Lecture Notes in Computer Science 3972

Jun Wang Zhang Yi Jacek M. Zurada
Bao-Liang Lu Hujun Yin (Eds.)

Advances in Neural Networks – ISNN 2006

Third International Symposium on Neural Networks
Chengdu, China, May 28 - June 1, 2006
Proceedings, Part II

Volume Editors

Jun Wang
The Chinese University of Hong Kong
Dept. of Automation and Computer-Aided Engineering
Shatin, New Territories, Hong Kong
E-mail: jwang@acae.cuhk.edu.hk

Zhang Yi
University of Electronic Science and Technology of China
School of Computer Science and Engineering
Chengdu, Sichuan, China
E-mail: zhangyi@uestc.edu.cn

Jacek M. Zurada
University of Louisville, Dept. of Electrical and Computer Engineering
Louisville, Kentucky, USA
E-mail: jacek.zurada@louisville.edu

Bao-Liang Lu
Shanghai Jiao Tong University, Dept. of Computer Science and Engineering
Shanghai, China
E-mail: blu@cs.sjtu.edu.cn

Hujun Yin
University of Manchester, School of Electrical and Electronic Engineering
Manchester M60 IQD, UK
E-mail: h.yin@manchester.ac.uk

Library of Congress Control Number: 2006925897

CR Subject Classification (1998): F.1, F.2, D.1, G.2, I.2, C.2, I.4-5, J.1-4

LNCS Sublibrary: SL 1 – Theoretical Computer Science and General Issues

ISSN 0302-9743
ISBN-10 3-540-34437-3 Springer Berlin Heidelberg New York
ISBN-13 978-3-540-34437-7 Springer Berlin Heidelberg New York

Springer is a part of Springer Science+Business Media

springer.com

Printed in Germany

Typesetting: Camera-ready by author, data conversion by Scientific Publishing Services, Chennai, India
Printed on acid-free paper SPIN: 11760023 06/3142 5 4 3 2 1 0

Preface

This book and its sister volumes constitute the Proceedings of the Third International Symposium on Neural Networks (ISNN 2006) held in Chengdu in southwestern China during May 28–31, 2006. After a successful ISNN 2004 in Dalian and ISNN 2005 in Chongqing, ISNN became a well-established series of conferences on neural computation in the region with growing popularity and improving quality. ISNN 2006 received 2472 submissions from authors in 43 countries and regions (mainland China, Hong Kong, Macao, Taiwan, South Korea, Japan, Singapore, Thailand, Malaysia, India, Pakistan, Iran, Qatar, Turkey, Greece, Romania, Lithuania, Slovakia, Poland, Finland, Norway, Sweden, Demark, Germany, France, Spain, Portugal, Belgium, Netherlands, UK, Ireland, Canada, USA, Mexico, Cuba, Venezuela, Brazil, Chile, Australia, New Zealand, South Africa, Nigeria, and Tunisia) across six continents (Asia, Europe, North America, South America, Africa, and Oceania). Based on rigorous reviews, 616 high-quality papers were selected for publication in the proceedings with the acceptance rate being less than 25%. The papers are organized in 27 cohesive sections covering all major topics of neural network research and development. In addition to the numerous contributed papers, ten distinguished scholars gave plenary speeches (Robert J. Marks II, Erkki Oja, Marios M. Polycarpou, Donald C. Wunsch II, Zongben Xu, and Bo Zhang) and tutorials (Walter J. Freeman, Derong Liu, Paul J. Werbos, and Jacek M. Zurada). ISNN 2006 provided an academic forum for the participants to disseminate their new research findings and discuss emerging areas of research. It also created a stimulating environment for the participants to interact and exchange information on future challenges and opportunities of neural network research.

Many volunteers and organizations made great contributions to ISNN 2006. The organizers are grateful to the University of Electronic Science and Technology of China and the Chinese University of Hong Kong for their sponsorship; to the National Natural Science Foundation of China and K.C. Wong Education Foundation of Hong Kong for their financial supports; and to the Asia Pacific Neural Network Assembly, European Neural Network Society, IEEE Computational Intelligence Society, IEEE Circuits and Systems Society, and International Neural Network Society for their technical cosponsorship. The organizers would like to thank the members of the Advisory Committee for their supports, the members of the International Program Committee for reviewing the papers and members of the Publications Committee for checking the accepted papers in a short period of time. Particularly, the organizers would like to thank the publisher, Springer, for publishing the proceedings in the prestigious series of

Lecture Notes in Computer Science. Last but not least, the organizers would like to thank all the speakers and authors for their active participation at ISNN 2006, which is essential for the success of the symposium.

May 2006

Jun Wang
Zhang Yi
Jacek M. Zurada
Bao-Liang Lu
Hujun Yin

ISNN 2006 Organization

ISNN 2006 was organized and sponsored by the University of Electronic Science and Technology of China and the Chinese University of Hong Kong. It was technically cosponsored by the Asia Pacific Neural Network Assembly, European Neural Network Society, IEEE Circuits and Systems Society, IEEE Computational Intelligence Society, and International Neural Network Society. It was financially supported by the National Natural Science Foundation of China and K.C. Wong Education Foundation of Hong Kong.

Jun Wang, Hong Kong (General Chair)
Zhang Yi, Chengdu, China (General Co-chair)
Jacek M. Zurada, Louisville, USA (General Co-chair)

Advisory Committee

Shun-ichi Amari, Tokyo, Japan (Chair)
Hojjat Adeli, Columbus, USA
Guoliang Chen, Hefei, China
Chunbo Feng, Nanjing, China
Kunihiko Fukushima, Tokyo, Japan
Okyay Kaynak, Istanbul, Turkey
Yanda Li, Beijing, China
Erkki Oja, Helsinki, Finland
Marios M. Polycarpou, Nicosia, Cyprus
Shoujue Wang, Beijing, China
Youlun Xiong, Wuhan, China
Shuzi Yang, Wuhan, China
Siying Zhang, Qingdao, China
Shoubin Zou, Chengdu, China
Walter J. Freeman, Berkeley, USA (Co-chair)
Zheng Bao, X'ian, China
Ruwei Dai, Beijing, China
Toshio Fukuda, Nagoya, Japan
Zhenya He, Nanjing, China
Frank L. Lewis, Fort Worth, USA
Ruqian Lu, Beijing, China
Nikhil R. Pal, Calcutta, India
Tzyh-Jong Tarn, St. Louis, USA
Paul J. Werbos, Washington, USA
Lei Xu, Hong Kong
Bo Zhang, Beijing, China
Nanning Zheng, Xi'an, China

Steering Committee

Zongben Xu, Xi'an, China (Chair)
Tianping Chen, Shanghai, China
Wlodzislaw Duch, Torun, Poland
Xiaoxin Liao, Wuhan, China
Zhiyong Liu, Beijing, China
Zhengqi Sun, Beijing, China
Donald C. Wunsch II, Rolla, USA
Fuliang Yin, Dalian, China
Liming Zhang, Shanghai, Chinac
Mingtian Zhou, Chengdu, China
Houjun Wang, Chengdu, China (Co-chair)
Andrzej Cichocki, Tokyo, Japan
Anthony Kuh, Honolulu, USA
Derong Liu, Chicago, USA
Leszek Rutkowski, Czestochowa, Poland
DeLiang Wang, Columbus, USA
Gary G. Yen, Stillwater, USA
Juebang Yu, Chengdu, China
Chunguang Zhou, Changchun, China

Program Committee

Bao-Liang Lu, Shanghai, China (Chair)
Shigeo Abe, Kobe, Japan
Khurshid Ahmad, Surrey, UK
A. Bouzerdoum, Wollongong, Australia
Jinde Cao, Nanjing, China
Matthew Casey, Surrey, UK
Luonan Chen, Osaka, Japan
Yen-Wei Chen, Kyoto, Japan
Yuehui Chen, Jinan, China
Yiu Ming Cheung, Hong Kong
Sungzoon Cho, Seoul, Korea
Emilio Corchado, Burgos, Spain
Shuxue Ding, Fukushima, Japan
Meng Joo Er, Singapore
Mauro Forti, Siena, Italy
Marcus Gallagher, Brisbane, Australia
Chengling Gou, Beijing, China
Lei Guo, Nanjing, China
Min Han, Dalian, China
Zhifeng Hao, Guangzhou, China
Zengguang Hou, Beijing, China
Jinglu Hu, Fukuoka, Japan
Guangbin Huang, Singapore
Marc van Hulle, Leuven, Belgium
Danchi Jiang, Hobart, Australia
Hoon Kang, Seoul, Korea
Samuel Kaski, Helsinki, Finland
Tai-hoon Kim, Seoul, Korea
Yean-Der Kuan, Taipei, Taiwan
James Lam, Hong Kong
Xiaoli Li, Birmingham, UK
Yuanqing Li, Singapore
Xun Liang, Beijing, China
Lizhi Liao, Hong Kong
Fei Liu, Wuxi, China
Ju Liu, Jinan, China
Hongtao Lu, Shanghai, China
Fa-Long Luo, San Jose, USA
Jinwen Ma, Beijing, China
Stanislaw Osowski, Warsaw, Poland
Ikuko Nishkawa, Kyoto, Japan
Paul S. Pang, Auckland, New Zealand
Yi Shen, Wuhan, China
Michael Small, Hong Kong
Ponnuthurai N. Suganthan, Singapore
Fuchun Sun, Beijing, China
Hujun Yin, Manchester, UK (Co-chair)
Ajith Abraham, Seoul, South Korea
Sabri Arik, Istanbul, Turkey
Jianting Cao, Saitama, Japan
Wenming Cao, Hangzhou, China
Liang Chen, Prince George, Canada
Songcan Chen, Nanjing, China
Xue-wen Chen, Kansas, USA
Xiaochun Cheng, Berkshire, UK
Zheru Chi, Hong Kong
Jin-Young Choi, Seoul, Korea
Chuanyin Dang, Hong Kong
Tom Downs, Brisbane, Australia
Shumin Fei, Nanjing, China
Wai Keung Fung, Winnipeg, Canada
John Qiang Gan, Essex, UK
Chengan Guo, Dalian, China
Ping Guo, Beijing, China
Qing-Long Han, Rockhampton, Australia
Daniel W.C. Ho, Hong Kong
Dewen Hu, Changsha, China
Sanqing Hu, Chicago, USA
Shunan Huang, Singapore
Malik Magdon Ismail, Troy, USA
Joarder Kamruzzaman, Melbourne, Australia
Nikola Kasabov, Auckland, New Zealand
Tae Seon Kim, Seoul, Korea
Hon Keung Kwan, Windsor, Canada
James Kwok, Hong Kong
Shaowen Li, Chengdu, China
Yangmin Li, Macao
Hualou Liang, Houston, USA
Yanchun Liang, Changchun, China
Meng-Hiot Lim, Singapore
Guoping Liu, Treforest, UK
Meiqin Liu, Hangzhou, China
Wenlian Lu, Leipzig, Germany
Zhiwei Luo, Nagoya, Japan
Qing Ma, Kyoto, Japan
Zhiqing Meng, Hangzhou, China
Seiichi Ozawa, Kobe, Japan
Jagath C. Rajapakse, Singapore
Daming Shi, Singapore
Jochen J. Steil, Bielefeld, Germany
Changyin Sun, Nanjing, China

Norikazu Takahashi, Fukuoka, Japan
Yu Tang, Mexico City, Mexico
Christos Tjortjis, Manchester, UK
Michel Verleysen, Louwain, Belgium
Dan Wang, Singapore
Si Wu, Brighton, UK
Cheng Xiang, Singapore
Simon X. Yang, Guelph, Canada
Yingjie Yang, Leicester, UK
Dingli Yu, Liverpool, UK
Gerson Zaverucha, Rio de Janeiro, Brazil
Huaguang Zhang, Shenyang, China
Liqing Zhang, Shanghai, China
Tao Zhang, Tianjin, China
Yanqing Zhang, Atlanta, USA
Jin Zhou, Shanghai, China
Pu Sun, Ann Arbor, USA
Ying Tan, Hefei, China
Peter Tino, Birmingham, UK
Dan Ventura, Provo, USA
Bing Wang, Hull, UK
Kesheng Wang, Trondheim, Norway
Wei Wu, Dalian, China
Daoyi Xu, Chengdu, China
Xiaosong Yang, Wuhan, China
Zhengrong Yang, Exeter, UK
Wen Yu, Mexico City, Mexico
Zhigang Zeng, Hefei, China
Jie Zhang, Newcastle, UK
Qingfu Zhang, Essex, UK
Ya Zhang, Kansas, USA
Yunong Zhang, Maynooth, Ireland

Organizing Committee
Yue Wu (Chair), Chengdu, China
Xiaofeng Liao (Co-chair), Chongqing, China

Publications Committee
Chuandong Li (Chair), Chongqing, China
Jianwei Zhang (Co-chair), Hamburg, Germany
Mao Ye (Co-chair), Chengdu, China

Publicity Committee
Bin Jiang (Chair), Chengdu, China
Zeng-Guang Hou (Co-chair), Beijing, China
Jennie Si (Co-chair), Tempe, USA

Registration Committee
Xiaorong Pu (Chair), Chengdu, China

Local Arrangements Committee
Hongli Zhang (Chair), Chengdu, China

Secretariats
Jiancheng Lv, Chengdu, China
Tao Xiang, Chongqing, China

Table of Contents – Part II

Pattern Classification

Computer Vision

Image Processing

Signal Processing

System Modeling

Control Systems

Robotic Systems

Power Systems

Design an Effective Pattern Classification Model

Do-Hyeon Kim[1], Eui-Young Cha[1], and Kwang-Baek Kim[2]

[1] Dept. of Computer Engineering, Pusan National University,
San-30, Jangjeon-dong, Keumjeong-gu,
Busan, 609-735, Korea
unlimmit@hanmail.net, eycha@pnu.edu
[2] Dept. of Computer Engineering,
Silla University, San1-1, Gwaebop-dong,
Sasang-gu, Busan, 617-736, Korea
gbkim@silla.ac.kr

Abstract. This paper presents an effective pattern classification model by designing an artificial neural network based pattern classifiers for face recognition. First, a RGB image inputted from a frame grabber is converted into a HSV image. Then, the coarse facial region is extracted using the hue(H) and saturation(S) components except intensity(V) component which is sensitive to the environmental illumination. Next, the fine facial region extraction process is performed by matching with the edge and gray based templates. To make a light-invariant and qualified facial image, histogram equalization and intensity compensation processing using illumination plane are performed. The finally extracted and enhanced facial images are used for training the pattern classification models. The proposed hierarchical ART2 pattern classification model which has the Max-Min cluster selection strategy makes it possible to search clustered reference patterns effectively. Experimental results show that the proposed face recognition system is as good as the SVM model which is famous for face recognition field in recognition rate and even better in classification speed.

1 Introduction

Recently, face recognition becomes an important research area for natural human-computer interaction. The process of face recognition is generally composed of facial region extraction, feature extraction from acquired facial images, and recognition from extracted features. There are various techniques of facial region extraction: skin color based method using color information [1], motion-information method, template matching, neural networks method [2], and snake method. The recognition methods of extracted facial images are largely classified into geometrical feature matching, statistical method [3], PCA(Principle Component Analysis) method [4], LFA (Local Feature Analysis) method, LDA(Linear Discriminant Analysis) method [5], neural networks method [6][7][8] and SVM (Support Vector Machine) method [1][9].

J. Wang et al. (Eds.): ISNN 2006, LNCS 3972, pp. 1 – 7, 2006.

This paper presents an effective pattern classification model by designing an artificial neural network based pattern classifiers for face recognition.

2 Preprocessing and Facial Region Extraction Technique

2.1 Facial Region Extraction

The RGB image captured from a frame grabber is converted into a HSV image which is similar to the human beings' vision system. Then, the region of which color is similar to facial color is extracted using the hue and saturation components except intensity component which is sensitive to the environmental illumination. And then, hue and saturation mask is made from the corresponding region of the image with a predefined threshold. After the common region is extracted from two masks, the final facial skin mask is generated by interpolation to obtain the complete facial skin mask.

50 x 50 gray-based and edge-based templates are used to minimize the interference of illumination. After facial regions are manually extracted from various human face images, both gray and edge templates are created from the average image of these facial regions. To minimize the interference of illumination and to get balanced templates, right-left mirror images are added. And the gray template is finally equalized.

To improve the matching speed, we execute the matching with only facial region which is acquired by color information. To eliminate the effect of template size, we use various scaled versions of original image. The matching original image with template is performed by correlation. Given template *T* and image window *R* in original, correlation coefficient $\gamma(T,R)$ is computed by equation (1).

$$\gamma(T,R) = \frac{\sum_{i=0}^{M-1}\sum_{j=0}^{N-1}(T[i][j]-\mu_T)(R[i][j]-\mu_R)}{M \bullet N \bullet \sigma_T \bullet \sigma_R} \tag{1}$$

In equation (1), M, N are the size of image window, μ_T, μ_R are average of T, R, and σ_T, σ_R are the standard deviation of T,R respectively.

After matching gray template and edge template with image window region in original image, the average of both correlation coefficients is acquired. The region with the largest average correlation coefficient is set up as the optimal facial region in the current scale. This process is repeated at various scales 0.6 – 1.5. Then, the region with the highest correlation coefficient is determined as the final facial region.

2.2 Image Enhancement by Illumination Plane and Equalization

When the effect of intensity is added to image, it may result in a maximizing illumination effect and have a bad influence on recognition performance. So we have designed illumination plane to compensate this effect of illumination. Illumination plane can divide the extracted facial region into blocks and adjust compensative amounts of illumination according to the size of blocks. In other words, after the representative intensity values of each block are extracted, illumination plane image can be generated by interpolating to the size of original image.

3 Hierarchical ART2 Model

3.1 Structure of Hierarchical ART2

The structure of ART2(Adaptive Resonance Theory) is determined by vigilance parameter ρ. It defines the similarity of the stored cluster patterns and a certain input pattern. As ρ is large, the diameter of a cluster is large and one cluster can accept many patterns. As ρ is small, the diameter of a cluster is small, many clusters can be generated, and various patterns can be stored. But too many clusters may result in lower recognition speed. In this paper, considering such a vigilance parameter, we proposed the Hierarchical ART2 which is consists of two layer clusters.

H-ART2 is consist of the first layer of clusters connected to the input layer and second layer of clusters connected to the clusters in the first layer. The clusters in the first layer as a centroid reference vector patterns are the representative patterns for representing input pattern. In this paper, the input patterns classified by each clusters in the first layer are more specifically re-classified, and are formed as clusters in the second layer.

3.2 Learning Algorithm of H-ART2

The clusters in the first layer are generated by unsupervised learning of the conventional ART2 method. The representative patterns in the first layer are more specifically re-classified into the clusters in the second layer by the supervised learning algorithm which is a modified ART2 learning algorithm. The supervised learning algorithm of ART2 is similar to unsupervised learning algorithm. But, although an input pattern may pass the vigilance test, a new cluster is generated when the class of the input pattern is different from the class of winner cluster.

3.3 Searching Strategy

The numbers of clusters in ART2 is determined by vigilance parameter ρ. If the number of clusters is too large, the speed of learning and recognition becomes slow because there are too many clusters to be compared. On the contrary, if too small, the recognition rate becomes low because there is not sufficient space enough to represent various types of input patterns. Considering this trade-off, we have designed a hierarchical neural network structure and have introduced an appropriate method for selecting reference clusters.

Fig. 1. (a) shows the method of selecting clusters only in the first layer and Fig. 1. (b) shows the method of selecting every clusters in the second layer as candidate clusters. Both methods, without searching strategy, select the generated clusters as candidate clusters, and then search for the reference cluster pattern, which is most similar to input pattern, out of them. Other cases in Fig. 1 have their own searching strategies.

3.3.1 Selection Strategy by Ranking and 2-WC(Winner Classifier)

In the way of Fig. 1. (c), an input pattern is classified into the class of the candidate cluster which has the shortest distance between the input pattern and the selected candidate clusters. In the manner of Fig. 1. (d), an input pattern is classified into the

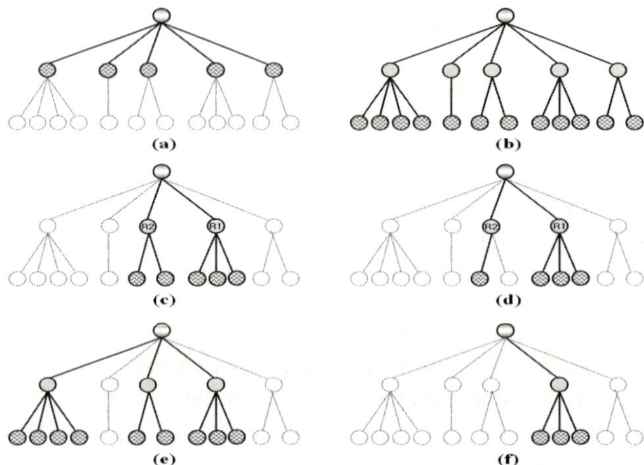

Fig. 1. Searching strategies of candidate clusters

class of candidate cluster which has the most frequent class out of the selected candidate clusters.

3.3.2. Proposed Selection Strategy by Relative Distance Ratio

Proposed searching strategy is shown in Fig. 1. (e) (f). In this method, you obtain the distances between the input pattern and every cluster in the first layer, calculate the maximum and minimum distances, and select only the clusters in the first layer, which has the ratio within the adaptively calculated distance range. Finally, you select clusters in the second layer, which are connected to the previously selected clusters, as candidate clusters. For instance, if cluster the selection ratio is p, the minimum distance is d_{min} and the maximum distance is d_{max}, only the clusters of first layer, which has the distance below $d_{threshold}$, are selected and finally each clusters connected to these selected cluster in the first layer become candidate clusters. Finally, an input pattern is classified into the class of the candidate cluster which has the shortest distance between the input pattern and the candidate clusters.

$$\| w_i - x_k \| \leq d_{threshold} \quad (2)$$

$$d_{threshold} = d_{\min} + (d_{\max} - d_{\min}) \times p \quad (3)$$

Fig. 1. (e) shows that three clusters in the first layer similar to the input pattern. Fig. 1 (f) shows that only one reference cluster is similar to the input pattern. This dynamic characteristic, selecting reference cluster according to the similarity to the input pattern based on selection ratio, not the fixed size of reference clusters, is the strong points of proposed searching strategy.

3.4. Pattern Classification of H-ART2

The clusters in the first layer selected by searching strategy are similar to input pattern and, in the same manner, the second cluster connected to the clusters in the first layer has similar pattern with input pattern. On the contrary, the unselected clusters in the first and second layer don't have similarity with the input pattern. Therefore fast recognition can be executed because a plenty of clusters has no possibility are pruned. Selecting candidate clusters by cluster selection ratio, not simple similarity with input pattern, can decrease the risk of misclassification.

4 Experimental Result

In this paper, facial data is acquired from the 320 x 240 x 24bit motion data of CCD camera. The number of people is 10, sex ratio is 6:4, and the ratio of people wearing glasses is 7:3. 1479 facial images of total 2215 are used to the training and the rest 736 is put to the test.

4.1 Performance Evaluation of the Proposed Pattern Classification Model

Facial data are normalized with 25 x 25 and are enhanced with illumination plane and equalization. For extract feature extraction, 3 kinds of image are prepared: original gray images, edge images for emphasizing the outline of face, and combined images from two previous images.

We compared the performance with respect to number of clusters in the first layer and the second layer changing vigilance parameter variously.

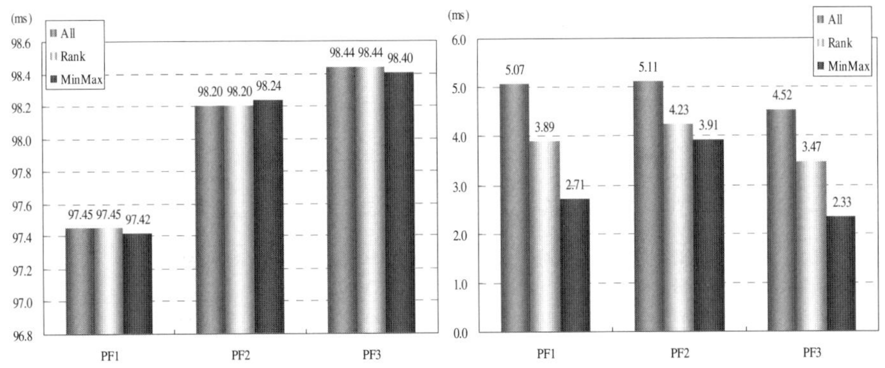

Fig. 2. Recognition accuracy and speed of H-ART2

According to searching strategies, the strategy of searching every candidate clusters(All), the strategy of searching only clusters in the first layer which have high similarity ranking with input pattern (Rank), and the proposed strategy of adaptively searching clusters in the first layer which have similarity ranking with input pattern by relative distance ratio between clusters (Max_Min) are experimented. The cluster selection ratio is set up to 0.3. In other words, in the strategy of searching by Rank and in the strategy of searching by Max_Min, the clusters in the first layer within 30% ranks are selected.

Fig. 2 shows the average recognition rate and the average recognition speed per one pattern according to the 3 kinds of preprocessing methods, the number of reference clusters, and searching strategies.

The combined image with gray image and edge image show better performance of recognition rate than gray image and edge image. As the strategy of searching by relative distance ratio reduces the search space, it raises recognition speed as double without lowering the recognition rate.

4.2 Performance Comparison with Other Pattern Classification Model

Table 1 shows comparison of proposed pattern classification performance with K-means clustering, Hierarchical clustering, and SVM. For the fairness of comparison, we make the number of reference patterns of K-means clustering to be equal to that of Hierarchical clustering. Three kinds of kernel functions, RBF, sigmoid, and polynomial, are applied to SVM. As the value of variable C is from 300 to 700, 600 ~700 SV(Support Vector) are generated. It shows variable C has little effect on SV.

Table 1. Comparison of pattern classification performance with others

Class	Model	Conditions	Recognition Accuracy(%)	Recognition Speed(ms)
K-means Clustering	K-1	#C=100	95.92	2.37
	K-2	#C=150	97.15	3.52
	K-3	#C=200	97.28	4.44
	K-4	#C=250	97.83	5.51
Hierarchical Clustering	H-1	#C=100	92.39	2.30
	H-2	#C=150	94.57	3.39
	H-3	#C=200	95.52	4.45
	H-4	#C=250	97.15	5.48
Support Vector Machine	SVM-r1	RBF, C=300 → SV=655	98.78	28.07
	SVM-r2	RBF, C=500 → SV=655	98.78	28.25
	SVM-r3	RBF, C=700 → SV=655	98.78	28.17
	SVM-s1	Sigmoid, C=300 → SV=624	98.64	19.67
	SVM-s2	Sigmoid, C=500 → SV=624	98.64	18.95
	SVM-s3	Sigmoid, C=700 → SV=624	98.64	18.94
	SVM-p1	Polynomial, C=300 → SV=772	98.91	22.36
	SVM-p2	Polynomial, C=500 → SV=745	98.91	21.70
	SVM-p3	Polynomial, C=700 → SV=733	98.91	21.62
H-ART2	HART-1	ρ=100 → #C1=116, #C2=192	98.23	3.11
	HART-2	ρ=105 → #C1=50, #C2=196	98.10	2.01
	HART-3	ρ=110 → #C1=38, #C2=206	98.64	2.12
	HART-4	ρ=115 → #C1=30, #C2=199	98.64	2.09

The proposed H-ART2 model shows better recognition rate and speed than K-means and Hierarchical clustering. Although SVM model is very stable without the influence of parameter and shows a good recognition rate, it is difficult to control the number of SV. Therefore recognition speed becomes slow because a lot of SV are generated.

5 Conclusion

In this paper, we have designed the efficient classification model for face recognition which is one of the biometric techniques. The proposed H-ART2 pattern classification model can reduce recognition time without lowering recognition rate because it selects a few clusters in the first layer, which is similar to the input pattern according to relative distance ratio, it can compare only the clusters in the second layer, which are connected to the previous clusters in the first layer. It can also improve recognition accuracy because it selects many clusters according to relative distance ratio if hard comparison is needed for an input pattern likely to be misclassified.

Experimental results with the facial data acquired from 10 male and female peoples at various distances and angles show that the proposed algorithm is efficient and effective for classification and the proposed face recognition system is as good as the SVM model which is famous for face recognition field in recognition rate and even better in classification speed.

References

1. Ai, H., Liang, L., Xu, G.: Face Detection based on Template Matching and Support Vector Machines, Proceedings of Image Processing 1 (2001) 1006-1009
2. Rowley, H. A., Baluja, S., Kanade,T.: Rotation Invariant Neural Network-Based Face Detection, Proceedings of Computer Vision and Pattern Recognition (1993) 38-44
3. Theodoridis, S., Koutroumbas, K.: Pattern Recognition, Academic Press (1999)
4. Romdhani, S.: Face Recognition using Principal Component Analysis, The MS Thesis, University of Glasgow (UK) (1997) 1-56
5. Belhumeur, P. N., Hespanha, J. P., Kriegman, D. J.: Recognition using Class Specific Linear Projection. IEEE Trans. Pattern Analysis and Machine Intelligence 19(7) (1997) 711-720
6. Kohonen, T.: The Self-Organizing Map, Proceedings of the IEEE, Vol.78, No.9, (1990) 1464-1480
7. Sanker, A., Mammone, R. J.: Neural Tree Networks in Neural Networks, Theory and Applications, Academic Press (1991)
8. Sanker, A., Mammone, R. J.: Growing and Pruning Neural Tree Networks, IEEE Trans. Computer 42(3) (1993) 291-299
9. Heisele, B., Ho, P., Poggio, T.: Face Recognition with Support Vector Machines - Global versus Component-Based Approach. 8th IEEE International Conference on Computer Vision, Vol.2, (2001) 688-694

Classifying Unbalanced Pattern Groups by Training Neural Network

Bo-Yu Li, Jing Peng, Yan-Qiu Chen, and Ya-Qiu Jin

Key Laboratory of Wave Scattering and Remote Sensing Information, (Ministry of Education), Fudan University, Shanghai 200433, China
liboyu@Fudan.edu.cn

Abstract. When training set is unbalanced, the conventional least square error (LSE) training strategy is less efficient to train neural network (NN) for classification because it often lead the NN to overcompensate for the dominant group. Therefore, in this paper a dynamic threshold learning algorithm (DTLA) is proposed as the substitute for the conventional LSE algorithm. This method uses multiple dynamic threshold parameters to gradually remove some training patterns that can be classified correctly by current Radial Basis Function (RBF) network out of the training set during training process, which changes the unbalanced training problem into a balanced training problem and improves the classification rate of the small group. Moreover, we use the dynamical threshold learning algorithm to classify the remote sensing images, when the unbalanced level of classes is high, a good effect is obtained.

1 Introduction

Unbalanced training set refers to number unbalance of training patterns or probability distribution unbalance of different groups [1]. In this paper, we will stress the problem of number unbalance of training patterns, we call the group that has more training patterns the dominant group (large class), similarly, we call the group that has few training patterns the non-dominant group(small class). Unbalanced training set can be observed in many applications. For example, in the task of image classification, we often obtain a training set that includes many positive examples and few negative examples because of different cost to obtain them; In defective product detection, most products will be good and only a few are defects. For these unbalanced data-sets, how to train ANN efficiently and impartially to classify every group will be immediate problem we must face. The conventional training method theoretically leads NN to ignore the small group because the conventional least square error (LSE) training algorithm will overcompensate for the dominant group, particularly when the separability of data-set is poor [1], [2]. As a consequence, although NN classifier probably achieves a good performance for the dominant group, it probably has a very low percentage of correct classification for small groups.

In this paper, the dynamic threshold learning algorithm (DTLA) is introduced to train neural network to solve this problem. In literatures [4], [5], a

J. Wang et al. (Eds.): ISNN 2006, LNCS 3972, pp. 8–13, 2006.

single threshold training method has been suggested to train RBF NN [3] for pattern recognition based on balanced data-set. However, it can not deal with the unbalanced data set because single threshold can not characterize the difference between large classes and small classes when there exist multiple classes in the data set. Therefore, Multi-threshold method is used to characterize these difference in the unbalanced data sets. Furthermore, new adjusting criteria are suggested to adjust the thresholds dynamically. Using this method, the number of training patterns that have large quantities could be reduced, consequently transform the unbalanced data set into a balanced one. Thus balances the training process. To prevent harming the performance of the NN classifier being trained and balance the training process, in the implementing we adjust the thresholds to remove some correctly recognized patterns out of the training set gradually during the training process. This method will result in the good results.

The DTLA method can make neural network classifier to care the small group effectively, which often includes some more important information of images, by changing the unbalanced situation into the balanced one. That is to say, we can use NN to capture the objective in which we are really interested even if its prototypes are very little. In this paper, the satellite images are especially focused on. Compared with the other training methods, experiment results show DTLA training method can stress the small group effectively.

2 The Dynamic Threshold Learning Algorithm (DTLA)

According to the analysis above, in order to train RBF NN impartially, the key problem is how to define suitable thresholds as controllers to remove recognized training patterns from the training set. Assuming each class corresponds to an output unit of the NN, during the training process, the recognized training patterns always have large output values (approaching to 1) on the corresponding output unit of the NN, while have small output values on the other output units(approaching to -1). Therefore, an upper threshold and a lower threshold should be considered to decide whether an input pattern should be ignored or not. Moreover, during the training iteration, large classes usually have more recognized patterns than that of small classes. Therefore, different thresholds are adopted for different output unit of the RBF NN, and the recognized training patterns of different groups are controlled to be removed from the training set sequentially. In most cases, large group have more recognized training patterns. If these patterns could be removed from the training set, it will help to restrain the overcompensation caused by large classes and force the RBF classifier to care the non-dominant class.

Assuming each output unit of RBF NN corresponds to a single group, the RBF NN has the same number of the output unit as the number of the groups. In order to simplify the presentation, in this paper, we do not distinguish the concept of the class and its corresponding output unit of NN. For example, the thresholds of the *ith* output unit of NN refers to the threshold of ω_i class.

For a training set that has two groups: ω_1 and ω_2. τ_1^u,τ_2^u are respectively the upper threshold of ω_1 and ω_2, and τ_1^L, τ_2^L are the lower threshold of ω_1 and ω_2 respectively($\tau_i^u \geq 0, \tau_i^L \geq 0, i = 1, 2$). The aim of this algorithm is to train RBF classifier to achieve the following goal for all training patterns of ω_1:

$$\begin{aligned} \forall x \in \omega_1 \Rightarrow f_1(x, w) = 1. \\ \forall x \in \omega_1 \Rightarrow f_2(x, w) = -1. \end{aligned} \quad (1)$$

Based on these observations, if $x \in \omega_1$ has been recognized by RBF NN and $f_1(x, w) > \tau_1^u \cap f_2(x, w) < -\tau_1^L$, then x will be removed from the training set in the next time of iteration, and will not make contribution to the error sum. The similar operation is done for $\forall x \in \omega_2$ if $f_2(x, w) > \tau_2^u \cap f_1(x, w) < -\tau_2^L$. We summarize the above presentation as the follows:

$$\begin{aligned} f_1(x, w) > f_2(x, w) \cap f_1(x, w) \geq \tau_1^u \cap f_2(x, w) < -\tau_1^L \cap \forall x \in \omega_1 = S - x, \\ f_2(x, w) > f_1(x, w) \cap f_2(x, w) \geq \tau_2^u \cap f_1(x, w) < -\tau_2^L \cap \forall x \in \omega_2 = S - x. \end{aligned} \quad (2)$$

where $f_1(x, w) > f_2(x, w)$ means that x can be recognized by RBF NN, $f_1(x,w) \geq \tau_1^u$ ensures the output of NN corresponding to x is larger than the threshold τ_1^u, $(S - x)$ means that in the next time of the training iteration we can remove the training pattern x out of training set S. In order to remove the recognized input patterns, a new error function (3) is proposed to train RBF NN classifier

$$\begin{aligned} E(w) = & \sum_{\forall x \in \omega_1} [max\{0, \tau_1^u - f_1(x, w)\}^2 + max\{0, \tau_1^L + f_2(x, w)\}^2] + \\ & \sum_{\forall x \in \omega_2} [max\{0, \tau_2^L + f_1(x, w)\}^2 + max\{0, \tau_2^u - f_2(x, w)\}^2]. \end{aligned} \quad (3)$$

In (3), for the training patterns that satisfy the constraint $f_1(x, w) > \tau_1^u \cap x \in \omega_1$ and $f_2(x, w) < -\tau_1^L$, the first sum term can ensure the patterns not to make any effect to the error sum $E(w)$. Similarly, the second sum term can ensure that the training patterns, which satisfy the constraint $f_1(x, w) < -\tau_2^L$ and $f_2(x, w) > \tau_2^u \cap x \in \omega_2$, can not make any contribution to the error sum $E(w)$. In order to balance the training process, more number of the training patterns in the large class must be ignored than that in the small group,which leads to define different upper thresholds and lower thresholds for different classes. For example, if $\tau_1^u < \tau_2^u$, more recognized patterns in ω_1 class should be ignored than that in ω_2 class. Using this method, the training process could be balanced. By decreasing both thresholds τ_1^u and τ_2^u, equation (3) can ignore more number of training patterns of both classes, so NN can focus on the difficult patterns.

For training the RBF NN to classify a training set with multi-groups, assuming $\omega_1, \cdots, \omega_{n_c}$ is n_c point sets defined on the real space R^n , and n_c is the number of the output units of the RBF NN. The equation (3) could be further extended as follows:

$$\begin{aligned} E(w) = & \sum_{\forall x} \{ \sum_{x \in \omega_i} (max\{0, \tau_i^u - f_i(x, w)\}^2 + \\ & \sum_{j=1 \cap j \neq i}^{n_c} [max\{0, \tau_i^L + f_j(x, w)\}^2])\}. \end{aligned} \quad (4)$$

where τ_i^u is the upper threshold of the ith neuron of RBF, and τ_i^L is the lower threshold of the ith neuron of RBF. The NN training process is also a process to minimize the $E(w)$. With the minimizing proceeding, the RBF NN classifier will recognize more number of the training patterns, and more training patterns x to meet the equation (2) will occur. In order to keep training process balanced and to avoid overcompensating for the dominant group, the thresholds should be readjusted according to these new situations. In order to achieve this target, an index is used to measure the unbalanced degree of classes.

$$p_i^{(t)} = \frac{n_i^c}{n_i}. \tag{5}$$

where n_i^c is the number of the current recognized training patterns of ω_i class, n_i is the number of training patterns of group ω_i. t is the epoch of the training, and $p_i^{(t)}$ is the current percentage of correct classification of the ω_i class. Let $p_{min}^{(t)} = min\{p_i^{(t)}|i = 1, ..., n_c\}$, then the step length of adjusting threshold can be defined as follows:

$$\kappa_i(t) = \frac{p_i^{(t)}}{p_{min}^{(t)}}. \tag{6}$$

Although a better performance can be achieved if $\tau_i^u \neq \tau_i^L$, in order to simplify this processing, here, let $\tau_i^u = \tau_i^L$. Then the criteria of adjusting thresholds is given as

$$\tau_i^u(t+1) = \tau_i^L(t+1) = \tau_i^u(t)/\kappa_i(t). \tag{7}$$

This criteria ensure to detect the dominant group during training process and curbs overcompensation for it. This rule implies that if the percentage of the correct classification (PCC) of the ith class is higher than the p_{min}, ω_i will probably becomes the dominant class and will dominate the training process. Therefore, the thresholds should be adjusted and dropped to remove more training patterns of this class out of training set. Moreover, for the large classes and small classes have different $\kappa_i(t)$ respectively, their corresponding thresholds τ_i^u(or τ_i^L)$(i = 1, ..., n_c)$ could be dynamically adjusted with different step length. Obviously, the largest class $p_j = max\{p_i|i = 1, ..., n_c\}$ can achieve the smallest thresholds because they have the largest adjusting step length: p_j/p_{min}.

In addition, in order to prevent over-fitting generalization and to further reduce the influence of the large class, some unidentified samples are deleted selectively at the last phase of the training process. This is because that if one pattern is identified wrongly and continuously over p times, it maybe noisy pattern and should be removed out of the current training set.

3 DTLA Application in Satellite Image Classification

NN plays very important role in image classification [6], [7]. In this experiments, we evaluate various training methods using the satimage database of UCI. We

Table 1. The averaged classification rate of task 2

Class	cotton crop	damp grey soil	other mixture class
DTLA	**95.5** %	**92.0** %	98.3 %
LSE	93.5 %	89.0 %	**99.4** %
Gaussian CPS [1]	95.0 %	90.6 %	98.0 %
RWTC[2]	95.0 %	90.0 %	98.5 %

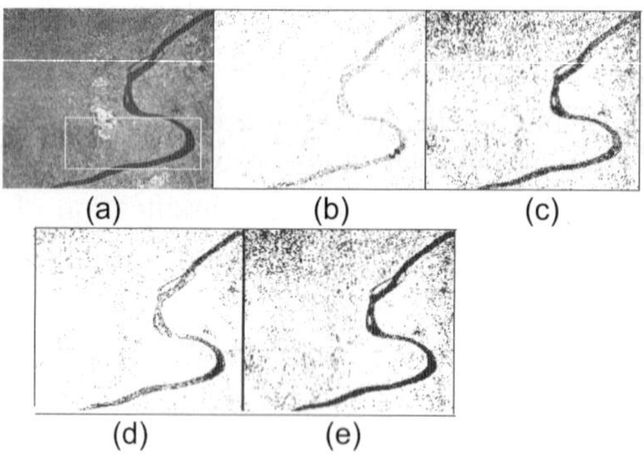

Fig. 1. (a) The original satellite image and pixels in box are used as training set. (b) Classification result using LSE method. (c) Classification result using Gaussian CPS [1]. (d) Classification result using RWTC [2]. (e) Classification result using DTLA method.

use the training/testing methodology to evaluate the performance of different training methods. The ten-fold cross validation is performed. In order to keep fair, all evaluations are performed on the same test set and on the same 5-10-2(3) network structure. For the sake of recurrence of the experiments results, the following conditions are also considered: we use the same initial weights in all experiments. For minimizing the error sum, the simple gradient descent method is employed for all training strategies with fixed learning rate($\eta = 0.02$). At the beginning of the training process, we randomly choose n_H training patterns as the initial centers of the activation function of hidden units and all width of hidden units are fixed to $\sigma_j = 10 (j = 1, \cdots, n_H)$. All initial thresholds are set according to the inverse proportion of samples number of each class. We list the experiment results in the table 1, from the table 1, we find DTLA algorithm is effective to deal with the multiple classes problem, where the small class (the damp grey soil) has a best classification rate than that obtained by other methods. Of course, in table 1, DTLA method achieves a high error on the large class, such as the mixture class, yet it should not be surprised because the improved performance on the small class is achieved at the cost of sacrificing the performance on the large class. An other experiment is to extract the river from the satellite image using pixel classification. The image, as shown in the figure 1(a), is a radar SAR remote sensing image, sized 298×291 pixels, photography in lower Mekong river in 1998, our aim is to detect the river correctly in this image. We can see the training set is seriously unbalanced, yet the segment result shows the DTLA achieve the best one, as shown in the figure 1 (e).

4 Conclusion

In this paper, we propose a new method to deal with the unbalanced problem of classification. This method can train the neural network to care the small classes in data set by transforming the unbalanced training into the balanced one. Compared with the other training schemes, the proposed method (DTLA) can improve the classification rate of NN on the small class. Experiments on classifying the satellite images show a good effect. Therefore, we think that although the data-set of the satellite image maybe be unbalanced, ambiguous and nonlinear, DTLA algorithm can make NN work well to detect the interesting objects.

Acknowledgements

The research work was supported by China State Major Basic Research Project (2001CB309401), the National Natural Science Foundation of China, project No. 60575022.

References

1. Murphey, Y.L., Guo, H., Feldkamp, L.A.: Neural Learning from Unbalanced Data. Applied Intelligence 21(2) (2004) 117–128
2. David, L., Andrew, R.W.: Optimized Feature Extraction and the Bayes Decision in Feed-Forward Classifier Networks. IEEE Trans on PAMI 13(4) (1991) 355–364
3. Friedhelm, S., Kestler, H.A., Palm, G.: Three Learning Phases for Radial-Basis-Function Networks. Neural Networks 14(4-5) (2001) 439–458
4. Sontag, E., Sussman, H.: Backpropagation Separates When Perceptrons Do. In: Proceeding IEEE International Joint Conference on Neural Networks (Washington DC) 1 (1989) 639–642
5. Francesco, L., Marco, S.: Efficient Training of RBF Neural Networks for Pattern Recognition. IEEE Trans on Neural Networks 12(5) (2001) 1235–1241
6. Sami, M.A., Linwood J., Park, J.D., Shannon, M.F.: A Neural Network Algorithm for Sea Ice Edge Classification. IEEE Trans on Geosciene and Remote Sensing 35(4) (1997) 817–826
7. James J.S., Timothy, J.M.: A Recurrent Neural Network Classifier for Improved Retrievals of Areal Extent of Snow Cover. IEEE Trans on Geoscience and Remote Sensing 39(10) (2001) 2135–2147

A Modified Constructive Fuzzy Neural Networks for Classification of Large-Scale and Complicated Data

Lunwen Wang[1], Yanhua Wu[1], Ying Tan[2], and Ling Zhang[3]

[1] 702 Research Division of Electronic Engineering Institute, Hefei, 230037, China
wanglunwen@163.com
[2] University of Science and Technology of China, Hefei, 230027, China
ytan@ustc.edu.cn
[3] Institute of Artificial intelligence, Anhui University, Hefei, 230039, China
Zling@ahu.edu.cn

Abstract. Constructive fuzzy neural networks (i.e., CFNN) proposed in [1] cannot be used for non-numerical data. In order to use CFNN to deal with non-numerical complicated data, rough set theory is adopted to improve the CFNN in this paper. First of all, we use rough set theory to extract core set of non-numerical attributes and decrease number of dimension of samples by reducing redundancy. Secondly, we can pre-classify the samples according to non-numerical attributes. Thirdly, we use CFNN to classify the samples according to numerical attributes. The proposed method not only increases classification accuracy but also speeds up classification process. Finally, the classification of wireless communication signals is given as an example to illustrate the validation of the proposed method in this paper.

1 Introduction

Constructive neural networks (i.e. CNN) from view of point of the geometrical meaning of models of M-P neurons were introduced in [2]. This approach gives the structure and parameters of the networks at the same time of processing given data. Based on the covering algorithms [3], this approach constructs a sphere neighborhood through a smart nonlinear transform, and converts a learning problem into a covering problem. It is characterized of little computation and is suitable for large-scale pattern classification. CNN has found a lot of practical applications, such as recognition of handwritten Chinese characters [4], and forecast of share index and price of stock market [5]. Recently, [1] adopts fuzzy theory to build an improved neural networks which was called as constructive fuzzy neural networks (CFNN), and makes it more powerful for classification. Unfortunately, CFNN is only suitable for the classification of numerical data and some non-numerical data that can be transferred to numerical data. However, complicated pattern usually includes both numerical data and non-numerical data, so CFNN is not suitable for classification of such complicated data.

Now, let's give a brief review of the preliminary concepts and notations of pattern classification [6]. The universe of objects can be regarded as 3-tuples as follows: *{U, C, V}*, where *U*, *C* and *V* are three finite, non-empty sets. *U* is a closed universe set that has *N* objects, *C* is a character set of the objects, *V* is a value set of

J. Wang et al. (Eds.): ISNN 2006, LNCS 3972, pp. 14 – 19, 2006.

the character. $u \in U$ is identification of the objects, $c \in C$ is special attribute of the objects. When U, normally having n types, is classified, an ideal classifier should has the property of the following:

$$y_i \cap y_j = \Phi, \ (i,j=1,2,\ldots,n \ i \neq j, \ \ y_i \neq \varphi) \tag{1}$$

$$y_1 \cup y_2 \cup \ldots \cup y_n = U \tag{2}$$

Eq. (1) shows the samples are classified correctly, eq. (2) shows every sample can be classified. However, in real problems, different objects have different characters, and furthermore the different aspect of an object has different characters. Basically, V can be divided into 2 types: numerical and non-numerical. Numerical data includes continuous values and discrete values, while non-numerical data includes characteristic, logical, and timely data etc. Scientists have put great efforts to investigate this kind of problem. However most of achievements can only classify one type or particular types of these data. In this paper, we want to introduce rough set theory to improve CFNN for constructing efficient classifiers.

2 Brief Description of CFNN

CFNN comes from the geometrical meaning of models of M-P neurons. In [1], we build CFNN from CNN by adopting fuzzy theory to improve the CNN, and give the membership functions of test samples, then determine all of the test samples. But there are 2 drawbacks in CFNN. One is that CFNN is only suitable for numerical data and for certain non-numerical data whose value can be mapped into numerical value. But most of objects have both numerical characters and non-numerical characters which cannot be changed into numerical value, so the application of CFNN is limited. Another shortcoming is that CFNN needs considerable running time to do the transformations, especially in the large-scale model classification, it will slow down the classification. The complexity of the algorithm is $O((K*S*W)^2)$ [3], where K is number of type, S is number of train samples of each kind and W is number of dimension of input vector. In large scale model classification, K、 S and W is usually very large, so the algorithm takes a long time.

There is a special model classification called layered classification in literature. It is also included in equations (1) and (2). Equations (1) and (2) can be regarded as first layer classification, and all samples can be classified into subset $U_k(i=1,2,\ldots,n)$. In the second layer classification, each set U_k can be classified into subset $U_{kj}(j=1, 2,\ldots,m)$ continually by equations (1) and (2). This process can continue if we need. On the other hand, there are many decompositions to the set U according to equations (1) and (2), but only small part of them meet our needs. In most cases, only one part meets our needs. So the part meeting our needs will be classified.

Currently, rough set theory is widely used in pattern recognition and data mining community. Rough set not only can be used to deal with non-numerical data, but also can be used to reduce attribute and eliminate redundancy of information according to non-numerical attribute. If we combine RS and CFNN, pre-classify the samples to sub-samples according to non-numerical character by RS, and re-classify the sub-samples according to numerical character by CFNN, the above problem will be solved easily. The classification of the first layer not only decreases number of dimensions of input (*W*), but also obtains subset that we need, and furthermore decrease the number of the samples (*S*) we have to deal with later. According to the complexity of the algorithm $O((K*S*W)^2)$, the complexity of CFNN will be reduced greatly. At the same time the second layered classification can maintain a high precision.

3 Reduction and Pre-classification Based on RS

The main idea of rough set is to deduce the decision and classification rules by reducing attributes on the basis of preserving the ability of classification. Reduction is one of the main tasks of rough set theory.

Definition 1. An information system is a quadruple: $S=(U,A,\{V_a\}, a)$. Where, U is a finite nonempty set of objects, A is a finite nonempty set of attributes, $\mathbf{V_a}$ is a nonempty set of values for $a \in A$, and $a: U \rightarrow V_{.a}$ is an information function.

Each information function a is a total function that maps an object of U to exactly one value in V_a. If A consists of conditional attributes set and conclusion attributes set, then $C \cup D=A$, $C \cap D=\phi$. S can be regarded as a decision system and denoted by $(U, C \cup D)$. When conclusion attributes set consists of only one element, the system is often denoted by $(U, C \cup \{d\})$.

Definition 2. Suppose P and Q are equivalence relation sets. $POS_P(Q)= \bigcup_{X \in U|Q} P_*(X)$ is called *positive region* of Q, denoted by $POS_P(Q)$. It is a set of all elements that can be classified to the equivalent by $U|P$. Relation r can be reduced in P respect to Q, if $r \in P$, and $POS_{IND(p-\{r\})}=(IND(Q))= POS_{IND(P)}(IND(Q))$, otherwise r can not be reduced. P is independent with respect to Q, if every relation r is indispensable, otherwise P relate to Q. The S is called a reduction of Q, if $S \subset P$ and $POS_S(Q)= POS_P(Q)$. The set of all the condition attributes indispensable in S is denoted by $CORE(A)$. CORE(A)= $\cap$ RED(A), where $RED(A)$ is the set of all *reductions* of A.

From definition 2, a set of the attribute A can be reduced to core(A). Classification by RS depends on equivalence relation. In fact, an attribute is an equivalence relation of U. Now, we can classify U according to core(A).

Definition 3. A decision system (U, $C \cup \{d\}$), $B \subseteq C$ is a subset of conditional attributes, IND(B, {d})= $(x,y) \in U \times U \mid d(x) = d(y)$ or $a \in B$, $a(x)=a(y)\}$ called binary relation or indiscernibility relation. Where $x \in U$ and $y \in U$. x and y can be called equivalence relation with respect to B, or x and y come from the same class with respect to B.

According to definition 3, U can be classified by equivalence relation on the basis of core(A). From the above, information system can be reduced and pre-classified. The attribute reduction eliminates redundant attributes on the basis of not losing information. The goal of pre-classification is to pick out the subset meeting our need from U according to non-numerical attributes. After it, not only number of the samples but also number of dimensions of input decreases both. From the above complexity $O((K*S*W)^2)$, the complexity of the algorithm reduces largely.

4 Method of Classification for Large-Scale and Complicated Data

Classification for large-scale and complicated data includes reduction of non-numerical attribute, getting core attributes, pre-classifying by RS, and re-classification by CFNN. The procedure of the proposed method can be described as follows.

(1) Reduction of non-numerical attribute

①Calculate positive region of D with respect to non-numerical attribute

②Calculate $POS_{C-\{c_i\}}(D)$ of non-numerical attribute c_i. If $POS_{C-\{c_i\}}(D) = POS_C(D)$, then c_i can be eliminated. Otherwise, c_i can not be eliminated. At the end, we will get core(C) of non- numerical attribute.

③ Calculate the minimal set P such that $POS_P(D)=POS_C(D)$, then P is minimal reduction of C.

(2) Delete the column in the information table which is not included in P, and get reduced train samples.

(3) Pre-classify by RS to pick out the subset meeting our need according to non-numerical attribute. If $P_i \in P$, $(i=1,2,...,m)$, $\{V_{Pi}\}$ is *set of values* of P_i, where $g_{Pi} \in \{V_{Pi}\}$ meet our need. We can get the subset according to $\cap (P_i=g_{Pi})$, $(i=1,2,...,m)$. The procedure is as follows:

① $Class \leftarrow \varnothing$, $Temp \leftarrow U$, $i=1$;

② Get the element $x(i)$ from $Temp$, where $i=1,2,...,n$, n is number of U.

③ If $((P_1==g_{P1})\&...\&(P_m==g_{Pm}))$, $Temp \leftarrow Temp - x(i)$, $Class \leftarrow Class \cup x(i)$;

④ Else $Temp \leftarrow Temp - x(i)$;

⑤ If ($Temp == \varnothing$) go to ⑥, otherwise, $i=i+1$, go to ②;

⑥ Output U$\leftarrow Class$. The non-numerical subset meeting our need is in U.

(4) The procedure of constructive fuzzy neural networks can be found in [1].

From above, steps (1) and (2) decrease dimensions of input samples, step (3) lessens samples which we will deal with later, step (4) increases classification accuracy. In summary, the proposed method is not only fast, but also accurate.

5 Experiments and Discussions

Communications signals were sampled from the receiver and the characters of the signals were extracted and listed in Table 1. Here, U is a set of signals, X_i $(i=1,2,...,n)$ are conditional attributes, where X_1 is modulation style, X_2 is the property of the signal, X_3 is the frequency of the signal, X_4 is the power of the signal, X_5 is the bandwidth of the frequency, X_6 is the parameter of modulation, X_7 is language of the signal, and Y is a decision attribute.

Table 1. Characters of communications signals

U	X_1	X_2	X_3	X_4	X_5	X_6	X_7	…	Y
1	AM	civilian	864	61	3.31	0.17	Chinese	…	a
2	AM	civilian	936	84	3.37	0.16	Chinese	…	b
3	AM	unknown	15500	96	3.51	0.19	Chinese	…	c
4	FSK	military	10500	43	1.42	1.2	Chinese	…	d
5	FSK	military	12300	51	1.16	1.0	Chinese	…	e

$C = \{X_1, X_2, \cdots, X_n\}$ is a set of conditional attributes, because $\cap(C - X_7) = \cap C$, X_7 can be eliminated from *C*, and other redundant attributes can be eliminated in the same way. Therefore, core attributes will be acquired as follows: $S_{core}=\{X_1, X_2, \cdots, X_6\}$.

Compared with Table 1, the number of attributes in Table 2 is reduced. Now we can pre-classify samples according to non-numerical attributes which meet our requirement. If X_1="AM" and X_2="civilian" are what we need, only signals 1 and 2 meet our need after pre-classification. Then we will deal with signal 1 and 2, and will not consider the others. So the number of input samples for CFNN is decreased, which plays an important role in the large-scale model classification task.

Table 2. Core attributes of signals

U	X_1	X_2	X_3	X_4	X_5	X_6	Y
1	AM	civilian	864	61	3.31	0.17	a
2	AM	civilian	936	84	3.37	0.16	b
3	AM	unknown	15500	96	3.51	0.19	c
4	FSK	military	10500	43	1.42	1.2	d
5	FSK	military	12300	51	1.16	1.0	e

Table 3. Results of the classifications

Method	Number of train	Number of test	Time(s)	Rate of right
1	500	500	2.1	85.2%
2	500	500	1.8	95.2%
3	500	500	1.6	98.5%

We have sampled many communications signals at different times, and extracted their characters accordingly. We have done experiments by using method 1(RS), method 2(CFNN) and method 3 (RS& CFNN). The results are shown in Table 3.

From Table 3, we can find that the time of classification by method 3 is less than that by method 1 and method 2, classification accuracy of method 3 is higher than that of method 1 and method 2. The reason is that the number of dimension of input is decreased by reduction, the samples processed later are lessened by pre-classification, and the accuracy is increased by CFNN. Therefore, we can conclude from our experiments that method 3 outperforms both method 1 and method 2.

6 Conclusions

In this paper, we successfully combine RS with CFNN to construct a two-layer pattern classifier. At the first layer, we use rough set theory to reduce redundancy of non-numerical attributes, extract core set of non-numerical attributes, decrease number of dimension of samples efficiently, pre-classify the samples according to non-numerical attributes which meet our need, eliminate the other samples which do not meet our need, lessen the samples size. At last, we decrease both dimensions of input and number of the samples we need to deal with later. At the second layer, we take advantage of CFNN to classify model precisely according to numerical attributes, because the CFNN has very explicit geometrical meaning such that it can deal with large-scale model and reach maximum precision of classification, comparing to other algorithms [1].

Our method combines RS with CFNN, which takes advantages of both RS and CFNN. It increases not only classification accuracy but also classification speed. From the experimental results of pattern recognition of wireless communications signals, the validity and efficiency of the proposed method have been demonstrated.

Acknowledgment

This work was supported by the Natural Science Foundation of China under Grant No.60135010; partially by the National Grand Fundamental Research 973 Program of China under Grant No. G1998030509.

References

1. Wang, L., Tan, Y., Zhang L.: Constructive Fuzzy Neural Networks and Its Application. In: J. Wang, X. Liao, Z. Yi (eds.): Advance in Neural Network-ISNN2005, Lecture Notes in Computer Science, Vol. 3497. Springer-Verlag, Berlin Heidelberg New York (2005) 440-445
2. Zhang, L., Zhang, B.: A Geometrical Representation of McCulloch-Pitts Neural Model and Its Applications. IEEE Transactions on Neural Networks10 (1999) 925-929
3. Zhang, L., Zhang, B., Ying, H.: An Alternative Covering Design Algorithm of Multilayer Neural Networks (in Chinese), Journal of Software (1999) 737742
4. Wu, M., Zhang, B.: A Neural Network Algorithm for Large Scale Pattern Recognition Problems (in Chinese). Journal of Software 12 (6) (2001) 851-855
5. Zhang Y.,Zhang, L.: A Structural Learning Algorithm Based on Covering Algorithm and Applying It to Stock Forecasting (in Chinese), Journal of Computer Research and Development41 (2004) 979-984
6. Niemann, H.: Pattern Classification,. Science Publishing Company (1988)
7. Pawlak Z.: Rough Scts Thcorctical Aspccts of Rcasoning about Data. Warsaw: Nowowiejskal, (1990)

A Hierarchical FloatBoost and MLP Classifier for Mobile Phone Embedded Eye Location System

Dan Chen[1,2], Xusheng Tang[3], Zongying Ou[3], and Ning Xi[2]

[1] College of Electrical Engineering and Automation, Fuzhou University, Fuzhou, China
[2] Shenyang Institution of Automation, Chinese Academy of Sciences, Shenyang, China
[3] School of Mechanical Engineering, Dalian University of Technology, Shenyang, China
chendan@sia.cn

Abstract. This paper is focused on cellular phone embedded eye location system. The proposed eye detection system is based on a hierarchy cascade FloatBoost classifier combined with an MLP neural net post classifier. The system firstly locates the face and eye candidates' areas in the whole image by a hierarchical FloatBoost classifier. Then geometrical and relative position information of eye-pair and the face are extracted. These features are input to a MLP neural net post classier to arrive at an eye/non-eye decision. Experimental results show that our cellular phone embedded eye detection system can accurately locate double eyes with less computational and memory cost. It runs at 400ms per image of size 256×256 pixels with high detection rates on a SANYO cellular phone with ARM926EJ-S processor that lacks floating-point hardware.

1 Introduction

The new-generation mobile phone offer a large set of functionalities, including user interfaces, and security access control etc. Robust eye detection is a crucial step towards these applications due to that they need to normalize faces or extract features according to eye positions [1]. A variety of eye detection algorithms have been developed in recent years such as grayscale projection [2], template matching [3], deformable template matching [4] and hierarchical Adaboost detector [5]. They can achieve good result under some constraints. Grayscale projection strongly depends on the lighting conditions. The drawback of template matching method is that their computation is complicated. Although the Adaboost eye detection systems [5] achieved pretty good results under the PC conditions, they require large resources. In practice, due to the limitation of hardware of the mobile phone (only a few KB memory and a processor with low frequency) and various factors, such as variation of size, pose, illumination conditions etc, robust eye detection on mobile phone is a challenging task.

In this paper, a novel approach for precisely and fast locating eyes on mobile phones is devised. The proposed approach locates face areas using FloatBoost face detector [6] prior to detecting eye location. Then eye candidates are extracted within the face areas by a followed FloatBoost eye detector. FloatBoost detector achieves the

J. Wang et al. (Eds.): ISNN 2006, LNCS 3972, pp. 20 – 25, 2006.

better performance with fewer features than AdaBoost. This is very important to the mobile phone because its memory is only a few KB. However, since the eye detector based on Haar-like feature only encode local gray intensive information of the eye, it is difficult to train an appropriate threshold to discriminate eye and eye-like images in the face regions such as eyebrows, thick frames of glasses etc. To solve this question, we train an eye-pair MLP classifier that is constructed using intrinsic geometrical and relative position information of eye-pair. Due to the non-accidental properties of geometrical relation information, the eye-like eye candidates can be excluded easily. The general process scheme for our proposed approach is shown in fig.1. The experimental results demonstrate that our eye location system can robustly cope with different light condition and achieve high detection rate on diverse test sets and can be implemented on a wide range of small resource-limit devices such as mobile phones.

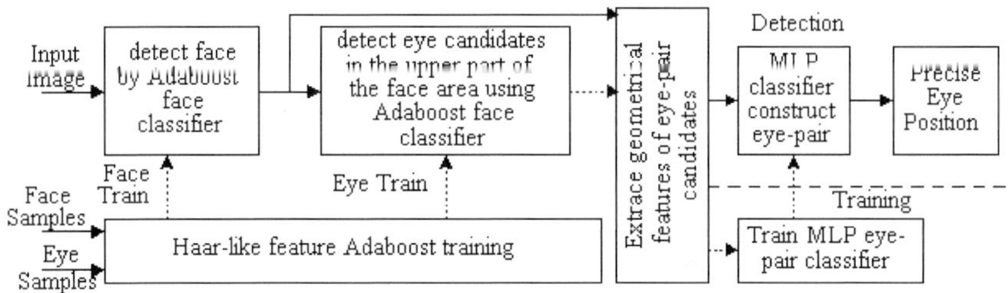

Fig. 1. Training and detection of Floatboost eye detector with MLP as post classifier

2 FloatBoost Classifiers

AdaBoost, introduced in [7], provides an effective approach to address nonlinear classification of high dimensional data. The AdaBoost constructs the strong classifier as a combination of weak classifiers with proper coefficients. Given a training set $\{x_i, y_i\}$, after T rounds of Adaboost training, T numbers of weak classifiers h_j and ensemble weights α_j are learned. Then a final strong classifier H (x) is defined as follow:

$$H(x) = \begin{cases} 1 & \sum_j^T \alpha_j h_j(x) \geq \theta \\ 0 & otherwise \end{cases} . \tag{1}$$

However, AdaBoost is sequential forward search procedures using the greedy selection strategy, neither the found weak classifiers nor their coefficients are optimal. Stan Li [6] proposed FloatBoost by incorporating the idea of Floating Search into AdaBoost for both feature selection and classifier design. The Floating Search method allows the number of backtracking steps to be controlled instead of being fixed beforehand. Specifically, it adds or deletes $l=1$ feature and backtracks r steps where r depends on the current situation. FloatBoost effectively improves the learning results. It needs fewer weaker classifiers than AdaBoost to achieve similar or higher performance. More details can be found in [6].

3 Eye Location System

3.1 Face Region Location and Detect Eye Candidates

If we can localize the face in the input image the problem of eye location can be greatly simplified due to the background being restricted to the face.

We adopted FloatBoost algorithm [6] combined with cascade approach [8] to build face detector. The cascade classifier combined successively more complex FloatBoost classifiers in a cascade structure. This structure results in extremely rapid object detector. The features used in FloatBoost training process are Haar-like features that can be calculated very fast by an integral image [8]. FloatBoost can effectively select fewer features than AdaBoost to achieve similar or higher performance. All of these ensure us to solve the problem of the resource-limitation of mobile phone.

Our face training set is drawn from FERET, ARData and ORL database. Total 7098 face images are processed with normalization to 20*20 pixels. Bootstrapping method is used for selecting negative samples. In detection process, face and non-face image regions can be discriminate according to Eq.(1).

We extend the above approach to build an eye candidates detector. There are total 7000 eye samples with 16*8 pixels. The negative samples are non-eye images, which are cropped from the face image due to eye searching area being restricted to the face.

After the face region is located, we only search the upper area of the face region according to the prior knowledge about face model. The reduction of searching area can save the detection time and dramatically reduce the possible false eye candidates. Since the eye detector based on Haar-like Feature only encode local gray intensive information of the eye, it is difficult to train an appropriate threshold to discriminate eye and eye-like images in the face regions such as eyebrows, thick frames of glasses, etc. So in this step we set low threshold to avoid false reject according to Eq. (1).

3.2 Precise Eyes Location Based on Eye-Pair MLP Post Classifier

Just as mentioned above, there are still some false eyes that cannot be discarded by hierarchical Floatboost classifier. Thus we propose a method that use MLP technology to construct the eye-pair classifier based on geometrical model of face. Firstly, eye candidates are grouped into eye-pair candidates. Secondly, the true eye-pair can be selected from the eye-pair candidates by using the MLP classifier. Finally, according to classifying results we can obtain the position of eye center. Due to the geometrical relation information encoded by MLP, the disadvantage of eye detector based on Haar-like features can be overcome.

3.2.1 Geometrical Features and Training Data

The results of prior detection process are a set of eye candidates' rectangles and associated face rectangles. We assume the detected face rectangle is $W_f \times H_f$ pixels and randomly selecting two detected eye candidate rectangles are $W_i \times H_i$ pixels and $W_j \times H_j$ pixels respectively. We define W_{ij}, H_{ij} as the horizontal and vertical distance of center of two eye candidates respectively. These distances can be easily calculated based on the prior results of detection. The geometrical relations are shown in Fig. 2.

As we know, the sizes of two eyes are similar in the front-view face. The positions of two eyes in the face are symmetry. The direction and distance of the line joining the centers of both eyes are also very useful cue to select the true eye-pair. According to these prior information about geometrical face model, we choose $\mathbf{L_i/H_f}$, $\mathbf{L_j/H_f}$, $\mathbf{W_{ij}/W_f}$, $\mathbf{H_{ij}/H_f}$, $\mathbf{H_{ij}/W_{ij}}$, $\mathbf{W_i/W_f}$, $\mathbf{H_i/H_f}$, $\mathbf{W_j/W_f}$, $\mathbf{H_j/H_f}$, $\mathbf{L_{if}/W_f}$, $\mathbf{L_{jf}/W_f}$, $\mathbf{|W_i-W_j|/W_f}$, $\mathbf{|H_i-H_j|/H_f}$ total 13 measurements for eye-pair classifying features.

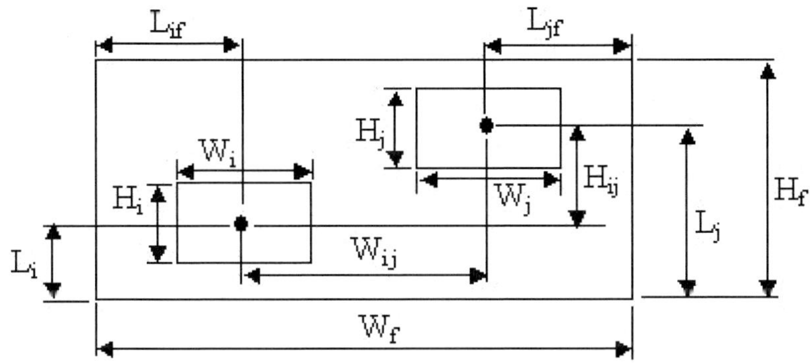

Fig. 2. Geometrical relation of Eye-pair Candidates

We will train an eye-pair classifier using MLP technology by supervised learning. We apply the following steps to get eye-pair MLP net training data:

For each image ($\mathbf{I_i}$) in the training set **(i=1…m)**

1. Applying Adaboost face and eye finder to $\mathbf{I_i}$, obtained a face rectangle $\mathbf{R_f}$ and a set of rectangles of eye candidates $\mathbf{R_{ec}}$ (assume k rectangles in the set totally)
2. Sorting ascending $\mathbf{R_{ec}}$ by x-coordinate value, selecting two eye candidates as eye-pair candidate $\mathbf{F_{eye\text{-}pair,\,j}}$ **(j=1…** C_k^2)
3. For each eye-pair candidates and associate $\mathbf{R_f}$, calculate 13 measurements of eye-pair classifying features and store them into a feature vector $\mathbf{x_{eye\text{-}pair,\,j}}$**(j=1…** C_k^2 **).** Manually labeling each $\mathbf{x_{eye\text{-}pair,\,j}}$,1 for true, -1 for false obtaining labeled samples ($\mathbf{x_{eye\text{-}pair,\,j}}$, $\mathbf{y_j}$) **(j=1…** C_k^2)

A training set of 2000 images has been selected from the above-mentioned database. Then 2000 position samples and 5672 negative samples are obtained using this method. The processed data was then random divided into training data and test data.

3.2.2 Multiplayer Neural Network Eye-Pair Classifier

The neural network eye-pair classifier, once trained, should be capable of selecting the true eye-pair much faster with less model size than the other forms of SVMs.

The standard back-propagation in the form of a three-layer neural network with one hidden layer is used to classify if a eye-pair candidates is true or not. The inputs to the network are the 13 eye-pair classifying features described above.

We tested various numbers of hidden units and found that 15 hidden units is enough to reach the best performance. In practical detection, after eye candidates have been extracted, any two-eye candidates are grouped as eye-pair candidates. Then eye-pair geometrical features vectors can be obtained by the method described in above section. These features are input to the trained neural net post classier to arrive at an eye-pair/non-eye-pair decision. The position of the eye-pair' classified as true is considered as the position of the eye center of the face. If more eye-pair candidates are classified as true, then the average position of theirs is considered as the eye center.

4 Experiment

An eye detection system on the cellular phone using our proposed algorithm is implemented. This system is implemented using C language. It is tested on a W21SA mobile phone (1MB of RAM and ARM926EJ-S processor). Two open face databases, BioID and JAFFE, are used in our experiments. The BioID database consists of 1520 images (384*286pixel). The JAFFE database consists of 213 images (256*256pixel).

A relative error measure [9] is used to evaluate the precision of eye localization. Let d_l , d_r be Euclidean distance between the detection left(right) eye position and the marked left(right) eye position respectively, d_{lr} be Euclidean distance between the marked left and right eye. Then the relative error is defined as follows:

$$err=max(d_l,d_r)/d_{lr}. \qquad (2)$$

Based on previous work by [9], if err<0.25, the eyes were deemed to found.

The results were list in Table. 1. We can see our eye detection system can accurately locate the eye position, even err<0.1, it has the 90% of hit rate on the test sets.

Table 1. The results of eye detection on different test database

Database	Hit rate (err<0.1)	Hit rate (err<0.25)	Time	Size(KB)
JAFFE	98.59%	98.59%	401ms	141
BioID	89.67%	95.72%	487ms	141

We also compared our eye detection system embedded on mobile phone with other newly published eye detection systems on PC. In Ref. [2], its hit rate was 97.18% and 94.18% on JAFFE and BioID respectively while err<0.25. Experiment results show the performances of our system embedded on mobile phone are equivalent to the previous published system on PC on both test sets.

The average time of detection and the model size of our system are also given in Table1. From Table1, we can see our system has achieved detection at two frames per second on mobile phone. The quick performance of our system is due to the efficiency of FloatBoost classifier and the hierarchical structure that allows the non-eyes of image areas to be quickly discarded while spending more computation on promising eye-like regions. Some real-life examples of eyes location are shown in Fig. 3.

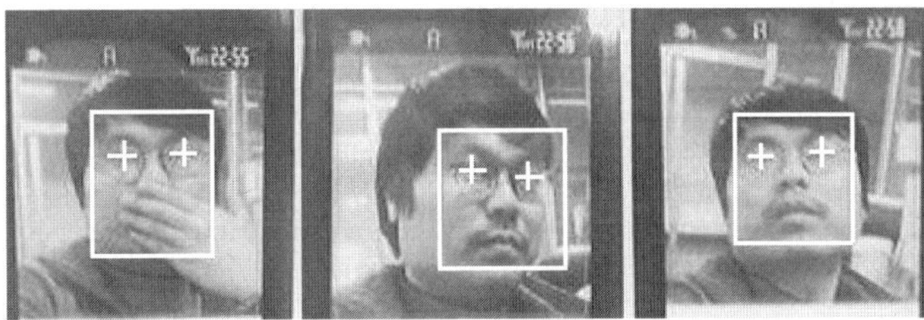

Fig. 3. Samples of our mobile phone embedded eye detection system on real-time detection

5 Conclusion

This paper has introduced a framework for locating eyes on mobile phone. The proposed method performs the eye location by means of FloatBoost classifiers and a MLP post eye-pair classifier. The FloatBoost classifiers can rapidly search the face and eye candidates. The eye pair MLP post classifier locates the true eye position finally. FloatBoost and MLP Neural Net technique can effectively select fewer features, which lead to the smaller size of eye classifier. Experiment results show that our method is fast, robust and reliable. Experimental results also show that our eye detection framework is very suitable for the resource-limit devices.

References

1. Riopka, T., Boult, T: The Eyes Have It. Proceedings of ACM SIGMM Multimedia Biometrics Methods and Applications Workshop. Berkeley, CA (2003) 9-16
2. Zhou, Z.-H., Geng, X.: Projection Functions for Eye Detection. Pattern Recognition 37(5) (2004) 1049-1056
3. Huang, W.J., Yin, B.C., Jiang, C.Y., Miao, J.: A New Approach for Eye Feature Extraction Using 3D Eye Template. ISIMP, (2001) 340 – 343
4. Huang, W.M., Mariani, R.: Face Detection and Precise Eyes Location. Proceedings of the 15th International Conference on Pattern Recognition, Vol. 4 (2000) 722–727
5. Yong, M., Xiaoqing, D., Zhenger, W., Ning, W.: Robust Precise Eye Location under Probabilistic Framework. Proc. Sixth IEEE Int. Conf. Autom. Face Gest. Recogn (2004) 339-344
6. Li, S.Z.: Floatboost Learning and Statistical Face Detection. IEEE Trans. on Pattern Analysis and Machine Intelligence 26(9) (2004) 1112-11123
7. Freund, Y., Schapire, R.: A Decision Theoretic Generalization of On-line Learning and an Application to Boosting. Journal of Computer and System Sciences. 55 (1) (1997) 119-139
8. Viola, P., Jones, M.: Robust Real-time Object Detection. Int. J. of Computer Vision 57(2) (2004) 137-154
9. Oliver J., Klaus J.K., Robert W.F.: Robust Face Detection Using the Hausdorff Distance. In: J. Bigun, F. Smeraldi (Eds.), Lecture Notes in Computer Science, Vol. 2091.Springer-Verlag, Berlin Heidelberg New York (2001) 90-95

Iris Recognition Using LVQ Neural Network

Seongwon Cho and Jaemin Kim

School of Electronic and Electrical Engineering, Hongik University,
72-1 Sangsu-dong, Mapo-gu, Seoul 121-791, Korea
swcho@hongik.ac.kr

Abstract. In this paper, we discuss human iris recognition, which is based on iris localization, feature extraction, and classification. The features for iris recognition are extracted from the segmented iris pattern using two-dimensional (2-D) wavelet transform based on Haar wavelet. We present an efficient initialization method of the weight vectors and a new method to determine the winner in LVQ neural network. The proposed methods have more accuracy than the conventional techniques.

1 Introduction

The human iris is the pigmented, round, contractile membrane of the eye, suspended between the cornea and lens and perforated by the pupil. The iris is composed of several layers: posterior surface, cooperative muscles, stromal layer, and anterior border layer. These multiplayer structures make the visual iris pattern. The visual pattern has two zones: the central pupillary zone with radiating spoke-like processes and frill and the surrounding ciliary zone with freckles, contractile lines and nevi. The zigzag circumferential collarette is placed between two zones [1].

The iris recognition is to analyze the iris pattern of an individual and to verify the individual's identity. Whereas most biological features (e.g., fingerprint, hand, and face) vary with age, the structure of iris is stabilized after two or three year old and is unique to an individual [2]. Due to high distinctiveness and stability, the iris pattern has been considered as the most reliable biometric measurement for the identification of individuals. Compared with the fingerprint recognition, the iris recognition does not require physical contact with a sensing device and is not invasive. However, the photo image of the iris pattern taken at a distance can have various artifacts due to long eyelashes and eyelid on the iris, reflection of illumination, and variation of the pupil size in different illumination. For reliable iris recognition, it is necessary to develop an efficient method for preprocessing the captured iris image, extracting the appropriate iris features, and classifying the extracted features.

In this paper, we address accurate localization of the iris pattern using a multiresolution active snake model, efficient extraction of features vector using Haar wavelet transform, and successful classification of the extracted feature vectors using neural network. This paper is organized as follows: In Section 2, we will review previous works. Section 3 describes the iris acquisition system used in our experiments. In Section 4, we present the accurate localization of the iris pattern by adapting the cost function of the snake model to the eye structure. Section 5 explains how to extract

J. Wang et al. (Eds.): ISNN 2006, LNCS 3972, pp. 26 – 33, 2006.

compact feature vectors using Haar wavelet transform. In Section 6, a new initialization method of the weight vectors and an efficient winner-selection method for competitive learning neural network are proposed. Experiment results are discussed in Section 7. Section 8 is the conclusion.

2 Previous Works

Researches on the iris recognition are classified into four categories: image acquisition, preprocessing, feature extraction, and verification [3], [4], [5], [6], [7], [8].

The image acquisition system consists of a simple lens, a monochrome CCD camera, and a frame grabbing board. A separate low-level tungsten halogen illuminator, an LED-based point light source, or infrared illuminators are used for the acquisition of a high-quality image of the iris. A beam splitter and a liquid crystal display are used to aid the user in alignment. To eliminate artifacts due to specular reflection, a polarizer is used [4], [6]. Once an iris image is captured, the image is analyzed to locate the pupil and the limbus. The iris is between the limbus and the pupillary boundary. The iris pattern is represented by the image intensity value at the polar coordinate (r, θ), where r is the distance from the center of the pupil [3], [4]. The localized iris pattern is then demodulated to extract features using a 2D Gabor filter and the extracted complex-valued coefficients are encoded into 256 bytes codes or less [3], [8]. The zero-crossing method based on 1D wavelet transform is also used to extract features [5]. The normalized Hamming distance between a pair of iris codes or the normalized correlation is used for verification or identification [3].

3 Iris Feature Extraction

An iris acquisition system to automatically capture an iris image while remaining noninvasive to the human operator. When we take a photo of the iris pattern in a normal illumination condition, the photo is dark and does not have good contrast. To have good contrast in the interior iris pattern without annoying the operator, we use two infrared lamps, which consists of many light-emitting diodes (LED). By careful positioning of the light source, the reflection of the source can be placed inside the pupil. In this case, the major problem is the reflections of the point source off eyeglasses. To eliminate these artifacts in the acquired images, we use directional infrared LEDs and adjust the directional angles of them. We also put an infrared-passing filter in front of the lens to eliminate reflection of various illuminants. To make iris images well-framed without unduly constraining the operator, we put half mirror in front of the camera. Human operator can see what the camera is capturing and adjust his position accordingly. The captured image size is 320x240, which is shown in Fig. 1. For the extraction of features from the iris pattern, Gabor transform has been widely used [3], [8]. Recently 1-D wavelet transform was also used [5]. In this section, we explain a new method for extracting feature from the iris pattern using 2-D Haar wavelet transform. We use Haar wavelet because it can be implemented easily and the computational cost can be saved.

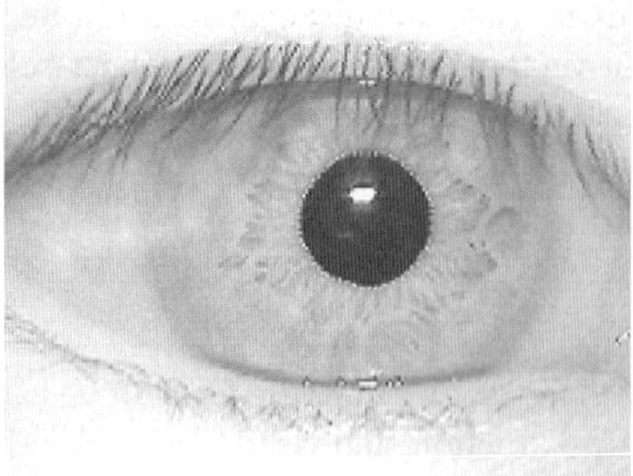

Fig. 1. Eye image with an iris pattern

3.1 Decomposition of 2-D Wavelet Transform

In this subsection, we briefly review a pyramid decomposition method [9]. To decompose a full-band 1-D signal into two subbands: L and H, a low-pass filter acts on the signal and the filtered signal is down sampled, which generates L subband, and a high-pass filter acts on the signal and the filtered signal is down sampled, which generates H subband. By using this decomposition in the x direction and in the y direction, a full-band 2-D signal is then decomposed into four bands: LL, LH, HL, and HH, where LH denotes one subband with L in the x direction and H in the y direction. For the pyramid decomposition, LL1 subband is decomposed into four bands: LL2, LH2, HL2, HH2. LL subbands are decomposed subsequently. This is shown in Fig. 2.

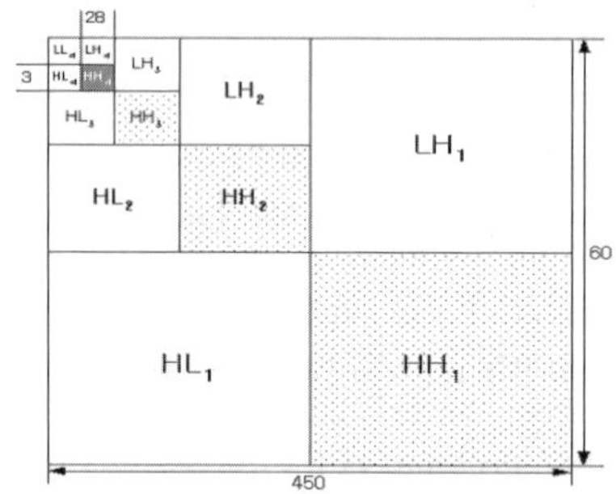

Fig. 2. Pyramid decomposition of the iris pattern

3.2 Feature Extraction Using 2-D Wavelet Transform

To analyze the iris image using Haar wavelet, the image in the polar coordinate is first transformed into the image in 2-D Cartesian coordinate and then is decomposed multiple subbands by the pyramid decomposition algorithm. Since sharp transition in the 2-D iris pattern is well represented by each HH subband, we can use each pixel intensity values of a HH subband as the feature vector for the classification. Since HH1 subband has too many pixels and is sensitive to the error of the iris location, we use a coarser HH subband such as HH4 or HH5. As the HH subband becomes coarser, the band becomes robust to the localization error but has less information.

4 Neural Network for Iris Recognition

Neural network is used to classify the extracted vectors. We use Learning Vector Quantization (LVQ) model due to its low complexity and high learning capability.

4.1 Initialization of the Weight Vectors

The LVQ is simple and faster than the error back-propagation algorithm, which is the most popular neural network model. However, the learning speed and classification performance of the LVQ are sensitive to the initial weight vectors. A simple method of initializing the weight vectors is to take the first m training vectors and use them as the initial weight vectors; the remaining vectors are then used for updating the weight vectors. Sometimes, the weight vectors are initialized to randomly selected training input vectors or the mean of the training vectors of each class. Other possible method of initializing the weights is to use K-mean clustering or CNN (condensed nearest neighbor) [10], [11], [12].

In the LVQ, an appropriate initial weight vector improves the learning time and classification performance. However, the distribution of initial weight vectors chosen by the previous initialization methods does not have large difference from the distribution of the training vectors. This is good for vector quantization. However, it is not appropriate for a pattern classifier based on a Nearest Neighbor method. When a nearest neighbor classifier is used, an input vector is classified to the class represented by the weight vector closest to it. Therefore, when we classify an input vector using reference (weight) vectors with uniform distribution, only reference (weight) vector close to decision boundaries between classes contribute to the classification performance. Accordingly, we can improve the performance of the LVQ by generating weight vectors close to decision boundaries and by removing unnecessary weight vectors.

Based on this fact, we propose a new weight initialization method for the LVQ, which generates the weight vectors close to decision boundaries. The proposed method is as follows:

[Step 1] Among training vectors of each class, take the first vector and use it as a weight vector for the class. The values of the remaining weight vectors for the classes are set to zero.

$$W_1^k = X_1^k \text{ for } k = 1,2,\ldots,M \tag{1}$$

where X_1^k is the first training pattern of the k-th class, W_1^k is the first weight vector for the k-th class, and M is the total number of classes.

[Step 2] Feed a new training vector as an input vector into the network.

[Step 3] Compute the distances between the input vector and the weight vectors.

$$d(k,j) = \sum_{i=0}^{N-1} \left(X_i - W_{i,j}^k\right)^2 \tag{2}$$

where X_i is the i-th element of the input vector, $W_{i,j}^k$ is the i-th element of the j-th weight vector for the k-th class, and N is the dimension of the input vector.

[Step 4] Determine whether the class represented by the weight vector with minimum distance is the same with the class of the input vector. Only if two classes are different, the input vector is assigned as a new weight vector for the class of the input.

[Step 5] Repeat Step 2-4 until all of the training vectors are processed.

Whereas for the original LVQ the number of the weight vectors for each class is to be predetermined, the proposed algorithm automatically determines the number of the weight vectors as the learning proceeds.

4.2 Dimensional Winner-Selection Method

As a distance measure, the Euclidean distance is used in the LVQ. However, when the dimension of the input vector is large like the iris feature vector, the winner selection based on the Euclidean distance can be wrong because of the loss of information about each dimension of the input vector. We propose a new winner selection method based on a new distance measure. For each training input vector, we do the following steps:

[Step 1] Compute the distance from an element of the input vector to the corresponding element of each weight vector.

[Step 2] Find the weight vector of which an element is closest to the corresponding element of the input vector, and then increase the winning count for the weight vector by one.

[Step 3] Repeat Step 1-2 until all elements of the input vector are processed.

[Step 4] After all elements are processed, the weight vector with the largest winning count is the winner. Once the winning weight vector is determined, the weight vectors are updated using the well-known LVQ updating rule [10].

In determining the winner, the proposed dimensional winner-selection method can consider information about each element of the input vector (such as quality), and its computational complexity is less than the conventional winner-selection methods using other distance measures such as the Euclidean distance and 1-norm.

5 Experimental Results

We use a database of 200 iris photographs for the experiments. The photographs were taken from 10 peoples in different hours and days. We used 100 iris photographs for learning and 100 for test. The parameter values for the LVQ are shown in Table 1.

Feature Extraction: We compared two different feature extraction methods based on Gabor transform and Wavelet transform. We used the same LVQ for two methods.

Table 1. Parameter values for LVQ

Learning rate	0.1
Update of learning rate	$\alpha(t)(1-1/\#\text{ of iteration})$
Total number of iteration	300

Table 2. Classification rates of two different feature extraction methods

	Gabor transform	Wavelet transform
Learning data	98%	98%
Test data	93%	94%

The performance of two methods is shown in Table 2. As for the learning data, two methods result in the same performance. As for the test data, feature extraction based on wavelet transform results in slightly better performance.

Initialization of Weight Vectors: We compared classification performance of two methods: one based on the proposed initialization method and the other based on a random initialization method. Wavelet transform was used for feature extraction. The performance of two methods is shown in Table 3. The proposed initialization method results in better performance.

Winner Selection: We compared the proposed winner-selection methods and other ones. We used wavelet transform for feature extraction and the proposed initialization methods for the initialization of LVQ. 2-D Euclidean norm is generally used as a

Table 3. Classification Rate of two different initialization methods

	Random initialization	The proposed one
Learning data	98%	100%
Test data	94%	98%

Table 4. Classification rates of two different winner-selection methods

Norm	Classification rate	
1-norm	98%	$\lvert\cdot\rvert + \lvert\cdot\rvert + \cdots$
2-norm	98%	$\sqrt{\lvert\cdot\rvert^2 + \lvert\cdot\rvert^2 + \cdots}$
∞ -norm	94%	$\max\{\lvert\cdot\rvert + \lvert\cdot\rvert + \cdots\}$

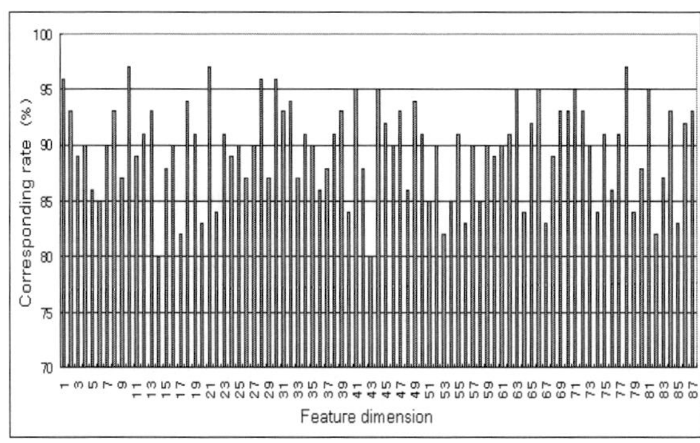

Fig. 3. Corresponding rate of each feature element

Table 5. Classification rates of a known winner-selection method using various distance measures

	Winner-selection method (Euclidean distance)	The proposed one
Learning data	100%	100%
Test data	98%	100%

distance measure in the LVQ but it does not guarantee better performance than other norms. For improving classification performance, we tested various norms. However, other norms do not improve the performance, which is shown in Table 4. To figure out the effect of each elements of the input vector on the classification performance, we computed the classification rate in each dimension of the input vector. This is shown in Fig. 3. If all elements have the similar contribution to the classification performance, we can use the Euclidean distance as a distance measure. However, some elements have more than 90% accuracy in the classification rate, and the other some elements have less than 85% accuracy in classification rate. This implies that the proposed dimensional-winner selection method is superior to the winner-selection method based on Euclidean norm, which is shown in Table 5. In the proposed winner-selection method, all test data are correctly classified at all tests. Even when we use only 10 elements of the input vector with the highest classification rate, misclassification was not found. This means that the proposed method is computationally effective.

6 Conclusions

In this paper, we presented an iris recognition system with high classification rate. For the accurate localization of the iris pattern, we used a multiresolution active snake model, in which each components of the energy function is adapted to the characteristics of the human eye. For the feature extraction, we used dyadic Haar wavelet transform, which is experimentally shown to be better than Gabor transform in classification performance. To improve classification rate, we used a neural network based on the LVQ. For the learning of the LVQ for the iris recognition, we proposed a new initialization method and dimensional winner-selection method, which were experimentally proven to results in better performance. We also reduced the size of feature vector without degrade of performance.

Acknowledgement

This work was supported by the Korea Research Foundation Grant funded by Korea Government (KRF-2004-042-D00132).

References

1. Adler, F.H.: Physiology of the Eye: Clinical Application. C.V. Mosby Company (1965)
2. Hallinan, P.W.: Recognizing Human Eyes. In: SPIE Proc. of Geometric Methods in Computer Vision 1570 (1991) 214-226

3. Daugman, J.G.: High Confidence Visual Recognition of Persons by a Test of Statistical Independence. IEEE Trans. Pattern Analysis and Machine Intelligence 15(11) (1993) 1148-1161
4. Williams, G.O.: Iris Recognition Technology. IEEE AES Systems Magazine (1997) 23-29
5. Boles, W.W., Boashash, B.: A Human Identification Technique Using Images of the Iris and Wavelet Transform. IEEE Trans. Signal Processing 46(4) (1998) 1185-1188
6. Wildes, R.P.: Iris Recognition: An Emerging Biometric Technology. Proceedings of the IEEE 85(9) (1997) 1348-1363
7. http://www.iriscan.com/html/irisrecogproduct.html.
8. Cho, S., Seong, H.: Iris Recognition Using Gabor Transform and Neural Net. Journal of Fuzzy Logic and Intelligent Systems 7(2) (1997) 397-401
9. Strang, G., Nguyen, T.: Wavelets and Filter Banks. Wellesley-Cambridge Press (1996)
10. Fausset, L.: Fundamentals of Neural Networks. Prentice Hall (1994)
11. Kohonen, T.: The Self-organization and Associate Memory. Springer-Verlag, Berlin Heidelberg New York (1985)
12. Yin F., Wang, J., Guo, C.: Advances in Neural Networks. Springer-Verlag, Berlin Heidelberg New York (2005)

Minimax Probability Machine for Iris Recognition

Yong Wang[1,2] and Jiu-qiang Han[1]

[1] School of Electronics and Information Engineering, Xi'an Jiaotong University, Xi'an, 710049, P.R. China
yongwang@126.com
[2] School of Electronics Engineering, Xidian University, Xi'an, 710071, P.R. China

Abstract. In this paper, a novel iris recognition method is proposed based on a state-of-the-art classification technique called minimax probability machine (MPM). Engaging the binary MPM technique, this work develops a multi-class MPM classification for reliable iris recognition with high accuracy. The experiments on iris database demonstrate that compared to the existent methods, the MPM-based iris recognition algorithm obtains better classification performance. It can significantly improve the recognition accuracy and has a competitive and promising performance.

1 Introduction

In recent years, with the increasing demands of security in our networked society, the technologies for personal identification work as the main solution to safeguard people's properties. Because traditional user authentication schemes are based on passwords, secret codes, identification cards or tokens. On one hand, schemes based only on passwords or secret codes can be cracked by intercepting the presentation of such a password, or even by counterfeiting it. On the other hand, an intruder can attack systems based on identification card or tokens by robbing, copying or simulation them. In that case, biometrics becomes an alternative recognition technique in aspect of protecting ourselves.

Among all biometrics (such as fingerprint, face, palmprint, gait, voice, iris, etc.), iris recognition is the most reliable one[1]. Because some properties of the human iris that enhance its suitability for use in automatic identification include: 1) its inherent isolation and protection from the external environment; 2) the possibility of surgically modifying it without high risk of damaging the user's vision; and 3) its physiological response to light, which provides the detection of a dead or plastic iris, avoiding this kind of counterfeit; also 4) its structure is genetically determined, the particulars of iris's minutiae are critically dependent on initial conditions in the embryonic mesoderm from which develops. Therefore, the above characteristics distribute to its application in airports, border crossings, banks, buildings, traveling, government and so on.

In iris recognition, how to improve robustness and increase recognition accuracy is a key problem. In the past, some iris feature dataset classification

J. Wang et al. (Eds.): ISNN 2006, LNCS 3972, pp. 34–39, 2006.

methods have been reported. Especially, Boles[2] uses dissimilarity function to distinguish different iris dataset and Ma[3] employs nearest feature linear method to realize iris feature classification. Nevertheless, these classification methods are limited to improve robustness and accuracy of iris recognition. Here, we study a novel classifier minimax probability machine (MPM),which provides a worst-case bound on the probability of misclassification of future data points ,and then propose an iris recognition based on MPM to increase accuracy and robustness of iris recognition.

This paper is organized as follows. Section 2 describes image preprocessing. Section 3 introduces iris dataset from iris images. In section 4, basic idea and formulation of binary MPM classification is introduced. Then multi-class MPM classification employing the binary MPM for iris recognition is discussed. Section 5 gives experimental results prior to conclusion in Section 6.

2 Image Preprocessing

An iris image contains not only the region of interest (ROI) but also some unuseful part such as eyelid, pupil etc. So, a captured iris image cannot be used directly. After the iris captures and before the feature extraction, three steps are performed: localization of the iris inside the image, image normalization and ROI enhancement. These procedures are called image preprocessing. Here, we adopt a proposed method based on three-step multi-scale strategy and histogram equalization. more details can be found in reference[4].

3 Iris Feature Forming Iris Dataset

From the above subsection, iris features are obtained by deploying iris ROI. For example, an iris ROI with length L and width W will become a $W \times L$ dimension feature vector. How to transform those features to iris dataset is an important step to make classification. Here, a new correlation strategy is used to describe iris dataset. The correlation strategy of two iris feature vectors is in details.

Assuming two observed feature vectors x and y respectively, an iris dataset $\{(E, \rho)\}$ can be represented by equation (1).

$$E = \Sigma_i (x_i - y_i)^2, \rho = \frac{\Sigma_i x_i y_i}{\sqrt{\sum_i {x_i}^2 \sum_i {y_i}^2}} \quad (1)$$

Thus, for n iris feature vectors, $(n-1)n/2$ data points form an iris dataset that describes relatives on different iris samples.

4 Iris Classification Based on MPM

4.1 MPM

MPM, a recently-proposed novel classifier, provides a worst-case bound on the probability of misclassification of future data points[5]. As discussed below, MPM minimizes directly an upper bound on the generalization error.

Suppose two random vectors x and y represent two classes of data with means and covariance matrices as $\{\overline{x}, \Sigma_x\}$,$\{\overline{y}, \Sigma_y\}$ respectively in a two-category classification task, where $x, y, \overline{x}, \overline{y} \in \Re^n$,and $\Sigma_x, \Sigma_y \in \Re^{n\times n}$ .With the reliable estimations of $\{\overline{x}, \Sigma_x\}$ and $\{\overline{y}, \Sigma_y\}$ for two classes of data, MPM attempts to determine the hyperplane $a^T z = b$ ($a \neq 0, z \in \Re^n, b \in \Re$, superscript T denotes the transpose) which can separate two classes data with a maximal probability. The minimax framework minimizes the generalization error by seeking the hyperplane for which the misclassification probabilities, $P_r\{a^T x \leq b\}$ and $P_r\{a^T y \geq b\}$, are low. Furthermore, the MPM problem can be transformed into a convex optimization problem:

$$\max_{\alpha,b,a\neq 0} \alpha \quad s.t. \quad supP_r\{a^T x \leq b\} \leq 1-\alpha, supP_r\{a^T y \geq b\} \leq 1-\alpha \qquad (2)$$

where α represents the lower bound of the accuracy for the future data, or the worst-case accuracy. Future points, z for which $a^T z > b$ are then classified as the class y; otherwise, they are judged as the class x. This derived decision hyperplane is claimed to minimize the worst-case probability of misclassification, or the error rate, for the classification of future data points.

The supremum in both constraints is computed via a theorem[6], that gives a formula for the supremum probability that a random vector lies on one side of a hyperplane:

$$\sup_{y\sim(\overline{y},\Sigma_y)} P_r\{a^T y \geq b\} = 1/(1+\delta^2), with \quad \delta^2 = \inf_{a^T w \geq b} (w-\overline{y})^T \Sigma_y{}^{-1}(w-\overline{y}) \quad (3)$$

Note that the infimum on the right side can be computed analytically from the hyperplane parameters a and b and the estimates $\overline{y}, \Sigma_y$. More specifically, this result can be used to simplify the optimization problem (3) to a second order cone programming problem[7]:

$$m = (\min_a \sqrt{a^T \Sigma_x a} + \sqrt{a^T \Sigma_y a}) \quad s.t. \quad a^T(\overline{x}-\overline{y}) = 1 \qquad (4)$$

The offset b of the hyperplane is uniquely determined for any a:

$$b = a^T\overline{x} - \sqrt{a^T \Sigma_x a}/m = a^T \overline{y} + \sqrt{a^T \Sigma_y a}/m \qquad (5)$$

Once a minimum value m is obtained for (4), the probability bound α can be computed as:

$$\alpha = 1/(1+m^2) \qquad (6)$$

Thus, we can obtain a hyperplane ${a^*}^T z > b^*$ from the above optimization processing. Classification of a new data point z_{new} is done by evaluating sign $({a^*}^T z_{new} - b^*)$: if this is $+1$,z_{new} is classified as from class x, otherwise z_{new} is from class y.

Moreover, for nonlinear classification, we can seek to map the problem to a higher dimensional feature space $\Re^f$ via a mapping $\varphi : \Re^n \rightarrow \Re^f$, such that a linear discriminant in the feature space corresponds to a nonlinear discriminant in the original space. The processing is called kernelization of MPM which can be solved by the same a second order cone programming problem in new space.

4.2 Iris Classification with Multi-class MPM

In the above subsection, basic theory of binary MPM classification has been introduced. Toward iris feature classification, we construct the multi-class classification technique by employing the binary MPM classifiers. Generally, there are two approaches for multi-class classification based on basic binary classifiers: one-against-one and one-against-all. In the mode of one-against-one, it needs to train $n(n-1)/2$ binary classifiers for n given classes while one-against-all mode needs to train n binary classifiers. The one-against-all approach means that we train n classifiers in which each MPM classifier is trained based on the data of one class against the other classes $(n-1)$. For each trained classifier, we can obtain its lower bound estimation of classification accuracy, denoted as α_i. In the predicting phase, we first perform a sorting of the n classifiers on a descending order of α_i. Then, the test data points are predicted based on the n sorted classifiers to find minimal lower bound of classification accuracy as unique claimed class. Here, we adopt the one-against-all approach since it only needs to train n binary classifiers for n given classes. These can help to reduce computation consuming.

5 Experiments

5.1 Iris Image Database

The performance of the proposed algorithm is evaluated with a common iris database named CASIA Iris Database[8]. The database is from 108 different eyes of 80 subjects. The images are acquired during different stages and the time interval between two collections is at least one month. The captured iris images are eight-bit gray images with a resolution of 320×280. These images make sure that there are enough iris image classes from different people to evaluate the proposed method.

5.2 Experimental Results

The results have been obtained using eyes from 108 people and, at least, 5 photos for each person (total 540 iris images including 108 classes). We select 216 samples (2 photos for each individual) for first session as the training set. The remaining 324 samples are used as the test set for second session. In the experiment, two types of kernel function are selected; they are linear and Gaussian kernel. The kernel parameter σ for the Gaussian kernel $(e^{-\|x-y\|^2/\sigma})$ is tuned using cross-validation over training set partitions. The algorithm is measured by correct classification rate (CCR: Correct classification sample numbers divided by all tested sample numbers).Fig.1 gives plot of CCR with Gaussian kernel parameter σ.

In the Fig.1, we obtain maximal classification rate 98.55% when $\sigma = 1.6$. To illustrate performance of Gaussian kernel function, linear kernel function is used in the experiment. Table 1 reports the results over test dataset for both the linear kernel and the Gaussian kernel with chosen as above in MPM classification.

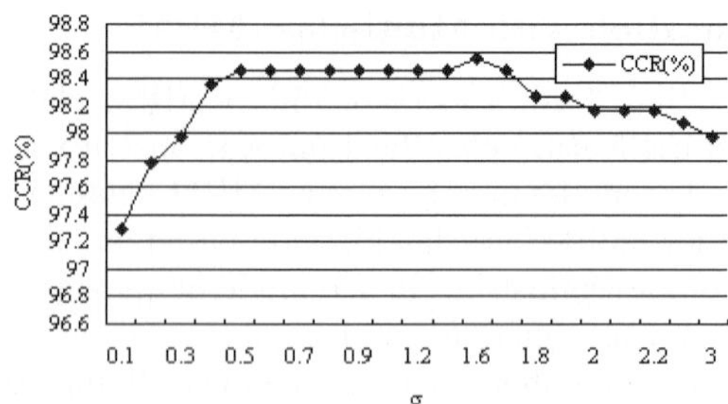

Fig. 1. Plot of CCR with σ

Table 1. Lower bound α and CCR compared to existing methods

Boles	Ma	Proposed method			
		Linear kernel		Gaussian kernel(σ=1.6)	
		α	CCR	α	CCR
92.61%	94.33%	99.20%	97.88%	99.39%	98.55%

As for the proposed iris database, several methods in published papers are listed in Table 1. These methods include Boles[2] and Ma[3]. Informally, Table 1 gives correct classification rate of these methods. From these data, we obtain 92.61%, 94.33%, 97.88% (MPM linear kernel) and 98.55% (MPM Gaussian kernel $\sigma = 1.6$) identification success rate respectively. On the guide of linear kernel and Gaussian kernel, MPM method increases 3.76% and 4.47% classification rate comparing with Ma method, 5.69% and 6.41% classification rate comparing with Boles method. Also, we notice lower bound α is smaller for a linear decision boundary than for the nonlinear decision boundary obtained via the Gaussian kernel. This clearly shows that Gaussian kernelization leads to more powerful decision boundaries. These comparisons show MPM method is more accurate in correction rate comparing with Boles and Ma methods, while keeping the stability of classification.

5.3 Results Analysis

From Table 1,we can find that the proposed MPM-based method have the best performance, followed by Ma and Boles et al. Compared with dissimilarity function as used by Boles and nearest feature linear method by Ma. Our method uses MPM for iris dataset classification. Importantly, MPM provides a lower bound estimation of classification accuracy of future data, which makes the scheme reliable and robust for better recognition performance.

Meanwhile, we observe that the accuracies of both kernels MPM for iris future data are different slightly in both linear and Gaussian kernel settings. Comparing the results of linear kernel with the results of Gaussian kernel, we note that both the worst-case bound and test accuracy for iris class in the Gaussian kernel are greater than those of the linear kernel. This also demonstrates the advantage of

Gaussian kernel setting. These comparisons indicate MPM-based approach has effective and emerging performance in iris recognition application.

6 Conclusion

In this paper, we present an iris recognition method employing a state-of-the-art classification technique called MPM. Different from other classification techniques, MPM provides a worst-case bound on the probability of misclassification of future data points and achieves good estimation on classification accuracy, which is important for classification of iris feature dataset and can help for reliability evaluation. From the experimental results, MPM could increase the recognition performance. These encouraging experimental results show that the MPM classification method is promising in various applications.

Acknowledgements

This work is supported by the Research Fund for the Doctoral Program of Higher Education under Grant No. 20050698025. We would like to thank National Laboratory of Pattern Recognition, Institute of Automation, Chinese Academy of Sciences for supporting this work.

References

1. Daugman, J.: High Confidence Visual Recognition of Persons by Test of Statistical Independence. IEEE Transactions on Pattern Analysis and Machine Intelligence 15(1993)1148–1161
2. Boles, W.W.: A Security System Based on Human Iris Identification Using Wavelet Transform. Engineering Applications of Artificial Intelligence 11(1998)77–85
3. Ma, L., Wang, Y.H., Tan, T.N.: Iris Rocognition Based on Multichannel Gabor Filtering. Proceedings of the Fifth Asian Conference on Computer Vision, Australia 1(2002)279–283
4. Wang, Y., Han, J.Q.: Three-stage Iris Localization with Multi-scale Approach (in Chinese). Microelectronics and Computer 22(2005)26–29
5. Lanckriet, G.R.G, Ghaoui, L.E., Bhattacharyya, C., Jordan, M.I.: Minimax Probability Machine. In Pro. Advances in Neural Information Processing System (2002)
6. Bertsimas, D., Sethuraman, J.: Moment Problems and Semidefinite Optimization. In: Handbook of Semidefinite Optimization. Kluwer Academic Publishers, Dodrecht (2000)469–509
7. Lobo, M., Vandenberghe, L., Boyd, S., Lebret, H.: An Analysis of Bayesian Classifiers: Applications of Second Order Cone Programming. Linear Algebra and Its Applications 284(1998)193–228
8. CASIA Iris Image Database. http://www.simobiometrics.com

Detecting Facial Features by Heteroassociative Memory Neural Network Utilizing Facial Statistics

Kyeong-Seop Kim, Tae-Ho Yoon, and Seung-Won Shin

School of Biomedical Engineering, College of Biomedical & Health Science, Konkuk University, Chungju, 380-701, Korea
kyeong@kku.ac.kr, {apple, sswcom98}@konkuk.ac.kr

Abstract. In this paper, we present an efficient algorithm of extracting the multiple facial features such as eyes, nose, and mouth. The face candidates are first obtained based on skin-color filtering in YC_bC_r color domain and skin-temperature values and then the elliptic measures are applied to extract a true face candidate and its boundary. A Sobel edge mask is performed and consequently horizontal projection operation is applied to locate the eyes referring to the maximum horizontal projection value in Y component. Once two eyes are located, the distance that crosses the center of eyes and extends to the face boundary, D_l is determined. A heteroassociative memory neural network model is utilized to find the facial features. An input neuron vector $\boldsymbol{X}$ accepts D_l and the output neurons vector $\boldsymbol{Y}$ maps it to the facial features such as eyes, nose and mouth.

1 Introduction

The Automatic extraction of human face and its relevant facial features such as head, eyes, nose, and mouth compromises quite important applications on surveillance and security. Birchfield[1] developed an elliptical head tracker algorithm to extract a face by utilizing the elliptic shape measure of a human face. This method maintains its robustness under the variations of illuminating light but it has a drawback that can not cope with the complicated background objects. Qian[8] and Yang[11] used the skin-color information with applying the skin-color filtering to extract a face. However, it would pick up the false regions if the background objects contain the color information similar to skin-color. Moreover, it is not immune to the effects of the varying illumination conditions. Kim[6] suggested a method that utilizes both skin-color information and the elliptic shape characteristics of a human face. The accuracy of locating a true face can be improved but it requires high amounts of computations and complexity. Genno[3] proposed non-contact imaging acquisition method for obtaining facial skin color information and simultaneously skin temperature. We use skin-color and its relevant skin temperature information to extract a face and apply image processing algorithms to obtain face boundary and the maximum horizontal projection value on it. Then, heteroassociative memory neural network model is used to extract eyes, nose, and mouth by supplying the relevant information related with the maximum projection value to the model.

J. Wang et al. (Eds.): ISNN 2006, LNCS 3972, pp. 40–48, 2006.

2 Face Detection

2.1 Extracting a Face Region

To extract the initial face candidates , we adopt the image acquisition system combining a CCD color imaging system with infrared imaging one as proposed by Genno[3] (Fig.1).

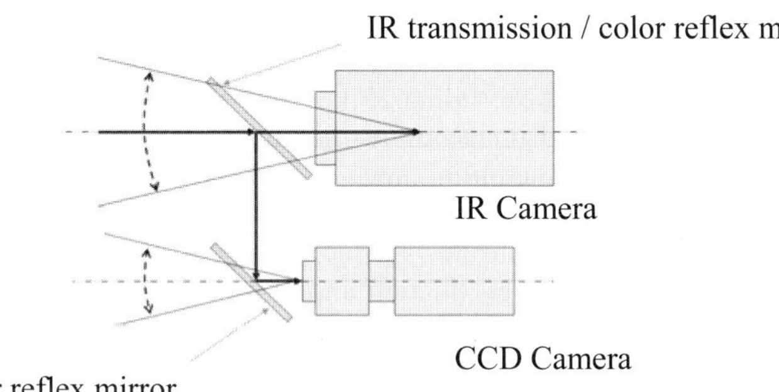

Fig. 1. The composite imaging acquisition system

We used a CCD Camera (Logitech V-UJ15, 320 pixels x 240 pixels) and IR camera (FLIR Therma CAM P20, 320 pixels x 240 pixels). Fig.2 shows an acquired sample CCD color image, and infrared one, respectively on a subject with mapping the identical focus on each imaging system.

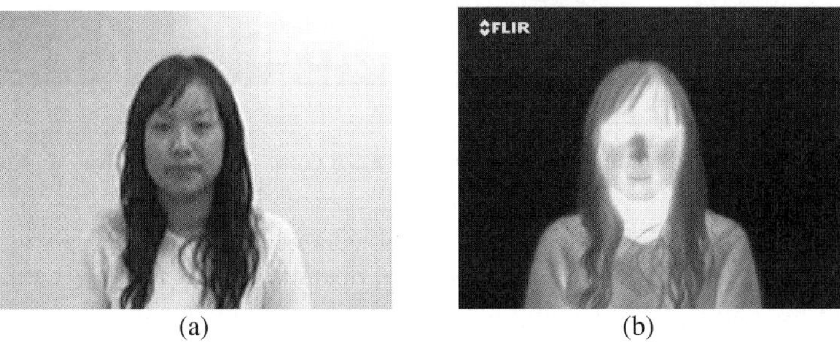

Fig. 2. (a) A captured CCD color image and (b) an infrared radiated one

To decorrelate the dependency of color features with respect to the illumination effects, a color plane transformation is performed to compress *RGB* color plane into YC_bC_r color domain as defined by Eq.(1) (Sharma[9]).

$$\begin{pmatrix} Y \\ C_b \\ C_r \end{pmatrix} = \begin{pmatrix} 0.257 & 0.504 & 0.098 \\ -0.148 & -0.291 & 0.439 \\ 0.439 & -0.368 & -0.071 \end{pmatrix} \begin{pmatrix} R \\ G \\ B \end{pmatrix} + \begin{pmatrix} 16 \\ 128 \\ 128 \end{pmatrix} . \tag{1}$$

The initial facial candidate regions $B(x,y)$ can be extracted by performing skin-color filtering with evaluating C_b and C_r color plane such as,

$$B(x, y) = \begin{cases} 1 & if\,(98 \le C_b \le 124) \cap (132 \le C_r \le 160) \\ 0 & Otherwise. \end{cases} \tag{2}$$

Here $B(x,y)$ is a binary image and '1' stands for the initial facial candidate regions in *RGB* color space. In similar fashion, we can also extract the facial candidate areas in infrared radiated image by utilizing the fact that the average human skin temperature ranges from 30°C to 34°C whereas the skin area covered with a cloth or hair has the scope of temperature value not exceeding 25°C. Thus, we can extract the initial facial regions in an infrared radiated image by restricting the temperature values between 30°C and 40°C. Fig. 3 shows the initial facial candidate regions in *RGB* color coordinates and infrared, respectively. In order to delete the isolated spurious regions, Morphological image processing (Gonzales[4]) such as opening and closing are applied. An opening is erosion followed by dilation and closing is dilation followed by erosion. Dilation is an operation that grows objects in a binary image controlled by a shape and a size of structuring element. In contrast with dilation, erosion shrinks objects in a binary image also controlled by a structuring element.

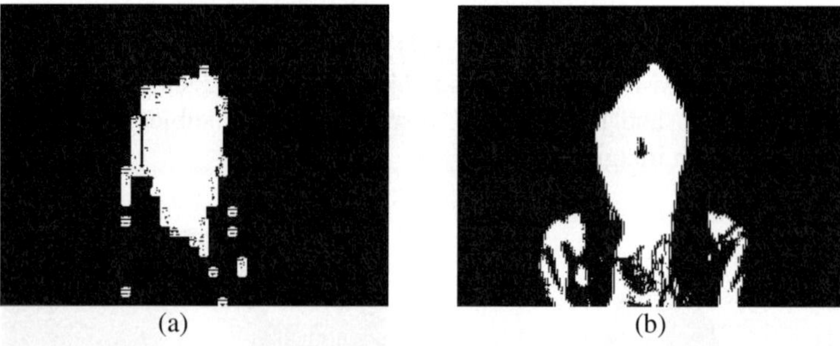

(a) (b)

Fig. 3. The initial facial candidate regions in (a) a *RGB* color and, (b) an infrared radiated one

In contrast with dilation, erosion shrinks objects in a binary image also controlled by a structuring element. Thus, Morphological opening deletes an object smaller than the structuring element and closing tends to join narrow breaks or fills holes smaller than the structuring element. We adopt an opening followed by a closing on Fig.3 (a) and (b), respectively. In this fashion, the facial candidate regions are refined by deleting the noisy objects that cannot be contained in the structuring element. Fig.4 shows the refined facial candidate regions by applying opening and closing successively. To merge the refined regions and delete the spurious regions further, the summation operation in terms of '1' pixel counts is performed in a line profile along the horizontal and vertical direction, respectively. Each count is compared with the threshold value and the pixels in the line profile are considered as a facial candidate region if the count exceeds the predefined threshold value. This operation is particularly useful to suppress the isolated neck regions if a subject's neck is exposed to the imaging acquisition system.

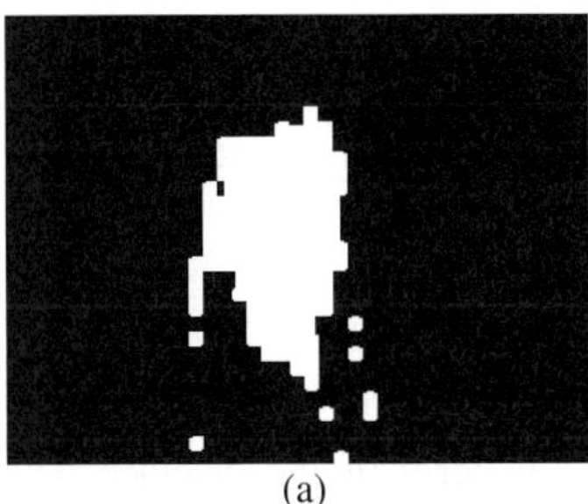
(a)

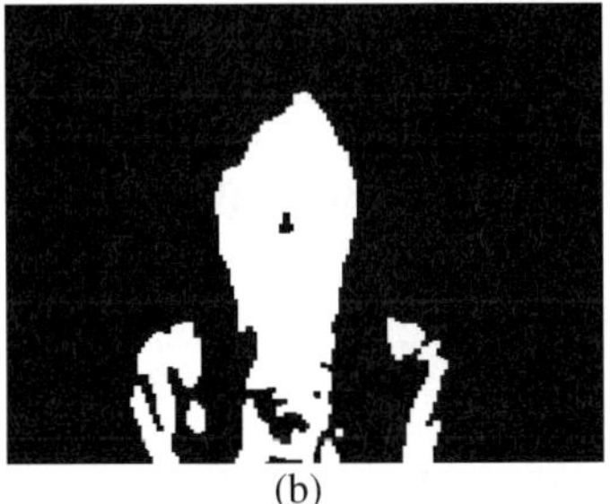
(b)

Fig. 4. Refined facial candidate regions by Morphological filtering in (a) a *RGB* color and (b) an infrared radiated image, respectively.

The final facial candidate regions can be determined by applying a logical 'AND' operation between refined facial candidate regions in *RGB* color and in infrared radiated image. Fig.5 shows the result of a logical 'AND' operation between the images depicted in Fig.4. (a) and in (b) (Kim[5]).

Fig. 5. The refined face region

The refined facial candidate regions are labeled by checking its 8-neighbor connectivity. The dimension of connectivity in each labeled blob is evaluated by sliding a 32 x 32 moving window and it is considered as a spurious region if the dimension of labeled object in terms of connectivity is smaller than that of the sliding window. After that, the features such as the orientation of the major axis of an ellipse fitted to the boundary of an object and a ratio of the major axis to the minor one are estimated to decide whether the labeled object is truly a facial object or not. The orientation θ and the ratio of the major axis to the minor one can be computed by evaluating a center point (x_c, y_c), 2nd moment features, μ_{xx}, μ_{xy}, μ_{yy} (Gonzales[4]). We assume that the ratio of the major axis to the minor one of a face object ranges from 1.0 to 1.8 and the amount of rotation ranges from 45° to 135° (a frontal-view face object has 90° rotation) if the labeled object is a face region.

2.2 Geometrical Model for Facial Features

After the elliptic features are applied to the refined face regions, the face boundary can be resolved and it is considered as a face mask to evaluate its confined *Y* component in YC_bC_r color domain. We used a vertical Sobel edge mask (Gonzales[4]) and applied horizontal projections to locate the eyes in terms of the maximum horizontal projection value in Y component. Once two eyes are located, the distance that crosses the center of eyes and extends to the face boundary, D_1 is extracted automatically. As Shih [10] suggested, we define the relative distances between the facial features such as eyes, nose, mouth, and its dimension, respectively, i.e.,

$$D_i / D_1, \ i = 2, 3, \ldots, 10. \tag{3}$$

Referring Fig.6, the geometrical properties in term of the distance between the facial features of a frontal face are defined as follows:

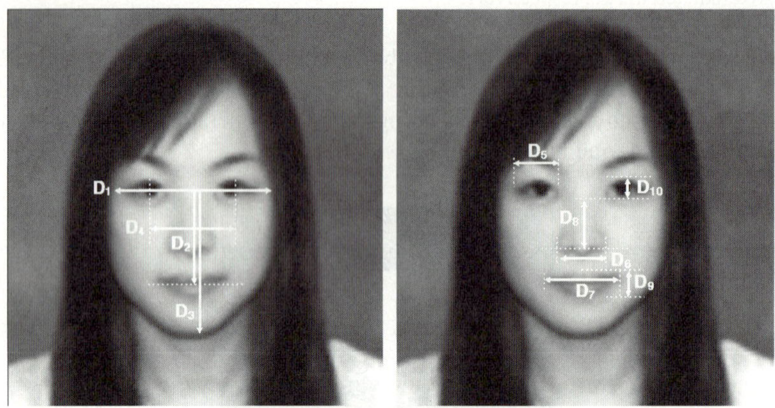

Fig. 6. The geometric facial features

(1) D_2: the vertical distance between the center of D_1 and the center of mouth.
(2) D_3: the vertical distance between the center of D_1 and the center of chin.
(3) D_4: the horizontal distance between the center of right and left eye.
(4) D_5: the horizontal width of eye.
(5) D_6: the horizontal width of nose.
(6) D_7: the horizontal width of mouth.
(7) D_8: the vertical width of nose.
(8) D_9: the vertical width of mouth.
(9) D_{10}: the vertical width of eye.

3 Heteroassociative Memory Neural Network Model for Extracting Facial Features

3.1 Training Stage

Once D_1 is automatically determined, we try to extract the facial features such as eyes, noses and mouth by utilizing heteroassociative memory neural network model (Fausett [2]). A input neuron vector X accepts a value of D_1 and the output neurons vector Y map it to D_i/D_1, $i = 2, 3, \ldots, 10$. In this case, a weight vector W is defined by

$$W = Y \cdot X^{-1} = [D_2, D_3, ..., D_{10}] \cdot D_1^{-1}. \quad (4)$$

To determine a weight vector *W*, we first obtain 300 Korean frontal-viewed face images (male: 150, female: 150, all subjects are in the mid-twenty) as suggested in Fig.1. The relative distance in CCD color images, D_i, i = 2, 3,...,10, is measured manually in order to decide a weight vector *W*. We randomly selected 30 males and 30 females as a training vector set for *W* and the rest of composite images are considered as a test vector set. To validate the similarity in the distance measures of between the randomly selected training vector set and the test vector set, we performed the independent samples t-test statistical analysis (Mould[7]) and Table 1 shows its result.

Table 1. The validation of a training vector set in terms of similarity in relative distance measures between a training vector set and a test vector set by independent samples t-test

		D_2/D_1	D_3/D_1	D_4/D_1	D_5/D_1	D_6/D_1	D_7/D_1	D_8/D_1	D_9/D_1	D_{10}/D_1
Mean	Training set	0.4892	0.7750	0.4860	0.2615	0.3174	0.3683	0.2755	0.1464	0.0980
	Test set	0.4861	0.7651	0.4770	0.2564	0.3257	0.3706	0.2790	0.1489	0.0982
Std.	Training set	0.0396	0.0614	0.0335	0.0194	0.0266	0.0313	0.0218	0.0130	0.0125
	Test set	0.0390	0.0541	0.0368	0.0177	0.0305	0.0355	0.0349	0.0188	0.0132
t	Training set v.s. Test set	0.576	1.324	-1.646	-1.909	1.889	0.451	0.735	0.976	1.248
Sig. (2-tailed)	Training set v.s. Test set	0.565	0.183	0.101	0.057	0.060	0.652	0.463	0.330	0.213

3.2 Test Stage

We tested 300 Korean frontal-viewed face images acquired by the composite imaging system as suggested in Fig.1. As we mentioned earlier, D_1 is automatically extracted for a given composite image by resolving face boundary and eyes locations in the horizontal projection direction. Then, D_1 is considered as the value of input neuron and the output neurons generate the normalized facial measures, D_i/D_1, i = 2, 3,...,10. Thus, D_i, i = 2, 3,...,10, are estimated and consequently the facial features such as eyes, nose and mouth can be described as shown in Fig.7.

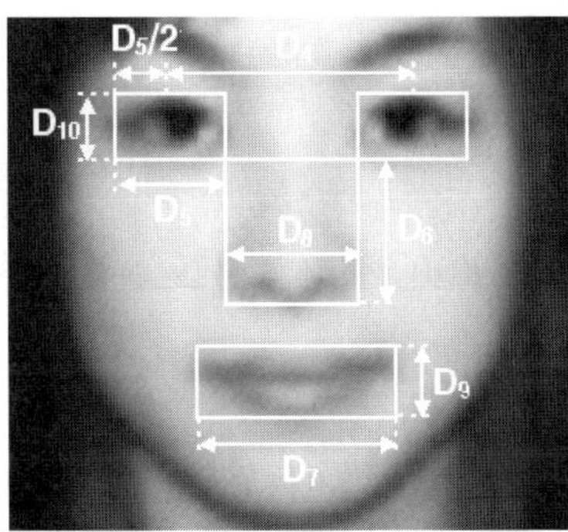

Fig. 7. The detected facial features: eyes, nose, and mouth

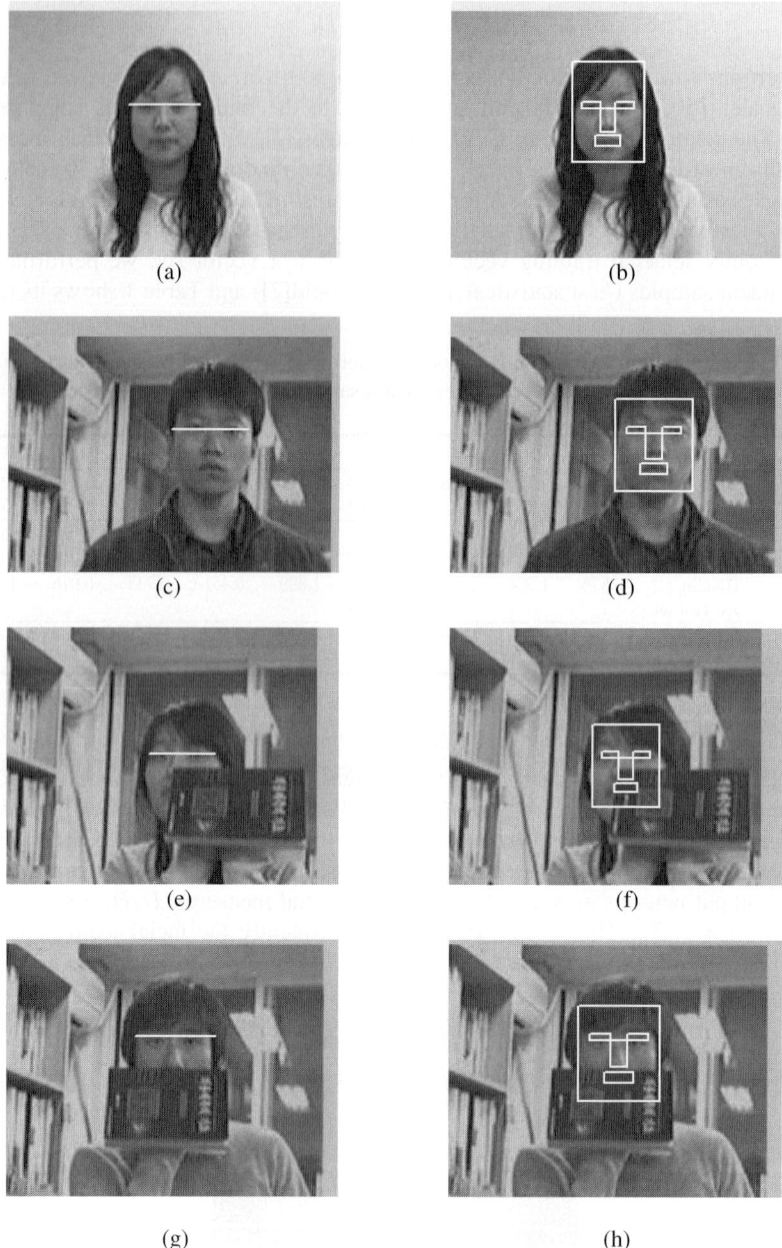

(g) (h)

Fig. 8. The experimental results of extracting the facial features by heteroassociative neural network model: (a) a face image superimposed by D_I, (b) the detected facial features on (a), (c) a face image in the complex background scene, (d) the detected facial feature on (c), (e) a partially blocked and slanted face image in the complex background scene, (f) the detected facial features on (e), (g) a partially blocked face image in the complex background scene, (h) the detected facial features on (g)

4 Experimental Results

Fig.8 shows the results of extracting the facial features. Fig.8 (a), (c), (e) and (g) are the original images in *Y* component with superimposed D_1. Here, D_1 is automatically extracted by detecting face boundary and eyes locations in the horizontal projection direction. Fig.8 (b), (d), (f) and (h) are the results of detecting the facial features: eyes, nose and mouth region which are depicted by its surrounding rectangular box, respectively.

Especially, Fig.8 (e) shows a partially blocked face image by a book in the complex background and the neural network model successfully detect the facial regions. However, if we observe Fig.8 (g) and (h), we can see that there exists a slight deviation in resolving the face facial region because a subject's forehead was covered completely with his hair. We believe that if a subject's forehead is significantly blocked by hair or some objects, the extraction algorithm of D_1 will slightly be erroneous and it consequently affects the final decision of neural network model.

5 Conclusions

We have proposed a novel approach for the extraction of human facial features such as eyes, nose and mouth based on a frontal-viewed CCD color image and its temperature values with utilizing a heteroassociative memory neural network model. The initial face candidates are determined by applying a logical 'AND' operation between the face candidate regions in YC_bC_r color domain and the corresponding infrared radiated image. In this way, we can avoid to extract false face regions especially a subject wears the clothes that has the similar color hue of his or her true face. Also the background objects which have the resembling color hue of human face would not be segmented into the face candidates. After the elliptic features are applied to get the face region, the face boundary and eyes positions can be resolved in terms of the maximum horizontal projection value in *Y* component. Once D_1 is extracted automatically as the distance that crosses the center of eyes and extends to the face boundary, this value is fed into the input neurons of heteroassocoative memory neural network model and the output neurons generates the relative measures to determine the facial features. Experimental results as depicted in Fig.8 show that the neural network model we used can successfully extract the facial features.

Acknowledgement

This work was supported by a grant of the "Development of Health Monitoring Technology for the Elderly," Ministry of Commerce, Industry and Energy, Republic of Korea.

References

1. Birchfield, S.: An Elliptical Head Tracker Proceeding of the 31st Asilmor, Conference on Signals, Systems and Computers (1997) 1710-1714
2. Fausett, L.: Fundamentals of Neural Networks, Architectures, Algorithms and Applications. Prentice Hall (1994) 101-125

3. Genno, H., Saijo, A., Yoshida, H., Suzuki, R., Osumi, M.: Non-Contact Method for Measuring Facial Skin Temperature, Industrial Ergonomics 19 (1997) 147-159
4. Gonzalez, R.C., Woods, R.E., Eddins, S.L.: Digital Image Processing Using MATLAB. Prentice Hall (2004) 334-358
5. Kim, K.S., Yoon, T.H., Han, M.H., Shin, S.W., Kim, I.Y., Lee, S.M.: Face Recognition and Thermal Feature Extraction for Health Monitoring. International Ubiquitous Healthcare Conference (2004) 85-86
6. Kim, Y.G., Han, J.H., Ahn, J.H.: Facial Regions Detection Using the Color and Shape Information in Color Still Images. Journal of Korea Multimedia Society 4 (2001) 67-74
7. Mould, R.F.: Introductory Medical Statistics, Institute of Physics Publishing Bristol and Philadelphia, 3rd Edition (1998) 115-128
8. Quian, R.J., Sezan, M.I., Matthews, K.E.: A Robust Real-Time Face Tracking Algorithms. International Conference on Image Processing (1998) 131-135
9. Sharma, G.: Digital Color Imaging Handbook. CRC Press, New York (2003) 574-587
10. Shih, F.Y., Chung, C.F.: Automatic Extraction of Head and Face Boundaries and Facial Features. Information Sciences 158 (2004) 117-130
11. Yang, J., Waibel, A.: A Real-Time Face Tracker. Proceeding of WACV, Proceedings 3rd IEEE Workshop (1996) 142-147

Recognizing Partially Damaged Facial Images by Subspace Auto-associative Memories

Xiaorong Pu, Zhang Yi, and Yue Wu

Computational Intelligence Laboratory, School of Computer Science and Engineering, University of Electronic Science and Technology of China, Chengdu 610054, China
puxiaor@uestc.edu.cn

Abstract. PCA and NMF subspace approaches have become the most representative methods in face recognition, which act in the similar way as a neural network auto-associative memory. By integrating with LDA subspace, in this paper, two subspace associative memories, PCA_{LDA} and NMF_{LDA}, are proposed, and how they recognize the partially damaged faces is presented. The theoretical expressions are plotted, and the comparative experiments are completed for the UMIST face database. It shows that NMF_{LDA} subspace associative memory outperform PCA_{LDA} subspace method significantly in recognizing partially damaged faces.

1 Introduction

Face recognition as a fundamental technology of biometrics has always been an attractive topic of research. The most representative recognition techniques frequently used in conjunction with face recognition are Principal Component Analysis (PCA) [1], Fishers Linear Discriminant Analysis (LDA) [2] and Non-negative Matrix Factorization with sparseness constraints ($NMFs$) [3], [8] subspace approaches, as well as neural network auto-associative memories [5]. $NMFs$ can be applied to recognize an individual based on the partial or local features of face images [3] by decomposing a non-negative face image matrix into both sparse and non-negative basis and encoding coefficient, while PCA is a classic method for face recognition based on global features by matrix factorizing to arbitrary sign 'eigenface' [1], [4]. In order to improve the performance in case of large variation in lighting direction and facial expression, LDA is applied to find an optimal subspace for classification by maximizing the ratio of the between-class scatter and the within-class scatter[2]. In this paper, by integrating LDA subspace, two reconstructed subspace PCA_{LDA} [6], [7], [8] and NMF_{LDA}[8] are adopted.

The goal of auto-associative memories is to find a set of connections between input units, so that when a portion of an input is presented as a memory key, the memory retrieves the complete pattern, filling in the missing components [9]. The fundamental problem in auto-associative memories is to find the suitable weights or connection strengths between the prototype vectors and the target outputs. Hebbian auto-associative memory sets up the connection weights as the sum of outer-product matrices from the prototype vectors[5]. The PCA_{LDA}

J. Wang et al. (Eds.): ISNN 2006, LNCS 3972, pp. 49–54, 2006.

subspace can be applied to recognize face images by mapping test face images into a specific subspace, which is equivalent to using a linear auto-associator to store and recall facial images[5], so does NMF_{LDA} subspace method.

The primary objective of this paper is to investigate, as subspace methods, why and how do PCA_{LDA} and NMF_{LDA} act as auto-associative memories. By using PCA_{LDA} subspace and NMF_{LDA} subspace associative memories, one of the significant advantages is that partially damaged facial images can be recognized easily without locate or detect the missing parts. We start our investigation by exploring the inherent connections between auto-associative memory and PCA_{LDA}/NMF_{LDA} subspace methods. The comparative experiments are carried out for recognizing partially damaged facial images on UMIST face database[10].

This paper is organized as follows. In Section 2, the relationship between Hebbian auto-associative memory and the PCA_{LDA}/NMF_{LDA} subspace auto-associators is discussed. Experiment results are summarized in Section 3 followed by conclusions in Section 4.

2 Review of Hebbian Auto-associative Memory and Subspace Auto-associators

In appearance-based approaches, a $2D$ facial image is viewed as an $i \times j$ vector in the image space. A set of face image samples can be represented as an $n \times m$ matrix $V = [V_1, V_2, \ldots, V_m]$, each column of which contains $n(n = i \times j)$ non-negative pixel values of one of the m facial images belonging to one of c classes.

2.1 Hebbian Auto-associative Memory

The weights or connection strengths of Hebbian auto-associative memory are represented by a matrix W_{Heb}, which is obtained by successively auto-associating each face vector V_k , and summing the resultant outer product matrices, with a formula [5],

$$W_{Heb} = \sum_{k=1}^{m} V_k V_k^T. \tag{1}$$

Recall of the k-th face from the memory is achieved by pre-multiplying the face vector V_k by the matrix W_{Heb} as

$$\hat{V_k} = W_{Heb} V_k, \tag{2}$$

where $\hat{V_k}$ represents the estimation of the k-th face by the memory. The quality of this estimation can be measured by computing the square coefficient of correlation between the vectors ${\epsilon_k}^2 = \|V_k - W_{Heb} V_k\|^2$.

2.2 Subspace Auto-associative Memories

Let us first look at the frameworks of a neural network auto-associative memory and a subspace auto-associator, as shown in Fig.1.

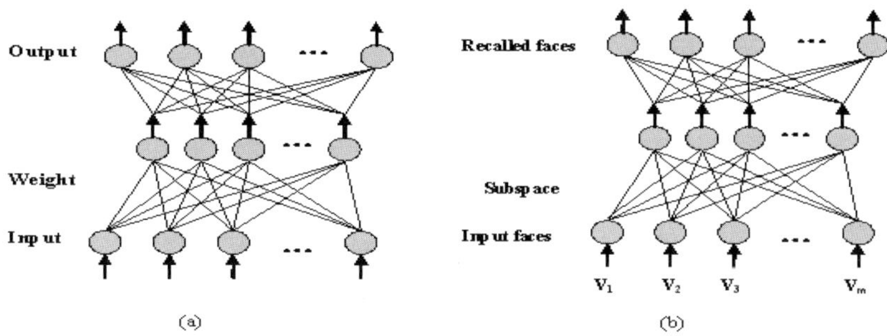

Fig. 1. (a)A Neural Network Auto-associator, (b) A Subspace Auto-associator

As shown in Fig. 1 (a), an association or weight is a link between the inputs and outputs of a network, which can be calculated by various rules, such as Hebbian rule, Kohonen rule. Obviously, As shown in Fig. 1 (b), both PCA_{LDA} subspace and NMF_{LDA} subspace perform as connect weights in auto-associative memories. Being similar to a neural network associative memory, a partially damaged facial image can be recognized by PCA_{LDA} subspace and NMF_{LDA} subspace auto-associators. As a classic neural network method, Hebbian auto-associative memory is able to fill the missing components when a portion of an input facial image is presented as a memory key, which is based on a set of connections between input units, while both PCA_{LDA} and NMF_{LDA} subspace auto-associators can recognize partially damaged faces based on their subspaces, i.e. 'eigen-fisher-faces' and NMF basis subspace. As a result, we argue that performing both PCA_{LDA} and NMF_{LDA} subspace auto-associators are equivalent to using Hebbian-type auto-associator because of their similar mathematical formulations (1)-(3) and (2)-(4)-(5), which indicate the inherent connections between a subspace and a neural network auto-associator.

As studied by Belhumeur [2] and Pu [8], the PCA_{LDA} subspace can be constructed by projecting PCA subspace into the LDA subspace as

$$W_{PL}^T = W_{LDA}^T W_{PCA}^T, \tag{3}$$

where $W_{PCA}(= [W_1\ W_2\ \dots\ W_r](r < m))$ is the set of the "r" orthonormal eigenvectors corresponding to the "r" largest orthogonal eigenvalues $\{\lambda_i | i = 1, 2, \dots, r\}$ of the covariance matrix C. The covariance matrix of a face set is calculated by $C = \frac{1}{m}\sum_{i=1}^{m}(V_i - \Psi)(V_i - \Psi)^T$ [1]. The LDA subspace is computed as $W_{LDA} = arg\max_W \frac{|W^T S_B W|}{|W^T S_W W|} = [W_1\ W_2\ \dots\ W_\mu]^T$, $\{W_i | i = 1, 2, \dots, \mu\}$ is the set of discriminant vectors of S_B and S_W corresponding to the c-1 largest generalized eigenvalues $\{\lambda_i | i = 1, 2, \dots, c-1\}$,i.e., $S_B W_i = \lambda_i S_W W_i$, S_B is the between-class scatter matrix calculated as $S_B = \sum_{i=1}^{c} N_i(b_i - \overline{b})(b_i - \overline{b})^T$, N_i is the number of samples in the i-th class C_i, and c is the number of classes, $\overline{b} = \frac{1}{m}\sum_{j=1}^{m} b_j$ is the mean image of the ensemble, and $b_i = \frac{1}{N_i}\sum_{j=1}^{N_i} b_j^i$ is the mean image of the i-th class C_i. The S_W is the within-class scatter matrix defined as $S_W = \sum_{i=1}^{c} S_{W_i}$, where S_{W_i} is the covariance matrix of class C_i.

Using these coefficients, we may recall a given face image V_k which may contain some noise as follows,

$$\hat{V_k} = W_{PL}F_k + \Psi, \tag{4}$$

where $F_k = W_{PL}^T(V_k - \Psi)$ is the feature vector of the image V_k, and $\Psi = \frac{1}{m}\sum_{i=1}^{m} V_i$ is the average face of the set. The $k-$nearest neighbor classifier is then used to calculate the Euclidian distance as $\varepsilon_k^2 = \|(F - F_k)\|^2$.

Non-negative matrix factorization(NMF) [4], [11] refers to the decomposition of a data matrix $V_{n\times m}$ into two matrices $W_{n\times r}(r < n)$ and $H_{r\times m}$, such that $V \approx W_{NMF}H_{NMF}$, where each column of V consists of n pixel values of one of the m facial images. The r columns of $W_{n\times r}$ are called basis images, and $H_{r\times m}$ is the matrix of encoding coefficients in one-to-one correspondence with a face in V. Unlike the orthogonality constraint of PCA, NMF does not allow negative entries in the matrix factors W_{NMF} and H_{NMF}.

Patrik [3] showed that NMF with sparseness constraints (NMFs) could lead to representations that were parts-based and match the intuitive features of facial images. The NMFs is defined as follows:

$$\begin{cases} \min_{W,H} \; D\left(V|WH\right) = \sum\limits_{i,j}\left(V_{ij} - (WH)_{ij}\right)^2 \\ \text{s.t.} \qquad W, H \geq 0, \text{ sparseness}(w_i) = S_w, \forall i, \text{ sparseness}(h_i) = S_h, \forall i \end{cases}$$

The sparseness measure based on the relationship between the L_1 norm and the L_2 norm is defined by $sparseness(x) = \frac{\sqrt{n} - ||x||_1/||x||_2}{\sqrt{n}-1}$, n is the dimensionality of x. Patrik proposed a projected gradient descent algorithm for NMF with sparseness constraints. For further details, refer to [3].

Pu [8] integrated LDA with the NMFs encoding coefficient vectors H_{NMF} rather than NMF basis, by $H_{NL}^T = W_{LDA}^T H_{NMF}^T = [H_1 \; H_2 \; \ldots \; H_t]^T (t \leq m)$. For a new face image V_k, the corresponding encoding coefficient is calculated based on the basis W_{NMF} by $H_k = W_{NMF}^{\dagger} V_k$. where $W_{NMF}^{\dagger}$ is the pseudoinverse of the basis matrix W_{NMF}.

In order to recall a given face V_k, its corresponding encoding coefficient H_k should be projected into the LDA subspace by $W_{LDA}^T H_k^T$ firstly. Then, a $k-$nearest neighbor classifier is used to calculate the Euclidian distance $\epsilon = [\epsilon_1, \; \epsilon_2, \; \ldots \; \epsilon_t]$ as $\epsilon_k^2 = \|H_{NL}^T - W_{LDA}^T H_k^T\|^2$.

After the H_k, corresponding to the minimum distance ϵ_k, was found out, the recalled face image can be calculated as

$$\hat{V_k} = W_{NMF}H_k. \tag{5}$$

Comparing Eq. 2 with Eq. 4 and Eq. 5, we can find that Hebbian-type associative memory model and PCA_{LDA}/NMF_{LDA} subspace associative memory share the same characteristics, even when they recognize partially damaged facial images.

3 Experiments

We have conducted experiments to compare the performance of PCA_{LDA} subspace auto-associative memory with that of NMF_{LDA} subspace auto-associative memory on the UMIST face database [10]. The UMIST database is a multiview database which consists of 575 gray-scale images of 20 subjects, each covering a wide range of poses from profile to frontal views as well as race, gender and appearance. We randomly selected 200 frontal facial images of 20 individuals among UMIST database, and resized them into 92×112.

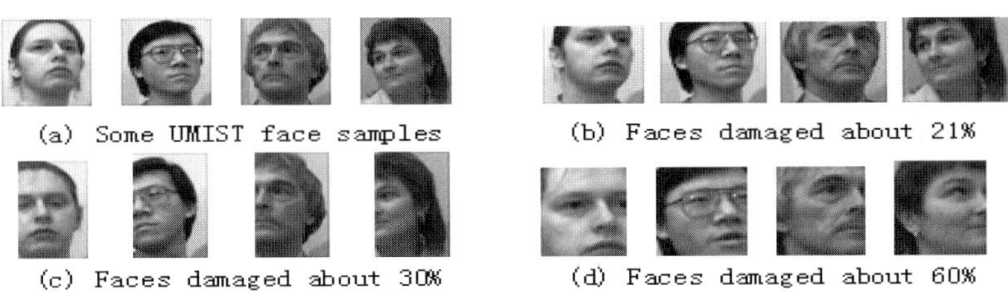
(a) Some UMIST face samples (b) Faces damaged about 21%
(c) Faces damaged about 30% (d) Faces damaged about 60%

Fig. 2. Some Prototype Facial Images in UMIST and Some Cropped Faces

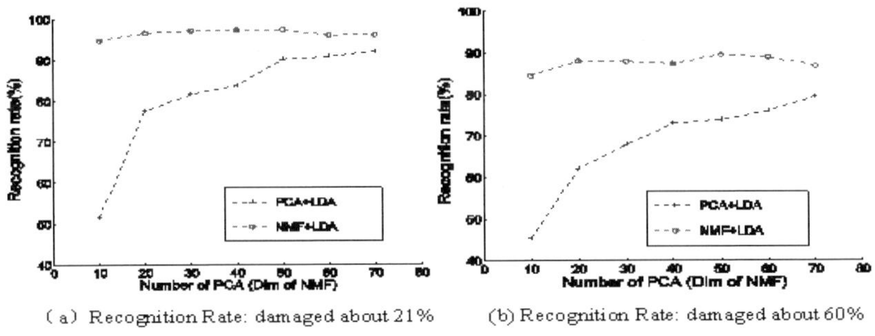

(a) Recognition Rate: damaged about 21% (b) Recognition Rate: damaged about 60%

Fig. 3. Average recognition rate of partially damaged facial images

There are 10 images for each person randomly partitioned into a training subset of 5 images and a test set of the other 5. All the facial images in test set are cropped partially, representing the percentage of information loss about 21%, 30% and 60%, respectively. Fig. 2 depicts some prototype UMIST facial images, and some cropped faces.The dimensionality $'r'$ of NMF and the number of principle components to represent the feature vectors were set to: $10, 20, 30, 40, 50, 60, 70$. The sparseness of the NMF basis was fixed at 0.75. An average of 800 NMF runs were executed.

The average recognition rates for differently damaged face images using PCA_{LDA} and NMF_{LDA} subspace auto-associative memories are shown in Fig. 3 (a) and (b), respectively. It can be seen that NMF_{LDA} subspace auto-associative memory achieved better recognition results than PCA_{LDA}.

4 Conclusions

Two subspace auto-associative memories are developed in this paper to recognize the partially damaged facial images. The experiments demonstrate the superiority of the parts-based subspace auto-associator,NMF_{LDA}, in partially damaged faces recognition in comparison with the global-based one,PCA_{LDA} subspace auto-associator. The experiments show that the NMF_{LDA} subspace auto-associative memory is less sensitive than the PCA_{LDA} subspace auto-associator to the information loss of testing patterns. And both of them can recall a stored facial image by specifying all or a portion of a key facial vector, as the Hebbian-type auto-associative memory does.

Acknowledgement

This work is supported by National Science Foundation of China under Grant 60471055 and the UESTC Youth Fund under Grant L08010601JX04030.

References

1. Turk, M.A., Pentland, A.: Eigenfaces for Recognition. Journal of Cognitive Neurosience 3 (1991) 71 - 86
2. Belhumeur, P.N., Hespanha, J.P., and Kriegman, D.J.: Eigenfaces versus Fisherfaces: Recognition Using Class Specific Linear Projection. IEEE Trans. Pattern Anal. Machine Intell. 19 (1997) 711 - 720
3. Patrik, O.H.: Non-negative Matrix Factorization with Sparseness Constraints, Journal of Machine Learning Research 5 (2004) 1457 - 1469
4. Lee, D. D., Seung, H. S.: Learning the Parts of Objects by Non-negative Matrix Factorization. Nature 401 (6755) (1999) 788 - 791
5. Valentin, D., Abdi, H., O'Toole, A.J.: Categorization and Identification of Human Face Images by Neural Networks: A Review of the Linear Autoassociative and Principle Component Approaches. Journal of Biological Systems 2(3) (1994) 413 - 429
6. Yambor, W.S.: Analysis of PCA-Based and Fisher Discriminant-based Image Recognition Algorithms. Technical Report CS-00-103, Colorado State University (2000)
7. Martinez, A.M., Kak, A.C.: PCA versus LDA. IEEE Transactions on Pattern Analysis and Machine Intelligence 23(2) (2001) 228 - 233
8. Pu, X., Zhang, Y. et al.: Face Recognition Using Fisher Non-negative Matrix Factorization with Sparseness Constraints, In: Wang, J., Liao, X., Zhang, Y. (eds.): Lecture Notes in Computer Science, vol. 3497. Springer-Verlag(2005) 112 - 117
9. Hagan, M. T. et al.: Neural Network Design, PWS Publishing Company (1996)
10. The UMIST database: http://images.ee.umist.ac.uk/danny/database.html
11. Paatero, P., Tapper, U.: Positive Matrix Factorization- A Non-negative Factor Model with Optimal Utilization of Error Estimates of Data Values. Environmetrics 5 (1994) 111 - 126

A Facial Expression Classification Algorithm Based on Principle Component Analysis

Qingzhang Chen[1,2], Weiyi Zhang[1] , Xiaoying Chen[1], and Jianghong Han[2]

[1] College of Information Engineering, Zhejiang University of Technology, Hangzhou 310032, China
qzchen@zjut.edu.cn
[2] School of Computer Science, Hefei University of Technology, Hefei 230009, China
qzchen@zjut.edu.cn

Abstract. In this paper, we try to develop an analytical framework for classifying human basic emotions. We try to find out what are the major components of each facial expression, what are the patterns that distinguish them from one another. We applied widely used pattern recognition technique-principle component analysis to characterize the feature point displacements of each basic human facial expression for each individual in the existing database. For faces not existent in the database, so called "novel face" in our experiment, we will first find the face in the database that has most likely neutral face to this individual, and base on an assumption that are widely accepted in cognitive science, we will classify this novel face to the category where the most similar one belongs, and classifying his/her facial expression using the so called "expression model" of the most similar individual. This kind of approach has never be exploited before, then we will examine its robustness in our experiment.

1 Introduction

With the quick development and popularization of computer and Internet, nowadays, more and more people resort to web-based/computer-based learning to acquire information or knowledge. But when you start your learning activity with curiosity and fascination, finally, as the learning task increase in difficult, you may experience confusion, frustration, or anxiety. Learning may be abandoned because of these negative feelings.

To solve this problem, we want computer to be intelligent and to interact naturally with us. Then, when people is likely to give up learning because of the negative feelings, a computer can detect user's feeling of frustration and redirect or motivate the user at such time. As a result, the learning activities will be greatly improved.

To fulfill this objective, what shall we do? The answer is that: we must give computer the ability to recognize, understand, even to have and express emotions. Scientists and experts had proposed various methods for this challenging task, but by far, no standard solution exists. In our project, we will choose one method, that is: we will teach computer how to recognize human basic emotions from their facial state. In this research, we will combine both theories in the arena of cognitive behavioral sciences and electrical engineering to target this problem. And do it under the condition

J. Wang et al. (Eds.): ISNN 2006, LNCS 3972, pp. 55 – 62, 2006.

that the coordinates of several feature points was previously obtained by some pre-processing techniques.

2 Related Work

When we begin our project, the first difficulty we must conquer is how to recognizing human expressions that relate to emotions since emotions are so subtle and so complex things. Fortunately, psychologist *Ekman* [1][2] has proposed six types of human basic emotions [3], which are *happiness, sadness, anger, surprise, disgust, contempt.* We try to develop an analytical framework for classifying these human basic emotions. Also, *Ekman* and *Friesen* have proposed that there are some prototypic facial expressions that are universal and correspond to basic human emotions.

Now, we can begin our project under these presuppositions above. To build a so called emotion-intelligent machine [4][5], people must design five functional elements, they are: *Input, Pattern Recognition, Reasoning, Learning* and *Output*. For our facial recognition system, we can classify these five elements into two functional parts. One is data extracts part, the other is reasoning and decision making part.

In this article, we focus on the reasoning and decision making part. So, for the data extracts part, we just have a glimpse.

Human body's raw data can be obtain by many kinds of sensors such as Video, EMG, BVP, GSR, Respiration and so on (Input). Then, the pattern analysis level extracts meaningful features in the row data (Pattern Recognition). For our project, we will use MPEG4 [6] (Motion Picture Expert Group) defined FDP (Facial definition parameter) points, or feature points, as a reference to classify the differences of facial features when expressing various emotions. We choose MPEG-4 FDP as our tools, because we expect that our proposed algorithm can be further integrated into an

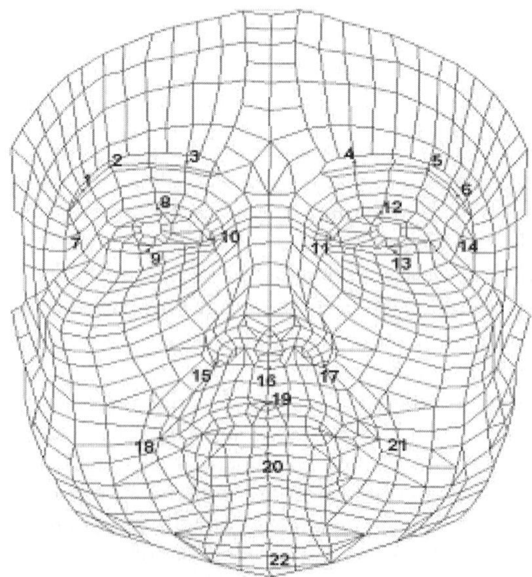

Fig. 1. 22 key MPEG4 FDP compatible feature points for our research

automatic system if automatic and accurate feature extraction. FDPs are used to customize a given face model to a particular face. The FDP set contains a 3D mesh, 3D feature points and optionally textures and other characteristics.

In our project, using the techniques above, we have extract 22 key MPEG4 FDP compatible feature points for our later research (Fig. 1).

3 System Overview

Now, we will talk about the reasoning and decision making part, which is our focused part. We will interpret this part with our facial expression system. And the facial emotions classification algorithm is the most important and most focused part in our project.

In this research we try to identify basic emotions such as neutral (no explicit emotion), happiness, anger, sadness, disgust, fear, surprise and contempt. Technically speaking, this task contains three stages[6][7], these stages show as follow:

(1) Extracting feature points coordinates in the face area;

(2) Using a reasoning algorithm to determine what expression the user is trying to express;

(3) Output a result claiming that certain expression is most likely being expressed, classifying into one of the basic emotions.

Since the first stage is not our focus in this work, we will manually extract 22 key MPEG-4 FDP compatible feature points for our later research. Then, put our attention on the second and third stages.

4 Human Facial Expression Transformation Model Building

In order to judge a particular facial expression that the observed individual might be expressing, we try to build a general and generic model for each of the basic emotion, then apply these models to each individual in the database to build so called "emotion model" of each basic emotion for every individual in the existing database, the resulting generic emotion model as a whole will be a reference when it comes to the even more challenging task of identifying the possible emotion of people that are not in the database. To build the model, we try to find that, among the 22 selected feature points, which ones are most important for certain emotion but not so important for another. We try to do build a quantitative analytical framework based on the results that we obtained by a statistical technique called PCA (principle component analysis) [8].

Now that PCA is a powerful and primary statistical tool for our project. Our final emotional transformation model is built based on PCA. We must understand some primary principle about it.

4.1 Principle Component Analysis as a Dimensional Reducing Technique

PCA is a statistical technique for forming new variables which are linear composites of the original variables and are orthogonal to each other. In most case, PCA provides us a means to represent a data set of p variables in a lower m-dimensional space where m is much less than p. while still reserving the critical information, the calculation is largely reduced. How well can we capture the configuration of the data in the reduced-dimensional space? To answer this, we have to know another property of PCA, which states that:

If a data set of p variables is represented by another set of variables in a lower m-dimensional space where m is much less than p

(1) The first new variable, or axis, obtained by PCA, accounts for the maximum variance of the data;

(2) The second new variable accounting for the maximum variance that has not been accounted by the first variable;

(3) The mth new variable accounts for the maximum variance that has not been accounted by the m-1 variables.

Typically the sum of variances of the new variables used to represent the data is used as a measure for the "information coverage" resulting from representing the data in a lower-dimensional space. So we will note that data reduction is not in terms of how much data has to be collected, as all the p variables are needed to form the principle components; rather it is in terms of how many nee variables are retained for further analysis. There is an important concept in PCA which is called "loading". It is defined as the simple correlations between the original and the new variables. Loadings give an indication of the extent to which the original variables are influential or important in forming new variables. That is, the higher the loading is the more influential the variable is in forming the principal components and vice versa. Since we have known some related techniques, we will start to build our emotion models.

4.2 Building Emotion Models

We try to rate the relative significance of the 22 points in performing certain expression. Since every point will have different movements, or motion vectors, when performing each specific expression, we can classify these patterns of FP movement to classify the basic emotions by applying statistical techniques on the collected motion vector information. We try to construct so-called "emotion equation" for each basic facial expression, all cross all individual in the database. The detailed steps are as the following:

4.2.1 Preprocessing

Firstly, we manually pinpoint all existing facial images in the database. For every neutral or expressive face, 22 Feature point coordinates are recorded. Then, we compute displacements in FPs for all possible pairs between neutral and expressive face of each individual

4.2.2 Principle Component Analysis for Each Basic Expression

Take a transformation from neutral to happiness for example, we now explain how to use the feature points coordinates can be manipulated to build a model for the recognizing basic emotions:

Firstly, we collect a 22*22 matrix, on each row we have 22 feature points motion vector of one person, and along 22 columns we have 22 two persons.

Secondly, we input this matrix into SAS, which is a commercially available statistical software package, to perform PCA.

Then, we will get the result which show the total variances covered by the principles and the respective loading of every feature on each of the new 3 dimension space in a component matrix, respectively. Also, we have seen the distribution of the new variables. We may further use these information revealed by variances and loadings of the PCA results for model building.

4.2.3 Emotion Model Building

Taken the example above, we have got the results after PCA analysis. From the result, we will find that they are three major components in this matrix, representing 87% of the variances of the matrix. And we can also know which point is most important to each component, by observing the loading of every feature on each principal component.

Then, we set a threshold of 0.8, only when one point has loading exceeding 0.8 on one component do we consider it has major contribution for that corresponding principal component. Since each component has different contribution to the total variances represented, we will also weight each component of the 22 points.

For every basic emotion, we repeat step 2 and 3, In the end, we can thus construct a "emotion equation" to model the transition from a neutral face to an expressive face. This generic model can be described as:

$$\begin{aligned} Ei &= SwiCi \\ &= w1\times(laFPa+lbFPb+\ldots)+w2\times(lcFPc+ldFPd+\ldots)+\ldots\ldots \end{aligned}$$

Where ***w*** is percentage of variances for each principal component(C), ***l*** is component loading for each feature point.

Taken the transformation from neutral to happiness for example, the resulting quantitative formula is as follows:

Happiness (FPd) =
68.099*(0.885*fid(3)+0.882*fid(1)+0.872*fid(2)+0.857*fid(8)+0.854*fid(7)+0.816*fid(9)+0.794*fid(22))+**11.048***(0.937*fid(16)+0.871*fid(19)+0.830*fid(17)+0.822*fid(15))

The other three formulas are:

Anger (FPd) =
63.914*(0.874*fid(17)+0.868*fid(21)+0.819*fid(13)+0.816*fid(5)+0.815*fid(12)+0.798*fid(6)+0.794*fid(11)+0.792*fid(14))+**17.284***(0.915*fid(8)+0.909*fid(9)+0.904*fid(1)+0.870*fid(3)+0.848*fid(2)+0.826*fid(10)+0.811*fid(7))+5.666*(0.878*fid(18)+0.769*fid(22))

Sad (FPd) =
78.027*(0.914*fid(11)+0.884*fid(15)+0.882*fid(4)+0.875*fid(10)+0.870*fid(8)+0.870*fid(3)+0.869*fid(9)+0.850*fid(2)+0.850*fid(16)+0.836*fid(5)+0.828*fid(12)+0.810*fid(13)+0.798*fid(17))+**7.803***(0.847*fid(18)+0.827*fid(20))

Surprise (FPd) =
72.881*(0.945*fid(7)+0.886*fid(9)+0.872*fid(10)+0.871*fid(8)+0.841*fid(15)+0.785*fid(2))+**10.398***(0.897*fid(6)+0.893*fid(14)+0.869*fid(12)+0.855*fid(5)+0.804*fid(4))+4.68*(0.807*fid(20)+0.790*fid(22));

4.2.4 Emotion Models for Each Individual in the Database

Following the steps above, we have had the generic model of each basic emotion in place. Then, we can input data gathered by the existing images in the database. That is, for every individual, since we already have four sets of motion vectors, each of them is composed of 22 feature point differences computed by comparing the relative movements of the corresponding points between expressive facial images and neutral facial images. We input these motion vectors into these four equations listed above, then for each individual, we have 4 numbers. At the end, we obtained a 4*22 matrix, which we call "Emotion Matrix", which can be used as the fundamental of our following expression recognition task, both for individuals inside or outside the existing database.

4.2.5 Facial Expression Recognition/Classification System

Now, we are going to state how to solve the most important problem in this article: recognizing basic emotion, the decision making flow is depicted below in Fig 2:

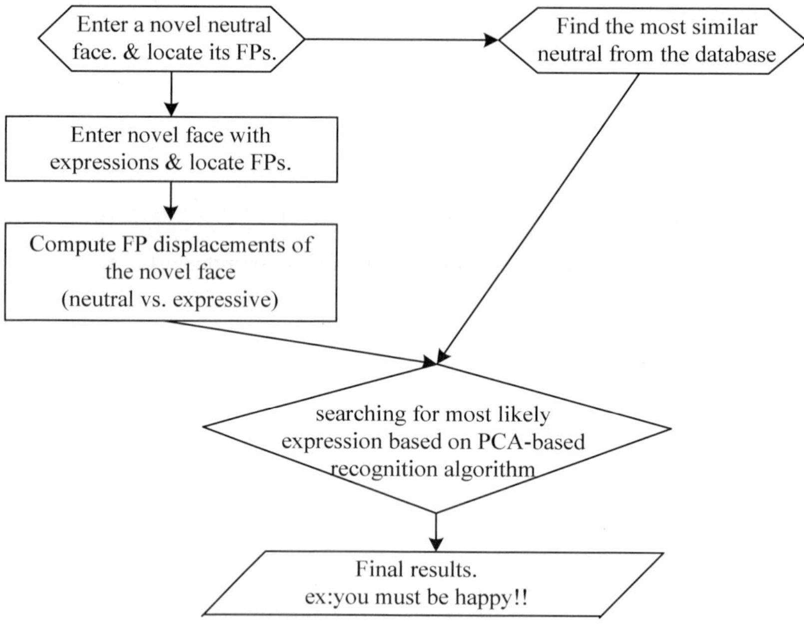

Fig. 2. Facial expression Classification

For example, we want to see if a user is angry or not, the recognition steps are as follows, the decision making task is composed of several steps:

(1) First we obtain the user's facial image expressing neutral emotion.

(2) Next, we compare the novel neutral face with neutral faces in the database, finding the most similar one. As a cognitive science theory states, people with similar neutral faces will also express their facial expression in a similar way[9]. Thus we use

the "emotion models" of the most similar existing individual as a reference when judging the possible basic emotion of the novel one, if he/she expresses any.

(3) The second angry picture of the novel person is taken, and we also extract its feature points. For this and the previously taken neutral face, feature point differences for every corresponding FDP points between these two pictures, or motion vectors, are calculated.

(4) We input the FP differences into the generic model obtained by PCA, and then we will get a number, say 100.

(5) We compare the number with 4 numbers; that is, 150, 178, 101, 451, standing for happiness, sadness, anger, and surprise of the most similar individual in the database, respectively. And since that (101-100) < (150-100) < (178-100) < (451-100), we claim that the user is most likely angry.

5 Test the Robustness of Our System

Now, our emotion models have been built. To text the robustness of our system, we design a mechanism to achieve this. The principle of mechanism is as follow:

We take human rating as a benchmark, and compare the machine recognition with it. If with the accuracy of human rating increasing, the accuracy of machine recognition also increases, and vice versa. In other words, the accuracy of these two methods forms almost a direct proportion. Then, our system is robust. So we do this examination, we invite a group of people to recognize the facial pictures, record the result, then, let the computer with our system to recognize these facial pictures. The outcomes answer for our assumption. There is a close relation between these two methods. By now, we have proved the robustness of or algorithm and our system.

6 Conclusion

Under the condition that the coordinates of several feature points was previously obtained by some pre-processing techniques. We founded out what are the major components of each facial expression that distinguish them from one another. We applied widely used pattern recognition technique-principle component analysis (PCA) to characterize the feature point displacements of each basic human facial expression for each individual in the existing database. For faces not being existent in the database, so called "novel face" in our experiment, we first find the face in the database that has most likely neutral face to this individual, and based on an assumption that are widely accepted in cognitive science, we will classify this novel face into the category where the most similar one belongs, and classifying his/her facial expression using the so called "expression model" of the most similar individual. Robustness of this approach is examined in our experiment, which obtained satisfactory results.

Acknowledgement

The Project Supported by Zhejiang Provincial Science Foundation of China (Y105388).

References

1. Ekman, P.: Facial Expression and Emotion. American Psychologist (1993) 384-392
2. Ekman, P.: Pictures of Facial Effect. Consulting Psychologist (1976)
3. Ekman, P.: Strong Evidence of Universals in Facial Expressions: A Reply to Russell's Mistaken Critique. Psychological Bulletin 115(2) (1994) 268-287
4. Picard, R.W., Vyzas, E., Healey, J.: Toward Machine Emotional Intelligence: Analysis of Affective Physiological State. IEEE Trans. Pattern Analysis and Machine Intelligence 23(10) (2001) 1175-1191
5. Picard, R.W.: Affective Computing. MIT Press (1997)
6. Pantic, M., Rothkrantz, L.J.M.: Automatic Analysis of Facial Expressions: The State of the Art. IEEE Trans. Pattern Analysis and Machine Intelligence 22(12) (2000)
7. Pantic, M., Rothkrantz, L.J.M.: An Expert System for Automatic Analysis of Facial Expressions. Imagine and Vision Computing 18(11) (2000) 881-905
8. Stork, D.G., Hart, P.E.: Pattern Classification. 2nd edn. Wiley Interscience, New York (1994)
9. Hong, H., Neven, H., von der Malsburg, C.: Online Facial Expression Recognition Based on Personalized Galleries. In: Third IEEE International Conference on Automatic Face and Gesture Recognition. IEEE Computer Society, Washington, DC, USA (1998) 354-359

Automatic Facial Expression Recognition

Huchuan Lu, Pei Wu, Hui Lin, and Deli Yang

School of Electronic and Information Engineering, Dalian University of Technology, Dalian, Liaoning Province, China
lhchuan@dlut.edu.cn

Abstract. We present a fully automatic real time system for face detection and basic facial expression recognition from video and images. The system automatically detects frontal faces in the video stream or images and classifies each of them into 7 expressions. Each video frame is first scanned in real time to detect upright-frontal faces. The faces found are scaled into image patches of equal size and sent downstream for further processing. Gabor energy filters are applied at the scaled image patches followed by a recognition engine. Best results are obtained by selecting a subset of Gabor features using AdaBoost and then training Support Vector Machines on the outputs of the features selected by AdaBoost.

1 Introduction

In this paper we present a fully automatic real time system for face detection and basic facial expression recognition from video and images. The system automatically detects frontal faces in the video stream or images and classifies each of them into 7 expressions: neutral, happiness, anger, fear, disgust, sadness and surprise. The method we present here is a combination of several typical methods. First, the images are converted into a Gabor magnitude representation, using a bank of Gabor filters. Then, we use AdaBoost to select a subset of Gabor features and train Support Vector Machine (SVM) on the outputs of the features selected by AdaBoost. This method was first researched by Bartlett [2]. Our report is based on a series of experiments comparing spatial frequency ranges, feature selection and recognition techniques. The combination of AdaBoost and SVM enhances both speed and accuracy of the system. It can reach a high level of accuracy above 90% in real time.

2 Facial Expression Analysis System

2.1 Preprocessing and Feature Extraction

The facial expression system is trained on Cohn Kanade's database [3]. It includes a series of facial expressions video sequences and each display is begun and ended with a neutral face. Two peak frames of the video sequences are selected per subject to form experiments database from which the training images are random selected, and the rest images are used for testing. We use a real time face detection system based on [4].

J. Wang et al. (Eds.): ISNN 2006, LNCS 3972, pp. 63 – 68, 2006.

In order to gain precise results, our experiments labeled face area manually for training images. The located faces are rescaled to 48*48 pixels. The images are converted into a Gabor magnitude representation, using a bank of Gabor filters at 8 orientations and 5 spatial frequencies. After that, an image is converted into 40 images with different scales and orientations and the features are the individual Gabor filters. So, there are 48*48*40=92160 possible features for a 48*48 pixels image.

2.2 Combination of SVM with AdaBoost

AdaBoost which is based on selecting a series of simple classifiers to form a strong classifier is a method to boost classification. In this system, we use AdaBoost to select features instead of being a classifier. A subset of these features is selected by AdaBoost from 92160 Gabor features. AdaBoost training continues until the classifier output distributions for the positive and negative samples are completely separated by a gap proportional. We select 100 features for each of the binary expression classifiers in our experiment and we prove that it gives better performance than others.

We explore training SVM classifiers on the features selected by AdaBoost. The combined classifiers are informally called AdaSVM in abbreviation of Adaptive Boosting Selected Feature representations in Support Vector Machines. We examine the effect of feature selection by AdaBoost on the number of support vectors. Smaller numbers of support vectors proffer two advantages: (1) the classification procedure is faster, and (2) the expected generalization error decreases as the number of support vectors decreases. Feature selection by AdaBoost reduces the number of support vectors employed by the nonlinear SVM in the AdaSVM system, to 12 to 26 percent, while linear SVM 10 to 33 percent and nonlinear SVM 14 to 43 percent [2]. We compare the number of support vectors for happy VS neutral using these different methods (see table 1). The number of training samples is 180.

Table 1. The number of support vectors of different methods

	Linear SVM	Nonlinear SVM(RBF)	Nonlinear SVM(RBF) in AdaSVM
Number of support vectors	59	78	33

3 Experiments

3.1 Experiments for Expressions Recognition

In order to solve multiclass facial expression recognition, we first train binary classifiers including one-versus-one (1 VS 1), one-versus-all (1 VS all). The simplest strategy is to train 1 VS all. The classifier is trained taking the examples from one kind of facial expression as positive and the examples from all other expressions as negative, i.e. happy VS nonhappy. 1 VS 1 method is pairwise partitioning that SVMs are trained to discriminate all pairs of emotions.

The database for experiments has been introduced as before, that the faces are labeled manually for training and scaled into 48*48 pixels. To reduce the unbalance of the expressions, we choose identify-matched pairs as the training set in 1 VS 1 method, where for example, the happy VS surprise SVM is trained on only those subjects who gave samples of both happiness and surprise. But in practice, we broaden the restriction as the images of some expression are too few. While in 1 VS all the strategy employs a larger number of training examples which dilutes identity effects.

Table 2 shows the results of 1 VS 1 recognition that uses identify-matched pairs as training set. Identify-matched pairs will help a lot to achieve higher recognition rate. According to the expression pairs presented here, we can get that some of them are difficult to classify directly by eyes too. That is the season why there still are several expression pairs with lower recognition rate than others in this table. With the same as people do, this system reaches different recognition rates for different facial expressions.

Table 2. 1 VS 1 recogniton rates (%)

	happy	fear	surprise	sad	disgust	anger	neutral
happy	------	91.49	96.30	100	98.08	95.09	100
fear	------	------	97.37	88.16	94.05	88.04	92.96
surprise	------	------	------	96.70	98.96	100	94.12
sad	------	------	------	------	90.79	90	85
disgust	------	------	------	------	------	80.68	96.30
anger	------	------	------	------	------	------	88.18
neutral	------	------	------	------	------	------	------

1 VS all method trains 7 SVMs with one kind of expression as positive and others as negative. The size of the training set or testing set is different for different expression pairs. The results (see table 3) show that all of them could reach a satisfying correct rate except neutral VS nonneutral and fear VS nonfear.

Table 3. 1 VS all recognition rates (%). Here the "non" is related to every expression, for example the first column is "hap. VS non" which means "happy VS nonhappy".

happy VS non	fear VS non	surprise VS non	sad VS non	disgust VS non	anger VS non	neutral VS non
95.798	86.554	96.638	91.176	95.798	93.277	81.512

In the system presented here, the SVM outputs are combined to make a 7 alternative forced choice. The binary classifiers used here are 21 SVMs that are got from 1 VS 1 experiment and they are combined by voting to get the output. The procedure counts the number of all classifiers aligned with each emotion. For example, if one SVM indicates happiness and not surprise, happiness gets +1 and surprise gets -1. These votes are summed over all of the SVMs. The expression getting the most votes is the recognized one. Here the testing set is 626 images. One experiment is to recognize the images, manually labeled face images, and the recognition rate is 88.6581%. The other experiment is to recognize original images with automatically face detection and the correct rate is 73.1629%. The recognition rate decreases using the automatic face detection because the face location is not as precise as that labeled manually.

3.2 Comparison

We also train SVM classifiers and AdaBoost classifiers directly on the output of the 96120 Gabor features. They perform no better than AdaSVM. The experiments use the same training set and testing set to compare the performances. We show 7 groups of expression pair here (see table 4). Here SVM uses RBF kernel function.

Table 4. Experiments results of comparing with SVM and AdaBoost (%). Here shows 7 groups of recognition rates for using AdaSVM, AdaBoost and SVM on a bank of Gabor features.

	AdaSVM	AdaBoost	SVM
happy vs fear	90.9091	90.9091	81.8182
happy vs sad	98.4848	96.9697	96.9697
happy vs neutral	99.1379	96.5517	93.9655
happy vs anger	94.6429	91.0714	85.7143
happy vs disgust	100	96.5517	96.5517
happy vs surprise	97.6744	100	97.6744
happy vs nonhappy	94.958	95.7983	94.5378

We also compare the time used by the three different methods for training and recognition on the expression pair happy VS neutral. Here the training set has 180 images and 116 images are in the testing set. The system presently operates on a computer of 2.0GHZ CPU and 512M memory. Table 5 shows the time used for every process.

Table 5. Comparison of the experimental speed. Here shows the time of every method used in training and testing process. The "Gabor" column shows the time for Gabor wavelet transforming of the input images and the following columns show the time of different methods working on the Gabor features.

	Gabor	AdaSVM	AdaBoost	SVM (RBF)
Training	4' 4''	206'	204'	338'
Testing	2'46''	54''	56''	84''

It is shown that, according to the recognition rates, AdaBoost+Gabor method performs better than SVM+Gabor method, but AdaSVM+Gabor method performs the best. Table 5 shows that SVM+Gabor method costs more time to calculate the results than any others. So, AdaSVM+Gabor method which we explore has the obvious superiority and worth doing.

3.3 Real Time Expression Recognition from Video

We combine the face detection and expression recognition into a system that operates on live digital video in real time. When the video is played, the system first detects face area automatically and then recognizes the expression and shows the result on the screen. We have done video tests on our own database and it performs very well. Our database consists of 7 expressions of 6 persons (3 are male and 3 are female), and every person has a video stream for each of the expression. The video begins with neutral expression and ends with the peak of target expression. The frame is 768*576 with uniform lighting. We do experiments on real time facial expression recognition using our database. The training images are all selected from Cohn Kanade's database and our database are only used for test. Fig.1 shows the output of the expression recognizer for a test video in which the subject poses a series of facial expressions from neutral to happy.

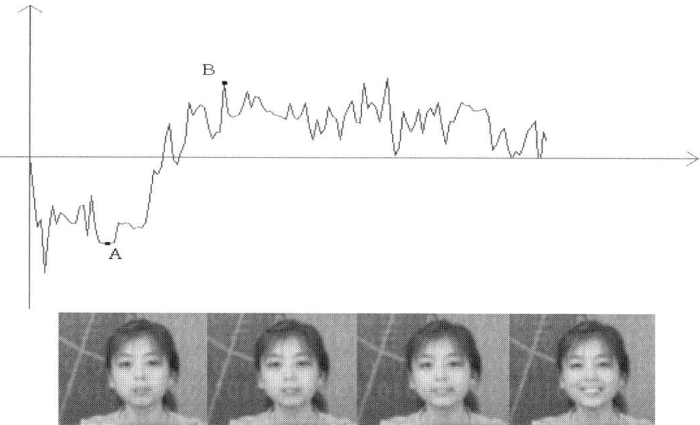

Fig. 1. Expression recognition from video. Dot A refers to the first image and dot B refers to the last image. The expression started with neutral and the curve is below the horizontal axis, when the expression changes the curve goes over the axis and finally reaches the peak when the expression reaches the target.

4 Conclusions

We present a system applied to the problem of fully automatic recognition of facial expressions, including Gabor wavelet transform, AdaBoost and support vector machines. We report results on a series of experiments comparing methods for multiclass decisions, spatial frequency ranges, feature selection methods, and recognition engines. Best results are obtained by selecting a subset of Gabor features using AdaBoost and then training Support Vector Machines on the outputs of the features selected by AdaBoost. The combination of AdaBoost and SVMs enhances both speed and accuracy of the system. Its performance in Cohn Kanade's database is averagely 93.4% for any expression pair recognition and 88.6% for 7-way forced choice.

Our results suggest that user independent fully automatic real time coding of basic expressions is an achievable goal with present computer power, at least for applications in which frontal views can be assumed. The following problem we will study on is how to reach high accuracy for 7 basic expressions recognition and how to solve the problem by a simple system, after the images are preprocessed by a bank of Gabor filters.

Acknowledgments

The Cohn Kanade database of facial images is used in portions of the research in this paper. So the authors would like to thank the Cohn Kanade Technical Agent. We also sincerely thank suggestions of Senior Research Scientist Wataru Ito and Dr. Yuanzhong Li at Imaging Software Technology Center, Research & Development Management Headquarters of FUJI PHOTO FILM CO., LTD.

References

1. Lyons, M., Budynek, J., Plante, A., Akamatsu, S.: Classifying Facial Attributes Using a 2-D Gabor Wavelet Representation and Discriminant Analysis. Proceedings of the 4th International conference on Automatic Face and Gesture Recognition (2000) 202–207
2. Littlewort, G., Bartlett, M. S., Fasel, I., Chenu, J., Movellan, J. R.: Analysis of Machine Learning Methods for Real-Time Recognition of Facial Expressions from Video. MPLab TR (2003)
3. Kanade, T., Cohn, J.F., Tian, Y.: Comprehensive Database for Facial Expression Analysis. Proceedings of the 4th IEEE International Conference on Automatic Face and Gesture Recognition (2000) 46–53
4. Viola, P., Jones, M.: Robust Real-time Object Detection. Technical Report CRL 2001/01, Cambridge Research-Laboratory (2001)
5. Osuna, E., Freund, R.: An Improved Training Algorithm for Support Vector Machine. Proceedings of 1997 IEEE Workshop on Neural Networks for Signal Processing (1997) 276-285
6. Platt, J.: Fast Training of Support Vector Machines using Sequential Minimal Optimization. MIT Press (1998)
7. Fan, R. E., Chen, P. H., Lin, C.J.: Working Set Selection Using the Second Order Information for Training SVM. http://www.csie.ntu.edu.tw/~cjlin/papers/quadworkset.pdf (2005)
8. Chen, P. H., Fan, R. E., Lin, C.J.: A Study on SMO-type Decomposition Methods for Support Vector Machines, Technical report. Department of Computer Science, National Taiwan University (2005), http://www.csie.ntu.edu.tw/~cjlin/papers/generalSMO.pdf

Facial Expression Recognition Using Active Appearance Model

Taehwa Hong[1], Yang-Bok Lee[2], Yong-Guk Kim[2], and Hagbae Kim[1]

[1] Department of Electrical and Electronic Engineering, Yonsei University, Seoul, Korea
{riccati, hbkim}@yonsei.ac.kr
[2] School of Computer Engineering, Sejong University, Seoul, Korea
yangbok@empal.com, ykim@sejong.ac.kr

Abstract. This paper describes a facial expression recognition system based upon Active Appearance Model (AAM), which has been typically used for the face recognition task. Given that AAM has been also used in tracking the moving object, we thought it could be effective in recognizing the facial expressions of humans. Our results show that the performance of the facial expression recognition using AAM is reliably high when it combined with an enhanced Fisher classification model.

1 Introduction

It is known that there are six prototypical (or basic) facial expressions, which are universal across human ethnicities and diverse cultures: surprise, fear, sadness, anger, disgust and happiness. Such foundation provides a convenient framework, by which we test and compare facial expression recognition systems.

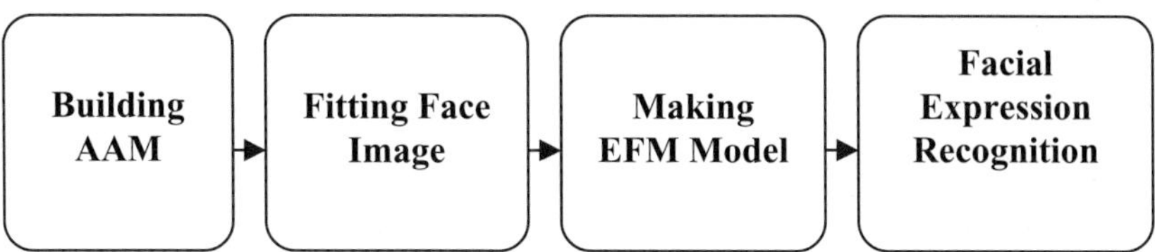

Fig. 1. Block diagram of facial expression recognition

2 Active Appearance Models (AAMs)

Active Appearance Models (AAMs) [1], [2], [3], [4], [5], [6], first proposed in [7], Active Blobs [8], [9] and Morphable Models [10], [11], [12] are nonlinear, generative, and parametric models of a certain visual phenomenon and commonly used for

J. Wang et al. (Eds.): ISNN 2006, LNCS 3972, pp. 69 – 76, 2006.

tracking faces in video or to recognize the face. AAMs are fundamentally constructed by applying the Procrustes analysis followed by the principal components analysis (PCA) to a collection of training images of faces with a mesh of canonical feature points on them. AAMs are then fit frame-by-frame to input video to track the face through the video. The most frequent application of AAMs to date has been face modeling [13]. However, AAMs may be useful for other phenomena too [9], [11].

In a typical application, the first step is to fit the AAM to an input image, i.e. model parameters are found to maximize the "match" between the model instance and the input image. The model parameters are then used in whatever the application is. For example, the parameters could be passed to a classifier to yield a face recognition algorithm. Many different classification tasks are possible. In [13], for example, the same model was used for face recognition, pose estimation, and expression recognition. Fitting an AAM to an image is a nonlinear optimization problem. The usual

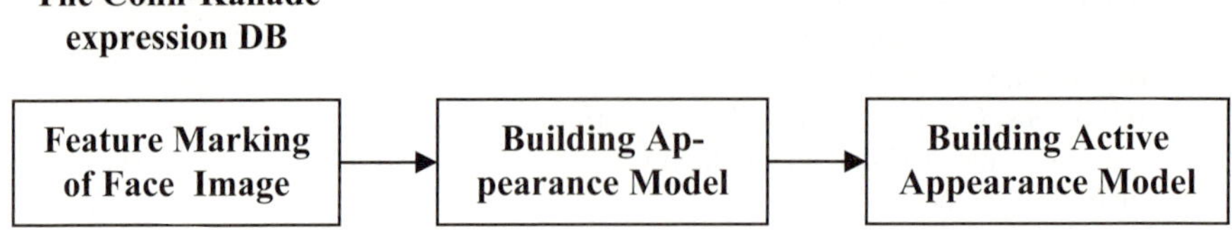

Fig. 2. Block diagram of building an AAM model

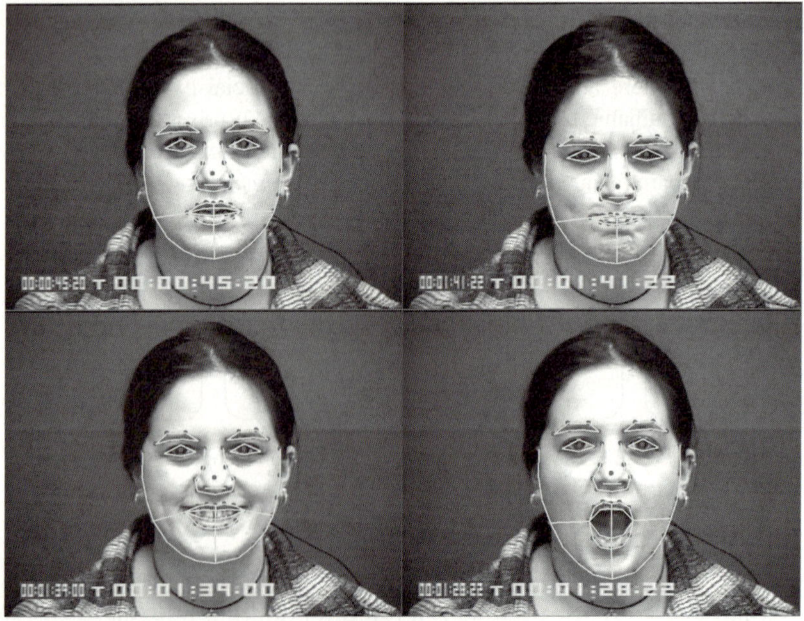

Fig. 3. Landmarking features on four facial expression images of a person

approach [1], [4], [5] is to iteratively solve for incremental additive updates to the parameters (the shape and appearance coefficients.) Given the current estimates of the shape parameters, it is possible to warp the input image onto the model coordinate frame and then compute an error image between the current model instance and theimage that the AAM is being fit to. In most previous algorithms, it is simply assumed that there is a constant linear relationship between this error image and the additive incremental updates to the parameters. The constant coefficients in this linear relationship can then be found either by linear regression [1], [6], [7] or by other numerical methods [4], [5].

2.1 Building Models

We build an AAM with 272 facial expression images from the Cohn-Kanade facial expression database. For each facial image, 68 feature points are manually marked. The AAM instance is then built by warping the appearance to each mesh in the shape model as shown in Fig. 2. The examples of landmarking features on four facial expression images of a person and the instances of the AAM are shown in Fig. 3 and Fig. 4, respectively.

2.2 AAM Fitting

AAM Fittings were performed for the remaining 172 images, which were not included in generating the AAM model. Both Figure 5 and 6 show reconstruction images and vertex points of the images taken through the fitting process.

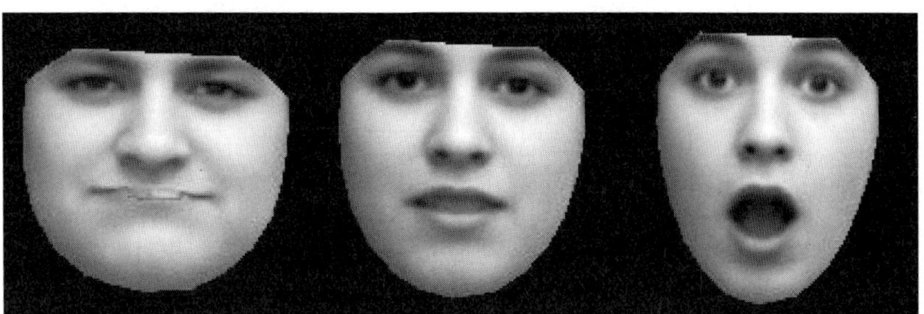

Fig. 4. The instances of the AAM

3 EFM Classifiers

We present in this section an EFM classifier that determines the discriminative features for the reduced image space. The EFM is introduced to improve on the generalization ability of the standard Fisher linear discriminant (FLD) based methods. The EFM firstly applies Principle component analysis (PCA) for dimensionality reduction before proceeding with FLD type of analysis. Then, it discriminate the reduced PCA subspaces.

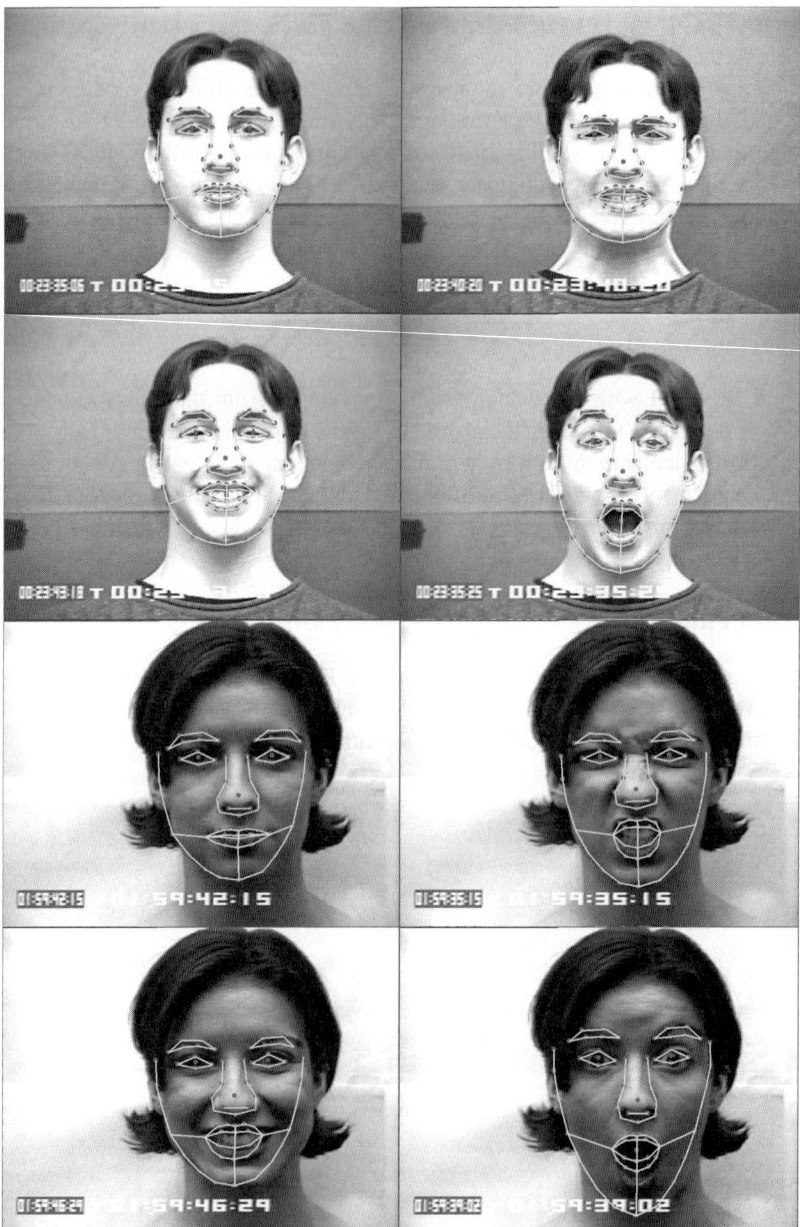

Fig. 5. The examples of AAM fitting

3.1 EFM Algorithm

Let Y be a random vector representing the lower dimensional feature. Let $w_1, w_2, ..., w_L$ and $N_1, N_2, ..., N_L$ denote the classes and the number of images within each class,

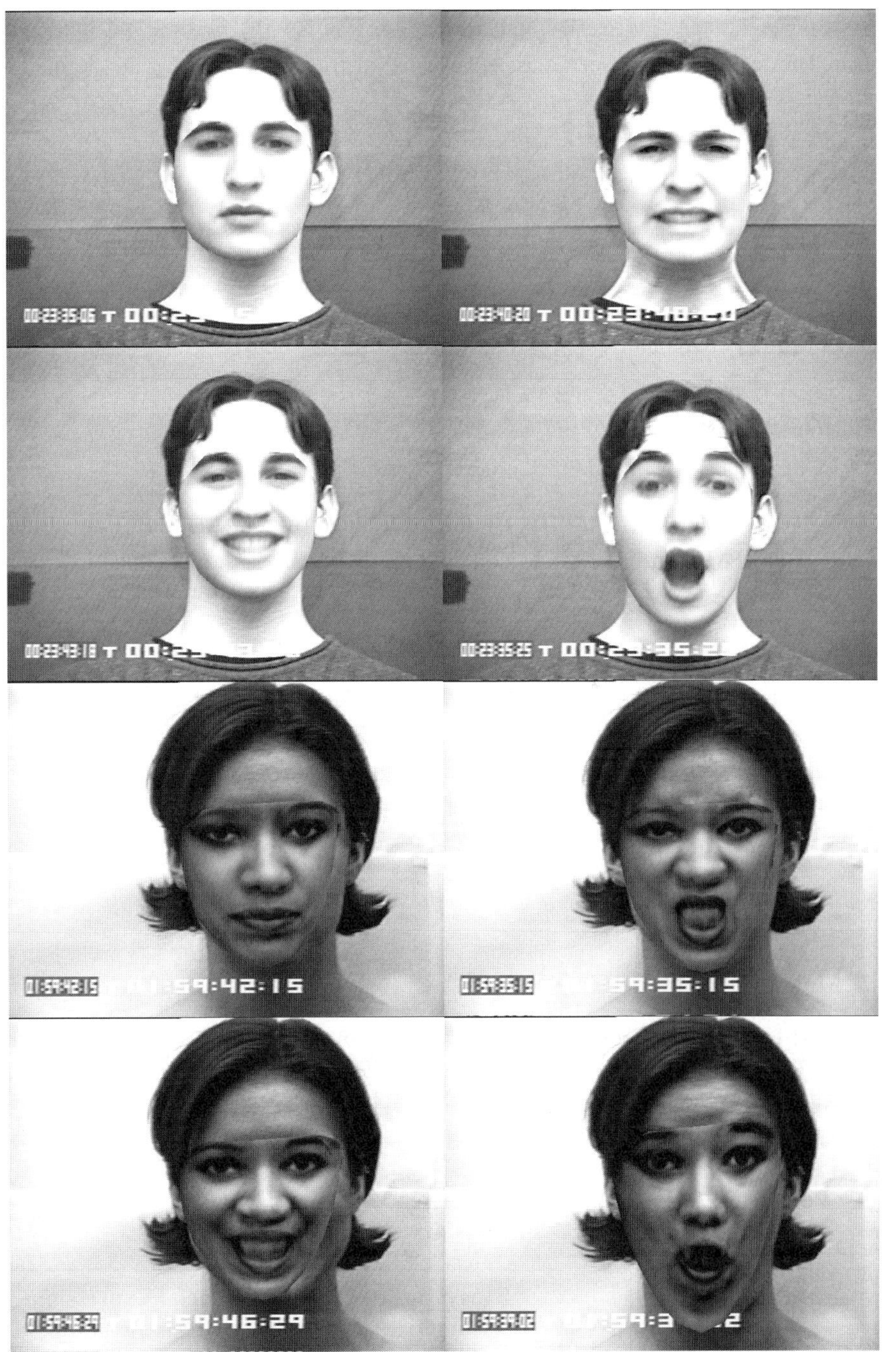

Fig. 6. The examples of reconstructed face images

respectively. Let $M_1, M_2, ..., M_L$ and M be the mean of the classes and the grand mean. The within-and between-class covariance matrices $\sum w$ and $\sum b$ are defined as follows:

$$\sum w = \sum_{i=1}^{L} P(w_i) E\{(Y - M_i)(Y - M_i)^t \mid w_i\}, \tag{1}$$

$$\sum b = \sum_{i=1}^{L} P(w_i)(M_i - M)(M_i - M)^t, \tag{2}$$

The EFM firstly diagonalizes the within-class covariance matrix $\sum w$.

$$\sum w \Xi = \Xi \Gamma \text{ and } \Xi' \Xi = I, \tag{3}$$

$$\Gamma^{-1/2} \Xi^t \sum w \Xi \Gamma^{-1/2} = I, \tag{4}$$

where Ξ, Γ are the eigenvector and the diagonal eigenvalue matrices of $\sum w$, respectively. The EFM then proceeds to compute the between-class covariance matrix as follows:

$$\Gamma^{-1/2} \Xi^t \sum b \Xi \Gamma^{-1/2} = K_b, \tag{5}$$

The EFM diagonalizes the new between-class covariance matrix K_b.

$$K_b \Theta = \Theta \Delta \text{ and } \Theta' \Theta = I, \tag{6}$$

where Θ, Δ are the eigenvector and the diagonal eigenvalue matrices of K_b, respectively. The overall transformation matrix of the EFM is finally defined as follows:

$$T = \Xi \Gamma^{-1/2} \Theta, \tag{7}$$

3.2 Making EFM Model

The EFM is generated through the information of 68 feature points that are manually marked to make an AMM model. In addition, the EFM expression classification is performed after geometrical normalization which is based on eyes and nose out of vertex made from AAM Fitting.

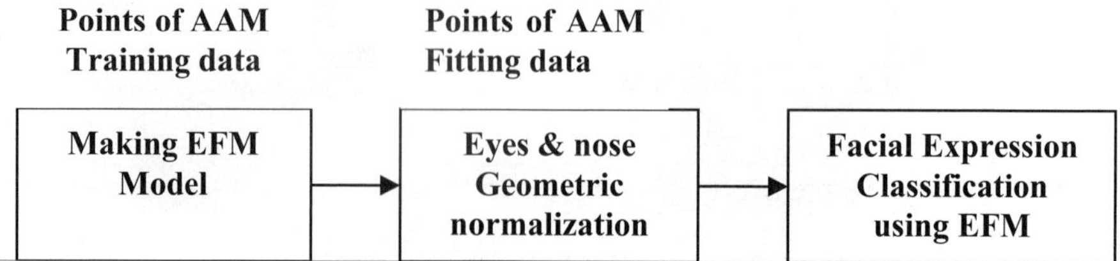

Fig. 7. Block diagram of EFM classifier

4 Performance Evaluation

In this research, recognition experiments were made for four expressions of neutral, disgust, happiness and surprise, adapting the EFM classifier. As shown in Table 1, average rate of recognition for all expressions was 88.9%, while 97.1% for surprise and 72.7%, the lowest, for disgust. In the case of disgust, eight out of 33 were classified into the wrong category of neutral expression as in the table 2.

Table 1. Recognition results

	Success No.	Test Image No.	Rate(%)
Neutral	46	51	90.2
Disgust	24	33	72.7
Happiness	41	44	93.2
Surprise	33	34	97.1
Total	144	162	88.9

Table 2. Confusion matrix

	Neutral	Disgust	Happiness	Surprise
Neutral	46	5	0	0
Disgust	8	24	1	0
Happiness	2	1	41	0
Surprise	1	0	0	33

5 Discussion and Conclusions

The work introduces Facial Expression Recognition using AAM. It was using the EFM for the facial expression recognition task. Our results suggest that it can be successfully applied to the facial expression case. We are developing an automatic facial expression system for interaction system, in which the whole processes will be carried out without any human intervention.

Acknowledgement

This research was performed for the Intelligent Robotics Development Program, one of the 21st Century Frontier R&D Programs funded by the Ministry of Commerce, Industry and Energy of Korea.

References

1. Cootes, T., Edwards, G., Taylor, C.: Active Appearance Models. In Proceedings of the European Conference on Computer Vision 2 (1998) 484–498
2. Cootes, T., Edwards, G., Taylor, C.: A Comparative Evaluation of Active Appearance Models Algorithms. In Proceedings of the British Machine Vision Conference (1998)

3. Cootes, T., Kittipanyangam, P.: Comparing Variations on the Active Appearance Model Algorithm. In Proceedings of the British Machine Vision Conf. 2 (2002) 837– 846
4. Cootes, T.F.: Statistical Models of Appearance for Computer Vision. Online technical report available from http://www.isbe.man.ac.uk/.bim/refs.html, Sept. (2001)
5. Cootes,T., Edwards, G., Taylor, C.: Active Appearance Models. IEEE Transactions on Pattern Analysis and Machine Intelligence 23(6) (2001) 681–685
6. Edwards, G.J.: Learning to Identify Faces in Images and Video Sequences. PhD thesis, University of Manchester, Division of Imaging Science and Biomedical Engineering (1999)
7. Edwards, G.J., Taylor, C.J., Cootes, T. F.: Interpreting Face Images Using Active Appearance Models. In Proc. Internat. Conf. on Autom. Face and Gesture Recog. (1998) 300–305
8. Sclaroff, S., Isidoro, J.: Active Blobs. In Proceedings of the IEEE International Conference on Computer Vision (1998) 1146–1153
9. Sclaroff, S., Isidoro, J.: Active Blobs: Region-Based, Deformable Appearance Models. Computer Vision and Image Understanding, 89(2/3) (2003) 197–225
10. Blanz, V., Vetter, T.: A Morphable Model for the Synthesis of 3D Faces. In Computer Graphics, Annual Conference Series (SIGGRAPH) (1999) 187–194
11. Jones, M., Poggio, T.: Multidimensional Morphable Models: A framework for representing and matching object classes. In Proceedings of the IEEE International Conference on Computer Vision (1998) 683–688
12. Vetter, T., Poggio, T.: Linear Object Classes and Image Synthesis from a Single Example Image. IEEE Trans. on Pattern Analysis and Machine Intelligence 19(7) (1997) 733–742
13. Lanitis, A., Taylor, C.J., Cootes, T.F.: Automatic Interpretation and Coding of Face Images Using Flexible Models. IEEE Transactions on Pattern Analysis and Machine Intelligence 19(7) (1997) 742–756

Facial Expression Recognition Based on BoostingTree

Ning Sun[1,2], Wenming Zheng[1], Changyin Sun[3], Cairong Zou[2], and Li Zhao[1,2]

[1] Research Center of Learning Science, Southeast University, Nanjing 210096, China
sunning@seu.edu.cn
[2] Department of Radio Engineering, Southeast University, Nanjing 210096, China
[3] College of Electrical Engineering, Hohai University, Nanjing, Jiangsu, 210098, China

Abstract. In recent years, facial expression recognition has become an active research area that finds potential applications in the fields such as images processing and pattern recognition, and it plays a very important role in the applications of human-computer interfaces and human emotion analysis. This paper proposes an algorithm called BoostingTree, which is based on the conventional Adaboost and uses tree-structure to convert seven facial expressions to six binary problems, and also presents a novel method to compute projection matrix based on Principal Component Analysis (PCA). In this novel method, a block-merger combination is designed to solve the "data disaster" problem due to the combination of eigenvectors. In the experiment, we construct the weak classifiers set based on this novel method. The weak classifiers selected from the above set by Adaboost are combined into strong classifier to be as node classifier of one level of the tree structure. N-level tree structure built by BoostingTree can effectively solve multiclass problem such as facial expression recognition

1 Introduction

In the early 70s of last century, Ekman[1] postulated seven primary emotions that comprise Anger, Disgust, Fear, Happiness, Sadness, Surprise and Neutral face. So, automatic facial expression recognition is a typical multiclass pattern classification problem. In the past, various approaches have been made towards facial expression recognition. Such as, Calder et al[2] discussed the performance of applying PCA method to recognize facial expression. Otsuka et al[3] recognized one of six facial expression using hidden Markov model (HMM). Zheng et al[4] addressed the facial expression recognition (FER) problem using kernel canonical correlation analysis (KCCA). Boosting[5] is a general algorithm of constructing accurate strong classifier by combining several weak classifiers. The combined strong classifier could achieve very high accuracy when the accuracy of each weak classifier is slightly better than random guess. In the last decade, Boosting has been widely applied to many practical pattern recognition problems, for example, Schapire and Singer[6] presented a boosting-based system for Text Categorization, and Yang et al[7] proposed a face recognition method using AdaBoosted Gabor features. Furthermore, most important application of boosting algorithm is face detection. Viola et al[8] developed a face detection system based on cascaded Adaboost that is capable of detecting face very rapidly and accurately. There are two contributions in this paper. Firstly we present a novel algorithm called BoostingTree for multiclass pattern classification, which

J. Wang et al. (Eds.): ISNN 2006, LNCS 3972, pp. 77 – 84, 2006.

recognizes the facial expressions by transforming the seven patterns into six level binary pattern classifications in the form of tree-structure. Conventional binary Adaboost algorithm is directly applied to classify the binary pattern in the every level of the tree structure. In addition, we make some modification to PCA method that using block merger strategy combines eigenvectors to solve the data disaster and data redundancy problem.

2 Weak Classifiers Construction Based on PCA

In the procedure of Boosting algorithm, the effective features are learned from a large feature set firstly, and weak classifiers are constructed based on one of the selected features. Finally, these weak classifiers are boosted into a strong classifier. In general, the training error of the strong classifier approaches zero exponentially in the number of rounds, if the accuracy of combined weak classifiers are slight better than 50%. But the performance of weak classifiers can affect the convergence speed of training process and the capability of whole method deeply. So, constructing a good weak classifier should keep the following two rules:

1. Simplicity, the training of Boosting algorithm is a procedure of selecting and re-weighting over the weak classifiers. It is highly time-consuming, if the size of weak classifier set is large. So, the complexity of weak classifiers affects significantly the speed of training.
2. Accuracy, precise weak classifier is capable of showing the key feature of the certain pattern and achieving the lower training error of the boosting algorithm, which will reduce the process of training and speed up the convergence of Boosting training.

However, the two rules are always contradictory. In general, Simple weak classifier has less computation-consuming but poor performance and accurate weak classifier is always complicated. So finding the trade-off of both sides is the key problem of constructing weak classifier. PCA provides an optimal linear transformation from the original image space to an orthogonal eigenspace with reduced dimensionality in the sense of the least mean square reconstruction error. In the conventional PCA method, the project matrix is the set of m-dimension eigenvectors corresponding to the m largest eigenvalues. According to the facial expression recognition, it is not likely that the project matrix constructed by conventional PCA method is the optimal one for discrimination. For the sake of building the optimal project matrix, we compute the m-dimension project matrix by randomly combining the eigenvectors corresponding to unordered m-eigenvalues, which is more appropriate to facial expression recognition than conventional PCA method. However, there are two main drawbacks in the random combination: (1) The possible number of combination is $C_{\phi}^{\gamma} = \phi!/(\gamma!*(\phi-\gamma)!)$ if we stochastically select γ eigenvectors from the ϕ dimension eigenvector matrix. When the value of ϕ and γ is very large, the possible number is astronomical figures, which causes the so called "data disaster". (2) There are plenty of similar results produced by combining the eigenvector randomly. Based on the physics significance of PCA method, these similar results affect to construct the project matrix slightly, which brings out much redundant data. This paper proposes a combination strategy called "block merger" to solve the above problems. This strategy merges several consecutive eigenvectors into a block, which reduces the total number of eigenvectors ϕ and the amount of selected eigenvectors γ remarkably and guarantees that the content of two blocks is different completely. Firstly, the block merger strategy solves the data disaster due to randomly combination through decreasing the ϕ and γ. Secondly, the discrepancy of two block combination is more than 50%, which can avoid large amounts of redundant data. For instance, suppose that the dimension of training samples D is 120×120, the number of training samples M is 140, and the category of training samples N is 7. Computed by the PCA method, the dimension of eigenvector matrix is 140×140. According to PCA theory, there are only $M-1$ meaningful eigenvectors in the matrix if the M is less than the D. Therefore, there are 139 eigenvectors to be chosen in this case. Undergoing the computation, the eigenvectors corresponding to the maximum 126 eigenvalues accounts for the 99.9% energy, which are treated as the selection set Consider the factors about data validity, data universality, and for the convenience of computation, we use three forms of combination to build the project matrix that 27, 54, and 90 eigenvectors is combined respectively. Provided that randomly combination is processed, the number of possible combination is $C_{126}^{27}+C_{126}^{54}+C_{126}^{90}=1.68\times10^{36}$. If block merger strategy is applied, we merge $\eta=9$ eigenvectors into one block. So, the selection set can be separated into $\mu=126/9=14$ blocks, and the three forms of combination are also separated into $\nu_{27}=27/9=3$, $\nu_{54}=54/9=6$, $\nu_{90}=90/9=10$ blocks respectively. As the result of the above computation, the number of possible combination is only $C_{14}^{3}+C_{14}^{6}+C_{14}^{10}=4368$. Therefore, block merger strategy has the ability to solve the two drawbacks in the random combination of eigenvectors.

3 BoostingTree Method

Adaboost is the adaptive boosting algorithm. In Adaboost learning, we are given a sequence of training samples $s=(x_1,y_1),...,(x_m,y_m)$ along with a distribution $\omega_{t,i}=(\omega_{t,1}...\omega_{t,m})$. The $\omega_{t,i}$ is reweighed after each learning round t, and the distribution of incorrectly classified examples is increased so that the weak learner is forced to focus on the hard examples in the training set. The binary classified version of Adaboost is described in[5]. In this section, we describe an algorithm for facial expression recognition called BoostingTree, which can solve multiclass classified problems using the binary classified version of Adaboost directly. BoostingTree method transforms the N-category classified problems into $N-1$ level tree-structure binary classified problems. In the first level, we use the "one against the rest" strategy transform the N-category training samples into N binary training data. To certain classifier, in the "one against the rest" strategy, assumed that N-category training samples are $C1, C2, \cdots, CN$, and the corresponding N-category training samples are $\{x_1^{C1},x_2^{C1},...,x_{m1}^{C1}\},\{x_1^{C2},x_2^{C2},...,x_{m2}^{C2}\},\cdots,\{x_1^{CN},x_2^{CN},...,x_{mN}^{CN}\}$. To perform the transformation, the strategy chooses one category samples to be positive training set and the samples of rest categories to be negative training set. As a result, N binary training data are obtained. After that, we apply Adaboost algorithm to respectively train N strong classifiers according to the N binary training data. Then the N strong classifiers are tested by the test samples of this level, and the classifier achieving the lowest testing error is selected to be as the node classifier of the first level. After the above steps, the positive samples of the selected classifier are discarded and the relevant negative samples are transmitted to the next level to be as (N−1)-categories training set. The BoostingTree algorithm run repeatedly as the above rules until the training of whole tree structure classifier is finished. The algorithm is shown in Fig. 1.

- Given initial images as:
  training images $(x_1, y_1), \ldots, (x_n, y_n)$ where $x_i \in X \quad y_i \in \{1, \ldots, N\}$.
  test images $(x'_1, y'_1), \ldots, (x'_m, y'_m)$ where $x'_j \in X' \quad y' \in \{1, \ldots, N\}$.
- Initial training set A_N is whole train images and Initial test set B_N is whole test images.
- For $t = N, \ldots, 2 \quad (step = 1)$ where N is the number of binary classifier in each level
  - For $u = 1, \ldots, t$
    √According to y_u, use "one against the rest" strategy to categorize the A_t into two classes: positive samples $C_{t,u}$,negative samples $D_{t,u}$.
    √Call Adaboost to train classifiers h_u.
    √Compute error ε_u by test database B_t.
  - Choose the classifier h_t with the lowest test error ε_t.
  - Update train set $A_{t-1} = A_t - C_{t,t}$, and test set $B_{t-1} = B_t - D_{t,t}$, where $C_{t,t}$ is the positive samples according to the h_t and $D_{t,t}$ is the test samples corresponding to the h_t

Fig. 1. The BoostingTree algorithm

4 Experiment and Discussion

Facial expression recognition is a seven pattern classification problem. So, we make an experiment for facial expression recognition by BoostingTree algorithm. In this experiment, we use the Japanese Female Facial Expression (JAFFE)[9] database to train classifiers. The database includes 213 images with seven expressions (Anger, Disgust, Fear, Happiness, Neutral, Sadness, and Surprise). In preprocessing, we use horizontal and vertical projection to crop the face area from original image. The cropped images are normalized to 120×120 pixel and processed by illumination compensation and histogram equalization. We select 140 images, 20 images per expression, from JAFFE database to be training set, and the rest of the JAFFE as test set.

Based on the BoostingTree algorithm described in section 3, we build six-level tree structure classifier, each of which is a binary classification problem. The positive and negative training data of the first level is separated by one against the rest strategy. Therefore, We get seven eigenvector matrices in the size of 140×140 by applying PCA method. Based on the "block merger" strategy described in the section 2, seven weak classifier sets are computed, and the size of each set is 4368. Adaboost algorithm is used to learn and select weak classifier from the set, and combine the learned weak classifiers into strong classifiers. Table 1 is the test error of the seven strong classifiers in the first level processing. Obviously, the test error of classifier corresponding to Surprise is the lowest. Thus the classifier corresponding to Surprise is selected as the first level node classifier. Investigating the seven expression samples of JAFFE, we find out that the mouth area of the Surprise is highly different from else in the shape, which is probably the reason why BoostingTree selects the classifier corresponding to Surprise to be the first level node classifier.

Table 1. The test error of seven strong classifiers

Expression	Test Error
Anger	8.23%
Disgust	11.51%
Fear	8.72%
Happiness	5.87%
Neutral	13.92%
Sadness	16.47%
Surprise	3.58%

After the selection of the first level node classifier, BoostingTree discards the 20 positive samples according to the Surprise expression of the first level, and transmits the negative samples to the next level as the whole training samples. The number of eigenvector in the matrix is 120,100,80,60,40 respectively in next five levels. The training samples of the rest five levels are combined by "block merger" strategy all the same. As the presentation of section 2, the detailed parameters of constructing the weak classifiers through block merger strategy are list in the Table 2. In the Table 2, L is the level number, n is the total number of the eigenvectors to be combined, $\xi_{l,i}, l \in \{1,2,\cdots 6\}, i \in \{1,2,3\}$ is the number of eigenvectors in the three combination forms, and the number of the eigenvector in certain block is denoted as η. In addition, $T = \sum_{l,i} C_{\mu}^{\xi_{l,i}}$, $\mu = n/\eta$, $v_{\xi_{l,i}} = \xi_{l,i}/\eta$.

Analyzing the node classifier of the first level, we find out that the first one of weak classifiers combined into strong classifier is a 54 eigenvectors combination, and is not corresponding to the largest 54 eigenvalues. They are respectively corresponding to the eigenvalues listed in the brace $\{10,\cdots,18,28,\cdots,45,64,\cdots 81,91,\cdots 99,\}$, see Fig. 2.

Table 2. The detailed parameters of "block merger" strategy

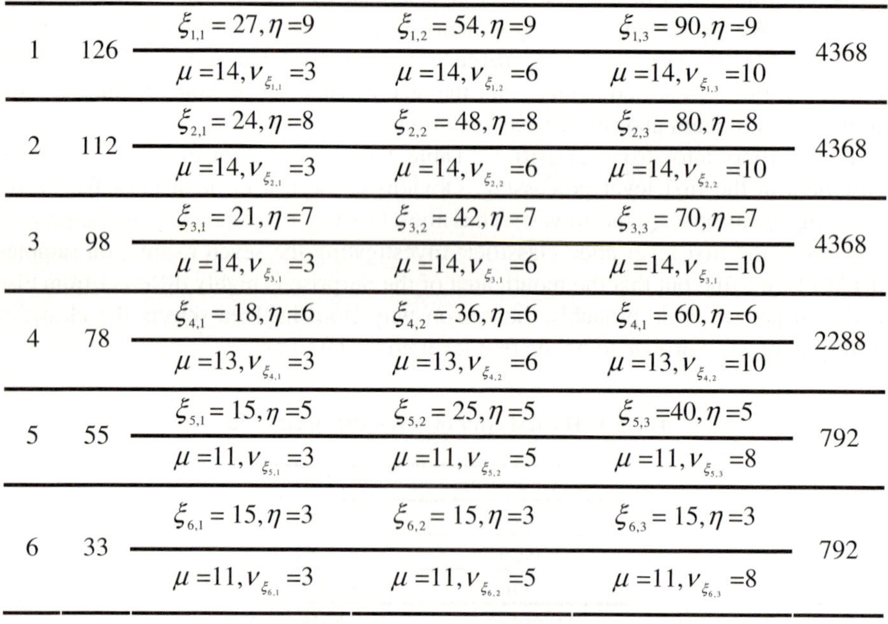

L	n				T
1	126	$\xi_{1,1}=27, \eta=9$	$\xi_{1,2}=54, \eta=9$	$\xi_{1,3}=90, \eta=9$	4368
		$\mu=14, \nu_{\xi_{1,1}}=3$	$\mu=14, \nu_{\xi_{1,2}}=6$	$\mu=14, \nu_{\xi_{1,3}}=10$	
2	112	$\xi_{2,1}=24, \eta=8$	$\xi_{2,2}=48, \eta=8$	$\xi_{2,3}=80, \eta=8$	4368
		$\mu=14, \nu_{\xi_{2,1}}=3$	$\mu=14, \nu_{\xi_{2,2}}=6$	$\mu=14, \nu_{\xi_{2,2}}=10$	
3	98	$\xi_{3,1}=21, \eta=7$	$\xi_{3,2}=42, \eta=7$	$\xi_{3,3}=70, \eta=7$	4368
		$\mu=14, \nu_{\xi_{3,1}}=3$	$\mu=14, \nu_{\xi_{3,1}}=6$	$\mu=14, \nu_{\xi_{3,1}}=10$	
4	78	$\xi_{4,1}=18, \eta=6$	$\xi_{4,2}=36, \eta=6$	$\xi_{4,1}=60, \eta=6$	2288
		$\mu=13, \nu_{\xi_{4,1}}=3$	$\mu=13, \nu_{\xi_{4,2}}=6$	$\mu=13, \nu_{\xi_{4,2}}=10$	
5	55	$\xi_{5,1}=15, \eta=5$	$\xi_{5,2}=25, \eta=5$	$\xi_{5,3}=40, \eta=5$	792
		$\mu=11, \nu_{\xi_{5,1}}=3$	$\mu=11, \nu_{\xi_{5,2}}=5$	$\mu=11, \nu_{\xi_{5,3}}=8$	
6	33	$\xi_{6,1}=15, \eta=3$	$\xi_{6,2}=15, \eta=3$	$\xi_{6,3}=15, \eta=3$	792
		$\mu=11, \nu_{\xi_{6,1}}=3$	$\mu=11, \nu_{\xi_{6,2}}=5$	$\mu=11, \nu_{\xi_{6,3}}=8$	

Fig. 2. 54 eigenfaces corresponding to the selected eigenvectors

To facial expression recognition, the above results show that the project matrix built by the conventional PCA method is not the best. The project matrix constructed by our proposed method is more suitable to the recognition of facial expression.

Fig. 3 shows the test error of each level. By selecting the strong classifier with the lowest test error in every level, the tree-structure classifier is combined, which is respectively corresponding to Surprise, Happiness, Anger, Disgust, Fear and Sadness from the first level to the sixth level. This kind of combination indicates that the most distinctive expression of each level is chosen by BoostingTree algorithm. The more distinctive the expressions are, the earlier they are chosen by the upper level, and thus the rest expression is more difficult to discriminate. As shown in Fig. 4, the lowest test error of each level is increased.

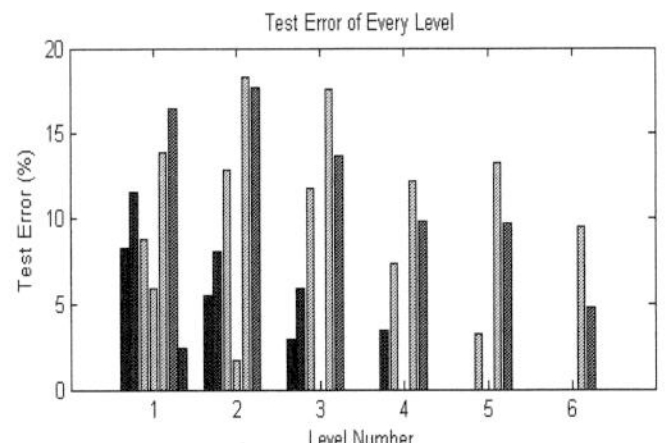

Fig. 3. Test error of each level

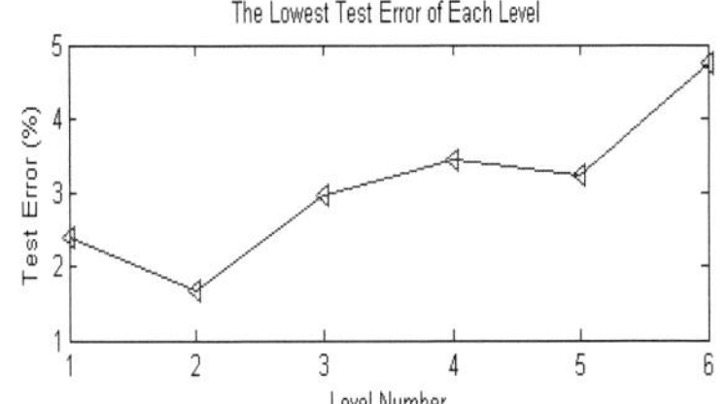

Fig. 4. The lowest test error of each level

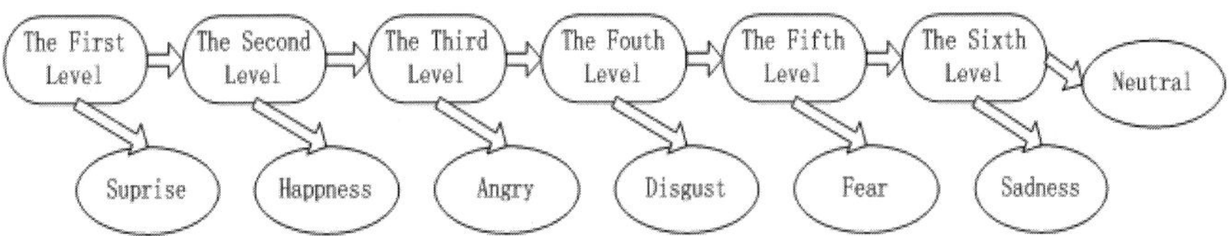

Fig. 5. Tree structure classifier

After the whole training process by BoostingTree, a tree-structure classifier is developed, see figure 5. We test the tree structure classifier using the rest samples of JAFFE, the correct rate of whole classifier is 86.9% Compared with other methods, such as Bayes, Adaboost, SVM algorithm described in paper[10], The performance of BoostingTree method is better than Bayes and Adaboost based method, and only slightly lower than the SVM based method. See table 3.

Table 3. Experiment result of several method for facial expression recognition

Method	Accuracy
BoostingTree	86. 9%
Bayes All	63.3%
Bayes FS	71.0%
AdaBoost	71.9%
L-SVM	92. 4%
NL-SVM	91.9%

5 Conclusion

In this paper, we have proposed a novel method called BoostingTree to solve the multi-pattern classification problems. In BoostingTree method, tree structure classifier is build to reduce the N-category classification problems into N−1 binary classification problems. Secondly, block merger strategy is designed to construct the weak classifier set, which can solve the drawbacks of "data disaster" and data redundancy caused by combining eigenvector randomly. In addition, the experiment result proves that the project matrix computed by the proposed method is much more appropriate to facial expression recognition than conventional PCA method. And our method can achieve better performance than the Bayes and Adaboost based facial expression recognition method

Acknowledgment

This work was partly supported by the national natural science foundations of China under grant 60503023, and partly supported by the natural science foundations of Jiangsu province under the grant BK2005407, partly supported by the key laboratory of image processing and image communication of Jiangsu province under the grant ZK205013, and partly supported by Program for New Century Excellent Talents in University (NCET).

References

1. Ekman, P., Friesen, W. V.: Constants Across Cultures in The Face And Emotion, J.Personality Social Psychol. 17(2) (1971) 124-129.
2. Calder, A. J., Burton, A. M., Miller P., Young A. W.: A Principal Component Analysis of Facial Expressions. Vision research 41(2001) 1179-1208.
3. Otsuka, T., Ohya, J.: Spotting Segments Displaying Facial Expressions from Image Sequences Using HMM. IEEE Proc. Int'l conf. On FG, Nara, Japan (1998) 442-447.
4. Zheng, W., Zhou, X., Zou, C., Zhao, L.: Facial Expression Recognition Using Kernel Canonical Correlation Analysis (KCCA). IEEE Transactions on Neural Networks (To appear).
5. Freund, Y.: Boosting A Weak Learning Algorithm by Majority. Information and Computation 121(2) (1995) 256-285.
6. Schapire, R.C., Singer Y.: BoosTexter: A Boosting-based System for Text Categorization. Machine Learning 39(2) (2000) 135-168.
7. Yang, P., Shan, S., Gao, W., Li, S. Z.,:Face Recognition Using Ada-Boosted Gabor Features. Proceedings - Sixth IEEE International Conference on Automatic Face and Gesture Recognition FGR (2004) 356-361.
8. Viola, P., Jones, M.: Robust Real Time Object Detection. 8th IEEE International Conference on Computer Vision (IC2CV). Vancouver,British Columbia (2001)
9. The Japanese Female Expression (JAFFE) Database http://ww.mis.atr.co.jp/~mlyons/jaffe.html.
10. Guo, G., Dyer, C. R.: Learning From Examples in the Small Sample Case: Face Expression Recognition. IEEE Trans. On System Man and Cybernetics 35(3) (2003) 477-488.

KDA Plus KPCA for Face Recognition

Wenming Zheng

Research Center for Science of Learning, Southeast University, Nanjing, 210096, China
wenming_zheng@seu.edu.cn

Abstract. Kernel discriminant analysis (KDA) and the kernel principal component analysis (KPCA), which are the extension of the linear discriminant analysis (LDA) and the principal component analysis (PCA), respectively, from linear domain to nonlinear domain via the kernel trick, are two very popular nonlinear feature extraction methods. In this paper, we present a new feature extraction algorithm by combing KDA and KPCA, and then apply it to the face recognition task. The experimental results on Yale face dataset show that the proposed method can significantly improve the performance both KDA and KPCA.

1 Introduction

Face recognition has been a very active research area in pattern recognition because of its numerous technical challenges and commercial applications. Within the last decade, numerous algorithms and methods for face recognition have been proposed. For detailed surveys, see [10], [11]. Among these methods, the principal component analysis (PCA) [12] and the Fisher linear discriminant analysis (LDA or FLD) [13] may be the two most successful methods. PCA is an orthogonal basis transform for extracting the structure from possible high-dimensional data sets [5], whereas LDA is a powerful technique used for extracting the discriminant features from the data sets. As for pattern classification, it was shown that the LDA based approaches outperforms the PCA based approaches [13]. The former realizes an optimal feature space on the Fisher's criterion function, subject to the uncorrelated constraints of the discriminant vectors. Although successful in most pattern recognition problems, both methods work better only for the linear patterns. As for nonlinear patterns, such as face pattern, which is subject to large variations in viewpoints, resulting in a highly nonconvex and complex distribution [14]. To overcome the limitation of PCA and LDA, Yang [9] proposed to use the kernel principal component analysis (KPCA) [5] and kernel discriminant analysis (KDA) [1], or equivalently, the generalized discriminant analysis (GDA) [2], for face recognition, and had received much better performance than PCA and LDA, respectively. Yang also showed that KDA method, or Kernel Fisherfaces, achieved the better results than KPCA [9].

The Kernel Fisherface method has proven to be a very powerful method for face recognition. However, further study shows that there are still drawbacks in this method. One of the major drawbacks of Kernel Fisherfaces method is that it will lose the within-class scatter information as for the so-called "small sample size" problem [8] because that all the optimal discriminant vectors in this case are limited in the null

J. Wang et al. (Eds.): ISNN 2006, LNCS 3972, pp. 85 – 92, 2006.

space of the within-class scatter matrix [8], and this information are also important for face recognition. To improve the performance of face recognition, we propose another learning algorithm combing the advantages of KDA and KPCA. Motivate by Duchene [6], our proposed algorithm can be divided into three steps:(1) compute the optimal discriminant vectors of KDA; (2) compute another projected vectors for extracting features based on a different criterion function; (3) using the two kinds of features for recognition.

2 Related Work

Assume that $\mathbf{X}$ is a n -dimensional sample set with N elements belonging to c classes. Let $\mathbf{X}_l$ denote the l th class sample set, and N_l be the cardinality of the $\mathbf{X}_l$. Thus we have $\mathbf{X}=\bigcup_{l=1}^{c}\mathbf{X}_l$, $N=\sum_{l=1}^{c}N_l$. Let $\mathbf{X}$ be mapped into a Hilbert space F through a nonlinear mapping function $\Phi:\mathbf{X}\rightarrow F$, $\mathbf{x}\rightarrow\Phi(\mathbf{x})$. The between-class scatter matrix $\mathbf{S}_B^\Phi$, the within-class scatter matrix $\mathbf{S}_W^\Phi$ and the total-scatter matrix $\mathbf{S}_T^\Phi$ in F are given as follows:

$$\mathbf{S}_B^\Phi=\sum_{i=1}^{c}N_i(\mathbf{u}_i^\Phi-\mathbf{u}^\Phi)(\mathbf{u}_i^\Phi-\mathbf{u}^\Phi)^T \tag{1}$$

$$\mathbf{S}_W^\Phi=\sum_{i=1}^{c}\sum_{j=1}^{N_i}(\Phi(\mathbf{x}_i^j)-\mathbf{u}_i^\Phi)(\Phi(\mathbf{x}_i^j)-\mathbf{u}_i^\Phi)^T \tag{2}$$

$$\mathbf{S}_T^\Phi=\sum_{i=1}^{c}\sum_{j=1}^{N_i}(\Phi(\mathbf{x}_i^j)-\mathbf{u}^\Phi)(\Phi(\mathbf{x}_i^j)-\mathbf{u}^\Phi)^T \tag{3}$$

where $\mathbf{u}_i^\Phi$ is the mean of the i th class and $\mathbf{u}^\Phi$ is the mean of all samples in F :

$$\mathbf{u}_i^\Phi=\frac{1}{N_i}\sum_{j=1}^{N_i}\Phi(\mathbf{x}_i^j),\ \mathbf{u}^\Phi=\frac{1}{N}\sum_{i=1}^{c}\sum_{j=1}^{N_i}\Phi(\mathbf{x}_i^j) \tag{4}$$

2.1 Kernel Discriminant Analysis (KDA)

KDA is a nonlinear extension of LDA via the kernel trick. The optimal discriminant vectors of KDA can be definded as follows: the first discriminant vector $\boldsymbol{\omega}_1$ is a unit eigenvector that maximizes the Fisher criterion $J_F(\boldsymbol{\omega})$ in the space F , where

$$J_F(\boldsymbol{\omega})=\frac{\boldsymbol{\omega}^T\mathbf{S}_B^\Phi\boldsymbol{\omega}}{\boldsymbol{\omega}^T\mathbf{S}_T^\Phi\boldsymbol{\omega}} \tag{5}$$

According to discriminant analysis [3], we obtain that $\boldsymbol{\omega}_1$ is the eigenvector corresponding to largest eigenvalue of the eigenequation

$$\mathbf{S}_B^\Phi\boldsymbol{\omega}=\lambda\mathbf{S}_T^\Phi\boldsymbol{\omega} \tag{6}$$

The i th ($i>1$) discriminant vector $\boldsymbol{\omega}_i$ is the eigenvector corresponding to the largest eigenvalue of the eigenequation (8) under the following constraints:

$$\boldsymbol{\omega}_i^T \mathbf{S}_T^{\Phi} \boldsymbol{\omega}_j = 0, \quad j = 1,2,\cdots,i-1 \tag{7}$$

2.2 Kernel Principal Component Analysis (KPCA)

KPCA is a nonlinear extension of the linear PCA via the kernel trick. The KPCA can be derived from the following criterion [7]:

$$J_P(\boldsymbol{\omega}) = \frac{\boldsymbol{\omega}^T \mathbf{S}_T^{\Phi} \boldsymbol{\omega}}{\boldsymbol{\omega}^T \mathbf{I}^{\Phi} \boldsymbol{\omega}} \tag{8}$$

where $\mathbf{I}^{\Phi}$ is a unit matrix in feature space F .
Note that the principal component vectors are statistically uncorrelated, thus the principal component vectors $\tilde{\boldsymbol{\omega}}_i$ ($i = 1,2,\cdots$) can be calculated as follows: $\tilde{\boldsymbol{\omega}}_1$ is the eigenvector corresponding to largest eigenvalue of the eigenequation

$$\lambda \boldsymbol{\omega} = \mathbf{S}_T^{\Phi} \boldsymbol{\omega} \tag{9}$$

The i th ($i>1$) discriminant vector $\tilde{\boldsymbol{\omega}}_i$ is the eigenvector corresponding to largest eigenvalue of the eigenequation (11) under the following constraints:

$$\tilde{\boldsymbol{\omega}}_i^T \tilde{\boldsymbol{\omega}}_j = 0, \quad j = 1,2,\cdots,i-1 \tag{10}$$

3 Feature Extraction Algorithm Combing KPCA and KDA

In this section, we will present another learning algorithm for feature extraction, which can be divided into three steps: (1) compute the optimal discriminant vectors of KDA; (2) compute another projected vectors based on the criterion $J_P(\boldsymbol{\omega})$ in equation (8), where the vectors are statistically uncorrelated with the KDA optimal discriminant vectors; (3) combine the two kinds of vectors for extracting features.

Suppose that $\boldsymbol{\omega}_i\,(i = 1,2,\cdots,c-1)$ are the $c-1$ optimal discriminant vectors of KDA using the MGDA algorithm [8]. Then the other c–1 optimal vectors in our proposed algorithm can be defined as: the $\boldsymbol{\omega}_j$ ($j = c, c+1,\cdots$) is the unit eigenvector that maximizes $J_P(\boldsymbol{\omega})$ under the following constraints:

$$\boldsymbol{\omega}_j^T \boldsymbol{\omega}_i = 0, \quad i = 1,2,\cdots,j-1 \tag{11}$$

3.1 Formulation of the Proposed Algorithm

Let

$$\Phi(\mathbf{X}) = [\Phi(\mathbf{x}_1^1) \ \cdots \ \Phi(\mathbf{x}_1^{N_1}) \ \cdots \ \Phi(\mathbf{x}_c^1) \ \cdots \ \Phi(\mathbf{x}_c^{N_c})] \tag{12}$$

From equations (1)-(4), we have

$$\mathbf{S}_B^\Phi = \Phi(\mathbf{X})(\mathbf{W}-\mathbf{M})(\mathbf{W}-\mathbf{M})^T(\Phi(\mathbf{X}))^T \tag{13}$$

$$\mathbf{S}_W^\Phi = \Phi(\mathbf{X})(\mathbf{I}-\mathbf{W})(\mathbf{I}-\mathbf{W})^T(\Phi(\mathbf{X}))^T \tag{14}$$

$$\mathbf{S}_T^\Phi = \Phi(\mathbf{X})(\mathbf{I}-\mathbf{M})(\mathbf{I}-\mathbf{M})^T(\Phi(\mathbf{X}))^T \tag{15}$$

where $\mathbf{W} = (\mathbf{W}_l)_{l=1,\cdots,c}$ is a $N \times N$ block diagonal matrix, $\mathbf{W}_l$ a $N_l \times N_l$ matrix with all terms equal to $1/N_l$, $\mathbf{I}$ the $N \times N$ identity matrix, and $\mathbf{M} = (m_{ij})_{i=1,\cdots,N;j=1,\cdots,N}$ a $N \times N$ matrix with all terms equal to $1/N$. Let

$$\mathbf{K} = (\Phi(\mathbf{X}))^T\Phi(\mathbf{X}) \tag{16}$$

be the gram matrix, where the i th row and j th column elements of $\mathbf{K}$ can be computed by the kernel function $k(\mathbf{x}_i, \mathbf{x}_j)$. From literatures [2], [8], [5], we know that the solutions $\boldsymbol{\omega}$ of KDA or KPCA can be written as

$$\boldsymbol{\omega} = \Phi(\mathbf{X})(\mathbf{I}-\mathbf{M})\boldsymbol{\alpha} \tag{17}$$

where $\boldsymbol{\alpha}$ is a $N \times 1$ vector.
Thus, the optimal discriminant vector of KDA can be expressed as:

$$\boldsymbol{\omega}_i = \Phi(\mathbf{X})(\mathbf{I}-\mathbf{M})\boldsymbol{\alpha}_i, \ i = 1,2,\cdots,c-1$$

The j th eigenvector $\boldsymbol{\omega}_j = \Phi(\mathbf{X})(\mathbf{I}-\mathbf{M})\boldsymbol{\alpha}_j$ ($j > c-1$) is the one that maximizes

$$J_P(\boldsymbol{\omega}) = \tilde{J}_P(\boldsymbol{\alpha}) = \frac{\boldsymbol{\alpha}^T\mathbf{T}\boldsymbol{\alpha}}{\boldsymbol{\alpha}^T\tilde{\mathbf{K}}\boldsymbol{\alpha}} \tag{18}$$

under the constraints

$$\boldsymbol{\omega}_j^T\boldsymbol{\omega}_i = \boldsymbol{\alpha}_j^T\tilde{\mathbf{K}}\boldsymbol{\alpha}_i = 0, \ i = 1,2,\cdots,j-1 \tag{19}$$

where

$$\mathbf{T} = (\mathbf{I}-\mathbf{M})^T\mathbf{K}(\mathbf{I}-\mathbf{M})(\mathbf{I}-\mathbf{M})^T\mathbf{K}(\mathbf{I}-\mathbf{M}) \tag{20}$$

$$\tilde{\mathbf{K}} = (\mathbf{I}-\mathbf{M})^T\mathbf{K}(\mathbf{I}-\mathbf{M}) \tag{21}$$

Moreover, we have the following theorem:
Theorem 1. *Suppose that* $\mathbf{B}$ *and* $\mathbf{R}$ *are positive semi-definite matrices,* $\mathbf{V}$ *is a positive matrix. The discriminant criterion function is defined as:*

$$F(\varphi) = \frac{\varphi^T\mathbf{B}\varphi}{\varphi^T\mathbf{V}\varphi}$$

Let φ_1 *be the discriminant eigenvector that maximizes* $F(\varphi)$ *and* φ_{r+1} *be the* $(r+1)$ *th* ($r \geq 1$) *discriminant eigenvector that maximizes* $F(\varphi)$ *under the following constraints:*

$$\varphi_{r+1}^T \mathbf{R} \varphi_i = 0, \ (i = 1, 2, \cdots, r)$$

Then φ_{r+1} *is the eigenvector corresponding to the largest eigenvalue of the following eigenequation:*

$$\mathbf{PB}\varphi = \lambda \mathbf{V} \varphi$$

where $\mathbf{P} = \mathbf{I} - \mathbf{RD}^T (\mathbf{DRV}^{-1}\mathbf{RD}^T)^{-1} \mathbf{DRV}^{-1}$, $\mathbf{D} = [\varphi_1 \quad \varphi_2 \quad \cdots \quad \varphi_r]^T$.

The proof of theorem can be found in [8].

From theorem 1, we obtain that $\boldsymbol{\alpha}_j$ is the eigenvector corresponding to the largest eigenvalue of the following eigenequation:

$$\mathbf{PT}\boldsymbol{\alpha} = \lambda \tilde{\mathbf{K}} \boldsymbol{\alpha} \tag{22}$$

where

$$\mathbf{P} = \mathbf{I} - \tilde{\mathbf{K}}\mathbf{D}^T (\mathbf{D}\tilde{\mathbf{K}}\mathbf{D}^T)^{-1} \mathbf{D} \tag{23}$$

$$\mathbf{D} = [\boldsymbol{\alpha}_1 \quad \boldsymbol{\alpha}_2 \quad \cdots \quad \boldsymbol{\alpha}_{j-1}]^T \tag{24}$$

The coefficients $\boldsymbol{\alpha}_j$ are normalized by requiring that the corresponding vectors $\boldsymbol{\omega}_j$ be normalized: $\boldsymbol{\omega}_j^T \boldsymbol{\omega}_j = 1$. Thus, the coefficients $\boldsymbol{\alpha}_j$ are divided by $\sqrt{\boldsymbol{\alpha}_j^T \mathbf{R} \boldsymbol{\alpha}_j}$ in order to get normalized vectors $\boldsymbol{\omega}_j$.

3.2 Features Combination for Recognition

Now let

$$\mathbf{E}_1 = [\boldsymbol{\omega}_1 \quad \boldsymbol{\omega}_2 \quad \cdots \quad \boldsymbol{\omega}_{c-1}] \tag{25}$$

$$\mathbf{E}_2 = [\boldsymbol{\omega}_c \quad \boldsymbol{\omega}_{c+1} \quad \cdots \quad \boldsymbol{\omega}_{2c-2}] \tag{26}$$

The projection of a test point $\mathbf{t}$ onto the discriminant transformation matrix $\mathbf{E}_1$ is computed as:

$$\mathbf{z}_i = \mathbf{E}_i^T (\Phi(\mathbf{t}) - \mathbf{u}^{\Phi}), \ (i = 1, 2) \tag{27}$$

Then $\mathbf{z}_i$ ($i = 1, 2$) are further normalized as follows

$$\mathbf{z}_i = \mathbf{z}_i / \|\mathbf{z}_i\|, \ (i = 1, 2) \tag{28}$$

The final feature $\mathbf{z}$ is consisted of $\mathbf{z}_1$ and $\mathbf{z}_2$:

$$\mathbf{z} = [\mathbf{z}_2^T, \mathbf{z}_2^T]^T \tag{29}$$

4 Experiments

In this section, we will use the proposed method (KDA+ KPCA) to perform the face recognition experiments on the Yale face database. We adopt the "leave-one-out " strategy to perform these experiments: To classify an face image, we remove it from the whole face image set, and the discriminant vectors are computed using the training set of the remainder images. The nearest neighbor classifier is used in both experiments. The Gaussian kernel defined as $k(\mathbf{x},\mathbf{y}) = \exp(-\|\mathbf{x}-\mathbf{y}\|^2/\sigma)$ is used in the experiment, where σ is the Gaussian kernel parameter.

The Yale face database contains 15 subjects, each of which contains 11 face images with variations in both facial expression and lighting condition. The original face images are sized 243×320 pixels with a 256-level gray scale. Figure 1 shows all the 11 images of one subject. To reduce the background part of each full image, each image is manually centered into a size of 190×170 pixels, and then down-sample into the size of 48×43. Moreover, note that the normal facial expression image and the without glasses image (or with glasses if subject normally wears glasses) in face subjects numbered 2, 3, 6, 7, 8, 12 and 14 are copies of each other. We remove the normal facial expression image, i.e., image numbered 6 in Figure 1, from every subject. After doing that, the total number of the data set is 150.

Since the number of the images in Yale face dataset is relatively small, we adopt "leave-one-out" strategy [4] to perform the experiments: to classify a face image, we remove it from the whole face image set, and the discriminant vectors are computed using the remainder images. Similar with the first experiment, we represent each image by a raster scan vector of the intensity values, and then normalize them within [-1,1] using a linear function. Table 1 shows the experimental results on Gaussian

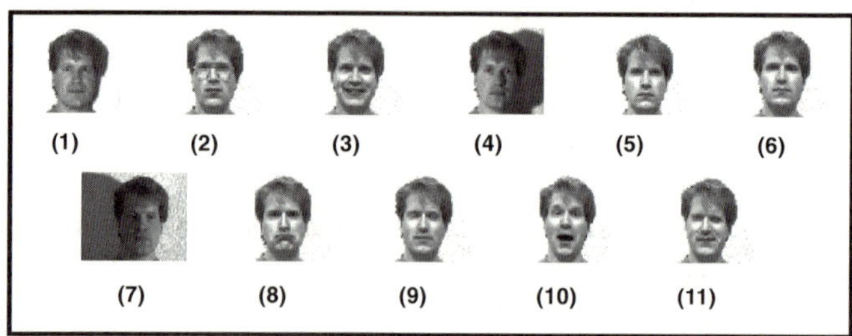

Fig. 1. Ten images for one subject in Yale face database

Table 2. Comparison of Average Recognition Rate On Yale face database

Method	Average Recognition Rate (%)
KDDA	86.00 (129/150)
KPCA	76.67 (115/150)
KDA	90.00 (135/150)
KDA+KPCA	95.33 (143/150)

kernel with parameter $\sigma = 1e6$, where the number of the discriminant vectors in the range space of the within-class scatter matrix is 15. From Table 1, we see that the proposed method achieves the highest recognition rate (95.33%), which is much better than the other methods.

5 Conclusion

In this paper, we have presented a new kernel-based learning algorithm by combing the advantages of KDA and KPCA. The proposed method fully utilizes the powerful discriminant ability of KDA whereas overcome its limitations as for the case of the small sample size problem. The experiment on Yale face database showed that the proposed method achieves better performance than both KDA method and KPCA method.

Acknowledgements

This work was partly supported by the national natural science foundations of China under grant 60503023, and partly supported by the natural science foundations of Jiangsu province under the grant BK2005407, partly supported by the key laboratory of image processing and image communication of Jiangsu province under the grant ZK205013, and partly supported by Program for New Century Excellent Talents in University (NCET).

References

1. Mika, S., Rätsch, G., Weston, J., Schölkopf, B., Müller, K.-R.: Fisher Discriminant Analysis with Kernels. Proceeding of IEEE Neural Networks for Signal Processing Workshop (1999) 41-48
2. Baudat, G., Anouar, F.: Generalized Discriminant Analysis Using a Kernel Approach. Neural Computation 12 (10) (2000) 2385-2404
3. Duda, R.O., Hart, P.E.: Pattern Classification and Scene Analysis, John Wiley & Sons Inc., New York (1973)
4. Fukunaga, K.: Introduction to Statistical Pattern Recognition. Academic Press Inc. (1990)
5. Schölkopf, B., Smola, A. J., Müller, K.-R.: Nonlinear Component Analysis as a Kernel Eigenvalue Problem. Neural Computation 10 (5) (1998) 1299-1319
6. Duchene, J.: A Significant Plane for Two-class Discrimination Problems. IEEE Transactions on Pattern Analysis and Machine Intelligence 8 (4) (1986) 557-559
7. Siedlecki, W., Siedlecka, K., Sklansky, J.: An Overview of Mapping Techniques for Exploratory Pattern Analysis. Pattern Recognition 21 (5) (1988) 411-429
8. Zheng, W., Zhao, L., Zou, C.: A Modified Algorithm for Generalized Discriminant Analysis. Neural Computation 16 (6) (2004) 1283-1297
9. Yang, M.H.: Kernel Eigenfaces vs. Kernel Fisherfaces: Face Recognition using Kernel Methods. Proceedings of the Fifth International conference on Automatic Face and Gesture Recognition (2002) 215-220

10. Samal, A., Iyengar, P.: Automatic Recognition and Analysis of Human Faces and Facial Expressions: a Survey. Pattern Recognition 25 (1) (1992) 65-77
11. Zhao, W., Chellappa, R., Phillips P.J., Rosenfeld, A.: Face Recognition, a Literature Survey. ACM Computing Surveys 35 (4) (2000) 399-458
12. Turk, M.A., Pentland, A.P.: Eigenfaces for Recognition. Journal of Cognitive Neuroscience 3 (1) (1991) 71-86
13. Belhumeur, P.N., Hespanha, J.P., Kriegman, D.J.: Eigenfaces vs. Fisherfaces: Recognition using Class Specific Linear Projection. IEEE Transactions on Pattern Analysis and Machine Intelligence 19 (7) (1997) 711-720
14. Lu, J., Plataniotis, K.N., Venetsanopoulos, A.N.: Face Recognition using Kernel Direct Discriminant Analysis Algorithms. IEEE Transactions on Neural Networks 14 (1) (2003) 117-126

Face Recognition Using a Neural Network Simulating Olfactory Systems

Guang Li[1], Jin Zhang[2,3], You Wang[2], and Walter J. Freeman[4]

[1] National Laboratory of Industrial Control Technology, Zhejiang University, Hangzhou 310027, China
[2] Department of Biomedical Engineering, Zhejiang University, Hangzhou, 310027, China
[3] Software College, Human University, Changsha, 410082, China
[4] Division of Neurobiology, University of California at Berkeley, Donner 101, Berkeley, CA, 94720-3206, USA
guangli@cbeis.zju.edu.cn, mail_zhangjin@163.com

Abstract. A novel chaotic neural network K-set has been constructed based in research on biological olfactory systems. This non-convergent neural network simulates the capacities of biological brains for signal processing in pattern recognition. Its accuracy and efficiency are demonstrated in this report on an application to human face recognition, with comparisons of performance with conventional pattern recognition algorithms.

1 Introduction

Face recognition (FC) plays an increasingly important role in a wide range of applications, such as criminal identification, credit card verification, security system, scene surveillance, etc. However, a straightforward implementation is difficult.

In order to optimize performance, many algorithms of feature extraction are proposed, such as PCA, ICA and so on. Upon the extraction of the proper set of features, a classifier such as nearest neighbor distances, Bayesian statistics, SVM, etc., is applied to collections of facial images. Biologically inspired neural networks are especially widely used, but most are deterministic and time-invariant, and they lack the speed, reliability, and capacities for abstraction and generalization that characterize biological neural systems.

The KIII set is a chaotic neural network mimicking the olfactory system of animals engaged in odorant identification, as revealed by research on the electrical activity recorded in animals trained to identify odors. The KIII model has static nonlinearities that support aperiodic broad-spectrum oscillations from nonconvergent 'chaotic' attractors. The multiple attractors form a landscape, each attractor in a basin that represents a class of input that the system is trained to identify and classify efficiently in image pattern recognition. In this paper, we introduce this novel chaotic neural network and its use for face recognition. First an algorithm for image partition is applied to extract the features of face images. Then the extracted feature vectors are given as inputs to train the KIII set by changing its connection weights, creating a landscape of chaotic attractors. Then independent inputs are given to demonstrate the efficiency of the KIII set as a classifier of facial images, with the potential for use in other types of pattern cognition.

J. Wang et al. (Eds.): ISNN 2006, LNCS 3972, pp. 93 – 97, 2006.

2 A Neural Network Mimicking Olfactory Systems — The KIII Set

2.1 K Set Hierarchy

In accordance with the anatomic architecture, KIII network is a multi-layer neural network model. The dynamics of every node is described with a second order differential equation (1), which is derived from measurement of open-loop impulse responses [1].

$$\frac{1}{a \cdot b}[x_i''(t) + (a+b)x_i'(t) + a \cdot b \cdot x_i(t)] = \sum_{j \neq i}^{N} \left[W_{ij} \cdot Q(x_j(t), q_j) \right] + I_i(t) \tag{1}$$

$$Q(x_i(t), q) = \begin{cases} q(1 - e^{-(\exp(x(t))-1)/q}) & x(t) > x_0 \\ -1 & x(t) < x_0 \end{cases} \tag{2}$$
$$x_0 = \ln(1 - q\ln(1 + 1/q))$$

In this equation $x_i(t)$ $(x_j(t))$ represents the state variable of ith (jth) neural population while W_{ij} indicates the connection strength between them. $I_i(t)$ is an input function. The parameter a, b reflect two rate constant. Q is a static nonlinear sigmoid function derived from Hodgkin-Huxley model and q represents the maximum asymptote of the sigmoid function.

The KIII network describes the whole olfactory neural system. After the parameter optimization, the KIII network generates EEG-like waveform with 1/f power spectra [2]. The KIII system [1] presents an aperiodic oscillation when there is no stimulus and the trajectory of the system soon goes to specific local basin and converges to an attractor when there is a stimulus.

2.2 Learning Rule

There are two kinds of learning rules: Hebbian learning reinforces the desired stimulus patterns while habituation decreases the impact of the background noise and the stimuli that are ambiguous, irrelevant or insignificant.

The algorithm is described as follows:

IF $P_M(i)>(1+K)*P_M$ AND $P_M(j)>(1+K)*P_M$ AND i != j *then*
$W`_{M(i)->M(j)} = h_{Heb}$ (i = 1,…,n)
Else IF i = = j
$W`_{M(i)->M(j)} = 0$
ELSE
$W`_{M(i)->M(j)} = h_{Hab}$ (i = 1,…,n)
ENDIF
ENDIF

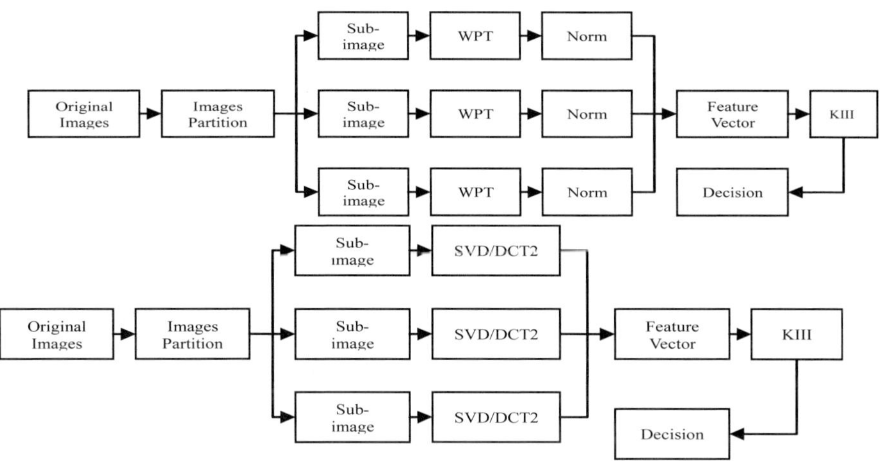

Fig. 1. The flow chart of face recognition

3 Application of KIII Model to Face Recognition

3.1 Different Recognition Process

In the process of recognition, 2 different flow charts (Fig.1) are used according to different the extracting methods, SVD (singular value decomposition), DCT (discrete cosine transform) and WPT (wavelet packet transform).

First, the original image is divided into n equal sub-images. In each sub-image, only one feature is extracted. When the feature of sub-image is decomposed, an n-dimension feature vector is generated as the input of KIII model. Fig.1 (upper) shows the flow chart based on SVD or DCT and Fig.1 (lower) shows the flow chart based on WPT.

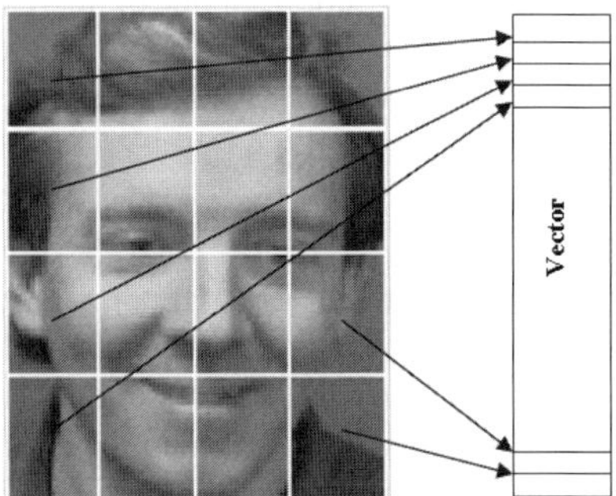

Fig. 2. Image partition

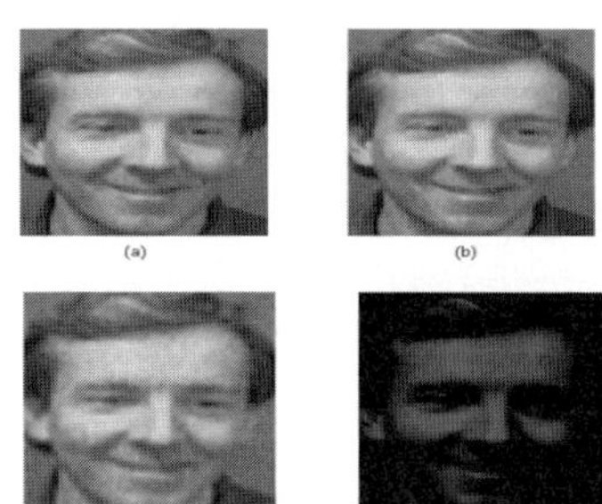

Fig. 3. Face image comparison processed by DCT

3.2 Feature Extraction Methods

In extracting a feature vector from the entire image, we divide the original image into sub-images and extract the feature of each sub-image from the whole sub-image (Fig.2). Then, the features are combined to form the whole feature vector of the original image. In our simulation, the face images are divided into 8, 16, 32, 64 and 80 sub-images individually. Either DCT, SVD, or WPT is used separately to extract the feature vector.

DCT can concentrate most of the signal energy effectively. In Fig.3, (a) is the original face image, (b) is the reconstructed image without discarding any coefficient, (c) is the reconstructed image discarding about 90% coefficient and (d) is the reconstructed image only discarding one maximal coefficient.

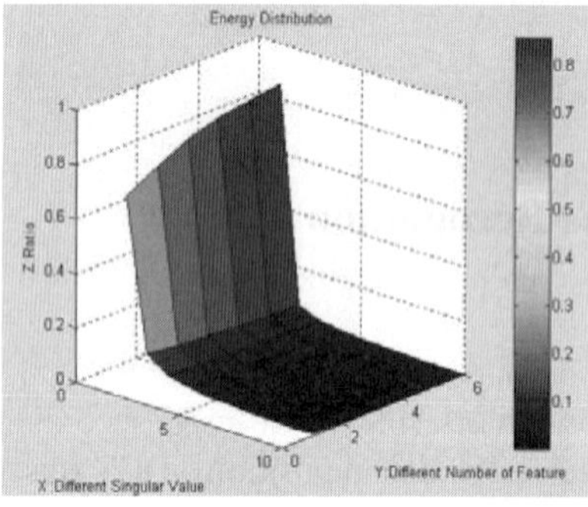

Fig. 4. Energy distribution in SVD

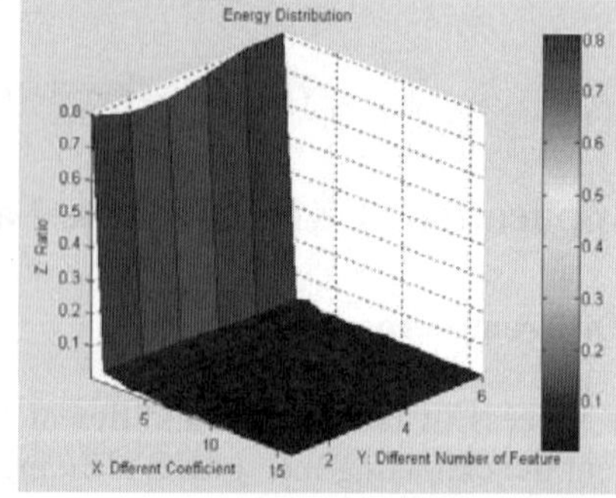

Fig. 5. Energy distribution in WPT

In SVD, the largest singular composes the n-dimension feature vector. In Fig.4, X axis denotes 10 large singular values in each sub-image. Y axis denotes the number of feature, 8*|y i|. Z axis denotes the ratio of singular value to the sum of singular value.

The WPT of image results in an array of wavelet coefficients. In this paper, we use 2 level WPT calculate the coefficients and the norm of all the coefficients of lowest frequency range in sub-bands as the feature. From Fig.5, it is shown that most of the energy of the image concentrates in the lowest frequency range.

3.3 Experiment Results

ORL face dataset is used to evaluate the performance. Five images of each person are selected for training and others are used to test. Compared with other ANN, each pattern is only learned 10 times in KIII. The output of OB layer is stored as cognition standard. And the nearest neighbor principle is used to classify new images. From Table 1, it is shown that, the higher is the dimension of the feature vector, the better is the performance. The DCT-based feature seems a little better, but SVD/WPT-based classifier seems more stable. From Table.2, KIII model shows better performance than all of the others.

Table 1. Average recognition accuracy for ORL database

Accuracy		Number of feature					
		8	16	32	48	64	80
Method	SVD	0.693	0.815	0.883	0.900	0.908	0.910
	DCT	0.670	0.820	0.885	0.900	0.915	0.910
	WPT	0.695	0.818	0.873	0.898	0.908	0.898

Table 2. Comparison the KIII method with other algorithm

Method	MLP[3]	HMM[4]	PCA[4]	WPT+KIII	SVD+KIII	DCT+KIII
Accuracy	84.0%	87.0%	90.0%	90.8%	91.0%	91.5%

4 Discussion

KIII network is a kind of chaotic neural network derived from biological neural systems. In this paper, the KIII model is used to face recognition. Based on the feature vectors extracted by DCT/SVD/WPT, the potential of the KIII set for pattern cognition is shown, especially to classify complex images.

Acknowledgements

This research is supported by the National Natural Science Foundation of China (No. 60421002) and the National Basic Research Program of China (973 Program, No. 2004CB720302).

References

1. Freeman, W.J., Kozma, R.: Biocomplexity: Adaptive Behavior in Complex Stochastic Dynamic Systems, Biosystems, 59 (2001), 109-123
2. Li, G., Lou, Z., Wang, L., Li, X., Freeman, W.J.: Application of Chaotic Neural Model Based on Olfactory System on Pattern Recognitions. In: Wang, L., Chen, K., Ong Y.S. (eds.): Advances in Natural Computation. Lecture Notes in Computer Science, Vol. 3610. Springer-Verlag, Berlin Heidelberg New York (2005) 378 - 381
3. Pan, Z., Adams, R. and Bolouri, H.: Dimensionality Reduction of Face Images Using Discrete Cosine Transforms for Recognition, IEEE Conference on Computer Vision and Pattern Recognition (2000)
4. Samaria, F.: Face Recognition Using Hidden Markov Models, PhD Thesis, Cambridge University (1994)

Face Recognition Using Neural Networks and Pattern Averaging

Adnan Khashman

Department of Electrical & Electronic Engineering,
Near East University, Lefkosa, North Cyprus, Turkey
amk@neu.edu.tr

Abstract. The human ability to recognize objects has not so far been matched by intelligent machines. This is more evident when it comes to recognizing faces, where a quick human "glance" is sufficient to recognize a "familiar" face. Face recognition has recently attracted more research aimed at developing reliable recognition by machines. Current face recognition methods rely on detecting certain features within a face and using these features for face recognition. This paper introduces a novel approach to face recognition by simulating our ability to recognize "familiar" faces after a quick "glance" using pattern averaging and neural networks. A real-life application will be presented throughout recognizing the faces of 30 persons. Time costs and the neural network parameters will be described, in addition to future work aimed at further improving the developed system.

1 Introduction

Many face recognition techniques have been developed over the past few decades. These techniques use different methods such as the appearance-based method [1], [2], [3]. Other methods were also developed that use dimensionality reduction techniques such as the Principal Component Analysis (PCA) [1], the Linear Discriminant Analysis (LDA) [4] and the Locality Preserving Projections (LPP) [3]. The main difference between PCA, LDA, and LPP is that PCA and LDA focus on the global structure of the Euclidean space, while LPP focuses on local structure of the manifold, but they are all considered as linear subspace learning algorithms [5]. Some nonlinear techniques have also been suggested to find the nonlinear structure of the manifold, such as Locally Linear Embedding (LLE) [6]. Known approaches that use the Eigenfaces method [1], the Fisherfaces method [4] and the Laplacianfaces method [3] have shown successful results in face recognition. However, these methods are appearance-based or feature-based methods that search for certain global or local representation of a face. None so far has considered modelling the way we humans recognize faces.

This paper presents a simple and efficient face recognition method that simulates our recognition of familiar faces using pattern averaging and neural networks. The human "glance" of a face can be approximated in machines using pattern averaging, whereas, the "familiarity" of a face can be simulated by a trained neural network. A real-life application will be presented throughout recognizing the faces of 30 persons to demonstrate this method. Successful results have been achieved.

J. Wang et al. (Eds.): ISNN 2006, LNCS 3972, pp. 98 – 103, 2006.

2 Image Pre-processing

The database used in developing face recognition systems relies on images of human faces captured and processed in preparation for implementing the recognition system. The variety of information in these face images makes face detection difficult due to various obstacles [7]. However, as our novel approach to face recognition is based on simulating a human glance and face familiarity, and since humans are able to recognize familiar faces despite the above problems, only face orientation is considered when developing the work presented within this paper, with the assumption that the input image contains only one face which is not occluded by other objects. Facial features, expressions, poses, structural components and imaging conditions are not considered as an obstacle in our work and are left to the neural network to learn via pattern averaging.

There are 30 persons whose faces are to be recognized and thus their face images would be used as the database for the work presented within this paper. Each face has three different projections, which were captured while looking: Left (LL), Straight (LS) and Right (LR) as shown in Figure 1 thus resulting in 90 images that can be used for developing and implementing this novel face recognition system. Figure 2 shows LS face images representing the 30 persons of various gender, ethnicity and age.

All original images are gray and of size (512x512) pixels. The images have been compressed and their size reduced to 128x128 pixels in order to meet one of the aims

Fig. 1. Face orientation, person 25 looking: a- left (LL) b- straight (LS) c- right (LR)

Fig. 2. Face image database of 30 persons looking straight (LS)

of this work, namely, providing a time and cost efficient system. A window of size 100x100 pixels; that contains the face and its background, is then extracted and the data within this relatively smaller size image is used for training and eventually testing the neural network.

The method used for presenting the images to the neural network uses pattern averaging, which in our hypothesis approximates the glance. The face image of size100x100 pixels is segmented and the values of the pixels within each segment are averaged. The result average values are then used as input data for the neural network. The averaging of the segments within an image reduces the amount of data required for neural network implementation thus providing a faster recognition system. Pattern averaging can be defined as follows:

$$PatAv_i = \frac{1}{s_k s_l} \sum_{l=1}^{s_l} \sum_{k=1}^{s_k} p_i(k,l), \tag{1}$$

where k and l are segment coordinates in the x and y directions respectively, i is the segment number, S_k and S_l are segment width and height respectively, $P_i(k,l)$ is pixel value at coordinates k and l in segment i, $PatAv_i$ is the average value of pattern in segment I , that is presented to neural network input layer neuron i. The number of segments in each window (of size XY pixels) containing a face, as well as the number of neurons in the input layer is i where

$$i = -\{0, 1, 2, \ldots, n\}, \tag{2}$$

and

$$n = (X/s_k)(Y/s_l). \tag{3}$$

Segment size of 10x10 pixels (S_k = S_l = 10) has been used and average values representing the image were obtained, thus resulting in 100 average values in total ($n = 100$) that were used as the input to the neural network for both training and testing.

In summary, each image has to undergo *compression, region extraction, segmentation* and *segment averaging* prior to presenting it to the neural network. Previous work using this pre-processing method showed sufficient representation of the objects within the images and meaningful data within the averaged patterns were obtained to aid the neural network learning and classification [8]. Figure 3 shows an

Fig. 3. Image pre-processing and averaging before neural network training / generalizing

example of this pre-processing phase. The original 512x512 pixel image is first compressed to 256x256 pixels and then to 128x128 pixels. This is followed by extracting a region of size 100x100 pixels that contains the face. The extracted region is then segmented and averaged yielding a 10x10 pixel pattern that represents the original image.

3 Neural Network Implementation

The neural network, which has been developed as part of this face recognition system, is based on the back propagation learning algorithm, with an input layer, one hidden layer and an output layer. The input layer has 100 neurons, each receiving an averaged value of the face image segments. The hidden layer consists of 99 neurons, whereas the output layer has 30 neurons according to the number of persons. Figure 4 shows the topology of this neural network and data presentation to the input layer.

The implementation of a neural network consists of training and testing. In this work a total of 90 face images (corresponding to 30 persons) are used. For training the neural network 60 face images (looking left *LL* and looking right *LR*) are used. The 30 remaining face images (looking straight *LS*) are used for testing purposes, where the system is expected to recognize the person looking straight at the camera by training it on face images looking left and right.

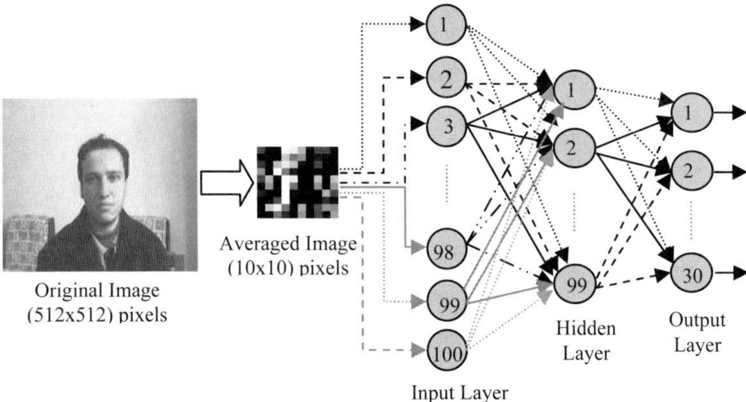

Fig. 4. Pattern averaging and neural network topology

This, in my hypothesis, simulates the familiarity of a face in machines, even though the test images (looking straight) present a neural network with different pixel values as a result of the difference in the orientation of the face.

4 Results and Discussion

The neural network learnt and converged after 4314 iterations and within 390 seconds, whereas the running time for the generalized neural network after training and using one forward pass was 0.21 seconds. These results were obtained using a 2.4

GHz PC with 256 MB of RAM, Windows XP OS and Borland C^{++} compiler. Table 1 lists the final parameters of the successfully trained neural network. The robustness, flexibility and speed of this novel face recognition system have been demonstrated through this application.

All training images (60 face images- looking left and right) were recognized when used for testing the trained neural network yielding 100% recognition rate. Testing the neural network using test images (30 face images – looking straight) that had not been presented to the network before, yielded a successful 96.67% recognition rate where 29 out of 30 face images were correctly recognized. The incorrectly recognized face, out of the testing image set, was identified as another person. Both persons have close face resemblance as shown in Figure 5. Table 2 shows the recognition rates where a total recognition rate of 98.89% has been achieved.

Table 1. Trained neural network final parameters

Input Nodes	Hidden nodes	Output Nodes	Learning rate	Momentum rate	Error	Iterations	Training time	Generalization time
100	99	30	0.008	0.32	0.002	4314	390 seconds	0.21 seconds

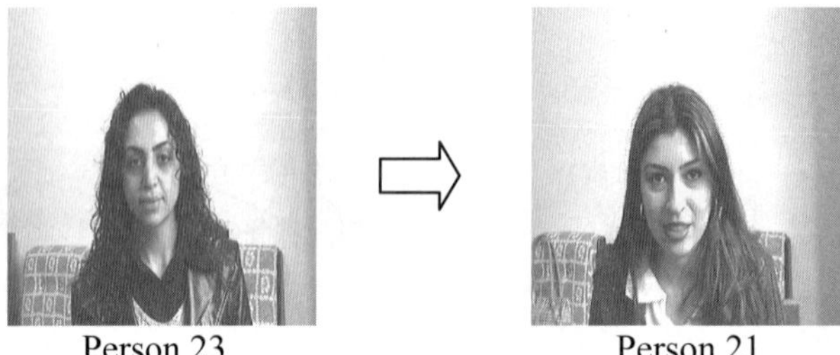

Fig. 5. Person 23 identified as person 21 and their close face resemblance

Table 2. Face recognition results for 30 persons

Image Set	Training (60 images)	Testing (30 images)	Total (90 images)
Recognition Rate	(60/60) 100%	(29/30) 96.67%	(89/90) 98.89%

5 Conclusions

A novel approach to face recognition, based on simulating the human "glance" and face "familiarity", is introduced in this paper. The glance effect is approximated via image pre-processing and pattern averaging. When we, humans, have a quick look (glance) at faces we do not observe the detailed features but rather a general global impression of a face. This in our hypothesis can be approximated by averaging the face image instead of searching for features within the face. The averaged patterns are representations of a face regardless of its expression or orientation. The quick glance is followed by familiarity with a face, which is simulated by training a neural network using face images with different orientations.

A real life application, using 90 face images of 30 persons of different gender, age and ethnicity, was implemented using this intelligent recognition system. A total recognition rate of 98.89% was obtained using 90 face images of the 30 persons in different orientations. The robustness and success of this face recognition system is further demonstrated by its quick run time (one neural network forward pass) of 0.21 seconds. Time cost was kept minimal through image-preprocessing and reduction of input/hidden layer neurons in the topology of neural network.

Future work includes enlarging the face database with more persons and more face orientations. Additionally, work will be carried out on further reducing neural network training and run times. This can be done by reducing the number of averaged segments in the face image while attempting to provide the neural network with meaningful representation of the face.

References

1. Turk, M., Pentland, A.P.: Face Recognition Using Eigenfaces. Proceedings of the IEEE Conference on Computer Vision and Pattern Recognition (1991) 586-591
2. Murase, H., Nayar, S.K.: Visual Learning and Recognition of 3-D Objects from Appearance. International Journal on Computer Vision 14 (1995) 5-24
3. He, X., Yan, S., Hu, Y., Niyogi, P., Zhang, H.J.: Face Recognition Using Laplacianfaces, IEEE Trans. PAMI 27(3) (2005) 328-340
4. Belhumeur, P.N., Hespanha, J.P., Kriegman, D.J.: Eigenfaces vs. Fisherfaces: Recognition Using Class Specific Linear Projection, IEEE Trans. PAMI 19(7) (1997) 711-720
5. Martinez, A.M., Kak, A.C.: PCA Versus LDA, IEEE Trans. PAMI, 23(2) (2001) 228-233
6. Roweis, S.T., Saul, L.K.: Nonlinear Dimensionality Reduction by Locally Linear Embedding, Science 290 (2000) 2323-2326
7. Yang, M.H., Kriegman, D.J., Ahuja, N.: Detecting Faces in Images: A Survey. IEEE Trans. PAMI 24(1) (2002) 34-58
8. Khashman, A., Sekeroglu, B.: Multi-Banknote Identification Using a Single Neural Network, Lecture Notes on Computer Science, Vol. 3708. Springer-Verlag, Berlin Heidelberg New York (2005) 123-129

Semi-supervised Support Vector Learning for Face Recognition

Ke Lu[1], Xiaofei He[2], and Jidong Zhao[1]

[1] School of Computer Science and Engineering, University of Electronic Science & Technology of China, Chengdu, Sichuan 610054, China
[2] Department of Computer Science, University of Chicago, 1100 E 58th Street, Chicago, IL 60637, USA
kel@uestc.edu.cn

Abstract. Recently semi-supervised learning has attracted a lot of attention. Different from traditional supervised learning, semi-supervised learning makes use of both labeled and unlabeled data. In face recognition, collecting labeled examples costs human effort, while vast amounts of unlabeled data are often readily available and offer some additional information. In this paper, based on Support Vector Machine (SVM), we introduce a novel semi-supervised learning method for face recognition. The basic idea of the method is that, if two data points are close to each other, they tend to share the same label. Therefore, it is reasonable to search a projection with maximal margin and locality preserving property. We compare our method to standard SVM and transductive SVM. Experimental results show efficiency and effectiveness of our method.

1 Introduction

Face recognition is a long standing research problem. It has received a lot of attentions during past decades. The typical face recognition methods include Eigenface, Fisherface, Laplacianface [1], Support Vector Machine, Bayesian, etc.

Most of traditional learning methods only make use of labeled data, while the unlabeled data is ignored. In recent years, much research has been done on semi-supervised learning [2]. Semi-supervised learning makes use of both labeled and unlabeled data. Generally, the unlabeled data can be used to better describe the intrinsic geometrical structure of the data space, and hence improve the classification performance. Also, most previous learning algorithms only consider the Euclidean structure of the data space. However, in many cases, the objects of interest might reside on a low-dimensional manifold which is nonlinearly embedded in the ambient space (data space). In such cases, the nonlinear manifold structure is much more important than the Euclidean structure. Specifically, the similarity between objects should be described by the geodesic distance rather than the Euclidean distance.

In this paper, we introduce a new semi-supervised algorithm for face recognition. Our algorithm is intrinsically based on Support Vector Machines and Locality Preserving Projections (LPP) [3]. LPP is a recently proposed algorithm for linear dimensionality reduction. We first build a nearest neighbor graph over all the data points (labeled and unlabelled) which models the local geometrical structure of the image

J. Wang et al. (Eds.): ISNN 2006, LNCS 3972, pp. 104 – 109, 2006.

space. By combining SVM and LPP, we can obtain a classifier which maximizes the margin and simultaneously preserves the local information.

2 Preliminaries

In this section, we provide a brief overview of Support Vector Machines [4] and Locality Preserving Projections [3].

2.1 Support Vector Machines

SVM training algorithms are based on the idea of structured risk minimization rather than empirical risk minimization, and give rise to new ways of training polynomial, neural network, and radial basis function (RBF) classifiers. SVMs make no assumptions on the distribution of the data and can, therefore, be applied even when we do not have enough knowledge to estimate the input data [4].

We shall consider SVM in the binary classification setting. We assume that we have a data set $D = \{x_i, y_i\}_{i=1}^{l}$ of labeled examples, where $y_i \in \{-1, 1\}$, and we wish to select, among the infinite number of linear classifiers that separate the data, one that minimizes the generalization error, or at least minimizes an upper bound on it. It is shown that the hyperplane with this property is the one that leaves the maximum margin between the two classes. Given a new data point x to classify, a label is assigned according to its relationship to the decision boundary, and the decision function is

$$f(x) = sign\left(\sum_{i=1}^{t} \alpha_i y_i \langle x_i, x \rangle - b\right)$$

The nonlinear SVM implicitly maps the input variable into a high-dimensional (often infinite-dimensional) space, and applies the linear SVM in the space. Computationally, this can be achieved by the application of a reproducing kernel. If define K as kernel function, the corresponding nonlinear decision function is

$$f(\mathrm{x}) = sign(\sum_{i=1}^{t} \alpha_i y_i \mathrm{K}(\mathrm{x}_i, \mathrm{x}) - b)$$

2.2 Locality Preserving Projections

LPP is a recently proposed linear dimensionality reduction algorithm [3]. For a data point x in the original space, we consider the projection $W \in R^n$ such that $y = W^T x$. LPP aims to discover the intrinsic manifold structure by Euclidean embedding. Give a local similarity matrix S, the optimal projections can be obtained by solving the following minimization problem:

$$\begin{aligned} W_{opt} &= \arg\min_W \sum_{i=1}^{m} \left(W^T x_i - W^T x_j\right)^2 S_{ij} \\ &= \arg\min_W \left(W^T X L X^T W\right) \end{aligned}$$

with the constraint

$$W^T XDX^T W = 1$$

where $L = D - S$ is the graph Laplacian [5] and $D_{ii} = \sum_j S_{ij}$. D_{ij} measures the local density around x_i. The bigger D_{ii} is, the more important x_i is. With simple algebra steps, we can finally get the following generalized eigenvector problem:

$$XLX^T W = \lambda XDX^T W$$

For the detailed derivation of LPP, please see [3].

3 Semi-supervised Support Vector Learning

Most existing semi-supervised learners do not readily generalize to new test data. A brute force approach is to incorporate the new test points and reestimate the function using semi-supervised learning, but this is very inefficient. Another problem of semi-supervised transduction is the computational complexity. In this section, we introduce a new semi-supervised learning algorithm which efficiently combines the characters of both SVMs and LPP.

Let $X = [x_1, x_2, \cdots, x_m]$ denote the set of data points, belonging to c classes. Let y_i be the label associated with x_i, $y_i \in \{-1, 1\}$. Suppose the first t data points are labeled, and the rest $m-t$ are unlabled. We construct a graph as follows: for any pair of data points, we put an edge between them if they are sufficiently close to each other. Correspondingly, the weights are defined as follows:

$$S_{ij} = \begin{cases} 1, & \text{if } x_i \text{ is among the } p \text{ nearest neighbors of } x_j \\ & \quad \text{or } x_j \text{ is among the } p \text{ nearest neighbors of } x_i; \\ 0, & \text{otherwise} \end{cases} \tag{1}$$

The above definition reflects the intrinsic manifold structure of the data space.

Recall that SVM aims to maximize the margin between two classes. The loss function of SVM can be described as follows:

$$\begin{aligned} L(x_i, y_i, w) &= \frac{2}{w} \\ y_i(x_i \cdot w + b) - 1 &\geq 0, \forall i \end{aligned} \tag{2}$$

One disadvantage of the above loss function is that it only takes into account the labeled data, while the unlabeled data is ignored. A natural extension is to incorporate the unlabeled data and preserve the graph structure. Specifically, we expect that if two data points are close to each other, they tend to be classified into the same class.

Let f be the classifier such that $f(x_i)$ is the estimated label for x_i. Thus, we expect to minimize the following loss function:

$$G(\{x\}_i, f) = \sum_{ij} \left\| f(x_i) - f(x_j) \right\|^2 S_{ij}$$

We define:

$$\mathrm{f} = (f_1, \ldots, f_m), \quad f_i = f(x_i), \quad D_{ii} = \sum_j S_{ij}$$

By simple algebra, we have

$$\begin{aligned} G(\{x_i\}, f) &= \sum_{ij} (f_i - f_j)^2 S_{ij} = \sum_{ij} f_i^{\,2} S_{ij} + \sum_{ij} f_j^{\,2} S_{ij} - 2\sum_{ij} S_{ij} f_i f_j \\ &= \mathrm{f}^{\mathrm{T}} S\mathrm{f} + \mathrm{f}^{\mathrm{T}} S\mathrm{f} - \mathrm{f}^{\mathrm{T}} D\mathrm{f} = 2\mathrm{f}^{\mathrm{T}} (D - S)\mathrm{f} \cong 2\mathrm{f}^{\mathrm{T}} L\mathrm{f} \end{aligned}$$

where $L = D - S$ is called graph Laplacian [5]. Suppose f is linear, $y = W^T x$. Thus, we have:

$$G(\{x\}_i, f) = W^T XLX^T W \tag{3}$$

By combing (2) and (3), the optimal classifier can be obtained as follows:

$$\begin{aligned} W^* &= \arg\max_W \left\| W \right\|^2 - \beta W^{\mathrm{T}} XLX^T W \\ y_i(x_i \cdot W + \mathrm{b}) - 1 &\geq 0,\, i = 1, \ldots, t \end{aligned} \tag{4}$$

The classifier obtained from (4) is a linear classifier. Suppose f is a nonlinear function in a Reproducing Kernel Hilbert Space (RKHS) H_K, where H_K is defined below:

$$H_K = \{ f \mid f(x) = \sum_i \alpha_i K(x_i, x), \alpha_i \in R \}$$

where K is a kernel function [4]. Thus, we get the following minimization problem in RKHS:

$$f^* = \arg\max_{f \in H_K} \left\| f \right\|^2 - \beta\, \mathrm{f}^{\mathrm{T}} L\mathrm{f}$$

4 Experimental Results

Two face databases were used to compare our algorithm with the standard SVM algorithm. The first one is the PIE (Pose, Illumination, and Experience) database from CMU, and the second one is the ORL database. In all the experiments, preprocessing to locate the faces was applied. Original images were normalized (in scale and orientation) such that the two eyes were aligned at the same position. Then, the facial areas were cropped into the final images for matching. The size of each cropped image in all the experiments is 32×32 pixels, with 256 gray levels per pixel. No further preprocessing is done. Thus, the image is represented as a 1024-dimensional vector. When applying SVM and our algorithm for face recognition, we use one vs. one and majority vote strategies for multi-class classification.

4.1 Experiments on PIE Database

The CMU PIE face database contains 68 subjects with 41,368 face images as a whole. The face images were captured by 13 synchronized cameras and 21 flashes, under varying pose, illumination and expression. We choose the five near frontal poses (C05, C07, C09, C27, C29) and use all the images under different illuminations and expressions, thus we get 170 images for each individual. Some sample images of one individual are shown in Figure 1. For each individual, l (= 5, 10, 20, 30) images are randomly selected for training and the rest are used for testing.

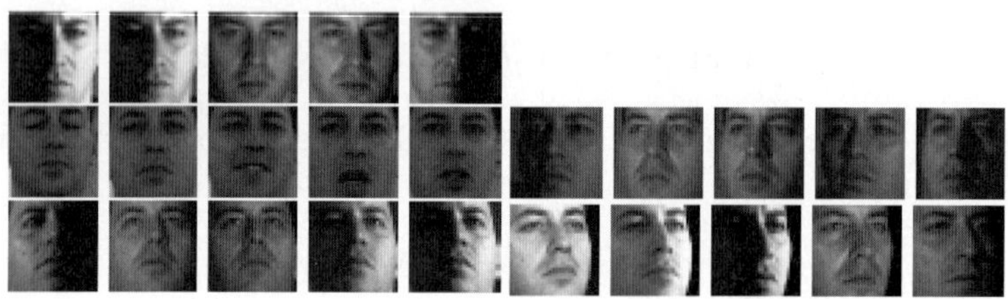

Fig. 1. Sample face images from the CMU PIE database. For each subject, there are 170 near frontal face images under varying pose, illumination, and expression.

Fig. 2. Sample face images from the ORL database. For each subject, there are 10 face images with different facial expression and details.

The training set is utilized to learn a classifier by using SVM and our algorithm. For each given l, we average the results over 10 random splits. Table1 shows the recognition rates. As can be seen, our method outperforms the standard SVM with different numbers of training samples (5, 10, 20, 30) per individual.

Table 1. Recognition error on PIE database

Method	5 Train	10 Train	20 Train	30 Train
SVM	39.9%	32.7%	26.5%	20.8%
Our method	29.7%	24.4%	20.4%	15.6%

It is found that our algorithm especially suitable when the number of training samples is extremely small. This is because that, our algorithm can efficiently makes use of unlabeled data to discover the intrinsic geometrical structure of the face manifold.

4.2 Experiments on ORL Database

The ORL (Olivetti Research Laboratory) face database is used in this test. It consists of a total of 400 face images, of a total of 40 people (10 samples per person). The images were captured at different times and have different variations including expressions (open or closed eyes, smiling or non-smiling) and facial details (glasses or no glasses). The images were taken with a tolerance for some tilting and rotation of the face up to 20 degrees. 10 sample images of one individual are displayed in Figure 2. For each individual, l (= 2, 3, 4, 5) images are randomly selected for training and the rest are used for testing.

Table 2. Recognition error on ORL database

Method	5 Train	10 Train	20 Train	30 Train
SVM	39.9%	32.7%	26.5%	20.8%
Our method	29.7%	24.4%	20.4%	15.6%

The experimental design is the same as that in the last subsection. Table 2 shows the recognition errors of SVM and our method. As can be seen, our method performs better than standard SVM. As before, our method is especially suitable when the number of training samples is extremely small.

5 Conclusions

In this paper, we introduce a new semi-supervised support vector learning algorithm which combines SVMs and LPP. The new algorithm maximizes the margin between two classes and simultaneously preserves the local information. We have applied the new method to face recognition. Experimental results show that the unlabeled data can be used to enhance the recognition rate.

There are still several questions that remain unclear. For example, it is unclear under what conditions, the local information is especially important. Also, it remains unclear how to define the locality. We are currently working on these problems.

References

1. He, Y., Yan, S., Hu, Y., Niyogi, P., and Zhang, H.-J.: Face Recognition Using Laplacianfaces. IEEE Trans. on Pattern Analysis and Machine Intelligence 27 (2005) 328-340
2. Belkin, M., and Niyogi, P.: Semi-supervised Learning on Riemannian Manifolds. Technical Report, TR-2001-30, Univ. of Chicago, Computer Science Dept. (2001)
3. He, X. and Niyogi, P.: Locality Preserving Projections. Advances in Neural Information Processing Systems 16 (2003)
4. Vapnik, V.N.: The Nature of Statistical Learning Theory. Springer-Verlag (1995)
5. Chung, F.R.K.: Spectral Graph Theory. Volume 92 of Regional Conference Series in Mathematics (1997)

Parts-Based Holistic Face Recognition with RBF Neural Networks

Wei Zhou, Xiaorong Pu, and Ziming Zheng

Computational Intelligence Laboratory, School of Computer Science and Engineering, University of Electronic Science and Technology of China, Chengdu 610054, P.R. China
zhouwei@uestc.edu.cn, puxiaor@uestc.edu.cn
http://cilab.uestc.edu.cn

Abstract. This paper proposes a method for face recognition by integrating non-negative matrix factorization with sparseness constraints (NMFs) and radial basis function (RBF) classifier. NMFs can represent a facial image based on either local or holistic features by constraining the sparseness of the basis images. The comparative experiments are carried out between NMFs with low or high sparseness and principle component analysis (PCA) for recognizing faces with or without occlusions. The simulation results show that RBF classifier outperforms $k-$nearest neighbor linear classifier significantly in recognizing faces with occlusions, and the holistic representations are generally less sensitive to occlusions or noise than parts-based representations.

1 Introduction

With the last decade, face recognition (FR) has become one of the most active and exciting research areas. One of the fundamental problems in face recognition is to find a suitable representation of the facial image. Most face recognition systems learn either holistic or parts-based representations. The principle component analysis (PCA), as a classical method for feature extraction, learns holistic representations of facial images [1], [2], while non-negative matrix factorization (NMF), a recently proposed approach, learns parts-based representations of faces [3]-[7]. However, we argue that NMF can not only learn parts-based representations but also holistic ones with different sparseness constraints.

Another issue in FR is to find an appropriate classifier. There are many conventional classifiers, which are generally divided into two categories: linear classifiers and nonlinear ones. Although successful in many cases, the linear classifier methods, such as $k-$nearest neighbor linear classifier, still fail to perform well when face images have larger variation in viewpoints, or with occlusions, which results in highly nonconvex and complex distribution [11]. The radial basis function (RBF) neural networks, which can be used as non-linear classifiers, have been applied extensively in pattern classification because of their salient features in being universal approximators, possing the best approximation property and the fast learning speed and having more compact topology than other neural networks [12].

J. Wang et al. (Eds.): ISNN 2006, LNCS 3972, pp. 110–115, 2006.

In this paper, we propose an efficient NMFs+RBF aggregate framework for FR, in which the non-negative matrix factorization with sparseness constraints (NMFs) is firstly applied to learn either the holistic representations or the parts-based ones, then the RBF classifier is adopted for pattern classification. The comparative performances are studied among the NMFs+RBF method, the PCA+RBF method, and the PCA+FLD (Fisher's Linear Discriminant) method [2], [8]. All simulations are carried out on the ORL face database [9].

This paper is organized as follows. The brief review of the NMFs method is introduced in Section 2. Section 3 describes RBF classifiers. Our proposed method is investigated in Section 4. Finally, section 5 and 6 provide comparative experimental studies and the conclusion.

2 Non-negative Matrix Factorization with Sparseness Constraints

Denote the training set of n face images by $V = (V_1, V_2, \dots, V_n) \subset \Re^{m\times n}$. Each column of V contains m pixel values (≥ 0) of one of the n facial images, and each image belongs to one of c classes. Non-negative matrix factorization (NMF) is a linear, non-negative approximate data representation[3], which refers to the decomposition of the matrix V into two matrices $W \subset \Re^{m\times r} (r < n)$(basis images) and $H \subset \Re^{r\times n}$(encoding coefficients), such that $V \approx WH$, $W, H \geq 0$. The NMF with sparseness constraints (NMFs) is defined as follows [10]:

$$\min_{W,H} \quad D(V|WH) = \sum_{i,j}(V_{ij} - (WH)_{ij})^2, \tag{1}$$

$$s.t. \quad W, H \geq 0, \quad sparseness(w_i) = S_w, \forall i, \quad sparseness(h_i) = S_h, \forall i, \tag{2}$$

where S_w and S_h are desired sparseness of W and H respectively, the sparseness level $sparseness(x) = \frac{\sqrt{n} - ||x||_1/||x||_2}{\sqrt{n}-1}$, n is the dimensionality of x. In [10], Patrik found the face representation switched from a global one to a local one when increasing the sparseness of the basis images. For further details, refer to [10].

3 Radial Basis Function Neural Networks

An RBF neural network, which involves an input layer, a hidden layer and an output layer, can be considered as a mapping: $\Re^r \rightarrow \Re^s$. The output of each hidden layer unit is of the following Gaussian form:

$$R_i(P) = \exp\left[-\frac{\|P - C_i\|^2}{\sigma_i^2}\right], \quad i = 1, \dots, u, \tag{3}$$

where $P \subset \Re^r$ is the input vector, $C_i \in \Re^r (1 \leq i \leq u)$ is the prototype of the input vectors, $\|\cdot\|$ indicates the Euclidean norm, and σ_i is the width of the ith RBF unit. The output is given as:

$$Y = GR, \tag{4}$$

where $G \subset \Re^{s \times u}$ is the connection weight and $R \subset \Re^{u \times n}$ is the matrix of the hidden units' output. Let the RBF center be defined as $C^k = \frac{1}{n^k} \sum_{i=1}^{n^k} P_i^k \quad k = 1, \ldots, u$, where P_i^k is the ith sample belonging to class k, n^k is the total number of training patterns in class k and u is the number of the RBF nodes.

The construction for the neural networks can be described as follows:

1. Initialize the number of hidden nodes: $u := s$, and set $k := 1$.
2. For each RBF node k, compute the RBF center C^k, the distances d_k, $dc(k, j)$ and $d_{\min}(k, l)$, where $d_k = \|P^k(f) - C^k\|(k = 1, \ldots, u)$, $dc(k, j) = \|C^k - C^j\|(j = 1, \ldots, u, j \neq k)$, $d_{\min}(k, l) = \arg\min_l (dc(k, j))(j = 1, \ldots, u, j \neq k)$, where $P^k(f)$ is the furthest point belongs to the class k.
   (a) If $d_k + d_c \leq d_{min}(k, l)$, which means no overlapping.
   (b) If $d_k + d_c > d_{\min}(k, l)$, find a split way for two overlapped classes k and l, which make the overlapping points between two classes is the least.
3. If $k \neq u$, set $k := k + 1$, go to (3);
4. Compute the width of class k: $\sigma^k = \max\left(\sigma_W^k, \sigma_B^k\right)$, where

$$\sigma_W^k = d_k / \sqrt{|\ln \beta|}, \sigma_B^k = \eta \times d_{min}(k, l), \eta \approx \sum_{i=1}^{c} \frac{d_k}{\sqrt{|\ln \beta|}} / \sum_{k=1}^{c} d_{min}(k, l)$$

   and β is the confidence coefficient, which lies in the range $0.5 \leq \beta < 1$. In our simulation,we set $\beta = 0.7$,
5. Employ the Hybrid Learning Algorithm (HLA) to estimate the σ^k and G. For further details, refer to [12].

4 Face Recognition Using NMFs + RBF

Let H_{train} be the matrix for training image set V_{train}, and it is calculated by

$$H_{train} = W^{\dagger} \cdot V_{train}, \tag{5}$$

where $W^{\dagger}$ is the pseudoinverse of the basis matrix W. During the training process, the basis features H_{train} are used as the input vectors for the RBF neural networks. During the testing process, the RBF neural networks inputs are H_{test}, where $H_{test} = W^{\dagger} \cdot V_{test} = (H_1, H_2, \ldots, H_n)$ is the matrix for testing image set, and the final output is

$$Y = G \cdot R(H_{test}), \tag{6}$$

where $Y = (Y_1, Y_2, \ldots, Y_n) \subset \Re^{c \times n}$, and $R(H_{test}) = (R(H_1), R(H_2), \ldots, R(H_n))$.

5 Experiments

Our experiments are carried out on the ORL face database, which contains 400 face images from 40 individuals, varying in position, rotation, scale and

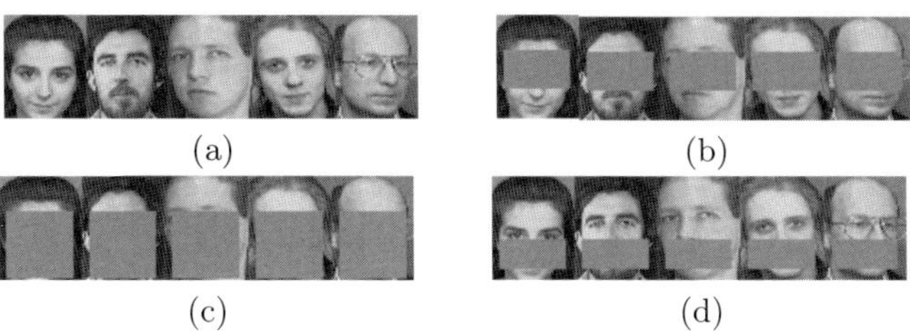

Fig. 1. Examples of test images. (a)Some ORL faces, (b)Eyes occluded faces, (c)Most parts occluded faces, (d)Mouth occluded faces

Table 1. Comparisons of PCA-based FR Methods and NMFs+RBF FR Method

Occlusion Condition	Number of features	PCA+FLD (%)	PCA+RBF (%)	NMFs+RBF (%)	
				0.45 (Holistic)	0.75 (Local)
Without Occlusions	5	72.54	69.04	70.72	43.80
	20	91.84	89.12	88.20	76.22
	80	94.06	94.44	91.76	93.58
Eyes Occluded Faces	5	43.42	50.88	46.84	11.32
	20	77.98	79.44	79.24	48.52
	80	84.58	89.92	88.16	89.14
Most Parts Occluded Faces	5	18.54	20.74	22.12	8.08
	20	47.12	49.40	61.20	37.32
	80	51.84	64.88	70.66	64.52
Mouth Occluded Faces	5	49.22	54.62	51.16	31.58
	20	85.72	83.08	84.46	64.68
	80	85.36	90.38	88.24	91.28

expression. Among total database, 200 images are randomly selected (5 face images per person) for training and another 200 for testing. Three experiments are employed in this paper. The first is the PCA + RBF method, which use the eigenface-based features as the input vectors of RBF neural network for training [1]. The second is PCA + FLD method, in which an optimal Fisher subspace is constructed and the k-nearest neighbor linear classifier is used for classification [2],[8]. The last one is NMF+RBF approach.

Tests are done with different number of features $(5, 10, 15, 20, 35, 50, 65, 80)$ and different sparseness levels for NMFs $(0.45, 0.55, 0.65, 0.75)$, with or without occlusions in the testing images. The occlusion is simulated in an image by using different grey patch of size $40 \times 72, 70 \times 72, 30 \times 72$ at the eye, face, mouth locations respectively; some samples are shown in Fig.1. 800 runs are executed for each PCA-based method, and 800 runs for NMFs+RBF method with different sparseness level. The number of features, basis vectors and the corresponding face recognition rates are shown in Table 1 (the sparseness levels are set to 0.45 and 0.75). Fig.2(a) indicates that NMFs+RBF method has better performance than the other two methods in recognizing faces with occlusions, due to its capability of learning parts-based representations. We also investigate how the sparseness

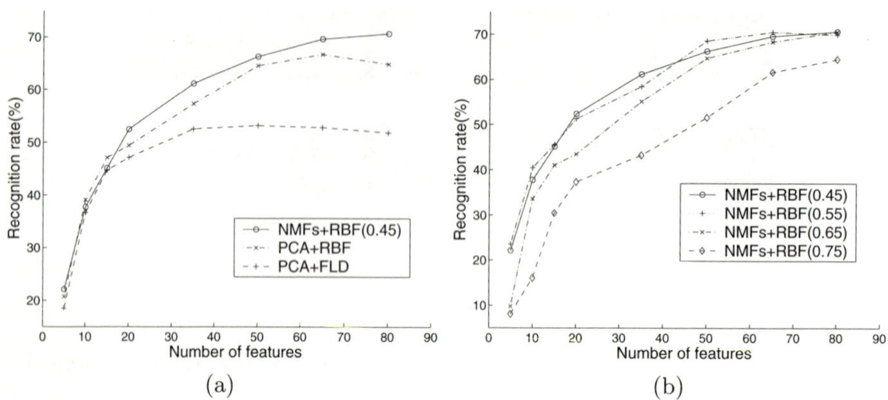

Fig. 2. The performance comparisons in face occlusion condition: (a) Average recognition rate for three FR approaches, (b) Average recognition rate for NMFs+RBF approaches with different levels of sparseness $(0.45, 0.55, 0.65, 0.75)$

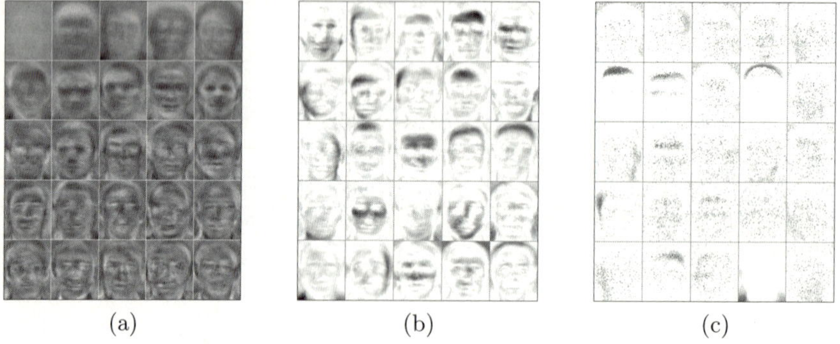

Fig. 3. Basis images learned form the ORL database: (a) PCA Eigenfaces, (b) NMFs-Faces (0.45), (c) NMFs-Faces (0.75). The trend can be seen from the holistic face representation to the parts-based face representation.

level affects the performance. Actually, the NMFs+RBF method with proper low sparseness can perform well in recognizing faces with or without occlusions (see Fig 2(b) and Table 1) and can learn holistic representations similar to PCA [1]; While it can learn parts-based representations by constraining its sparseness in a high level, as most researches studied [3], [5] (see Fig.3).

6 Conclusions

In this article, the NMFs+RBF method is developed for recognizing faces with or without occlusions. NMFs can learn both holistic and parts-based representations of facial images by constraining the sparseness of the basis images and an RBF classifier is designed for pattern classification. The comparative experiments are carried out among NMFs with low or high sparseness integrating with RBF, PCA integrating with RBF, and PCA integrating with FLD and a $k-$nearest neighbor linear classifier. The simulation results show that our proposed RBF nonlinear classifier outperforms $k-$nearest neighbor linear classifier significantly in recognizing faces with occlusions, and the methods of learning holistic representations, such as PCA or NMFs with low sparseness, are generally less sensitive to occlusions or noise than the methods of learning parts-based representations.

Acknowledgement

This work was supported by National Science Foundation of China under Grant 60471055, Specialized Research Fund for the Doctoral Program of Higher Education under Grant 20040614017, and the UESTC Youth Fund under Grant L08010601JX04030.

References

1. Turk, M.A., Pentland, A.: Eigenfaces for Recognition. Journal of Cognitive. Neurosience vol. **3** (1991) 71-86
2. Belhumeur, P.N., Hespanha, J.P., Kriegman, D.J.: Eigenfaces versus Fisherfaces: Recognition using Class Specific Linear Projection. IEEE Trans. Pattern Anal. Machine Intell. vol. **19** (1997) 711-720
3. Lee, D.D., Seung, H.S.: Learning the Parts of Objects by Non-negative Matrix Factorization. Nature 401(6755) (1999) 788-791
4. Guillamet, D., Vitri, J.: Non-negative Matrix Factorization for Face Recognition. LNCS vol. **2504** Springer-Verlag GmbH (2002) 336-344
5. Feng, T., Li, S.Z., Shum, H.Y., Zhang, H.J.: Local Non-negative Matrix Factorization as a Visual Representation. In: Proceedings of 2nd International Conference on Development and Learning (ICDL.02). IEEE Computer Society Cambridge MA (2002) 178-183
6. Foon, N.H., Jin, A.T.B., Ling, D.N.C.: Face Recognition Using Wavelet Transform and Non-negative Matrix Factorization. LNCS vol. **3339** Springer-Verlag GmbH (2004) 192-202
7. Pu, X.R., Yi, Z., Zheng, Z., et al.: Face Recognition Using Fisher Non-negative Matrix Factorization with Sparseness Constraints. LNCS vol. **3497** (2005) 112-117
8. Yambor W.S.: Analysis of PCA-Based and Fisher Discriminant-based Image Recognition Algorithms. Technical Report CS-00-103, Computer Science Department, Colorado State University (2000)
9. The ORL database is available from http://www.cam-orl.co.uk/facedatabase.html
10. Patrik O.H.: Non-negative Matrix Factorization with Sparseness Constraints. Journal of Machine Learning Research vol. **5** (2004) 1457-1469
11. Lu, J., Plataniotis, K.N., Venetsanopoulos, A.N.: Face Recognition using Kernel Direct Discriminant Analysis Algorithms. IEEE Trans. Neural Netw. 14(1) (2003) 117-126
12. Er, M.J., Wu, S., Lu J.: Face Recognition with Radial Basis Function (RBF) Neural Networks, IEEE Trans. Neural Netw. 13(3) (2002) 697-710

Combining Classifiers for Robust Face Detection

Lin-Lin Huang[1] and Akinobu Shimizu[2]

[1] Beijing University of Aeronautics and Astronautics,
37 Xueyuan Road, Haidian District, Beijing 100083, P.R. China
llhuang@buaa.edu.cn
[2] Tokyo University of Agriculture and Technology,
2-24-16 Naka-cho, Koganei-shi, Tokyo 184-8588, Japan

Abstract. In this paper, we propose a face detection method by combining classifiers. We apply two classifiers using features extracted from complementary feature subspaces learned by principal component analysis (PCA). The two classifiers employ the same classification model named a polynomial neural network (PNN). The outputs of the two classifiers are fused to make the final decision. The effectiveness of the proposed method has been demonstrated in experimentals.

1 Introduction

Face detection has become an active research topic due to its numerous applications. The methods proposed for face detection can be roughly divided into two categories: feature-based [1]and classification-based [2][3]. So far, the best results have been achieved by classification-based methods [4].

Face detection from cluttered images is challenging because faces are non-rigid objects. Also, variable lighting conditions, low image quality and cluttered backgrounds pose great difficulties. Different face/non-face classifiers can attain different degrees of success, but maybe none of them performs perfectly. Practical applications in pattern recognition have been shown to benefit from combining multiple classifiers since they potentially offer complementary information about the pattern to be classified [5].

The improvement of accuracy made from combining classifiers largely depends on the complementariness of participating classifiers. In this paper, we combine two classifiers using features extracted from complementary feature subspaces learned by principal component analysis (PCA). The two classifiers employ the same classification model named a polynomial neural network (PNN). The outputs of the two classifiers are fused to make the final decision of face detection.

Our face detection system has been described in [6]. The input image to detect faces from is re-scaled to multiple scales and each re-scaled image is scanned by a sliding window. Each local image in the window is assigned a face likelihood score by a classifier. The local maximums of face scores over different scales and locations are detected and each local maximum greater than a threshold signifies a detected face. In the proposed combining method, the local image is

J. Wang et al. (Eds.): ISNN 2006, LNCS 3972, pp. 116–121, 2006.

represented by two types of features, each feature is classified by the PNN to give a face score. The two scores are fused to give a combined score, and the local maximums of combined scores give detected faces.

2 Classification Method

2.1 Feature Extraction

Most classification-based methods have used the intensity values of window images as the input features of classifier. However, in a face image, the direction of edges is a prominent and stable feature. In [6], we extracted gradient direction feature from windowed local images and have achieved superior detection performance. After pre-processing the local image, the gradient vector $\mathbf{g}(x, y) = [g_x, g_y]^T$ is computed for each pixel using the Sobel operator. We then decompose the gradient vector into the components in eight chaincode directions (Fig. 1). The magnitude of $g_d(x, y)$ in d-th chaincode direction ($d = 1, \ldots, 8$) is assigned to the d-th directional plane (sub-image).

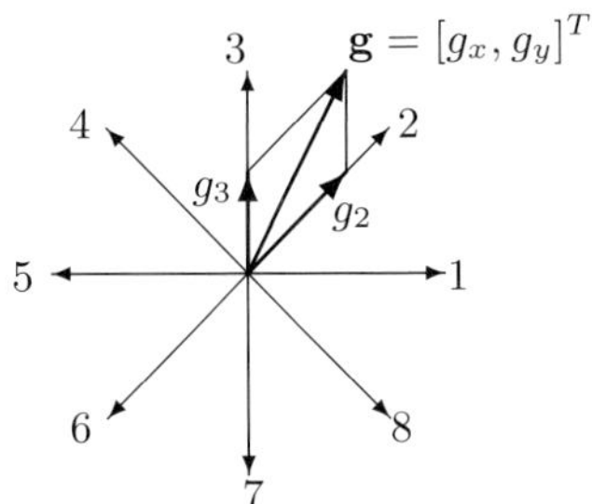

Fig. 1. Decomposition of gradient vector

To generate a feature vector of moderate dimensionality, each pair of planes of opposite directions are merged into a single orientation plane and each plane is compressed to 10×10 pixels by block averaging. We select three stable orientation planes (the orientation plane from directions 1 and 5 is abandoned) and the compressed intensity image to construct a feature vector. After removing some corner pixels in each compressed plane/image, we obtain 368 feature measurements.

2.2 Polynomial Neural Network

The two classifiers employ the same classification model, namely, polynomial neural network. The PNN is a single-layer network which uses as inputs not only the feature measurements of the input pattern but also the polynomial terms of the measurements [8].

The PNN has one output unit with the linear and binomial terms of feature values as inputs. To reduce the complexity, the dimensionality of input feature vector is reduced by principal component analysis (PCA). The subspace parameters are estimated on a sample set of face images. Denoting the principal components (projections on subspace) of input feature vector $\mathbf{x}$ by z_j, $j = 1, \ldots, m$, the outputs of PNN is computed by

$$y(\mathbf{x}) = s(\sum_{i=1}^{m} w_i z_i + \sum_{i=1}^{m}\sum_{j=i}^{m} w_{ij} z_i z_j + w^D D_f + w_0), \qquad (1)$$

where D_f is the residual of subspace projection, also called distance from feature subspace (DFFS). $s(\cdot)$ is the sigmoid function. The connecting weights are estimated on a sample set of face and non-face images by minimizing the mean square error (MSE) between 0-1 targets and actual sigmoid outputs.

Since the PNN is single layer network, the training process is not influenced by the random initialization of the weight.

2.3 Feature Subspace Learning and Combination Scheme

In study of classifier combination, various classifier combination schemes have been experimentally demonstrated that they consistently outperform a single classifier if they are different. The difference can be achieved by using different feature sets as well as different training sets [7]. We have achieved good detection accuracy using only one classifier [6]. Now we try to design a new classifier which could be complementary to the existent one so that the combination of them can result in improvement of detection performance.

As described in Section 2.2, the projection on the feature subspace is used as the input of the PNN for classification. It is desirable to construct two feature subspaces which could represent different types of face patterns. We then compute the feature subspace employed by Classifier2 on the set of face samples misclassified by Classifier1. In other words, the face samples misclassified by Classifier1(PNN1) are "difficult" patterns which could be covered by the second feature subspace used by Classifier2(PNN2).

Specifically, the subspace employed by the PNN1 is computed on the 29,900 face samples. Then the PNN1 is trained on the face samples and 44,644 non-face samples (collection method of training samples are explained in Section 3). The trained PNN1 is used to classify the 29,900 face samples while the samples with output value lower than a threshold are considered as misclassified face samples, from which the feature subspace employed by the PNN2 is learned.

The number of face samples misclassified by PNN1 changes with the threshold of output value. The number of misclassified face samples used to compute the second feature subspace heavily influences the effectiveness of combination. The larger the number, the more similar the two feature subspaces are. On the other hand, if the number is too small, the second feature subspace would be useless. We run several experiments to investigate the suitable value. The set of 2,299 samples was found to be the best.

The two classifiers can be combined either in cascade or in parallel. We tried three ways, namely, *cascade*, *parallel* and *cascade + parallel*.

In *cascade*, only if the test window passes through Classifier1, it is detected by Classifier2. If the output of Classifier2 is larger than a threshold, it is considered as a face candidate.In *parallel*, the two classifiers are directly combined using weighted sum rule. Denote the output of the Classifier1 as C_1, Classifier2 as C_2, the output of the combination is given by: $C_1+\alpha\cdot C_2$, where α is weight constant. Besides *cascade* and *parallel*, we also tried to organize the two classifiers in both parallel and cascade way, named *cascade* + *parallel*. The test window is firstly processed by Classifier1, only if it passes through, it is detected by Classifier2, then the output is computed as $C_1 + \alpha \cdot C_2$.

3 Experiments

3.1 Training

To train a neural network, a large number of training samples are needed. 29,900 face samples were generated from 2,990 real face images by varying the size, aspect ratio and reflection. Non-face samples were collected using the bootstrap strategy in three stages. The samples collected in the first stage were the non-face windowed local images that have small Euclidean distance from the mean vector of face samples. In the second and third stages, the non-face samples were collected by the PNN trained with the face samples and the already collected non-face samples. Using the PNN, the non-face local images that have face scores greater than a threshold are collected. The images for collecting non-face samples are totally different from the test images.

The two classifiers are trained independently. Except the first turn of training, the non-face samples are different with each other. Hence, the two classifiers can cover different patterns. They may reach a consensus on a face region, but probably not on a non-face region, which is very useful to reduce false positives.

3.2 Testing

The performance of the proposed method was tested on two sets of images, which are totally different with those used in training procedure. Test Set1 contains 270 images with clear faces and simple background. Each image has only one face, so totally, 270 faces in Set1. Test Set2 consists of 109 CMU [3] images with complex background, containing a total of 487 faces.

In testing, we set variable thresholds to the face likelihood score to investigate the tradeoff between the detection rate and the false positive rate (receiving operating characteristics, ROC). From the ROC curves shown in Fig.2, we can see that the combinations yield significant improvements compared to the original system, Classifier1. This justifies that the two feature subspaces are complementary to each other to certain degree and the combination of them is quite effective in reducing false positives. It is also shown that Classifier2 performs better than Classifier1. The reason could be explained as the feature subspace employed by PNN2 is learned on "difficult" face samples so that it has better generalization ability.

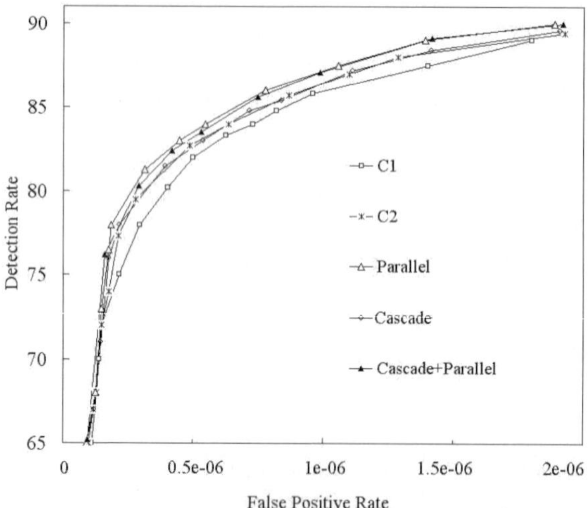

Fig. 2. ROC curves on 109 CMU images

Table 1. Detection results

	Test Set1		Test Set2	
	Det. rate	False rate	Det. rate	False rate
parallel	99.3%	0	89.3%	1.32×10^{-6}
cascade	99.3%	0	89.3%	1.80×10^{-6}
cascade + parallel	99.3%	0	89.3%	1.29×10^{-6}
C_1	99.3%	1.5×10^{-7}	89.3%	2.31×10^{-6}
C_2	99.3%	0.3×10^{-7}	89.3%	1.82×10^{-6}

Fig. 3. Detection examples of Test Set2

It is worthy to notice that both of *parallel* and *cascade+parallel* are superior to *cascade* while the performances of *parallel* and *cascade + parallel* are almost the same. This suggests that the final decision made from both classifiers is more accurate than that from one classifier as done in *cascade*. Moreover, in *cascade + parallel*, only the windows passing through Classifier1 are processed by Classifier2 so that *cascade + parallel* is faster than *parallel*.

Table 1 shows the detection rates. Since the images in Set1 have better quality and simpler background, the detection rate is higher than that of Set2. Some detection examples of Test Set2 are given in Fig.3.

4 Conclusion

In this paper, we propose a face detection method by combining classifiers. We apply two classifiers using the features extracted from complementary feature subspaces learned by PCA while a polynomial neural network is employed as classification model. The outputs of the two classifiers are fused to make the final decision of face detection. Experimental results and comparisons with other methods demonstrate the effectiveness of the proposed method.

References

1. Maio, D., Maltoni, D.: Real-Time Face Location on Gray-Level Static Images. Pattern Recognition **33** (2000) 1525–1539
2. Liu, C.: A Bayesian Discriminating Features Method for Face Detection. IEEE Trans. Pattern Anal. Mach. Intell. **25** (2003) 725–740
3. Rowley, H.A., Baluja, S., Kanade T.: Neural Network-Based Face Detection. IEEE Trans. Pattern Anal. Mach. Intell. **20** (1998) 23–38
4. Yang, M.H., Kriegman, D.J., Ahuja, N.: Detecting Faces in Images: A Survey. IEEE Trans. Pattern Anal. Mach. Intell. **24** (2002) 34–58
5. Kittler, J., Duin, P.W., Matas, J.: On Combining Classifiers. IEEE Trans. Pattern Anal. Mach. Intell. **20** (1998) 226–239
6. Huang, L.-L., Shimizu, A., Hagihara, Y., Kobatake, H.: Gradient Feature Extraction for Classification-Based Face Detection. Pattern Recognition **36** (2003) 2502–2511
7. Ali, K.M., Pazzani, M.J.: On the Link Between Error Correlation and Error Reduction in Decision Tree Ensembles. Technical Report, ICS-UCI (1995) 95–38
8. Schürmann, J.: Pattern Classification: A Unified View of Statistical Pattern Recognition and Neural Networks. Wiley Interscience (1996)
9. Robbins, H., Monro, S.: A Stochastic Approximation Method, Annals of Mathematics and Statistics **22** (1951) 400–407

Face Detection Method Based on Kernel Independent Component Analysis and Boosting Chain Algorithm

Yan Wu and Yin-Fang Zhuang

Department of Computer Science and Engineering, Tongji University,
Shanghai, 200092
yanwu@mail.tongji.edu.cn

Abstract. A face detection method based on Kernel Independent Component Analysis and Boosting Chain Algorithm was proposed. Moreover a linear optimization scheme was proposed to address the problems of redundancy in boosting learning and threshold adjusting in cascade coupling. Experiments were done to compare the performance of boosting chain with that of Adaboost and Floatboost and the results show the effectiveness of this new method.

1 Introduction

Face detection refers to finding out the positions, sizes and poses of faces (if they exist) in the input images. The paper [1] by P.Viola was regarded as a milestone in the improvement of detection speed. He brought Adaboost and Cascade together to implement the real-time face detection system [2]. After that, more scholars shifted their attention to Boosting algorithms. Stan Z.Li put forward a multi-view face detection algorithm based on FloatBoost [3].

Here a face detection method based on Kernel Independent Component Analysis (KICA) and Boosting Chain Algorithm was proposed. Moreover a linear optimization scheme was proposed to address the problems of redundancy in boosting learning and threshold adjusting in cascade coupling. Experiments were done to compare the performance of boosting chain with that of Adaboost and Floatboost.

2 Face Detection Method Based on KICA and Boosting Chain Algorithm

2.1 The Rebuilt Signal-Noise Ratio Threshold Value Function Constructed by KICA

Independent Component Analysis (ICA) put forward recently proves to be superior to eigenface method put forward by Turk M and Pentland A[4] in the aspect of identification. KICA is not a kernelized version of ICA but rather a new ICA algorithm. The use of a function space makes it possible to adapt to a variety of sources and thus makes our algorithms more robust to varying source distributions.

The contrast function adopted here is:

J. Wang et al. (Eds.): ISNN 2006, LNCS 3972, pp. 122 – 127, 2006.

$$C(\boldsymbol{W}) = I_{\delta_F}(\boldsymbol{K}_1,\ldots,\boldsymbol{K}_m) = -0.5 * \log \delta^{\kappa}(\boldsymbol{K}_1,\ldots,\boldsymbol{K}_m) . \tag{1}$$

Here $\delta^{\kappa}(\boldsymbol{K}_1,\ldots\ldots,\boldsymbol{K}_m) = \det \kappa_{\kappa} / \det \boldsymbol{D}_{\kappa}$ is kernel generalized variance. The diagonal elements in κ_{κ} are $\boldsymbol{K}_i + (N*\kappa/2)*\boldsymbol{I}$ and the other elements in κ_{κ} are $\kappa_{\kappa}(i,j) = \boldsymbol{K}_i * \boldsymbol{K}_j (i \neq j)$. $\boldsymbol{D}_{\kappa}$ is a diagonal matrix and its diagonal elements are $\boldsymbol{K}_i + (N*\kappa/2)*\boldsymbol{I}$. $\boldsymbol{K}_i$ is Gram matrix with $N*N$.

A naïve implementation of the KICA algorithm, based on the above contrast function, would scale as O (m^3N^3). As noted by several researchers, however, the spectrum of Gram matrices tends to show rapid decay and low-rank approximations of Gram matrices can therefore often provide sufficient fidelity. A positive semidefinite matrix $\boldsymbol{K}$ can always be factored as $\boldsymbol{G}*\boldsymbol{G}^T$, where $\boldsymbol{G}$ is an $N*N$ matrix. Our goal is to find a matrix $\boldsymbol{G}$ of size $N*M$, for small M (that is, $M<<N$), such that the difference $\boldsymbol{K} - \boldsymbol{G}*\boldsymbol{G}^T$ has norm less than a given value η. This can be achieved via incomplete Cholesky decomposition, whose overall complexity is O (M^2N).

Here we adopt the KICA algorithm put forward by [5], using the incomplete Cholesky decomposition to solve the computational issue.

Algorithm of KICA using Incomplete Cholesky Decomposition:

a. Whiten the input data so that $\boldsymbol{YY}^T = \boldsymbol{I}$, $\boldsymbol{Y} = [\boldsymbol{y}_1 | \ldots | \boldsymbol{y}_N]$, $\boldsymbol{I}$ is an identity matrix. Here use these column vectors $\boldsymbol{y}_1,\ldots,\boldsymbol{y}_N$ as the inputs to represent N images and adopt Gaussian Kernel.

b. Compute the contrast function $C(\boldsymbol{W})$.

  b1. Compute the centered Gram matrices $\boldsymbol{K}_1,\ldots\ldots,\boldsymbol{K}_m$ of the estimated sources ($\boldsymbol{x}_1,\ldots,\boldsymbol{x}_N$), where $\boldsymbol{x}_i = \boldsymbol{W} * \boldsymbol{y}_i$ and $\boldsymbol{W}$ is a parameter matrix.

  b2. Compute $\det \kappa_{\kappa}$ through Incomplete Cholesky Decomposition.

  b3. Compute $C(\boldsymbol{W}) = -0.5 * \log \delta^{\kappa}(\boldsymbol{K}_1,\ldots\ldots,\boldsymbol{K}_m)$ where $\delta^{\kappa}(\boldsymbol{K}_1,\ldots,\boldsymbol{K}_m) = \det \kappa_{\kappa} / \det \boldsymbol{D}_{\kappa}$.

c. Use the steepest descent algorithm to mini-mized $C(\boldsymbol{W})$.

  c1. Set k=0 and initiate $\boldsymbol{W}$ as $\boldsymbol{W}^0$.

  c2. Compute $\boldsymbol{d}^k = -\nabla C(\boldsymbol{W})$, If $\boldsymbol{d}^k$=0•then stop.

  c3. Choose α^k by executing imprecise linear searching

$$\alpha^k = \arg\min_{\alpha>0} C(\boldsymbol{W} + \alpha \boldsymbol{d}^k) . \tag{2}$$

  c4. Set $\boldsymbol{W}^k + \alpha^k \boldsymbol{d}^k \to \boldsymbol{W}^{k+1}, k+1 \to k$. Go to c2.

Orthogonalize each column of $\boldsymbol{W}$ and represent $\boldsymbol{W}$ as $[\boldsymbol{W}_1, \boldsymbol{W}_2, ..., \boldsymbol{W}_N]$. The projections of face vector $\boldsymbol{P}_i$ onto the subspace expanded by $\boldsymbol{W}_1, \boldsymbol{W}_2, ..., \boldsymbol{W}_N$ are $\boldsymbol{U} = \boldsymbol{W}^T \boldsymbol{P}$. When we rebuild face images, we use K independent components, where $K << N$, that is, $\boldsymbol{P}_i = \boldsymbol{W}\boldsymbol{U} \approx \sum_{i=1}^{K} \boldsymbol{W}_i \boldsymbol{U}_i$. Define the rebuilt signal-noise ratio:

$$\boldsymbol{R}_{SN}(\boldsymbol{P}) = 10 * \log\left(\frac{\|\boldsymbol{P}\|^2}{\|\boldsymbol{P} - \boldsymbol{U}\|^2}\right) . \quad (3)$$

If it is less than θ, we can say $\boldsymbol{P}$ is not a face image. By doing this, we can get several threshold value detection functions: $\text{sign}(R_{SN}(\boldsymbol{P}) - \theta)$.

2.2 Boosting Chain Algorithm Application to Face Detection

The main idea of Boosting Chain method [6] is to link each Boosting classifier to a chain structure and the successive Boosting classifiers use historical information of the previous Boosting classifiers. The algorithm could be improved with minor modifications on weighting schema and training strategy. Firstly, the positive sample weights are directly introduced into the substantial learning procedure. For negative samples, their weights are adjusted according to the classification errors of each previous weak classifier. Secondly, the successive training is directly based on the previous boosting classifier.

Boosting Chain Algorithm:

a. Initialize: i=0, $F_0 = 1$, $\Phi = \{\}$, $w_j = 1/p$ for all positive sample x_j, $w_j = 1/n_i$ for all positive sample x_j.

b. While $F_i > F$

b1. Set i=i+1.

b2. Train Φ_i to meet the f_i and d_i requirements on the validation set.

- Use initial weights w_j, and train set P and N_i.
- Train a node classifier Φ_i.

b3. Optimize the Node classifier(see Sect. 2.3).

b4. Set $F_i = F_{i-1} * f_i$, $\Phi = \Phi \cup \{\Phi_i\}$.

b5. Evaluate *boosting* chain Φ on non-face image set, and put false detections into the set N_{i+1}

b6. For each sample $\boldsymbol{x}_j$ in set N_{i+1}, update weight w_j for: $w_j^{i+1} \leftarrow w_j^0 \exp[-y_j \sum_{k=1}^{i} \sum_{t=1}^{m_k} \alpha_{k,t} h_{k,t}(\boldsymbol{x})]$.

Here Φ_i is the ith boosting classifier in the cascade. P is a positive training set and $p=|P|$. N_i is the ith negative training set and $n_i=|N_i|$. f_i is the maximum false positive rate of ith layer. d_i is the minimum detection rate of ith layer. $\boldsymbol{W}_j$ is the weighting of sample $\boldsymbol{x}_j$. F is the overall false positive rate. D is the overall detection rate. M is the number of classifiers used in the cascade.

2.3 Boosting Chain Optimization

α Optimization. The final decision function of AdaBoost could be regarded as the linear combination of weak learners $\{h_1(\boldsymbol{x}), h_2(\boldsymbol{x}), ..., h_t(\boldsymbol{x})\}$. When $h_i(\boldsymbol{x})$ is fixed, the weak learner maps the sample $\boldsymbol{x}_i$ from the original feature space F to a point in a new space $\boldsymbol{F}^*$ with new dimensionality T. Consequently, the optimization of α parameter can be regarded as finding an optimal separating hyperplane in the new space $\boldsymbol{F}^*$. The solution for finding optimized hyperplane can be obtained by resolving the following quadratic programming problem. Maximize:

$$\mathrm{L}(\beta) = \sum_{i=1}^{n} \beta_i - \frac{1}{2} \sum_{i,j=1}^{n} \beta_i \beta_j y_i y_j (h(\boldsymbol{x}_i) \bullet h(\boldsymbol{x}_j)) \ . \tag{4}$$

subject to the constraints $\sum_i^n \beta_i y_i = 0$ 和 $C_i \geq \beta_i \geq 0$, $i=1,\ldots,n$.

Boosting Redundancy Reduction

Algorithm for Boosting redundancy reduction

```
a. Train a linear SVM classifier over the set {h_i(x)},
   i=1,…,M and weight W.
b. Sort the classifier parameter vector α by value.
   Suppose the new index will be i_1,i_2,…,i_M .
c. k=1,…,N, N is the constant for feature elimination.
  c1. Remove the feature h_ik .
  c2. Compute current learning accuracy p_k .
  c3. Put back feature h_ik .
d. Remove the feature h_ik , with largest p_k .
e. Set M=M-1, and Go to a.
```

3 Experimental Results

3.1 Frontal Face Detection

About 3000 face examples are collected from various sources such as FERET and AR standard databases. For each aligned face example, a synthesized face example is generated by a random in-plane-rotation and out-of-plane-rotation in the range of $[-20^\circ,+20^\circ]$. Then they are re-scaled to the size of 20*20. Sufficient nonface examples are collected from images containing no faces.

The MIT+CMU test set is composed of 125 images containing 481 faces."N.Fea" stands for the number of features and "N.FA" for the number of false alarms. Our algorithm learned a total of 1223 features (weak classifiers) from the 20*20 training examples to achieve the performance. This is about 2/3 of 2546 computed the FloatBoost counterpart from the 20*20 training examples and about 4/9 of 3872 computed the AdaBoost counterpart from the 20*20 training examples.

Table 1. Comparisons of AdaBoost, FloatBoost and our method in frontal face detection

		AdaBoost	FloatBoost	Our method
N.Fea		3872	2546	1223
N.FA	10	82.7%	83.5%	85.9%
	31	89.1%	90.3%	94.1%

3.2 Multi-view Face Detection

The training set for multi-view face detection is created in the similar way to that for the frontal faces, except that the out-of-plane rotation covers the full range of $[-90^\circ,+90^\circ]$. The CMU profile face set is used to test the algorithm. From Table2 we can see that our method improves the detection rate by 2.7% and 4.6% respectively compared with that of AdaBoost and FloatBoost. Our method also gives fewer false alarms than AdaBoost and FloatBoost do.

Table 2. Comparisons of AdaBoost, FloatBoost and our method in multi-view face detection

Method	Detection rate (%)	False alarms
AdaBoost	92.1	50
FloatBoost	90.2	31
Our Method	94.8	20

4 Conclusions

Here KICA is first used to construct the rebuilt signal-noise ration threshold value functions. Then use historical information gained from Boosting learning to form a Boosting chain and apply history information to successive Boosting classifiers. Then use a linear optimization scheme to remove the redundancy in the Boosting learning and to adjust the threshold values between Boosting classifiers in the Boosting chain. The finally gained classifier was composed of fewer weak classifiers, which can achieve lower error rate compared with simple Boosting learning methods. Experimental results show that our method is superior to other traditional Boosting methods in the aspect of the number of features needed, the detection rate and the number of false alarms.

Acknowledgements

The National Natural Science Foundation of P. R. China under grant No.60475019 supported this research.

References

1. Viola, P.: Rapid Object Detection Using a Boosted Cascade of Simple Features. In: KasturiR, Medioni G (eds.): Proc. of the IEEE Computer Vision and Pattern Recognition. IEEE Computer Society, Vol. 1. (2001) 511-518.
2. Viola, P., Jones, M.: Robust Real Time Object Detection. In: Song-Chun Z., Alan Y., David M (eds.): Intl. Workshop on Statistical and Computational Theories of Vision. Int'l Journal of Computer Science. (2001)747-771.
3. Li, S.Z., Zhu, L., Zhang, Z.Q., Zhang, H.J.: Learning to Detect Multi-View Faces in Real-Time. In: James L. M., Alex P. P. (eds.): Proceedings of the 2nd International Conference on Development and Learning. IEEE Computer Society. (2002).
4. Yow, K. C., Cipolla, R.: Feature-Based Human Face Detection. Image and Vision Computing. 15(2) (1997) 713-735.
5. Bach, F.R., Jordan, M. I.: Kernel Independent Component Analysis. Journal of Machine Learning Research. 3(1) (2003) 1-48.
6. Xiao, R., Zhu L., Zhang, H. J.: Boosting Chain Learning for Object Detection. In: Triggs, Andrew Zisserman (eds.): Ninth IEEE International Conference on Computer Vision (ICCV'03). IEEE Computer Society 2003, Vol. 1. (2003) 709-715.

Recognition from a Single Sample per Person with Multiple SOM Fusion

Xiaoyang Tan[1,2], Jun Liu[1], and Songcan Chen[1,2]

[1] Department of Computer Science and Engineering,
Nanjing University of Aeronautics & Astronautics, Nanjing 210016, China
{x.tan, j.liu, s.chen}@nuaa.edu.cn
[2] Shanghai Key Laboratory of Intelligent Information Processing,
Fudan University, Shanghai 200433, China

Abstract. One of the main challenges faced by the current face recognition techniques lies in the difficulties of collecting samples, and many existing face recognition techniques rely heavily on the size and representative of training set. Those algorithms may suffer serious performance drop or even fail to work if *only one training sample per person* is available to the systems. In this paper, we present a multiple-SOMs-based fusion method to address this problem. Based on the localization of the face, multiple Self-Organizing Maps are constructed in different manners, and then fused to obtain a more compact and robust representation of the face, through which the discrimination and class-specific information can be easily explored from the single training image among a large number of classes. Experiments on the FERET face database show that the proposed fusion method can significantly improve the performance of the recognition system, achieving a top 1 matching rate of 90.0%.

1 Introduction

The aim of face recognition is to identify or verify one or more persons from still images or video images of a scene using a stored database of faces. Many research efforts [1] have been focused on how to improve the accuracy of a recognition system. However, it seems that most of them ignore the potential problem that may stem from the face database at hand, where there may be only one sample image per person stored, possibly due to the difficulties of collecting samples or the limitations of storage capability of the systems, etc.

Under this condition, most of the traditional methods such as eigenface [2] and fisherface [1] will suffer serious performance drop or even fail to work, due to the absence of enough samples for a reliable covariation estimation. This problem, called the *one sample per person problem* (or, *one sample problem* for short), is defined as follows: Given a stored database of faces with only one image per person, the goal is to identify a person from the database later in time in any different and unpredictable poses, lighting, etc from the individual image.

Due to its challenge and significance for real-world applications, several researchers have recently made attempts to face the challenge. The methods in

J. Wang et al. (Eds.): ISNN 2006, LNCS 3972, pp. 128–133, 2006.

literatures include synthesizing virtual samples [3, 4], probabilistic matching [4], class-specific subspace[4], neural network method [5], and so on.

In this paper, we attempt to address the above problems within a general framework based on the Self-Organization Map (SOM,[6]). The main idea behind is to extract latent local features that are invariant to appearance changes using the SOM network. Our previous work [5] shows that this strategy is very successful in handling large variations contained in the dataset,such as large expression changes and partial occlusions.

In this paper, we further extend the framework by incorporating a multiple classifier fusion technique. We use multiple SOM maps constructed in totally different manners to explore local discrimination information as diverse as possible. The local information collected are then fused for the subsequent recognition. Experimental results on the FERET dataset show that the proposed hybrid method significantly improves the performance of the recognition system with one sample per person.

The paper proceeds as follows. The proposed method is described in section 2. The experiments are reported in section 3. Finally, conclusions are drawn in section 4.

2 The Proposed Method

2.1 Multiple SOM-Based Representation

Although the effectiveness and robustness of the basic SOM-face model has been revealed in our previous work[5], there is much room left to improve the performance of recognition system by extending the single-map to multiple-maps.

The Single SOM Scheme. In the previous single map scheme, each face image is partitioned into M different local sub-blocks, each of which potentially represents specific local information of the image. A self-organizing map (SOM) neural network is then trained and used to project all the sub-blocks onto a quantized lower dimensional space so as to obtain a compact but robust representation.Such a representation of face image is called "SOM-face". Its main advantage lies in that, in the "SOM-face", the information contained in the face is distributed in an orderly way and represented by several neurons instead of only one vector, so the common features of different classes can be easily identified.

However, in this scheme, it is the overall distribution of local features from every class that is emphasized. Therefore, the salient class-specific statistical features may be submerged in the other features. In addition, when the training sets grow large, the resulting map should also grow large enough to decrease the degree of overlapping between classes, thus increasing the computational cost.

The Multiple SOMs Scheme. To overcome those problems, we develop two multiple-map schemes, in which multiple rather than single maps are constructed based on the localization of face images, so as to deliberately specialize the representation of classes and local feature clusters, respectively.

In the first scheme, a separate SOM map for each class is trained using only the sub-blocks from the corresponding class. That is, each map is trained using class-designated samples so that the distribution of local features within each class is approximated. Thus we name this scheme cMSOM (class-dependent Multiple SOMs). An advantage of this scheme is that the robustness of recognition system can be improved due to the characterization of salient local features within each class. The computational cost of this scheme, however, is linear in relation to the number of classes to be recognized. When the number of classes is very large, the number of maps to be trained will be very large as well. Another multiple-map scheme is thus proposed to overcome this problem and is described below.

In the second scheme, a partition mechanism is carefully designed to divide the sample space into multiple training sets, one for each map. Firstly, the training faces are divided into sub-blocks with equal size. Secondly, the sub-blocks at the same position of each face are collected to form a separate training set, which is then used to train a SOM map for that set of facial features. Since each map's training set is constructed in a lateral way, thus the name lMSOM (lateral Multiple SOM). This scheme aims to characterize the distribution of similar local features between classes. In contrast to the cMSOM scheme, the totality of maps in the lMSOM scheme relies only on the number of sub-blocks of each face.

In summary, the three schemes mentioned above (i.e. single SOM, cMSOM and lMSOM) are all trying to approximate the distribution of local features in sample space, based on the Kohonen maps, aiming to represent the salient statistical structures hidden in the high dimensional image data. We can visualize the original face image from the three different kinds of maps, and call the reconstructed face single-SOM-face, cMSOM-face and lMSOM-face respectively. See Fig.1 for an example.

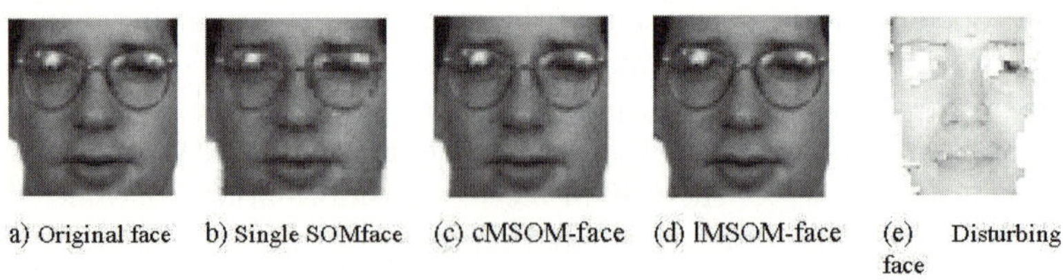

Fig. 1. Example of an original image and its variant representations

2.2 Multiple SOM-Faces Fusion

The high-level block diagram of the proposed method is shown in Fig.2. The multiple SOM-face representations of the testing face, together with a disturbing faces(Fig.1.e) constructed using E(PC)^{2}A technique [3], are fused first, then a soft k-NN ensemble decision scheme is employed to give the final class label.

Formally , to estimate the confidence c_{jk} that the j-th sub-block belongs to the k-th classes, we first calculate the L_2 pairwise distances between the j-th sub-block and all the "prototype sub-blocks" at the same position, according

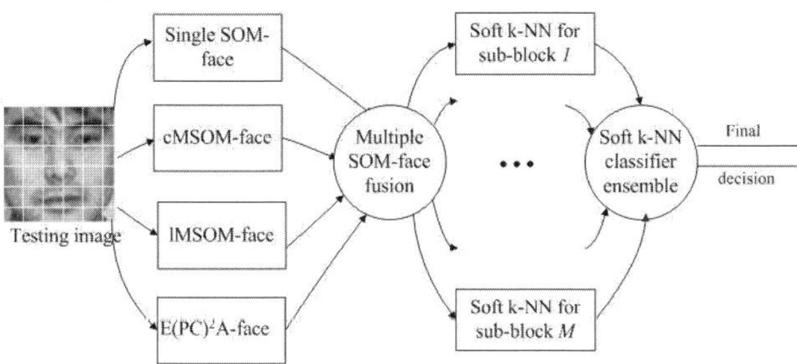

Fig. 2. Example of an original image and its variant representations

to the four models,respectively. Denote the obtained local pairwise distances as $\{d^i_{jk}\}_{k=1}^{k=N}, i = 1, 2, 3, 4$, where N is the total number of training samples. Note that in the situation of one sample per person, the number of training samples is equal to the number of training class.

To fuse them, a minimum aggregation rule [7] is employed , that is, the fused distance value d_{jk} is given by,

$$d_{jk} = \min_{i=1,2,3,4} (d^i_{jk}) \tag{1}$$

Then we have:

$$c_{jk} = \frac{log(\min_{k=1...N} \{d_{jk}\} + 1)}{log(d_{jk} + 1)} \tag{2}$$

Clearly, the class with minimum distance to the testing sub-block will yield a confidence value closer to one, while a large distance produces a very small confidence value, meaning that it is less likely for the testing sub-block to belong to that class.

In our implementation, only the first K pairwise distances in $\{d_{jk}\}_{k=1}^{k=N}$ where used for confidence estimation. This helps to reduce the influence of outliers.

Finally, the label of the test image can be obtained through a linearly weighted voting scheme, as follows,

$$Label = arg \max_{k=1...N} \sum_{j=1}^{M} C_{jk} \tag{3}$$

3 Experiments

The experimental face database used in this work comprises 400 gray-level frontal view face images from 200 persons, with the size of 256×384. There are 71 females and 129 males. Each person has two images (fa and fb) with different facial expressions. The fa images are used as gallery for training while the fb images as probes for testing. All the images are randomly selected from the FERET face database [8]. No special criterion is set forth for the selection. See [3] for some concrete face samples.

The feasibility and effectiveness of the proposed method is investigate by comparing the recognition performance between the proposed method and some other template-based approaches that deal with the *one image per person* problem, such as eigenface and $E(PC)^2A$.

The results are presented in Table 1. Table 1 reveals that the multiple SOM-face fusion method obtains the best performance among the compared methods. This promising result indicates that the fusion algorithm is able to combine information of different local features from different SOM-faces, resulting in a general performance improvement.

Next, we studied the behavior of the soft k-NN ensemble classifier constructed in the local feature spaces. Experimental results are presented in Figure 3. It is clear that the fusion algorithm outperforms any other individual schemes considered (i.e. single SOM, cMSOM and lMSOM schemes) consistently, concerning the top 1 matching rate.

Table 1. Comparison of recognition accuracies with different approaches

Methods	Accuracy(%)
Eigenface	83.0
EPC^2A	85.5
Single SOM-face	88.5
cMSOM-face	87.5
lSOM-face	89.0
MSOM-faces fusion	**90.5**

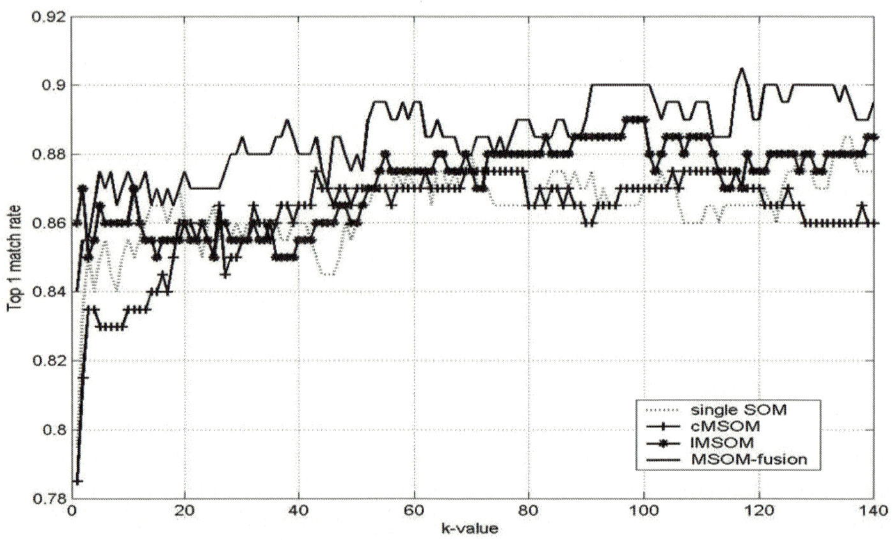

Fig. 3. Example of an original image and its variant representations

4 Conclusions

In this paper, a novel face representation and recognition approach is presented, where face images are first localized, then represented by their SOM-based proximities. Different ways of building SOM-faces and the possibility to fuse them are discussed. Experimental results on the FERET dataset demonstrate the effectiveness of the proposed method in dealing with the one sample problem.

Acknowledgement

We thank the support of Natural Science Foundation of China under Grant No. 60473035.

References

1. Zhao, W., Chellappa, R., Phillips, P.J., Rosenfeld, A.: Face Recognition: a Literature Survey. ACM Computing Survey, December Issue (2003) 399-458
2. Turk, M.A., Pentland, A.P.: Face Recognition Using Eigenfaces. In: Proc. of IEEE Conf. on Computer Vision and Pattern Recognition (1991) 586-591
3. Chen, S.C., Zhang, D.Q., Zhou, Z.H.: Enhanced $(PC)^2A$ for Face Recognition with One Training Image per Person. Pattern Recognition Letter 25(10) (2004) 1173-1181
4. Martinez, A.M.: Recognizing Imprecisely Localized, Partially Occluded, and Expression Variant Faces from a Single Sample per Class. IEEE Trans. Pattern Anal. Machine Intell. 24(6) (2002) 748-763
5. Tan, X.Y., Chen, S.C., Zhou, Z.H., Zhang, F.Y.: Recognizing Partially Occluded, Expression Variant Faces from Single Training Image per Person with SOM and Soft kNN Ensemble. IEEE Transactions on Neural Networks 16 (2005) 875-886
6. Kohonen, T. Self-Organizing Map, 2nd edition, Berlin: Springer-Verlag (1997)
7. Kittler, J., Hatef, M., Duin, R.P.W., Matas, J.: On Combining Classifiers. IEEE Trans. Pattern Anal. Machine Intell. 20(3) (1998) 226-239
8. Phillips, P.J., Wechsler, H., Huang, J., Rauss, P.J.: The FERET Database and Evaluation Procedure for Face Recognition Algorithms. Image and Vision Computing 16(5)(1998) 295-306

Investigating LLE Eigenface on Pose and Face Identification

Shaoning Pang and Nikola Kasabov

Knowledge Engineering & Discovery Research Institute,
Auckland University of Technology, Private Bag 92006,
Auckland 1020, New Zealand
spang@aut.ac.nz, nkasabov@aut.ac.nz

Abstract. This paper introduces a new concept of LLE eigenface modelled by local linear embedding (LLE), and compares it with the traditional PCA eigenface from principle component analysis (PCA) on pose identity and face identity recognition through face classification. LLE eigenface is found outperforming PCA eigenface on the discrimination/recogntion of both face identity and pose identity. The superiority on face identity recognition is own to a more balanced energy distribution on LLE eigenfaces, while the superiority on pose identity recognition is due to the fact that LLE preserves a better local neighborhood of face images.

1 Introduction

Eigenface technique so far has been extensively studied and widely used by researchers in face recognition and pose detection[1, 2]. The traditional Eigenface methods are a set of linear dimension reduction methods including principle component analysis (PCA) eigenface, Linear Discriminant Analysis(LDA) eigenface also known as fisherface or discriminant eigenface, and independent component analysis (ICA) eigenface often called ICA-face, etc. For face recognition, these Eigenface techniques are popularly used to compress a high dimensional face image to a low dimensional feature vector.

Local Linear Embedding (LLE) is a nonlinear dimension reduction introduced by LLE S. T. Roweis and L. Saul[2, 3, 4]. Motivated by the traditional eigenface techniques for face feature extraction, we introduce here a concept of LLE eigenface modelled by local linear embedding.

Because LLE is known for the superiority on multi-pose face synthesis [3], it is generally believed that LLE eigenfeatures are superior to PCA eigenfeature on pose recognition, but might not have better performance than PCA eigenfeature on typical face recognition. To test the properties of LLE eigenface and clarify the facts between LLE eigenface and PCA eigenface on face classification, we study in this paper two categories of face classification: pose recognition and face recognition.

J. Wang et al. (Eds.): ISNN 2006, LNCS 3972, pp. 134–139, 2006.

2 LLE Eigenface

Compared with the linear dimensionality reduction method, PCA(Principle Component Analysis), Locally Linear Embedding (LLE) is a method of nonlinear dimensionality reduction introduced by Sam T. Roweis and Lawrence K. Saul[2]. This method recovers global nonlinear structure from locally linear fits. It attempts to preserve as well as possible the local neighborhood of each object, while preserving the global distances through the rest of the objects.

Suppose the data consist of N real-valued vectors $\boldsymbol{x}$ from face features extraction. Each vector $\boldsymbol{x}$ is a high dimensional vector with dimensionality D, and $\boldsymbol{y}_i$ is the low dimensional vector embedded from $\boldsymbol{x}_i$ with embedding dimensionality d, where $D >> d$. The computation of LLE eigenface involves an optimal embedding procedure which reduces vector from high dimensional data $\boldsymbol{x}$ to low dimensional data $\boldsymbol{y}_i$ by minimizing the following cost function,

$$\Phi(\boldsymbol{y}) = \sum_i |\boldsymbol{y}_i - \sum_j \boldsymbol{W}_{ij}\boldsymbol{y}_j|^2. \quad (1)$$

This procedure consists of four steps as follows,

Step 1. Compute the neighbors of each data points,$\boldsymbol{x}_i$ by computing pairwise distances and finding neighbors. In the simplest formulation of LLE, one can identify K nearest neighbors per data point, as measured by Euclidean distance.

Step 2. Compute the weights W_{ij} that best reconstruct each data point $\boldsymbol{x}_i$ from its neighbors, minimizing the following cost function by constrained linear fits.

$$\varepsilon(\boldsymbol{W}) = \sum_i |\boldsymbol{x}_i - \sum_j \boldsymbol{W}_{ij}\boldsymbol{x}_j|^2, \quad (2)$$

Step 3. Compute the matrix M in terms of the above weights computation.

$$\boldsymbol{M} = (\boldsymbol{I} - \boldsymbol{W})^T(\boldsymbol{I} - \boldsymbol{W}), \quad (3)$$

where I is the $d \times d$ identity matrix.

Step 4. Calculate Embedding by conducting the eigenanalysis on weight matrix $\boldsymbol{M}$

$$\boldsymbol{M}\boldsymbol{u}_k = \lambda_k \boldsymbol{u}_k \quad (4)$$

Where $\boldsymbol{M}$ is the cost function matrix $\boldsymbol{\Sigma}_{nn}$ of LLE, and $\boldsymbol{u}_k$ is a LLE eigenvector. Since it is often represented as a different ghost face from the PCA eigenfaces, we call it LLE eigenface in this paper.

Note that the bottom $d+1$ eigenvectors of the matrix M are corresponding to its smallest $d+1$ eigenvalues. Thus, for fixed weights vectors $\boldsymbol{W}_{ij}$, the embedding vectors $\boldsymbol{y}_i$ are found by minimizing the cost function Eq.(1). That is, the optimal embedding can be found by computing the bottom $d+1$ eigenvector of the matrix, $\boldsymbol{M}$.

Therefore, for face image data, LLE eigenfaces are a subset of eigenvectors of matrix $\boldsymbol{M}$, which assumes that a facial image x from a training set $\{\boldsymbol{x}_i\}_{i=1}^N$ can be reconstructed from its neighbors with the lowest reconstruction error Eq.(1).

3 Experimental Results

As compare LLE eigenfeature with PCA eigenfeature on pose recognition and face recognition, we used the MPEG-7 face database [5], which consists of 1,355 face images of 271 persons (5 different face images per person are taken), each image has the size of 56×46. The database contains two parts: the first part includes 710 faces from 142 persons, which are all multi-pose face images. We denote this part of face images as *pose-faceset*. The second part contains the remaining 645 face images of 129 person, which are all front face (pose free face) images, but have light variations. We denote this part of face images as *light-faceset*.

For both LLE and PCA, when selecting features, we rank the eigenvectors by their energy, and select a set of top energy eignevectors. For the classification phase, we use a K-NN classifier, and evaluated classification accuracy in the way of Leave-one-out cross-validation.

3.1 Pose Recognition

The first experiment is to test the above pose encoding under five types of face pose: front, down, up, left and right. From the above database, we selected 150 images of 130 people, and manually extracted five subsets corresponding to the five types of pose as follows,

1. Front pose: The people look straight ahead.
2. Down pose: The people look down.
3. Up pose: The people look up.
4. Left pose: The people turn to the left side.
5. Right pose: The people turn to the right side.

To distinguish with other facesets, we denote this faceset for pose recognition as *5pose-faceset*.

In Fig 1., we have shown the results for pose recognition of *5pose-faceset*. The horizontal coordinate in this figure represents the dimensionality of eigenspace, and the vertical coordinate represents the recognition accuracy. Having the knowledge of LLE multi-pose face synthesis, it is not a surprise that LLE eigenfeature has a better performance than PCA eigenfeature under a varied dimensionality. But we notice that the superiority of LLE to PCA on pose recognition is decreasing with the increase of dimensionality.

To investigate the energy distributions on LLE eigenfaces and PCA eigenfaces, for each eigenface, we calculate the percentages of energy as the ratio of its eigenvalue to the sum of eigenvalues for all eigenfaces.

Fig. 2 shows the energy distribution of LLE eigenfaces and PCA eigenfaces, the left side is for PCA eigenface, and right side is for LLE eigenface. As seen in the figure, the energy of LLE eigenface has a clearly more balanced distribution than that of PCA. PCA eigenfaces sustains a set of the most top energy distribution, the imbalance is happened when dimensionality is 10, and become worse as the dimensionality grows to 150. Whereas the top energy of LLE eigenfaces is decreasing with the increase of dimensionality, and the histogram is seen becoming flatter as the dimensionality increases from 10 to 150.

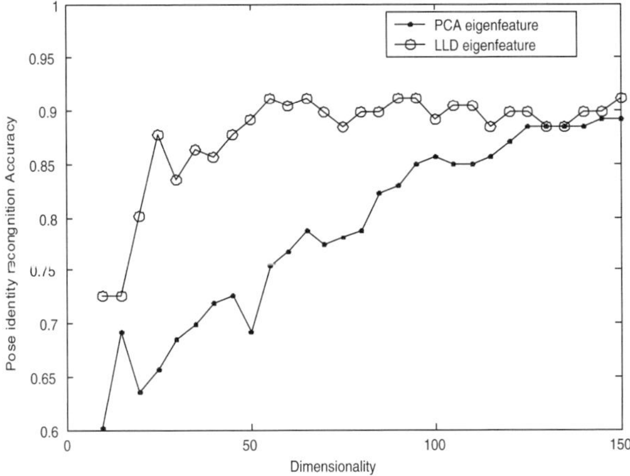

Fig. 1. Performance curves for pose recognition under different dimensionality

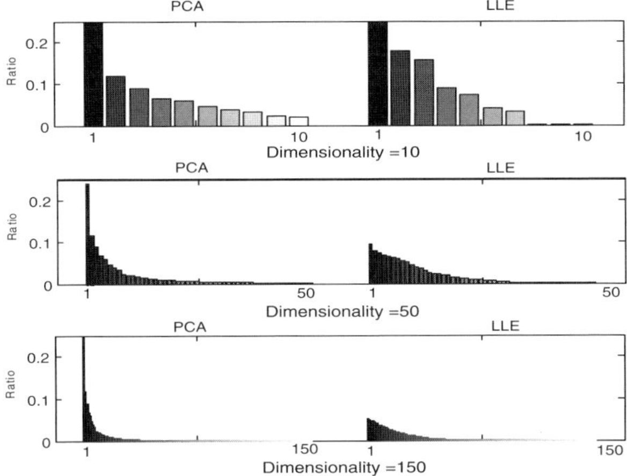

Fig. 2. Column illustration of the energy distribution on LLE and PCA eigenfaces for pose recognition over *5pose-faceset*, when the dimensionality is 10, 50, and 150

3.2 Face Recognition

Secondly, to test the response of LLE eigenfaces to a typical face recognition problem. We implemented LLE eigenfeature and PCA eigenfeature on face recognition over three face datasets: *light-set*, *pose-faceset*, *light-faceset* + *pose-faceset*, which is denoted as *whole-faceset*, and compare the face recognition performance under different number of eigenfeatures.

Table 1 presents the results of using LLE and PCA features on the face identity recognition over the three face datasets, and on the pose identity recognition over the *5pose-faceset* described in Section 3.1. The first 50 dimensions

Table 1. Performance of face recognition

Identity Type	faceset	PCA Accuracy	LLE Accuracy	Accuracy Difference
Face Id. Rec.	*light-faceset*	72.9%	78.3%	+5.4 %
	pose-faceset	73.5%	75.1%	+2.6%
	whole-faceset	70.7%	73.4%	+2.7%
Pose Id. Rec.	*5pose-faceset*	75.6%	92.5%	+16.9%

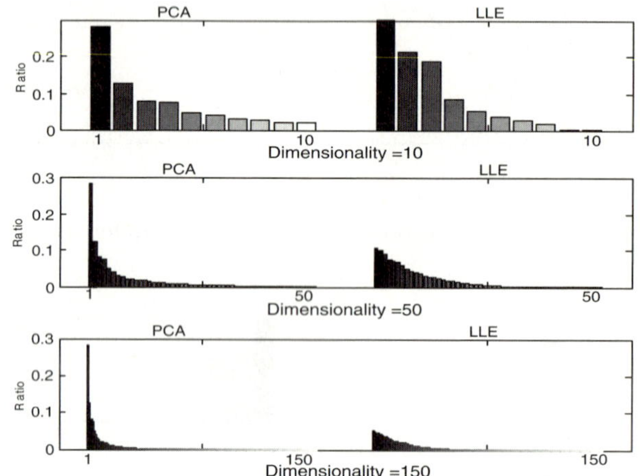

Fig. 3. Column illustration of the energy distribution on LLE and PCA eigenfaces for face recognition over *pose-faceset*, when the dimensionality is 10, 50, and 150

features are taken for K-NN classification, and the accuracy is evaluated under the policy of LOO cross-validation.

As can be seen, LLE outperforms PCA 17% on pose identity recognition performance, and about 3% on the performance of face identity recognition. The LLE eigenfeatues result in a better accuracy than PCA eigenfeatures, for pose free faceset, for pose affected faceset, and for light-and-pose affected faceset recognition. Obviously, the superiority of LLE on face identity recognition is less than that on pose identity recognition. Fig. 3 shows the same situation of the energy distribution on LLE and PCA eigenfaces for *pose-faceset* as in Fig. 2. It indicates that for face identity recognition, LLE eigenface also has a more balanced energy distribution than PCA eigenface.

4 Conclusions

LLE eigenfaces are projection directions that preserve as well as possible the local neighborhood of each class of faces. PCA eigenfaces are projection directions that maximize the total scatter across faces from all classes. Both can be employed to project high dimensional face images into a low dimensional feature space.

Yet, extensive experimental results have demonstrated that LLE eigenspace has a better dicriminablity than PCA eigenspace on face classification, and LLE is outperforming PCA not only on pose recognitions but on typical face recognition problems.

We investigated both face recognition and pose recognition problems by tracking the energy distribution on eigenfaces under different dimensionality. We found that LLE eigenface has a different disclamation for pose identity and face identity recognition. We can summarize the properties of the LLE eigenface on the recognition of pose identity and face identity as follows: (1) LLE eigenfaces manifest a better discrimination/recogniton of pose identity than the traditional PCA eigenfaces due to the fact that the LLE manifold learning preserves the local neighborhood of faces. (2) The superiority of LLE eigenface to PCA eigenface on face identity recognition is own to its more balanced energy distribution.

References

1. Ng, J. and Gong, S.: Composite Support Vector Machines for Detection of Faces Across Views and Pose Estimation. Image and Vision Computing 20 (2002) 359-368
2. Roweis, S. T. and Saul, L.: Nonlinear Dimensionality Reduction by Locally Linear Embedding. Science 290(5500) (2000) 2323–2326.
3. Saul, L. K. and Roweis, S. T.:Think Globally, Fit Locally: Unsupervised Learning of Low Dimensional Manifolds. Journal of Machine Learning Research 4 (2003) 119-155.
4. Saul, L. K. and Roweis, S. T.: An Introduction to Locally Linear Embedding. Report at AT&T labs -Research, 2000.
5. Kim, M., Kim, D., Bang, S. and Lee, S.: Face Recognition Descriptor Using the Embedded HMM with the 2nd-order Block-specific Eigenvectors.?ISO/IEC JTC1/SC21/WG11/M7997, Jeju, March 2002.

Multimodal Priority Verification of Face and Speech Using Momentum Back-Propagation Neural Network

Changhan Park[1], Myungseok Ki[2], Jaechan Namkung[3], and Joonki Paik[1]

[1] Image Processing and Intelligent Systems Laboratory, Department of Image Engineering, Graduate School of Advanced Imaging Science, Multimedia, and Film, Chung-Ang University, 221 Huksuk-dong, Tongjak-Ku, Seoul 156-756, Korea
initialchp@wm.cau.ac.kr
http://ipis.cau.ac.kr
[2] Broadcasting Media Research Group, Digital Broadcasting Research Division, ETRI, 161 Gajeong-dong, Yuseong-Gu, Daejeon 305-700, Korea
kkim@etri.re.kr
http://www.etri.re.kr
[3] Intelligent Image Communication Laboratory, Department of Computer Engineering, Kwangwoon University, 447-1 Wolge-dong, Nowon-Gu, Seoul 139-701, Korea
namjc@daisy.kw.ac.kr
http://vision.kw.ac.kr.

Abstract. In this paper, we propose a priority verification method for multimodal biometric features by using a momentum back-propagation artificial neural network (MBP-ANN). We also propose a personal verification method using both face and speech to improve the rate of single biometric verification. False acceptance rate (FAR) and false rejection rate (FRR) have been a fundamental bottleneck of real-time personal verification. The proposed multimodal biometric method is to improve both verification rate and reliability in real-time by overcoming technical limitations of single biometric verification methods. The proposed method uses principal component analysis (PCA) for face recognition and hidden markov model (HMM) for speech recognition. It also uses MBP-ANN for the final decision of personal verification. Based on experimental results, the proposed system can reduce FAR down to 0.0001%, which proves that the proposed method overcomes the limitation of single biometric system and proves stable personal verification in real-time.

1 Introduction

Human biometric characteristics are unique, so it can hardly be duplicated [1]. Such information includes; facial, speech, hands, body, fingerprints, and gesture to name a few. Face detection and recognition techniques are proven to be more popular than other biometric features based on efficiency and convenience [2], [3]. It can also use a low-cost personal computer (PC) camera instead of expensive equipments, and require minimal user interface. Face authentication has become a potential research field related to face recognition. Face recognition differs from face authentication because the former has to determine the identity of an object, while the latter needs

J. Wang et al. (Eds.): ISNN 2006, LNCS 3972, pp. 140 – 149, 2006.

to verify the claimed identity of a user. Speech [4] is the most basic and simplest communication media in the sense of efficiency and convenience. Each a single biometric information, however, has its own limitation. For this reason, we propose a multimodal biometric priority verification method to reduce false acceptance rate (FAR) and false rejection rate (FRR) in real-time.

An artificial neural network (ANN) provides a convenient tool for recognition and classification [5]. In an ANN information is accumulated in interconnected nodes, called neurons that are capable of learning by self-iterations. This form of artificial intelligence can handle very complex problems without known relationships, especially in the field of control [6].

There have been many approaches to extracting meaningful features for the recognition purpose. Those include principal component analysis (PCA) [7], neural networks (NN) [7], hidden markov models (HMM) [8], and linear discriminant analysis (LDA) [9]. In this paper, we use the PCA algorithm with unsupervised learning to extract face features. On the other hand we use the HMM algorithm with supervised learning for extracting speech feature. The proposed personal verification method accomplishes the higher priority of face and speech using a momentum back-propagation artificial neural network (MBP-ANN).

This paper is organized as follows: Sections 2 and 3 describe feature extraction of face and speech using the PCA and HMM algorithms, respectively. Sections 4 and 5 present the theory artificial neural network (ANN) and how to design the structure of the proposed system, respectively. Section 6 presents experimental results, and Section 7 concludes the paper with future research topics.

2 Face Extraction and Recognition

2.1 Face Feature Extraction and Recognition

The proposed face feature extraction and recognition method is shown in Fig. 1. The proposed method makes an edge-map image using a 13×9 template from the input face image. It then estimates the face poses and normalizes the size of the detected face to 60×60. The normalized image is stored in the multimodal database, and at the same time it trains the PCA module. The face recognition module distinguishes an input image from the trained images in the database.

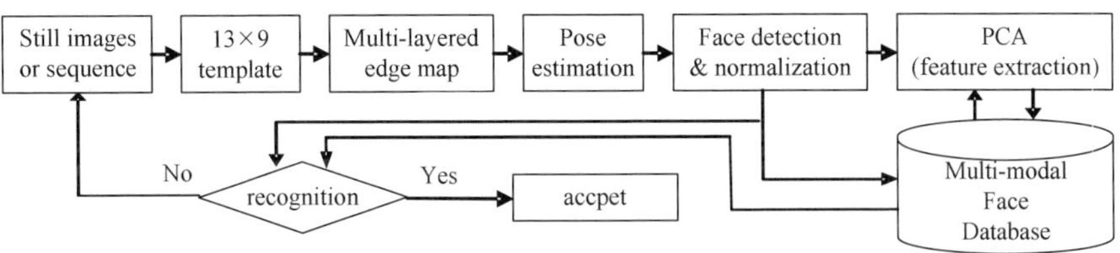

Fig. 1. Face feature extraction and recognition process

2.2 Face Detection and Building Database Using Multi-layered Relative Edge Map

In order to detect a face region and estimate face elements, we use the multi-layered relative edge map that can provide better result than just color-based methods [10]. A directional blob template can be selected according to the face size. More specifically, the template is defined so that the horizontal axis becomes longer than the vertical axis as shown in Fig. 2(a). The central pixel of a template in a $W \times H$ image is defined as $P_c=(x_c, y_c)$, which is created by averaging a 3×3 region. By using a $w_{ff} \times h_{ff}$ directional template for face components, the average intensity $\overline{I_{Dir}}$ of 8-neighborhood pixels is calculated on the central pixel, P_c. As a result, $\overline{I_c}$, the brightness value at P_c, and the brightness difference value can be obtained. The principal direction, $\overline{d_{pr}}$, and its magnitude, $|\overline{d_{pr}}|$, are also determined along the direction including the biggest brightness difference as shown in Fig. 2(b).

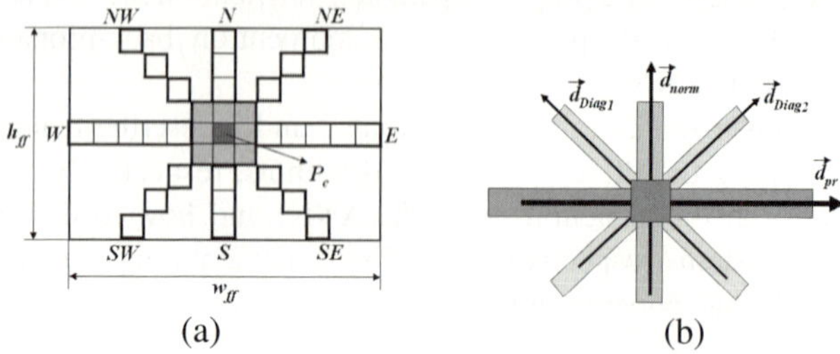

Fig. 2. (a) Directional template (b) New direction for edge map

2.3 Unsupervised PCA and Singular Value Decomposition (SVD)

In the process of PCA for pose estimation we compute covariance matrix C and its eigenvectors from training sets. Let $x_1, x_2 \cdots, x_N$ be N training face vectors. By definition, C can then be estimated as [11],

$$C = E\left[XX^T\right] = \frac{1}{N}\sum_{k=1}^{N} X_k X_k^T \cdot \tag{1}$$

The training data set are packed into the following matrix

$$X = \left[x_1, x_2 \cdots, x_N\right]. \tag{2}$$

The estimate of C can be approximately written as

$$C \cong \frac{1}{N} XX^T. \tag{3}$$

To estimate the eigenvectors of C, we only need to find the eigenvectors of XX^T. Even for images of moderate size, however, this is computationally complex. From the fundamental linear algebra [12], the eigenvectors of XX^T can be found from eigenvectors of $X^T X$. Suppose the rank of X is r, $r \le N$, then singular value decomposition of X is given as

$$X = \sum_{k=1}^{r} \sqrt{\lambda_k} u_k v_k^T , \tag{4}$$

where $\sqrt{\lambda_k}$, u_k , and v_k respectively represent singular values, left, and right singular vectors of *X*. u_k and v_k have the following relationship.

$$u_k = \frac{1}{\sqrt{\lambda_k}} X v_k . \tag{5}$$

Hence, we can easily find eigenface u_k after finding v_k . Recognized face classified using $d = \sum_{i=1}^{m} (r_i - t_i)^2$, where r_i and t_i represent input pattern, pattern of train face, respectively.

3 Speech Analysis and Feature Extraction

Speech recognition is classified into two categories in the sense of feature extraction method. One is to extract the linguistic information in speech signal, and the other is to extract an eigen specific of a speaker from speech signal [13]. The former performs extraction using the Mel-frequency cepstral coefficient (MFCC) based on the sense of hearing for human, and the latter extracts it by using the linear predictive coefficient (LPC) based on the sense of human speech. For this research we adopt the latter because an individual has its own sense of speech.

3.1 HMM for Speech Recognition and Verification

Speech verification calculates the cumulative distances with reference pattern when the test pattern is input. The reference patterns should be made in advance, and it can represent each speaker. This is classified by the pattern matching method that recognizes the pattern with calculated minimal cumulative distances and HMM. The HMM measures similarity with input pattern after modeling the speech signal statistically by extracting the feature from various speech waveforms. The training and verification process of speech is shown in Fig. 3.

We note that the proposed method can solve following three problems:

(i) Evaluation problem: Given an observation sequence $O = \{o_1, o_2, \cdots, o_T\}$ and the model $\lambda = (A, B, \pi)$, (where A represents the transition probability, B the output probability, and π the initial probability), $P(O \mid \lambda)$ can be calculated by using the forward and backward algorithms.

(ii) Learning problem: How to estimate the model parameter given $O = \{o_1, o_2, \cdots, o_T\}$ the model parameter can be by using Baum-Welch re-estimation.

(iii) Decoding (recognition) problem: Given a model, the best state sequence $q = \{q_1, q_2, \cdots, q_t\}$ of $O = \{o_1, o_2, \cdots, o_T\}$, can be estimated by using the Viterbi algorithm, where *q* represents the state sequence, *t* time, and *O* a specific vector for each frame.

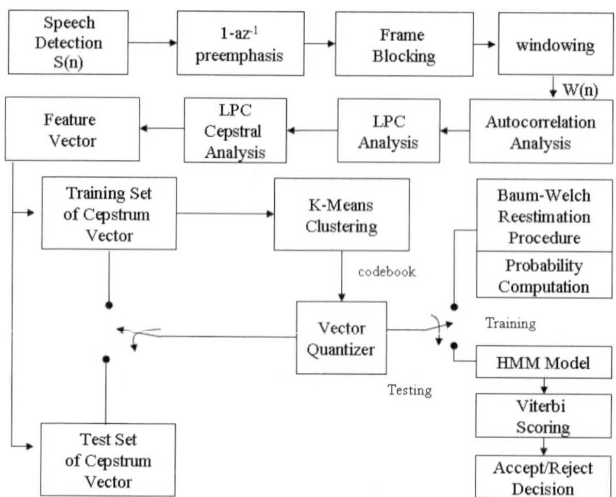

Fig. 3. Feature extraction and verification of speech

4 Artificial Neural Network

In this section, the proposed MBP-ANN method will be presented. An ANN is a computer model derived from a simplified concept of the brain [5]. It is a parallel distributed processing system composed of nodes, called neurons, and connections, called weights, and is based on the principle that a highly interconnected system of simple processing elements can learn complex interrelationships between independent and dependent variables. The most popular ANN is the back-propagation ANN (BP-ANN) [6].

4.1 Training Process of MBP-ANN

The BP-based training algorithm of the ANN uses change of weighted value between input layer, hidden layer, and output layer. Fig.4 shows the proposed MBP-ANN. Change of weighted value between input and hidden layers can be expressed as $\Delta V = \alpha\delta_x X$, and change of weighted value between hidden and output layers as $\Delta W = \alpha\delta_y Z$, where α represents the learning rate, δ_x and δ_y represent the error signals of the output of hidden and output layers, respectively. And X , Y, and Z represent the external input, the output of output layer, and the output of hidden layer, respectively.

The activation function of proposed method uses the sigmoid function, $f(x) = \frac{1-e^{-x}}{1+e^{-x}}$. The learning rate α initially has a small value because it can decrease with a change of weighted value at the learning step. In this case the learning becomes very slow. The MBP-ANN can accelerate the learning step of BP-ANN. The proposed method utilizes the weighted value of the previous learning step. The learning method of the MBP-ANN algorithm is the same to BP-ANN. But the change of weighted values ΔV and ΔW , only differ from additional momentum expression given in (6). The change ΔV_k and ΔW_k of weighted value at the $k-th$ learning step of MBP-ANN algorithm is given as:

$$\Delta V_k = \alpha\delta_z X + \beta\Delta V_{k-1}, \quad \Delta W_k = \alpha\delta_y Z + \beta\Delta W_{k-1}, \tag{6}$$

where α and β represent the learning rate, and a momentum constant, respectively. And δ_z and δ_y represent the error signal of the hidden and output layer, respectively.

Therefore, the weighted value V_{k+1} and W_{k+1} at the $k+1^{st}$ learning step is given as:

$$\Delta V_{k+1} = V_k + \Delta V_k = V_k + \alpha\delta_z X + \beta\Delta V_{k-1},$$

$$\Delta W_{k+1} = W_k + \Delta W_k = W_k + \alpha\delta_y Z + \beta\Delta W_{k-1}. \tag{7}$$

The square error $E = \frac{1}{2}(d-y)^2$ computes the target value d and the last output y.

Fig. 5 shows the dependencies of validating error and the optimal iterative number on a hidden node. Herein, the optimal iterative number means the minimal validating error (i.e., learning is stopped when the error in the validating set begins to increase). The proposed method has used 10 hidden layers.

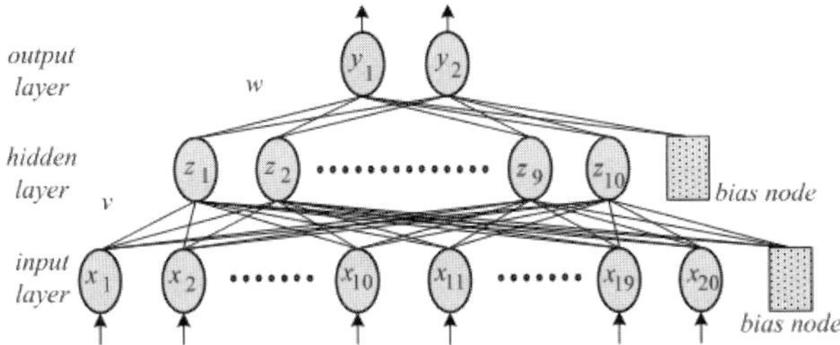

Fig. 4. Proposed momentum back-propagation artificial neural network (MBP-ANN)

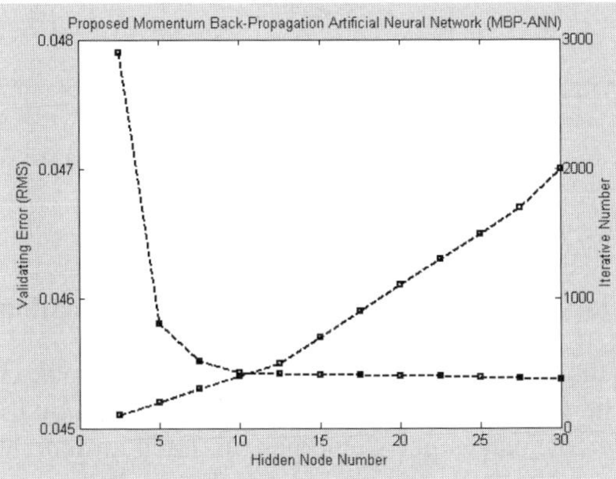

Fig. 5. The dependencies of validating error and the optimal iterative number on hidden node number

5 Proposed Multimodal Biometric Priority Verification System

The proposed multimodal biometric priority verification technique can solve the fundamental limitations inherent to a single biometric verification system. The proposed priority verification system consists of the input, the learning, and the verification modules. The input image of size 320×240 comes into the system in real-time together with the speech. In the learning modules, the face image is trained under the PCA framework, and the speech is trained with HMM. Feature extraction is also accomplished in the learning module. The verification module validates the recognized data from the image and speech by using the MBP-ANN algorithm. Personal information is saved in the form of a codebook class, and used for verification or rejection.

5.1 Personal Verification Using Multimodal Biometric

In this subsection, we present a personal priority verification method shown in Fig. 6. The proposed method first detects the face area from the input image. The face verification module compares the detected face with the pre-stored codebook class of personal information. The speech verification module extracts and recognizes the end-point of speech, and authenticates it after comparing with the codebook class. Decision processes of face and speech use the proposed MBP-ANN algorithm. If the face and speech verification results coincide, there in no further processing. Otherwise the MBP-ANN is used to solve the mismatch problem. Therefore, if the face and speech is same to the personal information of the codebook class, verification is accepted. Otherwise, it is rejected. The entire priority verification process is shown in Fig. 6.

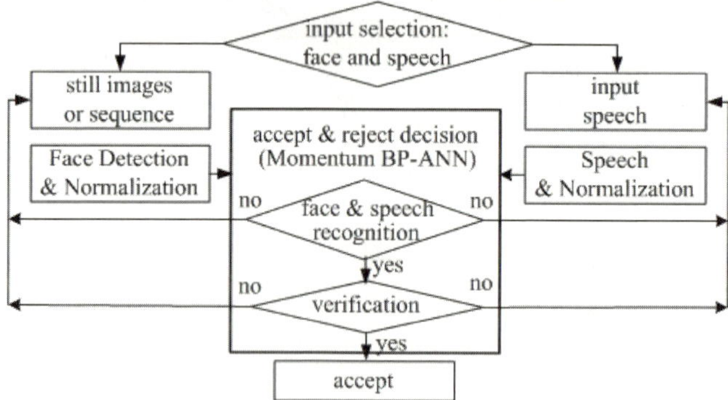

Fig. 6. The entire priority verification process

5.2 Code Book Class of Personal Face and Speech Information

In this subsection, the proposed personal information codebook class is described as shown in Fig. 7. The face feature extraction block is trained by using the PCA algorithm with ten different images per single person. Each an individual probability information projects the data to the original image. Fig. 8 shows a set of registered face images. The speech feature extraction block is trained by using the HMM algorithm with ten iterations per single person.

Person-1	*Face feature-1 class*	*Speech feature-1 class*
Person-2	*Face feature-2 class*	*Speech feature-2 class*
⋮	⋮	⋮
Person-n	*Face feature-3 class*	*Speech feature-3 class*

Fig. 7. Created personal code book class

Fig. 8. Some images of registered person

5.3 Proposed Priority MBP-ANN for Improved Verification

In this subsection, we propose a decision method for the face and speech to be certificated using the MBP-ANN. The proposed method learns the occurring number of a person with the recognized face and speech using the MBP-ANN. The proposed MBP-ANN algorithm is accomplished in 3 layers, and the input layer has 10 face images and 10 speeches. The hidden layer is composed of 10 outputs, and the output layer does two outputs to apply the priority of image and speech. We used initialization learning rate 0.01, error rate 0.03, and momentum constant 0.9. Fig 9 shows the tool of the MBP-ANN training.

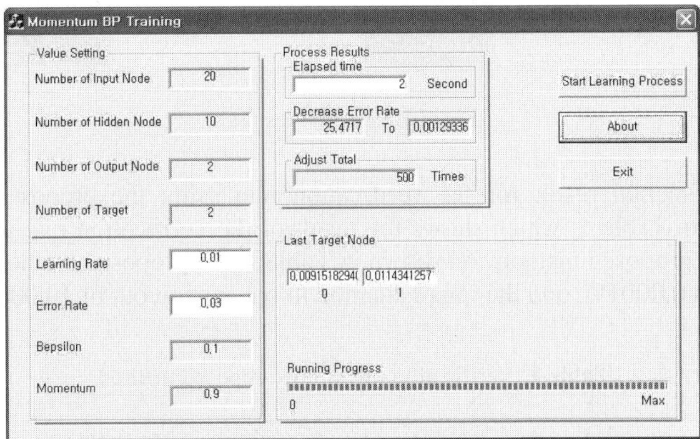

Fig. 9. Proposed momentum back-propagation training tool

6 Experimental results

The proposed multimodal biometric priority verification system is shown in Fig. 10, which shows the result of face and speech extraction. Fig. 11 shows the result of registered personal verification, and Fig. 12 shows the result of non registered person rejection.

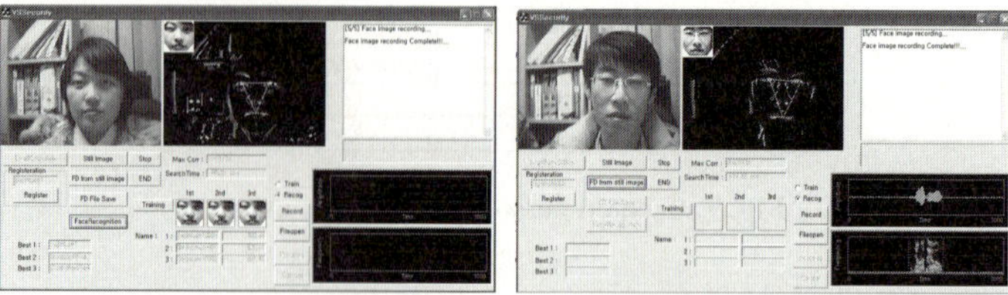

Fig. 10. The process to recognize face and speech

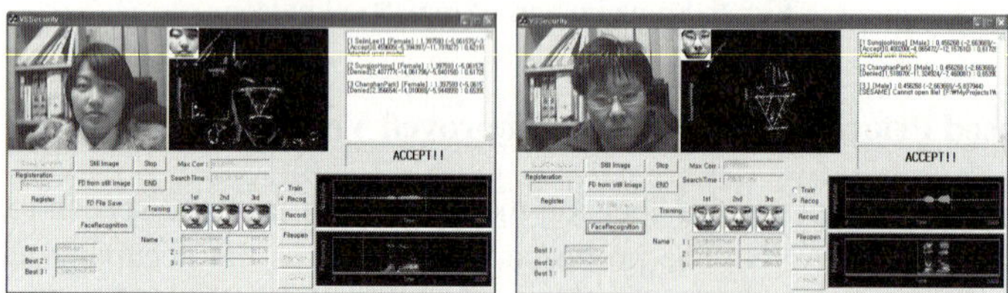

Fig. 11. Accepted results

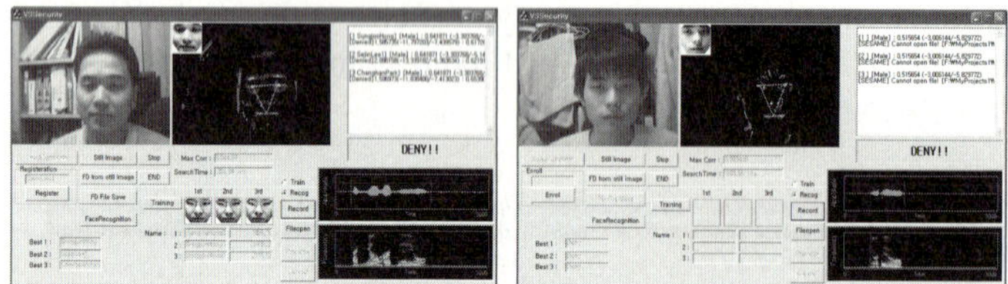

Fig. 12. Rejected results

The experimental result for the verification rate using the proposed method is summarized in Table 1, which shows the result of the verification rate and FAR obtained by the proposed method. As shown in Table 1, the proposed method can reduce FAR to down 0.0001%, and the impersonation to one person out of 10,000.

Table 1. Verification rate of the proposed method

Test DB	Verification Rate (%)	FAR (%)
face & speaker	99.99	0.0001

7 Conclusions

In this paper, we proposed a priority verification method for multi-modal biometric features by using the MBP-ANN. We also proposed a human verification method using combined face and speech information in order to improve the limitation of single biometric verification, which has the fundamental problems of high FAR and FRR. The proposed multimodal, biometric priority verification method improves the verification rate and reliability in real-time. We adopted the PCA framework for face recognition and HMM for speech recognition for real-time personal verification. As a result the proposed priority verification method can provides stable verification rate, and at the same time it overcomes the limitation of a single-mode system. Based on the experimental results, we show that FAR can be reduced down to 0.0001% in the human multimodal interface method using both face and speech information.

Acknowledgement

This research was supported by Korean Ministry of Science and Technology under the National Research Laboratory Project, Korean Ministry of Information and Communication under HNRC-ITRC program at Chung-Ang University supervised by IITA, and the Research Grant of Kwangwoon University in 2006.

References

1. Kong, S., Heo, J., Abidi, B., Paik, J., Abidi, M.: Recent Advances in Visual and Infrared Face Recognition - A review. Computer Vision, Image Understanding 97(1) (2005) 103-135
2. Kriegman, D., Yang, M., Ahuja, N.: Detecting Faces in Images: a Survey. IEEE Trans. Pattern Analysis, Machine Intelligence 24(1) (2002) 34-58
3. Liu, X., Chen, T., Kumar, V.: On Modeling Variations for Face Authentication. Proc. 2002 Int. Conf. Automatic Face, Gesture Recognition (2002) 369-374
4. Gu, Y., Thomas, T.: A Hybrid Score Measurement for HMM-based Speaker Verification. Proc. 1999 IEEE Int. Conf. Acoustics, Speech, Signal Processing 1 (1999) 317-320
5. Fu, R., Xu, T., Pan, Z.: Modeling of the Adsorption of Bovine Serum Albumin on Porous Polyethylene Membrane by Back-propagation Artificial Neural Network. Journal, Membrane Science 251 (2004) 137-144
6. Devillers, J.: Neural Network in QSAR and Drug Design. Academic Press, San Diego, CA (1996)
7. Rowley, H., Baluja, S., Kanade, T.: Neural Network-based Face Detection. IEEE Trans. Pattern Analysis, Machine Intelligence 20(1) (1998) 203-208
8. Samaria, F., Young, S.: HMM Based Architecture for Face Identification. Image, Vision Computing 12(8) (1994) 537-543
9. Belhumeur, P., Hespanha, J., Kriegman, D.: Eigenfaces vs Fisherfaces: Recognition Using Class Specification Linear Projection. IEEE Trans. Pattern Analysis, Machine Intelligence 19(7) (1997) 711-720
10. Kim, Y., Park, C., Paik, J.: A New 3D Active Camera System for Robust Face Recognition by Correcting Pose Variation. Proc. 2004 Int. Conf. Circuits, Systems (2004) 1482-1487
11. Zhang, J., Yan, Y., Lades, M.: Face Recognition: Eigenface, Elastic Matching, and Neural Nets. Proc. IEEE 85(9) (1997) 1423-1435
12. Sirivich, L., Kirby, M.: Low-dimensional Procedure for the Characterization of Human Faces. Journal, Optical Society of America A: Optics, Image Science, Vision 4(3) (1987) 519-524
13. Rabiner, L., Juang, B.: Fundamentals of Speech Recognition. Prentice-Hall (1993)

The Clustering Solution of Speech Recognition Models with SOM

Xiu-Ping Du and Pi-Lian He

School of Electronic & Information Engineering, Tianjin University, 300072 Tianjin, China
duyu@eyou.com

Abstract. This paper first introduces the system requirement and the system flow of the auto-plotting system. As the data points needed by the auto-plotting system coming from the remote speech signals, to reach high recognition accuracy, the Hidden Markov Model (HMM) approach was chosen as the speech recognition approach. Then the paper is detailed on the speaker dependent (SD), speaker independent (SI) and speaker adaptive (SA) speech recognition methods. We proposed the n-speech models SD system as the recognition system to gain the highest recognition performance in varying speech environments. However the system required that searching for the optimal model from the database should finish in 5 minutes, so the paper finally describes how the Self-Organizing Map (SOM) was used to pre clustering to the n-speech models, to decrease the time for speech recognition and results evaluation and decrease matching time, Experiments show the n-speech models SD system can select the best-matching model in the limited time and improve the average speech recognition accuracy to 97.2. It ideally suits the system requirements.

1 Introduction

In this paper, we developed an auto plotting system which needed to recognize the speech signal from remote place, produce series of number strings, then separate the number strings according to some certain rules, get series of data points with several attributes (such as symbol string, serial number, coordinate, distance, time, etc.), finally plot such data points on a special coordinate system.

The kernel task of the system was speech recognition. The speech signal content only consisted of numbers from 0 to 9, which was synthesized by computer, with a relatively strong stability. However, the speech speed could slow or quick in different time. In addition, the speech range could vary sometimes because of impact of transmission channel noise and the environment noise. The wireless signal was especially disturbed by environment factors, which made it much difficult to realize the system.

Speech can be considered as a sequence of words produced by a source affected by noise factors. Several techniques have been designed in order to correctly recast single spoken words from speech sounds, mostly relying on the HMM approach. An HMM is a statistical model that describes a probability distribution over a number of possible sequences. HTK (Hidden Markov Model Toolkit) was a major toolkit used to build HMM and to model time series and was widely used in speech recognition research world.

J. Wang et al. (Eds.): ISNN 2006, LNCS 3972, pp. 150 – 157, 2006.

The paper is arranged as follows. Firs, the system flow is introduced in section 2, then the speech recognition and SOM are described in section 3 and 4, respectively. Section 5 gives the experimental results and the conclusions are drawn in section 6.

2 System Flow

The system requirements presented as follows.

1) The system needed high recognition accuracy to the speech signals.

2) The system did not care the times of training speech model and the time consumption for training. However, while applying in a real task, it had only 5 minutes to prepare the system before starting the normal auto plotting process.

The system flow was shown as Fig.1

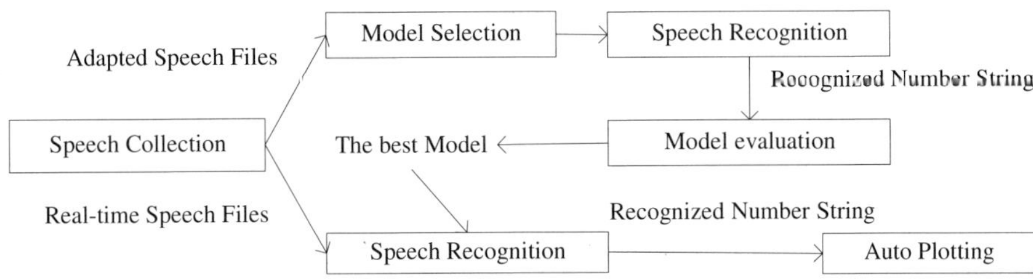

Fig. 1. The System Flow

1) Recording a 30 second adaptive speech file by the speech collection module and dictating the wave file to generate reference label sequences.

2) Repeatedly selecting the trained speech recognition models from the database, recognizing the speech file form step 1), getting the recognition result.

3) Evaluating the recognition result with reference label sequences, choosing the best–matching model.

4) Adopting the chosen model as the current speech recognition model, running the real-time speech recognition and auto plotting process.

Recording and scripting file could take 1 minute. A remarkable part of the preparing phase (clearly over 75%) was consumed in searching for the best-matching speech model from the database and for larger systems the portion increased even more. That was said, to finish prepare in 5 minutes, it should control the total time in 4 minutes, which included selecting all speech recognition models from the database, recognizing the adaptive file with each model, evaluating each recognition result, and picking out the optimal model.

3 Speech Recognition

In the system, each speech frame was represented by a 39-dimensional feature vector that consists of 12 mel frequency cepstral coefficients (MFCCs), normalised log energy along with the first and second differentials of these values. Cepsual mean normalization was applied. The speech data was pre-emphasized with

$$H(z) = 1 - 0.97z^{-1} \tag{1}$$

windowed to 25 ms frames with 10 ms frame shit. A left to right non skipping transitions topologies was used. Data preparation and training procedures were performed using the HTK 3.2 toolkit [1].

Percent Accuracy is a representative figure of recognizer performance in the system. HResults matches each of the recognized and reference label sequences by performing an optimal string match using dynamic programming. The optimal string match works by calculating a score for the match with respect to the reference such that identical labels match with score 0, a label insertion carries a score of 7, a deletion carries a score of 7 and a substitution carries a score of 100. The optimal string match is the label alignment that has the lowest possible score. Once the optimal alignment has been found, the number of substitution errors (S), deletion errors (D) and insertion errors (I) can be calculated. The percentage accuracy defined as

$$\text{Percent Accuracy} = \frac{N - D - S - I}{N} * 100\% \tag{2}$$

where N is the total number of labels in the reference transcriptions.

In the system, the percent accuracy was utilized to evaluate the model performance.

3.1 SI and SD

The speech recognition can divide into SI system and SD system. To set up a SD, it needs a lot of training speech samples. Moreover, when changing a new user, it needs a great of training speech samples to re-training. The consumed time of re-training for a new user is too long to be accepted in the system. The SI does not need to gather lots of speech samples of a new user to re-train and can provide a system to the new user in time. However to set up such a SI, it needs much more speech samples than SD. Above all, the speech recognition accuracy of SI is less than that of SD.

3.2 SA

In recent years, there is a growing interest in which has been shown to be an effective means of improving the performance of the large vocabulary continuous SI speech recognition system. To achieve the performance of a SD system for a SI system, the SA technique is an effect technique.

To SA technique, while many adaptation schemes have been proposed, MAP estimation [2] and MLLR [3] seem to be particularly promising. SA recognition method uses a few training speech sample of a new user to improve the performance of the original reference recognition system. If there exists a better model in the original reference recognition system, then the system can adapt to new user, so the system has a good recognition performance. But SA needed to train new model before real use, it could not meet the system object that reaching high recognition accuracy in 5 minutes.

3.3 n-Speech Models SD Recognition Solution

To gain the highest recognition accuracy, the SD recognition solution was adopted in the system. Because the features of the speech source could vary in different time, a single SD recognition model could not suit all instance. So it needed many speech models that could recognize in different speech environment. It was called n-speech models SD recognition solution. In the usual speech training, once a training model generated, we added it into the database. After training more and more times, n got more and more large.

Because there were many n-speech models, if we recognized and evaluated results with each model, it would take too long time to select the optimal model in 5 minutes. To decrease the searching time, we should decrease the counts of the model to search from. Therefore, we proposed to pre-cluster n-speech models to reduce the searching space. While recognizing, we could firstly find out the optimal recognition class, secondly found out the best-matching recognition model in the optimal class, then could decrease the prepare time.

The difference among n-speech models could be large or little. Clustering accord ing to the models was expected to make the least difference in a same class. If classing more detailed, namely the total speech cluster count m got larger, then the models in a same cluster got more similar. However if m was too large, then the counts in a cluster was relatively less, causing the matching time to select the optimal model increase. So we should setup a proper m value.

Assuming the total models count was n, the total clusters count was m, the average models count in each cluster was k, the average recognition time of each model was t, and the average time to select the optimal model was T, then T could be expressed as

$$T=(m+k)*t$$

In addition, we had

$$n=m*k$$

To get the least T, m should setup the minimum integer great than the square root of n, as pointed in (3).

$$m = [\sqrt{n}\ ] \tag{3}$$

As could reduce the computational complexity of the system from m2 to 2m.

4 Clustering by SOM

4.1 SOM

SOM proposed by Kohonen in 1982 that adopting unsupervised learning method to cluster. The algorithm can be represented in a simple computational and iterative form. Here gives a brief account of Kohonen's algorithm [4].

The algorithm describes a map from an input space $\boldsymbol{V}$ into an output space $\boldsymbol{A}$. The output space consists of nodes $\boldsymbol{y}_i$, which is usually arranged in vertices of a two-dimensional lattice. For each node $\boldsymbol{y}_i$ in $\boldsymbol{A}$, illustrated in Fig.2, a reference (weight) vector $\boldsymbol{y}_i$ is assigned, and the same input vector $\boldsymbol{x}$ in $\boldsymbol{V}$ is broadcasted to all of the nodes in $\boldsymbol{A}$. The best-matching node, the "winner" node $\boldsymbol{y}_c$. is defined according to (4).

$$\| x - y_c \| = \min\{\| x - y_i \|\} . \tag{4}$$

The winner nodes $\boldsymbol{y}_c$ are selected by referring to an arbitrary norm, here, the Euclidean norm in the input space $\boldsymbol{V}$. Updating of $\boldsymbol{y}_i$ is restricted to a topological neighborhood $\boldsymbol{N}_c$ of the winner node $\boldsymbol{y}_c$ and the weights are updated with (5).

$$\begin{array}{ll} y_i(t+1) = y_i(t) + \alpha\{x(t) - y_i(t)\} & \forall i \in N_c(t) \\ y_i(t+1) = y_i(t) & \forall i \notin N_c(t) \end{array} . \tag{5}$$

There were l nodes in the input layer, connected with m nodes in the output layer in the paper. The m value was defined at (3).

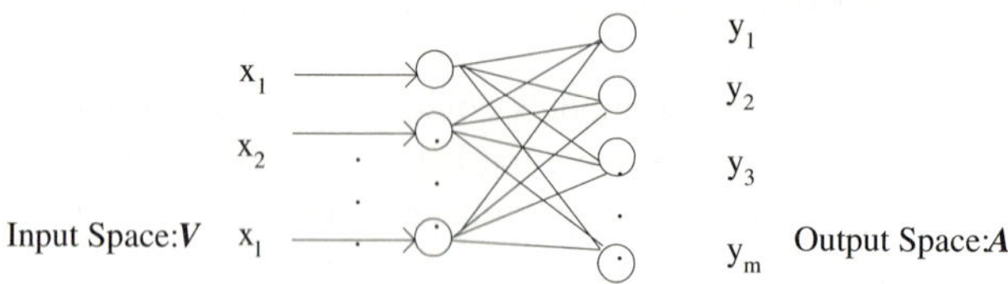

Fig. 2. SOM

4.2 Input Parameters

A part of parameter values of a speech recognition model was shown as Fig.3. We chose all MEAN, VARIANCE, MIXTURE, GCONST and TRANSP values in the file as the input vector. So the input nodes count l in the paper was 22374.

```
VARIANCE> 39
1.469456e+001 6.041367e+000 1.235755e+001 8.139022e+000 1.562447e+001 1.470236e+001 1.725561e+001 1.734955e+001
1.099444e+001 1.996369e+001 2.909897e+001 2.342045e+001 2.324371e+001 1.340878e+000 2.254325e+000 1.276471e+000
1.401697e+000 2.248518e+000 1.378227e+000 3.064810e+000 4.118820e+000 1.973377e+000 6.326252e+000 9.240440e+000
4.027103e+000 1.512562e+000 3.060281e-001 3.538994e-001 3.012249e-001 3.486748e-001 5.767410e-001 4.222811e-001
2.834813e-001 4.195176e-001 3.515428e-001 6.848796e-001 9.580414e-001 6.267970e-001 3.926489e-001
<GCONST> 1.079971e+002
~h "ou-g+u"
<BEGINHMM>
<NUMSTATES> 5
<STATE> 2
~s "g_s21"
<STATE> 3
~s "g_s31"
<STATE> 4
~s "g_s41"
<TRANSP> 5
0.000000e+000 1.000000e+000 0.000000e+000 0.000000e+000 0.000000e+000
0.000000e+000 6.814417e-001 3.185583e-001 0.000000e+000 0.000000e+000
0.000000e+000 0.000000e+000 5.166715e-001 4.833285e-001 0.000000e+000
0.000000e+000 0.000000e+000 0.000000e+000 6.290846e-001 3.709153e-001
0.000000e+000 0.000000e+000 0.000000e+000 0.000000e+000 0.000000e+000
<ENDHMM>
```

Fig. 3. Part Parameter Values of a Speech Recognition Model

4.3 Clustering Algorithm

The clustering algorithm to speech recognition models with SOM was as follows.

1) Selecting all models from the database.
2) Gaining all parameters from all speech recognition models.
3) Using the parameters from step 2) as input data, training and clustering with SOM.
4) Saving the clustering results into the database.

Using SOM to cluster, while there came a new training model, it needed to re-cluster. The former training results was no use any longer. There needed much time to get new cluster. As pointed in the system requirement in the paper, the time consumption in usual was no matter. So when meeting a new speech environment, we trained the speech model and saved the training results into the database. Because the speech was relatively stable in a certain time, it was not long from preparing the training files to finishing training.

4.4 The Hierarchical Clustering

To the models clustered by SOM, in some class, the counts of models could be much than a threshold that caused the too long time for searching through the models. The threshold was defined as 500 in the paper. When there was models count in a cluster going beyond the threshold, we clustered these models again with SOM to generate sub clusters. It was called hierarchical clustering. In this time, the output nodes count setup as 500. To different models in a same sub cluster, we could directly merge all training files, re-training to generate a new speech model. As could decrease the total searching space, then decreased the matching time.

5 Experiments and Results

The improved system flow was as follows.

1) Recording a 30 second adaptive speech file by the speech collection module and dictating the wave file to generate reference label sequences.
2) From the clusters of the speech recognition models in the database, selecting a model from each cluster, recognizing the file from step 1), getting recognition result, evaluating the result, then choosing out the best matching model cluster.
3) In the best matching model cluster, iteratively selecting the speech recognition model, recognizing the file from step 1), getting recognition result, evaluating the result with reference label sequences, choosing out the optimal model.
4) Adopting the chosen model as the current speech recognition model, running the real-time speech recognition and auto plotting.

The system ran in the PIV 1.5GCPU, 256MRAM, and 40G Hard disk computer. The operation system was Win2000.

The average performance of different speech recognition solution was shown as Table 1. From the table, we should choose the SD speech recognition solution in order to complete the prepare work in limited minutes.

The average time of searching for the optimal model was shown as Table 2. When the total speech recognition model count was 1000, the average time of searching for the optimal model was 191 second. However, while the count increasing to 2000, the average time rise to 382 second which beyond the limited time of system. After clustering, although the total speech recognition models count reached 5000, the count need to search from was only 142, and the average time was 27 second which less than the limited time.

Table 1. The average performance of different speech recognition solution

Speech recognition solution	Percent accuracy	Average time consumption for prepare (minute)
SD	97.2%	>1
SI	89.3%	0
SA	92.4%	>10

Table 2. Average time of searching for the optimal model

Model counts	The total models to search from		The average time of searching for the optimal model (second)	
	Before clustering	After clustering	Before clustering	After clustering
100	100	20	18	4
500	500	45	92	9
1000	1000	64	191	13
2000	2000	90	382	17
5000	5000	142	955	27

6 Conclusion

We presented the n-speech models system as the speech recognition approach that gave a significant performance improvement in the recognition accuracy. The hierarchical clustering solution with SOM obviously reduced time consumption on searching for the optimal speech recognition model. It suits the system requirements well.

References

1. Veeravalli, A.G., Pan, W.D., Adhami, R., Cox, P.G.: A Tutorial on Using Hidden Markov Models for Phoneme Recognition. In: SSST '05 (eds.): Proceedings of the Thirty-Seventh Southeastern Symposium on System Theory, IEEE, Tuskegee, AL (2005) 154-157
2. Legetter, C.J., Woodland, P.C.: Maximum Likelihood Linear Regression for Speaker Adaptation of Continuous Density Hidden Markov Models. Computer Speech and Language 9(2) (1995) 171-186

3. Gauvain, J.L., Lee, C.H.: Maximum a Posteriori Estimation for Multivariate Gaussian Mixture Observations of Markov Chains. IEEE Trans. Speech and Audio Processing 2(2) (1994) 291-298
4. Nishida, S., Ishii, K., Ura, T.: Adaptive Learning to Environment Using Self-Organizing Map and Its Application for Underwater Vehicles. In: UT '04 (eds.): Proceedings of the 2004 International Symposium on Underwater Technology, IEEE, Taipei, Taiwan (2004) 223–228

Study on Text-Dependent Speaker Recognition Based on Biomimetic Pattern Recognition

Shoujue Wang, Yi Huang, and Yu Cao

Lab of Artificial Neural Networks, Institute of Semiconductors,
Chinese Academy of Sciences, Beijing, 100083, China
camelyi@263.net, {wsjue, caoyu}@semi.ac.cn

Abstract. We studied the application of Biomimetic Pattern Recognition to speaker recognition. A speaker recognition neural network using *network matching degree* as criterion is proposed. It has been used in the system of text-dependent speaker recognition. Experimental results show that good effect could be obtained even with lesser samples. Furthermore, the misrecognition caused by untrained speakers occurring in testing could be controlled effectively. In addition, the basic idea "cognition" of Biomimetic Pattern Recognition results in no requirement of retraining the old system for enrolling new speakers.

1 Introduction

Automatic Speaker Recognition (ASR) means identifying or verifying the testing speaker's identity using the phonic features of speakers. It could be divided into two classes, according to whether the speech content has limitation: text-independent recognition and text-dependent recognition. Compared with the former, the latter might lead to some inconvenience for the speech content limitation; however, it could provide more accurate recognition results with less training samples [1]. Therefore, text-dependent recognition has more potential application fields including password entrance guard and access control, etc.

Artificial Neural Network (ANN) has been successfully applied in speaker recognition for its excellent ability on complex mapping. Traditional speaker recognition methods based on ANN use either Multilayer Perceptron (MLP) [2] which grounds on "division" or Radial Basis Function (RBF) [3] network on template matching. These methods deal with the distribution of the sample points in feature space relatively coarsely, hereby they have obvious limitations [4]. For instance, if new speaker occurs in the test, misrecognition will surely be resulted in; besides, adding new speakers leads to retraining the whole system. With the increasing amount of testing crowd, the overlapped regions of speaker feature space may rise so that the recognition performance may be lowered; furthermore, the training time of the network will increase rapidly. In theory, the retraining could not be achieved with high testing crowd number.

Considering the defects of traditional method based on "division", the author put forward Biomimetic Pattern Recognition (BPR) [5] theory. It has been applied successfully in omni-directionally oriented rigid objects recognition [5], and speaker-independent speech recognition [6], etc. The advantages of recognition results proved

J. Wang et al. (Eds.): ISNN 2006, LNCS 3972, pp. 158 – 164, 2006.

that BPR could resolve effectively the existing problems in traditional methods. In this paper, we proposed a speaker recognition neural network based on BPR and used it in the text-dependent speaker recognition system.

2 Theory of Biomimetic Pattern Recognition

Traditional pattern recognition methods rely on statistical decision theory, emphasizing on optimal separation of inhomogeneous samples in feature space. The most typical one is Support Vector Machine (SVM) [7]. Instead, Biomimetic Pattern Recognition bases on the Principle of Homology-Continuity (PHC) [5] in feature space. Guiding by the idea of "cognition", BPR could achieve the optimal covering of every class of samples in feature space using high-dimensional space geometry method [8] and multi-weight neuron theory [9].

In the BPR theory, the construction of the subspace of a type of samples only depends on the type of samples itself, which is completely different from those traditional methods depending on the relations among various types of samples. Particularly, the construction in BPR is based on the theory of high-dimensional manifold in point set topology, researching the topological characters of the samples belonging to the same type, and reasonably covering the distribution of the type of the samples in the feature space by high-dimensional geometric complex. Figure 1 shows the sample subspaces of the BPR and those of the traditional pattern recognition methods (MLP and RBF) in 2-dimensional space.

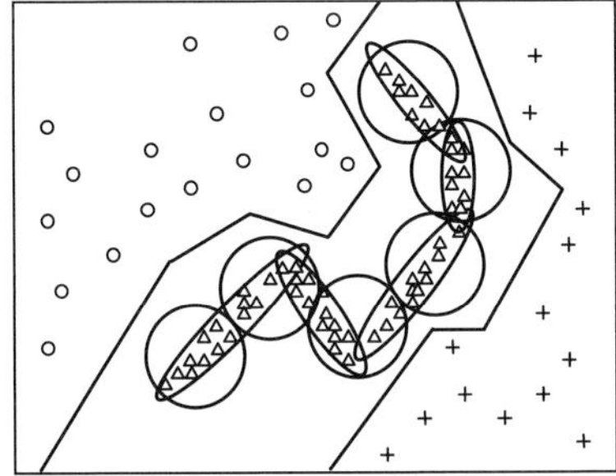

Fig. 1. Scheme of the pattern subspaces constructed by BPR and traditional methods

In the scheme, triangles are the samples to be recognized, circles and crosses are two kinds of samples different from triangles; fold lines are the partition modes of MLP method, great circles denote those of RBF network (equal to the recognition mode by template matching method), and curves consisted by Hyper-Sausages [10] stand for "cognition" modes of BPR.

3 Speaker Recognition Based on BPR

In the time domain, speech is a kind of short-time stationary signal. Therefore, the input speech samples are pretreated by adding windows and partitioning frames in common speech signal processing methods. Generally, human speech consists of many different speech phonemes such as surd, sonant, vowel, and so on, among which distinction exists. Therefore, the eigenvectors extracted from the partitioned speech signal frames might reveal some clustering quality in the feature space. This property is the basis of current main methods such as Vector Quantization (VQ) and Gaussian Mixture Model (GMM) [11].

Different from the idea of clustering, BPR takes the fact that "the whole of homologous samples is continuous in feature space" as the preknowledge of the distribution of the sample points. Every frame comes from the same speaker, the same vocal organs, the same oral cavity and nasal cavity, the same track and the same vocal cords. The physical parameters cannot saltate during the vocal course. Therefore, these speech frames accord with the Principle of Homologous Continuity (PHC). That is [5]: suppose that point set A includes all samples in the class A in feature space, x, y are two random elements in set A. If $x, y \in A$ and $\varepsilon > 0$ are given, there exists set B:

$$B = \left\{ x_1, x_2, x_3, \ldots, x_n \middle| \begin{array}{l} x_1 = x, x_n = y, n \in N, \rho(x_m, x_{m+1}) < \varepsilon, \\ \varepsilon > 0, 1 \le m \le n-1, m \in N \end{array} \right\}, B \subset A \tag{1}$$

Applied the above into learning process, it means choosing one or more than one appropriate close surface to build a continuous complex geometrical body in high-dimensional space to cover rightly the training samples. The choice depends on the distribution of the training samples in feature space for homologous things.

A universal expression for a multi-weight neuron is [8]:

$$Y = f[\Phi(W_1, W_2, \ldots, W_m, X) - \theta] \tag{2}$$

where W_i is the weight vector, X is the input vector, Φ denotes the computational function determined by the multi-weight neuron (multi input vector, single output scalar), θ represents the activation threshold of the neuron, and f is the nonlinear transfer function.

Assume that the feature space is n-dimensional real space R^n, namely, $X \in R^n$. Thus the following vector equation

$$\Phi(W_1, W_2, \ldots, W_m, X) = \theta \tag{3}$$

could be taken as a locus of the vector X in feature space R^n determined by a series of weight vectors $W_1, W_2, \ldots W_m$. The locus may be $n-1$ dimensional hypersurface or hyperplane in R^n which divides R^n into two parts. Selecting an appropriate Φ will obtain the close hypersurface. Thus, a finite covering region is formed in feature space. Changing the weight values and the threshold values will result in variation of the shape and the size of the hypersurface. In the learning process of the multi-weight neural networks for speaker recognition, we should choose appropriate multi-weight neurons according to the distribution of the speech eigenvector in feature space. The eigenvectors are extracted from the training speech frames of each speaker. Thus, the union of these neurons forms a connective region. For example, for a certain speaker A, there exists

$$Y_A = OR(Y_{Aj})_{j=1}^{p} \tag{4}$$

where p is the amount of required neurons for covering all the training frames of the speaker A. It relies on the training of the networks. The output of these neurons is

$$Y_{Aj} = f\left[\Phi(W_{Aj1}, W_{Aj2}, \ldots W_{Ajm}, X) - \theta_{Aj}\right] \tag{5}$$

We defined that if X is inside the close hypersurface, the output is *true* ; otherwise it is *false* .

The speech characteristics are not static and stationary. In one hand, even if the speaker and the text is steady, the speech signals usually have great variation for different voice speed or emotion, etc.; in another hand, since lesser speech samples could merely cover finite eigenvectors, the required information could not be adequate to cover all the speaker features. Besides, noise inevitably exists. Considering all the above, we do not require all the eigenvectors to be covered during the recognition process. A network matching degree of testing speech signal was proposed, which is called *Fitness*. *Fitness* was defined as the covering rate of the speech frames, namely, the ratio of the amount of covered frames to the total frame amount. In speaker identification, the samples to be tested are deduced into the network whose *Fitness* is maximal; in speaker verification, when *Fitness* is more than the predefined threshold value, the claimed identity is accepted, or else the claimed identity is rejected. Experiments proved that the use of *Fitness* greatly improved the robustness of the system. Furthermore, the false rejection and the false acceptance of the system can be well balanced by adjusting the lower limit of *Fitness*.

4 Experiments and Results of Text-Dependent Speaker Recognition

4.1 Feature Extraction

This paper utilized Mel-Frequency Cepstrum Coefficients (MFCC) as the speech frame feature in speaker recognition, which was confirmed to be one of the most successful descriptions for the characteristics in speech-related recognition tasks [11]. The brief approach is as follows: firstly, deal with the speech signal by voice activity detection and pre-strengthening; secondly, take every 25ms as one frame, while the relative displacement between two neighboring frames is 12.5ms; thirdly, extract 12 MFCC parameters from each frame; finally, raise the cepstrum parameters, thus a 12-dimensional eigenvector is obtained.

4.2 Construction of the Text-Dependent Speaker Recognition Neural Network Based on "Cognition"

Assume that n is the amount of training text-dependent speech samples for each speaker and $n \geq 2$. The former $n-1$ samples are used to train the neuron weight vectors by point distribution analysis in high-dimensional space [5, 6, 8-10] provided by the author; the n-th sample is used to train the activation threshold value under the stop condition of $Fitness > 0.85$. Hereby, recognition network is constructed for each speaker one by one.

4.3 Experiments and Results

Experimental samples include 25 persons (14 males, 11 females). Each speaker uttered "one, two, three, four" (named class A) and "seven, eight, nine, ten" (named class B) in Chinese respectively for 5 times. The total sample amount is 250. The sampling frequency is 22.05 kHz, mono, 16 bit precision.

In speaker identification, every speaker uttered 2 pieces of speech for training and the left 3 pieces for testing, thus 150 samples for testing totally including class A and B of 25 speakers; if 3 pieces for training and the left 2 for testing, thus 100 samples for testing totally. Table 1 gives the results of speaker identification.

Table 1. Results of speaker identification

Training speech samples	True identification samples	False identification samples	Correct rate
2×2×25	146	4	97.33%
3×2×25	100	0	100%

Table 2. Results of speaker verification with 2 training speech samples

Rejection condition	True acceptance samples	True rejection samples	Total correct rate	False acceptance samples	False rejection samples	Total false rate
$Fitness \leq 0.70$	137	3595	99.52%	5	13	0.48%
$Fitness \leq 0.75$	116	3597	99.01%	3	34	0.99%
$Fitness \leq 0.80$	89	3599	98.35%	1	61	1.65%

Table 3. Results of speaker verification with 3 training speech samples

Rejection condition	True acceptance samples	True rejection samples	Total correct rate	False acceptance samples	False rejection samples	Total false rate
$Fitness \leq 0.70$	100	2400	100%	0	0	0%
$Fitness \leq 0.75$	99	2400	99.96%	0	1	0.04%
$Fitness \leq 0.80$	96	2400	98.84%	1	61	0.16%

In speaker verification, when 2 speech samples of each speaker were used for training, 25 networks for speaker verification were tested by 150 samples in turn, namely, totally tested for 3750 times. Table 2 displayed the results under different rejection condition. When 3 speech samples were used for training, 25 networks verification were tested by 100 samples, totaled for 2500 times. From Table 3 we could see the results under different rejection condition.

5 Discussions and Conclusions

From the above experimental results, we conclude that:

(1) When the amount of training speech samples was 2, the tasks of identification and verification still performed high correct rate. The results confirmed the advantage of BPR with lesser training samples.

(2) When the training amount is 3, the speaker's information for network training was relatively adequate. Not only identification task but also verification task had fairly high correct rate.

(3) By the analysis about Tables 2 and 3, we know that in the task of verification, the threshold of rejection *Fitness* could be adjusted so as to control the security level according to various applications with different security requirement. The increase of the threshold of rejection might slightly raise the false rejection rate, however, re-verification could compensate for this problem in practical applications.

Our experimental results prove the excellent performance of BPR. Especially in text-dependent speaker verification, less sample amount for building a model, convenient system expansion, flexible security level control, high correct recognition rate and correct rejection rate show a very wide application prospect of this method.

References

1. Campbell, J.P.: Speaker Recognition: A Tutorial. Proceedings of the IEEE 85(9) (1997) 1437-1462
2. Farrel, K.R., Mammone, R.J., Assaleh, K.T.: Speaker Recognition Using Neural Networks and Conventional Classifiers. IEEE Transactions on Speech and Audio Processing 2(1) (1994) 194-205
3. Oglesby, J., Mason, J.S.: Radial Basis Function Networks for Speaker Recognition. Proceedings - ICASSP, IEEE International Conference on Acoustics, Speech and Signal Processing, Vol. 1 (1991) 393-396
4. Finan, R.A., Sapeluk, A.T., Damper, R.I.: Comparison of Multilayer and Radial Basis Function Neural Networks for Text-Dependent Speaker Recognition. IEEE International Conference on Neural Networks - Conference Proceedings, Vol. 4 (1996) 1992-1997
5. Wang, S.: Bionic (topological) Pattern Recognition - A New Model of Pattern Recognition Theory and Its Applications (in Chinese). Acta Electronica Sinica 30(10) (2002) 1417-1420
6. Qin, H., Wang, S.: Comparison of Biomimetic Pattern Recognition, HMM and DTW for Speaker-Independent Speech Recognition (in Chinese). Acta Electronica Sinica 33(5) (2005) 957-960

7. Vapnik, V.N.: The Nature of Statistic Learning Theory. Springer-Verlag, Berlin Heidelberg New York (1995)
8. Wang, S., Wang, B.: Analysis and Theory of High-Dimension Space Geometry for Artificial Neural Networks (in Chinese). Acta Electronica Sinica 30(1) (2002) 1-4
9. Wang, S., Li, Z., Chen, X., Wang, B.: Discussion on the Basic Mathematical Models of Neurons in General Purpose Neurocomputer (in Chinese). Acta Electronica Sinica 29(5) (2001) 577-580
10. Wang, S., Zhao, X.: Biomimetic Pattern Recognition Theory and Its Applications. Chinese Journal of Electronics 13(3) (2004) 373-377
11. Quatieri, T.F.: Discrete-Time Speech Signal Processing: Principles and Practice. Prentice Hall (2002)

A New Text-Independent Speaker Identification Using Vector Quantization and Multi-layer Perceptron

Ji-Soo Keum[1], Chan-Ho Park[2], and Hyon-Soo Lee[1]

[1] Dept. of Computer Engineering, Kyung Hee University, 1, Seocheon-dong, Giheung-gu, Yongin-si, Gyeonggi-do, Korea
{jskeum, leehs}@khu.ac.kr

[2] Dept. of Internet Information Science, Bucheon College, 424, Simgok-dong, Wonmi-gu, Bucheon-si, Gyeonggi-do, Korea
chpark@bc.ac.kr

Abstract. In this paper, we propose a new text-independent speaker identification method using VQ and MLP. It consists of three parts: a new spectral peak analysis based feature extraction, speaker clustering and model selection using VQ, and MLP based speaker identification. The feature vector reflects the speaker specific characteristics and has a long-term feature for which makes it text-independent. The proposed method has a computational efficient for feature extraction and identification. To evaluate the proposed method, we calculated the correct identification ratio (CIR), the average CIR of the proposed and GMM method was 92.27% and 85.78% for 5 seconds segments in 15-speaker identification. Experimental results, we have achieved a performance comparable to GMM-method.

1 Introduction

In order to identify the speaker or verify the speaker's voice, many kinds of methods have been proposed. In speaker identification, the purpose is to determine the voices that best match the enrolled speaker. It is divided into text-dependent and text-independent system by restriction of uttered speech data [1][3].

Current speaker identification models can be categorized into four main groups: Vector Quantization (VQ), Gaussian Mixture Model (GMM), Hidden Markov Model (HMM) and Neural Networks (NN) [4][5]. Widely used methods are the GMM and HMM. These methods have demonstrated excellent performance, but are computationally expensive.

For feature vectors, the Mel Frequency Cepstral Coefficient (MFCC) and Linear Prediction Coefficient (LPC) are used in most methods. This feature vectors includes both speech and speaker information, and so it is difficult to distinguish between the speech and speaker information. It is necessary to find the feature that has the speaking style for correct speaker identification.

In the NN method approach, Multi-layer Perceptron (MLP), Time Delay Neural Network (TDNN) and Radial Basis Function (RBF) apply speech and speaker recognition [6]. Also, MLP based Predictive Neural Network (PNN) is proposed for speaker recognition [7].

J. Wang et al. (Eds.): ISNN 2006, LNCS 3972, pp. 165 – 171, 2006.

To implement a NN method for speaker recognition, the problem of time alignment must first be solved. To accomplish this, normalized the speech length or speaking speed. In order to use the merit of NN as likes, the generalization using train, parallel processing and fault tolerance, it is necessary to extract the feature that has the speaker specific characteristics and which is suitable for input vector of NN.

In this paper, we propose new text-independent speaker identification that has a speaker specific feature and solves the time alignment problem. The proposed method uses the MLP as identification model because it exhibits excellent classification performance and the time alignment problem is solved during the feature extraction segment. And, before speaker identification the VQ methods are applied to cluster the similar speaker to improve the performance.

The rest of this paper is organized as follows. In section 2, we propose new text-independent speaker identification. Section 3 discusses the results of the experiment for the proposed method and GMM method. Section 4 presents a conclusion and summary of future research.

2 Proposed Speaker Identification System

Fig. 1 shows the block diagram for the proposed speaker identification system. It consists of three parts: (1) spectral peak analysis based feature extraction, (2) speaker clustering and model selection using VQ, and (3) MLP based speaker identification.

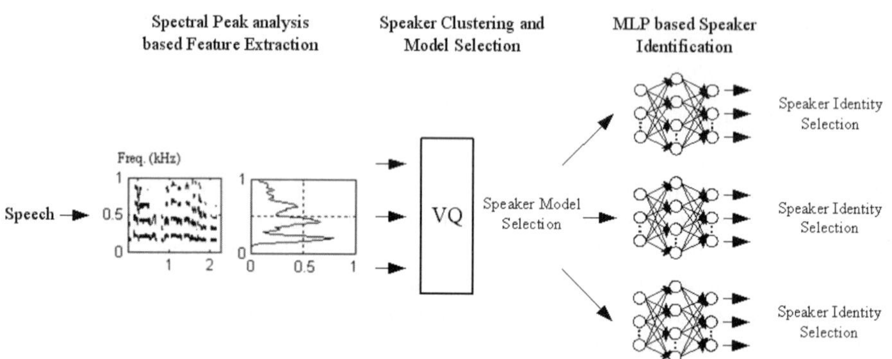

Fig. 1. Proposed speaker identification system

2.1 Spectral Peak Based Feature Extraction

The segmental information plays an important role in gathering information and our hearing systems are very sensitive to spectral transition [2]. When the speaker pronounces the sentence or words, it has a rhythm, and the intonation pattern varies according to the type of sentence and speaking style. In [10], a spectral peak track analysis based feature is used to speaker change detection and in [8][9] spectral peak track is used to characterizing sound of song and speech.

Fig. 2 shows that spectral peaks and histogram for 0~1000 Hz. Speech length is 5 seconds and sampled at 11025 Hz. A 512-point Fast Fourier Transform (FFT) was performed, and shifted window length was 256-point.

The spectral peak track exists in a specific frequency band and has duration time, and it differs from another speaker's voice. In Fig. 2(a), 2(b) the spectral peak track has a harmonic structure. Fig. 2(c), 2(d) presents a high peak track variation and Fig. 2(b) has a long duration compared to 2(a), 2(c) and 2(d). Therefore, when the spectral peak feature is applied it is possible to identify the speaker. The analysis and use of spectral peak track for speaker change detection is described in [10].

In (1), $X[k]$ is a FFT spectrum and $CX_n[k]$ is a clipped spectrum at n-th frame, it represents the spectral peaks. If the spectrum magnitude is below a threshold from spectrum peak then one must apply the clipping to remove the low magnitude.

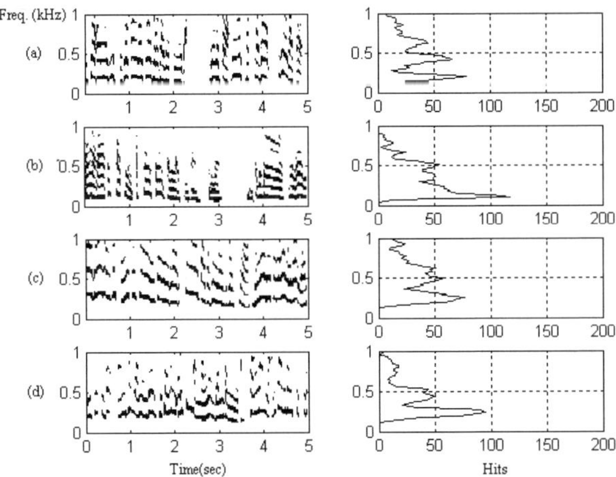

Fig. 2. Spectral peak track histogram for 5 seconds: (a) male 1, (b) male 2, (c) female 1, (d) female 2

From the clipped spectrogram, we were able to determine the continuity for time, which is similar to block matching. We define the $CT_k(n,n+i)$ as a connection from n frame to $n+i$ frame for k-th frequency. Where k denotes the frequency band index and n is frame index. During speech, even when the speaker does not change when speaking the same sentence or words, each speech length varies. It is necessary to normalize the speech length. In (3), $H_t(k)$ is normalized spectral peak feature vector by total frame number N. The $H_t(k)$ is 35-th dimensional vector, because we performed 512-point FFT and the spectral peaks exist mainly in 60~800 Hz. This normalized spectral peak feature is used input vector of the VQ and MLP.

$$CX_n[k] = \begin{cases} 1, & if \ |X[k]|^2 \geq threshold \\ 0, & if \ |X[k]|^2 < threshold \end{cases} \tag{1}$$

Where, $threshold = Max(X[k]) \times \alpha$

$$CT_k(n, n+i) = \begin{cases} 1, & if\ CX_n[k] = 1\ and\ \ldots\ CX_{n+i-1}[k] = 1\ and\ CX_{n+i}[k] = 1 \\ 0, & if\ CX_n[k] = 0\ or\ \ldots\ CX_{n+i-1}[k] = 0\ or\ CX_{n+i}[k] = 0 \end{cases} \tag{2}$$

$$H_l(k) = \frac{1}{N}\sum_{n=1}^{N}\sum_{i=1}^{I} CT_k(n, n+i) \tag{3}$$

2.2 Speaker Clustering and Identification Using VQ and MLP

In speaker identification, many errors occur in similar speaking style's voice and difference of train and test condition. Therefore, before speaker identification we applied the VQ to improve the performance.

The VQ maps k-dimensional vectors into a finite set of vectors $Y = \{y_i\text{: i} = 1, 2, \ldots, N\}$. Where the y_i is a codeword and the set of all the codewords are codebook [11]. The codeword is the centroid of the cluster and it is represents the average vector of each speaker. Using the normalized spectral peak feature, we perform the speaker clustering and construct the speaker model. Each centroid corresponds to the speaker's center and it gathers similar speaking style's speaker. In test phase, the proposed system is automatically selects the speaker model by VQ.

The MLP architecture consists of three layers: Input, Hidden and Output layer. The MLP is trained using Back Propagation (BP) algorithm [12]. The training of the MLP by BP involves three stages: the feed-forward of the input training vector, the calculation of error and the adjustment of the weights. The input vectors are the 35-dimensional normalized spectral peak feature and each the output neuron is correspondence to matched speaker's identity.

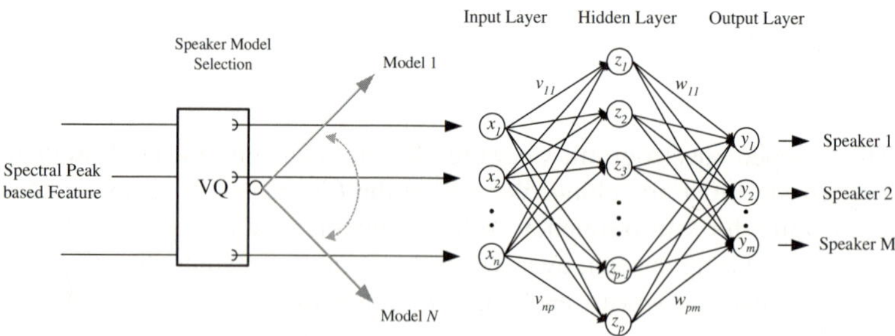

Fig. 3. MLP based speaker identification model

3 Experiments and Results

This section provides the results of experiments for the proposed VQ-MLP and GMM-based text-independent speaker identification.

3.1 Experimental Condition

Speech data sampled at 11025 Hz using general TV capturing board for Korean broadcast news. Captured speech data consists of 15 speakers (10 males, 5 females). To evaluate the proposed method, we calculated the correct identification ratio (CIR).

$$\mathrm{CIR}\ (\%) = \frac{\text{number of correctly identified segments}}{\text{total number of segments}} \times 100 \tag{4}$$

In 5 seconds segment, about 20 syllables exist in Korean sentences. The 5 seconds segment has sufficient information to represent the speaker's characteristic. Therefore, we use the 5 seconds for training and the length of test segment is 5, 10 and 15 seconds, and the segment moves 0.2 seconds.

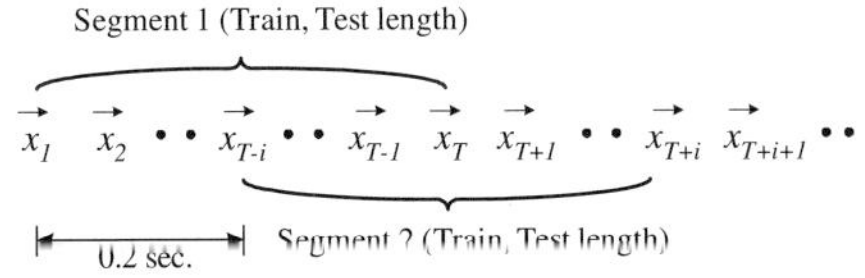

Segment 1 (Train, Test length)

$\vec{x}_1$ $\vec{x}_2$ • • $\vec{x}_{T-i}$ • • $\vec{x}_{T-1}$ $\vec{x}_T$ $\vec{x}_{T+1}$ • • $\vec{x}_{T+i}$ $\vec{x}_{T+i+1}$ • •

0.2 sec. Segment 2 (Train, Test length)

In the proposed method, the feature vector dimension is 35 for 1-segment. For (3), we setup the 3(about 92 ms) by value of I. Therefore, the normalized feature vector is the width is 21.5 Hz and the length (duration) is 92 ms block (21.5 Hz and 92 ms duration) that normalized by the total frame number of segment. The number of hidden neuron is selected by experiment results from 3 to 8 and used learning rate is 0.65. The number of training segment and test segment is 10 and 100 for each speaker respectively. Total test segments are 1,500 for 15 speakers.

The GMM is trained using the Expectation Maximization (EM) algorithm and the diagonal covariance matrix. The feature vectors for GMM method extracted by 13 dimensional MFCC every 20ms frame and the number of mixture is 2, 4, and 8. The number of test segments is 100 for each speaker.

3.2 Experimental Results

Table 1 shows the performance of the VQ based speaker clustering. Table 2 and Table 3 present the performance of the proposed method and GMM method respectively. We performed the experiment by changing the segment length of test data, and the number of cluster and mixture.

From the VQ experimental results, we have achieved best performance at the 5 clusters for 15 speaker. Using this result, we construct the 5 kinds of speaker model and the constructed each speaker models has 1 to 5 speakers.

Table 1. Results of the VQ using Spectral Peak Feature

Cluster No.	Train, Test (sec)		
	5	10	15
4	78.40	79.33	77.67
5	95.20	98.33	96.93
6	91.73	93.40	94.73

Table 2. Results of the MLP and Spectral Peak Feature

Class	Test (sec)		
	5	10	15
1	86.50	99.00	98.00
2	98.00	100.0	100.0
3	86.00	94.00	87.00
4	91.00	99.67	99.33
5	73.20	81.80	90.60
Avg.	86.94	94.89	94.98

Table 3. Results of the GMM and MFCC

Mixture	Test (sec)		
	5	10	15
2	84.80	88.40	89.40
4	83.13	83.27	84.80
8	87.13	86.60	84.53
Avg.	85.02	86.09	86.23

Based on the experimental results, we can see the merit of the proposed method using the spectral feature for speaker identification. The amounts of training patterns are very small, but the proposed method exhibited a comparable performance to GMM method. Also, we knows that the proposed spectral peak feature is suitable for small number of speaker identification problem as likes, News, debates and that we know the appearance of the speaker.

From the Table 2, we know that according to some speaker model has a lower performance compare to another speaker model. This is due to the spectral peak features is very similar with another speaker's feature. In experiment results shows that the spectral peak feature is sufficient to use for speaker identification.

In experimental results, the GMM-method has demonstrated excellent performance, but is computationally expensive. From the Table 3, we know that the GMM model requires large numbers of training data for high performance and the performance are relative to segments length and mixture number.

4 Conclusions

In this paper, we performed text-independent speaker identification based on the VQ and MLP. The spectral peak feature is applied it is possible to identify the speaker. This feature is suitable for input vector of the MLP, because of the time alignment problem is solved during the feature extraction segment. The proposed method uses the MLP as identification model, because it exhibits excellent classification performance and a trained network can produce its out very rapidly. Moreover, to reduce the errors in similar speaking style's voice and differences of train and test condition, we applied the VQ before speaker identification. The results experiment indicates that the proposed method is sufficient to speaker identification.

Future research will focus on spectral track feature extraction and robust speaker modeling based on neural network. Also, many kinds of speaker's voice will be investigated.

References

1. Joseph, P. Campbell, J.R.: Speaker Recognition A Tutorial. Proceeding of The IEEE 85(9) (1997) 1437-1462
2. Sadaoki, F.: Recent Advances in Speaker Recognition. Pattern Recognition Letter 18 (1997) 859-872
3. Herbert, G., Michael, S.: Text-independent Speaker Identification. IEEE Signal Processing Magazine (1994) 18-32
4. Reynolds, D.A.; Rose, R.C.: Robust Text-independent Speaker Identification using Gaussian Mixture Speaker Models. IEEE Trans. on Speech and Audio Processing 3(1) (1995) 72-83
5. Narayanaswamy, B., Gangadharaiah, R.: Extracting Additional Information from Gaussian Mixture Model Probabilities for Improved Text-Independent Speaker Identification. IEEE International Conference on Acoustics Speech and Signal Processing 1 (2005) 621-624
6. Farrell, K.R., Mammone, R.J., Assaleh, K.T.: Speaker Recognition using Neural Networks and Conventional Classifiers. IEEE Trans. on Speech and Audio Processing 2(1) (1994) 194-205
7. Hiroaki H.: Text-Independent Speaker Recognition using Neural Networks. IEICE Trans. INF. & SYST. E76-D(3) (1993) 345-351
8. Lu, L., Zhang, H.J., Jiang, H.: Content Analysis for Audio Classification and Segmentation. IEEE Trans. on Speech and Audio Processing 10(7) (2002) 504-516
9. Zhang, T., Kuo, J.: Audio Content Analysis for Online Audiovisual Data Segmentation and Classification. IEEE Trans. on Speech and Audio Processing 9(4) (2001) 441-457
10. Keum, J.S., Lee H.S.: Speaker Change Detection Based on Spectral Peak Track Analysis for Korean Broadcast News. The Fifth International Conference on Information Communications and Signal Processing (2005) 724-728
11. Mohamed, Q.: Vector Quantization. http://www.geocities.com/mohamedqasem/vector-quantization/vq.html
12. Laurene, F.: Fundamentals of Neural Networks. Prentice Hall (1994)

Neural Net Pattern Recognition Equations with Self-organization for Phoneme Recognition

Sung-Ill Kim

Division of Electronic and Electrical Engineering, Kyungnam University,
449 Wolyoung-dong, Masan City, 631-701 Korea
kimstar@kyungnam.ac.kr

Abstract. In this paper, the neural net pattern recognition equations were attempted to apply to speech recognition. The proposed method features a dynamic process of self-organization that has been proved to be successful in recognizing a depth perception in stereoscopic vision. This study showed that the dynamic process was also useful in recognizing human speech. In the processing, input vocal signals are first compared with standard models to measure similarities that are then given to the dynamic process of self-organization. The competitive and cooperative processes are conducted among neighboring input similarities, so that only one winner neuron is finally detected. In a comparative study, it showed that the proposed method outperformed the conventional Hidden Markov Models(HMM) speech recognizer under the same conditions.

1 Introduction

Many researches on hidden Markov models(HMM)[1] or artificial neural networks (ANN)[2,3,4] have been conducted in the field of speech recognition. In the case of HMM, it is still difficult to give a satisfactory explanation of humanlike speech understanding since it is originally based on the probabilistic modeling concepts. As the alternative approach, therefore, ANNs have been developed by modeling the information processing mechanism of physiological human brain. However, there are still demerits of dealing with too many parameters in both training and recognition processes.

In stereoscopic vision, the depth perception phenomenon has been successfully simulated by using the recently modified algorithms[5,6] of neural networks. It is assumed that the human brain system has complicated neural networks fusing disparities between the different images by a self-organizing process of competition and cooperation. The neural networks have even simpler architecture than ordinary ANN. Moreover, the dynamic process of self-organization makes it identify the most likely neuron among confusable candidates, resulting in a clear depth perception of specific objects.

In a similar way, it is assumed that speech recognition is conducted by the self-organization of neurons handling similarities between input speech signals and memorized patterns in our brain. In its application to speech recognition, the input similarities are competed and cooperated among neurons through the dynamic process. As a result, the so-called winner-take-all process plucks only one winner neuron out of candidate ones.

J. Wang et al. (Eds.): ISNN 2006, LNCS 3972, pp. 172 – 177, 2006.

The proposed approach presents the new mathematical algorithms of stereoscopic vision neural networks, namely, coupled pattern recognition(CPR)[6,7] equations with self-organization mentioned above. Moreover, the new approach would be investigated how well the neural net equations work in identifying specific speech among confusable hypotheses. In addition, the comparative study with the existing HMM would be made under the same condition.

2 Neural Net Pattern Recognition Equations

In the course of the process based on the neural net pattern recognition equations, the activity of a neuron with the highest similarity excels others that would vanish. Such a winner-take-all system is given by CPR equations as following:

$$\xi_u^a(t) = -\frac{dU}{d\xi_u^a(t)} \tag{1}$$

where ξ_u^a (t) is the time-dependent neural activity in which the amplitude of ξ_u^a means the projection of test patterns.

$$U(\xi_u^a(t)) = \frac{\alpha}{2}\xi_u^a(t)^2 - \frac{E}{3}\xi_u^a(t)^3 + \frac{C}{4}\xi_u^a(t)^4 \tag{2}$$

where $U(\xi_u^a(t))$ is the non-vanishing part of U under the derivative of equation (1). In this equation, α_u^a is given by

$$\alpha_u^a(t) = -\lambda_u^a + (B+C)\sum_{\substack{a'=a-a_s \\ a'\neq a}}^{a+a_s} \xi_u^{a'}(t)^2 - D\sum_{\substack{u'=n-l \\ u'\neq u}}^{n+l} \xi_{u'}^a(t)^2 \tag{3}$$

where B,C,D are positive constants which are chosen appropriately. In equation (3), λ_u^a, which is the similarity at u-th temporal frame to certain vocal signal /a/, can be defined as

$$\lambda_u^a = \frac{\log N(o_u;\mu_a,\Sigma_a) - <\log N>}{<\log N>} \tag{4}$$

where N is the Gaussian probability density function with input data o_u, mean μ_a, and covariance Σ_a. <log N> means the average over temporal frames. As shown in this equation, the similarity between two features can be obtained after normalization. From the equation, it is noticed that λ_u^a is the deviation of similarity measure from its mean or standard value.

In equation (3), particularly, the first term is the similarity of input data at u-th temporal frame of arbitrary hypothesis /a/. The second term means the competitive coupling among neighboring neural activities of all candidates, while the third term represents the cooperative coupling among neighboring frames. The summation index of competitive coupling runs over the disparity search area defined as $a - a_s \le a' \le a + a_s$ with the restriction of $a' \ne a$. The summation index of cooperative coupling, on the other hand, runs over the cooperation area defined as $u - l \le u' \le u + l$ with the restriction of $u' \ne u$. Accordingly, it is noticed that $\alpha_u^a(t)$ depends on the neural activities of both $\xi_u^{a'}(t)$ and $\xi_{u'}^a(t)$ as well as the input similarity λ_u^a.

In actual applications, it is dealt with only two cases of potential functions in which ξ_u^a converges to certain absolute minimum value. Figure 1 shows the typical potential functions where a stable solution is decided in each case by the minima of the potential depending on the value of α_u^a.

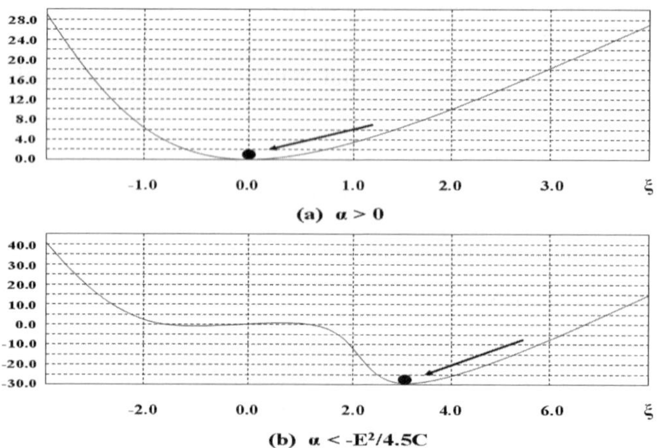

Fig. 1. Potential functions for loser neural activity(a) and winner neural activity(b)

In the case of $\alpha > 0$, ξ_u^a converges to a global minimum. In the case of $\alpha < -E^2/4.5C$, on the other hand, ξ_u^a saturates in certain positive value corresponding to absolute minimum in potential function. Therefore, it is noticed that only one neuron among candidates would win the competition through the dynamic process of self-organization based on the potential function.

3 Dynamic Process with Self-organization

The similarity map of all possible candidates is obtained as a result of similarity measure where each candidate phoneme, for example, has the similarity values in each frame. In the next step, the similarity map is given to CPR equations with self-organization in which the dynamic process with competitive and cooperative coupling is then performed among similarities.

α_u^a has the inhibitory coupling among neural activities in all candidates as well as the excitatory coupling among neighboring temporal frames. It is noticed in the above-mentioned process that the self-organization process exerts great influences on α_u^a as well as ξ_u^a. Figure 2 and 3 show the examples of time dependent behaviors of both α_u^a and ξ_u^a, respectively, at the fifth frame of each candidate phoneme.

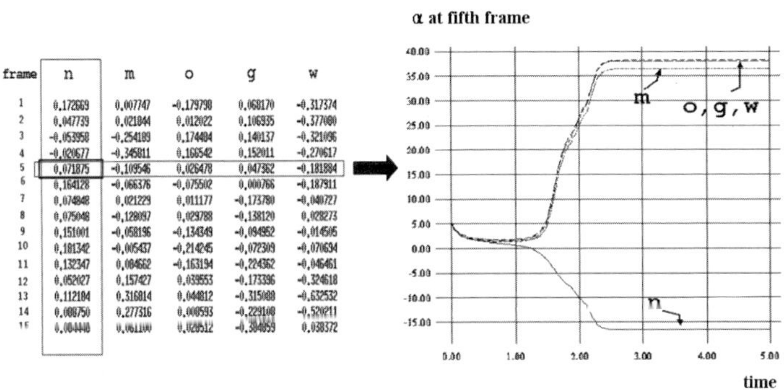

frame	n	m	o	g	w
1	0,172669	0,007747	-0,179798	0,068170	-0,317374
2	0,047739	0,021844	0,012022	0,106935	-0,377080
3	-0,053958	-0,254189	0,174484	0,140137	-0,321096
4	-0,020677	-0,345811	0,166542	0,152011	-0,270617
5	0,071875	-0,109546	0,026478	0,047362	-0,181884
6	0,164128	-0,066376	-0,075502	0,000766	-0,187911
7	0,074848	0,021229	0,011177	-0,173780	-0,040727
8	0,075048	-0,128097	0,029788	-0,138120	0,028273
9	0,151001	-0,058196	-0,134349	-0,094952	-0,014505
10	0,181342	-0,005437	-0,214245	-0,072309	-0,070634
11	0,132347	0,084662	-0,163194	-0,224362	-0,046461
12	0,052027	0,157427	0,039553	-0,173396	-0,324618
13	0,112184	0,316814	0,044812	-0,315088	-0,632532
14	0,088750	0,277316	0,008593	-0,229108	-0,520211
15	0,004440	0,061100	0,028512	-0,384859	0,038372

Fig. 2. Time dependent behaviors of α_u^a at the fifth frame of each candidate

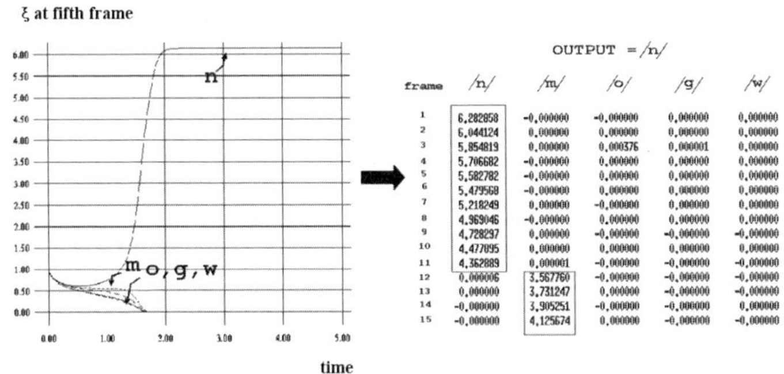

OUTPUT = /n/

frame	/n/	/m/	/o/	/g/	/w/
1	6,282858	-0,000000	-0,000000	0,000000	0,000000
2	6,044124	0,000000	0,000000	0,000000	0,000000
3	5,854819	0,000000	0,000376	0,000001	0,000000
4	5,706682	-0,000000	0,000000	0,000000	0,000000
5	5,582782	-0,000000	0,000000	0,000000	0,000000
6	5,479568	-0,000000	0,000000	0,000000	0,000000
7	5,218249	0,000000	-0,000000	0,000000	0,000000
8	4,969046	-0,000000	0,000000	0,000000	0,000000
9	4,728297	0,000000	-0,000000	-0,000000	-0,000000
10	4,477895	0,000000	0,000000	0,000000	0,000000
11	4,362889	0,000001	-0,000000	-0,000000	-0,000000
12	0,000006	3,567760	-0,000000	-0,000000	-0,000000
13	0,000000	3,731247	0,000000	-0,000000	-0,000000
14	-0,000000	3,905251	-0,000000	-0,000000	0,000000
15	-0,000000	4,125674	0,000000	-0,000000	-0,000000

Fig. 3. Time dependent behaviors of ξ_u^a at the fifth frame of each candidate

As a result, when ξ_u^a reaches certain saturated point through the cycles of recurrent networks, it is called a winner neuron, while it is called a loser neuron when it loses the whole activity to become close to 0. In this example, /n/ is recognized as the most likely candidate to input speech, as it has even more winner neurons than others.

4 Experimental Evaluation and Discussion

The phoneme recognition by the proposed method has been assessed on two kinds of test sets and compared with the conventional HMM recognizer under the same experimental condition. For the training of standard models, fist of all, each of recognition

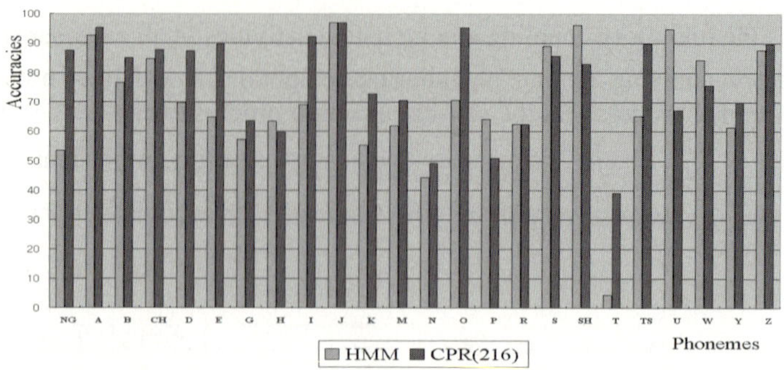

Fig. 4. Comparison of the proposed method(CPR) with HMM on 216 test sets

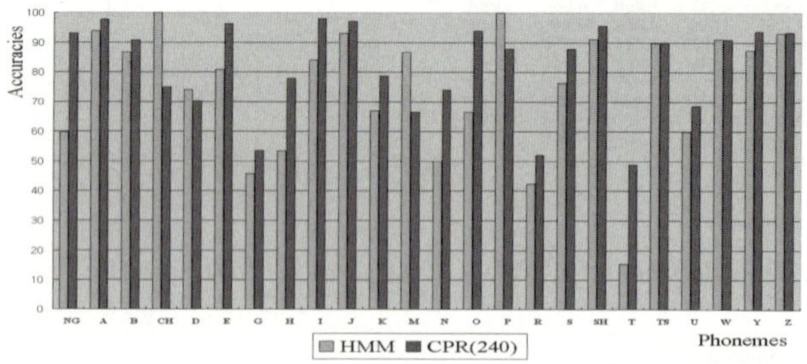

Fig. 5. Comparison of the proposed method(CPR) with HMM on 240 test sets

systems used two kinds of the phoneme-labeled training database. The labeled phonemes were extracted from ATR Japanese word speech database which was composed of 4000 words spoken by 10 male speakers, and from ASJ Japanese continuous speech database which was composed of 500 sentences by 6 male speakers.

For evaluation, test data consisted of two kinds, one from database of 216 words set and the other from 240 words set, each of which is the phoneme balanced ATR database spoken by 3 male speakers, respectively. For acoustic feature parameters, 10 dimensional Mel-frequency Cepstrum coefficients(MFCC) and their derivatives were used. Figure 4 and 5 show the average recognition accuracies in speaker independent experiments, in which the proposed method was compared with HMM on two kinds of test sets. The recognition accuracies using the proposed approach were 76.9% and 82.1% on 216 and 240 test sets, which were compared with 69.6% and 74.5% by HMM, respectively. Consequently, the performance of the proposed system was 7.5% higher than HMM in average. As a result, it was found that the proposed system outperformed the existing HMM recognizer. However, the accuracies of the system did not always show better performance than HMM in all phonemes. Since this study is restricted to phoneme recognition, therefore, we should conduct further experiments to word or continuous speech recognition as future works.

5 Conclusion

This study presented a new approach of speech recognition based on the neural net pattern recognition equations with self-organization. From the comparative study, it was shown that the proposed method outperformed the conventional HMM recognizer. Particularly, the dynamic process of self-organization makes it possible to enhance the discriminative capability in detecting the most likely candidate out of confused ones. Therefore, it is noticed that the visual cognitive mechanism, which has been used in the area of stereoscopic vision, might be also useful and compatible with the underlying principles of discriminating different sounds in the area of speech recognition.

Acknowledgement

This work is supported by the Kyungnam University Research Fund, 2005.

References

1. Woodland, P.C., Leggestter, C.J., Odell, J.J., et.al.: The 1994 HTK Large Vocabulary Speech Recognition System. Proc. IEEE Int. Con. on Acoustics, Speech, and Signal Processing 1 (1995) 73–76
2. Bourlard, H., Wellekens, C.J.: Links between Markov Models and Multi-layer Perceptrons. IEEE Tran. Patt. Anal. Machine Intell.12 (1990) 1167–1178
3. Lang, J., Waibel, A., Hinton, G.E.: A Time-Delay Neural Network Architecture for Isolated Word Recognition. Artificial Neural Networks, Paradigms, Applications and Hardware Implementations, IEEE press, New York (1992) 388–408
4. Martinelli, G.: Hidden Control Neural Network. IEEE Tran. on Circuits and Systems, Analog and Signal Processing 41(3) (1994) 245–247
5. Reinmann, D., Haken, H.: Stereo Vision by Self-organization. Biol. Cybern.71 (1994) 17–26
6. Kohonen, T.: Self-organizing map. Proc. IEEE, 78(9) (1990) 1464-1480
7. Yoshitomi, Y., Kanda, T., Kitazoe, T.: Neural Nets Pattern Recognition Equation for Stereo Vision. Trans. IPS (1998) 29–38

Music Genre Classification Using a Time-Delay Neural Network

Jae-Won Lee, Soo-Beom Park, and Sang-Kyoon Kim

Department of Computer Science, Inje University, Kimhae, 621-749, Korea
{jwviolet, park, skkim}@cs.inje.ac.kr

Abstract. A method is proposed for classifying music genre for audio retrieval systems using time-delay neural networks. The proposed classification method considers eight types of music genre: Blues, Country, Hard Core, Hard Rock, Jazz, R&B(Soul), Techno, and Trash Metal. The melody between bars in the music is used to distinguish the different genres. The melody pattern is extracted based on the sound of a snare drum, which is used to effectively represent the rhythm periodicity. Classification is based on a time-delay neural network that uses a Fourier transformed vector of the melody as an input pattern. This classification method was used to analyze 80 training data from ten different musical pieces for each genre and a further 40 test data from five additional musical pieces for each genre. The accuracy of the genre classifications that were obtained for the two sets of data was 92.5% and 60%, respectively.

1 Introduction

The practical use of multimedia data is increasing due to the rapid development of computer technology and the improved availability of the Internet. The vast amount of audio data that is available electronically as various multimedia data presents a considerable challenge for information retrieval. Content-based audio retrieval is an up-to-date overview of audio content analysis that can search for specific information from the content when users don't know the title of music. Using this method, users can acquire stored data from an audio database using descriptors as queries. However, since it is very inefficient to search for an entire tune among stored musical pieces, this method also requires that experts in the field extract descriptors from all of the music data in advance. These descriptors can then be used to investigate the audio database. Hence, there is a need to annotate music according to genre, atmosphere, tempo and so forth as descriptors in order to facilitate retrieval[1, 2, 3, 4]. Among the descriptors, music genres are categorical descriptors that are used to classify and describe music data. Traditionally, the classification of music genre has been determined by individuals. Research is needed to establish a means by which the increasing amounts of audio data that are becoming available on the Internet can be effectively searched. Recently, several different types of music genre have developed. A detailed list of the genres used by our classification method is provided in Chapter 2.

Today's well-known genres do not comply with any specific criteria, making it hard to classify music into particular genres. Even experts in the field have difficulty

J. Wang et al. (Eds.): ISNN 2006, LNCS 3972, pp. 178 – 187, 2006.

in classifying the music. The task of searching for audio data would be far more efficient if a method could be found to automatically classify music by genre.

Although there has been a considerable amount of research as regards matching audio data, little work has been done on determining the genre of music. Tzanetakis and Cook proposed a genre classification method that used 30 features from three feature sets: timbral texture, rhythmic content, and pitch content. They used a Gaussian classifier, a Gaussian mixture model, and a K-nearest neighbor classifier as a standard statistical pattern recognition classifier. They proved that using a large number of features is more effective than using only a few[5].

Preprocessing and feature extraction are essential processes in music genre classification. But effective feature extraction is very difficult, and the size and ambiguity of audio data make it difficult to develop comparative methods to investigate the extracted features. Tzanetakis and Cook used 30 seconds excerpts per musical piece for their datasets. If the data are extracted without considering the genre of the characteristic parts of the music, then the results could be adversely effected and could contain several extraneous and extreme values such as outliers. In addition, the results could be ambiguous when multi-genres such as Crossover are considered. A method is needed that can determine which extracts from a tune are representative of the whole. A means of extracting an area of interest from a tune and an audio matching algorithm are essential in order to develop an efficient audio retrieval system.

This paper proposes a music genre classification system that uses a time-delay neural network(TDNN) to compare the feature vector of a bar of musical data. Music consists of both melody and rhythm. Musical instruments express the rhythm by a bar. Rhythm has regularity and the drum or the bass are usually used as rhythm instruments. Therefore, we can extract bars as an area of interest in audio data by the regularity of the music's rhythm. Our method extracts bars based on the sound of a snare drum, as it has the most periodic occurrence and the most vivid timbre as compared to other drums.

A TDNN classifier based on the characteristics of a snare drum is used to extract bars. A TDNN is also used for music genre classification using the extracted bars. A TDNN is a neural network that incorporates a time-delay as a dynamic factor of multi-layered perception. Therefore, a TDNN can represent chronological relationships between events. A TDNN has a long training time but a short execution time. Consequently, it can be used effectively when retrieving data from very large audio datasets.

2 The Target Genres of Our Music Genre Classification Scheme

Recently, it has become popular to consider various types of music genre. New genres are continuously being developed while, at the same time, some genres fuse together. Usually, a new music genre is determined in accordance with some new music style that is currently in fashion. Sometimes, a part of the style of a genre disappears. However, these styles do not represent only styles of music. In a cultural context, each music genre connotes the intrinsic content or meaning of the lyrics, the musicians' behavior, and various other factors. That is, music genres may be created by a musician's ideas and by commercial intention. Recently, new genres have been created, such as Crossover genres that are a combination of more than two genres. A single album of a certain musician may include more than two genres. Therefore, because recent well-known genres have not been uniquely determined by any one particular set of criteria, it is hard to classify the music by the criteria of a single genre.

In previous studies on genre classification, Pachet and Cazaly used Internet music retailers' categories as target genres, and Tzanetakis and Cook proposed a total of 20 genres for a hierarchical genre classification[5, 6].

After due consideration of the properties of popular music described above, we selected eight generalized and fixed genres. We selected genres with pure musical properties, those that are not affected by the content of lyrics or cultural factors. The selected genres have uniform rhythm, a particular musical style and represent a fixed selection of instruments. The eight genres selected were: Blues, Country, Hard Core, Hard Rock, Jazz, R&B(Soul), Techno and Trash Metal. Experimental audio data were obtained from several pieces of music that uniquely represented each genre.

3 Time Delay Neural Network

The incorporation of a time-delay allows a TDNN to achieve a high degree of recognition with dynamic data, such as audio data. Thus, TDNNs have been widely used in the field of speech recognition[7]. A TDNN generally consists of an input layer, two hidden layers and an output layer. The hidden layers may detect local features of the data by using a time-delay unit. The output layer has an output value that is added to the square of the output value of the previous layer.

Fig. 1 shows a schematic diagram of a TDNN. The step size of Fig. 2 is the unit of time-delay. The kernel size is the number of nodes on the time axis when the present layer's nodes are combined and transferred to an upper-layer node.

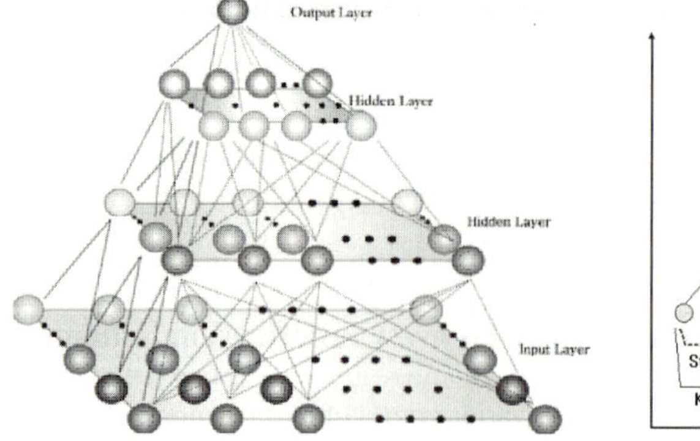

Fig. 1. Schematic diagram of a TDNN

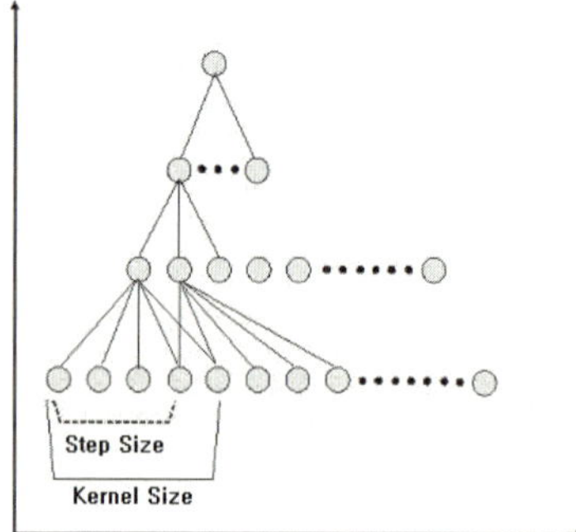

Fig. 2. Time axis section of a TDNN

4 Music Genre Classification System

Fig. 3 shows the main structure of our proposed classification system. A preprocessing step transforms stereo sound into mono. A fast Fourier transform(FFT) step determines the features of extracted bars, and then incorporates these features into a training pattern. The next step uses a TDNN classifier to extract bars. The classifier is trained with a training pattern that is based on the sound of a snare drum to extract a bar by detecting the beginning of a bar. Data normalization down-samples the size of the extracted bars and transforms the data into the same size. In the last step, the composition of a TDNN classifier for genre classification, the classifier is trained using the training pattern of the extracted bars.

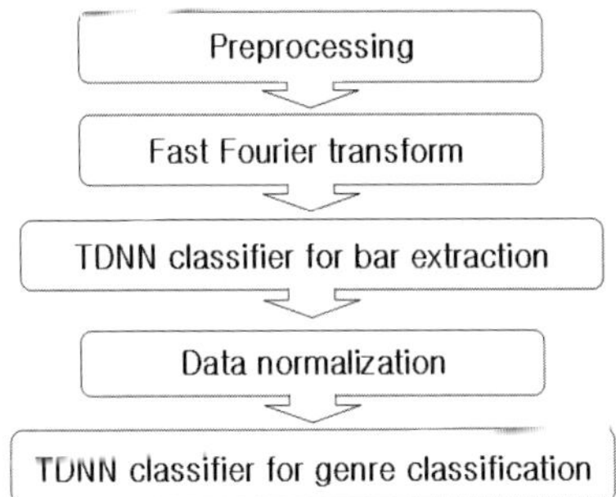

Fig. 3. Steps of the music genre classification system

4.1 Preprocessing

A wave file of CD-quality two channel stereo audio is generally sampled at a 44,100-Hz rate using 16-bit sampling. A wave file with a four-minute running time has a file size of approximately 40 Mbytes. To reduce the time required to process this large amount of data, a preprocessing step transforms stereo sound into mono.

4.2 Fast Fourier Transform through Considering Audio Data Properties

All music has a tempo and a beat. Because music has different tempos and beats, it is not reasonable to always extract the same size sample from the same position in the audio data. Our method searches for the beginning of bars and extracts the size of a bar from the music, which is then used as the size for the extraction feature. Most of our experimental data had four-quarters time or six-eighths time and had a regular tempo. Most of the audio data regularly consisted of a part for melody and a part for rhythm. Our method detected the start point of a bar by using the characteristics of rhythm regularity.

Rhythm is played by a drum, bass guitar, or similar instrument. For example, jazz uses a double bass instead of a bass guitar and other percussion instruments instead of drums. However, the music played with these different instruments sounds very similar. Our method focuses on the sound of the snare drum to detect the beginning of a bar. A snare drum is usually struck every second or fourth beat and has a uniform frequency.

In general, music has an introduction to establish the atmosphere desired by the composer. This introduction often does not have the properties of any specific music genre. Because the introduction can have an unrestricted tempo, it must be eliminated when selecting a sample bar. In tests, we found that all of the experimental data had an introduction that lasted less than 40 seconds. Therefore, the first 40 seconds of the wave file were not considered. The size of a bar is generally less than 220,000 samplings, so our method extracts 220,500 samplings, representing the five seconds following the musical introduction. Fig. 4 shows a sample of identical size.

To search for the sound of a snare drum, signals on the time axis are transformed into frequencies. A 256-point FFT is performed. Due to the symmetry of an FFT, only the first 128 frequency values are used. When the transformed data were analyzed, it was determined that the snare drum sound had frequency values from the 37th to the 43rd band inclusive. Fig. 5 shows an example of the 37th frequency band of the experimental data. The highest apexes represent the snare drum as part of the bar.

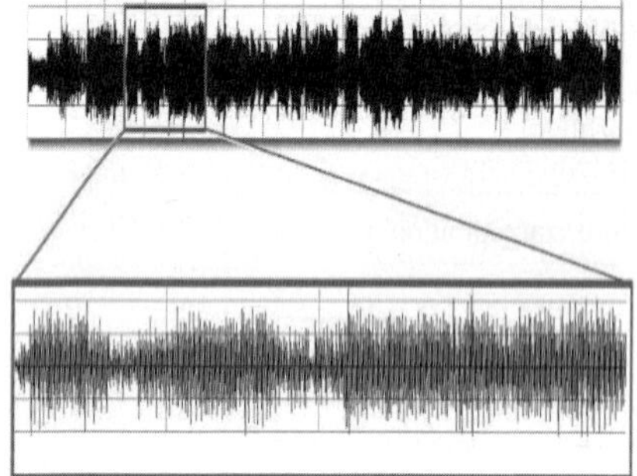

Fig. 4. A sample of identical size

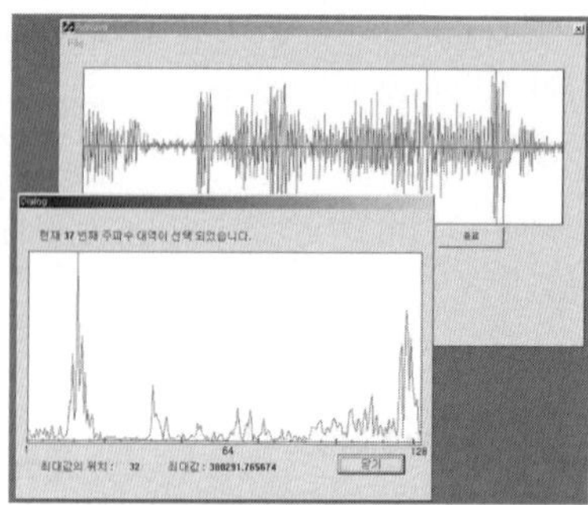

Fig. 5. The 37th frequency band

4.3 TDNN Classifier for Extracting a Bar

The Creation of a Training Pattern. In order to derive a training pattern that can be used as an input to the TDNN for extraction of a bar, a 256-point FFT is performed 16 times on 4096 data samples. This process yields 16×256 transformed values. From these values, our method extracts the seven values between the 37th and the 43rd frequency bands. The resulting 16×7 patterns are calculated by repeating the extraction of these seven values 16 times. To use the extracted pattern as a training pattern for our neural network, the extracted pattern values are normalized between 0 and 1.

TDNN Classifier Configuration. The proposed TDNN classifier for extracting a bar consists of an input layer, two hidden layers and an output layer. The input layer has 16×7 nodes, and each hidden layer consists of 7×4 and 4×2 nodes, respectively. The output layer has one node.

The kernel and step size of each layer are experimentally determined. Usually, kernel and step sizes of TDNN are made by the intuition of researcher. We coordinate according to increasing each size from a minimum size to the suitable size. The kernel sizes and step sizes of every layer are determined equally.

On the minimum spatial size of input layer, classifier misunderstands whenever music has a little similar frequency to snare drums. So, the spatial size of input layer is chosen as 7, and this is each value of 7 frequency domains. Also, its temporal size

Table 1. TDNN configuration for extracting a bar

	Spatial size	Kernel size	Step size	Nodes
Input layer	7	0	0	16×7
Hidden layer 1	4	4	2	7×4
Hidden layer 2	2	4	1	4×2
Output layer	1	4	0	1×1

is determined as 16. Next, the spatial size of hidden layer 1 was determined as four, the half of spatial size of input layer[8]. Similarly, the spatial size of hidden layer 2 is two as the half of spatial size of hidden layer 1. Table 1 shows the configuration of the layers of the TDNN classifier for extracting a bar in detail.

4.4 Data Normalization

The data that are acquired through the TDNN use bars of different sizes because of differences in the tempo of each musical sample. Therefore, the number of features extracted differs. In order to extract features of the same size from all the data, a normalizing process is required. Down-sampling is used to transform the extracted bar to a normalized format with 16,384 samplings.

4.5 The TDNN Classifier for Genre Classification

Creation of a Training Pattern. To compose a training pattern for use as an input to the TDNN classifier for genre classification, a 256-point FFT is performed 64 times on the normalized data. This process results in 64×256 transformed values. Only half of the 256 data points are used because of the symmetry of FFT. From these 128 data, we extract the 32 Mel-frequency cepstral coefficients (MFCCs), a set of perceptually motivated features that have been widely used in speech recognition[10]. Repeating the 64 frames for each of the 32 MFCCs results in 64×32 patterns. Finally, the extracted pattern values are normalized between 0 and 1.

Configuration of a TDNN Classifier for Genre Classification. The proposed TDNN classifier for genre classification consists of the same layers as those of the TDNN classifier for extracting a bar. Its input layer has 64×32 nodes, and each hidden layer consists of 31×16 and 10×8 nodes, respectively. The output layer has three nodes. Table 2 shows the configuration of the layers of the TDNN classifier for genre classification in detail.

Table 2. Configuration of a TDNN for Genre Classification

	Spatial size	Kernel size	Step size	Nodes
Input layer	32	0	0	64×32
Hidden layer 1	16	4	2	31×16
Hidden layer 2	8	4	1	10×8
Output layer	1	4	3	3×1

5 Experimental Results

Our music genre classification system was implemented using Visual C++ on a Pentium IV PC and Microsoft Windows operating system.

5.1 Experimental Results and Analysis of the TDNN Classifier for Extracting a Bar

On experiments of TDNN classifier for extracting a bar, training rate is 0.05, momentum is 0.5, target error value is 0.01, and the iteration limit is 40,000. To analyze the performance of the trained classifier, we experimented with training data and test data from a total of 120 musical pieces. The training data consisted of 80 data samples from ten musical pieces for each genre. The test data were 40 data samples from 5 musical pieces for each genre.

Fig. 6 shows the extraction of a bar after detecting candidates for the beginning of the bar. First, it extracts 220,500 samplings after the first 40 seconds of audio data. To reduce the error, the input patterns are overlapped by 5 samplings. Because the size of the input pattern is 4096, the number of input patterns per music piece is 34,100. Therefore, the bar extractor inputs nearly 68 patterns every 0.01 seconds. The bar extractor detects a candidate for a snare drum sound from the input patterns. Within the same time criteria, it determines the 500 candidates with the highest probability. The tolerance of time is 0.01 seconds. Time values are considered to be the same if they are identical down to two places of the values. In Fig. 6, the horizontal axis of the graph is time and the vertical axis is the number of candidates. Each division on the horizontal axis represents 0.1 seconds. If the time interval of the detected candidates is less than 0.05 seconds, the bar extractor chooses these candidates as the snare drum sound. Next, it chooses three points that are found to be the snare drum sound and extracts a bar according to both of the ends that were determined by the sets of 3 points. Fig. 6 shows the continuous and regular occurrence of a snare drum.

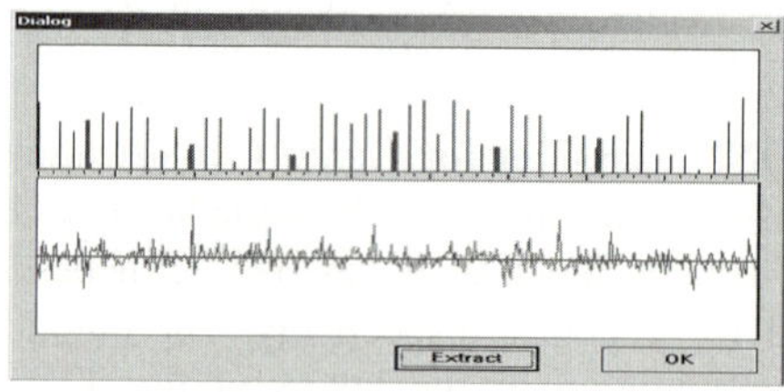

Fig. 6. Candidates for the beginning of a bar

According to our experimental results, the training data showed a 93.75% successful classification rate and the test data showed a 76.25% successful classification rate. The incorrect classification results arose from a shortage of training data for the various patterns of other harmony instruments used with a snare drum. Training the classifier for the various patterns of harmonizing instruments would improve the classification rate, as the training data usually included other instrumental sounds as well as the snare drum sound.

5.2 Experimental Results and Analysis of the TDNN Classifier for Genre Classification

The genre classifier has the same training rate, momentum, target error value, and iteration limit as the bar extractor. The training data consisted of a total of 80 data samples that were made up of ten data samples for each of the eight genres. The test data had a total of 40 data samples that were made up of five samples for each of the eight genres. Table 3 shows the classification rate of the experiment. The correct classification rates for the training and test data were 92.5% and 60%, respectively.

Table 3. Experimental results

Music genre	Training	Test	Music genre	Training	Test
Blues	9/10	4/5	Jazz	10/10	4/5
Country	9/10	2/5	R&B(soul)	9/10	2/5
Hard Core	8/10	2/5	Techno	10/10	4/5
Hard Rock	10/10	4/5	Trash Metal	9/10	2/5

Among the incorrect classification results, Hard Core and Trash Metal genres had the highest incorrect classification rates. The Hard Core genre was often incorrectly classified as Trash Metal, and Trash Metal was incorrectly classified as Hard Core or Hard Rock. These results indicate that these three genres use similar instruments and have similar rhythm, which is reasonable as Hard Core and Trash Metal are both derived from Hard-Rock. Also, by the same reasoning, the genres of Blues, Jazz and R&B have similar results. Even for a person, it is not easy to classify these genres.

The reason for incorrect classification of the other genres was that the test data had a remarkably different tempo from that of the training data. As differences in tempo affect bar extraction, this resulted in errors when creating the training pattern for genre classification. If the method of extracting a bar could be adjusted to allow flexibility for various tempos within each genre and the TDNN classifier was then trained with these patterns, the classification rates would improve. Table 4 is the confusion matrices for each training and test data.

Table 4. Confusion matrix of training and test data

	Training data									Test data							
	Bu	Co	HC	HR	Ja	RB	Te	TM		Bu	Co	HC	HR	Ja	RB	Te	TM
Bu	9								Bu	9							
Co		9							Co		9						
HC			8						HC			8					
HR		1	1	10		1		1	HR		1	1	10		1		1
Ja	1				10				Ja	1				10			
RB						9			RB						9		
Te			1				10		Te			1				10	
TM								9	TM								9

Table 5. Incorrect genre of training data

Genre	Title of music	Incorrect
Blues	Albert King "Answer to the Laundromat blues"	Jazz
Country	Garth Brooks "American honky-tonk bar association"	Hard Rock
Hard Core	Biohazard "Man with a promise"	Techno
	Nine Inch Nails "Mr. self destruct"	Hard Rock
Hard Rock	.	.
Jazz	.	.
R&B(soul)	Maxwell "Sumthin' sumthin'"	Hard Rock
Techno	.	.
Trash Metal	Sepultura "Sympthom of the universe"	Hard Rock

Table 6. Incorrect genre of test data

Genre	Title of music	Incorrect
Blues	BBKing "I'm gonna do what they do to me"	R&B(soul)
	Bob Dylan "Tangled up in blue"	Hard Rock
Country	Garth Brooks "Cowboy Cadillac"	Hard Rock
	John Denver "Rocky mountain high"	Techno
	Korn "Got the life"	Trash Metal
Hard Core	Marilyn Manson "School drop-outs"	Trash Metal
	Rage Against The Machine "Born of a broken man"	Techno
Hard Rock	Deep Purple "Mistreated"	Hard Core
Jazz	Duke Ellington "Little max"	R&B(soul)
	D' Angelo "Send it on"	Techno
R&B(soul)	D' Angelo "The root"	Hard Rock
	Maxwell "The urban theme"	Jazz
Techno	Moby "Have you seen my baby"	R&B(soul)
	Anthrax "In my world"	Hard Rock
Trash Metal	Metallica "Creeping death"	Hard Rock
	Slayer "Reborn"	Hard Core

6 Conclusions

We have proposed an audio segmentation method using bar extraction based on the sound of a snare drum and an audio classification method that uses a TDNN genre classifier. Although the performance of our proposed system was not altogether satisfactory, it was comparable to the ability of a person to classify musical genres. The system could be further improved by more effective feature extraction for the classification of similar genres and a more accurate version of the bar extractor to allow for various tempos.

Future work could also include the development of an extraction method for the climax of a musical piece. In addition, an effective querying and matching method for audio retrieval systems could be developed.

Acknowledgement

This work was supported by grant No. 2001-1-51200-007-1 from the Basic Research Program of the Korea Science & Engineering Foundation.

References

1. Wold, E., Blum, T., Keislar, D., Wheaton, J.: Content-based Classification, Search, and Retrieval of Audio. IEEE Multimedia 3(3) (1996) 27–36
2. Kosugi, N., Nishihara, Y., Kon'ya, S., Yamamuro, M., Kushima, K.: Music by Humming Using Similarity Retrieval over High Dimensional Feature Vector Space. IEEE Pacific Rim Conference on Communications, Computer and Signal Processing (1999) 404-407
3. Blackburn, S., DeRoure, D.: A Tool for Content-based Navigation of Music. Proceeding of the ACM Multimedia (1998) 361-368
4. Liu, C.C., Hsu, J.L., Chen, A.L.P.: An Approximate String Matching Algorithm for Content-based Music Data Retrieval. IEEE International Conference on Multimedia Computing and Systems, 1 (1999) 451-456
5. Tzanetakis, G., Cook, P.: Musical Genre Classification of Audio Signals. IEEE Transactions on Speech and Audio Processing 10(5) (2002) 293–302
6. Pachet, F., Roy, P., Cazaly, D.: A Combinatorial Approach to Content-based Music Selection. IEEE International Conference on Multimedia Computing and Systems, 1 (1999) 457–462
7. Waible, A., Hanazawa, T., Hinton, G., Shikano, K., Lang, K.J.: Phoneme Recognition Using Time-delay Neural Networks. IEEE Transactions on Acoustics, Speech, and Signal Processing 37 (1989) 328-339
8. Logan, B.: Mel Frequency Cepstral Coefficients for Music Modeling. International Symposium on Music Information Retrieval (2000)

Audio Signal Classification Using Support Vector Machines

Lei-Ting Chen[1], Ming-Jen Wang[2], Chia-Jiu Wang[2], and Heng-Ming Tai[3]

[1] School of Computer Science and Engineering,
University of Electronic Science and Technology of China, Chengdu, China
richardchen@uestc.edu.cn
[2] Department of Electrical and Computer Engineering, University of Colorado at Colorado, Springs, CO, USA
cwang@eas.uccs.edu
[3]Department of Electrical Engineering, University of Tulsa, Tulsa, OK, USA

Abstract. As the internet community grows larger, digital music distribution becomes widely available and is made easier than ever. Artists from all over the world can make their songs available by a single click. Websites, containing varieties of music style for download, charge only a fraction of the cost of a CD for the service. With the incredible amount of music pieces available, it is impossible to classify each piece by its style manually. A procedure is proposed using the support vector statistical learning algorithm to achieve the task autonomously. Digital music files are converted, partitioned and processed to obtain the desirable input vectors for the algorithm. As the machine learns the features of each music genre, it is capable of classifying input vectors from unknown pieces. A simulation was carried out to evaluate the efficiency of the algorithm. Results from the simulation are presented and discussed in this paper. Conclusions are drawn by comparing other algorithms against the proposed method.

1 SVM Introduction

The Support Vector Machine (SVM) is a statistical learning algorithm that can perform binary classification or real valued function approximations. The subject was brought up in the late seventies by V. Vapnik, but only receiving attentions in the recent years [1 - 4]. The algorithm has been applied to various area of interest such as hand writing recognition, voice recognition and many others. This paper will focus on the audio signal i.e., music classification. SVM will be used to classify different types of music genres such as Classical, Jazz and Rock. There are several reasons to build a music classification machine for both commercial and scientific purposes. Many music websites allow the users to search through their database based on the type of music that users prefer. For this particular purpose and the sheer volume of new pieces of music submitted everyday, the websites might want to have the ability to expeditiously classify a piece of new music so the entry can be added to the database correctly without human intervention. It is also interesting to see how humans can tell different genres of music by listening to a couple of seconds of a clip, yet there are still no reliable scientific theories on how this is done by the human brain.

J. Wang et al. (Eds.): ISNN 2006, LNCS 3972, pp. 188 – 193, 2006.

2 Using SVM for Classification Purposes

There have been many attempts to use learning algorithms to classify music styles. One of them is the Hidden Markov Models (HMM) [5] used to classify folk songs. HMM approach was able to achieve 75%, 77% and 66% for 2-way classification and 63% for 3-way classification using 6-state left-right HMM with the interval representation in the experiment. To rigorously examine the algorithm, one would have to collect way more than a handful of albums to generate both training and testing samples. In our work, we use the Matlab to convert the audio data into input vectors for SVM training and testing. Figure 1 shows the data preprocessing flow chart to produce input vectors for support vector machines.

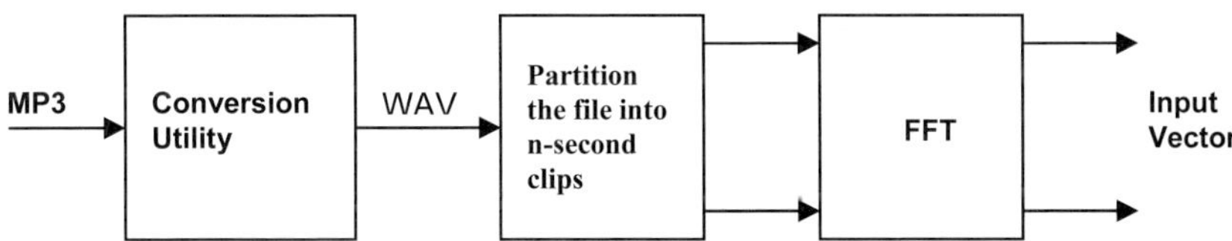

Fig. 1. Preprocessing Flow Chart

The most popular audio file format nowadays is the MP3 (MPEG Layer 3). This file format features high compression ratio yet with reasonably high audio quality. Music files from three categories will be used for the simulation. The categories aree Classical, Jazz and Rock. Several albums of each category were used to generate samples for the simulation.

2.1 Feature Extraction

How do we convert the MP3 files to sets of usable input vectors that SVM software can process? A lot of the preprocessing is done in Matlab. Since Matlab does not have a MP3 interface, these files will have to be converted to wav format. This is done by using any of the shareware [6] that converts MP3 to wav or vice versa. Once in wav format, we can see in Matlab that most files are in stereo format, which means the wav file read into the workspace is an M×2 matrix. As for our purpose, it would be easier to consolidate the matrix to the size of M×1. This can easily be done by adding

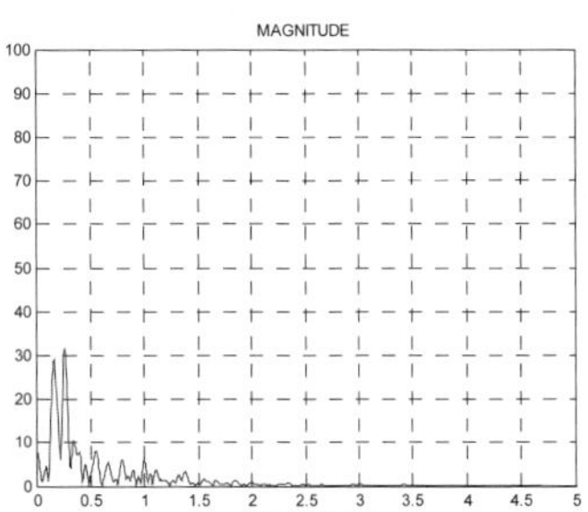

Fig. 2. Frequency spectrum of a Jazz

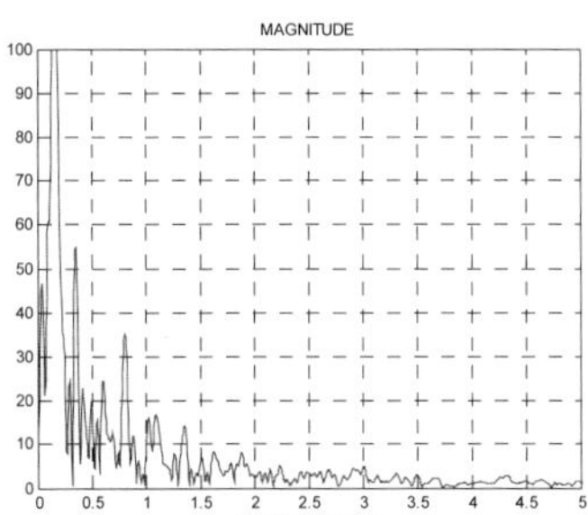

Fig. 3. Frequency spectrum of a Rock

each channel and average it out. Pauses and applause in the recordings may skew the sample accuracy. To avoid possible outliers, sections of the recording ranges from 5 to 10 seconds are removed from the front and the rear of the record. At this point, the raw data in the matrix is ready to be converted to feature set inputs. The feature used is the frequency spectrum of the music. Fast Fourier Transform (FFT) is used in obtaining the frequency spectrum. Figure 2 and Figure 3 show the frequency spectrum of a Jazz and a Rock sample respectively. The size of the FFT calculation is 1024 points.

2.2 Evaluation

With all preprocessing taken place, the samples are ready to be used to train Support Vector Machines. Samples from several unclassified songs are kept away separately from the pool. They are to be used as testing subjects later on. For the training file, one thousand samples were taken out of each pool of the respective category. Each sample is a vector of size 1×1024. Sample vectors are arranged into different combination of training files: Classical vs. Jazz, Jazz vs. Rock and Rock vs. Classical.

The training files are arranged such that both the false categories are in the same class (i.e., Classical "1" vs. Jazz and Rock "-1") in order to reduce the number of discrete machines needed to be trained to classify n categories. Unfortunately for the particular chosen feature set (frequency spectrum), simulations have shown that this method produces less than satisfying results. The reason behind this is that even though two categories of input may possess distinct spectrum, adding more categories will introduce more peaks at other frequencies (which may overlap the frequencies of the first two categories) and making it more difficult to differentiate the samples.

It is also important to note that sometimes certain combination of categories produces better contrast against each other and therefore generates better results in the simulation. Table 1 shows the results when grouping different categories together. Three SVMs called CvJ, JvR and RvC, are used to generate this table. The letters in the name can be interchanged (i.e., CvJ is the same machine as JvC). Each machine is trained by a 1v1 file consisting of two thousand samples, one thousand from each genre and each sample is a 1×1024 vector. Two hundred samples are then taken out of each experimental pool for testing. When the Classical samples are fed into the CvJ and RvC machines, about 97% of the samples are classified correctly. When Jazz samples are fed into the CvJ and JvR machines, the result is about 80%. And finally when Rock samples are used in RvC and JvR machines, the result diverges quite a bit. These numbers are obtained by doing three iterations of different sets of two hundred testing samples from the experimental pool. The results from all three runs are then averaged out.

The number of SVM needed in this case is dictated by the fact that 1v1 training produces better result then 1v2. To generalize the problem, for N categories, we can

Table 1. Sample Accuracy Comparison

	Classical		Jazz		Rock	
SVM	***CvJ***	RvC	***CvJ***	JvR	***RvC***	JvR
Accuracy (%)	***98***	97	***80.5***	79.5	***95***	48

use the N-choose-2 expression C(N, 2) to calculate how many SVMs that have to be trained. For our case three SVMs are sufficient. As the number of category increases, we can see the merit of the suggestion mentioned above to keep the number of SVM as small as possible.

Table 2. Confusion Matrix for 600 samples

600 Samples	Classical	Jazz	Rock
Classical	196	41	10
Jazz	4	159	0
Roc	0	0	190
Accuracy	98%	79.5%	95%

Table 2 shows the classification results for 600 individual samples vectors. Though the result seems quite good in one class, it may still be mediocre at another. The SVM used are CvJ, JvR and RvC. As we shall show that using sample-set based paradigm, instead of individual samples produces better overall result.

For an unknown piece of song, there could be thousands of sample vectors. When feeding the information to the machine, it most of the time will not be able to positively classify all samples. Figure 4 shows the flow chart on how to use the number produced by each machine to best determine the classification. As with all the statistical learning machines, the more it learns, the more accurate the results are.

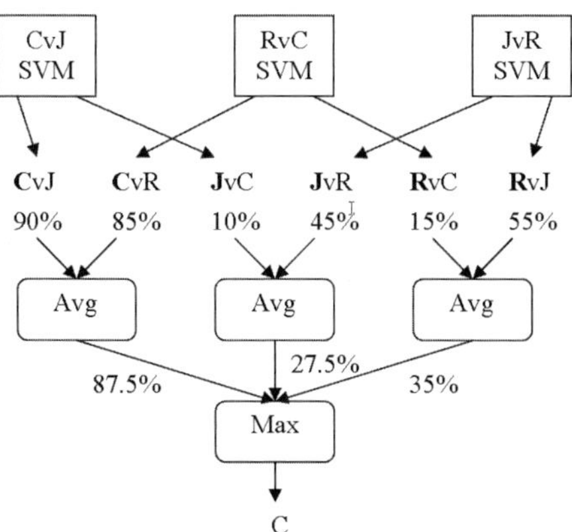

Fig. 4. An example of classification flow chart

In another experiment, we tried to evaluate sample-sets instead of individual samples. Let 1 sample-set = 100 samples. Each sample in the sample-set would be assigned a value by each of the SVMs. Each of the assigned value would then be evaluated by sign(x) = {1: x > 0; 0: x = 0; -1: x < 0}. After the sign(x) evaluation, it can be determined how many percent of the samples in the sample-set belong to each class. Since each sample-set is evaluated by three SVMs, six numbers would be generated to represent the percentage for each class, two percentage per class. Table 3 shows the confusion matrix for 300 sample-sets and each set has 100 samples. Table 4 shows a small section of the sample-set evaluation results.

With over 40000 individual samples from about 20 MP3s, 30000 samples are used to construct 300 sample-sets using the method described above, 100 sample-set from Classical genre, 100 from Jazz and 100 from Rock. The results are presented in Table 3.

Table 3. The confusion matrix for 300 samples

300 Sample-set	Classical	Jazz	Rock
Classical	99	0	0
Jazz	1	96	6
Rock	0	4	94
Accuracy	99%	96%	94%

Table 4. Evaluation results of some sample-sets

Set	Classical		Jazz		Rock		Classi.	Jazz	Rock	
	CvJ	CvR	JvC	JvR	RvC	RvJ	Ave.	Ave.	Ave.	Max.
1	97	100	3	100	0	0	98.5	51.5	0	C
2	96	100	4	100	0	0	98	52	0	C
3	99	100	1	100	0	0	99.5	50.5	0	C
4	99	100	1	100	0	0	99.5	50.5	0	C
5	89	100	11	100	0	0	94.5	55.5	0	C
6	91	100	9	100	0	0	95.5	54.5	0	C
7	87	100	13	100	0	0	93.5	56.5	0	C
8	96	100	4	100	0	0	98	52	0	C
9	83	100	17	100	0	0	91.5	58.5	0	C
10	90	100	10	100	0	0	95	55	0	C
11	91	100	9	100	0	0	95.5	54.5	0	C
12	92	100	8	99	0	1	96	53.5	0.5	C
13	77	100	23	100	0	0	88.5	61.5	0	C

Comparison is made between different classification methods used in other papers. Gaussian classifiers and Neural Networks are both discussed in paper [7] and [8] respectively. By combining multiple features together, the Neural Network method achieves results close to SVM with sample-set method. Table 5 shows the results achieved by each method. Considering that this simulation only utilizes one feature of the samples, the produced result is relatively decent. On the other hand, the good results obtained by this simulation could very well be biased by the album selected. Since the songs chosen from an album are all from the same artist, the same types of musical instruments are used. That might be why the spectral of the samples from these three genres are so very distinctive.

Table 5. Results comparison between different methods

Accuracy	Classical	Jazz	Rock
Gaussian Classifier [7]	86%	38%	49%
Neural Network [8]	97%	n/a	93%
SVM (individual sample)	98%	79.5%	95%
SVM (sample-set)	99%	96%	94%

3 Discussions

As shown in Table 5, this simulation results indicate that Support Vector Machine is a viable algorithm for music classification. There are many good algorithms such as Neural Networks, Hidden Markov Model and Gaussian Classifier that can do just as well or even better. But what makes SVM attractive is the simplicity of the algorithm and minimal sacrifice of the result. The classification accuracy of SVM is quite similar as the Neural Networks method in [8], a lot of improvements can still be done. Keep in mind that all the results that we've obtained so far is all based on one feature. There are many other features in audio spectrum that can used to classify a piece of music such as Mel-Frequency Ceptral Coefficient, Linear Predictive Coding Taps or Melodic Shape [9]. If incorporated, the result could be even much better.

References

1. Vapnik, V.: Statistical Learning Theory. Wiley, New York (1998)
2. 2.Burges, C.: A Tutorial on Support Vector Machines for PatternRecognition. http://citeseer.ist.psu.edu/burges98tutorial.html
3. Moore, A.: Support Vector Machines Tutorial. Carnegie Mellon University.http://www.cs.cmu.edu/~awm/tutorials
4. Cover, T.M.: Geometrical and Statistical Properties of Systems of Linear Inequality with Applications in Pattern Recognition. IEEE Trans. Electronic Computers 14 (1965) 326-334.
5. Chai, W., Vercoe, B.: Folk Music Classification Using Hidden Markov Models. Technical Report, MIT. http://web.media.mit.edu/~chaiwei/papers/chai_ICAI183.pdf
6. Easy MP3 Wav Converter. http://www.audiotool.net/
7. Tzanetakis, G., Essl, G., Cook, P.: Automatic Musical Genre Classification of Audio Signals. Computer Science and Music Department, Princeton University. http://ismir2001.ismir.net/pdf/tzanetakis.pdf
8. Scott, P.: Music Classification Using Neural Networks. http://www.stanford.edu/class/ee373a/ musicclassification.pdf
9. Deshpande, H., Singh, R., Nam, U.: Classification of Music Signals in the Visual Domain. Proceedings of the COST G-6 Conference on Digital Audio Effects (Page DAFX-01). Limerick, Ireland, December 6-8, 2001.

Gender Classification Based on Boosting Local Binary Pattern

Ning Sun[1,2], Wenming Zheng[2], Changyin Sun[3], Cairong Zou[2], and Li Zhao[1,2]

[1] Research Center of Learning Science, Southeast University, Nanjing 210096, China
sunning@seu.edu.cn
[2] Department of Radio Engineering, Southeast University, Nanjing 210096, China
[3] College of Electrical Engineering, Hohai University, Nanjing, Jiangsu, 210098, China

Abstract. This paper presents a novel approach for gender classification by boosting local binary pattern-based classifiers. The face area is scanned with scalable small windows from which Local Binary Pattern (LBP) histograms are obtained to effectively express the local feature of a face image. The Chi square distance between corresponding Local Binary Pattern histograms of sample image and template is used to construct weak classifiers pool. Adaboost algorithm is applied to build the final strong classifiers by selecting and combining the most useful weak classifiers. In addition, two experiments are made for classifying gender based on local binary pattern. The male and female images set are collected from FERET databases. In the first experiment, the features are extracted by LBP histograms from fixed sub windows. The second experiment is tested on our boosting LBP based method. Finally, the results of two experiments show that the features extracted by LBP operator are discriminative for gender classification and our proposed approach achieves better performance of classification than several others methods.

1 Introduction

Face is one of the most important biometric features of human. We can acquire much information of people naturally from observing their faces, such as identity, sexy, age or expression and so on. As a result, many researches achieved greatly remarkable success in the field of biometric person authentication such as face detection, face recognition, gesture recognition, in which the classification of gender is perhaps the most fundamental estimation problem. The earliest attempt to use computer vision techniques for gender classification was based on neural networks. Gollomb et al[1] trained a fully connected two-layer neural network, named SEXNET, to identify gender from 30×30 face images. Brunelli et al [2] used HyperBF networks to recognize male and female, in which two competing RBF networks are trained using several geometric features as inputs. Moghaddam et al [3] investigated to apply the Support Vector Machine (SVM) to classify gender with low-resolution "thumbnail" faces. And Wu et al [4] introduced an automatic real-time gender classification system based on Adaboost, in which the LUT–type weak classifiers are trained by the Simple Direct Appearance Model (SDAM) method.

J. Wang et al. (Eds.): ISNN 2006, LNCS 3972, pp. 194 – 201, 2006.

Local Binary Pattern is a powerful operator for texture description, proposed by Ojala[5] originally, which is defined as a grayscale invariant texture measure, derived from a general definition of texture in a local neighborhood. The LBP method has already been used in a large number of applications, including texture classification, image retrieval, face image analysis, and so on. Timo et al[6] presented a novel approach for face recognition, which takes advantage of the LBP histogram. In their method, the face area is equally divided into several sub windows from which the LBP features are extracted and concatenated to represent the local texture and global shape of face images. Recently, Li et al [7] proposed a systematic framework for fusing 2D and 3D information at both feature and decision levels. They used LBP as the representation of faces in 2D and 3D images, and applied the AdaBoost to selecting effective feature from a 2D+3D feature pool. In addition, the LBP operator is also used in the application of facial expression recognition [8, 9].

In this work, we present a novel approach for gender classification by boosting local binary pattern-based classifiers. Firstly, the training sample is scanned with the scalable sub windows, which is moved horizontally and vertically around the images. The histograms of Local Binary Pattern are extracted from the small windows to describe the local features. And the LBP histograms of each small window are averaged to generate a histogram template for the class of male or female. The Chi Square distance between histogram of samples and templates is computed as the features for discrimination. By mean of the Adaboost algorithm, the most useful features are selected, and the strong classifier is obtained in the form of linear combination of LBP feature based weak classifiers.

The rest of this paper is organized as follow: In Section 2, the LBP representation is introduced briefly. Section 3 describes AdaBoost algorithm for feature selection and strong classifier construction in detail. And two experiment results of gender classification based on LBP operator are given in the Section 4. At last, we make the conclusion in section 5.

2 Local Binary Pattern

The original version of the LBP operator labels the pixels of an image by thresholding the 3×3–neighbourhood of each pixel with the center value and considers the result as a binary number. Then the histogram of the labels is used as a texture descriptor. Just as described in Fig.1(a).

The major limitation of the original LBP operator is that it can not obtain dominant features with large scale structures. Thereby the primary LBP operator is extended to

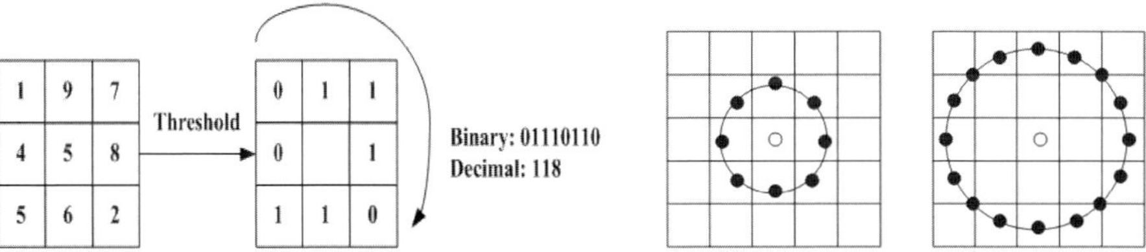

Fig. 1. (a) The basic LBP operator. (b) Two examples of the extended LBP: The circular (8, 1) neighborhood, and the circular (16, 2) neighborhood.

use neighborhoods of different size[10]. Using circular neighborhoods and bilinearly interpolating the pixel values allow any radius and number of pixels in the neighborhood. There are two samples of extended LBP in the Fig.1(b), where (*P,R*) means *P* sampling points on a circle of radius of *R*. As a result of the first extension, the size of feature vectors extracted by LBP operator with different number of neighborhoods is quite large, and the contribution of most feature vectors for texture description is limited. Hence, another extension of LBP is to use so called uniform patterns. A Local Binary Pattern is called uniform if it contains at most two bitwise transitions from 0 to 1 or vice versa when the binary string is considered circular. According to the texture image, the experiment in the paper [10] shows that the uniform patterns contribute 87.2% of the total pattern data when using the (8, 1) neighborhood and 70.7% in (16, 2) neighborhood. So, it is proved that uniform patterns appear to be fundamental properties of local image texture.

We use the notation $LBP_{P,R}^{u2}$ for operators, where the subscript shows that the operator is in a (P, R) neighborhood, and the superscript u2 stands for using uniform patterns and labeling all remaining patterns with a single label. A histogram of the labeled image $f_l(x, y)$ is defined as following:

$$H_i = \sum_{x,y} T\{f_l(x, y) = i\}, \quad i = 0, \cdots, n-1 \tag{1}$$

where n is the number of different labels produced by the LBP operator and

$$T(A) = \begin{cases} 1 & A \text{ is ture} \\ 0 & A \text{ is false} \end{cases} \tag{2}$$

This histogram contains information about the distribution of the local micro patterns, such as edges, spots and flat areas, over the whole image.

In former LBP based method for face analysis, the image is divided into several sub windows $W_0, W_1 \cdots W_{m-1}$ to achieve the goal of retaining the spatial information. The spatially enhanced histogram is defined as

$$H_{i,j} = \sum_{x,y} T\{f_l(x, y) = i\} T\{(x, y) \in W_j\} \quad i = 0, \cdots, n-1, j = 0, \cdots, m-1 \tag{3}$$

There are several possible dissimilarity measures proposed for histogram, such as histogram intersection, Log-likelihood statistic. In our work, Chi square statistic (x^2) is adopted:

$$x^2(S, M) = \sum_i \frac{(S_i - M_i)^2}{S_i + M_i} \tag{4}$$

where S and M respectively denote sample and template distributions.

3 Adaboost Learning

Boosting[11] is a general algorithm of constructing accurate strong classifier by combining several weak classifiers. The combined strong classifier could achieve very high accuracy when the accuracy of each weak classifier is slightly better than random guess. In the last decade, Boosting has been widely applied to many practical pattern recognition problems, for example, Schapire and Singer[12] presented a boosting-based system for Text Categorization, and Yang et al[13] proposed a face recognition method using AdaBoosted Gabor features. Furthermore, most important application of boosting algorithm is face detection. Viola et al[14] developed a face detection system based on cascaded Adaboost that is capable of detecting face very rapidly and accurately.

- Given example images (x_1, y_1), ..., (x_n, y_n) where $y_i = 0{,}1$ for negative and positive examples respectively.
- Initialize weights $\omega_{1,j} = 1/2m, 1/2l$ for $y_i = 0{,}1$ respectively, where m and l are the number of negatives and positives respectively.
- For $t = 1,...,T$:
  1. Normalize the weights, $\omega_{t,i} \leftarrow \dfrac{\omega_{t,i}}{\sum_{j=1}^{n} \omega_{t,j}}$ so that ω_t is a probability distribution.
  2. For each feature j, train a classifier h_j which is restricted to using a single feature. The error is evaluated with respect to $\varepsilon_j = \sum_i \omega_i |h_j(x_i) - y_i|$
  3. Choose the classifier, h_t, with the lowest error ε_t.
  4. Update the weights: $\omega_{t+1,i} = \omega_{t,i}\beta_t^{1-e_i}$, where $e_i = 0$ if example x_i is classified correctly, $e_i = 1$ otherwise, and $\beta_t = \dfrac{\varepsilon_t}{1-\varepsilon_t}$.
- The final strong classifier is: $H(x) = \begin{cases} 1 & \sum_{t=1}^{T} \alpha_t h_t(x) \geq 0.5 \sum_{t=1}^{T} \alpha_t \\ 0 & otherwise \end{cases}$

  where $\alpha_t = \log(1/\beta_t)$

Fig. 2. The binary-classified Adaboost algorithm [11]

In the procedure of Boosting algorithm, the effective features are learned from a large feature set firstly, and weak classifiers are constructed based on one of the selected features. Finally, these weak classifiers are boosted into a strong classifier. In general, the training error of the strong classifier approaches zero exponentially in the number of rounds, if the accuracy of combined weak classifiers are slight better than 50%. In Adaboost learning, we are given a sequence of training samples $s = (x_1, y_1),...,(x_m, y_m)$ along with a distribution $\omega_{t,i} = (\omega_{t,1}...\omega_{t,m})$. The $\omega_{t,i}$ is reweighed after each learning round t, and the distribution of incorrectly classified examples is increased so that the weak learner is forced to focus on the hard examples in the training set. The binary classified version of Adaboost is shown in Fig. 2.

4 Experiment and Discussion

We design two experiments for gender classification based on Local Binary Pattern operator. Two experiments are all made on the image sets collected from FERET database. The training set consists of 2000 images with 256 gray levels, 1200 of male subjects and 800 of female subjects, and the test set comprises others 400 images selected from FERET database, in which the subject of men and women are all 200. In preprocessing, we crop the face area from original image based on the two eyes location. The cropped images are scaled to 144 pixels high by 120 pixels wide and processed by illumination compensation and histogram equalization. There are several samples of the training set shown in the Fig.3.

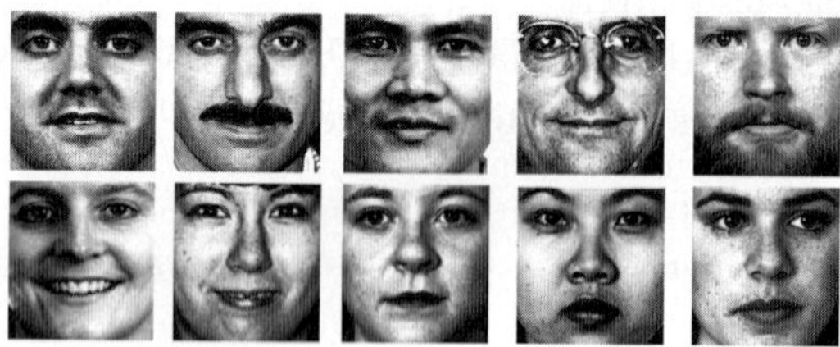

Fig. 3. Some samples of training set collected form FERET database

Experiment A – This experiment is performed to test the effectiveness of LBP feature for gender classification. Here a face image is equally divided into small sub windows from which LBP features are obtained and concatenated into a single, spatially enhanced feature histogram. We divide 144×120 pixels facial image into 24×20 pixels so that 36 (6×6) sub windows are given in total, see Fig.4. The 59-bin $LBP_{8,2}^{u2}$ operator is used for each sub window, and the length of the extracted histogram is 2124 (59×36).

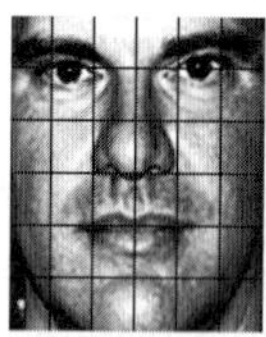

Fig. 4. An example of face image divided into 6×6 sub windows

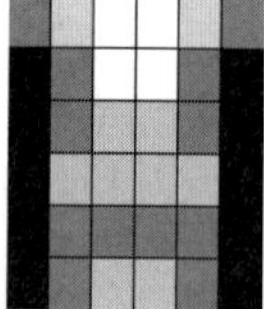

Fig. 5. The weights set corresponding to Chi Square distance

We apply the Self Organizing Maps (SOM) method to separate the training set into 10 classes, 5 for male images and 5 for female images. After the SOM training, the final weight vector for each node is the centroid of the class, i.e., the template vector, which corresponds to the template of each class. Investigating the male and female images from training set, we find out that the area of eyebrow, bridge of a nose, and chin contribute the most effective features to distinguish between men and women. As a result, a weights set corresponding to the divided face images is designed to improve the performance of the gender classification. The weights set is shown in Fig.5, where black squares mean weight is 0, dark grey is 1, light grey is 2 and white is 4.

In the classification, the weighted Chi Square distance is adopted to be as the measurement:

$$x_{\omega}^{2}(S,M)=\sum_{i,j}\omega_{j}\frac{\left(S_{i,j}-M_{i,j}\right)^{2}}{S_{i,j}+M_{i,j}} \tag{5}$$

where ω_j is the weight of sub windows j. Then, the histogram of the input image is matched with the closest template, and the input images are considered to be as male if their histograms are matched 1 to 5 classes, otherwise they are female.

Experiment B – This is an experiment of classifying gender based on boosting LBP. There are obviously two aspects that can be improved in experiment A: (1). The equal division to face image limits the variety of the size and position of the extracted features. By scaling and shifting the sub windows can capture much more features, which maintain the more abundant and detailed information of face images. (2). The weights set of Chi Square distance is predefined, which may be rough and subjective. It will be more rational that the weights of sub windows are computed by statistical learning such as Adaboost algorithm.

In experiment B, the scaling and shifting sub windows are used to obtain the features and Adaboost are applied to select the useful features and computed the weights set, which can overcome the above two drawbacks effectively. There are 12221 LBP features in total extracted by scanning the each face image with scalable windows. The LBP histograms in a given class are averaged to generate a histogram template for this class. The Chi square distance is applied to construct the weak classifier set by computing the dissimilarity measure of histogram between sample and template. Finally, Adaboost algorithm is used to combine the strong classifier for gender classification by learning features from the weak classifier set. The first three sub windows selected by Adaboost are shown in Fig.6.

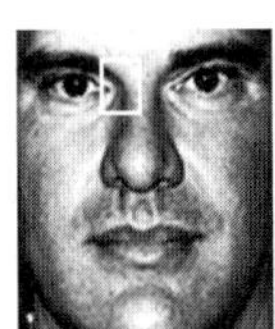 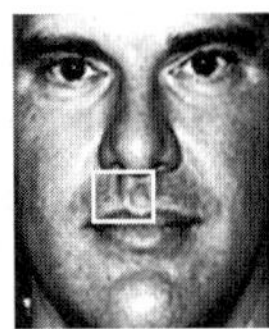 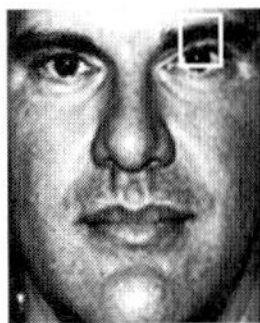

Fig. 6. The first three sub windows selected by Adaboost.

Result and Discussion – The performance of experiment A and experiment B is 82.75% and 95.75% respectively, and the experiment result compared with several gender classification methods also tested on the images set collected from FERET database are list in Table.1. The correct rate of experiment A and experiment B with other methods shows that the features extracted by LBP operator are discriminative for gender classification. And our simple template matching using LBP can achieve better result than the sixth method listed in the Table 1 also using nearest neighbor for discrimination. According to the experiment B, its accuracy is higher than ICA+LDA, ICA+SVM, only a bit lower than SVM+RBF kernel method. It is denoted that the boosting LBP approach can construct effective nonlinear classifier for gender classification. Furthermore, the comparison between our boosting LBP method and LUT-based Adaboost method indicates that the features obtained by LBP operator are more powerful to describe local texture than Haar like features for gender classification. Additionally, the accuracy of the two experiments in this section demonstrates that the LBP histograms of scalable and movable sub windows yield more complete and agile description of face images than the fixed one, and the weights set chosen by Adaboost algorithm is more reasonable than the predefined one.

Table 1. Comparisons of different methods for gender classification

Method	Accuracy
Experiment A	82.75%
Experiment B	95.75%
ICA+LDA in [3]	93.33%
ICA+SVM in [3]	95.67%
SVM+RBF kernel in [3]	96.62%
Nearest neighbor in [3]	72.84%
LUT-based Adaboost in [4]	85.46%

5 Conclusion

In this paper, we have proposed a novel method for gender classification by boosting statistical Local Binary Patterns based classifiers. LBP features extracted by scanning face images with scalable sub windows are introduced to be as a powerful texture description for classifying male and female. The Chi square distance between corresponding Local Binary Pattern histograms of sample image and template is used to construct weak classifiers pool. Adaboost algorithm is applied to build the final strong classifiers by selecting and combining the most useful weak classifiers. Two experiments illustrate that LBP features are effective for gender analysis, and the boosting LBP method can achieve better performance than several other methods.

Acknowledgment

This work was partly supported by the national natural science foundations of China under grant 60503023, and partly supported by the natural science foundations of Jiangsu province under the grant BK2005407, partly supported by the key laboratory of image processing and image communication of Jiangsu province under the grant ZK205013, and partly supported by Program for New Century Excellent Talents in University (NCET).

References

1. Golomb, B. A., Lawrence, D. T., Sejnowski, T. J.: SEXNET: A Neural Network Identifies Sex from Human Faces. In Advances in Neural Information Processing Systems (1991) 572-577.
2. Brunelli, R., Poggio, T.: Hyperbf Networks for Gender Classification. Proc. DARPA Image Understanding Workshop (1992)311-314.
3. Moghaddam, B., Yang, M. H.: Gender Classification with Support Vector Machines Proceedings of the International Conference on Automatic Face and Gesture Recognition (2000) 306-311.
4. Wu, B., Ai, H., Huang, C.: Real-time Gender Classification. In Proceedings of SPIE 5286 Third International Symposium on Multispectral Image Processing and Pattern Recognition (2003) 498-503.
5. Ojala, T., Pietikainen, M., Harwood, D.: A Comparative Study of Texture Measures with Classification Based on Feature Distributions. Pattern Recognition 29 (1996) 51–59.
6. Ahonen, T., Hadid, A., Pietikainen, M.: Face Recognition with Local Binary Patterns. In Proceedings of the European Conference on Computer Vision (2004) 469–481.
7. Li, S. Z., Zhao, C., Zhu, X., Lei, Z.: 3D+2D Face Recognition by Fusion at Both Feature and Decision Levels. In Proceedings of IEEE International Workshop on Analysis and Modeling of Faces and Gestures (2005).
8. Feng, X., Pietikäinen, M., Hadid, A.: Facial Expression Recognition with Local Binary Patterns and Linear Programming. Pattern Recognition and Image Analysis 15(2) (2005) 550-552.
9. Shan, C., Gong, S., McOwan, P.: Conditional Mutual Information Based Boosting for Facial Expression Recognition Proc. The 16th British Machine Vision Conference (BMVC 2005)
10. Ojala, T., Pietikänen, M., Mäenpää, T.: Multiresolution Gray-Scale and Rotation Invariant Texture Classification with Local Binary Patterns. IEEE Trans. on Pattern Analysis and Machine Intelligence 24 (2002) 971-987.
11. Freund, Y.: Boosting A Weak Learning Algorithm by Majority, Information and Computation 121(2) (1995) 256-285.
12. Schapire, R. C., Singer, Y.: BoosTexter: A Boosting-based System for Text Categorization. Machine Learning 39(2) (2000) 135-168.
13. Yang, P., Shan S., Gao, W., Li, S Z.: Face Recognition Using Ada-Boosted Gabor Features. Proceedings - Sixth IEEE International Conference on Automatic Face and Gesture Recognition (2004) 356-361.
14. Violas, P., Jones, M.: Robust Real Time Object Detection. 8th IEEE International Conference on Computer Vision (IC2CV) (2001).

Multi-view Gender Classification Using Local Binary Patterns and Support Vector Machines

Hui-Cheng Lian and Bao-Liang Lu

Department of Computer Science and Engineering, Shanghai Jiao Tong University,
800 Dong Chuan Rd., Shanghai 200240, China
{lianhc, blu}@cs.sjtu.edu.cn

Abstract. In this paper, we present a novel approach to multi-view gender classification considering both shape and texture information to represent facial image. The face area is divided into small regions, from which local binary pattern(LBP) histograms are extracted and concatenated into a single vector efficiently representing the facial image. The classification is performed by using support vector machines(SVMs), which had been shown to be superior to traditional pattern classifiers in gender classification problem. The experiments clearly show the superiority of the proposed method over support gray faces on the CAS-PEAL face database and a highest correct classification rate of 96.75% is obtained. In addition, the simplicity of the proposed method leads to very fast feature extraction, and the regional histograms and global description of the face allow for multi-view gender classification.

1 Introduction

Gender classification is one of the most challenging problems for face recognition researchers. Similar to any pattern classification problems, two key points for gender classification are feature extraction and pattern classification. From the view of feature extraction, the most simple method is to use gray-scale or color pixel vectors as features [1]. Another kind of methods comes from the theory of subspace transformation such as PCA, ICA and LDA, which project faces into a low-dimensional space and then recognize them [2]. This kind of method has been shown not very robust to variations of face orientation. The third kind of methods is using texture information such as winkle and complexion [3]. The last kind of methods is combining the facial feature detection with wavelet transform to extract the local facial feature for classification [4, 5], such as the analysis of facial winkles and shapes.

Traditional pattern classifiers such as k-nearest-neighbor, Fisher linear discriminant, neural network and SVMs are often employed to gender classification. SVMs seem to be superior to all other classifiers [1]. Moghaddam and Yang developed an appearance-based method to classify gender from facial images using nonlinear SVMs and compared their performance with traditional classifiers including Fisher linear discriminant, nearest-neighbor, RBF networks and large ensemble-RBF classifiers [1]. However, they focused their study on very

J. Wang et al. (Eds.): ISNN 2006, LNCS 3972, pp. 202–209, 2006.

low-resolution masked images, in which only the main frontal facial regions are visible and almost completely excluded hair information.

Recently a powerful way of texture description called local binary patterns (LBP) has been proposed for texture classification [6], face detection [7], and face recognition [8, 9]. An exciting recognition rate of 97.9% has been yielded via using this kind of texture description on the FERET FA/FB image sets. Considering the efficiency and effectiveness of LBP for face representation, we propose a method that combines LBPs and SVMs for gender classification. We evaluate the proposed method on the CAS-PEAL face database with comparison to [1]. The experimental results indicate that LBP-based methods can do well on both low degree faces and 30 degree faces.

2 Feature Extraction with Local Binary Patterns

The original LBP operator, introduced by Ojala *et al.* [6], is a powerful way of texture description. The operator labels the pixels of an image by thresholding the 3×3-neighbourhood of each pixel with the center value and considering the result as a binary number. Then the histogram of the labels can be used as a texture descriptor. The basic LBP operator is illustrated in Fig. 1 (a).

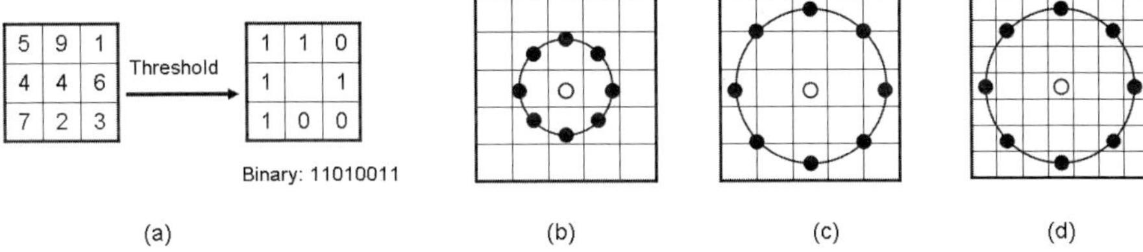

Fig. 1. The basic LBP operator and three examples of extended LBPs. (a) The basic LBP operator. (b) The circular (8,1) neighborhood. (c) The circular (8,2) neighborhood. (d) The circular (8,3) neighborhood.

The most prominent limitation of the LBP operator is its small spatial support area. Features calculated in a local 3×3 neighborhood cannot capture large scale structure that may be the dominant features of some textures. Later the operator was extended to use neighborhoods of different size [6]. Using circular neighborhoods and bilinearly interpolating the pixel values allow any radius and number of pixels in the neighborhood. Examples of these kinds of extended LBP are shown in Fig. 1(b), (c), (d).

Another extension to the original LBP operator is to use so called uniform patterns [6]. A local binary pattern is defined uniform if it contains at most two bitwise transitions from 0 to 1 or vice versa when the binary string is considered circular. For example, 00000000, 00011110, and 10000011 are uniform patterns. Ojala *et al.* [6] noticed that in their experiments with texture images, uniform patterns account for a bit less than 90% of all patterns when using the (8,1) neighborhood and for 70% in (16,2) neighborhood.

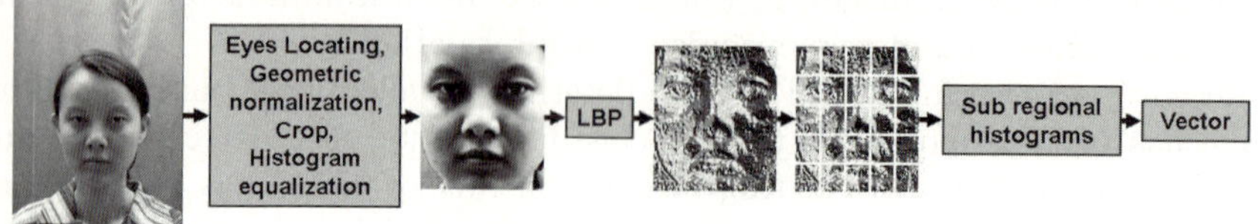

Fig. 2. Feature extraction. LBP transformation is done after locating eye positions from original images, geometric normalization, cropping and histogram normalization. The LBP face is divided into blocks and their histograms fitted together form a vector.

We use the notation $LBP^{u}_{P,R}$ for the uniform LBP operator. $LBP^{u}_{P,R}$ means using the LBP operator in a neighborhood of P sampling points on a circle of radius R. The superscript u stands for using uniform patterns and labelling all remaining patterns with a single label. The number of labels for a neighbourhood of 8 pixels is 256 for standard LBP and 59 for $LBP^{u}_{8,1}$.

A histogram of the labelled image $f_l(x, y)$ can be defined as

$$H_i = \sum_{x,y} I\{f_l(x,y) = i\}, i = 0, 1, ..., n-1,$$

where n is the number of different labels produced by the LBP operator and

$$I(A) = \begin{cases} 1, & \text{A is true} \\ 0, & \text{A is false.} \end{cases}$$

This histogram contains information about the distribution of the local micro-patterns over the whole image, such as edges, spots and flat areas. For efficient face representation, one should retain also spatial information. For this purpose, the image is divided into regions $R_0, R_1, ..., R_{m-1}$, as shown in Fig. 2 and the spatially enhanced histogram is defined as

$$H_{i,j} = \sum_{x,y} I\{f_l(x,y) = i\} I\{(x,y) \in R_j\}, \tag{1}$$

where $i = 0, 1, ..., n-1$ and $j = 0, 1, ..., m-1$.

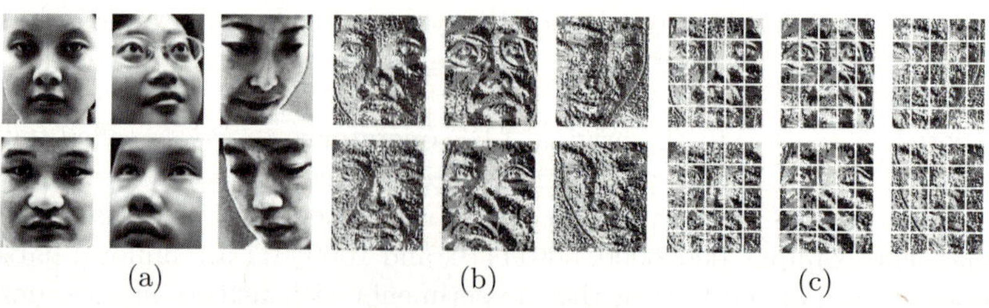

Fig. 3. Some of multi-view faces from the CAS-PEAL face database. (a) Preprocessed faces. (b) Corresponding LBP faces. (c) Corresponding 5×5 grid LBP faces.

The process of feature extraction for gender classification is illustrated in Fig. 3. An original image is processed by locating eye positions, geometric normalization, cropping and histogram normalization and a so called LBP face is obtained by performing LBP operator on this preprocessed facial image. $K \times K$ equal size blocks are divided from the LBP face with a grid on it and their histograms fitted together form a vector that will be fed into the gender classifier. Fig. 3 shows some of multi-view samples and their corresponding LBP faces from the CAS-PEAL face database.

3 Support Vector Machines

Support vector machine is a learning algorithm for pattern classification, regression and density estimation [10]. The basic training principle behind SVMs is finding the optimal linear hyperplane such that the expected classification error for unseen test samples is minimized. According to the structural risk minimization principle and VC dimension minimization principle [10], a linear SVM uses a systematic approach to find a linear function with the lowest capacity. For linearly nonseparable data, SVMs can nonlinearly map the input to a high-dimensional feature space where a linear hyperplane can be found.

Given a labelled set of M training samples $(\mathbf{x}_i, y_i)$ where $\mathbf{x}_i \in R^N$ and y_i is the associated label ($y_i \in \{-1, 1\}$), a SVM classifier finds the optimal hyperplane that correctly separates the training data while maximizing the margin. The discriminant hyperplane is defined by:

$$f(\mathbf{x}) = \sum_{i=1}^{M} y_i \alpha_i \cdot k(\mathbf{x}, \mathbf{x}_i) + b$$

where $k(\cdot, \cdot)$ is a kernel function, b is a bias and the sign of $f(\mathbf{x})$ determines the class membership of $\mathbf{x}$. To construct an optimal hyperplane is equivalent to finding all the nonzero α_i and is formulated as a quadratic programming (QP) problem with constraints.

For a linear SVM, the kernel function is just a simple dot product in the input space while for a nonlinear SVM the kernel function projects the samples to a higher dimension feature space via a nonlinear mapping function:

$$\Phi : R^N \rightarrow F^M,$$

where $M \gg N$, and then constructs a hyperplane in F. By using Mercer's theorem [10], the projecting samples into the high-dimensional feature space can be replaced by a simpler kernel function satisfying the condition

$$k(\mathbf{x}, \mathbf{x}_i) = \Phi(\mathbf{x}) \cdot \Phi(\mathbf{x}_i)$$

where Φ is the nonlinear projection function. Several kernel functions, such as, polynomials and radial basis functions, have been shown to satisfy Mercer's theorem and have been used successfully in nonlinear SVMs:

$$k(\mathbf{x}, \mathbf{x}_i) = ((\mathbf{x} \cdot \mathbf{x}_i) + 1)^d$$
$$k(\mathbf{x}, \mathbf{x}_i) = \exp(-\gamma |\mathbf{x} - \mathbf{x}_i|^2)$$

where d is the degree in a polynomial kernel and γ is the spread of a Gaussian cluster.

4 Experiments and Discussions

CAS-PEAL [11] is a large-scale face database that currently contains 14,384 pose images of 1,040 individuals. In this database there are thousands of samples with 9 different poses including looking up pose, looking middle pose and looking down pose with 0 degree, 15 degree and 30 degree, respectively. The eye coordinates are manually located. The image is scaled, lined up by the eye coordinates and cropped. The histogram of the image is equalized and the resolution of final image is 150×130 pixels. Table 1 shows the detailed contents about the database where 'Pxnn' gives all of the pose information about the data sets. The character 'P' represents pose variation. The 'x' (U, M, D) indicates the subject's pose (Up, Middle, Down). The 'nn' indicates the azimuth of the camera from which the image is obtained. Take the first row for example, it includes 1,040 images from 1,040 individuals (445 males and 595 females) and the first 400 images from 400 individuals (200 males and 200 females) are used as training samples and the rest 640 images from other 640 individuals are used as test samples. Consequently, the total number of training sample is 3,600 and the total number of test sample is 10,784.

Table 1. Description of training and test data sets from the CAS-PEAL face database for multi-view gender classification

set ID	Description	No. Male	No. Female	No. Total data	No. Training	No. Test
1	PD00	445	595	1,040	200×2	640
2	PD15	846	1,032	1,878	200×2	1,478
3	PD30	846	1,032	1,878	200×2	1,478
4	PM00	445	595	1,040	200×2	640
5	PM15	844	1,032	1,876	200×2	1,476
6	PM30	844	1,032	1,876	200×2	1,476
7	PU00	445	595	1,040	200×2	640
8	PU15	846	1,032	1,878	200×2	1,478
9	PU30	846	1,032	1,878	200×2	1,478
#	TOTAL	6,407	7,977	14,384	3,600	10,784

We use $LBP^{u}_{8,1}$ in all experiments and the 130×150 pixels image is divided into $K \times K$ equal size blocks, where K is ranged from 5 to 14. The LBP histograms of blocks are extracted and concatenated into a single, spatially enhanced feature histogram by using Equation (1). SVMs with linear kernel, polynomial kernel and RBF kernel are chosen to evaluate the performance of our

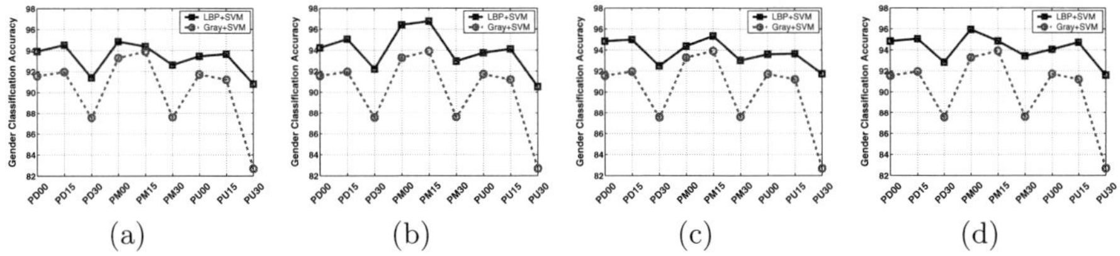

Fig. 4. Comparison of LBP method with the grey pixel approach using polynomial SVMs with degree 3 on the CAS-PEAL face database. (a) K=7. (b) K=8. (c) K=9. (d) K=10.

method. We also compare the proposed method with the existing approach [1]. The reasons why [1] is chosen are firstly this work of [1] is outstanding and secondly SVMs have been shown to be superior to traditional pattern classifiers for gender classification problem. The most important difference between our work and the existing approach [1] is that we use LBP histograms as feature extraction while the existing approach uses grey pixels. In addition to this difference, the simplicity of the proposed method allows for very fast feature extraction, and the regional histograms and global description of the face allow for efficiently multi-view gender classification. All of the SVMs in our experiments come from LibSVM [12] and all parameters are the default parameters of LibSVM except specially noting. To evaluate the efficiency of the LBP method, we firstly compare LBP method using K=7, 8, 9, 10 with the grey pixel approach the on data

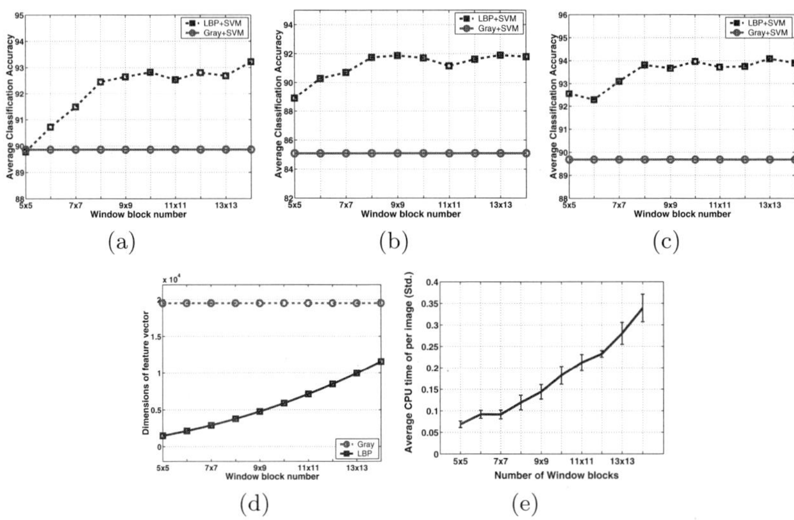

Fig. 5. Average classification accuracy comparison with three different kernels and the feature dimension comparison. (a) RBF kernel. (b) Linear kernel. (c) Polynomial kernel. (d) Feature dimension comparison. (e) CPU time (second) of LBP feature extraction.

sets of all degrees. The experimental results are shown in Fig. 4. From Fig. 4 we can see that the correct rates of LBP method are obviously higher than Moghaddam's method with a highest correct rate 96.75%. More particularly, on all 30 degree data sets (U, M, D), the grey pixel method almost fails with a lowest accuracy 82.68%, while LBP method can largely increase correct rates on these corresponding data sets, with a highest improvement from 82.68% to 91.75%.

To further show the performance, we compare LBP method with Moghaddam's method using three different SVM kernels and K ranged from 5 to 14 on average classification accuracy of all nine poses (see Fig. 5 (a), (b) and (c)). From them we can see that the average correct rate can even be highly improved from 89.69% to 94.08% meanwhile from Fig. 5 (d) we can see that the vector dimensions are much lower than that of the grey pixel approach. For example, when K is 5 or 6, the number of dimensions is 1,475(5×5×59) or 2,124(6×6×59), while the number of dimensions of the original grey pixel method is 19,500(150×130).

For there is no time needed for using the grey face directly, we do not compare CPU time of LBP feature extraction method with grey face method, while compare the LBP method with the Gabor method which is a very widely used method and can be found in [3, 4, 11]. From Fig. 5(e) we can see that the average CPU time of per image of LBP methods are only 0.07 seconds and 0.34 seconds for $K = 5$ and $K = 14$, while under the same conditions (3GHz PC/Matlab 6.1) the CPU times of Gabor methods are 6.31 seconds and 9.71 seconds for 4 scales with 6 frequencies and 5 scales with 8 frequencies.

The reason why LBP method can do well on both low degree faces and 30 degree faces is that this kind of method can describe face on three different levels of locality: the labels for the histogram contain information about the patterns on a pixel-level, the labels are summed over a small region to produce information on a regional level, and the reginal histograms are concatenated to build a global description of the face. And the reason why the feature dimension can be largely reduced is that facial image is first divided into small regions from which LBP histograms are extracted and concatenated into a single, spatially enhanced feature histogram efficiently representing the facial image. The simplicity and efficiency of uniform LBP operator and the superiority of SVMs over traditional pattern classifiers lead to a rapid and precise multi-view gender classification.

5 Conclusions

We have introduced a new method for multi-view gender classification by combining powerful LBP-based facial description with support vector machines. The efficiency and simplicity of LBP allow for very fast feature extraction, and the regional and global descriptions allow for capturing multi-view information of faces. Meanwhile the high generalization ability of SVMs allows for learning and classifying gender from a large set of multi-view faces. The experimental results show that classification accuracies are highly improved and a highest correct rate is 96.75% and a highest average correct rate is 94.08% on the CAS-PEAL face database.

Acknowledgment

This work was supported in part by the National Natural Science Foundation of China via the grants NSFC 60375022 and NSFC 60473040. Portions of the research in this paper use the CAS-PEAL face database collected under the sponsor of the Chinese National Hi-Tech Program and ISVISION Tech. Co. Ltd.

References

1. Moghaddam, B.,Yang, M.H.: Learning Gender with Support Faces. IEEE Transactions on Pattern Analysis and Machine Intelligence 24(5) (2002) 707-711
2. Balci, K., Atalay, V.: PCA for Gender Estimation: Which Eigenvectors Contribute? In: 16 th International Conference on Pattern Recognition (ICPR'02) Vol. 3. Quebec City, QC, Canada, (2002) 363-366
3. Iga, R.: Gender and Age Estimation System from Face Images. In: SICE Annual Conference in Fukui. August 4-6(2003) 756-761
4. Hosoi, S., Takikawa, E., Kawade M.: Ethnicity Estimation with Facial Images. In: Sixth IEEE International Conference on Automatic Face and Gesture Recognition May 17-19, 2004 Seoul, Korea (2004) 195-200
5. Lian, H. C., Lu, B. L.: Gender Recognition Using a Min-Max Modular Support Vector Machine. In: Proc. ICNC'05-FSKD'05, LNCS 3611 (2005) 433-436
6. Ojala, T., Pietikäinen, M., Mäenpää, T.: Multiresolution Gray-scale and Rotation Invariant Texture Classification with Local Binary Patterns. IEEE Transactions on Pattern Analysis and Machine Intelligence 24 (2002) 971-987
7. Jin, H.L, Liu, Q. S., Tong, X. F.: Face Detection Using Improved LBP Under Bayesian Framework. In: Proceedings of the Third International Conference on Image and Graphics (ICIG 04), Hong Kong, China, 18-20 Dec (2004) 306-309
8. Ahonen, T., Hadid, A., Pietikäinen, M.: Face Recognition with Local Binary Patterns. In: Proceedings of the European Conference on Computer Vision (2004) 469-481
9. Zhang, G.C.: Boosting Local Binary Pattern (LBP)-Based Face Recognition. In: Proceedings of the Sinobiometrics 2004, LNCS 3338 (2004) 179-186
10. Vapnik, V.: The Nature of Statistical Learning Theory. Springer, (1995)
11. Gao, W., Cao, B., Shan, S. G., *et al.*, The CAS-PEAL Large-Scale Chinese Face Database and Baseline Evaluations, technical report of JDL, (2004), avaible on http://www.jdl.ac.cn/~peal/peal_tr.pdf
12. Chang, C.C., Lin, C.J. LIBSVM : a Library for Support Vector Machines. http://www.csie.ntu.edu.tw/~cjlin/papers/libsvm.ps.gz

Gender Recognition Using a Min-Max Modular Support Vector Machine with Equal Clustering

Jun Luo and Bao-Liang Lu*

Department of Computer Science and Engineering, Shanghai Jiao Tong University, 800 Dong Chuan Rd., Shanghai 200240, China
{ljfootball, bllu}@sjtu.edu.cn

Abstract. Through task decomposition and module combination, min-max modular support vector machines (M^3-SVMs) can be successfully used for different pattern classification tasks. Based on an equal clustering algorithm, M^3-SVMs can divide the training data set of the original problem into several subsets with nearly equal number of samples, and combine them to a series of balanced subproblems which can be trained more efficiently and effectively. In this paper, we explore the use of M^3-SVMs with equal clustering method in gender recognition. The experimental results show that M^3-SVMs with equal clustering method can be successfully used for gender recognition and make the classification more efficient and accurate.

1 Introduction

Gender recognition is one of the most challenging problems among face recognition research areas. Nowadays, a lot of improvements have been achieved on this problem. As for classifiers, support vector machines (SVMs) have been successfully applied to solve the task and seem to be superior to other classifiers [1]. However, SVMs have to solve a quadratic optimization problem, and for a large-scale two-class problem such as gender classification it is rather difficult to improve classification accuracy.

In our previous work, a min-max modular support vector machine [2][3] is proposed to solve gender recognition problem [4] and also a novel clustering algorithm called equal clustering that can equally decompose training data sets is used to improve the performance of M^3-SVM [5]. In this paper, we explore the use of M^3-SVM with equal clustering method in gender recognition task. The gender recognition problem is a two-class pattern classification problem and the scale of training samples needs to be quite large in order to make the classification more accurate. Using M^3-SVM with equal clustering method, we can decompose the whole large problem of gender images into several balanced subproblems. Each individual subproblem becomes less complicated than the original problem and can be solved effectively and efficiently.

* To whome correspondence should be addressed. This work was supported by the National Natural Science Foundation of China under the grants NSFC 60375022 and NSFC 60473040.

J. Wang et al. (Eds.): ISNN 2006, LNCS 3972, pp. 210–215, 2006.

2 Min-Max Modular Support Vector Machine

The min-max modular neural network [6] is based on a 'divide-and-conquer' strategy and it can divide a large-scale problem into several independent smaller subproblems and then combine the results of each subproblem into a solution to the original problem. In essence, M^3-neural network is a general framework for machine learning.

M^3-SVM can easily solve multi-class problem and here for simplicity we just discuss the two-class problem. Given a training data set $\mathcal{S} = \mathcal{X}^+ \cup \mathcal{X}^-$, $\mathcal{X}^+$ and $\mathcal{X}^-$ denote the positive and negative training data sets for a two-class problem $\mathcal{T}$, respectively.

$$\mathcal{X}^+ = \{(x_i^+, +1)\}_{i=1}^{l^+}, \ \ \mathcal{X}^- = \{(x_i^-, -1)\}_{i=1}^{l^-} \tag{1}$$

where $x_i^+ \in \mathbf{R}^n$ and $x_i^- \in \mathbf{R}^n$ are the input vectors, and l^+ and l^- denote the total number of positive training data and the total number of negative training data of the two-class problem, respectively.

According to [2], M^3-SVMs consist of three steps. The first step is task decomposition. In this phase, $\mathcal{X}^+$ and $\mathcal{X}^-$ are decomposed into the following N^+ and N^- subsets, respectively.

$$\mathcal{X}^+ = \cup_{i=1}^{N^+} \mathcal{X}_i^+, \mathcal{X}_i^+ = \{(x_j^{+i}, +1)\}_{j=1}^{l_i^+}, i = 1, \ldots, N^+ \tag{2}$$

$$\mathcal{X}^- = \cup_{i=1}^{N^-} \mathcal{X}_i^-, \mathcal{X}_i^- = \{(x_j^{-i}, -1)\}_{j=1}^{l_i^-}, i = 1, \ldots, N^- \tag{3}$$

where $\cap_{i=1}^{N^+} \mathcal{X}_i^+ = \phi$, $\cap_{i=1}^{N^-} \mathcal{X}_i^- = \phi$, and ϕ denotes empty set as each subset is set to independent to others.

After the decomposition, every two subsets from $\mathcal{X}^+$ and $\mathcal{X}^-$ are chosen to form $N^+ \times N^-$ smaller subproblems as follows:

$$(\mathcal{T}^{(i,j)})^+ = \mathcal{X}_i^+, \ (\mathcal{T}^{(i,j)})^- = \mathcal{X}_j^- \tag{4}$$

where $(\mathcal{T}^{(i,j)})^+$ and $(\mathcal{T}^{(i,j)})^-$ denote the positive training data set and the negative training data set for the subproblem $\mathcal{T}^{(i,j)}$. And then all of the subproblems can be trained by SVMs in a massively parallel way.

The last step is to integrate all small trained modules to get a solution to the original problem using the minimization and maximization principles. The $N^+ \times N^-$ smaller SVMs are integrated into a M^3-SVM with N^+ MIN units and one MAX unit as follows,

$$\mathcal{T}^i(x) = \min_{j=1}^{N^-} \mathcal{T}^{(i,j)}(x) \ \text{ for } i = 1, \ldots, N^+ \text{ and } \mathcal{T}(x) = \max_{i=1}^{N^+} \mathcal{T}^i(x) \tag{5}$$

where $\mathcal{T}^{(i,j)}(x)$ denotes the transfer function of the trained SVM corresponding to the two-class subproblem $\mathcal{T}^{(i,j)}$, and $\mathcal{T}^i(x)$ denotes the transfer function of a combination of N^- SVMs integrated by the MIN unit.

3 Equal Clustering Task Decomposition Method

As mentioned above, the first step of M^3-SVM is task decomposition. Based on the previous work, several task decomposition strategies have been applied to solve gender recognition problem [4]. However, when training data is not identically distributed and no useful prior knowledge information can be found, the effectiveness of these two data partition methods will be unstable. In order to handle this problem, based on the algorithm "GeoClust" [7], we have proposed the equal clustering method in our previous work [5].

The way of generating reasonable clusters needs to solve the unconstrained nonlinear programming problem as follows:

$$\underset{c_1\, c_2\, \dots\, c_m}{\text{Minimize}} : h = \max_{i=1}^{m} |W_i - \overline{W}|,\ N = \sum_{i=1}^{m} W_i \tag{6}$$

where c_i and W_i denote the center vector and the number of samples in the ith cluster, respectively, m denotes the number of clusters, and $\overline{W}$ is the mean number of samples per cluster, i.e, $\overline{W} = \lfloor \frac{N}{m} \rfloor$.

Actually, the equal clustering algorithm is to generate spatially localized clusters that contain nearly equal number of samples to keep load balanced, so that it might catch local probability distribution of the training data set. This strategy has been evaluated on several benchmark data sets such as Banana data set and Letter Recognition data set [5], and its advantages have been demonstrated.

4 Experiments

4.1 Experimental Setup

In this section, we present experimental results on the CAS-PEAL face database [8] to compare M^3-SVMs using equal clustering method with the traditional SVMs and M^3-SVMs using random partition strategy. Here, there are several parameters for equal clustering algorithm and in our experiments we set *maxiter* $= 6000$, $\alpha = 0.01 \times 10^{-\lfloor \frac{(m-1)}{10} \rfloor}$, $l = 3$, and $\varepsilon = \lfloor \frac{N}{10m} \rfloor$. Here N denotes the number of training samples and m denotes the number of partition.

The training data sets include $2,670$ male and female samples respectively with 12 kinds of face poses. For each pose we select equal number of male and female samples. In order to enhance the performance, all images have been preprocessed and scaled into 65×75 images as shown in Fig.1. Here gray scale vector is chosen as feature and the dimension of each input vector is 4875. The probe sets represent face images with various pose degree ranged from $-30°$ to $30°$. In order to ensure the credibility of the conclusions, all experiments are repeated three times and the average is taken.

In all figures below, 'RP' means random partition method, and 'EC' means equal clustering method. For convenience, M^3-SVM-RP denotes min-max modular support vector machine using random partition strategy, while M^3-SVM-EC denotes the one using equal clustering strategy.

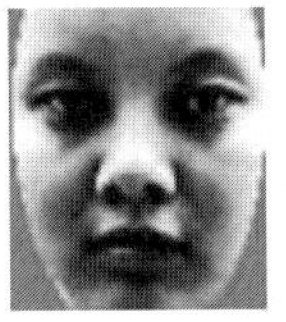 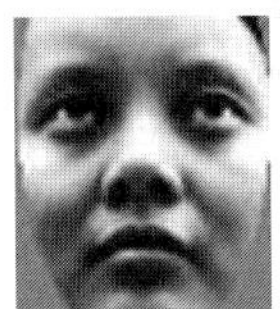 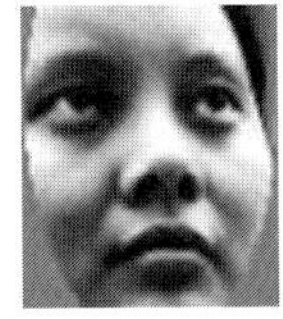 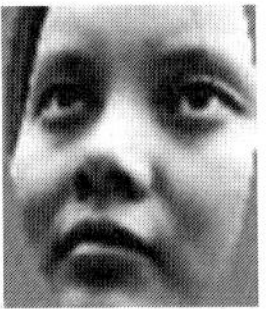 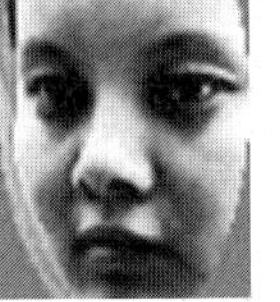 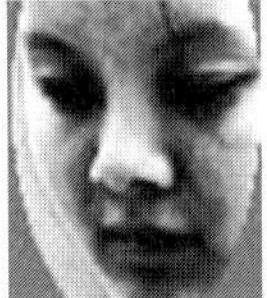

Fig. 1. A subject face instance. Six of poses including looking middle with 0 degree, looking up with 0 degree, looking up with left 22 degrees, looking up with right 22 degrees, looking middle with right 22 degrees and looking down with right 22 degrees.

4.2 Experimental Results

From Fig.2(a), we can see the advantage of M^3-SVM-EC over traditional SVM. First of all, M^3-SVM-EC can improve the classification accuracy by a high degree on gender recognition problem. In Table 1 we only present two division examples. The results of these two M^3-SVMs divided into 4 (M^3-SVM-EC-4) and 6 parts (M^3-SVM-EC-6) respectively, are both quite better than traditional SVM classifiers. Furthermore, from the results we can also see that the number of partition can't determine the final result. Meanwhile, Fig.2(b) illustrates that M^3-SVM with equal clustering method takes higher generalization accuracy than M^3-SVM with random partition method in most of time.

Moreover, as the equal clustering can make each modular have nearly same number of samples, M^3-SVM can run in a more efficient way than that with random partition method. From Fig.3(a) we can see that M^3-SVM with equal

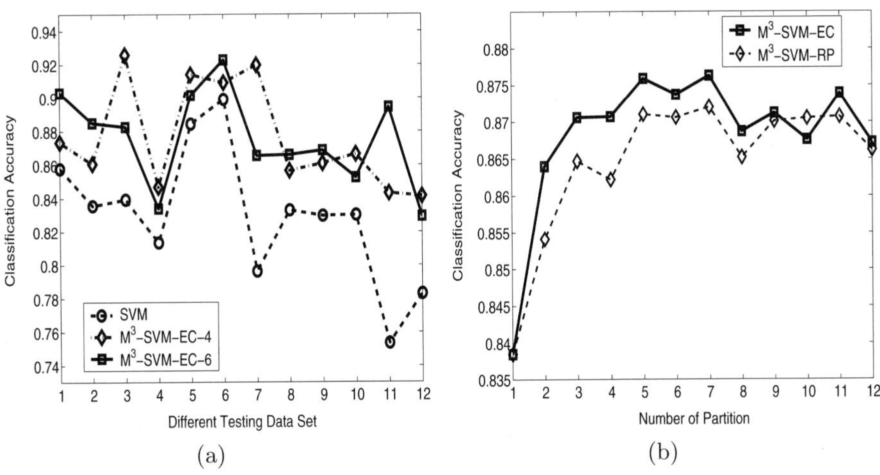

Fig. 2. The comparative results of M^3-SVMs based on equal clustering method and SVMs and M^3-SVMs based on random partition method. (a) Results of SVM and M^3-SVM divided into 4 and 6 parts with equal clustering method; (b) Results of M^3-SVMs with random partition and equal clustering method on total test data set.

Table 1. The problem statistics and the results based on different classifiers. The name of test data set is composed of three parts: 'POSE' denotes the face pose; 'PD', 'PM', and 'PU' denote person facing down, middle and up, respectively; the number denotes the pose angle.

Test Data	Female	Male	Total	Train	Test	SVM	M^3-SVM-EC-4	M^3-SVM-EC-6
POSE_PD00	445	595	1040	400	640	85.78%	87.34%	**90.31%**
POSE_PD15	846	1032	1878	600	1278	83.57%	86.07%	**88.50%**
POSE_PD22	44	158	202	40	162	83.95%	**92.59%**	88.27%
POSE_PD30	846	1032	1878	800	1078	81.35%	**84.69%**	83.40%
POSE_PM00	445	595	1040	400	640	88.44%	**91.41%**	90.16%
POSE_PM15	844	1032	1876	400	1476	89.91%	90.92%	**92.28%**
POSE_PM22	44	158	202	40	162	79.63%	**91.98%**	86.55%
POSE_PM30	844	1032	1876	600	1276	83.30%	85.66%	**86.60%**
POSE_PU00	445	595	1040	400	640	82.97%	86.09%	**86.87%**
POSE_PU15	445	595	1040	400	640	83.02%	**86.64%**	85.25%
POSE_PU22	44	158	202	60	142	75.35%	84.32%	**89.44%**
POSE_PU30	846	1032	1878	800	1078	78.29%	**84.14%**	82.93%

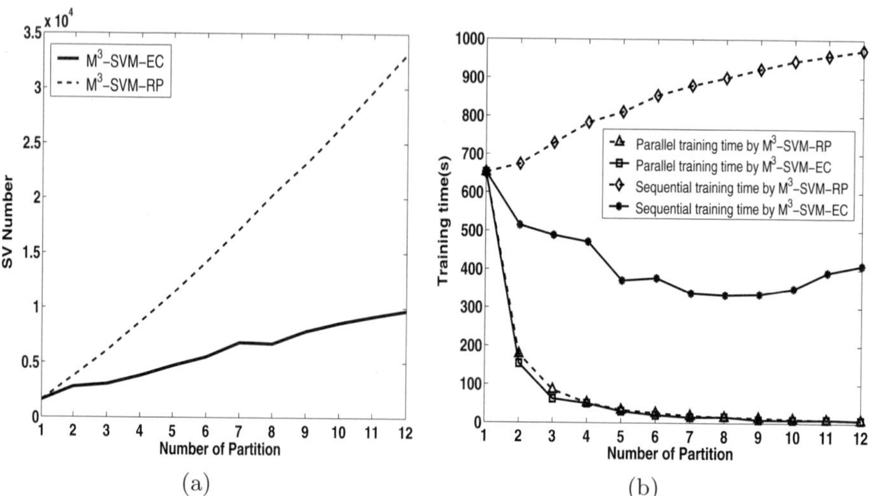

Fig. 3. The time performance of M^3-SVMs using equal clustering method and random partition method.(a)The number of SVs produced;(b)The training time of two task decomposition methods.

clustering method generates less support vectors than that with random partition method does, because it can make training data more separable. Actually, the smaller the number of support vectors is, the less the training time and the cost of realization of M^3-SVM are. Even though the data partitioning costs more preprocessing time, M^3-SVM-EC's less support vectors and more generalization accuracy can compensate for it and it takes less training time than M^3-SVM-RP does in both sequential and parallel modes as shown in Fig.3(b).

5 Conclusion

In this paper, M^3-SVMs with the equal clustering method are applied to solve gender recognition problem and some comparisons have been done to show the advantage of this method over traditional SVMs and M^3-SVMs with random partition method. From our experiments it can be seen that M^3-SVMs with equal clustering can improve the performance of gender recognition and be more efficient.

Acknowledgements

The authors thank Yi-Min Wen for providing the source programme of equal clustering algorithm.

References

1. Moghaddam, B., Yang, M.H.: Gender Classification with Support Vector Machines. In:Proceedings of the Fourth IEEE International Conference on Automatic Face and Gesture Recognition, (2000) 306-311
2. Lu, B.L., Wang, K.A., Utiyama, M., Isahara, H.: A Part-versus-part Method for Massively Parallel Training of Support Vector Machines. In: Proceedings of IJCNN'04, Budapast, (2004) 735-740
3. Wang, K.A., Lu, B.L.: Task Decomposition Using Geometric Relation for Min-Max Modular SVM. In: ISNN 2005: Lecture Notes in Computer Science, vol.3496. Springer-Verlag, Berlin Heidelberg (2005) 887-892
4. Lian, H.C., Lu, B.L., Takikawa, E., Hosoi, S.: Gender Recognition Using A Min-max Modular Support Vector Machine. In: ICNC 2005: Lecture Notes in Computer Science, vol.3611. Springer-Verlag, Berlin Heidelberg (2005) 438-441
5. Wen, Y.M., Lu, B.L., Zhao, H.: Equal Clustering Makes Min-Max Modular Support Vector Machines More Efficient. In: ICONIP'2005, Taipei, (2005)
6. Lu, B.L., Ito, M.: Task Decomposition and Module Combination Based on Class Relations: A Modular Neural Network for Pattern Classification. IEEE Transcations on Neural Network 10(5) (1999) 1244-1256
7. Choudhury, A., Nair, C.P., Keane, A.J.: A Data Parallel Approach for Large-Scale Gaussian Process Modeling. In: Proceedings of the Second SIAM International Conference on Data Mining, (2002)
8. Gao, W., Cao, B., Shan, S.G., Zhou, D.L., Zhang, X.H., Zhao, D.B.: The CAS-PEAL Large-Scale Chinese Face Database and Evaluation Protocols, Technical Report No. JDL_TR_04_FR_001, Joint Research & Development Laboratory, CAS, 2004

Palmprint Recognition Using ICA Based on Winner-Take-All Network and Radial Basis Probabilistic Neural Network

Li Shang[1,2], De-Shuang Huang, Ji-Xiang Du[2], and Zhi-Kai Huang [2]

[1] Department of Automation, University of Science and Technology of China, Hefei, Anhui 230026, China
shangli@iim.ac.cn
[2] Intelligent Computing Lab, Hefei Institute of Intelligent Machines, Chinese Academy of Sciences, P.O. Box 1130, Hefei, Anhui 230031, China
{Dshuang, du_jx, zhengch}@iim.ac.cn

Abstract. This paper proposes a novel method for recognizing palmprint using the winner-take-all (WTA) network based independent component analysis (ICA) algorithm and the radial basis probabilistic neural network (RBPNN) proposed by us. The WTA-ICA algorithm exploits the maximization of the sparse measure criterion as the cost function, and it extracts successfully palmprint features. The classification performance is implemented by the RBPNN. The RBPNN is trained by the orthogonal least square (OLS) algorithm and its structure is optimized by the recursive OLS (ROLS) algorithm. Experimental results show that the RBPNN achieves higher recognition rate and better classification efficiency with other usual classifiers.

1 Introduction

Biometric palmprint verification is such a technology, which recognizes a person based on unique features in his palm, such as the principal lines, wrinkles, ridges, and texture, etc. Palmprint recognition is a very challenging problem. Many recognition methods, especially the neural networks (NN) method [1], [2], etc., have been proposed. This paper focuses on using a novel radial basis probabilistic neural network (RBPNN) model [1] to perform the palmprint recognition task. Features of palmprint images can be extracted by certain transform, such as Fourier transform (FT) [3], the wavelet based transform [4], principal component analysis (PCA) and ICA [5], etc. Fourier and wavelet transforms have strong mathematical foundations and fast implementations, but they are not adaptive to particular data. While the significant advantage of the PCA and ICA is that they only depend on the statistic property of data. However, PCA is usually suitable for the second order accumulation variant, whereas ICA can be used for multi-dimensional data. Here, we utilize the winner-tall-all (WTA) network based ICA technique [6] to extract successfully features of palmprint images and use the RBPNN model as a classifier to implement the classification work.

J. Wang et al. (Eds.): ISNN 2006, LNCS 3972, pp. 216 – 221, 2006.

2 The Winner-Take-All (WTA) Based ICA Algorithm

2.1 Learning WTA-ICA Weights

Karvanen et al [7] generalized the sparseness measure of l^p norm criteria as follows:

$$Sparse(u) = -E\left\{\left(\sum_i |u_i^p|\right)^{\frac{1}{p}}\right\}. \tag{1}$$

Karvanen and Cichocki suggested that the range of p should be in (0,1), nevertheless, Zhang, et al [6] found the fact that $p \to \infty$ can also give a good sparseness measure. When $p \to \infty$, the limit of $\left(\sum_i |u_i^p|\right)^{\frac{1}{p}}$ is $\max_i \{|u_i|\}$, thus, the optimization process is relatively easy for the l^∞ norm sparseness measure function (i. e., the cost function), which is defined as follows:

$$J(\mathrm{u}) = E\left\{\lim_{p\to\infty}\left(\sum_i |u_i^p|\right)^{\frac{1}{p}}\right\} = E\left\{\max_i \{|u_i|\}\right\}. \tag{2}$$

Let w_j denotes the weight vector, u_j denotes the independent source vector. Replacing the $|\mathrm{u}_j|$ with $|\mathrm{u}_j|^2 = (\mathrm{w}_j^T\mathrm{x})^2$, Eqn. (4) can be rewritten as [6]:

$$J(W) = E\left\{\max\left[\left(\mathrm{w}_j^T\mathrm{x}\right)^2\right]\right\} = \int \max_j\left[\left(\mathrm{w}_j^T\mathrm{x}\right)^2\right] p(\mathrm{x})\,dx\,. \tag{3}$$

Let $c = \arg\max_j\left[\left(\mathrm{w}_j^T\mathrm{x}\right)^2\right]$ and $Q = \sum_j\left(\mathrm{w}_j^T\mathrm{x}\right)^{2r}$, the maximum part can be rewritten as:

$$\left(\mathrm{w}_c^T\mathrm{x}\right)^2 = \lim_{r\to\infty} Q^{\frac{1}{r}}\,. \tag{4}$$

The partial derivative of Eqn. (3) with respect to the jth column vector w_j is [6]:

$$\frac{\partial J}{\partial \mathrm{w}_j} = \int \lim_{r\to\infty} \frac{\partial Q^{\frac{1}{r}}}{\partial \mathrm{w}_j} p(\mathrm{x})d\mathrm{x}\,. \tag{5}$$

Assume that for w_j given, the term $\left(\mathrm{w}_c^T\mathrm{x}/\mathrm{w}_j^T\mathrm{x}\right)$ is the maximum. Combining $Q = \sum_j\left(\mathrm{w}_j^T\mathrm{x}\right)^{2r}$ into Eqn. (7), $\left(\partial Q^{\frac{1}{r}}/\partial \mathrm{w}_j\right)$ can be deduced as follows:

$$\frac{\partial Q^{\frac{1}{r}}}{\partial \mathrm{w}_j} = \frac{1}{r} Q^{\frac{r-1}{r}} \cdot \frac{\partial Q}{\partial \mathrm{w}_j} = 2Q^{\frac{1}{r}} \cdot \frac{\left(\mathrm{w}_j^T\mathrm{x}\right)^{2r-1}}{Q}\mathrm{x}\,. \tag{6}$$

where Eqn. (6) is the updating rule for the weight matrix W. Note that $\lim_{r\to\infty}\left[\mathrm{w}_j^T\mathrm{x}^{(2r-1)}/Q\right] = \delta_{cj}\left(\mathrm{w}_j^T\mathrm{x}\right)^{-1}$, where δ_{cj} is the Kronecker delta. If $c=j$, then $\delta_{cj}=1$, otherwise, $\delta_{cj}=0$. Thus, the limitation of Eqn. (6) was written as $\lim_{r\to\infty}\frac{\partial Q^{\frac{1}{r}}}{\partial \mathrm{w}_j} = 2\delta_{cj}\left(\mathrm{w}_c^T\mathrm{x}\right)\mathrm{x}\,$.

Here, we consider $(w_j^T x)x$ with $\|w_j\| = 1$ as an observation to get the mean of this observation, while w_j is estimated incrementally. For incremental estimation, where W is continuously improved, what is called an amnesic mean is defined as $\bar{x}^{(n)} = \alpha(n)\bar{x}^{(n-1)} + \beta(n)x_n$, where $\bar{x}^{(n)}$ is the mean at the *nth* iteration, x_n is the *nth* sample, and $\alpha(\cdot)$ and $\beta(\cdot)$ are respectively defined as $\alpha(n) = [n-1-\mu(n)]/n$ and $\beta(n) = [1+\mu(n)]/n$ [6], where $\mu(n)$ is a non-negative small function that discounts old estimate and gives more weights to the new x_n at time n. Thus, the weight updating rule is derived as:

$$w_j(n_j+1) = \alpha(n_j)w_j(n_j) + \beta(n_j)\frac{w_j(n_j)^T x_t}{\|w_j(n_j)\|}x_t . \tag{7}$$

where $w_j(n_j)$ is the component vector w_j after the $n_j th$ updating, x_t is the current whitened data input.

2.2 Two Architectures of Performing WTA-ICA

According to literature [6], there are two types of implementation architectures for WTA-ICA in the image recognition task. Architecture I treats images as random variables and pixels as observations, i. e., each row of the input data matrix denotes an image, and its goal is to find a set of statistically independent basis images. While architecture II utilizes pixels as random variables and images as observations, i. e., each column of the input data denotes an image, and its goal is to find a representation in which all coefficients are statistically independent. Convenient for calculating, we use PCA to realize data whitening and a dimensional reduction before performing WTA-ICA. Note that the basis images of architecture I show more localized features [2], see Fig. 1(a), and the basis generated by architecture II shows more globalized features, see Fig. 1(b).

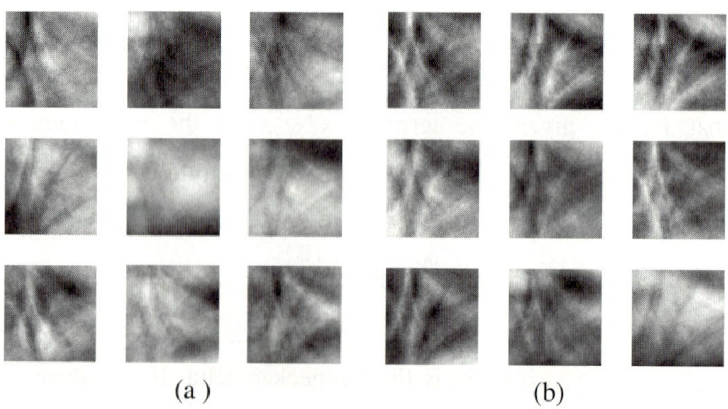

Fig. 1. Resulting bases of two WTA-ICA architectures. (a) The first 9 statistically independent basis images obtained by architecture I ; (b) The first 9 basis images (coefficients are statistically independent) obtained by architecture II .

3 The RBPNN Model and the Training Algorithm

The RBPNN model [1], [2]is shown in Fig 2. It can be seen that this network consists of four layers. The last layer for the RBPNN is just the output layer. In mathematics, for input vector x, the actual output value of the *ith* output neuron of the RBPNN, y_i^a, is written as:

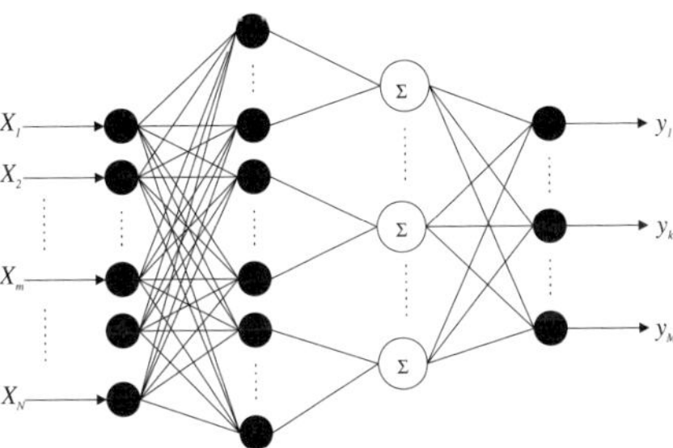

Fig. 2. The structure of radial basis probabilistic neural network

$$\mathrm{y}_i^a = \sum_{k=1}^{M} \mathrm{w}_{ik} h_k(\mathrm{x}) \,. \tag{8}$$

$$h_k(\mathrm{x}) = \sum_{i=1}^{n_k} \phi_i\left(\left\|\mathrm{x} - c_{ki}\right\|_2\right), \quad k = 1,2,3,\cdots, M \,. \tag{9}$$

where $h_k(\mathrm{x})$ is the *kth* output value of the second hidden layer of the RBPNN; w_{ik} is the synaptic weight between the *kth* neuron of the second hidden layer and the *ith* neuron of the output layer of the RBPNN; c_{ki} represents the *ith* hidden center vector for the *kth* pattern class of the first hidden layer; n_k represents the number of hidden center vector for the *kth* pattern class of the first hidden layer; M denotes the number of the neurons of the output layer and the second hidden layer, or the pattern class number for the training samples set; $\phi_i(\cdot)$ is the kernel function, which is generally Gaussian kernel function. $\phi_i\left(\left\|x - c_{ki}\right\|_2\right)$ can be written as:

$$\phi_i\left(\left\|\mathrm{x} - c_{ki}\right\|_2\right) = \exp\left[-\frac{\left\|\mathrm{x} - c_{ki}\right\|_2^2}{\sigma_i^2}\right]. \tag{10}$$

where σ_i is the shape parameter for Gaussian kernel function. Here, using the OLSA algorithm to train the RBPNN. Considering the form of matrix, Eqn. (8) can be written as [1] $Y^a = HW$, where Y^a and H are both an $N \times M$ matrix, W is a square matrix of $M \times M$. According to literature [1], it can be known that W has the form of $W = R^{-1}\hat{Y}$, where R is an $M \times M$ upper triangle matrix with the same rank as H, and $\hat{Y}$ is an $M \times M$ matrix. Both of them can be respectively written as $H = Q \times [R \quad \cdots \quad 0]^T$

and $Q^T \times Y = \begin{bmatrix} \hat{Y} & \tilde{Y} \end{bmatrix}^T$, where Q is an $N \times N$ orthogonal matrix with orthogonal columns satisfying $QQ^T = Q^T Q = I$, and $\tilde{Y}$ is an $(N-M) \times M$ matrix.

4 Experimental Results

We used the Hong Kong Polytechnic University (PolyU) palmprint database, available from http:// www.comp.polyu.edu.hk/~biometrics, to verify our RBPNN algorithm. This database includes 600 palmprint images with the size of 128×128 from 100 individuals, with 6 images from each. By PCA, the dimension of the training set is reduced to 196, i.e., the number of the fist k PCs is 196.

Using WTA-ICA architectures, basis vectors of palmprint images are extracted. Then we select all the 196 training samples as the hidden centers of the first hidden layer. The number of the second hidden neurons and the output layer neurons is set as 100 (class pattern), respectively. According to reference [1], the shape parameter σ_i is set as 650. Using the OLS algorithm to train the RBPNN, the recognition rate of the testing samples corresponding to two architectures is respectively 94.52% and 97.74%, as shown in Table 1. In order to optimize and prune the RBPNN, we use the ROLS algorithm to optimize the structure of RBPNN. As a result, the selected hidden centers number of the first hidden layer is reduced from 196 to 64 and the recognition rate of testing samples corresponding to each WTA-ICA architecture is invariant. It is clearly to see that the architecture II outperforms the architecture I in classification.

Table 1. Recognition rate of different classifiers

Recognition rate (%)	RBPNN	RBFNN	BPNN
WTA-ICA architecture I	94.52	91.93	89.61
WTA-ICA architecture II	97.74	93.76	92.53

Table 2. Training and classification CPU time for the PolyU Palmprint database

Methods	architecture I CPU time (sec.)		architecture II CPU time (sec.)	
	Training	Classification	Training	Classification
RBPNN	2.4	0.02	2.1	0.018
RBFNN	2.8	0.17	2.6	0.015
BPNN	197	0.26	192	0.22

With the same training and testing data, compared with the RBPNN and the BP to each architecture, the results were also shown in table 1. It can be clearly seen that the recognition rate of the RBPNN is the highest. On the other hand, for each architecture and classifier, the CPU time needed to recognize an image was shown in Table 2 (using the usual computer with 2.6GHz clock and 256Mb RAM). It was easily found that the training speed and testing speed with the RBPNN are also very fast.

5 Conclusions

In this paper, a novel palmprint recognition method was developed by using WTA-ICA algorithm and the RBPNN classifier. This algorithm exploits amnesic mean instead of conventional learning rate. So, it results in a faster convergence than the classical ICA method From the above experimental results, it can be concluded that our palmprint recognition method not only achieves higher statistical recognition rate, but also behaves faster training speed and testing speed. This method is indeed effective and efficient, which greatly support the claim that the RBPNN is a very promising neural network model in practical applications.

Acknowledgements

This work was supported by the National Science Foundation of China (Nos. 60472111, 30570368 and 60405002).

References

1. Huang, D. S.: Radial Basis Probabilistic Neural Networks: Model and Application. International Journal of Pattern Recognition and Artificial Intelligence 13(7) (1999) 1083-1101
2. Huang, D. S., Zhao, Wen-Bo: Determining The Centers of Radial Basis Probabilities Neural Networks by Recursive Orthogonal Least Square Algorithms. Applied Mathematics and Computation 162(1) (2005) 461-473
3. Li, W., David, Z., Xu, Z.: Palmprint Identification by Fourier Transform. Int. J. Pattern Recognition. Art. Intell. 16(4) (2002) 417-432
4. Kumar, A., Shen, H. C.: Recognition of Palmprints Using Wavelet-based Features. Proc. Intl. Conf. Sys., Cybern., SCI-2002, Vol. 3072. Orlando Florida (2002)
5. Bartlett, M. S., Movellan, J. R., Sejnowski, T. J.: Face Recognition by Independent Component Analysis. IEEE Transactions on Neural Networks 13 (2002) 1450-1464
6. Zhang, N., Weng, J.: Sparse Representation from A Winner-take-all Neural Network. IEEE Int. Joint Conf. on Neural Networks (IJCNN'04). Budapest Hungary (2004) 2209-2214
7. Karvanen, J., Cichocki, A.: Measuring Sparseness of Noise Signals. In Pro. of 4th Int. Sym. on Independent Component Analysis and Blind Signal Seperation (ICA'03). Nara New Public Hall Japan (2003) 125-130

An Implementation of the Korean Sign Language Recognizer Using Neural Network Based on the Post PC

Jung-Hyun Kim and Kwang-Seok Hong

School of Information and Communication Engineering, Sungkyunkwan University, 300, Chunchun-dong, Jangan-gu, Suwon, KyungKi-do, 440-746, Korea
kjh0328@skku.edu, kshong@skku.ac.kr
http://hci.skku.ac.kr

Abstract. A traditional studies about recognition and representation technology of sign language have several restrictions such as conditionality in space and limitation of motion according to the technology of wire communication, problem of image capture system or video processing system for an acquisition of sign language signals, and the sign language recognition system based on word and morpheme. In order to overcome these restrictions and problems, in this paper, we implement the Korean sign language recognizer in the shape of sentence using neural network based on the Post wearable PC platform. The advantages of our approach are as follows: 1) it improves efficiency of the sign language input module according to the technology of wireless communication, 2) it recognizes and represents continuous sign language of users with flexibility in real time, and 3) it is possible more effective and free interchange of ideas and information between deaf person and hearing person (the public). Experimental result shows the average recognition rate of 92.8% about significant, dynamic and continuous the Korean sign language.

1 Introduction

The sign language is defined any form of communication using gestures to represent words and ideas, especially an official system of hand gestures used by an aphasiac or a deaf person. That is, a sign language is a language which uses manual communication instead of sound to convey meaning - simultaneously combining hand shapes, orientation and movement of the hands, arms or body, and facial expressions to fluidly express a speaker's thoughts [1]. The related studies about recognition and representation technology of sign language is progressing actively in many different countries such as the Americas, Asia, Europe, South Korea etc. to a hand-signal recognition for crane operation of the construction field, the control system of mobile robot and avatar using hand action recognition, the communication with the hearing person (the public) through sign language recognition, the hand action recognition in virtual reality and so on. Especially, "standard sign language translation system" developed jointly by the professors of the KIST (The Korea Institute of Science and Technology) and the Samsung Electronics Co., Ltd. in the South Korea is sign language translation system that recognize and represents the sign language of elementary student level, and it is possible recognition about 20 basic sign language and 31

J. Wang et al. (Eds.): ISNN 2006, LNCS 3972, pp. 222 – 231, 2006.

finger-spelling. Also, the Hitachi Ltd. in the Japan announced "Japanese - sign language translation technology" that a sign language animation is created automatically after input Japanese sentence [2].

But, The general purposes and features of these studies are as follows: 1) it has general purposes to the control of optional hand signal and sign language recognition based on the desktop computer and the technology of wire communication, 2) it used image capture system or video processing system for an acquisition of sign language signals, and 3) it is pursuing free communication with hearing person (the public) through sign language recognition system that emphasize gesture recognition and representation based on word and morpheme.

Therefore, in this paper, in order to implement user interface to guarantee mobility in portable terminal, we proposes and implements real time the sign language recognizer in the shape of sentence based on the embedded-ubiquitous environment using the Post wearable PC platform(embedded i.MX21 board), blue-tooth module and neural network algorithm and Relational Database Management System(hereafter, RDBMS) module. The advantages of our approach are as follows: 1) it improves efficiency of the sign language input module according to the technology of wireless communication, 2) it contributes to user's the convenience, 3) it recognizes and represents continuous sign language of users with flexibility in real time, and 4) Because the ability of communication and representation of the sign language recognizer in the shape of sentence are very superior more than sign language recognizer based on word and morpheme, it is possible more effective and free interchange of ideas and information between deaf person(and an aphasiac) and hearing person(the public).

The proposed gesture recognition system consists of three modules: 1) the gesture input module that processes motion of dynamic hand to input data, 2) the RDBMS module to segment significant gestures from input data, and 3) the neural network module to recognize significant sign language of continuous, dynamic sign language and to enhance the extensibility of recognition.

2 The Regulation of the Korean Sign Language

Sign language differs from oral language in its relation to writing. The phonemic systems of oral languages are primarily sequential: that is, the majority of phonemes are produced in a sequence one after another, although many languages also have non-sequential aspects such as tone. As a consequence, traditional phonemic writing systems are also sequential, with at best diacritics for non-sequential aspects such as stress and tone. The signs are formed by 7 sign formation principles: indication; imitation; metonymy; metonymy-indication; metonymy-movement; blending; home-sign according to structure method of sign and gestures. Also, we prescribed basic elements (meaningless units) that correspond to phoneme of spoken language by “sign language morpheme” in Korean sign language. That is, There are 5 classes of cheremes (a motor analogue to a phoneme) in the Korean sign language: Configuration of the hand-DEZ (designator); position of the hand-TAB (tabulator); movement of the hand-SIG (signation); direction of the hand; and Non-manual markers (or Facial Expression). And there are 30 cheremes in the configuration, 23 cheremes in the position, 36 cheremes in the movement, and 20 cheremes in the direction and in the motion respectively [3]. In order to implement significant sign language recognition interface in real time, this study selected 25 basic sign language gesture through priority "Korean Standard Sign Language Tutor(hereafter, KSSLT)[4]" analysis and according to a classified standard of 'sign language morpheme'. And 23 hand gestures necessities of sign language gesture are classified as hand shapes, pitch and roll degree. Consequently, we developed 44 significant sign language recognition models according to associability and presentation of hand gestures and basic sign language gestures. The examples of sign language gesture and hand gesture about "the date (numbers-day-month)" in 44 significant sign language recognition models are shown in Table 1.

Table 1. The example of sign language about "the date (finger numbers-day-month)"

Object Language	Description of the Korean Sign Language
Finger-Number	1~30(or 31) : Signing of numbers
Month	Represent signing of numbers that correspond to "1(one)" with a left hand (indication of month), and draw crescent with thumb, index finger of a right hand
Day	Spread out thumb, index fingers of a both hands and rises on in front of chest.
Hand Gesture	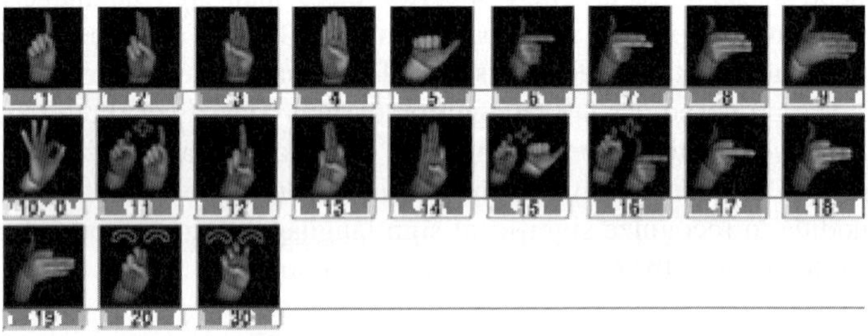

3 The Sign Language Input Module

In this paper, we use 5DT company's 5th data glove system (wireless) and fastrak® which is one of the most popular input devices in the haptic application field. As a traditional method to acquire significant dynamic gesture data of user from haptic devices, hand gesture input module using the technology of wire communication (based on desktop pc environment) has several restrictions such as conditionality on space, complexity between transmission mediums (cable elements) and limitation of motion. Also, in vision technology for image processing and recognition, recognition performance have the possibility of change and diversity according to the performance of the camera, change of background (colors) and illumination condition. In order to overcome these problems, in this paper, we implement sign language recognition system using 5th data glove system, fastrak®, neural network algorithm and RDBMS based on the wearable post PC (the embedded-ubiquitous environment using blue-tooth module, embedded i.MX21 board). 5th data glove system is basic gesture recognition equipment that can capture the degree of finger stooping using fiber-optic flex sensor and acquires data through this. Also, because it has pitch and roll sensor inside, the pitch and roll of wrist can be also recognized without other equipment possible. For motion information of 5th data glove system's each finger, user's dynamic gesture data is saved by f1=thumb, f2=index, f3=middle, f4=ring and f5=little in regular sequence. Each flexure value has a decimal range of 0 to 255, with a low value indicating an inflexed finger, and a high value indicating a flexed finger. Also, the most accurate electromagnetic motion tracking system available, fastrak® is the best solution for accurately computing position and orientation through space. With real time, six degrees of freedom tracking and virtually no latency, this award-winning system is ideal for head, hand, and instrument tracking. And fastrak® is a 3D digitizer and a quad receiver motion tracker, making it perfect for a wide range of applications requiring high resolution, accuracy, and range. By computing the position and orientation of a small receiver as it moves through space, it provides dynamic, real-time measurements of position (X, Y, and Z Cartesian coordinates) and orientation (azimuth, elevation, and roll) [6].

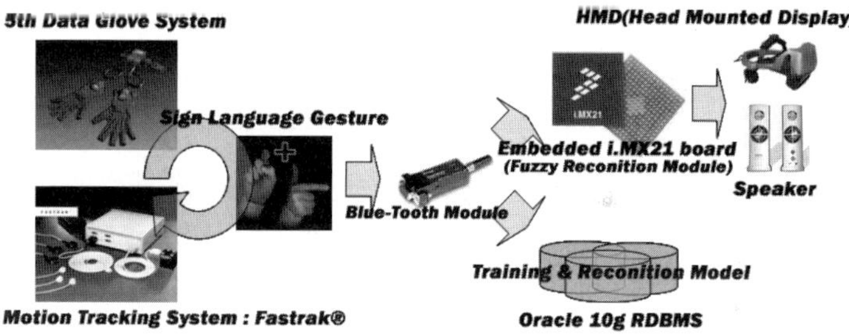

Fig. 1. The architecture of sign language input module

The captured dynamic gesture data of various users is transmitted to embedded i.MX21 board and server (Oracle 10g RDBMS) through blue-tooth module. The sign language data transmitted to server is used to as a training data for sign language recognition model by analytic function of Oracle 10g RDBMS SQL. And, sign language data that is transmitted to embedded i.MX21 board is used as the input to neural network recognition module for significant gestures recognition. The architecture of sign language input module is shown in Fig. 1.

4 The Training and Recognition Models Using RDBMS

The RDBMS is the main stream database management system that maintains data records and indices in tables and their relationships may be created and maintained across and among the data and tables. Also, often used for transaction processing and data warehouses and has the capability to recombine the data items from different tables, providing powerful tools for data usage [7], [8].

The RDBMS is used to classify saved gesture document data from the sign language input module (5th data glove system and fastrak®) into valid gesture record set and invalid record set (that is, status transition record set) and to efficiently analyze valid record set. The analytic function features, which are recently introduced in the SQL language, perfectly provide the analysis power for gesture validity analysis. According to the logic in source code, even though a record set is, based on the above process, decided as valid, the record set is regarded as a status transition gesture record set. If the valid record set contains less than 5 valid records in sequence. The continuous sign language from 5th data glove system and fastrak® are segmented into the significant record and transition record in 98% believability by the analytic function of SQL. A rule to segment valid gesture record set and invalid record-set (changing gesture set) is shown in Fig. 2.

1. Difference Between Preceding Average(preceding 3 and 1) and Current Row Value
2. Decide validity : Preceding Average - Current Value = 52-17 = 35

Average Between 3 Preceding and 1 Preceding from Current Row : 52

1. Difference between Preceding Average and Current Row Value > 5
2. 'X'(Invalidity Record) check

SerialNO	F1THUMB	F1INDEX	F1MIDDLE	F1RING	F1LITTLE	F1X-POSI.	F1Y-POSI.	F1Z-POSI.	F1_Px	F1_Py	GESTURE DATA...	VALIDITY
22	73	86	255	255	255	50.90	-14.53	-4.74	55.42	-5.71		X
23	51	48	255	255	255	52.05	-11.97	-2.13	53.29	-4.95		X
24	32	15	255	255	255	53.99	-9.50	0.01	50.02	-4.75		X
25	17	0	255	255	255	51.03	-10.50	-1.05	51.11	-4.83	Omission...	X
26	7	0	255	255	255	52.01	-9.85	-1.02	49.85	-5.23		
27	0	0	255	254	255	52.78	-8.77	-1.75	51.85	-6.02		
28	0	0	253	254	255	49.87	-10.18	-2.12	48.97	-5.27		
29	0	0	255	255	254	48.71	-9.43	-1.92	52.85	-4.89		
30	0	0	255	255	255	51.09	-8.75	-1.75	47.93	-4.94		

Fig. 2. The rule of segmentation and record set

- If the difference between preceding average (preceding 3 and 1) and current row value is over 5, the current value is regarded as transition sign language record.
- If one of 5th data glove system and fastrak® data value are over 5, current value data is also regarded as changing sign language record.

5 Neural Network Module for Sign Language Recognition

An Artificial Neural Network (hereinafter, ANN) is an information processing paradigm that is inspired by the way biological nervous systems, such as the brain, process information and is configured for a specific application, such as pattern recognition or data classification, through a learning process. Neural networks, with their remarkable ability to derive meaning from complicated or imprecise data, can be used to extract patterns and detect trends that are too complex to be noticed by either humans or other computer techniques [9]. Also, neural network is a powerful data modeling tool that is able to capture and represent complex input/output relationships. The true power and advantage of neural networks lies in their ability to represent both linear and non-linear relationships and in their ability to learn these relationships directly from the data being modeled.

With backpropagation, the input data is repeatedly presented to the neural network. With each presentation the output of the neural network is compared to the desired output and an error is computed. This error is then fed back (backpropagated) to the neural network and used to adjust the weights such that the error decreases with each iteration and the neural model gets closer and closer to producing the desired output. The proposed algorithm for perceptron training and Generalized Delta Rule (hereinafter, GDR) for weight conversion process are defined as in the following formula (1), and neural network training process that proposed in this paper is shown in Fig. 3.

$$y(t) = F(W_i \cdot A_i - \theta) \qquad W_i(t+1) = W_i(t) + @ \cdot (d_i - y_i) \cdot A_i(t)$$

$$\begin{bmatrix} \textit{where,} \\ y(t): \textit{output of R layer's PE} \\ F: \textit{fucntion that using in R layer's PE} \\ W_i: \textit{weights of A layer's \& R layer's PE} \\ A_i: \textit{input of A layer's PE} \\ \theta: \textit{critical value of R layer's PE} \end{bmatrix} \begin{bmatrix} \textit{where,} \\ W_i(t+1): \textit{weight after tarining} \\ W_i(t): \textit{weight from input i (time is t)} \\ @: \textit{training rate} \\ (d_i - y_i): \textit{error} \\ W_i(t) \cdot \textit{input pattern} \end{bmatrix} \quad (1)$$

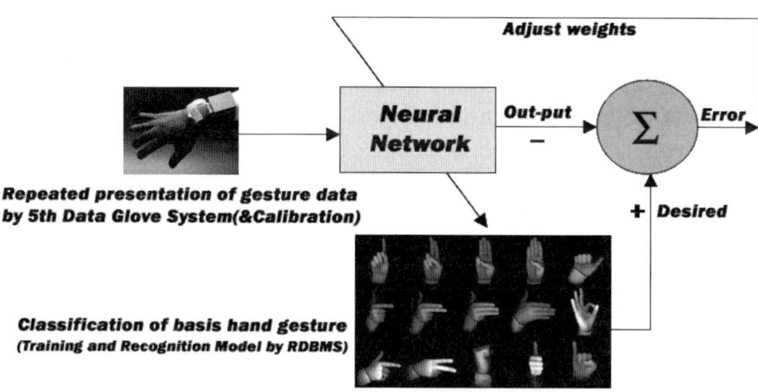

Fig. 3. The training process of neural network

The sign language is repeatedly presented to the neural network. With each presentation, the error between the network output and the desired output is computed and fed back to the neural network. The neural network uses this error to adjust its weights such that the error will be decreased. This sequence of events is usually repeated until an acceptable error has been reached or until the network no longer appears to be learning.

The captured dynamic sign language data of various users is transmitted to embedded i.MX21 board and saved as a document (text type). The significant sign language of users is then fed into a neural network that has been trained to make the association between the sign language data and a weight value that corresponds to the sign language recognition model by RDBMS. The output from the neural network is then

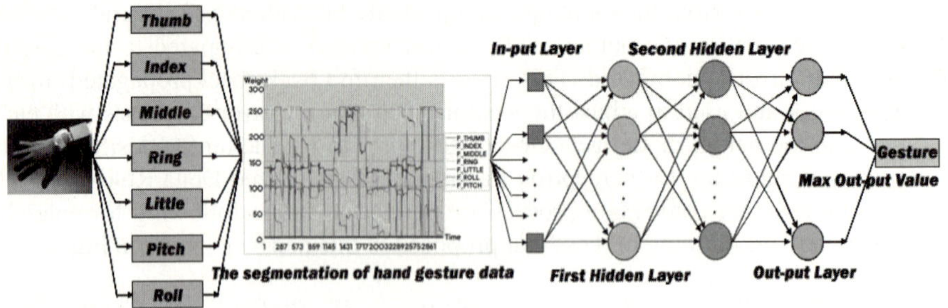

Fig. 4. The concept of sign language recognition system using neural network

calculated by weight value's summation and this is result of user's dynamic sign language recognition. The basic concept of sign language recognition system using neural network is shown in Fig. 4.

6 Experiments and Results

Experimental environment consists of blue-tooth module with i.MX21 board under embedded LINUX and implemented wireless-ubiquitous environment. In the proposed neural network sign language recognition system, sign language input module uses blue-tooth module for wireless communication. That is, 5th data glove system transmits 14kinds of hand gesture data (10 fingers gesture data, 4 pitch & roll data) and fastrak® transmits 12 kinds of sign language gesture data (position X, Y, and Z Cartesian coordinates and orientation-azimuth, elevation, and roll) in both hands to embedded i.MX21 board via blue-tooth module with transmission speeds of maximum 9,600 bps (right hand), 19,200 bps (left hand) and 115,200 bps in the case of fastrak®.

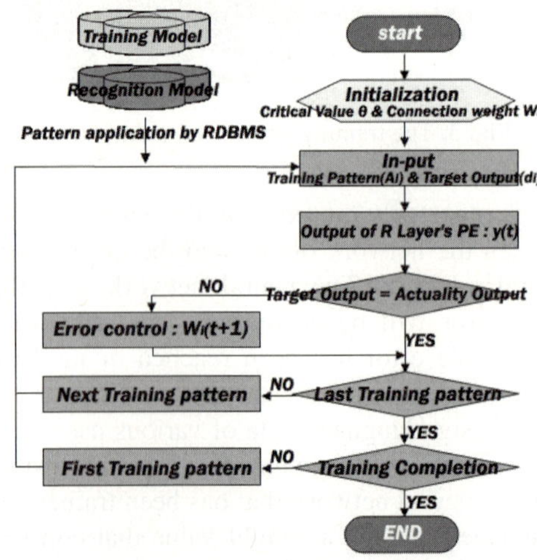

Fig. 5. The flow-chart of training module in neural network recognition system

The proposed neural network gesture recognition system's whole file size is 215 Kbytes and can calculate 175 samples per seconds on i.MX21 board. The neural network recognition system is consisted of 2 module of training module and recognition module. The overall system process of training module using neural network are as following: 1) After initialization of critical value and Connection weight, input training pattern and target pattern by RDBMS, 2) Calculate output of R layer's PE, and compare target output and actuality output and 3) If target output and actuality output are same, it end training. Otherwise, train last pattern after error control. Also, The overall system process of recognition module using neural network are as following: 1) After it loading connection weight, and input recognition pattern using 5th Data Glove System, 2) it calculate output results, and after compare with the recognition rate, remember maximum value, 3) If number of trained pattern conforms, it display recognition result . The process of training and recognition in neural network recognition system are shown in Fig. 5 and Fig. 6.

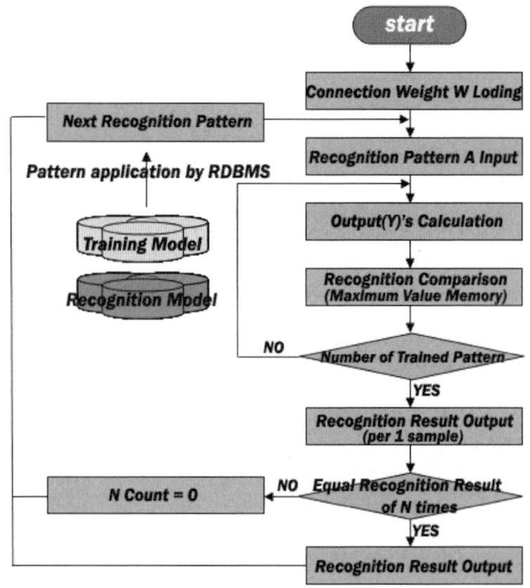

Fig. 6. The flow-chart of recognition module in neural network recognition system

Experimental set-up is as follows. The distance between sign language input module and embedded i.MX21 board for processing of sign language recognition is about radius 10M's ellipse form. As the gesture, we move the 5th data glove system and receivers of the fastrak® to prescribed position. For every 15 reagents, we repeat this action 10 times. Experimental result, Fig. 7 shows the average recognition rate of 92.8% in neural network max-min recognition module about significantly, dynamic sign language (44 recognition models). The root causes of errors between recognition of sign language are various: that is, imprecision of prescribed actions, user's action inexperience, and the changes in experiment environment of physical transformation at 5th data glove system fiber-optic flex sensor.

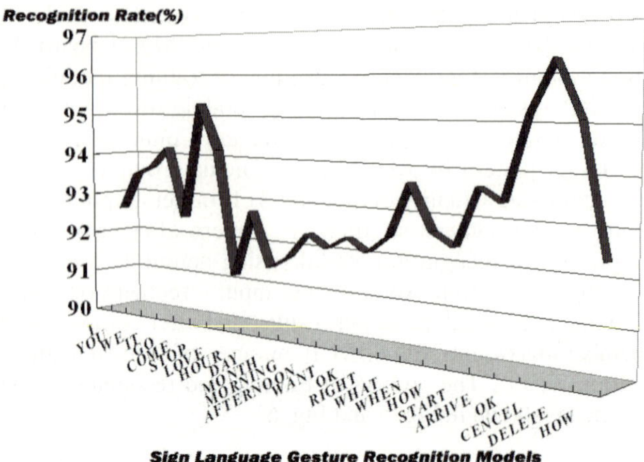

Fig. 7. The average recognition rate

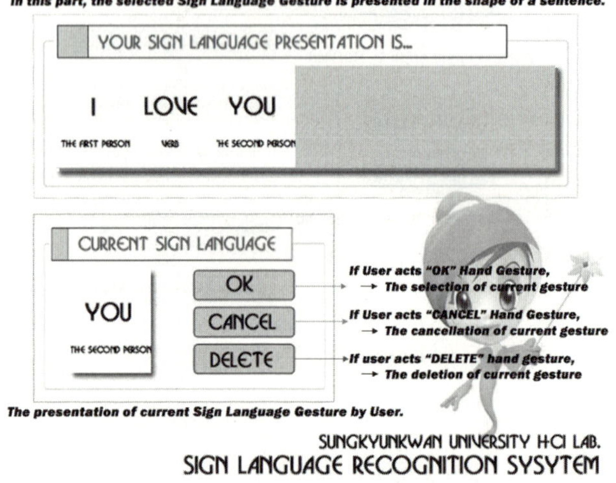

Fig. 8. The user interface for visual representation of the Korean sign language recognition

Also, the user interface for visual representation of the Korean sign language is shown in Fig. 8.

7 Conclusions

The Post-wearable PC is subset of ubiquitous computing that is the embedding of computers in the world around us. In this paper, we implemented sign language recognition system that analyzes user's intention more efficiently and more accurately can recognizes and represents continuous 44 significant, dynamic the Korean sign language of users with flexibility in real time using neural network recognition algorithm and RDBMS module in Post wearable PC platform. Also, because the sign language recognizer in the shape of sentence has very superior the ability of communication and representation, we can expect the function that understand life and culture of deaf person (and an aphasiac) and connect with modern society of hearing person through sign language recognizer in this study.

The result of the experiment for the Korean sign language recognition was successful and satisfactory, and in the future, by integrating sign language recognition system with other haptics such as smell, taste, hearing and sight, we would like to develop advanced multi-modal HCI technology.

Acknowledgement

This research was supported by the MIC (Ministry of Information and Communication), Korea, under the ITRC (Information Technology Research Center) support program supervised by the IITA (Institute of Information Technology Assessment)" (IITA-2005- C1090-0501-0019).

References

1. Use of Signs in Hearing Communities: http://en.wikipedia.org/wiki/Sign_language
2. Jang, H.Y., Kim, D.J., Kim, J.B., Bien, Z.N.: A Study on Hand-Signal Recognition System in 3-Dimensional Space. Journal of IEEK 41(2004) 103-114
3. Kim, S.G.: Standardization of Signed Korean, Journal of KSSE 9 (1992)
4. Kim, S.G.: Korean Standard Sign Language Tutor. Osung Publishing Company, Seoul (2000)
5. Kim, J.H., Kim, D.G., Shin, J.H., Lee, S.W., Hong, K.S.: Hand Gesture Recognition System using Fuzzy Algorithm and RDBMS for Post PC. FSKD2005. Lecture Notes in Artificial Intelligence, Vol. 3614. Springer-Verlag, Berlin Heidelberg New York (2005) 170-175
6. 5DT Data Glove 5 Manual and FASTRAK® Data Sheet: http://www.5dt.com
7. Relational DataBase Management System: http://www.auditmypc.com/acronym/RDBMS.asp
8. Oracle 10g DW Guide.: http://www.oracle.com
9. Simon, H.: Neural Network-A Comprehensive Foundation. Prentice-Hall,Inc., New Jersey (1999)

Gait Recognition Using Wavelet Descriptors and Independent Component Analysis

Jiwen Lu, Erhu Zhang, and Cuining Jing

Department of Information Science, Xi'an University of Technology,
Xi'an, Shaanxi, 710048, China
{lujiwen, eh-zhang, jcn_abela}@xaut.edu.cn

Abstract. This paper proposes an approach to automatic gait recognition based on wavelet descriptors and independent component analysis (ICA) for the purpose of human identification at a distance. Firstly, the background extraction method is applied to subtract the moving human figures accurately and to obtain binary silhouettes. Secondly, these silhouettes are described with wavelet descriptors and converted into one-dimensional signals to get the independent components (ICs) of these feature signals through ICA. Then, a fast and robust fixed-point algorithm for calculating the ICs is adopted and a selection criterion how to choose ICs is given. Lastly, the nearest neighbor and support vector machine classifiers are chosen for recognition and the method is tested on the XAUT and NLPR gait database. Experimental results show that our method has encouraging recognition accuracy with comparatively low computational cost.

1 Introduction

The demand for automatic human identification systems is strongly increasing in many important applications, especially at a great distance in many security-sensitive environments. Gait recognition aims at identifying person by the way he or she walk and it has some prominent advantages of being non-contact, non-invasive, unobvious, low resolution requirement and it is the only perceivable biometric feature for human identification at a great distance till now. Unlike face, gait is also difficult to conceal and has great potential applications in many situations especially for human identification at a great distance. Although gait recognition is a new research field, there have been some studies and researches in recent literatures [1], [2], [3], [4], [5], [6], [7], [8] and [9]. Currently, gait recognition approaches are classified two main classes, namely holistic-based methods [2], [3], [4], [5], [6], and [7] and model-based methods [8] and [9]. As the effectiveness of model-based techniques is still limited, most existing gait recognition methods are holistic-based. Hence, like other holistic-based algorithms, we can consider gait being composed of a sequence of body poses and recognize it by the similarity of these body poses and silhouettes and this paper proposes an automatic gait recognition method for human identification using wavelet descriptors and independent component analysis (ICA) based on this idea.

J. Wang et al. (Eds.): ISNN 2006, LNCS 3972, pp. 232 – 237, 2006.

2 Feature Extraction and Representation

Before training and recognition, motion segmentation, shadow elimination, feature representation and extraction should be processed at this preprocessing stage.

2.1 Segmentation of Human Motion

Human segmentation is the first step of our method and plays a key role in the whole gait recognition system. Here, let P represents a sequence including N frames. The resulting background $p(x, y)$ can be computed as formulas (1):

$$p(x, y) = median(p_1(x, y), p_2(x, y), \cdots, p_N(x, y)) . \tag{1}$$

For each image, the changing pixels can be detected by a suitable threshold T decided by traditional histogram and then we can easily obtain human silhouettes. After this procedure, there still exist some noises in the foreground, shadow elimination method [5] is used to eliminate the shadow and erosion, dilation and connected component analysis are adopted for further processing. Then, edge tracking is applied to obtain one connected edge. One example of background subtraction can be seen in Fig.1 from (a) to (h). Two gait databases, named NLPR and XAUT gait dataset, are selected here as the data set for our gait experiments.

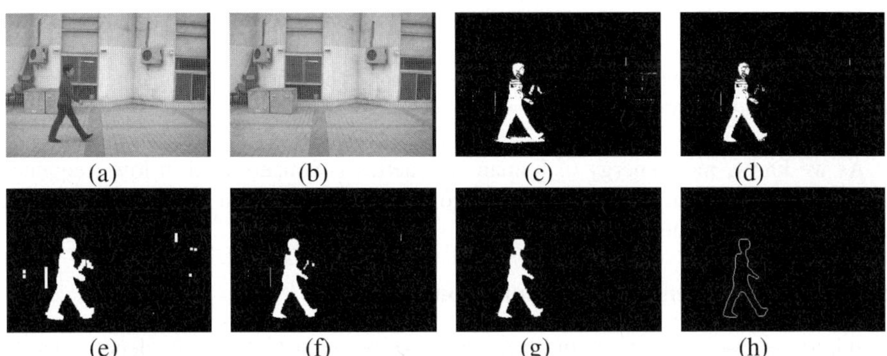

Fig. 1. One example of gait image preprocessing, here is one frame of gait image of NLPR gait database. (a) is one original image, (b) is the subtracted background, (c) is the differencing image and (d) is the image after shadow elimination, (e) and (f) are the image after dilation and erosion operation, (g) is the image after connected component analysis and (h) is the last edge.

2.2 Representation of Human Silhouette Using Wavelet Descriptors

Wavelet descriptors have been established recently and proved as a good method for representing a two-dimensional shape's boundary and its major advantage is that it has strong robust against rotation, scale and linear transformation. When one shape is represented in wavelet domain, its frequency component can be easily obtained. The general features of the shape are located in the lower frequencies while the details features are located in the higher frequencies. Now, we apply discrete wavelet descriptors to describe the human silhouettes. First, each counter is set in the complex plane and its centroid is set as the origin complex. Each point on the counter can be represented by a complex number $s_i = x_i + j \times y_i$, where N is the number of counter points. We select the same number of points to represent each counters and unwrap each counter counterclockwise from the top of the counter and converted into one dimension vector $[d_1, d_2, \cdots, d_N]$, d_i can be computed as formulas (2):

$$d_i = \sqrt{(x_i - x_c)^2 + (y_i - y_c)^2} \, . \tag{2}$$

Where (x_c, y_c) is the centroid of the human boundary. Therefore each gait sequence is transformed into a sequence of complex vectors with the same dimensions and the value of N in our experiment is 256 after re-sampling the boundary points. One example of wavelet descriptor describing human boundary can be seen as Fig.2:

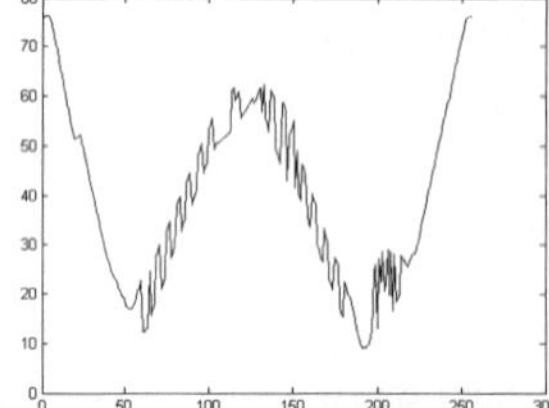

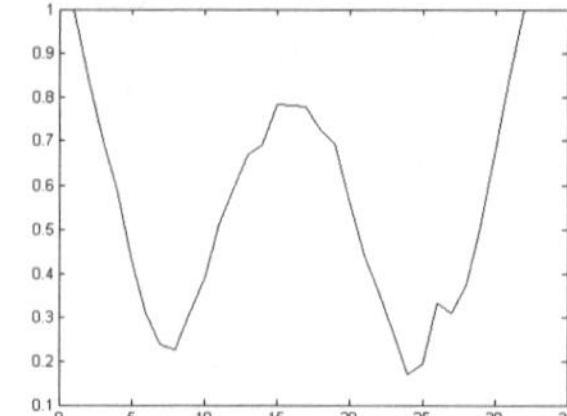

Fig. 2. Feature representation using wavelet descriptor. The left is the original boundary and the right is the boundary described with wavelet descriptor.

As we know, most energy of human silhouettes is concentrated in low frequency and only 64 points of lowest frequency components is selected for the least loss of silhouettes for computational convenience.

2.3 Feature Extraction and Compression Using ICA

At this stage, we will extract and train gait feature using ICA. Let us denote the observed variables x_i as a vector with zero-mean random variable $X = (x_1, x_2, \cdots, x_m)^T$, the component variables s_i as a vector $S = (s_1, s_2, \cdots, s_n)^T$ with the model AS

$$X = AS \, . \tag{3}$$

For reducing computational cost, an algorithm named FastICA [11] using a fix-point iteration algorithm finding the local extrema of the kurtosis of a linear combination of the observed variables is introduced. To reduce the computational cost, we select some ICs from A in the way that the ratio of the within-class scatter and between-class scatter is minimized [6], [7] and [12].

If the matrix X contains n individual persons and each person has m frames images, a_{ij} represents the entry at the ith row and the jth column. The value SB_j, which is called as the mean of within-class distance in the jth column, is then given by:

$$SB_j = \frac{1}{mn(m-1)} \sum_{i=1}^{n} \sum_{u=1}^{m} \sum_{v=1}^{m} (a_{(i-1)m+u,j} - a_{(i-1)m+v,j})^2 . \tag{4}$$

The value SI_j, which is called as the mean of between-class distance in the jth column:

$$SI_j = \frac{1}{n(m-1)} \sum_{s=1}^{n} \sum_{t=1}^{m} \rho(\bar{a}_{s,j} - \bar{a}_{t,j}) . \tag{5}$$

Where

$$\bar{a}_{i,j} = \sum_{u=1}^{m} a_{(i-1)m+u,j} . \tag{6}$$

In this paper, we employ the ratio of within-class distance and between-class distance to select stable mixing feature from A .The ratio γ_j is defined as:

$$\gamma_j = \frac{SB_j}{SI_j} . \tag{7}$$

From the definition γ_j, the smaller γ_j is, the better the classifier will be. Using (7), we choose the smallest γ_j and select the top k ($k < n$) column features from A and S .

3 Experimental Results and Analysis

The classification process is carried out through two different methods, namely the nearest neighbor (NN) and support vector machine (SVM) classifier derived from the ICs. NN classifier is a very simple classifier and the Euclidean distance is applied to evaluate the discriminatory of two gait sequences. SVM classifiers have high generalization capabilities in many tasks especially in pattern analysis and computer vision. SVM is based on structural risk minimization, which is the expectation of the test error for the trained machine. More details can be seen in [13].

There are several kernel functions used in SVM, here radial basis function (RBF) is adopted in our experiment and our SVM classifier is a 2-class classifier and there are two options for us: one is using N SVMs (N being the number of classes) while another is separating one class from the rest or using $N(N-1)/2$ SVMs one for each pair of class. We select the first option in our experiments because it is less complex.

Two public gait databases, namely Chinese National Laboratory of Pattern Recognition (NLPR) and Chinese Xi'an University of Technology (XAUT) gait database, are selected to test and evaluate the capability of the proposed method in this paper. Here NLPR database includes 20 subjects and four sequences for each views angle and have three angles, namely laterally (0°), obliquely (45°) and frontally (90°), XAUT database includes 10 subjects and four sequences for each views angle and have three angles, namely laterally (0°), obliquely (45°) and frontally (90°) Table 1 and 2 give the experiment results separately using different classifiers on the two datasets as follows:

Table 1. The recognition results using the NN classifier

	XAUT database (10 persons, 3 views)		NLPR database (20 persons, 3 views)	
	150 ICs selected	Using all ICs	300 ICs selected	Using all ICs
Rank 1	80.0%	80.0%	75.0%	75.0%
Rank 5	95.0%	92.5%	95.0%	90.0%
Rank 10	100%	100%	100%	100%

Table 2. The recognition results using the SVM classifier

	XAUT database (10 persons, 3 views)		NLPR database (20 persons, 3 views)	
	150 ICs selected	Using all ICs	150 ICs selected	Using all ICs
Rank 1	87.5%	85.0%	85.0%	82.5%
Rank 5	97.5%	92.5%	95.0%	90.0%
Rank 10	100%	100%	100%	100%

4 Conclusions and Future Work

This paper has proposed a simple gait recognition method based on human silhouettes using wavelet descriptors and ICA. From our analysis, we can make the conclusions that the ICs which are transformed from the frequency components have much better discriminatory capability than other gait feature. Although our recognition accuracy is comparatively high, much work still remains to be done in the future. Firstly, a much larger and most varied database with more subjects, more sequences with more different views and more variation in conditions such as the walkers wear different clothes in different seasons is strongly needed. Now we are building such a large gait databases consists of multi-view and work for solving this problem. Another is seeking better maturity measures, designing more effective classifiers, such as fusing NN and SVM into NN-SVM, extracting more effective gait feature, proposing better gait detection and segmentation algorithms and combination of holistic-based and model-based methods.

Acknowledgements

The authors would like to express their thanks to the Institute of Automatic, Chinese Academic of Science (CASIA) for Human ID image database. Portion of the research in this paper uses the CASIA Gait database are collected by Institute of Automatic, Chinese Academic of Science. This work is partly supported by the Science and Technology Innovation Foundation of Xi'an University of Technology (No. 104-210401, 104-210413, 104-210512).

References

1. http://sinobiometrics.com
2. Wang, L., Tan, T., Ning, H., Hu, W.: Silhouette Analysis-Based Gait Recognition for Human Identification. IEEE Transactions on Pattern Analysis and Machine Intelligence 25 (2003) 1505-1518
3. Mowbray, S.D., Nixon, M.S.: Automatic Gait Recognition via Fourier Descriptors of Deformable Objects. Proceedings of 4th International Conference on Audio and Video-based Biometrics Person Authentication (2003)
4. Yu, S., Wang, L., Hu, W., Tan T.: Gait Analysis for Human Identification in Frequency Domain. Proceedings of the Third International Conference on Image and Graphics. Hong Kong, China (2004)
5. Han, H.: Study on Gait Extraction and Human Identification Based on Computer Vision. Doctoral thesis. University of Science and Technology Beijing, Beijing (2003) (in Chinese)
6. Lu, J.W., Zhang, E.H, Zhang, Z.G., Xue, Y.X., Gait Recognition Using Independent Component Analysis, Proceedings of 2nd International Symposium on Neural Networks. Chongqing, China (2005).
7. Zhang, E.H., Lu, J.W., Duan, G.L.: Gait Recognition via Independent Component Analysis Based on Support Vector Machine and Neural Network, Proceedings of 1st International Conference on Natural Computation. Changsha, China (2005).
8. Lee, L.: Gait Analysis for Classification. Doctoral Thesis. Massachusetts Institute of Technology, USA (2002)
9. Wagg, D.K., Nixon, M.S.: An Automated Model-based Extraction and Analysis of Gait. IEEE Conference on Automatic Face and Gesture Recognition, Seoul, Korea, 5 (2004) 11-16
10. Hyvarinen, A.: Independent Component Analysis: Algorithm and Applications. Neural Networks 13 (2000) 411-430
11. Hyvarinen, A.: A Fast and Robust Fixed-Point Algorithm for Independent Component Analysis. IEEE Transaction on Neural Networks 3 (1999) 626-634
12. Yuen, P.C., Lai, J.H.: Face representation using independent component analysis. Pattern Recognition 35 (2002) 1247-1257
13. Deniz, O., Catrillon, M., Hernandez, M.: Face Recognition Using Independent Component Analysis and Support Vector Machines. Pattern Recognition Letters 24 (2003) 2153-2157

Gait Recognition Using Principal Curves and Neural Networks

Han Su[1,2] and Fenggang Huang[1]

[1] School of Computer Science and Technology, Harbin Engineering University, China
susuhan1016@yahoo.com.cn
[2] School of Computer Science, Sichuan Normal University, Chengdu 610066, China

Abstract. This paper presents a new method for human model-free gait recognition using principal curves analysis and neural networks. Principal curves are non-parametric, nonlinear generalizations of principal component analysis, and give a breakthrough to nonlinear principal component analysis. Different from the traditional statistical analysis methods, principal curve analysis seeks lower-dimensional manifolds for every class respectively, and forms the nonlinear summarization of the sample features and directions for each class. Neural network with the virtue of its universal approximation property is an outstanding method to model the nonlinear function of principal curve. Firstly, a background subtraction is used to separate objects from background. Secondly, we extract the contour of silhouettes and represent the spatio-temporal features. Finally, we use principal curves and neural networks to analyze the features to train and test gait sequences. Recognition results demonstrate that our method has encouraging recognition performance.

1 Introduction

Human gait recognition is the process of identifying individuals by their walking manners. As a biometric, gait has several particular advantages over other biometrics. It can be performed at a distance or at low resolution, while other biometric need higher resolution. Apart from this, it is difficult to disguise, and it requires no body-invading equipment to capture gait information. In these cases, gait recognition is becoming the second generation biometric recognition technology based on the vision movement. More and more researchers have devoted to this field. Current approaches to gait recognition can be broadly divided into two categories[1]: model-free ones[2,3,4,8,5,9] and model-based ones. The majority of the current approaches are model-free ones that deal directly with image statistics regardless of the underlying structure. Most recently approaches are based on analysis of silhouette. The methods of gait feature analysis include: Fourier analysis, HMM, PCA, ICA, LDA and CA etc. Different from them, principal curves analysis of this paper analyze gait data for each class directly. Principal curves(PC)[6,7] are smooth self-consistent curves that pass through the middle of the multidimensional data, and the theoretical foundation is to seek lower-dimensional non-Euclidean manifolds embedded multidimensional data space. They can reflect the inherent structure of the data and extract features effectively. According to these characteristics, this paper proposes a novel gait

J. Wang et al. (Eds.): ISNN 2006, LNCS 3972, pp. 238 – 243, 2006.

recognition method using principal curves and neural network(NN). We consider principal curves as the new method of nonlinear feature analysis and classification. PC seeks lower-dimensional manifolds for every classes respectively, forms the nonlinear summarization of the sample features and directions for each class. NN[8] are universal functional approximators and nonlinear models that can model real work complex relationships flexibly.

2 Gait Detection and Representation

2.1 Gait Detection

A background subtraction technique is used to segment and track the moving silhouette from the video sequence. Firstly, we model the background image. Then, the background subtraction is used to each frame and the common subset of background. Finally, the silhouettes are segmented, and we translate them into binary images. Note that the static background pixels sometimes include non-background pixels mainly due to the shadows, and the silhouettes can contain holes while directly subtracting the foreground image from original image. Then our foreground images are obtained by thresholding and applying morphology after background subtraction. Fig.1 is an example of silhouette extraction and contour extraction. And the silhouette contour is extracted by our preview method [9]. To get rid of the influence of the image size and different start, we normalized the sequences. We define that gait cycle is the time between the successive peak values of the width, and one stride includes three stance phases and two swing phases. Step length is one of our gait features by computing the width of the silhouette of lower limbs.

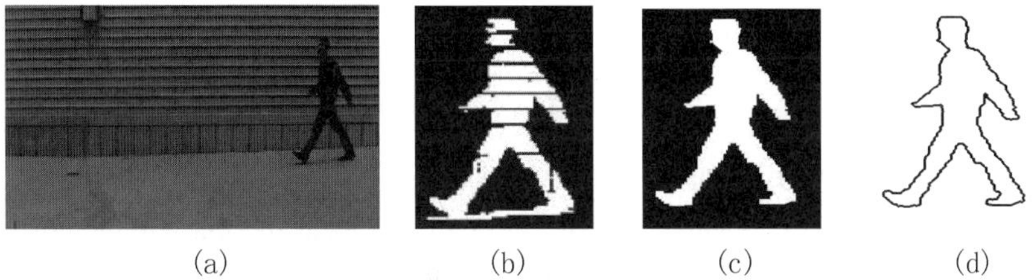

Fig. 1. An example of silhouette extraction and contour extraction. (a) is the original image, (b) is the difference image, (c) is the amended silhouette,(d) is the contour extracted from(c).

2.2 Spatiotemporal Gait Feature Extraction

The method of spatiotemporal analysis[9] is used to represent the contour of silhouette and extract the corresponding spatiotemporal gait feature. Before representing the contours, we have placed a window around the subject in order to reduce the computational complexity and the window has been normalized to be the same dimensions H×W. Fig.2 shows the representation of the silhouette's edge points. One point P_j on the boundary can be representing as $P_j = y_j + \theta_j i$, where y_j is the vertical distance between the origin and P_j, θ_j is the angle from P_j to x-axis in a plane. The variance of θ_j describes the variational orientation of one edge point. The set of boundary form the spatial contour matrix X . Through analyzing the shape contour, we extract the body width, height, step length and gait cycle to be our gait features.

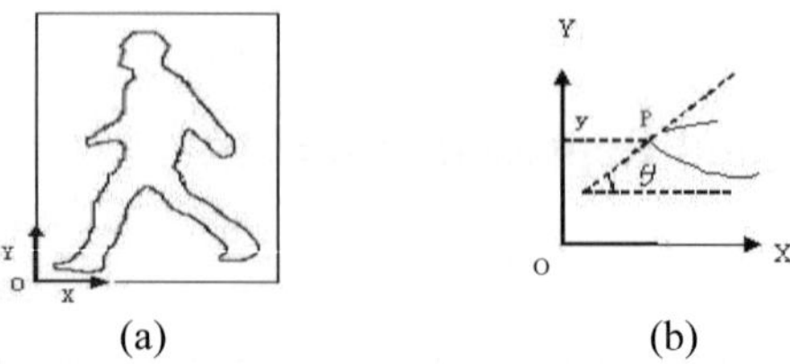

Fig. 2. Representation of the silhouette's boundary

The gait sequence is the pattern that be composed of a sequence of static body postures. The temporal variations of these postures are the keys for recognition. Recognizing people by gait depends greatly on how the silhouette shape of an individual changes over time in an image sequence[2]. We calculate all spatial contour matrixes for the bounding boxes, $X(t)$ is the spatial contour matrix of the tth frame in one's ith sequence. The temporal matrix is calculated as (1):

$$T(t1,t2) = X(t1) \otimes X(t2) = (t_{nm}) \in C^{hw}, 1 \le t1, t2 \le N . \tag{1}$$

where N is the number of spatial contour matrixes in a sequence,

$$X(t1) \otimes X(t2) = \begin{bmatrix} x_{11}(t1)X(t2) \cdots x_{1w}(t1)X(t2) \\ x_{21}(t1)X(t2) \cdots x_{2w}(t1)X(t2) \\ \vdots \\ x_{h1}(t1)X(t2) \cdots x_{hw}(t1)X(t2) \end{bmatrix} . \tag{2}$$

In (2), $x_{ij}(t1)X(t2)$ is calculated by $x_{nm}(t1) \cdot x_{pq}(t2) = | y_{t1} - y_{t2} | + | \theta_{t1} - \theta_{t2} | i$, and $t_{nm} = x_{nm}(t1)X(t2)$. In consideration of the relatively local motion within short time, we calculate $x_{nm}(t1)X'_{nm}(t2)$ instead of $x_{nm}(t1)X(t2)$. $X'_{nm}(t2)$ is a subset of $X(t2)$. It changes with the $x_{nm}(t1)$. $X'_{nm}(t2)$ is a $d \times d$, $d \ll H,W$ sub-matrix and center of $x_{nm}(t1)$. d is decided by interval between two frames. In additional we normalize t_{nm} to remove the disadvantage of unbalanced numerical values between $|\triangle y|$ and $|\triangle \theta|$.

For one image sequence, we can calculate the temporal matrix $T(ti,tj)$ between two consecutive frames and T that derives by summing all the temporal matrixes.

3 PC and NN Feature Analysis

Principal curves were defined by Hastie and Stuetzle[6] as "self-consistent" smooth curves which pass through the "middle" of a d-dimendional probability distribution or data cloud. And Principal curves are non-parametric, nonlinear generalizations of principal component analysis, and give a breakthrough as an approach to nonlinear principal component analysis.

Given a data set $X \in R^d$, a random vector $x_i \in X$ is defined by $x_i = f(\lambda_i) + e_i$ based on principal curves, where λ_i is the projection index, e_i is the noise which is IID and $f()$ denotes the vector function between the projection index and data.

Having extracted gait features by spatiotemporal representation, we analyze and train gait data using principal curves. Different from the traditional statistical analysis methods, we seek lower-dimensional manifolds for every classes respectively, forms the nonlinear summarization of the sample features and directions for each class[7].The essence of principal curves is 1D manifolds embedded Euclidean space, which describe not only the backbone of data set, but also the framework of it. Principal curves analysis can reserve the topological structure of data set. In accordance with the theory proposed by Chang[7], this paper uses principal curves for analyzing gait data.

For the gait feature set X of one person, $X = \{x_1, x_2, ... x_m\}, x_i \in R^d$, $i = 1...m$, a smooth curve $f(\lambda)$ is computed. x_i is gait feature vector for one sequence extracted in section 2.To form one person's principal curve template $f(\lambda)$, two steps are included, one is projection step, and the other is expectation step. First, we initialize the initial curve $f^{(0)}$ to be the first principal component. Then, all feature vectors are projected on the $f^{(j)}(\lambda)$, $\lambda_{f^{(j)}}(x)$ is calculated by:

$$\lambda_{f^{(j)}}(x) = \max\{\lambda : \| x - f(\lambda) \| = \min_{\omega} \| x - f(\omega) \|\} . \tag{3}$$

Expectation step computes the new principal curve $f^{(j+1)}(\lambda) = E(X \mid \lambda_f(X) = \lambda)$. Projection step and expectation iterate until $\mid D^2(x, f^{(j)}) - D^2(x, f^{(j+1)}) \mid / D^2(x, f^{(j)})$ is below some threshold, where $D^2(x, f) = E \| X - f(\lambda) \|^2$.

In this way, the principal curve $f(\lambda)$ for one person including all his(her) gait vectors is constructed. Considering $f(\lambda)$ is the curve parameterized by arc length and λ represents the ordering structure corresponding data points on the curve, we denote the gait feature set X to: $X = f(\boldsymbol{\lambda}) + E$, where $\boldsymbol{\lambda} = [\lambda(1), \lambda(2), ..., \lambda(n)]$, $\lambda(\mathrm{i})$ is an $m \times 1$ column vector that is composed of arc length of all m feature vectors along the curve, n equates to the number of nonlinear principal component, E is the residual.

Among the approaches for simulating $f()$, NN are universal functional approximators and NN are nonlinear models that can model real work complex relationships flexibly. So we use this outstanding method to model the arbitrary nonlinear representation $f()$ for analyzing gait feature. According to the method of Dong[8], we model $f()$ by two three-layer feedforward network with one sigmoid hidden layer and choose conjugate gradient learning method to train our neural networks because of the learning speed and the learning rate constants. The number of hidden-layer units is decided by cross-validation scheme.

Fig. 3 is the structure of our model including two three-layer feedforward networks, one maps any gait feature vector x_i to the projection index, and the other is the reverse process. We choose the dimension of λ according to need. The output of the first one is the input of the second one, the output of the second one is the corrected data for the first one. In essence, this is one autoassociative five-layer network.

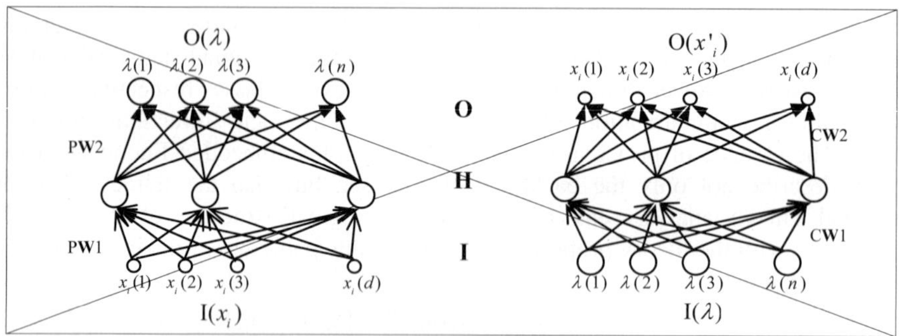

Fig. 3. The structure of the feedforward network for modeling the nonlinear function of principal curve

Thus, a principal curve for one person is formed in the training phase. For recognition, we compute k closest distances between the probe and the principal curves of different persons, and decide the class having most numbers of the nearest distances to be that of this probe.

4 Experiments and Analysis

To evaluate our method, we tested it on some gait databases using leave-one-out cross validation technique. First, we evaluate our approach for the data set of Little and Boyd. The database consists of 40 image sequences and 6 different subjects. The recognition rate on this database is 99.20%.Then, we train and test on the CMU MoBo and CASIA(NLPR)data set. CMU MoBo data set contain 6 simultaneous motion sequences of 25 subjects walking on a treadmill. Each subject is recorded performing four different types of walking: slow walk, fast walk, inclined walk and slow walk holding a ball. We test our method on the slow walk. The recognition rate is 88.00%. NLPR gait database include 80 sequences from 20 subjects and 4

Table 1. Recognition results on the NLPR database(0^0)

Methods	CCR(%)
BenAbdelkader[4]	72.50%
Lee[3]	87.50%
Phillips[5]	78.75%
Our method	89.31%

sequences per subject. Tab.1is the recognition rate on the NLPR database(0^0). Taking one with another, our method has an improvement in recognition performance.

The performances are owing to the spatiotemporal gait representation, the nonlinear analysis method and the universal functional approximators.

5 Conclusions

A new model-free gait recognition method based on PC and NN was presented in this paper. Principal curves analyzed the nonlinear gait data extracted by the spatiotemporal method, and neural networks modeled the nonlinear function of principal curve. Experimental results show that the method is efficient. In future, we will pay more attention to the feature space for describing, analyzing and recognizing the human gait on the larger database of subjects.

References

1. Nixon, M. S., Carter, J. N.: Advances in Automatic Gait Recognition. In: Proc. of IEEE Face and Gesture Analysis, Seoul, Korea (2004) 11-16
2. Wang, L., Tan, T.N., Ning, H.Z., Hu, W.M.: Silhouette Analysis-Based Gait Recognition for Human Identification. IEEE Transactions on Pattern Analysis and Machine Intelligence 25(12) (2003) 1505-1528
3. Lee, L., Grimson, W.: Gait Analysis for Recognition and Classification.In: Proc. of Intl. Conf. on Automatic Face and Gesture Recognition (2002) 155-162
4. Abdelkader, C. B., Cutler, R., Nanda, H., Davis, L.: EigenGait: Motion-Based Recognition Using Image Self- Similarity, In: AVBPA01, Lecture Notes in Computer Science, Vol. 2091. Springer-Verlag, Berlin Heidelberg New York (2001) 289-294
5. Phillips, P. J., Sakar, S., Robledo, I., Grother, P., Bowyer, K.: The Gait Identification Challenge Problem: Data Sets and Baseline Algorithm. In: Proc.of Intl. Conf. on Pattern Recognition, Québec, Canada (2002) 385-388
6. Hastie, T.: Principal Curves and Surfaces. USA, Stanford Univ. (1984)
7. Chang, K.Y.: Nonlinear Dimensionality Reduction Using Probabilistic Principal Surfaces. USA, Texas Univ. (2000)
8. Dong, D., McAvoy, T. J.: Nonlinear Principal Component Analysis Based on Principal Curves and Neural Networks. Computers Chemistry Engineering 20 (1) (1995) 65-78
9. Su, H., Huang, F.G.: A New Method for Human Gait Recognition Using Temporal Analysis. In: Yue Hao et al. (eds): Computational Intelligence and Security, Lecture Notes in Artificial Intelligence, Vol. 3801. Springer-Verlag, Berlin Heidelberg New York (2005) 1039-1044

An Adjacent Multiple Pedestrians Detection Based on ART2 Neural Network

Jong-Seok Lim, Woo-Beom Lee, and Wook-Hyun Kim

Department of Computer Engineering, Yeungnam University, Korea
{robertlim, beomlee, whkim}@yumail.ac.kr

Abstract. This paper presents a method to detect adjacent multiple pedestrians using the ART2 neural network from a moving camera image. A BMA(Block Matching Algorithm) is used to obtain a motion vector from two consecutive input frames. And a frame difference image is generated by the motion compensation with the motion vector. This image is transformed into binary image by the adapted threshold and a noise is also removed. To detect multiple pedestrians, a projection histogram is processed by the shape information of human being. However, in case that pedestrians exist adjacently each other, it is very different to separate them. So, we detect adjacent multiple pedestrians using the ART2 neural network. The experimental results on our test sequences will show the high efficiency of our method.

1 Introduction

In this paper we present a method to detect adjacent multiple pedestrians using the ART2 neural network. The class of applications we are interested in is the unmanned automobile system [1] with a car-mounted monocular color camera. In this system, a pedestrian detection is an essential part. That is, an unmanned car can head off the collision as previously perceiving a dangerous traffic situation. Therefore this can be used in various applications of a real world such as security or surveillance system and crosswalk signal control system [2].

The pedestrian detection is a previous step to track pedestrians. If pedestrians are detected correctly, the pedestrian tracking also can be achieved smoothly using all kinds of algorithms. Some of conventional pedestrian detection algorithms mainly detected one person and used a fixed camera. Generally, it is very ease to detect one pedestrian from a fixed camera. Because all moving objects can be detect by the frame difference from two consecutive input frames.

However, it is very difficult to detect multiple pedestrians from a moving camera. Because pedestrians are occluded or exist close at hand each other. Therefore, we classify adjacent pedestrians using the ART2 neural network clustering method [3].

There are many algorithms for moving object detection such as correlation based method, optical flow based method, shape based method, motion based method, frame difference based method, and model based method. The correlation based method [4] separates a local area of an image from that of the other

J. Wang et al. (Eds.): ISNN 2006, LNCS 3972, pp. 244–252, 2006.

image and detects a moving object by the correlation method. This method is operated in case of a moving camera but it is difficult to detect nonrigid objects. The optical flow based method [5] use an optical flow equation which modeled relation between speed and spatio-temporal gradient of a local illumination. But this method should be consistent a value of the brightness degree on the object surface and satisfy a speed smoothness constraint which adjacent neighboring pixels move with the equal shape. This method has drawbacks that obtains poor result in case that an interval between frames is wide, a illumination is change, a consistent shape can not keep it because of noise. The shape based method [6] depends on appearance features of an object. This method has a drawback that it should be consider conditions of a wide scope such as a shape of an object, a movement of flux, illumination, and a color. The motion based method [7] use an action pattern of a moving object in an image. The frame difference based method [8] use a difference of gray level from two consecutive input frames. But these methods do not detect pedestrians in case of having a subtle movement of background. The model based method [9] matches an input image with 2D model defined in advance. This method should be present an exact geometrical model.

In this paper, an efficient pedestrian detection algorithm that can handle situations with any object motion, multiple pedestrians and changing background by a moving camera is proposed. The rest of this paper is organized as follows. In the next section, the proposed algorithm is described in detail. Section 3 shows the experimental results. Finally, Section 4 concludes this paper.

2 The Proposed Approach

The proposed algorithm is divided into five major steps. The first step is to calculate the motion vector by BMA from two consecutive input frames. The second step is to calculate the frame difference using the motion vector of the previous step. In the third step, pre processing is carried out for the smooth work. The frame difference image is transformed into binary image and noises also eliminated by the proposed noise elimination method. Then, pedestrians are detected by the projection histogram[8]. However, if pedestrians exist adjacently each other, it is very difficult to detect them and can know only the location of pedestrians. Therefore, the pedestrians which is not detected in the fourth step is clustered by ART2 neural network in the fifth step.

The details of each step will be discussed in the following subsections.

2.1 Motion Vector

In case of a moving camera, the background in the image is change. In this status, it is very difficult to detect only moving objects using the frame difference between two consecutive input frames since a background region also is detected together with it. Therefore, we detect moving objects by the motion compensation using the motion vector which is calculated by the BMA.

The motion vector is founded by the direction dx, dy that is minimized the total sum D of the difference between the region of the current frame and that of the previous frame while scanning in the four direction. Their relation is shown as follows:

$$D = \Sigma\Sigma|f_n(x_i, y_j) - f_{n-1}(x_i + dx, y_j + dy)|. \quad (1)$$

where f_n is the current frame and f_{n-1} is the previous frame. x_i denotes the pixel location in the horizontal direction and y_j denotes the pixel location in the vertical direction. And the size of the region is 16×16 each and the maximum value of scanning in the region is eight pixels in the all direction.

Meanwhile, we calculate motion vector about the specific region in the image to reduce computing time. Once the motion vector is calculated, we find the representative motion vector to process the motion compensation. The representative motion vector is created by the motion vector that emerged most frequently among the entire motion vector. For example, if the representative motion vector calculated in an image is $x = 4$ and $y = 0$, it means that the previous frame was moved as the four pixels to the x direction and the zero pixel to the y direction.

2.2 Frame Difference

The frame difference between two consecutive input frames is widely used to detect moving objects due to a small computation time. However, this method does not detect moving objects in case that the background change by the moving camera. Therefore, we use the representative motion vector founded the previous subsection to detect moving objects.

General equation of the frame difference is shown as follows:

$$d(i, j) = |f_n(i, j) - f_{n-1}(i, j)|. \quad (2)$$

In this paper, we compute the frame difference by the motion compensation using the representative motion vector and the equation is shown as follows:

$$d(i, j) = |f_n(i + x, j + y) - f_{n-1}(i, j)|. \quad (3)$$

where x and y is a component of the representative motion vector.

Fig. 1 shows two examples of the frame difference image. Fig. 1(Left) is the image applied the motion compensation and Fig. 1(Right) is not. As shown in Fig. 1, Fig. 1(Left) is the clearer than Fig. 1(Right).

Fig. 1. Comparison of the frame difference image: motion compensation image(Left), general image(Right)

2.3 Pre Processing

After the frame difference step, moving objects is detected. However, due to the irregular camera movement, there exist some noise regions in the frame difference image. Therefore, a pre processing step to eliminate these noise regions is necessary.

The frame difference image should be transformed to a binary image before the noise regions are removed. The binary image is generated by thresholding the frame difference image obtained in previous subsection. The equation is shown as follows:

$$B(i,j) = \begin{cases} If \;\; d(i,j) \;\; \geq \;\; Th \;\; Then \;\; 255 \\ Otherwise \qquad\qquad\qquad 0. \end{cases} \tag{4}$$

where $B(i, j)$ denotes the binary image, $d(i, j)$ denotes the frame difference image and Th denotes the threshold value.

The transformed binary image has some noise regions. A traditional methods to remove noise regions are morphological operations. However, these operations can remove not only noise regions, but also object boundary. Therefore, we propose an effective noise elimination method instead of a traditional approaches. The method is to remove all of the pixels existing outside interested region. If the sum total of the pixels existing in the window which has an $n\times\, n$ size is smaller than a threshold value, all of the pixels in the window is deleted. The window consists of $10\times\, 20$ size. It is minimum size of pedestrian's face which we would like to detect it. And the threshold value consists of 44 pixels, which approaches a size of the circumference inscribed in the window.

Fig. 2 shows two examples for the pre processing. Fig. 2(Left) is the binary image transformed by the frame difference image(a threshold value is 40 pixels). Fig. 2(Right) is the image which noise regions are removed by our approach.

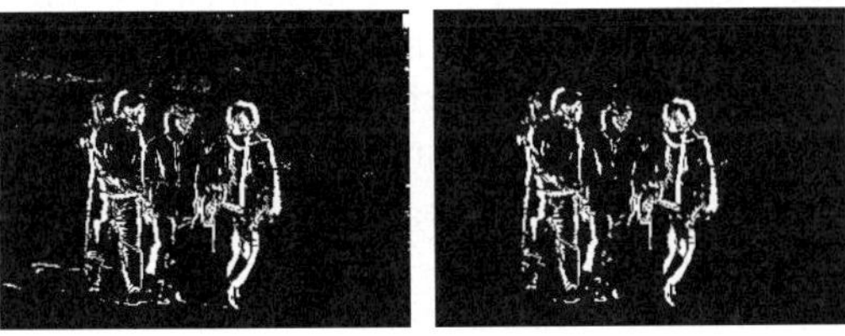

Fig. 2. Two example for the pre processing: binary image(Left), noise elimination image(Right)

2.4 Pedestrian Detection

The pedestrian detection step is to generate the bounding box around the respective pedestrians using the projection histogram. The projection histogram is made by counting pixels which is bigger than 0 existing to the horizontal or the vertical direction in the image. The equation is shown as follows:

$$H_i = \sum_{i=0}^{m-1} B'(i,j), \; V_j = \sum_{j=0}^{n-1} B'(i,j). \tag{5}$$

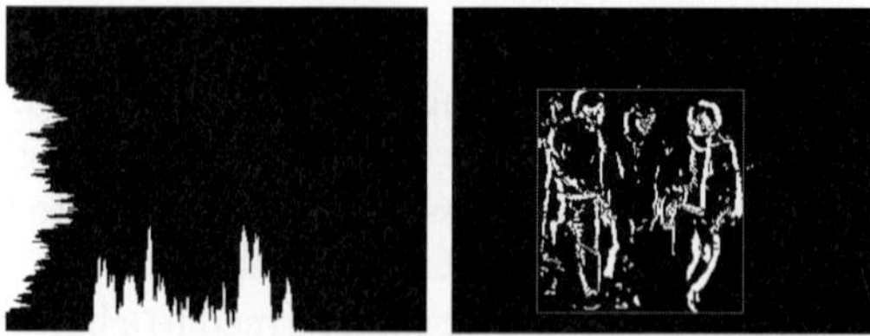

Fig. 3. Projection histogram and pedestrian detection result: projection histogram image(Left), pedestrian detection result(Right)

where H_i denotes the horizontal projection histogram , V_j denotes the vertical projection histogram and $B'(i,j)$ denotes the preprocessed binary image. Also, the range of i is from 0 to m-1, j is from 0 to n-1 in the $n\times\ m$ image size.

After generating the projection histogram, pedestrians are detected using human's shape information. The human being is very different from an automobile or an animal. Therefore it can distinguish human being from them using this information. However, this method can not detect pedestrians in case that multiple pedestrians exist adjacently each other or some occluded.

Fig. 3 shows the projection histogram and the image which pedestrians are detected. As shown in Fig. 3, pedestrians was not separated individually because pedestrians exist adjacently each other.

2.5 Classification Using ART2 Neural Network

If it failed to separate pedestrians by the projection histogram in the previous step, an ART2 neural network clustering method is used to classify pedestrians.

Originally, ART neural network model was proposed by Carpenter and Grossberg. There are ART1 and ART2. ART1 neural network can be worked with binary input and ART2 can work well with the analog input which has continuous values. And ART neural network has various characteristics such as unsupervised learning, real-time learning, dynamic structure, and stable-plastic network.

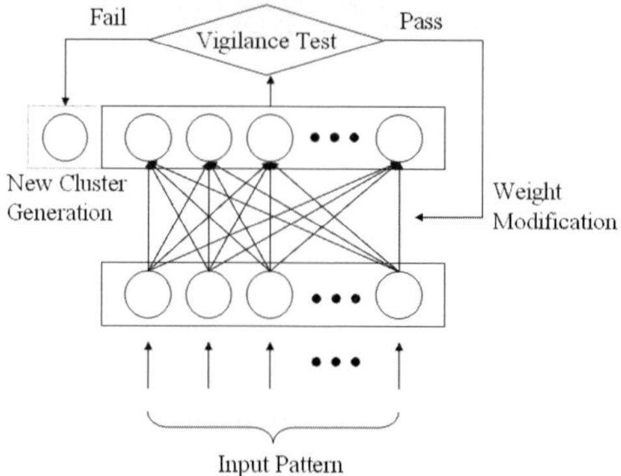

Fig. 4. Architecture of the ART2 neural network

Fig. 4 shows ART2 neural network structure. This consists of two layers. The first layer responds to an input pattern which is similar to stored one. If not, the second layer generates a new class after it closes the first layer.

The learning algorithm of the ART2 neural network is as follows:

1. Let x_k be the kth input vector and W_j be the jth weight vector of neural network(input and output node is $N\times N$).
2. Given a new input pattern, the second layer is adopted to select the winner cluster J, which yields the minimum distance.

$$d_j = ||W_j - x_k||, \ j = 1, 2, ..., M. \tag{6}$$

   where $||.||$ is an any norm, M is number of template pattern(weight of neural network).
3. The vigilance test is carried out for the input pattern. If $||W_j - x_k|| < \rho$, the input pattern is integrated into the winner cluster and the weight of the J is adjusted by

$$V_J^{new} = \frac{x_k + V_J^{old}||Cluster_J^{old}||}{||Cluster_J^{old}|| + 1}. \tag{7}$$

   where $||Cluster_J||$ denotes the number of patterns in cluster J. Otherwise, a new cluster is created with the input pattern.
4. Repeat (2) (3) for all of the input patterns. If the number of the iteration exceeds the predefined number or the cluster exemplar pattern is converged, then stop learning.

Generally, pedestrians who exist adjacently each other can detect as separating the head part of them. Therefore, we first divide the region of the pedestrians which was detected by the previous subsection into 4 equal parts vertically. Then, the top part feeds into input pattern of the ART2 neural network. If the number of clustered

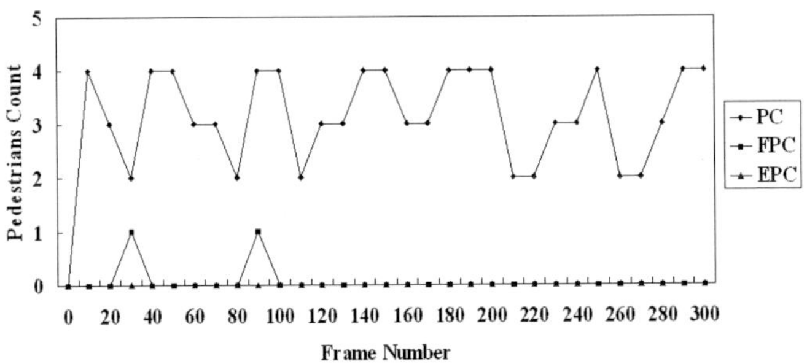

Fig. 5. The relation of PC, EPC and FPC

pedestrians is the smaller than the width of the region, the next part feeds into that of the ART2 neural network. Finally, it makes the final bounding box using clustered pedestrians and the projection histogram in the previous subsection.

3 Experimental Results

Simulations have been carried out on the video sequences captured in our city. And test platform is a personal computer with a 1.4 GHz Pentium 4 processor.

Fig. 6. Pedestrians detection results for six benchmark sequences

The test image is 300 sequences of 320×240 format, which is to take a picture with the cam in the city. In the experiment, the threshold value(Th) of the binary image set to 40 and the vigilance test value(ρ) set to 15, which is adaptively change according to an occlusion status of pedestrians. For example, if much occlusion occur, the value set to 10.

The quality evaluations for the pedestrian detection are applied on our algorithm. Both the error rate and the failure rate of the pedestrian detection are adopted to present the effectiveness of our algorithm. The error rate(e_r) and the failure rate(f_r) is defined as the following equation respectively:

$$e_r = \frac{N_{ep}}{N_p}, \; f_r = \frac{N_{fp}}{N_p}. \tag{8}$$

where N_{ep} denotes the number of none pedestrians who was detected in error and N_{fp} denotes the number of pedestrians who was not detected. And N_p denotes the number of pedestrians in the frame.

Fig. 5 shows the relation of the pedestrian count(PC), the error pedestrian count(EPC) and the failure pedestrian count(FPC) in the video sequence. In the experiment, both the error rate and the failure rate were found 0% and 0.6% respectively. We failed to detect some pedestrians because occlusion between them was much.

Fig. 6 shows the detection results for several benchmark sequences. The sequences are regular, slow, and fast pedestrian in JPG format. The detection results at frame #30, #60, and #90 of each sequence are shown in the figure. The regular and slow pedestrian sequences do not have occlusion so their detection results tend to be better than that of the fast sequences.

4 Conclusions

In this paper, we proposed an efficient method to detect adjacent multiple pedestrians in video sequences obtained from a moving camera. A BMA was used to compensate camera movement from the video sequence. And the moving object was detected by the frame difference and the pre processing was performed. Then, pedestrians were detected by the projection histogram. However, if pedestrians exist adjacently each other, we have clustered them using the ART2 neural network. The experimental results demonstrate the high efficiency of our approach as it have shown error rate of 0% and failure ratio of 0.6%.

References

1. Franke, U., Gavrila, D.M., Goerzig, S., Lindner, F., Paetzhold, F., Woehler, C.: Autonomous Driving Approaches Downtown. IEEE Intelligent Systems 13(6) (1998) 40-48
2. Agbinya, J.I., Rees, D.: Multi-Object Tracking in Video. Real-Time Imaging. Vol. 5. (1999) 295-304

3. Carpenter, G.A., Grossberg, S.: ART2:Self-Organization of Stable Category Recognition Codes for Analog Input Patterns. Applied Optics 26 (1987) 4919-4930
4. Inoue, H., Tachikawa, T., Inaba, M.: Robot Vision System with a Correlation Chip for Real-Time Tracking, Optical Flow and Depth Map Generation. Proc. of the IEEE International Conf. on Robotics and Automation, Vol. 2. (1992) 1621-1626
5. Yamamoto, S., Mae, Y., Shirai, Y., Miura, J.: Real-Time Multiple Object Tracking Based on Optical Flows. Proc. of the IEEE International Conf. on Robotics and Automation, Vol. 3. (1995) 2328-2333
6. Broggi, A., Bertozzi, M., Fascioli, A., Sechi, M.: Shape-Based Pedestrian Detection. Proc. of the IEEE Intelligent Vehicles Symposium 2000, (2000) 215-220
7. Mori, H., Charkari, N.M., Matsushita, T.: On-Line Vehicle and Pedestrian Detection Based on Sign Pattern. IEEE Trans. on Industrial Electronics 41(4) (1994) 384-391
8. Lim, J.S., Kim, W.H.: Multiple Pedestrians Tracking Using Difference Image and Projection Histogram. Proc. of the International Conf. on Imaging Science, Systems and Technology, Vol. 1. (2002) 329-334
9. Paragios, N., Deriche, R.: Geodesic Active Contours and Level Sets for the Detection and Tracking of Moving Objects. IEEE Trans. on PAMI. 22(3) (2000) 266-280

Recognition Method of Throwing Force of Athlete Based on Multi-class SVM

Jinghua Ma[1,2], Yunjian Ge[1], Jianhe Lei[1,2], Quanjun Song[1,2], Yu Ge[1], and Yong Yu[1]

[1] Robot Sensor Lab, Institute of Intelligent Machines, Chinese Academy of Sciences, P.O. Box 1130, Hefei, Anhui 230031, P.R. China
{jhma, yjge, geyu, yuyong}@iim.ac.cn, jhlei@yeah.net, qjsongrsl@yahoo.com.cn
[2] University of Science and Technology of China, 96 JinZhai Road, 230026 HeFei, Anhui Province, P. R. China

Abstract. A novel recognition method of throwing force of athlete combined with wavelet and multi-class support vector machine is introduced in the paper, which is based on the analysis of motion characters of gliding shot put. Utilizing the digital shot based on a three dimensional accelerometer, we get the three dimensional throwing forces in real time. Through wavelet transform, the general characteristics of force information are picked up. Then the general characteristics are input into the classifier for recognition of throwing force curves. The analysis provides the scientific basis for the motion training and instruction of shot put. The experiment shows that the method not only has high anti-noise ability and improves the recognition efficiency, but also decreases the burden of system and improves the recognition speed.

1 Introduction

The technical analysis of motion is an important component of sport training. Generally, coaches observe athletes' motion and then make instructions according to their sense and experience for most athletic sports. Obviously it is lack of scientific data and theoretical bases. The Institute of Intelligent Machines, Chinese Academy of Sciences and Hebei Normal School designed a kind of digital shot based on a three dimensional accelerometer which can detect the three dimensional throwing forces in real time[1]. According to the comprehension and analysis of the throwing force information, whether the motions are reasonable and accurate is judged and motions will be improved to conform to mechanics and technique principles.

Throwing force information shows the whole process of the motion, therefore, how to match and recognize the force information acquired from digital shot with motion phases correctly and find out the throwing curve corresponding to some motion phase is an important problem. Some researchers have utilized information processing techniques to analyze the motion of human body [2, 3, 4]. In this paper, we propose a recognition method based on wavelet and multi-class support vector machine. This method is utilized to recognize the shot athlete's throwing force curves and provide scientific basis for training and instructing system.

J. Wang et al. (Eds.): ISNN 2006, LNCS 3972, pp. 253 – 258, 2006.

2 The Analysis of the Motion of Shot Athlete

Generally, the motion of glide shot put is divided into five phases [5, 6]. In fig. 1, part 1 represents the phase of making ready. Part 2 is the phase of stretching the leg. Part 3 is the phase of gliding. Part 4 is the phase of transition and part 5 and 6 represents the phase of final release. The three curves represent Z, Y and X dimensional throwing force get from digital shot respectively. According to the expert knowledge, part A to E represents the five phases of the whole process respectively.

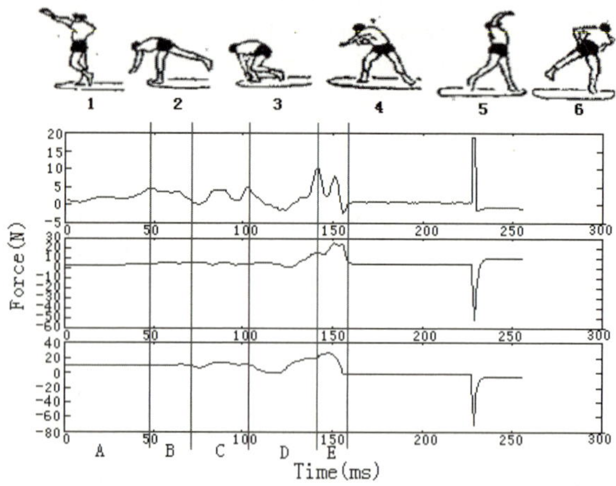

Fig. 1. The motion of putting the shot and the information of three dimensional throwing forces

The straight line after phase E shows that the shot is flying in air. At the end of this stage, the sudden change at the same time is the result of shot dropping on the ground. The stage before straight line is the whole process of putting the shot. In part B, the force of dimension Z increases at first, then tends to steady, while the forces of dimensions Y and X have few changes. In phase C, the athlete starts to glide and the force is getting to increase, but the change is not very large. At the end of this phase, that is the dropping instant of left foot, it comes to phase D. The forces of dimensions Y and X have a few changes, while the force of dimension Z increases suddenly because of the rotation from side to front and at the instant of right hand leaving right shoulder comes to the maximum. Phase E is the most important final release stage. The aim of this phase is to get the optimal parameters of shot put, including speed, height and angle of release. Because of the action of stretching the arm, the force of dimension Z begins to decrease, while that of dimension Y increases a little.

The technical motions of athletes are strengthened in training and their techniques have become relatively mature. Their motions have many resemblances. Besides, to an athlete, the motions of every shot put have strong repetitions. Therefore, the curves of throwing force have quite a few resemblances and repetitions. This principle is the basis to justify the efficiency of training and provides the theoretical basis to recognize the force information automatically. Utilizing the principle, we can find out the force curve which matches some phase of shot put and classify the patterns of the motion. Taking these for bases and combining them with expert knowledge, we will ultimately set up a scientific training and instructing system for shot put.

3 The Algorithm of SVM and its Application

Generally speaking, SVMs are the good tools to do the job of classification. Given a training set of instance-label pairs $(x_i, y_i), i = 1,...,l$ where $x_i \subset R^n$ and $y \in \{1,-1\}^l$, the SVMs require the solution of the following optimization problem:

$$\min_{w,b,\xi} \frac{1}{2} w^T w + C \sum_{i=1}^{l} \xi_i \tag{1}$$

Subject to

$$y_i(w^T \phi(x_i) + b) \geq 1 - \xi_i \qquad \xi_i \geq 0 \ , \ i = 1,...,l \tag{2}$$

Here training vectors x_i are mapped into a higher (maybe infinite) dimensional space by the function ϕ. Then SVM finds a linear separating hyper-plane with the maximal margin in this higher dimensional space. C > 0 is the penalty parameter of the error term. The SVM technique consists in constructing a decision surface f (x), which discriminates between two classes. $K(x_i, x_j) \equiv \phi(x_i)^T \phi(x_j)$ is called the kernel function.

The SVM has been extended to multi-class problems [7, 8]. There are mainly 2 different methods. The first considers the multi-class problem as a generalization of the binary classification scheme. The second divides the problem into a number of binary classifications [9]. The most widely used transformations are one-vs.-all, and one-vs.-one, in which each class is individually compared with all the others.

The one-vs.-all problem is introduced in this section. The k binary classifiers can be structured according to the binary classification algorithms. The m th class training sample is one class (Category $y_i^m = m$, m is positive integer), and all the other class training samples are regarded as the other class (Category $y_i^m = -1$). After optimization, the m th classification output function is:

$$f^m(x) = \sum_{x_i \in NSV} a_i^m y_i^m K(x_i, x) + b^m \tag{3}$$

Taking the test sample x as the input of the k binary classifiers, we can get k results from classification output function. The test sample belongs to the classifier which output is the maximum. The limitation of the method is that when constructing a binary classifier, all the k training samples are taken into computation. Whereas the classification of the curve of throwing force involves only five categories, it is appropriate to use one-vs.-all algorithm for this classification.

4 Recognition of Throwing Force Curve Based on Multi-class SVM

4.1 Preprocessing of the Throwing Force Curves

To classify the throwing force curve fast and accurately, the classification method must have the strong auto-adapted ability and the quick classification speed. In this paper, we preprocessed the throwing force curve by wavelet transform. The general characters, i.e. the approximation of the throwing force curve are extracted by the wavelet decomposition and the shape of throwing force curve is remained. This method decreases the influence of noise and taking the general characters as the input of recognition will reduce the computation and decrease the burden of the system.

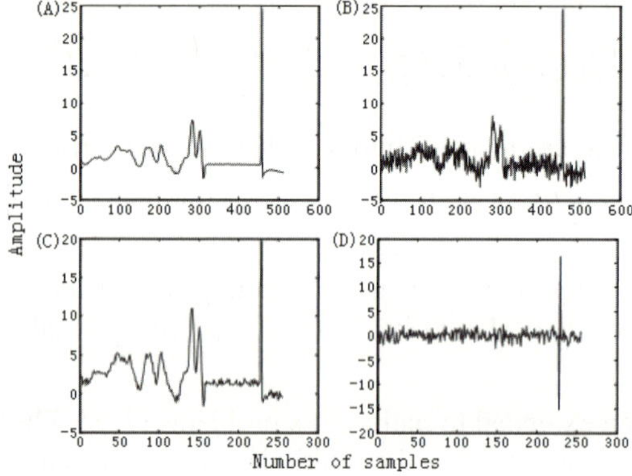

Fig. 2. The wavelet decomposition of the force signal (A) Original force signal (B) Force signal with noise (C) Approximation of force signal with noise (D) Detail of force signal with noise

Fig. 2 is the result of 1 scale db2 wavelet decomposition of throwing force information. The low frequency section of signal reflects the general characters, i.e. the approximation, and the high frequency section reflects some detail information and the noise signal. The general characters can reflect the main information of throwing force at large and will be utilized to be the input of next signal processing. It can not only reduce the complexity of computation, but also decrease the noise influence of the recognition result.

4.2 Experiment and Analysis

To justify the efficiency, we utilized the force data of dimension Z which are obtained from digital shot to do the comparison experiment. The steps are as follows.

1. Construct two training samples. Select two hundred force curves of dimension Z from the data obtained from digital shot. The number of corresponding curves of every phase is forty. Every curve has 512 sampling points. The data are normalized to construct sample A. After 2 scale wavelet decomposition of the above data, the low frequency section is constructed as sample B.

2. Select another two hundred force curves and normalize them to construct sample C for recognition. The number of corresponding curves of every phase is forty. Every curve has 512 sampling points. The sample C is with Gaussian white noise which variance is σ^2. The value of σ is 0, 0.05, 0.1, 0.15, 0.2, 0.25 and 0.3 respectively. Sample D is constructed by 2 scale wavelet decomposition of sample C.
3. Take sample A and B as the training set respectively and construct the binary classifiers which are corresponded to five phases. The classifiers are utilized to distinguish some phase with other phases. Thus the multi-class classifier is constructed. To compare with SVM, we take sample A as the input of BP neural network for training.
4. Recognize sample C and D which are influenced by Gaussian white noise with SVM classifier. Sample C is also recognized with BP neural network. The kernel of SVM is RBF, which parameter σ is 0.01 and penalty parameter C is 1000. Fig.3 is the comparison of recognition. A is the result of recognition result of the phase of stretching the leg while B is the result of recognition result of the phase of gliding. X-coordinate is the variance of Gaussian white noise, and y-coordinate is the percent of recognition. R1 is the SVM classifier trained by sample A, R2 is the SVM classifier trained by sample B, and R3 is the BP training neural network.

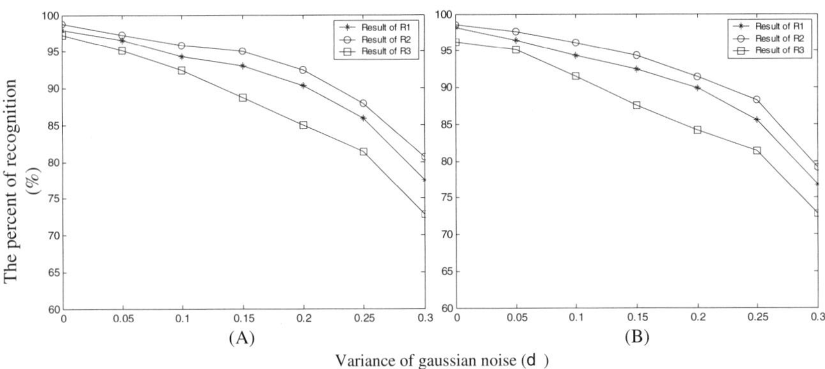

Fig. 3. The comparison of recognition (A) Recognition of the phase of stretching the leg (B) Recognition of the phase of gliding

From the result of experiment, we can see that comparing the SVM classifier with BP neural network, the recognition of the proposed method has improved greatly, especially to the signal with noise. The computation is reduced because of the wavelet decomposition. It is efficient to decrease the burden of the system and improve the speed of recognition.

5 Conclusion

The study of recognition method of throwing force of athlete is to analyze the throwing force of shot put and help coaches to instruct athletes on scientific view. In this paper, we analyzed the motion of shot put and its characteristics in five phases. The proposed method of recognition has the advantage of both wavelet transform and SVM. The general characters of force signal are input into classifier to recognize and match the throwing force curves with motion phases. The method improves the recognition efficiency, decreases the burden of system and improves the recognition speed. The experiment shows we get the satisfied result.

Acknowledgment

This research is supported by The National Natural Science Foundation of China, No.60343006 and No.60505012, and the Natural Science Foundation of Anhui Province, No.03042304.

References

1. Sun, W.Q., Song, G.M., Qiu, L.K., Shen, C.S., Tang, Y., Ge, Y.J.: Design and Implementation of Digital-Shot Based on Three-Axis Accelerometer. Computer Engineering and Applications 41(4) (2005) 102-104
2. Berbyuk, V., Lytvyn, B.A.: Mathematical Modeling of Human Walking on the Basis of Optimization of Controlled Processes in Biodynamical Systems. Journal of Mathematical Sciences 104(5) (2001) 1575-1586
3. Hahn, M.E., Farley, A.M., Lin, V., Chou, L.-S.: Neural Network Estimation of Balance Control during Locomotion. Journal of Biomechanics 38(4) (2005) 717-724
4. Tang, Y., Ge, Y.J., Yuan, H.Y., Wang, D.C.: Application of FNN in the Recognition of Force of Foot about Athlete. Journal of System Simulation 15(10) (2003) 1412-1417
5. Yuan, H.Y., Tang, Y., Song, G.M., Song, Q.J., Ge, Y.J., Liu, J.G.: Development and Application of Digital Shot Sensor. Journal of Transducer Technology 23(9) (2004) 61-63
6. Lu, D.M.: The Measurement of Sport Biomechanics. 1st edn. Beijing Sport University Press (2001)
7. An, J.L., Wang, Z.O., Ma, Z.P.: A New SVM Multiclass Classification Method. Information and Control 33(3) (2004) 262-267
8. Allwein, E.L., Schapire, R.E., Singer, Y.: Reducing Multiclass to Binary: A Unifying Approach for Margin Classifiers. Journal of Machine Learning 1(1) (2000) 113-141
9. Jeronimo, A.-G., Fernando, P.-C.: Multi-class Support Vector Machines: A New Approach. In: Digital Object Identifier. Vol. 2. IEEE ICASSP (2003) 781-784

A Constructive Learning Algorithm for Text Categorization

Weijun Chen[1] and Bo Zhang[2]

[1] School of Software, Tsinghua University, Beijing, 100084, P.R. China
cwj@tsinghua.edu.cn
[2] Department of Computer Science, Tsinghua University, Beijing, 100084, P.R. China
dczb@tsinghua.edu.cn

Abstract. The paper presents a new constructive learning algorithm CWSN (Covering With Sphere Neighborhoods) for three-layer neural networks, and uses it to solve the text categorization (TC) problem. The algorithm is based on a geometrical representation of M-P neuron, i.e., for each category, CWSN tries to find a set of sphere neighborhoods which cover as many positive documents as possible, and don't cover any negative documents. Each sphere neighborhood represents a covering area in the vector space and it also corresponds to a hidden neuron in the network. The experimental results show that CWSN demonstrates promising performance compared to other commonly used TC classifiers.

1 Introduction

In recent years, as the number of online text documents available on the Internet and corporate intranets continues to increase, there is a growing need for tools helping people better find, filter and manage these resources. Text categorization, which is defined as the activity of labelling natural language texts with thematic categories from a predefined set, is an important component in many information management tasks, such as automatic document indexing, document organization and filtering, etc.

Many approaches to text categorization and web page classification have been proposed [1]; for example, k-nearest neighbor (kNN), Bayesian probabilistic approaches, decision trees, inductive rule learning, support vector machine (SVM), and artificial neural networks (ANN). kNN is one of the most popular algorithms for text categorization, previous work has found that kNN achieves very good performance on different data sets [2]. However, kNN is a lazy learning method as no model needs to be built and nearly all the computation takes place at the classification stage. Furthermore, it generally has to store the whole data set in memory, resulting a high space complexity. The strength of neural networks is that they are robust, i.e., they have the ability to fit wide range of distribution accurately. However they also have some limits: (a) the network architecture must be fixed a priori; and (b) the computational cost increases very rapidly with the dimension of the input space. And to make things worse, a major characteristic, or difficulty, of text categorization problems is the high dimensionality of the feature space. The native feature space consists of the unique terms (words) that occur in documents, which can be tens of thousands of terms for

J. Wang et al. (Eds.): ISNN 2006, LNCS 3972, pp. 259 – 264, 2006.

even a moderate-sized text collection. This is prohibitively high for ordinary neural networks, which are not able to handle such a large number of input nodes.

In this paper, we present a novel algorithm named CWSN (Covering With Sphere Neighborhoods) to construct a three-layered neural network, and use it to solve the text categorization problem. The CWSN algorithm is inspired by a geometrical representation of McCulloch-Pitts neuron model, which was proposed by Zhang [3]. From the representation, a clear visual picture and interpretation of the model can be obtained, and the design of neural networks is transformed to a geometrical covering problem. So it provides us a direct and convenient way to construct neural classifiers.

2 The CWSN Algorithm

2.1 A Geometrical Representation of M-P Model

The geometrical representation of M-P model is summarized as follows, for more details see [3].

An M-P neuron is an element with n inputs and one output. The general form of its function is

$$y = \mathrm{sgn}(\boldsymbol{wx} - \varphi)$$

where

$\boldsymbol{x} = (x_1, x_2, x_3, \ldots, x_n)^T$ — an input vector

$\boldsymbol{w} = (w_1, w_2, w_3, \ldots, w_n)$ — a weight vector

φ — a threshold

It is well known that the function of an M-P neuron can geometrically be regarded as a spatial discriminator of an n-dimensional space divided by a hyperplane. In fact, note that $\boldsymbol{wx} - \varphi = 0$ can be interpreted as a hyperplane P in the n-dimensional space, when $(\boldsymbol{wx} - \varphi) > 0$, input vector $\boldsymbol{x}$ falls into the positive half-space of the hyperplane P. Meanwhile, $y = \mathrm{sgn}(\boldsymbol{wx} - \varphi) = 1$. When $(\boldsymbol{wx} - \varphi) < 0$, input vector $\boldsymbol{x}$ falls into the negative half-space of P, and $y = -1$.

Now assume that each input vector $\boldsymbol{x}$ has an equal length, which means all input vectors will be restricted to an n-dimensional sphere S^n (In more general cases, where the input vectors have different lengths, a transformation is needed). Then, $(\boldsymbol{wx} - \varphi) > 0$ represents the positive half-space partitioned by the hyperplane P and the intersection between the positive half-space and S^n is called a "sphere neighborhood" as shown in Fig.1. When an input vector $\boldsymbol{x}$ falls into this region, i.e. $\boldsymbol{x}$ is covered by the sphere neighborhood, then output $y = 1$, otherwise, $y = -1$.

If the weight vector $\boldsymbol{w}$ has the same length as $\boldsymbol{x}$ (which means $\boldsymbol{w}$ is also on the sphere), $\boldsymbol{w}$ becomes the center of the sphere neighborhood, and $r(\varphi) = \sqrt{R^2 - \varphi^2}$ (R is the radius of S^n), a monotonically decreasing function of φ, becomes the radius of the neighborhood. Now given an M-P neuron and its two parameters $(\boldsymbol{w}, \varphi)$, we can always find the corresponding sphere neighborhood on S^n. This clear visual picture of an M-P neuron is a great help to the analysis and construction of neural networks.

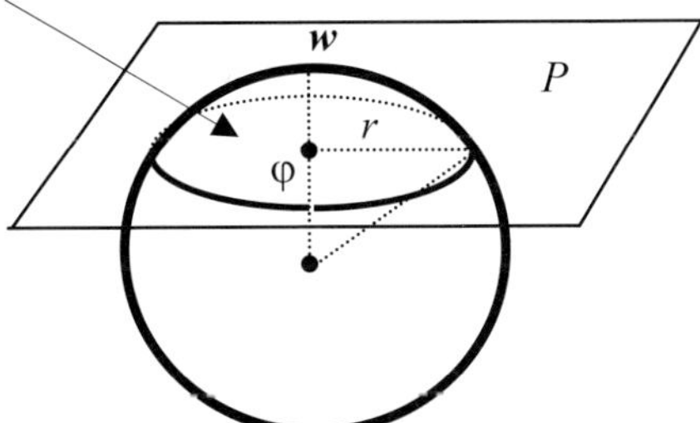

Fig. 1. A sphere neighborhood

2.2 Framework of CWSN Algorithm

First of all, we define our problem as follows:

Suppose the training set consists of N examples P_i (i = 1, 2, …, N), the i-th example P_i is an input-output pair ($\boldsymbol{x}^{(i)}$, $y^{(i)}$), the input vector $\boldsymbol{x}^{(i)}$ is identified by M values, i.e. $\boldsymbol{x}^{(i)} = (x_1^{(i)}, x_2^{(i)}, \cdots, x_M^{(i)})$, and the corresponding desired output $y^{(i)}$ has only K different candidate values, $y^{(i)} \in \{1, 2, \ldots, K\}$, i.e., the training examples will be classified into K classes. Now, the problem is how to construct a three-layer network: the input layer, the hidden layer and the output layer. Note that the number of hidden neurons is not determined before the training process and the number of output neurons is K. Our aim is that for each training example P_i, if the input vector $\boldsymbol{x}^{(i)}$ is presented to the network, then the $y^{(i)}$-th output unit is supposed to "fire".

From the above discussion, it can be seen that each M-P neuron corresponds to a sphere neighborhood on the sphere surface S^n. Inspired by this geometrical representation, we propose an algorithm framework CWSN to build the neural network constructively. The basic idea is to transform the design of a neural classifier to a training example covering problem. First, find K sets of sphere neighborhoods SET(k), k=1, 2,…,K, such that SET(k) covers only the training examples of class k and does not cover any examples of other classes. Next, for each sphere neighborhood in SET(k), construct a corresponding hidden neuron, and then connect these neurons to the k-th output neuron using the OR operation. The algorithm is described in detail as follows:

For any class k, $1 \le k \le K$, let POS(k) be the set of training examples belonging to the k-th class. And for an example $\boldsymbol{x}$ in POS(k), if COVER($\boldsymbol{x}$) = true, it means that $\boldsymbol{x}$ has been covered by some sphere neighborhood in SET(k).

Algorithm 1. CWSN learning algorithm

S1 $k \leftarrow 1$, SET(k) $\leftarrow \Phi$;

S2 POS(k) ← the set of training examples belonging to the k-th class;

S3 For each example $\boldsymbol{x}$ in POS(k), COVER($\boldsymbol{x}$) ← false;

S4 If all the examples in POS(k) have been covered, go to S9; otherwise randomly select an example ***sd*** from POS(k) which has not been covered yet;

S5 Execute the GetSphereNeighbor(***sd***, k) algorithm to find a sphere neighborhood SN which covers ***sd*** and some other examples of class k and does not cover any examples of other classes;

S6 SET(k) ← SET(k) ∪ { *SN* }, for each example $\boldsymbol{x}$ which is covered by *SN*, COVER($\boldsymbol{x}$) ← true;
S7 Suppose the center of *SN* is $\boldsymbol{w}$ and the threshold is φ, use these two parameters to construct a corresponding hidden neuron;
S8 Connect this new hidden neuron to the k-th output unit and set the weight to 1, then go to S4;
S9 $k \leftarrow k + 1$, if $k \leq K$, go to S2; otherwise return.

When the training process is completed, the CWSN algorithm will construct a three-layer neural network. Now for each test example, if its input vector $\boldsymbol{x}$ is presented to the network, the hidden neuron, whose corresponding sphere neighborhood covers the vector, will be trigered. Furthermore, the output unit which is connected to this hidden neuron will "fire" too, so the class of this example is determined.

2.3 GetSphereNeighbor Algorithm

Given a training example ***sd*** and its class k, the GetSphereNeighbor algorithm will find a sphere neighborhood *SN* which covers ***sd*** and some other examples of class k and does not cover any examples of other classes. Since our goal is to construct as few hidden units as possible, it seems natural that *SN* can cover as many examples of class k as possible. In GetSphereNeighbor, a self-adaptive and parameter-free technique is employed to find a sphere neighborhood of appropriate size.

The basic idea of the algorithm is that first construct an initial sphere neighborhood centered at the training example ***sd***. Then move its center (weight vector) and change its threshold such that more training examples of the k-th class can be covered. Consequently the number of sphere neighborhoods used to cover all the training examples of class k will not be too large.

Algorithm 2. GetSphereNeighbor (***sd***, k)
S1 Execute the ConstructSN (***sd***, k) algorithm to construct an initial sphere neighborhood *SN* which is centered at ***sd***, and the threshold of *SN* is adjusted to cover as many positive examples (i.e. examples of class k) as possible.
S2 NumExamples ← the number of positive examples covered by *SN*;
S3 MaxSN ← *SN*, MaxNumExamples ← NumExamples;
S4 ***cn*** ← the center of all the positive examples covered by MaxSN, i.e. the arithmetic mean vector of these examples. Let G be the set of positive examples covered by MaxSN, then

$$\boldsymbol{cn} \leftarrow \frac{\sum_{x \in G} \boldsymbol{x}}{|G|}$$

S5 Execute the ConstructSN (***cn***, k) algorithm to construct an new sphere neighborhood *SN* which is centered at ***cn***;
S6 NumExamples ← the number of positive examples covered by *SN*;
S7 If NumExamples > MaxNumExamples
MaxSN ← *SN*, MaxNumExamples ← NumExamples; go to S4;
S8 Return MaxSN;

Given a training example $\boldsymbol{cn}$ and its class k, the ConstructSN algorithm will find a sphere neighborhood SN, the center of SN is $\boldsymbol{cn}$, and the threshold φ is carefully adjusted to cover as many positive examples as possible.

Algorithm 3. ConstructSN ($\boldsymbol{cn}$, k)

S1 $d1 \leftarrow$ the "distance" between $\boldsymbol{cn}$ and the nearest negative example, i.e.

$$d1 \leftarrow \max_{x \notin POS(k)} \{\langle \boldsymbol{x}, \boldsymbol{cn} \rangle\}$$

note that <$\boldsymbol{x}$, $\boldsymbol{cn}$> denotes the dot product of $\boldsymbol{x}$ and $\boldsymbol{cn}$, and on a sphere surface, the closer two vectors are, the smaller the angle is between them and the larger their dot product will be;

S2 d2 $\leftarrow$ the "distance" between $\boldsymbol{cn}$ and the farest positive example whose "distance" is large than $d1$, i.e.:

$$d2 \leftarrow \min_{x \in POS(k)} \{\langle \boldsymbol{x}, \boldsymbol{cn} \rangle > d1\}$$

S3 $\varphi \leftarrow (d1 + d2) / 2$;

S4 Construct a sphere neighborhood SN, the center is $\boldsymbol{cn}$ and the threshold is φ;

S5 Return SN.

3 Text Categorization Experiments

We used the famous Reuters-21578 data in our experiments, it contains documents collected from the Reuters newswire in 1987 and has become a standard benchmark in text categorization evaluations. We adopted the 'Apte Split' in the corpus to obtain training and test data, then we discarded those documents with multiple category labels, and selected the categories which have at least one document in the training set and test set. It left us with 6559 training and 2570 test documents in 52 categories.

We represent documents using the well known vector space model. First, all the words in the document are tokenized, filtered using a stop list, and stemmed. Unique stems with their corresponding frequencies are kept. Second, each document $\boldsymbol{d}$ is represented as a vector of weighted terms (the standard *tf idf* weighting function is used): $\boldsymbol{d} = (x1, x2, \ldots, x_m)$, where $x_i = tf_i \times idf_i$. Here m is the total number of unique terms in the training collection, tf_i is the i-th term's frequency within $\boldsymbol{d}$, idf_i is its inverse document frequency. In order to make weights fall in the [0,1] interval and documents be represented by vectors of equal length, the cosine normalization is used.

In text categorization the high dimensionality of the term space may be problematic. In our experiments, there are 19824 unique terms in the training data. To select keyword features among them, the $\chi 2$ (CHI) statistic is adopted as the ranking metric.

To evaluate the system, we use the F1 measure which combines recall (r) and precision (p) with an equal weight: $F1 = 2pr / (p+r)$. We summarize the F1 scores over the different categories using both the macro and micro-averaging methods.

In our experiments, we compare the performance of different algorithms such as kNN, Rocchio algorithm, C4.5 decision tree and neural network (NN). Different values of parameters have been tried on each algorithm to ensure that the experimental results can reflect the best performance. The basic settings of each algorithm are:

(1) The k in kNN is set to 10. (2) The value of β is set to 1 and γ is set to 10 for the Rocchio algorithm. (3) For each category, a different C4.5 decision tree is built. (4) A typical three-layer NNet is trained on all the 52 categories, the number of hidden units is set to 64. The number of features is 500 for kNN, C4.5 and NN, 300 for Rocchio.

Table 1. Performance summary of different algorithms (%)

	kNN	Rocchio	C4.5	NN	CWSN
MicroF1	88.40	87.67	86.31	86.61	**89.17**
MacroF1	56.17	55.43	52.03	39.54	**57.22**

Table 1 shows that CWSN outperforms the linear classifier Rocchio, the decision tree algorithm C4.5, and the non-linear neural network NN in both the micro-averaged and the macro-averaged F1 scores, and is comparable to kNN which is the state-of-art method for text categorization. And CWSN is more efficient than kNN because it keeps fewer representatives for each category.

4 Conclusion

In this paper, we described a new constructive learning algorithm named CWSN, and used it to solve the text categorization problem. CWSN is based on a geometrical representation of M-P neuron, it uses a sphere neighborhood covering technique to dynamically construct a three-layer neural network. Extensive experiments have been conducted on a large-scale document corpus, the experimental results show the efficacy of our method. Ongoing and future work include further improvements of the GetSphereNeighbor and the ConstructSN algorithms to cover more positive examples.

References

1. Sebastiani, F.: Machine Learning in Automated Text Categorization. ACM Computing Surveys 34(1) (2002) 1–47
2. Yang, Y., Xin, L.: A Re-examination of Text Categorization Methods. In: 22th International Conference on Research and Development in Information Retrieval, SIGIR’99(1999) 42–49
3. Zhang, L., Zhang, B.: A Geometrical Representation of McCulloch-Pitts Neural Model and Its Applications. IEEE Transactions on Neural Networks 10(4) (1999) 925-929

Short-Text Classification Based on ICA and LSA

Qiang Pu and Guo-Wei Yang

School of Computer Science and Engineering,
University of Electronic Science and Technology of China,
Chengdu, 610054, China
{puqiang, gwyang}@uestc.edu.cn

Abstract. Many applications, such as word-sense disambiguation and information retrieval, can benefit from text classification. Text classifiers based on Independent Component Analysis (ICA) try to make the most of the independent components of text documents and give in many cases good classification effects. Short-text documents, however, usually have little overlap in their feature terms and, in this case, ICA can not work well. Our aim is to solve the short-text problem in text classification by using Latent Semantic Analysis (LSA) as a data preprocessing method, then employing ICA for the preprocessed data. The experiment shows that using ICA and LSA together rather than only using ICA in Chinese short-text classification can provide better classification effects.

1 Introduction

Text classification (TC) according to documents' topics is an application of automatic analysis of documents in modern information processing systems. ICA applied on word context data can give distinct features which reflect syntactic and semantic categories [1].The application of ICA in TC is based on the assumption that a document corpus is generated by a combination of several thematic topics. Thus, the independent components obtained by the ICA algorithms define statistically independent clusters of documents, allowing their thematic classification [2]. Text classifiers based on ICA, in many cases, give good classification effects. In short-text case, however, ICA can not work well because short-text documents usually have little overlap in their feature terms.

Our aim is to solve the short-text problem we called STP in text classification by using LSA as a data preprocessing method. ICA is then employed for the preprocessed data by utilizing its great classifying capability of identifying a generalizable low-dimensional basis set in the face of high-dimensional noisy data [3].

2 LSA and ICA Related to TC

TC methods are almost depended on document vector space model (VSM) [4] representation that our analyses and experiments are also based on.

Latent semantic analysis (LSA) [5] is a well-known simple approach to automatic generation of the concepts and analysis of terms co-occurrence that are useful in text classification and information retrieval purposes. LSA is based on singular value

J. Wang et al. (Eds.): ISNN 2006, LNCS 3972, pp. 265 – 270, 2006.

decomposition (SVD) [6] by which a smaller document matrix $\boldsymbol{A}$ that retains only the most important information from the original document matrix can be constructed.

$$\boldsymbol{A} = \boldsymbol{TSD}^T. \tag{1}$$

In (1), $\boldsymbol{T}$ and $\boldsymbol{D}$ contain orthogonal vectors and $\boldsymbol{S}$ is diagonal matrix of singular values. The eigenvectors with the largest singular value capture the axes of largest variation in data. So each document can be projected into a lower dimensional space, which is called dimensionality reduction often used to reduce data noise, avoid overlearning. After getting the reduced space, same co-occurrence terms are projected on the same dimensionality and documents are represented as linear combinations of orthogonal features. [4], [6] have introduced the applications of LSA in TC. [5] describes LSA in detail.

ICA is a method for presenting a set of multivariate observations as a linear combination of unknown latent variables that are statistically independent, giving no information on the values of the other components [7]. Lately it has been found out that it is a suitable method for analyzing text documents [3], [8].

We suppose that there are some main topics acting as a set of independent hidden random variables in each document. ICA as a technique that exploits the more effective higher order statistical features in documents can be employed to determine these independent random variables and extract the independent topic structure of documents. The classic ICA model can be expressed as following:

$$\boldsymbol{x} = \boldsymbol{As}. \tag{2}$$

In (2), $\boldsymbol{x}$ is the documents topic-mixed information, a $T \times D$ term by document matrix. $\boldsymbol{A}$ is a $T \times N$ mixing matrix and $\boldsymbol{S}$ is a $N \times D$ documents topic information matrix, N is the number of independent components representing main topics of a set of documents. If mixing matrix $\boldsymbol{A}$ is known, topics in a document corpus can be estimated as $\boldsymbol{y} = \boldsymbol{A}^{-1}\boldsymbol{x} = \boldsymbol{Wx}$, where separating matrix $\boldsymbol{W}$ can be determined by employing FastICA [7] algorithm. Consequently, new unseen documents can be classified under N topic categories by comparing their independent components with the ones of training documents, after projecting them to the ICA space by means of matrix $\boldsymbol{W}$ [2].

3 Short-Text Classification Problem

Each document has its special topic used for TC. Topics are considered as a set of independent hidden random variables. In VSM, suitable feature terms consisting of document vectors often implicate underlying topic of documents. Likewise, a topic can also determine which terms to appear in documents. If two documents, usually long-text documents, have some common feature terms, the same topic the two documents probably belong to. If regarding a set of documents as a topic-mixed corpus, TC based on ICA is then an application that ICA algorithm can maximize the statistical independence between documents from the observed topic-mixed document corpus and try to determine the documents' real topic categories.

But short-text documents usually have little overlap in their feature terms, especially after filtering stop-words or function-words. Fig.1 illustrates this case. This means that vectors of such documents are already independent of each other so that

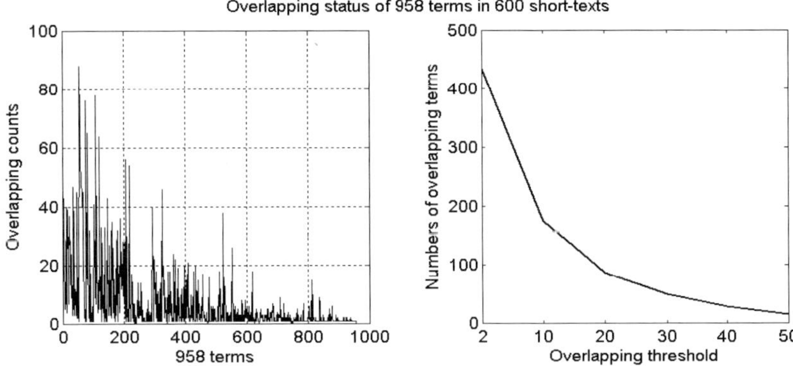

Fig. 1. Terms overlap in short-text documents. There are 958 terms in our 600 short-text documents including 6 topic categories. Left figure shows terms are in their overlap case and only a few terms are overlap terms. Right one clearly shows there are total 524 single terms (hapax legomena) if term overlap threshold is set greater than 2. In order to get an improved classification precision, the overlap threshold will be a larger number, e.g. 20. In this case, experiment shows only 87 terms among 958 are overlapped, that is, only 9.1% terms are overlapped.

ICA leads to nothing. Experiments in section 4 will show an unsatisfied classifying effect when only using ICA on Chinese short-text documents classification.

Fortunately, there is a possible solution to STP problem. Low terms overlap proportion is insufficient to determine that there is no similarity between documents because the co-occurrence terms may exist among different documents. LSA is just the right tool to find these co-occurrence terms by projecting them on the same dimensionality in reduced space. Co-occurrence terms can improve the similarity of short-text documents, that is, a better topic-mixed corpus is provided for ICA to extract the independent topic structure of documents.

4 Experiments

We randomly select 400 Chinese short-text documents downloaded from Chinese web sites (40930 terms after web-page preprocessing, Chinese terms splitting and out-of-vocabulary words recognizing by means of our programs) manually divided into 4 topic categories, such as *traffic* (TRA), *education* (EDU), *medicine* (MED) and *politics* (POL). Each topic contains 100 short documents.

Stop-words and function-words are removed. Jarque-Bera test shows that the remaining 660 terms are not normally distributed, so Chi-square [4] statistic is used to calculate these terms χ^2-value over the four thematic topics. Experiments show that topic category of each document can be better represented by each document's 12 terms with higher χ^2-value. The 12 terms are then selected as document feature terms, weighted by normalized $tf \times idf$ [4]. Each document is represented as 12-dimensional vector, $\boldsymbol{x}_j = (s_{1j}, \ldots, s_{Kj})$, $K = 12$, where s_{ij} is:

$$s_{ij} = round\left(10\times\frac{1+\lg\left(tf_{ij}\right)}{1+\lg\left(l_j\right)}\right) \tag{3}$$

In (3), tf_{ij} represents the occurrence of term i in document j and l_j is the length of document j. The calculation of s_{ij} employs the logarithm method, considering standardization of weights factor.

As a data preprocessing step to solve STP, LSA is employed to reduce the original 660 dimensionalities to 17. The number of independent components is set to 4 representing our four selected topic categories. The classifier is trained with the 80% of the short-text documents (320 documents randomly selected) in corpus by FastICA algorithm. Classifier obtained from FastICA determines projection directions into which short-text document must be projected to reveal the latent topic that the document belong to. As a contrast, a classifier only based on ICA is trained in the same corpus.

We use the remaining 80 documents in training set (denoted as *In_Topics_Test*) and 200 documents from other two topic categories (denoted as *Out_Topics_Test*): *military* (MIL) and *sports* (SPO) as testing set to verify classification effect of the classifier obtained in training phase. Fig. 2 shows the classifying results of the 80 documents in testing set by using LSA and ICA together (denoted as *Method_1*) and ICA only (denoted as *Method_2*) respectively.

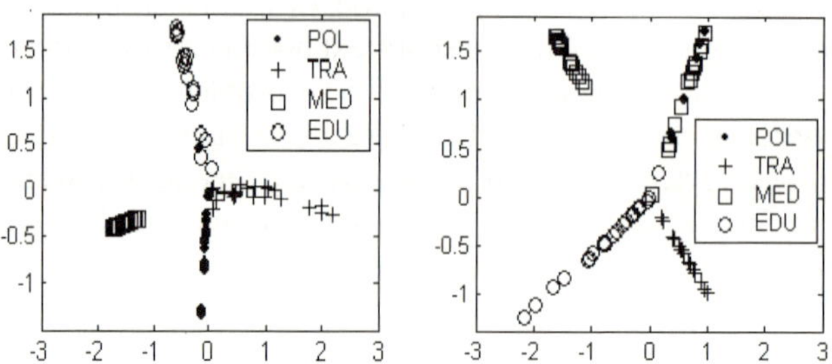

Fig. 2. The left figure is the classifying result obtained by *Method_1*, the right one shows the classifying result obtained by *Method_2*. In right figure, topics of *politics* and *medicine* are confused due to their little overlap feature terms. However, the left figure shows that four topic categories are clearly classified, proving that *Method_1* is suitable for short-text classification.

The detail classifying effects of Fig. 2 are illustrated in Table 1. According to the average within-class scatter (*Method_1*'s 0.99 vs. *Method_2*'s 3.75) and average between-class scatter (*Method_1*'s 2.35 vs. *Method_2*'s 1.23), *Method_1* in short-text classification makes a better classifying effect than that of *Method_2*. At the same time, *Method_1* gets a higher classifying precision than that of *Method_2*, such as in *politics* topic, classifying precision is *Method_1*'s 95% vs. *Method_2*'s 30%.

There are two classifiers learnt by means of *Method_1* and *Method_2*. Fig. 3 and Fig. 4 show the classifying capability of two classifiers learnt in training phase. There are two kinds of lines (real line and broken line) we called *topic status curve* (TSC) in these figures. One TSC is represented by average value of independent component of each document in one topic category. Real lines represent TSCs in training set, and broken lines represent TSCs in testing set.

Table 1. Comparison of classifying effects between *Method_1* and *Method_2*. C denotes four categories (*traffic*, *politics*, *education*, *medicine*), and M is the two classifying method. WS shows 'within-class scatter' in each topic category respectively. Each three values in BS corresponds to 'between-class scatter', such as topic *traffic*, the values (2.76, 0.86, 2.69) correspond to the 'between-class scatter' of *traffic* with *politics*, *medicine*, *education*, respectively. N is the number of documents classified in the corresponding topic category.

C \ M	TRA			POL			MED			EDU		
	WS	**BS**	**N**	**WS**	**BS**	**N**	**WS**	**BS**	**N**	**WS**	**BS**	**N**
ICA	0.96	2.76 0.86 2.69	20	5.83	2.76 0.54 0.003	6	4.60	0.86 0.54 0.51	33	4.61	2.69 0.003 0.51	21
LSA +ICA	0.52	0.57 2.85 1.14	20	5.02	0.57 0.87 0.10	18	0.21	2.85 0.87 0.39	20	3.66	1.14 0.10 0.39	22

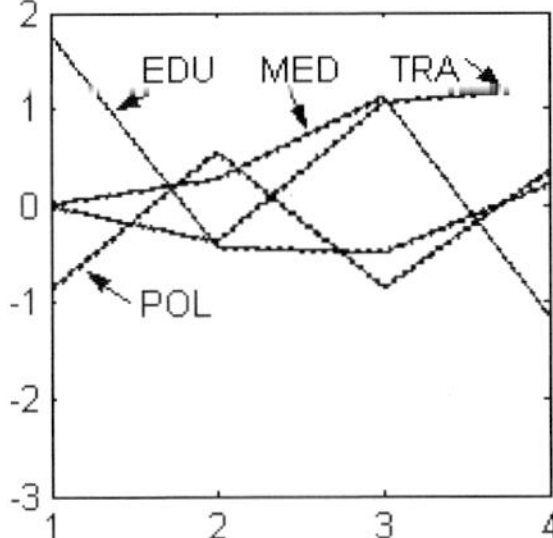

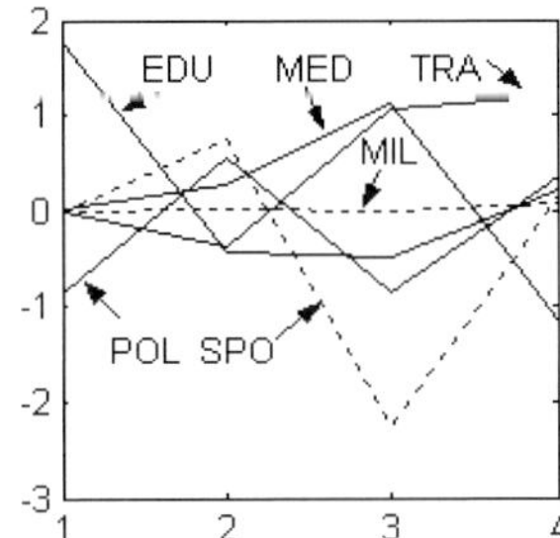

Fig. 3. The classifying effects when using classifier learnt by *Method_1*. Left figure shows the surprising similarity of the TSCs in training set and in *In_Topics_Test* set, the two kinds of TSCs are almost overlapped. That is, the classifying precision is perfect when documents are in *In_Topics_Test* set. In right figure, TSCs of *Out_Topics_Test* are very different from the four TSCs, which illustrates that the classifier learnt by *Method_1* can reject documents from *Out_Topics_Test* set and do not classify them into the four training topic categories.

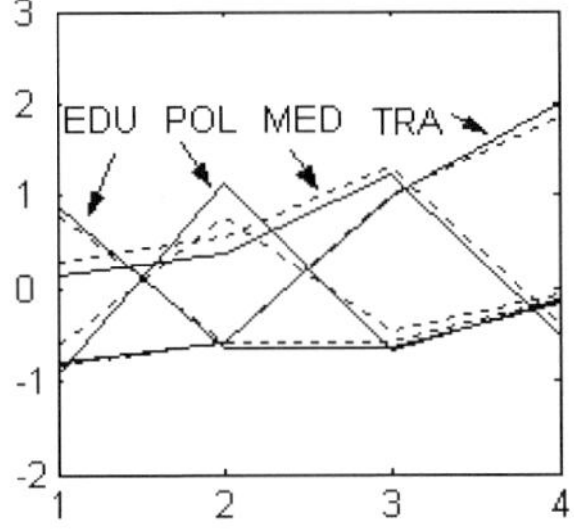

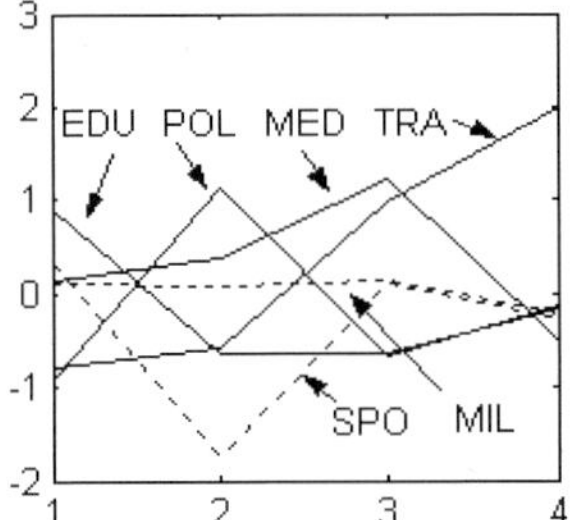

Fig. 4. The classifying effects when using classifier learnt by *Method_2*. Left figure shows the similarity of the TSCs in training set and in *In_Topics_Test* set. However, real lines and its corresponding broken line do not overlap well, that means, the classifying precision is unsatisfied when documents are in *In_Topics_Test* set. The rejection capability of this kind of classifier learnt by *Method_2* is almost as same as the classifier learnt by *Method_1*.

5 Conclusion

In this study, we have analyzed short-text problem in TC. This kind of problem can be solved without loss of generality by LSA projection. Experimental results illustrate that ICA and LSA together provide a satisfied classifying result in Chinese short-text classification, that is, LSA is good at analyzing terms co-occurrence, extracting and inferring relations of contextual usage of terms in documents, increasing the semantic terms overlapping, while ICA finds statistically independent topics, providing a great classifying capability. Using ICA only, however, can not obtain a satisfied classifying result. At the same time, experiments show that Chi-square statistic is a suitable method for extracting feature terms in short-text documents.

How many dimensionalities of the reduced dimensionality space should be determined when using LSA projection? In this study, the dimensionality of reduced space comes from experiment that makes ICA yield a good classifying effect. In future works, however, the dimensionality should be determined in manner of quantitative analysis and word-sense disambiguation will be taken into account in TC.

References

1. Honkela, T., Hyvärinen, A.: Linguistic Feature Extraction Using Independent Component Analysis. Proc. Int. Joint Conf. on Neural Networks (IJCNN), Budapest, Hungary (2004)
2. Sevillano, X., Alías, F., Socoró, J.C.: Reliability in ICA-based Text Classification. Fifth International Conference, ICA 2004, Granada, Spain (2004) 1213–1220
3. Kolenda, T., Hansen, L.K.: Independent Components in Text. In: Advances in Neural Information Processing Systems, volume 13 (2000) 235–256
4. Manning, C.D., Schütze, H.: Foundations of Statistical Natural Language Processing. MIT Press, Cambridge, MA (1999)
5. Landauer, T.K., Foltz, P.W., Laham, D.: Introduction to Latent Semantic Analysis. Discourse Processes 25 (1998) 259-284
6. Deerwester, S., Dumais, S.T., Furnas, G.W., Landauer, T.K., Harshman, R.: Indexing by Latent Semantic Analysis. Journal of the American Society of Information Science 41 (1990) 391–407
7. Hyvärinen, A.: Survey on Independent Component Analysis. Neural Computing Surveys 2 (1999) 94–128
8. Isbell, C.L., Viola, P.: Restructuring Sparse High Dimensional Data for Effective Retrieval. In: Advances in Neural Information Processing Systems, Vol. 11 (1998) 480–486

Writer Identification Using Modular MLP Classifier and Genetic Algorithm for Optimal Features Selection

Sami Gazzah[1] and Najoua Essoukri Ben Amara[2]

[1,2] Laboratoire des Systèmes et du Traitement de Signal (LSTS), Ecole National d'Ingénieurs de Tunis (ENIT)
sami.gazzah@laposte.net
[2] Ecole Nationale d'ingénieurs de Sousse 4000, Sousse - Tunisie
Najoua.BenAmara@enim.rnu.tn

Abstract. This paper describes the design and implementation of a system that identify the writer using off-line Arabic handwriting. Our approach is based on the combination of global and structural features. We used genetic algorithm for feature subset selection in order to eliminate the redundant and irrelevant ones. A modular Multilayer Perceptron (MLP) classifier was used. Experiments have shown writer identification accuracies reach acceptable performance levels with an average rate of 94.73% using optimal feature subset. Experiments are carried on a database of 180 text samples, whose text was made to ensure the involvement of the various internal shapes and letters locations within a word.

1 Introduction

As a signature, a person's handwriting has long been considered as a behavioural characteristic and accepted as a form of human identification. In spite of the existence of within-writer variation, we can notice that this variation is less significant than the between-writer variation. Writer identification can be divided into two categories: verification and identification of the writer. In writer verification, the object is to compare questioned handwriting with samples of handwriting obtained from sources for the purposes to determine whether the two samples were written by the same writer or by two different writers [1], [2]. On the other hand, writer identification aims to identify the writer as one of a group of N possible authors. Writer identity offers a valuable additional source of information to recognition system that allows transforming the complexity of the multi-writers system into mono-writer ones (Fig. 1). In spite of its usefulness and meaningful improvement in recognition for off-line handwriting, personal identification has been almost neglected [3]. The identification task includes a comparison of a set of handwriting samples provided from an unknown writer with each class reference pattern, and decides which reference pattern matches the unknown writer features best.

The recognition of the off-line multi-writers arabic handwriting is a non-trivial task due to the morphological characteristics of this script. It is assumed to be among the most difficult style due to the variations in character shape even if it is rewritten by the same person [4]. Our various work in recognition of the off-line Arabic Handwriting [5], shows the complexity of the task in particular in a multi-writer context For the handwriting, the variations within and between-writers are accentuated by

J. Wang et al. (Eds.): ISNN 2006, LNCS 3972, pp. 271–276, 2006.

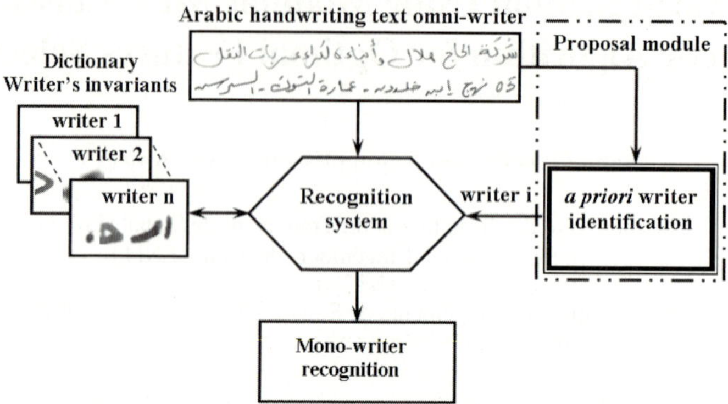

Fig. 1. Recognition system improved by adding a priori writer identification module

the calligraphic nature of Arabic script are more likely to increase the size and the complexity of the feature selection task. In this context, extract a meaningful characterisation of this cursive style, then finding a minimal set of features necessary for recognition and classification are a key part of designing efficient and implementable systems. The main purpose of feature selection is to reduce the number of features used in classification while maintaining acceptable classification accuracy. Less discriminatory features are eliminated, leaving a subset of the original features which retains sufficient information to discriminate well along classes [6].

In the present work we explain the importance of the feature selected set optimisation in section 2. In section 3 we will give an overview of our system, while in section 4 MLP classifiers design and the experimental results are given. Conclusions and future work are drawn in the end.

2 Features Selection

Features extraction is crucially important step in the construction of any pattern classification system and the selection of the most relevant ones offer a potential improvement to this system by: Simplifying features extracting task, optimizing recognition system complexity, reducing running time and improving the classification accuracy. However, it has been observed that beyond a certain point, the inclusion of additional features leads to a worse rather than better performance [7]. Research works in the feature subset selection have been based on two main models: the filter model and the wrapper model. In the first model, subset selection is performed as a pre-processing module before the learning algorithm. While the wrapper model, subset selection is performed as a cover around the learning algorithm in witch each subset possible solution evaluated using learning stage [8].

The main idea of features selection is to select a subset of input variables by cutout features with weakly or no predictive information while maintaining or performing classification accuracy.

In this context, Genetic algorithms have been successfully implemented to select feature subsets in classification problems. This paper focus on the feature subset selection for Writer Identification using arabic script.

3 Proposal Approach

The system is composed of the following five modules: text-image acquisition, pre-processing, features extraction, feature subset selection and classification. For the experiments reported in this paper, we have designed an Arabic letter source document which contains: 505 characters, 15 numerals and 6 punctuations. The choice of the letter contents was made to ensure the involvement of the various internal shapes of the letter within a sub-word (isolated, initial, middle and end). Handwriting samples of 60 individuals were collected. Each individual was required to copy the same letter three times, two samples were used for the training, the other for the tests which makes a total of 180 A4 format samples pages. A black pen, a white sheet and a plain were provided. These text images are initially digitized in grayscale at a resolution of 300 dpi.

3.1 Features Extraction

This step aims at the selection of relevant information that characterizes each class. Each selected feature should morphologically characterize the style of writing of a script writer (authorship) to dissociate it from to the other script writers (non-authorship).

We extracted two types of features: structural (height of the lines, inclination of the ascenders and characteristic of the diacritic points) and global (features resulting from the wavelets). Feature vectors are extracted from handwriting samples of 60 individuals at three levels: Ascender, sub-words and line.

- **Structural features**

Line height: We fixed spaces between the lines to avoid the overlaps between two succeeding lines. Under these conditions, the extraction of the lines of writings becomes easier by the method of the horizontal projection profile. However, diacritical signs are ignored by the extraction procedure. The height ' H' of the line is measured directly on the histogram.

Spaces between sub-words: An algorithm allowing to cross the baseline and to calculate the average of white spaces between sub-words.

Inclination of the ascenders: First, we extract the zone delimiting the ascenders (vertical distance from baseline to the character top edge) then the segmentation of the ascenders is operated with connected-component neighbourhoods. Finally, we measure the angle of inclination 'γ' on each ascender.

Dots boldness and shape: Diacritic dots shapes characterize the writer's style in Arabic handwriting. Our visual system detect the dot boldness by the density of black surface and their "ellipsoid" shape by comparing two constants, conventionally denoted major and minor axis.

Two features are extracted from isolated diacritic dots. First, the height and the width of each diacritic dot (isolated dots) are measured on each point's bounding box and the ratio of the sum of black pixels. Then we compute:

$$R_{h/l} = \frac{a}{b} \text{ and } R_{pixels} = \frac{P_n}{P_t} \times 100 \tag{1}$$

Where ‘b’ is the width, ‘a’ is the height, P_n is the sum of the black pixels and P_t is the sum of the total pixels in the bounding box.

- **Global features**

Wavelet transforms: We used Daubechies wavelet transforms to describe cursive Arabic word, the same feature set was successfully used in the case of Arabic Optical Font Recognition [9]. The 2D wavelet decomposition is applied until level three; yielding at this level four sub-bands images: approximation image, horizontal details, vertical details and diagonal details at level three. Therefore, we compute the mean and the standard deviation from each matrix subband.

Entropy: Entropy is a statistical measure of “randomness” (probability distribution) used to characterize the texture of the input image. It is defined as:

$$-\sum p(i).\log p(i) \tag{2}$$

Where *p(i)* is the probability of the ith bins level being used computed from the grayscale histogram.

3.2 Features Selection

The goal of dimension reduction is to present to the Multi-Layer Perceptron neural network the most relevant information. Consider the original feature set extracted in the previous step $F = \{f_1, f_2, ..., f_{20}\}$. Each individual in the population represents a candidate solution to the feature subset selection problem. We run our experiments using standard genetic algorithm for feature subset selection in order to cut out the redundant and irrelevant features. Filter approach is adopted for feature subset selection. Experiments held using the following parameter selected after several preliminary runs. The initial population: generated randomly. Population size: 30, Number of generation: 1000, Probability of crossover: 0.8, Probability of mutation: 0.002, Selection function: roulette wheel and as a Fitness function we used to minimize Mean Square Error. The first step was to find the optimal subset for each writer’s set of features by mean of a genetic algorithm. The second step consisted of cumulating the number of times each feature was selected for the sixty individuals representing the full dataset. Finally, after several experimental evaluations using neural networks, we retained the twelve feature’s first higher cumulated score [10].

4 Classification

This section briefly describes the techniques that we have used to implement a modular networks. The proposal architecture was tested with a Multi-Layer Perceptron network. Multi-Layer Perceptron (MLP) network has been successfully applied to many practical problems because of its non-linear mapping ability.

Class-modular classification architecture was adopted. It consists of 60 different classes MLP, each one representing the style of one writer which is trained sparely.

During the training stage, a sub-network learns the features of the authorship and non-authorship. Thus, each sub-network learns the good examples as well as the bad ones (among the whole training base). Finally, the pages of test are presented at each of the 60 trained sub-networks. We recorded an average rate of 94.73 %. Fig. 2 reports the results obtained by each subnetwork of the ensemble for both original feature set and optimal subset.

If we consider that the number of inputs for each network was associated with the size of the feature vector, the number of inputs for each subnetwork was reduced by 40% with the uses of the optimal subset. We have to compute less number of weights. The convergence speed of the networks is accelerated.

The results shown in Table 1 indicate that the network constructed using both original and optimal feature set deliver comparable performance level.

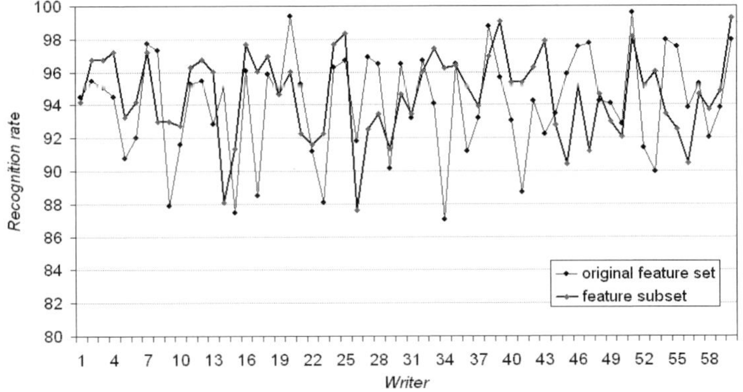

Fig. 2. Experimental results obtained with MLP modular network

Table 1. Size of the feature vector and average recognition rate for each set

Original network		Optimal network	
Number of inputs	Average rate	Number of inputs	Average rate
20	94.11	12	94.73

Optimal subset has the benefit of using much reduced complexity classifier when used less features. We can conclude that the genetic algorithm offers the advantage of eliminating the irrelevant and redundant features increasing the performance of the neural networks in both speed and global recognition rate.

5 Conclusion and Future Works

This paper has proposed an approach for writer identification by combining optimal local and global feature. The classification experiments were carried out using two features sets (original set Vs optimal subset) and Multi-Layer Perceptron for classification. Feature subset selection was performed using genetic algorithm to cutout features with weakly or no predictive information. Through experiments, we have found that the network constructed using both original and optimal feature set deliver comparable performance level on this particular task. Optimal subset has the benefits of using much reduced complexity classifier and making the step of features extraction easier when we have to extract only 12 features (compared to 20 features in the original set). Redundancy makes emphasis on less important characteristics when more than two inputs represent the same feature. In this case, it involves assigning a false valued weight associated to this characteristic and reflecting higher relevance. Additional works will be needed to increase the size of our database.

References

1. Srihari, S.N., Cha, S.H., Arora, H., Lee, S.: Individuality of Handwriting. Journal of Forensic Sciences 47(4) (2002) 856-872
2. Cha, S.H., Srihari, S.H.: Multiple Feature Integration for Writer Verification. Proceeding of the 7th International Workshop on Frontiers in Handwriting Recognition (IWFHR'00), Amsterdam, September (2000) 333-342
3. Writer Identification Using Arabic Script. The second International Conference on Machine Intelligence (ACIDCA-ICMI'05), Tozeur/Tunisia (2005) 1001-1005
4. Korsheed, M.S.M.: Automatic Recognition of words in Arabic manuscripts. PhD thesis, Churchill College University Of Cambridge (2000)
5. Amara, N., Ellouze, N.: Overview and Advances in Arabic Optical Character Recognition. Asian Journal on Information Technology, Grace Publications Network 3 (2004) 217-240
6. Raymer, M., Punch, W.F., Goodman, E.D., Kuhn, L.A., Jain, A.K.: Dimensionality Reduction Using Genetic Algorithms. IEEE Transactions on Evol. Comput.4(2000) 164-171
7. Oliveira, L.S.: A Methodology for Feature Selection Using Multi-objective Genetic Algorithms for Handwritten Digit String. International Journal of Pattern Recognition and Artificial intelligence 17 (6) (2003) 903-929
8. John, G.H., Kohavi, R., Pfleger, K.: Irrelevant Feature and the Subset Selection Problem. Eleventh International Conference on Machine learning (ML'94). New Brunswick (1994)
9. Gazzah, S., Amara, N.: Utilisation des Ondelettes et des Réseaux de Neurones en Identification des Fontes Arabes. Conférence Internationale Signaux, Circuits et Systèmes, (SCS'04), Monastir/Tunisia (2004) 498-502
10. Gazzah, S., Amara, N.: Writer Identification Using SVM Classifier and Genetic Algorithm for Optimal Futures Selection. International Arab Conference on Information Technology (ACIT'05). Amman / Jordan (2005) 461-466

Self-generation ART Neural Network for Character Recognition

Taekyung Kim[1], Seongwon Lee[2], and Joonki Paik[1]

[1] Image Processing and Intelligent Systems Laboratory, Department of Image Engineering, Graduate School of Advanced Imaging Science, Multimedia, and Film, Chung-Ang University, Seoul, Korea
kimktk@wm.cau.ac.kr, http://ipis.cau.ac.kr
[2] Department of Computer Engineering, College of Electronics and Information, Kwangwoon University, Seoul, Korea
swlee@kw.ac.kr

Abstract. In this paper, we present a novel self-generation, supervised character recognition algorithm based on adaptive resonance theory (ART) artificial neural network (ANN) and delta-bar-delta method. By combining two methods, the proposed algorithm can reduce noise problem in the ART ANN and the local minima problem in the delta-bar-delta method. The proposed method can extend itself based on new information contained in input patterns that require nodes of hidden layers in neural networks and effectively find characters. We experiment with various real-world documents such as a student ID and an identifier on a container. The experimental results show that the proposed self-generation. ART algorithm reduces the possibility of local minima and accelerates learning speed compared with existing.

1 Introduction

Artificial neural network (ANN) is a computing paradaim that originiates from biological nervous systems. It is composed of large number of interconnected neurons mapping a random N-dimensional input space to M-dimensional output space. Because of its learning ability ANN can be applied to associative memory, classification, pattern cognition and image processing to name a few [1-5].

The error back-propagation that is one of the supervised learning methods of ANN is the most widely used learning algorithm of a multilayer neural network. In this method the learning speed is determined by parameters for input patterns. The solutions of the error back-propagation algorithm may fall into a local minimum depending on the number of hidden layer nodes. That is, if a large number of hidden layers are required, there are extra nodes that have little effect on the performance of the decision, and slow the learning speed down. The smaller number of hidden layers is, the higher possibility of the result being located in local minima is. Although the number of required hidden nodes is esily found in simple problems such as an XOR problem, it is impractical to find the optimum number in 2-D pattern recognition. Thus, the number of hidden nodes is usually obtained by using heuristic methods.

J. Wang et al. (Eds.): ISNN 2006, LNCS 3972, pp. 277 – 286, 2006.

One of the other learning models is ART (Adaptive Resonance Theory) neural network by Carpenter [6]. ART has stability and adaptivity because it can classify and learn patterns [7]. Furthermore, ART doesn't have the local minima problem. ART, however, may classify the similar patterns into different clusters if the similarity is smaller than boundary condition [8]. In case of pattern recognition applications, ART algorithm often suffers from noise [9]. In this paper, we propose a novel character recognition algorithm that finds optimal result and is immune to noise by combining delta-bar-delta [10] and the ART neural network. We measure the performance of the proposed algorithm with several real-world documents such as student ID, resident ID and container identifier.

This paper is organized as follows. Section 2 presents background algorithms. In section 3 we describe the proposed Self-Generation ART Neural Network. The experimental results are presented in section 4. Finally, we conclude the paper in section 5.

2 ART Algorithm and Delta-Bar-Delta Method

2.1 ART Algorithm

One of sophisticated clustering technique for adaptively adjusting the number of clusters, called adaptive resonance theory (ART), was introduced by Carpenter and Grossberg [6]. In the ART algorithm, a backward network is adopted for vigilance test in addition to the conventional forward networks between the input neurons and output neurons, two widely used versions of ART algorithms are: ART-1 for binary-valued patterns and ART-2 for continuous-valued patterns. Figure 1 shows a general configuration of ART [12]. It can adaptively create a new neuron for an incoming input pattern if it is determined (by a vigilance test) to be sufficiently different from the existing clusters. Such a vigilance test is incorporated into the adaptive backward network. Suppose that w_{ij} and t_{ij} are the forward and backward weights between neuron j and input i, respectively. Note that x_i's and t_{ij}'s binary-valued while w_{ij}'s are real. The weight adjustments are applied to both the feed-forward weights w_{ij*} and the feedback weight t_{ij*} from winning neuron $j*$.

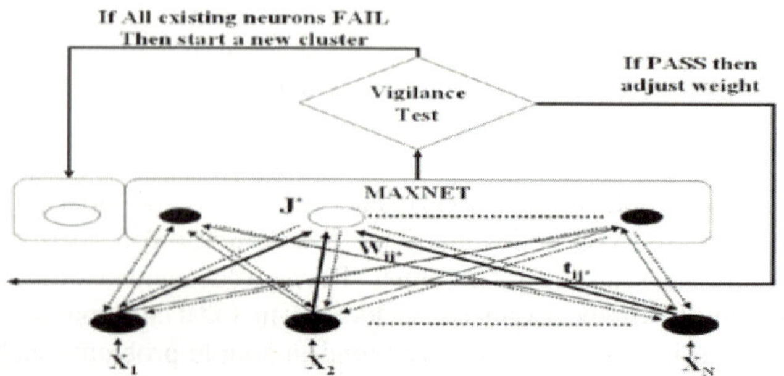

Fig. 1. The configuration of ART learning structure

The configuration of ART-2 is similar to ART-1 except that the backward weights $\|x - w_j\|$ are the same as the forward weights, and the MAXNET $\|x - w_{j*}\| < \rho$ is replaced by MINNET, so the minimum-distance solution is chosen as the winner.

2.2 Delta-Bar-Delta Algorithm

One of heuristics for back-propagation speed-up is the Delta-bar-Delta algorithm by Jacobs [11]. This algorithm is based on the idea that the slope of the error surface along different weight directions might differ considerably. Descending slopes along different directions can cause for minima across steep slopes to be jumped over and lead to very slow descent along shallow slopes. Simply trying to alleviate either of these problems exacerbates the other. In order to address both issues together, Jacobs suggests the following four heuristics, which form the basis for the delta-bar-delta algorithm:

$$w_{kj}(t+1) = w_{kj}(t-1) - \alpha_{kj}(t+1)\frac{\partial E}{\partial w_{kj}} = w_{kj}(t) + \alpha_{kj}(t+1)\delta_k Z_i, \tag{1}$$

where i th input layer, j th hidden layer, k th output layer when a multilayer network consists of weight change. The symbol α_{kj} is learning value, δ_{kj} output layer error value, Z_j and hidden layer output value. Delta rules to output layer formula (2) and hidden layer formula (3) are

$$\Delta_{kj} = \frac{\partial E}{\partial w_{kj}} = -\delta_k Z_j, \tag{2}$$

$$\Delta_{ji} = \frac{\partial E}{\partial w_{ji}} = -\delta_j x_i, \tag{3}$$

$$\overline{\Delta_{kj}}(t) = (1-\beta)\Delta_{kj}(t) + \beta\overline{\Delta_{kj}}(t-1), \tag{4}$$

where $0 < \beta < 1$. If weights change continuously learning step equal vector addition learning value, the next delta and delta-bar for learning value are

$$\begin{aligned}
&\alpha_{kj}(t+1) \\
&= \alpha_{kj}(t) + x \qquad && if \quad \overline{\Delta}(t-1) \bullet \Delta_{kj}(t) > 0 \\
&= (1-r)\alpha_{kj}(t) && if \quad \overline{\Delta}(t-1) \bullet \Delta_{kj}(t) < 0 \\
&= \alpha_{kj}(t) && if \quad \overline{\Delta}(t-1) \bullet \Delta_{kj}(t) = 0
\end{aligned}, \tag{5}$$

$$\begin{aligned}
&\alpha_{ji}(t+1) \\
&= \alpha_{ji}(t) + x \qquad && if \quad \overline{\Delta_{ji}}(t-1) \bullet \Delta_{ji}(t) > 0 \\
&= (1-r)\alpha_{ji}(t) && if \quad \overline{\Delta}_{ji}(t-1) \bullet \Delta_{ji}(t) < 0 \\
&= \alpha_{ji}(t) && if \quad \overline{\Delta_{ji}}(t-1) \bullet \Delta_{ji}(t) = 0
\end{aligned}, \tag{6}$$

where r is a fixed const value.

3 Proposed Self-generation ART Model

3.1 Comining Delta-Bar-Delta with ART-1

The error back-propagation learning algorithm determines the number of hidden layers with heuristically methods. In this paper, we propose self generation adaptive resonance theory (SG-ART) algorithm. The proposed self generation supervised learning algorithm that is a modified ART algorithm combined with delta-bar-delta method uses self-generating hidden nodes. The proposed SG-ART-1 is used to produce nodes from the input layer to the hidden layer. A winner-take-all method is used to the connection weight adaptation so that some stored patterns can be updated.

The proposed model skips forward activation processes in case of zero patterns, and affects all the weights for all clustering layers in the backward activation process. Analyzing this as a biological model, a winner neuron should be chosen by excitatory neurons. If the net value of the clustering layer node approaches 0, it has a forward inhibitory neuron structure and doesn't choose a winner node. In the proposed method, a winner node is chosen by a biological feedback inhibitory neuron structure.

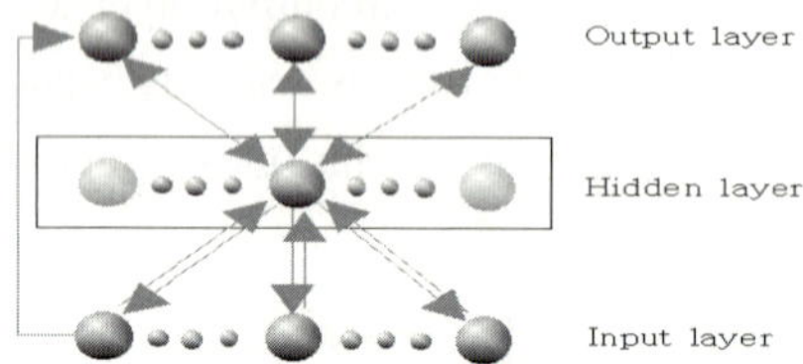

Fig. 2. Proposed self-generation learning supervised structure based on ART1

3.2 Proposed SG-ART-1 Algorithm

The proposed method adjusts the weight of established adaptive error back-propagation by winner-take-all method. A winner node from a hidden layer becomes representative class for the suggested patterns. In order to reform the representative class with the winner node the proposed algorithm adjusts the weights of synapses from the hidden layer to the input layer as shown in figure 3. The connection weight related to the output layer and its representative class is also adjusted for the target value to affect output value. It is shown in figure 4.

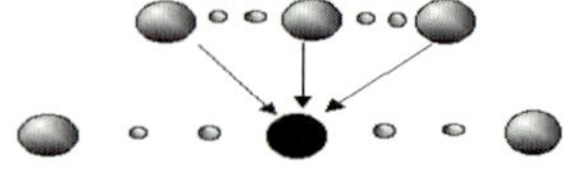

Fig. 3. An adjusting weight connecting hidden layer to input layer

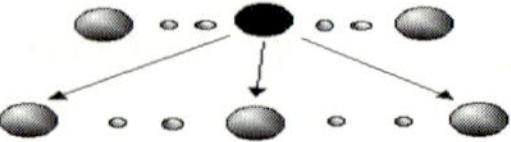

Fig. 4. An adjusting weight connecting output layer to hidden layer

In the proposed algorithm the forward activity process is skipped in case of zero patterns and weight adjustment affects all hidden node. It prevents the net value of hidden layer from being zero, which causes the failure of winner selection. The procedures of SG-ART-1 algorithm can be summarized as follows:

Step 1: Initialize top-down weight w_{ji} between input layer and hidden layer and bottom-up weight t_{ji} as

$$t_{ji} = 1, \qquad w_{ji} = \frac{1}{m+1}, \tag{7}$$

where $i(i=1,...,m)$ is input layer, $j(j=1,...,n)$ hidden layer, $k(k=1,...,p)$ output layer.
Step 2: suggest input vector X_i and target vector T_k
Step 3: Calculate output vector O_j of a hidden layer

$$O_j = \sum_{i=1}^{m} w_{ji} \times x_i \,, \tag{8}$$

Step 4: select winner node O_{j*}

$$O_{j*} = Max\lfloor O_j \rfloor, \tag{9}$$

Step 5: Calculate the similarity between bottom-up winner weight nodes and input patterns

$$S = \frac{\|T \bullet X\|}{\|X\|}, \tag{10}$$

Step 6: If the similarity is $S \geq \rho$ where Vigilance parameter ρ is between 0.5 with 1, select a winner node as a represented class and go to step 8. Otherwise, go to step 7.
Step 7: Reset the output vector of winner node ρ_j^* to zero, and select a new node for the representative class.
Step 8: Adjust the top-down and bottom-up weights of newly created representative class.

$$\begin{aligned} t_{j*i}(n+1) &= t_{j*i}(n) \times x_{i,} \\ w_{j*i}(n+1) &= \frac{t_{j*i}(n+1) \times x_i}{0.5 + \sum_{i-1}^{m} w_{j*i} \times x_i}, \end{aligned} \tag{11}$$

Step 9: Calculate NET and output vector O_k as

$$\begin{aligned} NET &= \sum_{j-1}^{n} w_{kj*} \times O_{j*} + \Theta_k, \\ w_{j*i}(n+1) &= \frac{1}{1+e^{-net}}, \end{aligned} \tag{12}$$

Step 10: Adjust weight w_{kj*}, bias Θ_k, learning rate α_{kj} as

$$
\begin{aligned}
& w_{kj*}(n+1) = w_{kj*}(t) + \alpha_{kj*}(t+1)\sigma_k O_j^*, \\
& \alpha_k = (T_k - O_k)O_k(1-O_k), \\
& \Delta_{kj} = \frac{\partial E}{\partial w_{kj}} = -\sigma_k Z_j, \\
& \overline{\Delta_{kj*}}(t) = (1-\beta)\Delta_{kj}(t) + \beta\overline{\Delta_{kj*}}(t)(t-1), \\
& if \quad \overline{\Delta_{kj*}}(t-1) \bullet \Delta_{kj*}(t) > 0, \\
& \qquad \alpha_{kj*}(t+1) = (1-\gamma)\alpha_{kj*}(t), \\
& if \quad \overline{\Delta_{kj*}}(t-1) \bullet \Delta_{kj*}(t) = 0, \\
& \qquad \alpha_{kj*}(t+1) = \alpha_{kj*}(t)
\end{aligned} \tag{13}
$$

where σ_k is output error and O_j^* winner nodes of hidden layers.

Step 11: If total sum of square (TSS) is greater than the limit of error ε , goto step 3. Otherwise, finish the learning procedure

3.3 Proposed SG-ART-2 Algorithm

The proposed SG-ART-2 algorithm does not take a binary pattern but an analog value. The proposed SG-ART-2 compute the output vector between input layer and hidden layer as

$$O_j = \sum_{i=1}^{N}(\|x_i - w_{ji}(t)\|), \tag{14}$$

where w_{ji} means the weight between an input layer and a hidden layer. An output vector of the hidden layer represents the error between clusters and input patterns. Therefore, the smallest output vector is the winner node as

$$O_j^* = Min[O_j], \tag{15}$$

where *Min* returns the minimum value. The verification of the winner node is calculated as

$$O_{j*} < \rho, \tag{16}$$

where ρ is vigilance parameter of ART-2 algorithm. An output vector of winner vector node is to be classified as a different pattern if the value of the winner node is smaller than the vigilance parameter. If it is classified as the same pattern, the weight is adjusted to reflect features of an input pattern. The adjustment weight of SG-ART-2 algorithm is

$$w_{j*i}(t+1) = \frac{w_{j*i}(t) \times cluster_j + x_i}{cluster_j + 1}, \tag{17}$$

where $cluster_j$ is the number of updated patters in the created cluster.

The structure of SG-ART-2 is shown in figure 5.

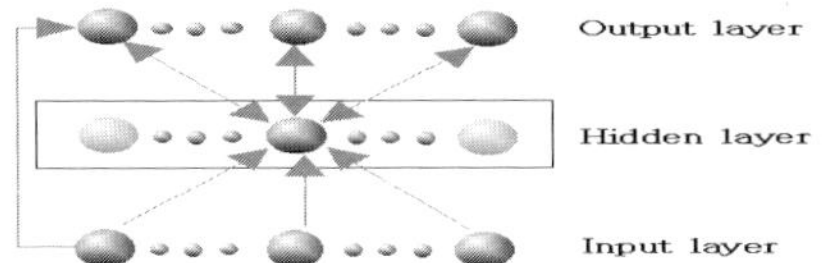

Fig. 5. Proposed self-generation learning supervised structure based on ART2

4 Environmental Results

The proposed algorithm was tested using character and number pattern found in student's identification card, citizen registration card, container image to size 800×600 is used in this paper. Performance comparison analyzed have been applied to 50th number patterns, and extracted from certificate of residence card and 50th character pattern extracted from container image from student's identification card in each algorithm. Figure 6 present the test images used for the proposed experiment.

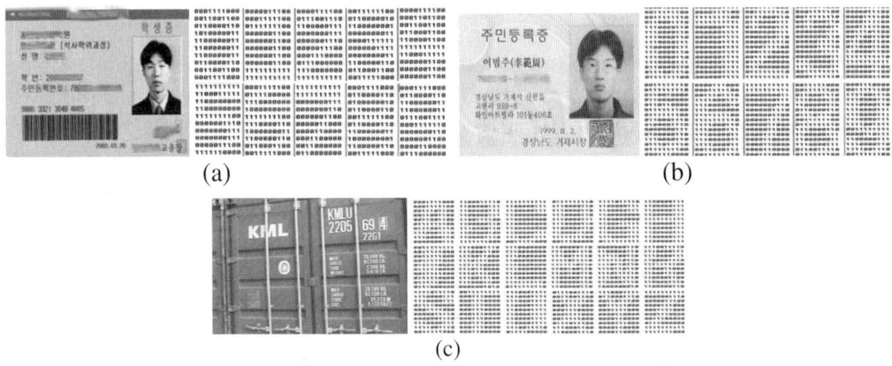

(a) Number of the pattern detected from student identification card, (b) Number of the pattern detected from certificate of residence card, (c) Number of the pattern detected from identifier of the container

Fig. 6. Test Images

The proposed algorithm is applied same using following parameters in [11]. In our simulation, we set up initial learning rate of 0.3 momentum value of 0.5 and delta-bar-delta algorithm parameter value set up by B (0, 7), K (0.005), and R (0.2), apply to vigilance parameter is 0.95 at ART algorithm. Learning conventional decision achieves iteration by 2000 times, existing method value by 0.04 total sum of square (TSS).

The performance of delta-bar-delta method about number pattern abstracted in student's identification card, running experiment, number of times for 10th iteration in momentum method, succeed to each of 4th learning, and average epoch number is delta-bar-delta method is 3952 times, average iteration number in momentum method is 2769 times. In a delta-bar-delta method and ART based on self-generation supervised learning method experiment number of times 10th repeatedly all of them. Also, proposed method use average number of iteration less than existing, confirms improvement in the learning speed and the convergence.

Table 1. Learning state comparison between proposed method with existing method

Method	Number of experiment	Number of Success	Number of Hidden layer	Number of Average iteration
Delta-bar-delta	10	4	10	3952
Momentum+BP	10	4	10	2769
ART-1 SGSL	10	10	23	670
ART-2 SGSL	10	10	18	411

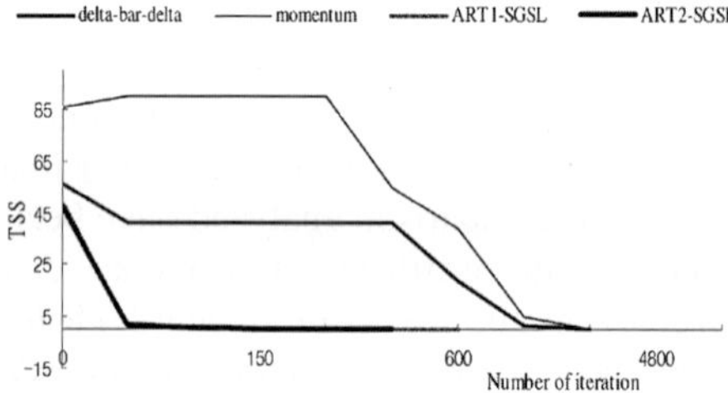

Fig. 7. Total sum of squares between proposed method with existing method change curved

Result using delta-bar-delta method that achieves 10th experiment about number pattern of existing method a certificate of residence card, momentum method was failed. In proposed experiment 10th succeed all, and showed more result by number of average iteration 679 times.

Result using delta-bar-delta method that achieves 10th experiment about character pattern than existing method in container image, momentum method didn't succeed to learning in once. In proposed 10th succeed all, and showed result by number of average iteration 939 times. Also, a proposed method then average number of iteration

Table 2. Learning state comparison between proposed method with existing method

Method	Number of experiment	Number of Success	Number of Hidden layer	Number of Average iteration
Delta-bar-delta	10	0	10	No learning
Momentum+BP	10	0	10	No learning
ART-1 SGSL	10	10	40	679
ART-2 SGSL	10	10	40	651

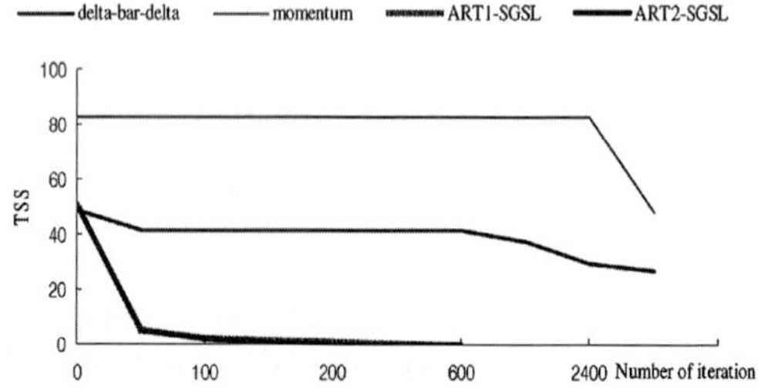

Fig. 8. Total sum of squares between proposed method with existing method change curved

Table 3. Learning state comparison between proposed method with existing method

Method	Number of experiment	Number of Success	Number of Hidden layer	Number of Average iteration
Delta-bar-delta	10	0	10	No learning
Momentum+BP	10	0	10	No learning
ART-1 SGSL	10	10	46	939
ART-2 SGSL	10	10	40	840

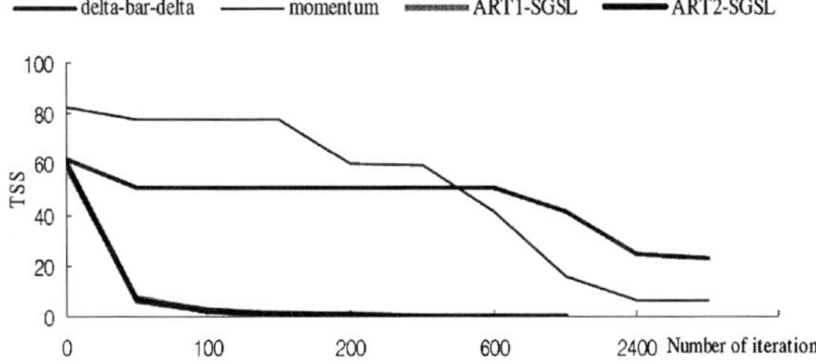

Fig. 9. Total sum of squares between proposed method and existing method change curved

compared to existing method confirmed that they are improved in the learning speed and the convergence.

Table 4 present the result of a learning difference by the existing method using a vigilance parameter which comes from the proposed method, and we got bad result of the existing method than p by number of hidden layer.

Table 4. Result of learning to the vigilance parameter by ART-2 based on SGSL

method	Vigilance parameter	Number of iteration	Created of number of hidden layer
ART-1 SGSL	0.85	No learning	34
	0.9	No learning	37
	0.95	939	46
ART-2 SGSL	0.15	No learning	33
	0.1	840	40
	0.05	970	48

5 Conclusions

In this paper we proposed a novel self-generation supervised learning method that combines delta-bar-delta method and ART artificial neural network. The proposed algorithm uses ART algorithm for connections between input layer and hidden layer and delta-bar-delta method for them between hidden layer and output layer. By combining advantages of two algorithms, the proposed algorithm can reduce noise problems in the ART algorithm and local minima problems in the delta-bar-delta method. It also improves convergence speed. Experimental results with some real-world images show that the proposed SG-ART algorithm can lessen the impact of problems in the existing algorithms and improve character recognition performance.

Acknowledgement

This research was supported by Korean Ministry of Science and Technology under the National Research Laboratory Project, Korean Ministry of Information and Communication under HNRC-ITRC program at Chung-Ang university supervised by IITA.

References

1. Rumelhart, D., Hinton, G., Williams, R.J.: Learning Representation by Back Propagation Errors. Nature (1986) 533-536
2. Rumelhart, D., McClelland, J.: The PDP Research Group. `Parallel Distributed Processing 1 318-536
3. James, A., David, M.: Neural Networks Algorithms, Applications, and Programming Techiques. Addison Wesley (1991) 89-125
4. Martin, T., Howard, B.: Neural Network Design. PWS Publishing Company (1996)
5. Carpenter, G., Grossberg,.S.: Neural Dynamics of Category Learning and Recognition, Attention, Memory Consolidation, and Amnesia. Brain Structure, Learning, and Memory, AAAS Symposium Series (1986)
6. Carpenter., G., Grossberg, S.: The ART of Adaptive Pattern Recognition by a Self-Organizing Neural networks. IEEE Computers (1988) 77-88
7. Kim, K., Kim, K.C.: A Study on Face Recognition using New Fuzzy ART. ITC-CSCC, 2 (1998) 1057-1060
8. Carpenter, G., Grossberg, S.: A Massively Parallel Architecture for a Self-organizing Neural Pattern Recognition Machine. Computer Vision, Graphics, and Image Processing 37 (1987) 54-115
9. Allen, R.: Several Studies on Natural Language and Back Propagation. IEEE ICNN 2 (1987) 335-342
10. Jacobs. R.A.: Increased Rates of Convergence Through Learning Rate Adaptation. Neural Networks 2 (1988) 295-308
11. Carpenter, G., Grossberg, S.: ART2 Self-Organization of Stable Category Recognition Codes for Analog Input Patterns. 1987 IEEE Int. Conf. Neural Networks (1987) 727-736

Handwritten Digit Recognition Using Low Rank Approximation Based Competitive Neural Network

Yafeng Hu, Feng Zhu, Hairong Lv, and Xianda Zhang

Department of Automation, Tsinghua University, Beijing 100084, China
yafh99@mails.tsinghua.edu.cn

Abstract. A novel approach for handwritten digit recognition is proposed in this paper, which combines the low rank approximation and the competitive neural network together. The images in each class are clustered into several subclasses by the competitive neural network, which is helpful for feature extraction. The low rank approximation is used for image feature extraction. Finally, the k-nearest neighbor classifier is applied to the classification. Experiment results on USPS dataset show the effectiveness of the proposed approach.

1 Introduction

Handwritten digit recognition plays an important role in pattern classification research, which is to automatically determine the digit based on the given image. It is widely used in OCR applications, such as the bank check reading, postal mail sorting and document processing. Although many efforts have been done to improve the performance of digital recognition [1], [2], [3], [4], it still lacks a technically satisfactory solution. In this paper, the proposed approach combines the low rank approximation with the competitive neural network. Training samples of each class are clustered into several subclasses by the competitive neural network, which is helpful for feature extraction. The low rank approximation is used for image feature extraction, which has been proved to be an effective method. The algorithm is tested on USPS handwritten digit dataset. The experiment results show that this approach has high performance.

2 Low Rank Approximation

The performance of digit recognition largely depends on the feature extraction approach. More recently, Ye [5] proposed a new approach for image feature extraction and image representation, which is called generalized low rank approximation (GLRAM). This method is proved to have superior performance over the SVD-based method in many cases.

Let $\mathbf{A}_i \in \Re^{m\times n}$, for $i = 1,2,\cdots,M$, be the M image matrixes. Denote $\mathbf{U} \in \Re^{m\times p}$ and $\mathbf{V} \in \Re^{n\times q}$ as two transform matrixes with orthogonal columns. The feature matrix of $\mathbf{A}_i$ is gotten as follows:

J. Wang et al. (Eds.): ISNN 2006, LNCS 3972, pp. 287 – 292, 2006.

$$\mathbf{D}_i = \mathbf{U}^T \mathbf{A}_i \mathbf{V} \tag{1}$$

The reconstructed image $\mathbf{U}\mathbf{D}_i\mathbf{V}^T$ is an approximation of $\mathbf{A}_i$. Then the optimization problem can be stated as

$$\min_{\substack{U^T U = I_p \\ V^T V = I_q}} \frac{1}{M} \sum_{i=1}^{M} \left\| \mathbf{A}_i - \mathbf{U}\mathbf{U}^T \mathbf{A}_i \mathbf{V}\mathbf{V}^T \right\|_F^2 \tag{2}$$

where the Frobenius norm is defined as $\| \mathbf{H} \|_F^2 = \sum_{i,j} h_{ij}^2$. Let

$$\begin{cases} \mathbf{M}_V = \dfrac{1}{M} \sum_{i=1}^{M} \mathbf{A}_i \mathbf{V}\mathbf{V}^T \mathbf{A}_i^T \\ \mathbf{M}_U = \dfrac{1}{M} \sum_{i=1}^{M} \mathbf{A}_i^T \mathbf{U}^T \mathbf{U} \mathbf{A}_i \end{cases} \tag{3}$$

Assume $\mathbf{U}$ and $\mathbf{V}$ be the optimal solution to the problem in Eq.(2), and they also solve the following optimization problem [5]:

$$\max_{\substack{\mathbf{U}^T \mathbf{U} = \mathbf{I}_p \\ \mathbf{V}^T \mathbf{V} = \mathbf{I}_q}} \frac{1}{M} \sum_{i=1}^{M} \left\| \mathbf{U}^T \mathbf{A}_i \mathbf{V} \right\|_F^2 \tag{4}$$

Eq.(4) can be further represented as

$$\begin{aligned} \frac{1}{M} \sum_{i=1}^{M} \left\| \mathbf{U}^T \mathbf{A}_i \mathbf{V} \right\|_F^2 &= \frac{1}{M} \sum_{i=1}^{M} trace(\mathbf{U}^T \mathbf{A}_i \mathbf{V}\mathbf{V}^T \mathbf{A}_i^T \mathbf{U}) \\ &= trace(\mathbf{U}^T \mathbf{M}_V \mathbf{U}) \\ &= trace(\mathbf{V}^T \mathbf{M}_U \mathbf{V}) \end{aligned} \tag{5}$$

Let Val_{U} be the sum of the first q largest eigenvalues of $\mathbf{M}_{\mathrm{U}}$. There has no closed form solution for this problem, and the algorithm is iterative, which is given as follows:

Step 1: Initializing. Assume that p and q are given, and compute the matrix

$$\mathbf{G}_p = \sum_{i=1}^{N} \mathbf{A}_i \mathbf{A}_i^T \tag{6}$$

Computer the eigenvectors of $\mathbf{G}_p$, and get $\mathbf{U}_0$.

Step 2: Updating. Let k denote the iterative times. Compute the eigenvectors of $\mathbf{M}_{\mathrm{U},k-1}$, and get $\mathbf{V}_k$ corresponding to $\mathbf{U}_{k-1}$.

Compute the eigenvectors of $\mathbf{M}_{\mathrm{V},k}$, and get $\mathbf{U}_k$ corresponding to $\mathbf{V}_k$. Compute $Val_{\mathrm{U},k}$ of $\mathbf{M}_{\mathrm{U},k-1}$.

Step 3: Stop. If $\left|Val_{\mathbf{U},k} - Val_{\mathbf{U},k-1}\right| < threshold\_value$, then the iterative procedure is stopped. Save $\mathbf{U}_k$ and $\mathbf{V}_k$ as the desired $\mathbf{U}$ and $\mathbf{V}$, respectively.

3 Competitive Neural Network

As a typical unsupervised learning network, competitive networks can group a set of vectors into clusters [6]. The network consists of three layers, which are the input layer, the competitive layer and output layer. Fig.1 illustrates the structure of a competitive neural network.

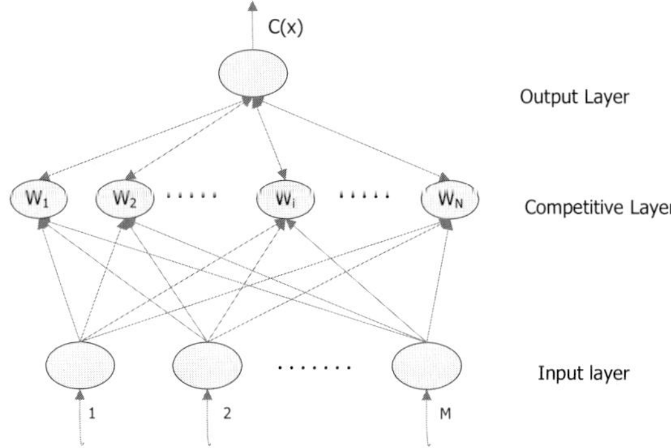

Fig. 1. Structure of the competitive neural network

When an input vector $\mathbf{x}$ is presented to the input layer, the neurons in competitive layer compete and the one whose weight vector is closest to $\mathbf{x}$ is chosen as the winner. The output layer picks up the wining neuron, and classifies the input vector to that class. During the learning process, the weight of the wining neuron is adjusted with the Kohenen learning rule stated below:

$$\mathbf{W}_i(t) = \mathbf{W}_i(t-1) + \alpha(\mathbf{x}(t) - \mathbf{W}_i(t-1)) \tag{7}$$

where i is the index of the wining neuron, α is the learning rata and t is the iteration number. Thus, the competitive network can learn to recognize groups of similar input vectors.

Some neurons may never win because their weights start out far from any input vectors. To keep this from happening, biases are added to these neurons and make them more likely to win. Eventually, every neuron in the competitive layer will respond to an equal number of training vectors.

4 Low Rank Approximation Based Competitive Neural Network

The algorithm of low rank approximation based competitive neural network has five steps: image preprocessing, feature extraction for clustering, clustering, feature extraction in every subclass and classification.

4.1 Image Preprocessing

Preprocessing is very important for the good performance of the system. Our algorithm includes size normalization and energy normalization. All original images are size normalized to fit in a 16×16 pixel box while preserving their aspect ratio. The value of every pixel is mapped into the [0, +1] range. Define the energy of an image $\mathbf{A}$ as $\mathrm{E}(\mathbf{A}) = \sum a_{ij}^2$, then the energy normalized image is obtained as follows:

$$\tilde{\mathbf{A}} = \frac{\mathbf{A}}{\sqrt{\mathrm{E}(\mathbf{A})}} \tag{8}$$

4.2 Feature Extraction for Clustering Using Low Rank Approximation

In this step, for the given p and q, get $\tilde{\mathbf{U}}_i \in \Re^{16\times(p+2)}$ and $\tilde{\mathbf{V}}_i \in \Re^{16\times(q+2)}$ for every class of the training images by the iterative low rank approximation algorithm, where $i = 1,2,\cdots 10$. By Eq.(1), obtain feature matrix sets $\tilde{\Phi}_i = \left\{\tilde{\mathbf{D}}_{i,1}, \tilde{\mathbf{D}}_{i,2}, \cdots, \tilde{\mathbf{D}}_{i,N_i}\right\}$.

4.3 Clustering in Every Class of Digit Images

Every feature matrix set $\tilde{\Phi}_i$ is clustered into L subsets, which is $\tilde{\Phi}_i = \left\{\tilde{\Phi}_{i,1}, \tilde{\Phi}_{i,2}, \cdots, \tilde{\Phi}_{i,L}\right\}$. This procedure is accomplished by the competitive neural network. Then, the training images of each class are divided into L subclasses, and the number of the total subclasses is $10L$.

4.4 Feature Extraction in Every Subclass Using Low Rank Approximation

Using the iterative low rank approximation algorithm for each subclass, and get $\mathbf{U}_{i,j}$ and $\mathbf{V}_{i,j}$,for $i = 1,2,\cdots 10$ and $j = 1,2,\cdots,L$. Through Eq.(1), obtain the final $10L$ feature matrix subsets $\Phi_{i,j} = \left\{\mathbf{D}_{i,j,1}, \mathbf{D}_{i,j,2}, \cdots, \mathbf{D}_{i,j,N_{ij}}\right\}$.

4.5 Classification with the K-Nearest Neighbor Classifier

For a testing image $\mathbf{A}_0$, which has been preprocessed, define the distance between $\mathbf{A}_0$ and the feature matrix $\mathbf{D}_{i,j,r}$ as

$$\mathrm{d}(\mathbf{A}_0, \mathbf{D}_{i,j,r}) = \left\| \mathbf{U}_{i,j}^T \mathbf{A}_0 \mathbf{V}_{i,j} - \mathbf{D}_{i,j,r} \right\|_F^2 \tag{9}$$

Then we use the k-nearest neighbor classifier for classification. Define the k-distance between $\mathbf{A}_0$ and $\Phi_{i,j}$ as follows:

$$\Psi_k(\mathbf{A}_0, \Phi_{i,j}) = \sum_{s=1}^{k} \mathrm{d}(\mathbf{A}_0, \mathbf{D}_{i,j,rs}) \tag{10}$$

where $\{d(\mathbf{A}_0, \mathbf{D}_{i,j,rs}) | s = 1, 2, \cdots, k\}$ are the first k minimums of $\{d(\mathbf{A}_0, \mathbf{D}_{i,j,r}) | r = 1, 2, \cdots, N_{i,j}\}$. If $\Psi_k(\mathbf{A}_0, \Phi_{l,v}) = \min_{i,j} \Psi_k(\mathbf{A}_0, \Phi_{i,j})$, the resulting decision is that $\mathbf{A}_0$ belongs to the l-th class. In the experiment, we set k=2.

5 Experiment Results

To test the performance of the proposed algorithm, the US Postal Service (USPS) handwritten digit dataset is used. It is derived from a project on recognizing handwritten digits on envelopes. The digits are downscaled to 16×16 pixels and 1:1 scaled. The training set has 7291 samples, and the test set has 2007 samples. A human error rate estimated to be 2.5% shows that it is a hard recognition task. Fig.2 shows samples from USPS.

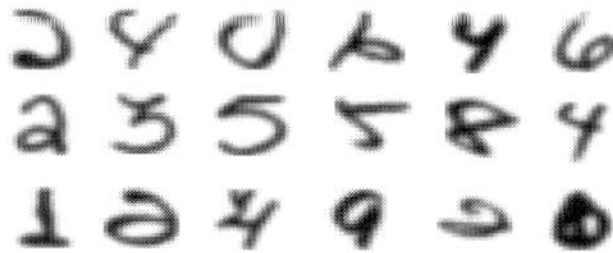

Fig. 2. Some samples in USPS

In the experiment, the number of subclasses is set as $L = 10$, and the network trains for 300 epochs. The algorithm is tested with the different dimensions of feature matrixes. The recognition error rates are listed in Table 1.

For comparison, Table 2 shows the best results obtained by other methods. It must be pointed out that all these methods are based on Euclidean distance. From Table 2, we can see that the proposed algorithm achieves excellent recognition performances.

Table 1. Experiment results vers the dimensions of feature matrixes on USPS dataset

Dimension(p*q)	3x3	4x4	5x5	6x6	7x7	8x8
Error Rate	10.4%	6.32%	4.38%	4.18%	4.38%	4.43%

Table 2. Error rates on the test data of USPS dataset

Method	Error Rate
Proposed Algorithm	4.18%
PCA	5.58%
Decision Tree C4.5 [7]	16.2%
Two-Layer Neural Net 7]	5.9%
5-Layer Neural Net [7]	4.2%
SVM with a polynomial kernel [7]	4.0%
Human Performance	2.5%

6 Conclusions

In this paper, we have presented an approach for handwritten digit recognition using low rank approximation based competitive neural network. By the competitive neural network, training images in each class are clustered into several subclasses. Then all images in a subclass are more similar to each other, which is very helpful for feature extraction. The low rank approximation is an effective method for image feature extraction, which is accomplished by an iterative procedure. At last, the k-nearest neighbor classifier is used for classification. The experiment results on USPS dataset show the effectiveness of the proposed method.

Acknowledgement

Research mainly supported by the Seed Foundation of Tsinghua University. Yafeng Hu and Feng Zhu are the first coauthors.

References

1. Liu, C.-L., Nakashima, K., Sako, H., Fujisawa, H.: Handwritten Digit Recognition: Benchmarking of State-of-the-Art Techniques. Pattern Recognition 36 (2003) 2271-2285
2. Kim, D., Bang, S.Y.: A Handwritten Numeral Character Classification Using Tolerant Rough Set. IEEE Trans. Pattern Anal. and Machine Intelligence 22 (2000) 923-937
3. Mayraz, G., Hinton, G.: Recognizing Handwritten Digits Using Hierarchical Products of Experts. IEEE Trans. Pattern Anal. and Machine Intelligence 24 (2002) 189-197
4. Zhang, B., Fu, M., Yan, H.: A Nonlinear Neural Network Model of Mixture of Local Principal Component Analysis: Appl. to Handwritten Digit Recognition. Pattern Recognition 34 (2001) 203-214
5. Ye, J.: Generalized Low Rank Approximations of Matrices. The 21st International Conference on Machine Learning (2004) 887–894
6. Kohonen, T.: Self-organization and Associative Memory. Springer-Verlag, Berlin (1987)
7. Vapnik, V.: The Nature of Statistical Learning Theory. Springer, New York (1995)
8. Zhang, X.D.: Matrix Analysis and Applications. Tsinghua University Press, Beijing (2004)

Multifont Arabic Characters Recognition Using HoughTransform and Neural Networks

Nadia Ben Amor[1] and Najoua Essoukri Ben Amara[2]

[1] Laboratory of Systems and Signal Processing (LSTS), National Engineering School of Tunis (ENIT), Tunisia
n.benamor@ttnet.tn
[2] Laboratory of Systems and Signal Processing (LSTS), National Engineering School of Sousse (ENISo), Tunisia
najoua.benamara@enim.rnu.tn

Abstract. Pattern recognition is a well-established field of study and Optical Character Recognition (OCR) has long been seen as one of its important contributions. However, Arabic has been one of the last major languages to receive attention. This paper describes the performance of an approach combining Hough transform in features extraction and Neural Networks in classification. Experimental tests have been carried out on a set of 85.000 samples of characters corresponding to5 different fonts. Some promising experimental results are reported.

1 Introduction

Arabic is a language spoken by Arabs in over 20 countries, and roughly associated with the geographic region of the Middle East and North Africa, but is also spoken as a second language by several Asian countries in which Islam is the principle religion (e.g. Indonesia). However, non-Semitic languages such as Farsi, Urdu, Malay, and some West African languages such as Hausa have adopted the Arabic alphabet for writing [1]. Due to the cursive nature of the script, there are several characteristics that make recognition of Arabic distinct from the recognition of Latin scripts or Chinese. The cursive nature of the language makes recognition more difficult especially when we deal with multifont characters. Many researchers have been working on cursive script recognition for more than three decades. Nevertheless, the field remains one of the most challenging problems in pattern recognition and all the existing systems are still limited to restricted applications. The following figure shows some of Arabic characters in the five considered fonts we have worked on so far. The work we present in this paper belongs to the general field of Arabic documents recognition exploring the use of multiple sources of information. In fact, several experimentations carried out in our laboratory had proved the importance of the cooperation of different types of information at different levels (features extraction, classification…) in order to overcome the variability of Arabic and especially multifont characters [2, 6, 9].

In spite of the different researches realised in the field of Arabic OCR (AOCR), we are not yet able to evaluate objectively the reached performances since the tests had

J. Wang et al. (Eds.): ISNN 2006, LNCS 3972, pp. 293 – 298, 2006.

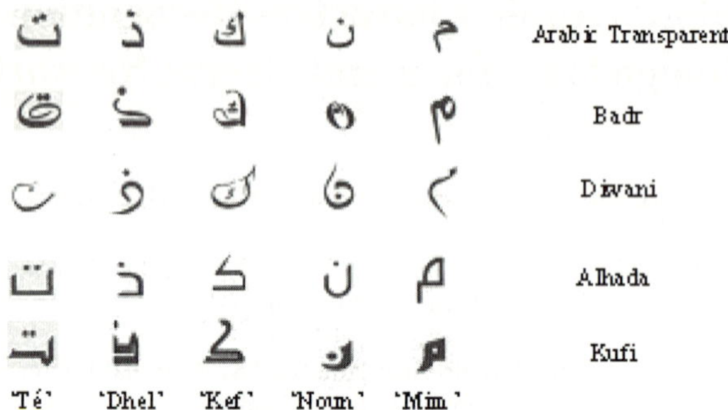

Fig. 1. Illustration of the five considered fonts

not been carried out on the same data base. Thus, the idea is to develop several single and hybrid approaches and to make tests on the same data base of multifont Arabic characters so that we can deduce the most suitable combination or method for Arabic Character Recognition.

2 Characters Recognition System

The main process of the AOCR system we developed can be presented by Fig.2.

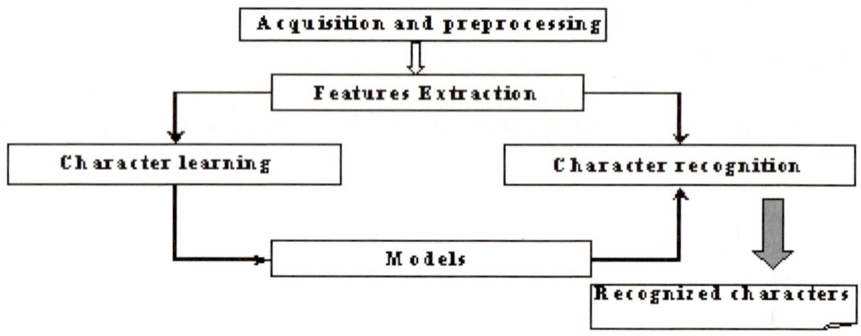

Fig. 2. Block diagram of the OCR

2.3 Pre-processing

Pre-processing covers all those functions carried out prior to features extraction to produce a cleaned up version of the original image so that it can be used directly and efficiently by the feature extraction components of the OCR. In our case, the goal of image pre-processing is to generate simple line-drawing image such as the one in Figure 3 which presents the edges detection of the character 'noun'. Our implementtation uses the canny edge detector for this extraction.

Fig. 3. Edges extraction using canny edge detector

2.2 Features Extraction

One of the two basic steps of pattern recognition is features selection. We quote from Lippman [4]: "Features should contain information required to distinguish between classes, be insensitive to irrelevant variability in the input, and also be limited in number to permit efficient computation of discriminant functions and to limit the amount of training data required." In fact, features extraction involves measuring those features of the input character that are relevant to classification. After features extraction, the character is represented by the set of extracted features. There are an infinite number of potential features that one can extract from a finite 2D pattern. However, only those features which are of possible relevance to classification need to be considered. Obviously, the extraction of suitable features helps the system reach the best recognition rate [5]. In a previous work, we have used wavelet transform in order to extract features and we have obtained very encouraging results [2]. In this paper, we present a Hough Transform based method for features extraction.

2.2.1 Hough Transform

The Hough Transform (HT) is known as the popular and powerful technique for finding multiple lines in a binary image, and has been used in various applications. The HT gathers evidence for the parameters of the equation that defines a shape, by mapping image points into the space defined by the parameters of the curve. After gathering evidence, shapes are extracted by finding local maxima in the parameter space (i.e., local peaks). The Hough technique is particularly useful for computing a global description of a feature (where the number of solution classes need not be known *a priori*), given local measurements. The motivating idea behind the Hough technique for line detection is that each input measurement (*e.g.* coordinate point) indicates its contribution to a globally consistent solution. Hough transform is used to identify features of a particular shape within a character image such as straight lines, curves and circles. When using the (HT) to detect straight lines, we rely on the fact that a line can be expressed in parametric format by the formula: $r = x\cos\theta + y\sin\theta$, where r is the length of a normal from the origin to the line and θ is the orientation of r with respect to the x-axis. To find all the lines within the character image we need to build up the Hough parameter space H. This is a two dimensional array that contains accumulator cells. These cells should be initialised with zero values and will be filled with line lengths for a particular θ and r. For our study the range of θ is usually from $0°$ to $180°$ although often we only need to consider a subset of these angles as we are usually only interested in lines that lie in particular direction. In our case, we have arbitrary chosen a step of $30°$ for the θ increment. Without using information from neighbouring pixels (which the Hough transform doesn't), each black pixel p(x,y) in the input image can possibly lie on a line of any angle . For each black pixel p(x,y) in the image, we take each angle along which we wish to find lines, calculate the value r as defined above and increment the value held in accumulator cell $H(r, \theta)$ by 1. The values in the resultant matrix will hold values that indicate the number of pixels that lie on a particular line $r=x\cos\theta+y\sin\theta$. These values don't represent actual lines within the source picture, merely a pixel count of points that lie upon a line of infinite length through the image. Lines passing through more pixels will have higher values than those lines passing through fewer pixels. The line can be plotted by substituting values for either x and y or r and θ and calculating the corresponding co-ordinates.

2.2.2 Line Extraction

To extract collinear point sets, one must first extract significant straight lines from the image. These lines correspond to major linear features. The advantage of the Hough transform[7] is the fact that it operates globally on the image rather than locally. The Hough transform works by allowing each edge point in the image to vote for all lines that pass through the point, and then selecting the lines with the most votes. After all edge points are considered, the peaks in the parameter space indicate which lines are supported by the most points from the image.

The first thing to understand about parameter space for line extraction is that there is no one-to-one relationship between pixels in the image and cells in the parameter space matrix. Rather, each cell in parameter space represents a line that spans across the entire image.

The transformation between feature space and parameter space is the following:

- Project a line through each edge pixel at every possible angle (you can also increment the angles at steps).
- For each line, calculate the minimum distance between the line and the origin.
- Increment the appropriate parameter space accumulator by one.

The resulting matrix: The x-axis of parameter space ranges from 1 to the square root of the sum of the squares of rows and columns from feature space. This number corresponds to the furthest possible minimum distance from the origin to a line passing through the image. The y-axis represents the angle of the line.

The larger the numbers in any given cell of the accumulator matrix, the larger the likelihood that a line exists at that angle and distance from the origin.

3 Artificial Neural Networks Classification

Artificial Neural Networks classifiers (ANN) have been used extensively in character recognition [8]. These networks can be used as a combined feature extractor and classifier, or as a "pure" classifier where the inputs are extracted features, which is the case in our system. We have implemented the Multilayer Perceptron (MLP), which is one of the well known and used ANN architecture in classification problem. We have tested a model presented by a network per character with only one output architecture.

3.1 Multilayer Perceptron Network Per Character

This structure implies the creation of twenty eight networks corresponding each to an Arabic character in its isolated form. Every network is characterized by:

- an input layer formed by two neurons
- only one hidden layer
- an output layer formed by a neuron corresponding to the considered character

During the learning phase, we present to the system not only the extracted features of the considered character but also those of others characters. Thus, it will not only learn good samples but also the bad ones.

4 Experimental Results

4.1 Test Vocabulary

The different tests have been carried out on isolated Arabic characters. Due to the absence in AOCR of a data base, we have created our own corpus which is formed by 85.000 samples in five different fonts among the most commonly used in Arabic writing which are: Arabic transparent, Badr, Alhada, Diwani, Koufi as shown in figure1.

4.2 Results of Using Hough Transform and MLP Network Per Character

In comparison of the achieved results with the wavelet/ANN based method we previously developed [2], we can say that wavelet transform is a little more efficient as far as features extraction is concerned. In fact, when using the Symmlet 4 wavelet to extract features, we obtained a rate of recognition of 98.63% with the same kind of neural network and on the same database.

Table 1. Recognition rate per character

Character	ا	ب	ت	ث	ج	ح	خ	د	ذ	ر	ز
Recognition rate HT/MLP	**99.10**	**95.13**	**97.16**	**98.86**	**97.02**	**96.80**	**96.26**	**95.69**	**96.73**	**96.73**	**96. 55**

Character	س	ش	ص	ض	ط	ظ	ع	غ	ف	ق	ك
Recognition rate HT/MLP	**94.65**	**96.16**	**96.24**	**94.62**	**94.64**	**95.24**	**96.76**	**96.22**	**95.50**	**95.34**	**95.16**

Character	ل	م	ن	ه	و	ى
Recognition rate HT/MLP	**98.38**	**97.02**	**97.18**	**98.55**	**97.31**	**97.98**

5 Conclusion

In this paper, a neural network classifier using features resulting from a Hough transform is proposed and evaluated for multifont Arabic characters. As results show, the achieved recognition rate is indebted to the selection of features especially when we deal with multifont characters. In our perspectives we are intending to optimise the step of features extraction especially by adapting the choice of the incrementing step of θ according to the character. Besides, we are intending to carry out other hybrid classifiers as well combining Hidden Markov Models and Artificial Neural Networks in order to take advantages of their different characteristics.

References

1. Amin, A.: Arabic Character Recognition. Handbook of Character Recognition and Document Image Analysis, World Scientific Publishing Company (1997) 397–420
2. Amor, N. B., Amara, N.: Applying Neural Networks and Wavelet Transform to Multifont Arabic Character Recognition. International Conference on Computing, Communications and Control Technologies (CCCT 2004), Austin (Texas), USA, on August 14-17, 2004.
3. Klassen, T.: Towards Neural Network Recognition Of Handwritten Arabic Letters. Dalhousie University (2001)
4. Lippmann, R.: Pattern Classification using Neural Networks. IEEE Communications Magazine (1989)
5. Brown, E. W.: Character Recognition by Feature Point Extraction. Northeastern University internal paper (1992)
6. Amor, N. B., Amara, N.: Hidden Markov Models and Wavelet Transform in Multifont Arabic Characters Recognition. International Conference on Computing, Communications and Control Technologies (CCCT 2005), July 24-27, in Austin, Texas, USA (Silicon Hills) 2005.
7. Illingworth, J., Kittler, J.: A Survey of the Hough Transform. Computer Vision, Graphics and Image Processing 44 (1988) 87-116
8. Altuwaijri, M. M., Bayoumi, M. A: Arabic Text Recognition Using Neural Network. ISCAS 94 IEEE International Symposium on Circuits and systems, Vol. 6, 1994
9. Amor, N. B., Amara, N.: Multifont Arabic Character Recognition Using Hough Transform and Hidden Markov Models. ISPA2005 IEEE 4th International Symposium on Image and Signal Processing and Analysis, September 15-17, 2005, Zagreb, Croatia

Recognition of English Calling Card by Using Multiresolution Images and Enhanced ART1-Based RBF Neural Networks

Kwang-Baek Kim[1] and Sungshin Kim[2]

[1] Department of Computer Eng., Silla University, Busan 617-736, Korea
gbkim@silla.ac.kr
[2] School of Electrical and Computer Eng., Pusan National University, Busan, Korea
sskim@pusan.ac.kr

Abstract. A novel hierarchical algorithm is proposed to recognize English calling cards. The algorithm processes multiresolution images of calling cards hierarchically to firstly extract individual characters and then to recognize the characters by using an enhanced neural network method. The horizontal smearing is applied to a 1/3 resolution image in order to extract the areas. The second vertical smearing and contour tracking masking is applied to a 1/2 resolution image to extract individual characters. And lastly, the original image is used in the recognition step because the image accurately includes the morphological information of the characters precisely. The enhanced RBF network is also proposed to recognize characters with diverse font types and sizes, by using the enhanced ART1 network adjusting the vigilance parameter dynamically according to the similarity between patterns. The results of experiments show that the proposed algorithm greatly improves the character extraction and recognition compared with traditional recognition algorithms.

1 Introduction

Mobile devices are often equipped with camera sensors, and various input devices, which can scan documents via a photo sensor. Work on calling card recognition by mobile devices is actively being pursued. The recognition of calling cards generally consists of two phases; the extraction of significant information from scanned images known as the extraction phase and the recognizing of the resulting information known as the recognition phase [1]. The extraction phase includes preprocessing to remove noise, the extraction of character strings and the identification of individual characters.

This paper proposed a new character extraction method using multiresolution images, which quickly and precisely extracts individual characters without interference from noise or the skew and the arrangement of character strings by using smearing and contour tracking algorithms on the original image and images with lower resolution. And this paper proposes an enhanced RBF(Radial Basis Function) network which is robust to noises and quickly adapts to new patterns.

J. Wang et al. (Eds.): ISNN 2006, LNCS 3972, pp. 299–305, 2006.

2 Related Works

Previous works have used a variety of methods to extract characters from calling card images, such as histograms [2], contour tracking algorithms [3,4], and smearing algorithms [5,6,7].

The histogram method [2] extracts character string areas by first scanning binarized images and extracts then locates individual characters. The success rate of this method degrades when the characters are heavily skewed in an unknown direction, have excessive noise or other problems. The extraction methods that use smearing algorithms [5,6,7] can be classified as RLSA(Run Length Smoothing Algorithm) and RXYC(Recursive X-Y Cuts) methods. The RXYC method is efficient for extracting large blocks, while the RLSA method is efficient for extracting small blocks. However, the RXYC method may incorrectly extract blocks under noisy conditions. A variety of methods have been proposed for recognizing individual characters such as the template matching method, statistical method and neural network method [8]. One is a RBF network to perform multivariate analysis and statistical interpolation to structure neural network models [9].

This paper proposes an enhanced RBF network, which uses an enhanced ART1 in the learning structure between the input and middle layers, and uses it to recognize the individual characters of calling cards.

3 Recognition System for Calling Cards

The proposed system consists of the extraction and the recognition phases. Multiresolution images are used to extract significant information from calling card images. The extraction of blocking areas and individual characters from low-resolution images reduces the processing time and it is also implement ably in

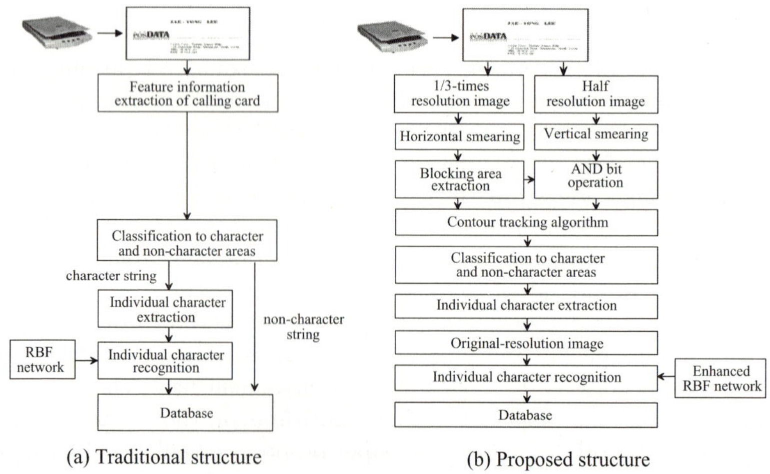

Fig. 1. The structure of the calling card recognition system

parallel processing in hardware. Also, contour tracking and smearing algorithms are used to prevent interference by skew, character arrangement and image noise for a robust extraction process. The whole structure of the proposed recognition system is based on multiresolution images, which is like the one in Fig. 1.

3.1 Image Preprocessing and Blocking Area Extraction

The proposed method eliminates noise from input images and extracts blocking areas including picture, logo, and character strings from the preprocessed image. The input image is binarized to precisely extract individual characters in the preprocessing, and the binary image is down-scaled to two low-resolution images.

The smearing algorithm on the 1/3 resolution binary image is employed to quickly extract blocking areas, and the blocking areas are extracted from the results by using the calling card feature of arranging characters on horizontal lines. The smearing algorithm expands black pixels in the direction in the image and removes the space between characters if it is below a given threshold. As individual characters are arranged in horizontal lines, horizontal smearing connects adjacent characters and generates a blocking area that includes a row of characters.

3.2 Individual Character Extraction

This method uses 1/2 – resolution images for individual character extraction to reduce processing time and to improve accuracy. In a image with half resolution, vertical smearing algorithm is applied to areas corresponding to blocking areas previously extracted, and the AND bit operation is executed between the result of vertical smearing and blocking areas with all black pixels.

Individual characters are able to be extracted by applying contour tracking algorithm to blocking areas generated by a sequence of pixel operations in the half-resolution image. Generally, contour tracking algorithm extracts individual characters along with its shapes accurately. But, in the case that a character like 'i' is separated to two parts, contour tracking algorithm may process small part of the character as noise and remove it. So, this paper makes characters like 'i' into connected areas by using vertical smearing algorithm and AND bit operation with black-pixel block.

3.3 Character Recognition Using an Enhanced RBF Networks

The RBF neural networks are mainly used to the classify learning data and to model nonlinear systems [9]. To improve the recognition success rate, this paper proposes enhanced RBF neural networks that adapts the ART1 network to the learning structure between the input layer and the middle layer and applies the output layer of the ART1 network to the middle layer. In the ART1 network, a vigilance parameter inversely determines the degree of mismatch between any input pattern and saved patterns that is allowed.

The enhanced ART1 network adjusts the vigilance parameter dynamically according to the homogeneity between the patterns using Yager's union operator.

The vigilance parameter is dynamically adjusted only in the case that the homogeneity between the stored pattern and the learning pattern is greater than or equal to the vigilance parameter. Let T^p and T^{p*} be the target value of the learning pattern and the stored pattern respectively. If T^p is equal to T^{p*}, the network decreases the vigilance parameter and adjusts the weight of connection between the input layer and the middle layer. Otherwise, the network increases the vigilance parameter and selects the next winner node. The algorithm dynamically adjusts the vigilance parameter as follows:

$$\begin{aligned} &if \quad (T^p_{j*} \neq T^{p*}_{j*}) \\ &then \quad \rho(t+1) = \wedge\left(1, \left((\rho(t))^{-2} + (\rho(t-1))^{-2}\right)^{-1/2}\right) \\ &else \quad \rho(t+1) = \wedge\left(1, \left((\rho(t))^{2} + (\rho(t-1))^{2}\right)^{1/2}\right) \end{aligned} \tag{1}$$

where ρ is the vigilance parameter.

The enhanced RBF network proposed in this paper adjusts dynamically to the vigilance parameter according to learning patterns by applying the enhanced ART1 to the middle layer. The learning process in the middle layer of the enhanced RBF network is as follows: In a similarity test, if the similarity between learning patterns and stored patterns is lower than the vigilance parameter, the network searches a new candidate winner node and executes the similarity test

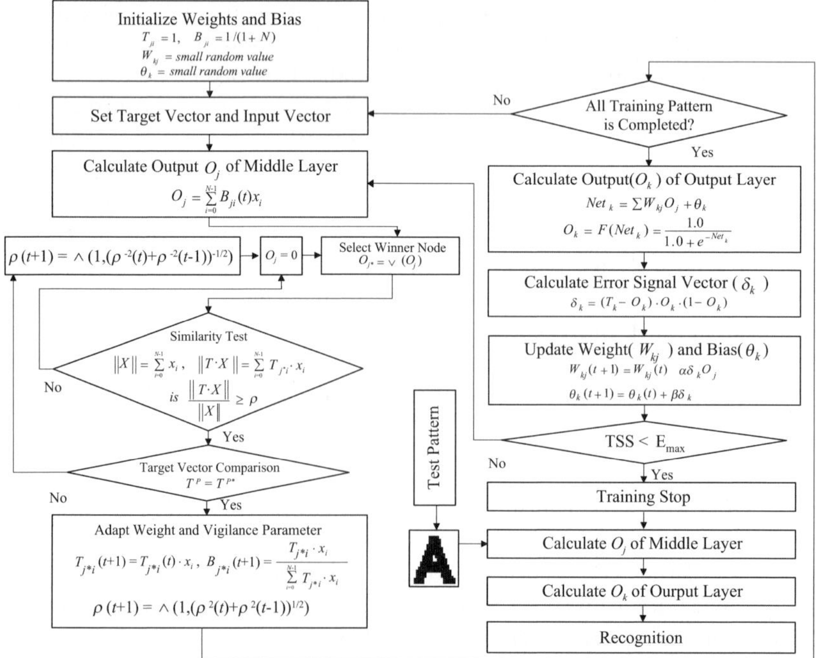

Fig. 2. The learning and recognition algorithm of the enhanced RBF networks

with the selected node. If no candidate winner node is selected, the network classifies learning pattern to a new cluster by creating a new node in the middle layer and adjusting the weight of connection between the input layer and the middle layer. This paper enhances the ART1-based RBF network by applying the enhanced ART1 algorithm to the middle layer, as shown in Fig. 2.

4 Performance Evaluations

The proposed recognition system is implemented using C++ Builder. 50 calling card images with a pixel size of 1500x800 are used to evaluate the performance of the proposed extraction algorithm. Table 1 compares the number of blocking areas and individual characters extracted by the proposed extraction algorithms in [1] and those from the proposed method. The first row in Table 1 indicates the total number of blocking areas and individual characters included in the 50 calling card images. Fig. 3(a) presents an example of successfully extracted individual characters and Fig. 3(b) show a failed extraction of an individual character. To evaluate the performance of the enhanced ART1-based RBF network, the proposed enhanced ART1 and the normal ART1 were used on the middle layer of the RBF network. Table 2, compares learning and recognition results obtained from recognition experiments using the two algorithms on previously extracted individual characters. For experiments presented in Table 2, all individual characters are separated into three character sets; alphabetic, numeric and special characters, and recognition experiments were performed on each character set. The initial vigilance parameter in the RBF network based on normal ART1 was fixed to 0.93, and in the enhanced RBF network, the vigi-

Table 1. The extraction result of blocking areas and individual characters with respect to extraction methods

	Blocking area	Character area	Character recognition
Total number of areas	629	7641	7641
# of extraction success by [1]	627	7353	7158
# of extraction failure by [1]	2	288	483
# of extraction success by the proposed method	627	7495	7428
# of extraction failure by the proposed method	2	146	213

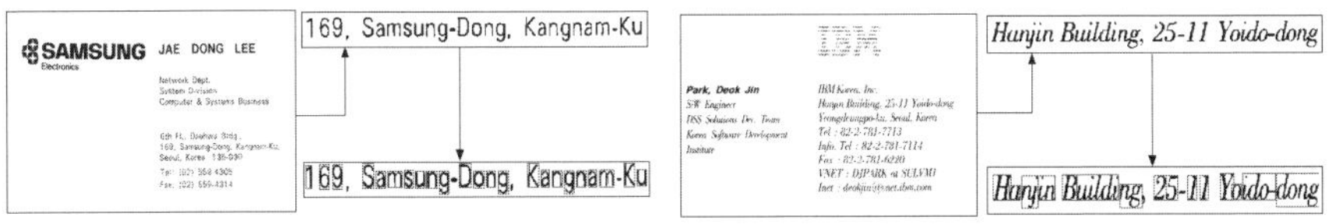

(a) An example of a successul individual character extraction (b) An example of a failed individual character extraction

Fig. 3. Examples of individual character extraction

Table 2. The results of learning and recognition using the enhanced RBF network

	# of hidden layer's nodes (# of nodes / # of input patterns)	Recognition performance (# of successes / # of input patterns)
The proposed enhanced RBF network		
Alphabetic char's	218 / 520	3716 / 3723
Numeric char's	91 / 200	2801 / 2821
Special char's	10 / 30	911 / 951
The RBF network based on normal ART1		
Alphabetic char's	496 / 520	3682 / 3723
Numeric char's	175 / 200	2773 / 2821
Special char's	21 / 30	889 / 951

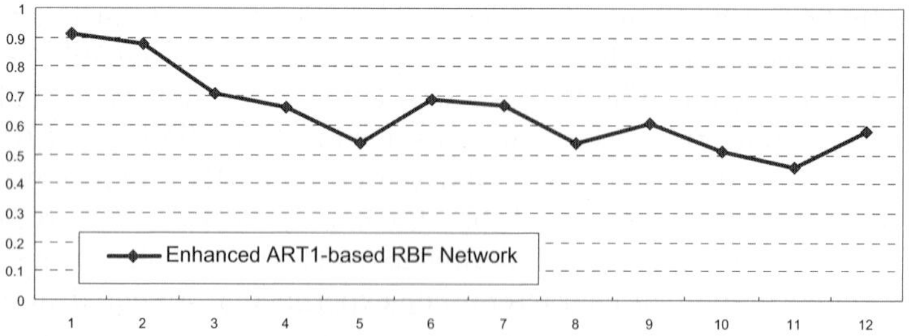

Fig. 4. Dynamical change of the vigilance parameter in the learning process for numeric characters

lance parameter was initially set at 0.93 and adapted dynamically. By comparing the number of recognition successes between the two algorithms in Table 2, the enhanced ART1-based RBF network performed better than the RBF network based on normal ART1.

Fig. 4 shows a trace of the dynamical change of the vigilance parameter appearing in the procedure that creates or modifies nodes at the middle layer of the enhanced ART1-based RBF network. In the enhanced ART1-based RBF network, the vigilance parameter is decreased if the target vector of the input pattern is equal to the one of winner node in the similarity test, or else the vigilance parameter is increased. Therefore, the enhanced ART1-based RBF network classifies similar patterns to the same cluster, improving the performance of learning and recognition.

5 Conclusions

This paper proposes a new character extraction algorithm based on multiresolution images that reduce preprocessing time and accuracy in the recognition of calling cards. To improve the recognition rate for individual characters, this paper proposed the enhanced ART1-based RBF network which constructs the middle layer by using the enhanced ART1 network, supporting the dynamical change of the vigilance parameter and applied it to the character recognition phase.

The results of the experiments showed that the proposed extraction algorithm and the enhanced RBF network performed better than previous algorithms used in calling card recognition. Moreover, the enhanced ART1-based RBF network recognized effectively the individual character thus it showed a high success rate of recognition. In future works, we will investigate and develop the preprocessing and recognition methods for the calling cards with hand-written characters.

References

1. Kim, K.B., Cho, J.H., Oh, A.S.: Recognition of an English Business Cards Using Enhanced Hybrid Network. In: Jun, W., Xiaofeng, L., Zhang Y. (eds.): Advances in Neural Networks – ISNN 2005. Lecture Notes in Computer Science, Vol. 3497. Springer-Verlag, Berlin Heideberg New York (2005) 209–215
2. Gonzalez, R.C, Woods, R.E.: Digital Iimage Processing. Second Edition. Prentice Hall. (2001)
3. Kim, K.B., Jang, S.W., Kim, C.K.: Recognition of Car License Plate by Using Dynamical Thresholding and Enhanced Neural Networks. In: Nicolai, P., Michel, A. W. (eds.): Computer Analysis of Images and Patterns. Lecture Notes in Computer Science. Vol. 2756. Springer-Verlag Berlin Heideberg New York (2003) 309–319
4. Kim, K.B., Kim, S.S., Ha, S.A.: Recognition of Passports Using a Hybrid Intelligent System. In: Mohamed, K., Aurelio, C. (eds.): Image Analysis and Recognition. Lecture Notes in Computer Science. Vol. 3314. Springer-Verlag Berlin Heideberg New York (2004) 762–767
5. Wang, K.Y., Casey, R.G., Wahl, F.M.J.: Document Analysis System. IBM J. Res. Develop. 26(6) (1982) 647–656
6. Nagy, G., Seth, S.C., Stoddard, S.D.: Document Analysis with an Expert System. Proceedings of Pattern Recognition in Practice II. Amsterdam. (1985) 19–21
7. Gonzalez, R.C., Wintz, P.: Digital Image Processing. Addison-Wesley Publishing Company Inc. (1977) 88–102
8. Schalkoff, R.: Pattern Recognition: Statistical, Structural and Neural Approaches. John Wiley & Sons. (1992)
9. Park, J., Sandberg, I.W.: Universal Approximation Using Radial Basis Function Network. Neural Comp. (1991) 246–257

A Method of Chinese Fax Recipient's Name Recognition Based on Hybrid Neural Networks

Zhou-Jing Wang, Kai-Biao Lin, and Wen-Lei Sun

Dept. of Automation, Xiamen University, Xiamen 361005, China
zjwang@public.xm.fj.cn, lkbiao208@163.com

Abstract. A professional Chinese fax information processing system is designed which has functions to automate incoming fax distribution in a company or institution, read an incoming fax cover sheet and route the fax to the receiver's email box. This paper reports our research as part of an effort to realize such a system and focuses on recognition of the handwritten recipient's on fax cover pages. We propose hybrid neural networks for large scale Chinese handwritten character recognition. The network is composed of the self-organizing competitive fuzzy layer and the multi-layer neural network using BP method, connected in cascade. The characteristic features of this network structure for Chinese handwritten character recognition are discussed and performances are evaluated on 8208 real world faxes which are taken from one company in 2004, the results of experiments compared to standard neural solutions based on MLP show that the whole system is of reasonable structure and satisfactory performance.

1 Introduction

Regardless of Internet's prevailing nowadays, fax still holds its own place as before for its unique characters, wins the satisfaction of both markets and users, and possesses a significant position in the field of office automation for the advancement of its technology. But documents transmitted through fax machines are received as images. Like many document images obtained from sources such as journals, maps, engineering drawings, fax images are desirable to be converted to computer editable form. A fax cover page usually is accompanied by the documents being transmitted and carries information referring to sender, recipient, comments, date, fax and phone numbers, etc. As use of fax machines is becoming popular, it calls for the need to process fax images more intelligently.

The work by Jenchyou, L. and Srihari, S.N. [1] may be the earliest open paper that reports their research to carry out an automatic fax routing system and focuses on locating the recipient's name and address on fax cover sheets. Likforman-Sulem L. [2] discusses the name block location process of a messaging system which deals with facsimile images including handwritten items, and proposed a method of name block location in facsimile images using spatial/visual cues. Azzabou, N. and Likforman-Sulem, L. [3] proposed a method to identify and extract a specific information (the name of the sender) from a fax cover page. For this purpose, methods based on image document analysis (OCR recognition, physical blocks selection), and text analysis

J. Wang et al. (Eds.): ISNN 2006, LNCS 3972, pp. 306 – 315, 2006.

methods (optimized dictionary lookup, local grammar rules), are implemented in parallel. Alam H. et al. [4] The Fax Assist routing system exploits textual analysis thanks to the recognition of all words: printed and handwritten. The image analysis stage is reduced to assigning each name a score according to its position in the image. Recipient names in faxes with common layouts are recognized with high precision (between 96 and 99.7 percentage) and a recall between 56% and 78% according to the type of the fax (only printed or including handwritten items).

As described above, there are many scholars plunging themselves into the research of automatic recognition of fax information because of the importance of fax and acquiring rather great production. However, the extraction and recognition of Chinese fax information can't apply those methods mechanically mentioned above for the specialty of Chinese characters. Thus, a professional Chinese fax information processing system is designed which has functions to automate incoming fax distribution in a company or institution, read an incoming fax cover sheet and route the fax to the receiver's email box. This paper reports our research as part of an effort to realize such a system and focuses on recognition of the handwritten recipient's on fax cover pages.

This paper is organized as following, i.e. the recognizing method of fax recipient is introduced in section 2, the recognizing method of handwritten Chinese characters based on hybrid neural networks is introduced in section 3, the experiments and results sampled in real world faxes are presented in section 4, and the conclusion is given in section 5.

2 Fax Recipient's Name Recognition System

We consider here images of fax front covers: field headers are printed but their content may be printed or handwritten. Fax images are of low quality that involves character recognition errors. Layouts are highly variable, as they are not normalized. There is a variety for field designation that involves the use of lexicon. Keywords themselves are homonymous to other words: many candidates will be generated which involves a process for selecting the right ones.

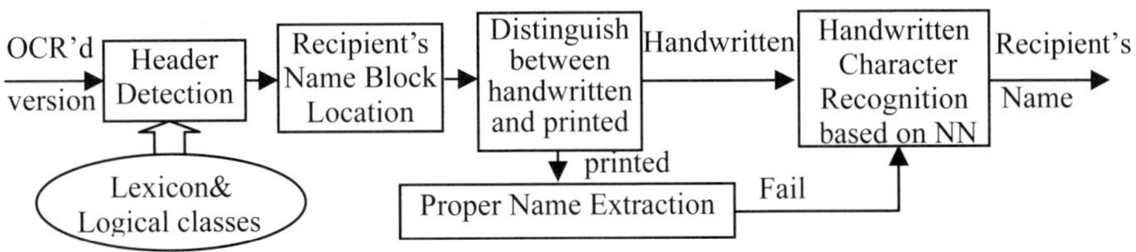

Fig. 1. The architecture of the system

The system architecture is shown in Fig.1. The start of the process is the binary image and its electronic version produced by an OCR system, which is a structured text file that contains the text strings of the printed text and the position of the words. Pseudo words including characteristic words related to the recipient are candidates for being potential field headers. These characteristic words belong to recipient-class of logical classes. For instance, recipient-class may include keywords such as '收件人', 'To', 'ATTN' and 'COMPANY' belonging to Chinese/English lexicons. Searching only for these words may induce errors as such general words do not always correspond to headers. Spatial constraints between hypothesized headers are added in order to select the best ones. Spatial relations between layout components enable to select components possibly including the recipient's name. These spatial relations and constraints are defined from very general cues such as proximity and alignment. The selection of potential recipient's name blocks is described in [3]. Before recognizing characters, it is necessary to distinguish that the recipient's name blocks are handwritten or printed characters. A neural network is in charge of classifying recipient's name block into 2 classes (printed or handwritten), and those contents about it will be introduced in other papers because of the length of this paper. If the name blocks are judged printed, then extract the name directly from OCR's Version. But if they are judged handwritten, send them to the handwritten characters recognition module of the system presented in this paper. Further, a special situation must be considered in this system, i.e. the name blocks are printed characters but have bad conglutinated underlines that make it difficult for OCR to recognize. The solution now is to regard printed characters as handwritten ones and then send them to handwritten characters recognition module.

The handwritten characters recognition module is the core module of our system, and a method of handwritten characters recognition based on hybrid neural networks is presented in this paper, which will be introduced in detail in section 3.

3 Handwritten Characters Recognition Based on Hybrid NN

Handwritten character recognition module includes three primary parts named image preprocessor, feature extractor and classifier based on hybrid neural networks. Image preprocessor part basically includes underline detection and cleaning, noise reduction, character isolation, and character dotted image normalization. The role of feature extractor is to process the isolated and normalized characters in order to extract features for classifier. Classifier is used for recognize characters, which is realized by hybrid neural networks. Feature extractor and classifier are the cores of our module. Here we concisely give an introduction to the module.

3.1 Preprocession

Preprocessing is one of the indispensable parts in character recognition. Firstly, adopt Urger smoothing method to denoise the character image in this paper, then implement horizontal projection to the character image denoised, and then detect underline by threshold method and clear it. More, divide up characters by histogram and unify them to the lattice of 64*64 pels. An example shown in Fig. 2 is used to illustrate our method. Part (a) is a part of primitive fax cover page. Part (b) is receiver name block intercepted after the process of recipient's name block location module. Part(c) is the figure after denoising to part (b). Part (d) is the horizontal projection aiming at removing the underline in part (c). Part (e) is the figure after removing underline. Part (f) is the vertical projection aiming at dividing characters. Part (g) is the figure that shows the positions of dividing lines. Part (h) is the figure after normalizing size of characters.

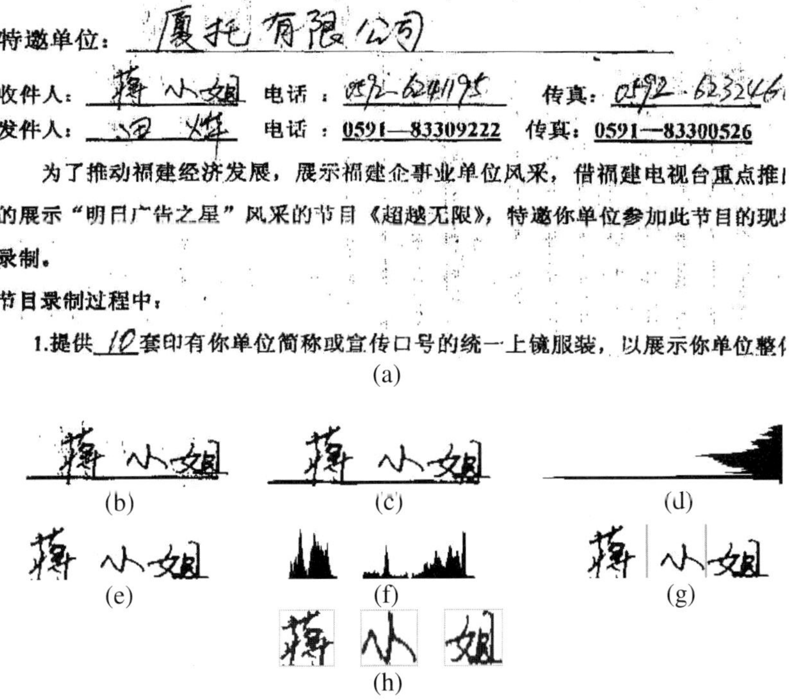

Fig. 2. An example of image preprocessing

3.2 Feature Extraction

Chinese characters recognition based on hybrid neural networks adopts different characteristics in different stages of recognition. Those characteristics are complementary in the description of characters information and the performance of anti-jamming, which can improve the effect of recognition evidently.

Classifying those characters roughly is the first stage of the recognition system, and generally, those characteristics adopted are the projection characteristics, such as horizontal and vertical projection of characters and the projection on 4 directions. But those projection characteristics are sensitive to the distortion of characters such as rotation. Once characters rotate, they would be misjudged by projection characteristics. Thus, a kind of characteristic that has strong ability of anti-jamming and anti-noise is proposed in this paper, and the extraction process is as the following:

Step1. Extract the contour line of characters by central projection transformation. This characteristic has such strong robustness that the selection of characters has no influence on the extraction of characteristics.

Step2. Unfold the contour line of characters as a line in one dimension, wavelet transform it to form a series of wavelet sub-template, and then get the curve of wavelet sub-template to compute sub-dimensions, which are used to be the characteristic, in order to reduce character dimension and analyze the contour characteristic of character image.

Step3. Compute sub-dimensions of contour line, and low and high frequency curves extracted from one-rank wavelet decomposition. And there are only 3 characters that are fit for rough classification of characters.

The second stage of the recognition system sub-classifies the results classified in the first stage. When sub-classifying, the difference among structures of different characters should be considered seriously. Therefore, the characteristic of local line fitting (LLF) strokes decomposition is adopted in the results from the first stage. This characteristic combines the characteristics of statistics and structures, divides character image into some square grids, and compute the ratio of pels in each grid to pels of the entire character. As grids divide strokes of character into some segments, compute the slope of segments after curve fitting and then compute the characteristic according to the slope. As we can get 3 character values in each grid, build character values in all grids into a character vector and that would be the LLF characteristic which describes those structure features such as the slope of character strokes and gets attention to statistical information through computing pels in grids at the same time. The size of character is unified to be 64*64, and in LLF, the size of grid is 16*16, the number of grids is 16 and there are 3 character values in each grid. So, as input vectors of network, the dimension of LLF projection is 48.

3.3 Classifier Based on Hybrid Neural Networks

The architecture of the classifier based on hybrid neural networks is shown in Fig. 3. It is an unbalanced tree structure that is generally divided into 2 hierarchical levels: level one is rough classification module and level two is recognition module.

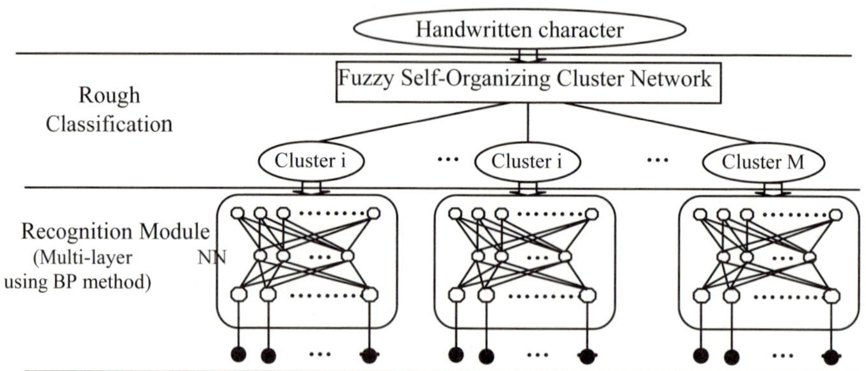

Fig. 3. The architecture of the classifier based on hybrid neural networks

The working flow of the hybrid neural networks classifier is as following. First, a normalized character is inputted into the module in the first level. This module of networks does feature extraction, and then classifies pattern by fuzzy self-organizing clustering candidates. The top 3 matched subsets are selected. Once a sample reaches a unitary subset, the recognition result is obtained. Unitary subset is a subset that has only one pattern category. In contrast, if a subset has more than one pattern category, it is called complex subset. If a given sample is classified into a complex subset, the recognition module in the succeeding level, which is corresponded with the best-matched subset, is active. The recognition module also performs feature extraction and then classifies pattern by multi-layer neural network using BP method.

3.3.1 Rough Classification

We use a fuzzy self-organizing clustering network as a tool to divide large-scale set of Chinese handwritten characters into many small subsets. The fuzzy neurons in the self-organizing layer perform the fuzzy partition of p input vectors x_j into M fuzzy groups and find a cluster center for each group in such away that a cost function of dissimilarity measure is minimized. Each input vector x_j belongs to the i th cluster with the membership grade μ_{ij} described by the equation (1).

$$\mu_{ij} = \mu_i(x_j) = \exp(\frac{\|x_j - c_i\|^2}{\sigma^2}) \tag{1}$$

The vector c_i is the center of i th cluster and σ characterizes the width of it.

The centers c_i are usually determined in the self-organizing competitive learning process. The most well known method is the fuzzy C-means clustering algorithm that can be formulated as follows [6]–[7].

Step1. Initialize the membership functions μ_{ij} with random values between 0 and 1 in such way, that constrains $\sum_{i=1}^{M} \mu_{ij} = 1$ for $j = 1,2,\cdots,p$ are satisfied.

Step2. Find M fuzzy cluster centers c_i using equation (2)

$$c_i = \sum_{j=1}^{p} \mu_{ij}^m x_j \Big/ \sum_{j=1}^{p} \mu_{ij}^m \tag{2}$$

Where m is the weighting exponent, $m \in [1, \infty]$.

Step3. Calculate the cost function $E = \sum_{i=1}^{M} \sum_{j}^{P} u_{ij}^m \|c_i - x_j\|$. If it is below the assumed tolerance value or its improvement over the previous iteration is negligible-stop, else go to the next step.

Step4. Calculate new entries of μ_{ij} using equation (3)

$$\mu_{ij} = \frac{1}{\sum_{k=1}^{M} (\frac{d_{ij}}{d_{kj}})^{1/(m-1)}} \tag{3}$$

Where d is the Euclidean distance between center and data vector, and go to step 2.

This iterated procedure repeated many times leads to some minimum of E, which however, is not necessarily the global minimum. The quality of solution is determined by the choice of initial cluster centers following from the random values of membership values μ_{ij}. The centers should be concentrated in these areas where most of the multidimensional data points are distributed. Special methods for measuring the density distribution of data should be applied. The most well known methods are the mountain clustering method and subtractive clustering. In our solution the subtractive clustering, since it is less sensitive to the dimensionality of the problem, has been used.

After this clustering, subset usually contains patterns from 10 to 30. A small percentage of subsets have only one pattern. The numbers in subset are relative to the characteristic extraction of rough classification. The function of this level is to lessen the burden of following recognition work.

3.3.2 Recognition Module

Recognition module is to further recognize the characters in the small scale subsets generated by level one. A multi-layer perceptrons (MLP) network is trained for the classification. The MLP network consists of several layers of neurons, in which the first one is the input layer, and the last one is the output layer, remaining layers are called hidden layers. There are complete connections between the nodes in successive layers but there is no connection within a layer. Every node, except the input layer nodes, computes the weighted sum of its inputs and applies a sigmoid function to compute its output, which is then transmitted to the nodes of the next layer. The objective of MLP learning is to set the connection weights such that the error between the network output and the target output is minimized. In this paper, we employed an MLP consisting of three layers: an input layer with the same number of nodes as the dimension of the feature set, and here, the dimension of LLF projection is 48, which is the number set in input layer, and the number of nodes in output layer is equal to the number of pattern in subsets.

Use experience function to compute the optimal nodes of hidden layer:

$$n1 = \sqrt{n+m} + a \tag{4}$$

Where, n---The number of neurons in input layer, m---the number of neurons in output layer, n1---The number of neurons in hidden layer, a ---the integer between 1 and 10. Set a as 8 in this paper, so the number of neurons in hide layer, $n1 = \sqrt{48+m} + 10$, for example, if the subsets have 20 characters, and the number of MLP in hidden layer is $n1 = \sqrt{48+20} + 8 = 16$, thus network is 48-16-20.For training weights in the network, a back-propagation algorithm is used. And, we set the learning rate and momentum as 0.1 and .08 respectively. This shows that the accuracy of hybrid neural networks in this paper is obviously higher than standard MLP network.

4 Experiments and Results

Those experimental samples in this paper are taken from 8208 faxes of one company in 2004, in which the recipient's names of 2605 faxes are handwritten character while others are printed characters. After the process of OCR recognition to those 8208 faxes, those names of 4258 faxes can be recognized and extracted directly by OCR tools with 100% accuracy, in which 246 faxes are handwritten characters and the names of them are written neatly and don't conglutinate with underlines. But 538 faxes are refused to recognize as the keyword before recipient's name block can't be found. Through analyzing the results, we know that the reason is because of the low quality of 323 fax images, while other 215 faxes are because of their handwritten keywords that influence the recognition result. Thus, there are totally 3412 faxes that need to be recognized by handwritten characters recognition module. And 2390 faxes in them are handwritten, and OCR can't recognize 1022 other faxes recipient's name blocks because of their conglutinated underlines or low image quality. From statistic analyses, we know that these 3412 faxes belong to 124 persons, and use 368 characters. But the frequencies of these 368 characters are totally different, as some appear about 10 times while some appear several hundred times, such as '先', '生','经', '理'. For convenience of sampling, we just take these characters to do experiments which frequency of appearance is more than 10 times in these 312 characters. We select 50 sets of characters sample, in which 30 sets are used to be training sets and others are testing sets. And as the frequency of some characters is not greater than sampling frequency, we implement different noise process on character samples to expand sampling sets.

The self-organizing clustering network is trained by 30 sets of training samples. These samples are clustered into 45 subsets that are not the same in size, i.e. the number of Chinese character category contained in each subset is not equal, that is, there are 5 subsets containing only 1 character, 18 subsets containing 2-10 characters, 12 subsets containing 10-20 characters, 8 subsets of 20-30 characters, and 2 subsets containing more than 30 characters. The largest subset has 36 kinds of Chinese category. This means that the large scale Chinese character set can be divided into many small subsets by use of the fuzzy self-organizing clustering network, among which there are 5 subsets need no further classification, for they have only one pattern category and thus can be used as preliminary recognition results. Therefore, this clustering module reduces the complexity of the recognition system very well.

The experiments on training and testing samples have been performed. The results show that all training samples are correctly clustered into 1st candidate subsets. For the testing samples, there are about 86% of samples dropped in the 1st candidate subsets, 92% of them in the top two candidate subsets and those which are not belong to top three candidate subsets is less than 4%. As it's the recognition for character sets of small size, the accuracy of the recognition module by multi-layer neural network using BP method reaches 100% for testing sets, and even reaches 96% for training sets. The overall recognition rate of the system is summarized as follows: The system gives the correct recognition rate of 100% for the 30 sets of training samples, and more than 95% for about 20 sets testing samples.

It is interesting to compare the behavior of our hybrid neural networks with the performance of the standard MLP network. To make a fair comparison we have applied the MLP network of approximately the same number of weights. The MLP network checked in the experiments was of such structure: 48-58-312. The network has been trained on the same learning data set and then tested on the data used for testing the hybrid neural networks. The results show that the correct recognition rate of 92% for the 30 sets of training samples, and only 81% for 20 sets testing samples. This shows that, the accuracy of hybrid neural networks in this paper is obviously higher than standard MLP network.

5 Conclusions

Based on research of other papers [1]-[5], a method of recognition for the name blocks of fax recipients according to the specialty of Chinese fax is proposed in this paper. First, fax image is sent into OCR recognition module, then keywords of recipient's name block are searched from the result of recognition, and the image of recipient's name block is extracted according to the location of keywords, and then we can get the recipient's name from characters recognition of the recipient's name block.

The hybrid neural networks proposed in the paper for the recognition of the Chinese handwritten character achieves good accuracy, better than obtained by using the classical MLP network. The distinct advantage of the proposed solution is association in one network of two kinds of sub-networks: The fuzzy self-organizing layer and the multi-layer neural network using BP method. The hybrid network outperforms the MLP in terms of the accuracy and also training time. The training time reduction is due to the very good starting condition for the proposed fuzzy structure. It was possible due to very strict connection of the parameters of the self-organizing neurons and the learning data. The well-initialized network means very quick learning, ending at the global minimum of the error function. Application of the fuzzy self-organizing neurons has another positive effect on the learning paradigms. It reduces the requirements for the learning data size, since the self-organizing layer is not sensitive to the size of learning patterns, assuming that they represent the typical ones. On the other hand due to the application of the self-organizing sub-network, the MLP needs a much smaller number of neurons, and this also reduces its requirements for the size of the learning data.

The system can be easily adopted for different companies by retraining the neural network using new data.

References

1. Jenchyou, L., Srihari, S.N.: Location of Name and Address on Fax Cover Pages. Proceedings of the Third International Conference on Document Analysis and Recognition 2 (1995) 756-759
2. Likforman-Sulem L.: Name Block Location in Facsimile Images Using Spatial/Visual Cues. Proceedings of the Sixth International Conference on Document Analysis and Recognition (2001) 680-684
3. Azzabou, N., Likforman-Sulem, L.; Neural Network-Based Proper Names Extraction in Fax Images. Proceedings of the 17th International Conference on Pattern Recognition 1 (2004) 421-424
4. Alam H. et al.: FaxAssist: An Automatic Routing of Unconstrained Fax to Email Location, IS&T/SPIE Conference on Document Recognition and Retrieval. San José (2000) 148–156

5. Likforman-Sulem, L., Vaillant, P., Yvon, F.: Proper Names Extraction from Fax Images Combining Textual and Image Features. Proceedings of Seventh International Conference on Document Analysis and Recognition 1 (2003) 545-549
6. Jang, J.S., Sun, C.T., Mizutani, E.: Neuro-Fuzzy and Soft Computing. NJ: Prentice-Hall, Englewood Cliffs (1997)
7. Pal, N., Bezdek, J.C., Tsao, E.C.: Generalized Clustering Networks and Kohonen Self-organizing Scheme. IEEE Trans. Neural Networks 4 (1993)
8. Pal, N.R, Pal, S., Das, J., Majumdar, K.: SOFM–MLP: A Hybrid Neural Network for Atmospheric Temperature Prediction. IEEE Transactions on Geoscience and Remote Sensing 41 (2003) 2783-2791

Fast Photo Time-Stamp Recognition Based on SGNN

Aiguo Li

Department of Computer Science and Technology,
Xi'an University of Science and Technology, 710054, Xi'an, China
liag@xust.edu.cn

Abstract. Photo time-stamp is a valuable information source for content-based retrieval of scanned photo databases. A fast photo-stamp recognizing approach based on Self-Generating Neural Networks (SGNN) is proposed in this paper. Network structures and parameters of SGNN needn't to be set by users, and their learning process needn't iteration, so SGNN can be trained on-line. Proposed method consists of three steps: A photo is roughly segmented to determine which corner of the photo contains time-stamp; The area which contains time-stamp of the photo is finely segmented, in order to locate each character in the time-stamp, projection technology is used to locate edges of these characters; The time-stamp is recognized based on SGNN. Experimental results show that proposed approach can achieve higher recognition accuracy and computing efficiency.

1 Introduction

Detecting and recognizing text information from image and video data has become a research point [1]. Time-stamp in video and image shows the time a video or image taken, which is important information for retrieval. So time-stamp recognition rouses interests of content-based image retrieval and overview researchers. In [2], automatic time-stamp extraction from family videotapes was discussed and good effect was obtained. A photo is a special type of image data. How to recognize the time-stamp in a photo and use it in photo retrieval is an interesting problem. The difficulty in photo time-stamp recognition is that time-stamp is always embedded in complicated background [3]. In [3], a template matching-based MAP (maximum posterior probability) approach is proposed to detect and recognize time-stamp of color photos. The advantage of this approach is that the mathematical concept is simple and clear. The shortcoming of this approach is that it needs large amount of prior knowledge and off-line learning. In [4], a template matching-based PSO (particle swarm optimization) approach is proposed to detect and recognize time-stamp of color photos. This approach uses a PSO algorithm to recognize time-stamps and need not large amount of prior knowledge and off-line learning, but time consumption of the approach is large. An SGNN based automatic photo time-stamp recognition approach is proposed in this paper. A photo is roughly segmented to determine which area contains the photo time-stamp. Four detecting areas are chosen in this step, which probably include the photo-stamp, and then their R

J. Wang et al. (Eds.): ISNN 2006, LNCS 3972, pp. 316–321, 2006.

component and G component are margin-detected using Sobel operator to determine whether the photo-stamp is in them. The area that contains time-stamp is finely segmented, in order to locate each character in the time stamp. Projection technology is used to locate edges of these characters. Then we use SGNN to do template matching on the determined area to complete the photo-stamp recognition. Experimental results show that the proposed approach can achieve high recognition accuracy.

2 SGNN

SGNN was first proposed by Wen *et al*[5]. Then Qin *et al*[6] made profound research in its applications. SGNN features simple network design and short learning time. The traditional neural network is a human-designed network structure and adjusts the connecting weight through learning the sample. It is a skillful job to select a good network structure for the current learning task, because only at a special application, can a certain network structure give full play of its capacity. But the SGNN generates a self-generating neural tree (SGNT) in learning the sample and its whole structure including its neurons, connection between the neurons and weights is generated on its own in the non-supervisory learning method during learning, so it has good adaptation and is applicable to classification and cluster. In this paper the SGNN is used as a clustering device for clustering images in pixels.

The following definitions are made before describing the generating algorithm of the SGNT.

1) The input sample vector e_i is the vector composed of its attributes, $e_i = (a_{i1}, a_{i2}, ..., a_{iq})$;

2) The neuron n_j is an ordered pair (w_j, c_j). w_j is weight vector of neuron n_j, $w_j = (w_{j1}, w_{j2}, ..., w_{jq})$; c_j is the aggregation of leaves n_j;

3) The SGNT is a tree <{ n_j }, { l_k }> generated automatically from the training samples in the following algorithm, where { n_j } the aggregation of neurons, { l_k } is the connecting aggregation of the tree. And only when at $n_j \in c_i$, n_i and n_j has direct connection.

4) For an input sample e_i, if $\forall j$, $d(n_k, e_i) \le d(n_j, e_i)$, then n_k is called winner. $d(n_j, e_i)$ is the distance between the neuron n_j and the sample e_j. The Euclidean distance is used to measure the distance between the sample and the neurons.

$$d(n_j, e_i) = \sqrt{\sum_{k=1}^{q} (w_{jk} - a_{ik})^2 / q} \tag{1}$$

The following pseudo code describes the SGNT generating algorithm: Input:

Given training sample set $E = \{e_i\}, i = 1,2,...,q$;

Output: one SGNT tree generated from E.

Program codes:

```
copy(n_1,e_1);          //n_1 is the root node.
for(i=2,j=2;i≤q;i++){
    n_win=choose(e_i,n_j);
   if(leaf(n_win)){
       copy(n_j,n_win);
       connect(n_j,n_win);
       j++;
   }
  copy(n_j,e_i);
connect(n_j,n_win);
j++;
update(n_win,e_i);
}.
```

The functions in the program are defined as:

copy(n_j , e_i / n_{win}): create n_j , reproduce the attributes of e_i or the weight of the winner n_{win} as the weight w_j of n_j .

choose(e_i ,n): find the winner of the sample e_i in the current SGNT.

leaf(n_{win}): judge whether the n_{win} is a leaf in the current SGNT.

connect(n_j , n_{win}): connect n_j and n_{win} , let n_j act as a subleaf node of n_{win} .

update(n_{win} , e_i): refresh the weight vector of n_{win} according to the attributes of e_i in the following formula:

$$w_{jk,i+1} = w_{jk,i} + \frac{1}{i+1}(a_{i+1,k} - w_{jk,i}) \tag{2}$$

In the final SGNT, the attributes of the training sample are the weight of the leaf neurons while the weights of all non-leaf neurons are the mean value of all its children as well as the mean value of all sample attributes covered by the sub-tree rooted in it.

3 Proposed Method

3.1 Rough Segmenting of an Image

Photo time-stamp always appears at specific position, that is, it can only appear at a certain corner. So, we select four detecting areas at four corners of a photo, and then judge which area the photo time-stamp is in. This is the purpose of rough segmenting.

According to imaging features of photo time-stamp, for R component and G component of each detecting area, Sobel operator is used for margin extraction.

$$E_R^i = Sobel(R^i), i = 1,2,...,4 \tag{3}$$

$$E_G^i = Sobel(G^i), i = 1,2,...,4 \quad (4)$$

$E_R^i, E_G^i, i = 1,2,...,4$ represent margin image obtained when R component and G component of i-th detecting area is done with Sobel operator.

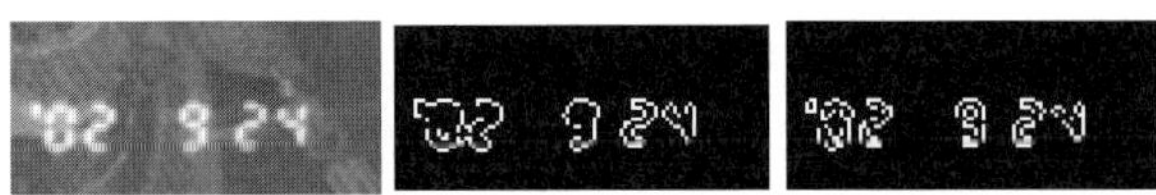

Fig. 1. Rough segmenting time-stamp (Left: original image; middle: E_R^i ; right: E_G^i)

Because time-stamp strokes are intensive margin in the margin image, $E_R^i, E_G^i, i = 1,2,...,4$ can be projected vertically and horizontally. According to the features of vertical and horizontal bar charts, it is easier to judge the area and approximate position of the time-stamp. Figure 1 shows an example of rough segmenting time-stamp of a color photo.

3.2 Fine Segmentation

R component and G component of T are written as T_R and T_G respectively. Background filtering is to be done to T_R and T_G respectively by the following method:

$$T_k^{'}(i,j) = \begin{cases} 0, & (q_1\overline{T}_k < T_k(i,j) < q_2\overline{T}_k) \vee \quad (T(i,j) < q_3\overline{T}_{k,\max}) \\ T(i,j), & otherwise \end{cases} \quad (5)$$

where, q_1, q_3 is constant in [0, 1], q_2 is constant in [1, 2]; $k = \{R, G\}$, i, j span are length and width of T respectively (with a pixel as a unit); $\overline{T}_k$ is average of pixel value of T_k ; $T_{k,\max}$ is maximum of pixel value of T_k .

T_R and T_G are changed into $T_R^{'}$ and $T_G^{'}$ respectively. If q_1, q_2, q_3 are selected properly, the background of $T_R^{'}$ and $T_G^{'}$ can be well filtered (shown in figure 2).

Fig. 2. Time-stamp background filtering (Left: original image; middle: $T_R^{'}$; right: $T_G^{'}$)

Process $T_R^{'}$ and $T_G^{'}$ with morphological operator open and close respectively:

$$T_R^{''} = T_R^{'} - close(open(T_R^{'})) \quad (6)$$

$$T_G^{''} = T_G^{'} - close(open(T_G^{'})) \quad (7)$$

Project $T_R^{''}$ and $T_G^{''}$ vertically and horizontally and segment each character of time-stamp according to features of horizontal and vertical bar charts. When segmenting each character, vertical and horizontal bar charts of $T_R^{''}$ and $T_G^{''}$ are taken into consideration. After fine segmenting of T, time-stamp template parameter is obtained.

3.3 Template Matching Based on SGNN

Assuming that C is a template set containing standard binary images for single character font of photo time stamps, including 12 characters: characters 0, 1,…,9, character marker for year, blank character. Before template matching, the character template set C is prepared in advance for backup. Size of these character templates should be selected as maximum possible size, such as 20×28 pixels. These standard binary images are then polluted by white noise in order to generate 12 train sample sets respectively. Each train set includes a lot of binary images with 20×28 pixels. 12 SGNTs are trained by the train sample sets respectively. Each SGNT represents a specify character template.

As soon as 12 character templates (i.e. 12 SGNTs) are generated, each time-stamp character S (a binary image) can be identified through inputting S into each SGNT respectively, and calculating minimum distance between S and output results of the SGNTs.

According to the result of fine segmenting, the size of photo time-stamp character w and h are obtained. Because it is possible for the size of photo time-stamp characters to be smaller than predefined character templates, the characters need to be converted so as to make character templates and actual time-stamp characters have the same size.

Assuming that size of character template is $w^{'} \times h^{'}$, whereas size of actual time-stamp character S is $w \times h$, and $w \leq w', h \leq h'$, each actual time-stamp character is enlarged according to the following formula:

$$S^{'}(i,j) = S(\left\lfloor i\frac{w}{w^{'}-1} \right\rfloor, \left\lfloor j\frac{h}{h^{'}-1} \right\rfloor), i = 0,1,...,w^{'}-1; j = 0,2,...,h^{'}-1 \quad (8)$$

where, $\lfloor \bullet \rfloor$ represents rounding operation.

4 Experiments

Our test set includes 320 pieces of color photos obtained from different camerae in JPEG format. Detecting area of time-stamp is set to rectangle area of 48×96.

Templates of 12 characters such as 0, 1,…,9, marker of year and blank character are prepared in advance. Size of the character templates is 20×28 pixels. These character templates are then polluted by white noise in order to generate 12 train sets, respectively. Each train set includes 300 binary images with 20×28 pixels. 12 SGNTs are trained by the train sets, respectively. Each SGNT represents a specify character of the character templates.

Proposed method and PSO based method [5] is used to recognize the time-stamp of color photos in test set respectively. Experimental results show in table 1. The experimental results show that proposed approach in this paper is effective.

Table 1. Experimental results of photo time stamp recognition

Methods	Correct recognition rate (%)	Average CPU time per photo (s)
Proposed method	93.75	1.5261
PSO based method	85.94	132.0056

5 Conclusions

Photo time-stamp is a valuable information source for photo retrieval. A fast photo time-stamp recognition approach based on SGNN is proposed in this paper. Given a photo to be recognized, we segment the photo roughly to determine which area contains the photo time-stamp. The area that contains time-stamp is finely segmented, in order to locate each character in the time-stamp. Then SGNN is used to do template matching on the determined area to complete the photo time-stamp recognition. Compared with the existing approaches, this approach needs no large amount of a prior knowledge and off-line learning processes. Experimental results show that the proposed approach can achieve high recognition accuracy.

References

1. Lienhart, A., Wernicke, A.: Localizing and Segmenting Text in Images and Videos. IEEE Transactions on Circuits and Systems for Video Technology 12(4) (2002) 256-268
2. Yin, P., Hua, X. S., Zhang H. J.: Automatic Time Stamp Extraction System for Home Videos. In: Proc. of IEEE Int. Symp. on Circuits and Systems, IEEE, New York (2002) 73-76
3. Chen, X, Zhang, H.: Photo Time-Stamp Detection and Recognition. In: Proc. of Int. Conf. on Document Analysis and Recognition, IEEE Computer Society, Los Alamitos (2003) 319-222
4. Bao, F.M., Li, A.G., Qin, Z.: Photo Time-Stamp Recognition Based on Particle Swarm Optimization. In: Zhong, N., Tirri, H., Yao, Y., *et al* (eds.): Proc. of IEEE/Wic/ACM Int. Conf. on Web Intelligence, IEEE Computer Society, Los Alamitos (2004) 529-532
5. Wen, W.X., Jennings A., Liu. H.: Learning A Neural Tree. In: Proc. of Int. Joint Conf. on Neural Networks, Beijing, China (1992) 751-756
6. Qin, Z., Bao, F.M., Li, A.G.: A Novel Image Fusion Method Based on SGNN. In: Wang, J., Liao, X., Yi, Z. (eds.): Advances in Neural Networks. Lecture Notes in Computer Science, Vol. 3497. Springer-Verlag, Berlin Heidelberg New York (2005) 747-752

Hierarchical Classification of Object Images Using Neural Networks

Jong-Ho Kim, Jae-Won Lee, Byoung-Doo Kang, O-Hwa Kwon, Chi-Young Seong, Sang-Kyoon Kim, and Se-Myung Park

Department of Computer Science, Inje University, Kimhae, 621-749, Korea
{lucky, jwviolet, dewey, eujin13, cyseong, skkim, smpark}@cs.inje.ac.kr

Abstract. We propose a hierarchical classifier of object images using neural networks for content-based image classification. The images for classification are object images that can be divided into foreground and background areas. In the preprocessing step, we extract the object region and shape-based texture features extracted from wavelet-transformed images. We group the image classes into clusters that have similar texture features using Principal Component Analysis (PCA) and K-means. The hierarchical classifier has five layers that combine the clusters. The hierarchical classifier consists of 59 neural network classifiers that were learned using the back-propagation algorithm. Of the various texture features, the diagonal moment was the most effective. A test showed classification rates of 81.5% correct with 1000 training images and of 75.1% correct with 1000 test images. The training and test sets each contained 10 images from each of 100 classes.

1 Introduction

Neural networks are derived from biological learning systems and consist of nets of neurons with complicated connections. This biological basis allows effective machine learning, and neural networks are used in a variety of fields. They have proved remarkably efficient in recognition and classification through learning.

Most neural network applications use a one-level neural network structure, which has only one set of input, output, and hidden layers. To obtain better performance from the one-level classifier, initial weights and the topological structure of the classifier are adjusted[1][2]. However, there is a drawback in the one-level neural network structure: as the number of classes to be classified increases, the number of nodes and learning time increase rapidly, and the classification efficiency rate decreases. It is therefore difficult to apply existing one-level neural networks to the classification of various kinds of images. To deal with this problem, we propose a hierarchical classifier of object images using neural networks. The images for classification are object images that can be divided into foreground and background. Our system extracts the object region by segmentation of the image. It extracts wavelet transform-based structural texture features from the extracted object region. The images are grouped into clusters that have similar texture features using Principal Component Analysis (PCA) and K-means. With the clustered classes, a hierarchical classifier is constructed using neural networks. The hierarchical classifier consists of 59 neural network classifiers

J. Wang et al. (Eds.): ISNN 2006, LNCS 3972, pp. 322 – 330, 2006.

for 100 kinds of image classes. It has five layers and the results from each node are combined in a top-down approach. In this way, we can reduce the number of image classes that should be allocated to one neural network classifier and improve classification performance.

2 Classification of the Object Image

2.1 Classification Objects

For image retrieval to become practical, it is necessary to limit the retrieval target to the object image in real-world images, as users generally want to extract the object image from a natural scene. Search sites such as AltaVista or Yahoo service text-based categories of object images for frequent retrieval. Therefore, we consider a method for the automatic classification of object images that uses a content-based method without manual processing.

2.2 Construction of Classifier

Fig. 1 shows the main structure of the image classifier. Our image classifier consists of three modules: a preprocessing module, a feature extraction module, and a classification module as follows:

1). The preprocessing module removes the background of an image and performs normalization. In the first step, it removes the background from the image using region segmentation. It also converts the JPEG image from the Internet into a BMP and normalizes the image size to 128 × 128 for the neural network classifier.

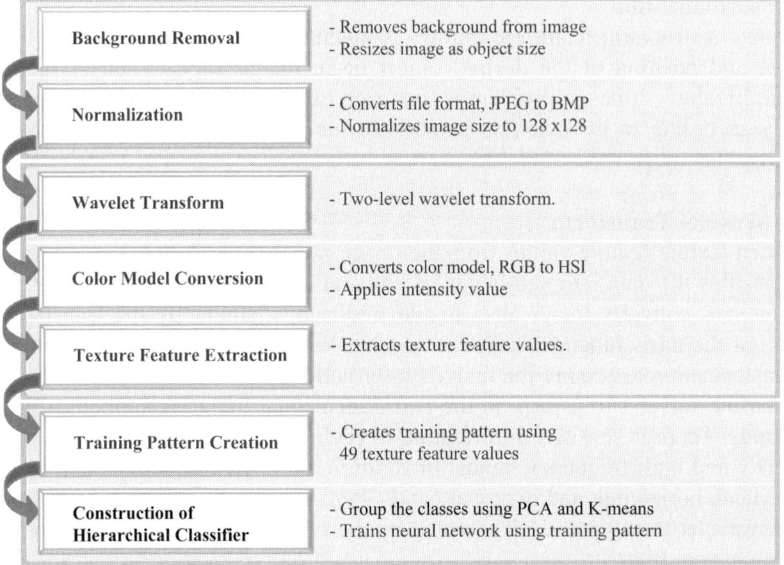

Fig. 1. The main structure of the classifier

2). The feature extraction module performs a wavelet transform and a color model conversion and extracts texture feature values. In the first step, it performs a one-level wavelet transform that by normalizing the image enables the acquisition of the shape and texture information from the image. In the second step, it converts the RGB color model to the HSI color model, so the intensity values can be used. In the third step, we choose a diagonal moment that gives the best classification rate for various texture features, and the module extracts a structural feature value from the diagonal moment. 3). The classes are grouped using PCA and K-means. Finally, the hierarchical neural network classifier is trained with input data using the learning pattern.

2.2.1 Background Removal

In the various images on the Internet, most object images contain a complex background. Therefore, our system performs preprocessing to remove the background from the original image data. We segment images using a JSEG image division algorithm[3]. After region segmentation, we extract the core object region from the center of the original image[4]. Fig. 2 shows an example of an image with a complex background. Fig. 3 shows the result of background regions removal.

Fig. 2. An example of a complex background image

Fig. 3. The result of background regions removal

2.2.2 Normalization

We isolate a structural feature to extract as much information as possible. However, the size and position of the desired object in an image vary. When extracting the structural feature values, a problem is that each of the feature values is extracted differently according to the size and position of the object. To solve this problem, we normalize the image to be 128×128.

2.2.3 Wavelet Transform

To extract texture feature values from the image, we first perform a wavelet transform that classifies the data into sub-bands of low and high frequency. The wavelet transform method converts image data to the frequency domain by the translation and dilation of the basis function, called a wavelet. We analyze the sub-bands created by the transformation to acquire the image's information.

When the wavelet transform in the two-dimensional image is applied, it produces four bands. There is sensitive information in each band, mainly divided into the low-frequency and high-frequency bands. In addition, the high-frequency band is divided into vertical, horizontal, and diagonal bands. Fig. 4 shows the convention for displaying the wavelet transform results created by the two-level wavelet transform through the two-channel filter.

Most related research has used either the HH2 domain acquired by the two-level wavelet transform or all six of the high-frequency bands. However, we use the HH1 domain, which contains more information than the others, because we have reduced the size of the image twice during the background removal and normalization steps.

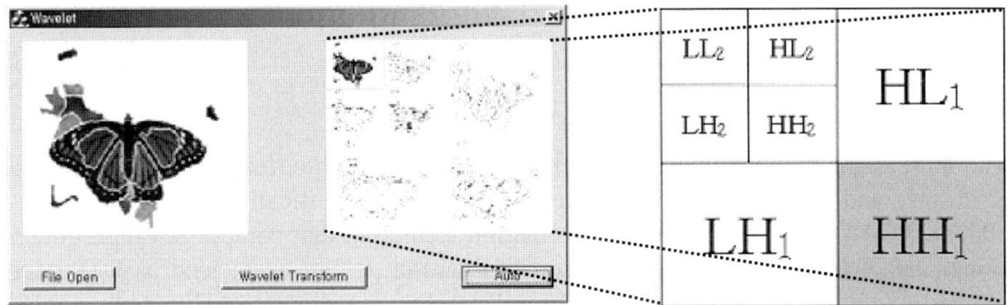

Fig. 4. The classification region by wavelet

2.2.4 Color Model Conversion

The red, green and blue (RGB) colors are mixed to produce the various colors of the image. Because of their strong mutual relation, the RGB components are difficult to handle in the image-processing algorithm. Consequently, many image-processing applications use the hue, saturation and intensity (HSI) color model, and we use the intensity of the HSI color model that is derived from the original RGB image.

2.2.5 Texture Feature Extraction

Usually, an important method for describing a region is to score the texture of the region quantitatively[5].

Texture is expressed using a statistical feature, a structural feature, and a spectrum feature. The statistical feature expresses the degree of roughness of the image. The structural feature expresses regularity, such as a horizon repetition. The spectrum feature expresses the periodicity by analyzing the high-energy component of the image.

The elements to express texture are coarseness, contrast, and directionality. Coarseness means the rough degree of texture. Contrast means the pitch distribution of brightness, and directionality means the specific direction of texture. Various texture feature values were acquired for our experiment.

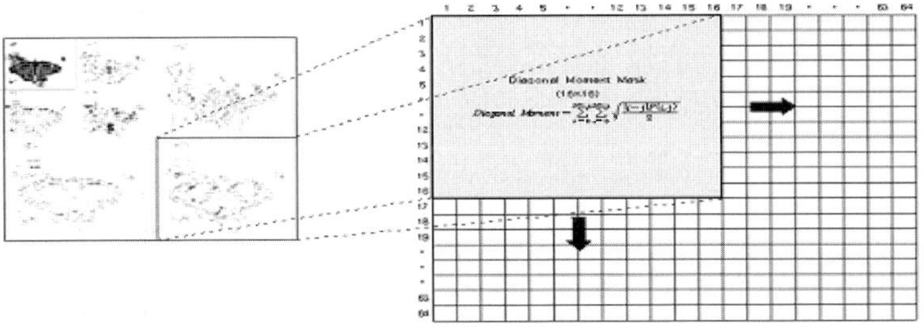

Fig. 5. The sliding window-based texture feature extraction

2.2.6 Training Pattern Creation

Texture feature values are extracted from the HH1 region, which is sized 64×64 after the wavelet transform. We acquire one texture feature value in a window of 16×16. This window is then moved to overlap from left to right and top to bottom by eight pixels in each direction. Finally, we acquire a total of 49 features of texture in the image. Fig. 5 shows how object shape-based texture feature values are extracted using the sliding windows method.

2.3 Hierarchical Classifier

The images are grouped into clusters that have similar texture features using PCA and K-means, and a hierarchical classifier is constructed vertically with the clustered classes. PCA is a simple linear transformation technique that compresses high-dimensional data with minimum loss of data information. It can be viewed as a classical method of multivariate statistical analysis for achieving dimensionality reduction[6], because a small set of uncorrelated variables is much easier to understand and use in further analysis than a larger set of correlated variables. The K-means clustering method is probably the best known[7]. The algorithm starts with k initial seeds of clustering, one for each cluster. All the n objects are then compared with each seed by means of the Euclidean distance and assigned to the closest cluster seed. The procedure is then many times[8].

The two steps for constructing the hierarchical classifier are as follows. First, extracted diagonal moment are transformed linearly to another feature space with PCA, and new values independent of each other called principal components are obtained. We choose the first four principal components patterns that explain features with about 90% of the total variance. Second, from each class, we choose one representative image with feature values that are closest to the average value for that class's distribution. Fig. 6 shows a feature value distribution of an image class that has 10 images. For this class, the fifth image was selected as the representative image. Fig. 7 shows the hierarchical classifier composition using PCA and K-means. Fig. 8 shows the hierarchical classifier composition prepared by human classifiers.

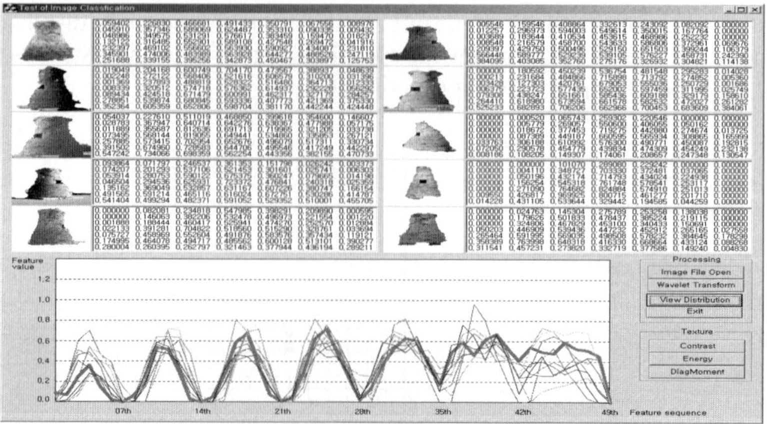

Fig. 6. The feature value distribution of an image

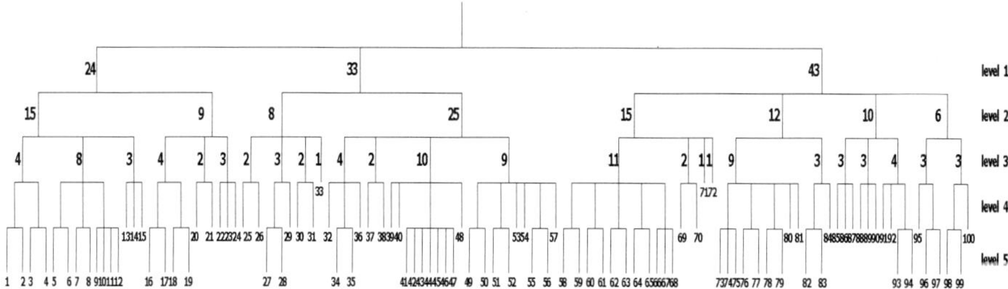

Fig. 7. The hierarchical classifier composition using PCA and K-means

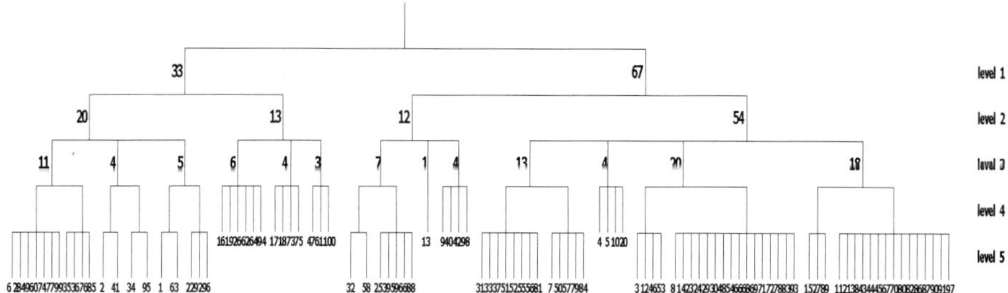

Fig. 8. The hierarchical classifier composition prepared by human classifiers

3 Experimental Results

We experimented with three different structures for the classifier : a one-level neural network classifier, a hierarchical neural network classifier prepared by human classifiers, and a hierarchical neural network classifier using PCA and K-means. We compared the performance of each classifier with various features. In addition, we analyzed the classes of images having higher and lower classification rates.

The environment for our experiment was Visual C++ on a Pentium IV PC and the Windows operating system. As experimental data for the various images from the Internet, we used object images that could be divided into foreground and background regions. We used 100 classes with 10 images per class. With 1000 images for neural network training, and 1000 images for testing on the trained neural networks, a total of 2000 images were used as experimental data. The texture features considered in the experiments were contrast, diagonal moment, energy, entropy, homogeneity, second diagonal moment, and uniformity.

Of the seven texture features, the diagonal moment shows the highest success rate. It proved impossible to train a one-level classifier, although we trained it for seven days with an error rate of 0.00001. When we reduced the data to 30 classes with 10 images per class, the training time was 123 hours for a one-level classifier and was reduced to one third using a hierarchical classifier. Table 1 shows the recognition rate for 30 classes using the one-level and hierarchical classifiers.

The one-level classifier achieved a classification rate of 72.3% on the training data and 67% on the test data. The hierarchical classifier achieved a classification rate of

Table 1. Recognition rate for 30 classes by the one-level and hierarchical classifiers

One-level classifier			
Training time	Training data	Test data	Total data
123h	217/300 (72.3%)	201/300 (67.0%)	418/600 (69.7%)
Hierarchical classifier			
Training time	Training data	Test data	Total data
23h	255/300 (85.0%)	228/300 (76.1%)	483/600 (80.5%)

85.0% on the training data and 76% on the test data. The hierarchical classifier by human classifiers achieved a classification rate of 53.8% on all 2000 data images. With the hierarchical neural network classifier using PCA and K-means, we achieve a rate of 81.5% on the training data and 75.1% on the test data.

The classification rate for the hierarchical classifier using PCA and K-means is higher than that of the hierarchical classifier prepared by human classifiers. When humans cluster images into classes, they tend to base the clusters on functions, usages and the kinds of objects in the images. However, this is a fault from the viewpoint of the computer, which constructs the classes with content information from the images such as texture features, rather than the functions, usages and kinds.

Table 2. Recognition rates for each classifier

(a) Hierarchical classifier prepared by human classifiers

Texture Feature	Training Data	Test Data	Total Data
Contrast	593/1000 (59.3%)	430/1000 (43.0%)	1023/2000 (51.2%)
Diagonal Moment	623/1000 (62.3%)	453/1000 (45.3%)	1076/2000 (53.8%)
Energy	543/1000 (54.3%)	444/1000 (44.4%)	987/2000 (49.4%)
Entropy	510/1000 (51.0%)	435/1000 (43.5%)	945/2000 (47.3%)
Homogeneity	420/1000 (42.0%)	410/1000 (41.0%)	830/2000 (41.5%)
Second DM	572/1000 (57.2%)	426/1000 (42.6%)	998/2000 (49.9%)
Uniformity	544/1000 (54.4%)	457/1000 (45.7%)	1001/2000 (50.1%)

(b) Hierarchical classifier by PCA and K-means

Texture Feature	Training Data	Test Data	Total Data
Contrast	680/1000 (68.0%)	627/1000 (62.7%)	1307/2000 (65.4%)
Diagonal Moment	815/1000 (81.5%)	751/1000 (75.1%)	1566/2000 (78.3%)
Energy	644/1000 (64.4%)	567/1000 (56.7%)	1211/2000 (60.6%)
Entropy	617/1000 (61.7%)	543/1000 (54.3%)	1160/2000 (58.0%)
Homogeneity	536/1000 (53.6%)	483/1000 (48.3%)	1019/2000 (51.0%)
Second DM	712/1000 (71.2%)	627/1000 (62.7%)	1339/2000 (67.0%)
Uniformity	701/1000 (70.1%)	621/1000 (62.1%)	1322/2000 (66.1%)

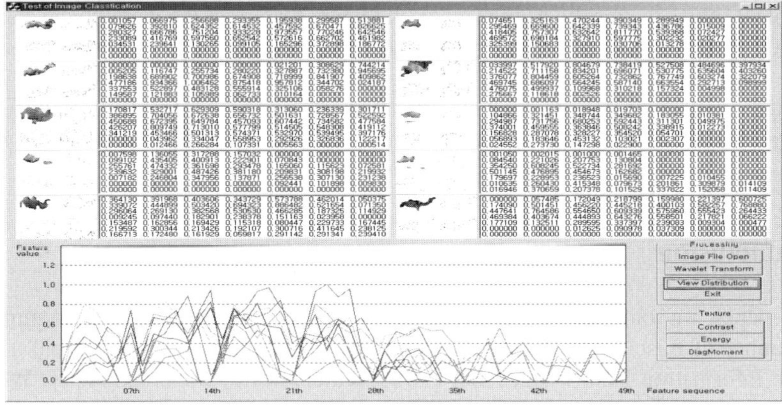

Fig. 9. The feature value distribution of a complex background image

Fig. 10. The classes of images for training and test

Table 2 shows the results from using each texture feature of the hierarchical classifier prepared by human classifiers (a) and hierarchical classifier by PCA and K-means (b). When the object was divided into the background clearly and was not complex, it was extracted easily and showed a high classification rate. However, Fig. 9 shows that when the object is not divided into the background and is complex, the classification rate is low. Fig. 10 shows the classes of images used in the training and test.

4 Conclusion

In this paper, we removed background regions from images in an attempt to minimize the number of misclassifications. By using the wavelet transform, we could acquire feature information from images and extract texture features hierarchically. We also transformed extracted features using PCA and grouped them by K-means, constructed a hierarchical structure and classified the images using neural network. We analyzed training and test classes on each level using a hierarchical neural classifier. As a result we were able to reduce training time. By constructing a hierarchical structure using computed features rather than human classifications, we achieved a higher classification rate. Our future work will focus on identifying better features for classifying images more effectively, and improving our system to extract objects from complex back-grounds.

Acknowledgement

This work was supported by the 2003 Inje University research grant.

References

1. Singh, V., Rao, S.: Application of Image Processing and Radial Basis Neural Network Techniques for Ore Sorting and Ore Classification. Minerals Engineering 18(15) (2005) 1412-1420
2. Giacinto, G., Roli, F., Didaci, L.: Fusion of Multiple Classifiers for Intrusion Detection in Computer Networks. Pattern Recognition Letters 24(12) (2003) 1795-1803
3. Deng, Y., Manjunath, B.S.: Unsupervised Segmentation of Color-Texture Regions in Images and Video. IEEE Transactions on pattern analysis and machine intelligence 23(8) (2001) 800-810
4. Park, S.B., Lee, J.W., Kim, S.K.: Content-based Image Classification using Neural Network. Pattern Recognition Letters 25(3) (2004) 287-300
5. Gonzalez, R.C., Woods, R.E.: Digital Image Processing. 2nd edn. Prentice Hall, Upper Saddle River New Jersey (2003)
6. Turk, M., Pentland, A.: Eigenfaces for Recognition. Journal of Cognitive Neuroscience 3(1) (1991) 71-86
7. Johnson, R.A., Wichern, D.W.: Applied Multivariate Statistical Analysis. 5th edn. Prentice Hall, Upper Saddle River New Jersey (2002)
8. Gose, E., Johnsonbaugh, R., Jost, S.: Pattern Recognition and Image Analysis. Prentice Hall, Upper Saddle River New Jersey (1996)

Structured-Based Neural Network Classification of Images Using Wavelet Coefficients

Weibao Zou, King Chuen Lo, and Zheru Chi

Centre for Multimedia Signal Processing,
Department of Electronic and Information Engineering,
The Hong Kong Polytechnic University, Hong Kong
enwzou@eie.polyu.edu.hk, enkclo@polyu.edu.hk,
enzheru@inet.polyu.edu.hk

Abstract. Image classification is a challenging problem in organizing a large image database. However, an effective method for such an objective is still under investigation. This paper presents a method based on wavelet for image classification with adaptive processing of data structures. After decomposed by wavelet, the features of an image can be reflected by the wavelet coefficients. Therefore, the nodes of tree representation of images are represented by distribution of histograms of wavelet coefficient projections. 2940 images derived from seven original categories are used in experiments. Half of the images are used for training neural network and the other images used for testing. The classification rate of training set is 90%, and the classification rate of testing set is 87%. The promising results prove the proposed method is very effective and reliable for image classification.

1 Introduction

Image content representation has been a popular research area in various applications of image processing in the past few years. Some systems attempted to represent image contents in terms of a series of keywords. However, keyword-based taxonomy reduces the system complexity while increasing the requirements of users' knowledge. On the other hand, much research has been conducted on representing image contents with visual features. Most of the applications represent images using low level visual features such as color, texture, shape and spatial layout in a very high dimensional feature space, either globally or locally. However, the most popular distance metrics such as Euclidean distance cannot guarantee that the contents are similar even their visual features are very close in the high dimensional space.

Most recently, with wavelet, feature-based representation of images potentially offers an attractive solution to this problem. It involves a number of features extracted from the raw image based on wavelet [1]. The features of images, such as edges of an object, can be reflected by the wavelet coefficients in low and high bands. Both features or objects and the spatial relationship among them play more important roles in characterizing image contents, because they convey more semantic meaning. By organizing these wavelet coefficients into tree representation, image contents can be represented more comprehensively at various details, which will be helpful for image classification.

J. Wang et al. (Eds.): ISNN 2006, LNCS 3972, pp. 331 – 336, 2006.

Neural Networks for adaptive processing of data structures are of paramount importance for structural pattern recognition and classification [2]. The main motivation of this adaptive processing is that neural networks are able to classify static information or temporal sequences and to perform automatic inferring or learning [3-5]. Sperduti and Starita have proposed supervised neural networks for classification of data structures [6]. This approach is based on using generalized recursive neurons and the Backpropagation Through Structure (BPTS) algorithm [2], [6]. In this study, BPTS algorithm is adopted in image classification using wavelet coefficients.

This introduction will be followed by a description of classification features. The adaptive processing of data structure is introduced in section 3. Then, design and results of experiments are reported respectively in sections 4 and 5. Some conclusions are drawn in the final section.

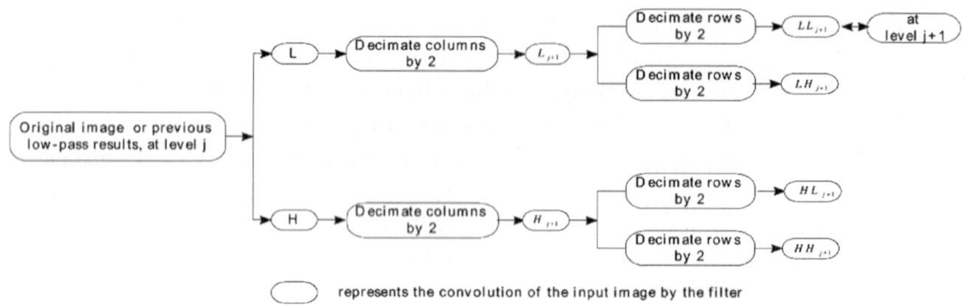

Fig. 1. Block chart of wavelet decomposition of an image

2 Description of Features for Images Classification

2.1 Extraction of Features from an Image with Wavelet

With the goal of extraction of features, as an application of the wavelet decomposition performed by a computer, a discrete wavelet is considered and is summarized in Fig. 1. It is noticed that the original image (or previous low-pass results at level j) can be decomposed into a low-pass band (L_{j+1}) and a high-pass band (H_{j+1}) through filters at level j+1. Through a low-pass filter and a high-pass filter, the low-pass band (L_{j+1}) is again decomposed into a low-pass band (LL_{j+1}) and a high-pass band (LH_{j+1}). The high-pass band (H_{j+1}) is decomposed into a low-pass band (HL_{j+1}) and a high-pass band (HH_{j+1}) through filters. Sub-bands LH_{j+1}, HL_{j+1} and HH_{j+1} respectively represent the characteristics of the image in the horizon, vertical, and diagonal views. The local details of image, such as edges of objects, are reserved in high bands. The basic energy of the image is reserved in LL_{j+1}. We will refer to the four sub-bands created at each level of the decomposition as LL_{j+1} (Low/Low), LH_{j+1} (Low/High), HL_{j+1} (High/Low), and HH_{j+1} (High/High).

2.2 Tree Structural Representations of an Image

A tree representation of an image decomposed by wavelet is shown in figure 2(a). At level one (L_1), the original image (the top image) is decomposed into four bands. The most left one in the line of L_1 is the low frequency band (LL_1), and the right three ones are the high frequency bands (LH_1, HL_1 and HH_1). At level tow (L_2), the previous low frequency band is decomposed into four bands. The most left one in the line of L_2 is the low frequency band (LL_2), and the other three ones are the high frequency bands (LH_2, HL_2 and HH_2). Obviously, from figure 2(a), it can be found that the basic energy and the edges of object can be respectively observed in low frequency band and high frequency bands at each level.

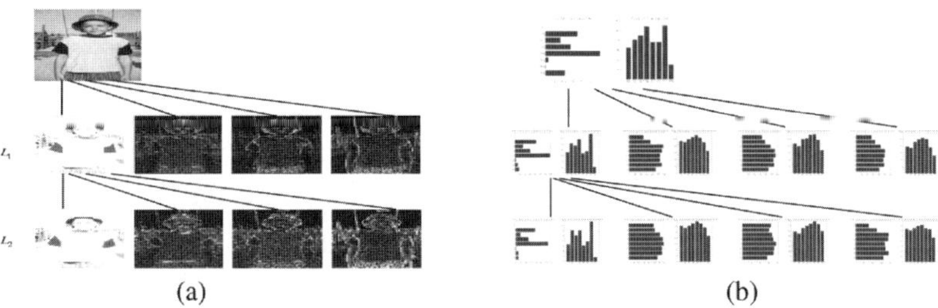

Fig. 2. A tree structural representation of an image and its wavelet coefficients: (a) A tree representation of a boy image decomposed by wavelet at two levels; (b) A tree representation of histograms of projections of wavelet coefficients corresponding to the images in (a)

2.3 Tree Structural Description of Histograms of Wavelet Coefficients

The features of an image can be reflected by wavelet coefficients. Therefore, the edge of an object can be described by the distribution of wavelet coefficients that can be respectively projected onto the x-axis and y-axis. The projections of wavelet coefficients are statistic by histograms with 8 bins in the x direction and 8 bins in the y direction. For example, for the low frequency band, its features are represented by the histograms in x-axis and y-axis after wavelet coefficients are projected. It is similar with the high frequency bands. However, for the original image, its histograms are the projection of gray value of pixels. Actually, such histograms can effectively represent the outline of objects. The tree representation of an image described by histograms of projections of wavelet coefficients is illustrated in figure 2(b).

3 Adaptive Processing of Data Structure

Connection list models have been successfully employed to solve learning tasks characterized by relatively poor representations in data structure such as static pattern or sequence. Most structured information presented in the real world, however, can hardly be represented by simple sequences. Although many early approaches based on syntactic pattern recognition were developed to learn structured information, devising a proper grammar is often a very difficult task because domain knowledge is incomplete or insufficient. On the contrary, the graph representation varies in the size of input units and can organize data flexibly. Its encoding process shown in Figure 3 is independent on the classification task and it is not necessary to pay attention specifically to their nodes. Recently, neural networks for processing data structures have been proposed by Sperduti [6]. It has been shown that they can be used to process data structures using BPTS. It extends the time unfolding carried out by back-propagation through time in the case of sequences. A general framework of adaptive processing of data structures was introduced by Tsoi [3] and Frasconi et al.[5][7][8].

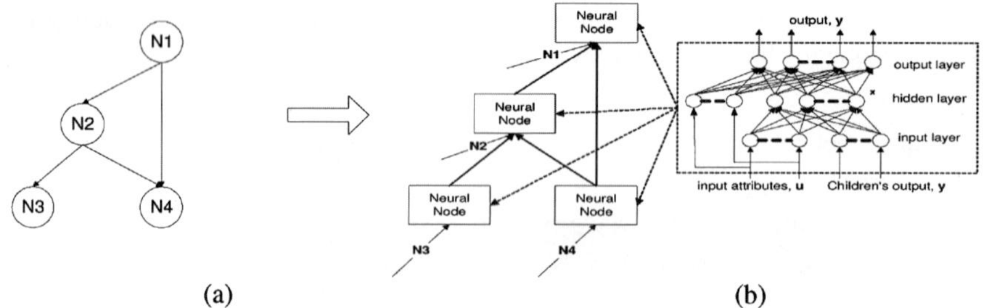

Fig. 3. An illustration of a data structure with its nodes encoded by a single-hidden-layer neural network.: (a) Directed Acyclic Graph (DAG), (b) The encoded data structure.

4 Design of Experiment

4.1 Data for the Experiment

Seven categories of images, building, beetle, fish, grass, people, warship and vehicle, are adopted in the experiments. In each category of image, there are ten independent images and two of them are shown in figure 4.

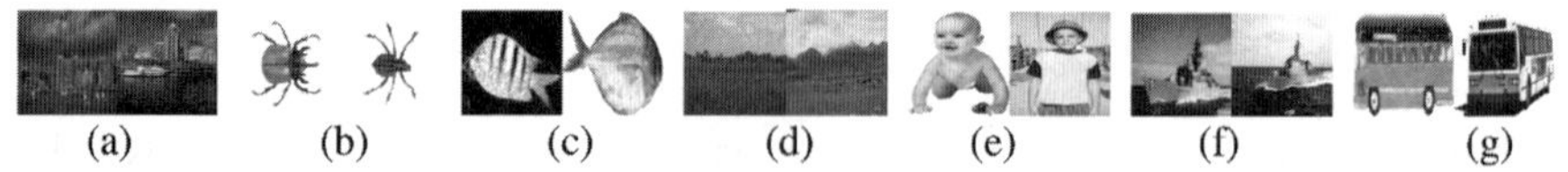

Fig. 4. Images for experiments: (a) Biulding; (b) Beetle; (c) Fish; (d) Grass; (e) People; (f) Warship; (g) Vehicle

4.2 Strategy for the Experiment

There are two parts in the experiments. One is training the neural network, and the other is testing. In order to make findings reliable, much more images are generated by translating the original images. For example, a window is set up in the original image, and it shifts from left to right, then top to bottom. Of course, the object must be contained in the window. However, the background is changing with the shifting window. Corresponding to each window, a new image is generated. Therefore, from each image, 40 new images are generated from this translation operation. The

Table 1. Image database for experiments

Category of image	Number of original image in each category	Number of derived images from an independent image	Number of images in each category after image translation
Building	10	40	400
Beetle	10	40	400
Fish	10	40	400
Grass	10	40	400
People	10	40	400
Warship	10	40	400
Vehicle	10	40	400
Total number	70	280	2800

information of images is tabulated in Table 1. Together with the original images, 1470 images are used for training the neural network and 1470 images are used for testing.

5 Experimental Results

This section reports the classification performance of our proposed method. The attributes of all the nodes in the tree are vectors of 16 inputs described by histograms according to the method in Section 2. In this investigation, a single hidden-layer is sufficient for the neural classifier, which has 16 input nodes, and 7 output nodes. The number of hidden nodes could obviously affect the result. Therefore, several different numbers of hidden nodes are adopted in the experiment. The results are shown in table 2. When the number of hidden nodes is 10, its classification accuracy rate for the training data set is 82%. With the trained neural network, the testing is carried out with another images not used in training set. The classification accuracy rate is 74% for the test data set. For 17 hidden nodes, its classification rate for training set is 90% and for testing set is 87%. For 23 hidden nodes, its classification for training set is 79% and for testing set is 71%. Obviously, based on the proposed method, the neural network is strong enough for classifying many categories of image. Especially, when the number of hidden nodes is 17, the experimental results are more reliable.

Table 2. Classification results with different hidden nodes

Number of hidden nodes	Classification rate on training set (%)	Classification rate on test set (%)
10	82	74
17	90	87
23	79	71

6 Conclusion

In this paper, a method based on wavelet coefficients is proposed for the image classification with adaptive processing of data structures. The classification features including extracting features from images with wavelet, histogram of projections of wavelet coefficients and tree structural image representation are first described. 16 bins of histograms are used as the input attributes for BPTS algorithm. Based on this, 1470 images are used for training neural network. The classification accuracy rate up to 90% is reached. Using such a trained neural network, with another 1470 images, the testing is carried out and its classification rate arrives at 87%. Clearly, better classification accuracy is achieved by our approach. It can be seen that the proposed method for image classification using wavelet coefficients is reliable and effective.

Acknowledgement

The work described in this paper was fully supported by a grant from the Research Grants Council of the Hong Kong Special Administrative Region, China (Project no. PolyU 5229/03E).

References

1. Cohen, A., Ryan, R. D.: Wavelets and Multiscale Signal Processing. Chapman & Hall Press (1995).
2. Giles, C. L., Gori, M.: Adaptive Processing of Sequences and Data Structures. Springer, Berlin New York (1998).
3. Tsoi, A. C.: Adaptive Processing of Data Structures: An Expository Overview and Comments. Technical report, Faculty of Informatics, University of Wollongong, Australia (1998).
4. Tsoi, A.C., Hangenbucnher, M.: Adaptive Processing of Data Structures. Keynote Speech, Proceedings of Third International Conference on Computational Intelligence and Multimedia Applications (ICCIMA '99) (1999) 2-2 (summary).
5. Frasconi, P., Gori, M., Sperduti, A.: A General Framework for Adaptive Processing of Data Structures. IEEE Trans. on Neural Networks 9 (1998) 768-785.
6. Sperduti, A., Starita, A.: Supervised Neural Networks for Classification of Structures. IEEE Trans. on Neural Networks 8(3) (1997) 714-735.
7. Goller, C., Kuchler, A.: Learning Task-dependent distributed representations by Back-propagation Through Structure. Proc. IEEE Int. Conf. Nerual Networks (1996) 347-352.
8. Cho, S., Chi, Z., Wang, Z., Siu, W.: An Efficient Learning Algorithm for Adaptive Processing of Data Structure. Neural Processing Letters 17 (2003) 175-190.

Remote Sensing Image Classification Algorithm Based on Hopfield Neural Network

Guang-jun Dong, Yong-sheng Zhang, and Chao-jie Zhu

Information Engineering University Institute of Surveying and Mapping,
Postfach 66, 450052 Zhengzhou, China
{topd, zhangys, zhuchaojie}@163.com

Abstract. Considering the feature of remote sensing images, we put forward a remote sensing image classification algorithm based on Hopfield neural network. First, the function and principle of Hopfield neural network is described in this paper. Then based on the common model of Hopfield neural network, the image classification algorithm using Hopfield neural network is realized and experimental results show that its precision is superior to that of the conventional maximum likelihood classification algorithm.

1 Introduction

Remote sensing image classification is one of the important application worlds for remote sensing technique, but also is a form of showing the property of image objects directly and vividly. For several decades a good harvest has been gained, but relative to multitudinous images and given application, few image classifications can obtain perfect precision and speed, which is also emphases and hotspot in the remote sensing world. On the hand, with more and more spatiotemporal dimensions of remote sensing data, classical algorithms have gradually exposed some weaknesses, for instance, multi-source and multi-dimensional data may not have normal distribution feature; establishing initial condition is difficult and in most cases discrete data (ground truth data) has no statistical meaning. So in recent years artificial neural network technique has being applied to Remote sensing image classification.

In 1980's, Hopfield put forward the Hopfield neural network, which has attracted many researchers' attention since its first appearance. By calculating a group of differential equations, Hopfield neural network can get an optimal or approximate optimal result. So Hopfield neural network has a wide range of applications, such as content addressable memory, pattern recognition, and combinatorial optimization. E.Oja processed data compression with advantage of simpler calculation, higher compression ratio, better fault-tolerant and in advance of setting signal-noise ratio over conventional coding technique [1]. This paper introduces a remote sensing image classification algorithm based on Hopfield Neural Network by combining Hopfield neural network and remote sensing image classification technique. The results of experiment show that this method has outstanding advantage of processing rapidity and classification precision compared to conventional methods.

J. Wang et al. (Eds.): ISNN 2006, LNCS 3972, pp. 337–342, 2006.

2 Hopfield Neural Network [2]

Hopfield network is a feedback neural network that is often used to solve the optimization problems. The dynamic functions of continuous Hopfield network are:

$$\frac{d\mu_i}{dt} = \frac{-\mu_i}{\tau_i} + \sum_{j=1}^{n} T_{ij}\upsilon_j + I_i \quad i = 1,2,\cdots,n$$
$$\upsilon_i = g(\mu_i) \qquad T_{ij} = T_{ji}, \qquad \forall i,j. \tag{1}$$

where n is the number of neurons, μ_i is the state of the i -th neuron, τ_i is the time constant of the i -th neuron, T_{ij} is the post—synaptic strength from the j -th neuron to the i -th neuron, υ_i is the output of the i -th neuron, I_i is the bias at the i -th neuron, $g(\cdot)$ is the hyperbolic function

$$g(x) = \tanh(\alpha x) = \frac{e^{\alpha x} - e^{-\alpha x}}{e^{\alpha x} + e^{-\alpha x}}. \tag{2}$$

Where α is a positive constant which affects the convergence rate of the network. The network has an energy function

$$E = -\frac{1}{2}\sum_{i=1}^{n}\sum_{j=1}^{n} T_{ij}\upsilon_i\upsilon_j - \sum_{i=1}^{n} I_i\upsilon_i + \sum_{i=1}^{n}\frac{1}{\tau_i}\int_0^{\upsilon_i} g_i^{-1}(\upsilon)d\upsilon. \tag{3}$$

if the slope of g_i or τ_i is big enough, the integral term in Eq.(3) can be neglected.

The steps of solving optimization problem are: Firstly, the energy function is constructed by optimization measure; Secondly, the weights and biases of network are computed according to the energy function, and then the network structure is produced; Finally, the stable point of network running is the local minimum point of energy function (if the local minimum point is unique, then it is the global optimum point), then it is the answer to optimization problem.

3 Remote Sensing Image Classification Model Based on Hopfield Neural Network

Optimization calculation models based on Hopfield Networks have been widely used in a number of specialized worlds, such as character and fingerprint recognition; at the same time can be applied to Geo-analysis field such as cell analysis of remote sensing image and noise detection. The basic springboard of Hopfield neural network application is through several iterations for single functions to realize complex map fitting and approach. Training rule does add-weight and sum for data processing using transform function, and training network system does pattern recognition and transforms it to potential output value.

On the base of universal model for Hopfield geo-analysis, combining processing flow for remote sensing image classification and according to the theory of Hopfield pattern recognition, we correspondingly form remote sensing image classification model based on Hopfield neural network as the following description:

1. Set control parameters listed as follows: Let k be the category count that is expected to be got, Let I be the effectual iteration time and Let L be the number of samples.
2. Set the estimation condition for Hopfield network. When ending classification every sample belongs to only a category.
3. Compute image probability histogram $h(j)$ and the center of initial category.

$$h(j) = (j_1, j_2, j_3 \ldots. j_k) . \tag{4}$$

j_i is the number of grey level value as i .

4. Do classification using Hopfield network. Assuming n sample vectors to cluster into m categories, we can use $m \times n$ cells to form feedback network arranged to array with m rows and n columns. There we construct Hopfield network S as binary matrix with I_i and k , where I_i is the distance between i and k center.

So $S(i,k)$ is thresholded as:

$$S(i,k) = \begin{cases} 1, & i \in k \\ 0 & i \notin k \end{cases} . \tag{5}$$

5. So the energy function can be expressed as:

$$E = \sum_{i=0}^{L-1} \sum_{j=0}^{L-1} \sum_{k=0}^{L-1} \frac{W_{ij} h\ (j) s(i,k) s(j,k)}{\sum_{j=0}^{L-1} h(j) s(j,k)} . \tag{6}$$

and total input of each neuron can be expressed as:

$$Net(i,k) = \frac{-1}{\sum_{j=0}^{L-1} h(j) s(j,k)} \sum_{j=0}^{L-1} W_{ij} h(j) s(j,k) . \tag{7}$$

where W_{ij} is the constraint function which often selected as Mahalanobis distance function.

6. Do iteration until convergence according to network dynamic equation. Every time the result of iteration is a new category center $Z_{j+1} = \frac{1}{Net(i)} \sum_{x \in i} x$, if the times of iteration are more than I , or parameter change is in the range of tolerance, iteration is terminated.
7. Evaluate whether in calculation result the cell state at k row and i column is 1, which means that i -th sample belongs to k -th category. Incorporate categories and evaluate classification precision.

In the fact of classification processing, we make sample gray level vector not some feature statistics as input value so as to conveniently form network structure, and at the same time do probability histogram statistics avoiding reiteration. Otherwise if we directly use multiphase feature statistics as input value, corresponding network structure will be dilatable, which leads to slow training speed and even difficult convergence because of drastic surge. In addition, we can form different weight functions using TM multi-spectral data. Such integration can adopt mean of different band histograms or choose maximum, namely using mean and maximum of several image histograms to form weight coefficient, respectively. The else procedure is unchanged corresponding to basic processing method.

4 Results of Experiment and Analysis

We select TM image as experiment data with size of 512*512. First obtain image gray level histograms and form networks according to Hopfield. Then through several iterative calculations, make energy function run to stability ultimately and get corresponding classification results. Simultaneously according to the original images or related map data we select samples to check precision. Those samples, which are at best uniform and stable, can be collected directly through given row and column interval in images, or collected with selectivity according to object complexity.

The procedure flow for remote sensing Image classification model based on Hopfield neural network is as follows:

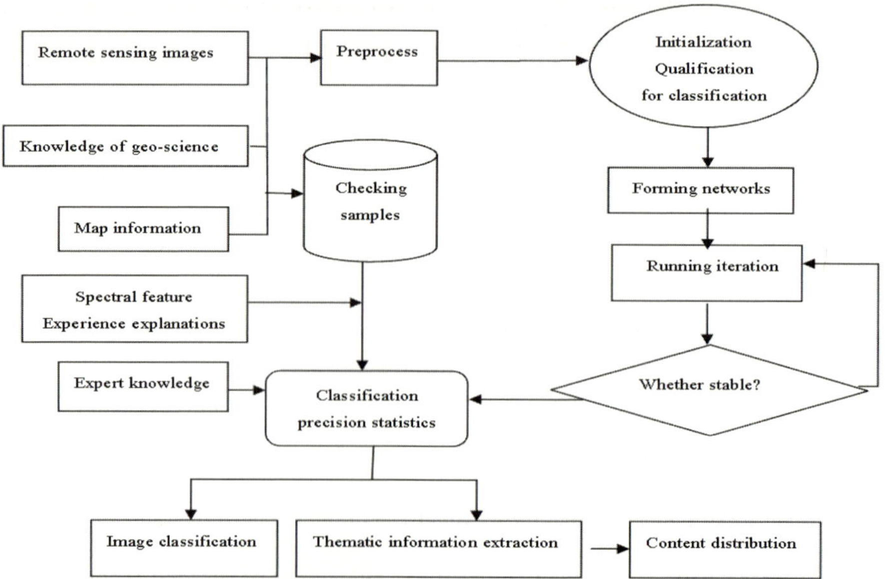

Fig. 1. The flow chart of Image classification model based on Hopfield neural network

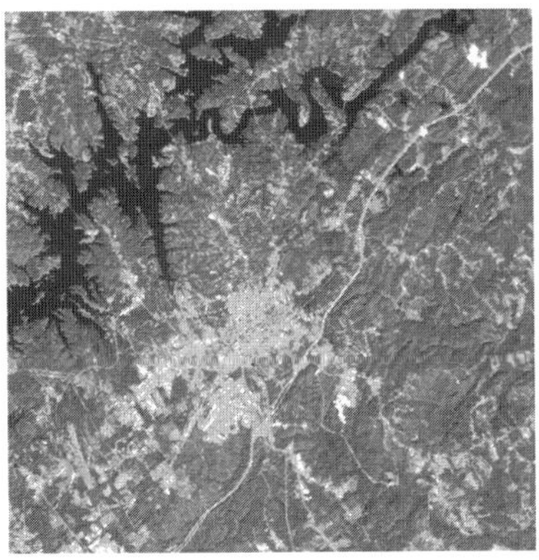

Fig. 2. The original image

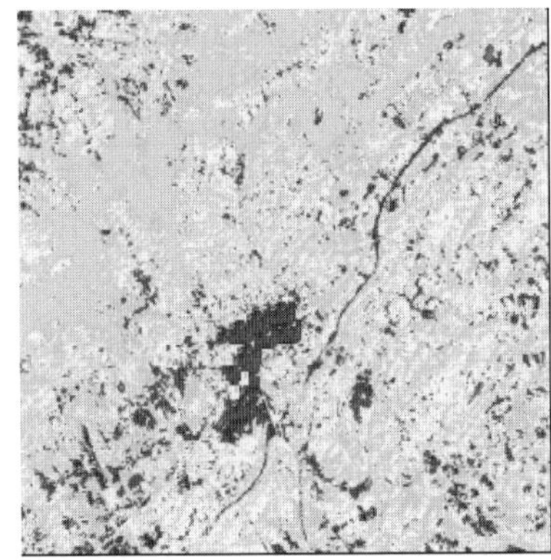

Fig. 3. Result of Hopfield classification

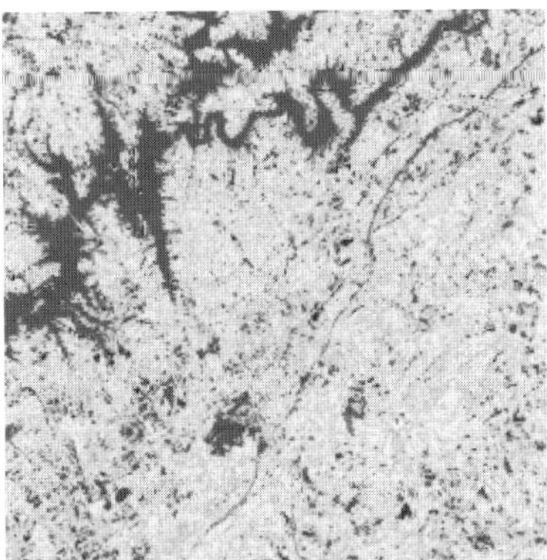

Fig. 4. Result of maximum likelihood classification

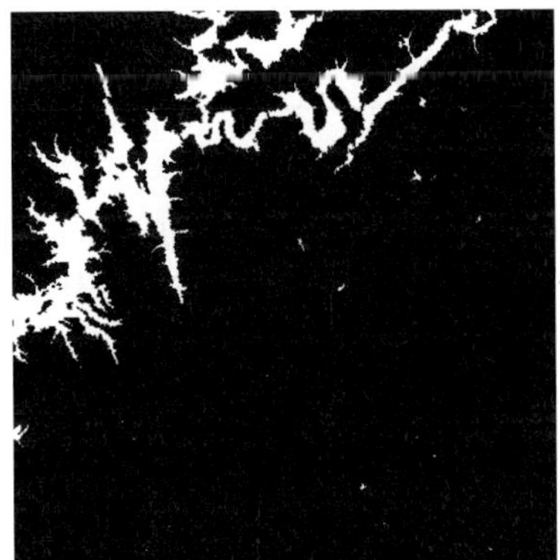

Fig. 5. Water bodies extraction

According to the above procedure flow, the results of experiment are shown:

With comparison to the original image, we select four kinds of prior samples which are load C1, water information C2, habitation C3, mountain C41, respectively, and compare to the result of post-classification processing getting the chart of error matrix as follows:

Table 1. Result comparison among different methods to classification

category / method	Time	C1	C2	C3	C4
Hopfield classification	2. 2s	83.4%	86.1%	85.6%	80.9
maximum likelihood classification	3. 6s	81.1%	85.1%	75.3%	76.5

The results of classification show that using maximum likelihood method has leaky and error cases, while using remote sensing image classification model based on Hopfield neural network greatly improves precision of classification with less error cases. So, this method is better to complicated data and processing time has advantage over maximum likelihood method.

5 Conclusion

According to the feature of remote sensing image classification, Hopfield neural network is introduced to image classification in this paper. The results of experiment suggest that this method has low demand of samples, high precision of classification and fast calculation, but also is helpful to extraction of theme information. Of course, this model is needed to further study. Researching on integration between Hopfield classification model and other models and extending knowledge applied in remote sensing processing is one of the main tasks in the following decades.

References

1. Bi, J., Fu, M., Zhang, Q.H..: Filtering and Estimation of Vehicular Dead Reckoning System Based on Hopfield Neural Network. Journal of Beijing Institute of Technology 12 (2003) 230-235
2. Haykin, S.: Neural Networks: A Comprehensive Foundation. Upper Saddle River, NJ: Prentice-Hall (1999)
3. Rosandich, R.G.: A New Network Architecture for Pattern Recognition. Neural Networks 10 (1996) 139-151
4. Seow, M.J., Asari, V.K.: High Storage Capacity Architecture for Pattern Recognition Using an Array of Hopfield Neural Networks. IEEE Computer Society Proceedings of 30th International Workshop on Applied Imagery and Pattern Recognition, (2001) 169-174
5. Hopfield, J.J., Tank, D.W.: Neural Computation of Decision in Optimization Problems. Biological Cybernetics 52 (1985)141-152

Tea Classification Based on Artificial Olfaction Using Bionic Olfactory Neural Network

Xinling Yang[1], Jun Fu[1], Zhengguo Lou[1], Liyu Wang[2], Guang Li[3], and Walter J. Freeman[4]

[1] Department of Biomedical Engineering, Zhejiang University, Hangzhou 310027, P.R. China
[2] Department of Optical Engineering, Zhejiang University, Hangzhou 310027, P.R. China
[3] National Laboratory of Industrial Control Technology, Zhejiang University, Hangzhou 310027, P.R. China
guangli@cbeis.zju.edu.cn
[4] Division of Neurobiology, University of California at Berkeley, LSA 142, Berkeley, CA, 94720-3200, USA

Abstract. Based on the research on mechanism of biological olfactory system, we constructed a K-set, which is a novel bionic neural network. Founded on the groundwork of K0, KI and KII sets, the KIII set in the K-set hierarchy simulates the whole olfactory neural system. In contrast to the conventional artificial neural networks, the KIII set operates in nonconvergent 'chaotic' dynamical modes similar to the biological olfactory system. In this paper, an application of electronic nose-brain for tea classification using the KIII set is presented and its performance is evaluated in comparison with other methods.

1 Introduction

The sense of smell is a chemical and neural process whereby odorant molecules stimulate the olfactory receptor cells that are located high up in the nose in the olfactory epithelium. Broad patterns of response are shown by the olfactory system consisting of a large number of nonspecific receptors [1]. The axons extended by these receptors converge synaptically and link to a limited number of secondary neurons that in turn drive the olfactory cortex of the brain [2]. To simulate the biological olfactory system, the concept of artificial olfaction, whose applicable product is called electronic nose-brain, is introduced.

Basically, an electronic nose-brain has the olfaction as a model and consists of a sensor array with partially overlapping selectivities and a pattern recognition algorithm. The sensor array simulates the receptors in the olfactory epithelium and the pattern recognition algorithm simulates the neural networks of the olfactory bulb, nucleus and cortex. The sensor with overlapping selectivities has broad responsiveness to different odorants as the odor receptor. Several kinds of sensors were selected to form the sensor array, such as metal oxide sensor, conducting organic polymer sensor, quartz crystal microbalance, etc. As stated above, the pattern recognition algorithm is a significant component in the electronic nose-brain system, which provides electronic nose-brain the capability in classifying a variety of odors. Derived from

J. Wang et al. (Eds.): ISNN 2006, LNCS 3972, pp. 343 – 348, 2006.

study on olfactory system, Freeman introduced a novel olfactory model called KIII [3]. Recently, some applications to bar code, figures and handwriting numbers recognition were performed using KIII model [4]. We built a preliminary prototype of electronic nose-brain using KIII model to separate three kinds of simple gases.

Traditionally, the classification of tea depends on human sense. However, it is inaccurate, laborious and time consuming owing to adaptation, fatigue and state of mind. Considering the wide variety of organic compounds in tea, it is really hard to hold out a common standard for tea classification [1]. One of those significant factors to distinguish different kind of tea is the aroma. At this point, we propose to explore whether the electronic nose-brain, which can avoid the limitations of the human sense, might offer a reliable alternative to traditional methods in tea classification.

2 Description of KIII Model

2.1 KIII Model

Generally, in conventional artificial neural network (ANN), chaos should be avoided for engineering purpose, because the trajectory of the system neither repeats nor converges and could not provide steady system output in chaotic state. However, in recent years, the theory of chaos is commonly used to understand the mesoscopic neural dynamics [5]. From recent research, it is believed that chaotic attractor is some kind of essential character of biological neural network [6]. The KIII network based on the olfactory neural system is a high dimensional chaotic network. In this model, the interaction of connected nodes leads to a high-dimensional chaotic attractor. After learning from different patterns, the system will form several low-dimensional local basins [7]. Therefore, the memory for different patterns might be regarded as the formation of local basins, while the recognition process refers to the transition from one basin to another. And the introduction of noise modeling the biological noise source made the KIII net work stable and robust [8].

From a standpoint of bionics, the olfactory neural system is composed of primary olfactory nerve (PON), olfactory bulb (OB), anterior nucleus (AON) and prepyriform cortex (PC). Fig. 1 [7] shows the topological structure of KIII network, in accordance with the anatomic architecture of olfactory neural system. In this model, PON is a KI [9] network; R represents the olfactory receptor, which offers input to the KIII network; the OB layer, AON and PC are composed of KII [9] units; The parameters in KIII network, such as connection strength values between different nodes, were optimized to fulfill features observed in lots of electro-physiological experiments [7].

Among the KIII models, every node is described as a second order differential equation as follows:

$$\begin{gathered}
\frac{1}{a\cdot b}[x_i''(t)+(a+b)x_i'(t)+a\cdot b\cdot x_i(t)]=\sum_{j\neq i}^{N}\left[W_{ij}\cdot Q(x_j(t),q_j)\right]+I_i(t) \\
Q(x,q)=\begin{cases} q(1-e^{-(e^x-1)/q}) & x>x_0 \\ -1 & x<x_0 \end{cases} \\
x_0=\ln(1-q\ln(1+1/q))
\end{gathered} \quad (1)$$

Here $x_i(t)$ represents the state variable of the *i*th node, while W_{ij} indicates the connection strength from *j* to *i*. $I_i(t)$ is external input to the *i*th node. The parameters *a*, *b* and *q* are constants derived from the electro-physiological experiments on biological olfactory system. $Q(\cdot)$ is a static sigmoid function derived from the Hodgkin-Huxley equation and evaluated by experiments.

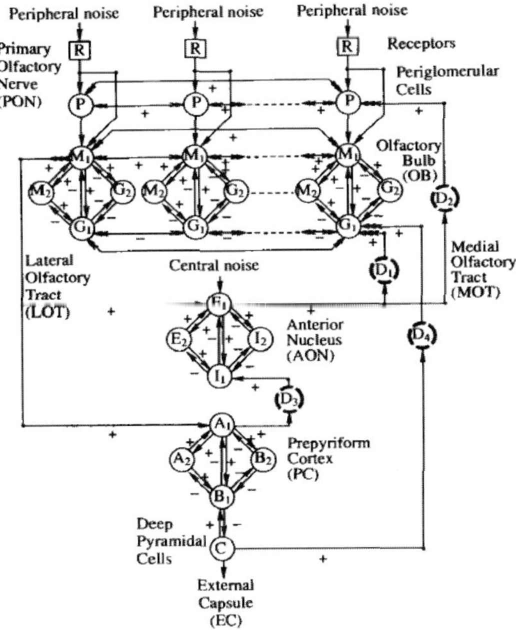

Fig. 1. The topological structure of the KIII network [7]

2.2 Learning Rules

The state of OB layer mitral level is used as the activity measure. The learning process only adjusts the connection strengths among the mitral level. A modified Hebbian learning rule and a habituation rule is employed to KIII model.

To measure the *ith* channel's activity, a value SD_i is extracted. The period with input patterns is divided into *S* segments and SD_i is the mean standard deviations of these segment. *SD*, composed of all the SD_i in the OB layer, depicts the activities of all the channels and SD^m is the mean activity measure of the whole OB layer.

$$SD_i = \frac{1}{S}\sum_{k=1}^{s} SD_{ik}, \quad S^m = \frac{1}{n}\sum_{i=1}^{n} SD_i, \quad SD = [SD_1, SD_2, ..., SD_n] \ . \tag{2}$$

The modified Hebbian learning rule in Equ.(3) means that each pair of *M* nodes co-activated should have their connection strengthened. *K* is inducted to avoid the saturation of weight space. And the habituation rule works at each node as in Equ.(4).

$$\begin{array}{llll} \text{if} & SD_i > (1+K)SD^m & \text{and} & SD_j > (1+K)SD^m \\ \text{then} & W_{ij} = R_{hebb} \times W_{ij} & \text{and} & W_{ji} = R_{hebb} \times W_{ji}, \quad (R_{hebb} > 1) \end{array} . \tag{3}$$

$$W_{ij} = W_{ij} \times r_{hab}, \quad (0 < r_{hab} < 1) . \tag{4}$$

During training, we acquire *SD* vectors with inputs of different patterns. After that, the cluster centers of *SD* in each pattern are calculated respectively. For classification, *SD* is obtained with inputs for classifying. The Euclidian distance from this *SD* to each cluster center is calculated. The minimum distance refers to the certain pattern.

3 Application in Tea Classification

Metal Oxide Semiconductor (MOS) sensors are commonly used in electronic nose-brain applications for its convenience in operating and steadiness in features. We made a sensor array to acquire the volatiles emitted by tea with seven metal oxide sensors of Figaro Co. (TGS2610, TGS2611, TGS800, TGS813, TGS822, TGS826, TGS880). A tea sample is heated before data acquirement. The mean value of the voltage signal during the steady state is acquired as the raw data of this sample. Sometimes there has some peak signal brought by noises. For this reason, a median filter must be added.

We firstly made a classification between green tea and black tea. To build up a testing set, thirty samples were acquired for each kind of tea while training set contains three samples of green tea and three samples of black tea.

Different from the application on classifying simple gases, the raw data of different kinds of tea are quite similar. Owing to this fact, four pre-processing methods, R_{odor}, $ln(R_{odor})$, R_{odor}/R_{air} and $ln(R_{odor})$-$ln(R_{air})$, were employed on the raw data. R_{air} and R_{odor} are the impedances of the sensor array during steady state phase in the air and exposed in the volatiles. The data, raw and pre-processed, should be normalized to avoid the influence of concentration. In the application, a seven-channel input KIII network is used with system parameters in reference [7]. All the data in the training set are used only once. The results are listed in Tab. 1.

The method using $ln(R_{odor})$ performs better. It is considered to be the most effective method. So in the later classification, this method is used as default. The result Euclidean distances to the cluster centers of the two patterns are provided in Fig. 2.

Fig. 3 shows the change of connection weight matrix in the mitral level. With the learning times increases, the difference of weight matrix between current and previous learning times descends rapidly. It is an important factor to scale learning speed. That means KIII network could be trained with a small quantity of learning times.

Table 1. Rate of correct recognition of different kinds of tea with five pre-processing method

	Raw Data	R_{odor}	$ln(R_{odor})$	R_{odor}/R_{air}	$ln(R_{odor})$-$ln(R_{air})$
Green tea	53.3%	76.7%	100%	53.3%	70%
Black tea	50%	60%	90%	46.7%	83.3%

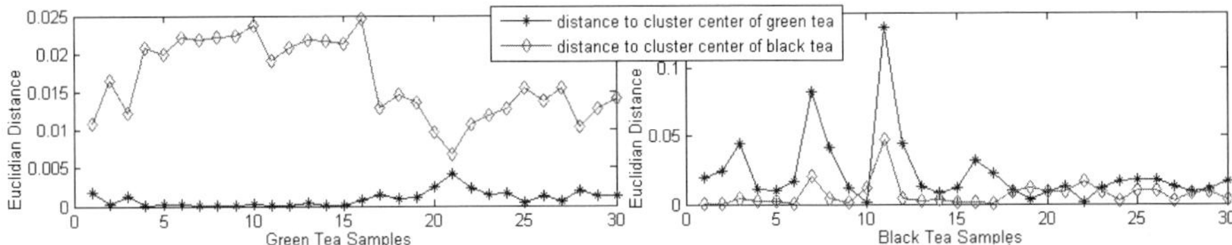

Fig. 2. The distances of different tea samples to the cluster centers of green tea and black tea

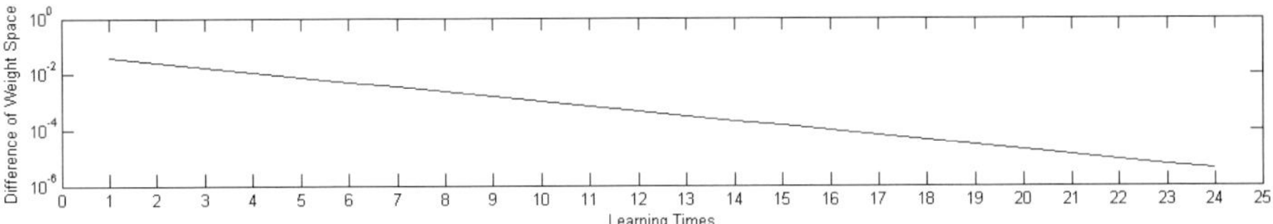

Fig. 3. Convergence curve of weight space in the OB layer mitral level of KIII network

To make a further step, we tried to classify more patterns using KIII model. The number of patterns increases to four, using data set composed of four kind of tea listed in Tab. 2. Fifteen samples of each kind of tea are acquired to build a testing set while 3 samples for each kind are introduced in the training set. At this time, a conventional artificial neural network, BP network, is carried out for comparing. And also, we invited 30 volunteers with normal olfaction to make the tea classification. All the volunteers were trained to remember the odor of each kind of tea. After that, they made the classification by smelling without seeing.

The results were recorded in Tab. 2. Obviously, BP and KIII are both efficient. However, the average classification rate of BP is a little lower. The maximum classification rate of BP is 100%, but the minimum goes down to 66.7%. While to the KIII network, it varies from 80% to 93.3%. The volunteers performed not so well as the electronic nose, because of some physiological and psychological factors [1].

Table 2. Rate of correct recognition of four kinds of tea

	Chinese Green Tea	Japanese Green Tea	Indian Black Tea	Chinese Black Tea	Average
KIII	86.7%	93.3%	93.3%	80%	88.3%
BP	100%	80%	66.7%	93.3%	85%
Human	46.7%	80%	83.3%	50%	65%

4 Discussion

In pattern recognition, KIII model shows good features. Compared with conventional artificial neural network, it is an accurate model in simulating the olfactory system. Fewer training times and less training sets are needed. Its weight matrix converges rapidly during learning. And the classification efficiency is relatively good. Different from the former work on KIII pattern recognition, which mostly used "0-1" digital data as input, a new way is provided to input with decimal. It is proved that decimal input also works effectively and indicates the possibility to reduce the required input channels contributed to pre-processing method. As a result, only a seven-channel KIII network is used instead of introducing more channels. However, it still has potential to be improved. In this work, the classification algorithm is quite simple. In fact, there are a lot of classification algorithms valid for KIII model. How to select a more effective algorithm that can be integrated with KIII model is part of our future works.

Acknowledgements

This research is partially supported by the National Natural Science Foundation of China (No. 60421002) and the National Basic Research Program of China (973 Program, No. 2004CB720302).

References

1. Dottie, R., Kashwanb, K.R., Bhuyanb, M., Hinesa, E.L., Gardner, J.W.: Electronic Nose Based Tea Quality Standardization. Neural Networks 16 (2003) 847–853
2. Persaud, K., Dodd, G.: Analysis of Discrimination Mechanisms in The Mammalian Olfactory System Using A Model Nose. Nature 299 (1982) 352–355
3. Freeman, W.J.: Neurodynamics. An Exploration in Mesoscopic Brain Dynamics. London UK: Springer-Verlag, Berlin Heidelberg New York (2000)
4. Li, G., Lou, Z., Wang, L., Li, X., Freeman, W.J.: Application of Chaotic Neural Model Based on Olfactory System on Pattern Recognitions. In: Wang, L., Chen, K., Ong Y.S. (eds.): Advances in Natural Computation. Lecture Notes in Computer Science, Vol. 3610. Springer-Verlag, Berlin Heidelberg New York (2005) 378–381
5. Freeman, W.J.: Mesoscopic Neurodynamics: From Neuron to Brain. Journal of Physiology-Paris (1994) 303–322
6. Kozma, R., Freeman, W.J.: Chaotic Resonance–Methods and Applications for Robust Classification of Noisy and Variable Patterns. Int. J. Bifurcation and Chaos. 11(6) (2001) 1607–1629
7. Chang, H., Freeman, W.J.: Biologically Modeled Noise Stabilizing Neurodynamics for Pattern Recognition. Int J of Bifurcation and Chaos, 8(2) (1998) 321–345
8. Chang, H.J., Freeman, W.J.: Local Homeostasis Stabilizes A Model of The Olfactory System Globally in Respect to Perturbations by Input During Pattern Classification. Int. J. Bifurcation and Chaos. 8(11) (1998) 2107–2123
9. Freeman, W.J.: Characteristics of the Synchronization of Brain Activity Imposed by Finite Conduction Velocities of Axons. Int J of Bifurcation and Chaos, 10 (2000) 2307–2322

Distinguishing Onion Leaves from Weed Leaves Based on Segmentation of Color Images and a BP Neural Network

Jun-Wei Lu[1], Pierre Gouton[2], and Yun-An Hu[1]

[1] Department of Control Engineering, Naval Aeronautical Engineering Institute, 264001 Yantai, China
[2] Laboratoire Electronique Informatique et Image (Le2i), University of Burgundy, 21078 Dijon, France

Abstract. A new algorithm to distinguish onion leaves from weed leaves in images is suggested. This algorithm is based on segmentation of color images and on BP neural network. It includes: discarding soil for conserving only plants in the image, color image segmentation, merging small regions by analyzing the frontier rates and the averages of color indices of the regions, at last a BP neural network is used to determine if the small regions belongs to onion leaf or not. The algorithm has been applied to many images and the correct identifiable percents for onion leaves are between 80%~ 90%.

1 Introduction

Decreasing the amount of agriculture chemical is one of the principal aims of Precision Agriculture. By using the analysis of image, one tempts to locate weeds or to measure the proportions of crop and weed leaves in field [1],[2]. In our work, Onions and weeds in field are chosen as objects to be studied. As the image acquisition is not controlled, difficulties appear when studying plants in field: (1) The overlapping of onion and weed leaves. (2) Onion and weed and leaves have nearly the same color. In this paper, a new algorithm to distinguish onion leaves from weed leaves in images is suggested. After the soil is discarded and the image conserves only the plants, the image is segmented into a lot of small regions, then these small regions are merged by analyzing the frontier rates and the averages of color indices of the regions, at last a BP neural network is used to determine if the small regions belongs to onion leaf or not.

2 Image Acquisition and Processing Equipment

A Nikon F.401X (Reflex 24×36mm automatic) camera was used for picture acquisition. All color pictures of onion and weeds were taken outdoors with sun illumination, at same time, in a field located at Quetigny in France, the light for all color picture changed little. The weeds include: Sharp- leaved fluellen, Perennial sow thistle, Creeping thistle, Medick sp., Common fumitory; Oxtonque sp., Charlock,

J. Wang et al. (Eds.): ISNN 2006, LNCS 3972, pp. 349 – 354, 2006.

False oat-grass and Dwarf spurge in the field. The vertical distance from the camera to ground is 1.35m. A scanner NIKON LS-30 digitized the roll films 35mm. After these films were scanned into computer, they were saved in a TIF format and then were processed with software VC++.

3 Discarding Soil for Conserving Only Plants in the Image

By analyzing our images in different color space (RGB, HSI, L*a*b, Ohta), we discover that our image in HSI space have better features than in other space, so HSI space is chosen to study our problems. As original images are represented by RGB format, a conversion from RGB to HSI is necessary, the formulas for converting are as following:

$$I = \frac{R+G+B}{3} \qquad S = 1 - \frac{3\min(R, G, B)}{R+G+B}$$

$$H = \begin{cases} \theta & G \geq B \\ 2\pi - \theta & G < B \end{cases} \qquad \theta = \cos^{-1}\left[\frac{\frac{1}{2}[(R-G)+(R-B)]}{\left[(R-G)^2+(R-B)(G-B)\right]^{\frac{1}{2}}}\right]$$

I: Intensity. S: Saturation. H: Hue. R: Red. G: Green. B: Blue

By comparing onion and weed leave characteristics with those of non-plants in different layers (hue, luminosity, saturation), we observe that there are some differences between values obtained from the living plants (onion and weeds) and those from the non-plants (ground, stones, and dead leaves) in layers of hue, saturation and luminosity.

The following algorithm is used to discard all non-plants and to conserve only living plants in the image:

If $\| H(i,j) - \mu_v(H) \| \leq K_h * \sigma_v(H)$
and $\| S(i,j) - \mu_v(S) \| \leq K_s * \sigma_v(S)$
and $\| I(i,j) - \mu_v(I) \| \leq K_i * \sigma_v(I)$
then $H(i,j)$, $S(i,j)$ *and* $I(i,j)$ *are maintained*
otherwise $H(i,j)=0$ *and* $S(i,j)=0$ *and* $I(i,j)=0$

$H(i,j)$, $S(i,j)$ and $I(i,j)$ represent respectively the values of hue, saturation and intensity of the pixels of coordinates i,j.

K_h, K_s, K_i -- constants, calibrated by test.

$\mu_v(H)$, $\mu_v(S)$ and $\mu_v(L)$ --averages of living plants in above-mentioned layers.

$\sigma_v(H)$, $\sigma_v(s)$ and $\sigma_v(L)$--standard deviations of living plants in also above-mentioned layers.

4 Segmentation of Images by Growth of Region

After the levels of hue, saturation and luminosity is divided in 10 classes, the image is scanned by a mask [3] of L inverted, from left to right and from top to bottom (Fig. 1).

If mod (x) represents classes (hue, saturation and luminosity) of a pixel x, and Reg (x) the region to which this pixel x belongs, four cases occur:

1.If {mod(F)=mod(G) and mod(F)≠mod(H) }, then Reg(F)=Reg(G).
2. If {mod(F) ≠mod(G) andt mod(F)=mod(H) }, then Reg(F)=Reg(H).
3.If {mod(F) ≠mod(G) and mod(F) ≠mod(H) }, then Reg(F)=K+1, where K+1 is the label of the new region.
4.if {mod(F)=mod(G) and mod(F) =mod(H) }, then Reg(F)=Reg(H)=Reg(G) and K=K-1 , two regions are merged: Reg(G) and Reg(H).

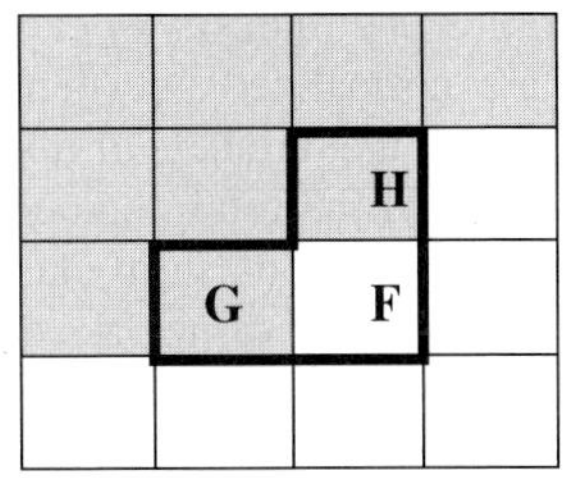

Fig. 1. Mask of inverted L

5 Merging Small Regions

When the image has been segmented by first stage, all lengths of frontier of each small region with its neighbors are memorized in a group.

If the region b has N neighbor regions, $l(b,x)$ is the length of frontier between the region b and the region x, and $v(b,x)$, the rate of frontier of the region x such that :

$$v(b,x)=l(b,x)/N$$

The following formula are used to calculate the differences between the region b with its neighbor region:

$$d(b,x) = \sqrt{(H_b - H_x)^2 + (I_b - I_x)^2}$$

H_b, H_x are the averages of the hue of all pixels belonging to region b and region x respectively.

I_b, I_x are the average of the intensity of all pixels belonging to region b and region x respectively .

We give the algorithm of merging small regions as following:
The region b is merged with its neighbor region x and the label k=k-1 only if :

$v(b,x)=1$
or
$0.5<v(b,x)<1$ and $K_d< d(b,x)=\min(d(b,x_1),d(b,x_2),d(b,x_3),d(b,x_4),....)$

where $x_1, x_2, x_3, x_4,...,$ *are all neighbor regions of the region b,* K_d *is constant and calibrated by test.*

6 Identifying Onion Leaves by a BP Neural Network

After these small regions are merged, many regions appear and are labeled in image. To decide witch of these regions belong to the onion leaves, the levels of hue, saturation and intensity *of* each region were examined, they were divided in 10 classes and are normalized between [0~1]. For each region, its total number of pixels is calculated; the numbers of pixels in each class are counted.

Supposed the total number of pixels of j region is N, the number of pixels in m class of hue is Hm, in m class of saturation is Sm, in m class of intensity *is* Im, the normalized value of m class of hue, saturation and intensity are:

$$h_m = \frac{H_m}{N} \qquad s_m = \frac{S_m}{N} \qquad i_m = \frac{I_m}{N}$$

h_m: the normalized value of m class of hue.
s_m: the normalized value of m class of saturation.
i_m: the normalized value of m class of saturation.
m: 1,2, ……, 10.

To identify these regions, which belong to the onion leaf, a rule of classification based on a neural network (Back propagation [4]) is established. This network includes one layer hidden with 5 neurons and one output layer with one neuron. In the hidden layer the Log-Sigmoid transfer function is used, the Tan-Sigmoid transfer function is used in the output layer. It uses one output parameter and 30 input parameters normalized between [0~1]. The output has two states: if the output is 1, then the object is an onion leaf; if the output is 0, then the object is the one no recognized. The diagrammatic sketch is shown as Fig.2.

The relation of inputs of this network with the normalized values of these classes of hue, saturation and intensity is listed in Table 1.

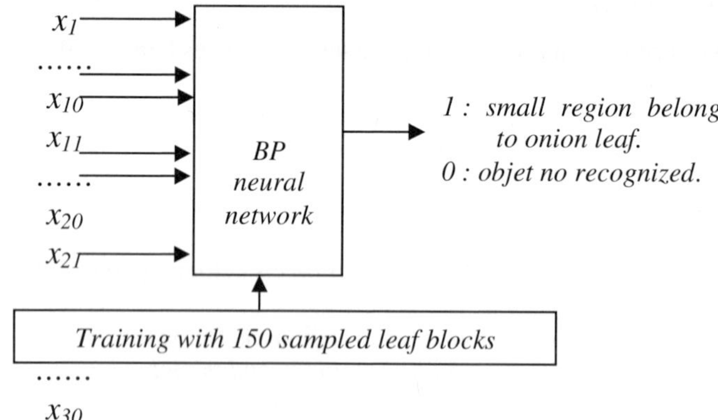

Fig. 2. Diagram of identifying the small onion region based on a neural network

Table 1. The normalized values of color and input of BP neural network

Hue	Saturation	Intensity
$h_1, h_2, \ldots\ldots, h_{10}$	$s_1, s_2, \ldots\ldots, s_{10}$	$i_1, i_2, \ldots\ldots, i_{10}$
$x_1, x_2, \ldots\ldots, x_{10}$	$x_{11}, x_{12}, \ldots\ldots, x_{20}$	$x_{21}, x_{22}, \ldots\ldots, x_{30}$

To train this BP neural network, 150 small regions were drawn precisely from onion leaf images and from weed leaf images by hand. As the color of onion and weed leaves varies in a great range, 150 small regions were chosen to represent all cases of onion and weed leaves. Some small regions, which have same color, were forsaken; another small regions, which have special color characteristics, were added. These 150 regions were utilized as samples to train this BP neural network. Their normalized values of hue, saturation and intensity were used as input vectors.

Among these small regions, the smallest region has only 100 pixels; the biggest region has 2500 pixels. After the training was finished, another 90 small regions of onion and weeds were chosen to test this network, all results were correct.

7 Results and Discussions

According to the algorithm mentioned in Sect.3~6, a program was written with VC++. Using this program to treat onion and weed images, many results were obtained.

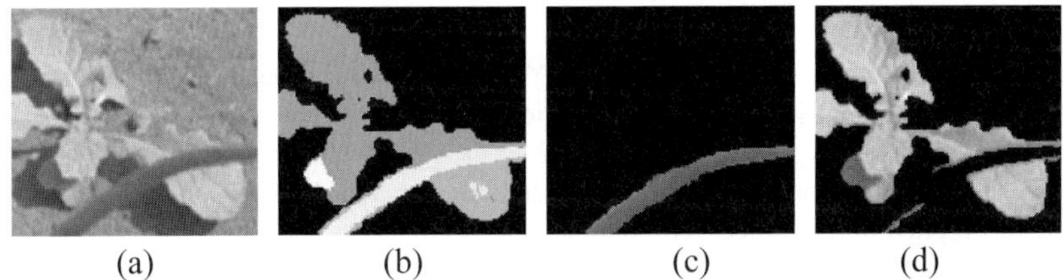

(a) (b) (c) (d)

Fig. 3. Some results in simple case of onion and weed leaves overlapping

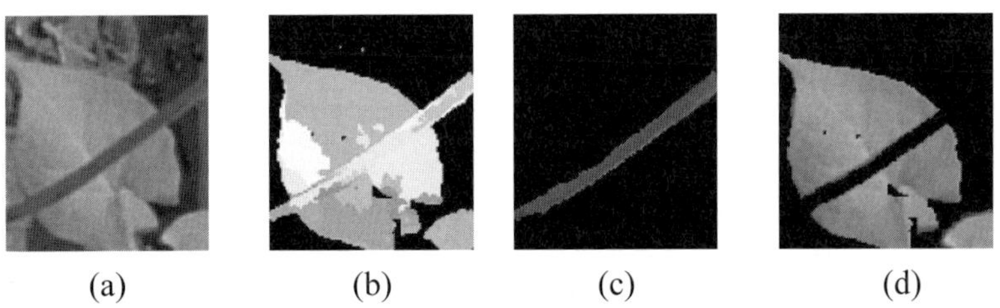

(a) (b) (c) (d)

Fig. 4. Some results in simple case of onion and weed leaves overlapping

Fig.3 and Fig.4 show two examples of the results in detail. In these two figures, (a) is original image. (b) is image after merging these regions. (c) is image of onion. (d) is image of weed. We can find that the onion leaves and weed leaves were entirely distinguished. After processing 40 complex images of onion and weed leaves, we have compared the results obtained by our algorithm with the ones identified by eye, the correct identifiable percents for onion leaves were enter 80%~ 90%. In general, these results are satisfied. Where the identifiable error comes from? Some results of complex images with many onion and weed leaves were analyzed. It was found that although a majority of onion and weed leaves were distinguished, a few onion leaves in original image are not identified as onion leaves by the algorithm. Analysis these onion leaves, we observed that their color is nearly same as the weed leaves. Modifying the parameters of the algorithms, the mistakes were not disappeared. That's to say: if one wants to increase the correct identifiable percents for onion leaves, another characteristics like form factors, multi-spectral, etc. must be added.

8 Conclusion

The algorithm proposed here has been verified on many images, the correct identifiable percents for onion leaves were enter 80%~ 90%, the results are satisfied and the algorithm is valid.

In our analysis, the information on the shape has possibilities to increase the quality of the identification. Another solution using multi-spectral information will make our algorithm robust.

References

1. Wang, N., Zhang, N., Dowell, F.E., Sun, Y., Peterson, D.E.: Design of an Optical Weed Sensor Using Plant Spectral Characteristics. Transactions of the ASAE 44(2) (2001) 409-419
2. Vioix, J.B., et al.: Spatial and Spectral Methods for Weed Detection and Localization. Eurasip Applied Signal Processing Journal 7 (2002) 679-685
3. Gouton, P.: Méthodes et Analyses des Images Multi-composantes: Application aux Images Couleur de Milieux Naturels. Habilitation à Diriger des Recherches, Université de Bourgogne, Dijon, France (2000)
4. Fausett, L.: Fundament of Neural Networks. Prentice Hall, Englewood Cliffs (1994)

Bark Classification Based on Textural Features Using Artificial Neural Networks

Zhi-Kai Huang[1,2], Chun-Hou Zheng[1,2], Ji-Xiang Du[1,2], and Yuan-yuan Wan[1]

[1] Intelligent Computing Lab, Hefei Institute of Intelligent Machines, Chinese Academy of Sciences, P.O. Box 1130, Hefei, Anhui 230031, China
[2] Department of Automation, University of Science and Technology of China, Hefei, China
huangzk@iim.ac.cn

Abstract. In this paper, a new method for bark classification based on textural and fractal dimension features using Artificial Neural Networks is presented. The approach involving the grey level co-occurrence matrices and fractal dimension is used for bark image analysis, which improves the accuracy of bark image classification by combining fractal dimension feature and structural texture features on bark image. Furthermore, we have investigated the relation between Artificial Neural Network (ANN) topologies and bark classification accuracy. Furthermore, the experimental results show the facts that this new approach can automaticly identify the plants categories and the classification accuracy of the new method is better than that of the method using the nearest neighbor classifier.

1 Introduction

Bark is the outer protective coating of the trunk and branches of trees and shrubs and includes all the tissues outside of the vascular cambium. The appearance of a bark depends on the type of cork cells produced by the cork cambium, the relative amount of cambial products, and the amount of secondary conducting tissue (phloem).The varied texture and thickness of bark are often functions of the environment in which the tree grows. The variation in the structure of bark often gives a tree its characteristic appearance, for example, the basswood's bark is brown/gray with deep vertical fissures and flat ridges, a crab apple's bark is reddish/brown, shallow fissures with broad flat topped scaly ridges, etc. A forester can recognize the species of trees by the differences in their bark either externally or by cutting a small slash to examine the inner structure. So, bark is a useful diagnostic feature for plant classification.

Recently, the automatic identification of the plant categories are widely use in agriculture, including plant diseases diagnosis, management of natural resources and plant production. The nearest neighbour classifier has been employed by several investigators to classify bark image, but results are not very satisfactory[1,2].

Application of Artificial Neural Networks (ANN) for bark recognition has not been reported in literatures. This led to high expectation of how neural networks can do in these fields where other approaches have not been successfully used. In this paper, multilayer perception (MLP) neural network with the back-propagation (BP) algorithm is chosen to be the classifier[3].

J. Wang et al. (Eds.): ISNN 2006, LNCS 3972, pp. 355 – 360, 2006.

2 Features for Bark

Architecture of typical pattern reorganization system has two major tasks. The first one is feature extraction, where a set of feature vectors is generated according to represent the content of each image in the database. The feature vectors should be smaller in size than the original image, if we take features directly from the raw image data, the number is too large for most classification tasks. The second task is to design classifiers to classify the image using feature vectors.

There has been many type of methods for texture features analysis. The feature has to represent statistical as well as structural characteristics of the texture using some mathematical measure or rule. The most popular texture feature extraction methods are based on gray level co-occurrence statistics, gray level run length statistics [4,5], probability density functions, fractal dimension, filtering methods like morphological filters, Fourier filters and Gabor filters, random field models, wavelet frames approaches [6,7,8].

2.1 Fractal Dimension Features

The complexity and self-similarity of texture at different scales are described by fractal dimension which was also computed in the this work. The fractal dimension can be used to quantify the texture information. A fractal set has its Hausdorff–Besicovitch dimension which is strictly greater than its topological dimension. Fractal-based texture analysis was introduced in literature [10] by Pentland, where a correlation between texture coarseness and fractal dimension of a texture was demonstrated. Fractal dimension is a defining property. The relationship between the ruler size r and the measured length L can be expressed as $L = cr^{1-D}$, where c is a scaling constant and D is the fractal dimension. Fractal dimension has been shown to correlate well with the function's intuitive roughness.

2.2 Brief Overview of GLCM

Gray-level co-occurrence matrix (GLCM) is traditional statistical method for second order texture analysis, also known as the gray-level spatial dependence matrix. The co-occurence matrix describes the relative frequencies Pij with which two neighboring pixels separated by distance d occur on the image, one with gray level i and the other with gray level j. The co-occurrence matrix is, therefore, a square matrix, that has the size of the largest pixel value in the image. Each element, $P[i, j]$, describes the number of occurrences of the following structure: pixel with the value i at a distance d from pixel of a value j. In the transformation from the image space into the co-occurrence matrix space, the neighboring pixels in one or some of the eight defined directions can be used; for $d = 1$, there are 4 possible directions, where the angles between the two pixels are: 0^0, 45^0, $90^0$and 135^0 , is initially regarded, and its reverse direction (negative direction) can be also counted into account. At last, the neighboring pixels have eight defined directions can be used.

Therefore, general GLCM texture measure is dependent on distance d and directionality, and known measures such as contrast, entropy, energy, dissimilarity, angular second moment (ASM) and homogeneity are expressed as follows:

Homogeneity:

$$Homogenneity = \sum_{i=1}^{Ng}\sum_{j=1}^{Ng}\frac{1}{1+(i-j)^2}P(i,j) \tag{1}$$

Homogeneity is a measure the closeness of the distribution of elements in the GLCM to the GLCM diagonal. This feature takes high values for low-contrast images due to the inverse $(i-j)^2$ dependence.

Contrast:

$$f_2 = \sum_{n=0}^{Ng-1} n^2 (\sum_{i=1}^{Ng}\sum_{j=1}^{Ng} P(i,j)) \tag{2}$$

where i, j subject to $|i-j|=d$.

This feature measures the local variations in the gray-level co-occurrence matrix. It takes high values for images of high contrast.

Correlation:

$$f_3 = \{\sum_{i=1}^{Ng}\sum_{j=1}^{Ng} i\cdot j\cdot P(i,j)) + \mu_x\mu_y\}/\delta_x\delta_y \tag{3}$$

where μ_x 、 δ_x is means and standard deviations.

Correlation measures the joint probability occurrence of the specified pixel pairs.

Energy:

$$Energy = \sqrt{\sum_{i=1}^{Ng}\sum_{j=1}^{Ng} P^2(i,j)} \tag{4}$$

This feature provides the sum of squared elements in the GLCM, and it also known as uniformity or the angular second moment.

Entropy:

$$Entropy = \sum_{i=1}^{Ng}\sum_{j=1}^{Ng} P(i,j)\log[P(i,j)] \tag{5}$$

Entropy is a measure of randomness and takes low values for smooth images.

In equation (1) ~ (5), N_g is dimension of the co-occurrence matrix, as grey value range of the input image. While, in GLCM texture measure, normalization of GLCM matrix, by each value dividing by the sum of element values, is applied, and then $P(i,j)$ is replaced to by the probability value.

3 Image Data and Experimental Results

We have collected more than 300 pictures of bark using digital camera. for developing plant leaves recognition software, where most of the features are extracted from the plant leaves. Concisely speaking, geometrical and morphological features are obtained. In order to improve the recognition accuracy, we tried to add the texture of bark as a supplementary characteristic. These images were recorded at a resolution of 640 x 480 pixels, with a bit depth of 24 bits/pixel. Thus, 256 levels were available for each R, G, and B color plane. The images were converted to jpeg format and grayscale before processing. Some bark images are shown in Fig.1.

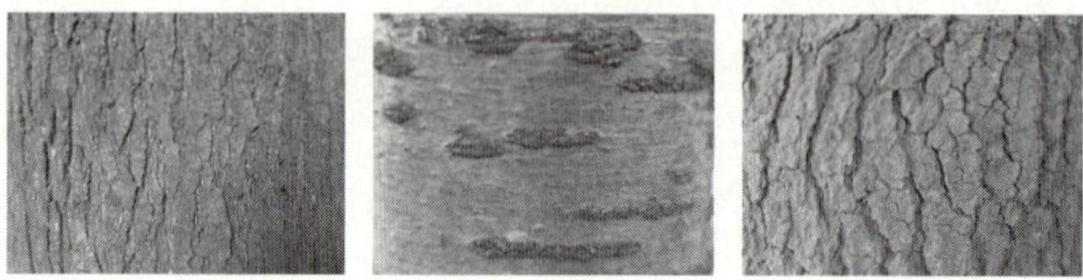

Fig. 1. Three kinds of original bark images

In fact it is so difficult for us to collect the data embodying its substantial features, because the same kind of trees living in different environment may have different bark texture. For example whether it has lichen, bark peeled, the photos we taken are the new ones or the old ones, etc.. In addition,because the trunk of the tree is cylinder and the two sides of the pictures are possibly blurred. So we must first segment on the pictures and get some useful texture features. As such, we selected 360 ones corresponding to twenty-four kinds of bark that have good shooting effects and the similar natural condition such as illumination. In addition, because of the particularity of interests (ROI), we must select a relatively bigger ROI with the size of 350×400 pixels, otherwise it will be very difficult to understand the texture even by our human's eyes.

In this study, first and second order statistical parameters and fractal dimension features are used to classify bark image, including sample mean, standard deviation, homogeneity, contrast, correlation, energy and entropy. We used distance $d = 1,2,3,4,5$ and the angles 0^0, 45^0, $90^0$and 135^0 between two pixels to extract GLCM texture features.

We used a three layer perceptions for experimenting. The number of input and output neurons required in the neural networks are dicided by the dimensionality of the input vectors and the number of categories, respectively, but the number of hidden layers is not related in a simple manner to such obvious properties of the classification problem. We need to seek some intermediate number units in hidden layer that will give a lower test error.

Training of the network is done using standard back propagation (BP) algorithm[9]. This algorithm consists of two passes: a forward pass, when an activity pattern is applied to the neurons of the network and its effect propagates through the network layer by layer, up to the last layer; and a backward pass, when the synaptic weights are adjusted in accordance with an error-correction rule. The image features were calculated using dedicated subroutines written in Matlab 7.0 language for a

512M memory, Pentium 2.4GHz PC computer. Totally twenty-four bark classes are used for identification. These were: maple, dogbane, trumpet creeper, osier, pine, phoenix tree, camphor, elm, etc. Every type of bark has at least 15 images for training, one image for testing. The obtained average recognition rates are shown in Table 1.

In order to compare average recognition with others feature extraction methods and different classifiers such as 1-NN, k-NN, and the corresponding results are shown in Table 2.

Table 1. Average Recognition Rates used Different Features and Different ANN topology

Features Used	ANN Topology	Number of Cases(Training/Test)	Training Epochs	Average Recognition Rates
d=1, D, θ= 0^0, 45^0, 90^0,135^0	20/12	180/24	500	73.91%
	24/12	180/24		78.26%
	28/12	180/24		73.91%
d=2, D,θ= 0^0, 45^0, 90^0,135^0	20/12	180/24	500	82.61%
	24/12	180/24		86.96%
	28/12	180/24		86.96%
d=3, D, θ= 0^0, 45^0, 90^0,135^0	20/12	180/24	500	86.96%
	24/12	180/24		82.61%
	28/12	180/24		86.96%
d=4, D, θ= 0^0, 45^0, 90^0,135^0	20/12	180/24	500	91.67%
	24/12	180/24		91.67%
	28/12	180/24		91.67%
d=4,θ= 0^0, 45^0, 90^0,135^0	20/12	180/24	500	78.26%
	24/12	180/24		82.61%
	28/12	180/24		86.96%

Table 2. Average Recognition Rates used other features and different classifiers

Texture methods	1-NN	k-NN
Auto-correlation method(ACM)	0.72	0.65
Co-occurrence matrices method (COMM)	0.77	0.75
Histogram method(HM)	0.65	0.62

First, the classification accuracy for the whole study area using different numbers of the GLCM texture features is also shown in Table 1. Second, average recognition rates can be enhanced much under the circumstances that the hidden neurons number is twice as the output neurons number. Third, recognition rates is also enhanced when the distance of GLCM is added to the textural features. Fourth, it can also be found that the classification accuracy is improved by combining the GLCM texture features and fractal dimension.. Finally, the textural and fractal dimension features can also be used to identify a bark type, and the best recognition rate can arrived to 91.67%.

4 Conclusions

In this paper, a new method for bark classification based on textural and fractal dimension features using ANN is presented. In this method, the fractal dimension features and grey level co-occurrence matrices features have been extract from bark image for plant classification. Moreover, the different ANN topologies are studied to obtain the best classification accuracy. Finally, the experimental results show that our method is more effective than others. The future study will focus on how to extract more efficient features from bark images, such as those derived from wavelet transforms and color information, to improve the classification accuracy futher.

References

1. Wan, Y. Y., Du J. X., Huang, D. S., Chi, Z.: Bark Texture Feature Extraction Based on Statistical Texture Analysis. International Symposium on Intelligent Multimedia, Video and Speech Processing, Hong Kong, (2004) 482-185
2. Chi, Z., Li, H. Q., and Wang, C.: Plant Species Recognition Based on Bark Patterns Using Novel Gabor Filter Banks. IEEE Int. Conf. Neural Networks & Signal Processing, Nanjing China (2003) 1035-1038
3. Huang, D.S.: Systematic Theory of Neural Networks for Pattern Recognition, Publishing House of Electronic Industry of China, Beijing (1996)
4. Loh, H. H., Leu, J. G., Luo, R. C.: The Analysis of Natural Textures Using Run Length Features. IEEE Trans. Industrial Electronics 35(2) (1988) 323 - 328
5. David, Clausi, A., Jernigan, M. Ed.: A Fast Method to Determine Co-Occurrence Texture Features, IEEE Trans. Geoscience and Remote Sensing 36(1) (1998) 298 - 300
6. Wojciech, Borkowski, Can.: Fractal Dimension Based Features Are Useful Descriptors of Leaf Complexity and Shape. J. For. Res. /Rev. can. rech. for. 29(9) (1999) 1301-1310
7. Randen, T., Husky, J.H.: Filtering for Texture Classification: A Comparative Study. IEEE Trans. Pattern Anal. Mach. Intel. 21 (4) (1999) 291 - 310
8. David, Clausi, A., Huang, D.: Design-Based Texture Feature Fusion Using Gabor Filters and Co-Occurrence Probabilities. IEEE Trans. Image Processing 14(7) (2005) 925-936
9. Huang, D.S., Ma, S.D.: Linear and Nonlinear Feedforward Neural Network Classifiers: A Comprehensive Understanding. Journal of Intelligent Systems 9(1) (1999) 1-38
10. Kube, P., Pentland, A. P.: On the Imaging of Fractal Surfaces. IEEE Trans. Pattern Anal. Mach. Intel. 10(5) (1988) 704–707.

Automated Spectral Classification of QSOs and Galaxies by Radial Basis Function Network with Dynamic Decay Adjustment

Mei-fang Zhao[1], Jin-fu Yang[1], Yue Wu[2], Fu-chao Wu[1], and Ali Luo[2]

[1] National Laboratory of Pattern Recognition, Institute of Automation, Chinese Academy of Sciences, Beijing 100080, China
mfzhao@nlpr.ia.ac.cn, yangjf@nlpr.ia.ac.cn, fcwu@nlpr.ia.ac.cn
[2] National Astronomical Observatories, Chinese Academy of Sciences, Beijing 100012, China
wuyue@lamost.org, lal@lamost.bao.ac.cn

Abstract. This paper presents a fast neural network method of radial basis function with dynamic decay adjustment (RBFN-DDA) to classify Quasi-Stellar Objects (QSOs) and galaxies automatically. The classification process is mainly comprised of three parts: (1) the dimensions of the normalized input spectra is reduced by the Principal Component Analysis (PCA); (2) the network is built from scratch: the number of required hidden units is determined during training and the individual radii of the Gaussians are adjusted dynamically until corresponding criterions are satisfied; (3) The trained network is used for the classification of the real spectra of QSOs and galaxies. The method of RBFN-DDA having constructive and fast training process solves the difficulty of selecting appropriate number of neurons before training in many methods of neural networks and achieves lower error rates of spectral classification. Besides, due to its efficiency, the proposed method would be particularly useful for the fast and automatic processing of voluminous spectra to be produced from the large-scale sky survey project.

1 Introduction

Celestial bodies in the universe can be divided by Milky Way galaxy into the objects within it and extragalactic objects. Objects within Milky Way galaxy are mainly composed of stars and ISM while components of extragalactic objects are more complicated. From morphology, QSOs are point source while galaxies belong to extended source. But both of them become point source if the observed objects are very far away. Then the spectra are used to distinguish them. Galaxies can be separated into normal and active galaxies according as active radiation regions exist. Active galaxies are mainly divided into active galactic nuclei (AGN) and starburst galaxies based on the existence of the activities of AGN or other violent activities. Since both AGNs and starburst galaxies have strong emission lines, especially Seyfert I galaxies in AGNs have rather broad emission lines and its spectra are similar to that of QSOs, the classifications of them are quite difficult. Furthermore, if the effects of redshifts are to be considered, the classifications would be even more uncertain. For projects of spectra survey, such as the Sloan Digital Sky Survey (SDSS) of the United States, the

J. Wang et al. (Eds.): ISNN 2006, LNCS 3972, pp. 361 – 368, 2006.

Two-Degree Field(2dF) Galaxy Redshift Survey of UK-Australia, the Large Sky Area Multi-object Fiber Spectroscopic Telescope(LAMOST) of China, the spectral classifications of different objects are groundwork for astronomers to carry out astrophysical researches.

The automated spectral classifications for extragalactic objects have been put emphasis upon internationally, i.e. [1], [2] and [3] have adopted principal component analysis(PCA) to classify the spectra of galaxies with known redshifts. As for the spectra with unknown redshifts, only [4] have made an attempt for the classification of galaxies with improved PCAZ method. In this paper, the method of Radial Basis Function Network with Dynamic Decay Adjustment (RBFN-DDA) is developed to classify the spectra of QSOs and galaxies with unknown redshifts and can become one part of the whole scheme of spectral classifications.

In order to check out the reliability of this method, both simulated LAMOST spectra and real spectra of SDSS are considered in the experiments of classification. And in comparisons to the classification results of the Nearest Neighbor algorithm、the Covering algorithm[5]、the Radial Basis Function Network (RBFN) and the Adaptive neuron-adding Radial Basis Function Network (ARBFN)[6], the RBFN-DDA has shown many superiorities. The plan of this paper is as follows. In § 2 the structure and algorithm of RBFN-DDA is introduced. Next, related parameters of this algorithm are analyzed and confirmed in experiments with simulated spectra. In § 4 the real observed spectra of SDSS are classified and related results of several different methods mentioned above are compared and discussed. Finally, in § 5 the important results of this work are summarized.

2 The Algorithm of RBFN-DDA

The Radial Basis Function Network was introduced by Moody and Darken[7] and has been used widely. However many training algorithms require a fixed structure and predetermined number of neurons in hidden layers before training. The RCE(Restricted Coulomb Energy) algorithm [8]and its probabilistic extension, the P-RCE algorithm take advantage of a growing structure in which the neurons in hidden layer are only introduced when necessary. It is faster for the training processes of the P-RCE algorithm to reach stability than that of the RBF network using gradient descent. The algorithm of Dynamic Decay Adjustment (DDA) inherits the constructive parts of P-RCE to dynamically construct a network that possesses appropriate numbers of RBF neurons, and the parameters of the network are calculated according to the information of neighbor points of the input data [9].

The model of RBFN-DDA and related algorithm are described as follows.

2.1 The Structure of RBFN-DDA

The RBFN-DDA is commonly comprised of three layers: input layer, hidden layer and output layer, as shown in Fig. 1. Input layer corresponds to input vector spaces, hidden layer represents the units of RBF and the activation function from input layer to hidden layer is similar to that of traditional RBF network. Generally the activation function is denoted as a Gaussian function. Each unit of output layer represents a potential class,

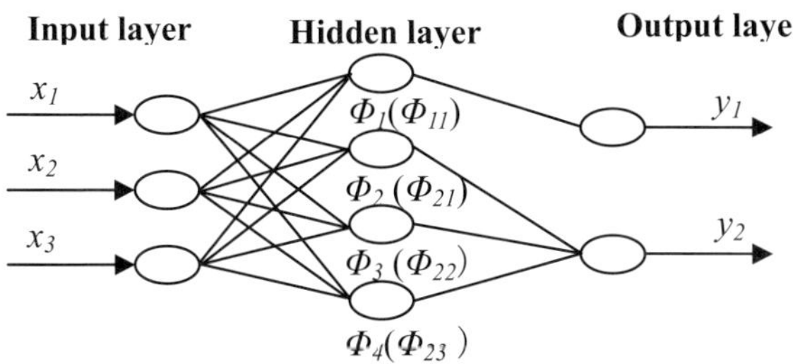

Fig. 1. The structure of RBFN-DDA

and every neuron is only connected with one output unit that just calculates the weighted sum of Gaussian functions of connected neurons.

2.2 The Algorithm of DDA

The training algorithm of P-RCE distributes an independent influencing area with same radius for each RBF unit. However, it is difficult to judge the class of new sample when it lies in the intersection of influencing areas of two neurons belonging to different classes. Therefore, the algorithm of DDA introduces two different thresholds[9]: θ^+(positive threshold) and θ^-(negative threshold) to define intersection of influencing area. As illustrated in Fig. 2, these two thresholds lead to an area of conflict where no neuron belonging to other classes is allowed to lie. And each training sample must be covered by at least one influencing area of a neuron of the same class. Then in Fig. 3, the new input sample has a higher activation value than θ^+ for the neuron of class B, but has a lower activation value than θ^- for the neuron of class A, so it is classified into class B. That is, θ^+ is used to control the lowest correct recognition rates of training samples and θ^- is used to avoid misclassification. The level of confidence of the algorithm of DDA is higher than that of P-RCE.

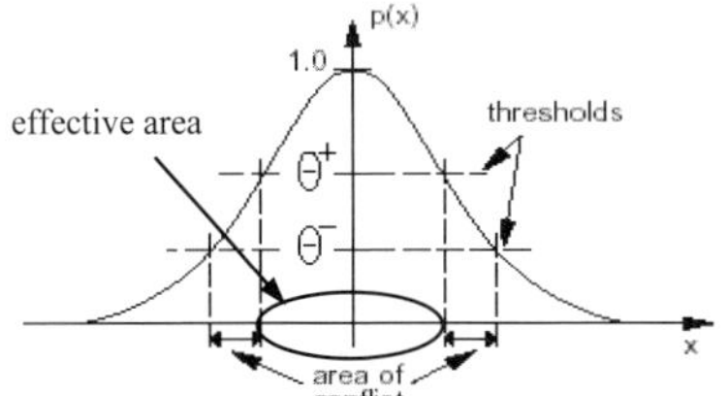

Fig. 2. One RBF unit in the DDA-algorithm

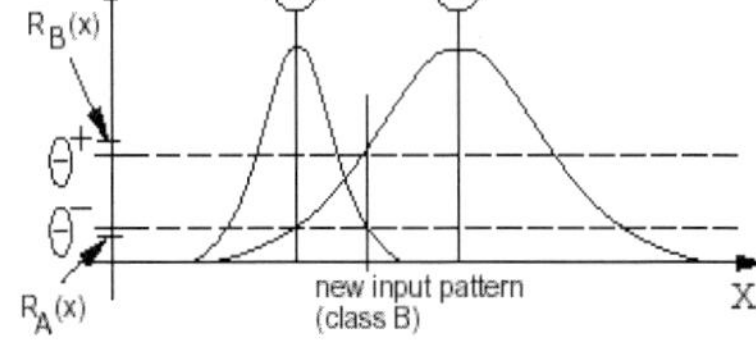

Fig. 3. Here the level of confidence is higher for the correct classification of the new pattern

An example of the algorithm of DDA is illustrated in Fig. 4: (1) given a sample of class A, it becomes the first new neuron of class A automatically; (2) likewise, the first training sample of class B is the first neuron of class B, and the radius of the neuron of class A is shrank; (3) the second sample of class B is recognized correctly by the first neuron of class B, so the radius of the first neuron of class A is shrank again; (4) the second sample of class A is incorrectly recognized, then it becomes the second neuron of class A and the radii of all existing neurons of class B will be adjusted based on it.

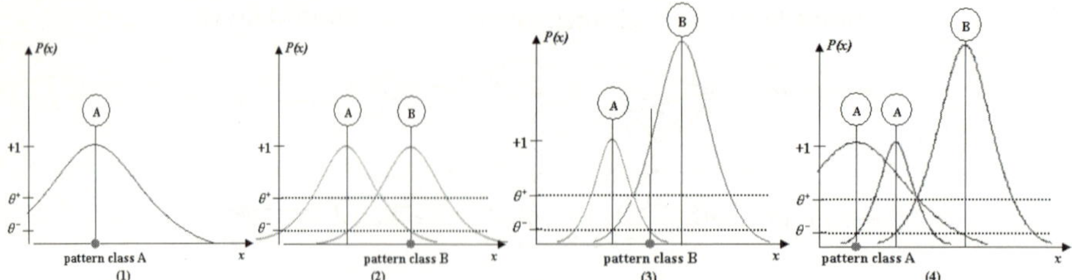

Fig. 4. An example of the first few steps of the DDA-algorithm

After training is finished, two conditions are true for all input-output pairs $(\vec{x}, k)$:

(1) at least one neuron of the correct class k has an activation value greater than or equal to θ^+:

$$\exists i : R_i^k(\vec{x}) \geq \theta^+, \tag{1}$$

(2) all neurons of conflicting classes have activations less than or equal to θ^-:

$$\forall l \neq k, 1 \leq j \leq m_l : R_j^l(\vec{x}) \leq \theta^-. \tag{2}$$

The detailed algorithm operates as follows:

- before training an epoch, all weights must be set to zero to avoid accumulation of duplicate information of the training samples;
- all training samples are presented to the network, if the new sample is classified correctly, the weight of the nearest neuron in the same class will be increased;
- otherwise a new neuron is introduced and its center and weight is confirmed;
- its initial radius is set as large as possible without misclassifying an already existing neuron of different class;
- the last step all the radii of neurons of different class are shrank so that they will not misclassify this new neuron.

This algorithm is run circularly for all training samples until no big change come into being basically and formula (1) and (2) are satisfied.

3 Confirmation of Parameters in RBFN-DDA

To observe the influences for classification under different thresholds, simulated spectra of galaxies are tested by the RBFN-DDA.

The simulated spectra are derived from four templates of normal galaxies and seven ones of starburst galaxies by Kinney & Calzetti[10] of which the redshifts all range from 0 to 1.2, and Seyfert I template by Rodriguez[11], Seyfert II template and LINER template by Calzetti and QSO template by Francis et al.[12] of which the ranges of redshifts are respectively 0.05~2.35, 0.01~2.07, 0.05~2.07 and 0.24~3.75. Generally the wave band of spectra is chosen as 3800~7420 angstroms with an interval of 5 angstroms and random Gaussian white-noise of which the Signal-to-Noise is ten are added. The step size of simulated redshifts is 0.01 for training spectra other than 0.001 for testing spectra in which the same spectra as training spectra will be eliminated. And starburst galaxies, normal galaxies, Seyfert I 、II and LINER galaxies are all set as class 1(galaxies) while QSOs are set as class 2. Finally there are 1,958 spectra of class 1 and 351 spectra of class 2 in the training samples, and 16,302 spectra of class 1 and 3159 spectra of class 2 for testing. To avoiding the differences in magnitude, the fluxes of all the real spectra are normalized. The principal component analysis (PCA) [13] is adopted here to reduce high dimensions of the spectra, the primary 19 components whose contribution rate of variance reaches 95% are selected.

Because of $0<\theta^-<\theta^+<1$, Fig. 5 a) demonstrates the correct classification rates and Fig. 5 b) shows the corresponding training time when θ^+ is valued from 0.3~0.9 and θ^- is valued from 0.001~θ^+. When θ^- is chosen from 0.001 to 0.4, all the correct rates are above 90% and that θ^+ still has many choices from θ^- to 1. However, when θ^+ and θ^- being farther, the training time increases with more neurons being added. Though the two thresholds need manual adjustment, it is not critical to choose their values and appropriate selection for the interval between them will result in fast training speed and high correct rates of classification. During the following experiments, $\theta^+=0.4$ and $\theta^-=0.2$ are chosen.

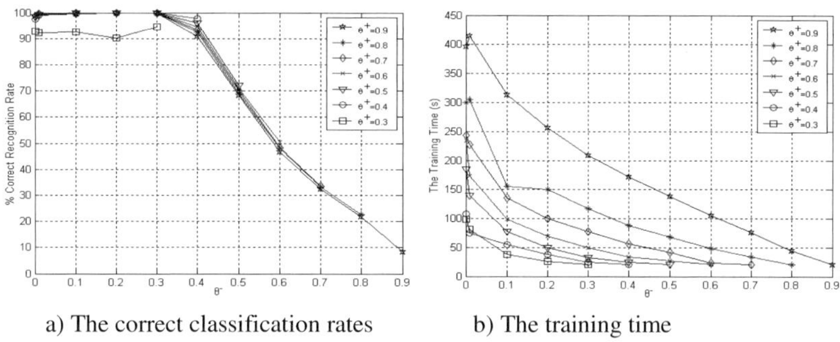

a) The correct classification rates b) The training time

Fig. 5. The correct classification rates and training time under different thresholds

4 Classification Results and Analysis

In the experiments of classification for the real spectra of QSOs and galaxies, there are 7,492 galaxies and 15,745 QSOs from SDSS Second Date Release. Only 1,038 galaxies and 1,000 QSOs are selected as training samples and the rest for testing.

In Table 1, the testing results of the Nearest Neighbor algorithm、the Covering algorithm、the Radial Basis Function Network (RBFN) , the Adaptive neuron-adding Radial Basis Function Network (ARBFN) and the RBFN-DDA are shown.

The Nearest Neighbor algorithm has the longest testing time, so its efficiency is the lowest. The covering algorithm adopts the biggest distance inside a class or the smallest distance between classes to find small supporting sets which substitute the original big training sets, so the testing speed is very fast that is the training time is just spent on

Table 1. The rejection and error rates of real spectral classification of Galaxies and QSOs

Method	Nearest Neighbor	Covering algorithm	RBFN	ARBFN	RBFN-DDA
Training samples	2038	2038	2038	2038	2038
Testing samples	21199	21199	21199	21199	21199
Training time/(s)	——	2003	371	15410	109
Testing time/(s)	10108	348	3668	1621	1594
Rejected number	——	——	——	——	437
Rejection rate/(%)	——	——	——	——	2.06
Misclassified number	931	2985	1165	701	586
Error rate/(%)	4.39	14.08	5.50	3.31	2.76

finding supporting sets. But the error rate is higher than that of the Nearest Neighbor algorithm, that is to say the supporting sets cannot yet replace the original sets. The RBFN shows fast training speed but slower testing speed. Commonly it is difficult to choose desirable numbers of neurons at the same time avoiding overtraining. The error rate of RBFN is also higher. The ARBFN introduces the mechanism of adding neurons adaptively during training based on traditional RBFN, resulting in avoiding overtraining to some extent, and the classification result is better. Nevertheless by reason of adding neurons and verifying the termination conditions continually, the training time is quite longer.

The RBFN-DDA demonstrates the shortest training and testing time. Furthermore only the algorithm of DDA has the rejection decision, only when all the neurons cannot identify a testing sample, the rejection decision is made, so this algorithm shows the smallest error classification rate. Among the 437 spectra rejected by the RBFN-DDA, there are 45 galaxies and 392 QSOs. And there are 93 galaxies and 493 QSOs misclassified. Fig. 6 and 7 respectively show the distribution of redshifts of these spectra, and Fig. 8 demonstrates the equivalent widths of Hα emission line of rejected and misclassified galaxies.

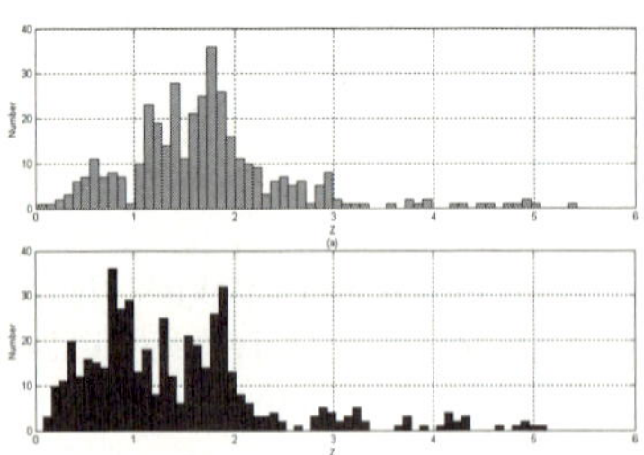

Fig. 6. Distribution of redshifts (a) Rejected QSOs (b) Misclassified QSOs

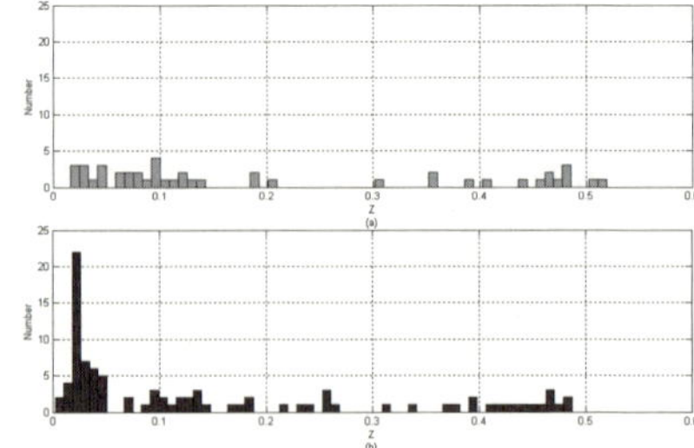

Fig. 7. Distribution of redshifts (a) Rejected galaxies (b) Misclassified galaxies

The spectra of QSOs have large fluxes in ultraviolet band and broad Hα emission line. As shown in Fig. 6, the spectra with redshifts less than 0.5 only account for a small proportion, and the great mass of rejected or misclassified spectra have large redshifts so that the broad feature goes far beyond the available wave band (3800~7420 angstroms). Furthermore the majority have redshifts close to 1 or 2, then those emission lines with large fluxes in ultraviolet band have not yet come into the available wave band. Under this condition, many QSO spectra with few features of emission lines would be misclassified into galaxies (class 1). And that the method of RBFN-DDA effectively refused a great deal of spectra with redshifts close to 1 or 2 so as to reduce a large number of misclassified spectra comparatively.

Fig. 7 shows that the rejected and misclassified galaxies (class 1) all cover the redshifts from 0 to 0.5, especially the majority of misclassified galaxies have small redshifts. Then the rejected and misclassified spectra with both small redshifts and the equivalent widths of broad Hα emission line within the available wave band are provided in Fig. 8. Most equivalent widths of Hα being positive, namely corresponding to emission lines, may be the possible reason of misclassification. Moreover the galaxies (class 1) include normal galaxies, starburst galaxies and active galactic nuclei, of them the spectra may have few features or big redshifts to be misclassified into QSOs. Still the RBFN-DDA has made rejection decisions resulting in fewer wrong recognized spectra.

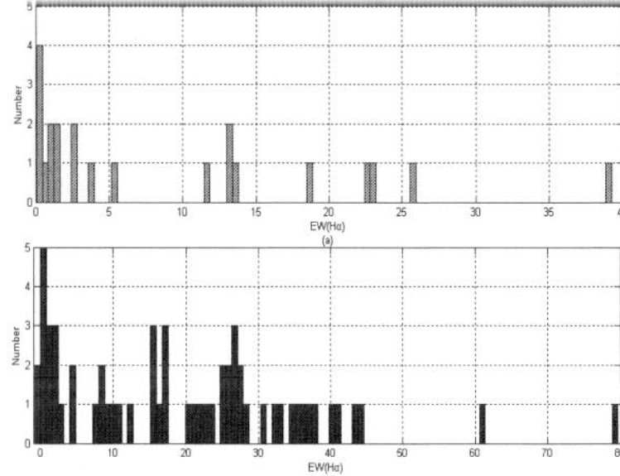

Fig. 8. The equivalent width of Hα emission line of galaxies(a) Rejected galaxies (b) Misclassified galaxies

Since the RBFN-DDA performs better than other methods with lowest error recognition rate and relatively small proportion of rejected spectra that can be identified further by other techniques, the conference level of this classification algorithm is higher comparatively.

5 Conclusion

A Radial Basis Function Neural Networks trained by Dynamic Decay Adjustment has been proposed. The main properties of it are: 1) constructive training: the network is built from scratch, the number of neurons is determined and the individual radii of the Gaussians are adjusted dynamically; 2) fast training: few epochs are needed to complete training; 3) guaranteed convergence: the algorithm can be proven to terminate when a finite number of training samples is used; 4) two uncritical manual parameters: only two thresholds are required to adjusted manually and the values of them are not critical; 5) distinct classification zones: after training terminates, correct classifications are above a threshold θ^+, wrong classification are below another threshold θ^-, and samples only residing in area of conflict have low classification probabilities, providing an additional "don't know" answer.

The DDA algorithm offers an easy to use methodology to train RBF network overcoming the difficulty of RBFN in selecting appropriate number of neurons, spending less time to achieve better classification performance than the Nearest Neighbor algorithm, the Covering algorithm, the RBFN and ARBFN. The RBFN-DDA classification method with higher confidence level will be of great benefit to the automated classifications and processes for voluminous spectra provided by the large-scale sky survey project.

Acknowledgements. This research was supported by National 863 Hi-Tech R & D Plan (project 863) under grant no. 2003AA133060, and by the National Natural Science Foundation, China, under grant no. 60202013.

References

1. Connolly, A.J., Szalay, A.S., Bershady, M.A. et al.: Spectral Classification of Galaxies: An Orthogonal Approach. Astronomical Journal 110(3) (1995) 1071-1082
2. Gaspar, G., Valerie, de L.: The ESO-Sculptor Survey: Spectral Classification of Galaxies with Z < 0.5. Astronomy and Astrophysics 332(4) (1998) 459-478
3. Zaritsky, D., Zabludoff, A.I., Jeffrey, A.W.: Spectral Classification of Galaxies along the Hubble Sequence. Astronomical Journal 110(5) (1995) 1602-1613
4. Zhu, G.H. et al.: Automated Measurement of Redshifts of Galaxy Spectra of LAMOST. Spectroscopy and Spectral Analysis 25(6) (2005) 1002-1005
5. Zhang, L., Zhang, B.: A Geometrical Representation of McCulloch-Pitts Neural Model and Its Applications. In: IEEE Transactions on Neural Networks 10(4) (1999) 925-929
6. Liu, H., Chandrasekar, V., Xu, G.: An Adaptive Neural Network Scheme for Radar Rainfall Estimation from WSR-88D Observations. J. Appl. Meteor. 40(11) (2001) 2038-2050
7. Moody, J. and Darken, C.J.: Fast Learning in Networks of Locally-Tuned Processing Units. Neural Computation 1(2) (1989) 281-294
8. Reilly, D.L., Cooper, L.N., Elbaum, C.: A Neural Model for Category Learning. Biol. Cybernet. 45(1) (1982) 35-41
9. Berthold, M.R., Diamond, J.: Boosting the Performance of RBF Networks with Dynamic Decay Adjustment. In: Proceedings of the Advances in Neural Information Processing Systems (NIPS), Vol. 7. MIT Press, Cambridge MA (1995) 521-528
10. Calzetti, D., Kinney, A.L., Storchi-Bergmann, T.: Dust Extinction of The Stellar Continua in Starburst Galaxies: The Ultraviolet and Optical Extinction Law. Astrophysical Journal 429(8) (1994) 582-601
11. Rodriguez-Pascual, P.M.: IUE Monitoring of the Bright Seyfert 1 Galaxy Fairall 9. In: Peterson, B.M., Cheng, F.-Z. and Wilson, A.S. (eds.): Emission Lines in Active Galaxies: New Methods and Techniques. ASP Conference Series, Vol. 113. (1997) 167-168
12. Francis, P.J. et al.: A High Signal-to-Noise Ratio Composite Quasar Spectrum. Astrophysical Journal 373(2) (1991) 465-470
13. Qin, D.M. et al.: A Fast Classification of Stellar Spectra Based on Principal Component Analysis. Spectroscopy and Spectral Analysis 23(1) (2003) 182-186

Feed-Forward Neural Network Using SARPROP Algorithm and Its Application in Radar Target Recognition

Zun-Hua Guo and Shao-Hong Li

Group 203, School of Electronics and Information Engineering,
BeiHang University, 100083 Beijing, China
grit2006@hotmail.com

Abstract. The feed-forward neural network using simulated annealing resilient propagation (SARPROP) algorithm was applied to the research community of radar target recognition in this paper. The high resolution radar range profiles were selected as the feature vectors for data representation, and the product spectrum based features were introduced to improve classification performance. Simulations are presented to identify the four different aircrafts. The results show that the SARPROP algorithm combined with product spectrum based features is effective and robust for the application of radar target recognition.

1 Introduction

Neural networks are a technique for pattern recognition and function approximation based originally on ideas from biology and the study of neurons. Neural networks are a very active research area and have been used to solve all sorts of problems including financial forecasting, internetwork routing, automatic target recognition, speech, vision, control systems and so on. Neural networks have achieved very significant success in various fields of application and proved to be effective and versatile [1-2].

Neural networks have been used in radar target recognition type problems before [3], and showed good classification performance. For example, the multi-layered feed-forward neural network, radial basis functions neural network, self-organized mapping neural network and fuzzy neural network have been used to classify the aircrafts, ships and the ground vehicles. This paper attempts to apply a multi-layered feed-forward neural network using the simulated annealing resilient propagation (SARPROP) algorithm [4-5] as the classifier in radar target recognition.

2 The Multi-layered Feed-Forward Network with SARPROP Algorithm

The feed-forward neural networks consist of units, corresponding to neurons. The values of each unit are fed forward via the weighted connections to other

J. Wang et al. (Eds.): ISNN 2006, LNCS 3972, pp. 369–374, 2006.

units. The most common transfer functions are linear and sigmoid functions. The backpropagation (BP) algorithm has proved successful at solving many complex problems. Standard backpropagation is a gradient descent algorithm, which leads to slow convergence, so a number of variations on the backpropagation algorithm that are based on other standard optimization techniques, have been proposed to improve the convergence properties, such as conjugate gradient methods, Levenberg-Marquardt algorithm and resilient propagation (RPROP) algorithm [1][6].

The SARPROP [4-5] algorithm is introduced for classifying the aircrafts in this paper. This algorithm applies the simulated annealing and weight decay techniques to the neural network training process, which can help the network escape from the local minima. This algorithm also shows good performance in the application of the pattern classification.

The sigmoid function is selected as the transfer function of the hidden and the output layer neurons:

$$f(x) = \frac{1}{1+\exp(-x)}. \tag{1}$$

The training rule is as the following:

(1) initialize the parameters: the weights of each neuron w_{ij},initial update value Δ_0, the upper limit of the update values Δ_{max}, the lower limit of the update values Δ_{min} , the increase factor η^+ , the decrease factor η^- and the temperature T;

(2) feed the training patterns set to the network, compute the input value and output value of each unit and the overall error function E;

(3) update the weights of each unit based on the following rules:

i) compute the SARPROP error gradient:

$$\partial E/\partial w_{ij}^{SARPROP} = \partial E/\partial w_{ij} - 0.01 * w_{ij}/(1 + w_{ij}^2) * SA. \tag{2}$$

where $SA = 2^{-T*epoch}$, *epoch* denotes the current training time,

ii) if $\partial E/\partial w_{ij}(t-1) * \partial E/\partial w_{ij}(t) > 0$, then

$$\begin{aligned} \Delta_{ij}(t) &= \min(\Delta_{ij}(t-1) * \eta^+, \Delta_{max}) \\ \Delta w_{ij}(t) &= -sign(\partial E/\partial w_{ij}(t)) * \Delta_{ij}(t) \\ w_{ij}(t+1) &= w_{ij}(t) + \Delta w_{ij}(t). \end{aligned} \tag{3}$$

iii) if $\partial E/\partial w_{ij}(t-1) * \partial E/\partial w_{ij}(t) < 0$, then

$$\begin{aligned} &\text{if } \Delta_{ij}(t-1) < 0.4 * SA^2, \text{then} \\ &\Delta_{ij}(t) = \Delta_{ij}(t-1) * \eta^- + 0.8 * r * SA^2 \\ &\text{else } \Delta_{ij}(t) = \Delta_{ij}(t-1) * \eta^- \\ &\Delta_{ij}(t) = \max(\Delta_{ij}(t), \Delta_{min}) \\ &\partial E/\partial w_{ij}(t-1) = 0. \end{aligned} \tag{4}$$

iv) if $\partial E/\partial w_{ij}(t-1) * \partial E/\partial w_{ij}(t) = 0$, then

$$\Delta w_{ij}(t) = -sign(\partial E/\partial w_{ij}(t)) * \Delta_{ij}(t)$$
$$w_{ij}(t+1) = w_{ij}(t) + \Delta w_{ij}(t). \tag{5}$$

where the $sign(\cdot)$ operator returns +1, if the argument is positive, returns -1, if the argument is negative and 0 otherwise. $\min(\cdot), \max(\cdot)$operator returns the minimum value and maximum value respectively, r is a random number in (0,1). The temperature T is problem dependent, in general the more complex the problem, the lower value is required;

(4) go to step (2), continue the training process, until the maximum number of training epochs is reached or the error goal is met;

(5) feed the test patterns set to the trained network and get the recognition result.

3 Range Profiles and Its Product Spectrum

When a target is illuminated by a high resolution radar, its electromagnetic features can be characterized with a backscatter-point model. A high resolution radar range profile can be viewed as a one-dimensional image of a target, where the parts of the target reflect the radar radiation, that is, the scatterers, are projected onto the line of sight. The range profiles provide potentially discriminative information on the overall size, shape and structure of the target. And it is much easier to obtain believable range profiles than to obtain focused microwave image [7-8]. So many efforts on the radar target recognition by range profiles have been published in recent years. It has been shown that the range profiles are promising candidate signatures for automatic target recognition in the modern war.

Feature extraction is a very important issue and a necessary step in the use of range profiles for classification purposes. Fourier transform, high-order spectral analysis and joint time-frequency analysis methods such as wavelet transform and Gabor transform are popular methods to obtain effective range profiles based features. The product spectrum based features which are originally used in speech signal processing community are introduced to improve the recognition rate in this paper.

Suppose that the range profiles sequence is $x(n), n = 0, 1, \cdots, N-1$, then its Fourier transform can be written as:

$$X(\omega) = \sum_{n=0}^{N-1} x(n) \exp(-j\omega n). \tag{6}$$

In general, $X(\omega)$ is complex and can be expressed in terms of its real and imaginary parts as:

$$X(\omega) = X_R(\omega) + jX_I(\omega). \tag{7}$$

or in terms of magnitude and phase as:

$$X(\omega) = |X(\omega)| \exp(j\theta(\omega)). \tag{8}$$

Sometimes it will be convenient to refer to the group delay rather than the phase [9]. The group delay is defined as the negative of the first derivative, with respect to ω, of the phase:

$$\tau_p(\omega) = -\frac{d\theta(\omega)}{d\omega}. \tag{9}$$

From the above expressions, we obtain:

$$\begin{aligned}\tau_p(\omega) &= -Im\frac{d\log(X(\omega))}{d\omega} \\ &= -\frac{X_R(\omega)X_I'(\omega) - X_I(\omega)X_R'(\omega)}{|X(\omega)|^2}.\end{aligned} \tag{10}$$

where the prime denotes differentiation with respect to ω.

Assuming that $Y(\omega)$ is the Fourier transform of the sequence $nx(n)$, then

$$Y(\omega) = \sum_{n=0}^{N-1} nx(n)\exp(-j\omega n) = jX'(\omega). \tag{11}$$

So the group delay can be computed by:

$$\tau_p(\omega) = \frac{X_R(\omega)Y_R(\omega) + X_I(\omega)Y_I(\omega)}{|X(\omega)|^2}. \tag{12}$$

The product spectrum $Q(\omega)$ is defined as the product of the power spectrum and the group delay [10]:

$$\begin{aligned}Q(\omega) &= |X(\omega)|^2\tau_p(\omega) \\ &= X_R(\omega)Y_R(\omega) + X_I(\omega)Y_I(\omega).\end{aligned} \tag{13}$$

Since some information about the target shape is more likely contained in the phase spectrum, and the product spectrum based features combine the information from the magnitude spectrum and phase spectrum, it will be helpful for radar target recognition.

4 Recognition Results

We use the two-dimension backscatters distribution data of four different scaled aircraft models provided by the laboratory for electromagnetism, Beijing University of Aeronautics and Astronautics to obtain the range profiles by step-frequency technique. The four aircrafts are: the stealth bomber B2 (scale 1:35), the stealth fighter plane F117 (scale 1:13), the supersonic fighter plane J6 (scale 1:8) and the fighter plane YF22 (scale 1:12).

The aspect angle of the range profiles change from 0° to 180° (0° is defined as the head direction) with an average interval of 0.6° , so the data set consists of 1200 patterns for four aircrafts. The data set for each aircraft model is divided into 6 subsets in the proportion of 6:1, that is, the first subset consists of the

1st, 7th, 13th range profile etc., and the rest can be deduced by analogy, one subset for training and the others for testing.

The structure of the feed-forward network is $256 \times 22 \times 2$, the corresponding parameters are set as follows: all weights are initialized to random values in the range(0,1), $\Delta_0 = 0.07, \Delta_{max} = 50, \Delta_{min} = 1.0e-6, \eta^+ = 1.2, \eta^- = 0.5, T = 0.013$. The clean patterns set is used to train the network and the patterns set with different level noise is used to have test and get the recognition results.

The recognition results are shown in Table 1. For comparison purposes, we present the recognition rates given by the range profiles and the product spectrum based features. We further examined the recognition rates in different Signal-to-Noise Ratio (SNR) conditions. The results show that the SARPROP algorithm combined with product spectrum based features is reliable and robust even in a noisy environment.

Table 1. Recognition rate (%)

Feature set	SNR/dB	B2	F117	J6	YF22
range profiles	clean	88.8	86.8	97.6	93.6
	20	81.6	71.2	95.6	88
	15	64.4	73.2	91.6	84
	10	65.6	52.8	93.2	87.2
	5	44.7	24.7	54	87.6
product spectrum	clean	96	94.8	96.8	97.2
	20	95.6	93.6	96	97.2
	15	95.4	93.6	95.6	96.8
	10	95.2	84.3	93.2	87.3
	5	95.6	79.6	89.4	55.3

5 Conclusions

This paper introduces the multi-layered feed-forward network using the SARPROP algorithm to the radar target recognition community. Since the product spectrum combines the information contained in the magnitude spectrum and the phase spectrum of the range profiles, it carries more information about the shape of the target and is helpful for target recognition. The simulation results show that the SARPROP algorithm combined with the product spectrum based features can yield good performance in various noise conditions. On the other hand, this technique is computational efficiency and suitable for real-time application.

Acknowledgement

The authors would like to thank the anonymous reviewers for their helpful comments.

References

1. Pingfan, Y., Changshui, Z.: Artificial Neural Network and Simulated Evolution Algorithm. Tsinghua University Press Beijing (2001)
2. Xingjun, Y., Junli, Z.: Artificial Neural Network and Blind Signal Processing. Tsinghua University Press Beijing (2003)
3. Michael, W.R.: Survey of Neural Network Technology for Automatic Target Recognition. IEEE Transactions on Neural Networks 1(1) (1990) 28–43
4. Nicholas, K. T., Tamas, D.G.: Simulated Annealing and Weight Decay in Adaptive Learning: The SARPROP Algorithm. IEEE Transactions on Neural Networks 9(4) (1998) 662–668
5. Nicholas, K. T., Tamas, D.G.: The SARPROP AlgorithmA Simulated Annealing Enhancement to Resilient Back Progagation. In: Proc. Int. Panel Conf. Soft and Intell. Comput. Budapest (1996) 289–293
6. Riedmiller, M., Braun, H.: A Direct Adaptive Method for Faster Backpropagation Learning: The RPROP Algorithm. In: Proceedings of the IEEE International Conference on Neural Networks. San Francisco (1993) 586–591
7. Donald, R.W.: High Resolution Radar. 2nd edn. Artech House Boston (1995)
8. Hsuehjyh, L., Shenghui, Y.: Using Range Profiles as Feature Vectors to Identify Aerospace Objects. IEEE Transactions on Antennas and Propagation 41(3) (1993) 261–268
9. Alan, V. O., Ronald, W.S.: Digital Signal Processing. Englewood Cliffs, Prentice-Hall New Jersey (1975)
10. Donglai, Z., Kuldip, K.P.: Product of Power Spectrum and Group Delay Function for Speech Recognition. In: Proceedings of the IEEE International Conference on Acoustics, Speech, and Signal Processing. Montreal (2004) 125–128

Camera Calibration and 3D Reconstruction Using RBF Network in Stereovision System

Hai-feng Hu

Department of Electronics and Communication Engineering, Sun Yat-sen University, Guangzhou, P.R. China 510275
isde35@zsu.edu.cn

Abstract. In this paper, RBF network (RBFN) is used to provide effective methodologies for solving difficult computational problems in camera calibration and 3D reconstruction process. RBFN works in three aspects: Firstly, a RBFN is adopted to learn and memorize the nonlinear relationship in stereovision system. Secondly, another RBFN is trained to search the correspondent lines in two images such that stereo matching is performed in one dimension. Finally, the trained network in the first stage is used to reconstruct the object's 3D figuration and surface. The technique avoids the complicated and large calculation in conventional methods. Experiments have been performed on common stereo pairs and the results are accurate and convincing.

1 Introduction

Camera calibration is a pre-requirement for most application in computer vision. Traditionally, it means the process of determining the internal camera geometric and optical characteristics (intrinsic parameters) and/or the 3-D position and orientation of the camera frame relative to a certain world coordinate system (extrinsic parameters). In many cases, the overall performance of the machine vision system strongly depends on the accuracy of the camera calibration.

Several methods for geometric camera calibration are presented in the literature. The classic approach that originates from the field of photogrammetry solves the problem through minimizing a nonlinear error function (e.g. [1], [4], [7], [12], [8]). However, due to slowness and computational burden of this technique, closed-form solutions have been also suggested (e.g. [6], [1]). In these methods, parameter values are computed directly through a non-iterative algorithm based on a closed-form solution. Since no iteration is required, the algorithms are fast. But there are two disadvantages. First, camera distortion cannot be incorporated, and therefore, distortion effects cannot be corrected. Second, the accuracy of the final solution is relatively poor. There are also calibration procedures where both nonlinear minimization and a closed form solution are used (e.g. [13], [14]). Tsai method [13] is the representative of this type, which uses a series of equations to determine camera parameters in two stages with a simplified model of radial lens distortion.

In this paper, RBF network is introduced to calibrate a stereovision system. A four-input-two-output-architecture network is adopted to directly map the stereo image points to their correspondent spatial points. So the nonlinear relationship of the

J. Wang et al. (Eds.): ISNN 2006, LNCS 3972, pp. 375 – 382, 2006.

stereovision system and physical parameters of the camera can be fully learned based on a training data set.

In stereovision system, traditional rectification (e.g. [2]) is realized through using some known physical camera parameters obtained from calibration. In this paper, we prove that some relationship between pairs of lines exists if some condition is satisfied. The new method is presented for searching correspondent lines utilizing a two-input-two-output RBF network.

The paper is organized as follows. Section 2 introduces how to use RBFN to calibrate a stereoscopic system. Section 3 proves that a mapping between pairs of lines exists if some condition is satisfied, and a RBFN is adopted to learn such relationship. Section 4 reports the tests and the last section is a brief discussion of our work.

2 Camera Calibration Using RBF Network

2.1 Stereovision System

Fig. 1 illustrates the basic geometry of stereovision system. p_1, p_2 are the projections of spatial point P onto the retinal plane π_1 and π_2 in respective. From literature [5], we have the following linear transformation equations in homogeneous coordinates

$$\begin{aligned} (u_1 m_{31}^1 - m_{11}^1)X + (u_1 m_{32}^1 - m_{12}^1)Y + (u_1 m_{33}^1 - m_{13}^1)Z = m_{14}^1 - u_1 m_{34}^1 \\ (v_1 m_{31}^1 - m_{21}^1)X + (v_1 m_{32}^1 - m_{22}^1)Y + (v_1 m_{33}^1 - m_{23}^1)Z = m_{24}^1 - v_1 m_{34}^1 \end{aligned} \tag{1}$$

$$\begin{aligned} (u_2 m_{31}^2 - m_{11}^2)X + (u_2 m_{32}^2 - m_{12}^2)Y + (u_2 m_{33}^2 - m_{13}^2)Z = m_{14}^2 - u_2 m_{34}^2 \\ (v_2 m_{31}^2 - m_{21}^2)X + (v_2 m_{32}^2 - m_{22}^2)Y + (v_2 m_{33}^2 - m_{23}^2)Z = m_{24}^2 - v_2 m_{34}^2 \end{aligned} \tag{2}$$

where $(u_1, v_1, 1)$ and $(u_2, v_2, 1)$ are the homogeneous coordinate of p_1 and p_2 in the π_1, π_2 respectively. $(X, Y, Z, 1)$ is the homogeneous coordinate of P in the world coordinate. $m_{ij}^k\,(k = 1, 2; i = 1..., 3; j = 1..., 4)$ are the camera calibration parameters.

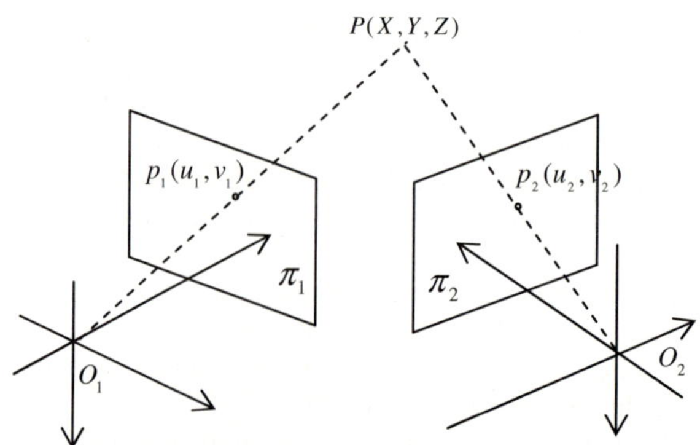

Fig. 1. Stereo imaging geometry

From the view of analytic geometry, we can determine the position of arbitrary spatial P with a least-square solution for P is the intersection of O_1p_1 and O_2p_2. In actual condition, camera distortion cannot be ignored and P should be determined through a very complex nonlinear mapping. If only one order radial distortion is considered, the mapping can be written as $f:(u,v,u^2,v^2,uv,u^3,v^3,u^2v,uv^2)\rightarrow$. (X,Y,Z). It is evident that great computation burden is requisite to get the exact solution of the function.

2.2 Direct Mapping by Neural Network

The learning and memory functions of the neural network are fully used in the calibration stage. The training samples are composed of the given spatial points and their images. Nonlinear relation of the stereovision system and the parameters of the camera are learnt by the neural network and they are stored in the weights for future needs.

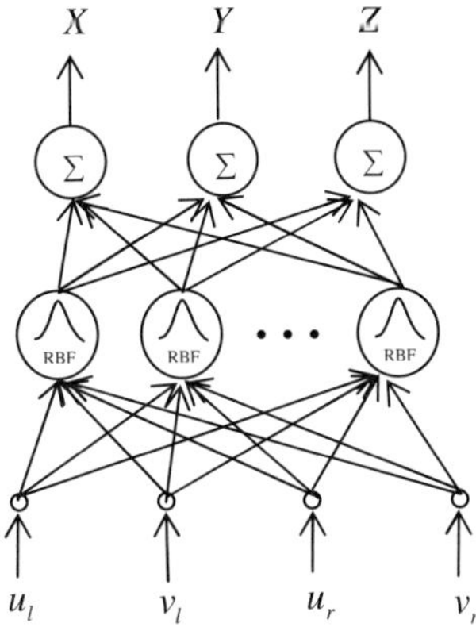

Fig. 2. RBF network used for calibration

The structure of the adopted neural network (denoted as network A) is illustrated in figure 2. It is 3-layer RBF neural network. The first layer is the input-layer with 4 neurons that correspond to the left image coordinate (u_l, v_l) and the right image coordinate (u_r, v_r). The last layer is the output layer with three neurons that correspond to the spatial point coordinate (X,Y,Z). The hidden layer is used to store the nonlinear relationship between the 2-D image and the 3-D object space and camera parameters.

3 Rectification of Stereo Pairs

The stereo matching problem can be solved much more efficiently if images are rectified, i.e. transforming the images so that the epipolar lines are aligned horizontally. In this case, stereo matching algorithm can easily take advantage of the epipolar constraint and reduce the search space to one dimension. In the following text, a new approach is proposed which can find conjugate epipolar lines through using RBFN.

3.1 The Relationship of the Correspondent Lines

Suppose there exists a line $l_1 : v_1 = k_1 u_1 + t_1$ in π_1, we modify the equations (1), and get the following equation:

$$u_1 = \frac{a_1^1 X + b_1^1 Y + c_1^1 Z + d_1^1}{a_3^1 X + b_3^1 Y + c_3^1 Z + d_3^1} \qquad v_1 = \frac{a_2^1 X + b_2^1 Y + c_2^1 Z + d_2^1}{a_3^1 X + b_3^1 Y + c_3^1 Z + d_3^1} \tag{3}$$

Then the line $l_1 : v_1 = k_1 u_1 + t_1$ could be modified as:

$$AX + BY + CZ + D = 0 \tag{4}$$

where A, B, C, D are the linear combination of k_1, t_1 and $a_i^1, b_i^1, c_i^1, d_i^1 (i = 1,2,3)$.

According to the basic knowledge of stereovision, a point (u_1, v_1) on l_1 must have a correspondent point (u_2, v_2) in π_2 which satisfies the following equation:

$$u_2 = \frac{a_1^2 X + b_1^2 Y + c_1^2 Z + d_1^2}{a_3^2 X + b_3^2 Y + c_3^2 Z + d_3^2} \qquad v_2 = \frac{a_2^2 X + b_2^2 Y + c_2^2 Z + d_2^2}{a_3^2 X + b_3^2 Y + c_3^2 Z + d_3^2} \tag{5}$$

Rearranging equation (5) and we can get

$$u_2 = \frac{P_1 Y + Q_1 Z + R_1}{P_3 Y + Q_3 Z + R_3} \qquad v_2 = \frac{P_2 Y + Q_2 Z + R_2}{P_3 Y + Q_3 Z + R_3} \tag{6}$$

where P_i, Q_i, R_i are the linear combination of $a_i^2, b_i^2, c_i^2, d_i^2 (i = 1,2,3)$ and A, B, C, D.

From Equation (4) and (6), we can eliminate Y and get the form of correspondent curve of l_1 in π_2

$$v_2 = \frac{T_1}{T_2} u_2 + \frac{W_1 - W_2 T_1 / T_2}{T_2 Z + W_2} u_2 + (S_1 - \frac{T_1}{T_2} S_2) + \frac{W_1 - W_1 T_1 / T_2}{T_2 Z + W_2} S_2 \tag{7}$$

Where

$$\begin{array}{lll} S_1 = P_2 / P_3 & S_2 = P_1 / P_3 & T_1 = Q_2 - Q_3 P_2 / P_3 \\ T_2 = Q_1 - Q_3 P_1 / P_3 & W_1 = R_2 - R_3 P_2 / P_3 & W_2 = R_1 - R_3 P_1 / P_3 \end{array} \tag{8}$$

If the object depth varies in a small range, i.e. $|Z - C_m| \le \delta$ (δ is small value, C_m is a constant) comes into existence under some condition, then equation (7) could be modified as the linear form, which means that l_2 in π_2 could be reduced as an epipolar line. A network can be used to learn the relationship between two l_1 and l_2.

3.2 Searching the Correspondent Line Using a RBF Network

We use a two-input-two-output RBF network to learn the relationship between the epipolar lines. Input vector is the slope and intercept of l_1, output vector is the slope and intercept of the correspondent line l_2. The new RBFN is denoted as network B.

4 Experimental Results

Fig. 3 shows the Schematic diagram of experimental setup for calibration. Two Panasonic TV CCD cameras with normal lenses were mounted on an ordinary z-raiser. The base line and vergence angle of the two cameras are about 150 mm and 20°, respectively. The focal length and the diagonal visual angle of the two lenses are 18 mm and about 30°, respectively. A calibration pattern produced by a laser printer is a normal chessboard map which contains many grid corners. A subtle method was used to get the exact position of each corner to serve as the calibration points ([14]).

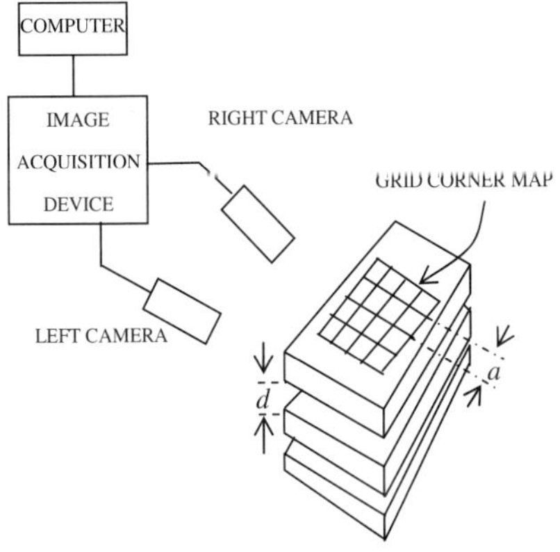

Fig. 3. Schematic diagram of experimental setup for calibration

Table 1. Average 3D position measurement errors by on different training data set (mm)

Data number	Training error	Sc=0.5	Sc=0.8	Sc=1.0	Sc=1.2
500	0.05	1.4051	1.4086	1.3970	1.3924
	0.02	0.9552	0.9512	0.9538	0.9543
	0.016	0.9705	0.9678	0.9570	0.9543
150	0.032	1.4933	1.5495	1.5231	1.4415
	0.016	1.4554	1.4467	1.4373	1.4087
	0.012	1.4422	1.3345	1.3126	1.3534
	0.008	1.1329	1.0784	0.9829	1.0309
	0.005	0.9984	1.0109	0.9842	0.9946
	0.002	1.4661	1.9983	2.3583	1.2058
50	0.02	10.7781	6.7019	2.2350	1.4637
	0.01	8.3398	6.4572	2.4528	1.5932
	0.005	6.2914	4.5785	2.0436	1.6810
	0.002	6.7811	4.5082	2.7297	3.7880
	0.001	6.6809	3.0409	3.7685	3.0149

4.1 Calibrating the Stereovision System Using Network A

As shown in Fig. 3, the distance between two adjacent corners is 10mm, i.e. a=10mm. Depth between two adjacent levels is 20mm, i.e. d=20mm. In our test, we get two different kinds of points. One is used for system calibration. The other is for testing the generalization ability of the network.

Table 1 illustrates the average 3D position measurement errorsusing different trained network. It should be noted that: (1) Sc is learning parameters (defined in [3]); (2) data number means the training data number, not the data number for generalization test. (3) The point number used for computing measurement errors is 500.

4.2 Stereo Image Rectification Using Network B

To learn the relationship between two correspondent lines in the stereo image planes, 180 lines on three different z planes are selected to train the network. To test the performance of the trained network, we select some correspondent image points from original stereo pairs (marked by crosses in fig. 4) and compute their difference in longitudinal direction after rectification. Table 2 shows the results of test errors.

Table 2. Average measurement errors and max error after rectification

Image pair	Data number	Average error (pix)	Max error (pix)
Phone	38	0.3585	1.294
Grid	96	0.6278	1.452

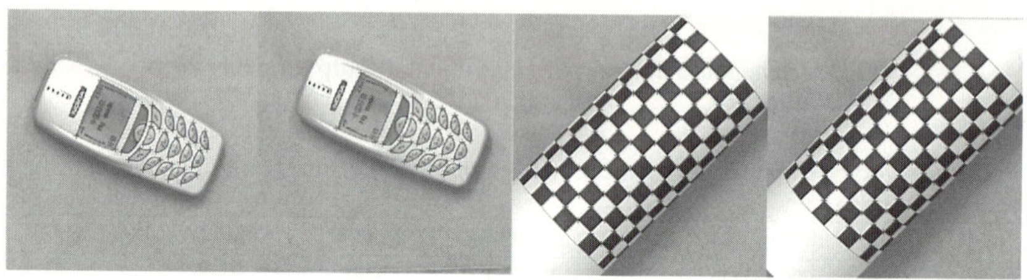

Fig. 4. The stereo images used for testing rectification

4.3 3D Reconstruction Using Network A

In this paper, we use two different stereo pairs to reconstruct the object's 3D figure. As illustrated in figure 5(a)-(b), breechblock and comb have been imaged by our lab's stereovision system. The original image size is 512×512. Figure 5(c)-(d) shows the image pairs that have been rectified and pruned such that the correspondent points lie on the same horizontal line. We performed the stereo matching on the stereo pairs based on the maximum flow algorithm [10]. Figure 5(e)-(f) are the depth map. Finally the 3D figure of the scene is reconstructed using the trained network in the first stage. The reconstructed configurations demonstrate the effectiveness of the RBFN in 3D reconstruction.

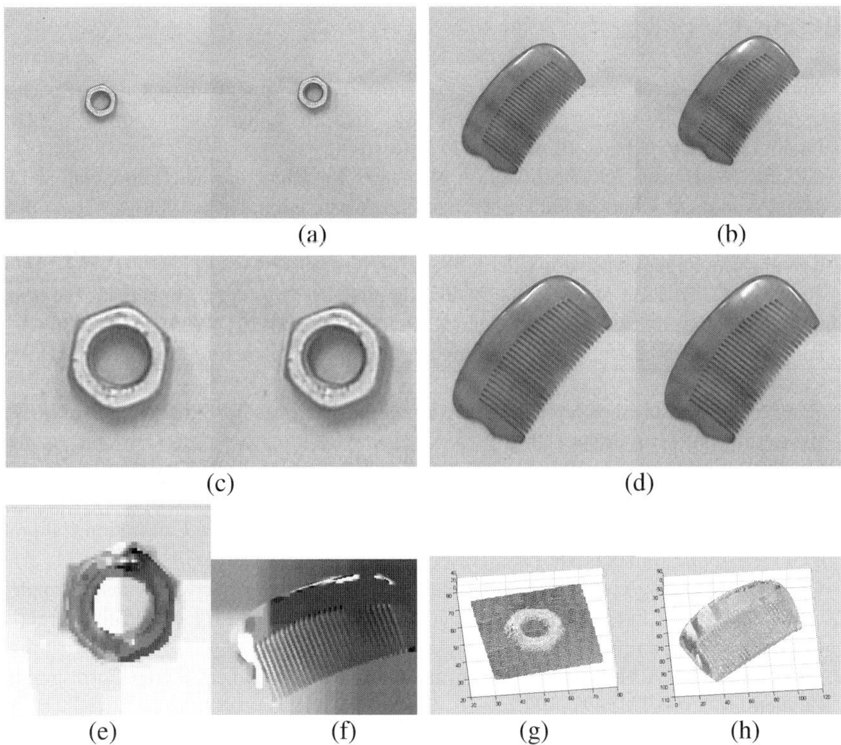

Fig. 5. Experiment results where (a)-(b) are the original stereo pair of nut, comb respectively which are captured by our stereovision system. (c)-(d) are the rectified and pruned images. (e)-(f) are the depth maps. (g)-(h) are the reconstruction configuration of two different stereo pairs.

5 Conclusion

This paper has demonstrated the effectiveness of RBF network in stereovision system. The RBF network works in three stages: firstly, it is adopted to learn the nonlinear relationship of vision system in the calibration stage; secondly, it is used to search the correspondent lines in rectification stage; finally, it is employed to reconstruct the 3D object figure in reconstruction stage. Through the practical application, it is shown that RBF network can be efficient, accurate and straightforward to implement in real stereovision system.

Acknowledgement

This work is supported by Nature Science Foundation of Guangdong (5300712) and Sun Yat-sen University Science Foundation.

References

1. Abdel-Aziz, Y.I., Karara, H.M.: Direct Linear Transformation into Object Space Coordinates in Close-Range Photogrammetry. In: Proc. Symposium on Close-Range Photogrammetry, Urbana, Illinois (1971) 1–18
2. Ayache, N., Hansen, C.: Rectification of Images for Binocular and Trinocular Stereovision. Proc. of Int. Conf. on Pattern Recognition, Washington, D.C. (1988) 11–16
3. Chen, S., Cowan, C.F.N., Grant, P.M.: Orthogonal Least Squares Learning Algorithm for Radial Basis Function Network. IEEE Trans. on Neural Network 2(2) (1991) 302–309
4. Faig, W.: Calibration of Close-Range Photogrammetric Systems: Mathematical Formulation. Photogrammetric Engineering and Remote Sensing 41(12) (1975) 1479–1486
5. Faugeras, O.: Three-Dimensional Computer Vision, a Geometric Viewpoint, MIT Press, Cambridge (1993)
6. Ganapathy S.: Decomposition of Transformation Matrices for Robot Vision. In: Proc. IEEE Int. Conf. RA (1984) 130–139
7. Gennery, D.B.: Stereo Camera Calibration. In: Proc. 10th Image Understanding Workshop (1979) 101–108
8. Isaguirre, A., Pu, P., Summers J.: A New Development in Camera Calibration Calibrating a Pair of Mobile Cameras. In: Proc. IEEE Int. Conf. RA (1985) 74–79
9. Kegl, B., Krzyak, A., Niemann, H.: Radial Basis Function Networks and Complexity Regularization in Function Learning and Classification. In: Proc. of 15th International Conference, Vol. 2 (2000) 81–86
10. Roy, S., Cox, I.: A Maximum-Flow Formulation of the N-Camera Stereo Correspondence Problem. ICCV (1998) 492–499
11. Schwenkar, F., Kestler, H.A., Palm, G.: Radial-Basis-Function Networks: Learning and Applications. In: Proc. of Fourth International Conference on Knowledge-Based Intelligent Engineering Systems and Allied Technologies, Vol. 1 (2000) 33–43
12. Slama, C.C.: Manual of Photogrammetry. 4th edn. American Society of Photogrammetry, Falls Church, Virginia (1980)
13. Tsai, R.Y.: A Versatile Camera Calibration Technique for High-Accuracy 3D Machine Vision Metrology Using Off-the-Shelf TV Cameras and Lenses. IEEE Trans. on RA 3(4) (1987) 323–344
14. Weng, J., Cohen, P., Herniou, M.: Camera Calibration with Distortion Models and Accuracy Evaluation. IEEE Trans. on PAMI 14(10) (1992) 965–980

A Versatile Method for Omnidirectional Stereo Camera Calibration Based on BP Algorithm

Chuanjiang Luo[1,2], Liancheng Su[1,2], Feng Zhu[1], and Zelin Shi[1]

[1] Optical-Electronic Information Laboratory, Shenyang Institute of Automation Chinese Academy of Sciences (CAS) ,114# Nanta Street, Shenyang, P.R. China
{lcj0603, lcsu, fzhu, zlshi}@sia.cn
[2] Graduate School, Chinese Academy of Sciences, Beijing, China

Abstract. This study describes a full model of calibrating an omnidirectional stereo vision system, which includes the rotation and translation between the camera and mirrors, and an algorithm implemented with a backpropagation technique of the neural network to determine this relative position from observations of known points in a single image. The system is composed of a perspective camera and two hyperbolic mirrors, which are configured to be separate and coaxial besides sharing one focus that coincides with the camera center for providing a single projection point. We divide the calibration into two steps. The first step we calibrate the camera's intrinsics without the mirrors in order to reduce computational complexity and in the second step we estimate the pose parameters of the CCD camera with respect to the mirrors based on a Levenberg-Marquart Backpropagation (LMBP) algorithm. The proposed technique can be easily applied to all kinds of catadioptric sensors and various amounts of misalignment between the mirrors and cameras.

1 Introduction

A catadioptric vision system using mirrors has been a popular means to get panoramic images [1], which contains a full horizontal field of view. This wide view is ideal for three-dimensional vision tasks such as motion estimation, localization, obstacle detection and mobile robots navigation. Omnidirectional stereo is a suitable sensing method for such tasks because it can acquire images and ranges of surrounding areas simultaneously.

Mobile robot navigation using binocular omnidirectional stereo vision has been reported in [2] and [3]. Such two-camera stereo systems are costly and complicated compared to single camera stereo systems. Omnidirectional stereo based on a double lobed mirror and a single camera was developed in [5], [6], and [7]. A double lobed mirror is a coaxial mirror pair, where the centers of both mirrors are collinear with the camera axis, and the mirrors have a profile radially symmetric around this axis. This arrangement has the advantage to produce two panoramic views of the scene in a single image. But the disadvantage of this method is the relatively small baseline it provides. Since the two mirrors are so close together, the effective baseline for stereo calculation is quite small.

J. Wang et al. (Eds.): ISNN 2006, LNCS 3972, pp. 383 – 389, 2006.

We have developed a large baseline omnidirectional stereo vision system based on a common perspective camera coupled with two hyperbolic mirrors [8], which separately fixed inside a glass cylinder. As the separation between the two mirrors provides much enlarged baseline, the precision of the system has improved correspondingly (see Fig. 1). The coaxial configuration of the camera and the two hyperbolic mirrors makes the epipolar line radially collinear, which makes the system free of the search process for complex epipolar curve in stereo matching (see Fig. 3).

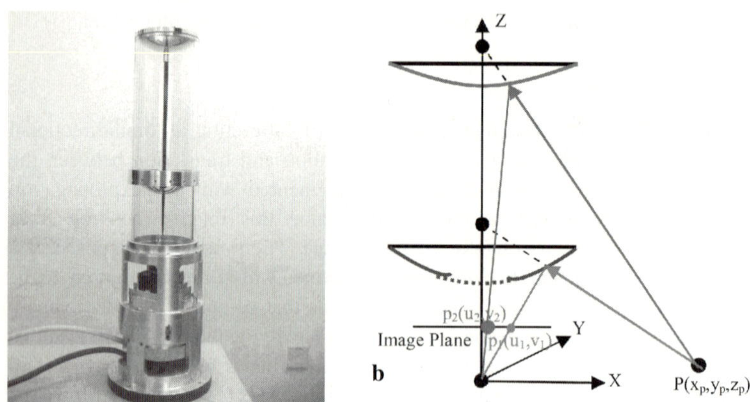

Fig. 1. The camera and the two hyperbolic mirrors are aligned vertically and coaxially. A hole in the mirror bellow permits imaging via the mirror above. **a:** The appearance of the stereo vision system. **b:** The configuration of the system.

In using the omnidirectional stereo vision system, its calibration is important, as in the case of conventional stereo systems [9]. We present a full model of the imaging process, which includes the rotation and translation between the camera and mirror, and an algorithm to determine this relative position from observations of known points in a single image.

Previous methods for omnidirectional camera calibration have required that the mirror and camera combination be single viewpoint, i.e., that rays reflected by the mirror onto the camera's center also intersect at a single point inside the mirror. Our calibration is performed within a general minimization framework, and easily accommodates any combination of mirror and camera. For single viewpoint combinations, the advantages of the single viewpoint can be exploited only if the camera and mirror are assumed to be properly aligned. So for these combinations, the simpler single viewpoint projection model, rather than the full model described here, should be adopted only if the misalignment between the mirror and camera is sufficiently small. In this case, the calibration algorithm that we describe is useful as a software verification of the alignment accuracy.

In this study, our projection model and calibration algorithm separate the conventional camera intrinsics (e.g., focal length, principal point) from the relative position between the mirrors and the camera (i.e., the camera-to-mirrors coordinate transformation) to reduce computational complexity and improve the calibration precision.

2 Method

2.1 Overview

The conventional camera intrinsics can be determined using any existing method, for the experiments described here, we have used the method proposed in http://www.vision.caltech.edu/bouguetj/calib_doc/. Once the camera intrinsics are known, the camera-to-mirrors transformation can be determined by obtaining an image of calibration targets whose three-dimensional positions are known, and then minimizing the difference between coordinates of the targets and the locations calculated from the targets' images through the projection model. Fig. 3 shows one example of calibration image used in our experiments. The locations of the three dimensional points have been surveyed with an accuracy of about one millimeter. Taking into account the inaccuracy of image point due to discrete distribution of pixels, the total measuring error is about five millimeters.

In the following subsections we describe the reprojection of points in the camera coordinate system assuming a general rotation and translation between the camera and mirrors (3.2) and the error function we minimize to determine the camera-to-mirrors transformation (3.3).

2.2 Projection Model

Fig 2 depicts the composition of a perspective camera with two hyperbolic mirrors. There are three essentially coordinate systems. (1) The camera coordinate system centered at the camera center O_c, the optical axis is aligned with the z-axis of the camera coordinate system; (2) The mirror system centered at common foci of the hyperbolic mirrors F_o, the mirrors axes is aligned with the z-axis of the mirror coordinate system (We assume that the axes of the mirrors are align well, and the common foci are coincident, from the mirrors manufacturing sheet we know it is reasonable); (3) The world system centered at O_w. The omnidirectional stereo vision system was placed on a plane desk. As both the base of vision system and desk surface are plane, the axis of the mirror is perpendicular to the base of the system and the surface of the desk feckly. We make the mirror system coincide with the world system to simplify the model and computations. So the equations of hyperboloid of two sheets in the system centered at O_w read as

$$(z-c_i)^2/a_i^2-(x^2+y^2)/b_i^2=1. \quad (i=1,2) \tag{1}$$

A known world point $P(x_w,y_w,z_w)$ in the world (or mirror) coordinate system whose projected points in the image plane are also known, $q_1(u_1,v_1)$ and $q_2(u_2,v_2)$ are respectively projected by the upper mirror and bellow mirror. Then we get their coordinates in the camera coordinate system:

$$\begin{bmatrix}x_i^c & y_i^c & z_i^c\end{bmatrix}^T=\begin{bmatrix}(u_i-u_0)k_u & (v_0-v_i)k_v & f\end{bmatrix}^T. \quad (i=1,2) \tag{2}$$

Where f is the focal length; ku, kv are the pixel scale factors; u_0, v_0 are the coordinates of the principal point, where the optical axis intersects the projection plane. They are intrinsic parameters of the perspective camera.

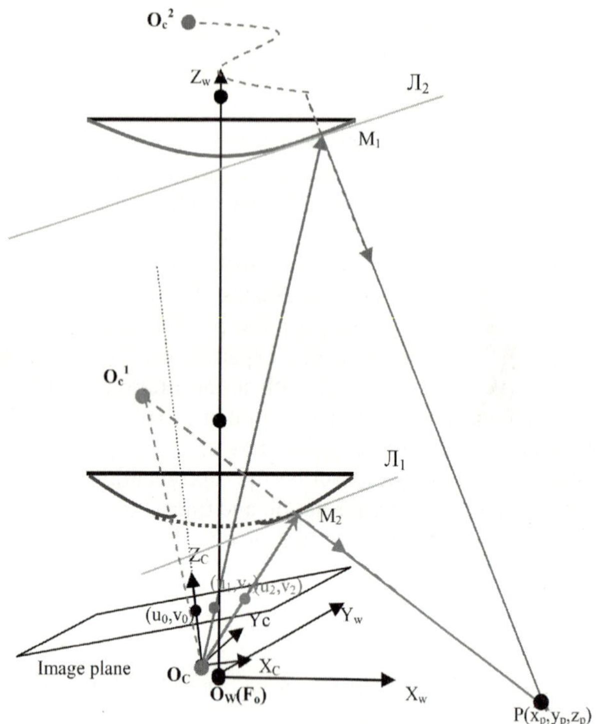

Fig. 2. The projection model of the omnidirectional stereo vision system. There are transformations between the camera coordinate system and the mirror (or world) coordinate system.

So the image points Pc (x_i^c, y_i^c, z_i^c) of the camera coordinate system can be expressed relative to the mirror coordinate system as:

$$\begin{bmatrix} x_i^m & y_i^m & z_i^m \end{bmatrix}^T = R\begin{bmatrix} x_i^c & y_i^c & z_i^c \end{bmatrix}^T + t . \qquad (i = 1,2) \tag{3}$$

Where R is a 3×3 rotation matrix with three rotation angles around the x-axis (pitch α), y-axis (yaw β) and z-axis (title γ) of the mirror coordinate system respectively; t= $[t_x\ t_y\ t_z]^T$ is the translation vector. So the origin O_c= $[0\ 0\ 0]^T$ of the camera coordinate system can be expressed in the world coordinate system O_m= $[tx\ ty\ tz]^T$, so the equations of lines O_cM_1 and O_cM_2 which intersect with the upper mirror and bellow mirror respectively at points M_1 and M_2, can be determined by solving simultaneous equations of the line O_cM_1 or O_cM_2 and the hyperboloid. Once the coordinates of the point M_1 and M_2 have been worked out, we can write out the equations of the tangent plane л$_1$ and л$_2$ which passes the upper and the bellow mirror at point M_1 and M_2 respectively. Then the symmetric points O_c^1 and O_c^2 of the origin of the camera coordinate system O_c relative to tangent plane л$_1$ and л$_2$ in the world coordinate system can be solved from the following simultaneous equations:

Fig. 3. A calibration image used in our experiments. The coaxial configuration of the camera and the two hyperbolic mirrors makes the epipolar line radially collinear, which makes the system free of the search process for complex epipolar curve in stereo matching.

$$\begin{cases} \dfrac{x_{o_c^i} - tx}{a_i^2 x_{M_i}} = \dfrac{y_{o_c^i} - ty}{a_i^2 y_{M_i}} = \dfrac{z_{o_c^i} - tz}{-b_i^2 z_{M_i} + b_i^2 c_i} \\ a_i^2 x_{M_i}(tx + x_{o_c^i}) + a_i^2 y_{M_i}(ty + y_{o_c^i}) - (-b_i^2 z_{M_i} + b_i^2 c_i)(tz + z_{o_c^i}) \\ + 2[-a_i^2 x_{M_i}^2 - a_i^2 y_{M_i}^2 - z_{M_i}(-b_i^2 z_{M_i} + b_i^2 c_i)] = 0 \end{cases} . \; (i = 1,2) \tag{4}$$

Hitherto the incident ray Oc_1M_2 and Oc_2M_1 can be written out to determine the world point $P(x_w,y_w,z_w)$. Generally, the two lines are non-co-plane due to various parameter errors and measuring errors, we solve out the midpoint G= $(\hat{x}_w, \hat{y}_w, \hat{z}_w)^T$ of the common perpendicular of the two lines by

$$\left.\begin{cases} [\overrightarrow{O_c^1 M_2} \times (\overrightarrow{O_c^1 M_2} \times \overrightarrow{O_c^2 M_1})] \bullet \overrightarrow{G_1 M_2} = 0 \\ \overrightarrow{G_1 M_1} = t\overrightarrow{G_1 O_c^2} \end{cases} \Rightarrow \overrightarrow{OG_1} \quad \begin{cases} [\overrightarrow{O_c^2 M_1} \times (\overrightarrow{O_c^1 M_2} \times \overrightarrow{O_c^2 M_1})] \bullet \overrightarrow{G_2 M_1} = 0 \\ \overrightarrow{G_2 M_2} = t\overrightarrow{G_2 O_c^1} \end{cases} \Rightarrow \overrightarrow{OG_2} \right. \quad \overrightarrow{OG} = (\overrightarrow{OG_1} + \overrightarrow{OG_2})/2 . \tag{5}$$

From all of them above, we finally come to the total expression to figure out the world point G= $(\hat{x}_w, \hat{y}_w, \hat{z}_w)^T$ from two image points respectively projected by the upper mirror and bellow mirror and six camera pose parameters left to be determined.

$$G(\alpha, \beta, \chi, t_x, t_y, t_z, u_1, v_1, u_2, v_2) = [\hat{x}_w \quad \hat{y}_w \quad \hat{y}_w]^T . \tag{6}$$

(6) is a very complex nonlinear equation with high power and six unknown parameters to determine. The artificial neural network trained with sets of image points of the calibration targets is used to estimate the camera-to-mirror transformation.

Taking advantage of the ANN capability, which adjusts the initial input camera-to-mirror transformations step by step to minimize the error function, the real transformations parameters of the camera-to-mirror can be identified precisely.

2.3 Error Function

Considering the world points with known coordinates, placed onto a calibration pattern, at the same time, their coordinates can be calculated using the equation (6) from back-projection of their image points. The difference between the positions of the real world coordinates and the calculated coordinates is the calibration error of the model. Minimizing the above error by means of an iterative algorithm such as Levenberg-Marquardt BP algorithm, the camera-to-mirror transformation is calibrated. The initial values for such algorithm are of consequence. In our system, we could assume the transformation between cameras and mirrors is quite small, as the calculation error without considering the camera-to-mirror transformation is not significant thus using R=I and T=0 as the initial values is a reasonable choice.

We minimize the following squared error ε^2:

$$\varepsilon^2 = \sum_{i=1}^{n} \left\| P_i - G_i\left(\alpha, \beta, \chi, t_x, t_y, t_z, u_1^i, v_1^i, u_2^i, v_2^i\right) \right\|^2 . \tag{7}$$

Where n is the number of the calibration points.

Because $G_i\left(\alpha, \beta, \chi, t_x, t_y, t_z, u_1^i, v_1^i, u_2^i, v_2^i\right)$ depends on the camera-to-mirror transformation, (7) is optimized with respect to the six camera-to-mirror parameters.

3 Experimental Result

The calibration was performed using a set of 81 points equally distributed on a desk with different heights from 0 to 122mm around the vision system.

The calibration results with real data are listed in Table 1.

Table 1. Calibration results with real data

	α	β	χ	t_x	t_y	t_z
value	-0.9539°	0.1366°	0.1436°	-0.0553mm	-0.1993mm	1.8717mm

The calibration error was estimated using a new set of 40 untrained points, the average square error of the set points is 34.24mm without considering the camera-to-mirror transformation. Then we calculate the error with the transformation values listed in Table 1, the average square error decrease to 12.57mm.

4 Conclusion

We have presented an imaging model for omnidirectional cameras that accounts for the full rotation and translation between the camera and mirror, and a LMBP method of the neural network for recovering the relative position from back–projection the point images in a single image. The method is general in that any combination of camera and mirror can be calibrated, included non-single-viewpoint combinations. For single viewpoint cameras, where the advantages of a single viewpoint can be exploited only if the camera and mirror is assumed to be perfectly aligned, this algorithm can be used to verify the alignment accuracy.

Acknowledgement

This works is supported by National Science Foundation of P.R. China (60575024).

References

1. Nayar, S.K.: Catadioptric Omnidirectional Camera. Proc. IEEE Conf. on Computer Vision and Pattern Recognition (1997) 482-488
2. Gluckman, J., Nayar, S. K., Thoresz, K. J.: Real-time Omnidirectional and Panoramic Stereo. Proc. DARPA Image Understanding Workshop (1998) 299-303
3. Koyasu, H., Miura, J., Shirai, Y.: Recognizing Moving Obstacles for Robot Navigation Using Real-time Omnidirectional Stereo Vision. Journal of Robotics and Mechatronics 14 (No.2) (2002) 147-156
4. Ma, J.: Omnidirectional Vision Based Localization and Path Planning for Mobile Robots. Doctor Thesis of Beijing Institute of Technology (2003)
5. Southwell, M.F.D., Basu, A., Reyda, J.: Panoramic stereo. Proc. International Conf. on Pattern Recognition (1996) 378-382
6. Conroy, T.L., Moore, J.B.: Resolution invariant surfaces for panoramic vision systems. Proc. IEEE International Conf. on Computer Vision (1999) 392-397
7. Cabral, E.L.L., de Souza Jr., J. C., Hunold, M.C.: Omnidirectional Stereo Vision with a Hyperbolic Double Lobed Mirror. Proc. International Conf. on Pattern recognition (2004) 1-4
8. Su, L., Zhu, F.: Design of a Novel Stereo Vision Navigation System for Mobile Robots. Proc. IEEE Conf. on Robotics and Biomimetics, Vol.1 (2005) 611-614
9. Luong, Q. T., Faugeras, O. D.: The Fundamental Matrix: Theory, Algorithms, and Stability Analysis. Int. J. of Computer Vision 17 (1) (1996) 43-76
10. Geyer, C., Daniilidis, K.: Conformal Rectification of Omnidirectional Stereo Pairs. Proc. IEEE Workshop on Omnidirectional Vision and Camera Networks 7 (2003) 1-6

Evolutionary Cellular Automata Based Neural Systems for Visual Servoing

Dong-Wook Lee[1], Chang-Hyun Park[2], and Kwee-Bo Sim[2]

[1] Korea Institute of Industrial Technology,
1271 Sa-1-dong, Sangrok-gu, Ansan 426-791, Korea
dwlee@kitech.re.kr
[2] Chung-Ang University, 221, Heukseok-dong, Dongjak-gu, Seoul 156-756, Korea
kbsim@cau.ac.kr

Abstract. This paper presents an evolutionary cellular automata based neural systems (Evolutionary CANS) for visual servoing of RV-M2 robot manipulator. The architecture of CANS consist of a two-dimensional (2-D) array of basic neurons. Each neuron of CANS has local connections only with contiguous neuron and acts as a form of pulse according to the dynamics of the chaotic neuron model. CANS are generated from initial cells according to the cellular automata (CA) rule. Therefore neural architecture is determined by both initial pattern of cells and production rule of CA. Production rules of CA are evolved based on a DNA coding. DNA coding has the redundancy and overlapping of gene and is apt for representation of the rule. In this paper we show the general expression of CA rule and propose translating method from DNA code to CA rule. In addition, we present visual servoing application using evolutionary CANS.

1 Introduction

The success of neural networks in information processing tasks depends mostly on well-designed structures of the systems. In general, its structures have to be defined before learning algorithms are executed. In practice, determination of structures takes most of effort in adjusting a neural system to particular tasks. In recent researches, the evolutionary approach is used for the structure and parameter optimization of neural networks. In addition, the resulting networks have usually been trained using back propagation. However, in this case, the computation cost can be so high as to make genetic algorithms impractical except for optimizing small topologies. In our systems, the mapping from genotype to phenotype uses the developmental model, which is CA. Since the chromosome is mapped the set of production rules, even if the size of neural networks are big, the chromosome is not so long. However, as the number of states of CA increases, the size of rule table also increases exponentially. So, it is difficult to evolve the rule table. In order to overcome this problem, we propose new encoding method based on biological DNA encoding. This method is an algorithm that is embodied by the characteristics of DNA. DNA coding [1] has many advantages.

J. Wang et al. (Eds.): ISNN 2006, LNCS 3972, pp. 390–397, 2006.

For example, it is suitable to represent the rule, and has good performance when the chromosome is long. Many researchers take developmental and evolutionary approach for designing a neural networks. Boers [2], Gruau [3] etc. proposed the design method of neural networks based on L-system and genetic algorithms. de Garis [4] has studied cellular automata and genetic programming based artificial brain, Sugisaka [5] also has developed artificial brain.

In this paper, we present a new type of neural architecture consisting of chaotic neurons. To construct CANS, we use the method of development and evolution. That is to say, we evolve the production rules of cellular automata (CA), which are the developmental model of neural systems. In section 2, we describe our neural systems, include neuron model and network architecture. Section 3 explains DNA coding for evolution of CA rules in detail. Section 4 shows the simulation result of visual servoing of RV-M2 robot manipulators.

2 An Architecture of Cellular Automata Based Neural Systems

2.1 Neuron Model

Nagumo-Sato proposed chaotic neuron model that is based on Hodgkin-Huxley equation that is deep consideration model of practical neuron membrane's characteristics. Thus, it has complex characteristics; nevertheless it has simple structure comparatively.

The equations of Nagumo-Sato's chaotic neuron model [6] are (1) and (3).

$$y(t+1) = u(x(t+1)) \tag{1}$$

$$y(t+1) = 2u(x(t+1)) - 1 \tag{2}$$

$$x(t+1) = S(t) - \alpha \sum_{d=0}^{t} k^d y(t-d) - \theta \tag{3}$$

where $y(t)$ is a output at time t, $x(t)$ is an inertia state at time t, $S(t)$ is a input at time t, $u(x)$ is a unit step function, k^d is a damping factor of refractoriness having values between 0 and 1, the constant α is a positive parameter, and θ is a threshold of neuron. Eq. (2) is the bipolar continuous version of (1). The neuron model which we use in order to construct the CANS, is (2) and (3).

2.2 Network Architecture

In this paper, we use CA model for development of neural networks. So, the production rules are encoded as a chromosome by DNA coding, and artificial neural system (CANS) emerges from initial cells and interaction between the rules. The network is obtained according to the development of 1-dimensional CA. Initial cells compose the first layer neurons of hidden layer. As the time step goes on the rest of the hidden layer is completed by CA rule and previous step patterns. The state of CA is a cell, which have various type of connection. Possible states for use are 27 (=3^3, 3 direction (right, left, down) and 3 connection

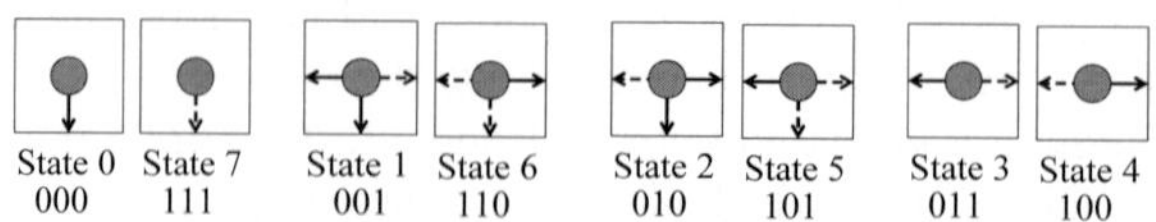

Fig. 1. State of the cell which is determined by its connection type (solid line: excitatory synapse, dashed line: inhibitory synapse)

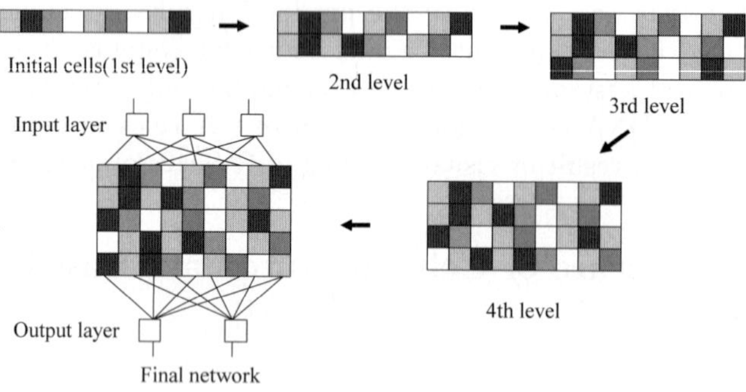

Fig. 2. Developmental level of CANS

(excitation, inhibition, no connection) is possible). In this paper, 8 states are selected by considering the symmetry (Fig. 1).

Fig. 2 shows the developmental process and how to connect the input and output cells to the network.

The characteristics of CANS are as follows.

1. Basic element (neuron): The characteristic equation of a neuron is chaotic neuron model which acts as a pulse type
2. Network architecture: The characteristics of the network are determined by not the connection strength (weights) but the connection types.
3. Generation method: CANS is produced based on development and evolution.

3 CA Rule Representation by DNA Coding

3.1 DNA Coding

Motivated by biological DNA, DNA coding [1] uses four symbols A (Adenine), G (Guanine), T (Tymine), and C (Cytocine) that denote nucleotide bases, not a binary representation as in the GA. A chromosome is represented by three successive symbols called a codon. A DNA code that begins from a Start codon (ATG) and ends at a Stop codon (TAA, TAG, TGA, or TGG) is translated into a meaningful code. This representation of chromosome can have multiple interpretations since the interpretations of Start and Stop codons allow overlaps as shown in Fig. 3. The length of chromosomes varies. DNA coding has floating

Table 1. The genetic code. Each codon represents one amino acid and termination sequence. U (Uracil) is replaced by T in DNA code

	T		C		A		G		
	UUU	Phe	UCU		UAU	Tyr	UGU	Cys	U
U	UUC		UCC	Ser	UAC		UGC		C
	UUA		UCA		UAA	Stop	UGA	Stop	A
	UUG		UGC		UAC		UGG	Trp	G
	CUU	Leu	CCU		CAU	His	CGU		U
C	CUC		CCC	Pro	CAC		CGC	Arg	C
	CUA		CCA		CAA	Gln	CGA		A
	CUG		CGG		CAG		CGG		G
	AUU		ACU		AAU	Asn	AGU	Ser	U
A	AUC	Ile	ACC	Thr	AAC		AGC		C
	AUA		ACA		AAA	Lys	AGA	Arg	A
	AUG	Met	AGG		AAG		AGG		G
	GUU		GCU		GAU	Asp	GGU		U
G	GUC	Val	GCC	Ala	GAC		GGC	Gly	C
	GUA		GCA		GAA	Glu	GGA		A
	GUG		GGG		GAG		GGG		G

representations without fixed crossover points. A codon is translated into an amino acid according to a translation table in Table 1.

3.2 CA Rule Representation

In this section, we propose CA rule representation by DNA coding. Eq. (4) shows the general representation of one dimensional CA rule ($\Phi_i(\sigma_i)$) [7].

$$\sigma_i^{new} = \Phi_i(\sigma_i) = \phi(\sigma_{i-r}, \cdots, \sigma_{i+r}) \tag{4}$$

where r is the radius of neighborhood.

We consider the case of $r = 1$. At this time, according to the state of i-th cell and its neighbor states, the next state of i-th cell can be represented as a function $\phi : (\sigma_{i-1}^j, \sigma_i^j, \sigma_{i+1}^j) \rightarrow \sigma_i^{j+1}$. And if the function ϕ consists of n subfunctions, then ϕ is represented in the form

$$\phi = \begin{cases} \phi_1 & \text{condition } 1 \\ \phi_2 & \text{condition } 2 \\ \cdots & \quad\cdots \\ \phi_n & \text{condition } n \end{cases} \tag{5}$$

where the maximum value of n is q^{2r+1} and q is total states.

Condition $1 \sim n$ in (5) is determined by its neighborhood, so we use (6) to (8) for representation of conditions in this paper.

$$\text{Type 1: If } \sigma_n = s_1 \tag{6}$$

$$\text{Type 2: If } \sigma_n = s_1 \text{ and } \sigma_m = s_2 \tag{7}$$

$$\text{Type 3: If } \sigma_n = s_1, \sigma_m = s_2 \text{ and } \sigma_l = s_3 \tag{8}$$

where s_1, s_2, s_3 are the states of the cell and $l, m, n \in \{i-1, i, i+1\}, l \neq m$.

Eq. (8) that is considered the 3 states is a specific rule and has highest priority when the conflict of the rule occurs. Eq. (6) (Eq. (7)) that is considered the 1 (2) state(s), is a general rule and has low priority than type 3. To consider the conflict between the rules, we assign the priority of the rule as follows. The rule of type 3 is highest priority, and type 2 is the next, and type 1 is the lowest priority. When the same priority rule is generated, early-translated rule has the higher priority. Sub-function consists of the operation as shown in table 2.

"Order" is the order of considering cell of states or operation. A/B is the abbreviation of And/Break for the control of what states are considered for condition term. Type 1, 2, and 3 are that the number of considering states is 1, 2, 3 respectively in the condition term. Fig 3 shows an example of the overlapped representation and translation of genetic code.

Table 2. Bit operation used in k

No.	Op. symbol	# of operand	Operation
0	3XOR	3	$\sigma_1 \wedge \sigma_2 \wedge \sigma_3$
1	AND	2	$\sigma_i \& \sigma_2$
2	OR	2	$\sigma_1 \| \sigma_2$
3	ADD	2	$(\sigma_1 + \sigma_2)\%q$
4	DIFFERENCE	2	$\|\sigma_1 - \sigma_2\|$
5	XOR	2	$\sigma_1 \wedge \sigma_2$
6	NOT	1	$\sim \sigma_1$
7	STATE	1	σ_4

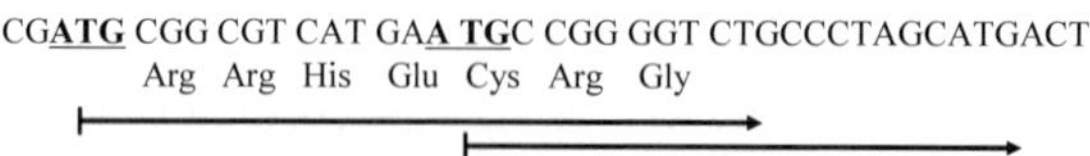

Fig. 3. Overlapping gene and translation of genetic code

Table 3. Translation table of codon

Amino acid	Phe	Leu	Val	Ser	Pro	Thr	Ala	Tyr	His	Gln	Asn	Lys	Asp	Glu	Cys	Arg	Gly
Order	0	2	1	3	4	1	5	1	3	1	5	4	4	2	5	0	4
State	0	0	1	3	7	2	7	4	5	5	2	5	1	6	3	4	6
And/Break	A	A	A	B	A	A	A	A	A	A	A	A	A	A	B	B	B
Operation	0	0	1	3	7	2	7	4	5	5	2	5	1	6	3	4	6

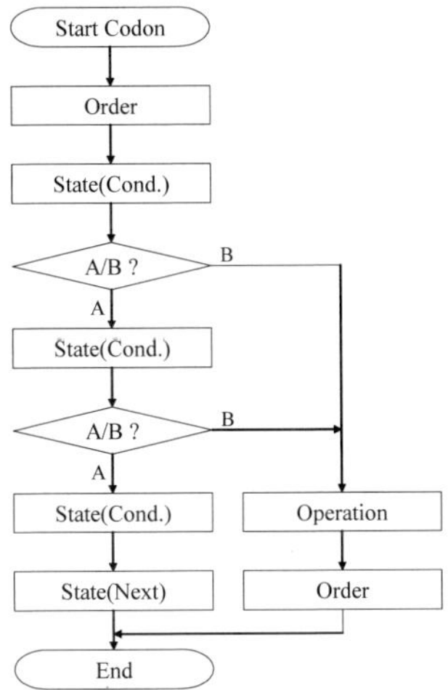

Fig. 4. Flowchart of translating DNA code (each block represents the codon)

Table 3 is the distribution of the condition to DNA codon. We design Table 3 by the considering the composition rate and three-dimensional geometry of the codon.

Fig. 4 shows the translation method of DNA code. The example of translation of the DNA code in Fig. 3 is as follows:

- ATG : Start
- Arg : Order (→ 0: $i-1$, i, $i+1$)
- Arg : State (→ 4)
- His : A/B (→ A)
- Glu : State (→ 6)
- Cys : A/B (→ B)
- Arg : Operation (→ 4: DIFFERENCE)
- Gly : Order (→ 4: $i+1$, $i-1$, i)

The complete rule is "if $\sigma_{i-1}^{j} = 4$ and $\sigma_{i}^{j} = 6$ then $\sigma_{i}^{j+1} = |\sigma_{i+1}^{j} - \sigma_{i-1}^{j}|$." This is type 2 rule.

4 Visual Servoing Experiments

4.1 Visual Servoing

In order to verify the effectiveness of the proposed scheme, we applied the visual servoing of RV-M2 robot manipulator. The object is 50×50×30mm cube and has four feature point in camera image. It assumes that the ideal image is already known. And the objective of the system is to confirm the actual image to the ideal image by moving robot manipulators. If the actual image corresponding

Table 4. Joint angle range(°) and maximum velocity(°/sec) of RV-M2 robot manipulator

Joint	θ_1	θ_2	θ_3	θ_4	θ_5
Angle	-150 ~ 150	-30 ~ 100	-120 ~ 0	-200 ~ 20	-180 ~ 180
Velocity	140	79	140	163	223

to the ideal image, a robot catch it using griper. Table 4 shows the joint angle range and the maximum velocity of RV-M2 robot manipulators.

The eight inputs of neural network controller are deviation of four feature point between ideal image and object image. And the outputs are variation of RV-M2's 5 joint angles.

4.2 Simulation Results

The number of input is eight deviations of four feature points and the number of output is five angle values. The fitness is determined as the inverse value of the error summation of each feature point. Fig 5 and Fig. 6 show the simulation results after 50 generation. As shown in Fig. 6, CANS controller shows good result when untrained pattern is given.

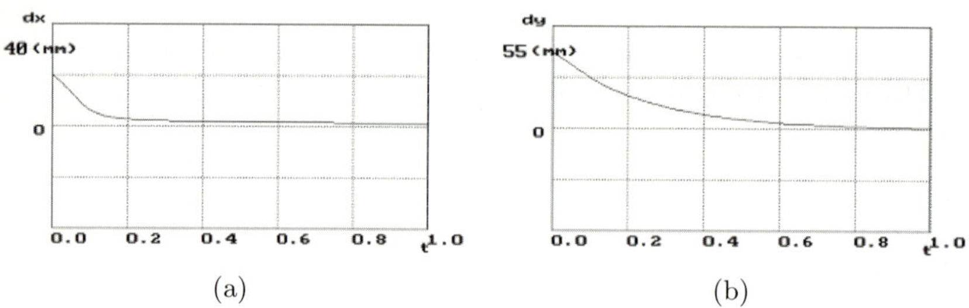

(a) (b)

Fig. 5. Deviation of x and y value after 50 generation. (a) Deviation of x value (dx) (b) Deviation of y value (dy).

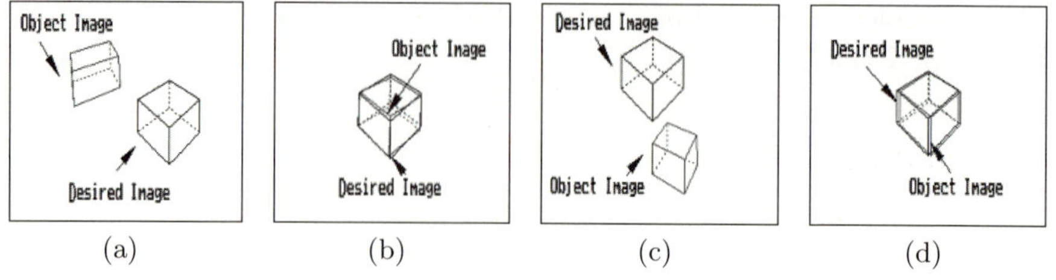

(a) (b) (c) (d)

Fig. 6. Simulation results of training pattern and new pattern. (a) Initial image (training) (b) Final image (training) (c) Initial image (new) (d) Final image (new).

5 Conclusion

In this paper, we proposed a new type of neural architecture and the evolution of its structure based on DNA coding. The CANS is based on the concept of development, learning, and evolution. To make CANS, we used chaotic neuron, cellular automata, and DNA coding. Also the proposed method is verified by simulation of visual servoing of robot manipulator.

The artificial neural networks have computational ability by interconnection of the artificial neuron that is simple component of networks. In conventional neural networks, weight between neurons is an important parameter for behavior of network, because it is the objective of learning. However, weight is the factor that evolution time gets long in evolution. Therefore, in this paper, only architecture of neural networks can display its ability by chaotic neuron model with neighborhood connection. In addition, these make realization of instinct behavior that is innate in birth by genetic instruction.

References

1. Yoshikawa, T., Furuhashi, T., Uchikawa, Y.: Effect of New Mechanismof Development from Artificial DNA and discovery of Fuzzy Control Rules. Proc. of IIZUKA '96 (1996) 498–501
2. Boers, E.J.W., Kuiper, H., Happel, B.L.M., Kuyper, S.: Designing Modular Artificial Neural Networks. Proc. of Computer Science in the Netherlands (1993) 87–96
3. Gruau, F., Whitley, D.: Adding Learning to the Cellular Development of Neural Networks: Evolution and the Baldwin Effect. Evolutionary Computation **1**(3), (1993) 213–233
4. de Garis, H.: CAM-BRAIN: The Genetic Programming of an Artificial Brain Which Grows/Evolves at Electronic Speeds in a Cellular Automata Machine. Proc. of The First Int Conf. on Evolutionary Computation **1**, (1994) 337–339b
5. Sugisaka, M.: Design of an Artificial Brain for Robots. Proc. of The Third Int Symp. on Artificial Life and Robotics. (1998) (I-2)–(I-11)
6. Nagumo, J., Sato, S.: Response Characteristic of a Mathematical Neuron Model. Kybernetik, **10**, (1972) 155–164
7. Sipper, M.: Non-Uniform Cellular Automata: Evaluation in Rule Space and Formation of Complex Structures. Artificial Life VI (1994) 394–399

Robust Visual Tracking Via Incremental Maximum Margin Criterion

Lu Wang, Ming Wen, Chong Wang, and Wenyuan Wang

Department of Automation, Tsinghua University, Beijing, 100084, P.R. China
{l-wang02, wenm03}@mails.tsinghua.edu.cn

Abstract. Robust visual object tracking is one of the key problems in computer vision. Subspace based tracking method is a promising approach in handling appearance variability. Linear Discriminant Analysis(LDA) has been applied to this problem, but LDA is not a stable algorithm especially for visual tracking. Maximum Margin Criterion(MMC) is a recently proposed discriminant criterion. Its promising specialities make it a better choice for the tracking problem. In this paper, we present a novel subspace tracking algorithm based on MMC. We also proposed an incremental version of the corresponding algorithm so that the tracker can update in realtime. Experiments show our tracking algorithm is able to track objects well under large lighting, pose and expression variation.

1 Introduction

Visual object tracking is an essential computer vision task and it has wide range of applications in robotics, video surveillance and human-machine interaction. The main challenge of visual tracking is the difficulty in handling appearance variability of the target object. Although many tracking algorithm have been proposed, these algorithms premised that the target object models do not change over time and will lose the object consequentially.

Object tracking with subspace model was first proposed in[4]. Their method used a pre-trained eigen subspace to represent the object and didn't update it during the tracking process. In[2], Ross presented a incremental PCA based tracker to handle large lighting and pose variation. However, PCA is an unsupervised learning algorithm and ignore the valuable class label information in the tracking procedure. In order to separate the foreground object and the background, Lin[7] proposed a LDA based tracker. And Shen[1] extended it with kernel LDA. As we all known, LDA requires computing the inverse of the within-class scatter matrix. But this matrix is often non-singular. In[7], Lin avoided this rank deficient problem just by adding a identity matrix to the scatter matrix, however it is not a stable way especially in the tracking circumstances, where the bias accumulates quickly. And the manipulation of matrix inverse also causes additional instability and computation burden.

Recently a novel efficient and robust subspace learning algorithm namely Maximum Margin Criterion (MMC)[3] was proposed. Some specialities of MMC make it a better choice for our tracking problem. Firstly, it is a discriminant criterion.

J. Wang et al. (Eds.): ISNN 2006, LNCS 3972, pp. 398–403, 2006.

Secondly, it avoid the singularity problem in nature. And thirdly, it does not need matrix inverse manipulations, making it efficient and stable. In this paper, we present a probabilistic tracking algorithm based on incremental MMC subspace learning. Our algorithm maintains a low dimensional subspace adaptively to represent the appearance of the object. With updating the subspace incrementally, the algorithm is robust to the appearance variation of the object.

The rest of this paper is organized as follows. Section 2 explains the details of our tracking algorithm. Section 3 presents the experiments. In section 4, we draw conclusions and our future work.

2 Tracking Algorithm Via Incremental MMC

This section describes the details of our tracking algorithm. First we introduce the probabilistic tracking model. Then the MMC algorithm is briefly reviewed. The following subsection shows how we use MMC to learn the appearance subspace and how to update it incrementally. Finally, the whole tracking algorithm is given.

2.1 Probabilistic Model

The probabilistic model used here is a simplified variant of the well-known CONDENSATION[5] algorithm. We use a rectangular window to enclose the target in the image frames. At time frame t, the window's location $\mathbf{x}_t$ is represented by five parameters $\mathbf{x}_t = (x, y, s_x, s_y, \theta)$, which denoting the centroid of the rectangle, the width and length, the orientation angle respectively. Then the observation $\mathbf{z}_t$ is referred to the selected rectangular image region.

At each time frame t, we have the observation $\mathbf{z}_t$, together with all previous object locations $\mathbf{x}_{1:t-1}$. Our goal is to infer the latent state $\mathbf{x}_t$. Suppose that the dynamics process is Markov and its distribution is $p(\mathbf{x}_t|\mathbf{x}_{t-1})$. Here we simply define it as a Gaussian random walk, then $p(\mathbf{x}_t|\mathbf{x}_{t-1})$ can be write to:

$$N(x_t|x_{t-1}, \sigma_x^2)N(y_t|y_{t-1}, \sigma_y^2)N(s_x^t|s_x^{t-1}, \sigma_{s_x}^2)N(s_y^t|s_y^{t-1}, \sigma_{s_y}^2)N(\theta^t|\theta^{t-1}, \sigma_\theta^2) \quad (1)$$

where $N(.;\mu, \sigma^2)$ denotes a Gaussian distribution function with mean μ and variance σ^2.

On the other hand, given an observation $\mathbf{z}_t$, we model the likelihood that the object is located at $\mathbf{x}_t$ by the distribution $p(\mathbf{z}_t|\mathbf{x}_t)$. Its concrete formulation will be defined in Section 2.3.

According to our probabilistic model, we begin by drawing N particles from the prior $p(\mathbf{x}_t|\mathbf{x}_{t-1})$. Then for each particle $\mathbf{x}_t^i$, we compute its $p(\mathbf{x}_t^i|\mathbf{x}_{t-1})$ and $p(\mathbf{z}_t|\mathbf{x}_t^i)$. With Bayes' rule, $p(\mathbf{x}_t|\mathbf{z}_t, \mathbf{x}_{t-1}) \propto p(\mathbf{z}_t|\mathbf{x}_t)p(\mathbf{x}_t|\mathbf{x}_{t-1})$, we can get the MAP $\mathbf{x}_t^*$ of the posterior probability by(2). Finally we select the $\mathbf{x}_t^*$ as the estimated lactation at frame t.

$$\mathbf{x}_t^* = \arg\max_{\mathbf{x}_t^i} p(\mathbf{x}_t^i|\mathbf{z}_t, \mathbf{x}_{t-1}) \quad (2)$$

2.2 Discriminative Subspace Via Maximum Margin Criterion

Maximum Margin Criterion (MMC)[3] is a recently proposed feature extraction criterion. Let $\boldsymbol{X}_i = \{x_1^i, ..., x_{N_i}^i\}$ be samples from class i. MMC computes the projection matrix W by maximizing the following objective function:

$$J(W) = W^T(S_b - S_w)W \tag{3}$$

where

$$S_b = \sum_{i=1}^{n} N_i(m_i - m)(m_i - m)^T$$

$$S_w = \sum_{i=1}^{n} \sum_{x \in X_i} (x - m_i)(x - m_i)^T$$

are the between and within-class scatter matrixes respectively; m_i is the mean of class i; N_i is the sample number in class-i; m is the mean of the overall samples. The projection matrix W can be obtained by solving the following eigenvector decomposition problem: $(S_b - S_w)W = \Lambda W$.

Starting at the first frame, we use a detector to locate the object or just do it manually. At the beginning of the tracking, we use MMC to learn an initial subspace. That is, some image patches near the object in the frame will be collected as the positive samples, and some patches far from the object as negative samples. From these two set of samples, we will get the initial subspace using MMC. During the tracking procedure, to address the various variation of the object's appearance, we need to update the subspace model periodically. This will be explained in details in the next subsection.

2.3 Incremental Updating the Subspace

At some frame t, when the location $\mathbf{x}_t^*$ is determined, we can collect samples for our subspace update in this frame.

Firstly we define the likelihood distribution $p(\mathbf{z}_t|\mathbf{x}_t)$. In our algorithm it is defined as a Gaussian distribution:

$$p(\mathbf{z}_t|\mathbf{x}_t) \propto \exp(-\lambda \left\|W^T z_{\mathbf{x}_t} - W^T \overline{z}_p\right\|^2) \tag{4}$$

where the $z_{\mathbf{x}_t}$ is the observation correspond to the sample $\mathbf{x}_t$ and $\overline{z}_p$ is the mean of the observations of all the positive samples. λ is a positive constant. $\|.\|$ denotes the L_2 distance and W is the projection matrix constructed by MMC.

Using the above definition, we can compute the likelihood of the N location particles. Then the particles with large likelihood and also near the location $\mathbf{x}_t^*$ will be selected as object locations. While the particles with large likelihood but away from $\mathbf{x}_t^*$ will be seen as the background. At last, their observations will be stored as positive and negative samples respectively.

When some new samples are collected, they are used to update the subspace. We proposed a method to update the subspace by incrementally updating the class scatter matrix and mean, so that the tracker does not need to store the previous samples and only store the mean and scatter matrix. We also found the similar method in [6].

Suppose that the old model was construct with N_p positive samples and N_n negative samples. The mean of the old positive and negative classes are m_p and m_n. And the within-class scatter matrixes of the two classes are S_p and S_n.

During the tracking process, we have newly collected L_p positive samples $\{x'_1, ..., x'_{L_p}\}$ and L_n negative samples $\{y'_1, ..., y'_{L_n}\}$. And the means and within-class scatter matrixes of these new samples are m'_p, m'_n, S'_p, S'_n respectively.

Now our goal is to compute the updated S^*_b and S^*_w using only the above parameters. As we can see easily, if we get the updated with-class scatter matrix S^*_p and S^*_n , it is easy to get S^*_b and S^*_w. And we show that:

$$S^*_p = S_p + \frac{L_p(L_p + 2N_p)}{(N_p + L_p)^2} S'_p + \frac{N_p L_p^2}{(N_p + L_p)^2}(D_p) + \frac{N_p^2}{(N_p + L_p)^2}(E_p) \tag{5}$$

where D_p and E_p are define below:

$$D_p = (m'_p - m_p)(m'_p - m_p)^T \tag{6}$$

$$E_p = \sum_{i=1}^{L_p} (x'_i - m_p)(x'_i - m_p)^T \tag{7}$$

Since S^*_n can be updated similarly, we omitted here.

2.4 Proposed Tracking Algorithm

Putting the inference, subspace modelling, incremental updating modules together, we present an adaptive tracking algorithm as follows:

1. **Initialization:** Decide the initial location $\mathbf{x}_0$ manually or by a detector. By applying small perturbations to $\mathbf{x}_0$ and crop the corresponding image regions as object samples. On the contrary, get the background samples by applying larger perturbations to $\mathbf{x}_0$. Then learn a initial subspace W_0 using MMC.
2. **Sampling:** Draw N location particles from prior distribution $p(\mathbf{x}_t|\mathbf{x}_{t-1})$ over locations. For each location, crop the corresponding image region as observation.
3. **Location Inference:** Compute $p(\mathbf{x}^i_t|\mathbf{x}_{t-1})$ and $p(\mathbf{z}_t|\mathbf{x}^i_t)$ of every particles using (1) and (4). Then find the MAP of $p(\mathbf{x}_t|\mathbf{z}_t, \mathbf{x}_{t-1})$ using (2). And set the new location at the MAP.
4. **Subspace Update:** For the interval of M frames, collect L_p positive samples and L_n negative samples. Update the S_b and S_w using (5) and (6), (7). Run MMC to update the subspace.
5. Go to step 3

3 Experiments

3.1 Experiment Settings

To test whether the proposed tracking algorithm with incremental MMC works well, we run some experiments on two well-known face tracking video sequences, which can be found at[1]. Both of the sequences were taken in office environment with resolution 320×240 and 30 frames per second.

For all these experiment, the affine parameters of prior dynamic distribution were: $\sigma_x = \sigma_y = 5$, $\sigma_{s_x} = \sigma_{s_y} = 2$, $\sigma_\theta = 0.1$. In the initialization, the prior MMC model was trained using 100 object samples and 100 background samples at the first frame. These samples are obtained by small perturbation and large perturbation to the initial location respectively. During the tracking procedure, we used 50 random particles at each frame to search the MAP. In each frame, after the location of object was decided, 5 new target object samples and background samples were selected and stored as the positive and negative samples. When every 10 frames passed, the subspace was incrementally updated.

The experiments contained two types for easy comparison: with and without the incremental subspace update. That is, in the latter experiments, the subspace model was not updated after the first frame.

3.2 Experiment Results

Firstly, we tested the algorithm with incremental MMC subspace updating. Figure 1 and Figure 2 show some snapshots of the tracking results in the two test video sequences. In the first sequence, the face undergoes large pose variation. And the face in the second sequence undergoes large illumination changes and facial expression variation. Our tracking algorithm is stable even if the face undergoes sharp illumination changes, pose and facial expression variation.

Fig. 1. Large pose variation. Frame number 2, 88, 391, 431.

We also tested the algorithm without subspace updating. In these experiments, the subspace model was fixed after the initialization. We found that the tracker lost the object after a short period time in both sequences, especially when the variation were large. The tracking snapshots are omitted here.

[1] http://www.cs.toronto.edu/ dross/ivt/

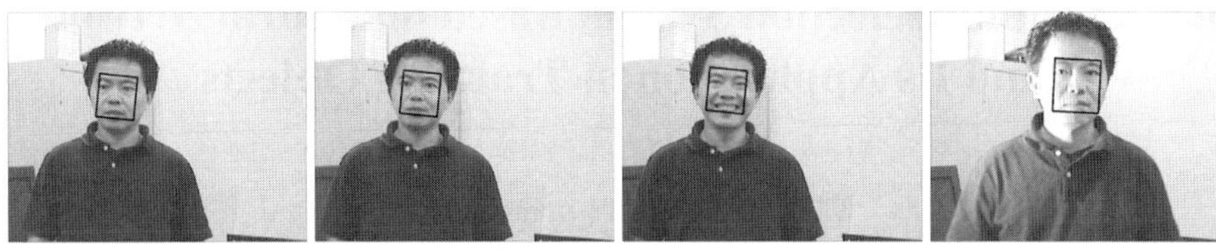

Fig. 2. Expression and illumination variation. Frame number 131, 185, 297, 1150.

4 Conclusions and Future Work

In this paper, we have presented a probabilistic object tracking algorithm with adaptive subspace learning. The proposed method is based on an incremental Maximum Margin Criterion algorithm. The experiments show the robustness of the tracking algorithm through large appearance variation. However, the problem of occlusion is not considered, and left for future work. In addition, we can also try better sampling and inference strategy.

References

1. Shen, C., Hengel, A.V.D., Brooks, M.J.: Visual Tracking via Efficient Kernel Discriminant Subspace Learning. In: Proceedings of IEEE Conference on Image Processing 2 (2005) 590 - 593
2. Ross, D., Lim, J., Yang, M.-H.: Adaptive Probabilistic Visual Tracking with Incremental Subspace Update. In: Proceedings of the Eighth European Conference on Computer Vision 2 (2004) 470 - 482
3. Li, H., Jiang, T., Zhang, K.: Efficient and Robust Feature Extraction by Maximum Margin Criterion. In: Proceedings of the Advances in Neural Information Processing Systems 16. MIT Press (2003)
4. Black, M.J., Jepson, A.D.: Eigentracking: Robust Matching and Tracking of Articulated Objects Using View-based Representation. International Journal of Computer Vision 26 (1998) 63 - 84
5. Isard, M., Blake, A.: CONDENSATION- Conditional Density Propagation for Visual Tracking. International Journal of Computer Vision 29 (1998) 5 - 28
6. Pang, S., Ozawa, S., Kasabov, N.: Incremental Linear Discriminant Analysis for Classification of Data Streams. IEEE Trans. on Systems, Man, and Cybernetics - Part B 35 (2005) 905 - 914
7. Lin, R., Yang, M.-H., Levinson, S.E.: Object Tracking Using Incremental Fisher Discriminant Analysis. In: Proceedings of the 17th International Conference on Pattern Recognition 2 (2004) 757 - 760

An Attention Selection System Based on Neural Network and Its Application in Tracking Objects

Chenlei Guo and Liming Zhang

Electronic Engineer Department, Fudan University, Shanghai, 200433, China
chenlei_guo@hotmail.com, lmzhang@fudan.edu.cn

Abstract. In this paper an attention selection system based on neural network is proposed, which combines supervised and unsupervised learning reasonably. A value system and memory tree with update ability are regarded as teachers to adjust the weights of neural network. Both bottom-up and top-down part are to simulate two-stage hypothesis of attention selection in biological vision. The system is able to track objects that it is interested in. Whenever it lost focus on tracked object, it can find the object again in a short time.

1 Introduction

Attention selection is an interesting topic. What can attract people's attention? Luminous color, moving objects, edges probably guide an unconscious attention. Another kind of attention is a conscious one such as audience attempting to find magician's secret. In 1980, Triesman put forward a feature integration theory [1] and a two-stage architecture: pre-attention stage for original image and decision stage made by the subjective information. In 1994, Wolfe proposed a Guided Search 2.0 model [2], which kept the two-stage architecture; he called the first stage as Bottom-up and the second one as Top-down. Based on these results done by these biologists, in 1999, C. Breazeal et al. proposed a kind of attention selection system [3] which implements Wolfe's model. Triesch, J. and Malsburg proposed a system [4] with strong adaptation to the tracking environment for detecting and tracking faces which integrates five different cues. However, their systems lack strong subjective guidance, thus when its attention is interfered, these systems will lose their object. This paper proposed a non task-specific attention selection system based on neural network, in which the combination of top-down and bottom-up information is achieved by Self-organization neural network.

2 System Architecture

Fig. 1 shows the architecture of attention selection system, which is divided into two parts: top-down and bottom-up from activation map. Activation map is constructed by neurons in 2-dimension array shown in Fig.2. The number of the neurons is the same as that of pixels of input image. Bottom-up includes motion, color and orientation filter, by which their saliency maps denoted by I_M , I_C and I_O, with the same size as

J. Wang et al. (Eds.): ISNN 2006, LNCS 3972, pp. 404 – 410, 2006.

input image I ,can be obtained. Their output units, denoted as M, C and O, are multiplied by weights w_1 , w_2 and w_3 as the input of corresponding neuron shown in Fig. 2. Top-down part consists of prediction module, value system and memory. Each unit of prediction map I_P is denoted as P, whose weight is w_4 . Note that the connection weights of all units for one saliency map are the same and they are adapted simultaneously in update. Moreover each neuron on activation map is also connected to its surrounding neurons in a $N \times N$ neighborhood N_i by connection weights S_{ik} (See Fig. 2). A self-organized unsupervised learning is used in the neural network. The position and output of winner neuron in the neural network are then sent to the value system. Its position is regarded as the candidate-area of attention at this time. Value system works as supervised learning's teacher of memory by judging whether the attention area matches the memory. At the same time, value system can adjust the connection weights w_1 , w_2 , w_3 and w_4 , so that the system can automatically adapt to the tracking environment in a short time.

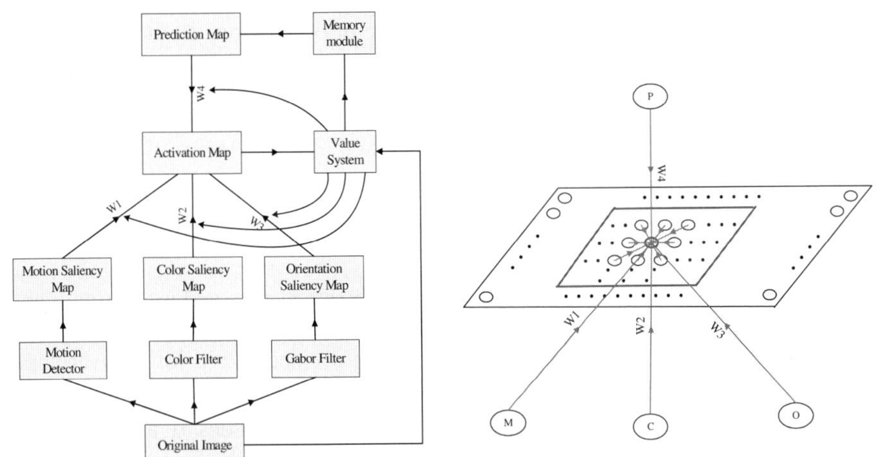

Fig. 1. System Architecture

Fig. 2. Neurons and connection on activation map

3 Implementation of Bottom-Up

Bottom-up of the system consists of 3 parts: motion, color and orientation. Motion saliency map[4] is realized by the differences between two consecutive frames in the video. Let $I(t)$ and $I(t-1)$ be input images at time t and $(t-1)$ respectively. The saliency map of motion is shown as

$$I_M(t) = \varphi(|I(t) - I(t-1)| - \eta) \tag{1}$$

where η is the threshold, typically $\eta = 15$. $\varphi(s)$ is step function: $\varphi(s) = 1$ for $s > 0$, otherwise $\varphi(s) = 0$.

Color saliency map is realized by the function below:

Let $I_R(\mathbf{x},t)$, $I_G(\mathbf{x},t)$ and $I_B(\mathbf{x},t)$ be the pixel's intensity corresponding value of RGB pathways. The color saliency map is given by

$$I_c(\mathbf{x},t) = \begin{cases} 1 & \begin{array}{l} \left|I_R(\mathbf{x},t) - \overline{I}_{R0}\right| < \sigma_R \cap \\ \left|I_G(\mathbf{x},t) - \overline{I}_{G0}\right| < \sigma_G \cap \left|I_B(\mathbf{x},t) - \overline{I}_{B0}\right| < \sigma_B \end{array} \\ 0 & otherwise \end{cases} \tag{2}$$

where $\overline{I}_{R0}$, $\overline{I}_{G0}$, $\overline{I}_{B0}$ are means and σ_R, σ_G, σ_B are variance corresponding to RGB channels, which are predefined. See [4] for relevant work.

Orientation saliency map is realized by Gabor filter. The coefficient of 2-D Gabor filter can be described as below:

$$f_o(x-x_0, y-y_0) = \exp\left[-\frac{1}{2}\left(\frac{(x-x_0)^2+(y-y_0)^2}{\sigma^2}\right)\right] \times \sin\left[\omega(x\cos\theta + y\sin\theta)\right] \tag{3}$$

where (x_0, y_0) is the centre point, θ is the angle of orientation, ω is the frequency of the filter, σ is special coefficient. Edges in different orientations are detected by Gabor filters. In this paper, coefficient of Gabor filter is $\omega\sigma = 1.8$, four angles of orientation are used $(\theta = -45°, 0°, 45°, 90°)$.

4 Implementation of Top-Down

Top-down of the system includes 3 parts: memory, value system and prediction map.

Memory is a search tree that use some predefined examples to learn by IHDR algorithm [5]. In training stage the example images are divided into a lot of overlapped sub-blocks with $l \times l$ size which belong to environment or interested objects. These sub-blocks are clustered into m classes just like m normal nodes If a normal node contains sub-blocks that only belong to pure environment or interested object, such node is called a leaf node. Otherwise, the node is continually split until leaf node. Every node is represented by the mean of its sub-blocks. In testing period, search the tree by comparison of testing block and node's mean to decide whether the block needs to be paid attention to or not. IHDR algorithm has short retrieval time and high accuracy. Moreover, the memory tree can be updated at real time, for example, when a new block is inputted, if the leaf node contains sub-blocks that belongs to both environment and objects, it can be split new leaf node in order to adjust its memory.

Prediction mechanism is described below. At time $(t-1)$ and $(t-2)$ the position of the object to be interested in is denoted as $\hat{p}_{t-1}(x_{t-1}, y_{t-1})$ and $\hat{p}_{t-2}(x_{t-2}, y_{t-2})$ respectively. According to the simple prediction mechanism, the position, at which the object maybe appears, can be calculated as:

$$\hat{p}_t(x_t, y_t) = \hat{p}_{t-1}(x_{t-1}, y_{t-1}) + \beta \cdot (\hat{p}_{t-1}(x_{t-1}, y_{t-1}) - \hat{p}_{t-2}(x_{t-2}, y_{t-2})) \tag{4}$$

where β is adaptation coefficient, typically $0.5 \le \beta \le 1.5$. β helps to adjust the accuracy of prediction smoothly. Extract the sub-block which is centered around $\hat{p}_t(x_t, y_t)$ by following equation.

$$R(x_t, y_t, t) = \{x, y / |x - x_t|, |y - y_t| \le l\} \tag{5}$$

Then input $R(x_t, y_t, t)$ into memory tree and decide whether it belongs to an interested object. If so, return $\hat{p}_t(x_t, y_t)$ as the result. Otherwise, set $\hat{p}_t(x_t, y_t) = \hat{p}_{t-1}(x_{t-1}, y_{t-1})$. Then search $N \times N$ area centered around $\hat{p}_t(x_t, y_t)$ and find the most familiar sub-block. Here $N > l$.

Value system acts as a guider to adjust the weights of each pathway. The input of Value system is from winner neuron of Activation map, which can be described as:

$$y_i = \sum_{j=1}^{4} w_j u_{ji} + \sum_{k \in N_i} s_{ik} y_k \quad \text{and} \quad \sum_{i=1}^{4} w_j = 1 \tag{6}$$

Where u_{ji} is the output of the i_{th} pixel in the j_{th} map, $j = 1, 2, 3, 4$ represent the index of M, C, O, P maps shown in Fig. 1, and the neuron i's neighborhood N_i is set as $N \times N$. S_{ik} is the connection weight between neuron i and neurons k within the neighborhood N_i, which likes a Gaussian shape. By competition, the winner neuron can be obtained by $p_{win} = \arg\max_i (y_i)$. Let $p_{win}(x_{win}, y_{win}, t)$ represent the coordinates and time of the winner neuron.

In Eq.(6) , For each y_k $k \in N_i$, has its input u_{jk} from saliency maps M, C, O, P. Thus, Eq.(6) can be described as:

$$y_i = \sum_{j=1}^{4} w_j u_{ji} + \sum_{k \in N_i} s_{ik} (\sum_{j=1}^{4} w_j u_{jk}) = \sum_{j=1}^{4} w_j (u_{ji} + \sum_{k \in N_i} s_{ik} u_{jk}) = \sum_{j=1}^{4} \tilde{D}_j \tag{7}$$

where $\tilde{D}_j, j = 1, 2, 3, 4$ is regarded as the absolute contribution corresponding to each saliency map, which is sent into value system.

Value system obtains two inputs from Activation map: the position of winner neuron and $\tilde{D}_j$. Normalize $\tilde{D}_j, j = 1, 2, 3, 4$ as relative contribution of saliency map:

$$D_j = \frac{\tilde{D}_j}{\sum_{j=1}^{4} \tilde{D}_j}, \; j = 1, 2, 3, 4 \tag{8}$$

Extract sub-block round winner position p_{win} and sent it into memory tree to test. If the object is to be interested in, the more contribution a saliency map makes, the larger connection weight activation map will be and vice versa. The weight adaptive formula is shown as follows:

$$\tilde{w}_j = w_j \cdot (1 + \alpha_j D_j), \qquad w_j = \frac{\tilde{w}_j}{\sum_{j=1}^{4} \tilde{w}_j}, \qquad j = 1, 2, 3, 4 \tag{10}$$

Here $\alpha_j, j = 1,2,3,4$ is coefficient of each pathway. If the attention is locked on the interested object, $\alpha_j > 0$, otherwise, $\alpha_j < 0$.

5 Experimental Results

There are two experiments: one is to test our idea with prediction mechanism based on memory; the other is to test the ability to lock on the lost object again and to adapt to the environment.

Exp 1. A video sequence with size 240×320 at 10f/S including similar color moving ball and cup is considered as a test sample. We hope to track the red ball using three different systems in two testing groups 100f/per group. System A is without prediction mechanism like that proposed by [2]; System B is with a simple prediction mechanism, like that proposed by [4]; System C is what we proposed. All the systems are in the same conditions. Fig. 3 shows the demo of our proposed system. Table 1 shows results of the experiment. Our System C performs best from Table 1.

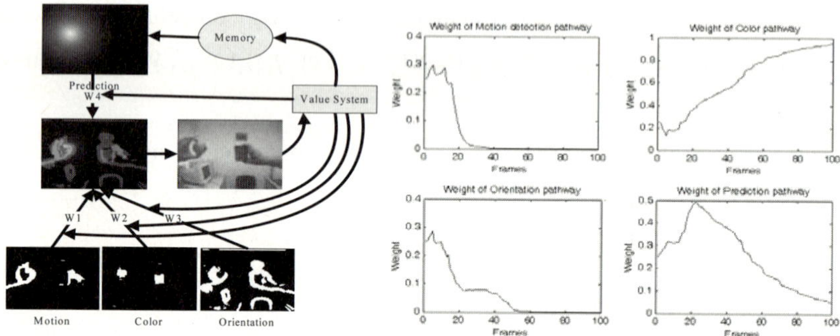

Fig. 3. Mechanism of system C

Fig. 4. Update process of connection weights while system C is tracking the ball

Table 1. Comparison among three different attention selection systems applied to tracking

System Type	Group A			Group B		
	Error	All	Accuracy	Error	All	Accuracy
System A	39	100	61.00%	32	100	68.00%
System B	11	100	89.00%	16	100	84.00%
System C	6	100	94.00%	4	100	96.00%

Fig. 4 shows the update process of weights when System C is tracking the red ball. In starting period, motion and orientation information play an important role in finding the object system interested in, so their weights are high at initial stage of tracking. However, when the object is locked, their weights is lower and lower, which means that their influence declines. Weight of color is always rising for the target is locked most of time. It has been shown by Table 1 that prediction mechanism based on memory is more effective and robust.

Exp2. shows the system's ability to lock on the lost object again. The first row in Fig. 5 illustrates tracking results from 1st to 40th frame. The dark cross is the focus of attention. At 3rd frame, system focused attention on the ball; however, attention was shifted to the hand at 6th frame. System found the ball again by adjusting weights at 11th frame. The second row in Fig. 5 shows that our system still can track object successfully when the light changes. The blue cross represents the focus of attention.

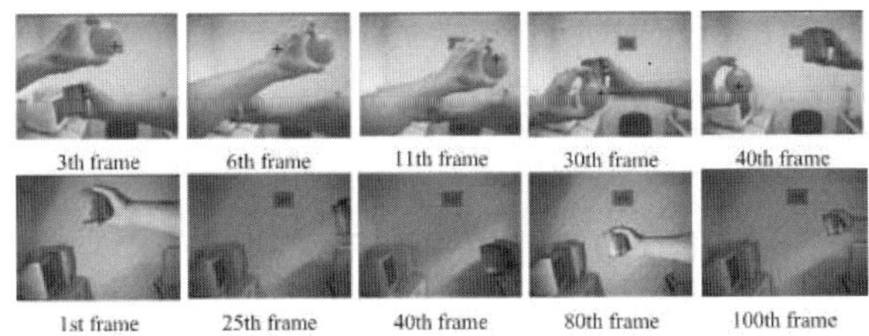

Fig. 5. The First row is tracking results of Exp.2. Video frames range from 1st to 40th. The second row is tracking results in an environment with light change.

6 Conclusions

In this paper, an attention selection system based on neural network is proposed on the basis of the results of biologists, which is applied to track objects. The system combines both supervised and unsupervised learning, and introduces HDR tree as the memory. Value system guide supervised learning and update of weights between neurons. Experimental results show that our proposed system performs better than other systems and has the ability to find the object again when losing focus on it.

Acknowledgements

This work was supported by NSF (No.60571052, 30370392) and Shanghai Science and Technology Committee (No.045115020).

References

1. Treisman, A., Gelade, G.: A Feature-Integration Theory of Attention. Cognit. Psychol. 12 (1980) 97–136
2. Wolfe, J.M.: Guided Search 2.0: A Revised Model of Visual Search. Psychonomic Bulletin & Review 1(2) (1994) 202-238

3. Breazeal, C., Scassellati, B.: A Context-Dependent Attention System for a Social Robot. In Proc. Int. Joint Conf. Artificial Intelligence (1999) 1146--1151
4. Triesch, J., von der Malsburg, C.: Democratic Integration: Self-organized Integration of Adaptive Cues. Neural Computation 13(9) (2001) 2049-2074.
5. Weng, J., Zhang, Y., Hwang, W.: Incremental Hierarchical Discriminant Regression for Online Image Classification, Document Analysis and Recognition, 2001. Proceedings of Sixth International Conference on 10-13 Sept. (2001) 476 – 480

Human Motion Tracking Based on Markov Random Field and Hopfield Neural Network

Zhihui Li and Fenggang Huang

School of Computer Science and Technology, Harbin Engineering University, Harbin 150001, China
lizhh@126.com

Abstract. This paper presents a method of human motion tracking based on Markov random field and Hopfield neural networks. The model of rigid body motion is first introduced in the MRF-based motion segmentation. The potential function in MRF is defined according to this motion model. The Hopfield neural network is first used in the implementation of MRF to take advantage of some mature Neural Network technique. After the introduction of the model of rigid body motion the joint angles of human body can be estimated .It is also helpful to the estimation of the proportions of human body, which is significant to the accurate estimation of human motion. Finally the experimental results are given and the existed problems in this method are pointed out.

1 Introduction

The visual analysis of human's motion is one of the most active research topics. It aims at analyzing and processing the image sequence of human motion and involves motion detection, moving objects classification, human tracking, and activity understanding and description. Tracking derived on the basis of movement detection and object classification may be considered to be equivalent to establishing coherent relations of image features between frames with respect to position, velocity, shape, texture, color, etc. After tracking the parameters of human motion changing with time can be derived. These parameters can be used in the further activity such as recognition.

Recent years some progress achieved in this domain and there are several surveys on this topic .See references [1]-[5]. The main present problem is how to segment the moving human body in fast and correct way. The present method can estimate the joint angle of human body but can not calculate the proportion of the parts of human body. Besides , most system can not solute the problem of occlusion in human motion effectively. This paper presents a method based on Markov random field (MRF) which can determine the joint positions as well as joint angles.

2 MRF-Based Tracking of Human Motion

Segmentation of moving object based on MRF mode is a very active topic .This approach describes the distribution of image data as 2-D random field. It can achieve the global coherence based on local correlation and the consistence of MRF and Gibbs

J. Wang et al. (Eds.): ISNN 2006, LNCS 3972, pp. 411 – 418, 2006.

distribution and by determining the conditional probability through potential functions. Motion segmentation by MRF has two advantages. One is its inherent local coherence, another is that the occlusion can be disposed by analysis several frames of images[6].

2.1 MRF Model

MRF model is the result that one dimension causal Markov chain extends into two dimensions. Hammersley-Clifford theorem assured the equivalence between MRF and Gibbs distribution [7][10].

MRF-based motion segmentation is to determine the regions of motions and the motion model of the regions under the prior observation field D and a series of constraints. That is to acquire the motion field which make the posterior probability maximum under the prior observation field D and a series of constraints. It is to obtain

$$v^* = \max_{v \in V} P(V = v \mid D = d) \tag{1}$$

According to Bayesian and Hammersley—Clifford theorem, the follow equation can be derived:

$$\begin{aligned} v^* &= \max_{v \in V} P(V = v \mid D = d) \propto \max_{v \in V} P(V = v) P(D = d \mid V = v) \\ &\propto \min_{v \in V} U(v \mid d) \propto \min_{v \in V} \{U(v) + U(d \mid v)\} \end{aligned} \tag{2}$$

2.2 Model of Human Body Motion

The majority researches on motion field estimation are based on the affine movement. The method in this paper is based on rotation of rigid body. For it is about pure silhouette of human body the 2-D rotational model is enough . Assume that the image function is f the motion model is as follow:

$$\hat{p}_2 = \phi(p_1, v) = \begin{bmatrix} \cos\theta & -\sin\theta \\ \sin\theta & \cos\theta \end{bmatrix} \cdot (p_1 - p_0) + p_0 + t_s \tag{3}$$

Where p_0 is the joint position vector or rotation axis, p_1 is the position vector of a pixel in current frame , $\hat{p}_2$ is the estimated position of p_1 which moved to next frame. Θ is the angle of rotation, t_s is shift, ϕ is motion model. Because p_0、t_s and θ determines an field uniquely, the motion field model parameter is defined as $v = (p_0, t_s, \theta)$.

The joint positions and angles can be conveniently derived by motion model of the rigid body.

2.3 Definition of Energy Functions

2.3.1 Definition of Clique Potential

U(v) is the function of clique potential which demonstrates the local relational constrains of a image. The second order neighborhood system(showing in figure 1) is chosen before the energy function is defined.

$$U(v) = \sum_{c \in C} V_c(v_s, v_n) \tag{4}$$

$V_c(v_s,v_n)$ is the clique potential, v_s,v_n are the parameters of the motion field of the neighboring pixel in the dualistic clique. To simplify the computation, we select c=(s,n) as the dualistic clique of a second order neighborhood system showing in Figure 1. s is the current pixel, n is a pixel in the second order neighborhood system N(I,j), and C is the set of all clique in the image. In order to introduce the Consistency constrains to the model and delete the isolated points caused by noise, the energy function is defined as the distance between v_s and v_n.

$$V_c(v_s, v_n) = \left\| v_s - v_n \right\| \tag{5}$$

(i-1,j-1)	(i,j-1)	(i+1,j-1)
(i-1,j)	(i,j)	(i+1,j)
(i-1,j+1)	(i,j+1)	(i+1,j+1)

Fig. 1. Second order neighborhood system N(i,j)

2.3.2 Conditional Probability Model

$p(d \mid v)$ can be regarded as the relationship between the observation fields and the motion fields while the motion fields are known. The observation field and the energy function in conditional probability model is defined by follow equation.

$$U(d \mid v) = d = \left\| f_1(p_1) - f_2(\hat{p}_2) \right\| \tag{6}$$

Where f_1 is the image function of current frame and f_2 is that of next frame. $p_1$and p_2 are same as in equation 3.Conditional probability model containing the motion model shows the relationship between the observation field and the motion field.

2.3.3 Total Potential Function

The total potential function is as follow.

$$U(v \mid d) = \beta_1 \cdot U(v) + \beta_2 \cdot U(d \mid v) \tag{7}$$

$\beta_1$and β_2 are the constants to adjust the ratio of the two type of potential functions. There are two parts in the definition of the MRF-based potential function. The first part contains the coherent constrains, and the second one includes the motion model. Since there is no coherent constrain in other approaches of the estimation of the motion field, it is the advantage of the MRF-based estimation.

3 Heteromorphic Hopfield Neural Network

MFR-based motion segmentation can be summed up to find the minimal solution of the energy function. Simulated annealing, Metroplis sampler and ICM(Iterated

Conditional Mode)algorithm are usually used to realize it. However, in this paper, the discrete Hopfield neural network are adopted to make use of some mature Neural Network technique. But the discrete Hopfield Neural Network must be improved in order to adapt the MRF algorithm—so we call it "Heteromorphic Hopfield neural network ".

3.1 Definition of Heteromorphic Hopfield Neural Network

Hopfield neural network is a kind of full connected Neural Network, which can be improved to realize the MRF algorithm. The neuron structure of heteromorphic Hopfield Neural Network is shown in figure 2. Each pixel of the objective region corresponds to a neuron, whose state is expressed by $u_{i.}$:$u_i \in \{0, 1, \cdots, k-1\}$.

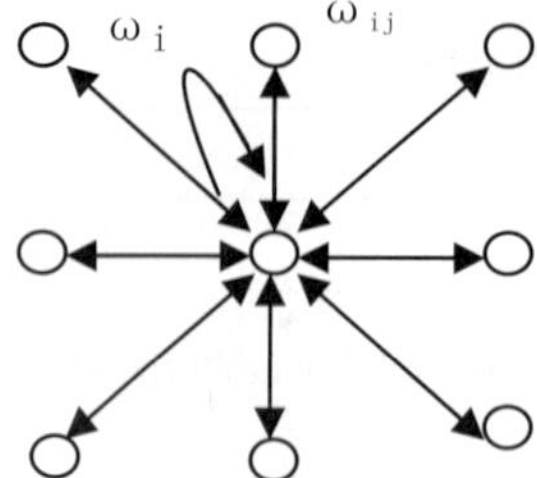

Fig. 2. Neuron structure of heteromorphic Hopfield Neural Network

Each state of the neuron corresponds to a class or a motion field and motion parameter v_i. Every neuron is connected with others in its neighborhood. Every neuron outputs data to others in its neighborhood by the connected weight and receives the information from others at the same time.

The energy of every neuron is defined as follow

$$H_i = \beta_1 \cdot \sum_{j=1}^{8} w_{ij} + \beta_2 \cdot w_i \tag{8}$$

$$w_{ij} = V_c(v_s, v_n) \quad w_i = U(d \mid v)$$

$$u_i(t+1) = \begin{cases} u_{in} & H_i(t+1) - H_i(t) < 0 \\ u_i(t) & otherwise \end{cases} \tag{9}$$

H_i is the energy and ω_i and ω_{ij} are the connected weight of each neuron. u_{in} is the possible new state of u_i.

3.2 Running Rule of Heteromorphic Hopfield Neural Network

(1) At time t select a neuron u_i from the network randomly. Let $u_{in}=u_l$,$u_l \in \{0, 1, \cdots, k-1\}$.$u_{in}$ is every state except u_i in turn.

(2) According to equition 8 calculate the weighted total of all inputs for the selected neuron.

(3) According to equition 9 compute the output of u_i at time t+1.
(4) All neurons' output keep invariable except u_i.
(5) Return (1) till the network enters steady state.

Network energy function can be defined as the sum of all neurons' energy(equation 8). Only when the energy reduces the neuron's state will change. So the network energy is always non-increased. So the network must be convergent and tend to the stable state of balance.

3.3 Difference Between Heteromorphic Hopfield and Classical Discrete Hopfield Neural Network

(1) The number of the state of each neuron in heteromorphic Hopfield neural network is the number of segmentation category, while that of classical discrete the Hopfield neural network is two(only 0 and 1).

(2) All neurons of classical discrete the Hopfield are connected to each other while only neighboring neurons of heteromorphic Hopfield neural network has the connections. The neighboring way of neurons are same as that of pixels in image.

(3) The weights of classical discrete the Hopfield neural network are constant while those in heteromorphic Hopfield neural network are not fixed and related to neuron's condition .

4 Algorithm and the Results of Experiment

The approach is applied to CASIA database [9] to estimate the motion of human body. In order to reduce the computational complexity the background subtraction is firstly used to obtain a rough body area on which the motion field will be estimated.

4.1 Evaluation of the Initial Value

The ICM algorithm is used to evaluate the initial value. For every pixel in target region calculate the motion parameter which make the error of every pixel minimum. The error is he sum of the error of each pixel and its neighboring pixels.

$$v^* = \arg\min_{p_0 \in R} \sum_{s \in \eta} [f_1(p_1) - f_2(\hat{p}_2)]^2 \tag{10}$$

The rotation axis p_0 of every pixel should be in the target area R. The meaning of equation 16 is that v is the number which makes error minimum under the constraint of $p_0 \in R$. After calculating the motion parameter of every pixel we use an unsupervised clustering method[8] to cluster all pixels into several categories and obtain the number of cluster-- k and the initial parameter v of every motion filed. This approach has the advantage of exactness but the computation is complex. In practical application we first calculate the error of shift for all pixel in objective region then compute the motion field parameters only for the pixels whose errors are large and cluster them finally. It is proved by the fact that this method reduced computation effectively and keeping the exactness at same time.

In figure 3 (a)(b) are the first and second frame in image sequence.(c) is the positions of estimated rotation axes. Table 1 includes the parameters of motion fields corresponding to the rotation axes.Column 6 are the parameters of shift motion.

Table 1. Parameters of motion fields in experiment

	1	2	3	4	5	6
p_0	(18,110)	(26,115)	(14,152)	(24,151)	(28,172)	(0,0)
θ	0.06	-0.12	-0.08	0.08	-0.04	0
t_s	(4,0)	(4,0)	(4,0)	(4,0)	(6,-2)	(4,0)

4.2 Determination of Motion Fields

The method in session 3 is used to determine the motion field. Each pixel in the target area corresponds to a neuron. Every time we choose one of neurons to disturb its state until the network enter stable state .Then the state of every neuron corresponds the motion field which the counterpart pixel of the neuron belongs to. The MRF algorithm considers both the motion model and local coherence. The ratio of the two parts can be adjusted by changing the values of β_1 and β_2. In this application, the proportion of β_1 and β_2 is 1:2.

(d)(e)in figure 3 show the pixels belonging to the first motion field. There are some isolated noise points before using MRF. MRF can reduce such noise points effectively for its local smoothing while the motion model works at same time. There are some errors in the results (the small region in (e)). Figure 4 shows the estimated motion fields.

There are some problems in the application of this method. The first one is that not all joint positions can be estimated (5 joints were obtained in this example). The second one is that there are some errors in application. The main cause of the error is that there are often some nonrigid deformation and occlusions in the human motion. These problems should be solved by estimation by several frames.

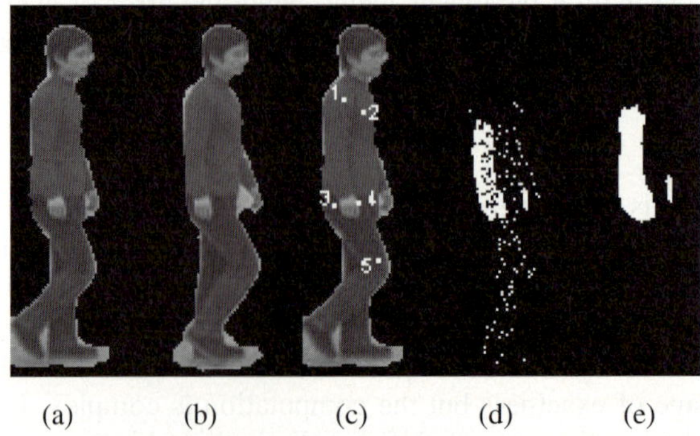

(a) (b) (c) (d) (e)

Fig. 3. Applicated example and experimental results. (a)(b) are the first and second frame in image sequence.(c) is the positions of estimated rotation axes. (d)(e) show the pixels belonging to the first motion field.

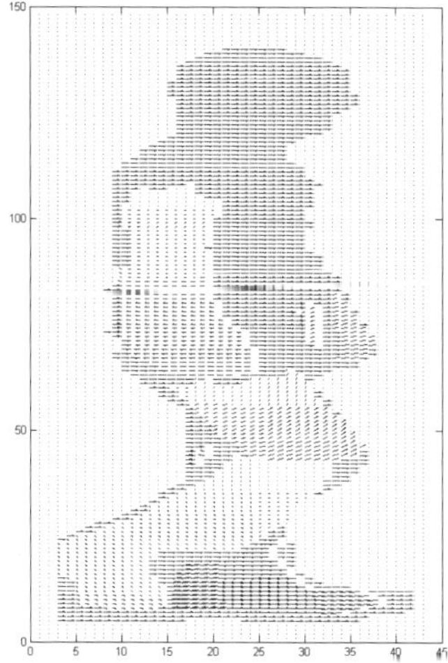

Fig. 4. The segmented motion fields

5 Conclusion

This paper presents a method of human motion tracking based on MRF and Hopfield neural networks. The model of rigid body motion is first introduced in the motion segmentation. The potential function in MRF is defined according to this motion model. The Hopfield neural network is first used in the implementation of MRF to take advantage of some mature Neural Network technique. After the introduction of the model of rigid body motion not only the joint angles of human body but also the joint positions of human body can be estimated, which is significant to the accurate estimation of human motion. Because this is the first time to introduce this motion model some problems happened in application. Further efforts should be made to improve the precision of estimation, realize the solution of occlusion and reduce the computation.

References

1. Aggarwal, J.K., Cai, Q., Liao, W., Sabata, B.: Articulated and Elastic Non-Rigid Motion: A Review. In: IEEE Workshop on Motion of Non-Rigid and Articulated Objects (1994) 2–14.
2. Cedras, C., Shah, M.: Motion-Based Recognition: A Survey. Image Vision Comput. 13 (2) (1995) 129–155.
3. Aggarwal, J.K., Cai, Q.: Human Motion Analysis: A Review. In: IEEE Workshop on Motion of Non-Rigid and Articulated Objects (1997) 90–102.
4. Aggarwal, J.K., Cai, Q.: Human Motion Analysis: A Review. Comput. Vision Image Understanding 73 (3) (1999) 428–440.
5. Wang, L., Hu, W., Tan, T.: Recent Developments In Human Motion Analysis. Pattern Recognition 36(3) (2003) 585 – 601.

6. Stiller, C., Konrad, J.: Estimating Motion in Image Sequences: A Tutorial on Modeling and Computation of 2D Motion. IEEE Signal Process. Mag.16(7)(1999) 70-91.
7. Konrad, J., Dubois, E.: Bayesian Estimation of Motion Vector Fields. IEEE Trans. Pattern Anal. Machine Intell. 14(9) (1999) 910-927
8. Figeneiredo, M.A.T., Jain, A.K.: Unsupervised Learning of Finite Mixture Models. IEEE Trans. Pattern Anal. Machine Intell. 24(3)(2002) 381-396
9. CASIA Gait Database. Http://Www.Sinobiometrics.Com.
10. Zhong, X.R., Xiao S.B.: Automatic Segmentation of Moving Object Based on Gaussian Markov Random Field Models. Journal of Suzhou University (Natural Science Edition) 19(3)(2003)78-83.
11. Kurita, N., Funahashi, K.L. On The Hopfield Neural Network and Mean Field Theory. Neural Networks 9(9) (1996)1531-1540.

Skin-Color Based Human Tracking Using a Probabilistic Noise Model Combined with Neural Network

Jin Young Kim[1], Min-Gyu Song, Seung You Na[1], Seong-Joon Baek[1], Seung Ho Choi[2], and Joohun Lee[3]

[1] Dept. of Electronics Eng., Chonnam National University, 300 Youngbong-Dong, Buk-Gu, Gwangjoo 500-757, South Korea
{beyondi,smg686,syna,tozero}@chonnam.ac.kr
[2] Dept. of Multimedia Eng., Dongshin University, 252 Daeho-dong, Najoo, Chollanamdo 520-714, South Korea
shchoi@dongshinu.ac.kr
[3] Dept. of Internet Broadcasting Dong-A Broadcasting College, 632-18 Samjook-Myeon, Ansung 456-717, South Korea
javacool@paran.com

Abstract. We develop a simple and fast human tracking system based on skin-color using Kalman filter for humanoid robots. For our human tracking system we propose a fuzzy and probabilistic model of observation noise, which is important in Kalman filter implementation. The uncertainty of the observed candidate region is estimated by neural network. Neural network is also used for the verification of face-like regions obtained from skin-color information. Then the probability of observation noise is controlled based on the uncertainty value of the observation. Through the real-human tracking experiments we compare the performance of the proposed model with the conventional Gaussian noise model. The experimental results show that the proposed model enhances the tracking performance and also can compensate the biased estimations of the baseline system.

1 Introduction

The trend in robot research has been changed to human friendly applications from industrial areas. Now, there are many robots applicable in home and office environments. Because the services of robots are targeted on human convenience and amusement, it is very important for robots to detect and track the person's existence or his activity [1-3].

In this paper we develop a simple and fast human tracking system using skin-color information and Kalman filter. The cardinal contribution is the proposition on observation noise model. In the implementation of Kalman filter it is important to model observation noise accurately. Generally the time-invariant and Gaussian noise model is used in Kalman filter application [4,5]. Real noises, however, have often time-varying characteristics. That is, when a person is moving here and there, the illumination condition on the person is not constant. However, it is not easy to detect the time-varying illumination condition directly with video information. Generally, we have only the

J. Wang et al. (Eds.): ISNN 2006, LNCS 3972, pp. 419–428, 2006.

estimated position information and this position signal, obtained using skin-color information, is perturbed randomly. From this observation we consider two sources of observation noises.

(1) The observed feature involves the uncertainty coming from inconstant lighting condition and other reasons.
(2) The probabilistic errors are added to the observed features. These errors are generated by algorithm's incompleteness.

So, the errors, occurring in real observations, are not only fuzzy but also probabilistic. To overcome this noise model problem in face detection and tracking, we devise an integrated noise model based on the fuzzy and probabilistic approaches. First, we measure the uncertainty value of the observed feature. And then, the probabilistic noise model parameters are controlled depending on the uncertainty value. In this paper we apply a feed-forward neural network for measuring the uncertainty values.

The rest of this paper is organized as follows. In Section2, we briefly introduce our proposed baseline system of human tracking. In Section 3, we describe the proposed fuzzy and probabilistic approach for observation noise modeling. In Section 4 we discuss the results of our experiments on human tracking. Finally in Section 5, we describe the conclusions of this study and possible future research.

2 The Proposed Baseline System for Human Tracking

Our main purpose of the human tracking system is providing an intelligent interface between human and robots. Especially, we apply our system to the embedded system based on the robot developed by our research group. The implemented baseline system of human tracking uses a skin-color information and Kalman filter.

Figure 1 shows the baseline human tracking system. The procedures for human detection and tracking are described as follows.

(1) All the pixels are transformed to skin-color probability values using the histogram-based skin-color model.
(2) The candidate regions of faces are determined based on x and y projections of the image.
(3) To validate the candidate regions, the neural network is applied to the regions. With some thresholds the candidate regions are rejected or accepted as faces.
(4) Using the information of the center positions of faces and Kalman filter, tracking a human is performed.

The above baseline system is developed under the concept of real time processing. So we don't use 2-dimensional signal processing if possible. For the detection of face regions we adopt 1-dimensional processing of x and y projections rather than image processing based approach. Also, we use neural networks for face validation, which can be implemented by parallel processing. In the following sections we describe the procedures more in detail.

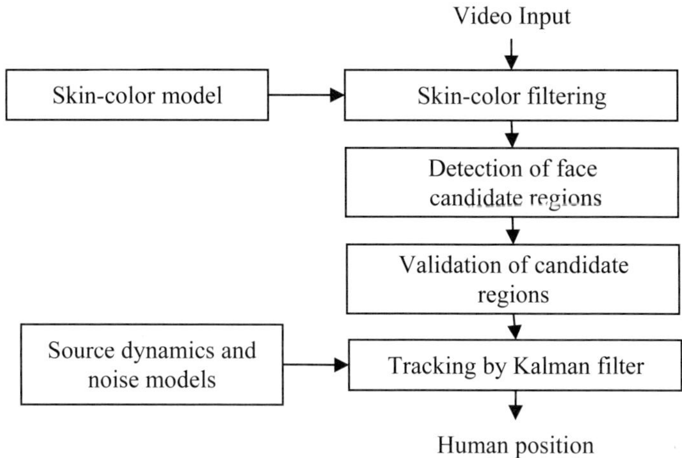

Fig. 1. Block diagram of the baseline system

2.1 Color Space and Skin-Color Model

Skin-color provides good information for extracting an area of the face. For description of skin color, several different kinds of color space, as like RGB, YCbCr, HSV and etc, have been proposed[6-9]. In this study we use *a-b* parameter proposed by S. Kawato [3]. This a-b color model is calculated from the normalized RGB using a simple transformation.

Chromatic colors (*r*,*g*), known as pure colors in the absence of brightness, are defined in a normalization process.

$$r = R/(R+G+B), \tag{1a}$$

$$g = G/(R+G+B). \tag{1b}$$

These values of r and g are modified to (a,b) space as follows.

$$a = r + g/2, \tag{2a}$$

$$b = \sqrt{3}/2g. \tag{2b}$$

This means that a color (R,G,B) is projected onto (a,b) plane. Any color can be projected onto the hatched regular triangular region, and the range of (a,b) is from (0,0) to $(1, \sqrt{3}/2)$.

There are some skin-color modeling approaches such as histogram, Gaussian mixture model, and fixed region method [7-9]. We use the histogram based probability model for skin-color modeling. We digitize (a,b) space into squares of 0.01size as a look-up table, and get a histogram for face images. Histogram is calculated with the face images, in which non-face regions are erased. Because we used a small number of images for skin-color modeling, the simple smoothing technique is applied for the histogram. And then the probability model is obtained by normalization process.

2.2 Detection of Face Candidate Regions

For the detection of face candidate regions the skin-color filtering is performed first. That is, each pixel's value is transformed to probability value in (a-b) color space with a threshold. Zero value is assigned to the pixels having low probability lower than a threshold. Figure 2 shows the results of skin-color filtering

Fig. 2. Extraction of pixels with high skin-color probabilities

After skin-color filtering x and y projections are performed for extracting candidate regions. The projections are accomplished with the following equations.

$$X(i) = \sum_{j=0}^{M-1} F(i,j), \tag{3a}$$

$$Y(j) = \sum_{i=0}^{N-1} F(i,j)\cdot \tag{3b}$$

where $F(i,j)$ is the probability of the (i,j) pixel, N is the width of the image, M is the height of the image, $X(i)$ is the x-projection and $Y(j)$ is the y-projection. In the study the basic face detection algorithm is as follows.

(1) Calculate x and y projections of $F(i,j)$.
(2) Smooth the x and y projections, X(i) and $Y(j)$, using a low-pass filter of Hamming window.
(3) Estimate candidate ranges in x-axis with some thresholds.
(4) Estimate candidate ranges in y-axis along the strip regions determined by the candidate ranges of x-axis.

By the above method the candidate regions of faces are determined. Figure 3 shows an example of face detection. Unfortunately, sometimes, the false-alarmed regions are detected.

So, we try a neural network for the rejection of the mis-detected regions. The neural network is a general feed-forward network. The inputs to the neural network are the 32x32 down-sampled images of the candidate regions. And the output is 1 for the correct regions and -1 for the false-alarmed regions. With a threshold we can discard the mis-detected regions. The thresholds are determined by empirical experiments.

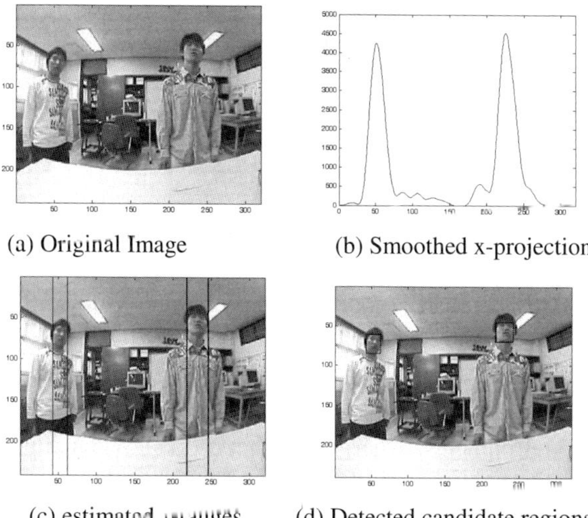

(a) Original Image (b) Smoothed x-projection

(c) estimated x-ranges (d) Detected candidate regions

Fig. 3. Detection of face regions

2.3 Human Tracking Based on Kalman Filter

The result of the face detection is the center (x,y) position of the face. In this study we only use x-position as a tracking parameter. In Kalman filter implementation the source dynamics model is very important. We adopt Langevin's model for the source dynamics [10].

$$\begin{aligned} \dot{x}_t &= a_x \dot{x}_{t-1} + b_x F_x, \\ x_t &= x_{t-1} + \Delta T \dot{x}_{t-1}, \\ a_x &= e^{-\beta_x \Delta T}, \\ b_x &= v_x \sqrt{1 - a_x^2}, \end{aligned} \tag{4}$$

where F_x is a normally distributed random variable, ΔT is $1/f_s$, $\beta_x = 10 s^{-1}$ and $v_x = 1 ms^{-1}$.

We assume a Gaussian random variable as the observation noise. In the baseline application the mean and variance of the noise model is fixed to the experimental values. The other things on the tracking implementation are same with the conventional Kalman filter implementations. We use a standard Kalman filter.

3 Fuzzy and Probabilistic Model of Observation Noise

In this chapter we describe a new idea on observation noise model applied to the human tracking Kalman filter. As simply explained in the introduction, we try to

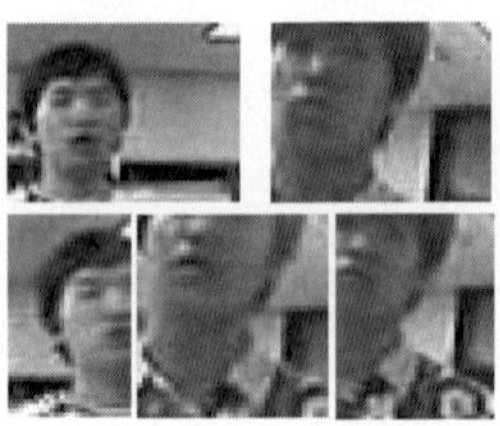

Fig. 4. Corrupted detections of face regions

adopt a fuzzy and probabilistic model to the observation noise. Our proposition is originated from the observations in human tracking experiments

(1) The lighting can degrade severely the performance of the fact detection. We can assume that this degradation may not be random, for the lighting is fixed in the fixed environments, where the robot works. When a person's activity is performed, the lighting condition under the fixed illumination still makes the observation uncertain. Although the degradation could be considered as random, the observation noises are correlated. Additionally there are consecutive biased observations.
(2) Also, there are errors made by the human tracking system, for its algorithms are not perfect. We consider these errors as probabilistic. Of course, these errors are generated by the observation itself. There are some objects having the skin-like colors. We assume that these additive errors have Gaussian probability density functions.

Figure 4 shows the corrupted detection results of the baseline system. From the figure we can see the biased detections of face regions. The main reason for the degradation is the lighting condition.

Thus the issue of this study is how the observation uncertainty is measured. In the section 2.2 we use a neural network for the rejection of the false-alarmed candidate regions. The output value of the neural network is the possibility measuring how much the candidate regions are shown like real faces. This possibility is the representative fuzzy value. Thus one method of uncertainty measure is to use the output of the neural network. Then the fuzzy and probabilistic model of the observation noises is constructed with the following steps.

(1) Measure the uncertainty value with the value of the neural network used for the rejection of the false-alarmed regions.
(2) Obtain the Gaussian statistics of the observation noises depending on the uncertainty value of the neural network output.

The above procedure is a basic approach for integrating fuzzy and probabilistic measure. In this study we modify the neural network to consider the biased estimation occurring due to the lighting condition. To estimate the bias sign we change the output to two nodes as shown in Figure 5. There are two outputs of the neural network. One is the possibility or uncertainty value of the input image. The other is the right or left deviation. We decide the bias of the detection as follows.

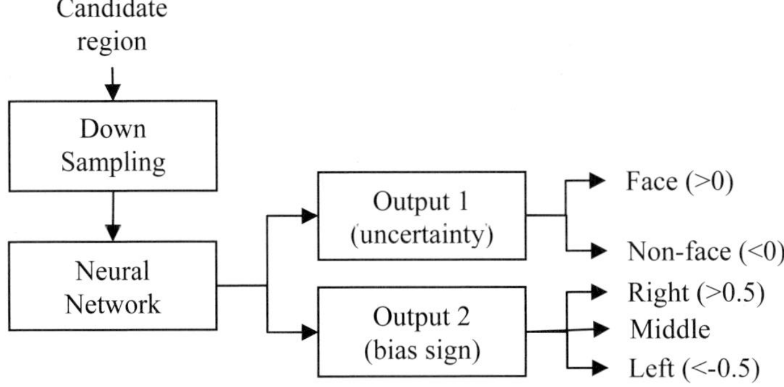

Fig. 5. Neural Network based uncertainty modeling

$$SOB = \begin{cases} left(-1), & \text{if } output2 < -0.5 \\ right(1), & \text{if } output2 > 0.5 \\ middle(0), & \text{else} \end{cases}, \tag{5}$$

where SOB is the sign of the bias. Using the uncertainty value and the bias information we can describe the Gaussian *pdf* as follows.

$$\mu(n) = f_1(output1, output2), \tag{6a}$$

$$\sigma(n) = f_2(output1, output2), \tag{6b}$$

where $\mu(n)$ and $\sigma(n)$ are the bias and standard deviation of the Gaussian noise model in the time index *n*. Output1 is the uncertainty and output2 is the bias information. Also, f_1 and f_2 are the control functions depending on the uncertainty value.

For training the neural networks we use 4,000 images. Table 1 shows the database used in training. In the table 'Biased Face' means that the image is filled with the partial face. And 'Face+Bacground' tell us that the candidate region includes background region and the face is biased to the left or the right. The error distribution depending on the uncertainty values are shown in Figure 6. From Figure 6 the following facts are observed.

(1) Using the uncertainty value it is possible to estimate the bias of the face detection.
(2) The noise distribution is symmetric along the *x*-axis. That is, the Gaussian *pdf*s have the same variance value and the same absolute mean value. Only the signs of mean values are different with the output2 of the neural networks.

We use a simple regression for representing the functions of f_1 and f_2. The order of regression function is three. On the other hand, the mean of the observation noise is not often zero. In that case we add the mean to the observation position vale *x*, for the observation noise assumed to be zero-mean in Kalman filter.

Table 1. Database used in training the neural network

	Number of Images		Number of Images
Correct Face	1,000	Non-face	2,000
Biased Face (Right)	500	Biased Face	500
Face(Right)+Background	500	Face(Left)+Background	500

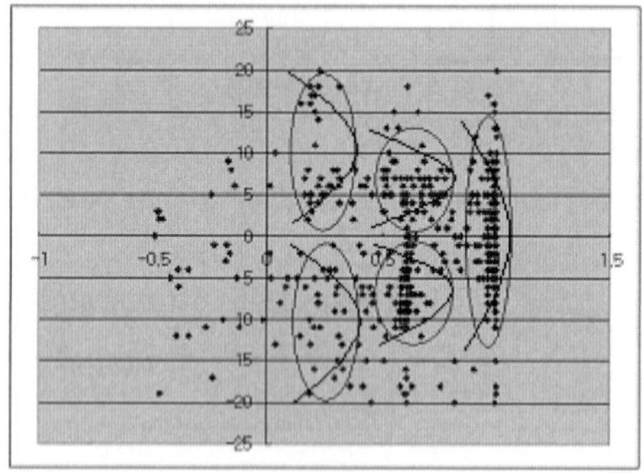

Fig. 6. Distribution of errors depending on uncertainty values(*x*-axis: uncertainty, *y*-axis: error pixel number)

4 Experimental Results and Discussions

We implement our human tracking system with Matlab. The video for evaluating the proposed algorithm is gathered in a common laboratory. The lighting is fluorescent lamps. And there is sun light from the windows. The frame rate is 30 Hz and the image size is 320x240. The recording is performed with a general CCD camera and the vision degree is 120°. In experiments two persons move along the *x*-axis. One person moves from the left to the right and moves back to the left. The other person moves vise versa.

We try two Kalman filter implementations. One, KF1, uses a common observation noise model, which use the fixed mean and standard deviation. The other, KF2, uses the proposed observation model, which is time-varying with the uncertainty value and a fuzzy and probabilistic model.

Figure 7 shows experimental results of the proposed method. From the figure we observe that the proposed method achieves a better performance compared with the baseline system.

(1) Generally, the performance of the KF2 is similar to or better than the original filter KF1. The average absolute error is reduced to 2.9 pixels from 4.1 pixels.
(2) Especially, the bias of the human tracking system is highly reduced when KF2 is used.

As shown in the figure the sun light is radiated from the left side window. Thus, in the region of (50,100) of x-position the degradation is severe as shown Figure 7 (b).

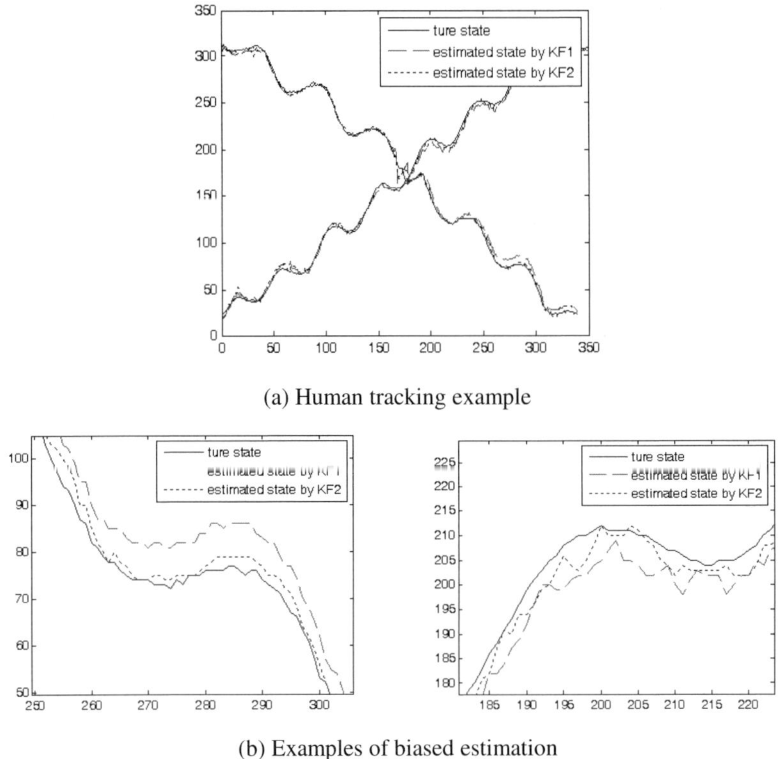

(a) Human tracking example

(b) Examples of biased estimation

Fig. 7. Human tracking results (x-axis : frame index, y-axis : estimated position)

With KF1, the original model, the bias problem cannot be overcome. But KF2, the proposed method, copes with the bias problem. The results, described here, are deduced from preliminary experiments. So, we have to construct more databases of video clips. We will evaluate more concretely in the recent future.

5 Concluding Remarks and Future Works

In this study we introduced a fuzzy and probabilistic approach for observation noise modeling. We implemented a simple human tracking system based on skin-color model and Kalman filter. Then we applied our noise model to the baseline system. Through the experiments with real video clips we showed the possibility of the proposed method in human tracking.

In the future we will perform more concrete research on noise model in human tracking for robots. We will construct enough databases for evaluating the proposed method and the enhancement algorithms of the current method. We expect the fuzzy and probabilistic approach can be successful in many other areas.

Acknowledgement

This work was supported by the Korea Research Foundation Grant funded by the Korea Government in 2004.

References

1. Perez, P., Hue, C., Vermaak, J., Gangent, M.: Color-based Probabilistic Tracking. ECCV02, (LNCS 2350) (2002) 661-675
2. Luo, R., Guo, Y.: Real-time Stereo Tracking of Multiple Moving Heads. IEEE ICCV Workshop (2001) 55-60
3. Kawato, S., Ohya, J.: Automatic Skin-color Distribution Extraction for Face Detection and Tracking. ICDSP2000 (2000) Vol. 2II, 1415-1418
4. Grewal, M.S., Andrews, A.P.: Kalman Filtering: Theory and Practice Using Matlab. John Wiley & Sons (2001)
5. Klein, L.A.: Sensor and Data Fusion: A Tool for Information Assessment and Decision Making. SPIE Press (2004)
6. Tomaz, F., Candeias T., Shahbazkia, H.: Improved Automatic Skin Detection in Color Images. VIIth International Conference on Digital Image Computing (2003) 10-12
7. Rowley, H.A., Baluja, S., Kanade, T.: Human Face Detection in Visual Scenes. CMU-CS-95-158, Carnegie Mellon University (1955)
8. Hsu, R.L.: Face Detection in Color Images. Department of Computer Science & Engineering, Michigan State University. MI 48824 (2004)
9. Vezhnevets, V., Andreeva, S.A.: A Survey on Pixel-Based Skin Color Detection Techniques. *Proc. Graphicon-2003*, Moscow State University (2003) 85-92
10. Ward, D.B., Lehmann, E.A., Williamson, R.R.: Particle Filtering Algorithms for Tracking an Acoustic Source in a Reverberant Environment. IEEE Trans. on Speech and Audio Processing 11(2003) 826-836

Object Detection Via Fusion of Global Classifier and Part-Based Classifier

Zhi Zeng, Shengjin Wang, and Xiaoqing Ding

Department of Electronic Engineering, Tsinghua University, Beiing, China
cengz03@mails.tsinghua.edu.cn

Abstract. We present a framework for object detection via fusion of global classifier and part-based classifier in this paper. The global classifier is built using a boosting cascade to eliminate most non-objects in the image and give a probabilistic confidence for the final fusion. In constructing the part-based classifier, we boost several neural networks to select the most effective object parts and combine the weak classifiers effectively. The fusion of these two classifiers generates a more powerful detector either on efficiency or accuracy. Our approach is evaluated on a database of real-world images containing rear-view cars. The fused classifier gives distinctively superior performance than traditional cascade classifiers.

1 Introduction

The main problem of object detection from still image is the contradiction between efficiency and accuracy when subjected to real-time application. In this paper we proposed an algorithm fusing the global classifier and part-based classifier to balance them. We make the following contributions in this algorithm: first, we describe the object part with Scale Invariant feature vector and use neural network to build weak classifiers on these features; second, we boost the weak classifiers to select the most effective parts in classification and combine them to form a strong part-based classifier; third, we present a general framework to fuse the global classifier and part-based classifier.

1.1 Previous Work

A number of statistical approaches have been proposed to detect object from image using a considerable set of samples. These methods differentiate either from types of features or learning methods. Rowley [1] use raw pixels to learn a neural network for detection, Papagorgicu [2] use an extensive Haar wavelet feature set and support vector machine to detect faces and pedestrians from still images, while Schneiderman [3] proposed a restricted Bayesian network to build object detector. Due to the complex representation of the feature or construction of classifier, these methods are difficult to implement in time-restrictive application.

Viola [4] described an effective object detection framework that reduces the processing time substantially while achieving almost the same accuracy as compared to

J. Wang et al. (Eds.): ISNN 2006, LNCS 3972, pp. 429 – 435, 2006.

a more complex single stage classifier. He proposed a simple but effective Haar-like feature set and cascade Adaboost classifier to achieve this performance. Lienhart extended the set of Haar-like features [5] and made a detector tree of boosted classifiers [6] to detect more complex and diverse object classes.

Instead of detecting object as a whole, Mohan [7] proposed an algorithm that firstly locates the constituent components and then combines the component detections with a classifier. This method need to pre-define several characteristic components and manually construct several component training set before hand. Compared with Mohan's work, Agarwal's approach [8] could automatically construct an object parts vocabulary using an interest operator to extract information-rich patches and then represent the object with the items in this vocabulary. However, since the size of the training samples is relative small and the interest operator is unstable in handling the large set of object samples, we make a full vocabulary of parts instead of doing interest pointer detection first and automatically select the best parts to represent the object via boosted neural networks.

1.2 Organization of This Paper

The rest of the paper is organized as follows: Section 2 make a brief introduction to the boosting cascade for rapid object detection. The detailed description of part-based classifier is provided in Section 3. In section 4, we present our approach for fusing the global cascade classifier and part-based classifier. Experimental result of our algorithm is listed in section 5 and the conclusion is made in section 6.

2 Global Object Detection Using Boosting Cascade

In this section, we give a brief description of Adaboost cascade classifier [4] for global object detection. In this method, Adaboost algorithm is used to automatically select effective features from an extensive Haar-like feature set and thus to construct a stage classifier; an cascade is then constructed to make the detection more effective with the main idea to eliminate most negative samples with very little processing.

Boosting cascade and Haar-like feature has been successfully validated for many detection tasks; however, as the stage number of the cascade increases, the Haar-like feature become less effective since it requires much more features in one stage than before. The restriction of Haar-like feature drives us to build a part-based classifier on more complex features after certain stages.

In order to fusing the result of the global object detection with part-based detection, we would rather describe the detection result of the cascade classifier as a probabilistic confidence than an absolute value. Firstly, we define the probabilistic confidence gotten from i-th stage classifier as:

$$P_i(O \mid x) = \frac{\exp(\sum_{i=1}^{T} \alpha_t h_t(x))}{\exp(\sum_{i=1}^{T} \alpha_t h_t(x)) + \exp(-\sum_{i=1}^{T} \alpha_t h_t(x))} \tag{1}$$

We then treat each stage as statistically independent. With this assumption and the definition above, we could get the probabilistic confidence of the final detection from the cascade classifier:

$$P(O \mid x) = \prod_i P_i(O \mid x) \tag{2}$$

This equation means that the probability of the candidate image being object equals the product of the confidence of each stage.

3 Part-Based Approach

In this section we describe the part-based object detection algorithm that is based on the representation of Scale Invariant feature, automatic selection and classification on boosted neural network classifiers.

3.1 Part Representation

A vocabulary of parts is first constructed from a set of sample object images. Because the object size is relative small (about 24x24 in our experiment) and the edge information may vary a lot even in the positive sample set so we can not detect a uniform interest point set, we make a full vocabulary of parts instead of doing interest pointer detection first. When d_x and d_y is the number of shift pixels between neighbor parts, to describe $M \times N$ sized object with $h \times w$ sized parts, we could make a parts vocabulary with K items, while

$$K = \frac{(M-h+1)\times(N-w+1)}{d_x \times d_y} \tag{3}$$

The vocabulary is described by the Scale Invariant feature [9], which represents image regions with multiple orientation planes. Each orientation plane contains only the gradient corresponding to that orientation. In implementation we first eliminate the pixels with gradient magnitude lower than certain threshold and then represent the part by the statistic of the gradient direction with an 8-element vector. For it represents regions only with normalized orientation statistics, Scale Invariant feature is robust to illumination change, scale variance and small rotation. It would be even more effective if the gradient distribution of the object is stable, such as the vehicles.

3.2 Boosting the Neural Network to Build Part-Based Classifier

Based on a Scale Invariant feature vector of each part in the vocabulary, we could firstly build a weak classifier to classify the samples. Adaboost, after that, could help us to select the most effective weak classifiers (It corresponds to the selection of effective parts in description the object) that make the smallest classification error and combine them to construct a stronger classifier.

In this part, we use backpropagation neural network as weak classifier. In the past years, many empirical studies [10], [11] show that decision tree cooperate well with

Adaboost, yet Schwenk [12] presents the advantage of neural network over decision trees in many cases and gives full analysis. Our experiments show that the combination of neural networks performs much better than decision trees in classifying the objects, for individual neural network fits Scale Invariant feature better than decision tree and Adaboost relaxes the overfitting problem of the individual neural networks.

Table 1. Boosting the neural network for automatic part selection and classifier combination

- Given example images $(x_1, y_1), \ldots, (x_n, y_n)$ where $y_i = 1, -1$ for positive and negative samples respectively.
- Build a full part vocabulary; train a weak classifier using backpropagation network for each part in the vocabulary.
- Initialize weights $w_{1,i} = \frac{1}{2m}, \frac{1}{2l}$ for $y_i = 1, -1$, m and l are the number of positives and negatives
- For t = 1, …, T
  1. Normalize the weights $w_{t,i} \leftarrow w_{t,i} / \sum_{j=1}^{n} w_{t,j}$.
  2. For each weak classifier h_j , compute the classification error $\varepsilon_j = \sum_i \frac{w_i \mid h_j(x_i) - y_i \mid}{2}$.
  3. Choose the classifier h_t with the lowest error ε_t , which correspond to the t-th effective part in the vocabulary.
  4. Let $\alpha_t = \frac{1}{2} \ln(\frac{1-\varepsilon_t}{\varepsilon_t})$, update the weights: $w_{t+1,i} = w_{t,i} \exp(-\alpha_t y_i h_t(x_i))$.
- The final strong classifier is $h(x) = sign(\sum_{t=1}^{T} a_t h_t(x))$

4 Fusion of Global Detector and Part-Based Classifier

In this section, we provide an approach for fusing together the global detector described in section 2 and the part-based classifier in section 3. We make use of the confidence of each scheme which is afterwards used to make combination decision.

We use the cascade classifier to detect the object globally, eliminating most easily-discriminated backgrounds and describing the left candidates by the probabilistic confidence presented in section 2. With the remaining candidates, we apply the part-based classifier to get another confidence by the similar method as formula (1). At last, the log-likelihood combination of the confidence is made to generate an overall confidence of the object. We define the decision rule as:

$$\text{If} \qquad \alpha \log P_g(O \mid x) + \beta \log P_l(O \mid x) > \lambda \tag{4}$$

We judge the candidate as positive; otherwise label it as negative.

$P_g(O \mid x)$ is the global confidence get from formula (2), while $P_l(O \mid x)$ is the local confidence computed from the similar method as formula (1).

5 Experimental Results

5.1 Detection Results of the Global Cascade Classifier

At first, we train the global cascade classifier with a rear-view car database which contains 1600 positive samples and 500 background images without objects; then test it on 178 images containing 200 cars. To evaluate the efficiency of the Haar-like feature, we test cascade classifiers of different stages.

Table 2. Classification results of cascade classifiers with different stages

Stage number	15	16	17	18	19	20
Correct detections	191	180	172	168	162	158
False positives	245	165	126	93	69	53

The test result above shows that Haar-like feature is less effective in more detailed classification after several stages, for the hit rate decline evidently when the stage order increases.

5.2 Construction of the Part-Based Classifier

Instead of making excessive use of Haar-like feature, we choose a certain number stages of cascade classifier, preserving the desired hit rate and allowing for some false alarms; then we apply the part-based detector to make careful classification. We evaluate our approaches in feature selection, weak classifier building and part-based classifier construction on a car image database which contain 1600 positive samples and 1600 negative samples.

Table 3 shows the advantage of neural network over decision trees in making weak classifier based on the Scale Invariant feature vector.

Table 3. Comparison of BP network and C4.5 in building the weak classifiers of parts (the best 5 parts)

Correct rate	Part 1	Part 2	Part 3	Part 4	Part 5
BP network	81.8%	81.5%	80.7%	80.6%	80.1%
C4.5	80.5%	80.1%	79.6%	79%	78.6%

To demonstrate the efficiency of the part-based approach in detailed classification of objects, we compared them with a more complex global classifier. The global classifier is trained with support vector machine on a dimension-reduced gabor feature vector. The results show the part-based approach (either by neural network or decision tree) achieves high accuracy than the global classifier, while the classifier based on neural network performs better than the one based on C4.5 decision trees.

Table 4. Comparison of part-based classifier and complex global classifier in detailed classification of the objects. We use a train set containing 500 positive samples and 500 negative samples in the experiment.

Classifier type	Part-Based Classifier(BP Network)	Part-Based Classifier (C4.5)	Global Classifier (SVM + Gabor)
FN	38	42	66
FP	52	57	84

5.3 Test Results of the Fused Detector

At last, we give an overall evaluation about the fused detector on 178 images (320x240) containing 200 cars. The classifier fusing algorithm whose part-based classifier is based on boosted neural networks achieves the highest detection rate and lowest false alarm rate on this test set.

Table 5. Comparison of different classifiers. (1. Cascade(20stages); 2. Cascade(15 sages); 3. Cascade(15 stages + Part-Based(C4.5); 4. Cascade(15 stages) + Part-Based(BP network).

Classifier type	1	2	3	4
Correct Detection	158 42	191	191	182
False Alarm	53	53	32	26
Time(s / image)	0.446	0.305	0.532	0.506

Fig. 1. Examples of test images on which our fused detector achieved perfect detection result

6 Conclusion

In this paper we present a novel framework for object detection by fusion of global cascade classifier and part-based classifier. The global detector makes its work fast and the part-based classifier keeps the high accuracy. In building the part-based classifier, we use Adaboost to automatically select and effectively combine the weak classifiers which was implemented by backpropagation neural networks. The experimental result has shown that, for classifying object by parts which are described by local Scale Invariant features, neural network perform better than traditional decision tree when combing together by Adaboost algorithm. Moreover, we have demonstrated the robustness and superiority of the proposed framework in detecting cars from a set of traffic scene images.

References

1. Rowley, H. A., Baluja, S., Kanade, T.: Neural Network-Based Face Detection. IEEE Trans. Pattern Analysis and Machine Intelligence 20(1) (1998) 23-28
2. Papageorgiou, C. F., Oren, M., Poggio, T.: A General Framework for Object Detection. International Conference on Computer Vision, Bombay, India (1998) 555-562
3. Schneiderman, H.: Learning a Restricted Bayesian Network for Object Detection. IEEE Conference on Computer Vision and Pattern Recognition 2 (2004) 639-646
4. Viola, O., Jones, M.: Rapid Object Detection Using a Boosted Cascade of Simple Features. IEEE Conference on Computer Vision and Pattern Recognition, Hawaii (2001) 905-910
5. LienHart, R., Maydt, J.: An Extended Set of Haar-like Features for Rapid Object Detection. IEEE International Conference on Image Processing 1 (2002) 900-903
6. LienHart, R., Liang, L., Kuranov, A.: A Detector Tree of Boosted Classifiers for Real-time Object Detection and Tracking. International Conference on Multimedia and Expo (2), New York, USA (2003)
7. Mohan, A., Papageorgiou, C., Poggio, T.: Example-Based Object Detection in Images by Components. IEEE Trans. Pattern Recognition and Machine Intelligence 23(4) (2001) 349-361
8. Agarwal, S., Awan, A., Roth, D.: Learning to Detect Objects in Images via a Sparse, Part-Based Representation. IEEE Trans. Pattern Recognition and Machine Intelligence 26(11) (2004) 1475-1490
9. Lowe, D. G.: Object Recognition from Local Scale-Invariant Features. International Conference on Computer Vision, Corfu, Greece (1999) 1150-1157.
10. Quinlan, J. R.: Bagging, Boosting and C4.5. International Conference on Machine Learning (1996) 725-730
11. Bauer, E., Kohavi, R.: An Empirical Comparison of Voting Classification Algorithms: Bagging, Boosting, and Variants. Machine Learning (36) (1999)105-139
12. Schwenk, H., Bengio, Y.: Boost Neural Networks. Neural Computation (12) (2000) 869-1887

A Cartoon Video Detection Method Based on Active Relevance Feedback and SVM

Xinbo Gao, Jie Li, and Na Zhang

School of Electronic Engineering, Xidian Univ., Xi'an 710071, P.R. China

Abstract. By analyzing the particular features of visual content for cartoon videos, 8 typical features of MPEG-7 descriptors are extracted to distinguish the cartoons from other videos. Then, a content-based video classifier is developed by combining the active relevance feedback technique and SVM for detecting the cartoon videos. The experimental results on the vast real video clips illustrate that compared with the classifier based on SVM and that based on traditional relevance feedback technique and SVM, the proposed classifier has a higher advantage of cartoon video detection.

1 Introduction

With the development of multimedia information, in order to search the multimedia of interest rapidly, the indexing and retrieval system emerges as the times require. The automatic multimedia processing and analysis technique has become a research focus. The video automatic classification technique is at the bottom of the interest and plays an important role in the development of VOD and HDTV as a kind of familiar multimedia [1]. Besides the function of classifying the enormous video data to reduce the uses' workload, the intelligent functions are expected such as recognizing the broadcast content and skipping the uninterested program automatically. Among these, the detection of cartoon videos automatically is a meaningful work for the children.

For this purpose, many content-based cartoon video detection methods have been put forward in recent years. For example, Matthew proposed a general video classification algorithm, but he did not make special analysis to the cartoon video [2]. Iuneua presented a detection algorithm aiming at the cartoon video classification, but only using several simple features such as color saturation [3]. With the constant increase of the cartoon types, the classification precision is more and more unsatisfactory. Afterward, Matthew also made analysis to the cartoon video based on the motion features [4]. The advantage is that the classification can be dealt with in real time, but the algorithm is limited to the discrimination of the traditional cartoon because of the great relation between the frame movement and the making tactic. So the method is difficult to generalize for wide applications. In sum, we make a full analysis to the cartoon videos first, and extract 8 MPEG-7 visual descriptors corresponding to the human visual perception characteristics to distinguish the cartoon from non-cartoon videos. Then

J. Wang et al. (Eds.): ISNN 2006, LNCS 3972, pp. 436–441, 2006.

a new classifier design method based on active relevance feedback and SVM is developed to classify them.

2 Cartoon Classification and Feature Analysis

2.1 Cartoons and Classification

In a nutshell, cartoon video is the unnatural video which is made by various means. At the point of view of the image processing, most cartoons have some of the following visual characteristics: (1) The color saturation is higher than that of the real video; (2) Tend to have regions of the same color; (3) Have obvious black edges; (4) The textures are simple, and seldom have the thin textures; (5) Seldom use the black, white and gray colors, so they are colorfulness.

Since the cartoons have many different types and the facture is also daedal, not all the cartoons have the above features completely. We cannot consider the cartoons as one class to distinguish from other videos. Based on the above definition, we divided the cartoons into three classes. One is the traditional cartoon, including the Chinese traditional cartoon and the Japanese traditional caricature, which has high color saturation and thin lines. Also the configuration of the setting and the characters are abstract. One is the Disney film cartoon. Being relative to the traditional cartoon, Disney cartoon has lower color saturation and richer lines, but the characters and setting are still abstract. The other is Hayao Miyazaki film cartoon which is popular in recent years. For the video uses the film facture technique, its color saturation is very low, even lower than the natural videos. But its color contrast is high and the facture of the character and setting is delicacy. At last, our classifier realizes the cartoon detection by classifying the video into three kinds of cartoon and non-cartoon videos.

2.2 Feature Extraction and Analysis

8 visual features are selected and extracted according to the analysis above to describe the cartoon video, such as HSV color histogram, the proportion of the multicolor region, color structure histogram, edge direction histogram descriptor, gradient histogram, edge intensity, homogenous texture histogram, and colorfulness metric [7][8].

After the feature extraction, a 405-bin histogram is obtained for each frame. To extract the features of the whole video, one can compute the median value of every frame aiming at a certain bin. After operating to the 405 bins, median histogram is produced to be the feature histogram of a video. For the high dimension of the feature, it is not feasible to perform any computation and analysis, and there exists considerable amount of spatial redundancy in the feature histogram. The Independence Component Analysis (ICA) is used to reduce the dimension and keep independent in the high order. Here, the fast and robust fixed-point algorithm (FastICA) [9] is adopted as analysis tool. The dimension is reduced from 405 to 20. At the same time, one can receive a separated matrix. All the videos are projected to the subspace by the separated matrix. The obtained ICA feature is used to be the direct feature of classification.

3 The Proposed Cartoon Detection Method

Because of the universality of the experimental material, the cartoon materials cannot be classified into one class. So in the experiment design, we classify the cartoon videos into three classes as described in section 2.1 to discriminate the cartoon and non-cartoon videos. The SVM [5] is selected for binary classification. Though it can be used to deal with the multiclass classification by the modified SVM, the performance is far from the binary case. For this end, we develop the following procedure shown in Fig.1. Three binary classifiers are integrated to implement the multiclass case.

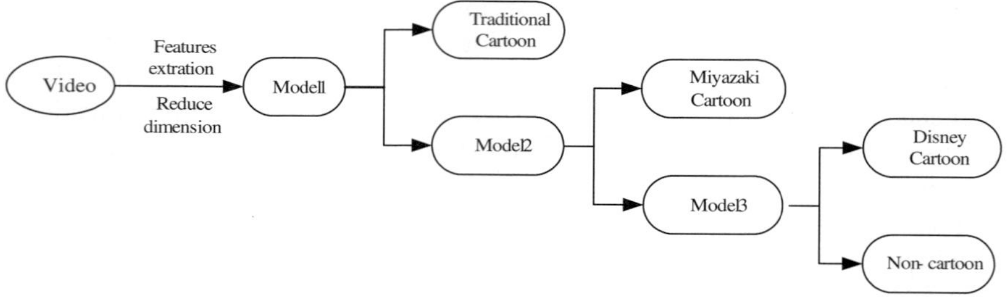

Fig. 1. Video classification flowchart

To improve the performance of cartoon detection, the active relevance feedback method [6] is employed as post-processing. The difference of relevant feedback and active relevant feedback lies in the matter that active relevant feedback algorithm does not feedback the stochastic results to users and search engine, but the most uncertain data. For SVM, the uncertain data almost distribute between two supporting vector hyperplanes. If the uncertain data can be marked by the users, the system will obtain more gains. How to measure the uncertainty is the key problem. Here, the sum of the distance of a test sample to two training sample sets is used as the metric.

$$S = D_c + D_{nc} = \frac{1}{N_c}\sum_{j=1}^{N_c} ||I_i - I_j||_2 + \frac{1}{N_{nc}}\sum_{j=1}^{N_{nc}} ||I_i - I_j||_2, \qquad (1)$$

where, D_c is the average distance between the feedback result and the cartoon training samples, and D_{nc} is the average distance between the feedback result and the non-cartoon training samples. I is the feature vector of a test sample.

In SVM, compared with the samples beyond two supporting vector hyperplanes, the samples between the supporting vector hyperplanes have a small distance to two training sample sets. So the smaller of S in (1), the more uncertain of the classification result. Here, we first use the initial classifier to do classification on the test data. Then, the uncertainty of the test data is computed. The first several test data whose S value are smaller than others are listed to users for re-labeling (Suppose the maximum of the S of the returned data is S_1). Furthermore, in SVM, the location of the separating hyperplane lies in the supporting vectors. It means that the vectors which are not supporting vectors do not play role in the classifier design. According to this, the classification label and the uncertainty of the non-supporting vectors are also computed. The non-supporting vectors whose S value do not exceed S_1 and the classification results are wrong remain in the training sample collections. Whereas the other non-supporting vectors are substituted by the test data whose result are marked to be wrong by users. After we get the new training sample collections, a new round begins. This new method can not only improve the performance of the classifier, but also avoid the traditional case that the number of the training samples increases sharply after the feedback at a time.

4 Experimental Results and Analysis

In our experiments the training and test video data are from the daily TV programs in order to illustrate the validity and the feasibility of the proposed cartoon video detection method. The cartoon materials come from 81 cartoon programs, many one minute sections are cut from them. The non-cartoon materials come from the news, films, soap opera, gymnasium program and advertisement. Also many one minute sections are extracted. With the classifier designed in Section 3, 40 traditional cartoon samples and 40 non-cartoon samples are selected as training samples for producing model1. Being similar with *model1*, *model2* and *model3* are also produced by 40 cartoon training samples of corresponding type and 40 non-cartoon training samples.

The experiments were carried out on the database of 237 cartoon test samples and 100 non-cartoon test samples. In the experiment, the kernel of SVM is RBF, and the parameter is set as follows: $gamma = 0.25, C = 1, cache\_size = 40$.

Corresponding to the same training and test database, and operating according to Fig.1, We used three algorithms as follows to conduct the comparison experiments, the only SVM algorithm, the combination of SVM and relevance feedback (RF), and the combination of SVM and active RF.

The experimental result is shown in Table 1, in which recall and precision defined in [10] is adopted as evaluation criteria. Through Table 1, we can draw the following conclusions.

(1) For the cartoon detection result with the SVM only, it is not satisfied. When the relevance feedback and the active relevance feedback are adopted, both the recall and the precision are increased than before.

Table 1. The detection results with the classifier trained by 40 samples

	SVM + RF		SVM + Active RF	
	Recall	Precision	Recall	Precision
SVM	84.0%	86.9%	84.0%	86.9%
The 1st round	84.0%	88.5%	84.2%	90.1%
The 2nd round	87.5%	87.2%	84.2%	93.0%
The 3rd round	84.9%	91.4%	87.8%	93.3%

(2) In the three rounds of feedback, the recall and the precision of the active RF ascend placidly. However, in the second round, the recall of the traditional RF rises sharply, whereas the precision descends obviously.

(3) Both the recall and precision of active RF case are higher than those of the RF case in the first and the third rounds. In the second round, the recall of the active RF is lower than that of the RF, but the high recall is at the cost of debasing the precision.

In order to illustrate the reliability of the proposed method, we reduce the number of training samples to 20. The experimental results are shown in Table 2. It can be found that both the recall and the precision of each round in Table 2 are lower than those in Table 1. It is the reason that the number of samples is only half of the first experiment. In addition, the initial samples are selected randomly. So, the classification results of SVM only have a great difference in different experiment. However, it is obvious that the trend in each round of the second experimental result is similar to the first one. In every round, both the recall and the precision of the active RF are higher than those of the RF, and in the first round, the recall of the RF declines again. However, in the second round, the recall of the active RF declines too. This is because not all the samples with smaller S value of (1) fall into the space between two supporting vector hyperplanes according to geometrical theory. We only can guarantee most of the data with smaller S value between them. But the amplitude of the declining is very small and does not affect the whole performance. On the whole, the performance of active RF is superior to that of the SVM and RF.

Table 2. The detection results with the classifier trained by 20 samples

	SVM + RF		SVM + Active RF	
	Recall	Precision	Recall	Precision
SVM	80.3%	78.6%	80.3%	78.6%
The 1st round	78.2%	85.5%	84.1%	86.1%
The 2nd round	81.5%	85.8%	83.8%	88.0%
The 3rd round	83.1%	87.1%	84.9%	88.2%

The two experimental results agree with the theory of section 2. For SVM is a small samples classifier, the random training samples cannot always determine an optimal separating hyperplane for classification on the abundant test samples. And a few training samples just can modify the location of separating hyperplane. So in the feedback, if the feedback samples are out of place, the separating hyperplan will lean to one side badly. The experimental results of the second round of the relevance feedback illustrate this issue. Also when the feedback samples cannot afford great information to the system, the difference of the new separating hyperplane and the old one will be little, and the performance will not be improved. So the result of active RF is better than the tradition RF case in the first round.

5 Conclusions

By analyzing the difference of visual content between the cartoons and non-cartoons videos, 8 kinds of features which are used to distinguish two kind videos are chosen. To detect the cartoons, a content-based video classifier is developed by combining the active RF technique and SVM. The experimental results on the daily TV program clips illustrate that the proposed detection scheme has the advantages of high precision and steady performance.

Acknowledgement

This work is supported by the Program for New Century Excellent Talents in University of China (NCET-04-0948), the National Natural Science Foundation of China (No.60202004) and the Key Project of the Ministry of Education of China (No.104173).

References

1. Zhang, N., Gao, X.B.: The Future Intelligent Television. Advanced Television Engineering **1** (2005) 134–136
2. Roach, M., Mason, J., Xu, L.Q.: Video Genre Verification Using both Acoustic and Visual Modes. IEEE Workshop on Multimedia Signal Processing, St. Thomas, Virgin Islands, USA (2002) 157–160
3. Ianeva, T.I., de Vries, A.P., Rohrig, H.: Detecting Cartoons: A Case Study in Automatic Video-genre Classification. International Conference on Multimedia and Expo, Baltimore, Maryland **(1)** (2003) 449–452
4. Roach, M., Mason, J.S., Pawlewski, M.: Motion-based Classification of Cartoons. International Conference on Intelligent Multimedia, Video and Speech Processing, Hong Kong, 2-4 May (2001) 146–149
5. Bian, Z., Zhang, X.G.: Pattern Recognition. The second Edition, Beijing, The Publishing House of Tsinghua, January (2000) 296–298
6. Chen, K., Jiang, Y., Zhou, Z.H.: The Active Relevance Feedback. Pattern Recognition and Artificial Intelligence **18(4)** (2005) 480–485
7. Manjunath, B.S. Ohm, J.-R., Vasudevan, V.V., Yamada, A.: Color and Texture Descriptors. IEEE Trans. on CSVT **11(6)** (2001) 703–715
8. Yong, M.R., Kim, M., Ho, K.K., Manjunath, B.S., and Kim, J.W.: MPEG-7 Homogeneous Texture Descriptor. ETRI Journal **23(2)** (2001) 41–51
9. Hyvarinen, A.: Fast and Robust Fixed-point Algorithms for Independent Component Analysis. IEEE Trans. on Neural Networks **10(3)** (1999) 626–634
10. Song, F., Gao, L.: Performance Evaluation Metric for Text Classifiers. Computer Engineering **30(3)** (2004) 107–127

Morphological Neural Networks of Background Clutter Adaptive Prediction for Detection of Small Targets in Image Data

Honggang Wu, Xiaofeng Li, Zaiming Li, and Yuebin Chen

School of Communication and Information Engineering,
University of Electronics Science and Technology of China, Chengdu, China

Abstract. An effective morphological neural network of background clutter prediction for detecting small targets in image data is proposed in this paper. The target of interest is assumed to have a very small spatial spread, and is obscured by heavy background clutter. The clutter is predicted exactly by morphological neural networks and subtracted from the input signal, leaving components of the target signal in the residual noise. Computer simulations of real infrared data show better performance compared with other traditional methods.

1 Introduction

This research aims to addresses the problem of detecting dim objects with very small spatial extent that are masked by spatially large background clutter in image data. Applications where this is of interest include the detection of tumors and other irregularities in medical images and target detection in infrared sensor data.

Since target of interest is obscured by around background clutter, no detector can find the target with very low false alarms in the raw image data. So the scheme of detection is divided into two steps. The first step is prediction and suppression of heavy background clutter, which enhance the detection to give a higher signal to noise ratio for the desired target. The second step is target detection with matched filter. This paper proposes morphological neural networks of background clutter prediction as the pre-processing algorithm for the first step in the scheme.

A number of models have been proposed for background clutter representation [6] in image data. A local demeaning filter has been used to track the non-stationary mean in an image by Reed *et al.* in [7]. Leonov *et al.* [8] summarized some kinds of nonparametric spatial filter for clutter estimation. Soni [9] used TDLMS adaptive filter to obtain a background estimate of spatially varying clutter.

However, these linear prediction methods are not of good performance for a complex background environment. So Yu *et al.* [2] applied morphological filter to track non-stationary and complex clutter. Morphological filter is a new nonlinear filter developed from mathematic morphology. It uses pre-defined SE (Structuring Element) to match and extract signal of desired figure. Morphological opening and closing operation can eliminate those bright and dark structures, which are smaller than defined SE. Thus the clutter is predicted and subtracted from input image, leaving components of the spatially small targets and residual noise in the output.

J. Wang et al. (Eds.): ISNN 2006, LNCS 3972, pp. 442 – 447, 2006.

Now adaptive design of SE has long been the hotspot and nodes in morphology domain [4,5]. Won *et al.* [3] proposed MSNN (Morphological Shared-Weight Neural Networks) algorithm to optimize SE of dilation and erosion operation. And Yu *et al.* [1] constructed neural network model of morphological opening and closing operation.

In this paper, a better 3-layer feed forward BP (Back Propagation) network model of opening and closing operation for clutter prediction is proposed. The network model is modified by extending the input layer data and the raw image is partitioned to some sub-blocks firstly for tracking non-stationary background. And simulations of real infrared data show better performance compared with other traditional methods.

2 Mathematical Model of Image Data

In this paper the infrared signal model developed by Chan *et al.* [6] was used for simulations. It's a two-dimensional Gaussian intensity function for the target

$$S(x,y)=\gamma\cdot e^{-[\frac{(x-x_0)^2}{\sigma_x^2}+\frac{(y-y_0)^2}{\sigma_y^2}]} \tag{1}$$

where γ is the maximum value of the target intensity function, (x_0, y_0) is the position of the center of the target, and (σ_x,σ_y) define the spatial spread of the target. And the input raw image $f(x,y)$ is formed as

$$f(x,y)=S(x,y)+B(x,y)+v(x,y) \tag{2}$$

where $B(x,y)$ represents the background clutter and $v(x,y)$ is noise component. If $B(x,y)$ is predicted accurately and subtracted from $f(x,y)$, $S(x,y)$ can be extracted easily in the residual output. And we define *the local signal to background ratio* (LSBR) as the signal to background ratio in a window around the region of interest.

$$LSBR=\frac{\sum_{x,y}[f(x,y)-m_w]^2}{\sigma_w^2}, \quad (x,y)\in\Omega_S \tag{3}$$

where m_w is the mean of the background in the window of interest, and σ_w^2 is the variance in the same window. Ω_S is defined as effected energy area of the target.

3 The Model of Morphological Neural Network

Morphological basic operations include dilation $\oplus$, erosion Θ, opening $\circ$, and closing $\bullet$. So a prediction of background by morphological filter can be represented as

$$\hat{B}=(f\circ b)\bullet b \tag{4}$$

where b is defined structuring element. Accordingly Yu *et al.* [1] proposed the three-layer feed forward BP neural network for optimizing the weights of SE. However, there are two concernful defects in [1]: The training samples of network are extracted from the whole frame, which leads to a bad performance for tracking clutter including different sub-structures; and the input layer data adopted by the node of output layer exceed the filter window of SE, which causes a not perfect model of the BP network. In this paper we will solve the two problems by the means as follows.

Firstly a raw image f is partitioned to some sub-blocks $A_n, n = 1,2,\cdots$. In every A_n the characteristics of intensity are so uniform that a fixed SE b_n can be effective in this sub-area. Therefore the training samples extracted from A_n are used for obtaining an optimal SE b_{nopt}. Thus the prediction of the whole background is obtained by

$$\hat{B} = \bigcup_n \{(A_n \circ b_{nopt}) \bullet b_{nopt}\} \tag{5}$$

Then a better neural network model will be built for optimizing b_n in the block A_n. Here for simplifying derivation we adopt a symmetric SE of $(2N+1)\times(2N+1)$. And what we need to do is optimizing the weights $b_{i,j}$ $(i, j \in [-N, N])$ by training neural networks in every A_n. And the neural networks structure of dilation and erosion can be easily found in [1]. Then we can design a 3-layer feed forward BP network model of opening and closing as Fig.1. If the hidden layer is erosion and the output layer is dilation, the network is an opening model; and if the hidden layer is dilation and the output layer is erosion, it's a closing network model. For avoiding the input layer data adopted by the node of output layer exceeding the filter window of SE, we extend data of the input layer. Suppose the numbers of training samples in every A_n is L, and the kth input exemplar is $F^k = \{f_{x_k-i,y_k-j} \mid i, j \in [-M, M]\}$, where $k \in [1, L], M = 2N$. So the input layer includes all data by which the node of the output layer can be calculated. Since a hidden layer node output is not from every input neuron, a binary function w for restriction is adopted in the network.

$$w(i,j,r,z) = \begin{cases} 1, (i-r, j-z) \in D_b \\ 0, otherwise \end{cases} \tag{6}$$

where D_b is the domain of b .Suppose the node output is Y^k when the input exemplar is F^k, and the expected output is d^k which is a blurring version of the input data or corresponding pixel of a sequential frame.

We define the square error square cost function as

$$J = \frac{1}{2L}\sum_{k=1}^{L}(Y^k - d^k)^2 \tag{7}$$

and the gradient of J with respect to the weight $b_{p,q}$ can be represented as

$$\delta_{p,q} = \frac{\partial J}{\partial b_{p,q}} = \frac{1}{L}\sum_{k=1}^{L}(d^k - Y^k)\cdot g(Y^k, b_{p,q}) \tag{8}$$

where $p,q \in [-N, N]$. and for opening operation network

$$g(Y^k,b_{p,q})=\begin{cases}1 & if\, Y^k=u_{-p,-q}+b_{p,q}, u_{-p,-q}\neq f_{x_k,y_k}-b_{p,q}\\ -1 & if\, Y^k\neq u_{-p,-q}+b_{p,q}, u_{r,z}=f_{x_k+r+p,y_k+z+q}-b_{p,q}, (r,z)\neq(-p,-q)\\ 0 & otherwise\end{cases} \quad (9)$$

for closing operation network

$$g(Y^k,b_{p,q})=\begin{cases}-1 & if\, Y^k=u_{p,q}-b_{p,q}, u_{p,q}\neq f_{x_k,y_k}+b_{p,q}\\ 1 & if\, Y^k\neq u_{p,q}-b_{p,q}, u_{r,z}=f_{x_k+r-p,y_k+z-q}+b_{p,q}, (r,z)\neq(p,q)\\ 0 & otherwise\end{cases} \quad (10)$$

Thus we can obtain the weights updating equation while η is the learning rate

$$b_{p,q}(t+1)=b_{p,q}(t)-\eta\delta_{p,q} \quad (11)$$

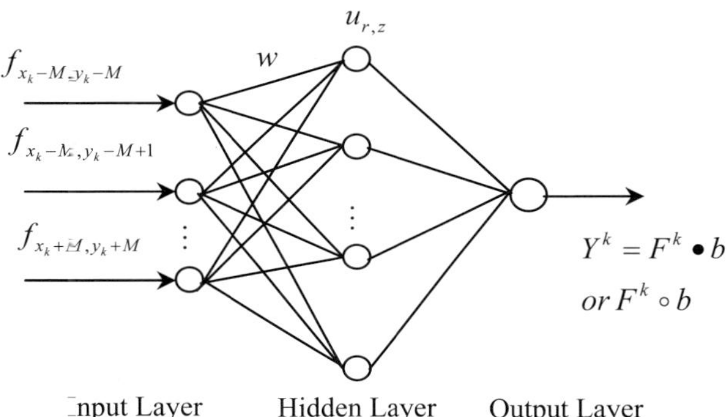

Fig. 1. Neural network model of morphological opening and closing operation

4 Simulations

To study the ability of the morphological neural network to separate targets in a heavy background clutter, simulations studies were conducted with some artificial signals inserted in the real infrared data. Fig.2 shows an infrared image of 240×240 with the 20 objects inserted in the background, and their spatial spread $\sigma_x=\sigma_y=0.5$. By (3) calculate their input LSBR in dB. Use TDLMS [9], morphological filter in [1] and the algorithm in this paper to suppress the background, respectively. And the LSBR gain can be obtained by

$$LSBR\, gain = Output\, LSBR - Input\, LSBR \quad \text{(dB)} \quad (12)$$

Table 1 details the signal to background ratio gains for the 20 targets of interest. Obviously the algorithm of this paper obtains more LSBR gain compared with other methods. A threshold for perfect detection can be defined as the highest threshold at which the weakest signal of interest is detected. The residual noise by clutter removal is generally supposed as a local Gaussian white model [10]. Then in a sub-window noise obeys the normal distribution of $N(\mu,\sigma^2)$. And by expected detection probability P_d and SNR set as minimum of output LSBR, the threshold of detection Th can be calculated by

$$Th = \sqrt{2}\sigma \cdot erfcinv(2P_d) + \sigma\sqrt{SNR} + \mu \tag{13}$$

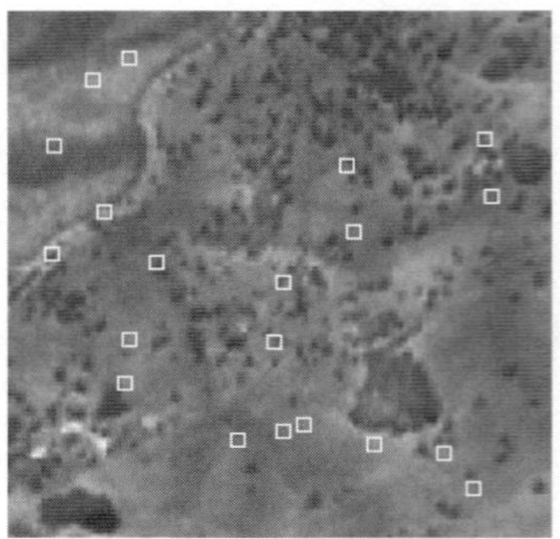

Fig. 2. The infrared image data with 20 targets

where $erfcinv(\cdot)$ is inverse complementary error function. Table 2 shows the detection results with different input intensity of signal, and P_d is set as 0.95. False alarms caused by the method in this paper are least; meanwhile detection rates are highest among the three methods.

Table 1. Comparison of LSBR gains

Position x	Position y	Input LSBR (dB)	LSBR Gain (dB) TDLMS In [9]	Morphological filter in [1]	Method of this paper
170	52	5.329	7.898	8.151	13.010
151	118	-8.131	17.217	19.084	24.090
196	102	-0.647	11.905	13.232	17.933
			• • •		
124	122	-9.345	21.859	22.778	27.542
Means		-1.338	12.593	13.913	18.446

Table 2. Comparison of detecting performance

γ	Means of input LSBR (dB)	Detection rates/ False alarm rates TDLMS In [9]	Morphological filter in [1]	Method of this paper
15	-3.498	$0.95/6.44\times10^{-2}$	$0.9/2.76\times10^{-2}$	$0.95/3.77\times10^{-3}$
18	-1.338	$0.95/2.57\times10^{-2}$	$0.95/1.4\times10^{-2}$	$1.00/1.55\times10^{-3}$
21	0.090	$0.95/1.09\times10^{-2}$	$0.9/6.49\times10^{-3}$	$0.95/6.42\times10^{-4}$

5 Conclusion

An effective morphological neural network of background clutter prediction for detecting small targets in image data is proposed in this paper. The traditional 3-layer feed forward BP network modal of opening and closing operation is modified by extending the input layer data. And for tracking complex background including different sub-structures, training samples extracted from the sub-window are used for optimizing the weights of SE in the corresponded window. Computer simulations of real infrared data show better performance compared with other traditional methods on LSBR gains and detecting effects.

References

1. Yu, N., Wu, C.Y., Li, F.M., et al.: Morphological Neural Networks for Automatic Target Detection by Simulated Annealing Learning Algorithm. Science in China 46(4) (2003) 262-288
2. Yu, N., Wu, C.Y, Tang, X.Y., *et al.*: Adaptive Background Perception Algorithm for Infrared Target Detection. Acta Electronica Sinica 33(2) (2005) 200-204
3. Won, Y. Gader, P.D., Coffield, P.C., *et al.*: Morphological Shared-Weight Networks with Applications to Automatic Target Recognition. IEEE Trans. on Neural Networks 8(5) (1997) 1195-1203
4. Davidson, J.L., Hummer, F.: Morphology Neural Network: An Introduction with Applications. IEEE Trans. on Signal Processing 12(2) (1993) 177-210
5. Gong, W., Shi, T.Y., Cheng, M.D.: Mathematical Morphology in Digital Space (in Chinese). Science Press, Beijing (1997)
6. Chan, D.S.K., Langan, D.A., Staver. D.A.: Spatial Processing Techniques for the Detection of Small Targets in IR Clutter. Proc. of SPIE, Vol. 1305. Technical Symp. Opt. Eng. And Phontonics in Aerospace Sensing, Orlando (1990) 53-62
7. Chen, J.Y., Reed, I.S.: A Detection Algorithm for Optical Target in Clutter. IEEE Trans. on Aerospace and Electronic Systems 23(1) (1987) 46-59
8. Leonov, S.: Nonparametric Methods for Clutter Removal. IEEE Trans. on Aerospace and Electronic Systems 37(3) (2001) 832-847
9. Soni, T., Zeidler, J.R., Ku, W.H.: Performance Evaluation of 2-D Adaptive Prediction Filters for Detection of Small Objects in Image Data. IEEE Trans. On Image Processing 2(3) (1993) 327-339
10. Hunt, B., Cannon, T.: Nonstationary Assumptions for Gaussian Models of Images. IEEE Trans. on Systems, Man and Cybernetics 6(12) 876-882

Two Important Action Scenes Detection Based on Probability Neural Networks

Yu-Liang Geng[1,2], De Xu[1], Jia-Zheng Yuan[1] and Song-He Feng[1,2]

[1] Institute of Computer Science and Technology, Beijing Jiaotong University, Beijing 100044, China
[2] Beijing Key Lab of Intelligent Telecommunications Software Multimedia, Beijing University of Posts and Communications, Beijing 100876, China
gengyuliang@hotmail.com

Abstract. In this paper, an effective classification approach for action scenes is proposed, which exploits the film grammar used by filmmakers as guideline to extract features, detect and classify action scenes. First, action scenes are detected by analyzing film rhythm of video sequence. Then four important features are extracted to characterize chase and fight scenes. After then the Probability Neural Networks is employed to classify the detected action scenes into fight, chase and uncertain scenes. Experimental results show that the proposed method works well over the real movie videos.

1 Introduction

Automated semantic scene, also called semantic event, detection is a crucial issue in content-based video retrieval. It can help viewers capture important and interesting content of video rapidly and attracts many researchers' attention.

There are two fundamentally different approaches for event detection. The first one is a bottom-up and content-independent approach, which employs low-level features to provide a feature curve whose maxima define the highlight parts of videos. In [1], the authors combine audio and video tempo to simulate human's emotion feeling and extract meaningful video segment. In [2], Ma *et al.* constructs a user attention curve based on visual, audio attention model to abstract significant parts of video. Due to the variety of video content, the content-independent approach can only estimate semantic scene content coarsely, so most of prior work is focused on the content-specific approach. This approach exploits domain knowledge and production rules to extract the predefined semantic scenes/events. Li *et al.* [3] extracts three types of events, namely, 2-speaker dialogs, multiple-speaker dialogs, and hybrid events, by analyzing the movie content structure and exploiting film's special editing features. Chu *et al.* [4] identifies four audio events, i.e. gunshot, explosion, car-braking, and engine sound by modelling audio events, and further detects gunplay and car-casing scenes by using HMM and SVM.

As mentioned above, exploiting domain knowledge and production rules involved in video is an effective way to detect semantic scenes. In this paper, we

J. Wang et al. (Eds.): ISNN 2006, LNCS 3972, pp. 448–453, 2006.

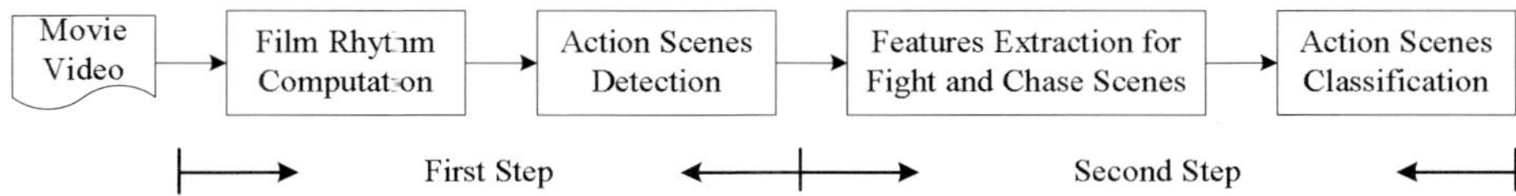

Fig. 1. Block diagram of the proposed approach

employ the film grammar as guideline to identify two important action scenes, namely, chase scene and fight scene in action movies. First, we exploit film rhythm to implement action scenes pre-detection. Then the action scenes are verified and further classified into chase scenes, fight scenes and uncertain scenes by using the probability neural networks (PNN). Experimental results validate the effectiveness of our proposed approach over real-world video data. The block diagram of our approach is shown in Fig. 1.

The organization of the paper is as follows. In Section 2, the action scenes are detected and verified under the guideline of film grammar. In Section 3, the action scenes are further classified into chase, fight and uncertain scenes by using the PNN. Section 4 and 5 give the experimental results and draw the conclusions.

2 Action Scenes Detection and Verification

In movie video, chase and fight scenes are two important semantic events, and often correspond to the plot climaxes [5], so extracting these events will facilitate movie content understanding, summarization and browsing.

In this section, We accomplish action scene detection, segmentation and verification by analyzing action scene characteristics [5]:

- In action scenes, the filmmaker often uses a series shots with high motion activity to create tense or strong atmosphere.
- Fast edits are frequently used to build a sense of kinetic action and speed, as you can see in any chase or fight scene.
- An action scene is often accompanied with high energy of audio effects.
- Two different scenes with high film rhythm may not be juxtaposed together.
- Dialog rarely occurs in action except for brief and terse words.

We can utilize the first three characteristics to compute film rhythm as discussed in our previous work [6], which will be used to detect action scenes. Then we use the last two characteristics to segment and verify the detected action scenes. The film rhythm of a given shot a is computed as follows

$$FR(a) = \alpha \times MIs(a) + \beta \times CFs(a) + \gamma \times ALs(a) \tag{1}$$

where $MIs(a)$ is the average motion intensity of shot a, $CFs(a)$ is the shot cut frequency that is defined as the inverse of shot length, and $ALs(a)$ is the average audio loudness of shot a. Before we compute the film rhythm, we normalize each component by using Gauss normalization formula. α, β, γ are the weight values.

As discussed above, two action scenes with high film rhythm may not be juxtaposed together. So we extract candidate action scenes by detecting the peaks of the film rhythm curve, which has been smoothed by a Gauss filter.

For each of candidate action scenes, we utilize the speech ratio in action scene, which is discussed in [6], to verify action scene. If the speech ratio is less than a given threshold, we regard it as a valid action scene. Otherwise, we regard it as other type, such as a quarrel scene.

3 Action Scenes Classification Based on PNN

In this section, we further classify the detected action scenes into chase scenes, fight scenes, and uncertain scenes by analyzing the intrinsical characteristics [5] involved in themselves.

3.1 Features Extraction for Chase and Fight Scenes

As a fight scene often happens in a limited space, so the shots filmed from the same camera have similar background. When the shots intercut back and forth between one opponent and another, a interleaving structure is generated. While a chase scene happens in open space, and the background change rapidly, so it does not have this attribute.

On the other hand, in the static shots, chase scene usually generates a relatively regular object motion trajectory for local object moving (e.g. running or flying through screen). While fight scene generates irregular motion trajectory for the physical fight. In the moving shots, the chase scene often generates a regular and continuous global motion when camera is tracking the moving objects. While fight scene often generates a uneven global motion because the camera jerking is often used to enforce tense atmosphere. With above analysis, we extract the following features to characterize the chase and fight scenes.

1) Average Inter-shot Similarity ($AISs$). For a given action scene, AS_k, we first compute the inter-shot similarity between shot a and shot b as follows.

$$ISs(a, b) = \max_{f_i \in KFS_a, f_j \in KFS_b} (D(f_i, f_j)) \tag{2}$$

where $D(f_i, f_j)$ is denoted as visual similarity of key frames f_i and f_j based on histograms intersection. And KFS_x is the key frame set of shot x. Here, we extract key frames of each shot by using divisive hierarchical clustering. The number of key frames depends on the visual content complexity of shot [7].

Then the average inter-shot similarity of action scene AS_k is computed as

$$AISs(k) = \sum_{a,b \in AS_i, a \neq b} ISs(a, b)/N_{AS_k} \tag{3}$$

where N_{AS_k} is the number of shots in action scene AS_k.

2) Scene Structure Feature (*SSF*). For a given action scene AS_k, we first cluster its shots into shot groups according to the inter-shot similarity [6]. Then the scene structure feature of action scene AS_k is defined as

$$SSF(k) = 1 - N_{SG_k}/N_{AS_k} \tag{4}$$

where N_{SG_k} is the number of shot groups in action scene AS_k.

3) Camera Motion Feature (*CMF*). We define a shot a has camera motion consistency if its dominant camera motion [6] is greater than a given threshold. Then we define the camera motion feature of action scene AS_k as

$$CMF(k) = N_{cs}/(N_{AS_k} - N_{us} - N_{ss}) \tag{5}$$

where N_{cs} is the number of the shots with camera motion consistency except the unknown and still motion types. N_{us}, N_{ss} is the number of unknown shots and still shots respectively.

4) Object Motion Feature (*OMF*). We track moving objects by computing salient regions in video. Since the same regions have similar color and motion information in the same shot, we utilize the average motion activity and color histogram to compute similarity between salient regions [8]. An approach similar to camera motion feature is adopted to describe the object motion feature. For a given action scene AS_k, the object motion feature computed as follows.

$$OMF(k) = N_{os}/N_{AS_k} \tag{6}$$

where N_{os} is the number of the shots with consistent motion trajectory.

Thus, a 4-Dimension vector, $\mathbf{F}$, is constructed, which is composed of the extracted features: $AISs$, SSF, CMF and OMF of action scene.

3.2 Action Scenes Classification Based on PNN

After event features are extracted, we employ a Probability Neural Networks (PNN) to implement action scenes classification. The PNN models the Bayesian classifier and provides a general solution to pattern classification [9].

A typical architecture of the PNN is composed of four layers. The input layer receives input vector from data set. In pattern layer, each pattern node represents a pattern of the training set. In category layer, the Bayesian posterior probability for each category is obtained as the output of the corresponding category node, which can be represented by

$$P(C_i|\mathbf{x}) \propto \sum_{j=1}^{k_i} \exp(-(\mathbf{x} - \mathbf{x}_{ij})^T(\mathbf{x} - \mathbf{x}_{ij})/2\sigma^2) \tag{7}$$

where $\mathbf{x}$ is the input vector, and $\mathbf{x}_{ij}$ is the ith sample of category C_i in training set. k_i is the sample size of category C_i. σ is a smooth parameter and will be determined in experiment. The output unit classifies the input vector into a specific category according to corresponding rules.

As mentioned in Section 3.1, each action scene was characterized by a 4-Dimension feature, **F**. The PNN receives **F** as its input vector, and the Bayesian posterior probabilities for each category: fight scene and chase scene, is obtained as the output of the corresponding category node. The classification rules are defined as: If two output probabilities have close values, $|P(C_1|\mathbf{x}) - P(C_2|\mathbf{x})| < 0.1$, we classify the input vector into the uncertain category. Otherwise, we classify the input vector into a specific category whose probability has the maximum.

4 Experimental Results

To evaluate the performance of proposed action scene classification algorithm, we collect test data from five action movies, namely, *Star Wars Episode* II, *The Bourne Supremacy*, *The Matrix Reloaded*, *Once a Thief*, and *I Robot*. For the subjectivity of the scene boundary and category, we employ a group of human observers to identify the action scene boundaries and categories as ground truth. There are 1305 shots, and 31 action scenes (18 fight scenes and 13 chase scenes) in total. We achieve 90.3% recall and 84.8% precision in the first step.

In training process of the PNN, only one parameter σ relies on the selection of users. Specht [9] suggests that the smooth parameter selection should rely on the dimension of the problem and the number of training samples. However, it is not difficult to find a suitable parameter σ because the experimental results do not change dramatically with small changes in σ. In experiment, we employ a linear search method to determine σ, and use the cross-validation to compute classification accuracy. Fig. 2 illustrates the effects of the parameter σ to the classification accuracy. The accuracy peak occurred at $\sigma = 0.3$ where it attained a classification accuracy of 0.71. As the semantic scene analysis is still a challenging issue, it is an encouraging experimental results.

The classification confusion matrix for the test data at $\sigma = 0.3$ is summarized in Fig. 3, where the PNN achieves the best classification performance. The numbers along the diagonal are the sample numbers classified by PNN correctly. As Fig. 3 shown, we achieve a satisfying results, which indicate that the features extracted can represent chase and fight scenes effectively. In experiment, the chase scene obtains the lower classification accuracy related to fight scene

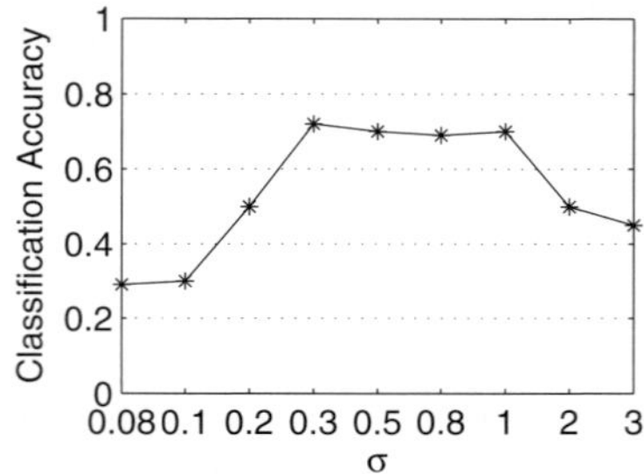

Fig. 2. The classification accuracy related to the parameter σ

PNN Classification Results

		Fight	Chase	Uncertain
True	Fight	14	0	4
	Chase	2	8	3

Fig. 3. The classification confusion matrix for the test data at $\sigma = 0.3$

because the chase scene is more complex, in which rapid camera motion, crowd scene are often used to enforce visual effect.

The feature selection is also an important step in action scene classification. In this paper, we only use the visual features to identify action scene categories. The audio feature may give us a useful clue, which is our further work.

5 Conclusions

We proposed an effective method for action scene classification using the PNN in this paper. The film grammar is used to extract features and identify scene categories. Experimental results show an encouraging classification performance. But semantic analysis is still a challenging task. Our future work will be focused on how to extract more useful audiovisual features to improve the performance of scene classification. Video summarization based on semantic event is another application of this work.

Acknowledgements

This research is supported by Science Foundation of Beijing Jiaotong University (Grant No. 2004SM013).

References

1. Lee, S.H., Yeh, C.H., Kuo, C.C.J.: Video Skimming Based on Story Units via General Tempo Analysis. In: Proc. of IEEE ICME, Vol. 2. (2004) 1099-1102
2. Ma, Y.F., Lu, L., Zhang, H.J., Li, M.J.: A User Attention Model for Video Summarization. In: Proc. of ACM Multimedia, (2002) 533-542
3. Li, Y., Narayanan, S., JayKuo, C.C.: Content-Based Movie Analysis and Indexing Based on AudioVisual Cues. IEEE Trans. on Circuits and Systems for Video Technology 14(8) (2004) 1073-1085
4. Chu, W.T., Cheng, W.H., Wu, J.L.: Generative and Discriminative Modeling toward Semantic Context Detection in Audio Tracks. In: Proc. of the 11th International Multimedia Modelling Conference, (2005) 38-45
5. Arijon, D.: Grammar of The Film Language. Silman-James Press, Los Angeles CA (1991)
6. Geng, Y.L., Xu, D., Wu, A.M.: Effective Video Scene Detection Approach Based on Cinematic Rules. In: Rajiv, K., Robert, J.H., Lakhmi, C.J. (eds.): Knowledge-Based Intelligent Information and Engineering Systems. Lecture Notes in Computer Science, Vol. 3682. Springer-Verlag, Berlin Heidelberg New York (2005) 1197-1204
7. Geng, Y.L., Xu, D.: Efficient Key Frames Extraction Based on Hierarchical Clustering. In: Proc. of IEEE TENCON, Vol. B. (2004) 77-80
8. Hu, Y.Q., Xie, X., *et al.*: Salient Region Detection Using Weighted Feature Maps Based on Human Visual Attention Model. In: Proc. of IEEE PCM, (2004) 993-1000
9. Specht, D.: Probablitic Neual Networks. Neual Networks 3(1) (1990) 109-118

Local Independent Factorization of Natural Scenes

Libo Ma, Liqing Zhang*, and Wenlu Yang

Department of Computer Science and Engineering, Shanghai Jiao Tong University,
1954 Hua Shan Road, Shanghai 200030, China
malibo@sjtu.edu.cn, zhang-lq@cs.sjtu.edu.cn

Abstract. In this paper, we study sparse representation of large-size natural scenes via local spatial dependency decomposition. We propose a local independent factorization model of natural scenes and develop a learning algorithm for adaptation of the synaptic weights. We investigate the dependency of neighboring location of the natural scene patches and derive learning algorithm to train the visual neural network. Some numerical experiments on natural scenes are performed to show the sparse representation of the visual sensory information.

1 Introduction

What is the internal representation of enormous amounts of information about the environment in the form of light, sound, and other senses? And how does our brain form useful internal representation and understand these information? These are some of basic issues in computational neuroscience and neural network modelling. In recent years, a general principle referred to as sparse coding [6] has been suggested to answer such questions. Furthermore, there are substantial evidences for sparse coding of sensory system occurring in the cortex of mammals [1, 7]. It has been postulated that the early visual system is adapted to the statistical properties of the environments. In recent research a number of models such as independent component analysis (ICA) [5] have been developed based on the statistical modelling of natural images. It has been shown that the statistically optimal linear features of natural images are very similar to those computed in simple cells or complex cells in V1.

In this paper, we propose a local independent factorization algorithm for internal neural representation of the visual sensory system. We investigate the dependency of neighboring location of the natural scenes patches and derive learning algorithm to train the visual neural network.

2 Local Independent Factorization Model and Algorithm

In this section, we propose a local independent factorization (LIF) model and develop a learning algorithm for training the model parameters. From this model,

* Corresponding author.

J. Wang et al. (Eds.): ISNN 2006, LNCS 3972, pp. 454–459, 2006.

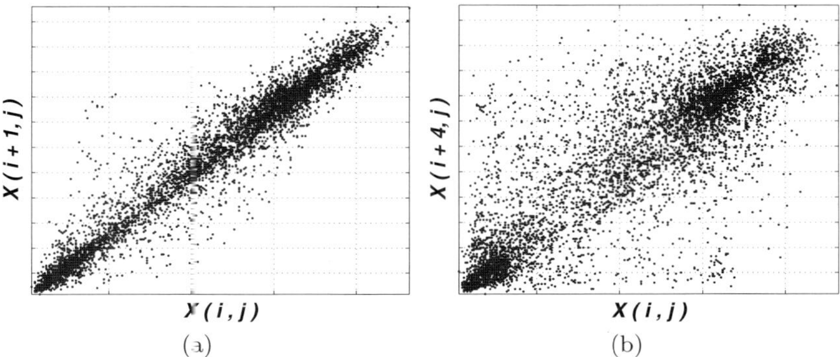

Fig. 1. Joint distribution of image pixel intensities of different distances. **(a)** scatterplots of pixel intensities of location (i, j) and neighbor location $(i+1, j)$. **(b)** scatterplots of pixel intensities of location (i, j) and far neighbor location $(i + 4, j)$.

we can decompose the internal neurons maximally independent by repeating the following two steps. One is updating local synaptic weights, by applying the natural gradient algorithm developed in [9]. During another phase, the whole natural images and samples are represented by the new synaptic weights matrix. It is well known that natural images are statistically redundant. To formulate local independent factorization algorithm, we consider spatial correlations of natural images. Even if we randomly choose one natural images, we can find that neighboring spatial locations are strongly correlated in intensity [8]. The scatterplots of two pairs of intensity value of neighboring points of different distances as illustrated in Fig. 1.

To investigate these spatial correlation, we consider a topography of input natural images, which is just based on the pixel lattice. Primarily we assume that the receptive field of one neuron is only relevant to corresponding local image patch of size $R \times R$ as illustrated in figure 2. That is, the image patches in the vicinity are strongly dependent and contribute to local neural sparse responses. In contrast, image patches out of the vicinity in the large image do not contribute to the corresponding local neural responses. Let $\boldsymbol{X}$ be an $M \times N$ given natural scene where M and N is the height and width respectively. Then $X(m, n)$ is defined as an image patch that the center is located at (m, n) and the size is $R \times R$. We assume that the size of receptive field is $R \times R$. Thus one patch $X(m, n)$ of $R \times R$ is projected onto a neuron (i, j) in the internal layer by connections $W(i, j, m, n)$ and the neural response is $Y(i, j)$. Whereas the whole synaptic connections $\boldsymbol{W}$ is an $M \times N \times L \times L$ four dimensional tensor which transform observed $\boldsymbol{X}$ to neural response $\boldsymbol{Y}$. Then the neighboring patch $X(m, n + 1)$ are projected onto neighboring neuron $(i, j + 1)$ subsequently. Furthermore, all the $L \times L$ neighboring patches which surround the center (m,n) are projected onto neurons of $L \times L$ in the internal layer in succession. Here, we choose corresponding 3×3 neurons locally as a unit to train the local synaptic weights $W(i, j, m, n)$. We concern with the local field of 3×3 and make the responses of this nine

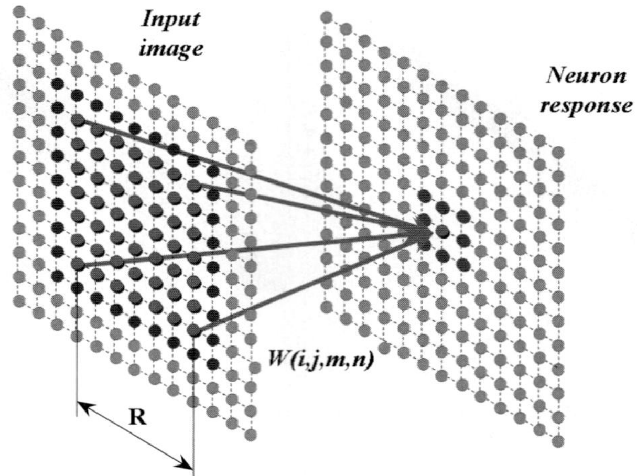

Fig. 2. An illustration of the local independent factorization model

neurons mutually independent. Through updating the synaptic connections of all the scattered 3×3 matrix, the global synaptic weights $\boldsymbol{W}$ are trained.

Now, the local independent factorization algorithm is described formally. The neural response in the internal layer can be formulated as follows

$$Y(i,j) = \sum_{|i-m| \le \frac{R-1}{2}, |j-n| \le \frac{R-1}{2}} W(i,j,m,n) X(m,n) \tag{1}$$

Where the neural response $Y(i,j)$ is only relevant to the image patch in size of $|i-m| \le \frac{R-1}{2}, |j-n| \le \frac{R-1}{2}$. The image pixels out side of this $R \times R$ receptive field is irrelative to the neural responses. $W(i,j,m,n)$ is the connecting synaptic weights between neuron (i,j) and the pixel (m,n). Now, We realign the nine neurons to a 9-dimensional vector in row wise and the local neural response of 3×3 is rewrite as

$$\begin{aligned} \mathcal{Y} = &(Y(i-1,j-1), Y(i-1,j), Y(i-1,j+1), \\ &Y(i,j-1), \quad Y(i,j), \quad Y(i,j+1), \\ &Y(i+1,j-1), Y(i+1,j), Y(i+1,j+1)) \end{aligned} \tag{2}$$

Suppose the joint probability density function of $\mathcal{Y}$ is $p_{\mathcal{Y}}(X)$, and the marginal probability density function of $Y(i,j)$ is $p_{Y(i,j)}$. In order to make the nine neurons in the internal layer maximally mutually independent, we use the mutual information as the dependency criterion. Then the mutual information of random variable $\mathcal{Y}$ is defined by the Kullback-Leibler divergence [2]

$$KL(p_{\mathcal{Y}}(X), \prod p_{Y(i,j)}(X)) = \int \log \frac{p_{\mathcal{Y}}(\xi)}{\prod p_{Y(i,j)}(\xi)} p_{\mathcal{Y}}(\xi) d\xi \tag{3}$$

Now, we can rewrite the equation (1) into the following ICA model $\mathcal{Y} = \mathcal{W}\mathcal{X}$, where $\mathcal{X}$ is R^2-dimensional vector. $\mathcal{W}$ is a $9 \times R^2$ matrix of synaptic weights. To minimize the mutual information, via simplifying the mutual information (3), we can derive a cost function [10] for learning the local synaptic weights as follows

$$\mathcal{L}(\mathcal{W}) = -\frac{1}{2}\log|\mathcal{W}\mathcal{W}^T| - \sum \log p_i(\mathcal{Y}(i)) \quad (4)$$

Whereas, there are unknown functions, the probability density functions $p_i(\mathcal{Y}(i))$ in this cost function. In order to find the minimum point of the cost function, we apply the natural gradient algorithm developed in [9]. The natural gradient algorithm given as

$$\Delta\mathcal{W} = \eta(\mathbf{I} - \varphi(\mathcal{Y})\mathcal{Y}^T)\mathcal{W} \quad (5)$$

where φ is a vector of activation function, which is related to the probability density functions of the neuron activities. However, the probability density functions of internal neurons are unknown during learning. In this paper we employ the generalized gaussian distribution [10] to approximate the probability density functions $p_i(\mathcal{Y}(i))$ and adapt the parameters in the model via maximum likelihood.

Applying the learning algorithm (5), the local synaptic weights are updated. Then the training process is performed in each scattered 3×3 field to obtain global synaptic weights $\boldsymbol{W}$.

3 Simulations

In this section, the local independent factorization model is applied to a video of natural scenes. In order to verify the validity of this algorithm, we also carry out two experiments. One is using FastICA [4] algorithm to estimate the basis vectors of the patches of the tree video. Another is using the Nonnegative Matrix Factorization (NMF) [3] algorithm to estimate the basis vectors as well as. We take a 35 seconds forest video with frame rate at 29 frames/s and transform it into avi format as the input training data. Then we downsample the resolution of this video fragment from 480×640 to 60×80 pixels and then we transform the color video to grey-level video. Furthermore, we choose 3×3 neurons and 17×17 receptive fields as the local training fields. We make the intensity of each pixel of time direction be zero-meaned. And then we apply the natural gradient algorithm to train the synaptic weights matrix $\boldsymbol{W}$. To implement this local independent factorization model, we use a Pentium IV 3.0 GHz CPU computer. Each iteration spends approximately 10 minutes.

The spatial distribution of the neural synaptic weights trained by the local independent factorization model are shown in Figure 3(a), which are different from the ICA bases obtained by ICA [6]. The local area shown by dashed square in figure 3(a) is enlarged to show the details of synaptic weights. From the distribution of the neural synaptic weights shown in figure 3(a), we can see that the local independent factorization algorithm tends to find out all the local

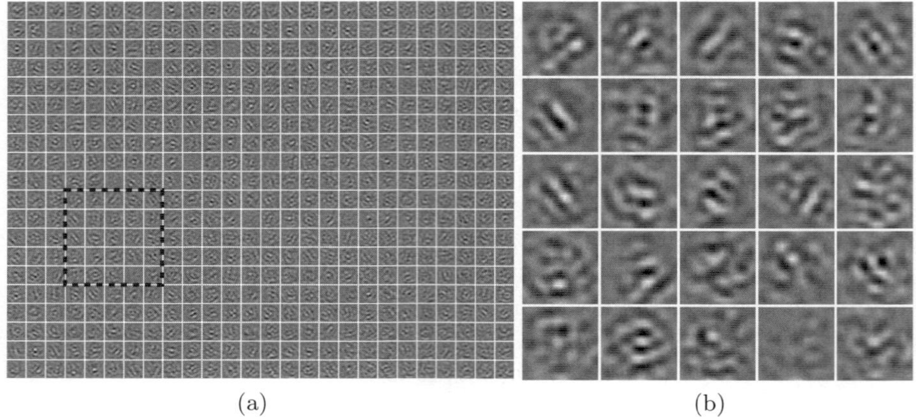

Fig. 3. (a) part of synaptic weights W by local independent factorization algorithm. **(b)** The enlarged synaptic weights that is shown by dashed square.

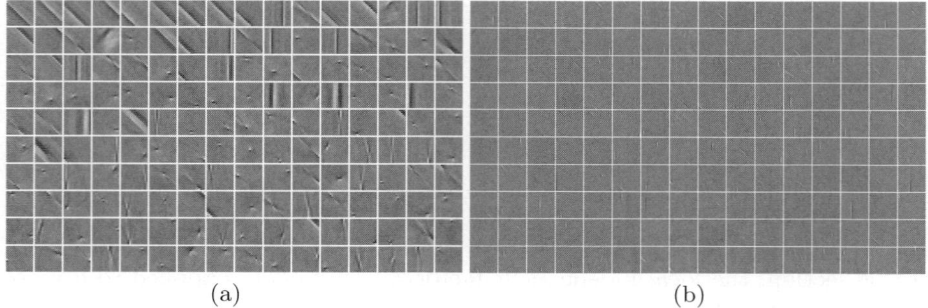

Fig. 4. The basis vectors estimated by FastICA and NMF algorithm from patches of natural scenes data respectively

independent components by applying sparse coding strategy to the small local neuronal area. The synaptic weights in our model seem to be selective tuning for location, orientation , and frequency. And it is similar to the Gabor-like classical receptive fields (CRFs) of simple cell. Furthermore, via extracting patches from this tree video as the input data, we also estimate the representation given by FastICA [4] and Nonnegative Matrix Factorization (NMF) [3] model respectively. A sample of 24000 patches of 32×32 pixels are sampled from the tree video. The simulation results are shown in figure 4(a) and figure 4(b).

4 Conclusions

In this paper, we investigate the relationship between natural image statistics and internal neural representation of the visual sensory system. We propose a local independent factorization algorithm via investigating the dependency of neighboring location of natural scene patches. Learning algorithm for training the neural network is derived by minimizing the mutual information of internal neurons. By using natural gradient algorithm to update the local synaptic weights we obtain global synaptic weights of natural scenes. Computer simulations are provided. It shows that the local independent factorization algorithm can elucidate the formulation of the internal sparse representation of the sensory system. We can see that the local independent factorization algorithm tends to find out all the local independent components by applying sparse coding strategy to the small local neuronal area. The synaptic weights in our model seem to be selective tuning for location, orientation, and frequency. And it is similar to the Gabor-like classical receptive fields (CRFs) of simple cell.

Acknowledgment

The work was supported by the National Basic Research Program of China (Grant No. 2005CB724301) and National Natural Science Foundation of China (Grant No.60375015).

References

1. Berry, M., Warland, D., Meister, M.: The Structure and Precision of Retinal Spike Trains. Proceedings of The National Academy of Sciences of The United States of America 94(10) (1997) 5411-5416
2. Haykin, S.: Neural Networks, A Comprehensive Foundation. Second Edition. Prentice-Hall (1999)
3. Hoyer, P.O.: Non-negative Matrix Factorization with Sparseness Constraints. Journal of Machine Learning Research 5 (2004) 1457-1469
4. Hyvärinen, A., Oja, E.: A Fast Fixed-point Algorithm for Independent Component Analysis. Neural Computation 9 (1997) 1483-1492
5. Jutten, C., Herault, J.: Blind Separation of Sources. Part I: An Adaptive Algorithm Based on Neuromimetic Architecture. Signal Process 24(1) (1991) 1-10
6. Olshausen, B., Field, D.: Emergence of Simple-cell Receptive Field Properties by Learning A Sparse Code for Natural Images. Nature 381 (1996) 607-609
7. Reinagel, P.: How Do Visual Neurons Respond in The Real World? Current Opinion in Neurobiology 11 (2001) 437-442
8. Simoncelli, E. P., Olshausen, B. A.: Natural Image Statistics and Neural Representation. Annual Review of Neuroscience 24 (2001) 1193-1216
9. Zhang, L.Q., Cichocki, A., Amari, S.: Natural Gradient Algorithm for Blind Source Separation of Overdetermined Mixture with Additive Noise. IEEE Signal Processing Letters 6(11) (1999) 293-295
10. Zhang, L.Q., Cichocki, A., Amari, S.: Self-adaptive Blind Source Separation Based on Activation Function Adaptation. IEEE Tansactions on Neural Networks 15(2) (2004) 233-244

Search Region Prediction for Motion Estimation Based on Neural Network Vector Quantization

DaeHyun Ryu and HyungJun Kim

Division of Information Technology, Hansei University, Korea
dhryu,harry@hansei.ac.kr

Abstract. We present a new search region prediction method using frequency sensitive competitive learning vector quantization for motion estimation of image sequences. The proposed method can decrease the computation time because of the smaller number of search points compared to other methods, and also reduces the bits required to represent motion vectors. The results of experiments show that the proposed method provides competitive PSNR values compared to other block matching algorithms while reducing the number of search points and minimizing the complexity of the search region prediction process.

1 Introduction

The goal of the temporal model for moving picture compression is to reduce redundancy between frames by forming a predicted frame and subtracting it from the current frame. The output of this process is a residual frame. The more accurate the prediction process is, the less energy is contained in the residual frame. The accuracy of the prediction can usually be improved by compensating for motion between the reference frame and the current frame. Since the temporal correlation as well as the spatial correlation is very high in moving pictures, a high compression ratio can be achieved using the Motion Compensated Coding(MCC) technology. MCC consists of a motion compensating by the precise motion estimation and prediction error encoding part[1, 2]. There are two categories of motion estimation techniques: Pel Recursive Algorithm(PRA) and Block Matching Algorithm(BMA). With BMA-based motion compensated prediction coding method, the amount of information for motion vectors and prediction error must be as small as possible[3].

In this paper, we propose a new method for estimating motion vectors in an image sequence. The proposed method predicts the search region by neural network vector quantization(VQ) and evaluates distortion for the predicted points. The paper is organized as follows: Section 2 reviews conventional block matching algorithms and their drawbacks. Section 3 describes the proposed method that predicts the motion region using neural network vector quantization and calculates motion vectors. Section 4 presents the results of the computer simulation for a sequence of images. It also compares the proposed method with other algorithms. Finally, Section 5 addresses the conclusions.

J. Wang et al. (Eds.): ISNN 2006, LNCS 3972, pp. 460–467, 2006.

2 Motion Vector Estimation Using Block Matching Algorithm

The simplest method of temporal prediction is to use the previous frame as the predictor for the current frame. Block matching algorithms are utilized to estimate motion at a block of pixels(size of $(M \times N)$), for example, the location of the block in the present frame compared to that of the block in the previous frame. This block of pixels is compared with a corresponding block within a search region in the previous frame as shown in Fig. 1. The process of BMA divides an image into fixed size subimages, and then finds one best match for the previous frame by maximizing cross correlation.

We define a function $D(\cdot)$ to locate the best match:

$$D(i,j) = \frac{1}{NM} \sum_{m=1}^{M} \sum_{n=1}^{N} G(U(m,n) - U_r(m+i, n+j)), \quad -p \leq i,j \leq p \qquad (1)$$

where $G(\cdot)$ is a nonlinear function to evaluate error power, U is a block of size $M \times N$ in the current frame, U_r is a search area of size $(M+2p) \times (N+2p)$ in the previous frame, and p is the maximum displacement allowed. The value (i,j) is displacement which minimizes the $D(i,j)$. Though motion vector detection schemes using BMA have been widely utilized, they have many drawbacks. For instance, they assume that all the pixels within the block have uniform motion because they detect motion vectors on a block-by-block basis. This assumption is acceptable for small block sizes (8×8 or 16×16). However, having a smaller block-size increases the number of blocks and requires higher transmission rate because of an increase in the amount of motion vectors to be transmitted[4, 5].

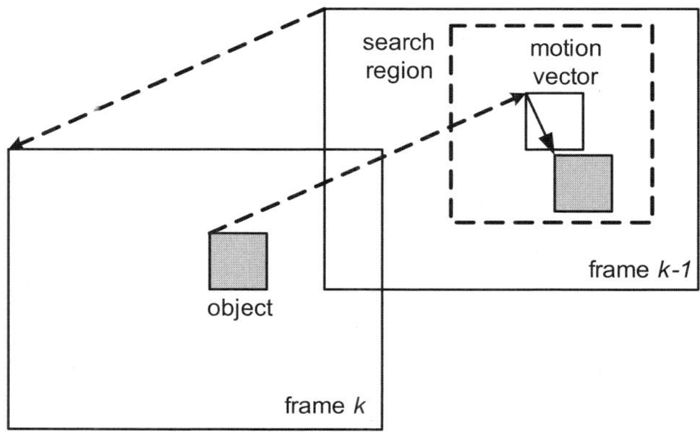

Fig. 1. Motion detection by block matching algorithm

3 Motion Vector Estimation Using Neural Network VQ

The performance of motion vector detection can be increased because motion vectors usually have high spatiotemporal correlation. We propose a new motion vector estimation technique using this correlation.

3.1 Search Region Prediction Using VQ

Vector quantization is a quantization technique which capitalizes any underlying structure in the data being quantized. The space of the vectors to be quantized is divided into a number of regions. A reproduction vector is calculated for each region. Given any data vector to be quantized, the region in which it lies is determined and the vector is represented by the reproduction vector for that region. More formally, vector quantization is defined as the mapping of arbitrary data vectors to an index m. Thus, the VQ is mapping of k-dimensional vector space to a finite set of symbols M.

$$VQ : \mathbf{x} = (x_1, x_1, \ldots, x_k) \longrightarrow m \tag{2}$$

where $m \in \{M\}$ and the set m has size M. Assuming a noiseless transmission or storage channel, m is decoded as $\mathbf{x}$. The collection of all possible production vectors is called the codebook. In general, this requires knowing the probability distribution of the input data. Typically, however, this distribution is not known, and the codebook is constructed through process called training. During the training, a set of data vectors that is representative of the data that will be encountered in practice is used to determine an optimal codebook[6].

3.2 Quantization of Motion Vectors Using Neural Networks

According to Section 3.1, the training and encoding processes are computationally expensive. Moreover, most of the algorithms currently used for VQ design, e.g., the LBG algorithm, are batch mode algorithms, and need to have access to the entire training data set during the training process. Also, in many communication applications, changes in the communication channel mean that a codebook designed under one condition is inappropriate for use in another condition. Under these circumstances, it is much more appropriate to work with adaptive VQ design methods, even if they are suboptimal in a theoretical sense. Another benefit of formulating vector quantization using a neural network is that a number of neural network training algorithms such as Competitive Learning(CL), Kohonen Self-organizing Feature Map(KSFM) and Frequency Sensitive Competitive Learning(FSCL) can be applied to VQ[7].

In this paper, we use FSCL to overcome a drawback of CL. Assume that the neural network VQ is to be trained on a large set of training data. Further assume that the weight vectors $\mathbf{w}_n(i)$ are initialized with random values. The algorithm for updating the weight vectors is as follows. The input vector is presented to all of the neural units and each unit computes the distortion between its weight and the input vector. The unit with the smallest distortion is designated as the

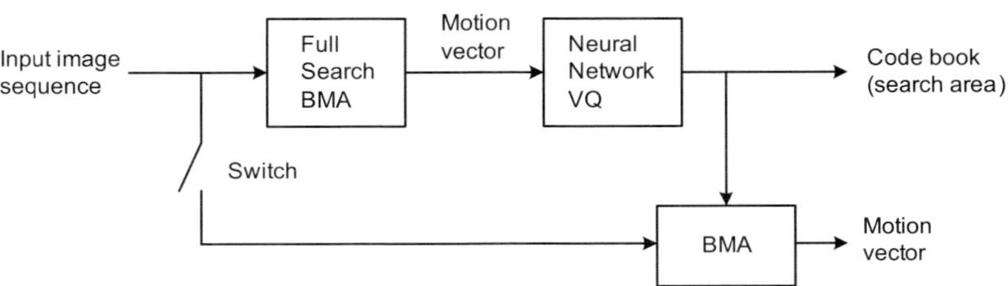

Fig. 2. Block diagram of the proposed motion vector estimation system

winner and its weight vector is adjusted towards the input vector. Let $\mathbf{w}_n(i)$ be the weight vector of neural unit i before the input is presented as follows:

$$z_i = \begin{cases} 1 & \text{if } d(\mathbf{x}, \mathbf{w}_i(n)) \leq d(\mathbf{x}, \mathbf{w}_i) \text{ for } i = 1, \ldots, M \\ 0 & \text{otherwise} \end{cases} \tag{3}$$

The new weight vectors $\mathbf{w}_i(n+1)$ are computed as

$$\mathbf{w}_i(n+1) = \mathbf{w}_i(n) + \varepsilon(\mathbf{x} - \mathbf{w}_i(n))z_i \tag{4}$$

In Eq.(4), the parameter is the learning rate, and is typically reduced monotonically to zero as the learning progresses. A problem with this training procedure is that it sometimes leads to neural units which are under utilized. To overcome this problem, FSCL algorithm has been suggested. In the FSCL network, each unit incorporates a count of the number of times it has been the winner. A modified distortion measure for the training process is defined as follows:

$$d^*(\mathbf{x}, \mathbf{w}_i) = d(\mathbf{x}, \mathbf{w}_i(n)) * u_i(n) \tag{5}$$

where $u_i(n)$ is the total number of times that neural unit i has been the winner during the training. The winning neural unit at each step of the training process is the unit with the minimum d^*.

3.3 Motion Vector Estimation by Search Region Prediction

Fig. 2 shows the block diagram of the proposed motion vector estimation method using a neural networks vector quantizer. First, we find motion vectors using the full search method from the training images and then, train the neural network vector quantizer codebook using these motion vectors. Second, a motion vector can be estimated using the codebook as a motion prediction region. The codewords in the codebook represent the motion vectors for the input image sequences.

Since the codebook is used as the search region for estimating the motion vectors, the search points and computation can be reduced compared with the full search BMA. In addition the information required to transmit the motion vectors can be reduced. For example, the number of possible motion vectors for BMA with a ± 7(the search region is 15 pixels in both horizontal and vertical

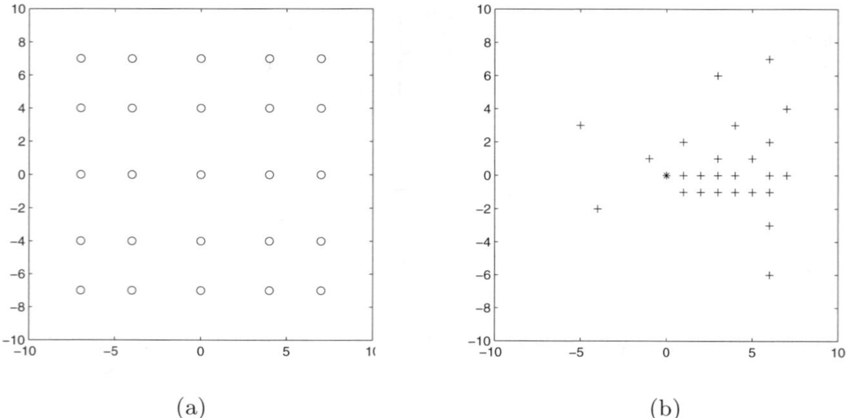

(a) (b)

Fig. 3. (a) Initial value of codebook and (b) Output value of codebook

directions) is 225, which requires about 8 bits per a motion vector for fixed length encoding. Therefore, we have compressed the number of motion vectors from 225 to 25 or from 8 bits to 5 bits per vector. The computational cost is also improved because the number of search point is reduced. The codebook is designed with the neural network vector quantizer utilizing the FSCL learning algorithm using the above motion vectors as the training input data. Fig. 3 shows the initial codebook and the output codebook that has 25 codewords.

4 Experimental Results and Discussions

The SIF version of Flower garden image sequence was used for the experiment. The size of a SIF sequence is half of its CCIR 601 version in both dimensions. The block size for BMA was set to 8×8. Since the MPEG recommended search region is 15 pixels in both horizontal and vertical directions, we choose a search region of ± 7 pixel in both spatial directions because the size of a SIF sequence is half of its CCIR 601 version in both dimensions. We also choose a codebook size of 64 motion vectors.

Table 1 shows the number of search points and the average PSNR of the first 30 frames. We compare the performance of the proposed method with that of BMA with ± 7 search region(15 pixels in both horizontal and vertical directions),

Table 1. Performance comparison

Method	No. of search pt.	Required bits/MV	Ave. PSNR(dB)
Full search(± 7)	225	8	22.6791
Full search(± 2)	25	5	17.3795
TSS	27	8	21.0388
Proposed	25	5	22.6388

±2 search region(5 pixels in both horizontal and vertical directions), and Three Step Search(TSS). As shown in Table 1, the number of possible motion vectors for TSS which is known to show better performance is 27, and the number of possible motion vectors for BMA is 225, which requires about 8 bits per motion vector for fixed length encoding. As mentioned in the previous section, the proposed method reduces not only the number of search points but also the computation time. For BMA with ±2 search region, the number of matches required is 25. The number of search points requires for the proposed method is the same as that of BMA with ±2 search region, but the average PSNR has a difference of about 5 dB. The proposed method achieves almost the same PSNR value as ±7 search region, and also requires the least number of search points and bits for a motion vector.

A common measure of distortion is the mean-square-error(MSE) method Which compares the reconstructed image I' and the original image I. The peak signal-to-noise ratio(PSNR) in decibel(dB) is used to measure the performance of algorithms which is defined by

$$PSNR = 20 \cdot \log_{10}\left(\frac{G_{max}}{RMSE}\right) \tag{6}$$

where G_{max} is the maximum pixel value(usually 255 for the 8-bit imagery), $RMSE$ is the root mean-squared error given as

$$RMSE = \sqrt{\frac{1}{NM}\sum_{i=0}^{N-1}\sum_{j=0}^{M-1}[I(i,j) - I'(i,j)]^2} \tag{7}$$

and N and M are the width and height, respectively, of the images. Note that the original image and the reconstructed image must be of the same size.

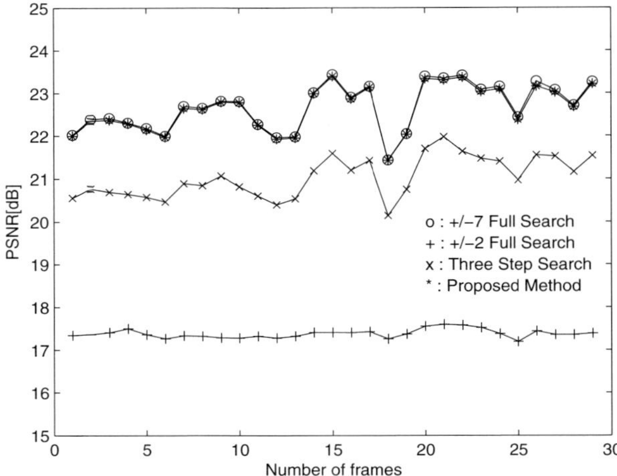

Fig. 4. PSNR values for the first 30 frames of Flower garden image sequence

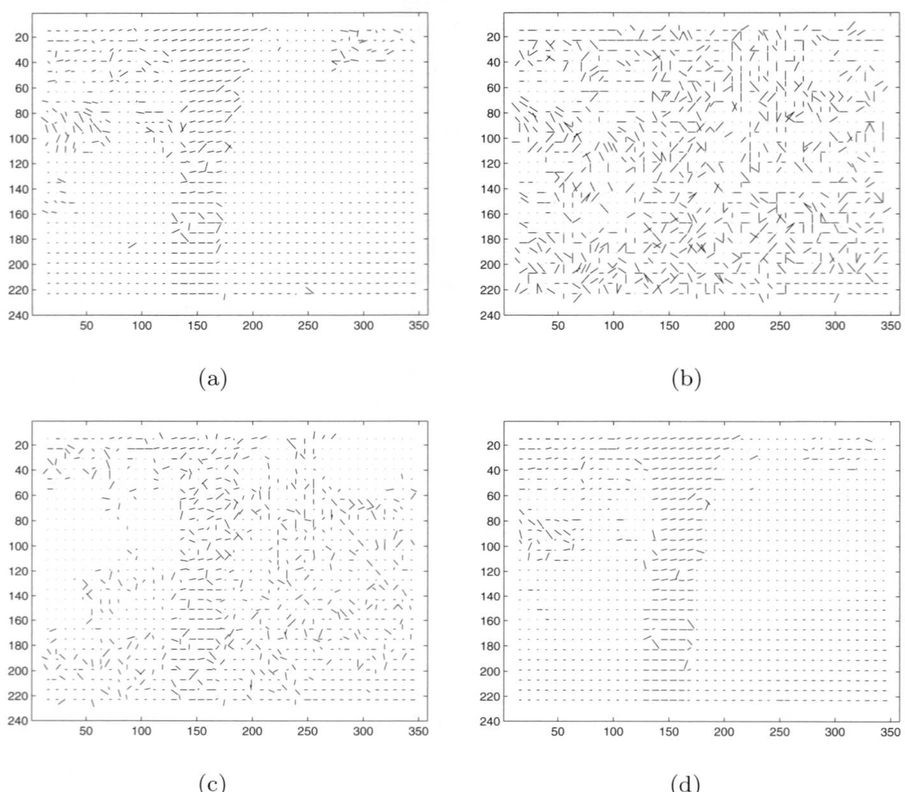

Fig. 5. Examples of motion vector map for the 16th frame of Flower garden image sequence: (a) full search, (b) search ±2 area, (c) TSS, and (d) the proposed method

Fig. 4 shows the PSNR values of 30 frames of which the smoothing effect of motion vectors has been calculated using four different methods. To make the effect more visible we set the block size to be 4 and the search area to be ±7, and exaggerate motion vectors. Fig. 5 shows examples of motion vector map for the 16th frame of Flower garden image sequence. It demonstrates that the smoothing effect of the proposed methods is superior to other methods. Using the smoothing effect, we can eliminate errors which may be caused by quantization process of motion vectors and we also can reduce the number of bits required to represent motion vectors.

5 Conclusions

The motion estimation method plays an important role in the moving image transmission systems. System performance depends on how accurately the motion vectors are estimated. Though number of motion estimation methods have been suggested, to detect the motion vectors more accurately the full search method which matches all point in the search area must be used, but it requires much computation and hardware complexity. In this paper, we found motion vectors using the full search BMA from the initial image sequences, and trained FSCL neural networks to design the codebook using the motion vectors. We used this codebook as the motion estimation region. This method uses the spatial correlation of motion vectors in image sequences, therefore reduces search area, bits required to transmit motion vectors and increased the compression rate. The computer simulations show that the proposed method is superior to the TSS about 1.5 dB. The proposed method is also robust to the noise because it has the motion vector smoothing effect during the vector quantization process. There is therefore a tradeoff in image quality associated with faster motion compensation schemes, since more accurate motion compensation requires longer processing time to find out the vector field but gets better image quality.

Acknowledgements

This work was supported by Hansei University.

References

1. Yao, W., Wenger, S., Jiantao, W., Katsaggelos, A. K.: Error Resilient Video Coding Techniques. IEEE Signal Processing Magazine 17(4) (2000) 61–82
2. Jain, A. K.: Image Data Compression: A Review. Proceeding IEEE 69(3) (1981) 384–389
3. Netravali, A. N., Limb, J. O.: Picture Coding: A Review. Proceedings of IEEE 63(3) (1980) 366–406
4. Iinuma, K., Koga, T., Niwa, K., Iijima, Y.: A Motion-Compensated Interframe Codec. in Proceeding Image Coding SPIE 594 (1985) 194–201
5. Lee, Y. Y., Woods, J. W.: Motion Vector Quantization for Video Coding. IEEE Trans. Image Processing 4(3) (1995)
6. Gersho, A., Gray, R. M.: Vector Quantization and Signal Compression. Kluwer Academic Publishers (1991)
7. Al Sayeed, C., Ishteak Hossain, A.: Image Compression Using Frequency-Sensitive Competitive Neural Network. In: Proceedings of the SPIE 5637 (2005) 611–618

Hierarchical Extraction of Remote Sensing Data Based on Support Vector Machines and Knowledge Processing

Chao-feng Li, Lei Xu, and Shi-tong Wang

School of Information Technology, Southern Yangtze University, 214122 Wuxi, China
chaofeng.li@163.com

Abstract. A new extraction method for remote sensing data is proposed by using both a support vector machine (SVM) and knowledge reasoning technique. The new method fulfils intelligent extraction of water, road and other plane-like objects from remote sensing images in a hierarchical manner. It firstly extracts water and road information by a SVM and pixel-based knowledge post-processing method, then removes them from original image, and then segments other plane-like objects using the SVM model and computes their features such as texture, elevation, slope, shape etc., finally extracts them by the polygon-based uncertain reasoning method. Experimental results indicate that the new method outperforms the single SVM and moreover avoids the complexity of single knowledge reasoning technique.

1 Introduction

In the last twenty years neural network models make great progress in remote sensing image classification [1-3]. But NN models are based on so-called Empirical Risk Minimization principle and don't guarantee good generalization ability. Recently Vapnik and his co-workers developed a kind of new classifier namely SVM [4-5]. SVM have been gained wide acceptance due to their high generalization ability and better performance than other traditional learning machines [6-7].

However remote sensing classification is a complicated task, and in order to gain better classification results, we have to consider much more features except spectral signatures, such as shape, size, texture, shadow etc. But SVM can only use measurable spectral signatures etc., and unable to use the rule-expressed knowledge. In this way, an intelligent knowledge processing is absolutely necessary. But single knowledge processing method needs a lot of geographical knowledge to infer, which will raise an issue of knowledge acquisition and knowledge reasoning [8-11]. So in this paper we integrate the SVM and knowledge processing together to extract remote sensing data.

2 Summary of the Integrated Intelligent Processing Procedure

The whole intelligent processing procedure consisting of three parts is shown in Fig.1.

J. Wang et al. (Eds.): ISNN 2006, LNCS 3972, pp. 468 – 473, 2006.

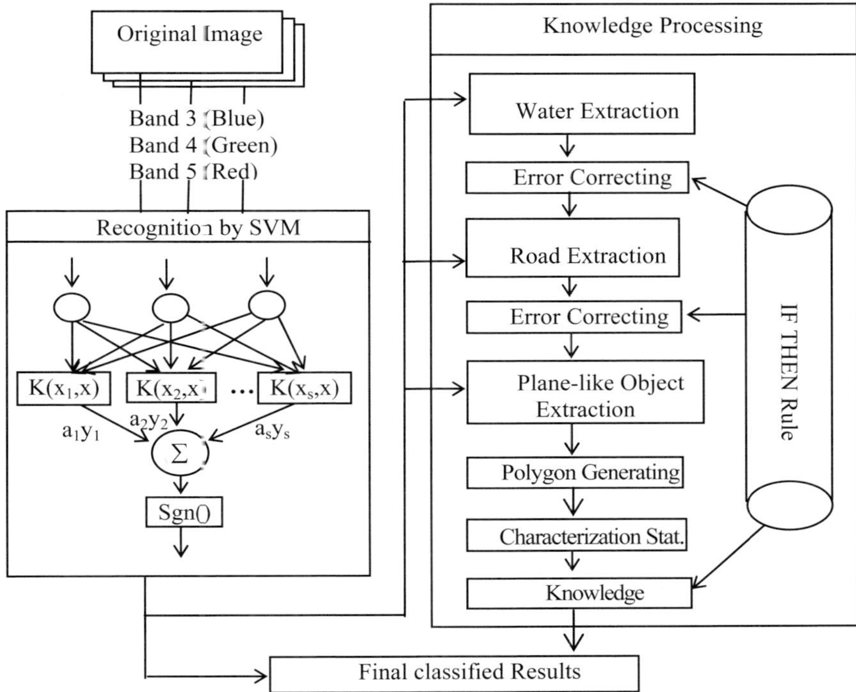

Fig. 1. Integrated intelligent processing flow

2.1 Preprocessor

An experimental image is TM image that covers Xuzhou city area in China. According to the analysis we only use band 3, band 4 and band 5 (Shown as Fig.2) to identify land-use categories. The size of the image is 540x486 pixels.

Fig. 2. The enhanced image of band 5

According to the terrain map analyzing the image visually, we divide it into 8 categories, namely city area, agriculture land, bare land, green land, hill, road, river & pond etc, lake. We will distinguish these categories by the method shown in Fig.1.

2.2 Brief Introduction of SVM Algorithm

A general non-linear SVM can be expressed as

$$f(x)=\sum_i a_i y_i K(\vec{x}_i,\vec{x})-b \tag{1}$$

Where $f(x)$ is the output of the SVM, K is a kernel function which measures the similarity of a stored training sample $\vec{x}_i$ to the input $\vec{x}$, $y\in\{-1,+1\}$ is the desired output of the classifier, b is a threshold, and a_i are weights which blend the different kernels [5]. Training of SVM consists of finding the parameter a_i, and the training is expressed as a minimization of a dual quadratic form:

$$\begin{aligned} &\mathit{Minimize}\quad W(\boldsymbol{\alpha})=\frac{1}{2}\sum_{i,j}^{n}\alpha_i\alpha_j y_i y_j K(\vec{\mathbf{x}}_i,\vec{\mathbf{x}})-\sum_{i=1}^{n}\alpha_i \\ &s.t.\quad \sum_{i=1}^{n}\alpha_i y_i=0 \\ &C\geq\alpha_i\geq 0 \qquad\qquad i=1,\ldots,l \end{aligned} \tag{2}$$

where α_i are Lagrange multipliers of a primal quadratic programming problem. There is a one-to-one correspondence between each α_i and each training sample $\vec{\mathbf{x}}_i$.

2.3 Intelligent Knowledge Processing

In this part the system mines geographical assistant knowledge to further analyze and correct these misclassified pixels. Remote sensing image processing knowledge is mainly from experts' experience and human common knowledge, and these knowledge is usually inaccurate and incomplete, and even fuzzy [8-9], so we use fuzzy rule to express them. Their basic forms are shown below.

IF (condition) THEN (conclusion) AND CF (certainty of conclusion)
An example is shown below.
IF DEM>60 THEN Hill CF 0.9

The above rule means that if elevation of the pixel is greater than 60, then the certainty that it is classified into Hill is 0.9.

In remote sensing image uncertain-reasoning classification, the pixel or polygon is classified into the category with the biggest certainty CF, namely $Class=\max[CF_{result}(1), CF_{result}(2),\cdots,CF_{result}(\omega)]$, where ω representing the number of categories.

3 Hierarchical Intelligent Extractions of Remote Sensing Special Data

3.1 Intelligent Extraction of Water Information

The extracted result of water information by a SVM is shown in figure 3. Comparing figure 3 with the original image visually, we can easily find there exist some misextracted hill-shadow (shown in figure 3 with line denoting). According to analyzing for water area, we construct two rules of "IF DEM>42 THEN water CF 0; IF Slope>4 THEN water CF 0." to remove other misclassified objects, and get final result of water information shown as figure 4.

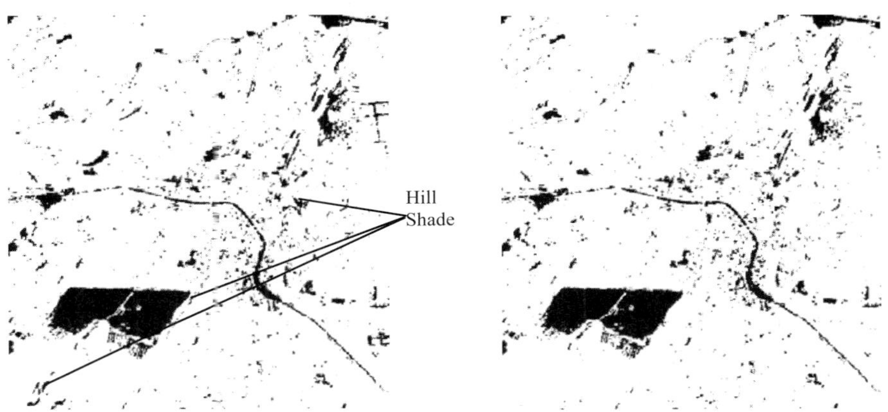

Fig. 3. Water extracted by SVM

Fig. 4. Water extracted by hybrid method

3.2 Intelligent Extraction of Road Information

In general roads in TM image are less than 3 pixels and even heavily influenced by their neighboring pixel. Considering these factors, we take a 3*3 window pixel gray value that reflects spatial structural information as the input of SVM to extract road object. At last the extracted result of road by SVM is shown as Fig. 5.

Comparing Fig. 5 with Fig. 1, we can find that there exist some misextracted objects such as sports ground and rare land, which may be distinguished with their shape information such as area and shape index $I = \sqrt{S} / P$ (where S and P are respectively area and perimeter of object). A sports ground usually has a bigger area, and show as rectangle in the image. After we remove the pixel whose texture is greater than 83, or the object whose area is less than 3 pixels, or the object whose area is more than 50 pixels and shape index is greater than 0.1, we can get an entire road-net image shown in Fig. 6. But there still exists some small rare land or residential area that is expected to be removed in future research.

Fig. 5. Road extracted by SVM

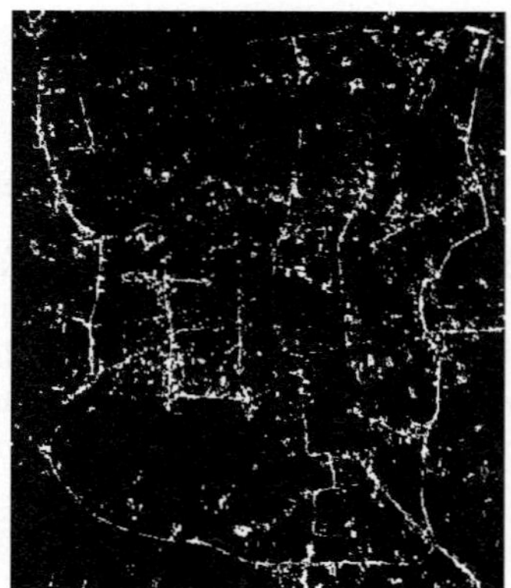

Fig. 6. Road extracted by hybrid method

3.3 Intelligent Extraction of Other Plane-Like Objects

Pixel-based classifiers cannot employ region feature of object to improve classification accuracy, so here we use the polygon-based knowledge reasoning method to extract plane-like objects. Firstly we remove extracted water and road information from the original image, and extract other plane-like objects by SVM, and then trace their edge to get edge image for each plane-like object (that of hill shown at left of Fig.7). Then tracing every polygon one by one in the edge image, and calculating their characteristics, such as categories classified by SVM, elevation, slope, area and shape index. At last we reassign these misclassified objects by the knowledge-reasoning method and obtain the final classified result.

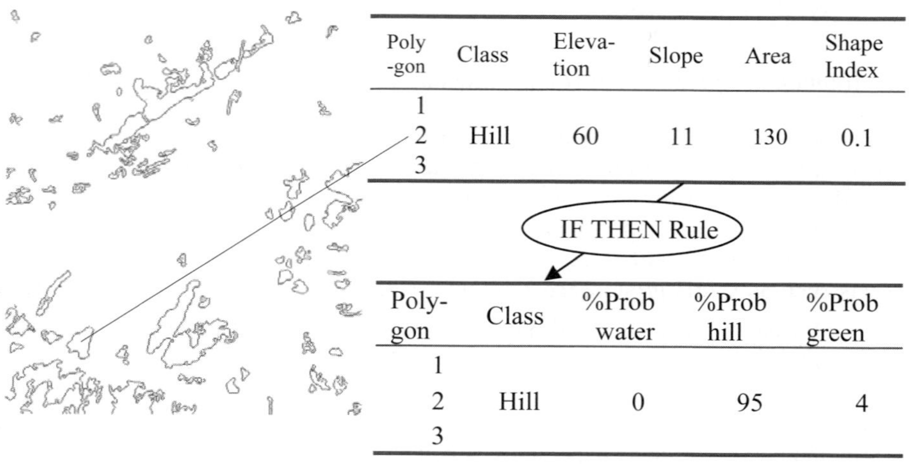

Fig. 7. Schematic representation of polygon-based classification

4 Experimental Results and Analysis

According to the land-use map and our visual observation, we randomly pick up a great deal of typical massive land as test samples, and calculate their respective precision of both the hybrid intelligent classifier and the single SVM classifier to compare their classification results. The overall accuracy and the kappa coefficient of single SVM classifier are respectively 89.8% and 0.87, while of hybrid intelligent classifier are respectively 92.9% and 0.91. Test samples are typical and easily classified, so test precision of two methods may be higher, and they cannot stand for realistic classification accuracy of the whole image. But the indexes reflect that the hybrid intelligent method is obviously superior to the single SVM classifier.

5 Conclusions

In this paper, we proposed a hybrid intelligent extraction method combining SVM with knowledge reasoning technique, and fulfilled hierarchical extraction for linear objects and plane-like information. Our experimental results indicate the success of the proposed method. However, road-object extraction model still needs to improve, and the rule base still needed to be extended in future research.

References

1. Atkinson, P.M., Tatnall, A.R.L.: Neural Networks in Remote Sensing. Int. J. Remote Sensing 18(4) (1997) 699–709
2. Giacinto, G., Roli, F., Bruzzone L.: Combination of Neural and Statistical Algorithms for Supervised Classification of Remote-Sensing Images. Pattern Recognition Letters 21(5) (2000) 385–397
3. Paola, J.D., Schowengerdt, R.A.: A Review and Analysis of Back-propagation Neural Networks for Classification of Remotely Sensed Multi-spectral Image. Int. J. GIS 16(16) (1995) 3033–3058
4. Vapnik, V.N.: Statistical Learning theory. Wiley, New York (1998)
5. Cortes, C., Vapnik, V.: Support Vector Networks. Machine Learning 20(3)(1995) 273–297
6. Burges, J.C.: A Tutorial on Support Vector Machines for Pattern Recognition. Knowledge Discovery and Data Mining 2(2) (1998) 121–167
7. Li, C.F., Wang, Z.Y., Remote Sensing Image Classification Method Based on Support Vector Machines and Fuzzy Membership Function. MIPPR 2005: SAR and Multispectral Image Processing, Vol.6043. SPIE, Wuhan (2005) 604324-1–7
8. Wharton, S.W.: A Spectral Knowledge-based Approach for Urban Land Cover Discrimination. IEEE Trans. on Geoscience and Remote Sensing 25(5) (1987) 272–282
9. Ton, J., et al.: Knowledge-based Segmentation of Landsat Images. IEEE Trans on Geoscience and Remote Sensing 29(3) (1991) 222–231
10. Palacio-prieto, J. L.: Improving Spectral Result in a GIS Context. Int. J. GIS 17(11) (1996) 2201–2209
11. Muvai, H.: Remote Sensing Image Analysis Using a Neural Network and Knowledge-based Processing. Int. J. Remote Sensing 18(4) (1997) 811–828

Eyes Location Using a Neural Network

Xiao-yi Feng[1], Li-ping Yang[1], Zhi Dang[1], and Matti Pietikäinen[2]

[1] College of Electronics and Information, Northwestern Polytechnic University, 710072 Xi'an, China
{Fengxiao, Liping, Dang}@nwpu.edu.cn
[2] Machine Vision Group, Infotech Oulu and Dept. of Electrical and Information Engineering, P.O. Box 4500 Fin-90014 University of Oulu, Finland
mkp@ee.oulu.fi

Abstract. This paper proposed a neural network based method for eyes location. In our work, face area is first located initially using an illumination invariant face skin model; Then, it is segmented by the combination of image transformation and a competitive Hopfield neural network (CHNN) and facial feature candidates such as eyes, eyebrows and mouth are obtained; Finally, eyes are located by facial features evaluation and validation, which is based on face's geometrical structures. Experimental results show that our system performs well under not good lighting conditions.

1 Introduction

Robust non-intrusive eyes location plays an important role in vision based man-machine interaction including automotive applications, such as driver inspection, face recognition and facial expression recognition, etc. In the past years, many works were addressed on this area. The most known methods include color-based approaches, neural network approaches, genetic algorithm approaches, and principle component analysis approaches [1-6].

In this paper, we propose a novel method for eyes location, which contains three steps: face detection, face segmentation and eyes validation. In the first step, an illumination invariant skin model suggested in [7] is applied to extract face area. In the second step, face image (color image) is first converted into a color-ratio image (gray image); then, both upper part and lower part of the color-ratio image are segmented separately, using the Hopfield neural network and we get the collection of facial features such as eyes, mouth, eyebrows candidates. These facial candidates are verified by face's geometric structure during eyes validation procedure.

The rest of the paper is organized as follows. Face detection method is first described in section 2 and face segmentation using CHNN is introduced in section 3. In section 4, Facial feature evaluation and eyes detection is presented. Experimental results are described in section 5. Finally in section 6 we conclude the paper.

2 Skin Locus Based Face Detection

In our work, face-like regions in an input image are detected using the skin detection method proposed by Martinkauppi et al. [7], who had found that the Normalized

J. Wang et al. (Eds.): ISNN 2006, LNCS 3972, pp. 474 – 479, 2006.

Color Coordinates (NCC) combined with skin locus most appropriate for skin detection under varying illumination. To detect face-like area, the image presented in RGB color space is converted to the NCC space *r, g* and *b*. Since $r+b+g = 1$, only two chromaticity *r* and *b* are used for detection. If *r* and *b* of a pixel fall into the area of the skin locus, the pixel belongs to skin. Skin detection result is shown in Fig.1 (b). This result is enhanced by morphological erode and dilate. Considering real application, the largest skin component is regarded as the most likely face area (see the part inside the green box in Fig.1 (b)).

(a) An input image

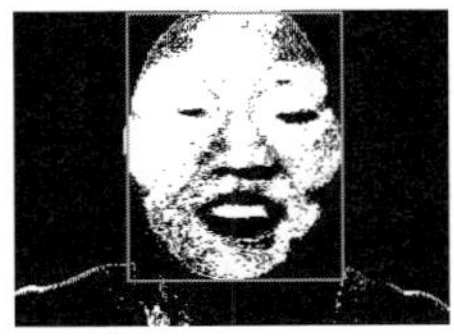

(b) Face area

Fig. 1. Face detection

3 Competitive Hopfield Neural Network Based Face Segmentation

3.1 Face Image Transformation

Based on the knowledge that facial features are darker than their surroundings, morphological valley detectors are usually used for eyes detection, while these methods usually fail under bad lighting conditions (See Fig.2).

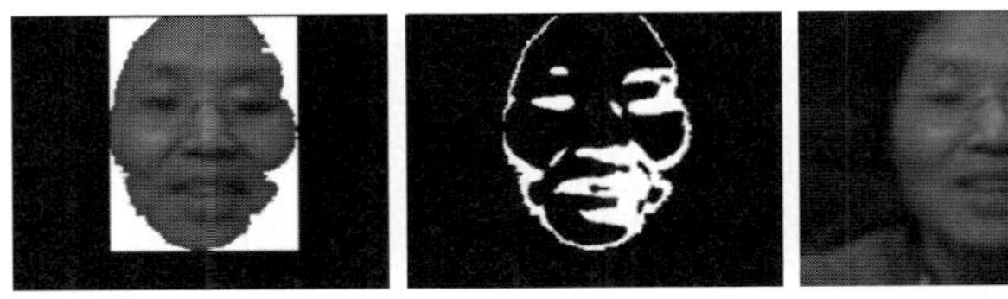

(a) Grey image (b) Valley detection (c) Eyes location

Fig. 2. Valley-based eyes detection

Based on the observation that eyes, eyebrows contain less red elements and lip contains more than the skin part, the color face region is converted to a gray level image named color-ratio image as follows:

$$f(x, y) = \min(255, b \times 255 / r) \tag{1}$$

Here $f(x, y)$ is the gray level of a pixel in position (x, y) in the color-ratio image, and *r* and *b* are chromaticity in the NCC space. The color-ratio image corresponding to the image in Fig. 1(a) is shown in Fig. 3(a).

3.2 Face Segmentation by the Hopfield Neural Network

It can be seen from the color-ratio image that eyes and eyebrows are with higher gray levels than their surroundings and lips are with lower gray levels. This knowledge is used for face segmentation.

To make our method insensitive to noise, both pixels' gray levels and their local mean gray values are taken into account during face segmentation, so a two-dimensional gray vector is composed of them in our work. Due to its fast convergence, the competitive Hopfield neural network is used for clustering.

Let L denote the total gray level and N denote the number of pixels in the image. The gray level of a pixel (suppose as i) and its mean gray value (suppose as j) forms a group of (i, j). Let f_{ij} and p_{ij} are the number and the associated probability density of group (i, j).

$$p_{ij} = f_{ij} / N , \sum_{i=0}^{L-1}\sum_{j=0}^{L-1} p_{ij} = 1 , p_{ij} \geq 0 , i, j = 0,1,2,3 \cdots L-1 \tag{2}$$

Distance between group (x_1, y_1) and (x_2, y_2) is defined as

$$d_{x_1 y_1 x_2 y_2} = \frac{1}{2} \times \left[(x_1 - x_2)^2 + (y_1 - y_2)^2 \right] \tag{3}$$

Suppose the two-dimensional vectors are classified into two kinds so CHNN consists of $L \times L \times 2$ neurons. Let V_{xyc} denote a neuron's state at (x, y, c), where x and y are the first and second part of a two-dimensional vector and c is its classification. The following conditions will be met when the system is steady.

$$\begin{gathered} V_{xyc} \in \{0,1\}, \ \sum_{c=0}^{1} V_{xyc} = 1, \\ 0 < \sum_{x=0}^{L-1}\sum_{y=0}^{L-1} V_{xyc} < L^2 , \sum_{x=0}^{L-1}\sum_{y=0}^{L-1}\sum_{c=0}^{1} V_{xyc} = L^2 \end{gathered} \tag{4}$$

The energy function is constructed as

$$\begin{aligned} E = {} & \frac{A}{2} \sum_{x_1=0}^{L-1}\sum_{y_1=0}^{L-1}\sum_{c-0}^{1}\sum_{x_2=0}^{L-1}\sum_{y_2=0}^{L-1} \frac{1}{\sum_{y=0}^{L-1}\sum_{x=0}^{L-1} p_{xy} V_{xyc}} t_{x_1 y_1 x_2 y_2 c} + \frac{B}{2} \sum_{x=0}^{L-1}\sum_{y=0}^{L-1}\sum_{c_1-0}^{1}\sum_{c_2=0}^{1} V_{xyc_1} V_{xyc_2} + , \\ & + \frac{C}{2} [\sum_{c=0}^{1}\sum_{y=0}^{L-1}\sum_{x=0}^{L-1} V_{xyc} - L^2] \end{aligned} \tag{5}$$

Here $t_{x_1y_1x_2y_2c} = V_{x_1y_1c} d_{x_1y_1x_2y_2} p_{x_2y_2} V_{x_2y_2c}$.

Introduce the winner-takes-all(WTA) learning rule that is

$$V_{xyc} = \begin{cases} 1 & if\ Net_{xyc} = \max\limits_{0<j<2}\{Net_{xyc}\} \\ 0 & otherwise \end{cases} \tag{6}$$

Formula (4) is reduced to

$$E = \frac{1}{2}\sum_{x_1=0}^{L-1}\sum_{y_1=0}^{L-1}\sum_{c=0}^{1}\sum_{x_2=0}^{L-1}\sum_{y_2=0}^{L-1}\frac{1}{\sum\limits_{y=0}^{L-1}\sum\limits_{x=0}^{L-1} p_{xy}V_{xyc}} t_{x_1y_1x_2y_2c} \tag{7}$$

The input of neuron (x, y, c) is

$$Net_{xyc} = \frac{-1}{\sum\limits_{x_1=0}^{L-1}\sum\limits_{y_1=0}^{L-1} p_{x_1y_1}V_{x_1y_1c}}\sum_{x_2=0}^{L-1}\sum_{y_2=0}^{L-1} tt_{xyx_2y_2} \tag{8}$$

Here $tt_{xyx_2y_2} = d_{xyx_2y_2} p_{x_2y_2} V_{x_2y_2c}$.

After network's initialization, formula (6) and (8) are used to update every neuron's state and all the pixels are classified by the clustering result.

The upper part and the lower part of the color-ratio image are segmented respectively using the network and the results are shown in Fig.3 (b) and (c).

4 Competitive Hopfield Neural Network Based Face Segmentation

After the possible facial features are detected, a similar method as proposed in [8] is applied to evaluate feature constellations, using a geometrical face model including eyes, eyebrows and mouth.

We first select two facial features locating at the upper half of face area to form a possible eye pair and evaluate each possible eye pair as follows:

$$E_{eyepair} = 0.5\exp(-10(\frac{D_{eyes} - 0.4B_{width}}{D_{eyes}})^2) + 0.25\left|\theta_{eyeleft} + \theta_{eyeright} - 2\times\theta\right| \tag{9}$$

Here B_{width} is the width of the face bounding box, D_{eyes} is distance between one possible eye pair and $D_{eyes} < 0.8B_{width}$. $\theta, \theta_{eyeleft}$ and $\theta_{eyeright}$ indicate directions of base line(The line passing through the centers of the eye candidates is called as base line) , left eye candidate and right eye candidate, respectively. The first item of this expression uses the distance between two eyes and the second item uses the direction of two eyes as eye pair constrains.

For each eye candidate pair, other facial features are evaluated as follows.

$$E_{feature} = \exp(-10(\frac{d_{feature} - D_{feature}}{D_{eyes}})^2) \quad (10)$$

Here $features = \{mouth, nostril, eyebrow\}$, $d_{feature}$ and $D_{feature}$ are real distance and reference distance from features to base line.

The total evaluation value is a weighted sum of values for each facial features. The weights for eye pair, mouth, nostril pair, and eyebrow pair are 0.4, 0.3, 0.1 and 0.05, respectively. The constellation which has the largest evaluation value and which is bigger than a valve (for example, 0.4) is assumed to be real facial features (if the largest evaluation is less than the valve, it is regarded that there is no face in the image or no eyes in the image). Fig.2 (c) and Fig.3 (d) are detection results.

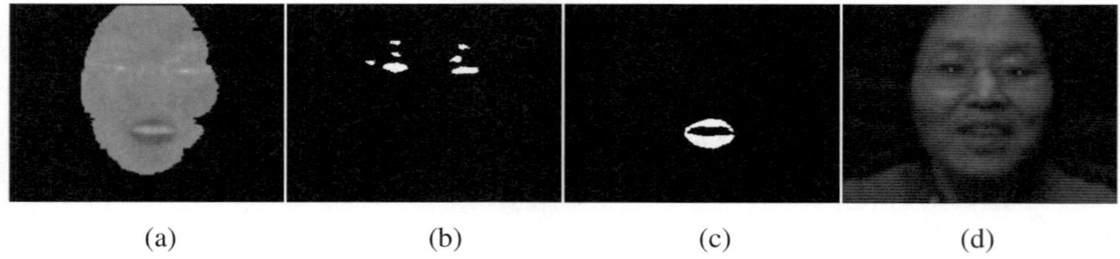

(a) (b) (c) (d)

Fig. 3. Neural-network based eyes detection. (a) Color-ratio image (b) Upper face segmentation (c) Lower face segmentation (d) Eyes detection.

5 Experimental Results

We evaluated the proposed eyes location under real lighting conditions. A digital video camera was placed above the computer. The image resolution is 320×240 and the frame rate is 15fps. Fig.4 shows some eyes tracking results. The located eyes are marked with red dots. It can be seen that the system works well in real conditions.

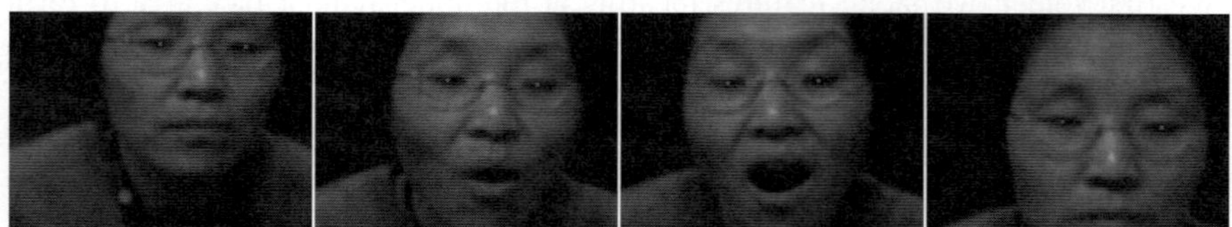

Fig. 4. Examples of eyes location result in a real image sequence

6 Conclusions

This paper presents a neural network based method for eyes location. After face detection, CHNN is used for detecting facial feature candidates. CHNN converges fast for its competitive learning mechanism, which assures its application in real time eyes location. Experimental results demonstrate that our system performs well even in not good lighting conditions.

Acknowledgement

This work was supported by the natural science foundation of Shaanxi province and by the NWPU talent program.

References

1. Wong, K.W., Lam, K.M., Siu W.C.: A Robust Scheme for Live Detection of Human Faces in Color Images. Signal Processing: Image Communication 18(2) (2003) 103-114
2. Yu, L., Eizenman, M.: A New Methodology for Determining Point-of-Gaze in Head-Mounted Eye Tracking Systems. IEEE Trans. Biomedical Engineering 51(1) (2004) 1765-1773
3. Paperno, E., Semyonov, D.: A New Method for Eye Location Tracking. IEEE Trans. Biomedical Engineering 50(1) (2003) 1174-1179
4. Kapoor, A., Picard, R.W.: Real-Time, Fully Automatic Upper Facial Feature Tracking. In: Williams, A. D. (eds.) Fifth IEEE Int. Conf. Automatic Face and Gesture Recognition (2002) 10-15
5. Huang, J., Wechsler, H.: Visual Routines for Eye Location Using Learning and Evolution. IEEE Trans. Evolutionary Computation 4(1) (2000) 73-82
6. Yoo, D.H., Chung, M.J.: Non-Intrusive Eye Gaze Estimation without Knowledge of Eye Pose. In: Proc. Sixth IEEE Int. Conf. Automatic Face and Gesture Recognition (2004) 785-790
7. Martinkauppi, B.: Face Color under Varying Illumination-Analysis and Applications. University of Oulu, Finland (2002)
8. Hannuksela, J.: Facial Feature based Head Tracking and Pose Estimation. University of Oulu, Finland (2003)

Gabor Neural Network for Endoscopic Image Registration

Vladimir Spinko[1], Daming Shi[1], Wan Sing Ng[2], and Jern-Lin Leong[3]

[1] School of Computer Engineering, Nanyang Technological University, Singapore 639798
{vlad0001, asdmshi}@ntu.edu.sg
[2] School of Mechanical and Aerospace Engineering, Nanyang Technological University, Singapore 639798
MWSNG@ntu.edu.sg
[3] Department of Otolaryngology, Singapore General Hospital, Singapore 169608
ljlent@gmail.com

Abstract. In this paper we present a Gabor Wavelet Network, a wavelet neural network based on Gabor functions, applied to image registration. Although wavelet network is time consuming technique, we decrease computational costs by incorporating three techniques: gradient-based feature selection, Gabor filtering, and wavelet neural network. Similarity criterion is built upon analyzing intensity function with Gabor Wavelet Network, which carries out the image registration by both gradient-based and texture features.

1 Introduction

Taking advantage of both the scaling property of wavelets and the effective learning mechanism of neural networks, wavelet networks have recently emerged as a powerful tool for many applications [8]. Gabor Wavelet Network has been used for facial feature localization [5], head pose estimation [6] etc. In stated works odd Gabor function is used as mother wavelet. Intensity pattern in the face region is approximated with number of Gabor functions. Odd Gabor function is written in the way to be encoded with translation, rotation, and dilation parameters, as it is defined in wavelet network theory. Distribution of the Gabor function displacement over the face region is defined by control points according to face features. In our work we use both even and odd Gabor functions in the conventional way. As obtained functions family is oversampled, we use uniform displacement of the Gabor functions over image region.

We propose a method for image registration by implying Gabor Wavelet Network (GWN), a wavelet network with combination of even and odd Gabor functions in wavelet family. For similarity estimation, we imply spatial information over the selected feature region, which is encoded by approximating the intensity function in the selected region with GWN. Despite the regions with well distinguished features are preferred for processing, we tried to relieve the constraints on the region pattern. Instead of introducing additional parameters for illumination changes, we propose to estimate variation of intensity function over its mean.

J. Wang et al. (Eds.): ISNN 2006, LNCS 3972, pp. 480–485, 2006.

The paper is organized as follows. In section 2, Gabor Wavelet Network is introduced as a combination of Gabor filters and wavelet neural network. Section 3 describes registration with similarity measure based on function analysis by Gabor Wavelet Network. Experimental results are presented in section 4. Section 5 concludes the paper.

2 Construction of Gabor Wavelet Network

Initially, Gabor analysis was introduced as the tool for signal processing and communications [2]. Gabor proposed expanding signal f as a series $f(x)=\sum c_{mn}\exp\{i2\pi mbx\}g(x-na)$, where $g(x-na)$ is the Gaussian. Since that, Gabor expansions were used for many different applications, and image processing is one of them.

Before using the Gabor function as a mother function for the network, frame property must be stated. Given parameters $a,b\in\Re$ and a function $h\in L(\Re)$, a frame for $L(\Re)$ of the form

$$\{h_{nm}\}=\left\{e^{i2\pi bmx}g(x-na)\right\}_{m,n\in Z} \tag{1}$$

is called a Gabor frame. A Gabor frame can only be a frame for $L(\Re)$ if $ab\le 1$ [3], and it is Reizs basis if and only if $ab=1$. To obtain the Gabor frame for two-dimensional case, we simply multiplied two one-dimensional Gabor functions:

$$\left\{e^{i2\pi b(mx+ny)}g(x-ka,y-la)\right\}_{k,l,m,n\in Z}. \tag{2}$$

We will define our Gabor frame based on functions used in Gabor filtering techniques. In [1] it is shown that if (2) is a frame for $L^2(\Re^2)$, than collection of functions

$$\{h_{nmkl}\}=\left\{\begin{Bmatrix}\cos(2\pi mbx)\cos(2\pi nby)\\ \cos(2\pi mbx)\sin(2\pi nby)\\ \sin(2\pi mbx)\cos(2\pi nby)\\ \sin(2\pi mbx)\sin(2\pi nby)\end{Bmatrix}g(x-ka,y-la)\right\}_{n,m,k,l\in Z} \tag{3}$$

is also a frame for $L^2(\Re^2)$ with same upper bound.

Now, as a frame of Gabor-like functions is obtained, GWN can be built. We will approximate functions with sum of the following structure:

$$\xi(x,y)=\sum_i w_i h_i(x,y)+\overline{g}\equiv\sum_{n,m,k,l} w_{nmkl}h_{nmkl}+\overline{g} \tag{4}$$

We need to select parameters a and b, constrained by inequality $ab<1$, and proceed to network training.

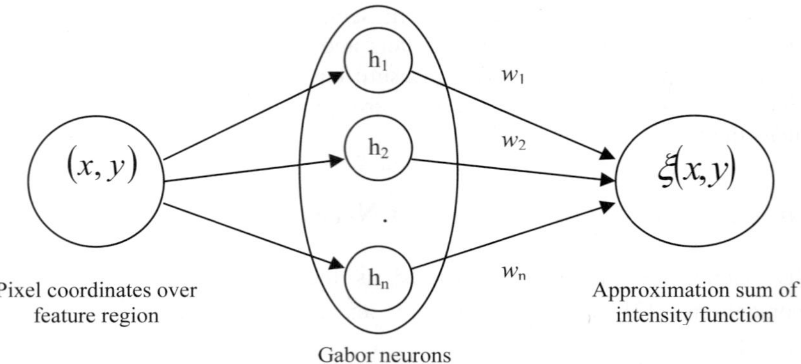

Fig. 1. Gabor Wavelet Network structure

Parameters selection is made as follows. To obtain frequencies proposed in Gabor filtering technique [4], we take $b = \sqrt{2}/L$, where L is image width, and $a = L/2$. Thus, frequencies in the family are $\left\{\sqrt{2}, 2\sqrt{2}, 3\sqrt{2}, \ldots\right\}$ cycles image-width^{-1}.

Next step in the GWN construction is to select the minimum number of wavelets from family (3), which still provide good approximation. Numerical criterion associated with such selection is Akaike's final prediction error (FPE) [9]:

$$FPE(f) = \frac{1 + n_p/N_t}{1 - n_p/N_t} \frac{1}{2N_t} \sum_{k=1}^{W} \sum_{l=1}^{W} \left[f(x_k, y_l) - I(x_k, y_l) \right]^2 \quad (5)$$

where n_p is the number of regressors in the estimator, $I(x, y)$ is the expected output of the training data (x, y), and $N_t = W^2$ is the length of training data, W is the window size.

Finally, we need to train our network. To do this we minimize the objective function:

$$\sum_{k=-\frac{M}{2}}^{\frac{M}{2}} \sum_{l=-\frac{M}{2}}^{\frac{M}{2}} \left[I(x+l, y+k) - \bar{I}_{Mask} - \left(\sum_{i=1}^{N} w_i g_i(x+l, y+k) \right) \right]^2 \quad (6)$$

with respect to weights w_i, where $\bar{I}_{\text{Mask}}$ is the mean of the $I(x, y)$ over the region of interest, and (x, y) is the center of the mask. When the minimization of (6) is complete, the approximation function is ready: $I(x, y) \approx \xi(x, y)$.

The structure of presented GWN is shown on Figure 1. Taking pixel coordinates over the selected window as an input, Gabor functions as neurons in hidden layer, we obtain a sum (4), an approximation of the intensity function, as output. Further we can proceed to describe using obtained structure for image registration.

3 Gabor Wavelet Network Applied to Image Registration

A gray scaled image is encoded as an array of integer values, where every pixel with its coordinates (x, y) has its intensity value $I(x, y)$. So, either the whole image or some part of the image can be described as two-dimensional discrete scalar-valued function. This discrete function can be approximated with two-dimensional continuous function $f_I(x, y)$. Moreover, it is reasonable to assume that $f_I(x, y) \in L^2(\Re)$. To find corresponding points on initial and a target images we suggest finding the image regions which can be approximated with same functions as a similarity measure.

Consider image region $M \times M$. The desired function in every pixel (x, y) takes value of its intensity $I(x, y)$. The problem is to find continuous function $f_I(x, y)$, such that $f_I(x, y) \approx I(x, y)$ for every pixel coordinates (x, y). To conduct this approximation we suggest using GWN. In other words, we obtain the source function from its given values on the coordinate lattice. In general case, such approximation can be made for any image region. But, for similarity measure we are interested in regions with high intensity contrast. Thus, pre-selection stage is made, which consists of selection regions with gradient-based technique [7].

Before stating similarity criterion, we describe the purpose of including parameter $\bar{g}$ into GWN. As we are interesting in two-dimensional function shape behavior, but not in absolute value, some method should be introduced to consider the functions which differ by a constant as the similar. That is the reason why parameter $\bar{g}$ was introduced. Weights in (4) are taken in the way to approximate deviation of the intensity function $I(x, y)$ from its mean over the region. So, this parameter is the mean of intensity over the processing region.

Now we will formulate the similarity criterion. As the intensity function is approximated, the second frame is analyzed to find the region where the intensity function is approximated with sum (6) with same parameters. Such regions are considered as corresponding regions and selected for image registration. To find the corresponding region on the target image, we need to find where the deviation of intensity function from its mean over the same mask can be approximated with the sample of weights and functions from the initial frame. Or, in other words, the objective function (6) is minimized with respect to (x, y) - image coordinates of the second frame, with given weights w_i.

4 Experimental Results

Experiments were conducted on endoscopic images. All images are 640×480 pixels, gray scaled with intensity normalized to $[0,1]$. Before Gabor Wavelet

Network was applied, pre-selection of the feature points were made with gradient based technique. The x and y coordinates of obtained 16×16 pixel region were scaled to $[0,1]\times[0,1]$. For every processed region network pruning and training (weights calibration) were conducted independently. After similarity calculation, six parameters model was applied to extract Affine transformation.

Apart registration with GWN, two other methods were selected for comparison: registration with feature points obtained by gradient-based technique, where similarity between two feature points is calculated by SSD minimization, and registration with Gabor filters.

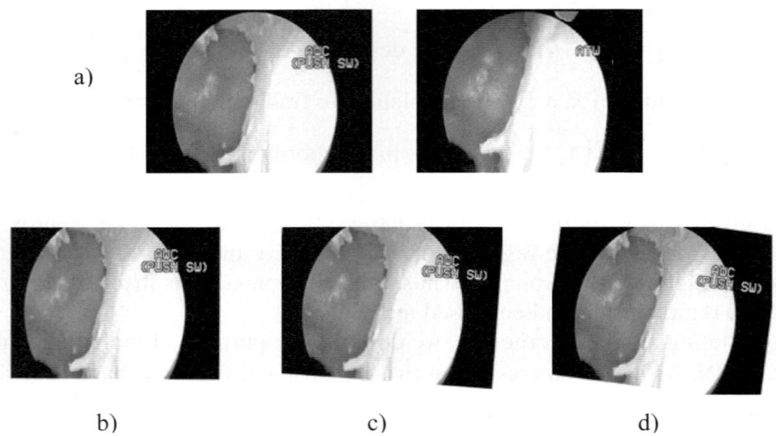

Fig. 2. Registered source images and estimated wrap: a) source and target images; b) gradient-based feature selection; c) registration with Gabor filters; d) registration with Gabor Wavelet Network

In Figure 2 the comparison of the three methods is presented. Experiments were conducted on two endoscopic images (Fig. 2a). Inaccuracy in the first method (Fig. 2b) can be explained by presence of light variation between two images, which causes inaccuracy in similarity measure. Performance of the second method (Fig. 2c) is better, but as in Gabor filtering technique whole images are processed, bright regions (especially left side of the target image) bring difficulties in registration. Our method (Fig. 2d) shows good accuracy and avoids influence of bright illumination, as pre-selection was made to cut off improper regions.

5 Conclusion and Future Work

A new method for image registration was presented in this paper. A similarity measure between feature regions in two images is built with GWN, a wavelet network based on Gabor functions. Gabor Wavelet Network is a computationally expensive tool, so pre-selection of regions of interest were made with gradient-based technique.

Further, pre-selection of frequencies is conducted to speed up the network training procedure. Function parameters are selected in such way that function family is dense in the space of continuous functions. Our experimental results show that our method is capable of dealing with image registration in the presence of illumination changes.

Future work can be divided into two directions. Firstly, introducing additional parameters (such as orientation) in the Gabor functions family, that can help dealing with object rotation. Furthermore, such parameters can provide more accurate transformation extraction after the similarity measure stage. Second part of the future work includes the application of different Gabor frames (i.e. irregular Gabor frames) for better intensity pattern function approximation.

References

1. Christensen, O.: Frames, Riesz Bases, and Discrete Gabor/Wavelet Expansions. Bull. Am. Math. Soc., New Ser. 38(3) (2001) 273-291
2. Gabor, D.: Thoery of Communications. IEE (London), 93(3) (1946) 429-457
3. Heil, C., Walnut, D.: Continiuous and Discrete Wavelet Transforms. SIAM Review 31(4) (1989) 628-666
4. Jain, A.K., Farrokhnia, F.: Unsupervised Texture Segmentation Using Gabor Filters. Pattern Recognition 24 (12) (1991) 1167-1186
5. Kruger, V., Sommer, G.: Gabor Wavelet Networks for Object Representation. Journal of the Optical Society of America (JOSA) 19(6) (2002) 1112-1119
6. Kruger, V., Sommer, G.: Gabor Wavelet Networks for Efficient Head Pose Estimation. Image and Vision Computing 20(9-10) (2002) 665-672
7. Shi, J., Tomasi, C.: Good Features to Track. IEEE Int. Conf. Computer Vision and Pattern Recognition (CVPR'94) (1994) 593-600
8. Zhang, Q., Benveniste, A.: Wavelet Network. IEEE Transactions on Neural Networks 3(6) (1992) 889-898
9. Zhang, Q.: Using Wavelet Network in Nonparametric Estimation. IEEE Trans. Neural Networks 8(2) (1997) 227-236

Isomap and Neural Networks Based Image Registration Scheme

Anbang Xu and Ping Guo*

Image Processing and Pattern Recognition Laboratory,
Beijing Normal University, Beijing, 100875, P.R. China
anbangxu@mail.bnu.edu.cn; pguo@ieee.org

Abstract. A novel image registration scheme is proposed. In the proposed scheme, the complete isometric mapping (Isomap) is used to extract features from the image sets, and these features are input vectors of feedforward neural networks. Neural network outputs are those translation, rotation and scaling parameters with respect to reference and observed image sets. Comparative experiments for Isomap based method, the discrete cosine transform (DCT) and Zernike moment are performed. The results show that the proposed scheme is not only accurate but also remarkably robust to noise.

1 Introduction

Image registration is the process of aligning two or more images of the same scene. Image registration techniques are embedded in a lot of visual intelligent systems, such as robotics, target recognition, remote medical treatment and autonomous navigation. The common image registration methods are divided into two types: intensity-based methods and feature-based methods. The analysis and evaluation for various techniques and methods of image registration are carried out on the basis of these two sorts, while the feature-based methods are more popular.

Recently, Itamar Ethanany[1] proposed to use feedforward neural network (FNN) to register an attacked image through 144 Discrete Cosine Transform (DCT) -base band coefficients as the feature vector. But this method has too large lumber of input feature vectors for the un-orthogonality of DCT based space thus expose low computational efficiency and high requirements on computer performance. Later, Wu and Xie[2] used low order Zernike moments instead of DCT coefficients to register affine transform parameters. Although the estimation accuracy and robustness toward noise are improved in some contents, the input vector dimension of neural networks is still high.

In this paper, we develop a novel method to image registration, which uses Isomap for feature extraction and then feed these feature vectors into a feedforward neural network to obtain register affine transform parameters. The input vector dimension is greatly reduced, consequently the network performance is improved. Experimental results show that the scheme we proposed is better than other methods in terms of accuracy and robustness to noise.

* Corresponding author.

J. Wang et al. (Eds.): ISNN 2006, LNCS 3972, pp. 486–491, 2006.

2 Isomap Based Image Registration Scheme

The proposed registration scheme needs to be performed with four steps as described following:

1. Synthesizing a set of training samples from the reference image.
2. Apply Isomap to extract features from the whole samples including training images and test images (registered images).
3. Train the feedforward neural network. The inputs of the neural network are the feature coefficients of training samples and its outputs are affine parameters.
4. Feed the feature coefficients of registered images to the trained network to get the affine parameters estimated. With obtained parameters, we can align images now.

2.1 Affine Transformation

Geometrical transformation can be represented in many different ways, affine transformation is one of the most common used transformations. It is a kind of rigid transformation, having many good mathematics characteristics. We can represent images with 2D matrices. Affine transformation can be described by the following matrix equation:

$$\begin{pmatrix} x_2 \\ y_2 \end{pmatrix} = \begin{pmatrix} t_x \\ t_y \end{pmatrix} + s \begin{pmatrix} cos\theta & -sin\theta \\ sin\theta & cos\theta \end{pmatrix} \begin{pmatrix} x_1 \\ y_1 \end{pmatrix}. \tag{1}$$

In the equation, there are four basic parameters for the transformation, where (x_1,y_1) denotes the original image coordinate, (x_2,y_2) denotes the transformed image coordinate in another image, t_x, t_y are the translation parameters, s is a scaling factor and θ is a rotation angle. In this paper, we will adopt this transformation model.

2.2 Isomap

Recently, a new dimensionality reduction technique has been introduced, namely isometric mapping[3]. The method attempts to best preserve the local neighborhood of each object, while preserving the global distances "through" the rest of the objects. It has been widely used for visualization purposes by mapping data into two or three dimensions. In this paper we will present how the method could be used for image registration purposes.

Isomap is a method that maps high-dimensional objects into a lower dimensional space, while preserving as much as possible the neighborhood of each object, as well as the 'geodesic' distances between all pairs of objects. Isomap works as follows:

1. Calculate the k closest neighbors of each object.
2. Create the Minimum Spanning Tree (MST) distances of the updated distance matrix.
3. Perform Multi-dimensional Scaling MDS on the new distance matrix.

Isomap aims at minimizing the least squares error of the geodesic distances, in the low dimension, of the neighbors' weights for every object.

2.3 Image Registration Scheme with FNN

The image registration scheme includes training the FNN to provide the required affine parameters. Each image in the training set is generated by applying an affine transformation. The affine parameters are randomly changed in a predefined range so as to reduce correlations among images. In order to improve the generalization and immunity of the FNN from over-sensitivity to distorted inputs, we introduce noise in the image synthesis. Then we employ Isomap as a feature extraction mechanism, the input dimension of the FNN is 10, which is greatly reduced compared to the 57-dimension of Zernike moment-based method and 144-diemnsion of DCT-based method.

Because there is no theoretical guidance to decide the number of neurons in hidden layer[4][5] and the computational efficiency of traditional methods such as cross validation and bootstrap is very low, for the FNN with three layer structure in this paper we can use the empirical formula[6] to compute the hidden neuron number:

$$p = \sqrt{0.43mn + 0.12n^2 + 2.54m + 0.77n + 0.35} + 0.5, \quad (2)$$

where p is the number of neurons in hidden layer, m is the number of input neurons and n is the number of output neurons. Here the structure of the FNN is that contains 10 inputs, 4 outputs and 6 hidden neurons. Sigmoid transfer functions are employed in the hidden layers while linear functions characterize the output-level neurons. The FNN is trained by using the Levenberg-Marquardt algorithm[7] based on gradient-descent learning technique.

3 Experiments

In the experiment, a pair of 256×256 resolution images was used. Fig. 1 shows one of the original images and a transformed image containing translation, rotation

(a)

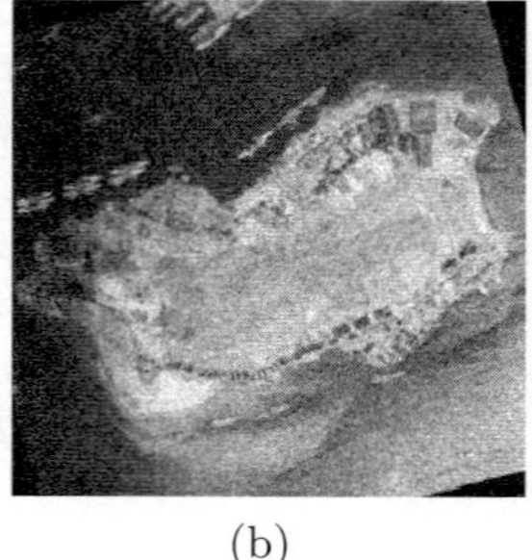

(b)

Fig. 1. (a)An original image and (b)a registered image in the training set with 15 degree rotation, 110% scaling, translation of 2 pixel and -3 pixel on X-axis and Y-axis respectively at a signal-to-noise ratio (SNR) of 15 dB and scaling.The training set consists of 300 images, each image is transformed from the reference image by translating, rotating and scaling randomly within a predefined range. Besides, additive Gaussian noise and Salt & Pepper type noise are applied on each image in various intensities. We can also generate some test samples to demonstrate the registration accuracy of the proposed method. We apply Isomap to the whole samples including training images and registered images, so the dimension of the sequence of vectors is reduced from 65536 to d ($d \ll 65536$). These feature coefficients of images are inputs of FNN, the FNN is trained and its outputs are affine parameters. After obtaining the trained FNN, the feature coefficients of test registered images are fed into the network to get the estimated affine parameters.

We use Isomap to reduce the sample dimension from 65536 to a lower d dimension, in Fig. 2 the relationship between residual variance and d is shown.

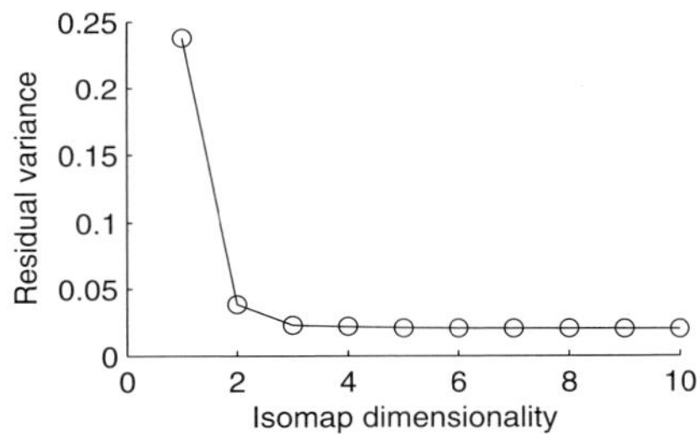

Fig. 2. The relationship between residual variance and dimensionality d

The intrinsic dimensionality of the data can be estimated by searching for the "elbow" at which this curve ceases to decrease significantly with increased dimensions. The residual variance of Isomap reaches the minimum at d=2, but in order to retain more information we can select d=10.

The accuracy of parameter estimation can be evaluated by root mean square error (RMSE) between the registered image and the original image, Fig. 3 shows the results with d=10 and k=12 under various test data set.

In another experiment, we compare the proposed method with the other methods in reference[1] and [2] under different noisy conditions. In the proposed method, we choose d=10 .While in Zernike moment-based method, d=58 and 17 hidden neurons, and in DCT-based method, d=64 and 18 hidden neurons. Same with the above experiment, a training set consisting of 300 images is synthesized. In order to evaluate registration performance with Gaussian noise, we take 40 images for each the evaluated SNR value. The test image as shown in Fig. 1 (a) is rotated 15 degree, 110% scaled, translated 2 pixels and -3 pixels on X-axis and Y-axis respectively. Fig. 4 depicts the results of estimating the affine transform parameters under different SNR values.

By analyzing the experiment results, it can be found that our method shows more accurate than the other two methods, especially when SNR is larger than 20 dB.

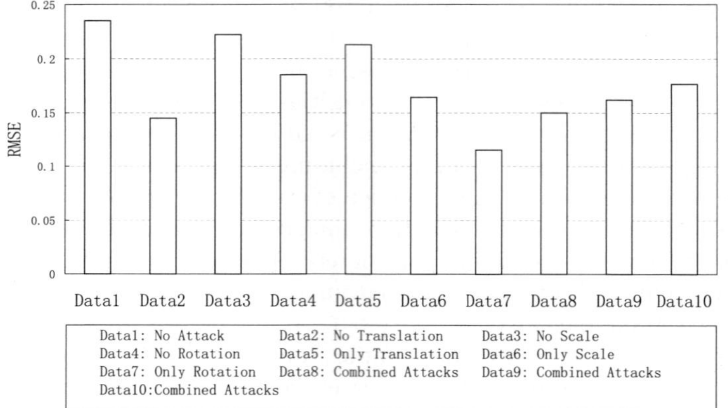

Fig. 3. RMSE between the registered image (under 15 db SNR) and the original image

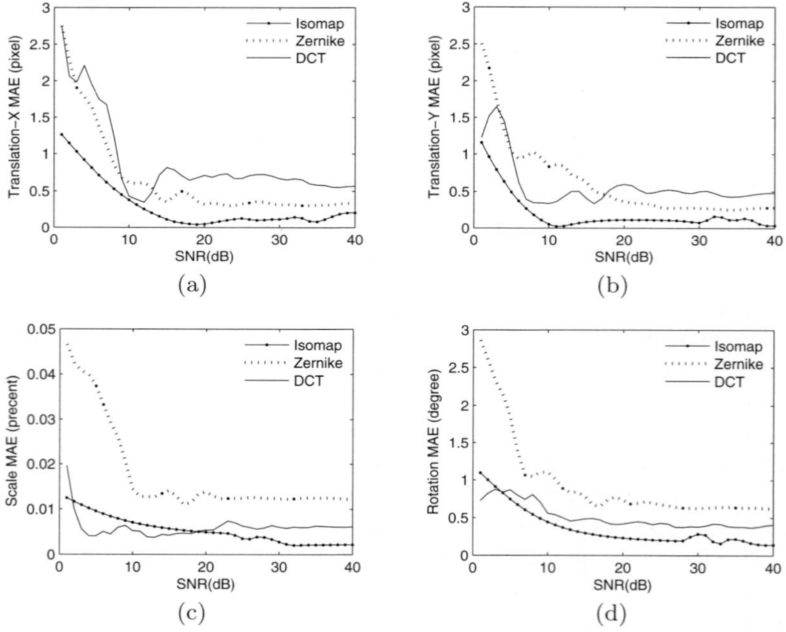

Fig. 4. Absolute registration error with respect to translation on X-axis (a), translation on Y-axis (b), scale (c) and rotation (d)

4 Conclusions

In this paper, a novel image registration scheme is proposed, which adopts the FNN and Isomap to register affine transform parameters. Experiment results show that the proposed scheme has more accurate registration performance and robust to noise than some other methods.

The proposed method still deserve further study. First, the performance of the method is not very satisfied when SNR is low. Second, there is no systematic method for selecting the parameter of Isomap in the proposed scheme. We will engage to find the solution for these problems in the further study. Third, in the Synthesizing a training set from the reference image,there is no image data due to rotation or scale. Although, as a feasible alternative to deal with this problem is to to cut the border of the images, the information is lost.

Acknowledgment

The research work described in this paper was fully supported by a grant from the National Natural Science Foundation of China (Project No. 60275002). The authors would like to thanks Prof. T.T. Wong of CUHK for his revision and constructive comments for this paper.

References

1. Elhanany, I., Sheinfeld, M., Beckl, A., *et al*: Robust Image Registration Based on Feedforward Neural Networks. In: Proceedings of IEEE International Conference on System, Man and Cybernetics, Vol. 2 (2000) 1507-1511
2. Wu, J., Xie, J.: Zernike Moment-based Image Registration Scheme Utilizing Feedforward Neural Networks. In: Proceedings of the 5th World Congress on Intelligent Control and Automation, Vol.5 (2004) 4046-4048
3. Tenenbaum, J.B., De Silva, V., Langford, J.C.: A Global Geometric Framework for Nonlinear Dimensionality Reduction. Science **290**(5500) (2000) 2319-2323
4. Ravdin P.M., Clark, G.M., Hilsenbeck, S.G., *et al*: A Demonstration That Breast Cancer Recurrence Can Be Predicted by Neural Network Analysis. Breast Cancer Research and Treatment **21**(1) (1992) 47-53
5. De Laurentiis, M. Ravdin, P.M.: Survival Analysis of Censored Data: Neural Network Analysis Detection of Complex Interaction between Variables. Breast Cancer Research and Treatment **32**(1) (1994) 113-118
6. White, H.: Connectionist Nonparametric Regression: Multilayer Feedforward Networks Can Learn Arbitrary Mapping. Neural Networks **3**(5) (1990) 535-549
7. Hagan, M.T., Menhaj, M.: Training Feedforward Networks with Marquardt Algorithm. IEEE Trans. Neural Networks **15**(6) (1994) 989-993

Unsupervised Image Segmentation Using an Iterative Entropy Regularized Likelihood Learning Algorithm

Zhiwu Lu

Institute of Computer Science & Technology of Peking University,
Beijing 100871, China
zhiwu.lu@yahoo.com.cn

Abstract. As for unsupervised image segmentation, one important application is content based image retrieval. In this context, the key problem is to automatically determine the number of regions(i.e., clusters) for each image so that we can then perform a query on the region of interest. This paper presents an iterative entropy regularized likelihood (ERL) learning algorithm for cluster analysis based on a mixture model to solve this problem. Several experiments have demonstrated that the iterative ERL learning algorithm can automatically detect the number of regions in a image and outperforms the generalized competitive clustering.

1 Introduction

Image segmentation is one of the basic problems of image processing. In general, there are two approaches to do such a task, i.e., region growing [1] and boundary detection [2]. In this paper, only the first kind of image segmentation is considered. Our study on unsupervised image segmentation was motivated by requirements and constraints in the context of image retrieval by content [3]. Most approaches use the query-by-example principle, performing queries such as "show me more images that look like this one". However, the user is often more particularly interested in specifying an object (or region) and in retrieving more images with similar objects (or regions), which is opposed to similar images as a whole. Our aim is to allow the user to perform a query on some parts (region of interest) of an image. Hence, we focus on the problem of clustering based segmentation of each image in the database to allow partial queries.

Although there have been various clustering methods, such as EM algorithm [4] and k-means algorithm [5], the number k of clusters in the data set is usually assumed to be pre-known. However, since image databases in the context of image retrieval are often huge, the prior setting of cluster number for each image is no longer feasible. Such requirement in this context motivates our interest to the idea of automatic model selection during clustering.

Inspired by the regularization theory [6, 7], we aim to solve the above problems, through implementing the entropy regularized likelihood (ERL) learning

J. Wang et al. (Eds.): ISNN 2006, LNCS 3972, pp. 492–497, 2006.

principle on a mixture model via an iterative algorithm for cluster analysis. Several experiments further demonstrate that the iterative ERL learning algorithm can automatically detect the number of regions for each image in the database during parameter learning. Moreover, the iterative algorithm outperforms the generalized competitive clustering [8] for color image segmentation and can often maintain more details of the color images.

2 Iterative Entropy Regularized Likelihood Learning Algorithm

We consider the following mixture model for cluster analysis:

$$p(x \mid \Theta) = \sum_{l=1}^{k} \alpha_l p(x \mid \theta_l), \sum_{l=1}^{k} \alpha_l = 1, \;\; \alpha_l \geq 0, \tag{1}$$

where $p(x \mid \theta_l)(l = 1, ..., k)$ are densities from the same parametric family, and k is the number of mixture components.

Given a sample data set $S = \{x_t\}_{t=1}^N$ generated from a mixture model with k^* actual components and $k \geq k^*$, the negative log-likelihood function on the mixture model $p(x \mid \Theta)$ is given by

$$L(\Theta) = -\frac{1}{N} \sum_{t=1}^{N} \ln(\sum_{l=1}^{k} (p(x_t \mid \theta_l)\alpha_l)). \tag{2}$$

The well-known maximum likelihood (ML) learning is just an implementation of minimizing $L(\Theta)$.

With the posterior probability that x_t arises from the l-th mixture component

$$P(l \mid x_t) = p(x_t \mid \theta_l)\alpha_l / \sum_{j=1}^{k} p(x_t \mid \theta_j)\alpha_j, \tag{3}$$

we have the discrete Shannon entropy of these posterior probabilities for the sample x_t

$$E(P(l \mid x_t)) = -\sum_{l=1}^{k} P(l \mid x_t) \ln P(l \mid x_t), \tag{4}$$

which can be made minimized when $P(l_0 \mid x_t) = 1, P(l \mid x_t) = 0(l \neq l_0)$, that is, the sample x_t is classified into the l_0-th cluster.

We now consider the average or mean entropy over the sample set S:

$$E(\Theta) = -\frac{1}{N} \sum_{t=1}^{N} \sum_{l=1}^{k} P(l \mid x_t) \ln P(l \mid x_t), \tag{5}$$

and use it to regularize the log likelihood function by

$$H(\Theta) = L(\Theta) + \gamma E(\Theta), \tag{6}$$

where $\gamma > 0$ is the regularization factor. That is, $E(\Theta)$ is a regularization term to reduce the model complexity such that the mixture model can be made as simple as possible by minimizing $H(\Theta)$.

We can then get the well-known Gaussian mixture model with

$$p(x \mid \theta_l) = \frac{1}{(2\pi)^{n/2}|\Sigma_l|^{1/2}} \exp(-(1/2)(x - m_l)^T \Sigma_l^{-1}(x - m_l)), \tag{7}$$

where n is the dimensionality of x, and $\theta_l = (m_l, \Sigma_l), l = 1, ..., k$ are the mean vectors and covariance matrices of the Gaussian distributions.

We now derive an iterative algorithm to solve the minimum of $H(\Theta)$ as follows. Let $\alpha_l = \exp(\beta_l)/\sum_{j=1}^{k} \exp(\beta_j)$, where $-\infty < \beta_l < \infty$. Using the general methods for matrix derivatives, we are led to the following series of equations:

$$U(l \mid x_t) = P(l \mid x_t)(1 + \gamma \sum_{j=1}^{k}(\delta_{jl} - P(j \mid x_t)) \ln(p(x_t \mid \theta_j)\alpha_j)), \tag{8}$$

$$\frac{\partial H(\Theta)}{\partial \beta_l} = -\frac{1}{N}\sum_{t=1}^{N}\sum_{j=1}^{k} U(j \mid x_t)(\delta_{jl} - \alpha_l) = 0, \tag{9}$$

$$\frac{\partial H(\Theta)}{\partial m_l} = -\frac{1}{N}\sum_{t=1}^{N} U(l \mid x_t)\Sigma_l^{-1}(x_t - m_l) = 0, \tag{10}$$

$$\frac{\partial H(\Theta)}{\partial \Sigma_l} = -\frac{1}{2N}\sum_{t=1}^{N} U(l \mid x_t)\Sigma_l^{-1}[(x_t - m_l)(x_t - m_l)^T - \Sigma_l]\Sigma_l^{-1} = 0, \tag{11}$$

where δ_{jl} is the Kronecker function. Then, the solution of those equations can be given explicitly as follows:

$$\hat{\alpha}_l = \frac{\sum_{t=1}^{N} U(l \mid x_t)}{\sum_{j=1}^{k}\sum_{t=1}^{N} U(j \mid x_t)}, \quad \hat{m}_l = \frac{1}{\sum_{t=1}^{N} U(l \mid x_t)}\sum_{t=1}^{N} U(l \mid x_t)x_t, \tag{12}$$

$$\hat{\Sigma}_l = \frac{1}{\sum_{t=1}^{N} U(l \mid x_t)}\sum_{t=1}^{N} U(l \mid x_t)(x_t - m_l)(x_t - m_l)^T. \tag{13}$$

These explicit expressions give us an iterative algorithm for the minimum of $H(\Theta)$. This iterative algorithm seems very similar to the EM algorithm for Gaussian mixture. Just like the EM algorithm, the iterative algorithm is usually faster than the general gradient algorithm, which is specially appropriate for image processing. However, it is different from the EM algorithm in that a regularization mechanism is implemented on the mixing proportions during the iterations, which leads to the automatic model selection.

3 Experimental Results

We further applied the iterative ERL learning algorithm to color image segmentation in the context of content based image retrieval. For each pixel of a color image, we consider these three color components red r, green g, and blue b as color feature, and a sample can be denoted as $x = (r, g, b), 0 \leq r, g, b \leq 255$. In order to make the iterative algorithm stable, we normalized three color components to $[0, 1]$, that is $x = (r/255, g/255, b/255)$. With the color images used in [8] for segmentation experiments, we then compared the iterative ERL learning algorithm with the generalized competitive clustering.

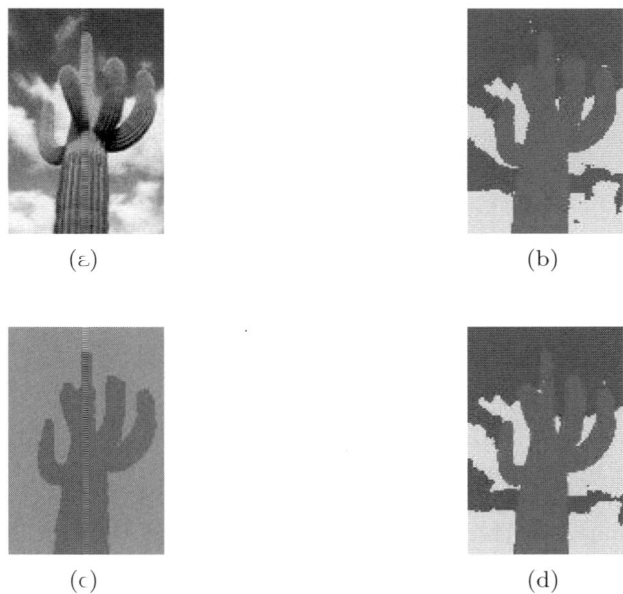

Fig. 1. The segmentation results on the cereus image by the iterative ERL learning algorithm and generalized competitive clustering. (a) The original color image; (b) The results by the iterative ERL learning algorithm with $\gamma = 0.3$; (c) The results by the iterative ERL learning algorithm with $\gamma = 0.8$; (d) The results by the generalized competitive clustering.

Using k^* to denote the true number of objects(or regions) in an image, we implement the iterative ERL learning algorithm always with $k \geq k^*$. Moreover, the other parameters are initialized randomly within certain intervals. In all the experiments, the learning is stopped when $|H(\hat{\Theta}) - H(\Theta)| < 10^{-7}$. During the iterative ERL learning process, all of the samples from an image are continuously classified into some clusters(i.e., regions), which can cause other clusters to have few samples. Hence, according to (13), the covariance matrices of some clusters may become singular after certain iterations, and we can then simply discard these clusters.

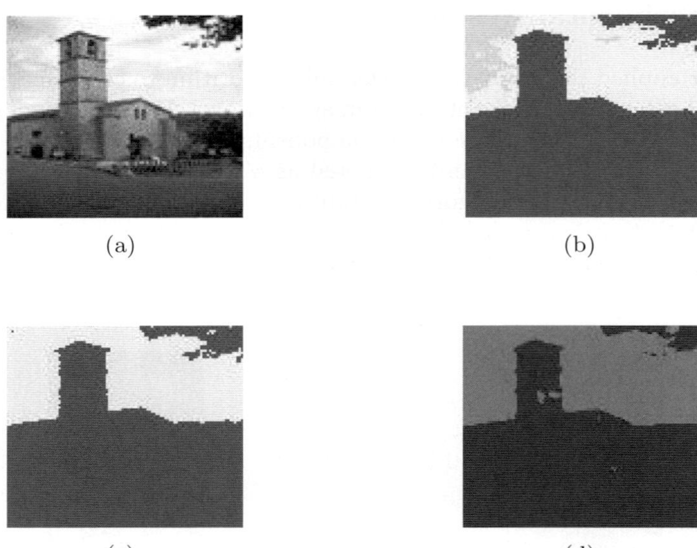

Fig. 2. The segmentation results on the building image by the iterative ERL learning algorithm and generalized competitive clustering. (a) The original color image; (b) The results by the iterative ERL learning algorithm with $\gamma = 0.3$; (c) The results by the iterative ERL learning algorithm with $\gamma = 0.8$; (d) The results by the generalized competitive clustering.

The iterative ERL learning algorithm with $k = 8$ is first implemented to make color image segmentation on the cereus image given by Fig. 1(a), and the results for different regularization factors are shown in Fig. 1(b) and Fig. 1(c). Comparing Fig.1(b) with Fig.1(d), we can find that the iterative ERL learning algorithm with $\gamma = 0.3$ can maintain the image details better than the generalized competitive clustering. However, with the regularization factor $\gamma = 0.8$, the iterative algorithm only obtains two clusters, that is, the regularization power is too strong to maintain some regions of the cereus image, and then only the cereus is extracted from the background.

Likewise, we implement the iterative ERL learning algorithm with $k = 8$ to make color image segmentation on the building image given by Fig. 2(a), and the results for different regularization factors are shown in Fig. 2(b) and Fig. 2(c). Compared with the generalized competitive clustering of the results shown in Fig.1(d), the iterative ERL learning algorithm with $\gamma = 0.3$ or $\gamma = 0.8$ can correctly detect the leaves and then maintain more details of the image. Moreover, with the regularization factor $\gamma = 0.3$, the iterative algorithm detects cloud from the sky and obtain one extra cluster.

The further experiments on the other images have also been made successfully for segmentation in the similar cases. Actually, in many experiments, the iterative ERL learning can automatically detect the number of regions for each image in the database and maintain the image details well at the same time.

4 Conclusions

We have investigated the unsupervised image segmentation in the context of content based image retrieval via an iterative ERL learning algorithm. With the regularization factor adjusted, our iterative ERL learning algorithm can detect the number of clusters automatically during parameter learning even on the sample set with a high degree of overlap. When applied to color image segmentation, the iterative algorithm can still get good results and outperforms the generalized competitive clustering.

References

1. Shih, F.Y., Cheng, S.X.: Automatic Seeded Region Growing for Color Image Segmentation. Image and Vision Computing, **23** (2005) 877–886
2. Dai, M., Baylou, P., Humbert, L., Najim, M.: Image Segmentation by a Dynamic Thresholding Using Edge Detection Based on Cascaded Uniform Filters. Signal Processing, **52** (1996) 49–63
3. Banerjee, M., Kundu, M.K.: Edge Based Features for Content Based Image Retrieval. Pattern Recognition, **36** (2003) 2649–2661
4. Render, R.A., Walker, H.F.: Mixture Densities, Maximum Likelihood and the EM Algorithm. SIAM Review, **26** (1984) 195–239
5. Chinrungrueng, C., Sequin, C.H.: Optimal Adaptive K-Means Algorithm with Dynamic Adjustment of Learning Rate. IEEE Transactions on Neural Networks, **6** (1995) 157–169
6. Dennis, D.C., Finbarr, O.S.: Asymptotic Analysis of Penalized Likelihood and Related Estimators. The Annals of Statistics, **18** (1990) 1676–1695
7. Vapnik, V.N.: An Overview of Statistical Learning Theory. IEEE Transactions on Neural Networks, **10** (1999) 988–999
8. Boujeman, N.: Generalized Competitive Clustering for Image Segmentation. 19th International Conference of the North American Fuzzy Information Processing Society, (2000) 133-137

An Improvement on Competitive Neural Networks Applied to Image Segmentation

Rui Yan[1], Meng Joo Er[1], and Huajin Tang[2]

[1] Intelligent Systems Center, Nanyang Technical University, 50, Nanyang Drive, 7th Storey, Research Techno Plaza, Border X Block, Singapore 637553
{yanrui, emjer}@ntu.edu.sg
[2] Department of Electrical and Computer Engineering, National University of Singapore, 4 Engineering Drive 3, Singapore 117576
engp1880@nus.edu.sg

Abstract. Image segmentation is a long existing problem and still regarded as unsolved to a large extent in computer vision. This letter describes the modeling method of competitive neural networks and elucidates its connection with the Hopfield type optimization network. A new algorithm to map the image segmentation problem onto competitive networks is proposed and its convergence is shown by the stability analysis. Finally, the improvement on the competitive neural networks based method is validated by the simulation results.

1 Introduction

Image segmentation is the partitioning of an image into meaningful sub-regions or grouping objects with the same attribution, which is often considered as an important yet a very difficult task in computer vision. Due to its critical importance in computer vision, image segmentation has been studied extensively and many techniques have been invented (see [1, 2] for a review of the subject). In the light of the excellent taxonomy work in [2], the algorithms can be broadly divided into three categories (similar to the classification adopted in [3]) according to different abstraction levels of input data, namely pixel level, local feature (or edge) level and object (or region) level. Pixel-based techniques use the intensities of individual pixels as input to the algorithm, while the feature level is based on the relative locations of perceptual features (e.g., edges, corners, junctions, etc.) and the object level is on properties of individual objects. Image segmentation has largely been performed by pixel-based or feature-based approaches.

A number of algorithms based on histogram analysis, region growing, edge detection or pixel classification have been proposed in [4, 5]. Among these, neural networks that fulfill unsupervised tasks are more attractive. In the pixel-level schemes [6, 7, 8, 9], the segmentation problem is often mapped onto a neural network (e.g., the well-known Hopfield network [10]) by means of an energy function, and thus the network converges to a (local) minimum of the energy function.

In this work, a new design algorithm of competitive neural networks is introduced for performing image segmentation. The previous design method of [6]

J. Wang et al. (Eds.): ISNN 2006, LNCS 3972, pp. 498–503, 2006.

is compared with the proposed method. The recently developed local minima escape (LME) algorithm [11] are incorporated to the competitive network.

2 The Competitive Model of Neural Networks

The image segmentation problem can be mapped into a neural network, with the cost function as the Lyapunov energy function of the network. Iterative updating of the neuron states will eventually force the network to converge to a stable state, preferably a valid one with the lowest energy. Hence, this can be seen as an optimization problem, and the best segmentation will be one with the lowest energy.

Let $g_x, x = 1, \cdots, n$ denote the intensity (gray level) values of a given image, and h_x be the number of pixels at the gray-level g_x. The network consists of c layers and each layer consists of n neurons. In each layer, the network has compatible vertical interactions W^i_{xy}, while between individual layers the lateral interactions are of the WTA. The task is to assign each gray level to an appropriate class. For every row of neurons, $x = 1, \cdots, n$, the output of neuron (x, i) is updated as follows:

$$\begin{aligned} v_{xi} &= \sigma\left(-\sum_y w^i_{xy} v_{yi} - J_{xi} + \max_j\left(\sum_y w^j_{xy} v_{yj} + J_{xj}\right)\right) \\ &= \sigma\left(Net_{xi} - \max_j(Net_{xj})\right) \end{aligned} \quad (1)$$

where Net_{xi} represents the total input to neuron (x, i) and $\sigma(\cdot)$ is a threshold function: $\sigma(\theta)$ is 1 if $\theta \geq 0$ and 0 otherwise.

The dynamics defined above performs the competition of the WTA in each row. The winner (the neuron with the largest input in a row) is set to 1 while all the others in the same row are set to 0.

The compatible interaction strength in layer i is written as

$$w^i_{x,y} = -h_x h_y d_{xy}, \quad (2)$$

where d_{xy} is the square of the Euclidean distance between two gray levels, $d_{xy} = (g_x - g_y)^2$.

The external input $J_{xi} \equiv J_x$ for all i and J_x is simply set to zero. It follows from the preceding equations that the input to neuron (x, i) is now computed as

$$Net_{xi} = -\sum_y^n h_x h_y d_{xy} v_{yi}. \quad (3)$$

3 Energy Function and Dynamics Analysis

The energy candidate function of the competitive network is formulated as

$$E = -\frac{1}{2}\sum_i \sum_{x,y} w^i_{xy} v_{xi} v_{yi} - \sum_{x,i} J_x v_{xi}. \quad (4)$$

By recalling the connection weights and external inputs in the above section, equation (4) becomes

$$E = \sum_{i}^{c} \sum_{x,y}^{n} v_{xi} v_{yi} h_x h_y d_{xy}. \tag{5}$$

It can be further written as $E = \sum_i^c E_i$, and $E_i = \sum_x^n \sum_y^n v_{xi} v_{yi} h_x h_y d_{xy}$ which represents the energy of all neurons in individual column.

Consider row r and let neuron (r, i) be the winning neuron before WTA updating. Upon updating, let neuron (r, j) be the winning neuron. Without loss of generality, let r be the first row. Suppose $V(k)$ and $V(k+1)$ be the network states before WTA updating and after WTA updating, respectively:

$$V(k) = \begin{pmatrix} 0 & 1 & \cdots & 0 & 0 \\ v_{21} & v_{2i} & \cdots & v_{2j} & v_{2c} \\ v_{31} & v_{3i} & \cdots & v_{3j} & v_{3c} \\ \vdots & \vdots & \vdots & \vdots & \vdots \\ v_{n1} & v_{ni} & \cdots & v_{nj} & v_{nc} \end{pmatrix}, V(k+1) = \begin{pmatrix} 0 & 0 & \cdots & 1 & 0 \\ v_{21} & v_{2i} & \cdots & v_{2j} & v_{2c} \\ v_{31} & v_{3i} & \cdots & v_{3j} & v_{3c} \\ \vdots & \vdots & \vdots & \vdots & \vdots \\ v_{n1} & v_{ni} & \cdots & v_{nj} & v_{nc} \end{pmatrix}.$$

At any time k, the energy of the network can be decomposed into three terms as below:

$$\begin{aligned} E(k) &= E_i(k) + E_j(k) + E_o \\ &= \sum_{x}^{n} \sum_{y}^{n} v_{xi}(k) v_{yi}(k) h_x h_y d_{xy} + \sum_{x}^{n} \sum_{y}^{n} v_{xj}(k) v_{yj}(k) h_x h_y d_{xy} + E_o, \end{aligned}$$

where E_o is the energy of all other neurons, not in column i or j. Note that E_o remains unchanged after updating. We have

$$\begin{aligned} E(k+1) - E(k) &= 2 \sum_{x=2}^{n} h_1 h_x d_{1x} v_{xj}(k+1) - 2 \sum_{x=2}^{n} h_1 h_x d_{1x} v_{xi}(k) \\ &= 2 \sum_{x=2}^{n} h_1 h_x d_{1x} v_{xj}(k) - 2 \sum_{x=2}^{n} h_1 h_x d_{1x} v_{xi}(k). \end{aligned}$$

Since neuron (r, j) is the winning neuron, $Net_{rj} > Net_{ri}$. It implies that $\sum_x^n h_r h_x d_{rx} v_{xj}(k) < \sum_x^n h_r h_x d_{rx} v_{xi}(k)$. Hence, we have $E(k+1) - E(k) < 0$. This shows that the energy of the network is decreasing and thus the convergence is ensured.

4 Simulation Studies

Case 1. Basic Simulation

It is noted that the dynamics defined by equations (1) can be simulated row-wise, i.e.,

$$v_{x,i} = \begin{cases} 1, & \text{if } Net_{x,i} = \max_j \{Net_{x,j}\}, \\ 0, & \text{otherwise.} \end{cases} \tag{6}$$

It is called WTA updating rule in [6] and [12].

The implementation of the competitive network is described as follows.

Algorithm 1. Competitive Model-Based Image Segmentation

1. *Input gray levels n of image and its associated histogram, as well as the number of classes c. Compute the distance matrix, according to the following equation, for all gray levels* $d_{xy} = (g_x - g_y)^2$.
2. *Initialize the neural network by randomly assigning one '1' per row, and setting the rest of neurons in the same row to '0'. Do the same for all other rows, while ensuring that there is at least one '1' per column.*
3. *Repeat n times: Choose one row randomly and calculate* $Net_{x,i}$ *to all neurons in the same row, according to equation (3). Apply the WTA mechanism (6) to update all the neuron states within the same row. This constitutes 1 epoch*
4. *Go to step 3) until convergence is reached, i.e. the network state,* $V = (v_{x,i})$, *for the previous epoch is the same as the current epoch.*

Case 2. Improving the Local Minima
Observed from the above discussion, the competitive network has a property of local search, similar to hill-climbing search. In the sequel, to improve the local minimum, the application of the recently proposed stochastic optimization technique, LME algorithm of [11] in the network is discussed.

The LME technique first disturbs its network parameters (W and J), using standard white Gaussian noise of varying weights, to obtain a new network $\hat{H}$.

$$\hat{w}_{xi,yj} = (1 - \alpha^w \eta^w_{xi,yj}) w_{xi,yj} + \beta^w \eta^w_{xi,yj}, \quad x < y \text{ or } (x = y, i \le j), \tag{7a}$$

$$\hat{w}_{xi,yj} = w_{yj,xi}, \quad x > y \text{ or } (x = y, i > j), \tag{7b}$$

$$\hat{J}_{x,i} = (1 - \alpha^i \eta^i_{xi}) J_{x,i} + \beta^i \eta^i_{x,i}, \tag{7c}$$

where $\eta^w_{xi,yj}, \eta^i_{x,i}$ are standard Gaussian noises, and $\alpha^w, \beta^w, \alpha^i, \beta^i$ are positive constants which control the strength of disturbance. Note that the connection matrix W is still symmetric after disturbance. The LME algorithm, applied to the WTA algorithm is summarized as follows:

Algorithm 2. Competitive Model with LME

1. *Carry out the Algorithm 1 for the network H.*
2. *This convergent state V is stored temporarily and set as the initial state for the disturbed network* $\hat{H}$. *Calculate the energy* $E_{previous}$ *of the network H.*
3. *Compute the disturbed distance matrix* $\hat{H}$ *using* $\hat{d}_{xy} = (g_x - g_y)^2 + \mathcal{G}(0, 1)$ *where* $\mathcal{G}$ *denotes a Gaussian noise with mean 0 and variance 1. Process the competitive dynamics of* $\hat{H}$ *as H until convergence is reached. Let this local minima be* $\hat{V}$.
4. *This new local minima* $\hat{V}$ *is now mapped back to H, as its new initial state. Perform the dynamics of H until convergence is reached. Let this new state be* $V_{current}$, *and calculate the current energy* $E_{current}$ *of the H network. If* $E_{current} < E_{previous}$, *accept the new state. Else retrieve its previous minimum state,* $V_{previous}$. *This constitutes 1 LME epoch.*

5. *Repeat the above procedure until the LME terminates, i.e., either (a) or (b) is satisfied:*
   (a) The total number of LME epochs reaches some preset number.
   (b) $V_{current} = V_{previous}$*, for some consecutive LME epochs.*

5 Simulation Results

The proposed network is applied to perform the image segmentation task and is compared with the results obtained from simulating the previous model [6]. For c classes, the lower energy value is, the better the segmentation will be. Since the energy functions are different, for an obtained state, the original energy value and its equivalent value calculated by our proposed energy function (see equation (5)) are calculated for all the simulation results.

Both energy functions were first tested using only the WTA algorithm, before incorporating the LME. Different number of classes c (6, 8, 10, 12) were preassigned to segment the image. A total of 20 simulation studies were carried out for each case, and the results shown in Table 1. The results show that the segmentation quality is improved as the number of classes rises.

From Table 1, the new mapping method gives more satisfactory results. Using the new algorithm, the energy of the network calculated using the proposed energy function is smaller than that of the original WTA algorithm, for all classes under simulation.

Using the LME algorithm, the energy of the network is reduced by 3.0×10^{10} on the average, as shown by Table 2. Though a large absolute value, this is however a small reduction percentage (approximately 0.9%), thus implying that the competitive network is by itself quite efficient.

Table 1. Image Segmentation using only WTA

Class	OE ($\times 10^4$)		EE($\times 10^{11}$)		PE ($\times 10^{11}$)	
	Min	Ave	Min	Ave	Min	Ave
6	8.2207	8.2207	36.486	36.486	32.558	32.869
8	4.4871	4.4871	14.676	14.676	13.711	13.864
10	3.0999	3.0999	8.3994	8.3994	6.8463	6.8869
12	2.3082	2.3322	5.6013	5.7445	4.0353	4.0891

OE–Original energy; EE–Equivalent energy and PE–Proposed energy.

Table 2. Image Segmentation with LME

Method	Class	Energy value	
		Min	Ave
WTA with LME	6	3.2538 e+12	3.2591 e+12
	8	1.3655 e+12	1.3723 e+12
	10	6.7925 e+11	6.8252 e+11
	12	4.0389 e+11	4.0584 e+11

6 Conclusions

A new algorithm based on competitive neural network was presented for image segmentation. The segmentation is treated as a constrained optimization problem and is based upon the global information of the gray level distribution. The new algorithm has higher computation efficiency and illustrates its advance in terms of reducing the segmentation energy, as compared to the existing approach. The stochastic optimization technique for improving the quality of solutions, LME, was adapted to the competitive network.

References

1. Pal, N.R., Pal, S. K.: A Review on Image Segmentation Techniques. Pattern Recognition **26** (1993) 1277–1294
2. Egmont-Pertersen, M., Ridder, D., Handels, H.: Image Processing with Neural Networks–a Review. Pattern Recognition **35** (2002) 2279–2301
3. Wang, D.L., Terman, D.: Image Segmentation based on Oscillatory Correlation. Neural Computation **9** (1997) 805–836
4. Pavlidis, T., Liow, Y.: Integrating Region Growing and Edge Detection. IEEE Trans. Pattern Analysis and Machine Intelligence **12** (1990) 225–233
5. Gonzalez, R., Woods, R,: Digital Image Processing. 2nd Edn. New Jersey: Prentice Hall (2002)
6. Cheng, K., Lin, J., Mao, C.: The Application of Competitive Hopfield Neural Network to Medical Image Segmentation. IEEE Trans. on Medical Imaging **15** (1996) 560–567
7. Wu, Y., Yang, C. Wang, T.: A New Approach of Color Quantization of Image based on Neural Network. International Joint Conference on Neural Networks (2001) 973–977
8. Gopal, S., Sahiner, B., Chan, H., Petrick, N.: Neural Network based Segmentation Using a Priori Image Models. International Conference on Neural Networks **4** (1997) 2455–2459
9. Phoha, V., Oldham, W.: Image Recovery and Segmentation Using Competitive Learning in a Layered Network. IEEE Trans. Neural Networks **7** (1996) 843–856
10. Hopfield, J., Tank, D.: Neural Computation of Decision in Optimization Problem. Biol. Cybern. **52** (1985) 141–152
11. Peng, M., Narendra, K, Gupta, A.: An Investigation into the Improvement of Local Minimum of the Hopfield Network. Neural Networks **9** (1996) 1241–1253
12. Tang, H., Tan, K.C., Yi, Z.: A Columnar Competitive Model for Solving Combinatorial Optimization Problems. IEEE Trans. Neural Networks **15** (2004) 1568–1573

Image Segmentation by Deterministic Annealing Algorithm with Adaptive Spatial Constraints

Xulei Yang[1], Aize Cao[2], and Qing Song[1]

[1] EEE School, Nanyang Technological University, Singapore
{yangxulei, eqsong}@pmail.ntu.edu.sg
[2] Medical Center, Vanderbilt University, USA
aizecao@vanderbilt.edu

Abstract. In this paper, we present an adaptive spatially-constrained deterministic annealing (ASDA) algorithm, which takes into account the spatial continuity constraints by using a dissimilarity index that allows spatial interactions between image pixels, for image segmentation. The local spatial continuity constraint reduces the noise effect and the classification ambiguity. More importantly, the strength of spatial constraint for each given image pixel is auto-selected by the scaled variance of its neighbor pixels, which results in the adaptiveness of the presented algorithm. The effectiveness and efficiency of the presented method for image segmentation are supported by experimental results on synthetic and MR images.

1 Introduction

Image segmentation plays an important role in a variety of applications such as robot vision, object recognition, and medical imaging. Fuzzy segmentation methods, especially the fuzzy c-means algorithm (FCM) [1], have been widely used in the image segmentation. Although the intensity-based FCM algorithm performs well on the segmentation of most noise-free images, it fails to classify the images corrupted by noise, outliers, and other imaging artifacts, and thus, makes accurate segmentation difficult. As observed in a real image, pixels of the same object usually form coherent patches. Thus, the incorporation of local spatial information in the clustering process could filter out noise and other imaging artifacts and reduce classification ambiguities such that yield a more accurate segmentation result [3]. In the last decades, many attempts, e.g. [2]–[5], have been made to introduce spatial contextual information into image segmentation procedures by modifying the objective function of conventional fuzzy clustering algorithm (mainly on FCM) to improve the segmentation performance, and promising results have been reported in the literature of image segmentation.

Spatially-constrained fuzzy clustering algorithms can raise insensitivity to noise to some extent, however, there remains several problems to be improved: first, it is well know that FCM kind algorithms are sensitive to initialization

J. Wang et al. (Eds.): ISNN 2006, LNCS 3972, pp. 504–510, 2006.

and converges to local optimal solutions; second, how to automatically determine (but not set as a prior) the strength of the spatial term is still a problem for most spatially-constrained fuzzy clustering algorithms. In this paper, we present an adaptive spatially-constrained deterministic annealing algorithm (ASDA) by adaptively incorporating the local spatial information into the formulation of standard deterministic annealing (DA) algorithm [6] [7] to deal with these existing problems. The proposed algorithm offers several improved features compared to existing image segmentation methods. First, it is less prone to noise and has better convergence due to the incorporation of spatial information. Second, it is independent of the data initialization and has the ability to avoid many poor local optima due to the mass-constrained DA process. Lastly, the strength of the spatial constraint for each pixel is auto-determined by the scaled variance of its neighbor pixels to improve the adaptiveness of the proposed algorithm.

The rest of this paper is organized as follows. The proposed ASDA algorithm is derived in section 2. In section 3, we show the experimental results of ASDA on synthetic and magnetic resonance (MR) data. Finally, conclusion and discussion are given in section 4.

2 Derivative and Implementation

2.1 The Proposed ASDA Algorithm

The deterministic annealing (DA) algorithm, in which the annealing process with its phase transitions leads to a natural hierarchical clustering, is independent of the choice of the initial data configuration [6] [7]. The standard DA to image segmentation, however, is sensitive to the noise and outliers. In this subsection, we modify the probability distribution of standard DA to incorporate the spatial contextual information by introducing a spatial penalty term. The penalty term acts as a regularizer and biases the solution toward piecewise-homogeneous labelling [5]. Such regularization is helpful in classifying images corrupted by noise. The spatial information is incorporated by considering the neighborhood pixels effect for the distance measure (dissimilarity) definition, that is

$$D_{jk} = d(x_j, v_k) + \frac{\alpha_j}{N_R} \sum_{r \in N_j} d(x_r, v_k) \tag{1}$$

where N_j stands for the set of neighbors falling into a (normally 3×3) window around x_j (not including x_j) and N_R is its cardinality. The spatial penalty term of image pixel j is regularized by the parameter α_j. Note unlike the standard spatially-constrained methods, where all the pixels are regularized by a constant parameter α. While they are regularized by dynamical values α_j in ASDA, which will be adaptively selected by the scaled variance of its neighbor pixels. The relative importance of the regularizing term α_j is proportional to the variance of a given pixel j against its neighbor pixels. We will discuss the adaptive selection of α_j later. Using the spatially-constrained distance measure (1), the probability distribution of the proposed ASDA is given by

$$p_s(v_k|x_j) = \frac{p(v_k)e^{-\left(d(x_j,v_k)+\frac{\alpha_j}{N_R}\sum\limits_{r\in N_j} d(x_r,v_k)\right)/T}}{\sum\limits_{i=1}^{c} p(v_i)e^{-\left(d(x_j,v_i)+\frac{\alpha_j}{N_R}\sum\limits_{r\in N_j} d(x_r,v_i)\right)/T}} \tag{2}$$

Accordingly, the free energy function becomes

$$F_s = -T\sum_{j=1}^{N} p(x_j)log\sum_{i=1}^{c} p(v_i)e^{\left(-d(x_j,v_i)+\frac{\alpha_j}{N_R}\sum\limits_{r\in N_j} d(x_r,v_i)\right)/T} \tag{3}$$

The cluster center v_k is obtained by minimizing (3) with respect to v_k as follows.

$$\frac{\partial(F_s)}{\partial(v_k)} = 0 \tag{4}$$

which leads to (detailed derivation is omitted)

$$v_k = \frac{\sum\limits_{j=1}^{N} p(x_j)p_s(v_k|x_j)(x_j + \frac{\alpha_j}{N_R}\sum\limits_{r\in N_j} x_r)}{\sum\limits_{j=1}^{N} (1+\alpha_j)p(x_j)p_s(v_k|x_j)} \tag{5}$$

Alternative updating (2) and (5) with the adaptive selection of α_j (discuss later) gives the proposed ASDA algorithm.

2.2 A Complexity Reduced Version of ASDA

As observed from (2) (5), the computation of the neighborhood terms takes much more time than the standard non-spatially-constrained algorithm. Here, we present a complexity reduced version of ASDA by equivalently replace $\sum\limits_{r\in N_j} x_r$ by $\bar{x_j}$ and approximately replace $\frac{1}{N_R}\sum\limits_{r\in N_j} d(x_r, v_i)$ by $d(\bar{x_j}, v_i)$, where $\bar{x_j}$ is the mean of neighborhood pixels at location j. This approximation still retains the spatial information such that obtains the similar segmentation results as the original version does. Unlike the original formulation, $\bar{x_j}$ can be computed in advance, thus the clustering time can be significantly reduced. To obtain the complexity reduced version of ASDA, the only modification is to replace (2) by

$$p_r(v_k|x_j) = \frac{p(v_k)e^{\{-d(x_j,v_k)+\alpha_j d(\bar{x_j},v_k)\}/T}}{\sum\limits_{i=1}^{c} p(v_i)e^{\{-d(x_j,v_i)+\alpha_j d(\bar{x_j},v_i)\}/T}} \tag{6}$$

and accordingly, (5) becomes

$$v_k = \frac{\sum_{j=1}^{N} p(x_j)p_r((v_k)|x_j)(x_j + \alpha\bar{x_j})}{\sum_{j=1}^{N}(1+\alpha_j)p(x_j)p_r(v_k|x_j)} \tag{7}$$

Equations (6) and (7) is just the complexity reduced version of ASDA algorithm, which is used in our experiments.

2.3 Adaptive Selection of Spatial Penalty α_j

Determination of an appropriate value of α is dependent on the image being clustered. Theoretically, the relative importance of the regularizing term (determined by α) is inversely proportional to the signal-to-noise (SNR) ratio of the observed image. Low SNR would require a higher value of α, and vice versa. However, in practical cases, SNR may not be known as a prior, which makes the selection of α difficult, in result, few attempts have been presented to deal with it. To our knowledge, only [2] so far discussed the adaptive selection of α by using a cross-validation method, however, that method suffers from extreme computation cost, which makes it impracticable in real applications. As observed from a real image, homogenous pixels are normally clustered together such that the intensity value x_j of a given image pixel j is normally similar to the values of its neighbor pixels, that means the variance of a given image pixel x_j against its neighbors is small. While if the given image pixel is contaminated, the variance becomes large. From this observation, we select the spatial strength α_j for each image pixel j by the scaled variance as follows,

$$\alpha_j = var(x_j)/v\bar{a}r \tag{8}$$

where $var(x_j)$ is the variance of x_j against its neighbors, which is defined by

$$var(x_j) = \frac{1}{N_R}\sum_{r\in N_j} d(x_r, x_j) \tag{9}$$

and $v\bar{a}r$ is the mean variance of all pixels, which is defined by

$$v\bar{a}r = \frac{1}{N}\sum_{j=1}^{N} var(x_j) \tag{10}$$

where N_j stands for the set of neighbors falling into a (normally 3×3) window around x_j (not including x_j) and N_R is its cardinality. From (8), the value of α_j can be calculated directly from the given image itself, which is computationally practical in really applications. It will be seen from the experiments that (8) is a reasonable estimation of the spatial strength, i.e., the ASDA with estimated α_j by (8) can produce satisfied segmentation results for the given images.

3 Experimental Results

To demonstrate and evaluate the performance of the proposed approach, we applied it to synthetic and magnetic resonance (MRI) images. To provide quantitative comparison, all the images were also performed by the conventional Fuzzy C-means (FCM) [1], mass-constrained deterministic annealing (DA) [7], and spatially-constrained FCM (SFCM) [5] [4] clustering algorithms. In all examples, we set $N_R = 9$ (a 3×3 window centered around each pixel), $m = 2$ (weighing exponent for FCM and SFCM), $\eta = 0.9$ (cooling rate for DA and ASDA), $\epsilon = 0.001$ (convergence threshold), the value of α_j for ASDA is determined by (8), while the value of α for SFCM is manually set as a prior.

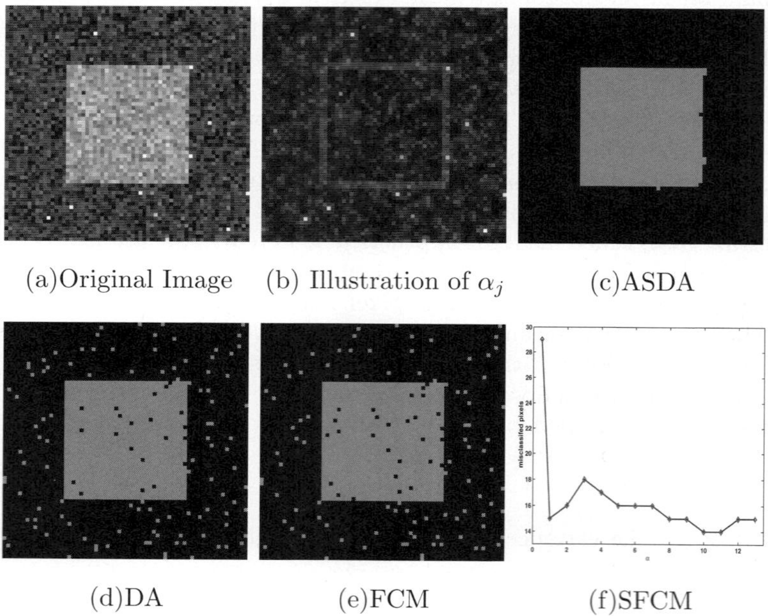

(a)Original Image (b) Illustration of α_j (c)ASDA

(d)DA (e)FCM (f)SFCM

Fig. 1. Segmentation results of different clustering algorithms on the synthetic image

3.1 Synthetic Image

This example simply demonstrate the effectiveness of (8) for determining the spatial strength α_j. Fig.1a shows a synthetic image. This image contains a two-class pattern corrupted by hybrid 1% "Gaussian" and 1% "pepper and salt" noise. The spatial strength α_j calculated from (8) is shown in Fig.1b, where the value of α_j has been scaled from 0 to 255 for visualization. It is obvious that the estimated spatial strength reflects the distribution of noise to some extent: the clear pixels have lower (darker) spatial penalty values and the noisy pixels have higher (lighter) spatial penalty values. This is in accordance with the

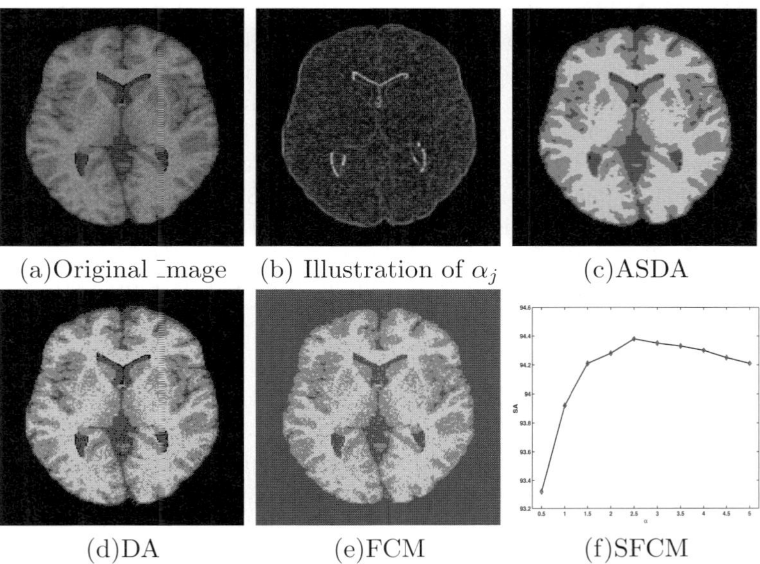

(a)Original Image (b) Illustration of α_j (c)ASDA

(d)DA (e)FCM (f)SFCM

Fig. 2. Segmentation results of different clustering algorithms on the MR image

statement that the relative importance of the spatial penalty term is inversely proportional to the signal-to-noise (SNR) ratio of the observed image. The segmented result of ASDA is shown in Fig.1c, which leads to 7 misclassified pixels. For comparison, the segmentation results of DA and FCM are shown in Fig.1d and Fig.1e respectively, where DA has 120 misclassified pixels and FCM leads to 127. The number of misclassified pixels of SFCM according to varying values of α is shown in Fig 1d, where the value of α is set as a prior. It can be seen that the minimum number 14 is reached by SFCM at $\alpha = 10$ and $\alpha = 11$. From this example we can see that because of the noise effect presents in the data, the performances of both FCM and DA are seriously deteriorated. In contrast, spatially constrained methods ASDA and SFCM obtain much better results, with ASDA achieving a little better performance than SFCM.

3.2 Simulated MR Images

The simulated MR images are obtained from the BrainWeb Simulated Brain Database [8]. Simulated brain data of varying noise are used to perform quantitative assessment of the proposed algorithm since ground truths are known for these data. Here in our experiments, we use a high-resolution T1-weighted phantom with slice thickness of $1mm$, 7% noise. The number of tissue classes in the segmentation was set to three, which corresponds to gray matter (GM), white matter (WM) and cerebrospinal fluid (CSF). Background and other tissues are ignored in the computation. Fig.2a shows the true partial volume model of the slice 75. Accordingly, Fig.2c-e show the segmented results obtained by applying

DA, FCM, and ASDA respectively. Fig.2f shows the results of SFCM with respect to the different given values of α. The DA and FCM have only 90.52% and 90.34% segmentation accuracy (SA) respectively, and the best result of SFCM leads to 94.38% SA at $\alpha = 2.5$, where SA is defined as the sum of the total number of pixels divided by the sum of number of correctly classified pixels [5] [2]. In contrast, the ASDA achieves an satisfied result with 94.31% SA, which is slightly worse than the best result of SFCM (94.38% SA at $\alpha = 2.5$). However, The SA obtained by ASDA is better than the average result of SFCM (94.13% SA at $\alpha = 0.5 : 0.5 : 5.0$). In addition, ASDA adaptively determine the values of α_j by (8) (the illustration is shown in Fig.2b), but SFCM requires α as a prior, which may be not practical for real applications.

4 Conclusion and Discussion

In this paper, we present an adaptive spatially-constrained DA algorithm for image segmentation. The key to the algorithm is a new dissimilarity measure that adaptively takes into account the influence of the neighboring pixels on the center pixel in a 3×3 window. The superiority of the proposed algorithm has been demonstrated by both synthetic and MR images. Typically, the standard DA runs $15 - 25$ times faster than the normal spatially-constrained ASDA, i.e., (2) and (5). Fortunately, the computation reduced version of ASDA, i.e., (6) and (7), significantly reduces the execution time, which typically runs $5 - 10$ faster than normal ASDA algorithm and $2 - 5$ slower than standard DA algorithm.

References

1. Bezdek, J.C.: Pattern Recognition with Fuzzy Objective Function ALgorithms. Plenum Press, New York (1981)
2. Pham, D.L.: Spatial Models for Fuzzy Clustering. Computer Vision and Image Understanding **84(2)** (2001) 285-297
3. Liew, A.W.C., Yan, H.: An Adaptive Spatial Fuzzy Clustering Algorithm for 3D MR Image Segmentation. IEEE Trans. on Medical Imaging **22(9)** (2003) 1063-1075
4. Liew, A.W.C., Leung, S.H., Lau, W.H.: Segmentation of Color Lip Images by Spatial Fuzzy Clustering. IEEE Trans. on Fuzzy Systems **11(4)** (2003) 542-549
5. Ahmed, M.N., Yamany, S.M., Mohamed, N., Farag, A.A., Moriarty, T.: A Modified Fuzzy C-means Algorithm for Bias Field Estimation and Segmentation of MRI Data. IEEE Trans. on Medical Imaging **21(3)** (2002) 193-199
6. Rose, K., Gurewitz, E., Fox, G.C.: Statistical Mechanics and Phase Transitions in Clustering. Physical Review letters **65(8)** (1990) 945-948
7. Rose, K.: Deterministic Annealing for Clustering, Compression, Classification, Regression, and Related Optimization Problems. Processings of the IEEE **86(11)** (1998) 2210-2239
8. McGill University, Canada, Available at http://www.bic.mni.mcgill.ca/brainweb

A Multi-scale Scheme for Image Segmentation Using Neuro-fuzzy Classification and Curve Evolution

Da Yuan[1,2], Hui Fan[1], and Fu-guo Dong[1]

[1] School of Information and Electronic Engineering,
Shandong Institute of Business and Technology, Yantai 264005, China
ydccec@126.com
[2] School of Computer Science and Technology, Shandong University,
Jinan 250100, China

Abstract. In this paper, we present a new scheme to segment a given image. This scheme utilizes neuro-fuzzy system to derive a proper set of contour pixels based on multi-scale images. We use these fuzzy derivatives to develop a new curve evolution model. The model automatically detect smooth boundaries, scaling the energy term, and change of topology according to the extracted contour pixels set. We present the numerical implementation and the experimental results based on the semi-implicit method. Experimental results show that one can obtains a high quality edge contour.

1 Introduction

In most computer vision applications, image segmentation constitutes a crucial initial step before performing the task of object recognition and representation. Segmented images or edge maps contain very useful information and are used very often to convey the essential content of an image. An edge in an image may be formulated using curve evolution model that in turn require solutions to PDEs. The basic idea in curve evolution models is to evolve a curve, subject to constraints from a given image, in order to detect objects in that image. For instance, staring with a curve around the object to be detected, the curve moves toward its interior normal and has to stop on the boundary of the object.

A geometric active contour model was introduced in [1][2], in which an evolving curve was formulated by the level set method [3]. The method works on a fixed grid, usually the image pixels grid, and automatically handles changes in the topology of the evolving contour. The approach based on Mumford-shah model and Vese-Chan model [4][5] was proposed without a stopping edges-function, which can detect contours both with or without gradient. Numerous models of linear and nonlinear diffusion have been proposed in literature for achieving image segmentation.

In this paper, we present a new image segmentation scheme that is based on curve evolution model and neuro-fuzzy Classification. In the next section we formulate the curve evolution model in terms of an energy minimization derived from neuro-fuzzy system. In section 3, we describe our numerical implementation of the model based on the semi-implicit scheme. Finally, we conclude the paper by giving a summary.

J. Wang et al. (Eds.): ISNN 2006, LNCS 3972, pp. 511 – 516, 2006.

2 Description of the Model

In [5] Chan T.F. and Vese L.A. proposed an active contour model using an energy minimization technique. Assume that the image u_0 is formed by two regions of approximately piecewise constant intensities u_0^i and u_0^o, and the object to be detected is represented by the region with value u_0^i. If the boundary is given by c_0 ,then $u_0 \approx u_0^i$ inside c_0 and $u_0 \approx u_0^o$ outside c_0.

$$F = F_1(C) + F_2(C) = \int_{inside(C)} |u_0 - c_1|^2 dx + \int_{outside(C)} |u_0 - c_2|^2 dx \ . \tag{1}$$

where c is any variable curve, c_1 and c_2 are constants depending on c. Therefore, The energy F is minimized when $c = c_0$:

$$F = \inf\{F_1(C) + F_2(C)\} \approx F_1(C_0) + F_2(C_0) \approx 0 \ .$$

In [3] C is represented by the zero level set of the function $\phi : R^N \to R$:

$$\begin{cases} C = \{x \in R^N : \phi(x) = 0\} \\ C_{inside} = \{x \in R^N : \phi(x) > 0\} \\ C_{outside} = \{x \in R^N : \phi(x) < 0\} \end{cases} \tag{2}$$

Using the standard definition for the Heaviside function H

$$H(z) = \begin{cases} 1 \text{ if } z \geq 0 \\ 0 \text{ if } z < 0 \end{cases}$$

Area inside $C = \int_\Omega H(\phi) dx$, thus

$$F(\phi, c_1, c_2) = \int_\Omega |u_0 - c_1|^2 H(\phi) dx + \int_\Omega |u_0 - c_1|^2 \{1 - H(\phi)\} dx \ . \tag{3}$$

Minimizing the energy function with respect to C_1 and C_2 gives

$$C_1(\phi) = \frac{\int_\Omega u_0 H(\phi) dx}{\int_\Omega H(\phi) dx} \ .$$

$$C_2(\phi) = \frac{\int_\Omega u_0 \{1 - H(\phi)\} dx}{\int_\Omega \{1 - H(\phi)\} dx} \ .$$

Because natural images are quite complex and can not always be well segmentation by the model. Even if they can be modeled, finding a solution that minimizes the function (3) could be very difficult. We combine neuro-fuzzy classification [10] with curve evolution technology for solving image segmentation problem. Our approach, instead, does not need to compute equation (3), extract nature contour pixels from images with neuro-fuzzy and formulate some attraction forces that be generated from this contour pixels information. The goal that we use neuro-fuzzy method is to classify a single pixel as contour or regular point according to its edge energy strength value $E_{i,j}$ in different scale images. Given an image I of L gray levels, we deal with image I with Gaussian function and obtain a series of images $I^{(r)}$ at different scale σ_r with $r = 1,...,R$. The $E_{i,j}^{(r)} = (E_{i,j}^{(1)},...,E_{i,j}^{(R)})$ are gradient values in different scale image $I^{(r)}$.

We use a three-layer feed-forward architecture to model the contour classification system. Nodes in the first layer evaluate the membership values for each input according to:

$$u_{rk}(E_{i,j}^{(r)}) = \exp(-(E_{i,j}^{(r)}/\sigma_{rk})^2) \qquad (4)$$

Where σ_{rk} is the width of the Gaussian function. The k nodes in the second layer perform precondition matching of rules by means of product operator. The k -th node computes the active strength of the k -th rule by:

$$u_{rk}(E_{i,j}) = \prod_{r=1}^{R} u_{rk}(E_{i,j}^{(r)}), k = 1,...,k \qquad (5)$$

The third layer supplies the class membership values by:

$$y = \sum_{k=1}^{k} u_k(E_{i,j}^{(r)} v_k) \Big/ \sum_{k=1}^{k} u_k(E_{i,j}) \qquad (6)$$

Where the value y obtained by (6) represents the fuzzy classification value. We select the contour class C_e corresponding to the maximum value of y. The k -th rule R_k is of the following form:

$$If\ E_{i,j}^{(1)} \in A_k^{(1)}\ and\ ...and\ E_{i,j}^{(R)} \in A_k^{(R)}\ then\ g_{i,j} \in C_e .$$

Where $A_k^{(r)}$ are the fuzzy sets defined over the variable $E_{i,j}^{(r)}$. Therefore, The energy F can be modified as:

$$F = \int_{C'} |\nabla u|^2 dx\ ,\ C' = \{x : x \in C_0\ and\ x \notin C_e\}. \qquad (7)$$

In the classical explicit snake model [6] the parametric curve is embedded into an energy minimization framework. However, the parametrization of the curve causes difficulties with respect to topological changes and numerical implementations. Thus, to prevent these difficulties, implicit curve evolution models have been developed. The basic ides is to represent the initial curve implicitly within a higher dimensional function, and to evolve this function under a partial differential equation. Our scheme is the minimization of an energy based on segmentation. In our implicit scheme the energy minimization parameter is embedded into the diffusion equation. The evolving contour model is given by the following evolution equation:

$$\frac{\partial u}{\partial t} = g(|\nabla G_\sigma * u|^2)|\nabla u| div\left(\frac{\nabla u}{|\nabla u|}\right) + \alpha |\nabla u| F . \qquad (8)$$

In order to reduce smoothing at edges, the diffusivity g is chosen as a decreasing function of the edge detector $|\nabla G_\sigma * u|$. Here, $\nabla G_\sigma * u$ is the gradient of a smoothed version of u which is obtained by convolving u with a Gaussian of standard deviation.

3 Numerical Implementation

A discrete m-dimensional image can be regarded as a vector $f \in R^N$, whose components $f_i, i \in \{1,....N\}$ display the grey values at the pixels. Pixel i represents the location x_i . Let h_l denote the grid size in the l direction. τ is the numerical time step. Thus u_i^k denotes the approximation of $u(x_i, t_k)$. Setting $a = 1/|\nabla u|$ the term $div(a\nabla u)$ may be approximated as follows:

$$div(a\nabla u) = \sum_{j \in N(i)} \frac{(a)_i^k + (a)_j^k}{2} \frac{u_j^{k+1} - u_i^{k+1}}{h^2} . \tag{9}$$

In matrix-vector notation, (9) becomes

$$div(a\nabla u) = \sum_{l \in \{x,y\}} A_l(u^k)u^{k+1} . \tag{10}$$

A semi-implicit discretization of the diffusion equation is given by

$$\frac{u^{k+1} - u^k}{\tau} = b\sum_{l=1}^{m} A_l(u^k)u^{k+1} + \alpha|\nabla u|F . \tag{11}$$

where b denotes the term $g\left(|\nabla u_\sigma * u|^2\right)$. Solving for u^{k+1} yields

$$u^{k+1} = \left(I - \tau b\sum_{l=1}^{m} A_l(u^k)\right)^{-1} \left(u^k + \alpha\tau|\nabla u|F\right). \tag{12}$$

where $I \in R^N$ is the unit matrix. Since the system matrix is strictly diagonally dominant. Additive operator splitting(AOS) method were introduced by Weickert et al.[7] as an unconditionally stable numerical scheme for non-linear diffusion in image processing. The AOS method lead to strictly diagonally dominant tridiagonal linear systems which has linear complexity and can be implemented very easily. The AOS scheme of (8) is given by

$$\begin{cases} u_x^{k+1} = \left(I - 2\tau b A_x(u^k)\right)^{-1}\left(u_x^k + \alpha\tau|\nabla u|F\right) \text{ in } x \text{ direction} \\ u_y^{k+1} = \left(I - 2\tau b A_y(u^k)\right)^{-1}\left(u_y^k + \alpha\tau|\nabla u|F\right) \text{ in } y \text{ direction} \end{cases} . \tag{13}$$

We can obtain the following general result:

$$u^{k+1} := 0.5(u_x^{k+1} + u_y^{k+1}) . \tag{14}$$

The AOS method has the same first-order Taylor expansion in τ as the semi-implicit scheme. Two methods are $O(\tau + h_1^2 + \cdots + h_m^2)$ approximations to the continuous equation.

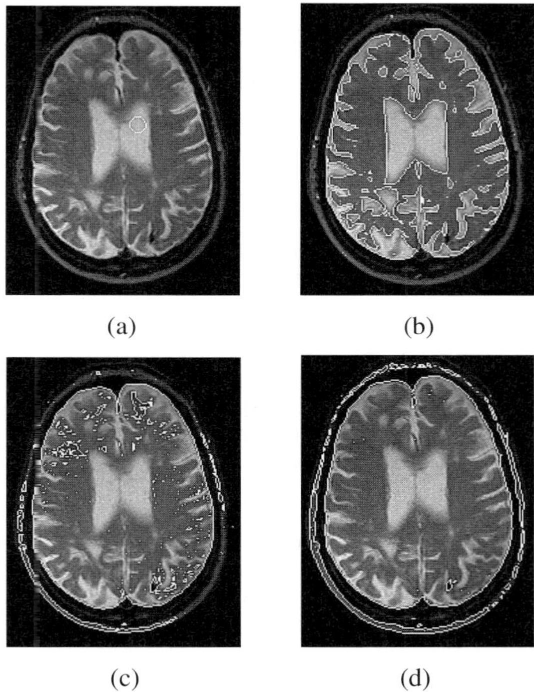

Fig. 1. Numerical result using the semi-implicit scheme (a) Original image and initial curve (b) after 2 iterations (c) after 4 iterations (d) after 10 iterations

Here, we present numerical results using our model. For the examples in Fig. 1, we show the convergence process of our algorithm applied to an image of the brain. In the energy minimization term, the parameters were chosen as follows: $h = 1$, $\tau = 0.25$ and $\alpha = 0.31$.

4 Conclusion

In this paper we proposed a curve evolution scheme based on neuro-fuzzy technology. The scheme automatically detect smooth boundaries, scaling the energy term, and change of topology according to the extracted contour pixels set. We derive suitable semi-implicit method for our active contour scheme. The semi-implicit method creates a discrete nonlinear diffusion scale space for arbitrarily large time steps. In particular, it is unconditionally stable and does not suffer from any time step size restriction. Unlike the explicit method, it can be fully adapted to the desired accuracy without the need to choose small time steps for stability reasons.

Acknowledgement

This work is supported by National Science Foundation of China 60403036 and National Science Foundation of Shandong Province Y2003G01. The authors would like to thank Prof. Cai-Ming Zhang.

References

1. Caselles, V., Catte, F., Coll, T., Dibos, F.: A Geometric Model for Active Contours in Image Processing. Nume. Math. 66(3)(1993) 1-31
2. Malladi, R., Sethian, J.A., Vemuri, B.C.: Shape Modeling with Front Propagation: A Level Set Approach. IEEE Trans. Patt. Analy. and Mach. Intell. 17(2)(1995) 158-175
3. Osher, S.J., Sethian, J.A.: Fronts Propagating with Curvature Dependent Speed: Algorithms Based on Hamilton-Jacobi Formulations. J. Comp. Phys. 79(1) (1988) 12-49
4. Mumford, D., Shah, J.: Optimal Approximation by Piecewise Smooth Functions and Associated Variational Problems. Comm. Pure Appl. Math. 42(5)(1989) 577-685
5. Chan, T.F., Vese, L.A.: Active Contours Without Edges. IEEE Trans. Imag. Proc. 10(2)(2001) 266-277
6. Kass, M., Witkin, A., Terzopoulos, D.: Snake: Active Contour Models. Int. J. Comp. Visi. 1(4)(1988) 321-331
7. Weickert, J., Kuhne, G.: Fast Methods for Implicit Active Contour Models. Geometric Level Set Methods in Imaging, Vision and Graphics, Springer-Verlag, New York(2003)
8. Weickert,J., Haar Romeny B.M., Viergever, M. A.: Image Processing Efficient and Reliable Scheme for Nonlinear Diffusion Filtering. IEEE Trans. Imag. Proc. 7(3)(1998) 389-410
9. Goldenberg, R., Kimmel, R., Rivlin, E., Rudzsky, M.: Fast Geodesic Active Contours. IEEE Trans. Imag. Proc. 10(10)(2001) 1467-1475
10. Castiello, C., Castellano, G., Caponetti, L., Fanelli, A.M.: Fuzzy Classification of Image Pixels. In: Ruspini, E.H., Várkonyi-Kóczy, A.R. (eds): Proceedings of the IEEE International Workshop on Intelligent Signal Processing. Vol. 1. IEEE I&M Society, Budapest(2003) 79-82

A Robust MR Image Segmentation Technique Using Spatial Information and Principle Component Analysis

Yen-Wei Chen[1,2] and Yuuta Iwasaki[2]

[1] College of Electronic and Information Engineering, Central South Forest University, Changsha 410004, China

[2] College of Information Science and Engineering, Ritsumeikan University, Nojihigashi, Kusatsu, Shiga 525-8577, Japan

chen@is.ritsumei.ac.jp

Abstract. Automated segmentation of MR images is a difficult problem due to the complexity of the images. Up to now, several approaches have been proposed based on spectral characteristics of MR images, but they are sensitive to the noise contained in the MR images. In this paper, we propose a robust method for noisy MR image segmentation. We use region-based features for a robust segmentation and use principle component analysis (PCA) to reduce the large dimensionality of feature space. Experimental results show that the proposed method is very tolerant to the noise and the segmentation performance is significantly improved.

1 Introduction

Automated segmentation of MR images is an important step for quantitative studies and 3-D visualization of anatomical structures. It has been studied from different viewpoints [1-4]. As a statistical classification task, the segmentation includes a strategy of feature extraction and classification [5]. We here concern the effect of feature extraction for robust segmentations.

For single spectral images, multiscale features obtained by local filtering have been used to represent the spatial features of MR images [6,7]. The effectiveness, however, is limited by the crucial choice of filter parameters. In MR analysis, we usually use several MR images (multi-spectral MR images) taken with different conditions, such as T1-weighted image, T2-weighted image, and proton density (PD) image. Most traditional segmentation methods based on spectral features use the spectral images independently [8,9]. They utilize the high contrast between different brain tissues in the three images, for instance, the CSF has uniquely bright intensity in the T2 image and the cerebrum region is relative bright in the PD image. C. Tsai et al. extract cerebrum region from PD image and the CSF from T2 image by thresholding the histograms. Although having different later classification method, many researches also obtain the CSF region by thresholding the image[10]. However the contrast in a MR image depends strongly upon the way the image is acquired. The intensity inhomogeneities in the RF field and the partial-volume effect cause the MR images to be very noisy. Since the multi-spectral features used in the conventional segmentation

J. Wang et al. (Eds.): ISNN 2006, LNCS 3972, pp. 517 – 522, 2006.

method are extracted from single pixel which is to be classified, it is very sensitive to the noise contained in the images. In this paper, we propose a robust method for noisy MR image segmentation. We use region-based multi-spectral features to combine the spectral features with spatial information so that the segmentation is very tolerant to the noise. We also use principle component analysis (PCA) [11] to reduce the large dimensionality of the region-based multi-spectral feature space. Experimental results show that the proposed method is very tolerant to the noise and the segmentation performance is significantly improved.

2 Robust Segmentation of MR Images Using Special Information and PCA

In MR image segmentations, T1, T2 and PD images are usually used. The T1, T2, PD images can be viewed as a multi-spectral images. Each pixel has three values: T1, T2, PD. In the conventional methods, since the multi-spectral features are extracted from single pixel to be classified as shown in Fig.1(a), it is sensitive to the noise contained in the images. In this paper, we propose to combine the multi-spectral features with spatial information for a robust segmentation. As shown in Fig.1(b), we use a region-based multi-spectral features for classifications instead of pixel-based features. In the proposed region-based method, since the spatial information is included in the classifications, the method is very tolerant and robust to the noise.

Though the proposed region-based method is robust to the noise, the limitation of the method is its large dimensionality of feature space. If we use a region of 3x3 for 2-D images, the dimensionality of feature space will be 27x1, while if we use a region of 3x3x3 for 3-D images, the dimensionality will be increased to 81x1. It should be noted that the dimensionality of conventional pixel-based feature space is only 3x1. In order to reduce the large dimensionality of feature space, we propose to use Principle Component Analysis (PCA) as a preprocessing of classifications to find an efficient representation of feature space and reduce the dimensionality of feature space as shown in Fig.1(b).

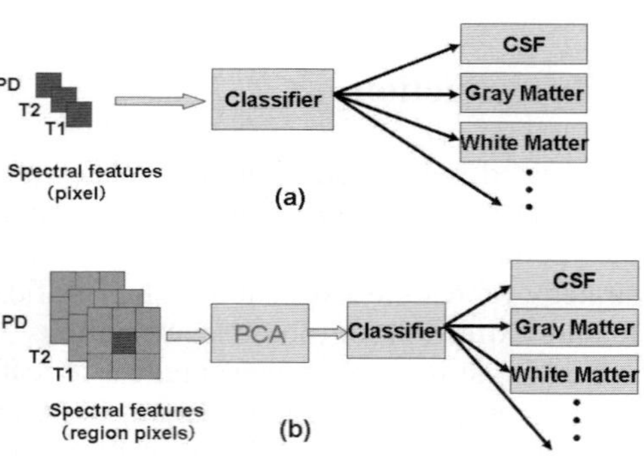

Fig. 1. (a) Conventional pixel-based method; (b) proposed region-based method

PCA is to linearly transform the data such that the outputs are uncorrelated. The general model can be described as follows. We start the basis assumption that a data vector $\mathbf{x} = (x_1, x_2, \cdots x_M)$ can be represented in terms of a linear superposition of basis functions,

$$\mathbf{x} = \mathbf{A}\mathbf{s} = \mathbf{a}_1 \cdot s_1 + \mathbf{a}_2 \cdot s_2 + \cdots + \mathbf{a}_N \cdot s_N \tag{1}$$

where $\mathbf{s} = (s_1, s_2 \cdots, s_N)$ is the coefficients, $\mathbf{A}$ is a $M \times N$ matrix and the columns $\mathbf{a}_1, \mathbf{a}_2, \cdots, \mathbf{a}_N$ are called basis functions. The basis functions are consistent and the coefficients vary with data. The goal of efficient coding is to find a set of $\mathbf{a}_i$ that result in the coefficient values being uncorrelated. Therefore, the linear image analysis process is to find a matrix $\mathbf{W}$, so that the resulting vector:

$$\mathbf{y} = \mathbf{W}\mathbf{x} \tag{2}$$

has uncorrelated components. $\mathbf{W}$ is constructed in such a way that $\mathbf{W}\mathbf{W}^T = \mathbf{I}$. Actually, the rows of $\mathbf{W}$ are the scaled versions of the basis functions (columns of $\mathbf{A}$), and they are the eigenvectors of the covariance matrix of $\mathbf{x}$. The PCA basis functions are obtained by analyzing sample data of MR brain images. Since the first several components will maintain most of information, we can easily reduce the dimensionality of features by choosing first several components as features

The unsupervised K-means algorithm [12] is used for clustering spectral features.

3 Segmentation of MR Brain Images

The proposed method has been applied to segment noisy MR brain images. The data is obtained from the Brain Web (http://www.bic.mni.mcgill.ca/brainweb) and the data included noise (9%). The typical multi-spectral MR images (2-D slice images) are shown in Fig.2.

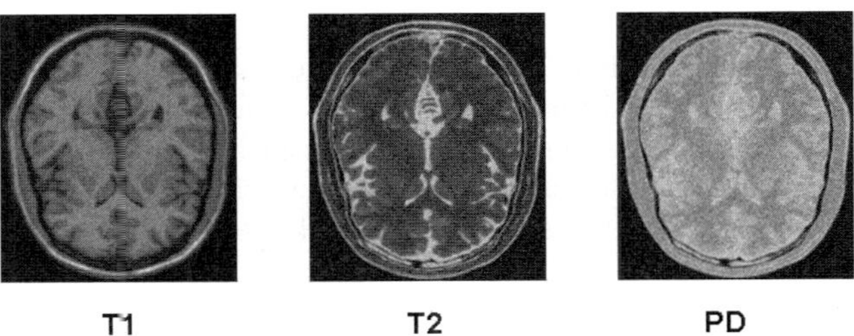

Fig. 2. Typical multi-spectral MR images

We use a 3x3 region to extract the multi-spectral features and the dimension of feature space is 27x1. By using PCA, we can project the 27-dimensional feature space into a 3-dimensional subspace, while maintaining 95% of information. Thus, we select the top three components as features for segmentation.

The segmentation results are show in Fig. 3. In order to make a comparison, the MR brain images are segmented by three methods: conventional pixel-based method (the dimension of feature vector is 3); proposed region-based method (the dimension of feature vector is 27) and proposed region-based method with PCA (the dimension of feature vector is 3). The segmentation results of white matter, gray matter and CSF are shown in Fig. 3. The ground truth shown in Fig. 3 is obtained from brain web database, which is labeled by experts manually.

As shown in Fig. 3, it can be seen that the noise significantly effect the segmentation results in the conventional pixel-based method, while the proposed region-based method is very tolerant to the noise. Especially, when we combine the region-based method with PCA, we can obtain a robust segmentation result with a small number of features as well as the conventional pixel-based method.

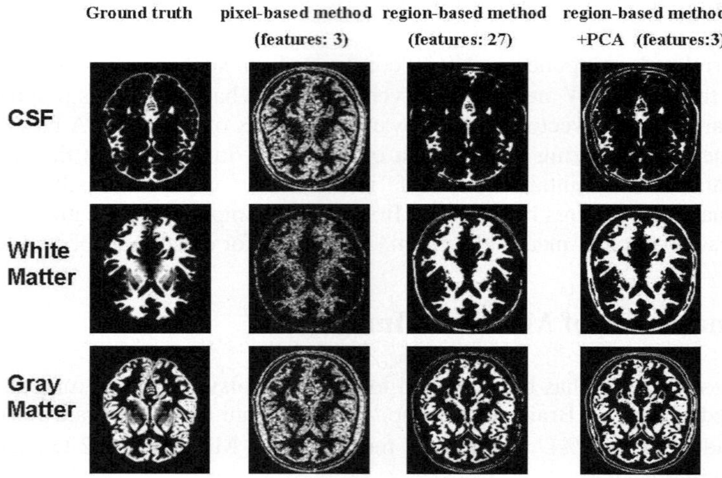

Fig. 3. Segmentation results of 2D MR brain images with noise

In order to make a quantitative evaluation, we compare the each segmented image with the corresponding ground truth. By taking the difference between the segmented image and the corresponding ground truth, we can easily count the number of pixels (N_p) that belong to the target class (white matter or gray matter or CSF) but are not segmented into the target class, and the number of pixels (N_n) that do not belong to the target class but are segmented into the target class. Segmentation performance is measured by two criteria of false positive rate δ_p and false negative rate δ_n, which are defined as

$$\delta_p = N_p / C$$

and

$$\delta_n = N_n / C$$

where C is the number of pixels belong to the target class in the ground truth.

The results are summarized in Table 1. It can be seen that the average false positive rate and average false negative rate for the conventional pixel-based method are 14.7% and 12.4%, respectively. On the other hand, by using the proposed region-based method, the average false positive rate and the average false negative rate are significantly decreased to about 4.5% and 2.5%, respectively, for both without PCA and with PCA. The significance of the proposed region-based method is obviously.

Table 1. Quantitative evaluation of segmentation performance

Method	Dimension of features	False positive rate (%)			False negative rate (%)		
		Gray Matter	White matter	CSF	Gray matter	White matter	CSF
Pixel-based method	**3x1**	**12.3**	**12.3**	**19.3**	**15.9**	**10.3**	**11.0**
Region-based method (w/o PCA)	**27x1**	**5.6**	**5.5**	**2.5**	**4.1**	**1.2**	**2.2**
Region-based method (with PCA)	**3x1**	**5.0**	**5.0**	**2.5**	**3.6**	**2.1**	**1.8**

The proposed method has also been used for segmentation of 3-D volume MR brain images. In the segmentation of 3-D images, a region of 3x3x3 is used for segmentations. Thus the dimension of feature vector is 81 and the dimension is reduced to 4 by PCA. The segmentation results are shown in Fig.4.

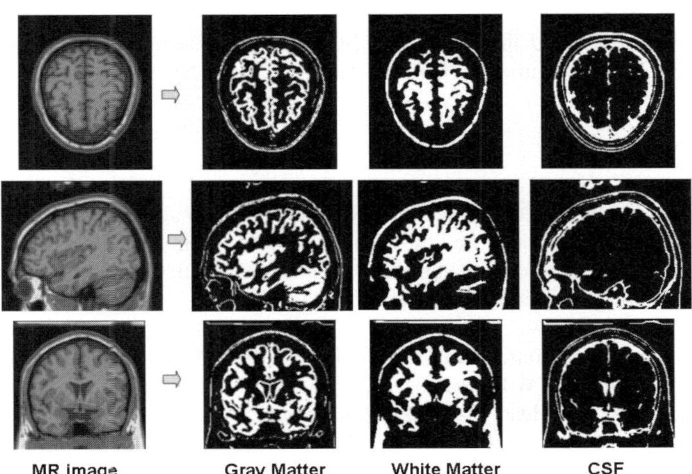

Fig. 4. Segmentation results of 3D MR brain images with noise

Table 2. Quantitative comparison of segmentation performance and computation time

method	Averaged false positive rate (%)	Averaged false negative rate (%)	Computation time (min)
Region-based method (w/o PCA)	5.33	3.47	300
Region-based method (with PCA)	5.67	3.62	3

The quantitative evaluation of segmentation performance and computation time are shown in Table 2. It can be seen that both false positive rate and false negative rate of the method with PCA are almost same with that of the method without PCA, while computation time is significantly reduced to 3 min from 300 min.

4 Conclusions

In this paper, we proposed a robust method for segmentation of noisy MR images using spatial information and PCA. We used region-based multi-spectral features for segmentation and used PCA for dimension reduction. The classification results of real MR images have shown that by using the proposed method, the segmentation performance can be significantly improved. Both false positive rate and false negative rate are reduced to less than 5%, while the computation time is almost same as the conventional pixel-based method.

Acknowledgement

This work was supported in part by the **S**trategic Information and **Co**mmunications R&D **P**romotion Programm**e** **(SCOPE)** under the Grand No. 051307017.

References

1. Bezdek, L.C., Hall, L.O., Clarke, L.P.: Review of MR Image Segmentation Techniques Using Pattern Recognition. Medical Physics, 20 (1993) 1033-1048.
2. Cohen, L.: On Active Contour Models and Ballons. Computer vision, Graphics and Image Processing: Image Understanding, 53(1991)211-218.
3. Hohne, K., Hanson, W.: Interactive 3d Segmentation of MRI and CT Volumes Using Morphological Operations. Journal of Computer Assisted Tomography, 16(1992) 285-294.
4. Cline, H.E., Lorensen, W.E., Kikinis, R., Jolesz, F.A.: Three-dimensional Segmentation of MR Images of the Head Using Probability and Connectivity. Journal of Computer Assisted Tomography, 14(1990)1037-1045.
5. Warfield, S.K., Kaus, M., Jolesz, F.A., Kikinis, R.: Adaptive, Template Moderated, Spatially Varying Statistical Classification. 4(2000)43-55.
6. Gerig, G., Kubler, O., Kikinis, R., Jolesz, F.A.: Nonlinear Anisotropic Filtering of MRI Data. IEEE Transactions on Medical Imaging, 11(1992)221-232.
7. Mohamed, N.A., Aly, A.F.: Volume Segmentation of CT/MRI Images Using Multiscale Features, Self-organizing Principal Component Analysis, and Self-organizing Feature Map. Proc. of Int. Conf. on Artificial Neural Networks, Houston, (1997).
8. Clark, M., Hall, L., Goldgof, D., etc.: MRI Segmentation Using Fuzzy Clustering Techniques: Integrating Knowledge. IEEE Eng. Med. & Biol. Mag., 13(1994)730-742.
9. Atkins, M.S., Mackiewich, M.T.: Fully Automatic Segmentation of the Brain in MRI. IEEE Trans. Med. Imagin, 17(1998)98-107.
10. Yoon, O.Y., Kwak, D.M., Kim, B.S., etc.: Automated Segmentation of MR Brain Image Using 3-Dimensional Clustering. IEICE Trains. Inf. & Syst., E85-D (2002)773-781.
11. Jolliffe, I.T.: Principal Component Analysis. Springer Verlag, New York, (1986).
12. Duda, R., Hart, P.: Pattern Classification and Scene Analysis. Join Wiley and Sons (1973)

Adaptive Segmentation of Color Image for Vision Navigation of Mobile Robots

Zeng-Shun Zhao, Zeng-Guang Hou, Min Tan, and Yong-Qian Zhang

Key Laboratory of Complex Systems and Intelligence Science,
Institute of Automation, Chinese Academy of Sciences,
P.O. Box 2728, Beijing 100080, China
{zengshun.zhao, zengguang.hou, min.tan}@mail.ia.ac.cn

Abstract. The self-localization problem is very important when the mobile robot has to move in autonomous way. Among techniques for self-localization, landmark-based approach is preferred for its simplicity and much less memory demanding for descriptions of robot surroundings. Door-plates are selected as visual landmarks. In this paper, we present an adaptive segmentation approach based on Principal Component Analysis (PCA) and scale-space filtering. To speed up the entire color segmentation and use the color information as a whole, PCA is implemented to project tristimulus R, G and B color space to the first principal component (1st PC) axis direction and scale-space filtering is used to get the centers of color classes. This method has been tested in the color segmentation of door-plate images captured by mobile robot CASIA-1. Experimental results are provided to demonstrate the effectiveness of this proposed method.

1 Introduction

Research on mobile robot navigation is an important issue which mainly includes: obstacle detection and avoidance, path planning, map building and self-localization. As a mobile robot moves in its environment, its actual position and orientation always differ from the position and orientation that it is commanded to hold [1]. So among tasks mentioned above, self-localization, which estimates in real time mobile robot's current location with respect to the immediate surroundings, is probably the most fundamental and difficult. Considering techniques for self-localization, landmark-based approach is preferred for its simplicity and much less memory demanding descriptions of robot surroundings [2,3]. This kind of navigation is heavily dependent on a powerful perception system to identify elements of the environment. So chosen elements should be simple enough in order to permit an easy identification from different view angles and distances [1]. A spatial navigation system based on visual templates is presented in [4]. Templates are created by selecting a number of high-contrast features in the image and storing them together with their relative spatial locations in the image. Inspired by this, we select door-plates as visual landmarks for our task.

J. Wang et al. (Eds.): ISNN 2006, LNCS 3972, pp. 523–532, 2006.

Color image segmentation is widely applied in fields of pattern recognition, robot vision, image retrieval, compression and face recognition [5,6]. An important first step in color vision tasks is to classify each pixel in an image into respective color classes. The leading approaches to accomplishing this task include color space threshold, linear color threshold, nearest neighbor classification and probabilistic methods. In this paper, nearest neighbor classification is employed in reduced color space obtained by PCA and centers of classes are obtained via scale-space filtering.

The rest of the paper is organized as follows. Classical color segmentation schemes are briefly described in the following section. Outline of proposed adaptive color segment method for Mobile Robots is presented in Section 3, which involves PCA and scale-space filtering based adaptive histogram analysis. Optimal scale selection and location of centers of color classes are also discussed in this section. Section 4 gives details about experiment and shows some segmentation results. At last, conclusions are drawn in Section 5.

2 Classical Schemes of Color Segmentation

For color image segmentation, selection of the initial number of classes and determination of their centers are crucial to clustering methods. Generally the numbers of pixels of the frequently appeared colors in an image are larger and can be considered as local summits in the 3-D color histogram of image [13]. The segmentation results of each color component can be combined in some way to obtain a final segmentation result [7].

Having the initial number of colors and their center, we can cluster pixels of image in 3-D color space, e.g., RGB color space, according to the rule of color similarity. Ref. [5] presented a segmentation algorithm for color images based on 1-D histogram thresholding and the fuzzy C-means (FCM) techniques. 1-D histogram thresholding is applied to one color component each time, then scale-space filtering is used to locate the thresholds in each color component's histogram. These thresholds are used to partition the color space into several hexahedra. The hexahedra that contain a number of pixels exceeding a predefined threshold are declared as valid clusters. In the step of determining the initial number of classes and their centers, similar method is also utilized in the 3-D color space [13]. First the classes of each color component are obtained by three histograms of color R, G and B, then these segmented parts are combined, false colors are removed. Finally rival penalized competitive learning algorithm (RPCL) is conducted. The primary drawback of this approach is relatively high memory cost. *Karhunen-Loeve* transformation is applied to RGB color space to extract features with large discriminant power to isolate the clusters in a given region [14]. At each step of segmentation, three new color features are adaptively calculated and used to compute the histograms for thresholding. However, performing K-L transformation using a recursive region splitting method at each segmentation step involves too much computational time for practical application.

3 An Adaptive Color Segmentation Method for Mobile Robots

3.1 Problem Presentation and Overview of the Proposed Algorithm

From analysis about classical schemes of color segmentation, it can be seen that color images segmentation by multi-thresholding still has three unsolved problems which are listed below: One is how to devise an automatic procedure for determining the optimal thresholds. Next, multiple histogram-based thresholding divides color space by thresholding each component histogram. Detecting the clusters of points within this space will be computationally expensive. The third problem is that thresholding is performed on only one color component at a time. Thus the regions extracted are not based on the information available from all three components simultaneously because the correlation among the three components is neglected. In another words, how to employ the color information as a whole for each pixel? We try to solve these three questions using our proposed method for our case.

The outline of our proposed segmentation scheme is illustrated in Fig. 1 which will be described in detail later. From Fig. 1 and the previous introduction, it can be seen that proposed method involves PCA and scale-space filtering.

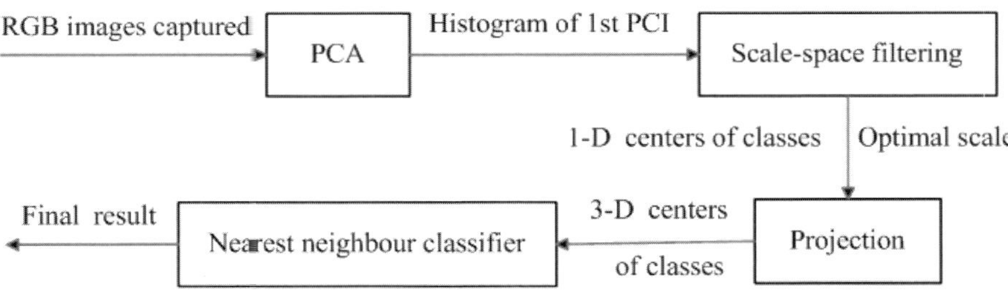

Fig. 1. Block diagram of the presented method

In this paper, image adopts RGB color space and each pixel is represented with R, G and B components. Instead of finding valleys of histogram as thresholds, peaks are located as centers of color classes for simplifying computation. Considering of the fact that regions of interest are just those digital numbers which could give much benefit, this paper proposes an adaptive segmentation approach based on PCA and scale-space filtering. Firstly, we employ PCA to obtain histogram of the first principal component of the captured image in RGB color space. Secondly, histogram is analyzed to track the optimal scale and locate 1-D peaks via scale-space filtering. Three dimensional peaks transformed from principal component space to RGB color space could be selected as centers of color classes. Finally, color image is segmented into several regions of different colors using classification based on the rule of minimal distance. Nearest neighbor classification is employed to classify each pixel in an image into some color classes, where we make one representative point in each class. Each image pixel converts its color to the color defined by the class.

3.2 Eigenspace Mapping to Obatain Principal Component Image

A Principal Cmponent Image(PCI) is obtained using PCA, which constitutes the foundation of our proposed transformation. PCA has many applications, for example, classification and recognition problems [9,10]. Considering a random vector $x = [x_1, x_2, ..., x_n]^T$, the mean vector and the covariance matrix of the vector population X are defined as

$$m_X = E(X) \tag{1}$$

$$C_X = E\{(X - m_X)(X - m_X)^T\} \tag{2}$$

where $E(\cdot)$ is the expected value of the argument, and T indicates vector transposition. For real vectors, the corresponding matrix is real and symmetric. Because C_X is real and symmetric, we can always find a set of orthonormal eigenvectors of this covariance matrix.

Given an image I in color space, $n = 3$. Let C_X be the covariance matrix of the distributions of R, G and B color components in I, and W_1, W_2, and W_3 be the eigenvectors of C_X with corresponding eigenvalues in decreasing order. Well-known methods of finding eigenvectors include the Jacobi method and the QR algorithm as well as neural networks [8,9]. We could obtain the first principal component image (1st PCI) by mapping images in RGB color space to eigenspace

$$Y_1 = W_1^T X. \tag{3}$$

The first principal component image, Y_1, has the largest variance, thus has the largest discriminant power. And we could also restore the image using the 1st PC as follows

$$\hat{X} = W_1 Y_1. \tag{4}$$

The goal of PCA is to perform dimensionality reduction. However, another reason for implementing PCA in the color segmentation scheme will be described later, which solves the problem how to make use of the color information of a pixel as a whole. At the same time, finding centers of color classes in principal component image are easier and more convenient than that in original color space.

3.3 Selection of Centers of Classes and Optimal Scale

Selection of centers of color classes is important to color segmentation by classifying, so is scale selection to linear scale- space analysis. We can do this via scale-space analysis about histogram of the first principal component image.

Description of a signal or an image is only meaningful over a limited range of resolutions. The interest in and perception of different features of a signal depends not only on the signal, but on the scale of observation [10,12]. The histogram of the first principal component image, $f(x)$, can be smoothed by convolution with a Gaussian function $g(x, \sigma)$ of space $\mathbf{x}$ with different standard deviations σ corresponding to different scales [11], which is showed as follows

$$F(x, \sigma) = f(x) * g(x, \sigma) = \frac{1}{\sqrt{2\pi\sigma}} \int_{-\infty}^{\infty} f(u) e^{(-\frac{(x-u)^2}{2\sigma^2})} du \tag{5}$$

where " $*$ " denotes convolution with respect to x; σ is the Gaussian's standard deviation and scale parameter. In the standard statistical way, we have defined the width of the Gaussian shape in terms of sigma. At any value of σ, the extrema in the nth derivative of the smoothed histogram are given by the zero-crossings in the $(n+1)$th derivative, computed using following relation:

$$\frac{\partial^n F}{\partial x^n} = f * \frac{\partial^n g}{\partial x^n} \tag{6}$$

So the 1st and 2nd of derivative of $F(x, \sigma)$ can be computed as follows:

$$F_x(x, \sigma) = \frac{\partial}{\partial x}[F(x, \sigma)] = f * \frac{\partial g}{\partial x} \tag{7}$$

$$F_{xx}(x, \sigma) = \frac{\partial}{\partial x}[F_x(x, \sigma)] = f * \frac{\partial^2 g}{\partial x^2}. \tag{8}$$

Zero-crossing point of smoothed histogram $F_x(x, \sigma)$ is thought of as the local extrema of the histogram, and at such a point, whether $F_{xx}(x, \sigma)$ is positive or negative could give us a criterion of peak or valley. So for our purpose we will restrict our attention to those in the first and second derivative of the filtered histogram.

Scale selection is one major problem of linear scale space analysis, mainly because the hypothesis that an extremum trajectory in scale-space arises from a single event is rarely true. Fortunately for our task, the regions of interest are just those digital numbers in door-plates having high-contrast to the background. Then we could define the number of color classes to 2 to speed up the computation. Of course, if we do not have this preset, proposed method can also give satisfactory color segment experimental results. With increasing scale, the number of zero-crossing points of smoothed histogram $F_x(x, \sigma)$ decreases gradually, so does those of $F(x, \sigma)$. We can track such a scale that number of peaks contained in 1st order derivative of histogram change from 3 to 2. Then the optimal scale is found. At such a scale, the 1-D centers of classes could be also determined, which is coarser than its true location but proved enough for practical real-time application.

3.4 Nearest Neighbor Classification with 3-D Color Center

After having acquired 1-D peaks via scale-space filtering,we could have 3-D peaks using inverse mapping from eigenspace to RGB color space. Such 3-D peaks in RGB color space can be selected as centers of color classes. Nearest neighbor classification is applied to classify each pixel in an image into some color classes, where we make one representative point in each class. Each image pixel converts its color to the color defined by the class. Then, color image is segmented into several regions of different colors using classification based on the rule of minimal distance.

4 Experimental Results

The experiment is conducted on mobile robot CASIA-1 shown as Fig. 2, which was developed by ourselves. The robot is equipped with infrared, ultrasonic and CCD sensors. The former two types of sensors provide distance information, and the latter is responsible for object recognition and tracking for robot localization and navigation. Color images captured by CCD use RGB color space, and the size is 360*288.

Fig. 2. CASIA-1

Fig. 3. Captured color image

Fig. 4. Image of 1st PC

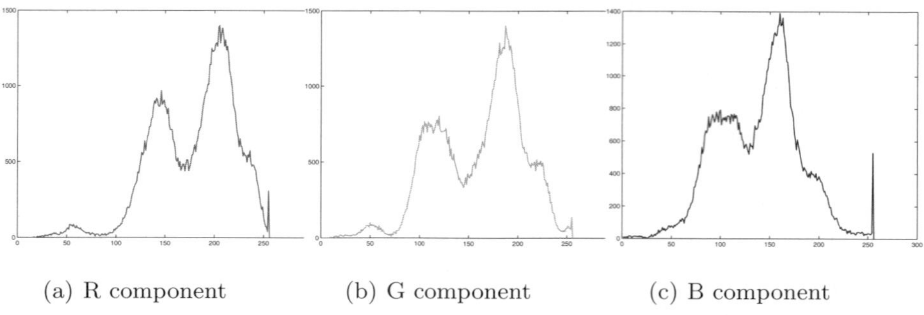

(a) R component (b) G component (c) B component

Fig. 5. Histograms of R, G, B components of Fig. 3

In the experiment, we could predefine the number of peaks as 2 or do not have this preset. Except this, no more parameters need to be prescribed as other methods do [5,13]. The following parts give segmentation result of Fig. 3.

Fig. 5 gives histograms of R, G, B components of Fig. 3. Fig. 4 shows the image of the first principal component (1st PCI) from which we can see that it

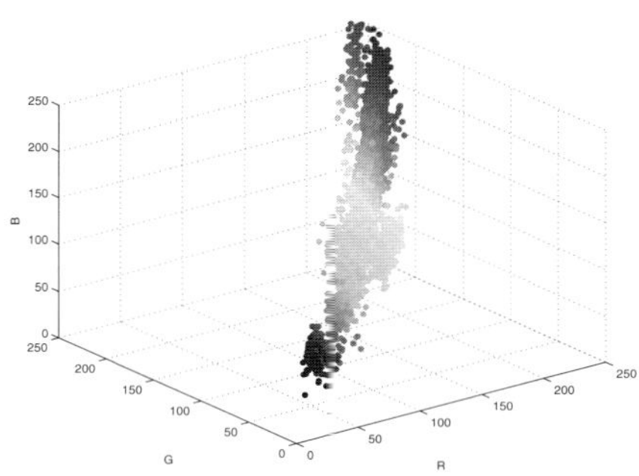

Fig. 6. Color distribution of original image

Fig. 7. Color distribution of image restored from 1st PC

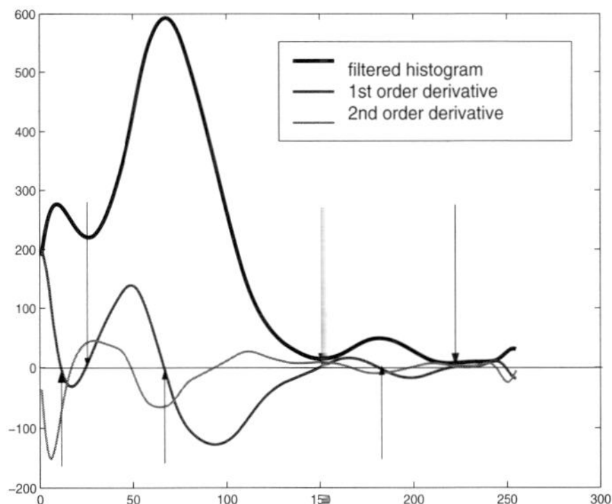

Fig. 8. Filtered histogram where $\sigma = 5.096$

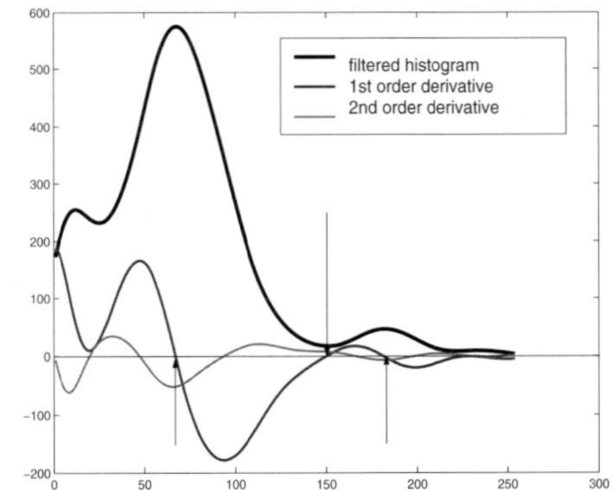

Fig. 9. Filtered histogram where $\sigma = 7.6439$

Fig. 10. Complementary image to Fig. 5

Fig. 11. Image restored from 1st PC

Fig. 12. Segmentation result with 2 classes

Fig. 13. Segmentation result with 4 classes

holds most information, especially our regions of interest for the first principal component retain the most energy. This can be observed more clearly from its complementary image showed as Fig. 10. Fig. 11 shows the image restored from the first principal component which has the least lose of information under the sense of least-mean-squares with $SNR = 27.342\ dB$. The effect of PCA also can be observed from the R, G, B color distribution of original image and image restored showed in Figs. 6 and 7.

With increasing scale, zero-crossing point of the first order derivative of filtered histogram decreases gradually. Thus the optimal scale could be tracked when the number of peaks change from 3 to 2. Fig. 8 shows the smoothed histogram where scale equals 5.096. Optimal scale automatically obtained in our experiment equals 7.6439 as showed in Fig. 9.Here we could get two 3-D centers in RGB color space equal (142, 119, 101) and (77, 50, 38) which are transformed from 1-D center 67 and 181 via PCA. This solves the problems described in Section 3. Combination of three component (e.g., R, G, B) segmentation results is not necessary any more, not having to consideration of false colors. But in Ref. [13], Three histograms of $R(I)$, $G(I)$, $B(I)$ components are divided into 5, 3, 4 parts, respectively. So 60 component combinations are obtained by component histogram segmentation. However, 44 centers are false colors among those 60 combinations. The left 16 centers are converged to 11 clusters after 20 times learning. This combination of each component segment is computationally expensive and could not use the color information of the pixel as a whole.

Final result by color segmentation method with two classes according to the rule of minimal distance is showed in Fig. 12. It can be seen that the digital numbers in the reduced color space image have a strong contrast to the background which is very helpful to real time navigation. Fig. 13 shows the segmentation result with four classes having more natural effect.

5 Conclusions

This paper focuses on the application of PCA and scale-space filtering in doorplate color image segmentation for vision navigation for mobile robots. Experiments on our developed mobile robot CASIA-1 show that under the circumstance of our task, the proposed approach can adaptively segmented captured color images into several main classes. It could be seen that digital numbers in the reduced color space image has a strong contrast to the background. Satisfactory experimental results show the effectiveness of this proposed method.

Acknowledgements

This research has been supported in part by the National Natural Science Foundation of China (Grant Nos. 60205004, 50475179 and 60334020), the National Basic Research Program (973) of China (Grant No. 2002CB312200), the Hi-Tech R&D Program (863) of China (Grant Nos. 2002AA423160 and 2005AA420040), and the Science and Technology New Star Program of Beijing (Grant No. H02082 0780130).

References

1. Mata M., Armingol J. M., de la Escalera A., Salichs M. A.: A Visual Landmark Recognition System for Topological Navigation of Mobile Robots. Proceedings of IEEE International Conference on Robotics and Automation, **2** (2001) 1124- 1129
2. Yoon K. J., Kweon I. S.: Artificial Landmark Tracking Based on the Color Histogram. Proceedings of the IEEE/RSJ International Conference on Intelligent Robots and Systems, (2001) 1918-1923
3. Moon I., Miura J., Shirai Y.: Automatic Extraction of Visual Landmarks for a Mobile Robot Under Uncertainty of Vision and Motion. Proceedings of IEEE International Conference on Robotics and Automation, (2001) 1188-1193
4. Balkenius C.: Spatial Learning with Perceptually Grounded Representations. Proceedings of The Second Euromicro Workshop on Advanced Mobile Robots, (1997) 16-21
5. Lim Y. W., Lee S. U.: On the Color Image Segmentation Algorithm Based on the Thresholding and the Fuzzy C-Means Techniques. Pattern Recognition **23** (1990) 935-952
6. Gonzale R. C., Woods R. E.: Digital Image Processing. 2nd Edn. Prentice Hall, Upper Saddle River, New Jersey (2002)
7. Yang C. K., Tsai W. H.: Reduction of Color Space Dimensionality by Moment-Preserving Thresholding and its Application for Edge Detection in Color Images. Pattern Recognition Letter **17** (5) (1996) 481-490
8. Diamantras K. I., Kung S. Y.: Principal Component Neural Networks: Theory and Applications. John Wiley and Sons, New York (1996)
9. Haykin S.: Neural Networks:A Comprehensive Foundation. 2nd Edn. Prentice-Hall, New Jersey (1999)
10. Witkin A. P.: Scale-Space Filtering. Proceedings of International Joint Conference on Artificial Intelligence, Karlsruhe, West Germany (1983) 1019-1021
11. Babaud J., Witkin A. P., Baudin M. and Duda R. 0.: Uniqueness of the Gaussian Kernel for Scale-Space Filtering. IEEE Transactions on Pattern Analysis and Machine Intelligence, **8** (1) (1986) 26-33

12. Dehghan H.: Zero-Crossing Contour Construction for Scale-Space Filtering. Conference Record of the Thirty-First Asilomar Signals, Systems and Computers, **2** (1997) 1479-1483
13. An Ch. W., Li G. ZH., Zhang Y. Q., Tan M.: Doorplate Adaptive Detection and Recognition for Indoor Mobile Robot Self-Localization. Proceedings of IEEE International Conference on Robotics and Biomimetics, (2004) 339-343
14. Ohlander R., Price K., Reddy D.R.: Picture Segmentation Using a Recursive Region Splitting Method. Computer Graphics and Image Process **8** (1978) 313-333

Image Filtering Using Support Vector Machine

Huaping Liu, Fuchun Sun, and Zengqi Sun

Department of Computer Science and Technology, Tsinghua University, P.R. China
State Key Laboratory of Intelligent Technology and Systems, Beijing, P.R. China

Abstract. In this paper, a support vector machine (SVM) approach for automatic impulsive noise detection in corrupted image is proposed. Once the noises are detected, a filtering action based on regularization can be taken to restore the image. Experimental results show that the proposed SVM-based approach provides excellent performance with respect to various percentages of impulse noise.

1 Introduction

Removal of noises while preserving the integrity of the image is an important problem in image processing. The median filters have attracted much attention due to their simplicity and capability of removing impulsive noises. However, it tends to blur fine details. To address the issue, modified median filters which incorporate some detectors have been proposed. The detector is used to discriminate the uncorrupted pixels from the corrupted ones. For example, [7] proposed a switching-based median filtering. [1] proposed a noise detector based on the rank-ordered mean(ROM). Recently, [2] proposed a new noise detector based on the center-weighted median information. Since the detector-based filter processes only the detected noises, it can preserve image details. The main issue of them lies in building a decision rule which can discriminate the uncorrupted pixels from the corrupted ones. In most cases, the noise detector typically employ a single threshold or a set of thresholds for local signal statistics.

In this paper, we proposed a new classification approach based on support vector machine(SVM) to address this problem. Support vector machines formulate these tasks in terms of convex optimization problems having a unique global optimum. The proposed classification approach can efficiently discriminate the noise pixels from the uncorrupted signal pixels. Furthermore, the determination of thresholds is no longer required.

2 Impulse Noise Detection and Filtering

In this work, the image model containing impulse noise with probability of occurrence p is represented as follows:

$$X_{ij} = \begin{cases} S_{ij} & \text{with probability } 1-p \\ I_p \text{ or } I_n & \text{with probability } p/2, \text{ respectively} \end{cases} \tag{1}$$

J. Wang et al. (Eds.): ISNN 2006, LNCS 3972, pp. 533–538, 2006.

where S_{ij} denotes the pixel values of the original image; X_{ij} denotes the pixel values of the degraded image. I_p and I_n denotes the erroneous point with positive and negative impulse values, respectively. In this paper, we assume that they vary in a small dynamic range, for example, $I_p \in [0, 10]$ and $I_n \in [245, 255]$.

In the following, we will introduce the extracted features using 3×3 sliding window. According to [2], we consider the outputs of CWM filters with weight value w: $Y_{ij}^w = median(X_{ij}^w)$, where $X_{ij}^w = \{X_{i+s,j+t}, w \diamond X_{ij} | -1 \leq s \leq 1, -1 \leq t \leq 1, (s,t) \neq (0,0)\}$, w denotes the center weight, which can take values from $\{1, 3, 5, 7\}$, $\diamond$ represents the repetition operation. For the pixel X_{ij}, the differences d_w can be defined as $d_w = |Y_{ij}^w - X_{ij}|$. These differences provide information about the likelihood of corruption for X_{ij}. To design a noise detector, [2] introduced four thresholds T_1, T_3,T_5 and T_7. Obviously, the thresholds are obtained through extensive experiments, which is not a trivial task. Since the differences d_1, d_3, d_5, d_7 have proven to be a set of useful features to detect the impulsive noises, we can use them to train the SVM to obtain a classifier. However, it is not necessary to incorporate all of $\{d_1, d_3, d_5, d_7\}$ into the feature space due to the information redundancy. Alternatively, since $d_1 \geq d_3 \geq d_5 \geq d_7$[2], we can select only d_1 and d_7 as the features. This selection will also speed the training of SVM. Furthermore, using only these differences is not enough because the line component whose width is just one pixel is difficult to distinguish from the noise. To address this issue, we also employ the features $u_1 = |X_{ij} - X_{c1}|, u_2 = |X_{ij} - X_{c2}|$, where X_{c1} and X_{c2} are pixel values closest to that of X_{ij} in its 3×3 neighborhood window. If a line component appears in the window, u_1 and u_2 will be rather small, if a noise appears, they will be large. Therefore, these two values supply a useful rule to distinguish the line component and the noise.

For every pixel, the corresponding feature vector can be obtained as $\mathbf{x} = \{d_1, d_7, u_1, u_2\}$ and the class label is 1(for uncorrupted signal) or -1 (for noise). The problem of classification is to assign each possible vector to a proper pattern class. To implement the pattern classification, the SVM is employed.

The architecture of an SVM noise detector involves three layers. The input layer is made up of source nodes that connect the SVM to the image. Its activation comes from a 3×3 window in the input image. Moreover, it calculates the feature vector. The hidden layer calculates the inner products between the input and support vectors $\mathbf{x}_1^*, \cdots, \mathbf{x}_v^*$, and applies an RBF function. The size of this layer v is determined as the number of support vectors identified during the training process. Taking every pixels in the image as the center of a 3×3 window for the classification, the detector searches the entire image to classify each pixel as uncorrupted pixel of noise. The output then represents the class of the the central pixel in the input window. For every pixel, if it is detected as uncorrupted signal, then no extra filtering action is needed. Otherwise, a filtering action is implemented. Here we mention a detail-preserving regularization filter[6], which was proposed recently. To use this filter, we first define $\mathcal{N}$ to be the set of of all noisy pixels, i.e., $\mathcal{N} = \{(i,j) | X_{ij}$ is detected as noise$\}$. Also, we define $\mathcal{N}^c = \{(i,j) | X_{ij}$ is detected as uncorrupted signal$\}$. Then the filtered output U_{ij} can be determined by minimizing the following performance

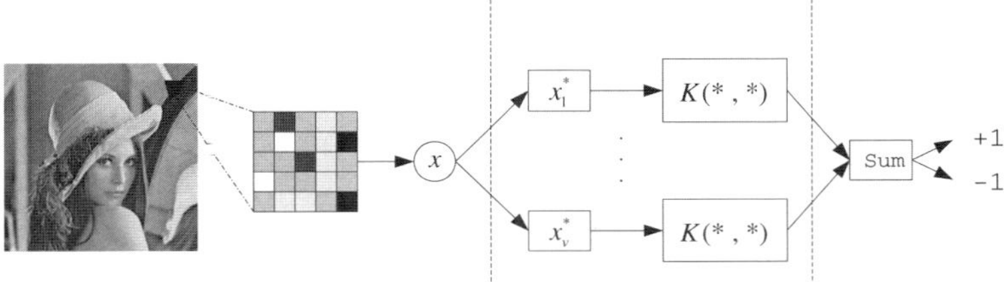

Fig. 1. Architecture of the noise detector, where the input vector $\mathbf{x} = \{d_1, d_7, u_1, u_2\}$

$$F = \sum_{(i,j)\in\mathcal{N}} [(U_{ij} - X_{ij})^2 + \beta(S_1 + S_2)/2] \tag{2}$$

where $S_1 = \sum_{(m,n)\in V_{ij}\cap\mathcal{N}^c} 2\psi(U_{ij} - X_{mn})$, $S_2 = \sum_{(m,n)\in V_{ij}\cap\mathcal{N}} \psi(U_{ij} - X_{mn})$, $\psi(\cdot)$ is a edge-preserving potential function, and β is a constant factor.

Remark 1. Since the regularization filtering approach requires solving a complicated optimization problem, we can also propose a simple alternative approach using the median value. That is to say, we can set $U_{ij} = Y_{ij}^1$. This will reduce the computation cost dramatically.

3 Experimental Results

Extensive experiments have been conducted on a variety of test images. The peak signal-to-noise ratio(PSNR) criterion is used to measure the restoration performance quantitatively.

First, the training data are obtained by the noisy 64 × 64 *Lenna* picture. We corrupted the *Lenna* by superimposing impulsive noise with probability 20% (Case I) or 30% (Case II) and select all of the pixel samples to form the training set, from which we calculate the feature vectors and train the SVMs. The paramters of interest in tuning an RBF kernel are γ and C, where C is the parameter used to penalize training errors during SVM estimation.

To see the influence of the parameters $(C; \gamma)$ on the filtering performance, we include some results for a wide range of parameter values in Table 1. Case I refers to using the filter whose SVM detector is trained by using the 20% corrupted *Lenna* as the training set; and Case II refers to using the filter whose SVM detector is trained by using the 30% corrupted *Lenna* as the training set. The two columns which are labelled as "nSV" are the numbers of the support vectors. The filters which are based on the trained SVM with different C and γ are utilized to restore the picture *Cameraman*, with 10%, 20% and 30% corrupted noises. The obtained PSNR are listed in Table 1 (see the 4-10*th* columns). All of the training process has been completed by using the software LIBSVM, which is available at `http://www.csie.ntu.edu.tw/~cjlin/libsvm`.

From the comparison results shown in Table 1, we observe that the SVM-based filters with different $(C; \gamma)$ always work well and the difference between the filtering performance of them is very little. For the purpose of comparison with

Table 1. PSNR for corrupted Cameraman

Parameters		Case I				Case II			
C	γ	nSV	10%	20%	30%	nSV	10%	20%	30%
1	0.0001	169	33.5536	27.8213	23.7721	312	33.5233	27.8282	23.7666
1	0.001	276	33.5504	27.9168	23.8285	381	33.5417	27.8395	23.7923
10	0.0001	128	33.5640	27.9054	23.8119	262	33.4107	27.7935	23.7257
10	0.001	239	33.3233	27.7754	23.7739	352	32.4508	27.4377	23.6451
100	0.0001	117	33.4674	27.8755	23.8148	245	32.5929	27.4627	23.6234
100	0.001	222	33.2990	27.7173	23.7458	322	32.2551	27.3339	23.5341
1000	0.0001	115	33.1123	27.7097	23.7653	229	31.3959	26.9449	23.3186
1000	0.001	192	33.1512	27.6591	23.6518	329	32.6405	27.3533	23.5642

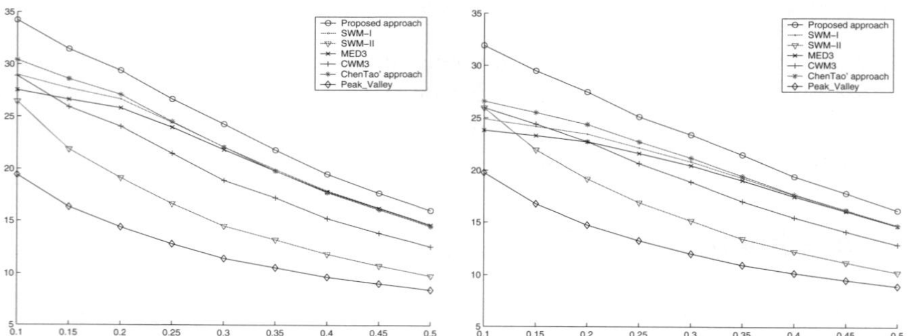

Fig. 2. LEFT: Airplane; RIGHT: Barbara

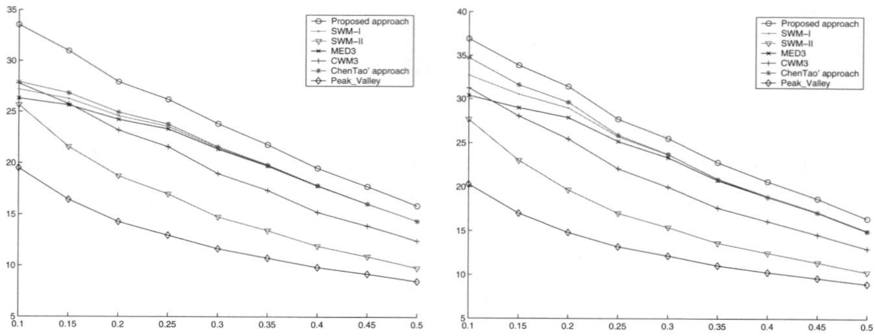

Fig. 3. LEFT: Cameraman; RIGHT: Lenna

other filters, we choose $C = 1$ and $\gamma = 0.001$ and also use the 64×64 *Lenna* which is corrupted by 20% noises as the training samples. Further, we have adopted the standard 3×3(MED3), the 3×3 weighted-center median filter whose weight is 3(CWM3). The 3×3 adaptive filter proposed by Chen et al [2](ChenTao), the

Fig. 4. LEFT: Original; RIGHT: Corrupted by 20% noises

Fig. 5. LEFT: Median filter; RIGHT: SWM-I filter

type I switching median (SWM-I) and the type II switching median (SWM-II)[7] have also been experimented.

Figs.2-3 give the PSNR curves under different noise densities. The proposed filter significantly outperforms other median filtering schemes. Subjective visual comparison using the image *Cameraman* are presented in Figs.4-6, which show that more details can be reserved using the proposed filter. Due to the space limitation, more experimental results cannot be included in this paper.

Remark 2. During the preparation of the final version of this paper, we found a related work in [5], which was published very recently. After a careful comparison, we point out there exist obvious differences between [5] and this paper. For example, though the features selected by [5] and by us are similar, the combination ways of them are rather different(we think using u_1 and u_2 separately will be more efficient than using their sum, and the feature d_7 was not considered

Fig. 6. LEFT: ChenTao's approach; RIGHT: Proposed approach

in [5]). In addition, in [5], a two-step approach is proposed, while in this paper, only one-step is needed. More differences lie in the experimental section.

Acknowledgements

This work was jointly supported by the National Key Project for Basic Research of China (Grant No: G2002cb312205), the National Science Foundation of China (Grants No: 60504003, 60474025, 60321002, 60334020), and the Basic Research Foundation of Tsinghua University (Grant No: JC2003028).

References

1. Abreu, E., Mitra, S.K.: A New Efficient Approach for Removal of Impulse Noise from Highly Corrupted Images, IEEE Trans. on Image Processing **5** (1996) 1012–1025
2. Chen, T., Wu, H.R.: Adaptive Impulse Detection Using Center-weighted Median Filters, IEEE Signal Processing Letters **8** (2001) 1–3
3. Eng, H.L., Ma, K.K.: Noise Adaptive Soft-swithing Median Filter, IEEE Trans. on Image Processing **10** (2001) 242–251
4. Lin, T.C., Yu, P.T.: Partition Fuzzy Median Filter Based on Fuzzy Rules for Image Restoration, Fuzzy Sets and Systems **147** (2004) 75–97
5. Lin, T.C., Yu, P.T.: Adaptive Two-pass Median Filter Based on Support Vector Machines for Image Restoration, Neural Computation **16** (2004) 333–354
6. Chan, R. H., Ho C. W., Nikolova, M., Salt-and-pepper Noise Removal by Median-type Noise Detectors and Detail-preserving Regularization, IEEE Trans. on Image Processing **14** (2005) 1479–1485
7. Sun, T., Neuvo, Y.: Detail-preserving Median Filters Based Filters in Image Processing, Pattern Recognition Letters **15** (1994) 341–347
8. Zhang, S., Karim, M.A.: A New Impulse Detector for Switching Median Filters, IEEE Signal Processing Letters **9** (2002) 360–363

The Application of Wavelet Neural Network with Orthonormal Bases in Digital Image Denoising

Deng-Chao Feng[1,2], Zhao-Xuan Yang[1], and Xiao-Jun Qiao[2]

[1] Institute of Electronic & Information Engineering, Tianjin University, Tianjin, 300072 P.R. China
[2] National Engineering Research Center for Information Technology in Agriculture, Beijing,100089 P.R. China
tyfdc001@163.com

Abstract. The resource of image noise is analysized in this paper. Considering the image fuzzy generated in the process of image denoising in spatial field, the image denoising method based on wavelet neural network with orthonormal bases is elaborated. The denoising principle and construction method of orthonormal wavelet network is described. In the simulation experiment, median filtering adaptive median filtering and sym wavelet neural network with orthonormal bases were used separately in the denoising for contaminated images. The experiment shows that, compared with traditional denoising method, image denoising method based on orthonormal wavelet neural network improves greatly the image quality and decreases the image ambiguity.

1 Introduction

Due to the limitation of devices and transmission channel being used, noise can be imported into an image in the acquiring、 transforming and transmitting process. Image without being processed is often disturbed by noise, which reduces the image quality and even shades the image characteristic. Therefore, image denoising has always been one of the important research topics in image processing.

Image denoising is to eliminate the noise and reconstruct the original image. Essentially, the characteristic of LP filter is used in all cases of noise elimination. While eliminating image noise, LP filter can eliminate some useful HP signals too. Traditional spatial field denoising methods such as average filtering and median filtering, consider different characteristics of different parts of the image as the same. This is not in accordance with original image, because while noise being eliminated, the edge and detail ambiguity of image is produced. Because of the great development and extensive application of wavelet theory, the image denoising algorithm based on wavelet is effectively implemented. [1] As a kind of noise-decorrelating transform, wavelet transform may, to a great extent, eliminate the correlation between image pixels. Wavelet neural network is the combination of wavelet analysis and neural network. It directly takes wavelet function or scaling function as the stimulation function of neural cell, and generates a network with similar structure with radial bases function. In traditional design of neural network, rational direction is a lack in determining the nodes number of hidden layer neural cell and initializing various parameters. However, wavelet neural network combines better time-frequency

J. Wang et al. (Eds.): ISNN 2006, LNCS 3972, pp. 539 – 544, 2006.

characteristic of wavelet and the adaptation of neural network. Therefore, with the development of wavelet, wavelet neural network is gradually applied in image denoising field.

2 Noise Model and Denoising Principle

2.1 Image Signal Model with Noise

A 2D signal model with noise may be described as

$$s(\mathrm{i,j})=f(\mathrm{i,j})+\delta e(\mathrm{i,j}),\ i,j=0,\ldots,m\text{-}1 \tag{1}$$

where f (i,j) is the origin signal, e(i,j) is the noise, δ is the noise intensity, and s(i,j) is the signal with noise.

In this paper, the neural network structure is optimized by wavelet multiresolution analysis theory and the adaptive image denoising is accomplished. According to the multiresolution analysis theory of wavelet, the information changing slightly corresponds to the LP part of signal, and that changing quickly responds to the HP part. According to the two scale recursive equation, the between scaling function and wavelet function can be expressed as Eq.(2) and Eq.(3)

$$\varphi(t)=\sqrt{2}\sum_{n=0}^{N}h_n\varphi(2t-n) \tag{2}$$

$$\psi(t)=\sqrt{2}\sum_{n\in z}g_n\varphi(2t-n) \tag{3}$$

Now suppose the 2D image is f (x,y) with the size of N*M. Then, the DWT (discrete wavelet transform) can be described as follows

$$\tilde{f}_{\varphi}(j_{,}m,n)=\frac{1}{\sqrt{MN}}\sum_{x=0}^{M-1}\sum_{y=0}^{N-1}f(x,y)\varphi_{j_0m,n}(x,y) \tag{4}$$

$$\tilde{f}_{\varphi}(j_{,}m,n)=\frac{1}{\sqrt{MN}}\sum_{x=0}^{M-1}\sum_{y=0}^{N-1}f(x,y)\varphi_{j,m,n}(x,y) \tag{5}$$

where j_o and j are resolution parameters and j_o is the initial scaling. $\tilde{f}_{\varphi}(j_{0,}m,n)$, which corresponds to LP, is the approximation of f (i,j) on scaling j_o. $\tilde{f}_{\varphi}(j_{,}m,n)$ corresponds to the HP part when $j \geqq j_o$, and it includes details in horizontal, vertical and diagonal directions. The wavelet coefficient obtained by analyzed wavelet may reserve the coefficient higher than threshold by threshold function. The denoised image can be obtained through wavelet construction.

2.2 Multiresolution Learning Algorithm

When selecting the original parameter of the wavelet network, multiresolution theory can be applied. Multiresolution is a series sets (V_m) which approximate $L^2(R)$ [2], that is,$0\subset V_1\subset\ldots V_m\subset V_{m+1}\subset\ldots\subset\infty$, and $\lim_{m=\infty} V_m=\{0\}$. Each V_m is composed of linear weighted series of $\Phi=\{\varphi_{m,n}(x),n\in Z\}$. Accordingly, the wavelet space is the difference between two neighboring spaces. It is known that $V_0=V_1\oplus W_1=V_2\oplus W_2\oplus W_1=\cdots$ from the definition of the multiresolution. The nesting of scales is equivalent to two-scale equation of Eq.(2) and Eq.(3),and $g_n=(-1)^n \bar{h}_{1-n}$ .The difference between two neighboring scale spaces is the detailed feature of f (x) in wavelet space. The multiresolution theory is the theoretical base of learning of wavelet network.

3 Wavelet Neural Network Denoising Method Based on Multiresolution Learning

In wavelet neural network, hidden layer function of general neural network is replaced with orthonormal wavelet function. The weight and threshold from corresponding input layer to hidden layer is replaced with the scaling and translation parameters separately.

The existing of compactly supported orthonormal wavelet and the study of its structure by Daubechies and Mallat multiresolution theory greatly ensure the feasibility of orthonormal bases wavelet network by multiresolution analysis. The node of hidden layer of network is composed of wavelet function node ψ and scaling function node φ.

In wavelet network composed of discrete wavelet function, to approximate f (x) with resolution m can described as

$$f_m(x)=C_m f(x)=c_{m,n}\varphi_{m,n}(x) \tag{6}$$

where f (x) $\in L^2(R)$, C_m is the orthonormal projection operator in spatial V_m, and $c_{m,n}$ is the weight of scaling cell of corresponding network which is the projection of scaling function $\varphi_{m,n}$(x).Owing to the multiresolution theory,W_m denotes orthogonal complement space of V_m on V_{m-1}, that is, $V_{m-1}=V_m\oplus W_m$, $V_m\perp W_m$, where D_m is the orthonormal projection operator of W_m. The approximation of f (x) at m-1 scale is

$$C_{m-1}f_{m-1}=[C_m\oplus D_m]f(x)=C_m f(x)\oplus D_m f(x)$$
$$D_m f(x)=\sum_{n=-\infty}^{+\infty} d_{m,n}\psi_{m,n}(x) \tag{7}$$

where $d_{m,n}$ is the projection of wavelet function $\psi_{m,n}$(x) of corresponding network, which is called the detail on this resolution.

The approximation of f (x) on resolution m-1 will be given by

$$f_{m-1}(x)=f_m(x)+\sum_{n=-\infty}^{+\infty} d_{m,n}\psi_{m,n}(x)=\sum_{n=-\infty}^{+\infty} c_{m,n}\varphi_{m,n}(x)+\sum_{n=-\infty}^{+\infty} d_{m,n}\psi_{m,n}(x) \tag{8}$$

When the wavelet network to be constructed is single hidden layer, the weight from input layer to hidden layer is fixed to be 1. Scaling function $\varphi_{m,n}$(x) in hidden layer constructs the approximation to signal on the most rough resolution and wavelet function $\psi_{m,n}$(x) constructs the gradual detail approximation to signal.

Being different of Fourier Transform, mother-wavelet is not unique. The selection of mother-wavelet is very important. Selection criterion mainly is compactly support, symmetry, vanishing moments and regularity. Length of compactly support usually is [5,9], and when it is much long, boundary problem occurs, and when too short, energy cannot be greatly converged. Symmetry in image processing is to avoid phase shift. The energy of wavelet transform with large vanishing moments is easily converged. Wavelet reconstructing process with good regularity is more smooth.

The orthonormal bases of wavelets have self-similarity and compact support, the wavelet coefficient has strong decorrelation and its decomposed coefficient range under each dimension is small. And the energy of high frequence coefficient is greatly converged. Therefore, the orthonormal bases of wavelets is widely applied in image denoising. In orthonormal bases of wavelets, the sym wavelet has more better symmetry than db wavelet, which can reserve the edge part of image in denoising process and reduce the phase shift of reconstruction. This paper took sym4 as the hidden layer of neural network.

From Reference [4], we can know that parameter of method of least square with random noise often can be drifted, and consequently unsteady phenomena occurs. On the bases of target function of RLS (recursive least square), damped least squares increases the damp to parameter variables and becomes a more steady distinguishing method. In the denoising processing network training, hidden layer node of wavelet network uses wavelet function. As wavelet network is initialized, parameter is near the optimized value. Damped least squares is used to do network training [3]. The recursive process is as follows:

$$W(t) = W(t-1) - \frac{H(t-1)\nabla^{t}\psi(t)\psi(t)}{r_0 + \nabla^{T}\psi(t)H(t-1)\nabla\psi(t)} + \mu r_0 H(t-1)[W(t-1) - W(t-2)] \tag{9}$$

$$H(t) = \frac{1}{r_o}\left[H(t-1) - \frac{H(t-1)\nabla^{t}\psi(t)\psi(t)}{r_0 + \nabla^{T}\psi(t)H(t-1)\nabla\psi(t)}\right]$$

$$\psi(t) = y(t) - \hat{y}(t), \quad \nabla\psi(t) = \frac{\partial\psi(t)}{\partial W}, \quad H(0) = \alpha^2 I \tag{10}$$

where r_o is the forgetting factor with range of [0.9,1], μ is damping factor with range of $[10^{-4},10^{-6}]$, W(t) is weight vector and a is a larger value.

In wavelet neural network, only the weight of the output layer needs to be trained.[4] The weight of input layer can be obtained only by the relation between intervals of sampling points and compactly supported fields, and then no training is required. Therefore, the training speed of wavelet neural network is greatly increased.

Training steps are as follows:

Select appropriately wavelet scaling function with generalized orthonormal bases.

(1) Construct a multi-ratio coefficient grid for each dimension of input.
(2) Train φ cell, when m=L.
(3) Add appropriate ψ cell when the error can not meet the demand,
(4) Delete ψ cell with smaller weight and test the network with new data.

4 Experiment and Conclusion

The above-mentioned denoising method is to be proved under the circumstance of MATLAB. Standard 512*512 image with gray-scale of 256, is adopted and tested.

The denoising performance is measured by PSNR (peak signal noise ratio)

$$R_{mse} = \sqrt{\frac{1}{N}\sum_{k}(s_k - d_k)^2}, \; P_{SNR} = 20\log_{10}\left(\frac{256}{Rmse}\right) \tag{11}$$

where s is the origin image, d is denoised image.

To compare perfomances of various algorithms, wavelet transform adopts the orthonormal base of compactly supported wavelet. Considering that sym wavelet has better symmetry than Daubechies wavelet, sym4 wavelet is adopted and the analyzed level is 3. In Fig.1, standard Gaussian white noise with σ_n =0.2 is added to the image. PSF is the 7*7 Gaussian point spread function when δ=20. Experimental results of experiments done with median filtering, adaptive median filtering and wavelet neural network separately are shown in Fig.1,

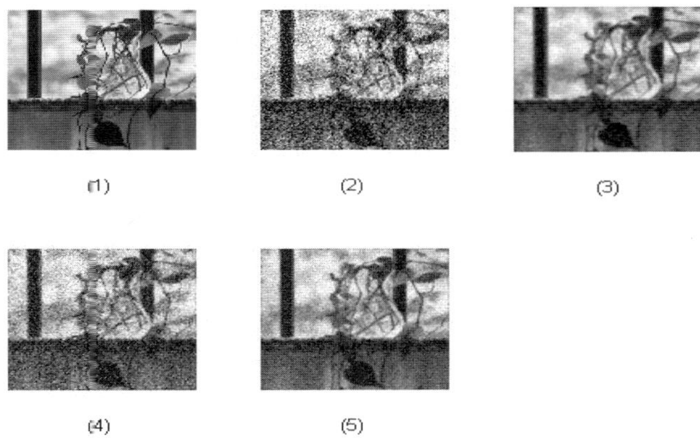

Fig. 1. Comparation of Image Processing Results

where *(1)*Original image; *(2)*Corrupted image,noise mean square errors equal to 0.2; *(3)*Reconstruction image by the median filter; *(4)*Reconstruction image by the adaptive median filter; *(5)*Reconstruction image by WNN. Corresponding histograms of images are showed in Fig.2.

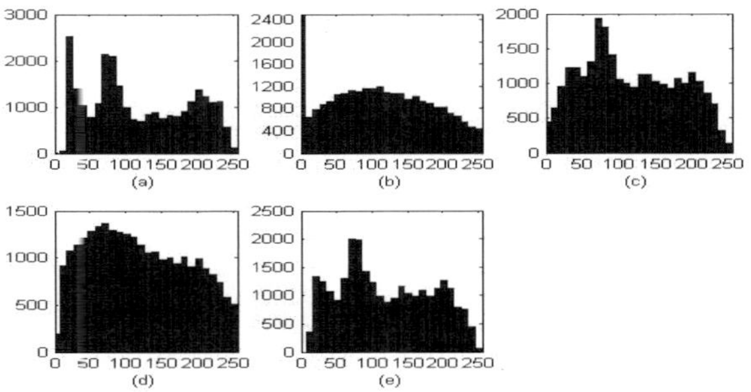

Fig. 2. Corresponding Histograms of Images

Image fuzzy exists through median and adaptive median filtering from the visual result. But the detailed feature is reserved when the noise is well removed through the approch of wavelet network.Tab.1 shows PSNR values of denoised images, which has been filtered by median filter, adaptive median filter and WNN(wavelet neural network).

Table 1. PSNR values of denoised images

Image size	PSF (gaussian)	σ_n	PSNR		
			Median filter	Adaptive median filter	Orthogonal bases WNN
Flower.JPG (512×512)	7×7, δ =10	0.1	46.5126	44.7434	47.8721
		0.2	45.7723	43.5748	46.0324
		0.3	45.5168	42.6588	45.8125
	7×7, δ =20	0.1	46.5077	44.0533	47.8947
		0.2	45.7747	43.5998	45.9146
		0.3	45.0880	42.3184	45.8763
	7×7, δ =30	0.1	46.5508	44.6006	46.8126
		0.2	45.6314	42.8280	45.8639
		0.3	44.9515	42.3640	44.9411

Simulation experiment results show that wavelet neural network with orthonormal bases makes a great improvement of the image quality. However, the selection of optimal wavelet bases function needs further research.

References

1. Zhang, H., Wu, B.: Research and Prospects of Wavelet Neural Networks. Journal of Southwest China Institute of Technology 17(1) (2002) 10–12
2. Cheng, Z.X.: Wavelet Analysis Algorithm and Application. 1st edn. Xi'an Jiaotong University Publishing House (2003)
3. Lin, M., Chen, Z., He, J., Yuan, Z.: The Neural Network Self-tuning One-step Predicitive Controller Based on Damped Least Square. Control and Decision 14(2) (1999) 69–72
4. Yan, P.F., Zhang, C.S.: Artificial Neural Networks and Evolutionary Computing. 2nd edn. Tsinghua University Publishing House (2005)

A Region-Based Image Enhancement Algorithm with the Grossberg Network

Bo Mi, Pengcheng Wei, and Yong Chen

Department of Computer Science and Engineering,
Chongqing University Chongqing, 400044, P.R. China
mi_bo@163.com

Abstract. In order to enhance the contrast of an image, histogram equalization is wildly used. With global histogram equalization (GHE), the image is enhanced as a whole, and this may induce some areas to be overenhanced or blurred. Although local histogram equalization (LHE) acts adaptively to overcome this problem, it brings noise and artifacts to image. In this paper, a region-based enhancement algorithm is proposed, in which Grossberg network is employed to generate histogram and extract regions. Simulation results show that the image is obviously improved with the advantage of both GHE and LHE.

1 Introduction

The purpose of image enhancement is to increase the visibility of an image. As we know, if an image is too dark or too light, the histogram always concentrates on a low or high band. And the details of the image are blurred. To effectively improve the image and enhance its contrast, one way is to equally distribute the grey levels to the whole range. By far, many histogram equalization methods have been proposed to improve image quality [1-3]. We classify them into two categories: global histogram equalization (GHE) and local histogram equalization (LHE or AHE).

Global histogram equalization is based on the statistical information of the whole image. Let $\Omega \subseteq IR^2$ be the image domain and $u:\Omega \to [a,b]$ be the given (low contrast) image. Let $f:[a,b] \to [c,d]$ be a given increasing function. Then $v := f(u)$ represents the stretched image. In order to distribute the grey levels equally to a new range from $[a,b]$, the probability density of each level must be obtained. Here, we set $a \le u \le b$, $h(u)$ is the normalized histogram (i.e. probability density) of *u*, then we can construct the transformation function *f* as follows:

$$H(u) = \sum h(\lambda), a \le \lambda \le u, \tag{1}$$

where $H(u)$ is the distribution density of *u*, and $0 \le H(u) \le 1$. Then the change of variables

$$f(u) = (d-c)H(u) + c. \tag{2}$$

Since $H(u)$ is a monotonic increasing function, the monotonicity of $f(u)$ can be guaranteed. This is an important property for global histogram equalization that

J. Wang et al. (Eds.): ISNN 2006, LNCS 3972, pp. 545 – 550, 2006.

guarantees the coherence of grey level order before and after transformation [2]. So the appearance of the enhanced image keeps similar to the original one. However, for global methods, it dose not consider the contents of an image except its grey level distribution. One transformation is applied to all the pixels of the input image so that some regions are over enhanced while some details are ignored.

To overcome such defect, many local methods have been proposed which are based on image features. Where a small-scale component is taken into account that an input-to-output mapping can be expressed as

$$I_O(x, y) = f[I(x, y), O_{\Omega'}(x, y)] \tag{3}$$

and $O_{\Omega'}(x, y)$ represents some local features within the neighborhood Ω' around the location (x, y).

In terms of the formula presented above, it is obvious that the transformation function changes adaptively according to the local contents. In fact, for each pixel, a local histogram of its neighborhood (i.e. Ω') is calculated. Then the transformation function is obtained in the same way as GHE. However, the computational burden is too high as the transformation for each pixel is performed independently. So Pizer *et al.* proposed interpolated AHE where the image is divided into fixed blocks and the histogram equalization mapping is calculated only for each block. Then the transformation function is assigned to pixels using interpolation algorithm [4]. Another disadvantage of LHE is the edge artifacts at sharp boundaries. It's mainly because the transformation function changes abruptly when the histograms of neighbor blocks are quite distinct. To solve this problem, Hui Zhu *et al.* suggested that the local transformation should be spatially continuous. They proposed a constrained local histogram equalization (CLHE) method where they applied a smoothness condition to constrain the local transformation [5]. Meanwhile, some other characteristic algorithms are also commendably performed [6-9].

Nevertheless, all the LHE methods mentioned above dose not adequately consider the contents of the whole image. So there are some defects can't be ignored: 1) The same grey level in the original image may be different and the grey order may even reverse. In this way, the enhanced image may appear quite different from the original one. 2) Owing to different background at different locations, pixels in similar objects may be transformed with distinct functions, so some unacceptable artificial effects emerge. 3) The transformation function of each pixel is determined by its location and the histogram of the sliding window (or fixed blocks). As the transformations formulae are obtained independently, the algorithms are time-consuming.

In order to effectively utilizing the features of the whole image for histogram equalization, we proposed a new HE method based on the relationship between image contents and its histogram. As we know, an image is always made up of a lot of regions that represent different objects. For a poor image, contents inside the regions are fogged in contrast to distinct boundaries. So if we enhance these regions separately, the image could be commendably enhanced while features are preserved.

2 Region-Based Image Enhancement Algorithm

Our method is inspired by Human visual property. It is known that two types of receptors on retina generate the vision: rods and cones [2]. Rods form basic information units in perceiving a visual image. They respond to edges patterns and acts at scotopic levels. When eyeballs are forced to focus on some areas by thews, scenes are projected onto retina center called centralfovea. Most cones lie on centralfovea and connect to teleneurons of their own. So the cones are sensitive to colors and details. Information obtained by the receptors is then integrated by a higher level of the central nervous system thus the image is perceived with both profiles and details. This motivates our model for equalizing histogram regionally. Here, we assume that one region is corresponding to one object. This assumption is in conformity with human visual habit, so it is reasonable. For each region Ω' we implement an independent transformation function that

$$f_{\Omega'}(u) = (L_{\max} - L_{\min})H_{\Omega'}(u) + L_{\min}, \tag{4}$$

where $L_{\min}$ and $L_{\max}$ represent the lowest and highest grey levels of Ω', which can also be rescaled to occupy a larger range. Note that the distribution density $H_{\Omega'}(u)$ is calculated only in region Ω'.

According to the formula (4), the first problem we face is how to determine different regions. Fortunately, we noticed that the histograms of the same region are similar with each other for most, especially medical, images. Figure 1 is a chest X-ray image, the histograms of chosen blocks illustrate that they are regionally dependent. So we divide the original image into several blocks as traditional methods, and then classify them into different categories according to the similarity of their histograms.

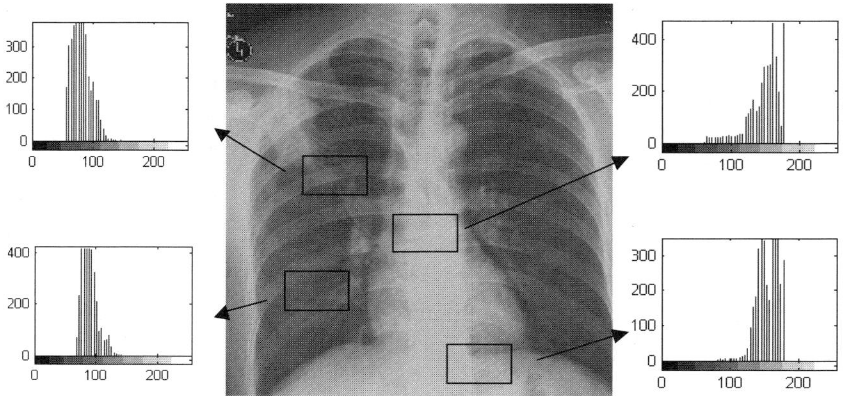

Fig. 1. A chest X-ray image

In order to realize such an algorithm, an artificial neural network called Grossberg network is employed [10]. Figure 2 illustrates a framework of Grossberg network and its property. First of all, a poor contrast image of size *N*×*N* is divided into blocks of size *n*×*n*. For each block, we count pixel quantities of different grey levels and arrange them in a vector form $p = (p_1 \quad p_2 \quad \cdots \quad p_M)^T$. We use p as the input of network and feed it into layer-1. The output of the layer-1 varies along with the time that

$$\varepsilon \frac{dn_i^1(t)}{dt} = -n_i^1(t) + ({}^{+}b_i^1 - n_i^1(t))p_i - ({}^{-}b_i^1 + n_i^1(t))\sum_{j \neq i} p_j, \tag{5}$$

where ${}^{+}b_i^1$ and ${}^{-}b_i^1$ represent the biases of nerve cells and determine the limits of output. Convergence speed is controlled by ε. Let ${}^{+}b_i^1 = 1$, ${}^{-}b_i^1 = 0$ and in stable states we have

$$n_i^1 = \frac{p_i}{1+P}, \text{ where } P = \sum_{j=1}^{M} p_j \tag{6}$$

If $P >> 1$, the normalized pixel quantity can be given by n_i^1 so the histogram of a block is obtained.

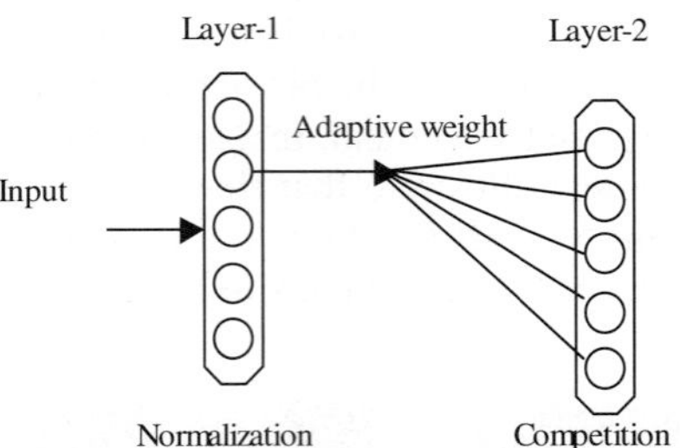

Fig. 2. Grossberg network

We use layer-2 to classify blocks into different categories. The operational equation is given by

$$\varepsilon \frac{dn^2(t)}{dt} = -n^2(t) + ({}^{+}b^2 - n^2(t))\left\{[{}^{+}W^2]f^2(n^2(t)) + W^2a^1\right\} - (n^2(t) + {}^{-}b^2)[{}^{-}W^2]f^2(n^2(t)), \tag{7}$$

where W^2 is adaptive weight, and a^1 represents the output of layer-1 which is equal to n^1. When $n^2(t)$ converged, the maximal inner product of prototypes (i.e. rows of matrix W^2) and a^1 would win the competition with supreme output $f_j^2(n_j^2(t))$. This also means the block is felicitously categorized according to the similarity between its histogram and some prototypes. In order to determine the adaptive weight properly, a learning rule of Grossberg network is employed as

$$\frac{dw_{i,j}^2(t)}{dt} = \alpha\left\{-w_{i,j}^2(t) + n_i^2(t)n_j^1(t)\right\}. \tag{8}$$

By far, all blocks can be accurately categorized. For each category, we construct an independent transformation function. Note that the histograms of blocks have already been worked out in layer-1 of Grossberg network. So we simply average them for each region and implement formula (4) to enhance the original image.

3 Experimental Results

In this section, the proposed method is applied to the chest X-ray image of Figure 1.

To illustrate the improvement of our method, the results of GHE and LHE are also demonstrated. The Grossberg network is simulated on IBM PC using MATLAB. Figure 3a is the global histogram equalization result. In figure 3a, it is obvious that some smooth areas are overenhanced while some details are even suppressed. The result of local histogram equalization method is shown in figure 3b. Pizer *et al.* consider LHE to be well-performed for medical image [4]. Comparing figure 3b with 3a, the spine and thorax can be seen much more clearly in LHE method so it is helpful for medical purpose. Nevertheless, because LHE enhances contrast locally, it brings in much noise and some distortions. This may lead to misdiagnosis, especially for inchoate pathological changes. Figure 3c and 3d are the results of the proposed method. The difference between 3c and 3d is that they employed six and ten nerve cells in layer-2 respectively. As a whole, the global appearance of original image is commendably preserved. Meanwhile, because the image is divided into dissimilar regions and enhanced severally, all grey levels have been adequately utilized and thus more details are visible. From figure 3c and 3d, it can be seen that the image is clearer than 3a and is cleaner than 3b as the trade-off between enhancement and preservation is well handled.

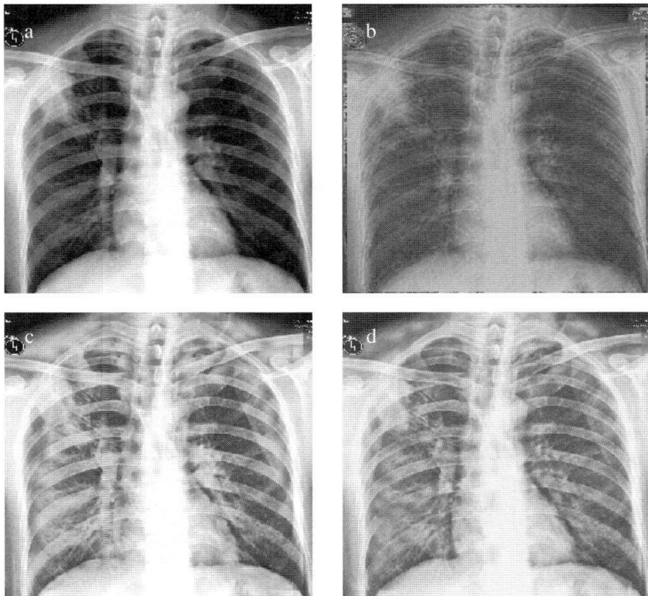

Fig. 3. (a) GHE result (b) LHE result (c) The proposed method (6 nerve cells in layer-2) (d) The proposed method (10 nerve cells in layer-2)

4 Conclusions

We have proposed a new histogram equalization method based on regions. This algorithm is inspired by the mechanism of visual perception on retina. Since Grossberg network is employed, the regions are extracted adaptively and the computational burden declines. With our method, the appearance of original image is preserved when enhancing details. Experiment results also illustrated that it achieves significant improvements in the quality of enhanced image.

References

1. Rosenfeld, A., Kak, A. C.: Digital Picture Processing, Vol. 1., Academic Press, San Diego (1982)
2. Gonzalez, R. C.: Woods, R. E., Digital Image Processing, Addison-Wesley, Reading, MA (1992)
3. Russ, J. C.: The Image Processing Handbook, 2nd ed., CRC Press, Boca Raton (1995)
4. Pizer, S. M., Amburn, E. P., Austin, J. D., Cromartie, R., Geselowitz, A., Greer, T., Romeny, B. H., Zimmerman, J. B., Zuiderveld, K.: Adaptive Histogram Equalization and Its Variations. Comput. Vision Graphics Image Process 39 (1987) 355–368
5. Zhu, H., Chan, F. H. Y., Lam, F. K.: Image Contrast Enhancement by Constrained Local Histogram Equalization. Computer Vision and Image Understanding 73 (1999) 281-290
6. Paranjape, R. P., Morrow, W. M., Rangayyan, R. M.: Adaptive-neighborhood Histogram Equalization for Image Enhancement. CVGIP: Graphical Models Image Process 54 (1992) 259–267
7. Rehm, K., Dallas, W. J.: Artifact Suppression in Digital Chest Radiographs Enhanced with Adaptive Histogram Equalization. Proc. SPIE 1092 (1989) 220–230
8. Cromartie, R., Pizer, S. M.: Structure-sensitive Adaptive Contrast Enhancement Methods and Their Evaluation. Image and Vision Comput (1993) 385
9. Rosenman, J., Roe, C. A., Muller, K. E., Pizer, S. M.: Portal film Enhancement: Technique and Clinical Utility. Int. J. Radiat. Oncol. Biol. Phys, 25 (1993) 333–338
10. Hagan, M., Demuth, H., Beale, M.: Neural Network Design, PWS Publishing, Boston (1996)

Contrast Enhancement for Image Based on Wavelet Neural Network and Stationary Wavelet Transform

Changjiang Zhang, Xiaodong Wang, and Haoran Zhang

College of Information Science and Engineering, Zhejiang Normal University, Jinhua 321004, China
{zcj74922, wxd, hylt}@zjnu.cn

Abstract. After performing discrete stationary wavelet transform (DSWT) to an image, local contrast is enhanced with non-linear operator in the high frequency sub-bands, which are at coarser resolution levels. In order to enhance global contrast for an infrared image, low frequency sub-band image is also enhanced employing non-incomplete Beta transform (IBT), simulated annealing algorithm (SA) and wavelet neural network (WNN). IBT is used to obtain non-linear gray transform curve. Transform parameters are determined by SA so as to obtain optimal non-linear gray transform parameters. Contrast type of original image is determined by a new criterion. Gray transform parameters space is determined respectively according to different contrast types. A kind of WNN is proposed to approximate the IBT in the whole low frequency sub-band image. The quality of enhanced image is evaluated by a total cost criterion. Experimental results show that the new algorithm can improve greatly the global and local contrast for images.

1 Introduction

A few algorithms are used widely to enhance contrast for images, such as histogram equalization, histogram matching, gray level transform and un-sharp mask algorithm. The common disadvantage of above algorithms is that the noise in the image is magnified while the contrast is enhanced. Recently, some new algorithms for image enhancement have been proposed. Shang-ming Zhou gave a kind of algorithm for contrast enhancement based on fuzzy operator [1]. However, the algorithm cannot be sure to be convergent. Ming Tang gave a kind of adaptive enhancement algorithm far infrared image sequences [2]. Performance of the algorithm is affected greatly by mathematic model. Thus improved histogram equalization algorithm was proposed to enhance contrast for kinds of image [3], however, the visual quality cannot be improved greatly with above algorithms. In addition, Tubbs gave a simple gray transform algorithm to enhance contrast for images [4]. However, the computation burden of the algorithm was large. Although many enhancing algorithms have been proposed, they still have some problems. For example, High image quality is demanded [5].

A new algorithm is proposed to solve above problems. The image is decomposed employing the DSWT [6]. Non-linear operator at the coarser resolution levels

J. Wang et al. (Eds.): ISNN 2006, LNCS 3972, pp. 551 – 556, 2006.

implements local contrast enhancement. In order to enhance the global contrast for an image, the low frequency sub-band image is also enhanced employing IBT, SA and WNN. Contrast type for original image is determined employing a new criterion. Contrast for original images are classified into seven types: particular dark (PD), medium dark (MD), medium dark slightly (MDS), medium bright slightly (MBS), medium bright (MB), particular bright (PB) and good gray level distribution (GGLD). IBT operator transforms original image to a new space. A certain criterion is used to optimize non-linear transform parameters. SA, which was given by William, is used to determine the optimal non-linear transform parameters [7]. In order to reduce the computation burden for calculating IBT, a new kind of WNN is proposed to approximate the IBT in the whole low frequency sub-band image.

2 Local Contrast Enhancement for Image

Based on DSWT, a kind of non-linear enhancement operator, which was proposed by A. Laine in 1994, is employed to enhance the local contrast for image [8]. Let $f_s^r[i,j]$ is the gray values of pixels in the **r***th* sub-band in the **s***th* decomposition level, where $s=1,2,\cdots,L$; $r=1,2,3$. $\max f_s^r$ is the maximum of gray value of all pixels in $f_s^r[i,j]$. Let $y_s^r[i,j]=f_s^r[i,j]/\max f_s^r$, $\mathrm{sigm}(x)=\dfrac{1}{1+e^{-x}}$, the contrast enhancement approach can be described by:

$$g_s^r[i,j]=\begin{cases} f_s^r[i,j], & \left|f_s^r[i,j]\right|<T_s^r \\ a\cdot\max f_s^r\{\mathrm{sigm}[c(y_s^r[i,j]-b)]-\mathrm{sigm}[-c(y_s^r[i,j]+b)]\}, & \left|f_s^r[i,j]\right|\geq T_s^r \end{cases} \tag{1}$$

Where T_s^r shows the enhancing threshold in the **r***th* sub-band in the **s***th* decomposition level, where $s=1,2,\cdots,L$; $r=1,2,3$.

3 Global Contrast Enhancement for Infrared Image

3.1 IBT

The incomplete Beta function can be written as following [4]:

$$F(u)=B^{-1}(\alpha,\beta)\times\int_0^u t^{\alpha-1}(1-t)^{\beta-1}dt,\quad 0<\alpha,\beta<10 \tag{2}$$

All the gray levels of original image have to be unitary before implementing IBT. All the gray levels of enhanced image have to be inverse-unitary after implementing IBT.

3.2 Objective Function for Image Contrast Enhancement

The objective function can be written as following [9]:

$$f = \frac{1}{MN}\sum_{i=1}^{M}\sum_{j=1}^{N} g'^2(i,j) - \left[\frac{1}{MN}\sum_{i=1}^{M}\sum_{j=1}^{N} g'(i,j)\right]^2 \tag{3}$$

where $g'(i,j)$ Shows gray level at (i,j) in enhanced image.

3.3 Contrast Classification for Image Based on Histogram

Contrast classification criterion can be described in Fig.1:

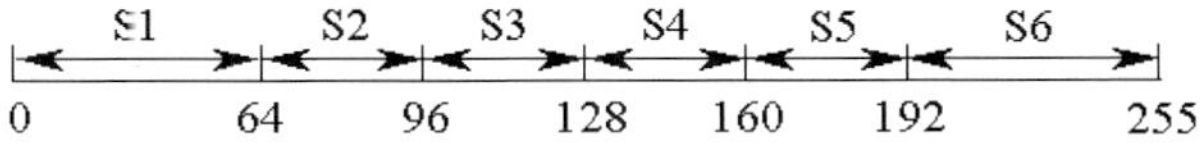

S1 S2 S3 S4 S5 S6
0 64 96 128 160 192 255

Fig. 1. Image classification sketch map based on gray level histogram

Given that original image has 256 gray levels (gray level ranges from 0 to 255), the whole gray level space is divided into six sub-spaces: S1, S2, S3 , S4 , S5 , S5。 Where Si（i=1, 2, ··· 6）is the number of all pixels which distribute in the ith sub-space. Let,

$S = \max_{i=1}^{6} S_i$, $S_1 = \sum_{k=2}^{6} \mathbf{Sk}$, $S_2 = \sum_{k=2}^{5} \mathbf{Sk}$, $S_3 = \sum_{k=1}^{5} \mathbf{Sk}$,
$S_4 = S1 + S6$, $S_5 = S2 + S3$, $S_6 = S4 + S5$,

Following classification criterion can be obtained:

$if\ S = S1\ \&\ S1 > S_1$
Image is PB;
$elseif\ S_2 > S_4\ \&\ S_5 > S_6\ \&\ S_5 > S1\ \&\ S_5 > S6\ \&\ S2 > S3$
Image is MD;
$elseif\ S_2 > S_4\ \&\ S_5 > S_6\ \&\ S_5 > S1\ \&\ S_5 > S6\ \&\ S2 < S3$
Image is MDS;
$elseif\ S_2 > S_4\ \&\ S_5 < S_6\ \&\ S1 < S_6\ \&\ S6 < S_6\ \&\ S4 > S5$
Image is MBS;
$elseif\ S_2 > S_4\ \&\ S_5 < S_6\ \&\ S1 < S_6\ \&\ S6 < S_6\ \&\ S4 < S5$
Image is MB;
$elseif\ S = S6\ \&\ S6 > S_3$
Image is PB;
$else$
Image is GGLD;
end

Where symbol & represents logic "and" operator.

3.4 Transform Parameters Optimization with SA

We will employ the SA to optimize transform parameters [7]. The range of α and β can be determined by Tab.1 so as to solve above problems.

Table 1. Range of α and β

Parameter	PD	MD	MDS	MBS	MB	PB
α	[0 , 2]	[0 , 2]	[0 , 2]	[1 , 3]	[1 , 4]	[7 , 9]
β	[7 , 9]	[1 , 4]	[1 , 3]	[0 , 2]	[0 , 2]	[0 , 2]

Let $\mathbf{x} = (\alpha, \beta)$, $F(\mathbf{x})$ is function to be minimized, corresponding to (3). Where $a_i < \alpha, \beta < b_i$ （$i = 1,2$ ）, a_i and b_i （$i = 1,2$ ）can be determined by Tab.1.

3.5 IBT Calculation with WNN

Let $f(x) \in L^2(R^n)$, WNN can be described approximately as follows:

$$Wf(x) = \sum_{i=1}^{N} w_i \psi[(a_i x - \tau_i)] \tag{4}$$

where τ_i is translation factor, a_i is scale factor, $Wf(x)$ shows the output of WNN. Parameters to be estimated are w_i , a_i , τ_i , $i = 1,2,\cdots,N$. "Forgetting factor" algorithm is used to train weight of WNN. Iterative prediction error algorithm is employed to train translation factors and scale factors. Weight, translation factors and scale factors are trained iteratively and mutually with above two algorithms [10]. Parameters α , β , g are input to trained WNN and output g' for IBT is obtained directly. Mexican hat wavelet is selected as mother wavelet:

$$\psi(x) = (1 - x^2) e^{-x^2/2} \tag{5}$$

4 Experimental Results

In experiments, three images are enhanced respectively. In order to prove the efficiency of the new algorithm, we will compare the performance between HE, USH, GWP [11] and WYQ [12]. Fig.2 shows IBT curve. The curves is employed to implement the global contrast enhancement to Fig.3 (a).

According to the experimental results, the clutter in the images is magnified greatly when USM is used to enhance the contrast for the images. This is obvious in Fig.3 (b). Local contrast is bad when HIS is used to enhance the contrast of the images. This is obvious in Fig.3 (c). The total contrast is good by GWP algorithm; however, it cannot reduce efficiently the clutter in the image. From Fig.3 (e), it is obvious that the clutter in the image is magnified greatly. According to Fig.3 (f), although WYQ algorithm can reduce the clutter well, the whole brightness of the image is too high so that some detail lost. Compared with the four algorithms, the new algorithm can reduce efficiently the clutter in the image while enhance the global and local contrast for the image well. According to Fig.3 (d), it is obvious that the new algorithm is more excellent in the total performance than USM, HIS, GWP and WYQ.

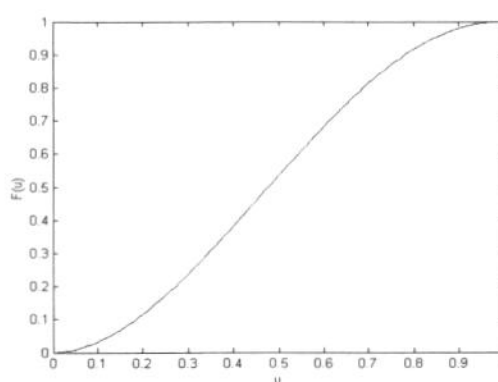

Fig. 2. IBT curves (α =2.1924, β =1.9542)

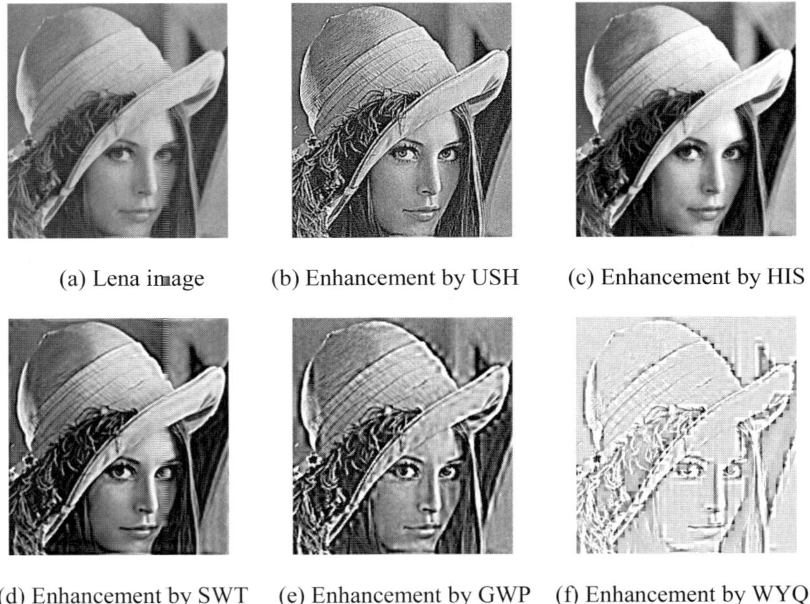

(a) Lena image (b) Enhancement by USH (c) Enhancement by HIS

(d) Enhancement by SWT (e) Enhancement by GWP (f) Enhancement by WYQ

Fig. 3. Enhancement results of Lena image

5 Conclusion

Experimental results show that the new algorithm can improve greatly the local and global contrast for the image while reduce well the clutter in the image. Employing the new algorithm, the detail in the image can be kept well. The visual quality of the new algorithm is more excellent than USM, HIS, GWP and WYQ.

Acknowledgements

This work is supported by Zhejiang Province Educational Office Foundation 20050292 and Zhejiang Normal University Foundation 20041076.

References

1. Zhou, S.M., Qiang G.: A New Fuzzy Relaxation Algorithm for Image Contrast Enhancement. Proceedings of the 3rd International Symposium on Image and Signal Processing and Analysis, Vol. 1024. Rome, Italy (2003) 11-16
2. Tang, M., Ma, S.D., Xiao, J.: Model-Based Adaptive Enhancement of Far Infrared Image Sequences. Pattern Recognition Letters 23(4) (2000) 827-835
3. Chen, S.D., Ramli, A.R.: Contrast Enhancement Using Recursive Mean-Separate Histogram Equalization for Scalable Brightness Preservation. IEEE Transactions on Consumer Electronics 49(4) (2003) 1301-1309
4. Tubbs, J.D.: A Note on Parametric Image Enhancement. Pattern Recogn. 30(1997) 616-621
5. Li, H.G., Li, X.G., Li, G.Z., Luo, Z.F.: A Method for Infrared Image Enhancement Based On Genitic Algorithm. Chinese J. Systems Engineering and Electronics 21(5) (1999) 44-46
6. Lang, M., Guo, H., Odegend, J.E.: Nonlinear Processing of a Shift-Invariant DWT For Noise Reduction. In SPIE Conference on Wavelet Applications. 2491 (1995) 1211-1214
7. William, L.G., Gary, D.F., John, R.: Global Optimization of Statistical Functions with Simulated Annealing. Journal of Econometrics 60 (7) (1994) 65-99
8. Laine, A., Schuler, S.: Hexagonal Wavelet Processing Of Digital Mammography. In Medical Imaging 1993, Part of SPIE's Thematic Applied Science and Engineering Series, Vol. 1898. Newport Beach, California (1993) 559-573
9. Rosenfield, A., Avinash, C.: Digital Picture Processing. Academic Press, New York (1982)
10. Wang, M.L., Zhang, C.J., Fu, M.Y.: Simulation Study of a Kind of Wavelet Neural Network Algorithm Used in Approaching Non-Linear Functions. Journal of Beijing Institute of Technology 22 (6) (2002) 274-278
11. Gong, W.P., Wang, Y.Z.: Contrast Enhancement Of Infrared Image via Wavelet Transforms. Chinese Journal of National University of Defense Technology 22 (6) (2000) 117-119
12. Wu, Y.Q., Shi, P.F.: Approach On Image Contrast Enhancement Based On Wavelet Transform. Chinese J. Infrared and Laser Engineering 32 (2) (2003) 4-7

Learning Image Distortion Using a GMDH Network

Yongtae Do and Myounghwan Kim

School of Electronic Engineering, Daegu University, Gyeongsan-City,
Gyeongbuk, 712-714, South Korea
ytdo@daegu.ac.kr, kjhgklk@orgio.net

Abstract. Using the Group Method of Data Handling (GMDH) a polynomial network is designed in this paper for learning the nonlinear image distortion of a camera. The GMDH network designed can effectively learn image distortion in various camera systems of different optical features unlike most existing techniques that assume a physical model explicitly. Compared to multilayer perceptrons (MLPs), which are popularly used to learn a nonlinear relation without modeling, a GMDH network is self-organizing and its learning is faster. We prove the advantages of the proposed technique with various simulated data sets and in a real experiment.

1 Introduction

The mapping from 3D world to 2D image plane in a camera is nonlinear. The nonlinearity is due mainly to camera lens distortion [1]. As the nonlinearity causes undesirable results in various applications of a camera, many techniques have been proposed to model and correct the nonlinear distortion. Particularly, in camera calibration where accuracy is of great importance, correcting image distortion due to lens has been a key issue [2-3].

The sophisticated physical modeling of nonlinear lens distortion, however, brings complications to camera calibration and image correction procedures. The worse is that the more complex model cannot guarantee the higher accuracy because optical characteristics of camera lenses are various. For example, Shih [4] reported that Tsai's technique [2], which uses a radial distortion model for camera calibration, was less accurate than a simple linear method when lens distortion was relatively small.

Using an artificial neural network (ANN) is an effective way to approximate a nonlinear system without an explicit physical model. It is well known that an MLP of sufficient hidden nodes is capable of approximating any continuous nonlinear function to desired accuracy [5]. Some thus employed MLPs to determine the nonlinear mapping between 3D space and image plane(s) [6], or to identify the part that could not be described by a physical camera model [7]. Although improved accuracy and robustness could be obtained, an MLP is still not a popular tool in image distortion modeling and camera calibration. For this fact, there are two important reasons: first, searching a good MLP structure often needs tedious

J. Wang et al. (Eds.): ISNN 2006, LNCS 3972, pp. 557 – 562, 2006.

trial-and-modifications; secondly, network learning may be slow especially when distortion pattern is complex.

This paper describes a network implementing the GMDH algorithm for learning the error of linear camera model, which is due mainly to nonlinear lens distortion. Since a GMDH network is self-organizing unlike MLP, there is no need to search network parameters intensively for an optimal structure. Its learning is fast and, in our tests, GMDH networks needed only several learning iterations.

2 Perspective Projection by Pin-Hole Camera Model

A simple way to relate a 3D point and its corresponding image point for a camera is using the pin-hole model as shown in Fig. 1. Let $\boldsymbol{P} = (x \ \ y \ \ z)^T$ be a 3D point in the world frame $\{\mathbf{W}\}$ and $\boldsymbol{I} = (i \ \ j)^T$ be its projected point on the image plane. Let $\boldsymbol{P_C} = (x_C \ \ y_C \ \ z_C)^T$ be the representation of $\boldsymbol{P}$ in the 3D camera frame $\{\mathbf{C}\}$ and $\boldsymbol{U} = (u \ \ v)^T$ be the representation of $\boldsymbol{I}$ in the row-column frame on the image plane. The following relationships are then given

$$u = i + i_o, \quad v = j + j_o. \tag{1}$$

$$\boldsymbol{P_C} = \boldsymbol{RP} + \boldsymbol{L}. \tag{2}$$

where $(i_o \ \ j_o)^T$ are the coordinates of optical image center in the row-column frame, $\boldsymbol{R}$ is a 3×3 rotation matrix, and $\boldsymbol{L}$ is a translation vector. The perspective projection gives the image coordinates by

$$i = f\, x_C / z_C, \quad j = f\, y_C / z_C. \tag{3}$$

where f is the focal length, the distance between an image plane and the pin-hole. Combining all above equations leads us to

$$u = \frac{m_{11}\, x + m_{12}\, y + m_{13}\, z + m_{14}}{m_{31}x + m_{32}y + m_{33}z + m_{34}}, \quad v = \frac{m_{21}\, x + m_{22}\, y + m_{23}\, z + m_{24}}{m_{31}x + m_{32}y + m_{33}z + m_{34}}. \tag{4}$$

where 12 unknowns, m_{11} , m_{12} ,..., m_{34} , can be determined using at least six control points.

The accuracy of the linear calibration method discussed above may not be high because lens distortion is not considered. So, the image coordinates estimated by the linear relations of Eq. (4), say (u_L, v_L), should be transformed to real distorted image coordinates (u_D, v_D) by

$$u_D = u_L + \delta_U(u,v), \quad v_D = v_L + \delta_V(u,v). \tag{5}$$

where $\delta_U(u,v)$ and $\delta_V(u,v)$ represent the amount of vertical and horizontal image distortion respectively.

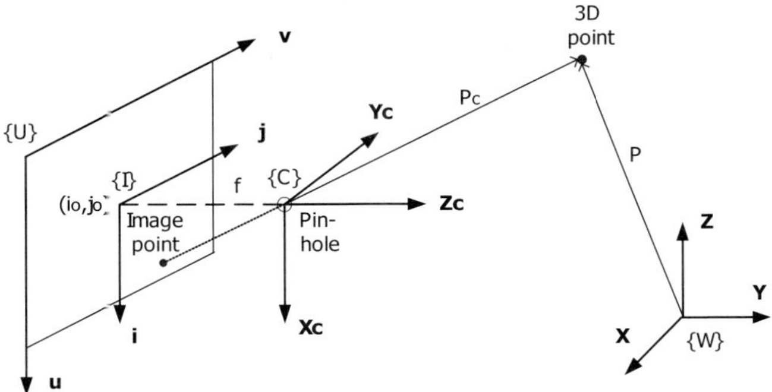

Fig. 1. Perspective projection using pin-hole camera model

3 GMDH Network for Learning Image Distortion

The GMDH is a heuristic self-organizing algorithm modeling the breed improvement of plants [8]. Each process in the breed improvement is comparable to the heuristic process in GMDH; sowing seed to selecting input variables, crossbreeding to structuring partial polynomials, and seed selection to self-selection of completed model [9]. A GMDH network is capable of approximating an arbitrary multi-variable function by constructing successive layers of second-order partial polynomials.

Fig. 2 shows a design of the GMDH network proposed to correct the projection error due to nonlinear lens distortion of a camera. Three input variables used for the network are image coordinates predicted by the perspective projection, *i.e.*, (u_L, v_L), and its distance from the optical image center, $d=((u_L-i_o)^2+(v_L-j_o)^2)^{1/2}$. The output variable is an actually measurable distorted image coordinate, u_D or v_D. The GMDH algorithm implemented for the system of Fig. 2 is outlined in the steps below:

Step 1: Collect N control points and their corresponding image points using a camera given. Divide them into N_T training data and N_E evaluation data. Compute ideal undistorted image points $\{ u_L(n), v_L(n) \mid n=1,2,\cdots,N \}$ from control points using Eq. (4).

Step 2: The output is set to a distorted real image coordinate u_D, which is represented by variable q in the figure (see Step 6 for v_D).

Step 3: In the first layer, a second-order polynomial is constituted by selecting two variables p_a and p_b from three input variables, $\{ p_1=u_L,\ p_2=v_L,\ p_3=d \}$,

$$ {}^1g = c_0 + c_1 p_a + c_2 p_b + c_3 p_a^2 + c_4 p_b^2 + c_5 p_a p_b \tag{6}$$

where 1 at the front superscript means the first layer. As there are three input variables, three polynomials can be composed like ${}^1g_1 = f(p_1,p_2)$, ${}^1g_2 = f(p_1,p_3)$, and ${}^1g_3 = f(p_2,p_3)$. The six coefficients $c_0,c_1,\cdots,c_5$ of the k 'th polynomial, where $k=1,2,3$, are determined by minimizing $\sum_{n=1}^{N_T}({}^1g_k(n)-q(n))^2$ using training data. Find then the partial polynomial resulting in the minimum estimate error, ${}^1e$, among $\{ {}^1e_k = \sum_{n=1}^{N_E}({}^1g_k(n)-q(n))^2 \mid k=1,2,3 \}$ using evaluation data.

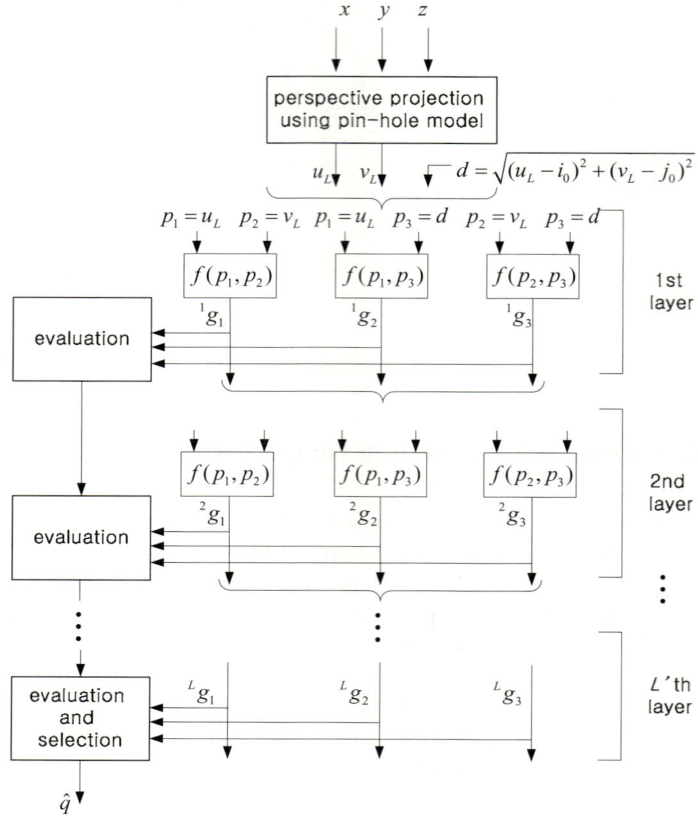

Fig. 2. GMDH network proposed for learning image distortion, where g is an estimate of real distorted image coordinate by a two-input second-order partial polynomial $f(\cdot)$, and $\hat{q}$ is the final estimate of the network

Step 4: In the L'th layer, where $L>1$, repeat the Step 3 to find the partial polynomial of the minimum error ^Le but the input variables are now defined as $\{ p_k = {}^{L-1}g_k \mid k=1,2,3 \}$.

Step 5: If $({}^{L-1}e - {}^Le) > \tau$, where τ is an arbitrary threshold, increase L constructing a new layer and go to Step 4. Otherwise, stop the iteration.

Step 6: Use v_D as the output variable q and repeat Step 3 to Step 5.

4 Experiments

The GMDH network designed was tested first using simulated data sets. The results were compared with those by Tsai's method [2], which is currently widely used and often employed for benchmarking camera calibration techniques. Results by the linear method of Eq. (4) were also compared to check the improvement by the GMDH network. Four hundred data were synthesized using Weng's model [3]. Half of them were used for training while the rest were used for testing. Figure 3 shows the test results. We could get comparatively good results steadily for different levels of radial and non-radial distortion using the network designed. It was also quite insensitive to the error of optical image center, which Tsai's technique assumed known in advance but actually difficult to know.

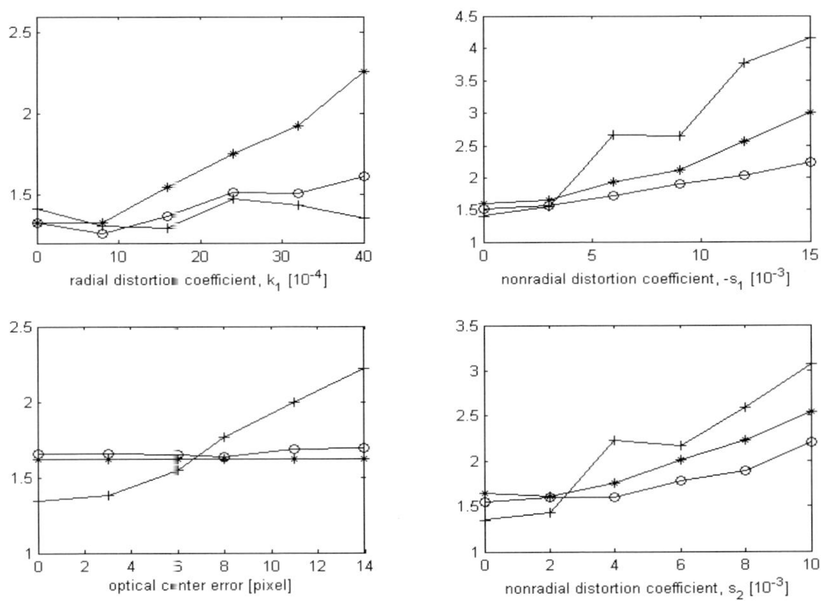

Fig. 3. Simulation results of various conditions. Lines of *, +, and o represent linear, Tsai, and GMDH network respectively. All vertical axes of figures are resultant error in [pixel].

Table 1. Result of real experiment [pixel]

Method	Training error	Testing error
Linear method	0.47	0.67
Tsai's method	0.72	1.86
GMDH network	0.38	0.59

For a real experiment, 400 control points and their corresponding image points were collected. To automate the data collection process, we employed a robotic arm that has an LED at its finger tip in the view of a camera. Half of the data collected were used for calibrating a 320×240 B/W RS-170 CCD camera with a 16 mm lens, and other data remained were used for testing. Table 1 shows the result obtained. Note that Tsai's technique was even worse than the linear method. This might be caused by the position error of the optical image center - we assumed the optical center at (320/2, 240/2) but it might be somewhere significant different.

5 Conclusions

We have described a design of GMDH network for learning the nonlinear lens distortion of a camera. Simulation studies and experimental result indicate that the network designed is effective for various types and levels of lens distortion unlike a conventional technique based on a physical model, which is good only when the model assumed corresponds to the real system used. Similar advantage can be expected when employing an MLP, but the GMDH network is self-organizing and needs much less iterations for learning. It, however, may require more memory for storing network weights, and its programming is rather complex as the size of a network is not fixed.

References

1. Brown, D.C.: Decentering Distortion of Lenses. Photogram. Eng. 32 (1966) 444-462
2. Tsai, R.Y.: An Efficient and Accurate Camera Calibration Technique for 3-D Machine Vision. In: Proc. IEEE Conf. Computer Vision and Pattern Recognition (1986) 364-374
3. Weng, J. et al.: Camera Calibration with Distortion Models and Accuracy Evaluation. IEEE Trans. Pattern Analysis and Machine Intelligence (1992) 965-980
4. Shih, S.–W, et al.: When Should We Consider Lens Distortion in Camera Calibration. Pattern Recognition 28 (1995) 447-461
5. Funahashi, K.–I.: On the Approximate Realization of Continuous Mapping by Neural Networks. Neural Networks 2 (1989) 183-192
6. Neubert, J., et al.: Automatic Training of a Neural Net for Active Stereo 3D Reconstruction. In: Proc. IEEE Conf. Robotics and Automation (2001) 2140-2146
7. Wen, J., Schweitzer, G.: Hybrid Calibration of CCD Cameras Using Artificial Neural Nets. In: Proc. IEEE-INNS Joint Conf. Neural Networks (1991) 337-342
8. Ivakhnenko, A.G.: Polynomial Theory of Complex Systems. IEEE Trans. Syst., Man, Cybern. 1 (1971) 364-378
9. Iwasaki, M.: GMDH-based Modeling and Feedforward Compensation for Nonlinear Friction in Table Drive Systems. IEEE Trans. Industrial Electronics 50 (2003) 1172-1178

An Edge Preserving Regularization Model for Image Restoration Based on Hopfield Neural Network

Jian Sun and Zongben Xu

Institute for Information and System Science,
Xi'an Jiaotong University, Xi'an 710049, China
sunjian@mailst.xjtu.edu.cn

Abstract. This paper designs an edge preserving regularization model for image restoration. First, we propose a generalized form of Digitized Total Variation (DTV), and then introduce it into restoration model as the regularization term. To minimize the proposed model, we map digital image onto network, and then develop energy descending schemes based on Hopfield neural network. Experiments show that our model can significantly better preserve the edges of image compared with the commonly used Laplacian regularization (with constant and adaptive coefficient). We also study the effects of neighborhood and gaussian parameter on the proposed model through experiments.

1 Introduction

Image restoration aims to recover the true image from its degraded version caused by blurring effect and noises. Many approaches have been proposed to solve this inverse problem. In these methods, restoration task is commonly modeled as mean square error, posterior probability, constrained optimization or regularization functional. Then transform or iteration based methods are used to minimize these models.

What we concern in this paper is regularization model and minimization by Hopfield neural network. This method was firstly investigated by Zhou et al.[1], in which objective functional was mapped onto a Hopfield neural network encoding each pixel by a set of binary elements. Later, Palik et al. [2] and Sun [3] improved the minimization algorithm based on Hopfield neural network. To better preserve the edges of image, papers [4]-[8] presented schemes by varying the regularization parameter to adaptively balance the smoothness and fidelity.

It's worth noticing that regularization terms are commonly chosen to be discrete Laplacian operator in most of the previous models, but image edges will be over-smoothed, no matter the regularization is constant or adaptive. To better preserve detailed information of image, we will discretize the continuous form of total variation in each pixel's local neighborhood, and then incorporate it into restoration model as the regularization term. We will also design sequential and parallel minimization algorithms based on Hopfield neural network to minimize

J. Wang et al. (Eds.): ISNN 2006, LNCS 3972, pp. 563–568, 2006.

the model. Experiments will show that our method can restore degraded images and preserve the edges effectively.

2 Regularization Model Based on Generalized Digitized Total Variation and Its Minimization by Hopfield NN

2.1 Restoration Model Based on Generalized Digitized TV

Total Variation $\int_{\Omega} |\nabla u|$ is a measurement of function smoothness, which was incoporated into image denoising model by Rudin et al.[9]. Due to its continuous formulation, it must be minimized by a PDE based dynamic system when removing noises.

Instead of using the continuous total variation, we discretize the form of total variation in each pixel's common neighborhood in this paper, which is more direct for digital image denoising or restoration. We discretize the form of total variation as

$$|D_{ij}u|_a = \sqrt{\sum_{(p,q)\in N(i,j)} \gamma_{ij}^{pq}(u_{pq} - u_{ij})^2 + a}$$

where

$$\gamma_{ij}^{pq} = d_{ij}^{pq} / \sum_{(p,q)\in N(i,j)} d_{ij}^{pq}$$

and

$$d_{ij}^{pq} = \frac{1}{2\pi s}\exp(-\frac{(i-p)^2 + (j-q)^2}{2s})$$

Obviously, $|D_{ij}u|_a$ measures grey variation of pixel (i,j) in its neighborhood $N(i,j)$. Normalized discrete gaussian value γ_{ij}^{pq} weights the intensity difference between pixels (i,j) and (p,q) in the summation. It means that farther pixels should be assigned lower weights in the summation because they are more possible to be different in intensity.

We find that the above definition is a generalization of Digitized Total Variation developed by Chan, T. F. et al.[10][11]. So we call our proposed term as Generalized Digitized Total Variation. Based on generalized DTV, we propose the following regularization model for image restoration

$$R(u) = \sum_{p\in\Omega} |D_p u|_a + \frac{1}{2}\lambda |Hu - f|^2$$

where u is the image vector and Ω is the image area. Through calculation and deduction similar to [11], we prove the following theorem.

Theorem 1. *Denote u as the image vector with length of MN (MN is the number of image pixels), then $R(u)$ is strictly convex, and its gradient is*

$$[\nabla R(u)]_p = \sum_{q\in N(p)} w_p^q \gamma_p^q (u_p - u_q) + \lambda [H^T(Hu - f)]_p$$

for any $1 \le p \le MN$, *where*

$$w_p^q = \frac{1}{|D_p u|_a} + \frac{1}{|D_q u|_a}$$

By theorem 1, we conclude that there exists only one optimum for our proposed model.

2.2 Minimization Algorithms Based on Hopfield NN

We will use Hopfield neural network to minimize our model in this section. First, map digital image with size of $M \times N$ onto a network with $M \times N$ cells, and each cell corresponds to a pixel. Then rewrite $\nabla R(u)$ as

$$\nabla R(u) = Wu + b$$

where

$$W_{pq}(u) = \begin{cases} \lambda[H^T H]_{pq} - \gamma_p^q(\frac{1}{|D_p u|_a} + \frac{1}{|D_q u|_a}) & q \in N(p) \\ \lambda[H^T H]_{pp} + \sum_{k\in N(p)} \gamma_p^k(\frac{1}{|D_p u|_a} + \frac{1}{|D_k u|_a}) & q = p \\ \lambda[H^T H]_{pq} & otherwise \end{cases}$$

and

$$b_p = -\lambda \sum_{q=1}^{MN} H_{qp} f_q$$

for $1 \le p, q \le MN$. Weights and bias of the network are chosen to be matrix W and vector b. Based on the above network structure, we propose two basic updating rules based on Hopfield neural network as follows.

Algorithm 1 (sequential updating)
Step 1 (Initialization): $t \leftarrow 0, u^{(t)} \leftarrow f$;
Step 2 (Updating): $v \leftarrow u^{(t)}$
For $p = 1, \cdots, MN$ do
{
Compute input of cell p

$$I_p^t = \sum_{q=1}^{MN} W_{pq}(v) v_q + b_p$$

Set

$$\triangle u_p = \begin{cases} -1 & I_p^t < 0 \\ 1 & I_p^t > 0 \\ 0 & otherwise \end{cases}$$

Then, compute $\eta = v + \triangle u_p \cdot e_p$, and

$$\Delta R = \sum_{q\in N(p), q=p} (|D_q(\eta)| - |D_q(u^t)|) + \frac{\lambda}{2}[(|H\eta|_p - f_p)^2 - (|Hu^t|_p - f_p)^2]$$

If $\Delta R < 0$, then $v \leftarrow \eta$.
}
Step 3: If $u^{(t)} = v$, then stop the algorithm; else $u^{(t+1)} \leftarrow v, t \leftarrow t+1$, return to step 2.

Algorithm 2 (parallel updating)

Step 1 (Initialization): set $t \leftarrow 1, \varepsilon \leftarrow 0.001, u^{(0)} \leftarrow f$;

Step 2 (Gradient descent search): use Goldstein or Wolfe rule to select step τ, and then update u according to

$$u^{(t)} \leftarrow u^{(t-1)} - \tau \nabla R(u^{t-1});$$

Step 3: If $|\nabla R(u)| < \varepsilon$, stop the algorithm; else $t \leftarrow t+1$, and return to step 2.

Algorithm 1 sequentially update image intensity, which is more flexible. Algorithm 2 updates image in parallel, which is more efficient. In the implementation, we combine the above two updating rules. The idea is, firstly update image intensity in parallel by algorithm 2 to approximately find the optimum image with an error ε, and then sequentially update by algorithm 1 to find the accurate optimum image.

3 Experiments

3.1 Model Comparison

We compare our proposed model with the classical Laplacian regularization models (constant regularization [1] and adaptive regularization [8] respectively) in this section.

To better compare these models, we select parameter λ corresponding to the same removed noise variance Γ, which means that the restored image u_0 by a given model with parameter λ satisfies $|Hu_0 - f|^2 = \Gamma$. Then we use local standard deviation mean square error (LSMSE) proposed by Perry and Guan [12] [4] to measure quality of the restored images, and smaller LSMSE corresponds to higher restoration quality.

Figure 1 shows the restoration results of degraded Lena image, which is blurred by discrete PSF matrix and guassian noises with mean 0 and variance 225. Table 1 gives LSMSE values of the restored images. We find that our model can significantly better remove noises in homogeneous area while preserving image edges, compared with Laplacian regularization.

3.2 Effect of Neighborhood and Gaussian Parameter on Restoration

From the third row of figure 1, the restored images are smoother when neighborhood is larger, but the edges are all well preserved through our observation. Table 2 gives the effect of gaussian parameter (in the definition of generalized DTV) on the restoration result. The neighborhood is set to be 3×3 window, and the test image is also the degraded Lena image. We find that LSMSE of the restored image is larger when gaussian parameter is larger, but the restoration result is not sensitive to gaussian parameter.

Fig. 1. Comparison of constant Laplacian regularization, adaptive Laplacian regularization and generalized DTV regularization on degraded Lena image, variance of removed noise is 200. From left to right are results by 4-neighborhood, 3-by-3 window and 5-by-5 window. From top to bottom are results by constant Laplacian regularization, adaptive Laplacian regularization and generalized DTV regularization.

Table 1. Comparison of LSMSE for restored Lena images by different models

	Constant Laplacian	Adaptive Laplacian	Generalized DTV
4-neighborhood	79.2285	78.4917	59.6865
3 by 3 window	83.1907	82.0541	64.7397
5 by 5 window	90.0558	88.5141	68.9711

Table 2. LSMSE for generalized DTV regularizaion with different gaussian parameters

Gaussian parameter (s)	0.2	0.5	1	2	4	10
LSMSE	63.0778	63.8942	64.7397	65.1509	65.3519	65.4707

4 Conclusion

Different from the previous work, we developed a regularization model for image restoration based on our proposed generalized DTV. To minimize this model, we developed an effective minimization algorithm using Hopfield neural network. Experiments showed that our restoration model can much better restore degraded images and preserve their edges, compared with constant and adaptive Laplacian regularization.

Acknowledgement

This work is supported by the NSFC key project under grand No. 70531030.

References

1. Zhou, Y.T., Chellappa, R., Vaid, A., Jenkins, B.K.: Image Restoration Using a Neural Network. IEEE Trans. on Acoustics, Speech, and Signal Processing **36**(7) (1988) 1141-1151
2. Palik, J.K., Katsaggelos, A.K.: Image Restoration Using a Modified Hopfield Network. IEEE Transactions on Image Processing **1**(1) (1992) 49-63
3. Sun, Y.: Hopfield Neural Netwoks Based Algorithms for Image Restoration and Reconstruction–Part I: Algorothms and Simulations. IEEE Trans. on Signal Processing **48**(7) (2000) 2105-2118
4. Perry, S.W., Guan, L.: Weight Assignment for Adaptive Image Restoration by Neural Networks. IEEE Trans. on Neural Network **11**(1) (2000) 156-170
5. Ghennam, S., Benmahammed, K.: Adaptive Image Restoration Using Hopfield Neural Network. In Proceedings of the 2001 IEEE Signal Processing Society Workshop (2001) 569 - 578
6. Palmer, A.S., Razaz, M., Mandic, D.P.: Spatially Adaptive Image Restoration by Neural Network Filtering. In Proceedings of VII Brazilian Symposium on Neural Networks (2002) 184 - 189
7. Wong, H.S., Guan, L.: Adaptive Regularization in Image Restoration Using Model-Based Neural Network. Optical Engineering **36**(12) (1997) 3297-3308
8. Kang, M.G., Katsaggelos, A.K.: General Choice of the Regularization Functional in Regularized Image Restoration. IEEE Trans. on Image Processing **4**(5) (1995) 594-602
9. Rudin, L.I., Osher, S., Fatemi, E.: Nonlinear Total Variation Based Noise Removal Algorithms. Physica D **60**(1-4) (2001) 259-268
10. Chan, T.F., Osher, S., Shen J.H.: The Digital TV Filter and Nonlinear Denoising. IEEE Trans. on Image Processing **10**(2) (2001) 231-241
11. Chan, T.F., Mulet, P.: On the Convergence of the Lagged Diffusivity Fixed Point Method in Total Vatiation Image Restoration. SIAM Journal of Numerical Analysis **36**(2) (1999) 354-367
12. Perry, S.W., Guan, L.: Perception-Based Adaptive Image Restoration. In Proc. Int. Conf. Acoust., Speech, Signal Processing (1998) 2893-2896

High-Dimensional Space Geometrical Informatics and Its Applications to Image Restoration

Shoujue Wang, Yu Cao, and Yi Huang

Laboratory of Artificial Neural Networks, Institute of Semiconductors
Chinese Academy of Sciences, Beijing 100083, China
{wsjue,caoyu}@semi.ac.cn, camelyi@263.net

Abstract. With a view to solve the problems in modern information science, we put forward a new subject named High-Dimensional Space Geometrical Informatics (HDSGI). It builds a bridge between information science and point distribution analysis in high-dimensional space. A good many experimental results certified the correctness and availability of the theory of HDSGI. The proposed method for image restoration is an instance of its application in signal processing. Using an iterative "further blurring-debluring-further blurring" algorithm, the deblured image could be obtained.

1 Introduction

In the remote past, human beings recognized the nature grounding on geometric concepts. Till thousands of years later, analytic methods occurred with the advances on mathematics, such as analytic geometry, differential calculus, etc. However, since we have stepped into information society, new problems and challenges to digital information processing with conventional mathematical tools emerged.

In current digital world, any information should be described as large amounts of digital numbers. How to deal with these numbers so as to find out or utilize the implied information becomes the main issue in information science. Here we quoted a statement from the literature of the mathematician Xingshen Chen [1]:

It might be interesting to characterize each period of mathematics using the number of the independent variables of a function or the dimension of the space which mathematics deals with. In this view, the mathematics in nineteen century is 1-dimensional, whereas in twenty century it is n-dimensional.

Modern information science and high-dimensional mathematics have consanguineous relationship. Each number containing certain information corresponds to a certain coordinate value; accordingly, each group of numbers corresponds to a certain point in high-dimensional space.

Considering this affinity, we bring forward a new subject-High-Dimensional Space Geometrical Informatics (HDSGI). In the second section, we provide its main features. The geometrical computation and analysis methods of HDSGI are described in the third section. At last, we give an instance of its application-image restoration.

J. Wang et al. (Eds.): ISNN 2006, LNCS 3972, pp. 569 – 574, 2006.

2 The Main Features of High-Dimensional Space Geometrical Informatics

Figure 1 shows the architecture of High-Dimensional Space Geometrical Informatics. That's also the main features of HDSGI. It could be summarized as the following:

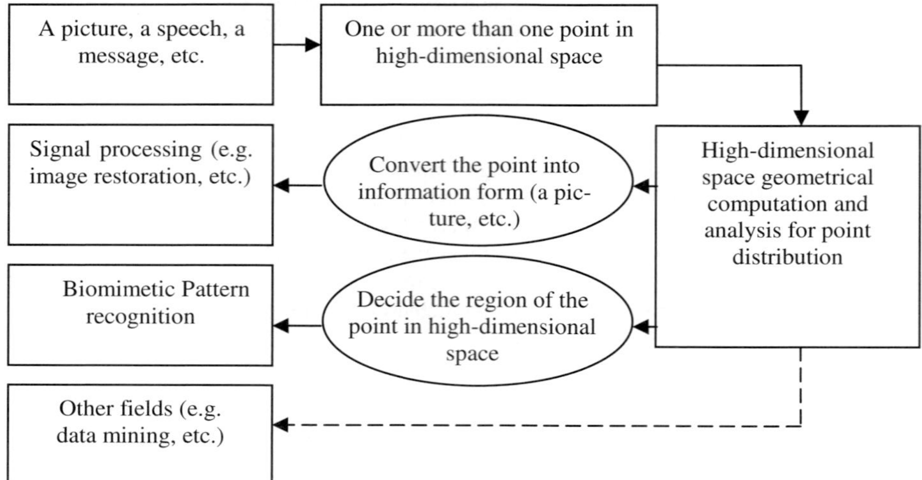

Fig. 1. Scheme of the Architecture of HDSGI

(1) Correspond all kinds of digital information (including a photo, a voice, a message, etc.) into one or more than one point in high-dimensional space.

(2) Use geometrical analysis methods to analyze the relationship among these points. During the research, we set up a series of analysis methods for point distribution in high-dimensional space [9].

(3) Study the connotative information (it might be a photo, a voice, a message, etc.) represented by the new subspaces which are constructed with the geometrical methods. During this course, we solved quite a few problems in information science with the geometrical computation and analysis methods of HDSGI. We have achieved converting the point into all kinds of information form such as an image, a speech. These are the problems of signal processing including image restoration and enhancement, etc. On the other hand, we could make a decision on the region of the point in high-dimensional space by the point distribution analysis. Thus, we proposed Biomimetic Pattern Recognition (BPR) [3, 4] which successfully resolved the problems in machine learning and pattern recognition, including human face recognition [5, 6], continuous speech recognition [7], etc. Besides, other fields of information science, such as data mining, are in our further research work.

3 The Geometrical Computation and Analysis Methods for High-Dimensional Space Geometrical Informatics

As a new proposed subject, High-Dimensional Space Geometrical Informatics has its special analysis methods. We displayed the most primary three points as follows.

1. All the operations are based on geometric figures. In lower-dimensional space, typical geometric figures involve points, lines, planes, etc. They are intuitive and ordinarily existing in our daily life. Note that the concepts of geometric figures in high-dimensional space are not entirely equal to those in lower-dimensional space [8].
2. All subspaces of the whole high-dimensional space are described with points. Modern information science cares about not only the discrete samples but also the intrinsic relation and the distribution of them in most cases. Thus, the subspaces should also be considered. We use the discrete points to describe the subspaces and simplify the complex continuous subspace during computation.
3. All computations in high-dimensional space are based on the combination of the basic simple operations in 2D planes. In high-dimensional space, some conceptions are only the extension of those in lower dimensional space. However, the computational complexity doesn't go up linearly with the increasing dimension. In the theory of HDSGI, we use a series of simple operations to implement the complex computation in high-dimensional space. Each simple operation is a basic computation in a true or artificial 2D plane in high-dimensional space.

The detailed geometrical analysis methods of HDSGI were provided in many of our literatures [2, 8, 9]. Limited to the length of the article, here we skip over them.

4 An Example of the Application-Image Restoration

Image restoration purposes to reconstruct the original clear image from the existing blurred image. According to the theory of High-Dimensional Space Geometrical Informatics, an image corresponds to a point in high-dimension space. Thus, the problem turns into point distribution analysis in high-dimensional space. Different from traditional methods, the emphasis of our method based on HDSGI diverts from merely hunting for the PSF to making full use of the intrinsic relationship between the images during the restoration process. That is the prominent property of HDSGI, whose original idea is not "isolated" but "connective". In further analysis, based on the Principle of Homologous Continuity (PHC) [3, 4] that "the whole of homologous samples is continuous in feature space", the change of neighboring points is gradual within certain short distance. Accordingly, the unknown points could be found by the information of known points and their intrinsic relation. This method is available to not only common image restoration but also blind image restoration [10] problem whose blur model (Point Spread Function, PSF) is unknown.

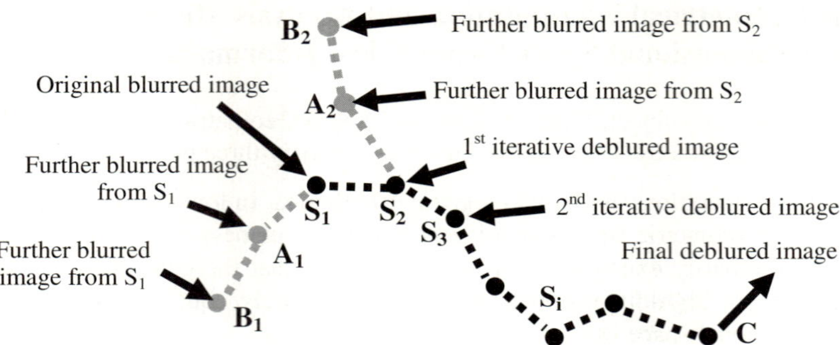

Fig. 2. Scheme of the image restoring process based on HDSGI

Figure 2 illustrates the proposed restoration method. By further blurred the original image, we could get two further blurred images. The coarse estimation of the blur from the observed information determines the trend of further blur. For an instance, if the observed image is estimated to be defocused blurred, we make further defocused blur on it twice. Hereby, we get three different blurred images. In the analysis methods of HDSGI, they correspond to three homologous points. In the first iteration, they are S_1, A_1, B_1, where S_1 is the original blurred point. The deblured point of the first iteration S_2 could be extrapolated from the arc fitted by these three points (S_1, A_1, B_1) with high-dimensional space geometry analysis methods [2, 8, 9]. In the same way, A_2, B_2 could be obtained from S_2, then S_3 is extrapolate by (S_2, A_2, B_2); the rest are deduced by analogy. This "further blurring-debluring-further blurring" process is performed iteratively till the final deblured image is reached.

The following equation group expresses the iterative algorithm:

$$\begin{aligned} C_i &= (1 - k_{i1} - k_{i2})S_i + k_{i1}A_i + k_{i2}B_i \\ S_{i+1} &= C_i \qquad\qquad i = 1,2,\cdots \end{aligned} \tag{1}$$

Where S_i and C_i denote the blurred image and the deblured image, A_i and B_i represents two further k_{i1}, k_{i2} are coefficients of A_i and B_i respectively; i is the iterative number; when $i = 1$, S_1 refers to the original blurred image.

Firstly, note that the coefficient of S_i is $(1 - k_{i1} - k_{i2})$. This constraint guarantees that in each current iteration, C_i is in the same 2D plane determined by (S_i, A_i, B_i) in high-dimensional space. Secondly, k_{i1} and k_{i2} are interrelated so as to fit an arc formed by (S_i, A_i, B_i) in this plane. C_i could be extrapolated in very short step on the arc of the current generation. It's certain that the deblured image C_{i+1} is clearer than C_i. Finally, C_i would be convergent to the true image. The value of the iteration number i depends on the degree of the blur and the practical requirement. In this way, we obtain the deblured image step by step.

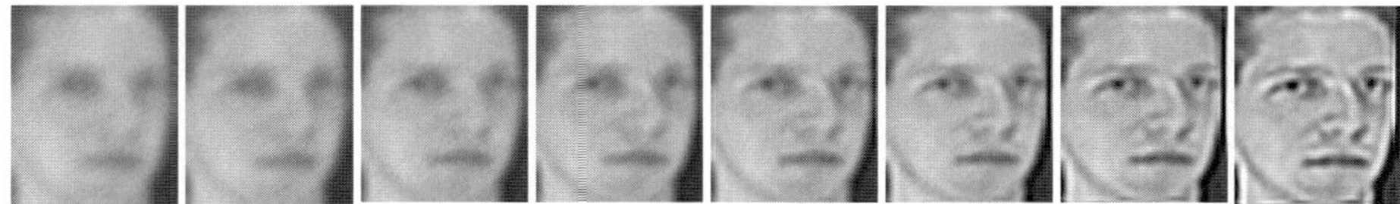

Fig. 3. One of the experimental results using the proposed restoration method

Figure 3 lists the restoration result of one image for seven generations. The first image was the original blurred image which was defocused blurred artificially. The other seven images were the debluread images from the first iteration to the seventh iteration respectively. We can see the good restoring effect by comparison.

Figure 4 gives partial restoration result of the blurred image but well embodied the "further blurring-debluring-further blurring" process. This astronomical photo was got from internet without any pre-knowledge about its blur. The first row is the restoration of the first iteration. We got the two further blurred images A_1, B_1 from S_1 by further defocused blurring. C_1 was extrapolated by the circular arc fitted by (S_1, A_1, B_1). The second row from S_2 to C_2 was deduced by analogy, in which S_2 got from the replacement of C_1 according to the iterative algorithm. This experiment shows that blind image restoration could also be achieved efficiently using this method.

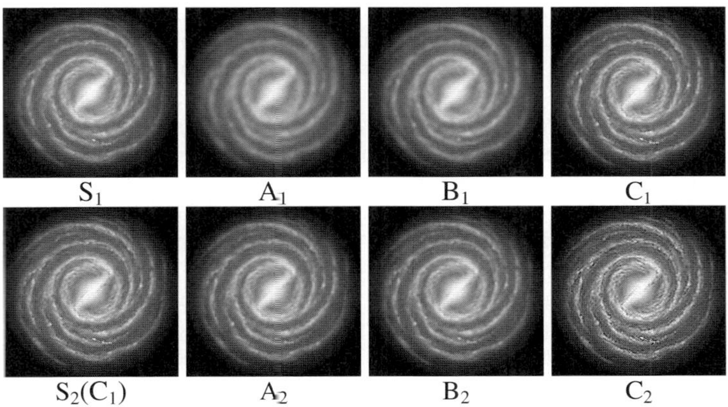

Fig. 4. Partial restoring process of one of our experiment

5 Conclusions

Information science is a rising and highly cross-disciplinary field, including computer science, cybernetics, bionics, artificial intelligence, and so on. In digital world, any information could be regarded as a point in high-dimensional space. Therefore, the tasks of information science turn into the analysis of point distribution in high-dimensional space. However, the current mathematical tools are not good at analyzing the distribution of these points in high-dimensional space. We put forward a new subject named High-Dimensional Space Geometrical Informatics (HDSGI), which also involves in a series of computational analysis methods. Great success in many fields strongly certified the correctness and availability of HDSGI.

The provided image restoration method is one instance of the applications of HDSGI in signal processing. Considering the intrinsic relation of the images during the restoring process, we used an iterative "further blurring-debluring-further blurring" method based on HDSGI. Experimental results showed good restoration effect and convergence property of the algorithm. Since HDSGI has exploited a new way of information science, we are confident of its good perspective in the future on further theory research and wider applications.

References

1. Chen, X.: Corpus of Chen Xingshen. East China Normal University Press, Shanghai (2002) 244
2. Wang, S.: Computational Information Geometry and Its Applications. Keynote Speech for ICNN&B'05: Second International Conference on Neural Networks and Brain 3(1) (2005) PL63-PL69
3. Wang, S.: Biomimetic Pattern Recognition. Neural Networks Society (INNS, ENNS, JNNS) Newsletter 1(1) (2003) 3-5
4. Wang, S., Zhao, X.: Biomimetic Pattern Recognition Theory and Its Applications. Chinese Journal of Electronics 13(3) (2004)
5. Wang, S., Xu, J., Wang, X., Qin, H.: Multi-Camera Human-Face Personal Identification System Based on the Biomimetic Pattern Recognition (in Chinese). Acta Electronica Sinica 31(1) (2003) 1-4
6. Wang, S., Qu, Y., Li, W., Qin, H.: Face Recognition: Biomimetic Pattern Recognition vs. Traditional Pattern Recognition (in Chinese). Acta Electronica Sinica 32(7) (2004) 1057-1061
7. Cao, W., Pan, X., Wang, S.: Continuous Speech Research Based on Two-Weight Neural Network. Advances in Neural Networks – ISNN 2005: Second International Symposium on Neural Networks, LNCS 3497 (2005) 345-350
8. Wang, S., Wang, B.: Analysis and Theory of High-Dimension Space Geometry for Artificial Neural Networks (in Chinese). Acta Electronica Sinica 30(1) (2002)
9. Wang, S., Lai, J.: Geometrical Learning, Descriptive Geometry, and Biomimetic Pattern Recognition. Neurocomputing 67(1-4) (2005) 9-28
10. Kundur, D, Hatzinakos, D.: Blind Image Deconvolution. IEEE Signal Processing Magazine 13(3) (1996) 43-64

Improved Variance-Based Fractal Image Compression Using Neural Networks

Yiming Zhou, Chao Zhang, and Zengke Zhang

Department of Automation, Tsinghua University, Beijing 100084, China
zhouym02@mails.tsinghua.edu.cn

Abstract. Although the baseline fractal image encoding algorithm could obtain very high compression ratio in contrast with other compression methods, it needs a great deal of encoding time, which limits it to widely practical applications. In recent years, an accelerating algorithm based on variance is addressed and has shortened the encoding time greatly; however, in the meantime, the image fidelity is obviously diminished. In this paper, a neural network is utilized to modify the variance-based encoding algorithm, which makes the quality of reconstructed images improved remarkably as the encoding time is significantly reduced. Experimental results show that the reconstructed images quality measured by peak-signal-to-noise-ratio is better than conventional variance-based algorithm, while the time consumption for encoding and the compression ratio are almost the same as the conventional variance-based algorithm.

1 Introduction

Compared with other image compression schemes, Fractal image compression could obtain high-resolution reconstructed images at very high compression ratio. However, a great deal of time cost during encoding limits it to practical applications. In order to solve this problem, some fast algorithms have been presented in [2]-[3], which present an assumption that two images can not be closely matched unless their variances are closely matched. The time cost for encoding can be shortened greatly. However, the quality of the reconstructed image may be diminished obviously.

Recently, some fractal encoding algorithms based on neural networks were addressed in [4]-[6]. These methods provide a more accurate description for the mapping between the matching subblocks and could operate in parallel; therefore, the encoding time complexity could be greatly reduced. However, the space complexity of the network is very high and the whole time cost for learning process is huge.

In this paper, we utilize a neural network to modify the variance-based fractal image compression. By introducing a variance threshold, a new fast fractal image encoding algorithm is proposed. According to different subblocks in the original image, some selected mapping would be constructed by neural network automatically. The simulation results demonstrate that the quality of the reconstructed images is enhanced evidently, furthermore, the space complexity and the whole time consumption are almost the same as the conventional variance-based algorithm.

J. Wang et al. (Eds.): ISNN 2006, LNCS 3972, pp. 575 – 580, 2006.

2 Overview of Fractal Image Encoding Algorithms

2.1 Baseline Fractal Image Encoding Algorithm

The baseline fractal image encoding algorithm is based on contractive transformations and partition iterated function system. During encoding, the original image is first partitioned into non-overlapping range blocks $\{R\}$, and overlapping domain blocks $\{D\}$. The size of domain blocks is always twice the size of range blocks [1].

Then the matching search between range blocks and domain blocks needs to minimize the following equation:

$$E(R,D) = \|R-(s\cdot D+oI)\| \tag{1}$$

where $\|\cdot\|$ is the two-norm, D is the contracted domain block under symmetric transformation, s controls the contrast, and o controls the brightness.

Finally, each range block in the original image would find its best matched domain block by (1), and coefficients of the symmetric transformation, location of the domain block, s and o could be obtained. Storing those coefficients of the entire range blocks would complete the image encoding.

2.2 Variance-Based Fractal Image Encoding

In order to accelerate the fractal image encoding, several fast algorithms are addressed in [2]-[3], which improve (1) to (2) and receive a conclusion that the variance of a range block and a domain block should not differ greatly if they are matched.

$$E^2(R,D) = \|R-\bar{r}I\|^2 - s^2\|D-\bar{d}I\|^2 \tag{2}$$

where $\bar{r}$ is the mean value of the range block R, and $\bar{d}$ is the mean value of the domain block D.

Then a pre-search can be introduced before the detailed search described by (1), and the detailed search space can be limited to a relatively small size domain pool. The encoding time cost can hence be greatly reduced.

3 Proposed Algorithm Using Neural Networks

From (1), the mapping is obviously a linear affine transformation. Furthermore, if the pixels of subblocks fluctuate little, the affine mapping would get a good result between the matched subblocks, and if the pixel values of subblocks fluctuate greatly, the affine mapping becomes unfit for the mapping and results in a great matching error.

In order to solve above problem, we utilize a neural network, which is very similar to the neural network shown in [6], to construct mapping for the subblocks, in which the pixels fluctuate greatly. In the network, each pixel in the subblocks is regarded as a neuron: pixel in range blocks as output neuron and pixel in its matched domain blocks as input neuron. Obviously there are four input neurons X_i, X_{i+1}, X_{i+2} and X_{i+3} according to each output neuron Y_j, and the corresponding weights are $W_{j,i}$, $W_{j,i+1}$, $W_{j,i+2}$ and $W_{j,i+3}$, and the threshold is β_j, the mapping structure is shown in Fig. 1.

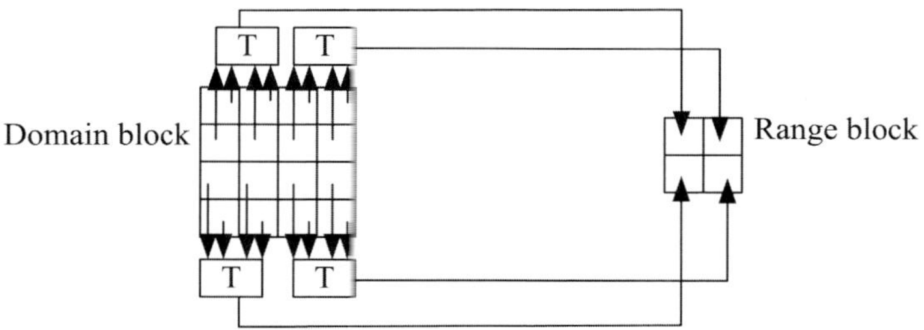

Fig. 1. The structure of the mapping by using neural network

The activation function, which is denoted as T above, is defined as follows:

$$Y_j = T(\sum_{k=i}^{i+3} W_{j,k} \times X_k - \beta_j)$$

$$T(x) = \begin{cases} x & if \quad 0 \le x \le 255 \\ 0 & otherwise \end{cases} \tag{3}$$

According to (3), the affine transformation mapping is just one possible situation when $W_{j,i} = W_{j,i+1} = W_{j,i+2} = W_{j,i+3} = \frac{s}{4}$ and $\beta_j = -o$, and this situation only happens if the pixels in the subblocks fluctuate little. Otherwise, those weights would not be close. Such conclusion demonstrates that the neural network could play a better role for constructing the matching mapping than affine transformation.

The error of neuron j is defined as follows:

$$e_j = d_j - Y_j \tag{4}$$

where d_j is the original pixel value, and Y_j is the corresponding output value of the proposed neural network.

Then the error-correction for weights and threshold is denoted as follows:

$$\Delta W_{j,k} = \eta \times \frac{e_j}{X_k}, \quad for\, k = i \sim i+3$$

$$\Delta \beta_j = -\eta \times e_j \tag{5}$$

where η is the rate of learning.

In addition, the initial values of all weights and thresholds are equal to $\frac{s}{4}$ and $-o$ respectively, so that learning process would be quickly. And all weights are set within -1 and 1; the threshold is set within -255 and 0. The learning process is repeated until the outputs of the proposed neural network are acceptable.

As we all know, the variance of a subblock, in which the pixel values fluctuate little, is generally small; and the variance of a subblock, in which the pixel values fluctuate greatly, is generally large. Therefore a variance threshold h is defined and can be utilized to construct a classification, which is shown in Fig. 2. After finding the best matched domain block for each range block in the variance-based encoding algorithm, if the variance of a range block is smaller than h, conventional affine transformation is used to construct the mapping between this range block and its matched domain block; otherwise neural network addressed above is used to construct the mapping.

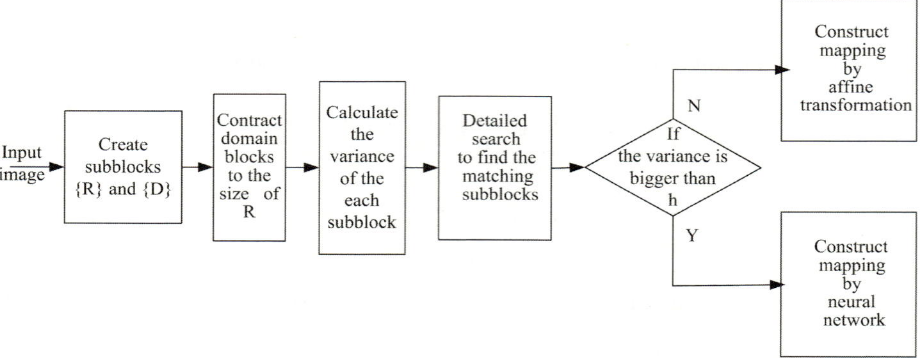

Fig. 2. The flow chart of the proposed fractal image algorithm

This novel algorithm has several advantages. Firstly, by using conventional affine transformation, the subblocks, in which the pixel values fluctuate greatly, play a dominant role to the error between the reconstructed image and original image. Consequently applying neural network to construct the mapping for such subblocks would improve the quality of the reconstructed images effectively. Moreover, although the new mapping produces a lot of weights, the majority will be just the initial value or zero, which doesn't affect the compression ratio greatly. Secondly, as to the most of the subblocks in the image, pixels are correlative in the neighborhood and fluctuate little. Applying conventional affine transformation rather than neural network to such subblocks could maintain the excellent performances of high compression ratio and fast speed stably. As a result, the proposed algorithm could obtain a better quality of the reconstructed images than variance-based algorithm, and the time consumption and the compression ratio are almost the same as the conventional algorithm.

4 Experimental Results

Two images with Peppers and Woman (256×256, 256 gray levels) were used to evaluate the performance of the proposed algorithm, and the encoding quality was estimated by PSNR. All experiments were run by Pentium 2.8GHz and VC++ 6.0.

There are four parameters affecting the encoding quality and encoding time in the proposed algorithm: threshold h for mapping classification, threshold η for learning rate, threshold Max for the biggest error-correction counts, and threshold k for the variance search window. Usually, Max is assigned to 100, and k is assigned to 5%.

As shown in Table 1, effects on reconstructed images with different h were addressed. Firstly, we could find that the variance-based algorithm speeds up encoding by almost 22 times, but the quality of the reconstructed images is lowered obviously. Secondly, in contrast with the variance-based algorithm, by utilizing the proposed algorithm, the quality of the reconstructed images is almost equal or much better than the baseline encoding algorithm, and the encoding time is almost the same as the variance-based algorithm. In this experiment, η is assigned to 2.0.

Table 1. Comparisons of the proposed algorithm and the VBFC, BFC algorithm

	h	300	250	200	150	120	VBFC	BFC
W	Time (s)	0.65	0.65	0.66	0.65	0.66	0.65	14.45
	PSNR(dB)	35.61	35.6	36.61	36.70	36.94	35.61	37.20
P	Time (s)	0.64	0.64	0.64	0.65	0.65	0.64	15.33
	PSNR(dB)	33.41	33.59	34.03	35.10	36.05	33.35	34.77

W is Woman, P is Peppers, VBFC denotes the variance-based fractal image algorithm, BFC denotes the baseline fractal image algorithm.

5 Conclusion

In this work, we utilize neural network to modify the variance-based fractal image compression. The encoding quality is improved greatly, while the encoding time is almost the same as the variance-based algorithm. Experimental results demonstrate that the proposed algorithm has a better performance than the conventional algorithm.

References

1. Fisher, F.: Fractal Image Compression: Theory and Application. Springer-Verlag, New York (1995)
2. Lee, C.K., Lee, W.K.: Fast Fractal Image Block Coding Based on Local Variances. IEEE Trans. Image Processing 7 (3) (1998) 888-891

3. He, C., Yang, S.X., Huang, X.: Variance-Based Accelerating Scheme for Fractal Image Encoding. Electronic Letters 40 (1) (2004) 115-116
4. Stark, J.: Iterated Function Systems as Neural Networks. Neural Networks 4 (1991) 679-690
5. Lee, S.J., Wu, P.Y., Sun, K.T.: Fractal Image Compression Using Neural Networks. Proceedings of the IJCNN'98, Anchorage, Alaska (1998) 613-618
6. Sun, K.T., Lee, S.J., Wu, P.Y.: Neural Network Approaches to Fractal Image Compression and Decompression. Neurocomputing 41 (2001) 91-107

Associative Cubes in Unsupervised Learning for Robust Gray-Scale Image Recognition

Hoon Kang

Intelligent Robot Vision Laboratory,
School of Electrical & Electronics Engineering, Chung-Ang University,
221 Huksuk-dong, Dongjak-gu, Seoul, 156-756, Korea
hkang@cau.ac.kr, http://sirius.cie.cau.ac.kr

Abstract. We consider a class of auto-associative memories, namely, "associative cubes" in which gray-level images and the hidden orthogonal basis functions such as Walsh-Hadamard or Fourier kernels, are mixed and updated in the weight cubes, ***C***. First, we develop an unsupervised learning procedure based upon the adaptive recursive algorithm. Here, each 2D training image is mapped into the associated 1D wavelet in the least-squares sense during the training phase. Second, we show how the recall procedure minimizes the recognition errors with a competitive network in the hidden layer. As the images corrupted by noises are applied to an associative cube, the nearest one among the original training images would be retrieved in the sense of the minimum Euclidean squared norm during the recall phase. The simulation results confirm the robustness of associative cubes even if the test data are heavily distorted by noises.

1 Introduction

Most of the associative memories deal with a pattern-matching problem. In this paper, we address a cubical structure of associative memories, an "**associative cube**" which stores the training images and retrieves those in an optimal sense. It is a class of auto-associative memories where training images have one-to-one correspondence to orthogonal basis functions such as the Walsh-Hadamard (Haar) bases or the sinusoidal Fourier kernels. The origin of unsupervised learning in associative memories stems from Hebbian learning [1]. In the Hopfield's auto-associative memory [2], the training data are stored near the vertices of a hypercube as the recurrent states. In the hetero-associative memory cases, the related works are Kosko's bidirectional associative memories (BAM) [3], Kohonen's optimal linear associative memories [4], the coding strategies in BAM [5], multilayer associative neural networks [6][7], and modified bidirectional decoding strategy [8]. However, in the case of continuous gray-level images, more cautious treatments are necessary in order to guarantee robust pattern recognition. As an extension, a continuous version is proposed in [9] where ten gray-scale is used. We propose an auto-associative memory with two layers in a cubical structure. Primarily, two phases of processes are involved in associative cubes, the training phase and the recalling phase. Here, a

J. Wang et al. (Eds.): ISNN 2006, LNCS 3972, pp. 581 – 588, 2006.

3D structure of an associative cube is formed where the depth of the z-axis stores the eigen-images by mixing the training images with the orthogonal discrete wavelets as shown in Figure 1. Also, we assume that the orientation and the scale of images are normalized and choose two kinds of orthogonal wavelets, the Haar wavelet and the Fourier kernels, with which the weight cube C is updated in the spatial domain. Therefore, the last-scale types of the wavelet packet decomposition [10] are adopted as the independent bases. The discrete basis functions have the following properties:

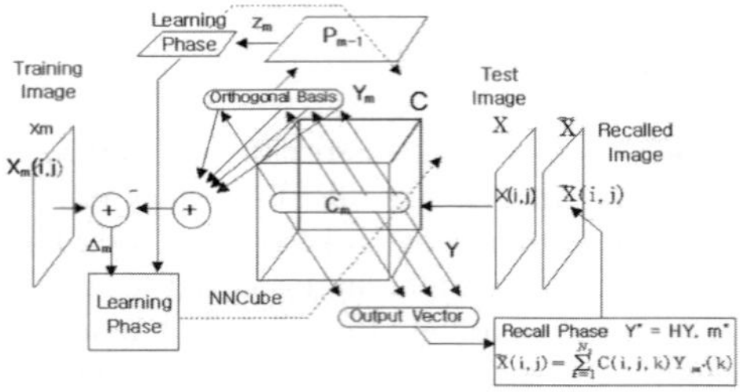

Fig. 1. The block diagram of an associative cube

- The basis vectors, Y_m, are orthogonal and the inner product between Y_i and Y_j is $\langle Y_i, Y_j \rangle = Y_i^T Y_j = K\delta_{ij}$ where δ_{ij} is the Kronecker delta.
- The basis matrix H consists of N_T rows and N_Z columns where N_T is the total number of the training images and N_Z is the depth of z-axis in an associative cube. Each column in H form the eigen-vector of the scaling and the rotating transformation with which any points in R^{N_Z} is mapped into the coupling-free coordinates in R^{N_T}. The transformation matrix H has the following characteristics:

$$H = \begin{bmatrix} Y_1^T & \cdots & Y_{N_T}^T \end{bmatrix}^T \in R^{N_T \times N_Z}, \quad H \cdot H^T = K \cdot I_{N_T} \in R^{N_T \times N_T} \tag{1}$$

where I_{N_T} is the $N_T \times N_T$ identity matrix. The matrices H and HH^T are represented for the Haar and the Fourier bases in Figure 1-(a) and Figure 1-(b), respectively.

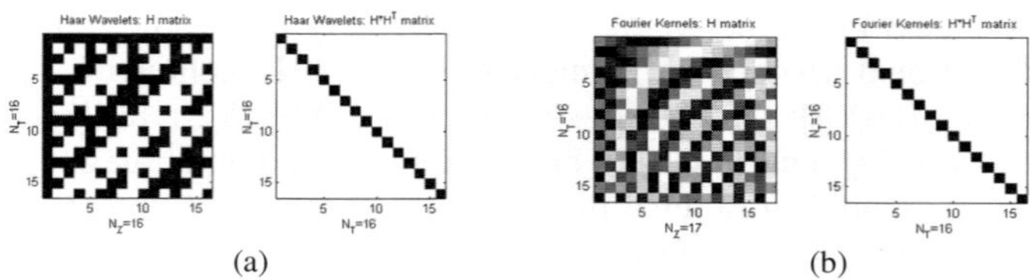

Fig. 2. (a) The matrices H and HH^T for the Haar basis vectors ($N_T = N_Z = 16$), (b) The matrices H and HH^T for the Fourier basis vectors ($N_T = 16$, $N_Z = 17$)

2 Unsupervised Learning & Retrieving Mechanism of Associative Cubes

The associative cubes are categorized as a neural network class of unsupervised learning by which we mean that the weight-updating algorithm is not based on the error signals between the desired pattern and the output pattern, but it is performed in terms of the desired ones only. We now investigate the encoding and the decoding schemes of the associative cubes for image recognition. First, the gray-scale images $X \in R^{N_X \times N_Y}$'s are converted from [0,1] to [-1,1] and we define the cubical memories $C \in R^{N_X \times N_Y \times N_Z}$ so that the recall process may have the form of

$$\mathrm{net}_Y(k) = \sum_{i=1}^{N_X} \sum_{j=1}^{N_Y} C(i,j,k) \cdot X(i,j), \; Y(k) = f\left(\mathrm{net}_Y(k)\right) \tag{2}$$

where $\mathrm{net}_Y \in R^{N_Z}$ is the net input vector; X and Y are the input image and the hidden node output, respectively; and C is the weight cube updated at each learning step. The activation function f is a normalization, $f(\mathrm{net}_Y(k)) = \mathrm{net}_Y(k) / \max\left|\mathrm{net}_Y(k)\right|$. Therefore, Y is within the range of [-1,1]. Second, the output Y of the hidden layer is applied to the transformation matrix defined in the previous section, in order to obtain the associated orthogonal basis vector or wavelet function for the recall process. Here, we utilize a competitive neural network of H which follows the winner-take-all rule defined by

$$Y^* = H \cdot Y = \left[Y_1^T Y \quad Y_2^T Y \quad \cdots \quad Y_{N_T}^T Y \right]^T, \; m^* = \arg\max_k \{Y^*(k)\} \tag{3}$$

Finally, the chosen orthogonal basis vector Y_{m^*} is then applied to the associative cube C resulting in the retrieved image $\tilde{X}$.

$$\tilde{X}(i,j) = \sum_{k=1}^{N_Z} C(i,j,k) \cdot Y_{m^*}(k) \tag{4}$$

2.1 Encoding or Learning Algorithm of Associative Cubes

The learning algorithm of the associative cube minimizes the quadratic energy function $E_{N_T}(C)$ in (5) which is a cubic version of the adaptive least squares algorithm. Here, the training image X_m, the hidden layer output, and the orthogonal wavelet Y_m take the values between -1 and +1. The weight cube C is updated as a new training input-output pair is stored in the associative cube.

$$\begin{aligned} E_{N_T}(C) = \frac{1}{2} \sum_{i=1}^{N_X} \sum_{j=1}^{N_Y} \Big[& \sum_{m=1}^{N_T} \Big\{ X_m(i,j) - \sum_{k=1}^{N_Z} C(i,j,k) Y_m(k) \Big\}^2 \\ & + \sum_{k_1=1}^{N_Z} \sum_{k_2=1}^{N_Z} (C(i,j,k_1) - C_0(i,j,k_1)) P_0^{-1}(k_1,k_2)(C(i,j,k_2) - C_0(i,j,k_2)) \Big] \end{aligned} \tag{5}$$

where C_0 is the initial guess of the weight cube C, and P_0 is a symmetric positive definite matrix such that $P_0 = I_{N_Z}$. For the (i,j)[th] pixel, (5) reduces to the local energy function in (6),

$$E_{N_T}(C)_{ij} = \frac{1}{2}\left[\sum_{m=1}^{N_T}\left\{X_m(i,j) - \sum_{k=1}^{N_Z} C(i,j,k)Y_m(k)\right\}^2 + \left\{(C-C_0)^T P_0^{-1}(C-C_0)\right\}_{ij}\right] \tag{6}$$

The first term of the local energy function on the right-hand side of (6) represents the difference between the actual observation of the gray-level training image X_m and the predicted model image $\langle C(i,j,\cdot), Y_m(\cdot)\rangle$ with the parameter C of the weight cube. The second term in (6) has been included to account for the initial conditions. Here, P_0 is a measure of confidence in the initial estimate C_0. Now, the following pseudo-coded algorithm shows the detailed learning or training procedure of the associative cubes:

Cubical Least Squares Algorithm for Associative Cube Learning

- initialize $m=1$, $C_0 = \square$, $P_0 = I_{N_Z}$ with $C_m \in R^{N_X \times N_Y \times N_Z}$, $P_m \in R^{N_Z \times N_Z}$;
- while ($m \le N_T$),
  - provide the mth training image $X_m \in [-1,1]^{N_X \times N_Y}$;
  - provide the associated orthogonal basis Y_m (the mth row of H);
  - define $Z_m = P_{m-1}Y_m$ and $\delta_m = 1 + Y_m^T Z_m$; and compute $P_m = P_{m-1} - Z_m Z_m^T / \delta_m$;
  - for all ($i=1,\ldots,N_X$, $j=1,\ldots,N_Y$, $k=1,\ldots,N_Z$), compute (7);

$$\Delta_m(i,j) = X_m(i,j) - \sum_{k=1}^{N_Z} C_{m-1}(i,j,k)\cdot Y_m(k),\ C_m(i,j,k) = C_{m-1}(i,j,k) + \Delta_m(i,j)\cdot Z_m(k)/\delta_m \tag{7}$$

  - $m = m+1$;
- end while;

Here, (7) is derived from the following procedure. First, we define the parameter vectors of $\phi_m, \theta_m, \Phi_M, \chi_M$ where $M = N_T$, and take just one (i,j)[th] pixel of the training pattern into account, such as

$$\begin{gathered}\phi_m = [Y_m(1) \ \cdots \ Y_m(N_Z)]^T = Y_m,\quad \theta_{m-1}^{(i,j)} = [C_{m-1}(i,j,1) \ \cdots \ C_{m-1}(i,j,N_Z)]^T \\ \chi_M^{(i,j)} = [X_1(i,j) \ \cdots \ X_M(i,j)]^T \in R^{M\times 1},\quad \Phi_M = \left[\phi_1^T \ \cdots \ \phi_M^T\right]^T \in R^{M \times N_Z}\end{gathered} \tag{8}$$

so that the second term of the squared summation in (6) may become

$$\sum_{k=1}^{N_Z} C_{m-1}(i,j,k)Y_m(k) = \phi_m^T \theta_{m-1}^{(i,j)} \tag{9}$$

The performance criterion in (6) is redefined as follows:

$$E_M(C)_{ij} = \frac{1}{2}\left[\chi_M^{(i,j)} - \Phi_M\theta^{(i,j)}\right]^T\left[\chi_M^{(i,j)} - \Phi_M\theta^{(i,j)}\right] + \frac{1}{2}\left(\theta^{(i,j)} - \theta_0^{(i,j)}\right)^T P_0^{-1}\left(\theta^{(i,j)} - \theta_0^{(i,j)}\right) \tag{10}$$

Now, we consider the following cubical least-squares learning theorem in C:

Theorem 2.1 (Cubical Least-Squares Learning). Given the quadratic local energy function defined in (10), let the learning algorithm in (7) be applied to the associative cube then for the (i,j)th pixel, the weight cube $C(i,j,k)$ minimizes the local energy function (10) and eventually the global energy function (5), in the least-squares sense.

Proof. Both for the cubic weights $C(i,j,k)$ to be updated and for the local energy function to be minimized, the condition $\partial E_M(\theta^{(i,j)})/\partial\theta^{(i,j)} = 0$ should be satisfied for the parameter vector θ. For the full proof, refer to [12].

2.2 Decoding or Recall Procedure of Associative Cubes

In the decoding phase, the matched output vector Y* and the associated orthogonal basis vector Y_{m^*} are found in order to retrieve the corresponding image with the minimum squared error. Since the encoding algorithm provides the weight cube C* minimizing the global energy function for the associative cube, the recall procedure is simple and straightforward as follows:

Recall Procedure of Associative Cubes

- provide the weight cube $C^* \equiv R^{N_x \times N_y \times N_z}$ from the learning algorithm;
- provide a test image $X \in [-1,1]^{N_x \times N_y}$ under the additive uniform noise;
- compute $net_Y(k) = \sum_{i=1}^{N_x}\sum_{j=1}^{N_y} C^*(\ ,j,k)\cdot X(i,j)$; normalize $Y \in [-1,1]^{N_z}$ with $Y(k) = net_Y(k)/\max_k |net_Y(k)|$; and compute $Y^* = H \cdot Y$;
- find an orthogonal index $m^* = \arg\ \max_k \{Y^*(k)\}$ with the competitive network; choose $Y_{m^*} = H(m^*,\cdot)$, the m*th row of the basis matrix H;
- compute the retrieved image $\tilde{X}(i,j) = \sum_{k=1}^{N_z} C^*(i,j,k)\cdot Y_{m^*}(k)$;

The block diagram of an associative cube is represented in Figure 1 where both structures of encoding and decoding are shown.

3 Experimental Results of Associative Cubes for Robust image Recognition

In the simulation, the resolution of 16 and 32 training images (N_T = 16, 32) is 64x64 with 256 gray levels and the size of the orthogonal basis function is varied. Each pixel in the 256 gray levels is transformed into the values between -1 and +1 both for learning and for recalling. Figure 3-(a) represents the 16 training images used for demonstrating the robustness of the associative cubes. First, we adopt the Haar wavelets for the orthogonal basis vectors. From the cubical learning algorithm, each slice in C^* is represented in Figure 3-(b) where the kth eigen-image is in $C^*(\cdot,\cdot,k)$. The test images of the 16th training pattern are blurred by the uniform noises in U[-0.5,0.5] in

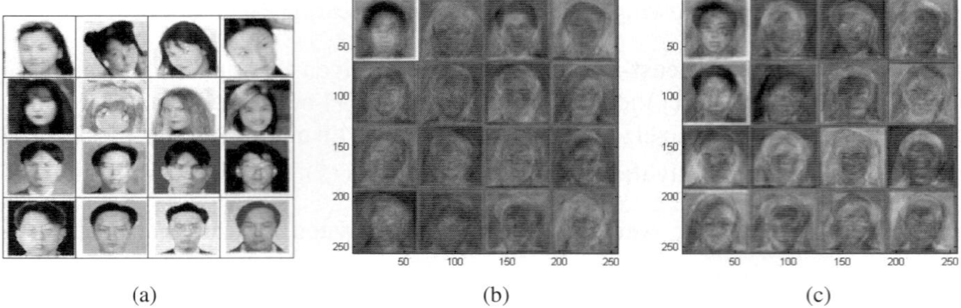

(a) (b) (c)

Fig. 3. (a) The original training images ($N_X = N_Y = 64$, $N_T = 16$), (b) The 16 eigen-images of the weight cube C^* based on the Haar wavelet, (c) those based on the sinusoidal Fourier kernels

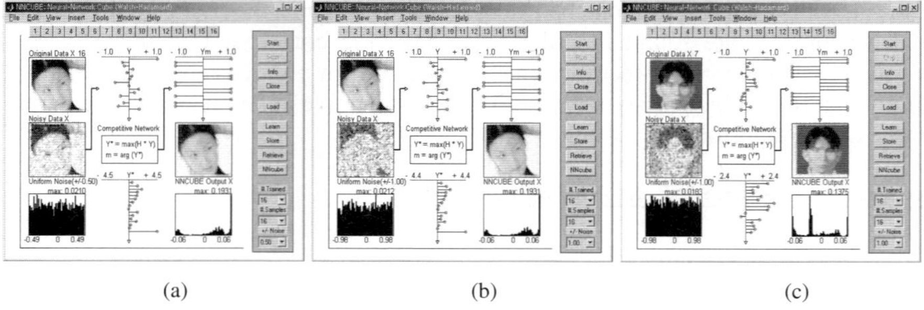

(a) (b) (c)

Fig. 4. The recognized images of the 16th test images (a) with uniform noise in U[-0.5,0.5], (b) with uniform noise in U[-1.0,1.0], (c) The result image of the 7th test image with uniform noise in U[-1.0,1.0] (Haar wavelets, $N_T = 16$, $N_Z = 16$)

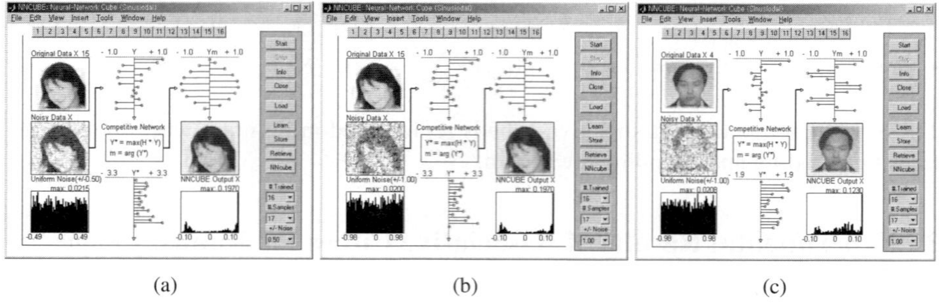

(a) (b) (c)

Fig. 5. The recognized images of the 15th test images (a) with uniform noise in U[-0.5,0.5] and (b) with uniform noise in U[-1.0,1.0], (c) The result image of the 4th test image with uniform noise in U[-1.0,1.0] (Fourier kernels, $N_T = 16$, $N_Z = 17$)

Figure 4-(a), and in U[-1,1] in Figure 4-(b), respectively. Here, the grades of each pixel in the test image with the additive noise are also bounded within [-1,1]. Figure 4-(c) is the recall result of the 7th test image under the noise U[-1,1]. Next, we choose the sinusoidal Fourier kernels and the eigen-image slices in C^* are shown in

Figure 3-(c). The test images of the 15th training pattern are distorted by the additive uniform noises in U[-0.5,0.5] in Figure 5-(a), and in U[-1,1] in Figure 5-(b), respectively. Similarly, Figure 5-(c) is the result of the 4th test image under the noise U[-1,1]. In Table 1, the average success rates of decoding for 16 and 32 test image sets are demonstrated for the two orthogonal bases in H under 1,000 random additive uniform noises per each test image. The results show that the average success rates of the two test images (pattern no.2 and no.4) tend to decrease for heavily disturbed uniform noises, due to similar patterns of the two blurred images. However, the rest of the test images reveal perfect average success rates of 100% even if they are heavily perturbed by the uniform noises.

Table 1. The performance comparison of the average success rates by using the two orthogonal basis vectors in the associative cubes (the comparisons between 16 and 32 image sets)

Orthogonal Bases		Haar Wavelets		Fourier Kernels	
No. of Test Images		16	32	16	32
H (HH^T)		16x16 ($16I_{16}$)	32x32 ($32I_{32}$)	16x17 ($8.5\ I_{16}$)	32x33 ($16.5\ I_{32}$)
Avg. Success Rates	Original Test Images	all: 100 %	all: 100 %	all: 100 %	all: 100 %
	Uniform Noise U[-0.25,0.25]	all: 100 %	all: 100 %	all: 100 %	all: 100 %
	Uniform Noise U[-0.50,0.50]	no.2: 96.1 % others: 100 %	no.31: 99.9 % others: 100 %	no.2: 95.7 % others: 100 %	no.31,32: 99.8 % others: 100 %
	Uniform Noise U[-0.75,0.75]	no.2: 87.2 % others: 100 %	no.16: 98.7 % no.31: 92.6 % no.32: 97.9 % others: 100 %	no.2: 85.9 % others: 100 %	no.16: 98.4 % no.31: 93.7 % no.32: 98.4 % others: 100 %
	Uniform Noise U[-1.00,1.00]	no.2: 79.3 % no.4: 99.9 % others: 100 %	no.2: 99.9 % no.16: 93.7 % no.22: 99.8 % no.31: 81.0 % no.32: 94.0 % others: 100 %	no.2: 81.1 % no.4: 99.8 % others: 100 %	no.2: 99.9 % no.16: 93.1 % no.22: 99.6 % no.31: 80.2 % no.32: 91.7 % others: 100 %

4 Conclusions and Discussion

We introduced a cubical neural network structure of the auto-associative memory, the associative cube, which performs learning and recalling in image recognition. The associative cubes are based upon a cubical least-squares algorithm with the energy function for learning and the competitive network for successful recall. The training patterns are mixed with the chosen orthogonal bases in the optimal weight cube C^* and the associative cube retrieves the nearest matched pattern among them in the optimal least-squares sense. Therefore, the associative cubes have the properties of robustness under noise, gray-level preserving within the minimum squared errors. Interestingly enough, if we add the eigen-images of the Haar wavelets together in pixel-by-pixel and normalize within [-1,1], we would get the 1st training pattern, and if we add and subtract those by taking turns based on the 2nd Haar wavelet elements, the 2nd training image could be obtained with the normalization, and so on. Similarly, for the eigen-images of the Fourier kernels, we could retrieve the mth training pattern by the sum-product operation with the mth sinusoidal harmonic elements. Therefore, all the training images are imbedded in one weight cube $C*$, not losing any information by interference of the cross-talks between similar training patterns nor encountering the storage-capacity problem in the learning algorithm. It is very promising that the associative cubes may be applied to a variety of applications such as image restoration, noise-free pattern recognition, control problems, and function approximation.

Acknowledgment

This paper was performed for the intelligent robotics development program, one of the 21st Century Frontier R&D Programs funded by the Ministry of Commerce, Industry and Energy of Korea, under the Grant No. 2003-0680-100.

References

1. Hebb, D.: Organization of Behavior. Science Edition Inc., New York (1961)
2. Hopfield, J. J.: Neural Networks and Physical Systems with Emergent Collective Computational Abilities. In: Proc. National Academic Sciences USA, Biophysics 79. Apr (1982) 2554-2558.
3. Kosko, B.: Bidirectional Associative Memories. IEEE Trans. Syst., Man, Cybern. 18 (1□ (1988) 49-60
4. Kohonen, T.: Self-Organization and Associative Memory. Springer-Verlag, Berlin Heidelberg, New York (1984)
5. Wang, Y. F., Cruz, J. B., Mulligan Jr., J. H.: Two Coding Strategies for Bi-directional Associative Memory. IEEE Trans. Neural Networks 1 (1990) 81-92
6. Kang, H.: Multilayered Associative Neural Networks (M.A.N.N.), Storage Capacity vs. Perfect Recall. IEEE Trans. Neural Networks 5 (1994) 812-822
7. Kang, H.: Multilayer Associative Neural Networks: Storage Capacity vs. Noise-free Recall. In: Simpson, P. K. (ed.): IEEE Trans. Neural Network Theory, Technology, and Applications, ser. IEEE Technology Update Series. IEEE, New Jersey. (1996) 215-221
8. Wang, Y. J., Lee, D. L.: A Modified Bi-directional Decoding Strategy based on the BAM Structure. IEEE Trans. Neural Networks 4 (1993) 710-717
9. Lee, D. L., Wang, Y. J.: Neighbor-layer Updating in MBDS for the Recall of Pure Bipolar Patterns in Gray-Scale Noise. IEEE Trans. Neural Networks 6 (1995) 1478-1489
10. Burrus, C. S., Gopinath, R. A., Guo, H.: Introduction to Wavelets and Wavelet Transforms. Prentice-Hall International Inc., New Jersey (1998)
11. Lewis, F. L., Optimal Estimation. John Wiley & Sons Inc., New York (1986)
12. Kang, H.: Unsupervised Learning with Associative Cubes for Robust Gray-Scale Image Recognition. In: Proc. ICNN&B'05, Beijing China (2005)

A Novel Graph Kernel Based SVM Algorithm for Image Semantic Retrieval

Songhe Feng[1,2], De Xu[1], Xu Yang[1], and Yuliang Geng[1]

[1] Dept. of Computer Science & Technology, Beijing Jiaotong Univ., Beijing, China 100044
[2] Beijing Key Lab of Intelligent Telecommunications Software and Multimedia, Beijing University of Posts and Communications, Beijing, China 100876
songhe_feng@163.com, xd@computer.njtu.edu.cn

Abstract. It has been shown that support vector machines (SVM) can be used in content-based image retrieval. Existing SVM based methods only extract low-level global or region-based features to form feature vectors and use traditional non-structured kernel function. However, these methods rarely consider the image structure or some new structured kernel types. In order to bridge the semantic gap between low-level features and high-level concepts, in this paper, a novel graph kernel based SVM method is proposed, which takes into account both low-level features and structural information of the image. Firstly, according to human selective visual attention model, for a given image, salient regions are extracted and the concept of Salient Region Adjacency Graph (SRAG) is proposed to represent the image semantics. Secondly, based on the SRAG, a novel graph kernel based SVM is constructed for image semantic retrieval. Experiments show that the proposed method shows better performance in image semantic retrieval than traditional method.

1 Introduction

Support Vector Machine (SVM) has shown its capacity in pattern recognition and recently been used for content-based image retrieval (CBIR) [1-3]. However, existing approaches using SVM were mostly focused on extracting effective global or region-based image features to form feature vectors as the training examples, which can not reflect the entire image semantics

In order to effectively represent the image semantics and narrow down the semantic gap, based on the human selective visual attention model, the concept of salient region adjacency graph (SRAG) is proposed to represent the image and further used for retrieval. Since traditional kernel function can not deal with structural data, here a novel graph kernel function is introduced and used for SVM.

Graph kernel has been created for prediction of the properties of chemical compounds and protein classification which have achieved good discrimination ability [5] [6]. Different from some existing kernel functions which can only be used in feature vector space, graph kernel can be used to deal with the structural data. Since SRAG is adopted to represent the image content, here, we propose an improved graph kernel for the image semantic retrieval. In contrast with previous methods, the advantage of the proposed graph kernel based SVM lies in that not only the low-level image

J. Wang et al. (Eds.): ISNN 2006, LNCS 3972, pp. 589 – 594, 2006.

features but also the structure features are considered for classification and retrieval. To the best of our knowledge, this graph kernel based SVM has not been used for content-based image retrieval systems.

The rest of this paper is organized as follows. In section 2, a novel method is introduced to extract salient regions and determine the salient region adjacency graph (SRAG). Our proposed graph kernel for SVM is explained and used for image semantic retrieval in Section 3. Then, Section 4 shows experimental results. Finally, conclusions are given in Section 5.

2 Salient Region Adjacency Graph Extraction Method

Region Adjacency Graph (RAG) is an effective way to represent an image because it provides a “spatial view” of the image. However, not all the segmented regions in the image are useful for image retrieval or further process, such as backgrounds or other unimportant regions. Here in order to reduce the computation cost and model human visual mechanism, we propose Salient Region Adjacency Graph (SRAG) to represent the image semantic content. In order to create Salient Region Adjacency Graph from the image, the salient regions should be firstly extracted. Then based on the salient regions, SRAG can be created subsequently.

2.1 Salient Region Detection

Selective visual attention model has been recognized that detection of region attracting user attention is much helpful in many applications such as scene analysis and object recognition [4]. In contrast to existing salient region detection methods which only consider the saliency map, in this paper, we combine the selective visual attention model with image segmentation method. The novelty of the combination lies in that it provides a robust and reliable method for salient region detection especially when there exists more than one salient region in the image.

In order to fulfill the proposed method, we first compute the saliency value by an improved selective visual attention model on pixel-level. Then, after combining image segmentation method, a maximum entropy-based algorithm is used to determine the salient regions. The details can be seen in our previous work [8].

2.2 Salient Region Adjacency Graph Construction

After salient regions are extracted from the image, the Salient Region Adjacency Graph (SRAG) can be created subsequently. Psychological experiments have shown that, human shift the attention from one salient object to another salient object. However, existing selective visual attention model can only simulate the shift from one salient region to another salient region. There exists the essential difference between object and region. Some existing methods [2] try to use so called attention window to represent the object or salient region, but how to define the size of the attention window has no general way. Although perfect object extraction is impossible from region-based segmentation, object is defined as a group of the related regions according to Gestalt rules, so here we propose the concept of Salient Region Adjacency Graph (SRAG) to denote the salient groups.

The Salient Region Adjacency Graph (SRAG) can be defined by two steps. Firstly, take each salient region as root vertex, and those regions that are adjacent to the root vertex are included to form the so called basic region adjacency graph (BRAG); Secondly, connect all the root vertices to group the BRAGs together, and form the final Salient Region Adjacency Graph (SRAG).

This method is based on two considerations: on one hand, each BRAG can denote the concept of "salient group" which is useful to simulate human's selective visual attention model; on the other hand, the final SRAG for image retrieval can effectively reduce the computation cost and avoid the NP-hard problem.

3 Graph Kernel Based SVM for Image Semantic Retrieval

Support Vector Machine (SVM) has been recently used for image retrieval [1] [2]. As mentioned above, existing methods using SVM only consider the low-level image features (color, texture, shape etc.) to form feature vectors, and use some typical non-structured kernel functions to deal with nonstructural data.

However, these methods rarely consider the image structure or some new kernel types. Therefore, they can't reflect the entire image semantics. Since image itself is a structural language and the input training examples are SRAGs, traditional kernel functions can't satisfy the structural data. In this paper, we consider not only the low-level image features, but also the image structure in order to extract high-level image semantics. Hence, we propose a novel graph kernel based SVM to handle structural data.

3.1 Preprocessing for SVM

Let the training set be $\Omega = \{(G_i, y_i) | i = 1 \cdots N\}$. Each labeled training sample is a pair (G_i, y_i) that consists of a SRAG $G_i = (V_i, E_i)$ and the class $y_i \in \{+1, -1\}$ that the SRAG belongs to. Here, $y_i = +1$ denotes G_i is similar to the query image and -1 otherwise. In each SRAG, $G_i = (V_i, E)$ where V_i denotes all the vertices in SRAG and E_i denotes the set of weighting value between vertices pairs.

As mentioned above, the proposed graph kernel takes into account both low-level visual features and image structure. So for each vertex in a given SRAG, according to [7], we extract some region-based low-level visual features which include 7-dimensional LUV dominant colors and color variances, 6-dimensional region locations to represent the vertex.

3.2 Graph Kernel Construction

Since the kernel function is crucial to the SVM classifier, here we concern with the proposed graph kernel construction. Suppose we want to calculate the kernel for a pair of SRAGs $G_1 = (V_1, E_1)$ and $G_2 = (V_2, E_2)$, according to the descriptions adopted by Kashima [7], the graph kernel can be defined as

$$K(G_1, G_2) = \frac{1}{|V_1| \bullet |V_2|} \sum_{v_1 \in V_1} \sum_{v_2 \in V_2} k(v_1, v_2) \tag{1}$$

Here, taking into account the structure of SRAG in each image, we define $k(v_1, v_2)$ as below

$$k(v_1, v_2) = s(v_1, v_2) \bullet \left\{ (1-\lambda) + \lambda \sum_{\substack{e_1 \in A(v_1) \\ e_2 \in A(v_2)}} \frac{s(e_1, e_2)}{|A(v_1)| \bullet |A(v_2)|} \bullet k(\delta(v_1, e_1), \delta(v_2, e_2)) \right\} \tag{2}$$

where $s(v_1, v_2)$ and $s(e_1, e_2)$ are indicator functions that returns 1 when the two vertices (edges) are similar and returns 0 otherwise. $A(v)$ is a set of edges adjacent to v, and $\delta(v, e)$ is a transition function that returns the vertex at the other side of e adjacent to v. $\lambda \in [0,1]$ is considered as a decaying constant. Here, we set $\lambda = 0.6$. The similarity functions for $s(v_1, v_2)$ and $s(e_1, e_2)$ can be separately defined as

$$D(v_1, v_2) = \sqrt{\sum_{i=1}^{n} (F_{v1}^{i} - F_{v2}^{i})^2} \tag{3}$$

$$D(e_1, e_2) = e^{\frac{-[D(v1, \delta(v1, e1))]^2}{\sigma}} - e^{\frac{-[D(v2, \delta(v2, e2))]^2}{\sigma}} \tag{4}$$

where F_{v_1} and F_{v_2} denotes the low-level feature vector of vertex v_1 and v_2, n is the dimension of the feature vector.

Here $D(v_1, v_2)$ and $D(e_1, e_2)$ are Euclidean distance functions between two vertices (edges), $s(v_1, v_2)$ (or $s(e_1, e_2)$) returns 1 when $D(v_1, v_2)$ (or $D(e_1, e_2)$) is lower than the threshold α (or β) and returns 0 otherwise. The threshold α and β are defined empirically and are set to 0.25 and 0.15 respectively in the experiment.

According to Kashima [5], the graph kernel $K(G_1, G_2)$ can be interpreted using a random walk on the vertex product graph $G_{1\times 2} = (V_1 \times V_2, E_{1\times 2} \subseteq E_1 \times E_2)$ of two graphs G_1 and G_2. However, in order to reduce the computational cost, we improve the random walks as follows. Unlike Kashima's method [5] where vertex is chosen randomly as the starting point, we only choose root vertices in the SRAG as the starting points. After the novel graph kernel is constructed, the SVM classifier can be used for image semantic retrieval.

4 Experimental Results

To evaluate the performance of the proposed method, we choose about 5000 images of 38 categories from the Corel Photo Gallery as our test image database. In current experiment, we choose a subset of the database which includes 2000 images from 10 selected categories. The selected ten categories are: flower, sunset, racecar, butterfly, forest, boat, building, horse, waterfall and factory.

In order to perform training, here we take sunset category as example. From the total 200 images of the sunset category, we choose 80 images as the training positive examples, and randomly choose 80 images from the other 9 categories as negative examples. The remaining 120 images of the sunset category are used for testing.

Firstly, for each of 160 training examples, according to the human selective visual attention model, the salient regions are extracted and the SRAG is created to represent the image semantics. Here, for each vertex in the SRAG, the 13-dimension low-level features are used to denote the vertex label. Secondly, based on the proposed graph kernel function, the extracted SRAGs are used for training. Finally, the trained SVM is used for image retrieval in the test image database.

We test the proposed graph kernel based SVM method using different query images and retrieval numbers. For comparison, here we use the traditional SVM classifier using Gaussian radial basis function (GRBF) as kernel function and use the proposed 13-dimension features to form feature vector. The comparison of average precision and recall rates are seen in Fig. 1.

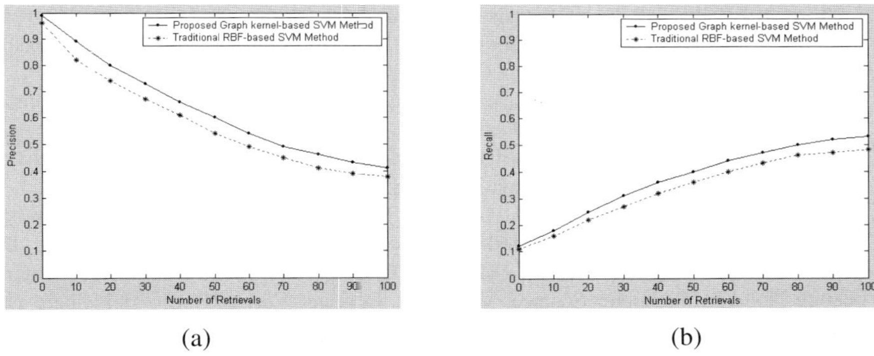

Fig. 1. (a) and (b) denote the average precision and recall rates of ten categories. (a) Precision comparisons. (b) Recall comparisons.

5 Conclusion

This paper presents a novel graph kernel based SVM algorithm for image semantic retrieval. The proposed method bridges the semantic gap by two steps. Firstly, according to human selective visual attention model, a novel salient region extraction method is introduced and so called salient region adjacency graph (SRAG) is created. Secondly, based on the SRAG, a novel graph kernel-based SVM is introduced for image semantic retrieval. In contrast with the traditional SVM kernels which only consider the global feature vectors, the advantage of the proposed graph kernel lies in that it considers both low-level features and image semantic structures. Although the experimental results have shown the encouraging performance, there still needs improvement. Future research work includes how to represent the semantic object effectively especially in the complicated image and how to design a better structured kernel to reflect the image semantic structure.

Acknowledgement

This work was supported by the Beijing Jiaotong University Research Project under Grant No.2004SM013.

References

1. Jing, F., Li, M., Zhang, H., Zhang, B.: An Efficient and Effective Region-Based Image Retrieval Framework. IEEE Trans. Image Processing 13(5) (2004) 699-709
2. Ko, B., Kwak, S., Byun, H.: SVM-based Salient Region(s) Extraction Method for Image Retrieval. in Proc. of IEEE International Conference on Pattern Recognition (2004)
3. Gondra, I., Heisterkamp, D.: Learning in Region-Based Image Retrieval with Generalized Support Vector Machines. In: Proc. of IEEE Computer Society Conference on Computer Vision and Pattern Recognition Workshops (CVPRW04) (2004)
4. Itti, L., Koch, C., Niebur, E.: A Model of Saliency-Based Visual Attention for Rapid Scene Analysis. IEEE Trans. Pattern Analysis and Machine Intelligence 20(11) (1998) 1254-1259
5. Kashima, H., Inokuchi, A.: Kernels for Graph Classification. In: Proc. of ICDM Workshop on Active Mining (2002)
6. Kashima, H., Tsuda, K., Inokuchi, A.: Marginalized kernels between labeled graphs. in Proc. of the Twentieth International Conference on Machine Learning (2003)
7. Fan, J., Gao, Y., Luo, H., Xu, G.: Statistical Modeling and Conceptualization of Natural Images. Journal of Pattern Recognition 38 (2005) 865-885
8. Feng, S., Xu, D., Yang, X., Wu, A.: A Novel Region-Based Image Retrieval Algorithm Using Selective Visual Attention Model. In: Jacques Blanc-Talon., Wilfried Philips., Dan Popescu, Paul Scheunders. (Eds.): Advanced Concepts for Intelligent Vision Systems. Lecture Notes in Computer Science, Vol. 3708. Springer-Verlag, Berlin Heidelberg New York (2005) 235-242

Content Based Image Retrieval Using a Bootstrapped SOM Network

Apostolos Georgakis* and Haibo Li

Digital Media Laboratory (DML), Department of Applied Physics and Electronics, Umeå University, SE-90187, Sweden
apostolos.georgakis@tfe.umu.se

Abstract. A modification of the well-known PicSOM retrieval system is presented. The algorithm is based on a variant of the self-organizing map algorithm that uses bootstrapping. In bootstrapping the feature space is randomly sampled and a series of subsets are created that are used during the training phase of the SOM algorithm. Afterwards, the resulting SOM networks are merged into one single network which is the final map of the training process. The experimental results have showed that the proposed system yields higher recall-precision rates over the PicSOM architecture.

1 Introduction

Image retrieval systems have become a necessity due to the mass availability of image information brought about by the spread of digital cameras and the Internet. The available systems today can be devised into two categories; the *keyword-based* systems and the *content-based* systems (CBIR). Keyword-based retrieval uses traditional database techniques to manage images. Using textual labels the images can be organized by topical or semantic hierarchies to facilitate easy navigation and browsing based on standard Boolean queries. Comprehensive surveys of early text-based image retrieval methods can be found in [1, 2]. Most text-based image retrieval systems require manual labeling which of course is a cumbersome and expensive task for large image databases. CBIR has been subjected to intensive research effort for more than two decade [3, 4, 5].

CBIR uses features related to the visual contents of an image such as color, shape, texture, and spatial layout to represent and index the images in the database (*training set*, $\mathcal{I}_{tr}$). Through this approach a typical image is described by a high-dimensional feature vector. The feature vectors corresponding to the images of the database form the feature space.

This paper provides a novel CBIR system which is based on the *self-organizing map* [6]. The proposed system is a variant of the *Picture SOM* (PicSOM) system which has been proposed by Laaksonen *et. al.* in [7]. The PicSOM system is a framework on which various algorithms and methods can be applied for content-based image retrieval. It relies on the so-called *Self-Organizing Map* (SOM).

* The work was supported by the Faculty of Science grand No. 541085100.

J. Wang et al. (Eds.): ISNN 2006, LNCS 3972, pp. 595–601, 2006.

In what follows, section 2 provides a brief description in feature extraction, section 3 covers the standard SOM algorithm while section 4 describes the presented variant. Finally, section 5 correspond to the experimental results related to the performance of the proposed system against the testbed system.

2 Feature Extraction

Let $\mathcal{I}_{tr}$ denote the image database. Let also $\mathbf{I}_i \in I\!R^{N_w}$, where N_w corresponds to the dimensionality of the feature vectors, denote the ith image in the database. In encoding each image into a numerical vector the following steps are taken:

- Each image is resized into a predetermined size.
- Each image is automatically segmented into a set of adjacent regions [8].
- Each region is encoded into a set of descriptors. The descriptors consist of property values which are defined over an entire region. The specific properties can be geometric, statistical or textural, or properties specified in some transform domain (*e.g.*, Fourier, Hough, Wavelet, Splines).
- All the descriptors corresponding to one image are packed together into on vector.

Since the PicSOM system corresponds to a framework for CBIR no particular preference is given on the descriptors employed. In this paper a wavelet transform is employed.

3 Self-organizing Maps

The SOM are feed-forward, competitive artificial neural networks that were invented by T. Kohonen [6]. The neurons on the computational layer are fully connected to the input layer and are arranged on a low-dimensional lattice. Grids with low dimensionality have prominent visualization properties, and therefore, are employed on the visualization of high-dimensional data.

Let $\mathcal{W}$ denote the set of reference vectors of the neurons, that is, $\mathcal{W} = \{\mathbf{w}_l(t) \in I\!R^{N_w}, l = 1, 2, \ldots, L\}$, where the parameter t denotes discrete time and L is the number of neurons on the lattice. Due to its competitive nature, the SOM algorithm identifies the best-matching, winning reference vector $\mathbf{w}_s(t)$ (or winner for short), to a specific feature vector $\mathbf{I}_j$ with respect to a certain distance metric. The index s of the winning reference vector is given by:

$$s = \arg \min_{l=1}^{L} \|\mathbf{I}_j - \mathbf{w}_l(t)\|, \tag{1}$$

where $\|\cdot\|$ denotes the Euclidean distance. The reference vector of the winner as well as the reference vectors of the neurons in its neighborhood are modified toward $\mathbf{I}_j$ using:

$$\mathbf{w}_i(t) = \begin{cases} \mathbf{w}_i(t) + a(t)\left[\mathbf{I}_j - \mathbf{w}_i(t)\right] & \forall i \in \mathcal{N}_s \\ \mathbf{w}_i(t) & \forall i \notin \mathcal{N}_s \end{cases} \tag{2}$$

where $a(t)$ is the learning rate and $\mathcal{N}_s$ denotes the neighborhood of the winner and the transition between the time instants t and $t+1$ is achieved whenever the entire feature space has been presented to the network.

3.1 Tree Structured SOM

The sequential search of the winner neuron in both the training as well as the testing phase of the SOM algorithm imposes a severe speed bottleneck in any SOM manifestation that deals with either large data sets or high-dimensional spaces. Among the various SOM speed-up approaches that can be found in the literature prominent position has the so-called *tree structured SOM* (TS-SOM) [6]. In TS-SOM the features are represented in hierarchical 2D or 3D grids of neurons where each grid is a standard SOM. The tree topology reduces the complexity (both time and computation) for the identification of the winner neuron.

4 Bagging SOM

From Eq. (2) is evident that the standard SOM algorithm performs an approximation of the unknown pdf of the feature space. This approximation is evident in the updating of the reference vectors of the neurons comprising the lattice. A logical question that arise here is how to boost the performance of the approximation. In doing so one probable answer is the ensemble of predictors.

Recently a number of predictor combining have been proposed [9, 10]. Perhaps the simplest approach is to bag the predictors. This paper proposes a variant of the standard SOM algorithm which relies in *bagging*, that is, on an ensemble or combination of predictors. Bagging works by applying a learning algorithm on a number of bootstrap samples of the feature space. Each of these applications yields a clustering or classification. The resulting ensemble of classifiers is combined by taking a uniform linear combination of all the constructed classifiers.

Bootstrapping is a simple but also effective approach of estimating a statistic of the feature space. The method consists of creating a number of pseudo data subsets, $\mathcal{I}^i, i = 1, 2, \ldots, D$, by sampling the set $\mathcal{I}_{tr}$ with uniform probability with replacement of each sample.

Each instance of the standard SOM algorithm is then trained separately using one of the $\mathcal{I}^i, \forall i$ data subset. Afterwards the networks are "merged" in order to create a final network in a process which will be explained in the subsection 4.1. Due to the fact that the SOM networks are trained on modified instances of the feature space (the density functions of the subsets $\mathcal{I}^i$ are expected to be different), then, with high probability we expect to get slightly different resulted network topologies.

Petrakieva *et. al.* in [11] are outlining the problems that arise with the SOM algorithm when the reference vectors of the neurons are randomly initialized. Namely, the resulted network topologies, even in the case of using the same feature space, are not completely identical. Since the SOM algorithm performs a dimensionality reduction through the mapping of the feature space on the 2D lattice, the maps are free to twist in any direction which oftenly offers local minimizations.

4.1 Network Merging

Let $\mathcal{L}_{ij}$ denote the set of feature vectors that are assigned to the neuron $\mathbf{w}_i^j$. The first step which generally admittedly is the most difficult one is the combination of the outputs of the several SOM networks toward the formation of one final SOM map, that is to find the associations between the neurons $\mathbf{w}_i^j, \forall i, j$.

For example, let suppose that there are only three SOM networks and each is assigned to extract five clusters from the space $\mathcal{I}_{tr}$. The goal afterwards is to combine the networks in a way so that features to be placed in a same neuron of the final map if and only if they were assigned to a same neuron in both of the networks. This task is not trivial because there is no guarantee that the ith cluster in the first network corresponds to the ith cluster in the second network. So, the networks must be aligned before they can be merged.

In this paper, the neurons are aligned according to the assumption that neurons that are "similar" should be close to each other also in the $I\!R^{N_w}$. In reality the order is reversed; neurons that are close to each other in the $I\!R^{N_w}$ should be similar. That is, lets $\mathbf{w}_{a_1}^{b_2}$ and $\mathbf{w}_{a_3}^{b_4}$ be the neurons whose sets $\mathcal{L}_{a_1}^{b_2}$ and $\mathcal{L}_{a_3}^{b_4}$ respectively contain more common features than any other neuron couple. Then, the reference vectors corresponding to these neurons will be, with higher probability, closer to each other than another possible neuron combination under the Euclidean distance.

In aligning the networks one should partition the LD neurons into L disjoint clusters with two constraints: a)each cluster will contain only D neurons, and b) each cluster will contain only one neuron from each of the D networks. The constrains simply state that each neuron of the output map is the union of just one neuron from each of the D constituent SOM networks. Under the above constrains is evident that a brute force approach to the global minimization problem has complexity $O\left(L^D\right)$[1] which obviously is unacceptable even for small values of the parameter D whereas the suboptimal solution described in subsection 4.2, which relies on dynamic programming, has complexity $O\left((D-1)\,L^2\right)$.

4.2 Neuron Alignment Through Dynamic Programming

In ordering the neurons according to dynamic programming the proposed approach uses the paired distance of neuron pairs in the $I\!R^{N_w}$ space. That is, if just two networks are to be merged then each neuron from the hypothetical network A should be matched with the neuron from the network B that would have been closer. Figure 1 depicts the above process. The first step is to merge networks A and B. In Fig. 1 it can be seen that the neuron $\mathbf{w}_1^1$ is closer to $\mathbf{w}_2^2$

[1] The reported computational complexity is due to the fact that we need to construct all the D-tuplets where each SOM network contributes with just one neuron per arrangement.

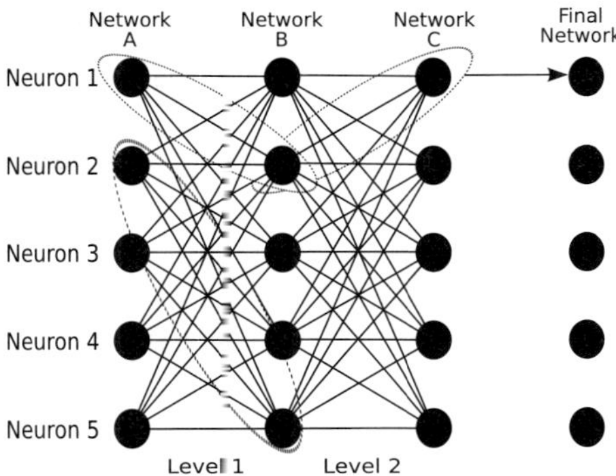

Fig. 1. The alignment of three networks and the subsequent merging into one final map

and further more $\mathbf{w}_2^1$ is closer to $\mathbf{w}_5^2$ and so on. The average vector between each pair of neurons will be used afterwards in the second level to merge the third map (network C) into the previous two networks.

In merging the third map onto the previous two, one need to compute the distances between the average vectors from the previous level and the reference vectors from network C. In that case the pair $\{\mathbf{w}_1^1, \mathbf{w}_2^2\}$ is closer to $\mathbf{w}_1^3$ and therefore these three neurons are grouped together. After the last network has been merged with the previous two maps we need to "build" the final network (see Fig. 1). In doing so the reference vector of each neuron is the average vector of the neurons grouped together in the previous step.

The last step towards the formation of the final map is the creation of the set $\mathcal{L}_{ifinal}$ which is the set of features assigned to the ith neuron of the final map. These sets are the unions of the sets corresponding to the clusters of neurons formed in the previous step. Let f_{ij} denote frequency of the jth image in the set $\mathcal{L}_{ifinal}$. If the frequency is close to the value D then more neurons in the constituent networks had that particular image assigned to them during the training phase. Therefore, the more frequent an image is the higher its importance to the particular neuron of the final map. That is, the images assigned to a particular neuron are ordered into descending order of frequency.

5 Evaluating Retrieval Performance

The proposed bootstrapped SOM approach is evaluated against the PicSOM architecture with a set of experimental settings using two image databases. The first one corresponds to the Corel Gallery [12] and the second collection corresponds to the SUNET image database [13].

Aiming at assessing the retrieval performance of the proposed SOM variant against that of the basic SOM method used in PicSOM two retrieval systems are trained using the two image databases. Afterwards, the systems are queried

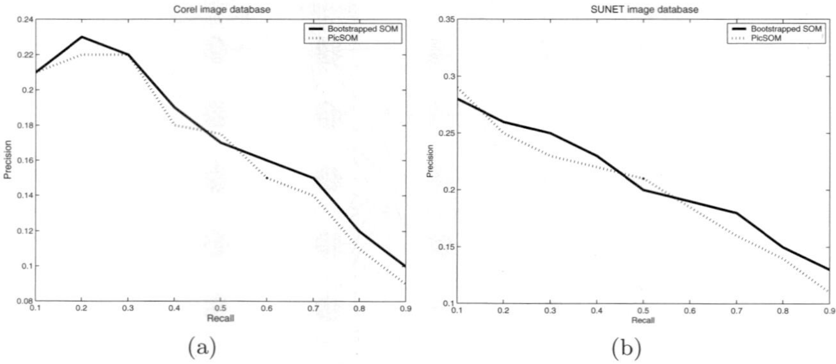

(a) (b)

Fig. 2. The average recall-precision curves for the PicSOM and the proposed SOM variant for: (a) the Corel image database and (b) the SUNET image database

using query-images randomly selected from the same datasets. The query-images undergo the same preprocessing steps as the images in the $\mathcal{I}_{tr}$.

For each image-based query, the system retrieves those training images that are represented by the best matching neuron of the final SOM map for both architectures. The retrieved images are by default ranked inside each neuron due to the process described in subsection 4.2. Finally, the retrieved images are labeled as either relevant or not to the query-images, with respect to the annotation category they bear. For each query, this classification leads to a different partition of the training set according to the retrieval volumes. The effectiveness of the proposed algorithm against the standard SOM is measured using the *precision* and the *recall* ratios [14, 15].

As the volume of retrieved images increases the above ratios are expected to change. The sequence of $(recall, precision)$ pairs obtained yields the so-called *recall-precision curve*. An average over all the curves corresponding to the same annotation categories that were obtained from the test set produces the average recall-precision curve [14].

Figures 2a and 2b depict the average recall-precision curves for the PicSOM architecture and the proposed SOM variant for all the annotation categories. It becomes evident that, in general, the bootstrapping of the feature space provides superior performance over the standard SOM algorithm with respect to volume of retrieved images despite the isolated lags.

6 Conclusions

This paper has provided a variant of the well-know PicSOM architecture for content-based image retrieval. The proposed modification relies on bootstrapping. In bootstrapping, the feature space is randomly sampled and a series of subsets are created that are used during the training phase of the SOM algorithm. Afterwards, the resulted SOM networks are merged into one single network which is the final map of the training process. The experimental results have showed that the proposed system yields higher recall-precision rates over the PicSOM architecture.

References

1. Chang, S.K., Hsu, A.: Image Information Systems: Where Do We Go from here? IEEE Trans. on Knowledge and Data Eng. **5** (1992) 431–442
2. Tamura, H., N.Yokoya: Image Database Systems: A Survey. Pattern Recognition **1** (1984) 29–43
3. Bimbo, A.D.: Visual Information Retrieval. San Mateo, CA: Morgan Kaufmann (1999)
4. Long, F., Zhang, H.J., Feng, D.: Fundamentals of Content-Based Image Retrieval. In Feng, D., Siu, W.C., Zhang, H.J., eds.: Multimedia Information Retrieval and Management - Technological Fundamentals and Applications. Springer (2002)
5. Lew, M.S.: Principles of Visual Information Retrieval. Springer Verlag, Heidelberg, Germany (2000)
6. Kohonen, T.: Self Organizing Maps. 3rd edn. Springer Verlag, Heidelberg, Germany (2001)
7. Laaksonen, J., Koskela, M., Oja E.: PicSOM - A Framework for Content-Based Image Database Retrieval Using Self-Organizing Maps. In: Proc. of 11th Scandinavian Conf. on Image Analysis (SCIA'99). (1999)
8. Pal, N.R., Pal, S.K.: A Review on Image Segmentation Techniques. Pattern Recognition **26** (1993) 1277–1294
9. Bakker, B., Heskes, T.: Clustering Ensembles of Neural Network Models. Neural Networks **12** (2003) 261–269
10. Breiman, L.: Using iterated Bagging to Debias Regressions. Machine Learning **45** (2001) 261–277
11. Petrakieva, L., Fyfe, C.: Bagging and Bumping Self-Organising Maps. Computing and Information Systems Journal (2003)
12. Corel: 1.300.000 Photo Gallery. ("http://www.corel.com")
13. SUNET: Image Database. ("ftp://ftp.sunet.se/pub/pictures")
14. Korfhage, R.R.: Information Storage and Retrieval. NY: J. Wiley (1997)
15. Sebastiani, F.: Machine Learning in Automated Text Categorization. ACM Computing Surveys **34** (2002) 1–47

Unsupervised Approach for Extracting the Textural Region of Interest from Real Image

Woo-Beom Lee, Jong-Seok Lim, and Wook-Hyun Kim

Department of Computer Engineering, Yeungnam University,
214-1 Dae-dong, Gyeongsan-si, Gyeongbuk 712-749, Republic of Korea
{beomlee, robertlim, whkim}@yumail.ac.kr

Abstract. Neural network is an important technique in many image understanding areas. Then the performance of neural network depends on the separative degree among the input vector extracted from an original image. However, most methods are not enough to understand the contents of a image. Accordingly, we propose a efficient method of extracting a spatial feature from a real image, and segmenting the TROI (: Textural Region Of Interest) from the clustered image without a pre-knowledge. Our approach presents the 2-passing k-means algorithm for extracting a spatial feature from image, and uses the unsupervised learning scheme for the block-based image clustering. Also, a segmentation of the clustered TROI is achieved by tuning 2D Gabor filter to the spatial frequency the clustered region. In order to evaluate the performance of the proposed method, the segmenting quality was measured according to the goodness based on the segmented shape of region. Our experimental results showed that the performance of the proposed method is very successful.

1 Introduction

Textures in image which described as fine, coarse, grained, smooth, etc. are a very effective cue for image segmentation, and neural network is an important technique in many image understanding areas. Then the performance of neural network for analyzing texture in image depends on the separative degree among the input vector extracted from an original image.

Although a numerous approach has been studied in this literature during the past decades[1, 2, 3, 4], most of these methods use a simple spatial feature having the low separative degree, plus are restricted to issue of supervised problems requiring a pre-knowledge. Thus, with respect to the separative degree of spatial feature and the unsupervised problem, previous methods are not enough to understand the contents of a image

Accordingly, we propose a efficient method of extracting a spatial feature from a real image, and segmenting the TROI from the clustered image without a pre-knowledge. The proposed method uses the 2-passing k-means algorithm for improving a separative degree of spatial feature. The extracted spatial feature corresponds to the input vector of SOM(: Self-Organizing Map). After the unsupervised learning of a SOM, the image is clustered into block-based type.

J. Wang et al. (Eds.): ISNN 2006, LNCS 3972, pp. 602–609, 2006.

In order to segment the clustered TROI, 2D Gabor filter is tuned to the spatial frequency of the clustered region. Unsupervised segmentation is achieved by applying zero-crossing algorithm to the thresholding image of the tuned Gabor filter. Then the thresholding value for segmentation is automatically determined by analyzing the Gabor filtering response.

2 The Unsupervised TROI Clustering

The unsupervised TORI clustering used in our approach consists of the following three stages: spatial feature extraction, the input vector production for a SOM, block-based clustering, and block-based merging.

2.1 Spatial Feature Extraction Using 2-Passing k-Means Algorithm

We define a pair of the spatial feature values for any sampling pixel S(x, y) in an image as follows

$$f = (f_A,\ f_M) = (\tan^{-1}(dx/dy),\ \sqrt{(dx^2 + dy^2)}). \tag{1}$$

where f_A and f_M are the spatial feature values of orientation angle and magnitude respectively. $dx = I(x+1, y) - I(x-1, y), dy = I(x, y+1) - I(x, y-1)$ is a horizontal and vertical gradient value for any $S(x, y)$ in an image respectively, and $I(x, y)$ is a gray level intensity in $S(x, y)$.

After completing the calculation for all sampling points in an image, the 2-passing k-means algorithm is performed for improving the separative degree of a spatial feature. Considering any block $B_{k,l}$, which is divided into equally $n \times n$ sized pixels from an image. A set $F_{k,l} = \{f_1, f_2, f_3, \cdots, f_{n\times n}\}$ of all f in the block $B_{k,l}$ are projected into the orientation features coordinate system that composed of the five regions, $R_{\{1,2,3,4,5\}}$ like shown Fig. 1(a). The 1-passing processing is then performed by optimizing the objective function as follows

$$K(F_{k,l}, C) = \sum_{i=1}^{n \times n} min_{j\in\{1,2\dots5\}} \parallel f_i - c_j \parallel^2, \tag{2}$$

and yields the 5 cluster centers $C_{k,l} = \{c_1, c_2, c_3, c_4, c_5\}$ corresponding to each region[5]. However, because the distribution of spatial feature is not uniform in the most case, several cluster centers become exist in the same region of the orientation features coordinate system(Fig. 1(c)). Thus, the cluster centers existing in the same region R_i are merged by 2-passing stage like form

$$z_i = \left(\sum_{c_j \in R_i}^{k=5} c_j \right) / N_i, \tag{3}$$

where N_i is the number of c that exists in R_i. As a results, the final centers $Z_{k,l} = \{z_1, z_2, z_3, z_4, z_5\}$ corresponding to spatial features of block can be acquired by this procedure.

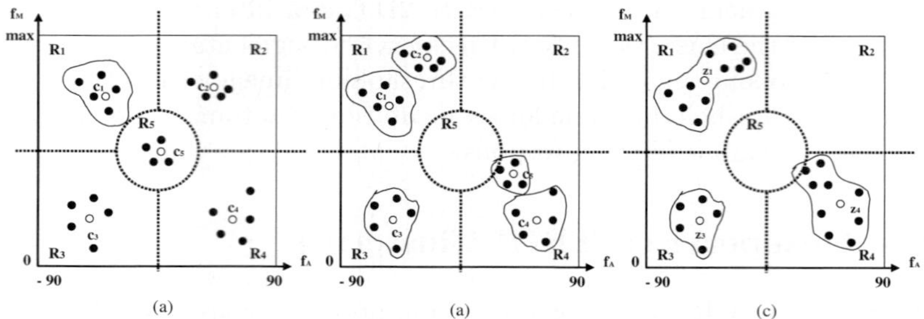

Fig. 1. Spatial feature extraction using 2-passing k-means algorithm ($k = 5$). (a) Center points C in the uniform case, (b) Center points C in the real case, (c) Merged center points Z by 2-passing k-means algorithm.

2.2 The Unsupervised Learning

We propose a novel approach for unsupervised block-based image clustering that uses self-organizing neural network schemes by Kohonen[5]. Fig. 2 shows the architecture of neural network used in our approach.

The input pattern vector $x_{k,l}$ for $B_{k,l}$ block may be defined as follows

$$x_{k,l} = Z_{k,l} = [z_1, z_2, z_3, z_4, z_5]^T \quad : \text{where, } \; z_i = [a_i, m_i, n_i] . \tag{4}$$

where a_i and m_i is the cluster centroid value of the region R_i respectively, and n_i is the total number of feature element that exists in the region R_i.

After yielding the input pattern vectors for all block in an image, the winner neuron T_j for unsupervised learning is determined by using the minimum-distance Euclidean criterion like form (5) and (6):

$$T_j = \begin{cases} \arg \{ \min \parallel y_{k,l} - w_j \parallel \} & \text{if } \; T_j \in T_{set} \\ \text{assign to new cluster, and append to } T_{set} & \text{otherwise} \end{cases} , \tag{5}$$

$$y_{k,l} = [i_1, i_2, i_3, i_4, i_5]^T \quad : \text{where, } \; i_i = \sum v_i c_i . \tag{6}$$

After the winner neuron T_j is selected, the block-based clustering is performed by using a competitive learning scheme like form (7) and (8):

$$w_j(n+1) = \begin{cases} w_j(n) + \alpha[y_{k,l} - w_j(n)] & \text{if } \; y_{k,l} \in T_j \\ w_j(n) & \text{otherwise} \end{cases} , \tag{7}$$

$$v_j(n+1) = v_j(n) + \delta_j[z_j - v_j(n)] \quad : \text{where, } \; \delta_j = \parallel i_j - w_j \parallel , \tag{8}$$

where α is the learning-rate parameter of w layer, and δ is the learning-rate parameter of v layer. It has a back-propagation effect by the learning of w layer.

During the unsupervised learning, each of blocks in an image is then clustered by comparing a similarity of between blocks. Consequently similar blocks

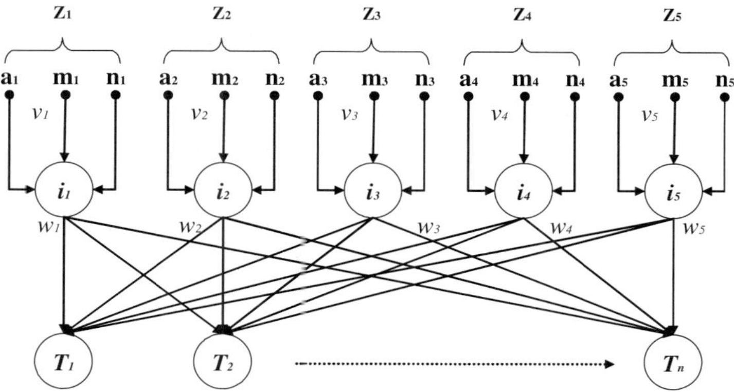

Fig. 2. The architecture of self-organizing neural network used in our approach

are assumed to belong to the same cluster, thus unique labels are assigned to each block. However, one textures may be split into several parts, causing a fragmentation problem despite a homogeneous texture region in image. Thus, to overcome this problem, a block-based merging procedure is proposed. If one block is identified as similar to a neighboring block, the same label is assigned and blocks are merged, yielding a number of distinct regions in the image.

3 The Unsupervised TROI Extraction

If one of the clustered TROIs is selected from the result of the block-based image, 2D Gabor filter for the unsupervised TROI extraction is designed by analyzing the spatial frequency of the clustered TROI. This is expected to extract from image.

3.1 Spatial Frequency Analysis of the Clustered TROIs

Each TROI, corresponding to the results of the block-based image clustering, are transformed into the frequency domain using the Fourier transformation method. As a result, each transformed region has a number of optimal frequency candidates, corresponding to the sorting-searching frequencies of the spectral peaks detected in the Fourier spectrum images of each regions.

To extract the optimal frequency for the t-th TROI, abbreviating to the TROI_t, the frequency distance $FD_t(k)$ can be defined as follows:

$$FD_t(k) = \sum_{i=1}^{n} \|OFC_{tk} - MF_i\|, \quad \text{where } (i \neq t,\ 1 \leqq k \leqq m)\,, \tag{9}$$

where OFC_{tk} is the k-th optimal frequency candidate in the TROI_t, which corresponds to the k-th center frequency that is ordered by the spatial frequency spectral magnitude of the TROI_t, and MF_i are the highest center frequencies in the TROI$_i$ that exists in the browsing results. Thus, $FD_t(k)$ is the sum of the Euclidean distance between the k-th optimal frequency candidate OFC_{tk} in the TROI$_t$ and the highest frequencies MF_i in the TROI$_i$. The optimal frequency OF_t for extracting the TROI$_t$ is determined by

$$OF_t = arg \left\{ max_{(1 \leqq k \leqq m)} \left\{ FD_t(k) \right\} \right\} = \{of_u, \ of_v\}, \tag{10}$$

where $\{of_u, of_v\}$ corresponds to the pixel location in the Fourier transformed image of the TROI$_t$. Here, OF_t does not means to the absolute optimal frequency, but the relative optimal frequency, which is the most distinct spatial frequency for segmenting when compared to the highest frequencies of the TROIs in the image.

3.2 2D Spatial Gabor Filtering

2D Gabor filters are more appropriates for analyzing image containing a very specific frequency and orientation characteristic, as they have a tunable orientation, center frequency, radial frequency bandwidth. The 2D Gabor function as a spatial filter in image processing is defined as

$$h(x,y) = g(x,y) \cdot \exp\Big(-2\pi i [u_0(x - x_0) + v_0(y - y_0)] \Big), \tag{11}$$

$$\text{where} \quad g(x,y) = \exp\big(-\pi[(x - x_0)^2 a^2 + (y - y_0)^2 b^2] \big).$$

(x_0, y_0) and (u_0, v_0) specify the center of the Gabor function in the spatial and frequency domain, respectively, as a and b denote the two scale parameters, $\lambda = b/a$ specifies the aspect ratio, and the radial center frequency F and orientation θ can be calculated as $F = \sqrt{u_0^2 + v_0^2}$ and $\theta = \tan^{-1}(v_0/u_0)$, respectively. Thus, a 2D Gabor filter is a complex sinusoidal function in the spatial domain, and acts as a local bandpass filter in the frequency domain.

Here the effective method is proposed for analyzing the filtering response. In a discrete spatial convolution, the Gabor function has real and imaginary components respectively given by

$$h_r(x,y) = g(x',y') \cdot \cos(2\pi F x'), \quad h_i(x,y) = g(x',y') \cdot \sin(2\pi F x'), \tag{12}$$

where $(x', y') = (x \cos\phi + y \sin\phi, -x \sin\phi + y \cos\phi)$ are rotated coordinates. However, for simplicity, ϕ is not considered with an aspect ratio $\lambda = 1$. Since the function $h_r(x,y)$ and $h_i(x,y)$ are symmetrically even and odd , respectively, along the preferred orientation direction, the convolution results are approximately identical, other than a difference of $\pi/2$ in the phase spectra. Therefore, a more uniform response can be obtained by considering the real and imaginary parts simultaneously. The response of the optimized Gabor filter, $m_t(x,y)$ can be defined in the form:

$$m_t(x,y) = \sqrt{of_r^2(x,y) + of_i^2(x,y)}, \tag{13}$$

$$\text{where} \quad of_r(x,y) = h_r(x,y) * t(x,y), \quad of_i(x,y) = h_i(x,y) * t(x,y).$$

$t(x, y)$ is regarded as the texture model corresponding to the TROI, * denotes the convolution operator, and $of_r(x, y)$ and $of_i(x, y)$ are the real and imaginary responses, respectively, of the optimized Gabor filter.

The optimized Gabor filtering then results in essentially uniform responses in the similar texture regions, which means that the proposed optimized Gabor filters can be effective for segmenting distinct texture regions in an image. Accordingly, after applying the optimized Gabor filter to the original image, the segmentation is then completed by extracting uniform regions from the response image.

4 The Unsupervised TROI Segmentation

The unsupervised segmentation of a TROI requires a threshold value for the binary transformation. Therefore, after applying the optimized Gabor filter to the corresponding TROI in the clustering results, the thresholding value for extracting the TROI can be created by

$$\psi\,[\ l : low\ bound,\ u : upper\ bound\ ] = \psi\left[\frac{\lfloor \omega \times L \rfloor}{\omega},\ \frac{\lceil \omega \times H \rceil}{\omega}\right], \tag{14}$$

where H and L are the highest and lowest response value, respectively, for $m_t(\cdot)$, in the TROI, ω is the precision coefficient, and $\lceil\cdot\rceil$ and $\lfloor\cdot\rfloor$ denote the $ceiling()$ function and $floor()$ function, respectively, for the truncation using the integer transformation. Thus, without any pre-knowledge or heuristic decision, the upper and lower bound for the binary image transformation can be automatically determined by using a threshold function in Eq. (14). Then, the binary image for extracting the TROI can be defined as below:

$$b_t(x, y) = 1 \quad \text{if } (l \leqq f(x, y) \leqq u),\ \text{otherwise } 0 \tag{15}$$

Finally, the TROI segmentation result $E_t()$ is achieved by using image AND-op. between the original image $O()$ and the thresholed binary image $b_t()$:

$$E_t(x, y) = O(x, y) \wedge b_t(x, y) \quad \text{for } \forall(x, y). \tag{16}$$

5 Experimental Results

In order to evaluate the segmentation quality of the proposed approach,the segmentation quality was measured by using two shape similarity[5]. One is the ratio of the area to the perimeter of a segmented region. The other is the ratio the width of the minior axis to the length of the major axis. Therefore, the shape measure function of a segmented region, SM can be defined as follows:

$$SM = \left(\frac{4\pi A}{P^2}\right) + \left(\frac{MaAL}{MiAW}\right), \tag{17}$$

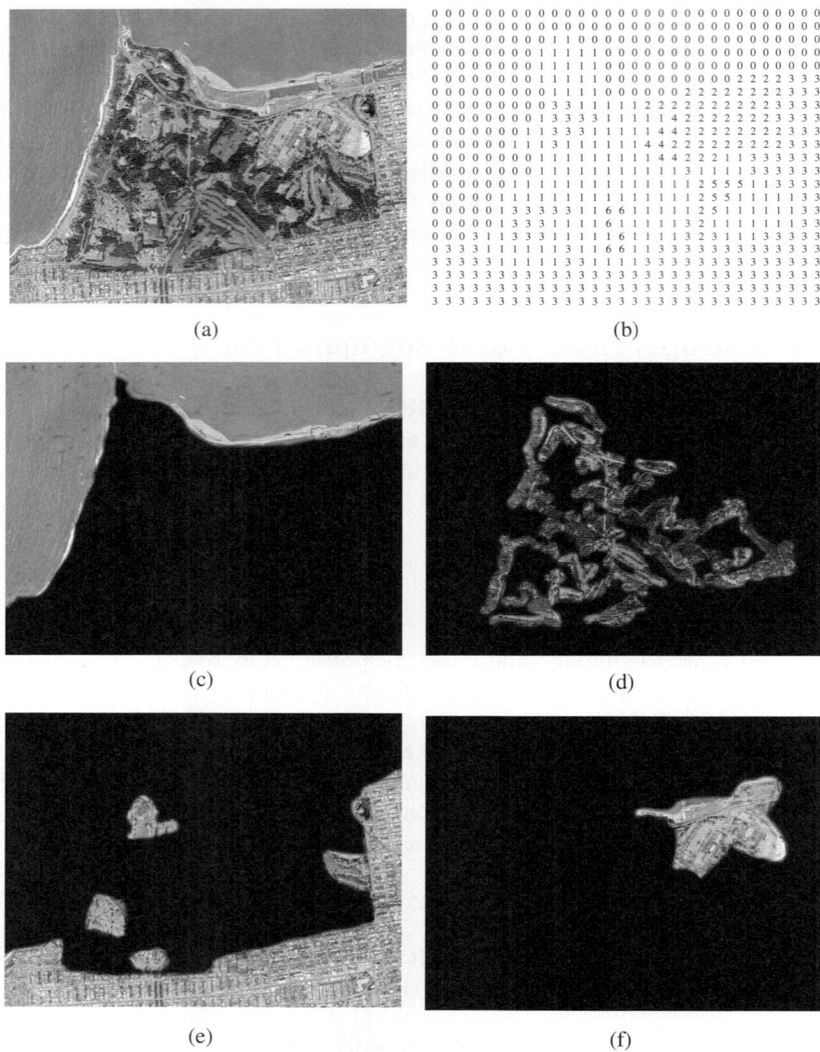

Fig. 3. Experimental results : (a) original satellite image: 600 × 460 pixels, (b) block-based clustering results: 20 × 20 pixels/block, (c) sea parts image, (d) forest parts image, (e) a housing complex parts image, (f) flat land parts image

where A is the total number of pixel inside the segmenting boundary, and P is the pixel distance around the circumference of the segmenting boundary. Also, $MaAL$ is the pixel distance length between the major axis endpoints, and $MiAW$ is the pixel distance length between the minor axis endpoints. As a results, the shape similarity between the ground truth and the segmented results were measured at the over to 85% about a various more than 100 real images. Fig. 3 shows the experimental result of a satellite image.

6 Conclusions

We propose a efficient method of extracting a spatial feature from a real image, and segmenting the TROI from the clustered image without a pre-knowledge. The proposed method uses the 2-passing k-means algorithm for improving a separative degree of spatial feature. Also, the focus of our approach is to reduce the constraint problems by the unsupervised learning schemes, and to improve the segmenting quality based on a shape goodness for recognition. However, some problems, such as block size, quantization, and the size of an 2D spatial filter, are left for future works. Nonetheless, the proposed method demonstrated a very successfully results in respect of the segmented shape similarity. This means that the proposed method guarantees the high quality in the shape similarity despite reducing the constraint problems. Therefore, our method has enough to detect and retrieve a partial image corresponding to a query image.

References

1. Randen, T., Husoy, J.: Filtering for Texure Classification: A Comparative Study. IEEE Trans. PAMI 21(4) (1999) 291-310
2. Idrissa, M., Acheroy, M.: Texture Classification using Gabor Filters. Pattern Recognition Letters 23 (2002) 1095-1102
3. Tsai, D., etc.: Optimal Gabor Filter Design for Texture Segmentation using Stochastic Optimazation. Image and Vision Computing 19 (2001) 299-316
4. Grigorescu, S., etc.: Comparesion of Texture Feature Based on Gabor Filters. IEEE Trans. Image Precessing 11(10) (2002) 1160-1167
5. Lee, W.B., Kim, W.H.: Texture Segmentation by Unsupervised Learning and Histogram Analysis using Boundary Tracing. In: Yue, H., et al (eds.): Computational Intelligence and Security. Lecture Notes in Artificial Intelligence, Vol. 3801. Springer-Verlag, Berlin Heidelberg New York (2005) 25-32

Image Fakery and Neural Network Based Detection

Wei Lu[1,2], Fu-Lai Chung[2], and Hongtao Lu[1]

[1] Department of Computer Science and Engineering,
Shanghai Jiao Tong University, Shanghai 200030, China
[2] Department of Computing, Hong Kong Polytechnic University,
Hung Hom, Kowloon, Hong Kong, China
luweicn@sjtu.edu.cn, cskchung@comp.polyu.edu.hk, lu-ht@cs.sjtu.edu.cn

Abstract. By right of the great convenience of computer graphics and digital imaging, it is much easier to alter the content of an image than before without any visually traces. Human has not believed what they see. Many digital images can not be judged whether they are real or feigned visually, i.e., many fake images are produced whose content is feigned. In this paper, firstly, image fakery is introduced, including how to produce fake images and its characters. Then, a fake image detection scheme is proposed, which uses radial basis function (RBF) neural network as a detector to make a binary decision on whether an image is fake or real. The experimental results also demonstrated the effectiveness of the proposed scheme.

1 Introduction

The development of digitization technology leads to the popularization of digital images. Nowadays, digital image editing softwares are cheap and convenient, thus, the content of images can be altered anywhere and anytime when it is needed to be what is wanted. What human saw has not been able to be believed. We see many images that are irrational and illogical with their content, where we can confirm the content of the images are altered. Unfortunately, all of the images can not be decided whether they are real or fake. Fake images are full of our lives. Fake images can be defined as the images that do not express the images imply originally. Fake images can be seen everywhere, such as movies, advertisements, etc. There is no general model for determining if an image is real or fake. However, some of fake images can be distinguished visually, since the content of these images are not natural. However, to those fake images that are produced for purpose of cheating, especially for political advantage, they can barely be distinguished.

Few contributions appeared about image fakery. Some similar concepts with image fakery have been introduced in the past ten years. One is digital forgery, the other is image splicing. In [1], digital forgeries refer to manipulate and alter digital images, also a method by detecting traces of resampling is proposed to exposing digital forgeries, and some other statistical tools for detecting digital

J. Wang et al. (Eds.): ISNN 2006, LNCS 3972, pp. 610–615, 2006.

(a) (b)

Fig. 1. Examples of fake images

forgeries are also proposed in [2]. In [3], a model for image splicing is proposed to combine difference objects of images into a new image. Aiming at how to detect spliced images, a statistics based model, bicoherence feature, is introduced [4]. Different from these image alteration concepts, fakery is to change the content or objects of images. Generally speaking, fake images are naturally in their signal character without any visual inconsistency. Figure 1 gives some examples of fake images, where a bomb is hung under a hawk in Fig. 1(a), and a girl stood on a crocodile in Fig. 1(b), which show that fake images are consistent in their signal character with real images.

Practically, it is hard to determine whether an image is real or fake from signal character, since image fakery is to altering the content of an image, i.e., adding, removing and replacing some objects into the image. Thus, it may be feasible to detect fake images from the relation and distribution of objects. From Fig. 1(a), we can see that it is illogical for a flying hawk hanging a bomb, so it can be concluded that this image is fake. Unfortunately, all of fake images can not be distinguished visually.

2 Image Fakery

In order to detect fake images, first of all, we need to know how fake images are created. Although early fake images are produced using darkroom photography techniques, it is obvious that almost all of fake images are generated using image editing software today. Here, we only discuss the common process of image fakery that combine different parts or objects of two images into one fake image aided by digital image editing tools. Generally speaking, there are four phases for image fakery that is shown in Fig. 2. The first phase is selecting fake object or area, where one of the main problems is to choose appropriate fake area, i.e., the fake object and the fake image should be similar with their lights, quality, tone, etc. The second phase is adjusting the fake object, where adjustment is to editing the fake object and make a proper object fakery, including scaling, rotation and lighting. The third phase is pasting the adjusted fake object into the original fake area to make a spliced image. The final phase is processing

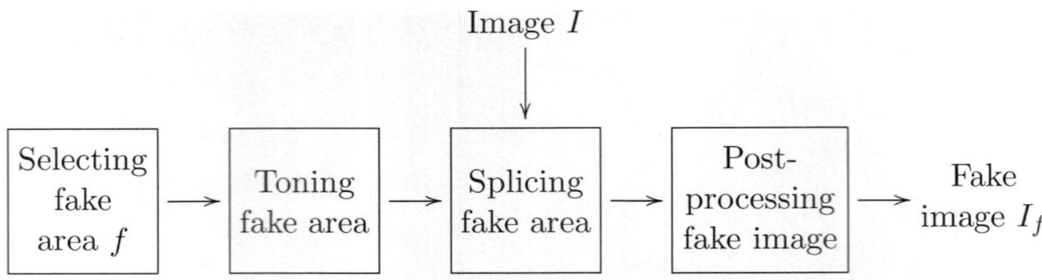

Fig. 2. A model for image fakery

the spliced image through image manipulation tools. Since the splicing is only a simple copy and paste process, there are some obvious artificial evidence, such as visible boundary between the fake object and un-fake area. In order to make a natural fake image, postprocessing is applied, such as toning, lighting, sharping, blurring, filtering, etc.

From Fig. 1 we can see that fake images are created through altering and replacing the content of the original image. So it is not any manipulated images which can be called fake images, i.e., fake images refer to those whose content is changed on condition that the image content is consistent in their signal characters. For example, filtering, noise addition and geometrical distortions can not be thought as fakery. Strictly speaking, only the images whose object or content is replaced with different object or content are fake images.

It is untouched how to exposing fake images. Although many fake images can be discerned to be inconsistent with human general knowledge. For example, in Fig. 1(a) it is impossible that a bomb is hung under a hawk. However, to the un-discerned fake images with human eyes, the methods should be developed to exposing them. In the rest parts of this paper, we introduced a fake image detection scheme based on neural network. Firstly, a watermark is inserted into the original image, and the altered area can be detected by extracting the watermark. Then, a radius basis function (RBF) neural network based detector is constructed to classify fake images and un-fake images, since there are many characters for RBF neural networks, such as simple network structure, fast learning ability, good classification ability and better approximation capability.

3 The Proposed Detection Scheme of Fake Images

In this section, the proposed method for detecting fake images is described in detail. Firstly, a fragile digital watermarking scheme is introduced as an assistant, which is used to find out the alteration of images. Then, the proposed neural network detector is trained using the constructed training set. Finally, the detection of fake images is performed by the output of the neural network.

3.1 Watermarking of Images

In our fake detection scheme, firstly, the altered area in an image should be able to be detected. Therefore, a LSB based fragile watermarking scheme [5] is used to watermark the original image, then the watermarked image is open to public.

In [5], a watermark W with the same size of the original image, i.e., $M \times N$, is constructed to embedded into the LSB plane of the original image I, where each watermark element is inserted into the LSB of each image pixel. A watermarked image I_w is obtained. When the watermark is detected, there will be difference between the extracted watermark W_e and the original watermark W, if the watermarked image is altered. Based on the difference a matrix can be obtained as follow:

$$A = \text{XOR}(W, W_e) \tag{1}$$

where A is the matrix composed of 0 and 1, the size of which is the same with the original image. Based on A, the alteration of the watermarked image can be concluded that when $A(i,j) = 1$, the pixel $I_w(i,j)$ in the watermarked image is altered, when $A(i,j) = 0$, it is not altered.

3.2 Training Neural Network

Once we get the matrix A, a training set for the proposed neural network can be constructed. Firstly, the matrix A is divide into blocks with size $m \times n$, i.e., $A = \{A_{xy} | x = 1, 2, \cdots, M/m,\ y = 1, 2, \cdots, N/n\}$. A new matrix B can be obtained as follow:

$$B(x,y) = \frac{1}{m \times n} \sum_{i=1}^{m} \sum_{j=1}^{n} A_{xy}(i,j) \tag{2}$$

Thus, the input set for the proposed neural network can be obtained, i.e., $B = \{B(x,y) | x = 1, 2, \cdots, M/m,\ y = 1, 2, \cdots, N/n\}$. To the output of the neural network δ, when the input image is fake, δ is set to 1, and when the input image is real, δ is set to 0. Thus, the training set is constructed as $T = \{B, \delta\}$, using which the proposed RBF neural network is trained which is a $(M/m \times N/n) - 8 - 1$ multiple layer perceptron.

3.3 Judgment of Image Fakery

In order to determine whether input image is fake, firstly, the watermark W_e is extracted from the input image, the matrices A and B are obtained one after the other. Then, the matrix B is input in the trained neural network, and an output δ is obtained. Finally, the decision on whether the input image is fake can be made that if $\delta \geq 0.5$, the input image is fake, and if $\delta < 0.5$, the input image is not fake.

4 Experimental Results

In this section, some experiments are carried out for proving the effectiveness of the proposed neural network based fake image detection scheme. Figure 3(a) shows the watermarked version of the popular image Lenna, its fake version is shown in Fig. 3(b), which is one of experimental image pair. At the same time, 250 fake images are used to train the RBF neural network.

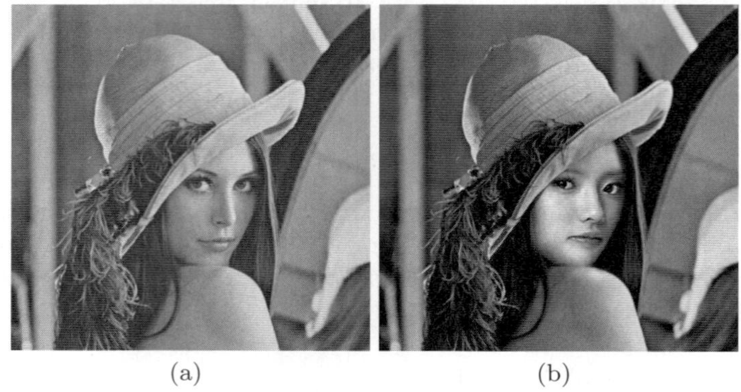

Fig. 3. (a) The open image. (b) The fake image.

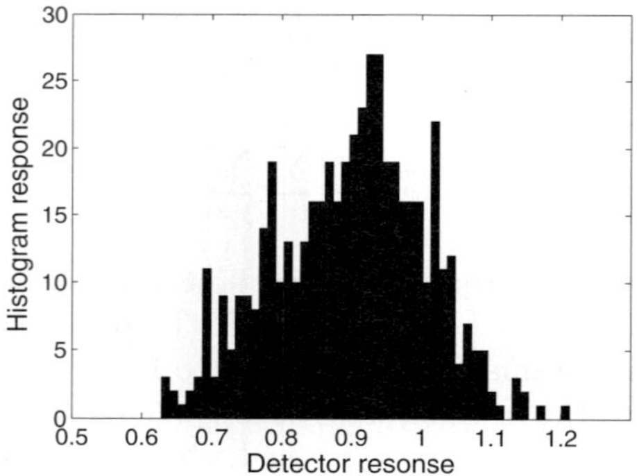

Fig. 4. The histogram response for 500 fake images (epochs=50)

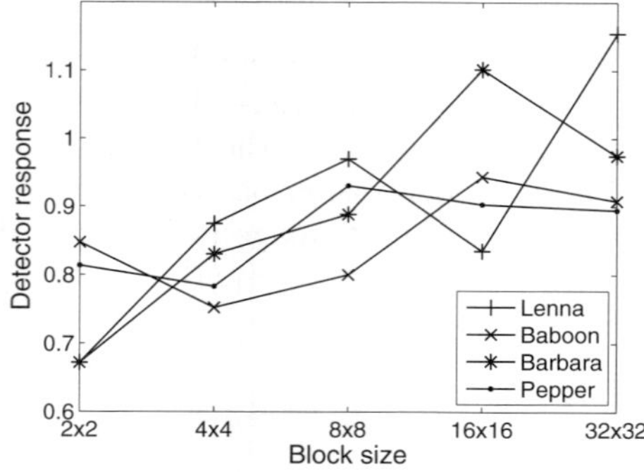

Fig. 5. The relation between the block size and the neural network detector

Figure 4 shows the histogram response of the neural network based detector's output under other 500 fake images. The low boundary value of the histogram is larger than 0.5, and most of the detection values locate in the range from 0.6 to 1.2, which shows that all of these tested fake images can be detected correctly.

In our scheme the size of blocks is also influential on the output of the neural network. Figure 5 gives some experimental results about the output δ of the neural network under different block size, which shows that the detector output increases with the increasing of the block size in the mass.

5 Conclusions

In this paper, we have proposed a detection scheme based on neural network for detecting fake images. Firstly a LSB-based watermarking is applied. Then, based on the detected altered area using the watermarking, the training set is constructed to train the proposed RBF neural network. Thus, fake images are detected based on the output of the proposed neural network. The purpose of watermarking is to detect the altered area. So the performance of the proposed watermarking influences directly the fake detection result. Furthermore, the security of the watermarking is also important, if it is attacked, the fake detection scheme will fail. Future work is to improve the proposed scheme without the assistant of digital watermarking or signature.

Acknowledgments

This work is supported by NSFC under project no. 60573033 and the ICRG grant under project A-PG49.

References

1. Popescu, A.C., Farid, H.: Exposing Digital Forgeries by Detecting Traces of Resampling. IEEE Transactions on Signal Processing **53** (2005) 758–767
2. Popescu, A., Farid, H.: Statistical Tools for Digital Forensics. In: 6th International Workshop on Information Hiding. Volume 3200., Toronto, Cananda (2004) 128–147
3. Ng, T.T., Chang, S.F.: A Model for Image Splicing. In: IEEE Int. Conf. on Image Processing (ICIP). Volume 2. (2004) 1169–1172
4. Ng, T.T., Chang, S.F., Sun, Q.: Blind Detection of Photomontage Using Higher Order Statistics. In: IEEE Int. Symposium on Circuits and Systems (ISCAS). Volume 5. (2004) 688–691
5. Lu, H., Shen, R., Chung, F.L.: Fragile Watermarking Scheme for Image Authentication. Electronics Letters **39** (2003) 898–900

Object Detection Using Unit-Linking PCNN Image Icons

Xiaodong Gu, Yuanyuan Wang, and Liming Zhang

Department of Electronic Engineering, Fudan University, Shanghai 200433, P.R. China
guxiaodong@263.net

Abstract. A new approach to object detection using image icons based on Unit-linking PCNN (Pulse Coupled Neural Network) is introduced in this paper. Unit-linking PCNN, which has been developed from PCNN exhibiting synchronous pulse bursts in cat and monkey visual cortexes, is a kind of time-space-coding SNN (Spiking Neural Network). We have used Unit-linking PCNN to produce the global image icons with translation and rotation invariance. Unit-linking PCNN image icon (namely global image icons) is the 1-dimentional time series, and is a kind of image feature extracted from the time information that Unit-linking PCNN code the 2-dimentional image into. Its translation and rotation invariance is a good property in object detection. In addition to translation, rotation invariance, the object detection approach in this paper is also independent of scale variation.

1 Introduction

Traditional AFRNNs (Average Firing Rate Neural Networks) are space-coding models. They only use the sum of pulse-average-firing-rates of signals during a short period of time. However, time plays an important role in information coding. The timing spikes have already been well established as a means for conveying information in the visual system in files. In 1995, Hopfield indicated that the sizes of variables are represented by the explicit time at which action potentials occur, rather than by the more usual 'firing rate' of neurons [1]. On the same issue, Sejnowski thought that time may be a new neural code. Lately SNNs(Spiking Neural Networks), pulse-emitting and space-time coding models, have attracted the attention of many researches. They mimic real neurons better and have more powerful computation performance than traditional AFRNNs[2]. PCNN (Pulse Coupled Neural Network) is a kind of SNN, which has the biological support. In 1990, Eckhorn introduced the linking field network [3] based on the phenomena of synchronous pulse bursts in the cat visual cortex. In 1993, Johnson developed PCNN [4] based on the linking model, which has been applied in many fields, such as image processing, object detection, optimization [5-7]. We introduced Unit-linking PCNN [8] based on PCNN and used it efficiently in image processing, such as image thinning [9], image shadow removal [10]. Using Unit-linking PCNN can conveniently change images into time series including the information of original images. Johnson indicated that under certain structure, the firing states of all neurons in PCNN had some invariance [11]. The desired image feature can be designed from these time series. We have introduced Unit-linking PCNN global image icons and local image icons based on these time

J. Wang et al. (Eds.): ISNN 2006, LNCS 3972, pp. 616 – 622, 2006.

series and have used them in navigation [12,13]. Especially Unit-linking PCNN global image icons have shown qualified performance in non-stationary-video navigation. In non-stationary-video navigation, the success rate of the navigation based on the global image time icon has been much better than based on CCIPCA(Candid Covariance-free Incremental Principal Component Analysis)[14]. For example [13], in the non-stationary-video AIBO (a SONY machine dog) navigation, the success rate of the navigation based on the global image time icon has been 90% , and increased by 35% than that of the navigation based on CCIPCA. The qualified performance of Unit-linking PCNN global image icon in non-stationary-video navigation is due to its translation and rotation invariance. In this paper, we use Unit-linking PCNN global image icon for object detection. Its translation, rotation invariance is also an advantage in object detection. In addition to translation and rotation invariance, the object detection approach in this paper is also independent of scale variation. For brevity, global image icon is called image icon in this paper.

2 Unit-Linking PCNN Image Icons

2.1 Unit-Linking PCNN

PCNN exhibits synchronous pulse bursts in cat and monkey visual cortexes and is a kind of SNN. We proposed the Unit-linking PCNN based on the PCNN. A Unit-linking Pulse Coupled Neuron (PCN) consists of three parts: the receptive field, the modulation field, and the pulse generator. Fig.1 illustrates a Unit-linking PCN j. It has two channels (F channel and L channel). F channel is feeding input (F_j); and L channel is linking input (L_j). In Fig.1, I_j , an input signal from the external source, is only inputted to F channel of j ($F_j = I_j$). $Y_1,\ldots,Y_k,\ldots$, output pulses emitted by neurons connected with j , are also input signals of neuron j, and only inputted to L channel of j. Y_j is the output signal of neuron j. In a Unit-linking PCN, we introduce the concept of Unit-linking. Namely as to neuron j, when one or more than one neurons fire in its neighbor field, L_j is equal to 1 (see Eq.(1)). In Eq.(1), $N(j)$ is the neighbor field of neuron j and j may belong or not belong to $N(j)$. In this paper, j does belong to $N(j)$. Unit-linking is the main difference between traditional PCNs and Unit-linking PCNs. Compared with traditional PCNs, this reduces parameters and makes the linking inputs of Unit-linking PCNs uniform so that the behaviors of networks consisting of Unit-linking neurons are easy to analyze and control. The uniform L_j is added a constant positive bias firstly. Then it is multiplied by F_j and the bias is taken

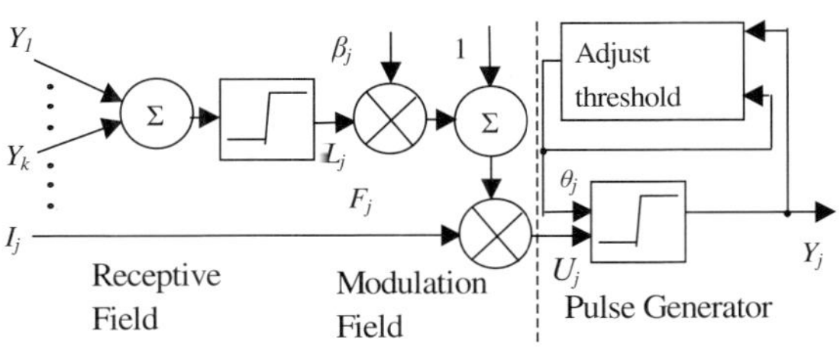

Fig. 1. A Unit-linking PCN

to be unity, and the modulation result is the total internal activity U_j (see Eq.(2), where β_j is the linking strength). Next, U_j is inputted to the pulse generator. If U_j is greater than the threshold θ_j, the neuron output Y_j turns into 1 (see Eq.(3)). Then Y_j feedbacks to make θ_j rises over U_j immediately so that Y_j turns into 0. Therefore, when U_j is greater than θ_j, neuron j outputs a pulse. On the other hand, θ_j drops lineally with time increasing (see Eq.(4)). In Eq.(4) V_j^T and α_j^T are the amplitude gain and the time constant of the threshold adjuster respectively.

$$L_j = \mathrm{Step}\big(\sum_{k\in N(j)} Y_k\big) = \begin{cases} 1, & if \quad \sum_{k\in N(j)} Y_k > 0 \\ 0, & else \end{cases} \tag{1}$$

$$U_j = F_j(1+\beta_j L_j) \tag{2}$$

$$Y_j = \mathrm{Step}(U_j - \theta_j) \tag{3}$$

$$\frac{d\theta_j(t)}{dt} = -\alpha_j^T + V_j^T Y_j(t), \quad \textit{the lower limit of integration is just before the last firing in solution} \tag{4}$$

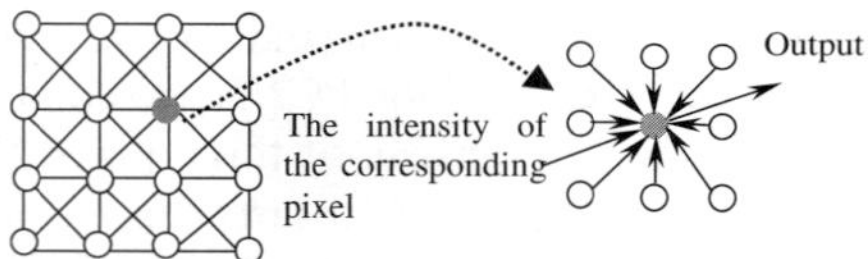

Fig. 2. The connection mode of each Unit-linking PCN for object detection

A Unit-linking PCNN consists of Unit-linking PCNs and one of those neurons is show in Fig. 1. For object detection in this paper, it is a single layer two-dimensional array of laterally linked neurons (see Fig. 2). The number of neurons in the network is equal to the number of pixels in the input image. One-to-one correspondence exists between image pixels and neurons. The F channel of each neuron receives the intensity of the corresponding pixel, and each neuron is connected with neurons in its 8-neighbor field by L channel. All neurons are identical.

2.2 Image Icons Based on Unit-Linking PCNN

Unit-linking PCNN image icon we introduced in [12,13]is the pulse train as a function of time, and is a kind of image feature. It is obtained from the time information which using Unit-linking PCNN changes the original image into. Unit-linking PCNN image icon is a 1-dimentional time series, each element of which is equal the sum of neurons of fired neurons at each iteration. Whether a neuron fires at each iteration is decided by 1) the intensity of the corresponding pixel and, 2) whether receiving the pulse waves emitted by the neighboring neurons or not. Therefore, for the grey image, Unit-linking PCNN image icon reflects the image intensity distribution pattern related to the geometry of the image. Obviously the image icon has the translation and rotation invariance because each neuron is identical and has the same connection mode. For the color image, first the original color image is divided into three channels (RGB or HSI). Then using Unit-linking PCNN to gain the Unit-linking PCNN image icon of each channel. Finally simply connecting these three image icons obtains the image icon of the color image. Obviously the color image icon also has the translation and rotation invariance. In this paper, the original color image is divided into HIS channels because the image icon of HIS channels is more stable to variation of the illumination than that of RGB channels. For object detection based on Unit-linking PCNN image icon, each one of three HSI channels is divided 20 levels to produce the image icon so the iteration times of each channel is 21 so Unit-linking PCNN color image icon including 3 channels is composed of 21*3=63 elements. By the way, sometimes for each channel we can use part elements to form image icons.

The algorithm of calculating Unit-linking PCNN image icon is described below.

(1) One channel (one of three HIS channels) of the original image is normalized between 0 and 1. β=0.2. Iteration times of each channel is 21, so total iteration times of three channels is 63 (21*3).
(2) Let pulse-waves spread fully at each level.
(3) Save the sum of neurons fired at each iteration to the image icon **T** in sequence.
(4) If a neuron fires, increase its threshold to prevent it from firing again.
(5) As to two other channels, repeat from (1) to (4).
(6) Obtain the HSI image icon **T** including 63 elements.

Here, **T**, a vector including 63 elements, saves the Unit-linking PCNN image icon based HSI. In **T,** elements from 1 to 21 form the Unit-linking PCNN image icon of I (Intensity) channel, elements from 22 to 42 form the image icon of H (Hue) channel, elements from 43 to 63 the image icon of S (Saturation) channel. More details have been shown in [12,13], where the image icon algorithm has been described by matrixes and is easy to implement by Matlab or C language.

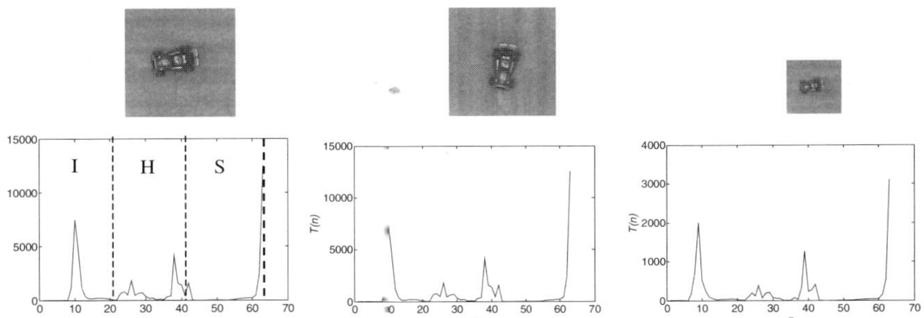

Fig. 3. Three color images in the 1st row and their HSI image icons are under them, and one-to-one correspondence exists between images and their HSI time icons

In the 1st row of Fig. 3, the left one is an image of a color truck tool (128*128), the middle one is its rotating result (128*128), and the right one (64*64) is the shrinking result of the left one. In the 2nd row of Fig.3, there are image icons under the corresponding color images. In image icons, the horizontal axis (n) indicates the iteration times and the vertical axis (T(n))indicates the sum of neurons fired at each iteration. Each HSI image icon consists of 63 elements. The image icons of the left one and the middle one are almost identical due to rotation invariance of image icons. In practice, the small rotation error caused by pixel discretization exists between the image icon of the left one and that of the middle one. The image icon of the right one is similar to that of the left one although they are not identical, because for the whole image, image icons are also independent of scale variation by using normalized Euclidean distance or vector angular distance. Note, scale invariance of image icon is some different from translation and rotation invariance of image icon. Scale invariance of image icon only exists for whole image. Translation and rotation invariance of image icon exists not only for the whole image, but also for the object in the same background.

3 Object Detection Based Unit-Linking PCNN Image Icons

The translation and rotation invariance of Unit-linking PCNN image icon is an advantage in object detection. In addition to translation and rotation invariance, the object detection approach in this paper is also independent of scale variation for the whole image.

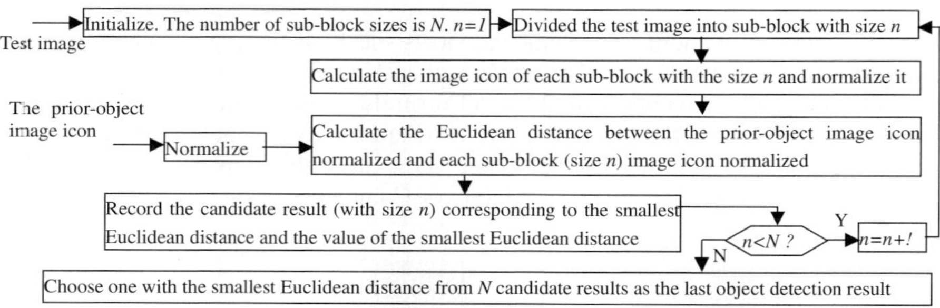

Fig. 4. Object detection based on Unit-linking PCNN image icons in this paper

In the object approach in this paper, a Unit-linking PCNN image icon of a block including the object is used as the object template, which is also called prior-object image icon. In detection (see Fig. 4), first the test image including the object is divided into sub-blocks overlapped and each image icon of each sub-block is calculated. Then calculate the Euclidean distance between the prior-object image icon normalized and each sub-block image icon normalized. The result of object detection is the sub-block corresponding to the smallest Euclidean distance. In order to be independent of scale variation, sub-blocks with different sizes are used. In this paper, three different sizes of sub-blocks are used. Each sub-block can be seen as a whole image, so can be the object block. Therefore, this object detection approach is also independent of scale variance besides translation and rotation invariance. Note, for the purpose of scale variation of object detection, before calculating the Euclidean distance between the prior-object image icon and each sub-block image icon, they all should be normalized; or vector-angular distance should be used. Namely normalized Euclidean distance or vector-angular distance is used to detect the similarity of the image icon in this approach. In order to detect the object correctly in different backgrounds, the prior object should occupy the prior block as large as possible because in this approach the prior block including the object is used to produce the object template. In detection, using the image icon of the prior block to march the image icons of different size sub-blocks in the test image detects the object, so object segmentation is avoided. According to the object detection approach in Fig. 4, even if the test image does not included the object, a sub-block with the smallest distance is still detected as the object. A distance threshold can be used to solve this problem.

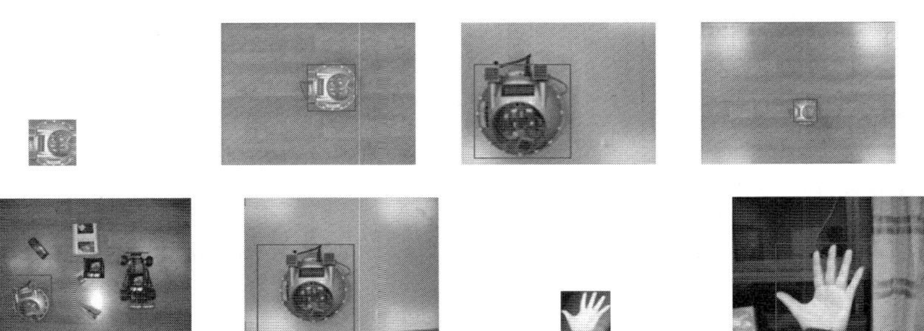

Fig. 5. The left one in the 1st row is the prior object (the robot Bread), and other images in the 1st row and 2 left ones in the 2nd row are test images, where objects with different sizes are in different backgrounds and the object detection results based on Unit-linking PCNN icon are enclosed by the red frames respectively. In the 2nd row, the 3rd one is the concaved prior object (the hand) and 4th one is the test image, where the red frame encloses the object hand.

In Fig. 5, the left one in the 1st row is the prior object (the robot Bread, 40*40), and other images in the 1st row and 2 left ones in the 2nd row are test images including objects with different sizes in different backgrounds. In the 2nd row, the 3rd one is the concaved prior object (the hand, 40*40) and 4th one is the test image including the hand. The size of each test image is 160*120. Three sizes (20*20,40*40,80*80) of sub-blocks are used. The move step of each sub-block is 5. The object detection results based on Unit-linking PCNN image icon are enclosed by the red frames in test images. Fig. 5 shows that the object robot Bread in different backgrounds and scales, and the concaved object hand are detected correctly.

4 Conclusions

This new approach to object detection based on Unit-linking PCNN image icon, is independent of translation, rotation, and scale variance. In detection, object segmentation is avoided because the whole prior block including the object is used to produce the prior image icon (namely to extract the image feather), and the prior image icon is used to match different size sub-blocks in the test image to search the object. For the complex object system, this approach may work together with other method to improve the performance of the system. Each image has only one Unit-linking PCNN image icon. However, one Unit-linking PCNN image icon corresponds many different images. Therefore, the relationship between invariance and separating capacity of the Unit-linking PCNN image icon should be researched further.

Acknowledgements

This research was supported by National Basic Research Program of China (2005CB724303), China Postdoctoral Science Foundation (No. 2003034282), Shanghai Postdoctoral Science Foundation, National Natural Science Foundation of China (Nos. 60571052, 30370392), and Key Project of Shanghai Science & Technology Committee (No. 045115020).

References

1. Hopfield J.J.: Pattern Recognition Computation Using Action Potential Timing for Stimulus Representation. Nature 376 (1995) 33-36
2. Maass W.: Fast Sigmoid Networks via Spiking Neurons. Neural Computation 9(1997) 279-304
3. Eckhorn, R., Reitboeck, H.J., Arndt, M., et al: Feature Linking via Synchronization among Distributed Assemblies: Simulation of Results from Cat Cortex. Neural Computation 2 (1990) 293-307
4. Johnson, J.L., Ritter, D.: Observation of Periodic Waves in a Pulse-coupled Neural Network. Opt.Lett. 18 (1993) 1253-1255
5. Johnson, J.L., Padgett, M.L.: PCNN Models and Applications. IEEE Trans. on Neural Networks 10 (1999) 480-498
6. Kuntimad, G., Ranganath, H.S.: Perfect Image Segmentation Using Pulse Coupled Neural Networks. IEEE Trans. on Neural Networks 10 (1999) 591-598
7. Caulfield, H.J., Kinser, J.M.: Finding Shortest Path in the Shortest Time Using PCNN's. IEEE Trans. on Neural Networks 10 (1999) 604-606
8. Gu, X.D., Guo, S. D., Yu, D. H.: A New Approach for Automated Image Segmentation Based on Unit-linking PCNN. Proceedings in IEEE International Conference on Machine learning and Cybernetics, Beijing, China (2002) 175-178
9. Gu, X.D., Yu, D.H., Zhang, L.M.: Image Thinning Using Pulse Coupled Neural Network, Pattern Recognition Letters 25 (2004) 1075-1084
10. Gu, X.D., Yu, D. H., Zhang, L.M.: Image Shadow Removal Using Pulse Coupled Neural Network. IEEE Trans. on Neural Networks 16(2005) 692-698
11. Johnson, J.L.: Pulse-coupled Neural Nets: Translation, Rotation, Scale, Distortion and Intensity Signal Invariance for Images. Applied Optics 33 (1994) 6239-6253
12. Gu, X.D., Yu, D.H., Zhang, L.M.: Global Icons and Local Icons of Images based Unit-linking PCNN and their Application to Robot Navigation. Lecture Notes In Computer Science, 3497(2005) 836-841
13. Gu, X.D.: Research on Several Theoretical and Applied Aspects on Unit-linking Pulse Coupled Neural Network. Post-doctoral Research Report, Fudan University (2005) 45-61
14. Weng, J., Zhang, Y., Hwang, W.: Candid Covariance-free Incremental Principal Component Analysis, IEEE Trans. on PAMI 25 (2003) 1034-1040

Robust Image Watermarking Using RBF Neural Network

Wei Lu[1,2], Hongtao Lu[1], and Fu-Lai Chung[2]

[1] Department of Computer Science and Engineering,
Shanghai Jiao Tong University, Shanghai 200030, China
[2] Department of Computing, Hong Kong Polytechnic University,
Hung Hom, Kowloon, Hong Kong, China
luweicn@sjtu.edu.cn, lu-ht@sjtu.edu.cn, cskchung@comp.polyu.edu.hk

Abstract. In recent years digital watermarking was developed significantly and applied broadly for copyright protection and authentication. In this paper, a digital image watermarking scheme is developed using neural network to embedded watermark into DCT domain of each subimage blocks obtained by subsampling, which achieves adaptively watermark embedding and stronger robustness. Furthermore, in order to improve the security of the proposed watermarking, a random permutation process is used in watermarking process. Experiments show that the proposed watermarking scheme is effect and encouraging.

1 Introduction

Digital watermarking is one of the ways for digital information copyright protection, authentication, tracing, etc., which is a two-step process. The first is to embed some information, i.e. watermark, into the protected digital media and open the watermarked media to public. The second is to detect the watermark or make a decision on whether the watermark exists when it is necessary. In recent years, many contributions about watermarking have been published, and it is still a hot domain now. To a watermarking system, there are some basic requirements, including robustness, security, imperceptibility.

Generally speaking, there are two types of watermarking, one type is spatial domain watermarking, the other is transformation domain watermarking. Because of good robustness and large capacity of information embedding, transformation domain watermarking is developed more, on which we focus. DCT domain watermarking is the most popular transformation domain watermarking, since its transformation coefficients are real number and easy for processing. In this paper, the constructed watermark is embedded into DCT domain.

As is well known, artificial neural networks (ANN) are mathematical tools, which is broadly applied in fitting nonlinear mapping relation. Some ANN based watermarking has been introduced [1, 2, 3], but not deeply, where back propagation neural networks (BPNN) and radial basis function neural networks (RBFNN) are often used, since they are simple and effective. In our proposed scheme, an RBF neural network is applied, since it is fast for convergent.

J. Wang et al. (Eds.): ISNN 2006, LNCS 3972, pp. 623–628, 2006.

In this paper, we develop a novel digital image watermarking scheme using RBF neural network to embed watermark into DCT domain of image blocks which are constructed using subsampling.

2 Related Theories

Artificial neural networks are broadly applied in fitting nonlinear mapping relation. Figure 1 shows an example of RBF neural network, which is a 3-5-1 3-layers structure. There are many characters for RBF neural networks, such as simple network structure, fast learning ability and better approximation capability, which is used in our watermarking scheme.

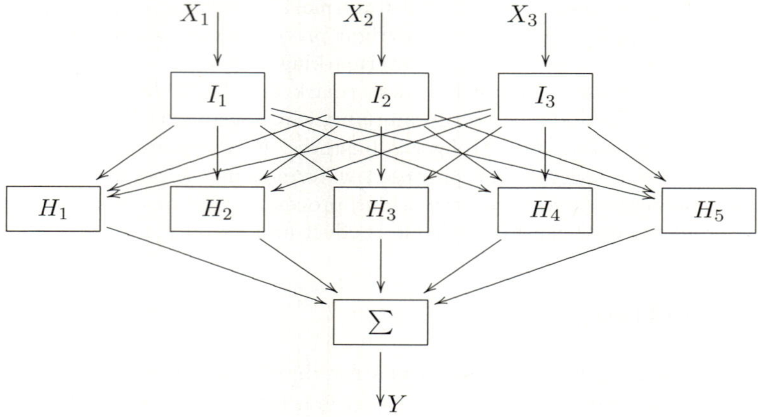

Fig. 1. The structure of RBF neural network

Subsampling is a common method in signal processing and analysis. To a digital image I with size $M \times N$, subsampling decomposes it into $m \times n$ subimages with size $M/m \times N/n$, where m and n are the subsampling intervals on the direction of row and column, which can be described as follows:

$$\begin{aligned}
I_1(i,j) &= I(m*(i-1)+1, n*(j-1)+1) \\
I_2(i,j) &= I(m*(i-1)+1, n*(j-1)+2) \\
I_3(i,j) &= I(m*(i-1)+1, n*(j-1)+3) \\
&\cdots\cdots \\
I_{n+1}(i,j) &= I(m*(i-1)+2, n*(j-1)+1) \\
I_{n+2}(i,j) &= I(m*(i-1)+2, n*(j-1)+2) \\
&\cdots\cdots \\
I_{m\times n}(i,j) &= I(m*i, n*j)
\end{aligned} \tag{1}$$

where $I_1, I_2, \cdots, I_{m\times n}$ denote the subimages.

3 The Proposed Watermarking Process

In this section, the proposed watermarking scheme is described in details. Firstly, the watermark embedding process is introduced, and an RBF neural network is used to achieve the improved capability of adaptively information embedding. Then, watermark detection is detailed.

3.1 Watermark Embedding

For the host image I with size $N \times N$, the subsampling process shown in Eq. (1) is used to construct a subimage set, i.e. $\{I_b | b = 1, 2, \cdots, m \times n\}$. Then, DCT is applied to these subimages, a coefficient set is obtained, i.e., $\{D_b | b = 1, 2, \cdots, m \times n\}$. In order to improve the security of the proposed watermarking scheme, the coefficient set is permuted randomly using a secret seed K. Thus, a encrypted coefficient sequence is obtained, i.e., $\{E_b | b = 1, 2, \cdots, m \times n\}$.

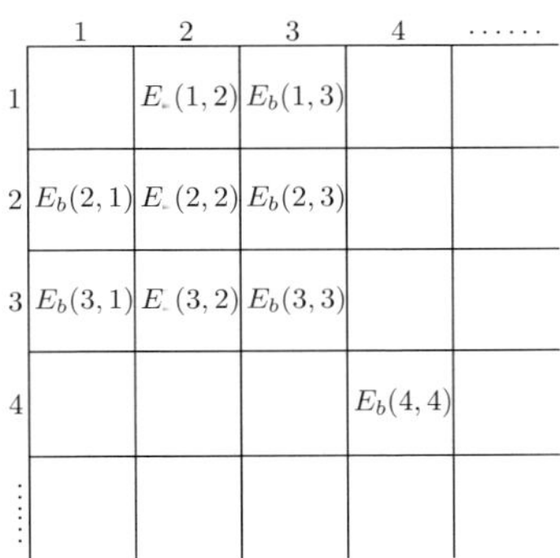

Fig. 2. The construction of the training set and the input set in RBFNN from DCT

For each DCT coefficient E_b, firstly, a coefficient subset is constructed, i.e., $S_b = \{E_b(1,2), E_b(1,3), E_b(2,1), E_b(2,2), E_b(2,3), E_b(3,1), E_b(3,2), E_b(3,3)\}$, which is shown in Fig. 2. Then, an RBF neural network with 8-5-1 structure is used for watermark embedding. In order to train the RBFNN, the first t coefficient elements in E_b are used to construct the training samples, i.e., $\{\{S_b, E_b(4,4)\} | b = 1, 2, \cdots, t\}$, where S_b is the input set, $E_b(4,4)$ is the out element, and t is the number of the training samples.

In our proposed scheme, the watermark is a random binary sequence $W = w_1, w_2, \cdots, w_l$, where $w_i \in \{-1, 1\}$ and $i = 1, 2, \cdots, l$. To the watermark embedding, the following l coefficient elements in S_b are used to construct the input set as $\{S_b | b = t+1, t+2, \cdots, t+l\}$. Suppose the watermark element w_i is to be embedded, the element S_{t+i} is fed to the trained RBFNN, and an output o_i is obtained. The watermark element w_i is embedded in o_i as follows:

$$o_i' = o_i + \delta * w_i \tag{2}$$

where δ is a scalar factor which controls the watermark embedding strength. Then, $E_{t+i}(4,4)$ is replaced with o'_i, the watermarked coefficient set E_b is obtained. Once each watermark element is embedded, through inverse random permutation, IDCT and composing of subimages, the watermarked image is obtained.

3.2 Watermark Extraction

The retrieve process of watermark is simple, which is a similar procedure with the watermark embedding. To the input image I^*, firstly, it is decomposed into subimages using subsampling as $\{I_i^*|i = 1, 2, \cdots, m \times n\}$. Then, DCT and random permutation is applied, we can get a permuted coefficient set $\{E_b^*|b = 1, 2, \cdots, m \times n\}$.

The same set is constructed as $S_b^* = \{E_b^*(1,2), E_b^*(1,3), E_b^*(2,1), E_b^*(2,2), E_b^*(2,3), E_b^*(3,1), E_b^*(3,2), E_b^*(3,3)\}$. Then, the training samples are constructed as $\{\{S_b^*, E_b^*(4,4)\}|b = 1, 2, \cdots, t\}$, which is used to train the RBFNN. In order to detect the watermark W^*, the following l elements in $\{S_b^*\}$ are input to the trained RBFNN, and the corresponding outputs are obtained as $\{o_i^*|i = 1, 2, \cdots, l\}$. Thus, W^* can be obtained as follow:

$$w_i = \begin{cases} 1 & \text{if } E_{i+t}^*(4,4) \geq o_i^* \\ -1 & \text{if } E_{i+t}^*(4,4) < o_i^* \end{cases} \tag{3}$$

In order to detect the similarity between the extracted watermark W^* and the original watermark W, the correlation coefficient is used as follow:

$$\rho = \frac{\sum_{i=1}^{l} w_i^* * w_i}{\sqrt{\sum_{i=1}^{l} (w_i^*)^2 * \sum_{i=1}^{l} (w_i)^2}} \tag{4}$$

If $\rho \geq 0.5$, the watermark W is considered to be exist in the image I^*, and if $\rho < 0.5$, the watermark W does not exist in the image I^*.

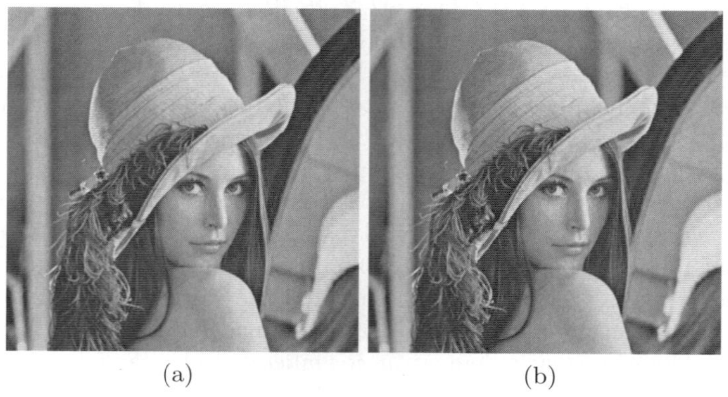

(a) (b)

Fig. 3. (a) The original image. (b) The watermarked image.

4 Simulation Results

The performance of the proposed watermarking scheme has been examined for invariant images. One of the tested images, Lenna, is shown in Fig. 3(a), and its watermarked version is shown in Fig. 3(b), where the watermark factor $\delta = 10$, $M = N = 512$, $m = n = 64$, $t = l = 1000$. The correlation coefficient ρ between the extract watermark from Fig. 3(b) and the original watermark is 1, which shows the watermark exists exactly.

Figure 4(a) shows the response of the correlation detector ρ under JPEG compression, and Fig. 4(b) shows the response under Gaussian lowpass filter and Gaussian noise addition using the image Lenna, which indicate that the proposed scheme is robust against these digital image manipulations. We also tested the proposed scheme using many other images. Figure 5 gives a histogram response of the correlation coefficient ρ under JPEG quality 80 using 500 images, which shows the strong robustness of the propose scheme. Table 1 shows some

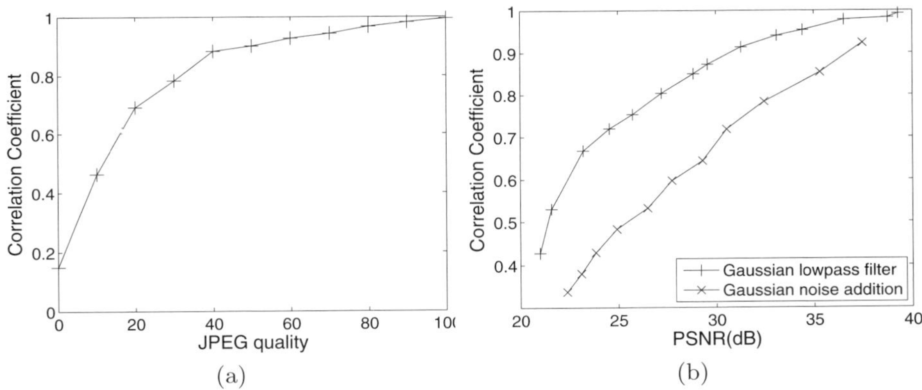

Fig. 4. (a) The detector response for JPEG encoding/decoding. (b) The detector response for lowpass filtering and noise addition.

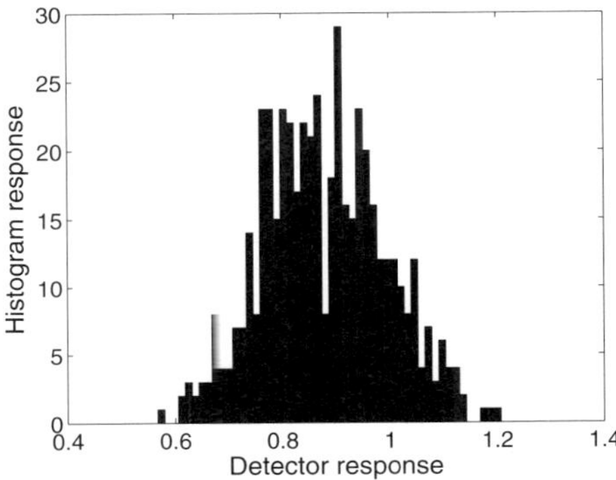

Fig. 5. The histogram response under JPEG quality 80 for 500 images

Table 1. The detector response under geometric distortions and multiple attacks

Attacks	ρ	
	Lenna	Pepper
Rotation (10°)	0.9201	0.9108
Rotation (45°)	0.9338	0.9195
Resizing (0.6)	0.8917	0.8740
Resizing (1.3)	0.9309	0.9083
Rotation (20°) + Resizing (0.8)	0.8773	0.8701
JPEG 80 + Rotation (60°)	0.8854	0.8593
JPEG 80 + Resizing (1.2)	0.8983	0.8772

other experiments under geometric distortions, including rotation and resizing, and multiple attacks. All these experiments show that our proposed scheme can achieve notable robustness against many watermarking attacks.

5 Conclusions

In this paper, we have introduced a robust digital image watermarking scheme using RBF neural network for watermark adaptively embedding in DCT domain. Simulations also proved that the proposed scheme is effective and robust against many watermark attacks. Neural network-based watermarking is still not developed broadly and deeply, and much more attentions need to be paid.

Acknowledgments

This work is supported by NSFC under project no.60573033 and the ICRG grant under project A-PG49.

References

1. Hwang, M.S., Chang, C.C., Hwang, K.F.: Digital Watermarking of Images Using Neural Networks. Journal of Electron. Imaging **9** (2000) 548–555
2. Yu, P.T., Tsai, H.H., Lin, J.S.: Digital Watermarking Based on Neural Networks for Color Images. Signal Processing **81** (2001) 663–671
3. Zhang, Z.M., Li, R.Y., Wang, L.: Adaptive Watermark Scheme with RBF Neural Networks. In: Proc. IEEE Int. Conf. Neural Networks & Signal Processing. Volume 8. (2003) 1517–1520

An Interactive Image Inpainting Method Based on RBF Networks

Peizhi Wen[1,3], Xiaojun Wu[2], and Chengke Wu[3]

[1] Department of Computer Science, Guilin University of Electronic Technology, Jinji Road 1, Guilin 5605683, P.R. China
wpzsia@163.com
[2] Harbin Institute of Technology Shenzhen Graduate School, HIT Campus, Shenzhen University Town, Xili 518055, P.R. China
wuxj@hitsz.edu.cn
[3] School of Telecommunication Engineering Xidian University, Taibai Road 2, Xi'an 710071, P.R. China
ckwu@xidian.edu.cn

Abstract. A simple and efficient inpaiting algorithm is proposed based on radial basis function network in this paper. Using the user defined areas, a neighborhood narrow band of the needing fixed pixels are computed by an erosion operator of mathematical morphology technique. Then the weights of RBF network are estimated and a continuous function is constructed, from which the needy inpainted pixels can be filled in.

1 Introduction

With the development of modern electronic devices, such as scanners, and let people scan old photos into digital archives. However, some old pictures would be damaged or corrupted for ill-considered conservation, shown in figure 1. In the ancient times, skilled artists used their experiences to fill the lost color information manually. Today we can computer and mathematical tools to finish this task automatically. Image inpainting, also called image retouching, means filling in the missing areas or modifying damaged ones in a non-detectable way by users who are not familiar with the original images [1]. The inpainting technique can be divided into three categories, that is films restoration, texture synthesis and image disocclusion [2]. Our work carried out in this paper belongs to the third scope.

The pioneering work of Bertalmio in image inpainting is on the basis of partial differential equations (PDEs). Colorful images are decomposed into three channels to deal with respectively. It fills the missing areas by propagating pixel value from outside of the masked region along isophotes whose directions are calculated at each pixel along the inpainting contour [2]. Chan and Shen proposed two inpainting methods: the Total Variational model (TVM) and Curvature-Driven Diffusion model (CDDM) [3, 4]. TVM uses the Euler-Lagrange equation and anisotropic diffusion process to deal with the small area inpainting. CDDM is an extension of TVM, where the geometric information of isophotes is taken into account. This algorithm allows to inpaint larger areas, but the result image is often blurry. Olivera offers a simple and

J. Wang et al. (Eds.): ISNN 2006, LNCS 3972, pp. 629 – 637, 2006.

fast inpainting algorithm based on an isotropic diffusion model extended with the notion of user-provided diffusion barriers [1]. But this method is limited to cope with small missing or damaged areas. Hirani and Totsuke combine global frequency and local spatial information for noise removal and use it special effects shots in post-production, which can produce nice inpainting results but it requires contents as prior knowledge of subimages.

Artificial neural networks have been applied in many fields, such as automotive, banking, electronics, financial, manufacturing and robotics etc [5-9]. In this paper, we use RBF network to estimate the weights to accomplish damaged image inpainting. Radial Basis Functions (RBFs) have been used in several fields as its accurate and stable interpolation properties. It is proved that any continuous function can be modeled up to a certain precision by RBFs [10].

In this paper we present a method to fill in the missing or damaged areas of images by RBF network. Usually, only a portion of an image needs to be inpainted instead of the whole image, so our method offers an interactive scheme to let users input processing regions. Numerical study shows it is efficient and robust. Paper is organized as follows. The radial basis functions network is introduced in section 2, and in section 3 we describe our inpainting method specifically. Some examples are presented in section 4. Part 5 is conclusion.

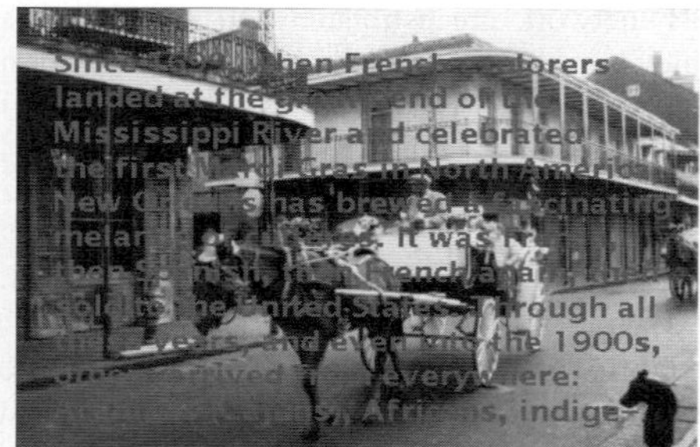

Fig. 1. Examples of corrupted and smeared pictures

2 Theories of Radial Basis Functions Network

RBFs network have traditionally been considered as a single-layer network with radial functions, as shown in figure 1. Samples $\{x_i\}_{i=1}^{N}$ are input of this neural network and to estimate the coefficients $\{\alpha_i\}_{i=1}^{N}$.

The problem of scattered data interpolation can be stated as given a set of fixed points $\boldsymbol{x}_1, \boldsymbol{x}_2, \cdots, \boldsymbol{x}_N \in R^n$ on a surface S in R^3 and a set of function values $f_1, f_2, \cdots, f_N \in R$, find an interpolant $\phi : R^3 \to R$ such that

$$\phi(\boldsymbol{x}_i) = f_i, \quad i = 1,2,\cdots,N. \tag{1}$$

where the set $\{x_i\}_{i=1}^{N}$ can be considered as the input of network and through a training algorithm to estimate the RBFs network parameters. So the training set can be represented as $T=\{(x_i, f_i)\}_{i=1}^{N}$. Most of these training algorithms are in accordance with the supervised training or to a joint unsupervised-supervised paradigm. The traditional RBFs has the following form

$$\phi(\boldsymbol{x})=\sum_{j=1}^{N}\alpha_j g_j\left(\left\|\boldsymbol{x}-\boldsymbol{x}_j\right\|\right)+p(\boldsymbol{x}) \tag{2}$$

where $p(\boldsymbol{x})$ is a polynomial, α_j are coefficients corresponding to each basis and $\|\cdot\|$ is the Euclidean norm on R^3. The basic function g is a real valued function on $[0,\infty)$, usually unbounded and of a global support. Using the training set and equation (2), we can get a linear system to estimate the unknown weights.

$$f(\boldsymbol{x}_i)=\sum_{j=1}^{N}\alpha_j g_j\left(\left\|\boldsymbol{x}_i-\boldsymbol{x}_j\right\|\right)+p(\boldsymbol{x}_i) \tag{3}$$

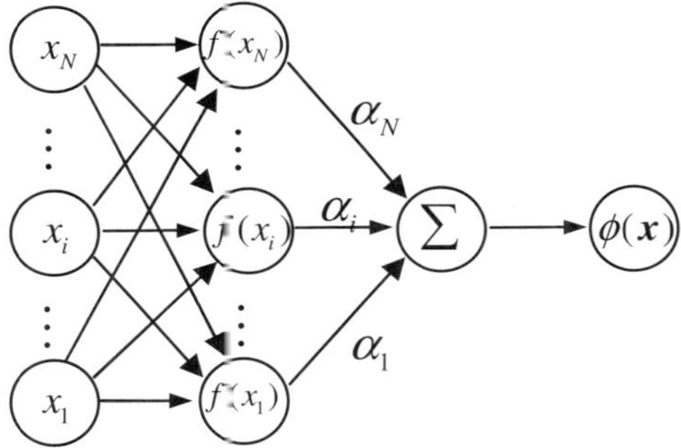

Fig. 2. RBF neural networks

There are many radial basis functions for use in (2). Some popular basis functions include thin-plate spline $g(r)=r^2\log(r)$, Gaussian $g(r)=\exp(-cr^2)$, multiquadric $g(r)=\sqrt{r^2+c^2}$, inverse multiquadric $g(r)=1/\sqrt{r^2+c^2}$, biharmonic $g(r)=r$, and triharmonic $g(r)=r^3$, where $r=\left\|\boldsymbol{x}-\boldsymbol{x}_j\right\|$. The polynomial $p(\boldsymbol{x})$ is appended for achieving polynomial precision according to the basis functions used. Additional so-called natural constraints are needed. For example, if $p(\boldsymbol{x})$ is a linear polynomial, the coefficients α must satisfy the following constraints:

$$\sum_{j=1}^{N}\alpha_j=\sum_{j=1}^{N}\alpha_j x_j=\sum_{j=1}^{N}\alpha_j y_j=\sum_{j=1}^{N}\alpha_j z_j=0 \tag{4}$$

The solution of the system composed of the weighting coefficients and the polynomial coefficients for the interpolation function $\phi(\boldsymbol{x})$. The output weights of the neural network can be determined by solving the linear system (3).

3 Image Inpainting Algorithm Description

3.1 Network Input Creation

As a matter of factor, a digital image is a definite sample of a continuous function of color field, texture and lighting, etc according to Nyquist sample theory. Some information lost in the smeared or damaged areas can be restored by the neighborhood pixels. All pixels locate on the some fixed regular grid. If all neighborhood pixels are used to estimate the coefficients of RBF networks, it would be very time consuming as the dense weight matrix. Therefore, we only employ a narrow band of neighborhood pixels of damaged areas as the input network samples. Figure 3 (a) is a portion of mask. These black lines and blocks are the damaged regions where the covered information should be filled in. This set of pixels are denoted as $\tilde{X}$.

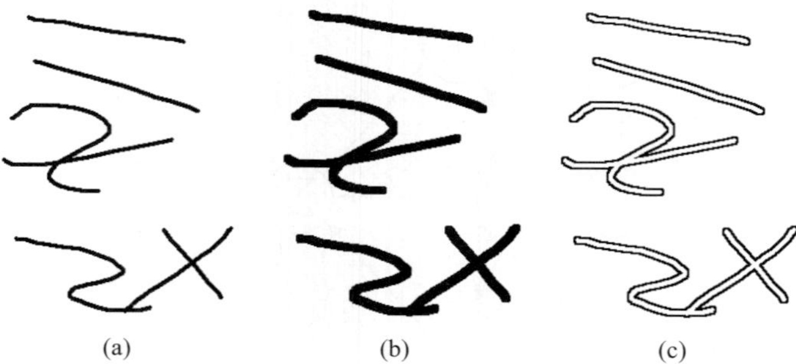

Fig. 3. Determining the RBF network input samples, (a) is the damaged areas, (b) is the eroded image and (c) is the narrow band of neighborhood of damaged areas

In order to obtain the neighborhood pixels of the smeared areas, we use erosion operator of morphological filter to extend the damaged areas to some pre-defined narrow band. The filed of mathematical morphology contributes a wide range of operators to image processing, all based on a few simple mathematical concepts from set theory. There are four often used morphological operators: Dilation, Erosion, Opening, and Closing [11]. The erosion operator is defined as the minimum of the difference of a local region of an image and a structuring element is generally chosen to emphasize or de-emphasize elements in the image. It can be formulated as:

$$(f \Theta b)(x) = \inf_{\substack{m \in S \\ x+m \in P}} \{f(x+m) - b(m)\} \tag{5}$$

where x is the location of pixel in image space, $f(x)$ is the gray scale values on the corresponding pixel x, and $b(x)$ is a structuring element. Non-zero pixels in eroded image are denoted as X. Then the desired narrow band can be calculated by

$$\bar{X} = X - \tilde{X} \tag{6}$$

Set $\{x_i\}_{i=1}^{N} \equiv \bar{X}$ is the input samples of RBF network, illustrated in figure 3(c). The size of the structuring element is determined by the size of the damaged regions. If a missing or damaged area is lager, it needs a larger structuring element correspondingly.

3.2 Inpainting Scheme

Our method allows users to select damaged areas to be inpainted since only a portion of image is missed or defiled. When the desired region is defined, we can use the method described in above subsection to estimate the RBF weights. Zhou [12] converts the inpainted area as an elevation field and takes the pixels gray values as the third dimension in RBF centers, plus the coordinates of each pixel it constitutes a problem of three dimensional surface interpolation. In that method, it is necessary to compute the normal of pixels in the fixing areas in order to avoid the trial solution of RBF, which is computationally expensive.

We adopt the fact that a digital image is a finite sample of a continuous function in our scheme. It is natural to utilize the pixel gray values as this function values on these samples. Therefore, we use the RBF network to estimate the weights of this unknown this continuous function. When this function is found, we can explicitly compute the function values at the pixels in the smeared areas. This process can be shown in figure 4. Figure (a) displays the area to be inpainted and the neighborhood narrow band pixels. Figure (b) is the continuous gray-scale function, but some information is missing for the block reason. An interpolated function is shown in figure (c), from which the damaged area can be estimated.

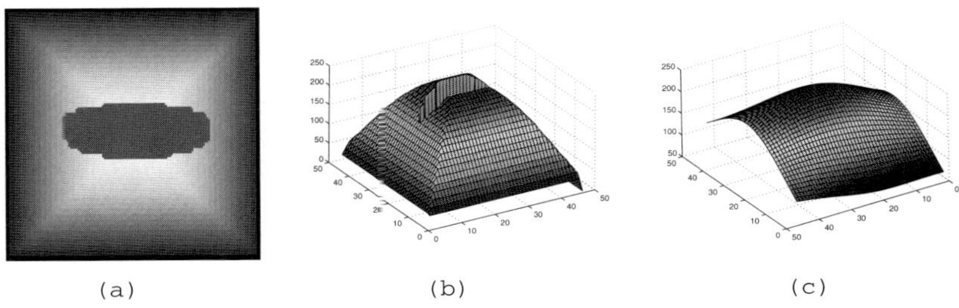

Fig. 4. The process of the image inpainting scheme, (a) a damaged a image and its neighborhood narrow band, (b) the continuous function with missing area, and (c) the interpolated function by RBF network

4 Experimental Examples

We implemented above described digital image inpainting algorithms in matlab. The colorful images are divided into three channels: red, green and blue to be handled respectively. The final output is obtained by combining these three channels. Users can specify particular area to be inpainted interactively. When the input narrow band is determined in accordance with the mask, the RBF network can be trained to get the weights. In our implementation, it is found that multi-quadric basis function has better performance than any others. We tested a lot of smeared images and the results are

Table 1. Mean square errors of exmaples

Image name	R. Channel	G. Channel	B. Channel	Inpainted pixels
Pavilion	19.27	22.506	12.629	6,877
Guohua I	19.176	25.827	23.251	12,871
Guohua II	51.545	57.192	51.668	9,363
Guohua III	8.614	9.428	7.072	24,072

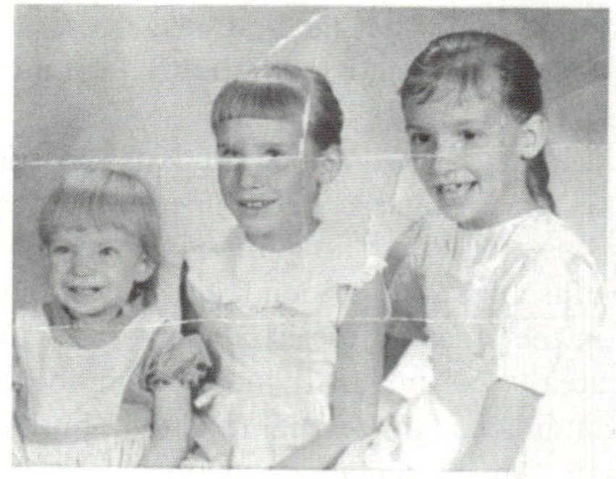

Fig. 5. Three-girl (courtesy of Marcelo Bertalmio [2]), left is an old picture and the right is the inpainted result

Fig. 6. New Orleans (courtesy of Marcelo Bertalmio [2]), left is an old picture and the right is the inpainted result

Fig. 7. Pavilion, left is an old picture and the right is the inpainted result

Fig. 8. Guohua I (courtesy of http://news.xinhuanet.com/misc), left is an old picture and the right is the inpainted result

Fig. 9. Guohua II (courtesy of http://www.tucoo.com/china_paint), left is an old picture and the right is the inpainted result

satisfactory. Figure 5 to figure 10 are the damaged images and the inpainted result images. They all look fine visually. We also use the mean square error (MSE) method to compute the reconstructed region for red, green and blue channels as a measure of the quality of the inpainting. This technique is often utilized in digital image

Fig. 10. GuohuaIII (courtesy of http://lib.hncj.cn/luntan/), left is an old picture and the right is the inpainted result

processing to assess errors. Table 1 shows the MSE of one of four examples as we have the original images and can compute the MSE with the results. From the table, we can see the MSEs are very small.

5 Conclusions

We have proposed a simple and efficient inpaiting algorithm based on radial basis function network. In our implementation, we offer to users an interactive function to select the damaged or smeared areas to eliminate the redundant pixels. In the selected areas, a neighborhood narrow band of the needing fixed pixels are computed by an erosion operator of mathematical morphology technique. Then the weights of RBF network are estimated and a continuous function is constructed. The missing area can be filled in by this function. The results produced by this method are comparable to previous RBF surface interpolation inpainting method.

References

1. Oliveira, M., Bowen, B., McKenna, R., Chang, Y.S.: Fast Digital Image Inpainting, In Proceedings of the Visualization, Imaging, and Image Processing IASTED Conference, Marbella, Spain, Sept. (2001), 261-266.
2. Bertalmio, M., Sapiro, G., Caselles, V., Ballester, C.: Image Inpainting. SIGGRAPH 2000, (2000), 417-424.
3. Chan, T., Shen, J.: Mathematical Models for Local Deterministic Inpaintings. UCLA CAM TR 00-11, March (2000).
4. Chan, T., Shen, J.: Non-texture Inpainting by Curvature-driven Diffusions (CCD). UCLA CAM TR 00-35, Sept. (2000).
5. Cairo, L., Nascimento, Jr.: Artificial Neural Networks for Control and Optimization, University of Manchester Institute of Science and Technology (UMIST), Control Systems Centre, Manchester, United Kingdom, (1994).

6. Sankar, K. Pal, Dinabandhu Bhandari, Harish, P., Kundu, M. K.: Cellular Neural Networks, Genetic Algorithms and Object Extraction. Far East Journal of Mathematical Sciences, (1993), 139-155.
7. Vellido, A., Lisboa, P.J.G., Vaughan, J.: Neural Networks in Business: a Survey of Applications (1992-1998), Expert Systems with Applications, (1999), (17), 51-70.
8. Sushmita Mitra, Sankar K. Pal and Malay K. Kundu: Finger Print Classification Using Fuzzy Multi-layer Perceptron, Neural Computing and Applications. (1994), 2, 227-233.
9. Kumaran, K., Geiger, D., Gurvits, L.: Illusory Surface Perception and Visual Organization, Network: Computation in Neural Systems, (1996), (7)1: 33-60.
10. Park, J., Sandberg, J.W.: Universal Approximation Using Radial Basis Functions Network. Neural Computation, (1991), (3), 246-257.
11. Edward R. Dougherty: Mathematical Morphology in Image Processing, Marcel Dekker, Inc. 270 Madison Avenue, New York, (1993).
12. Zhou, T., Tang, F., Wang, J., Wang, Z., Peng, Q.: Digital Image Inpainting with Radial Basis Functions. Journal of Image and Graphics (In Chinese), (2004), (9)10: 1190-1196.

No-Reference Perceptual Quality Assessment of JPEG Images Using General Regression Neural Network

Yanwei Yu, Zhengding Lu, Hefei Ling, and Fuhao Zou

College of Computer Science and Technology,
Huazhong University of Science & Technology, Wuhan 430074, China
yuyanwei198112@163.com, zdlu4409@public.wh.hb.cn,
lhefei@163.com, fuhao_zou@yahoo.com.cn

Abstract. No-reference perceptual quality assessment for JPEG images in real time is a critical requirement for some applications, such as in-service visual quality monitoring, where original information can not be available. This paper proposes a no-reference perceptual quality-assessment method based on a general regression neural network (GRNN). The three visual features of artifacts introduced in JPEG images are formulated block by block individually so that our method is computation-efficient and memory-efficient. The GRNN is used to realize the mapping of these visual features into a quality score in real time because of its excellent approximation and very short training time (one-pass learning). Experimental results on an on-line database show that our estimated scores have an excellent correlation with subjective MOS scores.

1 Introduction

Perceptual quality assessment of digital image and video is essential because they are subject to various distortions during acquisition, processing, or reproduction. Often, the subjective evaluation, for example, the mean opinion score (MOS), is the most reliable way of assessing the quality of an digital image or video, but the subjective evaluation method is not suitable for its inconvenience, slowness, high cost for most applications. Thus, in the past three to four decades, researchers turn to developing objective quality metrics. According to the availability of the original data, objective quality metrics are generally divided into three classes: full-reference (FR) where the reference signal is fully available, reduced-reference (RR) which requires certain features of the original signal, and no-reference (NR) in which a reference image or video is not available [1]. Most of current objective quality metrics belong to full-reference quality metrics, most of which try to simulate the characteristics of human visual error perception [1-5]. But in some applications, such as in-service visual quality monitoring, the reference data is not available. So it is urgent to develop a no-reference objective quality metric.

The present existing NR quality metrics mainly assessed the perceived quality by using the statistical features of distorted images [6-8]. Wang et al. [6] proposed a no-referenced perceptual quality assessment of JPEG compressed images. The metrics in [6] are computed locally, thus they are computation-efficient and memory- efficient. But their fatal shortcoming is that the model parameters are derived from the training data, so the generalization is poor. An objective quality-assessment method based on a

J. Wang et al. (Eds.): ISNN 2006, LNCS 3972, pp. 638 – 645, 2006.

Circular Back-Propagation (CBP) neural structure was proposed in [7]. Their method is unique that they aim at reproducing perceived image quality, rather than at defining a comprehensive model of the human visual system, but the method requires more storage space because the numerical features as CBPNN inputs derived from co-occurrence matrixes can not be computed locally.

In this paper, we propose a novel method (namely GRNN_NRQA for short) using general regression neural network (GRNN) for no-reference perceptual quality assessment of JPEG images. We quantify blocking artifacts and blurring artifacts as the most significant artifacts in JPEG images according to the visual characteristic of the distortion, such as loss of image details and edge smoothness. Then we emulate the human "understanding" mechanism to predict perceptual quality scores from numerical features by using GRNN because of its excellent approximation and very short training time (one-pass learning). The effectiveness of the proposed method has been verified by the experiments on the LIVE image quality assessment database [9]. Experimental results show that the proposed method outperforms the methods in [6-7] in terms of prediction accuracy, monotonicity and consistency in assessing the perceived visual quality of image and video sequence.

2 The Proposed Perceptual Quality Metric

The proposed perceptual quality evaluation scheme consists of two parts mainly. The first part describes blocking artifacts and blurring artifacts in JPEG distorted image using the mathematical formula. The second part estimates the JPEG distorted image quality from the numeric features extracted during the former phase by using GRNN. Fig. 1 gives the block diagram of the GRNN_NRQA scheme.

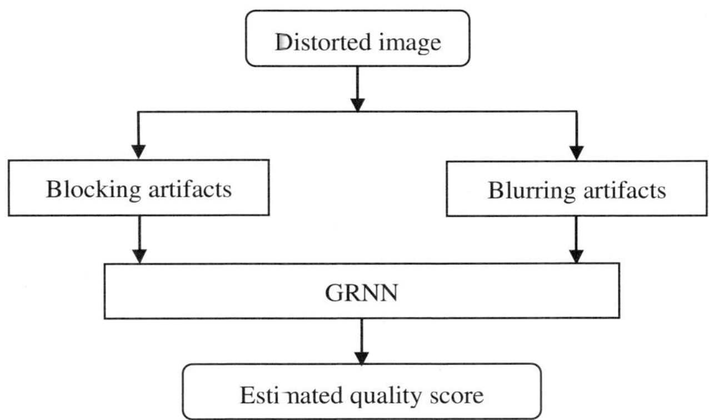

Fig. 1. Block diagram of the GRNN_NRQA scheme

2.1 The Formulation of Artifacts in JPEG Images

JPEG is a lossy image coding technique based on the block-based discrete cosine transform (BDCT) [10]. During the quantization of BDCT coefficients, blurring artifacts and blocking artifacts are introduced. The blocking artifact manifests an artificial discontinuity between adjacent blocks resulting from the independent quantization of the BDCT coefficients. Blurring artifacts are due to the loss of DCT high-frequent coefficients.

Because each macro block in the image may have different quantization steps and the same extent of blockiness may result in different degree of disturbance in human eyes due to the masking effect, we detect blocking artifacts locally block by block individually in this paper. Fig. 2 shows an example of 8×8 block and its neighboring blocks. It is enough for measuring blocking artifacts of Block A to consider its relationship with the neighboring blocks, i.e., Block B, C, D and E. However, at the time of processing Block A, only Block B and C are considered. The other two borders of Block A are to be considered when measuring horizontal blocking artifacts for Block D and vertical blocking artifacts for Block E. Note that we only process the 8×8 region near the block boundary to reduce computational complexity when measuring the artifacts at the block boundary because the artifacts at the block boundary is mainly affected by its neighboring region.

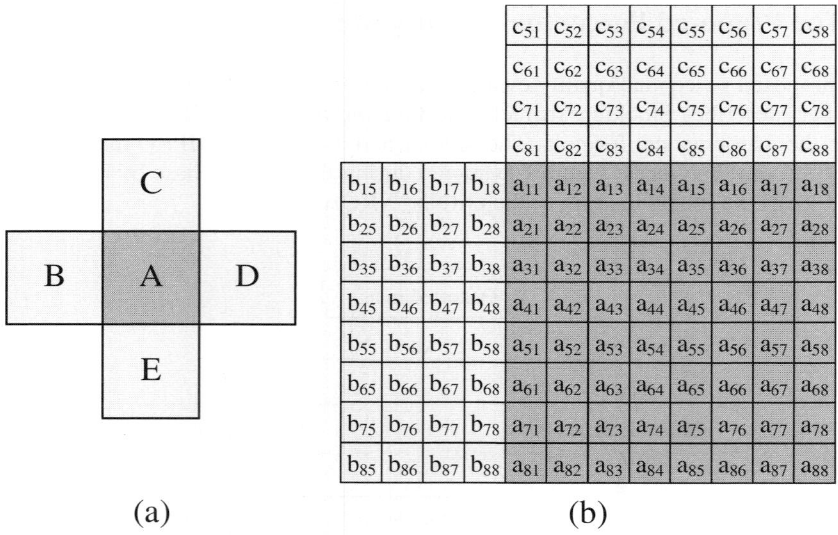

Fig. 2. An example of 8×8 block and its neighboring blocks

2.1.1 Blocking Artifacts Measure

Because the blocking artifact manifests an artificial discontinuity between adjacent blocks, it is reasonable to quantify the blocking artifacts of Block A by just examining the inter-pixel difference between Block A and its neighbors Block B and C.

By take into account the different masking effect of each block, the inter-pixel difference between Block A and its neighbors Block is divided by the sum of inter-pixel difference in the 8×8 region near the block boundary to denote the blocking artifact measure.

We define the horizontal inter-block difference $F1\_h$ between Block A and B as

$$F1\_h = \sum_{i=1}^{8} F1\_h(i) \tag{1}$$

$$F1\_h(i) = \begin{cases} \dfrac{|a_{i1} - b_{i8}|}{\sum_{j=5}^{7} |b_{i(j+1)} - b_{ij}| + \sum_{j=1}^{3} |a_{i(j+1)} - a_{ij}| + |a_{i1} - b_{i8}|} & a_{i1} \neq b_{i8} \\ 0 & a_{i1} = b_{i8} \end{cases} \tag{2}$$

Where $F1\_h(i)$ denotes the horizontal inter-block difference at the ith row between Block A and B.

The vertical inter-block difference $F1\_v$ between Block A and Block C are defined in the similar way. Assuming that the sensitivity of HVS to horizontal and vertical blocking artifacts are similar, then the blocking artifacts of an 8×8 block $F1_{blk}$ can be summarized as,

$$F1_{blk} = \frac{F1\_h + F1\_v}{2} \tag{3}$$

2.1.2 Blurring Artifacts Measure

The blurring artifacts manifest the loss of details in the block and "large splash" in the less coarse texture at the low bit-rate because of the unobviousness of the border between the neighboring blocks. So blurring artifacts can be measured by two aspects:

1) intra-block blurring artifacts measurement

The burring artifacts measure in the block can be measured by the descendant contrast between neighboring pixels in the block.

We define the horizontal contrast between neighboring pixels $F2\_h$ in the block A as,

$$F2\_h = \frac{1}{56} \sum_{i=1}^{8} \left(\sum_{j=1}^{7} |a_{i(j+1)} - a_{ij}| \right) \tag{4}$$

The vertical contrast between neighboring pixels $F2\_v$ in the block A is defined in the similar way. Then the blurring artifacts in an 8×8 block $F2_{blk}$ can be represented as,

$$F2_{blk} = \frac{F2\_h + F2\_v}{2} \tag{5}$$

2) inter-block blurring artifacts measurement

The burring artifacts at the border between the neighboring blocks can be measured in terms of the proportion of zero crossings from pixel to pixel locally. That means only the 8x8 region near the block boundary is considered, ranging from 4 pixels to the left of the boundary to 4 pixels to the right of the boundary. The total number of zero-crossings is divided by the total number of crossings to give the flatness measure in that particular region.

Defining a function for zero crossing as

$$ZC(x,y)=\begin{cases}\mathbf{1}, & |x-y|=\mathbf{0}\\ \mathbf{0}, & else\end{cases} \tag{6}$$

Then

$$F\mathbf{3}\_h=\frac{\mathbf{1}}{\mathbf{56}}\sum_{i=1}^{8}\left(\sum_{j=1}^{7}\left|b_{i(j+1)}-b_{i,j}\right|+\sum_{j=1}^{3}\left|a_{i(j+1)}-a_{i,j}\right|\right)+\frac{\mathbf{1}}{\mathbf{56}}\left|a_{i1}-b_{i8}\right| \tag{7}$$

Where $F\mathbf{3}\_h$ is the horizontal inter-block blurring artifacts measure across block A and block B.

The vertical inter-block blurring artifacts measure across block A and block C can be attained similarly. Then the inter-block blurring artifacts $F\mathbf{3}_{blk}$ near the regions of block A can be quantified as

$$F\mathbf{3}_{blk}=\frac{F\mathbf{3}\_h+F\mathbf{3}\_v}{\mathbf{2}} \tag{8}$$

Since all the blocks are subject to similar distortions and all the above artifacts for each single block have considered the masking effects, the overall measure of artifacts for an image block is attained by averaging the corresponding artifact of all blocks.

$$F\mathbf{1}=average(F\mathbf{1}_{blk}),\qquad F\mathbf{2}=average(F\mathbf{2}_{blk}),\qquad F\mathbf{3}=average(F\mathbf{3}_{blk}) \tag{9}$$

2.2 Perceptual Quality Assessment Using GRNN

The GRNN is then used to emulate the understanding system of human beings and predict the perceptual quality scores from the aforementioned visual features. The GRNN is a memory based neural network based on the estimation of a probability density function. The GRNN has the advantages of very short training time (one-pass learning) and guaranteed performance even with sparse data and does not need any priori information about the form of the regression functions in comparison to conventional regression techniques [11-12]. In mathematical terms, if we have a vector random variable x, a scalar random variable y, let X be a particular measured value of x, then the conditional mean of y given X can be represented as:

$$\hat{Y}(X)=\frac{\sum_{i=1}^{n}Y_i\exp(-\frac{D_i^{\mathbf{2}}}{\mathbf{2}\sigma^{\mathbf{2}}})}{\sum_{i=1}^{n}\exp(-\frac{D_i^{\mathbf{2}}}{\mathbf{2}\sigma^{\mathbf{2}}})} \tag{10}$$

Where $D_i^2 = (X - X_i)^T (X - X_i)$. X_i and Y_i are the sample values of the random x and y. n denotes the number of samples. σ is the width of the kernel which is the only unknown parameter in the above equation (10). After repeating experiments, we find the estimating performance is the best when σ is 0.018.

3 Experimental Results

The system for image quality assessment has been tested by using a database of JPEG images available at the LIVE Quality Assessment Database [9]. The database of JPEG images included 233 test images, 175 of which were generated by compressing (with JPEG) 29 high-resolution 24-bits/pixel RGB color images (typically 768 x 512) with different compression ratios. The Mean Opinion Score (MOS) of each image in the LIVE database is provided on a linear scale ranging from 1 ("bad") to 100 ("excellent"). In the present experiment, the subjective scores have been converted into the interval [-1,1] linearly.

The effectiveness of the GRNN_QA system is measured by using a conventional cross-validation approach. The available samples are randomly divided into a training set including 100 images and a test set including 133 images. During the training process of quality-assessment system only the training set is used, whereas the test set is applied only to predict the system generalization ability.

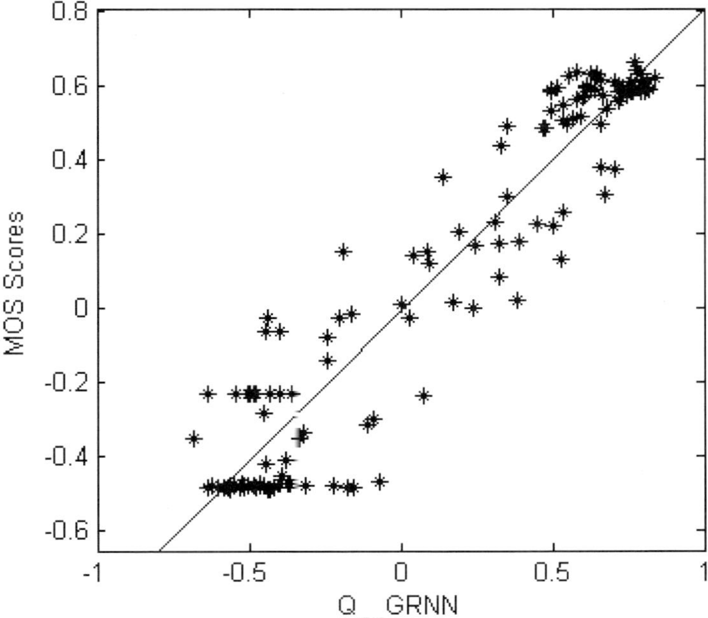

Fig. 3. Estimated perceptual quality scores Q_GRNN versus MOS Scores

The effectiveness of objective quality metrics is generally reflected by the correlation with subjective quality scores. The higher the correlation is, the more effective the objective quality metric is. In this paper, the correlation is quantified by the Pearson Correlation and the Root Mean Squared error (RMS).

The scatter plot in Fig. 3 shows the results obtained for the test set, with the estimated objective quality scores Q_GRNN as the x axis and the MOS scores as the y axis. It is obvious that the estimated scores correlate well with MOS scores and our proposed method is effective.

Table 1 shows the Pearson Correlation and the Root Mean Squared error (RMS) between various quality metrics and the subjective ratings of the JEPG database provided by LIVE [9]. It can be seen that our objective scores Q_GRNN have an excellent correlation with the subjective MOS scores and our method outperform the methods in [6-7].

Table 1. Pearson Correlation and RMS for LIVE database

algorithm	Pearson Correlation	RMS
Q_GRNN	0.954983	0.14053
S [6]	0.918092	0.7256
Q_CBP [7]	0.950152	0.14

4 Conclusions

In this paper, we proposed a no-reference perceptual quality assessment system GRNN_NRQA for JPEG images. Visual features effectively capturing the artifacts introduced by JPEG were formulated block by block individually and processed by GRNN which output the corresponding quality assessment. Experimental results confirmed that the system provided a satisfactory approximation of the subjective MOS scores.

The method is computationally efficient since no complicated transform is computed and the algorithm can be implemented without storing the entire image in memory, which makes embedded implementations easier. The basic methodology of the proposed method can also been used to develop NR quality assessment methods for MPEG compressed video. The next work is to apply our proposed method to our project to estimate the quality of watermarked video in real time.

Acknowledgement

This work is supported by the National Natural Science Foundation of China under Grant No. 60502024, the Natural Science Foundation of Hubei Province under Grant No. 2005ABA267, the Electronic Development Fund of Ministry of Information Industry of China and the Innovation Fund for Technology Based Firms of Ministry of Science and Technology of China under Grant No. 04C26214201284.

References

1. Wang, Z., Sheikh, H.R., and Bovik, A.C.: Objective Video Quality Assessment. The Handbook of Video Databases: Design and Applications (B. Furht and O. Marqure, eds.), CRC Press (2003) 1041–1078
2. Eckert, M.P., Bradley, A.P.: Perceptual Quality Metrics Applied to Still Image Compression. Signal Processing 70 (1998) 177–200
3. VQEG: Final Report from The Video Quality Experts Group on The Validation of Objective Models of Video Quality Assessment. http://www.vqeg.org/, 2000
4. Pappas, T.N., Safranek, R.J.: Perceptual Criteria for Image Quality Evaluation. Handbook of Image and Video Processing (A. Bovik, ed.), Academic Press (2000)
5. Wang, Z., Bovik, A. C.: Why Is Image Quality Assessment So Difficult? in Proc. IEEE Int. Conf.Acoust., Speech, and Signal Processing (2002) 3313–3316
6. Wang, Z., Sheikh, H.R., Bovik, A.C.: No-Reference Perceptual Quality Assessment of JPEG Compressed Images. IEEE International Conference on Image Processing (2002) 477–480
7. Gastaldo, P., Zunino, R.: No-reference Quality Assessment of JPEG Images by Using CBP Neural Networks. Proc. of the 2004 International Symposium on Circuits and Systems (ISCAS'04), 5 (2004) 772–775
8. Pan, F., Lin, X., Rahardja, S., Lin, W., Ong, E., Yao, S., Lu, Z, Yang, X.: A Locally-adaptive Algorithm for Measuring Blocking Artifacts in Images and Videos. Proc. of the 2004 International Symposium on Circuits and Systems (ISCAS'04), 3 (2004) 23–26
9. Sheikh, H.R., Wang, Z., Bovik, A.C., Cormack, L.K.: LIVE Image Quality Assessment Database. http://live.ece.utexas.edu/research/quality/
10. Shi, Y.Q., Sun, H.: Image and Video Compression for Multimedia Engineering: Fundamentals, Algorithms and Standards. CRC Press (1999)
11. Spetch, D.F.: A General Regression Neural Network. IEEE Transactions on Neural Netwolks 2 (1991) 568–576
12. Parzen, E.: On Estimation of a Probability Density Function and Mode. Ann. Math Statistics 33 (1962) 1065–1076

Minimum Description Length Shape Model Based on Elliptic Fourier Descriptors

Shaoyu Wang, Feihu Qi, and Huaqing Li

Department of Computer Science and Engineering, Shanghai Jiao Tong University, Shanghai, 200030, P.R. China
wang@sjtu.edu.cn

Abstract. This paper provides the construction of statistical shape model based on elliptic Fourier transformation and minimum description length (MDL). The method does not require manual identification of landmarks on training shapes. Each training shapes can be decomposed into a set of ellipse by elliptic Fourier transformation at a different frequency level. The MDL objective function is based on elliptic Fourier descriptors and principal component analysis (EF-PCA). Experiments show that our method can get better models.

1 Introduction

Statistical shape models have proven to be useful tools in image segmentation and interpretation. These models are usually created by manually indicating landmarks which defining the correspondences between similar structures on each training shape. However, establishing the correspondences by hand is tedious and prone to error.Taylor[1] described an approach where the best model was defined in terms of "compactness", as measured by the determinant of its covariance matrix. Although the method is workable, its objective function could not be rigorously justified and it was difficult to make the optimization converge to an acceptable solution. To solve these problems, Davies [2][3] defined a new objection function with a rigorous theoretical basis that is defined in an information theoretic framework. A minimum description length (MDL) objective function is used to measure model complexity, and optimised numerically with respect to the correspondences. One shape example was chosen as a reference shape and the positions of its correspondence points remained fixed throughout. The optimisation process involved perturbing the locations of the correspondence points of each shape in turn optimizing the MDL objective function. Although showed promise, the method is not suitable for shapes with too many landmarks because of using Cartesian coordinate points. For this reason, we use elliptic Fourier descriptors to describe shapes and use PCA to summarizing the information of the variations contained in the elliptic Fourier coefficients. Experiments proved our method could establish a novel MDL objective function and get more compact MDL shape model. The paper is structured as follows. Section 2 reviews the theory of Minimum Description Length. Section 3 introduces and discusses the

J. Wang et al. (Eds.): ISNN 2006, LNCS 3972, pp. 646–651, 2006.

reconstruction of shapes using EFT, and using it instead of Cartesian coordinates to build MDL model. Experimental results are shown and discussed in Section 4.

2 The Current MDL Objective Function

The key of MDL approach is to define an objective function to assess the "quality" of a model constructed from a given set of correspondences. Davies et al. [2] derived an object function with favours models that encode the training set most efficiently. This is motivated by an observation that simple explanations generalize best. This observation is formalized by minimizing description length, i.e. the information required to transmit the entire training set from sender to receiver. The correspondences problem is regarded as an optimization problem. Point samples are altered and refined across all training shapes until an objective function is minimized. The correspondences across the training shapes are optimised with respect to the description length cost (DL), node length cost (NodeCost) of landmark positions.

$$DL = \sum_{\lambda \geq \lambda_{cut}} (1 + \log \frac{\lambda_m}{\lambda_{cut}}) + \sum_{\lambda < \lambda_{cut}} \frac{\lambda_m}{\lambda_{cut}} \tag{1}$$

$$NodeCost = \sum (\alpha_i^{average} - \alpha_i^{target})^2 \tag{2}$$

λ_m is the eigenvalue of a principle components decomposition of landmark positions, λ_{cut} is is preset threshold; $\alpha_i^{average}$ and α_i^{target} are average and target parameters for the ith landmark.

H.H Thodberg [4]added the Curvature variation cost which expresses the mismatch of curvature features across the data set. The curvatures can be weighted with a factor and appended to the aligned position coordinates and included in the PCA:

$$Curvaturecost = c\frac{1}{N}\frac{1}{s}\sum_{i,r}(k_{i,r} - k_i^{mean})^2 \tag{3}$$

3 The Revised Objective Function Based EF-PCA

The previous MDL methods applied PCA directly on the x and y coordinates of landmarks to define objective function. When the number of landmarks is very large, the PCA processing is time consuming. To find a more effective method of establishing MDL objective function and get better MDL model, we use elliptic Fourier analysis to decompose each contour into a series of elliptical harmonics and arrange the normalized elliptic Fourier coefficients as a shape vector, then apply PCA on the variance-covariance matrix of the vectors of all training shapes.

Assume that N training shapes are available in the form of sets of np corresponding landmarks in the 2-dimensional space and standard Procrustes alignment has been applied on the training shapes to eliminate the variations introduced by translation, scaling and rotation. The goal of MDL is to place np points on each of shapes so that the statistical shape model is general, specific and compact to the largest possible extent.

Following the method of Kuhl et al. [5], a shape with a closed two-dimensional contour can be approximated as the sum of elliptical harmonics by elliptic Fourier analysis. We consider a closed contour which has k points is approximated by N harmonics, each harmonic has four Fourier coefficients a_n,b_n,c_n,d_n:

$$A_n = \frac{T}{2n^2\pi^2}\sum_{i=1}^{k}\frac{dx_i}{dt_i}(\cos\frac{2n\pi t_i}{T} - \cos\frac{2n\pi t_{i-1}}{T}) \tag{4}$$

$$B_n = \frac{T}{2n^2\pi^2}\sum_{i=1}^{k}\frac{dx_i}{dt_i}(\sin\frac{2n\pi t_i}{T} - \sin\frac{2n\pi t_{i-1}}{T}) \tag{5}$$

where n is the harmonic order of the quadruple of coefficients.

From geometry perspective, A_i and B_i represent the projection on the X axes of the semi-major and of semi-minor axis of the ith harmonic. The series of coefficient of the Y axes, C_i and D_i, are found in the same way. To make elliptic Fourier coefficients invariant to size, rotation and start point of contour traces, we use the following matrix transformation:

$$\begin{bmatrix} a_n \; b_n \\ c_n \; d_n \end{bmatrix} = \frac{1}{E^*}\begin{bmatrix} \cos\phi \; \sin\phi \\ \sin\phi \; \cos\phi \end{bmatrix}\begin{bmatrix} A_n \; B_n \\ C_n \; D_n \end{bmatrix}\begin{bmatrix} \text{cosn}\theta \; \text{-sinn}\theta \\ \text{sinn}\theta \; \text{cosn}\theta \end{bmatrix} \tag{6}$$

where E^* is magnitude of the semi-major axis of the best fitting ellipse:

$$E^* = \sqrt{(a^*)^2 + (b^*)^2} \tag{7}$$

ϕ is the orientation of this ellipse in radians

$$\phi = \tan^{-1}\frac{c^*}{a^*} \tag{8}$$

θ is the rotation of the starting point from the end of the ellipse

$$\theta = \frac{1}{2}\tan^{-1}\left[2\frac{(A_1B_1 + C_1D_1)}{(A_1^2 + C_1^2 - B_1^2 - D_1^2)}\right] \tag{9}$$

The values of a^* and b^* are given by

$$a^* = A_1\cos\theta + B_1\sin\theta \tag{10}$$

and

$$c^* = C_1\cos\theta + D_1\sin\theta \tag{11}$$

These normalizations cause the degeneration of the first three coefficients: $a_1 = 1;b_1 = c_1 = 0$. Thus for N harmonics there are 4N-3 nontrivial normalized coefficients. We arrange the normalized elliptic Fourier coefficients as a column vector that we then treated as a multivariate point representing the shape in a 4N-3 dimensional space of potential variation. The shape defined by elliptic Fourier coefficients is called as EFC shape, each EFC shape can be represented by a 4N-3 dimensional shape vector E_i

$$E_i = (d_1, a_2, b_2, d_2, \ldots, a_N, b_N, c_N, d_N)^T \tag{12}$$

Following the method of Cootes et al. [6], the set of all EFC shape vectors can be modeled by a multivariate Guassian distribution, PCA is then performed on the variance-covariance matrix of the vectors of all shapes to define a set of axes that are aligned with the principle directions of the data and to summarize the information contained in the coefficients of the elliptic Fourier descriptors. Starting with mean shape vector:

$$\mu = \frac{1}{n_s} \sum_{i=1}^{n_s} E_i \tag{13}$$

calculating the variance-covariance matrix:

$$D = \frac{1}{n_s} \sum_{i=1} n_s (E_i - \mu)(E_i - \mu)^T \tag{14}$$

solving for the eigenvaluesλ^m and eigenvectors $p^m$of D:

$$Dp^m = \lambda^m p^m \tag{15}$$

and sorting the eigenvalues(and the corresponding eigenvectors) in descending order:

$$\lambda_1 \geq \lambda_2 \geq \ldots \geq \lambda_{4N-4} \geq \lambda_{4N-3} \tag{16}$$

The above eigenvalues are then used to define our MDL's description length cost (DL):

$$DL = \sum_{\lambda_i \geq \lambda_{cut}} (1 + \lg \frac{\lambda_i}{\lambda_{cut}}) + \sum_{\lambda < \lambda_{cut}} \frac{\lambda_i}{\lambda_{cut}} \tag{17}$$

where λ_i is the cutoff constant and we set it to 10^{-5}

The DL describes the information needed to transmit the PCA representation of the EFC shapes. For a mode m with large eigenvalue the cost is $\lg(\lambda_m)$, while for smaller λ it should be tend to a constant.

4 Experimental Results

We tested our method on the training set of 24 box-bumps. The 9 training samples of Bump model are shown in figure. 1. These boxbumps are generated with a bump at a varying location and with varying aspect ratio of the box and have the similar gray-level distribution.

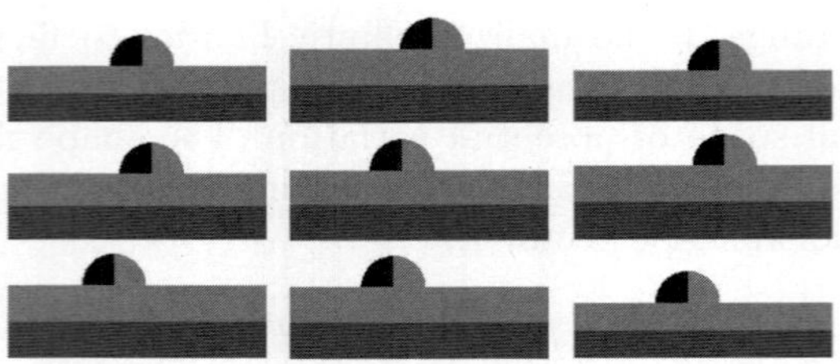

Fig. 1. The box-bumps used as training set

In our experiment, 8 control nodes have been used for the reparameterisation, 64 landmarks are sampled on each box-bump to evaluate the Description Length at the given parameterisation. To define the shape of a boxbump, previous MDL method need a shape vector has 128 dimensions. By using elliptic Fourier transform, a shape can be approximated by 20 harmonics and reduce the shape vector from 128 to 77. As for complex training shape, it needs more landmarks, and the dimension reduction ability of EF-PCA is more obvious. Figure. 2 shows the shape reconstruction by using 1 to 10 elliptic harmonics and proved 20 harmonics are sufficient to express fine scale shape structures.

We give quantitative results in table 1, tabulating the value of MDL objective function (F) and the variance explained by the first three modes, comparing the result with those for models built using the current MDL method.

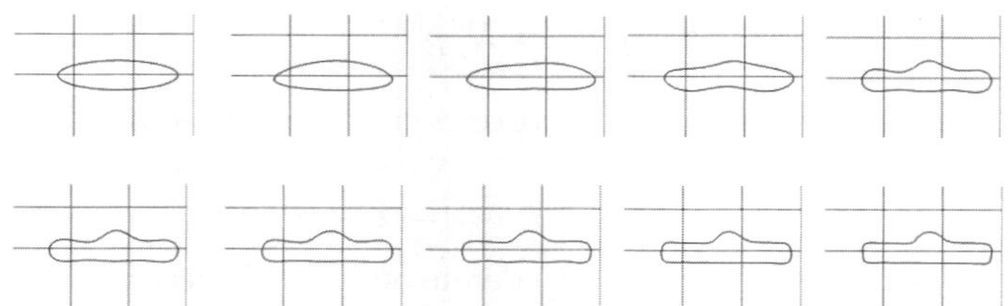

Fig. 2. The box-bumps reconstructed by the increasing number of harmonics from 1 to 10

Table 1. Eigenvalues and contribution of Principle Components of the first 3 modes of the current MDL method and our revised MDL method. V_T is the toatla variance and F is the value of the objective function.

	Current MDL method		Our revised MDL Method	
Component	Eigenvalue(10^{-4})	Proportion(%)	Eigenvalue(10^{-4})	Proportion(%)
1	45.953	49.690	43.515	51.247
2	23.326	25.223	22.092	26.018
3	6.353	6.870	6.237	7.80
V_T	92.479		84.912	
F	17.809		16.376	

Experiment results shows that the independent shape characteristics are well identified by PCA on the elliptic Fourier coefficients and MDL models based EF-PCA are more compact(have less variance per mode and less value of objective function) than the models built by the current MDL method.

5 Discussion and Conclusion

In this paper we present a more efficient way to define MDL's description length. We noticed that, to get the "best possible" model, the current MDL method needs to make the number of landmarks, n_p, very large and make the MDL optimisaion process very slow. To solve this problem, we describe overall training shapes mathematically by transforming x and y coordinates into elliptic Fourier descriptors, and apply PCA on these descriptors to identify the shape characteristics. By using EF-PCA, the length of shape vector could be reduced from $2 * n_p$ to $(4N - 3)$. Because the independent shape characteristics can be well identified, the revised Description Length function can detect more small shape variations. Experiments prove that our method can speed up the optimization process and quantitatively evaluate the DL objective function effectively.

Acknowledgements

This work was supported by the National Natural Science Foundation of China (No.60271033).

References

1. Kotcheff, A.C.W., Taylor, C.J.: Automatic Construction of Eigenshape Models by Direct Optimization. Medical Image Analysis 2(4) (1998) 303–314
2. Davies, R.H., Twining, C.J., Cootes, T.F., Waterton, J.C., Taylor, C.T.: A Minimum Description Length Approach to Statistical Shape Modelling. IEEE Trans. on Medical Imaging 21(5) (2002) 525–537
3. Davies, R.H., Twining, C.J., Cootes, T.F., Waterton, C.J., Taylor, C.J.: 3D Statistical Shape Models Using Direct Optimisation of Description Length. In: Heyden, A., Sparr, G.(eds.): ECCV(3). Lecture Notes in Computer Science, Vol. 2352. Springer-Verlag, Berlin Heidelberg Copenhagen (2002) 3–20
4. Thodberg, H.H., Olafsdottir, H.: Adding Curvature to Minimum Description Length Shape Models. In: 14th British Machine Vision Conference. Vol. 2. (2003) 251–260
5. Kuhl, F.P., Giardina, C.R.: Elliptic Fourier Features of a Closed Contour. Computer Graphics and Image Processing 18(3) (1982) 236–258
6. Coootes. T.F., Tayor, C.J., Cooper, D., Graham, J.: Training Models of Shape from Sets of Examples. In: 3rd British Machine Vision Conference. Springer-Verlag, (1992) 9–18

Neural Network Based Texture Segmentation Using a Markov Random Field Model

Tae Hyung Kim[1], Hyun Min Kang[1], Il Kyu Eom[1], and Yoo Shin Kim[2]

[1] Dept. Electronics Engineering, Pusan National University, 30 Jangjeon-dong, Geumjeong-gu, Busan 609-735, Republic of Korea
[2] Research Institute of Computer, Information and Communication, Pusan National Univesity, 30 Jangjeon-dong, Geumjeong-gu, Busan 609-735, Republic of Korea

Abstract. This paper presents a novel texture segmentation method using neural networks and a Markov random field (MRF) model. Multi-scale wavelet coefficients are used as input for the neural networks. The output of the neural network is modeled as *a posterior* probability. Initially, the multi-scale texture segmentation is performed by the posterior probabilities from the neural networks and MAP (maximum *a posterior*) classification. Then the MAP segmentation maps are produced at all scales. In order to obtain the more improved segmentation result at the finest scale, our proposed method fuses the multi-scale MAP segmentations sequentially from coarse to fine scales. This is done by computing the MAP segmentation given the segmentation map at one scale and *a priori* knowledge regarding contextual information which is extracted from the adjacent coarser scale segmentation. In this fusion process, the MRF prior distribution and Gibbs sampler are used, where the MRF model serves as the smoothness constraint and the Gibbs sampler acts as the MAP classifier.

1 Introduction

The choice of the texture segmentation method is influenced by the texture description or feature extraction method [1]. Of all methods for the texture feature extraction, signal processing methods are attractive due to their simplicity. Also, signal processing methods are supported by psychophysical research, which has given evidence that the human brain does a frequency analysis of the image [2]. Fourier [3] or wavelet transforms [4] can be used to extract texture image features. The resulting transformed features in a multi-scale wavelet domain are efficient for texture segmentation [4]. Therefore, the HMT (hidden Markov trees) model in a multi-scale wavelet-domain was used for texture segmentation [4]. Also, Kohonen neural network [5], or fuzzy c-means [6] with some features extracted from wavelet transforms, were also used for texture segmentation.

Neural networks can represent any distribution of inputs without complicated modeling methods [7]. An approach based on support vector machines and

J. Wang et al. (Eds.): ISNN 2006, LNCS 3972, pp. 652–660, 2006.

neural networks was investigated for texture segmentation [8], and a method by neural networks and HMTseg in the multi-scale wavelet domain was used for texture segmentation [9]. Another method for texture description is the model-based method. Model-based methods characterize texture images based on the construction of an image model. The model parameters capture the essential perceived qualities of texture. Markov random fields (MRFs) are widely used because they yield a local and economical texture description [10], [11]. A Gibbs sampler can be used as the MRF parameter esti-mator, or used as the texture synthesizer for generating a texture from an MRF texture model [12].

In this paper, we propose a novel method of supervised texture segmentation using neural networks in a multi-scale wavelet domain and MRF model. It is possible to compose neural networks so that their output represents *a posterior* probability when they are trained in a supervised training mode [13], [14]. In the proposed method, the input elements of neural networks consist of multi-scale wavelet coefficients with a quad tree structure; in addition, the texture segmentation maps by outputs of neural networks and MAP classification are initially produced at all scales. Texture segmentations in multi-scale images have contradictions between reliability and minuteness according to scale. The segmentation in coarse scales is accurate for large, homogeneous regions but poor along the boundaries between regions. The segmentation in fine scales is accurate near the boundaries between regions but has poor classification reliability due to the paucity of statistical information. A recent study demonstrated that texture segmentation in the finest scale is improved by using a multi-scale Bayesian image segmentation algorithm called HMTseg that fuses texture segmentations in multi-scale images [4]. In our proposed method, noises caused by classification errors in the initial segmentation maps are reduced by the MRF prior distribution and Gibbs sampler. Also, in order to obtain the more improved segmentation result at the finest scale, the multi-scale texture segmentation results are fused sequentially from coarse to fine scales. This fusion process is executed by computing the MAP segmentation, given the segmentation map at one scale and *a priori* knowledge (regarding contextual information which is extracted from the adjacent coarser scale segmentation), where the MRF model serves as the smoothness constraint and the Gibbs sampler acts as the MAP classifier. Texture segmentation by the proposed method performs better than the segmentation by HMT and HMTseg, and performs better than texture segmentation by neural networks and HMTseg.

2 Inputs and Outputs of Neural Networks

The elements of the input vector of the neural network consist of the multi-scale wavelet coefficients from the multi-level Haar wavelet transform. The multi-level Haar wavelet transform forms a pyramid structure through all scales as shown in Fig. 1-(a). The multi-scale wavelet coefficients, which analyze a common sub-region of an image, have persistence across the scale. The dependency between these coefficients can be represented as a wavelet quad-tree. The input vector of

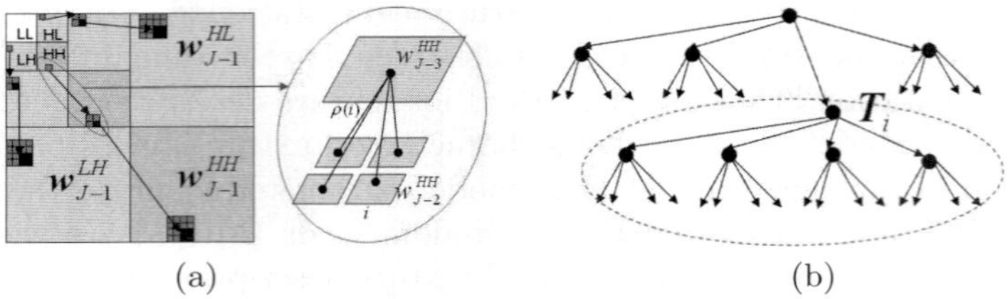

Fig. 1. (a) The Harr wavelet transform and the quad-tree structure of the wavelet coefficients; (b) The quad-tree structure of the wavelet coefficients in a wavelet sub-band and sub-tree $\boldsymbol{T}_i$ rooted at node i and its elements

the neural network is determined by considering the dependency of the wavelet coefficients across the scale. The coefficients of the sub-tree $\boldsymbol{T}_i$ rooted at location i (node i) of a subband wavelet quad-tree and the pixel intensities of the subimage (analyzed by node i), are used as inputs of neural networks for node i [see Fig. 1-(b)]. In Fig. 1-(b), a black node represents a wavelet coefficient, and $\boldsymbol{T}_i$ is a sub-tree rooted at node i. $\boldsymbol{T}_i$ is composed of $\boldsymbol{T}_i^{HL}$, $\boldsymbol{T}_i^{LH}$, and $\boldsymbol{T}_i^{HH}$ for each direction. By considering all sub-trees on node i of three wavelet subband quad-trees, the input vector $\boldsymbol{g}_i$ of the neural networks is defined as

$$\boldsymbol{g}_i \equiv \{\boldsymbol{T}_i, \boldsymbol{p}_i\}, \tag{1}$$

where $\boldsymbol{g}_i$ is for the region of the sub-image analyzed by sub-tree $\boldsymbol{T}_i$, and $\boldsymbol{p}_i$ represents the pixel intensities of the sub-image which are analyzed by node i.

The input and output structure of neural network can be made so that *a posterior* probability is estimated [13], [14]. The neural network structure in our algorithm has one group of multi-layer perceptron (MLP) networks for each wavelet scale. In each scale, one MLP network is assigned for each texture class. In addition, in pixel level, one group of MLP networks is also assigned. If wavelet decomposition level is L and the number of texture classes is C, then the number of MLP networks is $(L+1) \times C$. For the MLP network of texture class $c(c = 1, 2, , C)$, the target output for training is set up as follows:

$$t_i = \begin{cases} 1, & \boldsymbol{g}_i \in c \\ 0, & \boldsymbol{g}_i \notin c \end{cases}. \tag{2}$$

In this paper, the resilient back-propagation algorithm is used in MLP training [15]. The purpose of the resilient back-propagation training algorithm is to eliminate the faults of the back-propagation algorithm using the magnitudes of the partial derivatives of the cost function. After training the neural networks, the texture segmentation map is obtained by the following classification: Texture class, $\hat{c}$, by MLP is obtained from the MLP outputs.

$$\hat{c} \equiv arg \underset{c \in \{1,2,\ldots,C\}}{max} \frac{N(\boldsymbol{g}_i, \boldsymbol{w}_c)}{\sum_{k=1}^{C} N(\boldsymbol{g}_i, \boldsymbol{w}_k)}, \tag{3}$$

where $N(\boldsymbol{g}_i, \boldsymbol{w}_c)$ is the output of the MLP model, is composed of input vector $\boldsymbol{g}_i$ and weight vector $\boldsymbol{w}_c$ for class c. By using Eq. 3, the multi-scale texture segmentation maps are obtained. Multi-scale texture segmentations by Eq. 3 have contradictions between reliability and minuteness according to scale. In the proposed method, the multi-scale texture segmentation results by Eq. 3 are fused by using a Gibbs sampler and contextual information which is extracted from the adjacent coarser scale segmentation.

3 Texture Segmentation Using an MRF Model

The multi-scale texture segmentations by neural network have contradictions between reliability and minuteness according to scale. The initial texture segmentation maps by neural network have noises caused by classification errors. Here we will explain the reduction of noises in the segmentation maps by the MRF smoothness prior, the interscale decision fusion by defining the context vector v which has contextual information extracted from the adjacent coarser scale segmentation, and MAP classification by Gibbs sampler.

3.1 The Context Vector and the Multi-scale Decision Fusion

The multi-scale texture segmentation is used to assign the texture class labels to sites on a lattice of each scale. Node i of the wavelet subband quad-trees corresponds to site i of a scale. Let's define the context vector v_i^j for site i at scale j.

$$v_i^j \equiv \left[c_{\rho(i)}^{j-1}, c_{N1_{\rho(i)}}^{j-1}, c_{N2_{\rho(i)}}^{j-1}, c_{N3_{\rho(i)}}^{j-1}, c_{N4_{\rho(i)}}^{j-1}, c_{N5_{\rho(i)}}^{j-1}, c_{N6_{\rho(i)}}^{j-1}, c_{N7_{\rho(i)}}^{j-1}, c_{N8_{\rho(i)}}^{j-1} \right], \quad (4)$$

where $c_{\rho(i)}^{j-1}$ is the class label of parent site $\rho(i)$ of site i, $\left[c_{N1_{\rho(i)}}^{j-1}, \ldots, c_{N8_{\rho(i)}}^{j-1} \right]$ are class labels of neighbor sites of parent site $\rho(i)$. Then, the configuration $\boldsymbol{v}^j$ of context vectors at scale j is $\boldsymbol{v}^j = \{v_i^j | \forall i \in S, for\ the\ set\ S\ of\ sites\ at\ scale\ j\}$. The $\boldsymbol{v}^j$ contains classification information at the previous coarse scale.

We consider maximizing the posterior $p(\boldsymbol{c}^j|\boldsymbol{G}^j, \boldsymbol{v}^j)$, where $\boldsymbol{c}^j$ is the configuration of class labels on the lattice of scale j, and $\boldsymbol{G}^j \equiv \{\boldsymbol{g}_i^j\}$ is the collection of all $\boldsymbol{g}_i^j$ at scale j. Assuming that $\boldsymbol{G}^j$ and $\boldsymbol{v}^j$ are conditionally independent given the configuration $\boldsymbol{c}^j$, the posterior $p(\boldsymbol{c}^j|\boldsymbol{G}^j, \boldsymbol{v}^j)$ is rewritten as Eq. 5 by the naive Bayes' rule [16]. The naive Bayes' rule works quite well in practice, despite its manifest simplicity. Texture segmentation is performed by finding the optimal configuration $\boldsymbol{c}^*$ as follows:

$$\boldsymbol{c}^* = arg \max_{\boldsymbol{c}^j \in \boldsymbol{\Theta}^j} \{p(\boldsymbol{c}^j|\boldsymbol{G}^j, \boldsymbol{v}^j)\} = arg \max_{\boldsymbol{c}^j \in \boldsymbol{\Theta}^j} \{p(\boldsymbol{c}^j|\boldsymbol{G}^j) p(\boldsymbol{c}^j|\boldsymbol{v}^j)\}, \quad (5)$$

where $\boldsymbol{\Theta}^j$ is the configuration space (the set of all possible configurations) at scale j. The posterior distribution $p(\boldsymbol{c}^j|\boldsymbol{G}^j)$ has information obtained from all $\boldsymbol{g}_i$ of scale j for segmentation. The posterior distribution $p(\boldsymbol{c}^j|\boldsymbol{v}^j)$ has information obtained from all context vectors that are extracted from the previous coarse scale $j-1$. Considering Eq. 5, the interscale decision fusion for texture segmentation at scale j is performed.

3.2 The MRF Smoothness Prior and MAP Classification by Gibbs Sampler

Classification by Eq. 5 can be rewritten as follows:

$$\boldsymbol{c}^* = arg \underset{\boldsymbol{c}^j \in \boldsymbol{\Theta}^j}{max} \left\{p(\boldsymbol{c}^j|\boldsymbol{G}^j)p(\boldsymbol{c}^j|\boldsymbol{v}^j)\right\} = arg \underset{\boldsymbol{c}^j \in \boldsymbol{\Theta}^j}{max} \left\{p(\boldsymbol{c}^j|\boldsymbol{G}^j)p(\boldsymbol{v}^j|\boldsymbol{c}^j)p(\boldsymbol{c}^j)\right\} . \tag{6}$$

In Eq. 6, the $p(\boldsymbol{v}^j|\boldsymbol{c}^j)$ represents interscale dependency between the class labels, and can be rewritten as follows:

$$p(\boldsymbol{v}^j|\boldsymbol{c}^j) = \prod_{i \in S} p(v_i^j|c_i^j). \tag{7}$$

In Eq. 7, it is assumed that all v_i^j are independent given $\boldsymbol{c}^j$ and each v_i^j is distributed based on the distribution $p(v_i^j|c_i^j)$ independently of all other c_k^j and v_k^j, $k \neq i$. Here, we decide $p(v_i^j|c_i^j)$ as follows:

$$\begin{aligned} p(v_i^j|c_i^j) &\equiv \frac{the\ number\ of\ elements\ of\ vector\ v_i^j\ with\ same\ value\ as\ label\ c_i^j}{the\ number\ of\ all\ elements\ of\ vector\ v_i^j} \\ &= \frac{\delta_{c_i^j, c_{\rho(i)}^{j-1}} + \delta_{c_i^j, c_{N1_{\rho(i)}}^{j-1}} + \delta_{c_i^j, c_{N2_{\rho(i)}}^{j-1}} + \ldots + \delta_{c_i^j, c_{N7_{\rho(i)}}^{j-1}} + \delta_{c_i^j, c_{N8_{\rho(i)}}^{j-1}}}{9}, \end{aligned} \tag{8}$$

where $\delta_{m,n}$ is the Kronecker delta function.

In Eq. 6, a prior $p(\boldsymbol{c}^j)$ is modeled by an MRF model as follows:

$$p(\boldsymbol{c}^j) = \prod_{i \in S} p(c_i^j|c_{N_i}^j). \tag{9}$$

Here the MRF prior is characterized by the MLL (multi-level logistic) model for the local smoothness constraint [12] ($c_{N_i}^j$ represents class labels of neighbor sites of site i).

Assuming that all c_i^j are independent given $\boldsymbol{G}^j$ and each c_i^j is distributed based on the distribution $p(c_i^j|\boldsymbol{g}_i^j)$ independently of all other c_k^j and $\boldsymbol{g}_k^j$, $k \neq i$, the $p(\boldsymbol{c}^j|\boldsymbol{G}^j)$ of Eq. 6 can be rewritten as follows:

$$p(\boldsymbol{c}^j|\boldsymbol{G}^j) = \prod_{i \in S} p(c_i^j|\boldsymbol{g}_i^j). \tag{10}$$

Here the distributions $p(c_i^j|\boldsymbol{g}_i^j)$ are obtained from outputs of MLP as follows:

$$p(c_i^j|\boldsymbol{g}_i^j) = \frac{N(\boldsymbol{g}_i^j, \boldsymbol{w}_c^j)}{\sum_{k=1}^{C} N(\boldsymbol{g}_i^j, \boldsymbol{w}_k^j)}, \tag{11}$$

where $\boldsymbol{w}_c^j$ is the weight vector of MLP network for class c at scale j.

We can rewrite Eq. 5 as follows:

$$\boldsymbol{c}^* = arg \underset{\boldsymbol{c}^j \in \boldsymbol{\Theta}^j}{max} \left\{\prod_{i \in S} p(c_i^j|\boldsymbol{g}_i^j)p(v_i^j|c_i^j)p(c_i^j|c_{N_i}^j)\right\}, \tag{12}$$

Begin
  (1) Compute $p(c_i^j|\boldsymbol{g}_i^j)$ from MLP;
  (2) Initialize configuration $\boldsymbol{c}^j$ from $p(c_i^j|\boldsymbol{g}_i^j)$ and MAP classification (Eq.3);
  (3) For $i \in S$ do
    (3.1) If scale j is the initial coarse scale,
      Then compute $p_l = p(c_i^j = l|\boldsymbol{g}_i^j)p(v_i^j|c_i^j = l)p(c_i^j = l|c_{N_i}^j)$ for all $l \in L$;
      Else compute $p_l = p(c_i^j = l|\boldsymbol{g}_i^j)p(c_i^j = l|c_{N_i}^j)$ for all $i \in L$;
        where L is a set of all class labels;
    (3.2) Set c_i to l with probability p_l that has maximum value;
  (4) Repeat (3) for N times or until convergence;
End.

Fig. 2. Texture segmentation at scale j using a Gibbs sampler

Begin
  (1) Texture segmentation by using the algorithm of Fig.2 in the initial coarse scale $j = J$;
  (2) $j = j + 1$, and determine $p(v_i^j|c_i^j)$ for all site i of scale j from texture segmentation of scale $j - 1$;
  (3) Texture segmentation by using the algorithm of Fig.2 at scale j;
  (4) Repeat (3), (4) until j reaches at the finest scale;
End.

Fig. 3. The multi-scale decision fusion by sequential texture segmentation from coarse to fine scales

A Gibbs sampler can be used for MAP classification by Eq. 12. If texture segmentation is performed sequentially from coarse to fine scales by using Eq. 12, then texture segmentation results of each scale are fused sequentially from coarse to fine scales. In other words, the multi-scale decision fusion is performed. Finally, in the full resolution image (at the finest scale), the more improved texture segmentation result is obtained by the multi-scale decision fusion and the MRF smoothness prior that has the ability to reduce noises caused by classification errors. When Eq. 12 is used at the initial coarse scale, the context vector does not exist. But we can think that the configuration $\boldsymbol{c}^j$ is statistically independent of the context vectors at the initial coarse scale. In other words, at the initial coarse scale, $p(\boldsymbol{c}^j|\boldsymbol{v}^j) = p(\boldsymbol{c}^j)$.

Texture segmentation at scale j is shown in Fig. 2. The proposed segmentation algorithm of Fig. 2 uses a Gibbs sampler as the MAP classifier. Sequential texture segmentation from coarse to fine scale is shown in Fig. 3. The final texture segmentation result in the full resolution image is obtained by the proposed algorithm of Fig. 3.

4 Experiments and Results

In this paper, 17 Brodatz textures are used in the experiments. From each 512×512 Brodatz texture image, we randomly picked ten (overlapping) 64×64 blocks.

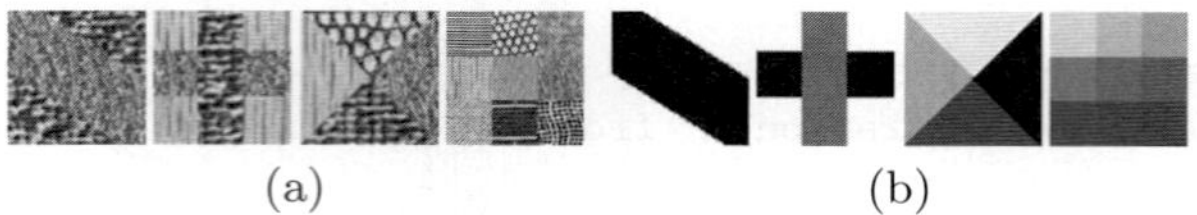

Fig. 4. (a) Test texture images; (b) ideal texture segmentations

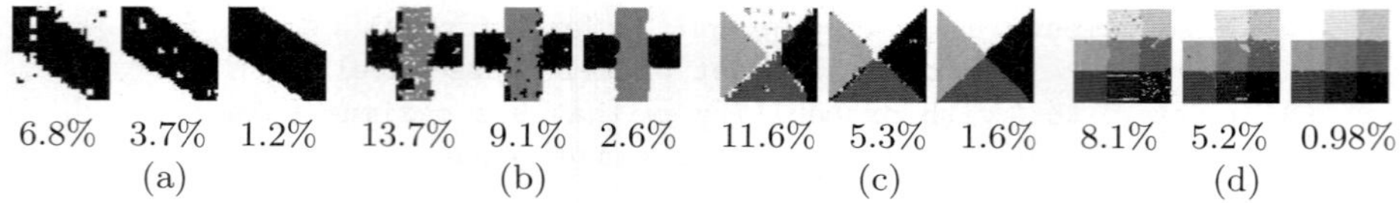

Fig. 5. Texture segmentation results and error rates by each segmentation method; (a) first column: by HMT and HMTseg; second column: by MLP and HMTseg; third column: by the proposed method (MLP and Gibbs sampler); (b) by the same method as (a); (c) by the same method as (a); (d) by the same method as (a)

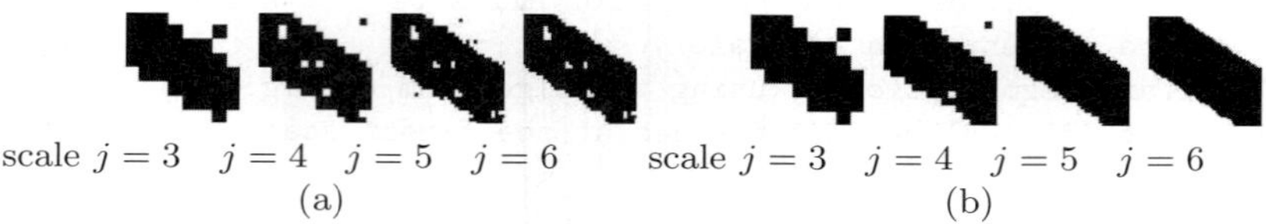

Fig. 6. The comparison of the texture segmentation results at each scale; (a) by MLP and HMTseg; (b) by the proposed method (MLP and Gibbs sampler)

Then, the multi-level wavelet transform (3 levels) of those blocks are used as training data. MLP networks have one hidden layer. The hidden layer of each MLP network has 20 hidden nodes. The image of scale j has $2^j \times 2^j$ pixels (If an image has 64×64 pixels, then the image has scale $j = 6$.). Therefore, when the full resolution image has 64×64 pixels, MLP networks of scale j have $(((4^{(6-j)} - 4)/3 + 1) \times 2 + ((4^{(7-j)} - 4)/3 + 1))$ input nodes (Refer Eq. 1.). The parameters of HMT were estimated using an EM algorithm with an intelligent parameter initialization [17]. Test images are shown in Fig. 4.

Segmentation results are in Fig. 5. Fig. 5 shows texture segmentation results by each segmentation method in the full resolution image. Under the picture of each segmentation result, the error rate between the ideal segmentation (Fig. 4-(b)) and the resulting segmentation (Fig. 5) is given. The error rate is the rate of the number of misclassified pixels to the total number of pixels in an image. The proposed method performs better than other methods.

To compare the proposed method with HMTseg, texture segmentation results at each scale are shown in Fig. 6 (In Fig. 6, scale $j = 6$ is finer than scale $j = 3$, and scale $j = 3$ is the initial coarse scale.). In the multi-scale decision fusion and the ability of reducing noises caused by classification error, the proposed method performs better than HMTseg.

5 Conclusion

In this paper, we proposed a segmentation method using MLP networks and a Gibbs sampler in a multi-scale wavelet domain. Multi-scale wavelet coefficients were used as input for MLP networks. Texture segmentation was performed by using outputs of MLP networks. The proposed method used the MLL model (an MRF model) for reducing noises caused by classification error, and defined the context vector for the multi-scale decision fusion. In addition, a Gibbs sampler was used as an MAP classifier. The results of texture segmentation by the proposed method are much better than those by other methods (HMT and HMTseg, MLP and HMTseg).

References

1. Chen, C.H., Pau, L.F., Wang, P.S.P. (eds.): The Handbook of Pattern Recognition and Computer Vision. 2nd edn. World Scientific Publishing Co. (1998) 207-248
2. Georgeson, M.A.: Spatial Fourier Analysis and Human Vision. In: Southland, N. S. (ed.): Tutorial Essays in Psychology. A Guide to Recent Advance, Vol. 2. Lawrence Earlbaum Associate, Hillsdale NJ (1979) Chapter 2
3. Jain, A.K., Farrokhnia, F.: Unsupervised Texture Segmentation Using Gabor Filters. Pattern Recognition 24 (1991) 1167-1186
4. Choi, H., Baraniuk, R.G.: Multiscale Image Segmentation Using Wavelet-Domain Hidden Markov Models. IEEE Trans. Image Processong 10(9) (September 2001)
5. Chen, Z., Feng, T.J., Houkes, Z.: Texture Segmentation Based on Wavelet and Kohonen Network for Remotely Sensed Images. IEEE Int. Conf. Systems, Man, and Cybernetics, Tokyo, Japan (1999) 816-821
6. Jiang, X., Zhao, R.: A New Method of Texture Segmentation. IEEE Int. Conf. Neural Networks and Signal Processing, Nanjing, China (December 14-17, 2003) 1083-1086
7. Duda, O.R., Hart, P.E., Stork, D.G.: Pattern Classification. 2nd edn. A Wiley Interscience Publication (2001) 161-192, 576-582
8. Kim, K.I., Jung, K., Park, S.H., Kim, H.J.: Support Vector Machines for Texture Classification. IEEE Trans. Pattern Analysis and Machine Intelligence 24(11) (November 2002) 1542-1550
9. Kim, T.H., Eom, I.K., Kim, Y.S.: Texture Segmentation Using Neural Networks and Multi-scale Wavelet Features. Lecture Note in Computer Science, Vol. 3611. Springer-Verlag, Berlin Heidelberg (2005) 400-409
10. Hu, R., Fahmy, M.M.: Texture Segmentation Based on A Hierarchical Markov Random Field Model. Signal Processing 26 (1992) 285-305
11. Li S.Z.: Markov Random Field Modeling in Computer Vision. Springer-Verlag, New York (1995)
12. Stan, Z.L.: Markov Random Field Modeling in Image Analysis. 2nd edn. Computer Science Workbench. Springer-Verlag (2001)
13. Richard, M.D., Lippmann, R.P.: Neural Network Classifiers Estimate Bayesian A Posteriori Probabilities. Neural Computation 3 (1991) 461-483

14. Rojas, R.: Short Proof of The Posterior Probability Property of Classifier Neural Networks. Neural Computation 8 (1996) 41-43
15. Reidmiller, M., Braun, H.: A Direct Adaptive Method for Faster Backpropagation Learning: The Rprop Algorithm. Proc. of ICNN, San Francisco (1993)
16. Duda, O.R., Hart, P.E., Stork, D.G.: Pattern Classification. 2nd edn. A Wiley Interscience Publication (2002) Revised Chapter Section 2.11
17. Fan, G.L., Xia, X.G.: Improved Hidden Markov Models in The Wavelet-Domain. IEEE Trans. Signal Processong 49(1) (January 2001)

Texture Segmentation Using SOM and Multi-scale Bayesian Estimation

Tae Hyung Kim[1], Il Kyu Eom[1], and Yoo Shin Kim[2]

[1] Dept. Electronics Engineering, Pusan National University,
30 Jangjeon-dong, Geumjeong-gu, Busan 609-735,
Republic of Korea

[2] Research Institute of Computer, Information and Communication, Pusan National Univesity, 30 Jangjeon-dong, Geumjeong-gu, Busan 609-735,
Republic of Korea

Abstract. This paper presents a likelihood estimation method from SOM (self organizing feature map), and texture segmentation is performed by using Bayesian estimation and SOM. Multi-scale wavelet coefficients are used as input for SOM, and likelihood probabilities for observations are obtained from trained SOMs. Texture segmentation is performed by the likelihood probability from trained SOMs and ML (maximum likelihood) classification. The result of texture segmentation is improved using contextual information. The proposed segmentation method performed better than segmentation method using HMT (hidden Markov trees) model. In addition, texture segmentation results by SOM and multi-scale Bayesian image segmentation technique called HMTseg also performed better than those by HMT and HMTseg.

1 Introduction

Of the many kinds of neural networks, SOM can be trained in an unsupervised training mode. SOM learns by self-organizing and competition, and learn both the distribution and topology of the input vectors they are trained on. In addition, a trained SOM forms *prototypes* for training data [6], [7]. Since SOM can represent any distribution of inputs without complicated modeling methods, it is easy to apply SOM to texture segmentation. Kohonen neural networks, with some features extracted from wavelet transforms, were used for texture segmentation [3]. An approach based on a support vector machines and neural networks has also been investigated for texture segmentation [9].

The texture description or feature extraction method is one of important factors for texture segmentation [2]. Of all methods for the texture feature extraction, signal processing methods are attractive due to their simplicity. Also, signal processing methods are supported by psychophysical research, which has given evidence that the human brain does a frequency analysis of the image [1]. Fourier [10] or wavelet transforms [5] can be used to extract texture image features. The resulting transformed features in multi-scale images are efficient for

J. Wang et al. (Eds.): ISNN 2006, LNCS 3972, pp. 661–668, 2006.

texture segmentation [5]. Therefore, the HMT model in a multi-scale wavelet-domain was used for texture segmentation [5]. Fuzzy c-means, with some features extracted from wavelet transforms, was used for texture segmentation [4].

In this paper, we propose a novel method of supervised texture segmentation using SOM in a multi-scale wavelet domain and a multi-scale Bayesian image segmentation technique called HMTseg. Firstly, a method for obtaining the likelihood probability of observations from SOM is proposed. Texture segmentation is performed by using likelihoods from SOM and ML classification. Secondly, a method using contextual information is proposed for improving texture segmentation by SOM. Finally, the proposed method uses HMTseg to fuse the multi-scale segmentations and to improve texture segmentation at the finest scale. The proposed method for texture segmentation performs better than the method of [5], which used HMT and HMTseg.

2 The Likelihood Estimation from SOM

In this section, we explain the method to estimate probability density from training samples or *prototypes* [6]. Then, from this estimation method we propose a method for estimating a distribution from SOM.

2.1 Statistical Characteristics of SOM

SOM has one layer, which is called the competitive layer, and neuron nodes in that layer are arranged in a lattice. The weight vectors are connected to the neuron nodes in the competitive layer and have identical dimensions as the input vectors. When an input vector is fed into SOM, the neuron whose weight vector is closest to the input vector according to some measure (e.g. Euclidean distance) is called the winning neuron for that input vector.

In SOM training, if a training vector $\boldsymbol{G}$ is fed into SOM, the weight vectors, which are connected to a winning neuron node i^* and all neuron nodes within a certain neighborhood $N_{i^*}(d)$ of the wining neuron, are updated using the Kohonen rule. Here, $N_{i^*}(d) = \{j, d_{i^*j} \leq d\}$, and d_{i^*j} is the distance between the winning neuron node i^* and the neuron node j in the competitive layer (the distance between positions of neuron nodes on the lattice). Thus, when a vector $\boldsymbol{G}$ is presented, the weights of the wining neuron and its close neighbors move toward $\boldsymbol{G}$. Consequently, after many presentations, weight vectors of neighboring neurons will be close to each other in a vector space. SOM allocates many neighboring neurons when recognizing a region which contains many training vectors in the training vector space, and relatively few neurons when recognizing a region which contains few training vectors. Thus, SOMs learn both the distribution and topology of the input vectors they are trained on. Then the weight vectors of the neurons in the layer become *prototypes* of the training vectors. SOM projects high dimensional training vectors on nodes which are arranged in a two-dimensional lattice.

2.2 Estimation of Probability Density from *Prototypes* [6]

The probability P that vector $\boldsymbol{x}$ will fall in a region R is given by

$$P = \int_R p(\boldsymbol{x}')d(\boldsymbol{x}'). \tag{1}$$

Thus P is an averaged version of the density function $p(\boldsymbol{x})$, and we can estimate the smoothed value of $p(\boldsymbol{x})$ by estimating the probability P. Suppose that n samples, $\boldsymbol{x}_1, \ldots, \boldsymbol{x}_n$, are drawn independently and identically distributed according to a probability law, $p(\boldsymbol{x})$. The probability that k of these n fall in R is given by the ratio k/n which will be a good estimate for the probability P. If we now assume that $p(\boldsymbol{x})$ is continuous and that the region R is so small that $p(\boldsymbol{x})$ does not vary appreciably within it, we can write

$$\int_R p(\boldsymbol{x}')d(\boldsymbol{x}') \approx p(\boldsymbol{x})V, \tag{2}$$

where V is the volume enclosed by R. Combining Eqs. 1 and 2, we arrive at the following obvious estimate for $p(\boldsymbol{x})$,

$$p(\boldsymbol{x}) \approx \frac{k/n}{V}. \tag{3}$$

If we want to obtain $p(\boldsymbol{x})$ rather than just an averaged version of it, we must be prepared to let V approach zero. From a practical standpoint, the number of samples is always limited. Thus, the volume V cannot be allowed to become arbitrarily small. To estimate $p(\boldsymbol{x})$ from n training samples or *prototypes* we can center a cell about $\boldsymbol{x}$ and let it grow until it captures k_n samples, where k_n is some specified function of n. These samples are the k_n nearest-neighbors of $\boldsymbol{x}$. If the density is high near $\boldsymbol{x}$, the cell will be relatively small (good resolution). If the density is low, the cell will grow large. In either case, we can take the estimation $p_n(\boldsymbol{x})$ of $p(\boldsymbol{x})$ for n training samples as follows:

$$p_n(\boldsymbol{x}) \approx \frac{k_n/n}{V_n}, \tag{4}$$

where V_n is the volume of the cell which contains k_n samples. If $p_n(\boldsymbol{x})$ is to converge to density $p(\boldsymbol{x})$, three conditions appear to be required as follows:

$$\lim_{n\to\infty} V_n = 0, \quad \lim_{n\to\infty} k_n = \infty, \quad \lim_{n\to\infty} k_n/n = 0. \tag{5}$$

2.3 The Likelihood Estimation from SOM

In section 2.1, it is presented that SOM projects high dimensional training vectors on nodes which are arranged in a two-dimensional lattice. Therefore we propose a method for estimating likelihood probability for an input vector $\boldsymbol{x}$ using SOM, which had been trained by training vectors of class c, as follows.

$$f(\boldsymbol{x}|c) \approx \frac{k_n/n}{V_n} = \frac{K/N}{V_n} = \frac{K/N}{\pi r_c(\boldsymbol{x})^2}, \tag{6}$$

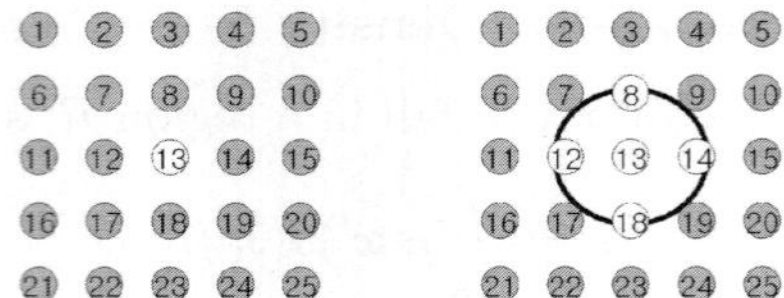

Fig. 1. (Elements of K neighboring neurons on the two-dimensional lattice of neurons; left: when $K = 1$; right: when $K = 5$

where, $c \in \{1, 2, \ldots, N_c\}$ is a class index, N is the number of all neurons in SOM, V_N is a circle with the radius $r_c(\boldsymbol{x})$. The radius $r_c(\boldsymbol{x})$ is the longest distance of all of the distances between input $\boldsymbol{x}$ and the weights of K neighboring neurons that are neighbors of winning neuron for the input $\boldsymbol{x}$ on the two-dimensional lattice of the competitive layer [see Fig. 1]. Thus, if the value of K is fixed, then $r_c(\boldsymbol{x})$ can be computed.

Estimating the likelihood probability more exactly from SOM, a proper K exists depending on the size of SOM and the amount of training data. Increasing the value of K results in an increment of the region R when estimating the distribution of training data. The smaller the amount of training data, the larger the region R grows according to increasing value of K. When the amount of training data is abundant, increasing the value of K improves the reliability of the estimation of the distribution of training data, and decreases the influence of noise. But, when there is few training data, increasing the value of K degrades the precision of the estimation of the distribution (Refer to Eqs. 3, 4, and 5). In this study, the values of K for SOMs are searched by experiments, and then the value of K for each SOM is fixed.

3 Texture Segmentation Using SOM

For texture segmentation we propose to use likelihood probability by SOM in a multi-scale wavelet domain. The multi-level Haar wavelet transform forms a pyramid structure through all scales. Haar wavelet transforms of three levels are shown in Fig. 2-(a). As can be seen, the coarse-scale coefficient w_{J-3}

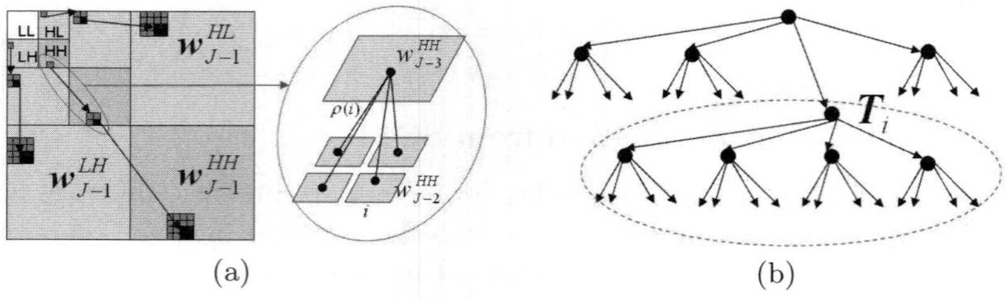

Fig. 2. (a) The Harr wavelet transform and the quad-tree structure of the wavelet coefficients; (b) The quad-tree structure of the wavelet coefficients in a wavelet subband and sub-tree $\boldsymbol{T}_i$ rooted at node i and its elements.

corresponds to four coefficients in the next finer scale and the dependency of these coefficients across scales has a quad-tree structure. The multi-scale wavelet coefficients, which analyze a common sub-region of an image, have persistence across the scale. The dependency between these coefficients can be represented as the dependency between parent and child nodes of a wavelet quad-tree [see Fig. 2-(b)]. In Fig. 2-(b), a black node represents a wavelet coefficient, and $\boldsymbol{T}_i$ is a sub-tree rooted at location i (node i). In this paper, we will often drop the scale $(J, J-1, \ldots)$ and the direction (LH, HL, HH), if they are not confused. To estimate likelihoods in multi-scale images, the proposed system has one group of SOMs for each wavelet scale. In each scale, one SOM is assigned for each texture class. In addition, in pixel level, one group of SOMs is also assigned. If wavelet decomposition level is L and the number of texture classes is C, then the number of SOMs is $(L+1) \times C$.

The input vector of SOM is determined by considering the dependency of the wavelet coefficients across the scale. The input vector $\boldsymbol{g}_i$ is defined as

$$\boldsymbol{g}_i \equiv \{\boldsymbol{T}_i^{HL}, \boldsymbol{T}_i^{LH}, \boldsymbol{T}_i^{HH}, \boldsymbol{I}_i\}, \tag{7}$$

where $\boldsymbol{g}_i$ is for the region of the sub-image analyzed by sub-tree $\boldsymbol{T}_i$ [see Fig. 2-(b)]; $\{\boldsymbol{T}_i^{HL}, \boldsymbol{T}_i^{LH}, \boldsymbol{T}_i^{HH}\}$ represents the wavelet coefficients of sub-trees rooted at node i in three wavelet sub-band (LH, HL, and HH bands) quad-trees (These coefficients of three sub-trees of three wavelet sub-bands analyze a common sub-region of an image.); $\boldsymbol{I}_i$ represents the pixel intensities of the sub-image analyzed by node i.

The classification of an input $\boldsymbol{g}_i$ in j'th wavelet scale is derived as follows:

$$\hat{c} = arg \overset{max}{c \in \{1, 2, \ldots, N_c\}} f(\boldsymbol{g}_i|c) \tag{8}$$

$$\approx arg \overset{max}{c \in \{1, 2, \ldots, N_c\}} \frac{K/N}{\pi r_c^i(\boldsymbol{g}_i)^2} = arg \overset{max}{c \in \{1, 2, \ldots, N_c\}} \frac{1}{r_c^i(\boldsymbol{g}_i)^2}, \tag{9}$$

where $r_c^i(\boldsymbol{g}_i)$ is a distance computed from SOM of scale j for class c by using the method explained in Eq. 6. In this study, the Euclidean distance is used as a distance measure. The multi-scale texture segments are obtained by Eq. 9.

If the contextual information of an image is used to classify the texture for one of the nodes of the wavelet quad-tree, then the multi-scale texture segmentations have been greatly improved. The multi-scale texture segments can be obtained by using contextual information as follows.

$$\hat{c}_i = arg \overset{max}{c_i \in \{1, 2, \ldots, N_c\}} f(\boldsymbol{D}_i|c_i), \tag{10}$$

where $\boldsymbol{D}_i = \{\boldsymbol{g}_i, \boldsymbol{D}_{N_i}, \boldsymbol{g}_{\rho(i)}, \boldsymbol{D}_{N_{\rho(i)}}\}$ is an observation that considers the context of node i of the wavelet quad-tree; $\boldsymbol{D}_{N_{\rho(i)}} \equiv \{\boldsymbol{g}_{N_{\rho(i)},1}, \boldsymbol{g}_{N_{\rho(i)},2}, \cdots, \boldsymbol{g}_{N_{\rho(i)},8}\}$ is an observation on the eight neighbors of the parent node $\rho(i)$ of node i; $\boldsymbol{D}_{N_i} \equiv \{\boldsymbol{g}_{N_i,1}, \boldsymbol{g}_{N_i,2}, \cdots \boldsymbol{g}_{N_i,8}\}$ is an observation on the eight neighbors of node i; and c_i is the class for node i of the wavelet quad-tree. If we assume that the wavelet coefficients in one scale are independent given a class, $f(\boldsymbol{D}_i|c_i)$ of Eq. 10 is rewritten as follows:

$$\begin{aligned} f(\boldsymbol{D}_i|c_i) &= f(\boldsymbol{g}_i|c_i)f(\boldsymbol{D}_{N_i}|c_i)f(\boldsymbol{g}_{\rho(i)}|c_i)f(\boldsymbol{D}_{N_{\rho(i)}}|c_i) \\ &\propto f(\boldsymbol{g}_i|c_i)f(\boldsymbol{D}_{N_i}|c_i)f(\boldsymbol{g}_{\rho(i)}|c_{\rho(i)})f(\boldsymbol{D}_{N_{\rho(i)}}|c_{\rho(i)}), \end{aligned} \tag{11}$$

$$\textit{where} \;\; f(\boldsymbol{D}_{N_i}|c_i) = \prod_{l=1}^{8} f(\boldsymbol{g}_{N_i,l}|c_i) \propto \prod_{l=1}^{8} \frac{1}{r_{c_i}(\boldsymbol{g}_{N_i,l})^2},$$

$$f(\boldsymbol{D}_{N_{\rho(i)}}|c_{\rho(i)}) = \prod_{l=1}^{8} f(\boldsymbol{g}_{N_{\rho(i)},l}|c_{\rho(i)}) \propto \prod_{l=1}^{8} \frac{1}{r_{c_{\rho(i)}}(\boldsymbol{g}_{N_{\rho(i)},l})^2}.$$

Texture segmentations in multi-scale images have contradictions between reliability and minuteness according to scale. The segmentation in coarse scales is accurate for large, homogeneous regions but poor along the boundaries between regions. The segmentation in fine scales is accurate near the boundaries between regions but has poor classification reliability due to the paucity of statistical information. A recent study modeled the distribution of the multi-scale wavelet coefficients in HMT, and performed texture segmentations by using HMT in a multi-scale wavelet domain, and showed that image segmentation in the finest scale is improved by using the HMTseg algorithm that fuses the multi-scale image segmentations obtained by HMT [5]. Therefore, to fuse multi-scale texture segments, we use the HMTseg algorithm. The HMTseg algorithm fuses the multi-scale segmentations one by one from coarse to fine scale and then finally improves texture segmentation at the finest scale.

4 Experiments and Results

In this paper, 16 Brodatz textures are used in the experiments. From each 512×512 Brodatz texture image, we randomly picked ten (overlapping) 64×64 blocks. Then, the multi-level wavelet transform (3 levels) of those blocks were used as training data. Neuron nodes in the competitive layer of SOM are arranged in a 7×7 planar square lattice. The image of scale j has $2^j \times 2^j$ pixels (If an image has 64×64 pixels, then the image has scale $j = 6$.). Therefore, when the full resolution image has 64×64 pixels, SOMs of scale j have $(((4^{(6-j)} - 4)/3 + 1) \times 2 + ((4^{(7-j)} - 4)/3 + 1))$ input nodes (Refer Eq. 7.). Euclidean distance is used to measure the distance between an input vector and the weight vector of

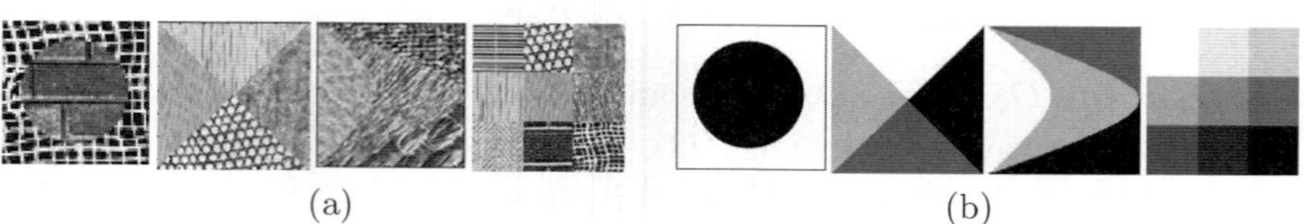
(a) (b)

Fig. 3. (a) Test texture images; (b) ideal texture segmentations

$j=3$ $j=4$ $j=5$ $j=6$ $j=3$ $j=4$ $j=5$ $j=6$ $j=3$ $j=4$ $j=5$ $j=6$
(a) (b) (c)

Fig. 4. Multi-scale texture segmentations by each system before applying HMTseg (j is scale index); (a) by SOMs with $K = 1$; (b) by SOMs, $K = 5$; (c) by SOMs, K values set according to each scale ($K = 1, 1, 1$, and 5 from left to right of Fig. (c))

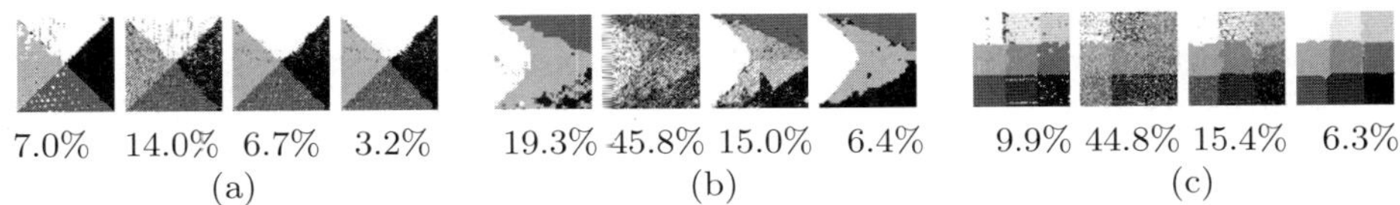

7.0% 14.0% 6.7% 3.2% 19.3% 45.8% 15.0% 6.4% 9.9% 44.8% 15.4% 6.3%
(a) (b) (c)

Fig. 5. Texture segmentation results and error rates by each segmentation method and HMTseg; (a) first column: by HMT and HMTseg; second column: by SOMs, $K = 1$ and HMTseg; third column: by SOMs, $K = 1, 1, 1$, and 5 from the coarsest scale to the finest scale, and HMTseg; fourth column: by SOMs, context $\boldsymbol{D}_i = \left\{\boldsymbol{g}_i, \boldsymbol{D}_{N_i}, \boldsymbol{g}_{\rho(i)}, \boldsymbol{D}_{N_{\rho(i)}}\right\}$ and using $K = 1, 1, 1$, and 5, and HMTseg; (b), (c) by the same method as (a).

a neuron. The parameters of HMT were estimated using an EM algorithm with an intelligent parameter initialization [8]. Test images are shown in Fig. 3.

A proper K in SOM will result in good segmentation performance. As shown on Fig. 4-(a) and 4-(b), SOM with $K = 5$ presents good performance in scale $j = 6$, and SOMs with $K = 1$ present good performance in scales $j = 3, 4$, and 5 (In Fig. 4, scale $j = 6$ is finer than scale $j = 3$.). The reason for these results is because the amount of training data is more abundant for the fine scales (especially, for scale $j = 6$) than for the coarse scales and segmentation for the fine scales is more sensitive to noise. Therefore, the segmentation performance by SOM with $K = 5$ is excellent at the finest scale $j = 6$. But, for the coarse scales which had less training data, increasing the value of K reduces the reliability of the likelihood estimation. Thus, SOMs with different values of K according to each scale have good texture segmentation performance. Texture segmentation results by SOMs with different values of K according to each scale are shown in Fig. 4-(c) (The multi-scale segmentation results by SOMs with $K = 1, 1, 1$, and 5 according to the scale, $j = 3, 4, 5$, and 6).

Segmentation results for other test images are in Fig. 5. Fig. 5 shows texture segmentation results by each segmentation method and HMTseg only in the final pixel domain (scale $j = 6$). Under the picture of each segmentation result, the error rate between the ideal segmentation (Fig. 3-(b)) and the resulting segmentation (Fig. 5) is given. The error rate is the rate of the number of misclassified pixels to the total number of pixels in an image. Let's consider Fig. 5-(a) and 5-(b) (the test image contains 4 texture classes). SOMs with using the proposed methods perform better than HMT except for SOM that uses $K = 1$ for all of the scales. However, Fig. 5-(c) indicates that SOMs using K

values that differed according to each scale did not perform as well as HMT. It seems that 7×7-sized SOM is more sensitive to noise than the Gaussian mixture model at the finest scale (scale $j = 6$) if the number of texture classes increases. However, when SOM uses contexts to reduce the influence of noise, it performs better than HMT. Comparing the second and third columns of Fig. 5 reveals that segmentation by SOM that uses proper values of K differing according to each scale is much better than SOM that uses $K = 1$ for all of the scales.

5 Conclusion

We proposed a novel method of supervised texture segmentation using SOM in a multi-scale wavelet domain and a multi-scale Bayesian image segmentation technique. A method for obtaining the likelihood probability of observations from SOM was proposed. To estimate the likelihood more exactly from SOM, the values of K for each SOM were searched by experiments. The performance of texture segmentation improved when K was set to different values for SOMs according to each scale or according to the amount of training data. We also propose a likelihood estimation method that used contextual information. This method improved texture segmentation. The results of texture segmentation by the proposed methods are much better than those by HMT and HMTseg.

References

1. Georgeson, M.A.: Spatial Fourier Analysis and Human Vision. In: Southland, N. S. (ed.): Tutorial Essays in Psychology. A Guide to Recent Advance, Vol. 2. Lawrence Earlbaum Associate, Hillsdale NJ (1979) Chapter 2
2. Chen, C.H., Pau, L.F., Wang, P.S.P. (eds.): The Handbook of Pattern Recognition and Computer Vision. 2nd edn. World Scientific Publishing Co. (1998) 207-248
3. Chen, Z., Feng, T.J., Houkes, Z.: Texture Segmentation Based on Wavelet and Kohonen Network for Remotely Sensed Images. IEEE Int. Conf. Systems, Man, and Cybernetics, Tokyo, Japan (1999) 816-821
4. Jiang, X., Zhao, R.: A New Method of Texture Segmentation. IEEE Int. Conf. Neural Networks and Signal Processing, Nanjing, China (December 14-17, 2003) 1083-1086
5. Choi, H., Baraniuk, R.G.: Multiscale Image Segmentation Using Wavelet-Domain Hidden Markov Models. IEEE Trans. Image Processong 10(9) (2001) 1309-1321
6. Duda, O.R., Hart, P.E., Stork, D.G.: Pattern Classification. 2nd edn. A Wiley Interscience Publication (2001) 161-192, 576-582
7. Howard, D., Beale, M.: Neural Network Toolbox for Use with MATLAB. User's Guide Version 4. The MathWorks Inc. 277-298
8. Fan, G., Xia, X.G.: Improved Hidden Markov Models in The Wavelet-Domain. IEEE Trans. Signal Processong 49(1) (2001) 115-120
9. Kim, K.I., Jung, K.C., Park, S.H., Kim, H.J.: Support Vector Machine for Texture Classification. IEEE Trans. Pattern Analysis and Machine Intelligence 24(11) (2002) 1542-1550
10. Jain, A.K., Farrokhnia, F.: Unsupervised Texture Segmentation Using Gabor Filters. Pattern Recognition 24 (1991) 1167-1186

Recognition of Concrete Surface Cracks Using the ART1-Based RBF Network

Kwang-Baek Kim[1], Kwee-Bo Sim[2], and Sang-Ho Ahn[3]

[1] Department of Computer Engineering, Silla University, Busan, Korea
gbkim@silla.ac.kr
[2] School of Electrical and Electronic Engineering, Chung-Ang Univ., Seoul, Korea
kbsim@cau.ac.kr
[3] Department of Architectural Engineering, Silla University, Busan, Korea
shahn@silla.ac.kr

Abstract. In this paper, we proposed the image processing techniques for extracting the cracks in a concrete surface crack image and the ART1-based RBF network for recognizing the directions of the extracted cracks. The image processing techniques used are the closing operation of morphological techniques, the Sobel masking used to extract edges of the cracks, and the iterated binarization for acquiring the binarized image from the crack image. The cracks are extracted from the concrete surface image after applying two times of noise reduction to the binarized image. We proposed the method for automatically recognizing the directions (horizontal, vertical, -45 degree, 45 direction degree) of the cracks with the ART1-based network. The proposed ART1-based RBF network applied ART1 to the learning between the input layer and the middle layer and the Delta learning method to the learning between the middle layer and the output layer. The experiments using real concrete crack images showed that the cracks in the concrete crack images were effectively extracted and the proposed ART1-based RBF network was effective in the recognition of the direction of extracted cracks.

1 Introduction

Because the cracks in concrete structures have bad effects on the tolerance, the durability, the waterproof and the appearance of the structures, they bring about some worse problems in the structures. Therefore the causes of cracks must be accurately examined and the durability and the safety of the structures must be evaluated. If necessary, the repair and rehabilitation works must be established. When we draw a deduction on the causes of cracks in the structures, the patterns and the distribution characteristics of cracks, these become important factors to judgment [1]. Because manual works by inspectors may be too subjective, techniques which enables objective examinations by computers is necessary [2], [3].

In this paper, we proposed the recognition method, which automatically extracts cracks from a surface image acquired by a digital camera, and it also recognizes the directions (horizontal, vertical, -45 degree, and 45 degree) of the cracks using the ART1-based RBF network. We compensate an effect of light on a concrete surface image by applying the closing operation, which is one of the morphological

J. Wang et al. (Eds.): ISNN 2006, LNCS 3972, pp. 669 – 675, 2006.

techniques, extract edges of cracks by Sobel masking, and binarize the image by applying the iterated binarization technique [4]. Two separate times of noise reduction are applied to the binarized image for effective noise elimination. After minute noises are eliminated by using the average of adjacent pixels corresponding to a 3x3 mask, more noises are eliminated by analyzing the regular ratio of length and width with Glassfire labeling algorithm. The specific region of cracks is extracted from the noise-eliminated binarized image. We proposed the method to automatically recognize the directions of cracks by applying the ART1-based RBF network. The proposed ART1-based RBF network applied ART1 to the learning between the input layer and the hidden layer, and the Delta learning method is used in learning between the hidden layer and the output layer.

2 Crack Detection Using Image Processing Techniques

The overall process for the crack detection and recognition algorithm using the proposed techniques in this paper is described in Fig. 1. At first, cracks are extracted from a concrete surface image using some image processing techniques, and then, the directions of cracks are automatically recognized by applying the ART1-based RBF network, which is proposed in this paper.

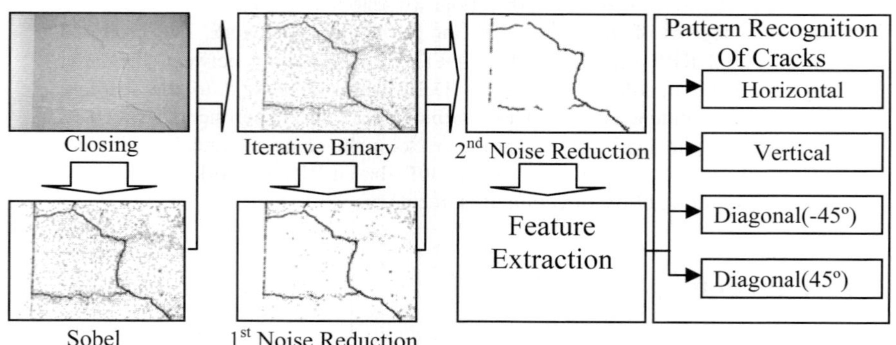

Fig. 1. Overview of the proposed crack detection and recognition algorithm

2.1 Compensation for an Effect of Light

The brightness of the background varies according to the direction and the amount of light in an environment when photographing concrete surfaces. Sobel masking, which is sensitive to a value of brightness, cannot extract edges in the dark regions due to the effect of light. Therefore, for compensating effectively an effect of light, we applied the closing operation being one of the morphological techniques. The closing operation performs the dilation operation after the erosion operation. The dilation operation and the erosion operation are as follows:

$$(f \odot g)(x) = \max\left\{y : g(z-x)+y \ll f(z)\right\} \tag{1}$$

$$(f \oplus g)(x) = \min\{y : -g(-(z-x)) + y \gg f(z)\} \tag{2}$$

In Fig. 2, (a) is the original image and (b) is the closing image, which is generated by applying the closing operation to (a) and this shows appreciable cracks.

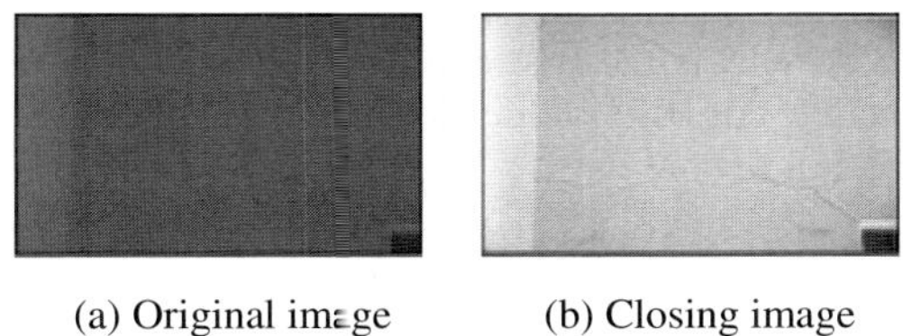

(a) Original image (b) Closing image

Fig. 2. Original image and closing image of a crack image

2.2 Crack Detection

Sobel masking is applied to the closing image for improving the performance of edge extraction based on features such as the great difference between the brightness of

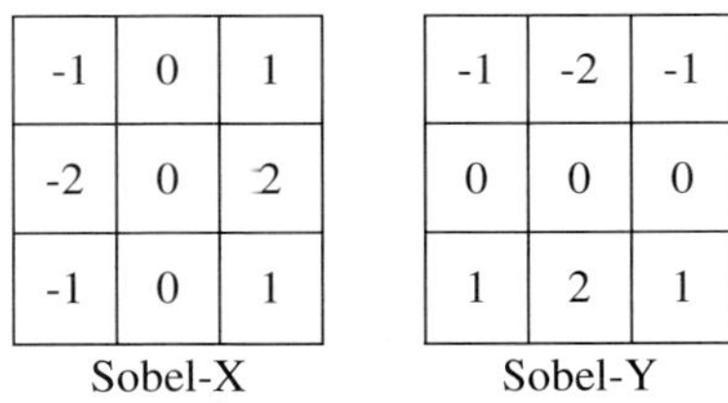

-1	0	1
-2	0	2
-1	0	1

Sobel-X

-1	-2	-1
0	0	0
1	2	1

Sobel-Y

Fig. 3. Sobel Mask

Step 1. Select the first estimated threshold value T^{t0}.

Step 2. Divide the image into two regions of R_1 and R_2 using the estimated threshold value T^t

Step 3. Calculate the average of the gray values, u_1 and u_2, for each region.

$$u_1 = \frac{\sum f(i,j)}{N_1} \qquad u_2 = \frac{\sum f(i,j)}{N_2}$$

N_1 and N_2 is the number of pixels in each region.

Step 4. Calculate the new threshold value.

$$T^{(t+1)} = \frac{u_1 + u_2}{2}$$

Step 5. Repeat Step 2 to Step 4 until the values of u_1 and u_2 cannot be changed.

Fig. 4. Iterated binarization algorithm

cracks and the brightness of the surface of a concrete structure. The edge extraction finds the change of brightness by the differential operator and the two masks shown in Fig. 3 is used for fast operation.

2.3 Binarization of a Crack Image

The iterated binarization selects the first estimated threshold value, then it iterates an update of threshold value until the value doesn't change, and this is followed by it selecting the final threshold value.

2.4 Noise Reduction

For eliminating noises without influencing the cracks and the background, noise reduction operation is applied twice. Firstly, after the 3x3 mask is applied to the binarized image, if 1's pixels are more than 0's ones among the adjacent 9 ones, the center pixel is set to 1. Otherwise, the center pixel is set to 0 as shown in Fig. 5. This process eliminates minute noises.

0	0	0
0	**0**	1
0	1	0

1	1	1
1	**1**	1
1	0	0

Fig. 5. 3x3 Mask for noise reduction

Secondly, the Glassfire labeling technique is applied for eliminating additional noises. Glassfire labeling is the labeling method examining the adjacent pixels of the current pixel one by one recursively until all the adjacent pixels are labeled [5]. In the labeled image, the area of each labeled region is calculated by using the first pixel and last pixel of the region. In this paper, through the experiment, the criterion of the area is set to 1.7. Therefore, if the ratio of the length and width is less than 1.7, they are determined as noises and are therefore eliminated.

3 Crack Recognition Using the ART1-Based RBF Network

The RBF network is a feed-forward neural network that consists of three layers, input, middle and output layer. In the RBF network, because the operations required between layers are different, learning algorithms between layers can be mutually different. So, the optimum organizations between layers can be separately constructed [6].

The middle layer of the RBF network executes the clustering operation, classifying an input data set to homogeneous clusters. The measurement of homogeneity in clusters is the distance between vectors in clusters. And the classification of input data to a cluster means that the distances between input data and each vector in the cluster are shorter than or equal to the fixed radius. But, the use of a fixed radius in clustering causes wrong classifications. Therefore the selection of the organization for the

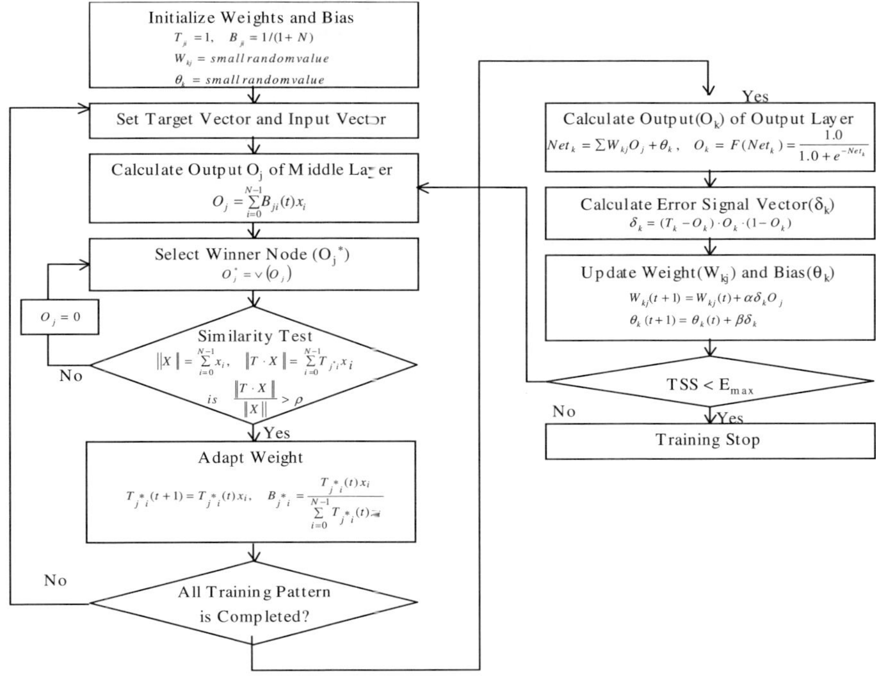

Fig. 6. The ART1-based RBF Network

middle layer determines the overall efficiency of the RBF network [7]. This paper proposes an enhanced RBF neural networks that adapts the ART1 network to the learning structure between the input layer and the middle layer and applies the output layer of the ART1 network to the middle layer. The learning of ART1-based RBF network is divided into two stages. In the first stage, competitive learning is applied as the learning structure between the input layer and the middle layer. And the supervised learning is accomplished between the middle layer and the output layer. Fig. 6 shows the learning process of the proposed ART1-based RBF network.

4 Experiments and Future Works

The crack images are acquired by Sony's Cyber-shot 5.0 digital camera. Fig. 7 is the result of the specific region extraction in a crack image and Fig. 8 is the result of the recognized directions of extracted cracks.

For analyzing the performance of the proposed ART1-based RBF network, the extracted 25 crack patterns are used as input patterns. The vigilance variable is set to 0.7, the number of output nodes to 5, and the learning rate to 0.9.

Table 1 summarizes the performance measurement of the ART1-based RBF network. In Table 1, the failure cases in crack recognition are the ones which uses enlarged or reduced images as input images, and the recognition of the non-directional cracks.

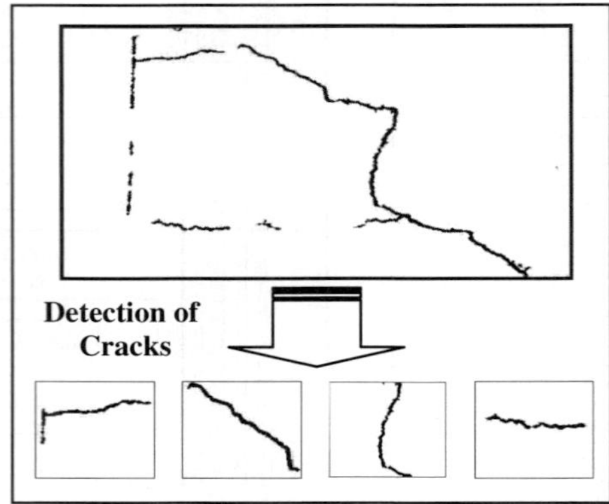

Fig. 7. Specific crack extraction in a crack image

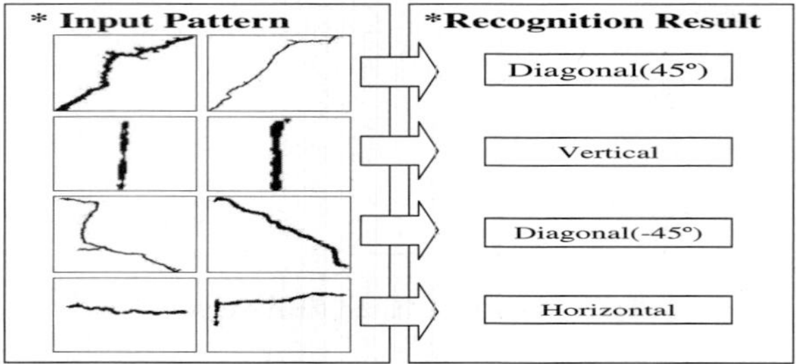

Fig. 8. Result of crack recognition

Table 1. Learning result of specific cracks

	ART1-based RBF Network
The number of node in the hidden layer	18
The number of Epoch	290
Recognition rate	24/25

5 Conclusion

In this paper, we proposed the recognition method to automatically extract the cracks of a concrete surface image acquired by the digital camera and to recognize the direction (horizontal, vertical, -45 degree, and 45 degree) of the specific cracks using the ART1-based RBF network. We compensate an effect of light on a concrete surface image by applying the closing operation, which is one of the morphological techniques, extract the edges of cracks by Sobel masking, and binarize the image by applying the iterated binarization technique. Noise reduction is applied twice to the binary image for effective noise elimination. After the specific regions of cracks are automatically extracted from the preprocessed image by applying Glassfire labeling algorithm to the extracted crack image, the cracks of the specific region are enlarged or reduced to 30x30 pixels and then used as input patterns to the proposed ART1-based RBF network. The ART1-based RBF network shows the effectiveness of learning and recognition for the directions of the extracted cracks.

In this paper, when the enlarged or reduced images as inputs are used, the recognition of non-directional cracks fails. In future studies, this will be examined by the new algorithm which will recognize the non-directional cracks by extracting parameters from the features discovered through the patterns of the cracks.

References

1. Lee, B. Y., Kim, Y. Y., Kim, J. K.: Development of Image Processing for Concrete Surface Cracks by Employing Enhanced Binarization and Shape Analysis Technique. Journal of the Korea Concrete Institute 17 (2005) 361-368
2. Lee, B.Y., Park, Y. D., Kim, J. K.: A Technique for Pattern Recognition of Concrete Surface Cracks. Journal of the Korea Concrete Institute 17 (2005) 369-374
3. Kim, Y. S., Haas and C. T.: An Algorithm for Automatic Crack Detection, Mapping and Representation. KSCE Journal of Civil Engineering. 4 (2000) 103-111
4. Gonzalez, R. C., Woods, R. E., and Eddins, S. L.: Digital Image Processing. Pearson Prentice Hall (2004)
5. Pitas, I.: Digital Image Processing Algorithms and Applications. John Wiley & Sons INC. (2000)
6. Kim, B. K., Lee, D. U., Sim, K. B.: Performance Improvement of Fuzzy RBF Networks. In: Lipo, W., Ke, C., Yew, S. O. (eds.): Advances in Natural Computation. Lecture Notes in Computer Science, Vol. 3610. Springer-Verlag, Berlin Heideberg New York (2005) 237-244
7. Kim, K. B., Joo, Y. H., Cho, J. H.: An Enhanced Fuzzy Neural Network. In: Kim, M. L., Hon, S., Simon, S., Wentong, C. Pingzhi, F. Susumu, H. (eds.): Parallel and Distributed Computing: Applications and Technologies. Lecture Notes in Computer Science, Vol. 3320. Springer-Verlag, Berlin Heideberg New York (2004) 176-179

SVM-Enabled Voice Activity Detection

Javier Ramírez[1], Pablo Yélamos[1], Juan Manuel Górriz[1], Carlos G. Puntonet[2], and José C. Segura[1]

[1] Dept. of Signal Theory, Networking and Communications, University of Granada, Spain
javierrp@ugr.es
[2] Dept. of Architecture and Computer Technology, University of Granada, Spain

Abstract. Detecting the presence of speech in a noisy signal is an unsolved problem affecting numerous speech processing applications. This paper shows an effective method employing support vector machines (SVM) for voice activity detection (VAD) in noisy environments. The use of kernels in SVM enables to map the data into some other dot product space (called feature space) via a nonlinear transformation. The feature vector includes the subband signal-to-noise ratios of the input speech and a radial basis function (RBF) kernel is used as SVM model. It is shown the ability of the proposed method to learn how the signal is masked by the acoustic noise and to define an effective non-linear decision rule. The proposed approach shows clear improvements over standardized VADs for discontinuous speech transmission and distributed speech recognition, and other recently reported VADs.

1 Introduction

Currently, there are technology barriers inhibiting speech processing systems that work in extremely noisy conditions from meeting the demands of modern applications. These systems often require a noise reduction system working in combination with a precise voice activity detector (VAD). The classification task is not as trivial as it appears and its performance is strongly affected by the increasing background noise level. Since their introduction in the late seventies [1], Support Vector Machines (SVMs) marked the beginning of a new era in the learning from examples paradigm. SVMs have attracted recent attention from the pattern recognition community due to a number of theoretical and computational merits derived from the Statistical Learning Theory [2] developed by Vladimir Vapnik at AT&T. This paper shows an effective SVM-based VAD for improving the performance of speech processing systems that need to operate in noisy environment. The proposed method combines a noise robust speech processing feature extraction process together with a trained SVM model for classification. The results are compared to standardized techniques and a representative set of VAD methods.

J. Wang et al. (Eds.): ISNN 2006, LNCS 3972, pp. 676–681, 2006.

2 Background on SVM Learning

SVMs have recently been proposed for pattern recognition in a wide range of applications by its ability for learning from experimental data. The reason is that SVMs are much more effective than other conventional parametric classifiers. In SVM-based pattern recognition, the objective is to build a function $f : R^N \longrightarrow \{\pm 1\}$ using training data that is, N-dimensional patterns $\mathbf{x}_i$ and class labels y_i:

$$(\mathbf{x}_1, y_1), (\mathbf{x}_2, y_2), ..., (\mathbf{x}_\ell, y_\ell) \in R^N \times \{\pm 1\} \tag{1}$$

so that f will correctly classify new examples $(\mathbf{x}, y)$.

Hyperplane classifiers are based on the class of decision functions:

$$f(\mathbf{x}) = \text{sign}\{(\mathbf{w} \cdot \mathbf{x}) + b\} \tag{2}$$

It can be shown that the optimal hyperplane is defined as the one with the maximal margin of separation between the two classes. The solution $\mathbf{w}$ of a constrained quadratic optimization process can be expanded in terms of a subset of the training patterns called support vectors that lie on the margin:

$$\mathbf{w} = \sum_{i=1}^{\ell} \nu_i \mathbf{x}_i \tag{3}$$

Thus, the decision rule depends only on dot products between patterns:

$$f(\mathbf{x}) = \text{sign}\{\sum_{i=1}^{\ell} \nu_i (\mathbf{x}_i \cdot \mathbf{x}) + b\} \tag{4}$$

The use of kernels in SVM enables to map the data into some other dot product space (called feature space) F via a nonlinear transformation $\Phi : R^N \longrightarrow F$ and perform the above linear algorithm in F. Fig. 1 illustrates this process where the 2-D input space is mapped to a 3-D feature space where the data is linearly separable. The kernel is related to the Φ function by $k(\mathbf{x}, \mathbf{y}) = (\Phi(\mathbf{x}) \cdot \Phi(\mathbf{y}))$. In the input space, the hyperplane corresponds to a nonlinear decision function whose form is determined by the kernel. There are three common kernels that are used by SVM practitioners for the nonlinear feature mapping: *i*) polynomial, *ii*)

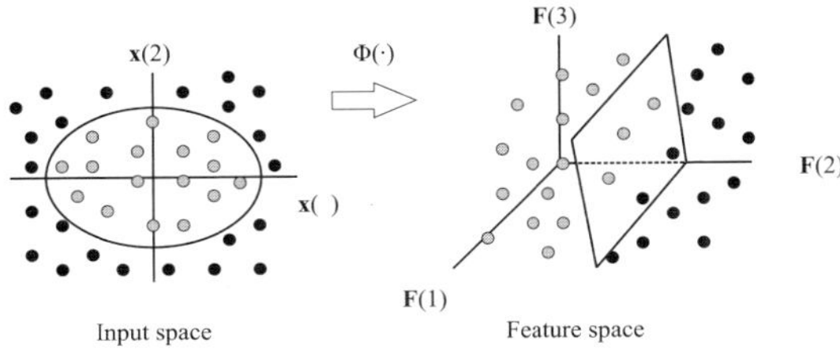

Fig. 1. Effect of the map from input to feature space where the separation boundary becomes linear

Radial basis function (RBF), and *iii*) sigmoid kernels. Thus, the decision function is nonlinear in the input space

$$f(\mathbf{x}) = \text{sign}\{\sum_{i=1}^{\ell} \nu_i k(\mathbf{x}_i, \mathbf{x}) + b\} \tag{5}$$

and the parameters ν_i are the solution of a quadratic programming problem that are usually determined by the well known Sequential Minimal Optimization (SMO) algorithm [3]. Many classification problems are always separable in the feature space and are able to obtain better results by using RBF kernels instead of linear and polynomial kernel functions [4, 5].

3 Proposed SVM-Based VAD

A block diagram of the proposed VAD is shown in Fig. 2. The fist step is the training process on the training data set and its associated class labels. The signal is preprocessed and a feature vector is extracted for training. Once the SVM model has been trained, the proposed SVM-based algorithm consists of the following stages: *i*) the input signal is decomposed into speech frames and feature extraction is conducted for classification, and *ii*) the speech features $\mathbf{x}$ are processed by the SVM decision function f defined in equation 5.

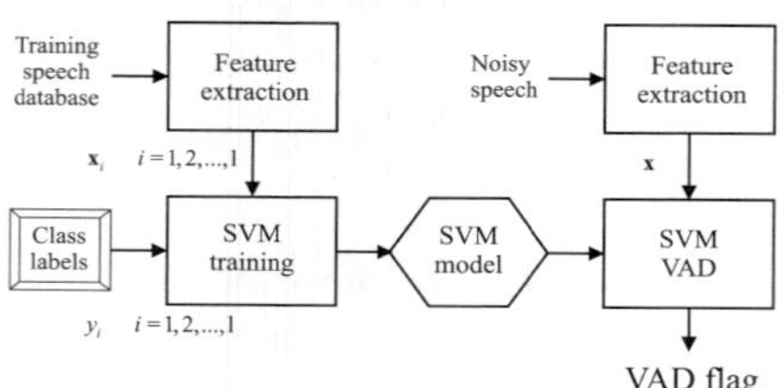

Fig. 2. Block diagram of the proposed SVM-based VAD

3.1 Preprocessing and Feature Extraction

The algorithm for feature extraction is stated as follows. The input signal $x(n)$ sampled at 8 kHz is decomposed into 25-ms overlapped frames with a 10-ms window shift. A denoising process based on a Wiener filter is applied to improve the performance of the VAD in high noise environments. Once the input signal has been denoised, a filterbank reduces the dimensionality of the feature vector to a representation including broadband spectral information suitable for detection.

3.2 Training the SVM Classification Rule

The SVM model has been trained using LIBSVM software tool [6]. A training set consisting of 12 utterances of the AURORA 3 Spanish SpeechDat-Car (SDC) was used. This database contains 4914 recordings using close-talking and distant

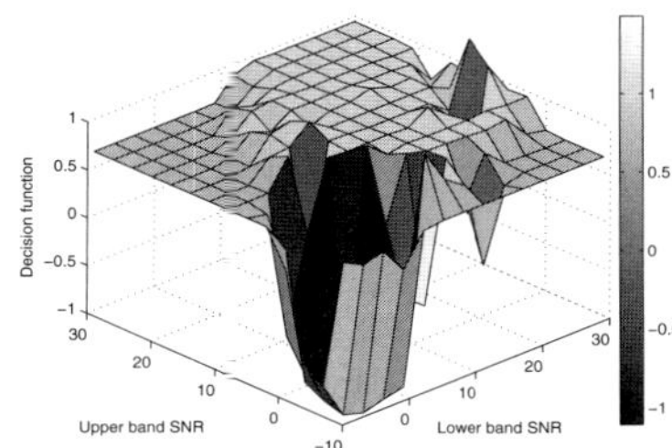

Fig. 3. Decision function of a 2-band trained SVM model

microphones from more than 160 speakers. The files are categorized into three noisy conditions: quiet, low noisy and highly noisy conditions, which represent different driving conditions with average SNR values between 25dB, and 5dB. The recordings used for training the SVM are selected to deal with different noisy conditions.

Fig. 3 shows the decision function of a 2-band trained SVM model. Note that, b can be used as a decision threshold for the VAD in the sense that the working point of the VAD can be shifted in order to meet the application requirements.

4 Experimental Framework

This section analyzes the proposed VAD and compares its performance to other algorithms used as a reference. The analysis is based on the ROC curves, a frequently used methodology to describe the VAD error rate. The AURORA subset of the original Spanish SDC database [7] was used again in this analysis. The non-speech hit rate (HR0) and the false alarm rate (FAR0= 100-HR1) were determined as a function of the decision threshold being the actual speech frames and actual speech pauses determined by hand-labelling the database on the close-talking microphone.

Before showing comparative results, the selection of the optimal number of subbands is addressed. Fig. 4 shows the influence of the noise reduction block and the number of subbands on the ROC curves in high noisy conditions. First, noise reduction is not carried to better show the influence of the number of subbands. Increasing the number of subbands improves the performance of the proposed VAD by shifting the ROC curves in the ROC space. For more than four subbands, the VAD reports no additional improvements. This value yields the best trade-off between computational cost and performance. On the other hand, the noise reduction block included in the proposed VAD reports an additional shift of the ROC curve as shown in Fig. 4.

Fig. 5 shows the ROC curves of the proposed VAD and other frequently referred algorithms [8, 9, 10, 11] for recordings from the distant microphone in high noisy conditions. The working points of the ITU-T G.729, ETSI AMR and AFE VADs are also included. The results show improvements in detection accuracy over standard VADs and over a representative set VAD algorithms

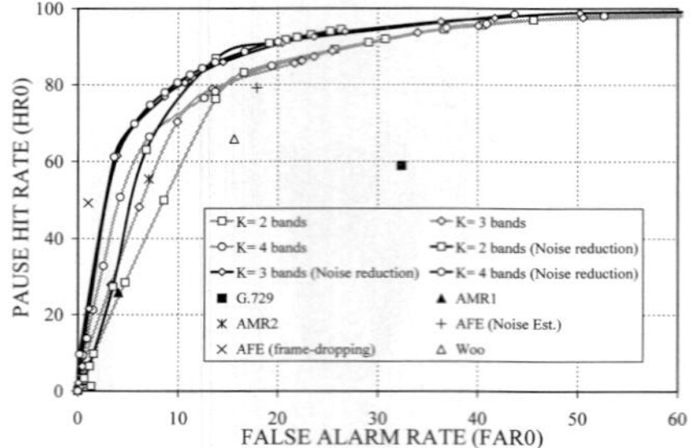

Fig. 4. Subband selection (High: high speed, good road, 5 dB average SNR)

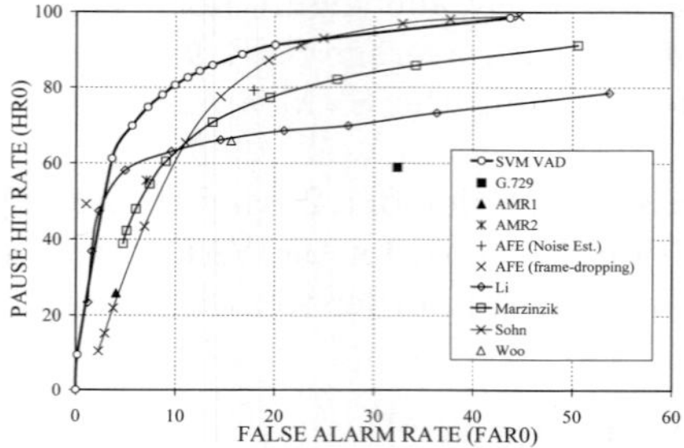

Fig. 5. Comparative results to other VAD methods

[8, 9, 10, 11]. Among all the VAD examined, our VAD yields the lowest false alarm rate for a fixed non-speech hit rate and also, the highest non-speech hit rate for a given false alarm rate. The benefits are especially important over ITU-T G.729, which is used along with a speech codec for discontinuous transmission, and over the Li's algorithm, that is based on an optimum linear filter for edge detection. The proposed VAD also improves Marzinzik's VAD [10] that tracks the power spectral envelopes, and the Sohn's VAD [11], that formulates the decision rule by means of a model-based statistical likelihood ratio test.

5 Conclusions

An effective algorithm for detecting presence of speech in a noisy signal is proposed in this paper. The proposed strategy combines spectral noise reduction techniques and support vector machine learning tools to derive a non-linear decision rule in the input space defined in terms of the subbands SNRs. With these and other innovations the proposed method has shown to be more effective than VADs that define the decision rule in terms of average SNR values. The non-speech and speech classes can be clearly distinguished in the 3-D space and that the SVM model learns how the signal is masked by the noise. On the other hand, increasing the number of subbands up to four improves the performance of the proposed VAD by shifting the ROC curve in the ROC space. Finally, the experiments conducted on the Spanish SpeechDat-Car database showed that the proposed algorithm outperforms ITU G.729, ETSI AMR1 and AMR2 and ETSI AFE standards as well as other recently reported VAD methods in speech/non-speech detection performance.

Acknowledgements

This work has received research funding from the EU 6th Framework Programme, under contract number IST-2002-507943 (HIWIRE, Human Input that Works in Real Environments) and SESIBONN and SR3-VoIP projects (TEC2004-06096-C03-00, TEC2004-03829/TCM) from the Spanish government. The views expressed here are those of the authors only. The Community is not liable for any use that may be made of the information contained therein.

References

1. Vapnik, V.: Estimation of Dependences Based on Empirical Data. Springer-Verlag, New York (1982)
2. Vapnik, V.: Statistical Learning Theory. John Wiley and Sons, Inc., New York (1998)
3. Platt, J.: Fast Training of Support Vector Machines Using Sequential Minimal Optimization. In: Advances in Kernel Methods - Support Vector Learning. MIT Press (1999) 185–208
4. Clarkson, P., Moreno, P.: On the Use of Support Vector Machines for Phonetic Classification. In: Proc. of the IEEE Int. Conference on Acoustics, Speech and Signal Processing. Volume 2. (1999) 585–588
5. Ganapathiraju, A., Hamaker, J. Picone, J.: Applications of Support Vector Machines to Speech Recognition. IEEE Transactions on Signal Processing **52** (2004) 2348–2355
6. Chang, C., Lin, C.J.: LIBSVM: A Library for Support Vector Machines. Technical report, Dept. of Computer Science and Information Engineering, National Taiwan University (2001)
7. Moreno, A., Borge, L., Christoph, D., Gael, R., Khalid, C., Stephan, E., Jeffrey, A.: SpeechDat-Car: A Large Speech Database for Automotive Environments. In: Proceedings of the II LREC Conference. (2000)
8. Woo, K., Yang, T., Park, K., Lee, C.: Robust Voice Activity Detection Algorithm for Estimating Noise Spectrum. Electronics Letters **36** (2000) 180–181
9. Li, Q., Zheng, J., Tsai, A., Zhou, Q.: Robust Endpoint Detection and Energy Normalization for Real-Time Speech and Speaker Recognition. IEEE Transactions on Speech and Audio Processing **10** (2002) 146–157
10. Marzinzik, M., Kollmeier, B.: Speech Pause Detection for Noise Spectrum Estimation by Tracking Power Envelope Dynamics. IEEE Transactions on Speech and Audio Processing **10** (2002) 341–351
11. Sohn, J., Kim, N.S., Sung, W.: A Statistical Model-Based Voice Activity Detection. IEEE Signal Processing Letters **16** (1999) 1–3

A Robust VAD Method for Array Signals

Xiaohong Ma, Jin Liu, and Fuliang Yin

School of Electronic and Information Engineering,
Dalian University of Technology,
Dalian 116023, China
maxh@dlut.edu.cn

Abstract. A new voice activity detection (VAD) method for microphone array signals is developed in this paper. A relatively pure speech signal can be obtained by applying noise canceling algorithms on some signals from microphone array. For suppressing correlated and uncorrelated noises, the proposed method doesn't perform the same processing, but analyze the natures of the background noises by calculating the correlation between the noisy signals during silence intervals firstly. If the additive noises are correlated, relatively pure speech component is separated by blind source separation (BSS) method. Otherwise, this speech component is estimated by beamforming and maximum a posterior (MAP) algorithm. Then, a voice activity detection method based on entropy is employed to determine whether this relatively pure speech signal is active or not. Finally, this VAD result is used as reference to produce those of all array signals. Simulation results illustrate the validity of the proposed method.

1 Introduction

Microphone array has increasingly been used in many systems for sound source localization [1] and speech enhancement [2]. Among these systems, sometimes more than one channel of array signals need to detect the presence of speech. Various types of VAD algorithms have been proposed, such as those based on zero crossing rates, cepstral features, adaptive noise modeling, source separation, periodicity estimates, decision-making based on a combination of different parameters, long-term speech information [3], and eigenspace-energy-entropy [4]. However, most of them deal with the single channel signal or signals in relatively mild noise environments, and their performance degrades severely under low signal-to-noise ratio (SNR) and non-stationary noise environments. To achieve the VAD results of all signals from microphone array in various noise environments, a new VAD method is proposed. There is a front-end processing of VAD, which aims to estimate relatively pure speech signal from the noisy signals using some signals from microphone array. Then, an entropy based voice activity detector is employed to determine whether this relatively pure speech signal is active or not. Finally, the VAD result is used as reference to produce those of all array signals.

J. Wang et al. (Eds.): ISNN 2006, LNCS 3972, pp. 682 – 687, 2006.

2 The Model of Array Signals

If the reverberation is ignored, a linear microphone array of N sensors is considered, which records the plane wave generated by a speaker in the presence of noises. The sensor signal $x_i(t)$ $(i = 1,2,\cdots,N)$ can be expressed as

$$x_i(t) = \alpha_i s(t - \tau_i) + n_i(t) \tag{1}$$

where $s(t)$ and $n_i(t)$ represent source speech signal and noise respectively. $s(t)$, $n_i(t)$ and $n_j(t)$ $(1 \le i, j \le N, i \ne j)$ are assumed uncorrelated with each other. α_i is attenuation factor due to propagation effect. τ_i denotes the propagation time of the direct path from the speech source to the i th microphone.

Sometimes, for the presence of air-condition or projector, the interference is directional. That is to say, $n_i(t)$ and $n_j(t)$ in Eq.(1) are correlated, and there is only a time delay between them. The sensor signal $x_i(t)$ can be expressed as

$$x_i(t) = \alpha_i s(t - \tau_i) + \beta_i n(t - \tau_i') \tag{2}$$

where $n(t)$ represents the directional noise, β_i is noise attenuation factor due to propagation effect, and τ_i' denotes the propagation time of the direct path from the noise source to the i th microphone.

3 The Proposed Method

The schematic diagram of the proposed method is shown in Fig.1. There are four main modules, the noise nature identifying and noise canceling module, the VAD module, the time delay estimation module, and the sequence shift module.

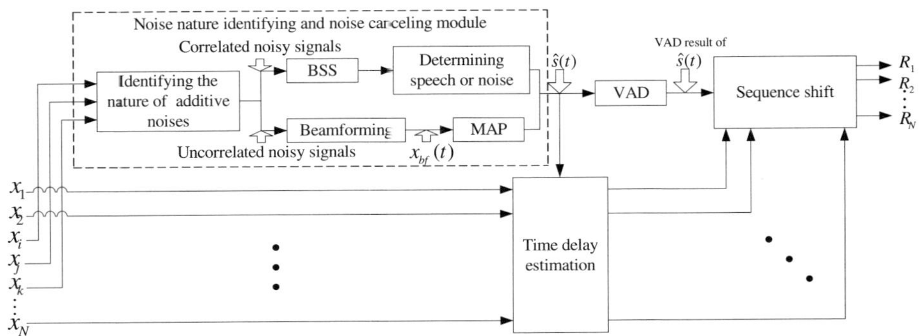

Fig. 1. Schematic diagram of the proposed method

3.1 Noise Nature Identifying and Noise Canceling

To be able to apply the most appropriate noise canceling algorithm afterwards, we need to identify the nature of noise of the array signals firstly.

To improve the accuracy and control the computation of the proposed method, we use any L-channel signals from microphone array to identify the nature of noises. Generally, the first 100ms of the noisy signal only includes noise. By calculating correlation coefficients for the first segment of the L-channel noisy signals, the nature of the noises can be determined.

Considering correlated noise environments, noise and speech signal are produced by different sources, so they are believed to be independent with each other. We can separate them by BSS. The BSS method we used here is the FastICA[5].The results of FastICA yield one output who mainly consists of the speech and the other output who mostly contains noise. Due to the arbitrary output orders in FastICA, the long-term spectral divergence (LTSD) are then calculated according to [3] to determine which one is the speech signal.

In uncorrelated noise environments, beamforming technique is firstly utilized to increase the SNR of signal before MAP. In the beamforming processing, to simplify the calculation, all the weights of $x_i(t)\ (i = 1, 2, \cdots, L)$ are set to be $\frac{1}{L}$ to produce one channel signal $x_{bf}(t)$. The output signal $x_{bf}(t)$ is still with noises. Then MAP algorithm is applied to cancel the noise further more. Relatively pure speech signal $\hat{s}(t)$ inferred by the gradient learning rule in [6] can be acquired.

3.2 Voice Activity Detection [4]

To achieve the satisfactory VAD results of N-channel signals in noise environments, VAD procedure is applied on the separated speech signal $\hat{s}(t)$ only once instead of on every N-channel signals. According to [4], noise and speech signal have different eigenspace-energy-entropy distributions. If we choose proper threshold of the entropy to the signal $\hat{s}(t)$, its presence can be detected. Obviously, signal $\hat{s}(t)$ is purer than the original array signal $x_i(t)$, so the VAD result of $\hat{s}(t)$ should be more accurate than that of $x_i(t)$.

3.3 Time Delay Estimation and Sequence Shift

According to Eq. (1) and (2), it is obviously that there exist time shifts between signal $\hat{s}(t)$ and the noisy signals from microphone array. Therefore, the VAD result of $\hat{s}(t)$ must have time shifts with that of N-channel signals. The time delay vector $\boldsymbol{\tau}_s$ between $\hat{s}(t)$ and array signals $\mathbf{x}(t)$ is estimated by resorting to the generalized cross correlation (GCC) method [7]. The beginning and ending tags of active speech segments in $\hat{s}(t)$ are shifted respectively according to $\boldsymbol{\tau}_s$, and the active speech segments $\mathbf{R} = (R_1, R_2, \cdots, R_N)^T$ of array signals $\mathbf{x} = (x_1, x_2, \cdots, x_N)^T$ can be achieved.

4 Experimental Results

Extensive experiments are carried out through a series trials in a simulated rectangular room ($4m\times 6m\times 3m$) to demonstrate the validity and robustness of the proposed method. The array of three microphones ($N=3$, $L=3$) is used in our simulations and the distance between adjacent microphones is set equally to 30cm. In each experiment, the speech signal, whose sampling rate is 8kHz and duration of time is 3.072s, is delayed for appropriate times according to the spatial locations. Various noises with appropriate time delays are added to each channel signal, and the average SNR is 5dB. As shown in the following figures, the horizontal axis represents samples whose magnitude is 10^4 , and the vertical axis represents the normalized amplitude of signals. The solid line represents the VAD results of the proposed method and the dashed line represents that of the conventional method [4]. The waveforms of pure speech source signal and its VAD result are shown in Fig.2.

Experiments with correlated noise are carried out firstly. The waveforms of three different noises are shown in Fig.3. Three noisy signals are shown in Fig.4 as well as the VAD results through two methods. The separated speech signals in each case are shown in Fig.5, with the addition of their VAD results. As shown in Fig.5, the separated speech signals are 'pure' enough for accurately detecting the active speech, in spite of the kinds and levels of noises. As shown in Fig.4, the results of the proposed method are affirmed to be more accurate than that of conventional method.

As for the experiments in uncorrelated noise environments, Fig.6 shows the waveforms of three uncorrelated white noises. Noisy signals and their VAD results through two methods are shown in Fig.7. The estimated speech signal and its VAD result are shown in Fig.10(a). Similarly, the experiment with uncorrelated color noises is achieved subsequently. The waveforms of three uncorrelated color noises are shown in Fig.8. Noisy signals and their VAD results through two methods are shown in Fig.9. The estimated speech signal is shown in Fig.10(b), with the addition of its VAD result. Clearly, relatively accurate VAD results are respectively obtained by using the proposed method, but the conventional method can not reach such satisfactory results.

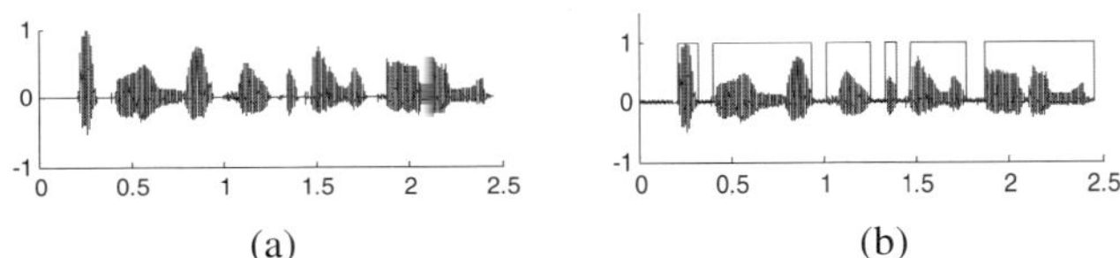

Fig. 2. (a) Waveform of pure speech. (b) waveform and VAD result of pure speech.

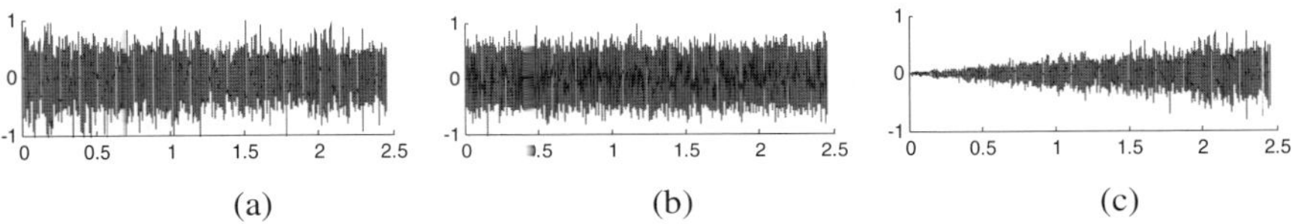

Fig. 3. Waveforms of noises. (a) color noise. (b) white noise. (c) color noise with gradually increasing intensity.

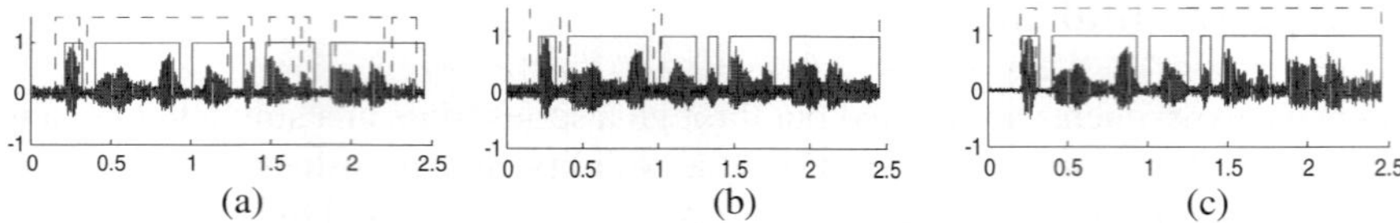

Fig. 4. Waveforms of the noisy signals (SNR=5dB) and the VAD results with two methods. (a) speech with color noise. (b) speech with white noise. (c) speech with color noise with gradually increasing intensity.

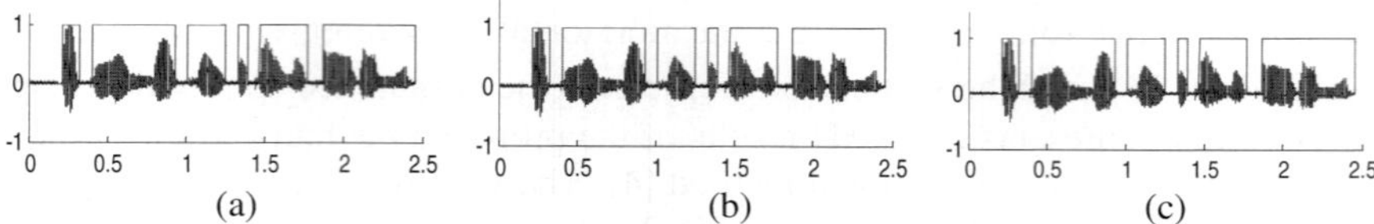

Fig. 5. Waveforms and VAD results of separated signals in each case. (a)-(c) separated speech signal.

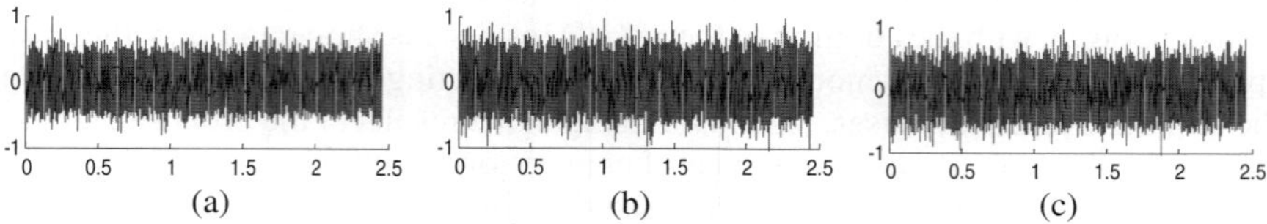

Fig. 6. Waveforms of white noises. (a) white noise1. (b) white noise2. (c) white noise3.

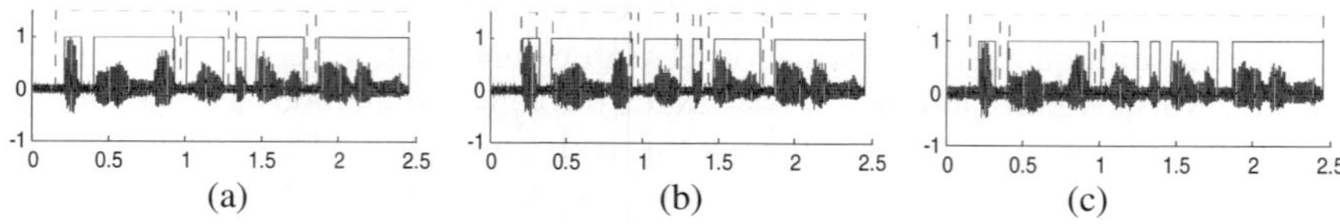

Fig. 7. Waveforms of the noisy signals (SNR=5dB) and the VAD results with two methods. (a) speech with white noise1. (b) speech with white noise2. (c) speech with white noise3.

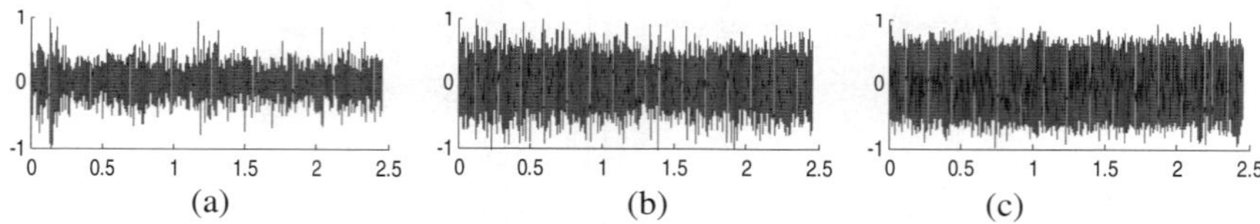

Fig. 8. Waveforms of color noises. (a) color noise1. (b) color noise2. (c) color noise3.

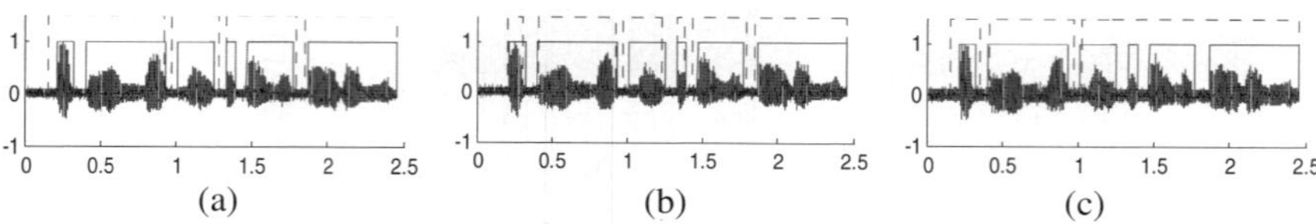

Fig. 9. Waveforms of the noisy signals (SNR=5dB) and the VAD results with two methods. (a) speech with color noise1. (b) speech with color noise2. (c) speech with color noise3.

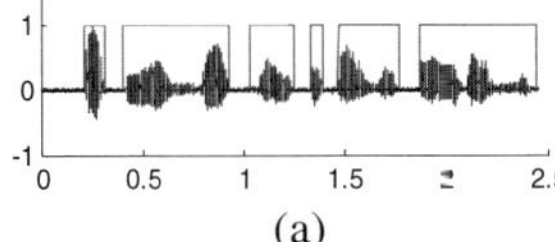

(a)

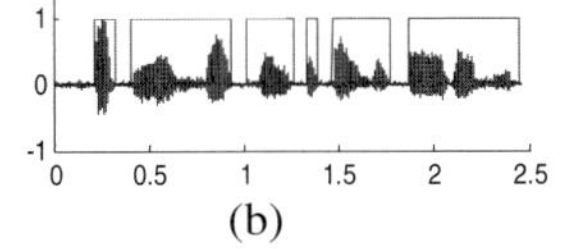

(b)

Fig. 10. Waveforms and VAD results of estimated signals in each case. (a)-(b) estimated speech signal.

5 Conclusion

A novel VAD strategy of array signals in correlated and uncorrelated noise environments is presented in this paper. By applying the BSS or beamforming and MAP algorithm on array signals before VAD, signal $\hat{s}(t)$ estimated from the mixed observed signals is pure enough to accurately detect the active speech, in spite of the kinds and the levels of noises. Using the VAD result of signal $\hat{s}(t)$ as reference for those of array signals, we can obtain VAD results of N -channel signals at the same time by using VAD method only once. Extensive experiments demonstrate that the proposed method is robust enough with various noises, and it enables us to detect the activity of speech more reliably.

Acknowledgement

This work was supported by the National and Liaoning Province Natural Science Foundation of China under Grant No. 60575011, No. 60372082, and No. 20052181.

References

1. Gustafsson, T., Rao, B. D., Trivedi, M.: Source Localization in Reverberant Environments: Modeling and Statistical Analysis. IEEE Trans. Speech and Audio Process 11 (2003) 791-803
2. Gannot, S., Cohen, I.: Speech Enhancement Based on the General Transfer Function GSC and Postfiltering. IEEE Trans. Speech and Audio Process 12 (2004) 561-571
3. Ramírez, J., Segura, J. C., Benítez, C., et. al.: Efficient Voice Activity Detection Algorithms Using Long-term Speech Information. Speech Communication 42 (2004) 271-287
4. Xu, W., Ding, Q., Wang, B. X.: A Speech Endpoint Detector Based on Eigenspace-energy-entropy (in Chinese). Journal of China Institute of Communications 24 (2003) 125-132
5. Hyvarinen, A., Oja, E.: A Fast Fixed-point Algorithm for Independent Component Analysis. Neural Computation, 9(1997) 1483-1492
6. Zhong, M. J., Tang H. W., Chen H. ., Tan Y. Y.: An EM Algorithm for Learning Sparse and Overcomplete Representation. Neurocomputing, 57 (2004) 469-476
7. Knapp, C., Carter, G.: The Generalized Correlation Method for Estimation of Time Delay. IEEE Trans. Acoustics, Speech, and Signal Process 24 (1976) 320-327

A Flexible Algorithm for Extracting Periodic Signals

Zhi-Lin Zhang[1] and Haitao Meng[2]

[1] School of Computer Science and Engineering,
University of Electronic Science and Technology of China,
Chengdu 610054, China
zlzhang@uestc.edu.cn

[2] School of Electric and Information Engineering,
Yancheng Institute of Technology,
Yancheng 224003, Jiangsu, China

Abstract. In this paper, we propose a flexible two-stage algorithm for extracting desired periodic signals. In the first stage, if the period and phase information of the desired signal is available (or can be estimated), a minimum mean square error approach is used to coarsely recover the desired source signal. If only the period information is available (or can be estimated), a robust correlation based method is proposed to achieve the same goal. The second stage uses a higher-order statistics based Newton-like algorithm, derived from a constrained maximum likelihood criteria, to process the extracted noisy signal as cleanly as possible. A parameterized nonlinearity is adopted in this stage, adapted according to the estimated statistics of the desired signal. Compared with many existing extraction algorithms, the proposed algorithm has better performance, which is confirmed by simulations.

1 Introduction

In real world [1], a large number of sensors are available but only a very few source signals are the desired ones. Thus it is essential to develop reliable, robust and effective source extraction algorithms which enable us to extract only the small number of desired source signals. In many applications, the desired source signals are periodic, such as many biomedical signals [3, 4, 5].

To extract periodic signals, many extraction algorithms have been proposed. Lu and Rajapakse [7] and Barros *et al.* [5] propose a class of algorithms based on both the second-order and the higher-order statistics. These algorithms need a reference signal, namely, they need the *a priori* information about the period and the phase of the desired signals. But in some applications, such as the fetal ECG extraction [3, 4], the accurate phase cannot be obtained.

Barros and Cichocki [6] propose a second-order statistics based algorithm, which only needs the period information of the desired signal. It is simple and runs quickly, but it cannot recover a clean desired signal, and its performance is sensitive to the estimate of the period.

J. Wang et al. (Eds.): ISNN 2006, LNCS 3972, pp. 688–693, 2006.

Recently, we propose an extraction algorithm [3], which also exploits both the second-order statistics and the higher-order statistics. But the algorithm only needs the period information. Thus it is very suitable to the case where phase information is unavailable. However, the algorithm is lacking robustness when the periods of some source signals are very close to each other.

Based on our previous work [2, 3, 4], in this paper we propose a novel algorithm, which has better performance, confirmed by computer simulations.

2 Problem Statement

Suppose one observes an n-dimentional stochastic signal vector $\mathbf{x}$ that is regarded as the linear transformation of an m-dimensional *mutually independent* zero-mean and unit-variance source vector $\mathbf{s}$, i.e., $\mathbf{x} = \mathbf{As}$, where $\mathbf{A}$ is an unknown mixing matrix. The goal of source extraction is to find a vector $\mathbf{w}$ such that $y = \mathbf{w}^T\mathbf{x} = \mathbf{w}^T\mathbf{As}$ is an estimated source signal up to a scalar. To cope with ill-conditioned cases and to make algorithms simpler and faster, whitening is often used to transform the observed signals $\mathbf{x}$ to $\tilde{\mathbf{x}} = \mathbf{Vx}$ such that $E\{\tilde{\mathbf{x}}\tilde{\mathbf{x}}^T\} = \mathbf{I}$, where $\mathbf{V}$ is a whitening matrix. For convenience, in the following we assume that $\mathbf{x}$ are the whitened observed signals and $n = m$.

Besides the independence assumption, we further assume the desired periodic signal s_i satisfies the following relations:

$$\begin{cases} E\{s_i(k)s_i(k-\tau^*)\} > 0 \\ E\{s_j(k)s_j(k-\tau^*)\} = 0 \quad \forall j \neq i \end{cases} \tag{1}$$

where s_j are other source signals, k is the time index, and τ^* is called the optimal time delay defined below:

Definition 1. *A non-zero τ^* is called the optimal time delay, if the delayed autocorrelation at τ^* of the desired source signal s_i is non-zero, while the delayed autocorrelations at τ^* of other source signals are zero. Here all of the source signals are supposed to be mutually independent.*

If there exists the optimal time delay τ^*, then the desired periodic source signal can be perfectly recovered by $y = (\mathbf{w}^*)^T\mathbf{x}$, where $\mathbf{w}^*$ is called the optimal solution, which is defined below:

Definition 2. *A column vector w^* is called the optimal solution, if the desired periodic source signal s_i is perfectly recovered by $y = (\boldsymbol{w}^*)^T\boldsymbol{x}$, where $y = cs_i$ and c is a non-zero scale. $\boldsymbol{x}$ are the observations or the whitened observations.*

Let us assume the optimal time delay τ^* is known. Obviously, under the constraint $\|\mathbf{w}\| = 1$, maximizing the objective function

$$J(\mathbf{w}) = E\{y(k)y(k-\tau^*)\} = \mathbf{w}^T E\{\mathbf{x}(k)\mathbf{x}(k-\tau^*)^T\}\mathbf{w} \tag{2}$$

leads to the desired periodic source signal s_i. The reason for this proposal is that for the desired periodic source signal, this delayed autocorrelation should have a positive larger value, while for other source signals this value should be zero.

Using the standard power method [1] and neglecting the small difference between $R_x(\tau^*)$ and $R_x(\tau^*)^T$, from the objective function (2) we can derive the Barros's algorithm [6]:

$$\begin{cases} \mathbf{w}^+ = R_x(\tau^*)\mathbf{w} \\ \mathbf{w} = \mathbf{w}^+/\|\mathbf{w}^+\| \end{cases} \tag{3}$$

where $R_x(\tau^*) = E\{\mathbf{x}(k)\mathbf{x}(k-\tau^*)^T\}$. Note here the Barros's algorithm is derived from a different aspect. Due to exploitation of second-order statistics, the algorithm is simple and fast. Since τ^* is the optimal time delay, the desired source signal is perfectly extracted.

But it is difficult to find such optimal time delay (in many cases it does not exist). A widely-used method is to take the desired signal's fundamental period as the time delay τ^* [6]. In this way, some practical issues should be considered:

1. The desired source signal s_i may weakly correlate with some of the other source signals, i.e., $E\{s_i(k)s_j(k-\tau^*)\} \neq 0, j \neq i$.
2. It is possible that several other source signals are autocorrelated at the time delay τ^*, i.e., $E\{s_j(k)s_j(k-\tau^*)\} \neq 0, j \neq i$.
3. Even if the source signals are strictly mutually uncorrelated, in fact the calculated correlations of source signals using limited samples are generally non-zero, due to the fact that the expectation operator is replaced by the mathematical average. That is to say, even if $E\{s_i(k)s_j(k-\tau^*)\} = 0$ and $E\{s_j(k)s_j(k-\tau^*)\} = 0, j \neq i$, it is very possible that $\sum_{k=\tau^*}^{N-1} s_i(k)s_j(k-\tau^*)/(N-\tau^*) \neq 0$ and $\sum_{k=\tau^*}^{N-1} s_j(k)s_j(k-\tau^*)/(N-\tau^*) \neq 0$.

We have shown [2] that due to the joint effect of the above issues, the performance of Barros's algorithm (3) is not good.

3 Proposed Algorithm

3.1 The First Stage: Coarse Recovery

In the case where the period of the desired source signal is known or can be estimated, we propose the following objective function:

$$J(\mathbf{w}) = \mathbf{w}^T \Big\{ \sum_{l=1}^{P} \big(\mathbf{R}_x(l\tau) + \mathbf{R}_x(l\tau)^T\big) \Big\} \mathbf{w}, \tag{4}$$

with the constraint $\|\mathbf{w}\| = 1$, where $\mathbf{R}_x(l\tau) = E\{\mathbf{x}(k)\mathbf{x}(k-l\tau)^T\}$, P is a positive integer, and τ is the fundamental period of the desired source signal.

Note that maximizing the objective function (4) is equivalent to finding the eigenvector corresponding to the maximal eigenvalue of $\sum_{l=1}^{P} \big(\mathbf{R}_x(l\tau) + \mathbf{R}_x(l\tau)^T\big)$. Thus we have the following algorithm for extracting the desired periodic source signal:

$$\mathbf{w} = EIG\Big(\sum_{l=1}^{P} \Big(\mathbf{R}_x(l\tau) + \mathbf{R}_x(l\tau)^T\Big)\Big), \tag{5}$$

where $EIG(\mathbf{Q})$ is the operator that calculates the normalized eigenvector corresponding to the maximal eigenvalue of the matrix $\mathbf{Q}$.

Since there are many fast algorithms that calculate the normalized eigenvector, the algorithm (5) is very fast. More importantly, since (4) is based on the averaged eigen-structure of the correlation matrix of source signals, the algorithm overcomes the three difficulties discussed in Section 2 [2]. Furthermore, we have the following theorem:

Theorem 1. *The cross-talk error of the extracted desired source signal* $y_1 = \boldsymbol{w}_1^T \boldsymbol{x}$ *tends to be smaller than that of the* $y_2 = \boldsymbol{w}_2^T \boldsymbol{x}$, *where* $\boldsymbol{w}_1$ *is obtained from the algorithm (5), and* $\boldsymbol{w}_2$ *is obtained from the algorithm (3).*

Due to the limit of space, we omit the proof. Note that, by adopting some techniques [2], the algorithm (5) is nonsensitive to the estimate errors of the fundamental period τ, unlike the algorithm (3).

In the case where both the period and the phase of the desired periodic source signal s_i are known or can be estimated, we can design a reference signal d [5, 7], satisfying that the desired source signal s_i is the one and only the one closest to d. In other words, $E\{ds_i\} > E\{ds_j\}$, where $j = 1, \cdots, n, j \neq i$. Accordingly, we can construct the following objective function:

$$J(\mathbf{w}) = E\{(d - \mathbf{w}^T \mathbf{x})^2\}. \tag{6}$$

By the minimum mean square error approaches, such as the LMS algorithm, the optimum Wiener solution is given by

$$\mathbf{w} = E\{d\mathbf{x}\}. \tag{7}$$

The above two algorithms (5) and (7) are based on second-order statistics. Although they can obtain good results, it can be shown that in the source extraction the algorithms based on second-order statistics are insufficient to recover the desired source signal, and that the solution $\mathbf{w}$ is just close to the optimal solution $\mathbf{w}^*$. To make the solution $\mathbf{w}$ closer to $\mathbf{w}^*$, we sort to the higher-order statistics in the following sub-section.

3.2 The Second Stage: Fine Extraction

Denote the solution from the previous stage by $\widehat{\mathbf{w}}$. In this stage we develop a new higher-order statistics based algorithm from a constrained maximum likelihood criteria, making the solution $\widehat{\mathbf{w}}$ further closer to the optimal solution $\mathbf{w}^*$, which implies the SNR of the extracted desired signal is higher.

Under the constraint $\|\mathbf{w}\| = 1$, the maximum likelihood criteria for extracting one source signal is given by

$$\begin{cases} \max & L(\mathbf{w}) = E\{\log p(\mathbf{w}^T \mathbf{x}(k))\} \\ s.t. & \|\mathbf{w}\| = 1 \end{cases} \tag{8}$$

where p denotes the probability density function of the source signal. Note that due to the existence of local maxima, maximizing (8) only leads to one source signal, but not necessarily the desired source signal s_i. However, if we take the $\widehat{\mathbf{w}}$ from the first stage as the initial value, we can necessarily obtain the s_i.

Using the Newton optimization method, we can easily obtain the following algorithm for extracting the desired source signal s_i:

$$\mathbf{w}^+(l+1) = \mathbf{w}(l) - \mu(l)E\{f\big(\mathbf{w}(l)^T\mathbf{x}\big)\mathbf{x}\}/E\{f'\big(\mathbf{w}(l)^T\mathbf{x}\big)\} \tag{9}$$

$$\mathbf{w}(l+1) = \mathbf{w}^+(l+1)/\|\mathbf{w}^+(l+1)\| \tag{10}$$

with the initial value $\mathbf{w}(0) = \widehat{\mathbf{w}}$. l denotes iterations, $\mu(l)$ is a time-dependent step-size, and the nonlinearity $f(y) = (\log p(y))' = p(y)'/p(y)$.

In general, p is unknown and should be estimated. We propose to use the parameterized generalized Gaussian distribution to model the probability distribution of the desired source signal:

$$p(y;\alpha) = \frac{\alpha}{2\lambda\Gamma(1/\alpha)}e^{-|\frac{y}{\lambda}|^\alpha} \tag{11}$$

where $\Gamma(x)$ is the Gamma function: $\Gamma(x) = \int_0^\infty t^{x-1}e^{-t}dt$. Since $E\{y^2\} = 1$, we have $\lambda = \sqrt{\Gamma(1/\alpha)/\Gamma(3/\alpha)}$. The Gaussian exponent α controls the peakness of the distribution. From the model, the nonlinearity can be derived by [1]

$$f(y) = \frac{d\log p(y)}{dy} = |y|^{\alpha-1}sgn(y). \tag{12}$$

In practice, two different values of Gaussian exponent α are used: $\alpha = 1.0$ when $\kappa(y)$ is positive; $\alpha = 4.0$ when $\kappa(y)$ is less than 0. Note that when $\alpha = 1$, we have $f'(y) = 2\delta(y)$, which can be implemented as:

$$f'(y) = 2, \quad \forall y \in [-\epsilon, \epsilon]; \quad f'(y) = 0, \quad otherwise \tag{13}$$

where ϵ is a very small positive number. The estimated kurtosis $\kappa(y)$ is given by

$$\kappa(y) = E\{(\widehat{\mathbf{w}}^T\mathbf{x})^4\}/(E\{(\widehat{\mathbf{w}}^T\mathbf{x})^2\})^2 - 3. \tag{14}$$

Since $\widehat{\mathbf{w}}$ is very close to the optimal solution $\mathbf{w}^*$, the estimated kurtosis $\kappa(y)$ is very close to its true value.

4 Simulation

The seven source signals were shown in Fig.1(a). The ECG signal s_2 was our desired source signal. These signals were randomly mixed and prewhitened. Then we easily estimated the desired source signal' period. We ran the algorithms in [2, 3, 5, 6, 7] and our proposed algorithm. Since the desired signal's phase information could not be estimated, the algorithms in [5] and in [7] did not work. Thus only the results by the remained algorithms are shown in Fig.1(b). Obviously, the algorithm in [6] and in [3] could not obtain the correct source signal. To make comparison between y_3 and y_4, the cross-talk error $PI = \frac{1}{N-1}(\sum_{i=1}^N \frac{e_i^2}{\max_i e_i^2} - 1)$ was used, where $[e_1, \cdots, e_N] = \mathbf{w}^T\mathbf{VA}$. The smaller the PI is, the better the performance is. The simulation was repeated 200 times independently. The averaged PI of y_3 was only 0.0345, while that of y_4 was 0.0001. The simulation showed our proposed algorithm has better extraction performance.

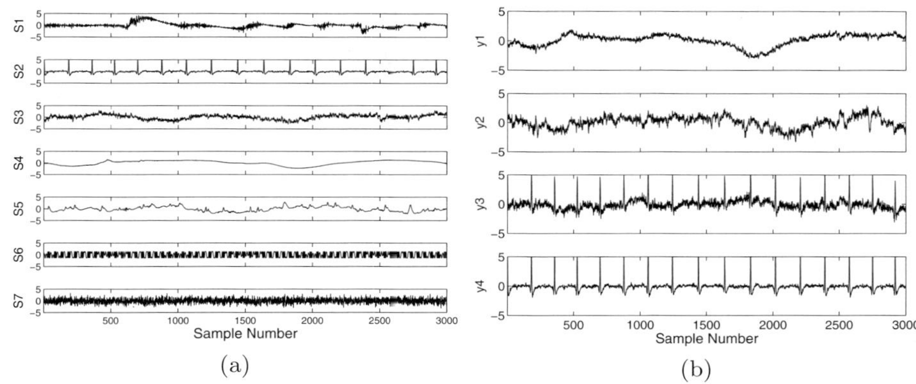

Fig. 1. Simulation. (a) Source signals. s_1-s_5 were all biomedical signals. They were, respectively, the muscle artifact, the ECG signal, the breathing artifact, the baseline wander and the electrode motion artifact. s_6 was a quasi-periodic signal. s_7 was a Gaussian noise. (b) The extracted signals, respectively by the algorithm in [3] (y_1), the one in [6] (y_2), the one in [2] (y_3) and our proposed algorithm (y_4).

5 Conclusion

In this paper, based on our previous work [4], we propose another novel two-stage algorithm for extracting periodic source signals. It converges quickly, and can achieve higher extraction performance, compared with the existing algorithms. In addition, it can be shown that it is nonsensitive to the estimate errors of the period of the desired source signal. Detailed discussion and analysis of the proposed algorithm will be given in a full separated paper in the near future.

References

1. Cichocki, A., Amari, S.: Adaptive Blind Signal and Image Processing: Learning Algorithms and Applications. John Wiley & Sons, New York (2002)
2. Zhang, Z.-L., Yi, Z.: Robust Extraction of Specific Signals with Temporal Structure. Neurocomputing (accepted)
3. Zhang, Z.-L., Yi, Z.: Extraction of Temporally Correlated Sources with Its Application to Non-invasive Fetal Electrocardiogram Extraction. Neurocomputing (accepted)
4. Zhang, Z.-L., Zhang, L.: A Two-Stage Based Approach for Extracting Periodic Signals. Proc. of the 6th International Conference on Independent Component Analysis and Blind Signal Separation (ICA 2006) (2006) (accepted)
5. Barros, A.K., Vigário, R., Jousmäki, V., Ohnishi, N.: Extraction of Event-related Signals from Multichannel Bioelectrical Measurements. IEEE Trans. Biomedical Engineering 47 (5) (2000) 583-588
6. Barros, A.K., Cichocki, A.: Extraction of Specific Signals with Temporal Structure. Neural Computation 13 (9) (2001) 1995-2003
7. Lu, W., Rajapakse, J.C.: Approach and Applications of Constrained ICA. IEEE Trans. Neural Networks 16 (1) (2005) 203-212

A Neural Network Method for Blind Signature Waveform Estimation of Synchronous CDMA Signals

Tianqi Zhang[1,2], Zengshan Tian[1], Zhengzhong Zhou[1,2], and Yujun Kuang[1,2]

[1] School of Communication and Information Engineering,
Chongqing University of Posts and Telecommunications (CQUPT),
Chongqing, 400065, China
[2] Research Centre for Optical Internet and Mobile Information Networks (COIMIN),
University of Electronic Science and Technology of China, Chengdu 610054, China
zhangtianqi@tsinghua.org.cn

Abstract. A principal component analysis (PCA) neural network (NN) based on signal eigen-analysis is proposed to blind signature waveform estimation in low signal to noise ratios (SNR) direct sequence synchronous code-division multiple-access (S-CDMA) signals. The received signal is firstly sampled and divided into non-overlapping signal vectors according to a temporal window, which duration is a period of signature waveform. Then an autocorrelation matrix is computed and accumulated by these signal vectors one by one. Since we have assumed that the synchronous point between the symbol waveform and observation window have been known, the signal vectors may be sampled and divided at the beginning of this synchronous point, therefore, each vector must contain all information of signature waveforms. In the end, the signature waveforms can be estimated by the principal eigenvectors of autocorrelation matrix blindly. Additionally, the eigen-analysis method becomes inefficiency when the estimated vector becomes longer. In this case, we can use the PCA NN method to realize the blind signature waveform estimation from low SNR input signals effectively.

1 Introduction

Spread spectrum signals have been used in military domain for a long time for secure communications. Nowadays, their field of application includes civilian transmissions, especially code-division multiple-access (CDMA) ones. Thanks to the orthogonal property of the signature waveform used, the CDMA technique allows one to solve the problem of growing number of users within a frequency band. Moreover, the performance of direct sequence synchronous CDMA (S-CDMA) systems is prior to the asynchronous ones in some cases, the S-CDMA systems are used widely in practice. We'll fix our attention on the case of S-CDMA systems in the following text. Since the signals of S-CDMA systems are often below the noise level, they have the distinguished capability of anti-jamming and lower probability of interception. Furthermore, the S-CDMA communication surveillance, management and military

J. Wang et al. (Eds.): ISNN 2006, LNCS 3972, pp. 694 – 699, 2006.

reconnaissance become a key and challenging problem now. Blind estimation of the signature waveforms from received S-CDMA signals is a vital step to the communication surveillance, management and military reconnaissance. In [1], a subspace method for blind signature waveform estimation in S-CDMA systems is proposed. However, the computation cost is relatively high due to complicated matrix manipulation. In [2], we have proposed an approach of neural network (NN) to realize the blind estimation of direct sequence spread spectrum (DS-SS, DS) signals, but it is still a problem for the case of S-CDMA signals.

This paper proposes an approach to the blind signature waveform estimation of S-CDMA signals. Unlike other methods, the estimator improves steadily with increasing of the number of code repeats (data group). The approach is applicable to arbitrary signature waveform, and message sequences, and can operate in environments containing arbitrary levels of additive white Gaussian noise. The method only requires the S-CDMA signal to have unrelated message and period of signature waveform. It can be further used in blind despreading of the underlying message sequence.

2 Signal Model

Let us consider a base-band direct sequence synchronous CDMA system with K users. The received signal is

$$x(t) = S(t) + n(t) . \tag{1}$$

where $n(t)$ is additive white Gaussian noise (AWGN) with variance of σ_n^2, $S(t)$ is the base-band signal of K users.

$$S(t) = \sum_{k=1}^{K} A_k \sum_{i=-M}^{M} m_k[i] p_k(t - iT_0 - \tau_k) . \tag{2}$$

where $2M+1$ is the number of data symbols per-user per-frame; A_k, $\tau_1 = \tau_2 = \cdots \tau_k = T_x \in [0, T_0]$, $\{m_k[i] \in \pm 1 : i = 0, \pm 1, \pm 2, \cdots, \pm M\}$ and $\{p_k(t) : 0 \le t \le T_0\}$ denote, respectively, the amplitude random time-delay, symbol stream, and signature waveform of the k-th user. $p_k(t) = \sum_{i=1}^{N} p_k[i] q(t - iT_c)$, $p_k[i] \in \{+1, -1\}$ is the signature (pseudo noise (PN)) sequence of the k-th user, $q(t)$ is the wave-form of one chip, where $T_0 = NT_c$, N is the bit number of signature sequence, T_0 is the periodic of signature sequence, T_c is the chip duration. The symbol is a random variable set with independent equal probability.

According to the above, the signature waveform and synchronization are required to de-spread the received direct sequence S-CDMA signals. But in some cases, we only have the received S-CDMA signals. We must estimate the signal parameters firstly (such as T_0, T_c), and then estimate the signature waveforms of all users.

3 Blind Estimation of Signature Waveforms

The received S-CDMA signal is sampled by period of T_c and divided into non-overlapping temporal windows, the duration of which is T_0. Then one of the received signal vector is

$$\mathbf{X}(l) = \mathbf{s}(l) + \mathbf{n}(l),\ l = 1,2,3,\cdots. \tag{3}$$

where $\mathbf{s}(l)$ is the l-th vector of useful signal, $\mathbf{n}(l)$ is the white Gaussian noise vector. The dimension of vector $\mathbf{X}(l)$ is $N = T_0 / T_c$. If the random time-delay is $T_x \in [0, T_0]$, $\mathbf{s}(l)$ may contain two consecutive symbol bits of each users, each modulated by a part of signature waveform of each users, i.e.

$$\mathbf{X}(l) = \sum_{k=1}^{K} A_k m_k^l \mathbf{p}_k^1 + \sum_{k=1}^{K} A_k m_k^{l+1} \mathbf{p}_k^2 + \mathbf{n}(l). \tag{4}$$

where m_k^l and m_k^{l+1} are two consecutive symbol bits of the k-th user, $\mathbf{p}_k^1$ ($\mathbf{p}_k^2$) is the right (left) part of signature waveform of the k-th user according to T_x.

According to signal eigen-analysis, we normalize the $\mathbf{u}_k^i$ by $\mathbf{u}_k^i = \mathbf{p}_k^i / \|\mathbf{p}_k^i\|$

$$\left(\mathbf{u}_k^i\right)^T \mathbf{u}_k^j = \delta(i-j),\ i,j = 1,2. \tag{5}$$

where $\mathbf{u}_k^i$ ($i = 1,2$) is the orthonormal version of $\mathbf{p}_k^i$, and $\delta(\square)$ is a Dirac function. From $\mathbf{u}_k^1$ and $\mathbf{u}_k^2$, we have

$$\mathbf{X}(l) = \sum_{k=1}^{K} A_k m_k^l \|\mathbf{p}_k^1\| \mathbf{u}_k^1 + \sum_{k=1}^{K} A_k m_k^{l+1} \|\mathbf{p}_k^2\| \mathbf{u}_k^2 + \mathbf{n}(l). \tag{6}$$

Assume $\mathbf{s}(l)$ and $\mathbf{n}(l)$ are mutually independent, we have

$$\mathbf{R}_X = E[\mathbf{X}\mathbf{X}^T] = \sigma_n^2 \left[\beta \cdot \left(\frac{T_0 - T_x}{T_c} \right) \cdot \sum_{k=1}^{K} A_k^2 \mathbf{u}_k^1 \left(\mathbf{u}_k^1\right)^T + \beta \cdot \left(\frac{T_x}{T_c} \right) \cdot \sum_{k=1}^{K} A_k^2 \mathbf{u}_k^2 \left(\mathbf{u}_k^2\right)^T + \mathbf{I} \right]. \tag{7}$$

where $\mathbf{I}$ is an identity matrix of dimension $N \times N$, the expectation of m_k is zero. The variance of m_k is σ_m^2, the symbol is uncorrelated from each other. The energy of signature waveform is $E_p = \int_{-\infty}^{+\infty} |\mathbf{p}_k|^2 dt \approx T_s \|\mathbf{p}_k\|^2$, $k = 1, 2, \cdots, K$, the variance of $\mathbf{s}(l)$ is $\sigma_s^2 = \sigma_m^2 E_p / T_0$, the β is $\beta = \sigma_s^2 / \sigma_n^2$. We can reconstruct a whole signature waveform by $\pm\mathbf{p}_k = \text{sign}(\mathbf{u}_k^1 \pm \mathbf{u}_k^2)$ ($\text{sign}(\cdot)$ is a sign function) and solve the problem of blind signature waveform estimation.

From Eq. (7), we can see there is $2K$ bigger eigenvalues in $\mathbf{R}_X$ associated with the useful S-CDMA signals, the others associated with the noise. The eigenvalues have relationship to both A_k and T_x. When synchronization is achieved (namely, $T_x = 0$), the energy in the useful signal will be contained in the first K eigenvalues of $\mathbf{R}_X$; otherwise, it will spill out to the rest of the other K eigenvalues. There is already a method in [4] to estimate T_x, so we'll focus on the case of $T_x = 0$ only. Furthermore, the source number K can be obtained using AIC or MDL criteria in [5].

In practice, the length of sample vector N may be so large that computation of the SVD on $N \times N$ matrix may not be feasible. There are usually two approaches to deal with this difficult problem. One approach is to subdivide the matrix into a set of overlapping matrices and the SVD applied to each of the smaller matrices individually. The other approach is the neural networks (NN) method. In the following text, we'll use a principal component analysis (PCA) NN to solve the estimation of signature waveforms.

4 Neural Network Implementation

As in [2], a two-layer PCA NN is used to estimate the signature waveforms in S-CDMA signals blindly. The number of input neurons is given by T_0/T_c.

Assume $T_x = 0$, then we can divide the received signal into non-overlapping temporal windows, which duration is $T_0 = NT_c$. Assume one of the vectors is

$$\mathbf{X}'(t) = \mathbf{X}'(l) = \left[x(t), x(t-T_c), \cdots, x[t-(N-1)T_c]\right]^T = \left[x_0(t), x_1(t), \cdots, x_{N-1}(t)\right]^T . \tag{8}$$

where $\{x_i(t) = x(t-iT_c), i = 0,1,\cdots,N-1\}$ are sampled by period of T_c. The synaptic weight vector of NN is

$$\mathbf{w}(t) = \left[w_0(t), w_1(t), \cdots, w_{N-1}(t)\right]^T . \tag{9}$$

where the sign of $\{ w_i(t),\ i = 0,1,\cdots,N-1 \}$ may denote the i-th bit of estimated signature waveform. Before $\mathbf{X}'(t)$ is put into the NN, we normalize it as follows.

$$\mathbf{X}(t) = \mathbf{X}'(t) / \left\| \mathbf{X}'(t) \right\| . \tag{10}$$

where $\left\| \mathbf{X}'(t) \right\| = \sqrt{\mathbf{X}'^T(t)\mathbf{X}'(t)}$. This will make the principal component vector to be estimated in a robust manner. The NN output layer has only one neuron, its output is

$$y(t) = \sum_{i=0}^{N-1} w_i(t-1)x_i(t) = \mathbf{w}^T(t-1)\mathbf{X}(t) = \mathbf{X}^T(t)\mathbf{w}(t-1) . \tag{11}$$

The original Hebbian algorithm is $\mathbf{w}(t) = \mathbf{w}(t-1) + \beta y(t)\mathbf{X}(t)$. However, we will modify this algorithm as follows

$$\mathbf{w}(t) = \mathbf{w}(t-1) + \beta y(t)\left[\mathbf{X}(t) - \mathbf{w}^T(t-1)\mathbf{X}(t)\mathbf{w}(t-1)\right] . \tag{12}$$

where β is a positive step-size parameter. In order to achieve good convergence performance, we express β as

$$\beta = 1/d(t)\,,\ d(t) = Bd(t-1) + y^2(t)\,. \tag{13}$$

The analysis of the PCA NN is detailed in [2]. According to the result of section 3, we'll have to extract the principal component vector one by one before complete all signature waveform estimation. When we extract the next principal component, we can use the same NN again, but its input signals are changed to $\tilde{\mathbf{X}}(t)$

$$\tilde{\mathbf{X}}(t) = \mathbf{X}(t) - y(t)\mathbf{w}(t)\,. \tag{14}$$

When we input $\tilde{\mathbf{X}}(t)$ to the NN, the synaptic weight vector will converge to the second principal component vector of $\mathbf{X}(t)$.

5 Simulations

We get principal eigenvectors and performance curves by neural computation. From Eq. (2), where $K = 7$, $\mathbf{A} = [A_1, A_2, \cdots, A_7]^T = [4.0, 3.5, 3.0, 2.5, 2.0, 1.5, 1.0]^T$, m_k is produced in uniform randomization, the length of $p_k(t)$ is *N=100bit* and *N=1000bit*, and $p_k(t),\ k = 1, \cdots, 7$ are randomly generated.

Fig. 1 shows the NN synaptic weights norm (||*W(t)*||) changes of NN synaptic weights vias time. It shows that the norm of NN synaptic weights are all arrived to a real numer of *1.0* when the time increased. Fig. 2 shows the convergence curves of NN. It shows that the averaged bit error rate changes with the input data group numbers. The curves from left to right associated with the signature waveform estimation of the first, second, third and fourth user. The curves show that the convergence property of NN is very good. Fig. 3 and Fig. 4 denote the performance curves. It shows the time taken for the NN to perfectly estimate the signature waveforms of the first, second, third and fourth user. Under the same conditions, the signature waveform estimation of the first user is very fast, this because the signal of the first user is strong, and the longer the PN

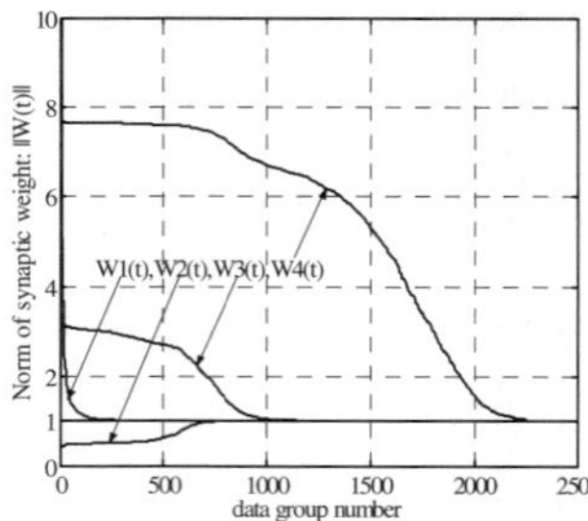

Fig. 1. The curves of ||*W(t)*||

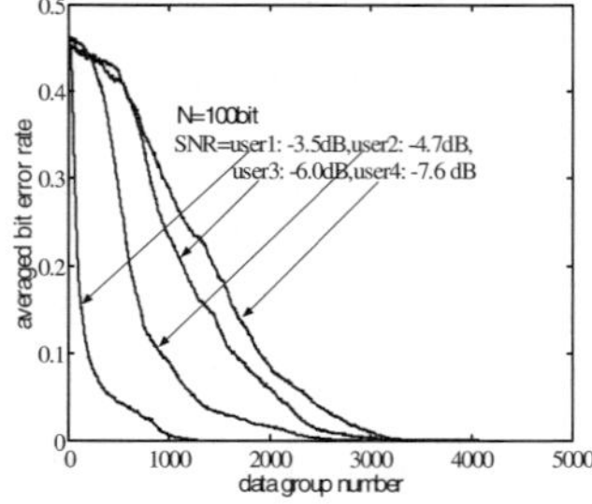

Fig. 2. Convergence curves

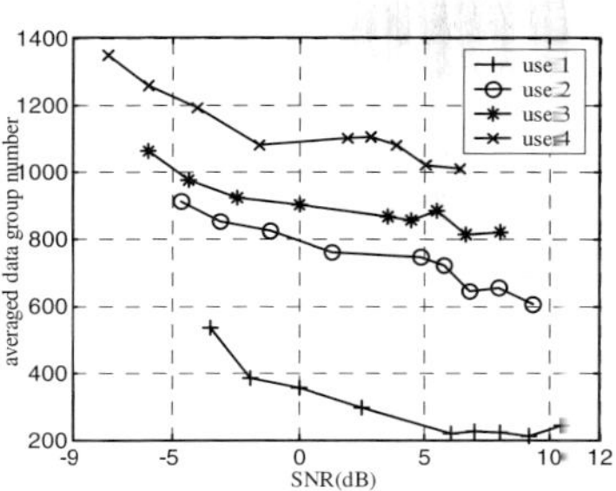

Fig. 3. Performance curves (*N=100bit*)

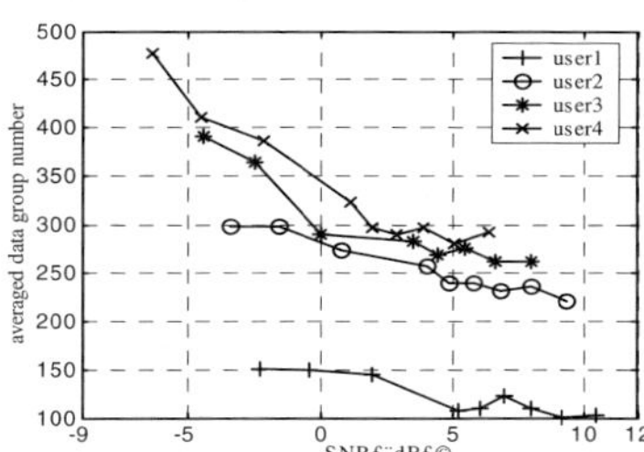

Fig. 4. Performance curves (*N=1000bit*)

sequence is, the better the performance is. We find the cross-correlation between every two users' signature waveforms may affect the estimation performance too.

6 Conclusions

A PCA NN approach for blind signature waveform estimation in direct sequence S-CDMA systems has been presented. The approach exploits cyclostationarity of communication signals adequately. Unlike the subspace-based method for blind signature waveform estimation in [1], the method of NN has low complexity and fast speed. The presented PCA NN has the advantage of dealing with longer received S-CDMA signal vectors. The method of blind signature waveform estimation is very robust and can work in low SNR conditions. Although this paper deals with the case of synchronous systems, it is straightforward to extend the method to asynchronous ones. It can be further used in multi-user detection, surveillance, management and reconnaissance of CDMA communications.

Acknowledgement

This work is supported by the Chongqing Economic Committee under Grant No. 04-2YZ-DZ107.

References

1. Liu, H., Xu, G.H.: A Subspace Method for Signature Waveform Estimation in Synchronous CDMA Systems. IEEE Trans. Com., 44(10) (1996) 1346-1354
2. Zhang, T.Q., Lin, X.K., Zhou, Z.Z.: Use PCA Neural Network to Extract the PN Sequence in Lower SNR DS/SS Signals. Lecture Notes in Computer Science, 3173(2004) 780-785
3. Chang, T.Q., Zhou, Z.Z.: A Spectral Method for Periodic Estimation of the PN Sequences in lower SNR DS-CDMA signals. 2002 IEEE International Conference on Communications, Circuits and Systems, Vol. 1, Chengdu, China (2002) 506-510
4. Nzoza, C.N., Gautier, R., Burel, G.: Blind Synchronization and Sequences Identification in CDMA Transmissions. IEEE Military Communications Conference, 3(2004) 1384-1390
5. Wax, M., Kailath, T.: Detection of Signals by Information Theoretic Criteria. IEEE Trans. Acoust., Speech, Signal Processing, 33(2) (1985) 387-392

A Signal-Dependent Quadratic Time Frequency Distribution for Neural Source Estimation

Pu Wang[1,2], Jianyu Yang[1], Zhi-Lin Zhang[2], Gang Wang[2], and Quanyi Mo[2]

[1] School of Electronic Engineering, University of Electronic Science and Technology of China, Chengdu 610054, P.R. China
pwang@ieee.org
[2] Blind Source Separation Research Group, University of Electronic Science and Technology of China, Chengdu 610054, P.R. China

Abstract. A novel method for kernel design of a quadratic time frequency distribution (TFD) as the initial step for neural source estimation is proposed. The kernel is constructed based on the product ambiguity function (AF), which efficiently suppresses cross terms and noise in the ambiguity domain. In order to reduce the influence from the strong signal to the weak signal, an iterative approach is implemented. Simulation results validate the method and demonstrate suppression of cross terms and noise, and high resolution in the time frequency domain.

1 Introduction

Neural activities in a human brain are basically nonstationary. And the source estimation needs to incorporate some prior knowledge regarding the source characteristics due to this problem is ill-posed. Such characteristics can include possible source locations, the total number of sources, or the frequency characteristics of source activities. In advance, time frequency (TF) analysis has played an important role in analyzing and characterizing neural data such as that from magnetoencephalography (MEG) or encephalography (EEG) [1-3]. Then the algorithms of source estimation are implemented from the region of interest (ROI) in the TF domain. In this paper, the initial step for neural source estimation, the TF analysis, is focused.

The TF analysis involved in neural source estimation mainly focuses on the quadratic class time frequency distribution (TFD) [1-3]. This kind of TFD can be categorized to two parts based on the characteristics of the kernels. The first type is about the fixed kernel, such as the spectrogram, Wigner-Ville distribution (WVD), the Choi-Williams distribution (CWD) [4], and the B-distribution [5]. These kernels can preserve the desirable resolution in the TF domain, but their application is bounded for the kernel fixation. The other type is signal-dependence, such as the kernel proposed in [6-7], which is adaptive to the signal under analysis. However, the performance of analysis of weak signal will dramatically degrade if unbalanced signals are sampled.

Due to analysis of the chirp signals judges the quadratic TFDs and the chirp is commonly encountered in the source estimation in the MEG analysis [1-3], the

J. Wang et al. (Eds.): ISNN 2006, LNCS 3972, pp. 700 – 705, 2006.

algorithm of kernel design for the chirp signals are first proposed. This algorithm is also extended for frequency modulated (FM) signals. The rest of this paper is organized as follows: Section 2 briefly reviews definition and main properties of the quadratic TFD and product ambiguity function (PAF); Section 3 refers to the kernel design for unbalanced signals. Section 4 provides simulation results and Section 5 ends this paper.

2 Preliminaries

2.1 Quadratic Time Frequency Distribution

The general form of the quadratic TFD [8] is given by

$$C_x(t,\omega) = \int_{-\infty}^{+\infty}\int_{-\infty}^{+\infty}\int_{-\infty}^{+\infty} x(u+\tau/2)x^*(u-\tau/2)\phi(\tau,v)e^{-j(tv+\omega\tau-uv)}du\,d\tau\,dv \tag{1}$$

where $\phi(\tau,v)$ is a weighting function, called the kernel, which determines the distribution and its properties. The quadratic TFD can be also obtained by employing a two dimensional Fourier transform on the weighted ambiguity function,

$$C_x(t,\omega) = \int_{-\infty}^{+\infty}\int_{-\infty}^{+\infty} AF(\tau,v)\phi(\tau,v)e^{-j(tv+\omega\tau)}d\tau\,dv \tag{2}$$

where $AF(\tau,v)$ is the ambiguity function (AF) of $x(t)$. Hence filtering the cross terms in the ambiguity domain will result in suppression of cross terms in the TF domain.

2.2 Product Ambiguity Function

Product ambiguity function (PAF) [9] is defined as the product of properly frequency scaled ambiguity functions. Mathematically, the PAF defined as

$$PAF(\tau_l,v) = \prod_{l=1}^{L} AF(\tau_l, v\tau_1/\tau_l), \tag{3}$$

will peak at $v = m_0\tau_1$. It is easy to find that the auto terms are amplified while the cross terms and noise are weakened. Obviously, the more sets of lags used, the better the noise suppression capability, but the higher the computational load.

3 Proposed Time Frequency Distribution

3.1 Proposed Kernel

This kernel retains as many auto terms energy as possible while filtering out as much cross terms energy as possible.

Lag Varying Filtering. This method implements filtering against the lag slice using a lag varying threshold $\eta(\tau)$. The threshold is obtained by

$$\eta(\tau) = k\times\max(PAF(\tau,v)), \qquad 0<k<1 \tag{4}$$

and this filtering operation can be shown by

$$\psi(\tau,v)=\begin{cases} PAF(\tau,v), & PAF(\tau,v)>=\eta(\tau) \\ 0, & PAF(\tau,v)<\eta(\tau) \end{cases}. \tag{5}$$

The threshold is related to the ratio of the auto terms energy and the noise floor energy. It controls the partition of the auto terms energy will be remained and of the noises will be filtered.

Another step, which is required to design the kernel, is the estimation of distance r of the significant radial contents. This estimation only implement at the distribution of auto terms which are located in lag varying filtering.

Generally, the distance r can be found directly from the modulus of the PAF:

$$\rho=\left\{\max r:\left|PAF(r,\phi)\right|\geq\varepsilon\left|PAF(0,0)\right|\right\}, \tag{6}$$

where λ controls the remains of the significant content. Therefore, the kernel form is given by the two steps:

$$\phi(r,\alpha)=\begin{cases} 1, & \text{if } \psi(\tau,v)\neq 0 \text{ and } r\in\rho \\ 0, & \text{otherwise} \end{cases}. \tag{7}$$

The performance of this method is influenced by the sets of used lags because the cross terms and noise suppressions are related to the sets of lags.

Residual Elimination. When the components have unbalanced or very different amplitudes, the kernel design is found difficult to efficiently characterize the components having low amplitudes. Hence, a recursive peeling technique is required to expose the weak components by removing the strong components.

This technique is required accurate parameter estimation, or two problems exist: (1) error propagation: the estimation of initial frequency and amplitude rely strongly on the accuracy of chirp rate estimation and the estimated error of chirp rate will pass on. (2) residual signals: the inaccuracy estimation will left the residual signal after removing the estimated signal from the observations.

According to the analysis, the key issue to the problem is the estimation of chirp rate. From this viewpoint, an iterative algorithm is introduced and the combination between the iterative algorithm [10] and the PAF is presented. This iterative algorithm makes use of the residuals to refine previously estimated chirp rate and brings about an estimation whose error can be confined within a preset threshold.

It's implemented as the following steps:

1. Initialization: $k=1$, $\hat{\mu}_0=0$, $\tau_{1,i}=0,\ 1\leq i\leq L$;
2. Define:

$$r_{k,j}(n)\triangleq x_{k-1}\left(n+\frac{\tau_{k-1,i}}{2}\right)x_{k-1}^{*}\left(n-\frac{\tau_{k-1,i}}{2}\right),\quad n=0,\cdots,N-\tau_{k-1,i}/2-1 \tag{8}$$

where $x_{k-1}(n)=x(n)\exp(-j\hat{\mu}_{k-1}n^2)$, k is the iterative number, $\tau_{k,i}$ is the ith lag used in the kth iteration;

3. Estimate $\Delta\mu_{k-1} = \mu - \hat{\mu}_{k-1}$, and define $e_{k-1} = |\Delta\mu_{k-1} / \mu|$ as the relative estimation error after $(k-1)$th iteration. Compare the error with a preset threshold, then switch to step 4 if $e_{k-1} < \varepsilon$, otherwise, repeat this iteration;
4. Demodulate $x(n)$ as $x(n)e^{-j\hat{\mu}n^2}$, where $\hat{\mu}$ is the chirp rate derived in step 3. Then estimate the harmonics whose frequency locates at $\hat{f}$;
5. Estimate the amplitude via least square:

$$\hat{A} = \arg\max_{a} \left| x(n) - ae^{j2\pi(\hat{f}n + \hat{\mu}n^2/2)} \right|. \tag{9}$$

The iterative principle of lags can be found in reference [10].

Derivation of the Kernel. The final step is to add the derived kernel derived in iterations and to adjust the amplitude to unit, since there are some areas jointly occupied by many kernels.

3.2 Extension to Nonlinear FM Signals

Due to a nonlinear FM signal can be treated as a set of chirp signals with time varying frequency rate, the modulus of the AF of a nonlinear FM signal would contain a collection of lines going through the origin with time varying angle. Therefore, the technique proposed in this paper could be applied to nonlinear FM signals. This will be illustrated by Example 2.

4 Simulation

We perform numerical experiments to test the validity of the proposed algorithm of kernel design for FM signal. The simulated signal refers to the magnetic filed in the MEG analysis. The simulated magnetic field is calculated at 512 time points. To generate the simulated neuromagnetic field, the ϕ components of the three sources $\omega_\phi^i(t)$, ($i = 1, 2, 3$) are FM, i.e., $\omega_\phi^i(t) = A_i \exp[j2\pi(f_i n + u_i n^2/2)]$. The values for { SNR_i, f_i, u_i } are set at {5dB, 0.3Hz, -0.3Hz/s} for the first source, {1dB, 0.1Hz, 0.2Hz/s} for the second source and {-5dB, 0 2Hz, 0.1Hz/s} for the third source, where $SNR_i = 10\log_{10}(A_i^2/\sigma^2)$. The θ components are set to zero. The Gaussian noise is uncorrelated between different sensor recordings.

Fig. 1 gives the kernel function derived using the lag varying filtering. The kernel is characterized by locating auto terms and excluding the cross terms and noise. It is observed that there are some undesirable areas, since filtering against the weak component is influenced by the less conspicuous peak. The TFD is derived by employing 2D Fourier transform to the weighted AF. In Fig.2, suppression of cross terms and noise, as well as good resolution, is achieved in the TF plane. Furthermore, the weak component is expressed distinctly.

Example 2 is to validate the proposed algorithm for the nonlinear FM signals. The simulated signals consist of the signal from first source and a cubic phase signal $s_c(t)$ $s_c(t) = A_c \exp[j2\pi(f_c n + \mu_c n^2/2 + \lambda_c n^3/3)]$, The values for { SNR_c, f_c, u_c, λ_c } are set at {1dB, 0.3Hz, -0.3Hz/s, -0.3Hz/s^2}.

The designed kernel is presented in Fig. 3. In this figure, a collection of lines go through the origin with time varying angle corresponding to the cubic phase signal, while a straight line is related to the chirp signal. In Fig. 4, the cross terms between two signals are almost eliminated while the cross terms introduced by nonlinear characteristics of cubic phase signal are not sufficiently suppressed.

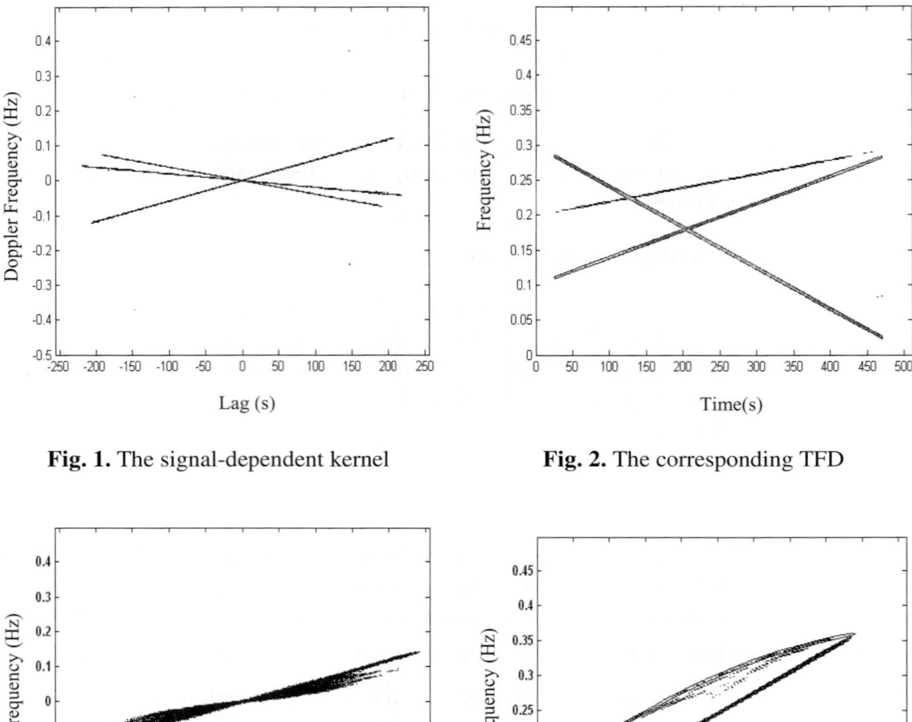

Fig. 1. The signal-dependent kernel

Fig. 2. The corresponding TFD

Fig. 3. The derived kernel based on PAF

Fig. 4. The corresponding TFD

5 Conclusion

The PAF-based kernel design and the resulting quadratic TFD are proposed for multi-component signals with unbalanced amplitudes. This TFD shows the ability to suppress cross terms and noise due to the kernel can discern the cross terms and noise in ambiguity function. The peeling technique along an iterative algorithm is implemented to reduce the error propagation and the residual signals. The simulation results verify the kernel design and the construction of quadratic TFD. The results demonstrate the suppression of cross terms and noise, and high resolution in the TF plain. It will provide convenience and benefit to neural source estimation.

References

1. Sun, M., Qian, S., Yan, X., Baumann, S.B., Xia, X.G., Dahl, R.E., Ryan, N.D., Sclabassi, R.J.: Localizing Functional Activity In The Brain Through Time-Frequency Analysis and Synthesis of The EEG. Proc. IEEE 84(9) (1996) 1302–1311
2. Sekihara, K., Nagarajan, S., Poeppel, D. Miyashita, Y.: Time-Frequency MEG-MUSIC Algorithm. IEEE Trans. Medical Imaging 18(1) (1999) 92–97
3. Sekihara, K., Nagarajan, S., Poeppel, D. Miyauchi, S, Fujimaki, N., Koizumi, H., Miyashita, Y.: Estimating Neural Sources From Each Time-Frequency Component of Magnetoencephalographic Data. IEEE Trans. Biomedical Engineering 47(5) (2000) 642–653
4. Choi, H.I., Williams W.J.: Improved Time-Frequency Representation of Multicomponent Signals Using the Exponential Kernels. IEEE Trans. Signal Processing 37(6) (1989) 862–871
5. Barkat, B., Boashash B.: A High-Resolution Quadratic Time-Frequency Distribution For Multicomponent Signal Analysis. IEEE Trans. Signal Processing 49(10) (2001) 2232–2239
6. Ristic, B., Boashash B.: Kernel Design For Time-Frequency Signal Analysis Using Radon Transform. IEEE Trans. Signal Processing 41(5) (1993) 1996–2008
7. Baraniuk, R.G., Jones D.L.: A Radially Gaussian Signal Dependent Time Frequency Representation. Proc. IEEE Int. Conf. Acoust. Speech, Signal Processing, Vol. 5, IEEE Press, (1991) 3181–3184
8. Cohen, L.: Time-Frequency Analysis. Englewood Cliffs, Prentice Hall New York (1995)
9. Barbarossa, S., Scaglione, A., Giannakis, G.B.: Product High-order Ambiguity Function For Multicomponent Polynomial-Phase Signal Modeling. IEEE Trans. Signal Processing 46(3) (1998) 691–708
10. Ikram, M.Z., Abed-Meraim, K., Hua, Y.: Estimating The Parameters of Chirp Signals: An Iterative Approach. IEEE Trans. Signal Processing 46(12) (1998) 3436–3441

Neural Network Channel Estimation Based on Least Mean Error Algorithm in the OFDM Systems

Jun Sun[1] and Dong-Feng Yuan[1,2]

[1] School of Information Science and Engineering, Shandong University, Jinan 250100, P.R. China
{freyja, dfyuan}@sdu.edu.cn
http://www.sdu.edu.cn
[2] State Key Lab. on Mobile Communications, Southeast University, Nanjing 210096, P.R. China

Abstract. We designed a new channel estimator including two parts of neural network to estimate the amplitude and the angle of the frequency domain channel coefficients, respectively. The least mean error (LSE) is used for training. This neural network channel estimator (NNCE) makes full use of the learning property of the neural network (NN). Once the NN was trained, it reflected the channel fading trait of the amplitude and the angle respectively. It was no need of any matrix computation and it can get any required accuracy. It has been validated that the estimator is available in the pilot-symbol-aided (PSA) OFDM system.

1 Introduction

Neural network has been pain more attention to in signal processing areas for its strong functions, especially for the function approximation, estimation, etc. At present, there are many methods to discuss the channel estimation method in the OFDM systems. All these methods can be mainly classified into two types. One is based on the PSA and the other is blind. In this paper, we will focus on the former method. One-dimensional (1-D), double 1-D, and two-dimensional (2-D) filtering algorithms have been proposed for PSA channel estimation in [1], [2], [3] and the references therein. In [4], the author proposed a minimum mean-square error (MMSE) PSA parameter estimator. It is robust to Doppler frequency for disperse fading channels with noise impairment. However, the implementation of the MMSE needs the matrix inverse and correlation computation. Besides, the additive noise information should also be known. These are troubled in some cases. As we know, because of the cycle prefix in the OFDM system, the received signal is considered as a product of a complex multiplicative component and an additional component. And in [5], the author has proposed a neural network channel estimator, which can estimate the amplitude and the angle of the channel respectively. From this motive, we proposed our new estimator structure using the Neural Network, which only need simply mathematic computation and no requirement of the noise information. The

J. Wang et al. (Eds.): ISNN 2006, LNCS 3972, pp. 706 – 711, 2006.

performance of our design is close to the ideal case in which the channel is perfectly known.

2 Systems and Channel Model

Fig.1 shows the system structure of the PSA OFDM system.

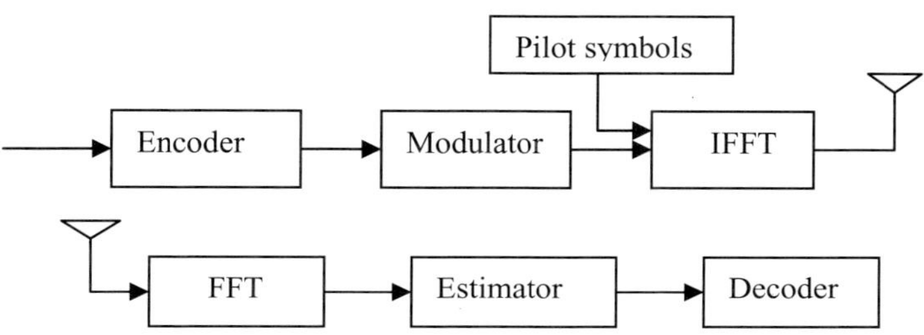

Fig. 1. PSA OFDM System with the transmitter (*top*) and the receiver (*underside*)

The signal at the *i*th tone of the *k*th block can be expressed as:

$$Y(i,k) = H(i,k)X(i,k) + W(i,k), 1 \le i \le N \tag{1}$$

The sign N is the total number of tones in one OFDM block. The transmitted modulated symbols are denoted by matrix X and the denotation $X(i,k)$ is an element in it. The matrix Y contains the received symbols. The W denotes the complex Gaussian process with zero mean and variance $\sigma_w^2 = 1$. The H is the frequency response of the multipath Rayleigh fading channel. In the following, we will omit k for simplicity by denoting $Y_i = H_i X_i + W_i$. The multipath Rayleigh fading channel with impulse response is (2). Where, the symbol L is the channel memory length, the set of { $h_l(t)$ } is correlated complex

$$g(t) = \sum_{l=0}^{L-1} h_l(t)\delta(t - \tau_l) \tag{2}$$

Gaussian processes with zero mean and variance σ_h^2 (assume $\sigma_h^2 = 1$ according to the Jake's model) and τ_l is the delay of the *l*th path. Here, we use the channel parameters of the typical urban (TU) model from the COST207 groups [6].

3 NNCE and the Algorithm

Fig.2 shows the structure of our channel estimator. It composes of a neural network consisting two subnetworks called NNA and NNG.

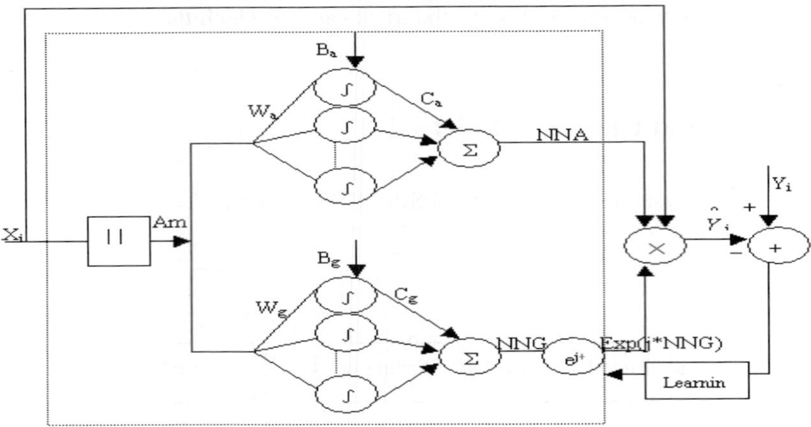

Fig. 2. NNCE Structure

Each subnetwork has n neurons in the first layer and a scalar output. The NNA aims at identifying the amplitude gain, while NNG aims at identifying the phase variant. Therefore, by using this structure, we aim at obtaining direct estimation of the amplitude and phase values of the complex channel coefficient at each tone from the pilot symbols. Then, we get the estimated output value of the system as:

$$\hat{Y}_i = X_i NNA(Am_i) e^{jNNG(Am_i)}, X_i = Am_i + An_i \tag{3}$$

Where,

$$NNA(Am_i) = \sum_{j=1}^{n} c_{aj} f(w_{aj} Am_i + b_{aj}), \quad NNG(Am_i) = \sum_{j=1}^{n} c_{gj} f(w_{gj} Am_i + b_{gj}) \tag{4}$$

The activation function $f(\cdot)$ is the hyperbolic tangent function, $\{waj\ , caj\ , baj\}$ and $\{wgj\ , cgj\ , bgj\}$ are the weights of subnetwork NNA and NNG, respectively. Here, we use vector to denote the weights as seen in Fig.2. For example, we have $Wa = \{w_{aj}, 1 \le j \le n\}$ .The system parameter vector will be denoted by θ, which includes all parameters to be updated:

$$\theta = [w_{a1}, \cdots w_{an}, b_{a1,\ldots} b_{an}, c_{a1,\ldots} c_{an}, w_{g1}, \cdots w_{gn}, b_{g1,\ldots} b_{gn}, c_{g1,\ldots} c_{gn}]$$

We use the LSE algorithm [7] to train the neural network and to update the system parameters in order to minimize the squared error:

$$J_i = \frac{1}{2}\left\| Y_i - \hat{Y}_i \right\|^2 = \frac{1}{2}\left[e_{Ri}^2 + e_{Ii}^2 \right] \tag{5}$$

Here, indexes R and I refer to the real and imaginary parts, respectively. Therefore, the neural network weights will be updated as follows:

$$\theta_i(l+1) = \theta_i(l) - \mu\nabla_{\theta_i(l)} J_i(l) \tag{6}$$

The sign l is the lth iteration. The symbol μ is a small positive constant.

$$\nabla_{\theta_i(l)} J(l) = e_{Ri}(l)\nabla_{\theta_i(l)} e_{Ri}(l) + e_{Ii}(l)\nabla_{\theta_i(l)} e_{Ii}(l) \tag{7}$$

Here, we omit the subscript i for simplicity. Then, we get the estimated frequency channel response:

$$\hat{H}_i = NNA(Am_i)e^{jNNG(Am_i)} \tag{8}$$

Where,

$$\nabla_{\theta(l)} e_R(l) = \begin{bmatrix} Am(l)^2 \cos(NNG(Am(l)) + An(l))c_{a1} f'(w_{a1}Am(l) + b_{a1}) \\ \vdots \\ Am(l)^2 \cos(NNG(Am(l)) + An(l))c_{an} f'(w_{an}Am(l) + b_{an}) \\ Am(l) \cos(NNG(Am(l)) + An(l))c_{a1} f'(w_{a1}Am(l) + b_{a1}) \\ \vdots \\ Am(l) \cos(NNG(Am(l)) + An(l))c_{an} f'(w_{an}Am(l) + b_{an}) \\ Am(l) \cos(NNG(Am(l)) + An(l)) f'(w_{a1}Am(l) + b_{a1}) \\ \vdots \\ Am(l) \cos(NNG(Am(l)) + An(l)) f'(w_{an}Am(l) + b_{an}) \\ -Am(l)^2 NNA(Am(l)) \sin(NNG(Am(l)) + An(l))c_{g1} f'(w_{g1}Am(l) + b_{g1}) \\ \vdots \\ -Am(l) NNA(Am(l)) \sin(NNG(Am(l)) + An(l)) f'(w_{g1}Am(l) + b_{g1}) \end{bmatrix} \tag{9}$$

The image part is similar to (9).

4 Results and Discussions

The transmitted signal was 1/2 rate convolution coded and16-QAM modulated. The TU channel has six paths with the propagation delay [0.0 0.2 0.6 1.6 2.4 5.0]μs, the path power [-3 0 –2 –6 –8 -10]dB. The Doppler power spectral density includes the Jakes and Gaussian models. The maximum Doppler frequency used here was 90Hz with the carrier frequency 2GHz and the sampling frequency was 0.2μs. We used 3 neurons in each subnetwork and μ=0.005. Fig.3 showed the BER performance using the neural network estimator in the OFDM system with 512 tones. And the number of the iteration was 400, 1000, 1500and 2000. It can be seen that the performance increased quickly with the iteration increases at first. When the iteration was up to 1500, the performance had little improvement. So, it can be concluded that the iteration has a limitation to get the optical performance. The final performance is close to the ideal one.

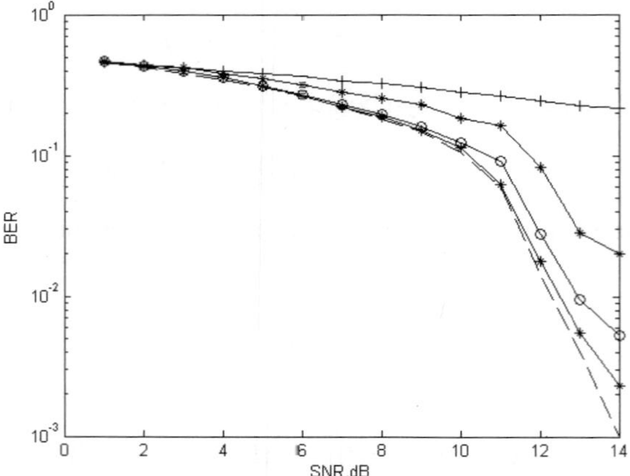

Fig. 3. The BER performance compares at different iterations: the system with the known channel information (*dashed line*), the NNCE system with 400 iterations (+ *line*), the NNCE system with 1000 iterations (* *line*), the NNCE system with 1500 iterations (*o line*) and the NNCE system with 2000 iterations ($\times$ *line*)

5 Conclusions

A new channel estimator structure based on the neural network is proposed in this paper. It can estimate the amplitude and the angle values of the frequency response of the channels by training the neural network with no need of any channel prior information. And, it provides more accurate estimation by continuous learning and updating. The results validated that our design was available. And the computation is less simple though it needs a little large iteration. In the time-invariance channel, only one pilot symbol at the beginning is needed to train the neural network. Once the NNCE has been trained, it can reflect the channel mechanics correctly. So, it is no need of additional pilot symbols and reduces the system spending. In time disperse channel, we can track the channel variance by inserting the pilot symbols at proper interval to train the NNCE again.

Acknowledgements

This work is supported by the foundations: National Scientific Foundation of China (No. 60372030); China Education Ministry Foundation for Visiting Scholar (No. [2003] 406); Key Project of Provincial Scientific Foundation of Shandong (No. Z2003G02); China Education Ministry Foundation for State key Lab. on Mobile Communications (No. A2005010); State Key Laboratory of Integrated Services Network, Xi'an, Shanxi, P. R. China (No. ISN7-05). The authors would like to thank all professors concerned the topic.

References

1. Mignone, V., Morello, A.: CD3-OFDM: A Novel Demodulation Scheme for Fixed and Mobile Receivers. IEEE Transactions on Communications. 44(9) (1996) 1144-1151
2. Hoeher, P., Kaiser, S., Robertson, P.: Two-dimensional Pilot-Symbol-Aided Channel Estimation by Wiener Filtering. In: IEEE Int. Conf. Acoustics, Speech, and Signal Processing, Vol.3. IEEE, Munich, Germany (1997) 1845-1848
3. Hoeher, X.P., Kaiser, S., Robertson, P.: Pilot-symbol-aided Channel Estimation in Time and Frequency. In: IEEE Global Telecommunication Conf., Vol.1. Phoenix (1997) 90 – 96
4. Li, Y.: Pilot-symbol-aided Channel Estimation for OFDM in Wireless Systems. IEEE Transactions on Vehicular Technology. 49(7) (2000) 1207-1215
5. Ibnkahla, M., Yuan, J.: A Neural Network MLSE Receiver Based on Natural Gradient Descent: Application to Satellite Communications. In: Seventh International Symposium on Signal Processing and Its Applications Proceedings. Vol. 1. IEEE, Paris, France (2003) 33 - 36
6. Pätzold, M.: Mobile Fading Channels. John Wiley & Sons, England (2002)
7. Proakis, J.G.: Digital Communications. 4rd edn. McGraw-Hill, New York (2001)

Higher-Order Feature Extraction of Non-Gaussian Acoustic Signals Using GGM-Based ICA

Wei Kong[1] and Bin Yang[2]

[1] Information Engineering College, Shanghai Maritime University, Shanghai 200135, China
weikong@cie.shmtu.edu.cn
[2] Logistics Research Center, Shanghai Maritime University, Shanghai 200135, China
binyang@cie.shmtu.edu.cn

Abstract. In this paper, independent component analysis (ICA) is applied for feature extraction of non-Gaussian acoustic signals. The generalized Gaussian model (GGM) is introduced as the p.d.f. estimator in ICA because it can provide a general method for modeling non-Gaussian statistical structure of univariate distributions. It is demonstrated that the proposed method can efficiently extract ICA features for not only sup-Gaussian but also sub-Gaussian signals. The basis vectors are localized in both time and frequency domain and the resulting coefficients are statistically independent and sparse. The experiments of Chinese speech and the underwater signals show that the proposed method is more efficient than conventional methods.

1 Introduction

Nowadays, many efforts have gone into finding learning algorithms to obtain the efficient statistical characteristics of sound signals. Jong-Hwan Lee and Gil-Jin Jang [1, 2] analysis the basis functions obtained for female and male speech signals, Anthoy J Bell and Terrence J Sejnowski[3] applied ICA method to extracting basis vectors from natural sounds, and Jong-Hawn Lee and Ho-Young Jung[4] provide methods to select dominant ICA feature vectors.

For speech and music signals, their distribution functions are always sup-Gaussian, however, some sound signals such as underwater signals, their distribution functions are sometimes sub-Gaussian. It is difficult for one ICA algorithm to extract features with different distributions. To solve the problem of probability density function (p.d.f.) mismatch in ICA, the generalized Gaussian model (GGM) is introduced to model the p.d.f. of the sources since it can provide a general structure of univariate distributions.

In this paper the proposed method is used to extract the higher-order structure features for Chinese speech and underwater signals, which are typical sup-Gaussian and sub-Gaussian signals respectively. The local characteristic in the time and frequency domain of the basis vectors and the sparseness of coefficients demonstrate that our method is efficient for feature extracting of non-Gaussian acoustic signals.

J. Wang et al. (Eds.): ISNN 2006, LNCS 3972, pp. 712 – 718, 2006.

2 ICA Feature Extraction Based on GGM

To extract independent feature vectors from sound signals, ICA assumes the observation $\boldsymbol{x}$ to be the linear mixture of the independent components $\boldsymbol{s}$ in the form as

$$x = As = \sum_{i=1}^{N} a_i s_i \tag{1}$$

ICA algorithm is performed to obtain the estimation of independent components $\boldsymbol{s}$ from signal segments $\boldsymbol{x}$ by the un-mixing matrix $\boldsymbol{W}$

$$u = \boldsymbol{W}\boldsymbol{x} \tag{2}$$

where $\boldsymbol{u}$ is the estimation of independent components $\boldsymbol{s}$. Basis vectors $\boldsymbol{A}$ can be calculated from the ICA algorithm by the relation $\boldsymbol{A}=\boldsymbol{W}^{-1}$.

The learning rules of *infomax* is represented as

$$\Delta W \propto \eta[I - \varphi(u)u^T]W \tag{3}$$

The vector $\varphi(u)$ is a function of the prior and is defined by $\varphi(u) = \frac{\partial \log p(u)}{\partial u}$, and $p(u)$ is the p.d.f. of u. In the *infomax* algorithm, Bell and Sejnowski[5] chose the nonlinearty to be a fixed *logistic* function: $g(u) = \frac{1}{1+e^{-u}}$, and the derivation of the logistic function $g'_i(u_i)$ is a suitable estimation for p.d.f. of speech and music signals. However, the fixed noninearity is not efficient when the p.d.f.s are widely different.

So the GGM is proposed in this paper as the p.d.f. estimator. The GGM can model a family of density functions that is peaked and symmetric at the mean, with a varying degree of normality in the following general form [6]

$$p_g(s \mid \theta) = \frac{\omega(q)}{\sigma}\exp[-c(q)\,|\frac{s-\mu}{\sigma}|^q], \qquad \theta = \{\mu, \theta, q\} \tag{4}$$

where

$$c(q) = [\frac{\Gamma[3/q]}{\Gamma[1/q]}]^{q/2} \tag{5}$$

and

$$\omega(q) = \frac{\Gamma[3/q]^{\frac{1}{2}}}{(2/q)\Gamma[1/q]^{\frac{3}{2}}} \tag{6}$$

$\mu = E[s], \sigma = \sqrt{E[(s-\mu)^2]}$ are the mean and standard deviation of the data respectively, and $\Gamma[\cdot]$ is the Gamma function. By inferring q, a wide class of statistical distributions can be characterized. The sub-Gaussian, Gaussian, Laplacian, and strong Laplacian (such as speech signal) distributions can be modeled by putting q = 2, q = 1, and q < 1 respectively.

For the purposes of finding the basis functions, the problem then becomes to estimate the value of q from the data. This can be accomplished by simply finding the maximum posteriori value q. The posterior distribution of q given the observations $\boldsymbol{x}=\{x_1,\ldots,x_n\}$ is

$$p(q \mid x) \propto p(x \mid q)\, p(q) \tag{7}$$

where the data likelihood is

$$p(x \mid q) = \prod_n \omega(q) \exp[\, -c(q) \mid x_n \mid^q] \tag{8}$$

and $p(q)$ defines the prior distribution for q, here Gamma function $\Gamma[\cdot]$ is used as $p(q)$.

In the case of the GGM, the vector $\varphi(s)$ in eq.3 can be derived as

$$\varphi_i(s_i) = -qc\sigma_i^{-q} \mid s_i - \mu_i \mid^{q-1} sign(s_i - \mu_i) \tag{9}$$

Using the learning rule eq. 3 the un-mixing matrix $\boldsymbol{W}$ is iterated by the natural gradient until convergence is achieved.

3 Feature Extraction of Chinese Speech Signal

Male and female Chinese speech signals of 130000 samples were used to test our proposed method. The sampling rates are both 8kHz. For each signal, the mean was subtracted, and 3250 samples of length 40 (5ms) were generated from the data. Each segment was pre-whitened to improve the convergence speed. The adaptation started from the 40×40 identity matrix and trained through the 3250 data vectors. The learning rate was gradually decreased from 0.2 to 0.05 during the iteration. Some of the basis functions of the male and female Chinese speech signals are shown in fig. 1.

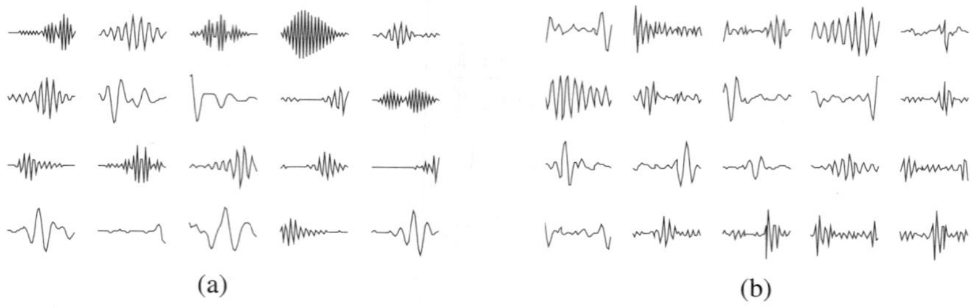

Fig. 1. (a)-(b) Basis vectors of male and female Chinese speech signals

Fig.2 shows the frequency spectrum of Fig.1 (a) and (b) respectively. It can be seen that the ICA basis vectors of Chinese speech signals are localized both in time and frequency domain.

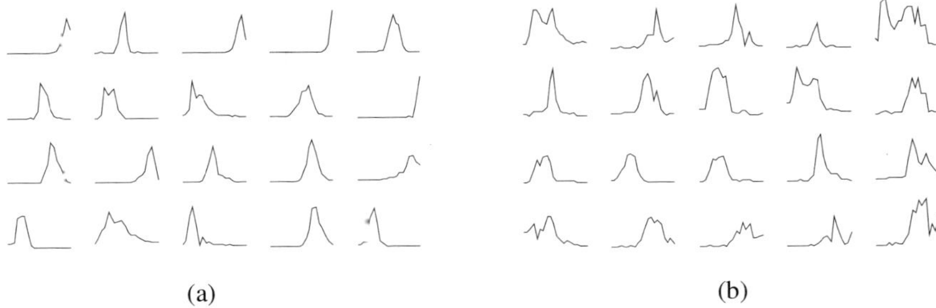

Fig. 2. (a)-(b) The frequency spectrum of Fig.1 (a) and (b)

Basis functions of male Chinese speech signals of the DFT, DCT, PCA and ICA basis are displayed in fig. 3. We can see that the DFT and DCT basis look similar and they are spread all over the time axis. For different signals the DFT and DCT basis are fixed. PCA basis is data driven and exhibits less regularity and global. However, the ICA basis functions are localized in time and frequency, thus they reflect both the phase and frequency information inherent in the data.

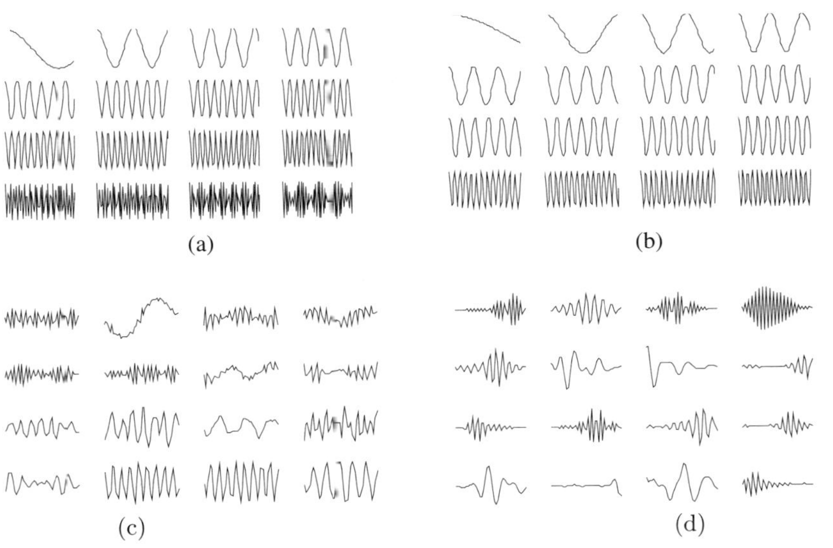

Fig. 3. Comparison of DFT, DCT, PCA and ICA basis vector of male Chinese speech signal, (a) DFT basis vector, (b) DCT basis vector, (c) PCA basis vector, (d) ICA basis vector

In addition, Fig. 4 compares the log-scaled histograms of the coefficients produced by the ICA basis filters with that by conventional methods. It can be seen that the distribution of the ICA coefficients is peakier than the others, and this characteristic yields greater sparseness in encode efficiency.

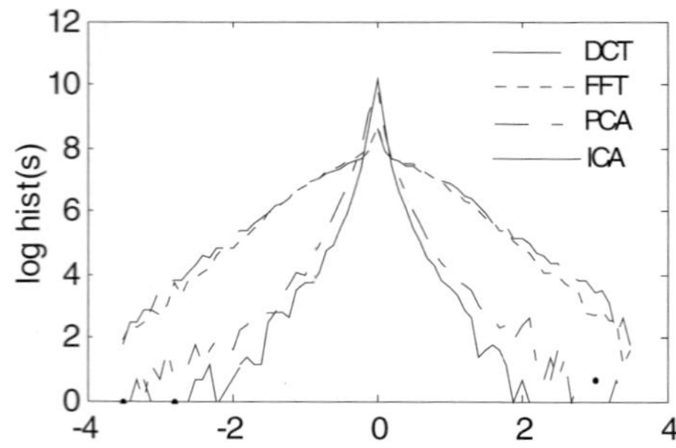

Fig. 4. The coefficients comparison of male Chinese speech signals in different methods

4 Feature Extraction of Underwater Signals

Different underwater acoustic signals have different kinds of statistic distributions. For example, the distributions of ship radiated signals are always sub-Gaussian, and the sea noises are nearly Gaussian. For this reason, the GGM-based ICA method is suitable for the feature extraction of underwater acoustic signals.

We down-sampled the originally ship radiated signals and sea noises data to 500Hz. For each signal, the mean was subtracted, and 1000 samples of length 64 were generated from random time-points of the data, and each segment was pre-whitened.

Using the GGM-based ICA algorithm, basis vectors of ship radiated signals and sea noises were extracted and some of them are shown in fig. 5. Fig.6 gives the frequency domain characteristic of fig.5.

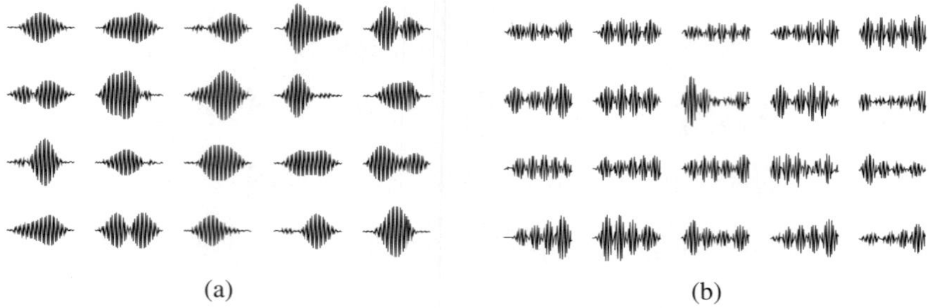

Fig. 5. (a)-(b) Some basis functions of ship radiated signals and sea noises

It can be seen that the ICA basis vectors of ship radiated signals and sea noise are obviously different. The ICA basis vectors of ship radiated signals are localized both in time and frequency domain, and not localized in sea noise.

It can be shown from fig.6 that the ICA feature of ship radiated signals are focus on low frequency, and that of sea noises are global in all frequency domain because sea noises are close to Gaussian distribution.

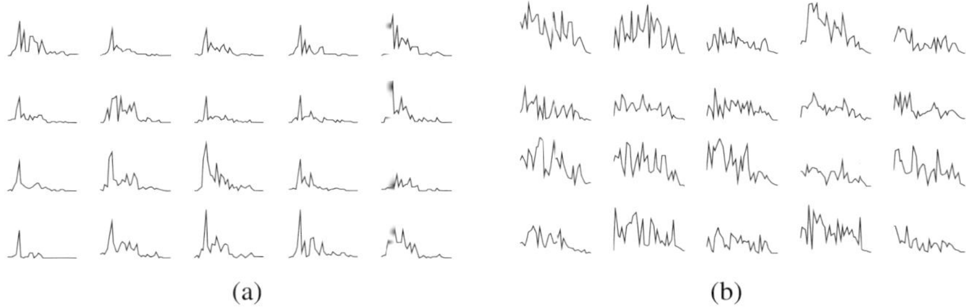

Fig. 6. (a)-(b) The frequency spectrum of fig.5 (a) and (b)

On the other hand, the log-scaled histograms of the coefficients of DFT, DCT, PCA and ICA for ship radiated signals are shown in fig.7. It can be seen that the distribution of the ICA coefficients is peakier than the others, and this characteristic yields greater sparseness in encode efficiency.

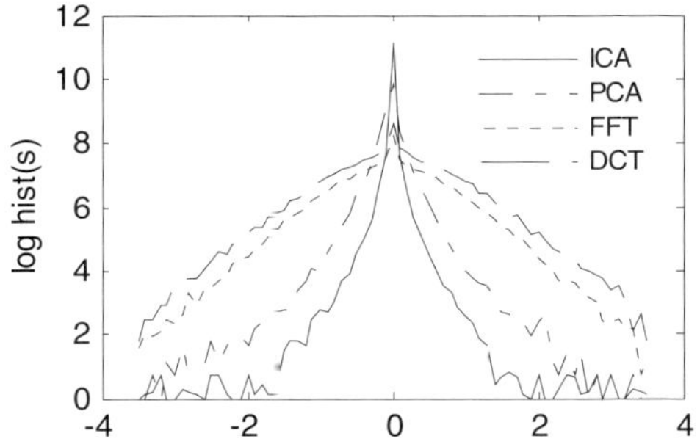

Fig. 7. The coefficients comparison of ship radiated signals in different methods

5 Conclusions

In this paper, we obtained an efficient feature extraction method for non-Gaussian acoustic signals by using GGM in ICA algorithm. By inferring the parameter of the GGM, different kinds of non-Gaussian distributions can be characterized. It is demonstrated that the presented method can extract the higher-order features of Chinese speech signals which are sup-Gaussian and the ship-radiated noises which are always sub-Gaussian. And the features are demonstrated localized both in time and frequency domain. The basis vectors as well as the coefficients are key factors in recognition and encoding the acoustic signals since they capture the higher-order structures of signals better than the conventional methods.

References

1. Lee, J.-H., Jung, H-J., Lee, T-W.: Speech Coding and Noise Reduction Using ICA-based Speech Features. In: International Workshop on Independent Component Analysis (ICA'00), Helsinki (2000) 417-422
2. Jang, G.-J., Lee, T.-W.: Learning Statistically Efficient Features for Speaker Recognition. Neurocomputing 49 (2002) 329-348
3. Bell, A. J., Sejnowski, T. J.: Learning the Higher-Order Structure of a Nature Sound. Network: Computation in Neural System 7 (1996) 261-266
4. Lee, J.-H., Jung, H.-Y.: Speech Feature Extraction Using Independent Component Analysis. In: Proc. ICASP. Vol. 3 (6). Istanbul, Turkey (2000) 1631-1634
5. Bell, A. J., Sejnowski, T. J.: An Information Maximization Approach to Blind Separation and Blind Deconvolution. Neural Computation 7(6) (1995) 1129-1159
6. Lee, T.-W., Lewicki, M. S.: The Generalized Gaussian Mixture Model Using ICA. International workshop on Independent Component Analysis (ICA'00). Helsinki, Finland, (2000) 239-244

Automatic Removal of Artifacts from EEG Data Using ICA and Exponential Analysis

Ning-Yan Bian[1], Bin Wang[1], Yang Cao[2], and Liming Zhang[1]

[1] Department of Electronics Engineering, Fudan University, Shanghai 200433, China
{042021040, wangbin, yang_cao, lmzhang}@fudan.edu.cn
[2] Department of Physiology and Biophysics, School of Life Science, Fudan University, Shanghai 200433, China

Abstract. Eye movements, cardiac signals, muscle noise and line noise, *etc.* present serious problems for the accuracy of Electroencephalographic (EEG) analysis. Some research results have shown that independent component analysis (ICA) can separate artifacts from multichannel EEG data. Further, considering the nonlinear dynamic properties of EEG signals, exponential analysis can be used to identify various artifacts and basic rhythms, such as α rhythm, *etc.*, from each independent component (IC). In this paper, we propose an automatic artifacts removal scheme for EEG data by combining ICA and exponential analysis. In addition, the proposed scheme can also be used to detect basic rhythms from EEG data. The experimental results on both the simulated data and the real EEG data demonstrate that the proposed scheme for artifacts removal has excellent performance.

1 Introduction

EEG recordings often contain some large and distracting artifacts, such as eye movements, muscle noise, cardiac signals, and linear noise, *etc.* which greatly influence the accuracy in EEG data analysis. Some research results have shown that Independent Component Analysis (ICA) can separate these artifacts (usually generated by non-cerebral sources) from the cerebral sources effectively [1]-[3]. The traditional methods to remove the artifacts are mainly semi-automatic, for example, scalp topographies combining visual inspection [1]. Recently, the proposed automatic identification methods mainly focus on the removal of eye movements, for example, ocular artifacts removal by calculating the correlation between independent components (ICs) and Electrooculography (EOG) signals [2]. Nevertheless, although such method is applicable with the precondition that there are EOG channels in the sources including EOG channels in the horizontal direction and vertical direction, it will not be applicable if the EOG channels are not available. In addition, there are perhaps basic rhythms in EEG recordings that should be identified, such as α rhythm and β rhythm.

We present here a generally applicable method for removing a wide variety of EEG artifacts by using exponents to quantitatively describe the nonlinear features of the ICs. Our method is able to identify various artifacts and basic rhythms in EEG recordings by a completely automatic way, without any human interference. The principle diagram of our method is given in Fig. 1.

J. Wang et al. (Eds.): ISNN 2006, LNCS 3972, pp. 719 – 726, 2006.

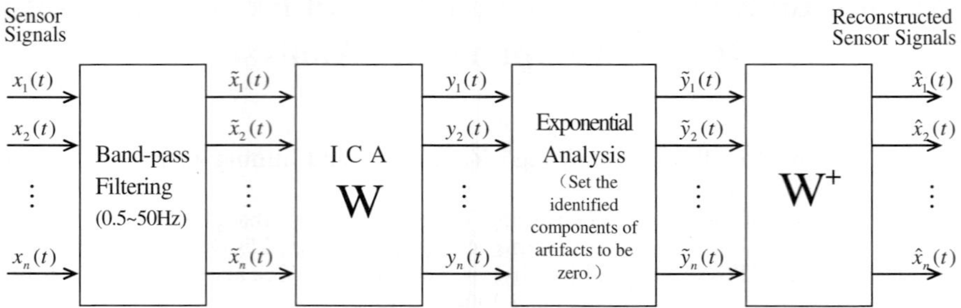

Fig. 1. Principle diagram

2 Independent Component Analysis

Independent Component Analysis (ICA) is an effective algorithm to realize blind source separation for EEG data. Some research has shown that ICA can separate the artifacts from multichannel EEG data and collect the basic rhythms in different ICs [1]-[3]. ICA can be described as follows: given the N dimension observed signals $\mathbf{x}(t)=[x_1(t),x_2(t),\cdots,x_N(t)]^T$, $\mathbf{s}(t)=[s_1(t),s_2(t),\cdots,s_N(t)]^T$ are the N unknown independent sources and $\mathbf{x}(t)=\mathbf{A}\mathbf{s}(t)$. The task is to separate the sources by a separating matrix **W**, making the outputs $\hat{\mathbf{s}}(t)$ close to the sources $\mathbf{s}(t)$:

$$\mathbf{y}(t) = \mathbf{W}\,\mathbf{x}(t) = \hat{\mathbf{s}}(t) \quad . \tag{1}$$

In this paper, we plan to use the "extended-Infomax" ICA algorithm proposed by Lee [3]. It uses the fact that maximizing the joint entropy $\mathbf{H}(\mathbf{y}_i)$ of the outputs of a neural processor minimizes the mutual information among the output components $\mathbf{y}_i=\mathbf{g}(u_i)$, where $\mathbf{g}(u_i)$ is a nonlinear function and $\mathbf{u}=\mathbf{W}\mathbf{x}$. The algorithm can separate the signal components with both super- and sub-Gaussian distributions by estimating the sign of the kurtosis of u_i .

3 Exponential Analysis

Many nonlinear analyses have revealed that the electrical activity of a brain measured by EEG exhibits a complex behavior with nonlinear dynamic properties. So the nonlinear dynamics theory may be an effective approach in characterizing the intrinsic nature of EEG. In the analysis of EEG data, different nonlinear measures are used in recent literature, such as correlation dimension, Lyapunov exponent, entropy and Hurst exponent [4][5].

We deem that the exponents mentioned above are useful to characterize certain nonlinear properties of EEG signals, but it is hard to identify various artifacts and basic rhythms with only one exponent. For example, Kurtosis values can detect the signals with peaked and sparse distribution, such as cardiac signals and eye movements, but they may not be well enough to detect muscle noise. Though Hurst exponents are good for identifying cardiac signals and some muscle artifacts, they may not detect eye movements effectively. Besides, both Kurtosis values and Hurst exponents are not good enough for recognizing some basic rhythms, but Lyapunov exponents can solve the problem. Therefore, we unite three exponents: Largest Lyapunov Exponent (LLE), Kurtosis and Hurst Exponent (HE), to analyze each IC, and present a completely automatic method for artifacts removal and basic rhythms identification in EEG data without any human interference. The procedures of the proposed method are given as follows:

Step 1: Calculate the LLE of each IC, and select periodic components by the threshold, and then analyze their spectrum by an automatic way.
Step 2: Calculate the Kurtosis value of each IC, and classify eye movements and cardiac signals by the threshold.
Step 3: Calculate the HE of each IC, and select other artifacts by the threshold.

In addition, it is worth pointing out that we use an automatic method to analyze the spectra of the ICs: detect the location of the largest peak value of the spectrum statistically, for example, if the location of the peak value is in the frequency range of α rhythm (8~13Hz) and there are no obvious peaks in other frequencies, then we consider this component is an α rhythm.

3.1 Largest Lyapunov Exponent

Lyapunov Exponent (LE) is a quantitative measure for distinguishing among various types of orbits based upon their sensitive dependence on the initial conditions, and is often used to discriminate between chaotic dynamics and periodic signals. A zero exponent means that the orbits maintain their relative positions, and a positive exponent indicates chaos, while a negative one corresponds to the periodic state.

To calculate the LE, we use the algorithm for Largest Lyapunov Exponent presented by Wolf [6]. For a given time series $\mathbf{x}(t)$ for m dimension phase space with delay $\tau : \{x(t), x(t+\tau), \cdots, x(t+(m-1)\tau)\}$, we locate the nearest neighbor to the initial point $\{x(t_0), x(t_0+\tau), \cdots, x(t_0+(m-1)\tau)\}$. And we denote the distance between these two points as $L(t_0)$. At time t_1, initial length will evolve to be length $L'(t_1)$. Then Lyapunov exponent is characterized by

$$\lambda = \frac{1}{t_M - t_0} \sum_{k=1}^{M} \log_2 \frac{L'(t_k)}{L'(t_{k-1})}, \tag{2}$$

where M is the iterative time. In one dimension projection, there is only one value of λ, but there are several values in multi-dimension phase space. And it will be chaotic with one positive exponent in a system. Therefore, it is useful to detect whether the largest LE is positive to decide chaos of a system. When calculating the LE of EEG data, an embedding dimension m between 5 to 20 and a delay of 1 should be chosen [4]. In this paper, we have chosen an embedding dimension of 10 and a delay of 1. If the LE is negative, the series can be characterized as periodic, so we relax the threshold to be -0.1.

3.2 Kurtosis

Kurtosis is the 4th cumulant of the data. For a time series, the Kurtosis is calculated using the following equations

$$\kappa = m_4 - 3m_2^2 \ , \tag{3}$$

$$m_n = E\{(x - m_1)^n\} \ , \tag{4}$$

where m_n is the nth central moment of the series and m_1 is the mean value. If the Kurtosis value is highly positive, the distribution of activity is peaked and sparse, and the identified data is likely to be an artifact, such as cardiac signals and eye movements. For an absolutely periodical series, the Kurtosis value is negative. Besides, if the Kurtosis value is close to zero, the series is likely to be a Gaussian noise.

In our method, we select a rejection threshold in terms of the number of standard deviation from the mean, *e.g.* 20%. It is likely to be an artifact when exceeding the threshold, such as eye movements and cardiac signals.

3.3 Hurst Exponent

Hurst Exponent (HE) is used to evaluate the presence or absence of long-range dependence and its degree in a time series. To calculate it, Hurst developed the rescaled range (*R/S*) analysis for time series [7]. Given a time series $\{\xi(t)\}$, $t = 1, 2, \cdots$, with the mean

$$< \xi >_\tau = \frac{1}{\tau}\sum_{t=1}^{\tau} \xi(t) \ . \tag{5}$$

The accumulated deviation from the mean can be defined as

$$X(t,\tau) = \sum_{t=1}^{\tau} \{\xi(t) - < \xi >_\tau\}, \quad 1 \le t \le \tau \ . \tag{6}$$

R is defined as a difference between the maximum and the minimum accumulated values

$$R(\tau) = \max_{1 \le t \le \tau} X(t,\tau) - \min_{1 \le t \le \tau} X(t,\tau) \ . \tag{7}$$

And standard deviation S estimated from the observed value is

$$S(\tau) = \left(\frac{1}{\tau}\sum_{t=1}^{\tau} \{\xi(t) - < \xi >_\tau\}^2\right)^{1/2} \ . \tag{8}$$

The Hurst exponent H is defined as

$$H = \log(R/S)/\log(cT) \ , \tag{9}$$

where T is the number of samples, c is the constant, typically $c = 1/2$, and H is in the range from 0 to 1. A value of H =0.5 corresponds to a standard Brownian activity, and 0< H <0.5 shows so called anti-persistent behavior, *e.g.* white uniform distributed noise has $H \cong 0.15$, while a value of 0.5< H <1 describes a temporally persistent or trend reinforcing time series. At limit a straight line with nonzero slope will have a HE of 1. It has been found by some research that the HE has a value of 0.70~0.76 for many human phenomena. Furthermore, some artifacts, such as cardiac signals, have a specific HE 0.64~0.69 [5]. Usually if HE is above 0.9, the signal is very likely to be a linear electrode artifact. Thus, we can automatically identify and remove the artifacts from sensor signals on the basis of the value of HE. In this paper, we set the threshold of HE in the range of 0.7 to 0.9.

4 Experimental Results

In this section, we will apply the proposed method to simulated data and real EEG data respectively, and give corresponding results with some analysis and discussions.

4.1 Experiments for Simulated Data

In order to validate the effectiveness of the proposed method, we first perform studies with known signals. The sources which consist of a simulated α rhythm (10Hz), a simulated β rhythm (25Hz), an eye blink signal, a cardiac signal, a Gaussian noise and a simulated high frequency interference (50Hz) are shown in Fig. 2(a) and the observed signals are shown in Fig. 2(b). Using the extended-Infomax algorithm, we obtain 6 ICs as shown in Fig. 2(c). By calculating the corresponding Lyapunov exponent, Kurtosis and Hurst exponent of each IC by the proposed method, the obtained results are shown in Table 1, where Kug means the global Kurtosis. LE detects IC2, IC4 and IC6 to be the periodic signals, and furthermore, we can easily identify IC2 as a simulated α rhythm and IC4 as a simulated β rhythm by spectral analysis of the three ICs by an automatic way. Kurtosis detects IC1, IC3 and IC5. Because of the

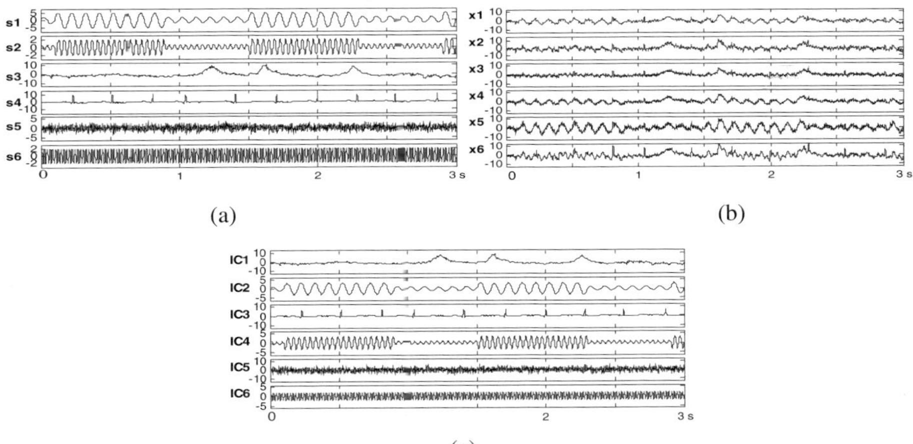

Fig. 2. Results of simulated data. (a) Simulated sources, (b) observed mixed signals, (c) estimated independent components.

Table 1. The three Exponents of each IC

	IC1	IC2	IC3	IC4	IC5	IC6
LLE	0.4342	-0.1267	0.0715	-0.4855	1.3136	-0.1664
Kug	6.0342	-0.8298	26.5441	-0.7071	0.0323	-1.5
HE	0.7925	0.6029	0.6647	0.5346	0.5643	0.4195

presence of low frequency activity corresponding to ocular, it is easy to recognize that IC1 is an eye movement component. And IC5 can be detected as a Gaussian noise with a Kurtosis value close to zero. Similarly, HE can further identify that IC3 is a cardiac signal.

4.2 Experiments for Real EEG Data

In this subsection, the performance of the proposed method is tested on EEG data of eye closed activities from the subject. The EEG recordings are collected from 26 electrodes with the sample frequency of 1000Hz and the test length of 5 seconds, as shown in Fig. 3(a). Firstly, we filter the EEG recordings by a bandpass filter from 0.5Hz to 50Hz, and Figure 3(b) shows the filtered EEG data. Because the EEG waves have a limited highest frequency and there is inevitably some noise in the recordings, we choose to remove the frequency above 50Hz. Meanwhile, we also eliminate the frequency below 0.5Hz, so as to remove the baseline drift and other slow interferences. We apply the exponential analysis to the estimated ICs after ICA processing. LE detects IC2, IC5, IC7, IC8, IC11, IC15 and IC18 to be the signals with periodic properties. And we can find IC5 has an obvious spectral peak at 10Hz by automatically analyzing its spectrum, so we consider IC5 as an α rhythm. Kurtosis detects two ICs with the larger Kurtosis values, IC3 and IC16, which exceed the mean value of 60% and 141%, respectively. Automatically, we can detect IC3 as an ocular movement because its spectral peak localizes at 0~1Hz.

Finally, HE further identifies that IC16 is a cardiac signal with the value H =0.6502, and IC23 is very likely to be a muscle movement with the value H =0.6295.

Figure 4 shows each IC, the corresponding spectra of detected artifacts and the scalp maps of these artifacts. From Fig. 4, we can see that most power spectrum of IC3 accumulates in the low frequency, and its scalp map indicates that the activity is in the frontal sites. For IC5, the activity is localized in the posterior areas of the scalp. We can also find that the wave of IC16 shows the striking cardiac characteristics. Finally, the spectrum range of IC23 is broad and disordered, and its activity is concentrated in the back temporal areas of the scalp map. All these results above also validate the effectiveness of the proposed method.

Using the proposed method in this paper, we can automatically identify IC3, IC5, IC16 and IC23. Among the four identified ICs, IC3, IC16 and IC23 should be removed, while whether IC5 should be kept or analyzed separately lies on later requirement. Figure 3(c) shows the reconstructed EEG data which we obtain by eliminating artifacts and α rhythm. Compared with Fig. 3(b), we can find the ocular artifacts in Fp1, Fp2, F3, Fz, F4, *etc.* and obvious cardiac artifacts in O1, Oz and O2 are well removed in Fig. 3(c). Besides, the obvious α rhythm in 1~2s in the original

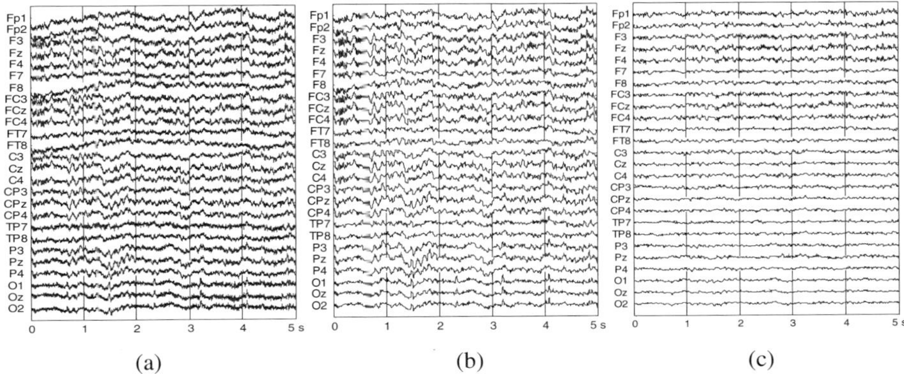

Fig. 3. Real EEG recordings. (a) The observed EEG signals of 26 channels, (b) the filtered EEG signals, (c) the EEG signals corrected by artifacts removal.

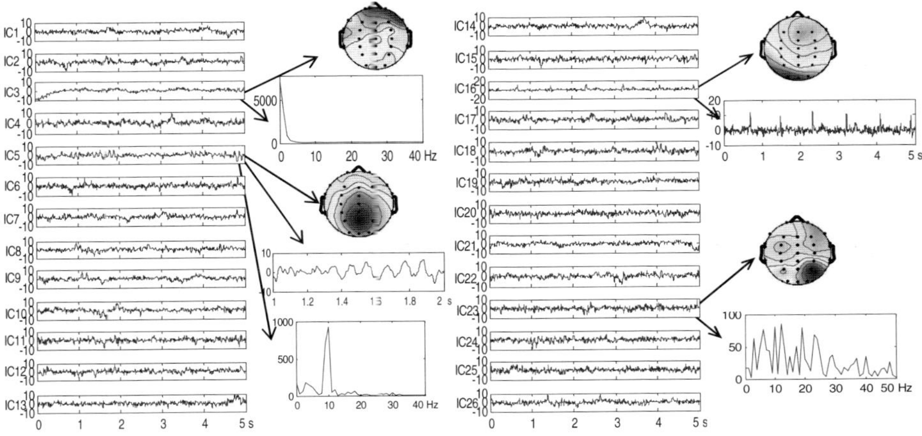

Fig. 4. Independent components. IC3: scalp map and spectrum pattern. IC5: scalp map, time series and spectrum pattern. IC16: scalp map and time series. IC23: scalp map and spectrum pattern.

EEG data is also removed in the corrected data in Fig. 3(c). It indicates that our method performs well in the automatic identification and removal of artifacts and some basic rhythms from EEG data.

5 Conclusions

In this paper, we have proposed a completely automatic scheme to identify artifacts and basic rhythms of EEG data by using exponents to analyze the nonlinearity of each

IC. This scheme is effective to automatically remove the artifacts and identify the basic rhythms of EEG data, and the whole process does not need any human interference. Then we reconstruct the data with the elimination of the automatically identified ICs by the proposed method, and the cleaned data can be obtained after artifacts removal. Finally, we point out that the proposed scheme can be considered as a kind of preprocessing for further quantitative study of the true EEG activity and should be very useful for many applications.

Acknowledgement

This research was supported by the grant from the National Natural Science Foundation of China (No. 30370392).

References

1. Jung, T.P., Makeig, S., Humphries, C., Lee, T.W., Sejnowski, T.J.: Removing Electroencephalographic Artifacts by Blind Source Separation. Psychophysiology 37 (2000) 163-178
2. Joyce, C.A., Gorodnitsky, I.F., Kutas, M.: Automatic Removal of Eye Movement and Blink Artifacts from EEG Data Using Blind Component Separation. Psychophysiology 41(2) (2004) 313-325
3. Lee, T.W.: Independent Component Analysis: Theory and Applications. Kluwer Academic, Boston (1998)
4. Natarajan, K., Acharya, R., Alias, F., Tiboleng, T., Puthusserypady, S.K.: Nonlinear Analysis of EEG Signals at Different Mental States. Biomedical Engineering Online (2004) 3:7
5. Vorobyov, S., Cichocki, A.: Blind Noise Reduction for Multisensory Signals Using ICA and Subspace Filtering, With Application to EEG Analysis. Biological Cybernetics 86 (2002) 293-303
6. Wolf, A., Swift, J.B., Swinney, L.H., Vastano, J.A.: Determining Lyapunov Exponent from a Time Series. Physica D 16 (1985) 285-317
7. Hurst, H.B., Black, R.P., Simaika, Y.M.: Long Term Storage. Constable, London (1965)

Identification of Vibrating Noise Signals of Electromotor Using Adaptive Wavelet Neural Network

Xue-Zhi Zhao and Bang-Yan Ye

College of Mechanical Engineering, South China University of Technology,
510640 Guangzhou, China
{mezhaoxz, byye}@scut.edu.cn

Abstract. Electromagnetic noise, unbalanced rotor noise and injuring bearing noise are three types of noise in faulting electromotor. An adaptive wavelet neural network is proposed to identify these noises. The process of wavelet-based feature extraction of signal is integrated into one part of neural network. During network's training course the wavelet's scale and shift parameters can be adaptively adjusted to fit input signal so that signal's feature could be extracted in maximum limit. The network's second part then uses these feature information to realize the identification of noise signal. The identification result of three types of noise of electromotor demonstrates that this neural network can give accurate identification result with high probability.

1 Introduction

There are general three types of noise in faulting electromotor, namely that electromagnetic noise, unbalanced rotor noise and injuring bearing noise[1]. It's important to identify the type of noise of a faulting electromotor so that the fault of electromotor can be correctly diagnosed by its noise.

An adaptive wavelet neural network is put forward to resolve this problem. In this neural network, the process of wavelet-based feature extraction of signal makes up the first part of the neural network, and the scale and shift parameters of wavelet will be adaptively adjusted to fit the input signals during the network's training course so that signals' feature information can be extracted in maximum limit. The learning algorithm of this neural network is given. The identification result of three types of noise signal of electromotor shows that this network can give accurate identification result with high probability.

2 The Architecture and Learning Algorithm of Adaptive Wavelet Neural Network

The architecture of adaptive wavelet neural network is illustrated in Fig.1, the idea of this architecture comes mainly from references[2][3] but its wavelet and learning algorithm are different. This network consists of two parts: the adaptive wavelet-based feature extraction part and identification part. The input of network is the noise signal of electromotor while its output is the identification result. Each output node stands for a type of noise.

J. Wang et al. (Eds.): ISNN 2006, LNCS 3972, pp. 727 – 734, 2006.

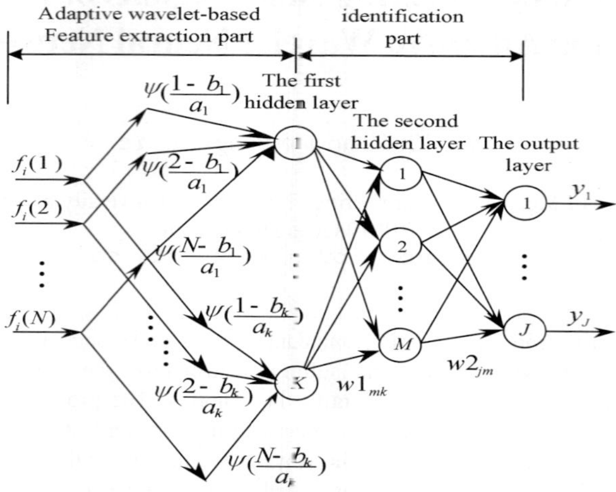

Fig. 1. The architecture of adaptive wavelet neural network

In the feature extract part, according to definition of wavelet transformation:

$$Wf(a,b) = |a|^{-1/2} \int_R f(t)\overline{\psi(\frac{t-b}{a})}dt \quad . \tag{1}$$

Where $\psi(t)$ is wavelet function, a and b are its scale and shift parameters, $f(t)$ is the input signal.

Equation (1) being referred to, the below discrete form of wavelet transformation is used to extract signal's feature in the neural network:

$$s_{ki} = \sum_{t=1}^{N} f_i(t) \cdot \psi(\frac{t-b_k}{a_k}) \quad . \quad (k=1,2,\cdots, K; i=1,2,\cdots, I) \tag{2}$$

Where $f_i(t)$ is the ith input signal to be identified (i=1,2,$\cdots$, I) and N is its length, a_k and b_k are scale and shift parameters of wavelet in the kth node, $s_{k\,i}$ is the feature information of $f_i\,(t)$ extracted by the wavelet in the kth node and $s_{k\,i}$ is also the output of the kth node in the first hidden layer (Fig.1).

In the second hidden layer (Fig.1), sigmoid function is used as neuron's transfer function, so output of the mth node in the second hidden layer is:

$$d_{mi} = \frac{1}{1+e^{-v_{mi}}} \quad . \quad (m=1,2,\cdots, M; i=1,2,\cdots, I) \tag{3}$$

Where $v_{mi} = \sum_{k=1}^{K} w1_{mk} \cdot s_{ki}$, $w1_{mk}$ is the weight to link the the first hidden layer with the second one (Fig.1).

In the output layer, sigmoid function is also used, so the output of the jth node is:

$$y_{ji} = \frac{1}{1+e^{-u_{ji}}} \ . \ (j=1,2,\cdots, J;\ i=1,2,\cdots, I) \tag{4}$$

Where $u_{ji} = \sum_{m=1}^{M} w2_{jm} \cdot d_{mi}$, $w2_{jm}$ is the weight to link the the second hidden layer with output layer (Fig.1).

Network's error function is:

$$E = \frac{1}{2}\sum_{i=1}^{I}\sum_{j=1}^{J}(\hat{y}_{ji} - y_{ji})^2 \ . \tag{5}$$

Where $\hat{y}_{ji}$ is the desired value of the jth node in the output layer corresponding to the ith signal $f_i(t)$. Each output node stands for one type of noise, if $f_i(t)$ is the jth type of noise signal, then let $\hat{y}_{ji}$ =1, as for other types of noise, let $\hat{y}_{ji}$ =0.

Except that weight $w1_{mk}$ and $w2_{jm}$ are corrected, a_k and b_k , the scale and shift parameters of wavelet, are also corrected in the network's training course so that wavelet can fit the input signal and $f_i(t)$'s feature can be extracted in maximum limit by this kind of adaptive wavelet transformation. For this reason, this network is called adaptive wavelet neural network.

Procedure of gradient descent is used to correct these parameters, according to (2)~(5), the gradient vectors of all parameters can be obtained as follows:

$$\delta w2_{jm} = \sum_{i=1}^{I}(\hat{y}_{ji} - y_{ji})\cdot y_{ji} \cdot (1-y_{ji})\cdot d_{mi} \ . \ (j=1,2,\cdots, J;\ m=1,2,\cdots, M)$$

$$\delta w1_{mk} = \sum_{i=1}^{I}\sum_{j=1}^{J}(\hat{y}_{ji} - y_{ji})\cdot y_{ji} \cdot (1-y_{ji})\cdot w2_{jm} \cdot d_{mi} \cdot (1-d_{mi})\cdot s_{ki} \ .$$

$$(m=1,2,\cdots, M;\ k=1,2,\cdots, K)$$

$$\delta a_k = \sum_{i=1}^{I}\sum_{m=1}^{M}\sum_{j=1}^{J}(\hat{y}_{ji} - y_{ji})\cdot y_{ji} \cdot (1-y_{ji})\cdot w2_{jm} \cdot d_{mi} \cdot (1-d_{mi})\cdot w1_{mk} \cdot \frac{\partial s_{ki}}{\partial a_k} \ .$$

$$(k=1,2,\cdots,K)$$

$$\delta b_k = \sum_{i=1}^{I}\sum_{m=1}^{M}\sum_{j=1}^{J}(\hat{y}_{ji} - y_{ji})\cdot y_j \cdot (1-y_{ji})\cdot w2_{jm} \cdot d_{mi} \cdot (1-d_{mi})\cdot w1_{mk} \cdot \frac{\partial s_{ki}}{\partial b_k} \ .$$

$$(k=1,2,\cdots,K)$$

where $\frac{\partial s_{ki}}{\partial a_k} = \sum_{t=1}^{N} f_i(t)\cdot \frac{\partial \psi(t')}{\partial t'} \cdot (-\frac{t-b_k}{a_k^2})$,

$\frac{\partial s_{ki}}{\partial b_k} = \sum_{t=1}^{N} f_i(t)\cdot \frac{\partial \psi(t')}{\partial t'} \cdot (\frac{-1}{a_k})$, and $t' = \frac{t-b_k}{a_k}$.

Mexican hat wavelet is an infinite smooth function and can easily highlight the sudden changing information of signal, so it's chosen to extract signal's feature in the neural network. Its equation is:

$$\psi(t) = (1 - t^2)e^{-\frac{t^2}{2}} . \tag{6}$$

Then

$$\frac{\partial \psi(t)}{\partial(t)} = (-3t + t^3)e^{-\frac{t^2}{2}} . \tag{7}$$

These gradient vectors being used, all parameters can be corrected by below:

$$x(t+1) = x(t) + \xi \cdot \delta x + \alpha[x(t) - x(t-1)] . \tag{8}$$

where ξ is learning constant; α is momentum factor; x stands for parameters of $w1_{mk}$, $w2_{jm}$, a_k, b_k; δx stands for gradient vector of these parameters.

3 Identification for Noise Signals of Electromotor

3.1 Training Vectors and Neural Network's Being Trained Process

The type of fault electromotor is YFK80/20-4, their power is 80W, number of pole is 4 and rotating speed is 730r/min. B&K precision sound meter is used to detect the noises of these electromotor. Sample frequency for electromagnetic and unbalanced rotor noise is 2500Hz. For injuring bearing will produce a wide band signal, so sample frequency for this noise is increase to 5000Hz. 2 signals are sampled for each type of noise, one signal is used to train neural network and the other one is used to test network. The data number of each signal is 512. The three types of noise signals used to train network is illustrated in Fig.2.

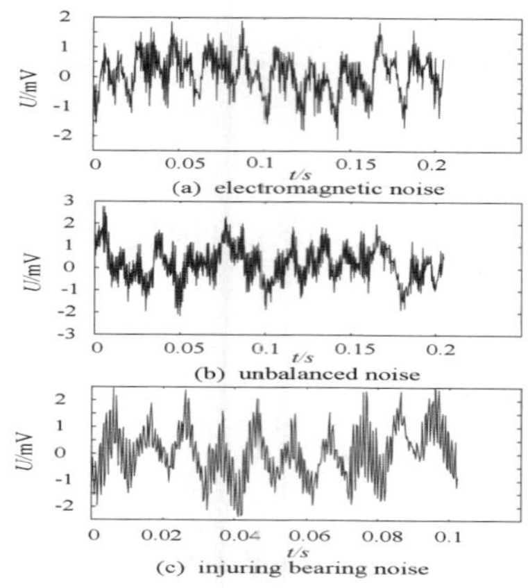

Fig. 2. The three types of noise signals of electromotor

Training vectors are obtained from these signals by the follow way: each noise signal is evenly divided into 5 segments, so 15 segments can be obtained and they are 15 input vectors, the data number of each segment is 100. The output layer of neural network has 3 nodes and each node stands for one type of noise. Output vectors are formed by the follow way: for 5 segments of electromagnetic noise, let desired value of the 1th node is 1, while the ones of the 2th and 3th node are 0. For 5 segments of unbalanced noise, let desired value of the 2th node is 1, while the ones of the 1th and 3th node are 0. For 5 segments of injuring bearing noise, let desired value of the 3th node is 1, while the ones of the 1th and 2th node are 0. The first hidden layer has 16 nodes and the second hidden layer has 10 nodes. In order to make initial wavelet cover the input signal evenly, the initial value of a_k =100/16=6.25, (k=1,2,···,16) and b_k is distributed in 0~100 evenly The initial weights are obtained according to reference[4]. ξ=0.01, α =0.9, E=0.02, After 162 iterations is trained, the network's error is 0.0199608. The error descent is illustrated in Fig. 3 and training results are illustrated in Fig. 4. From Fig.4 we can see that actual output values are very close to the desired ones.

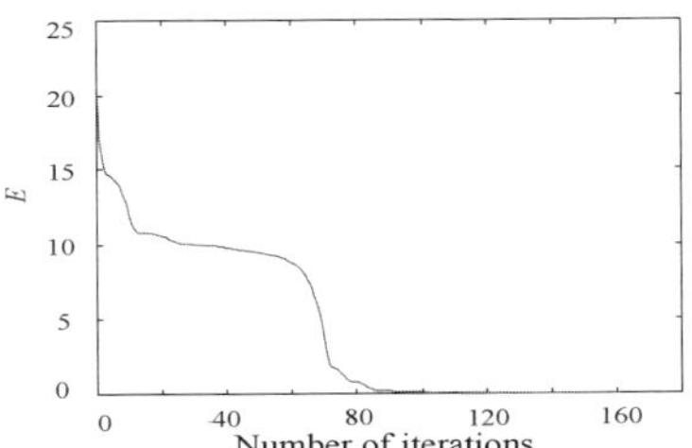

Fig. 3. The descent curve of error

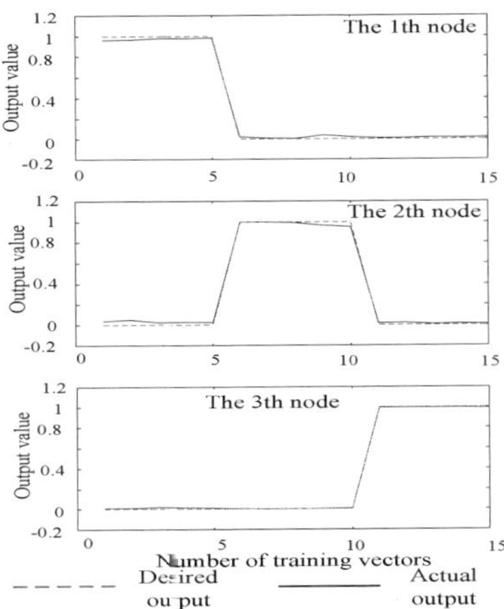

Fig. 4. The outputs of neural network after trained

The scale and shift parameters of wavelet are also adaptively adjusted during the training course. The comparison between their initial values and the final value is shown in Table 1. It's by this adaptive adjustment of a_k and b_k that signal's feature extraction effect is much better than classical wavelet transformation[5][6].

Table 1. The comparison of wavelet's scale and shift parameters between the initial values and the final values

a_k		b_k	
Initial value	Final value	Initial value	Final value
6.25	4.2041	0	0.0035
6.25	9.2718	6.25	6.2520
6.25	7.3285	12.5	12.4958
6.25	5.5617	18.75	18.7553
6.25	6.4980	25	24.9971
6.25	5.6923	31.25	31.2514
6.25	8.2397	37.5	37.4944
6.25	5.1038	43.75	43.7473
6.25	6.0655	50	49.9985
6.25	4.9513	56.25	56.2495
6.25	14.0224	62.5	62.5011
6.25	8.9391	68.75	68.7506
6.25	6.0727	75	75.0007
6.25	10.4517	81.25	81.2503
6.25	8.4348	87.5	87.5011
6.25	3.3482	93.75	93.7490

3.2 Neural Network's Identification for Three Types of Noise Signal of Electromotor

Another group of noise signals are used to test network (Fig.5). Similarly, each signal is evenly divided into 5 segments and also 15 input vectors are obtained. Corresponding to these 15 input vectors, the output results of the trained neural network are shown in table 2, 3 and 4.

Table 2 shows the output results for 5 segments of signal of Fig. 5(a), we can see that the output values of the 2th and 3th node are very close to 0, so unbalanced and injuring bearing noise can be excluded. As for the 5 output values of the 1th node, though the last value is 0.4564, it can't affect our judgment, for the other 4 outputs values are very close to 1, the ratio of 4:1 can make sure that this signal is electromagnetic noise.

Table 3 shows the output results for 5 segments of signal of Fig. 5(b). It's easy to see that output values of 2th node are all very close to 1 while ones of the 1th and 3th node are very close to 0, so we can decide that this signal is unbalanced noise.

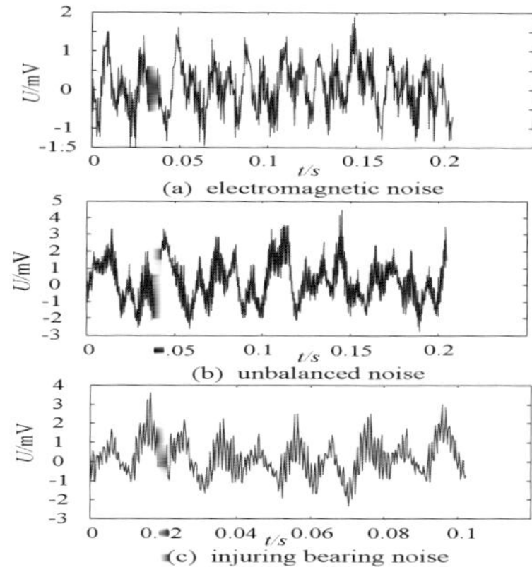

Fig. 5. The three noise signals used to test neural network

Table 2. The nodes outputs for the signal of Fig. 5(a)

Node 1	Node 2	Node 3
0.9784	0.0002	0.0035
0.9843	0.0026	0.0054
0.9891	0.0001	0.0102
0.9644	0.0009	0.0085
0.4564	0.0003	0.0035

Table 3. The nodes' outputs for the signal of Fig. 5(b)

Node 1	Node 2	Node 3
0.0026	0.9787	0.0430
0.0062	0.9985	0.0300
0.0020	0.9988	0.0299
0.0072	0.9756	0.0121
0.0073	0.9993	0.0186

Table 4 shows the output results for 5 segments of signal of Fig. 5(c). In the output values of the 1th node, the last one is 0.5982, but this can't bring trouble on our judgment, for the other 4 values of this node are very close to 0, the ratio of 4:1 can ensure that the signal isn't electromagnetic noise. Similarly, though the 3th output value of the 2th node is 0.2113, its other 4 values are very close to 0, so unbalanced noise can be excluded also. The key evidence is that output values of the 3th node are all very close to 1, so we can conclude that the signal is injuring bearing noise.

The above results show that the trained neural network can make an accurate and reliable identification for noise signals of electromotor. On the other hand, the output value of network can be regarded as the membership of noises, so this network also has nature of fuzzy judgment.

Table 4. The nodes' outputs for the signal of Fig.5(c)

Node 1	Node 2	Node 3
0.0190	0.0950	0.9975
0.1046	0.0018	0.9968
0.0110	0.2113	0.9994
0.0165	0.0015	0.9976
0.5982	0.0933	0.9956

4 Conclusion

(1) In the adaptive wavelet neural network, the process of adaptive wavelet-based feature extraction of signal is integrated into one part of neural network, so the feature extraction and identification of signal can be realized simultaneously.
(2) The scale and shift parameters of wavelet can be adaptively adjusted in the training course, by this adaptive wavelet parameters, feature of signal can be extracted much better than classical wavelet transformation.
(3) The training and identification process for three types of noise of electromotor shows that this network has fast learning speed and can give accurate and reliable identification result.
(4) Not only can this neural network be used to identify the noises of electromotor, but also it can be applied to identify and classify other fault signals.

Acknowledgement

This work is supported by National Natural Science Foundation of China (50305005). The authors would like to thank this support.

References

1. Zhao, X.Z., Chen, W.G., Ye, B.Y.: Probability Distribution Density of Vibrating Noises of Electromotor in Air-conditioner and Their Hypothesis Test. Proceedings of the CSEE. 24(6) (2004) 122–126
2. Brian, T., Szu, H.: Adaptive Wavelet Classification of Acoustic Backscatter and Imagery. Optical Engineering. 33(7) (1994) 2192–2202
3. Szu, H., Brian, T.: Neural Network Adaptive Wavelets for Signal Representation and Classification. Optical Engineering. 31(9) (1992) 1907–1916
4. Zhao, X.Z., Zou, C.H., Chen, T.J.: A Research on the Initialization of Parameters of Wavelet Neural Networks. Journal of South China University of Technology. 31(2) (2003) 77–79
5. Lepore, F.P., Santos, M.B., Barreto, R.G.: Identification of Rotary Machines Excitation Forces Using Wavelet Transform and Neural Networks. Shock and Vibration. 9(5) (2002) 235–244
6. Mahmoud, W.A., Amamra, I., Alsammarie, J.M.: Identification of Digital Modulation Techniques Using the Discrete Wavelet Transform and Neural Network. Advances in Modelling and Analysis. 46(2) (2003) 43–55

Fractional Order Digital Differentiators Design Using Exponential Basis Function Neural Network

Ke Liao, Xiao Yuan, Yi-Fei Pu, and Ji-Liu Zhou

School of Electronic and Information Engineering, Sichuan University, Chengdu, China
larklk@yahoo.com.cn

Abstract. In this paper, the topic of fractional order digital differentiators (FODD) is designed using neural networks approximation method. First, FODD amplitude response is given in the form of sum of exponential basis functions. Then, the exponential basis function neural network is used to approximate FODD amplitude response. Finally, some examples compared with others' method are given to illustrate the advantages of this paper approach.

1 Introduction

The Fractional calculus (FC) is widely used in engineering including modern control theory, kinetics, and electromagnetic theory [1]. It is an old issue in mathematic with more than three hundred years history, including fractional derivate and fractional integral. Fractional derivate means to generalize the integer order n derivative $(d^n f(x))/(dx^n)$ of function $f(x)$ into fractional order v derivative.

Design of fractional order digital differentiator filter is important for fractional calculus engineering applications. Recently, some methods have been developed to design finite impulse response (FIR) and infinite impulse response (IIR) filters, such as in time domain windowed method, frequency sampling method, CAD method [2]. In paper [3], the power function series are considered to design the FIR digital filters in time domain and this method is easily to compute. Above methods suffer from some kinds of drawbacks: they could not well approximate the ideal fractional differentiator's frequency response both in low frequency and in high frequency, when the derivative order v is less than one.

In this paper, we design a constant phase FIR filter to approximate the ideal fractional differentiator's frequency response. The exponential basis function neural network is used to compute the impulse response $h(n)$ of filters. Experimental results show that it gets a better approximation performance in all frequency bands, which overcomes above others method drawbacks.

2 Frequency Response of Fractional Differentiators

Ideal fractional differentiator's frequency response is [2] [3]

$$H_v(i\omega) = (i\omega)^v = H(\omega)\cdot e^{i\vartheta(\omega)} \qquad 0 \le \omega < \pi \tag{1}$$

J. Wang et al. (Eds.): ISNN 2006, LNCS 3972, pp. 735 – 740, 2006.

Where v is the order of the fractional order differentiators and is a noninteger, ω is the digital frequency, $H(\omega)$ is the ideal amplitude response, and $\phi(\omega)$ is the ideal phase response. They can be expressed as

$$H(\omega) = \omega^{v}, \; \phi(\omega) = \frac{v\pi}{2} \qquad 0 \le \omega < \pi \tag{2}$$

Using impulse response $h(n)(n \in Z)$, $H_v(i\omega)$ can be expressed as

$$H_v(i\omega) = \sum_{n=-\infty}^{+\infty} h(n) \cdot e^{-in\omega} \qquad -\infty \le n \le +\infty \tag{3}$$

The impulse response of ideal frequency response is infinite, which is impossible to be realized. In engineering design, finite N order impulse response $H_F(i\omega)$ is used to approximate the ideal $H_v(i\omega)$

$$H_F(i\omega) = \sum_{n=0}^{N-1} h(n) \cdot e^{-in\omega} \qquad n = 0, 1, \cdots N-1 \tag{4}$$

Because of filter length effect, the designing response $H_F(i\omega)$ can be realized in the form of FIR filter, and it is not identical with the ideal $H_v(i\omega)$.

3 Exponential Basis Function Neural Network

3.1 Neural Network Algorithm

Exponential basis function neural network is one kind of functional link artificial neural network (FLANN). In paper [4] [5] [6], authors used FLANN in some engineering applications and got good performance, here we generalize it into FODD design. From equation (4), the exponential basis function neural network is given in figure 1.

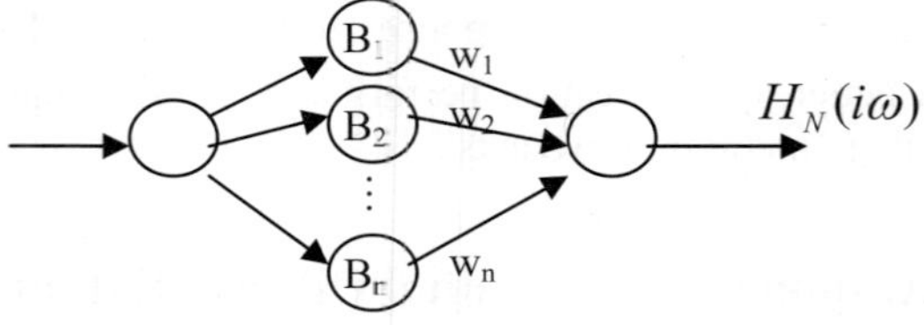

Fig. 1. Exponential basis function neural network

The exponential basis function neural network is a feed-forward network with three layers neurons.

The input layer only has one neuron, as well as the output layer. The hidden layer has $n = N$ neurons. The activation function of input layer and output layer is linear function. Exponential basis functions $B_1 = 1, B_2 = e^{-i\omega}, \cdots, B_n = e^{-i(N-1)\omega}$ are used as the activation functions of hidden layer. The weight coefficients of input layer to hidden layer are ones and the output weight coefficient is one too. The connection weights between output layer and hidden layer are w_n $(n = 1, 2 \cdots N)$.

So the output of neural network is

$$H_N(i\omega) = \sum_{n=1}^{N} w_n \cdot B_n \tag{5}$$

Network output (5) is similar to the fractional order differentiator's frequency response (4). Design of $h(n)$ can be achieved by way of design of w_n. Then $H_N(i\omega)$ is used to approximate the ideal $H(\omega)$.

Note $W = [w_1, w_2, \cdots, w_N]^T$ and $B(\omega) = [B_1, B_2, \cdots, B_N]^T$

Equation (5) can be expressed as

$$H_N(i\omega) = W^T \cdot B(\omega) \tag{6}$$

For M training samples, define the error function as

$$E(l) = H(\omega) - H_N(i\omega) \qquad l = 0, 1, \cdots M - 1 \tag{7}$$

Define the Lyapunov energy function as [4]

$$J = \frac{1}{2}\sum_{l=0}^{M-1} E^2(l) \tag{8}$$

The modification of weight coefficients is [4]

$$w_{k+1} = w_k + \Delta w_k \qquad k = 1, 2, \cdots, N \tag{9}$$

Where $\Delta w_k = \eta E(l) S(\omega)$ (10)

and η is the learning speed with $0 < \eta < 1$.

3.2 Neural Network Stable Condition

Theorem: When the learning speed η is $0 < \eta < 2/N$, the exponential basis function neural network is stable and will eventually converge to its minimum.

Proof: The theorem is easy to be proved refer to literatures [4] [6] and using the following expression:

$$\left\| \frac{\partial E(l)}{\partial \omega} \right\|_2^2 = \sum_{n=0}^{N-1} \left| Z^{-n} \right|^2 = \sum_{n=0}^{N-1} \left| e^{-in\omega} \right|^2 = N$$

In order to insure the neural network is convergent, the above expression gains its maximum value N . Here the verification is omitted.

3.3 Neural Network Simulation Procedure [4]

Step 1: For N -point FIR filter, randomly generate the initial weight coefficients $W = [w_1, w_2, \cdots w_n]^T$. Sampling M -points uniformly over $\omega \in [0, \pi]$ to get the ideal train set $\{\omega_l, H(\omega)\}$. The parameter η is chosen between $(0, 2/N)$.

Step 2: Count the output of neural network

$H_N(i\omega) = W^T \cdot B(\omega)$

Step 3: Count the error energy function between ideal output and the factual.

$E(l) = H(\omega) - H_N(i\omega)$

$J = J + \frac{1}{2} E^2(l)$

Step 4: Step 4: Modify the weight coefficients W

$W = W + \eta E(l) B(\omega)$

Step 5: If all the training samples are used, judge whether the error function J is less than a small prescribed error ξ or not. If J is satisfied, the training procedure ends, otherwise, return to step 2 to iterate.

Step 6: Count the output of stable neural network.

4 Design Examples

Examples based on exponential basis neural network approximation are given in this section with performance compared with others' method.

4.1 FODD Design Example

For instance, Fractional-Order Digital Differentiators design with orders $v = 0.4$ is shown in figure 2.

Where the designed FODD filter is 40 orders $(N = 40)$ and the hidden layer neurons of neural network are $N = 40$, the sampling frequency points are $M = 200$ points. Let the learning speed be $\eta = 0.3/N = 0.015$.

From Figure 2, actual neural network output is well approximated to the ideal value. For order $v = 0.4$, the absolute value of error is 0.0451 after the network is

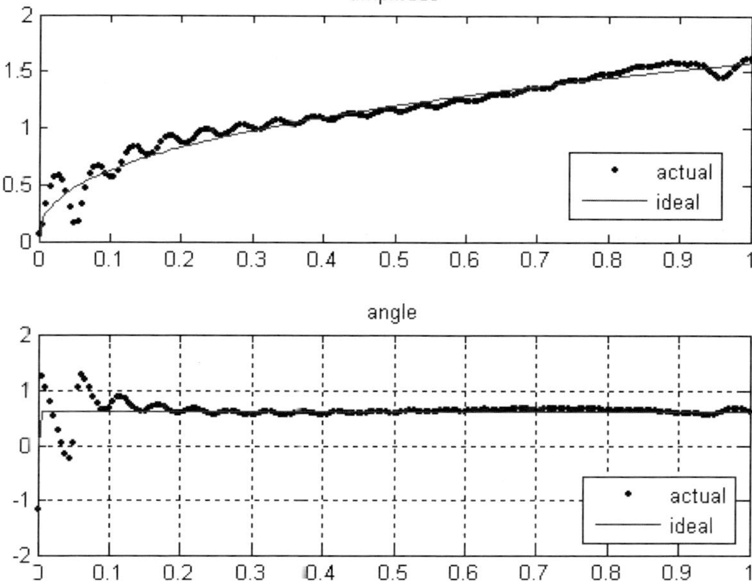

Fig. 2. The amplitude response and phase response of variant FODD with $v = 0.4$

convergent and the iteration time is 15. The phase response is nearly constant phase, which greatly satisfies the ideal FODD phase response.

A conclusion can be drawn that using exponential basis function neural network to design FODD gets good approximation performance, and the network convergence speed is fast.

4.2 Comparison with Others' Method

Here, we compare our method with method of literature [3].

We design exponential basis function neural network fractional order filter with order $N = 40$. Literature [3] method designs filter with order $N = 11$. The mean-square error (MSE) of these two methods between actual amplitude response and ideal value are listed in table 1, for different fractional orders v.

Table 1. Two methods' mean-square error between actual and ideal amplitude response

	Literature[3] method	This paper method
$v = 0.2$	3.8656	0.8434
$v = 0.5$	3.5556	0.4143
$v = 0.75$	0.1863	0.1366
$v = 1$	0.5469	0.0137
$v = 1.5$	1.5977	1.0581

Table 1 show that the exponential basis function neural network gains a better approximation performance than literature [3] method. Another merit is that FODD this paper realized has a constant value phase which is better than literature [3] with linear phase.

5 Conclusions and Discussions

Exponential basis function neural network optimization technique to design fractional order digital differentiator filter is investigated in this paper. The designed FIR filter is easy to implement. Choosing proper learning speed, the exponential basis function neural network iteration algorithm is convergent and its convergence speed is fast.

This approach has a good amplitude response approximation performance, and the filter's phase is almost constant value, which well satisfies the phase condition of FODD. Method of literature [3] doesn't have a good performance in the phase response approximation. Mean-square error comparison between this two papers are given, which shows this paper approach can reach better performance than literature [3]. Although the order of FIR filter this paper method is larger than that of literature [3], it is worthy in some sense.

References

1. Oldham, K.B. (ed.): The Fractional Calculus. Academic Press, NY and London. (1974).
2. Yuan, X., Zhang, H.Y., Yu, J.B.: Fractional-order Derivative and Design of Fractional Digital Differentiators. Acta Electronica Sinica, Vol.32. No.10. (2004) 1568-1665
3. Tseng, C.C.: Design of Fractional Order Digital FIR Differentiators. IEEE Signal Processing Letters 8 (2001) 77-79
4. Zeng, Z.Z., Tang, Z.: Study on the Optimization Design of the FIR with Linear Phase. Signal Processing 17 (2001) 296-301
5. Zeng, Z.Z., Liu, J.-H.: Study on the Optimization Design on the Differentiator. Systems Engineering and Electronics 14 (2002) 58-63
6. Zhou, H.: Optimal Design of the Type-three FIR Filter. Systems Engineering and Electronics. 26 (2004) 1892-1894

Multivariate Chaotic Time Series Prediction Based on Radial Basis Function Neural Network

Min Han, Wei Guo, and Mingming Fan

School of Electronic and Information Engineering,
Dalian University of Technology, Dalian, 116023, China
minhan@dlut.edu.cn

Abstract. In this paper, a new predictive algorithm for multivariate chaotic time series is proposed. Considering the correlations among time series, multivariate time series instead of univariate ones are taken as the inputs of predictive model. The model is implemented by a radial basis function neural network. To determine the number of model inputs, C-C method is applied to construct the embedding of the chaotic time series by choosing delay time window. The annual river runoff and annual sunspots are used in the simulation, and the proposed method is proven effective and valid.

1 Introduction

Chaotic phenomena frequently appear in economics, meteorology, chemical processes, biology, hydrodynamics and many other situations. Till now, most of the time series predictions are based on univariable time series and paid less attention to the correlation among different sources of information. In reality, there are always multiple variables in the complex systems. With multivariate time series, the large amount information and better exploitation are obtained and can allow a state space reconstruction, which is much less distorted than the univariate ones [1]. And adopting multivariables leads to improvements of prediction accuracy [2]. Thus, n this paper multivariate time series are predicted based on interrelations of different variables.

There are several techniques for modeling and forecasting nonlinear time series, such as the threshold model, exponential model, local linear model, nearest neighbor regression, feedforward neural network [3], recurrent neural network [4], and so on. Among these techniques, artificial neural network are the most recent trend. The advantage of the network models over the other methods of time series forecasting is that feedforward neural models use only linear parameters, whereas traditional polynomial, spline, and trigonometric extension use exponential parameters to achieve the same approximation rate [5]. Radial basis function neural network (RBFNN) is widely used for prediction and controlling because of its good generalization ability and a simple network that avoids necessary and length calculation. Therefore, the RBFNN is trained for the predictive model of time series.

The method which is presented in the present paper is investigated for multivariate chaotic time series prediction. For multivariate chaotic time series, the reconstruction of the equivalent attractor's state space by embedding the time series in a vector space is the first step of the analysis. To construct an embedding of a nonlinear time series,

J. Wang et al. (Eds.): ISNN 2006, LNCS 3972, pp. 741 – 746, 2006.

an appropriate delay time has to been determined. Delay time is usually estimated by the autocorrelation function. However, this does not treat the nonlinearity appropriately. And some researchers have suggested that the delay time should not be chosen independent of embedding dimension, and the appropriate value for delay time window should be chosen, which is the total time spanned by the components of each embedded point [6]. C-C method [7] is a technique for choosing either the delay time or the delay time window using the correlation integral. Therefore, the C-C method is applied to determine the input vector dimension of the RBFNN. In the simulation, time series of annual Yellow River runoff and annual sunspots is used to validate the proposed predictive algorithm. The results show that the prediction with multivariate time series is encouraging.

2 Multivariate Time Series Predictive Model

In this paper, a multivariate time series predictive model is proposed. The model is based on the RBFNN. The input of the model is determined based on C-C method by reconstructing the phase space of multivariable time series. And then the predictive model based on RBFNN is trained.

2.1 State Space Reconstruction Based on C-C Method

State Space reconstruction is the first step in nonlinear time series analysis of data from chaotic systems. The most common state space reconstruction method in the analysis of chaotic time series is the method of delay (MOD).

The algorithm is based on multivariate chaotic time series set. Assuming there are n chaotic time series concerned in this paper, they are $\{x_i(1), x_i(2), \cdots, x_i(N)\}$, $(i=1, 2, \cdots, n)$, where N is the length of the data set. According to MOD, the reconstructed vector of multivariate time series in the phase space could be denoted as Eq.(1).

$$\mathbf{x}_i(k) = [x_i(k), x_i(k+\tau_i), \cdots, x_i(k+(m_i-1)\tau_i)]^T, \quad i=1,2,\cdots,n;\ k=1,2,\cdots,M. \quad (1)$$

In Eq.(1), m_i and τ_i $(i=1, 2, \cdots, n)$ are the selected embedding dimension and the delay time of the ith chaotic time series, respectively. So there are M embedded points generated in the phase space, $M = N - \max_i[(m_i-1)\tau_i]$, where $(m_i\text{-}1)\tau_i$ is the delay time window of the ith time series, k=1,2,$\cdots$,M. And, the whole multivariate time series could be denoted as Eq.(2).

$$\mathbf{X}(k) = [\mathbf{x}_1^T(k), \mathbf{x}_2^T(k), \cdots, \mathbf{x}_n^T(k)]^T, \quad k=1,2,\cdots,M. \quad (2)$$

The C-C method could be used to estimate both the delay time window and delay time. Basically the delay time window is the optimal time for independence of the data, while delay time is the first locally optimal time. Compared with the use of the mutual information, the C-C method is easier to implement, useful for smaller data sets, and less demanding computationally. The details of the C-C method are described as below.

The correlation integral for the embedded time series is as Eq.(3).

$$C(m_i, N, r_i, t_i) = \frac{2}{M(M-1)} \sum_{1 \le p < q \le M} \theta(r_i - \|\mathbf{x}_i(p) - \mathbf{x}_i(q)\|), \quad r_i > 0;\ i = 1, 2, \cdots, n; \quad (3)$$

where, $\theta(a)$=0, if a<0, $\theta(a)$=1, if $a \ge 0$; r_i is the radius of the ith time series; and $\|\cdot\|$ denotes the sup-norm. $C(\cdot)$ measures the fraction of the pairs of points $\mathbf{x}_i(p)$ and $\mathbf{x}_i(q)$, i=1,2, ···, n, $1 \le p < q \le M$.

Subdivide time series $\{x_i(1), x_i(2), \cdots, x_i(N)\}$(i=1, 2, ···, n) into t_i disjoint time series as Eq. (4). The length of each disjoint time series is N/t_i.

$$\{x_i(l), x_i(t_i + l), x_i(2t_i + l), \cdots\}, \quad l = 1, \cdots, t_i; \quad i = 1, 2, \cdots, n. \quad (4)$$

In order to determine the nonlinear dependence and eliminate spurious temporal correlations, the statistic $S(\cdot)$ is defined based on correlation integral and computed from these disjoint time series as follows:

$$S(m_i, N, r_i, t_i) = \frac{1}{t_i} \sum_{l=1}^{t_i} [C_l(m_i, N/t_i, r_i, t_i) - (C_l(1, N/t_i, r_i, t_i))^{m_i}], \quad i = 1, 2, \cdots, n; \quad (5)$$

where C_l is the correlation integral for the lth disjoint time series.

For fixed m_i and τ_i, $S(\cdot)$ will be identically equal to 0 for all r_i, if the data are indentically distributed and $N \to \infty$. However, real data sets are finite, and the data may be serially correlated; in general, $S(\cdot) \ne 0$. Thus, the locally optimal times may be either the zero crossing of $S(\cdot)$ or the times at which $S(\cdot)$ shows the least variation with r_i. The measure of the variation is defined as:

$$\Delta S(m_i, N, t_i) = \max\{S(m_i, N, r_{i,j}, t_i)\} - \min\{S(m_i, N, r_{i,j}, t_i)\}, \ i = 1, \cdots, n;\ j = 1, 2, \cdots. \quad (6)$$

The studies led to the conclusions that m_i should be between 2 and 5, r_{ij} should be between $\sigma/2$ and 2σ. σ is the mean-square deviation of the time series. Then we define

$$\begin{cases} \bar{S}(t_i) = \frac{1}{16} \sum_{m_i=2}^{5} \sum_{j=1}^{4} S(m_i, N, r_{i,j}, t_i) \\ \Delta\bar{S}(t_i) = \frac{1}{4} \sum_{m_i=2}^{5} \Delta S(m_i, N, t_i) \\ S_{cor}(t_i) = \Delta\bar{S}(t_i) + |\bar{S}(t_i)| \end{cases} . \quad (7)$$

The optimal time is t_i for which $\bar{S}(t_i)$ and $\Delta\bar{S}(t_i)$ are both close to 0. These two quantities are equal important, so $S_{cor}(t_i)$ is defined as the sum of them. The first local minimum of $\Delta\bar{S}(t_i)$ is the delay time τ_i, which is the first locally optimal time for independence of the data. The minimum of $S_{cor}(t_i)$ is the delay time window.

2.2 Radial Basis Function Neural Network (RBFNN)

Radial basis function (RBF) is radial symmetric, which has only several connections that can affect output of network in some local range of input space. It has been proven that when enough units are provided, RBFNN can approximate any multivariate continuous function as much as desired. Thus, RBFNN has faster rates of convergence than other feedforward neural networks. It has been used in function approximation, pattern recognition and signal processing, etc. In this paper, RBFNN is applied to approximate a function representing multivariate time series.

RBFNN is a three-layer feedforward neural network. In the structure of RBFNN, input data $\mathbf{X}(k)$ (k=1,2, ⋯, M) is applied. Assumed that H neurons in the hidden layer and O neurons in the output layer. For a RBFNN, the activation function has different forms. In this paper, Gaussian function is chosen. The response of the jth hidden neurons for the kth input data has the following form:

$$\phi_j = \sum_{j=1}^{H} \exp\left(-\frac{\left\|\mathbf{X}(k)-\mathbf{C}_j\right\|^2}{2b_j^2}\right), \quad j=1,2,\cdots,H; \tag{8}$$

where $\mathbf{C}_j$ the jth basis function centers of neurons of hidden layer, b_j is called the scaling parameter or the width of activation function of the jth hidden neurons, $\|\cdot\|$ denotes the Euclidean norm.

Consequently, the output of RBF network is simply a weighted linear summation of the activation functions. The ith output of the network y_i can be described as:

$$y_i = \sum_{j=1}^{H} w_{ij}\phi_j, \quad i=1,2,\cdots,O;\ j=1,2,\cdots,H; \tag{9}$$

where w_{ij} is the weight from the jth hidden node to the ith output nodes. Thus, the error e_i between the network output y_i and the desired output d_i can be computed as the Eq.(10).

$$e_i = d_i - y_i, \quad i=1,2,\cdots,O. \tag{10}$$

And the cost function E is defined in terms of the error e_i.

$$E = \frac{1}{2}\sum_{i=1}^{O}[d_i - y_i]^2 = \frac{1}{2}\sum_{i=1}^{O} e_i^2, \quad i=1,2,\cdots,O. \tag{11}$$

The learning algorithm for RBFNN can be done in two stages:

1. Adjusting the parameters of hidden layer, including the centers $\mathbf{C}_j$ (j=1,2,⋯,H) and the scaling parameters b_j, respectively.
2. After confirming the centre of the neuron, calculation of the output weights $\{w_{ij}\}$ (i=1,2,⋯,O; j=1,2,⋯,H) of the network to minimize the cost values.

3 Simulation

To test the performance, the prediction of the annual sunspots and runoff of Yellow River in Northern China has been considered as an application of the previously described method. The Yellow River is the second longest river in China. Tracing to a source high up the majestic Yagradagze mountain in the nation's far west, it loops north, bends south, and flows east for 5,464 km until it empties into the sea, draining a basin of 745,000 km^2. Commonly, the river runoff is influenced by many factors, such as climate, sun activity, human activity, and so on. Among these variables, the sun affects the climate and consequently the runoff and the number of sunspots is a coefficient of sun activity. Thus, sun activity is a very important factor for the variation of runoff. And the activity of sunspots and the river runoff have some underlying correlations. In this paper, the time series of annual river runoff from year 1700 to 2003 is obtained from the Sanmenxia hydrologic station which located in Henan province of China. It should be noted that the noise of the runoff time series has been reduced before the prediction. In this simulation, the number of the sunspots as well as the historical properties of the Yellow River is applied to predict the present runoff of the Yellow River and the number of the sunspots, respectively.

In this work, the neural network is a multi-input single-output model. Hence, for one of multivariate time series, the predictive model will be provided more information, especially for the runoff time series. As the method mentioned above, N=304. Based on the C-C method, the parameters are chosen as follows: m_1=6, m_2=6, τ_1=1, τ_2=1, and M=299. That is to say, there are 299 state points in the reconstructed state space. Out of these 299 points, the first 289 points, which represent about 96% of the series, are selected as training set, whereas the remaining 10 points, accounting for about 4% of the series, are used for testing the forecasting performance. Fig.1 and Fig.2 show the predicted and observed values of the annual Yellow River runoff and the annual sunspots number.

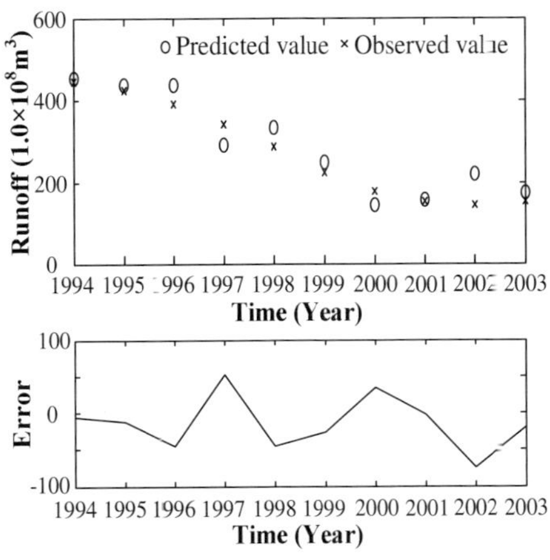

Fig. 1. Predicted and observed values of Yellow River runoff and predictive error

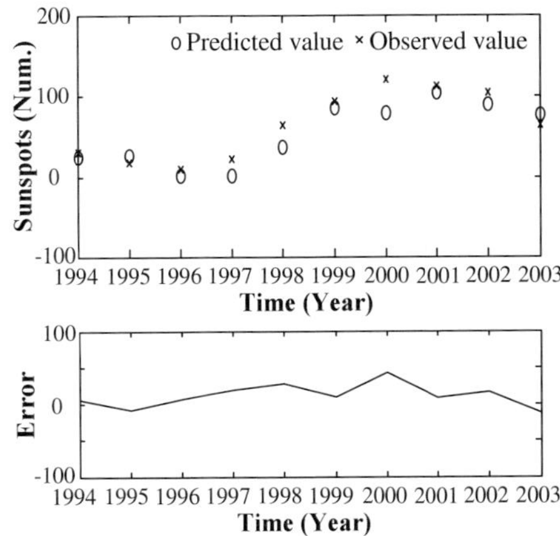

Fig. 2. Predicted and observed values of Sunspots and predictive error

To show the predict performance of the RBFNN for the two chaotic time series, the root mean square errors of predictive outputs are listed in Table 1. From the results, it can be seen that the predictive method proposed is an effective and available algorithm to predict nonlinear multivariate time series.

Table 1. RMSE of RBFNN Prediction

Time Series	RMSE
Yellow River Runoff	38.8351
Sunspots	19.1104

4 Conclusions

For multivariate time series, it is not an easy task to predict, whereas, there is the underlying correlation among these variables. In this paper, a new method based on RBFNN is proposed to predict the multivariate chaotic time series. The RBFNN is applied to model the complex system and C-C method is used to determine the number of the model input. The nonlinear multivariate time series river runoff and sunspots are predicted with the proposed method. And the simulation results show the validity of the method.

Acknowledgements

This research is supported by the project (60374064) of the National Natural Science Foundation of China. It is also supported by the project (50139020) of the National Natural Science Foundation of China. All of these supports are appreciated.

References

1. Porporato, A., Ridolfi, L.: Multivariate Nonlinear Prediction of River Flows. Journal of Hydrology, 248(1-4) (2001) 109-122
2. Suzuki, T., Ikeguchi, T., Suzuki, M.: Multivariable Nonlinear Analysis of Foreign Exchange Rates. Physica A-Statistical Mechanics and Its Applications, 323(5) (2003) 591-600
3. Tronci, S., Giona, M., Baratti R.: Reconstruction of Chaotic Time Series by Neural Models: A Case Study. Neurocomputing, 55(3-4) (2003) 581-591
4. Han, M., Xi, J., Xu, S., Yin F. L.: Prediction of Chaotic Time Series Based on the Recurrent Predictor Neural Network. IEEE Transaction on Signal Processing, 52(12) (2004) 3409-3416
5. Barron, A.: Approximation and Estimation Bounds for Artificial Neural Network, Technical Report 59, Department of Statistics, University of Illinois at Urbana-Champaign (1991)
6. Rosenstein, M.T., Collins, C.J., Luca, De: Reconstruction Expansion as a Geometry-Based Framework for Choosing Proper Delay Time. Physica D: Nonlinear Phenomena, 73(1-4) (1994) 82-98
7. Kim, H. S., Eykholt, R., Salas, J. D.: Nonlinear Dynamics, Delay Times, and Embedding Windows. Physica D: Nonlinear Phenomena, 127(1-2) (1999) 48-60

Time Series Prediction Using LS-SVM with Particle Swarm Optimization

Xiaodong Wang[1], Haoran Zhang[1], Changjiang Zhang[1], Xiushan Cai[1], Jinshan Wang[1], and Meiying Ye[2]

[1] College of Information Science and Engineering, Zhejiang Normal University, Jinhua, 321004, P.R. China
{Wxd, Hylt, Zcj74922, Xiushan, Dz66}@zjnu.cn
[2] College of Mathematics and Physics, Zhejiang Normal University, Jinhua, 321004, P.R. China
Ymy@zjnu.cn

Abstract. Time series analysis is an important and complex problem in machine learning. In this paper, least squares support vector machine (LS-SVM) combined with particle swarm optimization (PSO) is used to time series prediction. The LS-SVM can overcome some shortcoming in the multilayer perceptron (MLP) and the PSO is used to tune the LS-SVM parameters automatically. A benchmark problem, Hénon map time series, has been used as an example for demonstration. It is showed this approach can escape from the blindness of man-made choice of the LS-SVM parameters. It enhances the efficiency and the capability of prediction.

1 Introduction

Successful time series prediction is a major goal in many areas of research, e.g., biology, meteorology, physics, business and engineering. However, most time series of practical relevance are of nonlinear and chaotic nature that makes conventional linear prediction methods inapplicable. Hence, a number of nonlinear prediction methods have been developed including artificial neural networks, though, not initially proposed for time series prediction, exceed conventional methods by orders of magnitude in accuracy. One of the most common artificial neural networks in the area of time series prediction is the multilayer perceptron (MLP) with error back-propagation learning algorithm. The MLP employs gradient descent method to provide a suitable solution for network weights by minimizing the sum of squared errors. Training the artificial neural networks is usually done by iterative updating of the weights according to the error signal. Although the artificial neural network is developed in time series prediction, some inherent drawbacks, e.g., the multiple local minima problem, the choice of the number of hidden units and the danger of over fitting, etc., would make it difficult to put the MLP into some practice. In order to overcome the shortcoming in the MLP, the present study focuses on the problem of time series prediction using least squares support vector machine (LS-SVM) [1] [2] regression, whose parameters are automatically tuned using particle swarm optimization (PSO) [3] [4].

J. Wang et al. (Eds.): ISNN 2006, LNCS 3972 pp. 747 – 752, 2006.

2 Problem Description

The object of time series prediction is building an estimate function for the system's transfer function only using its output. Many conventional regression techniques can be used to solve problems of estimating function. In this investigation, we concentrate on the LS-SVM regression. Supposing that a time series $\{x_t\}$, $t = 1,\cdots,N$, is given, using the delays method, we represent the data in d -dimensional space by vectors

$$X_t = [x_t, x_{t+1}, \cdots, x_{t+(d-1)}], \tag{1}$$

where d is the embedding dimension.

Time series prediction is to deduce the future values of a time series according to its past values. We can use the map f to make the prediction. The prediction can be described as

$$X_{t+T} = f(X_t), \tag{2}$$

$$f : (x_t, x_{t+1}, \cdots, x_{t+(d-1)}) \rightarrow (x_{t+T}, x_{t+1+T}, \cdots, x_{t+(d-1)+T}), \tag{3}$$

where

$$x_{t+(d-1)+T} = g(x_t, x_{t+1}, \cdots, x_{t+(d-1)}), \tag{4}$$

g is an unknown function, and T is the prediction step. Here $T = 1$ means one-step-ahead prediction, and $T > 1$ means multi-step prediction. In this work, we try applying LS-SVM and PSO to estimate the unknown function g .

3 LS-SVM and Its Parameter Selection by PSO

In the following, we briefly introduce LS-SVM regression and PSO. For further details on them we refer to Refs. [1]-[4].

3.1 LS-SVM Regression for Time Series Prediction

Consider a given training set of N data points $\{x_t, y_t\}_{t=1}^{N}$ with input data $x_t \in R^d$ and output $y_t \in R$. In feature space LS-SVM regression models take the form:

$$y(x) = w^T \varphi(x) + b \tag{5}$$

where the nonlinear mapping $\varphi(\cdot)$ maps the input data into a higher dimensional feature space.

The solution of LS-SVM for function estimation is given by the following set of linear equations

$$\begin{bmatrix} 0 & \vec{1}^T \\ \vec{1} & \varphi(x_t)^T \varphi(x_l) + \gamma^{-1} I \end{bmatrix} \begin{bmatrix} b \\ \alpha \end{bmatrix} = \begin{bmatrix} 0 \\ y \end{bmatrix} \tag{6}$$

where $y = [y_1;\cdots;y_N]$, $\vec{1} = [1;\cdots;1]$, $\alpha = [\alpha_1;\cdots;\alpha_N]$, and the Mercer's condition

$$K(x_t, x_l) = \varphi(x_t)^T \varphi(x_l)\ ,\ t,l = 1,\cdots,N \tag{7}$$

has been applied. This finally results into the following LS-SVM model for function estimation

$$y(x) = \sum_{=1}^{N} \alpha_t K(x, x_t) + b \tag{8}$$

where α_t, b are the solution to the linear system, $K(\cdot,\cdot)$ represents the high dimensional feature spaces that is nonlinearly mapped from the input space x. The LS-SVM approximates the function using the Eq.(8).

In this work, the radial basis function (RBF) function is used as the kernel function

$$K(x_t, x_l) = \exp(-\|x - x_t\|^2 / \sigma^2) \tag{9}$$

Note that in the case of RBF kernels, one has only two additional tuning parameters σ in Eq.(9) and γ in Eq.(6).

From the training LS-SVM problem, one can see that there are two free parameters, viz. kernel width parameter sigma σ and regularization parameter γ, which may affect LS-SVM generalization performance. So these parameters need to be properly tuned/optimized to minimize the generalization error. In this paper, these parameters are automatically tuned using the PSO in the training phase.

3.2 PSO for Parameter Tuning of LS-SVM

Due to the simple concept, easy implementation and quick convergence, nowadays PSO has gained much attention and wide applications in different fields [5]. Also, it is promising to solve the parameter-tuning problem of LS-SVM by adopting PSO.

The PSO is initialized with a group of random particles (solutions) and then searches for optima by updating generations. Particles profit from the discoveries and previous experience of other particles during the exploration and search for higher objective function values. Let i indicate a particle's index in the swarm. Each of m particles fly through the d -dimensional search space R^d with a velocity v_i, which is dynamically adjusted according to its own previous best solution s_i and the previous best solution $\hat{s}$ of the entire swarm. The velocity updates are calculated as a linear combination of position and velocity vectors. The particles interact and move according to the following equations

$$v_i(j+1) = wv_i(j) + c_1 r_1(j)(s_i(j) - p_i(j)) + c_2 r_2(j)(\hat{s}_i(j) - p_i(j)) \tag{10}$$

$$p_i(j+1) = v_i(j+1) + p_i(j) \tag{11}$$

where $r_1(j)$ and $r_2(j)$ are random numbers between zero and one. $c_1(j)$ and $c_2(j)$ are learning factors, usually about $c_1 = c_2 = 2$. And w is an inertia weight, which is commonly taken as a decreasing linear function in index j from 0.9 to 0.6. It is possible to clamp the velocity vectors by specifying upper and lower bounds on v_i, to avoid too rapid movement of particles in the search space. Then we can use the standard procedure to find the optimum. The searching is a repeat process, and the stop criteria are that the maximum iteration number is reached or the minimum error condition is satisfied. The standard procedure is described as below:

(1) Set the iteration number j to zero. Initialize randomly the swarm (containing N particles) such that the position $p_i(0)$ of each particle to meet the prescribed conditions.
(2) Evaluate the fitness of each particle $F(p_i(j))$.
(3) Compare the personal best of each particle to its current fitness, and set $s_i(j)$ to the better performance, i.e.

$$s_i(j) = \begin{cases} s_i(j-1) & \text{if } f(s_i(j)) \le f(s_i(j-1) \\ s_i(j) & \text{if } f(s_i(j)) > f(s_i(j-1) \end{cases} \tag{12}$$

(4) Set the global best $\hat{s}(j)$ to the position of the particle with the best fitness within the swarm, i.e.

$$\hat{s}(j) \in \{s_1(j), s_2(j), \cdots, s_m(j)\} \big| F(\hat{s}(j)) = \max\{F(s_1(j)), F(s_2(j)), \cdots, F(s_m(j))\} \tag{13}$$

(5) Change the velocity vector for each particle according to Eq.(10).
(6) Move each particle to its new position, according to Eq.(11).
(7) Let $j = j+1$.
(8) Go to step (2), and repeat until meets the stop criteria.

It can be easily seen that there are two key steps when applying PSO to optimization problems: the representation of the solution and the fitness function. One of the desirable merits of PSO is that PSO takes real numbers as particles. It is not like genetic algorithms, where transformation of binary encoding and special genetic operators are needed. To build the relation between PSO and parameter optimization of LS-SVM in the time series prediction, PSO produces sets of particles, which represent all of the LS-SVM parameters (σ, γ). The fitness function is defined as:

$$fitness = \frac{1}{\text{RMSE}(\sigma, \gamma)} \tag{14}$$

where $\text{RMSE}(\sigma, \gamma)$ is the root-mean-square error of predicted results, which varies with the LS-SVM parameters (σ, γ). When the termination criterion is met, the individual with the biggest fitness corresponds to the optimal parameters of the LS-SVM.

4 Experiments

To illustrate the effectiveness of using the LS-SVM combined with PSO for time series prediction, the two-dimensional Hénon map is used as evaluation of the prediction power. As it is well known, the Hénon map is described by the following equations:

$$\begin{aligned} x_{t+1} &= 1 - Ax_t^2 + y_t \\ y_{t+1} &= Bx_t \end{aligned} \tag{15}$$

where A and B are the two bifurcation parameters fixed at 1.4 and 0.3, respectively. Here take its x component. We consider initial condition $x_0 = 0.01$, $y_0 = 0.04$. Following the conventional settings for predicting the Hénon map time series, we set the embedding dimension $d = 4$. The length of the series is $N = 900$. All the generated values are partitioned into three parts. The first 100 points are removed to eliminate the effect of the initial value, the second 400 points are used for training to build the model, and the last 400 points are used as the purpose of test.

In Fig.1 (a), one-step-ahead prediction results of the Hénon map time series by the LS-SVM regression, whose parameters are automatically tuned using the PSO. The RMSE(σ, γ) is 0.003. We can see that the prediction is quite accurate.

Cross-validation is a popular technique for estimating generalization performance. In order to better evaluate the performance of the proposed approach, we also use the k -folds cross validation [6] to chosen the appropriate parameters (σ and γ). In this example, we use $k = 5$ for the number of folds. Fig.1 (b) shows the dependence of prediction error with prediction step when LS-SVM parameters are automatically tuned using the PSO and Cross-validation, respectively. With the increasing prediction step T , the prediction errors increase. From Fig.1 (b), we compared tuning LS-SVM parameters by PSO with by 5-fold cross-validation, the PSO obtains lower prediction errors than the 5-folds cross validation. The results may be attributable to the fact that PSO is more likely to converge to the global.

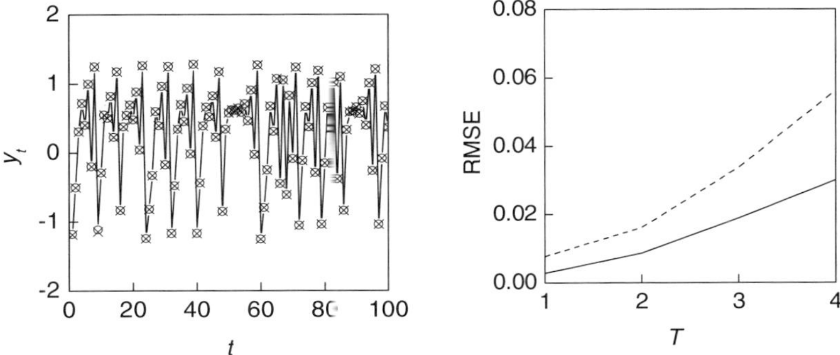

Fig. 1. (a) A piece of one-step-ahead prediction results of Hénon map time series by the proposed approach (prediction values '–', true values ' '); (b) Root-mean-square errors as a function of T. The solid line indicates the prediction errors using PSO and the dashed line indicates that using 5-folds cross validation.

5 Conclusions

In this paper, LS-SVM with PSO is used for time series prediction. The Hénon map time series has been used as examples for demonstration. The results demonstrate that the prediction method using the LS-SVM with PSO is suitable for the multi-step prediction. It is showed this approach can escape from the blindness of man-made choice of the LS-SVM parameters. It enhances the efficiency and the capability of prediction.

It should be pointed out that, although the processes are focused on the Hénon map time series, we believe that the proposed method can be used to many other complex time series.

Acknowledgements

The Project Supported by Zhejiang Provincial Natural Science Foundation of China (Y105281) and Key Construction Subject of Zhejiang Province.

References

1. Suykens, J.A.K., Vandewalle, J.: Least Squares Support Vector Machine Classifiers. Neural Processing Letters 9 (3) (1999) 293–300
2. Suykens, J.A.K., Gestel, T.V., Brabanter, J.D., Moor, B.D., Vandewalle, J.: Least Squares Support Vector Machines. World Scientific, Singapore (2002)
3. Kennedy, J., Eberhart, R.C.: Particle Swarm Optimization. In: Proc. IEEE Int. Conf. Neural Networks. Volume IV, IEEE Service Center (1995) 1942–1948
4. Kennedy, J., Eberhart, R.C., Shi, Y.: Swarm Intelligence. San Francisco: Morgan Kaufmann Publishers (2001)
5. Eberhart, R.C., Shi, Y.: Particle Swarm Optimization: Developments, Applications and Resources. In: Proc. Congress on Evolutionary Computation (2001) 81–86
6. Duan, K., Keerthi, S.S., Poo, A.N.: Evaluation of Simple Performance Measures for Tuning SVM Hyperparameters. Neurocomputing 51 (2003) 41–59

A Regularized Minimum Cross-Entropy Algorithm on Mixtures of Experts for Time Series Prediction

Zhiwu Lu

Institute of Computer Science & Technology of Peking University,
Beijing 100871, China
zhiwu.lu@yahoo.com.cn

Abstract. The well-known mixtures of experts(ME) model is usually trained by expectation maximization(EM) algorithm for maximum likelihood learning. However, we have to first determine the number of experts, which is often hardly known. Derived from regularization theory, a regularized minimum cross-entropy(RMCE) algorithm is proposed to train ME model, which can automatically make model selection. When time series is modeled by ME, it is demonstrated by some climate prediction experiments that RMCE algorithm outperforms EM algorithm. We also compare RMCE algorithm with other regression methods such as back-propagation(BP) algorithm and normalized radial basis function(NRBF) network, and find that RMCE algorithm still shows promising results.

1 Introduction

With an implementation of the so-called "divide-and-conquer" principle, the mixtures of experts(ME) model [1] consists of a set of expert networks and a gating network that cooperate with each other to solve a complex problem. Specifically, the expert networks are used to solve different input regions which are softly decomposed from the whole input space by a softmax based gating network. Then the outputs of the expert networks are combined by the softmax based gating network to obtain the solution of the problem. The motivation of the ME model is that individual expert networks can focus on specific regions and attack them well.

As for modeling time series, noise and non-stationarity are two key problems. The noisy characteristic refers to the unavailability of complete information from the past behavior of the time series to fully capture the dependency between the future and the past. The noise in the data may lead to the over-fitting problem. The non-stationarity implies that the time series switch their dynamics between different regions. In general, it is hard for a single model to capture such a dynamic input-output relationship inherent in the data. A potential solution to the above problems is to use the ME model [2]. The well-known expectation maximization(EM) algorithm [3] is carried out to train ME model for maximum likelihood learning, and performs well in nonlinear time series prediction.

J. Wang et al. (Eds.): ISNN 2006, LNCS 3972, pp. 753–758, 2006.

However, we have to first provide the number of experts, which has much effect on the predicting performance of ME model.

In light of both the minimum cross-entropy [4] principle and regularization theory [5], this paper aims to solve these problems, through implementing the regularized minimum cross-entropy(RMCE) principle on a typical architecture of the ME model via a gradient algorithm for time series prediction. When time series is modeled by ME, it is demonstrated by climate prediction experiments that RMCE algorithm outperforms EM algorithm. We also compare RMCE algorithm with other regression methods for time series prediction such as back-propagation(BP) algorithm [6] and normalized radial basis function(NRBF) network [7], and find that RMCE algorithm still shows promising results.

2 Regularized Minimum Cross-Entropy Algorithm

In order to give the regression of a mapping $x \to y : x \in R^n, y \in R$, we consider the following mixtures of experts [1] model:

$$q(y \mid x) = \sum_{l=1}^{k} q(y \mid x, l) P(l \mid x),\ q(y \mid x, l) = \frac{1}{\sqrt{2\pi}\tau_l} \exp(-\frac{(y - f_l(x, \theta_l))^2}{2\tau_l^2}), \quad (1)$$

$$P(l \mid x) = q(x \mid l)\alpha_l / \sum_{j=1}^{k} q(x \mid j)\alpha_j,\ q(x \mid l) = \frac{1}{(2\pi)^{\frac{n}{2}} \sigma_l^n} \exp(-\frac{\|x - m_l\|^2}{2\sigma_l^2}), \quad (2)$$

where $f_l(x, \theta_l)$ and $P(l \mid x)$ are the outputs of expert networks and the so-called softmax gating network, respectively. Moreover, the mixture proportions α_l are constrained by $\sum_{l=1}^{k} \alpha_l = 1,\ \alpha_l \geq 0$.

With the conditional density $q(y|x)$, we can get the following expected regression equation:

$$E(y \mid x) = \int y q(y \mid x) dy = \sum_{l=1}^{k} f_l(x, \theta_l) P(l \mid x). \quad (3)$$

That is, $E(y \mid x)$ is a sum of the outputs of experts weighted by the gate functions $P(l \mid x)$, respectively.

Given a data set of $\{x_t, y_t\}_{t=1}^{N}$, the joint density among x, y, and l can be estimated as $q(x, y, l) = q(y|x, l)q(x|l)\alpha_l$ for the above ME model. Since there are two types of Bayesian decomposition of a joint density, we can take into account another estimation $p(x, y, l) = p(l \mid x, y)p(y \mid x)p(x)$ of the joint density among x, y, and l. All the aspects of this estimation are further specified as [8]:

$$p(l \mid x, y) = \sum_{j=1}^{k} P(j \mid x, y)\delta(j - l),\ p(y|x) = \begin{cases} \delta(y - y_t), x = x_t \\ \text{not care, otherwise} \end{cases}, \quad (4)$$

$$p(x) = \frac{1}{N} \sum_{t=1}^{N} \delta(x - x_t),\ P(l \mid x, y) = \frac{q(y \mid x, l) q(x \mid l) \alpha_l}{\sum_{j=1}^{k} q(y \mid x, j) q(x \mid j) \alpha_j}, \quad (5)$$

where $\delta(x)$ is the δ-function.

Under general probability theory, the two density functions $q(x, y, l)$ and $p(x, y, l)$ should be the same. However, since we only give them certain estimation on the data set, there may be some difference between them. Our aim is to make their difference as small as possible during parameter learning. The problem of estimating all the parameters Θ of $p(x, y, l)$ and $q(x, y, l)$ can then be resolved through minimizing their cross-entropy regularized by the entropy of $p(x, y, l)$:

$$\min_{\Theta} H(p,\ q) = \min_{\Theta}(H_c(p,\ q) + \gamma H(p)), \tag{6}$$

where $H_c(p,\ q) = \int p\ ln(p/q)$ is the cross-entropy, $H(p) = -\int p\ ln\ p$ is the entropy of p, and γ is the regularization parameter.

We now substitute those components of $p(x, y, l)$ and $q(x, y, l)$ into $H_c(p,\ q)$ and $H(p)$, and have

$$H_c(p,\ q) = -\frac{1}{N}\sum_{t=1}^{N} \ln \sum_{l=1}^{k} (q(y_t \mid x_t, l) q(x_t \mid l)\alpha_l), \tag{7}$$

$$H(p) = -\frac{1}{N}\sum_{t=1}^{N}\sum_{l=1}^{k} P(l \mid x_t, y_t) \ln P(l \mid x_t, y_t). \tag{8}$$

Here, $H(p)$ is the regularization term which determines the model complexity, and the ME model can be made as simple as possible by minimizing $H(p)$ since the discrete entropy

$$H(P(l \mid x_t, y_t)) = -\sum_{l=1}^{k} P(l \mid x_t, y_t) \ln P(l \mid x_t, y_t) \tag{9}$$

is minimized when $P(l_0 \mid x_t, y_t) = 1, P(l \mid x_t, y_t) = 0 (l \neq l_0)$. Moreover, $H_c(p,\ q)$ is the empirical error [5] of learning on the data set $\{x_t, y_t\}_{t=1}^{N}$, and the maximum likelihood(ML) learning by minimizing $H_c(p,\ q)$ is only a special case of the RMCE learning with no regularized term.

In order to make the above minimum problem without constraint conditions, we can implement a substitution

$$\sigma_l = exp(d_l), \tau_l = exp(r_l), \alpha_l = exp(\beta_l)/\sum_{j=1}^{k} exp(\beta_j). \tag{10}$$

We also suppose each expert of the ME model has a linear output, that is, $f_l(x, \theta_l) = w_l^T x + b_l$. According to the derivatives of $H(p,\ q)$ with respect to the parameters w_l, b_l, r_l, m_l, d_l and β_l, respectively, we have the following gradient learning algorithm:

$$\Delta w_l = \frac{\eta}{N}\sum_{t=1}^{N} U(l \mid x_t, y_t) \frac{y_t - w_l^T x_t - b_l}{exp(2r_l)} x_t, \tag{11}$$

$$\Delta b_l = \frac{\eta}{N}\sum_{t=1}^{N} U(l \mid x_t, y_t) \frac{y_t - w_l^T x_t - b_l}{exp(2r_l)}, \tag{12}$$

$$\Delta r_l = \frac{\eta}{N}\sum_{t=1}^{N} U(l \mid x_t, y_t)\frac{(y_t - w_l^T x_t - b_l)^2 - exp(2r_l)}{exp(2r_l)}, \quad (13)$$

$$\Delta m_l = \frac{\eta}{N}\sum_{t=1}^{N} U(l \mid x_t, y_t)\frac{x_t - m_l}{exp(2d_l)}, \quad (14)$$

$$\Delta d_l = \frac{\eta}{N}\sum_{t=1}^{N} U(l \mid x_t, y_t)\frac{||x_t - m_l||^2 - n\ exp(2d_l)}{exp(2d_l)}, \quad (15)$$

$$\Delta \beta_l = \frac{\eta}{N}\sum_{t=1}^{N}\sum_{j=1}^{k} U(j \mid x_t, y_t)(\delta_{jl} - \alpha_l), \quad (16)$$

where

$$U(l \mid x_t, y_t) = P(l \mid x_t, y_t)(1 + \gamma\sum_{j=1}^{k}(\delta_{jl} - P(j \mid x_t, y_t))\ln(q(y_t \mid x_t, j)q(x_t \mid j)\alpha_j),$$

δ_{jl} is the Kronecker function, and η is the learning rate which is usually selected as a small positive constant.

The above gradient RMCE algorithm is designed for regression problem, and can be implemented directly to train ME model for time series prediction, which is just some kind of regression.

3 Results on Time Series Prediction

In this section, we apply the RMCE algorithm to time series prediction and then make a comparison with other prediction methods such as BP algorithm, NRBF network, and EM algorithm on the data set from the China Meteorological Administration, which consists of a total of 624 monthly averaged rain data recorded from January 1951 to December 2002 in Changchun, Beijing and Wuhan, respectively, denoted by $\{x_t\}_{t=1}^N$, $N = 624$. For each city, the data from January 1951 to December 1992 are used as the training set, and the remaining data are used as the test set. The performances of different prediction methods on the test set are measured by normalized mean square errors (NMSE).

The data set of each city can be rewritten as $\{x_{t(i,j)} : 1 \le i \le 52, 1 \le j \le 12\}$. The current rain data $x_{t(i,j)}$ is predicted by $y + m$ previous rain data as inputs, that is, $x_{t(i,j)} = f(x_{t(i,j)-1}, ..., x_{t(i,j)-m}, x_{t(i-1,j)}, ..., x_{t(i-y,j)})$, where $m \le 12$. If $y = 0$, the formation of input patterns is the same as most approaches in the literature. After certain experiments, our experience tells that it is the best to set $m = 1$ and $y = 3$. Hence, there are a total of 468 data patterns in the training set, and 120 data patterns in the test set.

Some parameters of the four prediction methods are set as follows. The gradient RMCE algorithm is implemented always with $k = 6$, η =0.1, and $\gamma \in [0, 1]$, while all the other parameters are initialized randomly within certain intervals. The EM algorithm has the same initialization as RMCE algorithm and the number of experts is also $k = 6$. The three-layer BP network is trained by the algorithm of Matlab 7.0, and there are six neurons in the hidden layer and

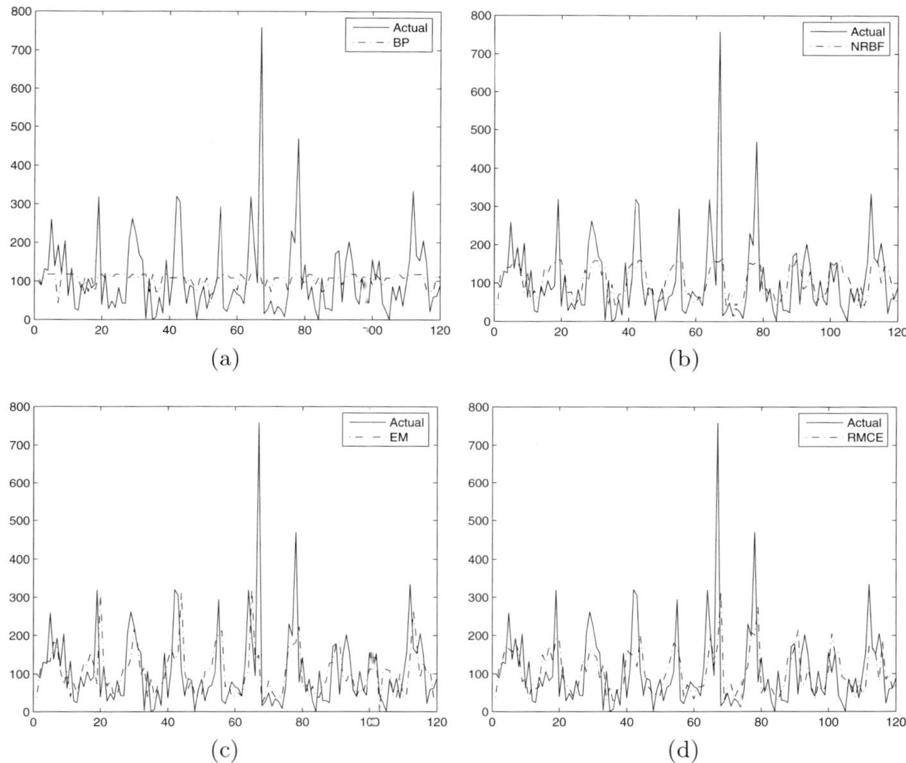

Fig. 1. The experimental results on Wuhan rain data by different time series prediction methods. (a) BP algorithm; (b) NRBF network; (c) EM algorithm; (d) RMCE algorithm.

Table 1. The test prediction errors (NMSE) on the rain data of three cities for different time series prediction methods

City	BP	NRBF	EM	RMCE
Changchun	0.293003	0.275631	0.303316	0.259814
Beijing	0.447599	0.440193	0.453667	0.373330
Wuhan	0.961792	0.816321	0.918462	0.863020

one neuron in the output layer. The NRBF network with six basis functions is trained by the conventional two-stage training algorithms [7].

Typically, we give the prediction results on Wuhan rain data by different prediction methods shown in Fig. 1. Moreover, the prediction errors for those prediction methods on the rain data of the three cities Changchun, Beijing, and Wuhan are listed in Table 1. The above experiments show that RMCE algorithm with automatic model selection outperforms both BP and EM algorithms. Though the prediction error on Wuhan rain data by RMCE algorithm is slightly more than the one by NRBF network, we can observe that RMCE algorithm tracks the change of rain data even better by comparing Fig. 1(b) with Fig. 1(d). Moreover, since we can get better results by RMCE algorithm on both Changchun and Beijing rain data, RMCE algorithm is at least thought to be more stable than NRBF network.

4 Conclusions

We have investigated time series prediction problems from the RMCE principle. When time series prediction is implemented based on ME model, RMCE algorithm outperforms EM algorithm in several climate prediction experiments due to automatic model selection. As compared with other methods such as BP algorithm and NRBF network, RMCE algorithm also shows promising results.

References

1. Lu, Z., Cheng, Q., Ma, J.: A Gradient BYY Harmony Learning Algorithm on Mixture of Experts for Curve Detection. Lecture Notes in Computer Science, Vol. **3578** (2005) 250–257
2. Weigend, A.S., Manageas, M., Srivastava, A.N.: Nonlinear Gated Experts for Time Series: Discovering Regimes and Avoiding Over-fitting. Int. J. Neural Systems **6** (1995) 373–399
3. Jordan, M.I., Xu, L.: Convergence Results for the EM Approach to Mixtures-of-Experts Architectures. Neural Networks **8** (1995) 1409–1431
4. Pal, N.R.: On Minimum Cross-Entropy Thresholding. Pattern Recognition **29** (1996) 575–580
5. Vapnik, V.N.: An Overview of Statistical Learning Theory. IEEE Trans. Neural Networks **10** (1999) 988–999
6. Lee, T.L.: Back-Propagation Neural Network for Long-Term Tidal Predictions. Ocean Engineering **31** (2004) 225–238
7. Moody, J., Darken, J.: Fast Learning in Networks of Locally-Tuned Processing Units. Neural Comput. **1** (1989) 281–294
8. Xu, L.: BYY Harmony Learning, Structural RPCL, and Topological Self-Organizing on Mixture Modes. Neural Networks **15** (2002) 1231–1237

Prediction for Chaotic Time Series Based on Discrete Volterra Neural Networks

Li-Sheng Yin, Xi-Yue Huang, Zu-Yuan Yang, and Chang-Cheng Xiang

College of Automation, University of Chongqing, Chongqing 400044, China
lishengyin@sohu.com

Abstract. In this paper, based on the Volterra expansion of nonlinear dynamical system functions and the deterministic and nonlinear characterization of chaotic time series, the discrete Volterra neural networks are proposed to make prediction of chaotic time series. The predictive model of chaotic time series is established with the discrete Volterra neural networks and the steps of the learning algorithm with discrete Volterra neural networks are expressed. The predictive model and the learning algorithm are more effective and reliable than the adaptive higher-order nonlinear FIR filter. The Experimental and simulating results show the discrete Volterra neural networks can be successfully used to predict chaotic time series.

Keywords: Chaotic time series; discrete Volterra neural networks; prediction.

1 Introduction

Prediction for chaotic time series is to approximate the unknown nonlinear functional mapping of a chaotic signal. The laws underlying the chaotic time series can be expressed as a deterministic dynamical system. Farmer and Sidorowich [1] suggest reconstructing the dynamics in phase space by choosing a suitable embedding dimension and time delay. Takens' theorem [2] ensures that the method is reliable.

In the last decade, particular interest has been put into predicting chaotic time series using Volterra series. Zhang Jiashu and Wu Weigen [3] proposed that the adaptive higher-order nonlinear FIR filter prediction of spatiotemporal chaotic time series. It usually cannot be realized in practical problem. So, advanced identification tools are badly needed for chaotic system.

In recent years, particular interest has been put into predicting chaotic time series using neural networks because of their universal approximation capabilities. Most applications in this field are based on feed-forward neural networks, such as the Back Propagation (BP) network [4], Radial Basis Function (RBF) network [5], Recurrent neural networks (RNNs) [6], and so on. It is widely used tool for the prediction of time series.

In this paper, the discrete Volterra neural networks are proposed to make prediction of chaotic time series. The predictive model of chaotic time series is established with the discrete Volterra neural networks and the steps of the learning algorithm with discrete Volterra neural networks are expressed. The Experimental and simulating results show the discrete Volterra neural networks can be successfully used to predict chaotic time series.

J. Wang et al. (Eds.): ISNN 2006, LNCS 3972, pp. 759–764, 2006.

2 The Equivalence of Three-layer Neural Networks and Volterra Series [7]

The $M+1$ input and the single output system corresponds to the three layer neural networks of Fig.1.

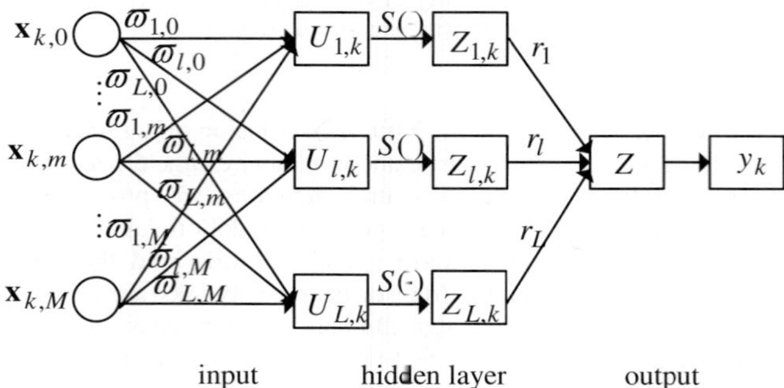

Fig. 1. The three layer neural networks inresponse to $M+1$ input and the single output system

The input vector $x_k^T=[x_{k,0},x_{k,1},\cdots x_{k,M}]$ at each time k is obtained by the time-delay input $x(k)$, where $x_{k,m}=x(k-m)$ for each index k .Thus, the output of the l th hidden unit ($l=1,2,\cdots,L$) for each k is $Z_{l,k}=S_l(u_{l,k})$, where $u_{l,k}=\sum_{m=0}^{M}\varpi_{l,m}x_{k,m}$ ..If a sigmoidal activation function is used, then another free parameter, θ_l is introduced as the characteristic "threshold" or "offset" of the j th unit. For instance, the "logistic" function $S_l(u_{l,k})=\dfrac{1}{1+\exp[-\lambda(u_{l,k}-\theta_l)]}$ is a commonly used sigmoidal activation function. Note that, in addition to the offset, the exponent contains another parameter, which is however fixed. For a linear output unit, there have $y_k=\sum_{l=1}^{L}r_lZ_{l,k}$ at each time k .Then, using the Taylor series expansion of each sigmoidal function about its offset value θ_l, the output is expressed as

$$y_k=\sum_{l=1}^{L}r_l\sum_{i=0}^{\infty}d_i(\theta_l)\cdot\sum_{m_1=0}^{M}\cdots\sum_{m_i=0}^{M}\varpi_{l,m_1}\cdots\varpi_{l,m_i}x_{k,m_1}\cdots x_{k,m_i}\,, \tag{1}$$

if the discrete Volterra series is expressed as

$$y(k)=\sum_{i}\sum_{m_1,\cdots,m_i}^{M}h_i(m_1,\cdots,m_i)x(k-m_1)\cdots x(k-m_i)\,, \tag{2}$$

if an expansion of the Volterra kernels $h_i(m_1,\cdots,m_i)$ is considered on a complete basis $\{b_l(m)\}$ of L basis functions defined over the system memory $[0, M]$, then

$$h_i(m_1,\cdots,m_i) = \sum_{l_1=1}^{L}\cdots\sum_{l_i=1}^{L} c_i(l_1,\cdots,l_i) b_{l_1}(m_1)\cdots b_{l_i}(m_i)\,, \tag{3}$$

where $y_k = y(k)$, $x_{k,m_i} = x(k-m_i)$ .The three-layer neural networks and Volterra series is equivalent by (1)、(2)、(3).

3 Establishment of Discrete Volterra Neural Networks for Chaotic Time Series Prediction and Chaotic Predictive Algorithm

3.1 Model of Chaotic Time Series Prediction

The chaotic time series prediction is based on the Takens' delay-coordinate phase reconstruct theory. If the time series of one of the variables is available, based on the fact that the interaction between the variables is such that every component contains information on the complex dynamics of the system, a smooth function can be found to model the portraits of time series. If the chaotic time series are $\{x(t), t = 1,2,\cdots,N\}$, then the reconstruct state vector is $x(t) = (x(t), x(t+\tau),\cdots, x(t+(m-1)\tau))$, where m ($m = 2,3,\cdots$) is called the embedding dimension($m = 2d+1$, d is called the freedom of dynamics of the system),and τ is the delay time. The predictive reconstruct of chaotic series is a inverse problem to the dynamics of the system essentially. There exists a smooth function defined on the reconstructed manifold in R^m to interpret the dynamics $x(t+T) = F(x(t))$,where T $(T>0)$ is forward predictive step length, and $F(\cdot)$ is the reconstructed predictive model.

3.2 Realized Architecture of Discrete Volterra Neural Networks and Chaotic Predictive Algorithm

There are many methods of constructing function to approximate the function F .The discrete-time Volterra models can be viewed as a multivariate power series expansion of a nonlinear function. Thus, the nonlinear predictive models F of chaotic time series can be constructed by the Volterra series expansion equation.

If the input of the discrete nonlinear dynamic system is $\vec{x}(n) = [x(n), x(n-1), \cdots, x(n-N+1)]^T$, and the output is $y(n) = \hat{x}(n+1)$, then the Volterra series of the nonlinear system function becomes:

$$\begin{aligned} x(n+1) = F(\vec{x}(n)) &= h_0 + \sum_{m=0}^{+\infty} h(m)x(n-m) + \sum_{m_1=0}^{+\infty}\sum_{m_2=0}^{+\infty} h_2(m_1,m_2)x(n-m_1)x(n-m_2) \\ &+\cdots\cdots \\ &+ \sum_{m_1=0}^{+\infty}\sum_{m_2=0}^{+\infty}\cdots\sum_{m_p=0}^{+\infty} h_p(m_1,m_2,\cdots,m_p)x(n-m_1)x(n-m_2)\cdots x(n-m_p) + \cdots \end{aligned}$$

,

where $h_p(m_1, m_2, \cdots, m_p)$ is the Volterra kernel function with p order. In the practical application, the infinite series are adopted the form of intercept summation with M order.

The steps of the chaotic time series prediction of the nonlinear Volterra neural networks are showed:

Step1) Based on the Takens' delay-coordinate phase reconstruct theory, the intercept order M of the discrete Volterra neural networks is determined.

The dimension m of chaotic time series is calculated by the way of G- P algorithms, and the delay time τ is calculated by the self-correlation method. For the overall description of the dynamics characteristic of the original system by the Takens' delay-coordinate phase reconstruct theory, a chaotic series demand $m \geq 2d + 1$ variances at least, so the intercept order of the discrete Volterra neural networks is $M = m$;

Step2) The discrete Volterra neural networks is structured in Fig.2, where $(x(n), x(n+\tau), \cdots, x(n+(m-1)\tau))$ is the input, $\hat{x}(n)$ is the output, $\{g_i\}$ $(i = 1,2, \cdots, N)$ are the polynomial activation functions, and $\varpi = (\varpi_{ij})_{N \times m}$ is the weight coefficient matrices of the hidden layer;

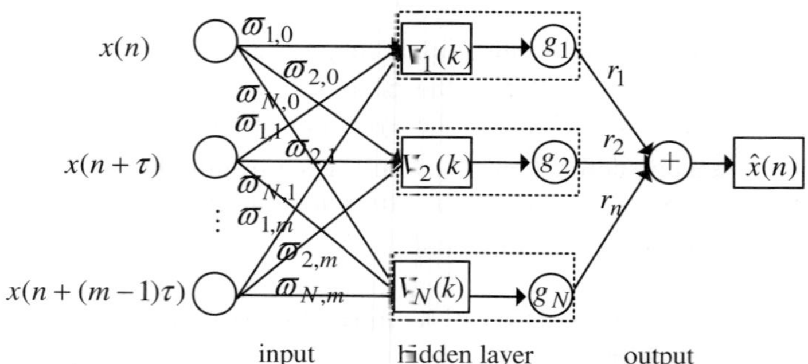

Fig. 2. The discrete Volterra Neural Networks for chaotic time series prediction

Step3) The neural networks are initialized and the weights are vested the initial values;

Step4) The neural networks are simulated and the output value is the predictive value at the time;

Step5) The error is calculated. If the error is in the scope of the permission, $\varpi = (\varpi_{ij})_{N \times m}$ is output and it turns into step 7), otherwise it continues;

Step6) The weights in the hidden layer are amended and it turns into step 4);

Step7) Using the Taylor series expansion of each function $\{g_i\}$ $(i = 1,2, \cdots, N)$ about its offset value θ_l, and the kernel functions are expressed.

4 Example Analysis and Conclusions

Considering Lorenz chaotic system

$$\text{Lorenz map:} \quad \begin{cases} \dot{x} = \sigma(y - x) \\ \dot{y} = rx - y - xz \\ \dot{z} = -bz + xy \end{cases}$$

Where $\sigma = 10$, $r = 28$, $b = 8/3$. The initial value is $x(0) = 0$, $y(0) = 5$, $z(0) = -5$; and the fixing step length of initial value is $0.05s$. Time series to the branch x with $60s$ is produced by the Runge-Kutta algorithms and the total data is 1200. The embedded dimension of the sampling chaotic time series m is 8 by the G- P algorithms. The delay time is τ=1 by the self-correlation function algorithms and the input dimension of the FIR neural networks adaptive predictive filters is 8.The former 1000 data is trained and other 200 data is predicted by the discrete nonlinear Volterra neural networks . That the experimental outcome of Lorenz chaotic sampling time series, the true value (real line) and the predictive value (star line) and the predictive error curve are showed in Fig.3.,Fig.4. and Fig.5.

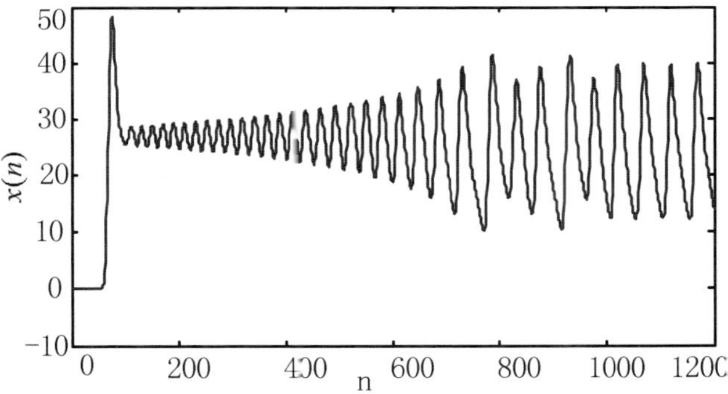

Fig. 3. Lorenz chaotic sampling time series

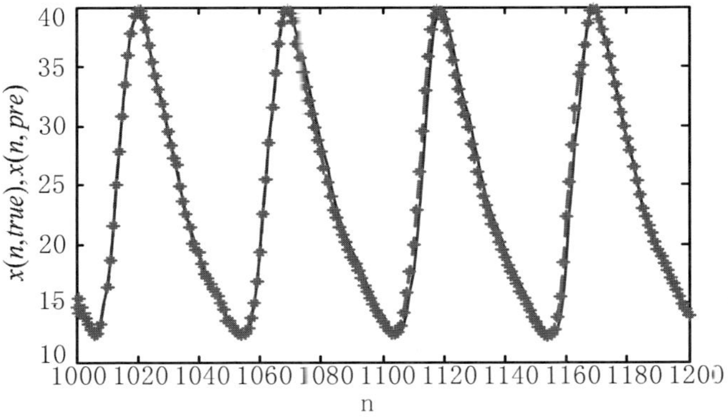

Fig. 4. True value (real line) and predictive value (star line)

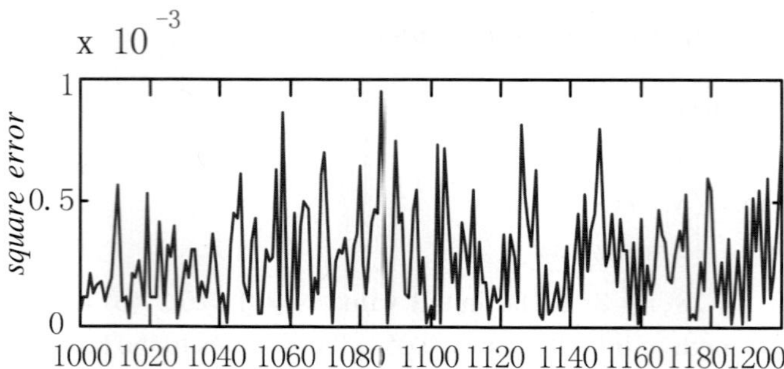

Fig. 5. predictive error curve

In Fig.3 the sampling chaotic time series number is 1200 by the Runge-Kutta algorithms.

The former 1000 datum are used to learn and train the discrete nonlinear Volterra neural networks every 8 datum. After the learned and trained stage, the true value (real line) and predictive value (star line) are shown in Fig.4.

The predictive error curve of the true value and the predictive value is very small in Fig.5.

The true value and the predictive value in the discrete nonlinear Volterra neural networks are to find a inner law in the series itself, which can avoid the disturbance of some subjective factors and enjoys higher reliability. In this study, the fusion of chaotic theory with the discrete nonlinear Volterra neural networks provide a new method for chaotic time series prediction. Simulation results for the modeling and prediction of chaotic time series show better predictive effectiveness and reliability.

References

1. Farmer, J.D., Sidorowich, J.J.: Predicting Chaotic Time Series. Phys.Rev. Lett. 59 (1987) 845–848
2. Takens, F.: On the Numerical Determination of the Dimension of an Attractor. Lecture Notes in Mathematics in Dynamical Systems and Turbulence, Vol. 898. Berlin Germany: Springer-Verlag (1981) 230–241
3. Jiashu, Z., Weigen, W., Xianci, X.: The Adaptive Higher-order Nonlinear FIR Filter Prediction of Spatiotemporal Chaotic Time Series. In Proceedings of the 3rd World Congress on Intelligent Control and Automation, Hefei China (2000)
4. Bakker, R., Schouten, J.C., Giles C. L.: Learning Chaotic Attractors by Neural Networks. Neural Computer 12 (2000) 2355–2383
5. Leung, H., Lo, T., Wang, S.: Prediction of Noisy Chaotic Time Series Using an Optimal Radial Basis Function Neural Network. IEEE Trans. Neural Networks 12 (2001) 1163–1172
6. Han, M., Jianhui, X., Shiguo, X., Yin, F.: Prediction of Time Series Based on the Recurrent Predictor Neural Network. IEEE Transactions on Signal Processing 52(12) (2004)
7. Vasilis, Z., Marmarelis, Xiao, Z.: Volterra Models and Three-layer Perceptrons. IEEE Transactions on Neural Networks 8(6) (1997)

A New Pre-processing Method for Regression

Wen-Feng Jing, De-Yu Meng, Ming-Wei Dai, and Zongben Xu

Institute for Information and System Science, Xi'an Jiaotong University, Xi'an, 710049, China
zbxu@mail.xjtu.edu.cn

Abstract. A new pre-processing method for regression is developed. The core idea is using three rules to clarify regression raw data. The rules are realized through introducing a judge function on the regression datum whose value determines the importance of the datum. By applying the rules, a new pre-processing method for regression is developed. Performance of the new method on a series of simulations demonstrate that it not only significantly increases computational efficiency and robustness, but also preserves generalization capability of a regression method. Incorporated with any regression method, the developed method then can be efficiently applied to regression of large data sets.

1 Introduction

With the rapid development of information acquisition technology, we can easily obtain a huge number of raw data in various areas. How to discover knowledge from these data naturally becomes a fundamental problem. Data mining (DM) is the main procedure of knowledge discovery from database, aiming at discovering underlying rules under large collections of raw data. The regression is one of the fundamental problems in DM. Its aim is to find the relationship between input and output data sets generated from two unknown but fixed distributions. The regression problem has attracted extensive attention in past decades due to its wide-scope applications in scientific research and engineering [1].

Various approaches for regression have been developed[2]. For any regression method, three features are normally used to assess its performance: computational efficiency, robustness and generalization capability. In these terms, each of the existing approaches has its own advantages and disadvantages. Nevertheless, the effectiveness of all the methods face challenges when the data set becomes huge.

There are many ways to tackle such a problem. Preprocessing the data is one of the useful ways. Most of the existing preprocessing methods have focused on dimension reduction techniques, such as PCA and ICA[3]. In this work, we propose a new preprocessing method, whose aim is to reduce the number of data instead of dimension. The main idea is to introduce a judge function that is constructed based on the information collected from neighborhood of a datum, and, to apply the judge function to measure the importance of each datum. Based on this measure, the less important data are deleted. In this way, the data size is reduced and any regression method can be more efficiently and robustly applied. A series of simulations are provided and they demonstrate

J. Wang et al. (Eds.): ISNN 2006, LNCS 3972, pp. 765–770, 2006.

the suggested new method not only significantly increase the computational efficiency and robustness, but also preserve the generalization capability of a regression method.

The paper is organized as follows. Section 2 provides three rules to guide the formulation of the new pre-processing method. The implementation strategies to realize the rules are presented in section 3. Section 4 gives a series of artificial and real world simulations. Finally we conclude the paper in section 5.

2 Rules for Preprocessing

A regression training datum is regarded to be "good" if it is in or nearby the regression surface. A regression method run on "good" training data set will have high efficiency, strong robustness and good generalization capability.

Although all training data of a regression problem are expected to be "good", there usually exist some kinds of unexpectations. The first unexpectation is that a training datum is obviously deviated from other neighborhood data. In this case the datum is usually a noise and should be deleted in order to improve robustness of the adopted regression method. Consequently, we suggest the following **Rule 1** of the new pre-processing method: *when a training datum deviates far from other neighborhood training data, it should be deleted.*

The second unexpectation is that in the neighborhood of one training datum (x_i, y_i) (in the sense of input metric space) there are few other training data. In this case the datum (x_i, y_i) contains almost all information of the underlying regression surface in the neighborhood of x_i and should be preserved to avoid the loss of generalization capability of the regression method adopted. Thus we suggest the use of following **Rule 2**: *when a training datum is isolated from most of the other training data, it should be preserved.*

The third unexpectation is that the outputs of the regression training data have a "thickness" in some input areas. Intuitively, the training data whose outputs are in the middle of the "thickness" are more important than the ones whose outputs are on the edge. Thus we can delete some training data which are the least important for the regression problem (i.e. the data whose outputs are on the edge of the "thickness") to increase computational efficiency and meanwhile preserve the generalization capability of the adopted regression method. So the suggested **Rule 3** is: *when the outputs of training data in some input area are "thick", the datum in the middle of the "thickness" should be preserved and ones on the edge should be deleted.*

The suggested **Rule 1-3** proposed above can guide us to make a pre-processing on the raw regression data set to reduce its size. In the next section three implement strategies to realize the three rules will be presented.

3 The New Pre-processing Method

For a practical regression problem with training data set $D = \{x_i, y_i\}_{i=1}^{l}$, we adopt three implement strategies to realize **Rules 1-3**.

The first strategy is guided by **Rule 2**. By applying a clustering method to the input of the training data $D_{in} = \{x_i\}_{i=1}^{l}$, a partition $C_1, C_2, ..., C_k$ is yielded, where $C_s = \{x_i^{(s)}\}_{i=1}^{n_s}$, $n_s = |C_s|$, $s = 1, 2, ..., k$. Since a valid clustering method obtains a good separation of D_{in}, i.e., the interior distance of a cluster is minimized and exterior distance between each clusters is maximized, the corollary can be deduced that if the number of elements in one cluster C_s is very small, the elements in this cluster are isolated from most of the others. According to **Rule 2**, the data $\{x_i, y_i\}_{i=1}^{n_s}$, whose x_i is in such a cluster, should be preserved.

The second strategy is guided by **Rule 1**. A datum $(x_i^{(s)}, y_i^{(s)})$ deviating from other neighborhood data $(x_j^{(s)}, y_j^{(s)})$, $j = 1, ..., n_s$, $j \neq i$ has the property that $d_x(x_i^{(j)}, x_s^{(j)})$ is of the small value while $d_y(y_i^{(j)}, y_s^{(j)})$ of the large value comparatively. The judge function can then be formulated as:

$$f(x_i^{(s)}) = \sum_{j=1}^{n_s} \exp(-d_x(x_i^{(s)}, x_j^{(s)}))(1 - \exp(-d_y(y_i^{(s)}, y_j^{(s)}))). \tag{1}$$

The function has the property that when it has a much larger value on $x_i^{(s)}$ than the values on others in the similar cluster, then the output $y_i^{(s)}$ possibly has a much larger or smaller value than other neighborhood outputs' values (since when $d_x(x_i^{(s)}, x_j^{(s)})$ is small, the larger $d_y(y_i^{(s)}, y_j^{(s)})$, the larger $f(x_i^{(s)})$ is likely to be). According to **Rule 1**, the data on which the value of judge function is very large should be deleted.

The third strategy is guided by **Rule 3**. The adopted window measure $d_y^{(s)}$ is formulated as follows:

$$d_y^{(s)}(y_i^{(s)}, y_j^{(s)}) = \begin{cases} d_y(y_i^{(s)}, y_j^{(s)}), \text{if } d_y(y_i^{(s)}, y_j^{(s)}) > h^{(s)}/2 \\ 0, \qquad\qquad\qquad \text{if } d_y(y_i^{(s)}, y_j^{(s)}) \leq h^{(s)}/2 \end{cases} \tag{2}$$

where

$$h^{(s)} = \max_{1 \leq i \leq n_s} (d_y(y_i^{(s)}, \mu^{(s)})), \ \mu^{(s)} = \frac{1}{n_s} \sum_{j=1}^{n_s} y_j^{(s)}. \tag{3}$$

Then the judge function is then transformed into

$$f(x_i^{(s)}) = \sum_{j=1}^{n_s} \exp(-d_x(x_i^{(s)}, x_j^{(s)}))(1 - \exp(-d_y^{(s)}(y_i^{(s)}, y_j^{(s)}))). \tag{4}$$

The function of the window measure $d_y^{(s)}$ is to enlarge the value of the judge function $f(x_i^{(s)})$ for the datum $(x_j^{(s)}, y_j^{(s)})$ on the edge of the "thickness" while to lessen its value for one in the middle. This is because usually for a datum $(x_j^{(s)}, y_j^{(s)})$ on the edge, $f(x_i^{(s)})$ has several times more non-zero positive additional elements than it has for the datum in the middle. Then according to **Rule 3,** the data with larger values of the judge function should be deleted.

Based on three strategies proposed above, the algorithm can be formally stated as follows:

Step I (*Initialization*): Set an appropriate natural number $0 < M \leq l$;
Step II (*Clustering*): Utilize an valid clustering method to cluster $D_{in} = \{x_i\}_{i=1}^{l}$ and a partition $C_1, C_2, ..., C_k$ is yielded, where $C_s = \{x_i^{(s)}\}_{i=1}^{n_s}$, $n_s = |C_s|$, $s = 1, 2, ..., k$;
Step III (*Preprocessing*): For $x_i^{(s)}$ in the cluster C_s satisfying $|C_s| > \frac{M}{k}$, compute $f(x_i^{(s)})$ (as (4)) and then rank $D^{(s)} = \{x_i^{(s)}, y_i^{(s)}\}_{i=1}^{n_s}, s = 1, 2, ..., k$ in the decreasing order of $f(x_i^{(s)})$; delete the first $|C_s| - \frac{M}{k}$ data and consequently get a new data set $D^{(s)\prime}$ with $\frac{M}{k}$ elements. For the cluster C_s satisfying $|C_s| \leq \frac{M}{k}$, set $D^{(s)\prime} = D^{(s)}$;
Step IV (*Regression*): The new training data set $D^{'} = \bigcup_{s=1}^{k} D^{(s)'}$ is obtained and any regression method can then be applied.

A prominent advantage of the new pre-processing method is that it can control the maximal size of the regression method through setting the value of M in **Step I.** Consequently incorporated with any regression method, the developed pre-processing method can be efficiently applied to regression of large data sets.

4 Simulations

In this section we will present artificial and real world simulation to verify the effectiveness of the new pre-processing method.

4.1 Artificial Data Sets

For demonstrating the performance of the pre-processing method visibly, two artificial regression data sets are formulated, .

The first artificial simulation is on *a linear regression data set*, formulated as $D = \{x_i, \zeta_i\}_{i=1}^{800}$ where x_i is generated from the uniform distribution on [0,3] and ζ_i is generated from the uniform distribution on the interval [-0.2,0.2]. The underlying optimal regression surface for this regression problem is the line $y = 0$. We apply the developed pre-processing method to the data set, in which set M as 400 and adopt fuzzy c-means clustering [4] as the clustering method (in the simulation we set the clustering number as 10). The computational result is shown as Figure 1. We can observe that although after preprocessing the reduced data set is only with half of the original number, almost all the data near the accurate regression surface are preserved. That is to say, this simulation demonstrates that the new pre-processing method can improve the computational efficiency of a regression method while preserving its generalization capability.

The second artificial simulation is on *a sinc regression data set*, formulated as:$D = \{x_i, y_i + \zeta_i\}_{i=1}^{720} \cup \{x_i, y_i\}_{i=721}^{800}$ where x_i is generated from the uniform distribution on the interval [0,3] and $y_i = -sin(x_i \times \pi), i = 1, 2, ..., 720$; $\zeta_i \in R, i = 1, 2, ..., 720$ is generated from the uniform distribution on the interval [-0.05,0.05]; $y_i, i = 721, ..., 800$ is generated from the uniform distribution on the interval [-1,1]. It can be seen that $\{x_i, y_i\}_{i=721}^{800}$ are noises and the real optimal

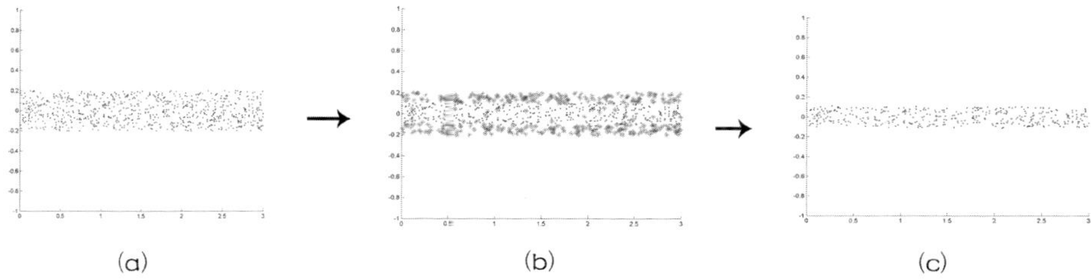

Fig. 1. (a) The original regression data set with number 800; (b) The effect of the pre-processing method. All the data which will be deleted are denoted as circles; (c) The reduced data set with number 400.

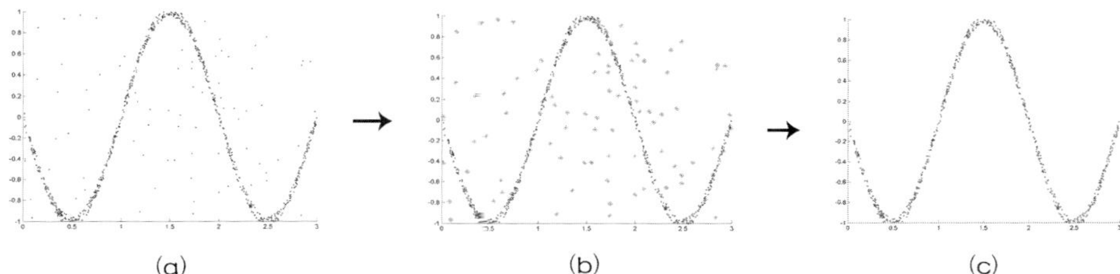

Fig. 2. (a) The original regression data set with number 800; (b) The effect of the pre-processing method. All the data which should be deleted are denoted as circles; (c) The reduced data set with number 700.

regression surface is $y = -sin(x \times \pi)$. The proposed pre-processing method is applied to D, in which set $M = 700$, adopt fuzzy c-means clustering as the clustering method, where the clustering number is set as 10. Figure 2 demonstrates the computational performance. Although only 100 data are reduced, all the noisy data (i.e. $\{x_i, y_i\}_{i=721}^{800}$) are eliminated and all the data near the accurate regression surface are preserved. That is to say, this simulation demonstrates that the new pre-processing method can improve the robustness of a regression method while preserving its generalization capability.

4.2 A Real World Data Set

The real world data set we selected is Boston Housing data, which is a fairly standard data set used for testing regression problems, the source of which is [5]. It contains 506 data with 13 attributes. The goal is to determine median value of owner-occupied homes. We set the first 400 data as the training data and the left as the test data. Also fuzzy c-means clustering is adopted as the clustering method and the support vector regression ([1]), which is known as one of the most promising regression method currently, is adopted as the regression method. In our simulations, Gaussian kernel was used and the parameters of SVR were estimated using 5-fold cross validation method [6]. The M in **Step I** of the pre-processing method is set to be 200. **Table 1** gives the performance of SVR run on the original data D and the transformed data D' after pre-processing

Table 1. Performance comparison of SVR on the data sets before preprocessing and after

	C	σ	ε	number of the training data	training time(s)	testing error
D	100	80	0.3	400	139.6	18.8919
D'	70	80	0.2	178	$1.9 + 19.5 = 21.4$	18.9540

(the training time of D' is the addition of the time spent by the pre-processing method and the SVR training time on the transformed data).From **Table 1** we can observe that the computational efficiency is greatly improved while the generalization capability is preserved. This further shows the high effectiveness of the presented pre-processing method.

5 Conclusions

In the paper we initiate a pre-processing method for regression to improve the computational efficiency and robustness of a regression method with preservation of its generalization capability. The core idea is formulating three reasonable rules to clarify the regression training data. To realize the rules, three corresponding implement strategies are constructed, respectively utilizing technologies of clustering, judge function formulation and window measure definition. By integrating the strategies new pre-processing method is proposed and a series of simulations results are presented. The performance of all the simulations shows the high effectiveness of the new pre-processing method.

Some future work include: (i) to give the theoretical evidence to support the pre-processing method; (ii) to utilize the method for more practical regression problems to further verify its effectiveness.

Acknowledgement

This research was supported by the NSFC project under contract 10371097.

References

1. Drucker, H., Burges, C.J.C., Kaufman, L., et al.: Support Vector Regression Machines. In: M.C. Mozer, M.I. Jordan and T. Petsche, Eds.: Advances in Neural Information Processing Systems, Vol. 9. MIT Press (1997) 155-200
2. George, A.F., Seber, C. J.: Wild Nonlinear Regression. Wiley (2005)
3. Gorriz, J.M., Segura-Lun, J.C.A.: A Survey of Orecasting Preprocessing Techniques Using RNs. Informatica 29 (2005) 13-32
4. Pal, N., Bezdek, J.: On Cluster Validity for The Fuzzy C-means Model. IEEE Trans. Fuzzy System 3(3) (1995) 370-379
5. Harrison, D., Rubinfeld, D.L.: Hedonic Prices and the Demand for Clean Air. J. Environ. Economics & Management 5 (1978) 81-102
6. Ratsch, G., Mūller, K.R.: Soft Margins for Adaboost. Machine Learning 42(3) (2001) 287-320

A New On-Line Modeling Approach to Nonlinear Dynamic Systems

Shirong Liu[1], Qijiang Yu[1,2], and Jinshou Yu[2]

[1] College of Automation, Hangzhou Dianzi University, Hangzhou, Zhejiang 310018, China
liushirong@hziee.edu.cn

[2] Research Institute of Automation, East China University of Science and Technology
Shanghai 200237, China
jshyu@ecust.edu.cn

Abstract. An improved radial basis function neural network (IRBFNN) with unsymmetrical Gaussian function is presented to simplify the structure of RBFNN. The improved resource allocating network (IRAN) is developed to design IRBFNN online for nonlinear dynamic system modeling, integrating the typical resource allocating network (RAN) with merging method for similar hidden units, deleting strategy for redundant hidden units, and LMS learning algorithm with moving data window for output link weights of network. The proposed approach can effectively improve the precision and generalization of IRBFNN. The combination of IRBFNN and IRAN is competent for the online modeling of nonlinear dynamic systems. The feasibility and effectiveness of the modeling method have been demonstrated by simulations.

1 Introduction

Radial basis function neural networks (RBFNN) have been widely used for nonlinear system modeling because of their simple topological structure and their ability to reveal the learning process in an explicit manner. Generally there have been many design methods of the network, such as the orthogonal least squares (OLS) or recursive orthogonal least squares (ROLS) algorithms [1], and dynamic clustering algorithm based on data samples [2]. These methods are difficult to model the complex time-varying dynamic systems online with the characteristics of dynamic data stream, such as communication systems and complicated dynamic processes.

Platt presented the sequential topology structure generating method for RBFNN [3], called resource allocating networks (RAN), which allocates the hidden units according to observed data novelty, distance and error criteria, and learns the link weights of network using LMS. Since RAN was presented, some improved algorithms have been developed by [4], [5]. In RBFNN, Gaussian function is usually used as the basis function, whose width is symmetric and usually proportional to the distance from the center to its nearest neighbor [6]. The precision and generalization of the network are determined by the number of hidden unit and output link weights.

In our study an improved radial basis function neural network (IRBFNN) with unsymmetrical Gaussian functions is proposed to simplify the structure of RBFNN, which based on the concept of hyper-rectangle partition of input space [7]. An

J. Wang et al. (Eds.): ISNN 2006, LNCS 3972, pp. 771 – 776, 2006.

improved resource allocating network (IRAN) is used to design the structure of IRBFNN online, which integrates the hidden unit creating with similar hidden unit merging, redundant hidden unit deleting, and LMS with moving data window. The combination of IRBFNN and IRAN is competent for the online modeling of nonlinear time-varying systems.

2 Improved Radial Basis Function Neural Networks

Considering MISO system, the output of RBFNN has the following form,

$$\hat{y}(n)=\sum_{i=1}^{M} w_i(n)\varphi_i(\| \mathbf{x}(n)-\mathbf{o_i}(n)\|), \tag{1}$$

where $\mathbf{x}(n)$ is the input vector, M the number of hidden units, $\varphi_i(\cdot)$ is the i-th radial basis function, $\mathbf{o_i}$ the center of the i-th radial basis function, $w_i(n)$ the output link weight. Let $\sigma_i(n)$ the width coefficient, and it is usually proportional to the distance from the center to its nearest neighbor. Radial basis function with Gaussian form is

$$\varphi_i(\mathbf{x}(n))=\exp\left(\frac{1}{2\sigma_i^2(n)}\left\|\mathbf{x}(n)-\mathbf{o}_i(n)\right\|\right). \tag{2}$$

Kobat proposed the hyper-rectangle based method to partition the input space into relatively homogeneous hyper-rectangle regions [7]. The center of Gaussian function depends on the center of hyper-rectangle region. At the border of hyper-rectangle region, the function value is equal to a pre-specified value z, as shown in Fig.1. We propose an improved RBFNN (IRBFNN) with unsymmetrical Gaussian function to simplify the structure of RBFNN and mitigate the impact of the different scale of each dimension variable. For instance, the function with one dimension is expressed as

$$\varphi_i(x)=\exp\left(\frac{|x-o_i|^2}{2\sigma_i^{\,2}}\right),\ \sigma_i=\begin{cases}\sigma_{il} & ,x\le o_i\\ \sigma_{ir} & ,x>o_i\end{cases} \tag{3}$$

where $\sigma_{il}=p\cdot|o_i-o_{i-1}|$ and $\sigma_{ir}=p\cdot|o_{i+1}-o_i|$, σ_{il} and σ_{ir} are the left and right widths of the i-th hidden unit respectively, σ_{il} and σ_{ir} are proportional to the distance from o_i to o_{i-1} and to o_{i+1} respectively, p is an overlap coefficient, $0<p<1$.

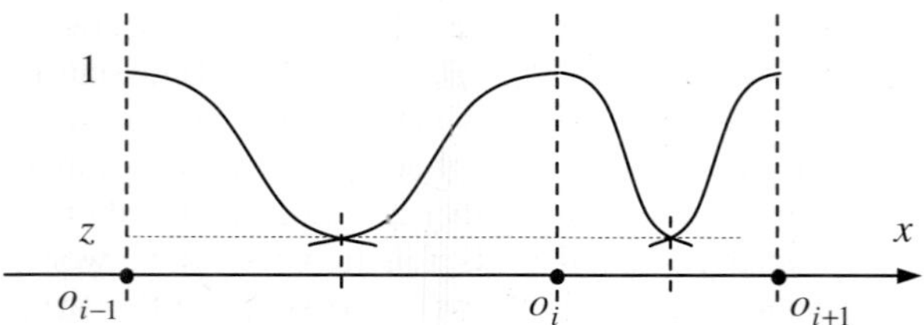

Fig. 1. Unsymmetrical Gaussian function

Let the input vector $\mathbf{x} = [x_1 \ x_2 \ \cdots \ x_k]^T$, the corresponding center of hidden units: $\mathbf{o_i} = [o_1(i_1) \ o_2(i_2) \cdots o_k(i_k)]^T$. The output of the i-th hidden unit is

$$\varphi_i(\mathbf{x}) = \exp\left\{-\frac{(x_1 - o_1(i_1))^2}{2\sigma_{i1}^2} - \frac{(x_2 - o_2(i_2))^2}{2\sigma_{i2}^2} - \cdots - \frac{(x_k - o_k(i_k))^2}{2\sigma_{ik}^2}\right\}, \tag{4}$$

where $\sigma_{i1},\cdots,\sigma_{ik}$ are the unsymmetrical widths of Gaussian function. Set one center array: $\mathbf{dimension}_i = [o_i(1) \ \cdots \ o_i(n_1)]^T, o_i(1) < \cdots < o_i(n_i)$, $i = 1,\cdots,k$ for each dimension. For the i-th hidden unit, the widths of the function are

$$\sigma_{i1} = \begin{cases} \sigma_{1l}(i_1), & x_1 \le o_1(i_1) \\ \sigma_{1r}(i_1), & x_1 > o_1(i_1) \end{cases}, \cdots, \sigma_{ik} = \begin{cases} \sigma_{kl}(i_k), & x_k \le o_k(i_k) \\ \sigma_{kr}(i_k), & x_k > o_k(i_k) \end{cases}, \tag{5}$$

where $\sigma_{jl}(i_j)$ and $\sigma_{jr}(i_j)$ are the left and right widths of the i-th hidden unit respectively, $j = 1,\cdots,k$. The left and right widths can be determined by

$$\sigma_{il}(j) = p \cdot | o_i(j) - o_i(j-1) |, \quad \sigma_{ir}(j) = p \cdot | o_i(j+1) - o_i(j) |, \tag{6}$$

where p is an overlap coefficient, $0 < p < 1$.

3 IRAN Algorithm

Resource allocating network (RAN) has been used as a tool for online modeling of nonlinear dynamic systems, but there exist the disadvantages such as the network's size too large and generalization poor. We developed an improved resource allocating network (IRAN) that integrates the criteria for creating hidden units, merging strategy for similar hidden units, pruning criterion of useless hidden units, and output link weight updating algorithm with moving data window.

After putting the new data sample $(\mathbf{x}(j), y(j))$ into the window, whether to create new hidden unit or not is determined by the following criteria

$$| e(j) | = | y(j) - \hat{y}(j) | > E_1, \tag{7}$$

$$|x_i(j) - o_{i,nearest}| \ge E_{2i}, i = 1,2,\cdots k, \tag{8}$$

where $o_{i,nearest}$ is the element of center array which is closest to $x_i(j)$, E_1 the error threshold, E_{2i} the distance threshold to be selected appropriately. Criterion (7) decides whether the existing hidden units are sufficient or not according to the error tolerance and criterion (8) ensures that the new hidden unit to be added is sufficiently far from all the existing units. When both criteria are met, the new $x_i(j)$ is added into the center array as a new element

There may be two or more hidden units whose centers are very close and their weights are almost similar during the network creating process, and they should be merged. Suppose that there are two hidden units p and q, the merging criteria are

$$\| \mathbf{o}_p - \mathbf{o}_q \| < E_3, \quad | w(p) - w(q) | < E_4 \tag{9}$$

where E_3 and E_4 are the merging thresholds. The new center and weight by merging operation are calculated by

$$\mathbf{o}_p = \frac{(\mathbf{o}_p + \mathbf{o}_q)}{2}, \quad w_p = \frac{w_p + w_q}{2}. \tag{10}$$

When a hidden unit is far away from work space, it is not activated and no contribution to the output in a continual period of time. It is shown that this hidden unit is a redundant unit and should be pruned. The pruning criterion of hidden unit is

$$n_{count} > C_{\max} \tag{11}$$

where n_{count} is the time of the hidden unit not activated continually and $C_{\max}$ is a sufficient large positive integer.

The RLMS algorithm with moving data window is adopted to improve output link weight learning. Let $(\mathbf{x}(L), y(L))$ the newest input-output data pair and $(\mathbf{x}(1), y(1))$ the oldest input-output data pair. Define the output matrix of hidden unit $\mathbf{O} = [o_{ik}]$, $\mathbf{O} \in \mathbf{R}^{M \times L}$, $o_{ik} = \varphi_i(\mathbf{x}_k)$. The output link weights are updated by

$$\mathbf{w}(j+1) = \mathbf{w}(j) + \eta \cdot \mathbf{O} \cdot \mathbf{E}, \tag{12}$$

where $\mathbf{E} = [\lambda^{L-1} e_1 \quad \lambda^{L-2} e_2 \quad \cdots \quad e_L]^T$ is the error matrix, e_k the output error at time k ($k = 1, 2, \cdots, L$), L the length of moving windows, λ ($0 < \lambda < 1$) the weighted coefficient, and η the learning rate.

The online design procedure of IRBFNN with IRAN is summarized below:

Step1 For $j = 1$, $(\mathbf{x}(1), y(1))$, set $\mathbf{o}_1 = \mathbf{x}(1)$, $n_1 = \cdots = n_k = 1$, $\text{dimension}_1 = [x_1(1)], \ldots$, $\text{dimension}_k = [x_k(1)]$, and $w_1(1) = y(1)$. Set $j = j+1$ and go to Step2.

Step2 For $j > 1$, $(\mathbf{x}(j), y(j))$, put it into the moving window and move the oldest data out from the window. Compute the output of IRBFNN by (1) and (2), and then go to Step3.

Step3 If (7) and (8) hold simultaneously, a new element, $x_i(j)$, is added to the corresponding center array, the weight of the new hidden unit is set to $e(j)$, and then go to Step4; or else go to Step4 directly.

Step4 If (9) hold, merge the corresponding hidden units using (10). If (11) hold, prune the corresponding hidden unit away from the network. Tune the width of each element of center array by (6). Go to Step 5.

Step5 Update the output link weights using (12). Set $j = j+1$ and go to Step2.

4 Simulation Study

In order to demonstrate the effectiveness of the proposed online modeling approach, the plant model adopted by [8] was used as an example in our study. It is described by

$$y(k+1)=\frac{y(k)[y(k)+0.3]}{1+y(k)^2}+\beta u(k)[u(k)+0.8][u(k)-0.5],\quad y(0)=0,$$

where $\beta=\begin{cases}1.0 & ,0<k\le 500\\ 1.5 & ,2250<k\le 1000\end{cases}$,

$$u(k)=\begin{cases}\sin\frac{2\pi k}{250} & ,\quad 0<k\le 250\\ 0.8\sin\frac{2\pi(k-250)}{250}+0.2\sin\frac{2\pi(k-250)}{25}, & 250<k\le 500\\ \sin\frac{2\pi(k-500)}{250} & ,\quad 500<k\le 750\\ 0.8\sin\frac{2\pi(k-750)}{250}-0.2\sin\frac{2\pi(k-750)}{25}, & 750<k\le 1000\end{cases}.$$

The parameters of IRAN and IRBFNN are taken as: $E_1=0.02$, $E_{21}=0.05$, $E_3=0.055$, $E_4=0.065$, $C_{\max}=800$, $L=20$, $\lambda=0.9$, $\eta=0.85$, and $p=0.7$.

The dynamic data stream is generated by the plant model in time sequence. At $k=500$, coefficient β in the plant model jumps from 1.0 to 1.5; it is used to test the adaptation of modeling approach. The outputs of the IRBFNN model and the plant and error of model are given in Fig. 2, respectively. Although β suddenly changes at $k=500$, the output of model tracks the output of plant rapidly and strongly.

The creating processes of hidden unit by IRAN and RAN are shown in Fig.3. It is obvious that the network structure obtained by IRAN is more parsimonious than that one by RAN. The learning process of the output link weight of the hidden unit created by IRAN at $k=44$ is shown in Fig. 4. It has been shown that the generalization of the model with IRAN is superior to the one with RAN. Because of the length limitation of paper, the simulation results are not given in this paper.

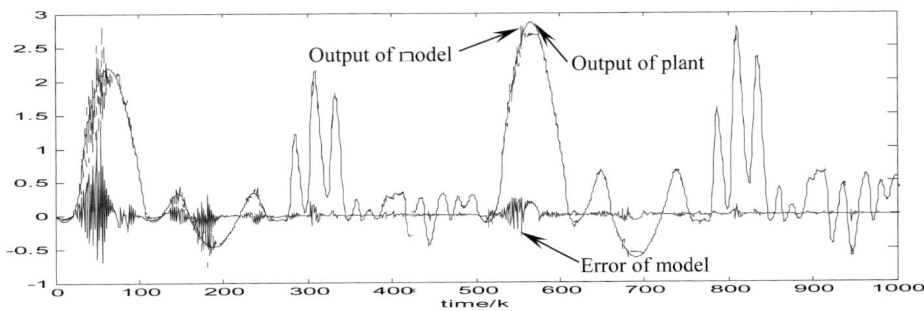

Fig. 2. Output of plant, output of model, and error of model

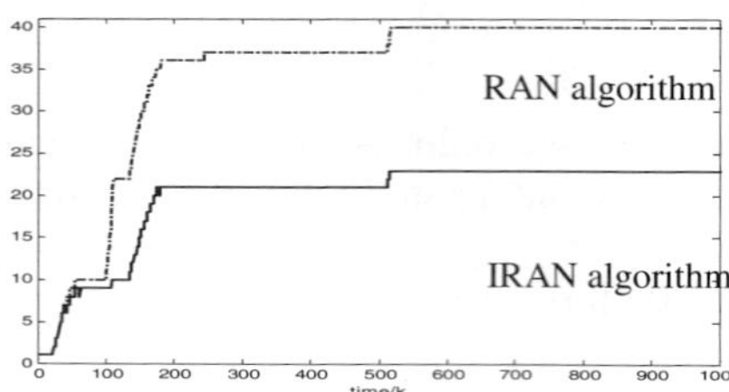

Fig. 3. Creating processes of hidden units

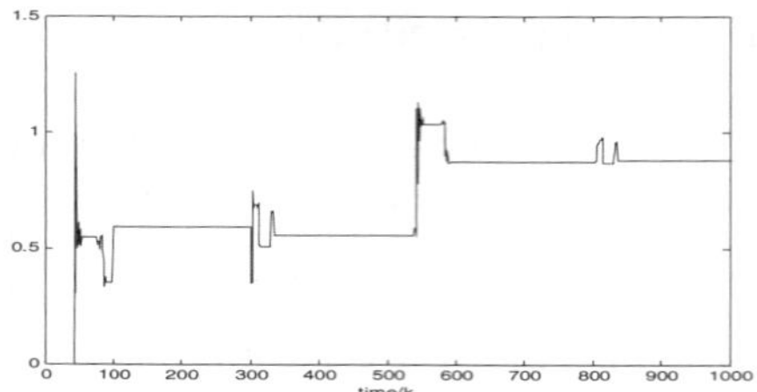

Fig. 4. Learning process of an output link weight

5 Conclusions

Nonlinear time-varying system can be modeled online by IRBFNN and IRAN successfully. Because the model designed by the proposed approach can track the dynamic characteristics of plant effectively and strongly, it can be applied to dynamic process monitoring, fault detecting, model-based self-adaptive controller, and data stream modeling for communication systems.

Acknowledgments

The work was supported by Natural Science Foundation of Zhejiang Province, China (Grant No. Y104560) and Research Start-up Foundation of Hangzhou Dianzi University, China (Grant No. KYS091505043).

References

1. Chen, S., Cowan, C. F. N., Grant, P. N.: Orthogonal Least Squares Learning Algorithms for Radial Basis Function Networks. IEEE Trans. Neural Networks **2** (1991) 302-309
2. Moody, J., Darken, C.: Fast Learning in Networks of Locally-Tuned Processing Units. Neural Computation **1** (1989) 281-294
3. Platt, C. J.: A Resource Allocating Network for Function Interpolation. Neural Computation **4** (1991) 473-493
4. Kadirkamanathan, V., Niranjan, M.: A Function Estimation Approach to Sequential Learning with Neural Networks. Neural Computation **5** (1993) 954-975
5. Yang, G., Lu, J., Liu, Z. Y.: A New Sequential Learning Algorithm for RBF Neural Networks. Science of China-Series E **47** (2004) 447-460
6. Saha, A., Keeler, J. D.: Algorithms for Better Representation and Faster Learning in Radial Basis Function Network. In: Touretzky, D. S. (ed.): Advances in Neural Information Processing Systems. Morgan Kaufmann, San Mateo, CA **2** (1990) 482-489
7. Kubat, M.: A Hyperrectangle-Based Method that Creates RBF Networks. In: Howlett, R. J., Jain, L. C. (eds.): Radial Basis Function Networks. Physica-Verlag, Heidelberg **1** (2001) 32-50
8. Patra, J. C.: Nonlinear Dynamic System Identification Using Chebyshev Functional Link Artificial Neural Networks. IEEE Trans SMC-Part B **32** (2002) 505-511

Online Modeling of Nonlinear Systems Using Improved Adaptive Kernel Methods

Xiaodong Wang[1], Haoran Zhang[1], Changjiang Zhang[1], Xiushan Cai[1], Jinshan Wang[1], and Meiying Ye[2]

[1] College of Information Science and Engineering,
Zhejiang Normal University,
Jinhua 321004, P.R. China
{Wxd, Hylt, Zcj74922, Xiushan, Dz66}@zjnu.cn
[2] College of Mathematics and Physics,
Zhejiang Normal University,
Jinhua 321004, P.R. China
Ymy@zjnu.cn

Abstract. The least squares support vector machines (LS-SVMs) is a kernel method. The training problem of LS-SVMs is solved by finding a solution to a set of linear equations. This makes online adaptive implementation of the algorithm feasible. An improved adaptive algorithm is proposed for training the LS-SVMs in this paper. This algorithm is especially useful on online nonlinear system modeling. The experiments with benchmark problem have shown the validity of the proposed method even in the case of additive noise to the system.

1 Introduction

Obtaining an accurate model of a complex, nonlinear, dynamic system is the first step towards the creation of high performance controllers. In system modeling field, researchers are very enthusiastic about the potential of neural networks especially regarding the multilayer perceptron (MLP) [1]~[3]. However, their performance is not always satisfactory. Some inherent drawbacks, e.g., the multiple local minima problem, the choice of the number of hidden units and the danger of over fitting, etc., would make it difficult to put the MLP into some practice.

The least squares support vector machine (LS-SVM) [4] achieves higher generalization performance than the MLP in solving the machine learning problem. Unlike MLP' training that requires nonlinear optimization with the danger of getting stuck into local minima, training LS-SVM is equivalent to solving a set of linear equations. Consequently, the solution of LS-SVM is always unique and globally optimal. In our previous work [5], LS-SVM has been successfully used for nonlinear system modeling. But existed LS-SVM algorithm is trained offline in batch way. Offline training algorithm is not fit for the practical applications such as online system modeling and control problems, where the data come is sequentially. Hence this paper proposes an improved adaptive kernel method for nonlinear system modeling.

J. Wang et al. (Eds.): ISNN 2006, LNCS 3972, pp. 777 – 782, 2006.

2 Problem Description

A wide class of nonlinear dynamic systems with an input u and an output y can be described in discrete time by the NARX (nonlinear autoregressive with exogenous input) input–output model

$$y(k+1) = f(\mathbf{x}(k)) \tag{1}$$

where $f(\cdot)$ is some nonlinear function, $y(k+1)$ denotes the output predicted at the future time instant $k+1$ and $\mathbf{x}(k)$ is the regressor vector, consisting of a finite number of past inputs and outputs:

$$\mathbf{x}(k) = [y(k), \cdots, y(k-n_y+1), u(k), \cdots, u(k-n_u+1)]^T \tag{2}$$

The dynamic order of the system is represented by the number of lags n_u and n_y.

The task of system modeling is essentially to find suitable mappings, which can approximate the mappings implied in a nonlinear dynamic system. The function $f(\cdot)$ can be approximated by some general function approximators such as neural networks, neuro-fuzzy systems, splines, interpolated look-up tables, etc. The aim of system modeling is only to obtain an accurate predictor for y. In this work, we use the adaptive LS-SVM regression mentioned above for nonlinear system modeling.

3 LS-SVM Regression Algorithm

Consider a given training set of N data points $\{x_i, y_i\}_{i=1}^N$ with input data $x_i \in R^n$ and output $y_i \in R$. In feature space LS-SVM models take the form:

$$y(x) = w^T \varphi(x) + b \tag{3}$$

where the nonlinear mapping $\varphi(\cdot)$ maps the input data into a higher dimensional feature space. In LS-SVM, the following optimization problem is formulated

$$\min \frac{1}{2} w^T w + \frac{C}{2} \sum_{k=1}^{N} e_i^2 \tag{4}$$

subject to the equality constrains

$$y_i = w \cdot \varphi(x_i) + b + e_i, \quad i = 1, \cdots, N \tag{5}$$

Using the Karush-Kuhn-Tucker (KKT) conditions we get the linear equations

$$\begin{bmatrix} 0 & e1^T \\ e1 & Q + C^{-1} I \end{bmatrix} \begin{bmatrix} b \\ \alpha \end{bmatrix} = \begin{bmatrix} 0 \\ y \end{bmatrix} \tag{6}$$

where $y = [y_1, \cdots, y_N]^T$, $e1 = [1, \cdots, 1]^T$, $\alpha = [\alpha_1, \cdots, \alpha_N]^T$, $Q_{ij} = \phi(x_i) \cdot \phi(x_j)$ $= K(x_i, x_j)$, $i, j = 1, \cdots, N$, and C is a regularization factor.

This finally results of LS-SVM model for function estimation is

$$y(x) = \sum_{i=}^{N} \alpha_i K(x_i, x_j) + b \tag{7}$$

where α_i and b are the solutions to the linear system, $K(x_i, x_j)$ represents the high dimensional feature space that is nonlinearly mapped from the input space x. The LS-SVM approximates the function using the equation (6).

In this work, the radial basis function (RBF) is used as kernel function. It takes the form

$$K(x_i, x_j) = \exp\left(-\|x_i - x_j\|^2 / 2\sigma^2\right) \tag{8}$$

where σ is a positive real constant.

Note that the LS-SVM is still found by working in the dual space, but unlike standard SVM solutions, the LS-SVM uses equality constraints. The equation (6) is solved by finding a solution to a set of linear equations. This makes online adaptive implementation of the algorithm feasible.

4 Improved Adaptive LS-SVM for System Modeling

In this section, we formulate an improved adaptive solution for the LS-SVM regression based on linear equation (6). We use a windows size of length L. The training data are described by $\{X(k), Y(k)\}$, the inputs $X(k) = [x_k, x_{k+1}, \cdots, x_{k+L-1}]$, and targets $Y(k) = [y_k, y_{k+1}, \cdots, y_{k+L-1}]^T$. At the k moment, $Q_{ij}(k) = K(x_{i+k-1}, x_{j+k-1})$, $i, j = 1, \cdots, L$, $\alpha(k) = [\alpha_k, \alpha_{k+1}, \cdots, \alpha_{k+L-1}]^T$, $b(k) = b_k$, $y(k) = y_k$, and then the equation (6) can be rewritten as

$$y(k) = \sum_{i=k}^{k+L-1} \alpha_i(k) K(x, x_i) + b(k) \tag{9}$$

Let $U(k) = Q_k + C^{-1} I$, where I is a one matrix, and then we get the matrix equation described by

$$\begin{bmatrix} 0 & e1^T \\ e1 & U(k) \end{bmatrix} \begin{bmatrix} b(k) \\ \alpha(k) \end{bmatrix} = \begin{bmatrix} 0 \\ y(k) \end{bmatrix} \tag{10}$$

Assuming $P(k) = U(k)^{-1}$, we can then use the equation (10) to compute $P(k)$ to get that

$$b(k) = \frac{e1^T P(k) y(k)}{e1^T P(k) e1} \tag{11}$$

$$\alpha(k) = P(k)\left(y(k) - \frac{e1 e1^T P(k) y(k)}{e1^T P(k) e1}\right) = P(k)(y(k) - e1 b(k)) \tag{12}$$

where

$$P(k)=U(k)^{-1}=\left[Q(k)+C^{-1}I\right]^{-1}$$
$$=\begin{bmatrix} h(k) & H(k)^T \\ H(k) & D(k) \end{bmatrix}^{-1}=\begin{bmatrix} 0 & 0 \\ 0 & D(k)^{-1} \end{bmatrix}+s_h(k)s_h(k)^T c_h(k) \tag{13}$$

$$h(k)=K(x_k,x_k)+C^{-1}, H(k)=[K(x_{k+1},x_k)+C^{-1},\cdots,K(x_{k+l-1},x_k)]^T,$$

$$D(k)=\begin{bmatrix} K(x_{k+1},x_{k+1})+C^{-1} & \cdots & K(x_{k+L-1},x_{k+1}) \\ \vdots & \ddots & \vdots \\ K(x_{k+1},x_{k+L-1}) & \cdots & K(x_{k+L-1},x_{k+L-1})+C^{-1} \end{bmatrix},$$

$$s_h(k)=[-1,H(k)^T D(k)^{-1}]^T, c_h(k)=1/(h(k)-H(k)^T D(k)^{-1}H(k)).$$

At the $k+1$ moment, the new data pair (x_{k+L},y_{k+L}) entered into the training data and the old data pair (x_k,y_k) is thrown away from the training data. The kernel function changes into $Q_{ij}(k+1)=K(x_{i+k},x_{j-k})$, $i,j=1,\cdots,L$, then

$$P(k+1)=U(k+1)^{-1}=\left[Q(k+1)+C^{-1}I\right]^{-1} \tag{14}$$

The improved adaptive algorithm of the LS-SVM for finding threshold value $b(k)$ and $\alpha(k)$ can be summarized by

(1) Initialization: $k=1$.
(2) The new data pair (x_{k+L},y_{k+L}) coming in and the old data pair (x_k,y_k) is thrown away in order to obtain the training data $(X(k),y(k))$.
(3) Computing the kernel function $Q(k)$ and $P(k)$.
(4) Computing the $b(k)$, $\alpha(k)$ and predicting $y(k)$.
(5) $k\leftarrow k+1$ go to step (2).

5 Experiments

To verify the proposed improved adaptive kernel methods in nonlinear system modeling, experiments with benchmark problem were presented. Two different models are used as evaluation of the modeling power of the improved adaptive LS-SVM regression, under Windows XP.

Example 1. The plant of the nonlinear system is given by the following difference equation:

$$x(k)=0.3y(k-1)+0.6y(k-2)+0.6\sin(\pi u(k-1)) + 0.3\sin(3\pi u(k-1))+0.1\sin(5\pi u(k-1)) \tag{15}$$

where the output at time t is a linear function of past output at times $k-1$ and $k-2$ plus a nonlinear function of the input at time $k-1$. The reference input $u(k-1)$ to the system is selected as

$$u(k-1) = \sin(2\pi(k-1)/250) \tag{16}$$

Example 2. The plant of the nonlinear system is given by

$$y(k) = \frac{0.2y(k-1)+0.6y(k-2)}{1+y(k-1)^2} + \sin(u(k-1)) \tag{17}$$

The reference input is selected as

$$u(k-1) = \sin(2\pi(k-1)/10) + \sin(2\pi u(k-1)/25) \tag{18}$$

Figs. 1 and 2 depict the modeling results using the proposed improved adaptive LS-SVM to examples 1 and 2, respectively. In the Figs. 1(b) and 2(b), the noise added to the systems is level of 2%.

As can be seen from Figs. 1(a) and 2(a), which demonstrate that the identified and actual points of the system are largely indistinguishable if the training data are noise-free, almost perfect modeling are achieved. In this situation, the root mean squared error (RMSE) is 0.0051 for example 1 and 0.096 for example 2.

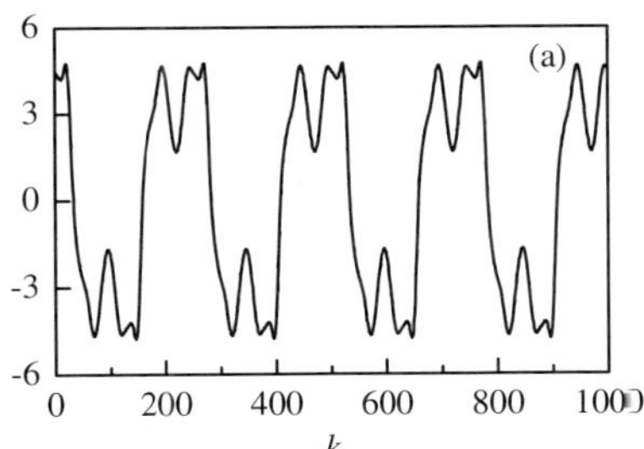

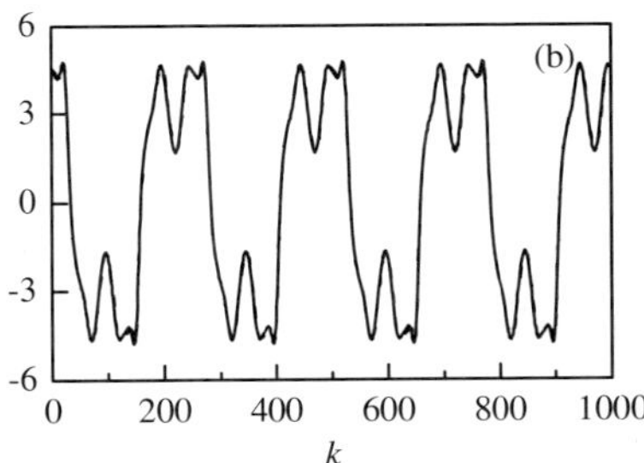

Fig. 1. The plant output (solid) and improved adaptive LS-SVM output (dashed) of Example 1 for (a) noise-free and (b) 2% noise level

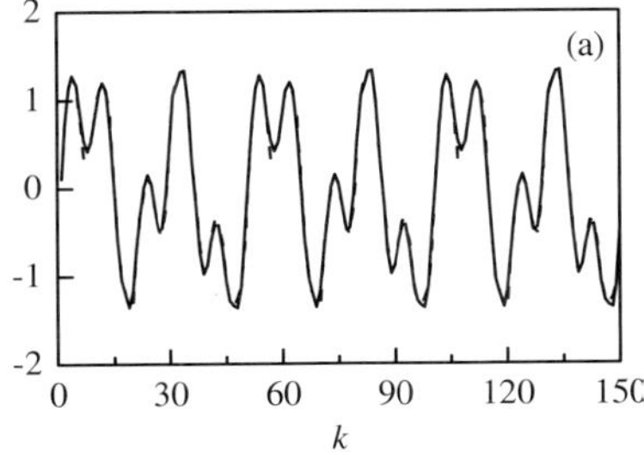

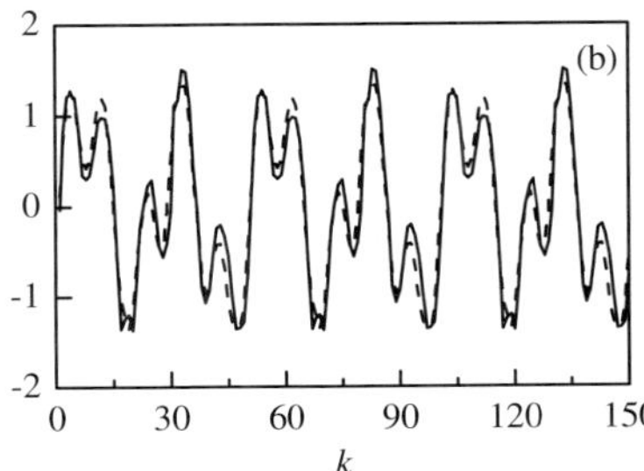

Fig. 2. The plant output (solid) and improved adaptive LS-SVM output (dashed) of Example 2 for (a) noise-free and (b) 2% noise level

From the Figs. 1(b) and 2(b), if the training data are corrupted with noise of 2% levels, it is verified that additive noise has an influence on the nonlinear system modeling. In this situation, the RMSE is 0.21 for example 1 and 0.85 for example 2. It may be seen that the modeling of the plant is satisfactory even when a training data are contaminated with additive noise.

6 Conclusions

The LS-SVM algorithm is trained offline in batch way. But offline training algorithm is not fit for the practical applications such as online system modeling, where the data come is sequentially. In this paper, an online nonlinear system modeling scheme based on improved adaptive kernel methods has been presented. Two different models are used as evaluation of the identifying power of the improved adaptive LS-SVM regression. The results indicate that this approach is effective even in the case of additive noise to the system.

Acknowledgements

The Project Supported by Zhejiang Provincial Natural Science Foundation of China (Y105281) and Key Construction Subject of Zhejiang Provincial.

References

1. Lu, S., Basar, T.: Robust Nonlinear System Identification Using Neural Network Models. IEEE Trans on Neural Networks 9 (1998) 407–429
2. Narendra, K.S., Parthasarathy, K.: Identification and Control of Dynamical Systems Using Neural Networks. IEEE Trans on Neural Networks 1 (1990) 4–26
3. Griñó, R., Cembrano, G., Torras, C.: Nonlinear System Identification Using Additive Dynamic Neural Networks--Two On-line Approaches. IEEE Trans on Circuits and Systems I: Fundamental Theory and Applications 47 (2000) 150–165
4. Suykens, J.A.K., Vandewalle, J.: Least Squares Support Vector Machine Classifiers. Neural Process Letter 9 (1999) 293–300
5. Ye, M.Y., Wang, X.D.: Chaotic Time Series Prediction Using Least Squares Support Vector Machines. Chinese Physics 13 (2004) 454–458

A Novel Multiple Neural Networks Modeling Method Based on FCM

Jian Cheng, Yi-Nan Guo, and Jian-Sheng Qian

School of Information and Electrical Engineering, China University of Mining and Technology, Xuzhou, Jiangsu, 221008, China
chjpaper@126.com

Abstract. A single neural network model developed from a limited amount of sample data usually lacks robustness and generalization. Neural network model robustness and prediction accuracy can be improved by combining multiple neural networks. In this paper a new method of the multiple neural networks for nonlinear modeling is proposed. A whole training sample data set is partitioned into several subsets with different centers using fuzzy c-means clustering algorithm (FCM), and the individual neural network is trained by each subset to construct the subnet respectively. The degrees of memberships are used for combining the outputs of subnets to obtain the final result, which are gained from the relationship between a new input sample data and each cluster center. This model has been evaluated and applied to estimate the status-of-loose of jig washer bed. Simulation results and actual application demonstrate that this model has better generalization, better prediction accuracy and wider potential application online.

1 Introduction

Single neural network has been increasingly used in building nonlinear models which utilizes sample data in industrial processes generally [1]. Even though neural networks have significant capability in representing nonlinear functions, inconsistency of accuracy still seems to be a problem where a neural network model cannot cope or perform well when it is applied to new unseen data. So robustness of model is one of the main criteria that need to be considered when judging the performance of neural network models [2]. Furthermore, advanced process control and supervision of industrial processes require accurate process models promoting investigations in the robustness of neural networks models. Therefore, the main problem in nonlinear modeling based on neural network is robustness and prediction accuracy. A lot of researchers have been concentrated on achieving better performance via optimal or suboptimum structure and optimum training parameters of neural network.

The multiple-modeling method, proposed by Bates and Granger [3], can enhance the robustness and generalization by combining several models. Now it is widely accepted that multiple models is a simple and effective method to gain better modeling performance [4]. In this method, several sub-models, which have different characteristics and the same objectives, are constructed according to training sample data

J. Wang et al. (Eds.): ISNN 2006, LNCS 3972, pp. 783 – 789, 2006.

subsets. Then outputs of sub-models are integrated to improve whole estimation performance. In this paper, the training sample data set is clustered into several subsets via fuzzy C-mean clustering algorithm, and subnets are trained by these subsets respectively. Then a novel fuzzy integrated method, in which degrees of membership are gained by the relationship of a new sample data and subsets, is proposed to combine the outputs of subnets to get the final output of multiple-model. Simulation results and application indicate performance of such multiple-model is better than it of single model. And this model has better generalization, predication accuracy and on-line application potential. This paper presents a novel multiple neural network modeling method based on FCM.

2 Fuzzy C-Means Clustering Algorithm

Fuzzy C-means clustering algorithm (FCM) is one of the most important and popular fuzzy clustering algorithm. Its main idea is to obtain a partition that minimizes the within-cluster scatter or maximizes the between-cluster scatter. The FCM was proposed first by Dunn in 1973, and generalized by Bezdek [5]. Sample data set $X=\{x_1,x_2,\cdots,x_n\}$ is classified into c clusters. Each sample data x_j includes p features, i.e. $x_j=\{x_{j1},x_{j2},\cdots,x_{jp}\}$, where x_j in the set X (X is a p-dimensional space). Because these features all can have different units in general, each of the features has to be normalized to unified scale before classification.

Objective function approach is adopted for classifying n data points to c clusters. In this approach, each cluster is considered as one hyper spherical shape with hypothetical geometric cluster center. The main aim of the objective function is to minimize the Euclidian distance between each data point within the cluster, and maximize the Euclidian distance between other cluster centers. The classification matrix $U=\{u_{ij}\}$ is a fuzzy matrix, also u_{ij} should satisfy the following condition:

$$u_{ij}\in[0,\quad 1],\ j=1,2,\cdots,n,\ i=1,2,\cdots,c,$$

$$\sum_{i=1}^{c}u_{ij}=1,\quad \forall j \tag{2.1}$$

where, u_{ij} is the degree of membership of x_j in the i-th cluster. The object function is defined as follows:

$$J_m(U,Z)=\sum_{i=1}^{c}J_i=\sum_{i=1}^{c}\sum_{j=1}^{n}u_{ij}^m d_{ij}^2 \tag{2.2}$$

where, $Z=\{z_1,z_2,\cdots,z_c\}$ is the cluster centers, $d_{ij}=\|z_i-x_j\|$ is the distance between x_j and the i-th cluster center z_i, u_{ij} is the membership function value of j-th sample data belongs to the i-th cluster, $m\in[1,\quad\infty)$ is the weighted exponent on each fuzzy membership. They are defined as follows:

$$z_i^{(k)} = \left(\sum_{j=1}^{n} u_{ij}^{m} x_j\right) \Bigg/ \left(\sum_{j=1}^{n} u_{ij}^{m}\right) \tag{2.3}$$

$$u_{ij} = \left[\sum_{k=1}^{c} \left(\frac{d_{ij}}{d_{kj}}\right)^{2/(m-1)}\right]^{-1} \tag{2.4}$$

If $d_{ij} = 0$, then $u_{ij} = 1$, and $u_{kj} = 0$, ($k = i$).

Because we do not have a clear idea on how many cluster should be for a given data set. The initial clusters are chosen randomly. The following algorithm can be used to minimize the objective function shown in formula (2.2).

Step 1. Choose an initial Z randomly, let $k = 0$.
Step 2. Calculate the initial membership function matrix $U^{(0)}$ using formula (2.4).
Step 3. Let $k = k + 1$, calculate the cluster center Z using formula (2.3).
Step 4. Calculate the matrix $U^{(k)}$, if $\left\|U^{(k)} - U^{(k-1)}\right\| < \varepsilon$, where ε is the convergent threshold value, then stop the iterative process, else go back to Step 3.

The weighted exponent m plays an important role in the FCM algorithm. Hence, choose a suitable weighted exponent m is very important when implementing FCM. For the best choice about m, the interval [1.5, 2.5] is most recommended. However, these recommendations were based on empirical studies and many not be appropriate for many real sets. Overall $m = 2$ is used most. Based on a local optimality test for FCM solutions, Yu et al. [6] created theoretical rules for selecting a suitable weighted exponent m depends on the data set itself.

3 Multiple Neural Networks (MNN) Architecture and Algorithm

3.1 Modeling Multiple Neural Networks

The basic idea of multiple neural networks is to build several independently trained neural networks with relevant features, and to estimate a given input sample data by obtaining a output from each individual neural network and then utilizing combination methods to decide the final output. Here, the FCM is employed as building blocks for MNN. The training sample data is clustered into c subsets via FCM, and then all subnets are trained by subsets respectively.

Methods for combining multiple neural networks can be divided into the following approaches [7]: simple averaging and weighted averaging, non-linear combining methods, supra bayesian, stacked generalization. In this paper, a new fuzzy integrated method is proposed, which is divided into two steps: firstly, degrees of membership are gained via the relationship between a new sample data and each subset; secondly, the outputs of subnets are synthesized to obtain the final output of MNN by the former degrees. Suppose that the MNN has c subnets, the output Y of the MNN is shown:

$$Y = \sum_{i=1}^{c} w_i f_i(X) \tag{3.1}$$

where, X is a new sample data, w_i is the degree of membership of a new X to the i-th subnet ($i = 1,2,\cdots,c$), $f_i(X)$ is the output of the subnet i. Estimating online, the degree $\{w_1, w_2, \cdots, w_c\}$ is implemented as the subnet weight, and the final output of the MNN is calculated by the formula (3.1).

w_i can be decided by the following method: w_i is related to the Euclidian distance between a new input sample data X and z_i which is the center of i-th cluster, and calculated by the formula as follows:

$$\begin{cases} w_i = 1, w_{j \neq i} = 0, & if \quad d_i = 0, \quad i = 1,\cdots,c \\ w_i = \left(\dfrac{1}{d_i}\right) \Big/ \left(\displaystyle\sum_{i=1}^{N} \dfrac{1}{d_i}\right), & otherwise \end{cases} \tag{3.2}$$

where, $d_i = \|X - z_i\|^2$ and $\sum_{i=1}^{c} w_i = 1$, $w_i \in [0, \ 1]$.

3.2 MNN Algorithm

MNN algorithm is shown as follows:

1) The training and the checking sample data set are chosen in the sample data set. The sample data is Smoothed and normalized before simulation. Then the training sample data set is classified into c clusters and corresponding cluster center z_i ($i = 1,2,\cdots,c$) by using FCM.
2) The structure of each subnet is built and trained by the cluster respectively.
3) $f_i(X)$,which is the output of the i-th subnet, is obtained, where X is a new input sample data.
4) The degree of membership of each subnet and weights of general output node w_i are updated by formula (3.2).
5) The final output of MNN is synthesized by formula (3.1).

4 Experimental Validation

In coal preparation plant with jig washer, the status- of-loose of jig washer bed, which describes how efficient the jig washer could be working, is the main factor affecting the separation effect and operating jig washer. Keeping good status-of-loose of jig washer bed, we can ensure the separation effect of jig washer, quality of products and the competitive advantages of the enterprise in global economy. But status-of-loose is difficult to measure on-line directly in industrial process situations. Therefore, status-of-loose estimation becomes one of key problems that we concern mostly. According to the analysis of processing mechanism, four key variables, which can be measured on-line by buoy sensor, are chosen to be input variables of the MNN. So the status-of-loose can be described by the nonlinear function as follows:

$$P = f(v_0, h_0, h_m, T) \tag{4.1}$$

where *P* is the status-of-loose of jig washer bed, v_0 is the beginning speed of up stream, h_0 is thickness of jig washer bed when it is compactness, h_m is the max range of buoy jumpiness, and *T* is the jigging period and $f(\ )$ represents complex nonlinear function. Data of input variables are collected from online sensor, and jigging separation effect according to former sampling time can be obtained, so the value of *P* can be analyzed offline using the approach that is proposed in the reference [8]. From the industrial field, we collect 400 samples. After smoothing and normalizing all samples, the first 350 samples are used for training the MNN, while the remaining 50 samples are used for verifying the identified MNN only. The training sample data is clustered by the FCM, here, weighted exponent $m = 2$ and cluster number $c = 4$ by several simulating test. And the results of clustering analysis are shown in Table 1.

Table 1. The results of clustering analysis

Cluster c	Clustering center					Sample number
	v_0	h_0	h_m	T	P	
1	39.27	17.42	21.48	0.98	0.17	76
2	45.09	20.74	26.26	1.05	0.43	71
3	42.75	19.60	24.64	0.97	0.74	131
4	40.80	18.43	22.92	1.03	0.53	72

Each subnet includes four inputs and one output with the same structure. Input variables include v_0、h_0、h_m、T and output variable is *P*. Three kinds of structures are adopted in subnets respectively.

① BP network: Each subnet includes one hidden layer, which has 14 nodes. Levenberg-Marquardt learning algorithm is adopted as training algorithm of subnet. This is a kind of improved BP algorithm which has rapid training rate;

② RBF network: Each subnet includes one hidden layer, which has 14 nodes. The center of the radial basis function is chosen randomly. All subnets are trained by the least squares estimate algorithm;

③ ANFIS network: Membership functions assigned to each input variable is bell-shaped function and the number of membership functions was arbitrarily set to 2. So the number of rules is 2^4=16. The number of rules is not too large to lead to "dimension curse". Parameters of each ANFIS are identified by hybrid learning algorithm [9]. The learning ability and the generalization of the model can be evaluated by the root mean squared error (RMSE):

$$RMSE = \sqrt{\frac{1}{N}\sum_{i=1}^{N}\left(P_i - \hat{P}_i\right)^2} \tag{4.2}$$

Where, N is the number of sample data; P_i and $\hat{P}_i$ are the i-th real sample value and model estimated value. These structures of network are chosen in single neural network and MNN respectively. Simulation results are shown in Table.2 (each value is the average of twenty runs), where $RMSE_{trn}$ is training RMSE, and $RMSE_{chk}$ is checking RMSE.

Table 2. Simulation error of three kind of neural networks

Single neural network			Multiple neural networks		
Model	$RMSE_{trn}$	$RMSE_{chk}$	Model	$RMSE_{trn}$	$RMSE_{chk}$
ANFIS	0.0154	0.0544	ANFIS-MNN	0.0027	0.0049
RBF	0.0214	0.0896	RBF-MNN	0.0034	0.0097
BP	0.0821	0.6487	BP-MNN	0.0103	0.0591

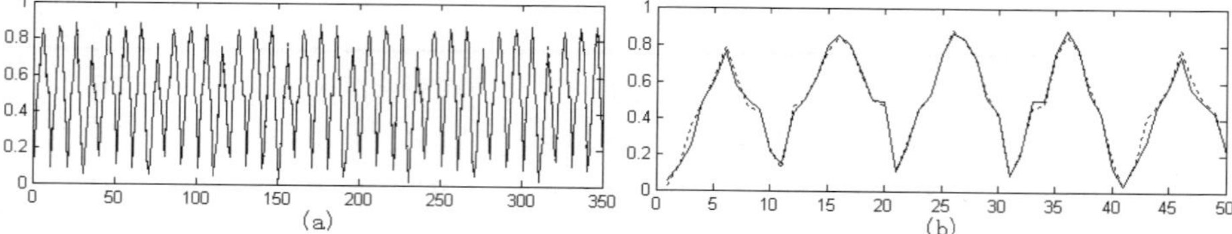

Fig. 1. The simulation results of ANFIS-MNN: (a) Results of training, (b) Results of checking. (solid line is real value and dashed line is estimated value)

Table 2 indicates that single neural networks have poor generalization. For single neural network, the best performance is a $RMSE_{chk}$ of 0.0544 from ANFIS network and the worst performance is a $RMSE_{chk}$ of 0.6487 from BP network. This demonstrates the variation in single neural network performance and the non-robust nature of single neural network. It is also shown in Table 2 that, for multiple neural networks, BP-MNN gives the worst performance and the best performance achieved by ANFIS-MNN, where the $RMSE_{chk}$ are 0.0591 and 0.0049 respectively. Training error of RBF-MNN and ANFIS-MNN are similar, but prediction accuracy of ANFIS-MNN is better than RBF-MNN obviously. Therefore, multiple neural networks can get better performance than the best of single neural network. And ANFIS-MNN has better robustness and generalization than other models in this paper. Limited by the paper space, Fig.1 only shows the results of ANFIS-MNN, and indicates the error is very little, the prediction accuracy is very high and the performance is quite well.

5 Conclusion

In the MNN, the generalizations of subnets are not the same and, therefore, different networks generate different errors. Combining these subnets can improve the robustness of this neural network model by sharing and averaging out these errors. A new fuzzy integrated multiple neural network to solve the case based nonlinear modeling problem is proposed in this paper. It is shown that fuzzy integrated combination and combining subnets of various structures generally improve model performance. Simulation results and actual application demonstrate that this model has better generalization, prediction accuracy and wider potential application online. Because support vector machines (SVM) can solve small-sample learning problems better, subnets should be built by SVM in the future.

Acknowledgement

This research is supported by National Natural Science Foundation of China under grant 70533050, Postdoctoral Science Foundation of China under grant 2005037225, Postdoctoral Science Foundation of Jiangsu Province under grant [2004]300 and Young Science Foundation of CUMT under grant 4465.

References

1. Yan, P.F.: Some Views on the Research of Multilayer Feed Forward Neural Networks. Acta Automatica Sinica 23 (1997) 129-135 (in Chinese)
2. Wolpert, D.H.: Stacked Generalization. Neural Networks 5 (1992) 241-259
3. Cho, S.B., Kim, J.H.: Combining Multiple Neural Networks by Fuzzy Integral for Recognition. IEEE Transactions on Systems, Man and Cybernetics 25 (1992) 380-384
4. Kegelmeyer, W.P., Bowyer, K.: Combination of Multiple Classifiers Using Local Accuracy Estimates. IEEE Transactions on Pattern Anal. and Machine Intelligence 19 (1997) 405-410
5. Bezdek, J.C.: Pattern Recognition with Fuzzy Objective Function Algorithms. Plenum Press, New York (1981)
6. Yu, J., Yang, M.S.: Optimality Test for Generalized FCM and Its Application to Parameter Selection. IEEE Transactions on Fuzzy Systems 13 (2005) 164-176
7. Ahmad, Z., Zhang, J.: A Comparison of Different Methods for Combining Multiple Neural Networks Models. Proceedings of International Joint Conference on Neural Networks, Vol. 1 (2002) 828-833
8. Cheng, J., Guo, Y.N., et al.: Status Evaluation of Loose of Jig Bed Based on Fuzzy Inference System. Journal of China University of Mining and Technology 13 (2003) 142-144
9. Jang, J.S.R.: ANFIS: Adaptive-network-based Fuzzy Inference System. IEEE Transactions on Systems, Man and Cybernetics 23 (1993) 665-685

Nonlinear System Identification Using Multi-resolution Reproducing Kernel Based Support Vector Regression

Hong Peng, Jun Wang, Min Tang, and Lichun Wan

School of Mathematics & Computer Science, School of Electric Information, Xihua University, Chengdu, Sichuan, 610039, China
ph66@tom.com

Abstract. A new reproducing kernel in reproducing kernel Hilbert space (RKHS), namely the multi-resolution reproducing kernel, is presented in this paper. The multi-resolution reproducing kernel is generated by scaling basis function at some scale and wavelet basis function with different resolution. Based on multi-resolution reproducing kernel and ν- support vector regression (ν-SVR) method, a new regression model is proposed. The regression model used to nonlinear system identification, incorporate the advantage of the support vector machines and the multi-resolution property of wavelet. Simulation examples are given to illustrate the feasibility and effectiveness of the method.

1 Introduction

Nonlinear system identification is a crucial but complex problem. There have been numerous recent papers in the area based on multi-layer perception (MLP), radial basis function (RBF) networks, fuzzy neural networks (FNN), wavelet networks (WNN), etc. The support vector machine (SVM) is a new universal learning machines proposed by Vapnik et al.[1], [2], which is successfully applied to function regression [3], [4]. SVM use support vector kernel to map the data from input space to a high-dimensional feature space in which the problem becomes linearly separable. The SVM based on structure risk minimum principle have better generalization capability than other approximate method of nonlinear function. In this paper, we present a new ν- support vector regression model based on multi-resolution reproducing kernel, which is an important spread to the standard support vector regression. In the regression model, the multi-resolution reproducing kernel, which is a reproducing kernel in reproducing kernel Hilbert space (RKHS), is generated by scaling basis function at some scale and wavelet basis function with different resolution. Thus, the regression model not only has the advantages of support vector machine, but also has the capability of multi-resolution which is useful to approximate nonlinear function. We use the regression model to nonlinear system identification problem. It demonstrates that the regression model is applicable to nonlinear system identification and outperforms many previous methods.

J. Wang et al. (Eds.): ISNN 2006, LNCS 3972, pp. 790–795, 2006.

2 Nonlinear System Identification

We consider the following nonlinear system model, which is a nonlinear discrete-time system:

$$y(k) = f(y(k-1), \cdots, y(k-n), u(k-1), \cdots, u(k-m)) + e(k), \tag{1}$$

where $y(k)$ and $u(k)$ are the input and output of system at time k respectively, $e(k)$ is Gaussian white noise, $e(k) \sim N(0, \sigma^2)$, $f(x)$ is an unknown nonlinear function belonging to some functional space (continuous, square integrable, etc.), m and n are two given positive integer. Let us define the following notation:

$$x(k) = [(y(k-1), \cdots, y(k-n), u(k-1), \cdots, u(k-m)]^T, \tag{2}$$

thus, Eq.(1) can be rewritten as follows:

$$y(k) = f(x(k)) + e(k). \tag{3}$$

Let $D = \{(x_i, y_i)|i = 1, \cdots, N, x_i \in R^d, y_i \in R\}$, in which $d = m + n$, be a sample of the input-output pair (x, y), referred to as the training data set. The identification problem of nonlinear system is to find a estimator $\hat{f}$ of f based on the data sample D.

Our approach to nonlinear system identification in this paper is to estimate function f by using ν- support vector regression (ν-SVR) with multi-resolution reproducing kernel.

3 Multi-resolution Reproducing Kernel

3.1 Multi-resolution Analysis

Here, we briefly review some results from the theory of wavelet that are relevant to this work. For more comprehensive discussion of wavelets, see e.g., [5]. Consider a sequence $V_i (i \in Z)$ of closed subspaces of $L^2(R^d)$. Based on multi-resolution analysis theory, the space V_0 can be decomposed as

$$\begin{aligned} V_0 = V_1 \bigoplus W_1 &= V_2 \bigoplus W_2 \bigoplus W_1 = V_3 \bigoplus W_3 \bigoplus W_2 \bigoplus W_1 \\ &= \cdots = V_L \bigoplus W_L \bigoplus \cdots \bigoplus W_2 \bigoplus W_1, \end{aligned} \tag{4}$$

and

$$V_m = span\{\varphi_{m,k} = 2^{-m/2}\varphi(2^{-m}x - k)\}, \tag{5}$$

$$W_m = span\{\psi_{m,k}(x) = 2^{-m/2}\psi(2^{-m}x - k)\}, \tag{6}$$

for $m \in Z$; $\varphi(x)$ and $\psi(x)$ are the scaling function and the wavelet function respectively. Using the scaling functions and the wavelets as basis functions, $f(x)$ can be expressed via following form:

$$f(x) = \sum_{m=1}^{L} \sum_{k=0}^{I_m} d_{m,k} \psi_{m,k}(x) + \sum_{k=0}^{I'_L} c_{L,k} \varphi_{L,k}(x). \tag{7}$$

3.2 Multi-resolution Reproducing Kernel

Here, we introduce some notations that will be used through the rest of the paper: let X be a domain included in R^d and R^X be the set of function $f: X \to R$.

Definition 1. *A Hilbert space H with inner product $<\cdot,\cdot>_H$ is a Reproducing Kernel Hilbert Space (RKHS) of R^X, if (i) H is a subspace of R^X. (ii) $\forall x \in X$, $\exists M_x > 0$ such that $\forall g \in H$, $|g(x)| \leq M_x\|g\|$. (see [6])*

Definition 2. *Let H is a reproducing kernel Hilbert space of R^X, there exists an unique application $K : X \times X \to R$, called Reproducing Kernel of H so that $\forall x \in X$, $\forall g \in H$, $g(x) =< g(\cdot), K(x,\cdot) >_H$. (see [6])*

Eq.(7) show that any $f(x) \in L^2(R)$ can be represented as a low-pass approximation at scale L plus the sum of L detail (wavelet) components at different resolutions. Let H be the set of functions so that:

$$H = \{f|\exists c_{m,k}, d_{m,k}, f = \sum_{m=1}^{L}\sum_{k=0}^{I_m} d_{m,k}\psi_{m,k}(x) + \sum_{k=0}^{I_L^{'}} c_{L,k}\varphi_{L,k}(x)\}. \tag{8}$$

According to wavelet theory, obviously, H is a Hilbert space, and closed subspace of V_0 as (4), thus, for $\forall f \in H$,

$$\begin{aligned} f(x) &= \sum_{m=1}^{L}\sum_{k=0}^{I_m} < f, \psi_{m,k} > \psi_{m,k}(x) + \sum_{k=0}^{I_L^{'}} < f, \varphi_{L,k} > \varphi_{L,k}(x) \\ &= < f(\cdot), \sum_{m=1}^{L}\sum_{k=0}^{I_m} \psi_{m,k}(\cdot)\psi_{m,k}(x) + \sum_{k=0}^{I_L^{'}} \varphi_{L,k}(\cdot)\varphi_{L,k}(x) > . \end{aligned} \tag{9}$$

Since R^X has a structure of vector space, any f in H belongs to R^X. Owing to (9), for $\forall g \in H$ and $\forall x \in X$, we have

$$\begin{aligned} |g(x)| &= |\sum_{m=1}^{L}\sum_{k=0}^{I_m} < g, \psi_{m,k} > \psi_{m,k}(x) + \sum_{k=0}^{I_L^{'}} < g, \varphi_{L,k} > \varphi_{L,k}(x)| \\ &\leq \|g\| \cdot (\sum_{m=1}^{L}\sum_{k=0}^{I_m} \|\psi_{m,k}\|^2 + \sum_{k=0}^{I_L^{'}} \|\varphi_{L,k}\|^2) = \|g\| \cdot M_x, \end{aligned} \tag{10}$$

since $\varphi(x), \psi(x) \in L^2(R)$, M_x is finite. Thus, according to $Definition1$ and $Definition2$, one can conclude that H is a Reproducing Kernel Hilbert Space, and further admits an unique reproducing kernel, *i.e.*, $\forall x \in X, \forall f \in H$, $f(x) =< f(\cdot), K(x,\cdot) >$. Due to the unicity of the functional evaluation in a reproducing kernel Hilbert space, we can obtain the reproducing kernel of H as follows:

$$K(x, x') = \sum_{m=1}^{L}\sum_{k=0}^{I_m} \psi_{m,k}(x)\psi_{m,k}(x') + \sum_{k=0}^{I_L^{'}} \varphi_{L,k}(x)\varphi_{L,k}(x'). \tag{11}$$

Eq.(11) show that the multi-resolution reproducing kernel $K(x', x)$ is a reproducing kernel in reproducing kernel Hilbert space, and is generated by scaling basis function at scale L and wavelet basis function with different resolution.

4 Support Vector Regression Model Based on Multi-resolution Reproducing Kernel

We now describe how ν- support vector regression (ν-SVR) approach based on multi-resolution reproducing kernel may be solve the system identification problem. Suppose training data D are independent identically distributed (*i.i.d.*), $D = \{(x_1, y_1), \ldots, (x_n, y_n)\} \subset X \times R$, where $x_i \in X \subset R^d$ denote input vector, and $y_i \in R$ is desired output. A suitable cost function should been chosen for support vector regression, such as the ε-insensitive loss $|y - f(x)|_\varepsilon = max\{0, |y - f(x)| - \varepsilon\}$. This does not penalize errors below some $\varepsilon > 0$, chosen a priori.

For ν-SVR [4], we first define a mapping $\Phi(x)$ from the space X of regressors to the possibly infinite dimensional hypothesis space H, in which an inner product $< \cdot, \cdot >_H$ is defined. We choose to limit our choice of regression function $f(\cdot)$ to the class of functions which can be expressed as inner products in H, taken between some weight vector w and the mapped regressor $\Phi(x)$, *i.e.*, $f(x) =< \omega, \Phi(x) >_H +b$. The regression function in the hypothesis space is consequently linear, and thus the nonlinear regression problem of estimating $f(x)$ has become a linear regression problem in the hypothesis space H. Note that the mapping $\Phi(\cdot)$ need never be computed explicitly; instead, we use the fact that if H is the reproducing kernel Hilbert space induced by $K(\cdot, \cdot)$, then writing $\Phi(x) = K(x, \cdot)$, we get $\Phi(x) = K(x, \cdot)$. The latter requirement is met for kernels fulfilling the Mercer condition.

We now describe the optimization problem to be undertaken in finding $f(\cdot)$. All support vector regression methods involve the minimization of a regularized risk functional, which represents a tradeoff between smoothness and training error (the latter is determined by the cost functional). In the case of the ν-SVR method, the regularized risk R_{reg} at the optimum is given by

$$\min_{w,b,\epsilon} R_{reg} = \frac{1}{2}\|\omega\|^2 + C(\nu\epsilon + \frac{1}{n}\sum_{i=1}^{n} |y - f(x)|_\varepsilon). \tag{12}$$

The terms C and ν in (12) specify the tradeoff between model simplicity. Solving (12) is equivalent to finding

$$min \quad \frac{1}{2}\sum_{i,j=1}^{n}(\alpha_i + \alpha_i^*)(\alpha_j - \alpha_j^*)K(x_i, x_j) + \sum_{i=1}^{n} y_i(\alpha_i - \alpha_i^*)$$

$$\sum_{i=1}^{n}(\alpha_i - \alpha_i^*) = 0, \sum_{i=1}^{n}(\alpha_i + \alpha_i^*) \leq C\nu, 0 \leq \alpha_i, \alpha_i^* \leq C/n. \tag{13}$$

Hence, regression estimation of f is obtained as follows:

$$f(x) = \sum_{i=1}^{n} (\alpha_i - \alpha_i^*) K(x, x_i) + b. \tag{14}$$

The kernel function $K(x', x)$ is a dot product in the hypothesis space, which must satisfy the Mercer condition. It is be proved easily that the multi-resolution reproducing kernel (11) satisfy the Mercer condition, and is a admissive SV kernel. In this work, the scaling reproducing kernel is used as support vector kernel of support vector regression model. According to (11) and (14), the support vector regression model base on the multi-resolution reproducing kernel is:

$$f(x) = \sum_{i=1}^{n} (\alpha_i - \alpha_i^*) [\sum_{m=1}^{L} \sum_{k=0}^{I_m} \psi_{m,k}(x) \psi_{m,k}(x_i) + \sum_{k=0}^{I_L'} \varphi_{L,k}(x) \varphi_{L,k}(x_i)] + b. \tag{15}$$

5 Simulation Result

In this Section, two examples of nonlinear system identification problem is given to demonstrate the learning capability of the presented support vector regression model based on multi-resolution reproducing kernel (MRK-SVR). In these examples, Meyer scaling function and wavelets are selected to generate the multi-resolution reproducing kernel. For the purpose of comparing the MRK-SVR with classical methods, the measure proposed in [7] is adopted as the performance index, *i.e.*,

$$J = \sqrt{\frac{\sum_{i=1}^{n} (\hat{y}_i - y_i)^2}{\sum_{i=1}^{n} (y_i - \bar{y})^2}} \qquad with \qquad \bar{y} = \frac{1}{n} \sum_{i=1}^{n} y_i. \tag{16}$$

In Eq.(17), y_i is the desired output and $\hat{y}_i$ is the estimated output.

Example 1. The follows piecewise function was studied in [7], *i.e.*,

$$f(x) = \begin{cases} -2.18x - 12.86 & -10 \le x < -2 \\ 4.246x & -2 \le x < 0 \\ 10e^{-0.05x-0.5} \sin[(0.03x + 0.7)x] & 0 \le x \le 10. \end{cases} \tag{17}$$

Here, we sampled 200 points distributed uniformly over $[-10, 10]$ as training data. The approximation errors J using our MRK-SVR, SVR with Meyer wavelet kernel, SVR with Gaussian kernel and WNN[7] are 0.04643, 0.06932 ,0.08467 and 0.05057, respectively. Obviously, the performance of our MRK-SVR is superior to that of other methods.

Example 2. The nonlinear dynamical system to be identified is defined as follows:

$$y(t+1) = \frac{[y(t)y(t-1)y(t-2)x(t-1)(y(t-2)-1) + x(t)]}{[1 + y^2(t-1) + y^2(t-2)]}. \tag{18}$$

The input signal $x(t)$ is selected as

$$x(t) = \begin{cases} \sin(2\pi t/250) & t \leq 500 \\ 0.8\sin(2\pi t/250) + 0.2\sin(2\pi t/25) & t > 500. \end{cases} \quad (19)$$

Data on-line for 800 time steps are collected to form input-output sample pairs for constructing the MRK-SVR. The approximation errors J using our MRK-SVR, SVR with Meyer wavelet kernel and SVR with Gaussian kernel are 0.05295, 0.07801 and 0.09463, respectively. The simulation results show that MR-SVR has good identification capability for nonlinear dynamic system.

6 Conclusions

A new support vector regression model for nonlinear system identification problem has been introduced. In the regression model, the multi-resolution reproducing kernel used to a admissive SV kernel is actually a reproducing kernel in reproducing kernel Hilbert space, and is generated by scaling basis function at some scale and wavelet basis function with different resolution. The regression model not only has the advantages of support vector machine, but also has the capability of multi-resolution which is useful to approximate nonlinear function. Simulation example have been used test the method and to compare it with classical approach, showing the feasibility and effectiveness of the method in nonlinear system identification problem.

Acknowledgment

This work is supported by the importance project foundation of the education department of Sichuan Province, China ($No.2005A117$).

References

1. Vapnik, V.: The Nature of Statistical Learning Theory. Springer-Verlag, New York (2001)
2. Cortes, C., Vapnik, V.: Support Vector Networks. Machine Learning **20** (1995) 273-297
3. Smola, A., Schölkopf, B.: A Tutorial on Support Vector Regression. NeuroCOLT Technical Report NC-TR-98-030, Royal Holloway College, University of London, UK, (1998)
4. Schölkopf, B., Smola, A., Williamson, R.C., Bartlett, P.L.: New Support Vector Algorithms. Neural Computation **20** (2000) 1207-1245
5. Mallat, S.G.: A Theory for Multiresolution Signal Decomposition: The Wavelets Representation. IEEE Trans. Pattern Anal. Machine Intell. **11** (1989) 674-693
6. Rakotomamonjy, A., Canu, S., Brucq, D.D.: Frame Kernels for Learning. ICANN, LNCS 245, (2002) 707-712
7. Zhang, Q., Benveniste, A.: Wavelet Networks. IEEE Trans. Neural Networks **3** (1992) 889-898

A New Recurrent Neurofuzzy Network for Identification of Dynamic Systems

Marcos A. Gonzalez-Olvera and Yu Tang

National Autonomous University of Mexico (UNAM), Av. Universidad 3000, CP 04150
Edificio Bernardo Quintana, Engineering Faculty, Mexico City, Mexico
mangel@verona.fi-p.unam.mx, tang@servidor.unam.mx

Abstract. In this paper a new structure of a recurrent neurofuzzy network is proposed. The network considers two cascade-interconnected Fuzzy Inference Systems (FISs), one recurrent and one static, that model the behaviour of a unknown dynamic system from input-output data. Each FIS's rule involves a linear system in a controllable canonical form. The training for the recurrent FIS is made by a gradient-based Real-Time Recurrent Learning Algorithm (RTRLA), while the training for the static FIS is based on a simple gradient method. The initial parameter conditions previous to training are obtained by extracting information from a static FISs trained with delayed input-output signals. To demonstrate its effectiveness, the identification of two non-linear dynamic systems is included.

1 Introduction

The problem of system identification is one of the most important aspects to solve not only in the control engineering, but in other knowledge areas. In general, the current output of a dynamic system is a function of its previous outputs and inputs, which makes the identification problem more complex than for static systems [1]. When linear models such as ARX or ARMAX turn to be insufficient, nonlinear structures should be considered. Some nonlinear models have been developed to deal with this problem [2], such as memory neural networks [3], recurrent neural networks [4], partially recurrent neural networks [5], fuzzy dynamic systems [6][7], and neuro-fuzzy networks [1]. There are several methods for training static neural and fuzzy structures, such as *ANFIS* [8] and [9], but when dealing with temporary properties the recurrency must be taken into account, which makes the training process not as straightforward as in the static ones. Some algorithms, such as Backpropagation-Through-Time (BTT) [10] have been developed, but they require a considerable computation effort and memory usage.

In this paper a new recurrent neurofuzzy network structure named *Controllable-Canonical-Form-Based Recurrent Neurofuzzy Network* (CReNN) is proposed. This structure depends on two Fuzzy Inference Systems (FISs), one recurrent and one static: the first one takes an external input and its delayed outputs (which are considered as internal variables or states to produce the internal state of the network); while the second one has a usual static Takagi-Sugeno-Kang (TSK) structure, that receives the delayed outputs from the first FIS and calculates the final output. To deal with simpler and faster algorithms, in this paper the Real-Time Recurrent Learning Algorithm (RTRLA), proposed by [4] is used, much faster and less demanding in online computation than BTT.

J. Wang et al. (Eds.): ISNN 2006, LNCS 3972, pp. 796–801, 2006.

The proposed structure has the following advantages: As the structure is given in a controllable form the on-line computational load is reduced, and the stability checking of the trained network is easier [11].

This paper is organized as follows: In Section 2 the proposed structure is presented. In Section 3 the training process for both obtaining the initial parameter conditions and training are discussed and described, and the results dealing with the identification of two nonlinear system are presented in Section 4. Conclusions are drawn in Section 5.

2 Structure of the Proposed Recurrent Neurofuzzy Network

2.1 Problem Statement

In general, a discrete nonlinear system can be seen as a dynamic mapping $\mathbf{g} : u(k) \rightarrow y(k)$, where $u(k)$ is the input and $y(k)$ the output of the system at sample time k. In the control engineering, two main structures are used for this mapping: Input-Output and State-Variable representation. The second representation is usually structured, in the SISO (Single-Input-Single Output) case, as $\mathbf{x}(k+1) = \mathbf{f}_d(\mathbf{x}(k), u(k))$, $y(k) = h_d(\mathbf{x}(k))$, where $\mathbf{x} \in \Re^n$, $u \in \Re$, $y \in \Re$ and $k \in \mathbb{Z}^+$ as the sample time. The objective then is to find a recurrent neurofuzzy structure with an input-output mapping $\hat{g} : u(k) \rightarrow \hat{y}(k)$ such that makes the distance $||\hat{g}(u) - g(u)||$ minimum, when only input-output data and some information about the system (such as the order, its equilibrium point) is available.

2.2 Proposed Network Structure

The CReNN is structured as a state-variable representation, with two FISs $\mathbf{f}(\cdot,\cdot)$ and $h(\cdot)$, one recurrent and one static, with states $\mathbf{z} \in \Re^{n_z}$, in the form

$$\hat{g} : \begin{cases} \mathbf{z}(k+1) = \mathbf{f}(\mathbf{z}(k), \mathbf{u}(k)) \\ \hat{y}(k) \quad = \quad \mathbf{h}(\mathbf{z}(k)) \end{cases} \tag{1}$$

The recurrent part $\mathbf{f}(\cdot,\cdot)$ is defined as a recurrent FIS, with as many rules as internal states considered. It is proposed, for a SISO case, that each rule takes the form:

$$\begin{aligned} \text{Rule i} \quad &: \ \text{IF } \mathbf{z}(k) \text{ is } A_i \text{ and } u(k) \text{ is } B_i \text{ THEN} \\ \mathbf{z}^i(k+1) &\stackrel{def}{=} \begin{pmatrix} & -C_i^T & \\ I_{(n_z-1)\times(n_z-1)} & & \mathbf{0}_{(n_z-1)\times 1} \end{pmatrix} \mathbf{z}(k) + \begin{pmatrix} d_i \\ \mathbf{0}_{(n_z-1)\times 1} \end{pmatrix} u(k) \\ &= \mathbf{C}\mathbf{z}(k) + Du(k) \end{aligned} \tag{2}$$

where $C_i^T = [c_{i1} \ c_{i2} \ \ldots \ c_{in_z}] \in \Re^{n_z}$, and A_i and B_i are fuzzy sets, whose membership functions are gaussian functions defined as $A_i(z_i) = \exp(-\sigma_i^2(z_i - \mu_i)^2)$ and $B_i(u) = \exp(-s_i^2(u - m_i)^2)$.

For the static FIS $h(\cdot)$ it is proposed that each rule takes usual TSK form, with as many rules as internal states, where its j-th rule is given as: IF $\mathbf{z}(k)$ is $\mathbf{H}_j$ THEN $\hat{y}^j(k) = \mathbf{g}_j^T\bar{\mathbf{z}}(k)$, where $\mathbf{g}_j^T = [g_{j0} \ g_{j1} \ g_{j2} \ \cdots \ g_{jn_z}]$ and $\bar{\mathbf{z}}^T(k) = [1 \ \mathbf{z}(k)^T]$. The membership functions for $\mathbf{H}_j$ are defined as gaussian functions $\mathbf{H}_j = \exp(-\sum_l \alpha_{jl}^2 (z_l - \beta_{jl})^2)$.

The conjunction operation is assumed to be taken as the arithmetic multiplication, so the firing strength for each FIS is, for $\mathbf{f}$, defined as

$$R_f^i(k) = \exp\left\{-\sigma_i^2(z_i(k) - \mu_i)^2 - s_i^2(u(k) - m_i)^2\right\}$$

while for h is $R_h^p(k) = \exp\left\{-\sum_q \alpha_{pq}^2(z_q(k) - \beta_{pq})^2\right\}$.

For both systems, the defuzzification process is calculated by a weighted averages, for $\mathbf{f}$, $\mathbf{z}(k+1) = \dfrac{\sum_r R_f^r(k)\mathbf{z}^r(k+1)}{\sum_r R_f^r(k)}$, and for h $\hat{y}(k) = \dfrac{\sum_r R_h^r(k)\hat{y}^r(k)}{\sum_r R_h^r(k)}$.

2.3 Stability Checking

As the recurrent part of the net is given as a dynamical fuzzy discrete system, its stability properties must be studied, where the stability checking can be done on-line (to ensure the stability of the resulted network) or at the very end of the training process. As (2) is given a form suitable for this purpose [12], it can be shown[12] that if the next condition is satisfied, the recurrent fuzzy system $\mathbf{f}(\cdot,\cdot)$ will be Lyapunov stable:

Condition 2.1 *Dynamic fuzzy system Lyapunov stability.*

Let $\mathbf{f}(\cdot,\cdot)$ be a recurrent fuzzy system with rules defined as in (2). This system is local and asymptotically Lyapunov stable if it exists a matrix $P > 0$ such that the set of inequalities $\mathbf{C}_j^T P\mathbf{C}_j - P < 0$ is satisfied $\forall\ j = 1,\ \ldots,\ n_z$

3 Training Algorithm for CReNN

This section presents a method to find the parameters of the CRFIS $\hat{g}$, such that minimizes the criteria $J(k) = \frac{1}{2}(\hat{y}(k) - y(k))^2$ over k. To tune the parameters of the CReNN a gradient-based training method is proposed; so, in order to initialize them and avoid getting trapped into a local minimum far from the global minimum, we propose an initialization procedure, where the number of internal variables n_z must be given in advance. For the sake of simplicity and without loss of generality, $n_z = 3$ and $n_u = 1$ are assumed.

3.1 Initialization Algorithm

Let $U \in \Re^N$ and $Y \in \Re^N$ be the training signals containing N points:

1. Normalize U and Y.
2. Train a static TSK-FIS $\mathbf{f}_s$ with gaussian membership functions using ANFIS [8] with n_z rules, such that its input is $U_{FIS} = [u(k)\ u(k-1)\ \ldots\ u(k-n_z+1)\ y(k)\ y(k-1)\ \ldots\ y(k-n_z+1)]$, and its desired output is $y(k+1)$; so each rule would have the structure, for the i-th rule: IF $y(k)$ is A_{i1} and $y(k-1)$ is A_{i2} and $y(k-2)$ is A_{i3} and $u(k)$ is B_{i1} and $u(k-1)$ is B_{i2} and $u(k-2)$ is B_{i3} THEN $\hat{y}^i(k+1) = a_{i0} + a_{i1}y(k) + a_{i2}y(k-1) + a_{i3}y(k-2) + b_{i1}u(k) + b_{i2}u(k-1) + b_{i3}u(k-2)$

3. The dynamics of each rule is contained in the parameters a_{ij}, $i \neq 0$, so make $C_i^T = [a_{i1}\ a_{i2}\ \dots\ a_{in_z}]$.
4. Initialize σ_{ij} and μ_{ij} by scaling those of A_{ij} in $\mathbf{f}_s$. If each rule of f_s is seen as an independent linear system, then under a step excitation $u(k) = 1,\ k \geq 0$ the final value of each discrete system and so the used scale, using the $\mathcal{Z}$-transform, will be $scale_i = \frac{Z_{final}}{Y_{final}} = \frac{1}{b_{i1}+b_{i2}+b_{i3}}$.
5. Equal the values of s_{ij} and m_{ij} of $\mathbf{f}$ with those of B_{ij}.
6. Generate an historical of $\mathbf{z}(k)$ using the initialized $\mathbf{f}$ and U, and with it train h using ANFIS. Consider the input as $\mathbf{z}(k)$ and the objective output as $y(k)$. Take $\mathbf{z}(0) = \mathbf{0}_{n_z \times 1}$.

3.2 Parameter Tuning of the CReNN

Once the CReNN has been initialized, the next step is to tune the parameters in order to move them towards the optimum. Considering the single output case, the goal is to minimize the given cost function. The gradient-based Real-Time Recurrent Learning Algorithm (RTRLA)[4] takes into account the recurrency of the network, so the parameter θ of a certain recurrent network is updated by $\theta(k+1) = \theta(k) - \eta_\theta \frac{\partial J(k)}{\partial \theta}$ where $\frac{\partial J(k)}{\partial (\theta}$ is itself a recurrent equation.

The parameters related with the static part of the CRFIS are trained using the simple Gradient-Based Algorithm. For example, the parameters $\mathbf{g}_j$ are updated by $\mathbf{g}_j(k+1) = \mathbf{g}_j(k) - \eta_g \frac{\partial J(k)}{\partial \mathbf{g}_j}$, where $\frac{\partial J(k)}{\partial \mathbf{g}_j(k)} = (\hat{y}(k) - y(k)) \frac{\partial \hat{y}(k)}{\partial \mathbf{g}_j}$ and $\frac{\partial \hat{y}(k)}{\partial \mathbf{g}_j(k)} = \frac{R_h^j(k)}{\sum_n R_h^n(k)} \bar{\mathbf{z}}^T(k)$. The update rules for α_{ij} and β_{ij} can be obtained in a similar way.

The parameters related to $\mathbf{f}$ must be updated thinking on its recurrency. As the other training rules can be derived in a similar way, only the update rules for C_i^T is shown. Each vector C_i^T is updated by $C_i^T(k+1) = C_j^T(k) - \eta_c(\hat{y}(k) - y(k)) \frac{\partial \hat{y}(k)}{\partial C_i^T}$, where $\frac{\partial \hat{y}(k)}{\partial C_i^T} = \frac{1}{\sum_r R_h^r(k)} \left(\sum_r \hat{y}_r(k) \frac{\partial R_h^r(k)}{\partial C_i^T} + R_h^r \bar{\mathbf{g}}_r^T \frac{\partial \mathbf{z}(k)}{\partial C_i^T} \right) - \frac{1}{(\sum_r R_h^r(k))^2} \left(\sum_r R_h^r(k) \hat{y}_r(k) \right)$, with $\bar{\mathbf{g}}_j = [g_{j1}\ g_{j2}\ \dots\ g_{jn_z}]$ and $\frac{\partial R_h^r(k)}{\partial C_i^T} = -2R_h^r \left(\sum_l \alpha_{rl}^2 (z_l(k) - \beta_{rl}) \frac{\partial z_l(k)}{\partial C_i^T} \right)$.

These equations are under the recurrent equation given by

$$\frac{\partial \mathbf{z}(k)}{\partial C_i^T} = \frac{\sum_r \left(R_f^r(k-1) \frac{\mathbf{z}^r(k)}{\partial C_i^T} + \mathbf{z}^r(k) \left(\frac{\partial R_f^r(k-1)}{\partial C_i^T} \right)^T \right)}{\sum_r R_f^r(k-1)} - \frac{\sum_r \left(R_f^r(k-1) \mathbf{z}^r(k) \left(\frac{\partial R_f^r(k-1)}{\partial C_i^T} \right)^T \right)}{\left(\sum_r R_f^r(k-1) \right)^2},$$

$\frac{\partial z_1^r(k)}{\partial C_i^T} = C_i^T \frac{\partial \mathbf{z}(k-1)}{\partial C_i^T} + \mathbf{z}^T(k-1)\delta_{ir}$, $\frac{\partial z_j^r(k)}{\partial C_i^T} = \frac{\partial z_{j-1}(k-1)}{\partial C_i^T}$, for $j = 2, 3 \dots n_z$; and finally $\frac{\partial R_f^r(k-1)}{\partial C_i^T} = -2R_f^r(k-1) \sum_s \sigma_{rs}^2 (z_s(k-1) - \mu_{rs}) \frac{\partial z_s(k-1)}{\partial C_i^T}$.

4 Results

4.1 Simulation Results

The first system to be identified is the same plant used by [13] (widely used as a benchmark), given is the difference equation $y(k) = f_N(y(k-1), y(k-2), y(k-3),$

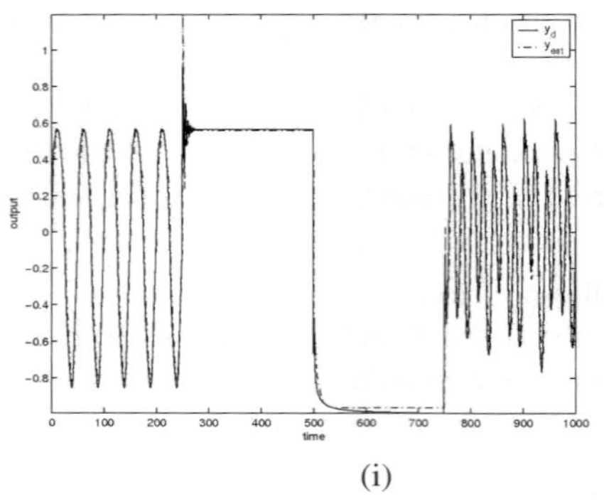

(i)

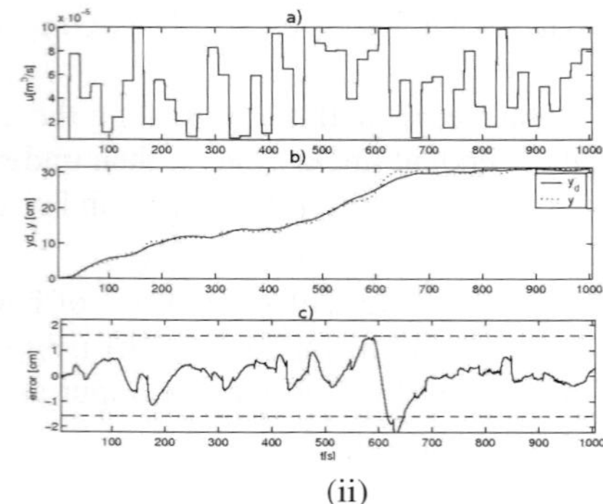

(ii)

Fig. 1. Results. (i) Simulated. The dashed line is the estimated output (ii) Experimental. a) Input signal, b) Output signal, the dashed line is the estimated output, c)Identification error within 5%.

$u(k), u(k-1))$, where $f(x_1, x_2, x_3, x_4, x_5) = \frac{x_1 x_2 x_3 x_5 (x_3 - 1) + x_4}{1 + x_3^2 + x_2^2}$. For this example, a recurrent network with 3 internal states ($n_z = 3$) and a single external input $n_u = 1$ is considered. After applying the initializing procedure described in Section 3, considering the training constants $\eta_\sigma = \eta_\mu = \eta_s = \eta_m = \eta_c = 10^{-6}$, $\eta_\alpha = \eta_\beta = 2 \cdot 10^{-3}$, $\eta_g = 3 \cdot 10^{-4}$, the values for C_i^T, $C_1^T = [0.7552\ -0.1209\ 0.1813]$, $C_2^T = [0.5979\ -0.2267\ -0.03667]$ and $C_3^T = [0.3887\ -0.5253\ -0.07293]$ were obtained (for the sake of space, the results for the other parameters σ, μ, etc. are not shown). For this results, a common positive definite matrix $P = [4.28\ -0.716\ 0.127;\ -0.716\ 2.54\ -0.173;\ 0.127\ -0.713\ 1.09]$ was found such that complies with Condition 2.1, so the obtained net is local and asymptotically stable.

In Figure 1.i it is shown the simulated result, where a MSE of 0.0418 is achieved. Whereas other structures such as RSONFIN[1], DFNN[14] and TRFN-S[1] shown minor identification identification error (0.0248, 0.0025 and 0.0084, with 30, 39 and 51 parameters, respectively), these structures do not consider a Lyapunov stability analysis.

4.2 Experimental Results

The second identification result is given by a nonlinear laboratory three water tank array DTS200 system is identified, which is known to be nonlinear. The water is fed through a voltage-controlled pump to a first tank, flows through a base connection to a second tank, which does so to the third tank. The last tank returns the water to the common reservoir. The system output is taken as the liquid level in the second tank. As it is known that the system has three water reservoirs, $n_z = 3$ is assigned. The results are shown in Figure 1.ii, where a normalized MSE of 0.02 is achieved.

The DTS200 system is known to be also asymptotically stable. After training the net, the values for C_i^T, $C_1^T = [1.260\ 0.1584\ -0.4194]$, $C_2^T = [1.047\ 0.2335\ -0.2809]$ and $C_3^T = [1.286\ 0.1884\ -0.4755]$ were obtained; and the found common positive definite matrix such that complies with Condition 2.1 was $P = [563.0\ -293.0\ -260.0;\ -293.0\ 269.0\ 21.8;\ -260.0\ 21.8\ 235.0]$, so the identified network is local and asymptotically stable.

5 Conclusions

This paper has presented a new structure for Recurrent Fuzzy Systems, a Controllable-Canonical-Form Based Recurrent Neurofuzzy Network (CReNN), whose the structure has two cascaded FIS: a recurrent one that handles the recurrency, trained using a Real-Time Gradient-Based Recurrent Learning Algorithm; and a second static one trained by a standard Gradient Algorithm. The parameters were initialized in order to avoid getting trapped into a local minima far from the global one. It was also presented a Lyapunov stability condition, based in the dynamic fuzzy systems. The method was applied to identify two nonlinear systems, one via simulation and the second one using real experimental data, having a performance near to other structures. Simulation and experiment results validated the proposed scheme.

Acknowledgements

The authors want to thank to UNAM program, PAPIIT-IN110402 and CONACYT for their support on this work.

References

1. Juang, C.F.: A TSK-Type Recurrent Fuzzy Network for Dynamic Systems Processing by Neural Network and Genetic Algorithms. IEEE Trans. Fuzzy Systems **10** (2002) 155–170
2. Nelles, O.: Nonlinear System Identification. Springer, Berlin (2001)
3. Sastry, P., Santharam, G., Innikrishnan, K.: Memory Neuron Networks for Identification and Control of Dynamic Systems. IEEE Trans. Neural Networks **5** (1994) 306–319
4. Williams, R.J., Zipser, D.: A Learning Algorithm for Continually Running Fully Recurrent Neural Networks. Neural Computation **2** (1989) 270–280
5. Elman, J.L.: Finding Structure in Time. Cognitive Science **14** (1990) 179–211
6. Yu, W., Li, X.: Fuzzy Identification Using Fuzzy Neural Network. IEEE Trans. on Fuzzy Systems **12** (2004)
7. Wang, L.X.: Adaptive Fuzzy Systems and Control. Prentice Hall (1994)
8. Jang, J.S.R.: ANFIS: Adaptive-Network-Based Fuzzy Inference System. IEEE Transactions on Systems, Man and Cybernetics **23** (1993) 665–685
9. Chiu, S.L.: A Cluster Eestimation Method with Extension to Fuzzy Model Identification. Proc. FUZZ-IEEE'94 (1994) 1240–1245
10. Rumelhart, D., Hinton, G., Williams, R.: 8. In: Learning Internal Representations by Error Propagation. MIT Press, Cambridge (1986)
11. Lin, C.T.: Neural Fuzzy Systems: a Neuro-Fuzzy Synergism to Intelligent Systems. 1st edn. Prentice Hall (1996)
12. Kosko, B.: Fuzzy Engineering. Prentice Hall (1992)
13. Narendra, K., Parthasarathy, K. Identification and Control of Dynamical Systems Using Neural Networks. IEEE Tras. Neural Networks **1** (1990) 4–27
14. Mastorocostas, P.A., Theocharis, J.B.: A Recurrent Fuzzy-Neural Model for Dynamic System Identification. IEEE Trans. onf Systems, Man and Cybernetics - Part B: Cybernetics **32** (2002) 176–190

Identification of Dynamic Systems Using Recurrent Fuzzy Wavelet Network

Jun Wang[1], Hong Peng[1], and Jian Xiao[2]

[1] School of Electrical Information, School of Mathematics & Computer Science, Xihua University, Chengdu, Sichuan, 610039, China
jiayu@mail.sc.cninfo.net
[2] School of Electrical Engineering, Southwest Jiaotong University, Chengdu , Sichuan, 610031, China

Abstract. This paper proposes a dynamic recurrent fuzzy wavelet network (RCFWN) for identified nonlinear dynamic systems. Temporary relations are embedded in the network by adding feedback connections in the second layer of the fuzzy wavelet network. In addition, the study algorithm of the RCFWN is introduced and its stability analysis is studied. Finally, the RCFWN is applied in several simulations. The results verify the effectiveness of the RCFWN.

1 Introduction

The combination of wavelet analysis, fuzzy system and neural network is a new idea and technology, which have the advantages of better accuracy, generalization capability, fast convergence and multiresolution. Fuzzy wavelet network (FWN) was proposed in [1] and [2]. However, the proposed FWN is static network due to the inherent feedforward network structure. Since for a dynamic system [4]-[5], the output is a function of past output, past input or both identification and control of this system is not as straightforward as a static system. Recently, dynamic neural network has been demonstrated to be successful [4], which include feedback neural network and dynamic recurrent neural network(DRNN).This paper will focus on the motivation and implementation of identification schemes for nonlinear system using dynamic recurrent fuzzy wavelet network(DRFWN). Based on our previous researches [2][3], a DRFWN model is presented to deal with the problems of dynamic system identification. Methods of model identification are described. The analysis of stability performance in learning algorithm is provided.

2 Structure of the Dynamic Recurrent Fuzzy Wavelet Network

The structure of the RCFWN is illustrated in Fig.1.

Layer 1. Layer 1 is input layer. The node number is n.

J. Wang et al. (Eds.): ISNN 2006, LNCS 3972, pp. 802–807, 2006.

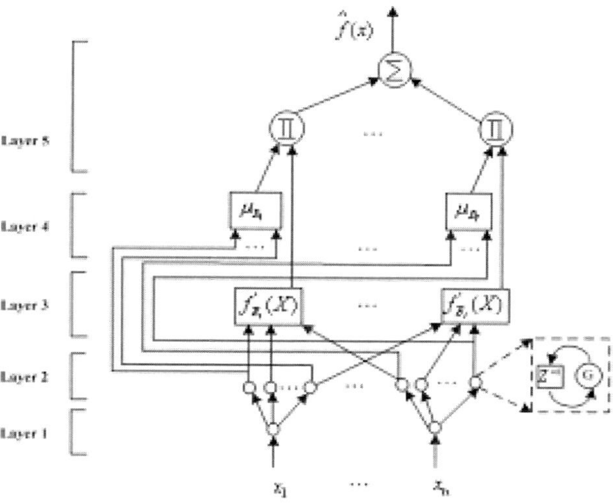

Fig. 1. The architecture of the RCFWN

Layer 2. In this layer, every node represents a linguistic variable, which compute fuzzy membership function. The B-spline wavelet membership function, a particular function, is adopted here as the membership function. Then

$$O_{ij}^{(2)} = \phi_{p_{i,j},l_{p_{i,j}+k}}\big(I_{ij}^{(2)}(k)\big) = 2^{-p_i/2}\phi\big(2^{-p_i}I_{ij}^{(2)}(k) - l_{p_{i,j}+k}\big) \tag{1}$$

where p_i and $l_{p_{i,j}+k}$ are dilation and translation of the scaling function, respectively. $i = 1, \ldots, M$, and M is the fuzzy rule number. For the approximation function network, the input of this layer is determined by

$$I_{ij}^{(2)}(k) = O_j^{(1)}(k) + d_{ij}O_{ij}^{(2)}(k-1) \tag{2}$$

where d_{ij} is the weight of the dynamic feedback. Clearly, the memory terms $O_{ij}^{(2)}(k-1)$ stores the past information of the network, so, it can realize the dynamic reflection.

Layer 3. Every node in this layer implements the fuzzy subsystem $f'_{E_i}(x)$ under the resolution factor p_i. Assuming the rule number of the fuzzy subsystem with resolution p_i is I. The output can be expressed by

$$O_i^{(3)} = f'_{E_i}(x) = \frac{\sum_{k=1}^{I_{p_i}} w_{ik} \prod_{j=1}^{n} \phi_{p_{i,j},l_{p_{i,j}+k}}\big(I_{ij}^{(2)}(k)\big)}{\sum_{k=1}^{I_{p_i}} \prod_{j=1}^{n} \phi_{p_{i,j},l_{p_{i,j}+k}}\big(I_{ij}^{(2)}(k)\big)} = \frac{\sum_{k=1}^{I_{p_i}} w_{ik}\Phi_{p_i,L_k}(X)}{\sum_{k=1}^{I_{p_i}} \Phi_{p_i,L_k}(X)} \tag{3}$$

Layer 4. In layer 4, the membership function in frequency domain can be denoted by

$$O_i^{(4)} = \mu_{B_i}(P_i, X) = \prod_{j=1}^{n} \mu_{B_i}(p_{i,j}, x_j) \tag{4}$$

Layer 5. This layer is output layer which denote fuzzy system with multiresolution capability. The output can be calculated by

$$O^{(5)} = \hat{f}(x) = \frac{\sum_{i=1}^{I} \mu_{B_i}(P_i, X) f'_{E_i}(x)}{\sum_{i=1}^{I} \mu_{B_i}(P_i, X)} = \sum_{i=1}^{I} \hat{\mu}_{B_i}(P_i, X) \frac{\sum_{k=1}^{I_{p_i}} w_{ik} \Phi_{p_i, L_k}(X)}{\sum_{k=1}^{I_{p_i}} \Phi_{p_i, L_k}(X)} \tag{5}$$

$\hat{\mu}_{B_i}(P_i, X)$ represents the when the membership function in frequency domain is the ratio of the corresponding scale function for x_j in time domain.

3 Online Learning Algorithm and Stability Analysis

In this paper, we adopt BP algorithm to train the free parameters of RCFWN network. The cost function E is defined as

$$E(k) = \frac{1}{2p} \sum_{l=1}^{p} \left[y_l^d(k) - \hat{y}_l(k) \right]^2 \tag{6}$$

where $y_l^d(k)$ is lth desired output and p denotes the number of output nodes. When the BP learning algorithm is used, the weighting vector of the RCFWN model is adjusted such that the error defined in (6) is less than the desired threshold value after a given number of training cycles.As a result, the weights d_{ij} of the dynamic feedback, the output weights w_{ik} and translation factors $l_{p_{i,j}+k}$ of RCFWN are updated by

$$W(k+1) = W(k) + \Delta W(k) = W(k) + \eta\left(-\frac{\partial E(k)}{\partial W}\right) \tag{7}$$

where η and W denote the learning rate and tuning parameters of the RCFWN. Let $W = [w, d, l]^T$. For single output system, let $e(k) = y^d(k) - \hat{y}(k)$.

Then the gradient of error $E(\cdot)$ in (6) with respect to an arbitrary weights vector W is

$$\frac{\partial E(k)}{\partial W} = -e(k) \frac{\partial \hat{y}(k)}{\partial W} = -e(k) \frac{\partial O^{(5)}(k)}{\partial W} \tag{8}$$

$\Delta W = [\Delta w, \Delta d, \Delta l]^T$ represents the change in training parameter vector. For BP algorithm, selecting an appropriate learning rate η is very importance. So, we develop some convergence theorems for selecting appropriate learning rates. Define a discrete Lyapunov function as follows:

$$V(k) = \frac{1}{2} \left[e(k) \right]^2 \tag{9}$$

where $e(k)$ denote the identification error. The change of the Lyapunov function is

$$\Delta V(k) = V(k+1) - V(k) = \frac{1}{2} \left[e^2(k+1) - e^2(k) \right] \tag{10}$$

During the training process, the error difference can be expressed by

$$\Delta e(k) \approx \frac{\partial e(k)}{\partial W} \cdot \Delta W = \left[\frac{\partial e(k)}{\partial w}, \frac{\partial e(k)}{\partial d}, \frac{\partial e(k)}{\partial l}\right] \times \left[\Delta w, \Delta d, \Delta l\right]^T \tag{11}$$

In this paper, $O^{(5)}(k)$ is the kth output system. Let $[\eta_1, \eta_2, \eta_3]^T = [\eta_w, \eta_d, \eta_l]^T$, we have the following convergence theorem.

Theorem 1. *Define*

$$D_{max} = \left[D_{1,max}, D_{2,max}, D_{3,max}\right]^T = \left[\max_k \left|\frac{\partial O(k)}{\partial w}\right|, \max_k \left|\frac{\partial O(k)}{\partial d}\right|, \max_k \left|\frac{\partial O(k)}{\partial l}\right|\right]^T \tag{12}$$

If η_i are chosen to satisfy

$$0 < \eta_i < \frac{2}{(D_{i,max})^2}, \quad i = 1, 2, 3 \tag{13}$$

Then asymptotic convergence is guarantee of the RCFWN2 identification system.

Proof. Due to

$$\Delta V(k) = \frac{1}{2}\left[e^2(k+1) - e^2(k)\right] = \Delta e(k)\left[e(k) + \frac{1}{2}\Delta e(k)\right] \tag{14}$$

Substitute (11) into (14),

$$\Delta V(k) = -\beta e^2(k) \tag{15}$$

where

$$\beta = \sum_{i=1}^{3}\left[\frac{1}{2}\eta_i\left(\frac{\partial O(k)}{\partial w_i}\right)^2\left(2 - \eta_i\left(\frac{\partial O(k)}{\partial w_i}\right)^2\right)\right] \tag{16}$$

If $\beta > 0$, $\Delta V(k) < 0$. So, the stability of the RCFWN identification system can be compact support. Then $2 - \eta_i(\frac{\partial O(k)}{\partial w_i})^2 > 0$ is required, that is

$$0 < \eta_i < \frac{2}{(D_{i,max})^2}, \quad i = 1, 2, 3 \tag{17}$$

Theorem 2. *If the learning rates $\eta = \eta_1 = \eta_2 = \eta_3$, the convergence condition of the RCFWN system is*

$$0 < \eta < \frac{2}{(D_{max})^2} \tag{18}$$

where

$$D_{max} = \max_k \|D(k)\| = \max_k \left\|\left[\left|\frac{\partial O(k)}{\partial w}\right|, \left|\frac{\partial O(k)}{\partial d}\right|, \left|\frac{\partial O(k)}{\partial l}\right|\right]^T\right\| \tag{19}$$

$\|\cdot\|$ is the usual Euclidean norm.

Proof. Since

$$\beta = \frac{1}{2}\eta\left\|\frac{\partial O(k)}{\partial W}\right\|^2\left(2-\eta\left\|\frac{\partial O(k)}{\partial W}\right\|^2\right) = \frac{1}{2}\left\|D(k)\right\|^2\eta\left(2-\eta\left\|D(k)\right\|^2\right) > 0 \quad (20)$$

We obtain $0 < \eta < \frac{2}{(D_{max})^2}$ that guarantees convergence.

Theorem 3. *If wavelet function is selected to be B-spline function, in order to guarantee the asymptotic stability of the RCFWN system, it is requested.*

$$0 < \eta_w < (2/R_I) \quad (21)$$

In which R_I *is wavelet node number of the layer* 4, $R_I = T_1 + \cdots + T_c$.

4 Simulation

Example 1. We select the following function

$$y(k+1) = \frac{y(k)y(k-1)y(k-2)u(k-1)(y(k-2)-1)+u(k)}{1+y^2(k-1)+y^2(k-2)} \quad (22)$$

Suppose $x_d = [u(k), y(k-1)]^T$ is input set. $y_d = y(k+1)$ is output. 1000 sample date are randomly selected. Defined index performance is 0.0285. Rule number is 10, study rate η_w, η_l and η_d is 0.15. Input $u(k)$ signal is random signal distributed uniformly over $[-2, 2]$. Because the wavelet membership function has compact support and the study algorithm is local, only finite scale variable and wavelet nodes are adopted in dynamic identification. After the training, the following input is added to the input of the system to verify the performance of the RCFWN.

$$u(k) = \begin{cases} \sin(2\pi k/250) & k \le 500 \\ 0.2\sin(2\pi k/40) + 0.8\sin(2\pi k/200) & k > 500 \end{cases} \quad (23)$$

Fig. 2(a) is the result of the identification and the actual output. Fig. 2(b) is the error curve to show that the two output result is very similar. The two figures are shown that the proposed RCFWN model has good generalization capability and fast convergence.

Example 2. The nonlinear dynamic system to be identified is defined as follows:

$$\begin{aligned} y(k) = {} & 0.5y(k-1) - u(k-2) + 0.1y^2(k-1) - 0.01y(k-1)y(k-2) \\ & + 0.4y(k-2)u(k-2) + 0.5e(k-1) + 0.2u(k-1)e(k-2) + e(k) \end{aligned} \quad (24)$$

where $e(k)$ is white noise signal, average is 0, variance is 0.2. Input is

$$u(k) = 0.6\sin(2\pi k/85) + 0.4\sin(2\pi k/165) \quad (25)$$

We collected data on-line for 200 time steps to form input-output sample pairs for constructing the RCFWN. The performance index is 0.08. Fig.2(c) shows that our RCFWN has good generalization capability and fast convergence.

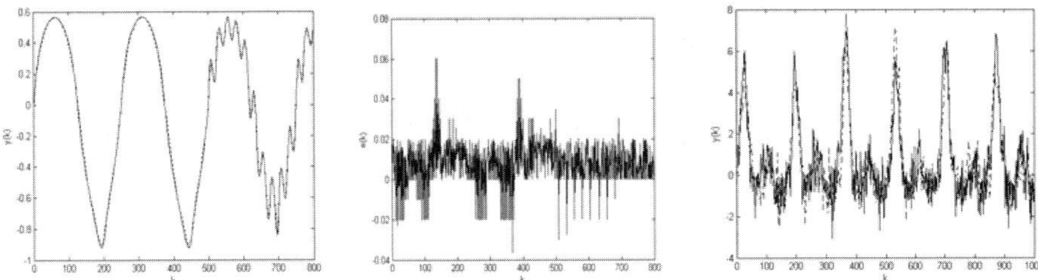

Fig. 2. (a) The identification of dynamic system using RCFWN(dotted line).(b)The error curve of the RCFWN. (c) The identification of dynamic system using RCFWN (dotted line).

5 Conclusions

Simulation results for this RCFWN show that the following advantages: (1) RCFWN has the capacities of attractor dynamics and temporary information storages. (2) The RCFWN has smaller network and a small number of tuning parameters than FNN and WNN. (3) The RCFWN has online adapting ability for dynamic system control. (4) The RCFWN has multi-resolution capability through B-spline membership function.

Acknowledgments

This paper is supported by the importance project fund of the education department of Sichuan Province, China (No. 0229957 and No. 2005A117).

References

1. Ho, D.W.C., Zhang, P.A., Xu, J.H.: Fuzzy Wavelet Networks for Function Learning. IEEE Trans. on Fuzzy Systems, Vol. 9. **(1)** (2001) 200-211
2. Wang, J., Xiao, J., Hu, D.: Fuzzy Wavelet Network Modeling with B-Spline Wavelet. The Fourth International Conference on Machine Learning and Cybernetics, (2005) 889-898
3. Wang, J., Xiao, J.: Constructing Fuzzy Wavelet Network Modeling. International Conference on Intelligent Computing, (2005) 169-172
4. Lee, C.H., Teng, C.C.: Identification and Control of Dynamic Systems Using Recurrent Fuzzy Neural Networks. IEEE Trans. Fuzzy Syst., Vol. 8. **(6)** (2000) 349–366
5. Zhang, J., Morris, A.J.: Recurrent Neuro-Fuzzy Networks for Nonlinear Process Modeling. IEEE Trans. Neural Networks, Vol. 10. (1999) 313–326

Simulation Studies of On-Line Identification of Complex Processes with Neural Networks

Francisco Cubillos and Gonzalo Acuña

Facultad de Ingeniería, Universidad de Santiago de Chile,USACH,
Avda. Ecuador 3659, Santiago, Chile
{fcubillo, gacuna}@lauca.usach.cl

Abstract. This paper analyzes various formulations for the recursive training of neural networks that can be used for identifying and optimizing nonlinear processes on line. The study considers feedforward type networks (FFNN) adapted by three different methods: inverse Hessian matrix approximation, calculation of the inverse Hessian matrix using a Gauss-Newton recursive sequential algorithm, and calculation of the inverse Hessian matrix in a recursive type Gauss-Newton algorithm. The study is completed using two network structures that are linear in the parameters: a radial basis network and a principal components network, both trained using a recursive least squares algorithm. The corresponding algorithms and a comparative test consisting of the on-line estimation of a reaction rate are detailed. The results indicate that all the structures were capable of converging satisfactorily in a few iteration cycles, FFNN type networks showing better prediction capacity, but the computational effort of the recursive algorithms is greater.

1 Introduction

Neural networks are parallel-processing nonparametric structures that replicate on a small scale some of the operations seen in biological systems, such as learning and adaptation. In process identification, which involves the functional approximation task, the most frequently used structure is the feedforward neural network (FFNN) with a hidden layer, using sigmoids as activation functions, because it has been shown that it is capable of approximating any continuous nonlinear function [1]. Traditionally, this kind of network is trained using the backpropagation method, which is coherent with the network's parallel structure. However, since it is a first order method, it converges slowly, which makes it prohibitive for real time control applications. In [2] and [3] neural controllers adapted recursively by backpropagation, are studied achieving good results only after a significant number of adaptations. Wilson and Martinez provide foundations, scope and solutions for the training of networks by means of first order methods [4]. Because of the above, the development of more powerful training algorithms for FFNN networks and the study of other more attractive neural structures is the subject of intensive studies. Among the former, the algorithms that use the Hessian matrix or its approximations have given excellent results and have proved to be much faster with respect to training time, especially for

J. Wang et al. (Eds.): ISNN 2006, LNCS 3972, pp. 808 – 814, 2006.

the case of the approximation of continuous functions [5]. In relation to other structures, networks whose parameters are linear with respect to their outputs have been proposed, and they can be trained using robust recursive least squares (RLS) techniques. In this paper we present a formulation for the recursive training of an FFNN neural network whose parameters can be adapted using three different methods: approximation of the inverse Hessian matrix using a BFGS (Broyden-Fletcher-Golgfarb-Shanno) algorithm; calculation of the inverse Hessian matrix in a Gauss-Newton recursive sequential algorithm; and calculation of the inverse Hessian matrix in a Gauss-Newton recursive sequential algorithm implemented in parallel with respect to the nodes. We also complete the study using two network structures that are linear in the parameters: a radial basis network and a principal component network, both of them trained using a least squares recursive algorithm.

2 Neural Structures

2.1 Feedfordward Neural Network

The most widely used training method for an FFNN neural network is the so-called backpropagation. To improve backpropagation, minimization methods that include or approximate the information from the second derivatives (Hessian matrix), such as the quasi-Newton or Gauss-Newton, have been proposed [2].

With the purpose of adapting recursively the weights of a feedforward neural network, a cost function that weights the past information and the new information incorporated into the system must be defined conveniently. If we assume that we have sampled the process for k instants, we can define a weighted cost function J between the values calculated by the neural network "y" and those desired "d" as the equation (1) shows.

$$J_k = \frac{1}{2}\sum_{i=1}^{k}\beta^{(k-1)} * \left((y-d)^t * (y-d)\right) \quad (1)$$

where $\beta \leq 1$ is a constant called the forgetting factor.

The way in which the inverse Hessian matrix is determined gives rise to a varied series of adaptation algorithms. In this work three different recursive type variants are studied: 1) Newton Recursive Sequential (NRS) Method, 2) Newton Recursive Parallel (NRP) Method, and 3) Quasi-Newton Methods.

2.2 Networks That Use Principal Component Analysis

The incorporation of principal component analysis (PCA) to neural networks of the multilayer type was introduced by Peel et al. in 1992 [6]. In their proposal they adopt a structure with a hidden layer in which the weights of the first layer are calculated in such a way that the input vector is transformed into principal components. The principal components analysis technique consists in decomposing the input vector **x** (with a mean of zero and variance of 1). Training of this kind of network may be summarized in the following steps:

i) Obtain the singular vectors of the input matrix [**X**], previously scaled to a mean of zero and a variance of 1.
ii) Calculate the corresponding principal components.
iii) Determine the matrix [ψ] as a function of the degree of the polynomial p and the number of principal components to be used.
iv) Calculate the parameters vector ω by some linear fitting method.

Because the determination of the weights of these kinds of networks is reduced to a linear regression problem, this same structure can be adapted recursively using the robust techniques of recursive least squares, provided the topology of the network and its singular vectors, which are considered constant during the adaptation, are specified beforehand. If a large variation in the structure of the input data is seen, a continuous updating of the singular vectors over a mobile data window can be implemented. This implementation can be developed as an unsupervised neural network [7]. Therefore, this kind of network can have a totally parallel character, with a first zone in charge of calculating the principal components and another in charge of generating the outputs.

2.3 Networks with Radial Basis Functions (RBF)

Networks with radial basis functions (RBF) are considered as a kind of neural network and have been used successfully for process identification and control [8]. These networks are a good alternative to multilayer type networks for identifying processes, since the outputs of an RBF network are linear with respect to their parameters. The RBF type networks consist of a layer of several nodes in which the activation function acts on the Euclidian norm of the difference between the input vector **x** and a vector c called the "center," which is specific for each node. The network's output is the result of a linear combination between the outputs of the nodes and a weight vector θ. The most widely used activation functions are: Gaussian, Multiquadratic, Reciprocal multiquadratic and Degree n spline. An interesting technique for choosing the centers was suggested by Moody and Darken [9] based on an algorithm that divides the set of data into n sectors, producing a set of n centers, minimizing the mean quadratic error incurred by representing the data set by the n centers. These centers can be part of the same data set or arbitrary values. One of the main characteristics of this way of choosing, called *n-mean clustering*, is that it can be implemented recursively allowing a continuous adaptation of the centers.

Chen *et al* [10], present an algorithm for recursive training of RBF networks consisting of two subalgorithms, one based on *n-mean clustering* to adapt the centers, and the other, a recursive least squares routine, to determine the network's output parameters. The *n-mean clustering* recursive algorithm (NMCR) is not supervised.

3 Application Example: Identification of a Kinetic Parameter

3.1 Definition of the Problem

The various algorithms and structures given above are used below for the recursive training of neural structures that are capable of predicting a kinetic parameter as a function of input and output variables obtained in a nonisothermal highly linear reactor CSTR in which the first order reversible reaction A $\Leftrightarrow$ R takes place. The neural structure must predict the kinetic parameter Υ when the actual concentration of product R and input temperature to the reactor Ti are known for present and past instants (equation 2).

$$\Upsilon = NN(\mathbf{R}_k, \mathbf{Ti}_k, \mathbf{Ti}_{k-1}) \tag{2}$$

The data set obtained by means of simulations using a phenomenological model of the process is shown in Figure 1, where Ti and Υ are between [-1 and 1].

A total of six recursively adapted neural structures were evaluated, whose main characteristics are given in Table 1. The structural parameters such as the number of nodes in the hidden layer, the order of the polynomial in the PCA network, and the number of bases in the RBF network were determined by the AGTSE cross-correlation criterion [11] using another data set.

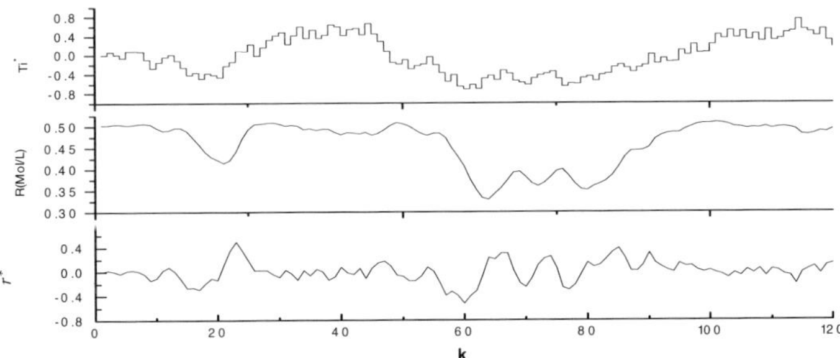

Fig. 1. Values of Ti (input var. of the process), R (output var. of the process) and Υ (kinetic parameter to be estimated) obtained in a nonisothermal CSTR reactor with reaction A $\Leftrightarrow$ R

Table 1. Used neural structures. Table contains structural parameters such as the number of nodes in the hidden layer, the order of the polynomial in the PCA network, and the number of bases in the RBF.

Network	Type	*F(v)*	#Nodes	#Parameters	Algorithm
SS/NRS	FFNN	Tanh/Tanh	4	21	NRS
SS/NRP	FFNN	Tanh/Tanh	4	21	NRP
SS/BFGS	FFNN	Tanh/Tanh	4	21	BFGS
SL/NRS	FFNN	Tanh/Lin.	4	20	NRS
PCA	PCA	Polyn. $p=3$	3	12	GSO / RLS
RBF	RBF	Spline n=2	12	12	NMCR/RLS

3.2 Implementation and Results

The FFNN networks were initialized with the following parameters: ß=0.99, $[\mathbf{H}]^{-1}$= 10.000*$\mathbf{I}$ and $\boldsymbol{\theta}$ with random values between -1 and 1. In the PCA network, the singular vectors were extracted from a DVS of the first 30 data using the orthonormalization method of Gram-Schmidt (GSO), and they were considered constant during the adaptation. The singular values were the following: $\boldsymbol{\sigma}$ = [3.9713 1.501 0.8035]. In the case of the RBF network the 12 centers were initialized with random values between -1 and 1 for Ti and between 0.2 and 0.6 for R. In the PCA and RBF networks the output parameters $\boldsymbol{\theta}$ were estimated with a robust algorithm of recursive least squares with variable forgetting factor RLS/VFF. The results obtained for each of the structures analyzed are shown in Figures 2 and 3, and it is seen that all of them converge satisfactorily. The residual modeling error and the robustness of the estimator were analyzed by presenting the training set repeatedly over a certain number of cycles.

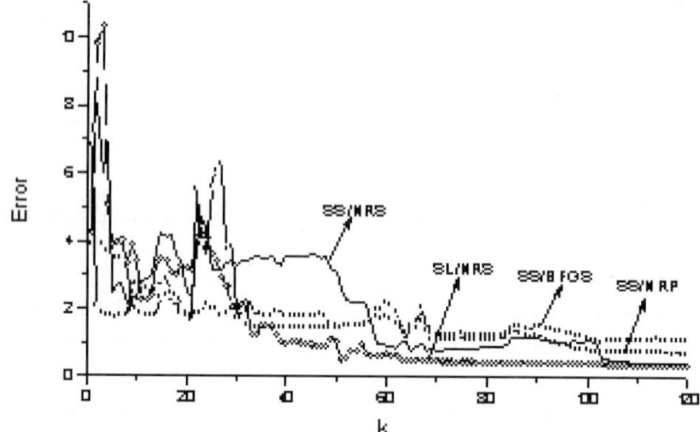

Fig. 2. Evolution of ê for *FFNN* networks

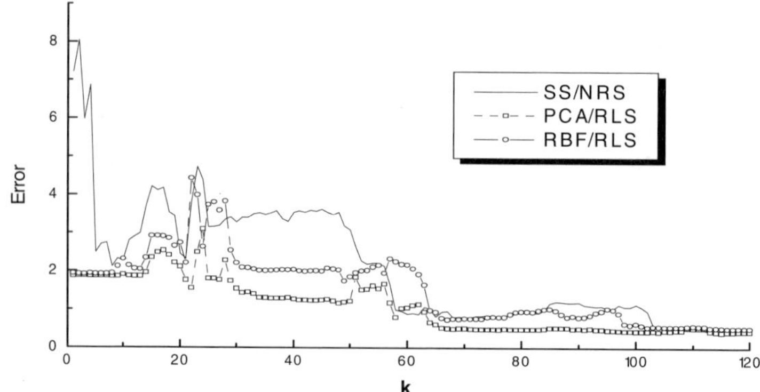

Fig. 3. Evolution of ê for PCA and RBF networks

4 Conclusions

From the results obtained it is seen that all the structures analyzed were capable of adapting satisfactorily to the proposed example, converging in less than one presentation cycle. Cyclic repetitions also showed that the algorithms are stable and tend to different residual errors, providing an indication of the modeling capacity of each structure in particular. In this sense, feedforward type networks show less residual error due to their great nonlinear adjustment flexibility, in contrast with the PCA and RBF networks, which are more rigid. The results also indicate differences between the convergence speeds, which are greater in the PCA and RBF networks - which practically converge for k>65- compared with the feedforward networks – which converge for k>100 (except SS/NRS), due to the RLS adaptation algorithm, which guarantees global optima, and to the smaller number of parameters used. Among the different feedforward network alternatives studied there are differences in both the convergence speed and the residual error, due basically to the simplifications introduced in the algorithm for adapting the inverse Hessian matrix. In fact, BFGS and NRP are approximations of NRS that imply some deterioration of the previously mentioned indicators. Other characteristics such as the computational load of the adaptation and the ease of the structured modeling can also be analyzed. In general, feedforward networks are simpler to specify and initialize than PCA and RBF networks, however the algorithms are computationally more costly in floating point operations and storage requirements. On the other hand, in PCA and RBF networks specification of the intermediate layer requires an additional effort, mainly in the RFB networks, in which the number of centers and the activation function has a marked influence on their performance.

Acknowledgements

This work thanks the partial financing provided by the Chilean Government through Fondecyt Grants 1040208 and 1020041.

References

1. Hornik, K., Stinchcombe, M., White, H.: Multilayer Feedforward Networks are Universal Approximators. Neural Network 2 (1989) 359-366
2. Ge, S.S., Hang, C.C., Lee, T.H.: Stable Adaptive Neural Network Control. Springer (2002)
3. Zhang, T., Ge, S.S., Hang, C.C.: Adaptive Neural Network Control for Strict Feed-back Nonlinear Systems using Backstepping Design. Automatica 36 (2000) 1835-1846
4. Wilson, R., Martinez, T.: The General Inefficiency of Batch Training for Gradient Descent Learning. Neural Networks 16 (2003) 1429-1451
5. Van der Smagt, P.: Minimisation Methods for Training Feedforward Neural Networks. Neural Networks 7 (1994) 1-11
6. Peel, C., Willis, M., Tham, M.: A Fast Procedure for Training NN. J. Proc. Contr. 2 (1992)
7. Oja, E.: Principal Components, Minor Components and Linear Neural Networks. Neural Networks 5 (1992) 927-935

8. Spooner, J.T., Passino, K.: Decentralized Adaptive Control of Nonlinear Systems Using Radial Basis Neural Networks. IEEE Transactions on Auto. Control 44 (1999) 2050-2057
9. Moody J., Darken, C.: Fast Learning in Networks of Locally Tuned Processing Units. Neural Computation 1 (1989) 281-294
10. Chen, S., Billings, S., Grant, P.: Recursive Hybrid Algorithm for Non Linear System Identification using Radial Basis Functions Networks. Int. J. Control 55 (1991) 1051-1070
11. Dhib, R., Hyson, W.: Neural Network Identification of Styrene Free Radical Polymerization. Polymer Reaction Engineering (USA) 10 (2002) 101-113

Consecutive Identification of ANFIS-Based Fuzzy Systems with the Aid of Genetic Data Granulation

Sung-Kwun Oh[1], Keon-Jun Park[1], and Witold Pedrycz[2]

[1] Department of Electrical Engineering, The University of Suwon, San 2-2 Wau-ri, Bongdam-eup, Hwaseong-si, Gyeonggi-do, 445-743, South Korea
ohsk@suwon.ac.kr

[2] Department of Electrical and Computer Engineering, University of Alberta, Edmonton, AB T6G 2G6, Canada and Systems Research Institute, Polish Academy of Sciences, Warsaw, Poland

Abstract. We introduce a consecutive identification of ANFIS-based fuzzy systems with the aid of genetic data granulation to carry out the model identification of complex and nonlinear systems. The proposed model implements system structure and parameter identification with the aid of information granulation and genetic algorithms. The design methodology emerges as a hybrid structural optimization and parametric optimization. Information granulation realized with HCM clustering algorithm help determine the initial parameters of fuzzy model such as the initial apexes of the membership functions in the premise and the initial values of polynomial functions in the consequence. And the structure and the parameters of fuzzy model are identified by GAs and the membership parameters are tuned by GAs. In this case we exploit a consecutive identification. The numerical example is included to evaluate the performance of the proposed model.

1 Introduction

In many other fields, the often-encountered problem is to understand an unknown system from its input output data set. The researches on the process have been exploited for a long time. Linguistic modeling [1] and fuzzy relation equation-based approach [2] were proposed as primordial identification methods for fuzzy models. The general class of Sugeno-Takagi models [3] gave rise to more sophisticated rule-based systems where the rules come with conclusions forming local regression models. While appealing with respect to the basic topology (a modular fuzzy model composed of a series of rules) [4], these models still await formal solutions as far as the structure optimization of the model is concerned, say a construction of the underlying fuzzy sets-information granules being viewed as basic building blocks. Some enhancements to the model have been proposed by Oh and Pedrycz [5], yet the problem of finding "good" initial parameters of the fuzzy sets in the rules remains open.

This study is aimed at a systematic approach to the identification procedure of ANFIS-based fuzzy model and proposes a hybrid algorithm for system identification. The design methodology emerges as a hybrid structural optimization and parametric

J. Wang et al. (Eds.): ISNN 2006, LNCS 3972, pp. 815 – 820, 2006.

optimization. To identify the structure and parameters of fuzzy model we exploit a consecutive identification using GAs. The proposed model is contrasted with the performance of conventional models in the literature.

2 Information Granulation (IG)

Roughly speaking, information granules [6], [7] are viewed as related collections of objects (data point, in particular) drawn together by the criteria of proximity, similarity, or functionality. Granulation of information is an inherent and omnipresent activity of human beings carried out with intent of gaining a better insight into a problem under consideration and arriving at its efficient solution. In particular, granulation of information is aimed at transforming the problem at hand into several smaller and therefore manageable tasks. In this way, we partition this problem into a series of well-defined subproblems (modules) of a far lower computational complexity than the original one. The form of information granulation themselves becomes an important design feature of the fuzzy model, which are geared toward capturing relationships between information granules.

It is worth emphasizing that the HCM clustering [8] has been used extensively not only to organize and categorize data by the criteria of similarity, but it becomes useful in data compression and model identification.

3 ANFIS-Based Fuzzy Systems with the Aid of IG

3.1 Premise Identification

In the premise part of the rules, we confine ourselves to a triangular type of membership functions whose parameters are subject to some optimization. The identification of the premise part is completed in the following manner.

Given is a set of data $\mathbf{U}=\{\mathbf{x}_1, \mathbf{x}_2, \ldots, \mathbf{x}_l ; \mathbf{y}\}$, where $\mathbf{x}_k=[x_{1k}, \ldots, x_{mk}]^T$, $\mathbf{y}=[y_1, \ldots, y_m]^T$, k=1, 2, …, l, and l is the number of variables and , m is the number of data.

[Step 1] Arrange a set of data **U** into data set $\mathbf{X}_k$.

$$\mathbf{X}_k=[\mathbf{x}_k ; \mathbf{y}] \tag{1}$$

[Step 2] Complete the HCM clustering to determine the centers (prototypes) $\mathbf{v}_{kg}$.

[Step 2-1] Classify data set $\mathbf{X}_k$ into c-clusters, which in essence leads to the granulation of information.

[Step 2-2] Calculate the center vectors $\mathbf{v}_{kg}$ of each cluster.

$$\mathrm{v}_{kg} = \{v_{k1}, v_{k2}, \ldots, v_{kc}\} \tag{2}$$

Where, g = 1, 2, …, c.

[Step 3] Partition the corresponding input space using the prototypes of the clusters $\mathbf{v}_{kg}$. Associate each clusters with some meaning (semantics), say Small, Big, etc.

[Step 4] Set the initial apexes of the membership functions using the prototypes $\mathbf{v}_{kg}$.

3.2 Consequence identification

We identify the structure considering the initial values of polynomial functions based upon the information granulation realized for the consequence.

[Step 1] Find a set of data included in the fuzzy space of the j-th rule.
[Step 2] Compute the prototypes $\mathbf{V}_j$ of the data set by taking the arithmetic mean of each rule.

$$\mathrm{V}_j = \{V_{1j}, V_{2j}, \ldots, V_{kj}; M_j\} \tag{3}$$

Where, j=1, 2, …, n. V_{kj} and M_j are prototypes of input and output data, respectively.
[Step 3] Set the initial values of polynomial functions with the center vectors $\mathbf{V}_j$.

4 Optimization of IG-Based FIS (ANFIS Type)

Genetic algorithms [9] have proven to be useful in optimization of such problems because of their ability to efficiently use historical information to obtain new solutions with enhanced performance and a global nature of search supported there. GAs are also theoretically and empirically proven to support robust searches in complex search spaces. Moreover, they do not get trapped in local minima, as opposed to gradient-descent techniques being quite susceptible to this shortcoming.

In this study, to consecutively identify the fuzzy model using genetic algorithms we determine such a structure as the number of input variables, input variables being selected and the number of membership functions standing in the premise and the type of polynomial in the conclusion. And also the membership parameters of the premise are genetically optimized. Figure 1 shows an arrangement of the content of the string to be used in genetic optimization to consecutively identify the fuzzy model. Here, parentheses mean the number of chromosomes for each parameter.

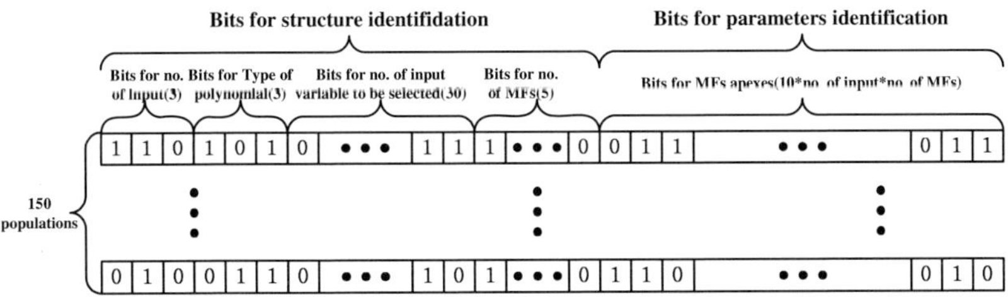

Fig. 1. The consecutive coding to identify the structure and parameters

5 Experimental Studies

In this section we consider comprehensive numeric studies illustrating the design of the fuzzy model. We demonstrate how IG-based FIS of ANFIS type can be utilized to predict future values of a chaotic time series. The performance of the proposed model is also contrasted with some other models existing in the literature. The time series is generated by the chaotic Mackey–Glass differential delay equation [10] of the form:

$$\dot{x}(t) = \frac{0.2x(t-\tau)}{1+x^{10}(t-\tau)} - 0.1x(t) \tag{4}$$

The prediction of future values of this series arises is a benchmark problem that has been used and reported by a number of researchers. From the Mackey–Glass time series $x(t)$, we extracted 1000 input–output data pairs for the type from the following the type of vector format such as: [$x(t$-30), $x(t$-24), $x(t$-18), $x(t$-12), $x(t$-6), $x(t)$; $x(t$ +6)] where t = 118–1117. The first 500 pairs were used as the training data set while the remaining 500 pairs were the testing data set for assessing the predictive performance. We carried out the identification on a basis of the experimental data to design Max_Min-based and IG-based fuzzy model of ANFIS type. The maximal number of input variables was set to be equal to four. For each model the corresponding input variables were picked up $x(t$-30), $x(t$-18), $x(t$-12), $x(t)$. The number of membership functions assigned to each input of Max_Min-based fuzzy model were set up to 2 each input variables, and the other hand, the number of membership functions of IG-based fuzzy model of ANFIS type was selected to 2, 2, 3 and 2. At the conclusion part, both models come with the consequence of type 3. For each model, we conducted consecutively the optimization of the parameters of the premise membership functions. Table 1 summarizes the performance index for Max_Min-based and IG-based fuzzy model of ANFIS type. It becomes evident that the performance of the IG-based fuzzy of ANFIS type is better than that of a Max_Min-based fuzzy model.

Table 1. Performance index of Max_Min-based and IG-based fuzzy model (ANFIS type)

Model	Input variable	No. Of MFs	Type	PI	E_PI
Max/Min_FIS	$x(t$-30), $x(t$-18), $x(t$-12), $x(t)$	2x2x2x2	Type 3	0.0001	0.0002
IG_FIS	$x(t$-30), $x(t$-18), $x(t$-12), $x(t)$	2x2x3x2	Type 3	0.00006	0.0001

Figure 2 depicts the values of the performance index produced in successive generation of the genetic optimization. We note that the performance of the IG-based fuzzy model of ANFIS type is good starting from initial generation; this could have been cause by the characteristics of the experimental data at hand. The resulting prediction error for the training and testing data for IG-based fuzzy model of ANFIS type are displayed in figure 3. The identification error (performance index) of the proposed model is also compared with the performance of some other models; refer to Table 2. Here the non-dimensional error index (NDEI) is defined as the root mean square errors divided by the standard deviation of the target series.

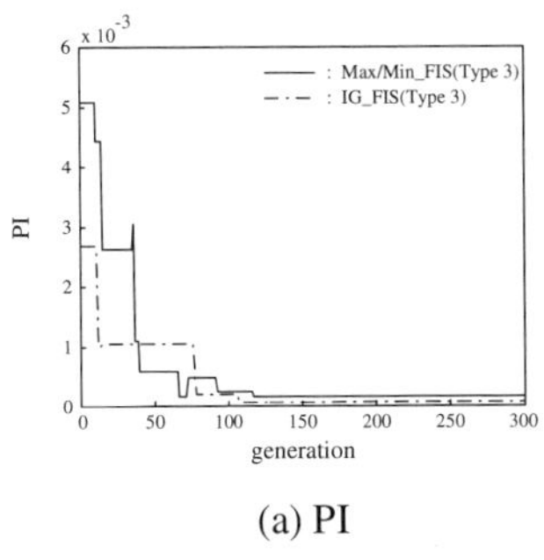

(a) PI

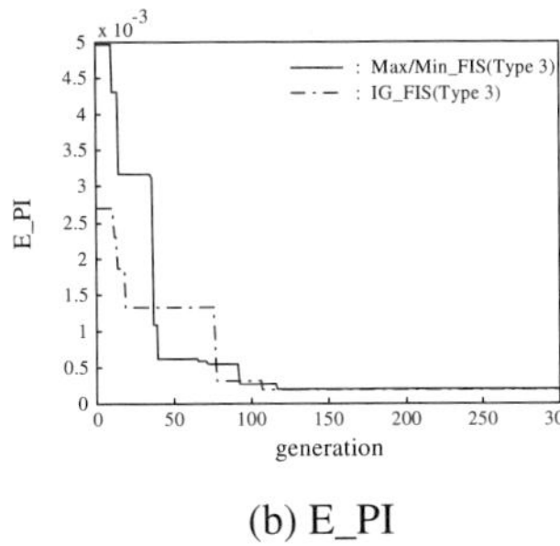

(b) E_PI

Fig. 2. Convergence process of performance index for Max_Min-based and IG-based fuzzy model of ANFIS type

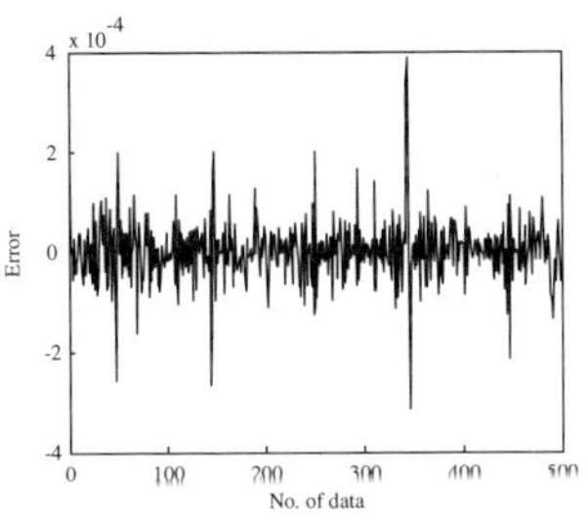

(a) Predicting error of training data

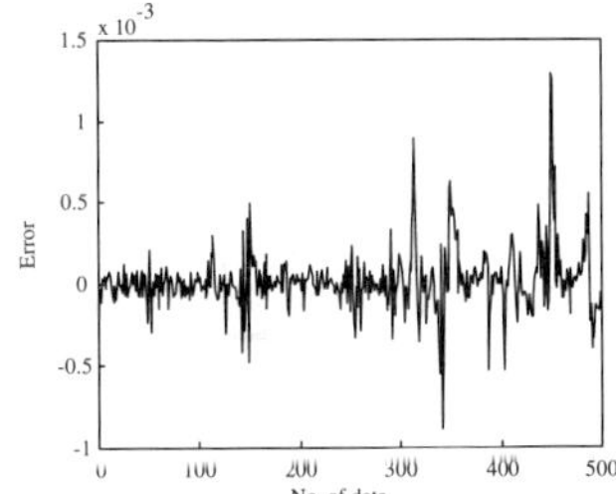

(b) Predicting error of testing data

Fig. 3. Predicting error of training and testing data for IG-based fuzzy model of ANFIS type

Table 2. Comparison of identification error with previous models

Model	No. of rules	PI_t	PI	E_PI	NDEI
Wang's model [11]	7	0.004			
	23	0.013			
	31	0.010			
Cascaded-correlation NN [12]					0.06
Backpropagation MLP [12]					0.02
6th-order polynomial [12]					0.04
ANFIS [13]	16		0.0016	0.0015	0.007
FNN model [14]			0.014	0.009	
Recurrent neural network [15]		0.0138			
Our model	24		0.00006	0.00001	0.00004

6 Conclusions

This study introduced a consecutive identification for ANFIS-based fuzzy systems with the aid of genetic data granulation. The main concept comes up with an optimization of information granules by exploiting techniques of clustering and genetic algorithms. We have reflected the characteristics of experimental data to the model and have consecutively identified the structure and parameters to optimally construct the model. The comprehensive experimental studies involving well-known data sets show a superb performance of the proposed model in comparison to the existing models. And we can construct a well-organized model. While the detailed discussion was focused on triangular fuzzy sets, the developed methodology applies equally well to any other class of fuzzy sets as well as a type of nonlinear local model. Moreover, the models scale up quite easily and do not suffer from the curse of dimensionality encountered in other identification techniques of rule-based systems.

Acknowledgements. This work has been supported by KESRI(R-2004-B-274), which is funded by MOCIE(Ministry of commerce, industry and energy).

References

1. Tong, R.M.: Synthesis of Fuzzy Models for Industrial Processes. Int. J Gen Syst. **4** (1978) 143-162
2. Pedrycz, W.: Numerical and Application Aspects of Fuzzy Relational Equations. Fuzzy Sets Syst. **11** (1983) 1-18
3. Takagi, T, Sugeno, M.: Fuzzy Identification of Systems and Its Applications to Modeling and Control. IEEE Trans Syst, Cybern. **SMC-15(1)** (1985) 116-132
4. Sugeno, M, Yasukawa, T.: Linguistic Modeling Based on Numerical Data. In: IFSA'91 Brussels, Computer, Management & System Science. (1991) 264-267
5. Oh, S.K., Pedrycz, W.: Identification of Fuzzy Systems By Means of an Auto-Tuning Algorithm and Its Application to Nonlinear Systems. Fuzzy Sets and Syst. **115(2)** (2000) 205-230
6. Zadeh, L.A.: Toward A Theory of Fuzzy Information Granulation and Its Centrality in Human Reasoning and Fuzzy Logic. Fuzzy Sets and Syst. **90** (1997) 111-117
7. Pderycz, W., Vukovich, G.: Granular Neural Networks. Neurocomputing **36** (2001) 205-224
8. Krishnaiah, P.R., Kanal, L.N., Editors: Classification, Pattern Recognition, and Reduction of Dimensionality. Volume 2 of Handbook of Statistics. North-Holland, Amsterdam (1982)
9. Golderg, D. E.: Genetic Algorithm in Search, Optimization & Machine Learning. Addison Wesley (1989)
10. Mackey, M.C., Glass, L.: Oscillation and Chaos in Physiological Control Systems. Science **197** (1977) 287-289
11. Wang, L.X., Mendel, J.M.: Generating Fuzzy Rules from Numerical Data with Applications. IEEE Trans. Systems, Man, Cybern. **22(6)** (1992) 1414-1427
12. Crowder, R.S. III.: Predicting the Mackey-Glass Time Series with Cascade-Correlation Learning. In D. Touretzky, G. Hinton, and T. Sejnowski, Editors, Proceedings of the 1990 Connectionist Models Summer School. Carnegie Mellon University (1990) 117-123
13. Jang J.R.: ANFIS: Adaptive-Network-Based Fuzzy Inference System. IEEE Trans. System, Man, and Cybern. **23(3)** (1993) 665-685
14. Maguire, L.P., Roche, B., Mcginnity, T.M., Mcdaid, L.J.: Predicting a Chaotic Time Series Using a Fuzzy Neural Network. Information Sciences **112** (1998) 125-136
15. Li, C.J., Huang, T.Y.: Automatic Structure and Parameter Training Methods for Modeling of Mechanical Systems by Recurrent Neural Networks. Applied Mathematical Modeling. **23** (1999) 933-944
16. Park, H.S., Oh, S.K.: Fuzzy Relation-based Fuzzy Neural-Networks Using a Hybrid Identification Algorithm. International Journal of Control, Automations, and Systems **1**(3) (2003) 289-300

Two-Phase Identification of ANFIS-Based Fuzzy Systems with Fuzzy Set by Means of Information Granulation and Genetic Optimization

Sung-Kwun Oh, Keon-Jun Park, and Hyun-Ki Kim

Department of Electrical Engineering, The University of Suwon, San 2-2 Wau-ri, Bongdam-eup, Hwaseong-si, Gyeonggi-do, 445-743, South Korea
ohsk@suwon.ac.kr

Abstract. In this study, we propose the consecutive optimization of ANFIS-based fuzzy systems with fuzzy set. The proposed model formed by using respective fuzzy spaces (fuzzy set) implements system structure and parameter identification with the aid of information granulation and genetic algorithms. Information granules are sought as associated collections of objects (data, in particular) drawn together by the criteria of proximity, similarity, or functionality. Information granulation realized with HCM clustering help determine the initial parameters of fuzzy model such as the initial apexes of the membership functions in the premise and the initial values of polynomial functions in the consequence. And the initial parameters are tuned with the aid of the genetic algorithms and the least square method. To optimally identify the structure and parameters we exploit the consecutive optimization of ANFIS-based fuzzy model by means of genetic algorithms. The proposed model is contrasted with the performance of conventional fuzzy models in the literature.

1 Introduction

Fuzzy modeling has been a focal point of the technology of fuzzy set from its very inception. Fuzzy modeling has been studied to deal with complex, ill-defined, and uncertain systems in may other avenues. Linguistic modeling [1] and fuzzy relation equation-based approach [2] were proposed as primordial identification methods for fuzzy models. The general class of Sugeno-Takagi models [3] gave rise to more sophisticated rule-based systems where the rules come with conclusions forming local regression models. While appealing with respect to the basic topology (a modular fuzzy model composed of a series of rules) [4], these models still await formal solutions as far as the structure optimization of the model is concerned, say a construction of the underlying fuzzy sets—information granules being viewed as basic building blocks of any fuzzy model. Some enhancements to the model have been proposed by Oh and Pedrycz [5], yet the problem of finding "good" initial parameters of the fuzzy sets in the rules remains open.

In this study we consider the problem of ANFIS-based fuzzy model that is a development of information granules-fuzzy sets. The design methodology emerges as a

J. Wang et al. (Eds.): ISNN 2006, LNCS 3972, pp. 821 – 826, 2006.

hybrid structural and parametric optimization. To optimally identify the structure and parameters we exploit the consecutive optimization of fuzzy model that is formed by using respective fuzzy spaces.

2 Information Granulation (IG)

Information granules [6], [7] are informally viewed as linked collections of objects (data point, in particular) drawn together by the criteria of proximity, similarity, or functionality. Granulation of information is an inherent and omnipresent activity of human beings carried out with intent of better understanding of the problem. In particular, granulation of information is aimed at splitting the problem into several manageable chunks. In this way, we partition this problem into a series of well-defined subproblems (modules) of a far lower computational complexity than the original one. The form of information granulation (IG) themselves becomes an important design feature of the fuzzy model, which are geared toward capturing relationships between information granules. It is worth emphasizing that the HCM clustering [8] has been used extensively not only to organize and categorize data, but it becomes useful in data compression and model identification.

3. ANFIS-Based Fuzzy Systems with Fuzzy Set by Means of IG

3.1 Premise Identification

The HCM clustering helps us organize the data into cluster, and in this way we take into account the characteristics of the experimental data. In the regions where some clusters of data have occurred, we end up with some fuzzy sets that help represent the specificity of the data set. In the sequel, the modal values of the clusters are refined.

The identification of the premise part of fuzzy rules that is formed by the isolated input space of each input variable is completed in the following manner.

Given is a set of data $\mathbf{U}=\{\mathbf{x}_1, \mathbf{x}_2, \ldots, \mathbf{x}_l ; \mathbf{y}\}$, where $\mathbf{x}_k=[x_{1k}, \ldots, x_{mk}]^T$, $\mathbf{y}=[y_1, \ldots, y_m]^T$, l is the number of variables and , m is the number of data.

[Step 1] Arrange a set of data $\mathbf{U}$ into data set $\mathbf{X}_k$.

$$\mathbf{X}_k=[\mathbf{x}_k ; \mathbf{y}] \quad (1)$$

[Step 2] Complete the HCM clustering to determine the centers (prototypes) $\mathbf{v}_{kg}$ with data set $\mathbf{X}_k$. and calculate the center vectors $\mathbf{v}_{kg}$ of each cluster.

$$\mathrm{v}_{kg}=\{v_{k1}, v_{k2}, \ldots, v_{kc}\} \quad (2)$$

Where, $g = 1, 2, \ldots, c$.

[Step 3] Partition the corresponding isolated input space using the prototypes of the clusters $\mathbf{v}_{kg}$. Associate each clusters with some meaning, say Small, Big, etc.

[Step 4] Set the initial apexes of the membership functions using the prototypes $\mathbf{v}_{kg}$.

3.2 Consequence Identification

The characteristics of input-output data is also involved in the conclusion parts as follows:

[Step 1] Find a set of data included in the isolated fuzzy space of the j-th rule.
[Step 2] Compute the prototypes $\mathbf{V}_j$ of the data set by taking the arithmetic mean of each rule.

$$\mathrm{V}_j = \{V_{1j}, V_{2j}, \ldots, V_{kj}; M_j\} \tag{3}$$

Where, j=1, 2, …, n. V_{kj} and M_j are prototypes of input and output data, respectively.
[Step 3] Set the initial values of polynomial functions with the center vectors $\mathbf{V}_j$.

4 Consecutive Optimization by Means of Genetic Algorithms

The optimization of fuzzy models may not be fully supported by the standard gradient-based optimization techniques, because of the nonlinearity of fuzzy models represented by rules based on linguistic levels. This forces us to explore other optimization techniques such as genetic algorithms. It has been demonstrated that genetic algorithms [9] are useful in a global optimization of such problems given their ability to efficiently use historical information to produce new improved solutions with enhanced performance.

To optimally identify the fuzzy model we determine such an initial structure as the number of input variables, input variables being selected and the number of membership functions standing in the premise and the type of polynomial in conclusion. The membership parameters of the premise are genetically optimized. Figure 1 shows an arrangement of the content of the string to be used in genetic optimization to consecutively identify the fuzzy model. Here, parentheses mean the number of chromosomes for each parameter.

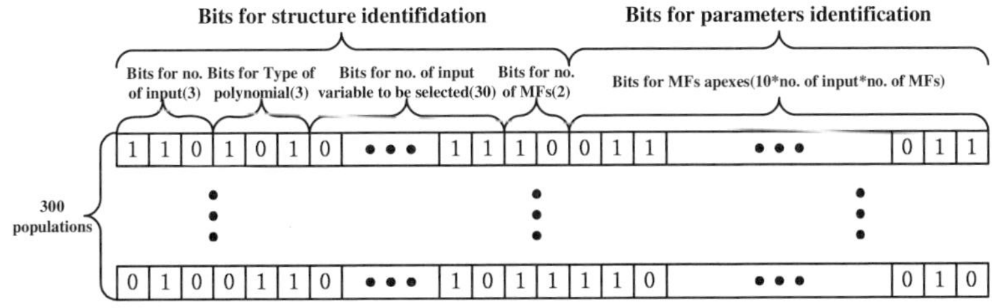

Fig. 1. The consecutive coding to identify the structure and parameters

5 Experimental Studies

5.1 Nonlinear Static System

In this section, we consider a nonlinear static system with two inputs, x_1, x_2, and a single output.

$$y = (1 + x_1^{-2} + x_2^{-1.5})^2, \quad 1 \le x_1, x_2 \le 5 \tag{4}$$

This system represents the nonlinear characteristic. Using Eq. (4), 50 input-output data are generated. To come up with a quantitative evaluation of the IG-based fuzzy model of ANFIS type, we use the standard MSE performance index. Because only two input variables are considered, we considered the number of membership functions and consequence type. Table 1 summarizes the performance index for Max_Min-based and IG-based fuzzy model of ANFIS type. It becomes evident that the performance of the IG-based fuzzy model of ANFIS type is better than that of the Max_Min-based fuzzy model. The identification error of the proposed model is also compared to the performance of some other models in table 2. Compared with approaches presented in previous literatures, our modeling method has much more accuracy.

Table 1. Performance index of Max_Min-based and IG-based fuzzy model (ANFIS type)

Model	No. Of MFs	Type	PI
Max/Min_FIS	5+5	Type 3	$5.037e^{-19}$
IG_FIS	5+5	Type 3	$6.273e^{-25}$

Table 2. Comparison of identification error with previous models

Model		No. of rules	PI
Sugeno and Yasukawa[10]		6	0.079
Gomez-Skarmeta et al[11]		5	0.070
Kim et al.[12]		3	0.019
Kim et al.[13]		3	0.0089
Oh et al.[14]	Basic PNN		0.0212
	Modified PNN		0.0041
Park et al.[15]	BFPNN	9	0.0033
	MFPNN	9	0.0023
Our Model		10	$6.273e^{-25}$

5.2 Gas Furnace Process

The time series data (296 input-output pairs) resulting from the gas furnace process has been intensively studied in the previous literatures [5,18,19]. The delayed terms of methane gas flow rate $u(t)$ and carbon dioxide density $y(t)$ are used as six input variables with vector formats such as [$u(t\text{-}3)$, $u(t\text{-}2)$, $u(t\text{-}1)$, $y(t\text{-}3)$, $y(t\text{-}2)$, $y(t\text{-}1)$]. And as output variable $y(t)$ is used. The first one (consisting of 148 pairs) was used for training. The remaining part of the series serves as a testing set. We consider the MSE being regarded here as a performance index. Table 3 summarizes the performance index for Max_Min-based and IG-based fuzzy model of ANFIS type. From the Table 3 we know that the performance of IG-based fuzzy model of ANFIS type is better than that of the Max_Min-based fuzzy model.

Table 3. Performance index of Max_Min-based and IG-based fuzzy model (ANFIS type)

Model	Input Variable	No. Of MFs	Type	PI	E_PI
Max/Min_FIS	$u(t\text{-}3)$, $y(t\text{-}1)$	4+4	Type 3	0.014	0.268
IG_FIS	$u(t\text{-}3)$, $y(t\text{-}1)$	5+5	Type 3	0.012	0.260

The identification error of the proposed model is also compared to the performance of some other models in Table 4. The performance of the proposed model is better in the sense of its approximation and prediction abilities than other works studied in the literatures as shown in Table 4.

Table 4. Comparison of identification error with previous models

Model		PI_t	PI	E_PI	No. of rules
Tong's model[16]		0.469			19
Pedrycz's model[2]		0.776			20
Xu's model[17]		0.328			25
Sugeno's model[4]		0.355			6
Oh et al.'s Model[5,18]	Simplified		0.024	0.328	4
	Linear		0.022	0.326	4
			0.021	0.364	6
HCM+GA[19]	Simplified		0.035	0.289	4
			0.022	0.333	6
	Linear		0.026	0.272	4
			0.020	0.264	6
Our model			0.012	0.260	10

6 Conclusions

We have developed a comprehensive identification for ANFIS-based fuzzy model with fuzzy set by means of information granulation realized with HCM clustering and consecutive optimization of the structure and parameters with the aid of gentic optimization. We used the isolated input space for each input variable and defined the fuzzy space by information granule. Information granulation realized with HCM clustering help determine the initial parameters of fuzzy model such as the initial apexes of the membership functions and the initial values of polynomial function being used in the premise and consequence part of the fuzzy rules. The membership parameters are tuned (adjusted) with the aid of the genetic algorithms and the least square method. Two examples are given to illustrate the validity and compare its performance with those of other models.

Acknowledgements. This work has been supported by KESRI (I-2004-0-074-0-00), which is funded by MOCIE (Ministry of Commerce, Industry And Energy).

References

1. Tong, R.M.: Synthesis of Fuzzy Models for Industrial Processes. Int. J. Gen Syst. **4** (1978) 143-162
2. Pedrycz, W.: An Identification Algorithm in Fuzzy Relational System. Fuzzy Sets Syst. **13** (1984) 153-167
3. Takagi, T, Sugeno, M.: Fuzzy Identification of Systems and Its Applications to Modeling and Control. IEEE Trans Syst, Cybern. **SMC-15(1)** (1985) 116-132
4. Sugeno, M, Yasukawa, T.: Linguistic Modeling Based on Numerical Data. In: IFSA'91 Brussels, Computer, Management & System Science. (1991) 264-267
5. Oh, S.K., Pedrycz, W.: Identification of Fuzzy Systems by Means of an Auto-tuning Algorithm and Its Application to Nonlinear Systems. Fuzzy Sets and Syst. **115(2)** (2000) 205-230
6. Zadeh, L.A.: Toward a Theory of Fuzzy Information Granulation and Its Centrality in Human Reasoning and Fuzzy Logic. Fuzzy Sets and Syst. **90** (1997) 111-117
7. Pderycz, W., Vukovich, G.: Granular Neural Networks. Neurocomputing. **36** (2001) 205-224
8. Krishnaiah, P.R., Kanal, L.N., editors.: Classification, Pattern Recognition, and Reduction of Dimensionality. Volume 2 of Handbook of Statistics. North-Holland Amsterdam (1982)
9. Golderg, D.E.: Genetic Algorithm in Search, Optimization & Machine Learning. Addison Wesley. (1989)
10. Sugeno, M., Yasukawa, T.: A Fuzzy-Logic-Based Approach to Qualitative Modeling. IEEE Trans. on Fuzzy systems. **1(1)** (1993) 7-13
11. Gomez Skarmeta, A.F., Delgado, M., Vila, M.A.: About the Use Of Fuzzy Clustering Techniques for Fuzzy Model Identification. Fuzzy Sets and Systems. **106** (1999) 179-188
12. Kim, E.T., Park, M.K., Ji, S.H., Park, M.N.: A New Approach to Fuzzy Modeling. IEEE Trans. on Fuzzy Systems. **5(3)** (1997) 328-337
13. Kim, E.T, Lee, H.J., Park, M.K., Park, M.N.: A Simply Identified Sugeno-type Fuzzy Model via Double Clustering. Information Sciences. **110** (1998) 25-39
14. Oh, S.K., Pedrycz, W., Park, B.J.: Polynomial Neural Networks Architecture: Analysis and Design. Computers and Electrical Engineering. **29(6)** (2003) 703-725
15. Park, B.J., Pedrycz, W., Oh, S.K.: Fuzzy Polynomial Neural Networks: Hybrid Architectures of Fuzzy Modeling. IEEE Trans. on Fuzzy Systems. **10(5)** (2002) 607-621
16. Tong, R.M.: The Evaluation of Fuzzy Models Derived from Experimental Data. Fuzzy Sets Syst. 13 (1980) 1-12
17. Xu, C.W., Zailu, Y.: Fuzzy Model Identification Self-learning for Dynamic System. IEEE Trans. on Syst. Man, Cybern., SMC-17(4) (1987) 683-689
18. Park, C.S., Oh, S.K., Pedrycz, W.: Fuzzy Identification by Means of Auto-Tuning Algorithm and Weighting Factor. The Third Asian Fuzzy Systems Symposium (AFSS) (1998) 701-706
19. Park, B.J., Pedrycz, W., Oh, S.K.: Identification of Fuzzy Models with the Aid of Evolutionary Data Granulation. IEE Proc.-Control Theory and Applications, **148(5)** (2001) 406-418
20. Park, H.S., Oh, S.K.: Fuzzy Relation-based Fuzzy Neural-Networks Using a Hybrid Identification Algorithm. Int. J. of Control, Automations, and Systems. **1**(3) (2003) 289-300

A New Modeling Approach of STLF with Integrated Dynamics Mechanism and Based on the Fusion of Dynamic Optimal Neighbor Phase Points and ICNN

Zhisheng Zhang[1], Yaming Sun[1], and Shiying Zhang[2]

[1] School of Electrical and Automation Engineering, Tianjin University, Tianjin 300072, People's Republic of China
[2] School of Management Engineering, Tianjin University, Tianjin 300072, People's Republic of China

Abstract. Based on the time evolution similarity principle of the topological neighbor phase points in the Phase Space Reconstruction (PSR), a new modeling approach of Short-Term Load Forecasting (STLF) with integrated dynamics mechanism and based on the fusion of the dynamic optimal neighbor phase points (DONP) and Improved Chaotic Neural Networks (ICNN) model was presented in this paper. The ICNN model can characterize complicated dynamics behavior. It possesses the sensitivity to the initial load value and to the walking of the whole chaotic track. The input dimension of ICNN is decided using PSRT, and the training samples are formed by means of the stepping dynamic space track on the basis of the DONP. So it can improve associative memory and generalization ability of ICNN model. The testing results show that proposed model and algorithm can enhance effectively the precision of STLF and its stability.

1 Introduction

The STLF plays an important role in power industry. Economic benefit and security operation and market business of power systems have very close relation with the accuracy of the load forecasting. A conservative estimate is that a 1% reduction in forecasting error for a 10,000 MW power department can save up to $1.6 million annually [1].

The research was shown that load time series of electric power systems possess complicated nonlinear and chaotic characteristic [2]. Therefore, in this paper presented study principle to STLF model is divided into two parts. Firstly, the basic chaotic load component (BCLC) that is determined by chaotic dynamics behavior of the load internal nonlinear factors must be studied based on the theory of Chaos and PSRT. Secondly, the external stochastic factors load component that is decided by climatic factors should be studied according to the indeterminacy theory. So the strategy to build and realize two forecasting models separately must be adopted by different theory. This problem has become the main respect in the research of STLF modeling. There are three kinds of questions in current correlative study in home and abroad. Firstly, the hybrid model including above two type factors is adopted, such as [3], the model can't

J. Wang et al. (Eds.): ISNN 2006, LNCS 3972, pp. 827 – 835, 2006.

embody the main chaotic dynamics behavior, and the ability to track the evolutional trajectory in PSR is lost. Secondly, dynamic recurrent neural networks are adopted, which own dynamic performance but isn't chaotic dynamic mechanism, such as [4]. Thirdly, the learning samples are obtained by PSRT, but the neural network is static and can't character chaotic dynamics mechanism effectively, such as [5]. Due to limited space, other questions aren't illustrated. Above questions ignore integrated chaotic dynamics mechanism, so the forecasting precision can't be satisfied. So the forecasting model with integrated chaotic dynamics mechanism is the key issue in this paper.

This paper presents the modeling approach with integrated dynamics mechanism and based on the fusion of DONPA in PSR and the ICNN. Its' key point is in the modeling process that the chaos dynamical performance can be shown perfectly to guarantee the unity of internal and external, and to guarantee the consistency of part and whole. For example, they are going as stepping dynamic tracking training. The learning samples of ICNN are obtained by DONPA to control the precision degree of approximating part trajectory and it is necessary to overcome the part instability of chaos characteristic in PSR. They embody the integrated dynamics mechanism and ensure the enhancement of forecasting precision and whole stability.

The Euclid-distance based K-nearest neighbor approach (NNA) has been used as a normal forecasting algorithm of chaotic time series [6], but the K-nearest neighbor approach isn't chaos dynamic concept, and the descent of forecasting precision with the increase of embedding dimension in the PSR has been proven in [7]. In this paper, the forecasting model based on the DONPA is presented, and it is a perfect chaos dynamic approach and can identify and eliminate the false neighbor points from the nearest neighbor points and can guarantee the whole forecasting precision and stability.

It is different with the chaotic learning algorithm of neural network [8], which is only the performance improvement of traditional neural network. In the paper, The ICNN is composed of modified chaotic neurons, and it embodies the sufficient chaos mechanism. So the ICNN possesses the sensitivity to the initial load value and to the walking of whole chaotic track. And it can characterize complicated dynamics behavior and has global searching optimal ability.

2 Chaotic Time Series Analysis and Forecasting Mechanism

The PSRT is used to analyze the chaotic time series. The evolution of any phasor is influenced by other interrelated phasors in PSR. So the evolution of every phasor implies the whole information in the chaos system. According to the embedding theorems of Takens and Saner et al., the topology of the attractor can be preserved by using delay-coordinate maps as long as m>2D is satisfied, where D is the fractal dimension of the system attractor. And this problem is a basic one of the prediction theory to chaotic time series. To time series $x_1, x_2, \ldots x_{n-1}, x_n \cdots$, if the embedding dimension is m and the delay time is τ, the phase space is reconstructed as following:

$$X(t_i) = [x(t_i), x(t_i + \tau), x(t_i + 2\tau), \cdots, x(t_i + (m-1)\tau)] \tag{1}$$

In which, each phase point $X(t_i)$ is represented as a column phasor of m sub-phasors (state points). The track of phase points describes the evolution of the system in m-dimension phase space.

This paper uses the G-P method [9] to acquire the embedding dimension, and uses multiple-autocorrelation method (MACM) [10] to extract the delay time.

The chaotic attractors have global stability, attractability and internal fractal ability in three aspects characteristic. The all the moving state will be attracted and close to the attractor when those moving are outside the attractor, contrarily the walking trajectories will exclude each other when those moving are inside the attractor. The trajectories can't always separate exponentially inside the attractor, and they will overlap and nest to form fractal construction in finite space. This dynamics law is the important basis of an argument to forecasting mechanism. The tracks will not cross if the space is reconstructed by PSRT. The forecasting phase point has the same moving trend with the neighbor phase points when they are outside the attractor. The forecasting phase points (FPP) have the trend to form fractal construction when they are inside the attractor. The nearest neighbor phase points (NNP) should be obtained. The function, which describes the relation of NNP and its successive series will represent approximately the relation of forecasting phase point and its successive series.

3 A New Approach of the Building DONP in PSR

3.1 K-Nearest Neighbor Approach in PSR

The K-nearest neighbor approach is a common forecasting approach of chaotic time series, and it is called as nearest neighbor phase points approach (NNPA). The approach computes the similarity of two state phasors that represent the phase points, according to Euclid-distance. In practice, some researchers have discovered that the forecasting precision will descend with increase of the embedding dimension [6], when the Euclid distance is used to select the NNP. The essence of the question is that the internal law of chaotic time series comes from the nonlinearity. The Fig.1 is the schematic diagram of affecting forecast on evolutionary track of NNP. It expresses the relativity and dynamics of chaotic time series in reconstructed phase space. Analyzing Fig.1, the Euclid distance is adopt, and $X_{sp}(t)$ and $X_{sq}(t)$ are the NNP of the FPP $X_s(t)$ in Euclid γ space, and X_s, X_{sp}, X_{sq} are denoted as NNP0,NNP1,NNP2 respectively in Fig.1. But $X_{sp}(t+1)$ of NNP1 is apart from $X_s(t+1)$ of the FPP at the time t+1, and $X_{sq}(t+1)$ of NNP2 has similar change with $X_s(t+1)$ of the FPP at the

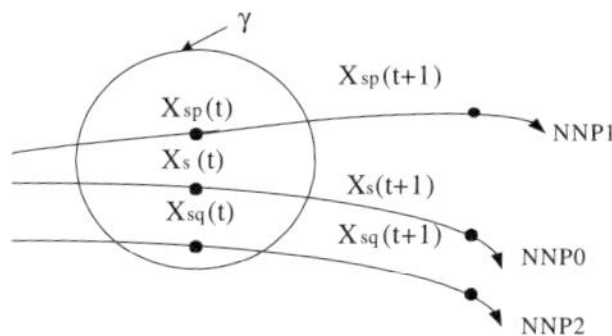

Fig. 1. A schematic diagram of affecting forecast on evolutionary track of NNP

time t+1. It is quite evident that the local dynamics of the trajectory of the NNP2 X_{sq} is fitter than the NNP1 X_{sp}. If the NNP1 X_{sp} is selected as the reference states of the FPP by the Euclid distance, the forecasting error will increase. The traditional Euclid distance approach can only consider the distance at the specified time, but it can't consider the moving trend of the trajectory. And this is the key of affecting the forecasting precision and can form the accumulated error.

3.2 Forecasting Approach Based on Dynamic Optimal Neighbor Phase Points

The geometrical notion of the NNPA is that the evolution trajectory of each dimensional sub-phasor in the NNP should be all near the FPP. Because the evolutional trajectories will overlap and nest continuously in the reconstructed phase space, and they evolve in the course of attraction and exclusion, so the convergence rate of the neighbor points is different. These neighbor points will shrink or expand with the time evolution. The trajectories of some points can depart generally from the evolutional trajectory of forecasting points. If these points are selected as reference states of the forecasting point, the accuracy of the local dynamic estimation will drop. These points are called as false neighbor points. The approach of this paper adopts the dynamic notion to eliminate effectively the false neighbor points.

The dynamic optimal neighbor phase points (DONP) are selected by the Euclid distance between the FPP and the NNP firstly, and then are identified by the exponential separating rate of these trajectories. In (2) can eliminate the false neighbor points, and it is easy to compute. In the m-dimension reconstructed phase space, the FPP is $X_s(t)$, $X_{sp}(t)$ is the pth neighbor phase point, p=1, 2, …k, k is the number of the neighbor phase points. $\hbar_p(t)$ denotes the Euclid distance between $X_s(t)$ and $X_{sp}(t)$, and it is described as $\hbar_p(t) = \|X_s(t) - X_{sp}(t)\|$. The exponential separating rate of the trajectory of the FPP and the trajectory of the pth NNP is as following:

$$\eta_p = \ln \left| \frac{\hbar_p(t)}{\hbar_p(t-1)} \right| \tag{2}$$

In which $\hbar_p(t-1)$ is the Euclid distance between the FPP and the pth NNP at time t-1. In this paper, the rate η_p is called as exponential separating rate of the local trajectories. The DONP should satisfy following conditions:

$$\text{(a) } \hbar_p(t) \le \delta \text{; (b) } \eta_p \le \varsigma \text{.}$$

δ ensures that the DONP are nearest authentically FPP, and ς ensures that the separating rate of the trajectories in these DONP is small. If the two conditions can't form the intersection or the number of the optimal neighbor points in the intersection can't satisfy the necessary number, then δ and ς can be enlarged. On the contrary, they can be reduced. It is obvious that the parameter δ and ς are adjusted according to different FPP on chaos trajectory position in phase space and the necessary number of DONP. This approach have a high approximating ability, corresponding sample sets is helpful to improve the generalization ability of ICNN.

4 Improved Chaotic Neural Networks

The notion that the chaotic character is embedded in NN is derived from the chaotic action of a biologic neuron unit. Biological neurons generally have chaotic characteristics permanently or transitorily, so chaotic character is the basic essence of chaotic neuron units. It is tested that chaotic dynamic characteristics can improve the ability of sympathetic vibration in the brain, but the simple periodic vibration can't accomplish complicated information processing.

Chaotic neurons are the basic units of building chaotic NNs. So the chaotic character of the single chaotic neuron is an important foundation study.

4.1 Modified Aihara Chaotic Neurons

Aihara chaotic neuron model [11] is common, but it is very complicated. This paper adopts the modified Aihara chaotic neurons, which own comparably chaotic character [12]. It is shown in Fig.2 (A). To the chaotic NNs, the dynamic action of the *i*th chaotic neuron at time *k*+1 is described as:

$$x_i(k+1) = Bx_i(k) + \sum_{j=1}^{n} w_{ij}^F I_j(k) + \sum_{j=1}^{m} w_{ij}^R y_j(k) \quad (3)$$

$$y_i(k+1) = f_N(x_i(k+1)) \quad (4)$$

$$f_N(x_i(k+1)) = \frac{1}{1+e^{-x_i(k+1)}} \quad (5)$$

Where $y_i(k+1)$ is the output of the *i*th chaotic neuron at time *k*+1, f_N is a sigmoid function, w_{ij}^F is connection weight from the jth external input neuron to the ith chaotic neurons. *F* is the symbol of weight space from external input neurons to chaotic neurons, w_{ij}^R is connection weight from the internal feedback neurons to the ith chaotic neurons, *R* is the symbol of weight space from feedback input neurons to chaotic neurons, $I_j(k)$ is the strength of the *j*th external input at time *k*, The *n* and *m* are the number

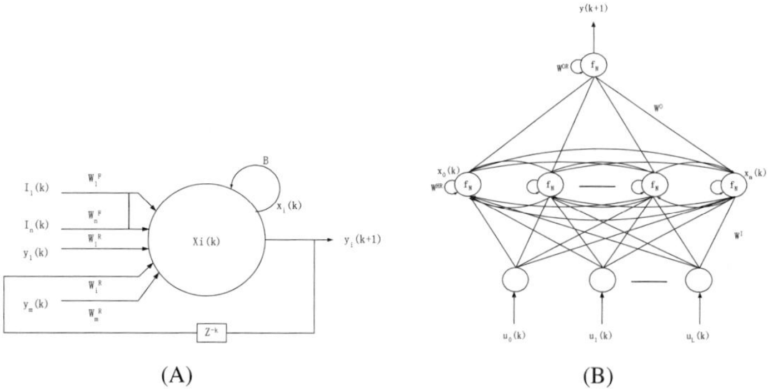

Fig. 2. Modified Aihara chaotic neuron unit (A) and its network architecture (B)

of external input neurons and internal feedback input neurons applied to the chaotic neuron, respectively. The damping factors of the external, feedback, and refractoriness are equal to *B*.

4.2 Improved Chaotic Neural Networks Model

In this paper the Improved Chaotic Neural Networks (ICNN) model is built based on modified Aihara chaotic neuron, as shown Fig.2 (B).

ICNN has two types of different coupling weights, and the first type is for directions among the neurons of interlayer, and the second type is for forward direction between layers. The first type includes self-feedback in the hidden layer and the output layer. For increasing the dynamic character of the ICNN, this connection weights in interlayer is defined as nonsymmetrical form, $w_{ij}^{OR} \neq w_{ji}^{OR}$, $w_{ii}^{OR} \neq 0$, and $w_{ij}^{HR} \neq w_{ji}^{HR}$, $w_{ii}^{HR} \neq 0$.

Consider Fig.2 (B), *l*, *m* is the number of the input nodes and the hidden nodes. The number of the output points is one, $u_i(k)$ is the *i*th input node at time *k*, $x_j(k)$ are the output of *j*th neuron in hidden layer. *B* is refractory parameter of chaotic neuron, W^I denotes the weight phasors between input and hidden layer, W^O represents the weight phasors between hidden and output layer, W^{HR} and W^{OR} denote inter-connecting weight space phasor among neurons in the hidden layer and among neurons in the output layer, $S_j^H(k)$ is the interior state of the *j*th chaotic neuron at time *k* in hidden layer. It is as follows:

$$S_j^H(k) = \sum_{i=1}^{l} w_{ij}^I u_i(k) + \sum_{q=1}^{m} w_{qj}^{HR} x_q(k-1) + B \cdot S_j^H(k-1) \tag{6}$$

$$x_j(k) = f_N[S_j^H(k)] \tag{7}$$

In (8) $S^O(k)$ is the interior state of the chaotic neuron at time *k* in output layer. In (9) $y(k)$ is the output of the neuron at time *k* in output layer. It is as follows:

$$S^O(k) = \sum_{j=1}^{m} w_j^O x_j(k) + w^{OR} y(k-1) + B \cdot S^O(k-1) \tag{8}$$

$$y(k) = f_N[S^O(k)] \tag{9}$$

5 STLF Model with Integrated Dynamics Mechanism and Based on the Fusion of DONPA and ICNN

This paper presents further the new STLF model and approach with integrated dynamics mechanism and this model based on the fusion of DONPA and the ICNN. The core of study is based on the PSRT and chaotic theory. ① The number of ICNN input nodes is decided by the embedding saturation dimension *m* in PSR. ② The *k* DONP are obtained by the Euclid distance and by seeking the least of exponential separating rate with the time evolutional trajectory, and the *m*+1 sub-phasors of each

DONP form the learning samples of ICNN. ③ They are gone as stepping dynamic tracking training. ④ ICNN has a high dynamic sensitivity which can guarantee to obtain the optimal forecasting phase point based on DONPA. So the forecasting function is of hidden and dynamic and nonlinear. The proposed forecasting approach can eliminate effectively the adverse effect of false neighbor points on the local dynamics. Because of the ICNN, model can characterize sensitively complicated dynamics behavior and has global searching optimal ability.

6 Forecasting Simulation and Error Analysis

6.1 Three Forecasting Models Used for Simulation and Test

This paper adopts three forecasting models, which are the normal K-nearest neighbor approach (model-I), the DONP approach (model-II), the forecasting model and approach based on the fusion of DONPA and the ICNN (model-III).

The model-I adopts the normal K-nearest neighbor points approach in PSR, and the model-II uses the DONPA presented in this paper. The two models both adopt local linear polynomial approach to solve the forecasting function f.

The model-III is built by the ICNN, and its number of input dimension is decided by the embedding dimension m. The structure of the ICNN modules is 7-8-1. The ant colony optimization algorithm (ACOA) is adopted. In our study, it is test that the ACOA is superior to genetic algorithm and the improved BP algorithm.

6.2 Forecasting Simulation in Power Network

The error results of three models to power network are analyzed and queued, and it is described in table 1. The error performance of the Model-III is the best in the all models and the forecasting precision can increase effectively. The load series possess chaotic character, so it should be analyzed by the dynamics behavior NN.

Table 1. The error index comparison of three models

Error% / Category	Model-I		Model-II		Model-III	
	E_{MAPE}	E_{MAX}	E_{MAPE}	E_{MAX}	E_{MAPE}	E_{MAX}
Power Network	2.584	5.726	1.909	3.782	1.193	2.762
Performance queue	□		□		□	

Table 2. The daily load forecasting errors of a calm temperature week in power network

Error% / Temp.℃		Power Network	
		E_{MAPE}	E_{MAX}
Mon.	4~13	1.193	2.762
Tues.	3~13	1.203	2.384
Wed.	5~16	1.216	2.659
Thurs.	4~17	1.235	2.681
Fri.	4~17	1.192	2.493
Sat.	5~13	1.258	2.812
Sun.	4~16	1.207	2.548

In order to analyze the forecasting performance deeply, the load systems are used to simulate and test in a week. The load forecasting errors are described in the table 2. It is shown that the proposed approach can enhance effectively the precision of STLF and high steadiness.

6.3 Forecasting Simulation Considering External Stochastic Factors

The forecasting precision of Model III isn't satisfied in the high temperature period in summer. Due to limited space, the model considering external stochastic factors can't be described in detail. The forecasting model of the external stochastic load component is constructed by the principle of pattern recognition which uses the load clustering algorithm based on ACOA considering the external meteorological factors and the load date type. The forecasting results of the BCLC and the external stochastic factors load component are synthesized to form total load forecasting result. The load series of actual power network is used as forecasting simulation, and the testing result show that the proposed STLF model can insure the forecasting precision and its stability in whole year, and the forecasting precision is enhanced evidently in high temperature period in summer and in special holiday period. The load forecasting errors of model III and model III considering external stochastic factors in high temperature week are compared in the table 3.

Table 3. the daily load forecasting errors of high temperature week in power network

Error% / Temp.℃		Model III		Model III considering external stochastic factors	
		E_{MAPE}	E_{MAX}	E_{MAPE}	E_{MAX}
Sat.	26~34	2.238	4.515	1.207	2.377
Sun.	26~37	2.522	5.463	1.230	2.542
Mon.	25~37	2.329	5.251	1.179	2.386
Tues.	24~35	2.365	4.694	1.052	2.290
Wed.	25~34	2.097	4.270	1.148	2.083
Thurs.	26~34	2.271	4.329	1.066	2.457
Fri.	24~29	2.360	5.518	1.293	2.808

7 Conclusions

This paper presents a new forecasting modeling approach based on the fusion of DONPA and the ICNN. According to the whole testing results show that the proposed approach can enhance effectively the precision of STLF and its stability, so the study of the integrated dynamics mechanism process theoretical significance.

References

1. Hobbs, B. F., Jitprapaikulsarn, S., Konda, S.: Analysis of the Value for Unit Commitment of Improved Load Forecasts. IEEE Trans on Power Systems 14(4) (1999) 1342-1348
2. Tsakoumis, A.C., Fessas, P., Mladenov, V.M.: Application of Chaotic Time Series for Short-term Load Prediction. WSEAS Transactions on Systems 2(3) (2003) 517-523

3. Drezga, Irislav, Rahman, Saifur: Phase-space Based Short-term Load Forecasting for Deregulated Electric Power Industry. In: David Brown, Mary Pastel Anderson (eds.): Proceedings of the International Joint Conference on Neural Networks. IEEE, Piscataway, NJ, USA (1999) 3405-3409
4. Senjyu, T., Mandal, P., Uezato, K., Funabashi, T.: Next Day Load Curve Forecasting Using Recurrent Neural Network Structure. IEE Proceedings-Generation, Transmission and Distribution 151(3) (2004) 388-394
5. Michanos, S.P., Tsakoumis, A.C., Fessas, P.: Short-term Load Forecasting Using a Chaotic Time Series. In: Liviu Goras (ed.): Proceedings of the International Symposium on Signals, Circuits and Systems. IEEE, Piscataway, NJ, USA. (2003) 437 -440
6. Drezga, I., Rahman, S.: Input Variable Selection for ANN-based Short-term Load Forecasting. IEEE Transactions on Power Systems 13(4)(1998) 1238-1244
7. Hou, Y.X., He, P.L., Wang, L.: Reconstructing High Dimension Phase Space: an Improved Approach to Chaotic Time Series Forecasting. Journal of Tianjin University 32(5) (1999) 594-598
8. Pan, S.T., Chen, S.-C., Chiu, S.-H.: A New Learning Algorithm of Neural Network for Identification of Chaotic Systems. In: Howard Eisner (ed.): Proceedings of the IEEE International Conference on Systems, Man and Cybernetics. IEEE, Piscataway, NJ, USA. (2003) 1316-1321
9. Grassberger, P., Procaccia, I.: Measuring the Strangeness of Strange Attractors. Physica D 29(9) (1983) 189-208
10. Lin, J.Y., Wang, Y.K., Huang, Z.P.: Selection of Proper Time-delay in Phase Space Reconstruction of Speech Signals. Signal Processing 15(3)(1999) 220-225
11. Aihara, K., Takabe, T., Toyoda, M.: Chaotic Neural Networks. Physics Letters A 144(6) (1990) 333-340
12. Kim, S.H., Hong, S.D., Park, W.W.: An Adaptive Neurocontroller with Modified Chaotic Neural Networks. In: Ken Marko, Paul Werbos (eds.): Proceedings of International Joint Conference on Neural Networks. IEEE, Piscataway, NJ, USA (2001) 509-514

Adaptive Neural Network Control for Nonlinear Systems Based on Approximation Errors

Yan-Jun Liu and Wei Wang

School of Electronic and Information Engineering, Dalian University of Technology, Dalian 116023, China
liuyjsir@163.com

Abstract. A stable adaptive neural network control approach is proposed in this paper for uncertain nonlinear strict-feedback systems based on backstepping. The key assumptions are that the neural network approximation errors satisfy certain bounding conditions. By a special scheme, the controller singularity problem is avoided perfectly. The proposed scheme improves the control performance of systems and extends the application scope of nonlinear systems. The overall neural network control systems guarantee that all the signals of the systems are uniformly ultimately bounded and the tracking error converges to a small neighborhood of zero by suitably choosing the design parameter.

1 Introduction

The research on the adaptive control of nonlinear systems with linearly parameterized uncertainty has advanced significantly since the late 1980s [1-4]. For systems with high uncertainty, for example, the uncertainty that cannot be linearly parameterized or is completely unknown, adaptive control approach obtained further development by means of neural network (NN) based on backstepping [5-8].

Although significant progress has been obtained by combining backstepping method and NN techniques, there still exist some problems that need to be solved in practical applications. An adaptive NN theorem with boundedness was proposed in literature [5,6]. Estimated values of unknown bounds of the NN approximation error were on-line adaptively turned. However, in order to avoid the controller singularity problem, the gain functions are assumed to be constants in [5,6]. This assumption cannot be satisfied in many cases. The situation that the gain functions are unknown has been studied in [7,8]. By introducing the integral-type Lyapunov functions, an adaptive backstepping NN approach was proposed in [7]. However, due to the integral operation, this approach is complicated and difficult to use in practice. An adaptive NN control scheme that does not requires the integral-type Lyapunov functions and guarantees stability of the closed-loop systems [8]. But, authors in [7,8] assumed that unknown bounds of the NN approximation error are less than bounded constants, if unknown bounds are larger than the assumed bounds, no performance of systems is guaranteed.

Considering above the disadvantage. A stable adaptive NN control approach for uncertain nonlinear strict-feedback systems is proposed in this paper. Gain functions are assumed unknown smooth functions and the approximation error satisfies certain

J. Wang et al. (Eds.): ISNN 2006, LNCS 3972, pp. 836 – 841, 2006.

bounding conditions. By combining backstepping and the approximation property of an RBF NN, and introducing a special-type Lyapunov functions, this approach avoids the controller singularity problem perfectly. Because unknown bounds of the NN approximation error are adaptively tuned online, the control performance of the closed-loop systems is improved and the application scope is extended.

2 Problem Description

The many practical nonlinear systems can be expressed in the following form

$$\begin{aligned} \dot{x}_i &= f_i(\bar{x}_i) + g_i(\bar{x}_i)x_{i+1}, i = 1,2,\cdots,n-1, \\ \dot{x}_n &= f_n(\bar{x}_n) + g_n(\bar{x}_n)u, \\ y &= x_1, \end{aligned} \tag{1}$$

where $\bar{x}_i = [x_1,\cdots,x_i]^T$, $u \in R$, $y \in R$ are system's state variables, input and output, respectively. Smooth functions $f_i(\bar{x}_i)$, $g_i(\bar{x}_i)$ are unknown. The reference signal y_d is bounded. The control object is to design an NN controller such that all the signals of systems are uniformly ultimately bounded and the tracking error converges to a small neighborhood of zero. We make following assumptions as commonly being done in the literature.

Assumption 1. The signs of $g_i(\bar{x}_i)$ are bounded. Without losing generality, we will assume $g_{i1} > g_i(\bar{x}_i) > g_{i0} > 0, \forall \bar{x}_i \in \Omega_i \in R^i, i = 1,2,\cdots,n$.

Assumption 2. There exist constants $g_{id} > 0$, such that $|\dot{g}_i(\bar{x}_i)| \le g_{id}$, $\forall \bar{x}_i \in \Omega_i \in R^i$.

3 Adaptive Neural Network Control Design

In this section, we will incorporate backstepping into adaptive NN control design. In the following, we assume that $u = x_{(n+1)d}$, $e_0 = e_{n+1} = 0$, $x_{1d} = y_d$ and $V_0 = 0$.

Step $i(1 \le i \le n)$. A virtual controller $x_{(i+1)d}$ can be designed to make the error $e_i = x_i - x_{id}$ as small as possible. The derivative of e_i is

$$\dot{e}_i = \dot{x}_i - \dot{x}_{id} = g_i(\bar{x}_i)[g_i^{-1}(\bar{x}_i)f_i(\bar{x}_i) - g_i^{-1}(\bar{x}_i)\dot{x}_{id} + x_{i+1}] . \tag{2}$$

From (2), it can be seen that $H_i(z_i) = g_i^{-1}(\bar{x}_i)[f_i(\bar{x}_i) - \dot{x}_{id}]$ is unknown smooth function, where $z_i = [\bar{x}_i^T, \partial x_{id}/\partial x_1,\cdots,\partial x_{id}/\partial x_{i-1},\Phi_{i-1}]^T \in R^{2i}$ are taken as inputs of the NN, $\dot{x}_{id} = \sum_{j=1}^{i-1}\dot{x}_j \partial x_{id}/\partial x_j + \Phi_{i-1}$, $\Phi_{i-1} = \sum_{j=1}^{i-1}(\dot{\theta}_j \partial x_{id}/\partial\theta_j + \dot{x}_{jd}\partial x_{id}/\partial x_{jd} + \dot{\beta}_j), i = 2,\cdots,n$ is introduced as a medium variable, which is computable.

Remark 1. The reason why vector $z_i \in R^{2i}$ is regarded as inputs of the NN is that make the inputs of the NN minimization. If we introduce a large amount of NN weights estimates θ_i to be regarded as inputs of the NN, it would lead to the dimension curse of RBF NN. Therefore, by introducing computable medium variable $(\partial x_{id}/\partial x_1),\cdots,(\partial x_{id}/\partial x_{i-1}),\Phi_{i-1}$, inputs dimension of the NN is reduced.

We make use of the NN to approximate the unknown function $H_i(z_i)$. The universal approximation property of an RBF NN is referred to [9]. We have

$$H_i(z_i)=\theta_i^{*T}\xi_i(z_i)+\delta_i(z_i), \tag{3}$$

where θ_i^* is the optimal weight vector, and $\delta_i(z_i)$ denotes the approximation error.

Throughout this paper, by introducing $\theta_i^T\xi_i(z_i)$ as the NN, we make the following assumption on the approximation error for each network.

Assumption3. On the compact set Ω_i, we assume $|\delta_i(z_i)|\le\psi_i^* s_i(z_i), i=1,2,\cdots,n$, where $\psi_i^*\ge 0$ is unknown and $s_i(z_i):\Omega_i\mapsto R^+$ is known smooth function.

Remark 2. Most of the analytical results in the adaptive NN control literature make the key assumptions that the approximation error is bounded by some constants [7,8], if the approximation error is larger than the assumed bounds, no performance of systems can be guaranteed. Assumption 3 relaxes these conditions by requiring only that the approximation error is bounded. Choosing the virtual control input, we have

$$x_{(i+1)d}=-e_{i-1}-k_ie_i-\theta_i^T\xi_i(z_i)+\beta_i, \tag{4}$$

where $\beta_i=-\psi_i\omega_i$, $\omega_i=s_i\tanh(e_is_i/\varepsilon)$, ψ_i will be defined later. For arbitrary $\varepsilon>0$ and $\Pi\in R$, the following inequality holds

$$0\le|\Pi|-\Pi\tanh(\Pi/\varepsilon)\le k\varepsilon, \tag{5}$$

where k is a constant that satisfies $k=\exp[-(k+1)]$, i.e., $k=0.2785$. By using (3) (4), and introducing the error variable $e_{i+1}=x_{i+1}-x_{(i+1)d}$, (2) becomes

$$\dot e_i=g_i(\bar x_i)[-\tilde\theta_i^T\xi_i(z_i)+\delta_i(z_i)+e_{i+1}-e_{i-1}-k_ie_i+\beta_i]. \tag{6}$$

Choosing the following adaptation law

$$\dot\theta_i=\Gamma_i[\xi_i(z_i)e_i-\sigma_i(\theta_i-\theta_i^0)], \tag{7}$$

$$\dot\psi_i=\gamma_i[e_i\omega_i-\eta_i(\psi_i-\psi_i^0)], \tag{8}$$

where $\Gamma_i=\Gamma_i^T>0$ is a gain matrix, $\gamma_i>0,\sigma_i,\eta_i>0$, θ_i^0,ψ_i^0 are the initial estimation of θ_i^*,ψ_i^*. Consider the following Lyapunov function candidate

$$V_i=V_{i-1}+e_i^2/2g_i(\bar x_i)+\tilde\theta_i^T\Gamma_i^{-1}\tilde\theta_i/2+\tilde\psi_i^2/2\gamma_i. \tag{9}$$

Using (6) (7), the time derivative of V_i is

$$\dot{V}_i = \dot{V}_{i-1} + (e_i e_{i+1} - e_{i-1} e_i) - [k_i + \dot{g}_i(\bar{x}_i)/2g_i^2(\bar{x}_i)]e_i^2 - \sigma_i \tilde{\theta}_i^T (\theta_i - \theta_i^0) + \Psi_i, \tag{10}$$

where $\Psi_i = e_i[\delta_i(z_i) + \beta_i] + \tilde{\psi}_i \dot{\psi}_i / \gamma_i$.

According to the assumptions 1 and 2, the following inequality holds

$$-[k_i + \dot{g}_i(\bar{x}_i)/2g_i^2(\bar{x}_i)]e_i^2 \le -[k_i - g_{id}/2g_i^2(\bar{x}_i)]e_i^2 \le -[k_i - g_{id}/2g_{i0}^2]e_i^2 \le -k_i^* e_i^2, \tag{11}$$

where k_i is chosen such that $k_i^* = k_i - g_{id}/2g_{i0}^2 > 0$.

By completion of squares, we obtain

$$-\sigma_i \tilde{\theta}_i^T (\theta_i - \theta_i^0) = -\sigma_i \tilde{\theta}_i^T \tilde{\theta}_i - \sigma_i \tilde{\theta}_i^T (\theta_i^* - \theta_i^0) \le -\frac{\sigma_i}{2} \tilde{\theta}_i^T \tilde{\theta}_i + \frac{\sigma_i}{2} \left\| \theta_i^* - \theta_i^0 \right\|^2. \tag{12}$$

According to the assumption 3 and using (5) (8), we obtain the following inequality

$$\begin{aligned}\Psi_i &= e_i[\delta_i(z_i) + \beta_i] + \tilde{\psi}_i \dot{\psi}_i / \gamma_i \le e_i \beta_i + \psi_i^* |e_i| s_i(z_i) + \tilde{\psi}_i \dot{\psi}_i / \gamma_i \\ &= -\psi_i e_i s_i \tanh(e_i s_i / \varepsilon) + \psi_i^* |e_i| s_i + \tilde{\psi}_i [e_i s_i \tanh(e_i s_i / \varepsilon) - \eta_i (\psi_i - \psi_i^0)] \\ &\le -\eta_i \tilde{\psi}_i^2 / 2 + \eta_i \left| \psi_i^* - \psi_i^0 \right|^2 / 2 + k \varepsilon \psi_i^*.\end{aligned} \tag{13}$$

In step i-1, it has been obtained that

$$\dot{V}_{i-1} \le e_{i-1} e_i - \sum_{j=1}^{i-1} k_j^* e_j^2 - \frac{1}{2} \sum_{j=1}^{i-1} \left(\sigma_j \tilde{\theta}_j^T \tilde{\theta}_j + \eta_j \tilde{\psi}_j^2 \right) + \frac{1}{2} \sum_{j=1}^{i-1} \left(\sigma_j \left\| \theta_j^* - \theta_j^0 \right\|^2 + \eta_j \left| \psi_j^* - \psi_j^0 \right|^2 \right) + k\varepsilon \sum_{j=1}^{i-1} \psi_j^* . \tag{14}$$

By using (11) - (14), we have the following inequality

$$\dot{V}_i \le e_i e_{i+1} - \sum_{j=1}^{i} k_j^* e_j^2 - \frac{1}{2} \sum_{j=1}^{i} \left(\sigma_j \tilde{\theta}_j^T \tilde{\theta}_j + \eta_j \tilde{\psi}_j^2 \right) + \frac{1}{2} \sum_{j=1}^{i} \left(\sigma_j \left\| \theta_j^* - \theta_j^0 \right\|^2 + \eta_j \left| \psi_j^* - \psi_j^0 \right|^2 \right) + k\varepsilon \sum_{j=1}^{i} \psi_j^* . \tag{15}$$

From (4), we obtain the practical control input

$$u = x_{(n+1)d} = -e_{n-1} - k_n e_n - \theta_n^T \xi_n (z_n) + \beta_n . \tag{16}$$

From (15), we have the following inequality

$$\dot{V}_n \le -\sum_{j=1}^{n} k_j^* e_j^2 - \frac{1}{2} \sum_{j=1}^{n} \left(\sigma_j \tilde{\theta}_j^T \tilde{\theta}_j + \eta_j \tilde{\psi}_j^2 \right) + \frac{1}{2} \sum_{j=1}^{n} \left(\sigma_j \left\| \theta_j^* - \theta_j^0 \right\|^2 + \eta_j \left| \psi_j^* - \psi_j^0 \right|^2 \right) + k\varepsilon \sum_{j=1}^{n} \psi_j^* . \tag{17}$$

Theorem 1. Under the assumptions 1, 2 and 3, let $y_d(t)$ be a bounded reference signal for all $t \ge 0$. Consider the systems consisting of (1). By choosing the virtual control inputs (4) and the practical control input (16), the NN weights adaptation law (7), and the parameters adaptation law (8), all the signals of the systems are uniformly ultimately bounded is guaranteed and the tracking error converges to a small neighborhood of zero by suitably choosing the design parameters.

Proof: Our proof will follow the quite standard stability analysis techniques of NN based on adaptive control as can be found in [7][8]. (17) can be rewritten as follows

$$\dot{V}_n \le -\mu V_n + \alpha , \tag{18}$$

where $\mu = \min_{1\le i\le n}\left(2k_i^* g_{i0}, \sigma_i \lambda_{\min}(\Gamma_i), \eta_i \gamma_i\right)$, $\lambda_{\min}(\Gamma_i)$ is minimal eigenvalue of the matrix $\Gamma_i, i = 1,\cdots,n$, $\alpha = nk\varepsilon\psi + \frac{1}{2}\sum_{j=1}^{n}\left(\sigma_j \left\|\theta_j^* - \theta_j^0\right\|^2 + \eta_j \left|\psi_j^* - \psi_j^0\right|^2\right)$, $\psi = \max_{1\le i\le n}\left(\psi_i^*\right)$.

The process of proof is similar to that of the theorem 1 in [8], using (18), we conclude that all signal of the systems are bounded.

Let $\beta = \alpha / \mu$, (18) satisfies $0 \le V_n(t) \le \beta + (V(0) - \beta)\exp(-\mu t)$. Therefore, we obtain $\sum_{i=1}^{n} e_i^2 / 2g_i(\bar{x}_i) \le \beta + (V(0) - \beta)\exp(-\mu t) < \beta + V(0)\exp(-\mu t)$. Let $g_{\max} = \max_{1\le i\le n}\{g_{i1}\}$, the inequality $e_1^2 < 2g_{\max}\beta + 2g_{\max}V(0)\exp(-\mu t)$ is true. It implies that given $\delta > \sqrt{2g_{\max}\beta}$, there exists T such that $t \ge T$, the tracking error satisfies $|e_1| < \delta$.

4 Simulation Experiment

In this section, simulation experiment is presented to demonstrate the feasibility of the proposed approach. The model of the strict-feedback systems is described as follows

$$\begin{aligned} \dot{x}_1 &= 0.5x_1 + (1 + 0.1x_1^2)x_2, \\ \dot{x}_2 &= x_1 x_2 + [2 + \cos(x_1)]u, \\ y &= x_1, \end{aligned} \tag{19}$$

where x_1 and x_2 are states, and y is the output of the systems. The initial conditions are $x_0 = [x_1(0), x_2(0)]^T = [1,0]^T$ and the reference signal is $y_d = \sin(t)$.

The system (19) is of second order, using the proposed approach, according to (16), we can obtain the practical input of the closed-loop system as follows

$$u = -e_1 - k_2 e_2 - \theta_2^T \xi_2(z_2) + \beta_2 ,$$

Where $e_1 = x_1 - x_{1d}$, $e_2 = x_2 - x_{2d}$, $x_{1d} = y_d$, $x_{2d} = -k_1 e_1 - \theta_1^T \xi_1(z_1) + \beta_1$, $z_1 = [x_1, \dot{x}_{1d}]^T$ $z_2 = \left[\bar{x}_2^T, \partial x_{2d} / \partial x_1, \Phi_1\right]^T$, $\Phi_1 = \partial x_{2d} / \partial x_{1d} \dot{x}_{1d} + \partial x_{2d} / \partial \theta_1 \dot{\theta}_1 + \dot{\beta}_1$, $\beta_1 = -\psi_1 \omega_1$, $\beta_2 = -\psi_2 \omega_2$, and the NN weights θ_1 and θ_2 are updated by (6). Equation (7) is chosen as adaptation law ψ_1 and ψ_2. The NN $\theta_1^T \xi_1(z_1)$ contains 20 nodes, with centers $\mu_j, j = 1,\cdots,20$ evenly spaced in $[-4,4]\times[-4,4]$, and widths $\sigma = 2$. The NN $\theta_2^T \xi_2(z_2)$ contains 125 nodes, with centers $\mu_j, j = 1,\cdots,125$ evenly spaced in $[-4,4]\times[-4,4]\times[-4,4]\times[-6,6]$, and width $\sigma = 2$. The design parameters of the proposed controller are $k_1 = k_2 = 3.5$, $\sigma_1 = \sigma_2 = 2, \gamma_1 = \gamma_2 = 2$, $\eta_1 = \eta_2 = 1$,

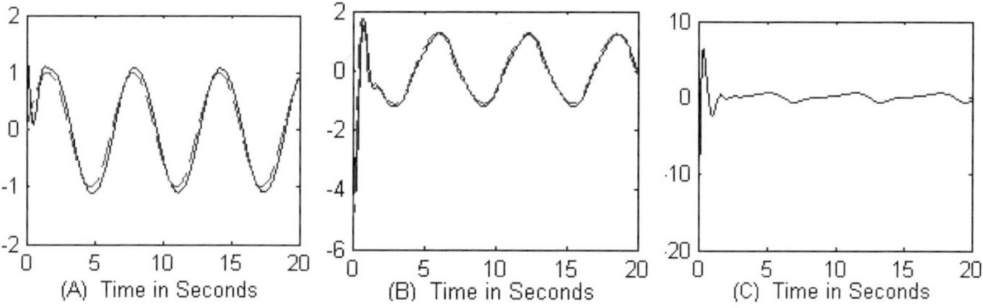

Fig. 1. (A) The output $y = x_1$ (solid line) and the reference signal $y_d = \sin(t)$ (dashed line); (B) The state of the system x_2 (solid line) and the virtual input x_{2d} (dashed line); (C) The practical control input u

$\Gamma_1 = \Gamma_2 = diag\{2\}, \varepsilon = 0.2$. The initial weights are $\theta_1(0) = \theta_2(0) = 0.5$ and the initial parameters are $\psi_1(0) = \psi_2(0) = 0$. For simplicity, let functions s_1, s_2 be up to 1.

Fig. 1 shows the simulation result of applying proposed approach to system (19) for tracking the reference signal $y_d = \sin(t)$. From the fig. (A) and (B), we can see that good tracking performance is obtained. The boundedness of u is shown in fig. (C).

5 Conclusion

A stable adaptive neural networks control approach for uncertain nonlinear strict-feedback systems is proposed. The control input and the adaptation law are obtained using Lyapunov synthesis approach. The proposed approach guarantees that all the signals of the resulting closed-loop systems are uniformly ultimately bounded and the tracking error converges to a small neighborhood of zero.

References

1. Isidori, A. (2nd): Nonlinear Control Systems. Spinger, Berlin (1989)
2. Nam, K., and Arapostations, A.: A Model-Reference Adaptive Control Scheme for Pure-Feedback Nonlinear Systems. IEEE Trans. Automat. Control 33 (1988) 803-811
3. Krstic, M., Kanellakopoulos, L., Kokotovic, P.V.: Nonlinear Adaptive Control Design. New York (1995)
4. Seto, D., Annaswamy, A.M., Baillieul, J.: Adaptive Control of Nonlinear Systems with A Triangular Structure. IEEE Trans. Automat. Control 39 (1994) 1411-1428
5. Polycarpou, M.M.: Stable Adaptive Neural Control Scheme for Nonlinear Systems. IEEE Trans. Automat. Control 41 (1996) 447-451
6. Polycarpou, M.M., Mears, M.J.: Stable Adaptive Tracking of Uncertain Systems Using Nonlinearly Parameterized On-Line Approximators. Int. J. Control 70 (1998) 363-384
7. Zhang, T., Ge, S.S., Hang, C.: Adaptive Neural Network Control for Strict-Feedback Nonlinear Systems Using Backstepping Design. Automatica 36 (2000) 1835–1846
8. Ge, S.S., Wang, C.: Direct Adaptive NN Control of a Class of Nonlinear Systems. IEEE Trans. Neural Network 13 (2002) 214-221
9. Park, J., Sandberg, I.W.: Universal Approximation Using Radial-Basis-Function Network. Neural Computation 3 (1991) 246-257

Adaptive Neural Network Control for Switched System with Unknown Nonlinear Part by Using Backstepping Approach: SISO Case

Fei Long[1,2], Shumin Fei[1], Zhumu Fu[1], and Shiyou Zheng[1]

[1] Department of Automatic Control, Southeast University, Nanjing, 210096, P.R. China
[2] School of Information Engineering, Guizhou University, Guiyang, 550000, P.R. China
flong1973@yahoo.com.cn

Abstract. In this paper, we address, in a backstepping way, stabilization problem for a class of switched nonlinear systems whose subsystem with trigonal structure by using neural network. An adaptive neural network switching control design is given. Backsteppping, domination and adaptive bounding design technique are combined to construct adaptive neural network stabilizer and switching law. Based on common Lyapunov function approach, the stabilization of the resulting closed-loop systems is proved.

1 Introduction

Switched dynamical systems have been attracting much attention because the study for these systems is not only academically challenging, but also of practice importance. For instance, for the switched linear system that consists of two stable subsystems, it would be unstable if we apply unsuitable switching rule to this system. Conversely, if its two subsystems are unstable and we adopt suitable switching path, the switched system would be stable. As such, how to design a switching law so that the switched system achieves certain performance is indeed an important and well-motivated problem. During the last decades, applications of neural network for non-switched nonlinear system have made great progress (see [1], [2], [3], [4], [5], [6], [7] and [8] for references). The study for non-switched nonlinear systems using universal function approximation has received much attention and many methods have been proposed. Typically, these methods use neural networks as approximation models for the unknown part of systems. One key advantage of these schemes is that the adaptive laws were derived based on Lyapunov synthesis and, therefore, guaranteed the stability of non-switched systems without the requirement for offline training. Due to the difference between non-switched systems and switched systems, a stable controller designed in non-switched system may become unstable in switched sys-tem via unsuitable switching law, thus we may run into troubles when we implement these networks controllers in switched system in which the data are typically available at switching time instants. Therefore, the study for switched system based on neural network is necessary and significant. In [9], the tracking control problem is studied for switched systems with unknown nonlinear parts by using RBF neural network. In

J. Wang et al. (Eds.): ISNN 2006, LNCS 3972, pp. 842 – 848, 2006.

[10] and [11], based on RBF neural networks, H-infinity control problem for similar switched systems in [9] is investigated in SISO case and MIMO case, respectively.

2 System Description and Preliminaries

Consider the following switched nonlinear system

$$\dot{x} = Ax + F_{\sigma(t)}(x) + G_{\sigma(t)}(x)u \tag{1}$$

where $x \in \mathbb{R}^n$ and $u \in \mathbb{R}$ is the system state and control input, respectively.

$$\bar{x}_i = (x_1, \cdots, x_i)^T,\ 1 \le i \le n;\ F_{\sigma(t)}(x) = \left(f_{1\sigma(t)}(x_1), \cdots, f_{i\sigma(t)}(\bar{x}_i), \cdots, f_{n\sigma(t)}(x)\right)^T;$$

$$G_{\sigma(t)}(x) = \left(0, \cdots, 0, g_{\sigma(t)}(x)\right)^T;\ A = \begin{bmatrix} 0 & I_{n-1} \\ 0 & 0 \end{bmatrix}.$$

$\sigma(\cdot) : [0, +\infty) \to \{1, 2, \cdots, N\} \overset{def}{=} \underline{\mathbb{N}}$ stands for the piecewise constant switching law to be designed. Moreover, $\sigma(t) = k$ implies that the kth subsystem is active. $g_k(x)\ (k \in \underline{\mathbb{N}})$ are known smooth function and $|g_k(x)| \ge l$ (l is a positive constant) for every $x \in \mathbb{R}^n$. $f_{ik}(x_i)\ (1 \le i \le n, 1 \le k \le N)$ are unknown smooth functions.

In this note, our control objective is to design a switching law $\sigma(t)$ and an associated adaptive neural network controller so that the resulting closed-loop system is stabilizable. On compact sets $\Omega_i \in \mathbb{R}^i\ (1 \le i \le n)$, $f_{ik}(\bar{x}_i)\ (1 \le i \le n, 1 \le k \le N)$ can be approximated by the following RBF neural networks, respectively.

$$\hat{f}_{ik}(\bar{x}_i, \theta_{ik}) = \theta_{ik}^T \phi_{ik}(\bar{x}_i) \tag{2}$$

where $\phi_{ik}(\cdot) : \Omega_i \to \mathbb{R}^{p_i}, k \in \underline{\mathbb{N}}$ are known Guassian basis function vectors, $p_i\ (1 \le i \le n)$ are the RBF neural networks node numbers. Defined $\theta_i \in \mathbb{R}^{p_i}$ $(i = 1, 2, \cdots, n)$ are the optimal weights as follows

$$\theta_i = \arg \min_{\hat{\theta}_{ik} \in \mathbb{R}^{p_i}} \left\{ \min_{k \in \underline{\mathbb{N}}} \left\{ \sup_{\bar{x}_i \in \Omega_i} | f_{ik}(\bar{x}_i) - \hat{\theta}_{ik}^T \phi_{ik}(\bar{x}_i) | \right\} \right\} \tag{3}$$

By means of (2) and (3), the system (1) can be written by

$$\dot{x} = Ax + F_{\sigma(t)}(x, \theta) + G_{\sigma(t)}(x)u + \upsilon_{\sigma(t)} \tag{4}$$

where

$$F_{\sigma(t)}(x, \theta) = \left(\varphi_{1\sigma(t)}^T(x_1)\theta, \cdots, \varphi_{i\sigma(t)}^T(\bar{x}_i)\theta, \cdots, \varphi_{n\sigma(t)}^T(x)\theta\right)^T,\ \theta^T = \left(\theta_1^T, \cdots, \theta_n^T\right),$$

$$\upsilon_{\sigma(t)} = \left(\upsilon_{1\sigma(t)}, \cdots, \upsilon_{n\sigma(t)}\right)^T,\ \upsilon_{ik} = f_{ik}(\bar{x}_i) - \theta_i^T \phi_{ik}(\bar{x}_i)\ (i = 1, 2, \cdots, n;\ k = 1, 2, \cdots, N)$$

$$\varphi_{ik}^{T}=\left(\underbrace{0,\cdots,0}_{\sum_{j=1}^{i-1}p_j},\phi_{fik}^{T},\underbrace{0,\cdots,0}_{\sum_{j=i+1}^{n}p_j}\right)\ (i=1,2,\cdots,n;\ k=1,2,\cdots,N).$$

On the compact subsets $\Omega_i\subset\mathbb{R}^i\ (i=1,2,\cdots,n)$, define

$$\omega=\max_{k\in\underline{\mathbb{N}}}\left\{\max_{1\le i\le n}\left\{\sup_{\overline{x}_i\in\Omega_i}\left|\upsilon_{ik}(\overline{x}_i)\right|\right\}\right\}. \tag{5}$$

where $\omega\ge 0$ is the unknown constant because the compact sets $\Omega_i\ (i=1,2,\cdots,n)$ are actually unknown. In fact, by virtue of neural network approximation theory [12], it is obvious that (5) is well-defined.

In this paper, we introduce the following Lemma.

Lemma 2.1 [1]: For every $\lambda\in\mathbb{R}$, there exists a constant $\varepsilon>0$ such that

$$|\lambda|\le\lambda\tanh(\lambda/\varepsilon)+0.2785\varepsilon.$$

3 Controller Design

In the section, we firstly employ backstepping technique to design the adaptive neural network stabilizer for each subsystem of switched system (4). Secondly, we utilize attenuation level of each subsystem to design switching law for switched system.

3.1 Adaptive NN Stabilizer of subsystem Design

For the kth subsystem of system (4),

$$\dot{x}=Ax+F_k(x,\theta)+G_k(x)u+\upsilon_k \tag{6}$$

we design the stabilizer and adaptive law of system (6) in a backstepping way. In this subsection, we assume that the number k be fixed.

Define a change of coordinates as follows

$$z_1=x_1;\ z_i=x_i-\alpha_{(i-1)k}(\overline{x}_{i-1},\hat{\theta},\hat{\omega}),\ 2\le i\le n \tag{7}$$

where $\hat{\theta}$ and $\hat{\omega}$ is the estimate of θ and ω, respectively.

From (6) and (7), we select the stabilizing functions

$$\alpha_{1k}=-nz_1-\hat{\theta}^T\beta_{\theta,1k}-\hat{\omega}\beta_{\omega,1k};\ \alpha_{ik}=-z_{i-1}-nz_i-\hat{\theta}^T\beta_{\theta,ik}-\hat{\omega}\beta_{\omega,ik}+\Lambda_{ik},\ 2\le i\le n \tag{8}$$

where;

$$\Lambda_{ik}=\sum_{j=1}^{i-1}\left(\frac{\partial\alpha_{(i-1)k}}{\partial x_j}x_{j+1}+\frac{\partial\alpha_{(i-1)k}}{\partial\hat{\theta}}\Gamma\beta_{\theta,jk}z_j+\frac{\partial\alpha_{(i-1)k}}{\partial\hat{\omega}_k}\rho\beta_{\omega,jk}z_j\right),\ 2\le i\le n;$$

$$\beta_{\theta,1k} = \varphi_{1k}(x_1);\ \beta_{\theta,ik} = \varphi_{ik}(\bar{x}_i) - \sum_{j=1}^{i-1} \frac{\partial \alpha_{(i-1)k}}{\partial x_j} \varphi_{jk}(\bar{x}_j), 2 \le i \le n;\ \beta_{\omega,ik} = \gamma_{ik} \tanh(\gamma_{ik} z_i / \varepsilon), 1 \le i \le n;$$

$$\gamma_{1k} = 1;\ \gamma_{ik} = 1 + \sum_{j=1}^{i-1} \left| \frac{\partial \alpha_{(i-1)k}}{\partial x_j} \right|,\ 2 \le i \le n .$$

We select, respectively, the adaptive law and the stabilizer as follows.

$$\dot{\hat{\theta}} = \sum_{i=1}^{n} \Gamma \beta_{\theta,ik} z_i,\ \dot{\hat{\omega}} = \sum_{i=1}^{n} \rho \beta_{\omega,ik} z_i \tag{9}$$

$$u = \alpha_{nk}(x, \hat{\theta}, \hat{\omega}) / g_k(x) \tag{10}$$

where $\Gamma = \Gamma^T > 0,\ \rho > 0$.

By means of Lemma 2.1, it follows from backstepping approach and $\omega < (\sum_{i=1}^{n} z_i^2)/(0.2785\varepsilon)$ that the system (6) is stabilizable by adaptive law (9), control law (10) and the following Lyapunov function

$$V = 2^{-1} \sum_{i=1}^{n} z_i^2 + 2^{-1} \tilde{\theta}^T \Gamma^{-1} \tilde{\theta} + (2\rho)^{-1} \tilde{\omega}^2 \tag{11}$$

where $\tilde{\theta} = \hat{\theta} - \theta,\ \tilde{\omega} = \hat{\omega} - \omega$.

3.2 Switching Law Design

In this subsection, we will employ attenuation level of each subsystem to design switching strategy for switched system (4). Preselecting the compact set $Z_0 \subset \mathbb{R}^n$ and contains the initial state $z(t_0) = z_0$. Preselecting a positive real number ε and d, moreover $d > \max_{z \in Z_0} z^T z$. Suppose that

$$\bigcup_{k=1}^{N} \left\{ z \in \mathbb{R}^n \ \middle|\ \max_{1 \le i \le n} (f_{ik}(\bar{x}_i) - \theta_i^T \phi_{ik}(\bar{x}_i)) \le d / (0.2785\varepsilon) \right\} \subseteq \left\{ z \in \mathbb{R}^n \ \middle|\ \|z\|^2 \le d \right\}$$

For initial state $x(t_0) = (x_1(t_0), x_2(t_0), \cdots, x_n(t_0))^T = (x_{10}, x_{20}, \cdots, x_{n0})^T$, set

$$\sigma(t_0) = \arg \min_{1 \le k \le N} \left\{ \max_{1 \le i \le n} (f_{ik}(\bar{x}_{i0}) - \hat{\theta}_i^T(t_0) \phi_{ik}(\bar{x}_{i0})) \right\}$$

where $\bar{x}_{i0} = (x_1(t_0), x_2(t_0), \cdots, x_i(t_0))^T = (x_{10}, x_{20}, \cdots, x_{i0})^T$, the symbol "arg min" denotes the index that attains the minimum. If there are one more than such index, we just pick the smallest one.

The first switching time instant is determined by

$$t_1 = \inf \left\{ t \ge t_0 \middle|\ \textit{there exists a } i \in \underline{n} \textit{ such that } f_{i\sigma(t_0)}(\bar{x}_i(t)) - \hat{\theta}_i^T(t) \phi_{i\sigma(t_0)}(\bar{x}_i(t) \ge (d / 0.2785\varepsilon) \right\}$$

where $\underline{n} \overset{def}{=} \{1, 2, \cdots, n\}$

The corresponding switching index is chosen to be

$$\sigma(t_1) = \arg\min_{1\le k\le N}\left\{\max_{1\le i\le n}(f_{ik}(\overline{x}_i(t_1)) - \hat{\theta}_i^T(t_1)\phi_{ik}(\overline{x}_i(t_1)))\right\}$$

Finally, we define the switching time/index sequences recursively by

$$t_{j+1} = \inf\left\{t \ge t_j \,\middle|\, \textit{there exists a } i \in \underline{n} \textit{ such that } f_{i\sigma(t_j)}(\overline{x}_i(t)) - \hat{\theta}_i^T(t)\phi_{i\sigma(tj)}(\overline{x}_i(t) > (d/0.2785\varepsilon)\right\}$$

The corresponding switching index is chosen to be

$$\sigma(t_{j+1}) = \arg\min_{1\le k\le N}\left\{\max_{1\le i\le n}(f_{ik}(\overline{x}_i(t_{j+1})) - \hat{\theta}_i^T(t_{j+1})\phi_{ik}(\overline{x}_i(t_{j+1})))\right\}$$

Theorem 3.1. Suppose the adaptive neural network controller and the above switching law are applied to switched system (1). For any given positive constant ε, the following set formula holds.

$$\bigcup_{k=1}^{N}\left\{z\in\mathbb{R}^n \,\middle|\, \max_{1\le i\le n}[f_{ik}(\overline{x}_i) - \theta_i^T\phi_{ik}(\overline{x}_i)] \le d/(0.2785\varepsilon)\right\} \subseteq \left\{z\in\mathbb{R}^n \,\middle|\, \|z\|^2 \le d\right\}. \quad (12)$$

Then, for any initial conditions, the resulting closed-loop system is asymptotically stable.

Proof. The adaptive neural network controller and switching law developed in this subsection are applied to switched system (1). Then the resulting closed-loop system can be written as

$$\begin{cases}\dot{z}_1 = -nz_1 + z_2 - \tilde{\theta}^T\beta_{\theta,1\sigma(t)} - \hat{\omega}_{\sigma(t)}\beta_{\omega,1\sigma(t)} + \delta_{1\sigma(t)} \\ \dot{z}_i = -z_{i-1} - nz_i + z_{i+1} - \tilde{\theta}^T\beta_{\theta,i\sigma(t)} - \hat{\omega}_{\sigma(t)}\beta_{\omega,i\sigma(t)} + \delta_{i\sigma(t)},\ 2\le i\le n-1 \\ \dot{z}_n = -z_{n-1} - nz_n - \tilde{\theta}^T\beta_{\theta,n\sigma(t)} - \hat{\omega}_{\sigma(t)}\beta_{\omega,n\sigma(t)} + \delta_{n\sigma(t)}\end{cases} \quad (13)$$

where $\delta_{1\sigma(t)} = \upsilon_{1\sigma(t)}$, $\delta_{i\sigma(t)} = \upsilon_{i\sigma(t)} - \sum_{j=1}^{i-1}\frac{\partial\alpha_{(i-1)\sigma(t)}}{\partial x_j}\upsilon_{j\sigma(t)}$.

For switched system (13), by Lemma 2.1, the network reconstruction terms can be dealt with as follows.

$$\begin{cases}\delta_{1k}z_1 \le |\upsilon_{1k}|\times|z_1| \le \omega\beta_{\omega,1k}z_1 + 0.2785\varepsilon\omega,\ 1\le k\le N \\ \delta_{ik}z_i \le \left|\upsilon_{ik} - \sum_{j=1}^{i-1}\frac{\partial\alpha_{(i-1)k}}{\partial x_j}\upsilon_{jk}\right|\times|z_i| \le \beta_{\omega,ik}\omega z_i + 0.2785\varepsilon\omega,\ 2\le i\le n-1, 1\le k\le N\end{cases} \quad (14)$$

Consider the Lyapunov function candidate (11), $\forall t\in[t_j, t_{j+1})$, the derivative of V along the trajectories of the system (13) is given by

$$\begin{aligned}\dot{V} &= \sum_{i=1}^{n} z_i\dot{z}_i + \tilde{\theta}^T\Gamma^{-1}\dot{\hat{\theta}} + \rho^{-1}\tilde{\omega}\dot{\hat{\omega}} \\ &= -n\|z\|^2 + \rho^{-1}\tilde{\omega}\dot{\hat{\omega}} + \tilde{\theta}^T\Gamma^{-1}(\dot{\hat{\theta}} - \sum_{i=1}^{n}\Gamma\beta_{\theta,ik}z_i) - \sum_{i=1}^{n}\hat{\omega}\beta_{\omega,ik}z_i + \sum_{i=1}^{n}\delta_{ik}z_i\end{aligned} \quad (15)$$

In view of (14) and (9), for every $z \in \left\{ z \in \mathbb{R}^n \,\middle|\, \|z\|^2 \le d \right\}$ we have

$$\dot{V} \le -nd + 0.2785n\varepsilon\omega \tag{16}$$

By means of (5) and the switching law developed in the subsection, the symbol ω of (16) stands for

$$\max_{1 \le i \le n} \left\{ \sup_{t \in [t_j, t_{j+1})} \left| \upsilon_{i\sigma(t_j)}(\overline{x}_i(t)) \right| \right\}$$

Again in view of (12) and the switching law developed in this subsection, $\dot{V}(t) < 0$, for every $t \in [t_0, \infty)$. Therefore by Lyapunov stability theorem, the conclusion holds. This is completes the proof. ◊

4 Conclusions

In this paper, an adaptive neural network stabilizer and a switching law were given for a class of switched nonlinear systems whose subsystem with trigonal structure. RBF neural networks are used to approximate unknown nonlinear terms. Combining with backsteppping approach and adaptive bounding design technique, an adaptive neural network switching control design is given. Furthermore, for above mention switched nonlinear system, an adaptive neural network stabilizer and switching law were constructed by using above control design technique. The stabilization of the resulting closed-loop system was proved based on common Lyapnov function approach. Future work includes its extension to the MIMO case of such switched nonlinear systems.

Acknowledgement

This work was partially supported by the NSFC under Grant 60574006.

References

1. Liu, G. P., et al: Variable Neural Networks for Adaptive Control of Nonlinear Systems. IEEE Trans. Systems, Man, Cybermetics-Part C 29 (1999) 34-43
2. Sridhar, S., Hassan, K. K.: Output Feedback Control of Nonlinear Systems Using RBF Neural Networks. IEEE Trans. Neural Network 11 (2000) 69-79
3. Levin, A. U., Narendra, K. S.: Control of Nonlinear Dynamical Systems Using Neural Networks-Part II: Observability, Identification, and Control. IEEE Trans. Neural Networks 7 (1996) 30-42
4. Cheng, W. S., Li, J. M.: Adaptive Backstepping Neural Network Control for Unknown Nonlinear Time-Delay Systems. In: Wang, J., Liao, X, Yi, Z. (eds.): Advance in Neural Networks---ISNN2005. Lecture Notes in Computer Science, Vol. 3498. Springer-Verlag, Berlin Heidelberg New York (2005) 30–35

5. Ge, S. S., et al: A Direct Method for Robust Adaptive Nonlinear Control with Guaranteed Transient Performance. System Control Letter 37 (1999) 275-284
6. Lewis, F. L., et al: Multilayer Neural-Net Robot Controller with Guaranteed Tracking Performance. IEEE Trans. Neural Networks 7 (1996) 388-398
7. Polycarpou, M. M.: Stable Adaptive Neural Control Scheme for Nonlinear Systems. IEEE Trans. Automat. Contr. 41 (1996) 447-450
8. Ge, S. S., et al: Stable Adaptive Neural Network Control. MA: Kluwer, Norwell (2001)
9. Long, F., Fei, S. M.: State Feedback Control for A Class of Switched Nonlinear Systems Based on RBF Neural Networks. In Proc. 23rd Chinese Control Conference 2 (2004) 1611-1614
10. Long, F., Fei, S. M., Zheng, S. Y.: H-infinity Control for Switched Nonlinear Systems Based on RBF Neural Networks. In: Wang, J., Liao, X, Yi, Z. (eds.): Advance in Neural Networks---ISNN2005. Lecture Notes in Computer Science, Vol. 3498. Springer-Verlag, Berlin Heidelberg New York (2005) 54–59
11. Long, F., Fei, S. M., Zheng, S. Y.: Adaptive H^{∞} Control for A Class of Switched Nonlinear Systems Based on RBF Neural Networks. Int. J. Hybrid Systems 4(4) (2004) 369-380
12. Haykin, S.: Neural Networks: A Comprehensive Foundation. 2nd edn. Prentice Hall, New York (1994)

Adaptive Neural Control for a Class of MIMO Non-linear Systems with Guaranteed Transient Performance

Tingliang Hu, Jihong Zhu, and Zengqi Sun

State Key Lab of Intelligent Technology and Systems,
Department of Computer Science and Technology,
Tsinghua University, Beijing 100084, China
htl02@mails.tsinghua.edu.cn

Abstract. A robust adaptive control scheme is presented for a class of uncertain continuous-time multi-input multi-output (MIMO) nonlinear systems. Within these schemes, multiple multi-layer neural networks are employed to approximate the uncertainties of the plant's nonlinear functions and robustifying control term is used to compensate for approximation errors. All parameter adaptive laws and robustifying control term are derived based on Lyapunov stability analysis so that all the signals in the closed loop are guaranteed to be semi-globally uniformly ultimately bounded and the tracking error of the output is proven to converge to a small neighborhood of zero. While the relationships among the control parameters, adaptive gains and robust gains are established to guarantee the transient performance of the closed loop system.

1 Introduction

The introduction of geometric techniques and in particular the feedback linearization methods have led to great success in the development of controllers for a class of nonlinear systems [1]. However, the plant to be controlled is too complex to find the exact system dynamics. In order to relax exact dynamics restrictions, several adaptive schemes have been introduced [2, 3, 4]. In such schemes, it is assumed that an accurate model of the plant is available, and the unknown parameters are assumed to appear linearly with respect to known nonlinear functions. This assumption is not sufficient for many practical situations, because it is difficult to describe an unknown nonlinear plant by a set of known nonlinear functions precisely.

Due to the universal approximators, we may substitute unknown system nonlinearities by neural networks, which is of known structure but contains a number of unknown parameters, plus a modeling error term, thus, transforming the original problem into a nonlinear robust adaptive control one. Therefore it is not necessary to spend much effort on system modeling which might be very difficult in some cases. There are a lot of control schemes [5-9] using linear neural networks, e.g., radial basis function, which need determine the centers and variances like the fuzzy space partition. Otherwise, there are a number of literatures using nonlinear neural networks [10-14], In [10] not only the output weights are adapted, but also the hidden-layer weights

J. Wang et al. (Eds.): ISNN 2006, LNCS 3972, pp. 849 – 858, 2006.

of the multilayer neural network are tuned, but this control scheme exploits the special properties of manipulators. So it can not be extend to general class of MIMO nonlinear systems. In [11] the result in [5] on radial basis function neural network was extended such that the variances of Gussian functions are also adapted in addition to the output weights. However, one limitation of these schemes is that they can only applied to nonlinear systems where certain types of matching conditions are required to be satisfied or to some classes of strict-feedback systems.

In this paper multiple multi-layer neural networks are used to approximate the uncertain nonlinearities in controlled plant. The on-line updating rules of the neural networks weights and the parameters in robustifying control terms are obtained by Lyapunov stability theory. Therefore all the signals in the closed loop are guaranteed to be uniformly ultimately bounded and the tracking error of the output is proven to converge to a small neighborhood of zero. The relationships among the control parameters, adaptive gains and robust gains are established to guarantee the transient performance of the closed loop system.

2 Problem Statement

Consider the following MIMO system that is described by differential equation:

$$y^{(n)} = f(x) + \Delta f^*(x) + g(x)\left(I + \Delta g^*(x)\right)u \tag{1}$$

where $y = [y_1, \cdots, y_m]^T \in R^m$ is output vector, $y^{(n)} = [y_1^{(n_1)}, \cdots, y_m^{(n_m)}]^T \in R^m$, $y_i^{(n_i)} = d^{n_i} y_i / dt^{n_i}$, $x = [y_1, \cdots, y_1^{(n_1-1)}, \cdots\cdots, y_m, \cdots, y_m^{(n_m-1)}] \in R^n$ with $n = n_1 + \cdots + n_m$ is the measurable state vector, $f(x) \in R^m$ and $g(x) \in R^{m\times m}$ are known continuous functions and $g(x)$ is nonsingular in the interest compact set $\Omega \subset R^n$. $\Delta f^*(x) \in R^m$ and $\Delta g^*(x) \in R^{m\times m}$ are unknown continuous functions, $u \in R^m$ is the input vector. Define the bounded reference signal vector Y_d and the tracking error vector E as:

$$Y_d = [y_{1d}, \cdots, y_{1d}^{(n_1-1)}, \cdots\cdots, y_{md}, \cdots, y_{md}^{(n_m-1)}] \in R^n \tag{2}$$

$$E = [e_1, \cdots, e_1^{(n_1-1)}, \cdots\cdots, e_m, \cdots, e_m^{(n_m-1)}] \in R^n \tag{3}$$

where $e_i = y_i - y_{id} \in R, \quad i = 1, \cdots, m$. The control objective is to make the system output asymptotically follows a given bounded reference signal.

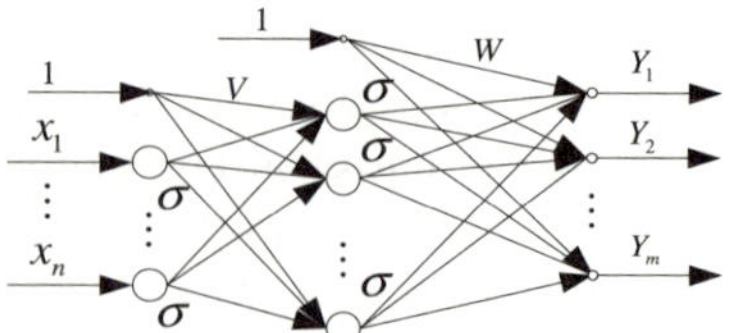

Fig. 1. Neural network structure

Next, we briefly summarize the neural network structure and properties required to derive the adaptive laws. The three-layer neural network shown in Fig. 1 with input $x \in R^{(n+1)}$ and output $Y \in R^m$ can be represented as

$$Y = W^T \sigma(V^T \bar{x}) \tag{4}$$

where $W \in R^{(N+1)m}$ is the weight matrix between the hidden layer and the output layer.

$V \in R^{(n+1)N}$ is the weight matrix between the input and hidden layer. The hidden layer output vector is $\sigma(z) = [1, \sigma(z_1), \sigma(z_2), \cdots, \sigma(z_N)]$ and $\bar{x} = [1, \sigma(x_1), \sigma(x_2), \cdots, \sigma(x_n)]$ stands for output vector of the input layer, $\sigma(\cdot)$ is activation function. n, N and m are the number of neurons in input layer, hidden layer and output layer. For any given continuous function $f(x)$, there exist hidden layer neurons N, and ideal bounded weight matrices W^* and V^* ($\|W^*\|_F \le W_m$, $\|V^*\|_F \le V_m$), such that

$$f(x) = W^{*T}\sigma(V^{*T}\bar{x}) + \varepsilon(x), \ \|\varepsilon(x)\| \le \varepsilon_m \quad \forall x \in \Omega \tag{5}$$

where $\Omega \subset R^n$ is a compact set. Now, let W and V be the estimated values of the ideal bounded weight matrices W^* and V^*, respectively. The weight estimation errors are defined as $\tilde{W} = W^* - W$ and $\tilde{V} = V^* - V$. Using the Taylor series expansion of $\sigma(V^{*T}\bar{x})$ about $\sigma(V^T\bar{x})$, the following properties can be obtained.

Property 1: For a neural network approximator (5), we have

$$W^{*T}\sigma(V^{*T}\bar{x}) - W^T\sigma(V^T\bar{x}) + \varepsilon(x) \tag{6}$$

$$= \tilde{W}^T\left(\sigma(V^T\bar{x}) - \dot{\sigma}(V^T\bar{x})V^T\bar{x}\right) + W^T\dot{\sigma}(V^T\bar{x})\tilde{V}^T\bar{x} + \theta \tag{7}$$

$$\theta = \tilde{W}^T\dot{\sigma}(V^T\bar{x})V^{*T}\bar{x} + W^{*T}O(\tilde{V}^T\bar{x})^2 + \varepsilon(x)$$

$$\|\theta\| \le c_1 + c_2\|W\|_F + c_3\|V\|_F \tag{8}$$

where $\dot{\sigma} = d\sigma(z)/dz\big|_{z=V^T\bar{x}}$. where c_1, c_2 and c_3 are positive constants.

Lemma: For matrices $A, B \in R^{n\times m}$

$$tr\left(A^T B\right) - tr\left(A^T A\right) \le tr\left(B^T B\right)/2 - tr\left(A^T A\right)/2 \tag{9}$$

Proof: Let $A = (a_{ij}), B = (b_{ij})$, $1 \le i \le n$, $1 \le j \le m$. Due to $ab \le \left(a^2 + b^2\right)/2, a, b \in R$, then

$$\mathrm{tr}\left(A^T B\right) - \mathrm{tr}\left(A^T A\right) = \sum_{\substack{1\le i\le n\\1\le j\le m}}\left(a_{ij}b_{ij} - a_{ij}^2\right) \le \sum_{\substack{1\le i\le n\\1\le j\le m}}\left(b_{ij}^2/2 - a_{ij}^2/2\right) = \mathrm{tr}\left(B^T B\right)/2 - \mathrm{tr}\left(A^T A\right)/2$$

3 Adaptive Controller

3.1 Controller Structure and the Error Dynamics

The controller structure is chosen as follows:

$$u = \alpha(x)\left(v + \beta(x) + \delta\right) \tag{10}$$

where $\alpha(x)$ and $\beta(x)$ are function matrix and vector of the state vector x , δ is robustifying control terms, v is a new control input. Let

$$v = y_d^{(n)} - \lambda^T E \tag{11}$$

where $\boldsymbol{\lambda}^T = \mathrm{diag}(\lambda_1^T, \lambda_2^T, \cdots, \lambda_m^T) \in R^{m\times n}$, $\boldsymbol{\lambda}_i = [\lambda_{i0}, \lambda_{i1}, \cdots, \lambda_{i,(n_i-1)}]^T \in R^n$, $i = 1, \cdots, m$ .such that $\boldsymbol{e}_i^{(n_i)} + \lambda_{i,(n_i-1)}\boldsymbol{e}_i^{(n_i-1)} + \cdots + \lambda_{i0}\boldsymbol{e}_i = 0,\ i = 1, \cdots, m$ is Hurwitz.

Applying the controller (10) into the system (1), we obtain

$$\begin{aligned}\boldsymbol{y}^{(n)} &= \boldsymbol{f}(x) + \Delta\boldsymbol{f}^*(x) + \boldsymbol{g}(\boldsymbol{x})\left(\mathrm{I} + \Delta\boldsymbol{g}^*(\boldsymbol{x})\right)\boldsymbol{\alpha}(\boldsymbol{x})\boldsymbol{v} \\ &+ \boldsymbol{g}(\boldsymbol{x})\left(\mathrm{I} + \Delta\boldsymbol{g}^*(\boldsymbol{x})\right)\boldsymbol{\alpha}(\boldsymbol{x})\boldsymbol{\beta}(x) + \boldsymbol{g}(\boldsymbol{x})\left(\mathrm{I} + \Delta\boldsymbol{g}^*(\boldsymbol{x})\right)\boldsymbol{\alpha}(\boldsymbol{x})\boldsymbol{\delta}\end{aligned} \tag{12}$$

To guarantee the asymptotic convergence of the output error $\boldsymbol{E}$, the following relationships should be satisfied:

$$\begin{aligned}&\boldsymbol{f}(x)\Delta\boldsymbol{f}^*(x) + \boldsymbol{g}(\boldsymbol{x})\left(\mathrm{I} + \Delta\boldsymbol{g}^*(\boldsymbol{x})\right)\boldsymbol{\alpha}^*(\boldsymbol{x})\boldsymbol{\beta}^*(\boldsymbol{x}) = 0 \\ &\boldsymbol{g}(\boldsymbol{x})\left(\mathrm{I} + \Delta\boldsymbol{g}^*(\boldsymbol{x})\right)\boldsymbol{\alpha}^*(\boldsymbol{x}) = 1, \quad \forall x \in \Omega\end{aligned} \tag{13}$$

where $\boldsymbol{\alpha}^*(x)$ and $\boldsymbol{\beta}^*(x)$ denote the ideal nonlinear function matrix and vectors.
Define following error matrices and error vector:

$$\tilde{\boldsymbol{\alpha}}(\boldsymbol{x}) = \boldsymbol{\alpha}^*(\boldsymbol{x}) - \boldsymbol{\alpha}(\boldsymbol{x}),\ \tilde{\boldsymbol{\psi}}(\boldsymbol{x}) = \boldsymbol{\alpha}^{*-1}(\boldsymbol{x}) - \boldsymbol{\alpha}^{-1}(\boldsymbol{x}),\ \tilde{\boldsymbol{\beta}}(\boldsymbol{x}) = \boldsymbol{\beta}^*(\boldsymbol{x}) - \boldsymbol{\beta}(\boldsymbol{x}) \tag{14}$$

From (12), (13) and (14), we have

$$\boldsymbol{y}^{(n)} = \boldsymbol{v} + \boldsymbol{\delta} - \tilde{\boldsymbol{\beta}}(x) + \tilde{\boldsymbol{\psi}}(\boldsymbol{x})\boldsymbol{u}$$

Substituting (11) into above equation, we obtain the error system dynamic and transform it to the state space form

$$\dot{\boldsymbol{E}} = \boldsymbol{A}\boldsymbol{E} + \boldsymbol{B}\left(\boldsymbol{\delta} + \tilde{\boldsymbol{\psi}}(x)u - \tilde{\boldsymbol{\beta}}(x)\right) \tag{15}$$

where $\boldsymbol{A} = \mathrm{diag}\{\boldsymbol{A}_1, \boldsymbol{A}_2, \cdots, \boldsymbol{A}_m\}, \boldsymbol{B} = [\boldsymbol{B}_1, \boldsymbol{B}_2, \cdots, \boldsymbol{B}_m]^T$

$$\boldsymbol{A}_i = \begin{bmatrix} 0 & 1 & 0 & \cdots & 0 \\ 0 & 0 & 1 & \cdots & 0 \\ \vdots & \vdots & \vdots & \vdots & \vdots \\ 0 & 0 & 0 & \cdots & 1 \\ -\lambda_{i0} & -\lambda_{i1} & -\lambda_{i2} & \cdots & -\lambda_{i,(n_i-1)} \end{bmatrix} \qquad \boldsymbol{B}_i = \begin{bmatrix} 0 & \cdots & 0 & \cdots & 0 \\ 0 & \cdots & 0 & \cdots & 0 \\ \cdots & \cdots & \cdots & \cdots & \cdots \\ 0 & \cdots & 1_i & \cdots & 0 \end{bmatrix}_{n_i\times m} \qquad i = 1, \cdots, m$$

Let

$$\boldsymbol{\alpha}^*(x) = \left(\boldsymbol{g}(x)\left(\mathrm{I} + \Delta\boldsymbol{g}^*(x)\right)\right)^{-1}, \quad \boldsymbol{\beta}^*(x) = -\boldsymbol{f}(x) - \Delta\boldsymbol{f}^*(x) \tag{16}$$

According to (14), we have

$$\tilde{\boldsymbol{\psi}}(x) = \boldsymbol{\alpha}^{*-1}(x) - \boldsymbol{\alpha}^{-1}(x) = \boldsymbol{g}(\boldsymbol{x})\tilde{\Delta}\boldsymbol{g}(\boldsymbol{x}),\ \tilde{\boldsymbol{\beta}}(x) = -\tilde{\Delta}\boldsymbol{f}(x) \tag{17}$$

where $\tilde{\Delta}\boldsymbol{g}(\boldsymbol{x}) = \Delta\boldsymbol{g}^*(\boldsymbol{x}) - \Delta\boldsymbol{g}(\boldsymbol{x})$ and $\tilde{\Delta}\boldsymbol{f}(x) = \Delta\boldsymbol{f}^*(x) - \Delta\boldsymbol{f}(x)$. Although there are $m(m+1)$ uncertain nonlinear functions in the controller, $m+1$ single-hidden-layer neural networks can be used to approximate $\Delta g^*(x)$ and $\Delta f^*(x)$. The outputs of the $m+1$ neural networks with ideal weights can be expressed as

$$\Delta g^*(x) = [W_1^{*T}\sigma(V_1^{*T}\bar{x}),\cdots,W_m^{*T}\sigma(V_m^{*T}\bar{x})]+\varepsilon_g(\bar{x}),\quad \Delta f^*(x) = W^{*T}\sigma(V^{*T}\bar{x})+\varepsilon_f(\bar{x}) \tag{18}$$

where $W^*, V^*, W_1^*, V_1^*, \cdots, W_m^*, V_m^*$ are the ideal bounded weights of the neural networks and $M_W, M_V, M_{W_1}, M_{V_1}, \cdots, M_{W_m}, M_{V_m}$ are their bounds respectively, $\|\varepsilon_g(\bar{x})\|_F \le \varepsilon_G$ and $\|\varepsilon_g(\bar{x})\| \le \varepsilon_F$ are the bounded approximated error matrix and vector.

The estimated outputs of the neural networks can be represented as:

$$\Delta g(x) = [W_1^T\sigma(V_1^T\bar{x});\cdots,W_m^T\sigma(V_m^T\bar{x})],\qquad \Delta f(x) = W^T\sigma(V^T\bar{x}) \tag{19}$$

where $W, V, W_1, V_1, \cdots, W_m, V_m$ denote the estimated value of the ideal weight matrices $W^*, V^*, W_1^*, V_1^*, \cdots, W_m^*, V_m^*$ respectively. In addition, define following error matrices

$$\tilde{W} = W^* - W, \tilde{V} = V^* - V,\ \tilde{W}_i = W_i^* - W_i, \tilde{V}_i = V_i^* - V_i,\ i = 1,\cdots,m \tag{20}$$

3.2 Stability Analysis and Adaptive Rules

Based on the error dynamic equation (15) and Taylor series expansion of ideal output of the neural networks about the estimated values, theorem of neural networks direct adaptive control is first given as follows:

Theorem: Consider system (1) and the reference signal (2), if the controller structure is (10), where $\alpha(x) = \left(g(x)\left(I + W_1^T\sigma(V_1^T\bar{x}),\cdots,W_m^T\sigma(V_m^T\bar{x})\right)\right)^{-1}$, $\beta(x) = -f(x) - W^T\sigma(V^T\bar{x})$ and the parameter adaptive rules are

$$\begin{aligned}
\dot{W} &= \gamma_1\left[\left(\sigma(V^T\bar{x}) - \dot{\sigma}\left(V_i^T\bar{x}\right)V^T\bar{x}\right)E^T PB - \kappa_1 W\right]\\
\dot{V} &= \gamma_2\left(\bar{x}E^T PBW^T\dot{\sigma}\left(V_i^T\bar{x}\right) - \kappa_2 V\right)\\
\dot{W}_i &= \gamma_W\left[\left(\sigma(V_i^T\bar{x}) - \dot{\sigma}\left(V_i^T\bar{x}\right)V_i^T\bar{x}\right)E^T PBg(x)u_i - \kappa_W W_i\right]\\
\dot{V}_i &= \gamma_V\left(\bar{x}E^T PBg(x)W_i^T\dot{\sigma}\left(V_i^T\bar{x}\right)u_i - \kappa_V V_i\right) \qquad i = 1,\cdots,m\\
\dot{D}_i &= \gamma_\delta\left|E^T PB_i\right|U - \kappa_\delta D_i,\ i = 1,\cdots,m\\
\delta_i &= -\mathrm{sgn}\left(E^T PB_i\right)D^T U,\ i = 1,\cdots,m\\
U &= \left[1,\|W\|_F,\|V\|_F,|u_1|,|u_1|\|W_1\|_F,|u_1|\|V_1\|_F,\cdots,|u_m|,|u_m|\|W_m\|_F,|u_m|\|V_m\|_F\right]^T
\end{aligned} \tag{21}$$

where $\gamma_1,\gamma_2,\gamma_W,\gamma_V,\gamma_\delta,\kappa_1,\kappa_2,\kappa_W,\kappa_V > 0$. Then for bounded initial values, all signals in the closed loop system remain bounded and the output tracking errors converge to a small neighborhood around zero by an appropriate choice of the design parameters.

Proof: Choosing the Lyapunov function as:

$$\begin{aligned}
V &= E^T PE/2 + \mathrm{tr}(\tilde{W}^T\tilde{W})/2\gamma_1 + \mathrm{tr}(\tilde{V}^T\tilde{V})/2\gamma_2\\
&+\sum_1^m\left(\mathrm{tr}(\tilde{W}_i^T\tilde{W}_i)/2\gamma_W + \mathrm{tr}(\tilde{V}_i^T\tilde{V}_i)/2\gamma_V\right) + \sum_1^m \tilde{D}_i^T\tilde{D}_i/2\gamma_\delta
\end{aligned} \tag{22}$$

where $\gamma_1,\gamma_2,\gamma_W,\gamma_V,\gamma_\delta>0$, P is a symmetric positive define matrix which satisfies the following Lyapunov equation $A^T P+PA=-Q$, $Q=Q^T>0$.

Differentiate (22) and substitute (15), (16), (17), (18), (19), (20) into it, we have

$$\begin{aligned}\dot V=&-\frac{1}{2}E^TQE+\sum_1^m\dot{\tilde D}_i^T\tilde D_i/\gamma_\delta+E^TPB\delta\\&+\mathrm{tr}\left(\tilde W^T\left(\sigma(V^T\bar x)-\dot\sigma V^T\bar x\right)E^TPB+\tilde W^T\dot{\tilde W}/\gamma_1\right)+\mathrm{tr}\left(\tilde V^T\bar xE^TPBW^T\dot\sigma+\tilde V^T\dot{\tilde V}/\gamma_2\right)\\&+\sum_1^m\left[\begin{array}{l}\mathrm{tr}\left(\tilde W_i^T\left(\sigma(V_i^T\bar x)-\dot\sigma V_i^T\bar x\right)E^TPBg(x)u_i+\tilde W_i^T\dot{\tilde W}_i/\gamma_W\right)\\+\mathrm{tr}\left(\tilde V_i^T\bar xE^TPBg(x)W_i^T\dot\sigma u_i+\tilde V_i^T\dot{\tilde V}_i/\gamma_V\right)\end{array}\right]\\&+E^TPB\left[\begin{array}{l}\tilde W^T\dot\sigma V^{*T}\bar x+W^{*T}O(\tilde V^T\bar x)^2+\varepsilon_f(x)\\+g(x)\sum_1^m\left(\tilde W_i^T\dot\sigma V_i^{*T}\bar x+W_i^{*T}O(\tilde V_i\bar x)^2+g(x)\varepsilon_{mg}(x)\right)u_i\end{array}\right]\end{aligned}\tag{23}$$

Since $g(x)$ is continuous on the interest compact set Ω, therefore $\|g(x)\|\le M_g$ is bounded. In terms of (20), we have

$$W=-\dot{\tilde W},\ V=-\dot{\tilde V},W_1=-\dot{\tilde W}_1,\ V_1=-\dot{\tilde V}_1,\cdots\cdots,W_m=-\dot{\tilde W}_m,\ V_m=-\dot{\tilde V}_m\tag{24}$$

Substituting (21), (24) into (23) and using the inequality (8), we have

$$\begin{aligned}\dot V\le&-E^TQE/2+\sum_1^m\dot{\tilde D}_i^T\tilde D_i/\gamma_\delta+E^TPB\delta\\&+\kappa_1\mathrm{tr}\left(\tilde W^TW\right)+\kappa_2\mathrm{tr}\left(\tilde V^TV\right)+\sum_1^m\left[\kappa_W\mathrm{tr}\left(\tilde W_i^TW_i\right)+\kappa_V\mathrm{tr}\left(\tilde V_i^TV_i\right)\right]/\gamma_\delta\\&+\left\|E^TPB\right\|\left[c_{01}+c_{02}\|W\|_F+c_{03}\|V\|_F+\left(c_{1i}+c_{2i}\|W_i\|_F+c_{3i}\|V_i\|_F\right)\|u_i\|\right]\\=&-\frac{1}{2}E^TQE+\sum_1^m\dot{\tilde D}_i^T\tilde D_i/\gamma_\delta+E^TPB\delta+\kappa_1\mathrm{tr}\left(\tilde W^TW\right)\\&+\kappa_2\mathrm{tr}\left(\tilde V^TV\right)+\sum_1^m\left[\kappa_W\mathrm{tr}\left(\tilde W_i^TW_i\right)+\kappa_V\mathrm{tr}\left(\tilde V_i^T\tilde V_i\right)\right]+\left\|E^TPB\right\|\zeta\end{aligned}$$

where $\zeta=c_{01}+c_{02}\|W\|_F+c_{03}\|V\|_F+\sum_1^m\left(c_{1i}+c_{2i}\|W_i\|_F+c_{3i}\|V_i\|_F\right)\|u_i\|$

Let

$$\Delta g(x)=[W_1^T\sigma(V_1^T\bar x);\cdots,W_m^T\sigma(V_m^T\bar x)],\qquad \Delta f(x)=W^T\sigma(V^T\bar x)\tag{25}$$

where

$$\begin{aligned}D^*&=\left[c_{01},c_{02},c_{03},c_{11},c_{12},c_{13},\cdots,c_{m1},c_{m2},c_{m3}\right]^T\\U&=\left[1,\|W\|_F,\|V\|_F,|u_1|,|u_1|\|W_1\|_F,|u_1|\|V_1\|_F,\cdots,|u_m|,|u_m|\|W_m\|_F,|u_m|\|V_m\|_F\right]\end{aligned}$$

Let

$$\delta_i=-\mathrm{sgn}\left(E^TPB_i\right)D_i^TU,\ i=1,\cdots,m\tag{26}$$

where $D_i^T=[\hat c_{i01},\hat c_{i02},\hat c_{i03},\hat c_{i11},\hat c_{i12},\hat c_{i13},\cdots,\hat c_{im1},\hat c_{im2},\hat c_{im3}]$, $i=1,\cdots,m$. Define following errors

$$\tilde D_i=D^*-D_i,\ i=1,\cdots,m\tag{27}$$

Then

$$\dot{V} \le -\boldsymbol{E}^T\boldsymbol{QE}/2 + \sum_1^m \tilde{\boldsymbol{D}}_i^T \dot{\tilde{\boldsymbol{D}}}_i/\gamma_\delta + \kappa_1 \mathrm{tr}\left(\tilde{\boldsymbol{W}}^T\boldsymbol{W}\right) + \kappa_2 \mathrm{tr}\left(\tilde{\boldsymbol{V}}^T\boldsymbol{V}\right) + \sum_1^m\left[\kappa_W \mathrm{tr}\left(\tilde{\boldsymbol{W}}_i^T\boldsymbol{W}_i\right) + \kappa_V \mathrm{tr}\left(\tilde{\boldsymbol{V}}_i^T\boldsymbol{V}_i\right)\right] + \left|\boldsymbol{E}^T\boldsymbol{PB}\right|_1 \tilde{\boldsymbol{D}}_1^T\boldsymbol{U} + \cdots + \left|\boldsymbol{E}^T\boldsymbol{PB}_m\right| \tilde{\boldsymbol{D}}_m^T\boldsymbol{U}$$

Substituting related parts of (21) into above equation, we have

$$\begin{aligned}\dot{V} &\le -\boldsymbol{E}^T\boldsymbol{QE}/2 + \kappa_1 \mathrm{tr}\left(\tilde{\boldsymbol{W}}^T\boldsymbol{W}\right) + \kappa_2 \mathrm{tr}\left(\tilde{\boldsymbol{V}}^T\boldsymbol{V}\right) + \sum_1^m\left[\kappa_W \mathrm{tr}\left(\tilde{\boldsymbol{W}}_i^T\boldsymbol{W}_i\right) + \kappa_V \mathrm{tr}\left(\tilde{\boldsymbol{V}}_i^T\boldsymbol{V}_i\right)\right] + \kappa_\delta \sum_1^m \tilde{\boldsymbol{D}}_i^T\boldsymbol{D} \\ &\le -\boldsymbol{E}^T\boldsymbol{QE}/2 - \kappa_1 \mathrm{tr}\left(\tilde{\boldsymbol{W}}^T\tilde{\boldsymbol{W}}\right) - \kappa_2 \mathrm{tr}\left(\tilde{\boldsymbol{V}}^T\tilde{\boldsymbol{V}}\right) - \sum_1^m\left[\kappa_W \mathrm{tr}\left(\tilde{\boldsymbol{W}}_i^T\tilde{\boldsymbol{W}}_i\right) + \kappa_V \mathrm{tr}\left(\tilde{\boldsymbol{V}}_i^T\boldsymbol{V}_i\right)\right] - \kappa_\delta \sum_1^m \tilde{\boldsymbol{D}}_i^T\tilde{\boldsymbol{D}}_i \\ &\quad + \kappa_1 \mathrm{tr}\left(\boldsymbol{W}^{*T}\boldsymbol{W}^*\right) + \kappa_2 \mathrm{tr}\left(\boldsymbol{V}^{*T}\boldsymbol{V}^*\right) + \sum_1^m\left[\kappa_W \mathrm{tr}\left(\tilde{\boldsymbol{W}}_i^{*T}\boldsymbol{W}_i^*\right) + \kappa_V \mathrm{tr}\left(\boldsymbol{V}_i^{*T}\boldsymbol{V}_i^*\right)\right] + \kappa_\delta \sum_1^m \boldsymbol{D}_i^{*T}\boldsymbol{D}_i^*\end{aligned} \tag{28}$$

Choose

$$\eta = \frac{\lambda_{min}(\boldsymbol{Q})}{\lambda_{max}(\boldsymbol{P})}, \kappa_1 = \frac{\eta}{2\gamma_1}, \ \kappa_2 = \frac{\eta}{2\gamma_2}, \kappa_W = \frac{\eta}{2\gamma_W}, \ \kappa_V = \frac{\eta}{2\gamma_V}, \ \kappa_\delta = \frac{\eta}{2\gamma_\delta} \tag{29}$$

where $\lambda_{min}(\cdot), \lambda_{max}(\cdot)$ are the smallest and the biggest e matrix igenvalues. We have

$$\dot{V} \le -\eta V + \eta_0 \tag{30}$$

where $\eta_0 = \kappa_1 \mathrm{tr}\left(\boldsymbol{W}^{*T}\boldsymbol{W}^*\right) + \kappa_2 \mathrm{tr}\left(\boldsymbol{V}^{*T}\boldsymbol{V}^*\right) + \sum_1^m\left[\kappa_W \mathrm{tr}\left(\tilde{\boldsymbol{W}}_i^{*T}\boldsymbol{W}_i^*\right) + \kappa_V \mathrm{tr}\left(\boldsymbol{V}_i^{*T}\boldsymbol{V}_i^*\right)\right] + \kappa_\delta \sum_1^m \boldsymbol{D}_i^{*T}\boldsymbol{D}_i^*$

From (29), we have

$$\dot{V} \le -\frac{1}{2}\eta\lambda\, min(\boldsymbol{P})\|\boldsymbol{E}\|^2 - \frac{\eta}{2\gamma_1}\|\tilde{\boldsymbol{W}}\|_F^2 - \frac{\eta}{2\gamma_2}\|\tilde{\boldsymbol{V}}\|_F^2 - \sum_{i=1}^m\left[\frac{\eta}{2\gamma_w}\|\tilde{\boldsymbol{W}}_i\|_F^2 + \frac{\eta}{2\gamma_v}\|\tilde{\boldsymbol{V}}_i\|_F^2\right] - \frac{\eta}{2\gamma_\delta}\sum_1^m\|\tilde{\boldsymbol{D}}_i\|^2 + \eta_0$$

Thus, the derivative of Lyapunov function is negative as long as one of following inequalities held.

$$\begin{aligned}&E \notin \Omega_e \triangleq \left\{E \,\middle|\, \|E\| \le \sqrt{\frac{2}{\lambda\, min(\boldsymbol{P})}\frac{\eta_0}{\eta}}\right\}, \quad \tilde{\boldsymbol{W}} \notin \Omega_w \triangleq \left\{\tilde{\boldsymbol{W}} \,\middle|\, \|\tilde{\boldsymbol{W}}\|_F \le \sqrt{2\gamma_1\frac{\eta_0}{\eta}}\right\} \\ &\tilde{\boldsymbol{V}} \notin \Omega_v \triangleq \left\{\tilde{\boldsymbol{V}} \,\middle|\, \|\tilde{\boldsymbol{V}}\|_F \le \sqrt{2\gamma_2\frac{\eta_0}{\eta}}\right\}, \quad \tilde{\boldsymbol{W}}_i \notin \Omega_{\tilde{W}_i} \triangleq \left\{\tilde{\boldsymbol{W}}_i \,\middle|\, \|\tilde{\boldsymbol{W}}_i\|_F \le \sqrt{2\gamma_w\frac{\eta_0}{\eta}}\right\} \\ &\tilde{\boldsymbol{V}}_i \notin \Omega_{\tilde{V}_i} \triangleq \left\{\tilde{\boldsymbol{V}}_i \,\middle|\, \|\tilde{\boldsymbol{V}}_i\|_F \le \sqrt{2\gamma_v\frac{\eta_0}{\eta}}\right\}, \quad \boldsymbol{D}_i \notin \Omega_{\tilde{D}_i} \triangleq \left\{\tilde{\boldsymbol{D}}_i \,\middle|\, \|\tilde{\boldsymbol{D}}_i\| \le \sqrt{2\gamma_\delta\frac{\eta_0}{\eta}}\right\}, \\ &i = 1,\cdots,m\end{aligned} \tag{31}$$

According to a standard Lyapunov theorem extension [15], these demonstrate the uniformly ultimately boundedness of $\boldsymbol{e}, \tilde{\boldsymbol{W}}, \tilde{\boldsymbol{V}}, \tilde{\boldsymbol{W}}_i, \tilde{\boldsymbol{V}}_i, \tilde{\boldsymbol{D}}_i, i = 1,\cdots,m$. In terms of (3), (20) and (27), we known that $\boldsymbol{x}, \boldsymbol{W}, \boldsymbol{V}, \boldsymbol{W}_i, \boldsymbol{V}_i, \boldsymbol{D}_i, i = 1,\cdots,m$ are bounded. Because of $\boldsymbol{\alpha}(\boldsymbol{x}) = \left(\boldsymbol{g}(\boldsymbol{x})\left(\mathrm{I} + \boldsymbol{W}_1^T\boldsymbol{\sigma}(\boldsymbol{V}_1^T\bar{\boldsymbol{x}}), \cdots, \boldsymbol{W}_m^T\boldsymbol{\sigma}(\boldsymbol{V}_m^T\bar{\boldsymbol{x}})\right)\right)^{-1}$, we must give the suitably initial values of neural networks to guarantee the existence of $\boldsymbol{\alpha}(\boldsymbol{x})$. In general, $\boldsymbol{\alpha}(\boldsymbol{x})$ is bounded if the initial values of neural networks weights are equal to zero. If the initial values of neural networks weights are given randomly, the following measures can be taken to guarantee the boundedness of $\boldsymbol{\alpha}(\boldsymbol{x})$. Let $A = \mathrm{I}$ and $B = \Delta g(x)$. According to (9)

$\alpha(x)=(I+\Delta g(x))^T/\left[g(x)\left(I+(\Delta g(x))^T\Delta g(x)+\gamma\left(\Delta g(x)+(\Delta g(x))^T\right)\right)\right]$ is nonsingular, then, u is bounded. Therefore all signals in the closed loop system are bounded.

Remark: In application, the symbol function $\operatorname{sgn}(\cdot)$ in (21) is replaced by a saturation function of the from:

$$sat\left(e^T PB_i\right)=\begin{cases} sgn\left(e^T PB_i\right) & if\ \left|e^T PB_i\right|>\varepsilon \\ \dfrac{e^T PB_i}{\varepsilon} & if\ \left|e^T PB_i\right|\le\varepsilon \end{cases}$$

where $\varepsilon>0$ is a small constant in order to remedy the control chattering.

4 Simulation Example

Since the dynamics of robot manipulators are highly nonlinear and may contain uncertain elements such as friction, we can use a two rigid-link robot manipulator to verify the effectiveness of the proposed control scheme. The dynamic model [16] of two rigid-link robot manipulator is as follows.

$$\boldsymbol{M}(\boldsymbol{q})\ddot{\boldsymbol{q}}+\boldsymbol{V}_m(\boldsymbol{q},\dot{\boldsymbol{q}})\dot{\boldsymbol{q}}+\boldsymbol{G}(\boldsymbol{q})+\boldsymbol{F}(\dot{\boldsymbol{q}})+\boldsymbol{T}_L(t)=\tau(t)$$

$$\boldsymbol{M}(\boldsymbol{q})=\begin{bmatrix} l_2^2m_2+l_1^2(m_1+m_2)+2l_1l_2m_2\cos(q_2) & l_2^2m_2+l_1l_2m_2\cos(q_2) \\ l_2^2m_2+l_1l_2m_2\cos(q_2) & l_2^2m_2 \end{bmatrix}$$

$$\boldsymbol{V}_m(\boldsymbol{q},\dot{\boldsymbol{q}})\dot{\boldsymbol{q}}=\begin{bmatrix} -2l_1l_2m_2\sin(q_2)(\dot{q}_1\dot{q}_2+0.5\dot{q}_2^2) \\ l_1l_2m_2\sin(q_2)\dot{q}_1^2 \end{bmatrix},\quad \boldsymbol{G}(\boldsymbol{q})=\begin{bmatrix} l_1(m_1+m_2)g\cos(q_1)+l_2m_2\cos(q_1+q_2) \\ l_2m_2 g\cos(q_1+q_2) \end{bmatrix}$$

$$\boldsymbol{F}(\dot{\boldsymbol{q}})=\left[12\dot{q}_1+0.5\,sgn(\dot{q}_1)\quad 12\dot{q}_2+0.5\,sgn(\dot{q}_2)\right]^T,\ \boldsymbol{T}_L(t)=\left[5\sin(5t)\quad 5\sin(5t)\right]^T$$

where q_1 and q_2 are the angle of joints 1 and 2; $m_1=0.8$ Kg and $m_2=2.3$ Kg are the mass of links 1 and 2; $l_1=1.0$ m and $l_2=1.0$ m are the length of links 1 and 2; g=9.8 m/s^2 is the gravity acceleration. The objective is to control the angles of joints 1 and 2 to track the desired command $q_{1d}=\sin t$ and $q_{2d}=\cos t$. First, the tracking response are given under the nominal condition ($\boldsymbol{F}(\dot{\boldsymbol{q}})=0$ and $\boldsymbol{T}_L(t)=0$). Secondly, the mass of the link 2 increases 1Kg and the friction forces $\boldsymbol{F}(\dot{\boldsymbol{q}})$ and external forces $\boldsymbol{T}_L(t)$ are considered at 5s. Finally, the manipulator should be settled at $q_{1d}=0.5\ rad$, $q_{2d}=-0.5\ rad$ from 15s. The reference model is given as follows

$$\dot{y}_m=A_m y_m+B_m r$$

$$A_m=\begin{bmatrix} 0 & 1 & 0 & 0 \\ -10 & -6.3 & 0 & 0 \\ 0 & 0 & 0 & 1 \\ 0 & 0 & -10 & -6.3 \end{bmatrix}\qquad B_m=\begin{bmatrix} 0 & 0 \\ 10 & 0 \\ 0 & 0 \\ 0 & 10 \end{bmatrix}$$

Adaptive gains: $\gamma_1=\gamma_2=\gamma_W=\gamma_V=0.1,\quad \gamma_\delta=0.1,\quad \kappa_1=\kappa_2=1, \kappa_w=\kappa_v=\kappa_\delta=5$

Control parameter: $\lambda=\operatorname{diag}([-100\ \ -63]\ \ [-100\ \ -63])^T\qquad Q=I$

Initial values of the weights are zero and number of hidden layer neutrons $N = 50$.
Initial value of system states: $x_1 = 0,\ x_2 = 0, x_3 = 0,\ x_4 = 0$

Fig. 2 is the response of link1, where solid line represents the actual response and dash line is the desired trajectory. Fig. 3 is the response of link 2, the denotation is same as link 1. Fig.4 is the tracking errors, and Fig. 5 is the control input torque.

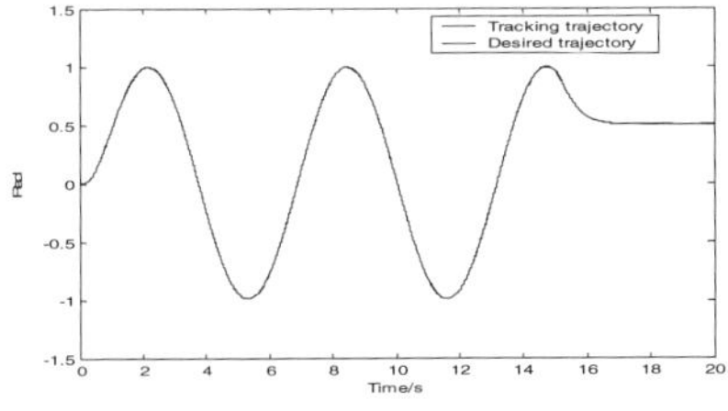

Fig. 2. Tracking response of link 1

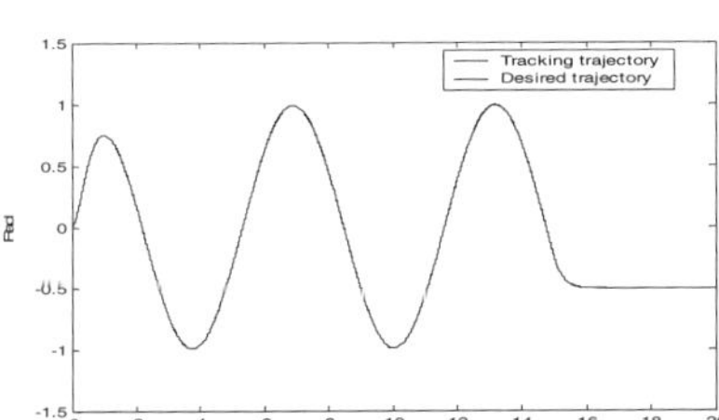

Fig. 3. Tracking response of link 2

Fig. 4. Tracking errors

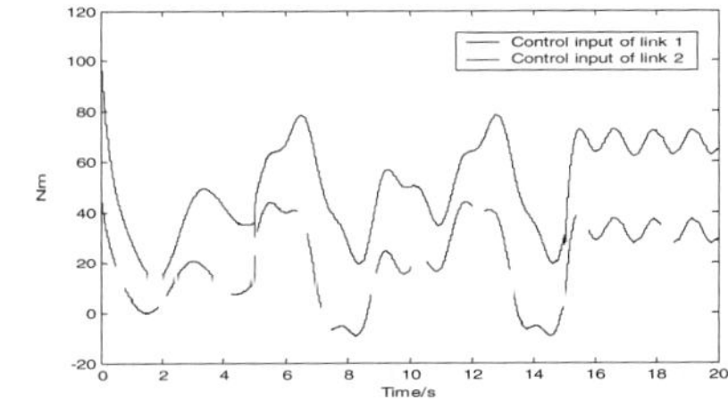

Fig. 5. Control inputs

5 Conclusion

A neural networks direct adaptive control scheme for a class of MIMO nonlinear system with unknown uncertainties is developed. Multiple multi-layer neural networks are employed to approximate the uncertainty of the plant's unknown nonlinear functions and robustifying control terms are used to compensate for approximation errors. All parameter adaptive laws and robustifying control terms are derived based on Lyapunov stability analysis so that all the signals in the closed loop are guaranteed to be semi-globally uniformly ultimately bounded and the tracking error of the output is proven to converge to a small neighborhood of zero. By suitably choosing the design parameters, the transient performance of the closed loop system is guaranteed.

References

1. Isidori, A.: Nonlinear Control System. Springer-Verlag, Berlin (1989)
2. Sastry, S.S., Isidori, A.: Adaptive Control of Linearizable Systems. IEEE Trans. Automat. Contr. 34 (1989) 1123–1131

3. Kanellakopoulos, I., Kokotovic, P.V., Morse, A.S.: Systematic Design of Adaptive Controllers for Feedback Linearizable Systems. IEEE Trans. Automat. Contr. 36 (1991) 1241–1253.
4. Slotine, J.-J.E, Li, W.: Applied Nonlinear Control. Prentice-Hall, Englewood Cliffs, NJ (1991)
5. Sanner, R., Slotine, J.-J.E.: Gaussian Networks for Direct Adaptive Control. IEEE Trans. Neural Networks 3 (1992) 837–863
6. Polycarpou, M. M.: Stable Adaptive Neural Control for Nonlinear Systems. IEEE Trans. Automat. Contr. 41 (1996) 447–450
7. Rovithakis, G.A., Christodulou, M.A.: Neural Adaptive Regulation of Unknown Nonlinear Dynamical Systems. IEEE Trans. Syst., Man, Cybern. B 27 (1997) 810–822
8. Xu, H., Ioannou, P.A.: Robust Adaptive Control for a Class of MIMO Nonlinear Systems With Guaranteed Error Bounds. IEEE Trans. Automat. Contr. 48 (2003) 728–742.
9. Ge, S.S., Wang, C.: Adaptive Neural Control of Uncertain MIMO Nonlinear Systems. IEEE Trans. Neural Networks 15 (2004) 674–692
10. Lewis, F.L., Liu, K., Yesildirek, A.: Neural Net Robot Controller with Guaranteed Tracking Performance. IEEE Trans. Neural Networks 6 (1995) 703–715
11. Cheny, Sh.-Ch., Chenz, W.-L.: Adaptive Radial Basis Function Neural Network Control with Variable Variance Parameters. Int. Journal of Systems Science 2 (2001) 413- 424
12. Spooner, J.T., Passino K.M.: Stable Adaptive Control Using Fuzzy Systems and Neural Networks. IEEE Trans. Fuzzy Syst. 4 (1996) 339–359
13. Zhang, T., Ge, S. S., Hang, C.C.: Design and Performance Analysis of a Direct Adaptive Controller for Nonlinear Systems. Automatica 35 (1999) 1809–1817
14. Park, J.H., Huh, S.H., Kim, S.H., Seo, S.J., Park, G.T.: Direct Adaptive Controller for Nonaffion Nonlinear Systems Using Self-Structuring Neural Networks. IEEE Trans. Neural Networks 16 (2005) 414-422
15. Narendra, K.S., Annaswamy, A.M.: A New Adaptive Law for Robust Adaptation without Persistent Excitation. IEEE Trans. Automat. Contr.32 (1987) 134-145
16. Taylor, D.: Composite Control of Direct-drive Robots. Proceedings of the IEEE Conference on Decision and Control (1989) 1670-1675

Adaptive Neural Compensation Control for Input-Delay Nonlinear Systems by Passive Approach

Zhandong Yu, Xiren Zhao, and Xiuyan Peng

School of Automatization, Harbin Engineering University,
150001 Harbin, China
zhandong_yu@163.com

Abstract. This paper focuses on the design of passive controller with adaptive neural compensation for uncertain strict-feedback nonlinear systems with input-delay. For local linearization model, the delay-dependent γ-passive control is presented. Then, γ-passive control law of local linear model is decomposed as the virtual control of sub-systems by backstepping. In order to compensate the nonlinear dynamics, the adaptive neural model is proposed. The NN weights are turned on-line by Lyapunov stability theory with no prior training. The design procedure of whole systems is a combination of local γ-passive control and adaptive neural network compensation techniques.

1 Introduction

The subject of the adaptive control for nonlinear systems with uncertainties has been attracting the attention of researchers for years. It is showed that for strict-feedback nonlinear systems, the Lyapunov function can be modified by backstepping [1]. Utilizing the differential geometry theory, we can judge whether the nonlinear systems are essentially in previous structure [2]. Recently, the researches about backstepping, which are reported in various publications, have rapidly developed [3-5]. However, it is worth noting that input delays are often encountered because of transmission of information in real control systems. The memoryless controller has been proposed for input delay systems. Although the approach is easy to implement, it tend to be more conservative when time delay is small [6, 7]. In this paper, we investigate the design of passive controller with adaptive neural compensation for a class of uncertain strict-feedback nonlinear systems with input-delay.

2 Problem Formulation

2.1 Model of the Nonlinear Systems

The uncertain nonlinear systems are described as follow:

$$\begin{aligned}\dot{x}_i(t) &= x_{i+1}(t) + f_i(x_{[i]}) + g_i(x_{[i]})w(t) \quad 1 \le i \le n-1 \\ \dot{x}_n(t) &= f_n(x_{[n]}) + g_n(x_{[n]})w(t) + k_1 u(t) + k_2 u(t-\tau)\end{aligned}$$

J. Wang et al. (Eds.): ISNN 2006, LNCS 3972, pp. 859–866, 2006.

It can be written more compactly as

$$\dot{x}(t) = f(x) + g(x)w(t) + B_2u(t) + B_3u(t-\tau) \tag{1}$$

where, $x=[x_1,...,x_n]^T \in R^n$ is the state vector, and $x_{[i]}=[x_1,...,x_i]^T \in R^i$; $u(t), w(t) \in R$ are input of system and unknown disturbance signal, respectively. $w(t) \in L_2$. $\tau>0$ is input delay. $f_i(x_{[i]})$, $g_i(x_{[i]})$ are smooth functions, which can be expanded as known linear term and unknown nonlinear term at equilibrium point $x=0$ [8]:

$$\begin{aligned}\dot{x}(t) &= Ax(t) + B_1w(t) + B_2u(t) + B_3u(t-\tau) + \begin{bmatrix}\tilde{f}_1(x_{[1]})\\ \vdots \\ \tilde{f}_n(x_{[n]})\end{bmatrix} + \begin{bmatrix}\tilde{g}_1(x_{[1]})\\ \vdots \\ \tilde{g}_n(x_{[n]})\end{bmatrix} w(t) \\ &= Ax(t) + B_1w(t) + B_2u(t) + B_3u(t-\tau) + \tilde{f}(x) + \tilde{g}(x)w(t)\end{aligned} \tag{2}$$

where, $\tilde{f}(x) = f(x) - Ax$, $B_1 = g(0)$, $\tilde{g}(x) = g(x) - B_1$

$$A = \left.\frac{\partial f}{\partial x}\right|_{x=0} = \begin{bmatrix} a_{11} & 1 & 0 & \cdots & 0 \\ a_{21} & a_{22} & 1 & \cdots & 0 \\ \vdots & \vdots & & \ddots & \vdots \\ & & & & 1 \\ a_{n1} & a_{n2} & a_{n3} & \cdots & a_{nn}\end{bmatrix}, \quad B_2 = \begin{bmatrix}0\\ \vdots \\ 0 \\ k_1\end{bmatrix}, \quad B_3 = \begin{bmatrix}0\\ \vdots \\ 0 \\ k_2\end{bmatrix}$$

For the local linearization model, the delay-dependent γ-passive controller is presented firstly. And then, we decompose the γ-passive control law by backstepping, which can guarantee global boundedness [8]. In the procedure, the linear virtual control law of sub-systems is obtained. In order to compensate nonlinear dynamics, the adaptive neural model is introduced. The whole design procedure is a combination of local γ-passive control and adaptive neural network compensation techniques.

2.2 Neural Network for Approximation

Neural network techniques are particularly useful for function approximation. The three-layer feed forward NN approximates smooth function in compact set as:

$$h(Z) = W^T \Psi(V^T Z) + \rho(Z) \tag{3}$$

where $Z=[x_1,\ldots,x_n,1]^T \in \Omega_z$ is the input vector of NN; $W=[w_1,\ldots,w_l]^T \in R^l$ and $V=[v_1,\ldots, v_l] \in R^{(n+1)\times l}$ are the 1-to-2 layer and the 2-to-3 layer weights respectively; the NN node number is l; $\rho(Z)$ is the NN approximation error. The activation function of hidden layer is Sigmoid: $\Psi(x)=(1+exp(-\gamma x))^{-1}$, $\gamma>0$. $\mathbf{\Psi}(V^T Z)=[\Psi(v_1{}^T Z),\ldots,\Psi(v_l{}^T Z), 1]^T$. According to the universal approximation theorem, there exist expected constant weights W^*, V^* in (3) such that $|\rho(Z)| \leq \rho_N<0$ for all $Z \in \Omega_z$. Unfortunately, W^* and V^* are unknown. Let $\hat{W}$ and $\hat{V}$ be the estimates of W^* and V^*. The estimation errors are $\tilde{W} = \hat{W} - W^*$ and $\tilde{V} = \hat{V} - V^*$. Taylor expansion of $\Psi(V^{*T}Z)$ at $\hat{V}^T Z$ is:

$$\Psi(V^{*T}Z) = \Psi(\hat{V}^T Z) + \hat{\Psi}' \cdot (V^{*T}Z - \hat{V}^T Z) + O(\tilde{V}^T Z)^2 = \Psi(\hat{V}^T Z) - \hat{\Psi}'\tilde{V}^T Z + O(\tilde{V}^T Z)^2 \tag{4}$$

where, $\hat{\Psi} = \Psi(\hat{V}^T Z)$, $\hat{\Psi}' = diag\{\hat{\psi}_1' \ \ \hat{\psi}_2' \ \ \ldots \ \ \hat{\psi}_l'\}$, $\hat{\psi}_i' = \psi'(\hat{v}_i^T Z) = d[\psi(z_a)]/dz_a|_{z_z=\hat{v}_i^T Z}$ $i=1,\ldots,l$; $O(\tilde{V}^T Z)^2$ is high-order remainder term. The estimation error can be expressed as:

$$\hat{W}^T\Psi(\hat{V}^TZ) - W^{*T}\Psi(V^{*T}Z) = (\tilde{W}+W^*)^T\Psi(\hat{V}^TZ) - W^{*T}[\Psi(\hat{V}^TZ) - \hat{\Psi}'\tilde{V}^TZ + O(\tilde{V}^TZ)^2] \\ = \tilde{W}^T(\hat{\Psi} - \hat{\Psi}'\hat{V}^TZ) + \hat{W}^T\hat{\Psi}'\tilde{V}^TZ + d_u \tag{5}$$

where, $d_u = \tilde{W}^T\hat{\Psi}'V^{*T}Z - W^{*T}O(\tilde{V}^TZ)^2$, $\psi(x)\in[0,\ 1]$, $\dot{\psi}(x)\in[0,\ 0.25\gamma]$, $|x\dot{\psi}(x)|\le 0.2239$, $\|\hat{\Psi}'\|_F \le \sum_{i=1}^{l}\psi'(\hat{v}_i^TZ)\le 0.25\gamma l$, $\|\hat{\Psi}'\hat{V}^TZ\| \le \sum_{i=1}^{l}|\hat{v}_i^TZ\psi'(\hat{v}_i^TZ)|\le 0.2239l$. We have

$$|d_u| \le \|\hat{W}^T\hat{\Psi}'V^{*T}Z\| + \|W^{*T}\hat{\Psi}'\hat{V}^TZ\| + \|W^*\|\|\Psi(V^{*T}Z) - \Psi(\hat{V}^TZ)\| \\ \le 0.25\gamma l\|V^*\|_F\|\hat{W}\|\|Z\| + 1.2239l\|W^*\| \tag{6}$$

where, $\|A\|$, $\|A\|_F$ are Euclid and Frobenius norms respectively; The update laws are

$$\dot{\hat{W}} = \Gamma_w[(\hat{\Psi} - \hat{\Psi}'\hat{V}^TZ)z_i - \sigma_w\hat{W}] \quad \sigma_w > 0 \\ \dot{\hat{V}} = \Gamma_v[(Z\hat{W}^T\hat{\Psi}'z_i - \sigma_v\hat{V}] \quad \sigma_v > 0 \tag{7}$$

3 Adaptive Neural Compensation Control by Passive Approach

3.1 γ-Passive Control with History Feedback for Linear Input-Delay Systems

The local linearization model of (2) is as follow:

$$\dot{x}(t) = Ax(t) + B_1w(t) + B_2u(t) + B_3u(t-\tau) \tag{8}$$

Theorem 1. System (8) with control law (9) is γ-passive, if there exist symmetric matrices $0 < X, N_1 \in R^{n\times n}$ and matrix $Y \in R^{1\times n}$ and γ>0 such that the LMI (10) hold:

$$u(t) = YX^{-1}\left[x(t) + \int_{-\tau}^{0} B_3u(t+\theta)d\theta\right] \tag{9}$$

$$\begin{bmatrix} M & B_1 & XC^T \\ B_1^T & -\gamma^2 I & D^T \\ CX & D & -I \end{bmatrix} < 0 \tag{10}$$

where, $M = XA^T + Y^T(B_2+B_3)^T + AX + (B_2+B_3)Y + \tau AN_1A^T + \tau Y^TB_3^TN_1^{-1}B_3Y$.

Proof: Define state transformation as $\xi(t) = x(t) + \int_{-\tau}^{0} B_3u(t+\theta)d\theta$ [6]. Taking time derivative of $\xi(t)$ and combining system (8) with the controller $u(t) = K\xi(t)$, where K will be chosen later, we obtain

$$\dot{\xi}(t) = A\xi(t) + (B_2+B_3)K\xi(t) + B_1w(t) - A\int_{-\tau}^{0} B_3K\xi(t+\theta)d\theta \tag{11}$$

Introduce the Lyapunov-Krasovskii function candidate as

$$V(\xi) = \xi^T(t)P\xi(t) + \int_{-\tau}^{0}\int_{t+\theta}\xi^T(\sigma)K^TB_3^TN_1^{-1}B_3K\xi(\sigma)d\sigma d\theta \tag{12}$$

where $0 < P = P^T \in R^{n\times n}$ and $0 < N_1 = N_1^T \in R^{n\times n}$. Its derivative along (8) is

$$\begin{aligned}\dot{V}(\xi) &= \xi^T(t)[(A+(B_2+B_3)K)^T P + P(A+(B_2+B_3)K)]\xi(t) + 2\xi^T(t)PB_1w(t) \\ &-2\xi^T(t)PA\int_{-\tau}^{0} B_3K\xi(t+\theta)d\theta + \tau\xi^T(t)K^T B_3^T N_1^{-1}B_3K\xi(t) - \int_{-\tau}^{0}\xi^T(t+\theta)K^T B_3^T N_1^{-1}B_3K\xi(t+\theta)d\theta \\ &\le \xi^T(t)[(A+(B_2+B_3)K)^T P + P(A+(B_2+B_3)K)]\xi(t) + 2\xi^T(t)PB_1w(t) \\ &+\tau\xi^T(t)PAN_1A^T P\xi(t) + \tau\xi^T(t)K^T B_3^T N_1^{-1}B_3K\xi(t) \\ &= \begin{bmatrix}\xi^T(t) & w^T(t)\end{bmatrix}\begin{bmatrix} S & PB_1 \\ B_1P & 0\end{bmatrix}\begin{bmatrix}\xi(t) \\ w(t)\end{bmatrix}\end{aligned} \tag{13}$$

where, $S = A^T P + PA + K^T(B_2+B_3)^T P + P(B_2+B_3)K + \tau PAN_1A^T P + \tau K^T B_3^T N_1^{-1}B_3K$. Let y as the output of system (Eq.(14)), and define the supply rate function as (15)

$$y = C\left[x(t) + \int_{-\tau}^{0} B_3u(t+\theta)d\theta\right] + Dw(t) \tag{14}$$

$$s(w, y) = \frac{1}{2}(\gamma^2 \| w \|^2 - \| y \|^2) \tag{15}$$

By evaluating the Hamiltonian $H(\xi) = -\dot{V}(\xi) + 2s(w, y)$ [7], it yields:

$$H(\xi) = -\begin{bmatrix}\xi^T(t) & w^T(t)\end{bmatrix}\begin{bmatrix} S + C^TC & PB_1 + C^TD \\ B_1^TP + D^TC & -\gamma^2I + D^TD\end{bmatrix}\begin{bmatrix}\xi(t) \\ w(t)\end{bmatrix} \tag{16}$$

If (17) is hold, $-\dot{V}(\xi) + 2s(w, y) < 0$, and from which it follows as (18).

$$\begin{bmatrix} S + C^TC & PB_1 + C^TD \\ B_1^TP + D^TC & -\gamma^2I + D^TD\end{bmatrix} < 0 \tag{17}$$

$$V(\xi(t_1)) - V(\xi(t_0)) < \int_{t_0}^{t_1}(\gamma^2 \| w \|^2 - \| y \|^2)dt \tag{18}$$

Since $V(\xi) > 0$, it follows that system (8) with (9) is γ-passive. Using the Schur complement to (17), we obtain (19), by multiplying $diag\{X, I, I\}$ at other side of matrix.

$$\begin{bmatrix} P^{-1}SP^{-1} & B_1 & P^{-1}C^T \\ B_1^T & -\gamma^2I & D^T \\ CP^{-1} & D & -I\end{bmatrix} \le \begin{bmatrix} M & B_1 & XC^T \\ B_1^T & -\gamma^2I & D^T \\ CX & D & -I\end{bmatrix} \tag{19}$$

where, $X = P^{-1}, Y = KP^{-1}$. Therefore, the conclusion can be made that system (11) is γ-passive, if there exist $X, N_1 \in R^{n\times n}, Y \in R^{1\times n}$ and $\gamma \in R^+$, such that the LMI (10) hold.

3.2 Decomposing of Local Linear Control Law by Backstepping

In the procedure, virtual control laws of linear sub-systems are obtained. A factorization of X^{-1} that makes the γ-passive design suitable for backstepping is as (20), where L is lower triangular matrix, and Λ>0 is diagonal matrix.

$$X^{-1} = P = L^T \Lambda L \tag{20}$$

$$L := \begin{bmatrix} 1 & 0 & \cdots & 0 \\ -\alpha_{11} & 1 & \ddots & \vdots \\ \vdots & \ddots & \ddots & 0 \\ -\alpha_{n-1,1} & \cdots & -\alpha_{n-1,n-1} & 1 \end{bmatrix} \quad \Lambda := \begin{bmatrix} \lambda_1 & & & \\ & \lambda_2 & & \\ & & \ddots & \\ & & & \lambda_n \end{bmatrix}$$

Furthermore, under the state transformation $z(t) = L\xi(t)$, (8) can be expressed as:

$$\dot{z}(t) = \bar{A}z(t) + \bar{B}_1 w(t) + (B_2 + B_3)u(t) - \bar{A}\int_{-\tau}^{0} B_3 u(t+\theta)d\theta \tag{21}$$

where, $\bar{A} = LAL^{-1}$, $\bar{B}_1 = LB_1$. The control is $u(t) = YX^{-1}L^{-1}z(t) = YL^T\Lambda z(t)$, and the virtual control law of sub-system is

$$\alpha_i(\xi_{[i]}) = \alpha_{[i]}\xi_{[i]} = \alpha_{[i]}L_{[i]}^{-1}z_{[i]}(t) = \bar{\alpha}_{[i]}z_{[i]}(t) = \bar{\alpha}_i(z_{[i]}) \tag{22}$$

where, $\alpha_{[i]} := [\alpha_{i1} \ \cdots \ \alpha_{ii}]$, $\bar{\alpha}_{[i]} := [\bar{\alpha}_{i1} \ \cdots \ \bar{\alpha}_{ii}]$, $1 \le i \le n$. $\alpha_{[0]} = 0$, $\alpha_{[n]} = 0_{[n]}^T$. And the stock functions of sub-systems are defined as:

$$\begin{aligned} V_{[i]} &= V_{[i-1]} + z_i\lambda_i z_i \quad i = 1,\dots,n-1 \\ V_{[n]} &= V_{[n-1]} + z_n\lambda_n z_n + \int_{-\tau}^{0}\int_{t+\theta} z^T(\sigma)(L^{-1})^T K^T B_3^T N_1^{-1} B_3 K L^{-1} z(\sigma) d\sigma d\theta \end{aligned} \tag{23}$$

where, $V_{[0]}=0$. The outputs are $y_i = (C_{[i]}L_{[i]}^{-1})z_{[i]} + Dw(t)$. The Hamiltonian functions are

$$\begin{aligned} H_i &= -\dot{V}_{[i]} + 2s_1(w, y_1) = -\dot{V}_{[i]} + \gamma^2 w^T w - y_i^T y_i = -\begin{bmatrix} z_{[i]}^T(t) & w(t) \end{bmatrix} T \begin{bmatrix} z_{[i]}(t) \\ w(t) \end{bmatrix} - 2z_{i+1}\lambda_i z_i \\ T &:= \begin{bmatrix} \bar{A}_{[i]}^T\Lambda_{[i]} + \Lambda_{[i]}\bar{A}_{[i]} + (L_{[i]}^{-1})^T C_{[i]}^T C_{[i]} L_{[i]}^{-1} & \Lambda_{[i]}\bar{B}_{1[i]} + (C_{[i]}L_{[i]}^{-1})^T D \\ \bar{B}_{1[i]}^T\Lambda_{[i]} + D^T C_{[i]}L_{[i]}^{-1} & -\gamma^2 + D^T D \end{bmatrix} \end{aligned} \tag{24}$$

If $z_{i+1}=0$, $H_i>0$ according to LMI (17). So the sub-systems with virtual control (22) are γ-passive. According to Theorem 1, the closed loop system is γ-passive.

3.3 Adaptive Neural Compensation

For the nonlinear terms in (2), three-layer feed forward neural networks are designed to compensate them. The design procedure is to add the compensation terms $\tilde{\alpha}_{i-1}$ to the linear virtual control law (22), while the global boundedness is guaranteed. Based on the Lyapunov's stability theory, the NN weights are turned on-line as (7) with no prior training needed. The new state transformation is:

$$z_i(t) = \xi_i(t) - \bar{\alpha}_{i-1}(z_{[i-1]}) - \tilde{\alpha}_{i-1} = \xi_i(t) - \bar{\alpha}_{i-1}(z_{[i-1]}) - \hat{W}_{i-1}^T\Psi(\hat{V}_{i-1}^T Z_{i-1}) + k_{i-1}(t)z_{i-1} \tag{25}$$

where, $\bar{\alpha}_0 = \tilde{\alpha}_0 = 0$, $Z_i = \left[z_{[i]}^T, \tilde{\alpha}_{i-1}, \frac{\partial \tilde{\alpha}_{i-1}}{\partial z_{[i]}}, \sum_{j=1}^{i-1}\left(\frac{\partial \tilde{\alpha}_{i-1}}{\partial \hat{W}_j}\dot{\hat{W}}_j + \sum_{r=1}^{N}\frac{\partial \tilde{\alpha}_{i-1}}{\partial \hat{V}_{jr}}\dot{\hat{V}}_{jr}\right), 1 \right]^T$, $k_i(t) = -\frac{1}{\varepsilon_i}\left[\frac{1}{2} + \|\hat{W}_i\|^2 \|Z_i\|^2\right]$.

4 Stability Analysis

In this section, we show that the control law and adaptive law presented by above procedure guarantee the uniformly ultimately boundedness of the closed-loop system.

Step 1: The $z_{[1]}$ sub-system is

$$\begin{aligned}\dot{z}_1(t) &= \dot{\xi}_1(t) = a_{[1]}\xi_1(t) + \xi_2(t) + b_1 w(t) + \tilde{f}_1(x_1) + \tilde{g}_1(x_1)w(t) \\ &= a_{[1]}L_{[1]}^{-1}z_1(t) + z_2(t) + \bar{\alpha}_1(z_{[1]}) + \tilde{\alpha}_1 + b_1 w(t) + \tilde{f}_1(x_1) + \tilde{g}_1(x_1)w(t)\end{aligned} \tag{26}$$

It is shown that neural networks approximates the nonlinear term in (26) such as $\tilde{f}_1(x_1)+\tilde{g}_1(x_1)w(t) = W_1^{*T}\Psi(V_1^{*T}Z_1)+\rho_1$. Define $V_{[1]} = z_{[1]}^T\Lambda_{[1]}z_{[1]} + \tilde{W}_1^T\Gamma_{w1}^{-1}\tilde{W}_1 + tr\{\tilde{V}_1^T\Gamma_{v1}^{-1}\tilde{V}_1\}$ as stock function. Using (7), inequalities $H_i>0$, $\xi(t)$ and (6), we obtain

$$\begin{aligned}\dot{V}_{[1]} &\le z_{[1]}^T(t)[\bar{a}_{[1]}{}^T\Lambda_{[1]} + \Lambda_{[1]}\bar{a}_{[1]}]z_{[1]}(t) + 2z_{[1]}^T\Lambda_{[1]}b_1 w(t) + 2z_2\lambda_1 z_1 - 2\sigma_{w1}\tilde{W}_1^T\hat{W}_1 - 2\sigma_{v1}tr\{\tilde{V}_1^T\hat{V}_1\} \\ &\quad + 2(\|\rho_{N1}\| + 0.25\gamma l\|V_1^*\|_F\|\hat{W}_1\|\|Z_1\| + 1.2239 l\|W_1^*\|)\lambda_1|z_1| + 2k_1(t)\lambda_1 z_1^2 \\ &\le z_{[1]}^T(t)[\bar{a}_{[1]}{}^T\Lambda_{[1]} + \Lambda_{[1]}\bar{a}_{[1]}]z_{[1]}(t) + 2z_{[1]}^T\Lambda_{[1]}b_1 w(t) + 2z_2\lambda_1 z_1 + N_1 - \sigma_{w1}\|\tilde{W}_1\|^2 - \sigma_{v1}\|\tilde{V}_1\|_F^2 \\ &= -[C_{[1]}L_{[1]}^{-1}z_{[1]}(t) + Dw(t)]^T[C_{[1]}L_{[1]}^{-1}z_{[1]}(t) + Dw(t)] - \sigma_{w1}\|\tilde{W}_1\|^2 - \sigma_{v1}\|\tilde{V}_1\|_F^2 + 2z_2\lambda_1 z_1 + N_1\end{aligned} \tag{27}$$

where, $N_1 = 2\lambda_1\left[\frac{\varepsilon_1\gamma^2 l^2}{64}\|V_1^*\|_F^2 + 1.5\varepsilon_1 l^2\|W_1^*\|^2 + \varepsilon_1\rho_{N1}^2 + \frac{\sigma_{w1}\|W_1^*\|^2}{2\lambda_1} + \frac{\sigma_{v1}\|V_1^*\|_F^2}{2\lambda_1}\right]$. In the procedure, the inequalities $2\tilde{W}_1^T\hat{W}_1 \ge \|\tilde{W}_1\|^2 - \|W_1^*\|^2$ and $2tr\{\tilde{V}_1^T\hat{V}_1\} \ge \|\tilde{V}_1\|_F^2 + \|V_1^*\|_F^2$ are used. If $z_2\equiv 0$, the subsystem is uniformly ultimately boundedness.

Step *i*: The derivative of z_i is

$$\begin{aligned}\dot{z}_i(t) &= a_{[i]}L_{[i]}^{-1}z_{[i]} + z_{i+1} + \bar{\alpha}_{[i]}z_{[i]} + \tilde{\alpha}_i + b_i w(t) + \tilde{f}_i(x_{[i]}) + \tilde{g}_i(x_{[i]})w(t) - \bar{\alpha}_{[i-1]}\left\{\bar{A}_{[i-1]}z_{[i-1]} + \begin{bmatrix}0_{i-1}\\ z_i\end{bmatrix} + \bar{B}_{1[i-1]}w\right\} - \dot{\tilde{\alpha}}_{i-1} \\ &= \left\{a_{[i]}L_{[i]}^{-1} + \bar{\alpha}_{[i]} - \begin{bmatrix}\bar{\alpha}_{[i-1]}\bar{A}_{[i-1]} & \bar{\alpha}_{i-1,i-1}\end{bmatrix}\right\}z_{[i]} + z_{i+1} + \left[b_i - \bar{\alpha}_{[i-1]}\bar{B}_{1[i-1]}\right]w + \tilde{\alpha}_i + \tilde{f}_i(x_{[i]}) + \tilde{g}_i(x_{[i]})w(t) - \dot{\tilde{\alpha}}_{i-1} \\ &= \bar{a}_{[i]}z_{[i]} + z_{i+1} + \bar{b}_i w + \tilde{\alpha}_i + \tilde{f}_i(x_{[i]}) + \tilde{g}_i(x_{[i]})w(t) - \dot{\tilde{\alpha}}_{i-1}\end{aligned}$$

The stock function $V_{[i]} = V_{[i-1]} + z_i\lambda_i z_i + \tilde{W}_i^T\Gamma_{wi}^{-1}\tilde{W}_i + tr\{\tilde{V}_i^T\Gamma_{vi}^{-1}\tilde{V}_i\}$. Similar as Step 1,

$$\begin{aligned}\dot{V}_{[i]} &\le \dot{V}_{[i-1]} + 2z_i\lambda_i[\bar{a}_{[i]}z_{[i]} + z_{i+1} + \bar{b}_i w] - \sigma_{wi}\|\tilde{W}_i\|^2 - \sigma_{vi}\|\tilde{V}_i\|_F^2 + N_i \\ &\le z_{[i]}^T[\bar{A}_{[i]}^T\Lambda_{[i]} + \Lambda_{[i]}\bar{A}_{[i]}]z_{[i]} + 2z_{[i]}^T\Lambda_{[i]}\bar{B}_{1[i]}w(t) - \sum_{j=1}^{i}\left[\sigma_{wj}\|\tilde{W}_j\|^2 + \sigma_{wj}\|\tilde{V}_j\|_F^2 - N_j\right] + 2z_i\lambda_i z_{i+1} \\ &\le -[C_{[i]}L_{[i]}^{-1}z_{[i]}(t) + Dw(t)]^T[C_{[i]}L_{[i]}^{-1}z_{[i]}(t) + Dw(t)] - \sum_{j=1}^{i}\left[\sigma_{wj}\|\tilde{W}_j\|^2 + \sigma_{wj}\|\tilde{V}_j\|_F^2 - N_j\right] + 2z_i\lambda_i z_{i+1}\end{aligned} \tag{28}$$

If $z_i\equiv 0$, the sub-system is uniformly ultimately boundeness because of the bounded N_i. N_i are chosen as N_1 expect that all subscripts are changed as i.

Step *n*: The dynamic of z_n is given by

$$\dot{z}_n(t) = \bar{a}_{[n]}z_{[n]} + \bar{b}_n w(t) + \tilde{f}_n(x_{[n]}) + \tilde{g}_n(x_{[n]})w(t) + (k_2 + k_3)u(t) - A\int_{-\tau}^{0}B_3 u(t+\theta)d\theta - \dot{\tilde{\alpha}}_{n-1}$$

The stock function of whole closed loop system is $V_{[n]}=V_{[n-1]}+z_n\lambda_n z_n+\tilde{W}_n^T\Gamma_{wn}^{-1}\tilde{W}_n$ $+tr\left\{\tilde{V}_n^T\Gamma_{vn}^{-1}\tilde{V}_n\right\}+\int_{-\tau}^{0}\int_{t+\theta} z^T(\sigma)(L^{-1})^T K^T B_3^T N_1^{-1}B_3KL^{-1}z(\sigma)d\sigma d\theta$. The derivative of $V_{[n]}$ is

$$\begin{aligned}
\dot{V}_{[n]} &= \dot{V}_{[n-1]}+2z_n\lambda_n\left[\bar{a}_{[n]}z_{[n]}+\bar{b}_n w(t)+(k_2+k_3)KL_{[n]}^{-1}z_{[n]}(t)-a_{[n]}\int_{-\tau}^{0}B_3KL_{[n]}^{-1}z_{[n]}(t)d\theta\right] \\
&+2z_n\lambda_n[-d_{un}+\rho_n]+2\lambda_n k_n(t)z_n^2-2\sigma_{wn}\tilde{W}_n^T\hat{W}_n-2\sigma_{vn}tr\left\{\tilde{V}_n^T\hat{V}_n\right\} \\
&\le z_{[n]}^T[\bar{A}_{[n]}^T\Lambda_{[n]}+\Lambda_{[n]}\bar{A}_{[n]}+(L^{-1})^T K^T(B_2+B_3)^T L^T\Lambda+\Lambda L(B_2+B_3)KL^{-1}+\tau\Lambda\bar{A}LN_1L^T\bar{A}^T\Lambda \\
&+\tau(L^{-1})^T K^T B_3^T N_1^{-1}B_3KL^{-1}]z_{[n]}+2z_{[n]}^T\Lambda_{[n]}\bar{B}_{1[n]}w(t)-\sum_{j=1}^{n}\left[\sigma_{wj}\|\tilde{W}_j\|^2+\sigma_{wj}\|\tilde{V}_j\|_F^2-N_j\right] \\
&\le -[C_{[n]}L_{[n]}^{-1}z_{[n]}(t)+Dw(t)]^T[C_{[n]}L_{[n]}^{-1}z_{[n]}(t)+Dw(t)]-\sum_{j=1}^{n}\left[\sigma_{wj}\|\tilde{W}_j\|^2+\sigma_{wj}\|\tilde{V}_j\|_F^2-N_j\right]
\end{aligned} \tag{29}$$

Define the compact set Ω. If $V_{[n]}$ is out of the Ω set, $\dot{V}\le 0$ in (29). So system convergences to Ω. Convergence radius depended on the values of ε, l, γ, σ.

$$\Omega=\left\{z_{[n]}^T,\tilde{W}_1^T,\dots,\tilde{W}_n^T,\tilde{V}_1^T,\dots,\tilde{V}_n^T\,\middle|\,\left[C_{[n]}L_{[n]}^{-1}z_{[n]}(t)+Dw(t)\right]^T\left[C_{[n]}L_{[n]}^{-1}z_{[n]}(t)+Dw(t)\right]-\sum_{j=1}^{n}\left[\sigma_{wj}\|\tilde{W}_j\|^2+\sigma_{wj}\|\tilde{V}_j\|_F^2\right]\le\sum_{j=1}^{n}N_j\right\}$$

5 Conclusion

A passive controller with adaptive neural compensation for a class of uncertain strict-feedback nonlinear systems with input-delay is proposed. The state-feedback control with input history feedback for linear input-delay system is obtained by γ-passive approach. In order to overcome the higher-order nonlinear terms, the adaptive neural model is introduced. In design procedure, the backstepping approach guarantees global uniformly ultimately boundedness.

Acknowledgement

The work is supported by Fundamental Research Fund No.HEUFT05062.

References

1. Kanellakopoulos, L, Kokotovic, P.V., Morse, A.S.: Systematic Design of Adaptive Controllers for Feedback Linearizable Systems. IEEE Trans on Aut. Control. 36(11) (1991) 1241-1253
2. Yeh, P.C., Kokotovic, P.V.: Adaptive Control of a Class of Nonlinear Discrete-time Systems. Int. J. Control. 62(2) (1995) 303-324
3. Jose, A.R., Rodolfo, S., America, M.: Cascade Control for a Class of Uncertain Nonlinear Systems: a Backstepping Approach. Chem Engineering Science 55(16) (2000) 3209-3221.
4. Gong, J. Q., Yao, B.: Neural Network Adaptive Robust Control of Nonlinear Systems in Semi-strict Feedback Form. Automatica 37(8) (2001) 1149-1160.

5. Zhang, T., Ge, S.S., Hang, C.C.: Adaptive Neural Network Control for Strict-feedback Nonlinear Systems Using Backstepping Design. Automatica 36(12) (2000) 1835-1846.
6. Yue, D.: Robust Stabilization of Uncertain Systems with Unknown Input Delay. Automatica 36(2) (2004), 331-336
7. Magdi, S.M., Mohamed, Z.: Passive Control Synthesis for Uncertain Systems with Multiple-state Delays. Computers and Electrical Engineering 28(3) (2002) 195-216
8. Kenan, E., Zigang, P., Kokotovic, P.V.: Locally Optimal and Robust Backstepping Design. IEEE Trans on Automatic Control 45(2) (2002) 260-271

Nonlinear System Adaptive Control by Using Multiple Neural Network Models

Xiao-Li Li[1], Yun-Feng Kang[1], and Wei Wang[2]

[1] Department of Automation, Information and Engineering School, University of Science and Technology Beijing, Beijing, 100083, P.R.China
[2] Research Center of Information and Control, Dalian University of Technology, Dalian, 116024, P.R. China

Abstract. Multiple radial based function (RBF)neural network models are used to cover the uncertainty of time variant nonlinear system, and multiple element controllers are set up based on the multiple RBF models. At every sample time, the closest model is selected by an index function which is formed by the integration of model output error. The element controller based on this model will be switched as the controller of the controlled system. This kind of multiple model adaptive controller (MMAC)is an extension of the MMAC in linear system, and it can improve the transient response and performance of the controlled system greatly.

1 Introduction

It is well known that the conventional adaptive control system based on a fixed or slowly adaptive model can get good performance. But when the parameters or structure of the system change abruptly from one context to another, the conventional adaptive control will react slowly, the output of the system will change abruptly and may be out of control at this time. One way to solve this problem is to use multi-model adaptive control (MMAC). From mid 1990's to now, a lot of MMAC algorithms combined with a switching index function have been given successively, and this kind of MMAC can guarantee the stability of closed-loop system. In recent ten years, The papers about switching MMAC have covered continuous time system[1], discrete time system[2, 3, 4], stochastic system[5, 6], etc, and there are also some practical implementations in this field.

In linear system MMAC, multiple adaptive and fix models are used to cover the uncertainty of parameters of the system, and the controller based on these models and an index function can improve the transient response. But the MMAC in linear system can not be extended to nonlinear system easily. In recent year, neural network has been used to set up multiple model adaptive controller of nonlinear system[7], but the structure of controller is still as linear forms, the nonlinear system always be represented by a combination of linear model with a unmodelled nonlinear dynamic part, and neural network is always used to identify the unmodelled dynamic part. The application of multiple model neural network adaptive controller in continuous time system can also be seen

J. Wang et al. (Eds.): ISNN 2006, LNCS 3972, pp. 867–874, 2006.

in[8], but the details of controller are not given, and the result of stability in linear system can not be extended to this kind of controller easily.

In this paper,multiple RBF neural network models can be used to approximate the nonlinear system in different local operation environment. Different weights of the different RBF models will be obtained according to different parameter of nonlinear system. The MMAC in linear system will be extended to nonlinear system by changing the identified parameters of linear model into the identified weights of RBF neural network. The property analysis in linear system can also be extended in this case. From the simulation, it can be seen that by using the multiple RBF model adaptive controller the control performance can be improved greatly.

2 Description of Plant and Adaptive Control

The system to be controlled is a single-input, single-output, discrete time system and described by the following equation.

$$y(t+1) = f(y(t), y(t-1), \cdots, y(t-n), a(t)) + B(z^{-1}, t)u(t) \tag{1}$$

where z^{-1} is a back-shift operator, $a(t)$ is parameter vector.

$$B(z^{-1}, t) = b_0(t) + b_1(t)z^{-1} + \cdots + b_m(t)z^{-m} \tag{2}$$

is known *in priori*.

When $f(y(t), y(t-1), \cdots, y(t-n), a(t))$ is a linear function, such as

$$f(y(t), y(t-1), \cdots, y(t-n), a(t)) = (a_1(t) + a_2(t)z^{-1} + \cdots + a_n(t)z^{-n})y(t) \tag{3}$$

the system (1) can be rewritten in the following regression form

$$y(t+1) = \phi^T(t)\theta(t) \tag{4}$$

$$\phi^T(t) = [y(t), \cdots, y(t-n), u(t), \cdots, u(t-m)] \tag{5}$$

$$\theta^T(t) = [a_1(t), \cdots, a_n(t), b_1(t), \cdots, b_m(t)] \tag{6}$$

Here, we assume that parameter $\theta(t)$ is the constant function or piecewise constant function of t. If $\theta(t)$ is unknown, the following adaptive control algorithm can be obtained by using the following recursive estimator.

$$\hat{\theta}(t) = \hat{\theta}(t-1) + \frac{\phi(t-1)e(t)}{1 + \phi(t-1)^T\phi(t-1)} \tag{7}$$

$$\hat{\theta}(t) = [\hat{a}_1(t), \cdots, \hat{a}_n(t), \hat{b}_1(t), \cdots, \hat{b}_m(t)]^T \tag{8}$$

$$e(t) = y(t) - \hat{y}(t) = y(t) - \phi^T(t-1)\hat{\theta}(t-1) \tag{9}$$

$$y^*(t+1) = \phi^T(t)\hat{\theta}(t) \tag{10}$$

where $y^*(t)$ is the setpoint value of output, and the control input can be obtained by (10).

When $f(y(t), y(t-1), \cdots, y(t-n), a(t))$ is a nonlinear function, the method used above can not be used again, and the MMAC based on (4)-(10) can not be used again. Compared the weights of RBF neural network with the parameter θ in linear system (4), it can be found that nonlinear system can have the same regression form as linear system if the nonlinear part of system is identified by RBF neural network. So the result in linear MMAC can be extended to nonlinear system by using the RBF neural network.

3 Nonlinear System Adaptive Control by Using Multiple RBF Neural Network Models

When $f(y(t), y(t-1), \cdots, y(t-n), a(t))$ is a nonlinear function, RBF neural network can be used to form an adaptive controller. In this case, the system (1) should satisfy the following assumption.

$\mathcal{A}1$: The parameter $a(t)$ in nonlinear function $f(\cdot)$ is the constant function or piecewise constant function of t, and the varying scope of parameter $a(t)$ is known *in priori*.
$\mathcal{A}2$: Polynomial $B(z^{-1})$ is known *in priori*,the roots of $B(z^{-1})$ lie inside the unit circle in complex plane (i.e. the system is minimum phase).

3.1 RBF Neural Network

Here a three layers RBF neural network with L input $x_i(i=1,\cdots,L)$ and M output $y_i(i=1,\cdots,M)$ is used to approximate the nonlinear function $f(\cdot)$. The structure of RBF neural network can be seen as below.

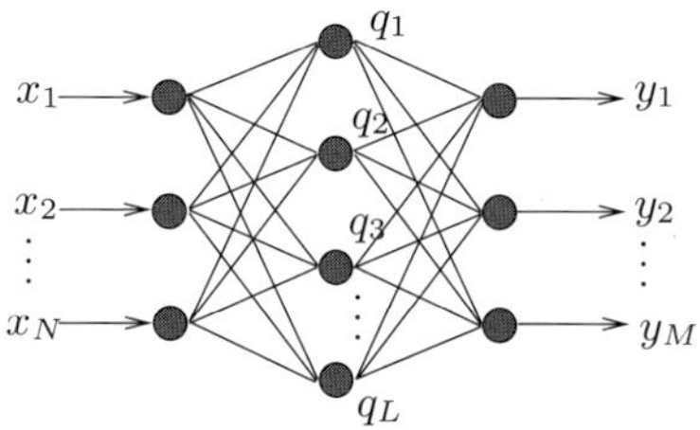

Fig. 1. the radial based function neural network

$$q_i(t) = K(|X(t) - S_i(t)|) = e^{-\frac{\sum_{j=1}^{N}(x_j(t)-s_{i,j}(t))^2}{2\alpha_i^2(t)}}, \quad 1 \le i \le L$$

$$y_j(t) = \sum_{i=1}^{L} w_{j,i}(t)q_i(t) + \lambda_j(t) = \sum_{i=0}^{L} w_{j,i}q_i(t) = Q^T(t)W_j(t); \quad 1 \le j \le M \quad (11)$$

$$w_{j,0}(t) = \lambda_j(t); \quad q_0(t) = 1 \quad (12)$$

$$Q(t) = [q_0(t), \cdots, q_L(t)]^T, \ \ W_j = [w_{j,0}, \cdots, w_{j,L}]^T \tag{13}$$

where

$q_i(t)$ is the output of the hidden unit (i.e. the output of radial based function).
$X(t)$ is the input vector.
$S_i(t)$ is the transforming center vector of the unit in hidden layer.
$\alpha_i(t)$ is the control parameter according to the central vecter.
The training step of the weights of network is as follows.

$$G(t) = \frac{P(t-1)Q(t)}{\delta + Q^T(t)P(t-1)Q(t)} \tag{14}$$

$$P(t) = P(t-1) - G(t)Q^T(t)P(t-1) \tag{15}$$

$$\hat{W}_i(t) = \hat{W}_i(t-1) + G(t)[d_i(t) - Q^T(t)\hat{W}_i(t-1)] \tag{16}$$

where
$P(t)$ is called inverse correlation matrix, δ is a forgetting factor,
$d_i(t)$ $(i = 1, 2, \cdots, M)$ is desired output of the unit of output layer.

3.2 Adaptive Control Using RBF Neural Network

Consider nonlinear system (1) satisfies the following assumption:

*A*3: The parameter $a(t)$ of $f(\cdot)$ is bounded, and lie in a close convex region Ω
*A*4: The nonlinear function $f(\cdot)$ is also bounded, if $y(t), \cdots, y(t{-}n)$, $(0 \le t \le \infty)$ are bounded.

The objective of the adaptive control problem is to determine a bounded control input $u(t)$ such that the output $y(t)$ of the system asymptotically tracks a specified arbitrary bounded reference output $y^*(t)$,i.e.

$$\lim_{t \to \infty} \mid y(t) - y^*(t) \mid = 0 \tag{17}$$

A three layers RBF neural network with $(n+1)$ input and one output can be used to identify the nonlinear part $f(\cdot)$, and the assumption below should be satisfied.
*A*5:For a certain value of $a(t)$ and over a certain compact set,$f(\cdot)$ can be approximated by a RBF neural network with proper choice of the structure and weights. i.e.

$$|f(\cdot) - \hat{f}(\cdot)| < \varepsilon \tag{18}$$

where $\hat{f}(\cdot) = Q^T(t)W(t)$, and ε can be any specified positive number.
In this case , the control input of $u(t)$ can be got as below

$$y^*(t+1) = Q^T(t)W(t) + B(z^{-1})u(t) \tag{19}$$

Because $a(t)$ is unknown *in priori*, so the initial value of weights of RBF neural network is difficult to select. When $a(t)$ is constant, if the initial value of weight is far from the weight according to $a(t)$, the recursive algorithm of RBF neural network needs long time to make the weights of RBF neural network converge to the weight according to $a(t)$, and the transient response of the system is also not very well. When $a(t)$ is a piecewise constant function of t, i.e the system (1) is nonlinear system with jumping parameters, the control performance will be worse if only one adaptive controller is used.

3.3 Multiple Neural Network Model Adaptive Control

To solve the problem mentioned above, multiple model adaptive control using neural network will be set up.

Firstly,decomposing the parameters set Ω to obtain a finite cover $\{\Omega_i\}_1^V$, which satisfies the following condition:

$\mathcal{C}1 : \Omega_i \subset \Omega, \ \ \Omega_i \neq \{\}, \ \ i = 1, \cdots, V$
$\mathcal{C}2 : \bigcup_i^V \Omega_i = \Omega$
$\mathcal{C}3$: for each $i = 1, \cdots, V$, let a_i and r_i denote the 'center' and 'radius' of Ω_i ,i.e. $a_i \in \Omega_i$,and $\|a - a_i\| \leq r_i$ for all $a \in \Omega_i$.

According to the V parameters of a, consider the V nonlinear systems of (1)

$$y(t+1) = f(y(t), y(t-1), \cdots, y(t-n), a_i) + B(z^{-1}, t)u(t), \ \ i = 1, 2, \cdots, V \quad (20)$$

Setting up V controller as (11)-(19) for a given output setpoint $y^*(t)$. After finite training steps, the weights of V RBF neural network will be convergent, the outputs of every system ,$i = 1, 2, .., V$ will trace the setpoint value $y^*(t)$ asymptotically. The V convergent weights W_{fi}, $i = 1, 2, \cdots, V$ will be kept to be used for multiple model adaptive controller in the following part.

Before the multiple model adaptive controller is set up, the following three steps should be done.

1. setting up V fix models with parameter W_{fi}, $i = 1, 2, \cdots, V$, and output

$$\hat{y}_{fi}(t+1) = Q^T W_{fi}(t) + B(z^{-1})u(t) \quad (21)$$

2. two adaptive models with weight W_{ai}, $i = 1, 2$ which will be updated online as (11)-(16) will be used for good steady state performance. The output of adaptive model is

$$\hat{y}_{ai}(t+1) = Q^T W_{ai}(t) + B(z^{-1})u(t) \quad (22)$$

3. two switching functions should be given

$$J_{ai}(t, t_0) = e_{ai}^2(t) + \beta e_{ai}^2(t-1) + \cdots + \beta^{t-t_0} e_{ai}^2(t_0), \ \ e_{ai}(t) = y(t) - \hat{y}_{ai}(t), \ \ i = 1, 2 \quad (23)$$

$$J_{fi}(t, t_0) = e_{fi}^2(t) + \beta e_{fi}^2(t-1) + \cdots + \beta^{t-t_0} e_{fi}^2(t_0),$$

$$e_{fi}(t) = y(t) - \hat{y}_{fi}(t), \ \ i = 1, 2, \cdots, V, \ \ 0 \leq \beta \leq 1 \quad (24)$$

By using (21)-(24), the multiple model adaptive control with structure as Fig.2 can be obtained as following.

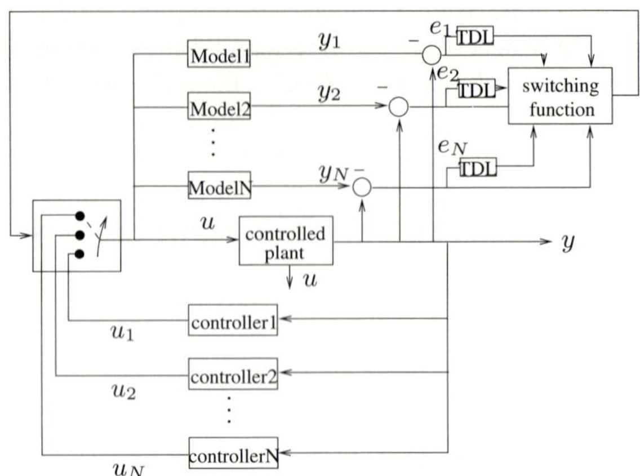

Fig. 2. Structure of MNNMAC

Multiple Neural Network Model Adaptive Control (MNNMAC)

1)$t = t_0$,let $W = W_0$,calculate the control input as (19)
2)$t > t_0$,calculate

$$i(t) = \arg_{k \in \{1,2,\cdots,V\}} min J_{fk}(t, t_0), j(t) = \arg_{k \in \{1,2\}} min J_{ak}(t, t_0)$$

if $J_{aj(t)}(t, t_0) \leq J_{fi(t)}(t, t_0)$, let $W = W_{aj(t)}$,calculate control input as (19), and let $W_{a2} = W_{aj(t)}$, $J_{a2}(t, t_0) = J_{aj(t)}(t, t_0)$, $t = t + 1$, go back to 2)

else,let $W = W_{fi(t)}$,calculate control input as (19),and let $W_{a2} = W_{fi(t)}$, $J_{a2}(t, t_0) = J_{fi(t)}(t, t_0)$, $t = t + 1$, go back to 2)

Remark 1. From the structure of MNNMAC, it can be seen the controller of the system can be switched to the controller based on the model with minimum index function value when the parameter of system change abruptly from one to another. The identified weights of the second adaptive model will be initiated to the weights of the model with the minimum index function value at every sample time. The existence of the two adaptive models will keep the steady property of conventional adaptive control and the control performance can be improved by using multiple models.

Remark 2. Using MNNMAC for nonlinear system as (1), the method of stability analysis in linear system can also be used by changing the identified parameters in linear MMAC into the identified weights of nonlinear MMAC.

4 Simulation Analysis

Consider the following nonlinear system with jumping parameters

$$y(t + 1) = f(y(t), y(t - 1), a(t)) + 0.5u(t) \tag{25}$$

$$f(t) = \frac{a(t)y(t)y(t-1)}{1 + y^2(t) + y^2(t-1)} \tag{26}$$

$$a(t) = \begin{cases} -5 & ; \quad 0 \leq t < 50 \\ 5 & ; \quad 50 \leq t < 100 \\ -1 & ; \quad 100 \leq t < 150 \\ 1 & ; \quad 150 \leq t < 200 \end{cases} \tag{27}$$

When one adaptive neural network model controller is used, the input and output can be seen in Fig.3 ,$r = y^*(t) = 2$.The overrun can be seen in Fig 3 at the time parameter $a(t)$ changes abruptly. When multiple model neural network

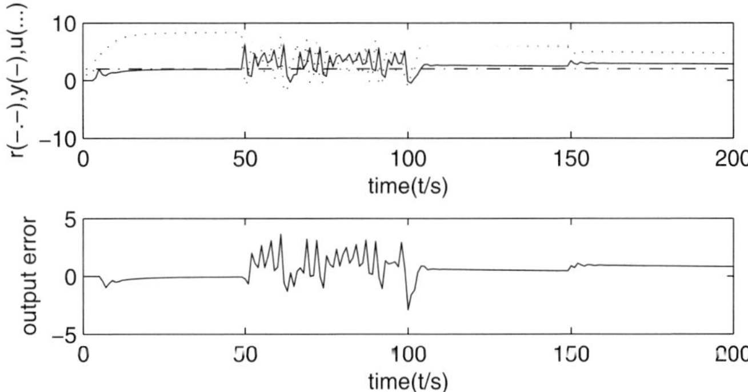

Fig. 3. System using one adaptive NN model

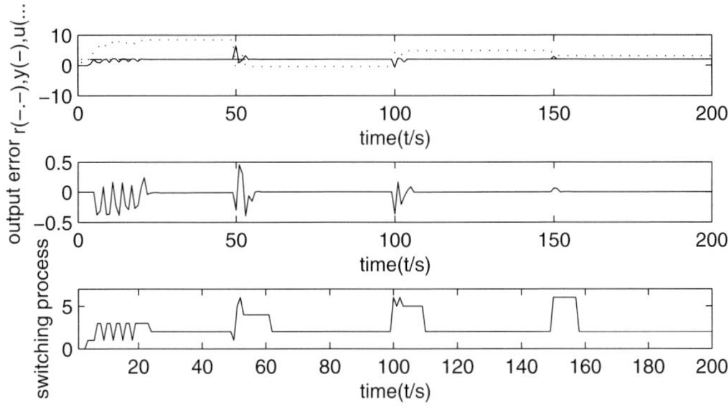

Fig. 4. System using MNNMAC

adaptive control is used, a three layers RBF neural network with 4 hidden units is used,and the result can be seen in Fig.4 ,$r = y^*(t) = 2$. It can be seen easily that control performance has been improved greatly. Four fix models with weight $w_1 = [10.4178, -3.1553, -3.1553, -3.1553, -3.1553]^T, w_2 = [-2.7671, 1.2472, 1.2472, 1.2472, 1.2472]^T, w_3 = [2.1094, -0.6376, -0.6376, -0.6376, -0.6376]^T$, $w_4 = [-2.1104, 0.6382, 0.6382, 0.6382, 0.6382]^T$ are used. Two adaptive models are also used to keep steady property of MNNMAC.

5 Conclusion

A nonlinear system multiple model adaptive control by using RBF neural network is proposed in this paper.By using MNNMAC, the control performance of the nonlinear system with jumping parameters has been improved greatly to parameter for the nonlinear system with jumping parameter, at the same time the steady property of conventional adaptive control has not been lost. From the simulation the effectiveness of MNNMAC can be tested easily. This kind of multiple model adaptive control also gives a way for better control of nonlinear system.

Acknowledgement

This work is partially supported by Innovation Talent Project Fund of University of Science and Technology Beijing,Fund of Key Discipline Project common constructed by Beijing Municipal Commission of Education and The Distinguish Youth Fund of National Natural Science Foundation of P.R.China (No.69825106).

References

1. Narendra, K.S., Balakrishnan, J.: Adaptive Control Using Multiple Models. IEEE Trans. Automatic Control 42(2) (1997) 171-187
2. Narendra, K.S., Xiang, C.: Adaptive Control of Discrete-time Systems Using Multiple Models. IEEE Trans. Automatic Control 45(9) (2000) 1669-1685
3. Li, X.L., Wang, W.: Minimum Variance Based Multi-model Adaptive Control. Proc. IFAC World Congress. Beijing. China (1999) 325-329
4. Li, X.L., Wang, W., Wang, S.N.: Multiple Model Adaptive Control for Discrete Time Systems. American Control Conference. Arlington. Virginia. USA (2001) 4820-4825
5. Chen, L.J., Narendra, K.S.: Nonlinear Adaptive Control Using Neural Networks and Multiple Models. Automatica 37(8) (2001) 1245-1255
6. Narendra, K.S., Driollet, O.: Stochastic Adaptive Control Using Multiple Estimation Models. Int. J. Adapt. Control Signal Process 15(3) (2001) 287-317
7. Li, X.L., Zhang, W.C., Wang, L.: Stochastic Adaptive Control Using Multiple Models for Improved Performance in the Presence of Bounded Disturbance.Impulsive Dynamical Systems and Applications.DCDIS Proceedings. Vol. 3 Wuxi China (2005) 1061-1068
8. Zufiria, P.J., Et,al.:Neural Adaptive Control of Non-linear Plants via a Multiple Inverse Models Approach. Int. J. Adapt. Control Signal Process 13(4) (1999) 219-239

Implementable Adaptive Backstepping Neural Control of Uncertain Strict-Feedback Nonlinear Systems

Dingguo Chen[1] and Jiaben Yang[2]

[1] Siemens Power Transmission and Distribution Inc., 10900 Wayzata Blvd., Minnetonka, Minnesota 55305, USA
[2] Department of Automation, Tsinghua University, Beijing, 100084, People's Republic of China

Abstract. Presented in this paper is neural network based adaptive control for a class of affine nonlinear systems in the strict-feedback form with unknown nonlinearities. A popular recursive design methodology - backstepping is employed to systematically construct feedback control laws and associated Lyapunov functions. The significance of this paper is to make best use of available signals, avoid unnecessary parameterization, and minimize the node number of neural networks as on-line approximators. The design assures that all the signals in the closed loop are semi-globally uniformly, ultimately bounded and the outputs of the system converges to a tunable small neighborhood of the desired trajectory. Novel parameter tuning algorithms are obtained on a more practical basis.

1 Introduction

Despite the systematic approach to address the adaptive control design, the backstepping design method needs to be enhanced to overcome the well-known control singularity problem in which the control has a factor that tends to infinity as the estimate of one of the nonlinear functions tends to zero, which makes the controller unrealizable and hence impractical. Motivated to achieve an adaptive control design that has the capability to overcome the possible control singularity problem and allows convenient implementation, neural control schemes were proposed in [1, 2] that completely avoid the control singularity problem and have better implementability. The problem with the control scheme in [2] is that the virtual controller makes use of the derivatives of some signals which are not available even though the signals themselves are available. Therefore, the control scheme is not theoretically justifiable even though it supports convenient implementation. The problem with the control scheme in [1] is that it does not distinguish the available signals from unavailable signals, leading to unnecessary approximation errors; in addition, it requires an increases in the dimension of inputs and thus requires a greater node number of neural networks by multiple folds. The above listed problems will all be addressed in this paper.

J. Wang et al. (Eds.): ISNN 2006, LNCS 3972, pp. 875–880, 2006.

The rest of the paper is organized as follows: The system under study is described in Section 2. With the employment of the backstepping design methodology, the adaptive neural control scheme is presented in Section 3. Finally, the concluding remarks are given in Section 4.

2 Problem Statement

A class of uncertain nonlinear strict-feedback systems takes the following form:

$$\begin{aligned} \dot{x}_i &= f_i(\overline{x}_i) + g_i(\overline{x}_i)x_{i+1}, 1 \leq i \leq n-1 \\ \dot{x}_n &= f_n(\overline{x}_n) + g_n(\overline{x}_n)u, n \geq 2 \\ y &= x_1 \end{aligned} \tag{1}$$

where $x_i \in R$, $i = 1, \cdots, n$, $u \in R$, and $y \in R$ are state variables, control input and system output, respectively; $\overline{x}_i = [x_1, \cdots, x_i]^\top \in R^i$, $i = 1, \cdots, n$; functions $f_i(\overline{x}_i)$ and $g_i(\overline{x}_i)$, $i = 1, \cdots, n$, are smooth but unknown.

The control objective is to design a neural controller such that

- All the signals in the resulting closed-loop system are semiglobally uniformly ultimately bounded;
- The system output y follows a desired trajectory y_d which has up to the $(n+1)$th order bounded derivatives.

To facilitate the theoretical development, we make the following conventional assumptions:

AS1: The signs of functions $g_i(\overline{x}_i)$'s are known and $|g_i(\overline{x}_i)|$'s are bounded between two positive numbers for $\forall \overline{x}_n \in \Omega \subset R^n$. Without loss of generality, assume that there exist g_{i1} and g_{i0} such that $g_{i1} \geq g_i(.) \geq g_{i0} > 0$.

AS2: There exist g_{id}'s such that $|\dot{g}_i(.)| \leq g_{id}$ for $\forall \overline{x}_n \in \Omega \subset R^n$.

AS3: For a well defined continuous function $F(.)$ on a compact domain and a collection of RBF NN's taking the general form of $\hat{F}_\theta(.) = \theta^\top \xi(.)$ with approximation error $\epsilon_\theta = \hat{F}_\theta(.) - F(.)$ where appropriate dimensions are assumed, there exists vector θ^* and ϵ such that $|\epsilon_{\theta^*}| \leq \epsilon$.

Remark 1: The optimal vector θ^* is defined as $\theta^* = arg\min_\theta\{\sup_{x\in\Omega} |\hat{F}_\theta(x) - F(x)|\}$.

3 Adaptive Neural Control

To derive a closed-form representation of the neural controller, we adopt the popular recursive backstepping design technique. At each recursive backstepping stage i $(1 \leq i \leq n)$, an appropriate function of the state variables is employed as a pseudo-control input for the corresponding lower dimension sub-system of the overall system. Each backstepping stage works out a virtual controller just to allow its following backstepping stage to continue in the same manner until the final stage arrives in which the true controller will be worked out. At each step i, the virtual control contains unknown functions that need to be approximated by RBF neural networks. By using the virtual control, each lower dimension sub-system is stabilized with respect to respective Lyapunov function. The true control is derived to stabilize the overall closed-loop system with respect to a final Lyapunov function, which is the sum of all the sub-Lyapunov functions. The same problem has been studied in [1, 2]. However, in [2] the derivative of the virtual control inputs derived at each backstepping stage is used as a known signal in respective parameter adaptation laws, which is technically incorrect. As a matter of fact, this kind of signal is available only in Step 1. In [1], no distinction is made between known signals and unknown signals when attempting to approximate them as a whole. This inevitably increases the dimension of the unknown expression when treated as a whole, and therefore increases the node number of RBF neural networks by exponential of dimensions, which is unacceptable when it comes to practical implementation. The proposed design illustrated in the following utilizes only known signals, distinguishes between known and unknown signals, uses the RBF neural networks to only approximate unknown signals, and thus minimizes the node number of RBF neural networks significantly and demonstrates the effective application of RBF neural networks. In what follows, the sketch of the proposed design scheme is presented due to the page limitation.

Step 1: Define $e_1 = x_1 - x_{1d}$ where $x_{1d} = y_d$ for convenient presentation. Since $g_1(.)$ and $f_1(.)$ are unknown and $\dot{x}_{1d}$ is known, RBF neural networks, $\hat{\theta}_{11}^T\xi_{11}(x_1)$ and $\hat{\theta}_{12}^T\xi_{12}(x_1)$ are used to approximate $g_1^{-1}(x_1)$ and $-g_1^{-1}(x_1)f_1(x_1)$, respectively. By choosing $x_{2d} = \hat{\theta}_{11}^T\xi_{11}(x_1)\dot{x}_{1d} + \hat{\theta}_{12}^T\xi_{12}(x_1) - k_1e_1$ where $k_1 > 0$, and defining $e_2 = x_2 - x_{2d}$, $\dot{e}_1 = g_1(e_2 - k_1e_1 + \tilde{\theta}_{11}^T\xi_1\phi_{11} + \tilde{\theta}_{12}^T\xi_{12} + \epsilon_1)$ where $\phi_{11} = \dot{x}_{1d}$, $\epsilon_1 = (g_1^{-1}(x_1) - \theta_{11}^{*T}\xi_{11}(x_1))\phi_{11} + (-g_1^{-1}(x_1)f_1(x_1) - \theta_{12}^{*T}\xi_{12}(x_1))$; and $\tilde{(.)} = \hat{(.)} - (.)^*$. By applying AS3 and noticing that $\dot{x}_{1d}$ is bounded, we conclude that ϵ_1 is bounded.

Consider the Lyapunov function candidate: $V_1 = \frac{1}{2g_1(x_1)}e_1^2 + \frac{1}{2}\tilde{\theta}_{11}^T\Gamma_{11}^{-1}\tilde{\theta}_{11} + \frac{1}{2}\tilde{\theta}_{12}^T\Gamma_{12}^{-1}\tilde{\theta}_{12}$ where Γ_{11} and Γ_{12} are symmetric and positive definite.

By employing the σ-modification type of adaptation laws, we obtain

$$\begin{aligned}\dot{\hat{\theta}}_{11} &= \Gamma_{11}(-e_1\xi_{11}(x_1)\phi_{11} - \sigma_{11}\hat{\theta}_{11})\\ \dot{\hat{\theta}}_{12} &= \Gamma_{12}(-e_1\xi_{12}(x_1) - \sigma_{12}\hat{\theta}_{12})\end{aligned} \tag{2}$$

where $\sigma_{11} > 0$ and $\sigma_{12} > 0$ are constants.

Let $k_1 = k_{10} + k_{11}$ with $k_{10} > 0$ and $k_{11} > 0$. By completion of squares, and choosing k_{10} such that $k_{10}^* = k_{10} - (g_{1d}/2g_{10}^2) > 0$, the following inequality can be obtained:

$$\dot{V}_1 \le e_1e_2 - k_{10}^*e_1^2 + \frac{\sigma_{11}}{2}||\theta_{11}^*||^2 - \frac{\sigma_{11}}{2}||\tilde{\theta}_{11}||^2 + \frac{\sigma_{12}}{2}||\theta_{12}^*||^2 - \frac{\sigma_{12}}{2}||\tilde{\theta}_{12}||^2 + \frac{\epsilon_1^{*2}}{4k_{11}} \tag{3}$$

Step 2: Define $e_2 = x_2 - x_{2d}$. $\dot{x}_{2d}$ involves unknown functions $f_1(x_1)$ and $g_1(x_1)$. Consequently, $\dot{x}_{2d}$ should not be treated as an available signal that can be put in the parameter adaptation laws. Therefore, the parameters adaptation laws presented in [2] for Steps 2 up to n are technically incorrect. Define: $\phi_{21} = (\partial x_{2d}/\partial x_{1d})\dot{x}_{1d} + \frac{\partial x_{2d}}{\partial \hat{\theta}_{11}}\Gamma_{11}(-e_1\xi_{11}(x_1)\phi_{11} - \sigma_{11}\hat{\theta}_{11}) + \frac{\partial x_{2d}}{\partial \hat{\theta}_{12}}\Gamma_{12}(-e_1\xi_{12}(x_1) - \sigma_{11}\hat{\theta}_{12})$; $\hat{\theta}_1 = [\hat{\theta}_{11}^\tau\ \hat{\theta}_{12}^\tau]^\tau$; $\phi_{22} = \frac{\partial x_{2d}}{\partial x_1}$, $\phi_{23} = \frac{\partial x_{2d}}{\partial x_1}x_2$. $\theta_{21}^{*\tau}\xi_{21}$, $\theta_{22}^{*\tau}\xi_{22}$, $\theta_{23}^{*\tau}\xi_{23}$ and $\theta_{24}^{*\tau}\xi_{24}$ are used to approximate $-g_2^{-1}$, $-g_2^{-1}f_1$, $-g_2^{-1}g_1$ and $g_2^{-1}f_2$, respectively; $e_3 = x_3 - x_{3d}$, $\epsilon_2 = -\sum_{i=1}^{i=3}\theta_{2i}^{*\tau}\xi_{2i}\phi_{1i} - \theta_{24}^{*\tau}\xi_{24} - g_2^{-1}\phi_{21} - g_2^{-1}f_1\phi_{22} - g_2^{-1}g_1\phi_{23} + g_2^{-1}f_2$. Again it can be readily shown that $|\epsilon_2|$ is bounded.

Choose $x_{3d} = -e_1 + \sum_{i=1}^{i=3}\hat{\theta}_{2i}^\tau\xi_{2i}\phi_{1i} + \hat{\theta}_{24}^\tau\xi_{24} - k_2e_2$. Consider the Lyapunov function candidate $V_2 = V_1 + \frac{1}{2g_2}e_2^2 + \sum_{i=1}^4 \frac{1}{2}\tilde{\theta}_{2i}^\tau\Gamma_{2i}^{-1}\tilde{\theta}_{2i}$ where $\Gamma_{2i} = \Gamma_{2i}^\tau > 0$, $1 \le i \le 4$. Considering the following adaptation laws:

$$\begin{aligned}\dot{\hat{\theta}}_{2i} &= \Gamma_{2i}(-e_2\xi_{2i}\phi_{2i} - \sigma_{2i}\hat{\theta}_{2i}),\ \ 1 \le i \le 3\\ \dot{\hat{\theta}}_{24} &= \Gamma_{24}(-e_2\xi_{24} - \sigma_{24}\hat{\theta}_{24})\end{aligned} \tag{4}$$

where $\sigma_{2i} > 0$, $1 \le i \le 4$; and letting $k_2 = k_{20} + k_{21}$ with $k_{20} > 0$, $k_{21} > 0$ and k_{20} chosen to satisfy $k_{20}^* = k_{20} - \frac{g_{2d}}{2g_{20}^2} > 0$, we obtain

$$\dot{V}_2 \le e_2e_3 - \sum_{i=1}^{2} k_{i0}^*e_i^2 - \sum_{i=1} 2\sum_{j=1}^{2^i}\frac{\sigma_{ij}}{2}||\tilde{\theta}_{ij}||^2 + \sum_{i=1} 2\sum_{j=1}^{2^i}\frac{\sigma_{ij}}{2}||\theta_{ij}^*||^2 + \sum_{i=1}^{2}\frac{\epsilon_i^{*2}}{4k_{i1}} \tag{5}$$

Step j $(3 \le j < n)$: Define $e_{(j+1)} = x_{(j+1)} - x_{(j+1)d}$. x_{jd} is a virtual control which is determined at Step $j-1$, $\dot{x}_{jd} = -\phi_{j1} - \cdots - g_{j-1}^{-1}\cdots g_2^{-1}g_1\phi_{j(2^j-1)}$ where ϕ_{ji}'s for $i = 1, \ldots, 2^j - 1$ can be appropriately derived and are all computable whereas g_j^{-1}, $\cdots$, $g_j^{-1}\cdots g_2^{-1}g_1$ are unknown smooth functions and need to be approximated by RBF neural networks $\theta_{ji*}\xi_{ji}$'s, and $g_j^{-1}f_j$ is approximated by $\theta_{j2^j*}\xi_{j2^j}$.

Choose $x_{(j+1)d} = -e_{j-1} - k_je_j + \sum_{i=1}^{2^j-1}\theta_{ji}\xi_{ji}\phi_{ji} + \theta_{j2^j}\xi_{j2^j}$. Consider the Lyapunov function candidate $V_j = V_{j-1} + \frac{1}{2g_j}e_j^2 + \sum_{i=1}^{j2^j}\frac{1}{2}\tilde{\theta}_{ji}^\tau\Gamma_{ji}^{-1}\tilde{\theta}_{ji}$ where $\Gamma_{ji} = \Gamma_{ji}^\tau > 0$, $1 \le i \le 2^j$.

Choosing the following parameter updating laws:

$$\begin{aligned}\dot{\hat{\theta}}_{ji} &= \Gamma_{ji}(-e_j\xi_{ji}\phi_{ji} - \sigma_{ji}\hat{\theta}_{ji}),\ \ 1 \le i \le 2^j - 1\\ \dot{\hat{\theta}}_{j2^j} &= \Gamma_{j2^j}(-e_j\xi_{j2^j} - \sigma_{j2^j}\hat{\theta}_{j2^j})\end{aligned} \tag{6}$$

where $\sigma_{ji} > 0$, $1 \le i \le 2^j$; and letting $k_j = k_{j0} + k_{j1}$ with $k_{j0} > 0$, $k_{j1} > 0$ and k_{j0} chosen to satisfy $k_{j0}^* = k_{j0} - \frac{g_{jd}}{2g_{j0}^2} > 0$, we obtain

$$\dot{V}_j \le e_je_{j+1} - \sum_{i=1}^{j} k_{i0}^*e_i^2 - \sum_{i=1} j\sum_{k=1}^{2^i}\frac{\sigma_{ik}}{2}||\tilde{\theta}_{ik}||^2 + \sum_{i=1} j\sum_{k=1}^{2^i}\frac{\sigma_{ik}}{2}||\theta_{ik}^*||^2 + \sum_{i=1}^{j}\frac{\epsilon_i^{*2}}{4k_{i1}} \tag{7}$$

Step n: Define $e_n = x_n - x_{nd}$. x_{nd} is a virtual control which is determined at Step $n-1$, $\dot{x}_{nd} = -\phi_{n1} - \cdots - g_{n-1}^{-1}\cdots g_2^{-1}g_1\phi_{n(2^n-1)}$ where ϕ_{ni}'s for $i =$

$1, \ldots, 2^n - 1$ can be appropriately derived and are all computable, whereas g_n^{-1}, $\cdots$, $g_n^{-1} \cdots g_2^{-1} g_1$ are unknown smooth functions and need to be approximated by RBF neural networks $\theta_{ni*}\xi_{ni}$'s, and $g_n^{-1} f_n$ is approximated by $\theta_{n2^n*}\xi_{n2^n}$.

Choose the actual controller u as

$$u = -e_{n-1} - k_n e_n + \sum_{i=1}^{2^n - 1} \theta_{ni}\xi_{ni}\phi_{ni} + \theta_{n2^n}\xi_{n2^n}. \tag{8}$$

Consider the Lyapunov function candidate $V_n = V_{n-1} + \frac{1}{2g_n} e_n^2 + \sum_{i=1}^{n2^n} \frac{1}{2}\tilde{\theta}_{ni}^{\tau} \Gamma_{ni}^{-1}$ $\tilde{\theta}_{ni}$ where $\Gamma_{ni} = \Gamma_{ni}^{\tau} > 0,\ 1 \le i \le 2^n$.

Choose the following parameter updating laws:

$$\begin{aligned} \dot{\hat{\theta}}_{ni} &= \Gamma_{ni}(-e_n \xi_{ni}\phi_{ni} - \sigma_{ni}\hat{\theta}_{ni}),\ \ 1 \le i \le 2^n - 1 \\ \dot{\hat{\theta}}_{n2^n} &= \Gamma_{n2^n}(-e_n \xi_{n2^n} - \sigma_{n2^n}\hat{\theta}_{n2^n}) \end{aligned} \tag{9}$$

where $\sigma_{ni} > 0,\ 1 \le i \le 2^n$.

Let $k_n = k_{n0} + k_{n1}$ with $k_{n0} > 0$, $k_{n1} > 0$ and k_{n0} chosen to satisfy $k_{n0}^* = k_{n0} - \frac{g_{nd}}{2g_{n0}^2} > 0$; defining $\phi = \sum_{i=1} n \sum_{k=1}^{2^n} \frac{\sigma_{ik}}{2} ||\theta_{ik}^*||^2 + \sum_{i=1}^{n} \frac{\epsilon_n^{*2}}{4k_{i1}}$; and properly choose σ_{ji} and μ such that $0 \le \mu \le \sigma_{ji}/\lambda_{max}\{\Gamma_{ji}^{-1}\}$ where $\lambda_{max}\{.\}$ is the largest eigenvalue of the underlying matrix and such that $k_{j0}^* \ge \mu/(2g_{j0}^2)$. Then we obtain

$$\dot{V}_n \le -\sum_{i=1}^{n} k_{i0}^* e_i^2 - \frac{\mu}{2} \sum_{i=1} n \sum_{k=1}^{2^n} \tilde{\theta}_{ik}^{\tau} \Gamma_{ik}^{-1} \tilde{\theta}_{ik} + \phi \le -\mu V_n + \phi \tag{10}$$

Theorem 1. *Consider the controller (8) and the parameters adaptation laws (2), (4), (6) and (9) for the system (1) with assumptions AS1, AS2 and AS3 made in Section 2. Then for bounded initial conditions, the following must hold:*

- *Overall Stability - All signals in the closed-loop system remain bounded;*
- *Tracking Convergence - The output tracking error $y(t) - y_d(t)$ converges to a pre-designated sufficiently small neighborhood.*

Proof. Based on Eq. (10), we can conclude from the employment of a standard Lyapunov theorem that e_j,$\hat{\theta}_{ji}$ for $j = 1, \cdots, n$ and $i = 1, \cdots, 2^j$ are uniformly ultimately bounded. According to the definitions of e_j and the virtual controllers x_{jd} for $j = 1, \cdots, n$, we conclude that all the virtual controllers are bounded, which in turn leads to the boundedness of x_j for $j = 1, \cdots, n$. Consequently, these lead to the boundedness of ϕ_{ji} for $j = 1, \cdots, n$ and $i = 1, \cdots, 2^j$, which together with the boundedness of ξ_{ji}, e_j and $\hat{\theta}_{ji}$ for $j = 1, \cdots, n$ and $i = 1, \cdots, 2^j$, guarantees the boundedness of the actual control u. Thus all the signals in the closed-loop system remain bounded.

From the inequality (10), the following holds: $V_n(t) \le \frac{\theta}{\mu} + (V_n(0) - \frac{\theta}{\mu}) \exp(-\mu t)$. Substituting the $V_n(t)$ expression into the above inequality yields $\sum_{j=1}^{n} \frac{1}{2g_j} e_j^2 \le$

$\frac{\theta}{\mu} + (Vn(0) - \frac{\theta}{\mu}) \exp(-\mu t) \leq \frac{\theta}{\mu} + Vn(0) \exp(-\mu t)$. Let $g_{max} = \max_{1 \leq j \leq n}\{g_{j1}\}$. Then we obtain $\sum_{j=1}^{n} e_j^2 \leq 2g_{max}(\frac{\theta}{\mu} + Vn(0) \exp(-\mu t))$.

This implies that for any given small number $\eta > \sqrt{2g_{max}\frac{\theta}{\mu}}$, there exists T such that for all $t \geq T$, the tracking error satisfies $|e_1(t)| = |x_1(t) - x_{1d}(t)| = |y(t) - y_d(t)| < \eta$. □

Remark 2: In [1], the intermediate signals such as ϕ_{ji}'s are used to calculate yet other intermediate signals which are used as additional inputs to RBF NN's. These present a practical difficulty as the range of these signals have to be determined before designing RBF NN's. In addition, the computable intermediate signals are not directly used as available signals but as part of the unknown functions, which is an over-parameterization that should be avoided in practice. In this paper, the intermediate signals such as ϕ_{ji}'s are directly used as available signals. Only the unknown functions g_j's and f_j's or their combinations are approximated by RBF NN's. The input domain of RBF NN's are just the ranges of individual states of the state vector $\overline{x}_n$. These features make the proposed approach practically implementable.

4 Conclusion

In this paper, a study has been made on a class of nonlinear uncertain systems. A stable neural controller has been proposed and developed based on the backstepping design procedures. The developed control design scheme avoids the control singularity problem and is also applicable to a larger class of nonlinear systems that can be transformed to the type of nonlinear systems studied in this paper. It can also be applied to the studied nonlinear systems but with stable zero dynamics. The design parameters can be tuned to achieve desired control performance. The stability and convergence of the system with the developed control laws and parameter updating laws is guaranteed. All the signals of the closed-loop system are semi-globally uniformly ultimately bounded, and the output of the system converges to a small neighborhood of the desired trajectory. The proposed control design scheme is practically implementable compared to the existing control designs that are practically non-implementable [1]. Future study includes extending the proposed controller design scheme to decentralized nonlinear systems such as power systems.

References

1. Ge, S., Wang, C.: Direct Adaptive NN Control of A Class of Nonlinear Systems. IEEE Trans. Neural Networks **13** (2002) 214 – 221
2. Li, Y., Qiang, S., Zhuang, X., Kaynak, O.: Robust and Adaptive Backstepping Control for Nonlinear Systems Using RBF Neural Networks. IEEE Trans. Neural Networks **15** (2004) 693 – 701

A Discrete-Time System Adaptive Control Using Multiple Models and RBF Neural Networks

Jun-Yong Zhai, Shu-Min Fei, and Kan-Jian Zhang

Research Institute of Automation, Southeast University,
Nanjing 210096, China

Abstract. A new control scheme using multiple models and RBF neural networks is developed in this paper. The proposed scheme consists of multiple feedback linearization controllers, which are based on the known nominal dynamics model and a compensating controller, which is based on RBF neural networks. The compensating controller is applied to improve the transient performance. The neural network is trained online based on Lyapunov theory and learning convergence is thus guaranteed. Simulation results are presented to demonstrate the validity of the proposed method.

1 Introduction

The multiple models approach to nonlinear systems has become an active research area in the recent years. Unlike the classical approach [1], the multiple models approach represents the system as an interpolation of simple local models. Each local model describes the behavior of the system in a limited part of the operating space. The concept of multiple models, switching, and tuning was applied to describe the dynamic of the system using different models for different operating conditions and devise a suitable strategy for finding the closest model to the current system dynamic in [2-4]. However, the approximated linearized system cannot cover the operating range of the system dynamics in high performance engineering applications, such as advanced aircraft control and position control of DC motors [5]. In this case, gain-scheduled controllers can be designed over several fixed operating points covering the system's operating range and controller gains interpolated over this range [6,7]. RBF neural network is applied to compensate for modeling error in [8]. In this paper we focus on the problem that if the modeling is not perfect or there exist uncertainties, which are always the case in practice, then lead to the significant degradation of the performance. In this case, we propose a new control scheme that includes local controllers and a compensating controller to improve the transient response.

The rest of the paper is organized as follows. Section 2 presents the problem statement. A compensating controller design using RBF neural networks is proposed and the neural network is trained based on Lyapunov theory and thus its convergence is guaranteed in Section 3. Switching control strategy is studied in Section 4. A simulation study is described in section 5, which shows the validity of the proposed method. Finally, some conclusions are drawn in section 6.

J. Wang et al. (Eds.): ISNN 2006, LNCS 3972, pp. 881 – 887, 2006.

2 Problem Statement

The major motivation for the multiple models approach is that local modeling is simpler than global modeling. The first step is to divide the system operating range into several regions. A local linear model is associated with each region, which describes the system behavior in that region. Let a nonlinear discrete-time system be described by

$$y(k+1) = g(y(k),\cdots, y(k-n),u(k),\cdots,u(k-m)) . \tag{1}$$

where $g(\cdot)$ is a nonlinear function. $u(k)$ and $y(k)$ are the system input and output of at time k respectively with $n \leq m$. If the system operates within the ith region, the local model is expressed as

$$x(k+1) = A_i x(k) + B_i u(k) , \quad y(k+1) = C_i x(k) . \tag{2}$$

The closed-loop system with the state feedback $u(k) = -K_i x(k)$, it becomes

$$x(k+1) = \bar{A}_i x(k) , \quad y(k+1) = C_i x(k) . \tag{3}$$

where $\bar{A}_i = A_i - B_i K_i$ is stable matrix and we can obtain a stable controller. This procedure is repeated for all of the local models and m state feedback controllers are designed. If there exist uncertainties, then the above design might lead to degrade the performance. In this paper, we propose a new control scheme including m models and corresponding controllers and a compensating controller. Actual systems will not be expressed as in (2). Instead, there will exist an uncertainty term $f(x)$ in (2). Assumed that $f(x)$ satisfies the matching condition, therefore in the ith region the system can be described as

$$x(k+1) = A_i x(k) + B_i (u(k) + f(x)) . \tag{4}$$

Due to this term $f(x)$, the performance of the system to be controlled will degrade. Our objective is to design a compensating controller to improve the system transient response. If the term $f(x)$ were known in advance, then a modified controller is

$$u(k) = u_0(k) + u_c(k) , \quad u_c(k) = -f(k) . \tag{5}$$

where $u_0 = -K_i x(k)$. Unfortunately, the term $f(x)$ is unknown in practice. A well-estimated function $\hat{f}(x)$ of the term $f(x)$ could be used to improve the system response. With the same control law (5) and a new compensating control law, which will be discussed subsequently, the closed loop control system can be expressed as

$$x(k+1) = \bar{A}_i x(k) + B_i (u_c(k) + f(x)) . \tag{6}$$

Owing to their approximation capability, artificial neural networks will be used to identify this nonlinear function. For this we make the following assumption: Given a positive constant ε_0, and a continuous function $f: C \to R$, where $C \subset R^n$ is a compact set, there exists a weight vector θ^* such that the output $\hat{f}(x,\theta)$ of the neural network satisfies $\max_{x\in C}\left|\hat{f}(x,\theta^*)-f(x)\right|\le\varepsilon_0$.

3 Compensating Controller Design Using RBF Neural Network

Assume the nonlinear function $f(x)$ be approximated by an RBF neural network with output $\hat{f}(x,\theta)$, where $\theta\in R^h$ is the adjustable weight, and h denotes the number of weight in the neural network approximation. Then (2) can be rewritten as

$$x(k+1)=\bar{A}_i x(k)+B_i(u_c(k)+\hat{f}(x,\theta^*)+f(x)-\hat{f}(x,\theta^*)). \tag{7}$$

where θ^* denotes the optimal weight value in the approximation for x belonging to a compact set $C(M_x)\subset R^n$. We are concerned with weights that belong to a large compact set $B(M_\theta)$, where M_θ is a design constant. In the design of adaptive law, a projection approach is applied to estimate θ^*. In this way, the optimal weight θ^* is defined as the element in $B(M_\theta)$ that minimizes the function $\left\|f(x)-\hat{f}(x,\theta)\right\|$ for $x\in C$. That is

$$\theta^*=\arg\min_{\theta\in B}\sup_{x\in C}\left\|f(x)-\hat{f}(x,\theta)\right\|. \tag{8}$$

Now (7) can be rewritten as

$$x(k+1)=\bar{A}_i x(k)+B_i(u_c(k)+\hat{f}(x,\theta^*)+\delta(x)). \tag{9}$$

where δ denotes the neural network error

$$\delta(x)=f(x)-\hat{f}(x,\theta^*). \tag{10}$$

with a bounded constant $\delta_0=\sup_{x\in C}\left\|f(x)-\hat{f}(x,\theta^*)\right\|$. The nonlinear function $\hat{f}(x,\theta^*)$ can be formulated as

$$\hat{f}(x,\theta^*)=\theta^{*T}\Phi(x). \tag{11}$$

where θ^* is the optimal weight value. The vector field $\Phi(x)\in R^h$ is Gaussian type of functions defined as

$$\Phi_i(x)=\exp\left(-\left\|x-c_i\right\|^2/2\sigma_i^2\right),\quad (i=1,2,\cdots,h). \tag{12}$$

For the sake of simplicity, c_i and σ_i are chosen to be constant. Then it can be observed that the nonlinear function Φ_i appears linearly with respect to the network weight matrix θ, which is now only adjustable parameter in this configuration. As

will be seen later, this configuration will greatly simplify our analysis for the system with use of neural networks.
Now (9) can be rewritten as

$$x(k+1)=\bar{A}_i x(k)+B_i(u_c(k)+(\theta^*)^T\Phi(x)+\delta(x)). \tag{13}$$

A compensating controller can be improved the transient response. The new compensating control law is

$$u_c(k)=-\hat{\theta}(k)^T\Phi(x). \tag{14}$$

where $\hat{\theta}$ is the estimation of parameter θ^*. So (13) can be rewritten as

$$x(k+1)=\bar{A}_i x(k)+B_i(-\tilde{\theta}(k)^T\Phi(x)+\delta(x)). \tag{15}$$

where $\tilde{\theta}(k)$ denotes the parameter estimation error.

We denote by $\|\cdot\|$ a suitable vector norm. Given a matrix $A=[a_{ij}], A\in R^{n\times m}$, the Frobenius norm is defined by $\|A\|_F^2=tr(A^TA)$, with $tr(\cdot)$ the trace operation. $\lambda_{\max}(P)$ and $\lambda_{\min}(Q)$ denote the maximum eigenvalue of matrix P and the minimum eigenvalue of matrix Q, respectively.

Consider the following Lyapunov function candidate

$$V(k)=x(k)^TPx(k)+\alpha^{-1}tr(\tilde{\theta}(k)^T\tilde{\theta}(k)). \tag{16}$$

where α is a positive constant and P is positive definite matrix satisfying the following condition

$$P-2\bar{A}_i^TP\bar{A}_i-2\alpha\|\Phi(x)\|^2\bar{A}_i^TPB_iB_i^TP\bar{A}_i>0. \quad \text{for all } i. \tag{17}$$

Taking the difference of V along the trajectories of (15), we have

$$\begin{aligned}\Delta V&=x(k)^T(\bar{A}_i^TP\bar{A}_i-P)x(k)-2x(k)^T\bar{A}_i^TPB_i\tilde{\theta}(k)^T\Phi(x)+2x(k)^T\bar{A}_i^TPB_i\delta(x)\\&-2\Phi(x)^T\tilde{\theta}(k)B_i^TPB_i\delta(x)+\Phi(x)^T\tilde{\theta}(k)B_i^TPB_i\tilde{\theta}(k)^T\Phi(x)+\delta(x)^TB_i^TPB_i\delta(x)\\&+\alpha^{-1}tr(2(\tilde{\theta}(k+1)-\tilde{\theta}(k))^T\tilde{\theta}(k)+(\tilde{\theta}(k+1)-\tilde{\theta}(k))^T(\tilde{\theta}(k+1)-\tilde{\theta}(k))).\end{aligned} \tag{18}$$

If we choose the parameter update law as

$$\tilde{\theta}(k+1)=\tilde{\theta}(k)+\alpha\Phi(x)(x(k)^T\bar{A}_i^TPB_i+\delta(x)^TB_i^TPB_i). \tag{19}$$

Then we have

$$\begin{aligned}\Delta V&=x(k)^T(\bar{A}_i^TP\bar{A}_i-P)x(k)+2x(k)^T\bar{A}_i^TPB_i\delta(x)+\delta(x)^TB_i^TPB_i\delta(x)\\&+\Phi(x)^T\tilde{\theta}(k)B_i^TPB_i\tilde{\theta}(k)^T\Phi(x)+2\alpha\|\Phi(x)\|^2\delta(x)^TB_i^TPB_iB_i^TP\bar{A}_ix(k)\\&+\alpha\|\Phi(x)\|^2x(k)^T\bar{A}_i^TPB_iB_i^TP\bar{A}_ix(k)+\alpha\|\Phi(x)\|^2\delta(x)^TB_i^TPB_iB_i^TPB_i\delta(x).\end{aligned} \tag{20}$$

Using the following matrix inequality for matrix $R>0$ and X,Y with appropriate dimensions:

$$X^T Y + Y^T X \le X^T RX + Y^T R^{-1} Y . \tag{21}$$

From (20), Then we can obtain

$$\begin{aligned}\Delta V \le & -x(k)^T Qx(k) + \Phi(x)^T \tilde{\theta}(k) B_i^T PB_i \tilde{\theta}(k)^T \Phi(x) \\ & + 2\delta(x)^T (B_i^T PB_i + \alpha \|\Phi(x)\|^2 B_i^T PB_i B_i^T PB_i)\delta(x) .\end{aligned} \tag{22}$$

where $Q = P - 2\bar{A}_i^T P\bar{A}_i - 2\alpha \|\Phi(x)\|^2 \bar{A}_i^T PB_i B_i^T P\bar{A}_i > 0$.

If the following inequality

$$\begin{aligned}\lambda_{\min}(Q)\|x(k)\|^2 > & \lambda_{\max}(\tilde{\theta}(k) B_i^T PB_i \tilde{\theta}(k)^T)\|\Phi(x)\|^2 \\ & + 2\lambda_{\max}(B_i^T PB_i + \alpha \|\Phi(x)\|^2 B_i^T PB_i B_i^T PB_i)\delta_0^2 .\end{aligned} \tag{23}$$

is satisfied, then we have $\Delta V < 0$.

4 Switching Control Strategy

Each model performance index is evaluated as

$$J_i(t) = \alpha_1 \|e_i(t)\|^2 + \alpha_2 \sum_{l=1}^{t-1} \lambda^{(t-l)} \|e_i(l)\|^2, \quad i \in \{1,2,\cdots,N\} . \tag{24}$$

where $e_i(t)$ is the identification error. $\alpha_1 \ge 0$ and $\alpha_2 > 0$ are design parameters. α_1 and α_2 define the weights given to the instant and long term errors respectively. $\lambda (0<\lambda<1)$ is a constant.

The performance indices (24) are calculated at every k instant, and the system switches to the controller corresponding to the minimum value of the performance indices. If the system exhibits rapid switching between controllers and indeed it is possible that degraded performance could occur. There can be very fast switching, known as chattering, between two or more controllers having the same value of their respective performance function. The hybrid controller is still stabilizing, but it may not lead to the desired behavior in a practical implementation. One way to avoid chattering is to add a constant $\rho>0$ to the performance functions that are switched out and subtract ρ from the performance functions that are switched in. this works as a hysteresis function. More information can be found in [9].

5 Simulation Study

In this section simulations are given to show the validity of the proposed method. A discrete-time nonlinear system can be represented as

$$y(k+1) = (0.52 y(k) + 0.2 y^2(k-1)) / (1 + y^2(k-1)) + u(k) .$$

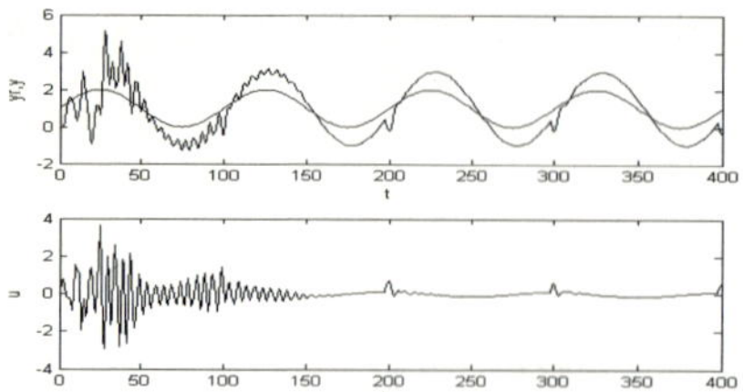

Fig. 1. The output response and the input of the plant using conventional method

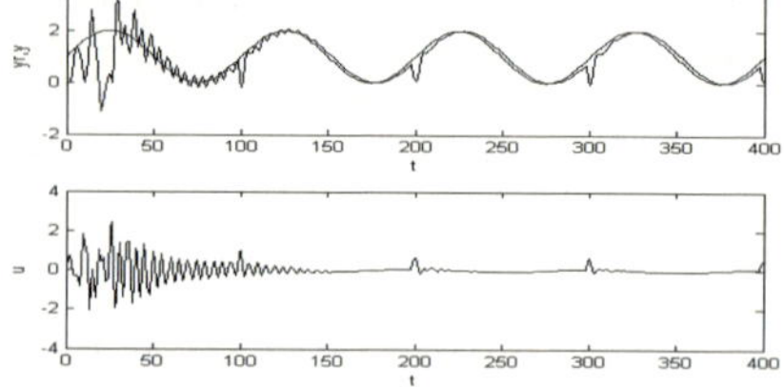

Fig. 2. The output response and the input of the plant using proposed method

Assuming there are four different regions that describe the nonlinear system. Our objective is to control the nonlinear plant to follow the reference signal $y_r = \sin(k)+1$. The simulation is carried out from $k=0$ to $k=400$. Fig.1 shows the output response and the input of the plant using conventionally adaptive control method, while that of the plant using multiple models and RBF neural networks in Fig.2. Apparently, the conventional adaptive control algorithm cannot deal with different modes well, while the method proposed in this paper improves the transient response performance.

6 Conclusion

In this paper a new control scheme is developed to improve the transient response of the controlled system. The proposed scheme consists of multiple feedback linearization controllers, which are based on the known local dynamics model, and a compensating controller, which is based on the RBF neural network. The neural network is trained on-line based on Lyapunov theory and learning convergence is thus guaranteed. Simulations are given to demonstrate the validity of the proposed method.

Acknowledgements

This work is supported by National Natural Science Foundation of China (60574006, 60404006), the Specialized Research Fund for Doctoral Program of Higher Education of China (20030286013), Natural Science Foundation of Jiangsu Province (BK2003405) and Graduate Innovative Project of Jiangsu Province (2005).

References

1. Chen, S., Billings, S.A.: Representation of Non-linear Systems: The NARMAX Model. Int. Journal of Control 49(3) (1989) 1013-1032
2. Narendra, K.S., Cheng, X.: Adaptive Control of Discrete-time Systems Using Multiple Models. IEEE Trans. Automatic Control 45(9) (2000) 1669-1686
3. Narendra, K.S., Balakrishnan, J., Ciliz, M.K.: Adaptation and Learning Using Multiple Models, Switching, and Tuning. IEEE Control Systems Magazine 15(3) (1995) 37-51

4. Narendra, K.S., Balakrishnan, J.: Adaptive Control Using Multiple Models. IEEE Trans. Automatic Control 42(2) (1997) 171-187
5. Gang, F.: Position Control of A PM Stepper Motor Using Neural Networks. Proceedings of the 39th Conference on Decision and Control, Sydney, Australia (2000) 1766-1769
6. Shamma, J.S., Athans, M.: Analysis of Gain Scheduled Control For Nonlinear Plants. IEEE Trans. Automatic Control 35(8) (1990) 898-907
7. Matausek, M.R., Jeftenic, B.I., Miljkovic, D.M., et al.: Gain Scheduling Control of DC Motor Drive With Field Weakening. IEEE Trans. Industrial Electronics 43(1) (1996) 153-162
8. Zhai, J.Y., Fei, S.M.: Multiple Models Adaptive Control Based on RBF Neural Network Dynamic Compensation. Lecture Notes in Computer Science, Vol. 3498. The second International Symposium on Neural Networks, Chongqing, China (2005) 36-41
9. Morse, A.S., Mayne, D.Q., Goodwin, G.C.: Applications of Hysteresis Switching in Parameter Adaptive Control. IEEE Trans. Automatic Control 37(9) (1992) 1343-1354

Robust Adaptive Neural Network Control for Strict-Feedback Nonlinear Systems Via Small-Gain Approaches

Yansheng Yang, Tieshan Li[1], and Xiaofeng Wang[2]

[1]Navigation College, Dalian Maritime University(DMU),
Dalian, 116026, P.R. China
tieshanli@126.com

[2] School of Finance, Dongbei University of Finance and Economics,
Dalian, 116024, P.R. China
Xfwang9796@163.com

Abstract. A novel robust adaptive neural network control (RANNC) is proposed for a class of strict-feedback nonlinear systems with both unknown system nonlinearities and unknown virtual control gain nonlinearities. The synthesis of RANNC is developed by use of the input-to-state stability (ISS), the backstepping technique, and generalized small gain approach. The key feature of RANNC is that the order of its dynamic compensator is only identical to the order n of controlled system, such that it can reduce the computation load and makes particularly suitable for parallel processing. In addition, the possible controller singularity problem can be removed elegantly. Finally, simulation results are presented to validate the effectiveness of the RANNC algorithm.

1 Introduction

In the past two decades, the adaptive tracking control of a class of nonlinear systems with parametric uncertainty has attracted a great deal of attention in the field of control and many approaches have been introduced (see [1, 2, 3, 4, 5] and references therein). However, the common feature of the adaptive control algorithms discussed in above literatures is to deal with the case of uncertainties in the linearly parameterized forms.Unfortunately, in industrial control environment, some systems are characterized by a wide of uncertainties referred to as unstructured ones, which cannot be modelled or repeatable. In order to cope with such kind of uncertainties, as an alternation, approximator-based control approaches have been studied for those systems in a Brunovsky form using Lyapunov stability theory, e.g. [6]- [9]. Recently, the developed approximator-based adaptive control approaches were extended to strict-feedback nonlinear systems with highly uncertain, nonlinear and unstructured systems using the idea of adaptive backstepping, including adaptive neural network control schemes [10]-[11] and adaptive fuzzy neural control schemes [12]. However, the number of hidden units becomes prohibitively large as we move to high dimensional systems, which imposes that there are many parameters need to be tuned in the

J. Wang et al. (Eds.): ISNN 2006, LNCS 3972, pp. 888–897, 2006.

approximator-based adaptive control schemes, such that the learning times tend to become unacceptably large and time-consuming process is unavoidable when the controllers are implemented, which restricts its applicability. This problem has been first pointed out and researched in [13] by using adaptive fuzzy control schemes.

In this paper, we will present a novel robust adaptive NN tracking approach to the aforementioned problem. We first give some preliminaries in the following.

2 Preliminaries

2.1 ISS and Small Gain Theorem

Definition 1. *[14] For system $\dot{x} = f(x, u)$, it is said to be input-to state practically stable (ISpS) if there exist a function γ of class K, called the nonlinear L_∞ gain, and a function β of class KL such that, for any initial condition $x(0)$, each measurable essentially bounded control $u(t)$ defined for all $t \geq 0$ and a nonnegative constant d, the associated solutions $x(t)$ defined on $[0, \infty)$ satisfy:*

$$\| x(t) \| \leq \beta(\| x(0) \|, t) + \gamma(\| u_t \|_\infty) + d.$$

When $d = 0$, the ISpS property reduces to the input-to-state stability (ISS) property introduced in [15].

Theorem 1. *Consider a system in composite feedback form of two ISpS systems*

$$\Sigma_{\tilde{z}\omega} : \begin{cases} \dot{x} = f(x, \omega) \\ \tilde{z} = H(x) \end{cases} \tag{1}$$

$$\Sigma_{\omega\tilde{z}} : \begin{cases} \dot{y} = g(y, \tilde{z}) \\ \omega = K(y, \tilde{z}) \end{cases} \tag{2}$$

If there exist two constants $d_1 > 0, d_2 > 0$, and let β_ω and β_ξ of class KL, and γ_z and γ_ω of class K be such that, for each ω in the L_∞ supremum norm, each $\tilde{z}$ in the L_∞ supremum norm, each $x \in R^n$ and each $y \in R^m$, all the solutions $X(x; \omega, t)$ and $Y(y; \tilde{z}, t)$ are defined on $[0, \infty)$ and satisfy, for almost all $t \geq 0$:

$$\| H(X(x; \omega, t)) \| \leq \beta_\omega(\| x \|, t) + \gamma_z(\| \omega_t \|_\infty) + d_1 \tag{3}$$

$$\| K(Y(y; \tilde{z}, t)) \| \leq \beta_\xi(\| y \|, t) + \gamma_\omega(\| \tilde{z}_t \|_\infty) + d_2 \tag{4}$$

Under these conditions, if

$$\gamma_z(\gamma_\omega(s)) < s \;\; (resp. \;\; \gamma_\omega(\gamma_z(s)) < s) \;\; \forall s > 0, \tag{5}$$

then the solution of the composite systems (1) and (2) is ISpS. [16].

2.2 RBF Neural Network

In this paper, we use the following RBF neural networks to approximate a smooth function $h(z) : R^q \rightarrow R$: $h_{nn}(z) = w^T S(z)$, where the input vector $z \in \Omega \subset R^n$, weight vector $w = [w_1, w_2, \ldots, w_l]^T \in R^l$, the neural network node number $l > 1$, and $S(z) = [s_1(z), s_2(z), \ldots, s_l(z)]^T$, with $s_i(z)$ being chosen as Gaussian functions, which have the form

$$s_i(z) = \exp\left[\frac{-(z-\mu_i)^T(z-\mu_i)}{\eta_i^2}\right], i = 1, 2, \ldots, l$$

where $\mu_i = [\mu_{i1}, \mu_{i2}, \ldots, \mu_{in}]^T$ is the center of the receptive field and η_i is the width of the Gaussian function.

For the unknown nonlinear function $f(x)$, we have the following approximation over the compact sets Ω

$$f(x) = w^{*T} S(x) + \varepsilon. \quad \forall x \in \Omega \subseteq R^n \tag{6}$$

where $S(x)$ is the basis function vector, ε is the approximation error with $|\ \varepsilon\ | \leq \varepsilon^*$, where $\varepsilon^* > 0$ is an unknown bound over a compact region $\Omega \in R^n$, and w^* is an unknown ideal constant weight vector, which is an "artificial" quantity required only for analytical purposes. Typically, w^* is chosen as the value of w that minimizes $|\ \varepsilon\ |$ for all $x \in \Omega$, where $\Omega \subseteq R^n$ is a compact set, i.e.,

$$w^* := \arg \min_{w \in R^n} \left\{ \sup_{x \in \Omega} |\ f(x) - w^T S(x)\ | \right\}.$$

Now, we introduce a key lemma which enables one to deal with nonlinear parameterization.

Lemma 1. *For any given real continuous function $f(x,\theta)$ with $f(0,\theta) = 0$, when the continuous function separation technique [17] and RBF NN approximation technique are used, then $f(x,\theta)$ can be denoted as follows*

$$f(x,\theta) = \bar{S}(x) A x \tag{7}$$

where $\bar{S}(x) = [1, S(x)] = [1, s_1(x), s_2(x), \cdots, s_l(x)]$, $s_i(x), i = 1, 2, \cdots, l$ are the RBF basis functions which are known$A^T = [\varepsilon, W^T]$, $\varepsilon^T = [\varepsilon_1, \varepsilon_2, \cdots, \varepsilon_n]$ is a vector of the approximation error and $W = \begin{bmatrix} w_{11}^* & w_{12}^* & \cdots & w_{1n}^* \\ w_{21}^* & w_{22}^* & \cdots & w_{2n}^* \\ \vdots & \vdots & \cdots & \vdots \\ w_{l1}^* & w_{l2}^* & \cdots & w_{ln}^* \end{bmatrix}$ is a weight matrix. Proof: (Omitted.)

3 Problem Formulation

Consider an uncertain nonlinear dynamic system in the following form

$$\begin{cases} \dot{x}_i = g_i(\bar{x}_i,\theta)x_{i+1} + f_i(\bar{x}_i,\theta) + \Delta_i(x,t), \quad 1 \leq i \leq n-1 \\ \dot{x}_n = f_n(x,\theta) + g_n(x,\theta)\, u + \Delta_n(x,t) \\ y = x_1 \end{cases} \tag{8}$$

where $x = [x_1, x_2, \cdots, x_n]^T \in R^n$ is the system state, $u \in R$ is the control input, $y \in R$ is the output of system and $\theta \in \Theta \subset R^q$ is an q-dimension of parameter uncertain vector, where Θ is a compact set. $\Delta_i(x,t)$'s are the disturbance uncertainties of the system. Let $\bar{x}_i = [x_1, x_2, \cdots, x_i]^T$. $f_i(\bar{x}_i, \theta)$'s are unknown smooth system functions with $f_i(0,\theta) = 0$ and $g_i(\bar{x}_i, \theta)$'s are unknown smooth functions which are referred to as the virtual control gain ones, all of which are continuous functions depending on the state x. The following assumptions are introduced.

Assumption 1. $g_i(\bar{x}_i, \theta)$, $i = 1, 2, \ldots, n$ are confined within a certain range such that $0 < b_{\min} \leq \mid g_i(\bar{x}_i, \theta) \mid \leq b_{\max}$, where $b_{\min}$ and $b_{\max}$ are the lower and upper bound parameters respectively.

The above assumption implies that $g_i(\bar{x}_i, \theta), i = 1, 2, \ldots, n$ are strictly either positive or negative. From now on, without loss of generality, we assume $g_i(\bar{x}_i, \theta) \geq b_{\min} > 0, i = 1, 2, \ldots, n, \forall (x, \theta) \in R^n \times \Theta$. Assumption 1 is reasonable because $g_i(\bar{x}_i, \theta)$ being away from zero is the controllable conditions of (8).

Assumption 2. For $1 \leq i \leq n$, there exist unknown positive constants p_i^* such that $\forall (x,t) \in R^n \times R_+$, $\mid \Delta_i(x,t) \mid \leq p_i^* \phi_i(\bar{x}_i)$, where $\phi_i(\cdot)$'s are known nonnegative smooth functions.

The primary goal of this paper is to track a given reference signal $y_d(t)$ while keeping the states and control bounded. $y_d(t)$ is assumed to be available together with its n time derivatives, and that $y_d^{(n)}(t)$ is piecewise continuous.

4 Design of Robust Adaptive NN Control

4.1 Control Design Procedure

We give the proceeding of the backstepping design as follows.

Step 1. Define the error variable $z_1 = x_1 - y_d$, then

$$\dot{z}_1 = g_1(\bar{x}_1, \theta) x_2 + f_1(x_1, \theta) + \Delta_1(x,t) - \dot{y}_d \tag{9}$$

Since $f_1(x_1, \theta)$ is an unknown continuous function with $f_1(0, \theta) = 0$, according to Lemma 1, $f_1(x_1, w)$ can be expressed as

$$f_1(x_1, \theta) = \bar{S}_1(x_1) A_1 x_1 = \bar{S}_1(x_1) A_1 z_1 + \bar{S}_1(x_1) A_1 y_d \tag{10}$$

Letting $c_{\theta 1} = \| A_1 \|$, we obtain $A_1^m = c_{\theta 1}^{-1} A_1$ and $\| A_1^m \| \leq 1$. Then, setting $\omega_1 = A_1^m z_1$. Defining a error variable $z_2 = x_2 - \alpha_1$ where α_1 is an intermediate stabilizing function and substituting (10) into (9), we get

$$\dot{z}_1 = g_1(\bar{x}_1, \theta)(z_2 + \alpha_1) + c_{\theta 1} \bar{S}_1(x_1) \omega_1 + \nu_1 \tag{11}$$

where $c_{\theta 1}$ is an unknown constant and $\nu_1 = \xi_1(x_1) A_1 y_d + \Delta_1(x,t) - \dot{y}_d$ is a bounded function.

Consider the stabilization of the subsystem (11) and the Lyapunov function candidate is given as follows

$$V_1(z_1, \hat{\lambda}_1) = \frac{1}{2}z_1^2 + \frac{1}{2}b_{\min}\Gamma_1^{-1}\widetilde{\lambda}_1^2 \tag{12}$$

where Γ_1 is a positive constant. The time derivative of V_1 is

$$\dot{V}_1(z_1, \hat{\lambda}_1) = z_1(g_1(\bar{x}_1, \theta)(z_2 + \alpha_1) + c_{\theta 1}\bar{S}_1(x_1)\omega_1 + \nu_1) - b_{\min}\Gamma_1^{-1}\widetilde{\lambda}_1\dot{\hat{\lambda}}_1 \tag{13}$$

We calculate some items in (13) first. Let $\gamma_1 > 0$, we can get

$$\begin{aligned} c_{\theta 1}\bar{S}_1(x_1)\omega_1 z_1 =& c_{\theta 1}\bar{S}_1(x_1)\omega_1 z_1 - \gamma_1^2\omega_1^T\omega_1 + \gamma_1^2\omega_1^T\omega_1 \\ =& -\gamma_1^2\left(\omega_1 - \frac{c_{\theta 1}}{2\gamma_1^2}\bar{S}_1 z_1\right)^2 + \frac{c_{\theta 1}^2}{4\gamma_1^2}\bar{S}_1\bar{S}_1^T z_1^2 + \gamma_1^2\omega_1^T\omega_1 \\ \leq& \frac{c_{\theta 1}^2}{4\gamma_1^2}\bar{S}_1\bar{S}_1^T z_1^2 + \gamma_1^2\omega_1^T\omega_1 \end{aligned} \tag{14}$$

note that, given any positive constant $\rho > 0$, we have

$$\nu_1 z_1 \leq \eta_1\psi_1(x_1) \parallel z_1 \parallel \leq \frac{\eta_1^2}{4\rho^2}\psi_1^2(x_1)z_1^2 + \rho^2 \tag{15}$$

where $\eta_1 = \max(\parallel A_1 y_d \parallel, p_1^*, \parallel \dot{y}_d \parallel)$ and $\psi_1(x_1) = 1 + \phi_1(\bar{x}_1) + \parallel \bar{S}_1 \parallel$. Noting (14) and (15), we can get

$$\begin{aligned} c_{\theta 1}\bar{S}_1(x_1)\omega_1 z_1 + \nu_1 z_1 \leq& \frac{c_{\theta 1}^2}{4\gamma_1^2}\bar{S}_1\bar{S}_1^T z_1^2 + \gamma_1^2\omega_1^T\omega_1 + \frac{\eta_1^2}{4\rho^2}\psi_1^2(x_1)z_1^2 + \rho^2 \\ \leq& b_{\min}\lambda_1\Phi_1(x_1)z_1^2 + \gamma_1^2\omega_1^T\omega_1 + \rho^2 \\ \leq& b_{\min}\hat{\lambda}_1\Phi_1(x_1)z_1^2 + b_{\min}\widetilde{\lambda}_1\Phi_1(x_1)z_1^2 + \gamma_1^2\omega_1^T\omega_1 + \rho^2 \end{aligned} \tag{16}$$

where $\Phi_1(x_1) = \frac{1}{4\gamma_1^2}\bar{S}_1\bar{S}_1^T + \frac{1}{4\rho^2}\psi_1^2$, $\lambda_1 = \max(b_{\min}^{-1}c_{\theta 1}^2, b_{\min}^{-1}\eta_1^2)$, $\widetilde{\lambda}_1 = (\lambda_1 - \hat{\lambda}_1)$ and $\hat{\lambda}_1$ is the estimate of λ_1.

Therefore, substituting (16) into (13), we can get

$$\begin{aligned} \dot{V}_1(z_1, \hat{\lambda}_1) \leq& g_1(\bar{x}_1, \theta)z_1 z_2 + g_1(\bar{x}_1, \theta)\alpha_1 z_1 + b_{\min}\hat{\lambda}_1\Phi_1(x_1)z_1^2 \\ &+ b_{\min}\Gamma_1^{-1}\widetilde{\lambda}_1\left(\Gamma_1\Phi_1(x_1)z_1^2 - \dot{\hat{\lambda}}_1\right) + \gamma_1^2\omega_1^T\omega_1 + \rho^2 \end{aligned} \tag{17}$$

Given a design constant $k_1 > 0$, we choose α_1 and the adaptive law for $\hat{\lambda}_1$ as

$$\alpha_1 = -\left(k_1 + \hat{\lambda}_1\Phi_1(x_1)\right)z_1 \tag{18}$$

$$\dot{\hat{\lambda}}_1 = \Gamma_1[\Phi_1(x_1)z_1^2 - \sigma_1(\hat{\lambda}_1 - \lambda_1^0)] \tag{19}$$

where λ_1^0 and σ_1 are design parameters. In light of Assumption 1, we can get

$$g_1\alpha_1 z_1 = g_1[-(k_1 + \hat{\lambda}_1\Phi_1(x_1))z_1^2] \le b_{\min}[-(k_1 + \hat{\lambda}_1\Phi_1(x_1))z_1^2] \quad (20)$$

note that,

$$g_1 z_1 z_2 \le \frac{1}{4}z_1^2 + g_1^2 z_2^2 \quad (21)$$

Using (18), (19), (20) and (21), $\dot{V}_1$ is converted into

$$\dot{V}_1(z_1, \hat{\lambda}_1) \le -\left(b_{\min}k_1 - \frac{1}{4}\right) z_1^2 - \frac{1}{2}b_{\min}\sigma_1\widetilde{\lambda}_1^2 + g_1^2 z_2^2 + \gamma_1^2\omega_1^T\omega_1 + \delta_1 \quad (22)$$

where $\delta_1 - \rho^2 + \frac{1}{2}b_{\min}\sigma_1 \mid \lambda_1 - \lambda_1^0 \mid^2$.

Step 2.

$$\dot{z}_2 = g_2(\bar{x}_2, \theta)x_3 + f_2(\bar{x}_2, \theta) + \Delta_2(x, t) - \dot{\alpha}_1 \quad (23)$$

Then the time derivative of α_1 is

$$\begin{aligned}\dot{\alpha}_1 =& \frac{\partial\alpha_1}{\partial x_1}\dot{x}_1 + \frac{\partial\alpha_1}{\partial\hat{\lambda}_1}\dot{\hat{\lambda}}_1 + \frac{\partial\alpha_1}{\partial y_d}\dot{y}_d \\ =& \frac{\partial\alpha_1}{\partial x_1}(g_1(\bar{x}_1, \theta)x_2 + f_1(\bar{x}_1, \theta) + \Delta_1) + \frac{\partial\alpha_1}{\partial\hat{\lambda}_1}\dot{\hat{\lambda}}_1 + \frac{\partial\alpha_1}{\partial y_d}\dot{y}_d \\ =& f_{12}(z_1, \bar{x}_2, \theta) + \frac{\partial\alpha_1}{\partial x_1}\Delta_1 + \frac{\partial\alpha_1}{\partial y_d}\dot{y}_d \end{aligned} \quad (24)$$

Substituting (24) into (23), we get

$$\dot{z}_2 = -g_1^2(\bar{x}_1, \theta)z_2 + g_2(\bar{x}_2, \theta)x_3 + f_2'(\bar{z}_2, y_d, w) + \Delta_2 - \frac{\partial\alpha_1}{\partial x_1}\Delta_1 - \frac{\partial\alpha_1}{\partial y_d}\dot{y}_d \quad (25)$$

where $f_2'(\bar{z}_2, y_d, \theta) = g_1^2(\bar{x}_1, \theta)z_2 + f_2(\bar{x}_2, \theta) - f_{12}(z_1, \bar{x}_2, \theta)$.

We also use Lemma 1 to treat the unknown function $f_2'(\bar{z}_2, y_d, \theta)$ and obtain

$$f_2'(\bar{z}_2, y_d, \theta) = \bar{S}_2(\bar{z}_2, y_d)A_2[\bar{z}_2, y_d]^T = \bar{S}_2 A_2^1\bar{z}_2^T + \bar{S}_2 A_2^2 y_d = c_{\theta 2}\bar{S}_2\omega_2 + \bar{S}_2 A_2^2 y_d$$

where $\omega_2 = A_2^m\bar{z}_2^T$ and $c_{\theta 2} = \| A_2^1 \| = \lambda_{\max}^{1/2}(A_2^{1T}A_2^1)$, such that $A_2^1 = c_{\theta 2}A_2^m$ and $\| A_2^m \| \le 1$. Defining the error variable $z_3 = x_3 - \alpha_2$, a direct substitution of above equation gives

$$\dot{z}_2 = -g_1^2(\bar{x}_2, \theta)z_2 + g_2(\bar{x}_2, \theta)(z_3 + \alpha_2) + c_{\theta 2}\bar{S}_2\omega_2 + \nu_2 \quad (26)$$

where $\nu_2 = \bar{S}_2 A_2^2 y_d + \Delta_2 - \frac{\partial\alpha_1}{\partial x_1}\Delta_1 - \frac{\partial\alpha_1}{\partial y_d}\dot{y}_d$.

Choosing Lyapunov function candidate

$$V_2 = V_1 + \frac{1}{2}z_2^2 + \frac{1}{2}b_{\min}\Gamma_2^{-1}\widetilde{\lambda}_2^2$$

where $\widetilde{\lambda}_2 = (\lambda_2 - \hat{\lambda}_2)$ and $\lambda_2 = \max(b_{\min}^{-1}c_{\theta 2}^2, b_{\min}^{-1}\eta_2^2)$.

A similar procedure with (14), (15) and (16) is used and the time derivative of V_2 becomes

$$\dot{V}_2 \leq -\left(b_{\min}k_1 - \frac{1}{4}\right)z_1^2 + \gamma_1^2\omega_1^T\omega_1 + \delta_1 + \gamma_2^2\omega_2^T\omega_2 + \rho^2 + z_2\left(g_2(z_3+\alpha_2) + b_{\min}\hat{\lambda}_2\Phi_2 z_2\right) + b_{\min}\Gamma_2^{-1}\widetilde{\lambda}_2(\Gamma_2\Phi_2 z_2^2 - \dot{\hat{\lambda}}_2)$$

where $\Phi_2 = \frac{1}{4\gamma_2^2}\bar{S}_2\bar{S}_2^T + \frac{1}{4\rho^2}\psi_2^2$, $\| \nu_2 \| \leq \eta_2\psi_2$ and $\psi_2 = 1 + \phi_2(\bar{x}_2) + \| \bar{S}_2 \| + \| \frac{\partial\alpha_1}{\partial x_1} \| \phi_1 + \| \frac{\partial\alpha_1}{\partial y_d} \|$.

Now, choose α_2 and adaptive law of $\hat{\lambda}_2$ as the forms of (18) and (19), using the same procedure as (20) and (21), then $\dot{V}_2$ is converted to

$$\dot{V}_2 \leq -\sum_{i=1}^{2}\left(b_{\min}k_i - \frac{1}{4}\right)z_i^2 - \frac{1}{2}b_{\min}\sum_{i=1}^{2}\sigma_i\widetilde{\lambda}_i^2 + g_2^2z_3^2 + \sum_{i=1}^{2}\gamma_i^2\omega_i^T\omega_i + \delta_2 \tag{27}$$

where $\delta_2 = 2\rho^2 + \frac{1}{2}b_{\min}\sum_{i=1}^{2}\sigma_i \mid \lambda_i - \lambda_i^0 \mid^2$.

A similar procedure is employed recursively for each step $k(3 \leq k \leq n-1)$. By considering the equation of system (8) for $i = k$, $\dot{x}_k = g_k(\bar{x}_k,\theta)x_{k+1} + f_k(\bar{x}_k,\theta) + \Delta_k(x,t)$, and the Lyapunov function candidate

$$V_k = V_{k-1} + \frac{1}{2}z_k^2 + \frac{1}{2}b_{\min}\Gamma_k^{-1}\widetilde{\lambda}_k^2$$

where $\widetilde{\lambda}_k = (\lambda_k - \hat{\lambda}_k)$.

We may design α_k and the learning law for $\hat{\lambda}_k$ with the similar forms of (18) and (19), respectively.

Step n: At this step, $u = \alpha_n$. Define $z_n = x_n - \alpha_{n-1}$, we have

$$\dot{z}_n = f_n(x,\theta) + g_n(x,\theta)u + \Delta_n - \dot{\alpha}_{n-1}$$

Taking the following Lyapunov function candidate

$$V_n = V_{n-1} + \frac{1}{2}z_n^2 + \frac{1}{2}b_{\min}\Gamma_n^{-1}\widetilde{\lambda}_n^2$$

where $\widetilde{\lambda}_n = (\lambda_n - \hat{\lambda}_n)$ and $\lambda_n = b_{\min}\max(b_{\min}^{-1}c_{\theta n}^2, b_{\min}^{-1}\eta_n^2)$.

And choosing the controller with adaptive law in step n as

$$u = \alpha_n = -\left(k_n + \hat{\lambda}_n\Phi_n\right)z_n \tag{28}$$

$$\dot{\hat{\lambda}}_n = \Gamma_n\left[\Phi_n z_n^2 - \sigma_n(\hat{\lambda}_n - \lambda_n^0)\right] \tag{29}$$

It follows from the recursive control design procedure similar to Step 2, so that

$$\begin{aligned}\dot{V}_n &\leq -\sum_{i=1}^{n-1}\left(b_{\min}k_i - \frac{1}{4}\right)z_i^2 - \frac{1}{2}b_{\min}\sum_{i=1}^{n}\sigma_i\widetilde{\lambda}_i^2 + \sum_{i=1}^{n}\gamma_i^2\omega_i^T\omega_i + \delta_n \\ &\leq -\sum_{i=1}^{n}\left(-b_{\min}k_i - \frac{1}{4}\right)z_i^2 - \frac{1}{2}b_{\min}\sum_{i=1}^{n}\sigma_i\widetilde{\lambda}_i^2 + \gamma^2 \| \omega \|^2 + \delta_n \end{aligned} \tag{30}$$

where $\delta_n = n\rho^2 + \frac{1}{2} b_{\min} \sum_{i=1}^{n} \sigma_i \mid \lambda_i - \lambda_i^0 \mid^2$, $\omega = [\omega_1, \omega_2, \ldots, \omega_n]^T$ and $\gamma = (\gamma_1^2 + \gamma_2^2 + \ldots + \gamma_n^2)^{1/2}$.

We are now in a position to state our main result in this paper.

Theorem 2. *Consider the system (8) and suppose that the packaged uncertain functions* $f_i'(\bar{z}_i, \bar{x}_{d(i)}, \theta), i = 1, 2, \cdots, n$ *can be dealt with by Lemma 1. If we pick* $\gamma < 1$, $k_i > \frac{5}{4} b_{\min}^{-1}, i = 1, 2, \ldots, n$ *in (30), then the robust adaptive NN tracking control* $u = \alpha_n$, *the intermediate stabilizing functions* α_i *and adaptive laws for* $\hat{\lambda}_i$ *can make all the solutions* $(z(t), \hat{\lambda})$ *of the derived closed loop system uniformly ultimately bounded. Furthermore, given any* $\mu_1 > 0$, *we can tune our design constants such that the output error* $z_1 = y(t) - y_d(t)$ *satisfies* $\lim_{t\to\infty} \mid z_1(t) \mid \leq \mu_1$. *Proof. (Omitted.)*

5 Simulation Example

In this section, we will discuss the following second-order plants as

$$\begin{aligned} \dot{x}_1 &= \ln(10 + x_1^2) x_2 + 0.5 x_1^3 + \Delta_1(x, t) \\ \dot{x}_2 &= (1 + e^{(-x_1^2 - x_2^2)}) u + \frac{x_1 x_2}{1 + x_1^2 + x_2^2} + \Delta_2(x, t) \end{aligned} \tag{31}$$

with the output $y = x_1$, $\Delta_1(x, t) = 0.6 \sin(x_2)$ and $\Delta_2(x, t) = 0.5(x_1^2 + x_2^2) \sin^3 t$. To select $p_1^* := 0.6$, $p_2^* := 0.5$, $\phi_1(x_1) = 1$ and $\phi_2(x_1, x_2) = x_1^2 + x_2^2$, and the desired signal is $y_d = \sin(0.5t) + 0.5 \sin(1.5t)$, we can use Theorem 2 to design

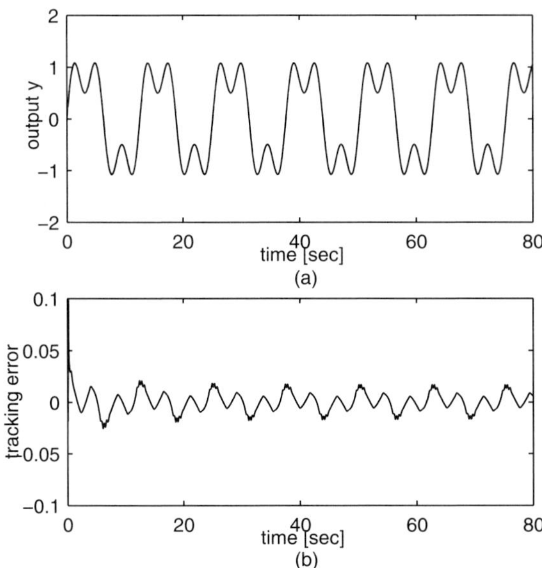

Fig. 1. Simulation results :(a) System output y and reference signal y_d (solid line: y and dashed line: y_d). (b) Tracking error z_1.

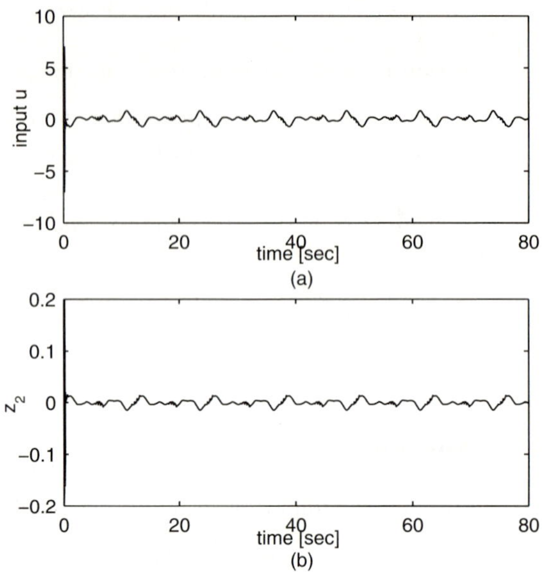

Fig. 2. Simulation results: (a) Control input u. (b) Intermediate error variable z_2.

the robust adaptive NN tracking controller. The simulation results are shown in Figs. 1 and 2.

6 Conclusion

We have considered the tracking control problem for a class of strict-feedback uncertain nonlinear systems. The systems may possess a wide class of uncertainties referred to as unstructured uncertainties, which are not linearly parameterized and have no prior knowledge of the bounding functions. We have incorporated the continuous function separation technique with RBF NN to model the unstructured uncertain functions in the systems and proposed a robust adaptive NN tracking control algorithm by combining backstepping technique with small-gain approach. The proposed algorithm can guarantee that the closed-loop system is semi-globally uniformly ultimately bounded. The main feature of the algorithm proposed is that the order of dynamic compensator of RANNC is only identical to the order n of controlled system, no matter how many hidden units in the neural networks are used. Then the computation load of the algorithm can be reduced, and it is a convenience to realize this algorithm for engineering. Finally, numerical simulation example is presented to illustrate the tracking performance of the closed-loop systems by use of the proposed algorithm.

Acknowledgements

This work was supported in part by the National Natural Science Foundation of China under Grant No. 60474014 and the Research Fund for the Doctoral Program of Higher Education of China under Grant No. 20020151005.

References

1. Krstic, M., Kanellakopoulos, I., Kokotovic, P. V.: Nonlinear and Adaptive Control Design. Wiley New York (1995)
2. Sastry, S. S., Isidori, A.: Adaptive Control of Linearization Systems. IEEE Trans. on Automatic Control 34 (1989) 1123-1131
3. Marino, R., Tomei, P.: Global Adaptive Output-feedback Control of Nonlinear Systems, Part I: Linear Parameterization; Part II: Nonlinear Parameterization. IEEE Trans. on Automatic Control 38 (1993) 33-49
4. Krstic, M., Kanellakopoulos, I., Kokotovic, P. V.: Adaptive Nonlinear Control without Overparametrization. Systems and Control Letters 19 (1992) 177-185
5. Kanellakopoulos, I.: Passive Adaptive Control of Nonlinear Systems. Int. J. of Adaptive Control and Signal Processing 7 (1993) 339-352
6. Yesidirek, A., Lewis, F. L.: Feedback Linearization Using Neural Networks. Automatica 31 (1995) 1659-1664
7. Wang, L. X.: Stable Adaptive Fuzzy Control of Nonlinear Systems. IEEE Trans. on Fuzzy Systems 1 (1993) 146-155
8. Yang, Y. S., Zhou, C. J., Jia, X. L.: Robust Adaptive Fuzzy Control and Its Application to Ship Roll Stabilization. Information Sciences 142(4) (2002) 177-194
9. Yang, Y. S., Ren, J. S: Adaptive Fuzzy Robust Tracking Controller Design via Small Gain Approach and Its Application. IEEE Trans. on Fuzzy Systems 11(6) (2003) 783-795
10. Yao, B., Tomizuka, M.: Adaptive Robust Control of SISO Nonlinear Systems in a Semi-strict Feedfack Form. Automatica 33 (1997) 893-900
11. Ge, S. S., Wang, J.: Robust Adaptive Neural Control for a Class of Perturbed Strict Feedback Nonlinear Systems. IEEE Trans. on Neural Networks 13 (2002) 1409-1419
12. Wang, W. Y., Chan, M. L., Lee, T. T., Liu, C. H.: Adaptive Fuzzy Control for Strict-feedback Cononical Nonlinear Systems with H_∞ Tracking Performance. IEEE Trans. Systems, Man, and Cybernetics-Part B: Cybernetics 30 (2000) 878-885
13. Yang, Y. S., Feng, G., Ren, J. S.: A Combined Backstepping and Small-gain Approach to Robust Adaptive Fuzzy Control for Strict-feedback Nonlinear Systems. IEEE Trans. on Systems, Man, and Cybernetics–Part A: Systems and Humans 34 (2004) 406-420
14. Sontag, E. D.: Smooth Stabilization Implies Coprime Factorization. IEEE Trans. on Automatic Control 34 (1989) 1411-1428
15. Sontag, E. D.: Further Results about Input-state Stabilization. IEEE Trans. on Automatic Control 35 (1990) 473-476
16. Jiang, Z. P., Mareels, I.: A Small Gain Control Method for Nonlinear Cascaded Systemns with Dynamic Uncertainties. IEEE Trans. on Automatic Control AC-42 (1997) 292-308
17. Lin, W., Qian, C.: Adaptive Control of Nonlinearly Parameterized Systems: The Smooth Feedbackcase. IEEE Trans. on Automatic Control 47(8) (2002) 1249-1266

Neural Network Based Robust Adaptive Control for a Class of Nonlinear Systems

Dan Wang[1] and Jin Wang[2]

[1] School of Automation and Electrical Engineering, Dalian Maritime University, Dalian 116026, P.R. China
dwangdl@gmail.com
[2] Dept. of Chemical Engineering, West Virginia University Institute of Technology, Montgomery, WV 25136, USA
jin.wang@mail.wvu.edu

Abstract. A neural network based robust adaptive control design scheme is developed for a class of nonlinear systems represented by input-output models with an unknown nonlinear function and unmodeled dynamics. By on-line approximating the unknown nonlinear functions and unmodeled dynamics by radial basis function (RBF) networks, the proposed approach does not require the unknown parameters to satisfy the linear dependence condition. It is proved that with the proposed control law, the closed-loop system is stable and the tracking error converges to zero in the presence of unmodeled dynamics and unknown nonlinearity.

1 Introduction

In the past two decades, a great deal of progress in the feedback control of nonlinear systems has been achieved. At the same time, Neural Network (NN) based adaptive control techniques are extensively studied. Motivated by the developments in the two areas, in this paper, a Neural Network based robust adaptive control approach for a class of nonlinear systems with uncertain nonlinearities and unmodeled dynamics is studied. Adaptive control of nonlinear systems with parameter uncertainty and uncertain nonlinearities are studied by many researchers [4]-[8]. Adaptive control for nonlinear systems with unmodeled dynamics has drawn great research attention [9] - [17]. In [4], adaptive control for a class of nonlinear systems is studied. The system under consideration is single-input-single-output, input-output linearizable, minimum phase, and modelled by an input-output model of the form of an nth-order differential equation. The uncertain nonlinear functions of the model depend linearly on constant unknown parameters. This is a wide class of nonlinear systems which includes as a special case the nonlinear systems treated in [6] and [18] for output feedback adaptive control and the linear systems treated in the traditional adaptive control literature, e.g., [19]-[21]. The dynamics of the system is extended by adding a series of integrator at the input side and the augmented system is represented by a state-space model, where the states are the input, the output, and a number of their derivatives. In the work, a semiglobal controller is designed that achieves

J. Wang et al. (Eds.): ISNN 2006, LNCS 3972, pp. 898–903, 2006.

asymptotic output tracking for reference signals which are bounded and have bounded derivatives up to the nth order. However, the adaptive controller is not robust to unmodeled dynamics. Improvements are reported in a later paper [5]. In which the results of [4] are extended to the case with bounded disturbance, but an upper bound on the disturbance must be known. In a recent paper [16], a robust adaptive controller for a class of nonlinear systems represented by input-output models containing unmodelled dynamics is presented. In the design of the adaptive controller, it is not necessary to know the upper bound of the disturbances. The systems considered in [4], [5], and [16] have a restriction of linear dependence on the unknown parameters. The linear dependence condition is removed in [17]. Instead, a smooth nonlinear function must be known such that an unknown nonlinear function in the system is bounded by the product of the known nonlinear function and an unknown constant.

In this work, inspired by above mentioned works, we further study the adaptive control problem for the same class of nonlinear systems. By on-line approximating the unmodeled dynamics by RBF networks, we will combine the feedback linearization technique and the neural network based adaptive control technique to develop a robust adaptive controller design method. The above mentioned restriction in [17] will be removed. In addition, it is proved that the closed-loop system is stable and the tracking error converges to zero.

2 Problem Formulation and Preliminaries

We consider a single-input single-output nonlinear system described by

$$\begin{aligned} y^{(n)} &= f(y, \dot{y}, \ldots, y^{(n-1)}, u, \dot{u}, \ldots, u^{(m-1)}) + \frac{1}{\gamma} u^{(m)} \\ &+ \Delta(y, \dot{y}, \ldots, y^{(n-1)}, u, \dot{u}, \ldots, u^{(m-1)}), \end{aligned} \tag{2.1}$$

where y is the output; u is the control; $y^{(i)}$ is the ith derivative of y; f is an unknown smooth nonlinear function; $\Delta(.)$ represents the uncertain nonlinearity and the unmodeled dynamics; and γ is an unknown constant parameter, but the sign of γ is known. Without loss of generality, we assume that $\gamma > 0$.

Let

$$\begin{aligned} x_1 = y, \;\; x_2 = y^{(1)}, \ldots, \;\; x_n = y^{(n-1)}, \\ z_1 = u, \;\; z_2 = u^{(1)}, \ldots, \;\; z_m = u^{(m-1)}. \end{aligned} \tag{2.2}$$

System (2.1) can be represented by

$$\begin{aligned} \dot{x}_i &= x_{i+1}, \quad 1 \le i \le n-1 \\ \dot{x}_n &= f(x, z) + \frac{1}{\gamma} v + \Delta(x, z), \\ \dot{z}_i &= z^{i+1}, 1 \le i \le m-1 \\ \dot{z}_m &= v, \end{aligned} \tag{2.3}$$

where $v = u^{(m)}$ is the control input for the augmented system (2.3), and $x = [x_1, ..., x_n]^T$, $z = [z_1, ..., z_m]^T$. We assume that the reference signal $y_r(t)$ is bounded with bounded derivatives up to the nth order and $y_r^{(n)}(t)$ is piecewise continuous. Denote

$$Y_r = [y_r, y_r^{(1)}, ..., y_r^{(n-1)}]^T. \tag{2.4}$$

Our objective is to design a robust adaptive state feedback controller for (2.3) such that the closed-loop system is stable and the output $y(t)$ of the system tracks the reference signal $y_r(t)$ with high quality in the presence of unmodeled dynamics and the unknown nonlinear function .

Remark 1: In [17], it is assumed that f is unknown but a smooth nonlinear function $\bar{f}$ must be known for control design which satisfies

$$|f(y, \dot{y}, \ldots, y^{(n-1)}, u, \dot{u}, \ldots, u^{(m-1)})| \le \theta \bar{f}(y, \dot{y}, \ldots, y^{(n-1)}, u, \dot{u}, \ldots, u^{(m-1)}), \tag{2.5}$$

where $\theta > 0$ is an unknown constant. In this paper, we do not need a known $\bar{f}$, *i. e.*, no prior knowledge of f can be used for control design. In the case, there is actually no difference between f and $\Delta(x, z)$ - both are totally unknown for control law design. So we will treat them together later. We use the system description here so that it is easy to see the relationship between this work and the existing results.

In this work, we assume that the state (x, z) of system (2.3) is available for feedback. Let $e_1 = x_1 - y_r$, $e_2 = x_2 - \dot{y}_r$, ..., $e_n = x_n - y^{(n-1)}$, $e = [e_1, e_2, ..., e_n]^T$. From (2.3), we get

$$\begin{aligned} \dot{e} &= Ae + b[f(e + Y_r, z) + \frac{1}{\gamma} v - y_r^{(n)} + \Delta(e + Y_r, z)], \\ \dot{z} &= \bar{A}z + \bar{b}v, \end{aligned} \tag{2.6}$$

where

$$A = \begin{bmatrix} 0 & 1 & \ldots & 0 \\ \vdots & \vdots & & \vdots \\ 0 & 0 & \ldots & 1 \\ 0 & 0 & \ldots & 0 \end{bmatrix}, \quad b = \begin{bmatrix} 0 \\ \vdots \\ 0 \\ 1 \end{bmatrix}. \tag{2.7}$$

$\bar{A}$ and $\bar{b}$ have the same forms as A and b but with different sizes. Let $A_c = A - bK$, where K is chosen so that A_c is Hurwitz. Then, we have

$$\dot{e} = A_c e + b[Ke + f(e + Y_r, z) + \frac{1}{\gamma} v - y_r^{(n)} + \Delta(e + Y_r, z)]. \tag{2.8}$$

Before introducing our control design method, let us first recall the approximation property of the RBF neural networks [1] [3]. The RBF neural networks take the form $\theta^T \xi(x)$ where $\theta \in R^N$ for some integer N is called weight vector, and

$\xi(x) \in R^n$ is a vector valued function defined in R^n. Denote the components of $\xi(x)$ by $\rho_i(x)$, $i = 1, \cdots, N$, then $\rho_i(x)$ is called a basis function. A commonly used basis function is the so-called Gaussian Function of the following form

$$\rho_j(x) = \frac{1}{\sqrt{2\pi}\sigma} exp(-\frac{\|x - \zeta_j\|^2}{2\sigma^2}), \quad \sigma \geq 0, \quad j = 1, \cdots, N \tag{2.9}$$

where $\zeta_j \in R^n, \quad j = 1, \cdots, N$, are constant vectors called the center of the basis function, and σ is a real number called the width of the basis function. According to the approximation property of the RBF networks [1] [2] [3], given a continuous real valued function $f : \Omega \to R$ with $\Omega \in R^n$ a compact set, and any $\delta_m > 0$, by appropriately choosing $\sigma, \quad \zeta_j \in R^n, \quad j = 1, \cdots, N$, for some sufficiently large integer N, there exists an ideal weight vector $\theta^* \in R^N$ such that the RBF network ${\theta^*}^T \xi(x)$ can approximate the given function f with the approximation error bounded by δ_m, *i.e.*,

$$f(x) = {\theta^*}^T \xi(x) + \delta^* \quad , \quad x \in \Omega \tag{2.10}$$

with $|\delta^*| \leq \delta_m$, where δ^* represents the network reconstruction error, i.e.,

$$\delta^* = f(x) - {\theta^*}^T \xi(x). \tag{2.11}$$

Since θ^* is unknown, we need to estimate θ^* online. We will use the notation $\hat{\theta}$ to denote the estimation of θ^* and develop an adaptive law to update the parameter $\hat{\theta}$.

3 Robust Adaptive Control Design

In this work, f in (2.1) and (2.3) is totally unknown. In this case, there is no difference between f and Δ. So, they will be treated as one unknown nonlinear function. Let

$$F = f(e + Y_r, z) + \Delta(e + Y_r, z). \tag{2.12}$$

Then, (2.8) becomes

$$\dot{e} = A_c e + b[Ke + F + \frac{1}{\gamma} v - y_r^{(n)}]. \tag{2.13}$$

Given a compact set $\Omega_{(e+Y_r,z)} \in R^{n+m}$, let θ^* and δ^* be such that for any $(e + Y_r, z) \in \Omega_{(e+Y_r,z)}$

$$F = {\theta^*}^T \xi(e + Y_r, z) + \delta^* \tag{2.14}$$

with $|\delta^*| \leq \delta_m$.

Following robust adaptive control law is proposed to solve the given tracking problem:

$$v = \hat{\gamma}[-Ke - \hat{\theta}^T \xi + y_r^{(n)} - Sign(e^T Pb)\delta_m], \tag{2.15}$$

where $\hat{\theta}$ is the estimate of θ^* and is updated as follows

$$\dot{\hat{\theta}} = \Gamma e^T Pb\xi \tag{2.16}$$

with any constant matrix $\Gamma = \Gamma^T > 0$. P is a matrix satisfying

$$PA_c + A_c^T P = -Q, \quad Q = Q^T > 0. \tag{2.17}$$

$\hat{\gamma}$ is the estimate of γ and is updated as follows

$$\dot{\hat{\gamma}} = -e^T Pb(-Ke - \hat{\theta}^T \xi + y_r^{(n)} - Sign(e^T Pb)\delta_m). \tag{2.18}$$

Theorem 1: For any $x(0)$, $z(0)$ and Y_r satisfying $(e+Y_r, z) \in \Omega_{(e+Y_r,z)}$, the proposed robust adaptive state feedback controller (2.15) and adaptive laws (2.16) and (2.18) guarantee that the closed-loop system is stable and the output $y(t)$ of the given system (2.3) converges to the reference signal $y_r(t)$ in the presence of unmodeled dynamics and unknown nonlinearity.

Proof: The proof is omitted due to the space limit. ♣

Remark 2: Due to the sign function used in the control law, there may be chattering phenomena in control input. However, since δ_m is very small in general, the Chattering cannot be significant. Chattering can be weaken by reducing the value of δ_m. In fact, it is easy to show that when the sign function term is dropped the closed-loop system is still stable with bounded tracking error.

Remark 3: In this paper, state feedback control is considered. Output feedback control is extensively studied in [4], [5], [16] and [17]. Interested readers may refer to the papers.

4 Conclusions

Combining the feedback linearization technique and the neural network based adaptive control technique, in this work, we developed a robust adaptive controller design method for a class of nonlinear systems represented by input-output models. Using RBF networks to on-line approximate the unmodeled dynamics, some restriction in existing results is removed. We proved that the closed-loop system is stable and the tracking error converges to zero.

Acknowledgement

This work is supported in part by the Scientific Research Foundation for the Returned Overseas Chinese Scholars, State Education Ministry.

References

1. Park, J., Sandberg, I. W.: Universal Approximation Using Radial-Basis-Function Networks. Neural Computation (3) (1991) 246-257
2. Polycarpou, M. M., Mears, Mark J.: Stable Adaptive Tracking of Uncertainty Systems Using Nonlinearly Parameterized On-line Approximators. Int. J. Control 70(3) (1998) 363-384
3. Sanner, R. M., Slotine, J. E.: Gaussian Networks for Direct Adaptive Control. IEEE Transactions on Neural Networks 3(6) (1992) 837-863
4. Khalil, H. K.: Adaptive Output Feedback Control of Nonlinear Systems Represented by Input-Output Models. IEEE Trans. Automat. Contr. 42(2) (1996) 177-188
5. Aloliwi, B., Khalil, H. K.: Robust Adaptive Output Feedback Control of Nonlinear Systems without Persistence of Excitation. Automatica 33 (1997) 2025-2032
6. Marino, R., Tomei, P.: Global Adaptive Output-Feedback Control of Nonlinear Systems, Part I: Linear Parameterization. Part II: Nonlinear Parameterization. IEEE Trans. Automat. Contr. 38(1) (1993) 17-49
7. Polycarpou, M. M., Ioannou, P. A.: A Robust Adaptive Nonlinear Control Design. Automatica 31 (1995) 423-427
8. Marino, R., Tomei, P.: Robust Adaptive State-Feedback Tracking for Nonlinear Systems. IEEE Trans. Automat. Contr. 43(1) (1998) 84-89
9. Taylor, D. G., Kokotovic, P. V., Marino, R., Kanellakopoulos, I.: Adaptive Regulation of Nonlinear Systems with Unmodeled Dynamics. IEEE Trans. Automat. Contr. 34(4) (1989) 405-412
10. Kanellakopoulos, I., Kokotovic, P. V., Marino, R.: An Extended Direct Scheme for Robust Adaptive Nonlinear Control. Automatica 27 (1991) 247-255
11. Jiang, Z. P., Praly, L.: Design of Robust Adaptive Controllers for Nonlinear Systems with Dynamic Uncertainties. Automatica 34 (1998) 825-840
12. Chang, Y.-C.: An Adaptive H_∞ Tracking Control for A Class of Nonlinear Multiple-Input Multiple-Output (MIMO) Systems. IEEE Trans. Automat. Contr. 46(9) (2001) 1432-1437
13. Jiang, Z. P.: A Combined Backstepping and Small-Gain Approach to Adaptive Output Feedback Control. Automatica 35 (1999) 1131-1139
14. Jiang, Z. P., Hill, D. J.: A Robust Adaptive Backstepping Scheme for Nonlinear Systems with Unmodeled Dynamics. IEEE Trans. Automat. Contr. 44(9) (1999) 1705-1711
15. Aloliwi, B., Khalil, H. K.: Robust Adaptive Control of Nonlinear Systems with Unmodeled Dynamics. Proc. IEEE Conf. Decision Control, Tampa, FL (1998) 2872-2873
16. Liu, Y., Li, X.-Y.: Robust Adaptive Control of Nonlinear Systems with Unmodeled Dynamics. Proc. Inst. Electr. Eng. Control Theory Applicat. 151(1) (2004) 83-88
17. Liu, Y., Li, X.-Y.: Robust Adaptive Control of Nonlinear Systems Represented by Input-Output Models. IEEE Transactions on Automatic Control 48(6) (2003) 1041-1045
18. Kanellakopoulos, I., Kokotovic, P. V., Morse, A. S.: Adaptive Output Feedback Control of Systems with Output Nonlinearities. IEEE Trans. Automat. Contr. 37(11) (1992) 1166-1182
19. Astrom, K. J., Wittenmark, B.: Adaptive Control. Reading, MA: Addison-Wesley (1989)
20. Narendra, K. S., Annaswamy, A. M.: Stable Adaptive Systems. Englewood Cliffs, NJ: Prentice-Hall (1989)
21. Sastry, S., Bodson, M.: Adaptive Control. Englewood Cliffs, NJ: Prentice-Hall (1989)

Robust H_∞ Control for Delayed Nonlinear Systems Based on Standard Neural Network Models

Mei-Qin Liu

College of Electrical Engineering, Zhejiang University, Hangzhou 310027, China
liumeiqin@zju.edu.cn

Abstract. A neural-network-based robust output feedback H_∞ control design is suggested for control of a class of nonlinear systems with time delays. The design approach employs a neural network, of which the activation functions satisfy the sector conditions, to approximate the delayed nonlinear system. A full-order dynamic output feedback controller is designed for the approximating neural network. The closed-loop neural control system is transformed into a novel neural network model termed standard neural network model (SNNM). Based on the robust H_∞ performance analysis of the SNNM, the parameters of output feedback controllers can be obtained by solving some lilinear matrix inequalities (LMIs). The well-designed controller ensures the asymptotic stability of the closed-loop system and guarantees an optimal H_∞ norm bound constraint on disturbance attenuation for all admissible uncertainties.

1 Introduction

Recently, the robustness issue has been an important focus of research in the neuro-control society, and several robust stability design approaches have been proposed [1]-[2]. However, they only deal with the problems of robust stability analysis and robust stabilization for neural control systems by considering modeling errors resulting from approximation of a plant with neural networks. It is well known that the H_∞ performance is closely relative to the capability of disturbance rejection. Regarding H_∞ control by neural networks, to the best of our knowledge, only a few results are published [3][4].

In biological and artificial neural networks, time delays arise in the processing of information storage and transmission. It is known that they can influence the stability of the entire network by creating oscillatory or unstable phenomena [5]. There are various research results about stability analysis of various time-delayed recurrent neural networks in many published literature [6]-[9]. However, we have noticed that, although some stability conditions obtained in some literature have explicit expressions and less conservativeness, there does not seem to be much (if any) study on the robust stabilization for delayed neural control systems influenced by external disturbances. To eliminate the effect of external disturbances, we must introduce the H_∞ robust technique to design the system. Furthermore, the problem of robust H_∞ control via output feedback controllers is still open and remains unsolved, which motivates the present study.

J. Wang et al. (Eds.): ISNN 2006, LNCS 3972, pp. 904–910, 2006.

In this paper, similar to the nominal models in linear robust control theory, we advance a standard neural network model (SNNM). Firstly, we analyze the robust H_∞ performance of the discrete-time SNNM with time delays. The optimal robust H_∞ performance problem is described as LMI eigenvalue minimization problem. On the other hand, based on the robust H_∞ performance analysis of the delayed SNNM, we will develop an output feedback control law for the delayed SNNM with inputs and outputs to ensure the asymptotic stability of the closed-loop system and optimize the H_∞ performance. Most delayed (or non-delayed) discrete-time nonlinear systems modeled by neural networks can be transformed into SNNMs to be robust H_∞ performance analyzed or robust H_∞ controller synthesized in a unified way.

Notation: Throughout this paper, $\mathfrak{R}^n$ denotes n dimensional Euclidean space, $\mathfrak{R}^{n\times m}$ is the set of all $n\times m$ real matrices, $\boldsymbol{I}$ denotes identity matrix of appropriate order, $*$ denotes the symmetric parts. The notations $\boldsymbol{X}>\boldsymbol{Y}$ and $\boldsymbol{X}\geq\boldsymbol{Y}$, respectively, where $\boldsymbol{X}$ and $\boldsymbol{Y}$ are matrices of same dimensions, mean that the matrix $\boldsymbol{X}-\boldsymbol{Y}$ is positive definite and positive semi-definite, respectively. If $\boldsymbol{X}\in\mathfrak{R}^p$ and $\boldsymbol{Y}\in\mathfrak{R}^q$, $\boldsymbol{C}(\boldsymbol{X};\boldsymbol{Y})$ denotes the space of all continuous functions mapping $\mathfrak{R}^p\rightarrow\mathfrak{R}^q$.

2 Problem Formulation

In linear robust control theory, systems with uncertainty can be transformed into a standard form known as linear fractional transformation (LFT). Similar to the LFT, we can analyze the performance or synthesize controllers for the nonlinear system composed of neural network by transforming them into SNNMs. The SNNM represents a neural network model as the interconnection of a linear dynamic system and static delayed (or non-delayed) nonlinear operators consisting of bounded activation functions. Here, we discuss only the discrete-time SNNM, though similar architecture and results for continuous-time SNNMs can also be achieved. A discrete-time SNNM with inputs and outputs is shown in Fig. 1. The block $\boldsymbol{\Phi}$ is a block diagonal operator

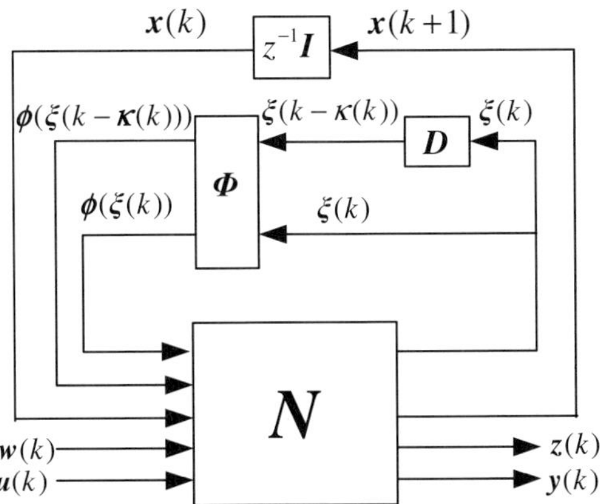

Fig. 1. Discrete-time standard neural network model (SNNM) with inputs and outputs

composed of nonlinear activation functions $\phi_i(\xi_i(\cdot))$, which are typically continuous, differentiable, monotonically increasing, slope-restricted, and bounded. The matrix $\boldsymbol{N}$ represents a linear mapping between the inputs and outputs of the time delay $z^{-1}\boldsymbol{I}$ in the discrete-time case (or integrator $\int$ in the continuous-time case) and the operator $\boldsymbol{\Phi}$. The vectors $\boldsymbol{\xi}(\cdot)$ and $\boldsymbol{\phi}(\boldsymbol{\xi}(\cdot))$ are the input and output of the nonlinear operator $\boldsymbol{\Phi}$, respectively. The block $\boldsymbol{D}$ represents the delayed element. $\kappa(\cdot)$ is the time-varying delay satisfying $0<\kappa(\cdot)\leq h$, where h is the maximal delay.

If $\boldsymbol{N}$ in Fig. 1 is partitioned as

$$\begin{bmatrix} \boldsymbol{A} & \boldsymbol{B}_p & \boldsymbol{B}_{pd} & \boldsymbol{B}_w & \boldsymbol{B}_u \\ \boldsymbol{C}_q & \boldsymbol{D}_p & \boldsymbol{D}_{pd} & \boldsymbol{D}_{qw} & \boldsymbol{D}_{qu} \\ \boldsymbol{C}_z & \boldsymbol{D}_{zp} & \boldsymbol{D}_{zpd} & \boldsymbol{D}_{zw} & \boldsymbol{D}_{zu} \\ \boldsymbol{C}_y & \boldsymbol{D}_{yp} & \boldsymbol{D}_{ypd} & \boldsymbol{D}_{yw} & \boldsymbol{D}_u \end{bmatrix}, \tag{1}$$

the input-output SNNM can be depicted as a linear difference inclusion (LDI):

$$\begin{cases} \boldsymbol{x}(k+1) = \boldsymbol{A}\boldsymbol{x}(k) + \boldsymbol{B}_p\boldsymbol{\phi}(\xi(k)) + \boldsymbol{B}_{pd}\boldsymbol{\phi}(\xi(k-\kappa(k))) + \boldsymbol{B}_w\boldsymbol{w}(k) + \boldsymbol{B}_u\boldsymbol{u}(k), \\ \xi(k) = \boldsymbol{C}_q\boldsymbol{x}(k) + \boldsymbol{D}_p\boldsymbol{\phi}(\xi(k)) + \boldsymbol{D}_{pd}\boldsymbol{\phi}(\xi(k-\kappa(k))) + \boldsymbol{D}_{qw}\boldsymbol{w}(k) + \boldsymbol{D}_{qu}\boldsymbol{u}(k), \\ z(k) = \boldsymbol{C}_z\boldsymbol{x}(k) + \boldsymbol{D}_{zp}\boldsymbol{\phi}(\xi(k)) + \boldsymbol{D}_{zpd}\boldsymbol{\phi}(\xi(k-\kappa(k))) + \boldsymbol{D}_{zw}\boldsymbol{w}(k) + \boldsymbol{D}_{zu}\boldsymbol{u}(k), \\ \boldsymbol{y}(k) = \boldsymbol{C}_y\boldsymbol{x}(k) + \boldsymbol{D}_{yp}\boldsymbol{\phi}(\xi(k)) + \boldsymbol{D}_{ypd}\boldsymbol{\phi}(\xi(k-\kappa(k))) + \boldsymbol{D}_{yw}\boldsymbol{w}(k) + \boldsymbol{D}_u\boldsymbol{u}(k), \end{cases} \tag{2}$$

with the initial condition function

$$\boldsymbol{\phi}(\xi(k_0+\theta)) = \boldsymbol{\phi}(\xi(k_0)),\ \forall\theta\in[-h,0], \tag{3}$$

where $\boldsymbol{x}\in\Re^n$ is the state vector, $\boldsymbol{A}\in\Re^{n\times n}$, $\boldsymbol{B}_p\in\Re^{n\times L}$, $\boldsymbol{B}_{pd}\in\Re^{n\times L}$, $\boldsymbol{B}_w\in\Re^{n\times r}$, $\boldsymbol{B}_u\in\Re^{n\times m}$, $\boldsymbol{C}_q\in\Re^{L\times n}$, $\boldsymbol{D}_p\in\Re^{L\times L}$, $\boldsymbol{D}_{pd}\in\Re^{L\times L}$, $\boldsymbol{D}_{qw}\in\Re^{L\times r}$, $\boldsymbol{D}_{qu}\in\Re^{L\times m}$, $\boldsymbol{C}_z\in\Re^{s\times n}$, $\boldsymbol{D}_{zp}\in\Re^{s\times L}$, $\boldsymbol{D}_{zpd}\in\Re^{s\times L}$, $\boldsymbol{D}_{zw}\in\Re^{s\times r}$, $\boldsymbol{D}_{zu}\in\Re^{s\times m}$, $\boldsymbol{C}_y\in\Re^{l\times n}$, $\boldsymbol{D}_{yp}\in\Re^{l\times L}$, $\boldsymbol{D}_{ypd}\in\Re^{l\times L}$, $\boldsymbol{D}_{yw}\in\Re^{l\times r}$ and $\boldsymbol{D}_u\in\Re^{l\times m}$ are the variable matrices, $\boldsymbol{\xi}\in\Re^L$ is the input of nonlinear operator $\boldsymbol{\Phi}$, $\boldsymbol{\phi}\in C(\Re^L;\Re^L)$ is the output of nonlinear operator $\boldsymbol{\Phi}$ satisfying $\boldsymbol{\phi}(0)=0$, $\boldsymbol{u}\in\Re^m$ is the control input, $\boldsymbol{w}\in\Re^r$ is the disturbance input, $\boldsymbol{y}\in\Re^l$ is the measured output, $z\in\Re^s$ is the performance output, and $L\in\Re$ is the number of nonlinear activation functions (that is, the total number of neurons in the hidden layers and output layer of the neural network).

3 Robust H$_\infty$ Controller Synthesis for the SNNM

In this paper, we assume that the activation functions in the SNNM (2) satisfy the sector conditions $\phi_i(\xi_i(k))/\xi_i(k)\in[q_i, u_i]$, i.e., $[\phi_i(\xi_i(k))-q_i\xi_i(k)]\cdot[\phi_i(\xi_i(k))-u_i\xi_i(k)]\leq 0$, $u_i>q_i\geq 0$ and the delays in the SNNM (2) are constant, i.e., $\kappa(\cdot)=h>0$. Here, we will first analyze the robust H$_\infty$ performance of the SNNM (2). Next based on the analysis results of the SNNM (2) we will design full-order dynamic output feedback controllers for the delayed nonlinear systems modeled by neural networks.

Theorem 1. If there exist a symmetric positive definite matrix $\boldsymbol{P}$ and $\boldsymbol{\Gamma}$, diagonal semi-positive definite matrices $\boldsymbol{\Lambda}$ and $\boldsymbol{T}$, and a positive scalar γ that satisfy

Minimize γ^2, (4)

$$\text{Subject to}\quad \begin{bmatrix} -P & 0 & PA & PB_p & PB_{pd} & PB_w \\ * & -I & C_z & D_{zp} & D_{zpd} & D_{zw} \\ * & * & -P & C_q^{\mathrm{T}} S & 0 & 0 \\ * & * & * & SD_p + D_p^{\mathrm{T}} S - 2T + \Gamma & SD_{pd} & SD_{qw} \\ * & * & * & * & -\Gamma & 0 \\ * & * & * & * & * & -\gamma^2 I \end{bmatrix} < 0, \tag{5}$$

where $S=\Lambda+(Q+U)T$, $Q=\mathrm{diag}(q_1, q_2, \ldots, q_L)$, $U=\mathrm{diag}(u_1, u_2, \ldots, u_L)$, then the origin of the SNNM (2) is asymptotically stable and the upper bound (i.e. γ) on the H$_\infty$ norm of the SNNM (2) for all admissible uncertainties is minimal.

The proof of Theorem 1 follows the same idea as that in the proof of Theorem 1 in [10], and is thus omitted here.

To significantly simplify the design procedure, we assume D_{yp}, D_{ypd} and D_{yw} are identically zeros. In most case, such assumption is satisfied, because the output of a lot of systems is either some states or linear combination of some states or inputs. Based on the above Theorem 1, we then discuss the design approach of the SNNM. The controller is of the form

$$\begin{cases} x_c(k+1) = A_c x_c(k) + B_c y(k), \\ u(k) = C_c x_c(k), \end{cases} \tag{6}$$

where $x_c \in \Re^n$ is the controller state, $A_c \in \Re^{n\times n}$, $B_c \in \Re^{n\times l}$, and $C_c \in \Re^{m\times n}$ are matrices of appropriate size. The overall closed-loop system of the SNNM (2) and the feedback controller (6) is described by

$$\begin{cases} x(k+1) = \tilde{A}\tilde{x}(k) + \tilde{B}_p \phi(\xi(k)) + \tilde{B}_{pd}\phi(\xi(k-\kappa(k))) + \tilde{B}_w w(k), \\ \xi(k) = \tilde{C}_q \tilde{x}(k) + D_p \phi(\xi(k)) + D_{pd}\phi(\xi(k-\kappa(k))) + D_{qw} w(k), \\ z(k) = \tilde{C}_z \tilde{x}(k) + D_{zp}\phi(\xi(k)) + D_{zpd}\phi(\xi(k-\kappa(k))) + D_{zw} w(k), \end{cases} \tag{7}$$

where $\tilde{x} = [x \quad x_c]^{\mathrm{T}}$, $\tilde{A} = \begin{bmatrix} A & B_u C_c \\ B_c C_y & A_c + B_c D_u C_c \end{bmatrix}$, $\tilde{B}_p = \begin{bmatrix} B_p \\ 0 \end{bmatrix}$, $\tilde{B}_{pd} = \begin{bmatrix} B_{pd} \\ 0 \end{bmatrix}$, $\tilde{B}_w = \begin{bmatrix} B_w \\ 0 \end{bmatrix}$, $\tilde{C}_q = [C_q \quad D_{qu} C_c]$, $\tilde{C}_z = [C_z \quad D_{zu} C_c]$.

Now, we are in a position to give the main result on the solvability of the dynamic output feedback control problem.

Theorem 2. Consider the discrete-time delayed SNNM (2) where $D_{yp}=0$, $D_{ypd}=0$, and $D_{yw}=0$. There exists a full-order dynamic output feedback controller (6) such that the closed-loop system (7) is globally asymptotically stable and the H$_\infty$ norm of the system (7) is minimal provided that there exist symmetric positive definite matrices X, Y and Ω, diagonal positive definite matrix Ψ and Σ, matrices $\hat{A}$ and $\hat{C}$, special structure matrix $\hat{B}$, and a positive scalar γ that satisfy the following optimization problem:

Minimize γ^2, (8)

Subject to

$$\begin{bmatrix} -X & -Y & 0 & AX+B_u\hat{C} & AY & B_p\Psi & B_{pd}\Psi & B_w \\ * & -Y & 0 & \hat{A} & AY+\hat{B} & B_p\Psi & B_{pd}\Psi & B_w \\ * & * & -I & C_zX+D_{zu}\hat{C} & C_zY & D_{zp}\Psi & D_{zpd}\Psi & D_{zw} \\ * & * & * & -X & -Y & (C_qX+D_{qu}\hat{C})^T & 0 & 0 \\ * & * & * & * & -Y & YC_q^T & 0 & 0 \\ * & * & * & * & * & \begin{pmatrix} D_p\Psi+\Psi D_p^T \\ +\Omega-2\Sigma \end{pmatrix} & D_{pd}\Psi & D_{qw} \\ * & * & * & * & * & * & -\Omega & 0 \\ * & * & * & * & * & * & * & -\gamma^2 I \end{bmatrix} < 0 \tag{9}$$

,

$$\boldsymbol{\Psi} \geq (Q+U)\boldsymbol{\Sigma}, \tag{10}$$

$$\begin{bmatrix} Y & Y \\ * & X \end{bmatrix} > 0, \tag{11}$$

where $\boldsymbol{Q}$=diag(q_1, q_2, …, q_L), $\boldsymbol{U}$=diag(u_1, u_2, …, u_L). Furthermore, the desired dynamic output feedback controller is given in the form of (6) with parameters as follow:

$$\begin{cases} C_c = \hat{C}(M^{\mathrm{T}})^{-1}, \\ B_cC_y = N^{-1}Y^{-1}\hat{B}Y^{-1}, \\ A_c = (YN)^{-1}(\hat{A} - AX - \hat{B}Y^{-1}X - B_u\hat{C})(M^{\mathrm{T}})^{-1} - B_cD_uC_c, \end{cases} \tag{12}$$

where $\boldsymbol{M}$ and $\boldsymbol{N}$ are any nonsingular matrices satisfying

$$MN^{\mathrm{T}} = I - XY^{-1}. \tag{13}$$

Proof: According to Theorem 1, if

$$\begin{bmatrix} -\boldsymbol{P} & 0 & \boldsymbol{P}\tilde{\boldsymbol{A}} & \boldsymbol{P}\tilde{\boldsymbol{B}}_p & \boldsymbol{P}\tilde{\boldsymbol{B}}_{pd} & \boldsymbol{P}\tilde{\boldsymbol{B}}_w \\ * & -\boldsymbol{I} & \tilde{\boldsymbol{C}}_z & \boldsymbol{D}_{zp} & \boldsymbol{D}_{zpd} & \boldsymbol{D}_{zw} \\ * & * & -\boldsymbol{P} & \tilde{\boldsymbol{C}}_q^{\mathrm{T}}\boldsymbol{S} & 0 & 0 \\ * & * & * & \boldsymbol{S}\boldsymbol{D}_p + \boldsymbol{D}_p^{\mathrm{T}}\boldsymbol{S} - 2\boldsymbol{T} + \boldsymbol{\Gamma} & \boldsymbol{S}\boldsymbol{D}_{pd} & \boldsymbol{S}\boldsymbol{D}_{qw} \\ * & * & * & * & -\boldsymbol{\Gamma} & 0 \\ * & * & * & * & * & -\gamma^2\boldsymbol{I} \end{bmatrix} < 0 \tag{14}$$

holds, the origin of the closed-loop system (7) is globally asymptotically stable and the H_∞ norm of the system (7) is bound. To proceed, we partition $\boldsymbol{P}$ and its inverse $\boldsymbol{P}^{-1}$ as

$$\boldsymbol{P} = \begin{bmatrix} \boldsymbol{Z} & \boldsymbol{N} \\ \boldsymbol{N}^{\mathrm{T}} & \boldsymbol{W} \end{bmatrix}, \quad \boldsymbol{P}^{-1} = \begin{bmatrix} X & M \\ M^{\mathrm{T}} & \boldsymbol{V} \end{bmatrix},$$

where $X \in \Re^{n\times n}$ and $Z \in \Re^{n\times n}$ are symmetric positive definite matrices. We define the following matrices

$$F_1 = \begin{bmatrix} X & I \\ M^{\mathrm{T}} & 0 \end{bmatrix},\ F_2 = \begin{bmatrix} I & Z \\ 0 & N^{\mathrm{T}} \end{bmatrix}.$$

The condition $P \cdot P^{-1}=I$ implies $PF_1=F_2$. Pre- and post-multiplying the left-hand side matrix of the Ieq.(14) by diag $(F_1^{\mathrm{T}}, I, F_1^{\mathrm{T}}, S^{-1}, S^{-1}, I)$ and diag $(F_1, I, F_1, S^{-1}, S^{-1}, I)$, respectively, and defining

$$\Psi = S^{-1}, \Sigma = S^{-1}TS^{-1}, \Omega = S^{-1}\Gamma S^{-1}, \tag{15}$$

$$\begin{cases} \hat{A} = AX + Z^{-1}NB_cC_yX + B_uC_cM^T + Z^{-1}N(A_c + B_cD_uC_c)M^{\mathrm{T}}, \\ \hat{B} = Z^{-1}NB_cC_yZ^{-1}, \\ \hat{C} = C_cM^T, \end{cases} \tag{16}$$

we rewrite the Ieq.(14) as

$$\begin{bmatrix} -X & -I & 0 & AX + B_u\hat{C} & A & B_p\Psi & B_{pd}\Psi & B_w \\ * & -Z & 0 & Z\hat{A} & ZA + Z\hat{B}Z & ZB_p\Psi & ZB_{pd}\Psi & ZB_w \\ * & * & -I & C_zX + D_{zu}\hat{C} & C_z & D_{zp}\Psi & D_{zpd}\Psi & D_{zw} \\ * & * & * & -X & -I & (C_qX + D_{qu}\hat{C})^{\mathrm{T}} & 0 & 0 \\ * & * & * & * & -Z & C_q^{\mathrm{T}} & 0 & 0 \\ * & * & * & * & * & \begin{pmatrix} D_p\Psi + \Psi D_p^{\mathrm{T}} \\ +\Omega - 2\Sigma \end{pmatrix} & D_{pd}\Psi & D_{qw} \\ * & * & * & * & * & * & -\Omega & 0 \\ * & * & * & * & * & * & * & -\gamma^2 I \end{bmatrix} < 0 \cdot \tag{17}$$

Pre- and post-multiplying the left-hand side matrix of the Ieq. (17) by diag$(I, Z^{-1}, I, I, Z^{-1}, I, I, I)$, and letting $Y=Z^{-1}$, we obtain the Ieq.(9). The controller parameters A_c, B_c, and C_c can be deduced from (16). It is worth noting that the structure of matrix $\hat{B}$ is determined by the form of C_y.

$T = \Psi^{-1}\Sigma\Psi^{-1}$ is derived from (15). Since $\Lambda \geq 0$, we obtain $S \geq (Q+U)T$, so the Ieq.(10) is necessary. Using the condition $P^{-1} \cdot P=I$, we have $MN^{\mathrm{T}} = I - XZ$ (i.e. $MN^{\mathrm{T}} = I - XY^{-1}$). By the Schur complement formula [11], the Ieq.(11) can be expressed as $Y - YX^{-1}Y > 0$, therefore $I - XY^{-1}$ is nonsingular. This ensures that there always exist nonsingular matrices M and N such that the Eq. (13) is satisfied. We thus complete the proof.

4 Conclusion

In this paper, we studied a control design algorithm for a class of discrete-time nonlinear systems with time delays based on neural-network approximation. Central to our design are the introduction of the SNNM, and the transformation of the closed-loop neuro-control system to the SNNM. A full-order dynamic output feedback controller

troller has been designed for the SNNM such that the closed-loop system is globally asymptotically stable and an optimal level on disturbance attenuation is guaranteed. The resulting design equations are described as a LMI optimization problem which can be solved by MATLAB LMI Control Toolbox [12] to determine control signals. The design approach can be extended to synthesize any nonlinear control systems as long as their equations can be transformed into the SNNM.

Acknowledgment

This work is supported by the National Natural Science Foundation of China under Grant 60504024, and the Research Project of Zhejiang Provincial Education Department under Grant 20050905.

References

1. Suykens, J. A. K., Vandewalle, J. P. L., De Moor, B. L. R.: Artificial Neural Networks for Modeling and Control of Non-linear Systems. Kluwer Academic Publishers, Norwell, MA (1996)
2. Xu H. J., Ioannou, P. A.: Robust Adaptive Control for a Class of MIMO Nonlinear Systems with Guaranteed Error Bounds. IEEE Trans. on Automatic Control 48(5) (2003) 728-742
3. Lin, C. L., Lin, T. Y.: An H_∞ Design Approach for Neural Net-based Control Schemes. IEEE Trans. on Automatic Control 46 (10) (2001) 1599-1605
4. Lin, F. J., Lee, T. S, Lin, C. H.: Robust H_∞ Controller Design with Recurrent Neural Network for Linear Synchronous Motor Drive. IEEE Trans. on Industrial Electronics 50 (3) (2003) 456-470
5. Baldi, P., Atiya, A. F.: How Delays Affect Neural Dynamics and Learning. IEEE Transactions on Neural Networks 5 (4) (1994) 612-621
6. Shen, Y., Zhao, G.Y., Jiang, M.H., Hu, S.G.: Stochastic High-order Hopfield Neural Networks. In: Wang, L.P., Chen, K., Ong, Y.S. (eds.): Advances in Natural Computation. Lecture Notes in Computer Science, Vol. 3610. Springer-Verlag, Berlin Heidelberg New York (2005) 740-749
7. Shen, Y., Jiang, M.H., Liao, X.X.: Global Exponential Stability of Cohen-Grossberg Neural Networks with Time-varying Delays and Continuously Distributed Delays. In: Wang, J., Liao, X.F., Zhang, Y. (eds.): Advances in Neural Networks. Lecture Notes in Computer Science, Vol. 3496. Springer-Verlag, Berlin Heidelberg New York (2005) 156-161
8. Shen, Y., Zhao, G.Y., Jiang, M.H., Mao, X.R.: Stochastic Lotka-Volterra Competitive Systems with Variable Delay. In: Huang, D.S., Zhang, X.P., Huang, G.B. (eds.): Advances in Intelligent Computing. Lecture Notes in Computer Science, Vol. 3645. Springer-Verlag, Berlin Heidelberg New York (2005) 238-247
9. Zeng, Z., Wang, J., Liao X.: Global Exponential Stability of a General Class of Recurrent Neural Networks with Time-Varying Delays. IEEE Trans. Circuits Syst.-I: Fundamental Theory and Applications 50 (10) (2003) 1353-1358
10. Liu, M., Yan, G., Wang, S.: Neural-model Based Robust H_∞ Controllers for Nonlinear Systems: an BMI Approach. In: International Conference on Systems, Man and Cybernetics, Vol. 6. Hague, Netherlands (2004) 5876-5881.
11. Boyd, S. P., Ghaoui, L. E., Feron, E., Balakrishnan, V.: Linear Matrix Inequalities in System and Control Theory. SIAM, Philadelphia (1994)
12. Gahinet, P., Nemirovski, A., Laub, A. J., Chilali, M.: LMI Control Toolbox- for Use with Matlab. The MATH Works, Inc., Natick, MA (1995)

SVM Based Nonlinear Self-tuning Control

Weimin Zhong[1,2], Daoying Pi[1], Chi Xu[3], and Sizhen Chu[3]

[1] National Laboratory of Industrial Control Technology,
Institute of Modern Control Engineering,
Zhejiang University, Hangzhou, 310027 P.R. China
[2] Automation Institute of East China University of Science
and Technology, Shanghai, 200237 P.R. China
{wmzhong, dypi, yxsun}@iipc.zju.edu.cn
[3] Hangzhou Automation Technology Institute, Hangzhou,
310027 P.R. China
{xch, csz}@tofine.net

Abstract. In this paper, a support vector machine (SVM) with polynomial kernel function enhanced nonlinear self-tuning controller is developed, which combines the SVM identifier and parameters' modifier together. The inverse model of a nonlinear system is achieved by off-line black-box identification according to input and output data. Then parameters of the model are modified online using gradient descent algorithm. Simulation results show that SVM based self-tuning control can be well applied to nonlinear uncertain system. And the SVM based self-tuning control of nonlinear system has good robustness performance in tracking reference input with good generalization ability.

1 Introduction

As a form of indirect adaptive control, self-tuning control has been of major interest for many years. It has been widely used in industrial process control system now. The basic idea of this adaptive control is that system identification is achieved through an identifier first, and then parameters are online self-tuning in order to generate the current control value [1]. Traditional self-tuning identification is based on the plant's linear or approximate linear model, and online parameters modification employs numeric methods such as iterative algorithm [2]. But in real industry case, most plants are complex and intrinsically nonlinear system, and traditional self-tuning control couldn't deal with them well. Against its drawbacks, some improvements have been done recently using new method.

Recently a new kind of learning machine called support vector machine is well used for classification, time series prediction, etc [3, 4]. SVM has excellent performance in function regression, so it can approximate nonlinear functions with arbitrary precision [3]. According to this, SVM can be used to do system identification and control [5,6]. This paper put forward a new self-tuning control architecture based on SVM with polynomial kernel function to deal with nonlinear system's adaptive control. It combines the SVM identifier and the parameters' modifier. The inverse model of a nonlinear system is achieved by off-line black-box identification according to

J. Wang et al. (Eds.): ISNN 2006, LNCS 3972, pp. 911–915, 2006.

input and output data. Then parameters of the model are modified online with gradient descent algorithm.

2 Support Vector Machine Based Self-tuning Control

The structure of SVM based self-tuning control can be seen in Fig. 1. u is the input, y is the corresponding output, and r is the system's reference input.

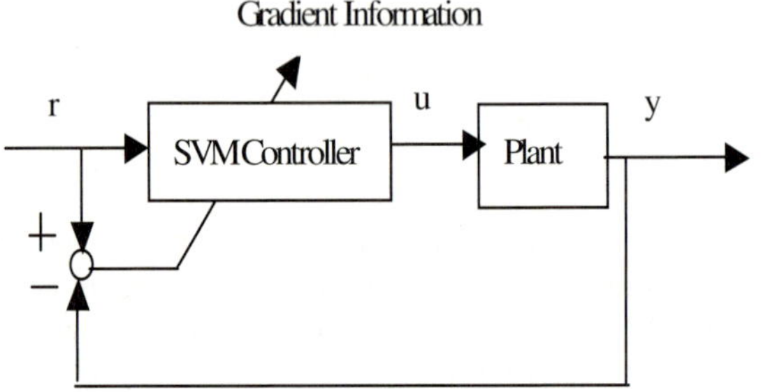

Fig. 1. The structure of SVM based self-tuning control

2.1 SVM Based Nonlinear Identification

Assume the discrete nonlinear system can be described as below:

$$y(k+1) = f[y(k), \cdots, y(k-n+1), u(k), \cdots, u(k-m+1)] \tag{1}$$

Where the structure information n and m may be unknown, and $f[\cdot]$ is an unknown nonzero nonlinear function. Control system's output $y(k+1)$ is required to track the reference input $r(k+1)$ precisely. So according to the certain equivalence principle, $y(k+1) = r(k+1)$ at sample time $k+1$, we can get the control algorithm:

$$\begin{aligned} u(k) = f^{-1}[&r(\mathrm{k}+1), y(k), \cdots, y(k-n+1), \\ &u(k-1), \cdots, u(k-m+1)] \end{aligned} \tag{2}$$

Where f^{-1} is inverse function of f. For linear systems, using linear control theory can get the control law easily. But when dealing with the complex and strong nonlinear systems, traditional control theory couldn't do it well. Here we introduce the SVM identifier to approximate the nonlinear plant's inverse model.

SVM strategy makes use of an approximation of the nonlinear regression constructed by kernel dot product. The inverse model is represented as follows:

$$\begin{aligned} u(k) = \hat{f}(\mathbf{I}_{\mathbf{k}}) = \hat{f}[&r(k+1), y(k), \cdots, y(k-n'+1), \\ &u(k-1), \cdots, u(k-m'+1)] \end{aligned} \tag{3}$$

$\hat{f}$ is another nonlinear function with SVM architecture. n' and m' are determined by approximation accuracy just like in neural network based system identification. According to training data pairs $\{\mathbf{I_s}, u_s\}(s=1,\cdots,d)$, $u_s=u(s)$, the inverse model is:

$$u(k)=\sum_{i=1}^{nsv} a_i K(\mathbf{I_i^{'}}\cdot\mathbf{I_k})+b=\sum_{i=1}^{nsv} a_i(\mathbf{I_i^{'}}\cdot\mathbf{I_k}+1)^2+b \tag{4}$$

Where $\mathbf{I_i^{'}}$ is the set of support vector from $\mathbf{I_s}$, a_i is corresponding support value, b is threshold value, $nsv\le d$ is the number of support vector. All above parameters are automatically obtained through SVM learning. And there is:

$$\begin{aligned}\mathbf{I_i^{'}}\cdot\mathbf{I_k}&=\mathbf{I_i^{'}}(1)r(k+1)+\mathbf{I_i^{'}}(2)y(k)+\cdots+\mathbf{I_i^{'}}(n'+1)y(k-n'+1)\\&\mathbf{I_i^{'}}(n'+2)u(k-1)+\cdots+\mathbf{I_i^{'}}(n'+m')u(k-m'+1)\end{aligned} \tag{5}$$

2.2 Parameter Modification

In order to overcome the disturbance and slow variation of plant, controller output should self-tuning online. In this paper, a gradient-based method is introduced. This is a simple but very efficient way. Consider (4), if plant's variation is small, we can assume support vectors $\mathbf{I_i^{'}}$ unchangeable, and so the modification can be focused on support values a_i. At time k, $r(k)$ and $y(k)$ are known, so we can use this information to update support values.

In order to determine the correction of support value, a quadratic objective function is introduced:

$$J(k)=\frac{1}{2}[r(k+1)-y(k+1)]^2 \tag{6}$$

So the modification algorithm is:

$$a_i(k+1)=a_i(k)+d(r(k+1)-y(k+1))\frac{\partial y(k+1)}{\partial a_i(k)} \qquad i=1,\cdots,nsv \tag{7}$$

Where d is the learning step and

$$\frac{\partial y(k+1)}{\partial a_i(k)}=\frac{\partial y(k+1)}{\partial u(k)}\cdot\frac{\partial u(k)}{\partial a_i(k)}=K(\mathbf{I_i^{'}}\cdot\mathbf{I_k})\cdot\frac{\partial y(k+1)}{\partial u(k)}$$

To an unknown plant, $\frac{\partial y(k+1)}{\partial u(k)}$ is unknown too. But through available information, we can get the variation direction of $y(k+1)$ when $u(k)$ is added to plant. So we can use SIGN function $sign\frac{\partial y(k+1)}{\partial u(k)}$ to substitute $\frac{\partial y(k+1)}{\partial u(k)}$, which can avoid the analytical solution of $\frac{\partial y(k+1)}{\partial u(k)}$.

$$sign\frac{\partial y(k+1)}{\partial u(k)} = sign\frac{y(k+1)-y(k)}{u(k)-u(k-1)} \tag{8}$$

The effect of this substitution is compensated by adjusting learning step d. And because $K(\mathbf{I}_{\mathbf{i}}^{'} \cdot \mathbf{I}_{\mathbf{k}}) > 0$, (14) can be written:

$$a_i(k+1) = a_i(k) + \eta(r(k+1) - y(k+1))sign\frac{y(k+1)-y(k)}{u(k)-u(k-1)} \qquad i = 1,\cdots,n \tag{9}$$

Where η is another learning step.

3 Simulation

A time-varying plant was investigated in [1],

$$\begin{cases} y(k+1) = 0.8\sin(y(k)) + 1.2u(k) & 0 < k < 200 \\ y(k+1) = 0.8\sin(y(k)) + \dfrac{y(k)}{7} + 1.2u(k) & k \geq 200 \end{cases} \tag{10}$$

With reference input:

$$\begin{cases} r(k) = 0.02 * k & k \leq 50 \\ r(k) = 1 & 50 < k \leq 100 \\ r(k) = 1 - 0.02 * (k - 100) & 100 < k \leq 200 \\ r(k) = -1 & 200 < k \leq 250 \\ r(k) = -1 + 0.02 * (k - 250) & k > 250 \end{cases} \tag{11}$$

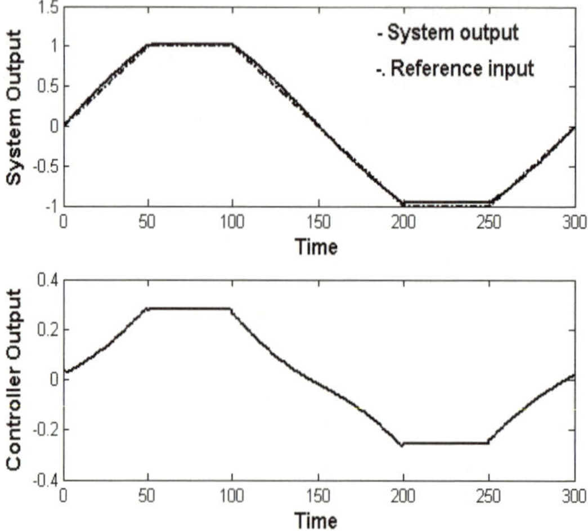

Fig. 2. Simulation results

And the perturbation is:

$$\begin{cases} v(k) = 0.1 & k = 70 \\ v(k) = -0.1 & k = 220 \\ v(k) = 0 & otherwise \end{cases} \tag{12}$$

Assume system is zero initial state. And select $\varepsilon' = 0.01$, $n' = 2$, $m' = 2$. Employ 250 random data pairs generated from $u \in [-1.5, 1.5]$ to train the SVM based identification. The simulation results are shown in Fig. 2. Solid line denotes system output, and dash-dot line denotes reference input. In lower figure, corresponding controller output is presented. We can see the SVM based self-control system can track the reference input (keystone wave) well.

4 Conclusion

SVM is an important new direction in the machine learning area. SVM based model is obtained by convex optimization problems with good performance in generalization. In this paper, a new SVM based self-tuning controller is developed for the control of nonlinear system. The analysis and experiment results show that this self-tuning control strategy will provide a robust stable control of nonlinear systems with good generalization ability. The effect of the controller is demonstrated by means of an example. Simulation results show that the control architecture gives good performance.

Acknowledgement

This work is sponsored by the 973 program of China under grant No.2002CB312200, the National Science Foundation of China under grant No.60574019 and No.60474045, the Scientific Research Foundation for the Returned Overseas Chinese Scholars, State Education Ministry, Zhejiang Province and Zhejiang University.

References

1. Xu, L.N.: Neural Network Control. Harbin Institute of Technology Press, Harbin (1999)
2. Han, Z.J.: Adaptive Control. Tsinghua University Press, Beijing (2000)
3. Vapnik, V.N.: The Nature of Statistical Learning Theory. Springer-Verlag, Berlin Heidelberg New York (1995)
4. Vapnik, V.N.: Statistical Learning Theory. John Wiley and Sons, New York (1998)
5. Zhong, W.M., Pi, D.Y., Sun, Y.X.: SVM with Quadratic Polynomial Kernel Function based Nonlinear Model One-step-ahead Predictive Control. Chinese Journal of Chemical Engineering 13(3) (2005) 373-379
6. Zhong, W.M., Pi, D.Y., Sun, Y.X.: Support Vector Machine based Nonlinear Multi-step-ahead Optimizing Predictive Control. Journal of Central South University of Technology 12(5) (2005) 591-595

SVM Based Internal Model Control for Nonlinear Systems

Weimin Zhong[1,2], Daoying Pi[1], Youxian Sun[1], Chi Xu[3], and Sizhen Chu[3]

[1] National Laboratory of Industrial Control Technology,
Institute of Modern Control Engineering,
Zhejiang University, Hangzhou, 310027 P.R. China
[2] Automation Institute of East China
University of Science and Technology, Shanghai, 200237 P.R. China
{wmzhong, dypi, yxsun}@iipc.zju.edu.cn
[3] Hangzhou Automation Technology Institute, Hangzhou,
310027 P.R. China
{xch, csz}@tofine.net

Abstract. In this paper, a design procedure of support vector machine (SVM) with RBF kernel function based internal model control (IMC) strategy for stable nonlinear systems with input-output form is proposed. The control scheme consists of two controllers: a SVM based controller which fulfils the direct inverse model control and a traditional controller which fulfils the close-loop control. And so the scheme can deal with the errors between the process and the SVM based internal model generated by model mismatch and additional disturbance. Simulations are given to illustrate the proposed design procedure and the properties of the SVM based internal model control scheme for unknown nonlinear systems with time delay.

1 Introduction

In nonlinear control literature, several techniques have been developed to control nonlinear plants subject to uncertainties and disturbance. Among them, nonlinear internal model control method has played a significant role due to its properties in good robustness against disturbances and model mismatch [1]. Developments of NIMC have been proposed for continuous-time systems [2] and for discrete-time systems [1, 3] using neural network and fuzzy theory. Generally, these methods have some drawbacks in modeling: training speed is slowly; generalization ability is not good; and prior knowledge is needed to some degree.

Recently a new kind of learning machine called SVM [4, 5] was presented and has been used for classification, function regression, and system identification and control, etc [6,7,8]. In [8], a simple SVM with RBF kernel function based internal model is built to control the greenhouse environment problem, in which time delay is not taken into account. The inverse model controller only generate response according to the current and past time's process outputs and controller outputs and a one-step-ahead value of the set point trajectory, not the feedback errors (including model mismatch and disturbance).

J. Wang et al. (Eds.): ISNN 2006, LNCS 3972, pp. 916–921, 2006.

In this paper, we detail the SVM based internal model control for nonlinear system with time delay whose structure is unknown. And another controller will be introduced in the feedback loop in order to fulfill the feedback to compensate the non-adaptive inverse model controller when the mismatch and additional disturbance exist.

2 SVM Based Nonlinear Modeling

In this paper, a stable single input/single output process is investigated. We consider the process to be described by the following discrete-time deterministic nonlinear input-output model [1]:

$$y(k+1) = h(y\{_{k-n+1}^{k}, u\{_{k-d-m+1}^{k-d}) + p(k) \quad (1)$$

Where $y\{_{k-n+1}^{k}$ denotes the known state $\{y(k),\cdots,y(k-n+1)\}$ and $u\{_{k-d-m+1}^{k-d}$ denotes the set of m steps past control inputs $\{u(k-d),\cdots,u(k-d-m+1)\}$; $d \in [0,1,2,\cdots]$ is the time delay; h is an unknown nonlinear function. And $p(k)$ is the effect of additional disturbance. And we assume that n , m and d are unknown.

Efficient black box modeling needs the training sequences are of sufficient size and distribution. Under the assumption that the system is zero initial, we use a random series of u_t (size is N_t , i.e., $u_t(1),...u_t(N_t)$) full of the input domain to generate the corresponding series of $y_{pt}(y_{pt}(1),\cdots,y_{pt}(N_t),y_{pt}(N_t+1),\cdots,y_{pt}(N_t+d),\cdots)$. We use n' and m' denote the truncation of y_{pt} and u_t respectively. Obviously, $m'>d$ should be guaranteed. The SVM based straight internal model will be represented as follow:

$$\begin{gathered} y_{sm}(k+1) = \phi_{sm}(y_{pt}\{_{i-n'+1}^{i}, u_t\{_{i-m'+1}^{i}, y\{_{k-n'+1}^{k}, u\{_{k-m'+1}^{k} : \boldsymbol{a}, b, K, C, \varepsilon) \\ i = \max(n',m'),\cdots,N_t \end{gathered} \quad (2)$$

Where ϕ_{sm} is a nonlinear function implemented by a feed-forward SVM. And here $\boldsymbol{a} = [a_1,\cdots a_{N_t-\max(n',m')+1}]'$.

In order to build the inverse model controller dealing with the unknown time delay, we will adopt more regression factors of y_{pt} before the time instant i , which will sufficiently take the time delay into account. So the SVM based inverse model controller will be represented as follow:

$$\begin{gathered} u(k) = \phi_{im}(y_{pt}\{_{i-n''+1}^{i+l}, u_t\{_{i-m''+1}^{i-1}, r\{_{k+1}^{k+l}, y\{_{k-n''+1}^{k}, u\{_{k-m''+1}^{k-1} : \boldsymbol{a}, b, K, C, \varepsilon) \\ i = \max(n'',m''),\cdots,N_t \end{gathered} \quad (3)$$

Where ϕ_{im} is another nonlinear function implemented by feed-forward SVM. $r\{_{k+1}^{k+l}$ is the l -step ahead tracking set point sequence at time k . And here $\boldsymbol{a} = [a_1,\cdots a_{N_t-\max(n'',m'')+1}]'$. n'' and m'' are the truncation number of past y_{pt} and u_t respectively. And in this case, $l \ge d$ should be guaranteed.

3 SVM Based Internal Model Control

In this paper, an internal model control structure is studied which is shown in Fig.1. Where dashed-line denotes past information is used, dash dotted line denotes the current and past information is used too. r is the tracking sequence of process output y. For the SVM based inverse model and internal model use the same set of training data, they have the same degree of accuracy. The feedback loop will function if there is model mismatch or additional disturbance.

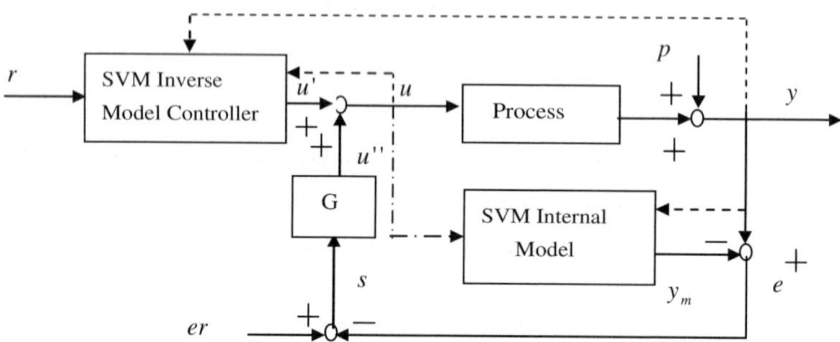

Fig.1. The block diagram of SVM based IMC

4 Simulation

In order to illustrate the results of SVM based internal model control, a nonlinear plant is selected as follows:

$$y(k+1) = 0.2y(k)^2 \sin u(k-1) + u(k-1) + 0.3u(k-1)u(k-2) + p(k) \tag{4}$$

Where y is the plant output, u is the input and p is the additional effect of disturbance and noise. The parameters of the true model are $n = 1, m = 2, d = 2$.

Set $n' = 3$, $m' = 5$, $n'' = 3$, $m'' = 3$, $l = 3$, $C = 10000$, $\varepsilon = 0.001$ and kernel function is RBF function $K(x_i \cdot x) = \frac{\|x_i - x\|^2}{2\alpha^2}$ with $\alpha^2 = 7.5$. And the feedback controller will be adopted the simple P-controller with the gain $K = 0.2$. Training data sets ($N_t = 150$) generated by applying by a series of random numbers between $u \in [-1.5, 1.5]$ are used to train the inverse model and straight model. And select the input reference trajectory is:

$$r(k) = \begin{cases} k/20 & k \le 20 \\ 1-(k-20)/20 & 20 < k \le 40 \\ \sin(0.1k - 4.1) & k > 40 \end{cases} \tag{5}$$

Fig. 2 shows the output of the internal model control system with additional noise between [-0.05, 0.05] and disturbance rejection $\Delta = -0.2$ at time $k = 30$ and $\Delta = 0.2$ at time $k = 60$.

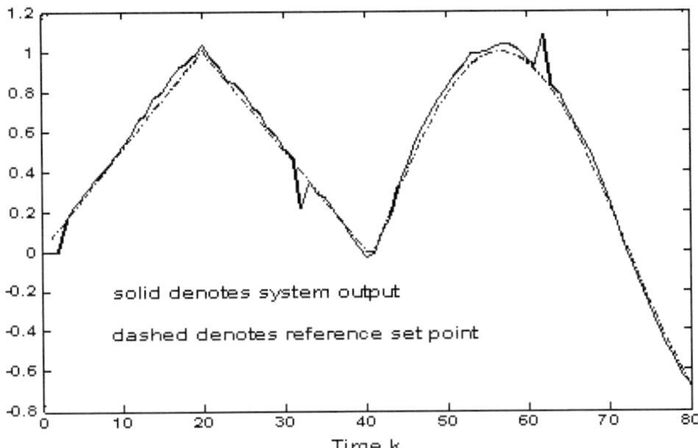

Fig. 2. Output under noise and disturbance

In most cases, the plant is slowly variable. So provided the both models are trained off-line perfectly (all parameters are the same as the above), assume (4) becomes:

$$y(k+1) = 0.2y(k)^2 \sin u(k-1) + u(k-1) + 0.6u(k-1)u(k-2) \tag{6}$$

Fig.3 shows the output response using method in [8] when system (4) is changed into (6). The performance is not good obviously. And Fig.4 shows the results of our method. The control performance is better than that of Fig.3.

Generally speaking, the black box-models are not guaranteed to be perfect and mismatch always exists. Fig.5 and Fig.6 show the output response of system (4) using

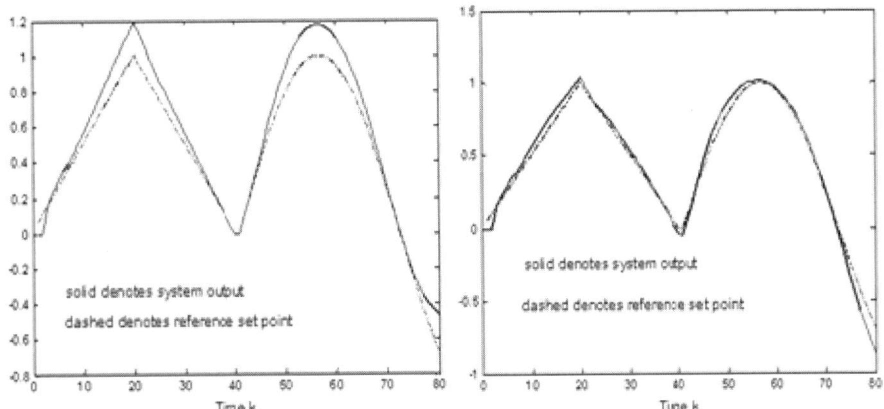

Fig. 3. Output response when system is changed using method in [8]

Fig. 4. Output response when system is changed using our method

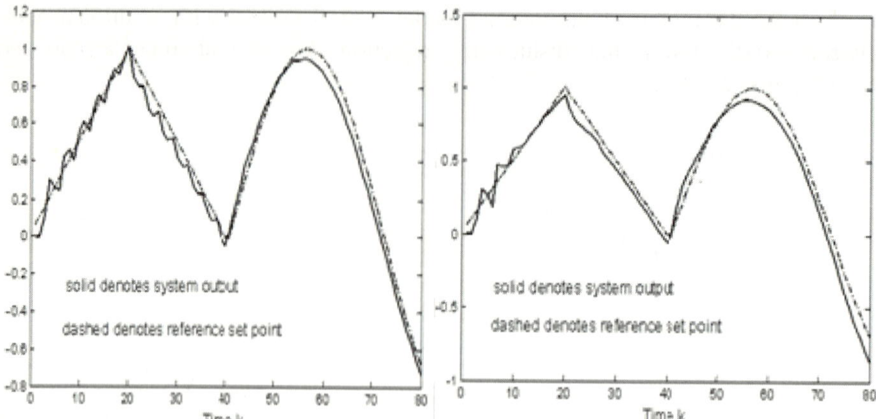

Fig. 5. Output response when mismatch exists using method in [8]

Fig. 6. Output response when mismatch exists using our method

method in [8] and our method when mismatch exists respectively. And we can see our performance is better than that of direct inverse control.

5 Conclusion

In this paper, a feed-forward and feedback SVM based internal model control structure is investigated. From [8], we can see that the identification and control performance of SVM based method is better than that of the neural network based. And some other conclusions are drawn from our simulation results:

(1) SVM is very suitable for building the model of nonlinear systems with unknown structure information including time delay.
(2) The SVM based internal model control could have good performance even when additional noise and disturbance exist.
(3) If the controlled process is varied slowly and model mismatch exists, the method of paper [8] will not work well. But our method will deal with this situation well because feedback controller is introduced.
(4) The control system has good robustness because the SVM based model can follow the variable input reference trajectory (ramp and sine functions) well with good generalization.

Acknowledgement

This work is sponsored by the 973 program of China under grant No.2002CB312200, the National Science Foundation of China under grant No.60574019 and No.60474045, the Scientific Research Foundation for the Returned Overseas Chinese Scholars, State Education Ministry, Zhejiang Province and Zhejiang University.

References

1. Reda, B., Sylvie, G., Laurent, F.: Nonlinear Internal Model Control: Application of Inverse Model based Fuzzy Control. IEEE Trans. Fuzzy Systems 11(6) (2003) 814-829
2. Calvet, J., Arkun, Y.: Feed Forward and Feedback Linearization of Nonlinear Systems and Its Implementation Using Internal Model Control. Ind.Eng.Chem.Res 27(10) (1998) 1822-1831
3. Isabelle, R., Léon, P.: Nonlinear Internal Model Control Using Neural Networks: Application to Processes with Delay and Design Issues. IEEE Trans. Neural Networks 11(1) (2000) 80-90
4. Vapnik, V.N.: The Nature of Statistical Learning Theory. Springer-Verlag, Berlin Heidelberg New York (1995)
5. Vapnik, V.N.: Statistical Learning Theory. John Wiley and Sons, New York (1998)
6. Zhong, W.M., Pi, D.Y., Sun, Y.X.: Support Vector Machine based Nonlinear Multi-step-ahead Optimizing Predictive Control. Journal of Central South University of Technology 12(5) (2005) 591-595
7. Zhong, W.M., Pi, D.Y., Sun, Y.X.: SVM with Quadratic Polynomial Kernel Function based Nonlinear Model One-step-ahead Predictive Control. Chinese Journal of Chemical Engineering 13(3) (2005) 373-379
8. Wang, D.C., Fang, T.J.: Internal Model Control Approach based on Support Vector Machines. Control Theory & Application 21(1) (2004) 85-88

Fast Online SVR Algorithm Based Adaptive Internal Model Control

Hui Wang[1], Daoying Pi[1], Youxian Sun[1], Chi Xu[2] and Sizhen Chu[2]

[1] The National Laboratory of Industrial Control Technology,
Institute of Modern Control Engineering, Zhejiang University,
Hangzhou, 310027 P.R. China

[2] Hangzhou Automation Technology Institute, Hangzhou 310012, China

huiwang@iipc.zju.edu.cn

Abstract. Based on fast online support vector regression (SVR) algorithm, reverse model of system model is constructed, and adaptive internal model controller is developed. First, SVR model and its online training algorithm are introduced. A kernel cache method is used to accelerate the online training algorithm, which makes it suitable for real-time control application. Then it is used in internal model control (IMC) for online constructing internal model and designing the internal model controller. Output errors of the system are used to control online SVR algorithm, which made the whole control system a closed-loop one. Last, the fast online SVM based adaptive internal model control was used to control a benchmark nonlinear system. Simulation results show that the controller has simple structure, good control performance and robustness.

1 Introduction

Internal model control (IMC) is a control method with simple designing, good adjusting characteristic and robustness [1-2]. Support vector machine (SVM) [3] is a new machine learning method and has been used for classification, function regression, and time series prediction, etc [4-5]. Compared with neural network SVM has well generalization ability, and is especially fit for machine learning in small sample condition. The training algorithm of SVM will not run into local minimum point. Also it can automatically construct the structure of system model. So this paper introduces SVM into internal model control to construct an adaptive controller. And a kernel cache method is used to accelerate the training algorithm, which makes it satisfy the need for real-time process in control application.

2 SVR Model and Its Fast Online Training Algorithm

SVM function regression can be expressed as: Given a training sample set $T=\{x_i, y_i, i=1,2,\cdots,l\}$, $x_i \in R^N$, $y_i \in R$, construct a linear regression function:

$$f(x) = W^T \Phi(x) + b \tag{1}$$

J. Wang et al. (Eds.): ISNN 2006, LNCS 3972, pp. 922–927, 2006.

on a feature space F. W is a vector in F, and $\Phi(x)$ maps the input x to a vector in F. W and b can be obtained by solving the following optimization problem:

$$\min_{W,b} T = \frac{1}{2}W^TW + C\sum_{i=1}^{l}(\xi_i + \xi_i^*) \text{ s.t. } \begin{array}{l} y_i - (W^T\Phi(x)+b) \le \varepsilon + \xi_i \\ (W^T\Phi(x)+b) - y_i \le \varepsilon + \xi_i^* \\ \xi_i, \xi_i^* \ge 0, i = 1\cdots l \end{array}. \quad (2)$$

Then define coefficient difference $\theta_i = \alpha_i - \alpha_i^*$, and define margin function

$$h(x_i) = f(x_i) - y_i = \sum_{j=1}^{l} K(x_i, x_j)\theta_j + b - y_i. \quad (3)$$

According to Lagrangian Multiplier Method and Karush-Kuhn-Tucker (KKT) conditions, training samples set can be separated into three subsets:

$$\begin{array}{l} \text{The Error Support Vectors Set: } E = \{i \mid |\theta_i| = C, |h(x_i)| \ge \varepsilon\} \\ \text{The Margin Support Vectors Set: } S = \{i \mid 0 < |\theta_i| < C, |h(x_i)| = \varepsilon\} \\ \text{The Remaining Samples Set: } R = \{i \mid |\theta_i| = 0, |h(x_i)| \le \varepsilon\} \end{array}. \quad (4)$$

According to quadratic optimization theory, if KKT condition is satisfied optimization problem (2) will get a unique resolution. So if we want the KKT conditions to be satisfied, every training sample must belong to one of the three sets. The destination of online SVM regression training algorithm is to add new samples into one of three sets or removes redundant samples out of three sets and in the same time keep any other sample belong to one of three sets. The change of θ_c may change the values of θ_i and $h(x_i)$ of other samples. The updating formulae see also reference [7].

We can use a cache to accelerate the algorithm. A cache is used to keep four kernel function matrixes $\boldsymbol{K}(x_n, x_E)$, $\boldsymbol{K}(x_n, x_R)$, $\boldsymbol{K}(x_n, x_S)$ and $\boldsymbol{K}(x_n, x_c)$.

Online SVM regression training algorithm is consisted of two sub-algorithms. One is the incremental algorithm, the other is the decremental algorithm. The incremental algorithm is used to add a new sample into one of the three sets, and the decremental algorithm is used to remove redundant samples in memory-limited condition.

The detailed process of incremental algorithm is as following:

1. Compute $\boldsymbol{K}(x_n, x_E)$, $\boldsymbol{K}(x_n, x_R)$, $\boldsymbol{K}(x_n, x_S)$, $\boldsymbol{K}(x_n, x_c)$, Set θ_c=0;
2. If $|h(x_c)| \le \varepsilon$, assign x_c to set R, update $\boldsymbol{K}(x_n, x_R)$, terminate.
3. Increase or decrease θ_c according to the sign of $-h(x_c)$, update $b, \theta_i, i \in S$, and $h(x_i)$, $i \in E \cup R \cup c$ until x_c enters into set S or E:

- If $h(x_c) = \varepsilon$, add x_c into set S, update $\boldsymbol{K}(x_n, x_S)$, terminate;
- If $\theta_c = C$ or $-C$, then add x_c into set E, compute $\boldsymbol{K}(x_n, x_E)$, terminate;

- If some sample changes from one of set *R*,*S*,*E* into another, update $\boldsymbol{K}(x_n, x_S)$, $\boldsymbol{K}(x_n, x_R)$ or $\boldsymbol{K}(x_n, x_E)$.

4. Go to step 3.

The detailed process of decremental algorithm is as following:

1. If $x_c \in R$, remove x_c out of training sample set, update $\boldsymbol{K}(x_n, x_E)$, $\boldsymbol{K}(x_n, x_R), \boldsymbol{K}(x_n, x_S)$, terminate.
2. If $x_c \in E$, Remove it out of set *E*, update $\boldsymbol{K}(x_n, x_E)$;
3. If $x_c \in S$, Remove it out of set *S*, update $\boldsymbol{K}(x_n, x_S)$, update matrix ***R***;
4. Increase or decrease θ_c according to the sign of $h(x_c)$, update $b, \theta_i, i \in S$, and $h(x_i)$, $i \in E \cup R \cup c$ until x_c enters into set *R*:

- If $\theta_c = 0$, remove x_c out of training sample set, update $\boldsymbol{K}(x_n, x_S)$, $\boldsymbol{K}(x_n, x_R)$ and $\boldsymbol{K}(x_n, x_E)$, terminate;
- If some sample changes from one of set *R*,*S*,*E* into another, update $\boldsymbol{K}(x_n, x_S)$, $\boldsymbol{K}(x_n, x_R)$ or $\boldsymbol{K}(x_n, x_E)$.

5. Go to step 4.

3 Fast Online SVR Algorithm Based Adaptive IMC

The system structure of adaptive inverse control is shown in fig 1. y_{sp} is reference input, *u* is output of SVM based Controller, *y* is system output, y_m is predictive output of SVM based internal model.

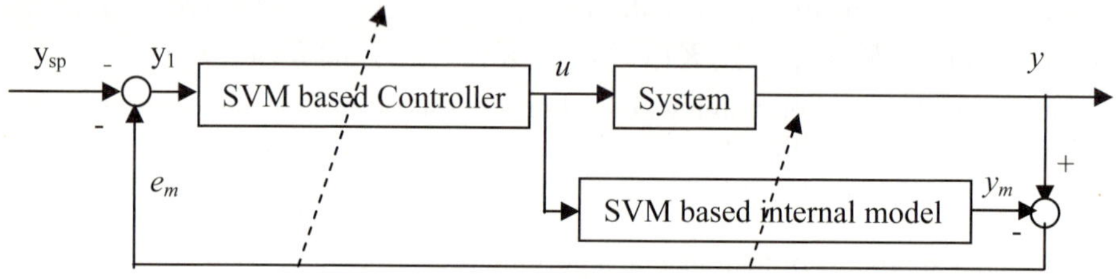

Fig. 1 Online SVM based IMC system

On-line SVM regression algorithm is used to train internal model and inverse model controller. Then internal model control algorithm is run to control the system output. Whenever error of output keeps large for some time, on-line SVM regression algorithm can be used to correct internal model and inverse model controller.

Assume that f_1 is SVM based internal model and f_2 is SVM based internal model controller, $y_m(k)$ is predictive output of f_1, and $u(k)$ is output of f_2. Then

According to principle of internal model control algorithm, during the control process input and output of f_1 can be expressed as:

$$y_m(k+1) = f_1[y(k), y(k-1), \cdots, y(k-n+1), u(k), u(k-1), \cdots, u(k-m+1)], \tag{5}$$

and input and output of f_2 can be expressed as

$$u(k) = f_2[y_1(k+1), y_1(k), \cdots, y_1(k-n+1), u(k-1), \cdots, u(k-m+1)], \tag{6}$$

$y_1(k+1)$ is computed by:

$$y_1(k+1) = y_{sp}(k) - (y(k) - y_m(k)), \tag{7}$$

The whole algorithm can be presented as follows:

1. Initialize parameters of the input and output of system; Construct initial training samples, use online SVM regression algorithm to train an initial internal model and internal model controller.
2. Construct a new training sample, use online SVM regression algorithm to online update internal model f_1 and inverse model controller f_2.
3. Compute $y_m(k+1)$ by (5), $y_1(k+1)$ by (7), and $u(k)$ by (6); Use $u(k)$ to control output of system; If it needs to stop algorithm, terminate; else go to step 4
4. Compute error of output: $y_m(k+1) - y(k+1)$, if it is large than the boundary value err_{mzx} , go to step 2;else go to step 3.

4 Simulation

A nonlinear benchmark problem [9] is described as follows :

$$y(k+1) = \frac{y(k)}{1 + y^2(k)} + u^3(k) + v, \tag{8}$$

where y is the plant output, $u \in [-1,1]$ is the input and v is a random disturb.

The training sample of internal model and internal model controller can be separately constructed as:

$$\begin{aligned} Y_i^1 &= \mathrm{y}(k+1), X_i^1 = [y(k), y(k-1), u(k), u(k-1)] \\ Y_i^2 &= \mathrm{u}(k), X_i^2 = [y_1(k+1), y_1(k), y(k-1)_1, u(k-1)] \end{aligned}, \tag{9}$$

The parameters of the algorithm is that : $C = 5$, $\varepsilon = 0.001$. Kernel function is RBF kernel. First 200 initial training samples constructed by randon input are used to train a initial internal model and internal model controller. In matlab environment the whole running time of training two SVR model with 200 steps is 45.312s, which is much faster than online SVR algorithm without kernel cache method.

Then internal model control algorithm runs 200 time step.Random disturb v has a positive step disturb with a amplitude of 0.1 in step 60 and a negative step disturb

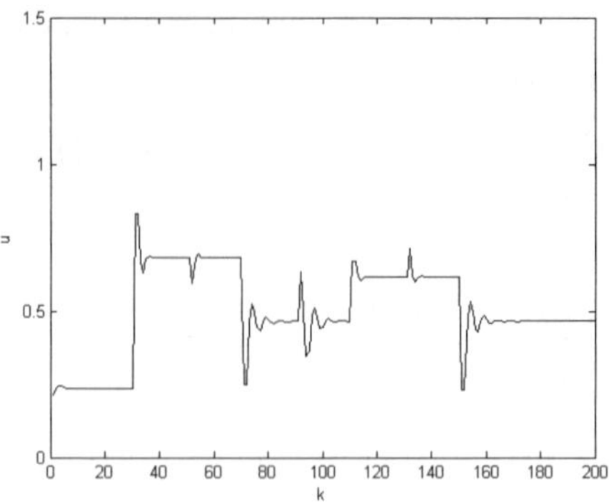

Fig. 2. Control input u

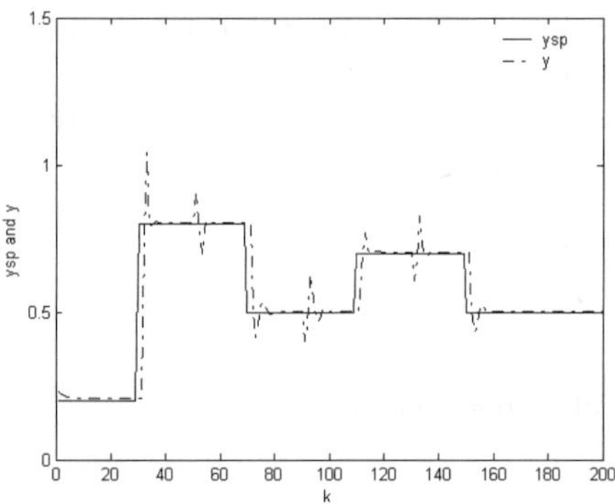

Fig. 3. Reference input ysp and system output y

with a amplitude of 0.1 in step 100 and 140. Sumulation results are shown in figure 2-3. From these figures we can see that on-line SVR algorithm can quickly approximate the system model with very high precision and the SVM based predictive controller can control the system response very well. Also when there are random step disturbs, the controller can rapidly counteract influance from disturbs and comeback system response in several time steps, which shows that the system has well robustness.

Then define e is error between reference input and system output.

$$e^2 = \frac{1}{200}\sum_{i=1}^{200} | y_{sp}(i) - y(i) | = 0.0302 . \tag{10}$$

It shows that the algorithm has high control accuracy.

5 Comparison and Conclusion

Here we compare online SVR algorithm with offline one and neural network (NN).

1. Usability. SVM has uniform structure of model, and only have several adjustable parameters. So it can be more easily used to construct system model than NN.

2. Speed. Speed of online SVM regression algorithms is faster than offline one and NN training algorithm.
3. Adaptivity. Online algorithm can automatically trace variance of model of identified object, so it can be used to construct inverse model of time-varying system.

So the new control algorithm is easy to use; with kernel cache method it has well real-time characteristic; it can control nonlinear system with good performance and robustness. Next work is to analyze the control performance quantitatively, and reduce the memory requirement of the algorithm.

Acknowledgements

This work is sponsored by the 973 program of China under grant No.2002CB312200, the National Natural Science Foundation of China under grant No.60574019 and No.60474045.

References

1. Hu, Q, Saha, P, Rangaiah, G.P.: Experimental Evaluation of An Augmented IMC for Nonlinear Systems. Control Engineering Pratice 8 (2000) 1167-1176
2. Hu, Q, Rangaiah, G.P.: Adaptive Internal Model Control of Nonlinear Process. Chemical Engineering Science 54 (1999) 1205-1220
3. Vapnik, V.N.: The Nature of Statistical Learning Theory. Springer-Verlag New York, USA (1995)
4. Smola, A.J., Schölkopf, B: A Tutorial on Support Vector Regression. Neurocolt Technical Report. Royal Holloway College, Unisversity of London (1998)
5. Müller, K.R., Smola, A.J.: Predicting Time Series with Support Vector Machines. Proceedings of ICANN'97, Lecture Notes in Computer Science, Vol.1327. Springer (1997) 999-1004
6. Cauwenberghs, G. and Poggio, T.: Incremental and Decremental Support Vector Machine. Advances in Neural Information Processing Systems 13 (2001) 409-415
7. Ma, J., Theiler, J., Perkins, S.: Accurate Online Support Vector Regression. Neural Computation 15(11) (2003) 2683-2704
8. Wang, Y., Tu, J.: Neuron Network Control. China Machine Press (1998)
9. Narendra, K.S., Parthasarathy, K.: Identification and Control of Dynamic Systems Using Neural Networks. IEEE Transactions on Neural Networks 1 (2000) 4-27
10. Zhong, W., Pi, D., Sun, Y.: Support Vector Machine Based Direct Inverse-model Identification. Control Theory & Applications 22 (2005) 307-310

A VSC Method for MIMO Systems Based on SVM

Yi-Bo Zhang[1], Dao-Ying Pi[1], Youxian Sun[1], Chi Xu[2], and Si-Zhen Chu[2]

[1] National Laboratory of Industrial Control Technology,
Zhejiang University, Hangzhou 310027, P.R. China
{ybzhang, dypi, yxsun}@iipc.zju.edu.cn
[2] Hangzhou Automation Technology Institute, Hangzhou 310012, China
{xch, csz}@tofine.net

Abstract. A variable structure control (VSC) scheme for linear black-box multi-input/multi-output (MIMO) systems based on support vector machine (SVM) is developed. After analyzing character of MIMO system, an additional control is designed to track trajectory. Then VSC algorithm is adopted to eliminate the difference. By estimating outputs of next step, VSC inputs and additional inputs are obtained directly by two kinds of trained SVMs, and so recognition of system parameters is avoided. A linear MIMO system is introduced to prove the scheme, and simulation shows that the high identification precision and quick training speed.

1 Introduction

The VSC system is a special kind of nonlinear controller characterized by a series of discontinuous control actions that change the control system structure upon reaching a set of switching surfaces [1-3]. The system has invariable properties, and the representative point of the system is constrained to move along a predetermined switching surface. Therefore, performance of system is determined after the switching surface is designed. But VSC must be applied in recognized systems.

Over the past few years, SVM has attracted a lot of researchers from the neural network and mathematical programming community [4-6]. SVM can track arbitrary curves with arbitrary precisions, which means it can be easily applied in recognition of linear and nonlinear systems. SVM have been used in SISO systems [5], but not in MIMO system yet. So research of MIMO system based on SVM is necessary.

A VSC scheme for MIMO systems based on SVM is proposed. After VSC is introduced, two kinds of SVMs are adopted to obtain control algorithm. Then a linear MIMO system is adopted to prove that the effectiveness of the scheme.

This paper is organized as follows: In section 2, SVM regression is mainly concerned, and VSC algorithm and Exponent Reaching Law is introduced in section 3. In the next section, the algorithm is proposed. In section 5, the scheme is proved.

2 Support Vector Machine Regression

Suppose that the training samples are (x_i, y_i), $i = 1,2,....,k$, $\{x_i \in R^n, y_i \in R\}$, the object is to solve the following regress problem:

J. Wang et al. (Eds.): ISNN 2006, LNCS 3972, pp. 928 – 933, 2006.

$$y = f(x) = \langle \omega \cdot x \rangle + b \tag{1}$$

where $\langle \cdot \rangle$ denotes inner product, and b is bias. Vapnik [4] suggested the use of ε − insensitive loss function where the error is not penalized if the loss function is less than ε. Using the error function together with a regularizing term, the optimization problem solved by the support vector machine can be formulated as:

$$\min \frac{1}{2}\|\omega\|^2 + C\sum_{i=1}(\zeta_i + \zeta_i^*) \qquad \text{s.t.} \quad \begin{cases} y_i - \langle \omega \cdot x \rangle - b \le \varepsilon + \xi_i \\ \langle \omega \cdot x_i \rangle + b - y_i \le \varepsilon + \xi_i^* \\ \xi_i, \xi_i^* \ge 0 \end{cases} \tag{2}$$

The constant $C > 0$ determines the tradeoff between the smoothness of f and the amount up to which deviations are larger than ε are tolerated. And ε-insensitive loss function is defined as:

$$|\xi^*|_\varepsilon = \begin{cases} |y_i - \langle \omega \cdot x_i \rangle| & |y_i - \langle \omega \cdot x_i \rangle| > \varepsilon \\ 0 & otherwise \end{cases} \tag{3}$$

Referring to Lagrange multipliers and KKT conditions, the optimal regression equation can be obtained as following:

$$f(x) = \sum_{i=1}^{nsv} \alpha_i x_i + b \tag{4}$$

where nsv is number of support vectors, while α_i are coefficients.

3 Variable Structure Control for MIMO Systems

Consider the following discrete controllable black-box MIMO system:

$$\begin{bmatrix} y_1(k+1) \\ y_2(k+1) \\ \cdots\cdots \\ y_n(k+1) \end{bmatrix} = \begin{bmatrix} a_{11} & a_{12} & \dots & a_{1n} \\ a_{21} & a_{22} & \dots & a_{2n} \\ \dots & \dots & \dots & \dots \\ a_{n1} & a_{n2} & \dots & a_{nn} \end{bmatrix} \begin{bmatrix} y_1(k) \\ y_2(k) \\ \dots \\ y_n(k) \end{bmatrix} + \begin{bmatrix} b_{11} & b_{12} & \dots & b_{1m} \\ b_{21} & b_{22} & \dots & b_{2m} \\ \dots & \dots & \dots & \dots \\ b_{n1} & b_{n2} & \dots & b_{nm} \end{bmatrix} \begin{bmatrix} x_1(k) \\ x_2(k) \\ \dots \\ x_m(k) \end{bmatrix} \tag{5}$$

where $x = [x_1(k) \quad x_2(k) \quad \dots \quad x_m(k)]^T$ and $y = [y_1(k) \quad y_2(k) \quad \dots \quad y_n(k)]^T$ denote input and output matrix respectively, while $A = \{a_{ij} \in R \quad i, j = 1,2,\dots,n\}$ and $B = \{b_{ij} \in R \quad i = 1,2,\dots,n \quad j = 1,2,\dots,m\}$. Suppose $d(k) = [d_1(k) \quad d_2(k) \quad \dots \quad d_n(k)]^T$ is desired output. Let $e(k) = d(k) - y(k)$, where $e(k) = [e_1(k) \quad e_2(k) \quad \dots \quad e_n(k)]^T \in R$ denotes difference. Define $x(k) = x_{vsc}(k) + x_d(k)$, then (5) becomes:

$$e(k+1) = \{Ae(k) - Bx_{vsc}(k)\} + \{-Bx_d(k) - Ad(k) + d(k+1)\} \tag{6}$$

where x_{vsc} denotes variable structure control to reduce effect of difference, while x_d denotes additional control to track $d(k)$. There must exist $x_d(k)$ that can satisfy

$$Bx_d(k) = -Ad(k) + d(k+1) \tag{7}$$

Then (6) becomes:

$$e(k+1) = Ae(k) - Bx_{vsc}(k) \tag{8}$$

Let

$$s(k) = c^T e(k) \tag{9}$$

where $s(k)$ is sliding mode function, and $c = [c_1 \quad c_2 \quad \dots \quad c_n]^T \in R$ is parameter of switching surface. So

$$\begin{aligned} s(k+1) &= c^T e(k+1) \\ &= c^T Ae(k) - c^T Bx_{vsc}(k) \end{aligned} \tag{10}$$

Exponent Reaching Law is adopted [1]:

$$s(k+1) = (1 - \delta\tau)s(k) - \varepsilon\tau \operatorname{sgn} s(k) \tag{11}$$

where $\tau > 0$, $\varepsilon > 0$ and $\delta > 0$ are sampling period, reaching speed and approaching speed respectively, and $1 - \delta\tau > 0$. Then (12) becomes:

$$c^T Ae(k) + c^T Bx_{vsc}(k) = (1 - \delta\tau)s(k) - \varepsilon\tau \operatorname{sgn} s(k)) \tag{12}$$

Because A, B are unknown, $x_{vsc}(k)$ and $x_d(k)$ cannot be obtained, which will be discussed in the next section.

4 The Scheme of VSC Based on SVM Regression

In order to obtain $x_{vsc}(k)$ and $x_d(k)$, two kind of SVMs are adopted.

4.1 SVM for Additional Control of the j th Input

Regard $x_d(k)$ as solution of (7), $x_d(k)$ can be expressed as linear combination of $d(k)$ and $d(k+1)$:

$$\begin{bmatrix} x_{d1}(k) \\ x_{d2}(k) \\ \dots \\ x_{dm}(k) \end{bmatrix} = \begin{bmatrix} e_{11} & e_{12} & \dots & e_{1,2n} \\ e_{21} & e_{22} & \dots & e_{2,2n} \\ \dots & \dots & \dots & \dots \\ e_{m1} & e_{m2} & \dots & e_{m,2n} \end{bmatrix} \begin{bmatrix} d_1(k) \\ \dots \\ d_n(k) \\ d_1(k+1) \\ \dots \\ d_n(k+1) \end{bmatrix} \tag{13}$$

where $\{e_{ij} \in R, i = 1,2,...m, j = 1,2,...,2n\}$. Suppose

$$\begin{aligned} U_{dj} &= [d_1(k) \quad \ldots \quad d_n(k) \quad d_1(k+1) \quad \ldots \quad d_n(k+1)]^T \\ V_{dj} &= x_{dj}(k) \end{aligned} \tag{14}$$

Training $\{U_{dj}, V_{dj}\}$ with SVM, then $x_{dj}(k)$ for the j th input can be obtained from output of this SVM. Because (6) has the same form with (1), training samples can be obtained from (1).

4.2 SVM for VSC of the j th Input

Similarly, regard $x_{vsc}(k)$ as solution for (10), then $x_{vsc}(k)$ can be expressed as linear combination of $s(k+1)$ and $e(k)$. Suppose

$$\begin{aligned} U_{vscj} &= [e_1(k) \quad \ldots \quad e_n(k) \quad s(k+1)]^T \\ V_{vscj} &= x_{vscj}(k) \end{aligned} \tag{15}$$

Training $\{U_{jvsc}, V_{jvsc}\}$ with SVM, then $x_{jvsc}(k)$ for the j th input can be obtained from output of this SVM. But training samples cannot be direct obtained. Let $d(k) = 0$ and consider (13), $x_d(k) = 0$. Then (10) becomes:

$$\begin{aligned} s(k+1) &= -c^T y(k+1) \\ &= c^T A[-y(k)] - c^T Bx(k) \end{aligned} \tag{16}$$

5 Simulation

Consider the discreted nominal linear MIMO system [3]:

$$\begin{bmatrix} y_1(k+1) \\ y_2(k+1) \\ y_3(k+1) \end{bmatrix} = \begin{bmatrix} 0.9658 & 0.0229 & 0.0133 \\ -0.0682 & 0.9021 & 0.0059 \\ 0.0425 & 0.0127 & 0.7066 \end{bmatrix} \begin{bmatrix} y_1(k) \\ y_2(k) \\ y_3(k) \end{bmatrix} + \begin{bmatrix} 8.678 & -0.961 & -2.755 \\ 0 & 2.861 & 0 \\ 0 & 0 & 8.8 \end{bmatrix} \begin{bmatrix} x_1(k) \\ x_2(k) \\ x_3(k) \end{bmatrix}$$

y_1, y_2 and y_3 represent the boiler temperature, the hot water tank temperature, and the space temperature respectively. The desired operating points of these states are 80,60 and 20 degrees. Sample time is $\tau = 0.01$. The initial condition for the state is $y(0) = y_0 = [79 \quad 59 \quad 19]^T$. Choose linear kernel, $C = 1000$, $\varepsilon = 0.01$, and inputs of the system are randomly created by computer within $[-5,5]$. Choose $\delta = 80$, $\varepsilon = 5$. Total training time for all three SVMs is 2.7 seconds. Simulating result is shown in Fig.1. It can be seen that the states converge to desired state in less than 30 sample times, and stable error is less than 0.02%. It is obviously that the scheme of this paper is much better than that of [3] What is more, the MIMO system in our scheme is considered as a black-box system while it is a known system in [3].

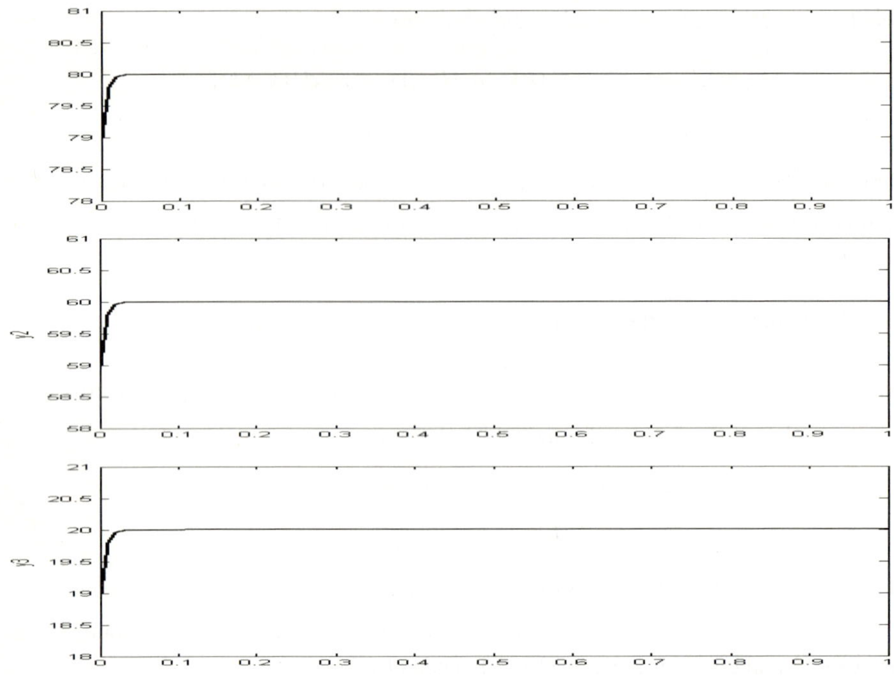

Fig. 1. These are the outputs of system by adopting scheme of [3]

6 Conclusion

Two kinds of SVMs are combined to obtain inputs. Then VSC is adopted to predict the output for the next step. Simulation shows that the scheme is effective.

Acknowledgements

This work is supported by 973 Program of China (Project No.2002CB312200), National Science Foundation of China (Project No.60474045 and No. 60574019), the scientific Research Foundation for the Returned Overseas Chinese Scholars, State Education Ministry, Zhejiang Province and Zhejiang University.

References

1. Gao, W.B.: Discrete-Time Variable Structure Control Systems. IEEE Transactions on Industrial Electronics 42(2) (1995) 117-121
2. Zhai, C.L.: Variable Structure Control Method for Discrete Time Systems. Journal of Shanghai Jiaotong University 34(1) (2000) 719-722
3. Tarek, M. M. N.: A New Variable Structure Control Design with Stability Analysis for MIMO Systems. Electronics, Circuits and Systems 2(2) (2000) 785-788

4. Vapnik, V.: The Nature of Statistical Learning Theory. Springer, NY (1995)
5. Zhong, W.M., Pi, D.Y., Sun, Y.X.: An Approach of Nonlinear Model Multi-step-ahead Predictive Control Based on SVM. Lecture Notes in Computer Science 35(16) (2005) 1036-1039
6. Sanchez Fernandez, M., Jeronimo Arenas, G: SVM Multi-Regression for Nonlinear Channel Estimation in Multiple-Input Multiple-Output Systems. IEEE Transactions on Signal Processing, 52(8) (2004)2298-2307
7. Drezet, P.M.L: Support Vector Machines for System Identification, UKACC International Conf. on Control 1(1) (1998) 668-692

Identification and Control of Dynamic Systems Based on Least Squares Wavelet Vector Machines

Jun Li and Jun-Hua Liu

School of Electrical Engineering, Xi'an Jiaotong University, Xi'an 710049, China
lijun691201@yahoo.com.cn

Abstract. A novel least squares support vector machines based on Mexican hat wavelet kernel is presented in the paper. The wavelet kernel which is admissible support vector kernel is characterized by its local analysis and approximate orthogonality, and we can well obtain estimates for regression by applying a least squares wavelet support vector machines (LS-WSVM). To test the validity of the proposed method, this paper demonstrates that LS-WSVM can be used effectively for the identification and adaptive control of nonlinear dynamical systems. Simulation results reveal that the identification and adaptive control schemes suggested based on LS-WSVM gives considerably better performance and show faster and stable learning in comparison to neural networks or fuzzy logic systems. LS-WSVM provides an attractive approach to study the properties of complex nonlinear system modeling and adaptive control.

1 Introduction

Multilayer Feedforward Neural networks (MFNN) obtain successful application in system identification and control because they provide good approximations to highly nonlinear and complex systems [1], [2]. However, MFNN has poor process interpretability and is hindered by problems associated with weight optimization such as slow learning and local minimization. In the last decade, support vector machines (SVM) and related methods have been introduced within the context of statistical learning theory and structural risk minimization [3]. SVM has been successfully applied for pattern recognition and regression [4]. In this method, one firstly maps the data into a high dimensional space via a nonlinear map, and in this high dimensional space, an optimal separating hyperplane or linear regression function is constructed. This will involve a quadratic programming problem, and will get a global optimal solution. It has been used as an alternative to conventional neural networks at present [3], [4].

Least squares support vector machines (LS-SVM), proposed by Suykens [5] have also been investigated for function estimation, in which the inequality constraints of original SVM are changed to equality constraints, and this leads to solving a linear system instead of a convex quadratic programming problem [6].

Since the wavelet technique shows promise for signal approximation [7], [8], a least squares wavelet support vector machines (LS-WSVM) approach is presented in this paper. That is, an admissible SV kernel, which is a wavelet kernel, may combine the wavelet technique with SVM [9].

J. Wang et al. (Eds.): ISNN 2006, LNCS 3972, pp. 934 – 942, 2006.

The effectiveness of the proposed method is verified by two simulation examples which include identification and adaptive control of nonlinear dynamical systems. Simulation results show that the identification and adaptive control schemes based on LS-WSVM gives considerably better performance and show faster and stable learning in comparison to previous neural networks or fuzzy logic systems methods.

This paper is organized as follows. In Section 2, a brief review of the LS-SVM formulation for function estimation and a novel wavelet kernel which is a kind of multidimensional wavelet function is presented. Simulation examples are provided in Section 3 to illustrate the performance of the LS-WSVM. These examples include modeling nonlinear components of multi-input multi-output (MIMO) dynamical systems, followed by tracking control based on a model reference adaptive scheme (MRAC) for nonlinear dynamical systems, respectively. Finally, Conclusions and suggested extensions to the research are presented in Section 4.

2 Least Squares Wavelet Support Vector Machines for Nonlinear Function Approximation

2.1 Least Squares Support Vector Machines for Function Approximation

Consider first a model in the prime weight space of the following form:

$$y(x) = w^T \varphi(x) + b . \tag{1}$$

where $x \in \mathrm{R}^n$, $y \in \mathrm{R}$ and $\varphi(\cdot): \mathbb{R}^n \to \mathbb{R}^{n_h}$ is the mapping to the high dimensional and potentially inifinite dimensional feature space. Given a training data set $\{\boldsymbol{x}_k, y_k\}_{k=1}^N$ we can formulate optimization problem in the prime weight space

$$\min_{w,b,e} J_P(w,b,e) = \frac{1}{2} w^T w + \frac{\gamma}{2} \sum_{k=1}^{N} e_k^2 . \tag{2}$$

such that

$$y_k = w^T \varphi(x_k) + b + e_k, k = 1, \ldots, N. \tag{3}$$

To solve the primal problem, we may construct the Lagrangian and derive the dual problem.

$$L(w,b,e,\alpha) = J(w,e) - \sum_{k=1}^{N} \alpha_k \left\{ w^T \varphi(x_k) + b + e_k - y_k \right\}. \tag{4}$$

where α_k are the Lagrangian multipliers. From the conditions for optimality and elimination of the variables w and e, we gets the following solution of KKT system

$$\begin{bmatrix} 0 & 1_N^T \\ 1_N & \Omega + \gamma^{-1} I_N \end{bmatrix} \begin{bmatrix} b \\ \alpha \end{bmatrix} = \begin{bmatrix} 0 \\ Y \end{bmatrix}. \tag{5}$$

with $Y=[y_1,\dots,y_N]^T$, $1_N=[1,\dots,1]^T$, $\alpha=[\alpha_1,\dots,\alpha_N]^T$, I_N is an $N\times N$ identity matrix, and the kernel matrix $\Omega\in\mathbb{R}^{N\times N}$ with $\Omega_{kj}=\phi(x_k)^T\phi(x_j)=K(x_k,x_j)$. In numerical linear algebra, efficient algorithms exist for solving large scale linear systems. The final LS-SVM model for function estimation becomes then.

$$y(x)=\sum_{k=1}^{N}\alpha_k K(x,x_k)+b. \tag{6}$$

where α_k, b are the solution to the linear system.

2.2 LS-SVM Based on Mexican Hat Wavelet Kernel

The formation of a support vector (SV) kernel is a kernel of dot-product type in some feature space. The chosen kernel function should be positive definite and satisfy the Mercer theorem [3], [10]. For SVM or LS-SVM, we usually use some kernel functions which include linear, Polynomials, Gaussian kernel functions etc.

There are two key properties that are required of a SV kernel function for an application. Firstly, it should capture the measure of similarity appropriate to the particular task and domain, and secondly, its evaluation should require significantly less computation than would be needed in an explicit evaluation of the corresponding feature mapping.

The Gaussian kernel $k(\boldsymbol{x},\boldsymbol{x}')=\exp\left(-\frac{\|\boldsymbol{x}-\boldsymbol{x}'\|^2}{2\sigma^2}\right)$ is one of the first support vector kernel studied for most of machine learning problems. $\sigma>0$ is a parameter chosen by user.

Wavelets usually form an orthonormal basis of $L_2(\mathbb{R})$ and are now commonly used for signal denoising or signal compression. In [9], ε-insensitive support vector machines based on wavelet kernels are presented and the existence of wavelet kernels is also proved by results of theoretic analysis. Since the wavelet technique shows promise for nonstationary signal approximation [11], it is valuable for us to study the problem of whether a better performance may be obtained if we combine the wavelet technique with LS-SVM.

In the paper, to construct wavelet kernel, without loss of generality, one select another translation-invariant wavelet kernel by a Mexican hat wavelet function adopted in [11].

$$h(x)=(1-x^2)\exp(-x^2/2). \tag{7}$$

Furthermore we may have the following conclusion.

Theorem: Given the mother wavelet (7) and the dilation a, $a, x\in R$, If $\boldsymbol{x}, \boldsymbol{x}'\in\mathbb{R}^n$, the wavelet kernel of this mother wavelet is

$$
\begin{aligned}
k(x,x') = k(x-x') &= \prod_{i=1}^{n} h\left(\frac{x_i - x_i'}{a}\right) \\
&= \prod_{i=1}^{n}\left(\left(1-\frac{\| x_i - x_i' \|^2}{a^2}\right)\exp\left(-\frac{\| x_i - x_i' \|^2}{2a^2}\right)\right)
\end{aligned}
\quad . \tag{8}
$$

which is an admissible support vector kernel. The proof of the theorem is similar to [9], so it is omitted here.

Now, the final least squares wavelet support vector machines (LS-WSVM) model for function estimation becomes

$$
y(x) = \sum_{k=1}^{N} \alpha_k \prod_{j=1}^{n} \left(1-\frac{\| x^j - x_k^j \|^2}{(a_k)^2}\right)\exp\left(-\frac{\| x^j - x_k^j \|^2}{2(a_k)^2}\right) + b \, . \tag{9}
$$

where the x_k^j denotes the jth component of the kth training sample.

Based on (6) and (9), LS-SVM can be expressed as the linear combination of wavelet kernel. The goal of LS-WSVM is to find the optimal approximation in the space spanned by multidimensional wavelets or wavelet kernels. Since LS-SVM cannot optimization the parameters of the kernels, it is different to determine N parameters a_k, k=1,..., N. For the sake of simplicity, let a_k=a such that the number of parameters becomes 1.

3 Application to Identification and Adaptive Control of Nonlinear Dynamic System

In this section, the LS-WSVM is applied to two illustrative examples. Note that in the case of RBF kernels or wavelet kernel, we all have only two additional tuning parameters (γ, σ) or (γ, a), which is less than for standard SVMs, and kernel parameters were tuned by 10-fold cross-validation or Bayesian Inference [6].

LS-WSVM is first used as identifier for multi-input multi-output dynamical process modeling. Furthermore, we expand on the comparison between our results and other landmark results related to the important identification problem. In the second example, LS-WSVM model is applied to design controller based on the model reference adaptive control (MRAC) scheme for dynamical systems.

Example 1: The identification of the nonlinear mulitivariable plant with two inputs and two outputs discussed in the example, it is shown that the LS-WSVM methods used to identify single-input single-output (SISO) plants can be used to identify MIMO plants as well.

The difference equation describing the plant was assumed to be of the form [12],

$$
\begin{bmatrix} y_{p1}(k+1) \\ y_{p2}(k+1) \end{bmatrix} = \begin{bmatrix} \dfrac{y_{p1}(k)}{1+y_{p2}^2(k)} \\ \dfrac{y_{p1}(k)\,y_{p2}(k)}{1+y_{p2}^2(k)} \end{bmatrix} + \begin{bmatrix} u_1(k) \\ u_2(k) \end{bmatrix} . \tag{10}
$$

A series-parallel identification model based on LS-WSVM is described by the equation

$$\begin{bmatrix} y_{p1}(k+1) \\ y_{p2}(k+1) \end{bmatrix} = \left[N\left[y_{p1}(k), y_{p1}(k) \right] \right] + \begin{bmatrix} u_1(k) \\ u_2(k) \end{bmatrix}. \tag{11}$$

The LS-WSVM model has two inputs variable and two outputs. 1600 simulated data points are generated from the plant model (10). The first 1500 data points are obtained by assuming random inputs $u_1(k)$ and $u_2(k)$ uniformly distributed in the interval [-3, 3], and the last 100 data points are obtained by a sinusoid vector input $[sin(2\pi k/25), cos(2\pi k/25)]^T$. We use the first 1500 data points to build a LS-WSVM model, and the performance of the LS-WSVM is tested using the remaining 100 data points. The (γ, a) parameters are set as (30, 0.2) by 10-fold cross-validation.

When the identification procedure was carried out for 1500 time steps, the responses of the plant and the identification model are shown in Fig.1. It can be seen

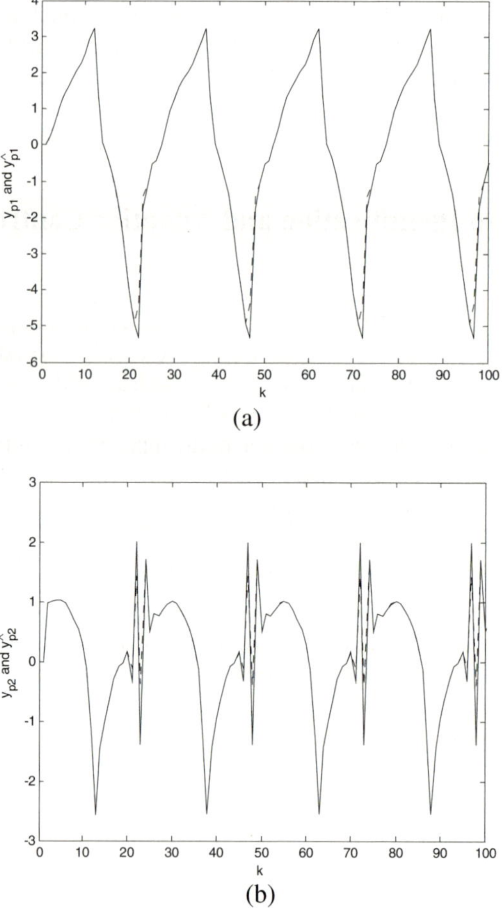

Fig. 1. Outputs of the identified model and the actual model

from Fig.1 (a) and Fig.1 (b) that the performance of LS-WSVM is very good. In Fig. 1, the dashed line represents the output of the identification model and the solid line represents the practical output of the plant.

The (γ, σ) parameters in RBF kernel is set to equal the (γ, a) parameters in wavelet kernel, the respective approximation results (MSE i.e. Mean Square Error) using LS-WSVM and LS-SVM based on RBF kernel are listed in Table 1. From Table 1, we can find the proposed wavelet kernel has better approximate results than the Gaussian kernel.

Furthermore, the approximation power of the LS-WSVM model can be appreciated if we compare the achieved modeling performance MSE with other results [12], [13]. In the examples, only 1500 data points have been used to identify the model in the training data region, while in [13], 5,000 data points have been used to identify corresponding two fuzzy systems model. In [12], 100,000 data points have been used to identify corresponding two neural networks model.

Table 1. Comparison results of multivariable nonlinear system identification

Kernel Name	MSE(training)	MSE (test)
Mexican Hat Wavelet Kernel	0.0125	0.0359
with γ=30□a=0.2	0.0256	0.0553
Gaussian RBF Kernel with	0.0197	0.0402
γ=30□σ=0.2	0.0315	0.0948

Example 2: We consider here the problem of controlling the three-order nonlinear plant which is described by the difference equation [12]

$$y(k+1)=f\left[y(k),y(k-1),y(k-2)\right]+u(k)+0.8u(k-1). \tag{12}$$

where the function

$$f\left[y(k),y(k-1),y(k-2)\right]=\frac{5y(k)y(k-1)}{1+y^2(k)+y^2(k-1)+y^2(k-2)}. \tag{13}$$

is assumed to be unknown. The aim of control is to determine a controller $u(k)$ which based on LS-WSVM model such that the output $y(k)$ of the closed-loop system follows the output $y_m(k)$ of the following reference model:

$$y_m(k+1)=0.32y_m(k)+0.64y_m(k-1)-0.5y_m(k-2)+r(k) \tag{14}$$

where $r(k)=sin(2\pi k/25)$. That is, we want the output error $e(k)=y(k)-y_m(k)$ converges to zero as k goes to infinite.

If the function f [*] of (13) was known, we would construct a controller based on the principle of inverse controls as follows:

$$\begin{aligned}u(k)=-f\left[y(k),y(k-1),y(k-2)\right]-0.8u(k-1)+0.32y(k)\\+0.64y(k-1)-0.5y(k-2)+r(k)\end{aligned}. \tag{15}$$

Resulting in the error equation $e(k+1)=0.32e(k)+0.64e(k-1)-0.5e(k-2)$. Since the reference model is asymptotically stable, it follows that $lim_{k\to\infty}e(k)=0$ for arbitrary initial conditions. However, since f [*] is unknown, the controller cannot be implemented. To solve this problem, we replace the f [*] in (13) by a LS-WSVM model N. That is, it is estimated on line as $\hat{f}$ by using a least squares wavelet support vector machines model N and the series-parallel method.

The following controller is as follows:

$$\begin{aligned} u(k) = -N\left[y(k), y(k-1), y(k-2)\right] - 0.8u(k-1) + 0.32y(k) \\ +0.64y(k-1) - 0.5y(k-2) + r(k) \end{aligned} \tag{16}$$

This results in the following nonlinear difference equation

$$\begin{aligned} y(k+1) = f\left[y(k), y(k-1), y(k-2)\right] + 0.32y(k) + 0.64y(k-1) \\ -0.5y(k-2) + r(k) - N\left[y(k), y(k-1), y(k-2)\right] \end{aligned} . \tag{17}$$

governing the behavior of the closed-loop system.

Fig.2 shows the structure of overall adaptive control systems which stands for a two-order nonlinear system. Likely, for the three-order nonlinear system in example 2, the basic configuration of the overall adaptive control system is also similar to Fig.2. From Fig.2 we see that the LS-WSVM consists of two parts: an identifier and a controller. The identifier uses the LS-WSVM model to approximate the unknown nonlinear function f in the plant, and this LS-WSVM model is then copied to the controller.

Hence, the control task is achieved by employing the inverse principle for feedback linearization within the framework of a model reference adaptive control (MRAC).

The simulation of overall closed-loop system proceeds as follows. First, the identification phase is carried out offline, the first 2500 data points are obtained by using the input $u(k)$ as a random signal uniformly distributed in the interval [-2, 2]. We use LS-WSVM model to the above three-order dynamic plant and the parameters in LS-WSVM are set as $\gamma=30$, $a=0.3$. After the identification procedure is terminated, (16) is used to generate the control input; that is, the controller in Fig.2 begins operating with N copied from the final N in the identifier.

Fig.3 shows the output $y(k)$ of closed-loop system with the controller together with the reference mode output $y_m(k)$. The control MSE using the above LS-WSVM model is 0.002. From the simulation results, we see that the proposed method works very well in producing accurate tracking control using the above scheme.

For the purpose of comparison, kernel parameters in LS-SVM based on RBF kernel are also set as $\gamma=30$, $\sigma=0.3$. The control MSE using the LS-SVM based on RBF kernel model is 0.012. Furthermore, the approximation power of the LS-WSVM model can be greatly appreciated if we compare the achieved modeling performance MSE with other results [12], [13], [14]. In [14], the best control MSE using adaptive fuzzy- neural model with 15 rules is 0.003. It is also inferior to our LS-WSVM model. Remember that in the example only 2500 data points are used to build the LS-WSVM model or LS-SVM model, while in [12], the neural controller achieved similar performance when the identification procedure was carried out 50,000 steps, i.e. 50,000 data points have been used to identify a neural network model.

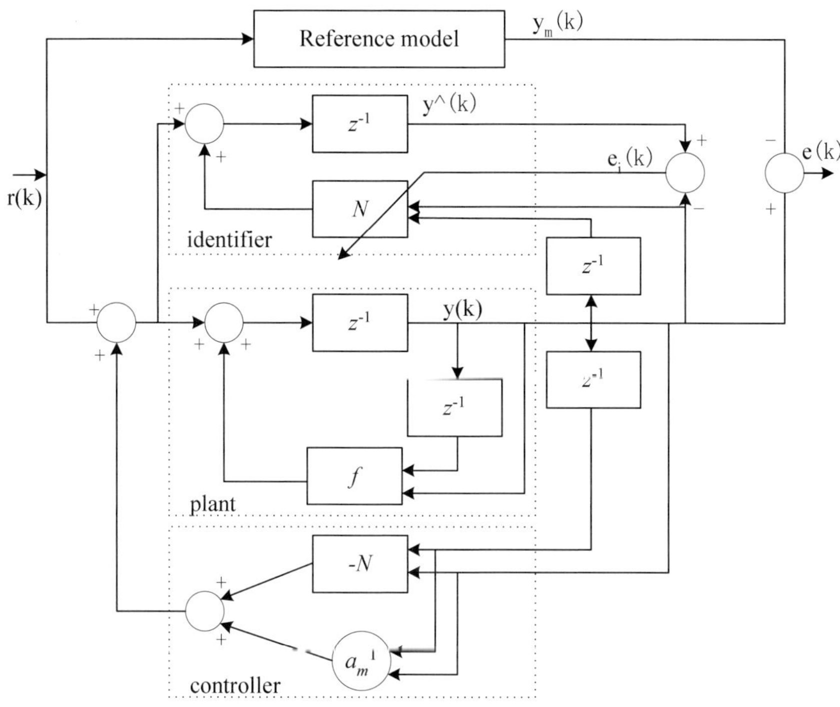

Fig. 2. Structure of the overall adaptive control systems

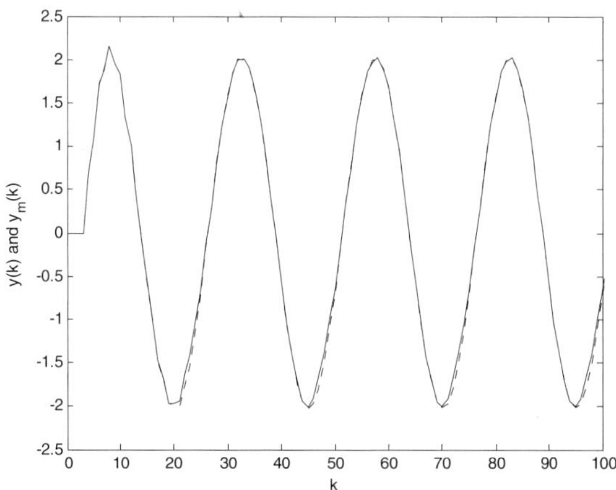

Fig. 3. The output $y(k)$ (solid line) of the closed-loop system and the reference trajectory $y_m(k)$ (dashed line) for Example 2

Notice that the wavelet kernel is orthonormal or orthonormal approximately, whereas the Gaussian kernel is correlative or even redundancy, which is the possible reason that the approximate accuracy is slightly higher than the Gaussian kernel LS-SVM in above examples.

4 Conclusion

The paper develops a novel least squares wavelet support vector machines approach to dynamical system identification and adaptive control, in which the wavelet kernel technique has played a central role. Compared with LS-SVM based on RBF kernel, LS-WSVM shows better performance with higher approximation accuracy. In the application examples, we have illustrated how LS-WSVM model can be successfully utilized for identification of MIMO nonlinear dynamical systems, and we also used the LS-WSVM to construct adaptive controllers for nonlinear dynamic systems. The simulation results demonstrate that the performances of identification and control scheme based on LS-WSVM are considerably satisfactory. It is an attractive new method for multivariable nonlinear system modeling and control in comparison to feedword neural network method or fuzzy logic systems method. MIMO systems identification and control using other types of LS-WSVM in which different kernel may be constructed based on wavelet frame may be exploited in the future.

Acknowledgement

This work is supported by the National Science Foundation of China (no.60276037).

References

1. Lin, C. T., Lee, C. S. G.: Neural Fuzzy System. Prentice-Hall, New Jersey (1996)
2. Haykin, S.: Neural Networks: A Comprehensive Foundation. 2rd edn. Prentice-Hall, New Jersey (2001)
3. Vapnik, V.: The Nature of Statistical Learning Theory. 2rd edn. Springer-Verlag, Berlin-Heidelberg New York (1998)
4. Burges, C. J. C.: A Tutorial on Support Vector Machines for Pattern Recognition. Data Mining Knowl. Disc, 2. (1998) 1–43
5. Suykens, J.A.K., Vandewalle, J.: Least Squares Support Vector Machine Classifiers. Neural Processing Letters, 9 (1999) 293–300
6. Suykens, J.A.K., Gestel, T.V., Brabanter, J.D., Moor, B.D., Vandewalle, J.: Least Squares Support Vector Machines. World Scientific, Singapore (2002)
7. Zhang,Q., Benveniste, A.: Wavelet Networks. IEEE Trans. Neural Networks 3 (1992) 889–898
8. Zhang, Q.: Using Wavelet Network in Nonparametric Estimation. IEEE Trans. Neural Networks 8 (1997) 227–236
9. Zhang, L., Zhou, W.D., Jiao, L.C.: Wavelet Support Vector Machine. IEEE Trans. Systems, Man, and Cybernetics-Part B: Cybernetics 34 (2003) 1–6
10. Smola, A., Scholkopf, B., Muller, K.R.: The Connection between Regularization Operators and Support Vector Kernels. Neural Networks 11 (1998) 637–649
11. Mallat, S.: A Wavelet Tour of Signal Processing. 2nd edn. Academic Press, San Diego (1999)
12. Narendra, K., Parthasarathy, K.: Identification and Control of Dynamical Systems Using Neural Networks. IEEE Trans. on Neural Networks 1 (1990) 4–27
13. Wang, L.X.: Adaptive Fuzzy Systems and Control: Design and Stability Analysis. Prentice Hall, Englewood Cliffs, NJ (1994)
14. Barada, S., Singh, H.: Generating Optimal Adaptive Fuzzy-Neural Models of Dynamical Systems with Applications to Control. IEEE Trans. Systems, Man, and Cybernetics-Part C: Applications and Reviews 28 (1998) 371–391

A Nonlinear Model Predictive Control Strategy Using Multiple Neural Network Models

Zainal Ahmad[1] and Jie Zhang[2]

[1] School of Chemical Engineering, University Sains Malaysia, Engineering Campus, Seri Ampangan, 14300, Nibong Tebal, Penang, Malaysia
chzahmad@eng.usm.my

[2] School of Chemical Engineering and Advanced Materials, University of Newcastle, Newcastle upon Tyne NE1 7RU, UK
jie.zhang@newcastle.ac.uk

Abstract. Combining multiple neural networks appears to be a very promising approach for improving neural network generalization since it is very difficult, if not impossible, to develop a perfect single neural network. Therefore in this paper, a nonlinear model predictive control (NMPC) strategy using multiple neural networks is proposed. Instead of using a single neural network as a model, multiple neural networks are developed and combined to model the nonlinear process and then used in NMPC. The proposed technique is applied to water level control in a conic water tank. Application results demonstrate that the proposed technique can significantly improve both setpoint tracking and disturbance rejection performance.

1 Introduction

Neural networks have been widely used not only in the engineering field but also in other applications like remote sensing, transportation, power systems, medicine, telecommunication, and banking. The growing interests in applying neural networks are due to the rapid development in computing power which enables neural networks being trained in short time durations when modeling the behavior of complex systems. Furthermore the characteristic of neural network models themselves that learn from examples rather than having to be programmed also contributed their increased applications. The architecture of single neural networks varies from multilayer perceptron to radial basis function and also recurrent neural networks. Currently, applications of single neural networks in process modeling and control are quite significant in industry especially in model based predictive control (MBPC) [1]. This is mainly due to the capability of neural networks in modeling nonlinear processes from process operation data. However, single neural networks usually lack generalization capability due to over-fitting, limitation of training data, and network training trapped in undesirable local minima. Recent studies have shown that this limitation can be overcome by combining multiple neural networks. Fig. 1 shows how multiple neural networks are combined. The individual networks in Fig. 1 model the same relationship and are developed from different data sets and/or different training algorithms. They can also have different structures. Instead of choosing the single "best"

J. Wang et al. (Eds.): ISNN 2006, LNCS 3972, pp. 943 – 948, 2006.

neural network model, all the individual neural networks are combined. Note here that if a single network is selected, then it is a common practice to select the best network on the training and/or testing data. However, this "best" network may not be the best when applied to unseen data. There are a number of methods in combining the networks like stacked neural network and bootstrap aggregated network where multiple networks are created on bootstrap re-samples of the original training data [2],[3].

A nonlinear model predictive control (NMPC) strategy using multiple neural networks is proposed in this paper. NMPC basically requires an accurate model of the nonlinear process to be controlled and predict the controlled variable over a future prediction horizon under a sequence of future control actions. An optimization procedure is then carried out to minimize the control errors over the prediction horizon by finding the appropriate sequence of control actions. Since a multiple neural network can offer enhanced prediction accuracy, it is expected that multiple neural network based NMPC can outperform single neural network based NMPC.

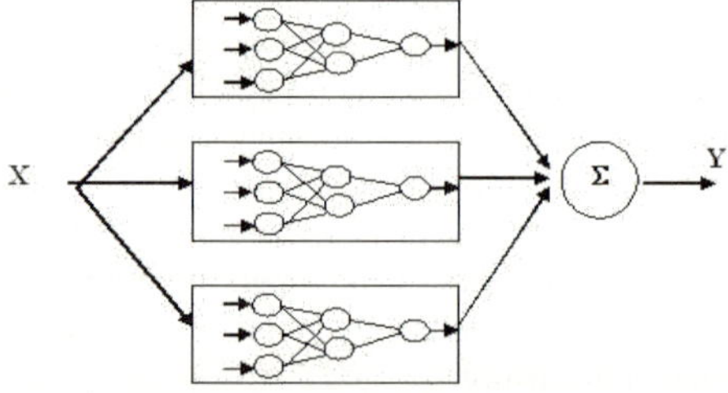

Fig. 1. Combining multiple neural networks

2 NMPC Using Multiple Neural Networks

Model Predictive Control (MPC) basically is a methodology that refers to a class of control algorithms in which a dynamic model of the plant is used to predict and optimize the future behavior of the process. At each control interval, the MPC algorithm computes a sequence of the manipulated variables in such a way to optimize the future behavior of the plant. MPC has been used in industry for nearly 30 years, and has become an industry standard (mainly in the petrochemical industry) due to its intrinsic capability for dealing with constraints and with multivariable systems. Most commercially available MPC technologies are based on a linear model of the controlled process. For processes that are highly nonlinear, the performance of linear model based MPC can be poor. This has motivated the development of Nonlinear Model Predictive Control (NMPC), where a more accurate nonlinear model of the plant is used for prediction and optimization. Many of the current NMPC schemes are based on physical models of the controlled processes. However, in many cases such models are difficult to obtain, and often not available at all. Neural network model based NMPC has been reported by Zhan and Ishida [4]. The basic structure of this predictive control scheme can be found in reference [5] which has been simplified by Hussain. In this paper we propose to use robust multiple neural network models in NMPC. Here instead of using single neural network model, multiple networks are used. The objective function of NMPC is given as

$$\min \sum_{k=1}^{N} \left(y_{sp}(t+k) - y_{predic}(t+k)\right)^2 + \sum_{k=1}^{M} \lambda\left(u(t+k-1) - u(t+k-2)\right)^2 \tag{1}$$

The optimization is subject to the following constraints

$$y_{\min} \le y_{\text{predict}}(t+k) \le y_{\max} \quad (k=1,..,N) \tag{2}$$

$$u_{\min} \le u(t+k) \le u_{\max} \quad (k=0,1,..,M-1) \tag{3}$$

$$\left|u(t+k) - u(t+k-1)\right| \le \Delta u_{\max} \quad (k=1,..,N) \tag{4}$$

where N is the predictive horizon, M is the control horizon, and λ is the control weight or a penalty term for large control action variations. The decision variable u(t+k) (k=0,..,M-1) are the control moves over a manipulated input horizon M (M<N) and are assumed to be kept constant for the remaining of the sampling intervals.

$$u(t+k) = u(t+M-1) \text{ (for } k=M,..,N-1) \tag{5}$$

Although the optimal control actions are obtained for a future control horizon, only the first control action is implemented. Then process output measurements are obtained and the process/model mismatch is compensated. The optimization is performed again at the next sampling interval, applying the following compensation for process (y(t))/model ($y^m(t)$) mismatch:

$$d(t) = c[y(t) - y^m(t)] \quad (0<c<1.0) \tag{6}$$

$$y_{predict}(t+k) = y^m(t+k) + d(t) \quad \text{(for all } k = 1,..,N) \tag{7}$$

where *c* is a tunable parameter.

3 Result and Discussion

Conic water tanks were used in this case study. There is an inlet stream to the tank and an outlet stream from the tank. Manipulating the inlet water flow rate will regulate the water tank level. A detailed schematics diagram can be found in reference [6] and mechanistic model is developed based on material balance and is used to simulate the process. The sampling time used is 10 seconds. The detailed mechanistic model indicates that the relationship between the inlet water flow rate and the water level in the tank is quite nonlinear. The outlet valve characteristic determines that the static gain increases with tank level. Because the tank is of a conical shape, the time constant of the processes increases with the tank level. Thus, both the static and dynamic characteristics of the process vary with the operating condition. All the network building data were generated from the simulation program and normally distributed noise with zero mean and a standard deviation of 0.7cm were added to the simulated tank level. A multiple neural network containing 20 individual networks with fixed structure (single hidden layer with 4 hidden neurons) and a multiple neural network containing 20 individual networks with various structures (single hidden layer with the number of hidden neurons ranging from 1 to 10) were developed to

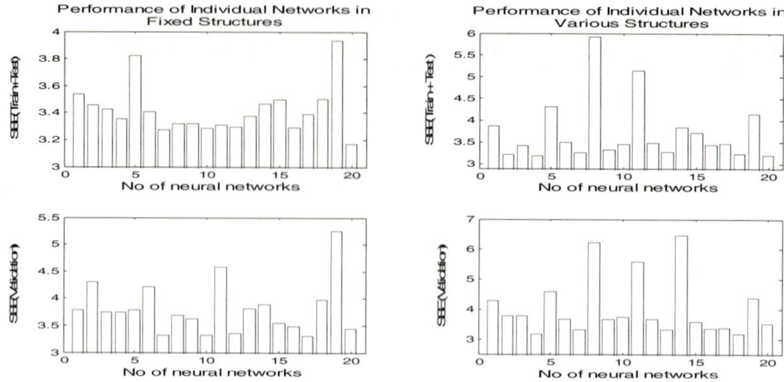

Fig. 2. SSE of one-step-ahead predictions from individual neural networks

model the nonlinear dynamic relationship between the inlet water flow rate and the water level in the tank from the process operating data.

In this study, multiple neural networks were combined using the simple averaging approach. The appropriate model structure was determined through cross validation. Fig. 2 shows the sum of squared errors (SSE) of one-step-ahead predictions from the individual networks on the training and testing data and on the unseen validation data. It can be seen that the individual networks give inconsistent performance on the training and testing data and on the unseen validation data. This indicated that non-robust nature of single neural networks. Neural network generalization capability can be combined by combining multiple neural networks. As shown in Fig. 3, the long range prediction performance of the combined neural network models with fixed and various structures is quite good, even though there are some errors in the predictions but the predictions still follow the trend in the actual data. Then NMPC is developed using the multiple neural networks to control the tank level.

The optimization method used is the sequential quadratic programming implemented as the *constr.m* function in the Matlab[TM] Optimisation Toolbox. The NMPC

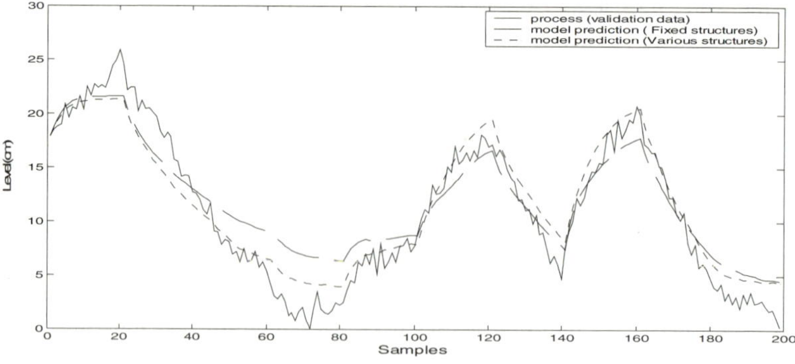

Fig. 3. Long range predictions on validation data from multiple neural network models with fixed and various structures

Table 1. Nominal values for simulation study in NMPC

Variables	Meanings	Nominal values
N	Prediction horizon	7
M	Control horizon	1
y_{min}	Minimum of the controlled variable	5 cm
y_{max}	Maximum of the controlled variable	30 cm
u_{max}	Maximum of the manipulated variable	250 cm^3/s
u_{min}	Minimum of the manipulated variable	70 cm^3/s
Δu_{max}	Maximum change in manipulated variable	180 cm^3/s
λ	Control weight	0.02
c	Integral term	0.1

parameters are listed in Table 1. To eliminate the offset of the prediction output, model plant mismatch is compensated using Eq (4) and Eq (5). An integral action was also added in the control action to eliminate any static offsets. Overall the setpoint tracking performance of NMPC based on multiple neural networks with various structures is better than based on multiple neural networks with fixed structures as shown in Fig. 4 and Table 2 especially in the high region of the operating condition. Even though the NMPC base on the "best" single network model gives quite poor performance in the

Table 2. Sum of squared control errors

	Setpoint tracking		Disturbance rejection	
	Fixed structures	Various structures	Fixed structures	Various structures
NMPC using 'best'	18495	8426	13.81	31.52
NMPC using multiple	5795	5419	14.60	7.89

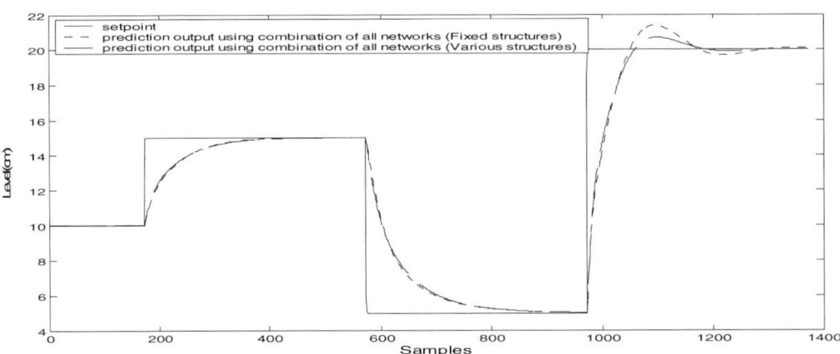

Fig. 4. Setpoint tracking performance for NMPC based on multiple neural networks with fixed and various structures

setpoint tracking as shown in Table 2 but by combining all these networks, better control performance is obtained. In disturbance rejection performance, multiple neural networks are superior to single neural networks as shown in Table 2. The sum of squared control errors from NMPC based on multiple neural networks is much lower than that from NMPC based on single neural networks.

4 Conclusions

Implementing NMPC using multiple neural network models is investigated in this paper. Instead of using the conventional single neural network models, multiple neural network models are used in order to take the advantage of their enhanced long range prediction performance. It is shown in the case study that NMPC based on multiple neural network models gives improved control performance in both setpoint tracking and disturbance rejection.

Acknowledgements

This work was supported by University Science Malaysia (for Z. Ahmad) and UK EPSRC through the Grant GR/R10875 (for J. Zhang).

References

1. Chen, J., Yea, Y.: Neural Network-Based Predictive Control for Multivariable Processes. Chemical Engineering and Communication 189 (2002) 866-894
2. Sridhar, D. V., Bartlett, E. B., Seagrave, R. C.: An Information Theoretic Approach for Combining Neural Network Process Models. Neural Networks 12 (1999) 915-926
3. Zhang, J.: Sequential Training of Bootstrap Aggregated Neural Networks for Nonlinear Systems Modeling. American Control Conference, Vol. 1 (2002) 531-536
4. Zhan, J., Ishida, M.: The Multi-step Predictive Control of Nonlinear SISO Processes with a Neural Model Predictive Control (NMPC) Method. Computers and Chemical Engineering 21(2) (1997) 201-210
5. Hussein, M. A.: Review of the Application of Neural Networks in Chemical Process Control-Simulation and Implementation. Artificial Intelligence in Engineering 13 (1999) 55-68
6. Zhang, J.: Developing Robust Neural Network Models by Using Both Dynamic and Static Process Operating Data. Ind. Eng. Chem. Res. 40 (2001) 234-241

Predictive Control Method of Improved Double-Controller Scheme Based on Neural Networks

Bing Han and Min Han

School of Electronic and Information Engineering, Dalian University of Technology, Dalian 116023, China
minhan@dlut.edu.cn

Abstract. This paper considers the problem of stabilizing a black-box plant with time delay using an improved double controller scheme. The PID parameters of the load controller of the double-controller scheme are obtained by a neural network controller with back propagation algorithm. Based on the adaptive algorithm of Universal Learning Network (ULN), ULN is adopted for modeling the plant and being a predictor of the control system. Simulation results prove the applicability and effectiveness of the improved double-controller scheme. ULN and the neural network controller give the double-controller scheme more representing abilities and robust ability.

1 Introduction

In modern industry processes, such as pH neutralization, temperature control, etc [1]. Requirements for the stability and robust ability of time delay control system increased significantly because of the increased complexity of the plants and the various disturbances. Due to the time delay part of the plant, it is highly difficult to control such process. Recent researches show that the key point to control the process with time delay is to predict the output in the future time, so as to the effect of the time delay can be restrained or eliminated in advance. The Smith predictor is an effective time delay compensator with a suitable process model [2]. Base on the idea of Smith, the double-controller scheme [3] is proposed for the time delay processes with strong robustness. The scheme consists of two controllers, a setpoint controller and a load controller, and the setpoint controller can be chosen to be PID controller simplicity. For the load controller, the theory on stabilizing PID controller [4] and internal model control principle [5] are adopted for turning the parameters. These methods are all based on the mathematical model of the plant, so they do not suit for the black-box process. For this problem, a self-tuning neural network controller is introduced for the load controller of double-controller scheme in this paper.

Since it is even impossible to obtain a suitable physically founded process model due to the complexity of the underlying processes or the lack of knowledge of critical parameter of the models, neural network, a nonlinear black-box model of the dynamic behavior, is seen as a promising way to overcome above problems [6]. So we employed a new type dynamical recurrent Universal Learning Network with an adaptive algorithm [7]. Universal Learning Network provides a generalized framework to model and

J. Wang et al. (Eds.): ISNN 2006, LNCS 3972, pp. 949–955, 2006.

control complex systems. Multiple branches and their arbitrary time delay can be effective to model the dynamic systems with long time delay in a very compact network. Because of the arbitrary time delay in the network, the high degree of freedom in their architectures gives ULN more flexibilities and representing abilities. Simulations show that the improved double-controller scheme cooperating with ULN predictive model can be effectively used in time delay system control.

In the following sections, section 2 gives the structure of ULN predictor; section 3 shows the control system design of double-controller scheme. In section 4, simulation results are presented. Finally, section 5 gives conclusions.

2 The Structure of Universal Learning Network Predictor

The structure of the ULN is showed in Fig. 1. In the Fig.1, i, j, k denote the node, $D_{ij}(p)$ is the time delay on the pth branch between the node i and the node j. It can be seen that the ULN belongs to recurrent neural network. The features in its structure include (1) all nodes are connected each other; (2) between any two nodes, there are multiple branches; (3) arbitrary time delay can be set on every branch.

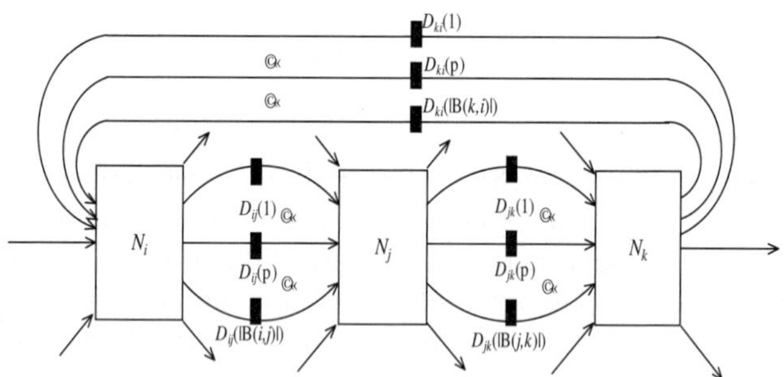

Fig. 1. Structure of the Universal Leaning Network

The generic equation that describes the behavior of ULN is expressed as Equation (1):

$$h_j(t) = f_j\Big(\{h_i(t - D_{ij}(p)) | i \in JF(j), p \in B(i,j)\},$$
$$\{\lambda_m(t) | m \in M(j)\}, \{r_n(t) | n \in N(j)\}\Big), (j \in J, t \in T) \tag{1}$$

where h_j (t) is the output value of node j at time t, $f_j(\ .\ )$ is the nonlinear function of node j, $\lambda_m(t)$ denotes value of m-th parameter at time t, r_n(t) can be seen as value of n-th external input variable at time t, $JF(j)$ is set of numbers of nodes whose outputs are connected to node j, $B(i,j)$ the set of numbers of branches from node i to node j, $M(j)$ the set of numbers of parameters, where output of node j can be partially differentiable with respect to these parameters, $N(j)$ the set of numbers of external input variables that are connected to j node, T the discrete set of sampling instants.

The criterion of Universal Learning Network is Equation (2).

$$E = \frac{1}{T}\sum_{t}^{T}\left[h(t) - \hat{h}(t)\right]^2 \tag{2}$$

Here $h(t)$ is the output network model, $\hat{h}(t)$ is the output of the teacher signal.

ULN adjusts weights by back propagation $\frac{\partial E}{\partial h_j(t)}$, compared to the general neural networks, the output of every node is not only connected with the output of every node at present time, but also connected with the output before.

$\lambda_m(t)$ is adjusted as follows

$$\lambda_m(t) \leftarrow \lambda_m(t) - \gamma \frac{\partial^+ E}{\partial \lambda_m(t)} \tag{3}$$

Where, $\frac{\partial^+ E}{\partial \lambda_m(t)}$ is the ordered derivative [8]. It takes into consideration both the direct and indirect paths that lead to the causal relationship.

The adaptive algorithm and time delay tuning of ULN is described in literature [7].

3 Control System Design

In this section, the double-controller scheme based on neural network method is introduced. The general structure of double-controller scheme is shown in Fig. 2. It is a modified Smith predictive control structure [4].

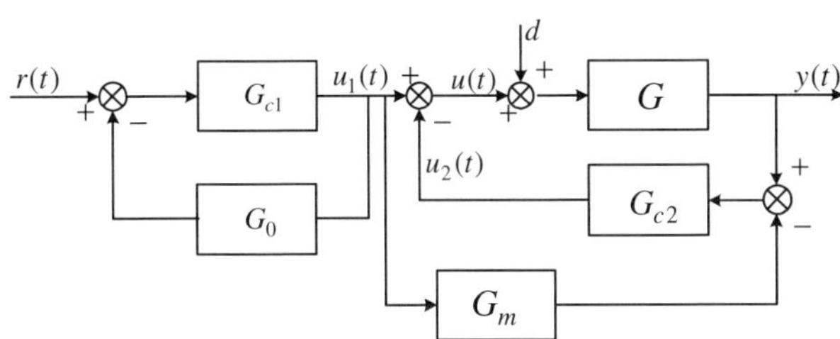

Fig. 2. Block diagram of the double-scheme controller

In Fig.2, G_{c1} is the PID type setpoint controller, G_{c2} is the load controller. The manipulated variable $u(t)=u_1(t)-u_2(t)$, is the difference between the outputs of G_{c1} and G_{c2}. G is the plant model, G_m is the ULN model that comes from identification of G, G_0 is the ULN predictor. $r(t)$ and $y(t)$ are the set value and output of the plant model, respectively. d is the disturbance of the control system. The setpoint controller G_{c1} determines the set point tracking of the plant, and the load controller is used for load rejection.

3.1 Design of Universal Learning Network Predictor

In Fig.2, ULN predictor G_0 is obtained by time delays on the branches that connect with the output node set one sampling time. The model is derived from measured input/output data of the plant, which is used to predict the output series of the plant. Comparing the prediction with the set point, the control series $u_1(t)$ is calculated by the setpoint controller. Supposing the time delay is t_d, at the same time, in order to eliminate the infection of long time delay, the output of system after the time delay is necessary to be predicted, so the criterion function is chosen as the following equation.

$$E = \sum_{k=1}^{n} [\hat{y}(t+t_d) - r(t+t_d)]^2 \tag{4}$$

Here, $\hat{y}(t+t_d)$ is the predicting output at $t + t_d$, $r(t + t_d)$ is the reference signal of the system at $t + t_d$.

The most expensive part of realization of a predictive control scheme is the derivation of the highly complex system model. Artificial neural network can deal with nonlinear problem, but the time delay in traditional neural network is not arbitrary, which limits the application of neural network in such case. However the time delay in ULN can be set arbitrarily based on the real system. So the ULN can be easily used as predictor.

3.2 Design of the Load Controller

For a complex plant with time delay, the performance of the double-controller is sensitive to the parameters of load controller. The traditional methods are disabled, when the plant is a black-box process. Another method to solve the problem is using BP neural network to adjust the parameters of a PID controller. The structure of the controller is showed in Fig.3.

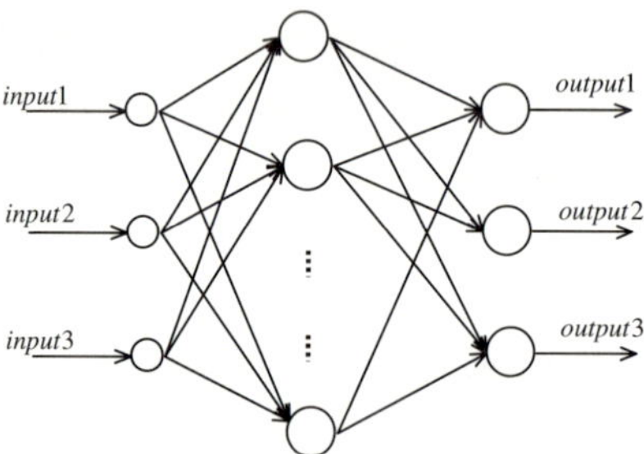

Fig. 3. Structure of neural network controller

In Fig.3, the input is the error at time $t + t_d$, $t + t_d - 1$, $t + t_d - 2$, and the output is the value of parameters: K_p, K_i, K_d. The value of the weight between input nodes and hidden nodes, output nodes are adjusted with BP algorithm. The criterion function of the network is described as Eq.(5).

$$J = \frac{1}{2}[y(t) - y_m(t)]^2 \tag{5}$$

Here, $y_m(t)$ is the output of ULN model G_m. The weights between node i and node j are adjusted as following Eq.(6):

$$w_{ij} \leftarrow w_{ij} - \eta \frac{\partial J}{\partial w_{ij}} \tag{6}$$

Then the parameters: K_p, K_i and K_d are adjusted.

4 Simulations

In order to validate the improved double-controller scheme mentioned in section 3, a typical process, a second order plant with time delay, is considered in the simulation. The process is described by equation (7).

$$G_2(s) = \frac{s+3}{(s+1)(s+2)} e^{-10s} \tag{7}$$

Form the equation, the process is described by a second order model which contains 10 steps time delay. It has been seen as a black-box plant during the simulation. The closed loop input and output serial of the plant can be obtained for providing the training data, by using PID controller to build the control system. The adaptive algorithm of ULN is adopted to identify the plant model. The simulation conditions are showed in Table 1. After 5,000 epochs learning, the time delay on the branches which connect with the output node is 10 sampling time, the learning error is 0.009209, test error is 0.016151. The training error and generalizing error curves are showed in Fig.4, and the generalizing curve is showed in Fig.5. It can be seen that ULN can embody the plant exactly.

The improved double-controller scheme showed in Fig.2 is driven to follow a step signal. In order to test the robustness of the system, add a step disturbance showed as Fig.2, d=0.1, after 200 iteration steps. From the curves in Fig.6, it can be concluded that controller with ULN predictor overcomes the effect of time delay well, for the ULN's universal approximation ability and the adaptability of neural network controller. It is an effectively control method for black-box process with time delay.

Table 1. Simulation conditions

Notations	Meaning	Values
N	Number of nodes	6
B	Number of branches	3
$\omega_{ij}(p)$	Initial weight	-0.1~0.1
γ	Learning coefficient	0.00001
$L_{\max}$	Learning epochs	10,000
$D_{6i}(p)$	Initial time delay	15
A	Breadth of node function	5

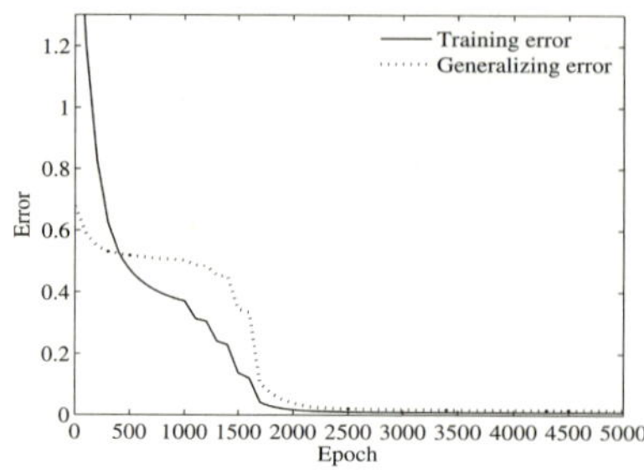

Fig. 4. Learning curve by universal learning network

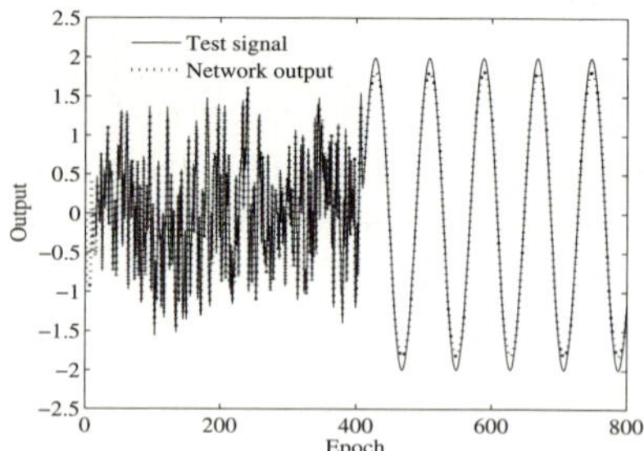

Fig. 5. Learning result by universal learning network

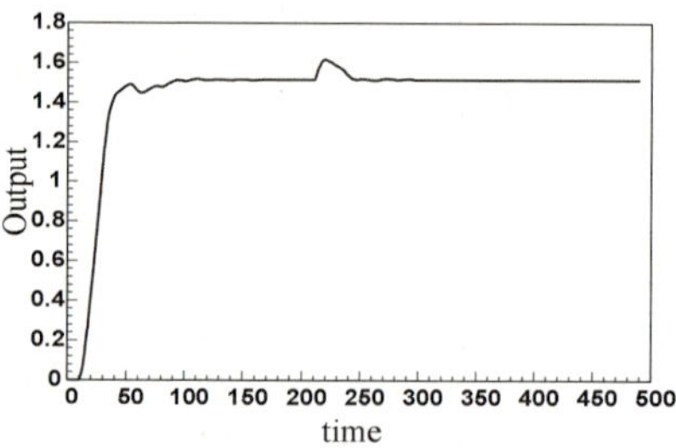

Fig. 6. Responses of the improved double-controller system

5 Conclusions

In this paper, a special type of the recurrent neural network—Universal Learning Network is used to identify the black-box system with time delay. The ULN is not only can embody the character of the system, but also can be used as a predictor in control system. Cooperated with ULN predictor, an improved double-controller scheme is discussed, results showed that the neural network controller can tune the parameters of the load controller adaptively. It is proved that the improved double controller scheme can be effectively used in black-box control system with time delay.

Acknowledgements

This research is supported by the project (60374064) of the National Nature Science Foundation of China; it is also supported by the project (50139020) of the National Natural Science Foundation of China. All of these supports are appreciated.

References

1. Faanes, A., Skogested, S.: PH-neutralization: Integrated Process and Control Design. Computers and Chemical Engineering **28** (8) (2004) 1475-1487
2. Huang, J.Q., Lewis. F.L.: Neural-network Predictive Control for Nonlinear Dynamic Systems with Time-delay. IEEE Trans. on Neural Networks **14** (2) (2003) 377-389

3. Tian, Y.C., Gao. F.: Double-controller Scheme for Control of Processes with Dominant Delay. IEE proceeding Control Theory and Applications **145** (5) (1998) 479-484
4. Guillermo, J.S., Aniruddha, D., Bhattacharyya, S.P.: New Results on the Synthesis of PID Controllers. IEEE Trans. on Automatic Control **47** (2) (2002) 241-252
5. Kaya, I: Two-degree-of-freedom IMC Structure and Controller Design for Integrating Process Based on Gain and Phase-margin Specifications. IEE Proceeding Control Theory and Applications **151** (4) (2004) 481-487
6. Piche, S., Sayyar-Rodsari, B., Johnson, D., Gerules, M.: Nonlinear Model Predictive Control Using Neural Networks. IEEE Control Systems Magazine **20** (3) (2000) 53-62
7. Han, M., Han, B., Guo, W.: Process Control of pH Neutralization Based on Adaptive Algorithm of Universal Learning Network. Journal of Process Control **16** (1) (2006)1-7
8. Werbos, P. J.: Backpropagation through Time: What It Dose and How to Do It. In Proceedings of the IEEE **78** (10) (1990) 1550-1560

Discrete-Time Sliding-Mode Control Based on Neural Networks

José de Jesús Rubio and Wen Yu

Departamento de Control Automático, CINVESTAV-IPN
A.P. 14-740, Av.IPN 2508, México D.F., 07360, México
yuw@ctrl.cinvestav.mx

Abstract. In this paper, we present a new sliding mode controller for a class of unknown nonlinear discrete-time systems. We make the following two modifications: 1) The neural identifier which is used to estimate the unknown nonlinear system, applies new learning algorithms. The stability and non-zero properties are proved by dead-zone and projection technique. 2) We propose a new sliding surface and give a necessary condition to assure exponential decrease of the sliding surface. The time-varying gain in the sliding mode produces a low-chattering control signal. The closed-loop system with sliding mode controller and neural identifier is proved to be stable by Lyapunov method.

1 Introduction

Nowadays, almost all of control schemes are implemented on digital devices. Some advantages of the continuous time controllers are lost, because the control signal can only be applied at fixed time steps. In order to overcome these problems, we may design discrete-time controllers directly instead of discretization of continuous-time controllers. Sliding-mode technique is very popular for this additional controller. Since the sliding-mode control switches the structure of the system during the evolution of the state vector to maintain the state trajectories in a predefined subspace, it is robust to parameter and structure uncertainties and invariance to unknown disturbances. However, the control chattering caused by the discontinuity of the control action is undesirable in most applications. In continuous-time, the sliding mode condition is [15], $s\dot{s} < 0$, where s is sliding surface, the switching can happen at any instant when the state trajectories cross the switching hyperplane.

But in discrete-time, the controller can only change the structure at discrete instants. Since the control input is applied in the sampling points, discrete-time sliding mode control is also called quasi-sliding mode control [1]. It is clear that the continuous-time sliding surface condition cannot be applied to discrete-time directly, for example, the existence of a quasi sliding regiment does not guarantee chattering motions close to the sliding manifold [13]. The simplest discrete-time sliding mode condition is to substitute the derivative by forward difference, so discrete-time sliding motion on the hyperplane is [5], $s(k)\left[s(k+1) - s(k)\right] < 0$.

J. Wang et al. (Eds.): ISNN 2006, LNCS 3972, pp. 956–961, 2006.

This condition is necessary but not sufficient for the existence of a quasi-sliding motion [4][13]. Furthermore, this condition cannot assure convergence of the trajectories onto the hyperplane, and it may increase amplitude chatter of the trajectories around the hyperplane [14]. A necessary and sufficient condition for discrete-time sliding motion is proposed in [12] and [6], it is, $|s(k+1)| < |s(k)|$. This condition can assure the sliding motion converge to the hyperplane, but is very difficult to design a sliding controller to satisfy this condition. So [1] gives another condition for discrete-time sliding mode, the sliding hyperplane $s(k) = 0$ should satisfy $|s(k)| \leq q$, where the parameters $q > 0$ is called the quasi-sliding mode band width. It does not require the trajectory to cross the sliding plane $s(k) = 0$ in each control step. Crossing the plane $s(k) = 0$ is permissible, but not required. The above sliding mode controllers require partial information of the nonlinear plant to design equivalent controller. When the equivalent controller is known, discrete-time sliding surface satisfies both [12] and [1] in [11].

If the plant is absolute unknown, artificial neural network can be used to estimate the plant. [3] proposes a sliding-mode control with discrete-time recurrent neural networks, the stability of the regulation is proved by Lyapunov technique. [9] gives a discrete time sliding mode controller with a modified switching function in order to make a small chattering, an adaptive term is added to improve transient performance. The gain of the sliding surface is constant. To improve robustness of discrete-time sliding mode control, [8] uses fuzzy neural networks to compensate the uncertainties. [14] considers regulation of state-space linear system with neural networks and sliding mode control. [4] presents a stable discrete-time sliding mode control which is not sensitive to sampling interval. There are two shortcomings, stability and non-zero properties do not discussed for the identifiers, the sliding mode gains are constants. To the best of our knowledge, sliding mode control with time varying gain and dead-zone identification has not yet been established in the literature.

In this paper we present a new neural sliding mode control, the neural estimator use dead-zone and project algorithm to assure the identification error is bounded and the equivalent controller exists. We propose a new sliding surface with time-varying gain, a necessary condition is given to avoids big chattering in the controller. The discrete-time sliding surface proposed in this paper can only satisfy the condition [1]. Lyapunov method is used to prove that the closed-loop system with sliding mode control and neural identifier is uniform stable.

2 Robust Identification Using Neural Networks

We consider a single-input single-output nonlinear system described as

$$y(k+1) = f\left[z(k)\right] + g\left[z(k)\right] u(k) \tag{1}$$

where $z(k) = \left[y(k), y(k-1)\ldots, u(k-1), u(k-2)\ldots\right]^T$, $m < n$, y is the output, u is the input, the arguments of $f(\cdot)$ and $g(\cdot)$ are unknown real functions. The nonlinear system is assumed to be controllable. We assume that $f\left(z(k)\right)$ and $g\left(z(k)\right)$ are smooth, $g\left(z(k)\right)$ is bounded and different from zero

$$0 < \underline{g} \leq |g(z(k))| \leq \overline{g} < \infty \tag{2}$$

(1) can be rewritten in state space form. If the nonlinear functions f and g in (1) are unknown, multilayer neural networks can be used to approximate them. We define $\widehat{f}$ and $\widehat{g}$ as the estimations of f and g, respectively, they can be expressed as

$$\widehat{f} = \sum_{j=1}^{q} w_j^f(k)\phi_j^f\left(\sum_{i=1}^{n+m} w_{ji}^f(k)z_i(k)\right), \widehat{g} = \sum_{j=1}^{q} w_j^g(k)\phi_j^g\left(\sum_{i=1}^{n+m} w_{ji}^g(k)z_i(k)\right) \tag{3}$$

where $w_j^f(k)$ and $w_{ji}^f(k)$ are the weights of the output and the hidden layer, respectively, q is hidden layer number. We select active functions of the neural networks as, $\phi_j^f(\cdot)$ is a hyperbolic tangent function, it can be positive and negative, $\phi_j^g(\cdot)$ is a sigmoid function, it is positive in nonlinear neurons, so $\phi_j^g(\cdot) \neq 0$. Since f and g are estimated by $\widehat{f}$ and $\widehat{g}$, the nonlinear system (1) can be modeled with a neural network model as

$$\beta\widehat{y}(k+1) = \widehat{f}[z(k)] + \widehat{g}[z(k)]\,u(k) \tag{4}$$

where β is a positive constant $\beta > 1$ which is a design parameter. We define the system identification error as

$$e_i(k+1) = \widehat{y}(k+1) - y(k+1) \tag{5}$$

The following theorem gives a stable learning algorithm for discrete-time nonlinear system identification via multilayer neural network.

Theorem 1. *If we use neural model (4) to identify nonlinear system (1), the following learning law is used for the weights updating,*

$$\begin{aligned} w_j^f(k+1) &= w_j^f(k) - \eta(k)\phi_j^f e_i(k) \\ w_{ji}^g(k+1) &= w_{ji}^g(k) - \eta(k)w_j^g(k)\dot{\phi}_j^g z_i(k)u(k)e_i(k) \end{aligned} \tag{6}$$

$$w_j^g(k+1) = \begin{cases} \overline{w}_j^g(k+1) & \text{if } \left|\overline{w}_j^g(k+1) - w_j^g(1)\right| \leq M \\ \frac{\overline{w}_j^g(k+1) - w_j^g(1)}{\left|\overline{w}_j^g(k+1) - w_j^g(1)\right|} M + w_j^g(1) & \text{if } \left|\overline{w}_j^g(k+1) - w_j^g(1)\right| > M \end{cases} \tag{7}$$

where $j = 1 \ldots q$, $i = 1 \ldots n+m$, $a(k) = \phi_j^{f2} + \left|\dot{\phi}_j^f w_j^f z_i\right|^2 + \left|u\phi_j^g\right|^2 + \left|u\dot{\phi}_j^g w_j^g z\right|^2$, $M > 0$, $\eta(k) = 0$ *when* $e_i^2(k) < \frac{\overline{\zeta}^2}{1-\eta}$ *or* $\beta e_i(k)\, e_i(k+1) < e_i^2(k)$. *We can obtain two conclusions 1) the identification error* $e_i(k)$ *is bounded and* $\widehat{g}[z(k)] \neq 0$, *2) The identification error* $e_i(k)$ *satisfies* $\lim_{k\to\infty} e_i^2(k) = \frac{\overline{\zeta}^2}{1-\eta}$, *where* $\overline{\zeta}$ *is upper bound of* $\zeta(k)$, $\zeta(k) = \zeta_f + \zeta_g u$.

3 Sliding Mode Control

In this section we will design a sliding mode control via neural identification to force the nonlinear system (1) to follow the trajectory generated by the following reference model

$$y_m(k+1) = a_0 y_m(k) + \ldots + a_n y_m(k-n) + b_0 v(k) + \ldots + b_m v(k-m) \quad (8)$$

where $v(k)$ is reference input. We define tracking error as

$$e_c(k+1) = y_m(k+1) - y(k+1) \quad (9)$$

Now we define a vector $e(k) = [e_c(k-n+1) \cdots e_c(k)]^T$. In this paper we propose a novel quasi-sliding mode controller as

$$u(k) = \frac{1}{\widehat{g}}\{-\widehat{f} + y_m(k+1) + K^T e(k) - \varepsilon(k) sign\left[s(k)\right]\} \quad (10)$$

where $\widehat{f}$ and $\widehat{g}$ are obtained from neural identifier, see (3) and (6), $K=[k_n \cdots k_1]^T \in R^{nx1}$, which is selected such that $s^n + \sqrt{2}k_1 s^{n-1} + \cdots + 2^{\frac{n}{2}} k_n$ is stable, $\varepsilon(k+1)$ is a time varying gain which satisfies

$$\varepsilon(k) = -e_i(k) sign\left[s(k)\right] + e^{-ck} \quad (11)$$

where $e_i(k)$ is the identification error which is obtained from (1), (4) and (5), c is a positive constant, $c \geq 1$. $s(k)$ is sliding surface which is defined as a function of tracking error (9)

$$s(k) = e_c(k) + K^T e(k-1) \quad (12)$$

From Theorem 1 we know the identification error $e_i(k)$ is bounded, *i.e.*

$$|e_i(k+1)| \leq H \quad |e_i(k)| \leq H \quad (13)$$

where H is a positive constant. The following lemma gives the boundedness of the sliding surface.

Lemma 1. *The sliding surface (12) in the discrete-time sliding mode controller (10) satisfies one sliding surface condition [1]*

$$|s(k+1)| \leq 1 + (1+\beta) H \quad (14)$$

where $e_c(k+1)$ and $e(k)$ are defined in (9) and (10), β and H are given in (4) and (13)

Remark 1. The sliding mode control proposed in this paper is similar with [9], the difference is that in [9] the gain is constant, in this paper the gain $\varepsilon(k)$ is time-varying by the identification error $e_i(k)$ and exponential decreased term e^{-ck}. The sliding mode gain decrease when the neural identifier converges, it produces less chattering than [9]. The chattering in this paper depends on two facts: identification error and constant c. A smaller c causes less chattering and bigger tracking error. A better c can be chosen by simulations.

The sliding surface can be rewritten as $e_c(k+1) = -K^T e(k) + s(k+1)$. By the definition $e(k) = [e_c(k-n+1) \cdots e_c(k)]^T$, so $e(k+1) = Ae(k) + bs(k+1)$. Because $\det(sI - \alpha A) = \alpha^n k_n + \alpha^{n-1} k_{n-1} s + \cdots + \alpha k_1 s^{n-1} + s^n$ [16], we select $K = [k_1 \cdots k_n]^T$ such that $\sqrt{2}A$ is stable ($\alpha = \sqrt{2}$). A stable $\sqrt{2}A$ can make the following Lyapunov equation have positive definite solutions for P and Q, $2A^T PA - P = -Q$, where $P = P^T > 0$, $Q = Q^T > 0$.

Theorem 2. *The close-loop system with sliding mode control (10) and neural identifier (3) and (6), is uniformly stable and the upper bound of the tracking error satisfies*

$$\lim_{k\to\infty} \frac{1}{T} \sum_{k=1}^{T} e^T(k)e(k) \le \frac{2b^T P b}{\lambda_{\min}(Q)} r^2 \tag{15}$$

where $r = (1+\beta) H$, $b = [0, \cdots 0, 1]^T$.

4 Simulation

In this section a tank model given by [10], is chosen to illustrate the new sliding-mode controller via neural identifier proposed in this paper. The dynamic of the tank can be represented by $h(k+1) = h(k) + \frac{-\sqrt{19.6h(k)}}{\sqrt{h(k)+3}} T + \frac{T}{\sqrt{h(k)+3}} u(k)$. We compare our neural sliding mode controller with normal neural controller [2] (layer number is also 1), and neural sliding mode controller with sliding mode control with fixed gain $\varepsilon = 0.05$ [9]. The control results and tracking error are shown in Fig.1 and Fig.2. We found that the new discrete-time sliding-mode control with neural networks proposed in this paper has a better tracking performance when the neural identification does not work well.

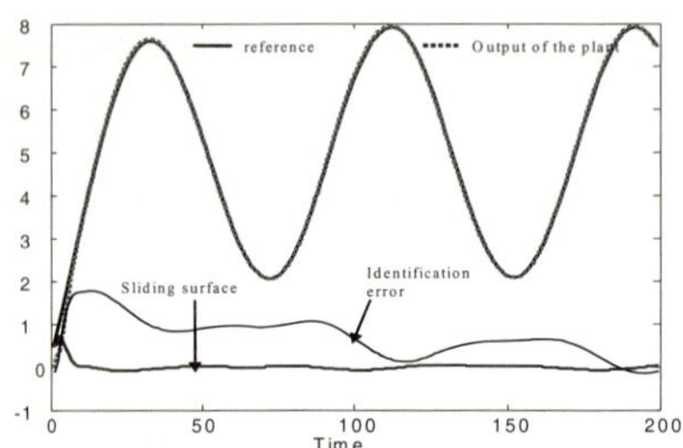

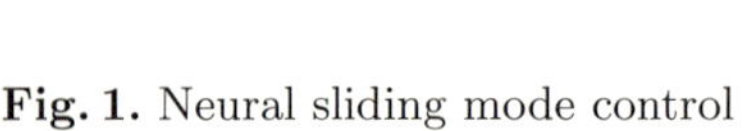

Fig. 1. Neural sliding mode control

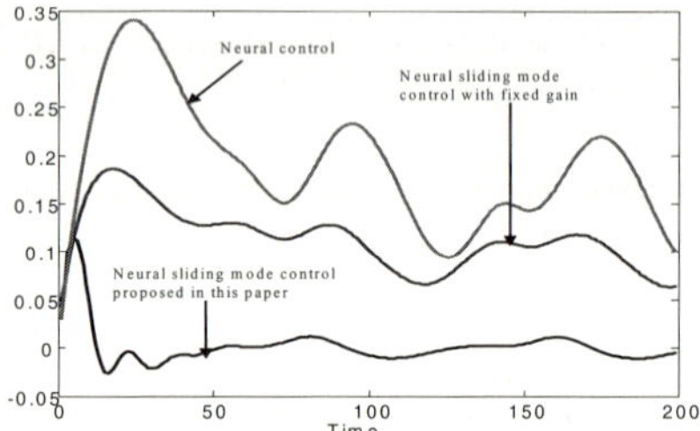

Fig. 2. Comparision for tracking errors

5 Conclusion

In this paper, a new sliding mode controller for a class of unknown nonlinear discrete-time systems is proposed. The main contributions are: 1) double dead-zones and projection technique guarantee the stability and non-zero properties of identification process, 2) time-varying gain in the sliding mode control assure less chattering compared to normal discrete-time sliding mode control. The closed-loop system with sliding mode controller and neural identifier is proved to be stable by Lyapunov method.

References

1. Bartoszewicz, A.: Discrete-Time Quasi-Sliding Mode Control Strategies. IEEE Transactions on Industrial Electronics 45(4) (1998) 633-637
2. Chen, F.C., Khalil, H. K.; Adaptive Control of a Class of Nonlinear Discrete Time Systems Using Neural Networks. IEEE Transactions on Automatic Control 40(5) (1995) 791-801
3. Fang, Y., Chow, T.W., Li, X.D.: Use of a Recurrent Neural Network in Discrete Sliding-Mode Control. IEE Proceeding-Control Theory Applications 146(1) (1999) 84-90
4. Furuta, K.; Sliding Mode Control of a Discrete System. Systems and Control Letters 14(2) (1990) 145-152
5. Gao,W., Wang, Y., Homaifa, A.: Discrete-Time Variable Structure Control Systems. IEEE Trans. Ind. Electron. 42(1) (1995) 117–122
6. S. Hui and H. Kak, On Discrete-Time Variable Structure Sliding Mode Control. Systems and Control Letters 38(3) (1999) 283-288
7. Jagannathan, S. Control of a Class of Nonlinear Discrete-Time Systems Using Multilayer Neural Networks. IEEE Transactions on Neural Networks 12(5) (2001) 1113-1120
8. Lin, F.J., Fung, R.F., Wai, R.-J.: Conparison of Sliding-Mode and Fuzzy Neural Network Control for Motor-Toggle Servomechanism. IEEE Transactions on Mechatronics 3(4) (1998) 302-318
9. Muñoz, D., Sbarbaro, D.: An Adaptive Sliding-Mode Controller for Discrete Nonlinear Systems. IEEE Transactions on Industrial Electronics 47(3) (2000) 574-581
10. Nounou, H.N., Passino, K.M.: Stable Auto-Tuning of Adaptive Fuzzy /Neural Controllers for Nonlinear Discrete-Time Systems. IEEE Transactions on Fuzzy Systems 12(1) (2004) 70-83
11. Saaj, M., Bandyopadhyay, C.B., Unbehauen, H.: A new Algorithm for Discrete-Time Sliding-Mode Control Using Fast output Sampling Feedback. IEEE Transactions on Industrial Electronics 49(3) (2002) 518-523
12. Sarpturk, S. Z., Istefanopolos, Y., Kaynak, O.: On the Stability of Discrete-Time Sliding Mode Control Systems. IEEE Transactions on Automatic Control 32(10) (1987) 930-932
13. Sira-Ramirez, H.: Nonlinear Discrete Variable Strusture Systems in Quasi-Sliding Mode. Int. J. Control 45(5) (1999) 1171-1187
14. Tsai, C.H., Chung, H.Y.: Neuro Sliding Mode Control With Its Appications to Seesaw Systems. IEEE Transactions on Neural Networks 15(1) (2004) 124-134
15. Utkin, V.I.: Sliding Modes in Optimization and Control. Springer-Verlag, (1992)
16. Yu,W., Moreno,M.A., Li, X.: Observer Based Neuro Identifier. IEE Proceedings - Control Theory and Applications 147(2) (2000) 145-152
17. Yu,W., Poznyak, A.S., Sanchez, E.N.: Neural Adaptive Control of Two-Link Manipulator with Sliding Mode. Proc. of IEEE International Conference on Robotics and Automation, Vol.4. (1999) 3122-3127
18. Zhu, Q., Guo, L.: Stable Adaptive Neurocontrol for Nonlinear Discrete- Time Systems. IEEE Transactions on Neural Networks 15(3) (2004) 653-662

Statistic Tracking Control: A Multi-objective Optimization Algorithm

Lei Guo

Research Institute of Automation,
Southeast University, Nanjing 210096, China
l.guo@seu.edu.cn

Abstract. This paper addresses a new type of control framework for dynamical stochastic systems, which is called *statistic tracking control* here. General non-Gaussian systems are considered and the tracked objective is the statistic information (including the moments and the entropy) of a given target probability density function (PDF), rather than a deterministic signal. The control is aiming at making the statistic information of the output PDFs to follow those of a target PDF. The B-spline neural network with modelling error is applied to approximate the corresponding dynamic functional. For the nonlinear weighting system with time delays in the presence of exogenous disturbances, the generalized H_2 and H_∞ optimization technique is then used to guarantee the tracking, robustness and transient performance simultaneously in terms of LMI formulations.

1 Introduction

Tracking control of dynamical stochastic systems has received much attention for Gaussian systems, where for a given reference input $r(t)$, the objective is to ensure the mean or variance of the difference between the output and $r(t)$ is convergent or minimized (see [1, 4]). Actually, in the classical stochastic theory, mean and variance are the objectives for control and filtering (e.g. LQG control and Kalman filtering) [4]. Two main confinements existed for the above so-called *"stochastic tracking"* problem: 1) Gaussian systems are considered and only the the mean or variance are concentrated; 2) the tracked target is supposed to be a deterministic signal.

However, non-Gaussian variables exist in many complex stochastic systems due to nonlinearity, which may even posses asymmetric and multiple-peak stochastic distributions (see [7, 10]). For non-Gaussian systems, mean and variance are insufficient to characterize the stochastic properties. On the other hand, along with the development of advanced instruments and data processing techniques, the measurement for feedback can be the stochastic distributions of the system output rather than the value of variables. Typical examples include the retention of paper making, particle size distribution, molecular weight distribution, and flame grey-level distribution processes [11]. Consequently, the controlled

J. Wang et al. (Eds.): ISNN 2006, LNCS 3972, pp. 962–967, 2006.

objective can be supposed as a target probability density functions (PDF) corresponding to a system output. To pursue a dynamic feedback control strategy for this purpose, new groups of strategies for general stochastic systems have been developed in the past a few years (see, e.g. [2, 3, 11]). This kind of problem has been called the PDF tracking or shape control, or the stochastic distribution control (SDC) problem.

It is noted that the PDF tracking problem is actually a dynamical tracking problem for a functional in the infinite dimensional space. The traditional performance index adopted in the previous works is the generalized distance between the output PDF and the target one. Since the controlled output is the shape of the conditional output PDF, the B-spline neural network has been widely used for the modelling of output PDFs so that the problem can be reduced to a tracking problem in finite-dimension for the weighting systems [6, 7]. It has been shown that this measure index leads to the nonlinearity and constraints for the weighting dynamic system. More importantly, some fundamental theoretical obstacles are required to discuss deeply, such as controllability, the control accuracy and the convergent rate. It is a challenging problem to reduce the problem complexity and to supply the control laws with feasible design algorithms.

In this paper, we present a new type of stochastic tracking control problem for the non-Gaussian systems, called *"statistic tracking control"* (STC) here. Differing from either the conventional stochastic tracking or the PDF tracking problems, the goal of control is to ensure the statistics information of the system output to follow that of a target PDF. Alternatively, every order moments and entropy can be selected as the tracking objectives based on the characteristics of different target PDFs. In this paper, as an application example, we adopt the mean and entropy as the tracked stochastic information. It can be shown that the main results of this paper have three advantages compared with the results on PDF tracking or the minimum entropy control in [6, 7, 11]. Firstly, technically the application of entropy tracking will eliminant the nonlinearity and the constraints resulting from the B-spline expansions for the output PDFs. Secondly, recalling that the mean and the variance are two objectives for the Gaussian systems, our control objective is a reasonable generalization for the non-Gaussian systems. Thirdly, noting that the functional tracking problem has been transferred into a finite-dimensional tracking one, analysis on controllability and performances turns to be possible. It is noted that STC also differs from the so-called minimum entropy control problems, where the reference input is still a deterministic signal and only optimization problem is discussed for the difference between the output and the given reference input [12].

After applying B-spline neural networks for the integrated functions in the definitions of the entropy and mean, it can be shown that the above STC problem can be transformed to a weighting tracking problem. We will consider the generalized mixed H_2 and $H\infty$ controller design for the tracking problem of the time delayed weighting models. The control objective is to find the gains of the mixed H_2 and $H\infty$ controllers such that the closed-loop systems is asymptotically stable, the weight dynamics can follow the desired set of weight, and the disturbance or un-modelling errors can be attenuated. LMI-based design approaches are provided, which also has independent significance for the tracking problem in multi-objective optimization setting for uncertain time delayed systems.

2 Main Results

2.1 Formulation of the STC Problem

For a dynamical stochastic system, denote $u(t) \in R^m$ as the input, $\eta(t) \in [a, b]$ as the stochastic output, whose conditional PDF is denoted by $\gamma(y, u(t))$, where y is the variable in the sample interval. It is noted that $\gamma(y, u(t))$ is a dynamic functional of y along with the time variable t. In the previous works, the PDF tracking problem has been widely studied with some effective design algorithms provided for various models, where the B-spline expansions are used to approximate $\gamma(y, u(t))$ or $\sqrt{\gamma(y, u(t))}$ and the control objective is to find $u(t)$ such that $\gamma(y, u(t))$ is convergent to the target PDF $g(y)$ (see *e.g.* [6, 7, 11]). For instance, in [7] the square root B-spline model is used and the control objective is that $e(y, t) = \sqrt{g(y)} - \sqrt{\gamma(y, u(t))} \to 0$. This procedure will lead to a nonlinear function in the B-spline model (see also the inequality in (3) and $h(V(t))$ in (8) of [5]), where the fault detection and diagnosis problem was investigated.

Our idea results form a simple observation. It is well known that mean and variance can characterize the stochastic property of a Gaussian variable. Generally, the moments from the lower order to higher order can decide the shape of a non-Gaussian PDF. In addition, entropy also has been widely used in communication and control theories as a measure for the average information contained in a given PDF [9, 12]. Thus, the PDF tracking can be achieved via the tracking of the above proper statistic information, which motivated the so called STC problem. It is noted that the STC approach will be a finite dimensional tracking problem, rather than the infinite one resulting from the previous PDF control theory.

To illustrate the design algorithm, in this paper we consider a special STC problem–*entropy tracking control* (ETC) problem. The following performance index is considered

$$\delta(\gamma, u(t)) = Q_1 \gamma(y, u(t)) ln(\gamma(y, u(t))) + Q_2 y \gamma(y, u(t)) \tag{1}$$

where Q_1 and Q_2 are two weighting parameters. In (1) the integral of the first term is the entropy and that of the second one is the mean of the output PDF respectively. It is noted that the performance index is reduced to the minimum variance control for the Gaussian system when $Q_2 = 0$ is assumed [10].

In this paper we construct the B-spline expansions to $\delta(\gamma, u(t))$ as follows

$$\delta(\gamma, u(t)) = C(y)V(t) + \varepsilon(y, t) \tag{2}$$

where

$$C(y) = \left[B_1(y) \ B_2(y) \cdots B_n(y) \right], V(t) = \left[v_1(t) \ v_2(t) \cdots v_n(t) \right]^T \tag{3}$$

and $B_i(y)$ $(i = 1, 2, \cdots, n)$ are pre-specified basis functions and $v_i(u(t)) := v_i(t), (i = 1, 2, \cdots, n)$ are the corresponding weights. $\varepsilon(y,t)$ represents the modelling error. Based on (1) and (2), for the target PDF, we can find the corresponding weights, which can be denoted as V_g. That is, $\delta(g, u(t)) = C(y)V_g(t)$. The tracking objective is to find $u(t)$ such that $\delta(\gamma, u(t)) - \delta(g, u(t)) = C(y)x(t) + \varepsilon(y,t)$ converges to 0 in the absence of $\varepsilon(y,t)$ and attenuate the influence of $\varepsilon(y,t)$, where $x(t) := V(t) - V_g(t)$.

2.2 Multi-objective Tracking and Optimization

The next step is to establish the dynamic model between the control input and the weights $x(t)$, this procedure has been widely used in PDF control and entropy optimization problems (see [7, 11]). To simply the design algorithm, originally only linear models were considered, where the shape of output PDFs cannot be changed [6]. In this paper, we adopt the following time delayed model with nonlinearity and the exogenous disturbances

$$\dot{x}(t) = Ax(t)+A_d x(t-d)+Ff(x(t))+B_{11}w_0(t)+B_{12}V_g+B_2u(t)+B_{2d}u(t-d) \quad (4)$$

where $u(t)$ and $w(t)$ represent the control and the exogenous vector respectively. $d(t)$ is the time delay satisfying $0 < \dot{d}(t) \leq \beta$. The initial condition is supposed as $x(t) = \phi(t), \quad t \in [-d(t), 0]$. $x(t-d)$ and $u(t-d)$ represent the delayed state and the control input respectively. It is noted that here uncertain vector $w_0(t)$ actually includes two parts. One part can be considered as the un-modelled dynamics of the weighting model (4), and another is resulted from $\varepsilon(y,t)$. In this paper, $w_0(t)$ is supposed to satisfy $\|w_0\|_2 < \infty$. A, F, A_d, B_2, B_{2d} and B_{1j} $(j = 1, 2)$ are known coefficient matrices. $f(x(t))$ is a nonlinear function satisfying

$$\|f(x_1(t)) - f(x_2(t))\| \leq \|U(x_1(t) - x_2(t))\| \quad (5)$$

for any $x_1(t)$ and $x_2(t)$, where U is a known matrix. It is noted that $f(W(t))$ can also be regarded as a kind of unknown modelling uncertainty. Actually, such a model for the weights can also be established by the neural network identifications.

For the context of tracking in the finite time, we use the truncated H_2 norm $\|x(t)\|_{2T}$ instead of the conventional H_2 norm $\|x(t)\|_2$. Generalization of the conventional H_∞ control is required for the tracking control problem since the reference input (e. g. V_g) does not satisfy $\|V_g\|_2 < \infty$ as long as V_g is non-zero. In order to guarantee the disturbance attenuation performance as $\|x(t)\|_{2T}^2 \leq \rho^2 \|w_0(t)\|_{2T}^2$, we select the first reference output as

$$z_\infty(t) = \begin{bmatrix} x(t) \\ \rho V_g \end{bmatrix} = \begin{bmatrix} I \\ 0 \end{bmatrix} x(t) + \begin{bmatrix} 0 & 0 \\ 0 & \rho I \end{bmatrix} \begin{bmatrix} w_0(t) \\ V_g \end{bmatrix} \quad (6)$$

where $w(t) = \begin{bmatrix} w_0(t) \\ V_g \end{bmatrix}$. It can be shown that $\|z_\infty(t)\|_{2T}^2 \leq \rho^2 \|w(t)\|_{2T}^2$ implies that $\|x(t)\|_{2T}^2 \leq \gamma \|w_0(t)\|_{2T}^2$.

H_2 performance is used to confine the transient response of the system and the high gain of the control input, which can be selected as $z_0(t) = C_0x(t) + D_0u(t)$. Thus, we can get the following augmented model for the future multi-objective control

$$\begin{cases} \dot{x}(t) = Ax(t) + A_dx(t-d) + Ff(x(t)) + B_1w(t) + B_2u(t) + B_{2d}u(t-d) \\ z_0(t) = C_0x(t) + D_0u(t) \\ z_\infty(t) = C_1x(t) + D_1w(t) \end{cases} \tag{7}$$

where $C_1 = \begin{bmatrix} I \\ 0 \end{bmatrix}$, $D_1 = \begin{bmatrix} 0 & 0 \\ 0 & \sqrt{\rho}I \end{bmatrix}$ and $B_1 = \begin{bmatrix} B_{11} \; B_{12} \end{bmatrix}$. It is noted that the initial conditions generally are non-zero.

At this stage, the considered PDF control problem has been formulated into the tracking problem for the above nonlinear weighting systems, and the control objective is to find $u(t)$ such that the tracking performance, disturbance attenuation performance and stability are guaranteed *simultaneously.*

It has been shown that tracking control can be solved via H_∞ setting. However, few tracking results have been provided for the time delayed systems with nonlinearity (see [6, 7] and references therein for the PI or PID tracking control problems). In this part, a new mixed H_2 and H_∞ optimization formulation will be applied to the above weighting tracking problem.

Denote $\delta(P_0, S_0) := \phi^T(0)P_0\phi(0) + \int_{-d}^0 \phi^T(\tau)S_0\phi(\tau)d\tau$, $sym(A) := A + A^T$ and $MM^T := \int_{-d(0)}^0 \phi^T(\tau)\phi(\tau)d\tau$. For the time delayed system with non zero initial conditions, we can get the following result by applying the approach in [8] to the case of the truncated norms.

Theorem 1. *For Σ with scalars $\gamma > 0$, μ and the weighting matrices T, if the optimization problem $\delta_1 := \min\{\alpha + tr(M^TTM)\}$ subject to*

$$\begin{bmatrix} \Phi_{11} & \Phi_{12} \\ \Phi_{12}^T & \Phi_{22} \end{bmatrix} < 0, \quad \begin{bmatrix} \alpha & \phi^T(0) \\ \phi(0) & Q \end{bmatrix} \geq 0 \tag{8}$$

is feasible with respect to $Q > I$ and L, where

$$\Phi_{11} := \begin{bmatrix} sym(AQ + B_2L) + T + \mu^2F^TF & A_dQ + B_{2d}L \\ (A_d^TQ + B_{2d}^TL)^T & -(1-\beta)T \end{bmatrix},$$

$$\Phi_{12} := \begin{bmatrix} B_1 & QC_1^T & QC_0^T + L^TD_0^T & QU^T \\ 0 & 0 & 0 & 0 \end{bmatrix}, \Phi_{22} := \begin{bmatrix} -\gamma^2I & D_1^T & 0 & 0 \\ D_1 & -I & 0 & 0 \\ 0 & 0 & -I & 0 \\ 0 & 0 & 0 & -\mu^2I \end{bmatrix}$$

then there exists a mixed guaranteed cost and generalized H_∞ state feedback controller $u = Kx$, where $K = LQ^{-1}$, such that the closed-loop system is asymptotically stable and satisfies $J_2 \leq \delta(Q^{-1}, Q^{-1}T_iQ^{-1}) \leq \delta_1$ and $(\|z\|_2^2 - \gamma^2\|w\|_2^2) < \delta_1$.

The proof of Theorem and simulations are omitted here to save space.

More generally, one can consider the state feedback controller with the form of $u(t) = Kx(t) + RV_g$ to guarantee the zero equilibrium if it is necessary.

3 Conclusions

This paper presents a new stochastic control setting where the tracking objective is some statistic information rather than a deterministic reference signal or a PDF. In this paper, it is called the *statistic tracking control* problem. B-spline neural networks are used to model the integrated function related to the tracked moments or entropy. By establishing the weighting systems, we introduce the generalized H_2 and H_∞ optimization problem to formulate the weighting tracking one. LMI-based approaches to the multi-objective optimal controller design where the initial condition are non-zero and the truncated norm is adopted. To achieve the enhanced tracking performance, one improvement is to use the different B-spline expansions acting on the mean and the entropy respectively. This may result in a simultaneous stabilization and optimization problem.

Acknowledgement. This work is supported by *National Science Foundation of China under Grant 60474050.*

References

1. Bar-Shalom, Y., Li, X. R. and Kirubarajan, T.: Estimation with Applications to Tracking and Navigation. John Wiley & Sons, London (2001)
2. Crespo, L. G. and Sun, J. Q.: Nonlinear Stochastic Control via Stationary Probability Density Functions. In Proc. ACC2002, Anchorage, AK, USA (2002) 2029-2033
3. Forbes, M. G., Forbes, J. F. and Guay, M.: Regulatory Control Design for Stochastic Processes: Shaping the Probability Density Function. In Proc. ACC2003, Denver, USA (2003) 3998-4003
4. Goodwin, G. C. and Sin, K. S.: Adaptive Filtering, Prediction and Control. Prentice- Hall, Englewood Cliffs, NJ (1984)
5. Guo, L. and Wang, H.: Fault Detection and Diagnosis for General Stochastic Systems Using B-Spline Expansions and Nonlinear Filters. IEEE Trans. on Circuits and Systems-I: Regular Papers. **52** (2005) 1644-1652
6. Guo, L. and Wang, H.: Generalized Discrete-time PI Control of Output PDFs Using Square Root B-Spline Expansion. Automatica. **41** (2005) 159-162
7. Guo, L. and Wang, H.: PID Controller Design for Output PDFs of Stochastic Systems Using Linear Matrix Inequalities. IEEE Trans. on Systems, Man and Cybernetics- Part B. **35** (2005) 65-71
8. Guo, L., Zhang, Y. M. and Feng, C. B.: Generalized H_∞ Performance and Mixed H_2/H_∞ Optimization for Time Delay Systems. In Proceedings of 8th Int. Conf. on Control, Automation, Robotics and Vision. Kunming (2004) 160-165
9. Guo, L. and Wang, H.: Minimum Entropy Filtering for Multivariate Stochastic Systems with Non-Gaussian Noises. IEEE Trans. on Automatic Control. **51** (2006) March, to appear
10. Papoulis, A.: Probability, Random Variables and Stochastic Processes, 3rd. McGraw- Hill, New York (1991)
11. Wang, H.: Bounded Dynamic Stochastic Systems: Modelling and Control. Springer-Verlag, London (2000)
12. Yue, H. and Wang, H.: Minimum Entropy Control of Closed Loop Tracking Errors for Dynamic Stochastic Systems. IEEE Trans. on Automatic Control. **48** (2003) 118-122

Minimum Entropy Control for Stochastic Systems Based on the Wavelet Neural Networks

Chengzhi Yang

Kunming University of Science and Technology 650051, China
jy5663@public.km.yn.cn

Abstract. The main idea of this paper is to characterize the uncertainties of control system base upon entropy concept. The wavelet neural networks is used to approach the nonlinear system through minimizing Renyi's entropy criterion of the system estimated error, and the controller design is based upon minimizing Renyi's entropy criterion of the system tracking errors. An illustrative example is utilized to demonstrate the effectiveness of this control solution, and satisfactory results have been obtained.

1 Introduction

In this note, the control system subjected to a non-Gaussian stochastic disturbance is considered, thus minimum variance of tracking error cannot be used to represent the performance of the closed loop system alone. This is simply because the spread area of a non-symmetric distribution cannot be fully described by the variance. In order to overcome these restricts, the high-order statistics to information is applied. The distinguished feature of our new control system design is based on the minimization of system Renyi's entropy H criterion where it can be described as $H(x) = \{\log \int_{-\infty}^{\infty} [f_x(x)]^{\alpha} dx\}/(1-\alpha)$ [1]. The wavelet neural networks identifier (WNNI) is used to approach the nonlinear system through minimizing entropy of the system estimated error, and the controller design method to minimize the system tracking error entropy is proposed.

2 Statement of the Problem

2.1 Control System

The control problem addressed in this paper is shown in Fig 1. The controlled system can be characterized by an unknown nonlinear function and the controller can be adjusted on line by the information $\partial y/\partial u$ of system sensitivity ($\partial y/\partial u$ can be approximated by $\partial\hat{y}/\partial u$ using the WNNI. $r(k)$ is the system desire output, $u(k)$ is the system control signal, $y(k)$ is the system measured output, $\hat{y}(k)$ is the system estimated output of the WNNI and $\omega(k)$ is the system arbitrary bounded independent noise term with a known or unknown probability density function $\gamma_{\omega}(x)$.

J. Wang et al. (Eds.): ISNN 2006, LNCS 3972, pp. 968 – 974, 2006.

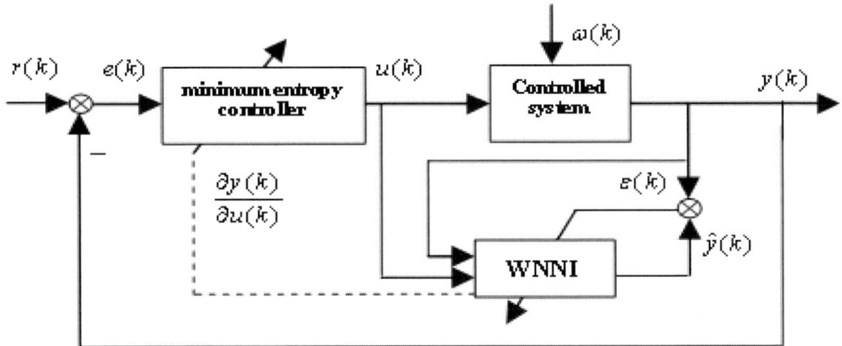

Fig. 1. the minimum entropy controller using the wavelet neural networks

From Fig 1, $e(k)$ is the system tracking error which can be defined as:

$$e(k) = r(k) - y(k) \tag{1}$$

$\varepsilon(k)$ is the system estimated error which can be defined as:

$$\varepsilon(k) = y(k) - \hat{y}(k) \tag{2}$$

2.2 Entropy Criterion of the Systems

The purpose of the controller design is to minimize the uncertainties or the randomness embedded in the system errors. For this purpose the control horizon will be divided into a number of equally lengthened subintervals as $\{T^1, T^2, \cdots, T^l, \cdots\}$. Within each subinterval there are a fixed number of sample points $\{\ 1, 2, \cdots, N\ \}$, where N is sampling numbers in a subinterval. In each subinterval a number of sampled signals can be collected and the probability density functions (PDFs) of samples is constructed. Then the control components $u(k)$ ($k = 1, 2, \cdots, N$) based on the shape of the system tracking errors PDFs is tuned so as to minimize the system tracking error entropy, and the coefficients θ of the WNNI based on the shape of the system estimated error PDFs is tuned either so as to minimize the system estimated error entropy. This process goes on until $u(k)$ (and θ) cannot be further tuned. The whole operational procedure is shown in Fig 2.

Assume that the current instant is at the beginning of the subinterval T^{l-1}, the system tracking error e and the system estimated error ε should have the samples value $\left\{e^{l-1}(1), e^{l-1}(2), \cdots, e^{l-1}(N)\right\}$ and $\left\{\varepsilon^{l-1}(1), \varepsilon^{l-1}(2), \cdots, \varepsilon^{l-1}(N)\right\}$ given by $e = r - y$ and $\varepsilon = y - \hat{y}$ respectively. Using the numerical methods given by Erdogmus and Principe in [2], a simple and useful plug-in estimator for the Renyi's entropy can be obtained by the Parzen window estimate[2] of the system PDFs in the entropy definition. Suppose the random variable e and ε are independent and identically distributed samples, the Parzen window estimate of the system tracking error PDFs and the system estimated error ε PDFs in the subintervals T^{l-1} can be calculated respectively by

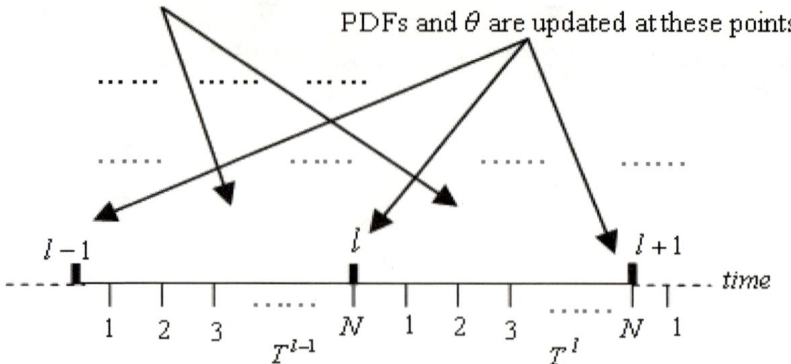

Fig. 2. The proposed closed loop operational procedure

$$\hat{f}_e^{l-1}(e) = \frac{1}{N}\sum_{i=1}^{N} k_\sigma(e - e^{l-1}(i)) \quad l = 1, 2, \cdots \tag{3}$$

$$\hat{f}_\varepsilon^{l-1}(\varepsilon) = \frac{1}{N}\sum_{i=1}^{N} k_\sigma(\varepsilon - \varepsilon^{l-1}(i)) \quad l = 1, 2, \cdots \tag{4}$$

where $k_\sigma(\cdot)$ is a real symmetric kernel function($k_\sigma(x) = k_\sigma(-x) \in (\infty, -\infty)$), satisfying $\int_{-\infty}^{\infty} k_\sigma(z)dz = 1$, $\sup| k_\sigma(z) | < \infty$ and $\lim_{|z|\to\infty} k_\sigma(z) = 0$. The bandwidth $\sigma(>0) = \sigma^N$ is a function of the sample size N [2]. Thus α -order Renyi's entropy of the system tracking error samples e and the system estimated error samples ε within the subinterval T^{l-1} can be given by respectively:

$$H_1^{l-1}(e) = \frac{1}{1-\alpha}\log\left\{\frac{1}{N^\alpha}\sum_{i=1}^{N}\left[\sum_{j=1}^{N} k_\sigma(e^{l-1}(i) - e^{l-1}(j))\right]^{\alpha-1}\right\} \tag{5}$$

$$H_2^{l-1}(\varepsilon) = \frac{1}{1-\alpha}\log\left\{\frac{1}{N^\alpha}\sum_{i=1}^{N}\left[\sum_{j=1}^{N} k_\sigma(\varepsilon^{l-1}(i) - \varepsilon^{l-1}(j))\right]^{\alpha-1}\right\} \tag{6}$$

Considered that the idea of the cost function of control system is described by the entropy, the criterion J_1 of the control system tracking error is performed by Renyi's entropy formula as follows:

$$J_1 = H_1(e) \tag{7}$$

Obviously the system output will approach the system desired output when the system estimated error entropy is minimized. The criterion J_2 of the system estimated error is performed by Renyi's entropy formula as follows:

$$J_2 = H_2(\varepsilon) \tag{8}$$

3 Minimizing Entropy Control Solution

Based on the analysis above, it generally leads to an optimal control gradient algorithm for the nonlinear dynamic stochastic system and the control signals components within the subinterval T^l can be defined as:

$$u^l(k) = u^{l-1}(k) - \lambda \bullet \frac{\partial J_1^{l-1}}{\partial u^{l-1}(k)} \quad k = 1, 2, \cdots, N; \tag{9}$$

where λ is the learning rate coefficient. In equation (9) minimum tracking error entropy control strategy of the closed-loop system has proposed, where the system tracking error is characterized on line by its PDFs using kernel function estimator (3). Obviously the key problem of controller (9) design is how to get $\partial J_1^{l-1} / \partial u^{l-1}(k)$ (or $\partial H_1^{l-1} / \partial u^{l-1}(k)$). In particular if Shannon entropy concept is used in this note (it means to make $\alpha \to 1$ [2]), the relationship $\partial J_1^{l-1} / \partial u^{l-1}(k)$ can be obtained as follows:

$$\frac{\partial J_1^{\,l-1}}{\partial u^{l-1}(k)} = \frac{1}{N} \sum_{i=1}^{N} \frac{\sum_{j=1}^{N} k'_\sigma (e^{l-1}(i) - e^{l-1}(j)) \left(\frac{\partial y^{l-1}(i)}{\partial u^{l-1}(k)} - \frac{\partial y^{l-1}(j)}{\partial u^{l-1}(k)} \right)}{\sum_{j=1}^{N} k_\sigma (e^{l-1}(i) - e^{l-1}(j))} \tag{10}$$

In equation (10) when the model of the controlled system is unknown, $\partial y / \partial u$ cannot be gotten directly. Here $\partial \hat{y} / \partial u$ is used to approximate $\partial y / \partial u$.

From paper [3] the output of the WNNI is defined as follows:

$$\hat{y}(k) = \sum_{p=1}^{n} \omega 2_q \psi_q \left(\frac{\sum_{q=1}^{m} \omega 1_{pq} x_p(k) - b_q}{a_q} \right) \tag{11}$$

The coefficients $\theta = \lfloor \omega 1_{pq}, \omega 2_q, a_q, b_q \rfloor$ of the WNNI within the subinterval T^{l-1} can be calculated by gradient algorithms:

$$\theta^l = \theta^{l-1} - \delta \frac{\partial J_2^{l-1}}{\partial \theta^{l-1}} + \eta \Delta \theta^{l-1} \tag{12}$$

where δ and η are the learning rate and the dynamic factor respectively. Alike to the results obtained above, the relationship $\partial J_1^{l-1} / \partial u^{l-1}(k)$ in equation (12) within the subinterval T^{l-1} can be obtained as following:

$$\frac{\partial J_2^{l-1}}{\partial \theta^{l-1}} = -\frac{1}{N}\sum_{i=1}^{N} \frac{\sum_{j=1}^{N} k'_\sigma(\varepsilon^{l-1}(i)-\varepsilon^{l-1}(j))\left(\frac{\partial \hat{y}^{l-1}(i)}{\partial \theta^{l-1}} - \frac{\partial \hat{y}^{l-1}(j)}{\partial \theta^{l-1}}\right)}{\sum_{j=1}^{N} k_\sigma(\varepsilon^{l-1}(i)-\varepsilon^{l-1}(j))} \tag{13}$$

The components in equation (13) can be expressed as following[3] ($k = 1, 2, \cdots, N;\ p = 1, 2, \cdots, n$):

$$\frac{\partial \hat{y}^{l-1}(k)}{\partial \omega 1_{pq}^{l-1}} = \sum_{q=1}^{m} \omega 2_q^{l-1} \bullet \psi_q^{l-1'}(k) \bullet x_p^{l-1}(k) \bullet \frac{1}{a_q^{l-1}} \tag{14}$$

$$\frac{\partial \hat{y}^{l-1}(k)}{\partial \omega 2_q^{l-1}} = \sum_{q=1}^{m} \psi_q^{l-1}(k) \tag{15}$$

$$\frac{\partial \hat{y}^{l-1}(k)}{\partial a_q^{l-1}} = \sum_{q=1}^{m} \omega 2_q^{l-1} \bullet \psi_q^{l-1'}(k) \bullet \frac{\varphi_q^{l-1}(k) - b_q^{l-1}}{(a_q^{l-1})^2} \tag{16}$$

$$\frac{\partial \hat{y}^{l-1}(k)}{\partial b_q^{l-1}} = -\sum_{q=1}^{m} \omega 2_q^{l-1} \bullet \psi_q^{l-1'}(k) \frac{1}{a_q^{l-1}} \tag{17}$$

From equation (12) the parameters of the WNNI can be obtained. Thus the system sensitivity information can be got through calculating $\partial \hat{y} / \partial u$ using equation (11) [3].

4 Experiment Results

To demonstrate the effectiveness of the proposed numerical solution, a simulation example should be studied using the proposed minimum entropy control algorithm in this paper. The WNNI use the 5-8-1 structure, that means $n = 5$ and $m = 8$. Thus the input vector $x(k)$ of the WNNI can be presented as:

$$x(k) = [y(k-2),\ y(k-1),\ y(k),\ u(k-2),\ u(k-1)]^T \tag{18}$$

Generally the following the wavelet function can be defined as follows [3]:

$$\psi_q(k) = x(k) \bullet \exp(-x^2(k)/2) \tag{19}$$

If a typical nonlinear system model is considered by following relation:

$$y(k)=\frac{[y(k-1)\bullet y(k-2)]\bullet[y(k-1)+3]}{\left[1+y\ (k-1)+y^2(k-2)\right]}+u(k-1)+0.2u(k-2)+0.6\omega(k-5) \tag{20}$$

where the desired output signal is given by $r(k)=1(k)$, the noise ω_n is a arbitrary random signal bounded in [-1, 1] and the bandwidth σ of Parzen window estimate is set to 0.25. The kernel function of random variable e is represented by a symmetric standard kernel function as:

$$k_\sigma(e)=\begin{cases}3/8(4-e^2)/4 & e^2\le 4\\ 0 & otherwise\end{cases} \tag{21}$$

In each simulating subinterval, there are 200 samples and the 100 subintervals have been used in total. The learning rate coefficient λ in equation (9) is set to -0.5, and the control signal initial value $u^0(1)$ and the system output initial value $y^0(1)$ are set to 5 and 4 respectively.

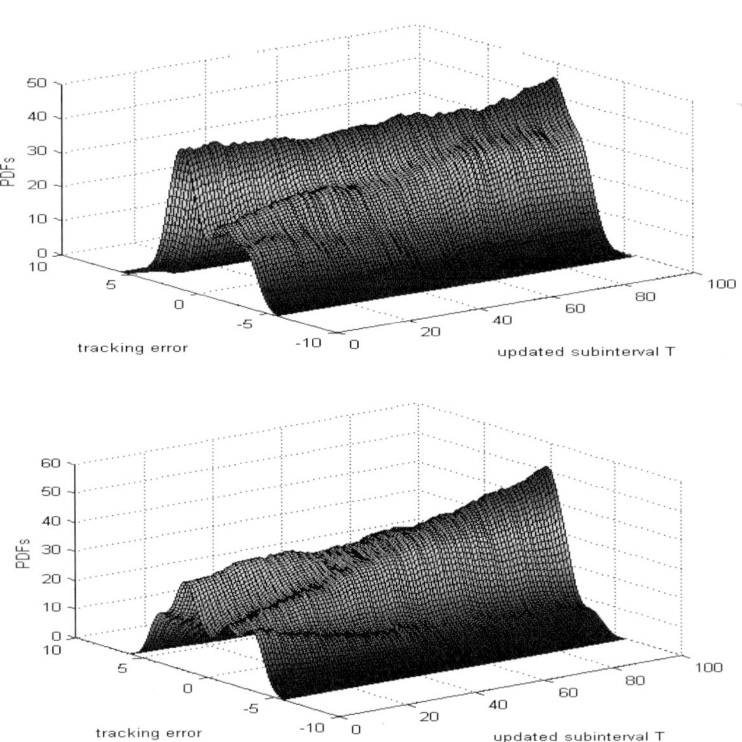

Fig. 3. a) PDFs of WNNI estimated error ε b) PDFs of tracking error e

The simulation results are shown in Fig 3 a) and Fig 3 b), where it can be seen that the control input is driving the system error signals toward a smaller randomness direction which it can be verified by the PDFs curves of system tracking error and the system estimated error. The shape of PDFs will become narrower and sharper when updated point l increases and the entropy values also have a well convergence properties accompanying with updated point l increases. All can indicate that the uncertainties of the system error are decreasing.

5 Conclusions

In this paper, the minimum entropy control problem for non-Gaussian stochastic system has been addressed. Based on Parzen window estimate method of the system error PDFs, the Renyi's entropy is used to represent the criterion of the system control and the system estimation using the WNNI. A numerical control strategy has been developed which controls the PDFs of the system output error so as to make it to track a given desired output. A simulated example is used to illustrate the effectiveness of the proposed algorithm and encouraging results have been obtained.

References

1. Erdogmus, D., Kenneth E.H.: Adaptive Blind Deconvolution of Linear Channels Using Renyi's Entropy with Parzen Window Estimation. IEEE Trans. Signal Processing 52(6) (2004) 1489-1498
2. Erdogmus, D., Principe, J.C.: An Error-Entropy Minimization Algorithm for Supervised Training of Nonlinear Adaptive Systems. IEEE Trans. Signal Processing 50(7) (2002) 1780-1786
3. Zhang, Q., Benveniate, A.: Wavelet Networks. IEEE Trans. Neural Networks 3(6) (1992) 889-898

Stochastic Optimal Control of Nonlinear Jump Systems Using Neural Networks

Fei Liu and Xiao-Li Luan

Institute of Automation, Southern Yangtze University
Wuxi, 214122, P.R. China
fliu@thmz.com

Abstract. For a class of nonlinear stochastic Markovian jump systems, a novel feedback control law design is presented, which includes three steps. Firstly, the multi-layer neural networks are used to approximate the nonlinearities in the different jump modes. Secondly, the overall system is represented by the mode-dependent linear difference inclusion, which is suitable for control synthesis based on Lyapunov stability. Finally, by introducing stochastic quadratic performance cost, the existence of feedback control law is transformed into the solvability of a set of linear matrix inequalities. And the optimal upper bound of stochastic cost can be efficiently searched by means of convex optimization with global convergence assured.

1 Introduction

Many practical systems experience randomicity, such as abrupt environment alterations, sudden operating condition variations, random component failures and repairs, etc. Compared with determinate systems, these complex dynamical systems possess not only determinate continuous states but also stochastic discrete events. While stochastic dynamic is modeled by Markovian chain taking values in a finite set, a class of so-called stochastic Markovian jump systems (MJS) has attracted a lot of research interest since 1961 [1], on which many modeling and control problems of engineering systems can be built, for examples, manufacturing systems, power systems, network systems, etc. For stochastic MJS, there have been a large amount of reported results addressed on linear cases, covering stochastic stability, stochastic controllability and observability, and quadratic control, see [2,3] and references therein.

However, almost all of engineering systems have inherent nonlinearities, to which linear stability theory and optimal control are difficult to be applied. Together with stochastic jump in their structure and parameters, the stability analysis and control synthesis become more complicated. Recently, a few papers have appeared to try on this topic. Some Hamilton-Jacobi-equation-based sufficient conditions have been derived for a class of nonlinear MJS [4]. Based on linear matrix inequality (LMI) approach, a robust state-feedback controller has been developed such that the closed-loop bilinear MJS is stochastically exponentially stable [5]. By combining local asymptotic stabilizing state feedback

J. Wang et al. (Eds.): ISNN 2006, LNCS 3972, pp. 975–980, 2006.

law and observer, a locally asymptotically stabilizing output feedback controller has been investigated for nonlinear jump system [6]. But, contrasting to linear case, in existing references, the control problems for nonlinear cases of MJS have not formed some systematic methods and remained as an open research area.

The aim of this paper focuses on stochastic quadratic optimal control for nonlinear MJS. Different from exiting results, artificial neural network is used as a main tool to deal with nonlinearities. Though neural network has been widely used in nonlinear area owing to its universal approximation capability, the farther issue concerned here is its compatibility to stability guaranteed control theory [7]. By means of linear differential inclusion (LDI) representation of multi-layer neural network [8], [9], a basic framework is build on stochastic Lyapunov theory [10], which gives a uniform way to analysis and control stochastic MJS with nonlinearities. The stochastic stability of nonlinear MJS can be guaranteed by searching a linear state feedback law in terms of linear matrix inequality (LMI). It is proved that if the approximation errors between neural networks and nonlinearities are norm-bounded, the state feedback law is also robust. Moreover, by introducing quadratic optimal performance cost to nonlinear MJS, an optimal controller is obtained with a guaranteed cost bound, which can be minimized by LMI optimization algorithm. To illustrate the presented approach, a simple nonlinear MJS is selected as an example, in which the cyclic frequency of nonlinear dynamic changes stochastically according to transition probability.Similar case may be often encountered in a real-time environment, and the application result is satisfied.

2 System Formulation and Analysis

Consider the nonlinear Markovian jump system, $\dot{x}(t) = f(x(t), r(t))$, where$f(\cdot)$ is a nonlinear function vector, x is state vector, $r(t)$ represents a continuous-time discrete-state Markovian mode taking values in a finite set $S = \{1, 2, \cdots, s\}$ with transition probability matrix $\Pi = [\pi_{rj}]$, in which π_{rj} is the transition rate from mode r to j. For the convenience of notations, while $r(t) = r$, $f(x(t), r(t))$ is represented by $f_r(x)$. Without loss of generality, in each mode r, nonlinear function $f_r(x)$ is approximated via a 3-layer neural network, in which only single input x and single output y_r are indicated for simplification, where y_r is the approximation of $f_r(x)$. Assume the hidden layer has h neutrons $Z_{ri}, i = 1, \cdots, h$,with weight vectors $W_{r1} = [w_{r11}, \cdots, w_{r1h}]^T, W_{r2} = [w_{r21}, \cdots, w_{r2h}]$. The activation function vector of the hidden layer is $\Psi_r[\varpi] = [\varphi_r(\varpi_1), \cdots, \varphi_r(\varpi_h)^T]$. With the output layer having the similar activation function, the neural network-based approximation of $f_r(x)$ is $y_r = \Psi_r[W_{r2}\Psi_r[W_{r1}x]]$. While the maximum and minimum derivatives of activation function $\varphi_r(\varpi_i)$ are respectively defined as $\xi_r(k_i = 1)$ and $\xi_r(k_i = 0)$, y_r can be represented by following max-min form,

$$y_r = \sum_{k_1=0}^{1} \sum_{k_2=0}^{1} \cdots \sum_{k_{h+1}=0}^{1} h_{r(h+1)k_{h+1}} \cdots h_{r2k_2} h_{r1k_1} \xi_r(k_{h+1}) W_{r2} \Xi W_{r1} x. \quad (1)$$

where real numbers $h_{rlk} > 0, l = 1, \cdots, h, h+1$, satisfying $\sum_{k=0}^{1} h_{rlk} = 1$, and $\Xi = diag[\xi_r(k_1), \xi_r(k_2), \cdots, \xi_r(k_h)]$.

Define two sets of index vector, respectively, $\Gamma_o = \{\nu \in \{0,1\}\}$ for output layer and $\Gamma_h = \{\nu \in R^h \mid \nu_j \in \{0,1\}, j = 1, \cdots, h\}$ for hidden layer. The total index vector ν of 3-layers neural network has 2×2^h combinations in the set $\Omega = \Gamma_o \bigoplus \Gamma_h$. For above representation (1), by introducing $A_{r\nu} = \xi_r(k_{h+1}) W_{r2} \Xi W_{r1}$, and $\sum_{\nu} \mu_{r\nu} = \sum_{k_1=0}^{1} \sum_{k_2=0}^{1} \cdots \sum_{k_{h+1}=0}^{1} h_{r(h+1)k_{h+1}} \cdots h_{r2k_2} h_{r1k_1}$, in which $\nu \in \Omega$, and obviously $\mu_{r\nu} > 0, \sum_{\nu} \mu_{r\nu} = 1$, the representation (1) is introducing rewritten as $y_r = \sum_{\nu} \mu_{r\nu} A_{r\nu} x$. Assuming the approximation error between $f_r(x)$ and y_r is specified by a set of error upper bound $\rho_r > 0$, there exist appropriate weights such that the approximation error $\triangle f_r(x)$ satisfying

$$\| \triangle f_r(x) \| = \| f_r(x) - y_r \| \leq \rho_r \| x \| . \tag{2}$$

Thus, by means of neural networks, nonlinear MJS is translated into a group of linear differential inclusions, in which the different inclusion is powered by stochastic Markovian process, i.e.

$$\dot{x}(t) = \sum_{\nu} \mu_{r\nu} A_{r\nu} x(t) + \triangle f_r(x). \tag{3}$$

In system (3), the detailed structure and quantitative size of error dynamics $\triangle f_r(x)$ is not needed, while the only requirement is norm-bounded. This condition is easily satisfied in practical cases. Also the bounds of norm may vary according to different nonlinearities in different modes.

In following section, benefited from basic stability theory of linear MJS [2], the quadratic optimal control problem of nonlinear MJS is investigated under mode-dependent LDI form. Moreover, with the bounded error dynamics, the presented results are also robust to the approximation error caused by neural network.

3 Feedback Control and Optimization

For a class of nonlinear stochastic MJS with control input $u(t)$, introduce stochastic quadratic performance cost as

$$J = E\{\int_0^{\infty} [x^T(t) Q x(t) + u^T(t) R u(t)] dt \mid x_0, r_0\}, \tag{4}$$

where Q and R are selected weight matrices respectively relating to state and control input. $u(t)$ may enter system through nonlinear functions $g_r(u)$, which can be approximated by neural networks in LDI form in a similar way as $f_r(x)$. For the sake of derivation simplification, thereinafter, $g_r(u)$ will be processed as $B_r u$, where linear inclusion is omitted without loss of generality.

Considering state feedback control law $u(t) = K_r x(t)$, the resulting closed-loop system is given by

$$\dot{x}(t) = (\sum_{\nu} \mu_{r\nu} A_{r\nu} + B_r K_r) x(t) + \triangle f_r(x). \tag{5}$$

Based on stochastic quadratic performance cost (4), the feedback control problem of nonlinear stochastic MJS is how to find a control law such that the closed-loop system (5) is asymptotically stable and satisfies $J \leq J_0$, where J_0 is cost upper bound. Furthermore, the optimal control problem is how to make the upper bound J_0 minimizing in the presence of stochastic jumps and nonlinearities. Following Theorem 1 and Theorem 2 will give the solutions for these problems respectively.

Theorem 1. There exist Markovian mode-dependent feedback control law $u(t) = Y_r X_r^{-1} x(t)$ such that nonlinear stochastic MJS is asymptotically stable and satisfies $J \leq J_0$ with $J_0 = x^T(0) X_r^{-1} x(0)$, if following coupling LMIs (6) is feasible about positive definite symmetric matrices X_r, matrices Y_r and positive scalar ε for all $r \in S$ and $\nu \in \Omega$.

$$\begin{bmatrix} M_r & X_r & Y_r^T & \rho_r X_r & N_r \\ X_r & -Q_r^{-1} & 0 & 0 & 0 \\ Y_r & 0 & -R_r^{-1} & 0 & 0 \\ \rho_r X_r & 0 & 0 & -\varepsilon I & 0 \\ N_r^T & 0 & 0 & 0 & L \end{bmatrix} < 0, \tag{6}$$

where

$$M_r = A_{r\nu} X_r + X_r A_{r\nu}^T + B_r Y_r + Y_r^T B_r^T + \pi_{rr} X_r + \varepsilon,$$

$$N_r = \left[\sqrt{\pi_{r1}} X_r \cdots \sqrt{\pi_{(r-1)r}} X_r \ \sqrt{\pi_{(r+1)r}} X_r \cdots \sqrt{\pi_{rS}} X_r \right],$$

$$L = -diag\{ X_1 \cdots X_{r-1} \ X_{r+1} \cdots X_S \}.$$

Theorem 2. The optimal upper bound J_{0min} of quadratic performance cost is obtained by convex optimization problem: minimizing J in variables X_r,$Y_r$and ε, subject to LMIs (6) and following (7)

$$\begin{bmatrix} -J & x^T(0) \\ x(0) & -X_r \end{bmatrix} \leq 0. \tag{7}$$

4 Numerical Example

Consider following nonlinear MJS with jump modes $r = \{1, 2, 3\}$

$$\dot{x}(t) = \begin{bmatrix} 0 & 1 \\ 0 & -1 \end{bmatrix} x(t) + \begin{bmatrix} 0 \\ sin(\gamma(r) x_1(t)) \end{bmatrix} + \begin{bmatrix} 0 \\ \lambda(r) \end{bmatrix} u(t),$$

where $\{\gamma(r), \lambda(r)\}$ is the mode-dependent parameters jumping within $\{0.5, 3\}_{r=1}$, $\{1.5, 0.5\}_{r=2}$, $\{2.5, 1.7\}_{r=3}$. The transition rate matrix is given by

$$\Pi = \begin{bmatrix} -3 \ 1.8 \ 1.2 \\ 0.3 \ -2 \ 1.7 \\ 0.3 \ 0.7 \ -1 \end{bmatrix}.$$

It is to be noted that the jumping $\gamma(r)$ appears in as cyclic frequency of nonlinear function. In this example, three 3-layer neural networks with $h = 2$ are chosen to respectively approximate three nonlinear functions $sin(\gamma(r)x_1(t))$ for each mode, and three LDI are obtained as

$$A_{11} = \begin{bmatrix} 0 & 0 \\ 0 & 0 \end{bmatrix}, A_{12} = \begin{bmatrix} 0 & 0 \\ 0.5152 & 0 \end{bmatrix}, A_{13} = \begin{bmatrix} 0 & 0 \\ -0.2257 & 0 \end{bmatrix}, A_{14} = \begin{bmatrix} 0 & 0 \\ 0.7409 & 0 \end{bmatrix},$$

$$A_{21} = \begin{bmatrix} 0 & 0 \\ 0 & 0 \end{bmatrix}, A_{22} = \begin{bmatrix} 0 & 0 \\ 1.2480 & 0 \end{bmatrix}, A_{23} = \begin{bmatrix} 0 & 0 \\ 4.4304 & 0 \end{bmatrix}, A_{24} = \begin{bmatrix} 0 & 0 \\ -3.1824 & 0 \end{bmatrix},$$

$$A_{31} = \begin{bmatrix} 0 & 0 \\ 0 & 0 \end{bmatrix}, A_{32} = \begin{bmatrix} 0 & 0 \\ 1.3656 & 0 \end{bmatrix}, A_{33} = \begin{bmatrix} 0 & 0 \\ 2.6068 & 0 \end{bmatrix}, A_{34} = \begin{bmatrix} 0 & 0 \\ -2.6065 & 0 \end{bmatrix},$$

The upper bounds of approximation errors are $\rho_1 = 0.18$, $\rho_2 = 0.09$, $\rho_3 = 0.34$, respectively.

Introducing weight matrices $Q = diag\{0.1, 0.5\}$, $R = 0.1$, assume the initial condition is $x_0 = \begin{bmatrix} 1.5 & 0 \end{bmatrix}^T$, apply Theorem 1, the mode-dependent gains of feedback control law are obtained as

$$K_1 = \begin{bmatrix} -1.4999 & -0.2032 \end{bmatrix}.$$

$$K_2 = \begin{bmatrix} -17.2227 & -7.2705 \end{bmatrix}.$$

$$K_3 = \begin{bmatrix} -2.2201 & -0.5583 \end{bmatrix}.$$

Furthermore, applying Theorem 2, the optimal control is achieved with minimum upper bound $J_{min} = 6.3070$, and optimal state feedback gains is as

$$K_1 = \begin{bmatrix} -18.2379 & -8.0189 \end{bmatrix}.$$

$$K_2 = \begin{bmatrix} -33.6998 & -14.7019 \end{bmatrix}.$$

$$K_3 = \begin{bmatrix} -11.9257 & -5.7706 \end{bmatrix}.$$

For addressed stochastic nonlinear MJS, thestate response under feedback control law and its optimization are shown in Fig. 1, contrastively.

5 Conclusion

While multi-layer neural networks being used to approximate the nonlinear functions according to different modes, the overall system can be represented by mode-dependent LDI. This procedure gives a uniform way to analysis and synthesis stochastic MJS with general nonlinearities. Benefited from LMI technique, stochastic optimal control can be achieved by means of convex optimization. The example reveals that the presented procedure is computationally flexible and efficient.

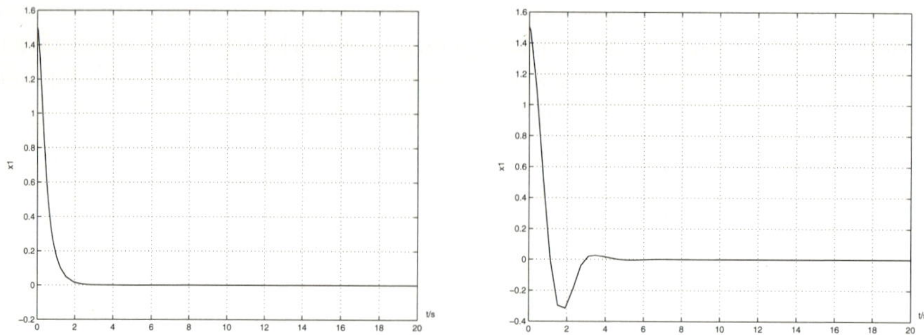

Fig. 1. State evolution by feedback control (Left) and optimization (Right)

Acknowledgements

This work is supported by The National Natural Science Foundation of China (NSFC: 60574001)and by Program for New Century Excellent Talents in University (NCET).

References

1. Krasovskii, N.N., Lidskii, E.A.: Analytical Design of Controllers in Systems with Random Attributes. Automat. Remote Control. 22(1961) 1021-1025
2. Ji, Y., Chizeck, H.J.: Controllability, Stability and Continuous-time Markovian Jump Linear Quadratic Control. IEEE Trans. Automat. Contr. 35(7) (1990) 777-788
3. Fang, Y., Loparo, K.A.: Stabilization of Continuous-Time Jump Linear Systems. IEEE Trans. Automat. Contr. 47(2002) 1590-603
4. Aliyu, M.D.S., Boukas, E.K.: Control for Markovian Jump Nonlinear Systems. Proc. IEEE Conf. Deci. Contr. (1998) 766-771
5. Wang, Z.D., Qiao, H., Burnham, K.J.: On Stabilization of Bilinear Uncertain Time-Delay Stochastic Systems with Markovian Jumping Parameters. IEEE Trans. Automat. Contr. 47(4) (2002) 640-646
6. Nakura, G., Ichikawa, A.: Stabilization of A Nonlinear Jump System. Systems and Control Letters 47(4) (2002) 79-85
7. Bass, E., Lee, K.Y.: Robust Control of Nonlinear System Using Norm-Bounded Neural Networks. IEEE Int. Conf. Neural Networks (1994) 2524-2529
8. Tanaka, K.: An Approach to Stability Criteria of Neural-Network Control Systems. IEEE Trans. Neural Networks 7(3) (1996) 629-642
9. Limanond, S., Si, J.: Neural-Network-Based Control Design: An LMI Approach. IEEE Trans. Neural Networks 9(6) (1998) 1422-1429
10. Wonham, W.M.: On A Matrix Riccati Equation of Stochastic Control. SIAM J. Contr. 6(1968) 681-697

Performance Estimation of a Neural Network-Based Controller

Johann Schumann[1] and Yan Liu[2]

[1] RIACS/NASA Ames, Moffett Field, CA 94035, USA
schumann@email.arc.nasa.gov
[2] Motorola Labs, Schaumburg, IL 60193, USA
yanliu@motorola.com

Abstract. Biologically inspired soft computing paradigms such as neural networks are popular learning models adopted in adaptive control systems for their ability to cope with a changing environment. However, continual changes induce uncertainty that limits the applicability of conventional validation techniques to assure a reliable system performance.

In this paper, we present a dynamic approach to estimate the performance of two types of neural networks employed in an adaptive flight controller: the validity index for the outputs of a Dynamic Cell Structure (DCS) network and confidence levels for the outputs of a Sigma-Pi (or MLP) network. Both tools provide statistical inference of the neural network predictions and an estimate of the current performance of the network. We further evaluate how the quality of each parameter of the network (e.g., weight) influences the output of the network by defining a metric for parameter sensitivity and parameter confidence for DCS and Sigma-Pi networks. Experimental results on the NASA F-15 flight control system demonstrate that our techniques effectively evaluate the network performance and provide validation inferences in a real-time manner.

1 Introduction

Adaptive Flight Control is considered as one of the most challenging real-time automation and control tasks as the system's functions are not static but evolve over time in a non-probabilistic manner. While these evolving functions, through judicious online learning, aid the adaptive controller to recuperate the system (aircraft) from an operational damage (sensor/actuator failure, changed aircraft dynamics: broken aileron or stabilator, etc.), they add an additional degree of complexity and system uncertainty. Since it is impossible to estimate and analyze all possible concerns relative to system safety beforehand, online adaptive systems require a non-conventional approach to verification and validation (V&V).

Neural networks are widely employed for function approximation, prediction and pattern recognition. The requirements on such models are usually described as satisfying certain criteria of precision and/or accuracy. Typical metrics used for performance evaluation of neural networks are Mean Square Error (MSE), Squared Error, etc. They are used to measure the learning performance of a

J. Wang et al. (Eds.): ISNN 2006, LNCS 3972, pp. 981–990, 2006.

neural network model. For prediction performance evaluation, the most popular metrics are prediction/confidence intervals defined to measure the reliability of network output. In the context of an online neural network based adaptive control system, the online neural network is expected to promptly respond to, adapt to and accommodate environmental changes. Therefore, within an online adaptive system, assuring the performance of the online neural network requires online evaluation of its adaptation performance.

In this paper, we present statistical approaches to estimate the performance of a neural network: the validity index (VI) for Dynamic Cell Structures (DCS) and confidence levels for the outputs of a Sigma-Pi (or MLP) network. Both tools provide error bars of the neural network outputs, given the current inputs and thus provide an estimate of the current performance of the network.

For safety-critical systems, a "black box" approach to network assessment is not sufficient. We therefore estimate how the quality of each parameter of the network (e.g., weight) influences the output of the network by calculating parameter sensitivity and parameters confidence for DCS and Sigma Pi networks.

There is only little related work on verification and validation of neural network-based controllers. [7, 18] discuss an approach toward V&V of such systems; NASA has developed a software verification process guide [12] addressing these issues. The use of Bayesian techniques to estimate neural network quality is presented in detail in [2]; another metric (validity index) for RBF has been introduced in [9]. Monitoring approaches for neuro-adaptive controllers, based on Lyapunov stability are discussed in [6].

2 Neural Network Based Flight Control

We illustrate our approach with the NASA F-15 Intelligent Flight Control System (IFCS) project. Its aim is to develop and test-fly a neuro-adaptive intelligent flight control system for a manned F-15 aircraft. Two principal architectures have been developed: the Gen-I architecture uses a DCS neural network as its online adaptive component, the Gen-II architecture a Sigma Pi network.

Figure 1 shows the basic architecture of the Gen-I and Gen-II controllers: pilot stick commands Θ_{cmd} are mixed with the current sensor readings Θ (e.g., airspeed, angle of attack, altitude) to form the desired behavior of the aircraft.

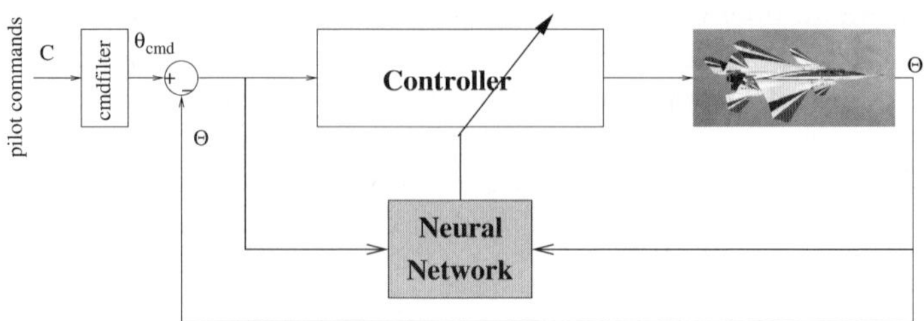

Fig. 1. IFCS Generic Adaptive Control Architecture

From these data, the PID controller calculates the necessary movements of the control surfaces (e.g., rudder, ailerons) and commands the actuators. The controller incorporates a model of the aircraft dynamics. If the aerodynamics of the aircraft changes (e.g., due to a damaged wing or a stuck rudder), there is a deviation between desired and actual state. The neural network is trained during operation to minimize this deviation. Whereas in the Gen-I architecture, the appropriate control derivatives are modified with a neural network, Gen-II uses a dynamic inverse controller with control augmentation, i.e., the neural network produces a control correction signal. For details on the control architecture see [16, 4, 17].

2.1 The Neural Networks

Dynamic Cell Structure Network. The Dynamic Cell Structures network is derived as a dynamically growing structure in order to achieve better adaptability. DCS can be seen as a special case of Self-Organizing Map (SOM) structures as introduced by Kohonen [8] and further improved to offer topology-preserving adaptive learning capabilities that can respond and learn to abstract from a much wider variety of complex data manifolds [13, 3]. In the IFCS Gen-I controller, the DCS provides derivative corrections during system operation.

The DCS network adopts the self-organizing structure and dynamically evol ves with respect to the learning data. It approximates the function that maps the input to the output space. At last, the input space is divided into different regions, referred to as the Voronoi regions [13, 3, 5]. Each Voronoi region is represented by its centroid, a neuron associated with its reference vector known as the "best matching unit" (bmu). Further, a "second best matching unit" (sbu) is defined as the neuron whose reference vector is the second closest to a particular input. An Euclidean distance metric is adopted for finding both units.

The training algorithm of the DCS network combines the competitive Hebbian learning rule and the Kohonen learning rule. The Hebbian learning rule is used to adjust the connection strength C_{ij} between two neurons. The Kohonen learning rule is used to adjust the weight representations of the neurons ($\boldsymbol{w_i}$), which are activated based on the best-matching methods during the learning. If needed, new neurons are inserted. After learning, when DCS is used for prediction (the recall mode), it will recall parameter values at any chosen dimension. It should be noted that the computation of an output is different from that during training. When DCS is in recall mode, the output is computed based on two neurons for a particular input. One is the bmu of the input; the other is the closest neighbor of the bmu other than the sbu of the input. In the absence of neighboring neurons of the bmu, the output value is calculated using the bmu only. Since our performance estimation does not depend on the specific learning algorithm, it will not be discussed in this paper. For details on DCS and the learning algorithm see [13, 3, 5, 10].

Sigma Pi Neural Network. The IFCS Gen-II controller uses a Sigma-Pi ($\Sigma\Pi$) neural network [15], where the inputs are $\mathbf{x}$ subjected to arbitrary basis functions (e.g., square, scaling, logistic function). Then Cartesian products (Π) of these function values are calculated. The final output of the network o is a weighted sum (Σ) of these products (Figure 2):

$$o = \sum_i w_i p_i \quad \text{where} \quad p_i = \prod_j \beta(\mathbf{x}_j)$$

with weights w_i and $\beta(\mathbf{x}_j)$ the basis functions. During the training, the weights w_i are modified as to minimize the tracking error of the controller. As our approach to confidence and sensitivity analysis does not depend on the specific training algorithm for the network, it will not be discussed here. For details see again [16]. Figure 2 (right) shows how the network weights w_i develop over time during an operational scenario. At $t = 1.5s$, a failure occurs, triggering adaptation of the neural network.

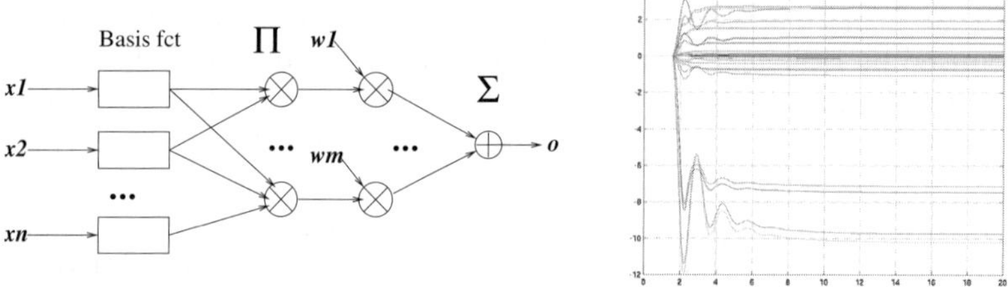

Fig. 2. Architecture of $\Sigma\Pi$ network (left). Development of the NN weights over time during adaptation (right). The failure occurred at $t = 1.5s$.

3 Estimating Network Performance

3.1 Data Confidence

Validity Index. Following the definition of Validity Index (VI) in RBF networks by Leonard et.al.[9], we define the validity index in DCS networks as an estimated confidence measure of a DCS output, given the current input. The VI can be used to measure the accuracy of the DCS network fitting and thus provide inferences for future validation activities. Based on the primary rules of DCS learning and properties of the network structure, we employ confidence intervals and variances to calculate the validity index in DCS. The computation of a validity index for a given input consists of two steps: (1) compute the local error associated with each neuron, and (2) estimate the standard error of the DCS output for the given input using information from step (1). Details can be found in [11, 10].

We have modified the DCS training algorithm to calculate the validity index. Because all needed information is present at the final step of each training cycle, we can simply calculate $s_i^{\prime 2}$ for each neuron after the learning stops. When the DCS is in recall mode, the validity index is computed based on the local errors

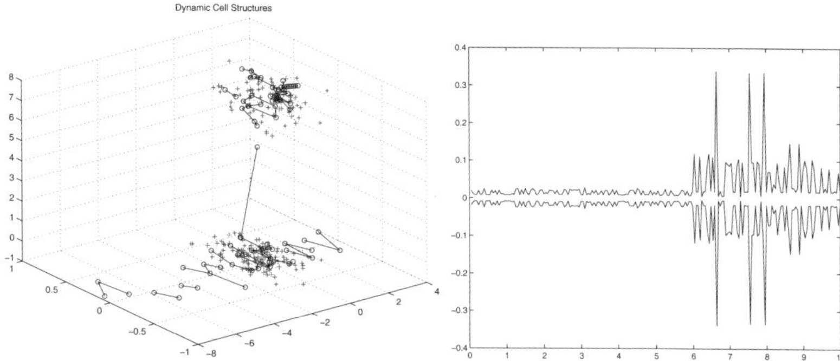

Fig. 3. Online operation of DCS VI on failure mode simulation data. Left: The final form of DCS network structures. Right: VI shown as error bars for each DCS output.

and then associated with every DCS output. We have simulated the online learning of the DCS network under a failure mode condition. Running at 20 Hz, the DCS network updates its learning data buffer (of size 200) at every second and learns on the up-to-date data set of size 200. We first start the DCS network under nominal flight conditions with 200 data points. After that, every second, we set the DCS network in recall mode and calculate the derivative corrections for the freshly generated 20 data points, as well as their validity index. Then we set the DCS network back to the learning mode and update the data buffer to contain the new data points.

Figure 3 shows the experimental results of our simulation on the failure mode condition. The left plot shows the final form of the DCS network structure at the end of the simulation. The 200 data points in the data buffer at the end of the simulation are shown as crosses in the 3-D space. The network structure is represented by circles (as neurons) connected by lines as a topological mapping to the learning data. The right plot presents the validity index, shown as error bars. The x-axis here represents the time frames. The failure occurs at $t = 5.0s$. We compute the validity index for the data points that are generated five seconds before and five seconds after the failure occurs.

A trend revealed by the validity index in our simulations is the increasingly larger error bars after the failure occurs. At $t = 6.0s$, the network has learned these 20 failure data points generated from $\Delta t = 5.0 \sim 6.0s$. The network performance became less stable. After that, the error bars start shrinking while the DCS network adapts to the new domain and accommodates the failure. After the failure occurs, the change (increase/decrease) of the validity index varies depending on the characteristics of the failure as well as the accommodation performance of the DCS network. Nevertheless, the validity index explicitly indicates how well and how fast the DCS network accommodates the failure.

Bayesian Confidence Tool. For the Gen-II architecture, the *Confidence Tool* (CT) [7] produces a quality measure of the neural network output. Our

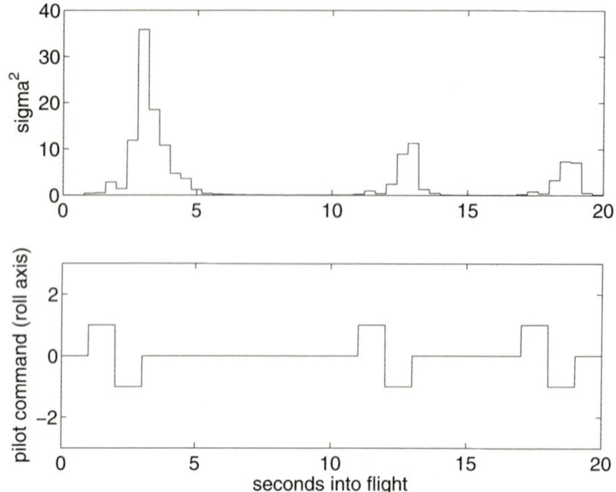

Fig. 4. Confidence value σ^2 over time *(top)* and pilot commands for roll axis (*bottom*). A failure has occurred at $t = 1.5s$.

performance measure is the probability density $p(o|\mathbf{x}, D)$ of the network output o given inputs $\mathbf{x}$, when the network has been trained with training data D. Assuming a Gaussian distribution, we use the standard deviation σ^2 as our performance measure. A small σ^2 (a narrow bell-shaped curve) means that, with a high probability, the actual value is close to the returned value. This indicates a good performance of the network. A large σ^2 corresponds to a shallow and wide curve. Here, a large deviation is probable, indicating poor performance.

Our confidence tool uses an algorithm, following the derivation in [2]. The CT has been implemented for Sigma-Pi and multi-layer perceptron (MLP) networks in Matlab (for a Simulink environment) and in C. For details see [7, 18]. The CT tool is currently implemented on the flight computer on a NASA F-15 aircraft and is scheduled for manned test-flights in the near future.

Figure 4 shows the results of a (Simulink) simulation experiment. In the top panel, σ^2 is shown over time. At time $t = 1.0s$, the pilot issues a doublet command (fast stick movement from neutral into positive, then negative and back to neutral position; Fig. 4*(lower panel)*). Shortly afterwards ($t = 1.5s$), one control surface of the aircraft (stabilizer) gets stuck at a fixed angle ("the failure"). Because the system dynamics and the model behavior do not match any more, the neural network has to produce an augmentation control signal to compensate for this deviation. The σ^2 of the network output increases substantially, indicating a large uncertainty in the network output. Due to the online training of the network, this uncertainty decreases very quickly.

A second and third pilot command (identical to the first one) is executed at $t = 11s$, and $t = 17s$, respectively. During that time, the network's confidence is still reduced, but much less than before. This is a clear indication that the network has successfully adapted to handle this failure situation.

3.2 Sensitivity and Confidence

For the analysis of any controller's behavior, it is important to estimate its sensitivity with respect to input perturbations. A badly designed controller might amplify the perturbations, which could lead to oscillations and instability. The higher the *robustness* of the controller, the less influence arises from input perturbations. It is obvious that such a metric (i.e., $\frac{\partial \mathbf{o}}{\partial \mathbf{x}}$ for outputs $\mathbf{o}$ and inputs $\mathbf{x}$) is also applicable to an adaptive control system. For an adaptive component, like a neural network, the estimation of the sensitivity is a "black box" method, i.e., no knowledge about the internal structure or parameters is necessary.

In this paper, however, we focus on *parameter sensitivity*. This means, we calculate $\frac{\partial \mathbf{o}}{\partial p}$ for each of the adjustable parameters $p \in \mathcal{P}$. For a neural network, $\mathcal{P}$ is comprised of the network weights w_i, for the DCS network, it is the reference vectors of the neurons $\boldsymbol{w_i}$. During training of the network, these parameters are adjusted to minimize the error. Depending on the architecture of the adaptive controller, the network can be pre-trained, i.e., the parameters are determined during the design phase ("system identification"), or the parameters are changing while the system is in operation ("online adaptation"). In both cases, one needs to know, which influence the actual values of the parameters have on the output of the neural network: if the influence of a parameter or neuron is negligible, then this neuron might be removed from the network. On the other hand, extremely high sensitivity might cause numerical problems.

Even more information can be obtained if we consider each parameter of the neural network not as a scalar value, but as a probability distribution. Then, we can formulate the sensitivity problem in a statistical way. The probability of the output of the neural network is $p(\mathbf{o}|\mathcal{P}, \mathbf{x})$ given parameters $\mathcal{P}$ and inputs $\mathbf{x}$. If we again assume a Gaussian probability distribution, we can define our parameter confidence as the variance $\sigma^2_{\mathcal{P}}$. In contrast to calculating the network output confidence value, we do not marginalize over the weights, but over the inputs.

A Sensitivity Metric for DCS Networks. Within the IFCS Gen-I, the DCS networks are employed for online adaptation/learning. Their parameters (connection strength C_{ij} and reference vectors $\boldsymbol{w}_i$) are updated during system operation. It should be noted that the connection strength C does not contribute to the network predictions while it is in recall mode. This implies that the sensitivity of the connection strength is merely a structure related parameter that influences the reference vectors instead of the network output. We therefore only measure the sensitivity of the reference vector of the DCS network. Using the simulation data obtained from the IFCS Gen-I simulator, we calculate the parameter sensitivity s and its confidence σ^2 after each learning epoch during a flight scenario. The sensitivity analysis has been conducted on a N-dimension space, where N is the number of dimensions of the input space.

Figure 5 shows two sensitivity snapshots at different times of the simulation where the network has been trained with 2-dimensional data. Each neuron is associated with a 2-dimensional sensitivity ellipse. At the beginning of the

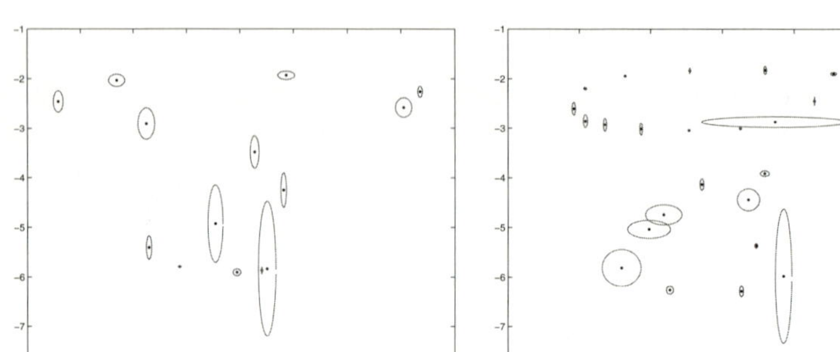

Fig. 5. Sensitivity analysis for DCS networks

simulation, the network is initialized with two neurons whose reference vectors represent two randomly selected training data points. The network continues learning and adjusts its own structure to adapt to the data. Figure 5 (left) shows the situation at $t = 5.0s$. Figure 5 (right) shows the situation at $t = 10.0s$. At $t = 5.0s$, most neurons exhibit relatively large sensitivity, while only a few (31%) neurons have small sensitivity values. However, at $t = 10.0s$, when the network has well adapted to the data, Figure 5 (right) clearly indicates that now most (78%) neurons have small sensitivity values.

A Sensitivity Metric for Sigma-Pi Networks. We have implemented the sensitivity analysis for the online adaptive Sigma-Pi network of the IFCS Gen-II controller. We calculate the parameter sensitivity s and its confidence σ^2 for the network parameters w_i at each point in time during a flight scenario. Figure 6 shows two sensitivity snapshots at various stages of the scenario. At the beginning of the scenario, all parameters of the network are set to zero, giving (trivially) in the same sensitivity. At $t = 1.5$, a failure is induced into the system. In order to compensate for the failure, the network weights adapt (see Fig. 2(right)). Figure 6(left) shows the situation at $t = 5.0s$. A considerable amount of adaptation and weight changes has taken place already. However, the confidence for each of the 60 neurons is still relatively small, as indicated by the large error bars. After approximately 20 seconds, the neural network is fully trained. Figure 6(right) now shows quite different values for the sensitivity. Whereas the sensitivity for most of the neurons is really small now, a few (here 7) neurons exhibit high sensitivity. Although their σ^2 is somewhat larger than that for the other neurons, a clear distinction between the different groups can be made.

Independently from this analysis, the network architecture had been modified several times during the design of the Gen-II controller. So, the number of weights in the network (for roll axis) was reduced from 60 (as shown) to 6. The results obtained with our Parameter Confidence tool (Figure 6) clearly demonstrate (after the fact) that this substantial reduction of the network size is justified.

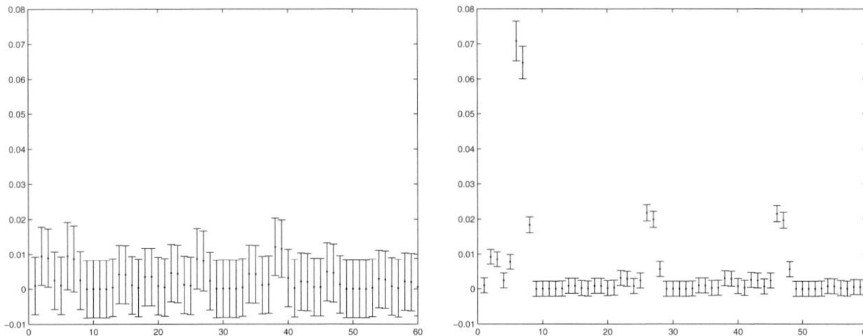

Fig. 6. Parameter sensitivity and confidence at $t = 5s$ (left) and $t = 20s$ (right)

4 Conclusions

We have presented tools for the estimation of the performance of a neural network used in an adaptive controller. For two, highly disjoint architectures, Dynamic Cell Structures (DCS), and Sigma Pi networks, we have shown how the network prediction performance in form of statistical error bars (validity index for DCS, network confidence for Sigma Pi) can be calculated. The online estimation is import in control applications, where the neural network is being trained during operation. The availability of this information plays an important role for verification and validation of such a system in a safety-critical application.

Further insight on the actual performance of the neural network can be gained by looking at the parameter sensitivity and parameter confidence. In this paper, we have presented tools to calculate the sensitivity of individual parameters (reference vectors of neurons in DCS; weights in Sigma Pi networks) and the parameters confidence (again an error-bar). Our approaches are based upon Bayesian statistics and thus provide a solid statistical background for the performance estimation.

Our tools are primarily designed to provide dynamic data on the performance of the networks, but can also be used during the early design phase of an adaptive controller, when the architecture and size of the network is determined. Our Bayesian approach allows different models (e.g. networks with different numbers of hidden units, or different network types such as multi-layer perceptrons, Sigma-Pi, RBF, or DCS) to be compared using only the training data. More generally, the Bayesian approach provides an objective and principled framework for dealing with the issues of model complexity.

Our tools are capable of calculating a performance index for the neural network. The actual performance of the entire system (in our case, the aircraft) also depends on a multitude of other parameters (e.g., robustness of controller, performance metric, type of failure). In aeronautics, the performance of an aircraft is defined in terms of its handling quality (e.g., the Cooper-Harper rating). Current research aims to relate our performance metric with the aircraft handling quality. With the real-time availability of handling quality estimates, our tools can be used to alert the pilot and provide assistance/support to decision making.

References

1. Ahrns, I., Bruske, J., Sommer, G.: On-line Learning with Dynamic Cell Structures. In: Fogelman-Soulié, F., Gallinari, P. (eds.): Proc. Int. Conf. Artificial Neural Networks, Vol. 2. EC2, Nanterre, France (1995) 141-146
2. Bishop, C.M.: Neural Networks for Pattern Recognition. Oxford University Press, Oxford, UK (1995)
3. Bruske, J., Sommer, G.: Dynamic Cell Structures. In: Proc. Neural Information Processing Systems, Vol. 7. (1995) 497-504
4. Calise A., Rysdyk. R.: Nonlinear Adaptive Flight Control Using Neural Networks. IEEE Control Systems Magazine 21(6) (1998) 14-26
5. Fritzke, B.: Growing Cell Structures - a Self-organizing Network for Unsupervised and Supervised Learning. Neural Networks 7(9) (1993) 1441-1460
6. Fuller, E., Yerramalla, S., Cukic, B.,Gururajan, S.: An Approach to Predicting Non-deterministic Neural Network Behavior. In: Proc. Intl. Joint Conference on Neural Networks (IJCNN) (2005)
7. Gupta, P., Schumann, J.: A Tool for Verification and Validation of Neural Network Based Adaptive Controllers for High Assurance Systems. In: Proc. High Assurance Software Engineering, IEEE Press (2004)
8. Kohonen, T.: Self-Organizing Maps. Springer-Verlag, New York (1997)
9. Leonard, J.A., Kramer, M.A., Ungar, L.H.: Using Radial Basis Functions to Approximate a Function and Its Error Bounds. IEEE Transactions on Neural Networks 3(4) (1992) 624-627
10. Liu, Y.: Validating A Neural Network-based Online Adaptive System. PhD thesis, West Virginia University, Morgantown (2005)
11. Liu, Y., Cukic, B., Jiang, M., Xu, Z.: Predicting with Confidence - An Improved Dynamic Cell Structure. In: Wang., L, Chen, K., Ong., Y.S. (eds.): Lecture Notes in Computer Science: Advances in Neural Computation, Vol. 1, Springer-Verlag, Berlin Heidelburg (2005) 750-759
12. Mackall, D., Nelson, S., Schumann, J.: Verification and Validation of Neural Networks of Aerospace Applications. Technical Report CR-211409, NASA (2002)
13. Martinez, T., Schulten, K.: Topology Representing Networks. Neural Networks 7(3) (1994) 507-522
14. Norgaard, M., Ravn O., Poulsen, N., Hansen, L.K.: Neural Networks for Modeling and Control of Dynamic Systems. Springer Verlag (2002)
15. Rumelhart, McClelland, and the PDP Research Group: Parallel Distributed Processing. MIT Press (1986)
16. Rysdyk R.,Calise, A.: Fault Tolerant Flight Control via Adaptive Neural Network Augmentation. AIAA-98-4483 (1998) 1722-1728
17. Schumann, J., Gupta, P.: Monitoring the Performance of A Neuro-adaptive Controller. In: Fischer, R., Preuss, R., and von Toussaint, U.: Proc. 24th International Workshop on Bayesian Inference and Maximum Entropy Methods in Sciences and Engineering (MAXENT), AIP (2004) 289-296
18. Schumann, J., Gupta, P., Jacklin. S.: Toward Verification and Validation of Adaptive Aircraft Controllers. In: Proc. IEEE Aerospace Conference. IEEE Press (2005)

Some Key Issues in the Design of Self-Organizing Fuzzy Control Systems

Xue-Feng Dai, Shu-Dong Liu, and Deng-Zhi Cui

Computer and Control Engineering Institute, Qiqihar University
161006 Qiqihar, Heilongjiang, China
xfdai@tom.com

Abstract. The design of self-organizing fuzzy control (SOFC) systems does not rely on experts' knowledge. The systems can establish fuzzy reasoning rules and adjust fuzzy parameters on-line automatically. So they are suitable for the control of plants for which we have no appropriate mathematical models and no other knowledge. There are three kinds of SOFCs, i.e., conventional, neural networks based and genetic algorithm based ones. New achievements of the above SOFC are summarized in this paper.

1 Introduction

Fuzzy logic control has exhibited great power for a kind of plants that are characterized by nonlinear, time varying and uncertainties dynamics. However, the precise knowledge that fuzzy reasoning is based on is hard to obtain. Therefore, self-organizing fuzzy control (SOFC) emerged as the times required. Roughly speaking, there are three kinds of SOFC systems, the first is the work in the framework initialized by Procyk and Mamdani, and is named as conventional SOFC here. The controller consists of a fuzzy-rule-based controller and a learning mechanism, the later adjusts fuzzy parameters based on the system performance evaluation. The second is the work based on neural networks. It combined the learning power of neural networks and the intuitive knowledge reasoning of fuzzy systems. The controller is implemented as a multi-layer neural network. The last is the genetic algorithms based SOFC. In this kind of SOFC, both the system structure and the parameters are optimized by genetic algorithms. The key issues in the design of SOFC systems are the selection of membership functions, the determination of rule base size and the learning algorithms. We will summarize the achievements along these aspects in this paper.

2 Conventional Self-Organizing Fuzzy Control

The main features of conventional SOFC are that no initial control rules are needed, and the systems have the on-line learning ability of deciding the proper parameters of fuzzy membership functions and regulating control actions. In general, the way the input space is partitioned determines the number of rules. The space may be partitioned regularly [1]-[7] in advance, or partitioned into a finite number of domains of different

J. Wang et al. (Eds.): ISNN 2006, LNCS 3972, pp. 991 – 997, 2006.

size [8]. Specially [4], the upper limit of the maximum number of rules maybe fixed based on NARX model and delay of the plant to be controlled.

The membership functions used most often in conventional SOFC are the Gaussian function or triangular functions. The Gaussian function [9]-[10] is given by

$$\mu_{xi} = \exp[-\frac{1}{2\sigma_i}(\alpha x(k) - x_i)^2] \tag{1}$$

where σ_i is the width, x_i the center, and α a constant. The triangular functions are given ([1]-[2], [4], [6]-[8]) by, respectively

$$\mu_{xi} = \begin{cases} 1 + \dfrac{x(k) - x_i}{b - a}, & a \le x(k) \le x_i, \\ 1 - \dfrac{x(k) - x_i}{b - a}, & x_i \le x(k) \le b, \\ 0, & otherwise\ . \end{cases} \tag{2}$$

$$\mu_{xi} = \begin{cases} \max(0, \dfrac{|x_i - x(k)|}{\alpha}), & x(k) \le x_i, \\ \max(0, \dfrac{|x(k) - x_i|}{\beta}), & x(k) \ge x_i, \end{cases} \tag{3}$$

$$\mu_{xi} = \begin{cases} \dfrac{1}{b-a}(-|x(k) - x_i| + b - a), & x_i - \frac{1}{2}(b-a) \le x(k) \le x_i + \frac{1}{2}(b-a), \\ 0, & otherwise. \end{cases} \tag{4}$$

In the learning stage, all the parameters in membership functions and the ones in consequent part of every rule are updated [11]. The model estimation is used to find the relationship between the system output and all the selected parameters. Differently [8], the consequent linguistic value or the net linguistic control action is calculated by taking the α-cut of the C_n, where α=max[μ (C_n)].

The performance measure [1], [7] or performance decision table [12] is employed to calculate parameter adjustments. The aim is minimize the following index in [1]

$$J = |y_r(k+1) - y(k+1)|, \tag{5}$$

where y_r(k+1) and y(k+1) is the reference output and real plant output, respectively. Also, a fuzzy performance decision maker (FPDM) may be used as in [13]. Lastly, the learning of controller can be made in consideration of the motional trend of the error state vector as well as its locational information [14].

A sliding mode self-organizing control scheme, which blending VSS with a self-organizing controller, had been developed [12], [16]. The system [16] is composed of two fuzzy inference engines. One is fuzzy controller and the other is to determine the modification of the fuzzy reasoning rules of output based on the sliding-mode function. Similiar result includes self-organizing fuzzy PID control [3]. Moreover, SOFC systems can be embedded in the feedforward and feedback control structure [5], [15].

Table 1. Comparison of conventional self-organizing fuzzy control

References	[1]	[3]	[4]	[5]	[11]
Number of rules	√	√	√	N_r	√
Learning parameters	F	√	F, D_{max}	C	F
Learning algorithm	√			Kohonen	
Membership functions	(4)	(4)	(2)	T,E	(4)
Objective function	(5)	PIT			$1/2(y-y_d)^2$
Discrete model	FARMA		NARX		

GD----gradient descent algorithm, F----fuzzy parameters, C----control, PIT---- performance index table, √----the method is used (blank for no use)

3 Neural Networks Based Self-Organizing Fuzzy Control

In general, it's unnecessary to determine the number of rules in advance for neural networks based SOFC [14], [18]-[19]. There are no hidden nodes (i.e., no fuzzy rules) in the controllers initially [14], [18]-[19], [26]. The number may be determined by clustering algorithms [20], or by adaptive scheme [10] dynamically. Occasionally, the number also has to be specified *a prior* [21]. The membership functions used here includes Gaussian [22], bell-shaped, trapezoidal [9] and triangular [21] ones.

It's impossible to write out all fuzzy rules sometimes. Neural network manifests great potential in fuzzy rule expression. The learning vector quantization (LVQ) network is used to provide information about the better locations of the IF-part membership functions through un-supervised learning in [24]. The premise membership function is given by eqn. (1), and the consequent membership function is give by

$$\mu_{ui}^{j}=\begin{cases}(1+((c_i^j-u_i)/a_{Li})^2)^{-1}, & u_i\le c_i^j,\\ (1+((u_i-c_i^j)/a_{Ri})^2)^{-1}, & u_i>c_i^j.\end{cases} \tag{6}$$

The rule base is partitioned by regular partition M and partition N near the trial trajectory.

One of the issues in the design of neural networks based SOFC is the selection of criterion that when to add a node (rule) and when to delete a node (rule) in the network. Euclidean distance is used often [10], [14], [18]. Another one is the minimum distance of current observation with existing center vector [14], [18].

A distributed control system is proposed in [25] based on temporal self-organizing neural network, called competitive Hebbian (CTH) network. The CTH network can learn and recall complex trajectories by means of two sets of synaptic weights, namely, competitive feedfordward weight that encode the individual states of the trajectory and Hebbian lateral weights that encode the temporal order of trajectory states. An approach utilizing a class of Fuzzified Basis Function Networks, which automatically organizes its network structure to the required size and parameters, is developed in [19]. The proposed similar degree is given by

$$dgr(i,u) = \Phi_{i=1}^{m} \mu_{xi}, \tag{7}$$

where Φ is a logical *AND* product, and μ_{xi} is given in eqn. (1). The learning phase is implemented by the modified Kohonen competitive learning strategy.

With a clustering algorithm, the input space is partitioned into appropriate clusters (regions) that sufficiently cover the training data [20]. Each generated cluster is treated as a antecedent of a fuzzy IF–THEN rule and is expressed by a multidimensional Gaussian function there.

Not only the existing neural networks models were utilized to realize self-organizing fuzzy control, but also a few of specific neural networks were proposed to implement SOFC systems. They are tree-searched neural networks (TNN) [9], recurrent self-organizing neural fuzzy inference network [26], and fuzzy adaptive learning control network [23]. A self-organizing adaptive fuzzy controller based on TNN is proposed in [9]. The rules are updated on-line by an adaptive law. If the maximal firing strength is less than a predefined constant, then a new rule is generated. In [23], the spatial firing strength is used as degree measure of incoming x belong to cluster i,

$$dgr_i(x) = \exp\{[(\sigma_i(x-m_i)]^T[\sigma_i(x-m_i)]\}. \tag{8}$$

If $dgr_J \leqslant F_{in}$, then a new rule is generated, where $J = arg \max_{1\le j\le c(t)} dgr_j(x)$. Let $M(m_i,\sigma_i)$ represents the bell-shaped membership function with center m_i and width σ_i, the following fuzzy similar measure is defined for structure learning

$$dgr(i,t) = \max_{1\le j\le k} E[M(m_{i-new}, \sigma_{i-new}), M(m_j, \sigma_j)]\cdot \tag{9}$$

Table 2. Comparison of neural networks based self-organizing fuzzy control

References	[9]	[10]	[23]	[24]	[25]
Neural networks	TNN	FINN	FALCON	LVQ	CTH
Number of rules		Adaptive		M+N+1	
Membership functions	G	G	√	G	√
Structure learning	FC	Kohonen	√		Hebbian
Parameter learning	AL	LMS	√	√	
On-line or offline	On-line	Offline	On-line	On-line	Offline

AL----adaptive law, FC----firing strength criterion, G----Gaussian function

A fuzzy inference neural network (FINN) which utilizes Kohonen algorithm is proposed in [10]. It can partition an input-output pattern space and can extract fuzzy rules automatically. A neural-network-based fuzzy system (NNFS) is proposed, which utilizes a nearest-neighborhood clustering scheme for both structure learning and initial parameters setting [22].

4 Genetic Algorithms Based Self-Organizing Fuzzy Control

There is quite a few work about the automatic generation of fuzzy rules using genetic algorithms [15], [17], [27]-[28]. However, sometimes the input space must be partitioned in advance [15], [27]. The feedforword module is offline trained and optimized by improved genetic algorithm in [15]. Both of the fuzzy parameters and structure is optimized simultaneously using genetic algorithm. The following performance index J and fitness function F is established,

$$J = \sum_{k=0}^{P} [\tau^d(k) - \tau(k)]^2 , \tag{10}$$

$$F = \frac{1}{1+J} . \tag{11}$$

A similarity measure eqn. (8) is used to determine if add a new chromosome. Paper [27] proposed a self-organized genetic algorithm-based rule generation method that is a three-stage hierarchical scheme. The first stage selects rules to control the system in the vicinity of the set point. The second extends the above rules span of operation to the entire input space. The last then refines the rulebase and reduces the number of rules. The following fitness function is proposed for the last aim,

$$f(R) = w_b * n_b + \frac{w_t}{t_{av}} + \frac{w_r}{|R|} , \tag{12}$$

where n_b is the number of controllable states, t_{av} the average settling time, $|R|$ the number of rules, and w_b, w_t, and w_r three nonnegative constants of the fitness function.

To overcome the nonholonomic constraint in controller design, a self-organized fuzzy controller using an evolutionary algorithm is proposed in [28]. Each input and output function is described by the form that slightly different with eqn. (1), that is

$$\mu_{xi} = \begin{cases} 1 + \dfrac{x(k) - x_i}{x_i - a}, & a \le x(k) \le x_i, \\ 1 - \dfrac{x(k) - x_i}{b - x_i}, & x_i \le x(k) \le b, \\ 0, & otherwise . \end{cases} \tag{13}$$

A Pittsburgh Fuzzy Classifier System #1(P-FCS1) is proposed in [11]. It is based on the Pittsburgh model of learning classifier systems and employs variable length rule-sets and simultaneously evolves fuzzy set membership functions and relations. A new crossover operator which respects the functional linkage between fuzzy rules with overlapping input fuzzy set membership functions is introduced too.

Table 3. Comparison of genetic algorithms based self-organizing fuzzy control

References	[15]	[27]	[28]	[17]
Membership functions	G	eqn.(4)	eqn. (11)	T
Number of rules	√	√	√	√
Parameter learning	√	√	√	√
Space partition			√	√
Encode	Binary	Binary	Binary	Real

5 Conclusions

SOFC can be applied to the control of a variety of plants [1], [18], [28]. But SOFC can not absolutely disaffiliate the mathematical model and input/output data. In the design of conventional SOFC systems, discrete time models are required sometimes to specify the dimension of fuzzy rule. And, the design procedure of both neural networks based and genetic algorithm based SOFC systems is a time consuming work.

Acknowledgement

This work is supported by Heilongjiang Oversea Scholars Fund.

References

1. Moon, U.-C., Lee, K. Y.: A Boiler-turbine System Control Using a Fuzzy Auto- Regressive Moving Average (FARMA) Model. IEEE Trans. Energy Conversion 18(1) (2003) 142-148
2. Singh, Y. O.: A Modified Self-organizing Controller for Real-time Process Control Applications. Fuzzy Sets and Systems 96(2) (1998) 147-159
3. Kazemian, H. B.: The SOF-PID Controller for the Control of a MIMO Robot Arm. IEEE Trans. Fuzzy Systems 10(4) (2002) 523-532
4. Dias, J. M., Dourado, A.: A Self-organizing Fuzzy Controller with a Fixed Maximum Number of Rules and an Adaptive Similarity Factor. Fuzzy Sets and Syst.103 (1999) 27-48
5. Nie, J. and Lee, T. H.: Self-organizing Rule-based Control of Multivariable Nonlinear Servomechanisms. Fuzzy Sets and Systems 91(3) (1997) 285-304
6. Chen, H.-Y., Huang, S.-J.: Ti6A14V Laser Alloying Process Control by Using a Self-organizing Fuzzy Controller. Int. J. Mech. Tools Manufact. 44(5) (2004) 1653–1665
7. Huang, S.-J., Lee, J.-S.: A Stable Self-organizing Fuzzy Controller for Robotic Motion Control. IEEE Trans. Industrial Electronics 47(2) (2000) 421-428
8. Abreu, G., Ribeiro, J.: A Self-organizing Fuzzy Logic Controller for the Active Control of Flexible Structures Using Piezoelectric Actuators. Appl. Soft Computing 1(2002) 271–283
9. Lee, C.-H., Wang, S.-D.: A Self-organizing Adaptive Fuzzy Controller. Fuzzy Sets and Systems 80(3) (1996) 295-313

10. Nishina, T., Higiwara, M.: Fuzzy Inference and NN. Neurocomputing 14 (1997) 223-239
11. Park, G. Y., Seong, P. H.: Application of a Self-organizing Fuzzy Logic Controller to Nuclear Steam Generator Level Control. Nuclear Engin. and Design 167 (1997) 345-356
12. Denker, A., Ohnish, K.: Robust Tracking Control of Mechatronic Arms. IEEE/ASME Trans. Mechatronics 1(2) (1996) 181-188
13. Kim, Y.-T., Bien, Z.: Robust Self-learning Fuzzy Controller Design for a Class of Nonlinear MIMO Systems. Fuzzy Sets and Systems 111(2) (2000) 117-135
14. Gao, Y., Er, M. J.: Online Adaptive Fuzzy Neural Identification and Control of a Class of MIMO Nonlinear Systems. IEEE Trans. Fuzzy Systems 11(4) (2003) 462-477
15. Jin, Y.: Decentralized Adaptive Fuzzy Control of Robot Manipulators. IEEE Trans. System, Man, and Cybernetics—Part B: Cybernetics 28(1) (1998) 47-57
16. Wai, R., Lin, C., Hsu, C.: Self-organizing Fuzzy Control for Motor-toggle Servomechanism via Sliding-mode Technique. Fuzzy Sets and Syst.131 (2002) 235- 249
17. Carse, B., Fogarty, T. C., Munro, A.: Evolving Fuzzy Rule Based Controllers Using Genetic Algorithms. Fuzzy Set and Systems 80(1) (1996) 273-293
18. Er, M. J., Gao, Y.: Robust Adaptive Control of Robot Manipulators Using Generalized Fuzzy Neural Networks. IEEE Trans. Industrial Electronics 50(3) (2003) 620-628
19. Lee, T. H., Nie, J. H., Tan, W. K.: A Self-organizing Fuzzified Basis Function Network Control System Applicable to Nonlinear Servomechanisms. Mechat.5(6) (1995) 695- 713
20. Li, C., Lee, C.-Y.: Self-organizing Neuro-fuzzy System for Control of Unknown Plants. IEEE Trans. Fuzzy Systems 11(1) (2003) 135-150
21. Yeh, Z.-M., Tarng, Y. S., Nian, C. Y.: A Self-organizing Neural Fuzzy Logic Controller for Turning Operations. Int. J. Mech. Tools Manufact. 35(10) (1995) 1363-1374
22. Wang, Y., Rong, G..: A Self-organizing Neural-network-based Fuzzy System. Fuzzy Sets and Systems 103(1) (1999) 1-11
23. Lin, C.-T., Lin. C.-J., Lee, C. S. G.: Fuzzy Adaptive Learning Control Network with On-line Neural Learning. Fuzzy Sets and Systems 71(1) (1995) 25-45
24. Lin, W.-S., Tsai, C.-H.: Self-organizing Fuzzy Control of Multi-variable Systems Using Learning Vector Quantization Network. Fuzzy Sets and Systems 124(2) (2001) 197–212
25. Barreto, G. A., Araujo, A. F. R., Ducker, C., Ritter, H.: A Distributed Robotic Control System Based on a Temporal Self-organizing Neural Network. IEEE Trans. System, Man, and Cybernetics—Part C: Applications and Reviews 32(4) (2002) 347-357
26. Juang, C.-F., Lin, C.-T.: A Recurrent Self-Organizing Neural Fuzzy Inference Network. IEEE Trans. Neural Networks 10(4) (1999) 828-845
27. Pal, T., Pal, N. R.: SOGARG: A Self-organized Genetic Algorithm- Based Rule Generation Scheme for Fuzzy Controllers. IEEE Trans. Evol. Comput.7(4) (2003) 397-415
28. Kim, S. H., Park, C., Harashima, F.: A Self-organized Fuzzy Controller for Wheeled Mobile Robot Using an Evolutionary Algorithm. IEEE Trans. Industrial Electronics 48(2) (2001) 467-474

Nonlinear System Stabilisation by an Evolutionary Neural Network

Wasan Srikasam, Nachol Chaiyaratana, and Suwat Kuntanapreeda

Research and Development Center for Intelligent Systems,
King Mongkut's Institute of Technology North Bangkok
1518 Piboolsongkram Road, Bangsue, Bangkok 10800, Thailand
sir_tum@yahoo.com, nchl@kmitnb.ac.th, suwat@kmitnb.ac.th

Abstract. This paper presents the application of an evolutionary neural network controller in a stabilisation problem involving an inverted pendulum. It is guaranteed that the resulting continuous closed-loop system is asymptotically stable. The process of training the neural network controller can be treated as a constrained optimisation problem where the equality constraint is derived from the Lyapunov stability criteria. The decision variables in this investigation are made up from the connection weights in the neural network, a positive definite matrix required for the Lyapunov function and a matrix for the stability constraint while the objective value is calculated from the closed-loop system performance. The optimisation technique chosen for the task is a variant of genetic algorithms called a cooperative coevolutionary genetic algorithm (CCGA). Two control strategies are explored: model-reference control and optimal control. In the model-reference control, the simulation results indicate that the tracking performance of the system stabilised by the evolutionary neural network is superior to that controlled by a neural network, which is trained via a neural network emulator. In addition, the system stabilised by the evolutionary neural network requires the energy in the level which is comparable to that found in the system that uses a linear quadratic regulator in optimal control. This confirms the usefulness of the CCGA in nonlinear system stabilisation applications.

1 Introduction

Neural network controllers have been successfully applied to various nonlinear systems in the past decades. Since the goals of designing a closed-loop system must always include stability assurance, the need to guarantee the stability of neural network control systems is inevitable. The study into the stabilisation capability of neural network controllers can be divided into three main categories. In the first category, a modification of learning algorithms is made where the search for appropriate connection weights is limited to the space at which the stability can be guaranteed. In the second category, a neural network is used to approximate the behaviour of the nonlinear plant where the network is subsequently used to generate control effort according to a predefined control law or

J. Wang et al. (Eds.): ISNN 2006, LNCS 3972, pp. 998–1006, 2006.

a stability condition. Finally, in the third category an additional stability condition is introduced where the learning process can be treated as constrained optimisation.

The modification of learning algorithms, which leads to the solution search in the feasible connection weight space only, was first introduced by Sadegh [1] where the stability condition is integrated directly into the weight adaptation algorithm. The concept proposed by Sadegh [1] is subsequently adapted for robotic system control [2]. In addition to the controller role, neural networks have also been used as both observer and controller in a robotic system [3] where a Lyapunov stability condition is utilised during the weight space restriction. Furthermore, the work in this category has also been extended to include the control of discrete systems [4, 5].

In the case where it is desirable to utilise an explicit control law or a stability condition, the training of a neural network to mimic the behaviour of the nonlinear plant is usually carried out. Since the neural network would certainly have a well-defined mathematical structure, the trained neural network can be regulated in a way that a specific control effort can be produced as the network output. The control effort can be generated according to a linear control law that leads to feedback linearisation [6, 7], a control Lyapunov function [8], or an input-to-state stability condition [9].

Another technique, which has also received much attention, is to introduce stability conditions as optimisation constraints during the search for appropriate connection weights. Tanaka [10] has shown that such a technique can be successfully applied in the case that the system can be explained in terms of linear differential inclusions. In later works, Suykens et al. [11] have proposed a stability constraint, which can be applied to any nonlinear systems, for use with recurrent network controllers. In contrast, Kuntanapreeda and Fullmer [12] have proposed a stability constraint for hermitian systems that are controlled by feedforward networks while Ekachaiworasin and Kuntanapreeda [13] have extended the work to cover nonhermitian systems.

From an optimisation viewpoint, the training of a neural network controller usually involves a search in connection weight space where the search objective is defined according to a specific system performance. Although the use of iterative learning algorithms has been successfully applied to the early works mentioned above, it has also been proven that evolutionary computation algorithms are stronger candidates for optimising neural networks [14]. Nonetheless, the application of evolutionary computation for optimising a neural network controller, while also satisfying stability constraints, has rarely been studied. In this paper, a cooperative coevolutionary genetic algorithm or CCGA [15] will be considered for the optimisation of the connection weights in a neural network controller. Specifically, the stability constraint developed by Ekachaiworasin and Kuntanapreeda [13] will act as the optimisation constraint. This condition is chosen for this paper since it can be easily integrated into any optimisation techniques.

The organisation of this paper is as follows. In section 2, the architecture of the neural network and the associated stability condition will be explained. In section 3, the description of the CCGA will be given. The application of the CCGA to the task of connection weight optimisation and the performance of neural network controller will be discussed in section 4. Finally, the conclusions will be drawn in section 5.

2 Neural Network and Stability Condition

The architecture of neural network controller, which is chosen for this investigation, is a multilayer perceptron. The network has one hidden layer where the activation function in the hidden node is a hyperbolic tangent function while that in the output node is a linear function. The schematic diagram of the neural network controller is given in Fig. 1. From Fig. 1, it can be seen that each neuron does not have a connection weight that represents a threshold. This means that the input to the activation function is simply a weighted-sum of neuron inputs. In Fig. 1, w_{1ij} represents an element in the connection weight matrix $\mathbf{W}_1$ and reflects the connection weight between the ith neuron in the hidden layer and the jth network input. Similarly, w_{2ij} denotes the connection weight in $\mathbf{W}_2$ and links between the ith output neuron and the jth hidden neuron. Ekachaiworasin and Kuntanapreeda [13] have proven that the network in Fig. 1 can be used as a full-state feedback controller where the associated stability condition, derived from the Lyapunov stability criteria, for the closed-loop system can be described by

$$\mathbf{C}_1 = (\mathbf{A} + a\mathbf{B}\mathbf{W}_2\mathbf{W}_1)^T\mathbf{P} + \mathbf{P}(\mathbf{A} + a\mathbf{B}\mathbf{W}_2\mathbf{W}_1) + \mathbf{q}\mathbf{q}^T + \mathbf{I} = \mathbf{0} \quad (1)$$

$$\mathbf{C}_2 = \mathbf{P}\mathbf{B}\mathbf{W}_2 + \mathbf{W}_1^T + \sqrt{2}\mathbf{q} = \mathbf{0} \quad (2)$$

where $\mathbf{A}$ and $\mathbf{B}$ respectively denote system and input matrices of the plant model, which has been linearised around the equilibrium state. In the stability condition, $\mathbf{P}$ is a positive definite symmetric matrix for the Lyapunov function $V = \mathbf{x}^T\mathbf{P}\mathbf{x}$ where $\mathbf{x}$ represents the state vector while $\mathbf{I}$, a and $\mathbf{q}$ are an identity matrix, a constant and a constant matrix, respectively. In this paper, a is preset to 0.5. With this condition, the system is guaranteed to be asymptotically stable.

Two types of control strategies are explored in this paper: model-reference control and optimal control. In the model-reference control, the closed-loop behaviour of the system is shaped such that the performances of the system and the reference model are similar. Generally, an error cost function, which is derived from the difference between the performance of the system and that of the reference model, can provide an indication on how the feedback controller should be adjusted. In this investigation, the squared tracking error between the system output and a linear reference model output will be the indicator. In contrast, the cost function for the optimal control is calculated from the energy utilised by the system during the execution of the control task. Here, the energy is calculated from a quadratic function, which can be described by

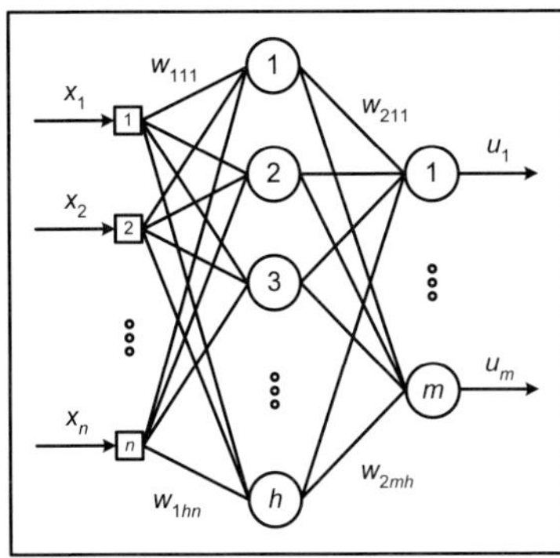

Fig. 1. Architecture of the neural network controller

$$J_e = \int_0^\infty (\mathbf{x}^T \mathbf{Q}\mathbf{x} + ru^2)dt \tag{3}$$

where J_e is the energy cost, $\mathbf{Q}$ is the weighting matrix for the state $\mathbf{x}$ and r is the weighting factor for the control effort u. Both cost functions explained will be parts of fitness functions during the evolution of the neural network controller using a cooperative coevolutionary genetic algorithm.

3 Cooperative Coevolutionary Genetic Algorithm

The cooperative coevolutionary genetic algorithm (CCGA) was first introduced by Potter and De Jong [15]. The CCGA functions by introducing an explicit notion of modularity to the optimisation process. This is done in order to provide reasonable opportunity for complex solutions to evolve in the form of interacting coadapted subcomponents. In brief, the CCGA explores the search space by utilising a population, which contains a number of species or subpopulations. In contrast to other types of genetic algorithm, each species in the CCGA represents a variable or a part of the problem, which requires to be optimised. A combination of an individual from each species will lead to a complete solution to the problem where the fitness value of the complete solution can then be identified. This value of fitness will be assigned to the individual of interest that participates in the formation of the solution. After the fitness values have been assigned to all individuals, the evolution of each species is then commenced using a standard genetic algorithm. The CCGA has been successfully used in a number of problem domains including multivariable function optimisation [15], sequential rule evolution [16], design of a cascade neural network [17] and container loading optimisation [18]. A comprehensive description of the CCGA and the summary of its applications can be found in Potter and De Jong [19].

4 Results and Discussions

The capability of the neural network controller in nonlinear control will be demonstrated in this section. Specifically, an inverted pendulum system will be investigated in the case studies. This system is chosen since it is a regulator control system that has only two states and a performance comparison against previously published results can be easily made. The model of the inverted pendulum is given by

$$\begin{aligned} \dot{x}_1 &= x_2 \\ \dot{x}_2 &= -x_2 + 2\sin x_1 + 10u \end{aligned} \tag{4}$$

where x_1 is the angular position, x_2 is the angular velocity and u is the control effort. The initial condition of the plant is $\left[x_1(0)\; x_2(0)\right]^T = \left[1\; 0\right]^T$ while the desired final condition is $\left[x_1(t_f)\; x_2(t_f)\right]^T = \left[0\; 0\right]^T$ where t_f is the trajectory time. As mentioned earlier in section 2, two control strategies will be explored: model-reference control and optimal control. Since the performance indices for both strategies are described in terms of cost functions, the search objective of the CCGA can be defined as

$$\begin{aligned} f = {} & \alpha_1 J_c + \alpha_2 \left(\sum_i \sum_j c_{1ij}^2\right)^{1/2} + \alpha_3 \left(\sum_i \sum_j c_{2ij}^2\right)^{1/2} \\ & + \alpha_4 \left| \left(\sum_i \sum_j c_{1ij}^2\right)^{1/2} - \left(\sum_i \sum_j c_{2ij}^2\right)^{1/2} \right| \end{aligned} \tag{5}$$

where f is the minimisation objective, J_c is the control cost which can be either the squared tracking error or the energy usage, c_{1ij} and c_{2ij} are elements from the stability constraint matrices $\mathbf{C}_1$ and $\mathbf{C}_2$ as given in equations (1) and (2), respectively and α_i is a weighting factor. It can be clearly seen that the search objective is a weighted-sum between the control cost and the constraint. A desired neural network controller would force the plant to perform according to the strategy employed while the stability condition is also satisfied.

The parameter settings for the multilayer perceptron and the CCGA are given in Table 1. From Table 1, it can be seen that the decision variables for the evolution of the neural network cover both connection weights and variables from the stability condition. In this work, each decision variable is treated as a species in the CCGA search. In addition, the evolution of the neural network is repeated ten times for each case study where the best result is selected for detailed analysis.

For the purpose of comparison, the performance of the evolutionary neural network for model-reference control will be compared with that of the neural network controller reported in Ekachaiworasin and Kuntanapreeda [13]. In the work by Ekachaiworasin and Kuntanapreeda [13], the model-reference control

Table 1. Parameter settings for the multilayer perceptron and the CCGA

Parameter	Setting and Value
Multilayer Perceptron	
Number of input nodes	2 (Number of feedback states)
Number of hidden nodes	4
Number of output nodes	1
CCGA	
Decision variables	$\mathbf{W}_1$ ($w_{1ij} \in [-2, 2]$), $\mathbf{W}_2$ ($w_{2ij} \in [-2, 2]$), $\mathbf{q}$ ($q_{ij} \in [-2, 2]$) and $\mathbf{P}$ ($p_{ij} \in [0, 2]$)
Number of variables (Number of species)	23
Chromosome representation	Binary chromosome
Chromosome length of a species member	20
Fitness scaling method	Linear scaling
Selection method	Stochastic universal sampling
Crossover method	Uniform crossover with probability = 0.8
Mutation method	Bit-flip mutation with probability = 0.3
Subpopulation size	100
Number of elitist individuals	1
Number of generations	300
Number of repeated runs	10

strategy is also explored. However, an additional neural network emulator is required for the controller training. In brief, a neural network is trained to emulate the behaviour of the plant. During the training of another neural network for use as the controller, the tracking error between the closed-loop system output and the reference model output is propagated backward through the neural network emulator where the resulting propagated signal provides an adjustment to the connection weights in the controller. Both evolutionary neural network and neural network controller reported in Ekachaiworasin and Kuntanapreeda [13] are designed such that the resulting closed-loop system behaviour is similar to that of the reference model, which in this study is a linear second order system with the damping ratio = 0.8 and the natural frequency = 3 rad/s. In contrast to the case study involving model-reference control, the performance of the evolutionary neural network for optimal control will be compared with that of a linear quadratic regulator (LQR). The regulator is designed according to the linearised model of the plant. The comparison will be carried out via the monitoring of the energy during the control task execution. The energy cost function is formulated such that the matrix $\mathbf{Q}$ is set to an identity matrix while the parameter r is set to 0.3 for both neural network evolution and linear regulator design.

The performances of the evolutionary neural networks and the techniques selected for comparison for the model-reference control and optimal control are displayed in Figs. 2 and 3, respectively. In addition, the corresponding control cost and constraint are also displayed in Table 2. Firstly, consider the results

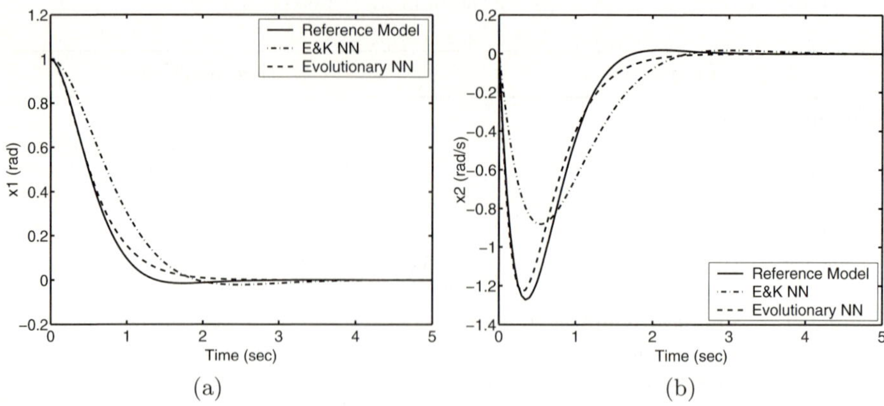

Fig. 2. State trajectory from model-reference control (a) angular position (b) angular velocity

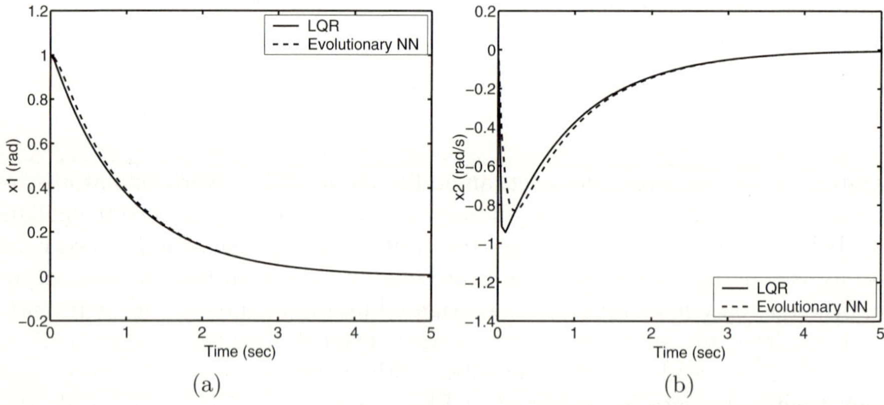

Fig. 3. State trajectory from optimal control (a) angular position (b) angular velocity

Table 2. Control cost and constraint

Control Strategy	Control Cost (J_c)	Constraint	
		$\left(\sum_i \sum_j c_{1ij}^2\right)^{1/2}$	$\left(\sum_i \sum_j c_{2ij}^2\right)^{1/2}$
Model-Reference Control			
E&K neural network	5.357349	0.000533	0.000380
Evolutionary neural network	0.254965	0.000033	0.000037
Optimal Control			
Linear quadratic regulator	20.329002	–	–
Evolutionary neural network	21.086844	0.000004	0.000005

from the first case study where the strategy employed is model-reference control. It can be clearly seen that the evolutionary neural network is superior to the neural network controller reported in Ekachaiworasin and Kuntanapreeda [13]. The tracking performance of the closed-loop system with the evolutionary neural network is very close to that of the reference model. On the other hand, a significant deviation from the desired response can be observed in the system using the previously reported neural network controller. The numerical results in Table 2 also support this observation where an improvement in both squared tracking error and constraint objective are noticeable. This improvement stems from the fact that the use of a neural network emulator, which is previously required during the training of the neural network controller, is no longer necessary in the current implementation. In the early work by Ekachaiworasin and Kuntanapreeda [13], the controller training capability is limited by the ability of the emulator at mimicking the behaviour of the plant. If the emulator cannot truly represent the plant, it would be impossible for the subsequent closed-loop system to behave like the reference model. Since the application of the CCGA as the training algorithm leads to a direct usage of the difference between the closed-loop system and the reference model as the training guidance, the improvement reported is hence not surprising.

The results from the optimal control are now considered. Figure 3 indicates that the state trajectory from the system with the evolutionary neural network controller and that from the system with the LQR are very similar. The energy usage reported in Table 2 also confirms this similarity where the energy required by the system with the linear controller is slightly lower. This implies that the evolved connection weights in the neural network must be close to the true optimal values since the state trajectory obtained approaches the ideally attainable one.

5 Conclusions

In this paper, the application of neural network controllers in the stabilisation of a nonlinear system is introduced. The Lyapunov-based stability condition imposed leads to the identification of the feasible connection weight space. As a result, the training of neural networks can be treated as constrained optimisation where the proposed technique for the task is a cooperative coevolutionary genetic algorithm (CCGA). Two control strategies, namely model-reference control and optimal control, are investigated where an inverted pendulum system is utilised in both case studies. The results from the model-reference control indicate that a significant improvement over the previous work by Ekachaiworasin and Kuntanapreeda [13] in terms of the tracking performance and the stability constraint handling is obtained. Furthermore, the results from the optimal control reveal that the energy usage from the system with the evolutionary neural network controller is similar to that from the system with a linear quadratic regulator. This clearly illustrates the suitability of the CCGA to the task.

References

1. Sadegh, N.: A Perceptron Network for Functional Identification and Control of Nonlinear Systems. IEEE Trans. Neural Networks **4**(6) (1993) 982–988
2. Fierro, R., Lewis, F.L.: Control of a Nonholonomic Mobile Robot Using Neural Networks. IEEE Trans. Neural Networks **9**(4) (1998) 589–600
3. Kim, Y.H., Lewis, F.L.: Neural Network Output Feedback Control of Robot Manipulators. IEEE Trans. Robot. Autom. **15**(2) (1999) 301–309
4. Jagannathan, S.: Control of a Class of Nonlinear Discrete-Time Systems Using Multilayer Neural Networks. IEEE Trans. Neural Networks **12**(5) (2001) 1113–1120
5. Zhu, Q., Guo, L.: Stable Adaptive Neurocontrol for Nonlinear Discrete-Time Systems. IEEE Trans. Neural Networks **15**(3) (2004) 653–662
6. He, S., Reif, K., Unbehauen, R.: A Neural Approach for Control of Nonlinear Systems with Feedback Linearization. IEEE Trans. Neural Networks **9**(6) (1998) 1409–1421
7. Nam, K.: Stabilization of Feedback Linearizable Systems Using a Radial Basis Function Network. IEEE Trans. Automat. Contr. **44**(5) (1999) 1026–1031
8. Kosmatopoulos, E.B.: Universal Stabilization Using Control Lyapunov Functions, Adaptive Derivative Feedback, and Neural Network Approximators. IEEE Trans. Syst. Man Cybern. B Cybern. **28**(3) (1998) 472–477
9. Sanchez, E.N., Perez, J.P.: Input-to-State Stabilization of Dynamic Neural Networks. IEEE Trans. Syst. Man Cybern. A Syst. Hum. **33**(4) (2003) 532–536
10. Tanaka, K.: An Approach to Stability Criteria of Neural-Network Control Systems. IEEE Trans. Neural Networks **7**(3) (1996) 629–642
11. Suykens, J.A.K., Vandewalle, J., De Moor, B.: Lur'e Systems with Multilayer Perceptron and Recurrent Neural Networks: Absolute Stability and Dissipativity. IEEE Trans. Automat. Contr. **44**(4) (1999) 770–774
12. Kuntanapreeda, S., Fullmer, R.R.: A Training Rule which Guarantees Finite-Region Stability for a Class of Closed-Loop Neural-Network Control Systems. IEEE Trans. Neural Networks **7**(3) (1996) 745–751
13. Ekachaiworasin, R., Kuntanapreeda, S.: A Training Rule which Guarantees Finite-Region Stability of Neural Network Closed-Loop Control: An Extension to Nonhermitian Systems. In: Amari, S.-I., Giles, C.L., Gori, M., Piuri, V. (eds.): Proceedings of the 2000 IEEE-INNS-ENNS International Joint Conference on Neural Networks. IEEE Computer Society, Los Alamitos (2000) 325–330
14. Yao, X.: Evolving Artificial Neural Networks. Proc. IEEE **87**(9) (1999) 1423–1447
15. Potter, M.A., De Jong, K.A.: A Cooperative Coevolutionary Approach to Function Optimization. In: Davidor, Y., Schwefel, H.-P., Männer, R. (eds.): Parallel Problem Solving from Nature–PPSN III. Lecture Notes in Computer Science, Vol. 866. Springer-Verlag, Berlin Heidelberg New York (1994) 249–257
16. De Jong, K.A., Potter, M.A.: Evolving Complex Structures via Cooperative Coevolution. In: McDonnell, J.R., Reynolds, R.G., Fogel, D.B. (eds.): Proceedings of the Fourth Annual Conference on Evolutionary Programming. MIT Press, Cambridge (1995) 307–318
17. Potter, M.A., De Jong, K.A.: Evolving Neural Networks with Collaborative Species. In: Ören, T.I., Birta, L.G. (eds.): Proceedings of the 1995 Summer Computer Simulation Conference. Society for Computer Simulation, San Diego (1995) 340–345
18. Pimpawat, C., Chaiyaratana, N.: Three-Dimensional Container Loading Using a Cooperative Co-evolutionary Genetic Algorithm. Appl. Artif. Intell. **18**(7) (2004) 581–601
19. Potter, M.A., De Jong, K.A.: Cooperative Coevolution: An Architecture for Evolving Coadapted Subcomponents. Evol. Comput. **8**(1) (2000) 1–29

Neural Network Control Design for Large-Scale Systems with Higher-Order Interconnections

Cong Ming and Sunan Huang

School of Mechanical Engineering, Dalian University of Technology,
Dalian 116024, China
congm@dlut.edu.cn

Abstract. A decentralized neural network controller for a class of large-scale nonlinear systems with the higher-order interconnections is proposed. The neural networks (NNs) are used to cancel the effects of unknown subsystems, while the robustifying terms are used to counter the effects of the interconnections. Semi-global asymptotic stability results are obtained and the tracking error converges to zero.

1 Introduction

In recent years, there has been an increased interest in the development theories for large-scale systems (see [1]). A significant proportion of these effects deals with the problem of control design methods implemented in a decentralized way. Earlier versions of the decentralized adaptive control methods were focused on control of large-scale linear systems. However, most physical systems are inherently nonlinear. Research on decentralized control for nonlinear systems was carried out in [2, 3, 4]. These previous works consider subsystems which are linear in a set of unknown parameters [2], or consider the isolated subsystems to be known [3, 4], such as input gain functions. Recently, Spooner and Passino [5] proposed a radial basis neural network control method to approximate unknown functions in nonlinear subsystems which may not be linearly parameterized. Specifically, the direct NN controller in [5] requires neither the knowledge of input gain functions, nor the direct estimation of the unknown input gain function. It thus avoids the control singularity problem. In additio, their result is focused on systems with first-order bounded interconnections. In this paper, we show that the decentralized semi-global stabilization can be achieved for the large-scale systems with the higher-order interconnections.

2 Preliminaries

We consider a large-scale nonlinear system comprised of n interconnected subsystems [5]. The ith subsystems to be single-input-single-output plant is given as

$$\left.\begin{array}{l} \dot{x}_i = f_i(x_1, ..., x_n) + g_i(x_i)u_i \\ y_i = \quad h_i(x_1, ..., x_n) \end{array}\right\}, \quad i = 1, 2, ...n, \tag{1}$$

J. Wang et al. (Eds.): ISNN 2006, LNCS 3972, pp. 1007–1012, 2006.

Using Lie derivative, we have

$$y_i^{(\rho_i)} = \alpha_i(x_i) + \beta_i(x_i)u_i + \Delta_i(x_1, ..., x_n), \tag{2}$$

where $y_i^{(\rho_i)}$ is the ρ_ith time derivative of y_i, and $L_{g_i}h_i(x_1, ..., x_n)$ is the Lie derivative of $h_i(x_1, ..., x_n)$. Here we are assuming that the influence of the other subsystems is represented by the $\Delta_i(x_1, ..., x_n)$ term in (2).

Assumption 1: The plant can be defined by (1) and transformed to (2) with input gain bounded by $0 < \beta_{0_i} \le \beta_i(x_i) \le \beta_{1i}$. The zero dynamics for each subsystem are exponentially attractive [6]. The ith subsystem input gain rate $\dot{\beta}_i(x_i)$ is bounded, i.e., $|\dot{\beta}_i(x_i)| \le B_i$.

Define a tracking error for the ith subsystem as $e_i(t) = y_i - y_{di}$ and thus $\dot{e}_i(t) = \dot{y}_i - \dot{y}_{di}, ..., e_i^{(\rho_i-1)}(t) = y_i^{(\rho_i-1)} - y_{di}^{(\rho_i-1)}]^T$, and a tracking error $e_i^{(\rho_i)}$ for the ith subsystem

$$e_i^{(\rho_i)} = -k_{i,1}e_i - k_{i,2}\dot{e}_i - ... - k_{i,\rho_i}e_i^{\rho_i-1} - \alpha_i(x_i) - \beta_i(x_i)u_i - v_i - \Delta_i, \tag{3}$$

where $k_{i,1}, k_{i,2}, ..., k_{i,\rho_i}$ are chosen such that the polynomial $k_{i,1} + k_{i,2}s + ... + s^{\rho_i}$ is strict Hurwitz, and $v_i = -k_{i,1}e_i - k_{i,2}\dot{e}_i - ... - k_{i,\rho_i}e_i^{(\rho_i-1)}$. Let $\mathbf{e}_i = [e_i, \dot{e}_i, ..., e_i^{(\rho_i-1)}]^T$ and $b_i = [0, 0, ..., 1]^T$. We have the following state equations

$$\dot{\mathbf{e}}_i = A_i\mathbf{e}_i - b_i[\alpha_i(x_i) + \beta_i(x_i)u_i + v_i + \Delta_i] \tag{4}$$

where

$$A_i = \begin{bmatrix} 0 & 1 & 0 & ... & 0 \\ 0 & 0 & 1 & ... & 0 \\ & & \vdots & & \\ -k_{i,1} & -k_{i,2} & -k_{i,3} & ... & -k_{i,\rho_i} \end{bmatrix}. \tag{5}$$

Since $k_{i,1} + k_{i,2}s + ... + k_{i,\rho_i}s^{\rho_i-1} + s^{\rho_i}$ is strict Hurwitz, A_i is an asymptotically stable matrix. For a given $Q_i > 0$, there exists a unique $P_i > 0$ satisfying

$$A_i^T P_i + P_i A_i = -Q_i. \tag{6}$$

Assumption 2: The desired trajectory $Y_{di} = [y_{di}, \dot{y}_{di}, ..., y_{di}^{(\rho_i)}]^T$ for the ith subsystem is continuous and available, and $||Y_{di}|| \le P_i$ with P_i being a known bound.

Assumption 3: The interconnections Δ_i are bounded by polynomial-type nonlinearities in $\mathbf{e}_l, 1 \le l \le n$, i.e.,

$$|\Delta_i(x_1, ..., x_n)| \le \sum_{j=1}^{n}(\zeta_{ij}^0 + \zeta_{ij}^1||\mathbf{e}_j||^1 + \zeta_{ij}^2||\mathbf{e}_j||^2 + ... + \zeta_{ij}^p||\mathbf{e}_j||^p) \tag{7}$$

with the unknown coefficients ζ_{ij}^k.

3 Main Result

In this section, a Lyapunov function is first proposed for developing a smooth local NN controller which is free from control singularity. To design the local controller, we consider an isolated subsystem

$$\dot{\mathbf{e}}_i = A_i\mathbf{e}_i - b_i[\alpha_i(x_i) + \beta_i(x_i)u_i + v_i]. \tag{8}$$

The following lemma is given to establish the existence of a desired control, that brings the filtered error approaches to zero.

Lemma 3.1. For the isolated subsystem of (8), if the desired local controller is chosen as

$$u^*_{di}(z_i) = -\frac{\alpha_i(x_i) + v_i}{\beta_i(x_i)}, \quad x_i \in \Omega_{x_i}, z_i \in \Omega_{z_i} \tag{9}$$

then the subsystem tracking error $\mathbf{e}_i$ converges to zero asymptotically.

Since u^*_{di} is a smooth function on a compact set Ω_{z_i} and unknown, and may be approximated by NNs

$$u^*_{di}(z_i) = -\frac{\alpha_i(x_i) + v_i}{\beta_i(x_i)} = W^{*T}_i\Phi_i(z_i) + \epsilon_i, \quad z_i = [x_i^T, v_i]^T, \tag{10}$$

with ϵ_i satisfying $|\epsilon_i| \leq \epsilon_{Mi}$ with ϵ_{Mi} constant. Let $\hat{W}_i$ be the estimate of W^* and the weight estimation error is $\tilde{W}_i = \hat{W}_i - W^*_i$.

An adaptive algorithm will be used to estimate $W^*{}_i$ with $\hat{W}_i$. These estimates are then used to define the decentralized control law as

$$u_i = \hat{W}_i^T\Phi_i(z_i) + u_{ci} \tag{11}$$

with

$$u_{ci} = \hat{\xi}_i sgn(\mathbf{e}_i^T P_i b_i) + \hat{\mu}_i(\mathbf{e}_i^T P_i b_i)[1 + ||\mathbf{e}_i||^{2(p-1)}], \tag{12}$$

where u_{ci} is introduced for countering uncertainties in the NN approximation error and system interconnections. Consider the following update laws:

$$\dot{\hat{W}}_i = \gamma_{1i}(\mathbf{e}_i^T P_i b_i)\Phi_i(z_i)\sum_{k=1}^{p} k(\mathbf{e}_i^T P_i\mathbf{e}_i)^{k-1} \tag{13}$$

$$\dot{\hat{\xi}}_i = \gamma_{2i}|\mathbf{e}_i^T P_i b_i|\sum_{k=1}^{p} k(\mathbf{e}_i^T P_i\mathbf{e}_i)^{k-1} \tag{14}$$

$$\dot{\hat{\mu}}_i = \gamma_{3i}(\mathbf{e}_i^T P_i b_i)^2[1 + ||\mathbf{e}_i||^{2(p-1)}]\sum_{k=1}^{p} k(\mathbf{e}_i^T P_i\mathbf{e}_i)^{k-1}, \tag{15}$$

where γ_{1i}, γ_{2i}, and γ_{3i} are positive constants.

Theorem 3.1. For the subsystems consisting of plant (2) satisfying Assumptions and the decentralized controller (11), if $B_i < \frac{\beta_{0i}\lambda_{min}(Q_i)}{2\lambda_{max}(P_i)}$, then all the signals in the closed-loop system are bounded, and the tracking error $\mathbf{e}_i$ decreases asymptotically to zero.

Proof. The actual value ξ_i^* of the control gain $\hat{\xi}_i$ will be given in (20). Let μ_i^* be the desired value of the control gain $\hat{\mu}_i$ to counter the effects of the interconnections. The true value μ_i^* will be obtained in the final step. Consider the Lyapunov function

$$V = \sum_{i=1}^{n} V_{1i} + \sum_{i=1}^{n} V_{2i}, \tag{16}$$

where $V_{1i} = \sum_{k=1}^{p} \frac{(\mathbf{e}_i P_i \mathbf{e}_i)^k}{\beta_i(x_i)}$, $V_{2i} = [\frac{1}{\gamma_{1i}} \tilde{W}_i^T \tilde{W}_i + \frac{1}{\gamma_{2i}} \tilde{\xi}^2 + \frac{1}{\gamma_{3i}} \tilde{\mu}_i^2]$, where $\tilde{\xi}_i = \hat{\xi}_i - \xi_i^*, \tilde{\mu}_i = \hat{\mu}_i - \mu_i^*$. Using the control of u_i, the tracking error dynamics may be expressed as

$$\dot{\mathbf{e}}_i = A_i \mathbf{e}_i - b_i[\beta_i(x_i)(\tilde{W}_i^T \Phi_i(z_i) - \epsilon_i + u_{ci}) + \Delta_i]. \tag{17}$$

By using Assumptions 1 and the Lyapunov equation (6), we have the following time derivative of $\dot{V}_{1i}$

$$\begin{aligned}\dot{V}_{1i} \le& \sum_{k=1}^{p} (\mathbf{e}_i^T P_i \mathbf{e}_i)^{k-1} [-\frac{k\mathbf{e}_i^T Q_i \mathbf{e}_i}{\beta_i} - 2k\mathbf{e}_i^T P_i b_i (\tilde{W}_i \Phi_i + u_{ci}) + 2k|\mathbf{e}_i^T P_i b_i| \epsilon_{Mi} \\ &+ \frac{B_i(\mathbf{e}_i^T P_i \mathbf{e}_i)}{\beta_i^2}] + \sum_{k=1}^{p} 2k(\mathbf{e}_i^T P_i \mathbf{e}_i)^{k-1} |\mathbf{e}_i^T P_i b_i| \frac{\sum_{k=0}^{p}\sum_{j=1}^{n} \zeta_{ij}^k ||\mathbf{e}_j||^k}{\beta_{0i}}.\end{aligned} \tag{18}$$

Since $||\mathbf{e}_j||^0 = 1$, the last term of (18) can be written as

$$\begin{aligned}&\sum_{k=1}^{p} 2k(\mathbf{e}_i^T P_i \mathbf{e}_i)^{k-1} |\mathbf{e}_i^T P_i b_i| \frac{\sum_{k=0}^{p}\sum_{j=1}^{n} \zeta_{ij}^k ||\mathbf{e}_j||^k}{\beta_{0i}} = \sum_{k=1}^{p} 2k(\mathbf{e}_i^T P_i \mathbf{e}_i)^{k-1} \\ &\times |\mathbf{e}_i^T P_i b_i| \sum_{j=1}^{n} \frac{\zeta_{ij}^0}{\beta_{0i}} + \sum_{k_1=1}^{p}\sum_{j=1}^{n}\sum_{k=1}^{p} \frac{2k(\mathbf{e}_i^T P_i \mathbf{e}_i)^{k-1} |\mathbf{e}_i^T P_i b_i| \zeta_{ij}^{k_1} ||\mathbf{e}_j||^{k_1}}{\beta_{0i}}.\end{aligned} \tag{19}$$

Utilizing inequality $2ab \le \delta a^2 + \delta^{-1} b^2$ (δ a positive constant) and $(\sum_{k=1}^{p} a_k b_k)^2 \le (\sum_{k=1}^{p} a_k^2)(\sum_{k=1}^{p} b_k^2)$, the terms in (19) can be written as

$$\begin{aligned}&\frac{1}{\beta_{0i}} \sum_{k_1=1}^{p}\sum_{j=1}^{n} 2[\sum_{k=1}^{p} k(\mathbf{e}_i^T P_i \mathbf{e}_i)^{k-1} |\mathbf{e}_i^T P_i b_i|] \zeta_{ij}^{k_1} ||\mathbf{e}_j||^{k_1} \\ &\le \sum_{j=1}^{n} [\frac{p^2(p+1)(2p+1)}{6\beta_{0i}} \sum_{k=1}^{p} (\mathbf{e}_i^T P_i \mathbf{e}_i)^{2(k-1)} |\mathbf{e}_i^T P_i b_i|^2 \delta_j + \frac{\zeta_{max}^2}{\beta_{0min}} \sum_{i=1}^{n} \delta_j^{-1} ||\mathbf{e}_j||^{2k}]\end{aligned}$$

where the constant δ_i is given later, $\zeta_{max} = max_{(1\leq i\leq n, 1\leq j\leq n, 1\leq k\leq p)}\{\zeta_{ij}^k\}$ and $\beta_{0min} = min_{(1\leq i\leq n)}\{\beta_{0i}\}$. For simplicity, we define $d_1 = \frac{p^2(p+1)(2p+1)}{6}$. Combining (19) and the above equation, $\sum_{i=1}^n \dot{V}_{1i}$ is given by

$$\sum_{i=1}^n \dot{V}_{1i} \leq \sum_{i=1}^n \sum_{k=1}^p (\mathbf{e}_i^T P_i \mathbf{e}_i)^{k-1}[-\frac{k\mathbf{e}_i^T Q_i \mathbf{e}_i}{\beta_i} - 2k\mathbf{e}_i^T P_i b_i(\tilde{W}_i \Phi_i + u_{ci}) + 2k|\mathbf{e}_i^T P_i b_i|\xi_i^*$$
$$+\frac{B_i(\mathbf{e}_i^T P_i \mathbf{e}_i)}{\beta_i^2}] + \sum_{i=1}^n \sum_{k=1}^p [\frac{d_1\sum_{j=1}^n \delta_j}{\beta_{0i}}(\mathbf{e}_i^T P_i \mathbf{e}_i)^{2(k-1)}|\mathbf{e}_i^T P_i b_i|^2 + \frac{n\zeta_{max}^2}{\beta_{0min}}\delta_i^{-1}||\mathbf{e}_i||^{2k}]$$

where

$$\xi_i^* = \epsilon_{Mi} + \frac{\sum_{j=1}^n \zeta_{ij}^0}{\beta_{0i}}. \tag{20}$$

With adaptations (13)-(15), $\dot{V}$ is given by

$$\dot{V} \leq \sum_{i=1}^n \sum_{k=1}^p (\mathbf{e}_i^T P_i \mathbf{e}_i)^{k-1}\{-\frac{k\mathbf{e}_i^T Q_i \mathbf{e}_i}{2\beta_i} + \frac{B_i(\mathbf{e}_i^T P_i \mathbf{e}_i)}{\beta_i^2} - 2\mu_i^* k(\mathbf{e}_i^T P_i b_i)^2$$
$$\times[1 + ||\mathbf{e}_i||^{2(p-1)}]\} + \sum_{i=1}^n \sum_{k=1}^p \frac{d_1\sum_{j=1}^n \delta_j}{\beta_{0i}}(\mathbf{e}_i^T P_i \mathbf{e}_i)^{2(k-1)}|\mathbf{e}_i^T P_i b_i|^2$$
$$-\sum_{i=1}^n \sum_{k=1}^p [\lambda_{min}^{k-1}(P_i)\lambda_{min}(Q_i)\frac{k}{2\beta_{1i}}||\mathbf{e}_i||^{2k} - \frac{n\zeta_{max}^2}{\beta_{0min}}\delta_i^{-1}||\mathbf{e}_i||^{2k}]. \tag{21}$$

We can determine a constant $\delta_i \geq max_{(1\leq k\leq p)}\{\frac{2n\beta_{1i}\zeta_{max}^2}{\lambda_{min}^{k-1}(P_i)\lambda_{min}(Q_i)k\beta_{0man}}\}$ such that the last two terms of (21) are negative. Then, (21) may be reduced to

$$\dot{V} \leq \sum_{i=1}^n \sum_{k=1}^p (\mathbf{e}_i^T P_i \mathbf{e}_i)^{k-1}[-\frac{k\mathbf{e}_i^T Q_i \mathbf{e}_i}{2\beta_i} + \frac{B_i(\mathbf{e}_i^T P_i \mathbf{e}_i)}{\beta_i^2}] - \sum_{i=1}^n \{2\mu_i^*[1 + ||\mathbf{e}_i||^{2(p-1)}]$$
$$\times \sum_{k=1}^p k\lambda_{min}^{k-1}(P_i)||\mathbf{e}_i||^{2(k-1)} - \frac{d_1\sum_{j=1}^n \delta_j}{\beta_{0i}}\sum_{k=1}^p \lambda_{max}^{2(k-1)}(P_i)||\mathbf{e}_i||^{4(k-1)}\}(\mathbf{e}_i^T P_i b_i)^2.$$

The last term arises due to the interconnection appearing in the state equations of the plant. However, this term can be countered using sufficiently high-control gain μ_i^*. For $k = 1, 3, 5...$, we choose

$$\mu_i^* \geq \mu_{1i}^* = max_{k=1,3,5,...\leq p}[\frac{d_1\sum_{j=1}^n \delta_j}{2k\beta_{0i}\lambda_{min}^{k-1}(P_i)}\lambda_{max}^{k-1}(P_i)], \tag{22}$$

while remaining parameters are chosen as

$$\mu_i^* \geq \mu_{2i}^* = max_{k=1+[p/2],2+[p/2],...,p}[\frac{d_1\sum_{j=1}^n \delta_j}{2\beta_{0i}(2k-p)\lambda_{min}^{2k-p-1}(P_i)}\lambda_{max}^{2(k-1)}(P_i)]. \tag{23}$$

Therefore, the last two terms of $\dot{V}$ can be made negative by choosing $\mu_i^* = max\{\mu_{1i}, \mu_{2i}\}$. This implies that

$$\dot{V} \leq -\sum_{i=1}^{n}\sum_{k=1}^{p} \frac{(\mathbf{e}_i^T P_i \mathbf{e}_i)^{k-1}}{\beta_i}[\frac{min\{k\}\lambda_{min}(Q_i)}{2} - \frac{B_i\lambda_{max}(P_i)}{\beta_{0i}}]||\mathbf{e}_i||^2. \quad (24)$$

Since we require that $B_i < \frac{\beta_{0i}\lambda_{min}(Q_i)}{2\lambda_{max}(P_i)}$ so that $\frac{\lambda_{min}(Q_i)}{2} - \frac{B_i\lambda_{max}(P_i)}{\beta_{0i}} > \varepsilon_i > 0$, where ε_i is a finite constant, this gives us $\dot{V} \leq -\sum_{i=1}^{n}\sum_{k=1}^{p} \frac{\varepsilon_i \lambda_{min}^{k-1}(P_i)}{\beta_{1i}}||\mathbf{e}_i||^{2k}$. Thus, the solution $\mathbf{e}_i, \hat{W}_i, \hat{\xi}_i, \hat{\mu}_i$ are bounded. Because of the boundedness of all the signals, from (17), $\dot{\mathbf{e}}_i$ is bounded. Moreover, since $V > 0$ and

$$\int_0^{\infty}\sum_{i=1}^{n}\sum_{k=1}^{p} \frac{\varepsilon_i\lambda_{min}^{k-1}(P_i)}{\beta_{1i}}||\mathbf{e}_i||^{2k} \leq -\int_0^{\infty}\dot{V}dt = V(0) - V(\infty) < \infty \quad (25)$$

this implies that $\mathbf{e}_i \in L_2$. Then by Barbalat's lemma, $\lim_{t\to\infty} \mathbf{e}_i(t) = 0$. Q.E.D.

4 Conclusion

In this paper, we have presented a decentralized NN control scheme for a class of large-scale nonlinear systems with higher-order interconnections. The main feature of the proposed approach is the application of the composite Lyapunov functions to construct the NN-based decentralized controller. Semi-global stability results are obtained and tracking error converges to zero.

References

1. Siljak, D.D.: Large-scale Dynamic Systems: Stability and Structure. North Holland (1978)
2. Fu, L.C.: Robust Adaptive Decentralized Control of Robot Manipulators. IEEE Trans. on Automat. Contr. 37(1992)106-110
3. Shi, L., Singh, S.K.: Decentralized Adaptive Controller Design for Large-Scale Systems with Higher Order Interconnections. IEEE Trans. on Automat. Contr. **37**(1992)1106-1118
4. Guo, Y., Jiang, Z.P., Hill, D.J.: Decentralized Robust Disturbance Attenuation for a Class of Large-Scale Nonlinear Systems. Systems & Control Letters **37**(1999)71-85
5. Spooner, J.T., Passino, K.M.: Decentralized Adaptive Control of Nonlinear Systems Using Radial Basis Neural Networks. IEEE Trans. on Automat. Contr. **44**(1999) 2050-2057
6. Sastry, S.S., Isidori, A.: Adaptive Control of Linearizable Systems. IEEE Trans. on Automat. Contr. **34**(1989) 1123-1131
7. Krstic, M., Kanellakopoulos, I., Kokotovic, P.: Nonlinear and Adaptive Control Design. Wiley, New York (1995)

Adaptive Pseudo Linear RBF Model for Process Control

Ding-Wen Yu[1] and Ding-Li Yu[2]

[1] Department of Automation, Northeast University at Qinhuangdao, China
[2] Control Systems Research Group, School of Engineering,
Liverpool John Moores University, Byrom Street, Liverpool L3 3A, UK
d.yu@livjm.ac.uk

Abstract. A pseudo-linear radial basis function (PLRBF) network is developed in this paper. This network is used to model a real process and its weights are on-line updated using a recursive orthogonal least squares (ROLS) algorithm. The developed adaptive model is then used in model predictive control strategy, which is applied to a pilot multivariable chemical reactor. The first stage of the project, simulation study, has been investigated and is presented. The effectiveness of the adaptive control in improving the closed-loop performance has been demonstrated for process time-varying dynamics and model-process mismatch.

1 Introduction

Different adaptive neural networks have been developed in recent years. For example, [1] proposed an adaptation algorithms for RBF network structure to recursively train the network model in off-line mode. An adaptive RBF network model was applied in [2] in the internal model control strategy to control an experimental process, and compared the performance with that achieved using a linear pole-placement controller. In this paper a pseudo-linear RBF network is on-line trained using ROLS [3-4] as the process model and is used in the model predictive control (MPC) of a laboratory-scaled chemical reactor. The reactor exhibits characteristics typical of many industrial processes, due to its non- linearity, coupling effects among the controlled variables and a long time-delay in heat exchange. The work in the first stage, simulation study, is described in the paper. It will be followed by the work in the second stage, real time application to the reactor and the application of the both weight and centre adapted RBF network model.

2 Process Descriptions

The schematic of the chemical reactor is shown in Fig.1. It consists of a continuously stirred tank (15 litres) to which the chemical solutions, NH_4OH, CH_3COOH and $N_{a2}SO_3$, and air are added. The liquid level in the tank is maintained at a pre-specified constant level by an outflow pump system. The concentrations and flow rates of solutions, CH_3COOH and $N_{a2}SO_3$, are constant except for some manual changes to mimic process disturbances. The concentration of NH_4OH is constant but

J. Wang et al. (Eds.): ISNN 2006, LNCS 3972, pp. 1013 – 1018, 2006.

the flow rate is adjustable by a servo-pump to regulate the pH value in the tank. The air-flow rate is also adjustable by a mass- flow meter connected to a compressing air network to regulate the percentage of the dissolved oxygen (pO_2) in the liquid in the tank. The tank is also equipped with an electric heating system to adjust the liquid temperature. The liquid in the tank is stirred continuously to make sure the pH, the dissolved oxygen and the temperature are consistent throughout the tank.

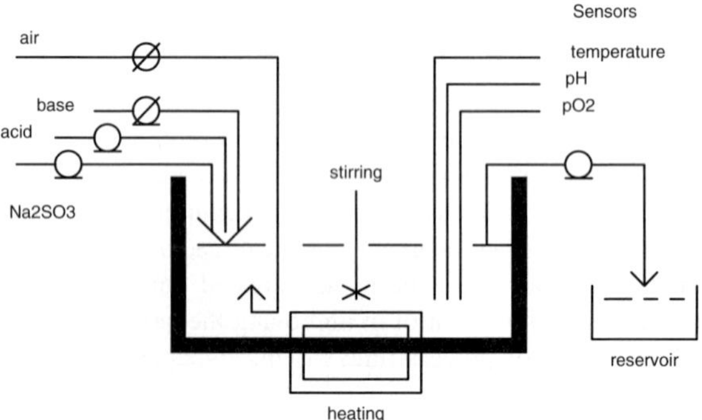

Fig. 1. The chemical reactor process

Process inputs and outputs are chosen as

$$u = \begin{bmatrix} Q \\ f_b \\ f_a \end{bmatrix}, \qquad y = \begin{bmatrix} T \\ pH \\ pO_2 \end{bmatrix} \tag{1}$$

where Q, f_b and f_a denote the heating power, the flow rate of the base and the flow rate of air respectively. The sample interval is selected to be 10 seconds.

3 Neural Network Models

The process is represented by the multivariable NARX model of the following form,

$$\hat{y}(k) = f(y(k-1),\cdots,y(k-n_y),u(k-1-d),\cdots,u(k-n_u-d)) + e(k) \tag{2}$$

where

$$y(k) = \begin{bmatrix} y_1(k) \\ \vdots \\ y_p(k) \end{bmatrix}, \quad u(k) = \begin{bmatrix} u_1(k) \\ \vdots \\ u_m(k) \end{bmatrix}, \quad e(k) = \begin{bmatrix} e_1(k) \\ \vdots \\ e_p(k) \end{bmatrix}$$

are the process output, input and noise respectively, p and m are the number of outputs and inputs respectively, n_y and n_u are the maximum lags in the outputs and inputs respectively, d is the maximum time delay in the inputs; and $f(*)$ is a vector-valued, non-linear function.

A PLRBF network is proposed in this work by augmenting the nominal RBF with the network input directly used as part of the regression signals. The PLRBF structure and operation are presented by the following equations.

$$\phi = \begin{bmatrix} z^T & x^T \end{bmatrix}^T \tag{3}$$

$$z_i = \exp\left(-\frac{\|x - c_i\|^2}{\sigma^2}\right), i = 1, \cdots, n_h - n \tag{4}$$

$$\hat{y} = W^T \phi \tag{5}$$

where $x, c_i \in \Re^n$ with $n = pn_y + mn_u$ are the input vector to the network and the centre vector in the hidden layer, $\phi \in \Re^{n_h}$ is the hidden layer output vector, $W \in \Re^{p \times n_h}$ is the output layer weight matrix and $\hat{y} \in \Re^p$ is the network output, or the prediction of the process output when the network is used as the process model. The network input vector is chosen as $x(k) = \begin{bmatrix} y^T(k-1), \cdots, y^T(k-n_y), u^T(n-1-d), \cdots, u^T(k-1-n_u) \end{bmatrix}^T$ according to (2). It is noted that such structured PLRBF is a combination of a linear part and a non-linear part in one model. This reduces the task of the hidden layer nodes to modelling only additional non-linear effects to a linear model throughout the whole operating region. Simulation studies have shown that much less hidden layer nodes are needed than a standard RBF to model a process to the same accuracy.

The weight matrix of the PLRBF is trained initially and then on-line updated using the ROLS algorithm. For the N input-output training data, it is formed according to (5),

$$Y = \hat{Y} + E = \Phi W + E \tag{6}$$

where $Y \in \Re^{N \times p}$ is the desired output matrix, $\hat{Y} \in \Re^{N \times p}$ is the neural network output matrix, $\Phi \in \Re^{N \times n_h}$ is the hidden layer output matrix, $E \in \Re^{N \times p}$ is the error matrix. $W(k)$ can be solved from

$$R(k) W(k) = \hat{Y}(k) \tag{7}$$

where R(k) is from the QR decomposition of Φ. Calculation of W can be achieved on-line using the following transformation (Bobrow and Murray, 1993),

$$\begin{bmatrix} \lambda R(k-1) & \lambda \hat{Y}(k-1) \\ \phi^T(k) & y^T(k) \end{bmatrix} \rightarrow \begin{bmatrix} R(k) & \hat{Y}(k) \\ 0 & \eta^T(k) \end{bmatrix} \tag{8}$$

The procedure of the ROLS algorithm is therefore the following: at stage t, calculate $R(t)$ and $\hat{Y}(t)$ according to (8), then solve $W(k)$ in (11). Initial values for $R(k)$ and $\hat{Y}(k)$ can be assigned as $R(0) = \alpha I$ and $\hat{Y}(0) = 0$, where α is a small positive number.

In this study, a PLRBF is trained as the process model and a MLP is trained as the process simulation for on-line control evaluation. The process time-delay and input-output orders were selected in the previous work using a linearized selection method [5]. The network input vector is chosen accordingly as,

$$x(k) = [T(k-1), T(k-2), pH(k-1), pH(k-2), pO_2(k-1), \\ pO_2(k-2), Q(k-1-d), f_b(k-1), f_a(k-1)]^T \tag{9}$$

where $d = 22$ is the time-delay from the heating power to the liquid temperature.

A training data set with 1800 samples is collected from the process. It is found that the data displays a time variant feature with the first 200 samples of dissolved oxygen behaving much different from the rest of the samples. It is thus decided the PLRBF model is trained using the last 1600 samples, while the MLP network is trained using the whole data set to present the process dynamics.

4 Model Predictive Control

For multivariable systems the neural network MPC strategy used to control this reactor the objective function used in the optimisation is

$$J = \sum_{i=1}^{p} (M_i - \hat{Y}_i)^T W_y^{(i)} (M_i - \hat{Y}_i) + \sum_{j=1}^{m} \xi(j) \Delta U_j^T W_u^{(j)} \Delta U_j \tag{10}$$

where

$$M_i^T = [m_i(t + N_1(i)), \cdots, m_i(t + N_2(i))]$$

$$\hat{Y}_i^T = [\hat{y}_i(t + N_1(i)), \cdots, \hat{y}_i(t + N_2(i))]$$

$$\Delta U_j^T = [u_j(t) - u_j(t-1), \cdots, u_j(t + N_u(j)) - u_j(t + N_u(j) - 1)]$$

N_1, N_2 are vectors specifying the prediction horizon, with the *i*th element specifying the parameter for the corresponding output, N_u is the control horizon vector, with the *j*th element specifying the parameter for the corresponding input.

A number of simulations have been tried and suitable control parameters for set point tracking are chosen as, $N_1 = [22 \quad 2 \quad 6]^T$, $N_2 = N_1 + [15 \quad 15 \quad 15]^T$, $N_u = [1 \quad 1 \quad 1]^T$, $\xi = [0.05 \quad 0.05 \quad 0.05]^T$.

The on-line adaptation of the model is performed by conducting (8) using the Givens rotation to the augmented matrix, then solve weight matrix W(k) in (7) by back substitution. A forgetting factor $\lambda = 0.98$ is found appropriate. To prevent the model losing useful past information in the period when the system dynamics does not change significantly, the following condition is applied to determine whether the weight is adapted at this sample instant.

$$\frac{1}{L} \sum_{i=1}^{L} \| y(k + L - i) - \hat{y}(k + L - i) \| > \delta \tag{11}$$

where L is the length of the window in which the mean modelling error norm is observed, δ is a pre-specified threshold which is given corresponding to the model accuracy. In this research, L=5 and δ =0.005 are used.

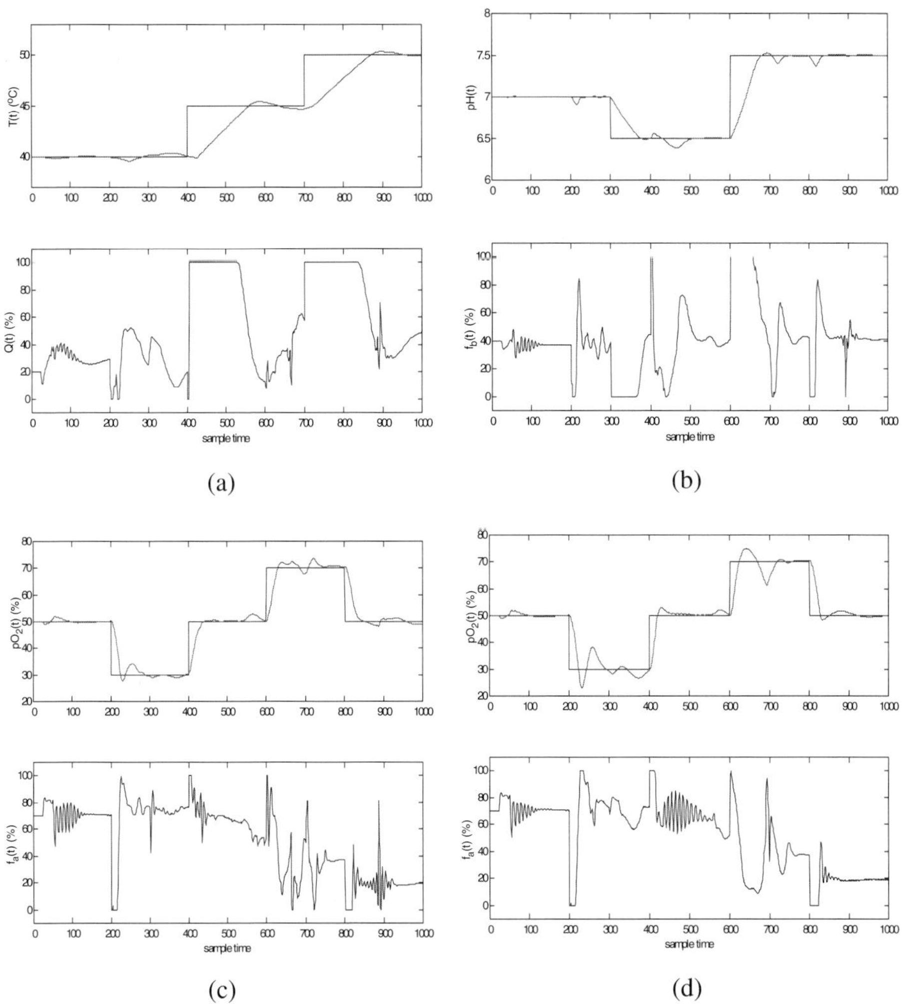

Fig. 3. Simulated response and control with adaptive and fixed model

In order to evaluate the performance of the adaptive model based MPC, a dynamics change of the process is simulated at sample time instant k=50 by changing some weights of the MLP model with 10%. This simulated model with the change is controlled twice, once by the adaptive model and once by the same model without adaptation, with all other control parameters being the same. The system response together with associated control variable for three variables are displayed in (a)-(c) of Fig.3. The response of the simulated process for the model without adaptation is displayed in (d) of Fig.3 for comparison. Due to the limited space only the dissolved oxygen is displayed.

The performance of the three variables is also compared numerically with measurement of the mean-squared-error of set-point tracking and on-line model prediction and displayed in Table 1.

Table 1. Comparison of control performance

	MSE for pO_2		MSE for all outputs	
	Tracking	Modelling	Tracking	Modelling
Adaptive model	0.0020	1.1324e-7	0.0095	2.6153e-7
Fixed model	0.0022	7.4740e-7	0.0101	2.3363e-6

The MSEs in the table show that the on-line modelling error is greatly reduced by on-line training. The tracking performance is also improved in the adaptive model case.

5 Conclusions

Model predictive control based on an adaptive PLRBF model is applied to a laboratory-scaled three-input three-output chemical reactor. The PLRBF uses much less hidden layer nodes while predicts much more accurately than a standard RBF network when they are used to model multivariable real processes. On-line update of RBF weights using the ROLS and implemented by applying the Givens rotation is numerically stable and computing efficiently. The tracking performance is greatly improved using the adaptive model than fixed model.

References

1. Yingwei, L., Sundararajan, N. and Saratchandran, P.: Identification of Time-Varying Nonlinear Systems Using Minimal Radial Basis Function Neural Networks. IEE Proc. Part.D 144 (2) (1997) 202-208.
2. Pereira, C., Henriques, J and Dourado, A.: Adaptive RBFNN Versus Conventional Self-tuning: Comparison of Two Parametric Model Approaches for Non-linear Control, Control Engn. Prac. 8 (1) (2000) 3-12.
3. Yu, D.L., Gomm, J.B. and Williams, D.: A Recursive Orthogonal Least Squares Algorithm for Training RBF Networks. Neural Processing Letters 5 (2) (1997) 167-176.
4. Gomm, J.B. and Yu, D.L.: Selecting Radial Basis Function Network Centres with Recursive Orthogonal Least Squares Training, IEEE Trans. Neural Networks 11 (2) (2000) 306-314.
5. Yu, D.L., Gomm, J.B. and Williams, D.: Neural Model Input Selection for A MIMO Chemical Process, Engn. Appl. of Artif. Intell. 13 (1) (2000) 15-23.

An Improved BP Algorithm Based on Global Revision Factor and Its Application to PID Control

Lin Lei, Houjun Wang, and Yufang Cheng

School of Automation and Engineering, University of Electronic Science and Technology of China, Chengdu, 610054, China
Science168@163.com

Abstract. To improve BP algorithm in overcoming the local minimum problem and accelerating the convergence speed, a new improved algorithm based on global revision factor of BP neural network is presented in this paper. The basic principle is to improve the formula of weight adjusting used momentum back propagation. A global revision factor is added in the weight value adjusting formula of momentum BP. Faster learning speed is obtained by adaptive adjusting this factor. The new BP algorithm is compared with other improved BP algorithms on many aspects. Simulation and applications for complex nonlinear function approximation, neural PID parameter tuning indicates that it has better training speed and precision than momentum back propagation and adaptive learning rate.

1 Introduction

BP neural network has favorable nonlinear approaching capability, generalization capability and easy application character. It was widely used in nonlinear optimization, pattern identification, intelligent control and so on. The BP neural networks are the first and most commonly used, since they are characterized by the output error signal which will be propagated backward to the input neurons from the successive layer to the preceding layer with strict learning schemes [1]. By training, BP neural network can give correct answers not only for learned examples, but also for the models similar to the learned examples, showing its strong associative ability and rational ability which are suitable for solving large, non-linear, and complex classification and function approximation problems. However general, BP network has some faults such as slow convergence speed, local minimum problem, difficulty of constructing network structure and so on [2], [3]. To overcome these problems, a new improved model namely Global Revision Factor BP based on Momentum Back Propagation (MOBP) is presented. In literature [2], its main modifying is accelerating the convergence. But, our method is accelerating the convergence of BP and precision. Feasibility and effectively of the new model is verified by applications for function approximation problem and neural PID parameters tuning compared with MOBP.

2 Weight Adjusting of MOBP

Generally, the typical BP neural network is three layers network. It is composed of input layer, output layer and several hidden layers. BP algorithm is a training

J. Wang et al. (Eds.): ISNN 2006, LNCS 3972, pp. 1019 – 1026, 2006.

algorithm with teachers, whose training procedures are divided into two parts: a forward propagation of information and a backward propagation (BP) of error. $\Delta W(n)$ denotes the weight increment in nth iteration. Weight value adjusting expressions of BP neural network is as follow [4]

$$\Delta W_{ji}(n) = \alpha \Delta W_{ji}(n-1) + \eta \delta_j(n) y_j(n) \tag{1}$$

Where, η is the learning rate. α is a momentum factor, $0 < \alpha < 1$. E is the error of nth iteration. $W(n)$ is the connection weight of nth iteration. Here m and N denote total number of neurons in input layer and in hidden layer. W_{jk} is the connection weight between hidden layer node j and input layer node k . θ_j is threshold of hidden layer node; W_{ij} is the connection weight between hidden layer node j and output layer node i . θ_i is gate value of output layer node. As provided sample in sequence, formula (1) can be expressed in the form of time sequence of t variable, $t \in [0, n]$. $\Delta W_{ji}(n)$ is obtained as following

$$\Delta W_{ji}(n) = \eta \sum_{t=0}^{n} \alpha^{n-1} \delta_j(t) y(y) = -\eta \sum_{t=0}^{n} \alpha^{n-1} \frac{\partial E(t)}{\partial W_{ji}(t)} \tag{2}$$

When sign $\frac{\partial E(t)}{\partial W_{ji}(t)}$ this time accords with that of last time. It means a greater $\Delta W_{ji}(n)$ and adjusting speed of W . When the sign of $\frac{\partial E(t)}{\partial W_{ji}(t)}$ this time is contrary to that of last time, some wave exists and the exponent weight sum result in the decrease of $\Delta W_{ji}(n)$, which has stabilization effect. S function or positive (negative) symmetry sigmoid function is selected as the activation function of hidden layers.

The new connection weight $W_{pq,k}(n)$ calculation formula is given [5], [6]

$$\Delta W_{pq,k}(n+1) = \eta \cdot \delta_{qk} \cdot O_{pj} + \alpha \cdot (\Delta W_{pq,k}(n)) \tag{3}$$

$$W_{pq,k}(n+1) = W_{pq,k}(n) + \Delta W_{pq,k} \tag{4}$$

Where $W_{pq,k}(n)$ is the connection weight between p node of hidden layer and q node in output layer k at n th iteration. δ_{qk} is the δ value of the q neuron in output layer k , O_{pj} is the output value of the p neuron in hidden layer j. The adjusting magnitude of weight value is affected directly by learning rate η , which is too small will lead to slow convergence speed and probably fall into local minimal points. But, when η is too large, it is will bring about swing near the minimal points in the course of convergence and even cause instability.

Define the output error function E is square sum of the difference between expected output and actual output.

$$E = \sum_{p=1}^{P} E^p = \sum_{p=1}^{P} \left| \frac{1}{2} \sum_{i=1}^{N} (D-O) \right| \tag{5}$$

Where, N is the total number of neurons in output layer. P is the total number of training samples. A global revision factor b is added in front of formula of traditional MOBP. Faster convergence speed and smaller iterative times are obtained by adaptive adjusting b. It is named Global Revision Factor BP (GRFBP) as follow

$$\Delta w_{pq,k}(n+1) = b(n+1) \cdot (b(n+1) \cdot \alpha \cdot (\Delta w_{pq,k}(n)) + \eta \delta_{qk} \cdot O_{pj}) \tag{6}$$

$$W_{pq,k}(n+1) = W_{pq,k}(n) + \Delta W_{pq,k}(n+1) \tag{7}$$

$b(n)$ is a varying parameter determined by nth iteration in training process. $W_{pq,k}(n)$ is the connection weight between node p in hidden layer and node q in output layer k at nth iteration, δ_{qk} is the δ value of the q neuron in output layer k, O_{pj} is the output value of the p neuron in hidden layer j.

$$E(n) = \sum_{m=1}^{N} E^m = \sum_{m=1}^{N} \left| \frac{1}{2} \sum_{q} (D-O) \right| \tag{8}$$

$$e_{(n)} = \frac{|E(n)| - |E(n-1)|}{|E(n)|} \tag{9}$$

$E(n)$ is the general error function at nth iteration of all output layer neurons for all samples, E^m is the error function produced by mth sample, $e(n)$ is the error variety rate.

If $e(n) < 0$, then $b(n+1) = b(n)[1 - \mu e^{-e(n)}]$ $\quad \mu \in (0,1]$ (10)

If $e(n) > 0$, then $b(n+1) = b(n)[1 - \mu e^{-e(n)}]$ $\quad \mu \in (0,1]$ (11)

μ is a small constant. The δ value of the p node in jth hidden layer can be calculated through (12)

$$\delta_{pj} = O_{pj}(1 - O_{pj}) \times \sum (\delta_{pj} \cdot W_{pq,k}) \tag{12}$$

Thereafter the weight adjusting formulas (8) to (12) are adopted to get weight of hidden nodes. In applications, a special instance is that global revision factor b takes constant satisfying with b > 1 during the all process. Generally, takes $1 < b < 2$. When b=1, the new algorithm is equal to traditional momentum BP model.

3 Simulation of GRFBP Algorithm

The GRFBP algorithm is compared with MOBP method and Levenberg-Marquart method by approximating the following complex nonlinear function using MATLAB.

$$y(k) = \frac{y(k-1)}{(1+y^2(k-1)) + u^3(k-1)} \tag{13}$$

The input for training of nonlinear system is

$$u(k) = 0.2\sin(2\pi k/25) + 0.3\sin(\pi k/15) + 0.3\sin(\pi k/75) \tag{14}$$

Table 1. Convergence speed and precision comparison for different BP

	Variable learning rate MOBP	GRFBP	Levenberg-Marquart
Iteration times	13454	181	9
Time (s)	145.1068	4.6150	0.5850
Mean square error (10-4)	9.996	8.109	7.621
Minimum	-0.7710	-0.7913	-0.7625

The training sample is made up of input or output pair where k ranges from 0 to 200. The output minimum is –0.7920. The training standard is the maximum of iteration times 15000 or the minimum mean square error 0.00001. Table 1 is the simulation result, in which each data is the mean of 10 trainings. The experiment results clearly show that performance of GRFBP is much faster with better precision than MOBP. Although Levenberg-Marquart method has much faster and much smaller Mean square error than GRFBP, precision of GRFBP is much better than Levenberg-Marquart method

4 PID Parameters Adaptive Tuning Using GRFBP

To obtain better result of PID controller, the best relationship is needed to found out from the variety nonlinear combine. Using GRFBP neural network, the PID controller can be setup, which its parameter is self-learning. The PID control carries out the best combination from the system study of the function. The structure of PID controller using GRFBP neural network is shown as figure 1. Where, $rin(\cdot)$ and $yout(\cdot)$ denote input and output of control system. k_p, k_i, k_d respectively denote proportion, integral and differential coefficient. The controller is composed of two parts. The one is the classic PID controller, directly carry out closeloop control towards control object, and three parameters are adjusted in on-line manner. The other part is the GRFBP neural network. According to the movement state of system, three parameters of PID controller are automatic regulated by GRFBP neural network, and let certain performance index attain to optimal value. We make outputs of output layer neuron corresponding to three adjustable parameters of PID controller. Making use of GRFBP neural

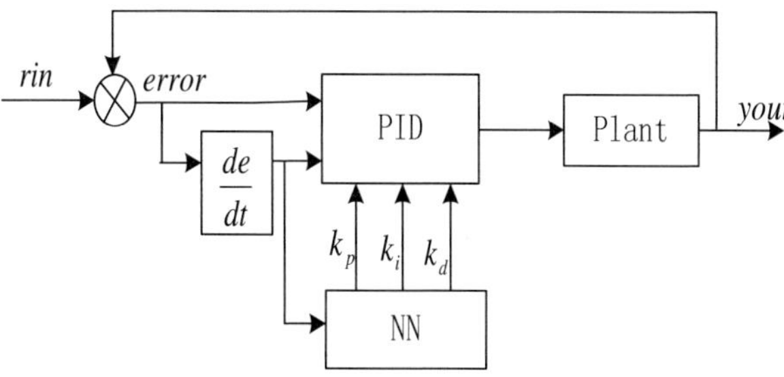

Fig. 1. The structure of GRFBP neural PID controller

network self-learning power coefficient, three parameters of PID controller can arrive at optimal value under certain optimal control rule.

The output $u(k)$ of PID controller in classical increment model is

$$u(k)=u(k-1)+k_p(error(k)-error(k-1))+k_i error(k)$$
$$+k_d(error(k)-2error(k-1)+error(k-2)) \tag{15}$$

Three layers BP network are adopted. The active function of Hidden layer neuron is symmetrical sigmoid function.

$$f_2(x)=\tanh(x)=\frac{1-e^{-x}}{1+e^{-x}} \tag{16}$$

The active function of output layer neurons is nonnegative Sigmoid

$$g(x)=1/(1+e^{-x}) \tag{17}$$

Take the index function as

$$E(k)=\frac{1}{2}(rin(k)-yout(k))^2 \tag{18}$$

Suppose the approximate mathematic model of controlled object is

$$yout(k)=\frac{a(k)yout(k-1)}{1+yout^2(k-1)}+u(k-1) \tag{19}$$

Table 2. Parameter selection of neural PID parameters tuning

	MOBP	GRFBP
lr	0.25	0.01
Mc	0.05	0.90
B	/	1.20
Goal	0.001	0.001
μ	/	0.80

where coefficient $a(k)$ is slowly varying with time k, $a(k)=1.2\left(1-0.8e^{-0.1k}\right)$. The structure of three layers BP network is selected, which has four inputs and three outputs. Take $a=2$. According to experience formula $h=\sqrt{m+n}+a$, hidden layer number $h=5$ is obtained. The structure of GRFBP network is $4-5-3$. Initial weight is a stochastic number in $[-0.5,0.5]$. The model parameters are given in table 2.

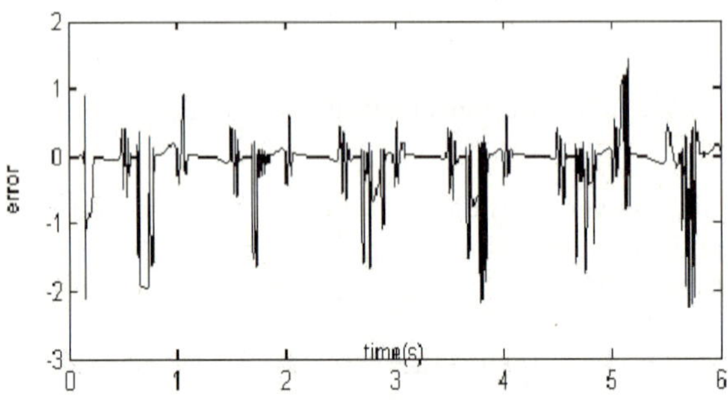

Fig. 2. Tracking error curve of MOBP

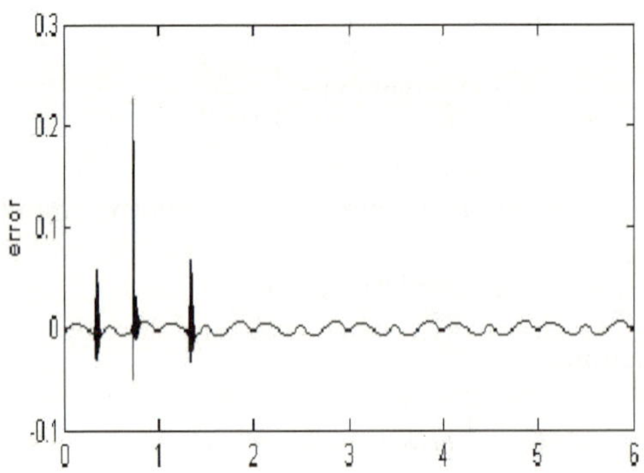

Fig. 3. Tracking error curve of GRFBP

In table 2, learning rate lr of MOBP is initial value, which is adjusted with adaptive adjusting rule during training. The magnitude parameters are selected as X_{iteinc}=1.05, X_{itedec}=0.7. Parameter b of GRFBP has also an initial value. Input instruction signal is sine signal. Its range is from -1 to +1. Simulations of two kinds of BP models are fulfilled through neural PID controller. Tracking error curve of MOBP and GRFBP are respectively shown in Fig. 2 and Fig.3. The adaptive tuning curve of parameters k_p, k_i, k_d using MOBP model are Fig.4. The simulation result of adaptive tuning curve of parameter k_p, k_i, k_d using GRFBP model are Fig.5. It can be seen that the control effect using GRFBP to implement PID parameters tuning and PID control are much better than that of variable learning rate MOBP. The details embodies

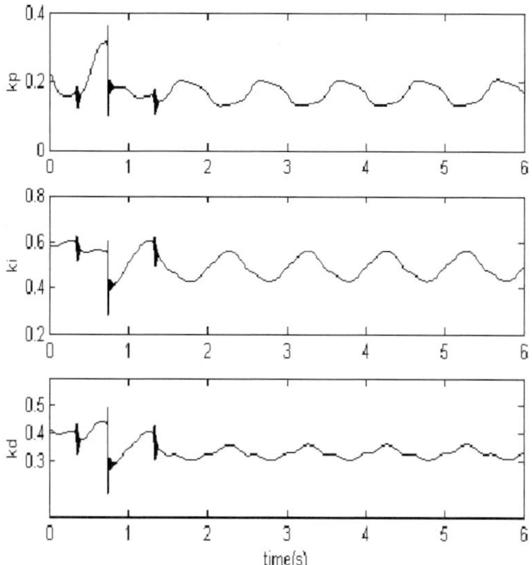

Fig. 4. Adaptive tuning curve of k_p, k_i, k_d using MOBP

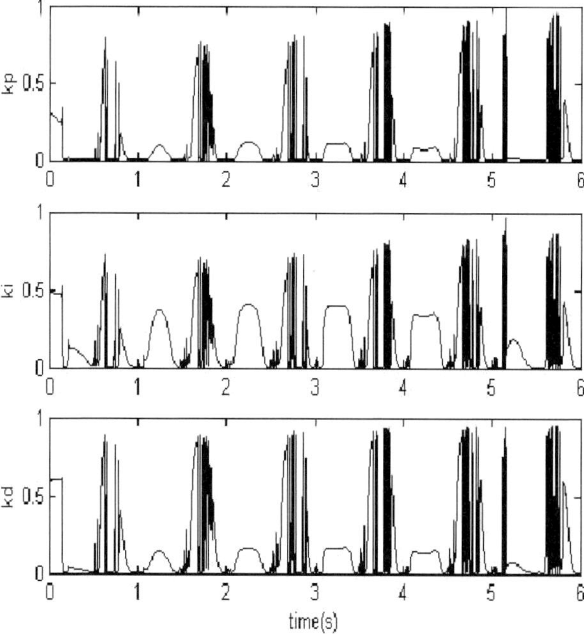

Fig. 5. Adaptive tuning curve of k_p, k_i, k_d using GRFBP

parameter k_p, k_i, k_d of PID controller and control variable u varying regularly, tracking curve smooth, no disturbance, and tracking error small and so on.

According to experiment and results we can obtain that apply improved BP algorithm to tune the parameters of neural PID controller is very effective, three control parameters k_p, k_i, k_d are easily tuned, this process do not need the man's experience and little depend on the system, and make the follow error to be diminished consumedly.

5 Conclusions

New improved learning algorithm of BP neural network namely GRFBP is presented in this paper. Experiments and actual applications indicate that GRFBP has some obvious superiority such as the training speed and training precision compared with MOBP and adaptive learning rate MOBP. It is also found that new learning algorithm can reduce the local minimum, to effectively obtain the global minimum, and can accelerate the convergence speed of the learning processes helpfully. And improved BP algorithm to tune the parameters of neural PID controller is very effective; three control parameters k_p, k_i, k_d are easily tuned. The simulation reveal that the improved BP network is used in tuning PID controller parameters has validity and value.

References

1. Yan, P., Zhang, C.: Artificial Neural Network and Evolution Simulation Calculation. Tsinghua Publication House Beijing (2000) 124-148
2. Kwon, T., and Chen, H.: Contrast Enhancement for Back Propagation. IEEE Transactions on Neural Networks 7(2) (1996) 515-524.
3. Yildirim, S., Sukkar, M. F., Demirci, R., and Aslantas, V.: Design of Adaptive NNs-Robust-PID Controller for A Robot Control. Proceedings of IEEE International Symposium on Intelligent Control (1996) 508-513
4. Li, J., Li, Y., Xun, J., Zhang, J.: Parallel Training Algorithm of BP Neural Algorithms. Proc. of the 3rd World Congress on Intelligent Control and Automation 2 (2000) 872-876
5. Wen, J., Zhao, J., Wei, L., Zhen, H.: The Improvement of BP Neural Network Learning Algorithm In: Signal Processing Proceedings. WCCC_ICSP 2000. 5th International Conference (2000) 1647-1649
6. Cheng, C., Tang, Y.: To Improve The Training Time of BP Neural Networks. Info-tech and Info-net. 2001 International Conferences. Vol.3 (2001) 473-479

Neuro-fuzzy Generalized Predictive Control of Boiler Steam Temperature

Xiang-Jie Liu and Ji-Zhen Liu

Department of Automation, North China Electric Power University,
Beijing, 102206, P.R. China
liuxj@ncepubj.edu.cn

Abstract. Reliable control of superheated steam temperature is necessary to ensure high efficiency and high load-following capability in the operation of modern power plant. A nonlinear generalized predictive controller based on neuro-fuzzy network (NFGPC) is proposed in this paper, which consists of local GPCs designed using the local linear models of the neuro-fuzzy network. The proposed nonlinear controller is applied to control the superheated steam temperature of a 200MW power plant, in which much better performance than the traditional cascade PI controller or the linear GPC is obtained.

1 Introduction

Continuous process in power plant and power station are complex systems characterized by nonlinearity, uncertainty and load disturbance[1]. The superheater is an important part of the steam generation process in the boiler-turbine system, where steam is superheated before entering the turbine that drives the generator.

From Fig. 1, the steam generated from the boiler drum passes through the low-temperature superheater before it enters the radiant-type platen superheater. Proper control of the superheated steam temperature by water spray is extremely important to ensure the overall efficiency and safety of the power plant. It is also important to reduce the temperature fluctuations inside the superheater, as it helps to minimize mechanical stress that causes micro-cracks in the unit, in order to prolong the life of the unit and to reduce maintenance costs. As the GPC is derived by minimizing these fluctuations, it is amongst the controllers that are most suitable for achieving this goal.

The multivariable multi-step adaptive regulator has been applied to control the superheated steam temperature in a 150t/h boiler[2], and generalized predictive control was proposed to control the steam temperature[3]. A nonlinear long-range predictive controller based on neural networks is developed in [4] to control the power plant process at several operating levels.

Fuzzy logic is capable of incorporating human experiences via the fuzzy rules. The neuro-fuzzy networks (NFNs) developed recently have the advantages of model transparency of fuzzy logic, and learning capability of neural networks [5]. The NFNs have been used to develop self-tuning control [6]. Since NFNs can be considered as a network that consists of several local regions, each of which contains a local linear model, nonlinear predictive control based on NFNs can be devised with the network incorporating all the local generalized predictive controllers (GPC) designed using the

J. Wang et al. (Eds.): ISNN 2006, LNCS 3972, pp. 1027 – 1032, 2006.

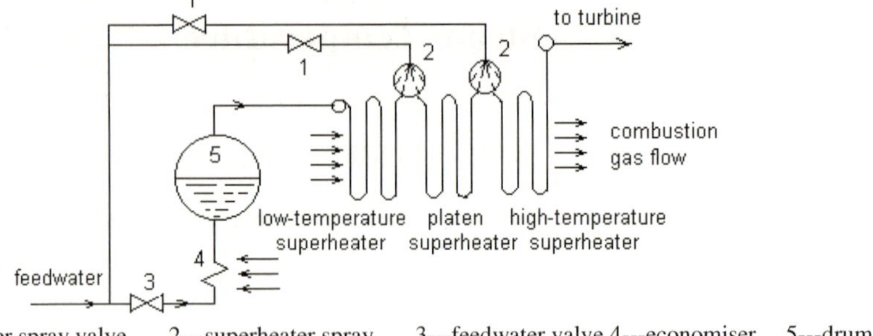

1--- superheater spray valve 2---superheater spray 3---feedwater valve 4---economiser 5---drum

Fig. 1. The boiler and superheater steam generation process

respective local linear models. The proposed controller is then applied to control the superheated steam temperature of the 200MW power unit. The proposed controller is tested first on the simulation of the process, before applying it to control the power plant.

2 Neuro-Fuzzy Network Modelling

2.1 Structure of the Nonlinear Predictive Model by Neuro-Fuzzy Network

Consider the following general single-input single-output nonlinear dynamic system:

$$\begin{aligned} y(t) = f[&y(t-1),\cdots,y(t-n_y'),u(t-d),\cdots,\\ &u(t-d-n_u'+1),e(t-1),\cdots,e(t-n_e')]+e(t)/\Delta \end{aligned} \tag{1}$$

where *f*[.] is a smooth nonlinear function such that a Taylor series expansion exists, *e*(*t*) is a zero mean white noise and Δ is the differencing operator, n_y', n_u', n_e' and *d* are respectively the known orders and time delay of the system. Let the local linear model of the nonlinear system (1) at the operating point O(*t*) be given by the following Controlled Auto-Regressive Integrated Moving Average (CARIMA) model:

$$\overline{A}(z^{-1})y(t) = z^{-d}B(z^{-1})\Delta u(t) + C(z^{-1})e(t) \tag{2}$$

where $\overline{A}(z^{-1}) = \Delta A(z^{-1})$, $B(z^{-1})$ and $C(z^{-1})$ are polynomials in z^{-1}, the backward shift operator. Note that the coefficients of these polynomials are a function of the operating point O(*t*). Since NFNs is a class of associative memory networks with knowledge stored locally [5], they can be applied to model this class of nonlinear systems. A schematic diagram of the NFN is shown in Fig. 2, where the membership functions are given by B-spline basis functions. The input of the network is the antecedent variable $[x_1, x_2 \cdots x_n]$, and the output, $\hat{y}(t)$, is a weighted sum of the output of the local linear models $\hat{y}_i(t)$.

The membership functions of the fuzzy variables can be obtained by:

$$a_i = \prod_{k=1}^{n} \mu_{A_k^i}(x_k) \ ; \quad \text{for } i = 1,2,\cdots,p \tag{3}$$

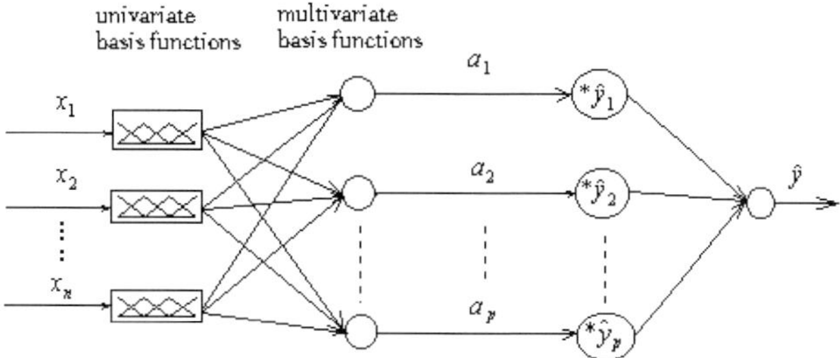

Fig. 2. Neuro-fuzzy network

where n is the dimension of the input vector x, and p, the total number of weights :

$$p = \prod_{i=1}^{n} (R_i + k_i) \tag{4}$$

where k_i and R_i are the order of the basis function and the number of inner knots respectively. The output of the NFN with p fuzzy rules is,

$$\hat{y} = \frac{\sum_{i=1}^{p} \hat{y}_i a_i}{\sum_{i=1}^{p} a_i} = \sum_{i=1}^{p} \hat{y}_i \alpha_i \tag{5}$$

3 Neuro-Fuzzy Network Generalized Predictive Control

The GPC is obtained by minimizing the following cost function [7],

$$J = E\{\sum_{j=d}^{N} q_j [\hat{y}(t+j) - y_r(t+j)]^2\} + \sum_{j=1}^{M} \lambda_j [\Delta u(t+j-1)]^2 \tag{6}$$

where q_j and λ_j are respectively the weighting factors for the prediction error and the control, $y_r(t+j)$ is the jth step ahead reference trajectory, d is the minimum costing horizon, N and M are respectively the maximum costing horizon for the prediction error and the control. The control computed from the NFGPC is the weighted sum of the control obtained from p local GPC controllers:

$$\Delta u(t) = \sum_{i=1}^{p} a_i \Delta u_i(t) \tag{7}$$

where $\Delta u_i(t)$ is the control in the ith region, $a_i(x)$ is defined previously in (3). From the NFN (5) and the control (7), J can be rewritten as:

$$J = E\{\sum_{j=d}^{N} q_j [\sum_{i=1}^{p} \alpha_i (\hat{y}_i(t+j) - y_r(t+j))]^2\} + \sum_{j=1}^{M} \lambda_j [\sum_{i=1}^{p} \alpha_i \Delta u_i(t+j-1)]^2 \tag{8}$$

The cost function is simplified first using the Cauchy inequality. Since

$$[\sum_{i=1}^{p}\alpha_i(\hat{y}_i(t+j)-y_r(t+j))]^2 \le p\sum_{i=1}^{p}[\alpha_i(\hat{y}_i(t+j)-y_r(t+j))]^2 ,$$

hence $$[\sum_{i=1}^{p}\alpha_i\Delta u_i(t+j-1)]^2 \le p\sum_{i=1}^{p}[\alpha_i\Delta u_i(t+j-1)]^2 \quad (9)$$

Rewriting (8) gives

$$E\{\sum_{j=d}^{N}\sum_{i=1}^{p}q_j[\alpha_i(\hat{y}(t+j)-y_r(t+j))]^2\}+\sum_{j=1}^{M}\sum_{i=1}^{p}\lambda_j[\alpha_i\Delta u_i(t+j-1)]^2$$

$$=\sum_{i=1}^{p}(\alpha_i)^2 J_i \quad (10)$$

Where $$J_i = E\{\sum_{j=d}^{N}q_j[\hat{y}_i(t+j)-y_r(t+j)]^2\}+\sum_{j=1}^{M}\lambda_j[\Delta u_i(t+j-1)]^2 \quad (11)$$

From (10), a set of local generalized predictive controllers is obtained, which forms part of the NFGPC.The local GPC in the unconstraint case is given by:

$$\Delta U_i(t) = (G_i^T QG_i+\lambda)^{-1}G_i^T Q[Y_r(t+1)-F_i\Delta U_i(t-1)-S_i(z^{-1})y_i(t)] \quad (12)$$

4 Neuro-Fuzzy Predictive Control of the Plant

Let θ be the superheated steam temperature, and μ_θ, the flow of spray water to the high temperature superheater. The response of θ can be approximated as[8]:

$$G(s)=\frac{\theta}{\mu_\theta(s)}=\frac{K_p}{(T_1 s+1)(T_2 s+1)}e^{-\tau s} \quad (13)$$

Table 1. Local CARIMA Models in Neuro-fuzzy Model

Load	$\overline{A}_i$	B_i
High	$1-2.8824z^{-1}+2.7682z^{-2}-0.8858z^{-3}$	0.0035
Medium high	$1-2.8940z^{-1}+2.7915z^{-2}-0.8975z^{-3}$	0.0028
Medium	$1-2.9000z^{-1}+2.8100z^{-2}-0.9100z^{-3}$	0.0023
Medium low	$1-2.9130z^{-1}+2.8279z^{-2}-0.9149z^{-3}$	0.0019
Low	$1-2.9200z^{-1}+2.8416z^{-2}-0.9216z^{-3}$	0.0016

The linear model (13) is, however, only a local model for the selected operating point. Since load is the unique antecedent variable, it is used to select the division between the local regions in the NFN. The load is divided into five regions, using also the experience of the operators, who regard a load of 200MW as high, 180MW as medium high, 160MW as medium, 140MW as medium low and 120MW as low. The NFN that models the power plant is then trained by the gradient method from the training data. For a sampling interval of 30s, the estimated linear local models $\overline{A}(z^{-1})$ used in the NFN are shown in Table 1.

For the NFGPC, the time delay *d*, which is also the minimum cost horizon, is set to 30secs, $Q = I$, and $\lambda = 0.1 \times I$, $N = 10$, $M = 6$. The local property of NFN implies only two local controllers are activated each time, determined by the load signal through five triangular membership functions. Consider the load changes, where the load increases from 140 MW to 195 MW at a rate of between 1% and 2.5%/min. around 20 minutes, and decreases gradually to 160MW from 60 minutes. From Fig. 3-a, good control of the superheated steam temperature is achieved by the NFGPC, as the temperature fluctuations for both the upward and downward load changes are within $\pm 7°C$. This result is comparable to that presented in [3], which is tested on a 380MW unit, under coal mill stop disturbance. In contrast, the fluctuations in the superheated steam temperature is much larger using the conventional cascade PI controller, as shown in Fig. 3-b.

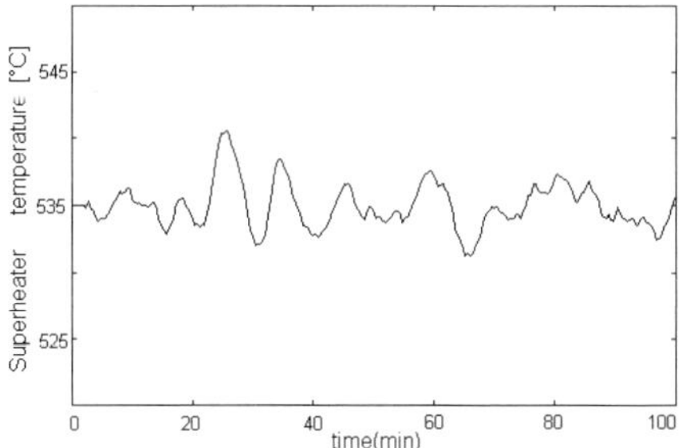

Fig. 3-a. Controlled by the NFGPC

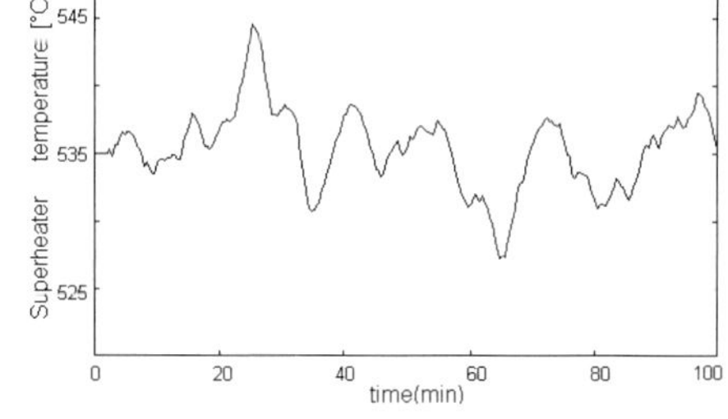

Fig. 3-b. Controlled by the cascade PI controller

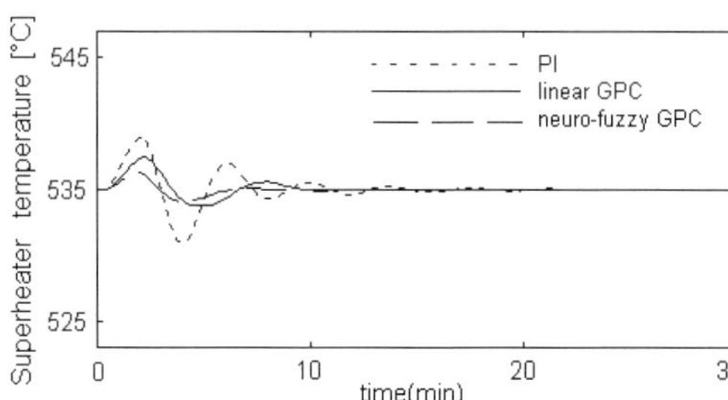

Fig. 4. Comparison of the NFGPC, conventional linear GPC, and cascade PI controller

As a further illustration, the power plant is simulated using the NFN model given in Table 1, and is controlled respectively by the NFGPC, the conventional linear GPC controller, and the cascaded PI controller while the load changes from 160MW to 200MW. The results in Fig.4 show that the best performance is obtained from the NFGPC as it is designed based on a more accurate process model. This is followed by the conventional linear GPC controller, as the plant is nonlinear. The performance of the conventional cascade PI controller is the worse, indicating that it is unable to control satisfactory the superheated steam temperature under large load changes.

5 Conclusion

The modeling and control of a 200MW power plant using the neuro-fuzzy approach is presented in this paper. The NFN consists of five local CARIMA models, and the output of the network is the interpolation of the local models using memberships given by the B-spline basis functions. The proposed NFGPC is similarly constructed, which consists of five GPCs, each of which is designed from the CARIMA models in the NFN. The NFGPC is most suitable for processes with smooth nonlinearity, such that its full operating range can be partitioned into several local linear operating regions. The proposed NFGPC therefore provides a useful alternative for controlling this class of nonlinear power plants, which are formerly difficult to be controlled using traditional methods.

Acknowledgment

This work is supported by Natural Science Foundation of Beijing(4062030), National Natural Science Foundation of China(50576022,69804003).

References

1. Liu, X.J., Lara-Rosano, F., Chan, C.W.: Neurofuzzy Network Modelling and Control of Steam Pressure in 300MW Steam-Boiler System. Engineering Applications of Artificial Intelligence 16(5) (2003) 431-440
2. Silva, R. N., Shirley, P. O., Lemos, J. M., Goncalves, A. C.: Adaptive Regulation of Superheated Steam Temperature: A Case Study in an Industrial Boiler. Control Engineering Practice 8(8) (2000) 1405-1415
3. Moelbak, T.: Advanced Control of Superheater Steam Temperatures - An Evaluation Based on Practical Applications. Control Engineering Practice 7(7) (1999) 1-10
4. Prasad, G., Swidenbank, E., Hogg, B.W.: A Neural Net Model-based Multivariable Long-range Predictive Control Strategy Applied Thermal Power Plant Control. IEEE Trans. Energy Conversion 13(2) (1998) 176-182
5. Brown, M., Harris, C. J.: Neurofuzzy Adaptive Modelling and Control. Englewood Cliffs, NJ: Prentice-Hall(1994)
6. Liu, X.J., Lara-Rosano, F., Chan, C.W.: Model-Reference Adaptive Control Based on Neurofuzzy Networks. IEEE Trans. Systems, Man and Cybernetics C. 34(3) (2004) 302-309
7. Clarke, D.W., Mohtadi, C., Tuffs, P.S.: Generalized Predictive Control, Parts 1 and 2, Automatica 23(2) (1987) 137-160
8. Liu, X.J., Zhou, X.X.: Identification of Boiler Models and its Fuzzy Logic Strategy. In Proc. the 14th IFAC World Congress, Beijing China, June, Vol. O (1999)149-154

Model-Free Control of a Nonlinear ANC System with a SPSA-Based Neural Network Controller

Yali Zhou[1], Qizhi Zhang[1], Xiaodong Li[2], and Woonseng Gan[3]

[1] Department of Computer Science and Automation, Beijing Institute of Machinery, P.O. Box 2865, Beijing, 100085, China
zhouyali@yahoo.com
[2] Institute of Acoustics, Academia Sinica, China
[3] School of EEE, Nanyang Technological University, Singapore

Abstract. In this paper, a feedforward active noise control (ANC) system using a mode-free neural network (MFNN) controller based on simultaneous perturbation stochastic approximation (SPSA) algorithm is considered. The SPSA-based MFNN control algorithm employed in the ANC system is first derived. Following this, computer simulations are carried out to verify that the SPSA-based MFNN control algorithm is effective for a nonlinear ANC system. Simulation results show that the proposed scheme is able to significantly reduce disturbances without the need to model the secondary-path and has better tracking ability under variable secondary-path. This observation implies that the SPSA-based MFNN controller frees the ANC system from the modeling of the secondary-path.

1 Introduction

The active noise control (ANC) using feedforward control techniques has attracted much research attention because it can complement traditional passive techniques and attain better performance on attenuation of low-frequency noises [1]. When the ANC system exhibits nonlinear response characteristics, the most common form of adaptive algorithm/architecture combination is the feedforward neural network (NN) using the gradient descent-based back-propagation (BP) algorithm [2-3], where the NN would be trained to derive an output signal to cancel the noise. But, in this control method, in order to update the weights of the NN, we need a gradient of the error function, namely, we must know the model of the secondary-path [4].

However, characteristics of the secondary-path usually vary with respect to temperature or other environments, that is, the secondary-path is time-variant. Therefore, it is difficult to estimate the exact characteristics of the secondary-path accurately. To solve the above problem, a model-free (MF) control scheme based on the simultaneous perturbation stochastic approximation (SPSA) algorithm is presented here [5]. This approach is based on the output error of the system to update the weights of the NN without the need to model the secondary-path [6].

In addition to being able to update the weights of the NN without the need to model the secondary-path, the presented algorithm can also give more simple formulae for updating the weights of the NN compared to the back-propagation neural network (BPNN) algorithm. Because the SPSA algorithm requires only two objective

J. Wang et al. (Eds.): ISNN 2006, LNCS 3972, pp. 1033 – 1038, 2006.

function measurements regardless of the number of weights being updated, it uses "simultaneous perturbation (SP)" to update all weights of the NN simultaneously and can simplify the derivation of the adaptation algorithm of weights greatly, this will be discussed in the following section.

2 Control Algorithm

The block diagram of a feedforward ANC system using the SPSA-based MFNN algorithm and the neural network controller with a single hidden layer are shown in Fig.1 and Fig.2, respectively. The primary-path *P(Z)* is from the noise source to the error microphone, and the secondary-path *S(Z)* is from the canceling loudspeaker to the error microphone. The SPSA-based NN controller is used to generate an anti-noise signal *u(t)*. Note that this algorithm does not use the estimator of the secondary-path [4].

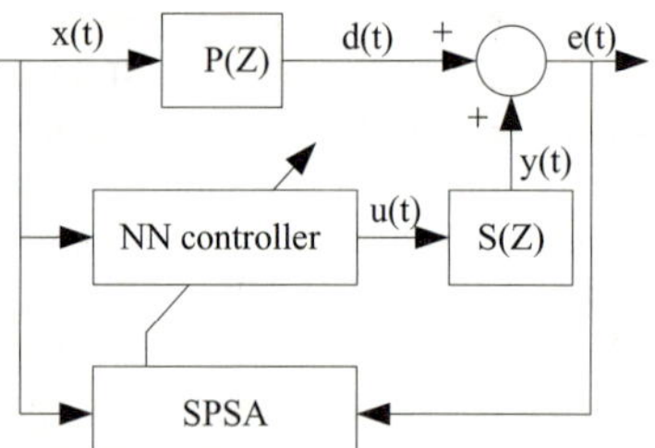

Fig. 1. The block diagram of an ANC system

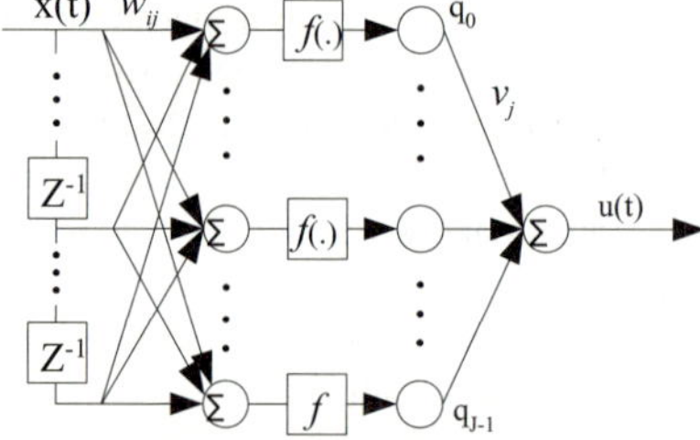

Fig. 2. The neural network controller

It is common knowledge that if the secondary-path of the ANC system is completely unknown, it is impossible to use usual gradient method as a learning rule to update the controller coefficients [3][7]. In this case, an estimator of the gradient of the error function is needed. The SPSA which was introduced by J. C. Spall [5] is a well-known gradient approximation approach that relies on measurements of the objective function, not on measurements of the gradient of the objective function.

The objective of the following analysis is to develop the SPSA-based MFNN algorithm to improve the noise cancellation capability of a nonlinear ANC system.

Step 1: Define the error function

Note that in ANC system, each sampling error signal does not contain enough information as an evaluation function to be optimized. That is, the expectation of the error signal has to be used as the evaluation function. For practicality, the sum of the error signal for a certain interval is used to approximate the expectation of the error signal. Thus, the error function is defined as:

$$J(u(t)) = \frac{1}{2}\sum_{t=1}^{\lambda} e^2(t) = \frac{1}{2}\sum_{t=1}^{\lambda}[y(t)+d(t)]^2 \ . \tag{1}$$

where t is the sampling number in a block interval, and λ is the total sampling number of one block interval.

Step 2: Compute the control signal $y(t)$
The output of the jth hidden layer neuron is

$$q_j(t) = f_j[net_j(t)] \ . \tag{2}$$

Where $net_j(t) = \sum_{i=0}^{I} w_{ij}(t)x(t-i)$,w_{ij} is the weights in the hidden layer, I is the number of the input layer neurons, $f(.)$ is a smoothing nonlinear activation function.

The output of the NN is

$$u(t) = \sum_{j=0}^{J}[v_j(t)q_j(t)] \ . \tag{3}$$

$v_j(t)$ is the jth weight in the output layer, J is the number of the hidden layer neurons.

Then the control signal $y(t)$ can be calculated using the following equation

$$y(t) = \sum_{j=0}^{M-1} s_j(t)u(t-j) = S(t)U(t) \ . \tag{4}$$

Where $S(t)=[s_0(t)\ s_1(t)\dots\ s_{M-1}(t)]$, is the secondary-path model, whose z-transform is $S(z)$, M is the length of the secondary-path.

Step 3: Define the weight vector
The weight vector of the NN including thresholds is defined as:

$$w = (w_{00},\cdots,w_{IJ-1},v_0,\cdots,v_J)^T = (w^1,\cdots,w^n)^T \ . \tag{5}$$

Where the superscript n denotes the number of weights to be estimated (including thresholds), as shown in Fig.2, $n=(I+1)J+(J+1)$, Superscript T is transpose of a vector:

Step 4: Generation of SP vector
The following perturbation vector Δ_k is generated as independent Bernoulli random variables with outcomes of ±1 that gives small disturbances to all weights [5]

$$\Delta_k = (\Delta_k^1,\cdots,\Delta_k^n)^T \ . \tag{6}$$

Where the subscript k denotes an iteration.

Step 5: Error function evaluations
Obtain two measurements of the error function $J(\square)$ based on the SP: $J(u(w))$ and $J(u(w+c_k\Delta_k))$ with the Δ_k from step 4.

Where c_k is a positive scalar and represents a magnitude of the perturbation.

Step 6: Gradient approximation

Generate the SP approximation to the unknown gradient $\frac{\partial J(u(w))}{\partial w}$ as [3][5]:

$$\Delta w(k) = \frac{J(u(w+c_k\Delta_k)) - J(u(w))}{c_k\Delta_k} \ . \tag{7}$$

Step 7: Update the weight vector w of the NN
Weights of the NN are updated in the following manner:

$$w(k+1) = w(k) - a_k \Delta w(k) \ . \tag{8}$$

Where a_k is a positive learning coefficient.

From Eqs. (7) and (8), it can be seen that the weights of the NN controller is updated without the need to model the secondary-path, so this algorithm is called MFNN control algorithm. At the same time, the conclusion can be derived that compared to the BPNN algorithm which were reported in Ref. [2] [4], the SPSA-based MFNN algorithm has more simple formulae for updating the weights of the NN.

3 Simulation Examples

Some simulations are presented to illustrate the noise-canceling performance of the SPSA-based MFNN algorithm on a nonlinear ANC system, and at the same time, a comparison between the SPSA-based MFNN algorithm and the filtered-X Least Mean Square (FXLMS) algorithm is made. A 300Hz sinusoidal signal is used to generate the primary disturbance signal and also used as the reference signal to the control algorithm. a 6-tap finite impulse response (FIR) filter is used in the FXLMS algorithm, the NN used in these simulations is three-layered feedforward network and the number of neurons is set as 15-10-1, the weights and the thresholds of the NN are randomly initialized in the interval [-1,1]. The inputs to the controller are the current and recent states of noise signal *x(t)*.The hidden layer nodes are sigmoid functions $f(net) = (1 - e^{-net})/(1 + e^{-net})$, and *net* is the weighted sum of input signals in the input layer. The output layer nodes are linear functions. The sampling frequency used is 3kHz, and the total sampling number of one block interval λ is set as 30. c_k and a_k are set as 0.01 and 0.001, respectively. The total simulation duration is 10 second.

The model used in this simulation has the following expressions with nonlinear terms:
The primary disturbance *d(t)* is expressed as [4]

$$\begin{aligned} d_{t+1} = {} & 0.8x_t + 0.6x_{t-1} - 0.2x_{t-2} - 0.5x_{t-3} - 0.1x_{t-4} \\ & + 0.4x_{t-5} - 0.05_{t-6} \end{aligned} \ . \tag{9}$$

The control signal *y(t)* is expressed as [4]

$$\begin{aligned} y_{t+1} = {} & 0.9u_t + 0.6u_{t-1}^3 + 0.1u_{t-2}^3 - 0.4u_{t-3}^3 - 0.1u_{t-4}^3 \\ & + 0.2u_{t-5}^3 + 0.1u_{t-6}^2 + 0.01u_{t-7}^2 + 0.001u_{t-8}^2 \end{aligned} \ . \tag{10}$$

Case 1: A simple static ANC example is first considered to illustrate the SPSA-based MFNN algorithm effectiveness by comparison with the result given by the FXLMS algorithm, the secondary-path is assumed to be time-invariant. Fig.3 presents the simulation result of the canceling errors in the frequency domain. The thin solid line shows the power spectrum of active noise canceling error when the ANC system is turned off, and the dashed-line curve shows the power spectrum of active noise canceling error when the FXLMS algorithm is used to adapt the coefficients of the controller, while the thick solid line shows the power spectrum of active noise canceling error when the SPSA-based MFNN algorithm is used to adapt the coefficients of the controller. From the results shown in Fig.3, it can be clearly seen that the SPSA-based MFNN algorithm is superior for a nonlinear problem compared to the linear FXLMS algorithm, the major disturbance frequency are attenuated by approximately 40 dB.

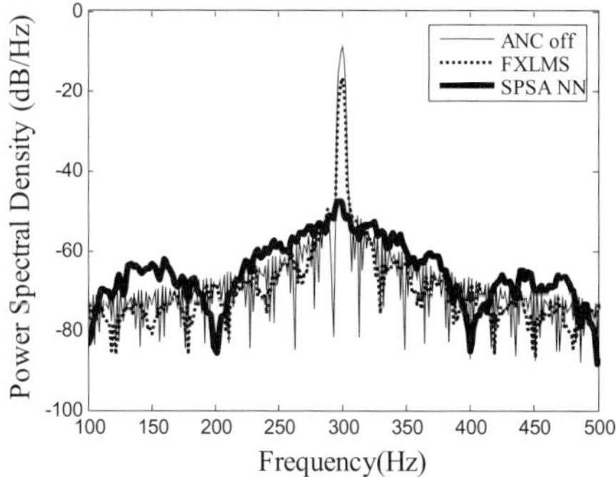

Fig. 3. The error signal spectrum for case 1

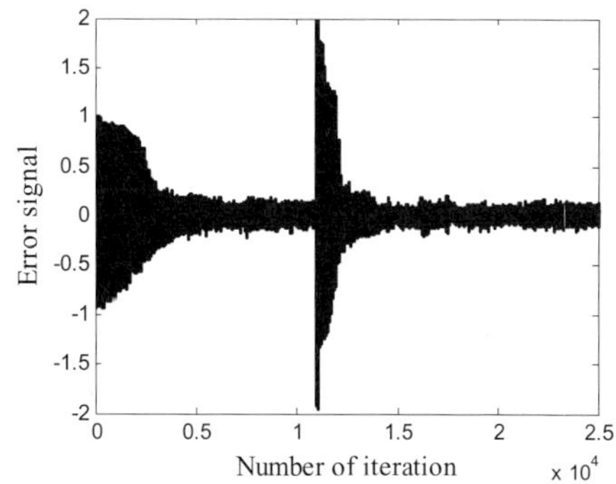

Fig. 4. The error signal versus number of iterations when the secondary-path is changed

Case 2: Next, we deal with a tracking problem. Using the same settings as in case 1, after the system has entered into steady-state phase, the secondary-path is altered by letting $S(z) = -S(z)$. Fig.4 shows the error signal in error microphone versus the number of iterations. When the number of iteration reaches 11,000, the secondary-path is changed. From the result shown in Fig.4, it can be seen that the system has a good tracking ability of the secondary-path. This simulation shows that the SPSA-based NN controller can free the ANC system from the modeling of the secondary-path.

4 Conclusions

The NN controller based on the SPSA algorithm has been developed for use in a nonlinear ANC system. This approach optimizes error function without using derivative of the error function. Therefore, the presented ANC algorithm does not require any estimation of the secondary-path. Some simulations were presented to verify that this algorithm is effective. The simulation results indicated that this algorithm was able to significantly reduce disturbances and an output error attenuation of approximately 40dB was achieved.

Acknowledgments

This research is supported by Scientific Research Common Program of Beijing Municipal Commission of Education (KM200511232008) and Training Funds for Elitist of Beijing (20051A0500603).

References

1. Nelson, P.A., Elliott, S.J.: Active Sound Control. Academic Press, London (1991)
2. Snyder, S.D., Tanaka, N.: Active Control of Vibration Using a Neural Network. IEEE Trans on Neural Networks, 6(4) (1995) 819-828
3. Maeda, Y., De Figueiredo, R.J.P.: Learning Rules for Neuro-Controller via Simultaneous Perturbation. IEEE Transactions on Neural Networks, 8(5) (1997) 1119-1130
4. Zhou, Y.L., Zhang, Q.Z., Li, X.D., Gan, W.S.: Analysis and DSP Implementation of an ANC System using a Filtered-Error Neural Network. Journal of Sound and Vibration, 285(1) (2005) 1-25
5. Spall, J.C.: Multivariate Stochastic Approximation Using Simultaneous Perturbation Gradient Approximation. IEEE Transactions on Automatic Control, 37(3) (1992) 332-341
6. Maeda, Y., Yoshida, T.: An Active Noise Control without Estimation of Secondary-Path. ACTIVE1999, USA (1999) 985-994
7. Spall, J.C., Cristion, J.A.: A Neural Network Controller for Systems with Unmodeled Dynamics with Applications to Wastewater Treatment. IEEE Transactions on Systems. Man. And Cybernetics, 27(3) (1997) 369-375

Robust Control for AC-Excited Hydrogenerator System Using Adaptive Fuzzy-Neural Network

Hui Li[1,2], Li Han[2], and Bei He[3]

[1] The Key Laboratory of High Voltage Engineering and Electrical New Technology, Ministry of Education, Chongqing University, Chongqing 400044, China
cqulh@163.com
[2] College of Electrical Engineering, Chongqing University, Chongqing 400044, China
hanli@cqu.edu.cn
[3] YangJiaPing Power Bureau, Chongqing Electrical Power Company, 400050, China
hb@cqep.com.cn

Abstract. The AC-excited hydrogenerator (ACEH) is a novel type of hydraulic generation system. Concern about its integrative control strategy is increasing, owing to the features of uncertain and nonlinear as well as parameters coupling and time-variation for three parts of water flux, hydroturbine and generator. A cascade-connected self-adaptive fuzzy-neural network control strategy is proposed, which the former controller uses a self-tuning fuzzy algorithm with the intelligent weight function rulers, the latter adopts a self-adaptive neural network controller based on dynamical coupling characteristics of controlled plants. By comparison with traditional PID control, simulation results have shown that this hydrogenerator system appears good robustness against load disturbance and system parameters uncertainty.

1 Introduction

The wide range of variable speed constant frequency (VSCF) operation, the capability of active and reactive power control make the AC-excited generator (ACEG) attractive for variable speed hydroelectric generators as well as wind power conversion system [1-4]. The stator of ACEG connects the grid directly and provides for variable speed operation by using a partially rated converter on the rotor side, which brings to some superior performances such as good power system stability, VSCF generation, stator active and reactive power regulation independently [1-6]. So the AC-excited hydrogenerator (ACEH) system can be operated round the optimal unit speed of hydroturbine by some suitable control strategies, when the water level or water flux is changed. The hydraulic efficiency and power system stabilization can be improved. The typical connection of ACEH system can be seen in [1,2,4].

The ACEH system is a more complex system including water, hydroturbine and generator portion as well as hydroturbine governing controller and generator excited system. Considering the water hammer effect, ACEH system has some distinct characteristics such as the system's nonlinear and great inertia, parameters variation, multi-variable characteristics. In order to develop the excellent operational performances of this hydrogenerator system, it is very important and necessary to study its comprehensive robust control strategies. Some methods and techniques have been

J. Wang et al. (Eds.): ISNN 2006, LNCS 3972, pp. 1039 – 1045, 2006.

researched to solve the ACEG excited control problem. Some of these methods are traditional vector control techniques based on stator flux or air-gap flux oriented frame [1-3], it is difficult to achieve the robust and stable control performances when system parameters are uncertain. To overcome the aforementioned drawbacks, the fuzzy logical control is also proposed and applied in [6], even though this kind of method is independent of the accurate plant models, it neglects the influence of the prime mover such as hydroturbine governing system.

To achieve excellent operational characteristics of ACEH system, a cascade-connected self-adaptive fuzzy-neural network (FNN) control strategy is proposed in this paper. Robust characteristics of ACEH system is studied and simulated with Matlab/Simulink.

2 Design of Fuzzy-Neural Network Controller

It is well known that the neural network technique has emerged as an attractive and powerful tool to control a wide class of complex nonlinear dynamic systems [7-13]. Considering the features of robustness and rapid convergence of fuzzy control [7,9], a novel approach of cascade–connected FNN controller is presented based on the dynamical coupling algorithm, the block diagram of which is shown in Fig. 1.

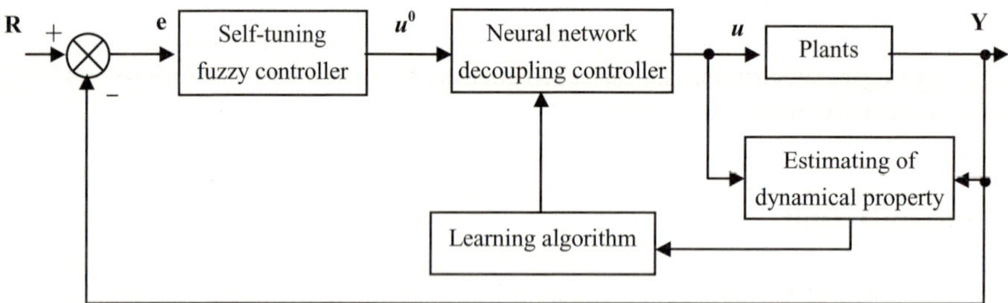

Fig. 1. Block diagram of adaptive FNN control system

This controller structure is made up of self-tuning fuzzy controller and self-adaptive neural network decoupling controller, the former is crucial to the system dynamical property, the latter is impact on the multivariable decoupling performances [8]. In the formerly fuzzy controller, the output vectors $\boldsymbol{u}^0=\{u_1^0, u_2^0,\ldots, u_n^0\}$are obtained by the error vectors $\boldsymbol{e}=\{e_1, e_2,\ldots, e_n\}$. In the neural network controller, the output vectors $\boldsymbol{u}=\{u_1, u_2,\ldots, u_n\}$are achieved by the weights adaptation law, which is based on estimating the dynamical multivariable coupling property of the controlled plants.

2.1 Algorithm of Self-tuning Fuzzy Controller

In the traditional fuzzy control, the control rules and fitness function play important role in the system performances; however, the appropriate control rules and fitness function are difficult to achieve for the system uncertainty and complexity [9,11,13]. Hence, the fuzzy algorithm is adopted based on continuous intelligent weight function, the structure block diagram of the single loop is shown in Fig. 2.

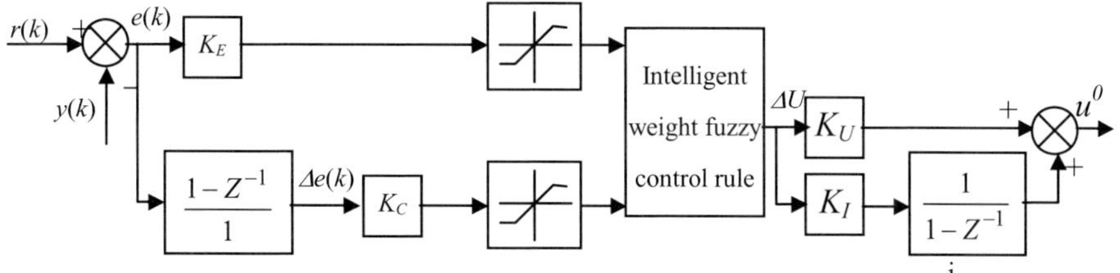

Fig. 2. Block of fuzzy controller with continuous intelligent weight functions

The fuzzy values of error $e(k)$ and differential error $\Delta e(k)$ are acquired as follows

$$\left.\begin{aligned} E &= K_E \cdot e(k) \\ E_C &= K_C \cdot \Delta e(k) \end{aligned}\right\} \tag{1}$$

Where K_E is error coefficient, K_C is differential error coefficient.

The self-tuning control rule is presented by the two-weight coefficients α_e, α_c

$$\left.\begin{aligned} \alpha_e &= |E| / (|E| + |E_C| + \varepsilon) \\ \alpha_C &= |E_C| / (|E| + |E_C| + \varepsilon) \end{aligned}\right\} \tag{2}$$

Where ε is a discretionary small positive constant.

The self-tuning rule of fuzzy controller is updated as follows [13]

$$\Delta U = \alpha_e E + \alpha_C E_C \tag{3}$$

The output of the self-tuning fuzzy controller is obtained as follows

$$u_i^{\ 0}(k) = K_U \Delta U + K_I \cdot \sum \Delta U \tag{4}$$

Where K_U is proportional coefficient, K_I is integral coefficient.

2.2 Self-adaptive Neural Network Controller

Three input-output structure of the self-adaptive neural network controller is shown in Fig. 3, the output vector of neural network controller is given as follows.

$$u_j(k) = \sum_{i=1}^{n} \omega_{ij}(k) u_i^0(k) \quad (j = 1, 2, 3, \cdots, n) \tag{5}$$

Where $\omega_{ij}(k)$ is weight value, $u_i^0(k)$ is output of the above fuzzy controller. Considering the minimization of the mean square error between the factual output and the desired output, the system cost function is defined as follows

$$E(k) = \frac{1}{2} \sum_{i=1}^{n} (r_i(k) - y_i(k))^2 \qquad i = 1, 2, \cdots, n \tag{6}$$

The weights of the neural network controller are updated as follows:

$$\omega_{ij}(k+1)=\omega_{ij}(k)-\eta\frac{\partial E(k)}{\partial\omega_{ij}(k)}=\omega_{ij}(k)+\eta\sum_{i=1}^{n}(r_i(k)-y_i(k))\frac{\partial y_i(k)}{\partial u_j(k)}u_i^0(k) \quad (7)$$

Where η is the learning-rate parameter; $\frac{\partial y_i(k)}{\partial u_j(k)}$ is dynamical coupling degree of the jth input to the ith output value.

3 FNN Control for AC-Excited Hydrogenertor System

3.1 Description of Control System for AC-Excited Hydrogenerators

In order to control independently active and reactive power for this hydrogenerator, the stator active power P_1, active power Q_1 and slip s are considered as input variables of the controller, respectively, the excited voltage components U_{rd}, U_{rq} based on synchronous rotational frame and hydrotubine regulation valve u are taken as output variables. The proposed FNN control system for ACEH system is shown as Fig. 3.

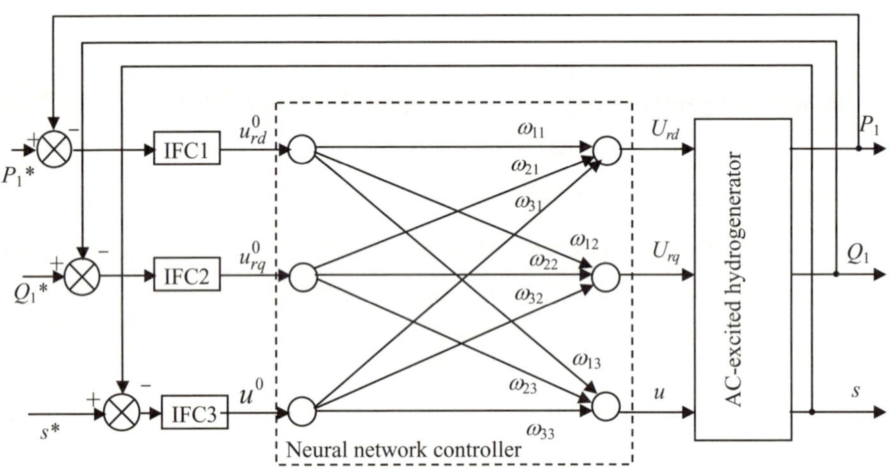

Fig. 3. Block of FNN control for AC-excited hydrogenerator system

As it can be seen, IFC1, IFC2 and IFC3 denote self-tuning fuzzy controller based on intelligent weight function, respectively, the mathematical models of ACEH are described in [2,4], where P_1*, Q_1* and s* denote the given value of stator active power, stator reactive power and slip, respectively.

3.2 Simulation

In order to testify the FNN control quality of the ACEH system, the operation performances of robustness against load disturbance and system parameters variation and uncertainty are simulated, respectively. The base region of controlled variables are set with [-1,1] in fuzzy controller, the initial learning-rate of neural network controller is set to 0.01, the initial weights are given as $\omega_{ij}(0)=1 \quad (i=j)$; $\omega_{ij}(0)=0 \quad (i \neq j)$. The main parameters of ACEH are described in Table 1.

Table 1. Main parameters of AC-excited hydrogenerator systems

Main parameters	Value
stator resistance R_s (p.u.)	0.00706
rotor resistance R_r (p.u.)	0.005
stator leakage inductance X_s (p.u.)	0.171
rotor leakage inductance X_r (p.u.)	0.156
mutual inductance X_m (p.u.)	2.9
system moment of inertia H (s)	5.04
time constant of hydraulic pressure driver system T_y (s)	5
time constant of water flux inertia T_w (s)	2

A. *Robustness against Load disturbance*

The water flux variation and grid power fluctuations can be considered as load disturbance ΔP_m. When ΔP_m is 0.3 p.u. from the time of 2 to 3 seconds, system operational performances are shown and compared as Fig. 4. (The thick real curve denotes the simulation result of the FNN control, while the thin real curve denotes it of the traditional PID). As it can be seen that the generating system has good robust ability with the FNN control, however it occurs more fluctuation of power and speed by using the conventional PID control.

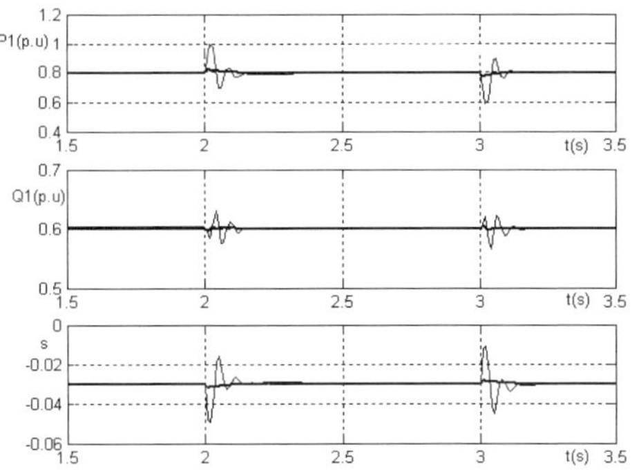

Fig. 4. Characteristics of against load disturbances

B. Robustness against parameters variation

The rotor resistance value is changed to 2 times with the original value in ACEG models, which is kept by original value in system control models, the active power regulation characteristics with rotor resistance variation is shown in Fig. 5. In the same way, when the constant time of water flux inertia T_w is set to 10 times with the original value, the active power regulation is also shown in Fig. 6. From the two figures, it can be seen that the ACEH system performance is seldom affected by the generator parameters variation or water flux time constant uncertainty by using FNN control, however the system occurs more fluctuation or unstable by using the conventional PID control, which is usually dependent on accurate plants models [13].

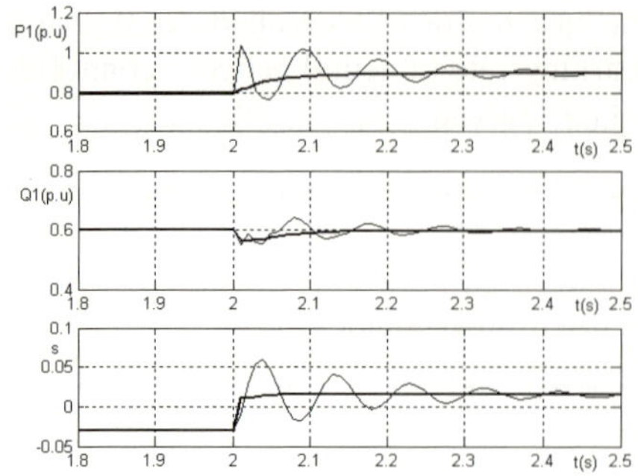

Fig. 5. Robustness when R_r is uncertain

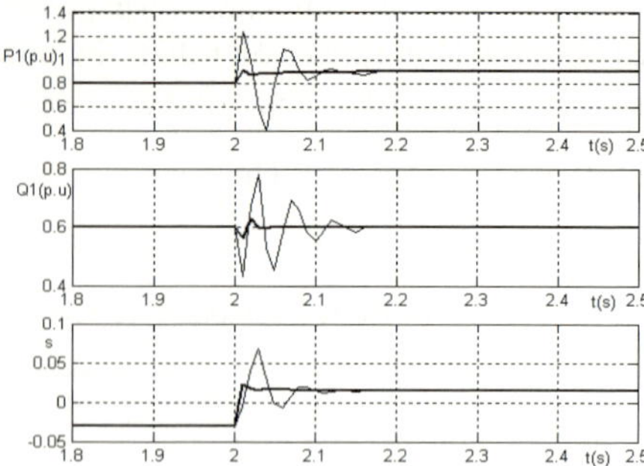

Fig. 6. Robustness when T_w is uncertain

4 Conclusion

Based on the complex characteristics of uncertain and nonlinear as well as parameters coupling and time-variation for ACEH system, the integrated control strategies of a cascade-connected self-adaptive FNN is proposed in this paper. The strong robustness is achieved by simulation, no matter what is the load disturbance and uncertainty of generator rotor resistance parameter or water flux time constant.

References

1. Yu, S., Yan, X., Liu, S., Li, H., Li, J.: Static State Stability Analysis of AC-excited Generators. The Fifth International Conference on Power Electronics and Drive Systems 2003 (PEDS 2003), Vol. 1 (2003) 733-736
2. Zhao, R., Chen, H.: A Control System for Hydro-generator with AC Excitation Running at Adjustable Speeds. Journal of Zhejiang University 33(6) (1999) 596-601
3. Okafor, F.N., Hofmann, W.: Modelling and Control of Slip Power Recovery Schemes for Small Hydro Power Stations. 7th AFRICON Conference in Africa, Vol.2 (2004) 1053-1058
4. Li, H., Yang, S.: Mathematical Models of Optimal Regulative Characteristics for Doubly Fed Hydrogenerator System. Power System Technology 29(9) (2005) 31-35
5. He, Y., Wang, W., Sun, D.: Adaptive Filtering of the Synchronizing Reference Signal and Its Application in an AC Excited Generation System. Proceedings of the Power Conversion Conference 2002. (PCC Osaka 2002) Vol. 3 (2002) 1411-1416
6. Kim, E. H., Kim, J.H., Lee, G.S.: Power Factor Control of a Doubly Fed Induction Machine Using Fuzzy Logic. Proceedings of the Fifth International Conference on Electrical Machines and Systems 2001. (ICEMS 01) (2001) 747-750
7. Lin, W. S., Chen, C.S.: Robust Neurofuzzy Controller Design of a Class of Uncertain Multivariable Nonlinear Systems. Proceedings of the 2001 IEEE International Conference on Control Applications 2001. (CCA '01) (2001) 902-907
8. Zhou, J., Han, Z.: A New Multivariable Fuzzy Self-Tuning Control System. Automatica Sinica 25 (1999) 215-219
9. Lin, C. M., Chen, C. H., Chin, W. L.: Adaptive Recurrent Fuzzy Neural Network Control for Linearized Multivariable Systems. Proceedings 2004 IEEE International Conference on Fuzzy Systems 2004 Vol. 2 (2004) 709-714

10. Pokorny, M., Rehberger, I., Cermak, P.: A Nonlinear Optimization and Fuzzy Modelling in Predictive Control Scheme. 26th Annual Conference of the IEEE Industrial Electronics Society, 2000. IECON 2000 Vol. 2 (2000) 1480-1484
11. Sadeghian, A.R.: Nonlinear Neuro-Fuzzy Prediction: Methodology, Design and Applications. The 10th IEEE International Conference on Fuzzy Systems 2001. Vol. 2 (2001) 1002-1026
12. Zhu, Q., Guo, L.: Stable Adaptive NeuroControl for Nonlinear Discrete Time Systems. IEEE Transactions on Neural Networks 15(3) (2004) 653-662
13. Li, S.: Fuzzy Control, NeuroControl and Intelligent Cybernetics. Haerbin Industrial University Press, China (1998)

Adaptive Fuzzy Neural Network Control for Transient Dynamics of Magneto-rheological Suspension with Time-Delay*

Xiaomin Dong, Miao Yu, Changrong Liao, Weimin Chen, Honghui Zhang, and Shanglian Huang

Center for Intelligent Structures, Dept. of Optoelectronic Engineering, Chongqing University, 400044, Chongqing, China
dongxiaomin99@tom.com

Abstract. Since Magneto-rheological (MR) suspension has nonlinearity and time-delay, the application of linear feedback strategy has been limited. This paper addresses the problem of control of MR suspension with time-delay when transient dynamics are presented. An adaptive Fuzzy-Neural Network Control (FNNC) scheme for the transient course is proposed using fuzzy logic control and artificial neural network methodologies. To attenuate the adverse effects of time-delay on control performance, a Time Delay Compensator (TDC) is established. Then, through a numerical example of a quarter car model and a real road test with a bump input, the comparison is made between passive suspension and semi-active suspension. The results show that the MR vehicle with FNNC strategy can depress the peak acceleration and shorten the setting time, and the effect of time-delay can be attenuated. The results of road test with the similarity of numerical study verify the feasibility of the control strategy.

1 Introduction

Recently, semi-active vehicle suspension with MR dampers has attracted more attention for its ability to improve the ride comfort [1]. However, MR suspension system always exhibits nonlinearity and time-delay like other semi-active suspension [2]. When MR vehicle runs across a bump, one may feel violent jerk. The time span of vehicle running across the bump is so short that it is necessary to consider the adverse effect of time-delay on control performance. Therefore, a Time Delay Compensator (TDC) is adopted in this paper to solve the problem. Some intelligent control strategies have been used to control the semi-active system [3]. Among them, the adaptive Fuzzy-Neural Network Control (FNNC) strategy has attracted increasing attention essentially because it can provide a powerful learning technique for complex unknown plants to perform complex tasks in highly nonlinear dynamical environment, and can also have available quantitative knowledge from repetitive adjustment of the system with better performance than those of fuzzy control with constant rules bases [4], [5]. Hence, the FNNC strategy is also proposed to control MR suspension system.

* This work was supported by the National Fund of Science (60574074 and 20048414).

J. Wang et al. (Eds.): ISNN 2006, LNCS 3972, pp. 1046 – 1051, 2006.

2 Adaptive Fuzzy-Neural Network Controller Design with Time-Delay Compensation for MR Suspension System

A schematic of the adaptive fuzzy-neural control system with time delay compensation is shown in Fig.1. The controller consists of two parts. One is the FNNC, which calculates the control force according to error and the change of the error; the other is the TDC, which is the neural network model of the predictor for the MR suspension without time-delay.

If state variables of the quarter car model in Fig.1 are defined as: $x = [x_1\ x_2\ x_3\ x_4]^T = [z_s - z_u\ \ \dot{z}_s\ \ z_u - w\ \ \dot{z}_u]^T$, then a standard state spaces form without time-delay can be written:

$$\dot{x} = Ax + Bu + L\dot{w} \tag{1}$$

where u is the control damping force of MR damper [6].

If considering the time-delay of MR suspension system, the control force is $u' = u(t - t_d)$, t_d is time-delay. Then Equation (1) can be written:

$$\dot{x} = Ax + Bu(t - t_d) + L\dot{w} \tag{2}$$

The MR suspension with time-delay can also be expressed as $G(s)e^{-\tau s}$, where $G(s)$ is a transfer function which can be transformed through Equation (1), and the input is control damping force of MR damper, the output is vertical vibration acceleration of car body, the disturbance is road input.

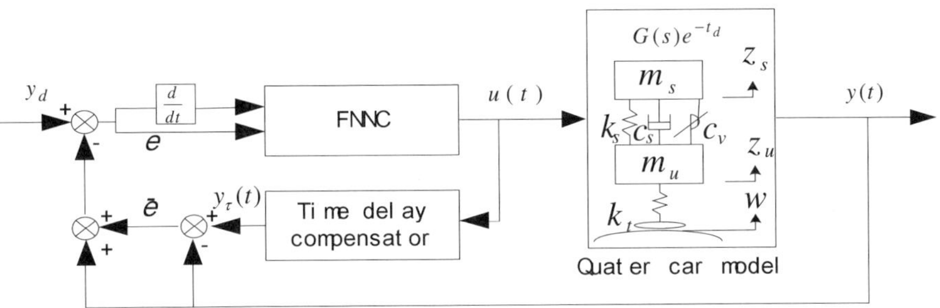

Fig. 1. Fuzzy-neural network control system with time-delay compensator

2.1 FNNC Design

As shown in Fig.2, the FNNC system is characterized by premise and consequence. The premise consists of two layers, the first layer is the fuzzy input linguistic vector $(e, \Delta e)^T$ and the second is to fuzz the input linguistic vector, which contains 14

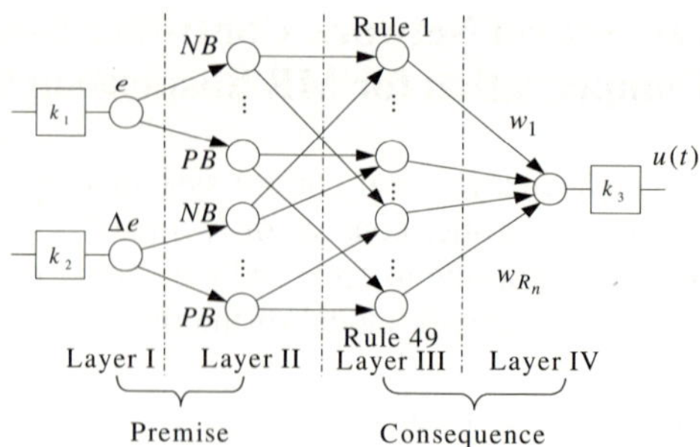

Fig. 2. Neural network model of FNNC

neurons corresponding to 14 fuzzy sets of two linguistic variables. Seven fuzzy sets of one input linguistic variable are $\{NB, NM, NS, ZE, PS, PM, PB\}$. The Gaussian membership functions with equal width intervals of the means are thus proposed to eliminate the sharp boundary and are defined:

$$A_{ij}(z) = \exp(-(z_k - \bar{z}_k)/2\sigma_k), k = 1,2...,2l \tag{3}$$

In which, $l = 7$, $\bar{z}_k$ and σ_k are the mean and variance of the Kth Gaussian membership function $\mu_k(z)$, respectively.

The third layer contains 49 neurons to realize fuzzy reasoning. The output from the FNNC system in the last layer can be expressed as:

$$u = \sum_{i=1}^{R_n} \mu_i w_i / \sum_{i=1}^{R_n} \mu_i = \sum \overline{\mu_i} w_i \tag{4}$$

where $R_n = 49$ and $\mu_i = \prod A_{i1}(e) * A_{i2}(\Delta e)$.

Weights of the last layer are tuned by back propagation algorithm. The error between expected output $y_d(t)$ and actual output $y(t)$ with time-delay is:

$$e = \frac{1}{2}\sum (y(t) - y_d(t))^2 \tag{5}$$

Then weights modified as:

$$w_i(t+1) = w_i(t) - \eta \frac{\partial e}{\partial w_i(t)} + \xi \Delta w_i(t) \tag{6}$$

in which η and ξ are learning factor and momentum factor.

To achieve good control performance, a Simulink model is formulated based on the equation (2) with a bump input, which is acquired by measuring the real bump road signal, and used in the teaching signal generation. The genetic algorithm is adopted to search out the best control force of MR damper that minimizes the fitness function (7).

$$Fit = a\sqrt{(\int_0^T \ddot{x}_s^2 dt)/T} + (1-a)\sqrt{(\int_0^T \ddot{x}_u^2 dt)/T} \tag{7}$$

where a represents the weighting factor, T is simulation time.

The control rules and the shape of each input/output membership function are tuned by learning from the teaching signal generated by the genetic algorithm. A trade-off is assumed between ride comfort and stability of the car body to choose an ideal control force.

2.2 Time-Delay Compensation

The principle of the TDC is based on the smith predictor [7]. The compensator can be realized using a back propagation network with four layers and nodes N_1-N_2-N_3-N_4 (In this paper, N_1=4, N_2=8, N_3=8, N_4=1). The mapping relationship of the model is described as:

$$y_M(t+1) = f_M(u(t-t_d), u(t-t_d-1), ..., u(t-t_d-m), y(t), ..., y(t-n)) \tag{8}$$

where m and n denote the orders of MR suspension system, and defining:

$$s^T = (s_1, ..., s_{N1})^T = (u(t-t_d), u(t-t_d-1), ... u(t-t_d-m), y(t), ..., y(t-n))^T$$

The activation functions in the last three layers of the TDC are respectively:

$$f_{1j} = 1/\{1+\exp[-(\sum_{j=1}^{N_1} \theta_{1ij} x_i + q_{1j})\} \text{ , } j = 1, ..., N_2 \tag{9}$$

$$f_{2k} = 1/\{1+\exp[-(\sum_{j=1}^{N_2} \theta_{2jk} f_{1j} + q_{2k})]\} \text{ , } k = 1, ..., N_3 \tag{10}$$

$$y_M(t+1) = 1/\{1+\exp[-(\sum_{k=1}^{N_3} \theta_{3kl} f_{2k} + q_{3l})]\} = f_M(f_{2k}(f_{1j})) \tag{11}$$

The output of conventional control system with the time-delay t_d is

$$y(t+1) = L^{-1}\{G(s)e^{-t_d s}\} = f(u(t-t_d), u(t-t_d-1), ... u(t-t_d-m), y(t), ... y(t-n)) \tag{12}$$

The output without time-delay is

$$y_\tau(t+1) = L^{-1}\{G(s)\} = f_\tau(u(t), ..., u(t-m), y_\tau(t), ..., y_\tau(t-n)) \tag{13}$$

where $L^{-1}\{\bullet\}$ is the inverse Laplace transform.

The predicted value of the output of the MR suspension system without time-delay is given by the neural network model and used to realize a compensating control. The network is trained by the sequence of the input-output samples. Using the same mapping network as (4), we can obtain the predicted value of the output without time-delay as follows:

$$y_{\tau}(t+1) = f_{M\tau}(u(t), u(t-1), ..., u(t-m), y_{\tau}(t), ..., y_{\tau}(t-n)) \tag{14}$$

The compensating error is

$$\bar{e} = y_{\tau}(t+1) - y(t+1) \tag{15}$$

Off-line and on-line learning algorithms can be used to modify the weights of TDC. The off-line learning results of the compensator can be used as a reference model of MR suspension system. The weights are modified by the index (15) using the principle of error gradient descent.

3 Simulation and Road Test

To verify our effort, the time response of passive suspension and MR suspension are compared when running across a bump through simulation and road test. The nominal parameters based on some saloon car are k_s=15000*N/m*, k_t=116900*N/m*, m_u=25.9*Kg*, m_s=264.2*Kg*,c_s=1000*Ns/m*. Other parameters used in the simulation of the genetic algorithm are population size=100, mutation probability=0.2, crossover probability=0.1. A control system based on dSPACE is fabricated. Four MR shock absorbers are used to replace the passive ones. Four accelerometers are placed on carriage's foursquare floor to record the vertical accelerate signal of sprung mass. Other four accelerometers are placed on two axis of vehicle to record the vertical acceleration signal of unsprung mass. Every seat has a passenger to simulate the condition of full load. The test car is driven straight down an arc road with the same dimension of the simulation at speed (20km/h).

The results of simulation and road test are shown in Fig.3-4 respectively. The result of road test is similar with the simulated. MR suspension employing FNNC with TDC or FNNC without TDC reduces the vertical vibration acceleration and adjusting time of car body compared to the passive suspension. The adjusting time of two strategies is almost the same. The FNNC with TDC is more effective than the FNNC without TDC in improving the ride comfort. The FNNC with TDC can reduce maximum peak-to peak acceleration 42.3%, but that without TDC can only reduce

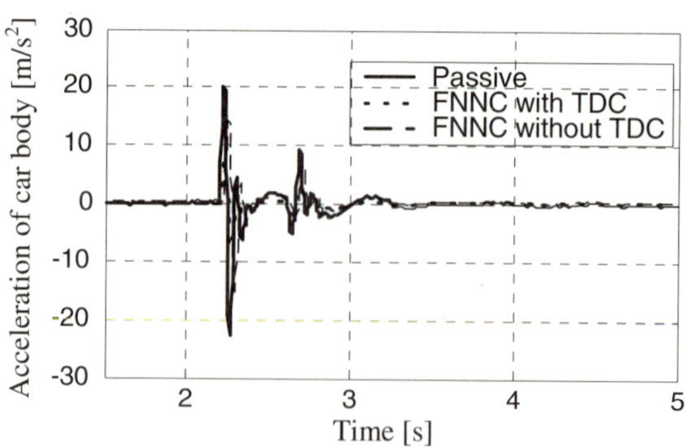

Fig. 3. Acceleration time history of car body (simulation)

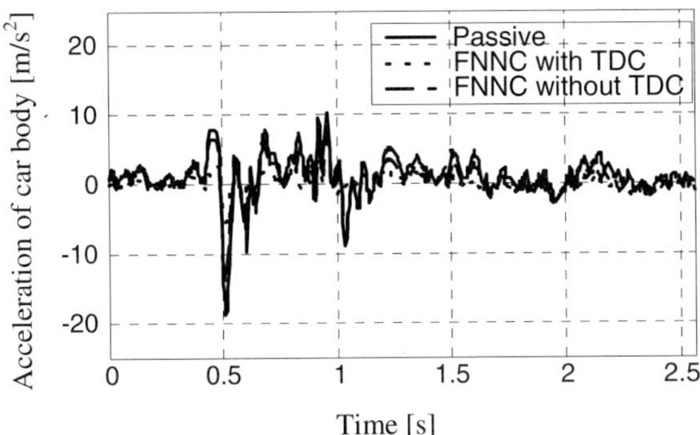

Fig. 4. Acceleration time history of car body (road test)

21.4%. Due to model error of simplification, it can also be seen that the experimental data is smaller than the simulated and experimental arc road is superposed on random road.

4 Conclusion

In this paper, a FNNC with TDC is proposed to control MR suspension with time-delay when running across a bump. The performances of the MR suspension system with FNNC strategy are evaluated through computer simulation and road test. Both the result of simulation and that of road test shows that the MR suspension system using FNNC can substantially reduce vertical peak acceleration of car body, shorten adjusting time and improve ride comfort. The MR suspension system using FNNC with TDC can achieve better control performance than that using FNNC without TDC.

References

1. Lee, H.S., Choi, S.B.: Control and Response Characteristics of a Magneto-rheological Fluid Damper for Passenger Vehicles. J. Intell. Materials Syst. and Structures 11 (2000) 80-87
2. Dong, X.M., Yu, M., et al.: Half Car Magneto-rheological Suspension System Accounting for Nonlinearity and Time-delay. Int'l J. of Modern Physics B 19(7-9) (2005) 1381-1387
3. Bani-Hani, K., Ghoboussi, L.: Nonlinear Structure Control Using Neural Networks. Journal of Engineering Mechanics, ASCE. 124 (1998) 319-327
4. Lin, F. J., Shyu, K. K., Wai, R. J.: Recurrent Fuzzy Neural Network Sliding Mode Controlled Motor-toggle Servomechanism. IEEE/ASME Trans. Mechanics 6(4) (2001) 453-466
5. Tong, S., Tong, J., Wang, T.: Fuzzy Adaptive Control of Multivariable Nonlinear System. Fuzzy Sets and Systems 111 (2000) 153-167
6. Yu, M., Huang, S.L., et al.: Research on Control Method for MR Damper. Chinese Journal of Chemical Physics, 14 (5) (2001) 606-612
7. Smith, O. H.: Closed Control of Loops with Dead Times. Chem. Eng. Prog. 53(6) (1957) 217-219

Adaptive Fuzzy Basis Function Network Based Fault-Tolerant Stable Control of Multi-machine Power Systems

Youping Fan[1], Yunping Chen[1], Shangsheng Li[1], Qingwu Gong[1], and Yi Chai[2]

[1] Faculty of Electrical Engineering, Wuhan University, Wuhan 430072, Hubei, China
Fyoupingnxinrong@yahoo.com.cn
[2] Automation College, Chongqing University, Chongqing 400044, China
chaiyi@cqu.edu.cn

Abstract. An approach base on an adaptive fuzzy basis function network (AFBFN) is presented for fault-tolerance treatment in uncertain power systems. The uncertain system is composed of unknown part and known part represented by a mathematical model. A fuzzy basis function network (FBFN) is trained offline to represent the model of unknown part. AFBFN is trained online to represent the unknown model included the unknown fault. The reference model is composed of the known mathematical model and FBFN. According to outputs of actual system, AFBFN and reference model, another AFBFN is adopted to complete the fault-tolerance process and obtain the feedback control input of the uncertain system, which makes the actual system to track output of the reference model. A simulation example of the multi-machine coupling power systems is given to validate the method. The result proved its effectiveness.

1 Introduction

In recent years, some nonlinear control methods using modern control theory such as adaptive control, sliding mode control and nonlinear H_∞ control have been introduced into power system control, but they are not effective for more complicated multi-machine coupling power system [1]. However, recent research indicates that the nonlinear system can be approached by fuzzy neural network, which has broad application in control area based on the ability of nonlinear processing and self-study, self-organization, self-adaptation. Therefore fuzzy neural network control of nonlinear system has aroused people's attention. Reference [2] applied the fuzzy relation matrix of neural network learning fuzzy inference system to reduce the dependence of fuzzy system on prior knowledge. Reference [3,4] adopted the network of Fuzzy ART map to isolate fault and identify the size of fault and its occurrence time. Reference [5] added fuzzy layer before neural network, which described the measured data with fuzzy set and improved the robustness of uncertainty and inaccurate measurement. Reference [6] put forward FBFN network based on T-S fuzzy model and applied it to unknown system to solve the problem of fault detection without taking the fault-tolerant problem into consideration. Reference [7,8] presented a fault detection method of dynamic system and implemented fault-tolerant using

J. Wang et al. (Eds.): ISNN 2006, LNCS 3972, pp. 1052 – 1061, 2006.

adaptive fuzzy neural network, which assumed the system model to be known, but it is difficult to construct exact model for most systems in practice.

This paper proposes a new fault-tolerant method to deal with uncertain system, which adopts FBFN and AFBFN to solve the fault-tolerance of uncertain systems with unknown mathematical model. The uncertain system can be divided into two parts: one part whose model is already known and the other unknown. The former is described by known mathematical model while the latter such as uncertainty of system and noise interference is described by unknown model. Here, FBFN is adopted to offline evaluate the model's unknown part when system is in normal condition and make the combination of mathematic model of known part and FBFN as the reference model of the fault-tolerance system. Online evaluate the unknown model of system using AFBFN and get the feedback input through fault-tolerance using another AFBFN network. Thus make the practical system to follow the expected output given by reference model. Finally, it is applied to the power system stability control with two machining coupling, validating its effectiveness.

2 FBFN Structure of Multi-machine Coupling Power System

The power system coupled by $\boldsymbol{m}$ subsystems can be described as:

$$x_{in_i} = F_i(\boldsymbol{X}, u_i) + d_{id}\,;\quad y_i = x_{i1} \qquad (i = 1,2,\cdots,m) \tag{1}$$

where $\boldsymbol{x}_i \in R^{n_i}$, $\boldsymbol{x}_i = [x_i, \dot{x}_i, \cdots x_i^{(n_i-1)}]^T = [x_{i1}, x_{i2}, \cdots, x_{in_i}]^T$ is the status vector of the i_{th} subsystem, u_i , y_i and d_{id} are the control input, control output and the bounded disturbances of the i_{th} subsystem separately.

FBFN is a type of neural network composed by fuzzy basis function based on ***T-S*** model, which has the ability to describe system fuzziness as T-S model and self-learning ability as neural network. FBFN adopting adaptive learning algorithm is called AFBFN. As $F_i(\boldsymbol{X}, u_i)$ is unknown in general, AFBFN is adopted to solve the fault-tolerant problem of uncertain system.

The ***T-S*** model is composed by a series of rules such as "If.... Then...." as follows:

$L^{(l)}$: If z_1 is A_1^l,..., z_p is A_p^l, Then y is y^l.

Where A_i^l is the fuzzy set at $U_i \subset R$, $i = 1,2,\cdots,p$, $z = (z_1, z_2, \cdots, z_p)^T \in U_1 \times U_2 \times \cdots \times U_p$ as antecedent variable $y^l = (y_1^l, y_2^l, \cdots, y_n^l)^T$ is the output of rule $L^{(l)}, l = 1,2,\cdots,m$, m is the total number of the rules.

The output **y** is the weighted mean of the output y^l in every rule:

$$y(z) = \sum_{l=1}^{m} \beta^l(z) \cdot y^l \tag{2}$$

where $\beta^l(z)$ should satisfy the following conditions:

$$\sum_{i=1}^{m} \beta^i(z) = 1; 0 \le \beta^i(z) \le 1, \quad i = 1, 2, \cdots, m \tag{3}$$

$$\beta^l(z) = \frac{1}{\sum_{l=1}^{m} \beta_l^*(z)} \cdot \beta_l^*(z) \tag{4}$$

$$\beta_l^*(z) = \prod_{j=1}^{n} \mu_{A_j^l}(z_j) \tag{5}$$

So the equation (2) can be described as:

$$y(z) = \frac{\sum_{l=1}^{m} y^l \left(\prod_{j=1}^{n} \mu_{A_j^l}(z_j)\right)}{\sum_{l=1}^{m} \left(\prod_{j=1}^{n} \mu_{A_j^l}(z_j)\right)} \tag{6}$$

Define the fuzzy basis function as:

$$F_l(z) = \frac{\prod_{j=1}^{n} \mu_{A_j^l}(z_j)}{\sum_{l=1}^{m} \left(\prod_{j=1}^{n} \mu_{A_j^l}(z_j)\right)} \qquad l = 1, 2, \cdots, m \tag{7}$$

Then the equation (6) is equivalent to following expansion of a fuzzy basis function:

$$y(z) = \sum_{l=1}^{m} F_l(z)\theta_l \tag{8}$$

where $\theta_l = y^l \in R^n$ is achieved through learning adjusting.

It is supposed that the membership function to be the Gaussian smoothing function:

$$\mu_{A_j^l}(z_j) = a_j^l \exp\left(-\frac{1}{2}\left(\frac{z_j - \bar{z}_j^l}{\sigma_j^l}\right)^2\right) \tag{9}$$

where a_j^l, $\bar{z}_j^l$ and σ_j^l are adjustable parameters. If we take a_j^l, $\bar{z}_j^l$ and σ_j^l as the design parameters of equation (8), the relationship between expansions of fuzzy basis function and parameters is nonlinear. If we make the above parameters fixed and take the parameter θ_l as adjustable parameter of equation (8), the relationship is linear.

3 Fault-Tolerant Method of Uncertain System

3.1 System Description

It is difficult or even impossible to construct exact mathematic model in actual project system. If the model between system input and output can be constructed, we can detect the system fault according to the information of input and output and then get the feedback input to enable the system to accommodate faults. System equation can be described as follows:

$$x(t)=\mathbf{h}_0[\mathbf{u}(t),\mathbf{x}(t-1),\cdots,\mathbf{x}(t-k)]+\mathbf{h}_1[\mathbf{u}(t),\mathbf{x}(t-1),\cdots,\mathbf{x}(t-k),\xi(t)] \tag{10}$$

where $\mathbf{x}(t)\in R^n$ represents system state, $\mathbf{u}(t)\subset R^p$ is the input of system, $\xi(t)$ is interference noise, and $\mathrm{h}_0(\cdot)$ is a known smooth nonlinear function which indicates the part can be modeled of system input and output. If there are faults in system, the model can be described as below:

$$x(t)=\mathbf{h}_0[\mathbf{u},\mathbf{x}]+\mathbf{h}_1[\mathbf{u},\mathbf{x},\xi(t)]+\mathbf{L}(t-t_0)\mathbf{f}(\mathbf{u},\mathbf{x}) \tag{11}$$

The notation $\mathbf{L}(t-t_0)\mathbf{f}(\mathbf{u},\mathbf{x})$ presents the unknown fault of system. $\mathbf{f}(\mathbf{u},\mathbf{x})$ is the abbreviation of $\mathbf{f}(\mathbf{u}(t),\mathbf{x}(t-1),\cdots,\mathbf{x}(t-k))$. $h_0(\mathbf{u},\mathbf{x})$ and $h_1(\mathbf{u},\mathbf{x},\xi(t))$ are the abbreviation of $h_0[\mathbf{u}(t), \mathbf{x}(t-1),\cdots,\mathbf{x}(t-k)]$ and $h_1[\mathbf{u}(t),\mathbf{x}(t-1),\cdots,\mathbf{x}(t-k),\xi(t)]$. $L(t-t_0)$ denotes system fault at the time of $t\geq t_0$:

$$\mathbf{L}(t-t_0)=\begin{cases}0 & t<t_0\\ 1 & t\geq t_0\end{cases} \tag{12}$$

When the system is completely unknown, the equation (11) can be rewritten:

$$x(t)=\mathbf{h}_1[\mathbf{u}(t),\mathbf{x}(t-1),\cdots,\mathbf{x}(t-k),\xi(t)]+\mathbf{L}(t-t_0)\mathbf{f}(\mathbf{u},\mathbf{x}) \tag{13}$$

3.2 The Fault-Tolerant System

The known model is the known mathematic model $\mathbf{h}_0(\mathbf{u},\mathbf{x})$ of practical system, while the estimated model is the estimation $\hat{\mathbf{h}}_1[\mathbf{u},\mathbf{x},\xi(t)]$ of system unknown model based on FBFN learning. The notation $\mathbf{u}_m$ is the reference input. $\mathbf{x}$ is the practical output. $\mathbf{u}_f$ is the feedback input. The known model and estimation model both constitute the reference model. We take the sum of $\mathbf{h}_0[\mathbf{u}_m,\mathbf{x}_m]$ and $\hat{\mathbf{h}}_1[\mathbf{u}_m,\mathbf{x}_m]$ under the reference input $\mathbf{u}_m$ as system reference output:

$$\mathbf{x}_m(t) = \mathbf{h}_0[\mathbf{u}_m, \mathbf{x}_m] + \hat{\mathbf{h}}_1[\mathbf{u}_m, \mathbf{x}_m] \tag{14}$$

The online estimation model is the unknown part model of system based on AFBFN estimation, which contains un-modeled part and unknown fault of system.

$$\hat{\mathbf{h}}(\mathbf{u}, \mathbf{x}) = \hat{\mathbf{h}}_1(\mathbf{u}, \mathbf{x}) + \hat{\mathbf{L}}(t - t_0)\hat{\mathbf{f}}(\mathbf{u}, \mathbf{x}) \tag{15}$$

where $\mathbf{u}=\mathbf{u}_m-\mathbf{u}_f$, $\hat{\mathbf{h}}_1(\mathbf{u},\mathbf{x})$ is the abbreviation of $\hat{\mathbf{h}}_1[\mathbf{u}(t),\mathbf{x}(t-1),\cdots,\mathbf{x}(t-k)]$. The system estimated output $\hat{\mathbf{x}}(t)$ is composed of the output of known model and AFBFN:

$$\hat{\mathbf{x}}(t) = \mathbf{h}_0(\mathbf{u}, \mathbf{x}) + \hat{\mathbf{h}}(\mathbf{u}, \mathbf{x}) \tag{16}$$

The feedback input $\mathbf{u}_f$ is gained through fault-tolerant AFBFN, which can be described as the mapping from network input to feedback control.

$$\mathbf{u}_f = \mathbf{g}(\mathbf{u}_m, \mathbf{x}_m, \hat{\mathbf{h}}, \mathbf{x}(t), \mathbf{x}(t-1), \cdots, \mathbf{x}(t-k)) \tag{17}$$

A. Off-line model estimation based on FBFN

The offline model estimation is the estimation of the part, which cannot be modeled when system is in norm condition, which can be acquired according to the sequence of input and output of norm system by applying FBFN to offline learning. We take the system input $\mathbf{u}=(u_1,u_2,\cdots;u_p)^T$ as antecedent variable of FBFN network. The network output is the estimated model output $\hat{\mathbf{h}}_1(\mathbf{u},\mathbf{x})$ and the least square multiplication is adopted as the learning method. From equation (10) we have:

$$\mathbf{h}_1(\mathbf{u}, \mathbf{x}) = x(t) - \mathbf{h}_0(\mathbf{u}, \mathbf{x}) \tag{18}$$

It is supposed there are N pairs of input-output data: $(u(k),x(k)),k=1,2,\cdots,N$, where $u(k)$ and $x(k)$ is the system input and output at the sampling time k. Supposing $y(t)=h_1(\mathbf{u},\mathbf{x})$. Taking the mode error into consideration, the output of equation (8) of FBFN can be rewritten as follows:

$$y(t) = \sum_{l=1}^{m} F_l(t)\theta_l + \xi(t) \tag{19}$$

where $\xi(t)$ indicates the error signal. $F_l(t)$ is taken as recession factor, weights θ_l of FBFN as adjustable parameter vector, and we get $\hat{\theta}_l$ by adopting the least square multiplication.

Then the equation (19) from $k=1$ to N can be rewritten as:

$$Y = F\theta + e \tag{20}$$

where $Y=[y(1),y(2),\cdots,y(N)]^T$, $F=[\phi_1,\phi_2,\cdots,\phi_m]$, $\phi_i=[F_i(1),F_i(2),\cdots,F_i(N)]^T$, $\theta=[\theta_1,\theta_2,\cdots,\theta_m]^T$, $e=[\xi(1),\xi(2),\cdots,\xi(N)]^T$.

After determination of the parameters a_j^l 、 $\bar{z}_j^l$ and σ_j^l of fuzzy basis function $F_l(t)$, a sequence of input-output pairs is obtained. Then we can acquire the parameter matrix $\hat{\theta}$ according to the principle of least square method. Thus we can get the output $\hat{\mathbf{h}}_1(\mathbf{u},\mathbf{x})$ of FBFN from the input:

$$\hat{\mathbf{h}}_1(\mathbf{u},\mathbf{x}) = \sum_{l=1}^{m} F_l(t)\hat{\theta}_l \tag{21}$$

B. On-line model estimation based on AFBFN

Online model estimation is to evaluate the unknown part (containing un-modeled part and unknown fault part) of system online by applying AFBFN. The AFBFN also takes the input $u = (u_1, u_2, \cdots, u_p)^T$ as antecedent variable, the output $\hat{\mathbf{h}}(\mathbf{u},\mathbf{x})$ is as follows:

$$\hat{\mathbf{h}}(\mathbf{u},\mathbf{x}) = \sum_{l=1}^{m} F_l \theta_L \tag{22}$$

Describe the parameter θ_l is described as the linear combination of system input $u(t)$:

$$\theta_l = Q^l u(t) \tag{23}$$

where Q^l is the $n \times p$ -dimensional parameter matrix corresponding to rule l. Then the equation (22) can be given as:

$$\hat{\mathbf{h}}(\mathbf{u},\mathbf{x}) = \sum_{l=1}^{m} F_l Q^l u(t) \tag{24}$$

The matrix Q is composed of parameters of all rules is given:

$$Q = \begin{bmatrix} Q^1 & Q^2 & \cdots & Q^m \end{bmatrix}^T \tag{25}$$

Cost function is defined as:

$$E = \frac{1}{2} r^T r \tag{26}$$

According to the equation (11) and (16), the following can be obtained:

$$r = x - \hat{x} = \mathbf{h}_1(\mathbf{u},\mathbf{x}) + L(t - t_0) f_1(\mathbf{u},\mathbf{x}) - \hat{\mathbf{h}}(\mathbf{u},\mathbf{x}) \tag{27}$$

The steepest descent method in parameter updating is adopted, which can be expressed as:

$$\dot{Q} = -\eta \frac{\partial E}{\partial Q} \tag{28}$$

Therefore

$$\dot{Q}_{ij}^{l} = -\eta \cdot \frac{\partial r^{T}}{\partial Q_{ij}^{l}} \cdot \frac{\partial E}{\partial r} \tag{29}$$

from equation （24） the following can be obtained

$$\hat{\mathbf{h}}(\mathbf{u},\mathbf{x}) = F_1 Q^1 u + F_2 Q^2 u + \cdots + F_m Q^m u \tag{30}$$

From equation （27） and （31） we get

$$\frac{\partial r^T}{\partial Q_{ij}^l} = -\frac{\partial \hat{h}^T}{\partial Q_{ij}^l} = \begin{bmatrix} 0 & \cdots & -F_l u_j & \cdots & 0 \end{bmatrix} \tag{31}$$

The right part is a n-dimensional row vector with $-F_l u_j$ as the i_{th} element while the others to be 0. So the equation (29) can be further expressed

$$\dot{Q}_{ij}^{l} = \eta . F_l u_j \cdot (x_i - \hat{x}_i) \tag{32}$$

where x_i is the i_{th} element of state vector. The parameter Q of fault model online is adjustable according to the equation (31). So the unknown model $\hat{\mathbf{h}}(\mathbf{u},\mathbf{x})$ of the system is established based on AFBFN according to equation (22).

C. Fault-tolerant AFBFN

Fault-tolerant AFBFN aims to design a AFBFN controller to make the practical system trace the output (expected output) of the reference model. The antecedent variable of fault-tolerant AFBFN represents reference input u_m, system output x, the output $\hat{\mathbf{h}}(\mathbf{u},\mathbf{x})$ of online model estimation AFBFN, and reference output x_m. The output is u_f. Similar to the online evaluation model AFBFN. The feedback control input is acquired according to the below equation:

$$u_f(t) = \sum_{l=1}^{m} F_l(Z)\theta_l^u \, , \; \theta_l^u \in R^p \tag{33}$$

The cost function is still $E = \frac{1}{2} r^T r$, where $r = x_m - x$.

Let $\theta^u = \begin{bmatrix} \theta_1^u & \theta_2^u & \cdots & \theta_m^u \end{bmatrix}^T$, update the parameter matrix based on below equation: where ψ indicates the learning rate. From above equation, giving

$$\dot{\theta}_{ij}^{u} = -\psi \frac{\partial E}{\partial \theta_{ij}^{u}} = -\psi \frac{\partial r^T}{\partial \theta_{ij}^{u}} \frac{\partial E}{\partial r} \tag{35}$$

where θ_{ij}^{u} is the i_{th} row and j_{th} list of element of matrix θ^{u}. The subscript i of θ_{ij}^{u} is the parameter vector θ_{i}^{u} corresponding to the i_{th} rule, and θ_{ij}^{u} represents the j_{th} element of parameter vector θ_{i}^{u}. From equation（35）,we get:

$$\frac{\partial u_{f}^{T}}{\partial \theta_{ij}^{u}} = \begin{bmatrix} 0 & \cdots & F_{i}(Z) & \cdots & 0 \end{bmatrix} \tag{36}$$

where $F_{i}(Z)$ is the j_{th} element of p-dimensional row vector in which the other elements are to be 0.

The following equation based on equation（36）and (35) is given:

$$\frac{\partial r^{T}}{\partial \theta_{ij}^{u}} = \frac{\partial u_{f}^{T}}{\partial \theta_{ij}^{u}}\left(\frac{\partial h_{0}^{T}}{\partial u} + \frac{\partial \hat{h}^{T}}{\partial u}\right) \tag{37}$$

Put equation $\frac{\partial x^{T}}{\partial u_{f}} = -\frac{\partial x^{T}}{\partial u}$ and $\frac{\partial x^{T}}{\partial u_{f}} = -\frac{\partial \hat{x}^{T}}{\partial u}$ into equation（34）, the parameter updating formula is expressed as follows:

$$\dot{\theta}_{ij}^{u} = -\psi \frac{\partial u_{f}^{T}}{\partial \theta_{ij}^{u}}\left(\frac{\partial h_{0}^{T}}{\partial u} + \frac{\partial \hat{h}^{T}}{\partial u}\right)(x_{m} - x) \tag{38}$$

where $\partial u_{f}^{T} / \partial \theta_{ij}^{u}$ is acquired from equation (36). The symbol $h_{0}(u,x)$ indicates the known mathematic model, and $\frac{\partial h_{0}^{T}}{\partial u}$ is easy to get. Equation（24）gives:

$$\frac{\partial \hat{h}^{T}}{\partial u} = \sum_{l=1}^{m} F_{l}(Q^{l})^{T} \tag{39}$$

The fault-tolerant AFBFN adjusts the network parameter online according to equation (16), and offer the feedback control u_{f} based on equation（32）.

4 Simulation Research

Taking the double-machine coupling power system $(i = 1,2)$ as example:

$$\begin{aligned}
\dot{x}_{i1} &= x_{i2} \\
\dot{x}_{i2} &= -\frac{T_{i}}{M_{i}}x_{i2} + \frac{1}{M_{i}}u_{i} + \sum_{j=1, j\neq i}^{2}\frac{E_{i}E_{j}Y_{ij}}{M_{i}}[\cos(\delta_{ij}^{0} - \theta_{ij}) - \cos(x_{i1} - x_{j1} + \delta_{ij}^{0} - \theta_{ij})] + d_{i} \\
y_{i1} &= x_{i1}
\end{aligned} \tag{40}$$

where $\boldsymbol{x}_i = [x_{i1}, x_{i2}]^T$ and y_i represents the state vector and output variable of subsystem separately, x_{i1} and x_{i2} indicates the rotor angle and rotor speed of the i_{th} generator, the other parameters are given in Reference [9]. Where,

$$F_i(\mathbf{X}, u_i) = f_i(\mathbf{X}) + g_i(\mathbf{X})u_i$$

$$f_i(X) = -\frac{T_i}{M_i}x_{i2} + \sum_{j=1, j\neq i}^{2} \frac{E_i E_j Y_{ij}}{M_i}[\cos(\delta_{ij}^0 - \theta_{ij}) - \cos(x_{i1} - x_{j1} + \delta_{ij}^0 - \theta_{ij})] \tag{41}$$

$$g_i(X) = \frac{1}{M_i}$$

5 fuzzy subsets for x_{i1} and x_{i2} separately, and so it needs 125 fuzzy rules to approach to $f_i(\mathbf{X})$. MATLAB is chosen as software in the simulation. Figure 1 shows the main outputs of generator 1and 2, that is, the output response curve for variation of frequency ($\Delta \mathrm{f}_{1,2}$), variation of power transfer ($\Delta \mathrm{Pe}$), and variation of turbine high-voltage output (ΔP_{t1}). They reflect the adjusting result of system after structure disturbance, hence demonstrate the effectiveness of the control.

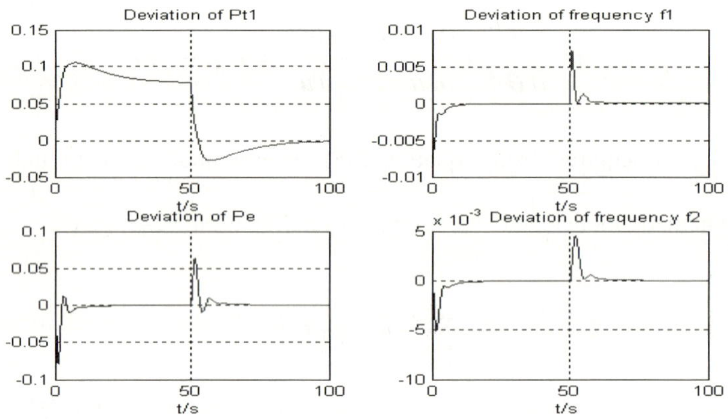

Fig. 1. Output response curves of the system to step load disturbance

5 Conclusion

An adaptive fuzzy neural network control scheme for multi-machine coupling power system with unknown model is presented. It can improve the robustness of control system by using adaptive compensator and meanwhile control the coupling system effectively based on the space-time approaching ability of fuzzy neural network. The simulation result shows that this method can process fault-tolerance of uncertain system, which enables the system to trace the output of reference model even in the case of fault. Adopting FBFN in the offline learning of system unknown part and utilizing AFBFN in online modeling of unknown system, another AFBFN is employed to solve the fault-tolerant problems of uncertainty system for the case unable to be modeled and partly unable to be modeled, which shows robustness for system uncertainty, imprecise measurement as well as system modeling error.

Acknowledgements

The research in this paper is supported by the National Natural Science Foundation of China (50477018) and the Post-doctoral Science Foundation of China (2004036135).

References

1. Xue, Y. J., Feng, R. P.: Adaptive Fuzzy Control of Space-Time Chaos in Continuous Coupling System. Acta Physica Sinica 50(3) (2001) 440-444
2. Wang, X. Z., Lu, M. L., McGreavy, C.: Learning Dynamic Fault Models Based on a Fuzzy Set Covering Method. Computers & Chemical Engineering 21(6) (1997) 621-630
3. Wang, Z. S.: Intelligent Fault-Tolerant Technology and Its Application. National Defense Industry Press, Peking China (2002) 10-24
4. Lin, C. J., Lin, C. T.: Reinforcement Learning for an ART-Based Fuzzy Adaptive Learning Control Network. IEEE Transactions on Neural Networks 7(3) (1996) 709-731
5. Wang, Y. H., Liu, Z. J., Li, H. G: Fault Diagnosis with Hybrid Integrated Neural Network. Journal of BeiJing University of Chemical Technology 30(1) (2003) 65-68
6. Wang, H., Mao, Z. Y.: Robot Intelligent Control. National Defense Industry Press, Peking China (2002) 1-34
7. AI-Jarrah, O. M., AI-Rousan, M.: Fault Detection and Accommodation in Dynamic Systems Using Adaptive Neurofuzzy Systems. IEE Proceedings-Control Theory and Applications 148(4) (2001) 283-290
8. Fan, Y. P., Huang, X.Y.: Adjusting Controlling on Gassing and Energy for High Power Excimer Laser Based on Intelligence. Control Theory and Application 19(4) (2002) 561-566
9. Lu, Q., Sun, Y., Mei, S.: Nonlinear Control Systems and Power System Dynamics. Kluwer Academic Publishers (2001)

Simulation Research on Applying Fault Tolerant Control to Marine Diesel Engine in Abnormal Operation

Xiao-Yan Xu[1], Min He[1], Wan-Neng Yu[2], and Hua-Yao Zheng[2]

[1] Electrical Engineering Department, Shanghai Maritime University 1066#, Shanghai 200135, China
{xyxu, minhe}@cen.shmtu.edu.cn
[2] Marine Simulation Center, Shanghai Maritime University, Shanghai 200135, China
{wnyu, hyzheng}@msc.shmtu.edu.cn

Abstract. Study of maintaining safe operation of marine main engine while one or several faults occur in marine systems has been put a high value recently in field of shipbuilding industry. The paper establishes a dynamic mathematical model of marine main engine system and analyses its characteristics while it is in abnormal conditions. Afterwards, a fault tolerant control system with artificial neural network algorithm is designed for improvement of operation of main engine that is in abnormal operation under consideration that a fault tolerant control system is able to fulfill function of fault toleration. The results of simulation experiments show that this fault tolerant control system is suitable for safe operation of marine main engine system.

1 Introduction

The operation state of marine main engine system has vital effect on determination of whether the ship is in safe navigation or not since the main engine is the main propulsion device in a ship. So, it is important to improve the reliability of main engine operation while the main engine is in abnormal condition. But marine main engine system is a nonlinear time-dependant system, and the parameters of its mathematical model vary in a wide range while its speed, load or other operational conditions vary. Furthermore, most traditional control systems are designed basing on the assumption that both sensors and actuators operate with proper function, and without consideration of whether a main engine system still operates properly or not when the main engine is in abnormal condition [1]. Under that consideration, it is difficult to control marine main engine system with traditional control system when the main engine operates in abnormal conditions. Recently, fault tolerant control theory has been developed quickly. If fault tolerant control method is applied, the system maintains the original performance, or if necessary, lowers the original performance in some measure to a certain degree, but operates without danger when one or more faults occur in the system since a fault tolerant control system has the ability of fault toleration [2].

In this paper, a fault tolerant control system based on ANN (artificial neural network) algorithm is applied to optimize the operation of marine main engine system when it is in abnormal conditions.

J. Wang et al. (Eds.): ISNN 2006, LNCS 3972, pp. 1062 – 1067, 2006.

2 Mathematical Model of Main Diesel Engine System

Figure 1 shows the block diagram of mathematical model of diesel engine [3].

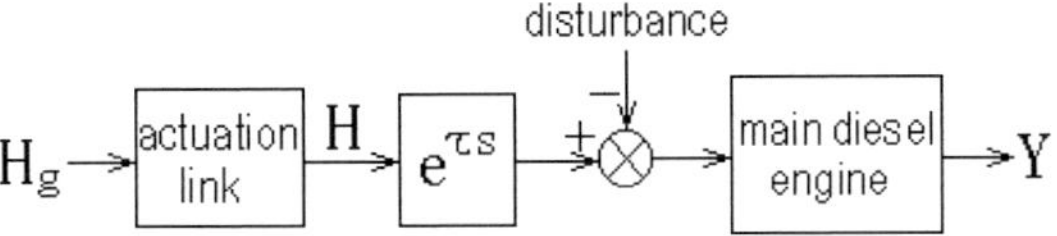

Fig. 1. Mathematical model of main diesel engine

Generally speaking, the actuator of marine main diesel engine is a DC servomotor which is used for the control of diesel engine governor, and it is a typical second order system whose transfer function is described as,

$$\frac{H(S)}{H_g(S)} = \frac{\varpi_{nd}^2}{S^2 + 2\xi_d \varpi_{nd} S + \varpi_{nd}^2} . \tag{1}$$

where, *Hg(S)* is the given extent of throttle opening of oil link; *H(S)* is the actual extent of throttle opening of oil link (that is, the output of the governor); ω_{nd} is the natural oscillation frequency of actuator; ξ_d is the damping coefficient of actuator.

Meanwhile, the differential equation of the main diesel engine if without turbocharger is,

$$T_a \frac{dy(t)}{dt} - y(t) = \eta(t-\tau) - \lambda(t) \tag{2}$$

where, $y(t)$ is the rotation speed of main diesel engine; $\eta(t)$ is the extent of throttle opening of oil link; $\lambda(t)$ is the disturbance; T_a is the time constant of main diesel engine; τ is the time delay of throttle opening of oil link. And the corresponding Laplace's equation is described as,

$$\frac{Y(S)}{e^{-\tau S} H(S) - \lambda(S)} = \frac{1}{T_a S + 1} . \tag{3}$$

where, $Y(S)$ is the Laplace transform of $y(t)$; $H(S)$ is the Laplace transform of $\eta(t)$; $\lambda(S)$ is the Laplace transform of $\lambda(t)$.

For the same reason, the transfer function of main diesel engine with turbocharger is described as,

$$\frac{Y(S)}{e^{-\tau S} H(S) - \lambda(S)} = \frac{\varpi_n^2}{S^2 + 2\xi \varpi_n S + \varpi_n^2} . \tag{4}$$

where, ω_n is the non-damping natural frequency of main diesel engine; ξ is the damping factor of main diesel engine.

Therefore, the transfer function of a whole main diesel engine system is obtained if without the consideration of disturbance. The transfer function of main diesel engine without turbocharger shows in figure 5, and with turbocharger in equation 6.

$$\frac{Y(S)}{H_g(S)} = \frac{\varpi_{nd}^2}{S^2 + 2\xi_{nd}\varpi_{nd}S + \varpi_{nd}^2} \cdot \frac{e^{-\tau S}}{T_a S + 1} \cdot \tag{5}$$

$$\frac{Y(S)}{H_g(S)} = \frac{\varpi_{nd}^2}{S^2 + 2\xi_{nd}\varpi_{nd}S + \varpi_{nd}^2} \cdot \frac{\varpi_n^2}{S^2 + 2\xi_n\varpi_n S + \omega_n^2} \cdot e^{-\tau S} \cdot \tag{6}$$

To improve the performance of main diesel engine system, a traditional PID (proportional-integrated-differential) controller is applied to regulate the rotation speed of main diesel engine. The block diagram of the main diesel engine system is shown in figure 2.

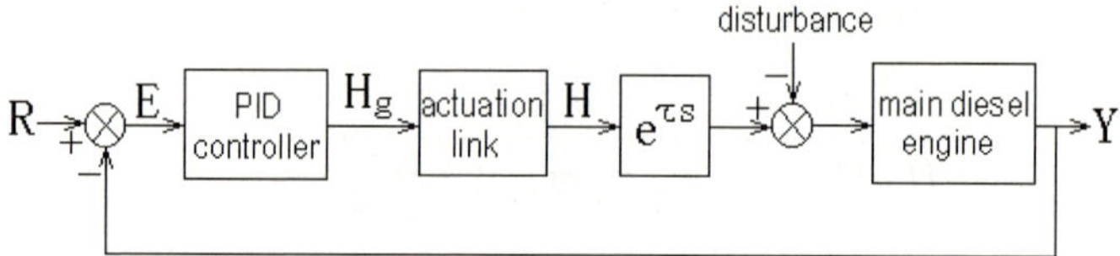

Fig. 2. Block diagram of main diesel engine system with PID controller

In figure 2, the transfer function of the PID controller is,

$$\frac{H_g(S)}{E(S)} = K_p\left(1 + \frac{1}{T_i S} + T_d S\right) \cdot \tag{7}$$

According to the rules of optimization design, parameters of the PID controller are chosen as: $K_p = 30$, $T_i = 1.5$s, $T_d = 0$. Figure 3 shows the speed response of the main diesel engine when a unit step input is applied on the system, and the parameters of the main diesel engine are: $T_a = 2.7$s, $\tau = 0.037$s, $\xi_{nd} = 0.707$, $\omega_{nd} = 35.4$. Where, curve 1 shows the speed response without the application of PID controller to the system and curve 2 shows the speed response after the application of PID controller.

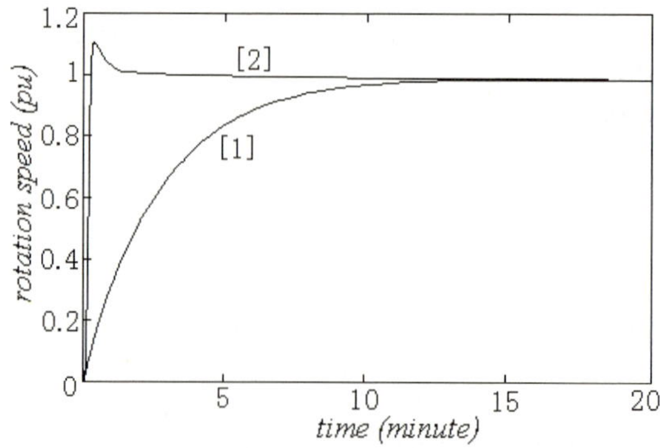

Fig. 3. Speed responses of main diesel engine before and after the application of PID controller

Obviously, the application of PID controller in engine system improves the performances of the main diesel engine.

3 Mathematical Model of Engine System in Abnormal Conditions

Like other industrial systems, faults sometimes occur in marine main diesel engine. For example, main engine starting failure, main engine fault stopping and auto-decreasing of the rotation speed, and so on.

In this paper, rotation speed auto-decreasing of marine main engine is selected for the study of the fault tolerant system of marine main engine. The rotation speed of main engine decreases under the following five conditions [4]: (1) cooling water inlet pressure low, (2) cooling water outlet temperature high, (3) lubricating oil inlet pressure low, (4) lubricating oil inlet temperature high, (5) fuel oil inlet pressure low.

When one or more than one of the abovementioned situations occur, the transfer function of the main diesel engine varies. In this simulation experiment, the transfer function of the diesel engine changes into,

$$\frac{Y(S)}{H_g(S)} = \frac{1250}{S^2 + 50S + 1250} \cdot \frac{0.7e^{-0.5S}}{0.4S^2 + 1.4S + 1} \ . \tag{8}$$

Consequently, the transient response of the main diesel engine system (including the PID controller) changes a lot, which is shown in figure 4. In figure 4, curve 1 shows the rotation speed response of the system while the fault occurs (only PID control system is applied), curve 2 shows the rotation speed response of a normal system, and curve 3 shows the oil throttle opening response of the main diesel engine. It is observed that the system is unstable because of the fault. Therefore, it is necessary to adopt the fault tolerant control to make the system stable in abnormal conditions.

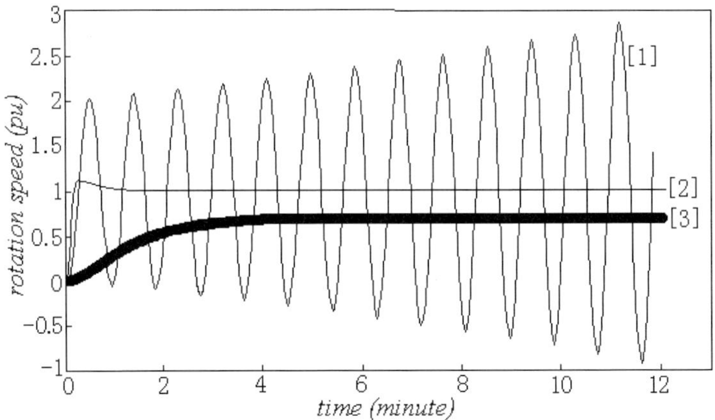

Fig. 4. Rotation speed response of main diesel engine while the main diesel engine system is in abnormal conditions

4 Fault Tolerant Control of Main Engine System

Due to the aforementioned analysis, a fault tolerant control system is applied to the marine main diesel engine system as shown in figure 5. Firstly, the fault-diagnosis with ANN algorithm detects the fault conditions. This link is an important one since the function of fault tolerant control is weak if not connected with fault detection and diagnosis unit [5, 6]. Secondary, fault tolerant control with ANN algorithm regulates the parameters of PID controller. To continue the aforementioned simulation experiment, the PID parameters are regulated to: K_p=14, T_i=1.5s, T_d=0.27s. The transient response after the application of fault tolerant control is shown in figure 6.

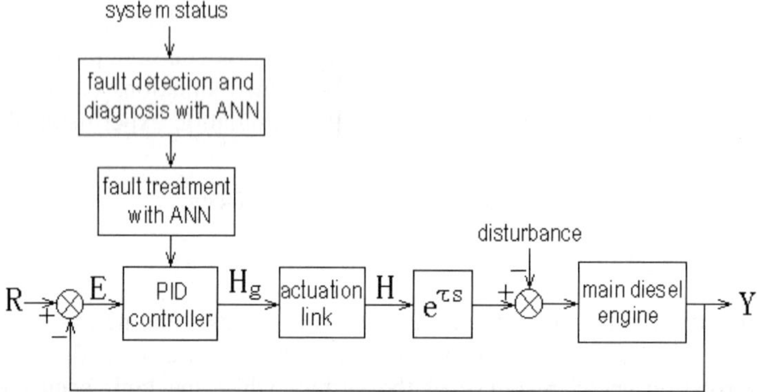

Fig. 5. Block diagram of marine main diesel engine system after the application of fault tolerant control system

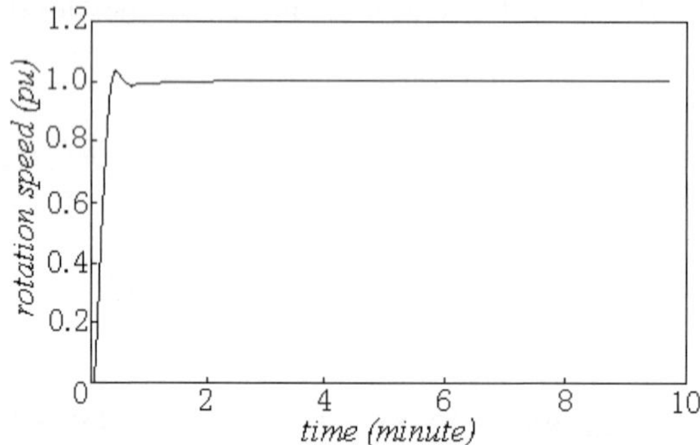

Fig. 6. Rotation speed response of marine main diesel engine system after the application of fault tolerant control system

5 Conclusion

To improve the operation of marine main diesel engine system, PID controller is able to optimize the performance of the system when the system is in normal conditions. But in abnormal conditions, that is, when faults occur, the performance of PID controller is not been satisfied, and it is necessary to adopt the fault tolerant control system to meet the requirements.

References

1. Chen, Y.: Marine Engine Room Intelligent Fault Tolerant Control System Based on Remote Communication. Master's Dissertation in Shanghai Maritime University. Shanghai, China (1999)
2. Liu, W.: Recent Development of Fault Tolerant Control of Dynamic System. J. Hebei Polytechnic Institute 19(1) (1997) 28-37
3. Zhao, G.: Research of Diesel Engine Electronic Speed Regulator Based on Generalized Predictive Control. Internal-Combustion Engine Engineering 19(3) (1998) 30 - 35
4. Zhu, Y.: Marine Engine Intelligent Fault Diagnosis System Based on FNN. Mater's Dissertation in Shanghai Maritime University. Shanghai, China (1997)
5. Patton, R. J.: Fault-Tolerant Control - the 1997 Situation. Proc. of IFAC/IMACs Symposium of Fault Detection - Supervision and Safety for Technical Process. Hull, England (1997) 1033 - 1055
6. Wen, X.: Fault Diagnosis and Fault Tolerant Control of Control System. Mechanical Industry Publishing House, Beijing (1998)

Hybrid Neural Network and Genetic Algorithms for Self-tuning of PI Controller in DSPM Motor Drive System

Rui-Ming Fang[1] and Qian Sun[2]

[1] Colleage of Information Science and Engineering,
National Huaqiao University, Quanzhou, Fujian, 362021, P.R. China
fangrm@hotmaill.com
[2] Department of Electrical Engineering, Hefei University, Hefei, Anhui, 230001, P.R. China

Abstract. Due to the nonlinear characteristics of Double Salient Permanent Magnet (DSPM) motor, the fixed-gain Proportional Integer (PI) controller can not perform well at all operating conditions. To increase the robustness of PI controllers, we present a self-tuning PI controller for speed control of DSPM motor drive system. The method is systematic and robust to parameter variations. We first treat the model of the DSPM motor drive. A well-trained multi-layer Neural Network (NN) is used to map the nonlinear relationship between the controller coefficients (Kp, Ki) and the control parameters (switching angles and current). Then we apply genetic algorithm to optimize the coefficients of the PI controller. A second NN is used to evaluate the fitness value of each chromosome in programming process of genetic algorithm. A main advantage of our method is that we do not require the accurate model of DSPM motor (which is always difficult to acquire), and the training process of NN can be done off-line through personnel computer, so that the controller can be implemented with a Digital Signal Processor (DSP-TMS320F2407). The experimental results illuminated that the proposed variable PI controller offers faster dynamic response and better adaptability over wider speed range.

1 Introduction

With the rapid progress of power electronics, digital signal processors (DSP), and control theory, doubly salient permanent magnet (DSPM) motors, which possess high efficiency, high power density and high energy conversion ratio [1], have been recognized as one of the key components in factory automation. The control of DPSM motors requires well performance for different operating conditions. The Proportional Integral (PI) controller is undoubtedly the most popular controller used in industrial control processes because of its simple structure and its reliability when used in a wide range of operating conditions [2]. However, traditional fixed-gain PI controllers are not robust enough to accommodate the variations of external disturbances and parameter and structural perturbations (friction) of DSPM motors during operation.

Much of the recent research in DSPM motors has been devoted to providing an intelligent PI-type controller that can self-tune its control gains according to the variations in operating conditions. This will take advantage of its simplicity and feasibility.

J. Wang et al. (Eds.): ISNN 2006, LNCS 3972, pp. 1068 – 1074, 2006.

Genetic algorithms (GAs) can be viewed as a general-purpose optimization method and have been successfully applied to search, optimization and machine learning tasks [3]. In [4], the authors proposed a GA-based self-tuning PI controller for DSPM motor drive systems, which improved the dynamic performance of the drive system. However, the system model is necessary for tuning the controller coefficients in an appropriate manner. Due to the much nonlinearity of the machines, power converter and controller, it is difficult to build an approximate mathematical model for DSPM motor drive system, the model in [4] also neglected some variations of parameters, which degraded the accuracy of the proposed method.

The neural network, which is well-known for its excellent learning ability and approximation to any arbitrary continuous function, is recently offering a new frontier in solving control problems of motor drives [5]. A well-trained feedforward multi-layer neural network can provide nonlinear input-output mapping with precise interpolation capability [6]. With appropriate training, this property permits a feedforward NN to identify the DSPM motor drive system.

In this paper, a self-tuning PI controller for speed control of DSPM motor drive system based on neural network and genetic algorithms is proposed. A multi-layer feedforward neural network is used to model the drive system, which learns its parameters with training patterns obtained from experiments. Genetic algorithms are adopted to optimize the coefficients of the PI controller, and a second NN model is used as the fitness value calculator of Gas that will output control performances, such as settling time t_s, maximum overshoot σ_p, and rising time t_r, according to the control coefficients (proportional coefficient K_p and integral coefficient K_i).

This paper is organized in the following manner. In Section 2, the DSPM motor drive system is introduced. A neural network based model of DSPM motor drive system is described in Section 2. In Section 4, the genetic algorithm based self-tuning PI control scheme is presented. The experimental setup and some experimental results are provided in Section 5 to verify methodology. And Section 6 concludes the paper.

2 System Modeling with Neural Network

The control scheme of DSPM motor drive with a PI controller is shown in Fig. 1.

The speed error between desired speed, n^*, and real speed, n, is converted into desired torque, T_e^*, through PI controller. This can be expressed as:

$$T^*(k) = K_p e(k) + K_i \sum_{j=0}^{k} e(j) \tag{1}$$

where $T^*(k)$ is the output of PI controller, K_p and K_i are proportional and integral gains, respectively, e is the speed error between desired speed, n^*, and real speed, n. k is the actual discrete time.

Then the real torque is achieved by changing control parameters, namely switched angles, θ_{on}^+, θ_{off}^+, θ_{on}^-, θ_{off}^-, as well as current reference I*. However, the relation between the torque and control angles is nonlinear and complex. So it is difficult to get definite expressions between the torque and control angles. Neural network mapping are usually good choices for such tasks. There is no criterion to select neuron numbers at every layer of the NN structure. Layer numbers and neuron numbers are determined with experiments. To map the relationship between $T^*(k)$, $n(k-1)$ and θ^+_{off}, θ^-_{on}, θ^-_{off}, I*, a multi-layer feedforward neural network with different number of layers, different number of neurons in each layer, and different transfer functions for each neuron, is tried. Finally the one shown in Fig.2 is used.

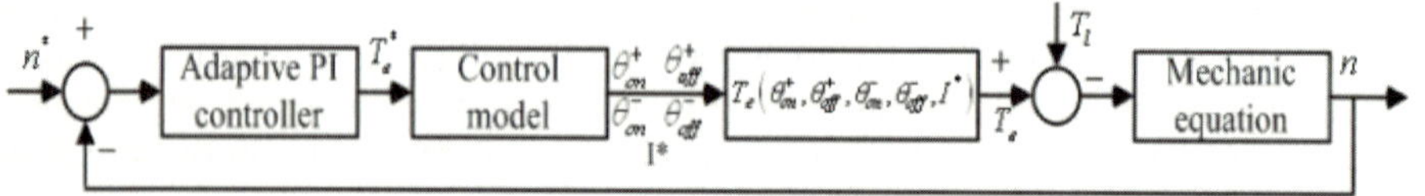

Fig. 1. Control scheme of DSPM motor drive

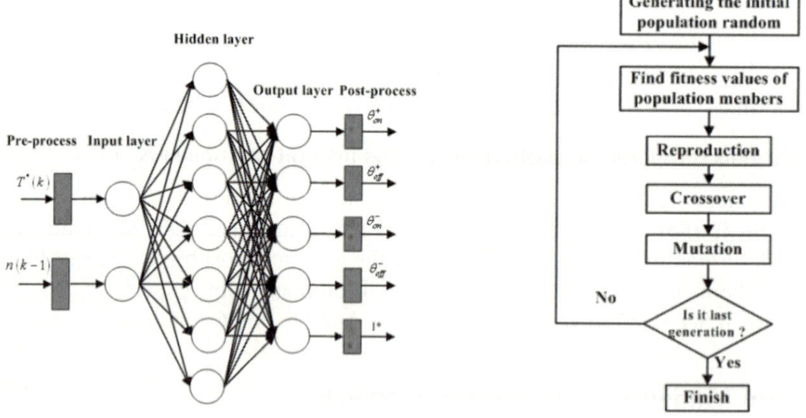

Fig. 2. Neural network structure for modeling of DSPM motor drive

Fig. 3. Flow diagram for the GA

To obtain the system model with NN, the input-output datasets of the system are required as training and testing data, which can be generated by experiments. Then we should choose a learning algorithm. In this paper, the back-propagation algorithm was used to train the NN, while the learning and momentum coefficients were determined as 0.65 and 0.85 by experiments, and a hyperbolic tangent sigmoid transfer function tansig() was chosen as the activation function, which gives the following relationship between its inputs and outputs:

$$\tan sig(x) = \frac{2}{1+e^{-2x}} - 1 \tag{2}$$

3 Optimization of PI Coefficients Using Genetic Algorithms

3.1 Genetic Algorithm

GAs are search algorithms based on the mechanism of genetic and natural selection developed by John Holland in 1975. GAs are widely applied to the optimization of an objective function in the learning of neural network, the tuning of fuzzy membership function, machine learning, system identification, control, etc, which were difficult to optimize. GA can obtain optimal solutions having no derivative information and without continuity of objective function. Since GAs are global search methods, which simultaneously search the solution population, the possibility of convergence to local minima will be reduced and solutions can be obtained by simple algorithms without complex procedure.

3.2 Optimization of PI Coefficients Using Genetic Algorithm

Gas are applied to obtain the optimum of K_p, K_i for the PI controller. The flow diagram of the GAs is shown in Fig. 3.

The parameters of PI controller are made as one string. In this paper, K_p and K_i consisted of 16 bits and 16 bits, and the solution population is composed as 30 strings. And the fitness function used for the each string evaluation of solution population is expressed as:

$$f = \frac{1}{\mathrm{T_s + T_r + s_p}} \tag{3}$$

where T_s is the settling time, T_r is the rising time, and σ_p is the maximum overshoot. T_s, T_r and σ_p all are the nonlinear functions of K_p, K_i. Nevertheless, the relationships are also difficult to express in mathematic forms. So we apply a second multi-layer feedforward neural network to map the nonlinear relationship between K_p, K_i and T_s, T_r, σ_p, as shown in Fig.4.

The optimum values for the K_p and K_i pairs were obtained using a computer program written in the C++ language for the GA. Firstly, 30 of the K_p and K_i pairs were generated randomly, and these strings were inputted into the ANN model as input. The settling time, rising time, and maximum overshoot were obtained from the ANN output. These values were then used as the fitness function. Then three different operators at bit level were executed. The one-point crossover method was used on the crossover operator (crossover rate was chosen 0.75). Mutual parameters of two random members on the crossover were divided into two parts and their positions were changed. A random bit of a random number on the mutation process was changed from 0 to 1 and 1 to 0(mutation rate was chosen 0.04). To avoid early convergence, the mutation rate was increased when convergence occurred in 5–10 generations. In addition, the member that had highest fitness values was directly inherited by the next generation, which can improve the performances of GA.

4 Experimental Results and Analysis

To testify the theoretical analysis and developed control strategy, we design a DSPM motor drive system, whose block diagram is shown in Fig.5, which typically consists of seven basic parts: DSPM motor, power converter, position transducer, current sensor, control circuit and microprocessor and personal computer. The processor used in this work was a TMS320LF2407 Digital Signal Processor (DSP). The processor communicated with the PC via a RS232 serial port.

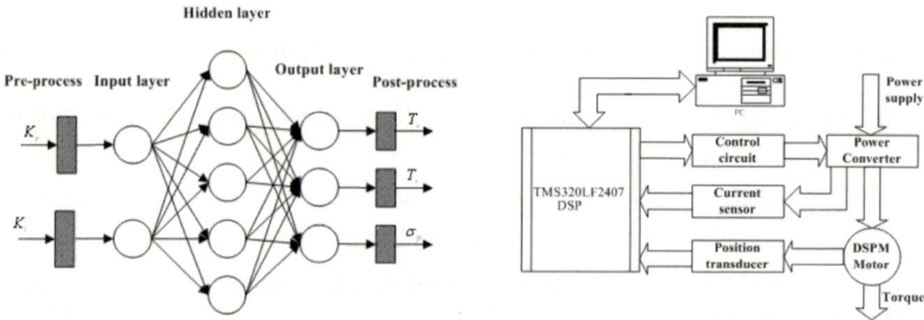

Fig. 4. The NN structure for calculator of fitness **Fig. 5.** Block diagram of DSPM motor drive

For the above system, given different switching angles and upper limit value of current, the output torques was measured. The dataset were used to train the NN model. Then given different PI coefficients, the resulted settling time, rising time, and maximum overshoot values were acquired, which can be used as the training data of NN fitness calculator. All the training processes were finished in personal computer with C++ program, and the results weights and biases were then transferred to DSP through RS232 port.

The change of mean squared error with the training process of the system neural network model and the fitness calculating neural network are shown in Fig. 6. The

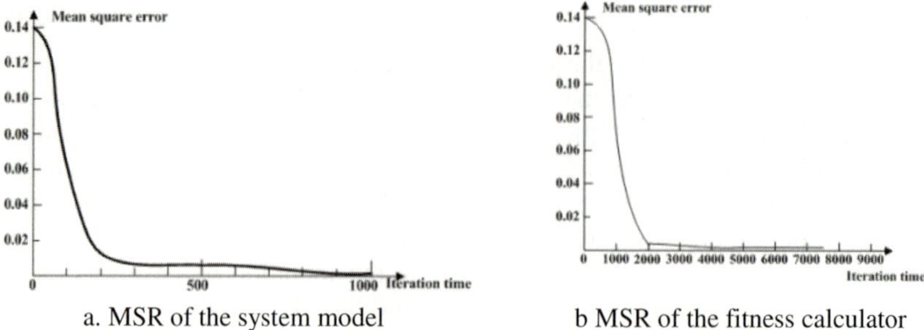

a. MSR of the system model b MSR of the fitness calculator

Fig. 6. The mean squared error values versus iteration number mean squared error reduced to lower than 0.001 by 1,000 iterations for the first NN, while 0.001 by 7500 iterations for the second NN.

In order to discern the appropriateness of the ANN model, the change in settling time with K_i for the actual system and the ANN model is shown in Fig. 7, where K_p is kept constant. It shows that the ANN model created for the system models it successfully.

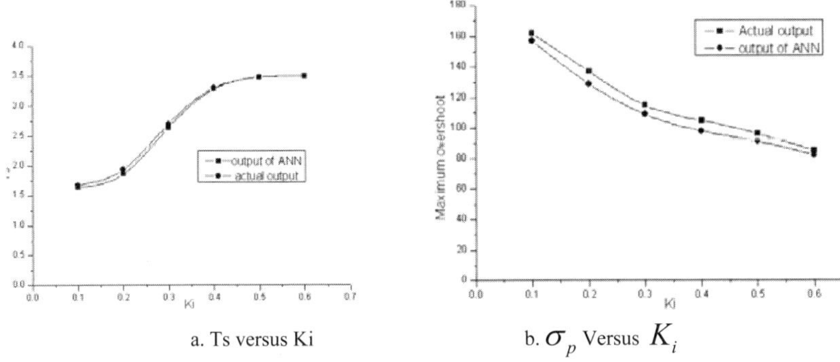

a. Ts versus Ki   b. σ_p Versus K_i

Fig. 7. The change of control performances with K_i

The transient responses of current and speed are recorded by an oscilloscope. Fig. 8 illustrated the transient response of current and speed of the DSPM motor drive with self-tuning PI controller and fix-gained PI controller when the speed decreased from 1200 rpm down to 600 rpm.

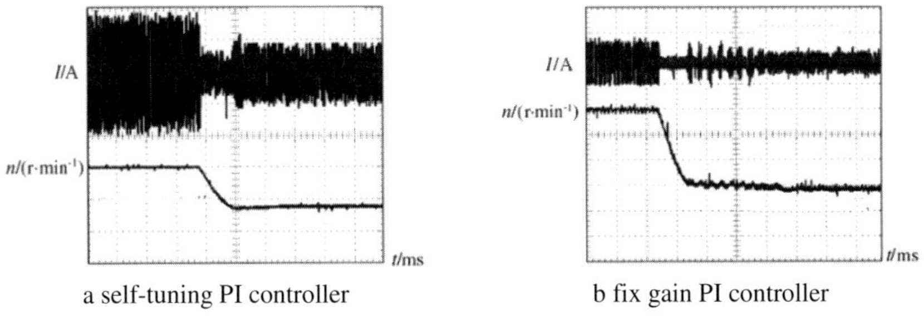

a self-tuning PI controller   b fix gain PI controller

Fig. 8. Measured dynamic responses when speed decreased from 1200RPM to 600RPM

It can be seen from the above figure that the DSPM motor drive with self-tuning PI controller exhibits fast responses and little overshoot.

5 Conclusions

In this study, a self-tuning PI controller for a DSPM motor drive based on neural network and genetic algorithms was presented. The multi-layer feedforward neural network was first used to model the drive system, and then used to model the fitness value calculator, which learned its parameters with training patterns obtained from experiments. Genetic algorithms were adopted to optimize the coefficients of the PI controller. The simulated and measured results illustrated that NN modeling can represent the physical system exactly, and the control performances can be improved if the system is controlled by control parameters obtained from the Gas-based optimization process.

References

1. Liao, Y., Liang, F., Lipo, T.A.: A Novel Permanent Motor with Doubly Salient Structure. IEEE Trans. Industrial Applications 31(5) (1995) 1069-1078
2. Astrom, K., Hagglumd, T.: PID Controllers: Theory, Design, and Tuning. 2nd edn. Instrument Society of America, New York (1995)
3. Goldberg, D.E.: Genetic Algorithms in Search, Optimization, and Machine Learning, Addison Wesley, New York (1989)
4. Sun, Q., Cheng, M., Zhou, E.: Variable PI Controller of a Novel Doubly Salient Permanent Magnet Motor Drive, Proceedings of CSEE. Libr. 23(6) (2003) 117-122
5. Simoes, M.G., Bose: Neural Network based Estimation of Feedback Signals for a Vector Controlled Induction Motor Drive. IEEE Trans. Industrial Applications 31(3) (1995) 620-629
6. Haykin, S.: Neural Networks: A Comprehensive Foundation, 2nd edn. Prentice Hall, Englewood Cliffs (1999)

An Efficient DC Servo Motor Control Based on Neural Noncausal Inverse Modeling of the Plant

H. Rıza Özçalık

Electrical Department of Engineering Faculty,
Kahramanmaraş Sütcü İmam University, 46001, Kahramanmaraş-Turkey
ozcalik@yahoo.com

Abstract. This study introduces an efficient speed controller for a DC servomotor based on neural noncausal inverse modeling of the motor. For this mission; first, motor mathematical model is obtained in digital form. Secondly, to be able to generate necessary inputs which drive the plant, open loop control signals, the inverse model of the system is identified by an ANN structure. Then, a neural controller is introduced immediately, which is trained by a composite error signal. During the identification and control process, an efficient numerical computing based on Newton-Raphson method simulates the dynamic of the motor. The success of the designed control system is tested by a simulation study considering real conditions to be able to occur in real-time running of the system.

1 Introduction

In this modern age, control phenomenon is an inevitable fact for the systems in the use of human being, not only to get enough satisfaction at the performance of them but also to protect them all along working-life of system. As time goes on, this necessity has been more important and popular.

In real life, still, conventional controllers are widespread. Especially, PID and PID like controllers are very popular. Whereas, these-kind controllers are always in need of tuning process manually due to their being inadaptable. Besides, if the PID controller is unable to deal with the complex process, no matter how tuned it, it will not work properly and perfectly. Therefore, intelligent-like controllers are compulsory and worth of introducing and taking their turns. Among them, self-tuning and other well-known adaptive controllers are already used in robotics, factories and other complex systems. In last years, artificial neural network (ANN) and fuzzy controllers have jumped into the real field and proved their dexterity of many respects. Neural network control is a control method using artificial neural networks. It has great potential since neural topology does not need the mathematical model of the system to be controlled. It is efficient for not only linear systems but also nonlinear ones [1], [2], [3]. In modeling process, various types of neural network structures are used. In this study multilayer perceptron(MLP) networks have been preferred together with back-propagation algorithm.

Obviously, the control of dynamical systems necessitates a special skillful method. Especially, the use of NN structures for the problem naturally involves an adaptive

J. Wang et al. (Eds.): ISNN 2006, LNCS 3972, pp. 1075 – 1083, 2006.

approach so that the identifier and controller parts are mostly constructed via the imitation of conventional adaptive methods. In general, modeling scheme makes use of discrete parametric models, ARX or ARMAX [1], [2], [4]. In fact, the construction of the models in this way ensures the convergence of controlling and identifying procedures, simply, within a small part of first epoch.

DC motor control has been studied by many researchers till this time. Among them, the works of Sharkawi et al. [5] and Chiasson [6] may be considerable regarding their novel and different view points. Sharkawi et al. used a direct simple neural controller supported by a reference model. Chiasson used a modern direct nonlinear-geometric technique.

In this work, a semi-explicit NN based control strategy is considered. First, the NN identifier estimates the motor structure and just after that, another NN controller is turned on in continuous way. This initial identification consumes very little time and it is, in fact, an on-line process all along control. This pure identification part has guaranteed proper initial weights affecting all performance. On the other hand, Newton-Raphson method has been preferred to solve the non-linear discrete model of the plant. This numerical solution has represented, in a way, system work and response to happen in real-time operation. As for the controller, it is another NN block trained by a composite error function.

2 Modeling of DC Servo Motor

There are some useful academic examples which demonstrate the success of the studies in Control Science. Obviously, among them, control studies for field or armature-controlled DC servo motor are rather popular. We have used a fan type DC servomotor given by Sharkawi et al. [5] with following rated values: 1 hp, 220 V, and 550 rpm. These values result in a nominal torque of 10 Nm. This machine has the ability of working with a wide speed range by tuning of the armature voltage.

2.1 Continuous Model

DC motor is an electro-mechanical system and can be easily modeled by the help of Newton and Kirchow laws describing electrical and mechanical parts of the plant dynamic:

$$\begin{aligned} V_a(t) &= Kw(t) + R_a i_a(t) + L_a\left[di_a(t)/dt\right] \\ J\left[dw(t)/dt\right] &= Ki_a(t) - Bw(t) - T_L(t) \end{aligned} \tag{1}$$

Where;

$w(t)$: motor speed (Rad/s)

$V_a(t)$: terminal voltage (V)

$i_a(t)$: armature current (A)

$T_L(t)$: load torque (Nm)

J : rotor and load equivalent inertia (Nm^2)

K : torque and back emf constant (NmA^{-1})

B : damping constant (Nms)

R_a : armature resistance (Ohm)

L_a : armature inductance (H)

Load is a fan-type one and can be expressed approximately by a torque, $T_L = \mu w^2$ [sign(w)]. In this formula, sign(.) function is for making the load effect, always, counter wise to motor original action. Load torque will change by speed and load coefficient naturally. Parameter values are, respectively; J = 0.068 Kg m^2, K = 3.475 Nm A^{-1}, R_a = 7.56 Ohm, L_a = 0.055 H, B = 0.03475 Nm s, μ = 0.0039 Nm s^2 and T = 40 ms.

2.2 Discrete Model

It is compulsory to construct a proper discrete model of the system for making a good simulation. For this task, a simple but efficient approximation is the first-forward difference or Euler technique [7].

$$D^r c(kT) \approx \frac{1}{T^r}\left\{\nabla^{r-1}c(kT) - \nabla^{r-1}c[(k-1)T]\right\} \tag{2}$$

$$= \frac{1}{T^r}\nabla^r c(kT)$$

By using this relation, continuous model can be easily converted into discrete form:

$$w^2(k)sign[w(k)] + a_1 w(k) + a_2 w^2(k-1)sign[w(k)] + a_3 w(k-1) + a_4 w(k-2) + bV_a(k) = 0 \tag{3}$$

Where;

$$a_1 = \frac{K^2T^2 + R_aJT + R_aBT^2 + L_aJ + L_aBT}{\mu.T(R_aT + L_a)}$$

$$a_2 = -\frac{L_a}{R_aT + L_a}$$

$$a_3 = -\frac{R_aJT + 2L_aJ + L_aBT}{\mu.T(R_aT + L_a)}$$

$$a_4 = \frac{L_aJ}{\mu.T(R_aT + L_a)}$$

$$b = \frac{KT}{\mu.(R_aT + L_a)}$$

Where, T is sampling time for control process. This value should be rather smaller than global time-constant of the system or, simply, rice time of the system response for a test step input. As seen in the relation (3), the digital model of the plant is strictly nonlinear so that the simulation study, in a way, reflects the real work of the DC motor. This nonlinear difference equation can be solved easily and efficiently by using Newton- Raphson method [8].

3 General Structure of Artificial Neural Network in Use

In this study, main tools for control are MLP type networks. Two-layer neural network structure has been chosen for both controller and inverse identifier. These networks are feed-forward structures with respect to signal flow. Figure 1 shows a general NN structure with one hidden layer. Our controller has got five inputs, two hidden layers so that each layer consists of five hidden neurons and one output. Inverse identifier block is another MLP structure and has five inputs, one hidden layer of five neurons and one output. Therefore, NN structures are rather simple for this control process.

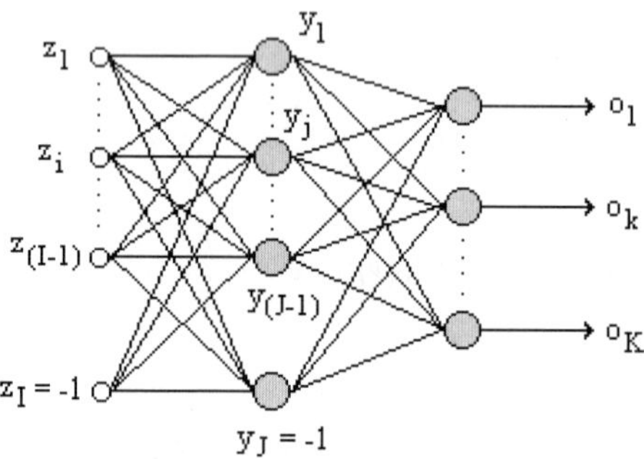

Fig. 1. Feed-forward neural network in use

The back-propagation algorithm based on delta learning method is employed to train the weights [9], [10]. The weights are updated as follows:

$$w_{ij}(k+1) = w_{ij}(k) + \eta \frac{\partial E}{\partial w_{ij}(k)} \tag{4}$$

Where, η is a prescribed learning rate. The cost function E for training the neural network is defined as

$$E = \frac{1}{2} \sum_{l=1}^{P} \sum_{r=1}^{K} (o_r^l - d_r^l)^2 \tag{5}$$

Where, o is neuron output, d is desired or set value, K is number of the neurons in the output layer and P is the number of the training patterns.

In all learning process bipolar sigmoid activation function has been used. This activation function is non-linear and very useful for continuous training scheme [8]. It is defined as:

$$o = f(net) = \frac{2}{1 + \exp(-\lambda net)} - 1 \tag{6}$$

Where, λ is a positive number called as steepness coefficient. This parameter is very important in learning procedure and lies down between one and infinity theoretically. Whereas, in application it extends, mostly, up to four.

4 Control Scheme

It is strictly true that the most important thing in a real system is to reach the desired outputs that are expected and prescribed beforehand. In this study, an efficient control system based on NN structures has been constructed as seen in Fig. 2.

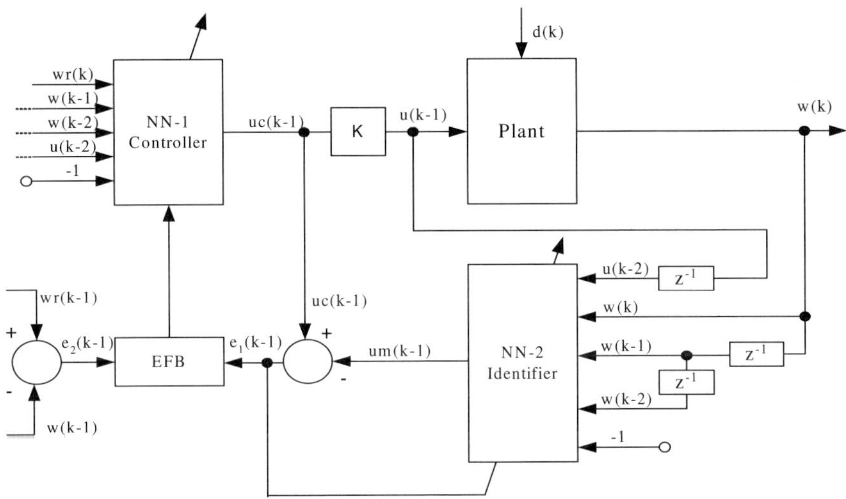

Fig. 2. A schematic diagram of the NN controller

The control system consists of two main parts: Identifying part and controller. In fact, both section are always on-line and work together.

4.1 Identifier

This part works as an inverse model of the plant. Considering relation (3), the motor can be seen as a system of two degrees for speed controlling process. Thus, it is possible to use a parametric model in the same degree. Therefore, ARX like parametric modeling may be a good choice for identifying process [11], [12].

$$w(k) + a_1 w(k-1) + a_2 w(k-2) = u(k-1) + b_2 u(k-2) + \varepsilon(k) \quad (7)$$

Where, u(k) is system input or control signal, ε(k) is noise-like disturbance or simulation equivalent of it. In general, this model is used to obtain system output but if used to find inputs, then, it will be an inverse model of the plant. In fact, to arrange the model for generating inputs results in a noncausal filter and, of course, to solve it very difficult by any conventional approximation. Consequently, to introduce an ANN filter is very useful and necessary. Therefore, ANN-2 is used as an efficient neural filter that is identifying actual control inputs. ANN-2 has five inputs, which are the back-shifted history of the actual speed and control inputs and has got one output, observed control signal. This filter can be written, briefly, as below:

$$u(k-1) = f\{w(k), w(k-1), w(k-2), u(k-2), -1\} \tag{8}$$

As easily seen in the equation (8), noise input has been ignored. Because, it is necessary to have it estimated beforehand. As an alternative view, the difficulty of the extraction of the noise from the complexity of whole process may be tolerated and, fortunately, compensated by the help of overall capability of ANN blocks.

4.2 Controller

Controller is another artificial neural network (ANN-1) and has been constructed by using general multi layer perceptron structure. It has got five inputs, reference speed, back-shifted history of actual speed and control signals and threshold; two hidden layers of five hidden neurons; and one output, predicted control signal. In the training of controller, an interesting way has been adopted so that complete error includes two kinds of components, modeling error e_1 and main system error (speed error) e. By using these errors together, a meaningful composite error function can be obtained.

$$e_c = g\left[|e_1| + \delta.|e|\right] \tag{9}$$

As seen in the relation, absolute values of modeling error and main error have composed a new error. Where, δ is a real number being between zero and one and can be named as “forming coefficient". On the other hand, g is a sign determiner and can be expressed as below.

$$g = sign(e_1) \quad \text{if} \quad |e_1| > |e| \quad ; \quad g = sign(e) \quad \text{if} \quad |e_1| < |e|$$

The logical base of this composition is the fact that this new error function would be able to take meaningful maximum value of the current errors in any case so that control scheme behaves to reduce all error components simultaneously by using the adaptation capability of NN structures. Adjusting parameter δ manually, controller weights can reach optimum values in time. To prefer this kind of error is to update controller weights so that both errors converge to zero simultaneously.

5 Simulation Results

Obviously, observing the results obtained in this study can only show the success of the control scheme. In a neural network based study, of course, the values of some parameters related with NN structures are very important and should have good values at the starting stage. For that reason, first, some parameters are determined: Learning rate can be taken as 0.075, steepness coefficient in activation function as 1.50, forming coefficient for composite error as 1.0, initial weights of NN structures as random values that lie between zero and one. To choose an appropriate work for the motor that can serve the aim of a good simulation. A control scheme is, of course, in need of the desired speed values which are wisely determined, in other words, the desired speed trajectory of the motor. This trajectory can be chosen as a composite sinusoidal function:

$$w_d(k-1) = 20\sin\{(2\pi/600)(k-1)\} + 10\sin\{(2\pi/300)(k-1)\} \tag{10}$$

This trajectory satisfies the real conditions that may happen in a typical servomotor work. At the beginning of the control, ANN-2 is trained by using a simple sinusoidal input, for example, like 0.5sin [(2π/600)(k-1)]. In this way, optimum initial weights can be generated. Obviously, another important input to the any system that is running would be disturbance of any kind. In this work, the disturbance consists of two parts: Load and noise-like disturbance. For this servomotor, main load is a fan and its weight related with load effect is represented by μ, load coefficient. In this study, fan is always on but to force the system and control scheme toward hard limits, load has been enhanced by changing μ as much as three times at sixth second of the trajectory and reduced it to original value at eighteenth second of it. As for the second disturbance, it is a white noise with zero mean value and rather big variance, 0.005. In these conditions, reference and actual angular speeds, tracking error for them and torque produced by the DC motor during all trajectory have come out as below.

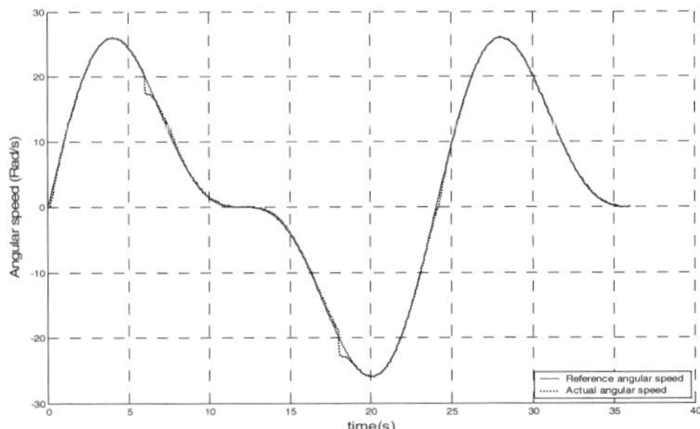

Fig. 3. Reference and actual speeds of the servo motor during control process

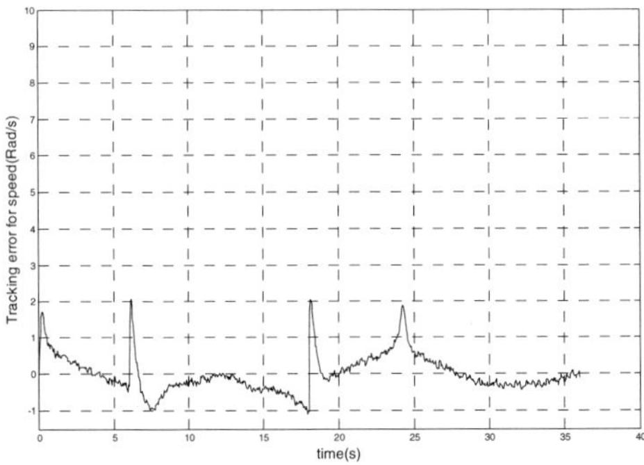

Fig. 4. Tracking error between reference and actual angular speeds

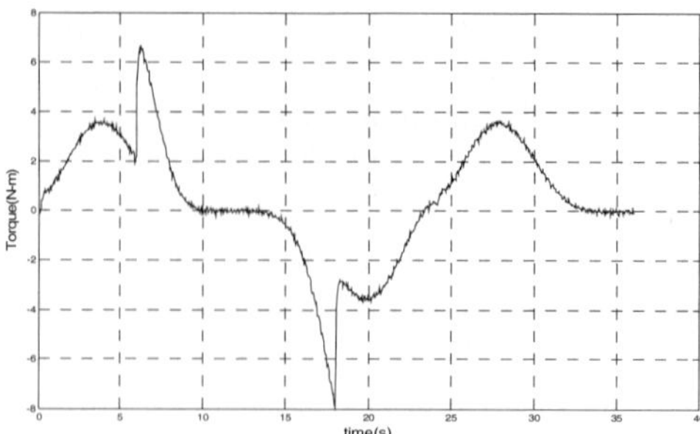

Fig. 5. Motor torque during control process (peaks show over-loading instants).

As seen in the figures, actual speed perfectly tracks the reference speed in all conditions even though desired speed varies quickly. All along trajectory, speed errors, mostly, remain under 1.5 rad/s or, approximately, at the scale of 5% with the exception of sudden variation points while the plant is under determined load. In the case of heavily loading conditions, deviations from set values have reached, only, up to a small value of 2 rad/s, rather small than 10 %. On the other hand, it is very important to observe the torque value. Considering motor ratings, torque has not to exceed rated value, 10 Nm. As seen in the figure 7, torque values have been remained in prescribed region.

6 Conclusions

An armature-controlled DC servomotor has been successfully controlled by the proposed control scheme based on ANN whereas high non-linearity and severe loading conditions exist. The original view-points can be expressed in this way: Using a neural noncausal inverse filter for getting good parameters and prediction of true control inputs without ceasing continuous process, just before the main control process and a composite error function for training the main controller. Actual speed has tracked the prescribed trajectory all along the process. In real time study, dynamic structure of the motor, by acting like a low-pass filter, will eliminate some undesired peaks and it will be able to arise much better results.

In similar work, Chiasson's use of the nonlinear geometric technique [6] produced generally good results, but did not consider enough severe nonlinear loading effects. On the other hand, Sharkawi et al. [5] used a good neural direct controller supported by a reference model and obtained good results but mostly studied by a certain load, so did not show enough the effects of severe torque variations.

In conclusion, it can be clamed that this work is highly successful in speed tracking under severe loading conditions and in recovering the working conditions according to prescribed trajectory.

References

1. Narendra, K.S., Parthasarathy, K.: Identification and Control of Dynamical System Using Neural Networks. IEEE Trans. on Neural Networks 3(1) (1990)
2. Hunt, K., Sbarbaro, D.: Neural Networks for Nonlinear Internal Model Control. IEE Proc.-D. 138(5) (1991) 431-438
3. Narendra, K.S., Mukhopadhyay, S.: Adaptive Control of Nonlinear Multivariable Systems Using Neural Networks. Neural Networks 7(5) (1994) 737-752
4. Watanabe, K., Fukuda, T., Tzafestas, S.G.: An Adaptive Control for CARMA Systems Using Linear Neural Networks. Int. J. Control 56(2) (1992) 483-497
5. Weerasooriya, S., El-Sharkawi, M.A.: Identification and Control of a DC Motor Using Back-Propagation Neural Networks. IEEE Trans. on Energy Conv. 6(4) (1991) 663-669
6. Chiasson, J.: Nonlinear Differential-Geometric Technique for Control of a Series DC Motor. IEEE Trans. Control Syst. Technol. 2(1) (1994) 35-42
7. Houpis, C.H.: Digital Control Systems. McGraw-Hill Inc., Singapore, 1992.
8. Rice, J.R.: Numerical Methods, Software, and Analysis. 2rd edn. Academic Press, New York (1992)
9. Zurada, J.M.: Introduction to Artificial Neural Systems. West Publishing Co., St. Paul (1992)
10. Haykin, S.: Neural Networks – A Comprehensive Foundation. Macmillan College Publishing Co., US (1994)
11. Soderstrom, T., Stoica, P.: System Identification. Prentice Hall, London (1989)
12. Isermann, R.: Parameter Adaptive Control Algorithms – A Tutorial. Automatica 18(5) (1982) 513-528

A Dynamic Time Delay Neural Network for Ultrasonic Motor Identification and Control

Yanchun Liang[1], Jie Zhang[1], Xu Xu [2,*], Xiaowei Yang[3], and Zhifeng Hao[3]

[1] College of Computer Science and Technology, Jilin University, Key Laboratory of Symbol Computation and Knowledge Engineering of the Ministry of Education, Changchun 130012, P.R. China
[2] College of Mathematics, Jilin University, Changchun 130012, P.R. China
[3] School of Mathematical Science, South China University of Technology, Guangzhou 510640, P.R. China
xuxu567@mail.jl.cn

Abstract A novel dynamic time delay neural network is proposed for ultrasonic motors identification and control in this paper. By introducing time delay neurons, the neural network identifier and controller of ultrasonic motors are constructed. Both of them are trained online by using an improved back-propagation algorithm. The usefulness and validity of the presented algorithm is examined by the experiments.

1 Introduction

An ultrasonic motor (USM) is a newly developed motor, which has some excellent performances and useful features [1]. However, it has heavy nonlinear features which vary with driving conditions. Therefore, precise control for ultrasonic motor is difficult. In recent years, control schemes based on artificial neural network [2] have become one of the main methods for the USM. However, the conventional neural control schemes based on static neurons network have some disadvantages. A neural network with time delay neurons is proposed in this paper for the ultrasonic motor control.

2 Time Delay Neural Network

A time delay neuron [3] and a three-layer neural network, which depicted in Figure 1, are proposed in this paper. In this figure, "•" represents dynamic time delay neuron, "O" represents linear neuron. The output of the neuron is governed by $y(t) = \sigma(\sum_{i=1}^{N} w_i x_i (t - \tau_i))$, where w_i are the neuron weights, τ_i are the delays, and $\sigma(\cdot)$ is the nonlinear activation function.

* Corresponding author.

J. Wang et al. (Eds.): ISNN 2006, LNCS 3972, pp. 1084 – 1089, 2006.

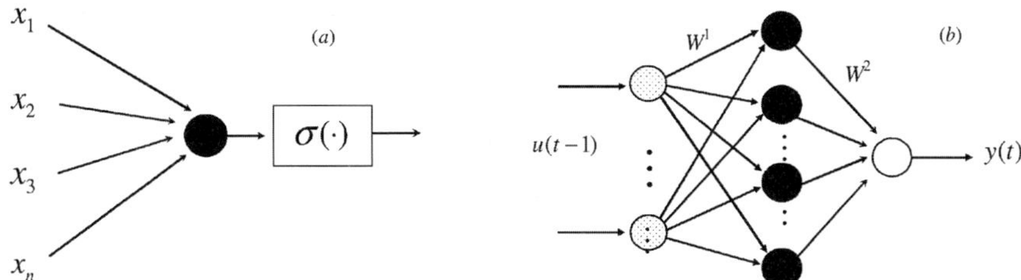

Fig. 1. (a) Time delay neuron; (b) Architecture of the DTDNN

The signal propagation in each layer is introduced as follows.

Layer 1: The input $x_i^1(t)$ and output $y_i^1(t)$ of neurons at time t are represented as

$$net_i^1(t) = x_i^1(t)\,,\; y_i^1(t) = f_1(net_i^1(t)) = net_i^1(t)\,. \qquad i = 1,2 \tag{1}$$

Layer 2: The output of the jth neuron in this layer is

$$net_j^2(t) = \sum_i w_{ij}^2 x_i^1(t-\tau_{ij})\,,\, o_j^2 = \sigma(net_j^2(t) - \theta_j^2)\,. \quad j = 1,\ldots,n \tag{2}$$

where $x_i^1(t-\tau_{ij})$ represents the ith input to the jth neuron of this layer at time $(t\text{-}\tau_{ij})$.

Layer 3: The node in this layer is linear neuron and its output is

$$net_k^3(t) = \sum\nolimits_{j=1}^{n} w_{jk}^3 o_j^2(t),\; y_I(t) = f_2(net_k^3(t) - \theta_k^3) = net_k^3(t) - \theta_k^3\,. \quad k = 1 \tag{3}$$

To obtain the learning algorithm, the energy function $E(t)$ is defined as $E(t) = 1/2(y_d(t) - y(t))^2$. Then the error term to be propagated is given by

$$\delta_k^3(t) = -\partial E(t)/\partial net_k^3(t) = -\partial E(t)/\partial y_I(t)\,.\partial y_I(t)/\partial net_k^3(t). \tag{4}$$

The weight adaptation law is governed by

$$\Delta w_{jk}^3 = -\alpha \partial E(t)/\partial w_{jk}^3 = \alpha \delta_k^3(t) o_j^2(t)\,. \tag{5}$$

The weights and threshold values of the output layer are updated as follows:

$$\begin{cases} w_{jk}^3(N+1) = w_{jk}^3(N) + \Delta w_{j,}^3 \\ \theta_k^3(N+1) = \theta_k^3(N) + \beta \delta_k^3(t). \end{cases} \tag{6}$$

For the hidden layers the δ terms, we have

$$\delta_j^2(t) = -\partial E(t)/\partial net_j^2(t) = \delta_j^2(t) = -\delta_k^3(t) w_{jk}^3 \sigma'(net_j^2(t)). \tag{7}$$

The weight is updated by

$$\Delta w_{ij}^2 = -\alpha \partial E(t)/\partial w_{ij}^2 = \alpha \delta_j^2(t) x_i^1(t-\tau_{ij}). \tag{8}$$

and the update law of τ_{ij} is

$$\Delta\tau_{ij} = -\alpha\partial E(t)/\partial\tau_{ij} = w_{ij}^2\,\partial x_i^1(t-\tau_{ij})/\partial\tau_{ij}\,. \quad (9)$$

In order to compute the value of $\partial x_i^1(t-\tau_{ij})/\partial\tau_{ij}$, we have

$$\partial x_i^1(t-\tau_{ij})/\partial\tau_{ij} \approx [x_i^1(t-\tau_{ij}) - x_i^1(t-\tau_{ij}-T)]/T\,. \quad (10)$$

where T is sample period. Then we have,

$$\Delta\tau_{ij} = \alpha\delta_j^2(t)w_{ij}^2[x_i^1(t-\tau_{ij}) - x_i^1(t-\tau_{ij}-T)]/T\,. \quad (11)$$

The weights and time delays of the hidden layer are updated as follows:

$$\begin{cases} w_{ij}^2(N+1) = w_{ij}^2(N) + \Delta w_{ij}^2, \\ \theta_j^2(N+1) = \theta_j^2(N) + \beta\delta_j^2(t), \\ \tau_{ij}(N+1) = \tau_{ij}(N) + \Delta\tau_{ij}. \end{cases} \quad (12)$$

3 DTDNN Control System

The block diagram of the DTDNN control system, as shown in Fig. 2, is adopted to control the speed of the USM. In this figure, u(t) and ω(t) represent the inputs of the DTDNNI, and $y_m(t)$ represents the desired speed.

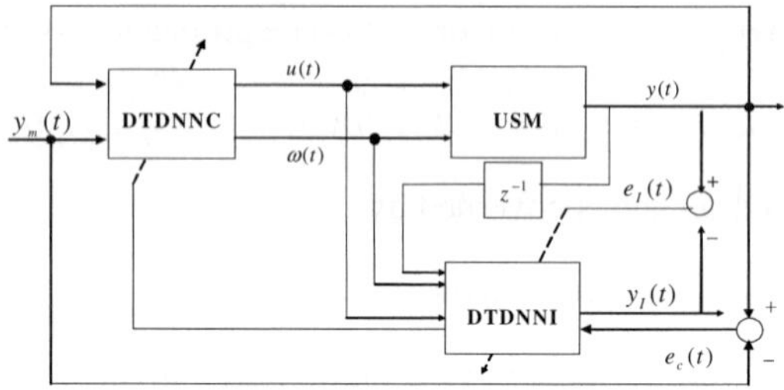

Fig. 2. DTDNN control system

To investigate the learning algorithms of DTDNNI, we define the energy function

$$E_I(t) = 1/2[y(t) - y_I(t)]^2 = 1/2e_I(t)^2. \quad (13)$$

During the learning process of the DTDNNI, the error term $\delta_k^3(t)$ in (4) becomes

$$\delta_k^3(t) = -\partial E_I(t)/\partial net_{I,k}^3(t) = e_I(t). \quad (14)$$

The sensitivity derivatives can be obtained by using the DTDNNI as follows:

$$\begin{cases} \dfrac{\partial y(t)}{\partial u(t)} \cong \dfrac{\partial y_I(t)}{\partial u(t)} = \dfrac{\partial y_{I,k}^3(t)}{\partial x_{I,1}^1(t)} = \sum_{k=1}^{l} w_{I,jk}^3 \sigma'(net_{I,j}^2(t)) \sum_{j=1}^{m} w_{I,1j}^2 = \varsigma, \\ \dfrac{\partial y(t)}{\partial \varpi(t)} \cong \dfrac{\partial y_I(t)}{\partial \varpi(t)} = \dfrac{\partial y_{I,k}^3(t)}{\partial x_{I,2}^1(t)} = \sum_{k=1}^{l} w_{I,jk}^3 \sigma'(net_{I,j}^2(t)) \sum_{j=1}^{m} w_{I,2j}^2 = \xi. \end{cases} \tag{15}$$

In addition, the energy function for the DTDNNC is defined as $E_c(t) = 1/2 e_c^2(t)$. Then the error term $\delta_k^3(t)$ in Eq. (4) becomes

$$\begin{cases} \delta_u^3(t) = -\dfrac{\partial E_C(t)}{\partial e_c(t)} \dfrac{\partial e_c(t)}{\partial y(t)} \dfrac{\partial y(t)}{\partial u(t)} \dfrac{\partial u(t)}{\partial y_{C,k}^3(t)} \dfrac{\partial y_{C,k}^3(t)}{\partial net_{C,k}^3(t)} = e_c \varsigma, \\ \delta_\varpi^3(t) = -\dfrac{\partial E_C(t)}{\partial e_c(t)} \dfrac{\partial e_c(t)}{\partial y(t)} \dfrac{\partial y(t)}{\partial \varpi(t)} \dfrac{\partial \varpi(t)}{\partial y_{C,k}^3(t)} \dfrac{\partial y_{C,k}^3(t)}{\partial net_{C,k}^3(t)} = e_c \xi. \end{cases} \tag{16}$$

4 Convergence of Algorithm

A Lyapunov function is defined as $V(k) = 1/2 e^2(k)$. Then we have

$$\Delta V(k) = V(k+1) - V(k) = 1/2[e^2(k+1) - e^2(k)]. \tag{17}$$

The error difference due to the learning can be represented by [4]

$$e(k+1) = e(k) + \Delta e = e(k) + [\partial e(k)/\partial W]^T \Delta W. \tag{18}$$

where ΔW represents a change in an arbitrary weight vector.

From the update rule of (5) and (14), we have

$$\Delta w_{jk}^3 = -\eta_I \, \partial E_I(t)/\partial w_{jk}^3 = -\eta_I e_I(t) \partial e_I(t)/\partial w_{jk}^3 = \eta_I e_I(t) o_j^2(t). \tag{19}$$

where η_I is the learning rate for weights in the DTDNNI. Then we have

Theorem 1: Let η_I be the learning-rate parameter of the connecting weights of the DTDNNI and $P_{w\max}$ be defined as $P_{w\max} \equiv \max_N \|P_w(N)\|$, and $P_w(N) = \partial y_k^3(t)/\partial w_{jk}^3$. Then, the convergence of the identification error is guaranteed if η_I is chosen as $0 < \eta_I < 1/H$, where H is the number of the hidden layer neuron in the DTDNNI.

Proof: Noting that $P_w(N) = \partial y_k^3(t)/\partial w_{jk}^3 = o_j^2(t)$ and the activation function of hidden layer in DTDNNI is hyperbolic tangential function, hence $|o_j^2(t)| \le 1, (j = 1,2,..,H)$ holds.

According to the definition of Euclidean norm one has $\left\|P_w(N)\right\| \le \sqrt{H}$.

Eq. (18) gives $e_I(N+1) = e_I(N) - \left[\partial y_k^3(N)/\partial w_{jk}^3\right]^T \Delta w_{jk}^3$, and

$$\left\|e_I(N+1)\right\| = \left\|e_I(N)[1-\eta_I P_w^T(N)P_w(N)]\right\| \le \left\|e_I(N)\right\| \left\|1-\eta_I P_w^T(N)P_w(N)\right\|. \tag{20}$$

If η_I is chosen as $0 < \eta_I \le 1/(P_{w\max})^2 = 1/H$, the term $\left\|1-\eta_I P_w^T(N)P_w(N)\right\|$ in (20) is less than 1. Therefore, the Lyapunov stability of $V > 0$ and $\Delta V < 0$ is guaranteed. The identification error will converge to zero as $t \to \infty$. This completes the proof of the theorem.

5 Experimental Results

Numerical simulations are performed using the proposed method for the identification of a longitudinal oscillation USM [5]. In order to show the effectiveness of the proposed scheme, the identify performance of the proposed identifier is compared

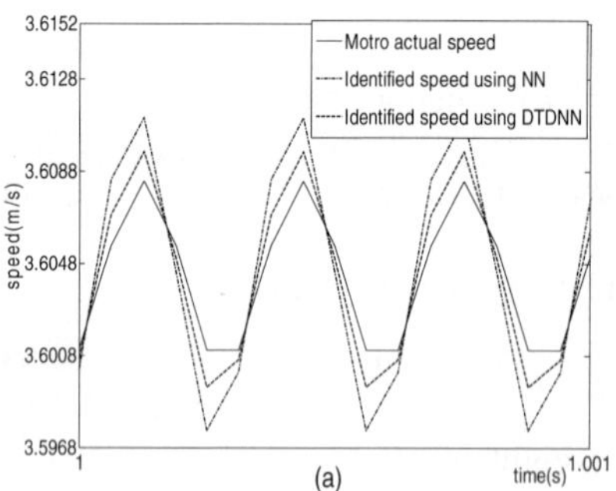

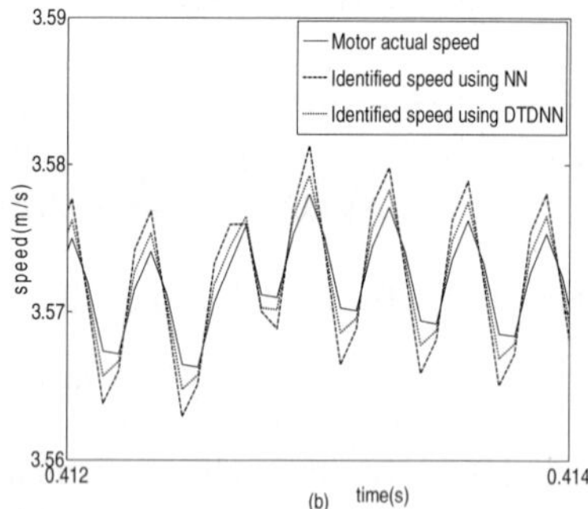

Fig. 3. The comparisons of identification results using the conventional NNI and the proposed method

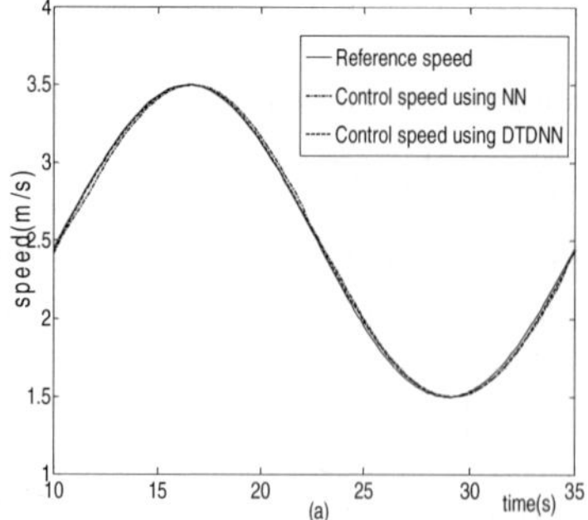

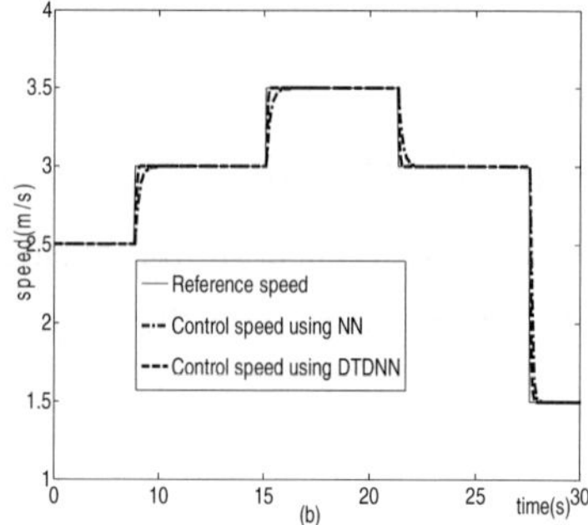

Fig. 4. Speed tracking characteristics. (a) Periodic sinusoidal command. (b) Desired velocity multiple stepwise changes.

with that of the conventional neural network identifier. From Fig. 3(a) and Fig. 3(b), it shows that the proposed identifier can approximate the nonlinear input-output mapping of ultrasonic motors quite well. The performance of the proposed control system is compared with that of the conventional neural network control system. From Fig. 4(a) and Fig. 4(b), the periodic desired output and discontinuously desired output demonstrate that the proposed control scheme has fairly adaptability.

6 Conclusions

Identification and control of ultrasonic motors based on a novel dynamic time delayed neural network are proposed in this paper. The results obtained in this paper show that the proposed identifier can approximate the nonlinear input–output mappings of the USM and can compensate the characteristic variations of the motor. And the high performance of speed identification of ultrasonic motors is also obtained when the external torque is applied. Good effectiveness of the proposed bimodal controller is also obtained for some different kinds of reference velocities.

Acknowledgment

The authors are grateful to the support of the National Natural Science Foundation of China (60433020, 10501017), the science-technology development project of Jilin Province of China (20050705-2), the doctoral funds of the Ministry of Education of China (20030183060), and “985” Project of Jilin University.

References

1. Sashida, T., Kenjo, T.: An Introduction to Ultrasonic Motors, Oxford: Clarendon, London (1993)
2. Xu, X., Liang, Y. C., Lee, H. P., Lin, W. Z., Lim, S. P., Lee, K. H., Shi, X. H.: Identification and Speed Control of Ultrasonic Motors Based on Neural Networks. Institute of Physics Publishing, Journal of Micromechanics and Microengineering. **13** (1) (2003) 104–114
3. Yazdizadeh, A., Khorasani, K.: Adaptive Time Delay Neural Network Structures for Nonlinear System Identification, Neurocomputing. **47** (1-4) (2002) 207-240
4. Yabuta, T., Yamada, T.: Learning Control Using Neural Networks. Proc. IEEE Int. Conf. Robotics and Automation, Sacramento, CA, Apr. (1991) 740-745
5. Xu, X., Liang, Y. C., Lee, H. P., Lin, W. Z., Lim, S. P., Lee, K. H.: Mechanical Modeling of a Longitudinal Oscillation Ultrasonic Motor and Temperature Effect Analysis. Smart Materials and Structures. **12** (4) (2003) 514-523

Application of PSO-Optimized Generalized CMAC Control on Linear Motor

Qiang Zhao and Shaoze Yan

Department of Precision Instruments and Mechanology,
Tsinghua University, Beijing 100084, China
account@hit.edu.cn, yansz@mail.tsinghua.edu.cn

Abstract. A Gaussian basis function based CMAC (GCMAC) is proposed for the feed-forward control of line motor. It has both the advantages of CMAC and GBF (Gaussian basis function), such as lower memory requirement, faster converging speed and more accurate approximation. Considering that the GCMAC's parameters selection is crucial for linear motor to get better control performance, we employ a particle swarm optimization algorithm to search for the optimal learning rate of the GCMAC. A numerical example of a linear motor model in wafer stage is preformed. The simulation results verify the effectiveness of the PSO-optimized GCMAC feed-forward controller.

1 Introduction

Linear motor is employed as the servo mechanism to provide linear motion in many mechanical systems, such as machine tools, precision instruments and integrated circuits manufacturing equipments [1-3]. Compared with conventional drives, linear motor has better dynamics, higher speed and longer achievable lengths of travel [4]. Using the traditional PID control alone cannot achieve satisfactory trajectory tracking, and many control algorithms were successfully designed to improve the performance of the linear motor, such as adaptive robust control (ARC) [5], model reference sliding-mode control [6], fuzzy sliding-mode control [7] and so on. However, the conventional control method based on state feedback provides limited control against disturbances and the time-variable property of the parameters. Neural networks are able to approximate any nonlinear continuous functions to an arbitrary accuracy, and it can be used in the control of linear motor.

The cerebellar model articulation controller (CMAC), originally proposed by Albus [8], has many advantages such as lower memory requirement, faster cycle time than other networks, so it is successfully applied in many real time occasions, but its drawback of low approximation precision restricts its application in precision tracking. On the other hand, the GBF has some advantages such as continuity and differentiability of the approximate function, and better accuracy. One possible difficulty is that the necessary number of basis functions may be large, which causes a heavy computational load. Here we employ the generalized CMAC neural network, namely Gaussian basis function based CMAC (GCMAC) [9], which has both the advantages of the CMAC and GBF, and meanwhile overcomes their respective drawbacks. Appropriate parameters, particularly the learning rate, are prerequisites for training the GCMAC and obtaining better performance. The particle swarm optimization (PSO)

J. Wang et al. (Eds.): ISNN 2006, LNCS 3972, pp. 1090–1095, 2006.

algorithm, proposed by Kennedy and Eberhart [10], has been used to solve many application problems [11], so we use this algorithm to find the optimal learning rate.

2 PSO-Optimized Generalized CMAC Neural Network Control

2.1 CMAC Feed-Forward Controller

The GCMAC is proposed by Chiang and Lin [9], a schematic representation of GCMAC is shown in Fig. 1, and its application as feed-forward controller for linear motor is shown in Fig. 2. The PD controller is responsible for stabilizing of the control system, and the GCMAC is used to decrease the disturbance and compensate for the system nonlinearity.

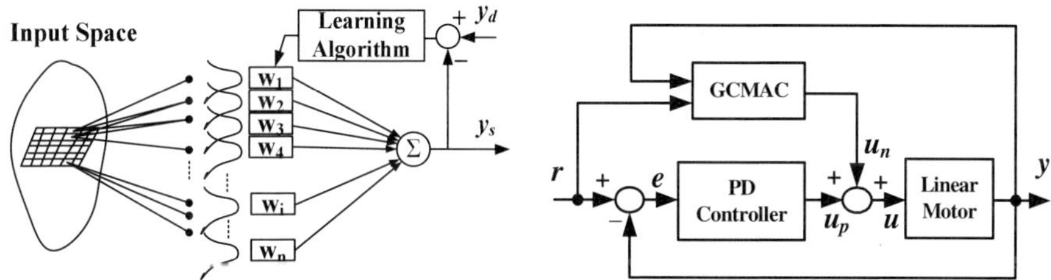

Fig. 1. Architecture of GCMAC

Fig. 2. Combined PD and GCMAC control

Both the CMAC and GCMAC use the basis functions that may dominate only in a local input space. When the constant basis in the traditional CMAC is replaced by a product of Gaussian functions, the output of GCMAC can be obtained as

$$y_s = a_s{}^T w(x_s) = \begin{bmatrix} a_{s,1} & a_{s,2} & \cdots & a_{s,N_h} \end{bmatrix} \begin{bmatrix} w_1 \\ w_2 \\ \vdots \\ w_{N_h} \end{bmatrix} = \sum_{i=1}^{N_h} a_{s,i} w_i \ . \tag{1}$$

where a_s is the basis-function selection vector and only consists of values of 0 and 1, x_s is the input vector of the s th pattern. N_h is the basis functions total number, and $w_i(s_x)$ is a vector with the i th element $w_i(x_s) = v_i b_i(x_s)$. $b_i(x)$ is Gaussian function:

$$b_i(x) = \prod_{k=1}^{N_x} \Phi_i(x_k) \ , \qquad \Phi_{i,k}(x_k) = \exp(-(x_k - u_{i,k})^2 / \sigma_{i,k}^2 \ , \tag{2}$$

where $u_{i,k}$ is the mean, $\sigma_{i,k}$ is the variance, and N_x is the dimension number of x. Usually, $u_{i,k}$ and $\sigma_{i,k}$ can be selected equably according to the range of x_k [12]: $u_{i,k} = i \cdot (x_{k\max} - x_{k\min}) / N_x$, $\sigma_{i,k} = (2 \sim 3)(x_{k\max} - x_{k\min}) / N_x$. v_i is the weight to be obtained through learning. First define the error function as

$$E = \frac{1}{2}(y_d - y_s)^2 = \frac{1}{2}(y_d - a_s^T w)^2, \tag{3}$$

where y_d is the desired output of the GCMAC.

For each input-output pair, the weight v_i is updated as the following equation:

$$\Delta v_i = -\frac{\alpha}{C}\frac{\partial E}{\partial v_i} = \frac{\alpha}{C}(y_d - a_s^T w(x_s))a_{s,i}b_i(x_s), \tag{4}$$

where C is generalization size and α is learning rate, $0 < \alpha < \min(2C / \sum_j b_j(x_j))$.

2.2 Model and Control for Linear Motor

Newton's law is introduced to derive the dynamical equation of a linear motor [11]:

$$m\frac{d^2 y}{dt^2} + c\frac{dy}{dt} + Ky = K_B I, \tag{5}$$

where y is output displacement of the linear motor, I is the input current to drive the linear motor. m is the total mass of motion component, c is the viscous friction coefficient. K is the elastic coefficient, K_B is the force constant.

With Laplace transformation, Eq. (5) can be rewritten in a transfer function form:

$$G(s) = \frac{K_B}{ms^2 + cs + K} = \frac{K_1\omega_n^2}{s^2 + 2\zeta\omega_n s + \omega_n^2}, \tag{6}$$

where damp ratio $\zeta = c/(2\sqrt{mK})$, frequency $\omega_n = \sqrt{K/m}$ and gain $K_1 = K_B / K$.

The algorithm to realize a combined PD and GCMAC control for the linear motor can be presented here. It is completed with Matlab in an M file form.

Algorithm 1:

```
1. setup the differential equation based on Eq.(6).
2. Set the reference signal.
3. Initialize N_h, C, α, v and other parameters.
4. Encode the input value of the GCMAC.
5. FOR    //each sampling moment
      Calculate output and error of linear motor.
      Calculate the output of GCMAC using Eq.(1).
      Compute the output of PD controller.
      Update weight of GCMAC Using Eq.(4).
      Update parameters of differential equation.
      Compute the differential of error.
   END
```

2.3 Optimal Parameter Selection Based on PSO Algorithm

The PSO algorithm first randomly initializes a swarm of particles, corresponding to individuals. Each particle is represented as $X_i = (x_{i,1}, x_{i,2}, ..., x_{i,n})$, $i = 1,2,..., N$, where N is the swarm size, and n is the total dimension number of each particle. Each particle adjusts its trajectory toward its own previous best position P_i and the previous best position P_g attained by the whole swarm. In each iteration k, the i th particle with respect to the j th dimension is updated by

$$v_{ij}^{(k+1)} = \omega v_{ij}^{(k)} + c_1 r_1 (P_{ij}^{(k)} - x_{ij}^{(k)}) + c_2 r_2 (P_{gj}^{(k)} - x_{ij}^{(k)}), \tag{7}$$

$$x_{ij}^{(k+1)} = x_{ij}^{(k)} + v_{ij}^{(k+1)} , \tag{8}$$

where $x_{i,j}^{(k)}$ and $v_{i,j}^{(k)}$ are current position and velocity. c_1 and c_2 are acceleration constants, r_1 and r_2 are random numbers within [0, 1]. ω is the inertial parameter.

For our problem, each particle can be encoded as $X = \{\alpha\}$, and the total sum of tracking error can be taken as the objective function to evaluate each particle:

$$f(X) = \sum_{k=0}^{K} \sqrt{(r(k) - y(k))^2} = \sum_{k=0}^{K} |e_k| . \tag{9}$$

Figure 3 depicts the block diagram of using PSO algorithm to optimize the learning rate of GCMAC. The parameter X is updated according the following Algorithm 2. The objective function $f(\cdot)$ needs to be computed by calling Algorithm 1.

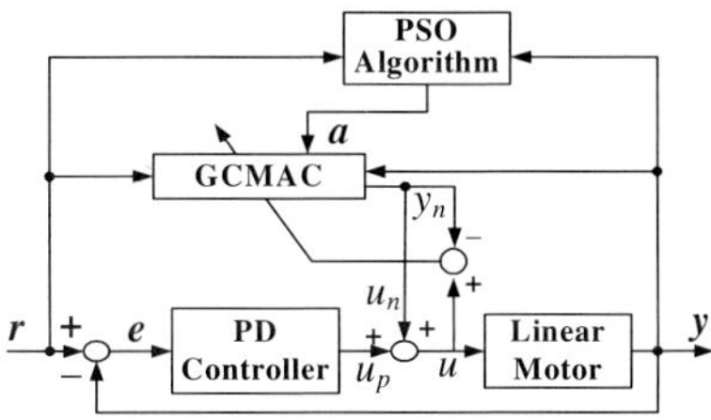

Fig. 3. System diagram of using PSO algorithm to optimize GCMAC

Algorithm 2:

1. Initializing: Generate randomly N particles X_i^0.
2. For each particle, set $P_i^0 = X_i^0$, and search for the minimum value $f_{\min}$ among $f(P_i^0)$, let $P_g^0 = X_{\min}$.
3. FOR //each iteration
   (1) Velocity and position updating: using Eq.(7) and (8).
   (2) Evaluate each particle: using Eq.(9).
   (3) Individual updating: If $f(P_i^k) > f(X_i^{k+1})$, then $P_i^{k+1} = X_i^{k+1}$.
   (4) Global updating: Search for the minimum value $f_{\min}$ among $f(P_i^{k+1})$, If $f(P_g^k) > f_{\min}$, then $P_g^{k+1} = X_{\min}$.

END

3 Numerical Simulation

A linear motor model used in wafer stage is taken from literature [13] as an example to verify the above control scheme. Its main parameters are set as shown in Table 1.

Table 1. Parameters of Linear Motor

m (kg)	c (N.s/m)	K (Kg/um)	K_B (N/A)	K_1	ζ	ω_n
10	0.0134	5	25	5×10^{-6}	0.3	707

The PD controller parameters are set as K_P=5 and K_D=3. Main parameters of the PSO algorithm are set as: N=10, $\omega = 0.7$, $c_1 = c_2 = 2$, Iterations times is set as 20. Parameters of GCMAC are set as: $C = 5$ and quantization size N_q =100.

Using the above algorithms, the optimization result is $\alpha = 0.2956$ with the objective value of 0.1762. The convergence process is shown in Fig. 4. Keeping $\alpha = 0.2956$ unchanged and making a fine tuning of the GCMAC's parameters with $C = 7$ and N_q =1000, Set the reference input displacement as $r = 0.2\sin(t)$ m, the linear motor's output displacement is shown in Figure 5. The corresponding error is also shown in Figure 5. It can be seen that the linear motor's displacement tracking error is within a range of -0.1mm<e<0.1mm. The GCMAC and PD controllers' outputs are shown in Figure 6. It is noted that the GCMAC take charge to the main control to restrain the disturbance and decrease the error.

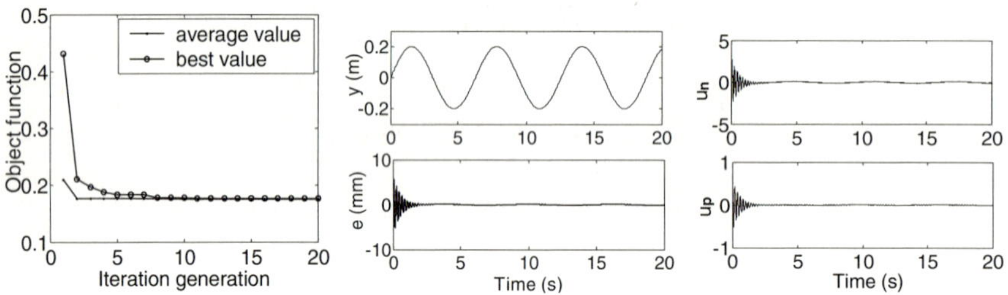

Fig. 4. Optimization process **Fig. 5.** Output and error **Fig. 6.** Controller output

Based on the above simulation model, a series of values of learning rates with uniform interval in range (0, 1) are taken to verify the previous optimization result, as shown in Table 2. it's reasonable to take $\alpha = 0.2956$ as the optimum.

Table 2. Tracking errors under different learning rates

Learning rate (α)	0.1	0.2	0.3	0.4	0.5	0.6	0.7	0.8~
Tracking error (e)	0.1127	0.0857	0.0780	0.0784	0.0864	0.1152	0.574	∞

4 Conclusions

In this paper, a combined PD feedback and GCMAC feed-forward control is presented for line motor. Introducing of particle swarm optimization can set the learning rate of the GCMAC in a best status, and enhance the tracking precision of the GCMAC controller. Simulation results show that the proposed control scheme is effective. Algorithm improvement and experiment will be discussed in the future.

Acknowledgement

This work is supported by National Science Foundation under grant No. 50390062.

References

1. Renton, D., Elbestawi, M.A.: Motion Control for Linear Motor Feed Drives in Advanced Machine Tools. International Journal of Machine Tools & Manufacture 41(4) (2001) 479-507
2. Dvoruk, S.K., Morozov, A.N., Pavlov, D.A., et al.: A Scanning System for the Michelson Interferometer with a Linear Motor Based on a Permanent Magnet. Instruments and Experimental Techniques 44 (3) (2001) 415-419
3. Spicer, D.F. Sackett, J.N. Black, C.T.: Precision X-Y Positioning Systems Using Linear Induction Motors. IEE Colloquium 494(1998) 1-3
4. Prischow, G.: A Comparison of Linear and Conventional Eletromechanical Drives. Annals of CIRP 47 (1998) 541-548
5. Li, X., Yao, B.: Output Feedback Adaptive Robust Precision Motion Control of Linear Motors. Automatica 37 (7) (2001) 1029-1039
6. Li, Y.F., Wikander, J.: Model Reference Discrete-Time Sliding Mode Control of Linear Motor Precision Servo Systems. Mechatronics 14 (7) (2004) 835-851
7. Hwang, C.L., Wei, M.H.: Improved Fuzzy Sliding-mode Control for a Linear Servo Motor System. Control Engineering Practice 5 (2) (1997) 219-227
8. Albus, J.S.: A New Approach to Manipulator Control: The Cerebellar Model Articulation Controller. Journal of Dynamic Systems, Measurement, and Control 97 (1975) 220-227
9. Chiang, C., Lin, C.: CMAC with General Basis Functions. Neural Network 9(7) (1996) 1199-1211
10. Kennedy, J.E., Eberhart, R.: Particle Swarm Optimization. In: Proceedings of the IEEE International Conference on Neural Networks. Vol. 4. Perth. Australia (1995) 1942-1948
11. Zhao, Q., Yan, S.: Collision-free Path Planning for Mobile Robots Using Chaotic Particle Swarm Optimization. In: Wang, L., Chen, K., Ong, Y. S.(eds.): Advances in Natural Computation. Lecture Notes in Computer Science, Vol. 3612. Springer-Verlag, Berlin Heidelberg New York (2005) 632-635
12. Qi, H., Li, X.: Improvement of Speedy Algorithm of CMAC Model for Gauss Basis Functions and its Application. Journal of Nanjing University of Science and Technology 29(2) (2005) 140-143
13. Song, W., Yu, G., Wang, Y., et al.: PID Control of Micro Feed Mechanism Based on Magnetic Levitation Technology. Journal of Harbin Institute of Technology 36 (1) (2004) 28-31

PID Control of Nonlinear Motor-Mechanism Coupling System Using Artificial Neural Network

Yi Zhang[1], Chun Feng[2], and Bailin Li[2]

[1] School of Electrical Engineering, Southwest Jiaotong University, Chengdu, Sichuan 610031, China
hi_zhangyi@163.com
[2] School of Mechanical Engineering, Southwest Jiaotong University, Chengdu, Sichuan 610031, China
ifengchun@163.com

Abstract. The basic assumption that the angular velocity of the input crank is constant in much mechanism synthesis and analysis may not be validated when an electric motor is connected to driven then mechanism. First, the controller-motor-mechanism coupling system is studied in this paper, numerically simulation result demonstrate the crank angular speed fluctuations for the case of a constant voltage supply to DC motor. Then a novel algorithm of motor-mechanism adaptive PID control with BP neural network is proposed, using the approximate ability to any nonlinear function of the neural network. The neural network are used to predicted models of the controlled variable, this information is transferred to PID controller, through the readjustment of the pre-established set. The simulation results show that the crank speed fluctuation can be reduced substantially by using feedback control.

1 Introduction

Planar four-bar mechanisms have a wide range of applications in industry, and the theory of analysis and synthesis is a well-developed subject in mechanics. One of the misleading assumptions in the analysis and synthesis of such mechanisms is that the angular velocity of the input link is constant. However, this may not be the case when an electric motor is connected to driven the mechanism through a gearbox. The changing inertia during the rotation of the mechanism produce varying external torque, the driving torque generated by the motor is no longer constant, then a periodically time changing behavior in crank angular speed is observed. Since the assumption of a constant crank speed is essential for the design of the mechanism to realize timing requirements, a PID controller with BP neural network is proposed in this paper to reduce the angular speed fluctuations introduced by the inertia effects rather than adding a mass flywheel, or adjusting the gear ratio.

Tao and Sadler [1] designed a PID controller and applied nonlinear programming techniques to determine the optimal controller gains presenting the best constant speed behavior for a four-bar mechanism. An experimentally validated simulation appeared in the paper by Dulger and Uyan [2], where they modeled and simulated a

J. Wang et al. (Eds.): ISNN 2006, LNCS 3972, pp. 1096 – 1103, 2006.

four-bar mechanism driven by a brushless servomotor. Other two applications of PD controller in closed-loop mechanisms were designed by Li et al. [3], Zhang and Chen [4]. They tried both a mass redistribution and a PD control to a four-bar mechanism to achieve a better system performance in their works.

In this paper, a proportional-plus-integral derivative (PID)-type controller incorporating neural network compensation is designed to regulate the crank angular speed fluctuations introduced by the inertia of the rotor and the rotating bars in a four-bar linkage when an electric motor is used to drive the mechanism. The popularity of PID controller can be attributed partly to their robust performance in a wide range of operating conditions and partly to their functional simplicity, which allows engineers to operate them in a simple, straightforward manner [5]. Therefore, the PID controller is widely used in industrial applications. However, the PID parameters are often tuned by experiences or simply by trial and error. In this paper, the PID parameters can be obtained systematically according to desired performances. Neural network control, simulating some human intelligence, is a nonlinear control basically independent of model and has the ability of learning and adaptive of environment. The neural network combining with PID control can partly solve the problem of PID controller parameter's real-time turning.

2 Motor-Mechanism Coupling System

2.1 Mathematical Model of a Four-Bar Mechanism

Fig.1. represents a general four-bar linkage. In the figure, the center of mass of each body is denoted by a dark circle and their locations are described by $\boldsymbol{r}_i$ and $\boldsymbol{\theta}_i$. For link $\boldsymbol{i}$, $\boldsymbol{m}_i$ denotes the mass of the link, $\boldsymbol{J}_i$ is the mass moment of inertia with respect to the centroid, and $\boldsymbol{L}_i$ is the length of the link. A torsional spring with a stiffness constant $\boldsymbol{k}$ and a torsional damper with a damping constant $\boldsymbol{C}$ are attached to the follower of the mechanism to represent a general loading situation. The crank angle$\boldsymbol{\varphi}_2$ is used as the generalized coordinate describing the motion of the linkage.

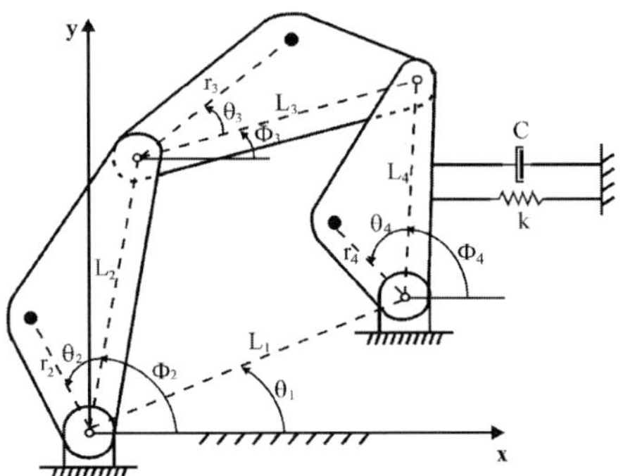

Fig. 1. Schematic of a four-bar Mechanism.

Lagrange's equation can be written to obtain the equations of motion of the mechanism as

$$\frac{d}{dt}(\frac{\partial K}{\partial \dot{\phi}_2})-\frac{\partial K}{\partial \phi_2}+\frac{\partial P}{\partial \phi_2}+\frac{\partial D}{\partial \dot{\phi}_2}=T \ . \tag{1}$$

where ***K*** denotes the kinetic energy, ***P*** denotes the potential energy, ***D*** denotes the dissipated energy, ***T*** denotes the externally applied torque.

Kinetic energy can be expressed as:

$$K=\sum_{2}^{4}(\frac{1}{2}m_i(V_{ix}^2+V_{iy}^2)+\frac{1}{2}J_i\dot{\phi}_i^2) \ . \tag{2}$$

where φ_i denotes the angular velocity of the link ***i***, V_{ix} and V_{iy} represent the ***x*** and ***y*** velocity components of the center of mass of link ***i*** . φ_3 and φ_4 can be expressed as the functions of φ_2 , V_{ix} and V_{iy} an be written as

$$V_{ix}=\alpha_i\dot{\phi}_2, \ V_{iy}=\beta_i\dot{\phi}_2, \ \dot{\phi}_i=\gamma_i\dot{\phi}, \ i=2,3,4 \ . \tag{3}$$

By substituting Eq.(3) into Eq. (2) , the kinetic energy expression is

$$K=\frac{1}{2}A(\phi_2)\dot{\phi}_2^2 \ . \tag{4}$$

where $A(\phi_2)=\sum_{2}^{4}(m_i(\alpha_i^2+\beta_i^2)+J_i\gamma_4^2)$.

Equation (1) becomes

$$A\ddot{\phi}_2+\frac{1}{2}\frac{dA}{d\phi_2}\dot{\phi}_2^2+k\gamma_4(\phi_4-\phi_{4.0})+C\gamma_4^2\dot{\phi}_2=T \ . \tag{5}$$

2.2 Mathematical Model of a Electric Motor

A schematic of the dc motor including a gearbox is demonstrated in Fig.2. The gear ratio is given as

$$n=T_b/T_a=\omega_a/\omega_b \ . \tag{6}$$

where ω_a and ω_b are the angular velocities of the shafts ***a*** and ***b***, respectively. The input to the motor-gear system is the voltage ***V***, and the output of the system is the torque T_b, which is equal to ***T*** defined in Eq.(1). ***R***, ***L*** and ***i*** are the armature resistance, inductance, and current, respectively. i_f, R_f and L_f are the field current, resistance, and inductance. ***J*** represents the moment of inertia of the rotor and ***B*** represents the viscous damping at the bearings, and T_L represents a constant mechanical load due to brush friction, gear friction, or dry bearing friction.

Using Kirchhoff's voltage law

$$V=Ri(t)+L\dot{i}(t)+e \ . \tag{7}$$

where ***e*** is the generated electromotive force of the motor. By using a Newtonian equation for the mechanism load, one can obtain

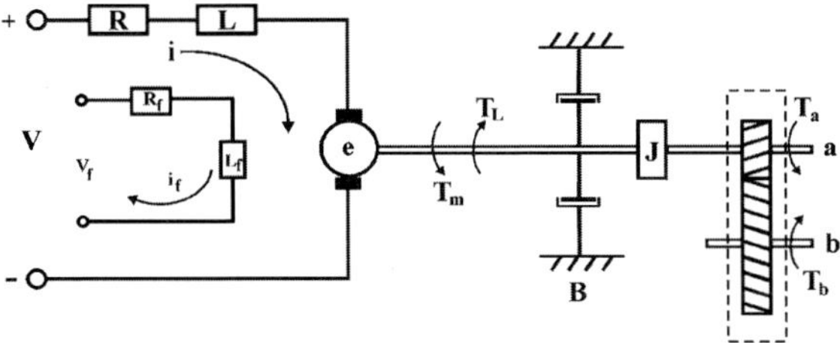

Fig. 2. Schematic of a DC motor and a gear box

$$T = n(T_m - T_L - B\omega_a - J\dot{\omega}_a) \ . \tag{8}$$

Where T_m in the equation represents the magnetic motor torque, and 'n'is the gear ratio defined in Eq.(1). The magnetic torque and the generated electromotive force are given by [1]

$$T_m = K_m i(t) \ \text{ and } \ e = K_g \omega_a \ . \tag{9}$$

where K_m and K_g represent the constants for motor torque and motor voltage, respectively. Because the shaft b is driving the crank, Eq. (6) can be changed

$$\omega_a = n\omega_b = n\dot{\phi}_2 \ . \tag{10}$$

The Eq.(9)–(10) are substituted into Eq.(7) and (8), one can obtain the mathematical model of the motor

$$\dot{i}(t) = \frac{1}{L}(V - Ri(t) - nK_g\dot{\phi}_2) \ . \tag{11}$$

$$\text{and } T = nK_m i(t) - nT_L - n^2 B\dot{\phi}_2 - n^2 J\ddot{\phi} \ . \tag{12}$$

2.3 State-Space Representation

Combining Eq.(5) and (12) , one can obtain the nonlinear equation of motion for the system as

$$A\ddot{\phi}_2 + \frac{1}{2}\frac{dA}{d\phi_2}\dot{\phi}_2^2 + k\gamma_4(\phi_4 - \phi_{4.0}) + C\gamma_4^2\dot{\phi}_2 = nK_m i(t) - nT_L - n^2 B\dot{\phi}_2 - n^2 J\ddot{\phi}_2 \ . \tag{13}$$

Let $x_1 = \phi_2, x_2 = \dot{x}_1 = \dot{\phi}_2, x_3 = i(t)$.

3 Motor-Mechanism System Simulation Result

The state-space representation of the motor-mechanism system (Eq.(13)) is simulated by using a Runge–Kutta fifth order numerical integration method for the open-loop step responses for the case tabulated in table 1. The initial conditions are taken as zero. For these simulations, the spring and damping constants are also taken as zero for simplicity. The results are presented in both graphical and tabular form for the step responses in Figs.3. and table 2.

Table 1. Mechanism and motor parameters

Mechanism parameter					Motor parameter		
Parameter	Link1	Link2	Link3	Link4	$R(\Omega)$	L(H)	K_m(N.m/a)
L(m)	0.5593	0.102	0.610	0.406	0.4	0.05	0.678
r(m)		0.0	0.305	0.203	K_g(v.s)	T_L(N.m)	J(kg.m2)
m(kg)		1.362	1.362	0.2041	0.678	0.0	0.056
J(kg.m2)		0.00071	0.0173	0.00509			B(N.m.s)
θ(rad)		0.0	0.0	0.0			0.026

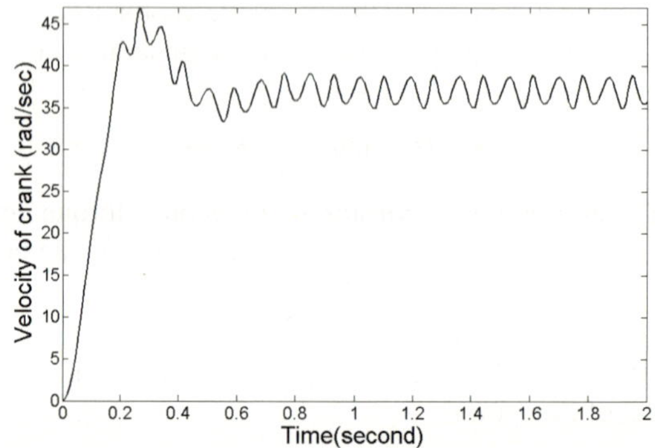

Fig. 3. Open-loop response of motor-mechanism system

Table 2. Summary of Open-loop response simulations

Speed Fl. (%)	Ave. speed(r/s)	Input Volts	Gear ratio
12.14	38.12	30	1:1

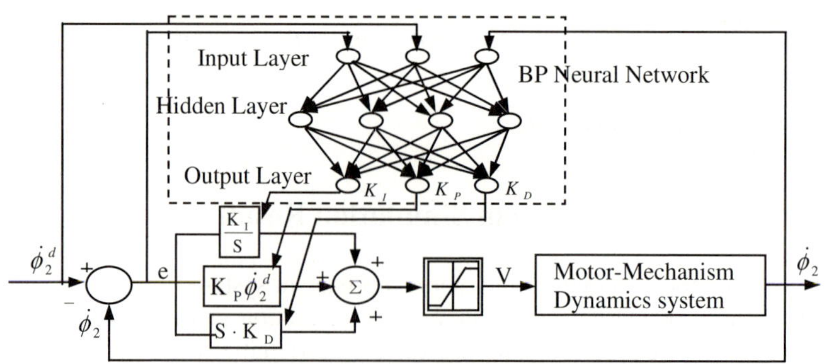

Fig. 4. Time varying PID control with BP neural network self-turning

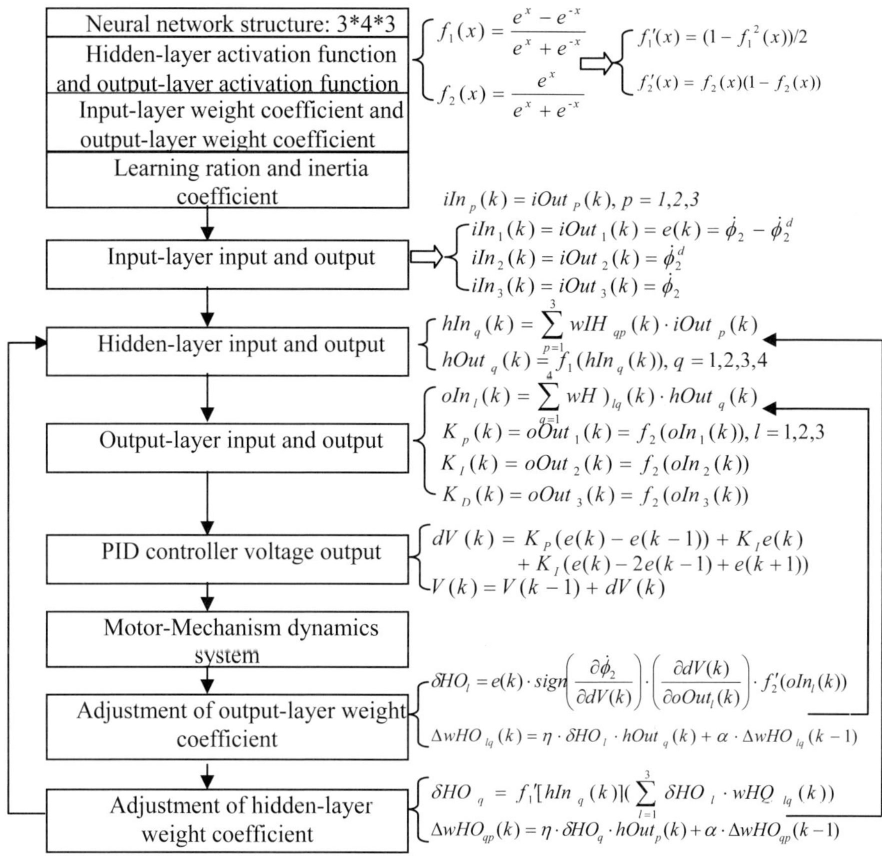

Fig. 5. Diagram of BP neural network self-turning PID controller

4 PID Neural Network Controller of Motor-Mechanism System

PID neural network is a new kind of networks. It utilizes the advantages of both PID control and neural structure. It consists of proportional (P), integral (I) and derivative (D) neurons and its weights are adjusted by the back-propagation algorithms. It can control different systems through quick learning process and has perfect performances. Fig.4. is the controller's scheme.

The whole BP algorithm is shown in Fig.5. Here we choose a three-layer BP neural network which are input-layer, hidden-layer and output layer. The input-layer has three neurons, the hidden layer has four and the output-layer has three. Considering the output junctions directly corresponding to three parameters can't be minus, the activation function of the output-layer neural network is defined as non-minus Sigmoid function, the activation function of the hidden layer neural network is defined as non-minus symmetrical Sigmoid function.

5 Experimental Results

The state-space representation of the motor-mechanism system (Eq.(13)) is integrated with the PID controller and without neural network. By trial and error, the values K_P=20, K_I=1, K_D=0.2 where selected, with a reference input angular velocity of 30(rad/s). The result for the PID controlled system using these gains is given in Fig.6. although the speed fluctuation was reduced quite well, there is a 13.1% overshoot because of the high proportional gain.

Using PID neural network controller we achieve a constant crack speed by adjusting the applied voltage accordingly. The simulation result for the cases tabulated in Tables 1 is presented in Fig.7. , the overshoot is 0.4232%, Rise time is 0.0427s, S.S. error is 0.0115%.

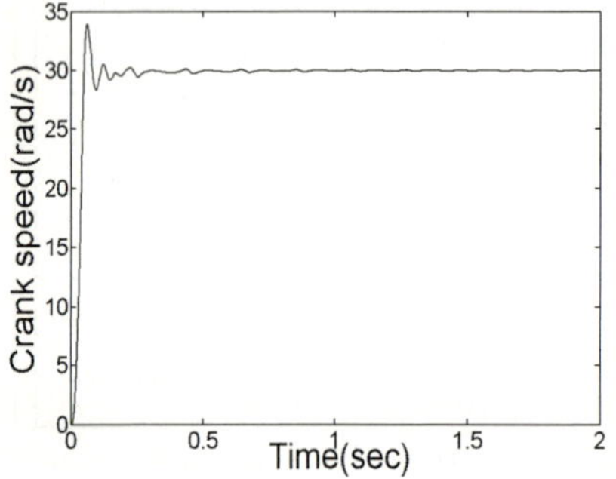

Fig. 6. PID Control result obtained result by trial and error

Fig. 7. PID neural network Control

6 Conclusions

PID neural network controller is a multilayered neural network and its structure is simple. PID neural network controller has abilities to regulate the crank speed of a four-bar mechanism and has perfect performance. Since the motor-mechanism system is nonlinear, BP neural networks are employed in order to identify controller gains which provide best performance in terms of steady speed fluctuation, steady state error, overshoot and rise time. The whole adjusting process is completed through self-learning and adaptive process. The examples considered illustrate the effectiveness of the approach, yielding significant reductions in speed fluctuation.

Acknowledgements

The authors would like to acknowledge the financial support from the Southwest Jiaotong University under grant 2005B11.

References

1. Tao, J., Sadler, J.P.: Constant Speed Control of a Motor Driven Mechanism System. Mech Mach Theory 30(5) (1995) 737–48
2. Dulger, L.C., Uyan, S.: Modeling, Simulation and Control of a Four-Bar Mechanism with Brushless Servomotor. Mechatronics 7(4) (1997) 369–83
3. Li, Q., Tso, S.K., Guo, L.S., Zhang, W.J.: Improving Motion Tracking of Servomotor-Driven Closed-Loop Mechanisms Using Mass-Redistribution. Mech Mach Theory 35(7) (2000) 1033–1045
4. Zhang, W.J., Chen, X.B.: Mechatronics Design for a Programmable Closed-Loop Mechanism. Proc Inst Mech Eng. 215(C) (2001) 365–75
5. Dorf, R.C., Bishop, R.H., Modern Control Systems, 8th edn. Reading, MA: Addison-Wesley (1998)
6. Gundogdu, O., Erenturk K.: Fuzzy Control of a DC Motor Driven Four-Bar Mechanism. Mechatronics 15 (2005) 423–438
7. Chu, S.Y., Teng, C.C.: Tuning of PID Controllers Based on Gain and Phase Margin Specifications Using Fuzzy Neural Network. Fuzzy Sets and Systems 101(1) (1999) 21–30

Design and Simulation of a Neural-PD Controller for Automatic Balancing of Rotor

Yuan Kang[1], Tsu-Wei Lin[1], Ming-Hui Chu[2], Yeon-Pun Chang[1] and Yea-Ping Wang[2]

[1] Department of Mechanical Engineering, Chung Yuan Christian University, Chung Li, Taiwan 32023, R.O.C.
yk@cycu.edu.tw
[2] Department of Automatic Engineering, Tung Nan Institute of Technology, Taipei, Taiwan 222, R.O.C.

Abstract. In this paper, the automatic balancing method is studied at constant speed or acceleration operation for rotor systems. The magnitude and phase of original imbalance is determined by the influence coefficient method through measurement of vibrations. In addition, a self-tuning neural-PD controller is designed to control the angular positions of correction masses for the automatic balancer. The equations of motion for a cylindrical rotor with radial imbalances are modeled to simulate two-plane balancing by means of axis-passed balancers. The dynamic responses before and after automatic balancing are investigated to justify the validity of the present method.

1 Introduction

The operation of rotating machinery is usually interrupted for the execution of balancing work. As a result, the efficiency, reliability and economy of the automatic balancer for rotating system are desired. The idea of a balancing system without interrupting the operation of rotating systems has received much attention. Goodman [1] extended the basic influence coefficient method to include the least squares and weighted least-square solutions. Van De Vegte and Lake [2] developed an automatic balancing system by measurement of the bearing vibrations. Sperling et al. [3] presented the numerical simulations and investigations for a rigid rotor to analyze the two-plane automatic balancing. Currently, the main process of automatic system in industry consists of try, comparison and correction by measuring vibrations and moving the correction masses to generate correction forces, which induces poor efficiency and low precision.

It is reasonable for an automatic balancing system to apply adaptive control to overcome the disturbance induced by varied operation speeds and mode shapes of a rotor. Akhyar and Omatu [4] proven that the parameters of a PID controller can be tuned by a neural network to control a nonlinear and time variant system. Omatu and Yoshioka [5] proposed a PID controller whose parameters are tuned by a neural network for the stable and fast response of an inverted pendulum, and the initial weights of the neural network can be tuned by genetic algorithm.

J. Wang et al. (Eds.): ISNN 2006, LNCS 3972, pp. 1104 – 1109, 2006.

2 On-Line Automatic Balancing for a Rotor System

The automatic balancing method based on the influence coefficient method is implemented by a digital computer. The imbalance of a rotating system is obtained by $u=A^{-1}r$, where A is denoted as the influence coefficient matrix representing the vibration due to the placement of a unit trial mass and r is original vibration measured.

The heavy rings of an automatic balancer are used as trial mass to produce correction masses by deriving DC motors to appropriate positions. In Fig.1, assume that the distance between the imbalance mass m and rotating center is r; the angle between r and X axis is θ; the correction masses are m_{b1} and m_{b2}; the radial distance of correction mass apart from the shaft center by R. According to the force equilibrium and vector decomposition, the objective angular positions of heavy rings are determined by

$$\phi_1^* = \theta + \pi - \frac{1}{2}\cos^{-1}\left(\frac{1}{2}\left(\frac{mr}{m_b R}\right)^2 - 1\right), \quad \phi_2^* = \theta + \pi + \frac{1}{2}\cos^{-1}\left(\frac{1}{2}\left(\frac{mr}{m_b R}\right)^2 - 1\right) \tag{1}$$

The automatic balancing system is shown in Fig. 2. The vibration signals are acquired in vibration measurement system by using eddy current proximity probes and data acquisition interfaces. A neural PD controller is designed to control the correction masses of automatic balancer moving to the objective positions. Fig. 3 shows the schematic diagram for a cylindrical rotor. The $OXYZ$ represents the inertial coordinate system. Some physical parameters are defined as follows. The mass, length and rotating

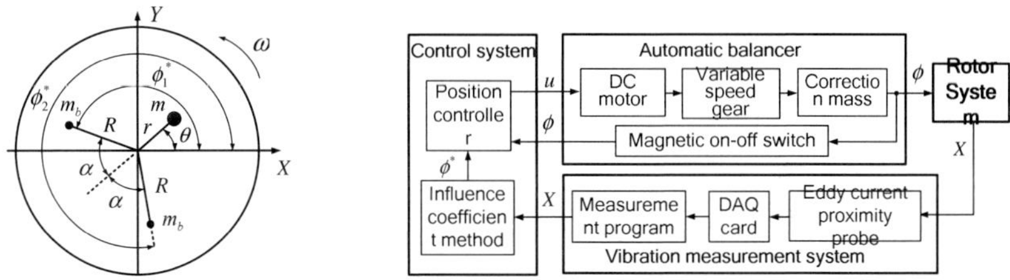

Fig. 1. Side view for automatic balancer **Fig. 2.** Block diagram of automatic balancing system

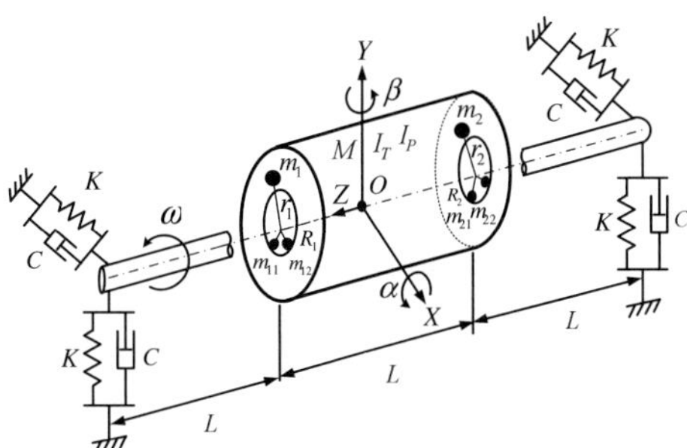

Fig. 3. Schematic diagram for cylindrical rotor

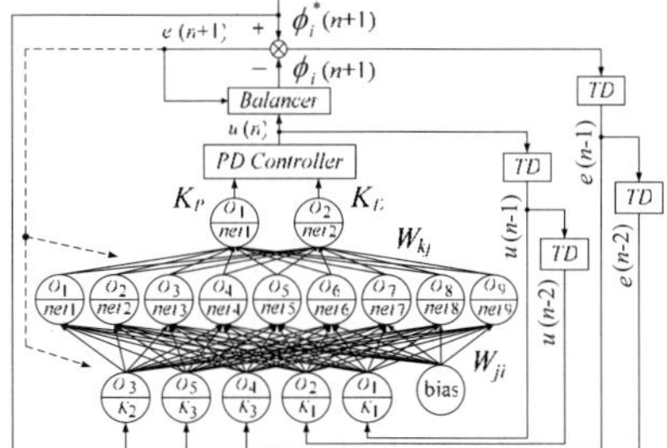

Fig. 4. The construction of neural tuner(TD: time delay)

speed of the rotor are denoted by M, L and ω, respectively; angular displacement about X and Y anises are α and β, respectively; polar and transverse moments of inertial are I_P and I_T, respectively; imbalance masses on the left and right side are m_1 and m_2, respectively, and the angles between X axis are θ_1 and θ_2; radial position of rotor on the both side are r_1 and r_1, respectively; the correction masses for the automatic balancers ar m_{11}, m_{12}, m_{21}, m_{22}; stiffness K and damping C for the both bearings. The dynamic equations for the cylindrical rotor with automatic balancers are expressed as

$$\begin{aligned}(M+m_1+m_2+4m_b)\ddot{X}+2C\dot{X}+2KX &= m_1r_1\omega^2\cos(\omega t+\theta_1)+m_2r_2\omega^2\cos(\omega t+\theta_2)\\ &+m_bR\omega^2[\cos(\omega t+\phi_{11})+\cos(\omega t+\phi_{12})]+m_bR\omega^2[\cos(\omega t+\phi_{21})+\cos(\omega t+\phi_{22})]\end{aligned} \tag{2}$$

$$\begin{aligned}(M+m_1+m_2+4m_b)\ddot{Y}+2C\dot{Y}+2KY &= m_1r_1\omega^2\sin(\omega t+\theta_1)+m_2r_2\omega^2\sin(\omega t+\theta_2)\\ &+m_bR\omega^2[\sin(\omega t+\phi_{11})+\sin(\omega t+\phi_{12})]+m_bR\omega^2[\sin(\omega t+\phi_{21})+\sin(\omega t+\phi_{22})]\end{aligned} \tag{3}$$

$$\begin{aligned}I_T\ddot{\alpha}+I_P\dot{\beta}+2CL^2\dot{\alpha}+2KL^2\alpha &= (L/2)\{2m_1r_1\omega^2\sin(\omega t+\theta_1)+(L/2)m_2r_2\omega^2\sin(\omega t+\theta_2)\\ &+m_bR\omega^2[\sin(\omega t+\phi_{11})+\sin(\omega t+\phi_{12})]+m_bR\omega^2[\sin(\omega t+\phi_{21})+\sin(\omega t+\phi_{22})]\}\end{aligned} \tag{4}$$

$$\begin{aligned}I_T\ddot{\beta}-I_P\omega\dot{\alpha}+2CL^2\dot{\beta}+2KL^2\beta &= (L/2)\{m_1\varepsilon_1\omega^2\cos(\omega t+\theta_1)+(L/2)m_2\varepsilon_2\omega^2\cos(\omega t+\theta_2)\\ &+m_bR\omega^2[\cos(\omega t+\phi_{11})+\cos(\omega t+\phi_{12})]+m_bR\omega^2[\cos(\omega t+\phi_{21})+\cos(\omega t+\phi_{22})]\}\end{aligned} \tag{5}$$

Two DC motors and gear mechanisms drive the correction masses. Also, the currents for driving the motor are delivered by brush and copper ring. A servo controller is designed to control the DC motors. The transfer function for angular displacement of correction masses and terminal voltage of motors is

$$\frac{\phi_i(s)}{E_a(s)}=\frac{N_TK_T}{R_aJ_ms^2+(R_aB_m+K_TK_b)s},\ i=1,2 \tag{6}$$

The parameters includes: the torque constant and back-emf coefficient of DC motor are $K_T=1.42\times10^{-2}$ N·m/A and $K_b=9.55\times10^{-3}$ V·s/rad, respectively; armature resistance R_a=18 Ohm; armature inertial $J_m=1\times10^{-6}$ kg·m^2; armature damper $B_m=5\times10^{-7}$ N·m; gear ratio $N_T=5.83\times10^{-3}$. Substituting above parameters into Eq. (6), one yields

$$\frac{\phi_i(s)}{E_a(s)}=\frac{0.459}{s^2+8s},\ i=1,2 \tag{7}$$

3 Position Controller Design for an Automatic Balancer

A self-tuning neural-PD controller is proposed to control the movement of correction masses fast and stably to the appropriate positions during balancing process. A discrete PD controller can be expressed by

$$u(n)=u(n-1)+K_P[e(n)-e(n-1)]+K_D[e(n)-2e(n-1)+e(n-2)] \tag{8}$$

The proportional and derivative parameters of PD controller can be on-line tuned by a neural network, which is called a neural tuner shown in Fig. 4. There are three layers with five, nine and two units in the input, hidden and output layers, respectively. The five inputs consist of the outputs of controller at previous sample times $u(n\text{-}1)$ and $u(n\text{-}2)$, the errors of objective position $e(n\text{-}1)$ and $e(n\text{-}2)$, and the objective position $\phi_i^*(n+1)$. The input signals are multiplied by gains K_1, K_2 and K_3 for normalization. The output signals are denoted as the proportional parameter K_P and the derivative parameter K_D. A simple sign function is used to approximate the Jacobin shown as

$$\frac{\partial \phi_i}{\partial u} = sign(\frac{\partial \phi_i}{\partial u_P}) \cdot \left| \frac{\partial \phi_i}{\partial u_P} \right| \tag{9}$$

Furthermore, the weights updated proposed by Omatu and Yoshioka [5].

4 Numerical Simulations

A magnetic on-off switch and a gear with 24 teeth are applied to measure the angular positions of the correction masses instead of the conventional rotary encoder due to the consideration of size and weight. The magnetic switch switches "on" when the tooth face of the gear closes to the head surface of the magnetic switch, and it switches "off" when the tooth face leaves the head surface. The on-off signals are converted to a series of pulse signals. By counting the number of pulses, the angular positions of correction masses can be evaluated. The position controls apply three different resolutions of automatic balancers with the speed ratio of 14/24000. The low- and high-resolution automatic balancers have the resolutions of 24 and 1000 pulse/rev, respectively; simultaneously, the maximum angular speed of correction masses is 7rpm. The ideal automatic balancer with the maximum speed of 140rpm has the possible minimum speed ratio fitting between the screw and screw washer.

The simulations of position control for the three balancers are shown in Fig. 5. The initial weights of the neural tuner are specified randomly between ±0.5; the normalized gains K_1= 0.075, K_2= 0.14 and K_3= 0.15; learning rate=0.1. The objective angular position ϕ_i^* is assumed as 120° and then changes to 60° in 15 seconds. For the three automatic

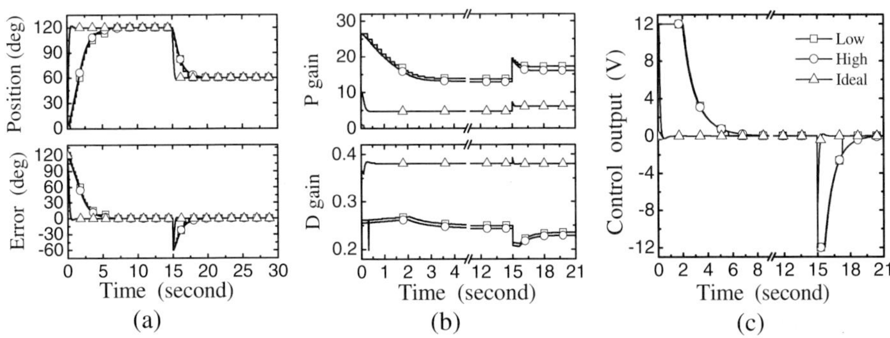

Fig. 5. Simulation results for position control of correction mass with neural-PD controller: (a) angular positions and errors, (b) P and D gains, (c) control voltage of DC motor

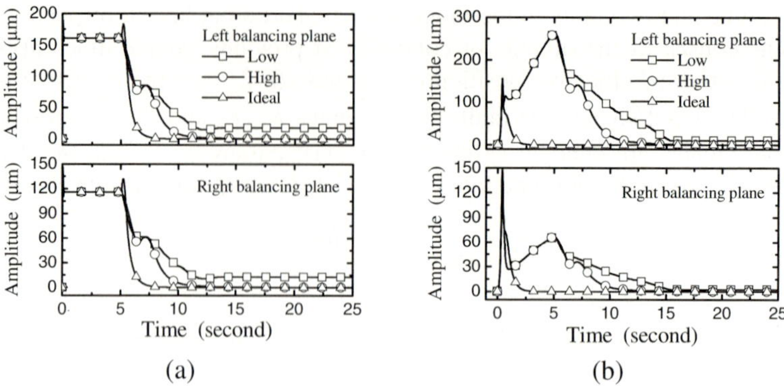

Fig. 6. Time responses of the rotor: (a) running at constant speed 1800 rpm operated with automatic balancers in 5 second, (b) accelerating up to 4800 rpm started with automatic balancers

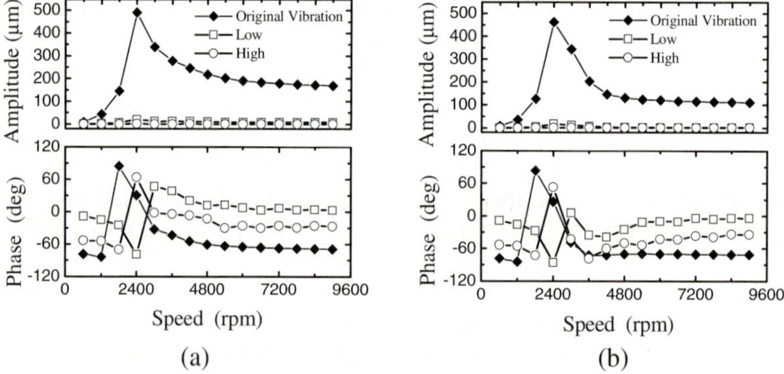

Fig. 7. The amplitudes and phases of whirls before and after balancing: (a) left balancing plane, (b) right balancing plane

balancers, the mean square errors (MSE) of the first position 120° are 6.3, 1.8, and 0.2, respectively; those in the second position 60° are 1.17, 0.02 and 0.06, respectively.

The parameters of automatic balancing system are specified as: M=3kg, I_P=0.04kg·m^2, I_P=0.0617kg·m^2, m_1=0.02kg, r_1=0.02m, θ_1=36°, m_2=0.015kg, r_2=0.03m, θ_2=162°, m_b=0.05kg, $\phi_{11}=\phi_{22}=0^{o}$, $\phi_{12}=\phi_{21}=180^{o}$, R_1=R_2=0.01m, L=0.2m, K=1×10^{5}N/m, and C=100N·s/m. The time responses of whirl for a rotor operate in two situations, which include running at constant speed 1800rpm and accelerating up to 4800rpm, as shown in Fig. 6. The vibrations, running at constant speed, in left balancing plane are decreased to 17.63, 0.44 and 0.42μm by using the three balancers, respectively; those in right balancing plane are decreased to 12.62, 0.45 and 0.46μm, respectively. The vibrations, during acceleration, in left balancing plane are decreased to 10.54, 1.54 and 1.35μm, respectively; those in right balancing plane are decreased to 7.22, 0.98 and 0.85μm, respectively. When rotating speed is over the critical speed 2400 rpm, the influence coefficient matrices obtained are varied as

$$A_1 = \begin{bmatrix} 0.0454\angle -118.5^\circ & 0.0381\angle -119.1^\circ \\ 0.0533\angle 113.4^\circ & 0.0464\angle 113.2^\circ \end{bmatrix}, \quad A_2 = \begin{bmatrix} 0.2489\angle -98.4^\circ & 0.1499\angle -108.2^\circ \\ 0.2608\angle 138.4^\circ & 0.1570\angle 128.7^\circ \end{bmatrix}$$

The amplitudes and phases of whirls before and after the automatic balancing under varied speed are shown in Fig. 7. The amplitudes of vibrations are decreased 95%, 99% and 99% by using the three different balancers, respectively.

5 Conclusions

In this paper, an on-line automatic balancing method based on influence coefficient method is proposed by designing a self-tuning neural-PD controller to control the correction masses of an automatic balancer. The simulation results show that the vibration can be decreased 90% during the transient state. Until it is operated in steady state, the vibration will be decreased 97%. The amplitude of residual whirling is below 1 μm. It can be proven that the proposed method is available for automatic balancing system and with high performance efficiency and stability.

Acknowledgement

This study was supported by the National Science Council of the R.O.C. with grant number NSC92-2623-7-033-003-ET.

References

1. Goodman, P.T.: A Least-Squares Method for Computing Balance Corrections. ASME J. Eng. Ind. 86 (1964) 273-279
2. Van de Vegte, J. and Lake, R.T.: Balancing of Rotating Systems Operation. J. Sound Vib. 57 (1978) 225-235
3. Sperling, L., Ryzhik, B., Linz, C. and Dukstein, H.: Simulation of Two-Plane Automatic Balancing of a Rigid Rotor. Math. Comput. Simul. 58 (2002) 351-365
4. Akhyar, S. and Omatu, S.: Self-Tuning PID Control by Neural Networks. Proceeding of International Joint Conference on Neural Networks (1993) 2749-2752
5. Omatu, S. and Yoshioka, M.: Self-Tuning Neural-PID Control and Applications. IEEE International Conference on Systems, Man, and Cybernetics, Computational Cybernetics and Simulation 3 (1997) 1985-1989

PD Control of Overhead Crane Systems with Neural Compensation

Rigoberto Toxqui Toxqui[1], Wen Yu[1], and Xiaoou Li[2]

[1] Departamento de Control Automático, CINVESTAV-IPN
A.P. 14-740, Av.IPN 2508, México D.F., 07360, México
yuw@ctrl.cinvestav.mx

[2] Sección de Computación, Departamento de Ingeniería Eléctrica
CINVESTAV-IPN
A.P. 14-740, Av.IPN 2508, México D.F., 07360, México

Abstract. This paper considers the problem of PD control of overhead crane in the presence of uncertainty associated with crane dynamics. By using radial basis function neural networks, these uncertainties can be compensated effectively. This new neural control can resolve the two problems for overhead crane control: 1) decrease steady-state error of normal PD control. 2) guarantee stability via neural compensation. By Lyapunov method and input-to-state stability technique, we prove that these robust controllers with neural compensators are stable. Real-time experiments are presented to show the applicability of the approach presented in this paper.

1 Introduction

Cranes are one of the most important systems for material handling of heavy goods. The need for faster cargo handling requires the control of the crane motion so that its dynamic performance is optimized. The control problem is that of reducing the swing of the payload while moving it to the desired position as fast as possible. Automatic cranes are comparatively rare in the industrial practice [15], because of the relation between investment costs and achievable cost saving. Since the swing of the payload depends on the acceleration of the trolley, minimizing the operation time and minimizing the payload swing are partially conflicting requirements.

It is a big challenging for control society. Several authors have considered control optimization techniques to be applied either the complete operation or to one of motions. The first idea in crane automation was proposed in [2], where time optimal control was considered by boundary conditions. This idea was further developed by [1] and [16]. In order to increase robustness, some requirements of time optimization have to be given up. One can use the condition of zero angular velocity at the target point [14]. To increase tracking accuracy, gain scheduling is a practicable method [4]. The nonlinearity of the cranes are viewed as the coupling of the actual rope length, rope angle and the crane trolley, model-based nonlinear control approach can improve dynamic behavior [19]. Observer-based feedback control was presented in [15]. The exact mathematical models for cranes

J. Wang et al. (Eds.): ISNN 2006, LNCS 3972, pp. 1110–1115, 2006.

systems are very important to controller design. Many attempts were made to introduce simplified models in order to construct "model-based" controller [15], for example planar operation [4], no-friction [14], etc.

The simplest stable controller for crane is proportional and derivative (PD) controller. Due to the existence of friction, gravity forces and the other uncertainties, the PD control cannot guarantee that the steady state error becomes zero. PID control does not require any component of robot dynamics in its control law, but it lacks global asymptotic stability [7]. Several efforts have been made to improve performance of PD controllers for robots. Global asymptotic stability PD control was realized by pulsing gravity compensation in [16]. If the friction and gravity are unknown, neural networks can be applied. In [11] the authors use neural networks to approximate the whole nonlinearity of robot dynamic. When the parameters in gravitational torque vector are unknown, adaptive PD control with gravity compensation was introduced by [12].

In this paper we make the following modifications: 1) we do not need structure information of uncertainties, neural networks are used to approximate 2) PD control with neural compensation is applied to overhead crane, 3) we give a new proof by means of Lyapunov and input-to-state stability analysis. The learning rules is obtained from the tracking error analysis.

2 PD Control of Overhead Crane

The overhead crane system described schematically in Fig.1 can be formed into a 3-inputs and 5-output systems. The dynamics of the overhead crane system can be obtained by Lagrangian method [10]

$$M(x)\ddot{x} + V(x,\dot{x})\dot{x} + G(x) + T\dot{x} = u \tag{1}$$

The control problem is to design a controller u to move the rail in such a way that the actual position of the payload reaches desired one. Form Fig.1 and (1)

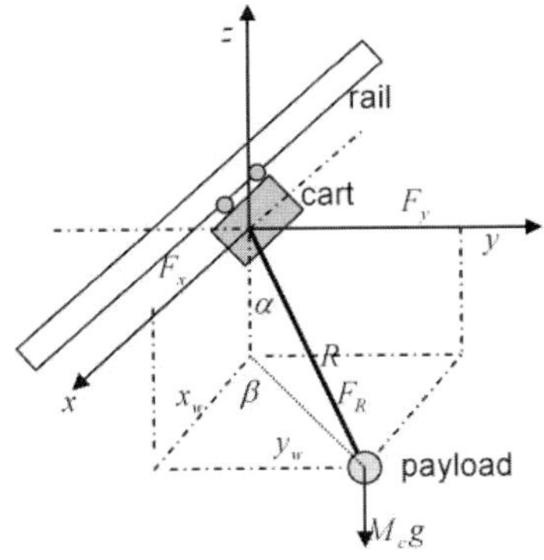

Fig. 1. Overhead crane

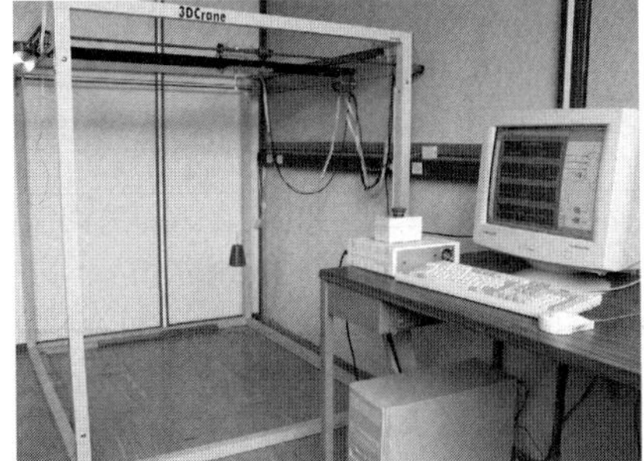

Fig. 2. 3D overhear crane

we find there are 3 control inputs $[F_x, F_y, F_R]$ which can force the crane to $[x_w, y_w, R]$, the swing angles $[\alpha, \beta]$ cannot be controlled by (1). In order to design an anti-swing control, let us analyze linezrization models for $[\alpha, \beta]$. Because the acceleration of the crane is much smaller than the gravitational acceleration, the rope length kept slowly varying and swing is not big, we have $|\ddot{x}_w| \ll g$, $|\ddot{y}_w| \ll g$, $\left|\ddot{R}\right| \ll g$, $\left|\dot{R}\right| \ll R$, $|\dot{\alpha}| \ll 1$, $\left|\dot{\beta}\right| \ll 1$, $s_1 = \sin\alpha \approx \alpha$, $c_1 = \cos\alpha \approx 1$. The approximated dynamic of $[\alpha, \beta]$ is $\ddot{\alpha} + \ddot{x}_w + g\alpha = 0$, $\ddot{\beta} + \ddot{y}_w + g\beta = 0$. Since $\ddot{x}_w = \frac{F_x}{M_r}$, $\ddot{y}_w = \frac{F_y}{M_m}$, the dynamic of swing angles are $\ddot{\alpha} + g\alpha = -\frac{F_x}{M_r}$, $\ddot{\beta} + g\beta = -\frac{F_y}{M_m}$. We design the control force F_x and F_y as

$$\begin{aligned} F_x &= A_1(x_w, \dot{x}_w) + A_2(\alpha, \dot{\alpha}) \\ F_y &= B_1(y_w, \dot{y}_w) + B_2\left(\beta, \dot{\beta}\right) \end{aligned} \tag{2}$$

Substitute these controllers into (1), the position control model is

$$M(x)\ddot{x} + V(x, \dot{x})\dot{x} + G(x) + T\dot{x} + D = u_1 \tag{3}$$

where $D = [A_2, B_2, 0, 0, 0]^T$, $u_1 = [A_1, B_1, 0, 0, F_R]^T$. The swing angles control model is $\ddot{\alpha} + g\alpha + \frac{A_1}{M_r} = -\frac{A_2}{M_r}$, $\ddot{\beta} + g\beta + \frac{B_1}{M_m} = -\frac{B_2}{M_m}$. If we regard $\frac{A_1}{M_r}$ and $\frac{B_1}{M_r}$ as disturbance, $\frac{A_2}{M_r}$ and $\frac{B_2}{M_m}$ as control inputs, (3) is second-order linear systems with disturbances. Standard PID control can be applied to regulate α and β to zero $A_2 = k_{p1}\alpha + k_{d1}\dot{\alpha} + k_{i1}\int_0^1 \alpha dt$, $B_2 = k_{p2}\beta + k_{d2}\dot{\beta} + k_{i2}\int_0^1 \beta dt$, where k_{pi}, k_{di} and k_{ii} are positive constants which are correspond to proportional, derivative and integral gains. For position control, we use PD control with neural networks compensator , which has the following form

$$u_1 = -K_p\left(x - x^d\right) - K_d\left(\dot{x} - \dot{x}^d\right) + \widehat{W}_t\sigma(\widehat{V}_t x) \tag{4}$$

where $x = [x_w, y_w, \alpha, \beta, R]^T$, $x^d = \left[x_w^d, y_w^d, 0, 0, R^d\right]^T$, x_w^d, y_w^d and R^d are the desired positions, in regulation case $\dot{x}_w^d = \dot{y}_w^d = \dot{R}^d = 0$. $K_p = diag\,[k_{p3}, k_{p4}, 0, 0, k_{p5}]$, $K_d = diag\,[k_{d3}, k_{d4}, 0, 0, k_{d5}]$. $\widehat{W}_t$ and $\widehat{V}_t$ are time-varying weights which are determined by the neural networks learning law. $\widehat{W}_t\sigma(\widehat{V}_t x)$ is a RBF neural network which can be expressed as $y = \widehat{W}_t\sigma(\widehat{V}_t x)$, where $y = [y_1, ..., y_5]^T$, $\widehat{W}_t \in R^{5\times n}$, $\widehat{V}_t \in R^{n\times 5}$, n is node number in hidden layer, $\sigma(x) = [\sigma_i...\sigma_n]^T$. The activation function used is $\sigma_i(s) = \exp\left(-\frac{(s-C_i)^2}{P_i}\right)$, where C_i represents the center of the basis function, P_i is variance representing the spread of the basis function. In this paper, we assume the Gaussian function is fixed, *i.e.* C_i and P_i are constants. We only adjust $\widehat{W}_t$ and $\widehat{V}_t$ in the RBF neural networks, our other paper [20] gives more detail on how to tune $\sigma(x)$. According to the Stone-Weierstrass theorem [5], a general nonlinear smooth function can be written as $f(x) = W^*\sigma(V^*x) + \mu(t)$, where W^* and V^* are optimal weights, $\mu(t)$ is the modeling error. First we consider RBF neural networks can approximate the friction $T\dot{x}$, gravity $G(x)$ and swing angles controller D exactly, *i.e.*, $\mu(t) = 0$

$$G(x) + T\dot{x} + D = W^*\sigma(V^*x) \tag{5}$$

We define position regulation error as $e = x - x^d$. The following theorem gives asymptotic stability of PD control with neural compensation for overhead crane systems.

Theorem 1. *If the neural networks can model the friction, gravity and swing angles controller exactly, the PD control with neural compensator* $u_1 = -K_p e - K_d \dot{e} + \widehat{W}_t \sigma(V^* x)$ *with following learning law*

$$\dot{\widehat{W}}_t^T = -K_w \sigma(V^* x) \dot{e}^T, \quad K_w \in R^{n \times n}, K_w > 0 \tag{6}$$

can make the position control loop asymptotic stability. If the derivative gain K_d *in the PD control (4) is selected such that* $K_d > \Lambda_\mu^{-1} + \Lambda_\sigma^{-1}$, *where* Λ_μ^{-1} *and* Λ_σ^{-1} *are given matrix, the updating laws for the weights of neural compensator in (4) are*

$$\dot{\widehat{W}}_t^T = -K_w \sigma(\widehat{V}_t x) \dot{e}^T, \dot{\widehat{V}}_t^T = -K_v x W^* D_\sigma \dot{e}^T \tag{7}$$

where K_w, K_v *are positive defined matrices, then PD control with neuro compensation (4) can make tracking error stable. The average tracking error* $\dot{e}$ *converges to*

$$\limsup_{T \to \infty} \frac{1}{T} \int_0^T \|\dot{e}\|_Q^2 \, dt \leq \overline{\eta}_\mu + \overline{\eta}_\sigma \tag{8}$$

where $Q = K_d - \Lambda_\mu^{-1} - \Lambda_\sigma^{-1}$.

Remark 1. The tracking error will converge to the ball radius the upper bounded of $\left(\overline{\eta}_\mu + \overline{\eta}_\sigma\right)$, and it is influenced by the prior known matrices W^*. Theorem 2 shows that W^* does not influence the stability property, we may select any value for W^* at first. From Theorem 2 we know the algorithm (6) can make the identification error convergent. W^* may be selected by following off-line steps: 1) Start from any initial value for W^*, 2)Do on-line identification with W^*, 3) Let W_t as new initial condition, *i.e.*, $W^* = W_t$, 4) If the identification error decreases, repeat the identification process, go to 2. Otherwise, stop off-line identification, now W_t is final value for W^*. From the definition of the Lyapunov function, we may see that the learning rules (6) will minimize the tacking error $\dot{e}$. This structure is different from normal neural networks which are used for approximation of nonlinear function [18]. The terms $-K_w \sigma(\widehat{V}_t x) \dot{e}^T$ and $-K_v x W^* D_\sigma \dot{e}^T$ are correspond to backpropagation scheme, only in backpropagation W^* is changed to W_t.

3 Experimental Case Study

The experiment is carried out on the Quanser overhead crane system (see Fig.1). The rail is 150cm long. The payload is about 500g. A/D-D/A board is based on a Xilinx FPGA microprocessor, which is a multifunction analog and digital timing I/O board dedicated to real-time data acquisition and control in the

Windows XP environment. The board is mounted in a PC Pentium-III 500MHz host computer. Because Xilinx FPGA chip supports real-time operations without introducing latencies caused by the Windows default timing system, the control program is operated in Windows XP with Matlab 6.5/Simulink. The sampling time is about 10ms. The responses of normal PD control $u_1 = -K_p\left(x - x^d\right) - K_d\left(\dot{x} - \dot{x}^d\right)$ with $K_p = 8$, $K_d = 0.1$, are shown in Fig.3. We can see that there are steady-state errors. Now we use PD control with neural compensation (4), the position regulation results are shown in Fig.4. We can also find the great differences between swing angle control (2) (Fig.5) and without swing angle control (Fig.6)

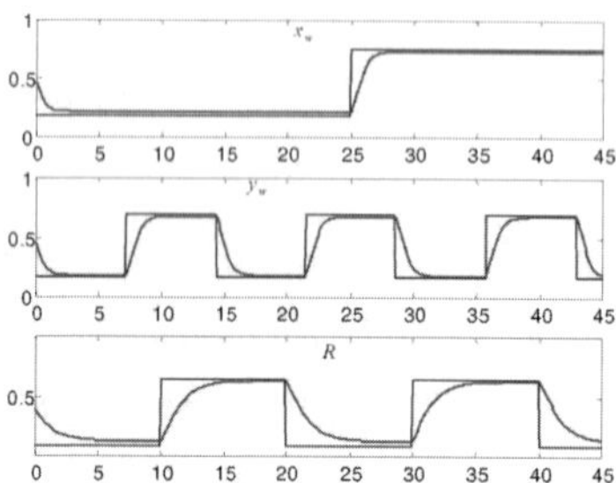

Fig. 3. Overhead crane

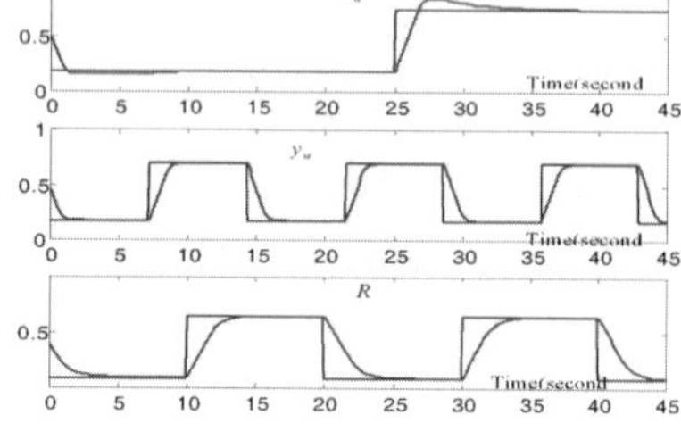

Fig. 4. 3D overhear crane

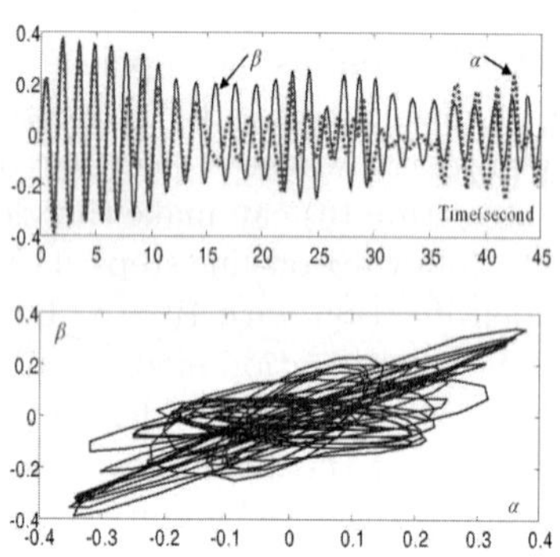

Fig. 5. Overhead crane

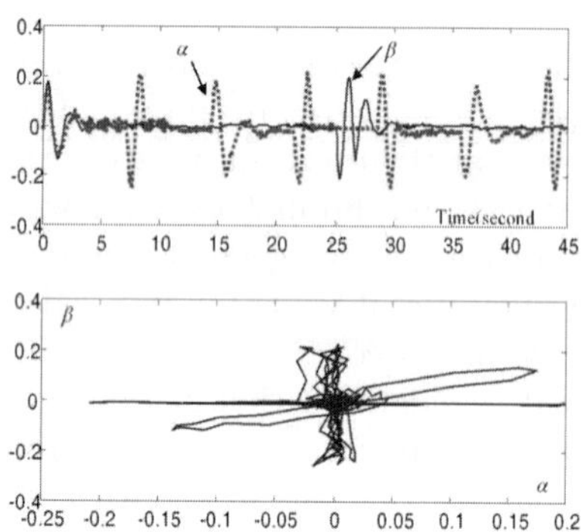

Fig. 6. 3D overhear crane

4 Conclusion

The main contributions of this paper are: 1) a new PD control scheme for overhead crane is proposed, which includes neural compensation and swing angles control, 2)by Lyapunov method and input-to-state stability technique, the stability of the closed-loop system are proved, 3) real-time experiments are presented. Future work will be carried out on improvement of the transient properties.

References

1. Auernig, J.W., Troger, H.: Time Optimal Control of Overhead Cranes with Hoisting of the Payload. Automatica 23(2) (1987) 437-447
2. Beeston, J.W.: Closed-Loop Time Optimatial Control of a Suspended Payload-a Design Study. Proc. of 4th IFAC World Congress, (1969) 85-99
3. Butler, H,. Honderd, G., Van Amerongen, J.: Model Reference Adaptive Control of a Gantry Crane Scale Model. IEEE Contr. Syst. Mag. 11(1) (1991) 57-62
4. Corriga, G., Giua, A., Usai, G.: An Implicit Gain-Scheduling Controller for Cranes. IEEE Trans. Control Systems Technology 6(1) (1998) 15-20
5. Cybenko, G.: Approximation by Superposition of Sigmoidal Activation Function. Math.Control, Sig Syst 2(2) (1989) 303-314
6. Fang, Y., Dixon,W.E., Dawson,D.M., Zergeroglu, E.: Nonlinear Coupling Control Laws for an Underactuated Overhead Crane System. IEEE/ASME TRANSACTIONS ON MECHATRONICS 8(2) (2003) 418-423
7. Kelly, R.: A Tuning Procedure for Stable PID Control of Robot Manipulators. Robotica 13(1) (1995) 141-148
8. Khalil, H.K:. Nonlinear Systems. 3rd Edition, Prentice Hall (2002)
9. Lee, H.H.: Modeling and Control of a Three-Dimensional Overhead Crane. Journal of Dynamic Systems, Measurement, and Control 120(2) (1998) 471-476
10. Lee, H.H.:A New Motion-Planning Scheme for Overhead Cranes With High-Speed Hoisting. Journal of Dynamic Systems, Measurement, and Control 126(2) (2004) 359-364
11. Lewis, F.L., Parisini, T.: Neural Network Feedback Control with Guaranteed Stability. Int. J. Control 70(1) (1998) 337-339
12. Moustafa, K.A., Ebeid, A.M.: Nonlinear Modeling and Control of Overhead Crane Load Sway. Journal of Dynamic Systems, Measurement, and Control 110(1) (1988) 266-271
13. Mohan, B.M., Patel, A.V.: Analytical Structures and Analysis of the Simplest Fuzzy PD Controller. IEEE Trans. on Syst., Man and Cybern. Part B 32(2) (2002) 239-248
14. Noakes, M.W., Jansen, J.F.: Generalized Input for Damped-Vibration Control of Suspended Payloads. Journal of Robotics and Autonomous Systems 10(2) (1992) 199-205
15. Sawodny, O., Aschemann, H., Lahres, S.: An Automated Gantry Crane as a Large Workspcae Robot. Control Engineering Practice 10(11) (2002) 1323-1338
16. Sakawa ,Y., Shindo, Y.: Optimal Control of Container Cranes. Automatica 18(1) (1982) 257-266
17. Sontag, E.D., Wang, Y.: On Characterization of the Input-to-State Stability Property. System & Control Letters 24(3) (1995) 351-359
18. Yu, W., Li, X.; Some New Results on System Identification with Dynamic Neural Networks. IEEE Trans. Neural Networks 12(2) (2001) 412-417
19. Yu, J., Lewis, F.L., Huang, T.: Nonlinear Feedback Control of a Gantry Crane. Proc. of 1995 American Control Conference, (1995) 4310-4315
20. Yu, W., Li, X.: System Identification Using Adjustable RBF Neural Network with Stable Learning Algorithms. Lecture Notes in Computer Science, Vol.3174. Srpinger-Verlgag, Berlin Heidelberg New York (2004) 212-217

A Study on Intelligent Control for Hybrid Actuator

Ke Zhang

School of Mechanical and Automation Engineering, Shanghai Institute of Technology, 120 Caobao Road, 200235 Shanghai, P.R. China
zkwy@hotmail.com

Abstract. A hybrid actuator is presented in this paper. Hybrid actuator is a new type of planar parallel robot, and requires precise control of the position of the mechanism. In order to achieve the desired accuracies, nonlinear factors as friction must be accurately compensated in the real-time servo control algorithm. According to the characteristics of the hybrid actuator, a hybrid intelligent control algorithm based on PID control and cerebellar model articulation control (CMAC) techniques was presented and used to perform control of hybrid actuator for the first time. Simulation results show that this method can improve the control effect remarkably compared with the traditional control strategy.

1 Introduction

The idea of hybrid machines was initially investigated by Tokuz and Jones [1], [2], and is a field of study with full potential. A hybrid mechanism is a configuration that combines the motions of two characteristically different electric motors by means of a mechanism to produce programmable output. Where one of the motions coming from a constant speed motor provides the main power, a small servo-motor introduces programmability to the resultant actuator. Demand for greater machine productivity with improved quality, diversity of product, competition on market, and industrial automation have accelerated needs for new alternative ideas like hybrid machines to generate programmable output motion [3]. Such machines will introduce to users greater flexibility with programmability option, and energy utilization will be realized at maximum.

Position control accuracy of motion determines the profile accuracy of hybrid actuator. For high precision hybrid actuator, factors such as temperature, vibration, friction, and backlash all have to be properly addressed. Temperature and vibration problems can be substantially reduced with proper mechanical design, environmental temperature regulation. Other nonlinear factors as friction, however, must be addressed by appropriate control of hybrid actuator in real time.

Standard proportional-integral-derivative (PID) type control algorithms are not capable of dealing with these phenomena within the required accuracy. Friction will cause limit cycle oscillations around the desired position under PID control or a steady state positioning error under PD control [4]. In order to have high precision motion control, the nonlinear factors must be accurately compensated in real-time control. Two different approaches can be taken to compensate the factors: 1. to compensate using mathematical models based on physical modeling [5], 2. to compensate using a

J. Wang et al. (Eds.): ISNN 2006, LNCS 3972, pp. 1116 – 1123, 2006.

learning algorithm controller [6]. Here we shall discuss the second approach, and compare results to the results of a standard PID controller, where the nonlinear factors were treated as disturbance of system.

2 Hybrid Actuator Description

Hybrid seven-bar actuator is the most representative planar parallel robot. Fig.1 represents seven link mechanism configuration having all revolute joints except one slider on output link. The hybrid mechanism has an adjustable link designed to include a power screw mechanism for converting rotary motion to linear motion by means of a small slider. The crank is driven by a DC motor (the main motor) through a reduction gearbox; the slider is driven by a lead screw coupled the second servomotor (the assist motor). Here the main motor is applied as a constant speed motor, and the constant speed motor profile is applied. Point-to-point positioning is certainly achieved for both motors, and the system output is taken from the last link.

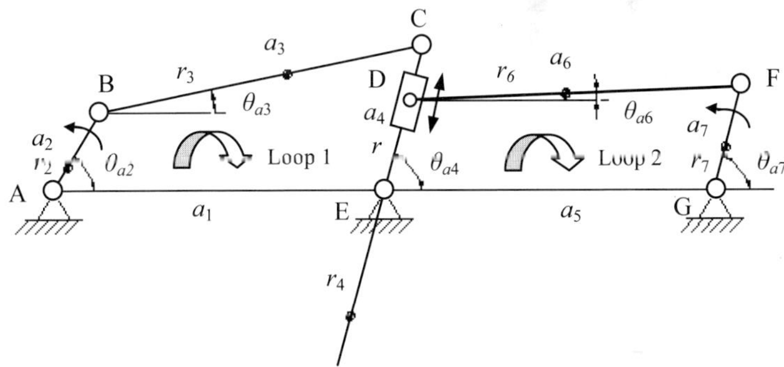

Fig. 1. Schematic diagram of seven link mechanism

$a_1, a_2, a_3, a_4, a_5, a_6, a_7$ link lengths of the seven bar mechanism (m)

$\theta_i, \dot{\theta}_i$ ($i = a_2, a_3, a_4, r, a_6, a_7$) angular displacement, angular velocity of the above links (rad, rad/s)

r_i positions to the centre of gravity in local coordinates (m)

r displacement of the slider on *CE* link relative to *E* point (m)

m_i masses of the links (kg)

G_i gravity of the links (N)

J_{is} link moment of inertias on the mass centre of the links(kgm^2)

M_0 the main driving torque (Nm)

M_ψ drag torque on output link (Nm)

The seven bar mechanism is shown with its position vectors in Fig. 1. By referring to Fig. 1, the loop closure equation is written as:

$$\overrightarrow{AB} + \overrightarrow{BC} = \overrightarrow{EC} + \overrightarrow{AE} \tag{1}$$

$$\overrightarrow{ED} + \overrightarrow{DF} = \overrightarrow{GF} + \overrightarrow{EG} \tag{2}$$

The above equations may be written in complex polar notation:

$$a_2 e^{i\theta_{a2}} + a_3 e^{i\theta_{a3}} = a_4 e^{i\theta_{a4}} + a_1 e^{i\theta_{a1}} \tag{3}$$

$$r e^{i\theta_{a4}} + a_6 e^{i\theta_{a6}} = a_7 e^{i\theta_{a7}} + a_5 e^{i\theta_{a5}} \tag{4}$$

Intermediate steps for the kinematic analysis are omitted herein. By solving first vector loop equation (3), angular positions of the link 3 and 4 are obtained. Similarly, solution of second loop equation has given the angular position of link 6 and link 7. They are definitely needed during the analysis of control model. Inputs necessary from the crank (θ_{a2}) and the slider (r) then calculated by using inverse kinematics. The motion of the slider is required following accurately with the crank motion.

3 Description of CMAC Algorithm

CMAC is essentially a table look up algorithm which is adaptive. It is adaptive since it can modify the table contents using a learning algorithm. It also has generalization capability due to the distributed storage of information. In CMAC, each state variable is quantized and the problem space is divided into discrete states. A vector of quantized input value specifies state and is used to generate addresses for retrieving information from memory for this state. Information is distributively stored. CMAC is composed of two basic components (Fig. 2):

1. Mapping an input space into a finite number of memory locations,
2. A learning algorithm based on the difference between a reference and the CMAC response.

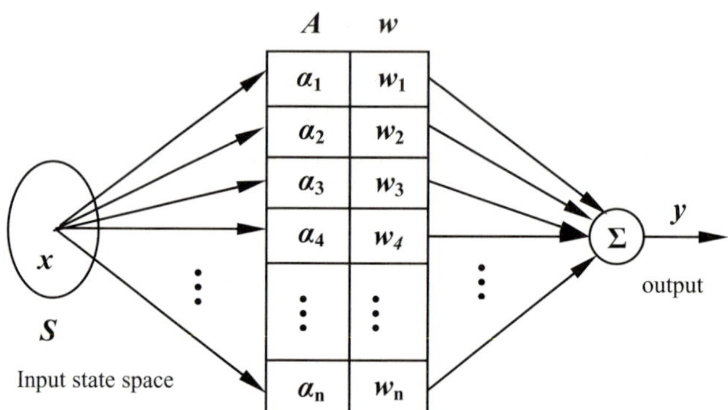

Fig. 2. Block diagram of the CMAC algorithm

The idea is that given a state in the input space, find the corresponding memory locations for that state, sum the content of these memory locations to get the response of the CMAC, compare it with the desired response, and modify the content of these active memory locations based on the learning algorithm. The main advantages of CMAC are:

1. Reduction in memory requirement,
2. Distributed storage of information,
3. Ability to learn nonlinear input-output mapping.

The mapping algorithm proposed by Albus [7] is defined a series of mapping from an input "state" vector, S, to a scalar output, y,

$$y = h(S) \tag{5}$$

where $h(S)$ represents the overall mapping of a particular state S to the scalar output y. this mapping can easily be extended for multidimensional I/O mapping, $Y = H(S)$ where $H(S)$ represents the multidimensional mapping of an input vector S, to an output vector Y. $H(S)$ consists of two smaller mappings in the follows order,

$$S \rightarrow A \rightarrow Y \tag{6}$$

Where S is sensory input vector, A is the binary association matrix, and Y is the output response.

4 Hybrid Control of Hybrid Actuator Based on CMAC

Hybrid control structure of hybrid actuator is shown in Fig. 3. Control system via hybrid controller based on CMAC and PID controller perform feedforward and feedback control respectively, where CMAC controller perform feedforward control to control inverse dynamic model of control object, PID controller perform closed-loop feedback control to ensure the stability of control system and restrain disturbance. The CMAC's learning ability of mapping functions make it well suited

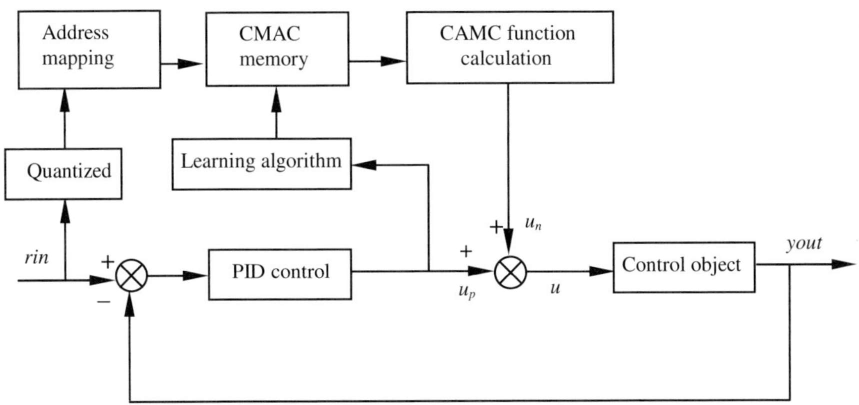

Fig. 3. Hybrid control structure based on CMAC and PID controller

for learning nonlinear factor characteristics. In the context of servo control, the CMAC and PID controller are added to the control system in parallel. A typical servo-control system normally has a process, actuators with appropriate power amplification sources, and sensors. The CMAC controller works in parallel with the servo loop controller.

The CMAC controller initialized to zero in all of its memory content. As the hybrid actuator starts the move, the CMAC contribution to the control action is zero. The learning algorithm compares the total control action and the CMAC control action, and modifies the content of active memory locations in the direction to minimize the difference. In other words, the objective function of the learning algorithm is to minimize the difference between the total control action and the CMAC control action contribution. The learning algorithm is as follows:

(1) Calculate CMAC control response,

$$u_n(k) = \sum_{i=1}^{c} w_i a_i \tag{7}$$

where w is the content of the active memory cells addressed for the current input state sample, a_i is binary choice vector, c is the generalization size, $u_n(k)$ is CMAC output.

(2) Compare it with the total control response from the CMAC and PID controllers, $u(k)$, and update the content of active memory locations by

$$\Delta w(k) = \eta \frac{u(k) - u_n(k)}{c} a_i = \eta \frac{u_p(k)}{c} a_i \tag{8}$$

$$w(k) = w(k-1) + \Delta w(k) + \alpha(w(k) - w(k-1)) \tag{9}$$

where η is the learning rate, $0 < \eta < 1$, $u_p(k)$ is PID output. In order to quicken convergence speed, inertia term is added to (8), α is inertia ($0 < \alpha < 1$).

(3) Repeat until the required accuracy criteria is satisfied. Criteria of weight-update is

$$E(k) = \frac{1}{2}(u(k) - u_n(k))^2 \cdot \frac{a_i}{c} \tag{10}$$

Control algorithm of hybrid controller is as follows:

$$u(k) = u_n(k) + u_p(k) \tag{11}$$

The learning stops when u and u_n are equal, the PID output is zero. The learning in CMAC continues as long as there is a non-zero output from the PID controller. The only way the learning stops is when all of the total control action is coming from the CMAC controller, none from the PID controller. Zero output from PID controller means perfect tracking. Therefore, the CMAC will continue to learn the necessary control law in order to achieve perfect tracking control. In practical applications, the learning may be stopped when u_p is less than a small threshold value compatible with the existing noise level in the system. It is important to notice that the CMAC does not copy PID function. It uses the PID as part of its learning algorithm to determine how well the control objectives are met. Other types of CMAC applications can be developed where the PID is replaced by some other block which provides a measure to tell the CMAC about the current performance.

5 Simulated Hybrid Actuator Motion Control

Mechanical properties of seven link mechanism; a_1=375mm, a_2=102mm, a_3=375mm, a_4=167mm, a_5=375mm, a_6=375mm, a_7=120mm. These link lengths and angle values for hybrid seven bar actuator were obtained in the studies of optimal kinematic design [8], [9]. We choose transfer function of control object as follows:

$$G(s)=\frac{20}{0.03s^2+s} \tag{12}$$

The CMAC parameters chosen in this work were N=300 memory cells, with a generalization size of c=5 and a learning rate η =0.1. The memory size, N, was chosen to be as small as possible while still giving a good resolution for the inputs. The generalization size and learning rate were chosen on a trial-and-error basis to determine what values give the best results. An increase in the generalization size increases the smoothness of the CMAC output. A generalization size of c=5 was found to be a good choice for both a smooth output and high accuracy. The learning rate of the CMAC affects the rate of change of the weights in the memory. A high learning rate, η, close to 1 was found to increase the overshoot and decrease the stability of the system. A low learning rate, $\eta \ll 1$, significantly reduced the speed at which the CMAC learned to control the system. A learning rate of η =0.1 was found to be a good choice.

Two different cases were simulated: the motion of the hybrid actuator under, 1. the PID controller alone (Fig. 3-CMAC is disconnected), 2. the PID plus the CMAC controller (Fig. 3-CMAC is connected). A disturbance was added to input of the control system. The results for the PID case alone are presented in Fig. 4. PID control would provide similar results except that instead of finite steady state error, there would be limit cycle oscillation about the desired position due to the integral action. The results for the hybrid control are shown in Fig. 5.

The PID algorithm, as expected, also had large steady state errors and large position errors. The CMAC, when it is connected shows significant improvement over the PID algorithm immediately. The steady-state position error is reduced almost to zero within the learning cycle. The CMAC behaves much like the addition of integral control. After the process has been repeated and the CMAC controller takes over, however, the errors are reduced by orders of magnitude and the system tracks almost perfectly.

In order to learn a function which is highly dependent on an input state component, the resolution of this input mapped into the CMAC memory must be properly adjusted. A function which varies little with a change in an input state component can be accurately represented with a low mapping resolution. It is also worth noting that, while PID controller alone act on control system, plus k_p play an important role of control effect, but while using hybrid controller based on PID and CMAC control, control effect is not dependent on k_p, k_p may be chosen in appropriate range.

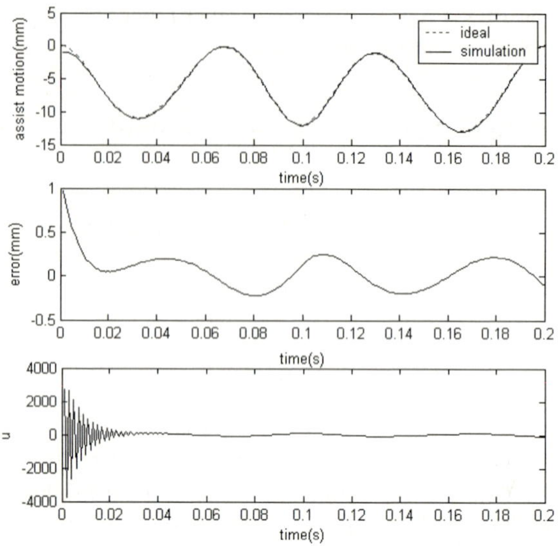

Fig. 4. Simulation results under PID controller

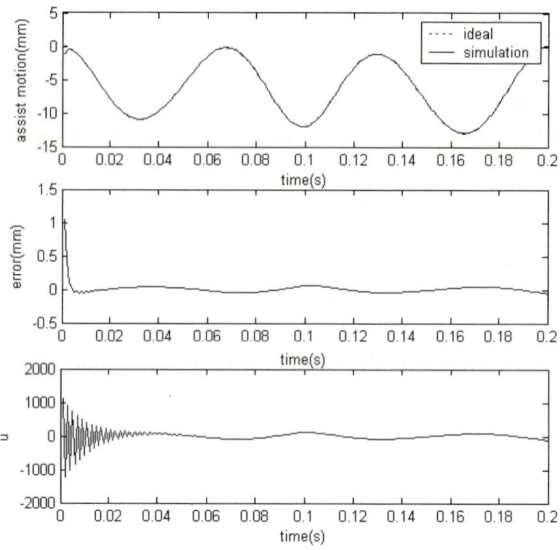

Fig. 5. Simulation results under hybrid controller

6 Conclusions

The goal of this work was to develop an improved servo-motion controller to meet the requirements of hybrid actuator. The standard PID controller is known to be inadequate for this goal. The hybrid controller proposed in this paper can provide extremely low position errors. Simulation results show that the CMAC learning controller was able to learn the disturbance in the hybrid actuator very quickly, and improve position control accuracy by at least one order of magnitude compared with PID controller.

References

1. Tokuz, L. C., Jones, J. R.: Programmable Modulation of Motion Using Hybrid Machines. Proceedings of ImechE, C414/071 (1991) 85-91
2. Tokuz, L. C., Jones, J. R.: Hybrid Machine Modelling and Control. PhD thesis, Liverpool Polytechnic (1992)
3. Shengze Wang, Yujuan Shen.: Dynamics Analysis and Design of a Varistructure Linkage. China Mechanical Engineering 13(3) (2002) 225-228
4. Yang, S., and Tomizuka, M.: Adaptive Pulse Width Control for Orecise Positioning Under The Influence of Stiction and Coulomb Friction. ASME Journal of Dynamic Systems, Measurement, and Control 110 (1988) 221-227
5. Cetinkunt, S., Yu, W. L., and Donmez, A.: Friction Characterization Experiments on A Single Point Deamond Turning Machine Tool. Int. J. Mach. Tool Manufact. 34(1) (1994) 19-32
6. Larsen, G. A., Cetinkunt, S., and Donmez, A.: CMAC Neural Network Control for High Precision Motion Control in The Presence of Large Friction. Journal of Dynamic Systems, Measurement, and Control 117 (1995) 415-420
7. Albus, J. S.: A New Approach to Manipulator Control: The Cerebellar Model Articulation Control (CMAC). Trans. ASME Journal of Dynamic Systems, Measurement, and Control 97(2) (1975) 220-227
8. Kireçci, A., Dülger, L. C.: A Study on a Hybrid Actuator. Mechanism and Machine Theory 35(8) (2000) 1141-1149
9. Dülger, L. C., Kireçci, A., and Topalbekiroglu, M.: Modelling and Simulation of Hybrid Actuator, Mechanism and Machine Theory 38(3) (2003) 395-407

Double Inverted Pendulum Control Based on Support Vector Machines and Fuzzy Inference

Han Liu, Haiyan Wu, and Fucai Qian

School of Automation and Information Engineering,
Xi'an University of Technology, 710048 Xi'an, China
liuhan@xaut.edu.cn

Abstract. In this paper, a fuzzy inference system based on support vector machines is proposed for nonlinear system control. Support vector machines provides a mechanism to extract support vectors for generating fuzzy if-then rules from the training data set, and a method to describe the fuzzy inference system in terms of kernel functions. Thus it has the inherent advantages that the model doesn't have to decide the number of fuzzy rules in advance, and has universal approximation ability and good generalization ability. The simulation results for stabilizing control of double inverted pendulum system are provided to show the validity and applicability of the proposed method.

1 Introduction

Nonlinear control is of great theoretical and practical importance in various fields of science and engineering. Although the framework of nonlinear control offers various solutions to the problems, the applications in real life still impose increasing demanding conditions, in the presence of which the problem become either mathematically intractable or a feasible solution unreachable. An alternative way is to develop a strategy that handles different operating conditions with different decision mechanisms, that is, to exploit the theory of fuzzy inference system. The main purpose of fuzzy inference system is to model human decision-making within the conceptual framework of fuzzy logic and approximate reasoning, and the traditional two methods to extract fuzzy rules have been discussed. One is cluster-partition [1], the major disadvantage of which is that the number of fuzzy rules increases exponentially with the number of input/output variables, and the noise patterns and data distribution have much affection to the cluster centers. The other is orthogonal least square-based approach (OLS) [2], and for its forward and greedy scheme to determine the optimal subset with the minimum empirical error, it could become trapped in a local minimum, so maybe has dissatisfaction generalization ability.

To overcome above drawbacks, support vector machines (SVM) [3] is proposed to find the reduced number of rules for the given precision. SVM can be seen as an approximate implementation of what Vapnik has defined as Structure Risk Minimization(SRM), and it has been widely used in pattern recognition, time series prediction, density estimation and so on[4,5].

In this paper, the inference framework realizes the task of function approximation between the input and output spaces in terms of a set of fuzzy if-then rules. Firstly, the

J. Wang et al. (Eds.): ISNN 2006, LNCS 3972, pp. 1124 – 1130, 2006.

properties of the SVM learning model and its utilization in fuzzy rules extraction are explored. And then the kernel function is related to the fuzzy basis function to mix the two mechanisms into a new fuzzy rule-based modeling method. Finally, the proposed method is applied to control double inverted pendulum system. Simulation results are provided to show the validity and applicability of proposed method.

2 Fuzzy Inference System Based on SVM

The SVM is derived from statistical learning theory. It determines support vectors and weights by minimizing an upper bound of generalization error. Assuming the given training data set $\{(x_i, y_i)\} i = 1,2,...,l$, where l is the number of data, x_i are input data, y_i are the corresponding target values, the task of support vector regression is to find the optimal $f(x, \mathbf{w}^*)$ from the linear function set $\{f(x, \mathbf{w})\}$ to approximate the dependent relationship between the input and output data. For the case decision surface, the original input data are firstly mapped into a high dimensional feature space, and the functional form of the mapping $\phi(x)$ does not need to be known since it is implicitly defined by the choice of kernel function which must satisfies Mercer's Theory [3] as (1)

$$K(x_i, \mathbf{x}) = < \phi(x_i) \cdot \phi(\mathbf{x}) > . \tag{1}$$

By introducing the e-insensitive loss function [6], the task can be transformed to minimize

$$\begin{aligned} &\min_{w,b,x_i,x_i^*} \|\mathbf{w}\|^2/2 + C\sum_{i=1}^{l}(\varepsilon_i + \varepsilon_i^*) \\ &s.t.\ \ y_i - (\mathbf{w}\cdot x_i + b) \le e - \varepsilon_i \\ &\qquad (\mathbf{w}\cdot x_i + b) - y_i \le e - \varepsilon_i^*,\ \ \varepsilon_i, \varepsilon_i^* \ge 0,\ \forall i, \end{aligned} \tag{2}$$

where e measures approximation precision, ε_i and ε_i^* are introduced as slackness variables when precision of f surpasses e, and the constant $C > 0$ determines the trade-off between the flatness of f and the amount up to which deviations larger than e are tolerated. Its dual quadratic programming-regression problem is

$$\begin{aligned} &\max_{w,b,x_i,x_i^*} \{-\frac{1}{2}\sum_{i,j=1}^{l}(\alpha_i - \alpha_i^*)(\alpha_j - \alpha_j^*)K(x_i, x_j) + \sum_{i=1}^{l}(\alpha_i - \alpha_i^*)y_i - \sum_{i=1}^{l}(\alpha_i + \alpha_i^*)e\} \\ &s.t.\ \sum_{i=1}^{l}(\alpha_i - \alpha_i^*) = 0,\quad 0 \le \alpha_i, \alpha_i^* \le C,\ i = 1,2,...,l, \end{aligned} \tag{3}$$

where α_i and α_i^* are Lagrange multipliers. In solving the above optimal problem, the vectors from the training set that associate with nonzero Lagrange multipliers are called the support vectors, and the decision function can be described as

$$f(x) = \sum_{i=1}^{l}(\alpha_i - \alpha_i^*)K(x_i, \mathbf{x}) + b, \tag{4}$$

where b is the bias value.

In the fuzzy inference system, the most common fuzzy rule base system consists of a set of linguistic rules in the following form

$$R_j\text{: If } x_1 \text{ is } A_1^j \text{ and } x_2 \text{ is } A_2^j \text{ and ... and } x_n \text{ is } A_n^j$$
$$\text{Then } z \text{ is } B_j, \qquad \text{for } j = 1, 2, \ldots, M,$$

where M is the number of fuzzy rules, $x_i\,(i = 1,2,\ldots,n)$ are the input variables, z is the output of the fuzzy system, and A_i^j and B_j are linguistic terms characterized by fuzzy membership functions $\mu_{A_i^j}(x_i)$ and $\mu_{B^j}(z)$, respectively. Let the fuzzy basis function be denoted as

$$p_j(x) = \prod_{i=1}^{n} \mu_{A_i^j}(x) \Big/ \sum_{j=1}^{M} \prod_{i=1}^{n} \mu_{A_i^j}(x), \tag{5}$$

then overall fuzzy inference function $f(x)$ can be viewed as a linear combination of the fuzzy basis function

$$f(x) = \sum_{i=1}^{M} p_j(x) z^j, \tag{6}$$

where z^j is the point in the output space at which $\mu_{B^j}(z)$ achieves its maximum value. And it has been proved that the fuzzy inference function is capable of uniformly approximating any real continuous function on a compact set to any arbitrary accuracy [7].

In this paper, the fuzzy basis function is chosen as the kernel function of SVM to mix those two mechanisms into a new fuzzy inference system. First, the denominator of the fuzzy basis function must be removed since the number of fuzzy rules is not available in advance. However, this scheme does not violate the spirit of fuzzy inference system, that is, the if-part determines the confidence values of the rules fired and the then-part aggregates the inference consequence by the confidence values of corresponding rule. It follows that the modified fuzzy basis function is

$$p_j(x) = \prod_{k=1}^{n} \left(-\frac{1}{2} \left(\frac{x_k - \overline{x}_k^{\,j}}{\sigma_k} \right)^2 \right), \tag{7}$$

where $\overline{x}_k^{\,j}$ and σ_k are real-valued parameters. The modified fuzzy basis function can be an eligible candidate of kernel function of the SVM since it satisfied the Mercer's Theorem. Using the support vectors obtained, a set of if-then fuzzy rules can be derived. One single linguistic fuzzy if-then rule is related to one fuzzy basis function naturally. Thus the overall fuzzy inference system can be represented as series expansion of fuzzy basis function, and this also makes the inference system itself to be interpretable. In order to obtain good approximated results, it needs to tune some parameters in SVM. At present, this procedure needs to try, that is, there is no unified method to determine the parameters of SVM. Here, cross-validation method is used. The SVM-based fuzzy basis function inference architecture is shown in Fig.1.

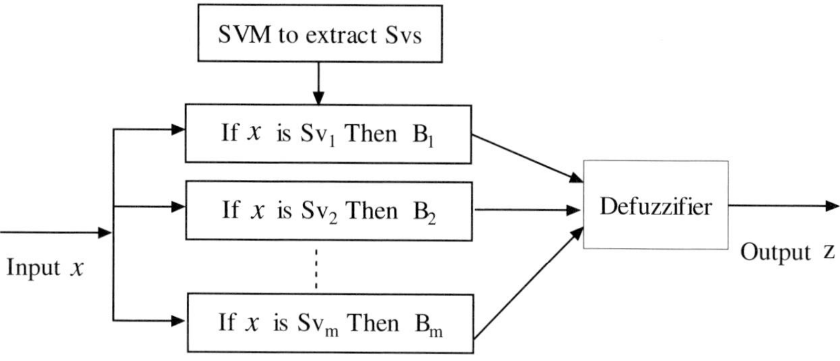

Fig. 1. SVM-based fuzzy basis function inference architecture

3 Simulation and Analysis

Double inverted pendulum system is nonlinear and well known for evaluating the validation of various kinds of control theories, as it is analogous to several real life systems. The structure of cart-type double inverted pendulum is shown in Fig.2. From the control engineering point of view, stabilizing control of pendulum at the upward unstable equilibrium position has attracted much attention from many researchers.

In this section, the SVM-based fuzzy inference system is implemented to approximate a controller of cart-type double inverted pendulum. LQR method is firstly used to generate a set of randomly sampled input-output pairs $(\boldsymbol{x},\boldsymbol{u})$ to construct the training data set, and then the proposed inference system which can approximate input-output pairs is applied to control the nonlinear system with arbitrarily chosen initial conditions.

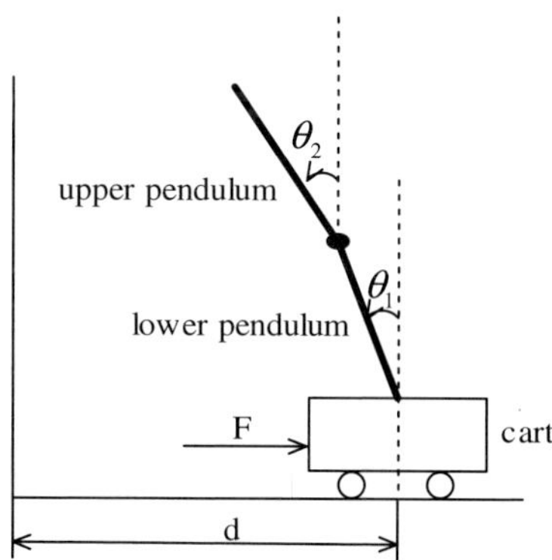

Fig. 2. The structure of cart-type double inverted pendulum

Let $\boldsymbol{x} = [r,\theta_1,\theta_2,\dot{r},\dot{\theta}_1,\dot{\theta}_2]$ be the states of the system, where r represents the cart's position from a zero reference point, θ_1 and θ_2 represent the angels from the vertical position of the lower and upper link, respectively. The system can be represented by the state-space model as

$$\begin{cases} \dot{\boldsymbol{x}} = \begin{bmatrix} \theta_{3\times3} & I_{3\times3} \\ M^{-1}(\theta_1,\theta_2)\cdot N(\theta_1,\theta_2) & \theta_{3\times3} \end{bmatrix} \boldsymbol{x} + \begin{bmatrix} \theta_{3\times3} \\ M^{-1}(\theta_1,\theta_2)\cdot G \end{bmatrix} u \\ y = [r,\theta_1,\theta_2]^T, \end{cases} \tag{8}$$

where u is the new input as the control power, matrixes $\boldsymbol{M}, \boldsymbol{N}, \boldsymbol{G}$ are defined as (9).

The descriptions and the values of the symbols are listed in Table.1.

$$\boldsymbol{M} = \begin{bmatrix} M_0 + M_1 + M_2 & (M_1 l_1 + M_2 L_1)\cos\theta_1 & M_2 l_2 \cos\theta_2 \\ (M_1 l_1 + M_2 L_1)\cos\theta_1 & J_1 + M_1 l_1^2 + M_2 L_1^2 & M_2 L_1 l_2 \cos(\theta_2 - \theta_1) \\ M_2 l_2 \cos\theta_2 & M_2 L_1 l_2 \cos(\theta_2 - \theta_1) & J_2 + M_2 l_2^2 \end{bmatrix},$$
$$\boldsymbol{N} = \begin{bmatrix} 0 & 0 & 0 \\ 0 & (M_1 l_1 + M_2 L_1) g \sin\theta_1 & 0 \\ 0 & 0 & M_2 l_2 g \sin\theta_2 \end{bmatrix}, \quad \boldsymbol{G} = \begin{bmatrix} G_0 \\ 0 \\ 0 \end{bmatrix}. \tag{9}$$

The purpose of control is to determine $u(\boldsymbol{x})$ such that the closed-loop system output y will converge to zero under certain initial conditions. First 200 samples distributing in $\boldsymbol{x} = [-2,2]\times[-\pi/10,\pi/10]\times[-\pi/10,\pi/10]\times[-2,2]\times[-\pi/10,\pi/10]\times[-\pi/10,\pi/10]$ are

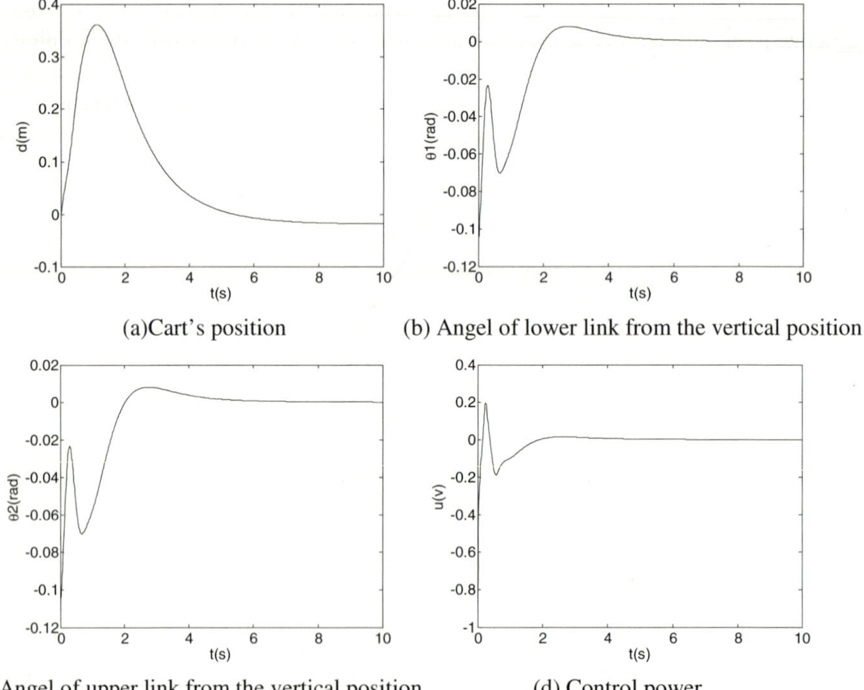

(a)Cart's position (b) Angel of lower link from the vertical position

(c) Angel of upper link from the vertical position (d) Control power

Fig. 3. Results for control of double inverted pendulum using SVM- based fuzzy inference randomly chosen. Assuming the initial state $\boldsymbol{x} = [0, -\pi/30, \pi/30, 0.6, -\pi/20, \pi/30]$, with the parameters are implemented as $C = 165.84$, $e = 0.05$ and $\sigma^2 = 134.62$, the simulation results are shown in Fig.3.

Table 1. The Descriptions and Values of Symbols

Symbol	Value	Descriptions
M_0	1.3280	Mass of the cart and the drive system(Kg)
M_1	0.22	Mass of the lower link (Kg)
M_2	0.187	Mass of the upper link (Kg)
J_1	0.004963	Mass moment of inertia of the lower link (Kgm^2)
J_2	0.004824	Mass moment of inertia of the upper link (Kgm^2)
L_1	0.49	Distance from the center of gravity of lower link to that of upper link (m)
l_1	0.304	Distance from center of gravity of lower link to hinge (m)
l_2	0.226	Distance from center of gravity of upper link to hinge (m)
G_0	11.887	The proportion of power to the voltage (Nms)

The proposed method automatically extracts the optimal number of fuzzy rules to approximate the original controller, and comparing with the method proposed in [8], it is obvious that the fuzzy inference system based on SVM can control the double inverted pendulum system to asymptotically stabilize to the upper equilibrium much faster. And it can be seen that the control power needed is very small, and the procedure of control is smooth, which are expected in the practical applications.

The optimum choice of fuzzy rules depends on the training data set, particularly on the learning process of support vector regression. By varying the insensitivity level, the controller can have better generalization ability or higher approximation accuracy. Besides, the original controller is based on sophisticated mathematical knowledge of the double inverted pendulum system, whereas the proposed approach needs only to have enough training patterns. This behavior shows that the proposed method is able to extract representative points from a training pattern pool to construct the final decision rules. Another important geometrical and algorithmic property is that usually, the support vectors are very few. Since the number of support vectors is the same as the number of fuzzy rules, the major advantage of this is that it is applicable to many practical and complicated situations, where we unable to determine the number of rules in advance.

4 Conclusions

For the fuzzy inference based on SVM network, SVM guarantees the systems have good generalization ability, and provides a mechanism to decide the number of fuzzy rules and membership functions, which is irrelevant to the dimension of the input space; while fuzzy logic techniques deal with reasoning, and doesn't have much learning capability. Therefore, mixing two mechanisms together can preserve the both advantages and create a new way for the nonlinear control.

Acknowledgement

Project supported by the Nature Science Specialties Foundation of Education Bureau, Shaanxi Province, China (Grant No. 05JK267).

References

1. Mendel, J.M.: Fuzzy logic systems for engineering: a tutorial. Proc. IEEE 83 (1995) 345-377
2. Wang, L.X., Jerry, M.: Fuzzy Basis Functions, Universal Approximation, and Orthogonal Least-squares Learning. IEEE Trans. on Neural Networks 3 (5) (1992) 807-814
3. Vapnik, V.N.: An Overview of Statistical Learning Theory. IEEE Trans. on Neural Networks 10 (5) (1999) 988-999
4. Chapelle, O., Haffner, P., Vapnik, V. N.: Support Vector Machines for Histogram-based Image Classification. IEEE Trans. on Neural Networks 10 (5) (1999) 1055-1064
5. Harris, D., Wu, D.H., Vapnik, V.N.: Support Vector Machines for Spam Categorization. IEEE Trans. on Neural Networks 10 (5) (1999) 1048-1054
6. Jacek, M.L.: TSK-Fuzzy Modeling Based on E-Insensitive Learning. IEEE Transactions on Fuzzy Systems 13(2) (2005)181-193
7. Lin, C.J., Lin C.T.: An ART-based Fuzzy Adaptive Learning Control Network. IEEE Trans. on Fuzzy Systems 5(4) (1997)477-496
8. Yang, Z.Q., Piao Y.G.: Control of Double Inverted Pendulum by Using State Variables Synthesis Fuzzy Neural Network. Control and Decision 17(1) (2002) 123-128

Adaptive Wavelet Neural Network Friction Compensation of Mechanical Systems

Shen-min Song, Zhuo-yi Song, Xing-lin Chen, and Guangren Duan

School of Astronautics, Department of Control Science and Engineering,
Harbin Institute of Technology, Harbin, China
{songshenmin, zysong, xlchen, grduan}@hit.edu.cn

Abstract. Recently, based on multi-resolution analysis, wavelet neural networks (WNN) have been proposed as an alternative to NN for approximating arbitrary nonlinear functions in $L^2(R)$. Discontinuous friction function is an unavoidable nonlinear effect that can limit control performance in mechanical systems. In this paper, adaptive WNN is used to design a friction compensator for a single joint mechanical system. Then asymptotically stability of the system is assured by adding a PD controller and adaptive robust terms. The simulation results show the validity of the control scheme.

1 Introduction

In control systems, NN is often used to design a controller based on the universal approximation property [1],[2], but unfortunately, this universal property is only proven for continuous functions[3]. Meanwhile, the functions to be approximated in control system are often discontinuous, such as friction functions. Although NN is always used to approximate discontinuous functions [3],[4],[6],[7], it is found that many NN nodes and many training iterations are required, and still do not yield very good results [5]. So it is necessary to modify traditional NN to provide the foundation of approximating discontinuous functions using NN. Recently, based on multi-resolution analysis (MRA), WNN is proposed as an alternative to NN to approximate functions in $L^2(R)$. So, in this paper, WNN is used to approximate friction functions of a mechanical system in real time to compensate it and adaptive law is used to adjust the parameters of WNN as existed friction models are imprecise.

2 WNN

Due to the similarity between wavelet decomposition and one-hidden-layer NN, the idea of combining wavelets and NN can hopefully remedy the weakness of each other, resulting in networks with theoretical foundation of approximating functions in $L^2(R)$ [9],[10],[11].

From orthogonal MRA point of view, $L^2(R)$ can be constructed by a sequence of orthogonal subspaces $\{W_j, j \in Z\}$ which is formed by dilation and integer translates of wavelet function ψ_j [12]. That is $L^2(R) = \sum_{j \in Z} \oplus W_j$ from which lemma1 is obtained.

J. Wang et al. (Eds.): ISNN 2006, LNCS 3972, pp. 1131 – 1139, 2006.

Lemma 1[13]**.** For $\forall f(x) \in L^2(R)$ and $\forall \varepsilon \geq 0$, there must exist a $f'(x) = \sum_{j\in Z}\sum_{k\in Z} c^{j,k}\psi^{j,k}(x)$, such that $\left\| f - f' \right\|_2 \leq \varepsilon$.

From multi-resolution theory, it is possible for the sum about j, k which denotes the dilation and translation of ψ_j contains only finite terms. Then lemma1 become the theoretical background of wavelet neural networks with one input. Simply speaking, WNN is feed-forward NN using wavelets as activation function.

Here Mexican-hat wavelet is used as activation function: $\psi(x) = (1 - x^2)e^{-x^2/2}$, and two-layer feed-forward (hidden layer and output layer) structure is used.

The NN equation with linear output activation function can be written in terms of vectors as $y = W^T\psi(V^T x + b)$,where W is hidden to output layer weight matrix and V is input to hidden layer weight matrix, b is the threshold vector. The accuracy analysis for WNN approximation in [15] has shown that WNN has the potential to achieve better approximation accuracy than NN.

3 Background

Consider a single joint mechanical system as studied in [8]. The mathematical model is formulated as:

$$M\ddot{\theta} + M_f' + M_\theta + \tau_d = u\,. \tag{1}$$

where $\ddot{\theta}(t)$, $\dot{\theta}(t)$, $\theta(t)$ represent the acceleration, velocity and position, respectively. M is the known constant inertia. M_f' is the friction and M_θ denotes the position dependency of friction and other friction modeling errors, τ_d reflects bounded unknown disturbances and $u(t)$ is actuation input.

Here Stribeck friction model is used which is formulated as:

$$M_f' = \left[M_c + M_s \exp(-M_\tau \dot{\theta}^2)\right]\mathrm{sgn}(\dot{\theta}) + B\dot{\theta}\,. \tag{2}$$

where M_c is Coulomb friction parameter and M_s static friction parameter. M_τ is corresponding to the Stribeck effect. B is constant viscous friction coefficient. Inertia M is assumed to be accurately known. Rewrite system dynamics as:

$$M\ddot{\theta} + B\dot{\theta} + M_f + M_\theta + \tau_d = u\,. \tag{3}$$

The non-parametric friction terms M_θ and τ_d are bounded as: $\left|M_\theta + \tau_d\right| \leq \tau_N$. where τ_N is an unknown constant bound for any position and velocity θ and $\dot{\theta}$.

Given a desired trajectory $\theta_d(t) \in R$, the tracking error is: $e(t) = \theta_d(t) - \theta(t)$, $\dot{e}(t) = \dot{\theta}_d(t) - \dot{\theta}(t)$.

The control objective is to force the plant state vector $\theta'(t) = [\theta(t) \ \ \dot{\theta}(t)]^T$ to follow desired trajectory $\theta_d'(t) = [\theta_d(t) \ \ \dot{\theta}_d(t)]^T$, so define filtered tracking error as: $r = \dot{e} + \lambda e$. And λ is a positive designed parameter.

Then the system dynamics (3) can be expressed in terms of $r(t)$ as:

$$M\dot{r} = M\ddot{\theta}_d + M\lambda(\dot{\theta}_d - \dot{\theta}) + f - u + M_\theta + \tau_d . \quad (4)$$

where the unknown nonlinear function $f(x) = B\dot{\theta} + M_f$, $x = \dot{\theta}$. Notice that there are no partially known information contained in $f(x)$, so that $f(x)$ is only related to $\dot{\theta}$ (the dependency on position is viewed as modeling error). In this way, not only the weights to be adjusted are reduced remarkably, multi-wavelets which is very complicated is not needed either.

Define a nominal control input torque as:

$$u_0 = M\ddot{\theta}_d + M\lambda(\dot{\theta}_d - \dot{\theta}) + k_v r + \hat{f} . \quad (5)$$

with gain $k_v > 0$ which determines the velocity of r running to zero. "$k_v r$" is in fact a PD controller, and $\hat{f}(x)$ obtained by WNN is an estimate of $f(x)$. The estimation error is $\tilde{f} = f - \hat{f}$. Then the closed-loop dynamics is given by:

$$M\dot{r} = -k_v r + \tilde{f} + M_\theta + \tau_d . \quad (6)$$

4 Control Scheme

4.1 WNN Friction Compensator

Some assumptions should be made before friction compensator designing.

Assumption 1 (Bounded Reference Trajectory and bounded WNN input): [5] The desired trajectory satisfy $\left\| [\theta_d \ \dot{\theta}_d]^T \right\|_F \le q$, the input to WNN $x(t) = \dot{\theta}$ is bounded by: $|x(t)| = |\dot{\theta}| \le c_0 + c_1 |r|$, where c_0, c_1 are constants.

From lemma1, there is a two layer WNN such that:

$$f(x) = W^T \psi(V^T x + b) + \varepsilon . \quad (7)$$

with the approximation error bounded: $\|\varepsilon\| \le \varepsilon_N$, and ε_N is unknown.

Assumption 2 (WNN Approximation Requirement):The selected WNN approximation property (11) holds for all x in the compact set we are interested in.

Assumption 3 (Bounded Ideal WNN Weights): On any compact subset of R^n , the ideal WNN weights (W , V , b) are bounded and the bounds are an unknown.

Define $\hat{W}$, $\hat{V}$, $\hat{b}$ as estimates of the ideal WNN weights and turned by some algorithms. Then the approximation of nonlinear function is:

$$\hat{f}(x) = \hat{W}^T \psi(\hat{V}^T x + \hat{b}) . \tag{8}$$

Define the estimation error as (9) and hidden layer error as (10):

$$\tilde{W} = W - \hat{W}, \quad \tilde{V} = V - \hat{V}, \quad \tilde{b} = b - \hat{b} . \tag{9}$$

$$\tilde{\psi} = \psi(V^T x + b) - \psi(\hat{V}^T x + \hat{b}) . \tag{10}$$

Use Taylor series we have:

$$\begin{aligned} \psi(V^T x + b) &= \psi(\hat{V}^T x + \hat{b}) + \psi'(\hat{V}^T x + \hat{b}) \times (\tilde{V}^T x + \tilde{b}) + O(\tilde{V}^T x + \tilde{b}) \\ &= \hat{\psi} + \hat{\psi}' \tilde{V}^T x + \hat{\psi}' \tilde{b} + O = \hat{\psi} + \tilde{\psi} \end{aligned} . \tag{11}$$

where $\hat{\psi}' \tilde{V}^T$ is first order term and O higher order terms. $\hat{\psi}'$ is a Jacobian matrix,

Assumption 4 (Bounded higher order terms): Suppose that higher order terms O is not larger than linear combination of first order term: $\left\| O(\tilde{V}^T x + \tilde{b}) \right\| \le c_2 (\tilde{V}^T x + \tilde{b}) + c_3$.

Notice that the higher order term is not needed to be bounded by a constant.

4.2 Controller Design

Here is presented a control scheme with a friction compensator, a traditional PD controller and adaptive robust terms to attenuate the approximation error, modeling error and disturbances. The control system diagram is shown in Fig.1.

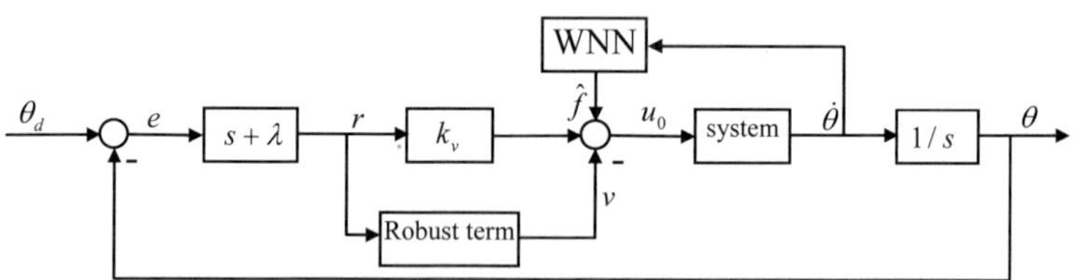

Fig. 1. Control diagram of system

Define the control input as:

$$u = u_0 - \upsilon + r \hat{C}^T \sigma(|r|) . \tag{12}$$

where u_0 is defined in (5), υ is an adaptive robust term to eliminate the effects of approximation error and external disturbances, $\hat{C}^T \sigma(|r|)$ is a NN(only the output layer weights are adjusted) which is used to compensate for the unknown terms in system dynamical equation(see following text). The input to this linear regression NN is $|r|$ and the weights from input layer to the hidden layer are all equal to 1, $\hat{C}$ is the weight matrix of the output layer. σ is sigmoid function.

The closed-loop error dynamics is:

$$M\dot{r} = -k_v r + W^T \psi(V^T x + b) - \hat{W}^T \psi(\hat{V}^T x + \hat{b}) + \varepsilon + \upsilon + \tau_d + M_\theta - r\hat{C}^T \sigma(|r|) . \tag{13}$$

Using (9) and substitute (11) in to (13), the closed-loop dynamics is

$$\begin{aligned} M\dot{r} &= -k_v r + \tilde{W}^T (\hat{\psi} - \hat{\psi}' \hat{V}^T x - \hat{\psi}' \hat{b}) - \hat{W}^T \hat{\psi}' \tilde{V}^T x \\ &- \hat{W}^T \hat{\psi}' \tilde{b} + \upsilon - r\hat{C}^T \sigma(|r|) + d_1 + d_2 \end{aligned} . \tag{14}$$

with unknown terms d_1 and d_2 given by:

$$d_1 = W^T \hat{\psi}' V^T x + W^T \hat{\psi}' b + W^T O, \quad d_2 = \varepsilon + \tau_d + M_\theta . \tag{15}$$

where d_2 is bounded by constant parameter and assumption 5 is made for d_1.

Assumption 5 (Bounded disturbance): The bound of d_1 in (15) satisfy (16), where ζ_0 is unknown constant, $\zeta_j(|r|)$ is unknown smooth function and $\zeta_j(0) = 0$.

$$|d_1| \le \zeta_0 + \sum_{j=1}^{n} \zeta_j(|r|) . \tag{16}$$

As $\zeta_j(|r|)$ is smooth function, there must exist smooth function $\xi_j(|r|)$ $(1 \le j \le n)$ which satisfy $\zeta_j(|r|) = |r| \xi_j(|r|)$.

Define positive definite function:

$$L = \frac{1}{2} Mr^2 + \frac{1}{2} tr(\tilde{W}^T S^{-1} \tilde{W}) + \frac{1}{2} tr(\hat{V}^T T^{-1} \hat{V}) + \frac{1}{2} \hat{b}^T Q^{-1} \hat{b} + \frac{1}{2} tr(\tilde{C}^T P^{-1} \tilde{C}) . \tag{17}$$

where the parameter matrixes: $S = S^T > 0, T = T^T > 0, Q = Q^T > 0, P = P^T > 0$.

Use (14) and the property $tr(ab^T) = a^T b$ with $a \in R^{n\times 1}, b \in R^{1\times n}$, the derivative of L is:

$$\dot{L}=-k_v r^2+tr\tilde{W}^T(S^{-1}\dot{\tilde{W}}+\hat{\psi}r-\hat{\psi}'\hat{V}^T xr-\hat{\psi}'\hat{b}r)+\tilde{b}^T(Q^{-1}\dot{\hat{b}}-\hat{\psi}'\hat{W}r)$$
$$+tr\tilde{V}^T(T^{-1}\dot{\hat{V}}-xr\hat{W}^T\hat{\psi}')+r(d_1+d_2+\upsilon)+tr(\tilde{C}^T P^{-1}\dot{\tilde{C}})-r^2 \hat{C}^T\sigma(|r|) \quad (18)$$

As $\dot{\tilde{W}}=-\dot{\hat{W}}$, let the WNN weight turning law be provided by:

$$\dot{\hat{W}}=S\hat{\psi}r-S\hat{\psi}'\hat{V}^T xr-S\hat{\psi}'\hat{b}r \quad \dot{\hat{V}}=TxrW^T\hat{\psi}' \quad \dot{\hat{b}}=Qr\hat{\psi}'\hat{W}. \quad (19)$$

Substitute (19) into (18) and use the property $2ab\le a^2+b^2$, we can get:

$$\dot{L}\le -k_v r^2-r^2\hat{C}^T\sigma(|r|)+tr(\tilde{C}^T P^{-1}\dot{\tilde{C}})+\upsilon r+|r|(\zeta_0+\varepsilon_N+\tau_N)+g(|r|)r^2. \quad (20)$$

where $g(|r|)=\frac{n}{2}+\sum_{j=1}^{n}\frac{1}{2}\xi_j^2(|r|)$ is a continuous function, so a NN is used to eliminate the effects d_1 brings to stability. The approximation property of the NN satisfies assumption 4:

$$g(|r|)=C^T\sigma(|r|)+\mu. \quad (21)$$

where C is unknown ideal bounded weight matrix of output layer, μ is bounded approximation error: $|\mu|\le\mu_M$, μ_M is unknown. Substitute (21) into (20):

$$\dot{L}\le -k_v r^2+r^2\tilde{C}^T\sigma(|r|)+tr(\tilde{C}^T P^{-1}\dot{\tilde{C}})+\upsilon r+|r|(\zeta_0+\varepsilon_N+\tau_N)+\mu_M r^2. \quad (22)$$

Define the adaptive robust term as:

$$\upsilon=-\hat{\beta}\,\mathrm{sgn}(r)-\hat{\mu}_M r. \quad (23)$$

Substitute (23) into (22):

$$\dot{L}\le -k_v r^2+\tilde{C}^T(r^2\sigma(|r|)-P^{-1}\dot{\hat{C}})+\tilde{\beta}|r|+\tilde{\mu}_M r^2. \quad (24)$$

Define positive function $L_1=\frac{1}{2k}\tilde{\beta}^2+\frac{1}{2\gamma}\tilde{\mu}_M{}^2$, where k,γ are all positive.

Let $\bar{L}=L+L_1$ as a Lyapunov candidate function, and the deviation of $\bar{L}$ is:

$$\dot{\bar{L}}=\dot{L}+\dot{L}_1\le -k_v r^2+\tilde{C}^T(r^2\sigma(|r|)-P^{-1}\dot{\hat{C}})+\tilde{\beta}(|r|+k^{-1}\dot{\tilde{\beta}})+\tilde{\mu}_M(r^2+\gamma^{-1}\dot{\tilde{\mu}}_M). \quad (25)$$

Let the weight turning law of NN be (26) and adaptive law of robust term be (27):

$$\dot{\hat{C}} = Pr^2\sigma(|r|) \ . \tag{26}$$

$$\dot{\hat{\beta}} = k|r| \, , \dot{\hat{\mu}}_M = \gamma r^2 \ . \tag{27}$$

Substitute (26) and (27) to (25), we can obtain $\dot{\bar{L}} \le -k_v r^2$. So the parameters of the system $r, \hat{W}_1, \hat{W}_2, \hat{V}_1, \hat{C}, \hat{\beta}, \hat{\mu}_M$ are all bounded and it can be known from (15) that $\dot{r}$ is bounded, so r is uniformly continuous. Moreover, since $\bar{L}$ is positive definite and we have $\int_0^\infty k_v r^2 \le -\int_0^\infty \dot{\bar{L}} = \bar{L}(0) - \bar{L}(\infty) < \infty$, then By Babalat lemma $\lim_{t\to\infty} r(t) = 0$, $\lim_{t\to\infty} e(t) = 0$

Theorem 1: For system (1) and controller (12), if assumptions hold, when the weight turning laws of NN are (19) and (26), the adaptive law of robust term is (27), all the parameters and signals in the closed-loop system are bounded and the tracking error decreases asymptotically to zero.

5 Simulation

The simulations conducted in this section are to investigate the effectiveness of the proposed WNN friction compensation scheme. The input signal is a low-velocity signal: $\theta_d = \sin(t) - 0.5\sin(2t)$.

In order to verify the robustness of the controller, simulation is conducted with external disturbance (simulated by white noise) and varied friction parameters: $M = 0.15kgm^2$, $B = (1.5 + 0.3\sin(10\theta)Nms/rad$, $M_c = (3.5 + 0.7\sin(10\theta)Nm$, $M_s = (5.0 + \sin(10\theta)Nm$, $M_\tau = (100 + 20\sin(10\theta)s^2/rad^2$. simulation time is 10 seconds and simulation step is 0.0001 second.

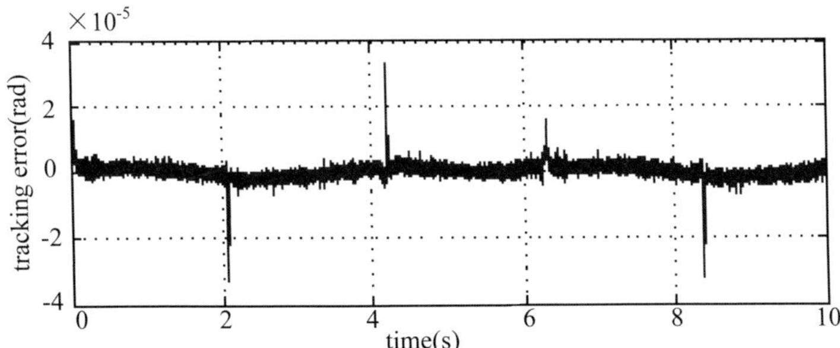

Fig. 2. Tracking error of system using control scheme of this paper

The control parameters are as follows: S=10×diag(5,5) , T=10×diag(5,5) , Q=diag(5,5) , P=5×diag(5,5) $k_v = 10$ $\lambda = 100$, $\gamma = 10, k = 10$. The are 5 wavelons in WNN and 5 nodes in the linear regression NN. Fig.2 depicts the tracking error of the system.

It can be seen from fig4 that the order of magnitude of the tracking error is 10^{-5} , so the control scheme is very successful.

Using traditional NN instead of WNN in this paper, the result is depicted in Fig.3. The magnitude of tracking error is 10^{-4} which is 5 times worse.

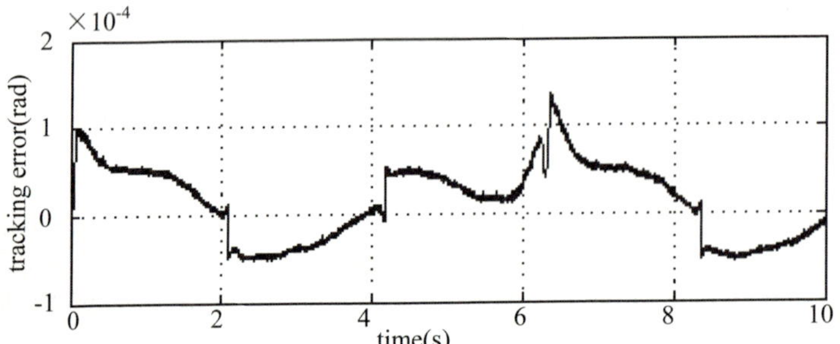

Fig. 3. Tracking error using NN instead of WNN

6 Conclusion

The WNN used to approximate friction functions is so simple that it has only one input and one hidden layer. The compensator and a PD controller plus adaptive robust terms construct the control scheme under which the system can achieve desired tracking performance. The controller is flexible because all the weights of WNN are updated by adaptive law. Using the updating rules, we can show that all the signals of the system are bounded and the tracking error decreased asymptotically to zero.

Acknowledgement

This work is supported by the Foundation HIT. 2002.12

References

1. Hagan, M. T., Demuth, H. B., de Jesus, O.: An Introduction to the Use of Neural Networks in Control Systems. Philadelphia, PA: Taylor and Francis (1999) 202-248
2. Chen, F. C., Khalil, H. K.: Adaptive Control of Nonlinear Systems Using Neural Networks. Int. J. Contr. 55(6) (1992) 1299–1317
3. Kim, Y. H., Lewis, F. L.: Reinforcement Adaptive Learning Neural Network Based Friction Compensation for High Speed and Precision. Proc. 37th IEEE Conf. Decision Contr. Tampa FL. (1998) 2231-2238

4. Qiang, S., Gao, X. Z., Zhuang, X.: Neural Networks in Friction Compensation, Velocity and Acceleration Measurement and PID Design. IEEE ICIT'02 Bangkok THAILAND (2002) 1145-1152
5. Selmic, R. R., Lewis, F. L.: Neural-Network Approximation of Piecewise Continuous Functions: Application to Friction Compensation. IEEE transactions on Neural Networks. 13(3) (2002) 745-751
6. Lewis, F. L., Jagannathan, S., Yesilidrek, A.: Neural Network Control of Robot Manipulators and Nonlinear Systems. Philadelphia, PA: Taylor and Francis (1999) 10-18
7. Lewis, F. L.: Neural Network Control of Robot Manipulators. IEEE Expert Special Track Intell. Contr. 64(1) (1996) 64–75
8. Armstrong-Hélouvry, B., Dupont, P., deWit, C. C.: A Survey of Models, Analysis Tools and Compensation Methods for the Control of Machines with Friction. Automatica 30(7) (1994) 1083–1138
9. Zhang, Q., Benveniste, A.: Wavelet Networks. IEEE Trans. Neural Networks. 3 (1992) 889-898
10. Pati, Y. C., Krishnaprasad, P. S.: Analysis and Synthesis of Feed-forward Neural Networks Using Discrete Affine Wavelet Transformations. IEEE Trans. Neural Networks. 3(4) (1993) 73-85
11. Zhang, J., Walter, G. G., Miao, Y., Lee, W. N. W.: Wavelet Neural Networks for Function Learning. IEEE Trans. Signal Processing. 43(4)(1995) 1485-1497
12. Mallat, S. G.: A Theory for Multi-resolution Signal Decomposition: the Wavelet Representation. IEEE Transactions on Pattern Analysis and Machine Intelligence. 11(7) (1989) 674~693
13. Gao, X.: A Comparative Research on Wavelet Neural Networks. Proceedings of the 9th International Conference on Neural Information Processing.(2002) 699~1703
14. Armstrong, B.: Friction: Experimental Determination, Modeling and Compensation. Proc. IEEE Int. Conf. Robot. Automat. Philadelphia PA (1988) 1422–1427
15. Delyon, B., Juditsky, A., Benveniste, A.: Accuracy Analysis for Wavelet Approximations. IEEE Trans. Neural Networks. 6(2) (1995) 332-348

Application of Collective Robotic Search Using Neural Network Based Dual Heuristic Programming (DHP)

Nian Zhang[1] and Donald C. Wunsch II[2]

[1] Dept. of Electrical and Computer Engineering,
South Dakota School of Mines & Technology,
501 E. St. Joseph Street, Rapid City, SD 57701, USA
nian.zhang@sdsmt.edu
[2] Dept. of Electrical and Computer Engineering,
University of Missouri-Rolla, 1870 Miner Circle, Rolla, MO 65409, USA
dwunsch@ece.umr.edu

Abstract. An important application of mobile robots is searching a region to locate the origin of a specific phenomenon. A variety of optimization algorithms can be employed to locate the target source, which has the maximum intensity of the distribution of some detected function. We propose a neural network based dual heuristic programming (DHP) algorithm to solve the collective robotic search problem. Experiments were carried out to investigate the effect of noise and the number of robots on the task performance, as well as the expenses. The experimental results were compared with those of stochastic optimization algorithm. It showed that the performance of the dual heuristic programming (DHP) is better than the stochastic optimization method.

1 Introduction

In recent years there has been growing interest in collective robotic search problem. The primary reason is that mobile robots can complete high-risk tasks. Mapping minefields, extraterrestrial and undersea exploration, detecting the location of chemical and biological weapons, and the location of explosive devices are its important applications. The goal of the team of robots is to find the origin of a specific phenomenon with the maximum intensity, by sharing information between robots, and to aggregate around the phenomenon [1]. Investigations of collective behavior are considerably rarefied, and studies involving collective search are rarer still. The foraging problem [2][3][4][5], in which robots collect objects scattered in the environment, is a canonical problem related to the source location problem. A decentralized alpha-beta coordination is proposed for an agent team searching for source targets [6]. Its simulations confirm the ability of the team to find a source and stabilize the steady-state mean squared error. It has been shown in [7] that how space-filling curves can enhance the efficiency and robustness of geographic search by robot collectives. Three neural networks algorithms for collective robotic search were investigated in [8]. We propose a neural network based dual heuristic programming algorithm to solve the collective robotic search problem. The performance is compared with that of the stochastic optimization algorithm in [8].

J. Wang et al. (Eds.): ISNN 2006, LNCS 3972, pp. 1140 – 1145, 2006.

2 Problem Description

Let's assume we have a two-dimensional bounded Euclidean space. The domain is shown in Fig. 1.

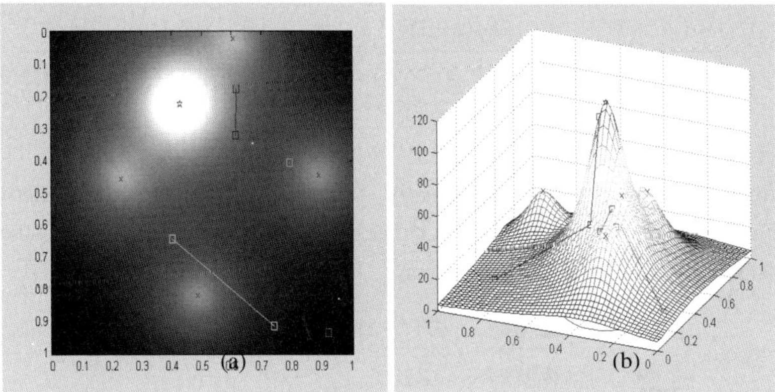

Fig. 1. Robot search region with five sources and four robots. Experiments were carried out on this 1 by 1 length unit square domain. The red star denotes the target, and the blue crosses denote the noises. The square blocks in different colors denote the robots. The lines indicate each step of an individual robot. In (a), x-axis and y-axis denotes the coordinates of sources and robots. The illumination of each source is shown. In (b), x-axis and y-axis denotes the coordinates of sources and robots. The z-axis denotes the illumination of each source.

3 Dual Heuristic Programming (DHP) Approach

DHP uses the critic network to estimate the derivatives of J function with respect to the state vector [9]. The cost-to-go function J in the Bellman equation of dynamic programming is expressed as follows:

$$J(t) = \sum_{k=0}^{\infty} \gamma^k U(t+k) \tag{1}$$

where γ is a discount factor for finite horizon problems ($0<\gamma<1$), and $U(\cdot)$ is the utility function. The critic is trained forward in time, which is of great importance of real-time operation. The critic network tries to minimize the following error measure over time:

$$\|E\| = \sum_t E^T(t)E(t) \tag{2}$$

$$\text{where } E(t) = \frac{\partial J[Y(t)]}{\partial Y(t)} - \gamma \frac{\partial J[Y(t+1)]}{\partial Y(t)} - \frac{\partial U(t)}{\partial Y(t)} \tag{3}$$

$\partial(\cdot)/\partial Y(t)$ is a vector containing partial derivatives of the scalar (.) with respect to the components of the state vector, Y(t). According to the chain rule, each of n components of the vector E (t) can be finally determined by

$$E_j(t) = \frac{\partial J(t)}{\partial R_j(t)} - \gamma \frac{\partial J(t+1)}{\partial R_j(t)} - \frac{\partial U(t)}{\partial R_j(t)} - \sum_{k=1}^{m} \frac{\partial U(t)}{\partial A_k(t)} \frac{\partial A_k(t)}{\partial R_j(t)} \tag{4}$$

where

$$\frac{\partial J(t+1)}{\partial R_j(t)} = \sum_{i=1}^{n} \lambda_i(t+1) \frac{\partial R_i(t+1)}{\partial R_j(t)} + \sum_{k=1}^{m} \sum_{i=1}^{n} \lambda_i(t+1) \frac{\partial R_i(t+1)}{\partial A_k(t)} \frac{\partial A_k(t)}{\partial R_j(t)} \tag{5}$$

$\lambda_i(t+1) = \partial J(t+1) / \partial R_i(t+1)$, and n, m are the numbers of outputs of the model and the action networks, respectively. The adaptation of action network is implemented by propagating $\lambda_i(t+1)$ back through the model down to the action and its goal can be expressed as equation:

$$\frac{\partial U(t)}{\partial A(t)} + \gamma \frac{\partial J(t+1)}{\partial A(t)} = 0, \forall t. \tag{6}$$

Adaptation of DHP is summarized in Fig. 2. In the figure, the discount factor is assumed to be equal to 1. The critic network is shown in two consecutive moments in time and pathways of backpropagation are depicted in dash lines.

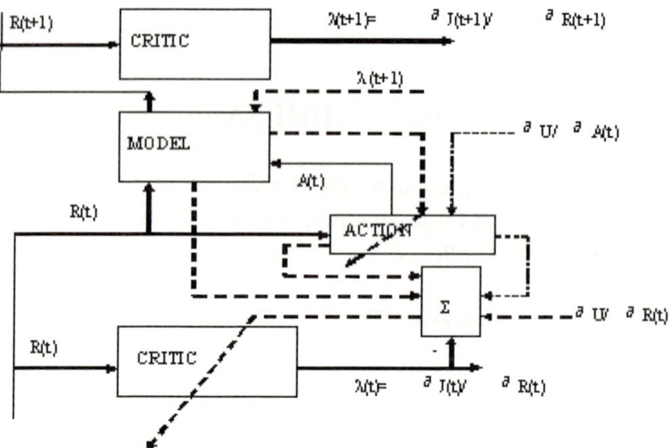

Fig. 2. Adaptation in DHP

3.1 The State Vector

In our case, the state vector $R(t)$ consists of five components at the time step t: $v_{lr}(t)$, $v_{mr}(t)$, $v_{sr}(t)$, $f_r(t)$, and $s_r(t)$, where $v_{lr}(t)$ stands for the direction of the center of mass for the large field strength robots relative to robot r, $v_{mr}(t)$ stands for the direction of the center of mass for the medium field strength robots relative to robot r, $v_{sr}(t)$ stands for the direction of the center of mass for the small field strength robots relative to robot r, $f_r(t)$ stands for the field intensity of the robot r [8], and $s_r(t)$ stands for its status. They are determined by the following expressions.

$$v_{cmr}(t) = \frac{\arctan\left(\dfrac{y_{cm}(t) - y_r(t)}{x_{cm}(t) - x_r(t)}\right)}{2\pi} \tag{7}$$

where cm stands for l, m or s, $x_{cm}(t)$ and $y_{cm}(t)$ are coordinates of the center of mass of *large*, *medium*, or *small* field strength robots, respectively, and v_{cmr} is the normalized direction of the corresponding center of mass relative to robot r.

$$x_{cm}(t) = \frac{\sum_{r \in C} x_r(t) f_r(t)}{\sum_{r \in C} f_r(t)}, \qquad y_{cm}(t) = \frac{\sum_{r \in C} y_r(t) f_r(t)}{\sum_{r \in C} f_r(t)} \tag{8}$$

where $x_r(t)$ and $y_r(t)$ are the coordinates of robot r at time step t. C is the class of robots having large, medium, or small field intensities.

$$x_r(t+1) = \cos(2\pi A_r(t)) + x_r(t) \tag{9}$$

$$y_r(t+1) = \sin(2\pi A_r(t)) + y_r(t) \tag{10}$$

where $A_r(t)$ is the output of the action network for robot r at time step t. The status $s_r(t)$ is calculated as follows:

$$s_r(t) = \frac{f_r(t)}{f_{\max}(t)} \tag{11}$$

where $f_{\max}(t)$ is the maximum field intensity that robots have ever sensed.

3.2 Utility Function

The utility function is designed as the sum of mean square of the distance of each robot from the source.

$$U(t) = \sum_r \frac{1}{2}[(x^* - x_r(t+1))^2 + (y^* - y_r(t+1))^2] \tag{12}$$

where x^* and y^* are the coordinates of the target source, $x_r(t+1)$ and $y_r(t+1)$ are the coordinates of robot r at time step $t+1$. The utility function makes use of the positions of the robots at time step $t+1$, but not t. The reason is that one may not know the cost that a robot incurs until it takes an action.

4 Training of Model, Critic, and Action Neural Networks

The critic network has two layers. There are 20 neurons in the hidden layer. The transfer function of the hidden layer is *logsig*. There are five neurons in the output layer. The transfer function of the output layer is *satlin*. Thus, the critic has five outputs, each of which is the derivative of the cost-to-go function with respect to the state vector component. The action network has two layers. There are 20 neurons in the hidden layer. The transfer function of the hidden layer is *logsig*. There is one neuron in the output layer. The transfer function of the output layer is *satlin*. Thus, the action network has only one output, the direction. Each robot can make a move of one grid square at each time step. The critical neural network builds a representation for the derivatives of J directly by being explicitly trained on them through $\partial U(t)/\partial R(t)$ and $\partial U(t)/\partial A(t)$, which are obtained by backpropagating the utility function, *U(t)* through the Model network.

5 Simulation and Experimental Results

We compare the performance of neural networks trained by dual heuristic programming (DHP) and stochastic optimization approaches [8]. The performance is evaluated by the average route length per robot. The less the route length per robot the robots take, the faster the robots converge at the target source. Two experiments were performed, with two sources and four sources, respectively. The route length comparisons are shown in Fig. 3 and Fig. 4, respectively.

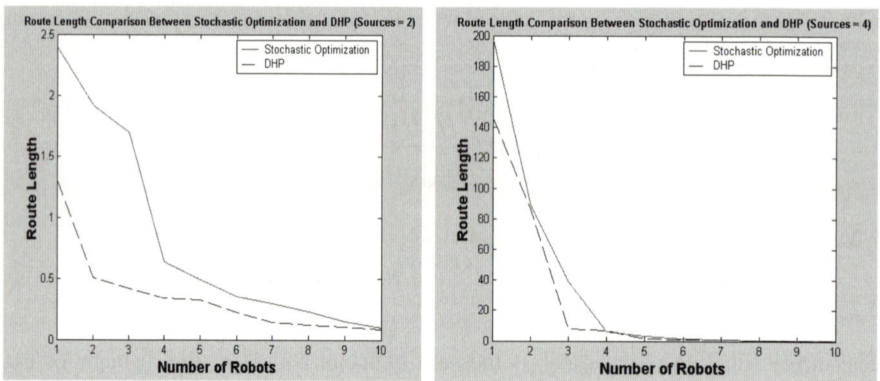

Fig. 3. Route length comparison (sources = 2) **Fig. 4.** Route length comparison (sources = 4)

From Fig. 3, we find that DHP (i.e. dashed line) has better results against stochastic optimization (i.e. solid line), especially when the number of robots is less than the number of sources. In addition, the more the robots, the less route length the robots take. Moreover, when the number of robots is much more than the sources (i.e. 10 to 2), the difference between the two approaches is small. From Fig. 4, we find that the route lengths of both of the two approaches are much greater than that in

Fig. 3. When the number of robots is more than the number of sources, the route length of two approaches decrease considerably. In addition, route length of DHP is less than the stochastic optimization approach when the number of robots is less than the number of sources. Moreover, when the number of robots exceeds the number of sources, the difference between the two approaches is very small.

6 Conclusions

This paper has presented Dual Heuristic Programming (DHP) approach to solve the collective robotic search problem. It is shown that the more the robots, the less the route length per robot. In addition, the more the noise, the more route length per robot. Furthermore, Dual Heuristic Programming (DHP) takes much less route length than the stochastic optimization approach, especially when the number of robots is less than the sources. Since the expenses are proportional to the route length, DHP proves to be an efficient approach to save more expenses in a very noisy environment. This is because DHP is a promising neural network design method to solve optimal control problem under the conditions of noise and uncertainty.

References

1. Gelenbe, E., Schmajuk, N., Staddon, J, and Reif, J.: Autonomous Search By Robots and Animals: A Survey. Robotics and Autonomous Systems 22 (1997) 23-34
2. Arkin, R. and Hobbins, J: Dimensions of Communication and Social Organization in Multi-agent Robotic Systems. Proceedings of Simulation of Adaptive Behavior (1994)
3. Goss, S. and Deneubourg, J.: Harvesting by a Group of Robots. Proceedings of European Conference of Artificial Life (1992)
4. Mataric, M: Interaction and Intelligent Behavior. MIT AI Lab Tech Report AITR-1495 (1994)
5. Steels, L.: Cooperation between Distributed Agents through Self-Organization. European Workshop on Modeling Autonomous Agents in a Multi-Agent World (1990)
6. Goldsmith, S.Y. and Robinett, R.: Collective Search by Mobile Robots Using Alpha-Beta Coordination. '98 Collective Robotics Workshop. Agent World, Paris (1998)
7. Spires, S.V. and Goldsmith, S.Y.: Exhaustive Geographic Search with Mobile Robots along Space-Filling Curves. Proceedings of the First International Workshop, CRW '98, Paris, France (1998)
8. Zhang, N., Novokhodko, A., Wunsch, D.C. II, and Dagli, C.H.: A Comparative Study of Neural Networks Based Learning Strategies on Robotic Search Problems. Proceedings of the SPIE Application and Science of Computational Intelligence Conference. Vol. 4390. Orlando, Florida (2001)
9. Prokhorov, D.V. and Wunsch, D.C. II: Adaptive Critic Designs. IEEE Trans. on Neural Networks 8(5) (1997) 997-1007

RBF Neural Network Based Shape Control of Hyper-redundant Manipulator with Constrained End-Effector

Jinguo Liu[1,3], Yuechao Wang[1], Shugen Ma[2,1], Bin Li[1]

[1] Robotics Laboratory of Chinese Academy of Sciences,
Shenyang Institute of Automation, Shenyang 110016, China
{liujinguo, ycwang, sgma, libin}@sia.cn
http://www.sia.ac.cn

[2] COE Research Institute, Ritsumeikan University, Shiga-ken 525-8577, Japan
shugen@fc.ritsumei.ac.jp

[3] Graduate School of Chinese Academy of Sciences, Bejing 100039, China

Abstract. Hyper-redundant manipulator has more degrees of freedom than the least necessary to perform a given task, thus it has the features of overcoming conventional industrial robot's limitation to carry out a designated difficult task. When the manipulator carries out the missions such as brushing or writing on a surface, drilling or inspection in a hole, the end-effector of the manipulator usually has both position and orientation requirement. Effective control of the hyper-redundant manipulator with such constrained end-effector is difficult for its redundancy. In this paper, a novel approach based on RBF neural network has been proposed to kinematically control the hyper-redundant manipulator. This technique, using variable regular polygon and RBF neural networks models, is completely capable of solving the control problem of a planar hyper-redundant manipulator with any number of links following any desired direction and path. With the shape transformation of variable regular polygon, the manipulator's configuration changes accordingly and moves actively to perform the tasks. Compared with other methods to our knowledge, this technique has such superiorities as fewer control parameters and higher precision. Simulations have demonstrated that this control technique is available and effective.

1 Introduction

The hyper-redundant manipulators have been widely used in different industries to perform many kinds of work such as welding, painting, assembling, space exploration, deep sea exploring, etc. The merits of hyper-redundant manipulator have attracted a lot attention [1~6]. The redundancy can be used to overcome the conventional industrial robot's limitation such as avoiding singularities, obstacle avoidance or joint limit avoidance. The control of a manipulator involves trajectory planning, inverse kinematics and inverse dynamics. In this paper, we mainly discuss the inverse kinematics of a hyper-redundant manipulator. Given a desired workspace trajectory, the task how to find out the corresponding joint space trajectory is a complex problem since redundant manipulators have more than necessary degrees of freedom (DOF)

J. Wang et al. (Eds.): ISNN 2006, LNCS 3972, pp. 1146 – 1152, 2006.

and multiple or infinite number of solutions. Moreover, the inverse kinematics equations are usually nonlinear and are difficult to find closed-form solutions. In the initial research, the Jacobian pseudo inverse approach had been widely applied [1]. Whereas in this technique there are a lot of matrix computations since the complexity increases exponentially with the redundancy increasing. Recent years, neural networks (NN) [2~3] have been successfully applied in robotic control, whose generalization capacities and structures make them robust and fault-tolerant in algorithms, which are able to solve a problem that was difficult to handle before, for instance, the highly nonlinear problems. The properties of NN make them so promising that they are widely applied to robotic control problems.

In this paper, a novel approach has been proposed to control the hyper-redundant manipulators. Using RBF neural networks controller and based on the concept of variable regular polygon technique, this approach has been developed to obtain the inverse kinematics solution of the planar modularization hyper-redundant with any number of links following any desired path and direction.

2 Shape Control Technique and Forward Kinematics

As shown in Fig.1, a hyper-redundant manipulator has a lot of links. It is composed of a serial chain of links with the same length where the internal variables $\theta_i (i = 1 \sim n)$ are the relative angles between adjacent links. The end effector's position is given by

$$\begin{pmatrix} x_e \\ y_e \end{pmatrix} = \left(\sum_{i=1}^{n} l \cos \psi_i , \sum_{i=1}^{n} l \sin \psi_i \right)^T \tag{1}$$

where ψ_i is the absolute link angle given by

$$\psi_i = \sum_{j=1}^{i} \theta_j \tag{2}$$

To control the hyper-redundant manipulator's end effector's position to realize the task as point to point, we have to calculate all the joints' relative angles, that is, to solve the inverse problem basically. It is apparent that the equation (1) has indeterminate solutions when $n > 2$. The algorithm's computation complexity increases when the number of the links increases. As they are analogous in shape and operation to

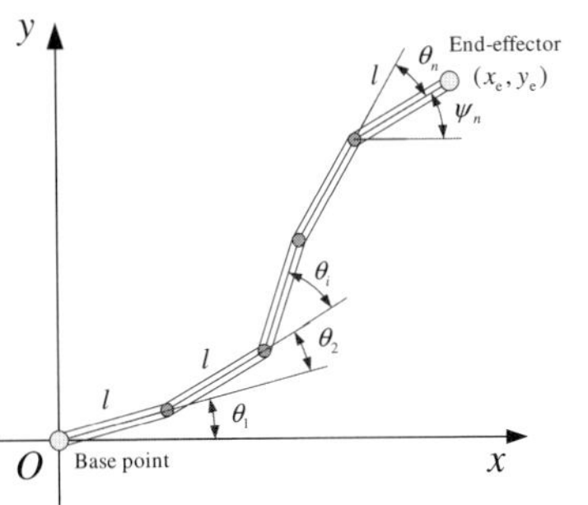

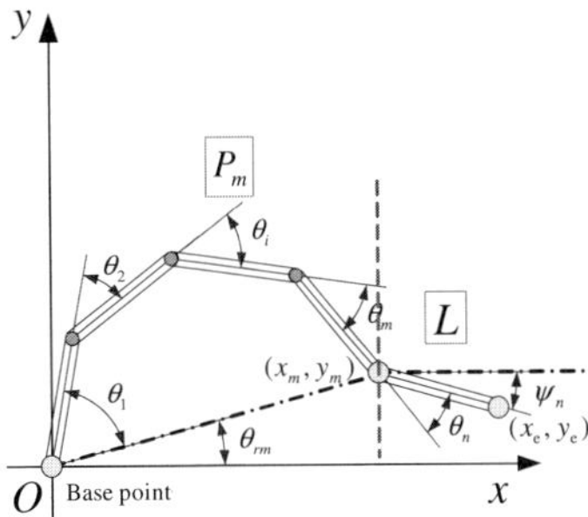

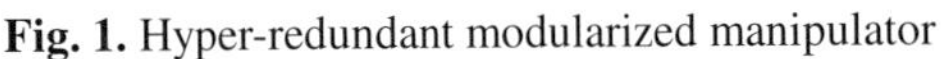

Fig. 1. Hyper-redundant modularized manipulator

Fig. 2. Shape controlled by *PmL*

snakes, elephant trunks and tentacles, Chirikjian and Burdick have proposed the backbone curve to control the hyper-redundant manipulator [4]. Ma and Konno have efficiently controlled the hyper-redundant manipulator by Serpenoid curve [5]. Kobyashi and Ohtake also adopted the shape control technique to control the continuous manipulator [6]. Whereas the shape control techniques mentioned above are always using continuous curve, on the contrary, most manipulators in application are in discrete curves so they are not precise enough to control the hyper-redundant manipulators since during dispersing the curve or integrating the curvature the error is unavoidable. So in this paper, variable regular polygon shape control approach has been proposed. Since the modularized manipulator has the links with the same length, it is easy for it to pose the configuration like a regular polygon. The hyper-redundant manipulator can be divided into end-effector part and the curve part. The shape of the curve part can be controlled by the variable regular polygon. The manipulator's shape changes according to the regular polygons' shape shifting and the end-effector' s motion. To make it easy to describe and calculate, we define an n-link hyper-redundant manipulator being composed of one arc subsystem P_m ($n=m+1$) and the end-effector L as shown in Fig.2. P_m represents a half part of a $2m$-link regular polygon and L represent the end-effector. The base part of manipulator is controlled by P_m's shape which comes from extension and rotation as shown in Fig.3.

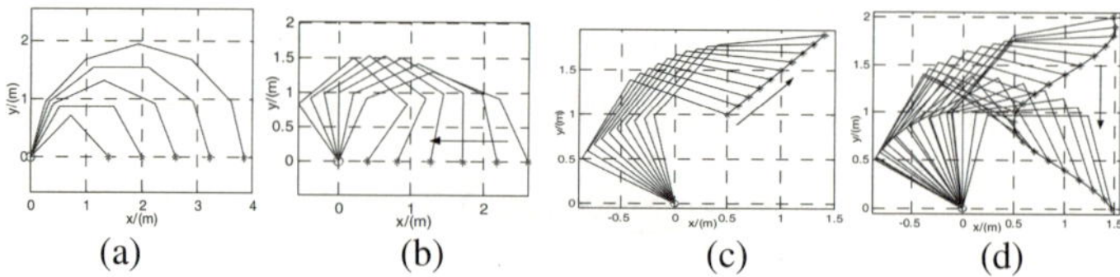

Fig. 3. Examples of *Pm*. (a) From standard *P2* to standard *P6*; (b) *P4*'s constriction (a inverse motion of extension); c) *P3*'s extension and rotation to follow a line; d) *P3*'s extension and rotation to follow a curve.

To guarantee that the relative angles between adjacent links have the same value, the joint's relative angle can be given by

$$\theta_1 = \frac{m-1}{2m}(1-\alpha)\pi \ \text{ And } \theta_i = \frac{\alpha-1}{m}\pi \ (i = 2 \sim m) \tag{3}$$

where α is the extension coefficient, $-1 \le \alpha \le 1$.

And the absolute angle for any joint can be calculated out as

$$\psi_1 = \theta_1 \tag{4}$$

$$\psi_i = \theta_1 + (i-1)\theta_i;\ (i = 2 \sim m) \tag{5}$$

According to θ_1 in (3) and θ_i in (4), we can get

$$\psi_i = (m/2 + i - 3/2)(\alpha - 1)\pi / m \tag{6}$$

If $\theta_{rm} = 0$, the end position of P_m is given by

$$\begin{pmatrix} x_m \\ y_m \end{pmatrix} = \left(\sum_{i=1}^{m} l\cos\psi_i , \sum_{i=1}^{m} l\sin\psi_i \right)^T \tag{7}$$

It can be obtained from (6) and (7) that

$$\begin{aligned} x_m &= \sum_{i=1}^{m} l\cos\psi_i = \begin{cases} ml;(\alpha=1) \\ = l\sin((\alpha-1)\pi/2)/\sin((\alpha-1)\pi/2/m);(\alpha\neq 1) \end{cases} \\ y_m &= \sum_{i=1}^{m} l\sin\psi_i = = \begin{cases} 0;(\alpha=1) \\ l\sin(m\theta_i/2)\sin((2\theta_1+(m-1)\theta_i)/2)/\sin(\theta_i/2);(\alpha\neq 1) \end{cases} \end{aligned} \tag{8}$$

If $\theta_{rm} \neq 0$, the end position of P_m is given by

$$\begin{pmatrix} x_{rm} \\ y_{rm} \end{pmatrix} = \begin{pmatrix} \cos\theta_{rm} & -\sin\theta_{rm} \\ \sin\theta_{rm} & \cos\theta_{rm} \end{pmatrix} \begin{pmatrix} x_m \\ y_m \end{pmatrix} \tag{9}$$

The end-effector's position is given by

$$\begin{pmatrix} x_e \\ y_e \end{pmatrix} = \begin{pmatrix} x_{rm} + \cos\psi_n \\ y_{rm} + \sin\psi_n \end{pmatrix} \tag{10}$$

where ψ_n comes from

$$\psi_n = \psi_m + \theta_n \tag{11}$$

To sum up, the end-effector's position and orientation are decided by three parameters: the extension coefficient α, the base rotation angle θ_{rm}, and the end-effector angle θ_n. To control the position and orientation of an n-link manipulator's end-effector, if n>3, this technique has the advantage of fewer control parameters.

3 RBF Neural Network Based Inverse Kinematics Solution of *PmL*

In the literature, more attention has been paid to the trajectory planning. When the manipulator carries out the missions such as painting or writing on a surface, the end-effector of the manipulator usually has been strictly constrained with both position and orientation requirement. According to the shape control technique presented above, the inverse kinematics of $P_m L$ can be given by

$$(\alpha, \theta_{rm}, \theta_n) = f(x_e, y_e, m, \psi_n) \tag{12}$$

where $f(.)$ represents the inverse kinematics function of the manipulator.

It can be seen from above forward kinematics discussion that the inverse kinematics solution of $P_m L$ exists and is sole. But we cannot yield the result directly from solving the equations from (1) to (12) because of their nonlinearity. In addition, trigonometric function often has a lot of available solutions, so it is hard to choose the satisfied one. For an n-link hyper-redundant modularized manipulator, there are three groups of nonlinear functions existing. Based on the biological receptive fields, Moody and Darken [7] proposed a network structure, namely, a radial basis function (RBF) neural network that employs local receptive fields to perform function mappings. The RBF neural network is well known in the field of approximation of nonlinear function and pattern recognition. In this paper, we construct a multiple RBF neural networks kinematically controller as shown in Fig.4. The principles of RBF neural networks have been introduced in [7] and [3] in detail.

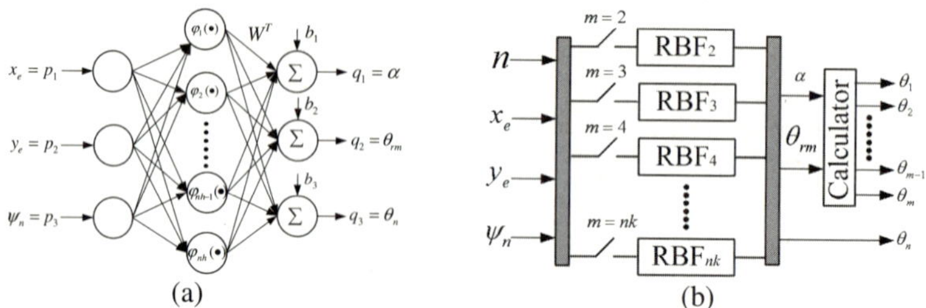

Fig. 4. RBF neural networks based kinematically controller. (a) A single neural network RBF*m*; (b) The multiple RBF neural networks based kinematically controller.

Fig.4a shows the single neural network model. The input ***p*** is a vector with 3 components that determine the end-effector's position and orientation. The output ***q*** is a vector with 3 components that determine the configuration of the manipulator. The output can be given by

$$q_i = \sum_{j=1}^{nh} [w_{ij}\varphi_j(p) + b_i] \tag{13}$$

$$\varphi_i(p) = \exp\left(-\frac{\left\|p - c_i\right\|^2}{2\sigma_i^2}\right) \tag{14}$$

where $\varphi(\cdot)$ are the Radial-Basis Function and *nh* is the number of hidden-layer neurons. The interconnection between hidden-layer and output weights by w_{ij} and the threshold value is b_i. Generally, the hidden-layer has Gaussian function for its activation function as (14) in which c_i is the center, and σ_i is the width.

For an *n*-link manipulator P_mL to carry out a designated mission, we use the forward kinematics data to train the neural networks RBF_m off-line, and then store it in the multiple RBF neural networks controller to control the correspondence manipulator on-line. The model includes a calculator which synthesizes the equations from (3) to (6) following the output of neural networks.

4 Simulations

As mentioned previously, hyper-redundant manipulator has several potential advantages over non-redundant manipulator. The extra degrees of the freedom can be used to achieve some special goals. In this paper, we mainly discussed the end-effector of the manipulator usually has both position and orientation requirement. Simulations of different operation with various requirements of the end-effector are given in Fig.5. It can be seen that the manipulator shows excellent flexibility of its end-effector and the RBF neural network controller has high precision in the simulation.

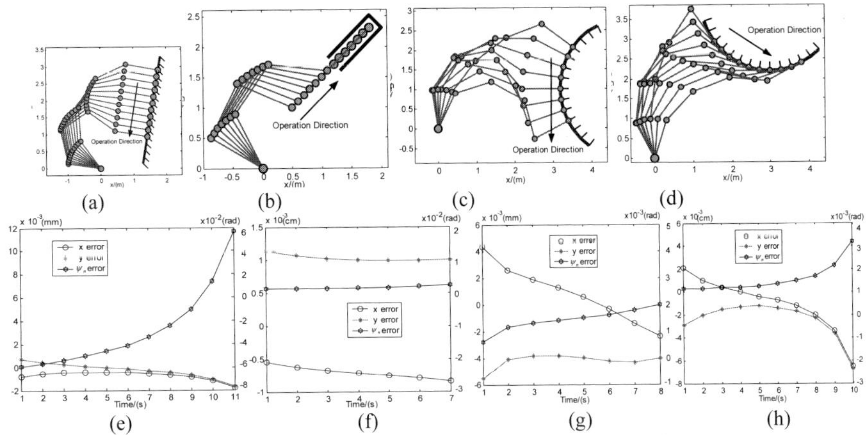

Fig. 5. Simulations of *PmL*'s operation in both path following and direction constrained of the end-effector. (a) *P3L* carries out a painting operation. (b) *P3L* carries out an inspection operation; (c) *P4L* carries out a painting operation; (d) *P4L* carries out a shaving operation; (e) Motion errors of (a); (f) Motion errors of (b); (g) Motion errors of (c); (h) Motion errors of (d).

5 Conclusion

Based on RBF neural networks and shape control technique, the approach presented in this paper can kinematically control the end-effector's position and direction of the hyper-redundant modularized manipulator with few control parameters and high precision. This approach has been demonstrated by several simulations and it has the potential application in more complicated environment. When the manipulator joint increases, joint torque will increase quickly. How to dynamic control it and optimize its motion are desired in the future.

References

1. Xiong, Y.: Robotics. China Mechanical Press (in Chinese) (1993)
2. Zhang, Y., Wang, J., Xu, Y.: A dual neural network for bi-criteria kinematic control redundant manipulators. IEEE Transactions on Robotics and Automation 18(6)(2002)923-931
3. Nanayakkara, T., Watanabe, K., Kiguchi, K., et al: Evolutionary structured RBF neural network based control of a seven-link redundant manipulator. In: IEEE Society (Eds.): Proceedings of the 39th SICE Annual Conference, Saga Univ. (2000)148-153
4. Chirikjian, G. S., Burdick, J.W.: The Kinematics of Hyper-Redundant Robot Locomotion. IEEE Transactions on Robotics and Automation 11(6)(1995)781-793

5. Ma, S., Konno, M.: An Obstacle Avoidance Scheme for Hyper-redundant Manipulators – Global Motion Planning in Posture Space. In: Harrigan R., and Jamshidi M. (Eds.): Proceedings of IEEE ICRA, Albuquerque (1997)161-166
6. Kobyashi, H., Ohtake, S.: Shape Control of Hyper Redundant Manipulator. In: Fukuda T., and Arimoto S. (Eds.): Proceedings of IEEE ICRA, Nagoya (1995)2803-2808
7. Moody, J., Darken, C.: Learning with Localized Receptive Fields. In: Touretzky, D., Hinton, G., and Sejnowski, T. (Eds.): Proceedings of Connectionist Models Summer School, San Mateo (1988) 133-143

Robust Adaptive Neural Networks with an Online Learning Technique for Robot Control

Zhi-gang Yu, Shen-min Song, Guang-ren Duan, and Run Pei

School of Aerospace, Harbin Institute of Technology,
Harbin 150001, China
yzglgy@163.com

Abstract. A new robust adaptive neural networks tracking control with online learning controller is proposed for robot systems. A learning strategy and robust adaptive neural networks are combined into a hybrid robust control scheme. The proposed controller deals mainly with external disturbances and nonlinear uncertainty in motion control. A neural network (NN) is used to approximate the uncertainties in a robotic system. Then the disadvantageous effects on tracking performance, due to the approximating error of the NN in robotic system, are attenuated to a prescribed level by an adaptive robust controller. The learning techniques of NN will improve robustness with respect to uncertainty of system, as a result, improving the dynamic performance of robot system. A simulation example demonstrates the effectiveness of the proposed control strategy.

1 Introduction

In the presence of uncertainties of the robot system, several robust control strategies [1],[2],[3] provide asymptotic motion tracking and an ultimate bounded force error. However, an accurate mathematical model for a complex nonlinear system is very difficult to be achieved and costly to be developed under realistic situations. In order to account for the system uncertainties and disturbance, many adaptive neural network controllers are proposed. Such as adaptive controllers for robots have been employed by Kung and Hwang [4]. Some robust adaptive neural networks controller have been applied to both of the problems indicated above [4],[5],[6],[7]. However, the dynamic performance does not be improved with time, since the controller lacks the capacity to compensate for the nonlinearity, such as, friction is an ubiquitous phenomenon in mechanical systems that is difficult to model and often requires to be compensated.

In this work, Lyapunov theory is used to provide a framework for stability. In conjunction with the Lyapunov theory, an adaptive learning scheme is introduced in order to improve the dynamic performances of robot system. Considering uncertainty in both environment and robot dynamics, the proposed robust controller with online learning technique deals mainly with external disturbances and nonlinear uncertainty in motion control for the tracking control of the robot system. According to the available system information, the learning scheme ensures that the tracking error asymptotically converges to zero. The simulations to a two-link robotic manipulator subject to uncertainties are performed to demonstrate the properties of the developed controller.

J. Wang et al. (Eds.): ISNN 2006, LNCS 3972, pp. 1153 – 1159, 2006.

2 The Dynamics of The Robot System and Problem Formulation

Consider the model describing the dynamics of an n-degrees-of-freedom rigid robot manipulator

$$\tau = M(q)\ddot{q} + C(q,\dot{q})\dot{q} + G(q) + \Delta(q,\dot{q}) + d . \tag{1}$$

where $q \in \mathbb{R}^n$, $\dot{q} \in \mathbb{R}^n$ and $\ddot{q} \in \mathbb{R}^n$ represent the vectors of joint position, velocity and acceleration for robot system, respectively, $\tau \in \mathbb{R}^n$ are the vector of joint input torques (control input), $M(q) \in \mathbb{R}^{n\times n}$ is the symmetric and positive-definite inertia matrix, $C(q,\dot{q}) \in \mathbb{R}^{n\times n}$ is the vector of gravitational torques. $\Delta(q,\dot{q})$ is the model error, d represents the uncertain torque caused by friction forces, vibration, backlash and elasticity.

In terms of the modeling uncertainties in a robotic system, firstly, we compensate the nonlinear component $\Delta(q,\dot{q}) + d$. Let $\tau = u + M(q)\ddot{q} + C(q,\dot{q})\dot{q} + G(q)$, where u is the output of the controller. Then, the tracking error dynamics is founded by

$$u = M(q)\dot{e} + C(q,\dot{q})\dot{e} + \Delta(q,\dot{q}) + d . \tag{2}$$

where $e(t) = q(t) - q_d(t)$, and $q_d(t)$ is the desired trajectory of the robot system.

In order to represent the ability of anti-disturbance, we define the following reference signal z_R. Thus, we consider an adaptive tracking performance criterion and formulate the robust tracking problem as follows: Given any smooth bounded desired motion trajectory, the robot system (1) uses a feedback controller $u = \varphi(q_d, \dot{q}_d, e, \dot{e})$, for given positive constant $\gamma > 0$, $p_1 \geq 0$ and $p_2 \geq 0$, so that the L_2 gain of the closed loop system with uncertainty $\Delta(q,\dot{q}) + d$ is as small as possible or less than a given constant γ, namely, satisfies $\|z_R\|_2 < \gamma \|\Delta(q,\dot{q}) + d\|_2$.

Based on the learning control strategy and the approximation capability of neural networks, a novel control scheme with learning strategy will be proposed. The adaptive law of the adjustable parameter vector in the neural networks adaptively compensates for the approximation error of neural network. As result in improving the transition performance of robot system described by (1). So that the ultimate result under the robust controller is that the tracking error $\underline{e}(t) = (e(t), \dot{e}(t))$ will be attenuated to an arbitrarily small residual tracking error set.

Lemma 1 (HJI inequation): Consider a nonlinear closed-loop system $\dot{x} = f(x) + g(x)\omega$, where $x \in \mathbb{R}^n$ denotes state variable, $f(x)$, $g(x)$ denotes a vector function, $z = h(x)$ is a reference signal. For a given positive constant $\gamma > 0$, if there exists a positive function $L(x) \geq 0$ that satisfy the following HJI inequation: $\dot{L} = \frac{\partial L}{\partial x}\dot{x} = \frac{\partial L}{\partial x} f(x) + \frac{\partial L}{\partial x} g(x)\omega \leq \frac{1}{2}\gamma^2 \|\omega\|^2 - \frac{1}{2}\|z_R\|^2$, such that the L_2 gain of the closed-loop system is less than γ.

3 Function Approximation by Neural Networks

The NN approximation error ε_f is a critical quantity, representing the minimum possible deviation of the ideal approximator $W_f^{*}\phi_f(\mathbf{x})$ from the unknown smooth function $f(\mathbf{x})$, where $\phi_f(\mathbf{x})$ denotes the vector of the radial basis function (RBF) of RBF neural networks, and W_f^{*} denotes the weight vector. Universal approximation results for radial basis function NN indicate that, if NN node number is sufficiently large, then ε_f can be made arbitrarily small on a compact region [8],[9]. The ideal weight vector W_f^{*} is an "artificial" quantity required for analytical purposes. It is assumption that W_f^{*} is defined as the value of W_f that minimizes $|\varepsilon_f|$ for all $\mathbf{x} \in V_B \subset \mathbb{R}^m$ in a compact region, i.e., $W_f^{*} := \arg\min_{W \in R^N} \left\{ \sup_{x \in \Omega_x} \left| f(\mathbf{x}) - W_f \phi_f(\mathbf{x}) \right| \right\}$. In general, the ideal NN weight W_f^{*} is unknown and needs to be estimated. In this paper, we shall consider $\hat{W}_f$ being the estimate of the ideal NN weight W_f^{*}.

4 Combined Robust Adaptive Neural Networks Tracking Control and Online Learning Technique

The control problem is to find a control law such that the state can track the desired trajectories in the presence of the uncertainties. Let $\Delta f(q,\dot{q}) = \Delta(q,\dot{q}) + d$ in robot system (1), then $\Delta f(q,\dot{q})$ can be approximated by a neural network. So we have $\Delta f(q,\dot{q}) = \hat{W}_f \phi_f + \varepsilon_f$. The input of neural network is joint positions and their derivative; the output of neural network is the uncertainty $\Delta f(q,\dot{q})$. If the parameter adjustment mechanism of the neural network reduces the approximation error, then the tracking error can go to zero. Substituting $\Delta f(q,\dot{q}) = \hat{W}_f \phi_f + \varepsilon_f$ into (2), then the dynamics equation of the closed-loop system becomes:

$$u = M(q)\dot{e} + C(q,\dot{q})\dot{e} + \hat{W}_f \phi_f + \varepsilon_f \,. \tag{3}$$

where the approximation ε_f is regard as the exterior disturbance, and we define the reference signal $z_R = pe$, where $p \geq 0$ are the given coefficient. So that L_2-gain is $J_R = \sup_{\|\varepsilon_f\| \neq 0} \|z_R\| / \|\varepsilon_f\|$. We define two state vectors $\begin{cases} x_1 = e \\ x_2 = \dot{e} + \alpha e \end{cases}$, where α is a given positive constant. So we rewrite (3) as

$$\begin{cases} \dot{x}_1 = x_2 - \alpha x_1 \\ M\dot{x}_2 = u - Cx_2 + \omega + \hat{W}_f \phi_f + \varepsilon_f \end{cases} . \tag{4}$$

where $\omega = M\alpha\dot{e} + C\alpha e$. For an n-joint robot system, NN have n output. Defining the torques error vectors $e_\tau(t) = (\tilde{\tau}_d - \tilde{\tau}) - W_f\phi_f$, where the desired nominal torque is $\tilde{\tau}_d = M(q_d)\ddot{q}_d + C(q_d, \dot{q}_d)\dot{q}_d + G(q_d)$ and real nominal torque is $\tilde{\tau} = M(q)\ddot{q} + C(q, \dot{q})\dot{q} + G(q)$.

Theorem 1. Consider the robot system (1) using the following controller and the robust adaptive law as follows:

$$u(t) = -x_1 - \omega + \hat{W}_f(t)\phi_f - (\varepsilon_2 + 1/2\gamma^2)x_2, \tag{5}$$

$$\dot{\hat{W}}_f(t) = -\eta \hat{W}_f(t). \tag{6}$$

the learnig algorithm for NN weights

$$\hat{W}_f(t+1) = \hat{W}_f(t) - \lambda e_\tau(t)\phi_f, \tag{7}$$

where $\varepsilon_2 > 0$ is the given constant, $\eta > 0$, $\lambda > 0$ is the learning-rate. The reference signal $z_R = px_1$ satisfies $\varepsilon_1 = \alpha - p^2/2$ ($\varepsilon_1 > 0$), such that L_2 -gain J_R of the closed-loop system (4) and (5) is less than γ.

Proof: Consider the positive-define Lyapunov candidate $L = \frac{1}{2}x_1^T x_1 + \frac{1}{2}x_2^T M x_2 + \frac{1}{2}W_f^T W_f$. We obtain the derivative of L

$$\begin{aligned}\dot{L} &= x_1^T \dot{x}_1 + x_2^T M\dot{x}_2 + \frac{1}{2}x_2^T \dot{M}x_2 + \hat{W}_f^T \dot{\hat{W}}_f \\ &= x_1^T(x_2 - \alpha x_1) + x_2^T(u - Cx_2 + \omega + \hat{W}_f\varphi_f + \varepsilon_f) + \frac{1}{2}x_2^T(\dot{M} - 2C)x_2 + \hat{W}_f^T\dot{\hat{W}}_f\end{aligned}$$

According to the property of the symmetry of robot system, namely, for an arbitrary vector x_2, satisfied $x_2^T(\dot{M} - 2C)x_2 = 0$. So we have

$$\dot{L} = -x_1^T \alpha x_1 + x_2^T(x_1 + u - Cx_2 + \omega + \hat{W}_f\phi_f + \varepsilon_f) + \hat{W}_f^T \dot{\hat{W}}_f.$$

Substituting (6) into the above equation, then

$$\dot{L} = -x_1^T \alpha x_1 + x_2^T(x_1 + u - Cx_2 + \omega + \hat{W}_f\phi_f + \varepsilon_f) - \eta\hat{W}_f^T \hat{W}_f.$$

Regarding the approximation error ε_f as the external disturbance, and let $H = \dot{L} - \frac{1}{2}\gamma^2\|\varepsilon_f\|^2 + \frac{1}{2}\|z_R\|^2$. Thus,

$$
\begin{aligned}
H &= -\alpha\|x_1\|^2 + x_2^T\left(x_1 + \omega - \hat{W}_f\phi_f + u\right) - x_2^T\varepsilon_f - \eta\|\hat{W}_f\|^2 - \frac{1}{2}\gamma^2\|\varepsilon_f\|^2 + \frac{1}{2}p^2\|x_1\|^2 \\
&= -\left(\alpha - \frac{1}{2}p^2\right)\|x_1\|^2 + x_2^T\left(x_1 + \omega - \hat{W}_f\phi_f + u\right) - \eta\|\hat{W}_f\|^2 - x_2^T\varepsilon_f - \frac{1}{2}\gamma^2\|\varepsilon_f\|^2
\end{aligned}.
$$

If we choose the inequation as follows:

$$
\begin{aligned}
-x_2^T\varepsilon_f - \frac{1}{2}\gamma^2\|\varepsilon_f\|^2 &= -\frac{1}{2}\left(\frac{1}{\gamma^2}\|x_2\|^2 + 2x_2^T\varepsilon_f + \gamma^2\|\varepsilon_f\|^2 - \frac{1}{\gamma^2}\|x_2\|^2\right) \\
&= -\frac{1}{2}\left\|\frac{1}{\gamma}x_2 + \gamma\varepsilon_f\right\|^2 + \frac{1}{2\gamma^2}\|x_2\|^2 \le \frac{1}{2\gamma^2}\|x_2\|^2
\end{aligned}.
$$

Then we have

$$
\begin{aligned}
H &\le -\left(\alpha - \frac{1}{2}p^2\right)\|x_1\|^2 + x_2^T\left(x_1 + \omega - \hat{W}_f\phi_f + u\right) - \eta\|\hat{W}_f\|^2 + \frac{1}{2\gamma^2}\|x_2\|^2 \\
&= -\left(\alpha - \frac{1}{2}p^2\right)\|x_1\|^2 + x_2^T\left(x_1 + \omega - \hat{W}_f\phi_f + u + \frac{1}{2\gamma^2}x_2\right) - \eta\|\hat{W}_f\|^2
\end{aligned}.
$$

Substituting (5) and $\varepsilon_1 = \alpha - p^2/2$ into the above equation, we obtain

$$
H \le -\varepsilon_1\|x_1\|^2 - \varepsilon_2\|x_2\|^2 - \eta\|W_f\|^2 \le 0.
$$

Therefore $\dot{L} \le 1/2\gamma^2\|\varepsilon_f\|^2 - 1/2\|z_R\|^2$ is achieved. According to the Lemma 1, the L_2 gain J_R of the closed loop system (4) is less than γ. The proof is complete.

The learnig algorithm (7) offers a new inititial value for the robust updating algorithm (6) at every discrete sampling time. Based on the theorem 1, we have the following important analytical results. NN has the ability of learning a nonlinear model without a prior knowledge of its structure. From a learning perspective, it is important not only to learn the unknown nonlinearity, but also to cause the tracking error to go to zero. The neural network controller is trained through the learning algorithm of gradient descent (7). The approximation error ε_f of neural network is to reduce. What is more important is that it can shorten the time of transition phase. Thus the dynamic performance of dynamic system will be greatly improved. In the efficient training process, this procedure is repeated as time progresses, then, $\|\varepsilon_f\|^2 < 1/\gamma^2\|z_R\|^2$ can be achieved. So we have $\dot{L} < 0$. Thus, the control error will eventually approach zero as the training process continues, i.e. the tracking error $\underline{e}(t) = \left(e(t), \dot{e}(t)\right)$ will be attenuated to an arbitrarily small residual tracking error set.

Let us consider the training process of NN in more detail. In the auxiliary neural learning algorithm, the gradient descent algorithm adopted can be readily extended to carry out the associated training process with robust updating algorithm. The experimental data of the uncertainty are directly obtained from the difference between desired nominal torque $\tilde{\tau}_d$ and real nominal torque $\tilde{\tau}$, where $\tilde{\tau}_d - \tilde{\tau} = \Delta f(q, \dot{q})$. The gradient decent algorithm is applied for neural network training. The objective is to find the parameter vector $\hat{W}$ to minimize: $\varepsilon(t)=1/2\left(\left|\Delta f(q,\dot{q})-\hat{W}_f\varphi_f\right|\right)^2$.

5 Simulation Study

We demonstrate the proposed control scheme by the tracking control of a two-link robotic manipulator with 2 degrees of freedom in the rotational angles described by angles q_1 and q_2, The dynamic equations describing the motion of the robotic system are of the following form [10]

$$\begin{bmatrix}\tau_1\\ \tau_2\end{bmatrix}=\begin{bmatrix}(m_1+m_2)r_1^2+m_2r_2^2+2m_2r_1r_2c_2 & m_2r_2^2+m_2r_1r_2c_2\\ m_2r_2^2+m_2r_1r_2c_2 & m_2r_2^2\end{bmatrix}\begin{bmatrix}\ddot{q}_1\\ \ddot{q}_2\end{bmatrix}+\begin{bmatrix}-m_2r_1r_2s_2\dot{q}_1(\dot{q}_1+\dot{q}_2)\\ m_2r_1r_2s_2\dot{q}_1^2\end{bmatrix}+\begin{bmatrix}(m_1+m_2)l_1gc_2+m_2l_2gc_1\\ m_2l_2gc_{12}\end{bmatrix}+\begin{bmatrix}d_1\\ d_2\end{bmatrix},$$

where m_1, m_2, $r_1=0.5l_1$ and $r_2=0.5l_2$, are the mass, the moment of inertia, the half-length of link 1 and 2, $g=9.8\,m/s$, and shorthand notations $c_1=\cos(q_1)$, $s_1=\sin(q_1)$, $c_{12}=\cos(q_1+q_2)$ etc. In the control experiments described below, the kinematics and inertial parameters of the arm are chosen as $l_1=1.2m$, $l_2=0.8m$, $m_1=4kg$, $m_2=2kg$, respectively. The initiation conditions are $q(0)=[0.7\quad 0.5]^{\mathrm{T}}$ rad, $\dot{q}(0)=[0\quad 0]^{\mathrm{T}}$ rad/s, $\ddot{q}(0)=[0\quad 0]^{\mathrm{T}}$ rad/s^2. In (5), (6) and (7), the design parameters are given by $\alpha=50$, $\varepsilon_1=0.1$, $\varepsilon_2=0.1$, $\gamma=0.06$, $\eta=0.2$, $p=20$. The combined effects of friction and the uncertainty of the robot system (The model error) are defined by $\Delta=\begin{bmatrix}0.5sign(\dot{q}_1)\left[0.1+\exp(-|\dot{q}_1|)\right]\\ sign(\dot{q}_2)\left[0.2+\exp(-|\dot{q}_2|)\right]\end{bmatrix}$ N×m.

The external torque disturbance $d=[d_1\quad d_2]^{\mathrm{T}}$ is random disturbance. The sampling time of robot system is 0.005s. The desired trajectories of the joint q_{1d}, q_{2d}, and the real trajectories q_1, q_2, are given in Fig. 1 and Fig. 2, respectively.

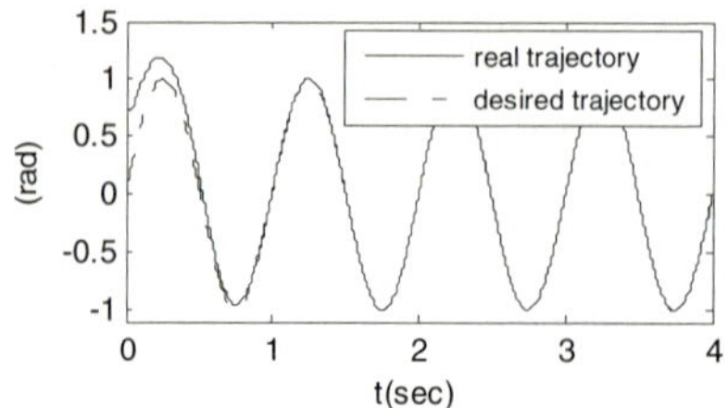

Fig. 1. The trajectories of the first joint

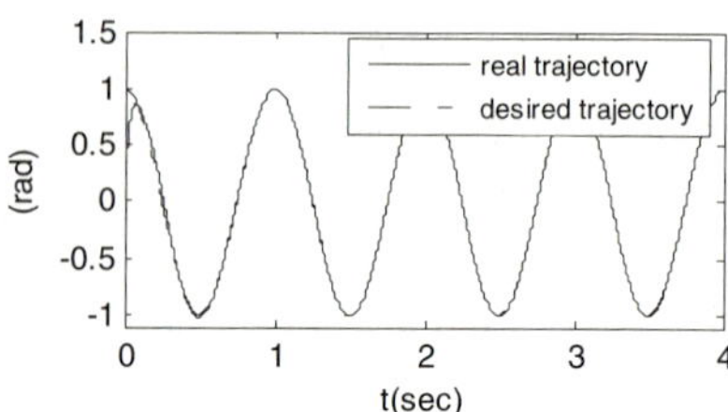

Fig. 2. The trajectories of the second joint

The simulation results have shown the expected performance. Thus the proposed robust adaptive neural network control with learning controller can control the robotic system to follow the desired trajectory without using any prior information about the uncertainty.

6 Conclusion

The robust adaptive neural network control integrates with the learning control of neural network to solve the tracking control problem of robot systems. The Lyapunov synthesis approach is used to develop robust adaptive control schemes based on neural network. A main feature of the adaptive neural network control law and parameter update algorithms is the stability properties of the algorithm being independent of the specific mechanism used to achieve the tracking performance with a prescribed attenuation.

Acknowledgement

This work is supported by foundation HIT2002.12.

References

1. Hwang, M.C., Hu, X.: A Robust Position/Force Linear Controller of Manipulators via Nonlinear H∞ Control and Neural Networks. IEEE Transactions on System, Man, Cybernetics–Part B 30(2) (2000) 310–321
2. Dixon, W.E., Zergeroglu, E., Dawson, D.M.: Global Robust Output Feedback Tracking Control of Robot Manipulators. Robotica 22 (4) (2004) 351–357
3. Wang, D., Soh, Y. C., Chean, C. C.: Robust Motion and Force Control of Constrained Manipulators by Learning. Automatica 31(2) (1995) 257–262
4. Kung, S.Y., Hwang, J.N.: Neural Network Architectures for Robotic Applications, IEEE Transactions on Robotics and Automation 5 (5) (1989) 641–657
5. Bai, P., Fang, T.J., Ge, Y.J.: Robust Neural-Network Compensating Control for Robot Manipulator Based on Computed Torque Control. Control Theory and Applications 18 (6) (2001) 895–903
6. Nasisi, O., Carelli, R.: Adaptive Servo Visual Robot Control, Robotics and Autonomous Systems 43 (1) (2003) 51–78
7. Wai, R.J.: Robust Control for Nonlinear Motor-Mechanism Coupling System Using Wavelet Neural Network. IEEE Transactions on Systems Man and Cybernetics-Part B: Cybernetics 33 (3)(2003) 489–497
8. Park J., Sandberg, I.W.: Universal Approximation Using Radial-Basis-Function Networks, Neural Computation 3 (2) (1991) 246–257
9. Liao, Y., Fang, S.C., Nuttle, H.L.W.: Relaxed Conditions for Radial-Basis Function Networks to Be Universal Approximators. Neural Networks 16(7) (2003) 1019–1028
10. Leung, T. P., Zhou, Q. J., Su, C. Y.: An Adaptive Variable Structure Model Following Control Design for Robot Manipulators, IEEE Trans. Automatic Control 36 (3) (1991) 347–353

A Particle Swarm Optimized Fuzzy Neural Network Control for Acrobot

Dong-bin Zhao and Jian-qiang Yi

Laboratory of Complex Systems and Intelligence Science, Institute of Automation, Chinese Academy of Sciences, Beijing 100080, P.R. China
{Dongbin.zhao, jianqiang.yi}@mail.ia.ac.cn

Abstract. This paper addresses the problem of controlling an acrobot, an under-actuated robotic systems, using fuzzy neural network approach. A five-layer Takagi-Sugeno fuzzy neural network control (TSFNNC) is proposed to swing up the acrobot from the low stable equilibrium to approach and balance around its top unstable equilibrium position. By analyzing the system dynamics, total energy and potential energy of the system are introduced in the second layer, with the system states as the inputs to the first layer. Fuzzy membership functions and rules are depicted in the third and fourth layers respectively. The fifth layer works as the final output. A modified particle swarm optimizer (PSO) is adopted to train the consequents in the fourth layer. Simulation results indicate that the integrated TSFNNC approach can control the acrobot system from upswing to balance process effectively. This approach provides an easy and feasible solution for similar control problems.

1 Introduction

Acrobot (**Acro**batic **Robot**) is a typical structure of two link under-actuated robotic systems, which are under extensive investigations recently [1-7]. The upswing control of the acrobot is to rotate the two links from the low stable to the top unstable equilibrium position. No smooth controller is proved to realize the whole process. A common adopted control scheme is to swing up the system close to the top unstable equilibrium with a nonlinear controller, and then switch to the other balance controller.

Spong [1] proposed the concept of partial feedback linearization, to drive the two links into the attractive basin of a linear balance controller. Smith, et al. [2] presented a dynamic fuzzy controller integrated with genetic algorithm and dynamic switching systems. Lai, et al. [3] developed a model-free controller for upswing and model-based controller for balancing. Xu, et al. [4] designed a time-optimal controller based on neural network reinforcement learning to reduce upswing time. Aiming at real application, Zhao and Yi [5] proposed a GA tuned fuzzy neural network control approach with limited torque output.

But during the switching process from upswing to balance, the control torque is usually not very smooth. Therefore, a further investigation concentrates on an integrated control approach for the whole process without control switching. Wang etc. [6] provided an adaptive sliding mode control. This paper is to develop a new integrated control approach based on computational intelligence. A five-layer Takagi-Sugeno fuzzy neural network control (TSFNNC) is proposed. Total energy and

J. Wang et al. (Eds.): ISNN 2006, LNCS 3972, pp. 1160 – 1165, 2006.

potential energy of the system are introduced in the second layer, with the system states as the inputs to the first layer. Fuzzy membership functions and rules are depicted in the third and fourth layers respectively. The fifth layer works as the final output. A modified particle swarm optimization (PSO) is adopted to train the consequents in the fourth layer.

2 System Dynamics

The model of the acrobot is shown in Fig. 1.

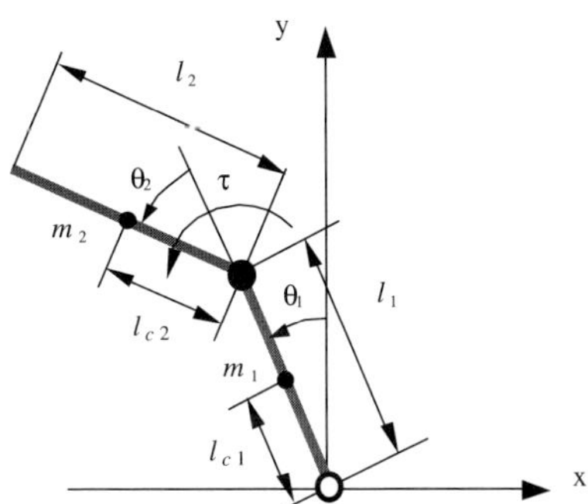

Fig. 1. Model of the acrobot system

Supposing there is no friction, the system dynamics equation is expressed by

$$\boldsymbol{\tau} = D(\boldsymbol{\theta})\boldsymbol{\alpha} + C(\boldsymbol{\theta}, \boldsymbol{\omega})\boldsymbol{\omega} + G(\boldsymbol{\theta}) \tag{1}$$

Where $\boldsymbol{\tau}$ is the external torque, which is only applied on the second joint. The variables $\boldsymbol{\theta}=[\theta_1\ \theta_2]^{\mathrm{T}}$, $\boldsymbol{\omega}=[\omega_1\ \omega_2]^{\mathrm{T}}$, and $\boldsymbol{\alpha}=[\alpha_1\ \alpha_2]^{\mathrm{T}}$ are angles, angular velocities and angular accelerations of the two links respectively. The terms D[$\boldsymbol{\theta}$], C[$\boldsymbol{\theta}, \boldsymbol{\omega}$] and G[$\boldsymbol{\theta}$] can be represented with five variables $\{q_1, q_2, q_3, q_4, q_5\}$. Detailed description of these parameters can be also seen in [5].

The kinetic energy E_k, potential energy E_p, and the total energy E are depicted by

$$\begin{aligned} E_k &= \frac{1}{2}\omega^T D(\theta)\omega \\ E_p &= q_4 g \cos\theta_1 + q_5 g \cos(\theta_1 + \theta_2) \\ E &= E_k + E_p \end{aligned} \tag{2}$$

One property of the two link under-actuated system is the passivity [7], which can be derived from equations 1 and 2, depicted as follows.

$$\dot{E} = \tau_2\,\omega_2 \tag{3}$$

It means that system energy will increase when the torque is exerted on the active joint in the same direction with its angular velocity, vice versa.

State vector **X** is selected as $[\theta_1, \omega_1, \theta_2, \omega_2]^T$. θ_1 and θ_2 vary within [-π , π]. From equ.2, four equilibrium positions of the system can be derived, of which the commonly reachable two are as follows.

(0, 0, 0, 0): Top unstable position with the top energy E_{top}, $E = E_{top} = q_4 g + q_5 g$

(π, 0, 0, 0): Low stable position with the low energy, $E = -E_{top} = -(q_4 g + q_5 g)$

The top unstable equilibrium position is the most difficult case for feedback stabilization among all the equilibriums. The integrated upswing control is to move the two links from the low stable to the top unstable position, and maintain there.

3 Particle Swarm Optimized Fuzzy Neural Network Control

From the analysis of the system dynamics, we can see that it will be a feasible way to increase the system energy when the energy is low by apply appropriate torque on the active joint. It is desirable to increase the total energy and potential energy simultaneously. When they both reach the top energy E_{top}, which is defined as an attractive region of linear controller [6], the system will be balanced easily.

To realize the above control scheme, a five-layer Takagi-Sugeno Fuzzy Neural Network is proposed, whose architecture is shown in Fig. 2. Each layer outputs serve as the inputs of its next layer. Layers functions are given as follows.

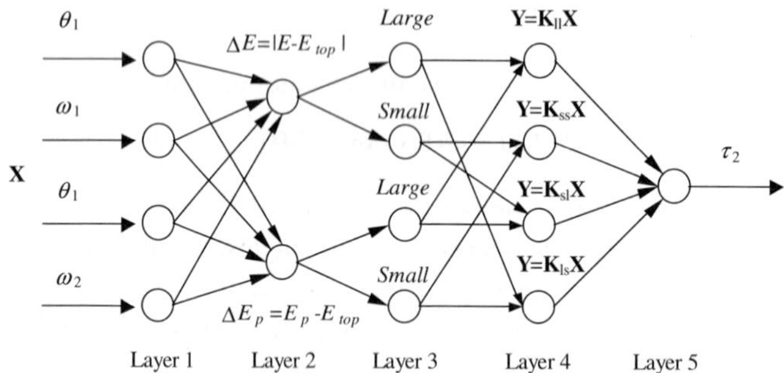

Fig. 2. TSFNNC Structure

Layer 1: Inputs

The state space variables θ_1, ω_1, θ_2, and ω_2 are as the inputs.

Layer 2: Energy error calculation

The absolute errors ΔE and ΔE_p are calculated. When they are small enough to zeros, the system approaches the top unstable position. When the errors are both large to E_{top}, the system is at the initial low stable position. When one of the errors is small and the other is large, they correspond to different system states, shown in Fig.3.

Layer 3: Membership definition

Two fuzzy subsets *Large* and *Small* are defined for each energy error. Triangle functions are adopted to describe the membership degree, shown in Fig. 4. The scope of

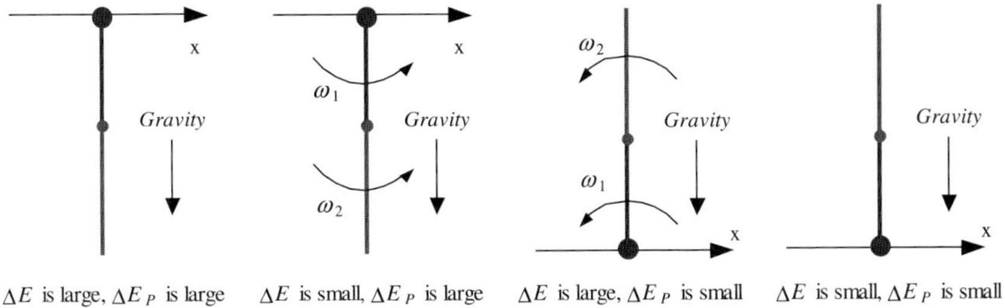

Fig. 3. Different state spaces of the acrobot system depicted by energy error

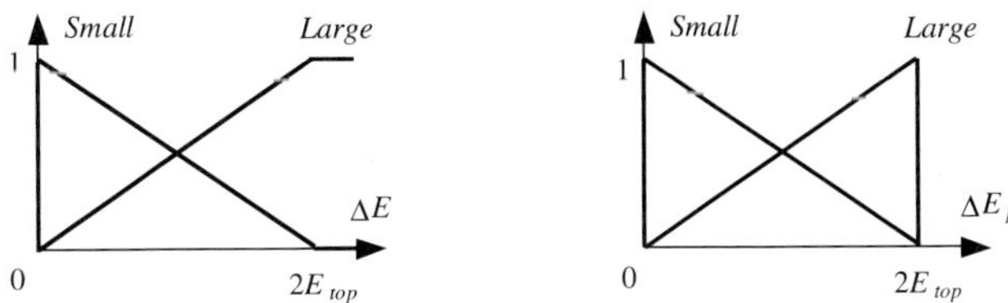

Fig. 4. Membership functions

the potential energy error ΔE_p is $[0, 2E_{top}]$. But the total energy error ΔE may be over $2E_{top}$, where the degree of the membership *Large* and *Small* is set to 1 and 0 respectively. For each energy error, the sum of the degrees is 1.

Layer 4: Fuzzy rules
There are only four fuzzy rules, corresponding to different sates.

$$\begin{aligned}
&\text{Rule 1: If } \Delta E \text{ is large and } \Delta E_p \text{ is large, then } Y_1 = \mathbf{K}_{\mathbf{ll}}\mathbf{X} = sign(x(4))\tau_{\max} \\
&\text{Rule 2: If } \Delta E \text{ is small and } \Delta E_p \text{ is large, then } Y_2 = \mathbf{K}_{\mathbf{ls}}\mathbf{X} \\
&\text{Rule 3: If } \Delta E \text{ is large and } \Delta E_p \text{ is small, then } Y_3 = \mathbf{K}_{\mathbf{sl}}\mathbf{X} = -sign(x(4))\tau_{\max} \\
&\text{Rule 4: If } \Delta E \text{ is small and } \Delta E_p \text{ is small, then } Y_4 = \mathbf{K}_{\mathbf{ss}}\mathbf{X}
\end{aligned} \tag{4}$$

where Rule 1 depicts the action on the initial low stable position, so the consequent is determined to the maximum output torque with the same direction as the angular velocity to increase the system energy. On the other side, Rule 3 depicts the action on the system with high energy, so it is just opposite to Rule 1. Rule 4 describes the system around the top unstable position, so the consequent can be derived with linear feedback controller. Rule 2 describes the system in upswing process, so only the feedback parameters of this consequent need to be determined. The optimum consequent will help to swing up the system quickly and smoothly.

Layer 5: Output
Due to the definition of the fuzzy membership, the final output becomes the summation of the outputs of layer 4. The output is saturated to the positive or negative maximum torque if it is beyond the torque limit.

Particle swarm optimization, proposed recently, shows good performance for fast speed and low error in neural network training and other application fields. Aiming at the consequents of the fuzzy neural network control to be determined, a modified particle swarm optimization algorithm [8] is adopted.

$$v_{id} = \omega v_{id} + c_1 r_1 (p_{id} - x_{id}) + c_2 r_2 (p_{gd} - x_{id}), x_{id} = x_{id} + v_{id} \tag{3}$$

where x_{id} is the current position of particle i, to represent the consequent $\mathbf{K}_{\mathbf{ls}}$. v_{id} is the velocity of the particle i. p_{id} is the best position of particle i, and p_{gd} is the best position of the whole particle swarm, to represent the final problem solution. c_1 and c_2 are learning factors. r_1 and r_2 are two random number. w is a velocity weight.

To evaluate the particle performance, the fitness is defined by

$$fitness = \exp\left(-\sqrt{\sum_{t=1}^{N} e(t)}\right), e(t) = \Delta E + \Delta E_p \tag{4}$$

where the total calculation step N is set to a large number, to provide enough time to swing up the system.

4 Simulation Results

The acrobot structure parameters are selected the same as [1] for simulation comparison. The coefficients $\{q_1, q_2, q_3, q_4, q_5\}$ are $\{1.3333, 1.3333, 1, 1.5, 1\}$. The maximum torque $\tau_{\max}$ is 4 Nm, and $\mathbf{K}_{\mathbf{ss}}$ is determined by LQR mechanism.

$$\mathbf{K}_{\mathbf{ss}} = [-245.9821, -106.2845, -98.4906, -50.0736] \tag{5}$$

The scope of the consequent $\mathbf{K}_{\mathbf{ls}}$ is [-10,10], among which the best is found by the modified particle swarm optimization. 50 particles are initialized, and 100 generations are calculated. The total calculation step N is 1000, and sampling time is 0.01s. The velocity weight w decreases from 0.9 to 0.4 with the generation increasing. The learning factors c_1 and c_2 are both 2.0. The maximum v_{id} is limited within [-2,2]. Several experiments all derive feasible $\mathbf{K}_{\mathbf{ls}}$ solutions, one of which is selected as

$$\mathbf{K}_{\mathbf{ls}} = [-1.6785, -2.0542, -1.8864, -1.9806] \tag{6}$$

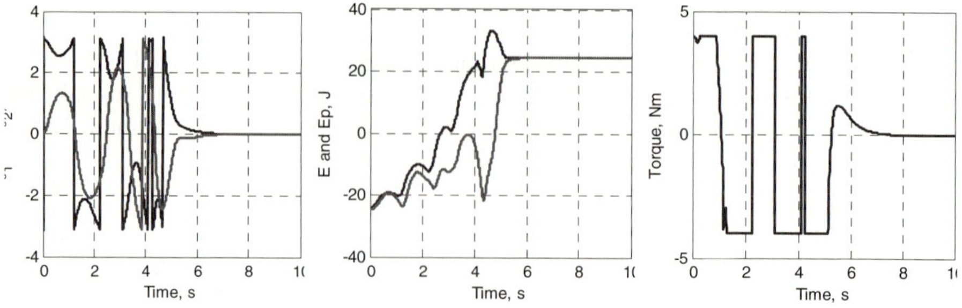

Fig. 5. Simulation results to swing up the acrobot. (*a*) θ_1 and θ_2; (*b*) E and E_p; (*c*) τ_2

Simulation results are shown in Fig. 5. It can be seen that system is swung up to its top unstable equilibrium position successfully. The torque is quite smaller compared to [1], which requires up to hundreds Nm. The torque also varies more smoothly compared to switch control method [5]. During several swings, the system energy E is pumped up effectively to E_{top}, which characterizes the method as an energy approach.

5 Conclusions

An integrated five-layer Takagi-Sugeno fuzzy neural network control is proposed. The consequent is learned with a modified particle swarm optimization algorithm. An acrobot is effectively swung up from the low stable equilibrium to approach and balance around its top unstable equilibrium with a limit torque for a short time. The control scheme can also be extended to other under-actuated systems. The control stability and other modified particle swarm optimizations will be further investigated in future research.

Acknowledgement

This work is supported by NSFC Projects (No.60475030 and No.60575047), and MOST Projects (No. 2003CB517106 and 2004DFB02100).

References

1. Spong, M.W.: The Swing up Control Problem for the Acrobot. IEEE Control Systems Magazine 15(1) (1995) 49-55
2. Smith, M.H., Zhang, T., Gruver, W.A.: Dynamic Fuzzy Control and System Stability for the Acrobot. IEEE International Conference on Fuzzy Systems (1998) 286-291
3. Lai, X., She, J.H. Ohyama, Y., Cai, Z.: Fuzzy Control Strategy for Acrobots Combining Model-free and Model-based Control. IEE Proceedings on Control Theory and Applications 146(6) (1999) 505-510
4. Xu, X., He, H.G.: Residual-gradient-based Neural Reinforcement Learning for the Optimal Control of an Acrobot. IEEE International Symposium on Intelligent Control (2002) 758-763
5. Zhao, D.B., Yi, J.Q.: GA-based Control to Swing up an Acrobot with Limited Torque. Proceedings of International Conference on Intelligent Computing (2005) 3358-3367
6. Wang, W., Yi, J.Q., Zhao, D.B., Liu, D.T.: Design of a New Stable Sliding-mode Controller for a Class of Second-order Underactuated Systems. IEE Proceedings – Control and Applications 151(6) (2004) 683-690
7. Fantoni, I., Lozano, R., Spong, M.W.: Energy Based Control of the Pendubot, IEEE Transactions on Automatic Control 45(4) (2000) 725-729
8. Shi, Y., Eberhart, R.: A Modified Particle Swarm Optimizer IEEE World Congress on Computational Intelligence (1998) 69-73

Adaptive Control Based on Recurrent Fuzzy Wavelet Neural Network and Its Application on Robotic Tracking Control

Wei Sun, Yaonan Wang, and Xiaohua Zhai

College of Electrical and Information Engineering,
Hunan University, Changsha, P.R. China
david-sun@126.com
Yaonan@mail.hunu.edu.cn
zhaixiaohua@163.com

Abstract. A kind of recurrent fuzzy wavelet neural network (RFWNN) is constructed by using recurrent wavelet neural network (RWNN) to realize fuzzy inference. In the network, temporal relations are embedded in the network by adding feedback connections on the first layer of the network, and wavelet basis function is used as fuzzy membership function. An adaptive control scheme based on RFWNN is proposed, in which, two RFWNNs are used to identify and control plant respectively. Simulation experiments are made by applying proposed adaptive control scheme on robotic tracking control problem to confirm its effectiveness.

1 Introduction

Recently, much research has been done on using neural networks (NN) to identify and control dynamic systems [1]-[3]. NN can be classified as feed forward neural networks and recurrent neural networks. Recurrent neural network [4]-[7] can capture the dynamical response of a system with its internal feedback loop. It is a dynamic mapping and demonstrates good performance in the presence of uncertainties, such as parameter variations, external disturbance, unmodeled and nonlinear dynamics.

Recurrent fuzzy neural network (RFNN) [8],[9] is a modified version of recurrent neural network, which uses recurrent network for realizing fuzzy inference. It is possible to train RFNN using the linguistic experience of human operators, and interpret the knowledge acquired from training data in linguistic form. And it is very easy to initialize the structure and parameters of RFNN from linguistic rules. Moreover, with its own internal feedback connections, RFNN can temporarily store dynamic information and cope with temporal problems efficiently.

In this paper, a RFWNN is proposed. In the network, the temporal relations are embedded by adding feedback connections on the first layer of fuzzy neural network, and wavelet basis function is used as fuzzy membership function. Back propagation algorithm is used to train the proposed RFWNN. For

J. Wang et al. (Eds.): ISNN 2006, LNCS 3972, pp. 1166–1171, 2006.

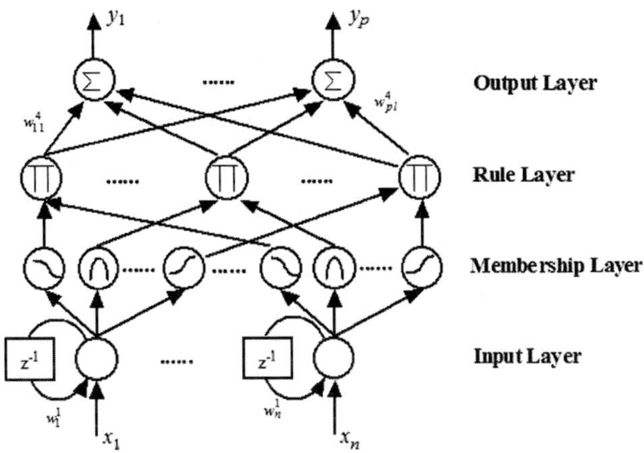

Fig. 1. Structure of four-layer RFWNN

control problem, an adaptive control scheme is developed, in which, two proposed RFWNNs are used to identify and control plant respectively. Finally, simulation experiments are made by applying proposed adaptive control scheme on robotic tracking control problem to confirm its effectiveness.

2 Construction of RFWNN

The structure of the proposed RFWNN is shown in Fig. 1, which comprises 4 layers. Using u_i^k and O_i^k to denote the input and output of the ith node in the kth layer separately, the operation functions of the nodes in each layer are introduced as follows.

Layer 1 (Input Layer): This layer accepts input variables. Its nodes transmit input values to the next layer. Feedback connections are added in this layer to embed temporal relations in the network.

$$u_i^1(k) = x_i^1(k) + w_i^1 O_i^1(k-1), O_i^1(k) = u_i^1(k),\ i = 1, 2, \ldots, n, \tag{1}$$

where k is the number of iterations, w_i^1 is the recurrent weights.

Layer 2 (Membership Layer): Nodes in this layer represent the terms of respective linguistic variables. Each node performs a wavelet basis membership function.

$$u_{ij}^2 = (O_i^1 - a_{ij})/b_{ij}, O_{ij}^2 = h(u_{ij}^2), i = 1, 2, \ldots, n, j = 1, 2, \ldots, m. \tag{2}$$

In (2),$h(\cdot)$is a mother wavelet used in this paper, which is defined as:

$$h(x) = cos(0.25x)exp(-x^2), \tag{3}$$

a_{ij} and $b_{ij}$are the dilation and translation parameters of the wavelet membership function, the subscript ij indicates the jth term of the ith input variable.

Layer 3(Rule Layer): This layer forms the fuzzy rule base and realizes the fuzzy inference. Each node is corresponding to a fuzzy rule. Links before each node represent the preconditions of the corresponding rule, and the node output represents the firing strength of corresponding rule.

The qth node of layer 3 performs the AND operation in qth rule. It multiplies the input signals and output the product. Using $O^2_{iq_i}, q_i = 1, 2, \ldots, m$ to denote the membership of x_i to its corresponding linguistic term in qth rule , then the input and output of qth node can be described as:

$$u^3_q = \prod_i O^2_{iq_i}, O^3_q = u^3_q, i = 1, 2, \ldots, n;\ q = 1, 2, \ldots, l. \tag{4}$$

Layer 4(Output Layer): Nodes in this layer performs the defuzzification operation. The input and output of sth node can be calculated by:

$$u^4_s = \sum_q w^4_{sq} O^3_q, O^4_s = u^4_s / \sum_q O^3_q, \tag{5}$$

where $s = 1, 2, \ldots, p; q = 1, 2, \ldots, l; w^4_{sq}$ is the weight, which represents the output action strength of the sth output associated with the qth rule.

From the above description, it is clear that the proposed RFWNN is a fuzzy logic system with memory elements in first layer. Since a fuzzy system has clear physical meaning, it is very easy to choose the number of nodes in each layer of RFWNN and determine the initial value of weights.

3 Adaptive Control Based on RFWNN

The block diagram of the adaptive control system based on RFWNN is shown in Fig. 2. In this scheme, two RFWNNs are used as controller (RFWNNC) and identifier (RFWNNI) separately. The plant is identified by RFWNNI, which provides the information about the plant to RFWNNC. The inputs of RFWNNC are $e(k)$ and $\dot{e}(k)$. $e(k)$ is the error between the desired output $r(t)$ and the actual system output $y(k)$. The output of RFWNNC is the control signal $u(k)$, which drives the plant such that $e(k)$ is minimized. Since the temporal relations are embedded in RFWNN, only $y(k-1)$ and $u(k)$ are need to be fed into RFWNNI for identifying the model of plant.

Both RFWNNI and RFWNNC are trained by BP algorithm. For training the RFWNNI in Fig. 2, the cost function is defined as follows:

$$J_I(k) = \frac{1}{2} \sum_{s=1}^{p} (e_{Is}(k))^2 = \frac{1}{2} \sum_{s=1}^{p} (y_s(k) - y_{Is}(k))^2, \tag{6}$$

where $y_s(k)$ is the sth output of the plant, $y_{Is}(k) = O^4_s$ is the sth output of RFWNNI, and $e_{Is}(k)$ is the error between $y_s(k)$ and $y_{Is}(k)$ for each discrete time k.

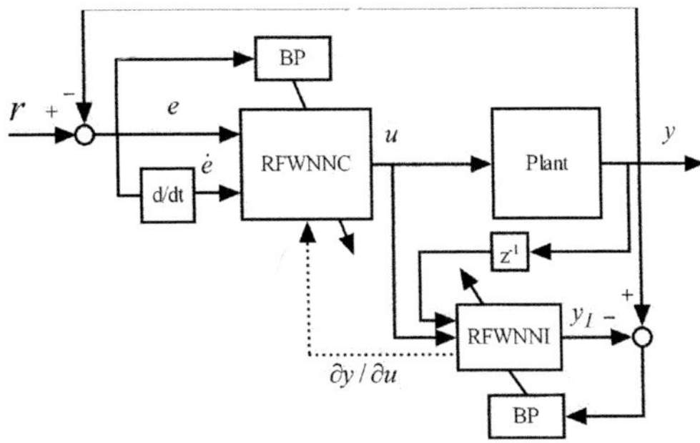

Fig. 2. Adaptive control system based on RFWNN

Then the weights of the RFWNNI can be adjusted by

$$W_I(k+1) = W_I(k) + \Delta W_I(k) = W_I(k) - \eta_I \frac{\partial J_I(k)}{\partial W_I(k)}, \tag{7}$$

where η_I represents the learning rate and W_I represents the tuning weights, in this case, which are w^4_{Isq}, a_{Iiq_i}, b_{Iiq_i}, and w^1_{Ii}.Subscript I represents RFWNNI.

For training RFWNNC in Fig. 2, the cost function is defined as

$$J_C(k) = \frac{1}{2}\sum_{s=1}^{h}(e_s(k))^2 - \frac{1}{2}\sum_{s=1}^{h}(r_s(k) - y_s(k))^2, \tag{8}$$

where $r_s(k)$ is the sth desired output, $y_s(k)$ is the sth actual system output and $e_s(k)$ is the error between $r_s(k)$ and $y_s(k)$.Thus the parameters of the RFWNNC can be adjusted by

$$W_C(k+1) = W_C(k) + \Delta W_C(k) = W_C(k) - \eta_C \frac{\partial J_C(k)}{\partial W_C(k)}. \tag{9}$$

In above equation,

$$\frac{\partial J_C(k)}{\partial W_C(k)} = \sum_s \frac{\partial J_C(k)}{\partial y_s(k)} \cdot \frac{\partial y_s(k)}{\partial W_C(k)} = \sum_s \{-e_s(k) \sum_t [\frac{\partial y_s(k)}{\partial u_t(k)} \cdot \frac{\partial u_t(k)}{\partial W_C(k)}]\}, \tag{10}$$

where u_t is the tth control signal, which is also the tth output of RFWNNC.

Note that the convergence of the RFWNNC cannot be guaranteed until $\partial y_s(k)/\partial u_t(k)$ is known. Obviously, the RFWNNI can provide this information to RFWNNC.

4 Simulation Experiments

Dynamics of robotic manipulators are highly nonlinear and may contain uncertain elements such as friction and load. Many efforts have been made in

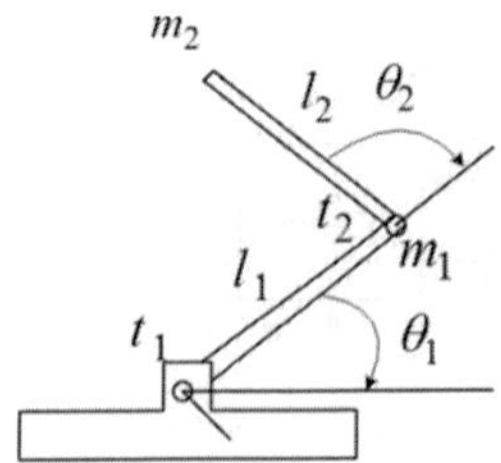

Fig. 3. Structure of two-joint robotic manipulator

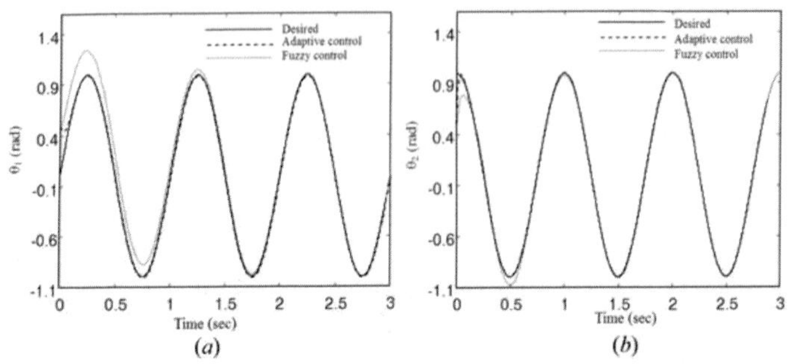

Fig. 4. Trajectories of robotic manipulator: (a) joint 1, (b) joint 2

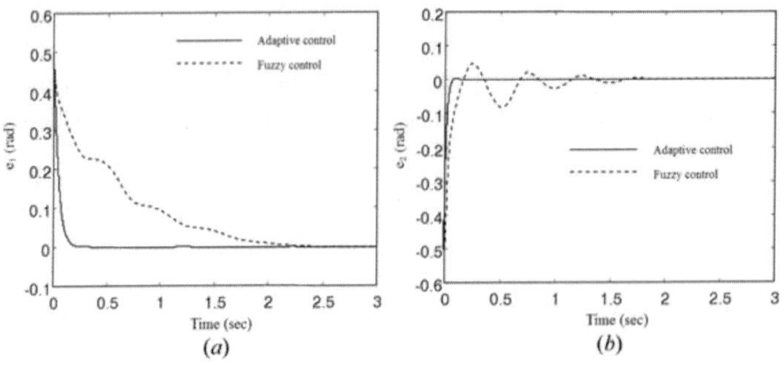

Fig. 5. Tracking errors of robotic manipulator: (a) joint 1, (b) joint 2

developing the precise tracking control of robot manipulators. In the simulation experiments of this paper, the proposed adaptive control scheme is applied on robotic tracking control.

For simplicity, the robotic manipulator used in the simulations just has two joints, see Fig. 3. In Fig. 3, m_1 and m_2 are masses of arm1 and arm2 respectively; l_1 and l_2 are lengths of arm1 and arm2; t_1 and t_2 are torques on arm1 and arm2; θ_1 and θ_2 are positions of arm1 and arm2. The system structure for the control of two-joint robotic manipulator is same as Fig. 2. In this case, the plant in Fig.2 represents the manipulator, the control signal is t_1 and t_2 , and the output of the system is θ_1 and θ_2.

In the simulation, the parameters of manipulator are defined as : $m_1 = 4kg$, $m_2 = 2kg$, $l_1 = 1m$, $l_2 = 0.5m$. Initial conditions are given as: $\theta_1(0) = 0rad$,$\theta_2(0) = 0rad$. The desired trajectories are $\hat{\theta}_1(t) = \sin(2\pi t)$, $\hat{\theta}_2(t)=\cos(2\pi t)$.

Simulation results of the proposed adaptive control method are shown in Fig.4 and Fig. 5, and compared with fuzzy control method. Fig.4 illustrates the trajectories of manipulator. Fig. 5 shows the tracking errors of manipulator.

5 Conclusions

This paper proposed an RFWNN for realizing fuzzy inference using the dynamic fuzzy rules. The proposed RFWNN consists of four layers and the feedback connections are added in first layer. Wavelet basis function is used as fuzzy membership function. The proposed RFWNN can be used for the identification and control of dynamic system. For identification, RFWNN only needs the current inputs and most recent outputs of plant as its inputs. For control, two RFWNNs are used to constitute an adaptive control system, one is used as identifier and another is used as controller. Finally, in this paper, the proposed adaptive control scheme based on RFWNN is used to control the robotic manipulator and simulation results verified its effectiveness.

References

1. Park, Y. M., Choi, M. S., Lee, K. Y.: An Optimal Tracking Neuro-controller for Nonlinear Dynamic Systems. IEEE Trans. on Neural Networks **7** (1996) 1099–1110
2. Narendra, K. S., Parthasarathy, K.: Identification and Control of Dynamical Systems Using Neural Networks. IEEE Trans. on Neural Networks **1** (1990) 4–27
3. Brdys, M. A., Kulawski, G. J.: Dynamic Neural Controllers for Induction Motor. IEEE Trans. on Neural Networks **10** (1999) 340–355
4. Ku, C. C., Lee, K. Y.: Diagonal Recurrent Neural Networks for Dynamic Systems Control. IEEE Trans. on Neural Networks **6** (1995) 144–156
5. Ma, S., Ji, C.: Fast Training of Recurrent Neural Networks Based on The EM Algorithm. IEEE Trans. on Neural Networks **9** (1998) 11–26
6. Sundareshan, M. K., Condarcure, T. A.: Recurrent Neural-network Training by A Learning Automaton Approach for Trajectory Learning and Control System Design. IEEE Trans. on Neural Networks **9** (1998) 354–368
7. Liang, X. B., Wang, J.: A Recurrent Neural Network for Nonlinear Optimization with A Continuously Differentiable Objective Function and Bound Constraints. IEEE Trans. on Neural Networks **11** (2000) 1251–1262
8. Lee, C. H., Teng,C. C.: Identification and Control of Dynamic Systems Using Recurrent Fuzzy Neural Networks. IEEE Trans. on Fuzzy Systems **8** (2000) 349–366
9. Lin, C. T., Chang, C. L., Cheng, W. C.: A Recurrent Fuzzy Cellular Neural Network System with Automatic Structure and Template Learning. IEEE Trans. on Circuits and Systems **51** (2004) 1024–1035

Dynamic Tracking Control of Mobile Robots Using an Improved Radial Basis Function Neural Network

Shirong Liu[1], Qijiang Yu[1,2], and Jinshou Yu[2]

[1] College of Automation, Hangzhou Dianzi University, Hangzhou, Zhejiang 310018, China
liushirong@hziee.edu.cn
[2] Research Institute of Automation, East China University of Science and Technology
Shanghai 200237, China
jshyu@ecust.edu.cn

Abstract. A novel dynamic control scheme for nonholonomic mobile robots is developed in this paper. The dynamics of mobile robot based on improved radial basis function neural network (IRBFNN) is modeled online by the improved algorithm of resource allocating network (IRAN). The control scheme of mobile robot integrates a velocity controller based on backstepping technology and a torque controller based on the IRBFNN and robust compensator. The simulations have shown that the control system is competent for the robust tracking control of mobile robot.

1 Introduction

In the tracking control of mobile robot, many researchers only considered the kinematics tracking control and ignored the dynamics of mobile robot. For example, Kanayama *et al* presented so called *perfect velocity tracking* using the velocity signals to control the actual vehicle motion, which does not hold in practice [1]. Fieero and Lewis integrated a velocity controller and a torque controller based on BP neural network, which was used to online model the dynamics of mobile robot, into the control system [2]. But BP neural network has the drawbacks of slow convergence easily resulting in the system tracking errors.

Radial basis function neural networks (RBFNN) have been widely used for nonlinear system modeling, and the basis function of the network is usually a Gaussian function in which the width is symmetric and always proportional to the distance between the center and its nearest neighbor [3]. The model precision is determined by the hidden units and output link weights of RBFNN. Grounded on the concept of hyper-rectangle based input space partition [4], we replace the typical Gaussian function with an unsymmetrical Gaussian function to decrease the hidden units of the network, and here the networks are called improved radial basis function neural networks (IRBFNN). A novel control scheme for mobile robots is proposed, including the backstepping based velocity controller and the torque controller with IRBFNN and the robust compensator. The dynamics of the robot based on IRBFNN is computed by the improved algorithm of resource allocating network [5]. The control system is competent for the robust tracking control of mobile robot.

J. Wang et al. (Eds.): ISNN 2006, LNCS 3972, pp. 1172 – 1177, 2006.

2 Model of a Nonholonomic Mobile Robot

The general kinematics and dynamic equations for mobile robots can be described by

$$\dot{\mathbf{q}} = \mathbf{S}(\mathbf{q})\mathbf{v}(t), \tag{1}$$

$$\mathbf{M}(\mathbf{q})\ddot{\mathbf{q}} + \mathbf{V}_m(\mathbf{q},\dot{\mathbf{q}})\dot{\mathbf{q}} + \mathbf{F}(\dot{\mathbf{q}}) + \mathbf{G}(\mathbf{q}) + \boldsymbol{\tau}_d = \mathbf{B}(\mathbf{q})\boldsymbol{\tau} - \mathbf{A}^T(\mathbf{q}), \tag{2}$$

where $\mathbf{q} = [x \quad y \quad \theta]^T$, $\mathbf{v} = [v \quad \omega]^T$, $\mathbf{M}(\mathbf{q})$ is a symmetric, positive definite mass matrix, $\mathbf{V}_m(\mathbf{q},\dot{\mathbf{q}})$ the centripetal and coriolis matrix, $\mathbf{F}(\dot{\mathbf{q}})$ the surface friction vector, $\mathbf{G}(\mathbf{q})$ the gravitational vector, $\boldsymbol{\tau}_d$ the bounded unknown disturbance including unstructured and unmodeled dynamics, $\mathbf{B}(\mathbf{q})$ the input transformation matrix, $\boldsymbol{\tau}$ the input vector, $\mathbf{A}(\mathbf{q})$ the Jacobian matrix associated with the constraints, and $\boldsymbol{\lambda}$ the constraint force vector. According to the constraint of pure rolling and no sliding, the velocity $\dot{\mathbf{q}}$ has the following constraint

$$\dot{y}\cos\theta - \dot{x}\sin\theta - d\dot{\theta} = 0. \tag{3}$$

It is obvious that $\mathbf{G}(\mathbf{q})$ is equal to zero in the mobile robot. For control purposes, the constraint matrix $\mathbf{A}^T(\mathbf{q})\boldsymbol{\lambda}$ in (2) is eliminated by an appropriate transform. The dynamic equation (2) can be expressed as

$$\overline{\mathbf{M}}\dot{\mathbf{v}} + \overline{\mathbf{V}}\mathbf{v} + \overline{\mathbf{F}} + \overline{\boldsymbol{\tau}}_d = \overline{\boldsymbol{\tau}}, \tag{4}$$

where $\overline{\mathbf{M}} = \mathbf{S}^\mathbf{T}\mathbf{M}\mathbf{S}$, $\overline{\mathbf{V}} = \mathbf{S}^\mathbf{T}(\mathbf{M}\dot{\mathbf{S}} + \mathbf{V}_m\mathbf{S})$, $\overline{\mathbf{F}} = \mathbf{S}\mathbf{F}$, $\overline{\boldsymbol{\tau}}_\mathbf{d} = \mathbf{S}^\mathbf{T}\boldsymbol{\tau}_d$, $\overline{\boldsymbol{\tau}} = \overline{\mathbf{B}}\boldsymbol{\tau}$, $\overline{\mathbf{B}} = \mathbf{S}^\mathbf{T}\mathbf{B}$. The parameter matrices or vectors in (4) usually contain some uncertainties such as mass, moment of inertia, friction coefficient, and so on.

3 IRBFNN and IRAN

For multi-input single-output (MISO) system, the output of RBFNN is represented as

$$\hat{y} = \sum_{i=1}^{M} w_i \varphi_i(\|\mathbf{x} - \mathbf{o_i}\|), \tag{5}$$

where $\mathbf{x}$ is the input vector, M is the number of hidden units, $\varphi_i(\cdot)$ ($i = 1,2,\cdots,M$) are the radial basis functions, $\mathbf{o_i}$ the centers of radial basis functions, w_i the output link weights. In general Gaussian function is selected as the radial basis function.

A hyper-rectangle based method was proposed to partition the input space to get relatively homogeneous hyper-rectangle regions [4]. The center of Gaussian function depends on the one of hyper-rectangle region. There is a certain value at the crossing border of two Gaussian functions. In our study, the unsymmetrical width is considered in Gaussian function (unsymmetrical Gaussian function) to mitigate the influence of the different scale of each dimension variable. This neural network is called improved radial basis function neural network (IRBFNN).

Let the input vector $\mathbf{x} = [x_1 \; x_2 \; \cdots \; x_k]^T$, the corresponding center of hidden units: $\mathbf{o_i} = [o_1(i_1) \; o_2(i_2) \cdots o_k(i_k)]^T$. The output of the i-th hidden unit is

$$\varphi_i(\mathbf{x}) = \exp\left\{-\frac{(x_1 - o_1(i_1))^2}{2\sigma_{i1}^{\;2}} - \frac{(x_2 - o_2(i_2))^2}{2\sigma_{i2}^{\;2}} - \cdots - \frac{(x_k - o_k(i_k))^2}{2\sigma_{ik}^{\;2}}\right\}. \tag{6}$$

where $\sigma_{i1},\cdots,\sigma_{ik}$ are the unsymmetrical widths of Gaussian function. Set one center array: $\mathbf{dimension}_i = [o_i(1) \quad \cdots \quad o_i(n_1)]^T, o_i(1) < \cdots < o_i(n_i)$, $i = 1,\cdots,k$ for each dimension. For the i-th hidden unit, the widths of the function are

$$\sigma_{i1} = \begin{cases} \sigma_{1l}(i_1), & x_1 \le o_1(i_1) \\ \sigma_{1r}(i_1), & x_1 > o_1(i_1) \end{cases}, \cdots, \sigma_{ik} = \begin{cases} \sigma_{kl}(i_k), & x_k \le o_k(i_k) \\ \sigma_{kr}(i_k), & x_k > o_k(i_k) \end{cases}, \tag{7}$$

where $\sigma_{jl}(i_j)$ and $\sigma_{jr}(i_j)$ are the left and right widths of the i-th hidden unit, $j = 1,\cdots,k$. The left and right widths can be determined by

$$\sigma_{il}(j) = p \cdot | o_i(j) - o_i(j-1) |, \quad \sigma_{ir}(j) = p \cdot | o_i(j+1) - o_i(j) |, \tag{8}$$

where p is an overlap coefficient, $0 < p < 1$.

Resource allocating network is used to create the hidden units of RBFNN sequentially [5], and it may be extended to online modeling of nonlinear dynamic systems. In creating RBFNN, it should be decided firstly whether to create new hidden unit or not after putting the new instance $\mathbf{x}(j), y(j)$ into the data window. A new hidden unit should be added if the following criteria are met

$$| e(j) | = | y(j) - \hat{y}(j) | > E_1, \tag{9}$$

$$|x_i(j) - o_{i,nearest}| \ge E_{2i}, i = 1,2,\cdots k, \tag{10}$$

where $o_{i,nearest}$ is the element of center array which is closest to $x_i(j)$, E_1 is the error threshold, E_{2i} is the distance threshold to be selected appropriately. Criterion (9) decides whether the existing units are sufficient or not for the permissible error, and criterion (10) ensures that the new hidden unit to be added is sufficiently far from all the existing units. When both are met, the current $x_i(j)$ is added to the center array as a new element.

There may be two or more hidden units whose centers and output link weights are all very close in the network creating process. These hidden units should be merged because they waste the resource of network. The hidden units that are not activated and become useless when they are far away from the work space should be pruned. The improved resource allocating network (IRAN) is a combination algorithm that includes criteria of adding new unit, merging strategy of similar units, pruning criterion of useless units, and output link weight learning with moving data window.

The detail presentation on IRBFNN and IRAN has been given by another paper in the proceedings of the conference.

4 Control System Design

The proposed control scheme for mobile robots is given in Fig. 1. In this scheme, the velocity command $\mathbf{v_c}$ is generated by the backstepping-based velocity controller, and the control torque $\boldsymbol{\tau}$ is computed by IRBFNN and a robust compensator. The pose and velocity of a reference robot are $\mathbf{q_r} = (x_r \quad y_r \quad \theta_r)^T$ and $\mathbf{v_r} = [v_r \quad \omega_r]^T$ respectively. According to the transformation between the local coordinate (built on the mobile robot) and the global coordinate, the equivalent trajectory tracking error in the mobile base can be expressed as

$$\mathbf{e}_M = \begin{bmatrix} x_e \\ y_e \\ \theta_e \end{bmatrix} = \begin{bmatrix} (x_r - x)\cos\theta + (y_r - y)\sin\theta \\ -(x_r - x)\sin\theta + (y_r - y)\cos\theta \\ \theta_r - \theta \end{bmatrix}, \dot{\mathbf{e}}_M = \begin{bmatrix} \omega y_e - v + v_r \cos\theta_e \\ -\omega x_e + v_r \sin\theta_e \\ \omega_r - \omega \end{bmatrix}. \tag{11}$$

The backstepping-based velocity controller proposed by [1] is

$$\mathbf{v_c} = \begin{bmatrix} v_c \\ \omega_c \end{bmatrix} = \begin{bmatrix} v_r \cos\theta_e + k_1 x_e \\ w_r + k_2 v_r y_e + k_3 v_r \sin\theta_o \end{bmatrix}, \tag{12}$$

where $k_1, k_2, k_3 > 0$ are the design parameters.

The torque controller is designed to convert the velocity command $\mathbf{v_c}$ into control torque $\boldsymbol{\tau}$ for the actual vehicle and make the actual velocity $\mathbf{v}$ of the mobile robot track the command $\mathbf{v_c}$. By substituting $\mathbf{v_c}$ for $\mathbf{v}$ in (4), the approximation of (4) is

$$\overline{\mathbf{M}}(\mathbf{q})\dot{\mathbf{v}}_\mathbf{c} + \overline{\mathbf{V}}(\mathbf{q},\dot{\mathbf{q}})\mathbf{v} + \overline{\mathbf{F}}(\mathbf{v}) + \overline{\boldsymbol{\tau}}_\mathbf{d} = f(\mathbf{v}, \dot{\mathbf{v}}_\mathbf{c}) \approx \overline{\boldsymbol{\tau}}. \tag{13}$$

We hope to use an appropriate nonlinear dynamic model $f(\mathbf{v}, \dot{\mathbf{v}}_\mathbf{c})$ to approximate $\overline{\boldsymbol{\tau}}$.

In the proposed control scheme, $f(\mathbf{v}, \dot{\mathbf{v}}_\mathbf{c})$ is generated by two IRBFNN models, $\dot{\mathbf{v}}_\mathbf{c} = [\dot{v}_c \quad \dot{\omega}_c]^T$ and $\mathbf{v} = [v \quad w]^T$ are used as the inputs of the networks. Two IRBFNN models are denoted RBF-NN1 and RBF-NN2, respectively. The input and output of RBF-NN1 are $\mathbf{x}_v = [v \quad \dot{v}_c]^T$ and $\overline{\tau}_v$ respectively, and the input and output of RBF-NN2 are $\mathbf{x}_\omega = [\omega \quad \dot{\omega}_c]^T$ and $\overline{\tau}_w$ respectively. Denote $\overline{\boldsymbol{\tau}}_n = [\overline{\tau}_v \quad \overline{\tau}_w]^T$, called virtual torques. The networks are learned online by IRAN. Obviously, if $\overline{\boldsymbol{\tau}}_n \to \overline{\boldsymbol{\tau}}$, then $\overline{\mathbf{B}}^{-1}\overline{\boldsymbol{\tau}} \to \boldsymbol{\tau}$.

In order to improve the robustness of control system, a robust compensator based on sliding mode is adopted in the torque controller. In Fig. 1, the velocity tracking error is defined as

$$\mathbf{e} = [e_v \quad e_\omega]^T = \mathbf{v_c} - \mathbf{v}. \tag{14}$$

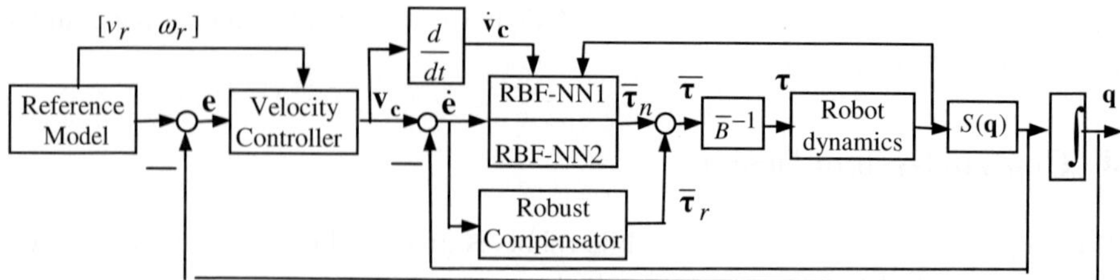

Fig. 1. Scheme of the control system of mobile robot

The sliding surface defined by [6] is

$$\mathbf{S(t)} = \begin{bmatrix} s_1(t) \\ s_2(t) \end{bmatrix} = (\frac{d}{dt} + \lambda)^2 \int_0^t \mathbf{e}\, d\tau = 0 . \tag{15}$$

The robust compensator can be derived from (15), and the compensated control is

$$\bar{\boldsymbol{\tau}}_r = \mathbf{k}_{r1}\,\mathrm{sgn}(\mathbf{S}(t)) + \mathbf{k}_{r2}\ddot{\mathbf{e}} + \mathbf{k}_{r3}\mathbf{e} , \tag{16}$$

where $\mathbf{k}_{r1}$, $\mathbf{k}_{r2}$ and $\mathbf{k}_{r3}$ are the design parameters.

5 Simulation Study

The model of mobile robot proposed by [2] is adopted in this paper, and the parameters of the robot are: $m = 10kg$, $I = 5kg.m^2$, $R = 0.5m$. The parameters of the velocity controller are: $k_1 = k_2 = k_3 = 1$ in (13). In the torque controller, the design parameters of RBF-NN1 are: $E_1 = 0.02$, $E_{21} = 0.3$ and $E_{22} = 0.2$; the one of RBF-NN2 are: $E_1 = 0.008$, $E_{21} = 0.03$ and $E_{22} = 0.3$; the overlap coefficient $p = 0.6$ and the learning rate $\eta = 0.8$; the design parameters of the robust controller are: $\mathbf{k}_{r1} = [10 \quad 10]^T$, $\mathbf{k}_{r2} = [1 \quad 0.5]^T$ and $\mathbf{k}_{r3} = [0.01 \quad 0.01]^T$. The parameters of merging and pruning strategies here are omitted for the length limitation of paper.

The straight line and circle tracking experiments are simulated respectively. Here only give the simulation results of the straight line tracking. Let motion trajectory of reference robot $y_r - x_r = 1$, with $\omega_r = 0\, rad/s$, $v_r = \sqrt{2}\, m/s$. Suppose the initial pose and velocity of the robot at $(x_0, y_0, \theta_0, v_0, w_0) = (0,0,0,0,0)$. The reference and tracking trajectories of mobile robot are shown in Fig. 2. It is shown that the mobile robot lands upon the reference path at $(2.2, 3.2)$ rapidly, and then tracks the reference trajectory smoothly. When the control torques of mobile robot enter into stable status, the hidden units of RBF-NN1 and RBF-NN2 converge to 48 and 60 respectively, the learning processes of hidden units given in Fig. 3.

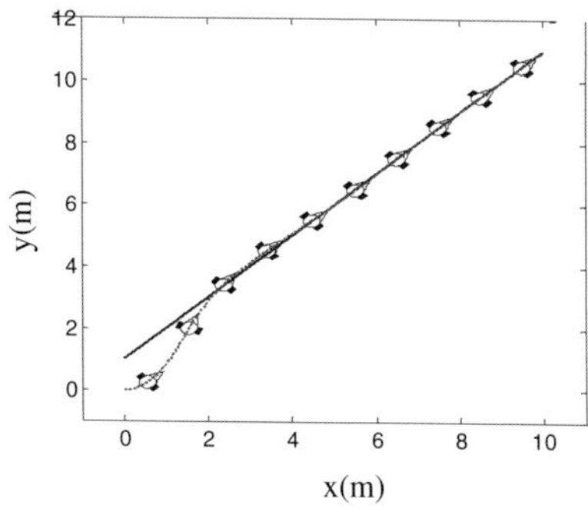

Fig. 2. Trajectory tracking of mobile robot

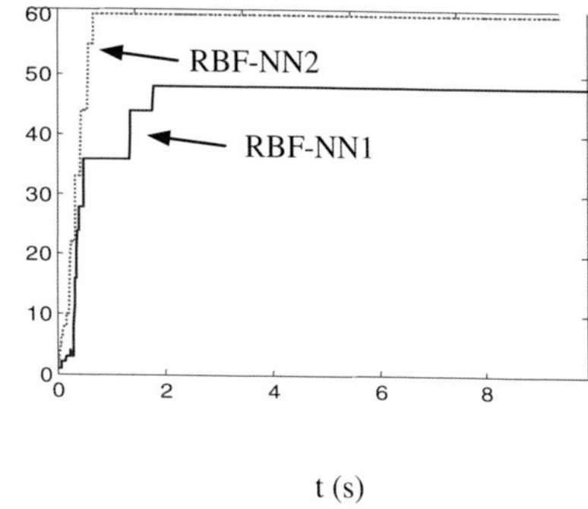

Fig. 3. Learning processes of hidden units

6 Conclusions

The control system of mobile robot with the backstepping-based velocity controller and torque controller using the IRBFNN and robust compensator is able to track the reference trajectory rapidly and smoothly with anti-disturbance. Because IRBFNN and IRAN can be used to model the dynamics of mobile robot online, they are competent for the design of adaptive torque controller of mobile robot system. The virtue of the control system has been demonstrated by various simulation examples.

Acknowledgment

The work was supported by Natural Science Foundation of Zhejiang Province, China (Grant No.Y104560) and Research Start-up Foundation of Hangzhou Dianzi University, China (Grant No.KYS091505043).

References

1. Kanayama, Y., Kimura,Y., Miyazaki, F., Noguchi, T.: A Stable Tracking Control Method for an Autonomous Mobile Robot. Proc. IEEE Conf. Robotics and Automa. (1990) 384-389
2. Fierro, R., Lewis, F. L.: Control of Nonholonomic Mobile Robot Using Neural Networks. IEEE Transactions on Neural Networks 9 (1998) 589-600
3. Saha, A., Keeler, J. D.: Algorithms for Better Representation and Faster Learning in Radial Basis Function Network. In: Touretzky, D. S. (ed.): Advances in Neural Information Processing Systems. Morgan Kaufmann, San Mateo, CA 2 (1990) 482-489
4. Kubat, M.: A Hyperrectangle-Based Method that Creates RBF Networks. In: Howlett, R. Jain, L. (eds.): Radial Basis Function Networks. Physica-Verlag, Heidelberg 1 (2001) 32-50
5. Platt, C. J.: A Resource Allocating Network for Function Interpolation. Neural Computation 4 (1991) 473-493
6. Slotine, J. J. E., Li, W.: Adaptive Nonlinear Control. Prentice-Hall, NJ (1991)

Grasping Control of Robot Hand Using Fuzzy Neural Network

Peng Chen, Yoshizo Hasegawa, and Mitushi Yamashita

Department of Environmental Science & Technology, Mie University,
1515 Kamihama, Tsu, Mie Pref., 514-8507, Japan
chen@bio.mie-u.ac.jp

Abstract. In this paper, we propose a grasping control method for robot hand using fuzzy theory and partially-linearized neural network. The robot hand has Double-Octagon Tactile Sensor (D.O.T.S), which has been proposed in our previous papers, to detect grasping force between the grasped object and the robot fingers. Because the measured forces are fluctuant due to the measuring error and vibration of the hand, the tactile information is ambiguous. In order to quickly control the grasping force to prevent the grasped object sliding out off the robot fingers, we apply the possibility theory to deal with the ambiguous problem of the tactile information, and use the partially-linearized neural network (P.L.N.N) to construct a fuzzy neural network. The method proposed in this paper is verified by applying it to practical grasping control of breakable objects, such as eggs, fruits, etc.

1 Introduction

When an object grasped in human's fingers starts to slip through the fingers, the minute partial-slip occurs first, and it spreads around the grasping area of the fingers[1]. Human can sense the partial-slip and control the grasp force quickly to prevent the object slip and slide out of his hand, because the tactile nerves in human's finger are highly denser. However, it is difficult to imitate the human's detection of the partial-slip using practical sized tactile sensor for robot hand with today's technology.

In this paper, we propose a grasping control method for slip prevention of grasped object in the robot hand using Double-Octagon Tactile Sensors (D.O.T.S)[2], possibility theory and neural network, to control the grasping force before the grasped object starts to slide in the robot hand. Because the conventional neural network (NN) cannot reflect the possibility of ambiguous problems[3], we propose a partially-linearized neural network (P.L.N.N) to realize a neuro-fuzzy for the grasping control. The contents in this paper are;

(1) The contracture of the system of grasping control and the measurement method of tactile information using the D.O.T.S;
(2) The calculation method of the dynamic and static friction coefficient between the hand fingers and the grasped object by the forces measured with the D.O.T.S for the judgment of slip state.
(3) The possibility function of the dynamic and static friction coefficient to deal with the fluctuant force measured by D.O.T.S due to the measuring error and vibration of the hand;

J. Wang et al. (Eds.): ISNN 2006, LNCS 3972, pp. 1178 – 1187, 2006.

(4) Partially-linearized neural network (P.L.N.N) to realize a fuzzy neural network for quickly control the grasping force to prevent the grasped object sliding out off the robot fingers.
(5) The results of verification for grasping breakable objects, such as eggs, fruits, etc.

2 Grasping Control System

The aim of this study is for robot hand to quickly detect the partial-slip and control the grasping force to prevent that the grasped object slip through the robot's finger. The grasping force must be applicable to grasp the object, and it will not too strong to damage the grasped object, such as eggs, fruits, etc.

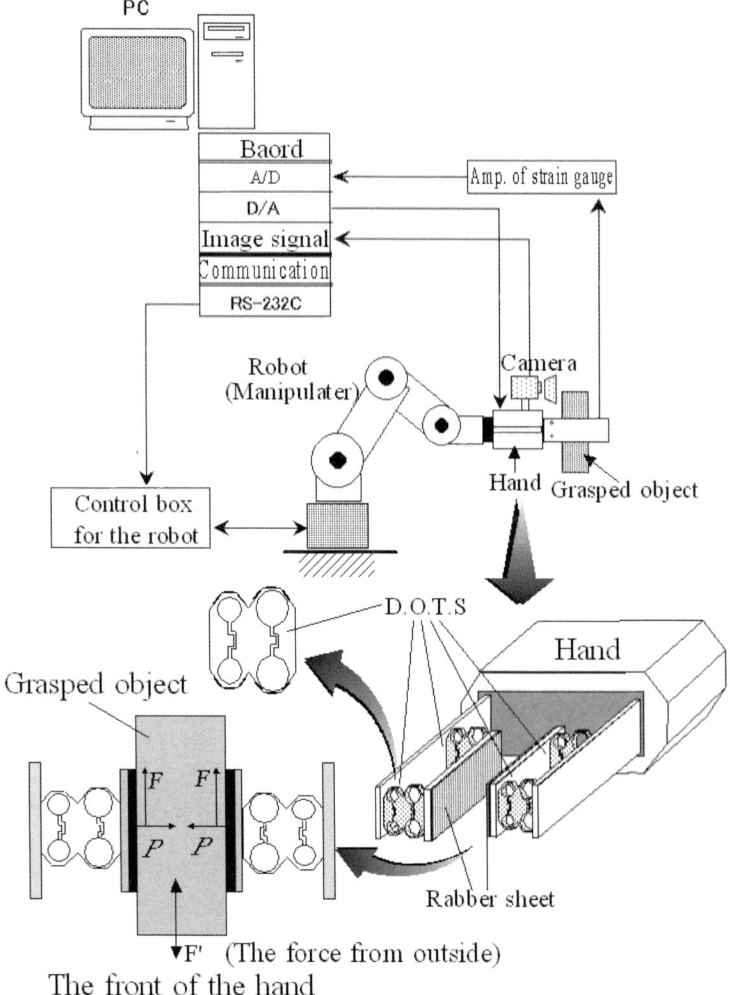

Fig. 1. The control system for robot hand

The construction of the system for this purpose is shown in Fig.1. The main parts of the system are;

tactile sensor: The Double-Octagon Tactile Sensors (D.O.T.S) are installed in the fingers.
camera: The camera is installed near the finger for easily taking a picture to detect the object.
manipulator: The manipulator has five degrees of freedom. The accuracy of the end position of the manipulator is $\pm 0.1mm$. It is controlled with the controller and the computer (PC). The detail characteristics of each part in the system are;

(a) The hand has two fingers. The surface of each finger is flat and parallel.

(b) The insides of the fingers are covered with rubber sheet, and the outsides of the fingers are rigidity body. The fingers is opened or closed by moving the rigid part of the finger.

(c) The static friction coefficient between the object and two fingers is the same. Rotating slip dose not occur between the fingers and the object.The elasticity of the fingers is caused by the rubber sheet and the fingers.

3 Slip and Damage Detection by Possibility Function

When an object grasped by the robot hand with the D.O.T.S starts to slip due to the force F' from external shown in Fig. 1 and Fig. 2, minute slip occurs first between the object and one finger. In this paper, the minute slip is called "partial-slip". It is important to detect the minute partial-slip by the D.O.T.S before the object starts to slide along the fingers.

In this chapter, we explain the partial-slip and clarify the mechanism of the partial-slip in robot hand, and give the condition equation for the judgment of the partial-slip. These are the basis for the control of grasping force.

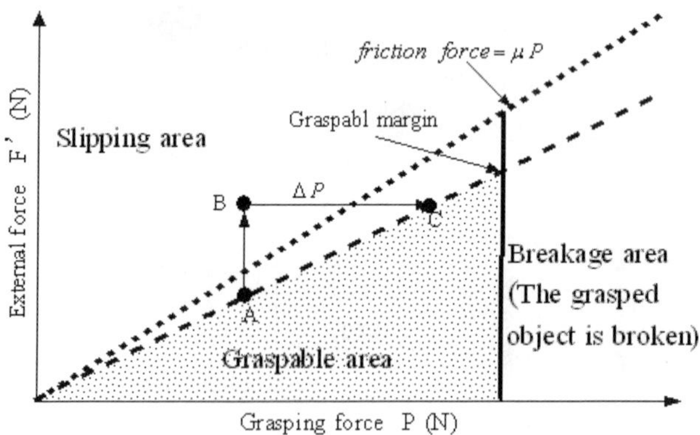

Fig. 2. The relationship between F' and P for the control of grasping force P

3.1 Judgment of the Partial-Slip

Fig 2 shows the relationship between the external force F' exerted on the grasped object and the grasping force P for the control. μ' is the maximum static friction coefficient between the grasped object and the fingers. The maximum static friction force $F_{\max}$ between the grasped object and the fingers

$$F_{\max} = \mu' P \tag{1}$$

can be measured by the D.O.T.S. If F'>F_{max}, the grasped object will slip out off the hand. In this case, we call that the object is in the "slipping area", such as the point "B" as show in Fig. 2, and we must add Δp to the grasping force P to move it to the "graspable area", such as the point "C" as show in Fig. 2. The grasping force has a limited value over which the grasped object will damaged, and the area is called "breakage area" as show in Fig. 2.

Fig. 3 shows an example of curves of grasping force P and shearing force (friction force) F which are measured by the D.O.T.S.

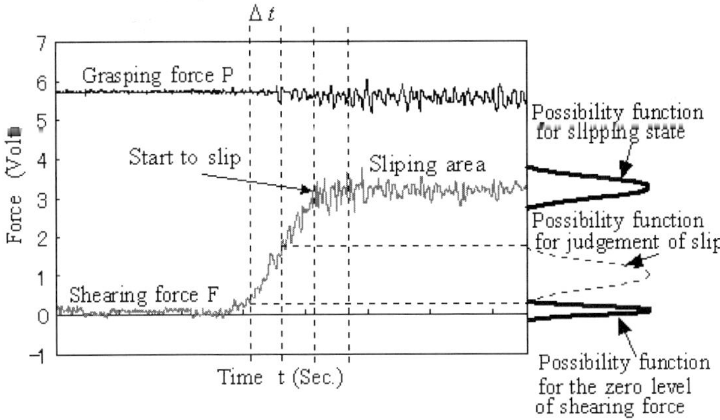

Fig. 3. Grasping force P and shearing force F

Because the forces are fluctuant due to the measuring error and vibration of the hand, we use the possibility theory[4] to resolve the ambiguous problem. If the current friction coefficient is μ , than,

$$\mu = \frac{F}{P} \tag{2}$$

The maximum static friction coefficient is μ'. IF $\mu \geq \mu'$, the grasped object starts to slip. The possibility function $p(\mu)$ of μ can be calculated by

$$p(\mu) = \sup_{\mu = F/P} p(F) \wedge p(P) \tag{3}$$

Here, $p(F)$ and $p(P)$ are the possibility function of shearing force F and grasping force P respectively. The possibility function can be obtained from probability density functions of the measured signal, as shown in Fig. 3, using possibility theory.

For example, when the probability density function conforms to the normal distribution, it can be changed to possibility distribution functions $p(x_i)$ by the following formulae.

$$P(x_i) = \sum_{k=1}^{N} \min\{\lambda_i , \lambda_k\} \tag{4}$$

λ_i and λ_k can be calculated as follows;

$$\lambda_i = \int_{x_{i-1}}^{x_i} \frac{1}{\sigma\sqrt{2\pi}} \exp\left\{-\frac{(x-\bar{x})^2}{2\sigma^2}\right\} dx, \quad \lambda_k = \int_{x_{k-1}}^{x_k} \frac{1}{\sigma\sqrt{2\pi}} \exp\left\{-\frac{(x-\bar{x})^2}{2\sigma^2}\right\} dx \tag{5}$$

Here, σ means the standard deviation, $\bar{x}$ means the mean value, and $x = \bar{x} - 3\sigma \sim \bar{x} + 3\sigma$. Fig.4 shows an example of the probability density function and the possibility function. The most merit of this method for deciding the membership function is the objectivity, because it is based on the mean value and the standard deviation of measured data.

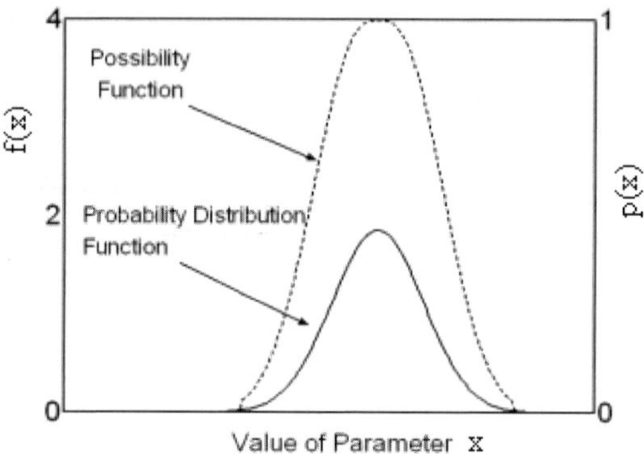

Fig. 4. Example of possibility function

Fig.5 shows examples of $p(\mu)$. In this figure, $p_f(\mu)$ and $p_d(\mu)$ mean the possibility function of slip and damage respectively. The "current state" means the static friction coefficient μ in rear time. If the possibility of the current state larger than 0.3 by matching with $p_f(\mu)$, the grasped object will slipped out off the hand. Conversely, if the possibility of the current state larger than 0.3 by matching with $p_d(\mu)$, the grasped object will be damaged.

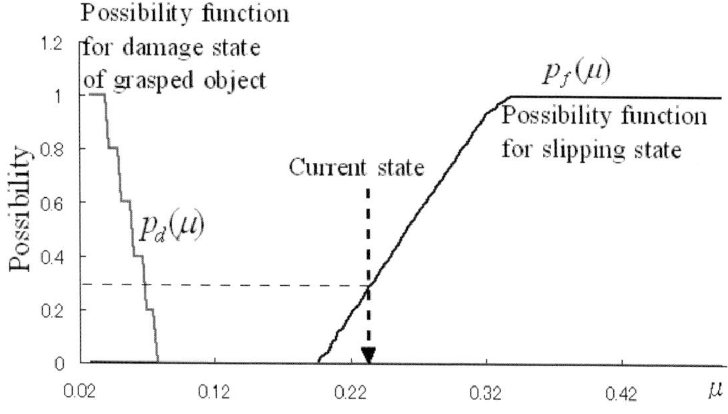

Fig. 5. An example of $p(\mu)$

4 Partially-Linearized Neural Network for Quick Control

The neuron numbers of m-th layer of a NN is N_m. The set $X^{(1)} = \{X_i^{(1,j)}\}$ expresses the pattern inputted to the 1st layer and the set $X^{(M)} = \{X_i^{(M,k)}\}$ is the teacher data to the last layer (M-th layer). Here, $i = 1\ to\ P,\ j = 1\ to\ N_1,\ k = 1\ to\ N_M$, and,

$X_i^{(1,j)}$: The value inputted to the j-th neuron in the input (1st) layer;

$X_i^{(M,k)}$: The output value of k-th neuron in the output (M-th) layer; $k = 1\ to\ N_M$

Even if the NN converge by learning $X^{(1)}$ and $X^{(M)}$, it cannot well deal with the ambiguous relationship between new $X^{(1)*}$ and $X_i^{(M)*}$, which have not been learnt. In order to predict $X_i^{(M)*}$ according to the probability distribution of $X^{(1)*}$, partially linear interpolation of the NN is introduced as Fig. 6, we called it "Partially-linearized Neural Network (P.L.N.N)".

In the NN which has converged by the data $X^{(1)}$ and $X^{(M)}$, the symbols are used as follows.

$X_i^{(m,t)}$: The value of t-th neuron in the hidden (m-th) layer; $t = 1\ to\ N_m$

$W_{uv}^{(m)}$: The weight between the u-th neuron in the m-th layer and the v-th neuron in the ($m+1$)-th layer; $m = 1\ to\ M; u = 1\ to\ N_m; v = 1\ to\ N_{m+1}$

If these values are all remembered by computer, when new values $X_j^{(1,u)*}$ ($X_j^{(1,u)} < X_j^{(1,u)*} < X_{j+1}^{(1,u)}$) are inputted to the first layer, the predicted value of v-th neuron ($v=1$ to N_m) in the ($m+1$)-th layer ($m=1$ to $M-1$) will estimated by

$$X_j^{(m+1,v)} = X_{i+1}^{(m+1,v)} - \frac{\{\sum_{\mu=0}^{Nm} W_{uv}^{(m)}(X_{i+1}^{(m,u)} - X_j^{(m,u)})\}(X_{i+1}^{(m+1,v)} - X_j^{(m+1,v)})}{\sum_{\mu=0}^{Nm} W_{uv}^{(m)}(X_{i+1}^{(m,u)} - X_i^{(m,u)})} \tag{6}$$

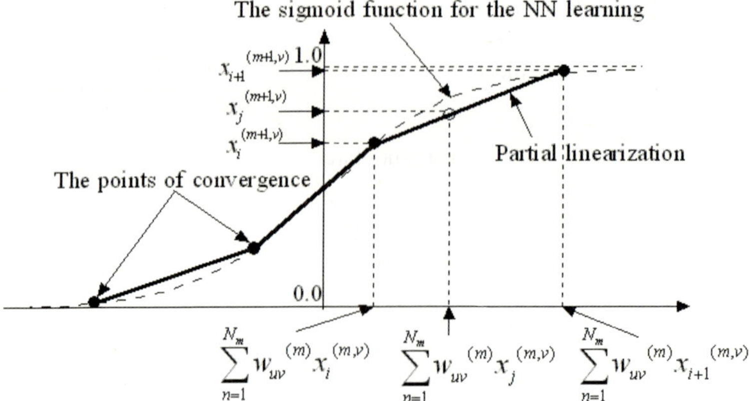

Fig. 6. The partial linearization of the sigmoid function

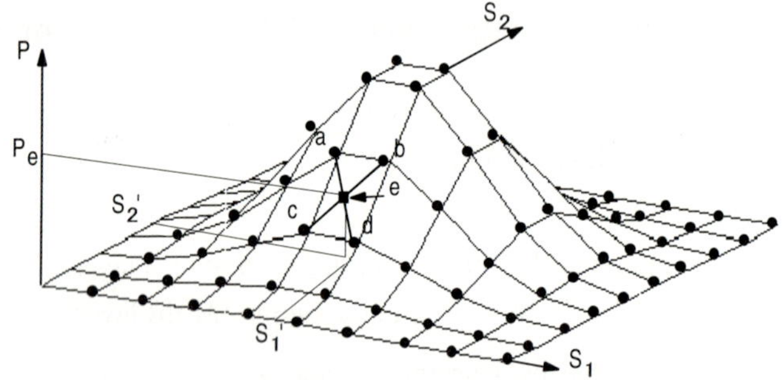

Fig. 7. Interpolation by the P.L.N.N

In the above way, the sigmoid function is partially linearized as shown in Fig. 6. If a function, such as Fig. 7, need to be learnt, the PNN will learn the points • shown in Fig. 7. When new data (s_1', s_2') are inputted into the converged P.L.N.N, the value symbolized by ■ correspond to the data (s_1', s_2') will be quickly identified as P_e shown in Fig. 7. So the P.L.N.N can be used to deal with ambiguous problems.

Four D.O.T.Ss, which can measure the shearing force F and grasping force P, are installed in the robot hand. We can obtained 8 forces from F_1 and P_1 to F_4 and P_4. Fig. 9 shows an example of the data for the learning of the partially-linearized neural network (P.N.N). The data are measured by the D.O.T.Ss when the hand grasped an egg and a man was pulling it out off the hand as shown in Photo.1.

The method proposed here was verified as shown in Photo.2. The P.L.N.N that finished the learning controls the grasping force to prevent the grasped egg slipping out off the hand.

Photo. 1 and 2 show the robot system and the experiments for learning and verification. In the case of Photo.1, we draw the grasped egg out off the hand and record the data of shearing force F and grasping force **P** for learning. In the case of Photo.2, we exerted an external force F' to the grasped egg, and the P.L.N.N controls the grasping force to prevent the egg slipping out off the hand.

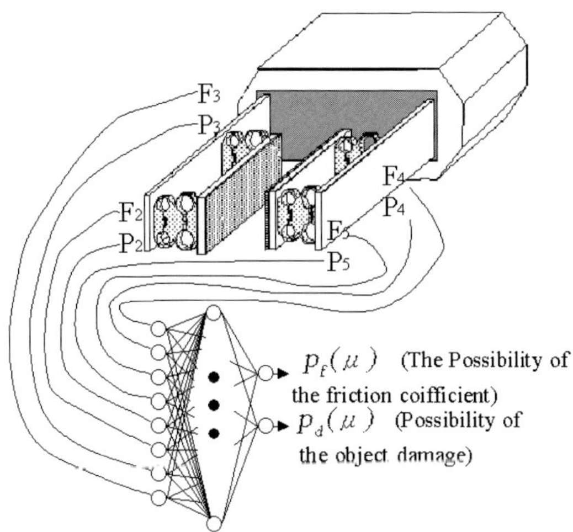

Fig. 8. The neural network for the control of the grasping force

P_1	F_1	P_2	F_2	P_3	F_3	P_4	F_4	$p_f(\mu)$	$p_d(\mu)$
10.00	0.80	10.00	0.88 …	0.43	0.38	0.973	0.31	0.00	1.00
8.00	0.76	8.00	0.86 …	0.55	0.39	1.018	0.33	0.00	0.80
6.00	0.74	6.00	0.93 …	0.64	0.41	0.959	0.35	0.00	0.60
4.00	0.76	4.00	0.92 …	0.64	0.44	0.917	0.39	0.00	0.40
2.00	0.78	2.00	0.91 …	0.43	0.36	0.935	0.29	0.00	0.20
1.70	0.80	0.89	0.10 …	0.60	0.51	1.011	0.50	0.00	0.00
1.73	0.76	1.11	0.08 …	0.40	0.45	1.102	0.40	0.20	0.00
1.71	0.78	1.04	0.09 …	0.45	0.39	1.364	0.31	0.50	0.00
1.67	0.73	0.90	0.09 …	0.88	0.28	1.445	0.18	0.70	0.00
1.81	0.75	0.99	0.09 …	0.39	0.37	1.616	0.29	0.90	0.00
1.74	0.75	0.94	0.09 …	0.47	0.32	1.553	0.22	1.00	0.00
1.68	0.78	0.98	0.10	0.54	0.32	1.672	0.21	1.00	0.00

Fig. 9. An example of data for the learning of the PNN

We have successfully made tests for verifying the method proposed in this paper. Fig. 10 shows a test result of verification. $P_f(\mu)$ and $P_d(\mu)$ are the possibility of slip and damage respectively that are the output value of the P.L.N.N. By the result, it is obvious that when the partial-slip between the grasped egg and the fingers accrues, the possibility $P_f(\mu)$ outputted by P.L.N.N became larger and the grasping force P

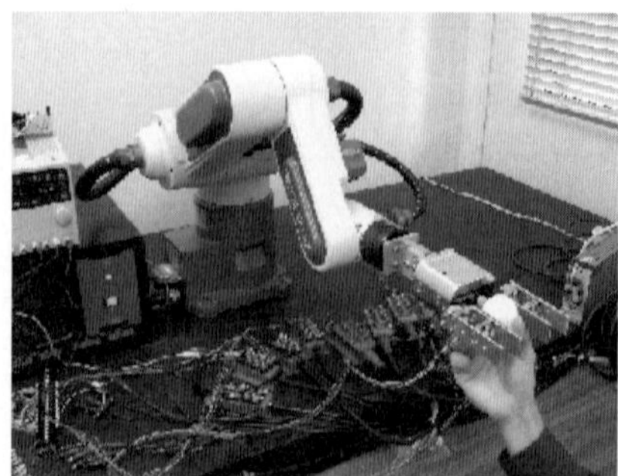

Photo. 1 Experiment for learning data

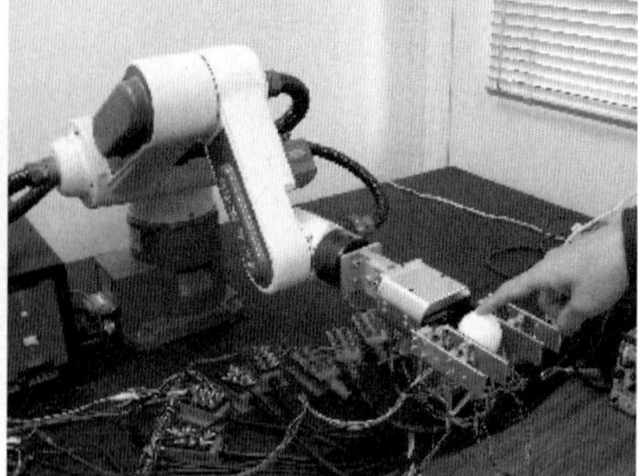

Photo. 2 Experiment for verification

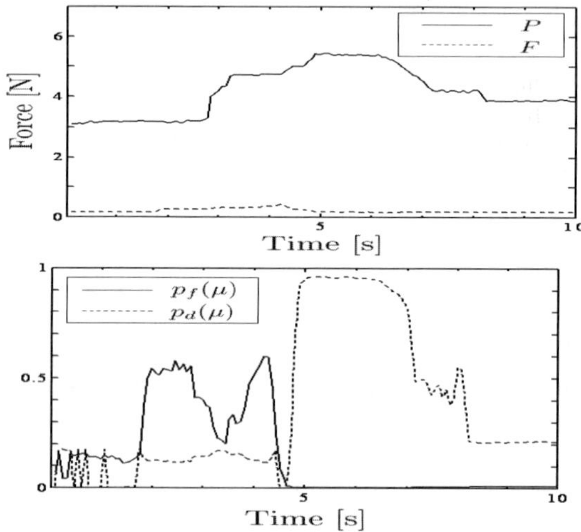

Fig. 10. A test result

will also be controlled enlarging gradually to decrease the $P_f(\mu)$ to prevent the grasped egg slipping out off the hand. Simultaneously, when the possibility $P_d(\mu)$ outputted by P.L.N.N became larger than 0.6 and the grasping force P will be controlled decreasing gradually to prevent the grasped egg to be damaged.

5 Conclusions

In this paper, we proposed a grasping control method of robot hand using possibility theory and partially-linearized neural network. Though the robot hand has special tactile sensor called Double-Octagon Tactile Sensor (D.O.T.S), the method proposed in this paper can also be used for other types of tactile installed in a robot hand, which can measure shearing force and grasping force.

In order to quickly control the grasping force to prevent the grasped object slipping out off the robot fingers due to some external force, we applied the possibility theory to deal with the ambiguous problem of the tactile information, and use the partially-linearized neural network (P.L.N.N) to construct a fuzzy neural network.

We showed the calculation method of the dynamic and static friction coefficient between the hand fingers and the grasped object by the forces measured with the D.O.T.S for the judgment of slip state. The possibility function of the dynamic and static friction coefficient have been established to deal with the fluctuant force measured by D.O.T.S due to the measuring error and vibration of the hand. Partially--linearized neural network (P.L.N.N) is used to realize a fuzzy neural network for quickly control the grasping force to prevent the grasped object sliding in the robot fingers. The method proposed in this paper was verified by applying it to practical grasping control of breakable objects, such as eggs, fruits, etc.

References

1. Maeno, T., Kobayashi, K., Kawai, T., Hirano, Y.: Grip Force Control by Detecting the Internal Strain Distribution Inside the Elastic Finger Having Curved Surface. Transactions of the Japan Society of Mechanical Engineers (C) 65 (633) (1999) 1907-1914
2. Nakayama, S., Chen, P., Matumiya, T., Toyota, T.: Development of Double-Octagon Tactile Sensor for Grasping Control. Proc. of 1999 IEEE/ASME International Conference on Advanced Intelligent Mechanics (1999) 218-223.
3. Bishop, C.: Neural Networks for Pattern Recognition. Oxford University Press (1995)
4. Dubois, D., Prade, H.: Possibility Theory, An approach to Computerized Processing. Plenum Press, New York (1988)

Position Control Based on Static Neural Networks of Anthropomorphic Robotic Fingers

Juan Ignacio Mulero-Martínez, Francisco García-Córdova, and Juan López-Coronado

Department of System Engineering and Automatic
Polytechnic University of Cartagena,
Campus Muralla del Mar, 30202, Cartagena, Murcia, Spain
{juan.mulero, francisco.garcia, jl.coronado}@upct.es

Abstract. A position neurocontroller for robot manipulators with a tendon-driven transmission system has been developed allowing to track desired trajectories and reject external disturbances. The main problem to control tendons proceeds from the different dimensions between the joint and the tendon spaces. In order to solve this problem we propose a static neural network in cascade with a torque resolutor. The position controller is built as a parametric neural network by using basis functions obtained directly from the finger structure. This controller insure that the tracking error converges to zero and the weights of the network are bounded. The implementation has been improved partitioning the neural network into subnets and using the Kronecker product. Both control and weight updating laws have been designed by means of a Lyapunov energy function. In order to improve the computational efficient of the neural network, this has been split up into subnets to compensate inertial, Coriolis/centrifugal and gravitational effects. The NN weights are initialised at zero and tuned on-line with no "off-line learning phase". This scheme has been applied to an anthropomorphic robotic finger with a transmission system based on tendons.

1 Introduction

Robot manipulators are mechanisms intended for transmission of motion. In many ocassions a transmission system is required to drive the robot manipulator. This motivates to design the system with a proper transmission ratio so that both the velocity and the force relations in the load shaft are achieved. There are many transmission systems such as gears, cam and followers, chains, pulleys and belts, timing belts, lead screws, et cetera. For example gearboxes permit to reduce the effect of non-linear terms from the dynamics of mechanisms. However, when the input and the output shafts are far away each to other or when a desired transmission ratio is not attained by using only two elements, it is necessary to use a train of mechanisms. Such a mechanical design has the advantage over other representations to decrease the volume and inertia of robotic manipulators locating actuators far away from the endpoint. Thus, the actuators can be controlled from the outset of the system allowing remote control and

J. Wang et al. (Eds.): ISNN 2006, LNCS 3972, pp. 1188–1197, 2006.

generally are attached to a supporting base. This provides light weight, small volume, compact and versatile manipulators.

The dynamic modelling of a tendon-driven manipulator was firstly developed by Lee et al., [1]. In [2] the dynamics of a tendon-driven manipulator with a completely decoupled transmission system has been developed using a Lagrangian approach. A systematic way to derive the dynamics equations for a tendon-driven robotic mechanism has been proposed recently by Lee, [3], on the basis of the recursive Newton-Euler equations. It may be observed that flexibility and elasticity play an important the role in the dynamics. This features have been studied by Kawamura, [4]. Other authors such as [5] includes more realistic models representing the tendon as a transmission line. In [6] it is claimed that the friction of the routing pulleys is not negligible and should be considered in the force control

Different kinds of remarks are to be made about the calculation of tendon forces. It may be observed that this computation represents an indeterminate problem since the different dimension of the joint space (n-dimensional) and the tendon space. (m-dimensional). The inputs of the dynamic model corresponds to the motor torques whereas the controller provides control actions based on the joint torques. As a consequence, a redundant system is addressed by resorting to the Morecki property, [7]. In this sense, this property can be conveniently applied to achieve total control of a system with more unknown variables than equations. A number of strategies have been provided in the literature to solve this problem. Amongst others, it is worth citing those based on the Penrose-Moore pseudoinverse, and those which apply the Jacobsen rectifiers, [8].

Linear techniques have been used to design position controllers insuring the tracking of reference trajectories and achieving a good performance for the whole system, [9]. These controllers were originally applied to the dexterous hand UTAH/MIT consisting of a completely decoupled transmission system with two tendons by joint according to an opponent system. In [8] the method of rectifiers was refined to be applied to hardly coupled tendon-driven transmission systems. Many control strategies for agonist/antagonist tendon-driven manipulators improving the performance of the system have been discussed in [10]. Proportional control generates some problems which can be solved by adding proximal stiffness and non-linear damping, or by using controllers which incorporate tendon force feedback, [11]. The control in the presence of nonlinear elasticity has been analyzed by Ozawa et al., [12], where the system is controlled in the actuator level in order to utilize efficiently the performance of the compliant elements. In [13] an adaptive control and an adaptive neural network control for tendon-driven robot manipulators with elastic tendons was developed. Recently, Mulero et al. proposed a parametric neural network to move an anthropomorfic finger with a transmission system based on tendons in an agonist/antagonist mode, [14] and [15].

The main contribution of this paper is the design and implementation of an adaptive neural network controller that tracks desired trajectories and reject disturbances with abilities of picking up parametric structured uncertainties in a tendon-driven transmission system. The neurocontroller is designed to learn the non-linear terms of the dynamics, such as inertial, coriolis/centripetal and gravitational terms, while Jacobsen rectifiers are applied to solve the redundancy problem. By exploiting the structural properties of the robot manipulators, a functional link neural network (FLNN) is developed to estimate the non-linear terms in the filtered error dynamics. As a result, the size of the neural network model of the robot manipulator becomes smaller than the usual dynamical ones. This network is characterised as being linear in the parameters and having only one hidden layer. The ideal approximation weights are assumed to be unknown and bounded. In order to illustrate the performance of the controller, a simulation of a finger is carried out while providing additional insight in the proposed control design

The paper is organized as follows. In section 2 the filtered error dynamics is introduced. In section 3, static neural networks are discussed on the basis of some interesting properties regarding with the Coriolis/centripetal matrix. Section 4 and 5 are devoted to describe the torque resolutor and the general control scheme which includes the update/control laws. A simulation of a four-link anthropomorfic finger is developed in section 6 in order to illustrate the performance of the system in terms of the tracking errors. Finally, the concluding remarks are discussed in the section 7.

2 Problem Statement

The dynamic equation of direct driven robot with n degrees of freedom has been widely referenced in the literature [16]

$$M(q)\ddot{q} + C(q,\dot{q}) + G(q) = \tau \tag{1}$$

where $q(t) \in \mathbb{R}^n$ represents the vector of generalized coordinates or joint variables, $M(q) \in \mathbb{R}^{n\times n}$ is the generalized inertia matrix, $C(q,\dot{q}) \in \mathbb{R}^{n\times n}$ the matrix of Coriolis/centripetal effects, $G(q) \in \mathbb{R}^n$ the vector of generalized gravitational forces and $\tau \in \mathbb{R}^n$ the vector of joint torques. Generally the nonlinear terms depending on positions and velocities, $C(q,\dot{q}) + G(q)$, are collected in a function $h(q,\dot{q})$ in order to clarify the model, [17]. The dynamics of the rotor consisting of a DC motor in series with a gear train directly connected to a link of the manipulator, is given by a second order differential equation. Taking account the effects of the inertia J_m, and the damping B_m of the rotors and the transmission system leads to

$$\left(M(q) + \tilde{M}\right)\ddot{q} + h(q,\dot{q}) + \tilde{B}\dot{q} = \tau_m \tag{2}$$

where $\tilde{M} = R_m^{-1}J_mR_m$ and $\tilde{B} = R_m^{-1}B_mR_m$ are the effect of inertia and damping to the system. A new matrix A called to as structural matrix collects information about both the routing of the tendons and the dimensions of the pulleys. If the element a_{ij} is non-null means that the tendon i is routed by the pulley rigidly attached to the joint shaft j. It is possible to get a decoupling in the transmission system if all the pulleys mounted on the same joint shaft have similar radii. In this case, the structural matrix is decompounded into a routing matrix $B \in R^{m\times n}$ and a dimension matrix $R \in \mathbb{R}^{n\times n}$, e.g. $A = BR$. The dimension matrix R is diagonal, $R = diag\left(r_1, ..., r_n\right)$, in such a way the element r_i represents the radius of the pulleys mounted on the shaft of the joint i. The routing matrix B has as much rows as tendons in the structure, so that the row i provides information about the pulleys through which the tendon i is routed. This equation can be expressed in terms of the motor torques, τ_m, considering the spooler pulleys rigidly attached to the shafts of the motors and pulling from the tendons, e.g. $\tau = A^T R_m^{-1} \tau_m$ where $R_m = diag\left(r_{m_1}, ..., r_{m_n}\right)$ is a diagonal matrix of radii of spooler pulleys. In a tendon-driven transmission system, the motor position vector Θ_m is related to the joint displacements q taking into account that the linear displacement of the tendon, $A\Theta$ is equal to the linear displacement caused by turning the motor and measured onto the output shaft of the gear train, $R_m \Theta_m = s = A\Theta$ where $R_m = diag\left(\frac{r_{m1}}{n_1}, ..., \frac{r_{mm}}{n_m}\right)$. Using the transformation of position coordinates and the joint torques the dynamics for the plant is concluded.

$$\left(M\left(\Theta\right) + \tilde{M}\right)\ddot{\Theta} + \tilde{B}_m \dot{\Theta} + h\left(\Theta, \dot{\Theta}\right) = R^T B^T R_m^{-1} F \tag{3}$$

where $\tilde{M}$ and $\tilde{B}_m$ represent respectively the rotor inertia and the motor damping matriz reflected into the joint space. These matrices are symmetric and definite-positive since $R_m^{-1} J_m R_m^{-1}$ and $R_m^{-1} B_m R_m^{-1}$ are diagonal matrices with positive eigenvalues and both $\tilde{M}$ and $\tilde{B}_m$ are congruent transformations.

$$\tilde{M} = R^T B^T R_m^{-1} J_m R_m^{-1} BR, \qquad \tilde{B}_m = R^T B^T R_m^{-1} B_m R_m^{-1} BR$$

The linear transformation $A^T \in R^{n\times m}$ causes a system of n equations and m unknown variables. It is necessary to make a coordinate transformation to move equilibrium points to the origin, so that a position error variable is considered, $e\left(t\right) = \Theta_d\left(t\right) - \Theta\left(t\right)$. Measurements of velocities are easy to get by tachometers, but sensors of acceleration are noisy and are not used for implementation in robotics field and for this reason a filtered error signal, $r\left(t\right)$, is defined as a derivative filter or PD so that no acceleration of errors $\ddot{e}\left(t\right)$ will appear in the error dynamic equation, [18], $r\left(t\right) = \dot{\Theta}_r\left(t\right) - \dot{\Theta}\left(t\right)$, where the reference velocity $\dot{\Theta}_r\left(t\right)$ is defined in terms of the position error $e\left(t\right) = \Theta_d - \Theta$ and the desired velocity $\dot{\Theta}_d\left(t\right)$ as $\dot{\Theta}_r\left(t\right) = \dot{\Theta}_d\left(t\right) + \Lambda e\left(t\right)$ being Λ a diagonal matrix of design parameters with big positive elements so that the system is Hurwitz. The definition of filtered error $r\left(t\right)$ in terms of position and velocity errors can be obtained from $\dot{\Theta}_r$

$$r\left(t\right) = \dot{e}\left(t\right) + \Lambda e\left(t\right) \tag{4}$$

The acceleration joint vector, $\ddot{q}$ can be carried out as $\ddot{q} = \ddot{q}_r - r$ and substituting into the plant given by equation (5), the error dynamic equation is derived.

$$-\bar{M}\left(\Theta\right)\dot{r} - \bar{C}\left(\Theta, \dot{\Theta}\right) r + R^T B^T R_m^{-1} f\left(x\right) + \tau_d = R^T B^T R_m^{-1} \tau_m \tag{5}$$

3 Static Neural Network

In this section, the problem of dynamic modelling of robots based on static neural networks is investigated. The controller implementation is streamlined by partitioning the neural network into several smaller subnets to achieve more efficient computation. The breakthrough of this partitioned scheme stands in its computational improvement. The linearity in the parameters is the major structural property of robot manipulators and has been analysed in [17], and [19]. The majority dynamical systems such as mechanical systems include matrix operations so that a matrix network emulator can facilitate the analysis of these systems. In the literature, two solutions have been proposed, those based on the operator and matrix GL, [20] and those based on the Kronecker product, [16] static neural networks are characterized by the fact that the inputs to the neural network are exclusively the joint positions and then are sufficient to approximate the non-linear terms $M(q)$ and $G(q)$. The columns of the matrices $M(q)$ and $G(q)$ are modelled as follows

$$\begin{aligned} M(q) &= M_{SNN}(q) + \varepsilon_m(q) \\ G(q) &= G_{SNN}(q) + \varepsilon_g(q) \end{aligned}$$

where $M_{SNN}(q)$ and $G_{SNN}(q)$ are static neural networks expressed as

$$M_{SNN}(q) = \left(W_{m_1}^T \phi_{m_1}(q) \cdots W_{m_n}^T \phi_{m_n}(q) \right), \qquad G_{SNN}(q) = W_g^T \phi_g(q) \quad (6)$$

with $\phi_{m_i}(q) \in \mathbb{R}^{L_i}$ and $\phi_g(q) \in \mathbb{R}^{L_g}$ being the vectors of activation functions; and $\varepsilon_m(q) \in \mathbb{R}^{n \times n}$ and $\varepsilon_g(q) \in \mathbb{R}^{n \times 1}$ being the approximation errors. Assuming that the columns of $M(q)$ can be expressed as in (6), for an arbitrary vector $x \in \mathbb{R}^n$, there follows the decomposition

$$M(q)x = W_m^T \gamma_m(q,x) + E_m(q,x)$$

where $\gamma_m(q,x) = x \otimes \phi_m(q)$ is the activation functions and $E_m(q,x)$ is the approximation error vector. It is important to take into account that the vectorial function $\gamma_m(q,x)$ only can be expressed by the Kronecker product if the hypothesis $\phi_{m_1} = \phi_{m_2} = ... = \phi_{m_n} = \phi_m \in R^{L_m}$ holds. Also in this is convenient to express the Coriolis/centripetal term in terms of neural networks as follows

$$\begin{aligned} C(q,x)y = {} & \frac{1}{2}\left(W_m^T \varphi(q,x) y + W_m^T \varphi(q,y) x - \varphi^T(q,x) W_m y \right) + && (7) \\ & + \frac{1}{2}\left(E_D(q,x) y + E_D(q,y) x - E_D^T(q,x) y \right) && (8) \end{aligned}$$

The Jacobian of $\gamma_m(q,x)$ is the matrix of activation functions $\varphi(q,x) = \frac{\partial \gamma_m(q,x)}{\partial q}$. The error term $E_D(q,\dot{q})$ can be seen as the approximation error associated with the fundamental matrix $M_D(q,\dot{q})$ while $E_V(q,\dot{q})$ regards with the velocity matrix $M_V(q,\dot{q})$. These errors will be expressed in terms of the gradient of the inertia error $\varepsilon_m(q)$ as follows

$$E_D(q,\dot{q}) = \sum_{i=1}^{n} \frac{\partial \varepsilon_m(q)}{\partial q_i} \dot{q} e_i^T, \qquad E_V(q,\dot{q}) = \sum_{i=1}^{n} \frac{\partial \varepsilon_m(q)}{\partial q_i} \dot{q}_i$$

Fig. 1. (a) Neural subnet for estimating $M(q)\ddot{q}_r$. (b) Neural subnet for estimating G(q)

As can be easily recognized, the obtained errors E_D and E_V depend on the partial derivative of the inertia error $\varepsilon_m(q)$ with respect to the joint coordinate q_i. An alternative way to generate these errors as a linear transformation, can be derived by resorting to the Kronecker product and the gradient of the inertia error $\frac{\partial \varepsilon_m(q)}{\partial q}$

$$E_D(q, \dot{q})\dot{q}_r = \frac{\partial \varepsilon_m(q)}{\partial q}(\dot{q}_r \otimes \dot{q}), \quad E_V(q, \dot{q})\dot{q}_r = \frac{\partial \varepsilon_m(q)}{\partial q}(\dot{q} \otimes \dot{q}_r) \tag{9}$$

The neural networks for $M(q)x$ and $G(q)$ are schematically shown in figure 1 while $C(q,x)y$ is schematically shown in figure 2.

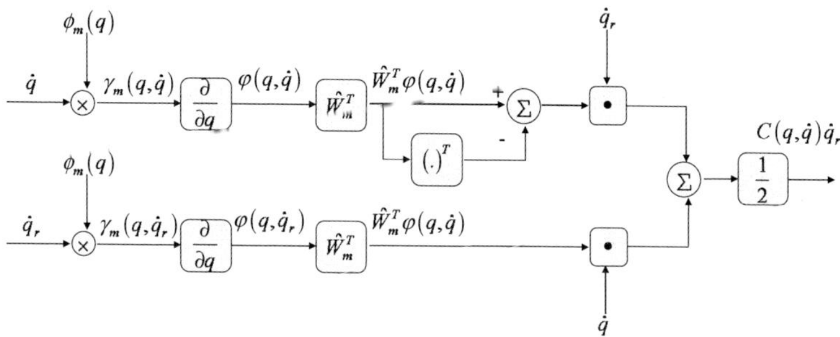

Fig. 2. Neural subnet for estimating $C(q, \dot{q})\dot{q}_r$

4 Design of the Torque Resolutor

This section is devoted to present the design of a torque resolutor to be applied to the robot manipulator. A tendon-driven system is an overactuated redundant system, i.e. the number of actuators is greater than the number of degrees of freedom. In this case, it is necessary to take into account not only the dynamics of the rotor but also the relation between the joint torques and motor torques by the transmission system, $\tau = A^T F$, where $F \in \mathbb{R}^m$ represents the forces pulling from the tendons and $A \in \mathbb{R}^{m \times n}$ is the structural matrix of the transmission system [21]. On the assumption that all the pulleys installed on the same joint axis have the same radius, the structural matrix can be decoupled, i.e. $A = B \cdot R$, where B is the routing matrix formed by elements -1, 0, or 1 and R is the matrix of dimensions of the pulleys are known.

$$B^T = \begin{pmatrix} 1 & 0 & 0 & 0 & 0 \\ 0 & 1 & -1 & -1 & 1 \\ 0 & -1 & 1 & -1 & 1 \\ 0 & 1 & 1 & 0 & 0 \end{pmatrix}$$

The design inspired in the Jacobsen rectifiers led to the following relations

$$\begin{aligned} F_1 &= \frac{\tau_1}{r_1} \\ F_3 &= O^+\left(\frac{\tau_3}{r_3}\right) + \delta_2,\ F_2 = O^-\left(\frac{\tau_3}{r_3}\right) + \delta_2 \\ F_4 &= \frac{1}{2}O^-\left(\frac{\tau_1}{r_1} + \frac{\tau_2}{r_2}\right) + \delta_1,\ F_5 = \frac{1}{2}O^+\left(\frac{\tau_1}{r_1} + \frac{\tau_2}{r_2}\right) + \delta_1 \end{aligned}$$

where δ_i is the pretensioning foce and O^+ and O^- are rectifiers designed to provide positive tendon forces. These rectifiers act as dual filters that enable the pass of a signal to the output depending on the sign of the input.

$$O^+(x) = \frac{|x| + x}{2}, \qquad O^-(x) = \frac{|x| - x}{2}$$

5 Control Architecture

In the figure 3 (a) it is shown the control system consisting of three blocks: i) a static parametric neural network to approximate the non-linearities while providing trajectory tracking, ii) a torque resolutor and iii) the dynamics of the robotic finger.

In view of the scheme 3 (a) it is worth remarking that $\ddot{\theta}_d(t)$ is necessary to build the filtered velocity error $\dot{r}(t)$. The main reason for using a torque resolutor is to be able to map between the tendon space and the joint space.

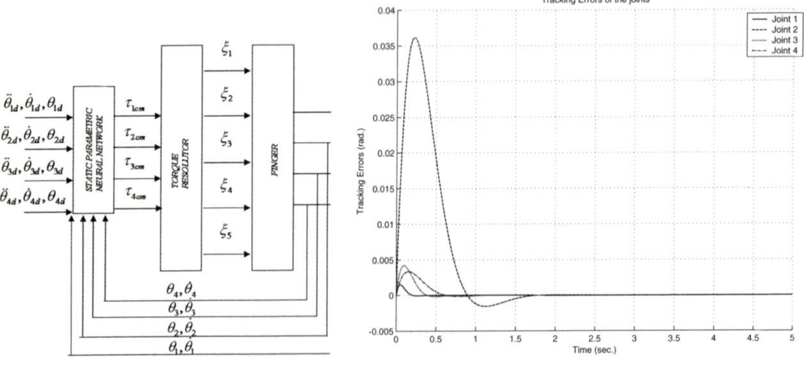

Fig. 3. (a) General scheme to control a robotic finger. ξ_i represents the tendon forces and τ_i the joint torques. (b) Tracking errors for the neurocontroller.

Furthermore, the use of Jacobsen rectifiers involves tendon working in a tension mode as opposed to the compression mode. This guarantees that the tension is always positive. Thus, the maximum velocity and acceleration of the finger can be used to determined the motor torques required to produce a suitable pretensioning of the tendons. The following theorem provides the control law and the updating law that stabilizes the system.

Theorem 1. *Let $q_d(t)$ bounded by q_B, i.e.* $\left\|\left(q_d^T(t), \dot{q}_d^T(t), \ddot{q}_d^T(t)\right)^T\right\| \leq q_B$ *and the initial condition for $r(t)$ satisfying* $\|r(0)\| < \frac{b_x - q_B}{c_0 + c_2}$ *where b_x is an upper bound for x,* $c_0 = \frac{1+\sigma_{\max}(\Lambda)}{\sigma_{\min}(\Lambda)}$ *and* $c_1 = (1 + \sigma_{\max}(\Lambda))\|e_0\| + q_B$
Suppose that the approximation error ε and unmodeled disturbances $\tau_d(t)$ are upper bounded by ε_N and d_B respectively. In the ideal case, $\varepsilon_N = d_B = 0$. The following weight updating laws are considered

$$\dot{\widehat{W}}_m = \Gamma_m\left(\left(\gamma_m(q, \ddot{q}_r) + \frac{1}{2}\varphi(q, \dot{q}_r)\dot{q}\right)r^T + \frac{1}{2}\varphi(q, \dot{q})\left(\dot{q}_r r^T - r\dot{q}_r^T\right)\right) \tag{10}$$

$$\dot{\widehat{W}}_g = \Gamma_g \phi_g(q) r^T, \quad \dot{\widehat{W}}_f = \Gamma_f \phi_f(q, \dot{q}) r^T$$

with $\Gamma_m = \Gamma_m^T > 0$, $\Gamma_g = \Gamma_g^T > 0$ and $\Gamma_f = \Gamma_f^T > 0$ symmetric positive-definite constant matrices. Then tracking error $r(t) \to 0$ as $t \to \infty$ and weights $\hat{W}_m$, $\hat{W}_g$ and $\hat{W}_f$ are bounded $\left(\left\|\hat{W}\right\|_F \leq W_B\right)$. *It can be concluded that $e \in L_2^n \cap L_\infty^n$, e is continuous, $e(t) \to 0$ and $\dot{e}(t) \to 0$ as $t \to \infty$; and τ is bounded.*

6 Results

The effectiveness of the static network based adaptive controller has been tested on a four-link robotic finger. This relatively simple example illustrates the essential features of the proposed control scheme as well as the sort of performance to be expected from the adaptive controller. The desired trajectory was chosen as follows

$$q_{d_k}(t) = a_k + b_k\left(1 - e^{-\alpha_k t^3}\right) + c_k\left(1 - e^{-\alpha_k t^3}\right)\sin(\omega_k t) \text{ for } k = 1 \ldots 4$$

where $a = \left(\frac{\pi}{45}, \frac{\pi}{8}, \frac{\pi}{30}, \frac{\pi}{60}\right)$, $b = \left(\frac{\pi}{4}, \frac{\pi}{8}, \frac{\pi}{16}, \frac{\pi}{8}\right)$, $c = \left(\frac{\pi}{18}, \frac{\pi}{4}, \frac{\pi}{16}, \frac{\pi}{8}\right)$, $\alpha_k = \left(2, \frac{9}{5}, \frac{9}{5}, \frac{9}{5}\right)$, $\omega_k = (15, 3.5, 8, 5)$. This provides bounded signals for position, velocity and acceleration. It is assumed that initially there is no knowledged about the system so that all the weights are initialized to zero. The system was simulated in SIMULINK using Runge-Kutta method with an integration fixed step of $\Delta t = 10^{-3}$ sec and a simulation range from 0 to 5 sec. The response of the controller with the proposed weight tuning appears in figure 3 (b) which shows the tracking errors for the four joints. It is possible to improve the tracking performance by increasing the gains of the controller. Therefore, it is concluded that the designed neurocontroller provides a good tracking of desired trajectories.

7 Conclusions

A neural network controller based on Christoffel symbols of first kind for a robotic finger has been presented. The non-linear compensation was developed by means of a neural network and both tuning and control laws were derived by a theorem that guarantees good tracking and boundedness for the weights and the signals. The neural network has been split up into subnets to improve and simplify the controller design and make faster the updating of weights. The weights are all initialized to zero so that no previous learning or training phase is needed. The tuning of the weights is carried out on-line, i.e. the weights are updated continuously as a function of time. The controller was connected to a torque resolutor based on the Jacobsen rectifiers. The main advantage of the proposed controller is that allows to derive tuning laws only for inertia, gravitational and frictional weights. The Coriolis/centripetal weights are not necessary to be used because of the approximation based on Christoffel symbols. This is very useful to implement neurocontrollers since the number of neurons diminishes and the computational performance improves. The activation functions for the neural networks depend on the non-linear functions associated with the inertia matrix, and therefore, a discretization of positions could be done for the inertia matrix. This is a very useful aspect because the position space for a revolute robot is compact and in consequence, the number of neurons is limited to approximate a non-linear function. The basic difference between this approach and that proposed by Li and Slotine, see [22] stems from the ability to combat the "curse of dimensionality".

Acknowledgement

This work is supported by the project TIC2003-08164-C03-03. The authors would like to thank the Spanish Research Foundation (CICYT) by the support given to this project.

References

1. Lee, J.J.: Tendon-Driven Manipulators: Analysis, Synthesis and Control. PhD thesis, Harvard University and Systems Research Center (1991)
2. Prisco, G.M.; Bergamasco, M.: Dynamic Modelling of a Class of Tendon Driven Manipulators. In: 8th International Conference on Advanced Robotics. ICAR '97. (1997) 893 – 899
3. Lee, J.J., Lee, Y.H.: Dynamic Analysis of Tendon Driven Robotic Mechanisms. Journal of Robotic Systems **20** (2003) 229–238
4. Kawamura, S., Kino, H., Won, C.: High-Speed Manipulation by Using Parallel Wire-Driven Robots. Robotica **18** (2000) 18–21
5. Johnstun, C., Smith, C.: Modeling and Design of a Mechanical Tendon Actuation Systems. ASME Journal of Dynamics systems. Measurement Control **114** (1992) 253

6. Nahvi, A., Hollerbach, J.M., Yangming, X., Hunter, I.: An Investigation of the Transmission System of a Tendon Driven Robot hand. In: IEEE/RSJ/GI International Conference on Intelligent Robots and Systems '94. 'Advanced Robotic Systems and the Real World', IROS '94. (1994)
7. Morecki, A., Busko, Z., Gasztold, H., Jaworek, K.: Synthesis and Control of the Anthropomorphic Two-Handed Manipulator, Proceedings of the 10th International Symposium on Industrial Robots (1980) 461–474
8. Lee, J.J.: Tendon-Driven Manipulators: Analysis, Synthesis and Control. PhD Thesis (1991)
9. Jacobsen, S., Wood, J., Knutti, D., Biggers, K.: The UTAH/M.I.T. Dextrous Hand: Work in Progress. The International Journal of Robotics Research **3** (1984) 21–50
10. Jacobsen, S.C., Ko, H., Iversen, E.K., Davis, C.C.: Control Strategies for Tendon-Driven Manipulators. IEEE Conference on Robotics and Automation (1989) 23–28
11. Jacobsen, S.C., Ko, H., Iversen, E.K., Davis, C.C.: Antagonistic Control of a Tendon Driven Manipulator. IEEE Proceedings of International Conference on Robotics and Automation (1989) 1334–1339
12. Ozawa, R., Kobayashi, H.: Set-Point Force Control of Tendon-Driven Robotic Mechanisms with Nonlinear Tendon Elasticity. Journal of the Robotics Society of Japan **19** (2001) 88–95
13. Kobayashi, H., Ozawa, R.: Adaptive Neural Network Control of Tendon-Driven Mechanisms with Elastic Tendons. Automatica **39** (2003) 1509–1519
14. Mulero-Martínez, J.I., Feliú-Batlle, J., López-Coronado, J.: Parametric Neurocontroller for Positioning of an Anthropomorfic Finger Based on an Oponent Driven-Tendon Transmission System. In: International Work-Conference on Artificial Neural Networks (IWANN). (2001)
15. Mulero-Martínez, J., López-Coronado, J., Feliu-Batlle, J., Calabozo-Morán, J.: Parametric Neurocontroller for Positioning of a Tendon-Driven Transmission System. In: IFAC. (2001)
16. Lewis, F.L., Jagannathan, S., Yesildirek, A.: Neural Network Control of Robot Manipulators and Nonlinear Systems. Taylor and Francis Ltd. (1999)
17. Sciavicco, L., Siciliano, B.: Modelling and Control of Robot Manipulators. Springer-Verlag, London (2002)
18. Slotine, J.J.E., Asada, H.: Robot Analysis and Control. John Wiley and Sons, Incorporated (1986)
19. Lewis, F.L., Abdallah, C.T., Dawson, D.M.: Control of Robot Manipulators. Macmillan Co., New York (1993)
20. Ge, S.S., Lee, T.H., Harris, C.J.: Adaptive Neural Network Control of Robotic Manipulators. World Scientific, London (1998)
21. Tsai, L.W.: Robot Analysis: The Mechanics of Serial and Parallel Manipulators. Interscience (1999)
22. Slotine, J.J., Li, W.: Applied Nonlinear Control. 1 Edition Prentice Hall (1990)

Control of Voluntary Movements in an Anthropomorphic Robot Finger by Using a Cortical Level Neural Controller

Francisco García-Córdova, Juan Ignacio Mulero-Martínez, and Juan López-Coronado

Department of System Engineering and Automatic
Polytechnic University of Cartagena,
Campus Muralla del Mar, 30202, Cartagena, Murcia, Spain
{francisco.garcia, juan.mulero, jl.coronado}@upct.es

Abstract. Biological control systems have long been studied as possible inspiration for the construction of robotic controllers. In this paper, we present a control of voluntary movements using a cortical network within constraints from neurophysiology. Neural controller is proposed to control desired joint trajectories for multi-joint opponent muscle control of a robot finger. Each joint is controlled by an agonist-antagonist muscle pair. Neural model proposes functional roles for pre-central cortical cell types in the computation of a descending command to spinal alpha and gamma motoneurons. Neurons in anterior area 5 are proposed to compute the position of the link in question using corollary discharges and feedback from muscles spindles. Neurons in posterior area 5 use this position perception to compute a desired movement direction. Through experimental results, we showed that neural controller exhibits key kinematic properties of human movements, dynamics compensation and including asymmetric bell-shaped velocity profiles. Neural controller suggests how the brain may set automatic and volitional gating mechanisms to vary the balance of static and dynamic feedback information to guide the movement command and to compensate for external forces.

1 Introduction

In recent years, the interface between biology and robotics is carried out by biorobotic researches. Biorobotics tries to emulate the very properties that allow humans to be successful. Each component of a biorobotic system must incorporate the knowledge of diverse areas as neuromuscular physiology, biomechanics, and neuroscience to name a few, into the design of sensors, actuators, circuits, processors, and control algorithms (see Fig. 1).

Modern robotics research is concerned with the control of complex plants. Such plants exhibit no-trivial dynamics and potentially long feedback delays. However, in order to be successful, many control techniques require accurate models of both the plant and the enviroment with which the plant interacts. In traditional form, the control of robots can be stably controlled due to fact that

J. Wang et al. (Eds.): ISNN 2006, LNCS 3972, pp. 1198–1207, 2006.

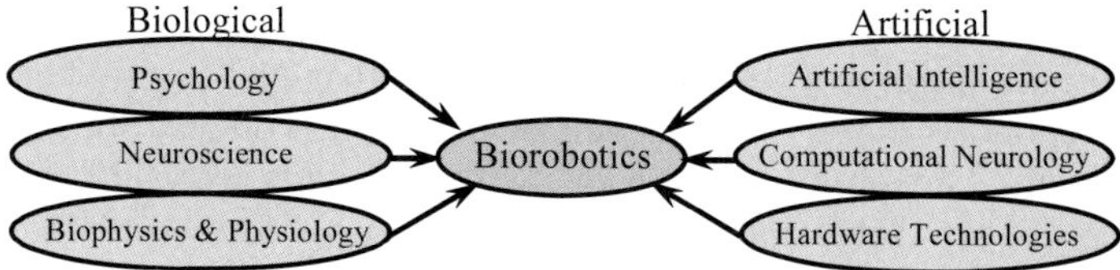

Fig. 1. Biological and artificial aspects of biorobotic systems (Adapted of [1])

each joint can be treated as an independent entity, stable high-speed control of a general dextrous robot finger using tendons (performance redundancy) or more strongly interdependent joints is, to date, highly problematic. In biology, both actuator and plant are in a constant state of flux, and behave in a complex and non-linear fashion. Delays of the sensory-motor loops are typically measured in tens or hundreds of milliseconds. Millions of years of evolutions have developed biological controllers that are very good at controlling these systems. These controllers do not rely on high-quality, pre-defined models of the plant. Rather, the control algorithm is tuned incrementally through experience with the environment. It is not surprising, then that we should turn to biology for inspiration.

It is now clear that natural movments such as reach and grasp are the results of a distributed control process, involving a complex array of sensorimotor structures. Emprirical experiments on the control of primate reaching movements have addressed issues such as the coordinate frames used in motor cortex and post-central areas, the relation of cell activity to movement variables [2], preparatory activity, load sensitivity, the latencies of various responses, and equilibrium point control [3]. Activity interpretable as motor command priming has been observed in areas 5 and 4 [2], and continuous, scalable and interruptible activities corresponding to evolving trajectory commands have been observed in area 4 [4]. Voluntary forelimb activity in primates is specialized for transporting and manipulating a wide range of objects of diverse mass. Controlling such movements requires accurate proprioception despite load variations, as well as finely graded force generation, in order to compensate for both inertial and static loads associated with the manipulated objects. Activity interpretable as static and inertial load compensation has long been associated with area 4, and a proprioceptive role for area 5 is also well established [2]. As part of an attempt to unify these diverse experimental data, Bullock et al. [5] proposed a computational model that incorporates model neurons corresponding to identified cortical cell types in a circuit that reflects known anatomical connectivity. Computer simulations in Bullock et al. [5] showed that properties of model elements correspond to the dynamic properties of many known cell types in area 4 and 5 of the cerebral cortex. Among these properties are delay period activation, response profiles during movement, kinematic and kinetic sensitivities, and latency of activity onset. In our earlier work, we proposed a biologically-inspired model of the voluntary movements for control of an one degree-of-freedom (DoF) system controlled by two antagonist artificial muscles [6], [1].

In this work, a control of voluntary movements for an antropomorphic finger is presented. The controller is developed of biologically inspired neural model [6]. A new aspect of this research work is to apply knowledge of human neuro-musculo-skeletal motion control to a robotic system, and to demonstrate that such system is able to respond voluntary similar human movements on an robot finger actuated by tendons. The robot finger used in our experiments is considered how a two DoF planar system. Actuation of the robot finger is performed by four tendons. Two pairs of artificial muscles (included its tendon) actuate on each joint. Each pair muscles pull againts one-another. The cortical level neural controller applies a strategy of trajectory control using an extension and revision of the Vector Integration to Endpoint (VITE) model [3], which exhibits key kinematic properties of human movements, including asymmetric bell-shaped velocity profiles. In the cortical network, control signals are not in form of torques to be applied to the joints, but instead control is performed directly in muscle space.

This paper is organized as follows. We first describe the neural controller of trajectory generation for vontuntary movements of a biomechanically-designed robotic system in Section II. Experimental results with the proposed scheme for control of voluntary reaching movements on an anthropomorphic finger are addressed in Section III. Finally, in Section IV, discussions based on experimental results are given.

2 Architecture of the Neural Control System

The proposed neural control system is a muti-channel central pattern generator capable of generating desired link movement trajectories by smoothly interpolating between initial and final muscle length commands for the agonist-antagonist muscles (DC-micromotors) involved in the movement. The rate of interpolation is controlled by the product of a difference vector which continuously computes the difference between the desired and present position of the joint. Basic properties of this circuit, notably its allowance for movement priming, and for performing the same movement at various speeds while maintaining synchrony among synergists.

Figure 2 depicts the neural controller, which uses several representations of neural variables postulated to be coded by activity levels distributed across cortical populations. Neural control system proposes that: (1) A finger movement difference vector (DV) is computed in parietal area 5 from a comparison of a target position vector (TPV) with a representation of current position called the perceived position vector (PPV). The DV command may be activated, or primed, prior to its overt performance. (2) The PPV is also computed in area 5, where it is derived by subtracting spindle-based feedback of position error, which is routed to area 5 via area 2, from an afference copy of an outflow position vector (OPV) from area 4. (3) The primed DV projects to a desired velocity vector (DVV) in area 4. A voluntarily scalable GO signal gates the DV input to the DVV in area 4. By virtue of the scaled gating signal, the phasic cell activity of the DVV serves as a volition-sensitive velocity command, which activates lower centers including gamma-dynamic motoneurons. (4) The DVV command is integrated by a tonic cell population in area 4, whose activity serves as an outflow position vector (OPV) to lower centers, including alpha

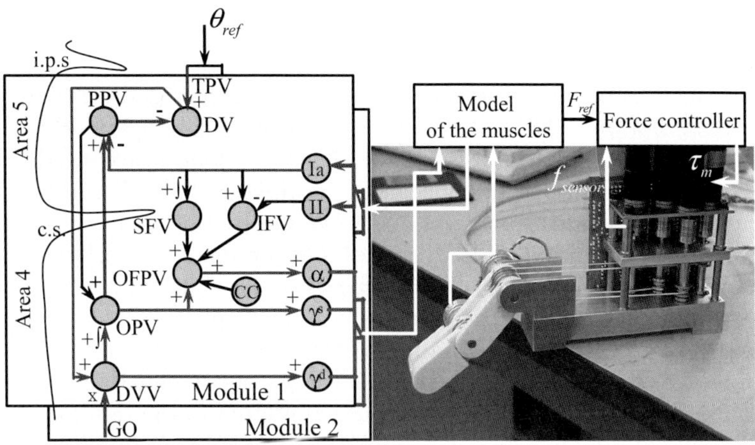

Fig. 2. Architecture of the neurobiological control system. GO, scalable gating signal; DVV, desired velocity vector; OPV, outflow position vector; OFPV, outflow force+position vector; SFV, static force vector; IFV, inertial force vector; PPV, perceived position vector; DV, difference vector; TPV, target position vector; γ^d, dynamic gamma motoneuron; γ^s, static gamma motoneuron; α, alpha motoneuron; Ia, type Ia afferent fiber; II, type II afferente fiber; CC, assumed cerebello-cortical input to the IFV stage; c.s., central sulcus; i.p.s., intraparietal sulcus. The symbol + represents excitation, - represents inhibition, x represents multiplicative gating, and $+\int$ represents integration.

and gamma-static motoneurons. This area 4 tonic cell pool serves as source of the efference copy signal used in area 5 to compute the perceived position vector (PPV). As the movement evolves, the difference vector (DV) activity in area 5 is driven toward baseline. This leads to termination of excitatory input to area 4 phasic cells, and thus to termination of the movement itself. (5) A reciprocal connection from the area 5 PPV cells to the motor-cortical tonic cells (OPV) enables the area 4 position command to track any movement imposed by external forces. This reciprocal connection also helps to keep spindles loaded and to avoid instabilities that would otherwise be associated with lags due to fine signal conduction rates and loads. (6) Phasic-tonic force-and-position-related (OFPV) cells in area 4 enable graded force recruitment to compensate for static and inertial loads, using inputs to area 4 from cerebellum and a center that integrates spindle feedback. These area 4 phasic-tonic cortico-motoneuronal cells enable force of a desired amount to be exerted against an obstacle without interfering with accurate proprioception (PPV), and while preserving a targe posture (TPV) should the obstacle give way. Finally these hypotheses were translated into a mathematical model.

2.1 Antropomorphic Finger Model

To explant the control of voluntary movements, a typical planar two-DoF finger was implemented. The detailed inverse dynamics equation of the finger are discribed as follows:

$$\boldsymbol{\tau}\left(\theta,\dot{\theta},u\right)=\mathbf{M}(\boldsymbol{\theta})\ddot{\boldsymbol{\theta}}+\mathbf{C}(\boldsymbol{\theta},\dot{\boldsymbol{\theta}})\dot{\boldsymbol{\theta}}+\mathbf{F}\dot{\boldsymbol{\theta}}+\mathbf{G}(\boldsymbol{\theta}), \tag{1}$$

where $\boldsymbol{\theta}\in\mathbb{R}^{n\times 1}$ is the vector of generalized joint coordinates describing the position of the robot finger, $\dot{\boldsymbol{\theta}}\in\mathbb{R}^{n\times 1}$ is the vector of joint velocities, $\ddot{\boldsymbol{\theta}}$ is the vector of joint accelerations, $\mathbf{M}\in\mathbb{R}^{n\times n}$ is the symmetric joint-space inertia matrix, $\mathbf{C}\in\mathbb{R}^{n\times n}$ describes Coriolis and non-lineal centripetal effects, $\mathbf{F}\in\mathbb{R}^{n\times n}$ is a diagonal matrix and describes viscous and Coulomb friction and is not generally considered part of the rigid body dynamics, $\mathbf{G}$ represents the aceleration of the gravity, $\boldsymbol{\tau}\in\mathbb{R}^{n\times 1}$is the vector of joint torques, u is the descending motor command, θ_1 and θ_2 represent the proximal and distal interphalangeal (PIP and DIP) joints, respectively. The movement ranges of the joints are $\boldsymbol{\theta}\in[-\pi/2,\,\pi/2]$.

It can be shown through the principle of energy conservation that the vector of forces, $\mathbf{F}_T\in\mathbb{R}^{m\times 1}$, exerted by the tendons are related to the resultant torques, $\boldsymbol{\tau}$, in the joints of the open-loop chain. The force and displacement transformations between the joint space and tendon space are discribed by the following relationship:

$$\boldsymbol{\tau}=\mathbf{A}^T\mathbf{F}_T, \tag{2}$$

$$\mathbf{L}=\mathbf{L}_0-\mathbf{A}\boldsymbol{\theta},\quad\text{where }\mathbf{A}^T=\begin{bmatrix} r_1 & -r_2 & r_3 & -r_4\\ 0 & 0 & r_5 & -r_6\end{bmatrix}, \tag{3}$$

where $\mathbf{L}\in\mathbb{R}^{m\times 1}$ is a vector representing the linear displacements of tendons, $\mathbf{L}_0\in\mathbb{R}^{m\times 1}$ is the vector of the passive resting lengths of the muscles, $\mathbf{A}$ is an $m\times n$ matrix which depends on the routing tendons called the structure matrix, r_i is the radii of the ith pulley, under the condition that all pulleys on the same joint axis and are of the same size, the elements of the matrix $\mathbf{A}^T$ are the radii of the pulleys installed at the consecutive axes of the equivalent open-loop chain. Notation $r_1=r_{2-4}$, and $r_5=r_6$. Actuaction of the finger is performed by two pairs of opposing muscles (see Figure 3). One pair of muscle actuates on the PIP joint, while the other on the DIP joint (see Figure 3b). Each muscle behaves as a non-linear visco-elastic (spring) element. Control of the robot finger is achieved by setting the rest length of each muscle.

2.2 Muscle Model

The actuation system of the robot finger consists of artificial muscles that use appropriate gearboxes, multi-radial flexible couplings, force and position sensors, and tendons (in order to pull and be pushed), in order to approach the behavior of the DC-motors like human or animal muscles. The artificial muscles introduce a PID force controller and the generation of desired force comes given by a mathematical model of the muscle in orden to represent the basic properties of the animal musculo-skeletal system. Essentially, muscle is springy tissue with a neurally controllabe contractile component, which gives it a neurally modifiable threshold length for force development. To simplify, we can assume that the force (F_i) development by a muscle is a threshold-linear function of its length (L_i) that depends upon its fixed resting length (Γ_i), its stiffness (K), and its neurally

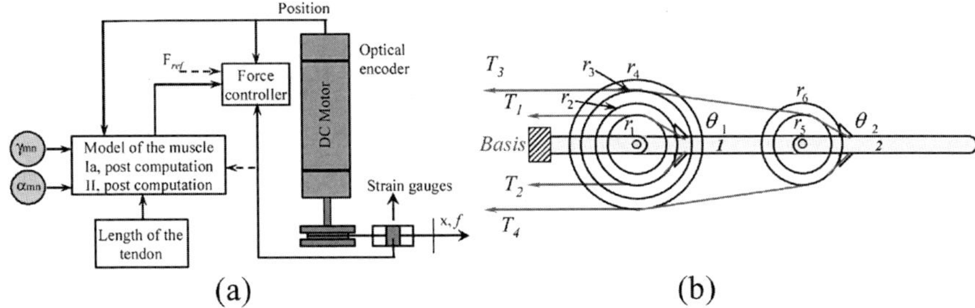

Fig. 3. Tendon-driven anthropomorphic robot finger with two-DoF and structurally isomorphic. (a) Schematic diagram of artificial muscle model using DC motors. (b) Planar schematic of the routing system of tendons.

modifiable contractile state (c_i). Then, a better approximation to real muscle, whose stiffness also varies with contractile state, is achieved by a quadratic force-length realtionship with nonlinear stiffness and viscosity. The dynamics of the muscle can be represented, such that:

$$F_{Ti} = K\left([L_{in} - (\Gamma_i - c_i)]^+\right)^2, \tag{4}$$

where the indices $i = \{1, 2\}$ designate antagonist muscle pairs, L_{in} is the normalized length of the muscle i and the threshold-linear function $[w_i]^+$ is defined as $\max(w_i, 0)$. Thus if $\omega_i = L_{in} - (\Gamma_i - c_i) > 0$, then $F_{Ti} = K.\omega_i$, whereas if $\omega_i \leq 0$, then $F_{Ti} = K.\omega_i = 0$. Equation (4) shows that a muscle is spring-like in that it develops a force only when stretched to a length L_{in} greater that the effective threshold length $\Gamma_i - c_i$. However, it also shows that muscle is more versatile than an ordinary spring because this threshold can be neurally adjusted by varying the muscle's state of contraction, c_i. The contractile state dynamics are defined by

$$\frac{d}{dt}c_i = \upsilon\left(-c_i + \alpha_i\right) - \left[F_i - \Gamma_{Fi}\right]^+, \tag{5}$$

where α_i represents alpha motor neuron pool activity in muscle control channel i, υ is contractile rate, and Γ_{Fi} is the force threshold in muscle control channel i. For the voluntary movements, the system operates via area 4. The process of assembling the net descending command to alpha motor neurons can be divided conceptually into kinematic and kinetic aspect. The kinematic aspect of trajectories control involves specifying the time series of position that the single link is intended to occupy between its initial and desired final position. The lengths of the muscles (L_i) are obtained through the equation (3) and normalized 0 to 1. The values of the parameters used in the experiments were $K = 2.5$, $\Gamma_i = 0.95$, $\Gamma_{Fi} = 10$, and $\upsilon = 0.25$. The initial lengths of the muscles were $L_{i0} = 0.5$.

2.3 Cortical Network

The system of equations of the cortical neural controller maintains a focus of neurophysiological data regarding temporal dynamics of cell types, it addresses both kinematic and kinetic aspects of reaching, it is capable of synchronizing movements among an arbitrary number of joints with complex musculature involving mono- and bi-articular muscles, to handle variable speeds and forces and shows the functional interpretation of properties of many types of identified cortical neurons. The interaction among the different neurons for a single link is described in the following way.

Tonic cells activity (OPV) is given by

$$\frac{d}{dt}y_i = (1 - y_i)\left(\eta x_i + [u_i - u_j]^+\right) - y_i\left(\eta x_j + [u_j - u_i]^+\right). \tag{6}$$

Area 5 cells activity is considered how a difference vector (DV) and can be described by

$$r_i = [T_i - x_i + B^r]^+. \tag{7}$$

The following equations describe the computation of a *Perceived Position Vector* (PPV) by anterior area 5 tonic cells that are assumed to receive an efference copy input from area 4 and position error feedback from muscle spindles:

$$\begin{aligned}\frac{d}{dt}x_i = (1 - x_i)&\left[\left(\Theta y_i + s_j^{(1)}(t-\tau) - s_i^{(1)}(t-\tau)\right)\right]^+ \\ -x_i&\left[\left(\Theta y_j + s_i^{(1)}(t-\tau) - s_j^{(1)}(t-\tau)\right)\right]^+,\end{aligned} \tag{8}$$

where $s_i^{(1)} = S\left(\varphi\left[\gamma_i^s - L_i\right]^+ + \phi\left[\gamma_i^d - \frac{d}{dt}L_i\right]^+\right)$, $\gamma_i^s = \chi y_i$, $\gamma_i^d = \rho u_i$, and $S(w) = \frac{w}{1+100w^2}$.

The *Desired Velocity Vector* (DVV) describes the area 4 phasic movement-time (MT) activity and is described by the following equation:

$$u_i = [g(r_i - r_j) + B^u]^+. \tag{9}$$

The sigmoidal GO signal is a input from a decision center in the brain and is described by

$$\begin{aligned}&\frac{d}{dt}g^{(1)} = \varepsilon\left[-g^{(1)} + \left(Cs - g^{(1)}\right)g^{(0)}\right], \quad \frac{d}{dt}g^{(2)} = \varepsilon\left[-g^{(2)} + \left(Cs - g^{(2)}\right)g^{(1)}\right], \\ &g = g^{(0)}\frac{g^{(2)}}{Cs}.\end{aligned} \tag{10}$$

The *Inertial Force Vector* (IFV) extracts velocity errors from the primary and secondary spindle feedback, and is described by the following equation:

$$q_i = \lambda\left[s_i^{(1)}(t-\tau) - s_i^{(2)}(t-\tau-\Lambda)\right], \text{ where } s_i^{(2)} = S\left(\varphi\left[\gamma_i^s - L_i\right]^+\right). \tag{11}$$

To compesate static loads, the neural controller integrates positional errors reported by the spindles and adds theses to the alpha motor neuron command. Spindle error integration is performed by a *Static Force Vector* (SFV), which is described by

$$\frac{d}{dt}f_i = (1 - f_i)h.k_i.s_i^{(1)}(t-\tau) - \psi f_i\left[f_j + s_j^{(2)}(t-\tau)\right]. \tag{12}$$

The activity of the phasic-tonic cells constitutes an *Outflow Force+Position Vector* (OFPV) and alpha-motor neurons (α-MN) are described by

$$a_i = y_i + q_i + f_i, \text{ and } \alpha_i = a_i + \delta s_i^{(1)}, \tag{13}$$

respectively. In cortical network, y_i is OPV; u_i is DVV; x_i is PPV; η is the gain on a pathway from the PPV to OPV; r_i is the activity of the DV cells; T_i is the target position vector; B^r is the baseline activity of the DV; γ_i^s and γ_i^d are the activities of static and dynamics gamma motor neurons; χ and ρ are gains of the γ_i^s and γ_i^d, respectively; $s_i^{(1)}$ is the activity of primary spindle afferents from muscle i; φ is the sensitivity to a stretch of the static nuclear bag and chain fibers; ϕ is the sensitivity of dynamic nuclear bag fibers to rate of stretch; Θ is the gain of the corollary discharge from area 4 tonic cells, calibrated such that $\Theta \approx \varphi$ to ensure accurate PPV calculation; variable t represents the time step; parameter τ is the delay on the feedback from spindles to central sites; the fuction $S(w)$ expresses the limited dynamic range of spindle afferent activity; g is the the GO signal that multiplies the DV; B^u is the baseline activity of the DVVV; $g^{(o)}$ is the step input from a forebrain decision center; ε is a slow integration rate; Cs is the value at which the GO cells sature; q_i is the IFV activity; λ is the feedback gain and Λ is a threshold; $s_i^{(2)}$ is secondary spindle afferent; f_i is the SFV activity; h is a gain that controls the strength and speed of load compensation (modulated by a specific muscle gain, k_i); ψ is a parameter scaling inhibition by the antagonist component of the SFV and by the antagonist spindle; a_i is the OFPV activity; α_i is the α-MN activity; and δ is the gain of the stretch reflex. The values of the parameters used in the experiments were $\Theta = \varphi = 0.7$, $\phi = 0.1$, $B^r = 0.001$, $B^u = 0.01$, $Cs = 25$, $\varepsilon = 1$, $\eta = 0.7$, $\psi = 15$, $\rho = 0.04$, $\Lambda = 0.003$, $\delta = 0.1$, $h = 0.025$, $k_i = 1$; $\tau = 0$, $\lambda = 8.5$, and $g^{(0)} = 2.6$.

3 Results

In the neural controller the desired angles of the joints becomes desired contraction of the artificial muscles. This way, the neural system controls lineal desplacement of tendons. The generated forces by the model of the muscle are introduced to the force controller as reference signals. We carried out experimental tests with the biomechanical system presented in the Figure 2 that emulates the musculo-skeletal system of a human finger. The finger initial configuration was from $\theta_1(0) = -\pi/2$, and $\theta_2(0) = 0$ corresponding to initial lengths of tendons $L_i = 0.5$. A step input as a voluntary movement and transformed to tondon space using the equation (3) was applied to DIP joint. Figure 4(a) shows experimental results of cell activities in cortical area 4 and 5 during the voluntary reaching movement. Activity of primary spindle afferents from muscle, i, is shown in the Figure 4(b). A movement sequence of robot finger is presented in Figure 5.

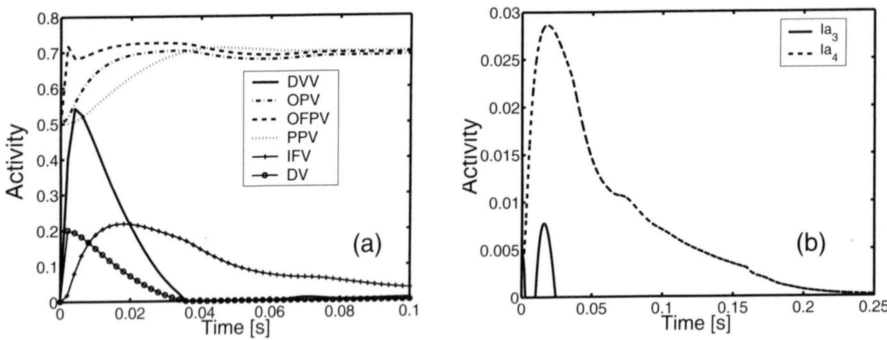

Fig. 4. Activity of the Cortical cells during a simple voluntary reaching task in the DIP joint. Desired angle was $\theta_{2d} = 36°$. (a) Cell responses. (b) Primary spindle activities during movement.

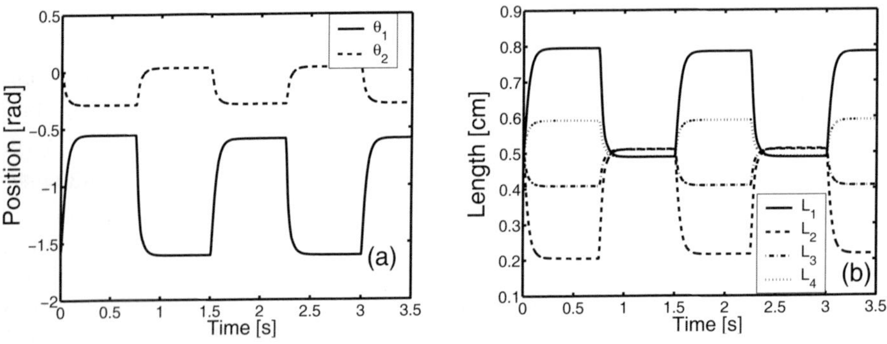

Fig. 5. Position control in a movement sequence with variation of $\theta_{1d} = -50°$ to $-90°$ and $\theta_{2d} = -36°$ to $0°$. (a) Response of the position in PIP and DIP joints. (b) Lenght of the tendons.

4 Discussions

In this paper, we have implemented an biologically inspired cortical network for the movement control of a robot finger. This proposed neural controller suggests how the brain may set automatic and volitional gating mechanism to the balance of static and dynamic feedback information to guide the movement command and to compensate for external forces. For example, with increasing movement speed, the system shifts from a feedback position controller to a feedforward trajectory generator with superimposed dynamics compensation. In particular, the spinal stretch reflex provides a component proportional to the position error, the SFV provides a component proportional to the integral of the position error, and the IFV provides a component proportional to time derivative of the position error. Here the errors in reference signals are provided by the OPV and DVV. By using the DVV, or desired velocity, as input to gamma-dynamic MNs allows substitution of a direct measurement of velocity error for a neural differentiation.

Modulation of the cortical network controller for trajectory generation and dynamics compensation with automatic and volitional mechanism allows the system to achieve a high degree of task sensitivity. The errors of the system by delays of the feedback signs can be corrected by making use of an adaptive cerebellar loop [7]. If the cerebellum learns to predict velocity errors based on movement context, it can automatically generate appropriate lauching pulses in area 4 before errors occur. Such a feedforward compensation circuit avoids oscillations due to feedback delays, and can operate at a much higher gain than a feedback circuit. In the experiments, we aproximate the pre-emptive function of such a cerebellar feedforward side-loop by reducing the delay (τ) on spindle feedback to zero. This is not meant to imply a non-physiological zero-delay in feedback, but is merely a way to mimic the availability of a calibrated feedforward compensation. The proposed control strategy can also compensate unexpected disturbances, which is as essential capability for compliance control and to exhibit key kinematic properties of human movements, including asymmetric bell-shaped velocity profiles.

Finally, to extend the proposed model in several directions, we propose (a) to extend the spinal cord inside the cortical network [7], as well as (b) to incorporate the control function of the cerebellum [8] in the presented neural architecture.

References

1. García-Córdova, F., Guerrero-González, A., Pedreño-Molina, J.L., Calabozo-Morán, J.: Emulation of the Animal Muscular Actuation System in an Experimental Platform. In: IEEE International Conference on Systems, Man and Cybernetics, Tucson, Arizona ,USA (2001)
2. Riehle, A., Mackay, W.A., Requin, J.: Are Extent and Force Independent Movement Parameters?. Preparation- and Movement-Related Neuronal Activity in the Monkey Cortex. Experimental Brain Research **99** (1994) 56–74
3. Cisek, P., Grossberg, S., Bullock, D.: A Cortico-spinal Model of Reaching and Proprioception under Multiple Task Constraints. Journal of Cognitive Neuroscience **10** (1998) 425–444
4. Scott, S.H., Kalaska, J.F.: Changes in Motor Cortex Activity During Reaching Movements with Similar Hand Paths but Different Arm Postures. Journal of Neurophysiology **73** (1995) 2563–2567
5. Bullock, D., Cisek, P., Grossberg, S.: Cortical Networks for Control Voluntary Arm Movements under Variable Force Conditions. Cerebral Cortex **8** (1998) 48–62
6. García-Córdova, F., Guerrero-González, A., Pedreño-Molina, J., López-Coronado, J.: A Cortical Network Controller for Control of Voluntary Movements in Robotics Systems. In: IEEE International Conference on Systems, Man and Cybernetics, Tucson, Arizona ,USA (2001)
7. Contreras-Vidal, J., Grossberg, S., Bullock, D.: Neural Model of Cerebellar Learning for Arm Movement Control: Cortico-Spino-Cerebellar Dynamics. Learn Memory **3** (1997) 475–502
8. Spoelstra, J., Schweighofer, N., Arbib, M.A.: Cerebellar Learning of Accurate Predictive Control for Fast-Reaching Movements. Biological Cybernetics **82** (2000) 321–333

Learning Control for Space Robotic Operation Using Support Vector Machines

Panfeng Huang[1,2], Wenfu Xu[3], Yangsheng Xu[2], and Bin Liang[3]

[1]College of Astronautics
Northwestern Polytechnical University, Xi'an, China
pfhuang@nwpu.edu.cn
[2] Department of Automation and Computer-Aided Engineering
The Chinese University of Hong Kong, Shatin, Hong Kong
ysxu@acae.cuhk.edu.hk
[3] Shenzhen Space Technology Center
Harbin Institute of Technology, Shenzhen, China
{wfxu, bliang}@robotsat.com

Abstract. Automatical operation of space robots is a challenging and ultimate goal of space servicing. In this paper, we present a novel approach for tracking and catching operation of space robots based on learning and transferring human control strategies (HCS). We firstly use an efficient support vector machine (SVM) to parameterize the model of HCS, and then develop a new SVM-based leaning structure to improve HCS in tracking and capturing control. The approach is fundamentally valuable in dealing with some problems such as small sample data and local minima, which makes it efficient in modeling, understanding and transferring its learning process. The simulation results demonstrate that the proposed method is useful and feasible in tracking trajectory and catching objects autonomously.

1 Introduction

It is very difficult mission for space robots to track and catch a free-flying object in space. In general, the model-based method was used in previous research and engineering work, which required the robot model in advance. However, it is difficult to accurately obtain the dynamic model of robot, especially for space robot due to its complicated multi-bodies construction. As well known, the robot obtains the object position and orientation information with respect to its end-effector or base using its vision sensors or other sensors, these parameters are in Cartesian space, the trajectory planning in Cartesian space is to plan the posture and velocities of end-effector in order to match the motion object and capture it completely. Moreover, a series of posture information must be converted to joint series of robot in Joint space by inverse kinematics solutions in order to realize the robot control. However, kinematics problems of space robot are different with terrestrial robots [1]. It is difficult and time-consuming to compute the inverse kinematics solutions for the computer installed on the robot.

J. Wang et al. (Eds.): ISNN 2006, LNCS 3972, pp. 1208–1217, 2006.

Learning HCS is to use past knowledge and experience in performing given tasks. Skill can be gained incrementally through learning and practicing. Acquiring, modeling, and transferring human skill has been an objective of research in the fields of artificial intelligence, robotics, and intelligent control for more than two decades. The main problem is not only important to the theory of machine intelligence, but also essential in practice for developing intelligent robotic and autonomous system. Most of space robots were operated to accomplish space missions by human's tele-operation. One of the most important characteristics of tele-operation is that a human operator is included in the control loop. A human operator gives the desired robot end-effector motion commands through a tele-operation device. For a given task, the operator's control inputs reflect his/her skill for performing the task. Allowing the robot system to learning human operator's skill is one major topic in this paper.

Learning by human skill has been studied from different disciplines with emphasis on different objectives about two decades. The idea of using learning method presented in [2], [3] is based on the observation that robot manipulators, being subject to playback control model, repeat their motions over and over in cycles. Tracking and capturing are defined herein as a robot approaching a moving object and match its location and velocity to be within pre-defined thresholds in shortest possible time. In an automatic tracking system, such as missiles intercepting aircrafts, efficient real-time autonomous tracking strategy is required when the motion of object is unknown beforehand. There are many strategies proposed to tackle such tracking and intercepting problem. These strategies deal with different motion type of the objects.

Considerable research efforts have been directed toward learning control architecture using artificial neural networks (ANN) [4],[5],[6] or Hidden Markov Model (HMM) [7]. Despite success, human-based control methods based on previous methods face several problems. For example, ANNs and HMMs have problems related to local minima. This drawback leads to more oscillations in experimental results - compared with the training state data - which sometimes causes failure. Moreover, another problem is that ANNs often require large amounts of training data to make the training process to reach satisfactory levels of precision. However, it is inconvenient and sometimes impractical to obtain large sets of training samples. Furthermore, robot behavior is difficult to analyze and interpret mathematically, and is highly dependent on their fixed architectures. From a mathematical point of view, Machine Learning and statistical learning estimate a function f mapping between inputs and outputs using n sets of training samples. Based on statistical learning theory, SVM is not only a tool for theoretical analysis but also a tool for creating practical algorithms for estimating multidimensional functions.

In recent several years, SVM has gained popularity for two reasons. First, it is satisfying from a theoretical point of view: support vector learning is based on some simple ideas, and provides an insight into the relationship between inputs and outputs. Second, it demonstrates high performance in practical applications. Thus, SVM has been applied to many areas, such as pattern identification, regression, and even nonlinear equalization, etc., [8], [9]. In this paper, we discuss how an SVM can be used for abstracting and transferring HCS problems.

The paper is organized as follows. Section 2 addresses the learning human control method and background of support vector machine. Section 3 introduces the SVM-based learning method, which mainly include problem description, training data collection and training process in Cartesian space and Joint space, respectively. In Section 4 simulation results attest and verify our learning method. Discussion and conclusion of this paper are presented in Section 5.

2 Learning Control Through Support Vector Machine

2.1 Support Vector Machine

Support vector machines provide a new approach to the problem of classification and regression with clear connections to the underlying statistical learning theory. They differ radically from comparable approaches such as neural networks: SVMs training always find a global minimum, and their simple geometric interpretation provides fertile ground for further investigation. An SVM is largely characterized by the choice of its kernel, and SVMs thus link the problem they are designed for with a large body of existing work on kernel based methods.

For classification problems, it was then, extended to regression estimation problems, i.e., to problems related to find the function $y = f(\tilde{\mathbf{x}})$ given by its measurements y_i with noise at some (usually random) vector $\tilde{\mathbf{x}}$

$$(y_1, \tilde{\mathbf{x}}_1), ..., (y_l, \tilde{\mathbf{x}}_l). \tag{1}$$

where each of them generated from an unknown probability distribution $P(\tilde{\mathbf{x}}, y)$ containing the underlying dependency. Here and below, bold face characters denote vectors.

In this paper, the term SVM will refer to both classification and regression methods, and the terms Support Vector Classification (SVC) and Support Vector Regression (SVR) will be used for specification. In SVR, the basic idea is to map the data X into a high-dimensional feature space $\mathcal{F}$ via a nonlinear mapping ϑ, and to do linear regression in this space [11].

$$f(\tilde{\mathbf{x}}) = (\mu \cdot \vartheta(\tilde{\mathbf{x}})) + \Delta \quad \text{with } \vartheta : R^n \rightarrow \mathcal{F},\ \mu \in \mathcal{F}, \tag{2}$$

where Δ is a threshold. Thus, linear regression in a high dimensional (feature) space corresponds to nonlinear regression in the low dimensional input space R^n. Note that the dot product in Equation (2) between μ and $\vartheta(\tilde{\mathbf{x}})$ would have to be computed in this high dimensional space (which is usually intractable), if we are not able to use the kernel that eventually leaves us with dot products that can be implicitly expressed in the low dimensional input space R^n. Since ϑ is fixed, we determine μ from the data by minimizing the sum of the empirical risk $R_{emp}[f]$ and a complexity term $\| \mu \|^2$, which enforces flatness in feature space

$$\begin{aligned} R_{reg}[f] = R_{emp}[f] + \lambda \| \mu \|^2 = \\ \sum_{i=1}^{l} M(f(\tilde{\mathbf{x}}_i) - y_i) + \lambda \| \mu \|^2 \end{aligned} \tag{3}$$

where l denotes the sample size $(\tilde{\mathbf{x}}_1, ..., \tilde{\mathbf{x}}_l)$, $M(.)$ is a loss function and λ is a regularization constant. For a large set of loss functions, Equation (3) can be minimized by solving a quadratic programming problem, which is uniquely solvable [10]. It can be shown that the vector μ can be written in terms of the data points

$$\mu = \sum_{i=1}^{l} (\beta_i - \beta_i^*) \vartheta(\tilde{\mathbf{x}}_i), \tag{4}$$

with β_i, β_i^* being the solution of the aforementioned quadratic programming problem [11]. The positive Lagrange multipliers β_i and β_i^* are called support values. They have an intuitive interpretation as forces pushing and pulling the estimate $f(\tilde{\mathbf{x}}_i)$ towards the measurements y_i [12]. Taking Equation (4) and Equation (2) into account, we are able to rewrite the whole problem in terms of dot products in the low dimensional input space

$$\begin{aligned} f(\tilde{\mathbf{x}}) &= \textstyle\sum_{l=1}^{l} (\beta_i - \beta_i^*)(\vartheta(\tilde{\mathbf{x}}_i) \cdot \vartheta(\tilde{\mathbf{x}})) + \Delta \\ &= \textstyle\sum_{l=1}^{l} (\beta_i - \beta_i^*) k(\tilde{\mathbf{x}}_i, \tilde{\mathbf{x}}) + \Delta. \end{aligned} \tag{5}$$

In Equation (5), we introduce a kernel function $k(\tilde{\mathbf{x}}_i, \tilde{\mathbf{x}}_j) = \vartheta(\tilde{\mathbf{x}}_i) \cdot \vartheta(\tilde{\mathbf{x}}_j)$. As explained in [13], any symmetric kernel function k satisfying Mercer's condition corresponds to a dot product in some feature space.

2.2 SVM Learning Approach

The skill that we are considering here is the control strategy demonstrated by a human expert to obtain a certain control target. For example, in controlling a robotic system, a human expert gives commands by a remote controller, such as a joystick, and the robot executes the task. The desired trajectory of the robot given by an expert through a joystick reflects the expert's control strategy. The goal of the human control strategy learning here is to model the expert control strategy and according to current states select one command that represents the most likely human expert strategy. Here, we consider the human expert actions as the measurable stochastic process and the strategy behind it as the mapping between the current states and commands. An SVR is employed to represent human expert strategy for the given task, and the model parameters are sought through an off-line learning process. This method allows human experts to transfer their control strategy to robots by the trained SVM model. The procedure for SVM learning approach can be summarized as follows:

1. Representing the control strategy by an SVM: Choosing a suitable kernel and structure of an SVM for characterizing the control strategy.
2. Collecting the training data: Obtaining the data representing the control strategy we want to model.
3. Training the model: Encoding the control strategy into an SVM.
4. Finding the best human performance: Learning/transferring the control strategy.

For training an SVM learning controller, the system states will usually be treated as the learning inputs and the control inputs/commands will be the learning outputs.

3 SVM Based Learning Control

To address the SVM based skill learning method in detail, it is necessary to describe the main tracking and catching problem. Herein, the trajectory of object and end-effector only contain the position information, and we define that the space manipulator has three degrees of freedom. Fig.1 shows the tracking and catching problem description, in which the desired trajectory is generated so as to approach the object trajectory, and the end-effector is controlled to track the desired trajectory.

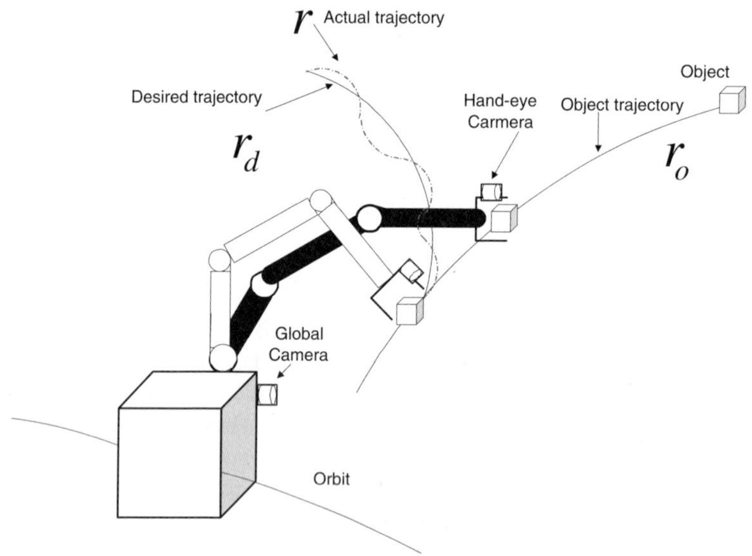

Fig. 1. Tracking and capturing problem description

Generally spoken, vision-servo system of robot system can guide the motion of end-effector to track the object. However, it is difficult to generate the tracking trajectory in time in complex space environment, which need the complicated algorithms to generate the corresponding trajectory online. Those algorithms are designed by human beings in advance, hence, we can consider that the originality of the algorithms derives from HCS. Fig. 2 shows the block diagram of the tracking process by HCS.

According to the proposed method, the "human control strategies" in Fig.2 can be substituted by the trained SVM which has learned and abstracted the human control strategies. Therefore, Fig.4 shows the block diagram of tracking process by SVM module.

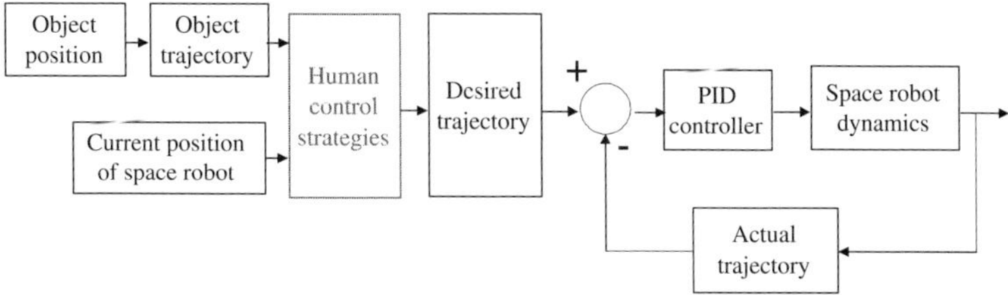

Fig. 2. The block diagram of tracking process by human control strategies

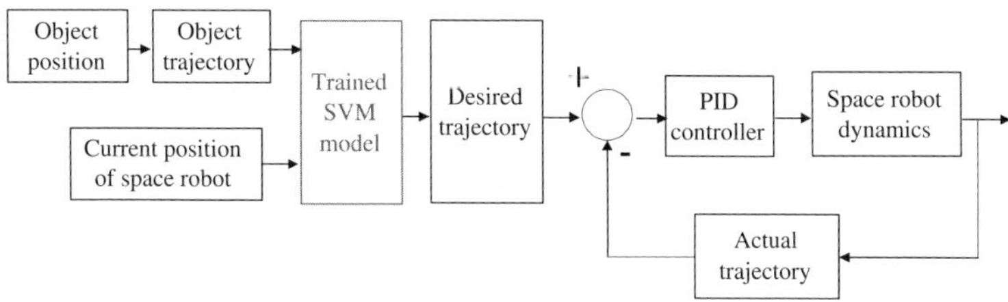

Fig. 3. The block diagram of tracking process by the trained SVM model

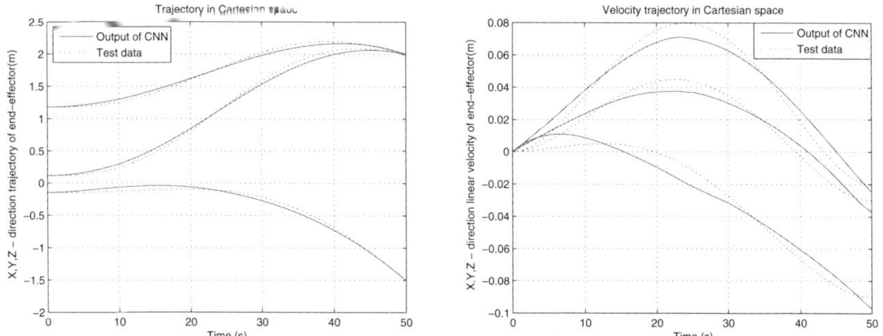

Fig. 4. Position trajectory of end-effector at X,Y,Z-direction

Fig. 5. Velocity trajectory of end-effector at X,Y,Z-direction

In order to collect data, we assume that this robot system is installed hand-eye camera and global camera which be used to measure and estimate the characteristics of the moving object, such as location and motion velocity. At the beginning, the space robot system can estimate the positions, velocities and accelerations of the object. Then, it can obtain the motion trajectory of the object. The operator can plan the dynamics-based trajectory of the end-effector to track and capture this object in its workspace. The control of robot manipulator uses the PID feedback controller.

To collect the training data, we assume that the object can enter into the workspace via some different trajectories. For different motion paths of the ob-

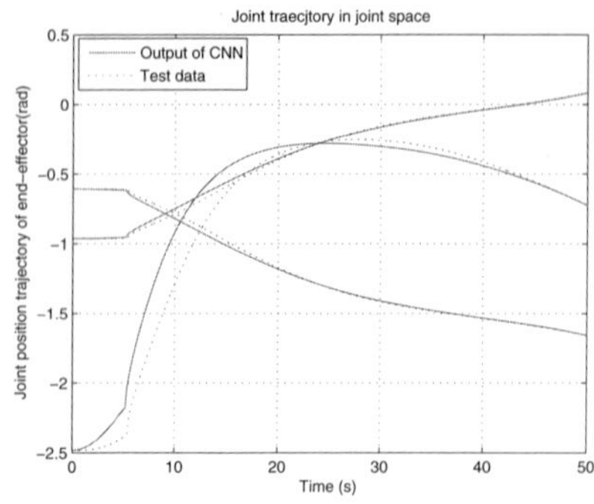

Fig. 6. Joint angle position trajectory

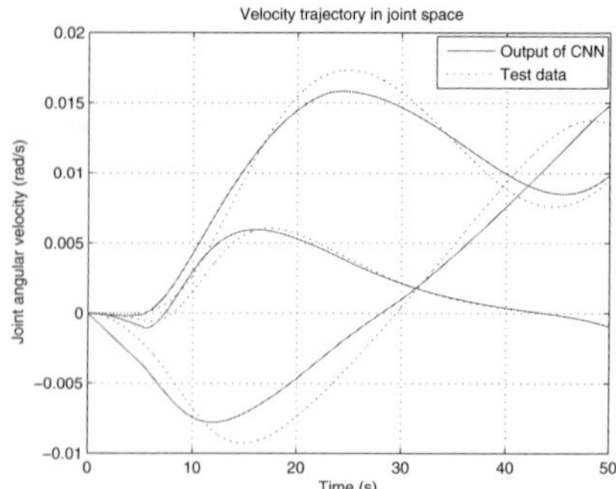

Fig. 7. Joint angular velocity trajectory

ject, the trajectory generator of space robot can generate the corresponding tracking trajectory with the help of human. Because an SVM does not require a large amount of training samples as do most ANNs, we try to let these trajectories distribute evenly in robot's workspace, thus, the SVMs can efficiently abstract human tracking trajectory planning skill.

Generally, it is desirable to employ support vector machine in order to represent human skill. This is because robot motion is generally multidimensional. And therefore, the skill learning using motion measurements should be multidimensional. For the learning trajectory, the learning procedure of multi-SVMs can be done in either Cartesian space or in Joint space. If it is done in Joint space, i.e., recording joint data and training the model in Joint space, the mapping from Cartesian space to Joint space is automatically avoided.

4 An Illustrative Example

4.1 System Description and Task

To evaluate the validity and effectiveness of the proposed scheme, we applied SVM based learning scheme to a space robot system which be operated to track and capture a free-flying object by tele-operation control. Moreover, this free-flying object can pass through the workspace of the robot. Skill learning is one of the most crucial issues for space robot now. To enable the reader fully understand our method. We define a space robot system as an illustrative example to verify our learning approach. The configuration of space robot is shown as Fig.2. We assume that this space robot comprise a three-link manipulator which has three revolution joint and a cubic base for simplification, each link of manipulator is cylindrical structure, and the radius of link r = 0.04m. Here, we only consider the position information of the object, Table I shows the dynamic parameters of robot system.

The task we investigated in this paper is to capture a free-flying object using a space manipulator. We will carry out two learning simulation in this study in

Table 1. Parameters of Space Robot System

	Space Base (Link0)	Space Manipulator		
		Link1	Link2	Link3
Mass(Kg)	200	5	10	7
Length(m)	1	0.25	0.8	0.4
$I_{xx}(Kg \cdot m^2)$	133	0.034	0.032	0.022
$I_{yy}(Kg \cdot m^2)$	133	0.034	0.549	0.149
$I_{zz}(Kg \cdot m^2)$	133	0.016	0.549	0.149

order to verify that the learning control approach for capturing object is valid and efficient. First experiment is to learn tracking trajectory in Cartesian space. Here, we select the motion trajectory of end-effector and object motion trajectory as the input sets of training data. Second experiment is the position trajectory learning in Joint space. For these experiments, the proposed approach is applied to learn operator skill form different angles. Leaning trajectory in Cartesian space is most intuitive, considering the human control hand by viewing hand position with respect to destination in Cartesian space. Learning trajectory in Joint space shows that the learning can be done in Joint space. Moreover, it is more desirable to learn trajectory in Joint space for a kinematically redundant robot to avoid the one-to-many mapping. On the other hand, it successfully avoids the complicated inverse kinematics problem, which is not easy sometimes.

4.2 Simulation Results

In our experiment, we simulate the learning position trajectory in Cartesian space and Joint space, respectively. We will introduce them in the following parts.

It is well known that the end-effector's motion is described in Cartesian space usually. Here, we only consider the position parameters of the object for simplicity, therefore, we assume that the object is moving along rectilineal trajectory and pass through the workspace of the space robot. During collecting data, we assume that the object pass through the workspace of the robot according to different speeds and different directions. We choose ten sets as samples to train the SVMs and four sets for testing the validation of the SVMs. With the trained SVMs, we test the learning performance, the experimental results are satisfactory and important to verify our learning approach for capturing object in space. Fig.4 and Fig.5 show the experimental results. In these figures, the real curves denote the output of SVMs, while dash curves denode real value by human tracking control. At the performance stage, the robot can track and capture the object quickly, and the tracking and capturing strategies are similar to its teacher's.

The tracking trajectory of robot is generally planned in Cartesian space. However, inverse kinematics problem is not avoided because the control process is executed in Joint space for robotic manipulator. However, it is difficult to calculate the inverse kinematics solutions in many cases since inverse kinematics is multi-solutions. Therefore, the robot system must firstly select appropriate solutions in order to satisfy the practical control requirement. Moreover, these problems are great payload for computer on space robot system. Therefore, the proposed method by learning joint trajectory can tackle these complicated problems. Especially, it is very useful to track and capture fast-maneuvering object.

The training idea is to use the joint trajectory as the training samples to train the SVMs. With the trained SVMs, we test the learning performance, the experimental results are satisfactory and important to verify our learning approach for capturing object in space. Fig.6 and Fig.7 show the experimental results. In these figures, the real curves denote the output of SVMs, while dash curves denode real value by human tracking control.

5 Conclusions

We have demonstrated that the SVM is a feasible parameter model for learning tracking skill in different domains. Although we only examined our proposed ideas by simulation in this paper, the training data form the model-based system can reflect the human skill. The SVM approach based on statistical learning theory overcomes the limitations of the general ANNs. This approach is fundamentally helpful in solving problems, such as small sample data and local minima, and therefore, is extremely useful in abstracting human skill for machine learning.

It is an important mission for tracking and capturing the object in space by space robots. The proposed method provides a feasible way to abstract human's tracking skill as a parametric model and transfer this skill to the robot system easily. It will be found useful in various other applications, aside from space robot, such as human action recognition and classification in man-machine interface, and pilot skill learning for unmanned helicopter.

References

1. Lindberg, R. E., Longman, R. W., Zedd, M. F.: Kinematic and Dynamic Properties of an Elbow Manipulator Mounted on a Satellite. Journal of the Astronautical Sciences 38(4) (1990) 397-421
2. Arimoto, S.: Learning Control Theory for Robot Motion. Int. J. Adaptive Contr. Signal Processing 4 (1990) 543-564
3. Moore, K., Dahleh, M., Bhattacharyya, S. P.: Interative Learning Control: A Survey and New Results. Journal of Robot System 9(5) (1992) 563-594
4. Antsaklis, P. J., Guest Editor: Special Issue on Neural Network in Control Systems. IEEE Control System Magazine 10(3) (1990) 3-87
5. Antsaklis, P. J., Guest Editor: Special Issue on Neural Network in Control Systems. IEEE Control System Magazine 12(2) (1992) 8-57
6. Zhang, Y., Wang, J., Xu, Y.: A Dual Neural Network for Bi-criteria Kinematic Control of Redundant Manipulators, IEEE Trans. Robotics and Automation 18(6) (2002) 923-931

7. Yang, J., Xu, Y., Chen, C. S.: Hidden Markov Model Approach to Skill Learning and its Application to Telerobotics. IEEE Trans. Robotics and Automation 10(5) (1994) 612-631
8. Sebald, D. J., Buckle, J. A.: Support Vector Machine Techniques for Nonlinear Equalization. IEEE Trans. Singal Processing 48 (2000) 3217-3226
9. Thawonmas, R., Abe, S.: Feature Reduction Based on Analysis of Fuzzy Regions. Proc. Int. Conf. on Artificial Neural Network Vol. 4 (1995) 2130-2133
10. Smola, A.: General Cost Function for Support Vector Regression. Proc. of the Ninth Australian Conf. on Neural Networks (1998) 79-83
11. Vapnik, V. N.: The Nature of Statistiacal Learning Theory. Springer-Verlag, New York (1995)
12. Cristianini, N., Shawe-Taylor, J.: An Introduction to Support Vector Machines and Other Kernel-based Learning Methods. Cambridge Unviersity Press, Cambridge (2000)
13. Bernhard, S.: Advanced in Kernel Mcthods Support Vector Learning. Cambridge, MA, MIT Press (1998)

Neural Networks for Mobile Robot Navigation: A Survey

An-Min Zou, Zeng-Guang Hou, Si-Yao Fu, and Min Tan

Key Laboratory of Complex Systems and Intelligence Science,
Institute of Automation, The Chinese Academy of Sciences,
P.O. Box 2728, Beijing 100080, China
{anmin.zou, zengguang.hou, siyao.fu, min.tan}@ia.ac.cn

Abstract. Nowadays, mobile robots have attracted more and more attention from researchers due to their extensive applications. Mobile robots need to have the capabilities of autonomy and intelligence, and they pose a challenge to researchers, which is to design algorithms that allow the robots to function autonomously in unstructured, dynamic, partially observable, and uncertain environments [1]. Navigation is the key to the relative technologies of mobile robots and neural networks are widely used in the field of mobile robot navigation due to their properties such as nonlinear mapping, ability to learn from examples, good generalization performance, massively parallel processing, and capability to approximate an arbitrary function given sufficient number of neurons. This paper surveys the developments in the last few years of the neural networks with applications to mobile robot navigation.

1 Introduction

In recent years, mobile robots have attracted more and more attention from researchers since they have been widely used in various fields, such as space exploration, under water survey, industrial and military industries, and service and medical applications, and so on. The robots need to have the capabilities of autonomy and intelligence, and they force the researches to deal with key issues such as uncertainty (in both sensing and action), reliability, and real-time response [2]. Therefore, a key challenge in robotics is to design algorithms that allow the robots to function autonomously in unstructured, dynamic, partially observable, and uncertain environments [1].

The problem of mobile robot navigation, which includes four fundamental matters of mobile robots: map building, localization, path planning, and obstacle avoidance, refers to plan a path with obstacle avoidance to a specified goal and to execute this plan based on sensor readings, and is the key to the robot to perform some designated tasks. Neural networks, motivated by how the human brain works, are increasingly being employed in various fields, including signal processing, pattern recognition, medicine, speech production and recognition, and business. In the last few years, neural networks including feedforward neural network, self-organizing neural network, principal component analysis

J. Wang et al. (Eds.): ISNN 2006, LNCS 3972, pp. 1218–1226, 2006.

(PCA), dynamic neural network, support vector machines (SVM), neuro-fuzzy approach, etc., have been widely used in the field of mobile robot navigation due to their properties such as nonlinear mapping, ability to learn from examples, good generalization performance, massively parallel processing, and capability to approximate an arbitrary function given sufficient number of neurons. The objective of this paper is to present the status of the applications of neural networks to mobile robot navigation. The rest of this paper is organized as follows: The methods of neural networks and hybrid approaches for mobile robot navigation are described in Sections 2 and 3, respectively, and conclusions are given in Section 4.

2 Neural Networks for Mobile Robot Navigation

2.1 Neural Networks for Interpretation of the Sensor Data

Sensors are necessary for a robot to know where it is or how it got there, or to be able to reason about where it has gone. The sensors can be flexible and mobile to measure the distance that wheels have traveled along the ground, to measure inertial changes and external structure in the environment. The sensors may be roughly divided into two classes: internal state sensors, such as accelerometers, gyroscope, which provide the internal information about the robot's movements, and external state sensors, such as laser, infrared sensors, sonar, and visual sensors, which provide the external information about the environment. The data from internal state sensors may provide position estimates of the robot in a 2D space; however, cumulative error is inevitable.The data from external state sensors may be used to directly recognize a place or a situation, or be converted to information in a map of the environment. The laser, infrared, and sonar sensors can provide distant and directional information about an object. Visual sensors can obtain rich-information of the environment that can be very expensive to process. In most cases, the sensor readings are imprecise and unreliable due to the noises. Therefore, it is important for the mobile robot navigation to process the sensor data with noises. Since neural networks have many processing nodes, each with primarily local connections, they may provide some degree of robustness or fault tolerance for interpretation of the sensor data.

Feedforward multi-layer perception neural network, which is trained by the back-propagation algorithm, has been used for classification, recognition and function approximation. Kim *et al.* proposed an approach to build environmental map where the nonlinearity error of range data from infrared sensors was corrected by using a feedforward neural network [3]. Thrun has used a feedforward neural network to "translate" the sonar readings of sonar sensors into occupancy values of each grid cell for building metric maps [4]. Meng and Kak presented a NEURO-NAV system for mobile robot navigation [5]. In the NEURO-NAV, a feedforward neural network, which is driven by the cells of the Hough transformation of the corridor guidelines in the camera image, is employed to obtain the approximate relative angles between the heading direction of the robot and the orientation of the hallway in order to drive the robot to move in the middle of the hallway [5]. In [6], a neural network based camera calibration method was presented for the global localization of mobile robots using monocular vision. Since every type of sensors have their own limitations for collecting the environmental information of a robot, sensor fusion is necessary for the mobile robot navigation. A sonar and infrared sensors fusion algorithm based on a feedforward neural network to obtain reliable data is studied in [7].

The self-organizing Kohonen neural network is known for its ability to perform classification, recognition, data compression and association in an unsupervised manner [8]. The self-organizing Kohonen neural networks are employed to recognition the landmarks using the measurements from laser sensors in order to provide coordinates of the landmarks for triangulation in [9]. Janet *et al.* proposed a global localization algorithm using self-organizing Kohonen neural networks [10]. By using the self-organizing Kohonen neural networks, the robot can determine the room where it is according to the sonar sensor data.

PCA, which has been applied to data compression, pattern recognition, and so on, is a statistical technique and is well known as one of the effective methods to extract the principal features from high-dimension data and decrease the dimension of the data. Crowley *et al.* presented an approach to estimate position of a mobile robot based on PCA of laser ranger sensor data [11]. Vlassis *et al.* proposed an approach for mobile robot localization where PCA was employed to decrease the dimensions of sonar sensor data [12]. PCA has been used to extract features of images for mobile robot localization in [13], [14]. Though PCA is an appropriate model for data generated by a Gaussian distribution, or data best described by a second order correlation; however, the distribution of natural images is highly non-Gaussian. Therefore, kernel PCA is used to extract features from image for mobile robot localization [15]. In the work reported in [15], kernel PCA has a higher localization rate than that of conventional PCA, whereas, the conventional PCA is faster than the kernel PCA.

SVM, which was proposed by Vapnik, is based on the statistical learning theory [16]. In [17], seat numbers were used as landmarks for mobile robot localization because the seat number could be employed to distinguish the landmarks, and SVM was adopted to segment number regions from images.

Hopfield neural network can be used as associative memory or to solve optimization problems [8]. In [18], [19], an improved neural network based on Lagrange programming method was presented for hierarchical optimization of nonlinear large-scale systems. Djekoune and Achour proposed a localization algorithm using stereo vision where the correspondence problem for a set of segments extracted from a pair of stereo images is formulated as minimization of a cost function that is performed by means of a two-dimensional Hopfield neural network [20].

2.2 Neural Networks for Obstacle Avoidance

In the environment, there are always static and non-static obstacles. Therefore, the robots need to autonomously navigate themselves in the environment with obstacle avoidance. The neural networks for obstacle avoidance of mobile robots should take the sensor data from the environment as the inputs, and output the direction for the robot to proceed. Fujii *et al.* proposed a multilayered model for collision avoidance of a mobile robot through reinforcement learning [21]. Silva *et al.* presented the MONODA (modular network for obstacle detection and avoidance) architecture for obstacle detection and avoidance of a mobile robot in an unknown environment [22]. This model consists of four modules that are three-layered feedforward neural networks (each detects the probability of obstacle in one direction of the robot). Ishii *et al.* developed an obstacle avoidance method for underwater vehicle based on self-organizing Kohonen neural networks [23]. Gaudiano and Chang studied an approach to avoid obstacle using a neural network model of classical and operant conditioning based on Grossberg's conditioning circuit [24] [25].

2.3 Neural Networks for Path Planning

The path planning problem, which may consist of two subproblems, path generation and path tracking, refers to determining a path between an initial pose of the robot and a final pose such that the robot does not collide with any obstacles in the environments and that the planned motion is consistent with the kinematic constraints of the vehicle. The existing path planning methods include A* algorithm [26], potential fields [27], BUG2 algorithm [28], and methods using intelligent control technique. The A* algorithm assumes that paths are made of a series of points in the free space. Each segments given a value that is the cost of that particular portion of the path. The drawback of the A* algorithm is that the generated path is made of a series of connected straight lines, which makes its curvature radius discontinuous, resulting in wheel slippage. Potential field methods were first introduced by Khatib [27]. The drawbacks of this approach are that convergence is not guaranteed and it requires very heavy calculation power. The A* algorithm and potential fields are employed in the static environment and assume that the map of the environment is known *a prior*. BUG2 algorithm is one of behavior based techniques, which divides the goal-seeking task into several dependent subtasks. Though it is simple and efficient, it does not always generate the optimal path. The methods using intelligent control such as neural networks, neuro-fuzzy, do not plan a global path for mobile robots and can be employed in an unknown environment. The input pattern of the neural network for path planning of mobile robots should consider the following data: robot's actual position and velocities; robot's previous positions and velocities; target position and sensor data, and then output the commands to drive the robot to follow a path towards the target with obstacle avoidance according to these data.

Kozakiewicz and Ejiri have used a human expert to train a feedforward neural network that reads inputs from a camera and outputs the appropriate commands actuators [29]. In [30], Sfeir *et al.* developed a path generation technique for mobile robot using memory neuron network proposed by Sastry *et al.* [31]. The memory neuron network is a feedforward neural network that uses memory neurons. A memory neuron is a combination of a classic perception and unit delays, which gives it memory abilities. Glasius, Komoda, and Gielen proposed a Hopfield-type neural network for dynamic trajectory formation without learning [32]. Fierro and Lewis studied a control structure that integrated a kinematic controller and a feedforward neural network computed-torque controller for non-holonomic mobile robot, and the neural network weights are tuned on-line, with no "off-line learning phase" needed [33], [34], [35]. Yang and Meng proposed a biologically inspired neural network approach for motion planning of mobile robots [36], [37], [38]. This model is inspired by Hodgkin and Huxley's membrane model [39] for a biological neural system and Grossberg's shutting model [40]. The proposed model for motion planning of mobile robots has the following properties: without any prior knowledge of the environment, without explicitly searching over the free workspace or the collision path, and without any learning procedure.

However, neural networks have also some drawbacks. For instance, a neural network can not explain its results explicitly and its learning is usually time-consuming. Further, the learning algorithm may not be able to guarantee the convergence to an optimal solution [41].

3 Hybrid Approaches for Mobile Robot Navigation

Though neural networks have some properties that are important for the mobile robot navigation, knowledge representation and extraction are difficult. Fuzzy systems are able to treat uncertain and imprecise information; they make use of knowledge in form of linguistic rules. Their main drawback is lack of systematic methodology for their design. The technology that combines or fuses the neural network with the fuzzy reasoning is being watched some very interesting architectures [42]. Several fuzzy neural networks have been presented and used for mobile robot navigation successfully [43-45, 47, 49-52].

Godjevac and Steele proposed a neuro-fuzzy controller based on Takagi-Sugeno design and a radial basis function for obstacle avoidance and wall following of a mobile robot [43]. Marichal *et al.* presented a neuro-fuzzy approach to guide a mobile robot with obstacle avoidance [44]. The proposed neuro-fuzzy strategy, which consists of a three-layer neural network along with a competitive learning algorithm, is able to extract the fuzzy rules and the membership functions according to the information provided by a set of trajectories that are obtained from a human guidance. Er and Deng studied a hybrid learning approach for obstacle avoidance of a mobile robot [45]. In [45], firstly, a neuro-fuzzy controller is developed from a pre-wired or innate controller based on supervised learning in a simulation environment. The fuzzy inference system has been constructed based on the Generalized Dynamic Fuzzy Neural Networks learning algorithm proposed by Wu and Er *et al.* [46], whereby structure identification and parameters estimation are performed automatically and simultaneously. Secondly, the controller is implemented on a real robot after the learning phase. A reinforcement learning algorithm based on the Fuzzy Actor-critic learning algorithm is employed so that the system can re-adapt to a new environment without human intervention.

Fuzzy Adaptive Resonance Theory (ART) and fuzzy ARTMAP were proposed by Carpenter and Grossberg *et al.* [47], [48]. Fuzzy ART is capable of learning stable recognition categories in response to both analog and binary input patterns, and fuzzy ARTMAP can rapidly learn stable categorical mapping between analog or binary input and output vectors. Araujo has used fuzzy ART neural network for on line map building from actual sensor data [49]. Later, this work has been extended. Prune-able Fuzzy ART neural network, which included the ability to selectively remove recognition categories, was introduced to build map of mobile robot in unknown environments [50]. The fuzzy ART based approach for map building of mobile robots has the following characteristics [50]: (1) Self-organization from perceived sensor data; (2) Multi-functionality for map building, motion planning; (3) Updatability: incremental, on-line update by learning separately each sensor data point, thus make the model available as soon as possible; (4) Compact geometric representation with small data requirements; (5) Low computational costs; (6) Possible application to higher dimensional space without adversely impacting on data size and complexity. Streilein *et al.* presented an approach to sonar-based object recognition using a fuzzy ARTMAP neural network for the mobile robot localization [51]. Azouaoui *et al.* proposed an approach for obstacle avoidance of mobile robot using fuzzy ARTMAP neural network [52]. This approach can provide robots with capability, after learning based on the supervised fast stable learning: Simplified fuzzy ARTMAP, to determine and use the rule allowing the robots to avoid collision.

4 Conclusions

In this paper, we have given a brief discussion on mobile robot navigation using neural networks. Although a great deal of progress has been made in the field of mobile robot navigation using neural networks, we have to go a long way to make the robot to have the capabilities of intelligence and autonomy truly, which will be possible when the neural hardware evolves and we get a better understanding of how the human brain works. In the mean time, perhaps the best approach is the hybrid approach that combines neural networks with other artificial intelligent algorithms such as fuzzy logic, knowledge-based systems and genetic algorithms.

Acknowledgements

This research has been supported in part by the National Natural Science Foundation of China (Grant Nos. 60205004, 50475179 and 60334020), the National Basic Research Program (973) of China (Grant No. 2002CB312200), the Hi-Tech R&D Program (863) of China (Grant Nos. 2002AA423160 and 2005AA420040), and the Science and Technology New Star Program of Beijing (Grant No. H02082 0780130).

References

1. Sukhatme, G. S., Mataric, M. J.: Robots: Intelligence, Versatility, Adaptivity. Communications of the ACM **45** (3) (2002) 30-32
2. Kortenkamp, D., Bonasso, R. P., Murphy, R. (Eds): Artificial Intelligence and Mobile Robots: Case Studies of Successful Robot Systems, AAAI press (1998)
3. Kim, H. H., Ha, Y. S., Jin, G. G.: A Study on the Environmental Map Building for a Mobile Robot Using Infrared Ranger-finder Sensors. Proceedings of the IEEE International Conference on Intelligent Robots and Systems, **1** (2003) 711-716
4. Thrun, S. B.: Exploration and Model Building in Mobile Robot Domains. Proceedings of the IEEE International Conference on Neural Networks, **1** (1993) 175-180
5. Meng, M., Kak, A. C.: Fast Vision-Guided Mobile Robot Navigation Using Neural Networks. Proceedings of the IEEE International Conference on Systems, Man and Cybernetics, **1** (1992) 111-116
6. Zou, A. , Hou, Z.G., Zhang, L., Tan, M.: A Neural Network-based Camera Calibration Method for Mobile Robot Localization Problems. Proceeding of the Second International Symposium on Neural Networks, Chongqing, China, Springer Lecture Notes in Computer Sciences (LNCS), **3498** (2005) 277-284
7. Barbera, H. M., Skarmeta, A. G., Izquierdo, M. Z., Blaya, J. B.: Neural Networks for Sonar and Infrared Sensors Fusion. Proceedings of the Third International Conference on Information Fusion, **2** (2000) WEB4/18 - WEB4/25
8. Lippman, R.: An Introduction to Computing with Neural Nets. IEEE ASSP Magazine **4** (1987) 4-22
9. Hu, H. S., Gu, D. B.: Landmark-Based Navigation of Mobile Robot in Manufacturing. Proceedings of the IEEE International Conference on Emerging Technologies and Factory Automation, **1** (1999) 114-121
10. Janet, J. A., Gutierrez-Osuna, R., Chase, T. A., White, M., Luo, R. C.: Global Self-localization for Autonomous Mobile Robots Using Self-organizing Kohonen Neural Networks. Proceedings of the IEEE International Conference on Intelligent Robotics and Systems, **3** (1995) 504-509
11. Crowley, J. L., Wallner, F., Schiele, B.: Position Estimation Using Principal Components of Range Data. Proceedings of the IEEE International Conference on Robotics and Automation, **4** (1998) 3121-3128
12. Vlassis, N., Motomura, Y., Krose, B.: Supervised Dimension Reduction of Intrinsically Low-dimensional Data. Neural Computation **14** (1) (2002) 191-215
13. Nayar, S. K., Murase, H., Nene, S. A.: Learning, Positioning, and Tracking Visual Appearance. Proceedings of the IEEE International Conference on Robotics and Automation, **4** (1994) 3237-3244
14. Artac, M., Jogan, M., Leonardis, A.: Mobile Robot Localization Using an Incremental Eigenspace Model. Proceedings of the IEEE International Conference on Robotics and Automation, **1** (2002) 1025-1030
15. Tamimi, H., Zell, A.: Global Visual Localization of Mobile Robots Using Kernel Principal Component Analysis. Proceedings of the IEEE International Conference on Intelligent Robots and Systems, **2** (2004) 1896-1901
16. Vapnik, V.: Statistical Learning Theory. John Willey & Sons (1998)
17. Zou, A. , Hou, Z.G., Tan, M.: Support Vector Machines (SVM) for Color Image Segmentation with Applications to Mobile Robot Localization Problems. Proceeding of the International Conference on Intelligent Computing, Hefei, China, Springer Lecture Notes in Computer Sciences (LNCS), **3645** (2005) 443-452

18. Hou, Z.G.: A Hierarchical Optimization Neural Network for Large-scale Dynamic Systems. Automatica **37** (12) (2001) 1931-1940
19. Hou, Z.G.,Wu, C.P., Bao.: A Neural Network for Hierarchical Optimization of Nonlinear Large-scale Systems. International Journal of Systems Science **29** (2) (1998) 159-166
20. Djekoune, O., Achour, K.: Vision-guided Mobile Robot Navigation Using Neural Network. Proceedings of the 2nd International Symposium on Image and Signal Processing and Analysis, (2001) 355-361
21. Fujii, T., Arai, Y., Asama, H., Endo, I.: Multilayered Reinforcement Learning for Complicated Collision Avoidance Problems. Proceedings of the IEEE International Conference on Robotics and Automation, **3** (1998) 2186-2191
22. Silva, C., Crisostomo, M., Ribeiro, B.: MONODA: A Neural Modular Architecture for Obstacle Avoidance without Knowledge of the Environment. Proceedings of the IEEE-INNS-ENNS International Joint Conference on Neural Networks, **6** (2000) 334-339
23. Ishii, K., Nishida, S., Watanabe, K., Ura, T.: A Collision Avoidance System based on Self-organizing Map and its Application to an Underwater Vehicle. 7th International Conference on Control, Automation, Robotics and Vision, **2** (2002) 602-607
24. Gaudiano, P., Chang, C.: Adaptive Obstacle Avoidance with a Neural Network for Operant Conditioning: Experiments with Real Robots. Proceedings of the IEEE International Symposium on Computational Intelligence in Robotics and Automation, (1997) 13-18
25. Grossberg, S., Levine, D. S.: Neural Dynamics of Attentionally Modulated Pavlovian Conditioning: Blocking, Inter-stimulus Interval, and Secondary Reinforcement. Applied Optics **26** (1987) 5015-5030
26. Nilsson, N. J.: Principles of Artificial Intelligence. Tioga Publishing Co., Palo Alto, CA (1980)
27. Khatib, O.: Real-time Obstacle Avoidance for Manipulators and Mobile Robots. The International Journal of Robotics Research **5** (1) (1986) 90-98
28. Prusky, A.: Robotique Mobile. La Planification de Trajectoire, Hermes (1996)
29. Kozakiewicz, C., Ejiri, M.: Neural Network Approach to Path Planning for Two Dimensional Robot Motion. International Workshop on Intelligent Robots and Systems, **2** (1991) 818-823
30. Sfeir, J., Kanaan, H. Y., Saad, M.: A Neural Network Based Path Generation Technique for Mobile Robots. Proceedings of the IEEE International Conference on Mechatronics, (2004) 176-181
31. Sastry, P. S., Santharam, G., Unnikrishnan, K. P.: Memory Neuron Networks for Identification and Control of Dynamic Systems. IEEE Transactions on Neural Networks, **5** (2) (1994) 306-319
32. Glasius, R., Komoda, A., Gielen, S. C. A. M.: Neural Network Dynamics for Path Planning and Obstacle Avoidance. Neural Networks **8** (1) (1995) 125-133
33. Fierro, R., Lewis, F. L.: Control of a Nonholonomic Mobile Robot Using Neural Networks. Proceedings of the IEEE International Symposium on Intelligent Control, (1995) 415-421
34. Fierro, R., Lewis, F. L.: Practical Point Stabilization of a Nonholonomic Mobile Robot Using Neural Networks. Proceedings of the 35th IEEE on Decision and Control, **2** (1996) 1722-1727
35. Fierro, R., Lewis, F. L.: Control of a Nonholonomic Mobile Robot Using Neural Networks. IEEE Transactions on Neural Networks **9** (4) (1998) 589-600
36. Yang, S. X., Meng, M.: An Efficient Neural Network Approach to Dynamic Robot Motion Planning. Neural Networks **13** (2) (2000) 143-148

37. Yang, S. X., Meng, M.: An Efficient Neural Network Method for Real-time Motion Planning with Safety Consideration. Robotics and Autonomous Systems **32** (2-3) (2000) 115-128
38. Yang, S. X., Meng, M. Q.-H.: Real-time Collision-free Motion Planning of a Mobile Robot Using a Neural Dynamic-based Approach. IEEE Transactions on Neural Networks **14** (6) (2003) 1541-1552
39. Hodgkin A. L., Huxley, A. F.: A Quantitative Description of Membrane Current and its Application to Conduction and Excitation in Nerve. Journey of Physiology (London) **117** (1952) 500-544
40. Grossberg, S.: Nonlinear Neural Networks: Principal, Mechanisms, and Architecture. Neural Networks **1** (1988) 17-61
41. Huang, S. H.: Artificial Neural Networks and its Manufacturing Application: Part 1. Online slides are available at : www.min.uc.edu/icams/resourses/ANN/ANNMaul.ppt
42. Jang, J.-S. R.: ANFIS: Adaptive-network-based Fuzzy Inference System. IEEE Transactions on Systems, Man, and Cybernetics **23** (3)(1993) 665-685
43. Godjevac, J., Steele, N.: Neuro-fuzzy Control of a Mobile Robot. Neurocomputing **28** (1999) 127-143
44. Marichal, G. N., Acosta, L., Moreno, L., Mendez, J. A., Rodrigo, J. J., Sigut, M.: Obstacle Avoidance for a Mobile Robot: A Neuro-fuzzy Approach. Fuzzy Sets and Systems **124** (2) (2001) 171-179
45. Er, M. J., Deng, C.: Obstacle Avoidance of a Mobile Robot Using Hybrid Learning Approach. IEEE Transactions on Industrial Electronics **52** (3) (2005) 898-905
46. Wu, S., Er, M. J., Gao, Y.: A Fast Approach for Automatic Generation of Fuzzy Rules by Generalized Dynamic Fuzzy Neural Networks. IEEE Transactions on Fuzzy Systems **9** (4) (2001) 578-594
47. Carpenter, G. A., Grossberg, S., Rosen, D. B.: Fuzzy ART: Fast Stable Learning and Categorization of Analog Patterns by an Adaptive Resonance System. Neural Networks **4** (1991) 759-771
48. Carpenter, G. A., Grossberg, S., Markuzon, N., Reynolds, J. H., Rosen, D. B.: Fuzzy ARTMAP: A Neural Network Architecture for Incremental Supervised Learning of Analog Multidimensional Maps. IEEE Transactions on Neural Networks **3** (1992) 698-713
49. Araujo, R., de Almeida, A. T.: Learning Sensor-based Navigation of a Real Mobile Robot in Unknown Worlds. IEEE Transactions on Systems, Man and Cybernetics, Part B 29 (2) (1999) 164-178
50. Araujo, R., Lourenco, D., Ferreira, G.: Integrating Laser and Infrared Sensors for Learning Dynamic Self-organizing World Maps. International Conference on Multisensor Fusion and Integration for Intelligent Systems, (2001) 293-298
51. Streilein, W. W., Gaudiano, P., Carpenter, G. A.: A Neural Network for Object Recognition through Sonar on a Mobile Robot. Proceedings of the IEEE ISIC/CIRA/ISAS Joint Conference, (1998) 271-276
52. Azouaoui, O., Ouaaz, M., Chohra, A., Farah, A., Achour, K.: Fuzzy ARTMap Neural Network Based Collision Avoidance Approach for Autonomous Robots Systems. Proceedings of the Second International Workshop on Robot Motion and Control, (2001) 285-290

Fault Diagnosis for Mobile Robots with Imperfect Models Based on Particle Filter and Neural Network

Zhuohua Duan[1,2], Zixing Cai[1], and Jinxia Yu[1,3]

[1] College of Information Science and Engineering, Central South University, Changsha 410083, Hunan, China
[2] Department of Computer, School of Information Engineering, Shaoguan University, Shaoguan 512003, Guangdong, China
duanzhuohua@163.com
[3] Department of Computer Science & Technology, Henan Polytechnic University, Jiaozuo 454003, Henan, China

Abstract. Fault detection and diagnosis (FDD) are increasingly important for wheeled mobile robots (WMRs), especially those in unknown environments such as planetary exploration. There are many kinds of fault diagnosis methods available for mobile robots, including multiple model-based approaches, particle filter based approaches, sensor fusion based approaches. Currently, all of these methods are designed for complete models. However, completely modeling a system is difficult, even impossible. In this paper, particle filter and neural network are integrated to diagnose complex systems with imperfect models. Two features are extracted from particles: the sum of sample weights, and the maximal a posteriori probability. These features are further feed to a neural network to decide whether the estimation given by the particle filter is credible or not. An incredible estimation indicates that the true state isn't included in the state space, i.e. it is a novel state (or an unknown fault). This method preserves the merits of particle filter and can diagnose known faults as well as detect unknown faults. It is testified on a real mobile robot.

1 Introduction

Wheeled mobile robots (WMRs) have been employed for applications including: military operations, surveillance, security, mining operations, and planetary exploration. In these application domains, WMRs operate in unknown environments where human intervention is expensive, slow, unreliable, or impossible [1]. There are many kinds of fault diagnosis methods available for mobile robots, including multiple model-based approaches, particle filter based approaches and sensor fusion based approaches [1-5]. Currently, all of these methods are designed for complete models. However, completely modeling a system is difficult, even impossible.

Some researches have developed several kinds of approaches for fault diagnosis with imperfect models. These include discrete model-based approach used in Livingstone [6], and hybrid model of discriminative and generative classifiers proposed by Symth [7]. The former is further extended to hybrid system with discrete and continuous models [8]. Both methods introduce an unknown fault for every component in the

J. Wang et al. (Eds.): ISNN 2006, LNCS 3972, pp. 1227 – 1232, 2006.

system. This may dramatically increase the state space size as well as the number of candidate diagnosis. Consequently, it increases the time expenditure and prevents its deep application to real time fault diagnosis for wheeled mobile robots.

In this paper, particle filter and neural network are integrated to deal with the problem of fault diagnosis for complex systems with imperfect models. Two parameters are extracted from sample-based expression for a posteriori probability density: the sum of sample weights, and the maximal a posteriori probability. A multi-layer perceptron is used to decide whether the estimation given by the particle filter is credible or not. An incredible estimation indicates the true state is not included in the state space, i.e. it is an unknown fault. This method preserves the merits of particle filter and can diagnose known faults as well as detect unknown faults. It is testified on a real mobile robot.

The rest of the paper is organized as follows: In section II, we briefly introduce the basic method of particle filter based fault diagnosis. Section III, we propose the integrated method for fault diagnosis of mobile robots with imperfect models. Experiments and analysis are given in section IV. Finally, the result is presented in section V.

2 Particle Filter Based Fault Diagnosis

The main idea for using PF as a fault diagnosis method is described as follows:

Let S represent the finite set of discrete fault and operational modes in the system; s_t represents the state of the system to be diagnosed at time t and $s_t \in S$; $\{s_t\}$ the discrete, first order Markov chain representing the evolution of the state over time. The problem of state monitoring and fault diagnosis consists of providing a belief (a distribution over the state set S) at each time step as it evolves according to the following transition model:

$$p(s_t=j|s_{t-1}=i) ,i,j\in S \tag{1}$$

Each of the discrete fault and operational modes changes the dynamics of the system. Let x_t denote multivariate continuous state of the system at time t. The non-linear conditional state transition models are denoted by $p(x_t|x_{t-1},s_t)$. The state of the system is observed through a sequence of measurements, $\{z_t\}$, based on the measurement model $p(z_t|x_t,s_t)$.

The problem of state monitoring and fault diagnosis consists of estimating the marginal distribution $p(s_t|z_{1..t})$ of the posterior distribution $p(x_t,s_t|z_{1..t})$. A recursive estimate of this posterior distribution may be obtained using the Bayes filter:

$$p(x_t,s_t \mid z_{1..t}) = \eta_t p(z_t \mid x_t,s_t) \int \sum_{s_{t-1}} p(x_t,s_t \mid x_{t-1},s_{t-1}) dx_{t-1} \tag{2}$$

There is no closed-form solution for this recursion. PFs appropriate the posterior with a set of N fully instantiated state samples or particles $\{(s_t^{[1]}, x_t^{[1]}),\ldots,(s_t^{[N]}, x_t^{[N]})\}$, and importance weights $\{w_t^{[i]}\}$:

$$\hat{P}_N(x_t,s_t \mid z_{1..t}) = \sum_{i=1}^{N} w_t^{[i]} \delta_{x_t^{[i]},s_t^{[i]}}(x_t,s_t) \tag{3}$$

where δ(.) denotes the Dirac delta function. The appropriation in equation (3) approaches the true posterior density as N→∞. Because it is difficult to draw samples from the true posterior, samples are drawn from a more tractable distribution *q(.)*, which is called the proposal (or importance) distribution. The most widely used proposal distribution is the transition distribution,

$$q(.)=p(x_t, s_t|x_{t-1}, s_{t-1}) \tag{4}$$

The importance weights are used to account for the discrepancy between the proposal distribution *q(.)* and the true distribution $p(x_t,s_t|z_{1..t})$. When the proposal distribution is given by equation (4), the importance weight of sample ($s_t^{[i]}$, $x_t^{[i]}$) is

$$w_t^{[i]} = p(z_t \mid x_t^{[i]}, s_t^{[i]}) \tag{5}$$

3 Fault Diagnosis for Mobile Robot with Imperfect Models Based on Particle Filter and Neural Networks

3.1 Two Typical Belief States

Particle filter infers according to the sample based expression of a posteriori probability density. The typical method is maximum a posteriori (MAP) estimation. Sometimes, the MAP estimation is incredible. For example, Fig. 1 shows two typical belief states obtained with the general particle filter applying to the models described in Section III. In Fig. 1(a), all states seem to be drawn from a uniform distribution and the belief state 'S20' happens to have the maximum belief, 0.052. Obviously, this kind of estimation is incredible. Another kind of belief state is shown in Fig. 1(b), in which all the samples are fallen into the state 'S1'.

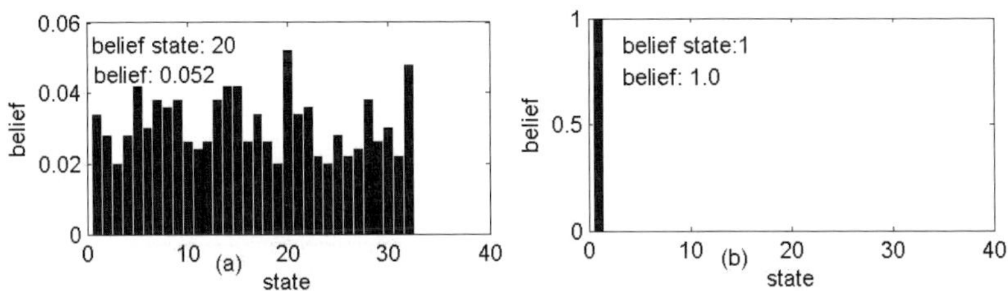

Fig. 1. Two typical belief state obtain from the general particle filter

In the first case, one would rather believe that none of the 32 states is believable, i.e. the true state is an unknown state that has not been included into the state space.

In fact, the former type of distribution shows that none of the states is fully superior to others, i.e. none of these state is true state. Further more, in the first case, the sum of the un-normalized weight of samples is negligible.

Based on this, we extracted two features from the weighted samples of particle filters during each time step *t*: the sum of sample weights (denoted with W_t), and the maximal a posteriori probability (denoted with B_t).

3.2 Fault Diagnosis for Mobile Robots with Imperfect Models Based on Particle Filter and Neural Network

The two features, W_t and B_t, contain the information of particles. Based on this, a multi layer perceptron (MLP) is employed to determine whether the estimation given by PF is credible or not.

The algorithm is shown as follows.

Algorithm 1. Fault diagnosis for mobile robots with imperfect models based on particle filter and neural network (PFNN)

1. **Initialize samples**
2. **for** each time step t **do**
3. **for i=1:N**
4. **prediction:**
5. $s_t^i \sim p(s_t \mid s_{t-1}^i)$
6. $x_t^i \sim p(x_t \mid x_{t-1}^i, s_t^i)$
7. **update:** $w_t^i = p(z_t \mid x_t^i, s_t^i)$
8. **end for**
9. $W_t = \sum_{i=1}^{N} w_t^i$
10. normalizing the weights: $\sum_{i=1}^{N} w_t^i = 1$
11. resampling
12. **estimation:**
13. $\hat{s}_t = \arg\max_{s_t} \hat{P}_N(s_t \mid z_{1..t})$
14. $B_t = \max \hat{P}_N(s_t \mid z_{1..t})$
15. credibility=MLP(W_t , B_t)
16. if credibility='incredible' $\hat{s}_t$ ='unknown fault'
17. **end for**

In algorithm 1, two parameters are collected based on particles. The first is the sum of the un-normalized weights of samples in step 9, and the second is the belief of the belief state in step 14. In step 15, a multi layer perceptron classifies the two features into two categories: 'credible' and 'incredible'. The MLP is trained with error back propagation algorithm. And it is trained off line with labeled training data.

4 Comparison of GPF and PFNN

The experiment platform is the robot shown in Fig. 2(a) [9]. We diagnose 32 kinds of faults of internal sensors, including four wheel-encoders and a gyroscope. The universal fault space is denoted with $S=\{S_i \mid i=1, 2, \ldots, 32\}$.

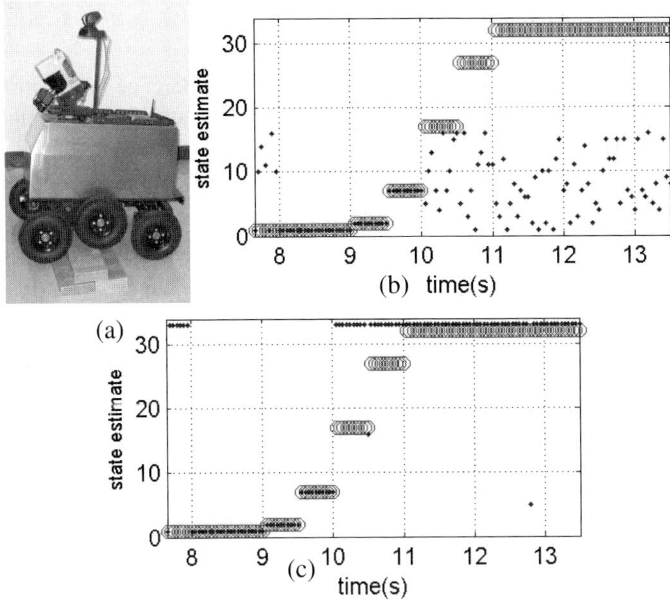

Fig. 2. The robot and state estimation with GPF and PFNN

We compared our algorithm (PFNN) with the general particle filter (GPF). In the experiment, the speeds of left and right wheels are set as +0.08rad/s and –0.08rad/s respectively, i.e. the robot is turning right, during 7.65-13.5s. The left front, left back, right front, right back and gyroscope fail at 9s, 9.5s, 10s, 10.5s and 11s respectively. So the 'true' states are 'S1' (7.65-9s), 'S2' (9-9.5s), 'S7' (9.5-10s), 'S17' (10-10.5s), 'S27' (10.5-11s), and 'S32' (11-13.5s).

Let S'={S1, S2, S3, S4, S5, S6, S7, S8, S9, S10, S11, S12, S13, S14, S15, S16} denote the actual fault space that we want to diagnose, i.e. we assume that at most two components fail simultaneously. The state estimation results of the GPF and PFNN are shown in fig. 2. Fig. 2 (b) shows that GPF misdiagnose when three or more components go wrong. It is understandable because any faults of three or more components are not included in the current state space S'. Conversely, fig. 2 (c) shows that the PFNN can classify most of these faults as unknown faults.

5 Conclusions

Particle filter and neural network are integrated for fault diagnosis of complex systems with imperfect models. It extracts two key features from the samples of particle filters, i.e. the sum of the un-normalized weights of samples and the maximal a posteriori probability. Then, an additional decision is made in the estimation step with a multi layer perceptron: when none of the state is probably to be true state, the 'unknown fault' state is assigned. The method is testified on a real mobile robot and the primary results are promising.

Acknowledgement

This work is partially supported by the National Natural Science Foundation of China (No. 60234030, 60404021) and National Foundation Project (No. A1420060159).

References

1. Verma, V., Gordon, G., Simmons, R., Thrun, S.: Real-time Fault Diagnosis [Robot Fault Diagnosis]. IEEE Robotics & Automation Magazine 11(2) (2004) 56-66
2. Roumeliotis, S. I., Sukhatme, G. S., Bekey, G. A.: Sensor Fault Detection and Identification in a Mobile Robot. IEEE/RSJ Int'l Conf. on Intelligent Robots and Systems. Victoria, BC, Canada (1998) 1383-1388
3. Goel, P., Dedeoglu, G., Roumeliotis, S. I., Sukhatme, G. S.: Fault Detection and Identification in a Mobile Robot Using Multiple Model Estimation and Neural Network. IEEE Int'l Conf. on Robotics & Automation. San Francisco, CA, USA (2000) 2302-2309
4. Hashimoto, M., Kawashima, H., Nakagami, T., Oba, F.: Sensor Fault Detection and Identification in Dead-Reckoning System of Mobile Robot: Interacting Multiple Model Approach. Int'l Conf. on Intelligent Robots and Systems. Maui, HI (2001) 1321-1326
5. Hashimoto, M., Kawashima, H., Oba, F.: A Multi-Model Based Fault Detection and Diagnosis of Internal Sensor for Mobile Robot. IEEE Int'l Conf. on Intelligent Robots and Systems. Las Vegas, NV, United States (2003) 3787-3792
6. Williams, B., Nayak, P.: A Model-based Approach to Reactive Self Configuring Systems. Proceedings of the 13th National Conference on Artificial Intelligence (AAAI-96). Vol. 2. Portland, OR, USA (1996) 971-978
7. Smyth, P.: Markov Monitoring With Unknown States. IEEE Journal on Selected Areas in Communications 12(9) (1994) 1600-1612
8. Hofbaur, M. W., Williams, B. C.: Hybrid Diagnosis with Unknown Behavioral Modes. in: Proceedings of the 13th International Workshop on Principles of Diagnosis (DX02) (2002)
9. Cai, Z. X., Zou, X. B., Wang, L., Duan, Z. H., Yu, J. X.: A Research on Mobile Robot Navigation Control in Unknown Environment: Objectives, Design and Experiences, Proceedings of Korea-Sino Symposium on Intelligent Systems. Busan, Korea (2004) 57-63

Adaptive Neural Network Path Tracking of Unmanned Ground Vehicle

Xiaohong Liao[1], Zhao Sun[1], Liguo Weng[1], Bin Li[1], Yongduan Song[1], and Yao Li[2]

[1] Department of Electrical Engineering, North Carolina A&T State University, Greensboro, NC 27411, USA
songyd@ncat.edu

[2] Department of Electrical and Computer Engineering, University of Maryland, College Park, MD 20742, USA
yaoli@umd.edu

Abstract. Unmanned ground vehicles (UGVs) play an increasingly important role in future space exploration and battlefield. This work is concerned with the automatic path tracking control of UGVs. By using the structure properties of the system, neuro-adaptive control algorithms are developed for high precision tracking without involving complex design procedures – the proposed control scheme only demands partial information of the system, no detail description of the system model is needed. Furthermore, uncertain effects such as external disturbance and uncertain parameters can easily be handled. In addition, all the internal signals are uniformly bounded and the control torque is smooth anywhere.

1 Modeling and Problem Formulation

As part of the effort in developing autonomous and cooperative control of multiple Unmanned Ground Vehicles (UGVs), this work explores the path tracking control problem of three-wheel mobile vehicle, as shown in Figure 1. While there exist various control schemes for the vehicle in the literature [1]-[13], our special interest lies in developing simple yet effective control algorithms for such vehicle.

We begin with the following dynamic equations, which are derived with the non-holonomic constraints [12]-[13],

$$M(q)\ddot{q}+V_m(q,\dot{q})\dot{q}+F(\dot{q})+G(q)+\tau_d=B(q)\tau+A^T(q)\lambda \tag{1}$$

where $q=\begin{pmatrix}x & y & \theta\end{pmatrix}^T$ is the generalized coordinates, $\tau=\begin{pmatrix}\tau_r & \tau_l\end{pmatrix}$ is a torque acting on the right wheel and left wheel, $\lambda\in R$ is the vector of constraint forces, $\tau_d\in R^3$ represents the unknown bounded disturbances including unmodeled dynamics, $F(\dot{q})\in R^3$ denotes the surface friction, $G(q)\in R^3$ is the gravitation vector, $M\in R^{3\times3}$ is a symmetric and positive definite inertia matrix, $V_m(q,\dot{q})\in R^{3\times3}$ is the centripetal and coriolis matrix, $B\in R^{3\times2}$ is an input transformation matrix defined by

J. Wang et al. (Eds.): ISNN 2006, LNCS 3972, pp. 1233–1238, 2006.

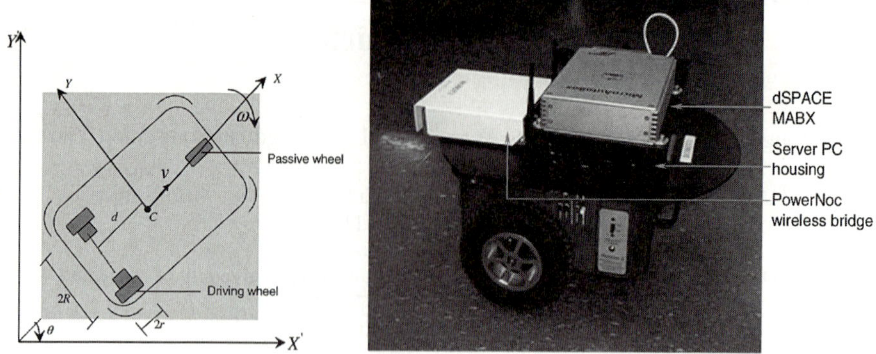

Fig. 1. Three-Wheel Mobile Vehicle under Consideration

$$B = \frac{1}{r}\begin{pmatrix} \cos\theta & \cos\theta \\ \sin\theta & \sin\theta \\ R & -R \end{pmatrix} \tag{2}$$

and $A(q) \in R^3$ is the matrix associated with the constraints given by

$$A(q) = [-\sin\theta \quad \cos\theta \quad -d]^T \tag{3}$$

Also, we have

$$\dot{q} = \begin{pmatrix} \cos\theta & -d\sin\theta \\ \sin\theta & d\cos\theta \\ 0 & 1 \end{pmatrix}\begin{bmatrix} v \\ \omega \end{bmatrix} \triangleq J\begin{bmatrix} v \\ \omega \end{bmatrix} \tag{4}$$

By proper manipulation, the following complete dynamic equation can be derived,

$$\tilde{M}\begin{pmatrix} \dot{v} \\ \dot{\omega} \end{pmatrix} + \tilde{F}(\cdot) = \tilde{B}\tau \quad \text{and} \quad \begin{pmatrix} \dot{x} \\ \dot{y} \end{pmatrix} = P\begin{pmatrix} v \\ \dot{\theta} \end{pmatrix} \tag{5}$$

where $\tilde{B} = J^T B$ is invertible, $\tilde{M} = (J^T MJ)$ is a symmetric and positive definite inertia matrix, and

$$\tilde{F}(\cdot) = \left\{ J^T [M\dot{J} + V_m J]\begin{bmatrix} v \\ \omega \end{bmatrix} + J^T G + J^T F + J^T \tau_d \right\} \tag{6}$$

where P is invertible given by

$$P(q) = \begin{pmatrix} \cos\theta & -d\sin\theta \\ \sin\theta & d\cos\theta \end{pmatrix} \tag{7}$$

The path control problem of UGV system is stated as: design a deriving torque for both left and right wheels such that the position tracking error vector $E = \left(e_x, e_y\right)^T$

and velocity vector $\dot{E} = \left(\dot{e}_x, \dot{e}_y\right)^T$ converge to a small compact set $\vartheta(E, \dot{E}) = \left\{E, \dot{E} / \|E\| \le p_1, \|\dot{E}\| \le p_2\right\}$ as time increases, where p_1 and p_2 are some constants associated with tracking precision, and the tracking errors e_x, e_y defined by $e_x = x - x_r$, $e_y = y - y_r$, where x_r, y_r are the desired trajectory. This is addressed in next section.

2 Neuro-Adaptive Path Tracking Control

Define the following variables

$$S = \begin{pmatrix} s_x \\ s_y \end{pmatrix} = \begin{pmatrix} \dot{e}_x + \beta_x e_x \\ \dot{e}_y + \beta_y e_y \end{pmatrix} \tag{8}$$

We then can express (7) in terms of S as follows,

$$\dot{S} = P\tilde{M}^{-1}\tilde{B}\tau + \tilde{F}_s(\cdot) \tag{9}$$

where

$$\tilde{F}_s(\cdot) = -P\tilde{M}^{-1}\tilde{F}(\cdot) + q_s + \dot{P}\begin{pmatrix} v \\ \omega \end{pmatrix} \tag{10}$$

with $q_s = -\begin{pmatrix} \ddot{x}_r \\ \ddot{y}_r \end{pmatrix} + \begin{pmatrix} \beta_x \dot{e}_x \\ \beta_y \dot{e}_y \end{pmatrix}$ .The proposed path tracking control is

$$\tau = (\tilde{B})^{-1} P^T (-KS + u_c) \tag{11}$$

where $K = diag(k_1, k_2)$ consists of two positive constants k_1, k_2, u_c is the compensating signal determined adaptively by algorithms to be specified later. The main idea behind the proposed control is to use a “PD like” control unit to achieve robustness while maintaining simplicity. Since PD control alone is not adaptive, an additional compensating signal is included in the strategy to make the whole system adaptive.

To complete the control design, we examine the closed-loop error dynamic equation derived from (9) and (11)

$$\dot{S} = Q(-KS + u_c) + \tilde{F}_s(\cdot) \tag{12}$$

where $Q = P\tilde{M}^{-1}P^T$. Now we define the lumped nonlinear vector

$$L = Q^{-1}\tilde{F}_s(\cdot) + \frac{1}{2}\frac{d}{dt}Q^{-1}S \tag{13}$$

Note that the quantity L represents the lumped nonlinearity and uncertainty of the system, which is the main source of complexity in the controller design and implementation. For this reason, our controller will not involve directly using L. From neural network theory, the lumped nonlinear function L can be approximated by a neural network with bounded approximation error as follows,

$$L = W^T \phi + \varepsilon \quad \|\varepsilon\| \le C_d < \infty \tag{14}$$

where $W \in R^{n_1 \times n}$ represents the ideal (optimal) NN weights and $\phi \in R^{n_1}$ is the basis function (n_1 is the number of the neurons), ε is the reconstruction error and C_d is an unknown constant. Now we choose the Lypunov candidate function

$$V = \frac{1}{2} S^T Q^{-1} S + \frac{1}{2} tr\left\{(W - \hat{W})^T (W - \hat{W})\right\} + \frac{1}{2}(C_d - \hat{C}_d)^2 \tag{15}$$

Which has the following time derivative

$$\begin{aligned} \dot{V}_2 &= -(\alpha - \hat{\alpha})\dot{\hat{\alpha}} \\ \dot{V}_1 &= S^T Q^{-1} \dot{S} + \frac{1}{2} S^T \frac{d}{dt} Q^{-1} S = S^T(-KS + u_c) + S^T L(\cdot) \end{aligned} \tag{16}$$

Now we design the compensating signal u_c as

$$u_c = -(\hat{W}^T \phi + \frac{S}{\|S\| + \mu} \hat{C}_d), \quad \dot{\hat{W}} = \phi S^T, \quad \dot{\hat{C}}_d = \frac{\|S\|^2}{\|S\| + \mu} \tag{17}$$

where $\mu > 0$ is chosen such that $\int_0^t \mu d\tau \le \rho < \infty$. It can be shown that

$$\begin{aligned} \dot{V}_1 &= -KS^T S + S^T u_c + S^T L = -KS^T S + S^T u_c + S^T (W^T \phi + \varepsilon) \\ &\le -KS^T S + S^T (W - \hat{W})^T \phi - \frac{\|S\|^2}{\|S\| + \mu} \hat{C}_d + \|S\| C_d \end{aligned}$$

Therefore, we arrive at:

$$\begin{aligned} \dot{V} &\le -KS^T S - tr\left\{(W - \hat{W})^T (\dot{\hat{W}} - \phi S^T)\right\} - (C_d - \hat{C}_d)\left\{\dot{\hat{C}}_d - \frac{\|S\|^2}{\|S\| + \mu}\right\} + \frac{\|S\| \mu C_d}{\|S\| + \mu} \\ &\le -KS^T S + \eta \end{aligned} \tag{18}$$

where $\eta = \mu C_d$. Since $\int_0^t \eta d\tau = \int_0^t \mu C_d d\tau \le \rho C_d < \infty$, we can show that $V(t) \in L_\infty$, which, in view of (15), implies that $S \in L_\infty$, $\hat{W} \in L_\infty$, $\hat{C}_d \in L_\infty$, which ensures that $e_x, e_y \in L_\infty$, $\dot{e}_x, \dot{e}_y \in L_\infty$, $u_c \in L_\infty$. Noting that $\dot{S} = P\tilde{M}^{-1} P^T (-KS + u_c) + \tilde{F}_s(\cdot)$, we see that $\dot{S}$ is bounded and therefore S is uniformly continuous. Hence, by Barbalat's lemma $\|S\| \to 0$ as $t \to \infty$. Consequently $\|e_x\| \to 0$ and $\|e_y\| \to 0$ as $t \to \infty$.

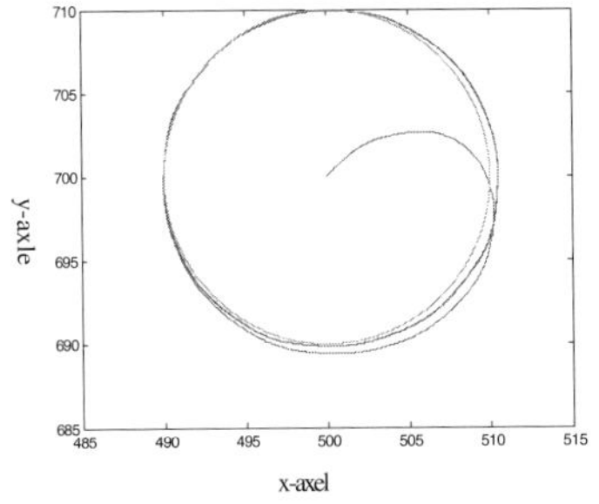

Fig. 2. Tracking Process

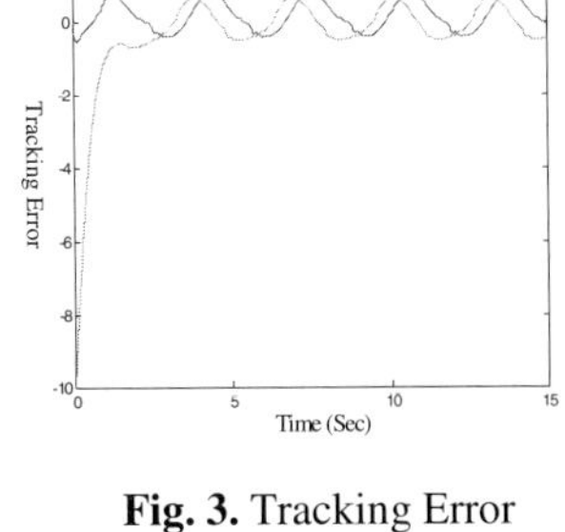

Fig. 3. Tracking Error

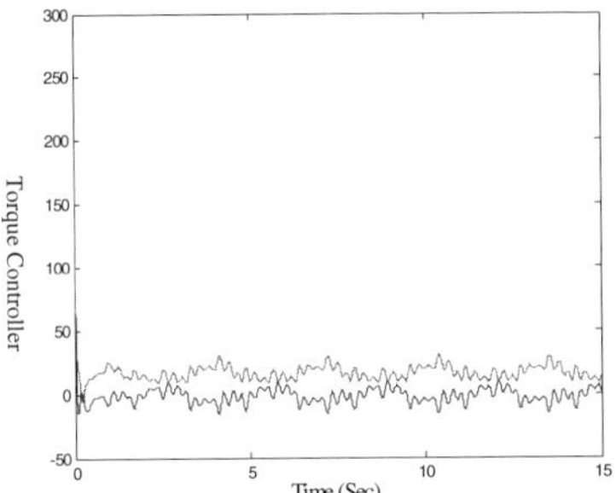

Fig. 4. Wheel Control Torques

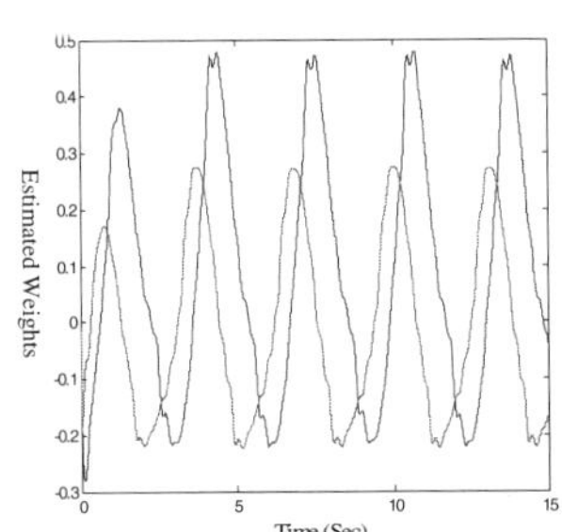

Fig. 5. Weights Updating

3 Simulation Results

To verify the effectiveness of the proposed control scheme, we conduct a series of computer simulation studies. The results related to the following case are presented:

The desired path is given as $\begin{cases} x_r = 500 + 20\sin(2t) \\ y_r = 700 + 20\sin(3t) \end{cases}$ The control parameters $K = diag(k_1, k_2)$ and $\beta = diag(\beta_1, \beta_2)$ are chosen as $k_1 = 300$, $k_2 = 200$, $\beta_1 = 3$, $\beta_2 = 2$, $\sigma = 0.1$, $\eta = 0.5$, and $\mu = 0.01$, which are fixed for all the test cases. For the given desired path, Figure 2 presents the tracking process and Figure 3 is the position tracking errors in *x/y* directions. Figure 4 is the control torque signal for both left and right wheels and Figure 5 depicts the updating of the NN weights. The results for other cases (not presented here due to page limit) are also satisfactory. As can be seen, fairly good tracking precision is achieved with the proposed control scheme for all the simulation cases.

4 Conclusion

Path tracking control is a fundamental issue in UGVs. The proposed control scheme has been shown effective in achieving high precision tracking with little vehicle dynamic information. The potential advantage of the proposed method, as compared with most other exiting methods, is that it can be easily designed and implemented without involving complicated procedures. Extension of the results to multi-UGVs formation control is under way.

Acknowledgements

This work was supported in part by ARO through the grant W911NF-04-1-0384 and NSF through the grant HRD-0450203.

References

1. Elfes, A., Bueno, S.S., Bergerman, M., de Paiva, E.C., Ramos, J.G.: Robotic Airships for Exploration of Planetary Bodies with An Atmosphere. Autonomy Challenges, Auto Robots 14(2003) 147-164
2. Hirose, S., Kuwahara, H., Wakabayashi, Y., Yoshioka, N.: The Mobility Design Concepts/characteristics and Ground Testing of an Offset-wheel Rover Vehicle. Space Tech. 17 (1997) 183-193
3. Ito, T., Mousavi Jahan Abadi,S.M.: Agent-based Material Handling and Inventory Planning in Warehouse. J. Intell. Manuf. 13(2002) 201-210
4. Letherwood, M.D., Gunter, D.D.: Ground Vehicle Modeling and Simulation of Military Vehicles Using High Performance Computing. Parallel Compute 27 (2001) 199-140.
5. Brockett, R.W.: Asymptotic Stability and Feedback Linearization in Differential Geometric Control Theory. Birkhauser, Cambridge MA (1983)
6. Rigatos, G.G., Tzafestas, C.S., Tzafestas, S. G.: Mobile Robot Motion Control in Partially Unknown Environments Using a Sliding Mode Fuzzy Logic Controller. Robotics and Autonomous Systems 2(2) (1994) 313-319
7. Yang, M., Kim, J.H.: Sliding Mode Control for Trajectory Tracking of Nonholonomic Wheeled Mobile Robots. IEEE Trans on Robotics and Automation. 15(3) (1999) 578-587
8. Fierro, R., Lewis, F. V.: Control of a Nonholonmic Mobile Robot: Backstepping Kinematics into Dynamics. J. Robotics Systems 14(3) (1997) 149-163
9. Zhang, Q., Shippen, J., Jones, B.: Robost Backstepping and Neural Network Control of a Low Quality Nonlonomic Mobile Robot. Int. J. Mach. To. Manuf. 39(7) (1999)1117-1134
10. Bolzern, P., Desantis, R., Locatelli, M., Masciouscchi, A. D.: Path Tracking for Articulated Vehicles with Off-axle Hitching. IEEE Trans on Control System Tech. 6(4) (1998) 515-523
11. Song, Y.D.: Adaptive Motion Tracking Control of Robot Manipulators-non-regressor Based Approach. INT. J. Control 63(1) (1996) 41-54.
12. Lewis, F.L., Yesildirek, A., Liu, K.: Multilayer Neural-net Robot Manipulators. MacMillan, New York (1993)
13. Sarkar, N., Yun, X., Kumar, V.: Control of Mechanical Systems with Rolling Constraints: Application to Dynamic Control of Mobile Robots. Int. J. Robot. Res.13(1) (1994) 55-69

A Nuclear Power Plant Expert System Using Artificial Neural Networks

Mal rey Lee[1], Hye-Jin Jeong[1], Young Joon Choi[2], and Thomas M. Gatton[3]

[1] School of Electronics & Information Engineering, ChonBuk National University, 664-14, 1Ga, DeokJin-Dong, JeonJu, ChonBuk, 561-756, Korea
mrlee@chonbuk.ac.kr
[2] Nuclear Safety Regulation Division, Korea Institute of Nuclear Safety, 19 Kusong-dong, Yusong-gu, Taejon, South Korea 305-338
[3] School of Engineering and Technology, National University, 11255 North Torrey Pines Road, La Jolla, CA, 92037 USA

Abstract. In this study, ANNs are introduced to act as a bridge between detailed computer codes and compact simulators with an aim to improve the capabilities of compact expert system. The ANNs compensate for the inaccuracies of a compact expert system occurring from simplified governing equations and a reduced number of physical control volumes, and predict the critical parameter usually calculated from the sophisticated computer code. To verify the applicability of the proposed methodology, computer simulations are undertaken for loss of flow accidents (LOFA).

1 Introduction

Nuclear power plant (NPP) expert systems are very important training tools for operators to understand the underlying principles of plant systems and components as well as the interrelationships between the behavior of various plant systems during transients or accidents. Operator training with expert system has received a high priority and the development of expert system with a much higher training capability than the previously existing expert system was performed as a result of the lessons learned from the accident at Three Mile Island Nuclear Station Unit 2 in 1979 [1]. Recently, operating NPPs have contained their own full scope expert system as a regulatory requirement. The full scope expert systems are complete replicas of the control room and are typically used for periodic operator training.

Scaled-down versions of the full scope expert system, so called compact expert system with less detail than the real control room but with sufficient simulation depth, have been developed to provide an optimized environment for operators or engineers to focus on understanding the physical phenomena and their interactions. Although the compact expert systems cannot completely replace the role of the full scope expert systems, they can be used as a cost-effective training alternative, analysis tool or full scope expert system. Generally, the compact expert systems are designed to demonstrate overall responses of normal operation and transients in real time or faster. In the thermal hydraulic models of compact expert systems, the governing

J. Wang et al. (Eds.): ISNN 2006, LNCS 3972, pp. 1239 – 1245, 2006.

equations, which translate physical phenomena into formal mathematical terms, are simplified with reasonable assumptions and empirical correlations and solutions are approximated using appropriate numerical schemes. In these processes, there always exists a trade-off between the accuracy of solutions and the computational efficiency. Moreover, many physical control volumes in plant modeling are lumped to reduce computing time. The simplification of equations and the reduction of control volume numbers usually degrade the accuracy of solutions. Therefore, while they have somewhat lower accuracy in the calculation results, they have the benefit of faster simulation calculation performance.

2 Accident Simulation Methodology

2.1 Conceptualization of the Proposed Methodology

The thermal hydraulic (TH) program playing the role as a compact simulator is designed to calculate the thermal hydraulic variables, such as pressure, temperature, and mass flow rate, at each node and junction during accident simulation. The Korean Standard Nuclear Power Plant (KSNP) combustion engineering type NPP is selected as the reference plant. The thermal hydraulic responses of the nuclear steam supply system of the KSNP to the selected accidents are simulated by RELAP5/MOD3 [6]. The minimum departure from nucleate boiling ratio (DNBR) is calculated by the COBRA-IV [7] for the LOFAs with the inputs obtained from RELAP5/MOD3. Each ANN is programmed to obtain the connection weights between the layers according to a determined procedure. The analysis results of RELAP5/MOD3 and COBRA-IV are used as the target values to train the ANNs. After training, connection weights between layers are saved as electronic files. The neural networks are again programmed as a subroutine in the TH program to perform their recall phases, in which the trained weight files are called by two neural networks. Variables calculated from the TH program are supplied as the input to the auto-associative neural network (AANN) after which the AANN produces the corrected version of the TH program output. Some of the AANN results are again fed to the back propagation neural network (BPN) after which the BPN predicts the minimum DNBR.

2.2 Thermal Hydraulic Program

Compact expert systems can be divided into three types [8]: (i) complex fast systems called plant analyzer, (ii) intermediate size modular systems and (iii) small simulators used for preliminary and conceptual studies. Plant analyzers are complete plant expert systems that can run interactively in real time, or faster, and can display the results dynamically using color coded graphics. Modular expert systems permit the user to select alternative models for different plant components from a library of stored programs. Small expert systems are useful for preliminary studies and to illustrate concepts and trends during normal operation or transients. The training cycles of operators are classified as follows [9]: (i) basic principle learning, (ii) specific functional training, (iii) full operating range training and (iv) detailed accident analysis. Expert systems relevant to the above training cycles are the basic principle, partial scope and full scope simulator. The basic principle expert system includes a physical model without the real operator interface. The partial scope expert system describes only a limited part of the process while the full scope simulator is an operator training tool. In this study, the TH program performs as a small expert system or a basic principle expert system as defined above. The essential objective of the TH program is to describe the overall responses of the NPP under various accident conditions within an appropriate time period. Therefore, it is necessary to adopt a coarse nodal representation of the reactor coolant system, which incorporates appropriate assumptions and excludes some components and systems. The TH program has been developed in reference to the plant model part of FISA-2/WS expert system [10], which uses the fully-implicit time scheme and the two-fluid model in the primary side and a two-region lumped parameter approach in the secondary side. In developing the TH program, major modifications from FISA-2/WS are as follows: (i) plant model change from Westinghouse type to Combustion Engineering type, (ii) nodal representation change for detailed simulation in the reactor vessel, (iii) elimination of some systems for simplified modeling, such as chemical and volume control system, plant control system, etc., (iv) simulation capabilities for accident simulation and (v) calculation algorithm such as steady state initialization, transient, etc. The TH program represents the whole plant as 18 nodes, which consists of 15 nodes for the primary side and 3 nodes for the secondary side. A schematic diagram of the TH program is shown in Fig. 2 below.

2.3 Training of the ANNs

The ANNs employed in this study are feed forward networks with an error back propagation algorithm [11]. The back propagation algorithm is the most widely used supervised training method and the best convergent method of computing the interconnection weights that minimize the total squared error between the network output and the target value. Because the algorithm is introduced in many papers, its mathematical formulations are not repeated in this paper. AANN is a feed forward network with multiple players being trained to produce an identity mapping between input and output. The relationships between the variables are embedded in the connection weights by the training process. As a result, the output is a correct version of the input pattern, although the input is distorted by noise and incomplete or missing data [12, 13]. This feature of the AANN is used to improve the accuracy of the TH program. Once the AANN is trained with the variables from RELAP5/MOD3, the output of the trained AANN will be a corrected version of the variables from the TH program. In this study, the AANN is composed of an input layer, three hidden layers and an output layer as shown in Fig. 1. Each layer of the AANN has 45, 54, 35, 54, and 45 nodes, respectively, and the activation function is a sigmoid function. The AANN has been trained with 45 variables, from RELAP5/MOD3. The training of a single AANN has been attempted by sequentially applying a (2,500 X 45) matrix for 10 LOFAs, while adjusting the network weights according to a predetermined procedure. However, a single AANN trained with 10 cases of the LOFAs has given unsatisfactory results. In general, when the trained space of the neural network is broad enough, the generality of neural network is selected Variables to Train the

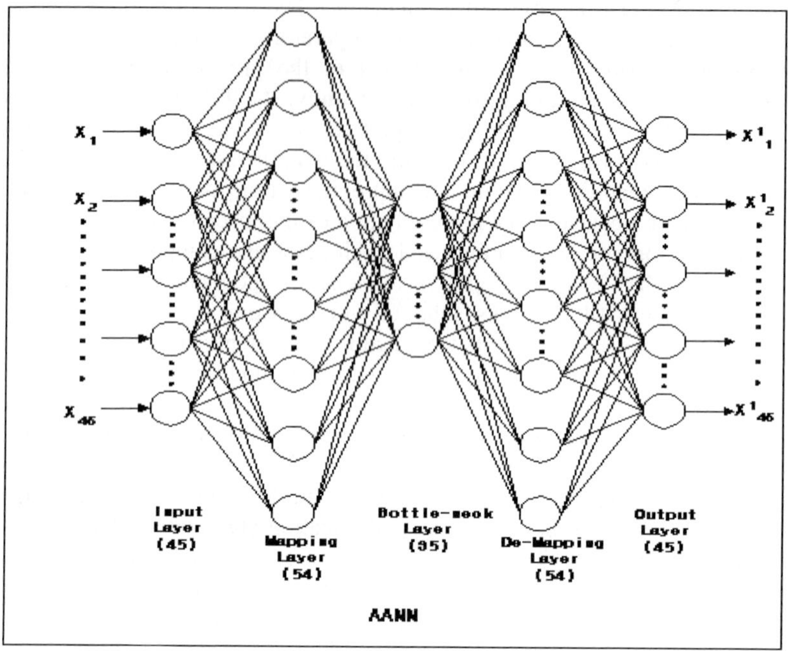

Fig. 1. Autoassociative Neural Network

AANNs and the Steady State. However, when many similarly changing states with many variables are combined together to train neural network, as performed in this study, there are limitations of generality for neural networks even when the total mean squared error is very low. To simplify the training and to improve accuracy, each accident case has been trained with a (250 X 45) matrix. The training of each AANN was terminated when the total mean squared error had reached the value of 1.0×10^{-6}. When each accident had been trained individually, better results were obtained. A three-layer BPN has been modeled for the minimum DNBR prediction. The input layer has four nodes for the same variables used in the COBRA-IV computer code. The hidden layer has nine nodes and the output layer has a single node representing the minimum DNBR. A sigmoid function is used as an activation function. To train the BPN, time-dependent inputs have been sequentially applied and training was terminated when the total mean squared error reached the value of 2.5×10^{-6}. To verify the trained results, input data (COBRA-IV) was reapplied to the trained BPN.

3 Application of the Proposed Methodology

3.1 Responses of the AANN to Total Loss of Flow Accident

A total loss of flow accident (TLOFA) has been selected as a sample case for illustrating the applicability of the proposed methodology because it produces a minimum DNBR that is more adverse than any partial loss of flow accidents. The dynamic behaviors of important parameters from the TH program and the AANN following the TLOFA and demonstrate how the results of TH program have been improved by the AANN. The results of RELAP5/MOD3 are also shown for comparison.

As can be seen in these figures, the results of AANN recall phase are almost the same as the target values obtained from RELAP5/MOD3. The variables from TH program are significantly improved by the AANN. The results of the TH program to the TLOFA show lower magnitudes or earlier responses than those of the RELAP5/MOD3 code due to its simplified governing equation and smaller node numbers. The increasing rates of pressure or temperature are lower and the responses are earlier than those of RELAP5/MOD3. These differences have been considerably reduced by the AANN. The responses of the AANN have taken place earlier than those of RELAP5/MOD3 as a result of the simulation method. The reactor trip signal is initiated with a 2.0 second time delay for conservative accident analysis after the reactor coolant pump flow rate reaches the reactor trip set point at 1.8 seconds in the RELAP5/MOD3 calculation, while the time to reach the reactor trip set point is 0.8 seconds, without considering the reactor trip delay in the TH program simulation. This means that the 3.0 second earlier response of the TH program makes the graphs of the AANN shift to the left as compared with those of RELAP5/MOD3. While some of the variables from the TH program show the same results or trends as those of RELAP5/MOD3, the other remaining variables also approach the target values. This is caused by the high degree of variable coherence which has been embedded in the weights of the AANN. As shown in Fig. 2, the decreasing trends of the mass flow rates (15 variables) of the TH program are similar to those of RELAP5/MOD3 and

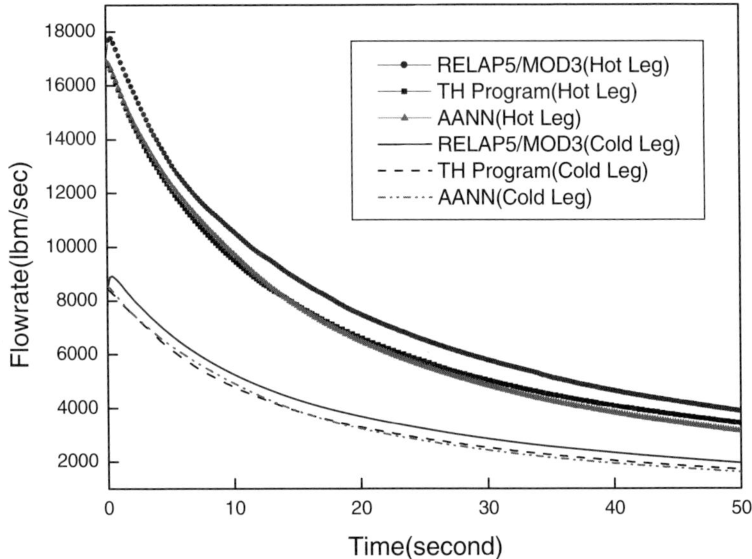

Fig. 2. Hot Leg and Cold Leg Flow Rate

play an important role as a contributor in directing the other variables, such as temperatures (14 variables) and pressures (15 variables), to the target values.

3.2 Prediction of the Minimum DNBR by the BPN

The four improved variables, core inlet pressure, core inlet temperature, core inlet flow rate and core power from the recall phase of the AANN have been used as the input to the BPN in order to predict the minimum DNBR. The minimum DNBR response to the TLOFA. The BPN gives a fairly accurate prediction of the minimum DNBR responses. The comparison of the earlier responses between the minimum DNBR from the BPN and the results of case 1 of the COBRA-IV code indicates the early reactor trip in the TH program. As can be seen from the above results, the proposed hybrid accident simulation methodology has shown to be practicable. The neural networks have played important roles as a complementary tool for improving the TH program which can be categorized as a small simulator or a basic principle simulator. More accidents may also need to be analyzed to fully illustrate the applicability of the proposed methodology. However, it is believed that the application to the TLOFA has illustrated the proposed approach and has verified its practicability.

4 Conclusions

An accident simulation methodology has been developed to improve the results of a compact expert system by introducing neural networks. The applicability of the proposed methodology has been illustrated with the analyses of the TLOFA. The results indicate that this methodology has the potential to enhance the performance of compact expert system in a cost-effective way. Therefore, more reliable overall responses of accident simulation can be presented to operators or engineers for training or education. Advantages of the proposed methodology can be summarized as follows: (i) more accurate computational results can be obtained from compact expert system while maintaining its merit, i.e., fast simulation capability: the AANN has improved the accuracy of compact expert system up to the level of detailed computer code, and (ii) multi-calculating stages can be integrated into one stage: the minimum DNBR has been calculated by the recall phase of the trained BPN without additional algorithmic calculation process.

References

1. Gonsalves, J. B.: Nuclear Plant Simulators: The Past and The Present. Nuclear News (1987) 4955
2. Hatice, A.: PWR System Simulation and Parameter Estimation with Neural Networks. Annals of Nuclear Energy 29 (2002) 20872103
3. Ricotti, M. E.: Neural Network Approach to Sensitivity and Uncertainty Analysis. Reliability Engineering & System Safety 64 (1999) 5971

4. Uhrig, R. E., Tsoukalas, L. H.: Soft Computing Technologies in Nuclear Engineering Applications. Progress in Nuclear Energy 34 (1999) 1375
5. Reifman, J.: Survey of Artificial Intelligence Methods for Detection and Identification of Component Faults in Nuclear Power Plants. Nuclear Technology 119 (1997) 7697
6. RELAP5 Development Team: RELAP5/MOD3 Code Manual, 1, NUREG/CR-5535, Idaho National Engineering Laboratory (1995)
7. Wheeler, C.: COBRA-IV-i : An Interim Version of COBRA for Thermal Hydraulic Analysis of Rod Bundle Nuclear Fuel Elements and Cores, BNWL-1962. Battelle Pacific Northwest Laboratory (1976)
8. Hetrick, D.: Power Plant Simulation and Reactor Safety. Nuclear Safety 28 (1987) 473486
9. Van Billoen, G.: Classification and Optimization of Training Tools for NPP Simulators. Proceeding of CSNI Specialist Meeting on Simulators and Plant Analyzers (1992) 1927
10. Jeong, J.: FISA-2/WS: A Compact Real-Time Simulator for Two-Loop PWR Plants. Nuclear Technology 90 (1990) 356370
11. Rumellhart, D.: Parallel Distributed Processing. MIT Press, Cambridge MA (1986)
12. Kramer, M.: Nonlinear Principal Component Analysis Using Auto-associative Neural Networks. AICHE Journal 37 (1991) 233243
13. Uhrig, R.: Intelligent Surveillance and Calibration Verification in Power System. International Conference on Intelligent System Application to Power Systems (1997) 5559

Short-Term Load Forecasting Based on Mutual Information and Artificial Neural Network

Zhiyong Wang and Yijia Cao

College of Electrical Engineering, Zhejiang University,
310027 Hangzhou, China
zjuwzy@yahoo.com.cn

Abstract. Short term load forecasting (STLF) has an essential role in the operation of electric power systems. Although artificial neural networks (ANN) based predictors are more widely used for STLF in recent years, there still exist some difficulties in choosing the proper input variables and selecting an appropriate architecture of the networks. A novel approach is proposed for STLF by combining mutual information (MI) and ANN. The MI theory is first briefly introduced and employed to perform input selection and determine the initial weights of ANN. Then ANN module is trained using historical daily load and weather data selected to perform the final forecast. To demonstrate the effectiveness of the approach, short-term load forecasting was performed on the Hang Zhou Electric Power Company in China, and the testing results show that the proposed model is feasible and promising for load forecasting.

1 Introduction

Short-term load forecasting (STLF) is aimed at predicting system load over an interval of a few minutes, hours or days ahead. It plays an important role in power system planning and operation. Nowadays, many countries have privatized and deregulated their power systems, and electricity has been turned into a commodity to be sold and bought at market prices. STLF becomes increasingly important since the rise of the competitive energy markets [1].

Recently Artificial Neural Network (ANN) has become a most promising approach for developing forecasting tools with enhanced performance capabilities [2-5]. The popularity of the ANN based predictive models can be owed to the highly nonlinear or even chaotic nature of electrical load over time and the availability of adequate load data [1]. But STLF is not an easy task. The load series is very complex and determined by varieties of factors such as weather-related variables and calendar date: the load at a given hour is dependent not only on the factors at the previous hour, but also on the factors on the previous days. And we could not take all the factors that may have influence on load as the input variables of ANN to avoid the "curse of dimensionality".

However there are very few theoretical considerations to help to select input variables for ANN predictors; usually, one must have some *a priori* knowledge about the behavior of the system under study, and of the factors that condition the output of that system. In this paper we present an approach that could select a priori inputs based only on the dataset. The remaining of this paper is organized as follows: Mutual

J. Wang et al. (Eds.): ISNN 2006, LNCS 3972, pp. 1246 – 1251, 2006.

Information, described in section 2 is used to select the inputs of ANN; next, section 3 details the proposed ANN based STLF model; section 4 illustrates the suitability of the proposed approach through an application to actual load data obtained from Hang Zhou Electric Power Company of China; and finally, section 5 presents the conclusions and further work.

2 Mutual Information Theory

Mutual Information (MI) is a basic concept in information theory. It is a measure of general interdependence between variables [6]. Let's consider two random variables X and Y, the MI $I(X,Y)$ is defined as follows:

$$I(X;Y) = H(X) + H(Y) - H(X,Y) \tag{1}$$

where $H(\bullet)$ computes the Shannon's entropy.

However, equation (1) leads to complicated integrations in the continuous entropy, and in this paper MI is estimated as [7]

$$I(X;Y) = \psi(k) - 1/k - \langle \psi(n_x) + \psi(n_y) \rangle + \psi(N) \tag{2}$$

where N is the size of dataset and $\psi(x)$ is the digamma function,

$$\psi(x) = \Gamma(x) - 1 - d\Gamma(x)/dx \tag{3}$$

$\psi(1) \approx -0.5772156$ and $\langle \cdots \rangle$ denotes average of $n_x(i)$ and $n_y(i)$ over all $i \in [1,2,...,N]$ and over all realizations of the samples. $n_x(i)$ and $n_y(i)$ are the number of points in the region $\|x_i - x_j\| \le \varepsilon_x(i)/2$ and $\|y_i - y_j\| \le \varepsilon_y(i)/2$, $\varepsilon_x(i)$ and $\varepsilon_y(i)$ are the projections of the distance from point i to its k-nearest neighbors. According to [7], k is set to be 6 in this paper.

The concept of MI can be easily expanded to include more than two variables and the joint mutual information (JMI) between a set of features $(X_1, X_2, ..., X_N)$ and the output Y is

$$I(X_1, X_2, ..., X_N; Y) = \sum_{i=1}^{N} I(X_i; Y | X_{i-1}, X_{i-2}, ..., X_1) \tag{4}$$

MI makes no assumptions about the nature of the relationship between variables, so it could be used to select reasonable features among a set of possible variables based only on the dataset [6]. In this paper MI was introduced to describe how much the information provided by the feature vector $(X_1, X_2, ..., X_N)$ decreases the uncertainty about the output Y. Given a large set of features, it is expected that some of them may be dependent on each other. Therefore, selecting feature subsets that could maximize MI with the output can provide an optimal subset that contains the most relevant features. The ideal greedy selection algorithm using MI is realized as follows:

1. $F \leftarrow \{X_1, X_2, ..., X_N\}, \ S \leftarrow \{\varnothing\}$;
2. $j = \arg\max_{X_j \in F} I(X_j;Y)$, $F \leftarrow F \setminus \{X_j\}, S \leftarrow \{X_j\}$;
3. $j = \arg\max_{X_j \in F} I(X_j, S;Y)$, $F \leftarrow F \setminus \{X_j\}, S \leftarrow S \cup \{X_j\}$;
4. IF $(I(S;Y) - I(S \setminus \{X_j\};Y)) > 0$, Goto step 3; ELSE $S \leftarrow S \setminus \{X_j\}$;
5. Output the set S, and the significance of the attributes γ $\forall \{a\} \in S, \gamma_{\{a\}} = (I(S;Y) - I(S \setminus \{a\};Y) / I(S;Y)$.

3 Load Forecasting Model

3.1 General ANN Model

In recent times, much research has been carried out on the application of artificial neural network to the load-forecasting problem [1]. The architecture of artificial neural network used in this paper consists of three layers: the input layer, the hidden layer and the output layer. The nodes in the input layer are relevant factors about load, and the node in the output layer provides the desired load. The number of nodes in the hidden layer is chosen by trial and error, selecting a few alternative number and then running simulations to find out one that gave the best predictive performance. The activation function for the input nodes and hidden nodes are *logistic* function, and for the output node is *linear* function. The adaptive back propagation algorithm [8] is used to adjust the weights a small amount at a time in a way that reduces the error. The principal statistics used to evaluate the performance of the network, mean absolute error (MAPE), is defined as

$$MAPE(\%) = \frac{1}{N} \sum_{i=1}^{N} \frac{\left| P_A^i - P_F^i \right|}{P_A^i} \times 100 \tag{5}$$

where P_A, P_F are the actual load and forecasted load, N is the number of data points.

3.2 Selection of ANN Inputs

One of the most important tasks in building an efficient ANN based forecasting model is the selection of the relevant input variables [1]. The input selection can be stated as follows: among a large set of potential input candidates, choose those variables that could maximize MI with the output. The suggested MI based approach provides a simple and reasonable procedure for determining the proper model inputs from an input candidate set. We obtain data on load domain from Hang Zhou Electric Power Company of China, which provides us with historical data about load between Jan. 1st 2000 to Jan. 1st 2002.

First, As we all know, weather factors have some accumulative effect which forces customers to keep the same behavior pattern as that of the previous days unless the environment drastically changes. Therefore, the selection of model inputs that considers a trend of weather data is performed. By this way, we can consider all the historical data that may have influence on the predicted day.

Table 1. Input candidate sets

Var. No.	Variable Meaning	Var. No.	Variable Meaning
$u_{1,\cdots,8}$	$S_{d-i}, i=0,...,7$	$u_{33,\cdots,40}$	$T_{\max.d-i}, i=0,...,7$
$u_{9,\cdots,16}$	$D_{d-i}, i=0,...,7$	$u_{41,\cdots,48}$	$T_{\min.d-i}, i=0,...,7$
$u_{17,\cdots,24}$	$H_{d-i}, i=0,...,7$	$u_{49,\cdots,56}$	$T_{ave.d-i}, i=0,...,7$
$u_{25,\cdots,32}$	$P_{d-i}, i=0,...,7$	$u_{57,\cdots,64}$	$L_{\max.d-i}, i=0,...,7$

Then, we select the data for the previous seven days $(d-1,...,d-7)$ of the day d to be forecasted. The input candidate sets can be found in table 1 where S is the season type; D is the day types including workday, weekend, and holiday; H is daily average relative humidity, P is daily amount of precipitation; $T_{\max}, T_{\min}$ and T_{ave} are maximum, minimum and average daily temperature, $L_{\max}$ is daily peak load. In this paper, the predicted peak load on d is regarded as output, while the other factors are regarded as the condition attributes.

Finally, by using the MI-based greedy selection algorithm proposed in section 2, the attributes space is reduced greatly. Only relevant factors and respective MI value are obtained, which can be used as the input nodes and initial weights of ANN. The relevant factors and their respective MI value to the output can be found in Table 2.

Table 2. Attribute sets after reduction

Var. No.	Variable Meaning	γ	Var. No.	Variable Meaning	γ
a_1	$L_{\max.d-1}$	0.179	a_7	$L_{\max.d-7}$	0.076
a_2	$L_{\max.d-2}$	0.128	a_8	$T_{mean.d-2}$	0.070
a_3	$T_{mean.d}$	0.107	a_9	$T_{\min.d}$	0.066
a_4	$L_{\max.d-3}$	0.088	a_{10}	$T_{\min.d-1}$	0.063
a_5	$T_{mean.d-1}$	0.083	a_{11}	$T_{\max.d-1}$	0.060
a_6	$T_{\max.d}$	0.080			

4 Test Results

In order to highlight the proposed method, using the actual load and temperature data (for the year 2000 to 2001) of Hang Zhou Electric Power Company, the prediction of the daily peak load for two month, e.g. July 2001 and December 2001 is presented, as an illustrative example.

The suggested method is now compared with traditional ANN and fuzzy neural networks (FNN) forecast techniques reported in the literature [2, 5]. For fair comparison, all the forecasting models are evaluated using the same data sets. The ANN STLF model consists of an input layer composed of 64 neurons, a hidden layer with 30 neurons and an output layer with one neuron representing the peak load forecast.

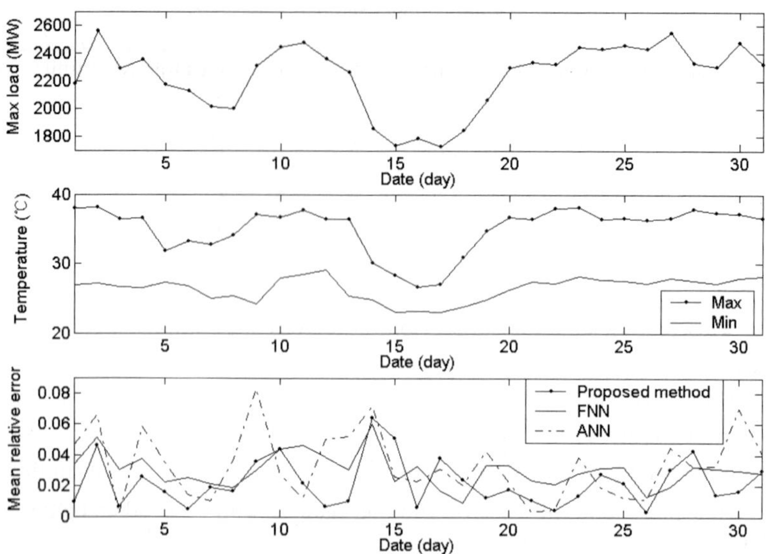

Fig. 1. Comparison of forecasting errors for July

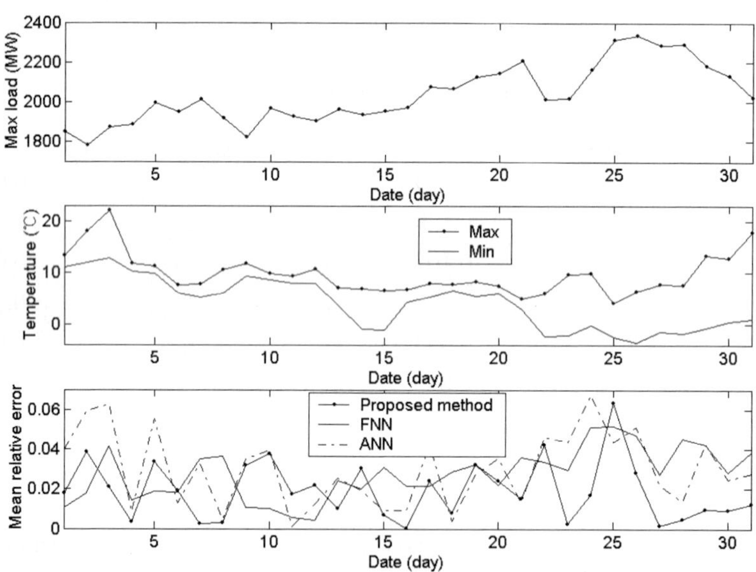

Fig. 2. Comparison of forecasting errors for December

Figures 1 and 2 show the results of monthly peak load forecasting test in different load types for the Hang Zhou power system. The test objects include the power system in winter and summer seasons. Daily peak loads in the winter reflect heating requirements and summer daily peak loads are the result of air conditioning demands. It is observed that the weather changes frequently in winter and summer, accordingly the load changes quickly. So it is hard for traditional ANN based forecasting method to obtain an accurate forecast and it is evident that the proposed approach exhibits a better forecasting performance than the FNN and conventional ANN method.

5 Conclusions

This paper has proposed a hybrid method of ANN and the mutual information theory for short-term load forecasting. A MI-based feature selection method was first used to obtain the most significant input variables and the initial weights of the ANN. Then, the ANN was used to predict the maximum daily load. To demonstrate the effectiveness of the proposed method, short-term load forecasting has been investigated on the Hang Zhou power system with different load types. The test results show the proposed forecasting method could provide an improvement of the forecasting accuracy for different day types and it could effectively support the conventional ANN based STLF.

References

1. Hippert, H.S., Pedreira, C.E., Souza, R.C.: Neural Networks for Short-term Load Forecasting: A Review and Evaluation. IEEE Trans. Power Syst. 16 (2001) 44–55
2. Park, D. C., El Sharkawi, M. A., Marks II, R. J., Atlas, L. E., Damborg, M. J.: Electric Load Forecasting Using An Artificial Neural Network. IEEE Trans. Power Syst. 6 (1991) 442–449
3. Khotanzad, A., Davis, M.H., Abaye, A., Maratukulam, D.J.: An Artificial Neural Network Hourly Temperature Forecaster with Applications in Load Forecasting. IEEE Trans. Power Syst. 11 (1996) 870–876
4. Taylor, J. W., Buizza, R.: Neural Network Load Forecasting with Weather Ensemble Predictions. IEEE Trans. Power Syst. 17 (2002) 626–632
5. Liao, G. C., Tsao, T. P.: Application of Fuzzy Neural Networks and Artificial Intelligence for Load Forecasting. EPSR. 70 (2004) 237–244
6. Cover, T. M., Thomas, J.A.: Elements of Information Theory. Wiley, New York (1991)
7. Harald, S., Alexander, K., Sergey, A., A., Peter, G.: Least Dependent Component Analysis Based on Mutual Information. Phys. Rev. E. 70 (2004) 066123
8. Magoulas, G. D., Vrahatis, M. N., Androulakis, G. S.: Improving the Convergence of the Backpropagation Algorithm Using Learning Rate Adaptation Methods. Neural Comp. 11 (1999) 1769–1796

Short Term Load Forecasting by Using Neural Networks with Variable Activation Functions and Embedded Chaos Algorithm

Qiyun Cheng and Xuelian Liu

Guiyang South Power Supply Bureau, Guizhou Power Grid Co.,
Guiyang 550002, Guizhou China
chqiyun@126.com

Abstract. In this paper a novel variant activation (transform) sigmoid function with three parameters is proposed, and then the improved BP algorithm based on it is educed and discussed, then Embedded Chaos-BP algorithm is proposed by means of combining the new fast BP algorithm and chaos optimization algorithm, Embedded chaos-BP algorithm converges fast and globally, and has no local minimum. The efficiency and advantage of our method is proved by simulation results of nonlinear function and prediction results of short-term load based on the improved and traditional BP ANNs.

1 Introduction

Load forecasting of electrical power system is one important job in electrical department and there are many mathematical ways of Short Term Load Forecasting [1]. Having the characteristics such as self-study and self-adaptation, Artificial Neural Network (ANN) is particularly suitable for dealing with nonlinear problems. Many examples have convinced that ANN can well solve the problem of Short Term Load Forecasting[2]. In the theory of neural network, the algorithm Back Propagation (BP) is most widely applied and it influences the world widely.

Though BP is very effective, two major problems exist in its utility. One is that its convergence is too slow, and the other is that E's local minimum usually stalemates the algorithm. Therefore, many scholars began to investigate the important factors of BP algorithm that impact on convergence [3, 4, 5]. In this paper, we take a further attempt to present three variable activation of varying parameter in order to obtain better nonlinear expression capability. Second, in order to overcome the problem of local minimum, many scholars' researches organically connect neural network and algorithm of better global convergence [6-9]. They have certain achievements. In this paper we present a modified searching algorithm combining chaos and BP so as to make searching most possible to find the optimum direction and accelerate the rate of convergence.

2 Activation Function

In this paper three parameters adjustable Sigmoid function is proposed as the activation function of nerve cells. It is named as BP-AA (Adjustable activation function algorithm):

J. Wang et al. (Eds.): ISNN 2006, LNCS 3972, pp. 1252 – 1258, 2006.

$$S_{a,b,\lambda}(x) = \frac{1}{1+e^{-\frac{x-b}{\lambda}}} + a \cdot \tag{1}$$

In the formula, a is a deviation parameter, b is a threshold (location parameter), λ is a gradient factor. Obviously, the function $S_{a,b,\lambda}(x)$ is richer in nonlinear expression capability than that of $S(x)$.

3 BP-AA Learning Algorithm

In this paper, the symbols applied in BP-AA algorithm formula are illustrated in the following contents:

① Input, hidden and output layers are respectively marked as $n_i(k)$, $n_h(k)$ and $n_o(k)$, the numbers of nerve cell in each layer are I, H and O;

② y_{ik} is $n_i(k)$ output value, X_{hk} is $n_h(k)$ total input, y_{hk} is the output value of $n_h(k)$, X_{ok} is the total input value of $n_o(k)$; y_{ok} is the actual output value of $n_o(k)$, y^*_{ok} is the expected output of $n_o(k)$;

③ a_{hk}, b_{hk} and λ_{hk} are respectively marked as three adjustable parameters of the activation function of hidden layer $n_h(k)$; a_{ok}, b_{ok} and λ_{ok} are respectively marked as three adjustable parameters of the activation function of output layer $n_o(k)$; w^I_{mn} is the weight between $n_i(m)$ and $n_h(n)$, w^O_{mn} is the weight between $n_h(m)$ and $n_o(n)$;

④ Superscript p stands for the p-th learning sample and the mean-square error in the p-th learning sample is $E^p = \frac{1}{2}\sum_{k=1}^{O}(y^p_{ok} - y^{p*}_{ok})^2$ while the sum of mean-square error of all the learning samples is $E = \sum_{p=1}^{P} E^p = \frac{1}{2}\sum_{p=1}^{P}\sum_{k=1}^{O}(y^p_{ok} - y^{p*}_{ok})^2$.

Then educes the results of the batch samples training model as follow:

① Adjustment of the weight w^O_{mn} between hidden layer and output layer:

$$\frac{\partial E}{\partial w^O_{mn}} = \frac{1}{\lambda_{on}}\sum_{p=1}^{P}(y^p_{on} - y^{p*}_{on})(y^p_{on} - a_{on})(1 + a_{on} - y^p_{on})y^p_{hm} \cdot \tag{2}$$

② Adjustment of the weight w^I_{mn} between input layer and hidden layer:

$$\frac{\partial E}{\partial w^I_{mn}} = \frac{1}{\lambda_{hn}}\sum_{p=1}^{P}\{(y^p_{hn} - a_{hn})(1 + a_{hn} - y^p_{hn})x^p_{im}\sum_{j=1}^{O}[\frac{1}{\lambda_{oj}}(y^p_{oj} - y^{p*}_{oj})(y^p_{oj} - a_{oj})(1 + a_{oj} - y^p_{oj})w^O_{nj}]\} \cdot \tag{3}$$

③ Adjustment of parameter of activation function of nerve unit in output layer:

$$\frac{\partial E}{\partial a_{ok}} = \sum_{p=1}^{P}(y^p_{ok} - y^{p*}_{ok}) \cdot \tag{4}$$

$$\frac{\partial E}{\partial b_{ok}} = -\frac{1}{\lambda_{ok}} \sum_{p=1}^{P} (y_{ok}^{p} - y_{ok}^{p*})(y_{ok}^{p} - a_{ok})(1 + a_{ok} - y_{ok}^{p}) \cdot \tag{5}$$

$$\frac{\partial E}{\partial \lambda_{ok}} = -\frac{1}{\lambda_{ok}^{2}} \sum_{p=1}^{P} (y_{ok}^{p} - y_{ok}^{p*})(y_{ok}^{p} - a_{ok})(1 + a_{ok} - y_{ok}^{p})(y_{ok}^{p} - b_{ok}) . \tag{6}$$

④ Adjustment of parameter of activation function of nerve unit in hidden layer:

$$\frac{\partial E}{\partial a_{ok}} = \sum_{p=1}^{P} [\sum_{j=1}^{O} \frac{1}{\lambda_{oj}} (y_{oj}^{p} - y_{oj}^{p*})(y_{oj}^{p} - a_{oj})(1 + a_{oj} - y_{oj}^{p}) w_{kj}^{O}] . \tag{7}$$

$$\frac{\partial E}{\partial b_{hk}} = -\frac{1}{\lambda_{hk}} \sum_{p=1}^{P} \{(y_{hk}^{p} - a_{hk})(1 + a_{hk} - y_{hk}^{p}) [\sum_{j=1}^{O} \frac{1}{\lambda_{oj}} (y_{oj}^{p} - y_{oj}^{p*})(y_{oj}^{p} - a_{oj})(1 + a_{oj} - y_{oj}^{p}) w_{kj}^{O}] . \tag{8}$$

$$\frac{\partial E^{p}}{\partial \lambda_{hk}^{p}} = \frac{1}{(\lambda_{hk}^{p})^{2}} \sum_{p=1}^{P} \{(y_{hk}^{p} - a_{hk}^{p})(1 + a_{hk}^{p} - y_{hk}^{p})(X_{hk}^{p} - b_{hk}^{p}) [\sum_{j=1}^{O} \frac{1}{\lambda_{oj}^{p}} (y_{oj}^{p} - y_{oj}^{p*})(y_{oj}^{p} - a_{oj}^{p})(1 + a_{oj}^{p} - y_{oj}^{p}) w_{kj}^{Op}] . \tag{9}$$

The same as traditional BP algorithm: $\Delta Y(t+1) = -\eta \frac{\partial E}{\partial Y} + \mu \, \Delta Y(t)$. In the formula: η is rate of learning, μ is momentum factor, t is loop count, Y represents variable w, a, b and λ.

4 Combination of Chaos Algorithm and BP-AA Algorithm

Accordingly, another big deficiency of BP algorithm is that objective function is easy to trap in local minimum. Therefore, it is proposed in this paper the searching algorithm by combining chaos system and improved BP algorithm. Then, two algorithms' predominance complements so as to not only overcome the deficiency of local minimum of objective function, but also that of big iteration amount of chaos algorithm.

4.1 Characteristics of Chaos System

Chaos system has some good characteristics such as state ergodicity. The well-known Logistic mapping is a chaos system model being widest researched recently. Its model is illustrated as follow:

$$x_{k+1} = f(\mu, x_k) = \mu x_k (1 - x_k) . \tag{10}$$

When μ=4,the system in the formula above is absolutely in chaos state. The formula is written as:

$$x_{k+1} = 4x_k (1 - x_k) . \tag{11}$$

4.2 Neural Network Learning Algorithm by Embedding Logistic Mapping

We embed the Logistic chaos mapping in the BP network learning so that make the algorithm efficient but uneasy to trap in local minimum. The probability that this algorithm gets to global optimum is 1. We call it BP-AAEC algorithm (Adjustable activation function and Embedding Chaos algorithm).

The algorithm is simply illustrated as follow:

Step1: The initialization of the algorithm

① Given deviation parameter of neural network a, threshold b, gradient factor λ and weight w is X_k. Set initial value of each parameter X_0, threshold of error and max loop count m and m1. Order k=0, k1=0, k2=0;

② Random initialization variable I, and chaos variable with same network parameter dimension, are marked as Y_0. Order k3=0;

③ Order X* as the current optimum network parameter, $X^* = X_0$;

Step2: By learning BP-AA algorithm, the parameter is X_k^*, k=k+1, if $E(X^*)<\varepsilon$, algorithm will end;

Step3: Compare current value with optimum value.

If $E(X^*) - E(X_k^*) > e$, order $X^* = X_k^*$, k1=0, k2=0, turn to step 2;

If $E(X^*) \geq E(X_k^*)$ and $E(X^*) - E(X_k^*) < e$, order $X^* = X_k^*$, k2=0,

If k1≤m, k1=k1+1, turn to step 2;

If k1>m, order k1=0, turn to step4 ;

If $E(X^*) < E(X_k^*)$, X^* keeps unchanged, k2=k2+1

If k2≤m1, turn to step1 ;

If k2>m1, algorithm ends ;

Step4: Enter chaos iterate, $Y_{k3+1} = 4Y_{k3}(1 - Y_{k3})$, k3=k3+1 ;

Step5: Generate new parameter from Y_{k3}, that is, $X_{k+1} = X^* + \left|X^*\right| \cdot (2Y_{k3} - 1) \cdot r$, turn to step2;

5 Simulation and Analysis of Result

In order to prove the efficiency of BP-AAEC improved learning algorithm, we will compare the regular BP algorithm with BP-AAEC algorithm proposed in this paper, respectively by the examples of nonlinear function approximation and actual load.

BP-AAEC network parameter's random initialization is illustrated as follow: weight w∈(−0. 3, 0. 3); deviation parameter a∈(-0.2, 0.2); Threshold value b∈(-0.3, 0.3); gradient factor λ∈(0.8,1.2). Two neural networks both adopt variable learning rate and momentum item, and batch training model.

5.1 Example 1: Nonlinear Function Approximation

We design a function with many local minimum points such as formula 13.

$$f(x) = \sin(0.5x)\log((x + 0.001)/40)\,. \tag{13}$$

This function has many extreme points in domain. Take 100 as the number of samples in this interval, 1-10-1 as network structure of the regular BP algorithm, 1-6-1 as network structure of BP-AAEC algorithm, and the result is illustrated in Table 1:

Table 1. Simulation result of nonlinear function approximation

Accuracy	Regular BP algorithm	BP-AAEC algorithm
0.012	3941	334
0.01	5665	527
0.005	No convergent	1032

Simulation result illustrates: compared with regular BP algorithm, BP-AAEC algorithm can reach the same function approximating accuracy with less number of times of iteration□though the number of cells in hidden layer decreases. It illustrates that BP-AAEC algorithm has faster convergence and stronger nonlinear expression capability. Furthermore, it can reach fairly high error accuracy and avoid trapping in local minimum when improper initial value of regular BP algorithm leads to trapping in local minimum and no convergence.

5.2 Example 2: Load Forecasting

This paper proceeds load forecasting simulation and checking calculation, exemplified by the load data of Jiangbei district of Chongqing in 2003.This paper divides the load's historical data into two types: workday load and weekend load. 24 neural networks of each type forecast the load value 24 hours each day. Its structures of input and output layer are illustrated as follow:

Table 2. Neural network structure

Input layer	L(d-7,t-1) L(d-7,t) Tmax(d-7) Tmin(d-7) L(d-2,t-1) L(d-2,t) Tmax(d-2) Tmin(d-2) L(d-1,t-1) L(d-1,t) Tmax(d-1) Tmin(d-1) Tmax(d) Tmin(d)
Output layer	L(d,t)

In table 2, d represents forecasting date, t represents forecasting time, L(d, t) represents the load forecasting at t time d day, Tmax(d) represents the highest temperature at the date of d, Tmin (d) represents the lowest temperature at the date of d.

Take 14-12-1 as regular BP network structure, and 5,000 as the number of times of training; take 14-8-1 as BP-AAEC network structure, and 1,000 as the number of times of training; and the training samples are totally the same. The training shows that BP-AAEC spends less time in 1,000 times' training than the time spent in 5,000 time's training by regular BP network but the network's error is smaller after training. It indicates that BP-AAEC algorithm is more predominant. Table 2 is the comparison of the results between regular BP network forecasting and BP-AAEC forecasting in one workday.

Table 3. Contrast table of load forecasting result and error by two methods in a workday

Time point	Actual load	Regular BP	BP-AAEC	Time point	Actual load	Regular BP	BP-AAEC
1	319.6	328.4	323.6	13	356.7	362.4	343
2	292	311.1	294.8	14	385.3	384.9	368.3
3	272.9	290.1	267.1	15	382.1	390.1	373
4	255.5	275.6	264	16	378	381.7	366.9
5	243.2	261.2	254	17	373.5	383.4	361.4
6	230.4	248.1	229.9	18	371.8	376.7	365
7	215.5	222.7	217.5	19	355.1	364.2	337.6
8	208.7	210.9	206.6	20	361.6	368.7	355
9	248.7	262	251.2	21	392.5	398.3	379.2
10	307.3	317.7	301	22	413	419	402
11	334	320	326.7	23	407	424	387.2
12	355.6	363.9	350.4	24	379.3	393.7	361
Average relative error						3.432	2.587

Forecasting results indicate that BP-AAEEC algorithm proposed in this paper can improve the accuracy of load forecasting and greatly decrease load-forecasting error.

6 Conclusion

In this paper a novel variant activation (transform) sigmoid function with three parameters is proposed. Compared with traditional neural network, this method has many characteristics such as strong nonlinear expression ability and fast convergence, which greatly improve the efficiency of learning. Embedded with global search capability, the improved chaos algorithm allows the neural network learning to reach optimum, and to improve the accuracy of load forecasting. It can reach better forecasting result to combine the algorithm proposed in this paper with other load forecasting algorithms (Fuzzy clustering, similar day, wavelet, etc.)

References

1. Niu, D., Cao, S.: Electric Load Forecasting Technology and Its Application. Chinese Electric Press (1998)
2. Ma, W., Bai, X.: Short Term Load Forecasting Using Artificial Neuron Network and Fuzzy Inference. Power System Technology 27(5) (2003) 29-32

3. Wu, Y., Zhao, M.: A New Neural Networks with Variable Activation Functions and Its Application, Science in China (Series E) 27(1) (1997) 55-60
4. Hu, Y., Yan, X., Zhang, X.: An Improved BP Algorithm Based on a Variant Sigmoid Function with Three Parameters. DCABES Proceedings (2004) 144-148
5. Long, X., Xiao, T.: Neural Networks with Variable Activation Functions. Computer Engineering 27(12) (2001) 71-73
6. Treadgold Nicholas K., Gedeon Tamas D.: Simulated Annealing and Weight Decay in Adaptive Learning: The Sarprop Algorithm. IEEE Trans on Neural Networks (1998) 662-668.
7. Li, X., Zhou, E.: Chaos BP Hybrid Learning Algorithm for Feedforward Neural Network. Control and Decision 19(4) (2004) 462-464
8. Jing, J., Sun, Y.: Short-Term Load Forecasting Using Chaotic Learning Algorithm for Neural Network. Automation Electric Power System 7(2004) 32-35
9. Wang, Y., Tan, W.: Genetic Based Neural Network Control for Chaotic System. ACTA Physica Sinica. 52 (11) (2003) 2723-2728

Short Term Load Forecasting Using Neural Network with Rough Set

Zhi Xiao[1], Shi-Jie Ye[2], Bo Zhong[3], and Cai-Xin Sun[4]

[1] College of Economics and Business Administration, Chongqing University, 400044 Chongqing, China
Xiaozhi@cqu.edu.cn
[2] College of Economics and Business Administration, Chongqing University, 400044 Chongqing, China
kg898972000@yahoo.com.cn
[3] College of Mathematics and Physics, Chongqing University, 400044 Chongqing, China
Zhongbo@cqu.edu.cn
[4] Key Laboratory of High Voltage Engineering and Electrical New Technology, Ministry of Education, Chongqing University, 400044 Chongqing, China
Suncaixin@cqu.edu.cn

Abstract. Accurate Short term load forecasting (STLF) plays a significant role in the management of power system of countries and regions on the grounds of insufficient electric energy for increased need. This paper presents an approach of back propagation neural network with rough set (RSBP) for complicated STLF with dynamic and non-linear factors in order to develop the accuracy of predictions. Through attribution reduction based on variable precision with rough set, the influence of noise data and weak interdependency data to BP is avoided so the time taken for training is decreased. Using load time series from a practical system, we tested the accuracy of forecasting in specific days with comparison.

1 Introduction

Load forecasting, aiming at making best use of electric energy and relieve the conflict between supply and need, is a very crucial issue for the operational planning of electric power systems and corresponding researches. Short term load forecasting (STLF) is a major part of load forecasting for its real-time, controllable and the random, dynamic and non-linear factors influencing it. According to the characteristics of STLF, neural network has received increasing attention as a non-linear and dynamic system with strong robustness and fault tolerance, and has combined with other advanced algorithms such as grey system, wavelet transform, support vector machines and fuzzy theory to face the need for higher accurate STLF.

Many researches have been proposed during the last decade regarding STLF. Otavio A.S. Carpinteiro provided a neural model made up of two self-organizing map (SOM) nets—one on top of the other. It has been successfully applied to domains in which the context information given by former events plays a primary role [1]. Ayca

J. Wang et al. (Eds.): ISNN 2006, LNCS 3972, pp. 1259 – 1268, 2006.

Kumluca Topallia and Ismet Erkmenb gave method based on hourly electric consumption which is proposed to cluster input data. In order to have an idea about the success of the model, several alternate models were formed and got considerably better results [2]. Liao and Tsao proposed an integrated evolving fuzzy neural network and simulated annealing for load forecasting method. First they used fuzzy hyper- rectangular composite neural networks for the initial load forecasting. Then used evolutionary programming and simulated annealing to find the optimal solution of the parameters of fuzzy hyper-rectangular composite neural networks [3]. Different types of neural network has combined with for taking advantages of each type to conquer their drawbacks in STLF [4][5]. In [6][14], fuzzy theory as an alternative way, has integrated with neural net work for its ability to fuzzy and uncertain data and conditions in order to make it possible neural network deal with unknown objects. Another technique, soft theory like support vector machines, have been successfully employed to solve nonlinear regression and time series problem and applied to help neural network to their common advantages for STLF under little amount of data[7][8]. Furthermore, Neural network combined with wavelet transform has presented for STLF with little influence from noise data and high accuracy through data discretization of wavelet transform [9][10]. Optimal combined forecasting and grey theory have also applied to develop neural network [13][15]. In short, the above neural based forecasting approaches can generally been classified into two categories in accordance with the employed techniques. The first category treats the load pattern as a time-series signal and predicts the future load using the already mentioned time-series analysis techniques. The other category, attempts to develop the initialization, learning and training process of neural network and get more precise predictions. In this paper, a multilayer back propagation network (BP), widely applied for STLF because of its ability to study and remember the relation between inputs and outputs as well as to approach any types of function, is combined with one tool of soft theory – rough set for reducing the influence due to the drawbacks of BP such as low training speed and easily affected by noise and weak interdependency data through attribution reduction with rough set.

2 Simple Intro on Rough Set

2.1 Process of Knowledge Abstraction

After attribution reduction with the significance, the condition attribution set is composed of members who have obvious effect to goal attributions. That is the main process of analysis rough set deals with dynamic and non-linear issues. In details, suppose goal attribution (d) relates several condition attributions(C_1,C_2 …C_n), and let the value of attributions in different period of time be u_t =($C_{1,t}$,$C_{2,t}$,……,$C_{n,t}$;d_t). Rough set attempts to find the relation between d and (C_1,C_1 …C_n) by calculating the dependence the goal to each condition and the significance each condition in the condition set with the value above. In rough set theory, we call U the universe set and D the set of decision attributes, and C is the set of condition attributes. So $C_{i,,t}$ is the value of condition attribute C_i corresponding to u_t, in other words that means $C_i(u_t)$= $C_{i,,t}$. Also d_t s the value of decision attribute d corresponding to u_t, which means

$C_i(u_t)= C_{i,,t}$ (t stands for the certain time). In one word, Rough set, avoiding the man-made function and weight of factors, obtains the importance of each condition attribute to decision attributes and the significance of each condition attribution in the set of conditions. In other words, we get a rule (IF…THEN…), which means what the decision attributes will be while what the values of the significant condition attributes are [11][12].

2.2 Process of Attributions Reduction

The process of attribution reduction with RS is shown as follows.

Step1. Data pretreatment. The data collected from fact is different from each other in unit and absolute value due to the measure and the criteria, so the difference must be eliminated before the datc is used. In this paper, the value of load time series is discrete so we use the order of spaces which each value of C_j ascends in to replace the original.

Step2. Characterization and the foundation of knowledge system. We draw the characteristic of each condition attribution with the corresponding data after pretreatment so as to make the calculation for the dependence the goal to each condition precise. The common characterization methods are Equal distance quartile, Naïve-Scaler algorithm, SsmiNaïve Scaler algorithm, and discretion basing on Boolean calculation or rough set theory. In this paper, Equal distance quartile is used. Through characterization we can found the knowledge system and equivalence relationship as follow.

$$R_C = \{(u,v) \in U \times U \mid C_i(u) = C_i(v), i = 1,2...m\} \tag{1}$$

$$R_{C_j} = \{(u,v) \in U \times U \mid C_i(u) = C_j(v), i \neq j, i = 1,2...m, j = 1,2...m\} \tag{2}$$

$$R_d = \{(u,v) \in U \times U \mid d(u) = d(v)\} \tag{3}$$

It's obvious that R_C、R_{Cj}、R_d are all equivalence relationship on U, and we mark the set of U as U/R_C、U/R_{Cj} (j=1,2,…,m)、U/R_d. The first two are knowledge system on the basis of condition attributes and the last is knowledge system on the basis of decision attributes.

Step3. Calculate the dependence degree of d to condition attributes by

$$H(R_d / R_c) = -\sum_{[x] \in U/R_c} p[x] \sum_{[y] \in U/R_d} p([y]/[x]) \ln(p([y]/[x])) \tag{4}$$

where the $p[x] = \dfrac{card[x]}{card[U]}, x \in U/R$ and card(x) is the influence of x which can be calculated by the number of elements of x. The bigger $H(R_d/R_c)$ is, the more dependence d to C_j is.

Step4. Calculate weight of conditions and reduce indexes. The significance of C_j in the set of C can be defined as:

$$\omega(C_j, C, d) = H(R_d / R_{C_j}) - H(R_d / R_C) \tag{5}$$

$$H(R_d / R_{C_j}) = -\sum_{[x]\in U/R_{cj}} p[x] \sum_{[y]\in U/R_d} p([y]/[x]) \ln(p([y]/[x])),\ j = 1,2......16 \quad (6)$$

The higher the value of $\omega(C_j, C, d)$ is, the more significant C_j is in the set of C. If the value is 0, C_j is thought as a redundant attribute and can be reduced from the set of C, and then we get a new set of C after reduction which can be expressed as C = C-C_j. Also the weight of each condition attribute remaining in C is given by as:

$$\lambda_j = \frac{\omega(C_j, C, d)}{\sum_{i=1}^{16} \omega(C_i, C, d)} \quad (7)$$

3 RSBP for STLF

3.1 The Structure of RSBP

A neural cell of BP is shown in Fig. 1. There are R inputs and each one has a appropriate weight so as to be connected with the next layer.

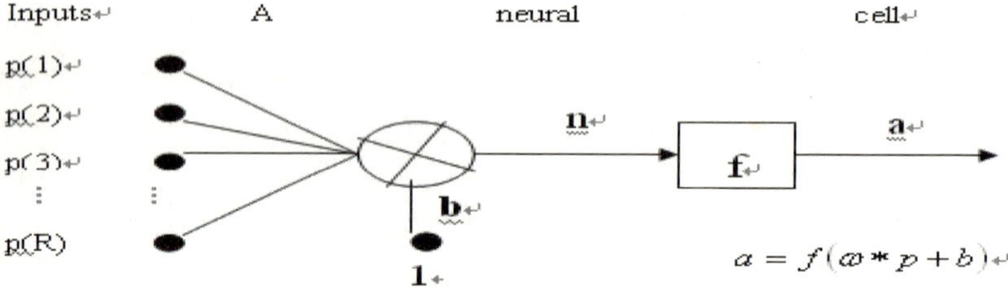

Fig. 1. A neural cell of BP

Here we apply a BP with an hidden layer in which the neural cells takes tan—sigmoid function for transform, and pruelin( ), a linear function, is used in output layer for transform in order to get values with a broad range. The whole structure is shown below:

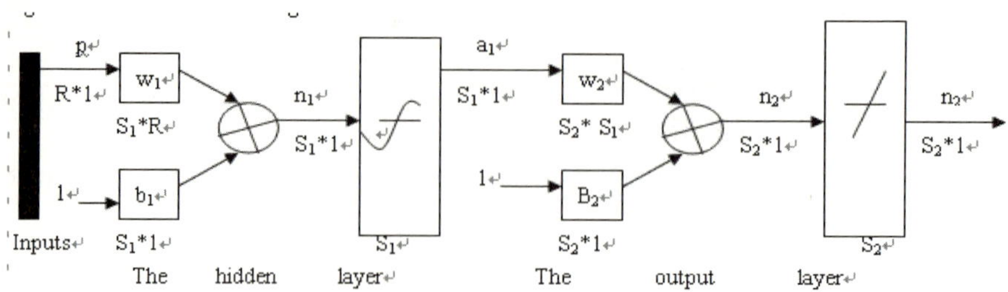

Fig. 2. The structure of BP in this paper

where $a_1 = \tan sig(\omega_1 * p + b_1)$ and $a_2 = purelin(\omega_2 * a_1 + b_2)$, besides there are 5 neural cells in hidden layer.

3.2 The Short Term Load Index System

The short term load index system is composed of goal attributions such as the maximum and the minimum of load of each day and condition attributions regarding weather. The values of attributions are from load time series of a practical power system from Jun 1st to Sep 30th as shown in Table 1.

Table 1. Actual load time series from Jun 1st to Sep 39th

Date	d_1	d_2	C_1	C_2	C_3	C_4	C_5	C_6	C_7
6.1	1916.852	1380.875	29.4	20.2	22.8	92	59	0.0	12
6.2	1992.921	1210.002	24.5	18.5	20.4	99	74	19.0	4
6.3	2283.564	1220.186	31.1	18.2	29.9	100	44	8.3	1
6.4	2419.789	1323.503	33.2	21.5	27.0	90	38	11.2	1
…	…	…	…	…		…	…	…	…
9.30	2305.640	1476.798	28.4	19.4	23.9	76	35	8.0	1

In Table 1, d_1 is the maximum of load in one day, d_2 is the maximum of load in one day, c_1 is the highest temperature, c_2 is the lowest temperature, c_3 is average temperature, c_4 is the highest relative humidity, c_5 is the lowest relative humidity, c_6 is sunlight time and c_7 is weather situation according to the situation of weather factors not mentioned above such as the speed of wind, rain, natural disasters, etc. It can be divided into 1 to 15, and the less is the better.

Suppose the goal attributions (d_1, d_2) relate the 7 condition attributions. We take the load time series of first 25 days as training samples and that of the last 5 days as forecasting samples. Now rough set is applied to find the dependency of each goal to each condition and the significance of each condition in the set of C with the corresponding data from training samples. Before that process, first we need to characterize the data with equal distance quartile, in other words, replace the data with the order of the space it lies in according to distance. The data after characterization is show in Table 2.

Table 2. The short term load index after characterization

d_1	d_2	C_1	C_2	C_3	C_4	C_5	C_6	C_7
1	1	2	1	3	3	1	3	1
1	1	1	1	1	3	3	1	3
1	1	1	2	3	2	3	1	3
2	1	1	2	2	3	3	2	3
…	…	…	…	…	…	…	…	…
2	3	2	3	3	1	2	2	3

We calculate the dependency each goal to each condition as the formula (4) given and the significance of each condition attribution in the set of C according to different goal attribution with formula (5) and formula (6). The results are shown in Table 3.

Table 3. The significance of each condition attribution according to different goal attribution

	C_1	C_2	C_3	C_4	C_5	C_6	C_7
d_1	0.70066	0.75129	0.63211	0.94428	0.93962	0.99962	1.00757
d_2	0.87240	0.87688	0.85741	0.88178	0.97178	1.05370	1.03245

It is obvious weather situation is the most significant for both goals, and the next 3 conditions are sunlight time, lowest humidity, highest humidity. Comparing with that, the effect the temperature conditions to load is inferior, especially the average temperature is the least significant of all 7 conditions so it can be reduced and the corresponding data and values should be deleted. And then the index system after attribution reduction and the new weight of each condition can be obtained with formula (7). So the following training process can be more effective and faster without the influence from noise and weak interdependent data.

Table 4. The index system after reduction and the new weight of each condition according to different goals

	C1	C2	C4	C5	C6	C7
d1	0.01172	0.01257	0.01580	0.01572	0.01672	0.01686
d2	0.01332	0.01339	0.01346	0.01484	0.01609	0.01577

3.3 The Training Process

In training process, we apply momentum method and adaptive learning rate to enhance the speed besides improve the credibility of algorithm. The sensitivity of local parts of error curved surface can be decreased with momentum method so that the trend that BP converges to local minimum can be restrained effectively. Using the function below, momentum method modifies w(k).

$$w(k+1) = w(k) + \alpha[(1-\eta)D(k) + \eta D(K+1)] \tag{8}$$

where w(k) is the weight of single cell, also it can be the weight vector, and $D(k) = \frac{-\partial E}{\partial w(k)}$ is the inverse gradient at time of k, so D(k-1) is the inverse gradient at time of k-1. α, a positive number, is the learning rate and η is the momentum gene between 0 and 1. With momentum the wave in process of study is alleviated and the constringency is improved. Adaptive learning rate is applied to save the time for study. A main reason for slow constringency speed of BP is the improper learning rate. The lower learning rate is the slower the constringency speed is, on the other hand it may cause excess modification resulting in oscillation or divergence. The algorithm of adaptive learning rate is provided as below.

$$w(k+1) = w(k) + \alpha(k)D(k) \tag{9}$$

$$\alpha(k) = 2^{\lambda}\alpha(k-1) \tag{10}$$

$$\lambda = sign[D(k)D(k-1)] \tag{11}$$

We take the maximum and the minimum of load as output vectors and the other attributions as input vectors, the value of data is consistent with that of short term load index system after reduction. There are 5 neural cells in hidden layer using tan-sigmoid function for transform, and the number of neural cells in output layer is the same as that of output vectors. After initialization, we start the training of network by letting error be 0.001, learning rate be 0.01 and training times be 3000. The training process is shown as:

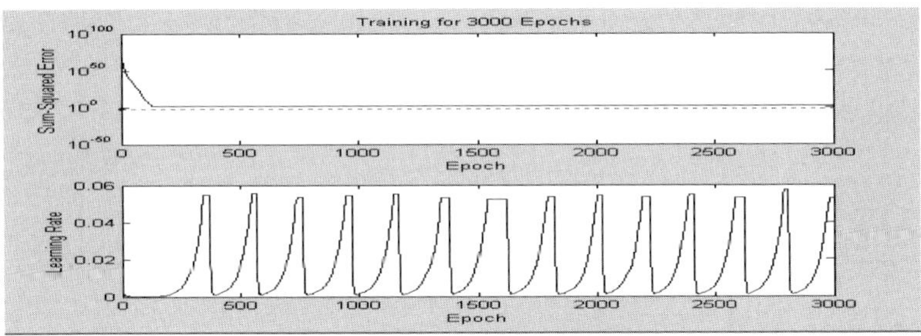

Fig. 3. The sum-squared errors and learning rate of each epoch

3.4 Forecasting

After training, the load time series of 7 condition attributions from forecasting samples are used as input vectors in the simulation of RSBP, and the predictions of each goal is obtained. At last, we compare the accuracy of predictions from RSBP and that of BP as below.

Table 5. The absolute errors between actual values and predictions

Date	E_1	E_2	E_3	E_4
9.26	4.9271	1.4202	8.3694	1.8721
9.27	1.4206	0.8709	8.8776	2.8208
9.28	30.6110	10.0383	30.6078	4.4624
9.29	3.1622	1.3312	1.8565	5.3820
9.30	2.0075	0.8389	2.2943	1.0896
SUM	42.1284	14.4996	52.0055	15.6270

In Table 5, $E_1$is the absolute errors of maximum from RSBP, E_2 is the absolute errors of minimum from RSBP, E_3 is the absolute errors of maximum from BP, E_4 is the absolute errors of minimum from BP.

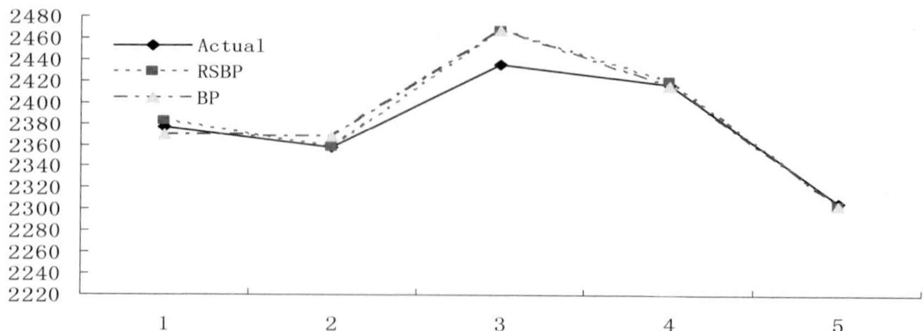

Fig. 4. The actual and predictions about the maximum of load

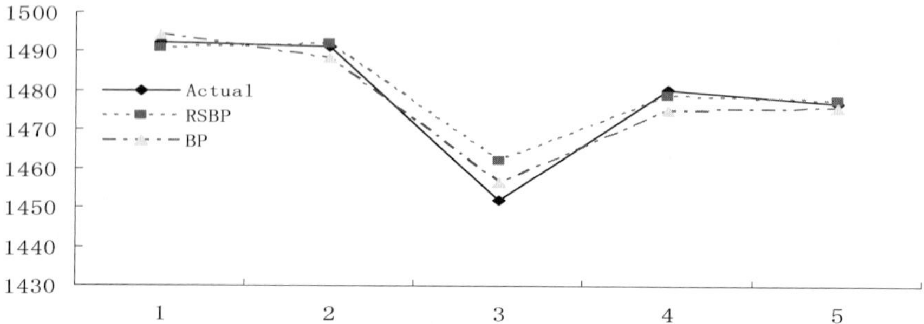

Fig. 5. The actual and predictions about the minimum of load

We can see, in Figs. 4-5, the predictions from both BPRS and BP approach close to the actual overall, especially there are some points at which the latter performs better than the former. However, it's undoubted that the accuracy of predictions from RSBP is better than that from BP when take the errors as a whole. And the former, with reference to the individuals, performs better at 7 items of 10 but the latter only performs better at 2 items.

4 Conclusion

Short term load forecasting (STLF) is important for the power system of countries and regions. Toward its non-linear, dynamic and random, neural network, as a dynamic system with strong robustness and fault tolerance has been applied with other algorithm for accurate STLF. Having reviewed the research recently, we attempted to combine BP with rough set. By taking advantage of soft computing of rough set to process the attribution reduction with the significance of each condition in the set of conditions, we got the short term load index system after reduction so as to reduce the influence from noise and weak interdependency data to the initialization, study and training of network. In training process, momentum method was applied to decrease the sensitivity of local parts of error curved surface so that the trend that BP converges to local minimum can be restrained effectively. Also adaptive learning rate was applied to reduce the time for study. By comparing the predictions from RSBP with that of BP, we testified the superiority of the former. Considering the factual circumstance of STLF, there are some issues need to be discussed and improved. First, the threshold for attribution deduction was manmade so there are subjective effects to the accuracy. Second, due to the limited origin of data, the index we used couldn't stand for all factual factors. Third, the credibility of model and the rationality of index system were not perfect because of the means by which we took for data characterization. Last, in order to save the time spent in initialization of network, the weights and the threshold of neural cells in hidden layer were determined automatically by program without optimization, it should be improved in prospective research.

Acknowledgements

The program we designed for the initialization, study, training and simulation of RSBP in this paper was written with the tool-box of MATLAB 6.1 produced by the MathWorks, Inc.

References

1. Carpinteiro, O.A.S., Reis, A.J.R., Alves da Silva, A.P.: A Hierarchical Neural Model in Short Term Load Forecasting. Applied Soft Computing 4 (2004) 405–412
2. Topallia, A.K., Erkmenb, I.: A Hybrid Learning for Neural Networks Applied to Short Term Load Forecasting. Neurocomputing 51 (2003) 495–500
3. Liao, G., Tsao, T.: Application of Fuzzy Neural Networks and Artificial Intelligence for Load Forecasting. Electric Power Systems Research 70 (2004) 237–244
4. Satish, B., Swarup, K.S., Srinivas, S., Rao, A.H.: Effect of Temperature on Short Term Load Forecasting Using an Integrated ANN. Electric Power Systems Research 72 (2004) 95–101
5. Hong, W.C.: Forecasting Regional Electricity Load Based on Recurrent Support Vector Machines with Genetic Algorithms. Electric Power Systems Research 74 (2005) 417–425
6. Al-Kandari, A.M., Soliman, S.A., El-Hawary, M.E.: Fuzzy Short-Term Electric Load Forecasting. International Journal of Electrical Power and Energy Systems 26 (2004) 111–122
7. Kodogiannis, V.S., Anagnostakis, E.M.: Soft Computing Based Techniques for Short-Term Load Forecasting. Fuzzy Sets and Systems 128 (2002) 413–426
8. Pai, P., Hong, W.: Support Vector Machines with Simulated Annealing Algorithms in Electricity Load Forecasting. Energy Conversion and Management 46 (2005) 2669–2688
9. Yao, S.J., Song, Y.H., Zhang, L.Z., Cheng, X.Y.: Wavelet Transform and Neural Networks for Short-Term Electrical Load Forecasting. Energy Conversion and Management 41 (2000) 1975–1988
10. Zhang, B., Dong, Z.: An Adaptive Neural-Wavelet Model for Short Term Load Forecasting. Electric Power Systems Research 59 (2001) 121–129
11. Rady, E.A., Kozae, A.M., Abd El-Monsef, M.M.E.: Generalized Rough Sets. Chaos Solitons and Fractals 21 (2004) 49–53
12. Hong, T., Tseng, L., Wang, S.: Learning Rules from Incomplete Training Examples by Rough Set. Expert Systems with Applications 22 (2002) 285-293

13. Zhang, D., Bi, Y. Q., Bi, Y. X., Bi, Y. M., Niu, Z., Luo, L.: Power Load Forecasting Method Base on Serial Grey Neural Network. Theory and Practice on Sysytem Engineering (2004) 128–132
14. Gao, J., Sun, H., Tang, G.: Application of Optimal Combined Forecasting Based on Fuzzy Synthetic Evaluation on Power Load Forecast. Journal of Systems Engineering 16 (2001) 106–110
15. Zhao, H.: The Application to Power Load Forecasting of an Optimization Combinatorial Prediction Modal. Operation Research and Management Science 14 (2005) 115–118

Application of Neural Network Based on Particle Swarm Optimization in Short-Term Load Forecasting

Dong-Xiao Niu[1], Bo Zhang[1], and Mian Xing[2]

[1] School of Business Administration, North China Electric Power University, 071003 Baoding, China
ndx@ncepu.edu.cn
[2] School of Mathematics and Physics, North China Electric Power University, 071003 Baoding, China
hdxm01@sohu.com

Abstract. To overcome the defects of neural network (NN) using back-propagation algorithm (BPNN) such as slow convergence rate and easy to fall into local minimum, the particle swarm optimization (PSO) algorithm was adopted to optimize BPNN model for short-term load forecasting (SLTF). Since those defects are partly caused by the random selection of network's initial values, PSO was used to optimize initial weights and thresholds of BPNN model, thus a novel model for STLF was built, namely PSO-BPNN model. The simulation results of daily and weekly loads forecasting for actual power system show that the proposed forecasting model can effectively improve the accuracy of SLTF and this model is stable and adaptable for both workday and rest-day. Furthermore, its forecasting performance is far better than that of simple BPNN model and BPNN model using genetic algorithm to determine the initial values (GA-BPNN).

1 Introduction

The Short-term load forecasting is one of the most crucial requirements for the power network's economic and stable running. Improving the accuracy of daily and weekly load forecasting is becoming even more significant than ever before due to the deregulation of the power utility worldwide. Recent studies have used artificial neural network for load forecasting due to its proven ability to be a universal approximate or for any non-linear continuous function with an arbitrary accuracy. Actually, a great number of promising results have been reported [1-4].

Concerning the training of NN, the BP algorithm is commonly adopted. However, Rmulhart proved that gradient-based BP algorithm relies on the selection of initial weights and thresholds, normally featured with random selection. Hence it will easily get stuck in a local minimum and converge slowly, which make the trained network inconsistent and the forecasting results less reliable.

Therefore some techniques are introduced in an attempt to resolve these drawbacks. Genetic algorithm (GA), featured with parallel search and global optimization, is becoming a popular NN training algorithm [5-8]. Ref. [6] showed that, compared

J. Wang et al. (Eds.): ISNN 2006, LNCS 3972, pp. 1269 – 1276, 2006.

with simple BP algorithm, using GA to determine the initial values of NN can largely improve the forecast precision as well as convergence speed. But involving complex genetic operations, GA makes the training time of NN to grow at exponential order along with the question scale [7]. Moreover, owing to the lack of effective local searching mechanism, the algorithm converges slowly or even ceases converging while approaching to optimal solution [8].

In this study PSO is adopted to determine the initial weights and thresholds of BPNN. As a result, a PSO-based NN model is developed for STLF, namely PSO-BPNN model, and is validated with daily and weekly loads forecasting for actual network. The simulation results show that this model is stable and adaptable for both workday and rest-day with better forecasting performance compared to that of BPNN model and GA-BPNN model.

2 PSO-BPNN Forecasting Model

2.1 The Principle of Particle Swarm Optimization

PSO is a population-based stochastic optimization technique developed by Dr. Eberhart and Dr. Kennedy in 1995, inspired by social behavior of bird flocking or fish schooling [9]. In a PSO system, the searching mechanism is similar to GA, but it doesn't concern such complex genetic operations as coding, crossover and mutation etc.. Instead, by cooperation and competition each particle adjusts flying according to its own flying experience and its companions' flying experience. Furthermore, this algorithm involves fewer parameters to adjust and is easy to implement and converge. Thus it has received much attention from various fields recently [10-13].

The PSO algorithm maintains a swarm of candidate solutions to the optimization problem under consideration. Each candidate solution is referred to as a particle. If the optimization problem has D variables, then each particle represents a D-dimensional point in the search space. The quality or fitness of a particle is measured using a fitness function that quantifies how close a particle is to the optimal solution. Each particle flies through the search space, having its position adjusted based on its distance from its own personal best position and the distance from the best particle of the swarm [13]. The performance of each particle, i.e. how close the particle is from the global optimum, is measured using a fitness function which depends on the optimization problem. Each particle i contains the following information: $X_i = (x_{i1}, x_{i2}, \cdots, x_{iD})$, the current position of the particle; $V_i = (v_{i1}, v_{i2}, \cdots, v_{iD})$, the current velocity of the particle; p_{best} , the personal best position of the particle, i.e. cognitive component; and g_{best} , the global best position of the particle swarm, i.e. social component.

Acting on the principle of pursing the optimal particle, in each iteration V_i and X_i are updated as follows:

$$V_{i+1} = w \cdot V_i + c_1 r_1 (P_i - X_i) + c_2 r_2 (P_g - X_i) \ , \tag{1}$$

$$X_{i+1} = X_i + V_i \ , \tag{2}$$

where w is the inertia weight, serving as a memory of previous velocities, and it usually linearly decreases from 0.9 to 0.4 throughout iteration; c_1 and c_2 are the acceleration constants, normally defined as 2; and r_1, r_2 are vectors with their elements sampled from a uniform distribution.

2.2 PSO Algorithm for Determining BPNN's Initial Values

The velocity--position search model of PSO algorithm is simple to implement and can coordinate global search and local search via inertia weight. In this way, it can guarantee optimal solution by bigger probability as well as increase the convergence speed in local regions, avoiding ceasing converging in local regions like GA [14]. Thereby PSO algorithm may theoretically optimize the initial weights and thresholds of BPNN and correspondingly build optimization NN model for STLF.

In order to design a PSO algorithm for determining initial values of BPNN, the first step is to establish reasonable particle model and search space. Since the PSO searching process is actually the adjustment of velocity and position of its different dimensions, the initial weights and thresholds of BPNN should correspond with particle position. Supposing the initial values of NN contain weights and thresholds altogether D parameters, D-dimensional X vectors are formed while each X vector represents a particle in PSO. The fitness degree can be measured by mean square error of NN's computed output and the expected output. Thus, the optimization process is essentially particles searching the best position in a D-dimensional space.

2.3 PSO-BPNN Forecasting Model

Since there have already been a universal computation program for BPNN in NN toolbox of Matlab which conveniences this paper greatly in network training. The main task for modeling PSO-BPNN focuses on the PSO algorithm for determining initial link weights and thresholds. Specific steps of modeling are summarized below:

Step1. Initialization of particles: assign randomly m groups of initial weights and thresholds within [-1, 1] under fixed network architecture and form initial set, where m equals the number of particles. The position vectors and velocity vectors are separately $X_i = (x_{i1}, x_{i2}, \cdots, x_{iD})$ and $V_i = (v_{i1}, v_{i2}, \cdots, v_{iD})$, where $i = 1,2,\cdots,m$, $-1 \le x_{ik} \le 1$, and $-V_{max} \le v_{ik} \le V_{max}$.

Step2. Calculation of fitness value F_i: set minimum mean square error of NN's actual output and expected output as the objective and calculate fitness value of X_i

$$f(X_i) = \frac{1}{n}\sum_{s=1}^{n} \frac{1}{1 + \sum_{k=1}^{j} (\hat{y}_{sk} - y_{sk})^2}, \tag{3}$$

where n denotes the number of samples, j denotes the number of nodes in output layer, $\hat{y}_{sk}$ is the actual output of the S th sample's k th node while y_{sk} is the corresponding expected output.

Step3. Comparison with current optimal values: if $f(X_i) > p_{best}$, then $p_{best} = f(X_i)$; and if $f(X_i) > g_{best}$, then $g_{best} = f(X_i)$.

Step4. Position Adjustment: change velocity and position using equation (1) and (2).

Step5. Stopping criterion: if iteration time is less than maximum iteration time $G_{\max}$, return to Step2; otherwise stop iterating.

Step6. Selection of NN's initial values: pick out the particle with the best fitness value from particle swarm and its position just represents the optimal initial weights and thresholds of NN under fixed network architecture.

Step7. Completion of NN training: input that optimal initial weights and thresholds into a NN with fixed network architecture and complete training in light of BP algorithm. Hence the PSO-BPNN model for STLF is completely built.

3 Simulation Results and Discussion

The performance validation of our proposed model is conducted using the same actual power load measurements recorded hourly over a period of five weeks (November 4–December 8, 2004) in certain region of Hebei Province.

In this section, PSO-BPNN model is applied to predict the one-day-ahead hourly load as well as one-week-ahead hourly load. A common baseline is also chosen for comparison with the simple BPNN model and GA-BPNN model by using the same data range of training (November 4–December 1, 2004) and testing (December 1–December 8, 2004) in the proposed model.

3.1 Basic Data of Forecasting Models

In order to design the network architecture, the changing rules of recorded hourly load require studying first. It is proved that the changing rules of hourly load in actual regions conform to the "weekly period" and "daily period" [15]. Thereby, for hourly load forecasting, we have selected one-day-ahead load, two-day-ahead load and one-week-ahead load at the same moment as well as the forecast day's average temperature as the network's input.

For illustrating the influence of PSO-BPNN model on forecasting performance, all NNs adopt 4-9-1 architecture and Sigmoid Function. The learning error ε is set as 2.0×10^{-5}. Other parameters involved in BP algorithm are set the same as that of NN toolbox of Matlab [16].

Especially for PSO-BPNN model: swarm scale m=50, inertia weight decreases linearly from 0.9 to 0.4 along with iteration, $c_1 = c_2$ =2, $V_{\max}$ =1.15, $G_{\max}$ =100.

For comparison GA-BPNN model: binary system coding method, probabilities of crossover and mutation are 0.90 and 0.03 respectively, and the group scale and iteration times are the same to that of PSO-BPNN model.

3.2 Daily Forecasting Performances of Three STLF Models

When testing a proposed model, it is necessary to quantify its prediction accuracy and validity in terms of meaningful measures. The following statistical metrics are used for comparing prediction performance, namely, Relative Error (RE), Maximum Relative Error (MRE), and Mean Absolute Relative Error (MARE). In addition, the whole experiment is carried out on a computer with the following configuration : CPU is SY1.8, EMS is 256MB, hard disk is 80G.

The hourly load data of testing sample are used to predict the 24-point load of December 2 2004, and the forecast errors of three models are drawn out in Fig.1 while the daily load forecasting performances of three models are shown in Table1.

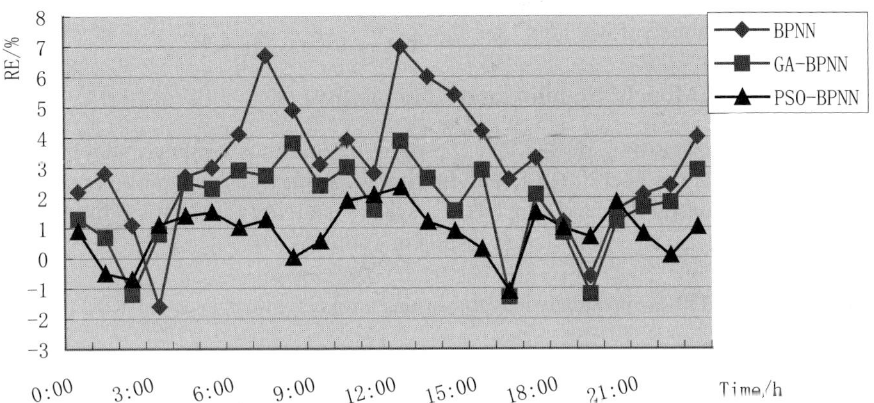

Fig. 1. Errors of forecasted load of actual network by three models

Table 1. Comparison of daily load forecasting performances of three models

Forecasting Model	MRE/%	MARE/%	Training Time/s
BPNN	7.02%	1.12%	4.9530
GA-BPNN	3.89%	2.11%	14.1210
PSO-BPNN	2.31%	1.02%	8.3021

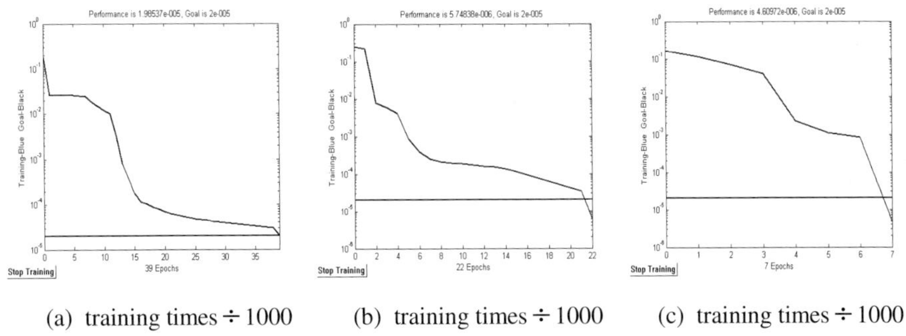

(a) training times ÷ 1000 (b) training times ÷ 1000 (c) training times ÷ 1000

Fig. 2. Training histories of learning error: (a) BPNN epoch and (b) GA-BPNN epoch and (c) PSO-BPNN epoch

Analysis: compared to simple BPNN forecasting model, both GA-BPNN model and PSO-BPNN model yield lower forecast errors. As for the optimized models, the error index values of PSO-BPNN model are better than that of GA-BPNN model, proving that PSO-BPNN model has higher stability and adaptability. Since optimized models involve determining initial values, their training times are longer than that of BPNN model. However, under the same learning scale, the training time of the proposed model decreases as much as 41.21% compared to GA-BPNN model. Thus the application of PSO algorithm proves superior to GA when applied for determining the initial values of NN. What's more, PSO algorithm is easy to implement and can improve the convergence of NN a lot. See Fig.2.

3.3 Analyses of Models' Stability and Adaptability

Only through forecasting results of a certain day cannot evaluate the proposed model's performance. Therefore the one-week forecasting is carried out to evaluate its forecast precision, stability and adaptability, which can testify its adaptability for both workday and rest-day. See simulation results in Table 2.

Table 2. Comparison of daily load forecasting errors of a week in actual network

Date (24 points)	BPNN		GA-BPNN		PSO-BPNN	
	MRE (%)	MARE (%)	MRE (%)	MARE (%)	MRE (%)	MARE (%)
12/02	6.23	2.81	3.29	2.21	2.56	1.06
12/03	7.11	3.18	4.83	2.64	2.67	1.16
12/04	6.58	2.64	4.17	2.15	2.06	1.23
12/05	6.02	2.34	3.66	1.83	2.37	1.12
12/06	6.84	2.87	3.94	1.99	2.18	1.21
12/07	5.75	3.06	3.82	1.87	1.79	1.05
12/08	7.06	3.17	4.01	2.30	2.05	1.20
Average	6.51	2.87	3.96	2.14	2.24	1.15

Analysis: compared to BPNN model, both GA-BPNN model and PSO-BPNN model receive obvious effect regarding two error indexes. When compared with each other, the MRE and MARE of GA-BPNN model are 3.96% and 2.14% respectively while those of PSO-BPNN model are 2.24% and 1.15%, a decrease of 43% and 46% separately. As for the variation scope within a week, the GA-BPNN model is 0.31% while PSO-BPNN model is only 0.18%. Therefore, the PSO-BPNN model shows better stability and adaptability for both workday and rest-day.

4 Conclusion

In this paper, PSO algorithm was adopted to determine the initial weights and thresholds of BPNN and a novel neural network model PSO-BPNN is proposed for STLF. Thus with better learning ability and generalization ability, PSO-BPNN model is relative stable and can converge globally faster. Based on historical power load, the proposed model was validated when existing neural networks are considered for comparison purposes.

Simulation results show that both GA-BPNN and PSO-BPNN models yield lower forecast errors compared with BPNN model. However, unlike GA which lacks effective searching mechanism, PSO algorithm concerns both local and global search, and can increase the convergence speed and guarantee optimal solution by bigger probability. The daily and weekly load forecasting proves that PSO-BPNN model can increase the forecast precision more effectively. Meanwhile it has better stability and adaptability for both workday and rest-day. This is beneficial to the power running with no need to model respectively.

Acknowledgment

This work is supported by National Science Foundation [Project No. 500077007] and Doctor Spot Special Foundation of NCEPU (North China Electric Power University) [Project No. 20040079008]. The provision of original load data from Hebei Provincial Power Corporation is also appreciated.

References

1. Metaxiotis, K., Kagiannas, A., Askounis, D. and Psarras, J.: Artificial Intelligence in Short Term Electric Load Forecasting: A State-of-the-Art Survey for the Researcher. Energy Conversion and Management 44 (2003) 1525–1534
2. Alfares, H.K., Nazeeruddin, M.: Electric Load Forecasting: Literature Survey and Classification of Methods. Int. J. Systems Science 33 (2002) 23–34
3. Li, Y.Y., Niu, D.X.: An Optimum Credibility Based Integrated Optimum Gray Neural Network of Monthly Power Load Forecasting. Power System Technology 29 (2005) 16–19
4. Tayati, W., Chankaipol, W.: Substation Short-Term Load Forecasting Using Neural Network with Genetic Algorithm. Proceedings of The 2002 IEEE Region 10 Conference on Computer, Communication, Control and Power Engineering, Beijing, China, (2002) 1787–1790
5. Leung, F.H.F., Lam, H.K., Ling, S.H., Tam, P.K.S.: Tuning of the Structure and Parameters of Neural Network Using an Improved Genetic Algorithm, IEEE Trans. Neural Networks 14 (2003) 79–88
6. Zhang, B.P.: Short Term Load Forecasting of Power System Based on GA-BP Neural Network. Ph.D. Thesis, Xi'an University of Technology (2003)
7. Yang, J.M., Kao, C.Y.: A Robust Evolutionary Algorithm for Training Networks. Neural Computing and Application 10 (2001) 214–230
8. Franchini, M.: Use of a Genetic Algorithm Combined with a Local Search Method for the Automatic Calibration of Conceptual Rainfall-Runoff Models. Hydrological Science Journal 41 (1996) 21–39
9. Kennedy, J., Eberhart, R.C: Particle Swarm Optimization. Proceedings of IEEE International Conference on Neutral Networks, Australia, Ⅳ (1995) 1942–1948
10. Parsopulos, K.E., Vrahatis, M.N.: Recent Approaches to Global Optimization Problems through Particle Swarm Optimization. Nature Computing, Kluwer Academic Publishers (2002) 235–306

11. Eberhart, R.C., Shi, Y.: Particle Swarm Optimization: Developments, Applications and Resources. Proc. IEEE Int. Conf. on Evolutionary Computation, Seoul (2001) 81–86
12. Shi, Y., Eberhart, R.C.: A Modified Particle Swarm Optimization. Proc. IEEE International Conference on Evolutionary Computation, Anchorage (1998) 69–73
13. Shi, Y., Eberhart, R.C.: Parameter Selection in Particle Swarm Optimization. Evolutionary Programming, VII, Proc. EP 98 (1998) 591–600
14. Dautenhahn, K.: Book Review: Swarm Intelligence. Genetic Programming and Evolvable Machines 3 (2002) 93–97
15. Niu, D.X., Cao, S.H., Zhao, L., et al.: Load Forecasting Technology and Its Application. Beijing, Electric Power Press of China (1998)
16. Liu, S., Yang, L.L., Wang, Z.G., et al.: Combined Power Load Forecast Model Based on Matlab Neural Network Toolbox. Electric Power Automation Equipment 23 (2003) 59–61

Study of Neural Networks for Electric Power Load Forecasting

Hui Wang[1,4], Bao-Sen Li[2], Xin-Yang Han[3], Dan-Li Wang[1], and Hong Jin[1]

[1] Intelligence Engineering Lab, Institute of Software, CAS, Beijing, China
Wanghui@ios.cn
[2] Jining Electric Power Company, Shandong, China
[3] Dynamic Economy Research Center, State Power Cooperation, Beijing, China
[4] School of Information Engineering, USTB, Beijing, China

Abstract. Electric Power Load Forecasting is important for the economic and secure operation of power system, and highly accurate forecasting result leads to substantial savings in operating cost and increased reliability of power supply. Conventional load forecasting techniques, including time series methods and stochastic methods, are widely used by electric power companies for forecasting load profiles. However, their accuracy is limited under some conditions. In this paper, neural networks have been successfully applied to load forecasting. Forecasting model with Neural Networks is set up based on the analysis of the characteristics of electric power load, and it works well even with rapidly changing weather conditions. This paper also proposes a novel method to improve the generalization ability of the Neural Networks, and this leads to further increasing accuracy of load forecasting.

1 Introduction

The objective of electric power company is to provide end users electricity with safety and stability, and electric load forecasting is very important for the operating of power systems. Short Term Load Forecasting is important for the economic and secure operation of power system. Fundamental operations functions such as unit commitment, hydro-thermal co-ordination, interchange evaluation and security assessment requires a reliable forecasting result. Many methods have been proposed for load forecasting, such as time series models, regression models, state space methods, etc. [1, 2, 3]. Recently, some news algorithms have been proposed, such as expert system models, neural networks models, fuzzy set models, etc. [4, 5, 6, 7].

With the rapid development of economy, the power consumption is growing fast in China. The load profile is quit different from that of years ago because it is becoming more sensitive to weather. Ordinary people are using more and more electric equipments, such as air conditioner, heater, etc. and these equipments are weather related load. Traditional methods don't work well under this new condition, especially when there is a sudden change of weather. Neural Networks is applicable to electric power load forecasting with advantages due to that the problems in load forecasting have the characteristics of nonlinearity, time-variation, vagueness and complexity that present difficulties to handle [7,8].

J. Wang et al. (Eds.): ISNN 2006, LNCS 3972, pp. 1277 – 1283, 2006.

This paper studies short term load forecasting models with Neural Networks based on the analysis of the characteristics of electric power load, and the practical experiments show that it works well even with rapidly changing weather conditions. The remainder of this paper is organized as follows. In section2, we describe the characteristics of electric load and study the relation of load and weather. Section 3 proposes forecasting models with neural networks. Section 4 proposes a novel method to improve the generalization ability of neural networks. Section 5 conducts experiments to examine the performance of the forecasting algorithm proposed in this paper. Finally some conclusions are drawn in section 6.

2 Analysis of Electric Power Load

There are many kinds of end users in electric power systems, and their electricity consumption patterns are quite different. As the first step to set up models to forecast the electricity demand, the characteristics of load should be studied and analyze the main factors related to load. Composition of load, time, weather, and random issues are the main factors that effect electric load. The purpose of short term load forecasting is to predict the load profile of the next day. Daily load profile includes two parts. One is the basic load, which is not related to weather; while the other part is highly related to the change of weather, such as temperature, humidity, wind speed, rain, etc. Temperature has a very big effect. The relation between load and temperature is a shape of saddle, as shown in Fig. 1.

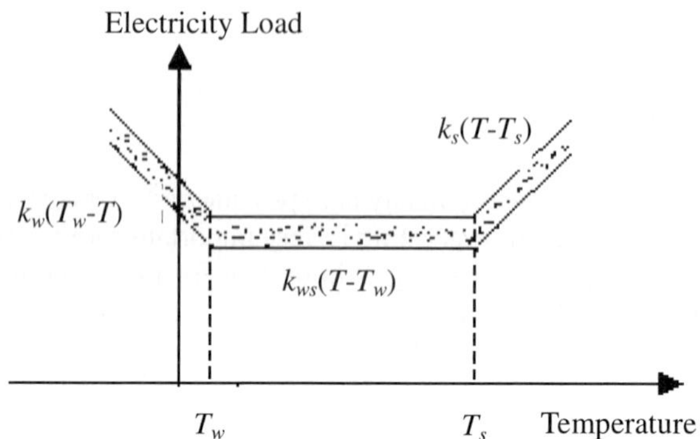

Fig. 1. Weather sensitive load changing with temperature

From Fig. 1 we can see that electricity load increases when the temperature is larger than a certain value; while it decreases when the temperature is lower than a certain value. k_w, k_{ws} and k_s are the corresponding slopes in figure 1. It's very important to get the value of changing points T_w *and* T_s. The plot might be a little different in different area, as shown in Fig. 2. In the figure, k_{w1}, k_{w2}, k_{w3}, k_{ws} and k_s are the corresponding slopes. This is because when the temperature is very low, some areas provide heat service through coal plants instead of electricity.

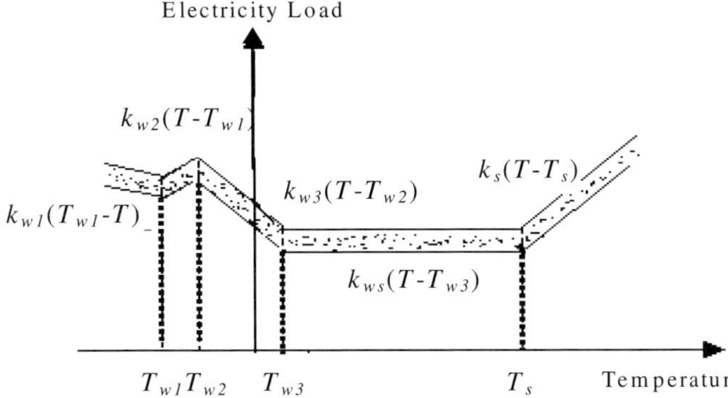

Fig. 2. Weather sensitive load changing with temperature in some areas

3 Short-Term Load Forecasting Model

Neural networks provide a flexible nonlinear modeling framework. Neural Networks is applicable to Electric Power Load Forecasting with advantages due to that the problems in load forecasting have the characteristics of non-linearity, time-variation,

Table 1. Inputs and Outputs of the forecasting model

	Node Number	Variables
Input	1, 2, …, 24	24 hour load of previous day
	25	Average temperature of previous day
	26	Average temperature of forecasting day
	27, 28, …, 33	Day type
Output	1, 2, …, 24	24 hour load of forecasting day

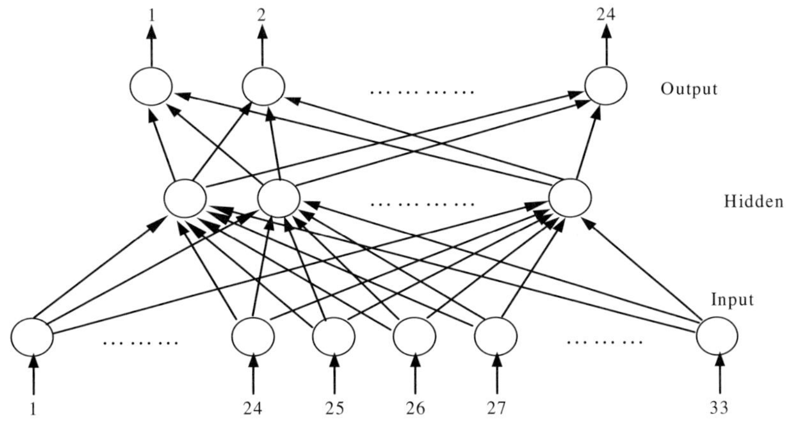

Fig. 3. Structure of the load-forecasting model

vagueness and complexity, which present difficulties to handle. Researchers have proposed forecasting models based on neural networks [7, 8, 9]. We designed a short term load forecasting model with neural networks as shown table 1. There are three layers in the neural network model. The output layer is the 24 hour load to be predicted, and the input layer are variables whose value can be got when forecasting, such as the 24 hour load of the previous day, temperatures, etc.

Based on the model described above, several other forecasting models were developed as shown in Table 2. These models have the same outputs which are the 24 hour load of the forecasting day, but the inputs are different.

Table 2. Inputs of other models based on neural networks

Model 1	01-24: 24 hour load of previous day 25-26: high and low temperature of previous day 27-28: high and low temperature of forecasting day 29-35: day type
Model 2	01-24: 24 hour load of previous day 25-48: 24 hour load of the day before previous day 49-50: high and low temperature of previous day 51-52: high and low temperature of forecasting day 53-59: day type
Model 3	01-24: 24 hour load of previous day 25-48: 24 hour load of the same day of last week 49-50: high and low temperature of previous day 51-52: high and low temperature of forecasting day 53-59: day type
Model 4	01-24: 24 hour load of previous day 25-48: 24 hour load of the day before previous day 49-72: 24 hour temperature of the previous day 73-96: 24 hour temperature of forecasting day 97-103: day type

4 A Novel Method to Improve Generalization Ability

When the neural networks are used after training, it's quite normal that the application is not good even though small error is reached during training. This is related to the generalization ability of neural networks. The generalization ability of neural networks is an important performance criterion of neural networks [10]. Many methods have been proposed to increase the generalization ability of neural networks, such as early stopping, regularization, neural network ensembles [11, 12], etc. In order to increase the forecasting accuracy, we developed a new method to improve the generalization ability of the neural networks from the point of training data set.

Training set has a big effect on the generalization ability of the neural networks. When collecting data for training, we consider two points. One is training data should be enough, and the other one is to minimize the redundancy of training data. When the training data is not enough, it's difficult for the neural networks to learn the hidden relation among the data. The neural networks just memorize the data in this situation. When the training data set is too large, it will take too much time to train the neural networks, and this does not lead to good generalization ability necessarily. We propose a new method for preparing the training data, and the steps are as following:

1) *Select data of the days which are similar to the day whose load is to going be predicted;*
2) *Select enough data of the curve area and guarantee the change of neighbor data is less than the training error;*
3) *Modify the error function of the training, as described below.*

Table 3. Short term load forecasting result (MW)

Hour	Actual value	Load forecasted	Error(%)
0	3585	3570	-0.42
1	3618	3522	-2.65
2	3477	3419	-1.67
3	3452	3384	-1.97
4	3388	3372	-0.47
5	3350	3405	1.64
6	3548	3624	2.14
7	3966	4000	0.86
8	3983	4031	1.21
9	4279	4261	-0.42
10	4250	4347	2.28
11	4500	4578	1.73
12	4476	4454	-0.49
13	4091	4040	-1.25
14	4218	4204	-0.33
15	4335	4307	-0.65
16	4411	4298	-2.56
17	4691	4555	-2.90
18	5108	5062	-0.90
19	5484	5424	-1.09
20	5264	5217	-0.89
21	4952	4847	-2.12
22	4522	4420	-2.26
23	4082	4016	-1.62
Root Mean-Square Error			1.76%

In the BP algorithm for neural network training, the error is formula 1.

$$E = \sum_{k=1}^{m} E_k \tag{1}$$

Where E_k is the error for each sample and m is the number of training sample. We modified the error function, as it is formula 2.

$$E = \sum_{k=1}^{m} \beta_k * E_k \tag{2}$$

β_k is a positive adjusting coefficient less than 1. It can be set by performing multi-regression analysis of the training samples. Firstly, sample error regression is computed, thenβ_k can be calculated through normalization.

5 Experiments

In order to test the proposed method for load forecasting, we used it in a provincial electric power company. The next day's 24-hour electric load was predicted and compared with the real values. The experimental results are as following:

6 Conclusions

Neural Networks is applicable to Electric Power Load Forecasting with advantages due to that the problems in load forecasting have the characteristics of nonlinearity, time-variation, vagueness and complexity which present difficulties to handle. Electric Power Load Forecasting model with Neural Networks is set up based on the analysis of the characteristics of electric power load, and it works well even with rapidly changing weather conditions. We use the method o forecast the 24 hours load and the average error is less than 2%. This paper also proposed a novel method to improve the generalization ability of the Neural Networks by modifying the error function for training the neural networks, and this set up a sound basis to improve the forecasting accuracy of the load forecasting models.

Acknowledgement

This work is supported by China NSF under Grant No. 60373056 and 60374058.

References

1. Song, K., Baek, Y., Hong, D., Jang, G.: Short-Term Load Forecasting for the Holidays Using Fuzzy Linear Regression Method. IEEE Transactions on Power Systems 20(1) (2005) 96–101
2. Charytoniuk, W., Chen, M.S., Van Olinda, P.: Nonparametric Regression Based Short-Term Load Forecasting. IEEE Transactions on Power Systems 13(3) (1998) 725–730

3. Fan, J.Y., McDonald, J.D.: A Real-Time Implementation of Short-Term Load Forecasting for Distribution Power Systems. IEEE Transactions on Power Systems, 9(2) (1994) 988–994
4. Rahman, S., Hazim, O.: Load Forecasting for Multiple Sites: Development of an Expert System-Based Technique. Electric Power Systems Research 39(3) (1996) 161–169
5. Papadakis, S. E., Theocharis, J. B., Bakirtzis, A. G.: A load curve based fuzzy modeling technique for short-term load forecasting, Fuzzy Sets and Systems 135(2) (2003) 279 – 303
6. Ruzic, S., Vuckovic, A., Nikolic, N.: Weather Sensitive Method for Short-Term Load Forecasting in Electric Power Utility of Serbia. IEEE Transactions on Power Systems 18(4) (2003) 1581–1586
7. Chen, H., Canizares, C.A., Singh, A.: ANN-Based Short-Term Load Forecasting in Electricity Markets. Proceedings of the IEEE Power Engineering Society Transmission and Distribution Conference, Vol. 2 (2001) 411–415
8. Khotanzad, A., Rohani, R.A., Maratukulam, D.: ANNSTLF– Artificial Neural Network Short-Term Load Forecaster–Generation Three. IEEE Transactions on Power Systems 13(4) (1998) 1413–1422
9. Taylor, J.W., Buizza, R.: Using Weather Ensemble Predictions in Electricity Demand Forecasting. International Journal of Forecasting 19(1) (2003) 57–70
10. Feng, N., Wang, F., Qiu, Y.: Novel Approach for Promoting the Generalization Ability of Neural Networks. International Journal of Signal Processing 2(2) (2005) 1304–4478
11. Inoue, H., Narihisa, H.,: Improving Generalization Ability of Self-Generating Neural Networks Through Ensemble Averaging. In: Terano, T., Liu, H., Chen, A. (Eds.): Proceedings of the 4th Pacific-Asia Conference on Knowledge Discovery and Data Mining. Lecture Notes in Computer Science, Vol. 1805. Springer-Verlag (2000) 177–180
12. Franco, L., Cannas, S. A.: Generalization and Selection of Examples in Feed forward Neural Networks. Neural Computation 12(10) (2000) 2405–2426

A Neural Network Approach to m-Daily-Ahead Electricity Price Prediction

Hsiao-Tien Pao

Department of Management Science, National Chiao Tung University, Taiwan
htpao@cc.nctu.edu.tw

Abstract. This paper proposes an artificial neural network (ANN) model to predict m-daily-ahead electricity price using direct forecasting approach on European Energy Exchange (EEX) market. The most important characteristic of this model is the single output node for m-period-ahead forecasts. The potentials of ANNs are investigated by employing cross-validation schemes. Out-of-sample performance evaluated with three criteria across five forecasting horizons shows that the proposed ANNs are more robust multi-step-ahead forecasting method than autoregressive error models (AUTOREG). Moreover, ANN predictions are quite accurate even when the length of forecast horizon is relatively short or long.

1 Introduction

Since the beginning of floating electricity prices, electricity price forecasting has become one of the main endeavors for researchers and practitioners in energy markets. Many of these problems can be modeled as mathematical programs. An overview of mathematical programming problems in electricity markets can be found in Conejo and Prieto [1]. Reported techniques to forecast day-ahead prices include ARIMA models [2] and [3], dynamic regression models [4], other time series techniques [5] and [6], neural network procedures [7]-[10], and wavelet transform models [11] and [12]. Recently, transfer function models have been proposed based on both past electricity prices and demands by Nogales and Conejo [13]. A data mining based electricity price forecast framework, which can predict the normal price as well as the price spikes have been proposed by Lu et al [14]. Finally, Conejo et al [15] present a wavelet transform and ARIMA hybrid model to forecast day-ahead electricity prices for the Spanish electricity markets.

Neural network applications for electricity price forecasting have yielded mixed results that may largely by attributed to problems in data selection and sampling variation. Cross-validation is a resampling technique that uses multiple training and test subsamples. Results from the cross-validation analysis will provide valuable insights on the reliability or robustness of neural networks with respect to sampling variation. A moving validation scheme with moving windows of fixed length provides an opportunity to investigate the effect of structural changes in a series on the performance of neural network forecasters [16].

This paper proposes an artificial neural network model to predict m-daily-ahead electricity price using direct forecasting approach on EEX market. The most important

J. Wang et al. (Eds.): ISNN 2006, LNCS 3972, pp. 1284 – 1289, 2006.

characteristic of this model is the single output node for m-period-ahead forecasts. In general, the proposed models are able to avoid too many rounds of errors. M-daily-ahead forecasting is useful for evaluating the robustness of a forecasting technique.

2 Neural Networks for Time Series Forecasting

Numerous neural network models have been proposed and used for forecasting [17]. They learn from examples or experiences and are particularly noted for their flexible function-mapping ability. Time series forecasting linear models assume that there is an underlying process from which data are generated and that the future values of a time series are solely determined by the past and current observations. Neural networks are able to capture the autocorrelation structure in a time series even if the underlying law governing the series is unknown or too complex to describe. In its applications, the data series is usually divided into a training set (in-sample data) and a test set (out-of-sample). The training set is used for the construction of the neural network whereas the test set is used for measuring the predictive ability of the model. Training process is used essentially to find the connection weights of the networks.

For an univariate time series forecasting problem, suppose we have N observations $y_1, y_2, \ldots, y_N$ in the training set, $y_{N+1}, y_{N+2}, \ldots, y_{N+m}$ in the test set, and we need the m-step-ahead forecasts. However, one or more output nodes can be used. If one output node is employed for multi-step-ahead forecasts, then the iterative forecasting approach is assumed and the forecast values are iteratively used as inputs for the next forecasts. On the other hand, if the number of output nodes is equal to the length of the forecasting horizon, then the direct forecasting approach is used in which we forecast the future values directly from the network outputs [16].

In order to avoid too many rounds of errors, this research proposes a network with single output node and p input nodes by using direct forecasting approach for m-step-ahead forecasting. The N-m-p+1 training patterns in the proposed network are

$$\begin{aligned} y_{p+m} &= f(y_p, y_{p-1}, \cdots, y_1) \\ y_{p+m+1} &= f(y_{p+1}, y_p, \cdots, y_2) \\ &\vdots \\ y_N &= f(y_{N-m}, y_{N-m-1}, \cdots, y_{N-m-p+1}) \end{aligned} \tag{1}$$

and the m testing patterns are

$$\begin{aligned} y_{N+1} &= f(y_{N+1-m}, y_{N-m}, \cdots, y_{N-m-p+2}) \\ y_{N+2} &= f(y_{N+2-m}, y_{N-m+1}, \cdots, y_{N-m-p+3}) \\ &\vdots \\ y_{N+m} &= f(y_N, y_{N-1}, \cdots, y_{N-p+1}). \end{aligned} \tag{2}$$

The training objective is to find the connection weights such that an overall predictive error means (SSE) is minimized. For this network structure, $SSE = \sum_{i=p+m}^{N}(y_i - \hat{y}_i)$ where $\hat{y}_i$ is the output from the network. The number of input nodes p corresponds to the number of lagged observations used to discover the underlying pattern in a time series. Too few or too many input nodes can effect either the learning or predictive capability of the network. Experimentation with a pilot sample is often used to find the appropriate numbers of hidden and input nodes.

3 Research Design and Methodology

This study focuses on the out-of-sample performance of the proposed neural networks in electricity prices at EEX. As discussed earlier, multi-step-ahead predictions are modeled. More specifically, the following research questions are addressed: 1) how robust is the neural network out-of-sample performance with respect to sampling variation and time frame? 2) what is the forecasting ability of neural networks in long and short forecast horizons? 3) what is the out-of-sample performance of neural networks relative to the linear models such as Box-Jenkins autoregressive error models?

To answer these questions, first, we employ a 20-fold cross-validation scheme with fixed length of moving series to deduct the sampling variation effects. Second, five different length forecast horizons with three performance measures are utilized in this study. Finally, modified Box-Jenkins models are applied to the data series and the out-of-sample results are compared to those of neural networks.

In Germany, the EEX based in Leipzig provides day-ahead prices for electricity and also forward contracts with varying maturities. Data on prices can be downloaded from their respective websites: www.eex.de. The electricity prices of the Phelix-Base at EEX applied here are 732 daily data recorded in December 1 2002 through December 1 2004. Fig. 1 plots this time series.

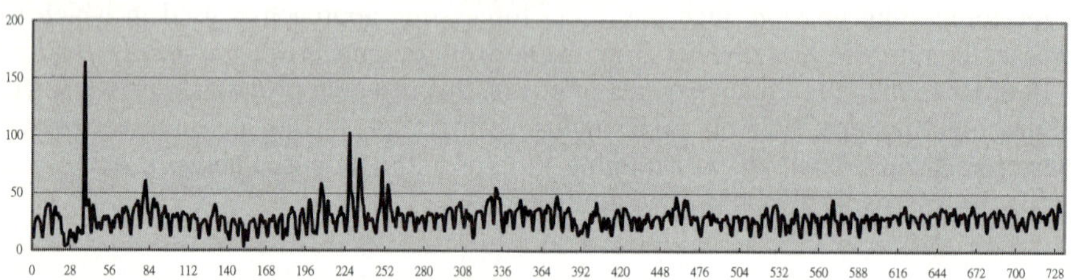

Fig. 1. Day-ahead prices for electricity from Dec-2002 to Dec-2004 at EEX

Looking at the autocorrelation function (ACF) for this working series, we see that the ACF has spikes at lags 1, 7, and multiple of 7. This implies that the time series observations separated by lags of multiple of 7 time units have a strong positive linear relationship. Thus, to build *m*-step-ahead forecasting models, we choose y_{t-m}, y_{t-m-7}, y_{t-m-14}, and y_{t-364} as the input nodes and y_t as the corresponding output node. The training and testing patterns for the *m*-step-ahead forecasting neural networks are proposed in the following equations:

$$\begin{aligned} &\textit{training} \quad \textit{pattern}: \quad y_t = f(y_{t-364}, y_{t-m-14}, y_{t-m-7}, y_{t-m}), \qquad t = 365, \cdots, N, \\ &\textit{testing} \quad \textit{pattern}: \quad y_p = f(y_{p-364}, y_{p-m-14}, y_{p-m-7}, y_{p-m}), \qquad p = N+1, \cdots, W \end{aligned} \tag{3}$$

where $W=N+m$ is the length of the time series in each subset and $N-364$ is the number of training data. Assume the length of available time series is always longer then the selected time period *W*. Then, we employ a cross validation scheme to evaluate the performance of the proposed *m*-step-ahead structure in Eq. (1)-(3). A moving cross-validation method with 20 folds is utilized. The walk-forward testing procedure uses multiple overlapping training and test sets as illustrated in Fig. 2. The number of training data and test data in each fold is *599-m-364* and *m*. Five forecasting horizons, *m*=7, 14, 21, 28, 91, are chosen in this study. The number of input nodes and the number of hidden nodes are not specified a priori. More than 50 experiments are conducted to determine the best combination of the learning rates, momentum, number of input nodes, and the number of hidden nodes. Throughout the training, the NeuralWare utility, 'SAVEBEST' is used to monitor and save the lowest root mean square (RMS) error from the training set. The best RMS error results are obtained by using a learning rate of 0.2, a momentum rate of 0.1, 4 input nodes: $y_{t-m}, y_{t-m-7}, y_{t-m-14}, y_{t-364}$, and 7 nodes in a single hidden layer that uses the generalized data learning rule and a sigmoid transfer function ($y=1/(1+e^{x})$). The best architecture of the networks is {4:7:1}.

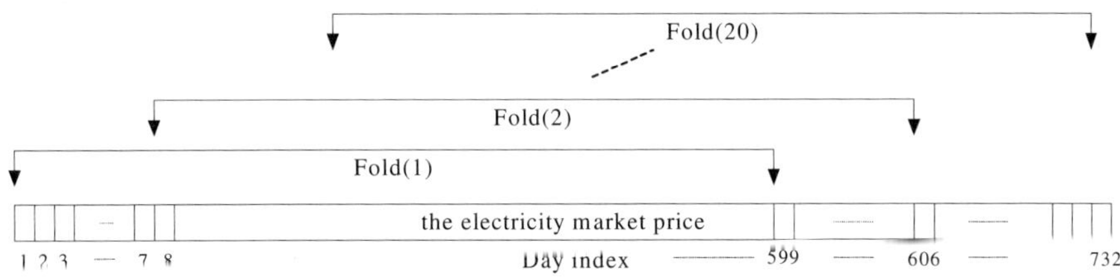

Fig. 2. Cross validation scheme

We also apply traditional autoregressive error models to the electricity price data. When time series data are used in regression analysis, often the error term is not independent through time. If the error term is autocorrelated, the efficiency of ordinary least-square parameter estimates is adversely affected and standard error estimates are biased. The following AUTOREG model corrects for serial correlation.

$$\begin{aligned} y_t &= c + v_t \\ v_t &= -\psi_1 v_{t-1} - \psi_2 v_{t-2} - \ldots - \psi_m v_{t-m} + \varepsilon_t \\ \varepsilon_t &\sim IN(0,\sigma^2). \end{aligned} \tag{4}$$

The notation $\varepsilon_t \sim IN(0,\sigma^2)$ indicates that each ε_t is normally and independently distributed with mean 0 and variance σ^2. The cross-validation method with 20 test sets is the same as the ANN models. To avoid the effects of sampling variation for the out-of-sample performance, the averages of RMSEs, MAEs, and MAPEs of 20 test sets for each forecasting horizon are compared with ANN models.

4 Empirical Results

This research proposes a neural network with single output node and *p* input nodes by using direct forecasting approach. The AUTOREG procedure is used to select available autoregressive error models to the 20 test sets for each horizon. It is not surprising to find that lags 1, 7, 12, 14, 21, 28 are included in all of the models. The averages of RMSEs, MAEs, and MAPEs of 20 test sets for each forecasting horizon are compared with ANN models. Table 1 shows the results. Several observations can be made from it. First, for each forecast horizon, the average values of three evaluation statistics of AUTOREG are larger than ANN models. Second, the variation in results is relatively large in short forecast horizon for each method. For example, in the one-week horizon, the standard errors of MAPE for 20 test sets are 2.28 and 3.96 respectively for both models, more than 2.16 (=(2.28+2.39+2.08+2.04+2.01)/5) and 3.54, the average values of standard errors for five forecast horizons of both models. As the forecast horizon increase, the observed variation decreases. For longer forecast horizon, ANNs still have smaller standard errors of three evaluation statistics, but AUTOREG is not. These results may be expected since the linear model is not available for too long prediction period. Finally, depend on the different input node $(y_{t-m}, y_{t-m-7,\dots,} y_{t-364})$, the proposed ANN with one output node model can be used to predict m period ahead using direct forecasting approach in which we forecast the future values directly from the network outputs. It is better than the iterative forecasting approach, because the forecast values are iteratively used as inputs for the next forecast in the iterative forecasting approach. Overall, depend on the cross-validation scheme, the proposed m-step-ahead forecasting ANNs are better than autoregressive error models.

Table 1. Out-of-sample comparison between neural networks and AUTOREG models

	Forecasting horizon									
	One Week		Two Weeks		Three Weeks		Four Weeks		Three Months	
	ANN	AUTO	ANN	AUTO	ANN	AUTO	ANN	AUTO	ANN	AUTO
	The averages of RMSE, MAE, and MAPE for 20 folds									
RMSE	**4.14**	6.34	**4.05**	5.89	**2.28**	4.57	**3.26**	5.25	**3.98**	5.48
MAE	**3.42**	5.26	**3.30**	4.85	**2.74**	4.01	**2.44**	4.10	**3.09**	5.20
MAPE(%)	**9.12**	14.6	**8.70**	14	**8.22**	11.5	**8.51**	13.2	**8.59**	14.3
	The standard errors of RMSE, MAE, and MAPE for 20 folds									
RMSE	**1.61**	2.10	**1.38**	2.07	**1.07**	1.62	**1.10**	1.71	**0.94**	1.84
MAE	**1.73**	2.54	**1.44**	2.23	**1.36**	1.90	**1.21**	1.86	**1.10**	1.92
MAPE(%)	**2.28**	3.96	**2.39**	3.50	**2.08**	2.94	**2.04**	3.40	**2.01**	3.88

5 Conclusions

In this study, we investigate the potential of neural network models in the prediction of daily electricity price utilizing cross-validation scheme. Our results clearly show that the accuracy of neural networks is not very sensitive to the length of forecast horizon, but autoregressive error models are. Neural network predictions are particularly superior when the forecast horizon is relatively short or long. Additionally, cross-validation results show that the sampling variation is relatively large in short forecasting horizons for both models. As the forecast horizon increase, the observed variation decreases. For longer forecast horizon, neural networks still have smaller observed variation, but AUTOREG is not. Furthermore, the proposed neural networks have a better ability to learn from data patterns in the training time period and successfully predict m-step-ahead outcomes for electricity prices on EEX.

References

1. Conejo, A.J., Prieto, F.J.: Mathematical Programming and Electricity Markets 9 (2001) 1-54
2. Contreras, J., Espinola, R., Nogales, F.J., and Conejo, A.J.: ARIMA Models to Predict Next-Day Electricity Prices. IEEE Trans. Power Syst. 18 (2003) 1014-1020
3. Fosso, O.B., Gjelsvik, A., Haugstad, A., Birger, M., and Wangensteen, I.: Generation Scheduling in a Deregulated System. IEEE Trans. Power Syst. 14 (1999) 75-81
4. Nogales, F.J., Contreras, J., Conejo, A.J., and Espinola, R.: Forecasting Next-Day Electricity Prices by Time Series Models. IEEE Trans. Power Syst. 17 (2002) 342-348
5. Obradovic, Z. and Tomsovic, K.: Time Series Methods for Forecasting Electricity Market Pricing. Proc. IEEE Power Eng. Soc. Summer Meeting 2 (1999) 1264-1265
6. Crespo, J., Hlouskova, J., Kossmeier, S., and Obersteiner, M.: Forecasting Electricity Spot Prices Using Linear Univariate Time Series Models. App. Energy 77 (2002) 87-106
7. Szkuta, B.R., Sanabria, L.A., and Dillon, T.S.: Electricity Price Short-Term Forecastin Using Artificial Neural Networks. IEEE Trans. Power Syst. 14 (1999) 851-857
8. Ramsay, B. and Wang, A.J.: A Neural Network Based Estimator for Electricity Spot-Pricing with Particular Reference to Weekend and Public Holidays. Neurocompute 23 (1998) 47-57
9. Zhang, L., Luh, P.B., and Kasiviswanathan, K.: Energy Clearing Price Prediction and Confidence Interval Estimation with Cascaded Neural Networks. IEEE Trans. Power Syst. 18 (2003) 99-105
10. Rodriguez, C.P., Anders G.J.: Energy Price Forecasting in the Ontario Competitive Power System Market. IEEE Trans. Power Syst. 19 (2004) 366-374
11. Yao, S.J., Song, Y.H.: Prediction of System Marginal Prices by Wavelet Transform and Neural Network. Elect. Mach. Power Syst. 19 (2004) 983-993
12. Kim, C.I., Yu, I.K., Song, Y.H.: Prediction of System Marginal Price of Electricity using Wavelet Transform Analysis. Energy Convers. Manage. 43 (2002) 1839-1851
13. Nogales, F.J., Conejo, A.J.: Electricity Price Forecasting Through Transfer Function Models. Journal of Operational Research Society (2005) 1-7
14. Lu, X., Dong, Z.Y., Li, X.: Electricity Market Price Spike Forecast with Data Mining Techniques. Electric Power Systems Research 73 (2005) 19-29
15. Conejo, A.J., Plazas, M.A., Espinola, R. Molina, A.B.: Day-Ahead Electricity Price Forecasting Using the Wavelet Transform and ARIMA Models. IEEE Trans. Power Syst. 20 (2005) 1035-1042
16. Hu, M.Y., Zhang, G., Jiang, C.X., Patuwo, B.E.: A Cross-Validation Analysis of Neural Network Out-of-Sample Performance in Exchange Rate Forecasting. Decision Science 30 (1999) 197-216
17. Zhang, G., Patuwo, B.E., Hu, M.Y.: Forecasting with Artificial Neural Networks: the State of the Art. International Journal of Forecasting 14 (1998) 35-62

Next-Day Power Market Clearing Price Forecasting Using Artificial Fish-Swarm Based Neural Network

Chuan Li and Shilong Wang

Faculty of Software Engineering, Chongqing University,
Chongqing 400044, China
chuanli@21cn.com, slwang@cqu.edu.cn

Abstract. Market clearing price (MCP) is one of the most important factors impacting on power system. Taking into account the features of deregulation and fluctuation, this paper uses artificial neural network to forecast next-day MCP, with period-decoupled data sequence and wavelet transform. For the purpose of better performance, an improved learning algorithm of artificial fish-swarm is proposed. By simulating fish-swarm actions, in random searching for foods, artificial fish-swarm based neural network (AFNN) achieves global optimum. Comparing with traditional next-day MCP forecasting methods, the suggested method could achieve better adaptability and greater predictive accuracy, which was proved by the experimental results.

1 Introduction

Market clearing price (MCP) is one of the most important factors for load controlling and market monitoring of power system. As its predictive accuracy is related to both sides' profits, next-day MCP forecasting method is essential for power market.

Many methods have been developed for next-day MCP forecasting recent years, such as Regressive Analyzing [1], Auto Regressive Integrated Moving Average [2] and Chaotic Time Series Analyzing [3]. With the successful applications in power load forecasting of NN [4], literature [5] forecasted short-term load using BPNN and literature [6] expanded the similarity of price time series into particular period. All of these studies have achieved good effects.

Affecting factors related to MCP include power load, demand, and other market conditions, such as power plants bidding strategies, market rules. So it is more difficult to forecast MCP. Using system load rate (SLR) instead of load to define relationship between supply capacity and market demand, using period-decoupled sequence instead of traditional 24-hour sequence, using wavelet transform to extract proper signals, using artificial fish-swarm algorithm (ASFA) based NN to map the influences of these nonlinear factors, the paper forecasts next-day MCP more effectively. BP learning, which is widely used in NN training, has shortages of long training period, slow convergence and probability to fall into local minimum point. By simulating fish-swarm actions, a NN training random optimizing method, artificial fish-swarm algorithm, is employed for next-day MCP forecasting.

J. Wang et al. (Eds.): ISNN 2006, LNCS 3972, pp. 1290 – 1295, 2006.

2 Forecasting Model

With the development of power market, electricity can be traded as commodities. Hence, MCP has the feature of fluctuation. Affected by supply and demand, short-term price also features particular periodicity at the same time in different days.

According to the study of literature [6], SLR data embody the effects of weather condition and historical price data including the impacts of power bidding strategies, system capacity and networks restrictions. So when forecasting next-day MCP, the historical price and SLR data are suggested as input variables of artificial fish-swarm based neural network (AFNN).

2.1 Extract Efficient Signal Using Wavelet Transform

As there are so many peaks, valleys and other noises in MCP curves, wavelet transform is employed to extract efficient signal from original signal function $f(t)$.

Let $\varphi(t) \in L^2(R)$ denote a wavelet base function, and sub-wavelet functions $\{\psi_{a,b}(t)\}$ can be inferred as

$$\psi_{a,b}(t) = |a|^{1/2} \varphi[(t-b)/a] \quad (a \neq 0, b \in R) \tag{1}$$

where a-scale factor; b-time shift factor. Approximated by multi-resolution, $f(t)$ is mapped into an orthogonal function space. Wavelet functions with the same scale form a base of the function space. With continual mapping, components in the same sub-spaces of $f(t)$ are grouped. Let J denote a scale, $k \in Z$, and $f(t)$ is divided into two parts:

$$f(t) = \sum_{j=-\infty}^{J} \sum_{k=-\infty}^{\infty} d_{j,k} \psi_{j,k}(t) + \sum_{k=-\infty}^{\infty} C_{J,k} \varphi_{J,k}(t) \tag{2}$$

where $d_{j,k}$-wavelet transform coefficient; $C_{J,k}$-scale expansion coefficient. According to equation (2), wavelet transform divides original signal into the first part (lower frequency components) and the second part (higher frequency components). The first part shows the profile of $f(t)$. If the signal fluctuation is regular, instead of using original signal, lower frequency components can reflect actual effects of input variables. Higher frequency components that are details of the signal reflect impacts of noises and other high frequency variables.

2.2 Forecasting Network

When forecasting next-day MCP, according to the price fluctuation at different hours, one-day price is decoupled into 24 different time series. If power market regulated half an hour clearing, one-day price could be decoupled into 48 time series.

Taking into consideration the periodicity of MCP, following factors are selected as input variables of NN: $S(d,t)$ - SLR of forecasting hour, $M(d,t-1)$ - price of one hour before forecasting, $S(d,t-1)$ - SLR of one hour before forecasting, $M(d-1,t)$ - price of forecasting day, $S(d-1,t)$ - SLR of forecasting day, $M(d-7,t)$ - price of 6 days before forecasting and $S(d-7,t)$ - SLR of 6 days before forecasting.

A three-layer full connected feed forward NN including a hidden layer can approximate any nonlinear functions arbitrarily well. Having considered all the above factors, the paper forecasts next-day MCP with a three-layer feed forward NN trained by ASFA. Fig. 1 shows the structure of the AFNN model. Arranging 24-hour $M(d,t)$, $(t=1,2,\ldots,24)$ price series, next-day MCP can be obtained.

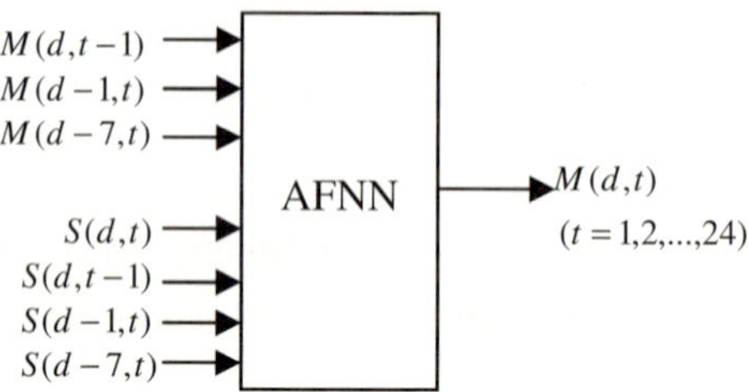

Fig. 1. AFNN for next-day MCP forecasting

3 NN Training Using ASFA

By simulating actions of the fish, a novel ASFA has been advanced recently [7],[8]. Deriving from basic actions of a single fish, ASFA reaches global optimum in each fish's local optimization for foods. As is researched, ASFA has the advantages of fast training, local extremum overcoming, possibility to reach global optimum, better adaptability and so on. So ASFA is suggested to train the weights of the NN.

When using ASFA to train a three-layer feed forward NN [9], let output vector of input neurons is $(x_1,\ldots,x_j,\ldots,x_n)$, input vector of hidden layer neurons is $(s_1,\ldots,s_i,\ldots,s_h)$, and output vector of output layer is $(y_1,\ldots,y_k,\ldots,y_m)$. Optimizing variables include the following parameters: $[\mathbf{w}_{\mathbf{ij}}]$-weight matrix of the input layer-hidden layer, $[\mathbf{v}_{\mathbf{ki}}]$-weight matrix of the hidden layer-output layer, $[\mathbf{w}_{\mathbf{i0}}]$-threshold vector of the hidden neurons, $[\mathbf{v}_{\mathbf{k0}}]$ - threshold vector of the output neurons. Natural fish swarm seeks for the most foods and AFNN seeks for the least error by adjusting weights. If F is food consistence, the error between actual output and expectation of NN is $E=1/F$.

Let V denote visual field of the artificial fish, S denote maximal step, δ denote swarm factor and each fish denote a NN. If the swarm consists of artificial fish vector $(X_1,\ldots,X_p,\ldots,X_q,\ldots,X_N)$, then distance between X_p and X_q is:

$$d_{pq}=\sum_{i=1}^{h}\sum_{j=1}^{n}\left[w_{ij}(p)-w_{ij}(q)\right]^2+\sum_{k=1}^{m}\sum_{i=1}^{h}\left[v_{ki}(p)-v_{ki}(q)\right]^2+\sum_{i=1}^{h}\left[w_{i0}(p)-w_{i0}(q)\right]^2 + \sum_{k=1}^{m}\left[w_{i0}(p)-w_{i0}(q)\right]^2 \quad (3)$$

Take example by w_{ij} of $[\mathbf{w_{ij}}]$ to analyze the optimizing actions of artificial fish. Optimizing processes of $[\mathbf{v_{ki}}]$, $[\mathbf{w_{i0}}]$ and $[\mathbf{v_{k0}}]$ are similar to w_{ij}.

1) Preying.

$$\begin{cases} w_{ij}(i+1) = w_{ij}(i) + Random(S)[w_{ij}(j) - w_{ij}(i)]/d_{ij} & (F_i < F_j) \\ w_{ij}(i+1) = w_{ij}(i) + Random(S) & (F_i \geq F_j) \end{cases} \tag{4}$$

where in its V, artificial fish $w_{ij}(i)$ compares F with another random artificial fish $w_{ij}(j)$. If $F_i < F_j$, then $w_{ij}(i)$ moves a random step $Random(S)$ in $[0\square\ S]$ towards $w_{ij}(j)$, or else $w_{ij}(i)$ moves in random direction to prey. Then set $w_{ij}(i+1) \leftarrow w_{ij}(i)$.

2) Swarming.

$$w_{ij}(i+1) = w_{ij}(i) + Random(S)[w_{ij}(c) - w_{ij}(i)]/d_{ic} \qquad (F_c / n_f > \delta.F_i) \tag{5}$$

where $w_{ij}(i)$ searches for fellow fish in its V. Let n_f donate the number of the fellow fish and $w_{ij}(l)$ denote one of the fellow fish. Suppose a virtual fish $w_{ij}(c)$ in the center of the fellow fish:

$$w_{ij}(c) = \sum_{l=1}^{n_f} w_{ij}(l)/n_f \tag{6}$$

If $F_c / n_f > \delta.F_i$, $w_{ij}(i)$ moves towards the center. Otherwise $w_{ij}(i)$ acts preying.

3) Following.

$$w_{ij}(i+1) = w_{ij}(i) + Random(S)[w_{ij}(m) - w_{ij}(i)]/d_{im} \qquad (F_m > \delta.F_i) \tag{7}$$

where $w_{ij}(i)$ seeks for the fellow fish in its V. Let $w_{ij}(m)$ denote an artificial fish that has the most F in the fellow fish. If $F_m > \delta.F_i$, $w_{ij}(i)$ moves towards $w_{ij}(m)$. Otherwise $w_{ij}(i)$ moves in accordance with prey action.

4) Bulletin.

$$w_{ij}(b) = w_{ij}(i) \qquad (F_i > F_b) \tag{8}$$

where set bulletin artificial fish $w_{ij}(b)$ to record the weight of optimized artificial fish. After each action, each fish $w_{ij}(i)$ compares F with $w_{ij}(b)$. If $F_i > F_b$, $F_i \rightarrow F_b$ and $w_{ij}(i) \rightarrow w_{ij}(b)$. So bulletin artificial fish $w_{ij}(b)$ is always the optimized fish.

Running actions of the artificial fish, each fish achieves local extremum. Let N_l donate iterative times, artificial fish run N_l times iteration to obtain optimum. Meanwhile, random moves of individuals overcome the limitations of long training period, slow convergence and possibility to abort optimization routine before it reaches optimum.

4 Experimental Results

In this section, we use historical price and SLR data of PJM Power Market (USA) to forecast next-day MCP. Decoupling historical data into 24-hour period, the Daubechies orthogonal wavelet transform is suggested to map time series. Fig. 2 shows MCP curves of June, 2004 transformed by MATLAB wavelet software package.

Using ASFA to train NN, the paper forecasts next-day MCP with network model of fig. 1. In this example, set N =100, V =0.5, S =0.2 and δ =0.25. Forecasting results and absolute percentage error (MAPE) of next-day MCP from June 21 to 27 (year 2004) are shown in fig. 3 and table 1. Meanwhile, the paper forecasts the same MCP using Regressive Analyzing (RA) and BPNN (no wavelet transform and 24-hour decoupling). Comparing next-day MCP with actual results, we can draw the conclusion that AFNN is more accuracy than traditional methods.

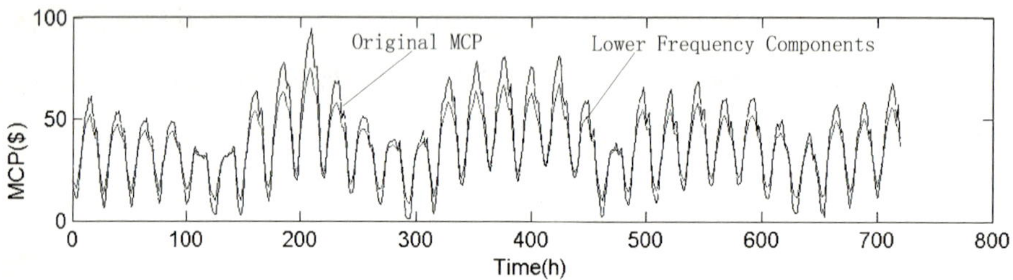

Fig. 2. Monthly MCP curves of PJM power market

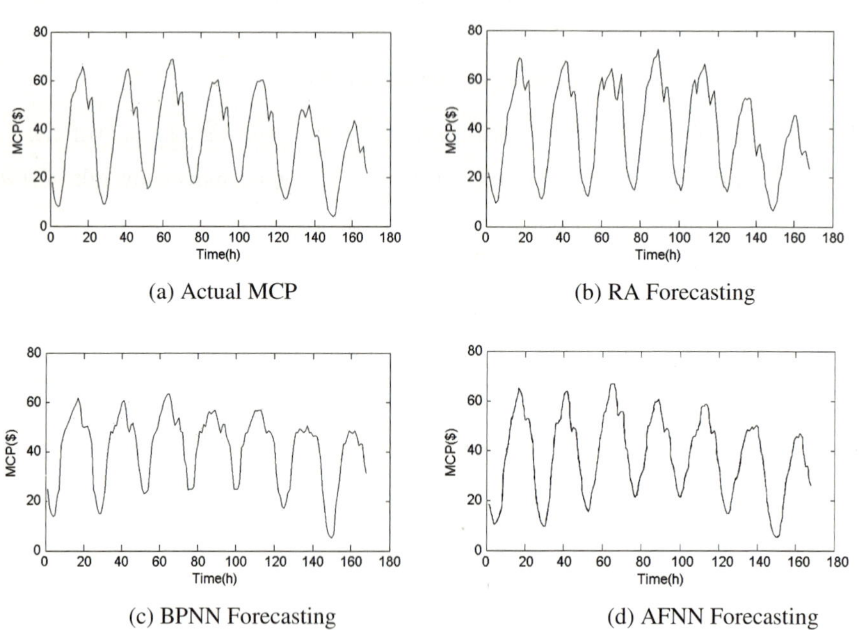

Fig. 3. Comparison of Actual MCP and forecasting MCP

Table 1. MAPE of next-day MCP forecasting using different methods (%)

Date	June 21	June 22	June 23	June 24	June 25	June 26	June 27
RA	13.7195	10.8565	11.1486	12.7778	11.8243	12.6226	14.3129
BPNN	7.3685	6.9371	7.8582	7.9237	8.3872	9.7126	9.6278
AFNN	4.3691	3.8692	3.4902	4.7840	3.9738	4.8235	5.1683

5 Conclusions

AFNN applied to forecast next-day MCP simulates swarm intelligence of animals. Deriving from basic actions of fish swarm, ASFA has the advantages of fast training, local extremum overcoming, swift global optimum accessing, and so on. AFNN forecasting method uses efficient 24 period-decoupled data extracted by wavelet transform. Comparing with traditional next-day MCP forecasting methods, the suggested method could achieve better adaptability and greater predictive accuracy which was proved by the experimental results.

References

1. Ni, E., Luh, P.B.: Forecasting Power Market Clearing Price and Its Discrete PDF Using a Bayesian-based Classification Method. IEEE Power Engineering Society Winter Meeting, Columbus (2001) 1518-1523
2. Zhou, M., Zheng, Y., Ni, Y.X., Li, G.Y.: An ARIMA Approach to Forecast Electricity Price with Accuracy Improvement by Predicted Errors. IEEE Power Engineering Society General Meeting, Tokyo (2004) 233 - 238
3. Nogales, F.J., Contreras, J., Conejo, A.J.: Forecasting Next-day Electricity Prices by Time Series Model. IEEE Transactions on Power System. **2** (2002) 342-348
4. Nagasaka, K., Al, M.M.: Long-term Peak Demand Prediction of 9 Japanese Power Utilities Using Radial Basis Function Networks. IEEE Power Engineering Society General Meeting, Tokyo, Japan (2004) 315-322
5. Hippert, H.S., Pedreira, C.E., Souza R.C.: Neural Networks for Short-Term Load Forecasting: A Review and Evaluation. IEEE Transaction on Power Systems. **16** (2001) 44-55
6. Zhang, X., Wang, X.F., Chen, F.H., Ye, B.: Short-term Electricity Price Forecasting Based on Period-decoupled Price Sequence. Proceedings of the CSEE. **15** (2005) 1-6
7. Li, X.L., Qian, J.X.: Studies on Artificial Fish Swarm Optimization Algorithm Based on Decomposition and Corrdination Techniques. Journal of Circuits and Systems. **1** (2003) 1-6
8. Li, X.L., Shao, Z.J., Qian, J.X.: An Optimizing Method Based on Autonomous Animates: Fish-swarm Algorithm. Methods and Practices of System Engineering. **11** (2002) 32-38
9. Ma, J.W., Zhang, G.L., Xie, H., Zhou, C.L.: Optimization of Feed-forward Neural Networks Based on Artificial Fish-swarm Algorithm. Computer Applications. **10** (2004) 21-23

Application of Evolutionary Neural Network to Power System Unit Commitment

Po-Hung Chen[1] and Hung-Cheng Chen[2]

[1] St. John's University, Department of Electrical Engineering, Taipei, Taiwan
[2] National Chin-Yi Institute of Technology, Institute of Information and Electrical Energy, Taichung, Taiwan
phchen@mail.sju.edu.tw, hcchen@chinyi.ncit.edu.tw

Abstract. This paper presents an evolutionary neural network (ENN) approach for solving the power system unit commitment problem. The proposed ENN approach combines a genetic algorithm (GA) with a back-propagation neural network (BPNN). The BPNN is first used as a dispatch tool to generate raw unit combinations for each hour temporarily ignoring time-dependent constraints. Then, the proposed decoding algorithm decodes the raw committed schedule of each unit into a feasible one. The GA is then used to find the finally optimal schedule. The most difficult time-dependent minimal uptime/downtime constraints are satisfied throughout the proposed encoding and decoding algorithm. Numerical results from a 10-unit example system indicate the attractive properties of the proposed ENN approach, which are a highly optimal solution and faster rate of computation.

1 Introduction

The UC problem is to determine the start up and shut down schedules of thermal units in order to meet forecasted load demand fluctuation over a future period [1]. Previous efforts at UC have applied various programming methods and optimization techniques to make the problem solvable [2]. Since even a small reduction in percentage fuel cost such as 1% may lead to large saving of cost for the electric utility companies, a complete and efficient algorithm is still in demand. In recent years, biologically artificial intelligence techniques, such as artificial neural network (ANN) and GA have emerged as candidates for the UC problem. Sasaki et al. [3] first presented the possibility of solving the UC problem applying a Hopfield neural network. The proposed ANN has solved a UC of 30 units over 24 periods, and the obtained results are encouraging. However, Wang et al. [4] found that UC problem cannot be handled accurately within the framework of the conventional Hopfield network. GA is a stochastic searching algorithm combining an artificial survival of the fittest with genetic operators that is suitable for a variety of optimization problems. In our previous work [5], a UC scheduling software using the GA method was completed and applied to the existing Taipower system of Taiwan. However, we found that there are numerous infeasible commitment schedules within the random-created initial population obviously slowing down the solution speed.

This paper develops an ENN approach, which combining a GA with a BPNN, to solve the UC problem. The BPNN is first used as a pre-dispatch tool to create raw

J. Wang et al. (Eds.): ISNN 2006, LNCS 3972, pp. 1296 – 1303, 2006.

unit combinations for each time interval ignoring time-dependent constraints temporarily. Then, a set of heuristic rules is judiciously applied to decode the raw operation schedule of each unit into a feasible one. Hence, the most difficult time-dependent constraints can be automatically satisfied. Finally, a GA is applied to find the optimal solution. Another attractive property of the ENN method is that it searches for many optimum points in parallel and obtains many feasible solutions rather than a single solution in one run.

2 Problem Description and Formulation

UC deals with the problem of obtaining the optimal generations for thermal units. It aims to minimize the system total fuel costs units while satisfying various constraints. With discretization of the total scheduling time into a set of shorter time intervals (say, one hour per interval), the UC problem can be mathematically formulated as a constrained nonlinear optimization problem as follows:

Problem

$$Minimize\ TF = \sum_{t=1}^{T}\sum_{i=1}^{N}[F_i^t(P_i^t)+SU_i^t(U_i^{t-1},U_i^t)+SD_i^t(U_i^{t-1},U_i^t)] \tag{1}$$

subject to the following constraints:

o System power balance

$$\sum_{i=1}^{N} P_i^t - P_D^t - P_{loss}^t = 0 \tag{2}$$

o Generation limits and ramp rate limits

$$Max(\underline{P_i}, P_i^{t-1} - DR_i) \le P_i^t \le Min(\overline{P_i}, P_i^{t-1} + UR_i) \tag{3}$$

o Minimal uptime/downtime

$$\tau_i^{t-1} \ge T_i^{up} \times U_i^{t-1}(1-U_i^t) \tag{4}$$

$$\left|\tau_i^{t-1}\right| \ge T_i^{down} \times (1-U_i^{t-1})U_i^t \tag{5}$$

where *TF* is the total fuel costs, P_i^t is the power generation of thermal unit *i* in hour *t*, $F_i^t(P_i^t)$ is the generation cost for P_i^t, SU_i^t / SD_i^t is the start-up/shut-down cost of unit *i*, U_i^t is the state of unit *i* in hour *t*, τ_i^{t-1} is the continuous on/off line hours till hour *t-1*, T_i^{up} / T_i^{down} is the minimal uptime/downtime of unit *i*, P_D^t is the system load demand in hour *t*, P_{loss}^t is the system transmission network losses in hour *t*, $\underline{P_i} / \overline{P_i}$ is the minimum/maximum generation of unit *i*, and UR_i / DR_i is the up/down ramp rate limit of thermal unit *i*.

3 ENN Solution Methodology

The problem of UC is essentially a non-linear, constrained, combinatorial optimization problem. In the proposed ENN method, each two-dimensional matrix represents a complete solution of the UC problem. The difficult minimal uptime/downtime constraints are embedded and satisfied throughout the proposed encoding and decoding algorithm. The solution methodology for solving the UC problem by the proposed ENN approach is outlined in the flowchart in Fig. 1 and will be described in detail later.

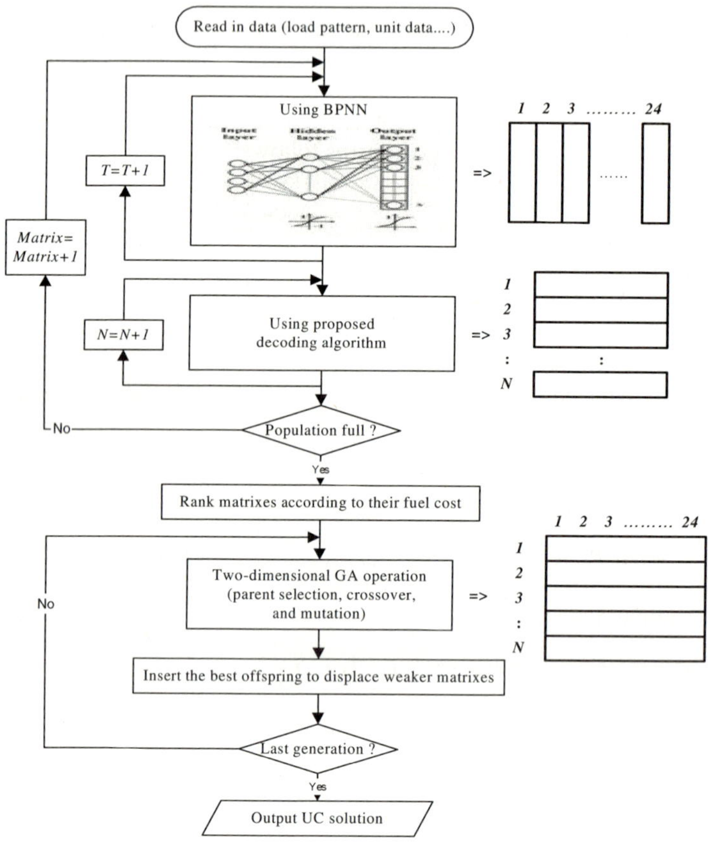

Fig. 1. General flow chart of the proposed ENN approach

3.1 Encoding

Implementation of a UC problem in the proposed ENN method starts from the parameter encoding, i.e., the representation of the problem. The encoding must be carefully designed to utilize the ability of both ANN and GA to efficiently transfer information between encoding strings and achieve the objective function of a problem. If N represents the number of thermal units, an $N \times 24$ matrix is needed to

		HOUR							
		1	*2*	*3*	.	.	.	.	*24*
	1	1	0	0	1	1	0	0	0
U	2	0	1	1	0	0	1	0	1
N	:			:			:		
I	:			:			:		
T	*N*	0	0	1	0	1	0	1	0

Fig. 2. The proposed two-dimensional encoding scheme

represent a complete operation schedule for a 24-hour UC problem. Fig. 2 shows the two-dimensional encoding scheme with raw binary representation. Each encoding matrix contains $N \times 24$ genes and each gene is assigned only one bit, where "0" indicates off-line, and "1" indicates on-line.

3.2 Back-Propagation Neural Network (BPNN)

The basic conception of ANN is intended to model the behavior of biological neural functions. The original desire for the development of ANN is intended to take advantage of parallel processors computing than traditional serial computation. From the literature survey, several models and learning algorithms of ANN have been proposed for solving combinatorial optimization problems [2]. In this paper, we establish a triple-layer feed-forward BPNN [6], as shown in Fig. 3, for solving the UC problem. The number of output layer neurons is set at N, where N is the number of thermal units. The input data for the BPNN include hourly load demand, load situation (in a load increasing or decreasing period), and a heuristic control variable. The output layer neurons generate raw unit combinations for each hour ignoring time-dependent constraints temporarily.

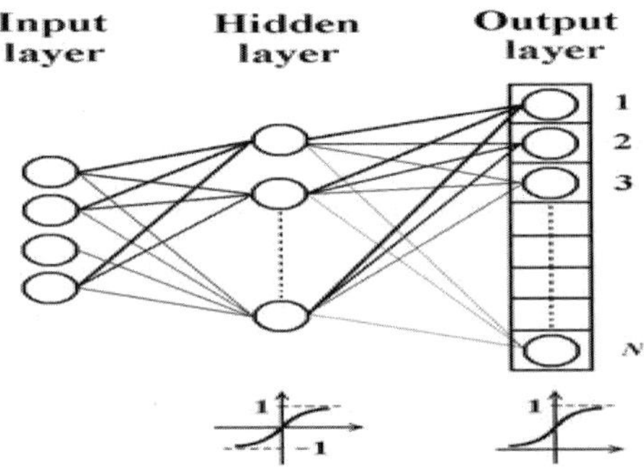

Fig. 3. Triple-layer feed-forward BPNN

In this paper, a faster back-propagation learning algorithm named "RPROP algorithm" is used as the learning rule. Riedmiller and Braun [7] showed that both convergence speed and memory requirement of the RPROP algorithm are better than traditional gradient-descent learning algorithms. In the RPROP algorithm, the update-values for each weight are modified according to the behavior of the sequence of signs of the partial derivatives in each dimension of the weight space, not according to the gradient value. The modified procedure of a weight of the RPROP algorithm can be mathematically formulated as follows:

$$\Delta_{ij}(t)=\begin{cases}\eta^{+}\times\Delta_{ij}(t-1)\ , & if\ \dfrac{\partial E}{\partial W_{ij}}(t-1)\times\dfrac{\partial E}{\partial W_{ij}}(t)>0\\ \eta^{-}\times\Delta_{ij}(t-1)\ , & if\ \dfrac{\partial E}{\partial W_{ij}}(t-1)\times\dfrac{\partial E}{\partial W_{ij}}(t)<0\\ \Delta_{ij}(t-1)\ , & \text{else}\end{cases} \tag{6}$$

$$\Delta W_{ij}(t)=\begin{cases}-\Delta_{ij}(t) & ,\quad if\ \dfrac{\partial E}{\partial w_{ij}}(t)>0\\ +\Delta_{ij}(t) & ,\quad if\ \dfrac{\partial E}{\partial w_{ij}}(t)<0\\ 0 & ,\quad \text{else}\end{cases} \tag{7}$$

$$\mathrm{W}_{ij}(t+1)=W_{ij}(t)+\Delta W_{ij}(t) \tag{8}$$

where $W_{ij}(t)$ is the weight back propagated from neuron j to neuron i harmonic order, η^{+}/η^{-} is the learning velocity, where $0<\eta^{-}<1<\eta^{+}$, $E(t)$ is the error function, and $\Delta_{ij}(t)$ is the update value of $W_{ij}(t)$

3.3 Decoding

Evaluation of an artificial chromosome is accomplished by decoding the encoded chromosome string and computing the corresponding chromosome's fitness value. For ease of exposition, consider a 24-hour case and assume the i-th unit has the following operation data:

o minimal uptime: 6 hours.
o minimal downtime: 5 hours.
o initial status (τ_i^0): -2 (i.e., the i-th unit has already been off line for 2 hours).

The detailed procedure that decodes the raw on/off status, u_i^t , into the actual on/off schedule, U_i^t , is illustrated in Fig. 4 and is summarized by the following rules:

(Rule A): Check if the initial status satisfies minimal uptime/downtime constraints. If not, continue the initial operation status from hour 1 until satisfied.

(Rule B): Check if $u_i^t = U_i^{t-1}$, then the i-th unit maintains the same operation status as the previous hour.

(Rule C): If the i-th unit starts from hour t, it must remain on-line at least T_i^{up} hours.

(Rule D): If the i-th unit shuts down from hour t, it must be kept off-line at least T_i^{down} hours.

HOUR

τ_i^0	1	2	3	4	5	6	7	8	9	10	11	12	13	14	15	16	17	18	19	20	21	22	23	24
-2	1	0	1	0	0	0	0	1	0	1	0	0	1	1	1	1	0	0	1	0	1	1	0	1

(a) a raw on/off status, u_i^t

HOUR

	1	2	3	4	5	6	7	8	9	10	11	12	13	14	15	16	17	18	19	20	21	22	23	24
	0	0	0	0	0	0	0	1	1	1	1	1	1	1	1	1	0	0	0	0	0	1	1	1
Rule:		A		B	B	B	B			C				B	B	B			D				C	

(b) the on/off schedule , U_i^t , after decoding

Fig. 4. An example of the proposed decoding algorithm

For a feasible schedule, the hourly generation cost is calculated as:

$$\sum_{i=1}^{N} F_i^t(P_i^t) = \sum_{i=1}^{N} (a_i P_i^{t^2} + b_i P_i^t + c_i) \tag{9}$$

where P_i^t is obtained by economic dispatch.

On the other hand, for a schedule that is not feasible, the hourly generation cost is calculated by including an additional penalty term:

a) $\text{if } \sum_{i=1}^{N} [\,Min(\overline{P_i}, P_i^{t-1} + UR_i)\,] < P_L^t \quad \text{then}$

$$\sum_{i=1}^{N} F_i^t(P_i^t) = [\sum_{i=1}^{N} (a_i P_i^{t^2} + b_i P_i^t + c_i)] + (P_L^t - \sum_{i=1}^{N} P_i^t) \times IC \tag{10}$$

where $P_i^t = Min(\overline{P_i}, P_i^{t-1} + UR_i)$ and IC is the incremental cost per MW of the most expensive unit.

b) $\text{if } \sum_{i=1}^{N} [\,Max(\underline{P_i}, P_i^{t-1} - DR_i)\,] > P_L^t \quad \text{then}$

$$\sum_{i=1}^{N} F_i^t(P_i^t) = [\sum_{i=1}^{N} (a_i P_i^{t^2} + b_i P_i^t + c_i)] + (\sum_{i=1}^{N} P_i^t - P_L^t) \times IC \tag{11}$$

where $P_i^t = Max(\underline{P_i}, P_i^{t-1} - DR_i)$.

Taking the penalty term into account, an infeasible schedule can be discarded in the solution procedure. The fitness function of the i-th matrix in the population is defined as:

$$FIT(i) = \frac{cost(max) - cost(i)}{cost(max) - cost(1)} \tag{12}$$

where *cost(i)* is the corresponding generation cost of the i-th matrix, and *cost(1)* is the cost of the highest ranking matrix, namely, the current best cost matrix, and *cost(max)* is the worst cost matrix.

4 Numerical Results

The proposed ENN solution method for UC was developed into software and tested on an example system to illustrate the capability of the proposed approach in practical applications. The test system consisted of 10 thermal units where the time period is 24 hours. The hourly MW load demand is shown in Table 1. The unit characteristics and the time-dependent parameters are given in Tables 2 and 3.

The test results including optimal on/off schedule, total fuel costs, and execution time created by the proposed approach are shown in Fig. 5. From the test results, we find that the proposed ENN approach could obtain global optimal solution same as the one obtained from a GA method [5]. Nevertheless, the execution time of the proposed approach is obviously less than the GA method. This makes the proposed ENN approach more attractive in practical applications.

Table 1. Hourly MW load demand

Hour	1	2	3	4	5	6	7	8	9	10	11	12
Demand	700	750	850	950	1000	1100	1150	1200	1300	1400	1450	1500
Hour	13	14	15	16	17	18	19	20	21	22	23	24
Demand	1400	1300	1200	1050	1000	1100	1200	1400	1300	1100	900	800

Table 2. Generation units' capacities and coefficients

Unit	$\underline{P_i}$	$\overline{P_i}$	$a_i(\$/MW^2)$	$b_i(\$/MW)$	$c_i(\$)$
1	150	455	0.00048	16.19	1000
2	150	455	0.00031	17.26	970
3	20	130	0.002	16.60	700
4	20	130	0.00211	16.50	680
5	25	162	0.00398	19.7	450
6	20	80	0.00712	22.26	370
7	25	85	0.00079	27.74	480
8	10	55	0.00413	25.92	660
9	10	55	0.00222	27.27	665
10	10	55	0.00173	27.79	670

Table 3. Generating units' time-dependent parameters

Unit	Min. Up Time	Min. Down Time	Hot Start Cost ($)	Cold Start Cost ($)	Cold Start Hours (hr)	Initial Status (hr)
1	8	8	4500	9000	5	8
2	8	8	5000	10000	5	8
3	5	5	550	1100	4	-5
4	5	5	560	1120	4	-5
5	6	6	900	1800	4	-6
6	3	3	170	340	2	-3
7	3	3	260	520	2	-3
8	1	1	30	60	0	-1
9	1	1	30	60	0	-1
10	1	1	30	60	0	-1

	1	2	3	4	5	6	7	8	9	10	11	12	13	14	15	16	17	18	19	20	21	22	23	24
1	1	1	1	1	1	1	1	1	1	1	1	1	1	1	1	1	1	1	1	1	1	1	1	1
2	1	1	1	1	1	1	1	1	1	1	1	1	1	1	1	1	1	1	1	1	1	1	1	1
3	0	0	0	0	0	1	1	1	1	1	1	1	1	1	0	0	0	0	0	1	1	1	1	1
4	0	0	0	0	0	0	0	0	1	1	1	1	1	1	1	1	1	1	1	1	1	0	0	0
5	0	0	0	1	1	1	1	1	1	1	1	1	1	1	1	1	1	1	1	1	1	1	0	0
6	0	0	0	0	0	0	0	0	0	1	1	1	1	0	0	0	0	0	0	1	1	1	0	0
7	0	0	0	0	0	0	0	0	0	0	0	0	0	0	0	0	0	0	0	0	0	0	0	0
8	0	0	0	0	0	0	0	0	0	0	0	0	0	0	0	0	0	0	0	0	0	0	0	0
9	0	0	0	0	0	0	0	0	0	0	0	1	0	0	0	0	0	0	0	0	0	0	0	0
10	0	0	0	0	0	0	0	0	0	0	1	1	0	0	0	0	0	0	0	0	0	0	0	0

o Start-up cost: 5,920 ($) o Generation cost: 545,219 ($)
o Total fuel costs: 551,139 ($) o CPU time: 16.35 sec.

Fig. 5. The optimal on/off schedule created by the proposed ENN approach

5 Conclusion

In this paper, a BPNN and a GA are respectively used as the pre- and post-dispatch tools to solve the UC problem. The proposed decoding algorithm involving a set of heuristic rules is applied to screen out the infeasible operation schedules. Hence, the difficult minimal uptime/downtime constraints can be automatically satisfied. This feature obviously speeds up solution time. In addition, the global optimum of the problem can be obtained with rather high probability.

Acknowledgment

Financial support given to this research by the National Science Council under Grant No. NSC 90-2213-E-129-005 is greatly appreciated.

References

1. Wood, A.J., Wollenberg, B.F.: Power Generation, Operation, and Control. 2nd edn. John Wiley & Sons, New York (1996)
2. Padhy, N.P.: Unit Commitment-A Bibliographical Survey. IEEE Trans. on Power Systems 19 (2004) 1196-1205
3. Sasaki, H., Watanabe, M., Yokoyama, R.: A Solution Method of Unit Commitment by Artificial Neural Network. IEEE Trans. on Power Systems 7 (1992) 974-981
4. Wang, C., Shahidehpour, S.M.: Effects of Ramp-rate Limits on Unit Commitment and Economic Dispatch. IEEE Trans. on Power Systems 8 (1993) 1341-1350
5. Chang, H.C., Chen, P.H.: A Genetic Algorithm for Solving the Unit Commitment Problem. Proceedings of the International Power Engineering Conference (1997) 831-836
6. Weerasooriya, S., El-Sharkawi, M.A.: Identification and Control of a DC Motor using Back-propagation Neural Networks. IEEE Trans. on Energy Conversion 6 (1991) 663-669
7. Riedmiller, M., Braun, H.: A Direct Adaptive Method for Faster Back Propagation Learning-The RPROP Algorithm. IEEE Int'l Conference on Neural Networks 1 (1993) 86-591
8. Davis, L.: Handbook of Genetic Algorithms, Van Nostrand Reinhold (1991)

Application of BP Neural Network in Power Load Simulator

Bing-Da Zhang and Ke Zhang

School of Electrical Engineering & Automation,
Tianjin University 300072, China
bdzhang@tju.edu.cn

Abstract. Adopting the fundamentals of PWM voltage source rectifier and the PID-control technology based on error-back-propagation neural network, an electronic power load simulator is designed in this paper, which can simulate the exact Volt-Ampere characteristics of power load and supply high quality feedback electric power. In order to enable the PID controller to perform better, genetic algorithms are used to accumulate the priori knowledge of the neural network's connection weights, and the system voltages and the controlled parameters are forecasted. The experimental results show that the electronic power load simulator runs well when the Volt-Ampere characteristics simulated are time-variable or voltage disturbances occur in power system.

1 Introduction

A power load simulator is one of the necessary equipments in the laboratory for dynamic simulation of power system. It is usually a bulb, a water-resistor, a motor-generator unit, etc. The scale of the laboratory is limited because this equipment occupies too much room, and consumes a mass of energy. With the development of Power Electronics Technology and PID-control technology, an electronic power load simulator comes to the laboratory. Its performance has close relation to the PID parameters. If the parameters are fixed, the load simulator can hardly satisfy the requests in experiment of power system. A PID controller based on error-back-propagation neural network (BP-NN), which is suitable for the electronic power load simulator, is put forward in this paper.

2 Power Load Simulator

The main circuit of the electronic power load simulator is similar to a VSR, as illustrated in Fig. 1. It is composed of three units used to absorb, store and feed back electrical power respectively.

The single-phase equivalent circuit to the main circuit of the absorption unit is illustrated in fig. 2, where e is the voltage of the simulated power system, and u is the modulating voltage of PWM.

The current absorbed by the load simulator from the simulated power system can be changed by adjusting the peak and the phase of the modulating voltage u. Thus,

J. Wang et al. (Eds.): ISNN 2006, LNCS 3972, pp. 1304 – 1310, 2006.

the characteristics of the simulated load can be controlled. The process is called AC-DC control. Regarding the feedback unit, the equivalent circuit is the same as shown in fig. 2. But, the controlled objects are the voltage of the storage unit and the phase difference between the feedback current and the voltage of the outer power system. The process is called DC-AC control. The whole control frame of the electronic power load simulator is depicted as Fig. 3.

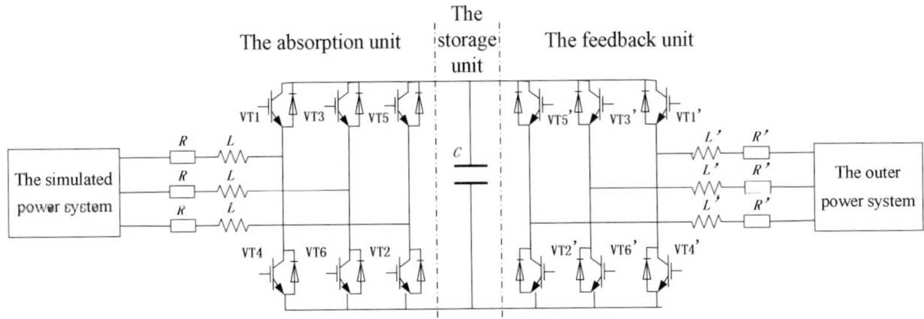

Fig. 1. The main circuit of the electronic power load simulator

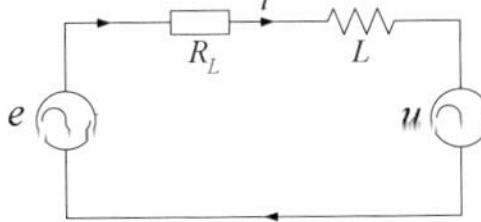

Fig. 2. The single-phase equivalent circuit to the main circuit of the absorption unit

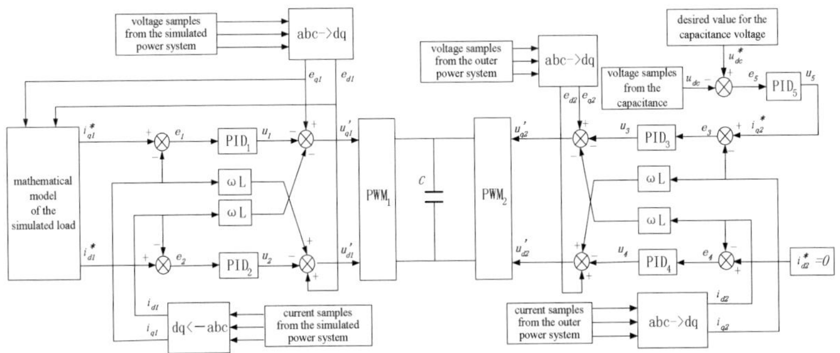

Fig. 3. The whole control frame of the electronic power load simulator

3 PID Control Based on BP Neural Network

The PID control law given by the following equation is considered:

$$\begin{aligned} u(k) &= u(k-1) + k_p[e(k) - e(k-1)] + k_i e(k) + k_d[e(k) - 2e(k-1) + e(k-2)] \\ &= u(k-1) + k_p \partial e(k) + k_i e(k) + k_d \partial^2 e(k) \quad , \end{aligned} \tag{1}$$

where $e(k) = r(k) - y(k)$, r is the desired output, and y is the actual output.

When the Volt-Ampere characteristics simulated are time-variable or voltage disturbances exist in power system, good control effect can hardly be achieved with a set of PID parameters k_p, k_i and k_d tuned beforehand. In practice, it has been proved that k_p, k_i and k_d should be adjusted according to the controlled object's error e, the error's first factorial increment ∂e and the error's second factorial increment $\partial^2 e$.

The BP-NN is a kind of multilayered forward feed network. A BP-NN which consists of M-inputs and L-outputs can be regarded as a nonlinear map from M dimensions to L dimensions in Euclidean Space. Furthermore, the BP-NN has the ability of self-study, self-organization and self-adaptation. Thus, a set of suitable PID parameters k_p, k_i and k_d can be adjusted according to $e(k)$, $\partial e(k)$ and $\partial^2 e(k)$ by BP-NN.

In the electronic power load simulator, though the controller PID_1, PID_2 modulate PWM_1 and PID_3, PID_4, PID_5 modulate PWM_2 respectively, each PID modulation effect is independent relatively and restricts mutually because DC of PWM_1 is the same as that of PWM_2. In order to make the PID controller based on BP-NN have better characteristics, it is regarded that the PID parameters are relative to $e(k)$, $\partial e(k)$ and $\partial^2 e(k)$ of the controlled objects. The BP-NN for all the PID controllers (NN-PID) in the load simulator is depicted as Fig. 4.

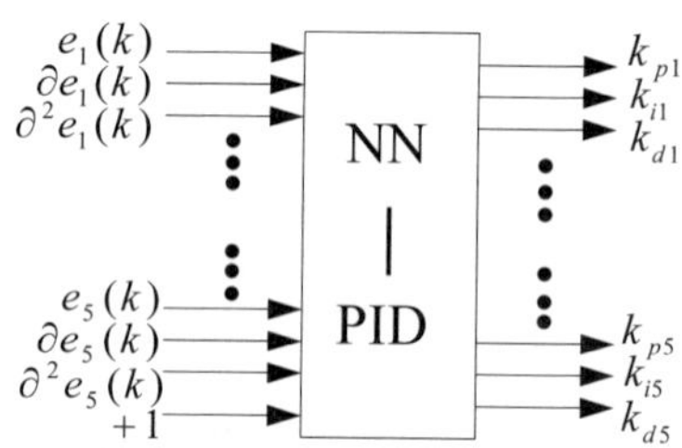

Fig. 4. The frame of NN-PID

The function of hidden nodes in BP-NN is to abstract and store the intrinsic rules in samples. A small number of hidden nodes are insufficient to embody and train the centralized sample rule, but a large amount of hidden nodes depress the ability of generalization. After tentative combination and optimal selection, 33 hidden nodes are set for NN-PID. Considering the inputs and outputs of NN-PID, the monopole activation function $g(x) = 1/(1+e^{-x})$ is selected for the output neurons, and the bipolar activation function $f(x) = (e^x - e^{-x})/(e^x + e^{-x})$ for the middle neurons.

4 Off-Line Study of NN-PID

The error function in the BP algorithms has a mass of semi-minimal ditch, so that the global minimal point is hardly obtained. Although the calculation of the genetic algorithms (GA) costs much, it can seek out the optimal or semi-optimal solution without any initial information. Thus, GA is employed to complete the off-line study of NN-PID in this paper.

There are many connection weights in NN-PID, and their value range is inestimable. Therefore, GA for seeking the ideal connection weights goes on a real code.

Regarding the load simulator, the main controlled objects of concern are the absorption currents i_{d1}, i_{q1} in the absorption unit, capacitance voltage u_{dc} in the storage unit, and the feedback currents i_{d2}, i_{q2} in the feedback unit. In order to guarantee the load simulator has good performance (stability, rapidity, static error), each PID input e, output u and the controlled object overshoot σ (If $r \geq 0$, $e < 0$, or $r < 0$, $e > 0$, then $\sigma = e$, otherwise, $\sigma = 0$) are associated with the GA fitness function. Hence, the fitness function is expressed as follows:

$$F = -\sum_{k=1}^{P}\sum_{i=1}^{5}(\alpha\left|e_i(k)\right| + \beta u_i^2(k) + \gamma\left|\sigma_i(k)\right|) , \tag{2}$$

where P is the amount of sampling; α, β, γ are weights of e, u, σ respectively, and $\alpha + \beta + \gamma = 1$.

A line-up proportion choice mechanism is adopted in GA. The higher adaption individual gets the greater scale factor from a set of given ones beforehand, which decide the individual selected probability. This method avoids the shortcoming that the choice relies too much on the fitness function. In the initial evolution period, the difference between the scale factors is big enough to speed up the GA convergence rate; in the later evolution period, a small difference will prevent the GA getting into local optimal points prematurely.

It is considered that the ideal connection weights have been found when some individual occupies a certain proportion in the solution community or the most of individuals approach some fitness.

5 On-Line Study of NN-PID

In order to make the electronic power load simulator work well, it is not enough to simply study off line. The NN-PID parameters should be consistent with the actual behavior in the run procedure. In other words, PID should adapt the changes of the interior parameters of the controlled system and the exterior condition automatically.

In order to enable NN-PID to perform well, the mathematical expression of the study goal is defined as

$$E(k) = \frac{1}{2}\sum_{i=1}^{5} e_i^2(k+1) = \frac{1}{2}\sum_{i=1}^{5}(r_i(k+1) - y_i(k+1))^2 . \tag{3}$$

The update rule for the NN-PID connection weights is derived as follows:

$$\Delta w(k+1) = -\eta\frac{\partial E(k)}{\partial w} + \alpha\Delta w(k) , \tag{4}$$

where η is the study rate; α is the momentum factor.

For the connection weights link between the output layer and the hidden layer,

$$\frac{\partial E(k)}{\partial w_3}=\frac{\partial E(k)}{\partial y(k+1)}\bullet\frac{\partial y(k+1)}{\partial u(k)}\bullet\frac{\partial u(k)}{\partial o_3(k)}\bullet\frac{\partial o_3(k)}{\partial net_3(k)}\bullet\frac{\partial net_3(k)}{\partial w_3}=\delta_3\bullet\frac{\partial net_3(k)}{\partial w_3}, \tag{5}$$

where $\frac{\partial u(k)}{\partial o_3(k)}=\begin{cases}\partial e(k), & o_3=k_p\\ e(k), & o_3=k_i\\ \partial^2 e(k), & o_3=k_d\end{cases}$;

$\frac{\partial o_3(k)}{\partial net_3(k)}=g(net_3(k))\bullet[1-g(net_3(k))]$; $\frac{\partial net_3(k)}{\partial w_3}=o_2(k)$.

And for the connection weights link between the hidden layer and the input layer,

$$\begin{aligned}\frac{\partial E(k)}{\partial w_2}&=(\sum\frac{\partial E(k)}{\partial y(k+1)}\bullet\frac{\partial y(k+1)}{\partial u(k)}\bullet\frac{\partial u(k)}{\partial o_3(k)}\bullet\frac{\partial o_3(k)}{\partial net_3(k)}\bullet\frac{\partial net_3(k)}{\partial o_2(k)})\bullet\frac{\partial o_2(k)}{\partial net_2(k)}\bullet\frac{\partial net_2(k)}{\partial w_2}\\&=(\sum\delta_3 w_3)\bullet\frac{\partial o_2(k)}{\partial net_2(k)}\bullet\frac{\partial net_3(k)}{\partial w_3},\end{aligned} \tag{6}$$

where $\frac{\partial o_2(k)}{\partial net_2(k)}=1-f^2(net_2(k))$; $\frac{\partial net_2(k)}{\partial w_2}=o_1(k)$.

$\partial E(k)/\partial y(k+1)$ and $\partial y(k+1)/\partial u(k)$ can't be computed because $r(k+1)$ and $y(k+1)$ are unknown. $e_{d1}(k+1)$, $e_{q1}(k+1)$, $e_{d2}(k+1)$ and $e_{q2}(k+1)$ must be forecasted before doping out $r(k+1)$ and $y(k+1)$. Generally speaking, the simulated power system is linked to the outer power system by a transformer in the laboratory for dynamic simulation of power system, and $e_{d1}(k+1)$, $e_{q1}(k+1)$, $e_{d2}(k+1)$ and $e_{q2}(k+1)$ change simultaneity. Moreover, the load simulator has little influence on the voltages both in the simulated system and the external system. Therefore, the model for forecasting the system voltages $e_{d1}(k+1)$, $e_{q1}(k+1)$, $e_{d2}(k+1)$ and $e_{q2}(k+1)$ based on BP-NN (NN-FV) is designed as Fig. 5.

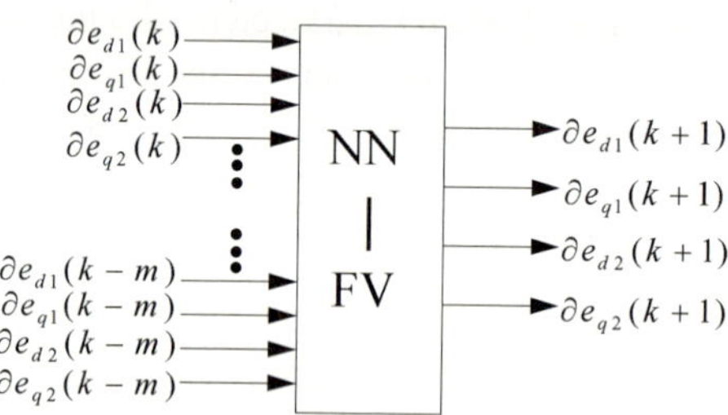

Fig. 5. The frame of NN-FV

For the desired outputs $r(k+1)$, $u^*_{dc}(k+1)$ and $i^*_{d2}(k+1)$ are constant; $i^*_{d1}(k+1)$, $i^*_{q1}(k+1)$ can be computed by $e_{d1}(k+1)$, $e_{q1}(k+1)$ and the mathematics model of the simulated object.

The actual outputs $y(k{+}1)$ relate to many factors such as $u^*_{dc}(k+1)$, $i^*_{d1}(k+1)$, $i^*_{q1}(k{+}1)$, $i^*_{d2}(k+1)$, $e_{q1}(k{+}1)$, $e_{d1}(k{+}1)$, $e_{q1}(k{+}1)$, $e_{d2}(k{+}1)$ and $e_{q2}(k{+}1)$. It is forecasted with the help of the ideal mathematics model of the load simulator in this paper.

6 Experimental Results

In the experiment, the parameters of the load simulator are as follows: filters of L=6 mH and R_L=0.1 Ω, a power storage capacitor of C=4200 μF. The simulated object is a resistor of R=50 Ω.

In order to exam the dynamic performance of the load simulator, the voltages of the simulated power system and the outer power system are set as shown in Fig. 6. The experimental results show that the simulated resistance is quite steady and its tracking error is within 0.2%. The absorbed current of the load simulator is shown as Fig. 7.

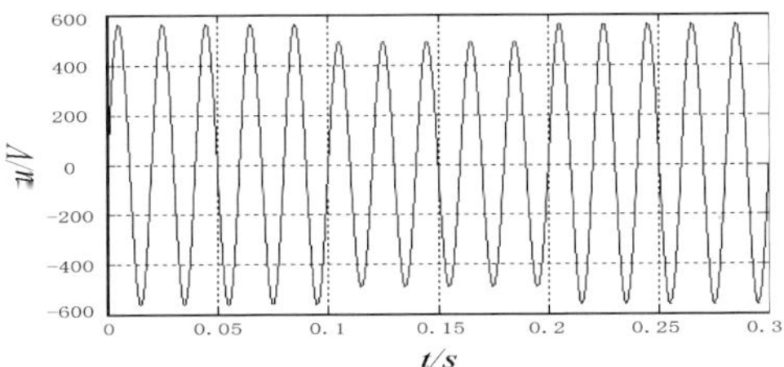

Fig. 6. The voltage of power system

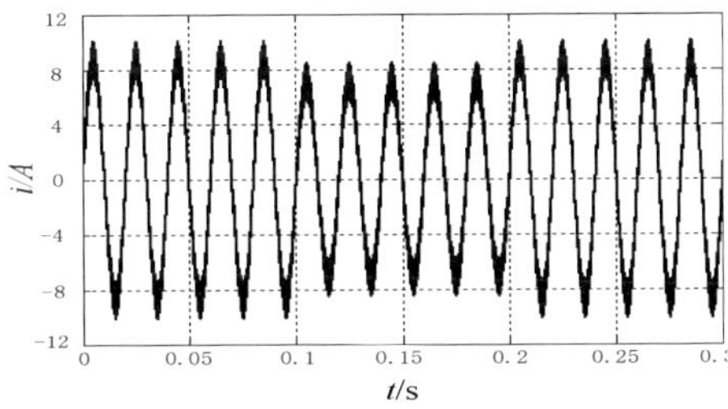

Fig. 7. The absorbed current of the load simulator

It is concerned that whether the feedback current of the load simulator has the same phase with the outer system voltage. Results from the analysis of the waveforms of the feedback current and the system voltage indicate that 1) when the simulated resistance is constant, the total harmonic content in the feedback current is less than 1.2%, and it will not exceed by 1.6% even if voltage disturbances exist in the power system; 2) both the phases of the fundamental wave components in the feedback current and the outer system voltage maintain consistent.

7 Conclusion

The electronic power load simulator performs well, when NN-PID has enough priori knowledge and the forecasting system voltages have high accuracy.

Seeking out the global optimal or semi-optimal connection weights of NN-PID by the genetic algorithms avoids the disadvantage that the BP algorithms need appropriate initial information and are easy to trap in the local minimal points.

The simulated power system is linked to the outer power system by a transformer in the laboratory for dynamic simulation of power system, so the forecasting system voltage based on NN-FV is reasonable.

The steady characteristics of power equipment have been well simulated by the load simulator, but there are many problems to be solved in simulating the transient characteristics.

References

1. Liu, J.: Advanced PID Control and its MATLAB Simulation. Publishing House of Electronics Industry, Beijing (2003)
2. Huang, Y.: Self-turning PID Controller Based on Genetic Neural Networks. Journal of System Simulation. 15(11) (2003) 1628–1630, 1641
3. Xu, L.: Neural Network Control. Publishing House of Electronics Industry, Beijing (2003)
4. Sakaguchi, A., Yamamoto, T.: A Design of Predictive PID Control Systems Using GA and GMDH Network. Lecture Notes in Proceedings of the 2002 IEEE International Conference on Control Applications, Sept. 18-20. Glasgow, Scotland, UK (2002) 266–271
5. Rech, C., Grundling, H. A., Hey, H. L., Pinheiro, J. R.: Analysis and Design of a Repetitive Predictive-PID Controller for PWM Inverters. IEEE (2001) 986–991
6. Tan, Y., Dang, X., Achiel, R. Van Cauwenberghe, Mehrdad, S.: PID Gradient Algorithm for Neural Network Based Generalised Nonlinear PID Controller. Control Theory and Applications 17(6) (2000) 861–867
7. Low, K.S., Chiun, K.Y., Ling, K.V.: Evaluating Generalized Predictive Control for a Brushless DC Drive. IEEE Transactions on Power Electronics 13 (1998) 1191–1198
8. Omatu, S., Yoshioka, M.: Self-tuning Neuro-PID Control and Applications. IEEE International Conference on Systems, Man, and Cybernetics (1997) 1985–1989
9. Martins, D.C., da Silva, P.A., Selinke, R.A.: PID Controller Implementation in Numerical Simulation Program Applied to the Power Static Converter. IEEE International Telecommunications Energy Conference (1996) 764–768
10. Bohn, C., Atherton, D.P.: An Analysis Package Comparing PID Antiwindup Strategies. IEEE Control Systems Magazine 15(2) (1995) 34–40

Feeder Load Balancing Using Neural Network

Abhisek Ukil, Willy Siti, and Jaco Jordaan

Tshwane University of Technology, Private Bag X680,
Pretoria, 0001, South Africa
{abhiukil, willysiti, jakop_s2003}@yahoo.com

Abstract. The distribution system problems, such as planning, loss minimization, and energy restoration, usually involve the phase balancing or network reconfiguration procedures. The determination of an optimal phase balance is, in general, a combinatorial optimization problem. This paper proposes optimal reconfiguration of the phase balancing using the neural network, to switch on and off the different switches, allowing the three phases supply by the transformer to the end-users to be balanced. This paper presents the application examples of the proposed method using the real and simulated test data.

1 Introduction

Phase balancing in the distribution system have different needs, from minimizing the losses in the system to relieving the transformer during the peak time and so forth. There are a number of normally closed and normally opened switches in a distribution system. By changing the open/close status of the feeder switches, load currents can be transferred from feeder to feeder, that is, from the heavily loaded to the less loaded feeders. In South Africa, to reduce the unbalance current in a feeder the connection phases of some feeders are changed manually after some field measurement and software analysis. This is, however, time-consuming and unsuccessful many times [1].

With the uses of the artificial intelligence, telecommunication and power electronics equipments in the power system, it is becoming easier to automate the phase balancing problem. The automation implementation will be technically advantageous as well as economical for the utilities and the customers, in terms of the variable costs reduction and better service quality, respectively.

The approach proposed here uses the neural network which will be able to switch on/off the different switches and keep the phases balanced. Each load will cater only one of the three phases following the constraint that for each load only one switch (to the phase) should be closed, while other two should remain open. For each loading condition, the neural network will be trained for the relevant minimum loss configuration. This can be applied to the small networks, for example, six to fifteen houses as the unbalanced loads.

2 Problem Description

To balance the three phase currents in every segment and then depressing the neutral line current is a very difficult task. Using the manual trial and error technique, used

J. Wang et al. (Eds.): ISNN 2006, LNCS 3972, pp. 1311 – 1316, 2006.

most of the time in South Africa for phase balancing, interruption of the service continuity is unavoidable when changing the connection phases of the distribution transformers to the feeder [1].

In South Africa, a distribution feeder is usually a three-phase, four-wire system. It can be radial or open loop structure [3]. The example feeder shown in Fig. 1 has three phase conductors for the section between the main transformer and the different load points. We limit our present study to six load points, as shown in Fig. 1. To improve the system phase voltage and current unbalances, the connections between the specific feeders and the distribution transformers should be suitably arranged.

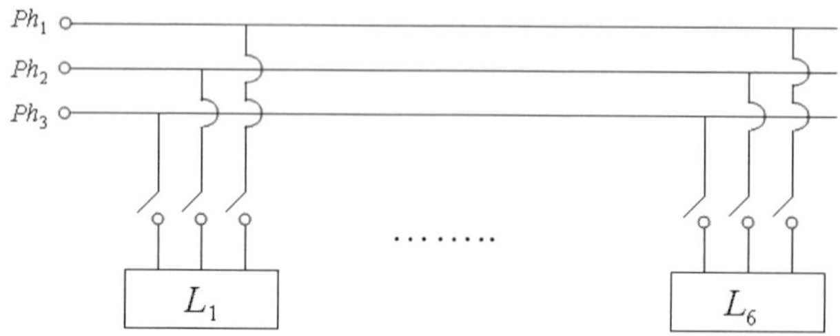

Fig. 1. Example Distribution Feeder

3 Feeder Reconfiguration Technique

In the case of a distribution system with some of the branches overloaded, and other branches lightly loaded, there is the need to reconfigure the networks such that loads are transferred from the heavily loaded feeder or transformers to the less loaded feeder (or transformers). The maximum load current the feeder conductor can take may be taken as the reference. The transfer of load must be done by satisfying the predefined objective to have minimum real power loss. Consequently, network reconfiguration may be redefined as the rearrangement of the network such as to minimize the total real power losses arising from the line branches. Mathematically, the total power loss can be expressed as follows [2]:

$$\sum_{i=1}^{n} r_i \frac{P_i^2 + Q_i^2}{|V_i|^2} , \tag{1}$$

where, r_i, P_i, Q_i, V_i are respectively the resistance, the active power, the reactive power and the voltage of the branch i, and n is the total number of branches in the system. Due to some practical considerations, there could be a constraint on the number of switch–on and off. Given a distribution system as shown in Fig. 1, a network with three phases with a known structure, the problem consists of finding a condition of balancing. The mathematical model [1] can be expressed as:

$$\mathbf{I}_{ph1k} = \sum_{i=1}^{3} sw_{k1i} \mathbf{I}_{ki} + \mathbf{I}_{ph1(k+1)} \tag{2}$$

$$\mathbf{I}_{ph2k} = \sum_{i=1}^{3} sw_{k2i}\mathbf{I}_{ki} + \mathbf{I}_{ph2(k+1)} \tag{3}$$

$$\mathbf{I}_{ph3k} = \sum_{i=1}^{3} sw_{k3i}\mathbf{I}_{ki} + \mathbf{I}_{ph3(k+1)} \ , \tag{4}$$

where, $\mathbf{I}_{ph1k}$, $\mathbf{I}_{ph2k}$ and $\mathbf{I}_{ph3k}$ represent the currents (phasors) per phase (1, 2 & 3) after the *k* point of connection; $sw_{k11},\ldots,sw_{k33}$ are different switches (the value of '1' means the switch is closed and '0' means it is open). Following the constraint of allowing only one breaker in each of the equations (2) – (4) to be closed, we can write the following set of modified constraints:

$$\sum_{i=1}^{3} sw_{k1i} - 1 = 0 \tag{5}$$

$$\sum_{i=1}^{3} sw_{k2i} - 1 = 0 \tag{6}$$

$$\sum_{i=1}^{3} sw_{k3i} - 1 = 0 \tag{7}$$

4 Neural Network-Based Approach

In the proposed strategy in this paper, the neural network must control the switch-closing sequence of each load for the minimum power loss which will lead to the optimal phase balance. The inputs to the neural network are the unbalanced load currents (six in the current study) and the outputs are the switch closing sequences for each load.

The input layer of the network has *N* input neurons, *N* being number of unbalanced load currents to be controlled. The following column vector has been assumed as the input

$$C_{Sw} = [I_{L1} \ldots\ldots\ldots\ldots I_{LN}] \ . \tag{8}$$

The output of the network is in the range {1, 2, 3} for each load, i.e., which switch (to the specific phase) should be closed for that specific load.

4.1 Neural Network Structure

For this application, we used the radial basis network [3]. Experimentations with the backpropagation and the radial basis network indicated faster training and better convergence for the latter. Radial basis networks may require more neurons than the standard feed-forward backpropagation networks, but often they can be designed in a fraction of the time needed to train the standard feed-forward networks. They work best when many training vectors are available [4]. Matlab® neural network toolbox [5] has been used for the implementation. We experimented with different kinds of radial basis networks, but generalized regression neural network ("GRNN") [5] produced the best result. A generalized regression neural network is often used for function approximation. It has a radial basis layer and a special linear layer [5].

4.2 Network Training

We have used the neural network-based operation for the test data in following structure: real and simulated data for six loads.

The real data set consisted of unbalanced load data from a South African city. The test data set had average load current values per houses in a specific locality of the city for the different times of each day in a month. We randomly selected six houses as our test data for each specific time, and we tested our result on 500 data. Simulated data were generated using the computer following the real load data structure.

First, we used the Matlab®-based fast heuristic method [1] for balancing the unbalanced load data. Details of the algorithm can be referred to in [1], but we explain the necessary part briefly below.

We consider the loads to be equally distributed per phase, i.e., we assume two loads to be connected per phase. So, the problem is to find the optimum three sets of two loads, with *minimum* differences among the individual sums of the three sets. To achieve this, first we calculate the ideal phase balance current value I_{ideal}, which is equal to the one-third of the sum of the all six load currents I_L.

$$I_{ideal} = \frac{1}{3}\sum_{j=1}^{6} I_{L_j} \ . \tag{9}$$

In the second step, we optimally select our 3 sets of currents for the three phase currents I_{ph}, each set comprising of two load currents $\{I_j, j = 1,2\}$.

$$I_{Load} = \{I_{L_j}, j = 1,...,6\} \ , \tag{10}$$

$$I_{ph} = \{I_j, j = 1,2\} \quad \text{where} \quad I_j \in I_{Load} \ . \tag{11}$$

Difference between the individual sum of these sets and the I_{ideal} should be *minimum*, ideally 0 for the perfect phase balance. So, we need to find three sets of $\{I_j, j = 1,2\}$, subject to the constraint,

$$\min \left| \sum_{j=1}^{2} I_j - I_{ideal} \right| \ , \ \text{where} \ \ I_j \in I_{Load} \ . \tag{12}$$

Following this, we obtain the output switching sequences as the target data set for training the networks. Using the output switching sequences and the input load currents, we calculate the balanced phase currents I_{ph1}, I_{ph2} and I_{ph3}. For example, I_{ph1} is calculated by adding the two load currents corresponding to the output switching sequences marked 1. Then we calculate the differences between I_{ph1}, I_{ph2} and I_{ph3}, which ideally should be zero. The differences indicate the quality of the phase balance [1].

Using the real and simulated unbalanced load as the input vector, and the output switching sequences as the target vector, we trained the above-mentioned neural network. Then, we tested the network with different unbalanced load data set. The output was the optimal switching sequences of {1, 2, 3} for the three-phases as explained above. Using the similar procedure as explained above, we computed the balanced phase currents and the differences between the phase currents, which indicate the quality of the balance.

5 Application Results

An Intel® Celeron® 1.9 GHz, 256 MB RAM computer was used for the test. Test results of the neural network-based approach for the simulated six load data format are shown in Table 1 to 4, for three different sample data. Table 1 shows the unbalanced load (current) data, Table 2 the output switching sequences, Table 3 the balanced phase currents and Table 4 the absolute differences between the balanced phase currents. In Table 2 to 4, 'NN' is the abbreviation for the Neural Network-based approach and 'HEU' is the abbreviation for the Heuristic Method [1]-based approach.

Table 1. Unbalanced Load (Current) Data

Data / I_L (A)	1	2	3
1	89	35	45
2	85	0	67
3	74	90	87
4	38	21	64
5	56	87	30
6	45	112	90

Table 2. Output Switching Sequences

	Data 1		Data 2		Data 3	
Switch Seq. for 6 Loads	NN	HEU	NN	HEU	NN	HEU
1	1	1	1	1	1	1
2	2	2	2	3	2	2
3	3	3	1	2	3	1
4	1	1	3	1	2	2
5	3	3	3	2	3	3
6	2	2	2	3	1	3

Table 3. Balanced Phase Currents

	Data 1		Data 2		Data 3	
I_{Phase} (A)	NN	HEU	NN	HEU	NN	HEU
Phase 1	127	127	125	56	135	132
Phase 2	130	130	112	177	131	131
Phase 3	130	130	108	112	117	120

Table 4. Differences between Phase Currents

	Data 1		Data 2		Data 3	
Difference (A)	NN	HEU	NN	HEU	NN	HEU
Phase 1-2	3	3	13	121	4	1
Phase 2-3	0	0	4	65	14	11
Phase 3-1	3	3	17	56	18	12

5.1 Comments on Application Results

- Summary of the neural network-based approach in comparison with the heuristic method [1]-based one is as follows.
  - Neural Network performs better than Heuristic Method: 14%
  - Neural Network performs same as Heuristic Method: 67%

  - Neural Network performs worse than Heuristic Method: 10%
  - Neural Network fails to converge or gives erroneous result: 9%
- From the above summary, it should be noted that the neural network-based approach mostly works similar to the heuristic method [1]-based approach. Deviation of the results in the 10% worse cases are not that severe, however, the 14% better performance is a significant improvement, as the heuristic method proved to be the most efficient [1].
- Speed of operation (average computation time 0.14 seconds) is similar with the heuristic method [1], once the network is trained. For this reason, once the network is suitably trained, we save and use it as a neural network object.
- This approach can be extended to any number of unbalanced load data. But as the training data construction depends on the heuristic method [1], at this stage, we have to limit our studies to the number of load data exactly divisible by 3 so that the loads can be equally distributed per phase.

6 Conclusion

The neural network-based approach for phase balancing for a small size network (six load data) proved the feasibility of the proposed control. The result achieved from the Matlab® implementation pertain to that obtained using the heuristic method. Besides, neural network-based approach gives better result in 14% cases, which is a significant improvement. This encourages the implementation of the neural network-based strategy on a large size network.

References

1. Ukil, A., Siti, M. and Jordaan, J.: MATLAB®-based Fast Load Balancing of Distribution System Using Heuristic Method. SAIEE Transactions (under review)
2. Chen, C.S. and Cho, M.Y.: Energy Loss Reduction by Critical Switches, of Distribution Feeders for Loss Minimization. IEEE Trans. Power Delivery 4 (1992)
3. Wasserman, P.D.: Advanced Methods in Neural Computing. Van Nostrand Reinhold, New York (1993)
4. Chen, S., Cowan, C.F.N. and Grant, P.M.: Orthogonal Least Squares Learning Algorithm for Radial Basis Function Networks. IEEE Trans. Neural Networks 2 (1991) 302-309
5. MATLAB® Documentation – Neural Network Toolbox, Version 6.5.0.180913a Release 13, Mathworks Inc., Natick, MA (2002)

A Neural Network Based Particle Swarm Optimization for the Transformers Connections of a Primary Feeder Considering Multi-objective Programming

Cheng Chien Kuo

Department of Electrical Engineering, Saint John's University,
499, Sec. 4, Tam King Road, Tamsui, Taipei, Taiwan, R.O.C
cckuo@mail.sju.edu.tw

Abstract. A new multi-objective formulation named normalized weighting method combined with particle swarm optimization for the connections between distribution transformers and a primary feeder problem is presented. The performance of Particle Swarm Optimization can be improved by strategically selecting the starting positions of the particles by back-propagation neural network. Six important objectives are considered in this problem. These six objectives are of equal important to electric utility companies, but they are somewhat non-commensurable with each other. In view of this, a normalized weighting method for the multi-objective problem is proposed. It can provide a set of flexible solutions using particle swarm optimization by following the intention of decision makers. To increase the realism, the load and operating constraints of the system are all considered. Comparative studies on actual Tai-power systems are given to demonstrate the effectiveness of the phase load balancing and the improvement of operation efficiency for the proposed method.

1 Introduction

The distribution feeders are designed to be balanced to perfect the utilization of three phase components and loads. However, it delivers the power to the customer by one single phase only in some cases. This usually will result in unequal phase loading such that the life time of devices will deteriorate [1]. There are many researches [2] focus on the improvement of system unbalance, especially the static reactive compensator technique [2]. All of these improvement methods should add more devices to either compensation or control the system reactive needed. In this paper, we develop a powerful tool to help system planner reduce feeder loss and maintain high power quality and efficiency. No more devices are needed in this suggestion such that the economic contribution is very important.

Recently, Eberhart and Kennedy proposed a particle swarm optimization (PSO) based on the analogy of swarm of bird and school of fish [3]. The main advantages of PSO algorithm are simple concept, easy implementation, robustness to control parameters, and computational efficiency when compared with mathematical algorithm and other heuristic optimization techniques. To ensure broad coverage of the search space, the particles should be initialized so that they are distributed as better quality as

J. Wang et al. (Eds.): ISNN 2006, LNCS 3972, pp. 1317 – 1323, 2006.

possible throughout the space. The standard method of initialization fails to accomplish this goal, especially in high-dimensional spaces. In this paper, using a starting configuration based on back-propagation neural network (BPN) can lead to improved performance.

Therefore, this paper aims to develop a unified approach to solve connections between distribution transformers and a primary feeder by taken into account the typical daily load patterns in distribution systems. A new multi-objective formulation combined with PSO and BPN for the above problems is presented.

2 Problem Description and Formulation

In this paper, [7] is used as the calculation method to transfer the loads of distribution transformers among different kinds of connection into simple three phases loads Sa Sb and Sc. Therefore, there are six candidate connections at most for each distribution transformers as shown in Fig. 1.

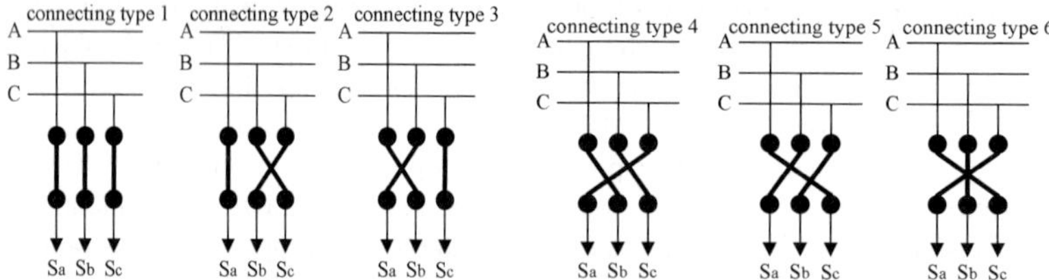

Fig. 1. Six candidate connection method

2.1 Objective Functions

The objective functions considered in this study are:

1.) Power balance index (PB): Power balance index of three phases that indicates the level of balance condition between three phases as stated below :

$$S^h_{i,AVG} = \frac{1}{3}(S^h_{i,a} + S^h_{i,b} + S^h_{i,c}) \quad ,i = 1 \sim N_l \qquad PB = \sum_{h=1}^{365\times 24}\sum_{i=1}^{N_l}\frac{\sum_{p=a,b,c}\left|S^h_{i,p} - S^h_{i,AVG}\right|}{\left|S^h_{i,AVG}\right|} \tag{1}$$

where:

$S^h_{i,p}$: complex power flowing of branch i of phase p at hour h,

$S^h_{i,AVG}$: average complex power flowing of branch I at hour h,

N_l : total number of branches.

2.) Total feeder loss (Ploss): Total feeder loss represents the economic condition of a system.

$$Ploss = \sum_{h=1}^{365\times 24}\sum_{p=a,b,c}\sum_{i=1}^{N_l} r_i \frac{(P^h_{i,p})^2 + (Q^h_{i,p})^2}{\left|V^h_{i,p}\right|^2} \tag{2}$$

where:

$Ploss$: total line losses of distribution feeder,

$P_{i,p}^{h}$: real power of phase p flowing out of bus i at hour h,

$Q_{i,p}^{h}$: reactive power of phase p flowing out of bus i at hour h,

$V_{i,p}^{h}$: voltage of phase p of bus i at hour h,

r_i : resistance between buses i and i+1.

3.) Voltage deviation (VD_SUM): Voltage deviation among buses of three phases that associates with the quality condition.

$$VD\_SUM = \sum_{h=1}^{365\times 24} \sum_{p=a,b,c} \sum_{i=1}^{N_b} \left| V^{ideal} - V_{i,p}^{h} \right| \tag{3}$$

where:

N_b : total number of buses,

4.) Zero unbalance factor (d0_SUM): Zero unbalance factor for all the buses of three phases that represents the power quality condition of a system

$$d0\_SUM = \sum_{h=1}^{365\times 24} \sum_{i=1}^{N_b} \frac{\left| V_{0,i}^{h} \right|}{\left| V_{1,i}^{h} \right|} \tag{4}$$

where

$V_{n,i}^{h}$: the n sequence voltage (including zero, positive and negative sequence) of bus i at hour h.

5.) Negative unbalance factor (d2_SUM): Negative unbalance factor for all the buses of three phases that also represents the power quality condition of a system

$$d2\_SUM = \sum_{h=1}^{365\times 24} \sum_{i=1}^{N_b} \frac{\left| V_{2,i}^{h} \right|}{\left| V_{1,i}^{h} \right|} \tag{5}$$

6.) LCO current (ILCO_SUM): The current passing through the LCO relay equipped in substation. Obviously, to minimize the LCO current can prevent the abnormal trip of the LCO relay

$$ILCO\_SUM = \sum_{h=1}^{365\times 24} ILCO^{h} = \sum_{h=1}^{365\times 24} \left| ICT_a^{h} + ICT_b^{h} + ICT_c^{h} \right| \tag{6}$$

where

$ILCO^{h}$: the LCO current of a feeder at hour h,

ICT_p^{h} : the secondary current of CT for phase p that flowing through the CB of feeder at hour h.

2.2 Operating Constraints

The voltage deviations, unbalance factors and LCO currents during each load period must lie within a permissible range for practical situation. Also, the current on each branch must stay within its ampacity limits for security reasons.
where

3 The Normalized Weighting Method

Most of the multi-objective programming techniques translate the objective functions into a combined single objective with weight values. This will arise one problem, the weight values are very difficult to determine because these objectives maybe vary widely in units. A better way to work with multi-objective problems is to provide a flexible NW method between the objective functions. The NW method presented here is based on this concept and will be described shortly.

3.1 Step 1

We first solve the single objective optimization problem expressed in (7). Due to its flexibility and efficiency mentioned above, a global optimization technique known as PSO [3] will be applied for such optimization applications throughout the study.

$$\min \ PB \qquad \text{subject to the constraints} \tag{7}$$

Assume the solution of (7) is S_{PB} , that is $PB(S_{PB}) = PB_{ideal}$. Then, treat all the other objectives as a single objective optimization problem one by one to find the ideal value for each objective. The nonideal values for each objective can be found by:

$$\begin{aligned} PB_{nonideal} = \max(&PB(S_{Ploss}), PB(S_{VD\_SUM}), \\ &PB(S_{d0\_SUM}), PB(S_{d2\_SUM}), PB(S_{ILCO\_SUM})) \end{aligned} \tag{8}$$

Due to its conflicting character, this single objective optimization problem can provide the best solution for the concerned objective, denoted as subscript ideal, while the worst value of objectives are denoted as nonideal. Also, the subscript ideal denotes the attainable goal value and nonideal denotes the worst value.

3.2 Step 2

A new single objective optimization problem is formulated as follows:

$$\min T = \sum_{\substack{Obj=PB,\, Plss, VD\_SUM, \\ d0\_SUM, d2\_SUM, \\ ILCO\_SUM}} W_{Obj} \frac{Obj - Obj_{ideal}}{Obj_{nonideal} - Obj_{ideal}} \quad \text{subject to the constraints} \tag{9}$$

Obviously, each objective value is normalized between (0,1) in (9), it is very easy to determine the weighting factor Wobj for each objective by direct thinking. Again, the GA is employed to find the optimal solution of (9).

4 Implementation of the Neural Network Based Particle Swarm Optimization Method

Conceptually, (7,9) belong to the class of problems known as combinatorial optimization with constraints. Possible combinations grow dramatically as the number of switches increases. Recently, the use of the global optimization technique called PSO [3], to solve real world problems have aroused researchers' interest due to its flexibility and efficiency. Therefore, PSO is used as the tool for solving Eqs. (7,9) in this paper.

4.1 Brief Review of the Particle Swarm

Particle swarm optimization (PSO), first introduced by Kennedy and Eberhart, is one of the modern heuristic algorithms. It was developed through simulation of a simplified social system, and has been found to be robust in solving continuous nonlinear optimization problems [3]. The PSO technique can generate a high-quality solution within shorter calculation time and stable convergence characteristic than other stochastic methods [4]. The feasibility of their method is compared with the reactive tabu system and enumeration method on practical power system, and has shown promising results [5]. Naka et al. have presented the use of a hybrid PSO method for solving efficiently the practical distribution state estimation problem [6].

4.2 Initial Population String Generated by Back-Propagation Neural Network

The particles should be initialized so that they are distributed as better as possible throughout the space. The standard method of initialization fails to accomplish this goal, especially in high-dimensional spaces. In this paper, using a starting configuration based on back-propagation neural network can lead to improved performance. Firstly, using random selection to generate some initial population and then put into the PSO to get the final solutions and evaluation values. Secondly, using these randomly guess data as the training set of BPN. The input and output layer are N_{tr} and six neurons respectively. The hidden layer is calculated by (N_{tr} +6)/2. Finally, randomly generate 10 times of the population size and put into the BPN to get the evaluation values. Take the better solutions as the initial generation.

5 Test Study

To illustrate the performance of the proposed solution methodology, consider a system that is part of the Taipower distribution system in Taiwan. It consists of 88 load centers and each load center equipped with one distribution transformer. To evaluate the P and Q loads for this feeder during whole year, the typical load pattern include three different types such as, weekday, weekend, and Sunday are considered. Also, the load forecasting is used to predict the load in the future. The solution space includes 6.664×10^{52} combinations such that PSO is used for better solution performance. The original configuration has many violations in constraints. After rearrange of the connection between distribution transformers and a primary feeder, the constraints are all satisfied. It shows that the PSO has the ability to solve the combinatorial optimization problem well.

Table 1. Numerical results for different strategies

WN37	Multi-objective programming					
	S1	S2	S3	S4	S5	S6
SA_{PB}	62.16	97.61	71.91	61.47	81.52	59.72
SA_{Ploss}	88.59	89.75	97.59	96.36	97.06	93.87
SA_{VD_SUM}	84.39	82.23	89.12	97.34	93.95	91.71
SA_{d0_SUM}	98.70	93.37	97.54	92.77	95.98	97.09
SA_{d2_SUM}	98.90	93.48	97.56	92.74	96.62	97.24
SA_{ILCO_SUM}	97.54	94.92	93.72	92.39	92.55	96.43

The multi-objective test results are summarized in Table 1 which according the different strategies. The satisfaction rates for each objective are defined in (10). It represents the level of satisfaction within the attainable search region for each objective. The strategy 1 denoted as S1 deal the different objectives as equal importance, and the simulation results reflect this decision correctly. All of the satisfaction rates are equally improved and the constraints are also met the requirement. The strategies S2, S3, and S4 treat the loss, voltage deviation, and LCO current as the primary objective, respectively. Also, the strategies S5 and S6 treat two of the six objectives as the primary concern. Obviously, according to the simulation results that shown in Table 1, the proposed method can solve the multi-objective problem according to the operator's will that represented as the weighting factor. Note that all the constraints are satisfied by the proposed method.

$$\begin{aligned} \textit{Satisfaction Rate of } PB &= SA_{PB} \\ &= (\max PB - PB)/(\max PB - \min PB) \end{aligned} \tag{10}$$

6 Conclusion

A new multi-objective formulation for the connections between distribution transformers and a primary feeder has been presented. A normalized weighting method combined with the particle swarm optimization for solving the multi-objective problem has been presented and tested on a Taipower distribution system. The performance of Particle Swarm Optimization can be improved by strategically selecting the starting positions of the particles by back-propagation neural network. The salient feature of the proposed method lies in that it can provide an easy way to determine the weighting factors between different objectives. Results obtained show that three phase voltage, currents and loads unbalance along a feeder as well as the voltage drop, line losses and the frequency of malfunction of grounding relay can be effectively reduced.

Acknowledgment

Support for this research by the National Science Council of the Republic of China under Grant No. NSC93-2622-E-129-006-CC3 is gratefully acknowledged.

References

1. Willians, J. E.: Operation of 3-Phase Induction Motors on Unbalanced Voltages. IEEE Transactions on Power Apparatus and Systems 5 (1954) 125-133
2. Miller, T. J. E.: Reactive Power Control in Electric Systems. John Wiley & Sons, New York, (1982)
3. Kennedy, J., Eberhart, R.: Particle Swarm Optimization. IEEE International Conference Neural Networks (ICNN'95), Perth, Australia (1995)1942–1948
4. Eberhart, R. C., Shi, Y.: Comparison between Genetic Algorithms and Particle Swarm Optimization. IEEE International Conference Evolution Computing (1998) 611–616
5. Yoshida, H., Kawata, K. Fukuyama, Y., Takayama, S., Nakanishi, Y.: A Particle Swarm Optimization for Reactive Power and Voltage Control Considering Voltage Security Assessment. IEEE Transactions on Power Systems 15 (2000) 1232–1239
6. Naka, S., Genji, T., Yura, T., Fukuyama, Y.: Practical Distribution State Estimation Using Hybrid Particle Swarm Optimization. Proceedings of the IEEE Power Engineering Society Winter Meeting 2 (2001) 815–820
7. Chen, T. H., Chang, Y. L.: Integrated Models of Distribution Transformers and Their Loads for Three-Phase Power Flow Analysis. IEEE Transactions on Power Delivery 11 (1996) 507-513

3-D Partial Discharge Patterns Recognition of Power Transformers Using Neural Networks

Hung-Cheng Chen[1], Po-Hung Chen[2], and Chien-Ming Chou[1]

[1] National Chin-Yi Institute of Technology,
Institute of Information and Electrical Energy,
Taiping, Taichung, 411, Taiwan, R.O.C.
{hcchen, s49312008}@chinyi.ncit.edu.tw
[2] St. John's University,
Department of Electrical Engineering,
Taipei, Taiwan, R.O.C.
phchen@mail.sju.edu.tw

Abstract. Partial discharge (PD) pattern recognition is an important tool in HV insulation diagnosis. A PD pattern recognition approach of HV power transformers based on a neural network is proposed in this paper. A commercial PD detector is firstly used to measure the 3-D PD patterns of epoxy resin power transformers. Then, two fractal features (fractal dimension and lacunarity) extracted from the raw 3-D PD patterns are presented for the neural-network-based (NN-based) recognition system. The system can quickly and stably learn to categorize input patterns and permit adaptive processes to access significant new information. To demonstrate the effectiveness of the proposed method, the recognition ability is investigated on 150 sets of field tested PD patterns of epoxy resin power transformers. Different types of PD within power transformers are identified with rather encouraged results.

1 Introduction

Power transformers play a crucial role in operation of transmission and distribution systems. A dielectric failure in a power transformer could result in unplanned outages of power systems, which affects a large number of customers [1]. Therefore, it is of great importance to detect incipient failures in power transformers as early as possible, so that they can be switched safely and improve the reliability of the power systems. Partial discharges phenomenon usually originates from insulation defects and is an important symptom to detect incipient failures in power transformers. Classification of different types of PDs is of importance for the diagnosis of the quality of HV power transformers. PD behavior can be represented in various ways. Because of the randomization of PD activity, one of the most popular representations is the statistics-based φ-Q-N distribution, i.e., the PD pattern is described using a pulse count N versus pulse height Q and phase angle φ diagram. Previous experimental results have adequately demonstrated that φ-Q-N distributions are strongly dependent upon PD

J. Wang et al. (Eds.): ISNN 2006, LNCS 3972, pp. 1324 – 1331, 2006.

sources, therefore the 3-D patterns can be used to characterize insulation defects [2]. This provides the basis for pattern recognition techniques that can identify the different types of defects.

The automated recognition of PD patterns has been widely studied recently. Various pattern recognition techniques have been proposed, including expert systems [3], fuzzy clustering [4], and neural networks (NNs) [5], [6]. The expert system and fuzzy approaches require human expertise, and have been successfully applied to this field. However, there are some difficulties in acquiring knowledge and in maintaining the database. NNs can directly acquire experience from the training data, and overcome some of the shortcomings of the expert system. However, the raw values of 3-D patterns were used with the NN for PD recognition in previous studies [7], the main drawbacks are that the structure of the NN has a great number of neurons with connections, and time-consuming in training. To improve the performance, two fractal features that extract relevant characteristics from the raw 3-D PD patterns are presented for the proposed NN-based classifier. It can quickly and stably learn to categorize input patterns and permit adaptive processes to access significant new information. To demonstrate the effectiveness of the proposed method, 150 sets of field-test PD patterns from HV epoxy resin power transformers are tested. Results of the studied cases show that different types of PD within power transformers are identified with rather encouraged results.

2 Fractal Features of 3-D PD Patterns for Recognition Purposes

2.1 Fractal Theory

Fractals have been very successfully used to address the problem of modeling and to provide a description of naturally occurring phenomena and shapes, wherein conventional and existing mathematical methods were found to be inadequate. In recent years, this technique has attracted increased attention for classification of textures and objects present in images and natural scenes, and for modeling complex physical processes. In this theory, fractal dimensions are allowed to depict surface asperity of complicated geometric things. Therefore, it's possible to study complex objects with simplified formulas and fewer parameters [8]. PD also is a natural phenomenon occurring in electrical insulation systems, which invariably contain tiny defects and non-uniformities, and gives rise to a variety of complex shapes and surfaces, both in a physical sense as well as in the shape of 3-D PD patterns acquired using digital PD detector. This complex nature of the PD pattern shapes and the ability of fractal geometry to model complex shapes, is the main reason which encouraged the authors to make an attempt to study its feasibility for PD pattern interpretation.

2.2 Calculation of Fractal Dimension

While the definition of fractal dimension by self-similarity is straightforward, it is often difficult to estimate/compute for a given image data. However, a related measure of fractal dimension, the box dimension, can be computed more easily. In this work, the method suggested by Voss [9] for the computation of fractal dimension D from the image data has been followed. Let *p(m,L)* define the probability that there are *m* points within a box of size *L* (i.e. cube of side *L*), which is centered about a point on the image surface. *P(m,L)* is normalized, as below, for all *L*.

$$\sum_{m=1}^{N} p(m,L) = 1 \tag{1}$$

where *N* is the number of possible points within the box. Let *S* be the number of image points (i.e. pixels in an image). If one overlays the image with boxes of side *L*, then the number of boxes with *m* points inside the box is *(S/m)p(m,L)*. Therefore, the expected total number of boxes needed to cover the whole image is

$$\langle N(L) \rangle = \sum_{m=1}^{N} \frac{S}{m} p(m,L) = S \sum_{m=1}^{N} \frac{1}{m} p(m,L) \tag{2}$$

Hence, if we let

$$N(L) = \sum_{m=1}^{N} \frac{1}{m} p(m,L) \tag{3}$$

this value is also proportional to L^{-D} and the box dimension can be estimated by calculating *p(m,L)* and *N(L)* for various values of *L*, and by doing a least square fit on [log(*L*), - log(*N*(*L*))]. To estimate *p(m,L)*, one must center the cube of size *L* around an image point and count the number of neighboring points *m*, that fall within the cube. Accumulating the occurrences of each number of neighboring points over the image gives the frequency of occurrence of *m*. This is normalized to obtain *p(m,L)*. Values of *L* are chosen to be odd to simplify the centering process. Also, the centering and counting activity is restricted to pixels having all their neighbors inside the image. This obviously will leave out image portions of width = (*L* – 1)/2 on the borders. This reduced image is then considered for the counting process. As is seen, large values of *L* results in increased image areas from being excluded during the counting process, thereby increasing uncertainty about counts near border areas of the image. This is one of the sources of errors for the estimation of *p(m,L)* and thereby *D*. Additionally, the computation time grows with the *L* value. Hence, *L* = 3, 5, 7, 9 and 11 were chosen for this work.

2.3 Calculation of Lacunarity

Theoretically, ideal fractal could confirm to statistical similarity for all scales. In other words, fractal dimensions are independent of scales. However, it has been observed that fractal dimension alone is insufficient for purposes of discrimination, since two differently appearing surfaces could have the same value of ***D***. To overcome this,

Mandelbrot [10] introduced the term called lacunarity *Λ*, which quantifies the denseness of an image surface. Many definitions of this term have been proposed and the basic idea in all these is to quantify the 'gaps or lacunae' present in a given surface. One of the useful definitions of this term as suggested by Mandelbrot [10] is

$$\Lambda(L)=\frac{M^2(L)-(M(L))^2}{((M(L))^2} \tag{4}$$

where

$$M(L)=\sum_{m=1}^{N} m\, p(m,L) \tag{5}$$

$$M^2(L)=\sum_{m=1}^{N} m^2\, p(m,L) \tag{6}$$

From the definition, we can obtain the idea that lacunarity reflects the density of fractal surfaces, namely the extent to which the density is. The smoother the surfaces, the less the lacunarity $\Lambda(L)$.

3 Discharge Experiments

In this paper, the tested object is an cast-resin HV power transformers that uses epoxy resin for HV insulation. The rated voltage and capacity of the tested HV power transformers are 12 kV and 2kVA, respectively. For testing purposes, four experimental models of power transformers with artificial insulation defects were purposely manufactured by an electrical manufacturer. The four PD models, including no defect, HV corona discharge, low voltage (LV) coil PD, and high voltage (HV) coil PD, are named Type I, II, III, and IV, respectively. In the testing process, all of the measuring data are digitally converted in order to store them in the computer. Then, the PD pattern classifier can automatically recognize the different defect types of the testing objects.

The individual 3-D PD patterns (stored as a 256x256 matrix) are plotted. The x and y axes correspond to the phase and amplitude (or charge), respectively. The matrix elements correspond to the pulse count data (or the z axis of the 3-D pattern). An example 3-D plot of the pattern from each one of the four types is given in Fig. 1. In order to simplify the calculation of both fractal dimension and lacunarity, a real gray-scaled image would be utilized instead of 3-D patterns. The amplitude values are linearly mapped to the varying intensities of the white color (uniformly mapped to one of the 16 gray colors in this work). This gray image is the input for computing the fractal features.

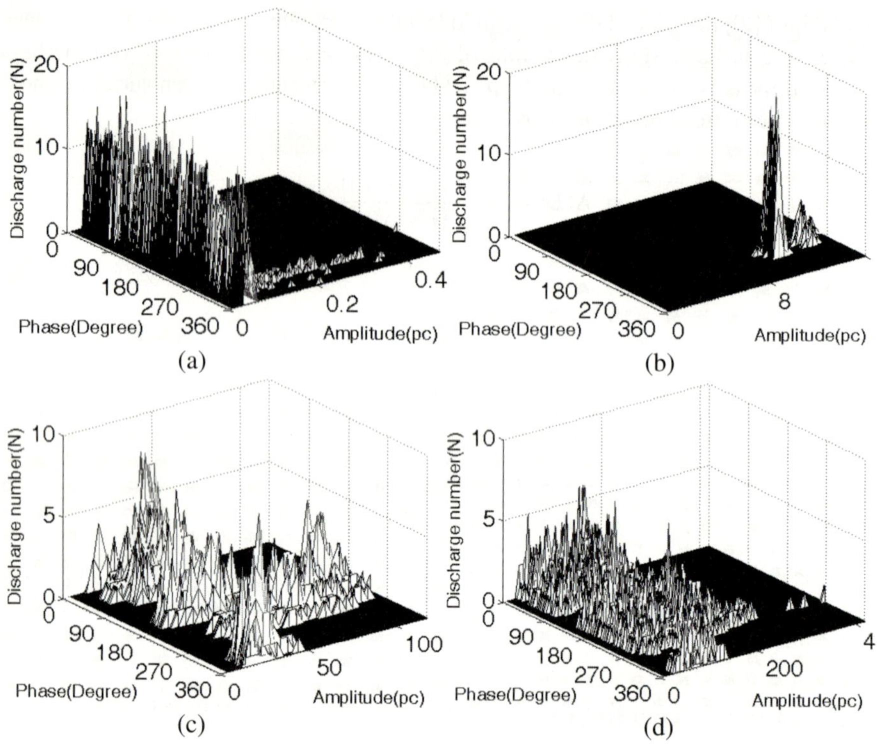

Fig. 1. Four typical defect types of PD pattern. (a) No defect (Type I). (b) HV corona discharge (Type II). (c) LV coil PD (Type III). (d) HV coil PD (Type IV).

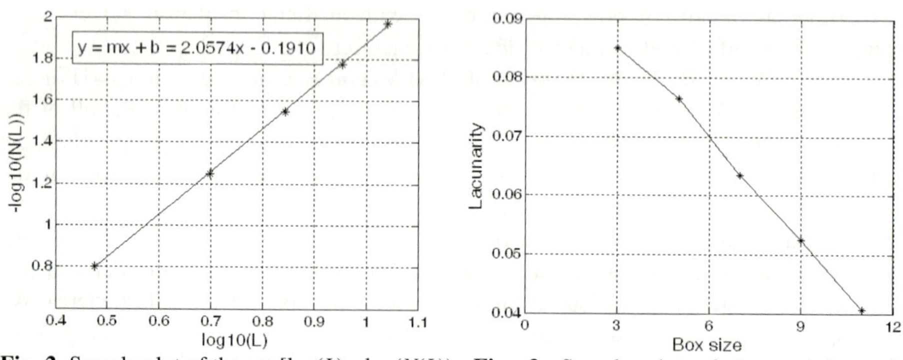

Fig. 2. Sample plot of the set [log(*L*), -log(*N*(*L*)] for different value of box size *L*

Fig. 3. Sample plot of the variation of lacunarity with respect to box size *L*

Fig. 2 is a sample plot of the set [log(*L*), -log(*N*(*L*))] for the five chosen values of *L* (computed for one of the pattern examples from Type III). A least square fit to this data set is performed to obtain the fractal dimension *D*. The corresponding lacunarity is also computed for each value of *L*. Fig. 3 shows its variation with respect to *L*. These two fractal features are computed for all the available patterns. Fig. 4 is a plot of fractal dimension and lacunarity of different discharge models. Lacunarity was found to be maximum for all the pattern examples considered, at $L = 3$ and so, this L value was chosen for convenience.

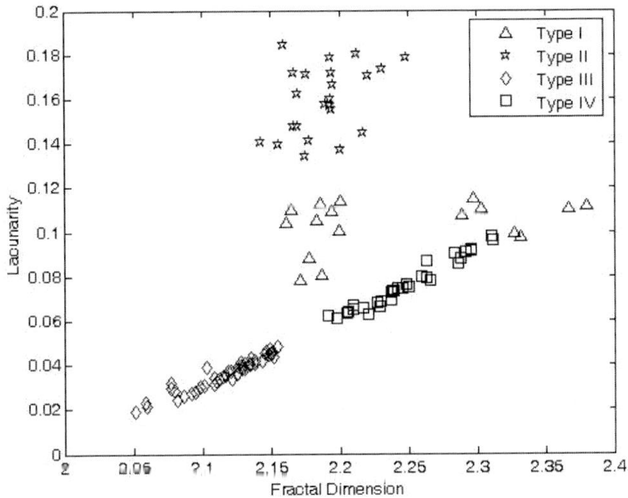

Fig. 4. Fractal dimension and lacunarity of different discharge models

4 Recognition Results and Discussion

Three neural network paradigms, back propagation network (BPN), probabilistic neural network (PNN), and learning vector quantization (LVQ), are utilized to classify PD pattern of the models. Four layers feed forward structure is used for the pattern recognition system. Its topological structure is shown in Fig. 5. The neuron number of its input is determined by the number of fractal features, viz., fractal dimension and lacunarity. The neuron number of both hidden layers is 6. The neuron number of output layer is determined by the number of patterns to be identified, which is 4 in this study. To demonstrate the recognition ability, 150 sets of field test PD patterns are used to test the proposed PD recognition system. The four defect models of 12-kV epoxy resin power transformers include the no-defect, HV corona discharge, LV coil PD, and HV coil PD, respectively. The NN-based PD recognition system randomly chooses 80 instances from the field test data as the training data set, and the rest of the instances of the field test data are the testing data set. Table 1 shows the recognized results of the proposed system with different input patterns. The recognition rates of the proposed system are quite high with about 100%, 94% and 98% for BPN, PNN, and LVQ, respectively. It is obvious that the NN-based PD recognition system has strong generalized capability. The recognized results of the three neural networks are almost of the same accuracy.

The field test data would unavoidably contain some noise and uncertainties which originate in environmental noise, transducers, or human mistakes. To evaluate the fault tolerance ability, total 150 sets of noise-contained testing data are generated by adding ±5% to ±30% of random, uniformly distributed, noise to the training data to take into account the noise and uncertainties. The test results with different amounts of noise added are also shown in Table 1 for the different neural networks. Usually, the noise-contained data indeed degrade the recognition abilities in proportion to the amounts of noise added. This table shows that all these neural networks rather withstand remarkable tolerance to the noise contained in the data. The proposed recognition systems show good tolerance to added noise, and have high accuracy rates of 78%, 72% and 70% in extreme noise of 30%.

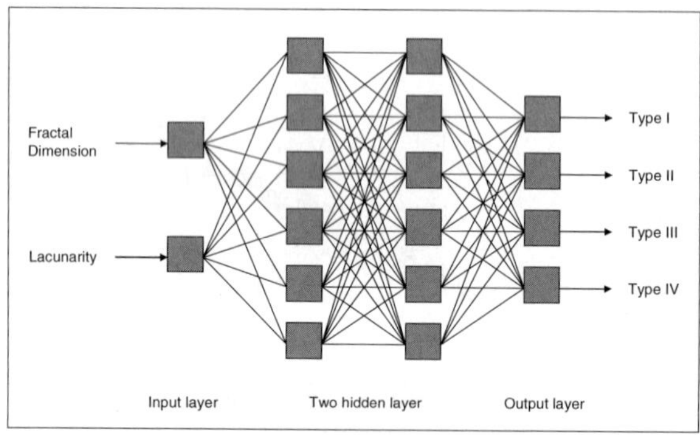

Fig. 5. Topology structure of NN-based pattern recognition system

Table 1. Recognized performances of different neural networks with various noises added

Proportion of noise	Recognition rate (%)		
	Back Propagation Network (BPN)	Probabilistic Neural network (PNN)	Learning Vector Quantization (LVQ)
±0%	100%	94%	98%
±5%	92%	92%	98%
±10%	88%	90%	94%
±15%	86%	90%	94%
±20%	80%	88%	90%
±25%	80%	78%	78%
±30%	78%	72%	70%

5 Conclusions

A method to analyze a PD pattern and identify the type of discharge source is an important tool for the diagnosis of HV insulation system. A NN-based PD pattern recognition method for HV power transformers, that uses fractal features to highlight the more detailed characteristics of the raw 3-D PD patterns, is proposed. This uses a fractal theory to extract the fractal dimension and lacunarity from the raw 3-D PD patterns. These fractal features are then applied to a neural network that performs the classification. The recognition rates of the proposed system are quite high with about

100%, 94% and 98%, and 78%, 72% and 70% in extreme noise of 30%, for BPN, PNN, and LVQ, respectively. The present experimental results indicate that this approach is able to implement an efficient classification with a very high recognition rate.

Acknowledgments

The research was supported in part by the National Science Council of the Republic of China, under Grant No. NSC93-2213-E-167-021.

Reference

1. Feinberg, R.: Modern Power Transformer Practice. John Wiley & Sons, New York (1979)
2. Krivda, A.: Automated Recognition of Partial Discharge. IEEE Trans. on Dielectrics and Electrical Insulation 2 (1995) 796-821
3. Satish, L., Gururaj, B.I.: Application of Expert System to Partial Discharge Pattern Recognition. in CIGRE Study Committee 33 Colloquium, Leningrad, Russia, (1991) Paper GIGRE SC 33.91
4. Tomsovic, K., Tapper, M., Ingvarsson, T.T.: A Fuzzy Information Approach to Integrating Different Transformer Diagnostic Methods. IEEE Trans. on Power Delivery 8 (1993) 1638-1643
5. Cho, K.B., Oh, J.Y.: An Overview of Application of Artificial Neural Network to Partial Discharge Pattern Classification. Proc. of the 5th International Conference on Properties and Applications of Dielectric Materials 1 (1997) 326-330
6. Zhang, H., Lee, W.J., Kwan, C., Ren, Z., Chen, H., Sheeley, J.: Artificial Neural Network Based On-Line Partial Discharge Monitoring System for Motors. IEEE Conference of Industrial and Commercial Power Systems Technical (2005) 125-132
7. Salama, M.M.A., Bartnikas, R.: Determination of Neural Network Topology for Partial Discharge Pulse Pattern Recognition. IEEE Trans. on Neural Networks 13 (2002) 446-456
8. Satish, L. Zaengl, W.S.: Can Fractal Features Be Used for Recognizing 3-D Partial Discharge Patterns. IEEE Trans. on Dielectrics and Electrical Insulation 3 (1995) 352-359
9. Voss, R.F.: Random Fractal: Characterization and Measurement. Plenum Press, New York (1985)
10. Mandelbrot, B.B.: The Fractal Geometry of Nature. Freeman, New York (1983)

Design of Self-adaptive Single Neuron Facts Controllers Based on Genetic Algorithm

Quan-Yuan Jiang, Chuang-Xin Guo, and Yi-Jia Cao

College of Electrical Engineering, Zhejiang University, Hangzhou 310027, China
jqy@zju.edu.cn

Abstract. With the growing application of Static Var Compensator (SVC) and Static Synchronous Compensator (STATCOM), the coordinating problem of SVC and STATCOM controllers in joint operation must be considered in modern power systems. This paper firstly establishes the nonlinear differential-algebra equations model of a single-machine infinite-bus (SMIB) power system installed with a SVC and a STATCOM and points out the possibility of the negative interactions between SVC and STATCOM controllers in this SMIB power system. Hence, a self-adaptive single neuron (SSN) control approach based on genetic algorithm is designed to eliminate the negative interactions and improve the stability of the closed-loop SMIB power system. The detailed simulation results demonstrate the effectiveness of the proposed SSN controllers.

1 Introduction

With the rapid development of power electronics over the past decades, FACTS (Flexible Alternating-Current Transmission Systems) is widely applied in modern power system [1]. SVC (Static Var Compensator) and STATCOM (Static Synchronous Compensator) are the common devices in FACTS family [2, 3]. A great deal of discussions has been seen about the design of SVC or STACOM controller [4, 5]. However, dynamic interactions and coordination of multiple FACTS controllers are not considered. This paper has been about to overcome this interaction by their co-ordinated design. A single-machine infinite-bus power system installed with a SVC and a STATCOM is investigated. The possibility of the negative interactions between SVC and STATCOM is pointed out. Hence, a self-adaptive single neuron controller based on genetic algorithm is designed to eliminate the negative interaction.

2 Interactions Between SVC and STATCOM

A single-machine-infinite-bus power system is adopted, and its model is given in reference [6]. One SVC and one STATCOM are installed in the middle of the transmission line as Fig.1. The dynamic model of the STATCOM is given in reference [7].

The model of SVC can be described with the following differential equation:

$$\dot{B}_{SVC} = \frac{K_{SVC}(V_{sref} - V_s) - B_{SVC}}{T_{SVC}} \tag{1}$$

J. Wang et al. (Eds.): ISNN 2006, LNCS 3972, pp. 1332 – 1337, 2006.

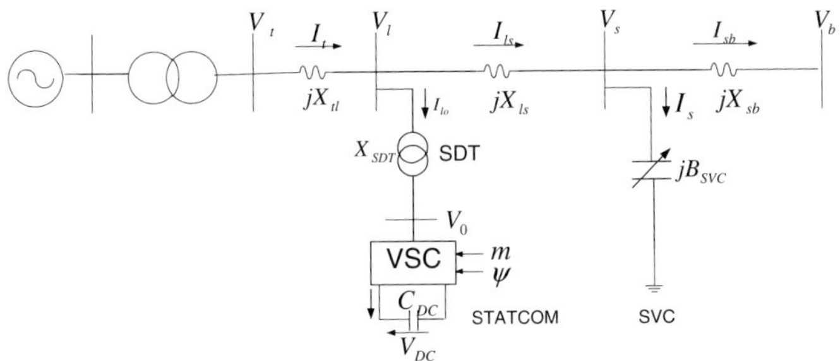

Fig. 1. A single-machine infinite-bus power system installed with SVC and STATCOM

where K_{SVC} and T_{SVC} are the gain and time constant of the SVC, B_{SVC} is the equivalent reactance of SVC.

According to multivariable control theory, for such a multi-input-multi-output (MIMO) system, the design of individual controllers to achieve a satisfactory joint operation of the controllers can only always be guaranteed when the interactions among the different control channels are small and can be ignored. Unfortunately, the detrimental interactions between different control channels of a FACTS device are reported in recent years [8]. In this paper, we focus on the potential detrimental interaction between SVC and STATCOM.

3 Coordinate Design of Self-adaptive Single Neuron FACTS Controller

In order to eliminate the negative interaction between SVC and STATCOM, a coordinated controller should be designed. In this paper, we propose a self-adaptive single-neuron controller to eliminate the negative interactions of the investigated power system.

3.1 Design of Self-adaptive Single-Neuron Controller

Single neuron is a basic element of neural networks. Generally, single neuron has multiple inputs and only one output. A typical self-adaptive single neuron (SSN) control system is shown in Fig.2 and Fig.3. From Fig.2 and Fig.3, we can see that a SSN control system consists of state converter, single neuron, and plant. The converter transfers the information acquired from the environment into the input information necessary for SSN. The variables in Fig.2 are explained as follows:

x_i $(i = 1, 2, \ldots, n)$	outputs of state converter and inputs of the single neuron
w_i $(i = 1, 2, \ldots, n)$	weights corresponding to inputs desired input
y	output of the system
u	output of the single neuron controller
K	gain of single neuron
r	reference input of system
e	error of the system

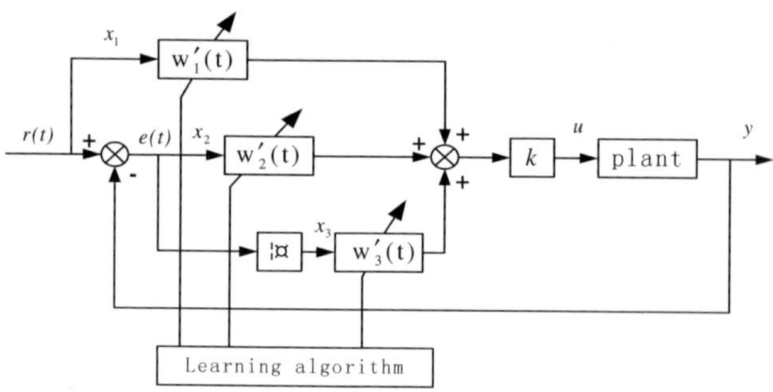

Fig. 2. Self-adaptive single neuron control system

Through adjusting weights w_i , the single neuron controller can achieve self-adaptive and self-organized capacity. In this paper, the weights w_i is adjusted based on the normal learning method:

$$
\begin{aligned}
&u(t) = k\sum_{i=1}^{n} w_i\,'(t)x_i(t) \\
&w_i\,'(t) = w_i(t)/(\sum_{i=1}^{n}|w_i(t)|) \\
&w_i(t+1) = w_i(t) + d\left[r(t) - y(t)\right]x_i(t) \quad (i = 1,2,\cdots n)
\end{aligned}
\tag{2}
$$

where d is the learning ratio. In this paper, the inputs of SSN, x_i $(i = 1, 2, \ldots, n)$ are set as follows:

$$
\begin{cases}
x_1(t) = r(t) \\
x_2(t) = r(t) - y(t) = e(t) \\
x_3(t) = \Delta e(t) = e(t) - e(t-1) \\
x_4(t) = \Delta^2 e(t) = e(t) - 2e(t-1) + e(t-2) \\
\cdots
\end{cases}
\tag{3}
$$

The online tuning of weights w_i mainly relates to the error $e(t)$ and its changing ratio.

There are three single-input single-output controllers to be designed, i. e. AC voltage controller of SVC, AC and DC voltage controllers of STATCOM. Assuming that the DC voltage controller of STATCOM adopts proportional and integral structure, and the other two controllers are SSN controllers. The entire control system is given in Fig.3. From Fig.2 and Fig.3, we can see that there are two three-input-single-output neurons, SSN1 and SSN2, which control the AC voltage of SVC and STATCOM respectively. u_1, u_2 are output signals of SSN1 and SSN2 controllers respectively. The feedback output signals y_1, y_2 and state variables x^i_j (i=1, 2; j=1,2, 3) of SSN1 and SSN2 are selected as follows:

(1) SSN1: Feedback signal $y_1 = V_s$. State variables x_j^1 (j=1,2, 3):

$$\begin{cases} x_1^1(t) = V_{sref} \\ x_2^1(t) = V_{sref} - V_s(t) = e_1(t) \\ x_3^1(t) = x_2^1(t) - x_2^1(t-1) = e_1(t) - e_1(t-1) \end{cases} \tag{4}$$

(2) SSN2: Feedback signal $y_2 = V_l$. State variables x_j^2 (j=1,2, 3):

$$\begin{cases} x_1^2(t) = V_{lref} \\ x_2^2(t) = V_{lref} - V_l(t) = e_2(t) \\ x_3^2(t) = x_2^2(t) - x_2^2(t-1) = e_2(t) - e_2(t-1) \end{cases} \tag{5}$$

According to the tuning algorithm of weights w_i described as equations (6), the control signals u_1, u_2 can be derived as follows:

$$\begin{aligned} u_i(t) &= \frac{k_i \sum_{j=1}^{3} w_j^i(t) x_j^i(t)}{\sum_{j=1}^{3} | w_j^i(t) |} \\ w_j^i(t+1) &= w_j^i(t) + d_i[r_i(t) - y_i(t)]x_j^i(t) \qquad (i = 1, 2; j = 1, 2, 3) \\ x_1^i(t) &= r_i(t) \end{aligned} \tag{6}$$

The equation (6) gives an iterative algorithm to compute the weight $w_j^i(t)$ and control signal $u_i(t)$ step by step. In order to startup the iterative process, the initial values of weight $x_j^i(0)$ should be determined firstly. Obviously, when $k_i = 0$ and $d_i = 0$, there are no control signals to effect the plant, i.e. $u_i(t) = 0$. In order to assure the plant normally operating when $u_i(t) = 0$, we must set $x_1^i(0) = 0$. The other initial weights $x_2^i(0)$, $x_3^i(0)$, and proportional gain k_i, learning ratio d_i can be optimized by genetic algorithm (GA).

3.2 Genetic Algorithms

GA is a stochastic parallel search technique and has been proven very effective for a wide variety of complex engineering problems, especially for the non-convex, randomness, non-differentiable and nonlinear complex problems. For the present application, a standard genetic algorithm (SGA) is proposed. It will result in a set of optimal parameters of the SSN controllers: $k_1 = 0.35$, $d_1 = 0.92$, $w_1^1(0) = 0$, $w_2^1(0) = 0.51$, $w_3^1(0) = 1.54$, $k_2 = 0.36$, $d_2 = 0.89$, $w_1^2(0) = 0$, $w_2^2(0) = 0.48$, $w_3^2(0) = 1.48$.

3.3 Simulation Verification

A full time domain simulation is carried out to verify the effectiveness of the derived SSN controllers. The dynamic responses of the SMIB power system with SVC and STATCOM to a step-change of the reference value of AC or DC voltage are investigated.

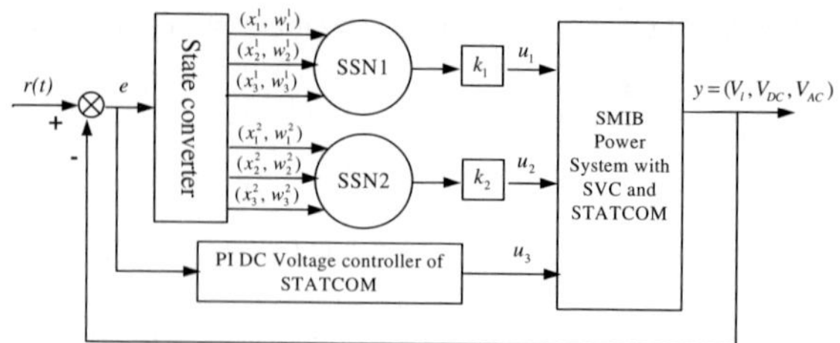

Fig. 3. Structure diagram of SMIB power system with SVC and STATCOM

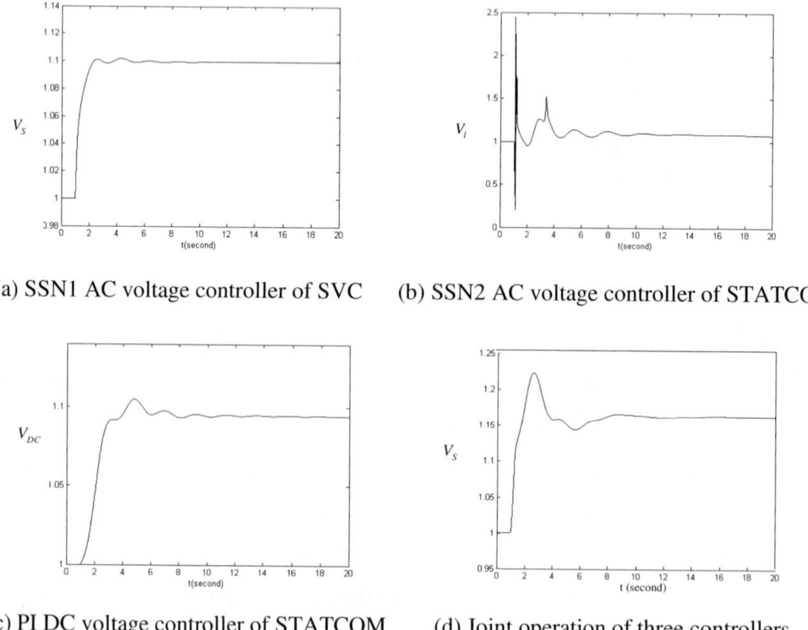

(a) SSN1 AC voltage controller of SVC (b) SSN2 AC voltage controller of STATCOM

(c) PI DC voltage controller of STATCOM (d) Joint operation of three controllers

Fig. 4. Time-domain simulation to evaluate the control performances

In the simulation, the reference value of AC or DC voltage is changed at 1.0 s of the simulation from 1.0 to 1.1 p.u. Results are given in Fig.4. From Fig.4, we can see that these three controllers, PI DC voltage controller of STATCOM, SSN1 and SSN2 controllers, can assure the SMIB power system stable not only in separate operating conditions but also in joint operating conditions.

4 Conclusions

In this paper, a self-adaptive single-neuron technique based on genetic algorithm is proposed for the coordinate design to eliminate the negative interactions among multiple control channels of SVC and STATCOM. The simulation results show the effectiveness of the derived controller. Comparing with the linear analysis technique and traditional multi-layer-neural-network-based control methods, the proposed single neuron controller also has the capability of self-adaptation to nonlinear and time-varying circumstance. But its structure is simple and concise, and it is therefore more efficient and easy to realize by software in real-time cases.

Acknowledgement

This work is supported by Natural Science Fund of China (#50507018), and by the Fund of Education Department of Zhejiang (20050908).

References

1. Hingorani, N.G., Narain, G.: Power Electronics in Electric Utilities: Role of Power Electronics in Future Power Systems. Proceedings of the IEEE 76 (1988) 481-482
2. Mathur, R.M., Varma, R.: Thyristor-Based Facts Controllers for Electrical Transmission Systems. IEEE Press 22 (2002) 42-43
3. Leha, P.W., Iravani, M.R.: Experimental Evaluation of Statcom Closed Loop Dynamics. IEEE Trans. On Power Delivery 13 (1998) 1378-1374
4. Zhou, E.Z.: Application of Satic Var Compensators to Increase Power System Damping. IEEE Transactions on Power System 8 (1993) 655-663
5. Rao, P., Crow, M.L., Yang, Z.: Statcom Control for Power System Voltage Control Applications. IEEE Trans. on Power Delivery 15 (2000) 1311-1317
6. Kundur, P.: Power System Stability and Control. 1st edn. McGraw-Hill New York (1994)
7. Wang, H.F.: Phillips-Heffron Model of Power System Installed with Statcom and Applications. IEE Proc Gener Transm Distribution 145 (1998) 111–118
8. Wang, H.F.: Interactions and Multivariable Design of Statcom AC and DC Voltage Control. International Journal of Electrical Power and Energy System 25 (2003) 387-394

Generalized Minimum Variance Neuro Controller for Power System Stabilization

Hee-Sang Ko[1], Kwang Y. Lee[2],
Min-Jae Kang[3], and Ho-Chan Kim[3]

[1] Dept. of Electrical and Computer Engineering,
The University of British Columbia, Vancouver, BC V6T 1Z4, Cananda
Hee-sang@ece.ubc.ca
[2] Dept. of Electrical Engineering, The Pennsylvania State University,
University Park, PA, 16802, USA
kwanglee@psu.edu
[3] Faculty of Electrical and Electronic Engineering, Cheju National University,
Cheju, 690-756, Korea
{minjk, hckim}@cheju.ac.kr

Abstract. This paper presents a power system stabilizer design that uses a generalized minimum variance-inverse dynamic neuro controller, which is the combination of the inverse dynamic neural model, the generalized minimum variance, and the neuro compensator. An inverse dynamic neural model represents the inverse dynamics of the system. The inverse dynamic neural model is trained to provide control input into the system, which makes the plant output reach the target value at the next sampling time. Once the inverse dynamic neural model is trained, it does not require retuning for cases with other types of disturbances. In this paper, a generalized minimum variance control scheme is adapted to prevent unstable system performance caused by non-minimum phase characteristics. In addition, a neural compensator is designed to compensate for modeling errors. The proposed control scheme is tested in a multimachine power system.

1 Introduction

Power system stabilization problem provides a typical example of designing a large-scale nonlinear and time varying control system. Power system is typically characterized by geographical separation because often generators are several hundreds of kilometers apart; hence, a decentralized control scheme is preferred to control a power system. Local controller is designed for each subsystem using local information. The excitation control is the most common tool used to enhance the damping of low frequency oscillations in a power system.

Considerable efforts have been made to design power system stabilizers (PSSs) for power systems, most of which is based on deMello and Concordia's pioneering work [1,2]. They used a linearized model of power system to find a proper set of parameters in the PSS. Linear optimal control and modern control theories are also introduced to improve the dynamic performance of power systems by accounting for the uncertainty of power system models [3,4]. These techniques, however, depend on the accuracy of

J. Wang et al. (Eds.): ISNN 2006, LNCS 3972, pp. 1338 – 1347, 2006.

the model, which is less reliable as the power system becomes larger. Adaptive techniques are also employed in the PSS design for a wide range of operations [5,6].

Recently, artificial neural networks (ANN) have attracted the attention of power system engineers. There has been a great deal of research that reports on ANN and its application to control and power systems [7]-[9]. Zhang and coworkers suggested a PSS that uses an inverse input-output mapped ANN [7]. Hsu and Chen [8] proposed a real time self-tuning approach where they use ANN to tune the parameters of a conventional proportional-integral type PSS. However, since their approach requires a mathematical model of the controlled system for a wide range of operating conditions, it bears a shortcoming in its large-scale system application. Wu, Hogg and Irwin [9] presented a hierarchically structured neural-PSS. Their neural-PSS consists of two subnets: one for input-output mapping, and the other for control. Most of the neural controllers, with few exceptions, have a common feature, that is, the use of two ANNs: one for system identification and the other for the controller.

In this paper, an inverse input-output relationship of the controlled system is identified by a neural network, which is based on the free-model concept [10], and such a neural network is said to be an inverse dynamic neural model (IDNM). Before training IDNM, the input/output data sets are collected by disturbing the system with various random step signals. After training, the IDNM is used as an inverse controller and remain fixed. The modeling error and disturbances resulting from the fixed IDNM are compensated by the neural compensator. The generalized minimum variance (GMV) control [11] is adapted to prevent excessive control input, which can be caused by a non-minimum phase characteristics. The combined scheme of the IDNM, the GMV, and the neuro compensator is called the generalized minimum variance - inverse dynamic neuro controller (GMV-IDNC) in this paper.

2 Generalized Minimum Variance-Inverse Dynamic Neuro Controller

2.1 Inverse Dynamic Neural Model (IDNM)

A function or a mapping of the plant input-output history can describe the future output of a system, which can be written in the form of a nonlinear function as follows:

$$y(k+1) = f(y(k), y(k-1), \ldots, y(k-n), u(k), u(k-1), \ldots, u(k-m)), \tag{1}$$

where n and m are the order of output $y(k)$ and input $u(k)$, respectively.

Using the free-model concept [10], (1) can be equivalently represented as the following:

$$y(k+1) = f(y(k), \Delta y(k), \ldots, \Delta^n y(k), u(k), u(k-1), \Delta u(k-1), \ldots, \Delta^m u(k-1)), \tag{2}$$

where Δ^i is the backward difference operator, defined as below:

$$\Delta^i x(k) = \Delta^{i-1} x(k) - \Delta^{i-1} x(k-1), \quad \Delta^0 x(k) = x(k)\,. \tag{3}$$

Now, let's assume that there exists the inverse relationship between input and output in (2), then

$$u(k) = g(\overline{\varphi}(k)), \tag{4}$$

$$\overline{\varphi}(k) = [y(k+1), y(k), \Delta y(k), \cdots, \Delta^n y(k), u(k-1), \Delta u(k-1), \cdots, \Delta^m u(k-1)) . \tag{5}$$

Since power system is a highly nonlinear and complex system, a method that is solely based on the input/output data is required to deal with the nonlinearity efficiently. From this point of view, the ANN is a good candidate for control since ANNs have the ability to learn and construct a complex nonlinear mapping through a set of input/output data. In this paper, the inverse relationship is obtained by using a general feedforward neural network [12] as shown in Fig. 1, named as IDNM.

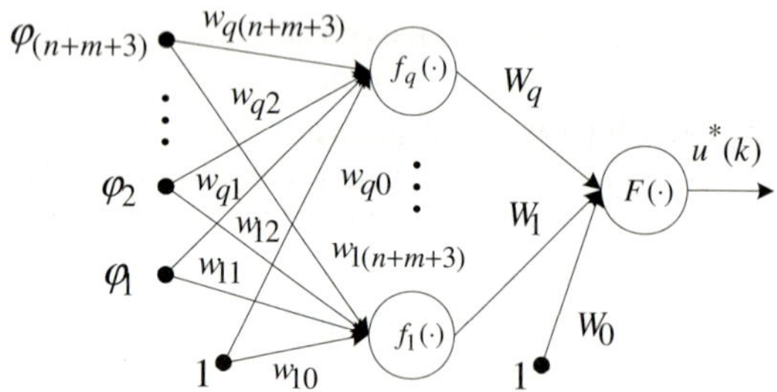

Fig. 1. A feedforward neural network

The three layer neural network is used: the first layer is for inputs. The second layer is called the hidden layer whose neurons are a hyperbolic tangent function. The third layer is the output layer whose neuron is a linear function. The output of the neural network, $u^*(k)$, is given by (6).

$$u^*(k) = F\left(\sum_{j=1}^{n_h} W_j f_j \left(\sum_{l=1}^{n_\varphi} w_{jl}\varphi_l + w_{j0} \right) + W_0 \right), \tag{6}$$

where φ_l is the lth component of input vector $\overline{\varphi}$, w_{jl} is the weights between input and hidden layer, w_{j0} (W_0) is the bias weights to hidden layer (or output layer), $n_h(n_\varphi)$ is the numbers of hidden neurons (or external input), W_j is the weights between hidden and output layer, f is a hyperbolic tangent function ($f(z) = \tanh(z)$) and F is a linear function ($F(z) = z$).

The proper weights are obtained by minimizing the following mean square error criterion:

$$J = \frac{1}{2N} \sum_{k=1}^{N} [u(k) - u^*(k)]^2 , \tag{7}$$

where N is the total number of patterns.

The training of IDNM is performed by the Levenberg-Marquart method, which is fast and robust convergence for off-line training of neural network [12].

Assuming such a network (6) has been obtained through a training process described above, it can then be used for controlling the system by replacing the future output $y(k+1)$ in the input vector (5) with the desired output, y_{ref}. If the network represents the exact inverse, the control input produced by the network will drive the system output $y(k+1)$ to y_{ref}. Once an IDNM is trained, (4) can be rewritten for a controller:

$$u_r(k) = g(y_{ref}, y(k), \Delta y(k), \cdots, \Delta^n y(k), u(k-1), \Delta u(k-1), \cdots, \Delta^m u(k-1)). \tag{8}$$

The output of IDNM, $u_r(k)$, can be very excessive or too active due to the non-minimum phase characteristics of a plant. Smooth control signal is preferred in reducing the stress on the equipment. For this reason, the generalized minimum variance control is introduced in the following section.

2.2 Generalized Minimum Variance Control

A generalized minimum variance (GMV) control is also known as a weight one-step ahead predictive control, which is one of the adaptive control techniques [11]. This is a modified control scheme of minimum variance (MV) control [11]. The objective of the MV control is to obtain the control input, which minimizes the difference between the plant output and the target value at the very next sampling time. The MV method does not place any limitation on the control input implicitly, which is the reason that MV control cannot prevent the excessive control input or instability. In the GMV control method however, a penalty is added to limit the control effort by modifying the objective function.

In this paper, the following objective function is defined:

$$J = \frac{1}{2}(y_{ref} - y(k+1))^2 + \frac{1}{2} w \cdot u(k)^2 \tag{9}$$

where the first term is the usual minimum variance objective function and w is the penalty on the control effort.

It can be shown that, by minimizing the objective function (9), the GMV based IDNM controller (GMV-IDNM) is obtained in the following form:

$$u(k) = \alpha \cdot u_r(k) \tag{10}$$

where α is a scalar as a function of the weight w and $u_r(k)$ is the output of the IDNM. It is shown that α ranges in an interval [0, 1]; α is 1 when w is 0, which corresponds to the MV control, and α decreases toward 0 as w increases toward infinity. Fig. 2 shows the control scheme when the GMV is implemented. In this case, $\bar{\Delta}_D$ is a vector backward difference operator, which generates output vector $[y(k), \Delta y(k), \cdots, \Delta^n y(k)]$ and input vector $[u(k-1), \Delta u(k-1), \cdots, \Delta^m u(k-1)]$.

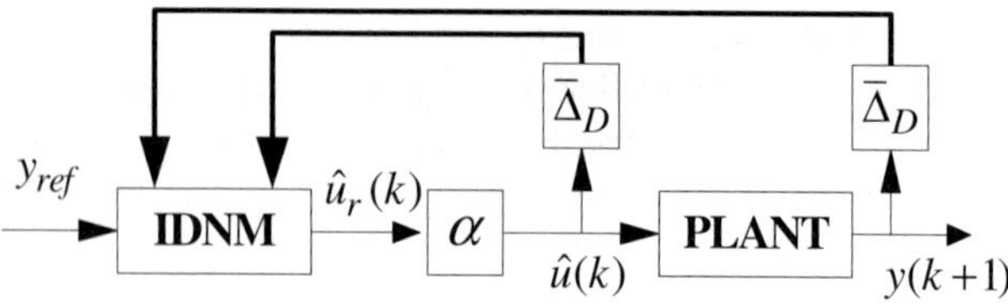

Fig. 2. GMV- IDNM

2.3 Neuro Compensator

To increase the robustness of the controller, it is necessary to consider a wide range of operating conditions and disturbances while designing the controller. In real applications, however, it is impractical to consider all of the operating conditions. Therefore, when the system is operating under conditions that the IDNM has never learned, error inevitably exists between the IDNM and real inverse dynamics even though the IDNM training may have been completed for previously given data sets.

A neuro compensator is constructed based on the pre-trained IDNM. In the application phase of IDNM, y_{ref} is not the same as $y(k+1)$, which was used in the training phase of the IDNM. Therefore, there exists a modeling error of $E(k)$ at time k. The neuro compensator is designed to compensate for this modeling error.

The error at time k can be estimated by extrapolating the history of previous errors using the Newton-backward-difference formula (NBDF) [14], as follows:

$$\hat{E}(k) = \sum_{r=1}^{l} (-1)^r \binom{r}{l+1} \Delta^r E(k-1), \tag{11}$$

where $E(k-1) = u(k-1) - \hat{u}(k-1)$ is the modeling error at time k-1, Δ^r is the backward difference operator defined in (2), l is the extrapolation order, and the binomial-coefficient is defined as $\binom{r}{l+1} = \frac{r(r-1)\cdots(r-l)}{(l+1)!}$.

Fig. 3 shows the design of the neuro compensator and the NBDF. The symbol D stands for a delay operator and $\hat{u}(k-1)$ is the output of IDNM based on the given

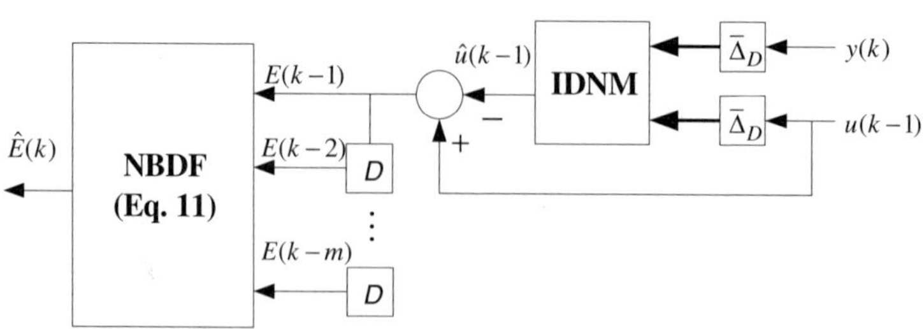

Fig. 3. The neuro compensator

input and output data given up to time k-2. An error at time k-1 between the estimated control input $\hat{u}(k-1)$ and real control input $u(k$-1) can be obtained. Finally, the compensating error for the control input at time k can be provided by the linear extrapolation by NBDF.

The compensated control input becomes as follows:

$$u(k) = \alpha(u_r(k) + \hat{E}(k)) \tag{12}$$

where $\hat{E}(k)$ is the estimate of the modeling error in the IDNM.

Fig. 4 presents the overall proposed control scheme of the GMV-IDNC.

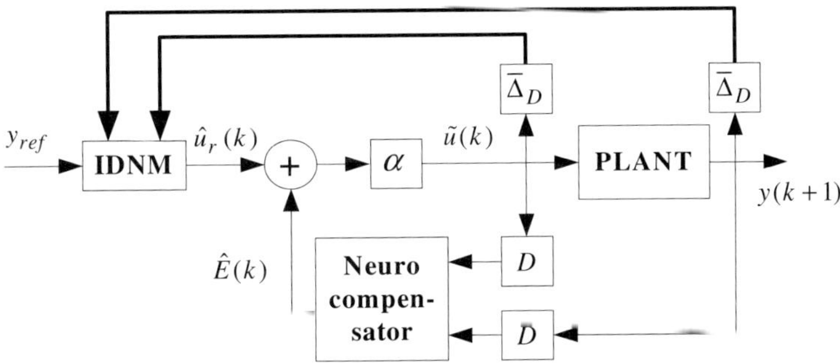

Fig. 4. The overall control scheme of the GMV-IDNC

3 Simulation

The proposed control scheme is tested in a three-machine nine-bus power system shown in Fig. 5. The mathematical models and parameters are given in [14]. A conventional PSS, tuned by a genetic algorithm [15], is used for comparison with the

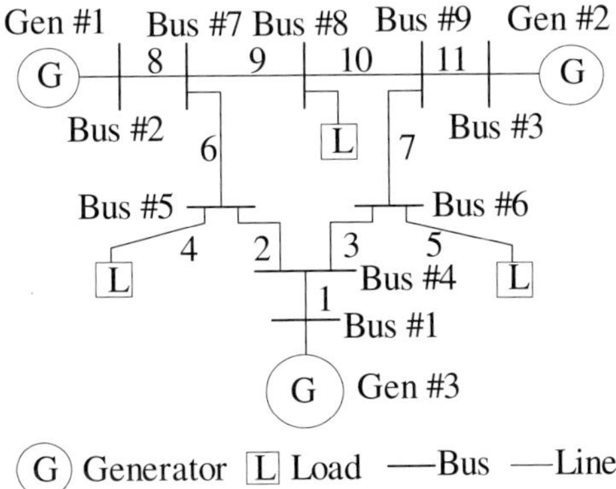

Fig. 5. Three-machine nine-bus power system

proposed control scheme in this paper. The proposed controller is designed for achnes 2 and 3, which were the optimal places determined for installing the power system stabilizer in [15].

The procedure is as follows: first, each machine is disturbed by adding various random step signals as control input. Then, the input-output data set is collected and chosen for the inputs to the IDNM based on the free-model concept. Second, the GMV controller gain is tuned relative to the weight on the control effort. Finally, the neuro compensator is implemented to complete the proposed control scheme.

The trained IDNM provides too active control input when it is used as a controller. Thus, the output of the IDNM is adjusted by the GMV gain α. In the power system under study, if α is set to 1, which is the MV control, when the active control input is applied, the system becomes unstable.

3.1 Identification of the IDNM

The proposed controller is designed for machines 2 and 3. The orders of input and output data in the inverse dynamics (4) are chosen as 3 since the higher order values beyond $\Delta^3 y(k)$, $\Delta^3 u(k-1)$ become very small in both machines. The number of neurons is selected as 14 in the hidden layer and the output layer has a single neuron. Also, in (11), the interpolation order l is set to 3 based on the above choice of the order of the input/output data set. Various random step signals are applied to disturb the system. To see the accuracy of the model of the IDNM, the following percentage of the output variance is used:

$$\varepsilon = \left(1 - \frac{\left\| u^* - u \right\|_2}{\left\| u - mean(u) \right\|_2}\right) \times 100 \quad \% \tag{13}$$

where u and u^* are the system input and the IDNM output, respectively, *mean* is a median value of the input vector and $\|\cdot\|_2$ is the 2-norm. The maximum iteration for the training of the IDNM is set as 1000 in this paper. The accuracy of the IDNM is 81.8392% and 87.0322% for machine 2 and machine 3, respectively. The gain of the GMV controller, α, is set as 0.2 and 0.5, for machine 2 and machine 3, respectively. In the following, a three-phase fault is introduced to test the proposed GMV-IDNC control scheme since it is the most severe disturbance in the power system.

3.2 Three-Phase Fault

Fault occurs at time 0 and lasts 0.09 seconds at near bus 7 on the line between buses 5 and 7 in Fig. 5. The faulted line is removed until 1 second and then reclosed.

For this network disturbance, the GMV-IDNC was tested and compared with a CPSS, and the GMV-IDNM. Fig. 6 shows the comparison of the three controllers in the system output (angular speed deviation) in machines 1, 2 and 3. The GMV-IDNC performs much better than the CPSS and GMV-IDNM, even if the IDNM was not

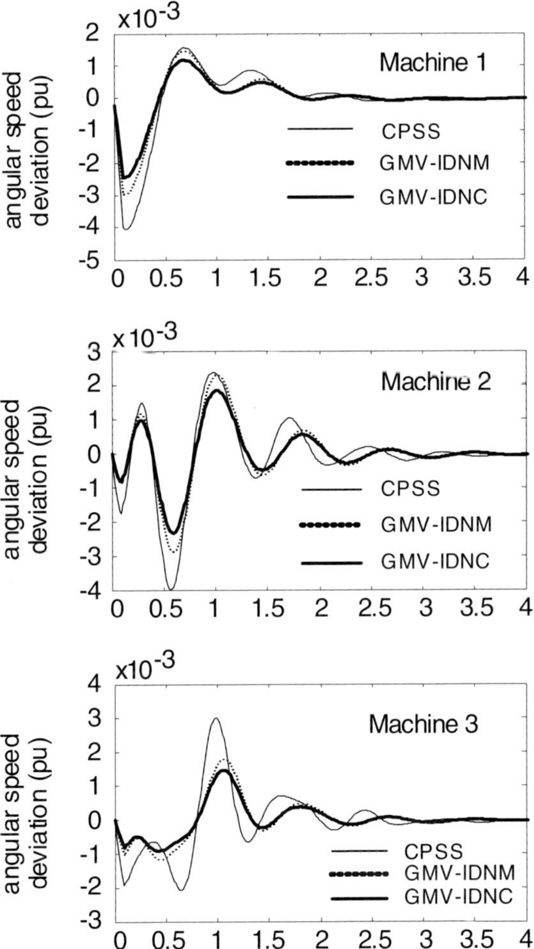

Fig. 6. The comparison of system outputs among the CPSS, GMV-IDNM, and GMV-IDNC in machines 1, 2, and 3

Table 1. RMS error of angular speed deviation

RMS error ($\times 10^{-3}$)			
Type	Machine		
	1	2	3
CPSS	0.5831	0.6230	0.5749
GMV-IDNM	0.5011	0.5805	0.3704
GMV-IDNC	0.4057	0.4720	0.2606

trained for this type of disturbance. Both GMV-IDNC and GMV-IDNM performs better than the CPSS. However, because of the compensation by the neuro compensator the GMV-IDNC performs better than the GMV-IDNM.

Table 1 shows the comparison of the three controllers measured in RMS values of the angular speed deviations. It is shown that the GMV improves the control performance and the neuro compensator further improves the performance by providing compensation control when a modeling error occurs due to the unlearned disturbances of the IDNM.

4 Conclusions

In this paper, a generalized minimum variance inverse dynamic neuro controller (GMV-IDNC) was presented and applied for the damping of low frequency oscillations in a multimachine power system. There are four main advantages of the proposed GMV-IDNC. First, the necessary input-output data sets for training and controlling the system are from the previously observed data, which has incremental forms using a backward difference operator. Second, the IDNM is trained once during a nomal operation and is used as an inverse controller; it is not required to retrain the IDNM for other disturbances. Third, the phenomenon of the excessive or too active control input, caused by a non-minimum phase characteristics, can be prevented by the GMV. The gain of the GMV, α, enhances the stability of the closed-loop system. Fourth, the proposed scheme has the neuro compensator to reduce the modeling error for various operating conditions because of using the same pre-trained IDNM. The GMV-IDNC was tested in a three-machine nine-bus power system under system fault conditions and compared with the conventional PSS and the GMV-IDNM without the neuro compensator. The GMV-IDNC out-performed the CPSS and the GMV-IDNM.

References

1. deMello, F.P., Concordia, C.A.: Concept of Synchronous Machine Stability as Affected by Excitation Control. IEEE Trans. on Power Apparatus and Systems 88(4) (1969) 316-329
2. deMello, F.P., Nolan, P.J., Laskowski, T.F., Undrill, J.M.: Coordinate Application of Stabilizers in Multimachine Power System. IEEE Trans. on Power Apparatus and Systems 99 (1980) 892-901
3. Doi, S.A.: Coordinated Synthesis of Power System Stabilizers in Multimachine Power Systems. IEEE Trans. on Power Apparatus and Systems 103 (1984) 1473-1479
4. Hwang, T.L., Hwang, T.Y., Yang, W.T.: Two-level Optimal Output Feedback Stabilizer Design. IEEE Trans. on Power Systems 6(3) (1991) 1042-1048
5. Ghosh, A., Ledwich, G., Malik, O.P., Hoper, G.S.: Power System Stabilizer Based on Adaptive Control Techniques. IEEE Trans. on Power Apparatus and Systems 103 (1984) 1983-1989
6. Pierre, D.A.: A Perspective on Adaptive Control of Power Systems. IEEE Trans. on Power Systems 2 (1987) 387-396
7. Zhang, Y., Malik, O.P., Hope, G.S., Chen, G.P.: Application of an Inverse Input/Output Mapped ANN as a Power System Stabilizer. IEEE Trans. on Energy Conversion 9(3) (1994) 433-441
8. Hsu, Y.Y., Chen, C.R.: Tuning of Power System Stabilizers Using an ANN. IEEE Trans. on Energy Conversion 6(4) (1991) 612-619

9. Wu, Q.H., Hogg, B.W., Irwin, G.W.: A Neural Network Regulator for Turbogenerators. IEEE Trans. on Neural Networks 3(1) (1992) 95-100
10. Ko, H.-S., Lee, K.Y., Kim, H.-C.: An Intelligent Based LQR Controller Design to Power System Stabilization. Electrical Power System Research 71(1) (2004) 1-9
11. Astrom, K., Wittenmark, B.: Adaptive Control, Addison Wesley (1995)
12. Haykin, S.: Neural Networks: A Comprehensive Foundation, Prentice Hall, New Jersey (1999)
13. Burden, R.L., Faires, J.D.: Numerical Analysis, PWS-KENT (1989)
14. Sauer, P.W., Pai, M. A.: Power System Dynamics and Stability. Prentice Hall, New Jersey (1998)
15. Abido, M. A.: Particle Swarm Optimization for Multimachine Power System Stabilizer Design. IEEE Power Engineering Society Summer Meeting 3 (2001) 1346-1351

Adaptive Control for Synchronous Generator Based on Pseudolinear Neural Networks

Youping Fan[1], Yunping Chen[1], Shangsheng Li[1], Dong Liu[1], and Yi Chai[2]

[1] Faculty of Electrical Engineering, Wuhan University, Wuhan 430072, Hubei, China
Fyoupingnxinrong@yahoo.com.cn
[2] Automation College, Chongqing University, Chongqing 400044, China
chaiyi@cqu.edu.cn

Abstract. Artificial neural networks can be used as intelligent controllers to control non-linear, dynamic systems through learning, which can easily accommodate the non-linearities and time dependencies. However, they require large training time and large number of neurons to deal with complex problems. Taking benefit of the characteristics of a Generalized Neuron that requires much smaller training data and shorter training time, the pseudo-linear neural network (PNN) based model predictive approach used in the single and multi-machine power system studies is proposed in this paper. A simulation is carried out. It is demonstrated that the proposed control strategy is applicable to some of nonlinear systems.

1 Introduction

Synchronous Generator in a power system is a non-linear, fast acting, multi-input multi-output (MIMO) device [1]. Conventional linear controllers (CONVC) for the synchronous generator consist of an automatic voltage regulator (AVR) to maintain constant terminal voltage, and a speed governor to maintain constant power and constant speed at some set point. They are designed to control, in some optimal fashion, the generator around one particular operating point; and because of nonlinearities, at any other point the generator's damping performance is degraded. More recently, artificial neural networks (ANNs) offer an alternative as intelligent nonlinear adaptive controllers, called neuro-controllers. i.e., a multilayer perceptron neuro-controllers (MLPNC) [2]. Because of the universal approximation ability of neural networks, the problems of neural network based model predictive control (NNMPC) can be solved by either direct optimization to train an NNMPC controller with optimized data, or solving a universal dynamic matrix control problem [3, 4].

The pseudo-linear neural network based model predictive approach widely used in the single and multi-machine power system studies is proposed in this paper. The organization is as follows. In Section 2, Pseudo-linear neural networks and training is described; PNN based model predictive control, the asymptotical stability of PNN based MPC and its algorithm is given. In Section 3, a simulation test result is given. Finally conclusions are drawn in Section 4.

J. Wang et al. (Eds.): ISNN 2006, LNCS 3972, pp. 1348 – 1353, 2006.

2 Pseudo-Linear Neural Networks and Training

A. ***Pseudo-linear Neural Networks and Training***

The pseudo-linear neural network (PNN), proposed by [5], [6], [7], and [8], is adopted for modeling the nonlinear process. Its architecture is shown in Fig.1.

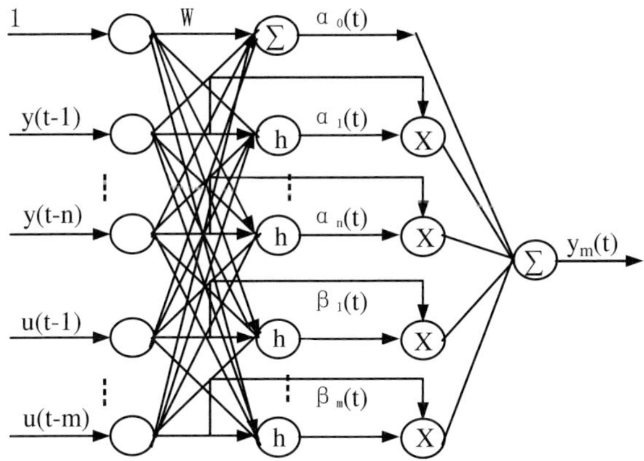

Fig. 1. The architecture of pseudo-linear neural networks

In Fig.1, the X circles symbolize multiplication neurons; the $\sum$ ones addition, and the ***h*** ones nonlinear neurons. At any discrete time ***t***, the inputs of PNN are $I(t)=[y(t-1),\ldots,y(t-n),u(t-1),\ldots,u(t-m)^T]$; $\alpha_j(t)$ and $\beta_j(t)$ are the output of the ***j***-th of the first hidden neuron, and $y_m(t)$ is the output of the network, ***W*** represents the weightings from the input layer to the first hidden layer, θ_i is the threshold. The input-output relationships of the PNN are

$$y(t)=\alpha_1(t)y(t-1)+\ldots+\alpha_n(t)y(t-n)+\ldots+\beta_1(t)u(t-1)+\ldots+\beta_m(t)u(t-m) \tag{1}$$

It is obviously from (1) that PNN model is a quasi-linear one as to the input and output of system, on which the controller design and system analysis will be convenient under some conditions.

B. ***Contingency Feature Space Decreasing Dimensions***

The MPC problem is formulated as follows, which in every sample period, should be solved.

$$\min J = \tfrac{1}{2} E_{t,P}^T \dot{E}_{t,P} + \tfrac{1}{2}\lambda \Delta U_{t,M}^T \Delta U_{t,M} \tag{2a}$$

S. t.

$$U_{\min} \le U(t) \le U_{\max} \tag{2b}$$

where $E_{t,p}$, is the error vector, $\Delta U_{t,M}$ is the increment of control, $\Delta U_{t,M} = [\Delta u(t), \Delta u(t+1), \ldots, \Delta u(t+M-1)]^T$, **P** is the length of prediction, **M** is the length of the control.

$$E_{t,P} = R_{t,P} - \widehat{Y}_{t,P} \tag{3}$$

where $R_{t,P}$ and $\widehat{Y}_{t,P}$, are set-point vector and predictive vector. And the control vector is

$$U_{t,M} = [u(t+1), u(t+2), \ldots, u(t+M)]^T \tag{4}$$

If the constraints are violated, then the $\boldsymbol{U(t)}$ takes the bounded $\boldsymbol{U_{\min}}$ or $\boldsymbol{U_{\max}}$. In the case that control $\boldsymbol{U}$ is within the constraints, an iteration of calculating $\boldsymbol{U_{t,M}}$ is necessary.

$$U_{t,M}^T(k+1) = U_{t,M}^T(k) + \frac{\eta}{1+\lambda\eta}\left(\frac{\partial \widehat{Y}_{t,P}}{\partial U_{t,M}(k)}\right)^T E_{t,P} \tag{5}$$

where the initial $U_{t,M}^T(0) = U_{t-1,M}^T$; and $\dfrac{\partial \widehat{Y}_{t,P}}{\partial U_{t,M}(k)}$ is the Jacobian matrix .

In order to simplify the problem, Consider only the unconstrained MPC Problem(2a). When the second order based optimization is applied, it can be obtained that

$$\Delta U_{t,M}^T(k+1) = -\eta(J^T J + \beta I_M)^{-1} J^T E_{t,P} \tag{6}$$

where $\beta = \lambda(1+\eta)$. When control length $\boldsymbol{M}$ is set to $\boldsymbol{1}$, the predictive control is of a simple form as

$$\Delta U(t) = \frac{\eta J^T E_{t,P}}{\beta + J^T J} \tag{7}$$

To calculate Jacobian matrix, let the neural network be

$$\widehat{y}(t+1) = f[y(t), \ldots, y(t-n+1), u(t), \ldots, u(t-m+1)] \tag{8}$$

$$I(t) = [y(t), \ldots, y(t-n+1), u(t), \ldots, u(t-m+1)]^T \tag{9}$$

Applying the chain rule to Eq. (8), the element of Jacobian matrix $\dfrac{\partial \widehat{Y}_{t,P}}{\partial U_{t,M}(k)}$ can be calculated recursively,

$$\frac{\partial \hat{y}(t+p)}{\partial u(t+l)} = \frac{\partial f(I(t+p-1))}{\partial u(t+l)} + \sum_{i=1}^{p-1} \frac{\partial \hat{y}(t+p)}{\partial \hat{y}(t+l)} \frac{\partial \hat{y}(t+l)}{\partial u(t+l)} \tag{10}$$

Denote the non-zero element of (10) be

$$a_l(I_j) = \frac{\partial \hat{y}(t+p)}{\partial \hat{y}(t+l)}, 1 \le l \le n \tag{11}$$

$$b_l(I_j) = \frac{\partial \hat{y}(t+l)}{\partial u(t+l)}, 1 \le l \le m \tag{12}$$

$$c_l(I_j) = \frac{\partial \hat{y}(t+p)}{\partial \hat{y}(t+l)} \tag{13}$$

Eq.(13) can be rewritten as

$$c_{p,l} = b_{p-l+1}(I_p) + \sum_{i=1}^{p-l} a_i(I_p) c_{p-i,l}, 1 \le p \le P, 0 \le l \le M-1 \tag{14}$$

When the neural network takes the PNN forms (1), there are

$$a_l(I_j) = \frac{\partial \hat{y}(t+p)}{\partial \hat{y}(t+l)} = a_l(I_j) + \sum_{i=1}^{n+m+1} h'(\bullet, I_j) w_{i,l} I_j, 1 \le l \le n \tag{15}$$

$$b_l(I_j) = \frac{\partial \hat{y}(t+l)}{\partial u(t+l)} = \beta_l(I_j) + \sum_{i=1}^{n+m+1} h'(\bullet, I_j) w_{i,l} I_j, 1 \le l \le m \tag{16}$$

Because the vector $\boldsymbol{I_j}$ may include the future control $\boldsymbol{u(t+l)}$, $\boldsymbol{0<l<M}$, the iteration values of Eq. (5) should be used.

3 Case Studies in Time-Domain Simulation

3.1 Plant Modeling[9]

In Fig. 2, the synchronous generator, turbine, exciter and transmission system connected to an infinite bus form the plant (dotted block in Fig. 2) that has to be controlled. Nonlinear equations are used to describe the dynamics of the plant in order to generate the data for the NN controllers and identifiers.

Where, $\boldsymbol{p_t}$ and $\boldsymbol{Q_t}$ are the real and reactive power at the generator terminals, respectively. $\boldsymbol{Z_e}$ is the transmission line impedance, $\boldsymbol{P_e}$ is the mechanical input power to the generator, $\boldsymbol{V_{fd}}$ is the generator field voltage, $\boldsymbol{V_b}$ is the infinite bus voltage, $\Delta\boldsymbol{\omega}$ is the generator speed deviation, is the terminal voltage, $\boldsymbol{V_{ref}}$ is the reference exciter voltage, and $\boldsymbol{P_{in}}$ is the turbine input power. In Fig.2 , there are four types of controllers: namely the neuro-controller (NNMPC or MLPNC), or the CONVC consisting of governor and AVR.

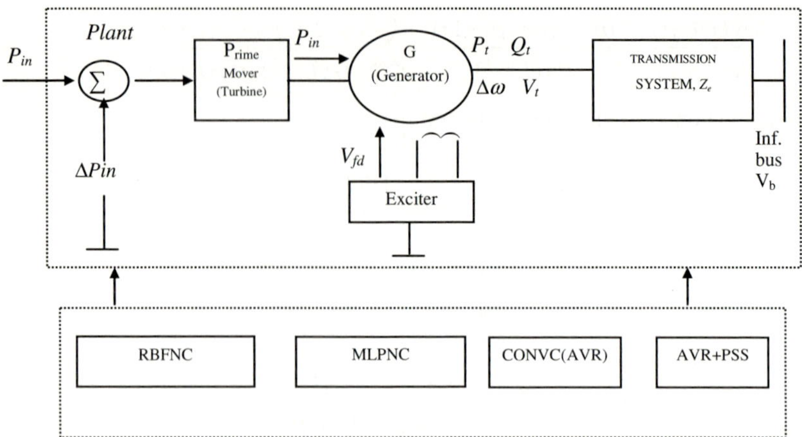

Fig. 2. Plant model used for control of a synchronous generator connected to an infinite bus

3.2 ±5%Step Changes in the Reference Voltage of the Exciter

First, the plant is operating in a steady-state condition ($\boldsymbol{P_t}$ **=1** pu, $\boldsymbol{Q_t}$ = 0.234 pu). At $\boldsymbol{t}$ =1s, a step increase in the reference voltage of the exciter $\boldsymbol{V_{ref}}$ is applied, resulting in a 5% step increase from the nominal value of the terminal voltage. At $\boldsymbol{t}$ =12s, the change in $\boldsymbol{V_{ref}}$ is removed, and the system returns to the initial steady-state condition. The rotor angle (δ) and terminal voltage (V_t) of the generator are indicators of how well the controllers are able to damp the system after this disturbance [10].

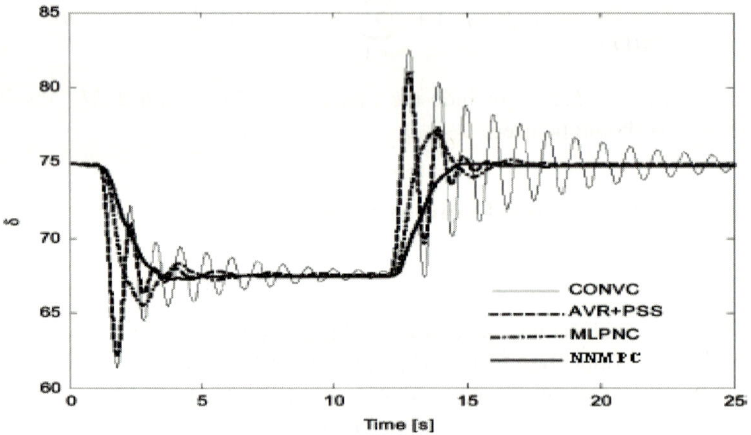

Fig. 3. Step changes in reference voltage of exciter ($\boldsymbol{P_t}$ **=1** pu, $\boldsymbol{Q_t}$ = 0.234 pu): Rotor angle (δ)

The results in Fig. 3 show that the neuro-controllers improve the transient system damping compared to CONVC and AVR PSS, and also that the NNMPC outperforms the MLPNC, i.e., the desired point is reached faster.

4 Conclusion

The proposed pseudo-linear neural network based model predictive approach employing the second order on-line optimization is demonstrated applicable to a synchronous generator connected to a power system. The results show that the MLPNC and NNMPC provide more damping than the CONVC and AVR + PSS. The NNMPC provides a better damping performance than the MLPNC. Moreover, the synchronous generator controlled by the CONVC goes unstable after the disturbance. In contrast though, the AVR PSS, MLPNC and NNMPC still control the generator effectively in a stable mode. Moreover, the RBFNC once again provides a better damping performance than the MLPNC and AVR PSS for this test.

Acknowledgements

This work is supported by the National Natural Science Foundation of China (50477018) and the Post-doctoral Science Foundation of China (2004036135).

References

1. 1.Anderson, P. M., Fouad, A. A.: Power System Control and Stability. IEEE Press, New York (1994)
2. Park, J.-W., Harley, R. G., Venayagamoorthy, G. K.: Comparison of MLP and RBF Neural Networks Using Deviation Signals for Indirect Adaptive Control of a Synchronous Generator. In Proc. Int. Joint Conf. Neural Networks, IJCNN02, Hawaii (2002) 919-924
3. Draeger, A., Engell, S., Ranke, H.: Model Predictive Control Using Neural Networks. IEEE Control System 15(1) (1995) 61-66
4. Wang, Y.J., Wang, H.: A Nonlinear Model Predictive Control Based on Pseudo-Linear Neural Networks. Proc. of 1999 European Control Conference, Germany (1999) 1057-1062
5. Wang, Y.J., Wang, H.: Nonlinear Model Predictive Control with Guaranteed Stability Based on Pseudo-Linear Neural Networks. J. Chongqing Univ.-Eng. Ed. 3(1) (1999) 26-29
6. Fan, Y.P., Huang, X.Y.: Adjusting-Controlling on Gassing and Energy for High Power Excimer Laser Based Intelligence. Control Theory and Applications 19(4) (2002) 561-566
7. Deif, A. S.: Advanced Matrix Theory for Scientists and Engineers. Abacus Press, London (1982)
8. Venayagamoorthy, G. K., Harley, R. G.: A Continually Online Trained Neuro-Controller for Excitation and Turbine Control of a Turbo-Generator. IEEE Trans. Energy Conv. 16(9) (2001) 261-269
9. Park, J.-W., Harley, R.G., Venayagamoorthy, G.K.: Indirect Adaptive Control for Synchronous Generator: Comparison of MLP/RBF Neural Networks Approach with Lyapunov Stability Analysis. IEEE Transactions on Neural Networks 15(2) (2004) 460-464
10. Kundur, P., Klein, M., Rogers, G.J., Zywno, M.S.: Application of Power System Stabilizers for Enhancement of Overall System Stability. IEEE Trans. Power Syst. 4(3) (1989) 614-626

A Research and Application of Chaotic Neural Network for Marine Generator Modeling

Wei-Feng Shi

Department of Electrical Automation,
Shanghai Maritime University, Shanghai 200135, China
wfshi@shmtu.edu.cn

Abstract. For enhancing approximation ability of chaotic neural network to nonlinear system, some characteristics are researched about neuron algorithm, architecture of network and learning rule of neural network. A local recurrent chaotic neural network is constructed based on Aihara chaotic neuron. A heuristic modified improved BP algorithm is applied in the chaotic neural network training with well ability of convergence and stability. The chaotic neural network is applied in marine generator modeling for a real time simulator. The application indicates that the chaotic neural network can be applied to build marine generator with ideal ergodicity and few number of neuron. There are relationships between value of mean square error and chaotic characteristic of neuron in marine generator modeling. When the neuron is in chaotic state, the minimum value of mean square error will be acquired.

1 Introduction

Artificial neural network (ANN) is provided with approximation ability to nonlinear system [1], but this ability is affected by factors of neuron algorithm, architecture of network and learning rule. There are some influences between ANN ergodicity and neuron algorithm. We think about chaotic characteristic. There are characteristics of quite centralization, dispersion with chaotic system and strong ergodicity with chaotic algorithm. Combining chaotic ergodicity with ANN to advance ergodicity for ANN. So exploring the ergodicity of chaotic neural network (CNN) is a worth problem. There are many questions need be solved with CNN, such as what kinds of chaotic algorithm can be used, what kinds of architecture of network will be constructed, what kinds of learning rule will be applied for network training.

2 Aihara Chaotic Neuron and Its Dynamic Characteristic

Aihara chaotic neuron comes from research of Nagumo-Sato chaotic neuron [2]. The Nagumo-Sato chaotic neuron can be described as equations (1).

$$\begin{cases} o(t+1) = H\left[A(t) - \alpha \sum_{r=0}^{t} k^{r} o(t-r) - \theta \right] \\ x(t+1) = A(t) - \alpha \sum_{r=0}^{t} k^{r} o(t-r) - \theta \end{cases} \tag{1}$$

J. Wang et al. (Eds.): ISNN 2006, LNCS 3972, pp. 1354 – 1362, 2006.

Here: $A(t)$ is input intensity on discrete time t. $w_{ij}^r = -\alpha k^r$. α is a positive parameter. k is rate of decay of non-response quality. $0 \leq k \leq 1$. r is time. $x(t+1)$ is state of neuron. $o(t+1)$ is output of neuron. H is hard limit transfer function.

When k is changing in definite range, the relationship of neuron between $x(t)$ and $a(t)$ is analyzed. When k is equal to 0.7, 0.85 or 0.95. The bifurcation characteristic of neuron is shown as Fig.1 (a), (b) or (c). It displays chaotic characteristic with the neural algorithm.

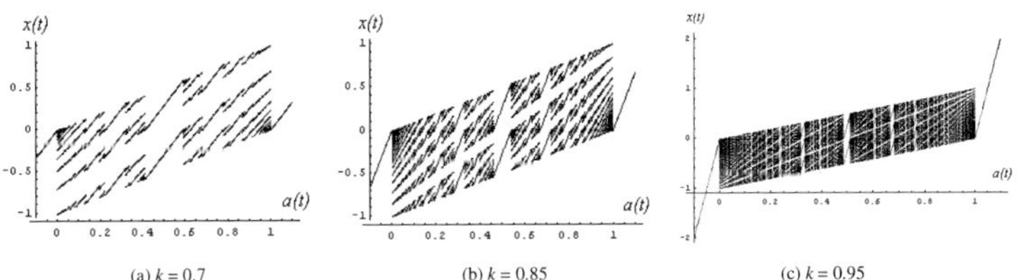

Fig. 1. Bifurcation characteristic of Nagumo-Sato neuron

Aihara brought forward using continuous function to replace hard limit transfer function of Nagurno-Sato neuron [3]. The neuron is shown in Fig.2. $f(\bullet)$ is a transfer function of neuron. It is a Logistic function with abrupt coefficient ε□0.04. The chaotic neuron is as hidden layer neuron. When it is affected by all neurons of input layer and hidden layer, the algorithm is as equation (2).

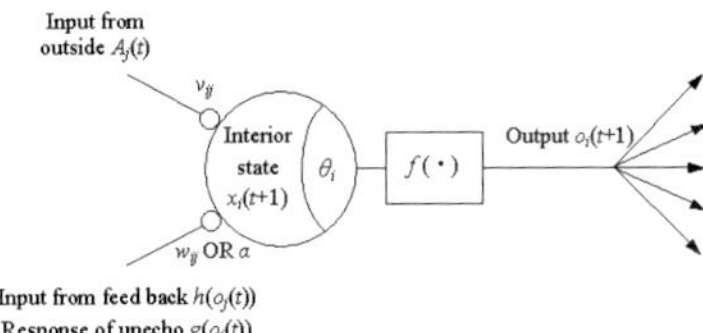

Fig. 2. The architecture of chaotic neuron

$$\begin{cases} o_i(t+1) = f\left\{\left[\sum_{j=1}^{M} w_{ij} \sum_{r=o}^{t} k^r h[o_j(t-r)]\right] + \sum_{j=1}^{N} v_{ij} \sum_{r=o}^{t} k^r A_j(t-r) - \alpha \sum_{r=o}^{t} k^r g[o_i(t-r)] - \theta_i\right\} \\ f[x_i(t)] = 1/\{1+\exp[-x_i(t)/\varepsilon]\} \end{cases} \quad (2)$$

Here: M is number of chaotic neuron. N is inputs number from outside. w_{ij} is weight between the jth neuron and the ith neuron. v_{ij} is weight of the jth input from outside to the ith neuron. $f(\bullet)$ is Logistic transfer function of neuron. h is interior feedback function of the jth neuron. $A_j(t-r)$ is the jth input intensity from outside in disperse time $(t-r)$, it is a known number $a(t)$ of input in certain time. $g(\bullet)$ is non-response function of neuron. θ_i is input bias of the ith neuron. i=1, 2, ···, N. The chaotic characteristic is produced by the neuron [4]. When parameter $a(t)$ changes, analyzing the time sequence, cobweb, bifurcation and Lyapunov exponent from neuron equation (2) (abrupt coefficient ε= 0.04). The characteristic of chaotic neuron is analyzed. The time sequence of chaotic neuron output is shown as Fig.3 when $a(t)$ equal to 0.04, 0.18 and 0.26. It is make out that CNN neuron loses the rule of period movement and enter a chaotic movement when $a(t)$ equal 0.26. Analyzing the cobweb figures, the figures of CNN neuron output are shown as Fig.4 when $a(t)$ equal to 1, 0.04, 0.18 and 0.28. The system enters chaos when $a(t)$ equal to 0.28. Fig.5 (a) to (c) are bifurcation figures of chaotic neuron with abrupt coefficient ε equal to 0.01, 0.04 and 0.06. The process of double period bifurcation can be seen clearly. Fig.5 (d) is graph of transfer function of neuron (ε = 0.04). Fig.5 (e) is graph of Lyapunov exponent of the chaotic neuron. The neuron produces chaos when the Lyapunov exponent is large than zero.

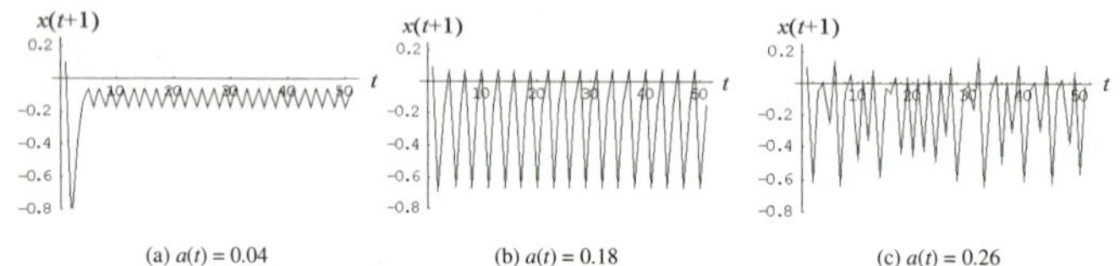

(a) $a(t)$ = 0.04 (b) $a(t)$ = 0.18 (c) $a(t)$ = 0.26

Fig. 3. The time sequence of CNN neuron output with ε = 0.04

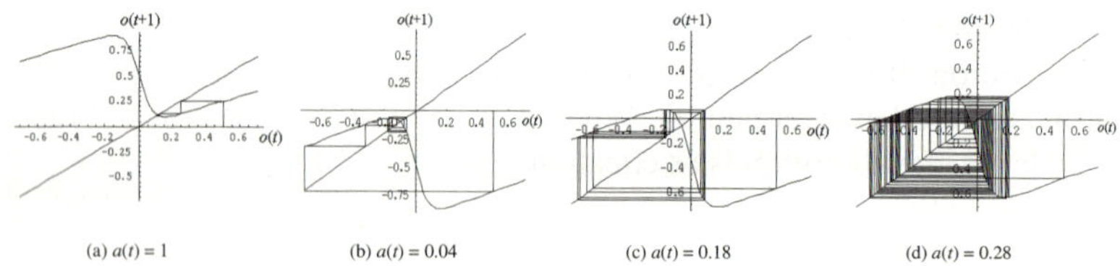

(a) $a(t)$ = 1 (b) $a(t)$ = 0.04 (c) $a(t)$ = 0.18 (d) $a(t)$ = 0.28

Fig. 4. The cobweb figures of CNN neuron with ε = 0.04

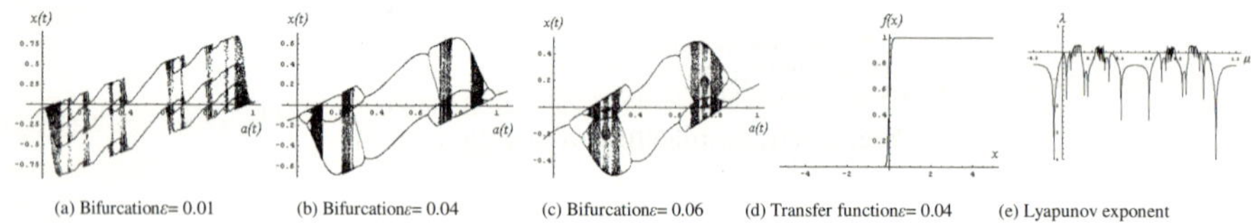

(a) Bifurcation ε= 0.01 (b) Bifurcation ε= 0.04 (c) Bifurcation ε= 0.06 (d) Transfer function ε= 0.04 (e) Lyapunov exponent

Fig. 5. Chaotic characteristic of CNN neuron

3 Construct an Aihara Local Recurrent Chaotic Neural Network

Recurrent network is provided with ability of powerful information process [5]. So, it can be used for information process with high denseness in marine generator modeling. Considering the generator ANN model should be fit for digital signal process chips in real time calculation, the architecture of network could not be too complex. So a local recurrent network form is applied with inner feedback delay. An Aihara local recurrent chaotic neural network (ARCNN) is constructed. The chaotic neuron and feedback network are combined for well global approach to nonlinear system.

The local recurrent network is inner delay feedback network with three layers. It is including one input layer, one hidden layer and one output layer. The architecture of network is shown as Fig.6. If input time sequence from outside is $\mathbf{u}(t)$, output of hidden layer neuron is $\mathbf{o}(t)$, output of network is $\mathbf{y}(t)$, then the network algorithm can be described as equations (3).

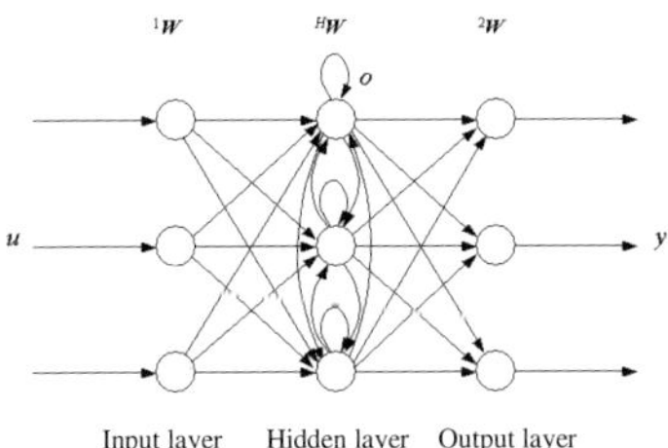

Fig. 6. The architecture of local recurrent network

$$\begin{cases} \mathbf{x}_o(t+1) = {}^{\mathrm{H}}\mathbf{W}\mathbf{o}(t) + {}^{1}\mathbf{W}\mathbf{u}(t) + {}^{1}\boldsymbol{\theta} \\ \mathbf{o}(t) = f_1(\mathbf{x}_o(t)) \\ \mathbf{y}(t) = f_2(\mathbf{x}(t)) = f_2({}^{2}\mathbf{W}\mathbf{o}(t) + {}^{2}\boldsymbol{\theta}) \end{cases} \tag{3}$$

Here, $f_1(\bullet)$ is Logistic function as equations (2), $f_2(\bullet)$ is linearity function. ${}^{1}\mathbf{W}$, ${}^{\mathrm{H}}\mathbf{W}$ and ${}^{2}\mathbf{W}$ are weight matrix of input layer to hidden layer, hidden layer to hidden layer and hidden layer to output layer respectively.

4 Applying BP Algorithm for ARCNN Training

BP gradient descent algorithm with momentum and adaptive learning method is adopted for ARCNN training. The algorithm process is as follow.

(1) Set up cost function for network is as equations (4).

$$E = \frac{1}{2}\sum_{j=t_0+1}^{t}(d_j - y)^2 = \frac{1}{2}\sum_{j=t_0+1}^{t} e_j^2 \tag{4}$$

Here, d_j and y are generator output sampling parameter and ARCNN output parameter.

(2) Iteratively adjusted algorithm of weights.

According to equations (2) and (3), an iteratively adjusted algorithm of weights can be determined. The weight of hidden layer to output layer ${}^{2}\mathbf{W}$ is as equations (5).

$$\frac{\partial E(n)}{\partial^2 w_i} = \frac{\partial E(n)}{\partial y(n)} \cdot \frac{\partial y(n)}{\partial x(n)} \cdot \frac{\partial x(n)}{\partial^2 w_i} = -e(n) f_2'(x(n)) o_i(n) = \delta(n) o_i(n) \tag{5}$$

Here, $\delta(n) = \dfrac{\partial E(n)}{\partial x(n)} = -e(n) f_2'(x(n))$

The weight of hidden layer ${}^{\mathrm{H}}\mathbf{W}$ is as equations (6).

$$\frac{\partial E(n)}{\partial^{\mathrm{H}} w_{ij}} = \frac{\partial E(n)}{\partial o_i(n)} \cdot \frac{\partial o_i(n)}{\partial x_{o,i}(n)} \cdot \frac{\partial x_{o,i}(n)}{\partial^{\mathrm{H}} w_{ij}} \tag{6}$$

Here, $\frac{\partial E(n)}{\partial o_i(n)} = \frac{\partial E(n)}{\partial x(n)} \cdot \frac{\partial x(n)}{\partial o_i(n)} = \delta(n) \cdot \frac{\partial \left[\sum_i {}^2 w_i o_i(n) \right]}{\partial o_i(n)} = \delta(n)^2 w_i(n)$

$$\frac{\partial o_i(n)}{\partial x_{o,i}(n)} = f_1'(x_{o,i}(n))$$

$$\frac{\partial o_i(n)}{\partial^{\mathrm{H}} w_{ij}} = \frac{\partial o_i(n)}{\partial x_{o,i}(n)} \cdot \frac{\partial x_{o,i}(n)}{\partial^{\mathrm{H}} w_{ij}} = f_1'(x_{o,i}(n)) \frac{\partial \left[\sum_j {}^{\mathrm{H}} w_{ij} o_j(n-1) + \sum_q {}^{\mathrm{I}} w_{iq} u_q(n-1) \right]}{\partial^{\mathrm{H}} w_{ij}}$$

$$= f_1'(x_{o,i}(n)) \left[o_j(n-1) + \sum_j {}^{\mathrm{H}} w_{ij} \frac{\partial o_j(n-1)}{\partial^{\mathrm{H}} w_{ij}} \right]$$

Order: $\delta_{o,i}(n) = \frac{\partial E(n)}{\partial x_{o,i}(n)} = \delta(n)^2 w_i(n) f_1'(x_{o,i}(n))$

For the qth input, the weight of ${}^{\mathrm{I}}\mathbf{W}$ is as equations (7).

$$\frac{\partial E(n)}{\partial^{\mathrm{I}} w_{iq}} = \frac{\partial E(n)}{\partial x_{o,i}(n)} \cdot \frac{\partial x_{o,i}(n)}{\partial^{\mathrm{I}} w_{ij}} = \delta_{o,i}(n) u_q(n-1) \tag{7}$$

The equation (7) indicates a process of dynamic recurrence equation. According to above equations, all the weights online iteratively from output to input adjusted equations of dynamic BP learning algorithm are as follow equations (8).

$$\begin{cases} \Delta^2 w_i = -\eta e(n) f_2'(x(n)) o_i(n) = \eta \delta(n) o_i(n) \\ \Delta^{\mathrm{H}} w_i = -\eta \delta(n)^2 w_i(n) f_1'(x_{o,i}(n)) \dfrac{\partial x_{o,i}(n)}{\partial^{\mathrm{H}} w_{ij}} = -\eta \delta_{o,i}(n) \dfrac{\partial x_{o,i}(n)}{\partial^{\mathrm{H}} w_{ij}} \\ \Delta^{\mathrm{I}} w_{iq} = -\eta \delta(n)^2 w_i(n) f_1'(x_{o,i}(n)) u_q(n-1) = -\eta \delta_{o,i}(n) u_q(n-1) \end{cases} \tag{8}$$

The feedback can enhance ability of information process in ARCNN if it is applied well. Otherwise feedback will produce instability with network [5]. For enhancing stability and ability of convergence, heuristic modification is used. We are considering use of momentum and variable learning rate. In ARCNN marine generator modeling, the learning algorithm is dynamic BP gradient descent algorithm with momentum and adaptive learning. The momentum method corresponds to add momentum filter in weights and biases changes. Thereby, the ARCNN algorithm accelerates convergence characteristic under the stable condition. The momentum modifies to weights and biases of back propagation algorithm equations are as follow equations (9).

$$\begin{cases} \Delta^m \mathbf{W}(n) = \gamma \Delta^m \mathbf{W}(n-1) - (1-\gamma) \alpha^m \mathbf{S} \; [{}^{(m-1)} \mathbf{o}(n)]^{\mathrm{T}} \\ \Delta^m \boldsymbol{\theta}(n) = \gamma \Delta^m \boldsymbol{\theta}(n-1) - (1-\gamma) \alpha^m \mathbf{S} \end{cases} \tag{9}$$

Here, $\Delta^m \mathbf{W}(n)$ and $\Delta^m \boldsymbol{\theta}(n)$ are the nth iteration momentum increment of weight and bias of the mth layer respective. γ is momentum coefficient. α is learning rate. ${}^m\mathbf{S}$ is vector of sensitivity. It is sensitivity between mean square error (MSE) and changes of the mth layer input element. ${}^{(m-1)}\mathbf{o}(n)$ is output of neuron.

The performance of the algorithm is sensitive to the proper setting of the learning rate. If the learning rate is set too large, the algorithm may oscillate and become unstable. If the learning rate is too small, the algorithm will take too long time to converge [1]. The principle of variable learning rate of ARCNN is: If the MSE increases by more than some set percentage after a weight update, then the weight update is discarded, the learning rate is multiplied by a factor less than one, and the momentum coefficient is set to zero. If the MSE increases by less than this percentage, then the weight update is accepted but the learning rate and the momentum coefficient are unchanged. If the MSE decreases after a weight update, then the weight update is accepted and learning rate is multiplied by some factor large than one. If momentum coefficient has been previously set to zero, it is reset to its original value.

5 Marine Generator Modeling Based on ARCNN

Considering in aspect of real time calculation of simulator system, the architecture and algorithm of built synchronous generator model must in a simply form under certain precision. ARCNN is constructed as three layers chaotic neural network for marine synchronization generator modeling. When neural network is training, we set momentum coefficient is 0.9, learning rate is 0.01, learning increasing rate is 1.05, learning decreasing rate is 0.7. The number of hidden layer choose a minimum value (such as 16) while the learning periodicity is large than 500 or MSE is less than 0.5×10^{-3}. A serial and parallel connection form is used in generator model identification. The input sample data of generator modeling are mechanical torque and excitation current. The output sample data are rotate speed and voltage. A supervised training method is used for neural network off line training [6]. The ARCNN is combined with tapped delay line (TDL) to build marine generator dynamic model.

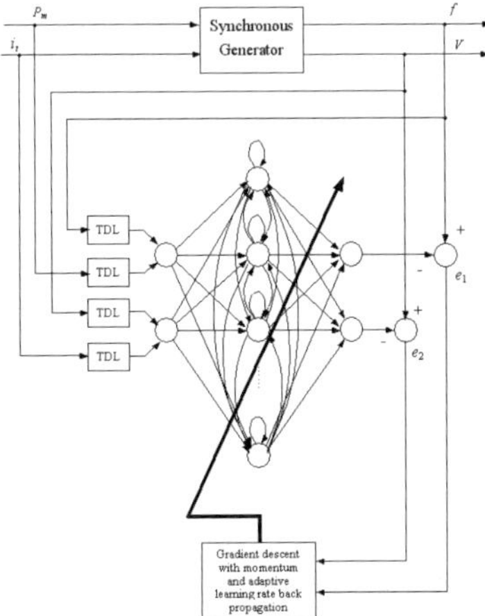

Fig. 7. The configuration of generator modeling based on ARCNN

The configuration of generator modeling based on ARCNN is shown in Fig.7. The response characteristic of generator is measured and shown in Fig.8 between input mechanical power and output frequency when a large load is starting. The response characteristic of generator is also measured between input excitation current and output voltage. The number of sampling data must cover with entire dynamic process of generator. There are four different generator running states sampling data sets for generator modeling. They are generator starting with zero load, large load and the largest load running on electric power network and three phases short circuit fault.

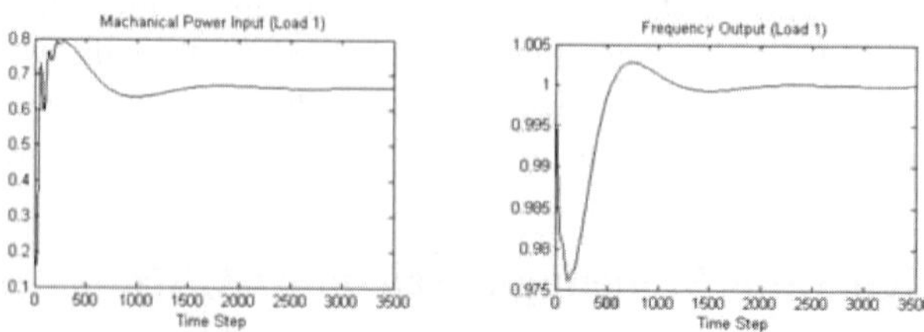

Fig. 8. Characteristic of mechanical power and frequency of marine synchronous generator

ARCNN is local feedback network. There are feedback values with network output. The output of ARCNN identification is as equations (10):

$$y^*(x_{k+1}) = N\phi[d_k,\ y^*(x_k),\ x_k;\ \mathbf{W}] \tag{10}$$

One axis configuration is used for generator modeling, two input/output of generator are selected with two networks respective. The outputs of ARCNN are as:

$$y^*[(x_{P_m})_{k+1}] = Ng\{(d_f)_k,\ y^*[(x_{P_m})_k],\ (x_{pm})_k;\ \mathbf{W}_1\}$$

$$y^*[(x_i)_{k+1}] = Nq\{(d_v)_k,\ y^*[(x_i)_{k+1}],\ (x_i)_k;\ \mathbf{W}_2\}$$

Here, $\mathbf{W}_1$ and $\mathbf{W}_2$ are ARCNN respective. The difference equations as:

$$y^*[(x_{P_m})_{k+1}] = Ng\{(d_f)_k,\ (d_f)_{k-1},\cdots,\ (d_f)_{k-n};\ y^*[(x_{P_m})_k],\ y^*[(x_{P_m})_{k-1}]\cdots,$$
$$y^*[(x_{P_m})_{k-n}];\ (x_{pm})_k,\ (x_{pm})_{k-1},\cdots,\ (x_{pm})_{k-m};\ \mathbf{W}_1\}$$

$$y^*[(x_i)_{k+1}] = Nq\{(d_v)_k,\ (d_v)_{k-1},\cdots,\ (d_v)_{k-n};\ y^*[(x_i)_k],\ y^*[(x_i)_{k-1}]\cdots,$$
$$y^*[(x_i)_{k-n}];\ (x_i)_{k-1},\ (x_i)_{k-2},\cdots,\ (x_i)_{k-m};\ \mathbf{W}_2\}$$

The y^* is feedback value of local recurrent network. The system built a one step forecast model based on current measurement data relative to system output. The network composes a dynamic system. The input of network is combined with input and preview output of process linearly according to autoregressive moving average model. The combined part of preview input is moving average part. The combined part of preview output is autoregressive part. The MSE between output of generator and ARCNN model are shown as Fig.9 (e) and (f). When the MSE reaches to 0.5×10^3,

ARCNN stops training. Then, the marine generator model is built by the ARCNN. The convergence of network is not ideal when a standard BP or other BP algorithm is used for network training. The MSE convergence characteristic is shown as Fig.9 (a) to (d). An improvement must be done as above for BP algorithm. All so it is discover that the MSE will be a minimum value when the abrupt coefficient of transfer function of ARCNN equal to 0.04. This means that there is affinity between ergodicity of network and chaotic characteristic of neuron.

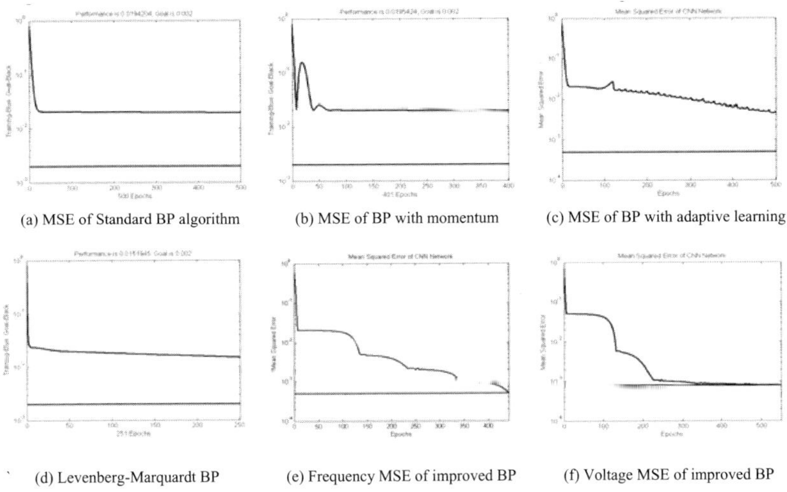

(a) MSE of Standard BP algorithm (b) MSE of BP with momentum (c) MSE of BP with adaptive learning

(d) Levenberg-Marquardt BP (e) Frequency MSE of improved BP (f) Voltage MSE of improved BP

Fig. 9. Some curves of MSE variation when ARCNN is training

6 The Results of Marine Generator Modeling Based on ARCNN

There are 16 chaotic neurons of hidden layer with each ARCNN for generator modeling. As one example, the compares of frequency between ARCNN model and sampling data of generator are shown in Fig.10. The left of figure shows dynamic process of ARCNN model and sampling data. Here, the real line is output value of ARCNN model. The broken line is output value of generator. The right figure shows error between two output values. The error is about 10^{-3} quantity class. The ARCNN is also provided with ability of generalization well in electric network short circuit fault.

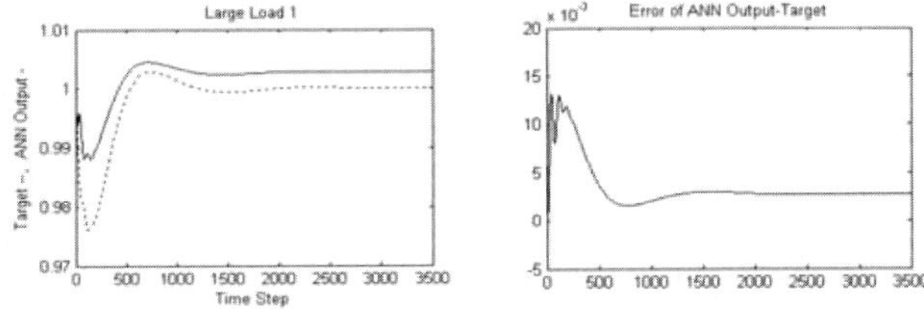

Fig. 10. The compares of rotation speed between ARCNN model and sampling data

7 Conclusion

In generator modeling, the generator system should be excited fully for obtaining sample data. The sample data should be chosen for ARCNN learning rightly. The selection fundamental is that the sample data should display the generator running in all states. The sample data should be converted in normalization form. If there are noises with the sample data collection, the sample data should be treated by a filter. The neural network astringency is well when system adopts BP gradient descent algorithm with momentum and adaptive learning for ARCNN training. If the system is changing with time, iterative algorithm should be used for ARCNN to adjust weights and biases. The precision of model is sensitive with abrupt coefficient (ε) of neuron transfer function. The MSE value will be a minimum value in modeling when the coefficient equal to 0.04. It is illuminated that the network ergodicity will be enhanced because of ANN neuron chaotic characteristic.

Acknowledgements

This project is supported by the fund of science and technology of Shanghai municipal education committee (05FZ09) and by key subject of SMU (XL0105).

References

1. Martin, T. H., Demuch, H.B., Beale, M.: Neural Network Design. PWS Publishing Company, Boston MA (1996)
2. Ruan, J., Gu, F., Chai, Z.: The Method and Application of Neural Dynamics Model. Science Publishing House, Beijing (2002)
3. Aihara, K., Takabe, T., Toyoda, M.: Chaotic NN. Physics Lett. A 144 (1990) 333–340
4. Adachi, M., Aihara, K.: Associative Dynamics in Chaotic Neural Network. Neural Networks,. 10 (1997) 83–98
5. Haykin, S.: Neural Networks: A Comprehensive Foundation. 2nd edn. Prentice-Hall, Inc., London (1999)
6. Fujita, H., Ohsawa,Y.: Modeling of Power System Dynamics Using Neural Network. Electrical Engineering in Japan 125 (1998) 10–18

Ship Synchronous Generator Modeling Based on RST and RBF Neural Networks

Xihuai Wang, Tengfei Zhang, and Jianmei Xiao

Department of Electrical and Automation, Shanghai Maritime University, Shanghai 200135, China
wxh@shmtu.edu.cn

Abstract. Ship synchronous generator modeling is the basis of control, analysis and design in the ship power systems. According to the strong non-linear relation characteristics of ship synchronous generator, a dynamic modeling method based on rough set theory (RST) and radial basis function (RBF) neural networks is presented in this paper. With the advantage of finding useful and minimal hidden patterns in data, RST is first applied to intelligent data analysis in this algorithm, including incompatible data elimination, important input nodes selection and radial basis function centers initialization, followed by a second stage adjusting the network parameters and training the weights of hidden nodes. The experimental results proved that this method could achieve greater accuracy and generalization ability.

1 Introduction

With the rapid development of modern ships, the demand for automation of ship power system to be much higher. Synchronous generators play a very important role in ship power systems. Analysis of the dynamic performance of ship synchronous generators has become very important; therefore, a valid dynamic model with high precision is a basic requirement. Up to date, many modeling methods have been proposed and implemented, which can be divided into two categories [1]. The first category assumes a known structure for the synchronous generator, using well-established theories like Park transformation. The second category deals with black-box modeling of synchronous generators using input-output data. In the black-box modeling, the structure of the model is not assumed to known. The only concern is to map the input data set to the output data set. Among these methods, the neural networks, which have the capabilities of self-learning and approaching any nonlinear functions, are paid great attention to the system dynamic modeling, and have obtained definite success [1-3]. However, due to the highly non-linear characteristics of synchronous generator, how to configure the initial network structure and to select the important input nodes is still a difficult problem before network training. Some input nodes, may not be significant and should be deleted because they mislead local learning and produce deterioration in the expected fit to future data. Therefore, if no preprocessing is carried out before constructing neural networks, the computation required may be too heavy with a poor generalization capability. Rough set theory (RST) is a relatively new mathematical and AI technique introduced in the early

J. Wang et al. (Eds.): ISNN 2006, LNCS 3972, pp. 1363 – 1369, 2006.

1980's by Z.Pawlak [4]. The technique is particularly suitable for dealing with imprecise, incomplete or inconsistent data, and discovering relationships in this data. It has been successfully applied to solve many real-life problems in medicine learning, data mining, etc [5-6]. In recent years, Radial basis function (RBF) neural networks have been successfully applied to many practical problems, especially for system dynamic modeling, with the good properties such as simple network structure, strong nonlinear approximation ability, rapid convergence speed and global convergence property, and so on. In this paper, we present a new dynamic modeling method based on RST and RBF neural networks

2 RST Data Analysis

Rough set theory has proved its usefulness as a new mathematical tool for data analysis and has been successfully applied in many real-life problems. The main advantage of rough set theory is that it does not need any preliminary or additional information about data.

2.1 To Establish the Information Table

Knowledge representation in rough set theory is done via information system. An information system can be characterized as a decision table, where the information is stored in a table. Each row in the table represents an individual record. Each column represents some an attribute of the records or a field.

Definition 1. An information system
$S = <U, A, V, f>$
U is a finite set of objects; R is a finite set of attributes that are further classified into disjoint sets of condition attributes C and decision attributes D ($A = C \cup D, C \cap D = \Phi$). V is a set of attributes values, $f : U \times A \to V$ is an information function, which appoint the value of each object x in U.

An equivalence relation is a reflexive, symmetric, and transitive binary relation on U. With each subset $R \subseteq A$, we associate an equivalence relation $IND(R)$ on U by setting $IND(R) = \{(x, y) \in U \times U : \forall a \in R, a(x) = a(y)\}$. The partition associated with $IND(R)$ is denoted as $U / IND(R)$.

2.2 To Be Eiscrete for Continuous Attributes

It is a pity that rough set theory can't deal with continuous attributes. When the value set of any attribute in a decision table is continuous valued or real numbers, then it is likely that there will be very few objects that will have the same value of the corresponding attribute. In such a situation the number of equivalence classes based on that attribute will be large and there will be very few elements in each of such equivalence class. This leads to the generation of a large numbers of antecedents in the classification rule, thereby making rough set theoretic classifiers inefficient. Therefore, to be discrete is necessary for continuous attributes. A rough set theory based software *RSES* [7], is a toolkit for analysis of information table data. With use of *Discretize/Generate cuts* from data table context menu we may generate decompositions of attribute value sets. With these descriptions, further discretization of continuous attributes or grouping (quantization) of nominal attributes referred to as cuts we may perform.

2.3 To Use RST to Get Suitable Input Nodes

It is well known that an information system or a decision system may usually have irrelevant or superfluous knowledge (attributes), which is inconvenient for us to get concise and meaningful decision. Since then, the reduction of attributes is demanded greatly. In rough set theory, the reduction is defined as a minimal set of attributes that enables the same classification of elements of the universe as the whole set of attributes.

Definition 2. Let P and Q be equivalence relations over U, then the set $POS_P(Q) = \bigcup_{X \subseteq U/Q} \underline{PX}$ is called the P-positive region of Q, which is the set of all objects of the universe U that can be properly classified to classes of U / Q employing knowledge U / P .

Definition 3. Let P and Q be equivalence relations over U, we say that r$\in$ P is Q-dispensable in P, if $POS_P(Q) = POS_{(P-\{r\})}(Q)$. Otherwise r is Q-indispensable in P If every r$\in$ P is Q-indispensable, P is Q-independent.

The attributes subset $R \subseteq P$ will be called a *Q*-reduction of *P*, if and only if *R* is the *Q*-independent subset of *P* and $POS_R(Q) = POS_P(Q)$.

Then, we can use the RST to get the key attributes. After this operation, we get *n* factors as the input of the networks.

2.4 To Eliminate Inconsistent Knowledge of the Training Data

Each row in decision table can regard as a decision rule. If the condition attributes of any two rules are equal whereas the decision attributes are not, we say that the two rules are inconsistent. When this happens both rules should be removed from the training data because they will mislead the local leaning.

If two or more rules represent the same class, all but one of the rules should be eliminated. This reduces computational time.

3 RBFNN Configuring and Training for Dynamic Modeling

RBF neural network is a three-architecture with no feedback, which has only one hidden layer, can approach any nonlinear functions precisely. The input layer units and the output units are determined by the practical problems and only the number of hidden layer units can be variable. A typical RBF neural network has *H* processing nodes in the hidden layer and *M* summing nodes in the output layer. The input sample is a *N*-dimensional vector.

The performance of RBF neural network is decided by the hidden layer, which consists of H hidden neurons (radial basis units) with radial activation functions, and each neuron only responds to an input that is close to its center. Different basis functions $\varphi(\cdot)$ can be adopted; a typical choice for this function is the Gaussian function, which is given by the following equation:

$$\varphi(x) = \exp(-\|x - c_i\|^2 / 2\sigma^2) \tag{1}$$

The detailed description of the network structure can be found in many inferences and can be considered as a multidimensional interpolation technique implementing general mappings $f : R^N \to R$ according to:

$$f(x) = w_0 + \sum_{i=1}^{H} w_i \varphi(\|x - c_i\|) \tag{2}$$

where $x = [x_1, x_2, \cdots, x_n]^T$ is the input vector and $c_i = [c_{i1}, c_{i2}, \cdots, c_{in}]^T$ are the center vectors; w_i are the weights of the output layer, $\varphi(\cdot)$ is the basis function, and $\|*\|$ denotes the Euclidean norm.

Using the advantage of RST in data processing, we can extract the connotative rules from the training data, and each rule is compact formally. Such every rule represents a certain class of the batch of data. If regarding the condition of the rule as input and the decision as output, each rule can be seen as a certain input-output sample. Because the decision rules set is minimum and has covered all relations in initial data, the condition of rules is exactly the ideal center vector space in RBF that we should seeking for based on clustering method. So, it is reasonable to regard each rule's condition as a hidden neuron center vector in RBF neural networks.

After having determined the hidden units, in the last step of the training process the weight matrix w is evaluated. The objective is to find the weight w which minimize the squared norm of the residuals:

$$E = \|Y - wz\|^2 \tag{3}$$

where Y is the $M \times K$ matrix of training targets, $z=(z_1, z_2, \ldots, z_k)$ is a $H \times K$ matrix, and w is the $M \times H$ weight matrix.

The weights w are initialized randomly and are updated by gradient descent method:

$$w(w+1) = w(n) - \eta \frac{\partial E^n}{\partial w} \tag{4}$$

Where η is the learning rate, which is small enough and decreases gradually.

4 Simulation

The system under consideration is a 2850kVA, 3657A ship synchronous generator, driven by a diesel engine. Electric loads including side-thruster (2200kW, 3300V,

465A), lubrication pump (240kW, 440V), steering gear, refrigerated container and so on [3]. The dynamic characteristic of ship power system depends on the diesel engine generator. According to working principles of diesel generator, we select excitation voltage (v_t) and diesel engine mechanical torque (P_m) as the input parameters; and the output parameters are terminal voltage (v), current (i) and frequency (f) of the ship synchronous generator. The parameters of input and output were measured as sampling data, and used for neural network training and testing. We selected some different running states of ship power system, such as diesel engine generator starting, lubrication pump running, side-thruster starting and three-phase ground fault. It is important that proper selection of input variables for building the dynamic network model to meet the highly non-linear characteristics of ship synchronous generator. Initially, the following notation is used for creating the input variables candidate set:

$$X(t) = [v_t(t), v_t(t-1), P_m(t), P_m(t-1), y(t-1), y(t-2), \cdots, y(t-n_y)]^T$$
$$y(t) = f[X(t)] \tag{5}$$

where $y(t\text{-}1)$, $y(t\text{-}2)$,…, $y(t\text{-}n_y)$ are the actual terminal output values when the time t-1,t-2, t-n_y. While $y(t)$ is the output of model at the moment t.

RST is used first to determine the significant lag n_y and then to select the most appropriate regress which make up the RBF network. At the beginning, the maximum lag was set to be 4. Thus, the establishing information table consists of 16 condition attributes. After analysis by RST, the 9 input nodes [$v_t(t)$, $v_t(t\text{-}1)$, $P_m(t)$, $P_m(t\text{-}1)$, $v(t\text{-}1)$, $v(t\text{-}2)$, $i(t\text{-}1)$, $i(t\text{-}2)$, $f(t\text{-}1)$] were selected as the best subset from the 16 input nodes.

The centers of RBF network were then initialized from the extracted 214 rules to configure the hidden units of networks. As an example, terminal voltage parameter of generator is selected to demonstrate the precision of model presented in this paper. In order to verify this method being superior, conventional RBF neural network model was presented using all the training data without pretreatment. The simulation results of generator starting, lubrication pump starting, side-thruster starting and three-phase grounded fault happened at t=0, 10, 20, 30s respectively are shown in Fig.1 and Fig.2. The relative errors between tests with actual terminal voltage values by both two methods are shown in Fig.3. Here, the relative error defined as:

$$RE = |\, y_i - \hat{y}_i \,| / y_i \tag{6}$$

where y_i is actual voltage values, and $\hat{y}_i$ is the test value.

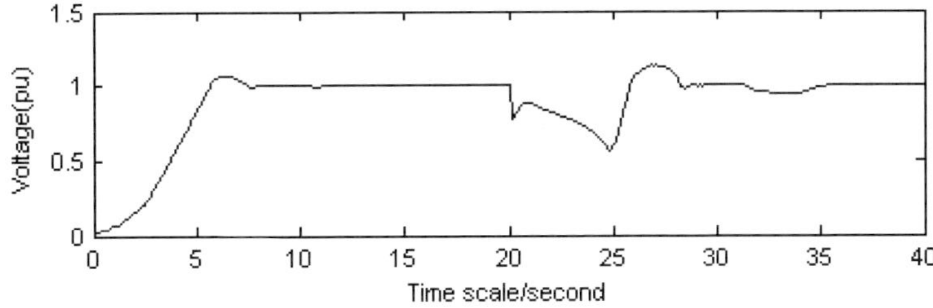

Fig. 1. The simulation results with RST+RBF neural network

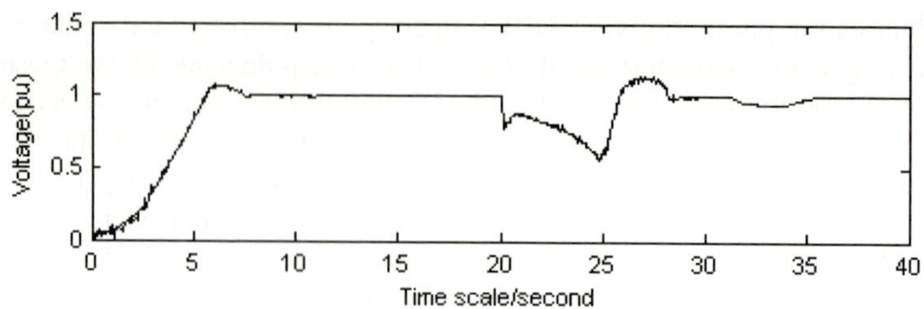

Fig. 2. The test results by conventional RBF neural network

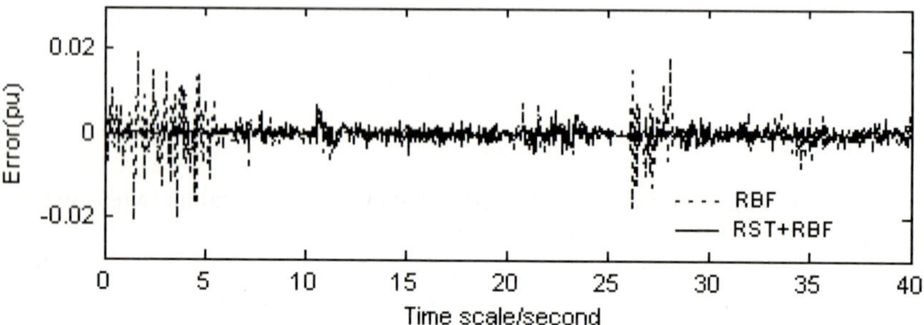

Fig. 3. The errors between model outputs with actual terminal voltage values

The root mean square error (*RMSE*) and the mean absolute error (*MAE*) are selected to evaluate the model performance, and the test results are shown in Table.1.

$$RMSE = \sqrt{\frac{1}{N}\sum_{i=1}^{N}(y_i - \hat{y}_i)^2} \qquad MAE = \frac{1}{N}\sum_{i=1}^{n}\left|y_i - \hat{y}_i\right| \tag{7}$$

Table 1. The test results error of both methods

Network model		RST+RBF	RBF
Training	RMSE	0.0013	0.0127
	MAE	0.0008	0.0087
Prediction	RMSE	0.0040	0.0706
	MAE	0.0025	0.0372

From Table.1 and simulation result all above, we can see that proper selection of input variables and preprocessing the training data could achieve the dynamic synchronous generator model with sufficient accuracy and better generalization.

5 Conclusions

Being the highly non-linear characteristics of the ship synchronous generators, building an excellent dynamic model mapping the complex non-linear relationships between generator parameters cannot be easily. A novel ship synchronous generator modeling approach based on RST and RBF neural network is presented in this paper. This method succeeds in combining the RBF neural network study and analysis ability with RST. The simulation results show the superiority of this method.

Acknowledgement

This work was supported by Key Science Project of Shanghai Education (04FA02) and Shanghai Leading Academic Discipline Project (T0602).

References

1. Karrari, M., Malik, O. P.: Identification of Synchronous Generators using Adaptive Wavelet Networks. Electrical Power and Energy Systems 27 (2005) 113-120
2. Zhang, T., Xiao, J.: Dynamic System Modeling Based on Rough Sets and RBF Neural Networks. In: Proceedings of the 5th World Congress on Intelligent Control and Automation (2004) 185-189
3. Shi, W., Yang, J., Tang, T.: RBF NN Based Marine Diesel Engine Generator Modeling. In: Proceedings of American Control Conference (2005) 2745-2749
4. Pawlak, Z.: Rough sets. International Journal of Computer and Information Science 11 (1982) 341-356
5. Pawlak, Z.: Some Issues on Rough Sets. In: Peters, J.F. et al. (Eds.): Transactions on Rough Sets I. Lecture Notes in Computer Science, Vol. 3100. Springer-Verlag, Berlin Heidelberg New York (2004) 1-58
6. Pawlak, Z.: Rough Sets and Intelligent Data Analysis. Information Sciences 147 (2002) 1-12
7. Bazan, J. G., Szczuka, M.: The Rough Set Exploration System. In: J.F. Peters et al. (Eds.): Transactions on Rough Sets III, Lecture Notes in Computer Science, Vol. 3400. Springer-Verlag, Berlin Heidelberg New York (2005) 37-56

A New Control Strategy of a Wind Power Generation and Flywheel Energy Storage Combined System

Jian Wang, Long-yun Kang, and Bing-gang Cao

Department of Mechatronics Engineering,
Xian JiaoTong University, Xi'an, 710049, China
wangjianxjtu@mail.xjtu.edu.cn

Abstract. The paper analyzes the structure character of a wind power generation and flywheel energy storage combined system, and presents a new control strategy—fuzzy neural network (FNN) based on genetic arithmetic (GA) for the nonlinear problem of the control system. The control strategy realizes automatic regulation of direct current (DC) bus voltage in the system, so it stabilizes the DC bus voltage of the system. The experimental results show that the controller has better self-learning and robustness, and realizes the satisfactory operation of the proposed system.

1 Introduction

In recent years, researchers have reported about the studies of the wind power generation and flywheel combined system. Boutot Tobie et.al introduced a low speed flywheel system for wind energy [1]. Ludovic Leclercq et.al proposed a control based on fuzzy logic for a flywheel energy storage system associated with wind and diesel generators [2]. Cárdenas Roberto *et al.* improved the control strategy based on fuzzy logic for enhanced power smoothing in wind energy systems [3][4].

In such systems, wind speed fluctuations produce fluctuations in the generator output voltage. The global scheme of the system under study is shown in Fig.1. An equivalent continuous model of this system is proposed in this paper [5][6]. When the DC bus voltage E_{dc} decreases, the induction machine is controlled to operate as a generator, transforming the inertial energy stored in the flywheel into electrical energy supplied to the DC Bus. When the DC bus voltage E_{dc} increases, the induction machine motors, transferring energy from the DC Bus to the flywheel. With reference to Fig. 1, the control plant is the DC bus voltage against fluctuations in this system.

Though the controller with fuzzy logic method could regulate the DC bus voltage of the wind power generation and flywheel combined system, the fuzzy logic rules of the controller built were required a lot of experimental data. To intelligentize the control of the system, this paper proposed a new control strategy of a wind power generation and flywheel combined system. Firstly, the parameters of fuzzy member function were regulated off-line by the GA; secondly, the DC bus voltage of the system was controlled online by the FNN.

J. Wang et al. (Eds.): ISNN 2006, LNCS 3972, pp. 1370 – 1377, 2006.

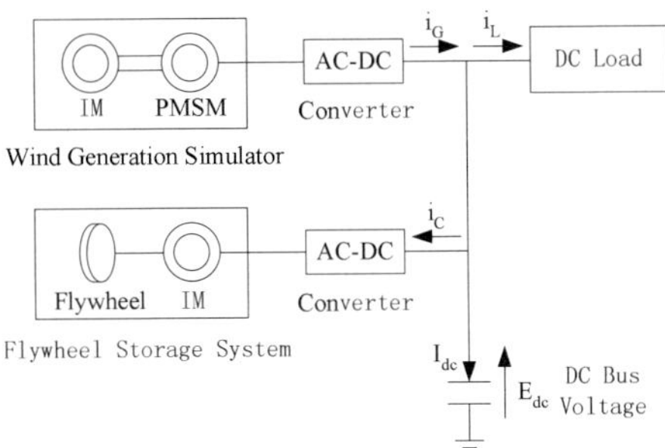

Fig. 1. Global scheme of the wind power generation and flywheel combined system

2 System Modeling

The energy control system based on vector-controlled induction machine is shown in Fig. 2. In the system, it is based on a standard indirect-rotor-flux-orientated (IRFO) [7] control of the induction machine driving the flywheel [8][9][10]. The current and voltage values are referred to the reference frame aligned to the rotor flux and take DC values in steady state. The torque current reference is derived from the FNN controller based on GA.

To obtain a relationship between the DC bus voltage and the current the power balance is used. In Fig. 1, the power balance between the DC bus side and the induction machine side is expressed as

$$E_{dc}\left((i_G - i_L) - C\frac{dE_{dc}}{dt}\right) = P_{loss} + k(V_d i_d + V_q i_q) \tag{1}$$

Where E_{dc} is the DC bus voltage, i_G is the output current of wind generation simulator, i_L is the current of DC load, P_{loss} is the inverter and iron power losses, k is the coefficient of the 2-3 axes scaling, V_d and V_q are the dq voltage of the stator respectively, i_d and i_q are the dq current of the stator respectively, C is the total capacitance of the DC bus.

According to the mathematical model of rotor-flux orientation of induction machine, V_d , i_d , V_q and i_q are given by

$$V_d = (R_s + \sigma L_s p)i_d - \omega_s \sigma L_s i_q \tag{2}$$

$$V_q = \omega_s \sigma L_s i_d + (R_s + \sigma L_s p)i_q + \omega_s \frac{L_m}{L_r}\varphi_r \tag{3}$$

Where R_s and R_r are the stator and rotor resistance respectively, L_s and L_r are the stator and rotor inductance respectively, L_m is the magnetizing inductance, φ_r is the rotor flux, ω_s is the rotational speed of stator, p is the differential operator,

$\sigma = 1 - \frac{L_m^2}{L_s L_r}$. Using (1)-(3), it can be shown that the power balance can be de rived as (4)

$$E_{dc}(i_G - i_L) - \frac{1}{2}C\frac{dE_{dc}^2}{dt} = P_{loss} + k\left[R_s(i_d^2 + i_q^2) + \left(\frac{L_m}{L_r}\right)^2 \cdot R_r i_q^2 + \omega_r \frac{L_m^2 i_d}{L_r} i_q + \frac{1}{2}\sigma L_s\left(\frac{di_q^2}{dt} + \frac{di_d^2}{dt}\right)\right] \quad (4)$$

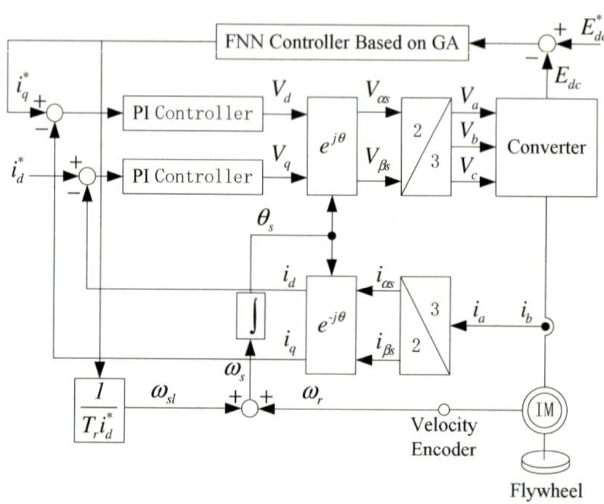

Fig. 2. Energy control system based on vector-controlled induction machine

Since the flywheel inertia will be large (the speed dynamics will be slow), and neglecting the variation in the energy stored in σL_s, C and P_{loss}, then the steady-state torque current i_q is expressed as

$$i_q = \frac{-\omega_r\left[\frac{L_m^2 i_d}{L_r}\right]}{2\left(R_s + R_r\left(\frac{L_m}{L_r}\right)^2\right)} + \frac{\sqrt{\left(\frac{\omega_r L_m^2 i_d}{L_r}\right)^2 - 4\left(R_s + R_r\left(\frac{L_m}{L_r}\right)^2\right)\left(R_s i_d^2 - \frac{1}{k}(E_{dc}(i_G - i_L))\right)}}{2\left(R_s + R_r\left(\frac{L_m}{L_r}\right)^2\right)} \quad (5)$$

3 Controller Constructing

The FNN control system based on GA is shown in Fig. 3. The principle of FNN is shown in Fig. 4. The FNN includes one input layer, two hidden layers, and one output layer. The input layer is composed of error e and differential coefficient of error de/dt. The output layer is u^*. The fuzzy variable of e is {N,Z,P}, where N,Z,P are abbreviations for the commonly used names "Negative", "Zero", "Positive". The fuzzy variable of de/dt is {NB,NS,Z,PS,PB}, where NB,NS,…, are abbreviations for the commonly used names "Negative Big", "Negative Small", and so on. The fuzzy member functions are expressed as

$$A_{ij}(x_i) = \exp\left[-\left(\frac{x_i - c_{ij}}{b_j}\right)^2\right] \tag{6}$$

Where x_i is the input variable, c_{ij} is the center of the function, b_j is the width of the function.

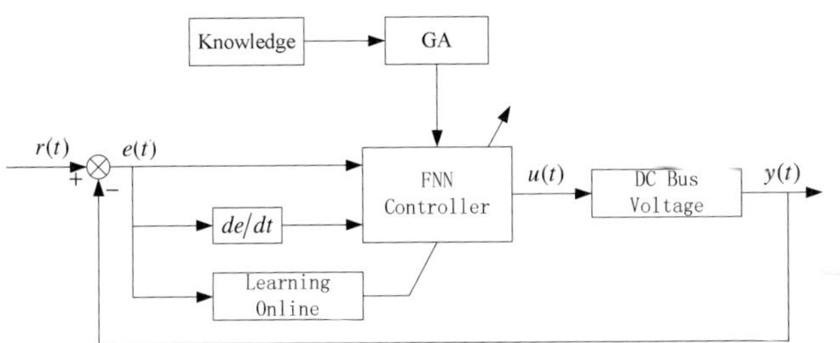

Fig. 3. FNN control system based on GA

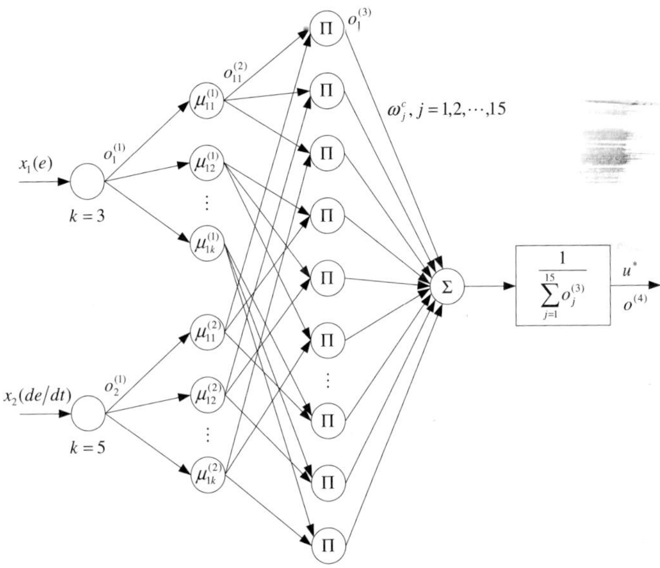

Fig. 4. Principle scheme of FNN

In Fig. 4, the input layer and first hidden layer correspond to the antecedent part of the fuzzy control rules, and the second hidden layer and output layer correspond to the conclusion part. The input-output relationships of units in the FNN are defined as

(1) Input layer:

$$o_i^{(1)} = x_i \qquad (i=1,2) \tag{7}$$

(2) First hidden layer:

$$net_{ik}^{(2)} = o_i^{(1)} \qquad (i=1,\ k=1,2,3;\ i=2,\ k=1,2,3,4,5) \tag{8}$$

$$o_{ik}^{(2)} = \mu(net_{ik}^{(2)}) \qquad (i=1,\ k=1,2,3;\ i=2,\ k=1,2,3,4,5) \tag{9}$$

(3) Second hidden layer:

$$net_{kl}^{(3)} = o_{1k}^{(2)} \cdot o_{2l}^{(2)} \qquad (k=1,2,3 \qquad l=1,2,3,4,5) \tag{10}$$

$$o_j^{(3)} = net_j^{(3)} \qquad (j=1,2,\cdots,15) \tag{11}$$

(4) Output layer:

$$o^{(4)} = u^* = \frac{net^{(4)}}{\sum_{j=1}^{15} o_j^{(3)}} \tag{12}$$

To regulate the parameters (c_{ij}, b_j) of fuzzy member function by the GA, the optimal problem should be described by the algorithmic language. If there are some data obtained $(e_i, de_i/dt, u_i)$, $1 \le i \le m$, the problem is described as when the inputs are $e = e_i$ and $de/dt = de_i/dt$, the output is $u_i^* = u_i$, searching the optimal parameters $\{c_{ij}, b_j\}$ of fuzzy member function, bringing on

$$\min E = \frac{1}{2}\sum_{i=1}^{m}\left(u_i - u_i^*\right)^2 \tag{13}$$

Where u_i is the demanded output, u_i^* is the output of FNN controller.

The optimal parameters are used in the FNN controller. The weight ω_j^c of FNN is regulated online by the BP gradient algorithm.

The performance index is defined as

$$J_c = \frac{1}{2}[r(t) - y(t)]^2 \tag{14}$$

Where $r(t)$ is the demanded input, $y(t)$ is the output of control object.

Then, the weight ω_j^c in the FNN is changed as

$$\omega_j^c(t+1) = \omega_j^c(t) - \eta\frac{\partial J}{\partial \omega_j^c} + \alpha\Delta\omega_j^c(t) \tag{15}$$

Where η is an adaptive learning rate, α is a momentum constant.

4 Experimental Results

The 15 fuzzy control rules as shown in Table 1 are seen as the learning sample. The number of population is 500. The parameters $\{c_{ij}, b_j\}$ are change at $[-1,1]$. The crossover rate p_c is 0.8. The mutation rate p_m is 0.07. The number of the generation is 30.

The optimal parameters $\{c_{ij}, b_j\}$ of FNN are obtained through 27 generation. The initial fuzzy member functions of error and differential coefficient of error are shown in Fig. 5 and Fig. 7. The fuzzy member functions of error and differential coefficient of error optimized are shown in Fig. 6 and Fig. 8.

Table 1. Fuzzy function rules

e \ de/dt	NB	NS	ZE	PS	PB
N	PB	PM	PS	ZE	NS
Z	PM	PS	ZE	NS	NM
P	PS	ZE	NS	NM	NB

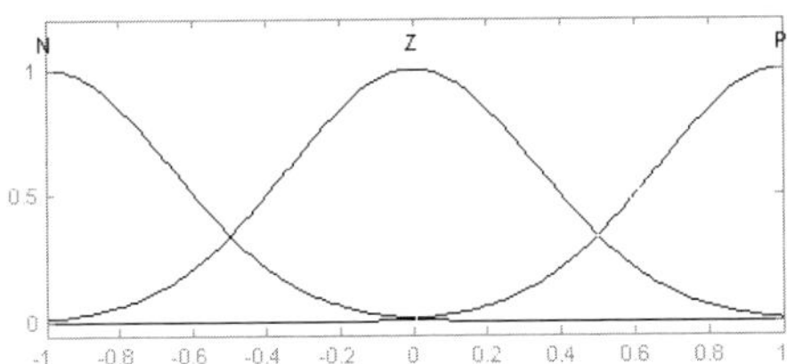

Fig. 5. The initial fuzzy member function of error

Fig. 6. The optimal fuzzy member function of error

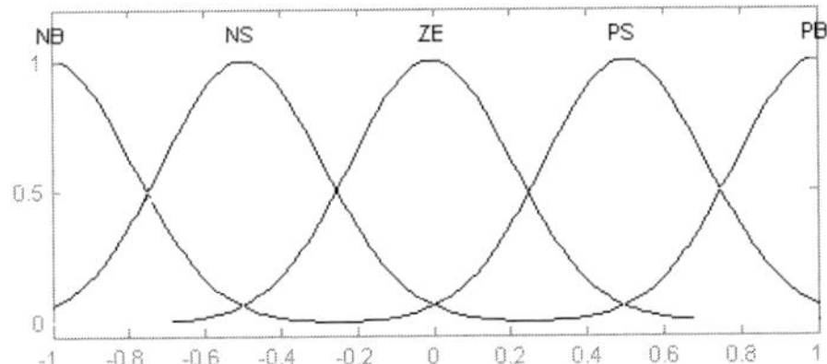

Fig. 7. The initial fuzzy member function of differential coefficient of error

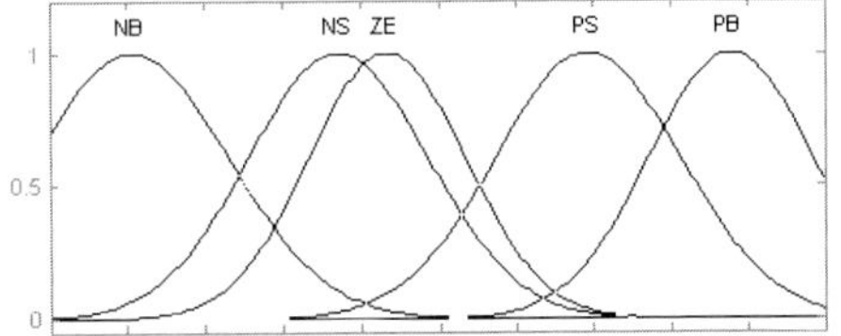

Fig. 8. The optimal fuzzy member function of differential coefficient of error

In the paper, induction machine of rating power 550W is used for the actuator of the flywheel energy storage system. PMSM of rating power 2kW and induction motor of rating power 2.2kW are combined for the actuator of the wind generator simulator.

The demanded DC bus voltage is 120V. The demanded DC load is 1500W. Fig. 9 shows the response of current $i_G - i_L$ and DC bus voltage E_{dc}. Fig. 10 shows the response of DC bus current I_{dc} and DC bus power P. The DC load is disconnected at $t = 0.135s$ and applied at $t = 0.625s$. When the load is connected, the overshoot of DC bus voltage E_{dc} is 14V. When the load is disconnected, the overshoot of DC bus voltage E_{dc} is 12V. The stable value of the DC bus voltage is 120V. The fluctuations of the DC bus voltage are very small.

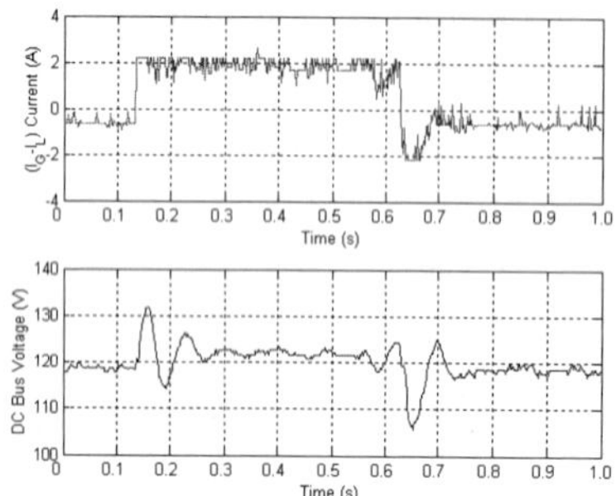

Fig. 9. The response of current $i_G - i_L$ and DC bus voltage E_{dc}

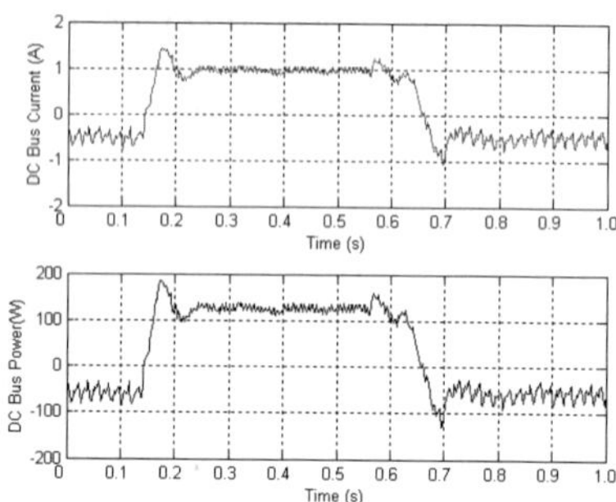

Fig. 10. The response of DC bus current I_{dc} and DC bus power P

5 Conclusions

This paper has proposed a novel energy complementary control strategy of a distributed power generation system. The control strategy for regulating the DC bus voltage is performed in an experimental wind flywheel hybrid energy system. The system uses a metal flywheel for power smoothing employing an inverter-fed vector-controlled induction machine. The FNN controller based on GA embedded in the system is used for the object of energy complementary control. The experimental results confirmed the satisfactory operation of the proposed system.

References

1. Boutot, T., Chang, L., Luke, D.: A Low Speed Flywheel System For Wind Energy conversion. In: Proceedings of th 2002 IEEE Canadian Conference on Electrical & Computer Engineering, Winnipeg, Manitoba, Canada (2002) 251-256
2. Leclercq, L., Robyns, B., Grave, J.M.: Control Based on Fuzzy Logic of a Flywheel Energy Storage System Associated with Wind and Diesel Generators. Mathematics and Computers in Simulation 63 (2003) 271-280
3. Cárdenas, R., Peña, R., Asher, G., Clare, J.: Control Strategies for Enhanced Power Smoothing in Wind Energy Systems Using a Flywheel Driven by a Vector-Controlled Induction Machine. IEEE Trans. Industrial Electronics 48 (2001) 625-635

4. Cárdenas, R., Peña, R., Asher, G., Clare, J.: Power Smoothing in Wind Generation Systems Using a Sensorless Vector Controlled Induction Machine Driving a Flywheel. IEEE Trans. Energy Conversion 19 (2004) 206-216
5. Marmin, F., Grave, J.M.: Etude de Faisabilité D'un Stabilisateur Cinétique de Réseau Pour Génératrice Éolienne. In: Proceedings of Innovelect'99, Mons, Belgium (1999) 41-46
6. Labrique, F., Buyse, H., Séguier, G., Bausière, R.: Les Convertisseurs de L'électronique de Puissance, Commande et Comportement Dynamique, Lavoisier 1998
7. Leonhard, W.: Control of Electrical Drives. Berlin, Germany: Springer-Verlag, 1985
8. Agaki, H., Sato, H.: Control and Performance of a Double-fed Induction Machine Intended for Flywheel Energy Storage System. IEEE Trans. Power Electron. 17 (2002) 109-116
9. Rudell, A. J., Bleij, J. A. M., Freris, L.: A Wind Diesel System with Variable Speed Fly-Wheel Storage. Wind Eng. 17 (1993) 129-145
10. Iqbal, M.T.: Simulation of a Wind Fuel Cell Hybrid Energy System. Renewable Energy 28 (2003) 511-522

Wavelet-Based Intelligent System for Recognition of Power Quality Disturbance Signals

Suriya Kaewarsa[1], Kitti Attakitmongcol[1], and Wichai Krongkitsiri[2]

[1] School of Electrical Engineering, Suranaree University of Technology, 111, Muang District, Nakhon Ratchasima, 30000, Thailand
suriya_ka@hotmail.com

[2] School of Electrical Engineering, Rajamangala University of Technology Isan, Sakon Nakhon Campus, Phangkhon District, Sakon Nakhon, 47160, Thailand
wichai_k@rmuti.ac.th

Abstract. Recognition of power quality events by analyzing the voltage and current waveform disturbances is a very important task for the power system monitoring. This paper presents a new approach for the recognition of power quality disturbances using wavelet transform and neural networks. The proposed method employs the wavelet transform using multiresolution signal decomposition techniques working together with multiple neural networks using a learning vector quantization network as a powerful classifier. Various transient events are tested, such as voltage sag, swell, interruption, notching, impulsive transient, and harmonic distortion show that the classifier can detect and classify different power quality signal types efficiency.

1 Introduction

Power quality (PQ) issues have been important for industrial facilities with sensitive loads. The demand for clean power has been increasing in the past several years. The reason is mainly due to the increased use of microelectronic processors in various types of equipment, such as computer terminals, programmable logic controller, diagnostic systems, etc. Most of these devices are quite susceptible to disturbances of the incoming alternating voltage waveform [1]. Poor power quality (PQ) may cause many problems for affected loads, such as malfunction, instabilities, short lifetime, and so on. Poor power quality is normally caused by power-line disturbances, such as impulses, notches, momentary interruptions, waveshape faults, voltage sag, swell, harmonic distortion, and flicker, resulting in failure of end-use equipment. In order to improve power quality, the sources and causes of such disturbances must be known before appropriate mitigating actions can be taken. A feasible approach to achieve this goal is to incorporate detection capabilities into monitoring equipment so that events of interest will be recognized, captured, and classified automatically. Thus, good performance monitoring equipment must have functions which involve the detection, localization, and classification of transient events. To monitor electrical power quality disturbances, short time discrete Fourier transform (STFT) is most often used. But for non-stationary signals, the STFT does not track the signal dynamics properly due to the limitations of a fixed window width chosen a priori. Thus, STFT cannot be used

J. Wang et al. (Eds.): ISNN 2006, LNCS 3972, pp. 1378 – 1385, 2006.

successfully to analyze transient signals comprising both high and low frequency components. On the other hand, wavelet transform (WT) uses short windows at high frequencies and long windows at low frequencies, thus closely monitoring the characteristics of non-stationary signals. The wavelet transform decomposes transients into a series of wavelet components, each of which corresponds to a time domain signal that covers a specific octave frequency band containing more detailed information. Such wavelet components appear to be useful for detecting, localizing, and classifying the sources of transients [2]. Hence, the wavelet transform is feasible and practical for analyzing power quality disturbances.

Santoso *et al.* [3] proposed to extract the features of power quality signals in terms of wavelet coefficients using the multiresolution analysis (MRA) as inputs of the neural network for identifying impulses, voltage sags, and transient oscillations. The detection, localization, and classification processes were performed by visual inspection. It yields low accuracy. Angrisani *et al.* [4] proposed to employ the continuous wavelet transform (CWT) to estimate the disturbance time duration and the discrete wavelet transform (DWT) to estimate the disturbance amplitude. The two features thus obtained are then used to classify the transient disturbance type. It gives medium accuracy. Santoso *et al.* [5] presented a wavelet-based neural classifier integrating the DWT, and learning vector quantization (LVQ) neural network to become an actual power disturbance classifier. The classifier employed the DWT coefficients as inputs to multiple LVQ neural networks to train and perform waveform recognition but the detection and localization processes were performed by visual inspection.

In this paper used wavelet-based neural classifier to automatically detect, localize, and classify the transient disturbance type, for high accuracy and low usage time. The underlying approach of the proposed method is to carry out waveform recognition in the wavelet domain using multiple neural networks. A final decision about the disturbance type is made by combining the outcomes of the networks using decision-making schemes.

2 Wavelet Transform

The wavelet transform (WT) is a mathematical tool, which allows us to view a time history of a signal in terms of its frequency components. A wavelet expansion coefficient represents a component that is itself local and is easier to interpret. The wavelet expansion may allow a separation of components that overlap in both time and frequency. The wavelet transform is with good performance in extracting features of the power quality disturbances because it is very sensitive to irregular signals and insensitive to regular signals. For WT, the original signal transforms from time domain to the time-scale domain. The transformation process decomposes the origin signal into several other signals with different resolution. The original time domain signal can be recovered by the decomposed signals without losing any information and the recover process is called the inverse wavelet transform.

Wavelet transform of sampled waveforms can be obtained by implementing the discrete wavelet transform (DWT) which is given by

$$\mathrm{DWT}(f,m,n)=\frac{1}{\sqrt{a_0^m}}\sum_k f(k)h^*\left(\frac{n-ka_0^m}{a_0^m}\right). \quad (1)$$

Where, the parameters a_0^m and ka_0^m are the scaling (dilation) and translation (time shift) constant respectively, k and m being integer variables and h is the wavelet function which may not be real, as assumed in the above equation for simplicity. In standard discrete wavelet transform the coefficients are sampled from the continuous wavelet transform (CWT) on a dyadic grid, $a_0 = 2$, yielding $a_0^0 = 1$, $a_0^{-1} = ½$, etc.

The choice of mother wavelet plays a significant role in detecting and localizing various types of disturbances. Daubechies' wavelets with 4, 6, 8, and 10 filter coefficients work well in most disturbance detection cases. At the lowest scale (scale 1), the mother wavelet is most localized in time and oscillates most rapidly within a very short period of time. As the wavelet goes to higher scales, the analyzing wavelets become less localized in time and it oscillate less due to the dilation nature of the wavelet transform analysis. As a result of higher scale signal decomposition, fast and short power quality disturbances will be detected at lower scales, whereas slow and long power quality disturbances will be detected at higher scales. Hence, we can detect both fast and slow waveshape faults with a signal type. Since Daub4 has the least number of filter coefficients and it gives the shortest support in the family, we use Daub4 in our algorithm [2].

3 Learning Vector Quantization

Artificial neural network is a sophisticated networks system that is made of many neurons connected with each other. In this study, the proposed classification is carried out in sets of multiple neural networks using a learning vector quantization network (LVQ). The LVQ network is a hybrid network which uses both unsupervised and supervised learning to form classifications [6].

In the LVQ network, each neuron in the first layer is assigned to a class and several other neurons are often assigned to the same class. Each class is then assigned to one neuron in the second layer. The number of neurons in the first layer (S^1) will therefore always be at least the number of neurons in the second layer (S^2) and will usually be larger. As with the competitive network, each neuron in the first layer of the LVQ network learns a prototype vector, which allows it to classify a region of the input space. Instead of computing the proximity of the input and weight vectors by using the inner product, the net input of the first layer can be obtained by

$$n_i^1 = -\left\|{}_i w^1 - p\right\|. \quad (2)$$

The output of the first layer of the LVQ network is

$$a^1 = \mathrm{compet}\left(n^1\right) \quad (3)$$

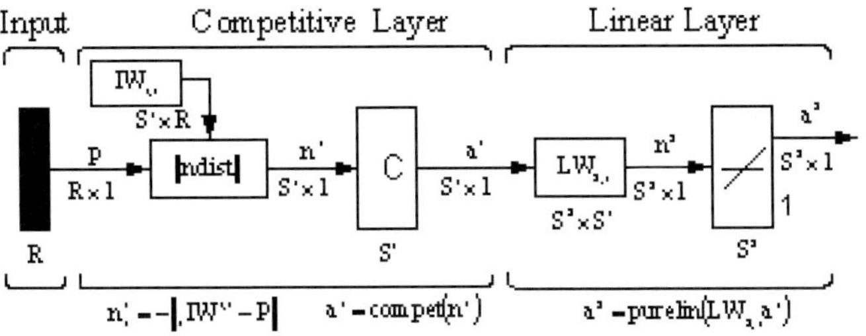

Fig. 1. Learning vector quantization network structure

Therefore the neuron whose weighting vector is closest to the input vector will output a 1, and the other neurons will output 0. In the LVQ network, the winning neuron indicates a subclass, rather than a class. There may be several different neurons (subclasses) that make up each class. The second layer of the LVQ network is used to combine subclasses into a single class. This is done with the W^2 matrix. The columns of W^2 represent subclasses, and the rows represent classes. W^2 has a single 1 in each column, with the other elements set to zero. The row in which the 1 occurs indicates which class the appropriate subclass belongs to.

$$\left(w_{ki}^2 = 1\right) \Rightarrow \text{Subclass is a part of class} \tag{4}$$

The LVQ learning rule proceeds as follows. At each iteration, an input vector p is presented to the network, and the distance from p to each prototype vector is computed. Then, the hidden neurons compete. If neuron i^* wins the competition, the i^*th element of a^1 is set to 1. Next, a^1 is multiplied by W^2 to get the final output a^2, which also has only one nonzero element, k^*, indicating that p is being assigned to class k^*. The Kohonen rule is used to improve the hidden layer of the LVQ network in two ways. First, if p is classified correctly, then the weights ${}_{i^*}w^1$ of the winning hidden neuron move toward p. This can be expressed in (5).

$${}_{i^*}w^1(q) = {}_{i^*}w^1(q-1) + \alpha\left(p(q) - {}_{i^*}w^1(q-1)\right) \tag{5}$$

Second, if p was classified incorrectly, then the weights ${}_{i^*}w^1$ move away from p. This can be expressed in (6).

$${}_{i^*}w^1(q) = {}_{i^*}w^1(q-1) - \alpha\left(p(q) - {}_{i^*}w^1(q-1)\right) \tag{6}$$

The result will be that each hidden neuron moves toward vectors that fall into the class for which it forms a subclass and away from vectors that fall into other classes.

4 Wavelet-Based Neural Classification Structure

The basic idea of the wavelet-based neural classifier is to perform waveform recognition in the wavelet domain using multiple neural networks. Fig. 2 shows the schematic block diagram of the wavelet-based neural classifier which consists of preprocessing, processing, and post-processing. The input of the neural network is a preprocessing signal. In this case, the time domain of power quality disturbance waveform is transformed into the wavelet domain before being fed to the neural network. This transformation detects and extracts disturbance features in the form of simultaneous time and frequency information, and gradient or slope of the disturbance signal using the dyadic orthonormal wavelet transform. The extracted features help the neural network in distinguishing one disturbance event from another. The processing phase contains a set of multiple artificial neural networks with wavelet transform coefficients as input signals. This processing phase performs waveform recognition in the wavelet domain since all input signals are in the wavelet domain. The output of the processing phase is the type of the disturbances. Since multiple neural networks are utilized, a post–processing phase is required to combine the outcomes of the multiple neural networks in order to make a decision

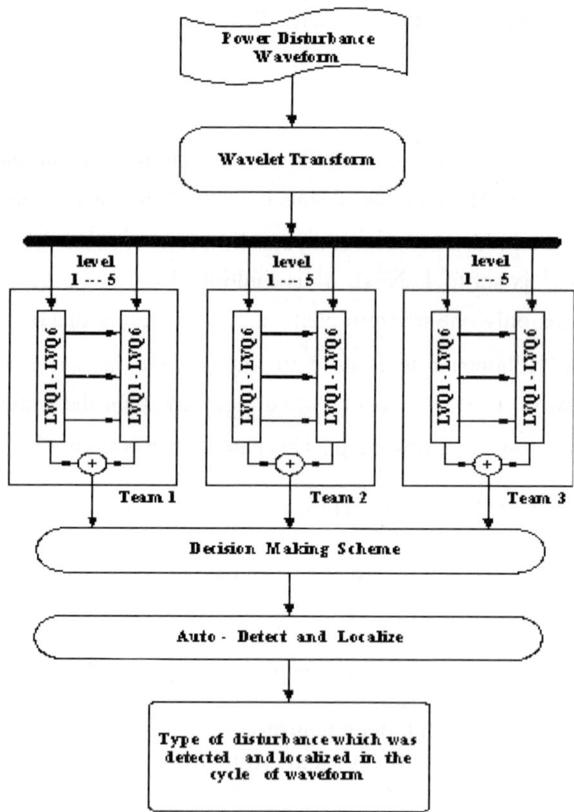

Fig. 2. Schematic block diagram of the proposed method

about the disturbance type and to provide a level of confidence for the decision made. The output of the classifier declared that it is a disturbance with belief interval of 90-95% as the degree as of belief.

The entire disturbance record (1000 sampling points) is used for this purpose. The disturbance features reside in five scales of decomposed signals. Teams of artificial neural networks which each team consists of 30 learning vector quantization networks are applied. The output of each team is then combined to produce a final decision about the disturbance with one of the decision making schemes. The LVQ must be trained using known disturbance waveforms before they can be used as a part of the classifier. Each of the LVQ is trained separately and their weight vectors are initialized independently. Thus, after training, the weight vectors are different from one another. In the testing phase, these disturbances are tested along with all other pre-specified disturbances. The schematic diagram for the testing phase is the same as the one shown in Fig. 2. The use of multiple set of neural networks arises from the need for achieving a higher accuracy rate. This is normally achieved by rejecting ambiguous patterns which cannot be recognized by a neural network. The use of multiple neural networks also provides a means of determining a degree of belief for each identified disturbance waveform. The voting scheme is the simplest method of combining the output of multiple neural networks. A decision is made based on which type of disturbance waveform receives the most votes.

5 Disturbance Data Set

The wavelet-based neural classifier presented in this work is designed to recognize 6 types of power quality disturbances including of type A voltage sag, type B voltage swell, type C interruption, type D impulsive transient, type E notching, and type F harmonic distortion as described in [7, 8]. Typical disturbance waveforms of these kinds are shown in Fig. 3. The power quality disturbance data set are split into the training data set and testing data set. Table 1 shows the number of disturbance records required for each type to train and test the classifier. The total number of disturbance

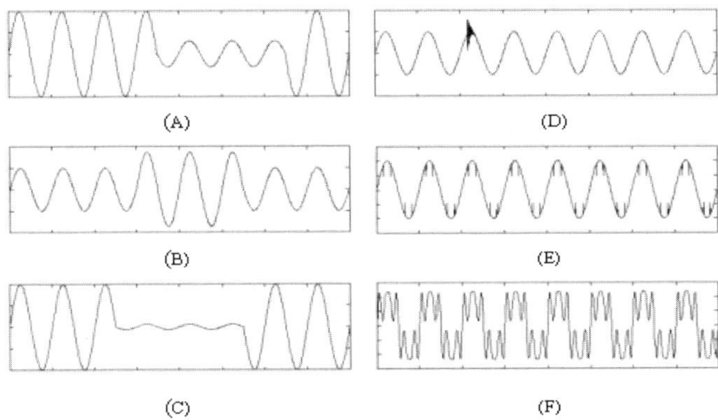

Fig. 3. Typical power quality disturbance categories in this research

records to train and test the classifier are 780 and 660 records, respectively. The sampling frequency for all types of disturbance considered here is 20 kHz.

Table 1. Power quality disturbance data set

Type	A	B	C	D	E	F	Total
Training	130	130	130	130	130	130	780
Testing	110	110	110	110	110	110	660

6 Simulation Results

This section discusses the simulation of the wavelet-based neural network classifier for recognizing power quality disturbance types. The proposed method is run by using MATLAB program. The random selected signal from 110 signals of each disturbance type is used to test neural networks. The proposed method is able to detect and classify all 6 types for power quality disturbances as shown in Table 2. From Table 2, all disturbance types tested are differentiated from pure sinusoids. Impulsive transient (type D), notching (type E), harmonic distortion (type F) and pure sinusoid are identified with 100 % accuracy. Voltage sag (type A), voltage swell (type B), interruption (type C) are classified with 96.40 %, 95.50 %, and 93.60 % accuracy, respectively.

Table 2. Results of classifying the power quality disturbances

Type	Correctly	Incorrect	Accuracy rate (%)
pure sine	110	-	100.00
A	106	4	96.40
B	105	5	95.50
C	103	7	93.60
D	110	-	100.00
E	110	-	100.00
F	110	-	100.00
Total of accuracy rate = 97.90 %			

After testing all power quality disturbances type A, B, C, D, E, F, and pure sinusoid, the proposed method is able to detect and classify the disturbance types with 97.90 % accuracy.

7 Conclusions

This paper proposed a prototype of wavelet-based neural network classifier for power quality disturbance recognition and classification. The experimental results showed that the proposed method has the ability of recognizing and classifying different power disturbance types efficiently. This work leads us to believe that wavelet analysis together with neural structure, as a new tool, offers a great potential for diagnosis of electrical power systems in the area of power quality problems.

Acknowledgments

The authors gratefully acknowledge the financial support from Rajamangala University of Technology Isan (RMUTI), Sakon Nakhon Campus, Thailand, during a period of this work.

References

1. Santoso, S., Powers, E.J., Grady, W. M.: Power Quality Disturbance Waveform Detection Using Wavelet Transform Analysis. Proc. IEEE Conf. Time-Frequency and Time- Scale Analysis (1994) 166-169
2. Kaewarsa, S., Attakitmongcol, K.: Diagnostic of Power Quality Disturbance Using Wavelet- Based Neural Network. Proc. IASTED Int.Conf. Energy and Power Systems (2005) 245-250
3. Santoso, S., Powers, E. J., Grady, W. M., Hofmann, P.: Power Quality Assessment via Wavelet Transform. IEEE Trans. Power Delivery 14(2) (1996) 924-930
4. Angrisani, L., Daponte, P., Apuzzo, M D., Testa, T.: A Measurement Method Based on the Wavelet Transform for Power Quality Analysis. IEEE Trans. Power Delivery 13(4) (1998) 990-998
5. Santoso, S., Powers, E. J., Grady, W. M., Parsons, A. C.: Power Quality Disturbance Waveform Recognition Using Wavelet-Based Neural Classifier-part 1: Theoretical Foundation. IEEE Trans. Power Delivery 15(1) (2000) 222-228
6. Hagan, M. T., Demuth, H. B., Beale, M.: Neural Network Design. Boston. PWS Publishing Company. (1996)
7. McEachern, A.: Handbook of Power Signatures. Basic Measuring Instruments. Foster city, CA (1988)
8. Dugan, R. C., McGranaghan, M. F., Santoso, S., Beaty, H. W.: Electrical Power System Quality, McGraw-Hill New York (2003)

Recognition and Classification of Power Quality Disturbances Based on Self-adaptive Wavelet Neural Network

Wei-Ming Tong[1], Xue-Lei Song[1], and Dong-Zhong Zhang[2]

[1] Harbin Institute of Technology, Harbin 150001, China
Dianqi@hit.edu.cn
[2] Heilongjiang University, Harbin 150080, China
zhangdzpop@163.com

Abstract. This paper presents a novel self-adaptive wavelet neural network method for automatic recognition and classification of power quality disturbances. The types of disturbances include harmonic distortions, flickers, voltage sags, voltage swells, voltage interruptions, voltage notches, voltage impulses and voltage transients. The self-adaptive wavelet neural network model constructed consists of four layers: input layer, preprocessing layer, hidden layer and output layer. The preprocessing layer is also called wavelet layer whose function is to extract features of power quality disturbances for recognition and classification; the other three layers just constitute the feedforward neural network whose function is to recognize and classify the types of power quality disturbances. The self-adaptive wavelet neural network has a good anti-interference performance, and the test and evaluation results demonstrate that utilizing it power quality disturbances can be recognized and classified effectively, accurately and reliably.

1 Introduction

In recent years, the power quality (PQ) has become a major concern to both electric utilities and electric customers, and has drawn a wide range of attention of the electric power researchers of the world. One reason is that the PQ has been being disturbed severely with the increasing number of polluting loads (such as non-linear loads, time-variant loads, fluctuating loads, unbalanced loads, and so on) connected to the electric power network [1], [2]; the other is that intelligent electrical devices have put forward more rigorous requirement for PQ of Electric power system. Therefore, the PQ imminently needs to be monitored and improved.

When monitoring and improving, recognition and classification of power quality is an important and complex task. Traditionally, carried out manually on collected data, recognition and classification is a costly and inefficient task. Therefore, recently several automatic recognition and classification methods based on artificial intelligence, such as artificial neural network (ANN) [3], were studied. ANN is a new and advanced pattern recognition technology and has been widely utilized in a lot of pattern recognition problems. Although ANN has the ability of self-organization,

J. Wang et al. (Eds.): ISNN 2006, LNCS 3972, pp. 1386 – 1394, 2006.

self-learning and self-adaptive, it is limited either merely in time domain or merely in frequency domain. Features of PQ mainly include two types which are time domain features and frequency domain features. Wavelet transform method may provide time information and frequency information simultaneously. So methods based on wavelet and ANN for recognition and classification of PQ were also studied [4], [5], [6]. Wavelet neural network (WNN) is a new technology which has the advantages of both wavelet transform and ANN. WNN has two types. One is that the neural cell of ANN is substituted with wavelet cell, that is, the wavelet function acts as the motivation function of ANN. The other is that wavelet transform acts as the preprocessing unit of ANN, that is, wavelet transform is used to extract features which are whereafter processed as inputs of ANN.

Self-adaptive WNN has been utilized in recognition of voice and image [7], [8]. Self-adaptive WNN can effectively and accurately extract features because it can adaptively adjust the scale parameter, the shift parameter and the weights during learning process. In addition, it has some other excellent performances which include good anti-interference ability, brief learning algorithm, rapid convergence rapidity, and so on. Therefore, this paper presents a novel self-adaptive wavelet neural network method for automatic recognition and classification of power quality disturbances.

2 Structure of Self-adaptive WNN

Generally, the self-adaptive wavelet neural network model constructed consists of four layers: input layer, preprocessing layer, hidden layer and output layer. The preprocessing layer is also called wavelet layer whose function is to extract features of power quality disturbances for recognition and classification over wavelet decomposition and whose outputs, of which each includes N-dimension features, are used as the inputs of the hidden layer. The number of wavelet layer nodes is defined as 16 here. The input layer, hidden layer and output layer just constitute the feedforward neural network whose function is to classify types of power quality disturbances. The inputs of the input layer are digital signals which need to be recognized and classified. Assuming 20ms as the period of the data window, then the number of input layer nodes is 64 when the sampling frequency is 3200Hz. The hidden layer varies with the wavelet layer. It receives the outputs of the wavelet layer and calculates them over the Sigmoid function. There are no definite methods to specify the number of hidden layer nodes. Increasing the number of hidden layer nodes can expand the search space and benefit convergence, but can also increase the calculation quantity. The output layer is similar with that of BP network. It receives the outputs of the hidden layer, and then calculates them over the Sigmoid function, and finally shows the results of recognition and classification. Here, the number of input layer nodes, wavelet layer nodes, hidden layer nodes and output layer nodes is defined as 64, 16, 16 and 3 separately.

The structure of self-adaptive WNN constructed is shown in Fig.1. The structure in which the hidden layer is added between the input layer and the output layer is the expansion of the self-adaptive WNN model brought forward by Szu. Here, $x(k)$ denotes the kth input of the input layer; $y(m)$ denotes the mth output of the output layer; w_{ji} denotes the weighting coefficient connecting the wavelet layer and the hidden layer; w_{im} denotes the weighting coefficient connecting the hidden layer and the output layer; a_j and b_j denote the scale parameter and the shift parameter of the jth node of the wavelet layer separately; K denotes the number of input layer nodes; J denotes the number of wavelet layer nodes; I denotes the number of hidden layer nodes; M denotes the number of output layer nodes. $\psi(x)$ denotes the mother wavelet function of the wavelet layer, $\sigma(x)$ denotes Sigmoid function which is used here as the transfer function both between the hidden layer and the output layer and between the wavelet layer and the hidden layer. So the self-adaptive WNN model constructed can be described by the following equation (1).

$$y(m) = \sigma\left(\sum_{i=1}^{I} w_{im} \cdot \sigma\left(\sum_{j=1}^{J} w_{ji} \cdot \sum_{k=1}^{K} x(k) \cdot \psi_{a_j,b_j}(k)\right)\right) \tag{1}$$

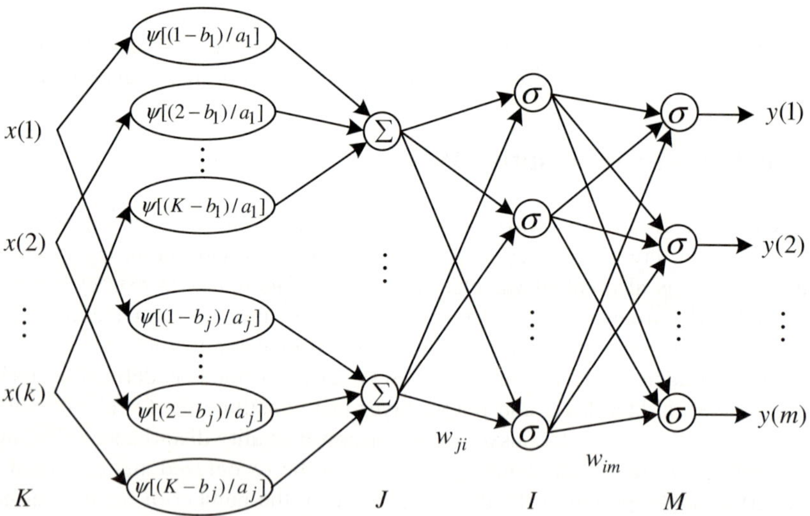

Fig. 1. Structure of self-adaptive WNN

2.1 Wavelet Layer

As the preprocessing layer, the wavelet layer is mainly utilized to extract features of PQ disturbances over wavelet transform. Like Fourier transform, the wavelet transform also consists of a pair of transform between time domain and frequency domain. For wavelet transform, the original signal transforms from time domain to time-frequency domain. The transform process decomposes the original signal into several other signals with different resolution. The original signal can be reconstructed by the decomposed signals without losing information. The reconstruction process is called inverse wavelet transform. Let $f(t)$ denotes the original time domain signal. The continuous wavelet transform (CWT) is defined as follows:

$$\mathrm{CWT}_f(a,b) = \frac{1}{\sqrt{a}} \int_{-\infty}^{+\infty} f(t)\psi\left(\frac{t-b}{a}\right)\mathrm{d}t \tag{2}$$

Where $\psi(t)$ represents the mother wavelet, a and b are the scale parameter and the shift parameter separately.

In order to make CWT be able to be implemented on a digital system, the scale parameter a and the shift parameter b are transformed from continuous to discrete as follow:

$$a = a_0^j, b = ka_0^j b_0 \tag{3}$$

Where j is the discrete scale parameter, k is the discrete shift parameter, and $a_0>1$, $b_0 \neq 0$. Therefore, the equation (2) can be rewritten as follows:

$$\mathrm{DWT}_f(j,k) = a_0^{-j/2} \int_{-\infty}^{+\infty} f(t)\psi\left(a_0^{-j}t - kb_0\right)\mathrm{d}t \tag{4}$$

The above equation (4) is just the discrete wavelet transform (DWT).

The choice of the mother wavelet is a key problem in constructing the self-adaptive WNN. Daubechies wavelet is orthogonal, biorthogonal, compactly supported and highly regular, and is able to utilize Mallat fast wavelet algorithm. Therefore, Daubechies wavelet has been widely used in extracting of features of singular signals. Therefore, Daubechies-8 wavelet is adopted to extract features of PQ disturbances in the wavelet layer of the self-adaptive WNN.

Different from classical wavelets, orthogonal wavelets, such as Daubechies wavelet, are unable to be represented by specific mathematic expressions. Generally, they are generated from the weighting combination of scale function. The wavelet function

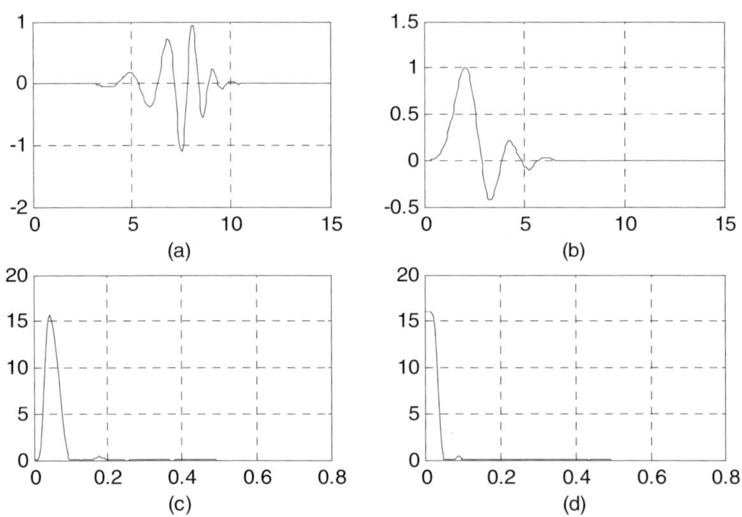

Fig. 2. Waveforms of $\psi(t)$, $\phi(t)$, $\Psi(\Omega)$ and $\Phi(\Omega)$ for Daubechies-8 period wavelet (a) $\psi(t)$; (b) $\phi(t)$; (c) $\Psi(\Omega)$; (d) $\Phi(\Omega)$

$\psi(t)$ and the scale function $\phi(t)$ are related with lowpass filter $H_0(z)$ and bandpass filter $H_1(z)$ which just constitute a dual channel filter group for multiresolution analysis. The relation between the wavelet function $\psi(t)$ and the scale function $\phi(t)$ is usually described by the dual-scale equation as follows:

$$\begin{cases} \phi(\frac{t}{2^j}) = \sqrt{2}\sum_{k=-\infty}^{+\infty} h_0(k)\phi(\frac{t}{2^{j-1}} - k) \\ \psi(\frac{t}{2^j}) = \sqrt{2}\sum_{k=-\infty}^{+\infty} h_1(k)\phi(\frac{t}{2^{j-1}} - k) \end{cases} \tag{5}$$

Where $h_0(k)$ and $h_1(k)$ are weighing coefficients separately.

For Daubechies-8 wavelet, the waveforms of $\psi(t)$, $\phi(t)$ and their Fourier transform $\Psi(\Omega)$, $\Phi(\Omega)$ are shown in Fig.2. And the weighting coefficients of Daubechies-8 mother wavelet is shown in Table 1.

Table 1. Weighting coefficients of Daubechies-8 mother wavelet

n	coefficient	n	coefficient
h_0	-0.00011748	h_8	0.12875
h_1	0.00067545	h_9	0.00047248
h_2	-0.0039174	h_{10}	-0.28402
h_3	-0.0048704	h_{11}	-0.015829
h_4	0.0087461	h_{12}	0.58535
h_5	0.13981	h_{13}	0.67563
h_6	-0.044088	h_{14}	0.31287
h_7	-0.17369	h_{15}	0.054416

2.2 Learning Algorithm

For learning of the self-adaptive WNN, a modified BP algorithm is adopted. The learning algorithm for batch processing is shown as follows.

Assume that $z^{(n)}(m)$ is the expected network output for nth input mode, and $y^n(m)$, $p^{(n)}(i)$ and $q^{(n)}(j)$ are the practical network output, hidden layer node output and wavelet layer node output for nth input mode separately, and then the error function can be defined as follows:

$$\xi = \frac{1}{2N}\sum_{n=1}^{N}\sum_{m=1}^{M}\left(y^{(n)}(m) - z^{(n)}(m)\right)^2 \tag{6}$$

From the above equation, the amendment values for network weights and wavelet parameters of self-adaptive WNN can be deduced as follow:

$$\Delta w_{im} = \frac{\partial \xi}{\partial w_{im}} = -\frac{1}{N}\sum_{n=1}^{N}\left(z^{(n)}(m) - y^{(n)}(m)\right)y^{(n)}(m)\left(1 - y^{(n)}(m)\right)p^{(n)}(i) \tag{7}$$

$$\Delta w_{ji} = \frac{\partial \xi}{\partial w_{ji}} = -\frac{1}{N}\sum_{n=1}^{N}\sum_{m=1}^{M}\left(z^{(n)}(m) - y^{(n)}(m)\right)y^{(n)}(m)\cdot \\ w_{im}p^{(n)}(i)\left(1 - p^{(n)}(i)\right)q^{(n)}(j) \tag{8}$$

$$\Delta a_j = \frac{\partial \xi}{\partial a_j} = -\frac{1}{N}\sum_{n=1}^{N}\sum_{m=1}^{M}\left(z^{(n)}(m) - y^{(n)}(m)\right)y^{(n)}(m)\left(1 - y^{(n)}(m)\right)\cdot \\ \sum_{i=1}^{I} w_{im}p^{(n)}(i)w_{ji}\sum_{k=1}^{K} x^{(n)}(k)\cdot\frac{\partial h_{a_j,b_j}}{\partial a_j} \tag{9}$$

$$\Delta b_j = \frac{\partial \xi}{\partial b_j} = -\frac{1}{N}\sum_{n=1}^{N}\sum_{m=1}^{M}\left(z^{(n)}(m) - y^{(n)}(m)\right)y^{(n)}(m)\left(1 - y^{(n)}(m)\right)\cdot \\ \sum_{i=1}^{I} w_{im}p^{(n)}(i)w_{ji}\sum_{k=1}^{K} x^{(n)}(k)\cdot\frac{\partial h_{a_j,b_j}}{\partial b_j} \tag{10}$$

2.3 Recognition and Classification Process

The block diagram of recognition and classification process of PQ disturbances based on self-adaptive WNN is shown in Fig.3. The self-adaptive WNN mainly consists of two parts: wavelet layer and ANN. Wavelet transform method may provide time information and frequency information simultaneously which are main feature types of PQ disturbances. So the recognition and classification process of PQ disturbances is as follows. Feed with PQ disturbance data, the wavelet layer preprocesses them and extracts the features of PQ disturbances over wavelet transform which are feed into the ANN at the same time. According to the disturbance database, then ANN recognizes and classifies the features extracted and finally shows the results (PQ disturbance types).

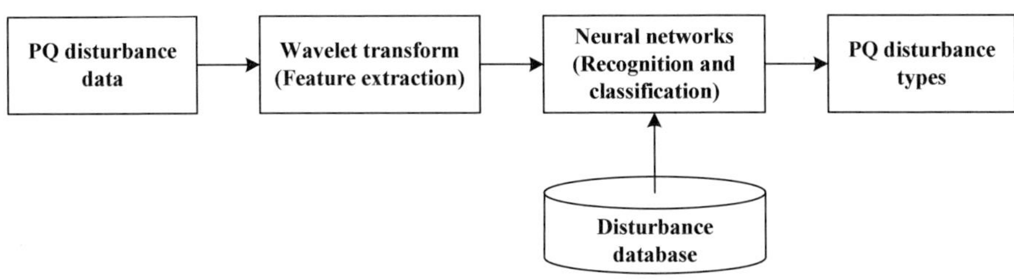

Fig. 3. Block diagram of recognition and classification process of PQ disturbances based on self-adaptive WNN

3 Training of Self-adaptive WNN

For a self-adaptive WNN, the convergence ability is very important. After the self-adaptive WNN model is constructed, training is needed to certify its performance such as convergence performance. During training process, it is an important task to collect sample data sets. Usually, the types of PQ disturbances include harmonic distortions, flickers, voltage sags, voltage swells, voltage interruptions, voltage notches, voltage impulses and voltage transients. Here, sample data sets are just PQ disturbance data sets including the above 8 PQ disturbance types. But obtaining a broad range of PQ disturbance data is very difficult because only a few PQ disturbance types can be obtained by monitoring electric power system. So Electromagnetic Transient Program (EMTP) is utilized to generate PQ disturbance data sets. Then PQ disturbance data sets generated are used to train the self-adaptive WNN over MATLAB software. The training target is defined as that the error between practical outputs and expected outputs of sample data sets must be less than the set value which is set as 0.1 here. Fig.4 shows the learning error curve. From Fig.4, it is obvious that the self-adaptive WNN has the ability of rapid convergence.

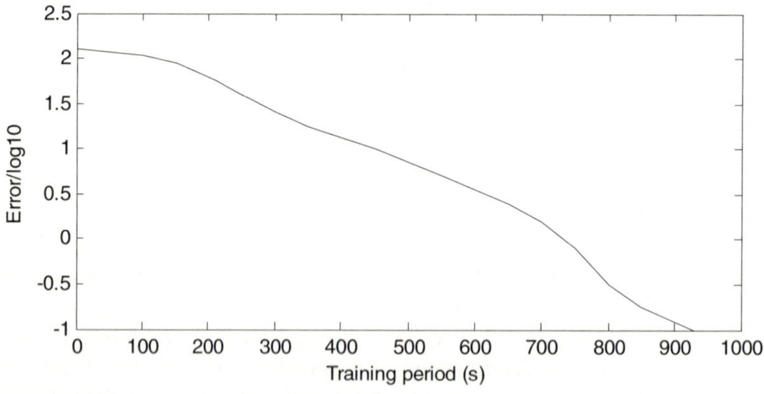

Fig. 4. Learning error curve of self-adaptive WNN

4 Test and Evaluation of Self-adaptive WNN

For trained self-adaptive WNN, it needs to be tested and evaluated by test data sets. The types of disturbances include harmonic distortions, flickers, voltage sags, voltage swells, voltage interruptions, voltage notches, voltage impulses and voltage transients. Similar to sample data sets, test data sets including 8 types of PQ disturbances are also able to be generated by EMTP. In this work, 800 test data sets are obtained, and then they are used to test and evaluate the constructed and trained self-adaptive WNN over MATLAB software. Table 2 shows the test and evaluation results. It can be seen that the constructed and trained self-adaptive WNN results in a correct identification and classification rate of 97.75% which shows that the proposed self-adaptive WNN method for automatic recognition and classification of PQ disturbances is effective, accurate and reliable.

Table 2. Test and evaluation results of self-adaptive WNN

Type of PQ disturbances	Number of test data sets	Number of sets correctly identified	Number of sets mistakenly identified	Correct identification rate (%)
Harmonic	50	50	0	100
flicker	50	50	0	100
Sag	50	49	1	98
Swell	50	48	2	96
Interruption	50	50	0	100
Notch	50	48	2	96
Impulse	50	49	1	98
Transient	50	47	3	94
Sum	400	391	9	97.75

5 Conclusion

In this paper a novel self-adaptive wavelet neural network method is proposed to automatically recognize and classify power quality disturbances. The self-adaptive wavelet neural network model constructed consists of two parts: wavelet layer and feedforward neural network. The function of the former is to extract features of power quality disturbances, and the function of the latter is to recognize and classify types of power quality disturbances. Firstly the structure of self-adaptive WNN is presented, and then the training process of self-adaptive WNN is introduced, and finally 800 test data sets of PQ disturbances are generated by EMTP to test and evaluate the performance of self-adaptive WNN over MATLAB software. The test and evaluation results demonstrate that the self-adaptive WNN method is able to recognize and classify PQ disturbances effectively, accurately and reliably.

References

1. Loredana, C., Alessandro, F., Simona, S.: A Distributed System for Electric Power Quality Measurement. IEEE Transactions on Instrumentation and Measurement 51(4) (2002) 776-781
2. Alessandro, F., Massimo, L., Simona, S.: A Calibration Procedure for a Digital Instrument for Electric Power Quality Measurement. IEEE Transactions on Instrumentation and Measurement 51(4) (2002) 716-722
3. Ghosh, A.K., Lubkeman, D.L.: The Classification of Power System Disturbance Waveforms Using A Neural Network Approach. IEEE Transactions on Power Delivery 10(1) (1995) 109-115
4. Perunicic, B., Malini, M., Wang, Z., Liu, Y.: Power Quality Disturbance Detection and Classification Using Wavelets and Artificial Neural Networks. Proceedings of the 8th ICHQP 24 (1998) 77-82

5. Santoso, S., Grady, M.W., Powers, J.E.: Power Quality Disturbance Waveform Recognition Using Wavelet-based Neural Classifier-Part 2: Application. IEEE Transactions on Power Delivery 15(1) (2000) 229-235
6. Borras, D., Castilla, M., Moreno, N.: Wavelet and Neural Structure: A New Tool for Diagnostic of Power System Disturbances. IEEE Transactions on Industry Application 37(1) (2001) 184-190
7. Szu, H.H., Telfer, B.A., Kadambe, S.: Neural Network Adaptive Wavelets for Signal Representation and Classification. Opt. Eng. 31(9) (1992) 1907-1906
8. Telfer, B.A., Szu, H.H., Dobeck, G.J.: Adaptive Wavelet Classification of Acoustic Backscatter and Imagery. Opt. Eng. 33(7) (1994) 2190-2203

Vibration Fault Diagnosis of Large Generator Sets Using Extension Neural Network-Type 1

Meng-hui Wang

Department of Electrical Engineering, National Chin-Yi Institute of Technology, 35, 215 Lane, Sec. 1, Chung Shan Road, Taiping, Taichung, Taiwan, ROC
wangmh@chinyi.ncit.edu.tw

Abstract. This paper proposes a novel neural network called Extension Neural Network-Type 1 (ENN1) for vibration fault recognition according to generator vibration characteristic spectra. The proposed ENN1 has a very simple structure and permits fast adaptive processes for new training data. Moreover, the learning speed of the proposed ENN1 is shown to be faster than the previous approaches. The proposed method has been tested on practical diagnostic records in China with rather encouraging results.

1 Introduction

A generator set fault not only damages the generator itself but also causes a break in the power system and loss of benefits. Generator fault diagnosis can produce significant cost saving by scheduling preventive maintenance and preventing extensive downtime periods caused by extensive failure [1]-[2], so that maintenance engineers can switch them safety and improve the reliability of power supplies. The possibility of incipient vibration faults is inherent due to stresses involved in the conversion of mechanical to electrical energy in generator set. A vibration fault is any form of periodic motion; its frequency can be discrete, or represented by a combination of fundamental and harmonic frequency components. Therefore, the information supplied by the frequency spectra of the vibration signals can be a valuable source to diagnose the fault condition of the generators. In the past, various vibration diagnosis tools have been proposed, including multilayer neural networks (MNN) [1], adaptive wavelets networks (AWN) [2], and fuzzy-based neural networks (FNN) [3]. However, the training data must be sufficient and compatible to ensure proper training in traditional NN, and there are time-consuming in training processes. In this paper, a novel neural network topology, ENN1 is proposed for vibration fault diagnosis in steam turbine generator sets. The proposed ENN1-based fault diagnosis method (EBFDM) permits fast adaptive processes for accessed significant and new information, and gives shorter learning times than previously approaches. Moreover, this EBFDM has shown higher accuracy, less memory consumption, and better noise rejection abilities in application.

2 The Proposed ENN1

In vibration diagnosis, the corresponding frequency spectra of the vibration signals with the fault type of the generator sets cover a range of values. Therefore, a new

J. Wang et al. (Eds.): ISNN 2006, LNCS 3972, pp. 1395 – 1401, 2006.

neural network topology, called the extension neural network (ENN1) is proposed to solve these problems in this paper.

2.1 Structure of ENN1

The schematic structure of the ENN1 is depicted in Fig. 1. There are two connection values (weights) between input nodes and output nodes; one connection represents the low bound w_{kj}^L, and the other connection represents the upper bound w_{kj}^U for this classical domain of the input features. The output layer (called the extension distance layer) is a strong lateral inhibition network, only one output node in the output layer remains active to indicate a classification of the input pattern.

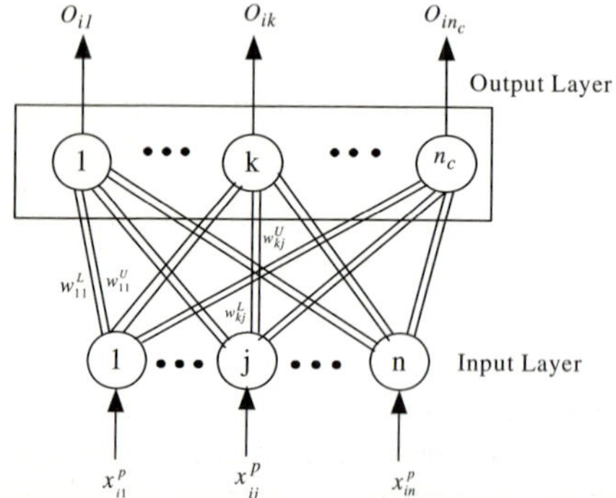

Fig. 1. The structure of extension neural network

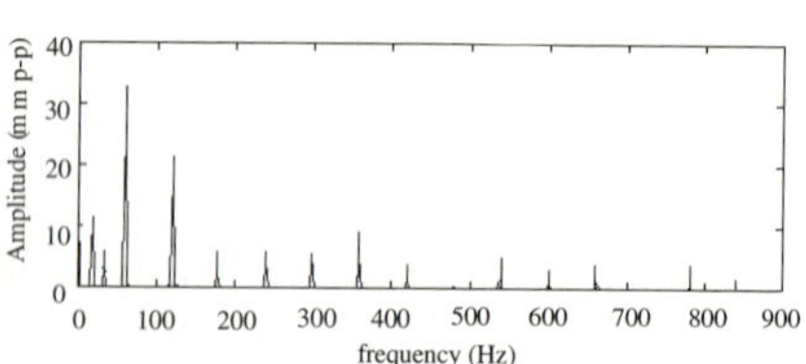

Fig. 2. The typical spectrum of vibration signals

2.2 Learning Algorithm of the ENN1

The learning of the ENN1 can be seen as supervised learning. Before the learning, several variables have to be defined. Let training pattern set be $X \equiv \{X_1, X_2, ..., X_{N_p}\}$, where N_p is the total number of training patterns. The i-th pattern is $X_i^p \equiv \{x_{i1}^p, x_{i2}^p, ..., x_{in}^p\}$, where n is the total number of the features, and the category of the i-th pattern is p. To evaluate the learning performance, the error function is defined below:

$$E_t = \frac{1}{2}\sum_{i=1}^{N_p}\sum_{j=1}^{n_c}\left(d_{ij} - o_{ij}\right)^2 \tag{1}$$

Where d_{ij} represents the desired j-th output for the i-th input pattern. The detailed supervised learning algorithm can be described as follows:

Step 1: Set the connection weights between input nodes and output nodes by the matter-element model [4]-[5]:

$$R_k = \begin{bmatrix} F_k, & c_1, V_{k1} \\ & c_2, V_{k2} \\ & \vdots \quad \vdots \\ & c_n, V_{kn} \end{bmatrix} \quad k = 1,2,\ldots,n_c \tag{2}$$

Where R_k : k-th cluster (matter-element); F_k : Name (or fault type) of the cluster R_k ; c_j : j-th characteristic (feature) of F_k ; n_c : Number of clusters, $V_{kj} = \left\langle w_{kj}^L, w_{kj}^U \right\rangle$ is the classical domain (boundary) of the k-th cluster (N_k) about the j-th feature c_j. The range of classical domains can be directly obtained from previous experience, or determined from training data as follows:

$$w_{kj}^L = Min\left\{x_{ij}^k\right\} ; \; w_{kj}^U = Max\left\{x_{ij}^k\right\} \tag{3}$$

It should be noted that the initial weights of the proposed ENN1 are directly determined from training data, which is different from the traditional neural networks.

Step 2: Calculate the initial cluster center of every cluster

$$Z_k = \left\{z_{k1}, z_{k2}, \ldots, z_{kn}\right\} \tag{4}$$

$$z_{kj} = (w_{kj}^L + w_{kj}^U) / 2 \quad \text{for } k=1,2\ldots n_c \, ; j=1,2\ldots n \tag{5}$$

Step 3: Read the i-th training pattern and its cluster number p

$$X_i^p = \left\{x_{i1}^p, x_{i2}^P, \ldots x_{in}^P\right\} \tag{6}$$

Step 4: Use the proposed extension distance (ED) to calculate the distance between the training pattern X_i^p and the k-th cluster as follows:

$$ED_{ik} = \sum_{j=1}^{n} \left(\frac{\left| x_{ij}^p - z_{kj} \right| - (w_{kj}^U - w_{kj}^L) / 2}{\left| (w_{kj}^U - w_{kj}^L) / 2 \right|} + 1 \right) \text{for } k=1,2\ldots n_c \tag{7}$$

The proposed extension distance is a new distance measurement; the proposed ED can describe the distance between the x and a range $\left\langle w^L, w^U \right\rangle$, which is different from the traditional Euclidean distance.

Step 5: Find the k^*, such that $ED_{ik^*} = Min\{ED_{ik}\}$. If $k^* = p$ then go to Step 7; otherwise Step 6.

Step 6: Update the weights of the p-th and the k^*-th clusters as follows:

$$\begin{cases} w_{pj}^{L(new)} = w_{pj}^{L(old)} + \eta(x_{ij}^{p} - z_{pj}^{old}) \\ w_{pj}^{U(new)} = w_{pj}^{U(old)} + \eta(x_{ij}^{p} - z_{pj}^{old}) \end{cases} \tag{8}$$

$$\begin{cases} w_{k^*j}^{L(new)} = w_{k^*j}^{L(old)} - \eta(x_{ij}^{p} - z_{k^*j}^{old}) \\ w_{k^*j}^{U(new)} = w_{k^*j}^{U(old)} - \eta(x_{ij}^{p} - z_{k^*j}^{old}) \end{cases} \tag{9}$$

Where η is a learning rate, it set to 0.9 based on the learning experience in this paper. From this step, we can clearly see that the learning process is only to adjust the weights of the p-th and the k^*-th clusters. Therefore, the proposed ENN1 has a speed advantage over other traditional neural networks, and can quickly adapt to new and important information.

Step 7: Repeat Step 2 to Step 6, if all patterns have been classified, then a learning epoch is finished.

Step 8: Stop, if the clustering process has converged, or the total error has arrived at a preset value, otherwise, return to Step 3.

It should be noted that the proposed ENN1 can take human expertise before the learning, because the initial weights can directly take according to previous experience, or determined from training data; and it can also produce meaningful weights after the learning, because the classified boundaries of the features are clearly determined by both the upper and lower bounds of weights.

3 The Proposed ENN1-Based Vibration Fault Diagnosis Method

3.1 Selecting of the Fault Features

The vibration frequency can be represented by a combination of fundamental and harmonic frequency components. A typical spectrum of vibration signal is shown in Fig. 2, the diagnostic information can be supplied by the spectrum of the vibration signal. In agreement with past studies [1]-[3], the six typical values (amplitude of <0.4f, 0.4f~0.5f, f, 2f, 3f, and >3f) are selected for inputs of the proposed ENN1-based diagnosis system, where f is the rotational frequency of the generator sets. There are four typical vibration faults with the features of the vibration signal. The four typical vibration faults in the tested records include oil-resonance, Imbalance misalignment and no faults.

3.2 ENN1-Based Fault Diagnosis Method (EBFDM)

Generally, fault records are rare at general power companies, so the historical database is very limited. The traditional MNN-based methods are not appropriate for application in this field due to needing a large amount of input data. In contrast, the proposed ENN1-based fault diagnosis method is only to find the range of fault features; it can be directly obtained from previous experience, or determined from limited training data. To demonstrate the effectiveness of the proposed fault diagnosis method, twenty sets of field-test data from steam-turbine generator sets in China were tested [1]-[3]. The proposed EBFDM has been successfully implemented using PC based software for vibration fault diagnosis of generator sets. Using the proposed EBFDM can be simply described as follows:

Step 1: Input the training patterns,
Step 2: Build the matter-elements model of every fault type as Eq.(2), and set-up the ENN1 according to the matter-elements models,
Step 3: Train the ENN1 using the proposed learning algorithm in Section 2.2,
Step 4: Go to step 3, if training process is not finished; otherwise go to Step 5,
Step 5: Save the weight vector of the trained ENN1,
Step 6: Use the trained ENN1 to diagnose fault.

Basically, the learning time of the EBFDM is shorter than the traditional neural-based methods due to the fact that knowledge from experience according to the matter-elements of every fault type is incorporated into the training process, and the learning process is only to adjust the weights of the changed clusters.

4 Test Results and Discuss

Table 1 compares the learning performance of the proposed ENN1 with other existing methods. The results show that the proposed ENN1 has the shortest training time and highest accuracy of all methods. Although the diagnosis system is trained off-line, the training time is not a critical point to be evaluated. It is an index, however, implying in some degree the efficiency of the algorithm developed. As exhibited in Table 1, it should be noted that the structure of the proposed ENN1 is the simpler than the other neural networks, 10 nodes and 48 connections are needed. Moreover, the proposed ENN1-based method also permits fast adaptive processes for a large amount of training data or new information, because the learning of ENN1 is only to tuning low bounds and upper bounds of excited connections, which is rather beneficial when implementing the ENN1-based diagnosis method in a Microcomputer for a real-time fault detecting device or a portable instrument.

The input data of a vibration diagnosis system will unavoidably contain some amounts of errors or uncertainties. To take into account the errors and uncertainties, a set of the testing data are created by adding *±5%* to *±30%* random uniformly

Table 1. Comparison of the classification performance of various methods

Compare	Diagnosis methods					
Item	MNN[1]	IBP[1]	AWN[2]	FNN[3]	ENN1	SOM[6]
Structure	6-16-4	6-16-4	6-16-4	6-16-4	6-4	6-4
No. of connections	160	160	160	160	48	24
Learning times (epochs)	2600	1000	1000	1000	2	100
Learning Error E_t	0.03	0.03	0.004	0.001	0.0	0.0

Table 2. Diagnosis performances of different methods with different percentages of errors added

Noise	Accuracy of different diagnosis methods						
Percentage (%)	MNN[1]	IBP[1]	AWN[2]	FNN[3]	ENN1	SOM	K-mean
± 0%	100%	100%	100%	100%	100%	100%	82%
± 5%	100%	100%	100%	87%	100%	90%	82%
± 10%	93%	93%	93%	81%	100%	90%	77%
± 15%	93%	93%	93%	73%	93%	87.%	73%
± 20%	92%	93%	93%	72%	93%	87.%	73%
± 25%	86%	87%	87%	65%	92%	82.%	72%
± 30%	78%	80%	85%	60%	87%	82.%	72%

distributed samples to the training data to appraise the fault-tolerant abilities of the proposed diagnosis method. The test results using different amounts of errors added are given in Table 2 for the seven diagnosis methods. The proposed method has a significantly higher diagnosis accuracy of 100% with ± 10% errors added. Moreover, the proposed method shows good tolerance to added errors, and has a high accuracy of 87% in extreme error of $\pm 30\%$.

5 Conclusions

This paper presents a novel vibration fault diagnosis method based on the ENN1 for steam turbine generators. Compared with other existing methods, the structure of the proposed ENN1 is simpler and the learning time is the faster than other methods. Moreover, the proposed ENN1-based diagnosis method also permits fast adaptive processes for the new data, which only tune the boundaries of classified features or add a new neural node. It is feasible to implement the proposed method in a Micro-computer for portable fault detecting devices. We hope this paper will lead to further investigation for industrial applications.

Acknowledgments

We gratefully acknowledge the support of the National Science Council, Taiwan, ROC, for financial support under grant NSC-92-2213-E-167-015.

References

1. Li, H., Sun, C.X., Liao, R.J., Chen, W.G., Hu, X.S.: Improved BP Algorithm in Vibration Fault Diagnosis of Steam Turbine Generator Set. Journal of Chongqing University. (1999)47-52
2. Li, H., Sun, C.X., Hu, X.S., Yue, G., Tang, N.F., and Wang, K.: Study of Method on Adaptive Wavelets Network for Vibration Fault Diagnosis of Steam Turbine Generation Set. Transactions of China Electrotechnics Society (2000) 52-57

3. Li, H., Sun, C.X., Hu, X.S., Yue, G., Wang, K.: The Fuzzy Inputting and Outputting Method in Vibration Fault Diagnosis of Steam Turbine Generator Set. Journal of Chongqing University (1999) 36-41
4. Cai, W.: The Extension Set and Incompatibility Problem. Journal of Scientific Exploration. (1983) 81-93
5. Wang, M.H.: A Novel Extension Neural Network for Power Transformer Fault Diagnosis. IEE Proceedings - Generation, Transmission and Distribution. 150 (6) (2003) 669-685
6. Kohonen, T.: Self-Organization and Associative memory. Vol. 3, Springer- Verlag, Berlin Heidelberg New York(1988)

Fault Data Compression of Power System with Wavelet Neural Network Based on Wavelet Entropy

Zhigang Liu and Dabo Zhang

Institute of Electrification & Automation, Southwest Jiaotong University, Chengdu, Sichuan 610031, China
zhigang_liu457@sohu.com

Abstract. Through the analysis of function approximation with wavelet transformation, an adaptive wavelet neural network is introduced in the paper, which is applied in data compression of fault data in power system. In addition, the wavelet entropy is adopted to choose the hidden nodes in the wavelet neural network. The learning algorithm of the wavelet neural network based on wavelet entropy is proposed and discussed for data compression of fault data in power system. The simulation results show that it is feasible and valid in the end.

1 Introduction

The notion of wavelet neural network was proposed by Q. H. Zhang and A. Benveniste in 1992 as an alternative to feedforward neural network for approximating arbitrary nonlinear functions[1]. Y. C. Pati and P. S. Krishnaprasad had presented the analysis and synthesis of feedforward neural network using discrete affine wavelet transformations [2]. The idea of combining both wavelet and neural network has been discussed in various works [3-8]. From activation function in neural networks, wavelet neural networks can be constructed through continuous wavelet transformation, orthogonal wavelet transformation and wavelet frame. In these wavelet neural networks, the activation functions are scale or wavelet functions generally.

It is known that there is a mass of transient messages in fault signals of power system. These messages have probably different each other, which depend on the fault type, the fault locations and other conditions in power system. That is to say, the component and content of transient messages are changeful. Hence, it is not enough to approximate the fault signals only with scale functions or wavelet functions in wavelet neural networks. Based on the model of wavelet neural network proposed in [8], an adaptive wavelet neural network is applied in fault data compression of power system in the paper.

There are different methods about the initialization of parameters and number choice of hidden layer nodes in different wavelet neural networks. We mainly discuss the number choice of hidden layer nodes in wavelet neural network here. If the scale parameter M has been confirmed, the number of hidden layer nodes can be determined through $2^M + p, p \geq 1$, p is a small integer in [4]. In discrete affine wavelet neural network, the number of hidden layer nodes is equal to group number of parameter (m, n) in [2]. Thinking about the importance of number choice of hidden layer nodes, and decreasing the complexity and computation of choice algorithm, we

J. Wang et al. (Eds.): ISNN 2006, LNCS 3972, pp. 1402 – 1408, 2006.

adopt wavelet entropy to decide the node number of hidden layers in wavelet neural network for the fault data compression of power system in the paper.

2 Wavelet Neural Network Model

The orthogonal wavelets are functions whose dilations and translations form an orthonormal basis of $L^2(R)$, the space of all square integrable functions on *R*. There exists a function $\psi(t)$ such that

$$\psi_{j,k}(t) = 2^{j/2}\psi(2^j t - k) \tag{1}$$

Forming an orthonormal basis of $L^2(R)$. Therefore, the wavelet basis induces an orthogonal decomposition of $L^2(R)$. The wavelet $\psi(t)$ is often generated from a companion $\varphi(t)$, known as the scale function. The dilations and translations of the scaling function induce a multi-resolution analysis of $L^2(R)$.

Two schemes for decomposing a $L^2(R)$ function $f(t)$ are listed below:

$$f(t) = \sum_{j,k\in Z} < f, \psi_{j,k} > \psi_{j,k}(t) \tag{2}$$

$$f(t) = \sum_{k\in Z} < f, \varphi_{J_0,k} > \varphi_{J_0,k}(t) + \sum_{j\geq J_0, k\in Z} < f, \psi_{j,k} > \psi_{j,k}(t) \tag{3}$$

Where $<\cdot,\cdot>$ represents the inner product and J_0 is an arbitrary integer, representing the lower resolution or scale in the decomposition. In (3), the detail part of function $f(t)$ is neglected through the function approximation. For the fault transient signals in power system, the detail part of signals have much useful message, hence it is improper to only make use of the wavelet neural network in [3] or in [4] for the processing of fault signals in power system.

Function $f(t) \in L^2(R)$, can be approximated arbitrarily closely, for some integer J_0. That is, for any $\varepsilon > 0$, there exists a J_0 sufficiently large such that

$$\left\| f(t) - \sum_{k\in Z} < f, \varphi_{J_0,k} > \varphi_{J_0,k} - \sum_{k\in Z} < f, \psi_{J_0,k} > \psi_{J_0,k} \right\| < \varepsilon \tag{4}$$

Namely

$$f(t) \approx \sum_{k\in Z} < f, \varphi_{J_0,k} > \varphi_{J_0,k} + \sum_{k\in Z} < f, \psi_{J_0,k} > \psi_{J_0,k} \tag{5}$$

The corresponding wavelet neural network can be constructed from (5), if a given set of J_0 and *K*.

$$\hat{f}(t) = \sum_{k=-K}^{K} c_k \varphi_{J_0,k}(t) + \sum_{k=-K}^{K} d_k \psi_{J_0,k}(t) \tag{6}$$

Based on (6), the wavelet neural network model is shown in Fig. 1.

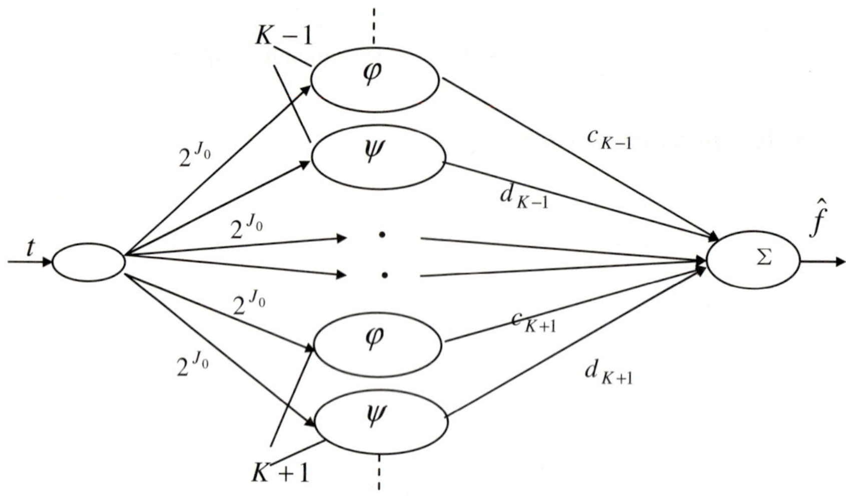

Fig. 1. Wavelet neural network in the paper

In Fig. 1, the function $f(t) \in L^2(R)$, can be approximated with both dilation functions and wavelet functions at same time, which can improve the approximation effect of signals or functions at their detail part.

3 Wavelet Entropy and Learning Algorithm

In wavelet neural network of Fig. 1, four parameters including K, J_0, c_k and d_k must be estimated or trained, where k=[-K, K]. Among the four parameters, K and J_0 can be estimated before the training of wavelet neural network. In fact, only c_k and d_k must be obtained through the training algorithm of wavelet neural network. We will discuss the choice and estimation of parameters K and J_0, firstly.

Similar to the multiple perception models, the choice of hidden layer is one of important problems in wavelet neural network. Because the scale and wavelet functions are directly considered as the neurons, the problem becomes the choice of neurons number. In fact, it corresponds to the decision of discrete translation parameter in wavelet transformation. For the stationary or slightly changeful signals, the problem is simple. Through the comparison between length of signal and support length of wavelet, the number can be obtained by virtue of even distribution.

But for the non-stationary signals or variable signals such as fault transient signals, the random and statistical knowledge must be adopted. In order to solve the problem, wavelet entropy [9,10] is introduced. In the uncertain system and procession, the information measurement can be obtained by entropy. Shannon entropy theory means: for an uncertain system, if a finite length random variable X denotes the system state characteristic, the probability of x_i is $p_i = P\{X = x_i\}$, $i = 1, \cdots, n$ and $\sum_{i=1}^{n} p_i = 1$, the information entropy of X is defined

$$H(X) = \sum_{i=1}^{n} p_i \ln(1/p_i) \tag{7}$$

Information entropy can be considered as a tool that measures the complexity of random signals. In wavelet transformation, once the mean coefficients $c_{j,k}$ of wavelet transformations are known, the energy for each time k and scale j can be calculated as

$$E_{j,k} = c_{j,k}^{\ 2} \tag{8}$$

Since the number of coefficients for each resolution level is different, the energy by calculating can be redefined, for each level, its mean value in successive time windows Δt denoted by the index i, which will now give the time evolution. Then, the mean energy is

$$E^{(i)}{}_j = \frac{1}{N} \sum_{k=i_0}^{i_0+\Delta t} E_{j,k} \tag{9}$$

where i_0 is the starting value of the time window ($i_0 = 1, 1+\Delta t, 1+2\Delta t, \ldots$) and N is the number of wavelet coefficients in the time window for each resolution level. For every time window i, the total mean energy can be evaluated as

$$F^{(i)}{}_{total} - \sum_j E_j^{\ 2} \tag{10}$$

and the probability distribution for each level can be define as

$$p^{(i)}{}_j = \frac{E_j^{(i)}}{E_{total}^{(i)}} \tag{11}$$

Wavelet entropy is defined as

$$WE^{(i)} = -\sum_j p_j^{(i)} \cdot \log_2 p_j^{(i)} \tag{12}$$

For the choice of parameter K in the proposed wavelet neural network, the calculation steps are listed below:

① Calculating the coefficients $c_{j,k}$ of wavelet transformation.
② Choosing time windows Δt and calculating the mean energy at resolution j.
③ Calculating wavelet entropy WE in different time section.
④ Deciding K in each time section according to value of wavelet entropy.
⑤ Calculating the total number of K in the whole signal time area.

For the decision of weigh value c_k and d_k, we can obtain through the learning of wavelet neural network. The algorithm below is presented:

Given training samples, the best weight value c_k and d_k can be obtained through the minimal square error:

$$(c_{-K},\cdots,c_K)=\min e_N(f,\hat{f}) \tag{13}$$

$$(d_{-K},\cdots,d_K)=\min e_N(f,\hat{f}) \tag{14}$$

Where

$$e_N(f,\hat{f})=1/N\sum_{i=1}^{N}(f(t_i)-\hat{f}(t_i))^2 \tag{15}$$

It is obvious that (13) and (14) can be solved though the partial derivative of (15):

$$\frac{\partial e_N(f,\hat{f})}{\partial c_k}=0 \tag{16}$$

$$\frac{\partial e_N(f,\hat{f})}{\partial d_k}=0 \tag{17}$$

The other steps of learning algorithm are similar to those of traditional BP algorithm in feedforward neural networks. There we pass over the content.

4 Data Compression in Power System

The principle of data compression with neural network is based on the map between input nodes and output original data, which saves the construction parameters of neural network, and realize the comeback of data though these parameters. The data compression procedure with neural network is equal to the function approximation.

Given 1000 original voltage data samples of single-phase short circuit in the end of 500kv power transmission line, the sample time is 0.05~0.1s. The voltage figure is drawn in Fig. 2(a). From the Fig. 2 (a), a great deal of transient message is found. In order to eliminate the useless transient message and keep the useful transient message, the detail of fault signal should be not neglected. We will adopt the wavelet neural network in Fig. 1 for the fault data compression. The whole procedure of data compression is put forward below:

① Choosing the base function in wavelet neural network. Considering the demand of orthogonality for activation functions in the wavelet neural network, scale functions and wavelet functions of wavelet should be orthogonal. In addition, the comparability between signal characteristic and wavelet should be thought, such as the length of support partition, the figure, the number of vanish moments of wavelets and so on. We adopt the scale function and wavelet function of DB4 wavelet as the activation functions in the wavelet neural network.

② Estimating parameters J_0 in the wavelet neural network. For parameter J_0, considering the complexity of wavelet decomposition for different resolution and the calculation of wavelet entropy, let $J_0 = 6$.

③ Estimating parameters K with the method proposed in previous section. During the computation procedure of wavelet entropy, let $\Delta t = 0.1ms$ shown in Fig. 2 (b), then the wavelet entropy values at $J_0 = 6$ resolution are 0.4101, 0.3976, 0.2186, 0.1132,0.0682 in corresponding domain. The number of parameter K is listed in Fig. 2 (c).

④ Training parameters c_k and d_k. We can obtain their values through the learning of wavelet neural network. The corresponding formulas are (13)~(17). The traditional BP algorithm in feedforward neural networks is adopted. The compressed fault voltage figure is shown in Fig. 3.

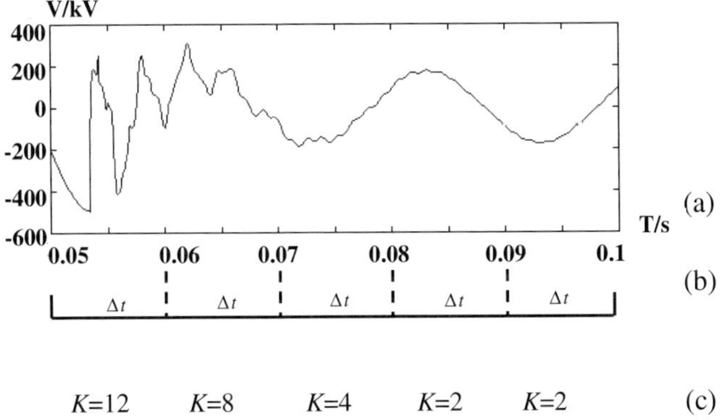

K=12 K=8 K=4 K=2 K=2 (c)

Fig. 2. (a) Fault voltage figure (b) Time Section (c) Number of parameter K

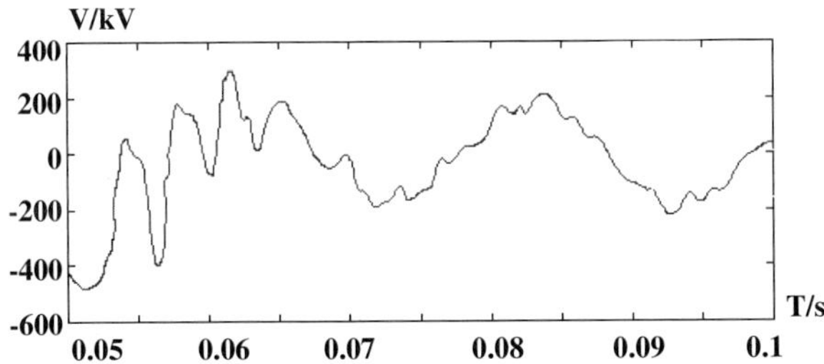

Fig. 3. Compressed fault voltage figure

The number of hidden layer nodes is 12+8+4+2+2=28. When the training demand is meet through the wavelet neural network training, the parameter K, the number of decomposition level, and parameters c_k and d_k can be saved. The construction of wavelet neural network is 1-28-1, thus the compression rate is about 10:1.

5 Conclusion

Thinking about data compression of fault signal in power system, we adopt an adaptive wavelet neural network based on the analysis of function approximation with wavelet transformation in the paper. In the wavelet neural network, scale functions and wavelet functions are adopted at the same time, which is more suitable to process the transient signals. The learning algorithm of the wavelet neural network based on wavelet entropy is proposed. The data compression of fault signal in power system is simulated and good compression effect is obtained in the end.

Acknowledgements

This work is supported by Foundation Science Research Fund of Southwest Jiaotong University (No.2004B01). The authors would like to thank Southwest Jiaotong University.

References

1. Zhang, Q.H., Benveniste, A.: Wavelet Network. IEEE Trans. on Neural Networks 3(6) (1992) 889-898
2. Pati, Y. C., Krishnaprasad, P. S.: Analysis and Synthesis of Feedforward Neural Network Using Discrete Affine Wavelet Transformations. IEEE Trans. on Neural Networks 4(1) (1993) 73-85
3. Szu, H. H., Telfer, B., Kadambe, B.: Neural Network Adaptive Wavelets for Signal Representation and Classification. Optical Engineering 31(9) (1992) 1907-1906
4. Zhang, J., Walter, G. G., Miao, Y. B.: Wavelet Neural Network for Function Learning. IEEE Trans. on Signal Processing 43(6) (1995) 1485-1497
5. Jiao, L.C., Pan, J., Fang, Y. W.: Multiwavelet Neural Network and Its Approximation Properties. IEEE Trans. on Neural Networks 12(5) (2001) 1060-1066
6. Zhang, Q.H.: Using Wavelet Networks in Nonparametric Estimation. IEEE Trans. on Neural Networks 8(2) (1997) 227-236
7. Delyon, B., Juditsky, A., Benveniste, A.: Accuracy Analysis for Wavelet Approximations. IEEE Trans. on Neural Networks 6(2) (1995) 332-348
8. Bakshi, B.R., Stepphanopoulous, B.R.: Wave-net: A Multi-resolution, Hierarchical Neural Network with Location Learning. AIChE Journal 39(1) (1993) 57-81
9. Quiroga, R.Q., Rosso, O.A., Basar, E.: Wavelet Entropy in Event-related Potential: A New Method Shows Ordering of EED Oscillations. Biological Cybernetics 84(4) (2001) 291-299
10. Sello, S.: Wavelet Entropy as a Measure of Solar Cycle Complexity. Astron. Astrophys 363(5) (2000) 311-315

Intelligent Built-in Test (BIT) for More-Electric Aircraft Power System Based on Hybrid Generalized LVQ Neural Network

Zhen Liu, Hui Lin, and Xin Luo

College of Automation, Northwestern Polytechnical University, Xi'an 710072, China
scdliu@tom.com, lhqr@nwpu.edu.cn, llxx028@tom.com

Abstract. This paper proposes a hybrid neural network model based on the Generalized Learning Vector Quantization(GLVQ) learning algorithm and applies this proposed method to the BIT system of More-Electric Aircraft Electrical Power System (MEAEPS). This paper first discusses the feasibility of application unsupervised neural networks to the BIT system and the representative Generalized LVQ (GLVQ) neural network is selected due to its good performance in clustering analysis. Next, we adopt a new form of loss factor to modify the original GLVQ algorithm in order to make it more suitable for our application. Since unsupervised networks cannot distinguish the similar classes, we add a LVQ layer to the GLVQ network to construct a hybrid neural network model. Finally, the proposed method has been applied to the intelligent BIT system of the MEAEPS, and the results show that the proposed method is promising to improve the performance of the BIT system.

1 Introduction

More-Electric Aircraft (MEA) utilizes electric power to drive aircraft subsystems currently powered by hydraulic, pneumatic or mechanical means to optimize the performance and life cycle cost of the aircraft [1]. Due to the electric power widespread using in the MEA, the reliability and fault tolerance capability of More-Electric Aircraft Electrical Power System (MEAEPS) must be significantly higher than conventional aircrafts. Furthermore, the MEA requires that more effective and efficient testing and fault diagnosis techniques be developed to improve the reliability of the EPS. An important means is to add a built-in test (BIT) technique in EPS, which is a kind of ability to identify its operation condition by itself, through the testing and diagnosis capabilities built into its own structure. BIT technique is an effective approach to improving testability and maintainability of a complex system. However, in practice, the conventional BIT technique suffers from some problems, such as notorious False Alarm (FA), Cannot Duplicate (CND) and Retest OK (RTOK), which influence the readiness of military aircraft. And with the increasing requirements of fault detection and the maintenance time, the conventional BIT technique becomes unsuitable to the MEA. To avoid these common problems, many Artificial Intelligence (AI) technologies (e.g. Expert System and Artificial Neural Networks) have been added

J. Wang et al. (Eds.): ISNN 2006, LNCS 3972, pp. 1409 – 1415, 2006.

to BIT system, which is called intelligent BIT [2]. The earlier intelligent BIT employed the Back-Propagation neural network, which is one of the most utilized supervised network models. However, in practice, not always it is easy to obtained the adequate training data from a complex system. Especially, it has the local minima problem and the learning time is long. Therefore, many unsupervised neural networks have received considerable attention recently [3][4]. These networks can perform through the adaptive learning process to classify the unlabeled data, so they have the good performance in clustering analysis. In this paper, another representative unsupervised neural network, Generalized LVQ(GLVQ) neural network, is applied in the BIT system of the MEAEPS. We present an analysis of the feasibility of using GLVQ network model for the development of intelligent BIT system, modify the original GLVQ learning algorithm and propose a hybrid neural network model. Experiments have shown that the proposed method improves the BIT performance of identifying the fault and false alarm, and it can overcome many drawbacks of the conventional BIT system.

2 The Feasibility Study

Compared with the supervised learning algorithms, the unsupervised method has the special property of categorizing the vast input data into a number of meaningful classes by means of the similarity or correlation between individual samples, and they are successful in various classification tasks involving very noisy signals, which is useful to eliminate the effect of BIT noise data. Four unsupervised network models are considered for the BIT system: Adaptive Resonance Theory II (ART-II), Kohonen's Self-Organizing Map (SOM), Generalized Learning Vector Quantization (GLVQ) and REINFORCE network. And the BIT system of MEAEPS employs six types of BIT techniques, they are analog voltage measurement, power level detector, statistical threshold, transition count, processor functional BIT routines, and activity detector. We apply the four neural networks respectively to the BIT system and evaluate the overall performance of intelligent BIT system through the simulation approach. We select 240 typical testing data and input them into each network model. The data is classified as 4 classes with 60 of each class: No Fault (NF), Armature Short Circuit (AS), Booster

	NF	AS	BO	CG
NF	57	3		
AS		51	5	4
BO		1	52	7
CG			5	55

	NF	AS	BO	CG
NF	49	5	6	
AS		45	7	8
BO		2	52	6
CG	9		8	43

	NF	AS	BO	CG
NF	59	1		
AS	3	50	7	
BO		3	51	6
CG		8		52

	NF	AS	BO	CG
NF	49	6	5	
AS	4	46	1	9
BO			52	8
CG		6	7	47

Fig. 1. Classification results using four neural networks. The matrix rows contain the desired classifications and the columns contain the actual classifications. The diagonal of the matrix contains the correct classifications for each class. These entries which are not on this diagonal are misclassifications. So the rate of correct recognition of ART II network is 89.6%, SOM's is 78.8%, GLVQ's is 88.3%, and REINFORCE's is 80.8%.

Detector Open Circuit (BO) and Converter Grounding (CG). The classification results are presented using confusion matrices as shown in figure 1.

From figure 1 we can calculate the percent correctly classification of each network. In order to test the network performance further, we use another data set with 240 testing data and input them into each network. The percent correctly classification are respectively 88.7%, 79.1%, 88.2% and 81.6%. As is seen form these results, the ART II network and the GLVQ network perform better than the other two networks, and they are the viable approaches for our study in this problem domain based on these factors which should be considered: BIT system configuration management, frequency of fault report, flexibility to future system modifications, convergence rate of networks, excessive training data, computational request, complexity, latency and reliability of neural network model.

Compared with the GLVQ network, the ART II network is very complex and its computational requirement is higher. Furthermore, the convergence rate of ART network is slower than GLVQ's. Considering these causes, we finally select the GLVQ network and apply it to the intelligent BIT system.

3 Algorithmic Defects of GLVQ and the Proposed Method

Being an unsupervised network, GLVQ cannot use the class information of the input samples, which is the inherent defect of unsupervised network. If two classes are very close to each other, they may be clustered together by this network. So we add a supervised LVQ layer [5] to the GLVQ network to avoid this problem. This method makes the class boundary be clearer than that only using the GLVQ network. Figure 2 shows the hybrid network architecture.

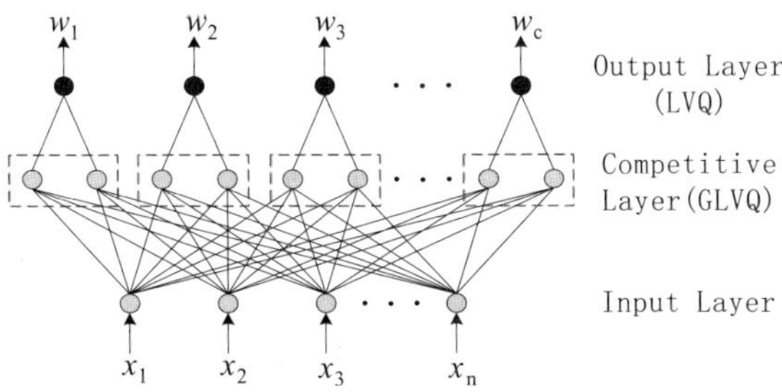

Fig. 2. The architecture of the hybrid network

GLVQ learning rules are derived from an optimization problem, and Pal et al. describe the algorithm in literature [6]. Let $\boldsymbol{X}=\{\boldsymbol{x}_1,\boldsymbol{x}_2,\cdots\boldsymbol{x}_n\}\in\Re^p$ be a set of samples and let $\boldsymbol{V}=\{\boldsymbol{v}_1,\boldsymbol{v}_2,\cdots\boldsymbol{v}_c\}$ be a set of codebook vectors. The subscript i denotes the best matching node according to the magnitude of Euclidean distance. A loss function L can be defined as follows [6]:

$$L = L(X, V) = \sum_{t=1}^{n}\sum_{r=1}^{c} g_r(\boldsymbol{x}_t, V)\|\boldsymbol{x}_t - \boldsymbol{v}_r\|^2, \tag{1}$$

where $g_r(\boldsymbol{x}, V) = 1$ if $r = i$, and $g_r(\boldsymbol{x}, V) = (\sum_{j=1}^{c}\|\boldsymbol{x} - \boldsymbol{v}_j\|^2)^{-1} = 1/D$ otherwise.

Through minimizing the objective function (1) using the gradient descent search approach, the GLVQ learning rules is developed as follows [6]:

$$\Delta\boldsymbol{v}_{i,t} = \alpha_t(\boldsymbol{x} - \boldsymbol{v}_{i,t-1})\frac{D^2 - D + \|\boldsymbol{x} - \boldsymbol{v}_{i,t-1}\|^2}{D^2} \quad \text{if } \boldsymbol{v}_j = \boldsymbol{v}_i, \tag{2}$$

$$\Delta\boldsymbol{v}_{j,t} = \alpha_t(\boldsymbol{x} - \boldsymbol{v}_{j,t-1})\frac{\|\boldsymbol{x} - \boldsymbol{v}_{i,t-1}\|^2}{D^2} \quad \text{if } \boldsymbol{v}_j \neq \boldsymbol{v}_i. \tag{3}$$

Nevertheless, Karayiannis et al. [7] and Gonzalez et al. [8] performed a different analysis and pointed out the defects of GLVQ, that is, the algorithmic sensitivity to the input data scaling and the number of prototypes being sought. Therefore, in literature [9], Karayiannis adopted the form of ratio $z_r = \|\boldsymbol{x} - \boldsymbol{v}_i\|^2 / \|\boldsymbol{x} - \boldsymbol{v}_r\|^2$ to avoid these limitations. Then the membership $u_r, (r = 1, 2, \cdots c)$, can be written as [9]

$$u_r = \begin{cases} 1 & \text{if } r = i \\ u\left(\dfrac{\|\boldsymbol{x} - \boldsymbol{v}_i\|^2}{\|\boldsymbol{x} - \boldsymbol{v}_r\|^2}\right) = u(z_r) & \text{if } r \neq i. \end{cases} \tag{4}$$

The obvious advantage of this form is that the learning rate is not affected by the input data scaling since z_r is the relative contribution of the prototype $\boldsymbol{v}_r$ to the loss function $L_{\boldsymbol{x}}$, and every non-winning prototype has the different contribution to the loss function according to the membership function $u(z_r)$.

The g_r of (1) is based on the sum of the distance of the input to the input vectors. Similarly, we adopt the total effect of all non-winning nodes to the winner node, but we use the relative contribution z_r instead of $\|\boldsymbol{x} - \boldsymbol{v}_r\|^2$. So the membership function has the form as follows:

$$u_r = \begin{cases} 1 & \text{if } r = i \\ u(z_1, z_2, \cdots z_c) & \text{if } r \neq i. \end{cases} \tag{5}$$

Then the loss function is

$$L_{\boldsymbol{x}} = \|\boldsymbol{x} - \boldsymbol{v}_i\|^2 + u\sum_{\substack{r=1 \\ r\neq i}}^{c}\|\boldsymbol{x} - \boldsymbol{v}_r\|^2. \tag{6}$$

The gradient of $L_{\boldsymbol{x}}$ with respect to $\boldsymbol{v}_i$ and $\boldsymbol{v}_j$ is

$$\nabla\boldsymbol{v}_i = (-2)(\boldsymbol{x} - \boldsymbol{v}_i)[1 + (\sum_{\substack{r=1 \\ r\neq i}}^{c}\frac{1}{z_r})(\sum_{\substack{k=1 \\ k\neq i}}^{c}\frac{\partial u}{\partial z_k}z_k)], \tag{7}$$

$$\nabla \boldsymbol{v}_j = (-2)(\boldsymbol{x} - \boldsymbol{v}_j)(u - z_j^{\ 2} \frac{\partial u}{\partial z_j} \sum_{\substack{r=1 \\ r \neq i}}^{c} \frac{1}{z_r}). \tag{8}$$

Let $S = \sum_{r=1, r \neq i}^{c} \frac{1}{z_r} > c - 1$ and $u = 1/(S+1) < 1/c$. Substituting S and u in (7) and (8) yields

$$\nabla \boldsymbol{v}_i = (-2)(\boldsymbol{x} - \boldsymbol{v}_i)[1 + \frac{S^2}{(S+1)^2}] \quad \mathit{if}\ \boldsymbol{v}_j = \boldsymbol{v}_i, \tag{9}$$

$$\nabla \boldsymbol{v}_j = (-2)(\boldsymbol{x} - \boldsymbol{v}_j)\frac{1}{(S+1)^2} \quad \mathit{if}\ \boldsymbol{v}_j \neq \boldsymbol{v}_i. \tag{10}$$

(9) and (10) are the modified learning rules, which eliminate these limitations mentioned earlier. So the whole learning algorithm can be described as follows:

Step 1: Given a training sample set $\boldsymbol{X}$ and target vector set $\boldsymbol{T}$. Initialize weight vector set $\boldsymbol{V}$. Choose the suitable values for the maximum iteration number N_1, N_2, the minimum error bound ε, and the learning rate α_0

Step 2: Unsupervised learning phase.

Step 2.1: For each training sample, find the corresponding winner node, update the winner by (9) and the non-winner by (10).

Step 2.2: Compare the total error E with ε. If $E < \varepsilon$ or the iteration number growing to N_1, go to step 3. Otherwise, change the learning rate and go to step 2.1.

Step 3: Relabel the weight vectors, and get a relabeling matrix $\boldsymbol{L}$.

Step 4: Supervised learning phase.

Step 4.1: For each training sample, find the corresponding winner node.

Step 4.2: If the winner represents the correct class, update its weight by (9) and others by (10). Otherwise, find the node that represents the correct class and update its weight by (9) and others by (10).

Step 4.3: Repeat from step 4.1 to step 4.2 until the status of cluster is stable.

4 Application of the Hybrid Network to Intelligent BIT System

The electrical system of MEA consists of generating system and power distribution system, which can supply 270 VDC, 28 VDC and 115 VAC/400Hz. The generating system employs switched reluctance starter/generator with the 270VDC output voltage, and it is the primary power generation.

According to BIT design rules and the characteristics of the electrical system architecture, the following Line Replaceable Units (LRUs) are tested by intelligent BIT system: Switched Reluctance Generator, Generator Control Unit, Power Control Unit, Power Interface I/O, and Power Supply Unit. In order to test the performance of the BIT system, we select six typical electrical faults. In our experiment, each fault mode contains 50 samples, and each input sample contains 6 items: the voltage of the regulating point (U_{por}), the current of the regulating point(I_{por}), the ripple of output voltage(U_r), and the excitation current of each phase(I_a, I_b, I_c). The normalized samples are shown in table 1.

Table 1. Normalized fault samples of BIT system(partial)

Fault Mode	U_{por}	I_{por}	U_r	I_a	I_b	I_c
No fault	0.904	0.501	0.078	0.487	0.474	0.480
Armature open	0.907	0.502	0.151	0.000	0.733	0.738
Armature short to ground	0.605	0.336	0.386	0.000	1.000	0.996
Converter short to ground	0.152	0.085	0.032	0.978	0.953	0.978
Power output short	0.188	1.000	1.000	0.243	0.220	0.249
Detecting circuit open	1.000	0.555	0.105	1.000	0.953	1.000
Booster inner fault	0.962	0.552	0.139	0.805	0.802	0.790

The output layer (LVQ layer) contains seven cells in order that it can distinguish seven different fault modes. Since the transit of dynamic state to steady state when one fault occurred, each fault mode is partitioned into two subclasses. Thus there should be thirteen cells in competitive layer (GLVQ layer). We input these training samples to this network with the initialize parameter $\alpha_0 = 0.6$, $\varepsilon = 0.0001$, and $N_1 = N_2 = 500$. After the training process finishing, we get the final weight matrix $\boldsymbol{V}$ and the relabeling matrix $\boldsymbol{L}$. So the output of this network can be computed by (11) as follows:

$$\boldsymbol{P} = \boldsymbol{L} \times compet(\boldsymbol{V} \times \boldsymbol{X}) . \tag{11}$$

In order to test the performance of this network, we first input the original training samples, and it can distinguish the all fault modes correctly. Then we input respectively several sets of testing data each containing 210 testing data into four neural networks: BP network, LVQ network, modified GLVQ network, and hybrid network. The rates of fault recognition are given as shown in table 2.

Table 2. The rates of correct fault recognition

	The rates of correct fault recognition			
	BP	LVQ	Modified GLVQ	Hybrid Network
Best	81.2%	89.6%	93.7%	100%
Average	68.5%	82.8%	89.5%	97.5%
Worst	53.7%	67.3%	83.7%	91.8%

5 Conclusions

This paper proposes a hybrid neural network model based on the GLVQ learning algorithm and applies this proposed method to the BIT system of the More-Electric Aircraft Electrical Power System. This hybrid network adopts the modified GLVQ algorithm and adds a supervised LVQ layer to the GLVQ network, which makes the boundaries among the fault classes more discriminative than using the GLVQ network alone. Experiments have shown that the hybrid network improves the performance of BIT, and overcomes many drawbacks of conventional BIT system.

Acknowledgement

This work is partially supported by Aeronautical Science Foundation of China under Grant No.04F53036.

References

1. Maldonado, M.A., Korba, G.J.: Power Management and Distribution System for a More Electric Aircraft (MADMEL). IEEE AES Magazine 14(12) (1999) 3-8
2. Richards, D.W.: Smart BIT: A Plan for Intelligent Built-In Test. IEEE AES Magazine 4(1) (1989) 26-29
3. Xu, Y.C., Wen, X.S., Yi, X.S., Tao, L.M.: New ART-2A Unsupervised Clustering Algorithm and Its Application on BIT Fault Diagnosis. Journal of Vibration Engineering 5(2) China (2002) 167-172
4. Wu, H.Q., Liu, Y., Ding, Y.L., Zhang X.W.: Application Study of SOM Artificial Neural Net in Airliner Fault Diagnosis. Journal of Nanjing University of Aeronautics & Astronautics 34(1) (2002) 31-34
5. Kohonen, T.: Self-Organization and Associative Memory. 3rd edn. Springer-Verlag, Berlin Heidelberg Germany (1989)
6. Pal, N.R., Bezdek, J.C., Tsao, E.C.-K.: Generalized Clustering Networks and Kohonen's Self-organizing Scheme. IEEE Transactions on Neural Networks 4(4) (1993) 549-557
7. Karayiannis, N.B., Pai, Pin-I: Fuzzy Algorithm for Learning Vector Quantization. IEEE Transactions on Neural Networks 7(5) (1996) 1196-1211
8. Gonzalez, A.I., Grana, M., Anjou, A.D': An Analysis of the GLVQ Algorithm. IEEE Transactions on Neural Networks 6(4) (1995) 1012-1016
9. Karayiannis, N.B.: A Methodology for Constructing Fuzzy Algorithms for Learning Vector Quantization. IEEE Transactions on Neural Networks 8(3) (1997) 505-518

Low Voltage Risk Assessment in Power System Using Neural Network Ensemble

Wei-Hua Chen, Quan-Yuan Jiang, and Yi-Jia Cao

College of Electrical Engineering, Zhejiang University, Hangzhou 310027, China
cwhwc@163.com

Abstract. Static voltage security is one of the important items of power system security. This paper provides an approach to calculate risk of low voltage in power system using neural network ensemble. Risk is defined as a condition under which there is a possibility of an adverse deviation from a desired outcome that is expected or hoped for. Risk index is used as an indicator of the low voltage security. It is calculated as the product of the probability of contingency and the impact of low voltage. Neural network ensemble (NNE) is used for the low voltage risk assessment to get the desired speed, accuracy and efficiency. The New England 39-bus test system is used as an example to demonstrate the efficiency of the proposed algorithm.

1 Introduction

In the new open access market environment, the operating conditions of power system are much closer to security boundaries. Moreover, operators do not know how safe the current system's operating conditions are and how the security changes as the operating conditions are relieved or stressed. The important of adopting efficient assessment approach has been verified by several major blackouts [1].

The deterministic approach [2] is prevailing when power system is in regulated monopoly structure. Yet there has been an enormous price to pay for using this approach due to the emphasis of the most severe and credible event. To satisfy the demand of power market, risk assessment approach is used to power system [3-4]. Risk index is the indicator of system security.

In power system risk assessment, the relationship between system status and risk index is nonlinear. So, neural network is often used to solve this problem. However, the generalization ability of single neural network is always not powerful in low voltage risk assessment [5]. In this problem, neural network ensemble is a successful alternative [6].

Based on risk theory and neural network ensemble, this paper presents a novel approach to assess the low voltage risk. Firstly, the probability model of system contingency and the impact model of low voltage are built. Based on these models, corresponding risk index is calculated. Then, the neural network ensemble system is formed. It is used for fast risk recognition and assessment. The risk index supports operators to make decision. The New England 39-bus test system is used as an example to demonstrate the efficiency of the proposed approach.

J. Wang et al. (Eds.): ISNN 2006, LNCS 3972, pp. 1416–1421, 2006.

2 Low Voltage Risk Assessment

It is defined that risk is a condition under which there is a possibility of an adverse deviation from a desired outcome that is expected or hoped for. Risk index can quantitatively capture the factors that determine security level: likelihood and severity of events.

Low voltage risk reflects the probability and impact of the low bus voltage induced by system contingencies. The probability model of system contingency and the impact model of low voltage are the basis of system low voltage risk assessment.

2.1 Probability Model

From the statistical material it can be seen that the fault process on a component is a homogeneous Possion process. The failure rate of component is assumed as λ_i. Then the fault probability of the component $\Pr(F_i)$ in unit time is

$$\Pr(F_i) = 1 - e^{-\lambda_i}, \qquad i = 1, \cdots, N_c \tag{1}$$

where F_i is the failure of component i, N_c is the number of critical components.

Power system contingency E_i is assumed to be caused by the failure of one component. Since $F_1, F_2, \cdots, F_{N_c}$ are independent of each other, then

$$\begin{aligned} \Pr(E_i) &= \Pr(\overline{F}_i \cap \overline{F}_2 \cdots \overline{F}_{i-1} \cap F_i \cap \overline{F}_{i+1} \cap \overline{F}_{N_c}) \\ &= \Pr(F_i) \prod_{j \neq i} \Pr(\overline{F}_j) = (1 - e^{-\lambda_i}) e^{-\sum_{j \neq i} \lambda_j} \end{aligned} \tag{2}$$

The equation (2) describes the probability model of power system contingency.

2.2 Impact Model

In this paper, severity function of post-contingency bus low voltage is used to reflect the impact of contingency. The low voltage severity function of bus j after contingency i is

$$I_{ij} = \frac{W_{V,j}}{N} \left| \frac{V_{ij} - V_j^{sche}}{\Delta V_{j,\lim}} \right|^N, \qquad j = 1, \cdots, N_b \tag{3}$$

where I_{ij} is the low voltage risk of bus j, $W_{V,j}$ is the weight factor for low voltage risk of bus j, V_{ij} is the bus j voltage magnitude, V_j^{sche} is the scheduled bus j voltage magnitude, $\Delta V_{j,\lim}$ is the bus j voltage variance limit, N_b is the number of bus, N is 1 in general.

The low voltage severity function of system after contingency i is

$$I_i = \sum_{j=1}^{N_b} I_{ij} \tag{4}$$

Equation (3) and (4) describe the impact model of power system contingency from the views of local and whole system respectively.

2.3 Low Voltage Risk Index

Low voltage risk index is used to quantify system voltage security level. It can provide evidences for operates to make decision. Low voltage risk index can be calculated as

$$R(V \mid X) = \sum_{i=1}^{N_c} \sum_{j=1}^{N_b} \Pr(E_i \mid X) \times \Pr(V_{ij} \mid E_i) \times I_{ij}(V_{ij}) \tag{5}$$

where X is system status.

3 Neural Network Ensemble

Theory and application have clearly demonstrated that the generalization ability of neural network ensemble improves its generalization ability at expense of little calculation. In this paper, neural network ensemble system is built for fast low voltage risk recognition and assessment.

The adopted architecture of neural network ensemble comprises the following subsystems: input subsystem, neural networks subsystem, combination subsystem, and output subsystem. The relationship of these subsystems is described as Fig.1.

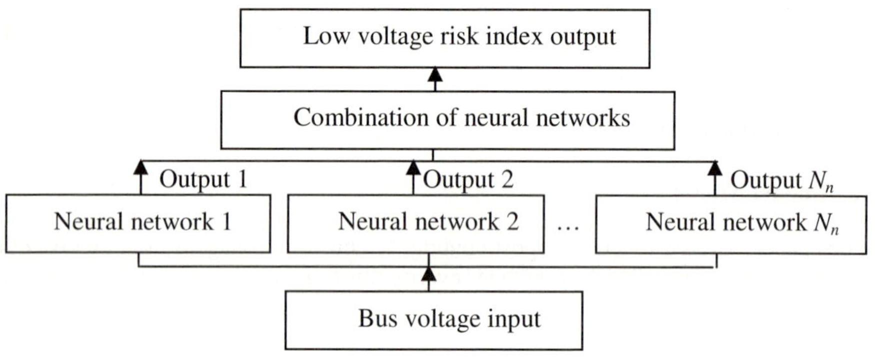

Fig. 1. Structure of neural network ensemble

The dimension of input vector $\boldsymbol{R}$ is equal to the member N_b of the system bus.

Neural networks subsystem is made of some BP networks. The number N_n of BP networks depends on the complexity of the problem. BP network is a multi-layered neural network with back propagation algorithm. To improve the generalization ability, different architectures of BP networks are selected.

In combination subsystem, the results of individual neural networks are combined by some methods. The common methods are majority voting method, simple averaging method, weighting averaging method and nonlinear method. Weighting averaging method is adopted in this paper.

Power system low voltage risk index is obtained in output subsystem. It provides the operators the intuitive information of the system security.

4 Case Study

In order to demonstrate the proposed approach, a case study is given based on the New England 39-bus test system [7].

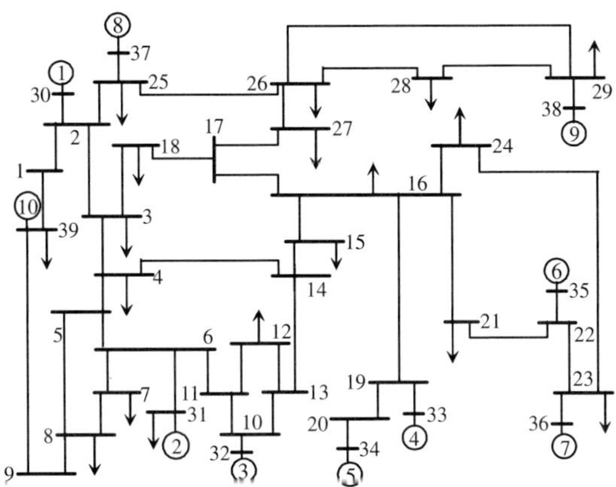

Fig. 2. New England 39-bus test system

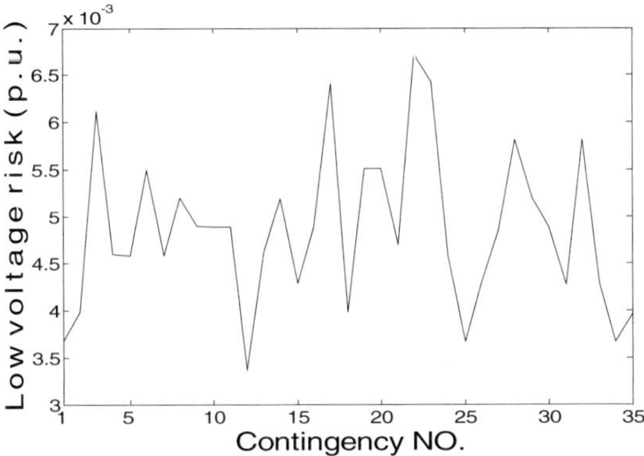

Fig. 3. Low voltage risk of *N*-1 line outage contingency

The system low voltage risk index is evaluated by N-1 static contingency (line outage) analysis. Assign all weights as 1, bus voltage magnitude limits as 1.0 p.u., voltage variance 0.075 p.u.. The system low voltage risk index responding to every line outage is described as Fig.3.

The dimension of input vector of the neural network ensemble is 39. It is equal to the number of system bus. The neural network ensemble is made of 6 different BP neural networks. Weighting averaging method is used for combination of the outputs of neural networks. The output of the neural network ensemble is power system low voltage risk index. Its dimension is one.

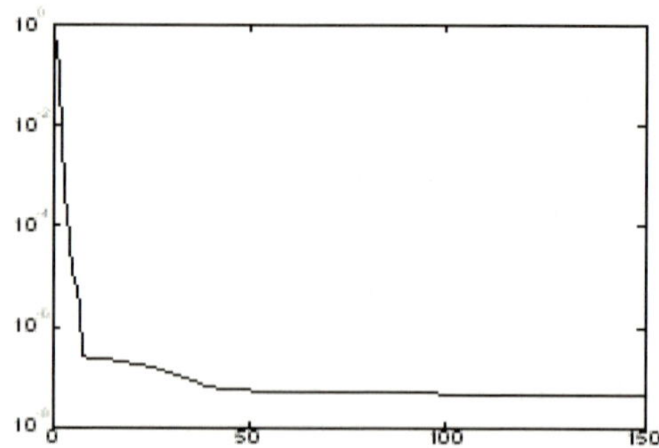

Fig. 4. Training of BP neural network

The system bus voltages and corresponding low voltage risk index after a contingency are selected as a group of data. There are 30 groups of data in the data set. 60 percent data of the set are used as training set. 30 percent data of the set are used as test set. Every neural network is trained and tested respectively. One training course is described as Fig.4.

Another 5 groups of data are used as test set for the neural network ensemble. A single BP network is also trained and tested by the same set. The results are given in Table 1.

Table 1. Test compare of NNE and BP network

NO.	Desired output(10^{-3})	NNE			Single BP network		
		Output(10^{-3})	Error		Output(10^{-3})	Error	
			p.u.(10^{-6})	%		p.u. (10^{-6})	%
1	3.676	3.674	1.249	0.034	3.683	7.131	0.194
2	3.981	3.979	1.035	0.026	3.973	7.086	0.178
3	6.118	6.121	2.814	0.046	6.129	11.318	0.185
4	4.592	4.593	1.331	0.029	4.583	8.862	0.193
5	4.586	4.588	2.384	0.052	4.593	7.383	0.161

It is clear that the generalization ability of neural network is superior to the single neural network. More analysis of error is given in Table 2.

Table 2. Error compare of NNE and BP network

	NNE		Single BP network	
	p.u. (10^{-6})	%	p.u. (10^{-6})	%
Maximal error	2.814	0.052	11.318	0.194
Minimal error	1.035	0.026	7.086	0.161
Average error	1.763	0.037	8.356	0.180

It is clear that every type of error of the neural network ensemble is less than that of the single neural network. Above all, the neural network ensemble is of the better generalization ability.

5 Conclusion

This paper provides a novel approach for risk assessment of power system low voltage using neural network ensemble. This approach uses risk index to quantitatively capture the likelihood of contingency and the impact of the low voltage. It is more adaptive for power market compared to traditional approach. This approach uses neural network ensemble to realize fast recognition and assessment of system low voltage risk. The virtues of this approach are less error and better generalization ability. The risk index supports operators to make decision. The New England 39-bus test system is used as an example to demonstrate the efficiency of the proposed approach.

Acknowledgement

This work is supported by the Special Fund for National Basic Research Program of China (2004CB217902), and by the Fund of Education Department of Zhejiang (20050908).

References

1. Qiang, W., Popovic, D. H., Hill, D.J.: Voltage Security Enhancement via Coordinated Control. IEEE Trans. on Power Systems 1 (2001) 127-135
2. Capitanescu, F., Van, Cutsem, T.: Unified Sensitivity Analysis of Unstable or Low Voltages Caused by Load Increases or Contigencies. IEEE Trans. on Power Systems 1 (2005) 321-329
3. McCalley, J., Found, A., Vittal, V.: A Risk-based Security Index for Determining Operating Limits in Stability-limited Electric Power System. IEEE Trans. on Power Systems 3 (1997) 1210-1220
4. Ming, Ni and McCalley, J.D.: Online Risk-Based Security Assessment. IEEE Transactions on Power Systems 1 (2003) 258-265
5. Nicolas, G. P., Cesar, H. M., Domingo, O. B.: Cooperative Coevolution of Artificial Neural Network Ensembles for Pattern Classification. IEEE Transactions on Evolutionary Computation 3 (2005) 271-302
6. Hansen, L.K., Salamon, P.: Neural Network Ensembles. IEEE Transactions on Pattern Analysis and Machine Intelligence 10 (1990) 993-1001.
7. Pai, M. A.: Energy Function Analysis for Power System Stability. Kluwer Academic, MA: Norwell Press (1989) 222-227

Risk Assessment of Cascading Outages in Power Systems Using Fuzzy Neural Network

Wei-Hua Chen, Quan-Yuan Jiang, Zhi-Yong Wang, and Yi-Jia Cao

College of Electrical Engineering, Zhejiang University, Hangzhou 310027, China
cwhwc@163.com

Abstract. This paper provides a strategy at the system level to assess and mitigate power system cascading outages considering the probabilities of hidden failure in protection system, which affect the risk of cascading outages. Some risk indices are used to assess the risk of cascading outages. The fuzzy neural network (FNN) is used to obtain the risk indices and to propose a solution that can decrease the system cascading outage risk under limited budget. The IEEE 118-bus system is used to illustrate the methodology.

1 Introduction

Cascading Outages become a major concern in the power industry recently as a result of some large-scale blackouts all over the world [1]. A cascading outage refers to a series of trips initiated by one component failure in the system. Risk assessment is an ideal method for cascading outages. It can quantitatively capture the factors that determine security level: likelihood and severity of events [2].

Many studies have been conducted on risk assessment. Energy function methods are used in reference [3]. However, it is difficult to construct an appropriate Lyapunov function and to achieve the level of accuracy desired. Reference [4] presents expert system, which relies on decision trees to assess the system stability and tend to be less robust to changes in power system state. Consequently, a more dependable and efficient approach is needed for risk assessment under a variety of time-varying network configurations and events. Using neural networks is an ideal choice for risk assessment. Neural network methods rely on reducing the computation at the expense of intensive studies. By performing training of a neural network using results obtained from history data or simulator, accuracy can be achieved within the computation and time constraints [5].

This paper provides a new strategy at the system level to assess and mitigate the cascading outages that involve protection system hidden failures. The strategy includes three parts: a) Improving assessment method. Risk assessment approach is used in this paper to overcome weakness in traditional methods. b) Valuing the risk index. Fuzzy neural network is used to improve accuracy and efficiency. c) Mitigating the risk of cascading outages. Some high-risk relays are replaced according to proposed upgrading solution based on relay sensitivities.

J. Wang et al. (Eds.): ISNN 2006, LNCS 3972, pp. 1422 – 1427, 2006.

2 Risk Assessment

Reference [6] built up hidden failure models and cascading outages model, and shows that probability P_W of hidden failures in the protective devices have a great impact on cascading outages. In this paper, the risk of cascading outages is assessed by some risk indices.

(1) Load Dropped Risk (R_{LD})

$$R_{LD} = P_{LD} \times I_{LD} = \frac{1}{N}\sum_{i=1}^{N} L(i) \times \frac{1}{P_S \times N}\sum_{i=1}^{N} P_L(i) \tag{1}$$

where P_{LD} is Load Dropped Probability, I_{LD} is Load Dropped Impact, $L(i)$ is 1 if there is the load dropped in events i, otherwise it is 0, N is the total number of event, $P_L(i)$ is load loss in events i, P_S is the system capacity, which is used to normalize load loss to account for the difference among various power systems.

(2) Generator Isolated Risk (R_{GI})

$$R_{GI} = P_{GI} \times I_{GI} = \frac{1}{N}\sum_{i=1}^{N} G(i) \times \frac{1}{P_S \times N}\sum_{i=1}^{N} P_G(i) \tag{2}$$

where $G(i)$ is 1 if there is generator isolated in events i, otherwise it is 0, $P_G(i)$ is load loss in events i.

(3) Network Split Risk (R_{NS})

$$R_{NS} = P_{NS} \times I_{NS} = \frac{1}{N}\sum_{i=1}^{N} S(i) \times \frac{1}{P_S \times N}\sum_{i=1}^{N} P_N(i) \tag{3}$$

where $S(i)$ is 1 if there is network split in events i, otherwise it is 0, $P_N(i)$ is load loss in events i.

(4) Integrated System Risk (R_{IS})

$$R_{IS} = w_{LD} \times R_{LD} + w_{GI} \times R_{GI} + w_{NS} \times R_{NS} \tag{4}$$

where $w_{LD} + w_{GI} + w_{NS} = 1$, w_{LD}, w_{GI}, w_{NS} are weighting factors.

3 Risk Assessment Using FNN

Fig.1 is a schematic diagram of the fuzzy neural network used for risk assessment of cascading outages in power systems. Two subsections are separated from the dashed line: data preprocessing subsection and the network subsection.

The parameter vector $\boldsymbol{P_W}$, $\boldsymbol{P_W} = [P_{W1}, P_{W2}, \cdots, P_{Wm}]$, where m is the number of relays, has a close relationship with cascading outages risk, so it is used as input. Application to real systems, the dimension of input is always high. To avoid "dimension exploding", fuzzy C-mean clustering approach is used to preprocess the input data. The result of preprocessing is cluster vector $\boldsymbol{x}$, $\boldsymbol{x} = [x_1, x_2, \cdots, x_n]$, where the value of n ranges from 1 to 5 according to scale of the system.

The input of layer 1 is $\boldsymbol{x} = [x_1, x_2, \cdots, x_n]$. The nodes of this layer is connected directly with x_i, and transmit $\boldsymbol{x}$ to the next layer. The number of nodes in layer 1 is $N_1 = n$.

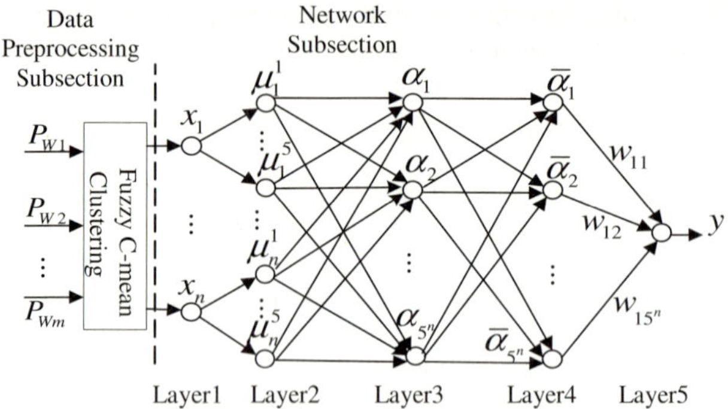

Fig. 1. Structure of fuzzy neural network

Every node in layer 2 represents a linguistic variable. There are generally five values of the linguistic variable according to the parameters of relays. They are Negative Big (NB), Negative Small (NS), Zero (ZE), Positive Small (PS), and Positive Big (PB). The nodes in this layer is used to calculate the membership function μ_i^j of input, where i = 1, 2, ⋯, n, j = 1, 2, ⋯, 5. In this paper, Gauss function is used as the membership function. The number of nodes in layer 2 is $N_2 = 5n$.

Every node in layer 3 represents a fuzzy rule. The using degree of every rule is

$$a_j = FUN(\mu_1^{i_1}, \mu_2^{i_2}, \cdots \mu_n^{i_n}) \tag{5}$$

where $i_1, i_2, \cdots, i_n \in [1, 2, \cdots, 5]$, $j = 1, 2, \cdots, 5^n$, FUN stands for multiplication. The number of nodes in layer 3 is $N_3 = 5^n$.

The layer 4 is used to normalizing by

$$\bar{\alpha}_j = \alpha_j / \sum_{i=1}^{5^n} \alpha_i \tag{6}$$

where $j = 1, 2, \cdots, 5^n$. The number of nodes in layer 4 is $N_4 = 5^n$.

The layer 5 is output layer. The output of this approach is risk index, and it is a more accurate result after defuzzyfication as following formula. The number of nodes in layer 5 is $N_5 = 1$.

$$y = \sum_{j=1}^{5^n} w_j \bar{\alpha}_j \tag{7}$$

where w_j is weighting factor.

This paper uses the sensitivity of the error to modify the weight factor and fuzzy membership parameters. Adjustment is made in successive steps in response to the training data. The process is repeated until an acceptably low error obtains in the output.

4 Case Study

In order to demonstrate the proposed strategy, a case study is given based on the IEEE 118-bus system [7]. The dimension of fuzzy neural network input is 186. Three clusters are obtained using fuzzy C-mean clustering approach. The number of nodes in layer 1 and 2 is 3 and 15. The initial membership functions of relay parameters are shown in Fig.3 (a). Here are 125 nodes in layer 3 and layer 4 respectively. There is one node in layer5.

Monte Carlo method is used for simulation. As a example, load dropped probability (P_{LD}) is shown in Fig.2. According to risk formula, the cascading outages risk on this system state is 5.04×10^{-4}.

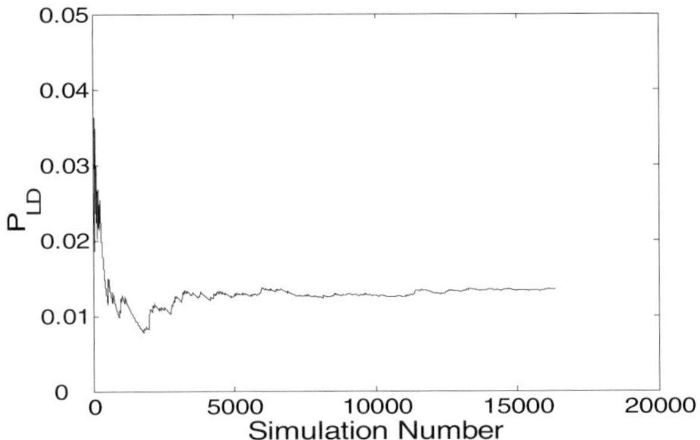

Fig. 2. Load dropped probability in one Monte Carlo simulation

Based on the relay parameters on this system state, 40% relay are selected randomly. The parameters P_W of the selected relays increase or decrease 20% randomly. Another 20 simulations are done again to obtain the training data in Table 1.

Table 1. Training data

No.	$R_{IS}(10^{-4})$	No.	$R_{IS}(10^{-4})$	No.	$R_{IS}(10^{-4})$	No.	$R_{IS}(10^{-4})$
1	5.04	6	4.75	11	4.99	16	5.83
2	5.94	7	6.44	12	6.35	17	3.11
3	4.02	8	4.34	13	4.16	18	4.58
4	3.66	9	4.12	14	4.87	19	6.88
5	4.34	10	3.02	15	2.54	20	3.54

Test scenarios are selected to be statistically independent from the training scenarios. Based on the relay parameters on this system state, 30% relay are selected random. The parameters of the selected relays increase or decrease 30% randomly. Another 10 simulations are done again to obtain the test data in Fig.3.

The test data are used to verify the fuzzy neural network. The results clearly show that the training of the fuzzy neural network has been successful for a variety of system state and disturbances for the IEEE 118-bus system, shown in Fig.3.

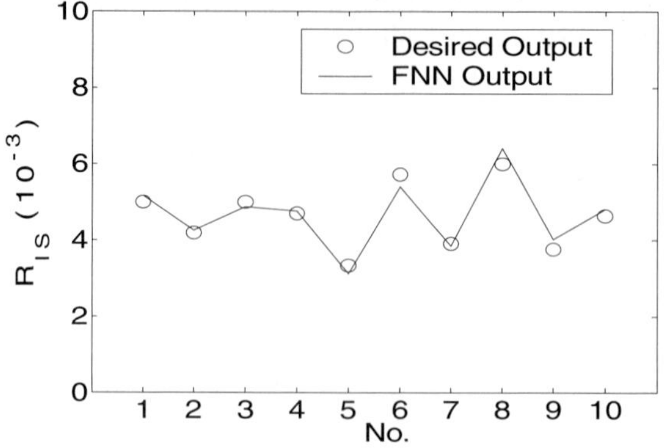

Fig. 3. Testing results

The supplemental function of fuzzy neural network is to calculate the relay parameter sensitivity to system cascading outage risk. The top 10 sensitivities are listed in Table 2. They should gain more attentions than other relays when planning a protection system upgrade.

Table 2. Top 10 relays of sensitivity

No.	Relay#	Sensitivity(10^{-3})	No.	Relay#	Sensitivity(10^{-3})
1	115	8.29	6	45	7.63
2	11	8.17	7	59	7.43
3	178	8.11	8	151	6.76
4	95	8.06	9	108	6.57
5	161	7.68	10	131	6.19

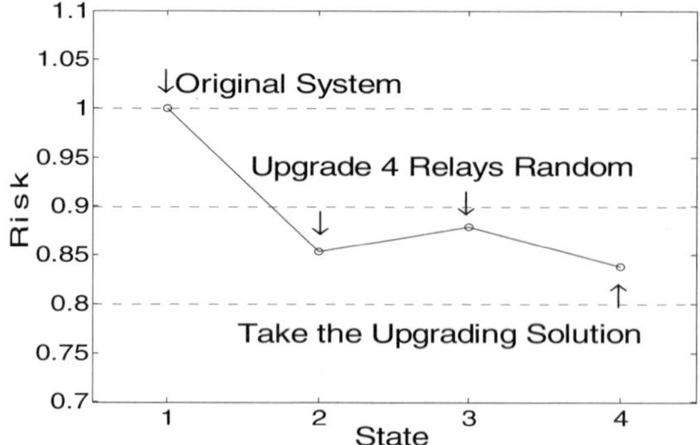

Fig. 4. Result of upgrading

If the number of relays can be upgraded is 4, relay 115, 11, 178 and 95 are selected to upgrade, whose replacement can decrease cascading outages risk effectively. Their improvements over the original system and replacement of another 4 relays random are compared in Fig.4. The relative cascading outages risk is used, and the risk of the original system is defined as base value one.

5 Conclusion

This paper provides a strategy to assess and mitigate power system cascading outages risk considering protection system hidden failure based on fuzzy neural network. Risk theory is used in the cascading outages security analysis and some risk indices are used to assess cascading outages risk. The fuzzy neural network is used to obtain the risk indices and to give a solution that can decrease the system cascading outage risk. The result of a case study illustrates the methodology.

Acknowledgement

This work is supported by the Special Fund for National Basic Research Program of China (2004CB217902), and by the Fund of Education Department of Zhejiang (20050908).

References

1. Yu, X. B. and Singh, C.: A Practical Approach for Integrated Power System Vulnerability Analysis with Protection Failures. IEEE Transactions on Power Systems 4 (2004) 1811-1820
2. Ming, Ni and McCalley, J.D.: Online Risk-Based Security Assessment. IEEE Transactions on Power Systems 1 (2003) 258-265
3. Abd, M. S., Moteleb, E.: Development of Energy Function Algorithm for Improving the Stability Assessment Criterion of Power Systems. Micromechatronics and Human Science, Proceedings of 2001 International Symposium (2001) 189-192
4. El-Arroudi, K.: Comprehensive Transmission Distance Protection Settings Using an Intelligent-Based Analysis of Events and Consequences. IEEE Transactions on Power Delivery 3 (2005) 1817-1824
5. Zhou, Q.: Application of Artificial Neural Networks in Power System Security and Vulnerability Assessment. IEEE Transactions on Power Systems 1 (1994) 525-532
6. Bae, K. and Thorp, J.S.: A Stochastic Study of Hidden Failures in Power System Protection. Decision Support System 3 (1999) 259-268
7. The Reliability Test System Task Force of the Application of Probability Methods Subcommittee. The IEEE reliability test system-1996. IEEE Transactions on Power Systems 3 (1999) 1010-1018

If the number of [illegible] can be upgraded [illegible] [illegible] and [illegible] are selected to upgrade, whose replacement [illegible] decrease cascading [illegible] risk effectively. Their improvements over the original system and [illegible] of another [illegible] randomly are compared in Fig. [illegible]. The relative cascading outages risk is used, and the risk of the original system is defined as base value one.

[illegible] Conclusion

This paper provides a strategy to assess and mitigate power system cascading outages [illegible]

Acknowledgement

This work is supported by the Special Fund for National Basic Research Program of China [illegible] and by the Fund of Education Department of Zhejiang [illegible]

References

1. [illegible] N. and Singh, [illegible] Approach to Integrated Power System Vulnerability Analysis [illegible] IEEE Transactions on Power Systems [illegible]
2. Ming, [illegible] and McCalley, J.D.: Online Risk-Based Security Assessment. IEEE Transactions on Power Systems [illegible]
3. [illegible] Development of [illegible] Function [illegible] for [illegible] the [illegible] Proceedings [illegible] International Symposium [illegible]
4. [illegible]

Author Index

Lecture Notes in Computer Science

For information about Vols. 1–3901

please contact your bookseller or Springer

Vol. 4004: S. Vaudenay (Ed.), Advances in Cryptology - EUROCRYPT 2006. XIV, 613 pages. 2006.

Vol. 3998: T. Calamoneri, I. Finocchi, G.F. Italiano (Eds.), Algorithms and Complexity. XII, 394 pages. 2006.

Vol. 3994: V.N. Alexandrov, G.D. van Albada, P.M.A. Sloot, J. Dongarra (Eds.), Computational Science – ICCS 2006, Part IV. XXIX, 1094 pages. 2006.

Vol. 3993: V.N. Alexandrov, G.D. van Albada, P.M.A. Sloot, J. Dongarra (Eds.), Computational Science – ICCS 2006, Part III. XXX, 1138 pages. 2006.

Vol. 3992: V.N. Alexandrov, G.D. van Albada, P.M.A. Sloot, J. Dongarra (Eds.), Computational Science – ICCS 2006, Part II. XXIX, 1121 pages. 2006.

Vol. 3991: V.N. Alexandrov, G.D. van Albada, P.M.A. Sloot, J. Dongarra (Eds.), Computational Science – ICCS 2006, Part I. CCXX, 1090 pages. 2006.

Vol. 3990: J. C. Beck, B.M. Smith (Eds.), Integration of AI and OR Techniques in Constraint Programming for Combinatorial Optimization Problems. X, 301 pages. 2006.

Vol. 3987: M. Hazas, J. Krumm, T. Strang (Eds.), Location- and Context-Awareness. X, 289 pages. 2006.

Vol. 3986: K. Stølen, W.H. Winsborough, F. Martinelli, F. Massacci (Eds.), Trust Management. XIV, 474 pages. 2006.

Vol. 3984: M. Gavrilova, O. Gervasi, V. Kumar, C.J. K. Tan, D. Taniar, A. Laganà, Y. Mun, H. Choo (Eds.), Computational Science and Its Applications - ICCSA 2006, Part V. XXV, 1045 pages. 2006.

Vol. 3983: M. Gavrilova, O. Gervasi, V. Kumar, C.J. K. Tan, D. Taniar, A. Laganà, Y. Mun, H. Choo (Eds.), Computational Science and Its Applications - ICCSA 2006, Part IV. XXVI, 1191 pages. 2006.

Vol. 3982: M. Gavrilova, O. Gervasi, V. Kumar, C.J. K. Tan, D. Taniar, A. Laganà, Y. Mun, H. Choo (Eds.), Computational Science and Its Applications - ICCSA 2006, Part III. XXV, 1243 pages. 2006.

Vol. 3981: M. Gavrilova, O. Gervasi, V. Kumar, C.J. K. Tan, D. Taniar, A. Laganà, Y. Mun, H. Choo (Eds.), Computational Science and Its Applications - ICCSA 2006, Part II. XXVI, 1255 pages. 2006.

Vol. 3980: M. Gavrilova, O. Gervasi, V. Kumar, C.J. K. Tan, D. Taniar, A. Laganà, Y. Mun, H. Choo (Eds.), Computational Science and Its Applications - ICCSA 2006, Part I. LXXV, 1199 pages. 2006.

Vol. 3979: T.S. Huang, N. Sebe, M.S. Lew, V. Pavlović, M. Kölsch, A. Galata, B. Kisačanin (Eds.), Computer Vision in Human-Computer Interaction. XII, 121 pages. 2006.

Vol. 3978: B. Hnich, M. Carlsson, F. Fages, F. Rossi (Eds.), Recent Advances in Constraints. VIII, 179 pages. 2006. (Sublibrary LNAI).

Vol. 3976: F. Boavida, T. Plagemann, B. Stiller, C. Westphal, E. Monteiro (Eds.), Networking 2006. Networking Technologies, Services, and Protocols; Performance of Computer and Communication Networks; Mobile and Wireless Communications Systems. XXVI, 1276 pages. 2006.

Vol. 3975: S. Mehrotra, D.D. Zeng, H. Chen, B. Thuraisingham, F.Y. Wang (Eds.), Intelligence and Security Informatics. XXII, 772 pages. 2006.

Vol. 3973: J. Wang, Z. Yi, J.M. Zurada, B.-L. Lu, H. Yin (Eds.), Advances in Neural Networks - ISNN 2006, Part III. XXIX, 1402 pages. 2006.

Vol. 3972: J. Wang, Z. Yi, J.M. Zurada, B.-L. Lu, H. Yin (Eds.), Advances in Neural Networks - ISNN 2006, Part II. XXVIII, 1444 pages. 2006.

Vol. 3971: J. Wang, Z. Yi, J.M. Zurada, B.-L. Lu, H. Yin (Eds.), Advances in Neural Networks - ISNN 2006, Part I. LXVII, 1442 pages. 2006.

Vol. 3970: T. Braun, G. Carle, S. Fahmy, Y. Koucheryavy (Eds.), Wired/Wireless Internet Communications. XIV, 350 pages. 2006.

Vol. 3968: K.P. Fishkin, B. Schiele, P. Nixon, A. Quigley (Eds.), Pervasive Computing. XV, 402 pages. 2006.

Vol. 3967: D. Grigoriev, J. Harrison, E.A. Hirsch (Eds.), Computer Science – Theory and Applications. XVI, 684 pages. 2006.

Vol. 3966: Q. Wang, D. Pfahl, D.M. Raffo, P. Wernick (Eds.), Software Process Change. XIV, 356 pages. 2006.

Vol. 3965: M. Bernardo, A. Cimatti (Eds.), Formal Methods for Hardware Verification. VII, 243 pages. 2006.

Vol. 3964: M. Ü. Uyar, A.Y. Duale, M.A. Fecko (Eds.), Testing of Communicating Systems. XI, 373 pages. 2006.

Vol. 3962: W. IJsselsteijn, Y. de Kort, C. Midden, B. Eggen, E. van den Hoven (Eds.), Persuasive Technology. XII, 216 pages. 2006.

Vol. 3960: R. Vieira, P. Quaresma, M.d.G.V. Nunes, N.J. Mamede, C. Oliveira, M.C. Dias (Eds.), Computational Processing of the Portuguese Language. XII, 274 pages. 2006. (Sublibrary LNAI).

Vol. 3959: J.-Y. Cai, S. B. Cooper, A. Li (Eds.), Theory and Applications of Models of Computation. XV, 794 pages. 2006.

Vol. 3958: M. Yung, Y. Dodis, A. Kiayias, T. Malkin (Eds.), Public Key Cryptography - PKC 2006. XIV, 543 pages. 2006.

Vol. 3956: G. Barthe, B. Gregoire, M. Huisman, J.-L. Lanet (Eds.), Construction and Analysis of Safe, Secure, and Interoperable Smart Devices. IX, 175 pages. 2006.

Vol. 3955: G. Antoniou, G. Potamias, C. Spyropoulos, D. Plexousakis (Eds.), Advances in Artificial Intelligence. XVII, 611 pages. 2006. (Sublibrary LNAI).

Vol. 3954: A. Leonardis, H. Bischof, A. Pinz (Eds.), Computer Vision – ECCV 2006, Part IV. XVII, 613 pages. 2006.

Vol. 3953: A. Leonardis, H. Bischof, A. Pinz (Eds.), Computer Vision – ECCV 2006, Part III. XVII, 649 pages. 2006.

Vol. 3952: A. Leonardis, H. Bischof, A. Pinz (Eds.), Computer Vision – ECCV 2006, Part II. XVII, 661 pages. 2006.

Vol. 3951: A. Leonardis, H. Bischof, A. Pinz (Eds.), Computer Vision – ECCV 2006, Part I. XXXV, 639 pages. 2006.

Vol. 3950: J.P. Müller, F. Zambonelli (Eds.), Agent-Oriented Software Engineering VI. XVI, 249 pages. 2006.

Vol. 3947: Y.-C. Chung, J.E. Moreira (Eds.), Advances in Grid and Pervasive Computing. XXI, 667 pages. 2006.

Vol. 3946: T.R. Roth-Berghofer, S. Schulz, D.B. Leake (Eds.), Modeling and Retrieval of Context. XI, 149 pages. 2006. (Sublibrary LNAI).

Vol. 3945: M. Hagiya, P. Wadler (Eds.), Functional and Logic Programming. X, 295 pages. 2006.

Vol. 3944: J. Quiñonero-Candela, I. Dagan, B. Magnini, F. d'Alché-Buc (Eds.), Machine Learning Challenges. XIII, 462 pages. 2006. (Sublibrary LNAI).

Vol. 3943: N. Guelfi, A. Savidis (Eds.), Rapid Integration of Software Engineering Techniques. X, 289 pages. 2006.

Vol. 3942: Z. Pan, R. Aylett, H. Diener, X. Jin, S. Göbel, L. Li (Eds.), Technologies for E-Learning and Digital Entertainment. XXV, 1396 pages. 2006.

Vol. 3941: S.W. Gilroy, M.D. Harrison (Eds.), Interactive Systems. XI, 267 pages. 2006.

Vol. 3940: C. Saunders, M. Grobelnik, S. Gunn, J. Shawe-Taylor (Eds.), Subspace, Latent Structure and Feature Selection. X, 209 pages. 2006.

Vol. 3939: C. Priami, L. Cardelli, S. Emmott (Eds.), Transactions on Computational Systems Biology IV. VII, 141 pages. 2006. (Sublibrary LNBI).

Vol. 3936: M. Lalmas, A. MacFarlane, S. Rüger, A. Tombros, T. Tsikrika, A. Yavlinsky (Eds.), Advances in Information Retrieval. XIX, 584 pages. 2006.

Vol. 3935: D. Won, S. Kim (Eds.), Information Security and Cryptology - ICISC 2005. XIV, 458 pages. 2006.

Vol. 3934: J.A. Clark, R.F. Paige, F.A. C. Polack, P.J. Brooke (Eds.), Security in Pervasive Computing. X, 243 pages. 2006.

Vol. 3933: F. Bonchi, J.-F. Boulicaut (Eds.), Knowledge Discovery in Inductive Databases. VIII, 251 pages. 2006.

Vol. 3931: B. Apolloni, M. Marinaro, G. Nicosia, R. Tagliaferri (Eds.), Neural Nets. XIII, 370 pages. 2006.

Vol. 3930: D.S. Yeung, Z.-Q. Liu, X.-Z. Wang, H. Yan (Eds.), Advances in Machine Learning and Cybernetics. XXI, 1110 pages. 2006. (Sublibrary LNAI).

Vol. 3929: W. MacCaull, M. Winter, I. Düntsch (Eds.), Relational Methods in Computer Science. VIII, 263 pages. 2006.

Vol. 3928: J. Domingo-Ferrer, J. Posegga, D. Schreckling (Eds.), Smart Card Research and Advanced Applications. XI, 359 pages. 2006.

Vol. 3927: J. Hespanha, A. Tiwari (Eds.), Hybrid Systems: Computation and Control. XII, 584 pages. 2006.

Vol. 3925: A. Valmari (Ed.), Model Checking Software. X, 307 pages. 2006.

Vol. 3924: P. Sestoft (Ed.), Programming Languages and Systems. XII, 343 pages. 2006.

Vol. 3923: A. Mycroft, A. Zeller (Eds.), Compiler Construction. XIII, 277 pages. 2006.

Vol. 3922: L. Baresi, R. Heckel (Eds.), Fundamental Approaches to Software Engineering. XIII, 427 pages. 2006.

Vol. 3921: L. Aceto, A. Ingólfsdóttir (Eds.), Foundations of Software Science and Computation Structures. XV, 447 pages. 2006.

Vol. 3920: H. Hermanns, J. Palsberg (Eds.), Tools and Algorithms for the Construction and Analysis of Systems. XIV, 506 pages. 2006.

Vol. 3918: W.K. Ng, M. Kitsuregawa, J. Li, K. Chang (Eds.), Advances in Knowledge Discovery and Data Mining. XXIV, 879 pages. 2006. (Sublibrary LNAI).

Vol. 3917: H. Chen, F.Y. Wang, C.C. Yang, D. Zeng, M. Chau, K. Chang (Eds.), Intelligence and Security Informatics. XII, 186 pages. 2006.

Vol. 3916: J. Li, Q. Yang, A.-H. Tan (Eds.), Data Mining for Biomedical Applications. VIII, 155 pages. 2006. (Sublibrary LNBI).

Vol. 3915: R. Nayak, M.J. Zaki (Eds.), Knowledge Discovery from XML Documents. VIII, 105 pages. 2006.

Vol. 3914: A. Garcia, R. Choren, C. Lucena, P. Giorgini, T. Holvoet, A. Romanovsky (Eds.), Software Engineering for Multi-Agent Systems IV. XIV, 255 pages. 2006.

Vol. 3911: R. Wyrzykowski, J. Dongarra, N. Meyer, J. Waśniewski (Eds.), Parallel Processing and Applied Mathematics. XXIII, 1126 pages. 2006.

Vol. 3910: S.A. Brueckner, G.D.M. Serugendo, D. Hales, F. Zambonelli (Eds.), Engineering Self-Organising Systems. XII, 245 pages. 2006. (Sublibrary LNAI).

Vol. 3909: A. Apostolico, C. Guerra, S. Istrail, P. Pevzner, M. Waterman (Eds.), Research in Computational Molecular Biology. XVII, 612 pages. 2006. (Sublibrary LNBI).

Vol. 3908: A. Bui, M. Bui, T. Böhme, H. Unger (Eds.), Innovative Internet Community Systems. VIII, 207 pages. 2006.

Vol. 3907: F. Rothlauf, J. Branke, S. Cagnoni, E. Costa, C. Cotta, R. Drechsler, E. Lutton, P. Machado, J.H. Moore, J. Romero, G.D. Smith, G. Squillero, H. Takagi (Eds.), Applications of Evolutionary Computing. XXIV, 813 pages. 2006.

Vol. 3906: J. Gottlieb, G.R. Raidl (Eds.), Evolutionary Computation in Combinatorial Optimization. XI, 293 pages. 2006.

Vol. 3905: P. Collet, M. Tomassini, M. Ebner, S. Gustafson, A. Ekárt (Eds.), Genetic Programming. XI, 361 pages. 2006.

Vol. 3904: M. Baldoni, U. Endriss, A. Omicini, P. Torroni (Eds.), Declarative Agent Languages and Technologies III. XII, 245 pages. 2006. (Sublibrary LNAI).

Vol. 3903: K. Chen, R. Deng, X. Lai, J. Zhou (Eds.), Information Security Practice and Experience. XIV, 392 pages. 2006.

Vol. 3902: R. Kronland-Martinet, T. Voinier, S. Ystad (Eds.), Computer Music Modeling and Retrieval. XI, 275 pages. 2006.